BOTANY FOR DEGREE STUDENTS

B.Sc. First Year

As per UGC Model Curriculum

Course: I
DIVERSITY OF MICROBES AND CRYPTOGAMS

Course: II
CELL BIOLOGY AND GENETICS

Dr. B.P. PANDEY
MSc, PhD, FPSI
Former Head
Department of Botany
J.V. College, Baraut, Uttar Pradesh

S. CHAND PUBLISHING

S Chand And Company Limited
(ISO 9001 Certified Company)

S. Chand And Company Limited

(ISO 9001 Certified Company)

Head Office: D-92, Sector–2, Noida – 201301, U.P. (India), Ph. 91-120-4682700

Registered Office: A-27, 2nd Floor, Mohan Co-operative Industrial Estate, New Delhi – 110 044, Phone: 011-49731800

www.**schandpublishing.com;** e-mail: **info@schandpublishing.com**

Marketing Offices:

Chennai	:	Ph: 23632120; chennai@schandpublishing.com
Guwahati	:	Ph: 2738811, 2735640; guwahati@schandpublishing.com
Hyderabad	:	Ph: 40186018; hyderabad@schandpublishing.com
Jalandhar	:	Ph: 4645630; jalandhar@schandpublishing.com
Kolkata	:	Ph: 23357458, 23353914; kolkata@schandpublishing.com
Lucknow	:	Ph: 4003633; lucknow@schandpublishing.com
Mumbai	:	Ph: 25000297; mumbai@schandpublishing.com
Patna	:	Ph: 2260011; patna@schandpublishing.com

First Edition 2007
Reprint 2008
Second Edition 2009
Reprints 2011, 2012 (Twice), 2015, 2016, 2018 (Thrice), 2019, 2020 (Twice), 2021 (Twice), 2022 (Twice)

Reprint 2023

ISBN: 978-81-219-2809-0 **Product Code:** H6BTY68BOTN10ENAB09R

PRINTED IN INDIA

By Vikas Publishing House Private Limited, Plot 20/4, Site-IV, Industrial Area Sahibabad, Ghaziabad – 201 010 and Published by S. Chand And Company Limited, A-27, 2nd Floor, Mohan Co-operative Industrial Estate, New Delhi – 110 044.

PREFACE

The present book 'Botany for Degree Students' B.Sc., First Year, has been written strictly according to **UGC Model Syllabus** for all Indian Universities. The syllabus of B.Sc. I has been divided into **Course-I** and **Course II.** Course I contains **'Diversity of Microbes and Cryptogams'** where subject matter is divided into *five* units, and expanded into *thirty two* chapters; while course II contains **'Cell Biology and Genetics,** having *two* units and *seven* chapters.

Thus, whole subject matter of B.Sc. I (Botany) has been expanded into *seven* units and *forty nine* chapters. The whole subject matter has been written in simple understandable language that can easily be grasped by students and other readers alike. To make subject easy and understandable, self-explanatory labelled diagrams have been added. Each unit or chapter as the case may be is followed by various types of questions, such as very short answer questions, short answer questions, long answer questions, diagrammatic questions and multiple choice questions, asked repeatedly in various Indian Universities. Every effort has been put up to provide latest and up-to date information of subject. I think, this book would be helpful a lot for whom, this has been written.

I am thankful to Mrs. Nirmala Gupta (CMD), Shri Navin Joshi, Vice-President (Publishing) and Shri R. S. Saxena, Advisor, S. Chand & Company Ltd., who have assigned this work to me to complete.

I am also thankful to those to whom I know or do not know but who have helped me a lot directly or indirectly, I thank all of them.

Suggestions from all corners would be welcomed and incorporated in subsequent editions.

'Neelkanth Kuteer'
Subhasnagar, BARAUT,
250 611.

Dr. B. P. Pandey

UGC MODEL SYLLABUS

B.Sc. First Year

BOTANY

Course-I. Diversity of Microbes and Cryptogams

Viruses and Bacteria: General account of viruses and mycoplasma; bacteria - structure, nutrition, reproduction and economic importance; general account of cyanobacteria.

Algae: General characters, classification and economic importance; important features and life history of Chlorophyceae - *Volvox, Oedogonium, Coleochaete*; Xanthophyceae – *Vaucheria;* Phaeophyceae - *Ectocarpus, Sargassum;* Rhodophyceae - *Polysiphonia.*

Fungi: General characters, classification and economic importance; important features and life history of Mastigomycotina - *Pythium, Phytophthora;* Zygomycotina – *Mucor,* Ascomycotina - *Saccharomyces, Eurotium, Chaetomium, Peziza;.* Basidiomycotina – *Puccinia, Agaricus;* Deuteromycotina - *Cercospora, Colletotrichum;* general account of Lichens.

Bryophyta: Amphibians of plant kingdom displaying alternation of generations; structure, reproduction and classification of Hepaticopsida (e.g. *Marchantia*); Anthocerotopsida (e.g. *Anthoceros*), Bryopsida (e.g *Funaria).*

Pteriodophyta: The first vascular plants; important characteristics of Psilopsida, Lycopsida, Sphenopsida and Pteropsida; structure, reproduction in *Rhynia, Lycopodium, Selaginella, Equisetum, Pteris* and *Marsilea.*

Course II. Cell Biology and Genetics

Structure and function of nucleus: Ultrastructure; nuclear membrane; nucleolus.

Chromosome organization: Morphology; centromere and telomere; chromosome alterations; deletions, duplications, translocations, inversions; variations in chromosome number, aneuploidy, polyploidy; sex chromosomes.

DNA the genetic material: DNA structure; replication, DNA-protein interaction; the nucleosome model; genetic code; satellite and repetitive DNA.

Cell division: Mitosis; meiosis.

Genetic inheritance: Mendelism; laws of segregation and independent assortment; linkage analysis; allelic and non-allelic interactions.

Gene expression: Structure of gene; transfer of genetic information; transcription, translation, protein synthesis; tRNA; ribosomes; regulation of gene expression in prokaryotes and eukaryotes; proteins, 1D, 2D and 3D structure.

Genetic variations: Mutations, spontaneous and induced; transposable genetic elements; DNA damage and repair.

Extranuclear genome: Presence and function of mitochondrial and plastid DNA; plasmids.

Structure and function of other organelles: Golgi, ER, peroxisomes, vacuoles.

The cell envelopes: Plasma membrane; bilayer lipid structure; functions; the cell wall.

CONTENTS

COURSE : I
DIVERSITY OF MICROBES AND CRYPTOGAMS

COURSE : II
CELL BIOLOGY AND GENETICS

Unit–6 : Cell Biology

Unit–7 : Genetics–Continuity of Life

COURSE - I

Diversity of Microbes and Cryptogams

[illegible] Hashimoto, who worked in 1894 with the dwarf disease of rice and the leaf hopper *Nephotettix* [illegible] [illegible]. Again in 1907 three workers in America, [illegible] Adams and Shaw suggested that there was some connection between curly top of sugarbeet and the leaf hopper *Eutettix tenella*. Some years later Smith and Boncquet (1915) confirmed this and showed that a single insect from an infected plant placed on a healthy plant for 5 minutes would produce the disease. The original criterion of a virus was an infectious entity that could pass through a filter of a pore size small enough to hold back all known cellular agents of diseases. However [illegible] were found that had virus like symptoms not associated with any pathogen visible [illegible] light microscope, but which could not be transmitted by mechanical inoculation. With such [illegible] the criterion of filterability could not be applied. The infectious nature was established by graft transmission and sometimes by insect vectors. Thus, it came about that certain diseases of the yellows and witches broom type such as aster yellows, came to be considered as due to viruses on quite inadequate grounds. Later on, critical examination by electron microscopy and the use of antibiotics, there have been strong indications that a number of so called virus diseases are really caused by mycoplasma like agents (Doi *et al.* 1967).

During most of the period between 1900 and 1935 attention was focussed on description of diseases. Ineffective attempts were made to refine filtration methods in order to define the size of viruses more closely. There were almost the only aspects of virus diseases that could be studied with the techniques that were available. The influence of various physical and chemical agents on virus infectivity was investigated but methods for assay of infective material were not refined. Holmes (1929) showed the local lesions produced in some hosts following mechanical inoculation could be used for the [illegible] [illegible] [illegible]. This technique enabled properties of viruses to be studied much more readily and paved the way for the isolation and purification of viruses a few years later.

In 1926 the first enzyme urease was isolated, crystallised and identified as a protein (Sumner, 1926). The isolation of others soon followed. In the early 1930's workers in various countries began attempting to isolate and purify plant viruses using methods similar to those that had been used for enzymes. Following detailed chemical work suggesting that the infectious agent of TMV might be protein, Stanley (1935) announced the isolation of this virus in an apparently crystalline form. [illegible] (1935, 1936) considered the [illegible] was a globulin containing no phosphorus. Bawden *et al.* (1936) described the isolation from TMV infected plants of liquid crystalline nucleoprotein containing nucleic acid of the pentose type. The present concept of the virus particle is that it consists of a protein shell within which is contained a nucleic acid molecule which by its own structure and coding property determined the arrangement of the amino acid sequence in the subunits of the protein cell. The nucleic acid is also the infective agent which as it passes to the healthy cell, dictates a change in the protein metabolism which results in the build up of virus protein usually at the expense of the normal protein.

Introduction. When Louis Pasteur (1822-1893) and Robert Koch (1843-1910) discovered that microbes were the cause of many diseases they opened up a new field of experiment and research. The word 'virus' which before their discoveries was applied to any poisonous substance came to be used for all kinds of infection agents.

[illegible] it was found that certain agents were so fine that they could pass through exceedingly [illegible] filters which did not allow even bacteria and other known microorganisms [illegible] [illegible] scientists, unable to conceive their extreme minuteness, [illegible] the opinion that these agents were [illegible] fluids and do not consist of discrete organisms. Finally, when it was established that they consist of sub-microscopic organisms, which were invisible to the most powerful microscopes of the conventional type, the term virus was restricted to apply only to these filter-passing or filterable sub-micro-organisms.

UNIT - 1

Diversity of Microbes
Viruses and Bacteria
Including Mycoplasma and Cyanobacteria

1
CHAPTER

Viruses

Historical. The virus comes from Latin language and means poison. This use of the word goes back to many hundreds of years, long before any one really knew what a virus was, or that it even existed as we know it today. It was generally believed that these 'viruses', or poisons, were carried in the night air and could cause many unexplained diseases. In 1770, a virus disease 'leaf roll' of potato was observed for the first time in England. In 1785, the same disease was reported from Germany. Virus diseases of plants have been described for many years. "Breaking" of tulips, which is now known to be a virus disease was first as early as 1576. Degeneration of potatoes was known in Europe reported in the eighteenth century. Tobacco mosaic virus was recognized by Swieten in Holland in 1887. Adloph Mayer a Dutch investigator working with this disease in Holland was first to point out in 1886 that it was readily transmissible and infectious. Iwanowski a Russian investigator (1892), demonstrated that tobacco mosaic virus would pass through a bacteria proof filter and thereby he distinguished this type of infectious agency from bacteria and from fungi.

This work was confirmed by Beizerink in 1898, who showed further that tobacco mosaic virus will diffuse through an agar agar layer suggested that it was a "*contagium vivum fluidium*". In 1898, Loeffler and Frosch were the first to show virus of animal origin to be filterable when they demonstrated that the foot and mouth disease of cattle was incited by an entity which passed through bacteria proof filter. The 'Sirah' disease of sugarcane caused by a virus was observed in Jawa in 1882, and the methods of its control were also investigated. Mosaic disease of sugarcane, which was then known as 'yellow stripe' disease was also discovered in Jawa in 1890. In 1887, the 'mosaic' of tomato was observed in England, and the disease was studied in detail in America. Mayer (1886), discovered tobacco 'mosaic' in north Europe. He proved that this disease can be transmitted to healthy plants by mechanical means. Iwanowsky (1892), discovered the first virus and thought it to be the small bacterium causing tobacco mosaic. Beijerinck (1898), discarded the Iwanowsky's statement and proposed his theory of *cantagium vivum fludium*. In 1901, Takami of Japan, transmitted the dwarf or stunt disease of paddy by an insect *Nephotettix apicalis*, and it was known for the first time that viruses could be transmitted by insects. For the first time in 1935, W.M. Stanley isolated the 'Tobacco Mosaic Virus' in paracrystalline form. Green (1935) and Laidlaw (1938) independently put forward the theory that viruses are obligate parasites which have developed parasitism to the highest degree. Bawden and Pirie (1938) isolated a fully crystalline form of the tomato bushy stunt virus. Stanley (1936) and Northrop (1938) suggested that the viruses are autocatalysts and multiply in susceptible cells by activating previously formed inactive precursors. According to Darlington (1944), the viruses are derived from cell proteins. In his view these undifferentiated proteins may come to acquire the properties of infection.

The discovery of the relationship between viruses and insects was not made in a day and a period of years elapsed between the time when insects were first suspected to transmitting plant viruses and the actual demonstration of this method of transmission. The first to prove experimentally the relationship between an insect and a plant virus seems to have been a Japanese farmer, Hashimoto, who worked in 1894 with the dwarf disease of rice and the leaf hopper *Nephotettix apicalis* var. *cincticeps*. About 1907, three workers in America, Ball, Adams and Shaw suggested that there was some connection between curly top of sugarbeet and the leaf hopper *Eutettix tenella*. Some years later, Smith and Boncquet (1915) confirmed this and showed that a single insect from an infected plant placed on a healthy plant for 5 minutes would produce the disease. The original criterion of a virus was an infectious entity that could pass through a filter with a pore size small enough to hold back all known cellular agents of diseases. However, diseases were soon found that had virus like symptoms not associated with any pathogen visible in the light microscope, but which could not be transmitted by mechanical inoculation. With such disease, the criterion of filterability could not be applied. The infectious nature was established by graft transmission and sometimes by insect vectors. Thus, it came about that certain diseases of the yellows and witches broom type such as aster yellows, came to be considered as due to viruses on quite inadequate grounds. Later on, critical examination by electron microscopy and the use of inhibitory drugs have been strong indications that a number of so called virus diseases are really caused by mycoplasma like agents (Doi *et al*, 1967).

During most of the period between 1900 and 1935 attention was focussed on description of diseases. Ineffective attempts were made to refine filtration methods in order to define the size of viruses more closely. There were almost the only aspects of virus diseases that could be studied with the techniques that were available. The influence of various physical and chemical agents on virus infectivity was investigated but methods for assay of infective material were not refined. Holmes (1929) showed the local lesions produced in some hosts following mechanical inoculation could be used for the rapid quantitative assay of infective virus. This technique enabled properties of viruses to be studied much more readily and paved the way for the isolation and purification of viruses a few years later.

In 1926 the first enzyme urease was isolated, crystallised and identified as a protein (Sumner, 1926). The isolation of others soon followed. In the early 1930's workers in various countries began attempting to isolate and purify plant viruses using methods similar to those that had been used for enzymes. Following detailed chemical studies suggesting that the infectious agent of TMV might be protein. Stanley (1935) announced the isolation of this virus in an apparently crystalline form. At first Stanley (1935, 1936) considered that the virus was a globulin containing no phosphorus. Bawden *et al* (1936) described the isolation from TMV infected plants of liquid crystalline nucleoprotein containing nucleic acid of the pentose type. The present concept of the virus particle is that it consists of a protein shell within which is contained a nucleic acid molecule which by its own structure and coding property determined the arrangement of the amino acid sequences in the subunits of the protein cell. The nucleic acid is also the infective agent which as it passes into a healthy cell, dictates a change in the protein metabolism which results in the build up of virus protein usually at the expense of the normal protein.

Introduction. When Louis Pasteur (1822-1895) and Robert Koch (1843-1910) discovered that microbes were the cause of many diseases, they opened up a new field of experiment and research. The word 'virus' which, before their discoveries, was applied to any poisonous substance came to be used for all kinds of infection agents.

As research progressed further, it was found that certain agents were so fine that they could pass through specially designed filters which did not allow even bacteria and other known microorganisms to pass through. Some scientists, unable to conceive their extreme minuteness, hazarded the opinion that these agents were contagious fluids and did not consist of discrete organisms. Finally, when it was established that they consist of sub-microscopic organisms, which were invisible to the most powerful microscopes of the conventional type, the term virus was restricted to apply only to these 'filter-passing' or filterable sub-micro-organisms.

It was the invention of electron microscope in recent years that has enabled these tiniest of organisms to be identified and photographed. The best of the light microscopes can magnify an object 2,000 times its original size. But the electron microscope gives a direct magnification of 10,000 and can enlarge photographically up to 200,000.

Some of the diseases caused by viruses have been known for a long time. Small-pox for instance, has been known for ages and has caused innumerable deaths. Even in these days it is responsible for thousands of deaths every year. But the nature and cause of these diseases have been understood only recently.

The behaviour of viruses has been a matter of great surprise and speculation: they possess certain characteristics of inorganic substances and they also behave like living organisms. This has led scientists to call them the missing links between dead and living matter.

In 1935, W.M. Stanley, an American biochemist, isolated the tobacco, mosaic virus and was surprised to find that it had all the properties of a crystalline solid. Its chemical constitution lent itself to analysis and produced regular shaped crystals like many chemical salts. In its isolated condition it was incapable of reproduction which is typical of all living organisms. Thus it was not alive in the accepted sense of the word. But as soon as it was brought into contact with a healthy tobacco leaf, it appeared to spring into life, flourishing and multiplying like any primitive living thing.

Chemically, viruses consist of nucleoproteins — a combination of nucleic acid and protein essential constituents of living matter. The more complex viruses contain some fats in addition. Their extreme simplicity leads one to hope that some day it may be possible to produce them in the laboratory if only for the purpose of producing vaccines.

There are a number of diseases caused by virus infection. Some of these are small-pox, measles, mumps, yellow fever, poliomycilitis (infantile paralysis), usually called polio, chicken pox, rabies, and influenza. It is believed that even common cold is caused by virus infection. In addition there are a number of animal and plant diseases caused by viruses.

The control of virus diseases is not adequate. Usually viruses are not affected by antibiotics. It is easier to control pneumonia than the common cold. Viruses can mutate just like other living organisms. By mutation is meant a sudden variation in some well-marked character. This gives rise to new types of viruses. The sudden outbreaks of some epidemics in unexpectedly virulent forms is attributed to such mutations. Naturally it takes some time before the new character is studied and brought under control. The deadly influenza epidemic of 1918 has been cited as a possible example.

Definition. According to Green (1935), 'that they are the smallest units showing the reproductive property considered typical of life.'

Stanely (1938) says, "*We are forced to conclude therefore, that although tobacco mosaic virus (T.M.V.) protein has the ordinary properties of molecules, it also has the ability to reproduce and to mutate, properties not ordinarily ascribed to molecules and hence that T.M.V. protein represents an entity unfamiliar to us.*"

The most accepted definition proposed by Bawden (1943), "*the virus is on obligatory parasitic pathogen with at least one dimension of less than 200 millimicrons mμ.*"

Characteristic features.

1. All the viruses are ultramicroscopic.

2. All viruses are obligatory intracellular parasites.

3. The viruses cannot be grown in artificial media in any case and the living cell seems essential for multiplication of viruses.

4. They are crystalline nucleo-proteins of very high molecular weight.

Nature of virus. Viruses are so small that they can not be seen even with the highest magnification of the microscope using visible light. They are recognizable only by their biological behaviour, such as, by the disease they cause. Their exact nature was a mystery for long. They were variously regarded as invisible form of bacteria, protozoa, enzymes, toxins or as unusual products of metabolism of the cells in which they were found. An American biochemist, Stanley in 1935, however isolated by chemical means from the diseased tobacco leaves a material which appeared to be a protein of high molecular weight. He studied this material in detail, and found that it possessed all the properties of a tobacco mosaic virus, which has already been discovered. There are a number of direct, indirect and circumstantial, evidences to show that the material itself is the virus. Stanley, on the basis of the chemical properties, identified the material as **autocatalytic proteins** which can multiply within a living cell only. Bawden purified the tobacco mosaic virus, and found it to be a crystalline nucleo-protein of very high molecular weight, retaining its infectivity even when diluted to a concentration of 1/1,000,000. This virus when examined with the electron microscope using X-rays, is found to be in the form of bundles of rod-like protein. Such nucleoprotein can not be obtained from healthy plants and is the virus itself. In 1938 the study of potato virus confirmed this. Since then, proteins of high molecular weight possessing all the properties of the respective viruses have been isolated and studied.

Most of the viruses are crystalline and rod-like, but different in size. Tobacco mosaic virus is rod-like in shape, 280 mμ in length and 18 mμ in breadth. 'Bushy stunt' virus of tomato is 274 mμ in diameter and its molecular weight is 8,800,000–12,800,000. Stanley studied the nucleo-proteins of the 'ring spot' virus of tobacco. The virus is rounded with 19 mμ diameter and 3,400,000 molecular weight. Tobacco 'necrosis' virus is smallest of all the plant viruses known so far. It is rounded with 13-20 mμ diameter.

Before 1935 they were believed to be living on the basis of few following properties:

1. That they can live only in a living cell.
2. They can infect healthy plants just like bacteria and fungi.
3. They multiply in number and grow in size, as the living organisms reproduce and grow.
4. They have physiologic specialization in relation to the insect vectors and the plants.
5. They respond to stimuli, such as acids, alkalies, light and temperature.

The supporters of the non-living nature of viruses based their views on the following few properties:

1. They are too small to be observed under visible light.
2. They retain infectivity even in very low concentrations.
3. They can be crystallized like a chemical substance.
4. They can be sedimented like proteins.
5. They can be precipitated by a number of chemical substances.
6. They retain the power of infection even after 31 years in non-living tobacco leaves.

Some workers did not agree with any of the above extreme views and adopted a middle course. They regarded viruses, representing a stage in between living and non-living with the acquired property of multiplication.

At present most of the virologists have agreed that viruses are nucleo-proteins of high molecular weight and have the power of multiplication.

Plant viruses. Plant viruses occur either as spheres or elongated rods. In both cases there is some sort of a ribonucleic acid RNA core surrounded by a protein sheath. The rod forms of tobacco mosaic also contain a central core of RNA as shown in the figure. The elongated rod of tobacco mosaic virus is twisted into uniform spiral to form what appears, in gross form, as a hollow rod.

The life-cycle of a plant virus. There is an incubation period in the body of an insect after it picks up virus particles from an infected plant. There also appears to be an incubation period in the plant cell after its infection with virus particles. Infection may be with viral RNA or with whole virus particles, but in the case of plant body it is always in the form of RNA. The changes of viral RNA reduce or change the course of viral infection. There is some evidence that RNA appears in the host nucleus before it appears in the host cytoplasm and that viral RNA is elaborated by the host cytoplasm before the viral protein is formed. The replication of the viral RNA is indeed somewhat of an enigma because current ideas of RNA application and protein synthesis are generaly related to information carried in the deoxyribonucleic acid DNA molecule.

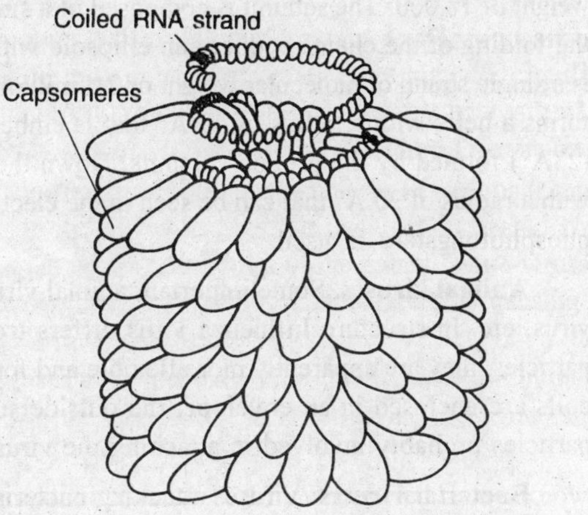

Fig. 1.1. Virus. Structure of tobacco mosaic virus (T.M.V.) a portion of rod of tobacco mosaic virus showing RNA core and protein sheath.

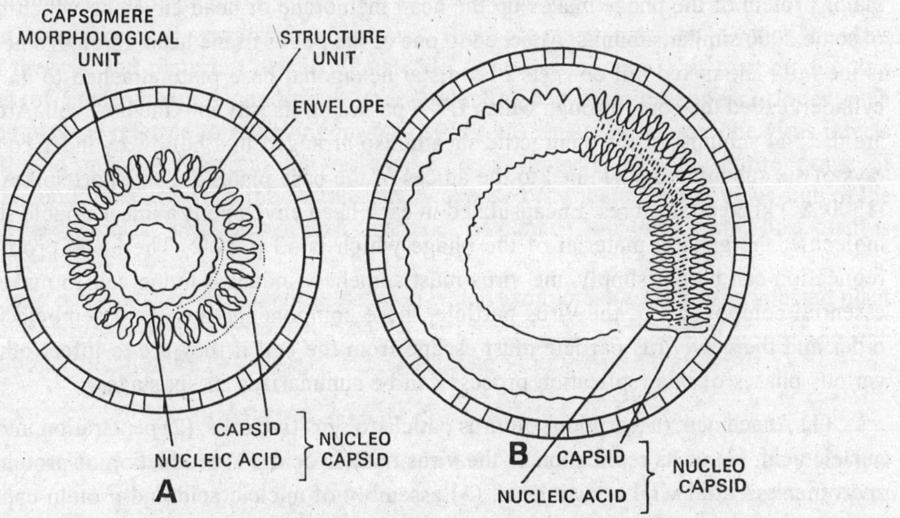

Fig. 1.2. Virus. Diagram showing detailed structure of virus particle

Viruses may be divided into three main classes according to the nature of their hosts: Plant viruses, Animal viruses and Bacterial viruses.

The important plant viruses are Tobacco mosaic virus, Tobacco satellite virus, Turnip yellow mosaic, Rugose mosaic virus. Generally all crystallized plant viruses have been found to consist mainly of ribonucleoproteins.

TMV is the extensively studied virus in plants, discovered by Iwanowski (1892) and obtained in pure state by Stanley (1935). It has a helical symmetry. Tobacco mosaic virus is a cylindrical structure with a molecular weight of 40 millions and dimensions of 160 × 3000 Å. Extensive structural studies of the virus—through X-ray diffraction, electron microscopy at various stages of dissociation, solubility properties as well as studies of the separate protein and RNA components, have yielded the following remakably detailed and precise model of the virus particle.

The protein coat of the virus consists of 2130 identical protein subunits each with a molecular weight of 18,000. The subunit is composed of a single chain of 158 amino acids of known sequence, the folding of the chain produces an ellipsoid with an axial rotation of 3:1. The RNA of the virus is a single strand of molecular weight of 2.4 million consisting of 6500 nucleotides. The RNA strand forms a helix with a radius of 40 A° that is embedded in a helix (radius 85 A°) of identical pitch (23A°) formed by the protein subunits. Down the middle of the molecule runs a cylindrical hole with a radius of 20 A° that can be seen in the electron microscope if special staining technique with phosphotungstate is used.

Animal viruses. Some important animal viruses are Influenza virus, Mumps, Pox virus, Reo virus, etc. In structure Influenza virus differs from TMV in that instead of being straight rigid particles, they are apparently more flexible and form loose coils. A further characteristic is that the coils are enclosed in an envelope, the outside surface of which is covered with minute protein particles probably involved in attaching the virus to the surface of host cells.

Bacterial viruses. Viruses attacking bacteria are known as bacteriophages or phages. These are tadpole like in shape T_4 phage attacking the bacterium *Escherichia coli*, is among the most complex of viruses. It consists of a head and a rather complicated tail. The shape of the head has been described as an elongated icosahedron having a length of 1250 A° and width of 850 A°. The major protein of the phage makes up the head membrane or head envelope which is constructed of some 2000 similar subunits. Attached to one of the points of the head, through a neck and collar, is the tail. The naked tail consists of a distal hexagonal base plate attached to an inner hollow cylinder, called the core, through which DNA passes on its way into the host cell. Around the core are the 144 subunits of the contractile sheath also arranged in a hollow cylinder consisting of 24 rays of six subunits each. Joined to the apices of the base plates are six short spikes and six long (1300 A°) kinked tail fibres. Encapsulated in each head envelope is a linear double stranded DNA molecule, the genetic material of the phage which is 53 μ by 2. The basic problems of virus replication can be put simply, the virus must somehow induce a living cell to make more of the essential components of the virus particle, these components must be assembeld in the proper order and the new virus particle must escape from the cell if they are to infect other cells. The various phases of this replication process can be summarized in six stages.

(1) Attachment (adsorption) of virus paticle to sensitive cells, (2) penetration into cell of virus nucleic acid, (3) of its replication of the virus nucleic acid, (4) production of protein capsomeres and other essential viral constituents, (5) assembly of nucleic acid and protein capsomeres into new virus particles, (6) release of mature virus particles from the cell.

Properties of viruses. Properties of various plant viruses have been studied in detail. Since the development of virology (science of viruses), tobacco mosaic virus has been the main source for investigations, and most of the virus properties described below are based on it.

1. Viruses are highly infectious. Bawden working on tobacco mosaic virus found it to be infectious even in 1/1,000,000 dilution.

2. Viruses are highly resistant to acids, alkalies and salts. Tobacco mosaic virus is affected by hydrochloric and nitric acids only when 1 gm. of acid is added to 100 c.c. virus solution. They are more affected by alkalies, such as sodium hydroxide and sodium carbonate. Salts like sodium chloride, silver nitrate, and mercuric chloride however do not affect them much, 1 gm copper sulphate in 100 c.c. virus solution, and 4% formaldehyde solution destroy tobacco mosaic virus completely. According to Vinson, acetone and alcohol have no effect on viruses at low temperatures.

3. Viruses are resistant to high temperatures. Tobacco mosaic virus can infect healthy plant even after being kept at 80°C for two days. However, it loses its power of infection within 10 minutes at 85°C–90°C temperatures.

4. Direct sunlight has no effect on viruses. Tobacco mosaic virus is not affected by X-rays and mercury vapours, till the exposures do not exceed half an hour and one hour respectively.

5. Viruses can be filtered through **Brefeld** and **Chamberlain** filters, but according to Mulvania, they can be filtered through collodian membrane also.

6. Viruses can retain the power of infection for long periods, even out of the living cells. Tobacco mosaic virus retains its infectivity in extracted cell sap for 5 years (Dickson, 1925) and according to Valleau and Johns, it remains infectious in dried tobacco leaves for 31 years.

7. Vinson and Petre precipitated viruses with the solution of safranin solution, acetone and ethyl alcohol. The precipitate obtained by safranin solution, has no power of infection but it regains it when ethyl alcohol added to the precipitate.

8. When the cell-sap contaminated with virus is subjected to a high speed centrifuge, viruses get sedimented like proteins.

9. Viruses increase in number and size within the living protoplasm of cell.

10. They can be transmitted from infected plant to a healthy plant by mechanical and biological means.

11. Insects which transmit viruses are known as vectors. A particular virus can be transmitted only by a particular species of insect and on a definite species of host. This sort of selection of the vector and the host is known as **physiologic specialization.**

The general properties described above are more or less common and the differences are of degree only.

Transmission of viruses. Viruses are transferred from infected plants to healthy ones by a number of agencies described below, and this is known as **transmission** of viruses.

1. **By grafting.** A large number of plants are vegetatively propagated. Grafting is the most common method used in propagating fruit and ornamental plants. When an infected plant is used for grafting with a healthy plant, the virus is transmitted through the cell solutions flowing from the infected part into the healthy parts.

2. **By seeds.** Seed transmission of viruses is not common. Mosaic viruses of cucurbits and legumes are transmitted by seeds. But in the crops which are vegetatively propagated by the use of setts, tubers, rhizomes, bulbs, and corms, etc., viruses are transmitted by these organs.

3. **By contact.** Viruses can be transmitted merely by contact or slight rubbing of the infected and healthy plant organs. Such transmission is quite easy in a thickly populated field where one plant is always in close contact with the other and even the slightest movement of the wind can help in rubbing the organs of one with the other. Viruses usually gain entry through the injuries caused on the plant surfaces.

4. **By air and water.** Unlike fungi and bacteria viruses are not easily transmitted by air and water. The tobacco necrosis virus has been observed to be transmitted by both air and water.

5. **By soil.** Tobacco mosaic virus lies in the soil with the plant debris after the harvest of the crop, and infects the new crop, when sown in the same field.

6. **By tools and agricultural operations.** At the time of agricultural operations such as topping, pruning, weeding, rogueing or irrigation, the agricultural tools and the hands of the worker may get contaminated by the juice of an infected plant if it gets injured. These tools and the hands of the workers transfer the viruses to the healthy plants, the contact with which cannot be avoided ordinarily.

7. **By smokers.** Tobacco mosaic virus remains infective in dry tobacco leaves for as long as 31 years. This virus is very resistant to high temperatures. It can spread by the fingers of the smokers, by the smoke and the unburnt pieces of cigarettes, cigars and biris. Tobacco chewers sputum can also transmit virus if it is there in the tobacco leaves.

8. **By store house.** Viruses, such as tobacco mosaic virus is trasmitted to the new stock of tobacco, when it is stored in the same store house, where the infected leaves were stored.

9. **By insects.** In nature, viruses are ordinarily transmitted by aphids, jassids and white flies which are the sucking insects. Most of the crop diseases viruses are transmitted by insects, and for this simple reason some virologists believe that there would have been no virus diseases if there were no insects. One virus can be transmitted by several species of insects and *vice versa.*

Insect vectors taking part in transmission, first suck juice from the infected plant. Now if it visits a healthy plant, the virus may be transferred and the new plant becomes infected. But in some cases a vector fails to infect a healthy plant soon after it has fed upon a diseased plant but it can infect the plant after some time. This period taken by the virus in developing infectivity within the vector is known as **incubation period.** Incubation period is different for different viruses and varies from hours to days. In such cases the insect vectors act as alternate hosts, in which the viruses undergo some changes before infecting the real or say, the primary host.

All the insect vectors cannot transmit different viruses indiscriminately. Curly top virus of beet-root is transmitted only by a leaf hopper *Eutettix tenellus*, and cannot be transmitted by any other sucking insects found on beet-root leaves. Mosaic virus of beet-root is transmitted only by 'peach aphid'. Stunt virus of paddy is transmitted only by a leaf hopper *Nephotettix apicalis*. From the plants, simultaneously infected with a mixture of tobacco mosaic and cucumber mosaic viruses, only tobacco mosaic virus is transmitted by needle prick, whereas cucumber mosaic virus is transmitted by aphids alone.

The above-mentioned examples show that there exists some close relationship between the virus and its insect vector. Though this relationship is not properly understood in some cases, it is not very difficult to comment upon it. All the species of insect vectors differ to a lesser or greater extent in their habit, morphology, anatomy and physiology. An insect which feeds on the pollen grains will not suck the sap, and one feeding on the leaves may not feed on the stems, or other organs. Viruses are sometimes present only in the cells of deeper tissues and as such they can be sucked up only by the insects having long proboscis. Surface feeders cannot transmit such viruses. Similarly the vectors feeding on the deeper tissues do not feed on the surface tissues.

In case, where there is a definite incubation period, the virus passes through the entire alimentary canal and coelome of the vector, then enters into the salivery glands and can go out with the saliva through the proboscis of the insect when it punctures into the host tissue. In passing through this long way it has to come in contact with several intestinal secretions. Some viruses multiply and grow while passing through the body of the vectors and others pass out as such without undergoing any change. Such viruses are favoured by the physiological activities of the vectors and pass out of it, retaining their infectivity. Physiological activities of a particular vector cannot suit all the viruses and therefore this differential adaptation of viruses to the vector is also related with the transmission of a particular virus and the reason for this is also related to their physiological behaviour.

Viruses can be recognized easily by the symptoms they produce on the host. Sometimes the same virus can cause widely different symptoms on different host plants, and the symptoms may vary in the same plant according to its age, nutrition and other environmental factors. Symptoms may sometimes be produced by a mixture of two or more viruses on the same plant. A few more common and easily detectable symptoms are given here:

1. **Chlorosis.** Due to the presence of viruses the chlorophyll of the green organs disappears at places, leaving yellowish spots, this is known as chlorosis. The presence of yellow spot at places, in the green tissue appears like a 'mosaic' pattern and therefore, the diseases with such symptoms are known as 'mosaic' diseases.

2. **Yellows.** When the chlorophyll disappears completely from the host tissue, the organs turn yellow and the symptom is known as 'yellows'.

3. **Vein clearing and vein banding.** The disappearance of chlorophyll along the veins of the leaves is known as **vein clearing** and when chlorophyll surrounding the veins disappears the symptom is known as **vein banding.**

4. **Necrosis.** The brownish spots due to the death and ultimate drying of the tissue are known as necrotic spots and the phenomenon is known as necrosis. Necrotic spots are also known as lesions.

Ring spots, bunchy top, galls, hypertrophy, atrophy, rolling, curling, crinkling of leaves, stunting and dwarfing of plants are various other symptoms usually produced on the hosts by viruses.

Control. The easy methods of preventing disease spread and losses due to them are as follows:

1. **Eradication.** The destruction of infected plants and susceptible weeds, lessen the possibilities of the spread of disease.

2. **Elimination of insects.** The use of insecticidal dusts and sprays reduces the chances of insect transmission.

3. **Selection of seeds.** It should be made from such fields which were free from infection. Seeds of cucurbits and legumes, setts of sugarcane, tubers, rhizomes and bulbs, etc., should be carefully selected for seed purposes.

4. **Tuber indexing.** It is done at the time of digging potatoes. Tubers taken from healthy plants are marked with ink and sown in small insect proof plots. The suspected diseased plants are removed carefully leaving only the healthy plants to grow. Thus the healthy seed is multiplied.

5. **Resistant varieties.** They are evolved at various research centres, and offer the best method for growing healthy crops.

TOBACCO MOSAIC VIRUS (TMV)

Symptoms. On the younger leaves of tobacco plants the veins may show a clearing and later this may be followed by mottling. As the leaves enlarge, abnormally dark green spots appear which develop into irregular crumpled blister-areas while the remainder of the tissue becomes more and more chlorotic. Plants are stunted in various degrees. The diseases characterized by mottling or variegation of leaves are referred to as 'mosaics'.

The virus. Typical tobacco mosaic virus.
Tobacco virus I.
Nicotiana virus 1 K.M. Smith.
Marmor tabaci Holmes.

The rod forms as illustrated by tobacco mosaic contain a central core of RNA, this elongated rod is twisted into a uniform spiral to form, in gross form, as a hollow rod. The RNA core remains surrounded by a protein sheath. Under appropriate conditions the viral particles induce the formation of a crystalline structure known as an inclusion body. The crystalline inclusion body of tobacco mosaic is known to contain true virus particles. (see fig. 1.1).

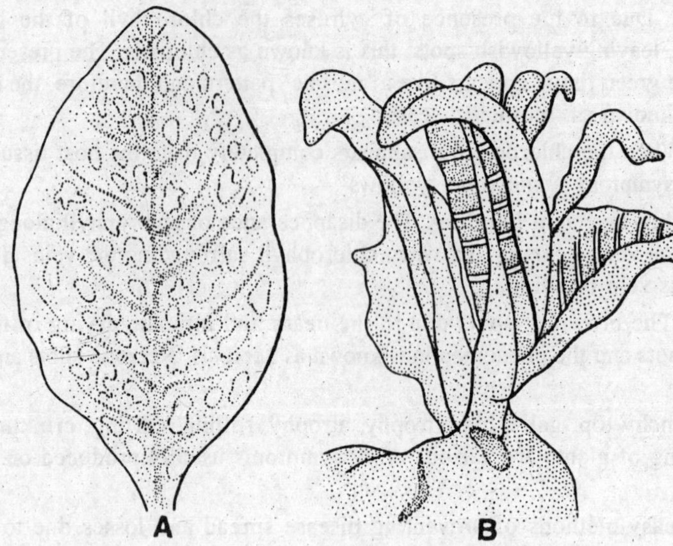

Fig. 1.3. Virus. A, leaf showing mosaic pattern (Tobacco Mosaic Virus T.M.V.), B, curling of leaves of tobacco.

The TMV remains active in extracted host plant juice upto 25 years. It is infectious in dilutions upto 1: 1,000,000 and withstands heating upto 90°C for 10 minutes.

Transmission. Dissemination of TMV from plant to plant takes place through mechanical transmission. This virus is so very infectious that this may be carried from one plant to another by almost any sort of contact. Workers in the tobacco field may carry tobacco virus from one plant to another on their hands or clothing. The TMV is not destroyed in the manufacture of smoking tobacco. Thus, should cigarettes or biris made from virus infected leaves be smoked by a field worker, he may easily carry the virus on his hands to healthy plants.

Control measures. (1) One should avoid infested soil and the use of tobacco refuge on land on which tobacco is to be grown.

(2) Seed beds should be steam sterilized and well removed from tobacco ware-houses.

(3) One should avoid contamination of hands with the virus from tobacco products. Though washing with soap is an effective practical means of cleaning hands between operations.

Virus Symptoms in Plants

Plant viruses on being introduced into the appropriate host produce characteristic symptoms. The presence of viruses can easily be recognized by the symptom picture they produce on the host, but symptoms as a basis of identifying specific viruses are not reliable. The same virus can cause widely different symptoms in different host plants, and symptoms may vary in the same plant with variation in age, nutrition and environmental conditions of growth. A host plant may also be attacked by a large number of viruses and similar symptoms can be caused by different viruses.

The most striking effects of virus infection are general reduction in growth, vigour and cropping power of the plant. Diseased growth is usually accompanied by certain distinct changes in the leaves and other organs of the plant. The most common symptom is change in colour of the leaves-chlorosis, and it is so pronounced that at one time this kind of virus disease was known as infectious yellow. In some type of chlorosis, light green or yellow patches of varying sizes and shape are irregularly distributed among normal green tissues giving a characteristic mosaic pattern.

Vein clearing and vein banding. These commonly appear in systemically infected leaves before mottle or mosaics, a clearing or chlorosis of the tissue in or immediately adjacent to the veins. This pattern is referred to as vein clearing. Vein banding consists of a broader band of chlorosis or necrosis in the interveinal parenchyma.

Ring spots. On inoculated and systemically infected leaves the symptoms appear in localized spots. These spots consist of various types of chlorosis and necrosis. They may be circular chlorotic areas in which case they are called as chlorotic ring spots. In the cases the necrosis may appear in rings alternating with normal green areas. Such spots are called as necrotic ring spots.

Necrosis. In diseases of both yellows and mosaic groups, necrosis appears in various forms. When necrosis involves parenchyma and veins of the leaves and stem it is called as streak. Top necrosis refers to relatively rapid killing of bud or branch or entire top of the palnts. Some viruses do not infect a given host systemically but affect the tissue at points of inoculation by causing a localized breakdown. This type of reaction is known as local necrosis.

Stunting and premature death. Virus diseases with few exceptions are hypoplastic in their general effect. This is shown in many cases by shorter internodes, smaller leaves, and fruits, and reduction of size of various other parts.

Excessive growth. Some viruses incite formation of masses of hypertrophied tissue, known as enation of surfaces of leaf and stem. In other cases virus infection results in stimulation of dormant buds, hyperplastic tissue and unusual differentiation. Spindle tuber of potato is example of such distorted growth. An extreme example of a typical growth is found in the extensive stimulation of numbers of buds in the potato witches broom disease.

Masked symptoms and symptomless carriers. There are many instances in which the virus is present in the infected plant but no distinguishable symptoms appear. This may be due to a particular set of environmental conditions under which the symptoms appear. Under such conditions the plant is referred to as having masked symptoms. In some diseases the virus infects a green host and increases therein but no visible signs appear over the entire range of environmental conditions to which the host is usually exposed and there is no obvious sign of the diseases as in latent mosaic of potato in many varieties of potato. Such an infected host plant is known as symptomless carrier.

Synergistic effect. There are numerous cases in which symptoms shown by the infected plant are due to the synergistic or combined action of two or more viruses. For example rugose mosaic of potato is due to infection by two viruses, namely, Potato virus X and Potato virus Y.

CLASSIFICATION OF VIRUSES

There are two general ways in which viruses may be classified. One is the classical monothetic hierarchial system applied by Linnaeus to plants and animals. This is a logical system in which divisions are made as to the relative importance of different properties which are then used to place a taxon in a particular phylum, order, family, genus, etc. An alternative system was proposed by Adanson (1763). He suggested that all known information should be used and that all characters should be considered equally important. There are at present proposals and proponents for both these approaches to virus classification (Lwoff, 1967, Gibbs and Harrison, 1968).

Lwoff *et al,* (1962) proposed a system in which the kind of nucleic acid in the particle (DNA or RNA) the architecture and symmetry of the virus particles, the presence or absence of an envelope, the number of morphological subunits (inicosahedral viruses), or the diameter of rod-shaped viruses were considered the most competent characters (in descending order) in making major subdivisions. Their system was supported in the proposals and recommendations of the Provisional Committee for Nomenclature of various groupings down to family level for the first three groupings.

Phylum	*subphylum*	*Class*
Vira	Deoxyvira	Deoxyhelica
		Deoxycubica
		Deoxybinala
	Ribovira	Ribohelica
		Ribocubica

Adansonial classification. Adanson (1763), considered that taxa were best derived by considering all available characters and giving equal weight to each. The method is labourious and has not been much used until recently. The availability of computer has renewed interest in this kind of classification.

Nomenclature. One of the main reasons why a Latin binomial system of nomenclature for viruses is inappropriate at the present time is that such names attempt to provide both a label and some information about the virus being labelled. In the present stage of knowledge or lack of it for most viruses, such names would be found to require changing frequently in the future. In an attempt to overcome this problem, Gibbs and his colleagues (Gibbs *et al,* 1966, Gibbs and Harrison, 1968) have suggested a two part system in which the first part of virus name is an unchanging label (the current vernacular name) and the other is a codified information store which can be readily adopted and changed as facts accumulate about the viruses. Each virus consists of four pairs of symbols as follows:

1st pair	Type of nucleic acid/Strand of nucleic acid.
2nd pair	Molecular weight of nucleic acid/Percent of nucleic acid in infective particle.
3rd pair	Outline of particle/Outline of nucleo capsid.
4th pair	Kinds of host infected/Kinds of vector.

For example, tobacco mosaic virus (R/1 : 2/5: E/E : S*)

*This property of the virus is not known.

BACTERIOPHAGES

The group of viruses which attacks bacteria was first described in 1915 by the British scientist Twort and more fully studied about in 1917 by the French investigator De Herelle. De Herelle noticed that some invisible agent was destroying his cultures of dysentery bacilli. Like other viruses, bacteriophages are filterable and will grow only in the presence of living cells in bacteria, which they cause to swell and dissolve. The bacteriophages (viruses) are found in nature wherever bacteria occur. They are especially found in abundance in the intestines of man and other animals.

There are several varieties of bacteriophages. Usually each kind of bacteriophage will attack only one species or even only one strain of bacteria.

Electron micrographs revealed that the bacteriophages may be spherical or comma shaped or may have a tail. They are about 5 mμ in diameter.

The initial step in viral reproduction is the attachment of the virus to the bacterial cell by means of its protein tail. An enzyme is found in the tail digests part of the bacterial cell wall and the deoxyribonucleic acid (DNA) is found in the core of virus passes into the bacterial cell. Hershey and Chase have shown that only the DNA of the virus is injected into the bacterium while the protein coat of the head and tail does not enter.

The inference is that the DNA contains all the genetic information for the synthesis of the complete virus particle. The DNA directs the biosynthetic system of the host cell to produce viral DNA and viral protein.

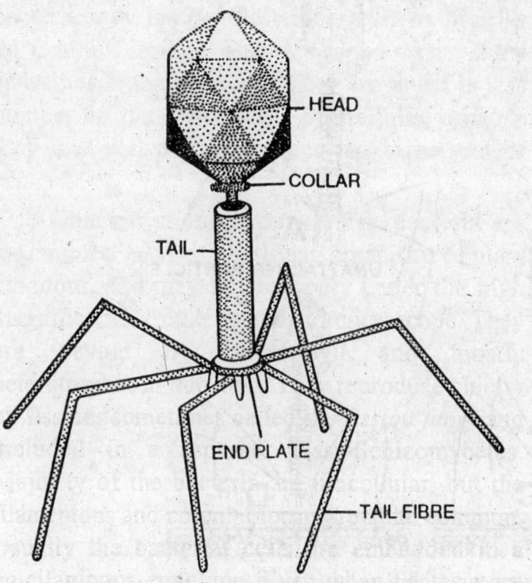

Fig. 1.4. Bacteriophage. Structure.

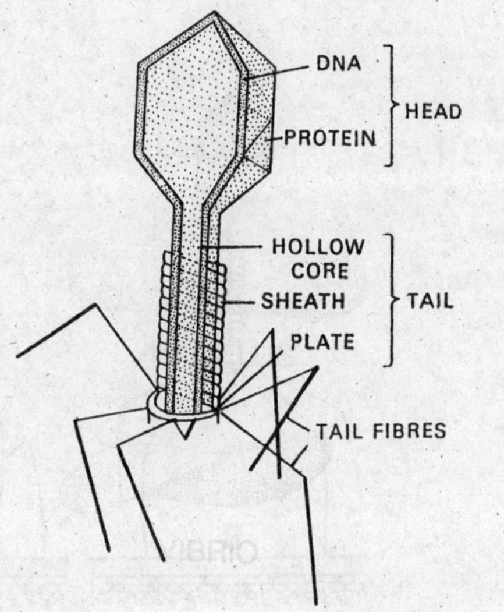

Fig. 1.5. Bacteriophage. Details of a phage particle.

In the first 10-15 minutes increasing number of viral particles accumulate, and some 30 minutes after infection the bacterial cell bursts and several hundred newly formed viral particles are released, ready to attack new bacterial cells.

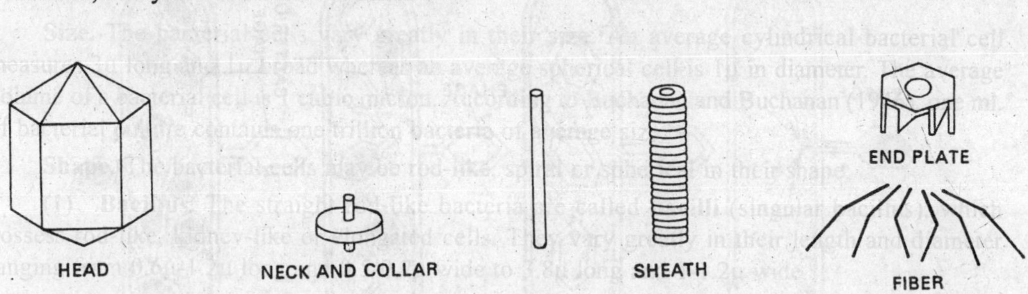

Fig. 1.6. Bacteriophage. Different components of a phage particle.

Multiplication of viruses. Viruses are regarded as "regressed parasites", that must take advantage of the biochemical machinery of the host cell within which it reproduces. In viruses most of the work has been done on bacterial viruses (bacteriophages). The bacteriophage attaches to the bacterial surface by its tail and then releases its nucleic acid into the bacterial cytoplasm. An independent synthesis of the viral components occurs within the cell followed by an assembly of those components into complete virus particles. After a well defined period the bacterial cell bursts and releases the newly formed virus particles. The middle part of this sequence appears to be general for all types of virus infections.

An interesting modification of this life cycle occurs when a certain type of bacteriophage, termed temperate, infects certain types of bacteria. In this instance genetic material of the temperate viruses attaches itself to a bacterial chromosome in a manner which is not understood. The bacterium shows no immediate sign of infection and continues to grow and divide nothing had happened. Such a bacterium is termed lysogenic. Occasionally this bacterium will spontaneously start making viral components and these components will be assembled into mature virus particles.

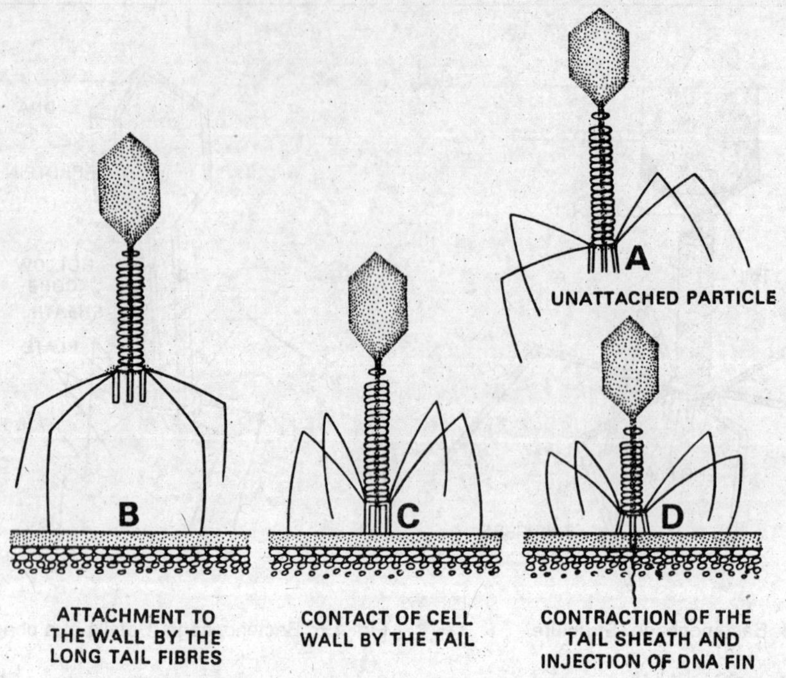

Fig. 1.7. Bacteriophage. Diagram showing adsorption and penetration of virus (phage) particle.

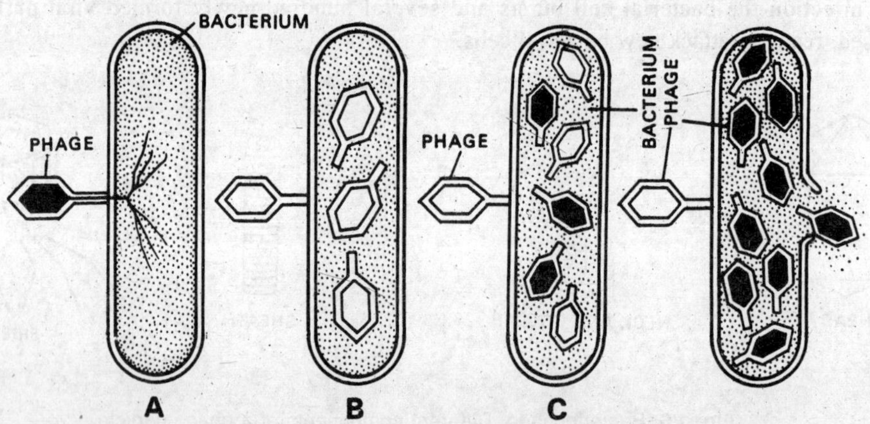

Fig. 1.8. Bacteriophage. Multiplication. A, entry of nucleic acid into the bacterial cell; B-C synthesis of viral components in the cell; D, lysis of bacterial cell and liberation of phage particle.

Biological importance of bactertiophages. The fact that bacteriophages can destroy bacteria suggest of course, that they might be used to combat bacterial diseases. Many bacteriophage preparations have been given, without much effect, to patients suffering from dysentery and staphylococcus infections. But the bacteriophages are ineffective in the presence of blood, pus or fecal matter and this fact has led to the abandonment of their use for therapeutic purposes.

Viral Genes

Viruses are submicroscopic entities which multiply only inside living cells and possess the ability to cause diseases to all forms of living organisms, from single-celled plants or animals to large trees and mammals.

The viruses are completely inactive outside of the host's living cells. Once inside the host's cells, however, the virus DNA or RNA (depending on the kind of virus) assumes control of the cell's biochemical activities. As soon as the virus enters the host cell, it changes the activities of the infected cell and makes it to the needs of virus. In this changed cell environment, the virus replicates. Hundreds of virus particles may be produced, all exactly like the one that infected the host cell.

Tobacco mosaic virus (TMV) contains RNA as the genetic material. Other viruses may contain DNA as the genetic material. There are also viruses that contain single staranded DNA as the genetic material. Viruses are diverse in size and shape but basically contain the genetic material wrapped around by a protein coat, generally known as the **capsid.**

Viral genomes are small and possess the information to code for a few proteins only. Thus, viruses use the host machinery such as enzymes, ribosomes and other components to replicate, to form the coat and to form more virus particles. The viruses infecting bacteria are known as **bacteriophages.** The expression of viral genes is regulated in such a way that the lysis of the host cells is caused. Thereafter many virus particles (**lytic**) are liberated which can infect new host cells to repeat the **lytic-cycle.** In some viruses after infection of the host cells viral particles get integrated with the host chromosome (DNA) and replicate along with it (**lysogeny**).

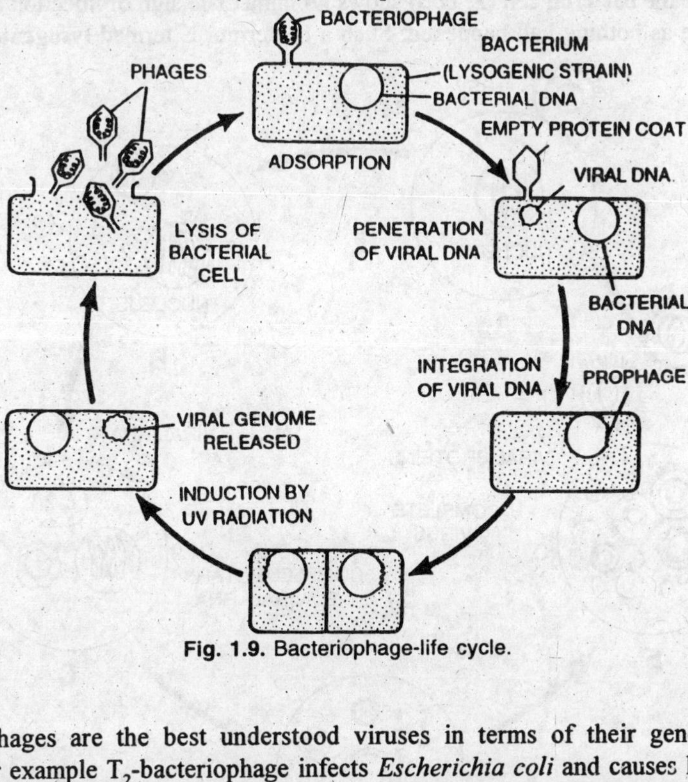

Fig. 1.9. Bacteriophage-life cycle.

Lytic-cycle

Bacteriophages are the best understood viruses in terms of their gene structure and expression. For example T_2-bacteriophage infects *Escherichia coli* and causes lytic cycle. This phage has a double stranded circular DNA enclosed in a hexagonal proteinaceous head, cylindrical hollow tail and six tentacles for attachment to the host. The phage gets attached to bacterial wall by its tentacles. The lysozyme found at the tip of the tail dissolves the host cell wall to form a pore. The tail now contracts and injects the viral DNA into the host cell. Thereafter the viral DNA synthesises an enzyme known as nuclease which degrades the host DNA. Only after few minutes the host DNA is destroyed by nuclease. On the other hand, the viral DNA is resistant to nuclease because its cytosine residues are methylated. Now the viral genes make use of host ribosomes for multiplication. First of all the replication of viral DNA takes places. Thereafter, the proteins of head tail and tentacles are synthesized. An independent synthesis of viral components occurs within the cell followed by an assembly of those components into complete virus particles. After a well defined period the bacterial cell bursts and releases the newly formed virus particles. The liberated virus particles again infect the bacterial cells and the lytic cycle is repeated.

Lysogenic Cycle

An interesting modification of lytic cycle occurs when a certain type of bacteriophage, termed temperate, infects certain types of bacteria. In this instance genetic material of the temperate viruses attaches itself to a bacterial chromosome in a different manner. Here the viral DNA gets attached to bacterial DNA, becomes inactive and is known as **prophage** or **provirus.** The inactivity of phage DNA is because of the synthesis of repressor protein by phage DNA that causes repression of all the phage-genes. Thereafter as the bacterial cell divides, phage DNA also replicates with it and is inherited by all the bacterial cells of following generations. In the other option, DNA or viral gene integrates with the bacterial (*E. coli*) chromosomal DNA and gives rise to a prophage. Here, the phage replicates along with the host. At times, the prophage may be activated due to altered environmental conditions. This inhibits the repressor protein synthesis, and results in the expression of lytic genes. Now the prophage becomes a lytic phage and resumes the lytic cycle. Here the bacterial cell (*E. coli*) shows no immediate sign of infection and continues to grow and divide as nothing had happened. Such a bacterium is termed **lysogenic**, and the cycle

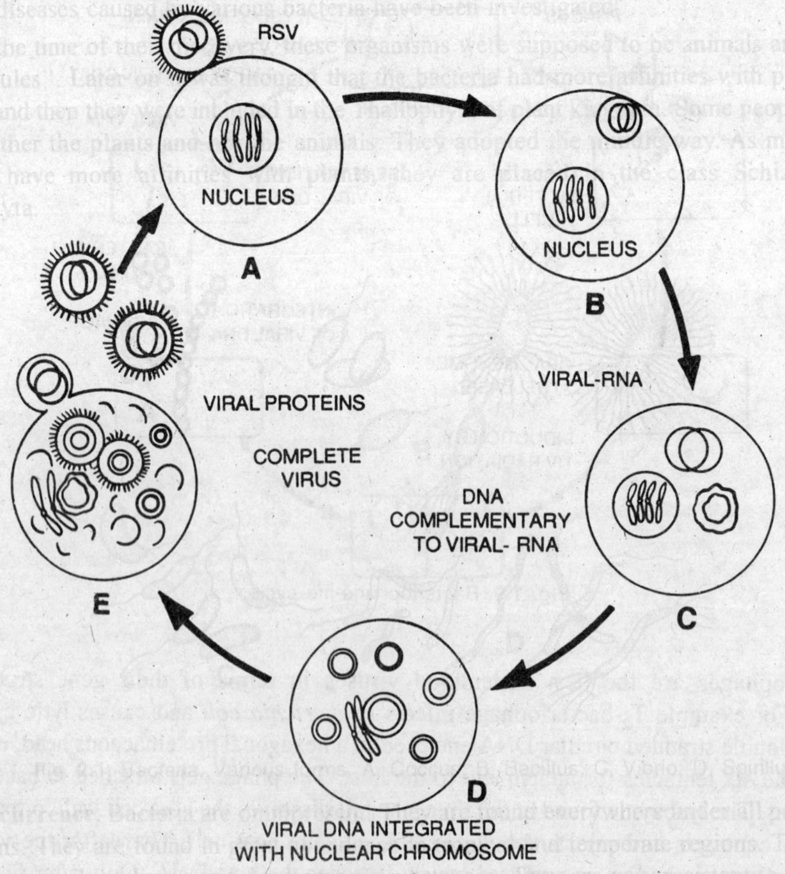

Fig. 1.10. Multiplication of Rous Sarcoma Virus (RSV). A-E, showing different stages.

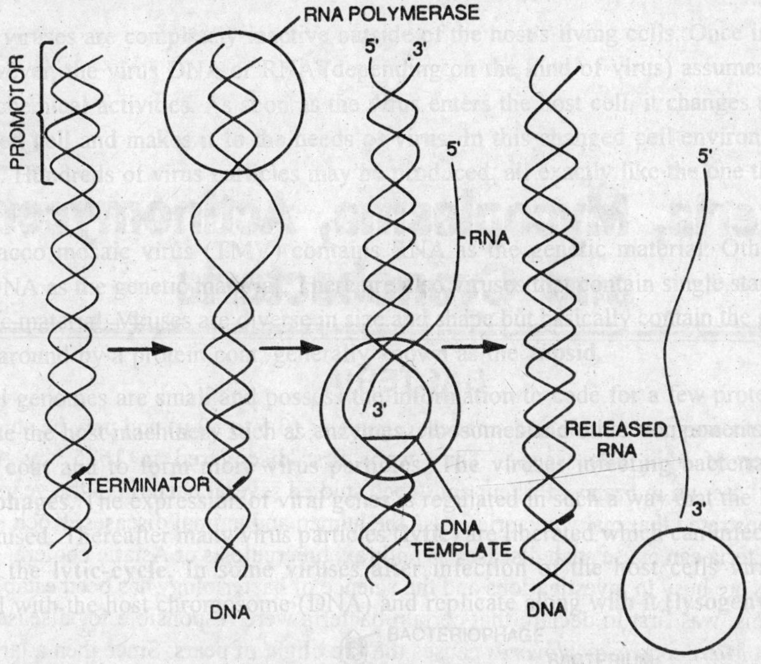

Fig. 1.11. Process of transcription. RNA polymerase binds to the promoter region of the DNA and starts the transcription process. The DNA unwinds and one of the two strands serves as a template for RNA synthesis, and a complementary copy of RNA is formed on the template strand.

involved, the **lysogenic cycle.** Occasionally this bacterium will spontaneously start making viral components and these components will be assembled into mature virus particles.

Reverse Transcription

This phenomenon was studied by Temin and Baltimore (1972), and thus known as Teminism. While studying the activity of double stranded RNA containing Rous sarcoma virus (RSV) in the cancerous tissue, they found that viral RNA synthesized DNA complementary to its strands. As soon as the RNA of RSV comes in contact of the cytoplasm of host cell, it synthesizes the enzymes **reverse transcriptase** that catalyses the conversion of RNA to DNA copy and exhibits a reverse phenomenon of the normal process. Such phenomenon is known as reverse transcription or Teminism. Once the DNA complementary to viral RNA is synthesized it follows the normal *rule and synthesizes viral* RNA and viral proteins as depicted in the figure. The central dogma says that the flow of information is unidirectional, *i.e.,* DNA → RNA → protein. However, with the discovery of enzyme reverse transcriptase, it becomes clear that RNA can also go back to DNA as represented here :

$$\text{DNA} \rightleftharpoons \text{RNA} \rightarrow \text{Protein.}$$

This is a characteristic of **retroviruses**, and this course is not followed by other RNA viruses. The virus HIV-III (human immuno deficiency virus III) which is responsible for causing AIDS disease is also a retrovirus. About twenty viral genes have been identified which are responsible for triggering cancer in cells. On the other hand, slightly modified versions of these genes are also present in normal cells which are known as **protooncognes** or **cellular oncogenes.** Most of the cellular oncogenes code for growth factors or for protein required for growth factor action such as receptors. The origin of cancer is a complex phenomenon, where viral oncogenes or activated cellular oncogene products are responsible for uncontrolled growth of cells. (see figs. 1.10 and 1.11).

2

CHAPTER

Bacteria, Mycoplasma, Actinomycetes and Cyanobacteria

BACTERIA

The bacteria are most simple and smallest organisms (0.5 to 50 μ) and may be studied under the high power of light microscope. They were first discovered in 1675, by Anton von Leeuwenhoek. Later on in France, Pasteur and Koch studied several bacteria in the years 1870-76; they also demonstrated that many bacteria cause the human and animal diseases. Koch (1876) had proved that bacteria can cause such diseases as anthrax, tuberculosis or Asiatic cholera. Since then several workers are busy in investigations and the science of bacteriology has been established. T.J. Burrill of Illinois, was first to declare that certain bacteria were responsible for disease in plants. He showed that *Micrococcus amylovorus* causes the fire blight of pears. Since then a large number of plant diseases caused by various bacteria have been investigated.

At the time of their discovery, these organisms were supposed to be animals and referred as "animalcules". Later on it was thought that the bacteria had more affinities with plants than the animals and then they were included in the Thallophyta of plant kingdom. Some people considered them neither the plants and nor the animals. They adopted the middle way. As majority of the bacteria have more affinities with plants, they are placed in the class Schizomycetes of Thallophyta.

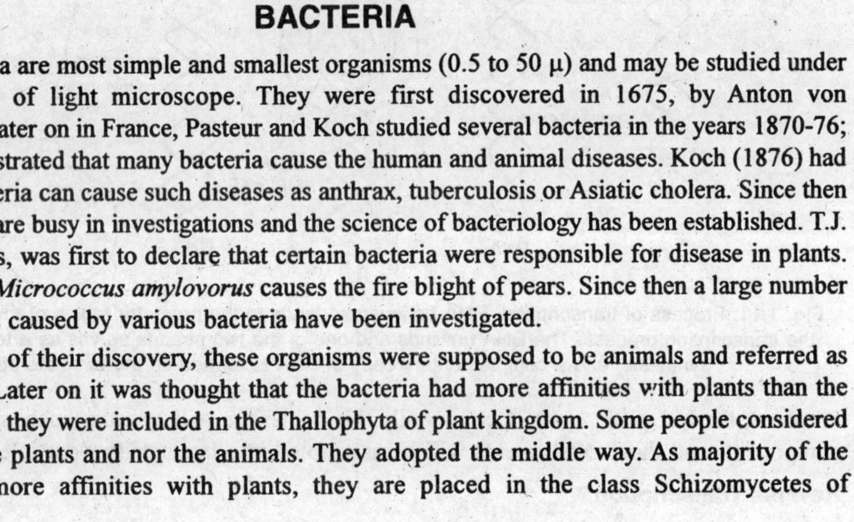

Fig. 2.1. Bacteria. Various forms. A, Coccus; B, Bacillus; C, Vibrio; D, Spirillum.

Occurrence. Bacteria are omnipresent. They are found everywhere under all possible natural conditions. They are found in great abundance in tropical and temperate regions. They are found in water, air, soil, on living and dead organic substances. They are very resistant to heat, cold and certain chemicals. They are also found in the intestines of human beings and animals. Certain bacteria may live in the temperature as high as 78°C in hot springs. They are found on the snow as well as in the deep soil. They are found in less number on the snow peaks of the hills, because they lack soil particles which make the seat of bacteria.

General characteristics. The bacteria are the simplest and smallest living organisms of plant kingdom, and may be seen only under the high magnification of the compound microscope. They are devoid of chlorophyll and mostly heterotrophic in nutrition. They reproduce chiefly by fission, sometimes called the *fission fungi* and included in a separate class-Schizomycetes. Majority of the bacteria are unicellular, but the filamentous and colonial forms are quite common. Usually the bacterial cells are embedded in a mucilaginous envelope. Even when bacteria are held together in groups or chains, each individual cell carries on its own metabolic process quite independently; there is no division of labour.

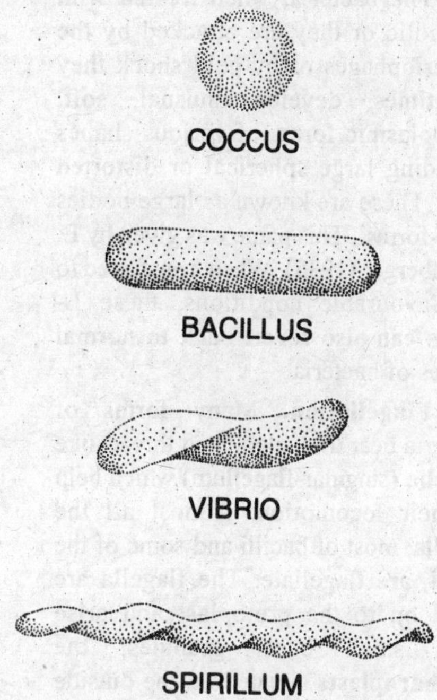

Fig. 2.2. Bacteria. Different types.

MORPHOLOGY OF BACTERIA

Size. The bacterial cells vary greatly in their size. An average cylindrical bacterial cell measures 3μ long and 1μ broad whereas an average spherical cell is 1μ in diameter. The average volume of a bacterial cell is 1 cubic micron. According to Buchanan and Buchanan (1946), one ml. of bacterial culture contains one trillion bacteria of average size.

Shape. The bacterial cells may be rod-like, spiral or spherical in their shape.

(1) Bacillus. The straight rod-like bacteria are called **bacilli** (singular bacillus), which possess rod-like, kidney-like or elongated cells. They vary greatly in their length and diameter ranging from 0.6μ-1.2μ long and 0.5-0.7μ wide to 3.8μ long and 1-1.2μ wide.

(2) Spirillum. The spiral or curved forms are called **spirilla** (singular spirillum), in which the cells are spirally coiled or coma-shaped or variously curved. They vary in size from 1.5μ to 4μ long and 0.2μ-0.4μ wide in *Vibrio* to upto 50μ long in *Spirillum.*

(3) Coccus. The rounded or spherical forms are called **cocci** (singular coccus) in which the cells are more or less spherical. They are smallest forms among bacteria. After division the cells may either separate from each other or may remain joined together to form groups of two cells in *Diplococcus*, a tetrad of four cells in *Micrococcus tetragenus* and a chain of cells in *Streptococcus.* The cocci range in diameter from 0.5μ-1.25μ.

Under favourable conditions of growth and development, the shape of one species remains constant, whereas under unfavourable conditions, some bacteria, such as the nitrogen fixing species change into at least three distinct forms during a complete life cycle. Under certain conditions some bacteria become very irregular in shape and sometimes abnormally enlarged. These modified cells are kown as "**involution forms**". Many bacteria in young cultures, show a mixture of several integrating forms, besides the original structure of the parent type. This phenomenon is known as **pleomorphism.**

The bacteria, when treated with antibiotic or they are attacked by the bacteriophages or by cold shock they sometimes, develop unusual, soft, protoplasmic forms of various shapes including large spherical or distorted cells. These are known as large bodies or 'L-forms'. This name was given by E. Klienberger (1935). When transferred to the favourable conditions, these L-forms can also revert back to normal shapes of bacteria.

Flagellation. Many forms of bacteria bear thin, elongated thread-like **flagella** (singular-flagellum) which help in their locomotion. Almost all the spirilla, most of bacilli and some of the cocci, are flagellate. The flagella are fixed on to the protoplast and arise from small basal granules, the **blepharoplasts** situated on the outside of the cytoplasm. The flagella may consist of either a simple filament or 2-3 fibrils coiled helically. In some forms (*e.g., Spirillum*) they twist to form a thick bundle. Bacteria having no flagella are termed **atrichous,** those with only one polar flagellum. **monotrichous** *(e.g., Vibrio)*, such a group of flagella at one end **lophotrichous** (*e.g., Spirillum*), with a group of flugella at both ends, **amphitrichous** and with flagella uniformly distributed all over the body are termed **peritrichous** (*e.g., Salmonella*).

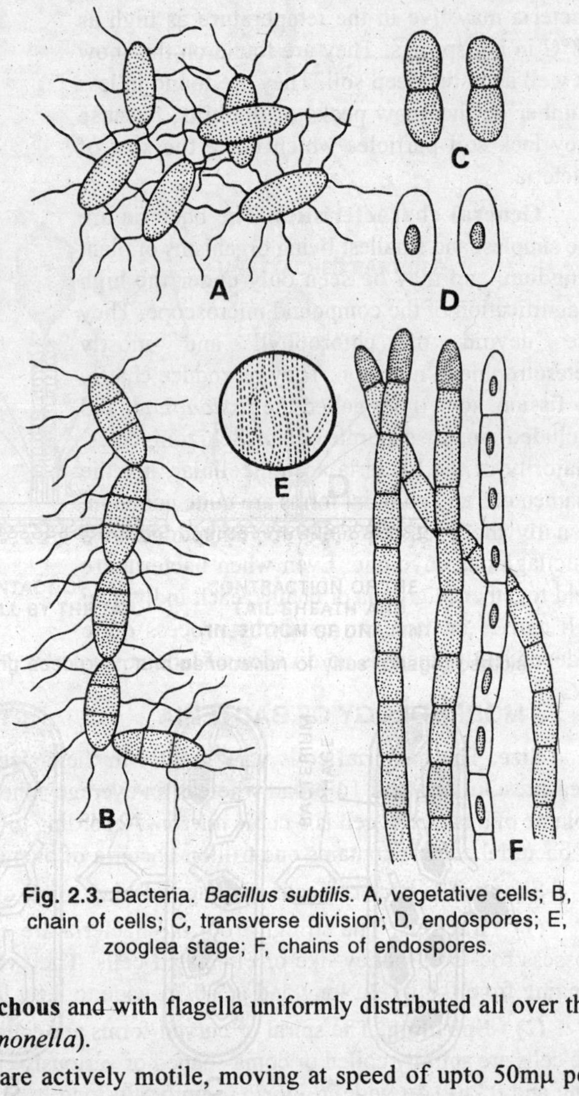

Fig. 2.3. Bacteria. *Bacillus subtilis.* A, vegetative cells; B, chain of cells; C, transverse division; D, endospores; E, zooglea stage; F, chains of endospores.

Flagella and pili. Many bacteria are actively motile, moving at speed of upto 50mμ per second. Such bacteria are normally found to possess flagella.

Bacterial flagella are much smaller than eukaryotic flagella, being about 120-180A° in diameter, they do not show the 9 plus 2 pattern of fibrils, are not enclosed in a membrane and show no ATPase activity.

Bacterial flagella consist essentially of a pure protein called "flagellin". The flagella of motile bacteria are distributed over the surface of the cell in characteristic fashion as described in the preceding paragraph. The flagellum pierces the cell wall and arises from a specialized area of the cytoplasm, sometimes called a blepharoplast by analogy with of higher organisms flagella. Depending upon the arrangement of flagella, they may be atrichous (non-flagellate), monotrichous (single polar flagellum), lophotrichous (a tuft of flagella), amphitrichous (flagella at both the poles) and finally peritrichous (*i.e.,* flagella all round).

The final structures to be considered are the pili or fimbriae. These are straight hair like structures concerned with attachment of the bacterial cell to solid surface and in some cases are known to act as conjugation canals through which DNA passes from one cell to another. Pili are composed of pure protein called 'pilin' and these subunits are helically arranged.

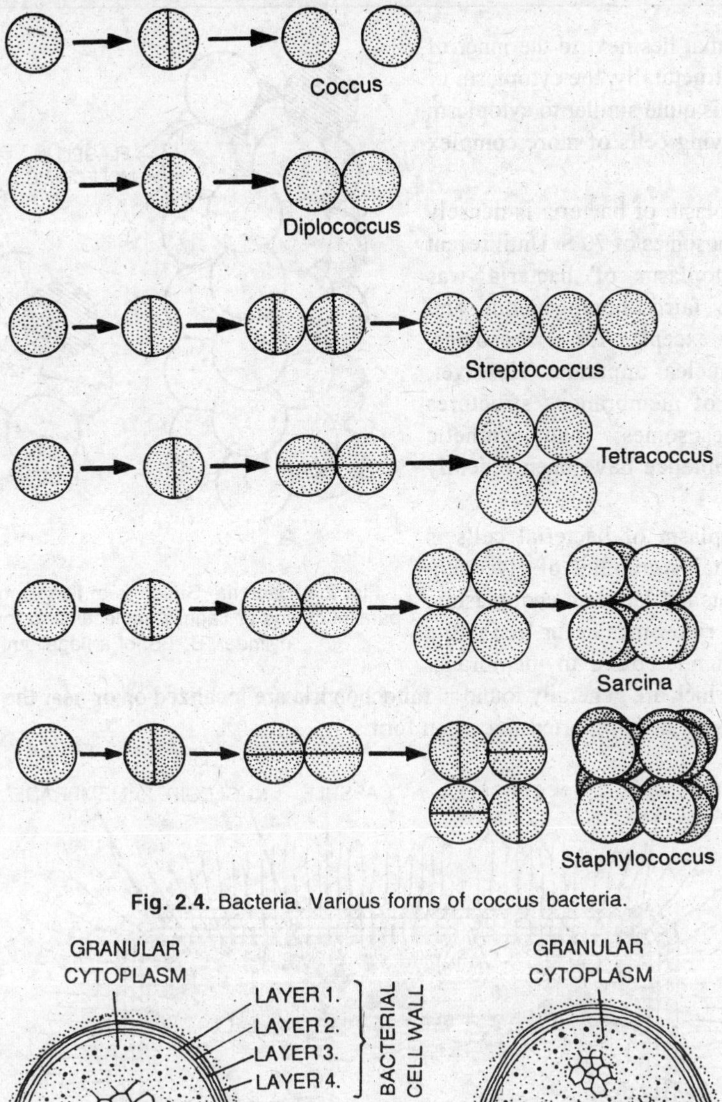

Fig. 2.4. Bacteria. Various forms of coccus bacteria.

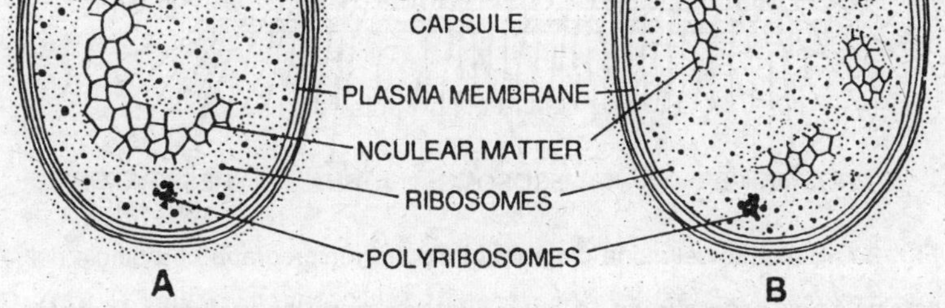

Fig. 2.5. Bacteria. Ultrastructure of cocci cells (A–B).

Cell structure. The bacterial cell contains vacuoles, ribosomes and granules of stored food. Water is an important constituent of bacterial cells. As much as 90 per cent of the cell is water. The movement of dissolved materials in and out of the cell is regulated by a cell membrane, formed by the cytoplasm that lies next to the inner of the cell wall. Structurally, the cytoplasm of a bacterial cell is quite similar to cytoplasm found in the living cells of more complex organisms.

The cytoplasm of bacteria is densely packed with ribosomes of 70 S. Until recent years the cytoplasm of bacteria was thought to be fairly homogeneous and structureless except for ribosomes, granules and nuclear apparatus. However, various types of membranous structures such as mesosomes, photosynthetic vesicles and lamellae have been recently demonstrated.

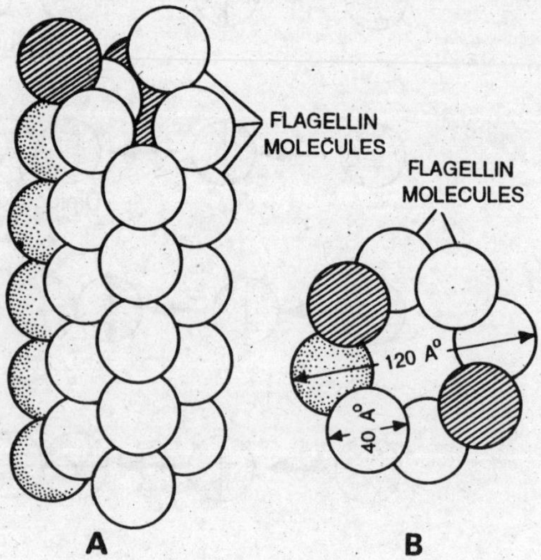

Fig. 2.6. Bacteria. Structure of flagellum. A, eight parallel chains of flagellin molecules to form a hollow cylinder; B, T.S. of a flagellum.

The cytoplasm of bacterial cells is dense and contains granules of glycogen, proteins and fats but lacks an endoplasmic reticulum. The ribosomes occur free in the cytoplasm and not bound to membrane. The enzymes which are generally found in mitochondria are localized on or near the cell wall. The mitochondria have been reported in certain forms.

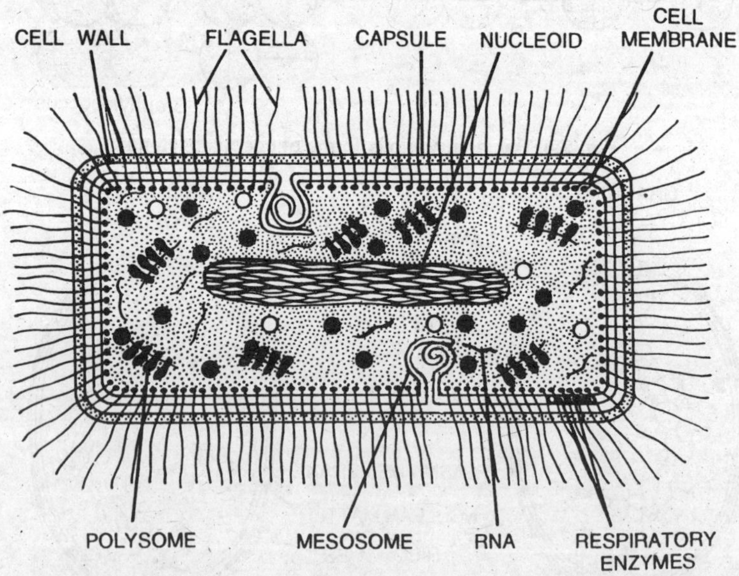

Fig. 2.7. Bacterial cell. Line diagram of electron micrograph of a single cell.

A few bacteria have been shown to have another granular inclusion, **volutin,** an inorganic metaphosphate related to ribonucleic acid, in the cytoplasm and some of the larger sulphur oxidising bacteria contain sulphur.

The cytoplasm contains many types of inclusions:

(*a*) **Vacuoles.** The vacuoles may or may not present. It depends upon the physiological state of the cell. The vacuoles, if present, remain surrounded by a membrane and may contain cell sap, volutin or sometimes lipid and sulphur.

(*b*) **Liquid globules.** The droplets of liquid occur as refractile globules. Sometimes it constitutes of 25% of the total weight.

(*c*) **Metachromatic granules.** They are also known as 'volutin granules'. They occur widespread in the cytoplasm. They occupy about 40% volume of the cell. They are composed of polymetaphosphate and increase in number towards late in their growth.

(*d*) **Glycogen granules.** The glycogen and other polysaccharide granules occur in some forms towards the polar region of the cell.

(*e*) **Mitochondria.** The mitochondria have been recorded from *Mycobacterium, Streptomyces* and others in the form of lamellated structures. They have been found to be centres of oxidation and reduction.

(*f*) **Chromatophores.** The chromatophores have been reported from the cells of autotrophic bacteria. These photosynthetic pigments are certainly different from those of higher plants.

Mesosomes. The mesosomes arise as invagination of plasmalemma and may become quite a complex whorl of convoluted membranes. The function of mesosomes is a matter of conjecture. It is believed that the mesosomes are active in cell wall synthesis and in the secretion of extracellular substance. It has been shown that transforming DNA taken up by whole cells of *Bacillus subtilis* apparently enters the cell *via* the mesosomes. There is considerable evidence that the bacterial nuclear body is attached to a mesosome.

Nucleus of bacteria (electron microscopic structure). Biologists generally agree that bacteria do not have nuclei of the type found in cells of higher plants and animals. In the cells of these more complex organisms, the nucleus contains a nucleolus or two–is bounded by a visible membrane, and divides by mitosis. Yet, it has been demonstrated repeatedly that there is some kind of mechanism in bacteria for the transmission of hereditary characteristics.

Most of the bacteriologists now believe that the nucleus in the bacteria is present in the cell as a discrete body. Feulgen tests applied in the case of bacteria have revealed the presence of purple stained bodies in a large number of cases. In bacteria nuclear apparatus is a more rudimentary structure. This incipient form of nucleus can be demonstrated by Feulgen stain and electron microscopy. It does not have nuclear envelope and it divides amitotically. In electron microscope the prokaryotic nuclear region appears as an electron translucent area and can be shown to contain very fine fibrils. These fibrils are molecular strands of DNA. The DNA is certainly restricted to this area of the cell. Using also radiographic technique, Cairns (1963) has demonstrated the circular DNA of *Escherichia coli.*

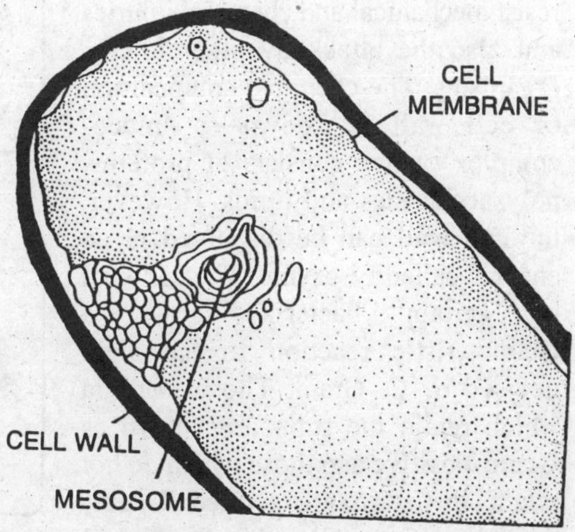

Fig. 2.8. Bacteria. Structure and position of mesosome in bacterial cell.

It is known that DNA comprises the genetic material of living cells. Some forms of bacteria have their DNA concentrated in one or two deeply staining bodies. Shortly before a bacterium divides, these DNA bodies divide and are equally distributed to the daugher cells. In this way the DNA bodies resemble chromosomes replicating in a dividing cell. More critical observations of bacterial 'chromosomes' have been made by studying ultrathin sections of bacteria under the electron microscope, in which the strands of DNA can be seen packed into a bacterial chromosme. It has been determined for some species of bacteria that the DNA molecule is about 3 millimicrons (mμ) wide and as much 1,200 (μ), or 1.2 mm long. A bacterial DNA molecule is thus more than 500 times as long as the bacterial cell that contains it. To make these measurements, bacterial chromosomes have been removed from the cell. In this condition the replicating strand of DNA has been observed to be in the form of a circle. Within the cell the circle of DNA appears to be twisted and folded.

The nuclear material of bacteria therefore is similar to and apparently has the same properties as all other organisms. The DNA is present in a distinct nuclear region but there is no nuclear membrane surrounding it. Cocci usually have one nuclear region per cell; rod-like bacilli usually have two or more per cell.

According to Lewis (1941), Robinow (1944) and others it has been advocated for many bacteria that nucleus is present in endospores and vegetative cells. Delameter and Hunter (1951), have demonstrated true mitosis in *Bacillus megatherium.* Knaysi (1951) supported in his statement that nucleus is present in many species of bacteria, *e.g.,* in *Staphylococcus flavo-cyaneus, Bacillus mycoides* and other species.

Cell wall. There is a definite thin cell wall, consisting of a single layer. It is made of a nitrogenous substance **chitin** resembling carbohydrates. The cell wall becomes gelatinous in some cases and appear-like a sheath or capsule. The sheath holds the cells together to form colonies. On stagnant water surface these colonies appear like scum and this scum-like resting stage in the life cycle of bacteria is known as **zooglea stage.** The slimy capsule composed largely of polysaccharides outside the cell wall serves as an additional protective layer.

The cell wall accounts for 20% of the dry weight of the cell. It gives shape and firmness to the cell. It can resist mechanical and chemical injuries and also the attack by other micro-organisms. The chemical composition of cell wall is known as 'muco-complex' which consists of proteins, polysaccharides and lipids. However, the cell wall can be dissolved in an enzyme called lysozyme. The cell wall of certain bacteria shows a characteristic reaction to the stain devised by C. Gram. Those bacteria which retain the stain are known as **Gram-positive** and those which do not retain the stain are termed **Gram-negative.** Generally the flagellated bacteria remain covered by a slimy layer whereas, the unflagellated bacteria under certain conditions of growth remain covered by a distinct capsule which is composed of polysaccharides or polypeptides.

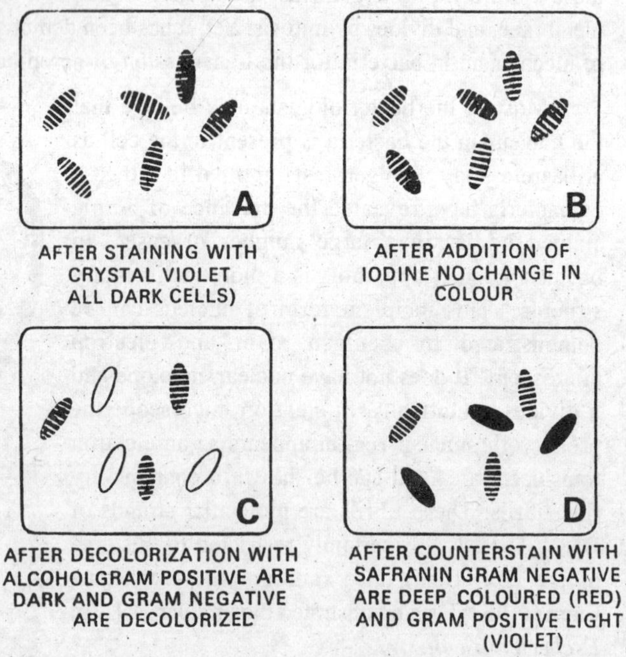

Fig. 2.9. Bacteria. Gram staining

Just beneath the cell wall there occurs a thin delicate, permeable cell membrane. This membrane surrounds the protoplast which consists of cytoplasm and the nucleus in the form of chromatin body. There is no cytoplasmic streaming in the bacterial cells as it is found in the cells of other higher plants.

The Gram reaction. One of the most important cytological features of bacteria is their reaction to a simple procedure called, after its discoverer, the Gram stain. The procedure involves staining the cells with the dye crystal violet and all bacteria will be stained blue. The bacteria are then treated with an iodine solution and then decolourized with alcohol. Gram positive bacteria retain the stain crystal violet. Gram negative are decolorized. This is a fundamental difference between Gram positive and Gram negative bacteria.

Table 2.1. Differences Between Gram Positive and Gram Negative Bacteria

Gram positive bacteria	*Gram negative bacteria*
Cell wall thicker contains only traces of lipids.	Cell wall thinner may contain upto 20% lipids.
Striking simplicity of amino acids. Glutamic acid and alanine occur in large amounts.	All the amino acids are present.
Diaminopimelic acid in some.	Diaminopimelic acid in some
Muramic acid more.	Muramic acid less.
Teichoic acid present	Absent.

Acid fast stain. Some bacteria are surrounded by a covering composed of fatty or waxy substances. These organisms are not readily stained, but when stained are able to retain the colour even after treatment with drastic decolorizing agents. They are called acid fast because the stained bacteria are resistant to decolorization with acid alcohol. Two well known members of this group are *Mycobacterium tuberculosis,* the etiological agent of tuberculosis in man and *M. leprae* believed to be the cause of leprosy in man. In the Ziehl-Neilsen method the primary stain is carbol fuschsin (basic fuschsin dissolved in a dilute solution of carbolic acid or phenol).

PHYSIOLOGY OF BACTERIA

Most of bacteria do not contain chlorophyll and unable to synthesize their own food. They are parasitic, saprophytic or symbiotic. The parasitic and saprophytic bacteria obtain the necessary energy for their growth and other metabolic activities from the tissues of the host or the dead decaying substrata on which they are found, by decomposing complex organic compound like carbohydrates and proteins present in the tissues. Such bacteria are **heterotrophic.**

Some bacteria obtain energy for metabolic activities, from the chemical reactions among inorganic substances. This phenomenon is known as **'chemosynthesis'** and the organisms are called **'chemosynthetic'** or **'autotrophic'** bacteria.

Heterotrophic bacteria. These bacteria may be parasitic, saprophytic and symbiotic in nature.

1. **Parasitic bacteria.** They derive their nutrition from the plants and animals on which they grow. With the result certain enzymes are produced which decompose or kill the protoplasm of the host cells. Such effects of the parasites on the host become visible to naked eye as disease symptoms. Many well known diseases of human beings like typhoid, tetanus, tuberculosis, pneumonia and many others and of plants like 'citrus canker', 'yellow rot of wheat' and 'ring rot of potato' are due to parasitic bacteria. They are also known **pathogenic** bacteria. Some bacteria grow well only in the presence of oxygen, while others grow well in absence of oxygen. The former are known as **aerobes** and the latter **anaerobes.**

2. **Saprophytic bacteria.** They grow on dead and decaying plants and animals, dung, rotten wood, stagnant water and many other decaying substances rich in organic matter. Certain enzymes secreted by the bacteria decompose the complex organic substances of the substrate, converting them into simpler ammonium compounds. They cause decay, and therefore also known as **putrefying** bacteria. The souring of milk, manufacture of cheese, preparation of butter from milk and vinegar from sugarcane juice, are various processes completed by the action of certain specific saprophytic bacteria. *Zygomonas* ferments glucose producing alcohol, lactic acid and carbon dioxide and plays significant role in wine industry. *Acetobacter* oxidises organic compounds to organic acids such as lactic acid thus having a significant role in vinegar industry. *Clostridium aceto-butylicum* forms butyl alcohol from carbohydrates. *Lactobacillus* converts sugars into lactic acid. Canned food is spoiled by *Bacillus stearuothermophilus* and *Clostridium thermosaccharolyticium.*

3. **Symbiotic bacteria.** Some organisms live in close association of other organisms and both partners do mutual benefit. This is termed as **symbiosis.** *Rhizobium* spp., is a striking example of this type. They occur in root nodules of leguminous plants and help in fixing the free nitrogen of the atmosphere in the soil for the plants which in return provide carbohydrates and protection to the bacteria. They are also called **nitrogen fixing bacteria** and add to fertility of the soil.

Azotobacter and *Clostridium* are other examples of nitrogen fixing bacteria. They are found in chalky soil, and obtain energy from the carbohydrates present in the soil The energy so obtained is used in fixing atmospheric nitrogen into aminoacids in the soil which react with the calcium salts, forming nitrites and thereafter nitrates.

Autotrophic bacteria. They may be chemosynthetic and photosynthetic.

A. **Chemosynthetic bacteria.** They oxidise inorganic substances and utilize the obtained energy converting the carbon of the carbon dioxide into organic material. On the basis of different inorganic substances oxidised by them, the following forms are recognized.

1. **Nitrifying bacteria.** They occur freely in soil and are of two types. *Nitrosomonas* oxidise the ammonium salts into nitrites and *Nitrobacter* oxidise nitrites into nitrates. They are anaerobic and utilize free oxygen for oxidation.

2. **Denitrifying bacteria.** They are found in oxygen deficient soils such as marshy places. These convert nitrates into nitrites and free nitrogen is liberated. They are anaerobic bacteria (*e.g., Pseudomonas aeruginosa, Micrococcus denitrificans*).

$$NO_3 \rightarrow NO_2 \rightarrow N_2O \rightarrow N_2$$

3. **Sulphur bacteria.** They oxidise sulphurated hydrogen (H_2S-Hydrogen sulphide) to water and free sulphur. This free sulphur is deposited as granules in the cytoplasm. They are aerobic (*e.g., Thiobacillus*).

4. **Iron bacteria.** These oxidise ferrous carbonate to ferric carbonate and the liberated energy is utilized. They are aerobic (*e.g., Ferrobacillus*).

5. **Hydrogen bacteria.** These oxidise hydrogen to water with the help of some oxides of nitrogen. They are aerobic (*e.g., Hydrogenomonas*).

6. **Carbon bacteria.** These oxidise charcoal and damp-coal into CO_2 in presence of free oxygen. They are aerobic (*e.g., Carboxydomonas*).

7. **Methane bacteria.** These oxidise methane (CH_4) to (CO_2) and H_2O. They are aerobic (*e.g., Methanomonas*).

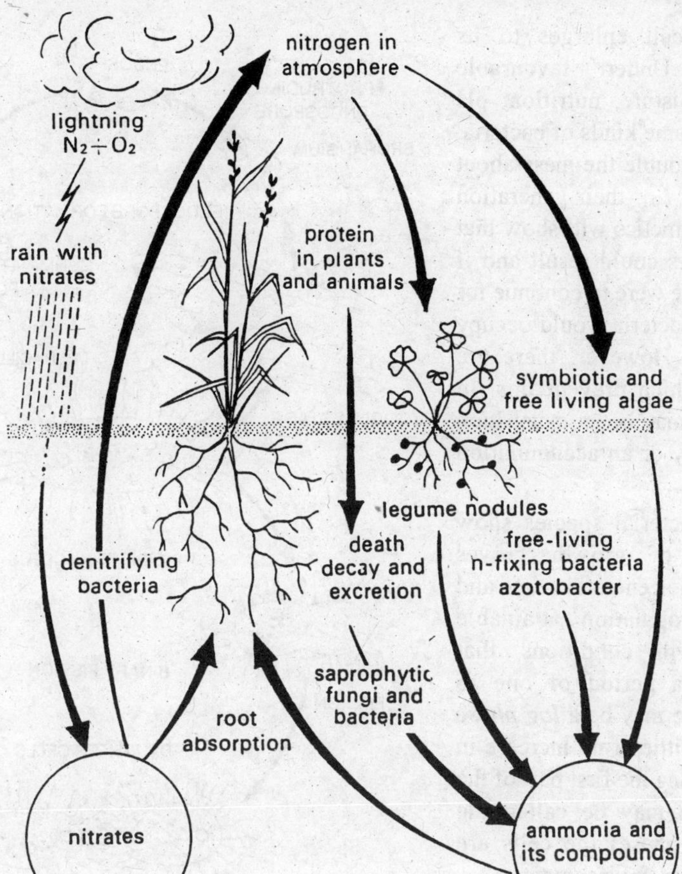

Fig. 2.10. Bacteria. The Nitrogen cycle

B. Photosynthetic bacteria. Very few bacteria contain a red pigment, which help in the synthesis of carbohydrates in presence of sunlight and are therefore known as **photosynthetic bacteria.**

Recently in some bacteria **bacteriochlorophyll** has been observed. It appears to be evenly distributed throughout the cytoplasm. With the aid of the electron microscope the pigment appears to be in minute grana somewhat different from those of the chloroplasts in the higher plants. These bacteria may well be considered independent.

Respiration. Resiration is a process in which energy is liberated as a result of oxidation of organic or inorganic substances. Bacteria oxidise organic and inorganic substances and utilize the liberated energy. Bacteria which can oxidise substances in complete absence of oxygen are known as **anaerobes** while others which require free oxygen for oxidation are known as **aerobes.**

GROWTH AND REPRODUCTION IN BACTERIA

Cell growth. If conditions are favourable, cell division is normally followed by a period of enlargement and growth. Through the absorption of water and food manufactured by the protoplasm the cell enlarges to its original size. Under favourable conditions of moisture, nutrition, pH and temperature some kinds of bacteria, *e.g., E.coli* may double the mass about every 20 minutes, *i.e.,* their generation time. Simple arithmetics will show that enormous numbers could result and if this rate of increase were to continue for even a few days bacteria would occupy the entire earth? However, there are several factors which prevent this. In some cases there may be an exhaustion of the food supply or an accumulation of waste products.

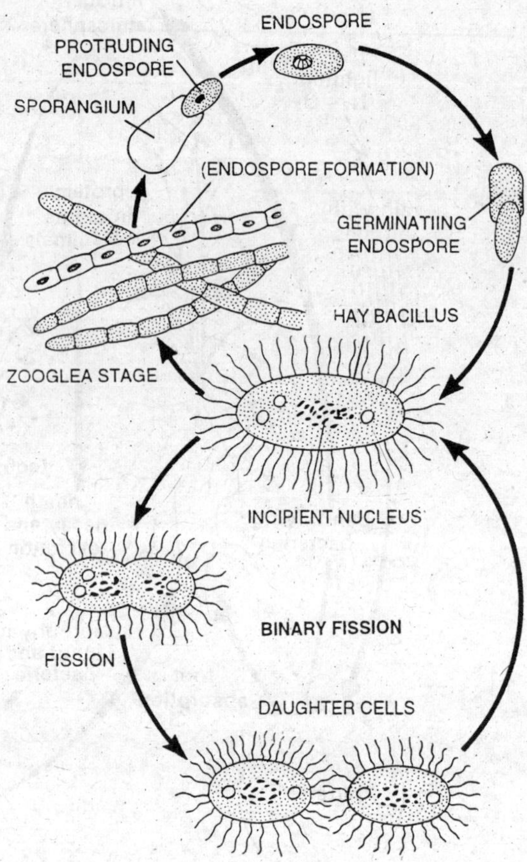

Fig. 2.11. Bacteria. Hay bacillus. Reproduction by endospore formation and binary fission.

Different bacterial species show various shapes of growth curves depending upon the generation time and the maximum population attainable under the growth conditions that prevail. During a period of one to several hours there may be a *lag phase* in which there is little or no increase in cell numbers. During the first part of the lag phase (which may be called the initial stationary phase) the cells are adapting to the new environment.

When the growing cells begin to divide they usually continue to do so at regular intervals until the maximum growth that can be supported by their environment is approached. This period of rapid cell division is known as the logarithmic phase of growth. The stationary phase occurs when rapid growth is halted by the depletion of nutrients, accumulation of waste products, or other factors. Unless the cells are transformed to new environment capable of supporting continuing growth they will eventually die.

Cell division. Cell division requires the doubling of all cell constituents and their orderly partitioning into the daughter cells. The first step in cell division is the doubling of the DNA complement of the nuclear body, which occurs, by the separation of the two strands and replication along each strand of a new complementary strand. Because of the enormous length of the DNA, the cell faces a problem in partitioning the doubled DNA between the two daughter cells. It is thought that the DNA molecule is attached at some point to the cell membrane and after replication, each one of the double strands while still attached, swings toward one of the two cell halves. Only after this slip has been accomplished can the new cross wall form. At the time the new cross wall, a mesosome of the membrane can be seen attached to the growing cross wall, and it is likely that this mesosome plays some role in the synthesis of cross wall. The DNA is probably attached to the other end of the same mesosome.

The highly coordinated process, DNA doubling/DNA partitioning/cross wall formation, is the equivalent of the mitotic process in eucaryotic organisms.

Reproduction. In bacteria, asexual and sexual methods of reproduction are known. Sexual reproduction has been investigated recently.

Asexual reproduction. A. Binary Fission. It is the commonest method of reproduction, found in bacteria. The binary fission is simply the multiplication of cells by division. The dividing cell elongates, and if the cell is spherical the elongation is very much restricted, and a rod-like bacterium elongates almost double its size. The protoplasmic mass then divides into two equal halves by a transverse wall or constriction. The two daughter cells soon grow to maturity and again divide within 20 minutes of their formation. This method is very quick and, if the favourable conditions prevail a single organism may produce 16,777,216 cells within 12 hours. But such favourable conditions never exist for a long period in nature.

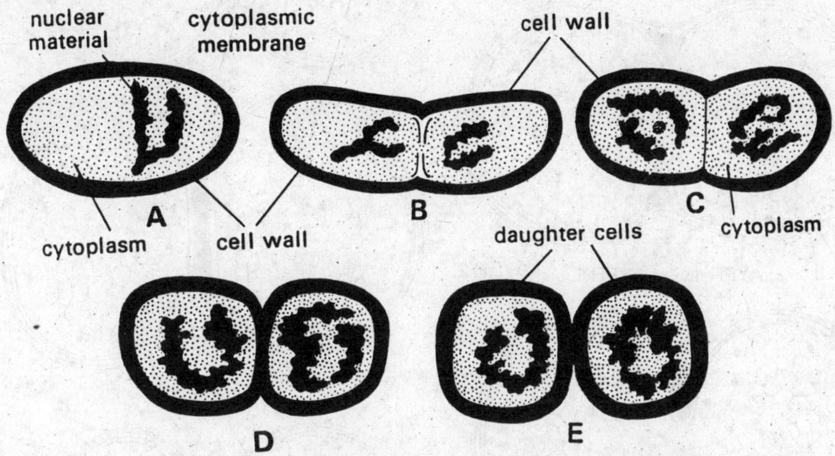

Fig. 2.12. Bacteria. Reproduction by binary fission (A–E)

If the repeated division of a bacterial cell continues to be in one plane only, chains are formed. When the cell divides in two planes, at right angles to each other, flat plate-like colonies are formed, and if the divisions occur in three planes, irregular colonies are formed.

B. Endospore formation. Endospores are formed only in few bacteria in adverse conditions. The endospores are known as **resting spores** and represent the resting stages of bacteria. The protoplast becomes rounded and secretes a hard, resistant wall around it. Each cell forms a single spore known as **endospore.** They are very resistant to temperatures as low as ice and as high as boiling water. They withstand dry conditions and remain alive in poisonous chemical substances. During unfavourable conditions they remain dormant for a long time and are set free from the parent cell when the conditions are favourable. On germination the outer membrane gets ruptured and the protoplasmic contents escape as a bacterial cell. Sometimes more than one endospore may be formed in the parent cell as in *Metabacterium polyspora.*

Endospores. Endospores are bodies produced inside the cells of considerable number of bacterial species. They are more resistant to unfavourable environmental conditions, such as heat, cold, desiccation, osmosis, and chemicals, than the vegetative cells, producing them. The bulk of evidence indicates the existence of a close relationship between spore formation and the exhaustion of nutrients essential for vegetative growth. Sporulation is a defence mechanism to protect the cell when the occasion arises. Spore formation is limited to two genera *Bacillus* and *Clostridium.* The spore structure is much more complex than that of vegetative cells in that it has many layers. The outermost is the exosporium, a thin delicate covering and within this is the spore coat, which is composed of a layer or layers of wall-like materials; below the spore coat is the cortex, a region consisting of many concentric rings like the cell wall, the cortex is composed of

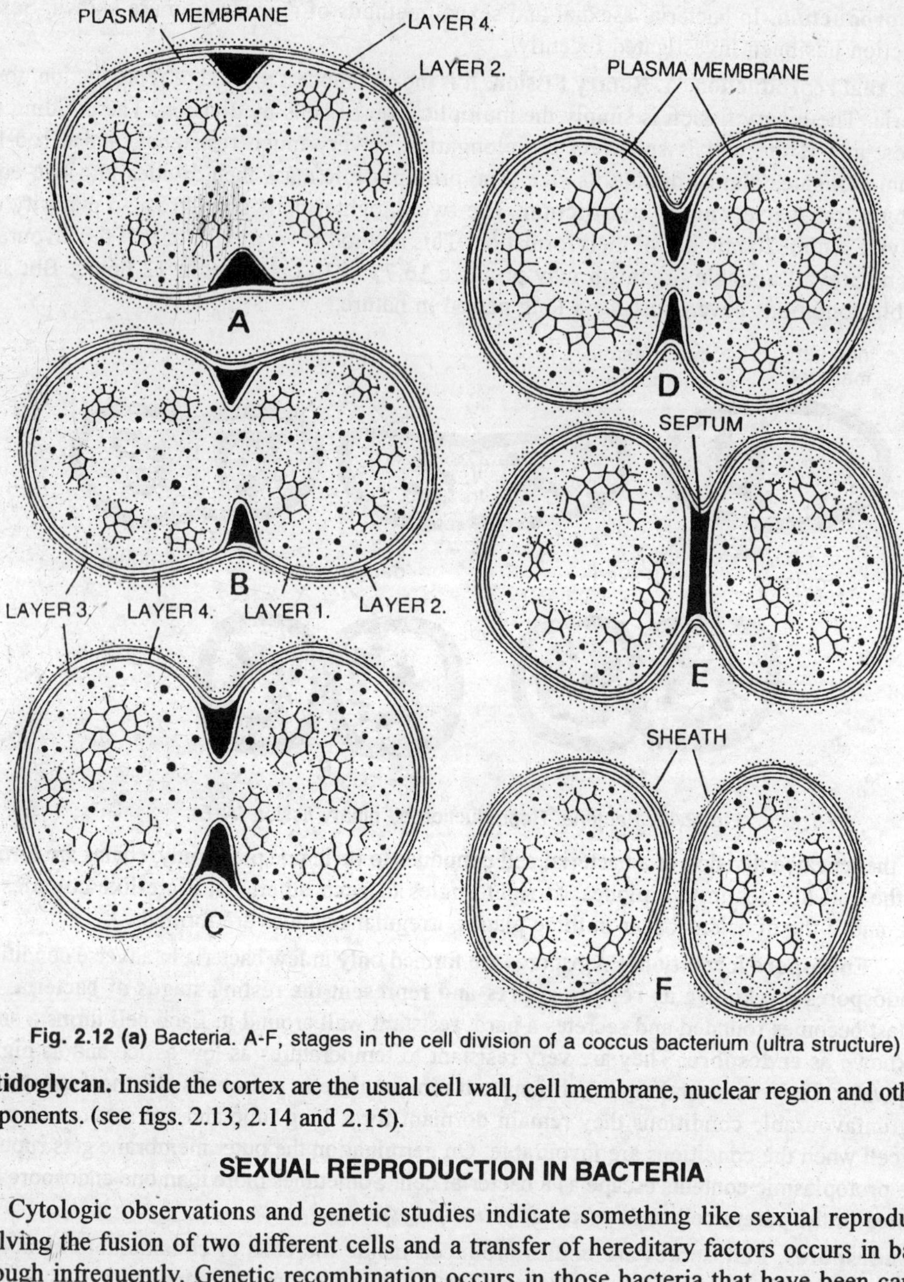

Fig. 2.12 (a) Bacteria. A-F, stages in the cell division of a coccus bacterium (ultra structure)

peptidoglycan. Inside the cortex are the usual cell wall, cell membrane, nuclear region and other cell components. (see figs. 2.13, 2.14 and 2.15).

SEXUAL REPRODUCTION IN BACTERIA

Cytologic observations and genetic studies indicate something like sexual reproduction, involving the fusion of two different cells and a transfer of hereditary factors occurs in bacteria although infrequently. Genetic recombination occurs in those bacteria that have been carefully studied and presumably occurs in other species as well. One of the most intensively studied species of bacteria, *Escherichia coli* has been shown to have sex-some act as males and transfer genetic information by direct contact with females. This ability to transfer genes is regulated by a fertility factor F+ which can itself be transferred to a female, thereby converting her into a male.

The usual vegetative bacterial cells are haploid and in sexual reproduction part or all of the chromosome passes from the male cell to the female cell, yielding a cell, *i.e.,* partly or completely diploid. Crossing over then occurs between the female chromosome and the male chromosome or fragment, followed by a process of segregation that yields haploid progeny cells.

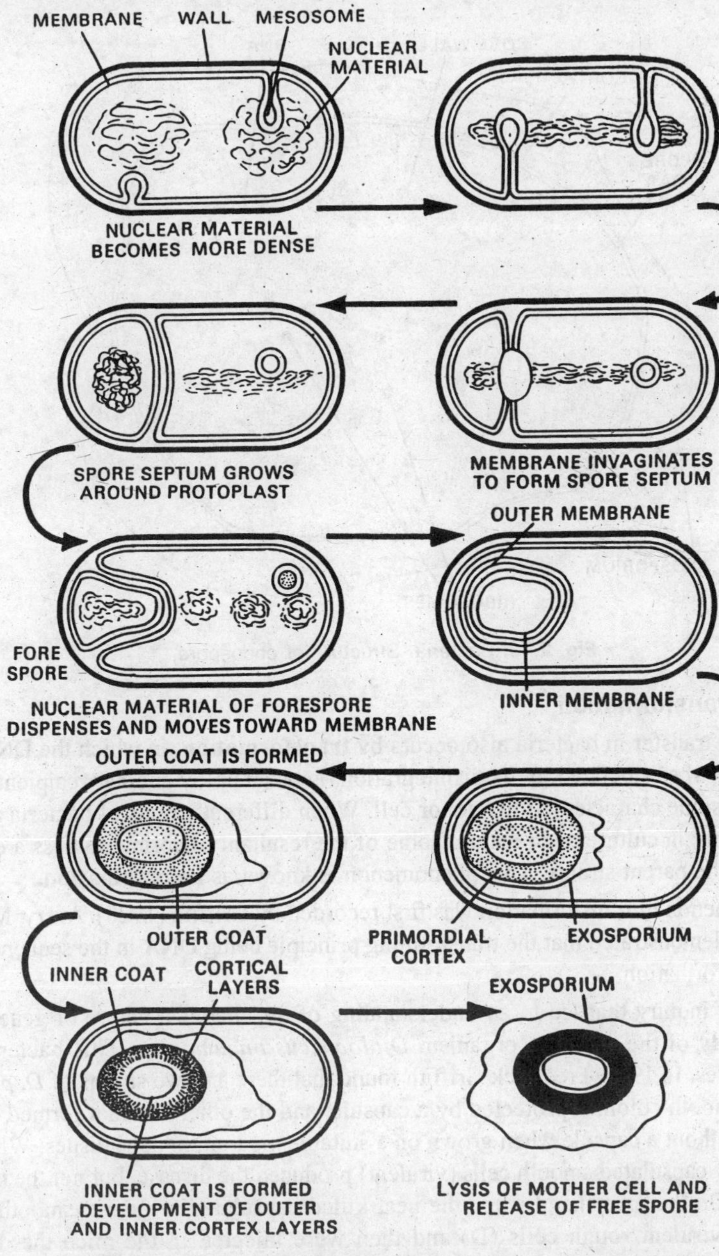

Fig. 2.13. Bacteria. Formation of endospore in a bacterial cell.

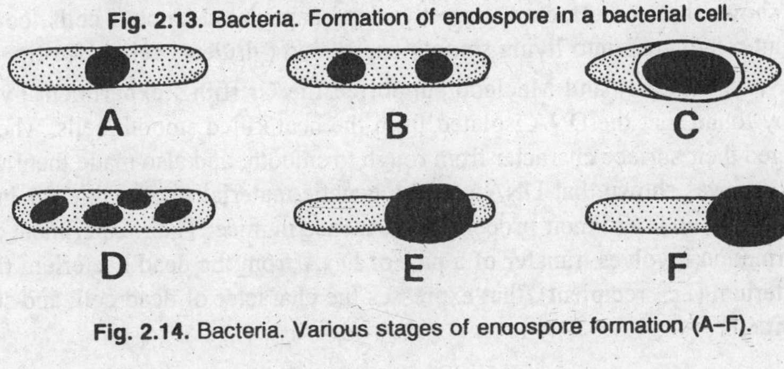

Fig. 2.14. Bacteria. Various stages of endospore formation (A–F).

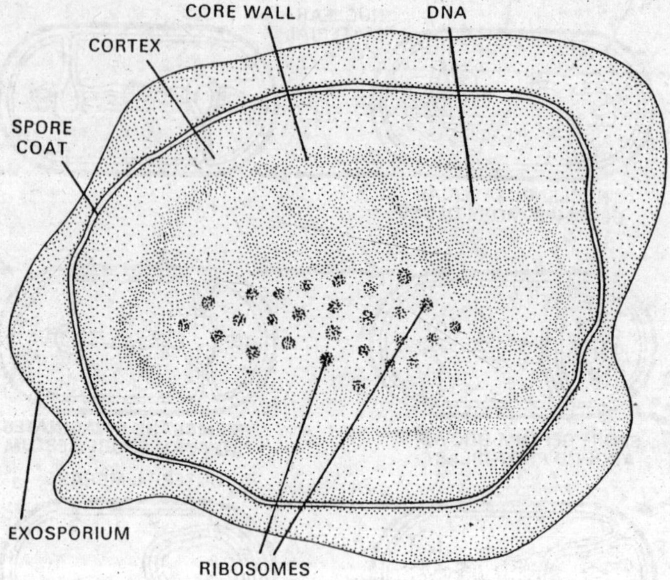

Fig. 2.15. Bacteria. Structure of endospore.

1. Bacterial transformation

The genetic transfer in bacteria also occurs by **transformation,** in which the DNA molecule of the donor cell, when liberated by its disintegration, is taken up by another recipient cell and its offspring inherit some characters of the donor cell. When different strains of bacteria are found in a mixed state either in culture or in nature, some of the resultant offspring possess a combination of characters of the parent strains. This phenomenon is known as **recombination.**

The phenomenon of transformation was first recorded by Griffith (1928). Avery, Macleod and McCarty (1944) demonstrated that the transforming principle being DNA in the sequence of events in bacterial transformation.

The lines of inquiry that led to an understanding of the chemical nature of genetic material arose from a study of the pestilent organism *Diplococcus pneumoniae*. This bacterium causes pneumonia in males. In 1928, Frederick Griffith found that there are two strains of *D. pneumoniae,* one that forms smooth colonies protected by a capsule, and the other one that formed irregular or rough colonies without a capsule when grown on a suitable medium in petri dishes. When injected into mice (A) only capsulated smooth cells (virulent) produced the disease, but not the non-virulent rough cells (B). On the other hand when the heat killed capsulated (virulent) smooth cells were mixed with non-virulent rough cells (D) and then were injected in the mice the disease was produced. This shows that some factors from the dead capsulated smooth cells, converted the living non-virulent rough cells into living smooth capsulated (virulent) cells. (see fig. 2.16).

In 1944, Avery, McCarty and Macleod supported the Griffith's experiment by molecular explanation. They found that the DNA isolated from the heat killed smooth cells, when added to rough cells changed their surface character from rough to smooth, and also made them virulent. By this experiment, this was shown that DNA was the genetic material responsible for inducing the smooth character of the cells and their property of virulence in mice. Their experiment proved that bacterial transformation involves transfer of a part of DNA from the dead bacterium (*i.e.,* donor) to the living bacterium (*i.e.,* recipient), that expresses the character of dead cell, and so is known as a **recombinant.**

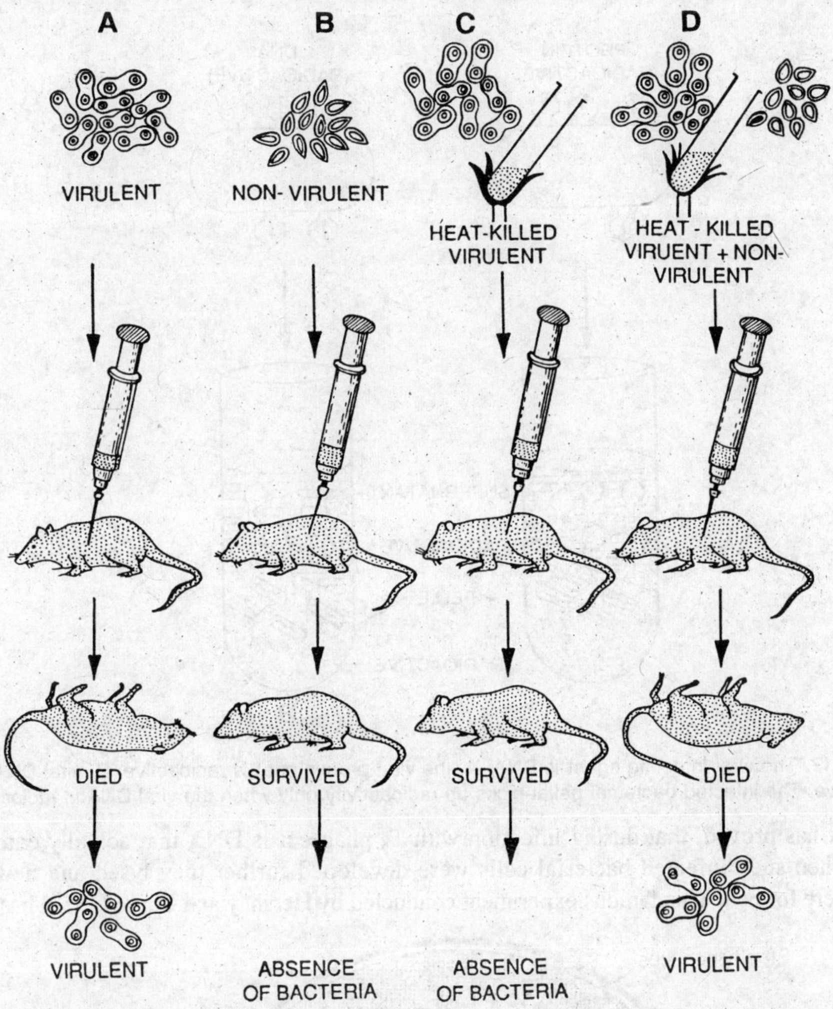

Fig. 2.16. Griffith's experiments with *Diplococcus pneumoniae* and mice.

Viral-infecting agent is DNA

A bacteriophage (T_2 virus) infects the bacterium *Escherichia coli.* After infection, the virus multiplies and T_2 phages are released with the lysis of the bacterial cells. As we know, the T_2 phage contains both DNA and proteins. Now the question arises, which of the two components has the information to programme for the multiplication of more viral particles.

To solve this problem Hershey and Chase (1952) devised an experiment with two different preparations of T_2 phage. In one preparation they made protein part radioactive and in the other preparation the DNA was made radioactive. Thereafter a culture of *E. coli* was made infected by these two phage preparations. Immediately after infection, and before lysis of bacteria the *E. coli* cells were gently agitated in a mixer so that the adhering phage particles were loosened and then the culture was centrifuged. With the result the heavier pellets of infected bacterial cells were settled down in the bottom of tube. The lighter viral particles and those particles which did not enter the bacterial cells were found in the supernatant. It was found that when T_2 phage with radioactive DNA was used to infect *E. coli* in the experiment, the heavier bacterial pellet was also radioactive. On the other hand when T_2 phage with radioactive protein was used, the bacterial pellet had very little radioactivity and most of the radioactivity was found in supernatant. This

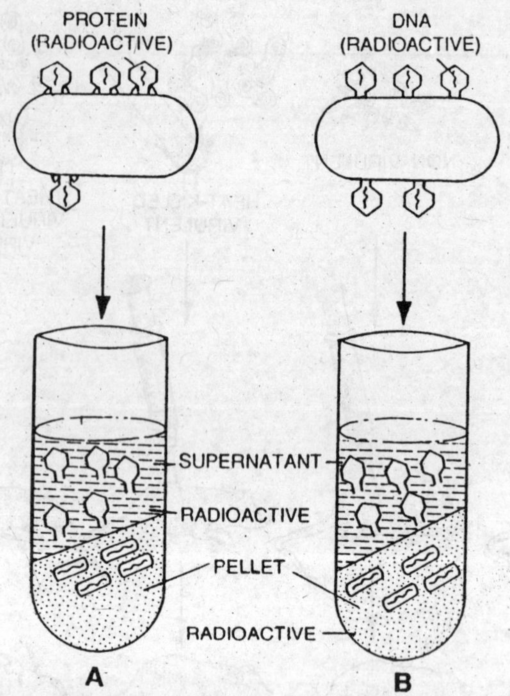

Fig. 2.17. The viral infecting agent is DNA. A, the viral protein coat is radioactive; B, viral DNA is radioactive. The infected bactetrial pellet picks up radioactivity only when the viral DNA is radioactive.

experiment has proved, that during infection with T_2 phage, it is DNA that actually entered the bacteria when such infected bacterial cells were developed further they lysed and new phage particles were formed. This famous experiment conducted by Hershey and Chase (1952) has proved

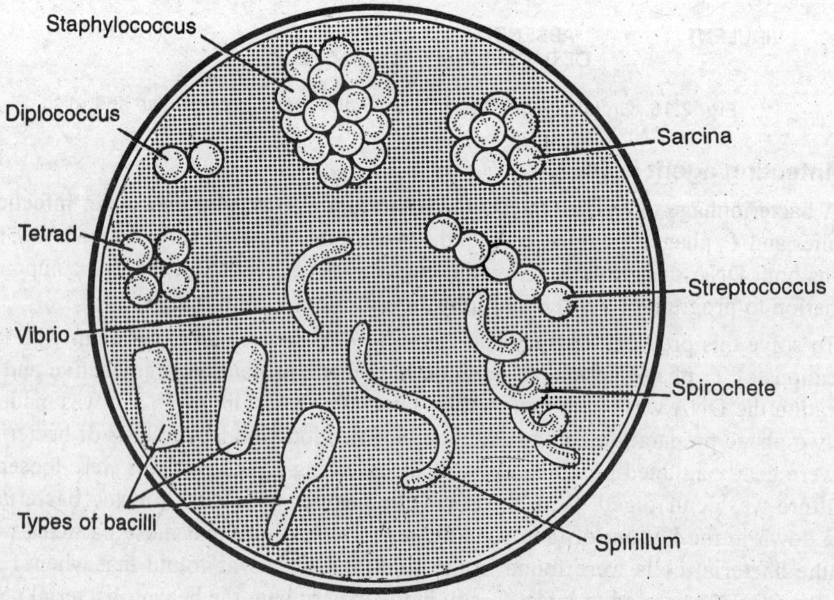

Fig. 2.18. Different types of bacteria.

that it is the viral DNA and not protein that contains information for the production of more T_2 phage particles hence DNA is genetic material. However, in some viruses (*e.g.*, TMV, influenza virus and polio virus) RNA serves as genetic material. (see fig. 2.17).

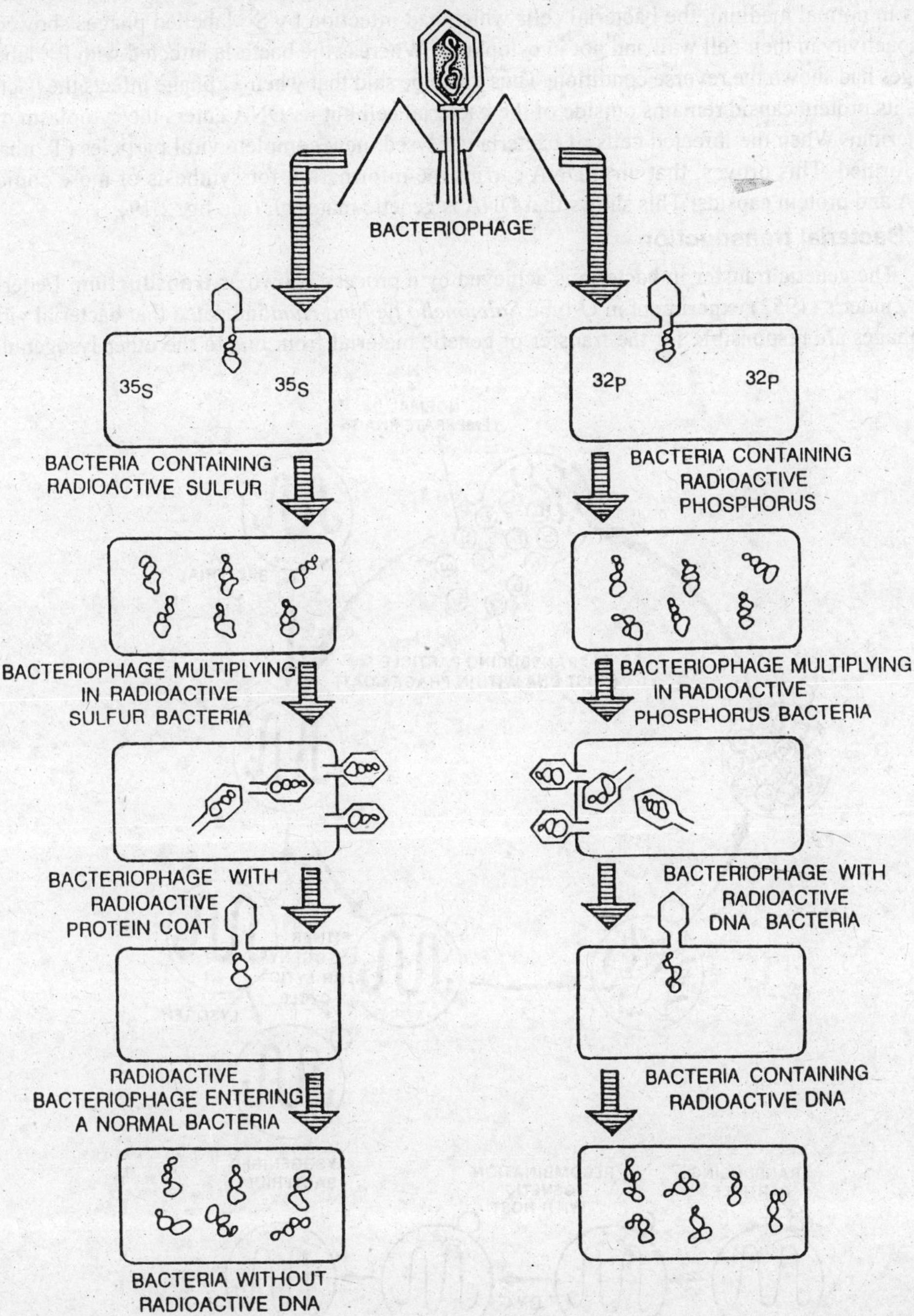

Fig. 2.19. Hershey and Chase's experiments show that DNA is hereditary material in bacteriophages.

Hershey and Chase conducted two experiments. In one experiment *E. coli* was given in a medium containing the radio-isotope S^{35} and in the other experiment *E. coli* was grown in a medium containing the radio-istope P^{32}. In these experiments *E. coli* cells were made infected with T_2 phage released from *E. coli* cells grown in S^{35} medium have S^{35} in their protein capsid, and those from P^{32} medium had P^{32} in their DNA. When these phages were used to infect new *E. coli* cells in normal medium, the bacterial cells which had infection by S^{35} labelled phages showed the radioactivity in their cell wall and not in cytoplasm. Whereas the bacteria infected with P^{32} labelled phages had shown the reverse condition. Thus it can be said that when T_2 phage infects the bacterial cell, its protein capsid remains outside of the bacterial cell but its DNA enters the cytoplasm of the bacterium. When the infected cells of bacteria get lysed, new complete viral particles (T_2 phages) are formed. This proves, that viral DNA carries the information for synthesis of more copies of DNA and protein capsids. This shows that DNA is genetic material. (see fig. 2.19).

2. Bacterial transduction

The genetic transfer in bacteria is achieved by a process known as **transduction.** Lederberg and Zinder's (1952) experiment in U-tube *Salmonella typhimurium* indicated that bacterial viruses or phages are responsible for the transfer of genetic material from one to the other lysogenic and

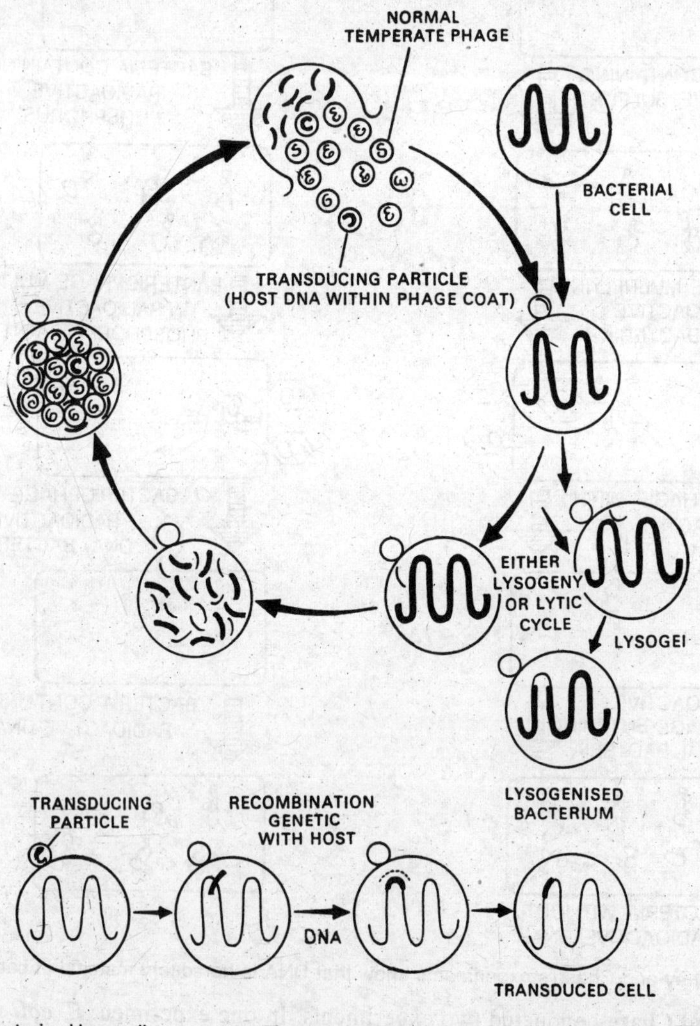

Fig. 2.20. Bacteria. Upper diagram: transduction mechanism where phage particles containing host can be formed. Lower diagram: genetic recombination with transducing particles.

lytic phages. Thus the host acquires a new genotype. Transduction has been demonstrated in many bacteria.

In this process, the DNA molecule that carries the hereditary characters of the donor bacterium, is being transferred to the recipient cell through the agency of the phage particle. In this process very few closely linked characters can be transferred by each particle. Thus the bacteriophage brings about genetic changes in those bacteria which survive the phage attack.

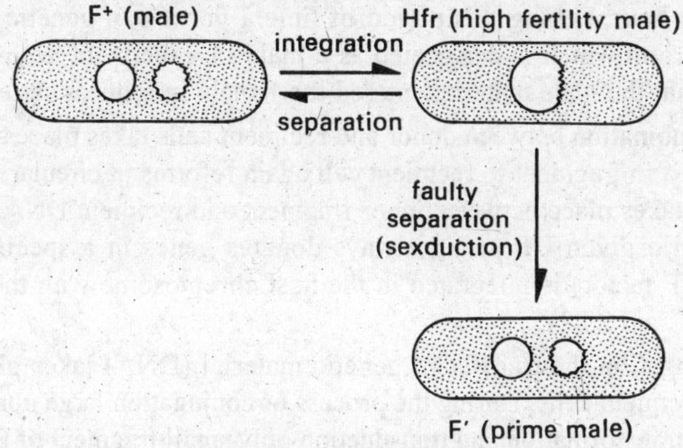

Fig. 2.21. Bacteria. Transduction.

When a bacterial cell is being infected with a temperate virus either lytic-cycle or lysogeny starts. Thereafter, host DNA breaks down into small fragments along with the multiplication of virus. Some of these DNA fragments are incorporated with the virus particles becoming transducing one. When bacteria lyse these particles along with normal virus particles are released

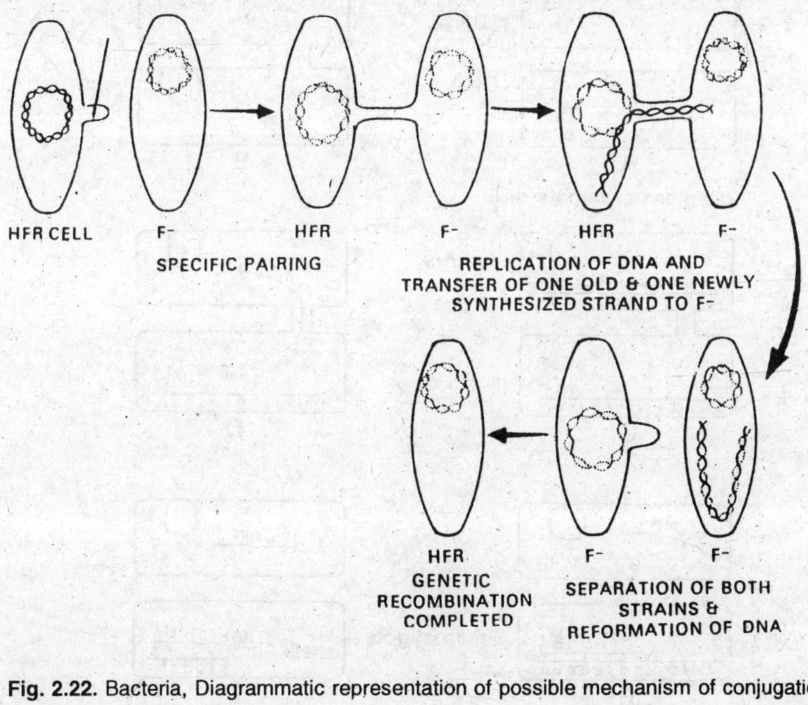

Fig. 2.22. Bacteria, Diagrammatic representation of possible mechanism of conjugation.

when this mixture of transducing and normal virus particles is allowed to infect the population of recipient cells, most of the bacteria are infected with normal virus particles and with the result lysogeny or lytic-cycle occurs again. A few bacteria are infected with transducing particles, transduction takes place and the DNA of virus particles undergo genetic recombinations with the bacterial DNA. (see figs. 2.20 and 2.21).

3. Bacterial conjugation

Wollman and Jacob (1956) have described conjugation in which two bacteria lie side by side for as much as half an hour. During this period of time a portion of genetic material is slowly passed from one bacterium which is designated as a male to a recipient designated as a female. This was established that the male material entered the female in a linear series.

The genetic recombination between donor and recipient cells takes place as follows: The Hfr DNA after leaving a part in fragment to recipient cell again reforms in circular manner. In F strain genetic recombination takes place between donor fragment and recipient DNA. Gene transfer is a sequential process and a given Hfr strain always donates genes in a specific order. A single stranded donor DNA (F factor) is integrated in the host chromosome with the help of **nuclease enzyme.** (see figs. 2.21 and 2.22).

In bacterial conjugation the transfer of genetic material (DNA) takes place by cell to cell contact of donor and recipient cells. During the process of conjugation large portion of the genome is transferred, while in transformation and transduction only small fragment of DNA is transferred. The process of conjugation was discovered by Lederberg and Tatum (1944) in a single strain of *Escherichia coli.* Conjugation has also been demonstrated in *Salmonella, Pseudomonas* and *Vibrio.*

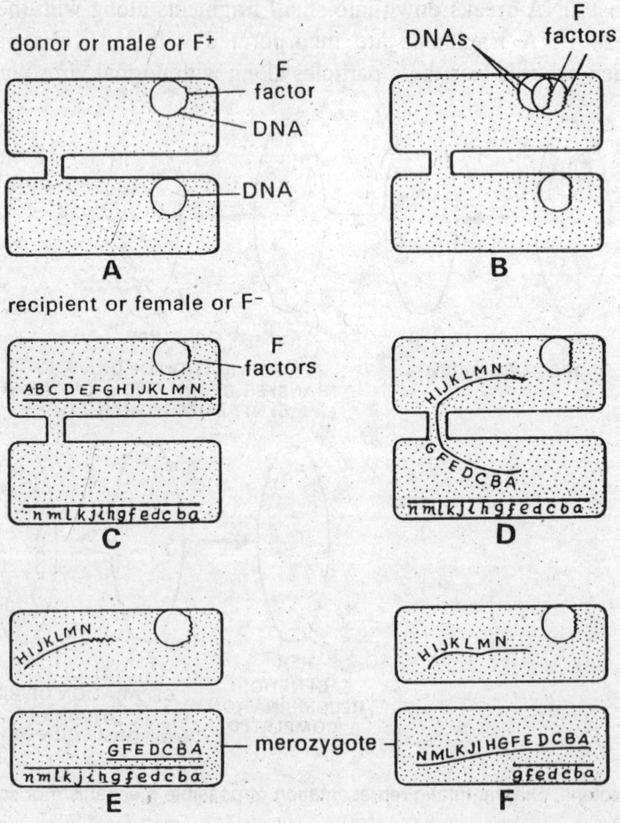

Fig. 2.23. Bacteria. Conjugation between sterile male and female of *Escherichia coli.*

In conjugation one way transfer of genetic material takes place from donor to recipient strain. The donor and recipient strains are always determined genetically. Recipient strain is designated as F, while donor strains are of two kinds and are designated as F^+ and H fr (high frequency of recombination). If the strain donates only a small portion of its genome it is called F^+, and if it donates large amount of genome it is called H fr. These F^+ and H fr factors are called **episomes.**

Strains F^+ and Hfr are characterised by the presence of specific flagellum like structures, the so called **sex pilus.** The sex pilus is absent in F^- strains, and is responsible for bacterial mating. Sex pili of F^+ and H fr touch the opposite mating type of cells specifically to transfer the genetic material.

Sex pilus has a hole of 2.5μm diameter which is large enough for a DNA molecule to pass through it lengthwise. At the time of pairing DNA of H fr strain (donor) is transferred to F^- strain (recipient) immediately. The circular DNA of H fr cells opens and replicates but during transfer, one strand of DNA is newly synthesized, whereas the other strand is derived from a pre-existing strand of H fr strain. After transfer of DNA both the cells are separated from each other.

The H fr DNA after leaving apart its fragment to recipient cell again reforms in circular manner. In F^- strain genetic recombination takes place between donor fragment and recipient DNA. Gene transfer is a sequential process, a given H fr strain always donates genes in a specific order. If F^- and H fr strains are allowed to mix in a suspension, different genes in a sequence of time are transferred from the genome of H fr to F^- strain. Genes that enter early, always appear in larger percentage of the recombinations than do genes that enter late. (see figs. 2.22, 2.23 and 2.24).

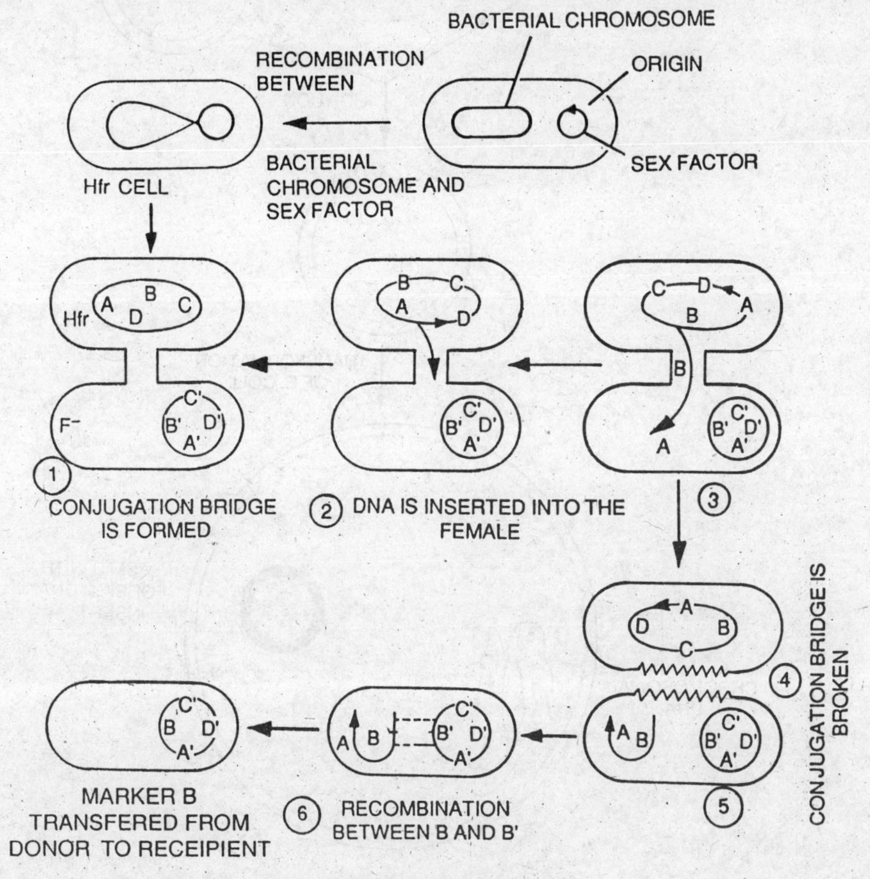

Fig. 2.24. Bacterial conjugation

Conjugation results in a number of recombinants in a suspension of F^- and H fr cells. These recombinants are variable in their genotypic constitution and so also in their phenotypic expression. These recombinants are entirely new and different from their parents.

PROKARYOTIC CHROMOSOMES

The genetic information of a prokaryotic cell is carried in the nucleoplasm on the structure termed **bacterial chromosome.** It consists of a double helical DNA (deoxyribonucleic acid) molecule, never associated with basic proteins, and has been shown in some prokaryotes to be

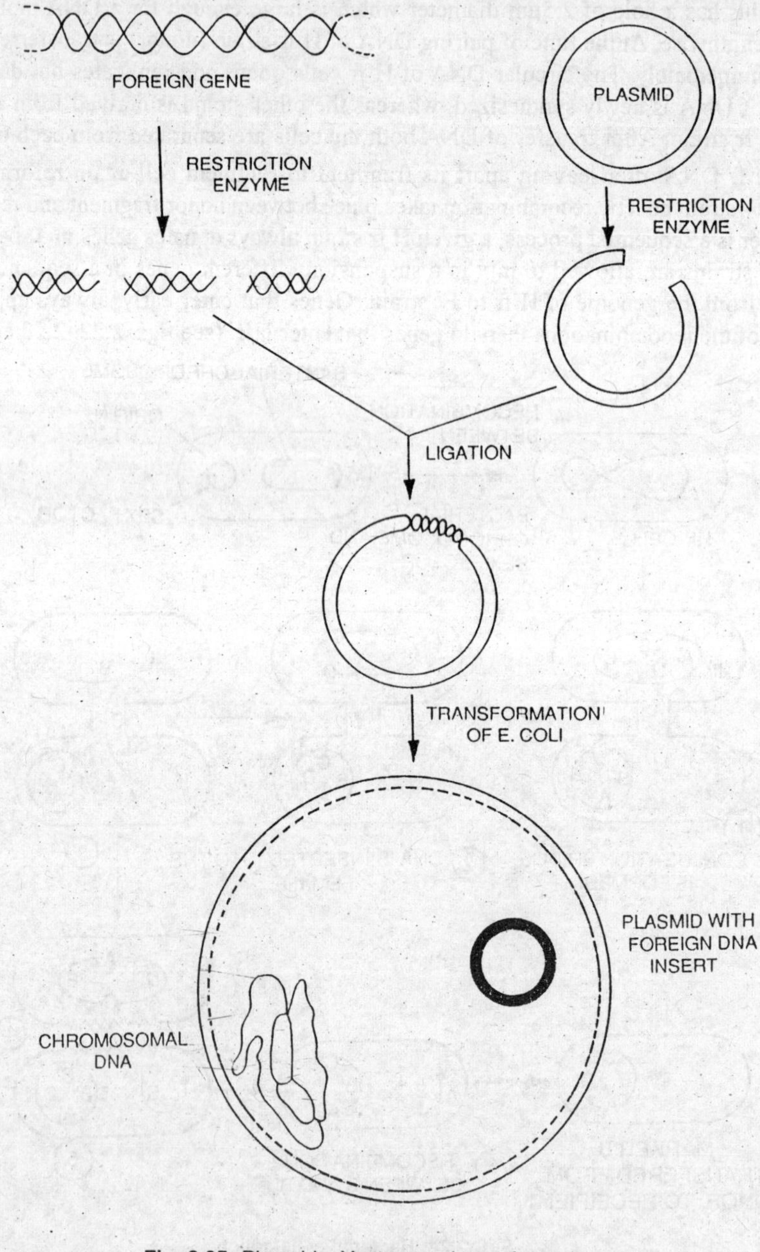

Fig. 2.25. Plasmids. Vectors and transformation.

circular. The bacterial chromosome is consequently not structurally homologous with the nuclear chromosomes of the eukaryotic cell, but rather with the organellar DNA present in the eukaryotic mitochondria and chloroplasts. It is probable, that a single bacterial chromosome (*viz.,* one very long DNA molecule) carries all the genetic information necessary to specify the essential properties of the prokaryotic cell.

Plasmids. Many bacteria can also harbour small, extrachromosomal circular DNA molecules capable of autonomous replication, which are known as **plasmids** so far investigated carry the determinants for such phenotypic characters as resistance to drugs and other antibacterial substances, and for the enzymes which mediate certain metabolic pathways. The amount of DNA in a plasmid is from 0.1 to 5 percent of that in the bacterial chromosome.

This was discovered in *Escherichia coli* that there are two mating types, and during conjugation one partner acts only as **genetic donor,** or **male,** and the other only as **genetic recipient** or **female.** Since the only function of the male is to transfer some of its DNA, it need not remain viable, whereas the female cell must remain viable, so that, the zygote may develop in it. This was discovered that maleness in bacteria is determined by a transmissible genetic element when male and female bacteria conjugate, every female cell is converted to a male. The genetic element governing the inherited property of maleness is called the **F factor** (F = fertility), it is transmitted only by direct cell to cell contact. In 1952 Lederberg coined the term **plasmid** as a genetic name for all extrachromosomal hereditary determinants, of which F is an example. It is now known that bacterial plasmids are small, circular molecules of DNA which carry the genes for their own replication. In many cases, they also carry genes which confer new properties on the host cell, such as resistance to drugs or the production of toxins. Many plasmids carry genes that govern the process of conjugation. Thus, conjugation is a mechanism imposed on the bacterial cell by a plasmid, the normal result of which is the transfer of plasmid DNA.

Types of plasmids. Most plasmids have been classified on the basis of host properties. Thus, there are the *R* **factor** (R = resistance) and the *COI* **factors** (COI = colicinogeny) of the gram-negative bacteria, the **penicillinase plasmids** of *Staphylococcus aureus,* the degradative plasmids of *Pseudomonas,* the **cryptic plasmids** etc.

Properties of plasmids. The sturcture of all known plasmids consists of circular, double stranded DNA molecules. Several of them have molecular weights in the range 5×10^7 to 7×10^7. One (an R factor) has a molecular weight of only 1×10^7, and some of the cryptic plasmids are even smaller. Since the amount of DNA required to code for an average polypeptide with a molecular weight of 40,000 is about 6×10^7, F_1 and other plasmids of similar size may contain as many as 100 genes.

Gene Expression in Prokaryotes

Bacterial chromosomes. It has been established now that a bacterial chromosome is always present in the centre of bacterial cell which is not capable of mitotic and meiotic divisions and is not enclosed within a definite nuclear membrane. An average bacterial cell contains one thousandth the DNA content of a typical eucaryotic cell. The bacterial chromosome is a thin elongated flexible and circular filament of DNA molecule associated with few proteins. The bacterial cells can divide very rapidly. The cell division is completed by the doubling of all the cell constituents followed by partition of the cell into two daughter cells. The first step in cell division is DNA duplication.

Genetic Recombination in Bacteria

This is a process where genetic materials, contained in two separate genomes, are brought together within one unit. In bacteria the recombination takes place by (1) **transformation,** (2) **transduction** and (3) **conjugation.** These phenomena have already been discussed in detail in preceding paragraphs, under the heading 'sexual reproduction in bacteria'.

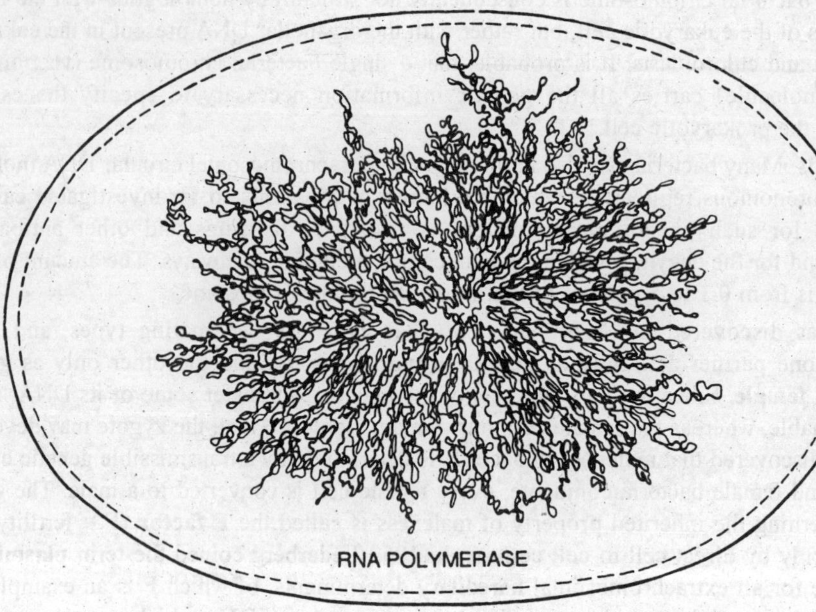

Fig. 2.26. Bacterial chromosome, chromosome of *Escherichia coli.*

Induction and Repression

The genetic potential of a virus is limited and it essentially uses the host machinery for its replication. On the other hand, the bacteria can synthesize their constituents from simple salts and sugar. The bacterium, *Escherichia coli* has about 2500 genes on its chromosome. The bacterium maintains its internal economy only by allowing the action of few genes at a time. This means that genes will be turned on or off as per requirement. A set of genes will be switched on when there is necessity to handle and metabolise a new substrate. When the set of these genes is switched on, enzymes are produced, which metabolise the new substrate. This phenomenon is called **induction** and the molecules (metabolites) evoking this phenomenon are **inducers.** On the other hand, when a metabolite needed by the bacterium is supplied in excess from outside, the bacterium inhibits making it and this stops the further production of the metabolite by the bacterium. These inactivated genes are thus named **repressible,** and the phenomenon is known as **repression** or **feed-back repression.** However, some of the genes are constantly expressed to take care of normal cellular activity such as glycolysis. These genes are known as **constitutive,** and the enzymes produced by them are known as **constitutive enzymes,** such as dehydrogenases.

Operon Concept

Francois Jacob and Jacques Monod (1961), while working at the Pasteur Institute, Paris, proposed that metabolic pathways are regulated as a unit. They studied the change in the nature of enzymes of *E. coli* when provided with different kinds of sugars. They found, when the sugar lactose is added to the cultures of *E. coli,* it induces three enzymes necessary to break down the lactose into glucose and galactose. These newly synthesized enzymes are–(i) β-galactosidase, (ii) permease and (iii) transacetylase. According to them, the synthesis of these three enzymes is controlled by a long DNA segment known as **operon** which is divisible into a operator site O and three structural genes Z, Y and A. The action of structural genes is regulated by operator site with the help of a **repressor** protein produced by the action of gene 'i' known as the **regulator gene.** The genes are expressed or not expressed that depends on whether the operator switch is on or off. When the switch is on, the three genes are transcribed by RNA polymerase into a single stretch of messenger RNA covering all the three genes. Each structural gene is generally known as **cistron** and the transcribed long mRNA covering all the cistrons is known as **polycistronic.**

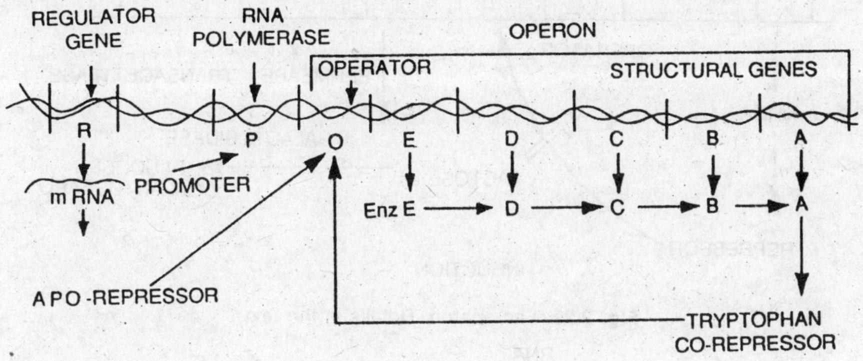

Fig. 2.27. Tryptophan. Operon, showing synthesis of tryptophan.

The switching on or off of the operator switch is achieved by a protein known as **repressor.** When this protein ties together to the operator (O) and blocks it, the switch is turned off and the three genes (Z, Y, A) are not expressed. (see fig. 2.28).

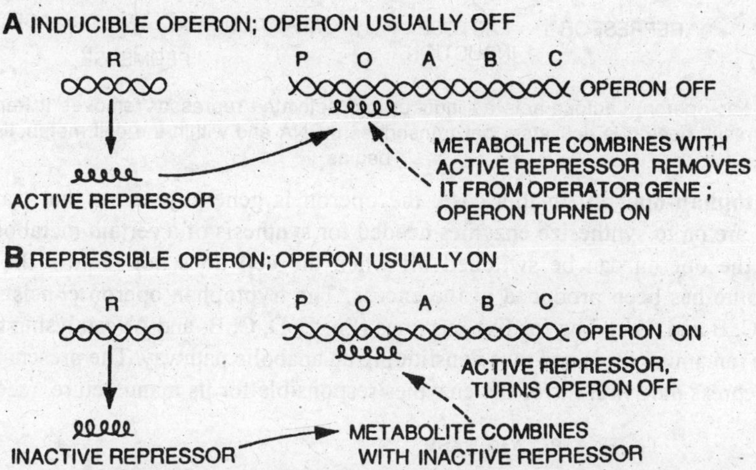

Fig. 2.28. Operons. A, inducible B, repressible. Both the operons become off when operator site is bound by the repressor formed by regulator gene R.

Lac-operon. This is an example of **inducible operon.** A few molecules of lactose are added into the cell by the action of the enzyme permease, a small quantity of which is found even under repressed conditions. These few molecules are then converted into an active form of lactose which binds to the repressor (i gene product) and thus, the repressor cannot bind to the operator switch any more. When the operator is free of repressor RNA polymerase starts he transcription of the operon by binding the promotor site P. Messenger RNA corresponding to all the three enzymes is synthesized which gets translated to produce three enzymes-β-galactosidase, permease and transacetylase. With the production of these three enzymes metabolism of lactose begins. The synthesis of enzymes is continued unless and until all lactose molecules are consumed by the cell. When the last molecules of lactose, bound to repressor are also consumed, the inactive repressor becomes active and binds to operator site (O) to switch off the operon as normal. (see figs. 2.29 and 2.30).

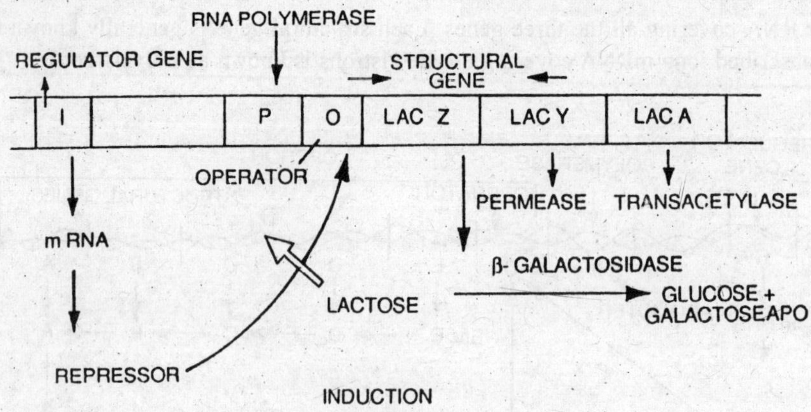

Fig. 2.29. Lac operon. Details in the text

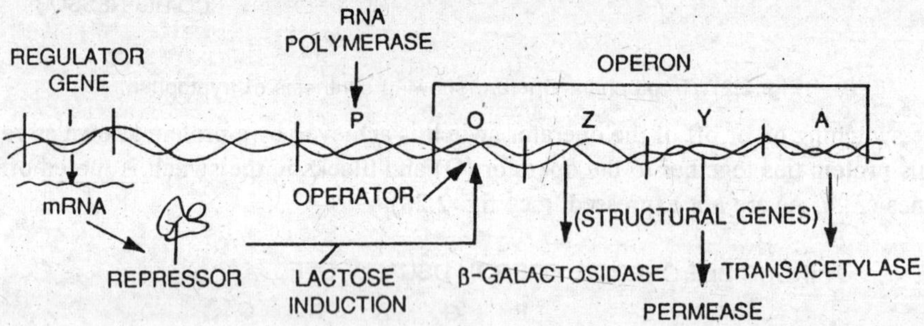

Fig. 2.30. Lac-operon. Lactose acts as inducer and activated repressor removes it from operator site with the result operon is activated that transcribes mRNA and within the cell metabolism of lactose begins.

Tryptophan-operon. In this case, the operon is generally on so that transcription and translation are on to synthesize enzymes needed for synthesis of a certain metabolite by the cell. However, the operon can be switched off when the cell does not require the metabolite or the metabolite has been produced in the excess. The tryptophan operon consists of five genes (trp E, D, C, B and A) coding for five enzymes (En_z E, D, C, B, and A) catalysing the synthesis of tryptophan (an amino acid) and thus constituting an anabolic pathway. The presence of tryptophan serves to repress the synthesis of the enzymes responsible for its manufacture. (see figs. 2.27 and 2.31).

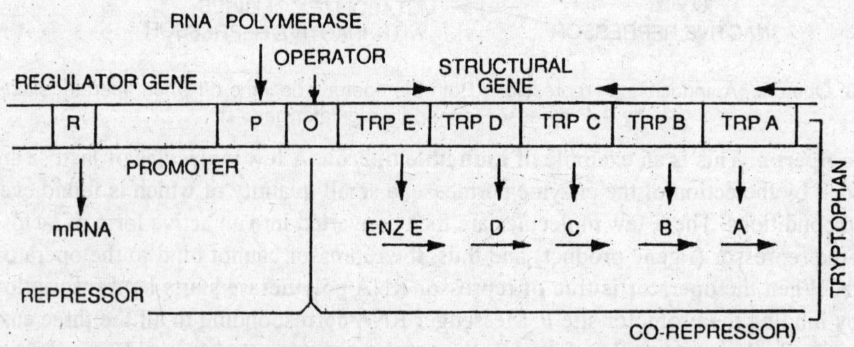

Fig. 2.31. Tryptophan operon. Details in the text.

The five structural genes coding for these five enzymes are regulated by a single operator switch. Here, the repressor protein produced by the regulatory genes does not bind to the operator by itself. However, when tryptophan is added, it binds to the repressor and is known as the **co-repressor.** The repressor-corepressor complex binds to the operator and blocks the expression of this operon.

The structure of the tryptophan operon is more or less similar to that of the lac operon, but has a functional variation. In this case the R gene product is equivalent to the i gene of lac operon. The R gene product produces protein which by itself is unable to bind to operator. This is referred to as **apo-repressor.** In the presence of tryptophan a co-repressor, the functional repressor is formed which now binds to the operator and prevents the transcription of the operon and production of tryptophan. Here the operon model explains both the induction and repression phenomenon in procaryotes.

CLASSIFICATION OF BACTERIA

Because of their simple morphology, bacteria canot be divided into orders, families, genera and species on the basis of their structure alone. The biochemical, physiological and cultural characters are to be considered for their classification. The scientific world of today accepts the Bergey's classification of Bacteria (1957).

Erwin F. Smith, a pioneer worker in this field at first accepted a system of classification proposed by Migula which was based on the number and arrangement of flagella. It was as follows:

(1) **Bacterium** included all atrichous rod-like forms.

(2) **Pseudomonas** included polar flagellate forms.

(3) **Bacillus** included peritrichous rod-like forms.

In 1905, Smith revised this system with slight modifications in the name of the divisions. Thus the name (i) **aplanobacter** was given for atrichous, rod-like bacteria, (ii) **bacterium** for polar flagellate bacteria, and (iii) **bacillus** for peritrichous bacteria.

The commonly used and more widely accepted divisions of bacteria are:

(1) **Bacterium** or **Bacillus** included all rod-like and kidney-shaped forms.

(2) **Coccus** includes all spherical bacteria.

(3) **Spirillum** includes spiral, comma like and variously curved bacteria.

Table 2.2

Author	*Non-motile*	*Motile with polar flagella*	*Motile with per-intrichous flagella*
Migula (1895-1900)	*Bacterium*	*Pseudomonas*	*Bacillus*
Lehmann and Neu-mann (1897-1927)	Non-spiral Motile or nonmotile *Bacterium*	Sporing motile or non- motile	*Bacillus*
Smith (1905)	*Aplanobacter*	*Bacterium*	*Bacillus*
Bergy (1923-1939)	*Phytomonas*		*Erwinia*
Dowson (1939)		*Pseudomonas* and *Xanthomonas*	*Bacterium*
Bergey (1948)	*Pseudomonas*	*Pseudomonas* *Yanthomonas*	*Bacterium*

BERGEY'S CLASSIFICATION (1957)

Modern schemes for classification of bacteria use a number of properties from, arrangement, Gram stain, motility, various enzymic properties and any other sort of information that can serve to delimit an organism of group of organisms.

A system of classification of Schizomycetes that is used by all bacteriologists and that has an international standing is "Bergey's Manual of Determinative Bacteriology". It represents the collaborative efforts of over 100 of the best qualified microbiologists at the time it was brought together and is a mounumental work. It is based on the International Rules of Nomenclature of the Bacteria and Viruses established by the International Committee of Bacteriological Nomenclature in 1947.

As outlined in Bergey's manual the entire group of bacteria (Class-Schizomycetes) comprises some 1500 species. These are divided into 10 orders differentiated from one another primarily on the basis of morphological characters and type of motility.

I. **Pseudomonadales.** Cells rigid, spheroidal or rod-like, straight curved, or spiral, some groups from trichomes; motile species have polar flagella.

II. **Chlamydobacteriales.** Rod-like cells in trichomes, often sheathed, deposit iron hydroxide in sheath; flagella sub-polar when present.

III. **Hyphomicrobiales.** Cells spheroidal or ovoid, connected on stalk or threads, no trichomes; exhibits budding and longitudinal fission; flagella polar when present.

IV. **Eubacteriales.** Typically unicellular spheroidal or rod-like cells, no trichomes, sheath or other accessory structures, motile species have peritrichous flagella.

V. **Actinomycetales.** Cells branch, many species form mycelia and mold like conidiophores and sporangiophores, polar flagellate, sporangiospores in only one species.

VI. **Caryophanales.** Trichomes often very long, peritrichous flagella.

VII. **Beggiatales.** Alga like trichome or coccoid cell, accumulate elemental S, gliding and oscillatory or rolling motion; no flagella.

VIII. **Myxobacteriales.** Cells coccoid, rod-like or fusiform, communal slime, fruiting bodies, cells flexuous, gliding motility in contact with solid surface; no flagella.

IX. **Spirochaetales.** Elongated, spiral cells rotatory and flexing motions and translatory motility; no flagella.

X. **Mycoplasmatales.** PPLO extremely pleomorphic and easily distorted cells without cell walls; complex life cycle; non-motile.

Phylogenetic relationships. The phylogenetic relationships of bacteria are not clear and because of this, their classification has undergone a number of changes. According to recent scheme, the bacteria are placed in the phylum Schizomycophyta in the Sub-kingdom Thallophyta, which include all plants that do not form embryos during development. According to other modern taxonomic systems these organisms have been placed in the kingdom Protista along with algae, fungi and protozoa, or in the kingdom Monera together with blue-green algae.

According to one school of thought, the bacteria have been descended from the blue-green algae, after becoming adapted to a saprophytic or parasitic existence losing their chlorophyll. This view is based on the general similarity of the cell structure of these two forms. According to other investigators, the fact that many bacteria possess flagella indicates that these organisms descended from simple flagellated forms, and perhaps these forms have also given rise to the green algae. There are others who believe that the heterotrophic bacteria of today have been evolved from autotrophic ones. According to these workers, the autotrophic bacteria may have appeared before any of the chlorophyll containing plants. It is also suggested that different groups of bacteria have descended independently from different ancestors. There are still others who believe that bacteria may be a terminal group in evolution that has given rise to no other forms or if they have not descended from the blue-green algae, perhaps the blue-green algae have descended from them.

ECONOMIC IMPORTANCE OF BACTERIA

Bacteria are of great economic value in our daily life. They are useful as well as harmful.

A. Useful bacteria. They do great benefit to the agriculture and industries.

1. Agriculture. Many species of saprophytic and symbiotic bacteria add to the fertility of the soil and provide nitrogen to the plants.

(*a*) **Ammonifying bacteria.** *Bacillus subtilis, B. mycoides, B. ramosus*, etc., act upon the dead animal and plant tissues and decompose their complex organic compounds like proteins into ammonium compounds. They are also known as **putrefying bacteria.**

(*b*) **Nitrifying bacteria.** *Nitrosomonas* oxidise the ammonium compounds into nitrites in presence of free oxygen and *Nitrobacter* oxidise nitrites into nitrates in the presence of free oxygen. Thus ammonifying and nitrifying bacteria increase the amount of nitrogenous compounds in the soil. Dead plants, animals and dung, etc., are converted into humus by the action of putrefying bacteria. This humus itself acts as fertilizer for plants.

(*c*) **Nitrogen fixing bacteria.** They are *Azotobacter, Clostridium* and *Rhizobium* spp. They fix free nitrogen of the soil and make it available to the plants. The first two bacteria live freely in soil and fix the atmospheric nitrogen in the form of nitrogenous compounds in the soil. The third one is a symbiotic type. They live in the root nodules of leguminous plants, take the free atmospheric nitrogen and fix it within its tissues. These bacteria enable plants to grow in soil where no nitrogenous fertilizers are available. The leguminous plants make the soil rich in nitrogen, and therefore used as green manures.

Nitrogen fixation. The phenomenon of nitrogen fixation takes place by special type of bacteria which fix free atmospheric nitrogen gas into ammonia by means of symbiosis with leguminous plants. The bacteria taking part in this process are *Rhizobium leguminosarum* (Rhizobiaceae) which live in soil. These bacteria produce IAA (Indol-Acetic Acid) due to which the root hairs curl. These rod-like bacteria penetrate through the tip of the root hair forming a continuous 'infection thread' that enters the cortical region within twentyfour hours. During its passage through the root hair, the infection thread gets surrounded by a cellulose wall. This wall is secreted by the host as a reaction to the infection. The infection thread ramifies in the cortical region and the bacterial rods are released in the cytoplasm of the cells which are stimulated. These cells enlarge and multiply to form the characteristic nodules all over the root system. On the outside, the root nodule possesses a cortical layer which is followed by an actively proliferating meristematical region, then the vascular system enclosing in the centre the bacterial zone possessing abundantly the branched rods of *Rhizobium leguminosarum.* These bacteria absorb atmospheric nitrogen and make it available to the host plant in the form of ammonia which is being converted into nitrates. In turn, the bacteria get shelter and carbohydrate-nutrition from the leguminous plant. On the death and decay of root nodules the rhizobia are again set free in the soil; the decomposition of roots adds nitrates into the soil thus increasing fertility of the soil.

Azotobacter is also found in the soil; this fixes the nitrogen gas of the atmosphere in the presence of carbohydrates. This fixation of free nitrogen from the atmosphere through ammonia into free nitrates and again their conversion into ammonia and free nitrogen takes place by means of nitrifying and denitrifying bacteria, along with other organisms. This process is termed nitrogen cycle.

2. Industry. A large number of saprophytic bacteria are employed in the manufacture of various industrial products.

(*a*) **Butter making industry.** Saprophytic bacteria such as *Lacto bacilli* popularly known as *starters* make the milk sour and produce various flavours. These bacteria are largely employed in butter industry for ripening milk and producing flavours in butter.

(*b*) **Cheese making industry.** Bacteria are employed in this industry. First the casein of milk is coagulated and then it is ripened by certain bacteria. Bacteria make the casein spongy, soft and give it characteristic taste and flavour.

Pasteurization. Heating milk at 62°C for 30 minutes or at 71°C for 15 seconds.

(*c*) **Vinegar making industry.** *Bacillus aceti* convert the sugar solution into vinegar.

(*d*) **Alcohol and acetone manufacture.** Butyl alcohol and acetone are manufactured by the action of bacteria on molasses.

(*e*) **Tobacco curing.** Crude dry tobacco leaves pass through curing and ripening processes before they are ready for use. Bacteria are employed in both these processes and the peculiar taste and smell in the tobacco is due to the bacterial activity. For this purpose molasses and alcohol are added to tobacco.

(*f*) **Tea curing.** Crude tea leaves are acted upon by certain bacteria. The process is known as curing, which is employed to impart a peculiar taste and flavour to the leaves. For this purpose alcohol is added to tea leaves.

(*g*) **Leather tanning.** The hides and skins after drying, salting and clearing are steeped in fluids containing specific bacteria. The process of fermentation goes on for some time and then they are transferred to tan-pits and are further allowed to be fermented. This whole process is known as tanning and the bacteria employed in the process are obtained from cowdung and the excreta of dogs and poultry.

(*h*) **Fibre retting.** Retting is the process of separating fibres from the plant tissues. Bacteria are employed in this industry, which cause decay of the softer tissues and render fibres easily separable mechanically. Fibres of flax, hemp, jute, coconut and other fibrous plants are obtained by immersing the specific plant organs in stagnant pond water where bacteria develop and cause retting.

(*i*) **The sewage work.** In order to remove solid and semi-solid constituents of sewage it is allowed to putrify. Putrifying bacteria are allowed to act upon sewage under anaerobic conditions. It gets decayed and liquified. It is now filtered and the liquid is either drained out to the river or used as manure in fields. For this purpose, in the soak pits the horse dung is filled up.

(*j*) **Ensilage.** It is the process of preserving green fodder in pits. Certain bacteria help in the preservation of fodder.

(*k*) **Medicines.** Antitoxins are the chemical substances produced in the host tissues in response to the attack of parasitic bacteria. Different vaccines and serums now prepared from these antitoxins are used in the treatment of specific ailments. The antibiotics such as streptomycin, aureomycin, chloromycetin, etc., are obtained from certain actinomycetous bacteria.

B. Harmful bacteria. Many bacteria are harmful to human affairs in many ways.

1. Pathogenic bacteria. These bacteria cause great losses to animal and plant life by causing various diseases in them. Cholera, typhoid, pneumonia, dysentery, tuberculosis, tetanus, etc., are more common human diseases. 'Ring disease of potato', 'yellowing rot of wheat,' 'Citrus canker,' 'wilt of cucumber' and 'crown gall' are the common bacterial diseases of plants.

Some common human and plant diseases. *Diplococcus pneumoniae* causes pneumonia. *Streptococcus phylgens* cause sore throat, *Staphylococcus aureus* causes boils. *Vibrio comma* causes asiatic cholera, *Mycobacterium tuberculosis* causes tuberculosis, *Corynebacterium*

Table 2.3. DISEASES CAUSED BY BACTERIA

In Plants

Name of disease and host	*Causal organisms*
Wilt of potato	*Pseudomonas solanacearum*
Wilt of tomato	*Pseudomonas solanacearum*
Wilt of tobacco	*Pseudomonas solanacearum*
Wilt of pepper	*Pseudomonas solanacearum*
Blight of beans	*Xanthomonas phaseoli*
Black rot of crucifers	*Xanthomonas campestris*
Citrus canker	*Xanthomonas citri*
Ring rot of potato	*Corynebacterium sepedonicum*
Canker of tomato	*Corynebacterium michiganese*
Crown gall of *Rubus* spp.	*Agrobacterium tumefaciens*
Crown gall of apple	*Agrobacterium tumefaciens*
Crown gall of rose	*Agrobacterium tumefaciens*
Crown gall of tomato	*Agrobacterium tumefaciens*
Scab of potato	*Streptomyces scabies*
Fire blight of apple and pear	*Erwinia amylovora*
Soft rot of vegetables	*Erwinia carotovora*
Wilt of cucurbits	*Erwinia tracheiphila*

In Man and Animals

Name of disease	*Causal organisms*
Cholera	*Vibrio cholerae*
Rat-bite fever	*Spirillum minus*
Enteritis in the intestine	*Escherichia coli*
Urinogenital infections	*Klebsiella* spp.
Wound infections	*Proteus* spp.
Enteric fever	*Salmonella* spp.
Bacillary dysentery	*Shigella* spp.
T.B.	*Mycobacterium tuberculosis*
Plague	*Pasteurella pestis*
Whooping cough	*Bordetella* spp.
Undulant fever	*Brucella* spp.
Influenza	*Haemophilis influenzae*
Abscesses	*Staphylococcus* spp.
Gonorrhoea	*Neisseria gonorrhoeae*
Meningitis	*Neissaria meningitides*
Diphtheria	*Corynebacterium diphtheriae*
Pneumonia	*Diplococcus pneumoniae*
Tetanus	*Clostridium tetani*
Anthrax	*Bacillus anthracis*

diptheriae causes diptheria, *Salmonella typhosa* causes typhoid fever, *Heamophilius pertusis* causes whooping cough, *Pastuerella pestis* causes plague, *Pseudomonas solanacearum* causes bacterial wilt of tomato and brown rot of potato, *Xanthomonas citri* causes citrus canker, *Erwinia amylovora* causes fireblight of apples and pears and *Xanthomonas malvacearum* causes angular leaf spot of cotton.

2. Food spoilage. Some saprophytic bacteria grow on unprotected foodstuffs like fruits, pickles, jams, jellies, bread, etc., and spoil them by causing decay. The use of salt, sugar and oil, etc., in preservation of pickles and jams, checks the growth of such bacteria. *Clostridium botulinum* produces a very virulent poison in canned food and many deaths occur due to it.

3. Loss of fertility. Anaerobic bacteria such as *Bacillus denitrificans* reduce the nitrates of the poorly aerated soil to nitrites and then to ammonium compounds, and free nitrogen is liberated. Thus the amount of readily available nitrogen goes to the atmosphere. They are known as denitrifying bacteria.

MYCOPLASMA

(PPLO-Pleuropneumonia like organisms and L-forms)

It has been established now that smallest cell, or free living organisms, is not a bacterium, it is so minute and could not be seen except through an electron microscope. Louis Pasteur discovered the existence of this organism, which is the causative agent of pleuropneumonia, a contagious disease of cattle. But Pasteur could not isolate, culture or see it under a microscope

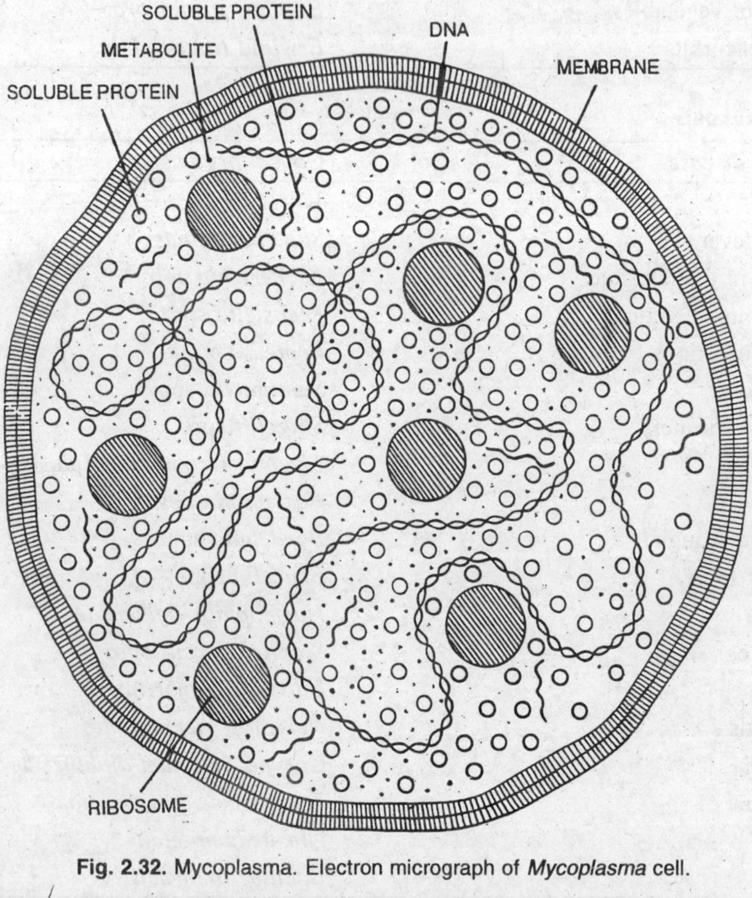

Fig. 2.32. Mycoplasma. Electron micrograph of *Mycoplasma* cell.

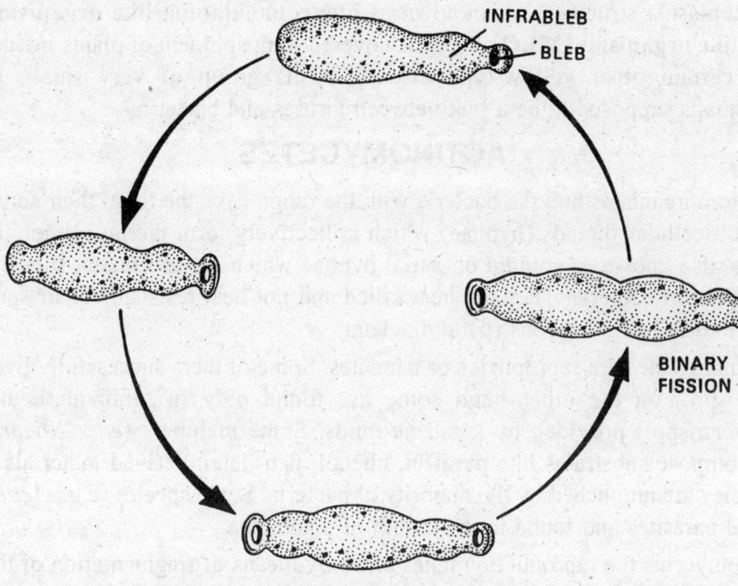

Fig. 2.33. Mycoplasma. Binary fission in *Mycoplasma gallisepticum.*

because it was too small. There are several kinds of this organism, but the smallest is about 0.1 micron in diameter and is known as PPLO (Pleuropneumonia like organisms) or *Mycoplasma.* In size these cells are like viruses in that they can pass through a fine-pored filter; then can however, grow in a nonliving medium, as do bacteria. Their internal organization is also bacterialike. They contain plasma membrane, ribosomes, complete metabolic machinery with some 40 known enzymes and an ill-defined nuclear apparatus lacking a nuclear membrane.

Mycoplasma and plant Pathology

Mycoplasma is a new discovery in plant pathology. Before 1976 this was thought that plant diseases were caused by fungi, bacteria, nematodes and viruses. However, in 1967 a team of Japanese workers proved that the plant diseases under the *yellow* group of viruses have been caused by *Mycoplasma,* which possess different characteristics from those of viruses.

For the first time in 1898, Nocard and Roux had discovered *Mycoplasma* as a disease incitant in animals. They also recorded that mycoplasma was similar to virus in size and could be grown in artificial culture. *Mycoplasma mycoides* (causal agent of pneumonia) was cultured by that time.

As mentioned above, mycoplasma causes *yellow* disease in plants. Since 1967 most of the plant diseases of yellow group have been reported to be caused by mycoplasma. Some important mycoplasma diseases occurring in this country have been reported from Indian Agricultural Research Institute, New Delhi. These diseases are: **Little leaf of brinjal; Citrus greening; Sandal spike; Grassy shoot of sugarcane; Rice yellow dwarf; Cotton little leaf** or **Cotton stenosis; Sesamum phyllody** and some others.

Mycoplasma has been isolated in pure cultures from insects, diseased mammals and infected plants. Mycoplasma is grown in highly specific culture medium containing sterol.

The mycoplasma is non-motile, gram-negative and contains both DNA and RNA. The respective cells of mycoplasma remain surrounded by proteinaceous membranes. Mycoplasma reproduces by budding and binary fission. Mycoplasma varies in its shape. The spherical forms measure 80-150mμ. The pleomorphic forms measure 210-435mμ and the large pleomorphic forms measure upto 1000mμ. The filamentous forms with branching structures have also been reported.

The mycoplasma structures also known as **pleuropneumonia like organisms** (PPLO) or **mycoplasma-like organisms** (**MLO**), were discovered in the phloem of plants infected with aster yellows and certain other yellow-type diseases. This group of very small, bacteria like microoraganisms is supposed to be a link between viruses and bacteria.

ACTINOMYCETES

These microorganisms link the bacteria with the fungi. Like the fungi their somatic structure consists of multicellular threads (hyphae) which collectively form the mycelium. They produce asexual spores also known as conidia on aerial hyphae which act as asexual spores of fungi and unlike endospores of bacteria. They are heat killed and not heat resistant. Their spores and cells are never more than 2 or 3 microns (μ) in diameter.

The Actinomycetes are saprophytes or parasites. Some of them successfully live in soil (*e.g., Streptomyces* spp.) on the other hand some are found only in semiaquatic habitats (*e.g., Micromonospora* spp.) provided by lacatrine muds. Some members (*e.g., Nocardia* spp.) are known to decompose substrates like paraffin, phenol, petroleum derived materials which are so resistant that they are untouched by the majority of bacteria. Some species (*e.g., Actinomyces* spp.) are specialized parasites and found in the mouth of mammals.

In Actinomycetes the reproduction takes place by means of fragmentation of the hyphae, by conidia produced singly or in chains. In certain Actinomycetes (family-Actinoplanaceae) the reproduction takes place by minute sporangia producing motile (Actinoplanes) or non-motile (Streptosporangium) spores.

Antibiotics. The genus *Streptomyces* of this group is most important, some species of which produce antibiotic substances of great medicinal value like streptomycin, chlorotetracyline, oxytetracycline, tetracyline, chloramphenicol and erythromycin.

CYANOBACTERIA

There are two main patterns of cellular organization–*prokaryotic* and *eukaryotic.* On the prokaryotic side, there are diverse forms of bacteria and a group generally termed *blue-green algae.* The term algae was applied to these organisms on the basis of their photosynthetic activities before their structural relationship to bacteria was uncovered with the electron microscope; they are, more properly referred to as *blue-green bacteria* or *cyanobacteria.* The cyanobacteria have been included in Volume 3 of Bergey's Manual. According to Bergey's classification they are *oxygenic phototrophic bacteria.*

In cyanobacteria their nuclear material deoxyribo-nucleic acid (DNA), is not delimited from the remainder of the protoplasm by a nuclear membrane, but rather it is dispersed to some degree throughout the cell. The membrane bounded plastids are absent. Large aqueous vacuoles, like those which occur in many green algae are absent from the cells of cyanobacteria or blue green algae. The cell walls of cyanobacteria show some chemical similarity to those of bacteria. Certain cyanobacteria may be infected with viruses which resemble bacteriophages advocates further similarity, between cyanobacteria and bacteria.

Characteristic Features of Cyanobacteria

1. They are omnipresent, and occur in all possible kinds of habitats.
2. The pigments found in this group are-chlorophyll *a,* β–carotene, Antheraxanthin, Aphanicin, Aphanizophyll, Flavacin, Lutein, Myxoxanthophyll, Oscilloxanthin, Zeaxanthin, Allophycocyanin, Phycocyanin, Phycoerythrin.
3. The storage products are cyanophycean starch and protein.
4. The flagella are absent.
5. They may be unicellular (*e.g., Chroococcus, Tetrapedia, Gloeocapsa*), colonial (*e.g., Aphanocapsa, Nostoc, Aphanothece*) and *filamentous* (*e.g., Oscillatoria*).

There are about 150 genera and 1,500 species of fresh water, marine and terrestrial blue green algae (cyanobacteria) in this class. The synonyms of the class are Cyanophyceae, Schizophyceae, and Phycochromophyceae. Since they are considered as bacteria (procaryotic), they are called **Cyanobacteria.**

Occurrence and distribution. The members of this class are worldwide in distribution and occur in all possible kinds of habitats. Majority of the Chamaesiphonales are marine occurring in the littoral and sub-littoral zones, while the majority of Chroococcales, and Hormogonales are fresh water. The red colour of the red sea is due to a blue green alga (Cyanobacteria), *i.e., Trichodesmium erythrium.* This form is found in the bottom of the sea. There are several terrestrial forms which occur on rocks and walls, *e.g., Porphyrosiphon, Scytonema,* etc. Some are found in the soil upto the depth of 8 feet. Some members of Chroococcaceae, Stigonemataceae and species of *Phormidium* inhabit in hot springs even upto 85°C temperature. The salts of calcium and magnesium are precipitated in the bottom by the thermal blue green algae and are called 'travertine'. Calcium carbonate is deposited by *Schizothrix, Lyngbya* and *Rivularia* species. Several Cyanobacteria, live in symbiosis. Some forms are endophytic. *Nostoc* colonies are found inside the tissues of *Anthoceros* and *Notothylas. Anabaena cycadeae* is found inside the coralloid roots of *Cycas. Anabaena azollae* is found in the tissues of *Azolla.* Many unicellular Chroococcaless are found in the association of unicellular animals such as Rhizopods and Cryptomonas, etc. Several Oscillatoriaceae are parasites. About forty species belonging to four genera of Oscillatoriaceae are parasitic inside the alimentary canal of human beings and other animals. Majority of cyanobacteria like soft water and sometimes found in such an abundance that the whole surface of the water becomes blue green in colour. A large part of the free floating population of microscopic organisms which constitute the **plankton** of lakes and oceans is made up of members of cyanobacteria.

Fossils. The fossils of cyanobacteria are not definitely known from the oldest geological era, *i.e.,* the *Archeozoic,* but immense graphite deposits indicate the presence of organisms whose remains were decomposed by the activity of bacteria. Many impressions have been of cyanobacteria.

Range of structure. The forms may be found as (*i*) unicellular individuals, (*ii*) non-filamentous colonies and (*iii*) filamentous forms.

1. **Unicellular forms.** Many members of Chroococcales are truly unicellular in structure, such as *Chroococcus, Tetrapedia* and *Gloeocapsa* are widespread unicellular cyanobacteria.

2. **Non-filamentous colonial forms.** The colonies are amorphous. Such colonies are resulted from the confluence of the gelatinous envelopes surrounding the individual cells, *e.g., Aphanocapsa, Aphanothece,* etc.

The shape of colony depends upon the plane of division of the cells forming the colony. When the cells divide in one plane the colony becomes plate-like, *e. g., Merismopedia.* When the cells divide in many planes; cubical or three dimensional colonies are formed. Irregular colonies are developed by the irregular divisions of the cells.

3. **Filamentous forms.** When a single cell divides continuously in one plane the ultimate result is a filament. Such types are commonly known in Oscillatoriaceae. In *Oscillatoria* the filament consists of a trichome surrounded by a sheath. In *Schizothirx* several trichomes are surrounded by a common sheath.

In *Nostochopsis,* the filaments are truly branched.

Leaving aside Oscillatoriaceae, in other filamentous forms the 'heterocysts' are found. The heterocysts may be intercalary or terminal.

Structure of the cell. The cell wall is two layered. The outer layer consists of pectic substance and the inner of cellulose.

The complete cell is filled up with cytoplasm. The protoplast may be differentiated into two regions. The outer peripheral region is called the **'chromoplasm'** and the central colourless region is **'centroplasm'** or **cental body**.

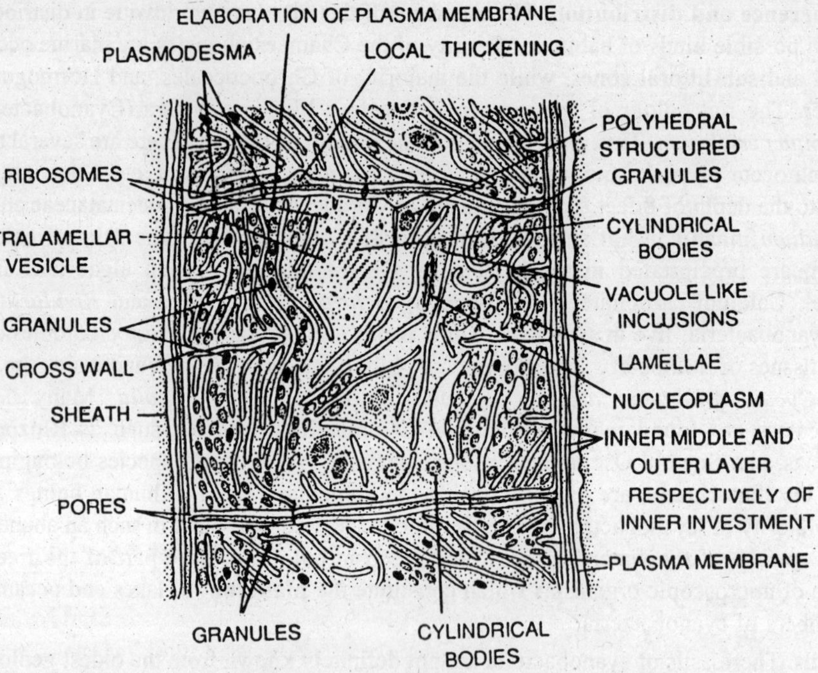

Fig. 2.34. Cynobacteria. Ultrastructure of a cell; a filamentous form

The coloured peripheral chromoplasm contains many pigments, *e.g.,* phycocyanin, phycoerythrin, carotenes, xanthophylls and chlorophylls in dispersed condition. These pigments are not found in chromatophores as in algae. In addition to these pigments oil drops and glycogen granules are also found in this region.

The central body or the centroplasm of the cell is colourless. According to one view, the central body is regarded as primitive nucleus having no nucleoli and nuclear membrane. There is no true nucleus. Several stainable chromatin grains are dispersed in the centroplasm. According to another view, the central body is a store-house of food material. According to West, the central body is an **incipient nucleus.**

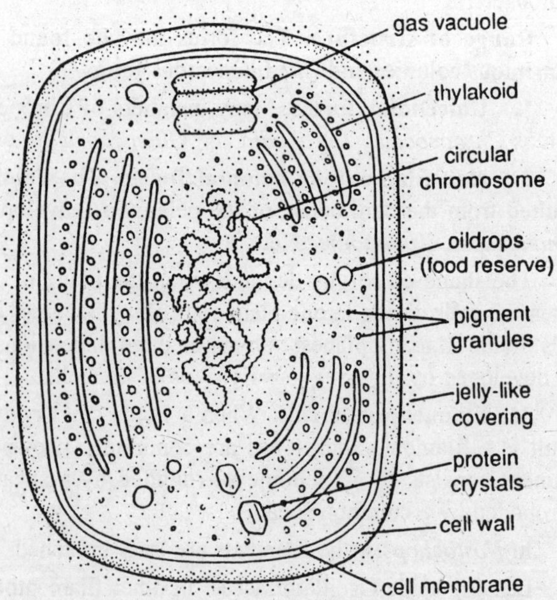

Fig. 2.35. A cyanobacterial cell. Ultra-structure.

In whole of the cytoplasm there are no vacuoles. It is viscous and gel like. Special pseudovaculoes have been reported from the cells. They are small, irregular bodies and commonly found in chromatoplasm. According to one view these pseudovacuoles are gas filled. According to another view, they are filled with viscous liquid.

Cytological work. The chromatin found in the central body has been demonstrated to be composed of DNA (deoxyribonucleic acid), RNA (ribonucleic acid), and protein. The DNA molecule determines the structure and heredity of the cell. The isolation of the radicals of nucleic acids (RNA and DNA) from cyanobacteria gives some substance to the view that a nucleus is present.

Movements of cyanobacteria. In some of the species the trichomes move forward and backward within the sheath, this movement is called 'gliding movement'.

The second type of movement is called the 'oscillation'. This movement is pendulumlike and found in many Oscillatoriaceae. The increase of temperature and illumination accelerates oscillating movement.

Photosynthesis and food products. The process of photosynthesis goes on in cyanobacteria in the same manner as in plants. The storage food products in cyanobacteria are somewhat different from those of green plants. The storage product is usually a starch like substance similar to glycogen and is different from the starch produced in plants. This is known as **Cyanophycean starch.** In many of the cyanobacteria the reserve foods are stored in the form of oils. An amino acid known as diaminopimelic acid is found in the proteins of cyanobacteria and bacteria but never found in higher plants or animals. This characteristic feature supports the view that both bacteria and cyanobacteria have common ancestry.

Nitrogen fixation by cyanobacteria. Many members of cyanobacteria have the ability to fix the atmospheric nitrogen in the soil. Soil is a living mass and apart from soil particles there are in it a number of bacteria, fungi, algae and protozoa. According to Russel, cyanobacteria occupy a volume three times that of the bacteria. For a long time it was felt that these are the bacteria in the sheath which fix nitrogen. P.K. De (1939) conclusively proved that cyanobacteria are the main agents for nitrogen fixation in rice fields, and the part played by bacteria is relatively unimportant. Watanabe (1951) made a far reaching research and proved that *Tolypothrix tenuis* is a strong nitrogen fixer, and it was also reported that it could fix as much as 780 lbs. of nitrogen per acre per year. Allen (1955) found that *Anabaena cylindrica* fixes 2,900 lbs. of nitrogen per acre per year.

Reclamation of alkaline usar soils by cyanobacteria. It has been found that some cyanobacteria form a thick stratum on the surface of saline **usar** soils during the rainy season when other plants including crop fail to grow on such soils. Various species of *Nostoc, Scytonema, Anabaena,* etc., have been found to be plentifully growing on **usar** soils during the rains. According to Dr. R.N. Singh (1950), they can be of use in the reclamation of usar lands. After a year of such reclamation, it has been possible to grow transplanted paddy crop with a yield of as much as 1576-2000 lbs. per acre. It has been found that the pH of the soil experimented upon in this manner fell from 9.7 to 7.6. Moreover, there was an improvement in the tilth and exchangeable calcium by about 20 to 33 percent, the water holding capacity of about 40 percent.

Reproduction. The sexual reproduction is altogether absent. The forms reproduce by means of vegetative and asexual methods.

1. Vegetative reproduction. The vegetative reproduction takes place by several methods.

***(i)* By cell division.** This is only the known method of propagation in the order Chroococcales. The cell becomes constricted in the middle and ultimately divides giving rise to two individuals.

***(ii)* Fragmentation of colony.** The colonies are fragmented into small bits. Each such bit develops into a colony by division of the cells in different planes.

***(iii)* Fragmentation of filaments.** In the family Oscillatoriaceae the filaments break into small fragments. The fragmentation may be accidental or otherwise.

***(iv)* By hormogones.** Inside the sheath, the trichomes are fragmented and hormogonia or hormogones are formed. Each such hormogone or hormogonium may develop into a new individual, *e.g.,* many Nostocales.

***(v)* Hormospores or hormocysts.** During the formation of hormospores, a large quantity of food material is accumulated in the cells of hormogones and they become thick walled, these structures are called the hormospores or hormocysts. Each such structure develops in a new plant on the approach of favourable conditions, *e.g., Westiella.*

***(vi)* Akinete formation.** Sometimes some of the cells of filamentous genera accumulate much of food, become thick-walled and are called the 'akinetes'. These are produced in unfavourable conditions, each such vegetative structure may develop into a new filament, *e.g., Nostoc.* On the approach of favourable conditions the akinetes develop into new filaments; the akinetes are also called the **'arthrospores or resting spores'.**

2. Asexual reproduction. The asexual reproduction takes place by means of (*i*) endospores and (*ii*) exospores.

***(i)* Endospores.** The endospores are produced inside the cell. During the formation of these spores the cytoplasm of the cell becomes cleaved into several bits. Thease bits later become endospores and are liberated. Each spore germinates into a new plant, *e.g., Dermocapsa.*

When the endospores are comparatively smaller in size and larger in number in the cell, they are called **'nannospores'.** Such spores develop individually into new individuals, *e.g., Dermocapsa.*

***(ii)* Exospores.** Such spores are produced outside the cell by constriction. In the case of *Chamaesiphon,* the exospores are developed from the terminal end of the plant in continuous succession. The cyanobacterium is epiphytic upon *Oedogonium.*

Heterocysts. In majority of filaments of cyanobacteria the button-like heterocysts are found. However, they are not found in Oscillatoriaceae. Each heterocyst is a spherical oval structure, with a thick wall of two layers. The inner thin layer is called the inner investment and the outer sheath is called the outer investment. The heterocysts may be intercalary or terminal in position. At the point of contact to the filament there is a pore in the heterocyst. The intercalary heterocyst is bipored and the terminal one is monopored. In the case of *Gloeotrichia, Rivularia, Cylindrospermum,* etc., the heterocysts are terminal and monopored. The filaments of *Anabaenopsis* bear heterocysts at their both the terminal ends and are monopored. In *Nostoc* they may be mono- or bipored.

Origin and affinities of cyanobacteria. The group is considered to be most ancient because of the presence of many special features such as (*a*) absence of the chromatophores, (*b*) presence of incipient nucleus, (*c*) lack of flagella, (*d*) lack of reproductive cells and (*e*) total lack of sexual reproduction.

They are worldwide in distribution and found in every possible type of habitat. An independent origin for the group is almost certain. But the ancestry is not known. The total absence of any swarmer rules out the possibility of flagellate ancestry.

Economic importance of cyanobacteria. Some species with gelatinous matrix including *Nostoc,* are boiled in soups and utilized as food. These form much of the plankton eaten by fish and the human beings in trun consume the fish. The growth of these in abundance makes the water of lakes or rivers unsuitable for human consumption. Excessive growth of some cyanobacteria such as *Aphanizomenon,* in lakes resulted in the death of innumerable fish from a depletion of the oxygen supply. The species of *Microcystis* and *Anabaena* may even kill the cattle which drink the water polluted with them. They cause various liver, heart and lung diseases to the cattle which consume the contaminated water, and ultimately they die. The active principle in *Microcystis* is described as a potent neurotoxin. If the species of *Oscillatoria* are consumed by human beings along with water, they cause the symptoms of itching, conjunctivitis, blocking of nasal passages and bronchial asthma.

The dominance of the cyanobacteria possessing the ability for nitrogen fixation particularly in the tropics suggests that these organisms are of considerable importance in maintaining soil fertility.

DESCRIPTION OF SOME IMPORTANT GENERA CYANOBACTERIA

Family–Chroococcaceae

The majority of the genera of this family are colonial and a few are unicellular. There is no tendency towards the formation of pseudofilament, when the cells are united in colonies.

CHROOCOCCUS

(*chroo,* colour; *coccus,* berry)

The genus is very widespread, the species being either free floating or soil inhabitants. The cells are single or else united into spherical or flattened colonies each containing a small number of cells, the individual sheaths being either homogeneous or lamellated. The plants which grow in water produce a concentric envelope but when they grow on damp soil, the sheath is often asymmetrical. The outer integument is not very gelatinous and indeed is quite thin in some species. The cell contents are variously coloured and either homogeneous or granular.

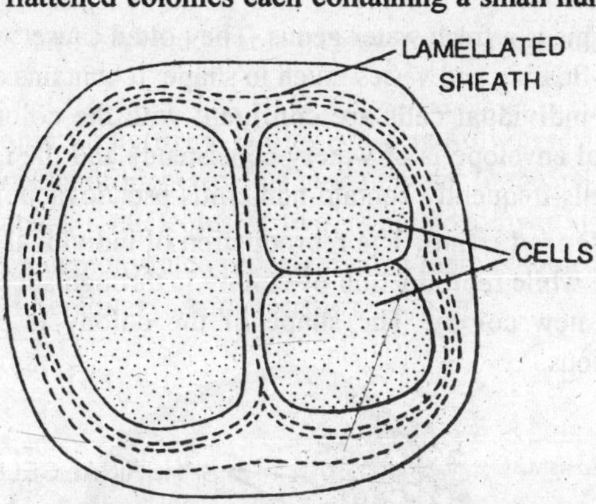

Fig. 2.36. Cyanobacteria. *Chroococcus turgidus.* A nonfilamentous species showing a colony of three cells.

Cytological work. Cytologically *Chroococcus turgidus* represents the simplest condition with the metachromatin granules only just differentiated. *Chroococcus macrococcus* is a more complex type. It contains a central body which is said to contain a fine reticulum with chromatin at the nodal points. During cell division the reticulum divides by simple constriction. There is no evidence of mitosis.

GLOEOCAPSA (Family-Chroococcaceae)

This is found growing in extensive strata on damp rocks. This genus resembles *Chroococcus* in having colonies composed of a few spherical cells, each surrounded by a homogeneous to lamellated sheath. The sheath is coloured yellow, brown, red, blue or violet.

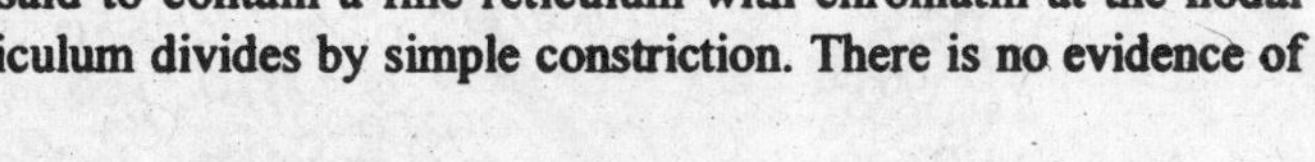

Fig. 2.37. Cyanobacteria. *Gloeocapsa.* Non-filamentous species. A-B, individuals; C, cell division.

Cell structure. Each cell is surrounded by a two layered cell wall. Each cell possesses a colourless central portion, the so-called **central body,** surrounded by an exterior portion, in which the pigment is diffused. The central body seems to represent an incipient nucleus. The central body is not separated from the rest of the cell by a membrane; it has no nucleus; it does not divide mitotically. The colourless granules found in the peripheral region of the cell are said to be of glycogen.

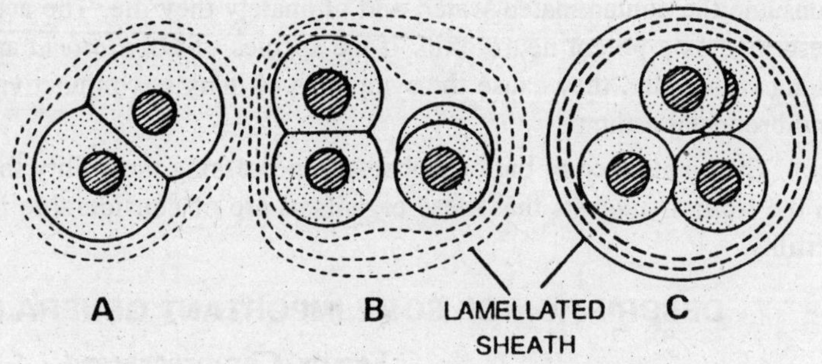

Fig. 2.38. Cyanobacteria. *Gloeocapsa*. Non-filamentous species A-C, various stages of cell division and colony formation.

Reproduction. Only vegetative method of reproduction takes place. The cell becomes constricted in the centre and ultimately divides giving rise to two individuals.

MICROCYSTIS (*micro,* small; *cystis,* bladder)

This is a fresh water genus. They often cause 'water blooms' in hard water lakes. The thallus is free-floating and varies much in shape. It contains a mass of single spherical cells but the sheaths of the individual cells are confluent with the colonial envelope. The gelatinous matrix of the colonial envelope is of watery consistency and the margins of the matrix are usually not evident. The cells frequently contain numerous pseudovacuoles.

Reproduction. The reproduction of individual cells takes place by means of fission in three planes, while reproduction of colony is through successive disintegrations, each portion develops into a new colony. The shape of the colony is primarily determined by the environmental conditions.

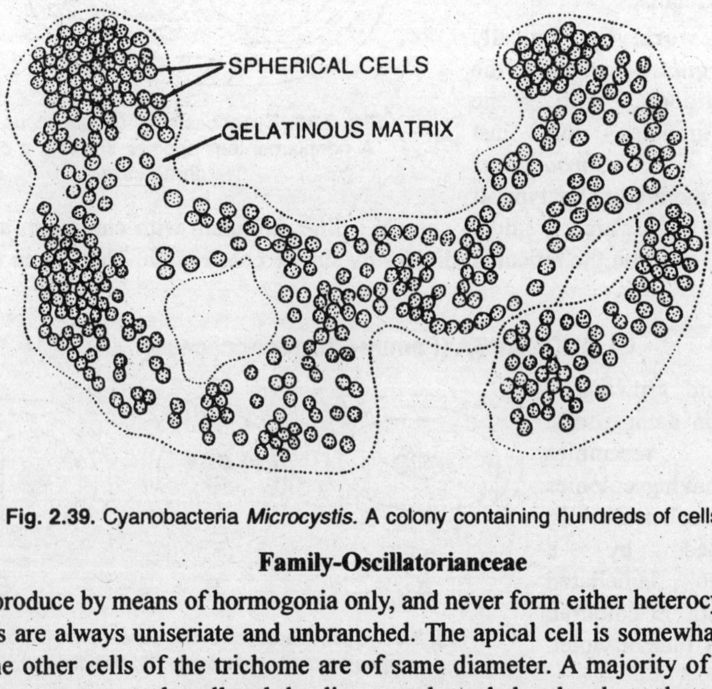

Fig. 2.39. Cyanobacteria *Microcystis.* A colony containing hundreds of cells.

Family-Oscillatorianceae

They reproduce by means of hormogonia only, and never form either heterocysts or akinetes. The trichomes are always uniseriate and unbranched. The apical cell is somewhat rounded at its apex while the other cells of the trichome are of same diameter. A majority of genera possess gelatinous, homogeneous or lamellated, hyaline or coloured sheaths about the trichomes. Some

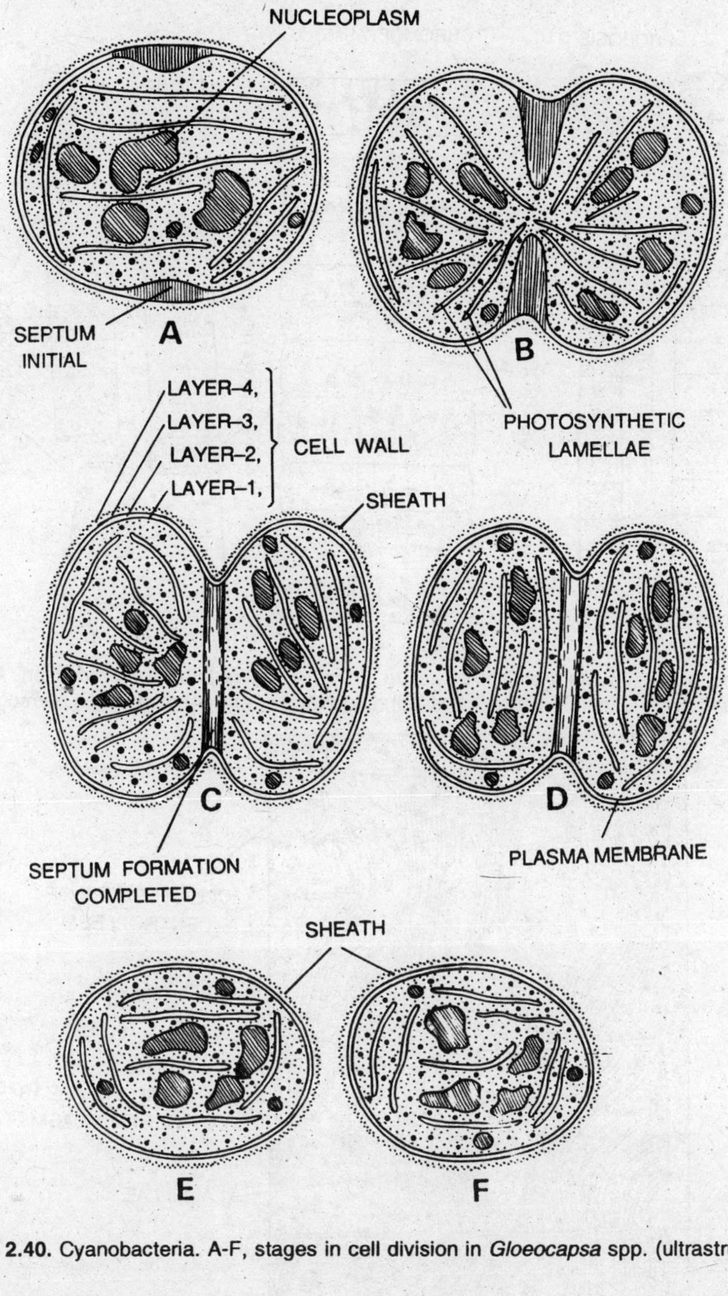

Fig. 2.40. Cyanobacteria. A-F, stages in cell division in *Gloeocapsa* spp. (ultrastructure).

genera possess several trichomes within a common sheath; others have one trichome within a sheath.

Genus-OSCILLATORIA

Occurrence. The species of *Oscillatoria* are commonly found in fresh water. They are found in ditches, ponds, pools and even the drains. Some species are also found on damp soil and damp rocks. They are found in the form of scums either in free floating condition on the surface of water or in the bottoms of pools and puddles. Some species are commonly found in polluted water.

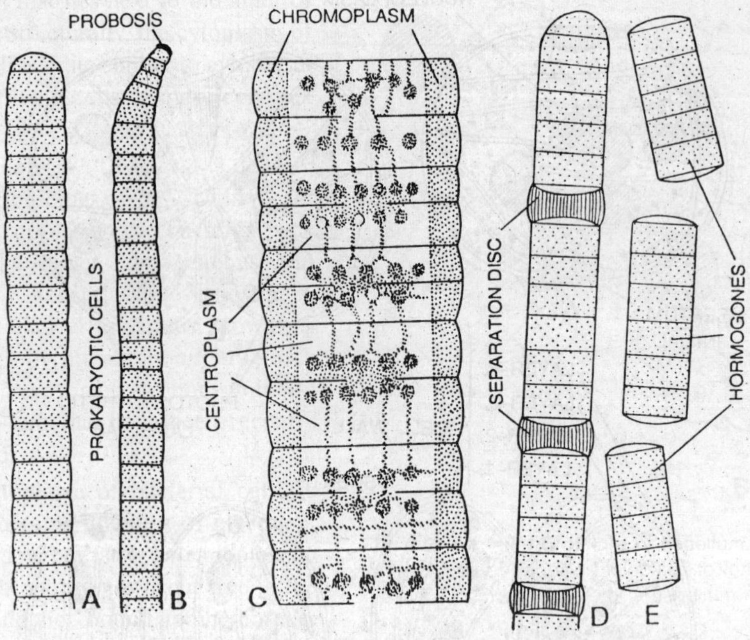

Fig. 2.41. Cyanobacteria. *Oscillatoria* sp. Vegetative structure and vegetative reproduction. A, *O. limosa*; B, *O. proboscidea*; C, cell structure (detailed); D, formation of hormogones; E, hormogones.

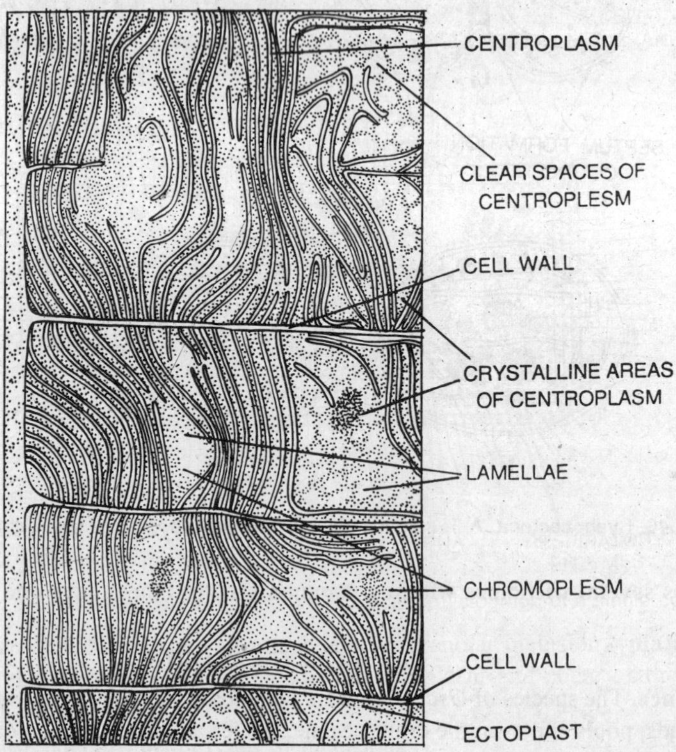

Fig. 2.42. Line diagram of the electron micrograph of *Oscillatoria chalybia* showing membranous nature of the pigment bearing chromoplasm, the centroplasm probably contains DNA is irregular in shape and not separated by a membrane from the chromoplsam. The centroplsam possesses granular background in which are embedded crystalline-like areas and clear spaces.

Structure. The body is filamentous. The filaments are uniseriate and unbranched. Each filament consists of a trichome ensheathed by a gelatinous sheath. The filaments are found either singly or in masses. The cells of the filament are placed end to end like coins. The cells of the broad species are broader than their length, but the cells of narrow species are much longer than their breadth. The terminal cell of filament may be somewhat round at its apex. In *Oscillatoria proboscidea* the apex of the filament forms a proboscis like structure.

The cells of the trichome are somewhat rectangular. The protoplast of each cell may easily be differentiated into two regions. The peripheral coloured region is called the chromatoplasm. This region contains many pigments such as phycocyanin, phycoerythrin, carotene, xanthophyll, chlorophyll and food reserves such as glycogen granules, etc. In addition to pigments pseudovacuoles are also found in the region. The central colourless region of the cells is called the centroplasm or central body. True nucleus is absent. The central body contains chromatin granules in dispersed condition. The central body represents the very primitive form of nucleus. According to West, this is an **incipient nucleus.** There are no nucleoli and nuclear membrane.

The cell wall of the cell is two layered, the outer layer consists of pectic substances whereas the layer next to protoplast consists of cellulose.

Movement of the filament. The filaments of *Oscillatoria* oscillate in pendulum-like fashion. The trichomes are ensheathed with gelatinous matrix and because of the swelling and expansion of this matrix the filaments move like the pendulum of the clock.

Reproduction. The reproduction takes place by vegetative method.

The vegetative reproduction takes place by means of **hormogones.** At various places of the filament colourless biconcave separation discs are formed. The filament breaks at these points giving rise to hormogones. Each hormogone consists of two or more cells. Each hormogone develops into a new plant.

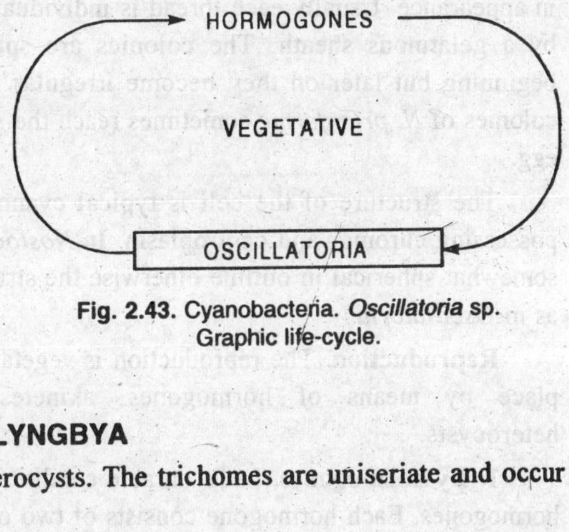

Fig. 2.43. Cyanobacteria. *Oscillatoria* sp. Graphic life-cycle.

Genus-LYNGBYA

The members do not form spores or heterocysts. The trichomes are uniseriate and occur singly in the enveloping sheath.

All cells are capable of division, and reproduction is by hormogones. The enveloping sheath is often thick and coloured.

Family-Nostocaceae

The trichomes of this family are always unbranched, uniseriate and without attenuation at their apices. The trichomes remain always surrounded by a sheath. They may be straight, spiral or twisted. The sheaths surrounding the trichomes are homogeneous. The cells are rounded or barrel-shaped and with or without constrictions at cross walls. The protoplasts of the vegetative cells may possess a homogeneous or granular structure and their colour may be blue green. All genera produce heterocysts, which may be terminal, intercalary, solitary or catenulate. Akinetes are also formed in several genera. They may be adjacent to or away from the heterocysts. The akinetes are longer than ordinary vegetative cells. They may be of different shapes.

Genus–NOSTOC (Family-Nostocaceae)

Occurrence. The species of *Nostoc* occur in fresh water ponds, pools, puddles and ditches. they are found in the form of mucilaginous masses free floating or attached among the grasses or other aquatic plants growing in the ponds. Some species are found on damp soil and some on the wet bark of the trees in irregular mucilaginous masses. Many species grow on the damp soil among mosses. Some species are sub-terranean and found in the soil even in the depth of three feet or more than that. A few species are found in the running streams of mountains. Some species are found in the bottom of the ponds attached to some substratum.

Certain species are endophytic. The colonies of *Nostoc* are found inside the tissues of *Anthoceros.* Certain species are found in the tissues of lichens, *e.g., Nostoc punctiforme* is found in a lichen *Peltigera canina.* Certain species of *Nostoc* are found in paddy fields where they fix atmospheric nitrogen in the soil and enrich the fertility of the soil of the fields.

Structure. Innumerable contorted filaments of *Nostoc* are found embedded in the mucilaginous envelope. The mucilaginous envelope forms more or less a firm boundary. Each thread possesses many spherical cells and it becomes moniliform in appearance. Usually each thread is individually ensheathed by a gelatinous sheath. The colonies are spherical in the beginning but later on they become irregular in shape. The colonies of *N. pruniforme* sometimes reach the size of a hen's egg.

The structure of the cell is typical cyanophycean type possessing chromo- and centroplasm. In *Nostoc* the cells are somewhat spherical in outline otherwise the structure is same as in *Oscillatoria.*

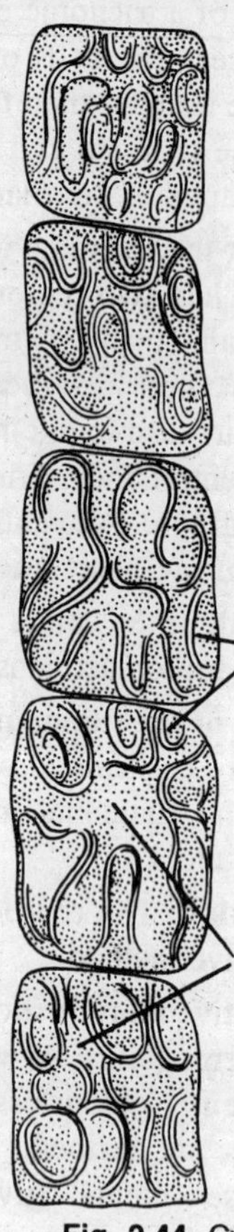

Fig. 2.44. Cyanobacteria. *Nostoc muscorum.* Enlargement of several cells showing membranous nature of chromoplasm.

Reproduction. The reproduction is vegetative and takes place by means of hormogones, akinetes and rarely heterocysts.

1. By hormogones. In favourable conditions the filaments break in small pieces called the hormogones. Each hormogone consists of two or more cells. Usually the filaments break at the heterocysts. Each hormogone is capable to give rise to a new plant. The hormogones may germinate after liberation from the gelatinous matrix or still within it.

2. Akinetes or arthrospores. The akinetes are produced in mature colonies. They are formed in unfavourable conditions. The akinetes are developed in between the heterocysts of the filament. Each cell develops into a single akinete. There are much food reserves in each cell and the walls are being thickened. They are also called arthrospores or resting spores. They are perennating bodies. They survive in adverse conditions even for the years. On the approach of favourable conditions they germinate directly or indirectly giving rise to new filaments. The contents of the akinete divide into small bits prior to germination. In many species of *Nostoc* a primary hormogone is produced from the germinating akinete.

3. Heterocysts. The reproduction by heterocysts is very rare. The heterocysts germinate in exceptional cases. The contents of the heterocyst of *N. commune* divide giving rise to filament. The newly developed filaments liberate from the heterocysts by breaking their thick walls.

ANABAENA
(Family Nostocaceae)

The filaments of this cyanobacterium occur either singly or in floccose colonies and freefloating or in a delicate mucous stratum in permanent and semipermanent pools. Some species of *Anabaena* are endophytic and live within the roots of *Cycas* and leaves of *Azolla.*

The trichomes are of the same thickness throughout and sometimes slightly attenuated at their apical ends. They are straight, circinate or irregularly contorted and occur singly within a sheath. The sheaths which surround the trichomes are hyaline and generally of watery nature. the sheaths may be broad or narrow, and many planktonic species possess sheaths several times broader than the vegetative cells.

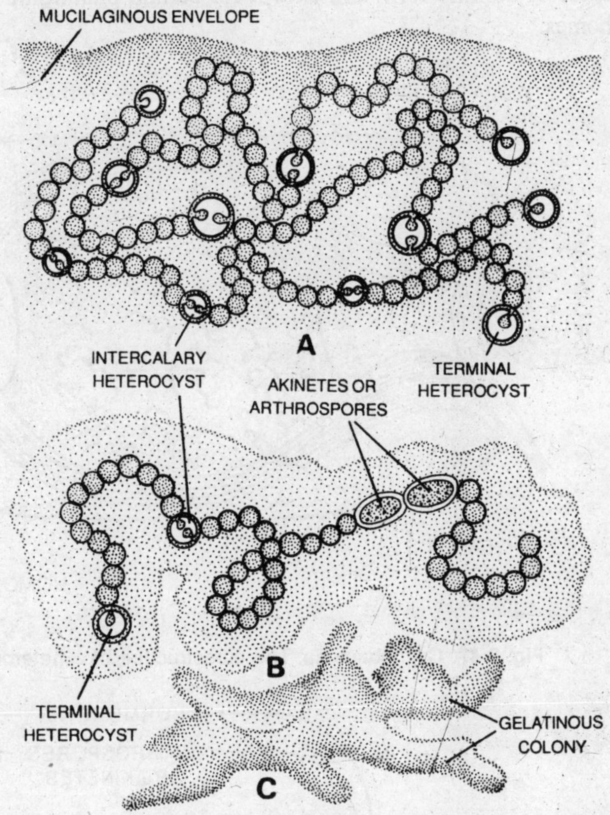

Fig. 2.45. Cyanobacteria. *Nostoc* sp. A, colony within mucilaginous envelope; B, a colony with terminal and intercalary heterocysts and akinetes (arthrospores); C, gelatinous colony in natural form.

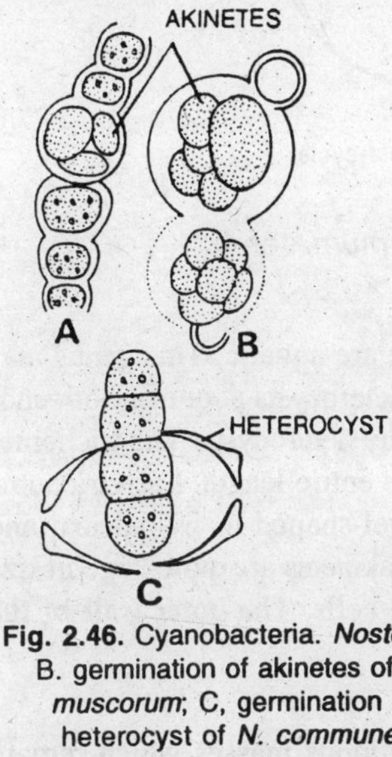

Fig. 2.46. Cyanobacteria. *Nostoc.* A-B. germination of akinetes of *N. muscorum*; C, germination of heterocyst of *N. commune*.

The cells are usually, spherical or barrel-shaped, rarely cylindrical and never discoid. The protoplasts of vegetative cell are usually filled with numerous pseudovacuoles. The intercalary heterocysts are of the same shape as vegetative cells, though slightly larger are present in any trichome. they are generally solitary and may be present in any trichome.

Akinetes develop only next to heterocysts, only remote from them or in both positions. The akinetes may be found singly or in very short catenuate series. They are always larger than the vegetative cells and generally cylindrical and with rounded ends.

Reproduction is similar to that of *Nostoc.*

The descriptions of *Anabaena* and of *Nostoc* are quite identical and it seems difficult to distinguish between the two. In actual practice there is no such difficulty, since

Nostoc possesses a firm gelatinous envelope in which the trichomes are always much contorted, on the other hand *Anabaena* possesses an extremely watery gelatinous sheath and never forms colonies of definite form and except for certain planktonic species, never possesses contorted trichomes.

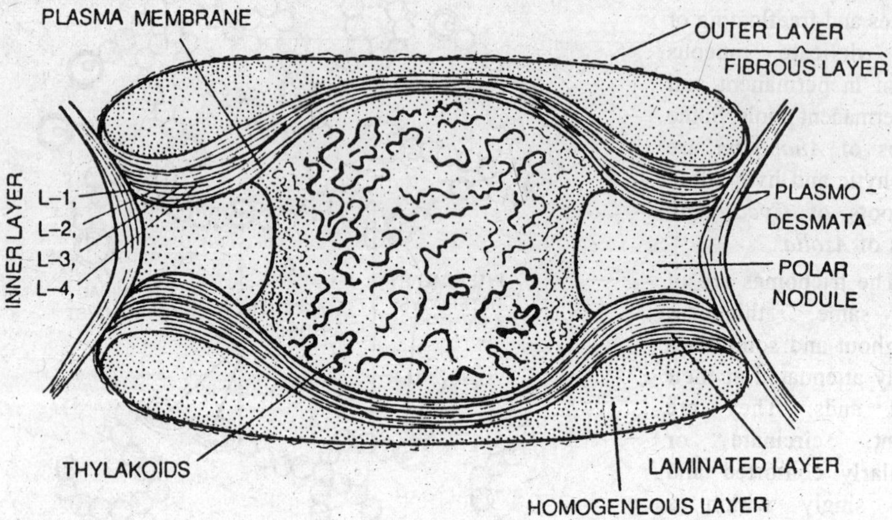

Fig. 2.47. Cyanobacteria. The ultrastructure of a heterocyst in L.S. (diagrammatic)

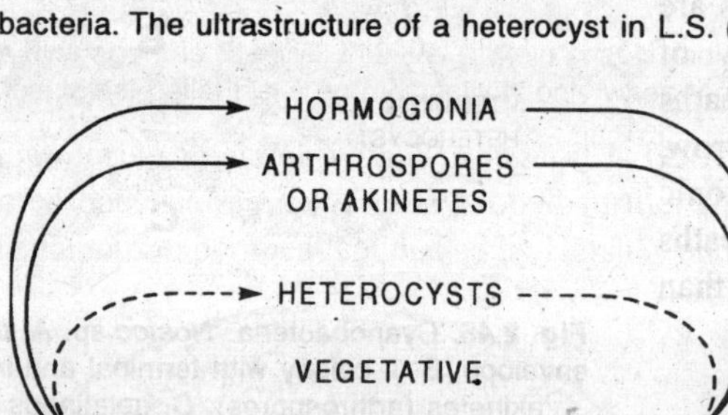

Fig. 2.48. Cyanobacteria *Nostoc* sp. Graphic life-cycle.

CYLINDROSPERMUM (*cylindro*, cylinder; *spermum*, seed)

(Family - Nostocaceae)

Some species of this genus are terrestrial while the remaining are aquatic. This genus may sharply be differentiated from other genera of the family by having heterocysts at one or both ends of the trichomes and by the formation of the akinetes only next to the heterocysts. The trichomes are generally short, straight or curved and of the same diameter of its entire length. Each trichome is surrounded by an extremely mucous sheath. The cells are barrel-shaped or cylindrical and possess rounded ends. Each cell is about twice as long as broad. The akinetes are quite large in size but the heterocysts are approximately the same size as vegetative cells. The outer wall of the akinete is often papillate.

Family-Rivulariaceae

The colonies form spherical, hemispherical or irregular gelatinous masses which remain attached to plants or stones or they occur on the soil. The colonies of *Rivularia atra* are very

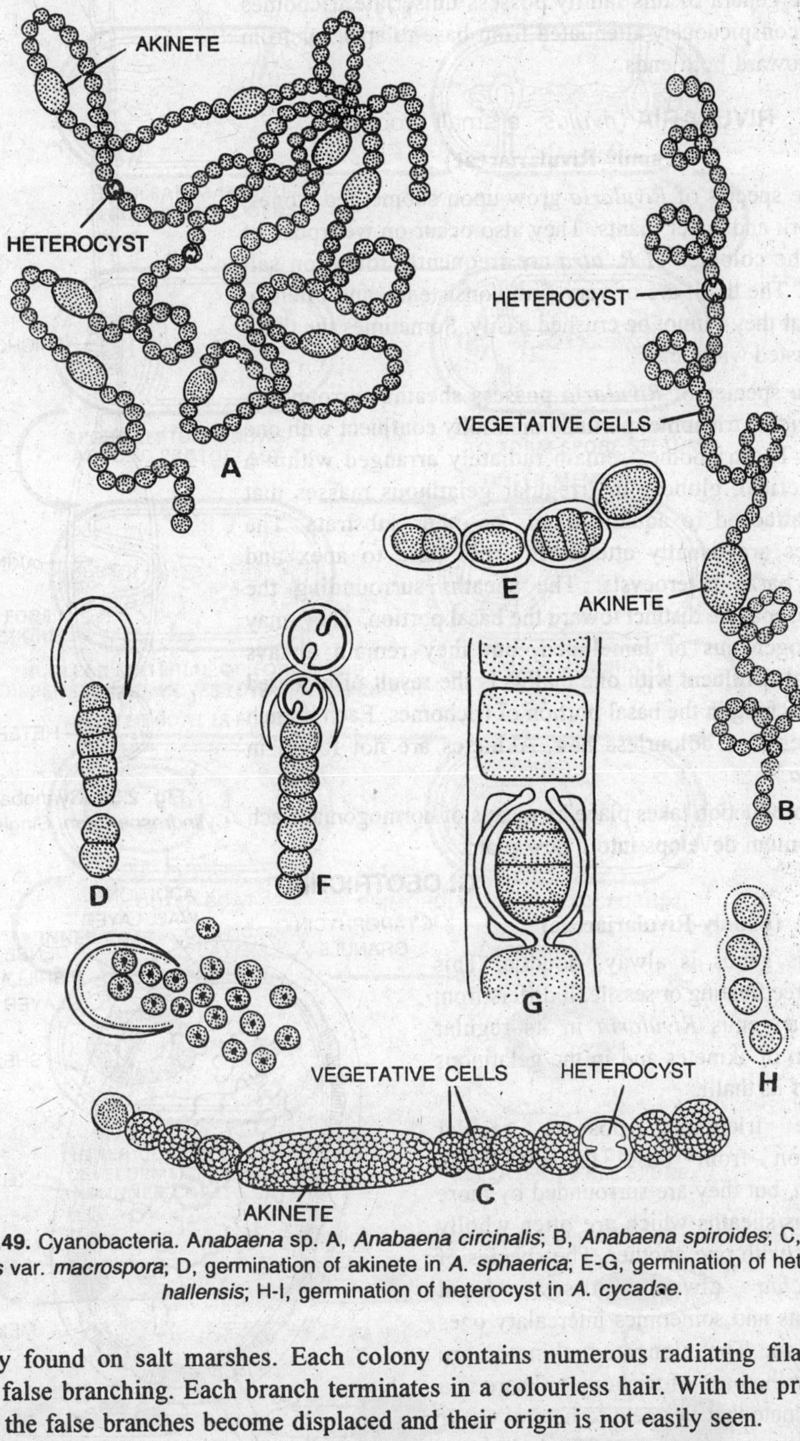

Fig. 2.49. Cyanobacteria. *Anabaena* sp. A, *Anabaena circinalis*; B, *Anabaena spiroides*; C, *Anabaena circinalis* var. *macrospora*; D, germination of akinete in *A. sphaerica*; E-G, germination of heterocyst in *A. hallensis*; H-I, germination of heterocyst in *A. cycadae*.

frequently found on salt marshes. Each colony contains numerous radiating filaments with repeated false branching. Each branch terminates in a colourless hair. With the production of mucilage the false branches become displaced and their origin is not easily seen.

The individual sheaths are visible near the base of the trichomes, but they are different farther up. The heterocysts are basal and in the genus *Gloeotrichia* akinetes or spores are produced next to them.

The genera of this family possess uniseriate trichomes that are conspicuously attenuated from base to apex, or from middle toward both ends.

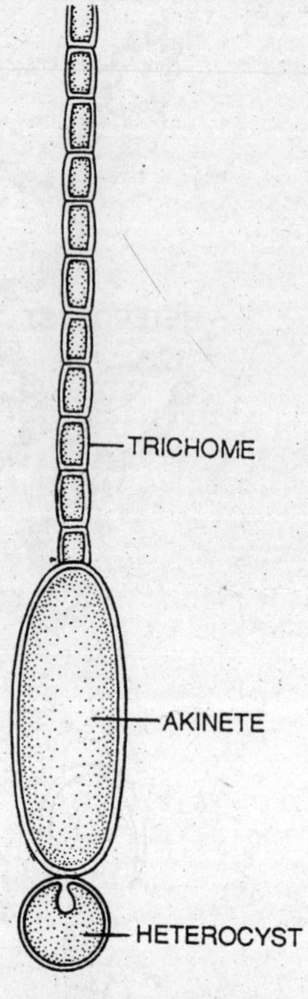

Fig. 2.50. Cyanobacteria. *Cylindrospermum*. Single trichome.

RIVULARIA (*rivulus,* a small brook)

(Family-Rivulariaceae)

The species of *Rivularia* grow upon submerged stones, woodwork and water plants. They also occur on wet rocks of cliffs. The colonies of *R. atra* are frequently found on salt marshes. The thalli are of very firm consistency and often so tough that they cannot be crushed easily. Sometimes the thalli are incrusted with lime.

The species of *Rivularia* possess sheaths surrounding the individual trichomes partially or wholly confluent with one another. The trichomes remain radiately arranged within a hemispherical, globose or irregular gelatinous masses that remain attached to aquatic plants or other substrata. The trichomes are usually attenuated from base to apex and possess basal heterocysts. The sheaths surrounding the trichomes may be distinct toward the basal portion. They may be homogeneous or lamellated, but they remain always somewhat confluent with one thallus is the result of repeated false branching in the basal portion of trichomes. Each branch terminates in a colourless hair. Akinetes are not found in *Rivularia.*

Reproduction takes place by means of hormogonia. Each hormogonium develops into a new plant.

GLOEOTRICHIA

(Family-Rivulariaceae)

This genus is always aquatic. This may be free floating or sessile. It differs from the allied genus *Rivularia* in its regular formation of akinetes and in the gelatinous texture of its thalli.

The trichomes possess regular attenuation from base to apex like *Rivularia,* but they are surrounded by more gelatinous sheaths which are often wholly confluent with one another. The species of *Gloeotrichia* always possess basal heterocysts and sometimes intercalary ones in addition. The elongated akinetes are present at the base of trichomes. There is a single akinete that lies next to heterocyst. If more than one akinetes are found, they are arranged in series of twos or threes.

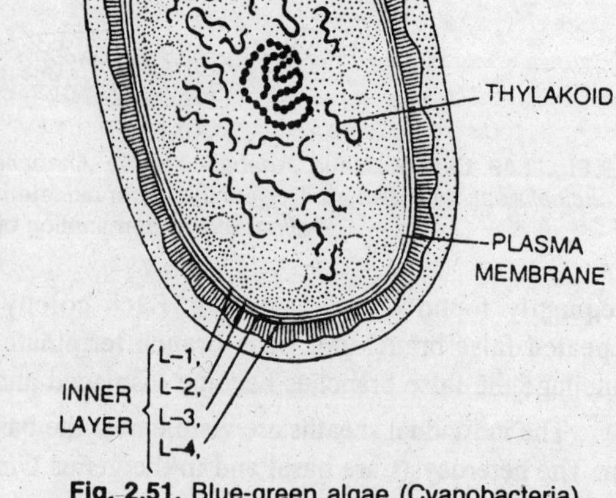

Fig. 2.51. Blue-green algae (Cyanobacteria). *Cylindrospermum.* Ultra structure of an akinete (diagrammatic)

Reproduction takes place by hormogonia and akinetes.

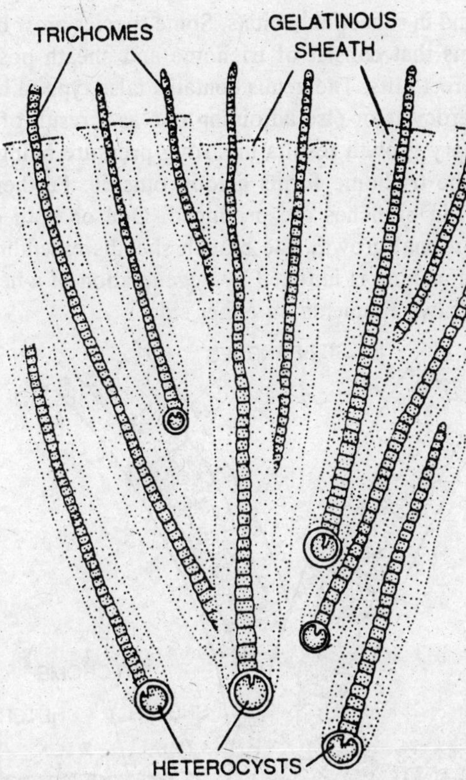

Fig. 2.52. Cyanobacteria. *Rivularia.* Trichome attenuated from base to apex and possess basal heterocysts.

Family-Scytonemataceae

The genera of this family possess uniseriate, falsely branched trichomes of the same diameter throughout their length. The filaments remain surrounded by a firm, well defined, hyaline or coloured and homogeneous or lamellated, sheath. With the result of the segmentation of a trichome into hormogonia, without their liberation from the sheath, the false branching takes place. The ends of the trichomes that develop from hormogonia grow through the old sheath of the parent filament, either singly or in pairs and then secrete a sheath of their own. Akinetes are not found. Heterocysts are always present. In many genera the region of the false branches is correlated with the position of heterocysts.

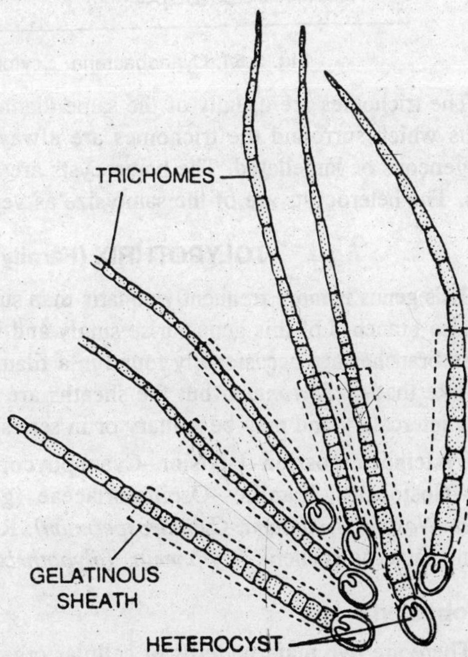

Fig. 2.53. Cyanobacteria. *Gloeotrichia echinulata.* Portion of a sterile colony.

SCYTONEMA (*scyto*, leather; *nema*, thread)

The genus is usually found in subaerial habitats. Some species grow best on damp soil, others on rocky cliffs. The filaments that consist of trichome and sheath possess distinct basal and apical regions forming little erect tufts. The genus contains false type of branching. The branches arise either between two heterocysts or else adjoining one as a result of the degeneration of an intercalary cell. The intercalary growth results in strong pressure being applied to the sheath, which finally ruptures and the trichome forms a loop outside. Further growth results in the breaking of this loop and twin branches are produced. One or both of these branches may subsequently proceed to additional growth, the branch sheaths extending back into the parent sheath. sometimes the false branching is initiated by degeneration of a heterocyst or a vegetative cell and subsequent growth of two filaments on either side.

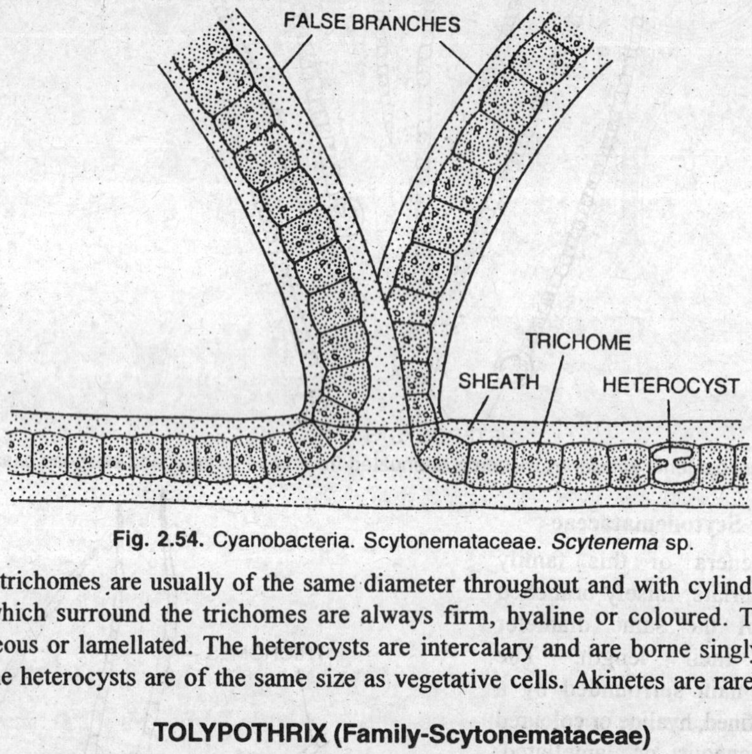

Fig. 2.54. Cyanobacteria. Scytonemataceae. *Scytenema* sp.

The trichomes are usually of the same diameter throughout and with cylindrical cells. The sheaths which surround the trichomes are always firm, hyaline or coloured. The sheaths are homogeneous or lamellated. The heterocysts are intercalary and are borne singly or in twos or threes. The heterocysts are of the same size as vegetative cells. Akinetes are rare.

TOLYPOTHRIX (Family-Scytonemataceae)

This genus is more frequently aquatic than subaerial, usually grows in small clumps or tufts. The false branches of this genus arise singly and immediately adjacent to heterocysts. The pairs of false branches are occasionally found in a filament. The general appearance of the filament is quite like that of *Scytonema* but the sheaths are narrower in *Tolypothrix*. The heterocysts are always intercalary and may be solitary or in series of two to six.

Systematic postion–Division–Cyanophycophyta; Class–Myxophyceae or Cyanophyceae; Order–Nostocales; Families–Oscillatoriaceae (genera: *Oscillatoria, Lyngbya*); Nostocaceae (genera: *Nostoc, Anabaena, Cylindrospermum);* Rivulariaceae-(genera: *Rivularia, Gloeotrichia),* Scytonemataceae-(genera: *Scytonema, Tolypothrix)*

Cyanobacteria

There are two main patterns of cellular organization-**prokaryotic** and **eukaryatic.** On the prokaryotic side, there are diverse forms of bacteria and a group generally termed **blue-green algae.** The term algae was applied to these organisms on the basis of their photosynthetic activities before their structural relationship to bacteria was uncovered with the electron microscope; they are, more popularly referred to as **blue-green bacteria** or **cyanobacteria.** The cyanobacteria have been included in Volume 3 of Bergey's Manual. According to Bergey's classification they are **oxygenic phototrophic bacteria.**

In cyanobacteria their nuclear material deoxyribo-nucleic acid (DNA), is not delimited from the remainder of the protoplasm by a nuclear membrane, but rather it is dispersed to some degree throughout the cell. The membrane bounded plastids are absent. Large aqueous vacuoles, like those which occur in many green algae are absent from the cells of cyanobacteria or bluegreen algae. The cell walls of cyanobacteria show some chemical similarity to those of bacteria. Certain cyanobacteria may be infected with viruses which resemble bacteriophages advocates further similarity, between cyanobacteria and bacteria.

QUESTION BANK

N.B. *Undermentioned questions are asked repeatedly in various Indian Universities and competetive examinations.*

(Cyanobacteria)

Long Answer Type

1. Give the characteristics of cyanobacteria.
2. Write an account of Cyanobacteria and their economic importance.
3. Write a comprehensive account of the methods of reproduction found in Cyanobacteria.
4. How the members of Chroococcaceae reproduce? Write a detailed note.
5. Write a comprehensive note on the occurrence, structure and reproduction of *Nostoc.*
6. Write an account of occurrence, structure, ultra structure and reproduction of *Oscillatoria.*
7. Give the important methods of reproduction in Cyanobacteria giving suitable examples.

Short Answer Type

1. What are Cyanobacteria?
2. Give characteristic features of Cyanobacteria.
3. Write a note on the reclamation of alkaline usar soils by Cyanobacteria.
4. Write a note on origin and affinities of blue-green algae (Cyanobacteria).
5. Write notes on: (a) economic importance of blue-green algae, (b) heterocysts, (c) hormogones, (d) hormospores.
6. Write notes on: (a) *Chroococcus,* (b) *Gloeocapsa,* (c) *Microcystis.*
7. Give characterstic features of *Nostoc*, and its economic importance.
8. Write short notes on: (a) *Oscillatoria,* (b) *Lyngbya.*
9. Write a note on oxygenic phototrophic bacteria.
10. Differentiate between Cyanobacteria and Bacteria.
11. What pigments are found in Cyanobacteria?
12. Point out the main differences between *Anabaena* and *Nostoc.*
13. Write notes on: (a) *Cylindrospermum,* (b) *Rivularia,* (c) *Gloeocapsa,* (d) *Scytonema,* (e) *Tolypothrix.*
14. Write notes on: (a) movements of filaments in *Oscillatoria*, (b) hormogones, (c) chromoplasm.
15. Give the important methods of reproduction in *Nostoc.*

Diagrammatic Type

1. Give the ultrastructure of a typical cyanobacterial cell.
2. Give the ultrastructure of a cell of *Oscillatoria.*
3. Give the ultrasturcture of a part of *Nostoc.*

Multiple Choice Type

1. The Cyanobacteria are:

   (i) eukaryotic, (ii) prokaryotic, (iii) eu- and prokaryotic, (iv) none of above.

2. In Cyanobacteria the nucleus is surrounded by
   (i) single-layered membrane, (ii) double-layered membrane, (iii) no membrane, (iv) none of above.
3. The storage products in blue-green algae are:
   (i) mannitol, (ii) cyanophycean starch and proteins, (iii) algin, (iv) agar-agar.
4. Flagellation is not found in:
   (i) Chlorophyceae, (ii) Phaeophyceae, (iii) Charophyceae, (iv) Myxophyceae.
5. The heterocysts are not found in:
   (i) *Oscillatoria,* (ii) *Nostoc,* (iii) *Rivularia,* (iv) *Gloeotrichia.*
6. The pendulum-like movement is found in:
   (i) *Nostoc,* (ii) *Anabaena;* (iii) *Oscillatoria,* (iv) *Rivularia.*
7. The tissue of *Anthoceros* contains:
   (i) *Nostoc,* (ii) *Cylindrospermum;* (iii) *Oscillatoria;* (iv) *Lyngbya.*
8. The roots of *Cycas* contain:
   (i) *Nostoc,* (ii) *Anabaena;* (iii) *Chroococcus,* (iv) *Oscillatoria.*
9. The leaves of *Azolla* contain:
   (i) *Oscillatoria,* (ii) *Anabaena,* (*iii*) *Rivularia,* (iv) *Gloeocapsa.*
10. Akinete formed only next to heterocyst:
    (i) *Cylindrospermum,* (ii) *Nostoc,* (iii) *Rivularia,* (iv) *Anabaena.*
11. The false branching is found in:
    (i) *Oscillatoria,* (ii) *Scytonema,* (iii) *Cylindrospermum,* (iv) none of above.
12. R. N. singh is related to:
    (i) Bryophytes, (ii) Brown algae, (iii) Red algae, (iv) Blue-green algae.

ANSWERS

1. (ii), 2. (iii), 3. (ii), 4. (iv), 5. (i), 6. (iii), 7. (i), 8. (ii), 9. (ii), 10. (i), 11. (ii), 12. (iv).

BACTERIA, VIRUSES, LICHENS* AND MYCOPLASMA

Long Answer Type

1. Describe the habit, structure and reproduction of a lichen collected by you. State the relationship between its components.
2. What is symbiosis? 'Lichens are the best example of symbiosis'. Justify this statement.
3. Give the distinguishing features of fungi, bacteria, virus, lichens and mycoplasma.
4. "Bacteria are both good and bad associates of human civilisation". Justify the statement.
5. Discuss briefly what you know about PPLO and bacteriophage.
6. Explain the various methods of reproduction found in bacteria.
7. Write an essay on food value of fungi, bacteria and lichens.
8. Briefly describe the ultra-structure of a fungal cell, a bacterium, TMV and bacteriophage.
9. Describe the characteristics of mycoplasma. How are they different from bacteria and viruses? Mention some of the plant diseases caused by PPLO.
10. Give the structure and reproduction of lichens.
11. What are lichens? Describe their methods of reproduction and economic importance.
12. Describe the ultra-structure of tobacco mosaic virus (TMV) and bacteriophage.
13. Describe the forms and internal features of lichen. Explain, how are they ecologically and economically important.
14. "Lichens are the best example of symbiosis". Justify this statement. Give an account of various methods of reproduction in lichens.
15. Describe the structure and economic importance of bacteria.

---

* Lichens described after Fungi

16. Give an account of the structure and economic importance of lichens.
17. Give an account of the structure and nature of plant viruses.
18. Write an essay on the economic importance of bacteria.
19. Give an account of symptoms of viruses. Describe their nature also.
20. Write about the methods of reproduction in bacteria.
21. Describe internal structure of lichen thallus. Also describe structures attached with the thallus.
22. Describe the method of reproduction in lichens or bacteria.
23. What are viruses? Are they living or non-living agents? Give the methods of their transmission.
24. What are lichens? Give habit and habitat, structure, vegetative and asexual reproduction and economic importance of lichens.
25. What are the procaryotic and eucaryotic cells? Which of the two is more primitive cells type and why?
26. Give an account of multiplication of plant viruses.
27. "The beneficial effects of bacteria outweigh the harmful ones". Justify the above statement and mention the part played by bacteria in the economy of nature.
28. Describe the structure and reproduction in a typical becterium. How does it differ from a blue-green alga?
29. Give an account of structure and reproduction of lichens with special reference to *Parmelia.*
30. How are plant viruses transmitted? Give the control measures for a typical plant viral disease.

Short Answer Type

1. In what ways do viruses differ from fission plants and fungi.
2. Give an apothecium, how will you identify whether it belongs to Ascomycetes or lichen.
3. Comment on the relationship between alga and fungus in a lichen.
4. Write the important differences between a procaryote and a eukaryote. Name one example of each.
5. Which algae and fungi are usually associated in lichens?
6. Why are not lichens found in Delhi?
7. Lichens are rare or absent in cities. What conclusions can you draw from this as to the physiology of lichens.
8. Write note on economic importance of lichens.
9. Write a note on organisation of thallus in lichen.
10. Write critical notes on fruticose lichen.
11. Write explanatory notes on: reproduction in lichen.
12. Discuss the comparative advantages gained by the fungus and the alga in a lichen.
13. Write critical notes on the isidia of lichens.
14. Write illustrated notes on the following : (i) nature of viruses, (ii) mycoplasma.
15. Write short notes on the following : (i) Tobacco Mosaic Virus (TMV), (ii) bacterial cell, (iii) fruticose lichens.
16. Describe the following : (i) nitrogen fixing bacteria, (ii) economic importance of lichens.
17. Write short notes on the following : (i) life-cycle of a bacteriophage, (ii) conjugation in bacteria.
18. Give the distinguishing features of fungi, bacteria, virus, lichens and mycoplasma.
19. Write short notes on the following : (a) economic importance of lichens, (ii) antibiotics from bacteria.
20. Write notes on: (a) nitrogen fixing bacteria, (b) virus disease problems in India.
21. Write notes on the following : (i) mycoplasma, (ii) antibiotics.
22. Give the symptoms of (i) citrus canker, (ii) mosaic disease of tobacco.
23. Write notes on the following : (i) bacteriophage, (ii) citrus canker.
24. Describe the following : (i) Tobacco mosaic virus, (ii) role of bacteria in industry.
25. Write notes on the following : (i) nitrogen fixing bacteria, (ii) nature of virus.

26. (a) What is the biological status of virus? (b) How do plant viruses multiply in a cell?
27. Write notes on the following : (a) tobacco mosaic virus, (ii) reproduction in bacteria.
28. Write notes on : (i) nitrogen fixing bacteria, (ii) bacteriophage.
29. Describe the following : (i) reproduction and economic importance of bacteria, (ii) structure of thallus in lichens.
30. Explain the following : (a) structure of virus, (b) role of bacteria, (c) symbiosis.
31. Write about the following : (a) that the viruses are living organisms, (ii) when a bacteriophage infects a bacterial cell, what part of it is left outside, (iii) what type of nucleic acid is present in a phage.
32. Write short notes on the following : (a) bacteriophage, (b) nature of plant viruses.
33. Write short notes on : (i) symbiotic relationship in lichens, (ii) the lichen acids.
34. Give the distinguished characteristics of the following types of lichens : (a) crustose lichen, (b) foliose lichen, (c) fruticose lichen.
35. Classify the bacteria on the basis of their flagellation.
36. Give the ultra-structure of a bacterial cell.
37. Write a note on chemosynthetic bacteria.

Multiple Choice

1. The bacterial cell wall consists of : (i) chitin, (ii) lignin, (iii) cellulose, (iv) pectose.
2. Bacteria were first discovered by (i) Pasteur, (ii) Koch, (iii) T.J. Burill, (iv) Anton von Leeuwenhoek.
3. Bacteria were first discovered in : (i) 1575, (ii) 1675, (iii) 1775, (iv) 1875.
4. The bacteria are : (i) unicellular, (ii) bicellular, (iii) multicellular, (iv) filamentous.
5. The bacterium with a group of flagella at both ends is : (i) lophotrichous, (ii) peritrichous, (iii) amphitrichous, (iv) none of above.
6. Which of the following is not found in bacterial cell : (i) glycogen, (ii) proteins, (iii) fats, (iv) endoplasmic reticulum.
7. Sort out nitrogen fixing bacteria : (i) *Azotobacter,* (ii) *Pseudomonas,* (iii) *Xanthomonas,* (iv) *Salmonella.*
8. Scientist concerned with classification of bacteria : (i) Alexopoulos, (ii) Hutchinson, (iii Bergey, (iv) Rendle.
9. Typhoid fever is : (i) viral disease, (ii) bacterial disease, (iii) mycoplasma disease, (iv) none of above.
10. Tetanus is : (i) viral disease, (ii) bacterial disease, (iii) mycoplasma disease, (iv) none of above.
11. The bacteria are : (i) eukaryotic, (ii) prokaryotic, (iii) mesokaryotic, (iv) none of above.
12. The microorganisms found in the leguminous root are : (i) *Nitrosomonas,* (ii) Azotobacter, (iii) *Rhizobium,* (iv) *Xanthomonas.*
13. Sort out the bacterial disease : (i) typhoid, (ii) cholera, (iii) tuberculosis, (iv) all the above.
14. A virulent poison is produced in canned foods by : (i) *Xanthomonas,* (ii) *Pseudomonas,* (iii) *Rhizobium* sp., (iv) *Clostridium* sp.
15. The nitrogen-fixing bacteria are : (i) *Azotobacter,* (ii) *Rhizobium,* (iii) *Clostridium,* (iv) all the above.
16. Sort out the ammonifying bacteria: (i) *Bacillus subtilis,* (ii) *Bacillus mycodes,* (iii) Bacillus ramosus, (iv) all the above.
17. Nitrogen fixation is done by : (i) soil fungi, (ii) bacteria, (iii) green algae, (iv) virus.
18. The bacteria belong to : (i) Bacillariophyceaė, (ii) Myxophyceae, (iii) Schizomycetes, (iv) Myxomycetes.
19. The common method of reproduction in bacteria is : (i) budding, (ii) fragmentation, (iii) fission, (iv) hormogone formation.
20. The bacteria-like algae are : (i) Chlorophyceae, (ii) Phaeophyceae, (iii) Myxophyceae, (iv) Rhodophyceae.
21. Sort out the sulphur bacteria : (i) *Ferrobacillus,* (ii) *Thiobacillus,* (iii) *Hydrogenomonas,* (iv) *Carboxylomonas.*

22. Sort out the bacterial disease : (i) late blight, (ii) early blight, (iii) leaf curl of potato, (iv) ring rot of potato.
23. Pleuropneumonia of cattle is caused by : (i) virus, (ii) fungus, (iii) mycoplasma, (iv) bacteria.
24. The mycoplasma is : (i) eukaryotic, (ii) prokaryotic, (iii) multicellular, (iv) none of above.
25. The membrane found around mycoplasma is : (i) single-layered, (ii) double-layered, (iii) three-layered, (iv) none of above.
26. The nuclear member one in mycoplasma is : (i) single-layered, (ii) double-layered, (iii) nuclear membrane lacking, (iv) nuclear membrane well-defined.
27. The diameter of smallest mycoplasma is : (i) 1 μ, (ii) 5 μ, (iii) .05 μ, (iv) .1 μ.
28. Little leaf of brinjal is caused by : (i) mycoplasma, (ii) fungi, (iii) virus, (iv) bacteria.
29. The mycoplasma contains : (i) only DNA, (ii) only RNA, (iii) both RNA and DNA, (iv) none of above.
30. Little leaf symptom is produced by : (i) fungi, (ii) algae, (iii) mycoplasma, (iv) virus.
31. Milk spoilage is due to : (i) *Aspergillus*, (ii) *Peudomonas*, (iii) *Lactobacillus*, (iv) *Staphylococcus*.
32. Tobacco mosaic virus was first crystallized by : (i) F.C. Bawden, (ii) K.M. smith, (iii) W.M. Stanley, (iv) V. Ivanowski.
33. A free living bacterium capable of fixing atmospheric nitrogen is : (i) *Clostridium*, (ii) *Rhizobium*, (iii) *Staphylococcus*, (iv) *Streptococcus*.
34. Sort out the bacterial disease, (i) red rust of tea, (ii) white rust, (iii) citrus canker, (iv) athlete's foot.
35. In nitrogen-cycle the bacteria which change, proteins to NH_2 are known as : (i) bacteria of decay, (ii) denitrifying bacteria, (iii) nitrate bacteria, (iv) nitrogen fixing bacteria.
36. The leguminous plants are important in agriculture because : (i) they are disease resistant (ii) four crops of legumes can be produced in a year, (iii) they require very little irrigation, (iv) they help in nitrogen economy of nature.
37. Small pox is caused by : (i) virus, (ii) bacteria, (iii) housefly, (iv) mosquito.
38. Bacteria that are responsible for fermentation of dairy milk and plant product : (i) *Lactobacillus*, (ii) *hay bacillus*, (iii) *Acetobacter*, (iv) *Rhizobium*.
39. Nitrogen-fixing bacteria are concerned with family : (i) Leguminosae, (ii) Cruciferae, (iii) Gramineae, (iv) Malvaceae.
40. Bacteriophage is made up of : (i) protein, (ii) DNA, (iii) nucleoprotein, (iv) lipid and protein.
41. In nitrogen-cycle, which of the following play an important role : (i) *Rhizopus*, (ii) *Mucor*, (iii) *Nitrobacter*, (iv) *Spirogyra*.
42. Bacteria cannot survive in a highly salted pickle because : (i) they become plasmolyed and consequently killed, (ii) they do anaerobic respiration, (iii) water is not available to them, (iv) all the above.
43. The organisms which participate most actively in nitrogen cycle in nature are : (i) saprophytic angiosperms, (ii) parasitic fungi, (iii) bacteria, (iv) legumes.
44. Which of the following is bacterial disease : (i) small pox, (ii) influenza, (iii) T.B., (iv) rabies.
45. A bacterial cell divides once every minute. It takes an hour to fill a cup. How many minutes will it take to fillhalf the cup : (i) 59, (ii) 49, (iii) 30, (iv) 29.
46. A free-living anaerobic bacterium capable of fixing nitrogen is : (i) *Azotobacter*, (ii) *Clostridium*, (iii) *Rhizobium*, (iv) *Streptococcus*.
47. A bacterial cell differs from eukaryotic cell due to (i) green algae, (ii) red algae, (iii) heterochromatin, (iv) rigid cell wall.
48. Which of the following disease is caused by bacteria : (i) arthritis, (ii) amoebic dysentery, (iii) beri-beri, (iv) diptheria.
49. Lichens are composite organs, consisting of algae and : (i) mosses, (ii) protozoa, (iii) fungi, (iv) bacteria.

50. The bulk of nitrogen in nature is fixed by : (i) lightning, (ii) symbiotic bacteria, (iii) denitryfying, (iv) chemical industries.
51. True nucleus is absent in : (i) mosses, (ii) green algae, (iii) soil fungi, (iv) bacteria.
52. Nodules with nitrogen fixing bacteria are present in : (i) mustard, (ii) wheat, (iii) gram, (iv) cotton.
53. Bacteria have cell membrane made up of (i) protein, (ii) cellulose, (iii) fat, (iv) chitin.
54. Bacteriophage consists of : (i) carbon and nitrogen, (ii) DNA, (iii) nucleoproteins, (iv) proteins only.
55. Bacteria cannot survive in a highly salted pickle because : (i) salts inhibit reproduction, (ii) bacteria do not get enough light for photosynthesis, (iii) they become plasmoslysed and killed, (iv) pickle does not contain nutrients necessary for bacteria to live.
56. One of the interesting features of the viruses is that they : (i) multiply only in host cytoplasm, (ii) behave as if they were plants. (iii) are made of proteins only, (iv) occur only inside bacteria.
57. Virology is the study of : (i) viruses, (ii) bacteria, (iii) fungi, (iv) algae.
58. Microbiology is the study of : (i) lower organisms with microscope, (ii) micro-organisms, (iii) bacteria, (iv) protozoa.
59. The free living nitrogen fixing bacteria belong to the genus : (i) *Pseudomonas,* (ii) *Azotobacter,* (iii) *Rhizobium,* (iv) *Xanthomonas.*
60. The production of nitrates from ammonia through *Nitrosomonas* is called : (i) nitrification, (ii) ammonification, (iii) nitrogen fixation, (iv) denitrification.
61. The bacteria which in association with some plant roots fix atmospheric nitrogen are called : (i) *Pseudomonas* (ii) *Bacillus,* (iii) *Salmonella,* (iv) Rhizobium.
62. *Escherichia coli* is the common inhabitant of : (i) human intestine, (ii) soil, (iii) water, (iv) decaying fruits.
63. Name of bacteria is first proposed by : (i) Anton von Leeuwenhock, (ii) Linnaeus, (iii) Watson, (iv) Ehrenberg.
64. Polymorphism is found in : (i) *Micrococcus,* (ii) *Diplococcus,* (iii) *Azotobacter,* (iv) *Clostridium.*
65. In nitrogen cycle which of the following plays an important role : (i) *Rhizopus,* (ii) *Mucor,* (iii) *Nitrobacter,* (iv) *Spirogyra.*
66. The nucleus absent in : (i) green algae, (ii) fungi, (iii) lichens, (iv) bacteria.
67. The nitrogen present in atmosphere : (i) is utilized by plants through micro-organisms, (ii) no use to plants, (iii) it is injurious to plants, (iv) none
68. Viruses cause : (i) dysentery, (ii) malaria, (iii) pneumonia, (iv) small pox.
69. Mitochondria are absent in : (i) yeast, (ii) fungi, (iii) bacteria, (iv) green algae.
70. Transfer of genetic material from one bacterium to another by virus is called : (i) transduction, (ii) transformation, (iii) conjugation, (iv) lysogeny.
71. In the nitrogen cycle, nitrite is converted to nitrate by : (i) *Azotobacter,* (ii) *Rhizobium,* (iii) *Nitrosomonas,* (iv) *Nitrobacter.*
72. The most abundant protein in the plant world is found in : (i) root hairs, (ii) mitochondria, (iii) chloroplasts, (iv) viruses.
73. Some diseases caused by bacteria are : (i) measles, mumps, malaria, (ii) tetanus, typhoid, tuberculosis, (iii) small pox, sleeping sickness, syphilis, (iv) pneumonia, polio, ring worm.
74. Citrus canker caused by : (i) *Xanthomonas citri* (ii) *Azotobacter,* (iii) *Pseudomonas,* (iv) *Erwinia.*
75. Virus is first isolated by : (i) Stanley, (ii) Miller, (iii) Iwanowsky, (iv) Schwann.
76. Vibrio is : (i) virus, (ii) bacteria, (iii) lichen, (iv) angiosperm.
77. In majority of lichens the fungal partner is : (i) member of Ascomycetes, (ii) member of Basidiomycetes, (iii) member of Oomycetes, (iv) none of above.
78. A group of algal cells imprisoned in a fungal mycelium and which is used for vegetative multiplication of a lichen is : (i) isidium, (ii) soredium, (iii) cephalodium, (iv) helotism.
79. The mode of life of lichen is : (i) saprophytic, (ii) parasitic, (iii) symbiotic, (iv) autotrophic.

80. A bacteriophage virus can be recognized by its : (i) extremely irregular shape, (ii) beautifully rounded shape, (iii) tadpole shape, (iv) rhomboidal shape.
81. Viruses are found to be consisting of : (i) purely proteins, (ii) purely nucleic acids, (iii) nucleoproteins, (iv) phosphates and sugars.
82. Among different kinds of water which of the following is free from bacteria : (i) deep well water, (ii) rain water as it falls down, (iii) sea water, (iv) water of hot springs.
83. A viral disease of potato : (i) brown rot, (ii) late blight, (iii) early blight, (iv) leaf roll.
84. A disease in cattle caused by viruses : (i) yellow fever, (ii) foot and mouth disease, (iii) anthrax, (iv) brucellosis.
85. The first person to use the Latin term virus in its current sense : (i) D. Iwanowski, (ii) M. Beijirinck, (iii) A. Meyer, (iv) W. M. Stanley.
86. The condition where the flagella are distributed all over the body of the bacterium is : (i) atrichous, (ii) monotrichous, (iii) amphitrichous, (iv) peritrichous.
87. A term meaning 'bacterium eater' : (i) cyanophage, (ii) coliphage, (iii) bacteriocide, iv) bacteriophage.
88. A virologist who discovered that T.M.V. can be crystallized : (i) Beijerinck, (ii) W. M. Stanley, (iii) Iwanowski, (iv) Adolf Meyer.
89. That part of the virus which gives it hereditary integrity : (i) capsomere, (ii) capsid, (iii) nucleic acid, (iv) nucleotide.
90. A viral disease of plants : (i) citrus canker, (ii) brown rot of potato, (iii) pea mosaic, (iv) angular leaf spot of cotton.
91. Cocci arranged in cubes with eight or more cells : (i) tetrads, (ii) streptococci, (iii) Diplococci, (iv) Sarcinae.
92. In bacteria the respiratory enzymes are situated in the : (i) cell membrane, (ii) cytoplasm, (iii) mitochondria, (iv) ribosomes.
93. One of the following bacterial characters is a plant character (i) presence of flagella, (ii) prokaryotic nucleus, (iii) rigid cell wall, (iv) heterotrophic nutrition.
94. The first bacteriologist : (i) Pasteur, (ii) Leeuwenhoek, (iii) Jenner, (iv) Robert Koch.

ANSWERS

1 (i), 2 (iv), 3 (ii), 4 (i), 5 (iii), 6 (iv), 7 (i), 8 (iii), 9 (ii), 10 (ii), 11 (ii), 12 (iii), 13 (iv), 14 (iv), 15 (iv), 16 (iv), 17 (ii), 18 (iii), 19 (iii), 20 (iii), 21 (ii), 22 (iv), 23 (iii), 24 (ii), 25 (iii), 26 (iii), 27 (iv), 28 (i), 29 (iii), 30 (iii), 31 (iii), 32 (iv), 33 (i), 34 (iii), 35 (ii), 36 (iv), 37 (i), 38 (i), 39 (i), 40 (iii), 41 (iii), 42 (i), 43 (iii), 44 (iii) 45 (i), 46 (ii), 47 (iii), 48 (iv), 49 (iii), 50 (ii), 51 (iv), 52 (iii), 53 (iv), 54 (iii), 55 (iii), 56 (i), 57 (i), 58 (ii), 59 (ii), 60 (i), 61 (iv), 62 (i), 63 (i), 64 (ii), 65 (iii), 66 (iv), 67 (i), 68 (iv), 69 (iii) 70 (i), 71 (iii), 72 (iii), 73 (ii), 74 (i), 75 (iii), 76 (ii), 77 (i), 78 (ii), 79 (iii), 80 (iii), 81 (iii), 82 (ii), 83 (iv), 84 (ii), 85 (i), 86 (iv), 87 (iv), 88 (iii) 89 (iii), 90 (iii), 91 (ii), 92 (i), 93 (iii), 94 (ii).

UNIT - 2
Diversity of Cryptogams
Algae

3
CHAPTER

Algae: General Topics

INTRODUCTION

"The algae are chlorophyll bearing organisms which possess unicellular sex organs or multicellular ones in which every cell forms a gamete." In this respect, the algae differ from all other green plants.

They are aquatic, both marine and fresh water, and occur on and within soil and on moist stones and wood as well as in association with fungi and certain animals. The algae are of great importance as primary producers of energy rich compounds which form the basis of the food cycle of all aquatic animal life. For this purpose, the planktonic algae are of special importance, since they serve as food for many animals. It is thought that 90 percent of the photosynthesis on earth is carried on by aquatic plants; the planktonic (suspended) algae are chiefly responsible for this. While photosynthesizing, they oxygenate their habitat, thus increasing the level of dissolved oxygen in their environment. Certain blue-green algae, like some bacteria, can use gaseous nitrogen from the atmosphere in building their protoplasm, and in this way they increase the nitrogenous compounds in water and soils of their habitat. This activity is called *nitrogen fixation.*

There are approximately 1800 genera with 21,000 species which are highly diverse with respect to habitat, size, organization, physiology, biochemistry and reproduction. **Phycology** is the study of algae and those who pursue such a study seriously are **phycologists.**

HABIT AND HABITAT

Most of the part of the land is covered over either by fresh water or sea water. Besides, several other algae are found in somewhat drier conditions. They are found on the trunks of trees, on telephone wires, on rocks, on walls, in hot springs and in several other unusual habitats. Here some of the algae have been classified according to their habitats. Special emphasis has been given on the occurrence of fresh water algae.

1. Hydrophytes. They are more or less completely submerged or free floating on the surface of the water. The hydrophytes may be subdivided into following heads.

***(i)* Benthophytes.** Several fresh water and marine algae are found in attached condition. The fresh water such as *Chara, Nitella, Cladophora, Gongrosira, Chaemosiphon,* etc., are found attached to some substratum in the bottom of the water. Almost all of brown algae (Phaeophyceae) are found in attached condition to some substrata in the sea.

***(ii)* Epactiphytes.** Such algae grow along the shores of lakes and ponds, and may be delimited from benthophytes with some difficulty. The most important fresh water forms are – *Oedogonium, Chaetophora,* some species of *Spirogyra, Mougeotia,* some diatoms, *Scytonema* and *Rivularia.*

***(iii)* Thermophytes.** Many algae are reported from hot springs. These algae may tolerate the temperature upto 70°C or more than that. According to Copeland, 53 genera and 153 species of Chroococcaceae may survive upto 84°C. Some Oscillatoriaceae may survive upto 85°C. This supports that Myxophyceae (blue-green algae) are primitive.

***(iv)* Planktophytes.** The algae which float on the surface of the water are called 'planktophytes'. They may be of two types, *'i.e.,* (*a*) euplanktophytes (*b*) tychoplanktophytes.

(*a*) Euplanktophytes. They are never attached, and from the very beginning are free floating, *e.g.,* diatoms, *Cosmarium, Closterium, Microcystis, Sphaeroplea, Scenedesmus, Pediastrum, Chlamydomonas, Volvox,* other Volvocales and some members of Chroococcales. The above given forms are fresh water in habit.

(*b*) Tychoplanktophytes. In the beginning, such algae are attached, but later on they become detached and free floating, *e.g.,* some species of *Spirogyra, Zygnema, Cladophora, Oedogonium, Rhizoclonium, Mougeotia, Tribonema, Microspora, Cylindrospermum, Tetraspora, Rivularia, Nostoc, Gloeotrichia, Sargassum,* etc.

***(v)* Halophytes.** The algae occur in saline waters are known as 'halophytes'. The most striking examples are *Dunaliella* and *Chlamydomonas* which occur in salt lakes, the species of *Scenedesmus, Aphanocapsa, Pediastrum, Aphanothece, Oscillatoria* are found in saline waters; the species of *Enteromorpha* are found in inland astuaries; many species of Ulvales, Ulotrichales, Conjugales, and Myxophyceae are found near the sea in astuaries.

***(vi)* Epiphytes.** Many algae are found upon other living plants and bigger species of algae. *Aphanochaete, Bulbochaete, Oedogonium* and *Microspora,* are found as epiphytes upon larger species of *Oedogonium, Cladophora, Rhizoclonium, Vaucheria* and *Hydrodictyon* species. *Coleochaete nitellarum* is epiphytic upon species of *Chara* and *Nitella.* Some of the species of *Coleochaete* are epiphytic upon some grasses grown on the banks of the ponds and the hydrophytes such as—*Vallisneria, Typha, Ipomoea* and several other aquatic plants. *Chaetonema* is found epiphytic on the mucilaginous masses of *Tetraspora* and *Batrachospermum.*

***(vii)* Epizoophytes.** Certain algae are found on living aquatic animals such as turtles, mollusc shells, fishes, etc. Species of *Cladophora* grow upon mollusc shells. *Protoderma* and *Basicladia* occur on the back of turtles. *Characiopsis* and *Characium* occur on the posterior and anterior legs of Branchipus respectively.

2. Edaphophytes. Such algae are also called terrestrial algae. They are found upon or inside the surface of the earth. They can be (*i*) saphophytes and (*ii*) cryptophytes.

***(i)* Saphophytes.** They are surface algae. Most of the species of Myxophyceae are found upon the surface of the soil. Besides, *Mesotaenium, Botrydium, Protosiphon, Oedocladium, Vaucheria, Fritschiella* and many others are met with upon the surface of the wet soil.

***(ii)* Cryptophytes.** Such algae are subterranean in habit and occur inside the soil. The species of Myxophyceae are found in the soil. The species of *Nostoc, Anabaena* and *Euglena* have been reported from the paddy fields, where they also fix the atmospheric nitrogen in the soil to enrich the fertility of the fields.

3. Aerophytes. Such algae are aerial in habitat. They are found upon the trunks of trees, walls, fencing wires,rocks, and animals and so many other aerial substrata.

***(i)* Epiphyllophytes.** Such algae are epiphytic upon leaves of trees. Species of *Trentepohlia* are commonly found upon the bark of trees. They also occur upon rocks and fencing wires. They are abundantly found on the fencing wires of Calcutta botanical gardens. *Phycopeltis* occurs upon *Rubus*; *Phyllosiphon* on *Arisaema* ; *Rhodochytrium* on *Asclepias* and *Solidago.*

***(ii)* Epiphloephytes.** These algae grow on the bark of trees mixed with many mosses and liverworts. *Phormidium, Scytonema, Haplosiphon* and *Schizothrix* grow on the bark of trees mixed with liverworts.

***(iii)* Epizoophytes.** These algae are found even on the bodies of land animals. Certain Chaetophorales are found even on the hairs of sloth.

(iv) **Lithophytes.** Many algae grow on the rocks and walls. The species of *Scytonema* grow on the walls in rainy season and the whole wall becomes black spotted. *Vaucheria, Nostoc* and many other algae are also found on wet rocks.

4. Cryophytes. These algae are found on ice and snow. These algal forms cause red snow, green snow, yellow snow, yellowish green snow and violet snow. In European countries, especially in arctic region the green snow is caused by *Chlamydomonas, Ankistrodesmus* and *Mesotaenium;* red snow is caused by species of *Chlamydomonas. Scotiella, Gloecapsa* and diatoms.

5. Symbionts or endophytes. Many algae grow in symbiotic association of other plants. The most striking example of symbiosis are lichens, here the algae are found in symbiotic association of fungi. Various Myxophyceae, *e.g., Chroococcus, Nostoc, Microcystis, Gloeocapsa, Scytonema, Rivularia,* etc., have been separated from lichens. Some green algae, *e.g., Coccomyxa, Chlorella, Protococcus, Palmella,* etc., are also found as symbionts in lichens.

Besides, several algae are endophytes in the tissue of other plants. *Anabaena azollae* is found inside the leaves of *Azolla* (a pteridophyte). *Anabaena cycadae* is found in the coralloid roots of *Cycas. Nostoc* has been reported from the tissues of *Anthoceros* and *Notothylas. Nostoc* is found in the leaves of *Sphagnum* (Bryophyta) and several angiosperms. *Chlorochytrium* is endophytic inside *Lemna, Ceratophyllum* and certain mosses.

6. Endozoophytes. Certain algae occur inside the body of animals. *Zooxanthella* is found inside fresh water sponges; *Zoochlorella* is found inside *Hydra viridis.* According to Langeron about 14 species of Oscillatoriaceae are found in the digestive and respiratory tracts of various vertebrates.

7. Parasites. Certain algae are parasites upon other plants. The most striking example is *Cephaleuros virescens* which causes the havoc of tea foliage in Assam and neighbouring areas. called 'red rust of tea'.

8. Fluviatile algae. Such algae are found in rapidly flowing waters; *Ulothrix* occurs in mountain falls. *Stigeoclonium, Batrachospermum* are reported from the swift running streams of Dehradun and other hilly tracts.

NUTRITION

From the view point of their nutrition the algae are **autotrophic**. They synthesize their food from inorganic materials such as carbon dioxide, water and minerals by means of photosynthesis. Chlorophyll is the most common pigment in all the algae, though in many, the green colour of the plastids is masked by other pigments, such as, **fucoxanthin** a yellow pigment which dominates in brown algae whereas **phycoerythrin** and **phycocyanin** pigments are found in red and blue green algae respectively. The algae also synthesize oil and proteins from the carbohydrates which they manufacture and soluble forms of nitrogen and other minerals available in solution in the water in which they are found. The aquatic species of algae obtain water and carbon dioxide by osmosis and diffusion processes respectively from the water in which they grow. This way, the process of nutrition in algae is quite similar to that of ordinary green plants.

The algae, like other chlorophyllous plants, require C, H, O, P, K, N, S, Ca, Fe and Mg and also traces of Mn, Bo, Zn, Cu and Co. For certain algae, additional elements are required such as Si for diatoms and Mo for *Scenedesmus.*

The algae which grow in an entirely inorganic medium in the presence of light are known as **photoautotrophic.** In other words, using light energy they synthesize their protoplasm from exclusively inorganic sources. Several other algae require in addition certain vitamins, usually B-12, thiamine or biotin and such algae are known as **photoauxotrophic**. A number of algae are **heterotrophic**. The algae which do not synthesize their protoplasm solely from inorganic sources but require some of the essential elements, usually carbon and nitrogen, are known as

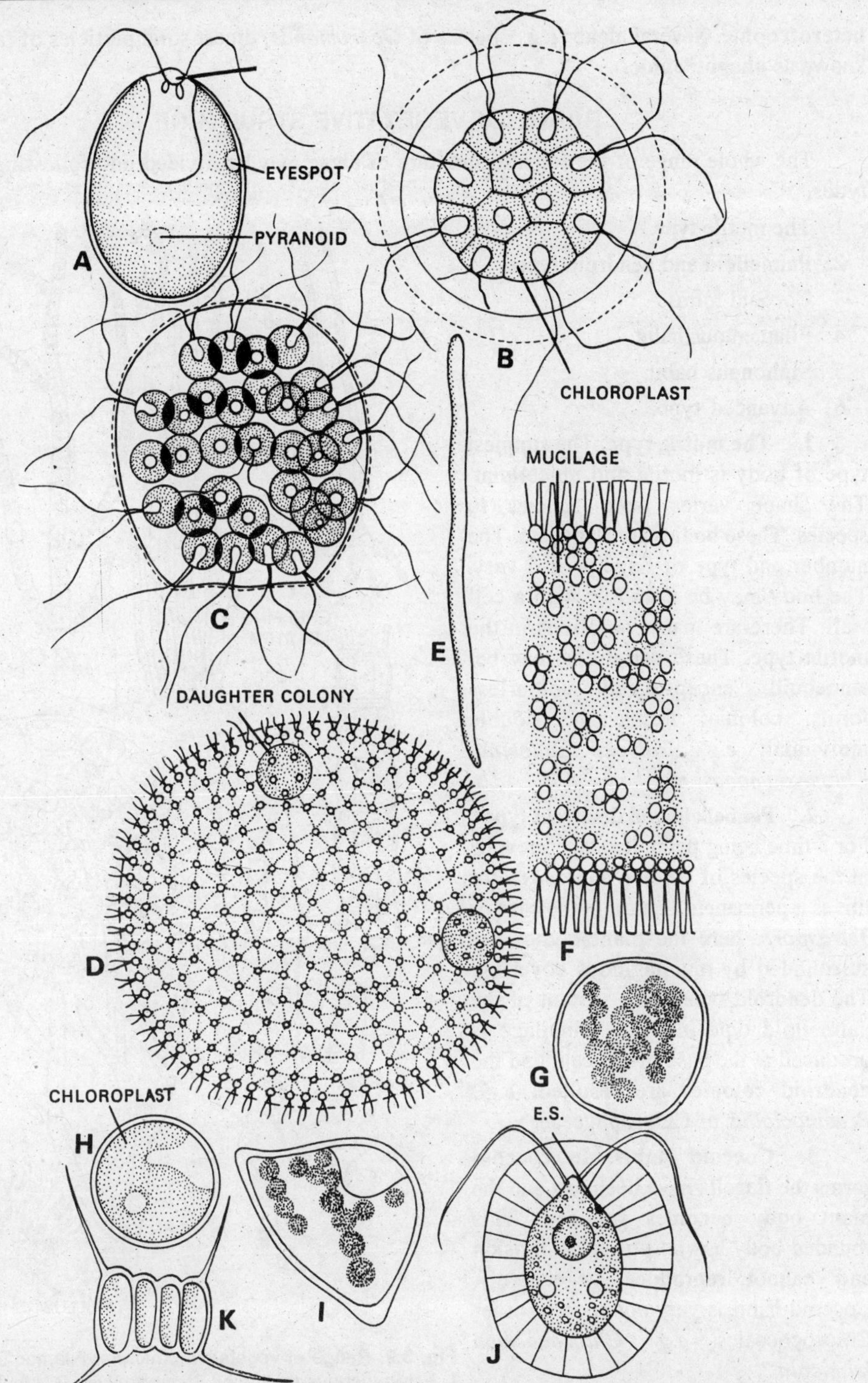

Fig. 3.1. Range of vegetative structure. The motile type A-D and J;A motile vegetative cell of *Chlamydomonas*; B, vegetative colony of *Pandorina* sp; C, vegetative colony of *Eudorina unicocca*; D, vegetative colony of *Volvox*; J, Motile cell of *Sphaerella*. The palmelloid type E and F; E, colony of *Tetraspora cylindrica*; F, portion of a colony of *T. cylindrica*; the coccoid type – G-I and K; G, H and I, vegetative cells of *Chlorococcum humicola*; K, colony of *Scenedesmus* sp.

heterotrophic. Several algae (*e.g.*, species of *Ocaromonas*) digest solid particles of food and are known as **phagotrophic.**

RANGE OF VEGETATIVE STRUCTURE

The whole range of the somatic structure of algae may be divided into following important types.

1. The motile type.
2. Palmelloid and dendroid types.
3. Coccoid forms.
4. Filamentous habit.
5. Siphonous habit.
6. Advanced type.

1. The motile type. The simplest type of body is motile and unicellular. The shape varies from species to species. These bodies are flagellate. The number and type of flagella also vary. The body may be naked or with a cell wall. There are many variations in the motile type. These variations may be, amoeboid, encapsulated colourless forms, colonial forms and double individual; *e.g., . Volvox, Eudorina, Chlamydomonas,* etc.

2. Palmelloid and dendroid types. For a time being palmella stage prevails in the species of *Chlamydomonas.* But this is a permanent feature in the case of *Tetraspora,* here the plant bodies are surrounded by mucilaginous covering. The dendroid type is the variant of the palmelloid type in which mucilage is produced at the base of the cell, and the dendroid colonies are resulted; *e.g., Prasinocladus* of Chlorophyceae.

3. Coccoid habit. In coccoid forms the flagella have been lost and the plant body becomes rounded. This rounded body has no power of division and cannot reproduce vegetatively. Coccoid habit is very common in order Chlorococcales, *e.g., Chlorococcum humicola.*

4. Filamentous habit. According to Blackman and Smith, the filaments originate from motile unicellular forms. According to these authors, the filamentous habit has been originated from

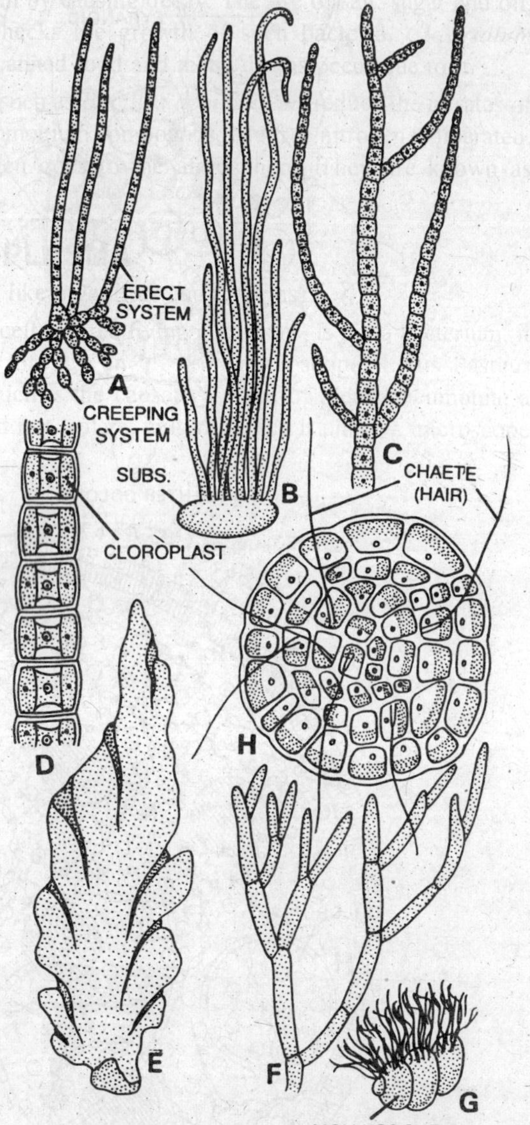

Fig. 3.2. Range of vegetative structure. Filamentous habit. A, heterotrichous thallus of *Stigeoclonium*; B, tubular thallus of *Enteromorpha*; C, portion of *Stigeoclonium*; D, portion of the filament of *Ulothrix;* E, foliaceous thallus of *Ulva*; F, a part of *Cladophora* thallus; G, habit of *Cladophora* (on mollusc shell); H, discoid thallus of *Coleochaete*.

motile unicellular bodies by the loss of motility and the restriction of vegetative division in one plane only. In the beginning, the cells were united by mucilaginous discs (false septation) but later the transverse septa of adjacent cells decayed and union between the cells of the filament took place. The view was supported by Dr. F.E. Fritsch. The filamentous forms possess several variations. The filaments may be simple, *e.g.*, *Ultothrix*, *Spirogyra*; heterotrichous, *e.g.*, *Stigeoclonium*; foliaceous, *e.g.*, *Ulva*; tubular, *e.g.*, *Enteromorpha*; discoid, *e.g.*, *Coleochaete*. The filaments of *Ulothrix* and *Spirogyra* are simple and unbranched. The filaments of *Stigeoclonium* are heterotrichous, *i.e.*, the plant body consists of two systems, one erect and the other prostrate. The prostrate system is comparatively smaller and possesses many erect branches. *Ulva* is a good example of foliaceous variant The divisions take place in all the planes and a foliaceous structure is developed. The thallus is two celled thick. In tubular variant, the thallus is a hollow tube with a wall one cell in thickness, *e.g.*, *Enteromorpha*. The discoid type has been evolved from heterotrichous filament. *Coleochaete scutata* is the well known example of this type. The cells of

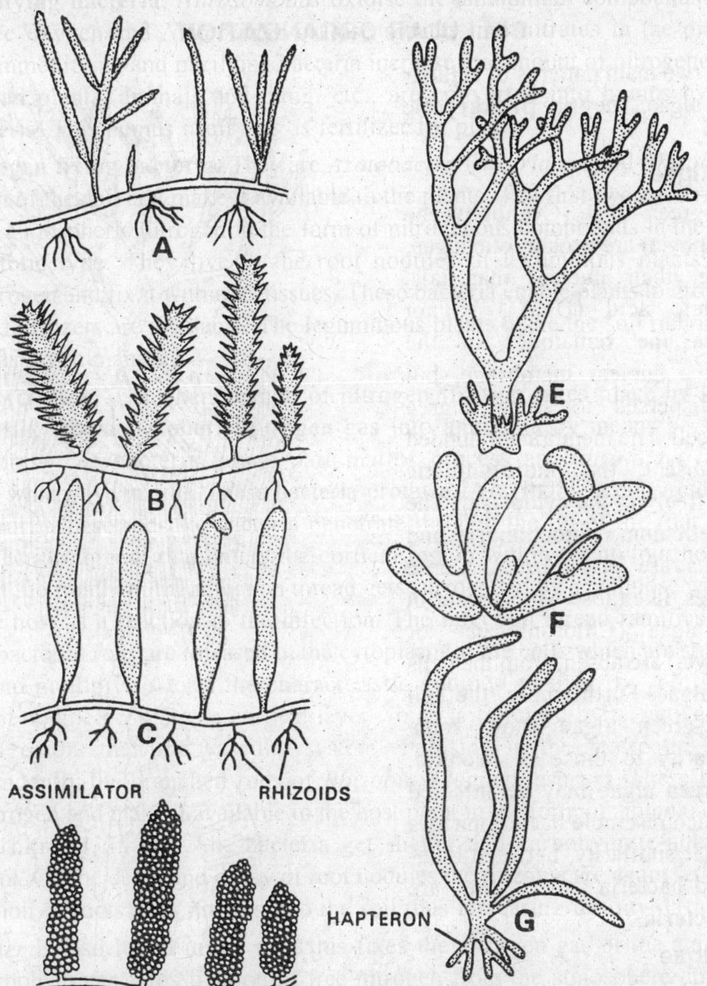

Fig. 3.3. Range of vegetative structure. Siphonous type. A, thallus of *Caulerpa cupressoides;* B, thallus of *C. crassifolia;* C, thallus of C. *prolifera*; D, thallus of *Caulerpa* sp.; F. thallus of *Valonia utricularis;* G, thallus of *Vaucheria* sp.

Coleochaete are joined end to end in branching filaments and a discoid habit is attained. The other species *Coleochaete pulvinata* possesses free branches.

5. Siphonous habit. The plant body consists of branching filaments having many nuclei and no partition walls. The cross walls appear only at the time of the formation of reproductive bodies. The well known examples of this type are – *Vaucheria, Codium, Caulerpa,* etc., of order Siphonales.

6. Advanced type. The advanced types, may be uniaxial filamentous type, multiaxial filamentous type and parenchymatous type. In uniaxial filamentous type, the single main axial thread has close branch systems to form more or less compact pseudoparenchymatous thallus, *e.g., Dumontia* of Rhodophyceae. In the multiaxial filamentous type, there is a number of axial threads and the branches and all of them form a compact cortex, *e.g., Scinaia furcellata* of Rhodophyceae. Many advanced types are true parenchymatous, *e.g., Laminaria, Fucus, Sargassum* of Phaeophyceae.

CELLULAR ORGANIZATION

There are two main patterns of cellular organization in algae. They are **prokaryotic** and **eukaryotic.**

Prokaryotic Algae

The blue-green algae (Cyanophyceae or Cyanophycophyta) are prokaryotic algae. In these algae, their nuclear materials, deoxyribo-nucleic acid (DNA), is not delimited from the remainder of the protoplasm by a nuclear membrane, but rather it is dispersed to some degree throughout the cell. The membrane-bounded plastids are absent; the photosynthetic lamellae occur freely in the cytoplasm. The endoplasmic reticulum, mitochondria and Golgi apparatus are also absent. Large aqueous vacuoles, like those which occur in many green algae (Chlorophyceae or Chlorophycophyta) are absent from the cells of blue-green algae. Furthermore, the cell walls of blue-green algae show some chemical similarity to those of bacteria. Certain blue-green algae may be infected with viruses which resemble bacteriophages advocates further similarity, between blue-green algae and bacteria and so, they are called **cyanobacteria.**

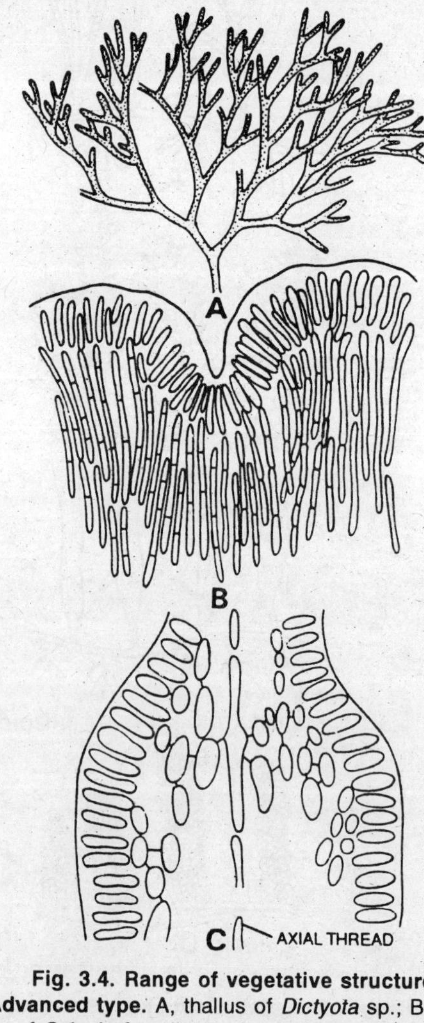

Fig. 3.4. Range of vegetative structure. Advanced type. A, thallus of *Dictyota* sp.; B, apex of *Scinaia furcellata* in longitudinal section; C, *Dumontia incrassata* (Rhodophyceae).

Eukaryotic Algae

On the other hand, the cells of eukaryotic algae have their DNA localized within a minutely perforated nuclear membrane, their nuclei being essentially similar to those of higher plants. These photosynthetic lamellae are confined within membranes as well defined chloroplasts. The chloroplasts may be massive, parietal or star-like structures, occurring singly within each cell, or they may be ribbon-like, bar-like, net-like or in the form of discoid structures as in non-algal green plants. In many green algae (Chlorophyceae), golden algae (Chrysophyceae), brown algae (Phaeophyceae) and red algae (Rhodophyceae), specialized regions, the **pyrenoids**, are present within the plastids. In the green algae, the pyrenoids are usually surrounded by starch grains. It is thought that pyrenoids are centres for enzymatic condensation of glucose into starch.

Endoplasmic reticulum, mitochondria and Golgi bodies are present in the cells of all eukaryotic algae so far studied.

The motile cells of algae may be flagellate. Associated with motility are such strucutres as contractile vacuoles, flagella and stigmas.

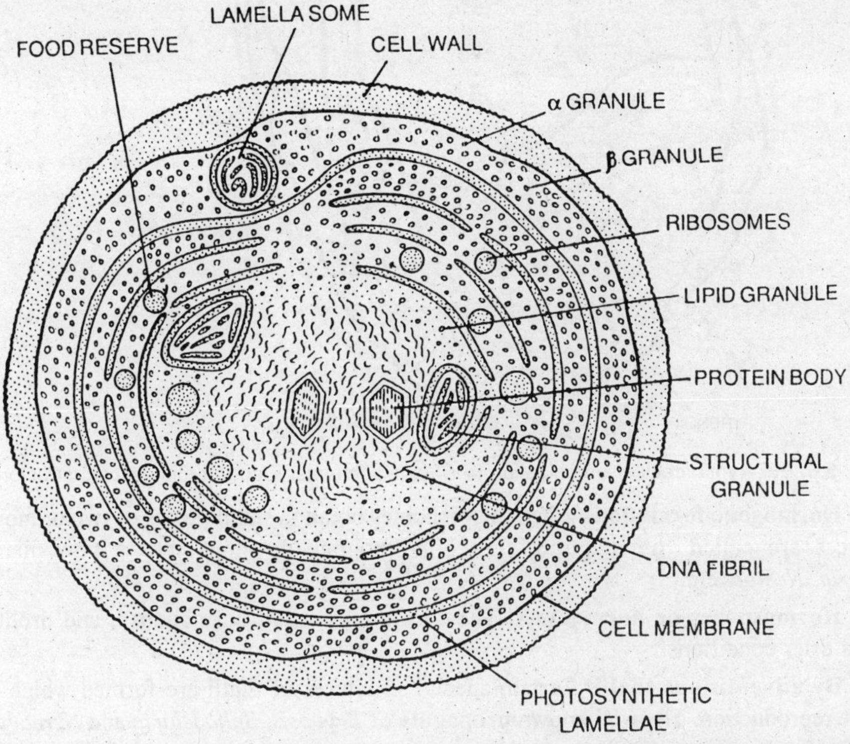

Fig. 3.5. Prokaryotic cellular organization in Myxophyceae. Line diagram of an electron micrograph of typical Myxophycean cell.

REPRODUCTION AND METHODS OF PERENNATION

There are three common methods of reproduction found in algae – (*i*) vegetative, (*ii*) asexual, and (*iii*) sexual. In addition to these methods, several perennating bodies also develop which face the adverse conditions.

1. Vegetative reproduction. This may be of several types.

(*i*) By cell division. The mother cells divide and the daughter cells are produced, which become new plants. This is exclusive type of reproduction in *Pleurococcus,* some desmids, diatoms, *Euglena,* etc.

(*ii*) Fragmentation. The plant body breaks into several parts or fragments and each such fragment develops into an individual. This type of vegetative reproduction is commonly met within filamentous forms, *e.g., Ulothrix, Spirogyra,* etc. The fragmentation of colonies also takes place in several blue green algae, *e.g., Aphanocapsa, Aphanothece, Nostoc,* etc.

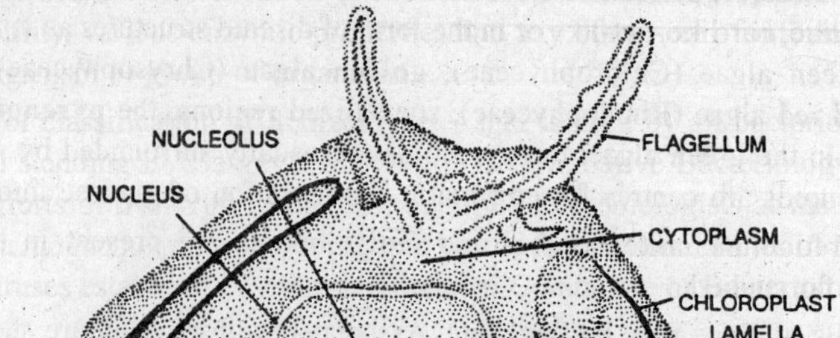

Fig. 3.6. Eukaryotic cellular organization in *Chlamydomonas*. Electron microscopic structure.

***(iii)* Hormogone formation.** When the trichomes break in small pieces of two or more cells, such pieces are called '**hormogones**'. Each hormogone develops into a new plant, *e.g., Oscillatoria, Nostoc,* etc.

***(iv)* Hormospores or hormocysts.** They are thick-walled hormogones, and produced in somewhat drier conditions.

***(v)* By adventitious thalli.** Certain special structures of thalli are formed which help in vegetative reproduction. The well known **propagula** of *Bryopsis, Sphacelaria* and *Nereocystis* are good examples.

***(vi)* By primary secondary protonema.** Such thread-like vegetative bodies develop in the case of *Chara,* which help in reproduction.

***(vii)* Tubers.** Usually these bodies are rounded and filled up with abundance of starch. Each body may give rise to a new plant, *e.g., Chara.*

***(viii)* Starch or amylum stars.** Such special star-shaped, starch filled bodies give rise to new plants frequently reported from *Chara.*

***(ix)* Bulbils.** Small bud-like structures. Usually develop on the rhizoids of *Chara* are called bulbils. Each such bulbil may develop into a new plant.

***(x)* Akinetes.** In most of the Chlorophyceae members, the akinetes are developed. Usually the protoplast of each cell converts in a single akinete. Sometimes they are formed in chains. Each akinete may develop into a new plant, *e.g., Oedogonium, Ulothrix,* etc.

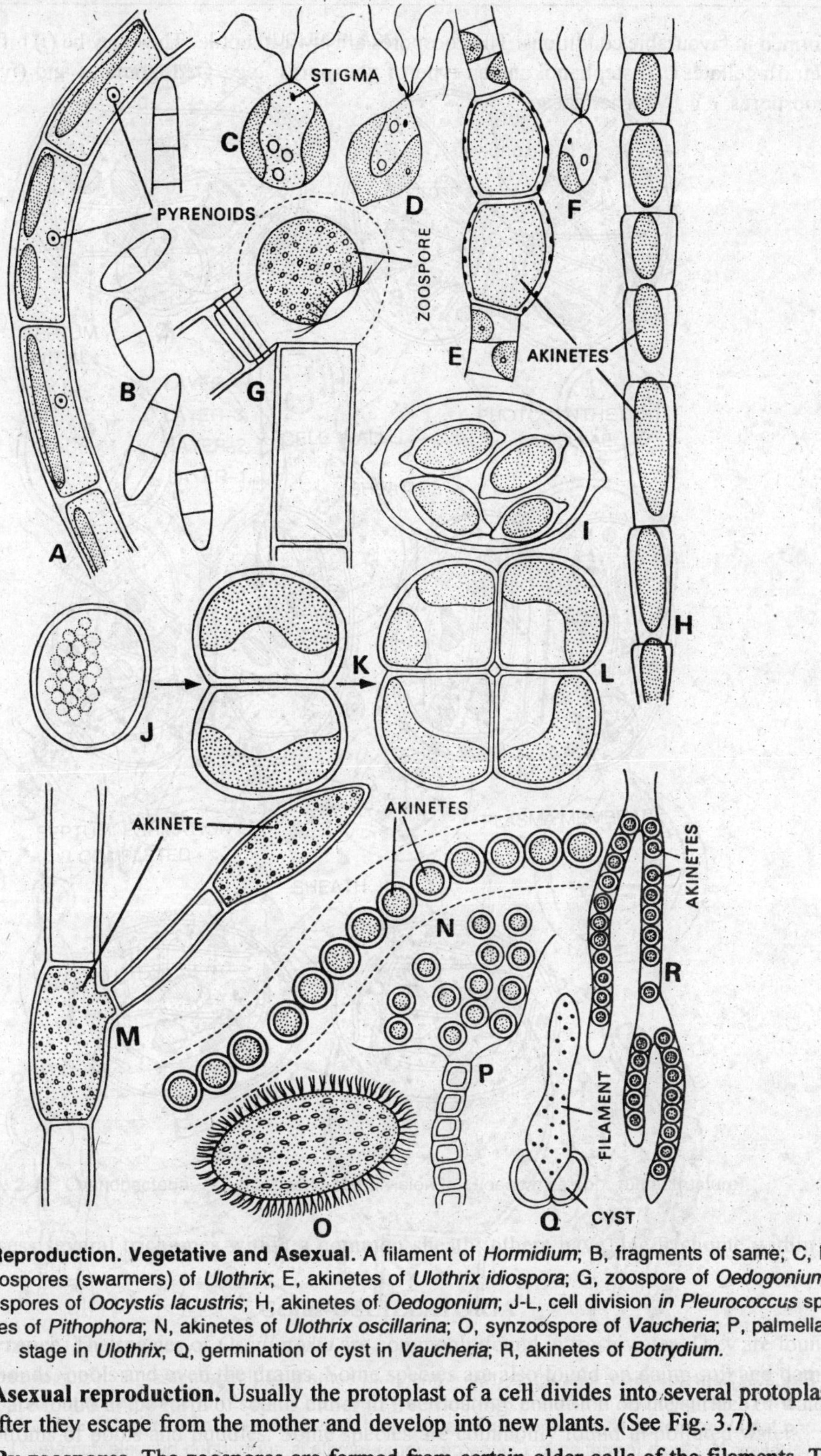

Fig. 3.7. Reproduction. Vegetative and Asexual. A filament of *Hormidium*; B, fragments of same; C, D and F, zoospores (swarmers) of *Ulothrix*; E, akinetes of *Ulothrix idiospora*; G, zoospore of *Oedogonium* sp.; I, autospores of *Oocystis lacustris*; H, akinetes of *Oedogonium*; J-L, cell division *in Pleurococcus* sp.; M, akinetes of *Pithophora*; N, akinetes of *Ulothrix oscillarina*; O, synzoospore of *Vaucheria*; P, palmella stage in *Ulothrix*; Q, germination of cyst in *Vaucheria*; R, akinetes of *Botrydium*.

2. Asexual reproduction. Usually the protoplast of a cell divides into several protoplasts and thereafter they escape from the mother and develop into new plants. (See Fig. 3.7).

***(i)* By zoospores.** The zoospores are formed from certain older cells of the filaments. The cytoplasm divides to form zoospores which are escaped from the mother cell. They are always formed in favourable conditions. The zoospores are always motile. They may be (*i*) biflagellate, (*ii*) tetraflagellate, (*iii*) stephanokontean type of zoospores, *e.g.*, Oedogoniales and (*iv*) compound zoospores, *e.g.*, Vaucheriaceae.

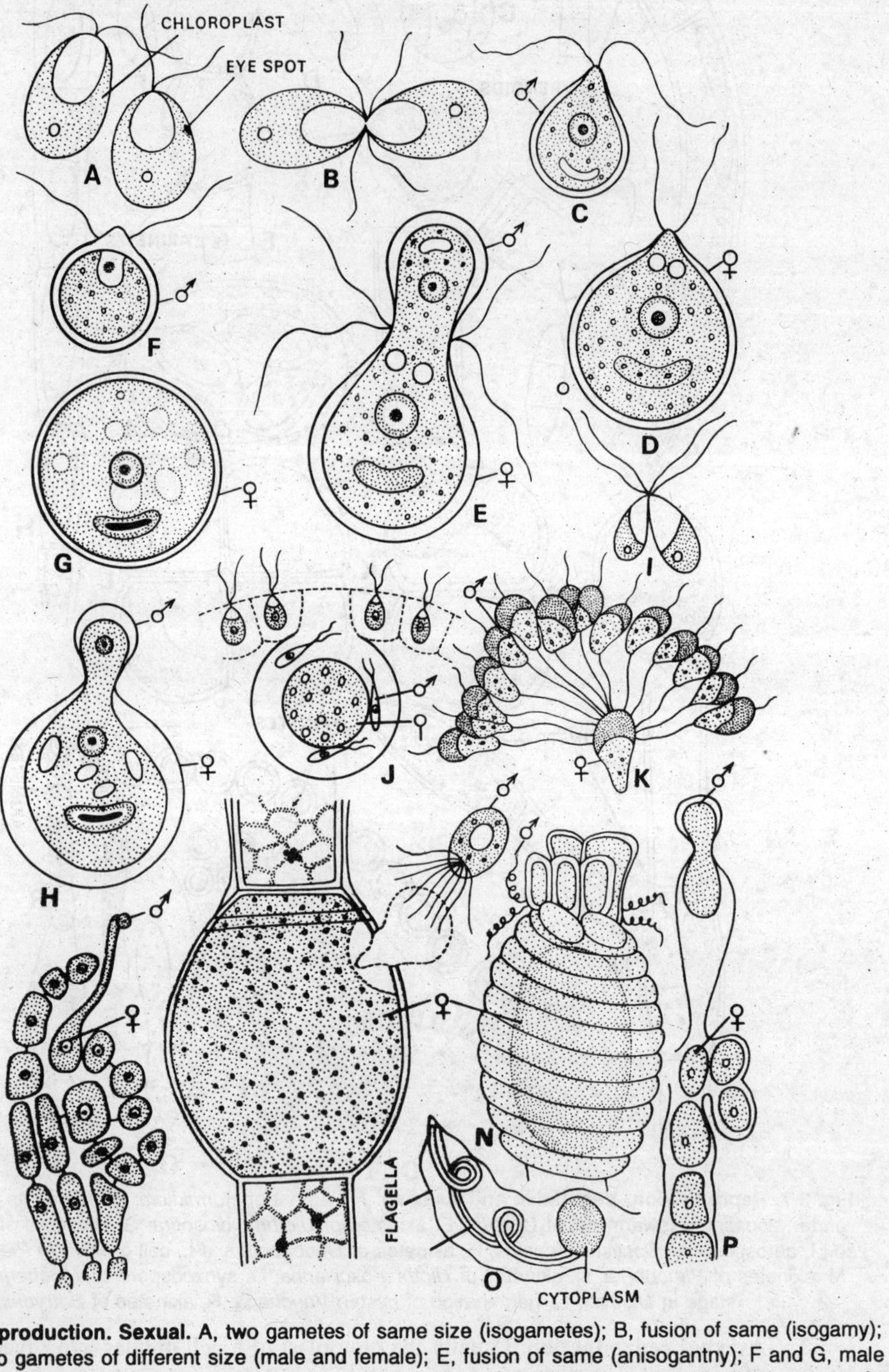

Fig. 3.8. Reproduction. Sexual. A, two gametes of same size (isogametes); B, fusion of same (isogamy); C and D, two gametes of different size (male and female); E, fusion of same (anisogantny); F and G, male and female gametes; H, fusion of same (oogamy); I, anisogamy in *Enteromorpha inteslinalis*; J. oogamy in *Volvox*; K, clump formation in *Ectocarpus*, large female is surrounded by male gametes; L, oogamy in *Oedogonium*; M, advanced oogamy in *Polysiphonia*; N, oogamy, in *Chara*; O, antherozoid of *Chara*; P, oogamy in *Batrachospermum*.

***(ii)* By aplanospores.** When motile phase of zoospores is eliminated, the bodies are called aplanospores. The aplanospores develop in unfavourable conditions. Each such spore is surrounded by a wall.

***(iii)* By hypnospores.** Actually they are very thick-walled aplanospores and develop only in adverse conditions, *e.g., Pediastrum, Vaucheria.*

***(iv)* Palmella stage.** Here the successive generations of divided cells are gelatinized and a thick mucilaginous envelope develops, *e.g., Chlamydomonas, Ulothrix,* etc.

***(v)* Autospores.** They are just like aplanospores except that they are smaller in size. They resemble in shape to mother cell except in size. Each autospore gives rise to a new plant. Such autospores are reported from many Chlorococcales.

***(vi)* Endospores.** In many blue green algae and Bacillariophyceae, the endospores are formed within the cells. On the approach of favourable conditions, each endospore develops in a new individual.

***(vii)* Auxospores.** In many members of Bacillariophyceae, such auxospores are produced. Each develops in a new plant.

***(viii)* Carpospores.** They are found in the carposporophytes of red algae (Rhodophyceae). Each such spore develops in a new individual.

***(ix)* Neutral spores.** These spores are not formed within the sporangia. They are found in Rhodophyceae.

***(x)* Monospores.** These spores develop within monosporangia. Each spore gives rise to a new plant, *e.g.*, many members of Rhodophyceae (*Bangia, Porphyra, Porphyridium,* etc).

***(xi)* Paraspores.** Such spores are reported from many members of Rhodophyceae. Each spore develops into a new plant.

***(xii)* Statospores.** They are found in Xanthophyceae and Bacillariophyceae where they act as perennating bodies.

***(xiii)* Daughter colonies.** In many Volvocales and Chlorococcales, the daughter colonies are developed asexually, *e.g., Volvox, Hydrodictyon, Pediastrum,* etc.

***(xiv)* Gongrosira stage of Vaucheria.** In the aseptate filaments of *Vaucheria,* the protoplast divides into several parts, several hypnospores or cysts are produced and the whole filament looks like an algal form '*Gongrosira*'.

***(xv)* Microspores.** They are produced in many Bacillariophyceae.

3. Sexual reproduction. It is greatly advanced method of reproduction and not known in Myxophyceae (blue green algae). There are two main types, *i.e.*, (*i*) isogamy and (*ii*) heterogamy.

***(i)* Isogamy.** The fusion of similar motile gametes is found in many species. Usually the gametes taking part in fusion come from two different individuals or filaments, sometimes these gametes come from two different cells of the same filament. Thousands of gametes come and aggregate in clumps. (See Fig. 3.8).

***(ii)* Heterogamy.** The fusion of dissimilar gametes is called heterogamy. There are variations of it.

***(a)* Anisogamy.** The motile gametes taking part in fusion may either differ in size (morphological anisogamy) or physiological behaviour (physiological anisogamy).

***(b)* Oogamy.** In this case, the male antherozoid fuses with the female egg. This fusion may be of primitive type as found in *Cylindrocapsa,* or advanced type as in *Oedogonium, Vaucheria, Chara, Polysiphonia,* etc.

(*iii*) **Aplanogamy or conjugation.** It implies the fusion of two non-flagellate amoeboid gametes (aplanogametes). They are morphologically similar but physiologically dissimilar, *e.g.*, order Conjugales.

In fresh water algae, the sexual reproduction is best means of perennation because it is followed by the formation of thick-walled zygote or oospore.

Conditions for sexual reproduction. (*a*) The sexual reproduction takes place after considerable accumulation of food material and the climax of vegetative activity is over.

(*b*) The bright light is the major factor for the production of the gametes.

(*c*) A suitable pH value is required.

(*d*) The optimum temperature is necessary.

Parthenogenesis. The female gametes convert into zygotes without fusion. The resultants are called **azygospores** or **parthenospores** and the phenomenon 'parthenogenesis', *e.g., Spirogyra, Oedogonium* and many others.

Autogamy. In this phenomenon, the fusion of the daughter protoplasts or of the divided nuclei of a cell without liberation takes place. This process is known in many diatoms and colourless dinoflagellates.

PIGMENTATION

Various pigments, such as red, yellow, green and blue have been found in various marine and fresh water algae. Usually these pigments are found in special plastids called chromatophores, however, in Myxophyceae the chromatophores are not found, and the pigments are dispersed in peripheral chromoplasm. Only, because of the presence of these conspicuous pigments the algae look variously coloured in their habitats. Besides, several algae are colourless and more or less saprophytic in nature. Colourless algae have been reported from several major groups. There are colourless diatoms, colourless dinoflagellates, colourless Chlorophyceae and colourless Rhodophyceae. The principal pigment chlorophyll has been reported from all major groups of algae. Besides, carotenoids, xanthophylls and phycobilins are also found in various algae.

Principal kinds of algal pigments. There are three main groups of principal algal pigments. They are as follows :

1. **The chlorophylls.** They are green fat soluble pigments and universally present in green and other algae.

2. **The carotenoids.** They are fat soluble yellow pigments and present in various groups of algae. They are subdivided into three groups: (*i*) carotenes; (*ii*) xanthophylls or oxycarotenes; and (*iii*) carotenoid acids.

3. **The phycobilins.** They are water soluble blue and red pigments. The phycobilins are subdivided into two groups, *i.e.*, (*i*) the phycocyanins and (*ii*) phycoerythrins.

Properties of chlorophylls. The algal chlorophylls are (*i*) green, (*ii*) in solution the chlorophylls are fluorescent and emit red light, (*iii*) they are also distinguished by strong absorption of the blue-green and green colours.

Properties or carotenoids. As stated, the carotenoids are subdivided into three groups, *i.e.*, (*i*) carotenes; (*ii*) xanthophylls and (*iii*) carotenoid acids. The properties are as follows:

1. **Carotenes.** (*i*) They are unsaturated hydrocarbons; (*ii*) they absorb blue and green light and transmit the yellow and red; (*iii*) they are fat soluble and yellow in colour.

2. **Xanthophylls.** They are also named as oxycarotenes. They are oxygen derivatives of carotenes and resemble the properties of carotenes.

3. **Carotenoid acids.** They consist of a chain of carbon atoms. Their properties resemble the properties of other two carotenoids.

Table 3.1

Pigments	*Chloro-phyceae*	*Xan-thophyceae*	*Bacillario-phyceae*	*Phaeo-phyceae*	*Rhodo-phyceae*	*Myxo-phyceae*
Chlorophylls :						
Chlorophyll a	AAA	AAA	AAA	AAA	AAA	AAA
Chlorophyll b	AA	O	O	O	O	O
Chlorophyll c	O	O	A	A	O	O
Chlorophyll d	O	O	O	O	A	O
Chlorophyll e	O	A	O	O	O	O
Carotenes :						
α Carotene	A	O	O	O	A	O
β Carotene	AAA	AAA	AAA	AAA	AAA	AAA
e Carotene	O	O	A	O	O	O
Flavicin	O	O	O	O	O	A
Xanthophyllus :						
Lutein	AAA	O	O	O	AA	O
Zeaxanthin	A	O	O	O	O	O
Violaxanthin	A	O	O	A	O	O
Flavoxanthin	O	O	O	A	O	O
Neoxanthin	A	O	O	A	O	O
Fucoxanthin	O	O	AA	AA	O	O
Neofucoxanthin A	O	O	A	A	O	O
Neofucoxanthin B	O	O	A	A	O	O
Diatoxanthin	O	O	A	O	O	O
Diadinoxanthin	O	O	A	O	O	O
Dinoxanthin	O	O	O	O	O	O
Neodinoxanthin	O	O	O	O	O	O
Peridinin	O	O	O	O	O	O
Myxoxanthin	O	O	O	O	O	A
Myxoxanthophyll	O	O	O	O	O	AA
Unnamed		AA		A		AAA
Phycobilins :						
r-Phycoerythrin	O	O	O	O	AAA	O
r-Phycocyanin	O	O	O	O	A	O
c-Phycoeythrin	O	O	O	O	O	A
r-Phycocyanin	O	O	O	O	O	AAA

AAA-Indicates the principal pigment in each of the four groups of pigments.
AA-Indicates a pigment comprising less than half of the total pigments of the group.
A-Indicates a pigment comprising a small fraction of the total pigments of the group.
D-Indicates absence of a pigment.

Properties of phycobilins. The phycobilins are subdivided into two groups. *i.e.*, (*i*) phycocyanins and (*ii*) phycoerythrins.

The phycobilins are proteins. They are soluble in water and insoluble in the fat solvents. They are strongly fluorescent and emit orange or red light.

(1) **Phycocyanins.** They absorb green, yellow, and red light and trasmit the blue.

(2) **Phycoerythrins.** They absorb blue green, green and yellow light and transmit the red.

All the phycobilins are strongly fluorescent and emit orange or red light.

Occurrence of principal pigments in algae. All autotrophic algae contain chlorophylls, carotenoids and xanthophylls. The Myxophyceae and Rhodophyceae contain phycobilins which overmask the other pigments. Phycobilins are always associated with other pigments such as chlorophylls and carotenoids.

Occurrence of individual pigments in algae. In **Chlorophyceae**, chlorophyll (α chlorophylls), β carotene (carotenoids) and lutein (xanthophylls) are present as principal pigments. Besides, other chlorophylls, carotenoids and xanthophylls are also found.

Xanthophyceae. Chlorophyll, α and β-carotene are present as principal pigments. Xanthophylls are also found.

Bacillariophyceae. Chlorophyll, α and β-carotene are found as principal pigments. In addition to these, several xanthophylls have also been reported.

Phaeophyceae. Chlorophyll, α, β-carotene and several xanthophylls are found.

Rhodophyceae. Chlorophyll, α, β-carotene and *r*-phycoerythrin are found as principal pigments. In addition to these chlorophyll *d,* α carotene, lutein and *r*-phycocyanin are also found.

Myxophyceae. Chlorophyll, α, β-carotene and *c*-phycocyanin are principal pigments. Besides, *c*-phycoerythrin and xanthophylls are also found.

ORIGIN AND EVOLUTION OF SEX IN ALGAE

Primarily there are two methods of reproduction in plant, *i.e.*, **asexual** and **sexual**. In asexual type of reproduction, no sex is involved and there is no fusion of any kind of gametes or cells. This type of reproduction is being effected by special cells known as **spores**. Each such spore is capable to develop into a new plant. On the other hand, in sexual reproduction the two cells or the gametes unite together, and the zygote is resulted, which develops into a new plant. Normally, the individual gametes are incapable to produce new plants without fusion.

Origin of sex. The most primitive algae, *i.e.*, the members of Cyanophyceae (Myxophyceae), *e.g., Gloeocapsa, Chroococcus,* etc., reproduce by means of fusion, whereas *Oscillatoria, Nostoc, Anabaena,* etc., reproduce vegetatively, by means of a group of few cells called **hormogonia,** which later on give rise to new plants by further division. Some members of Cyanophyceae reproduce by means of **arthrospores.** This shows that all the blue-green algae reproduce asexually and the sexual reproduction is altogether absent. In rest of algae, whether unicellular or filamentous, both asexual and sexual methods of reproduction prevail.

It is thought that probably the gametes have been originated from the motile asexual spores or zoospores. Except few cases, the sexual reproduction has not replaced the asexual method but has been added as a supplementary method. The zoospores usually resemble the gametes except for their size. For example, in algae such as *Chlamydomonas*, *Ulothrix*, *Cladophora,* etc., it is thought that the zoospores were probably produced before the gametes originated. In above mentioned and many other filamentous forms, the asexual reproduction takes place by means of uninucleate and biflagellate zoospores which are formed by repeated division contents of the ordinary vegetative cell. The zoospores are produced, in favourable conditions. In adverse conditions, the gametes are formed in the vegetative cells which earlier produced zoospores in favourable conditions. The gametes are quite similar to zoospores but for their size and behaviour. In forms like *Chlamydomonas debaryanum* the zoospores and gametes are quite similar. In *Ulothrix*, two types of zoospores, *i.e.*, the micro- and the macrozoospores, are produced and the gametes are produced in the same manner as the zoospores are formed. The microzoospores are quite similar to the gametes. At the end of the season, when the plants run in the shortage of food and other favourable conditions, the cells begin to produce gametes. Cholnosky, however, advocates that the gametes in *Ulothrix variabilis* develop from potential zoospores which fail to escape and undergo more divisions to give rise to the gametes. According to him, the gametes are quite similar except their size and number of flagella. In *Oedogonium*, the gametes and the zoospores are quite similar except their size.

One can conclude, that the gametes are formed as a result of aging of cells and unfavourable conditions. It may be said that the gametes are the reduced type of zoospores. Such reduced zoospores cannot reproduce asexually. They can give rise to a new offspring only with the result of fusion and formation of the zygote. All these examples support that the gametes have been originated from zoospores by means of reduction.

Evolution of sex. The members of Myxophyceae do not show any trace of sexuality. In other classes of algae, the sexuality has been established. Now to trace out the evolution of sex in algae, it becomes necessary to know about the nature of the gametes in beginning. It is thought, that in the beginning the gametes were of similar shape and size. These gametes were morphologically identical, but physiologically different, *e.g.*, in *Ulothrix* and certain species of *Chlamydomonas,* the gametes are of plus and minus strains. This is known as **isogamy**. Thereafter **anisogamy developed.** In this process, the two uniting gametes are of different sizes, *e.g.*, in *Chlamydomonas braunii* and *Pandorina.* In this condition, the number of divisions in one cell is more or less than the other cell, and therefore, the uniting gametes are unequal in size. Thereafter developed oogamy. The oogamy is the highly evolved condition of heterogamy. In oogamy, one cell does not divide at all, it simply increases in size, accumulates sufficient food material and acts as female cell or egg. The other cell divides repeatedly producing small motile cells, acting as male gametes. However, in *Polysiphonia* the resultant male gametes are non-motile. Less evolved oogamy is found in *Chlamydomonas coccifera, Volvox* and *Oedogonium.*

The evolution of sex in algae has not taken place in any one phylogenetic line. This has taken place along several independent lines. The example of this statement is found in Volvocales, where *Gonium* is isogamous, *Pandorina* slightly anisogamous, *Eudorina* and *Pleodorina* marked by anisogamy and *Volvox* oogamous. This *Gonium-Pandorina-Eudorina-Pleodorina-Volvox* series also shows progressive somatic differentiation. It has been assumed that it is correlated with the evolution from isogamy to oogamy through anisogamy. Whereas on the other hand, in another genus of Volvocales, *i.e., Chlamydomonas,* where species within the genus show all gradations from isogamy to oogamy, *e.g., Chlamydomonas snowiae* is isogamous, *C. braunii* is anisogamous and *C. coccifera* is oogamous.

In a few green algae, such as *Spirogyra, Vaucheria, Chara* and in nearly all the brown and red algae a far more advanced condition has been reached. In *Spirogyra,* the two uniting gametes are produced inside the ordinary vegetative cells known as gametangia. The gametangia are not differentiated into male and female structures. Those which receive the contents are supposed to be female ones while those from which the contents pass out are supposed to be male ones. In *Vaucheria,* the gametes are borne in well-differentiated gametangia, known as antheridia and oogonia. The antheridia produce antherozoids and the oogonium contains an egg. In *Chara,* there is further advancement. Here the sex organs are not developed all over the plant body, but are confined to nodes on the branches of limited growth. The male and female sex organs are highly developed and specialized organs known as globule and nucule respectively. The spermatogenous filaments develop within globule while the nucule possesses an egg.

In brown algae, *Fucus* and *Sargassum* possess fertile conceptacles on the receptacles which contain male and female sex organs. The oogonia may be liberated even before fertilization, and the fertilization takes place in the water. *Fucus* contains eight eggs in its oogonium whereas in

Sargassum seven eggs degenerate but one. The highest evolution of sex in algae has been seen in the red algae. In *Polysiphonia,* the male spermatia are nonmotile. After their liberation, the spermatia are carried away to the female by means of water currents. From these examples, it becomes quite clear, that there is no evolutionary sequence in the evolution of sex in algae. It is thought that it developed and evolved independently in green, brown and red algae.

In conclusion, we can say that the sex has been developed in response to circumstances and gametes have been developed from zoospores by means of reduction. The isogamy gave rise to oogamy through anisogamy. During oogamy, the female sex organ increases in size and stores food material to face adverse weather conditions, whereas, the male sex organ gives rise to numerous male gametes which ensure fertilization. The most important feature of sexual reproduction is the union of two gametic nuclei, which results in the formation of a diploid zygote nucleus. Sooner or later reduction division takes place which provides a means of reshuffling of paternal and maternal chromosomes brought together during the act of fertilization.

Asexual reproduction results in organic similarity while sexual reproduction in diversity. The sexual reproduction is not primarily meant for the multiplication of individuals. It is actually meant for reproduction of heritable variations that accelerate the process of evolution.

LIFE-CYCLE PATTERNS IN ALGAE

Many life-cycle patterns are found in algae. However, there is no regular and fixed alternation of generations, as found in higher plants. In blue-green algae, and certain Chlorophyceae (*e.g., Protococcus, Scenedesmus,* etc.) which reproduce asexually, there is no alternation of generations.

The representative life-cycle patterns in algae are as follows :

1. Haplontic type. This is the simplest and most primitive type of life-cycle. The other patterns of life-cycle have originated from this type. This type is found in all Chlorophyceae except a few. Sometimes this is called *Ulothrix* or Chlamydomonad type.

In such cases, the somatic phase (plant) is haploid (gametophyte) while the diploid phase (sporophyte) is represented by zygote. During germination, the zygote (2n) divides meiotically producing haploid (n) zoospores, which develop into individual plants. Here the unicellular (*e.g., Chlamydomonas*) or filamentous (*e.g., Ulothrix, Spirogyra, Oedogonium, Chara,* etc.) gametophyte (n) alternates with a one-celled zygote or sporophyte (2n). The haploid filamentous plants are known as **haplonts** which reproduce asexually by zoospores or aplanospores producing the individuals like parents.

2. Diplontic type. This pattern is reverse of haplontic type. In this case, somatic phase (plant) is diploid (sporophyte 2n) while the haploid phase (gametophyte n) is restricted to gametes which are produced by meiotic division. After gametic union, a diploid zygote is formed, which develops into a diploid (sporophyte 2n) plant by mitotic division. The well known examples of this pattern are – *Fucus, Sargassum, Codium, Bryopsis,* etc.

3. Isomorphic type. In this type, there are two exactly similar (**morphologically identical**) somatic phases (plants) showing alternation of generations. Here the one phase is diploid (sporophyte 2n) while the other haploid (gametophyte n). Among Chlorophyceae, this is found in Ulvaceae, Chaetophoraceae and Cladophoraceae. The orders Ectocarpales, Cutleriales, Tilopteridales, Sphacelariales and Dictyotales of Phaeophyceae also exhibit this pattern of life-cycle.

In such cases, the zygote develops into a diploid multicellular plant (sporophyte) by postponement of meiosis. Prior to zoospore (meiospore) formation there is meiosis. These zoospores (n) develop into haploid plants (gametophyte n). The haploid plants produce gametes (n) which after fusion develop into zygotes (2n).

4. Heteromorphic type. This pattern of life-cycle is exactly like that of preceding one (isomorphic type) only with the difference that the alternating haploid (n) and diploid (2n) somatic phases (plant) are, **morphologically different**. In such cases, the diploid multicellular sporophytic

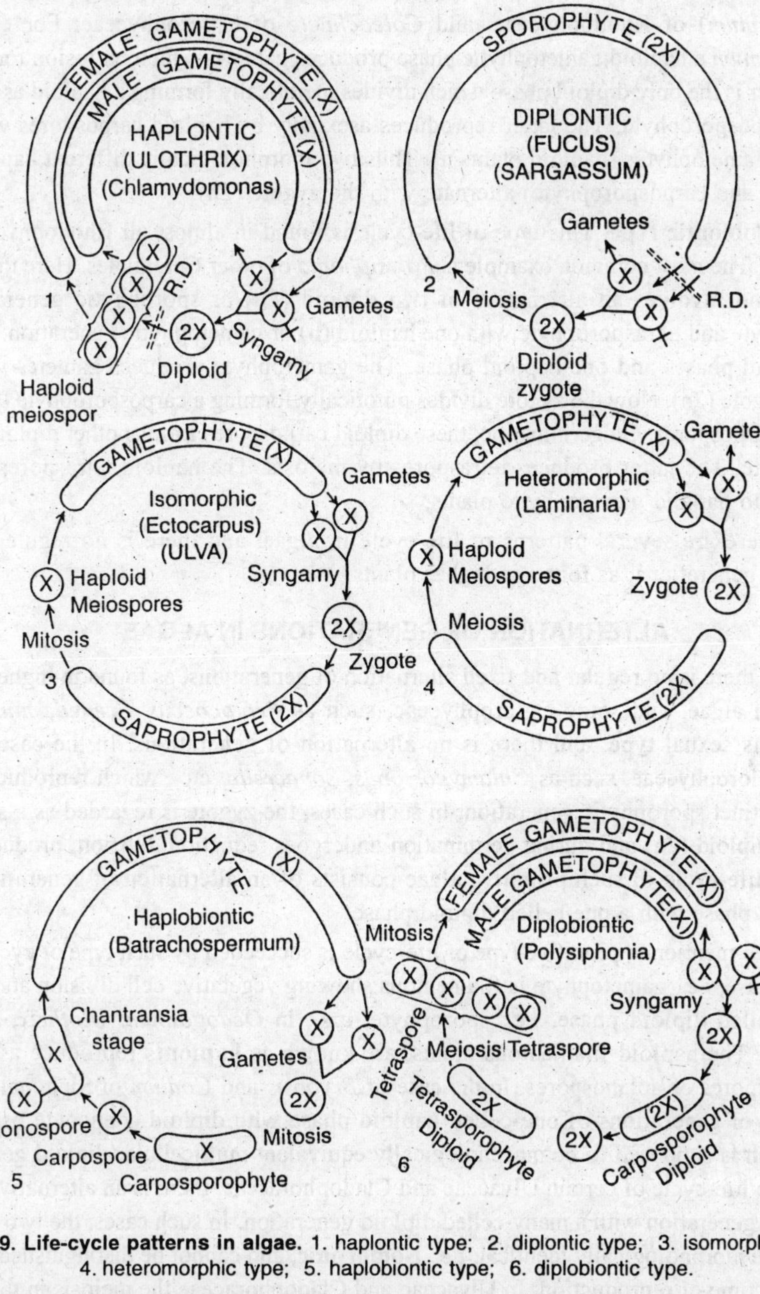

Fig. 3.9. Life-cycle patterns in algae. 1. haplontic type; 2. diplontic type; 3. isomorphic type; 4. heteromorphic type; 5. haplobiontic type; 6. diplobiontic type.

plant produces haploid zoospores (meiospores) by meiosis. These zoospores develop into gametophytes. Each gametophytic plant (n) produces gametes which after their union form a zygote and the latter develops into a diploid sporophytic plant by mitotic divisions. This pattern is found in Laminariales, Sporochnales, Desmarestiales, etc., of Phaeophyceae and *Urospora* of Chlorophyceae.

5. Haplobiontic type. In this pattern, there are three phases in the life-cycle. Out of three, two phases are haploid (n) and one diploid (2n). The examples are found among Nemalionales (*e.g.*, *Batrachospermum*) of Rhodophyceae, and *Coleochaete* of Chlorophyceae. For example, in *Batrachospermum* a haploid gametophytic phase produces gametes which on fusion form a zygote (2n). The latter is the only diploid phase which divides meiotically forming a haploid asexual phase known as carposporophyte. The latter reproduces asexually by haploid carpospores which again develop into gametophytes (haploid plants n). Thus two morphologically different haploid phases (gametophyte and carposporophyte) alternate with the zygote (2n).

6. Diplobiontic type. This type of life-cycle is found in almost all Rhodophyceae except Nemalionales. The most common example is *Polysiphonia* of order Ceramiales. Here the life-cycle is triphasic and involves an alternation of two diploid (2n) or sporophytic generations, *i.e.*, carposporophyte and tetrasporophyte with one haploid (n) or gametophytic generation. Thus there are two diploid phases and one haploid phase. The gametophyte produces gametes which unite and form a zygote (2n). Now the zygote divides mitotically forming a carposporophyte (2n) bearing diploid (2n) carpospores. On germination, these diploid carpospores form another diploid plant, the tetrasporophyte. The latter produces tetraspores by meiosis. The haploid tetraspores germinate and give rise to haploid gametophytic plants.

Thus, there are several patterns of life-cycle in algae, and there is no regular and fixed alternation of generations, as found in higher plants.

ALTERNATION OF GENERATIONS IN ALGAE

In algae, there is no regular and fixed alternation of generations, as found in higher plants. In all blue green algae, and some Chlorophyceae, such as *Protococcus*, *Scenedesmus,* etc., the reproduction is sexual type, and there is no alternation of generations. In the case of simple unicellular Chlorophyceae, such as *Chlamydomonas*, *Sphaerella,* etc., which reproduce sexually, there is no distinct sporophytic generation. In such cases, the zygote is regarded as a sporophyte, because it is diploid (2n) and during germination undergoes reduction division; producing gametophyte. The life-cycle of such primitive algae consists of an alternation of generations of one celled haploid phase with a one-celled diploid phase.

The above-mentiored primitive type of life-cycle is succeeded by such type of cycle in which the haploid phase, *i.e.,* gametophyte is filamentous showing vegetative cell division and alternates with a one-celled diploid phase, *i.e.*, sporophyte, *e.g.*, in *Oedogonium, Ulothrix*, *Spirogyra*, *Zygnema,* etc. The haploid filamentous plants, are known as **haplonts** reproduce asexually by means of zoospores or aplanospores. In the case of *Bryopsis* and *Codium* of Siphonales, there is an alternation of generations of one-celled haploid phase with diploid coenocyte or siphonous filament, which is supposed to be morphologically equivalent multicellular diploid generation or **diplont**. In the life-cycle of certain Ulvaceae and Cladophoraceae, there is an alternative of many-celled haploid generation with a many-celled diploid generation. In such cases, the two alternating generations are morphologically identical, *i.e.*, **isomorphic**, and cannot be distinguished from each other until the time of reproduction. In Ulvaceae and Cladophoraceae the meiosis in the zygote is delayed and the nuclei in the cell of the diploid plants usually do not divide reductionally until the plant is fully mature. Meiosis may take place in the nuclei of .all the cells of diploid plant, such as, in *Ulva* or may be restricted to the terminal actively growing cell, at the branch apices as in *Cladophora*. In one species of *Cladophora, i.e.*, *Cladophora glomerata*, reduction division does not occur in the development of the zoospores. The zoospores remain diploid like the plant that produces them. They are true asexual spores. In *C. glomerata,* meiosis occurs in the development of isogametes, which are consequently haploid. The zygote formed with the result of conjugation is diploid so that the haploid generation is represented only by the isogametes. In this way, two types of life-cycles occur in different species of *Cladophora.*

In one of the brown algae, *i.e., Dictyota* the two generations are morphologically identical but physiologically different. In this case, a flat ribbon-like sexual plant is identical with an asexual plant. The sexual plants which are haploid and bear antheridia or oogonia, whereas, the asexual plant which is produced from the oospore is diploid and shows reduction division (meiosis) of the spore mother cell during tetraspore formation. The tetraspores are haploid and produce the haploid sexual plants which bear antheridia or oogonia on their germination. Such type of life-cycle bears only two phases, *i.e.*, diploid and haploid and is known as biphasic life-cycle. This type of alternation of generations is known as **isomorphic** or **homologous**. The life-cycle of *Ectocarpus* is also of isomorphic type, where the gametophyte and sporophyte are alike.

In other brown algae, the two alternating generations are quite dissimilar. For example, in *Laminaria,* the sporophyte is a large complex plant and male and female gametophytes are simple and small. The type of life-cycle is also biphasic but the type of alternation of generations is heteromorphic, as the two phases are morphologically dissimilar. In *Fucus*, reduction division (meiosis) occurs during the gamete formation in the antheridia and oogonia and therefore, the eggs and sperms become haploid. Such type of life-cycle is found in animals as well as in higher seed plants where the gametophytes are reduced to a few nuclei.

In the red algae, there is a great variation in the life-histories. In these algae, the sexual reproduction and the female sex organs are more complicated. The alternation of generations in the lower genera of red algae such as *Batrachospermum* and *Nemalion* is more or less like that of green algae where the plant body is haploid and, therefore, gametophyte. The zygote formed by the fusion of egg and sperm is only diploid structure. The cystocarp or the fruit body of these algae produced from the oospore, though apparently sporophytic, is haploid because during its formation the zygote nucleus undergoes reduction division. In these lower red algae, there is a definite alternation of haploid to diploid number of chromosomes and back to haploid. In these cases, there is no alternation of gametophytic and sporophytic generations with different functions. Such life-cycle is triphasic, possessing three phases, *e.g., Batrachospermum* plant, cystocarp and zygote. The first two phases are haploid and the last phase is diploid one. This way, there is alternation of two haploid generations (gametophyte) with one diploid (sporophyte) generation. Such life-cycle is known as **haplobiontic** type of life-cycle.

There is distinct alternation of generations in the higher red algae, *e.g., Polysiphonia.* The plants are different in their chromosome numbers as well as in functions. The sexual plants bear male and female sex organs on them, and the asexual plants bear tetraspores. The tetraspores produce male and female plants (gametophytes) on germination with the result of fertilization the number of chromosomes being doubled, *i.e.*, diploid. The diploid zygote produces the asexual plant, *i.e.,* tetraspores, bearing tetrasporangia on germination. The tetrasporangia produce tetraspores. The reduction division takes place during tetraspore formation and therefore, the spores become haploid. Such life-cycle is also triphasic involves an alternation of two diploid or sporophytic generations with one haploid or gametophytic generation. Here, the diploid phases are represented by tetrasporophyte and carposporophyte, and the haploid phase by sexual plants. Such life-cycle is known as **diplobiontic** type of life-cycle.

ECONOMIC IMPORTANCE

The importance of algae has not been much realized by a layman but a deep study of the subject shows immense value to human beings. Two aspects of economic importance of algae have been realized, *i.e.*, (A) Harmful aspects and (B) Useful aspects.

A. Harmful Aspects

1. Harmful to living stock. The algae are harmful to humans in several ways. Volvocales, Chlorococcales, Myxophyceae and several others occur in such a great abundance in water, that they colour the whole water either green or blue green and cause the death of fishes. The algae block the gills of the fishes and they respire during night and make the respiration of fishes difficult by complete depletion of oxygen.

Sometimes the algae are found in water so abundantly that they make difficult to drink water of livestock. Some blue green algae have been reported poisonous and they directly cause the death of living stock who drink this contaminated water.

2. Blocking of photosynthesis. The epiphytic algae which are found upon other plants and trees block photosynthesis and indirectly harm the trees and plants.

3. Parasitic algae. The well known disease 'red rust of tea' is not caused by any parasitic fungus but an algal form *Cephaleuros virescens*. This causes a havoc to tea plants in Assam tea gardens. Besides, this parasitic form attacks several other plants, *e.g., Mangifera, Rhododendron. Coffea,* etc. The heavy losses are caused to tea and coffee by this parasitic algal form.

4. Mechanical injury. Sometimes the filamentous forms of algae are found in such a great abundance and net-like behaviour, that many fishes and other aquatic animals may get perish in these tangles, and direct death is inflicted upon them.

5. Contamination of water supply. Many blue-green, green and other algae contaminate the water of city reservoirs. This contamination develops a foul odour in the water and makes the water unhygienic. The algae also form some mucilaginous secretions which are the seats of harmful bacteria and other pathogens causing several human and animal diseases.

6. Fouling of ships. Some algae are attached to the ships, and this is called fouling of ship. The fouling retards the speed of the ship. To avoid this nuisance the ships are periodically dried up and painted with copper paint.

7. Deterioration of exposed fabrics. Commonly in rainy season, if the wet fabrics are exposed, within a few days a blue green alga appears on it and makes the cloth black spotted and weak. This was a serious problem during the Second World War. This algal growth is usually followed by bacterial infection and the fibres are completely destroyed.

B. Useful Aspects

1. Food for sea animals and fishes. The algae are used as a direct source of food by several sea animals and fishes. The marine algae are rich in iodine and several other important minerals. This makes the fundamental source of food for all marine animals and in this respect sea is the richest food producing area. Marine planktonic diatoms together with Dinoflagellata are of fundamental biological importance since all life of sea is dependent upon them.

2. Mineral contents. High mineral content, upto five percent of the wet material, in which all the mineral elements important in human and animal physiology are found, makes sea weeds a unique supplement for a well balanced diet. Potassium, sodium and chloride are found in the ionic form in sea weeds (Pillai 1956). According to Black (1953), the significance of iodine, as a constituent of food, is that besides being present in organic combination it is also available in part in the readily available form of the precursor of thyroxine, and hence this source of iodine surpasses mineral iodine in drinking water and iodized table salt. The another micronutrients besides iodine which are important in human metabolism are iron, copper, manganese and zinc, and all of them are present as the trace elements of sea weeds. The highest copper content is found in *Sarconema furcellatum* and *Acanthophora spicifera.*

At the present time, iodine is produced from brown seaweeds in Japan, France, Norway and Java. However, in Russia it is obtained from a red alga of the Black Sea, *Phyllophora nervosa* which contains from 0.2 to 0.5 percent on the dry matter. About 80 percent of the world's supply comes from nitrate mines in Chile.

3. Direct use of algae as food for man. Since the pre-historic times, several sea weeds have been used as direct source of food to human beings. Several fresh water algae have also been utilised in the preparation of various kinds of vitaminized food. As we know well that the fundamental food of sea living stock are algae and they are used as food by human beings. Since the algae are rich in vitamins and minerals, all the deficiencies are over run by the use of algae as food. The algae (sea weeds) form the most important part of the diet of Japan and China. And some people think that the artistic taste and cultural development of the people of Japan is because of the use of the sea weeds as food. In our country, a few species of *Spirogyra* and *Oedogonium* are utilised as food in South India.

Several preparations of algae are used in various countries. Of course Japan tops the list. **Suimono** is a Japanese preparation of dried fish and several sea weeds. **Mitsu** is another Japanese preparation which contains sea weeds, fruits, sugars and dried kidney beans. **Dulse** is an English dish of algae prepared from *Rhodymenia.* **Seatron** is prepared from *Nereocystis* in United States. **Laver** or **Norl** is prepared in Japan from *Porphyra.* **Green Laver** is prepared in South India from *Spirogyra* and *Oedogonium.* **Kompu** is the product of many Laminariales.

Apart from China and Japan, in Malaya, Indonesia, Myanmar and Thailand, the sea weeds are used as food. *Ulva lactuca* was used in Scotland for the preparation of salad and soups. *Laminaria saccharina, Rhodymenia palmata* have been used as food in parts of Scotland and Ireland and *Porphyra* is considered to be a tasteful culinary dish in many parts of England.

Among the more important algal food industries may be mentioned **carrageen**. This is a product of several sea weeds but principally *Chondrus crispus.* Carrageen is used by soaking it in water and mixing it with milk. Carrageen possesses the properties of gelling which is one of the factors which enhances its usefulness. Fruit juice may also be mixed with it forming fruit jelly. It is also employed in the preparation of ice creams and in the confectionery industry.

The sea weeds are also used as food in the regions of Far East and Australia. The inhabitants of the Hawaii island consume large quantities of sea weeds. The indigenous people of Chile use large quantities of *Durvillea antarctica* and some species of *Ulva.* The natives of New Zealand use certain green sea weeds in preparation of salad and soups.

The people of China and Japan consume the sea weeds on large scale. The people living on the sea coasts in these countries commonly use fresh sea weeds as food. In Japan, *Porphyra tenera* happens to be one of the most important edible alga and a product by the name of **Amanori** and **Asakusa-Nori** are made from it.

4. As a source of vitamins. The marine algae are the richest source of vitamins. The vitamins A, B and E are found abundantly in sea weeds. The vitamin B essentially required for the development of human body is found in great abundance in almost all Phaeophyceae. The cod liver oil is the rich source of vitamin A, which is acquired from sea weeds. Vitamin E is equally important for human beings which is found in many marine algae.

According to Lundin and Ericson (1956), in the sea weeds of Sweden maximum amount of vitamin B_{12} and folic acid are found in spring and summer and niacin and folic acid in winter. Vitamin B_{12} content and also that of B_1 are higher in green and red algae than in brown algae and that the niacin and vitamin C content appear to be about the same in the above three groups of marine algae. Several vitamins except ascorbic acid have been reported from *Chlorella.* The vitamins found in *Chlorella* are – thiamin, riboflavin, niacin, pyridoxine, pantothenic acid, chlorine, biotin, vitamin B_{12} and lipoic acid.

5. As a source of agar. The best agar is manufactured from *Gelidium* of Rhodophyceae, which is also called vegetative agar, Japan produces the largest quantity of agar. It produces 95% of the World production. Agar is also obtained from several other marine algae, the yield of agar, setting temperature and gel strength of the product from ten species belonging to *Gelidium, Sarconema, Hypnea* and *Gracilaria* were obtained by Thivy (1951). Japan is the chief agar producing country and it exports agar to most of the countries of the world.

The agar is used in several ways. It is employed in the preparation of ice cream, jellies, desserts, etc., in sizing the textiles and clearing many liquids. It is also used in preparing shaving creams, cosmetics and shoe polishes. The agar has constantly been used in biological laboratories for media preparation.

In India, agar resources, as annual yield of dry sea weeds of Chilka Lake have been estimated by Mitra (1946) to be about 4.06-5.08 metric tons, of Cape Comorin by Koshy and John (1948). Thivy (1957) about one metric ton, and of the Pamban area as estimated by Thivy (1957) about seven metric tons. Other large quantities are in Kathiawar Peninsula end estuaries, the resources of the Andamans are believed to be considerable.

6. Medicines and minerals. Several diseases caused by vitamin deficiency such as vitex, asthma, tooth decay, etc., may be eradicated., if flour of the sea weeds is added to the food. According to Dr. Weston, iodine is the most important element to enable the thyroid glands to secrete the thyrosin which contains 60% iodine. It controls the general development of the animal. Sea weeds are the best source of iodine for human beings. Several important sea weed medicinal preparations are prepared in various countries, *i.e.*, Kelpeck is prepared from kelps in Chicago; Burbank Vegetable tablets are sea weed preparations from United States. Kelpamalt is a sea weed medicinal preparation from New York (U.S.A.); Isokelp is prepared in California; Parakelp and Manamar are other medicinal sea weed American preparations. An antibiotic drug Chlorellum is also obtained from algae.

About fortyfive elements are found in a sea weed *Macrocystis pyrifera.* In addition to these elements, vitamins are also found. No other food contains such a great abundance of minerals and vitamins.

Marine algae are specially rich in vitamins A, B, C and E. The diatom *Nitzschia* is very rich in vitamin A and possibly this is the main source of vitamin A that is found in liver oils of many fishes. Vitamin B is also found in certain algae, for example, *Ulva, Porphyra,* etc., and *Alaria valida* is rich in Vitamin C, while some Fucoids and *Porphyra* are even richer.

Besides these several other marine forms are quite rich in minerals and vitamins. Several tribes using the sea weeds directly as food have their people handsome and sturdy.

7. Manufacture of iodine. The World's iodine supply is fulfilled from the sea weeds. Since a century back the iodine manufacture is in progress. Iodine is used in several ways.

8. Alginic acid, algin and mannitol. The alginic acid is manufactured from the cell wall of Phaeophyceae. It is insoluble in water and hard when dry. Sodium alginate is used in sizing material for water proof material, dyes, buttons, handles, combs and many of such things. This is also used as a sterilizer in daily use.

The algin is found in the form of calcium alginate and alginic acid. The Fucaceae are the chief source of algin in India. Yields of algin varying from 15.6 to 19.2 percent on air dry matter were estimated for Fucaceae and 10.4 percent for *Padina.*

A yield of 9.4 percent of mannitol from *Sargassum tenerrimum* and 73% from *S. wightii* have been reported.

9. Manufacture of soaps and alums. By burning sea weeds on the sea coast, the alkalies are prepared from sea weed ashes. These alkalies are employed in the manufacture of soaps and alums.

10. As a fodder for hens and milk cattle. By feeding the milk cattle and hens with sea weeds, iodine quantity of the milk and eggs may sufficiently be increased. In Scotland, New Zealand, Norway, France and United States several sea weeds are utilized for feeding hens and young livestock.

Some countries have even factories to process sea weeds into suitable cattle feed. The manufacture of cattle feed from sea weeds is made principally from brown algae and the processed food is fed to cattle, poultry and even pigs. It has been recorded that dried sea weeds served as cattle food have enhanced the milk-yielding and egg-laying capacity of cattle and poultry respectively.

11. Manufacture of potash. Species of *Macrocystis* and *Nereocystis* (Phaeophyceae) possess 30% of potash in their dry weight.

12. Used as fertilizers. Due to the presence of potassium chloride (KCl) in sea weeds, they are used as fertilizers in many countries, such as Japan, France, United States, England and South India.

Sea weeds are a store-house of the important potash, ionic sulphate, trace elements and growth substances, besides having every other element and radical required by plants. Sea weed manure seems to increase resistance to disease. Most of the nutrients including nitrogen compounds are in ionic form and a quick absorption by crops takes place and relatively little is left to be broken down by soil microflora, thus preventing acid conditions of the soil arising from the fermentation. In general, the minerals diffuse out from the sea weed thallus rapidly. Yet another. feature is that sea weed manure holds water and air at the same time and improves the soil in both respects. Like other manures sea weeds have a similar role but also contribute the required potassium, sulphur, phosphorus and calcium.

13. Manufacture of light weight buildings. Germany has discovered a process, in which the sea weeds are mixed with cement to make buildings light in weight and good heat resistant.

14. Manufacture of paper. It is probably thought that a rough quality of paper may be manufactured from sea weeds, as yet it is not practised.

15. Ornamental uses. Some algae like *Botrydium* and *Spirogyra* are grown in the garden ponds for their good looking habit.

16. Diatomin earth. In the beds of oceans, diatom earth is found. During millions of years, immense rocks of diatomaceous earth have been formed. At Lompoc and California, the deposits are several hundred feet deep, *i.e.*, 700 feet. At St. Maria oil fields California, the deposits are 3,000 feet deep and miles in length. Diatomaceous earth is very important commercial product and used in several ways. America is the largest producer of diatomaceous earth. This is employed as an absorbent in the manufacture of dynamite. It is used in the filtration of liquid in sugar industry. This earth can resist very high temperatures upto 1,500°C, and used in the manufacture of fire bricks. These bricks are used in the inner lining of the blast furnaces. This is also added in the cement to increase its cementing power. It is used as a mild abrasive for metal polishes. It is also used in the manufacture of tooth pastes and car polishes. It acts as a catalyst carrier in the hydrogenation of vegetable oil. In the manufacture of paints, lipsticks and other cosmetics, it is used in many ways.

17. Nitrogen fixation by blue green algae (cyanobacteria). Many members of blue green algae have the ability to fix the atmospheric nitrogen in the soil. Soil is a living mass and apart from soil particles there are in it a number of bacteria, fungi, algae and protozoa. According to Russel, algae occupy a volume three times that of the bacteria. For a long time, it was felt that these are the bacteria in the sheath of blue-green algae which fix nitrogen. P. K. De (1939) conclusively proved that blue-green algae (now called cyanobacteria) are the main agents for nitrogen fixation in rice fields, and the part played by other bacteria is relatively unimportant. Watanabe (1951) made a far reaching research and proved that *Tolyporthrix tenuis* is a strong nitrogen fixer, and it was also reported that it could fix as much as 780 lbs.of nitrogen per acre per year. Allen (1955) found that *Anabaena cylindrica* fixes 2,900 lb. of nitrogen per acre per year.

The work on this problem has been undertaken at I. A. R.I. New Delhi by G.S. Venkataraman and N. Dutta (1958). They have concluded that many blue green algae (cynobacteria) are capable of fixing atmospheric nitrogen in the soil. The dominance of the blue green algae, possessing the ability for nitrogen fixation particularly in the tropics, suggests that these organisms are of considerable importance in maintaining soil fertility. By increasing and strengthening the local blue green algal flora of a given habit, it may be possible to increase the fertility of these soils.

18. Reclamation of alkaline usar soils by blue-green algae (cyanobacteria). It has been found that some blue green algae form a thick stratum on the surface of saline usar soils during the rainy season when other plants including crop fail to grow. Various species of *Nostoc, Scytonema, Anabaena,* etc., have been found to be plentifully growing on usar soils during the rains. According to Dr. R.N. Singh (1950), these algae (cynobacteria) can be of use in the reclamation of usar lands. There is successive growth of the algal crop on such soils in a water-logged condition. After a year of such reclamation, it has been possible to grow transplanted paddy crop with a yield of as much as 1576-2000 lb. per acre. It has been found that the pH of the soil experimented upon in this matter fell from 9.7 to 7.6. Moreover, there was an improvement in the tilth and exchangeable calcium by about 20 to 33 percent. The water holding capacity increased by about 40 percent.

19. Utilization or *Chlorella.* In recent years, the alga *Chlorella* has assumed great importance and it is likely to become even more important in the future. *Chlorella* is rich in proteins, fats and vitamins, and in the presence of nutrients, CO_2 and sunlight, multiplies at an enormous rate. Several countries have been experimenting with the mass production of *Chlorella.* Pilot scale *Chlorella* farms have already been established in America, Japan, Holland, Germany and Israel. Such farms do not require excessive amounts of water and hence would be invaluable to augment food supplies in arid areas. Such 'farms' on a smaller scale could also be used to supply food in space stations of the future and even in the large ships.

Chlorella is also being increasingly used as means of purifying sewage when it is present in large, shallow tanks of effluent (after primary sedimentation of solids) the rapid photosynthesis produces abundant oxygen which is then used by the bacteria responsible for destroying the remaining organic matter. A number of sewage works using *Chlorella* now exist in the U.S.A. and at present largest in the world is opening at Auckland in New Zealand. It would seem that this cheap and effective means of sewage purification is likely to extend in the future.

Chlorella has the distinction of being the most widely studied alga as a potential food source for the world's expanding population. Each quart of a moderately thin suspension of *Chlorella* contains twenty billion cells of the alga. Arthur D. Little has produced *Chlorella* at the rate of 20 grams per square metre per day in his pilot plant established at Massachusetts. This way 17.5 tons of *Chlorella* may be produced per acre per year.

Chlorella has been largely used for studies on photosynthesis because it is easy to grow in various, environments and because the chlorophyll and end products, obtained during photosynthesis are quite similar to those found in higher plants. In newly grown cultures of *Chlorella,* large amount of protein is formed, but as the culture gets older the fats and carbohydrates are produced abundantly. The nutritional value of *Chlorella* is not only based on its high protein content but also on the fact that it contains the essential amino acids.

Chlorella contains most of the known vitamins such as thiamin, riboflavin, niacin, pyridoxine, pantothenic acid, choline, biotin, vitamin B_{12} and lipoic acid. Phytoplankton is the basic source of vitamins for fish, and makes the basis of food and vitamins for human beings. This has been calculated that a quarter pound of *Chlorella* will fulfil the need of all the vitamins except ascorbic acid, of a man for a day.

Conclusively, Sir John Murray is of opinion that the sea is as important as the land in the production of food materials. Marine Biology is a new field of study and may lead to the scientists to a new science 'mariculture' parallel to 'agriculture'. Certainly we have not realized the importance of sea vegetation.

CLASSIFICATION

For the first time, Aristotle (382-322 B.C.) and his pupil Theophrastus (372-287 B. C.), the father of Botany, classified the plants into three groups, *i.e.*, **trees**, **shrubs**, and **herbs**. This classification was based on the form and texture of the plants. Here, they considered the trees to be highest evolved.

In eighteenth century, Linnaeus (1707-1778) proposed his artificial sexual system of classification. He divided the plant kingdom into twenty five orders or classes. In his last order **Cryptogamia**, he placed the plants with concealed reproductive organs (flower, etc.). Linnaeus divided his order Cryptogamia into four suborders, *i.e.*, **(1) Filices** (pteridophytes), **(2) Musci** (mosses and leafy liverworts), **(3) Algae**, (algae, lichens and thallose liverworts), **(4) Fungi**.

The first natural system of classification of the plants was proposed by A. L. de Jussieu (1748-1836) in the end of eighteenth century, who divided the plants into three major groups, *i.e.*, **(1) Acotyledones**, **(2) Monocotyledones** and **(3) Dicotyledones**. His classification was based upon the sexual system of Linnaeus but in improved form. His Acotyledones were like Cryptogamia of Linnaeus.

In 1880, the cryptogamic plants were divided into three major groups, *i.e.*, **(1) Thallophyta** (algae, fungi and bacteria), **(2) Bryophyta**, and **(3) Pteridophyta**.

Before twentieth century only four classes were recognized among algae, *i.e.*, **(1) Chlorophyceae (2) Phaeophyceae**, **(3) Rhodophyceae** and **(4) Myxophyceae** or **Cyanophyceae**. Diatoms were then included in Phaeophyceae.

Formerly many zoologists placed many motile (flagellate) algae of today in the class Mastigophora of phylum Protozoa.

Rabenhorst in 1863, for the first time placed the *Chlamydomonas-Volvox* series in the Chlorophyceae of algae.

In the beginning of the twentieth century, the class Xanthophyceae was separated from Chlorophyceae and certain pigmented flagellate types were included in the class. According to Dr. F.E. Fritsch (1935, 1944, 1945), the algae have been divided into following eleven classes:

1. Chlorophyceae
2. Xanthophyceae
3. Chrysophyceae
4. Bacillariophyceae
5. Cryptophyceae
6. Dinophyceae
7. Chloromonadineae
8. Euglenophyceae or Euglinieae
9. Phaeophyceae
10. Rhodophyceae
11. Myxophyceae

FRITSCH'S SYSTEM

Dr. Fritsch's classification is mainly based on the following facts:

1. Pigmentation
2. The assimilatory food products or metabolic products.
3. Type of flagella.

In addition to above given facts, a few minor characters may also be taken into consideration. Here we will consider the characters of few important classes, *i.e.,* 1. Chlorophyceae, 2. Xanthophyceae, 3. Bacillariophyceae, 4. Phaeophyceae, 5. Rhodophyceae and 6. Myxophyceae.

1. Chlorophyceae (360 genera; 5,000 species).

Major Charactersitic Features

(*i*) The pigments are just like in higher green plants, *i.e.*, chlorophyll *a*, chlorophyll *b*, xanthophyll and carotenes. These pigments are localized in definite plastids or chromatophores.

(*ii*) The photosynthetic food product is starch, rarely oil as in *Vaucheria*. Usually in the chromatophores the pyrenoids are present. They are organellae for the formation of starch. A part of pyrenoid converts into starch.

(*iii*) The flagellation is of isokonate type, *i.e.,* both of the flagella are equal in length.

Minor Characteristic Features

(*i*) Vegetative body may be one to many celled.

(*ii*) The cell consists of cellulose.

(*iii*) Sexual reproduction ranges from isogamy to oogamy.

(*iv*) The life-cycle is mostly of haplontic type.

(v) The parenchymatous cells do not occur.

(*vi*) Most of the species are fresh water and few are marine.

2. Xanthophyceae (75 genera; 675 species)

Major Characteristic Features

(*i*) The chromatophores are yellow green, containing chlorophyll *a,* carotenes (β carotene) and xanthophylls in them.

(*ii*) The starch is not found. The pyrenoids are absent. The chief food products are oils.

(*iii*) The flagellation is of heterokontae type, *i.e.*, one flagellum is short and other long.

Minor Characteristic Features

(*i*) The cell wall consists of pectin, and in majority of cases two overlapping halves are present, *e.g.*, *Tribonema.*

(*ii*) The sexual reproduction is rarely found and is doubtfully established for only one species.

(*iii*) The resting stages possess two pieced membranes.

(*iv*) The walls are silicified.

(*v*) The life-cycle is mostly of haplontic type.

(*vi*) Majority of the species are fresh water, some are marine.

3. Bacillariophyceae (170 genera; 5,300 species)

Major Characteristic Features

(*i*) The chromatophores are golden brown or yellow, due to the presence of a pigment called **diatomin**. The other pigments are chlorophyll *a,* carotenes (β-Carotene) and xanthophylls.

(*ii*) The food products are fats or volutins.

(*iii*) The flagellate bodies are 1 to 2 flagellate.

Minor Characteristic Features

(*i*) Majority of them are unicellular and, some are colonial.

(*ii*) The sexual reproduction is of special type resulting in the formation of auxospores.

(*iii*) They are diplontic.

(*iv*) They are widely distributed in sea and fresh waters.

4. Phaephyceae (195 genera; 1,000 species)

Major Characteristic Features

(*i*) The chromatophores are yellowish brown in colour possessing xanthophylls in abundance in them, chlorophyll *a,* and carotenes (β carotene) are also found.

(*ii*) The reserve food products are alcohols, mannitol and laminarin.

(*iii*) The motile reproductive cells are pyriform with two laterally inserted flagella; one of which is of tinsel type. In all except Fucales, the anterior flagellum is longer.

Minor Characteristic Features

(*i*) The unicellular bodies are unknown.

(*ii*) The special type of granules, which are excretory in nature are vesicles.

(*iii*) The sexual reproduction ranges from isogamy to oogamy.

(*iv*) The zygote has no resting period.

(*v*) The great diversity prevails in the life cycles, possessing clear alternation of generations.

(*vi*) With the exception of three species, rest are marine.

5. Rhodophyceae (400 genera; 2,500 species)

Major Characteristic Features

(*i*) The chromatophores possess r-phycoerythrin and c-phycocyanin pigments. The other pigments are chlorophyll *a,* carotenes (β carotene) and the xanthophylls. The plants look reddish in colour.

(*ii*) The reseve food products are polysaccharides, floridean starch and a soluble sugar called floridoside.

(*iii*) The flagellation is altogether absent.

Major Characteristic Features

(*i*) The prominent plasmodesmata are present.

(*ii*) Sexual reproduction is special and advanced.

(*iii*) The life-cycle clearly shows alternation of generations.

(*iv*) About 50 species belonging to near about a dozen genera are fresh water. The rest are marine.

6. Myxophyceae (150 genera; 1,500 species)

Major Characteristic Features

(*i*) The pigments are not localized in definite chromatophores. The pigments are localized in the peripheral portion of the protoplast. The main pigments are chlorophyll *a,* carotenes (β carotene) and xanthophylls. The overmasking blue pigment is c-phycocyanin. The red pigment c-phycoerythrin is also found in peripheral regions in dispersed condition. This peripheral region with pigments is called chromatoplasm.

(*ii*) The reserve food products are sugars and glycogen-like compounds called the cyanophycean starch.

(*iii*) The flagella are altogether absent.

Minor Characteristic Features

(*i*) Sexual reproduction is unknown.

(*ii*) Majority of them are fresh water forms. Some are found in sea water.

PASCHER'S AND SMITH'S SYSTEM

Gilbert M. Smith (1955) followed A. Pascher's (1914, 1921, 1927) classification with slight modifications. This is as follows —

Division 1. **Chlorophyta**
- Class 1. **Chlorophyceae** (green algae)
- Class 2. **Charophyceae** (stoneworts)

Division 2. **Euglenophyta**
- Class 1. **Euglenophyceae** (euglenoids)

Division 3. **Phyrrophyta**
- Class 1. **Desmophyceae** (dinophysids)
- Class 2. **Dinophyceae** (dinoflagelloids)

Division 4. **Chrysophyta**
- Class 1. **Chrysophyceae** (golden brown algae)
- Class 2. **Xanthophyceae** (yellow green algae)
- Class 3. **Bacillariophyceae** (diatoms)

Division 5. **Phaeophyta** (brown algae)
- Class 1. **Isogenerateae**
- Class 2. **Heterogenerateae**
- Class 3. **Cyclosporeae**

Division 6. **Cyanophyta** (blue green algae)
- Class 1. **Myxophyceae**

Division 7. **Rhodophyta** (red algae)
- Class 1. **Rhodophyceae**

Algae of uncertain systematic position —
1. **Chloromonadales**
2. **Cryptophyceae**

PAPENFUSS SYSTEM

According to Papenfuss (1953), the algae have been divided into eight phyla. These phyla and classes are as below:

Phylum 1. CHLOROPHYCOPHYTA
- Class (a) Chlorophyceae

Phylum 2. CHAROPHYCOPHYTA
- Class (a) Charophyceae

Phylum 3. EUGLENOPHYCOPHYTA
- Class (a) Euglenophyceae

Phylum 4. CHRYSOPHYCOPHYTA
- Class (a) Xanthophyceae
- (b) Chrysophyceae
- (c) Bacillariophyceae.

Phylum 5. PYRROPHYCOPHYTA
- Class (a) Dinophyceae
- (b) Cryptophyceae
- (c) Chloromonadophyceae

Phylum 6. PHAEOPHYCOPHYTA
- Class (a) Phaeophyceae

Phylum 7. SCHIZOPHYCOPHYTA
- Class (a) Schizophyceae

Phylum 8. RHODOPHYCOPHYTA
- Class (a) Rhodophyceae

Chapman (1962) has also proposed a system of classification based on morphological characters, pigments and biochemical differences. He has placed, Charophyceae, Chlorophyceae, Phaeophyceae and Rhodophyceae under a large group EUPHYCOPHYTA. According to him, these classes of algae are interrelated, and therefore, have been grouped under Euphycophyta. The classification proposed by Chapman (1962) is as follows:

A. EUPHYCOPHYTA
- classes
  1. Charophyceae
  2. Chlorophyceae
  3. Phaeophyceae
  4. Rhodophyceae

B. MYXOPHYCOPHYTA
- class Myxophyceae

C. CHRYSOPHYCOPHYTA
- classes
  1. Chrysophyceae
  2. Xanthophyceae
  3. Bacillariophyceae

D. PYRROPHYCOPHYTA
- classes
  1. Cryptophyceae
  2. Dinophyceae.

MODERN TRENDS IN CLASSIFICATION

The classification of algae has continually been modified since its beginning in the 1753 edition of Linnaeus **Species Plantarum**. In **natural** or **phylogenetic** systems (which represent real evolutionary relationships) of classification, alterations in the classification of algae are made as a result of augmentation of our knowledge. This has resulted in the present, largely **polyphyletic** system (the system having many distinct lines of evolutionary history) in which the algae are grouped in seven to eleven categories of divisional rank. The criteria on which the classification is based are **pigments, reserve food** and **flagellation**. However, our knowlege of the comparative chemistry of the several pigments and storage products is fragmentary, especially in relation to the genetics and evolution of the organisms, that it is quite possible our current polyphyletic groupings may be modified into a more monophyletic system in the future.

At the levels of orders, families and genera it has been noted that (1) In the Chlorophycophyta and Phaeophycophyta, the orders are delimited largely on vegetative or somatic characters, variation in reproduction being treated as of secondary importance. In the division Rhodophycophyta, in contrast, the orders are at present delimited on the basis of the organization of the female reproductive branch, vegetative attributes being secondary. (2) Among the Chrysophycophyta, the families of certain orders are at present delimited on the basis of flagellar numbers while the same criterion is used at the generic level in the unicellular motile Chlorophycophyta. These examples indicate that modern classifications are, at least approximations of relationship.

Like all plants, the algae are classified in accordance with the recommendations and prescriptions of the International Code of Botanical Nomenclature. This code recognizes the individual organism as belonging to a species, the species to a genus, the genus to a family, the family to an order, the order to a class and the class to a division.

The following suffixes for use in the classification of algae have been recommended by the International Code of Botanical Nomenclature;

Division–**phyta**	Family–**aceae**
Class–**phyceae**	Sub-family–**oideae**
Sub-class–**phycideae**	Genus–**a Greek name**
Order–**ales**	Species–**a Latin name**
Sub-order–**inales**	

The algae have been divided into 11 divisions. They are as follows.

1. Cyanophycophyta (Blue-green algae)
2. Chlorophycophyta (Green algae)
3. Charophyta (Stoneworts)
4. Euglenophycophyta (Euglenoids)
5. Xanthophycophyta (Yellow-green algae)
6. Chrysophycophyta (Golden algae)
7. Bacillariophycophyta (Diatoms)
8. Phaeophycophyta (Brown algae)
9. Pyrrophycophyta
10. Cryptophycophyta
11. Rhodophycophyta (Red algae)

Representatives belonging to few important divisions— Chlorophycophyta, Xanthophycophyta, Phaeophycophyta and Rhodophycophyta have been discussed in detail in the present text.

QUESTION BANK

N.B. *Undermentioned questions are asked repeatedly in various Indian Universities and competitive examinations.*

Long Answer Type

1. How would you define algae ? Give characteristic features of this diversified group of plant kingdom.
2. Trace out the history of phycology. Also mention the Indian work.
3. Write a comprehensive note on habit and habitat of algae.
4. What type of nutrition is found in various groups of algae ?
5. Describe the range of vegetative structure in various groups of algae.
6. Give the various modes of reproduction found in algae.
7. Write a detailed note on the methods of perennation in algae.
8. What type of life-cycles are met within algae ?
9. Write a detailed note on pigmentation of algae.
10. Write an essay on the economic importance of algae.
11. Give the modem trends of classification of algae.
12. What criteria have been adopted for classifying algae ?
13. Give the outline of Fritsch's system of classification of algae.
14. Give a comprehensive account of origin and evolution of sex in algae.

Short Answer Type

1. Write a note on history of phycology.
2. Write a note on Algology in India.
3. Write notes on — Planktophytes, Cryophytes and Fluviatile algae.
4. Write a brief account of nutrition in algae.
5. Write notes on palmelloid and dendroid types, siphonous type, and advanced type of algae.
6. Write a note on evolution of sex in algae.
7. Write notes on: Heteromorphic and haplobiontic types of life-cycle in algae.

Diagrammatic Type

1. Give the cellular structure of a procaryotic alga.
2. Give the structure of a cell of eucaryotic alga.

Multiple Choice Type

1. Nitrogen fixation in soil is caused by:
   (i) Chlorophyceae, (ii) Phaeophyceae, (iii) Myxophyceae, (iv) Rhodophyceae.
2. Nitrogen fixation in soil is caused by:
   (i) green algae, (ii) brown algae, (iii) blue-green algae, (iv) red algae.
3. R.N. Singh worked on:
   (i) red algae, (ii) brown algae, (iii) blue-green algae, (iv) green algae.
4. Cyanophyta is written by: (i) T.V. Desikachary, (ii) M.S. Randhawa, (iii) R.N. Singh, (iv) J.N. Misra.
5. Zygnemaceae is written by:
   (i) T.V. Desikachary, (ii) M.S. Randhawa, (iii) R.N. Singh, (iv) J.N. Misra.
6. Phaeophyceae in India, is written by:
   (i) T.V. Desikachary, (ii) M.S. Randhawa, (iii) R.N. Singh, (iv) J.N. Misra.
7. Role of blue-green algae in Nitrogen economy of Indian Agriculture is written by:
   (i) T.V. Desikachary, (ii) M.S. Randhawa, (iii) R.N. Singh, (iv) J.N. Misra.
8. Cryophytes are found:
   (i) in deep sea, (ii) on sea shore, (iii) on ice and snow, (iv) on humus soil.

9. Chlorophyll is found in:
   (i) blue-green algae, (ii) green algae, (iii) brown algae, (iv) all the above.
10. Amylum stars are found in:
    (i) *Vaucheria,* (ii) *Ulothrix,* (iii) *Chara,* (iv) *Polysiphonia.*
11. Heterocysts are found in:
    (i) Chlorophyceae, (ii) Phaeophyceae, (iii) Rhodophyceae, (iv) Myxophyceae.
12. Diplobiontic type of life-cycle is found in:
    (i) *Vaucheria,* (ii) *Oedogonium,* (iii) *Polysiphonia,* (iv) *Ectocarpus.*
13. Carrageen is obtained from:
    (i) *Chara,* (ii) *Polysiphonia,*(iii) *Sargassum,* (iv) *Chondrus*
14. Agar-agar is manufactured from:
    (i) *Ectocarpus.* (ii) *Polysiphonia,* (iii) *Spirogyra,* (iv) *Gelidium*
15. Alginic; acid is obtained from:
    (i) Chlorophyceae, (ii) Phaeophyceae, (iii) Rhodophyceae, (iv) Myxophyceae.
16. Diatomin earth is a product of:
    (i) Chlorophyceae, (ii) Bacillariophyceae, (iii) Xanthophyceae, (iv) Rhodophyceae.
17. High protein contents are found in:
    (i) *Chlorella,* (ii) *Chlamydomomas,* (iii) *Spirogyra,* (iv) *Ulothrix.*
18. Red rust of tea is caused by:
    (i) *Puccinia,* (ii) *Ustilago,* (iii) *Melampsora,* (iv) *Cephaleuros.*
19. Red rust of tea is caused by:
    (i) Algae, (ii) Fungi, (iii) *Cuscuta,* (iv) *Balanophora.*
20. Point out the algologists:
    (i) M.O.P. Iyengar; (ii) Y. Bharadwaja (iii) R.N. Singh, (iv) all the above.

ANSWERS

1 (iii), 2 (iii), 3 (iii), 4 (i), 5 (ii), 6 (iv), 7 (iii), 8 (iii), 9 (iv), 10 (iii), 11 (iv), 12 (iii), 13 (iv), 14 (iv), 15 (ii), 16 (ii), 17 (i), 18 (iv) , 19 (i), 20 (iv).

4 CHAPTER

Chlorophycophyta–Chlorophyceae

(The Green Algae)

Characteristic features

Pigmentation. Chlorophylls *a* and *b*, α and β-carotene, Astaxanthin, Lutein, Neoxanthin, Siphonein, Siphonoxanthin, Violaxanthin, Zeaxanthin.

Storage products. Starch, oils.

Flagellation. 1, 2, 4 to many, equal apical or subapical insertion.

There are about 360 genera and 5,700 species.

Occurrence and distribution. About 90% of Chlorophyceae are fresh water and the remaining 10% are marine. Order Conjugales is exclusively fresh water. The algae are common both in quiet and running waters. The species of *Coleochaete* grow on aquatic plants and grasses. The species of *Cladophora* grow even upon mollusc shells. *Trentepohlia* is an aerial alga and grows on rocks and fencing wires. Volvocales and Chlorococcales occur in planktonic habit, *i.e.*, they are free floating. The cryo algae are found upon ice and snow. *Chlamydomonas nivalis* is the main cause of red snow. *Chlorella* is endophytic and grows inside the tissue of *Hydrilla. Cephaleuros* is a parasitic form and causes 'red rust of tea.'

Several Chlorophyceae are marine and confined in shallow waters. Many Siphonales are marine. Conclusively, it can be said that the forms of Chlorophyceae are widely distributed and may be expected anywhere.

Range of structure. A sufficiently large range of vegetative or somatic structure is found in Chlorophyceae.

1. **Motile forms.** There may be motile unicellcular forms, *e.g.*, *Chlamydomonas.* The species of *Phacotus* are motile and unicellular but encapsulated. Several forms are motile colonies, *e.g.*, *Gonium sociale* is a 4-celled colony and *Gonium pectorale* is 16-celled colony. *Pandorina* 8-celled; *Eudorina* 32-celled and *Pleodorina* is 32-128 celled colony. *Volvox* is the highest evolved colonial form and consists of 500-50,000 cells. Only a few cells are reproductive and the rest are vegetative. In one species of *Chlamydomonas,* double individuals are found.

2. **Palmelloid form.** All Tetrasporineae are palmelloid. In the species of *Chlamydomonas,* the palmelloid phase prevails temporarily. In *Tetraspora,* this is the permanent phase.

3. **Coccoid form.** Most of Chlorococcales are coccoid.

4. **Filamentous form.** Several Chlorophyceae are filamentous. The filaments may be (*i*) unbranched or (*ii*) branched. The typical examples of unbranched filamentous forms are Ulotrichales, *Spirogyra, Oedogonium,* etc. Branched filaments are found in Cladophorales. Heterotrichous filamentous forms are found in Chaetophorales.

5. **Siphonous habit.** In such algae, the filaments or thalli are branched, multinucleate but without septa. Such types are found in order Siphonles, *e.g.*, *Valonia, Caulerpa, Codium,* etc.

6. **Uniaxial type.** *Chara* is the typical example of this type. Here the ensheathing layer of axis is always one-celled thick.

7. **Parenchmatous type.** The expanded sheets, which look like parenchyma are also found within Chlorophyceae. *Ulva* possesses two-celled thick expanded sheet.

The cell wall. In Chlorophyceae, the cell wall usually consists of 2 or 3 layers. The innermost layer is of cellulose and encircles the protoplast. The second layer consists of pectic compounds of pectose sometimes associated with cellulose. In the case of *Oedogonium* and *Cladophora,* the outermost layer of the cell wall is chitinous in nature. In the cell wall of Siphonales and Conjugales, the outermost wall layer is covered over with mucilaginous substance and becomes slimy, *e.g., Spirogyra.* Usually calcium silicates and carbonates are deposited on the mucilaginous layer. Some of the more primitive Chlorophyceae lack cell walls and the protoplasts are naked.

The protoplast. In motile cell, two or more contractile vacuoles are found which pulsate alternately and help in excretion. Usually the filamentous forms possess a single large central vacuole in each cell. In some Chlorophyceae, *e.g., Prasiola,* the vacuoles are not found. The vacuole is filled up with cell sap rich in tannins. The cytoplasmic connections are clearly seen in colonial forms such as *Volvox.* Such plasmodesmata are not found in filamentous forms but in *Trentepohlia.*

The chloroplast. In Chlorophyceae, a variety of chloroplasts is known. In the genus *Chlamydomonas,* various types of chloroplasts are found. The detailed account of the chloroplasts of *Chlamydomonas* will be given in the description of the genus. The typical 'chlamydomonad type' of chloroplast is basin-shaped.

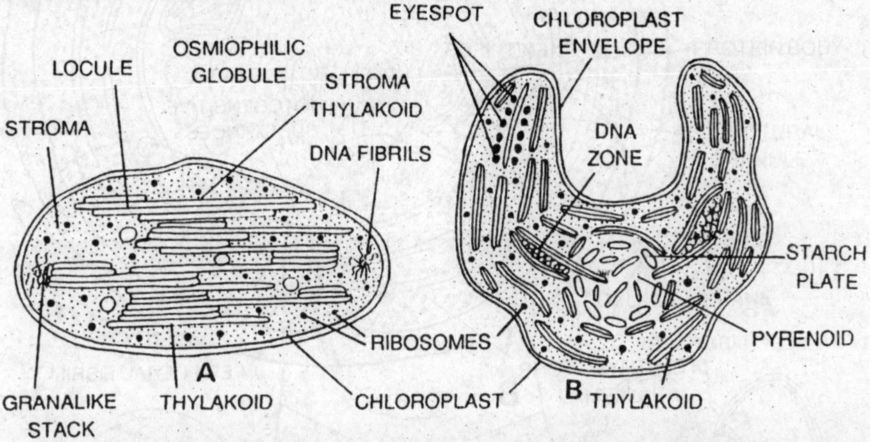

Fig. 4.1. Ultra structures of the chloroplasts of Chlorophyceae (green algae). A, discoid chloroplast showing grana-like arrangement of thylakoid cup-shaped chloroplast of *Chlamydomonas.*

Stellate chloroplasts are found in the cells of *Zygnema* and thallus of *Prasiola.* In Oedogoniales and Cladophorales, the chloroplasts are reticulate having many pyrenoids in them at anastamoses. In *Spirogyra,* the chloroplasts are ribbon-like and 1-14 in number in each cell. They are spiral and sometimes cut on the margins. In most of the Chlorophyceae, one chloroplast is found in each cell but exceptions are there. In *Chlamydomonas alpina,* several discoid chloroplasts are developed by breaking a single cup-shaped chloroplast.

In *Mougeotia,* special type of movement of chloroplast is found. Each chloroplast has independent capacity of its movement. The chloroplast rotates to a 90° and exposes its edge to the light.

The pyrenoids. The pyrenoids are found in the chloroplasts of Chlorophyceae with few exceptions such as *Chara, Vaucheria, Microspora,* etc.

There are different views about the functions of pyrenoids. According to one view, the pyrenoids are regarded as reserve proteins. According to other view, they are regarded as special organellae of the cells associated with metabolic activities. According to one rejected view, they are secreting receptacles.

The pyrenoids are crystalloid in structure and sometimes called 'pyrenocrystals'. According to some algologists, the crystals are embedded in a substance.

The compound pyrenoids have been reported from the species of *Tetraspora* and *Enteromorpha.* The compound pyrenoid is surrounded by a starch sheath. There are as many starch grains deposited around the pyrenoid and the whole pyrenoid is surrounded by a starch sheath. Whenever, the starch sheath of starch grains is incomplete around the pyrenoid, the structure is called the 'polar pyrenoid'.

In majority of Chlorophyceae, the pyrenoids are responsible for starch formation. In order Siphonales where the pyrenoids are lacking, the starch is formed by leucoplasts in the cytoplasm.

The pigments. The most important pigment of Chlorophyceae is chlorophyll *a.* In addition to it chlorophyll *b,* carotenoids and xanthophylls are also present. Haematochrome is reported from *Trentepohlia* and because of this pigment this looks orange red in colour. Phycoporphyrin is recorded from several Conjugales. Fucoxanthin has been found in *Zygnema pectinatum.* The lutein is among principal xanthophylls.

The nucleus. In majority of Chlorophyceae, the cells are uninucleate except Cladophorales where the condition is multinucleate. The multinucleate condition is also seen in Siphonales where the filaments are without septa. Fundamentally the nucleus of Chlorophyceae resembles with the nucleus of higher plants. The nucleolus is deeply stainable and conspicuous. It appears as a central

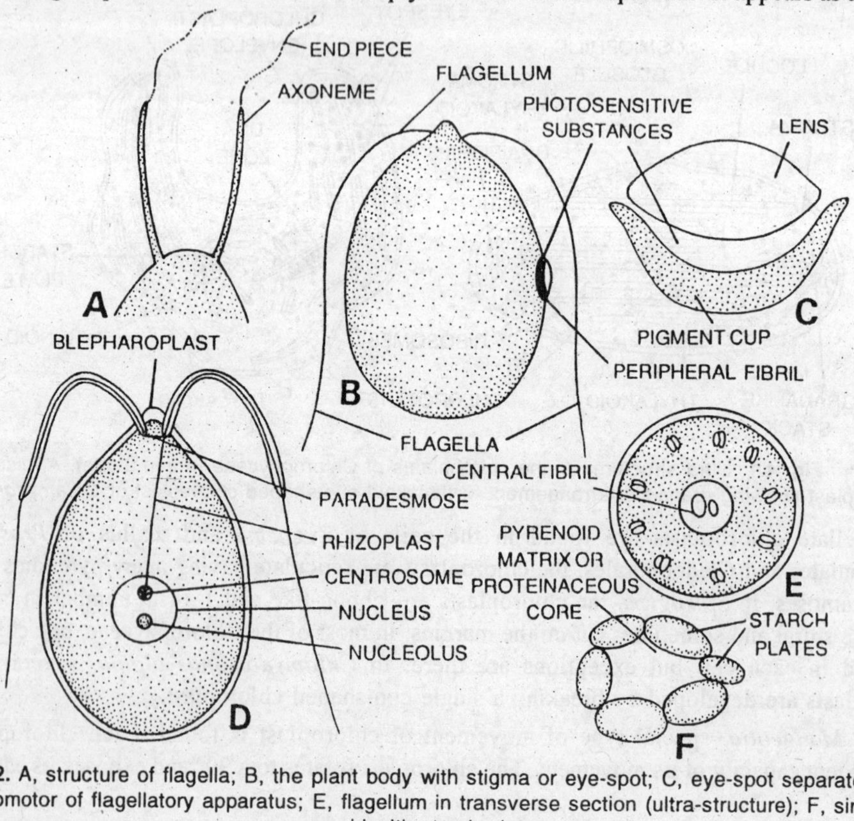

Fig. 4.2. A, structure of flagella; B, the plant body with stigma or eye-spot; C, eye-spot separated; D, neuromotor of flagellatory apparatus; E, flagellum in transverse section (ultra-structure); F, single pyrenoid with starch plates.

caryosome. The centrosomes have been recorded from Volvocales and *Cladophora.* Mitosis and meiosis occur in all types.

The flagella. The flagella of Chlorophyceae and especially of order Volvocales are produced by an apparatus called **'Neuromotor** or **flagellar apparatus'**. Usually two flagella occur at the anterior end of the motile body. At the base of each flagellum, there is a flagellum producing, **granule** called **'blepharoplast'**. These two blepharoplasts are connected to each other by a thin fibril called **paradesmose.** One of the blepharoplasts is connected by a long fibre called **rhizoplast** to the centrosome situated inside or outside the nucleus. The whole apparatus is called 'flagellar or neuromotor apparatus'.

In Chlorophyceae, both the flagella are of whiplash or acronematic type. The terminal end of the flagellum is thin and uncovered called the **end piece**, whereas the basal part of the flagellum, is comparatively longer and covered by a cytoplasmic coating. This part is called **axoneme**.

Under the electron microscope, the flagellum is seen to consist of a peripheral cylinder of 9 doublet fibrils surrounding a central pair of singlet fibrils. This type of arrangement is known as 9+2 arrangement. The entire group of fibrils is enclosed in a double sheath and the two central fibrils have subsidiary (single) sheath of their own.

The eye spot or stigma. Such spots are universally found on the motile organisms of the Chlorophyceae. They are lens-like and sensitive to light. The details of the stigma may be read from order Volvocales.

Contractile vacuoles. The motile bodies of Chlorophyceae possess contractile vacuoles on them. These vacuoles are present on the anterior of the body. They are 2-6 in number. They are pulsating in nature and excretory in function.

Reproduction. Three main types of reproduction are found in Chlorophyceae, *i.e.*, (1) Vegetative, (2) Asexual and (3) Sexual.

1. Vegetative reproduction. This type of reproduction takes place vegetatively by several means.

(i) **By fragmentation,** *e.g.,* in *Spirogyra, Ulothrix, Oedogonium* and several others.

(ii) **By amylum stars.** Such structures are found in *Chara* which help in vegetative propagation.

(iii) **By tubers.** They are found in *Chara.* Such bodies are full of starch.

(iv) **By secondary protonema.** Such structures are found in *Chara* which help in vegetative propagation.

(v) **By bulbils.** Such vegetative structures develop in *Chara* which develop into new plants.

(vi) **By cell division.** Here the algal form divides vegetatively, *e.g.*, *Protococcus*.

(vii) **By akinetes.** The akinetes are vegetative structures. Here the vegetative cells develop into spore-like stages with much thicker walls and more food material. Akinetes have been reported from several Chlorophyceae, *e.g., Oedogonium, Ulothrix,* etc.

2. Asexual reproduction. This type of reproduction is a common feature in Chlorophyceae and takes place in several ways.

(i) **By zoospores.** This is the most important type of asexual reproduction. The zoospores are motile and naked protoplasts. They may be biflagellate or quadriflagellate. In *Oedogonium,* stephanokontean type of zoospores are found. In such cases, many flagella are present at the anterior end of the zoospore in circle.

***(ii)* By aplanospores.** They are non-flagellate (non-motile) spores and sometimes interpreted as abortive zoospores. They are found in many Chlorophyceae and usually develop in unfavourable conditions.

***(iii)* Palmella stage.** Hundreds and thousands of cells are embedded within a common gelatinous matrix. In *Chlamydomonas,* the palmella stage occurs temporarily whereas this is permanent feature in *Palmella.*

***(iv)* Hypnospores.** They are very thick-walled aplanospores. They are perennating bodies. They are found in many green algae, *e.g., Pediastrum.*

(v) Autospores. The autospores are aplanospores with the shape of mother cells. They are present in order Chlorococcales. When autospores are found in the colonies they are called autocolonies.

***(vi)* Daughter colonies.** In *Volvox,* daughter colonies and sometimes even grand daughter colonies are developed asexually.

3. Sexual reproduction. This may be divided into four main sub-types, *i.e.,* (*i*) isogamy, (*ii*) anisogamy, (*iii*) aplanogamy and (*iv*) oogamy.

***(i)* Isogamy.** This is the simplest and primitive type of sexual reproduction. Two morphologically identical flagellated zoogametes take part in fusion. Usually such gametes come from two different individuals. The resultants are zygospores. Isogamy is found in many Chlorophyceae, *e.g.,* species of *Chlamydomonas, Ulothrix,* etc.

***(ii)* Anisogamy.** Here the gametes taking part in fusion are not identical. The gametes are flagellate. One gamete is smaller and the other larger in size. The smaller one is supposed to be male and the larger female. The resultants are zygospores. The examples of anisogamy are found in several Chlorophyceae, *e.g., Chlamydomonas braunii.*

***(iii)* Oogamy.** This is the highest evolved type. There is union of small flagellate antherozoid with a large non-motile egg. The resultants are called oospores, *e.g., Volvox, Oedogonium,* etc. and several other Chlorophyceae.

The species may be homothallic (monoecious), *i.e.,* both male and female sex organs are present on the same plant, or heterothallic (dioecious), *i.e.,* male and female sex organs develop on two different thalli.

***(iv)* Aplanogamy.** This is an unusual type found in order Conjugales. Here the amoeboid gametes (aplanogametes) fuse to each other and zygospores are resulted, *e.g., Zygnema, Spirogyra,* etc.

Parthenogenesis. Sometimes the perfect spores are attained without fusion and are called azygospores or parthenospores. Here the female gamete may give rise to a new plant without fusing with male gamete. The process is called parthenogenesis. This process has been recorded from several Chlorophyceae. Sometimes one species of the genus reproduces parthenogenetically and the other species does not, *e.g., Ulva lactuca* is parthenogenetic species and *U. lobata* is not.

Zygote and its germination. The zygote is a very resistant perennating body which usually faces the adverse environmental conditions for its survival. The zygotes of Chlorophyceae may be thin-walled or thick-walled. The thin-walled zygotes germinate within a day or two after their formation. The thick-walled zygotes remain dormant for a short or long period and then germinate.

Iso - and anisogamous species form thick-walled zygotes which possess flagella on them for a short time and the zygote remains motile during this period. Later on, these flagella are withdrawn and the zygote secretes a thick wall around it.

All oogamous species possess non-motile zygotes from the very beginning.

The rest period of the zygotes of various Chlorophyceae varies from genus to genus. The zygotes of *Chlamydomonas* germinate in ten days; those of *Ulothrix* in 5 to 9 months, and the zygotes of *Oedogonium* take the largest period in their germination, *i.e.*, 12 to 14 months.

At the time of the germination of the zygote, meiosis occurs to bring the haploid phase again.

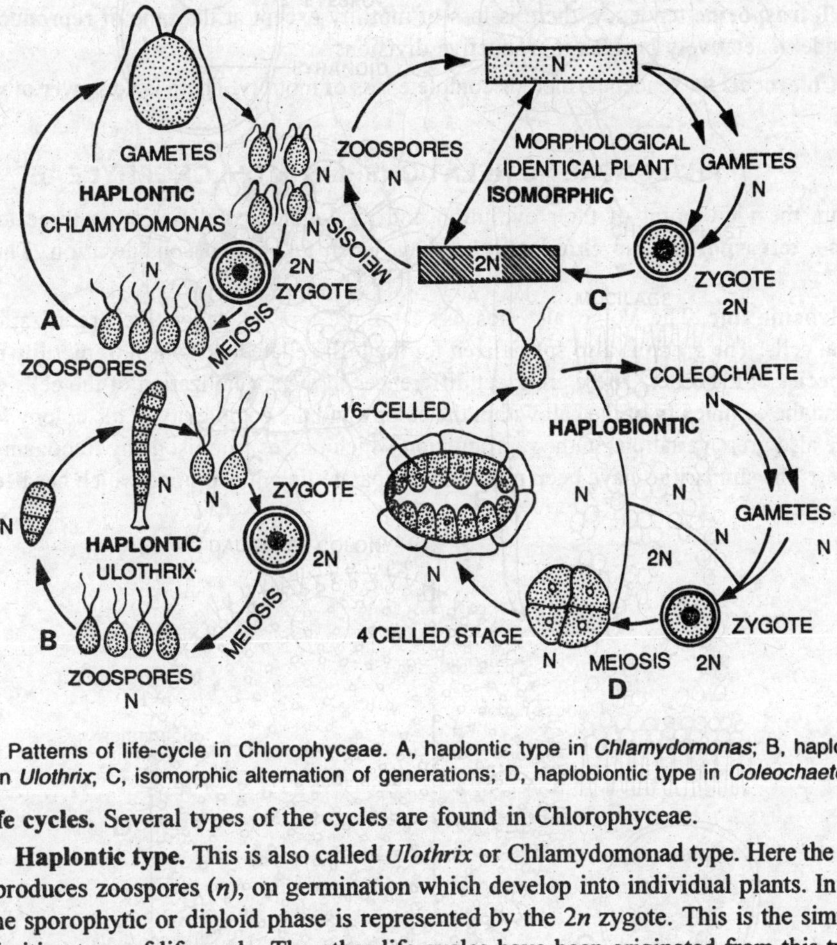

Fig. 4.3. Patterns of life-cycle in Chlorophyceae. A, haplontic type in *Chlamydomonas*; B, haplontic type in *Ulothrix*; C, isomorphic alternation of generations; D, haplobiontic type in *Coleochaete*.

Life cycles. Several types of the cycles are found in Chlorophyceae.

1. **Haplontic type.** This is also called *Ulothrix* or Chlamydomonad type. Here the zygote is $2n$ and produces zoospores (n), on germination which develop into individual plants. In this life-cycle, the sporophytic or diploid phase is represented by the $2n$ zygote. This is the simplest and most primitive type of life-cycle. The other life cycles have been originated from this type. This type is found in all Chlorophyceae except few.

2. **Isomorphic alternation of generations.** In this type, there are two exactly similar alternating phases, but one diploid ($2n$) and the other haploid (n). The haploid plants produce the gametes which after fusion produce zygotes ($2n$). The zygote is a new plant, which germinates and produces zoospores. Prior to zoospore formation there is meiosis, *e.g.*, Cladophoraceae, Ulvaceae, Chaetophoraceae.

3. **Heteromorphic alternation of generations.** In *Draparnaldiopsis indica* there are two alternating multicellular phases which are haploid (n) and diploid ($2n$) respectively. These phases are dissimilar in structure.

4. **Haplobiontic type.** Here there are three phases in the life cycle. Out of three, two phases are haploid and one diploid. The example of this type is *Coleochaete.* Details may be consulted from the description of the genus.

Evolution in chlorophyceae. F.F. Blackman in 1900, reported about evolutionary tendencies in Chlorophyceae. These tendencies are **(1) Volvocine, (2) Tetrasporine and (3) Chlorococcine** tendencies.

In **Volvocine** evolutionary tendency, there is an intention of motility by flagella. They do not divide vegetatively.

In **Tetrasporine** tendency, there is loss of motility except at the time of reproduction. The cells divide vegetatively besides reproductive divisions.

In **Chlorococcine** tendency, there is complete loss of motility. There is no power of vegetative division.

PHYLOGENETIC RELATIONSHIPS IN CHLOROPHYCEAE

From the view point of their evolution and phylogenetic relationships three tendencies **volvocine, tetrasporine** and **chlorococcine** have been taken into consideration. They are as follows.

Volvocine line. The Volvocales possess a unique type of vegetative organization of the individual cells. The group is also specialized for the biflagellate mechanism of motility possessed by the species of the order. There are great differences in the specialization of the cells within the colony and the complexity of the cell within the colony and the complexity of the colony formation. There are also great variations in the reproduction which ranges from isogamy to oogamy through anisogamy. The similarities have been placed in a separate line of evolution which has been known as **volvocine line**.

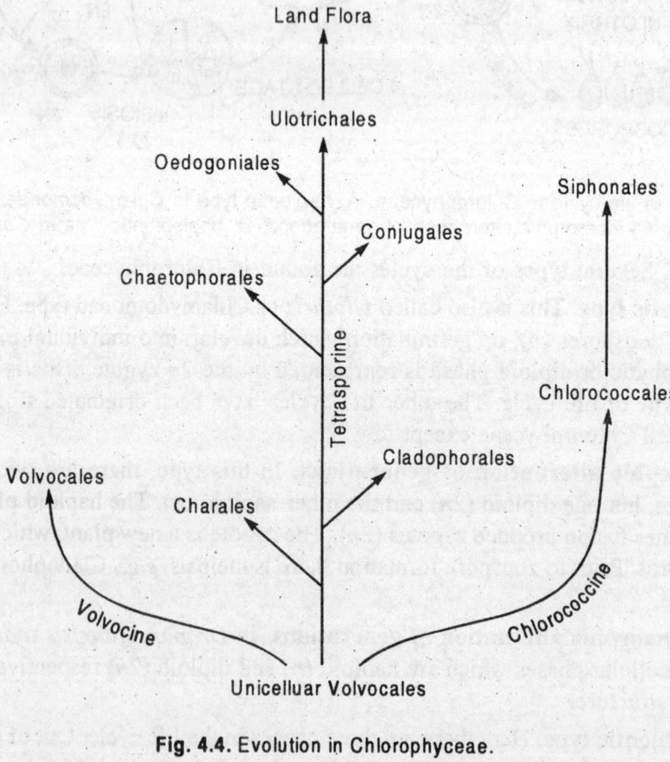

Fig. 4.4. Evolution in Chlorophyceae.

In the order Conjugales, the protoplasts taking part in conjugation are quite similar, and this method of reproduction has been treated as unique among the different groups. The Charales also constitute a unique group on the basis of the complexity of antheridium (globule) and oogonium (nucule), which makes a definite classification of the group difficult. In the order Charales, the vegetative organization is also unique, and because of the presence of all these characteristics sometimes the order is included in a separate class, the Charophyceae. But the cellulose walls, pigmentation and food reserves are clearly of chlorophycean type and therefore, they have been placed in the class Chlorophyceae.

Tetrasporine line. In this line of evolution, the organisms probably began with a unicellular form much like *Chlamydomonas,* but motility was soon restricted to reproductive cells and vegetative body became an unbranched filament like *Ulothrix.* In the tetrasporine line of evolution, the members of Ulotrichales play an important role. According to this tendency, the species of *Tetraspora* are supposed to be unicellular members of the Volvocales in which the cells are in an immobile cytoplasmic pseudocilia on the anterior surface. The cell also contains a single nucleus, a cup-shaped chloropolast containing single pyrenoid. The cellular contents of *Tetraspora* indicate a relationship to the Volvocales, and the lack of motility indicates that they have given rise to an unbranched filamentous genus such as *Ulothrix.* The evolution of unbranched filamentous species was followed by the branching filaments. The branching of the filaments becomes more and more complex in the genera such as *Draparnaldia* and *Chaetophora.* In this line of evolution, the simple isogamy also terminates into oogamy. The tetrasporine line also gives rise to another tendency where the filamentous condition changes into leaf-like thallus in the development of *Ulva* and *Enteromorpha.* The gametes of these genera are of primitive type but they show a type of alternation of generations. It is also thought that in this line of evolution a form similar to *Coleochaete* could have given rise to the liverworts.

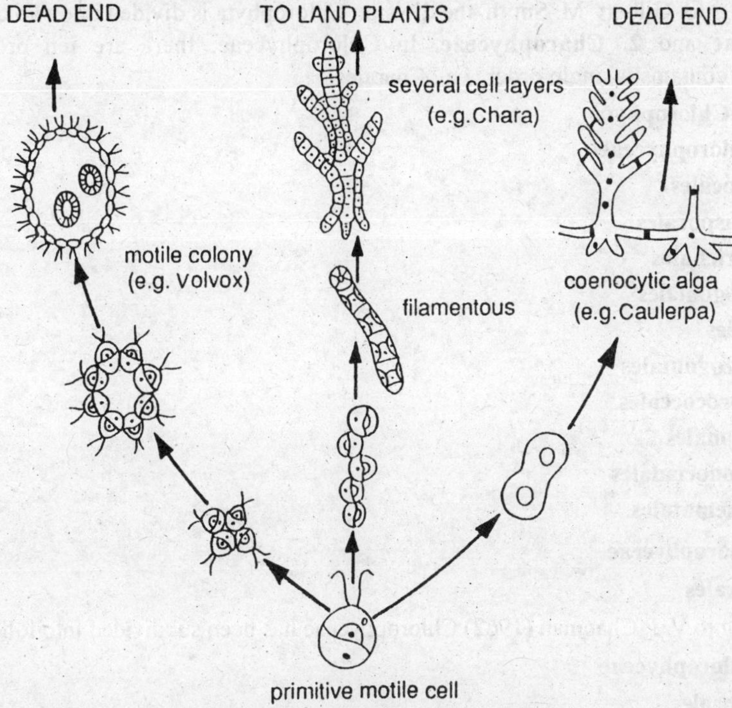

Fig. 4.5. Evolution in green algae

Chlorococcine line. This line of evolution has been named after *Chlorococcum* and Chlorococcales. In this line, the simple chlamydomonad flagellate type of organism has been thought to be ancestor. Here the line of evolution has proceeded in more than one direction. The motility of the vegetative parts was soon lost and *Chlorella* is an example of this type. The species of *Pediastrum* and *Hydrodictyon* demonstrated the characteristic of being siphonaceous as they possess coenocytic cells and the formation of daughter colonies within the individual cells and the mother colonies. The second divergence of this tendency is shown in *Codium* where the coenocytic cells are found in branched forms. If such a form could have developed a vascular cylinder in its central axis and the sporangia at the apex of its branches, it could have given rise to a plant quite similar to simplest known Tracheophyta. It is believed that the chlorococcine line of evolution is the most logical solution of the relationship of Chlorococcales with the Tracheophyta. They have been treated to be Chlorococcales the ancestors of simplest known Tracheophyta.

Classification of Chlorophyceae. This classification is propounded by Dr. F.E. Fritsch (1944), and based upon the structure of vegetative and reproductive bodies.

There are nine orders in class Chlorophyceae.

1. **Volvocales**
2. **Chlorococcales**
3. **Ulotrichales**
4. **Cladophorales**
5. **Oedogoniales**
7. **Conjugales**
8. **Siphonales**
9. **Charales.**

According to Gilbert M. Smith, the division Chlorophyta is divided into two classes, *i.e.*, **1. Chlorophyceae** and **2. Charophyceae.** In Chlorophyceae, there are ten orders whereas Charophyceae contains a single order, *i.e.*, Charales.

Division–**Chlorophyta**

Class–**Chlorophyceae**

1. **Volvocales**
2. **Tetrasporales**
3. **Ulotrichales**
4. **Oedogoniales**
5. **Ulvales**
6. **Schizogoniales**
7. **Chlorococcales**
8. **Siphonales**
9. **Siphonocladales**
10. **Zygnematales.**

Class–**Charophyceae**

1. **Charales**

According to V. J. Chapman (1962) Chlorophyceae has been subdivided into following orders.

Class–**Chlorophyceae**

1. **Volvocales**
2. **Chlorococcales**
3. **Ulotrichales**
4. **Oedogoniales**

5. **Chaetophorales**
6. **Siphonocladales**
7. **Dasycladales**
8. **Conjugales**

Class–**Charophyceae**

9. **Charales.**

He has separated the Characeae which comprises of one fresh water order the Charales; from other green algae (Chlorophyceae) because of their remarkable morphological and reproductive features.

Economic importance. The role of green algae is of minor importance as far as the human food is concerned. Occasionally the filamentous genera such as *Spirogyra* and *Oedogonium* are being used in the preparation of soups. The most important green algae with a use as a food today is sea lettuce, *i.e., Ulva lactuca.* This species of *Ulva* and few other species of *Enteromorpha* are used as salads and as an ingredient in soups. Algae may not appear to be of great food value to us at present, but the future of these algae in the human diet seems promising. An adequate world supply of food is becoming a problem every day, and laboratories throughout the world are experimenting with algae as part of diet. In modem days, the work *on Chloroella* is in progress. Many countries such as Thailand, Japan, Israel are harvesting thousands of tons of plankton from their nearby seas. This alga has been proved to be substitute for the proteins in human diets.

Order VOLVOCALES

There are about 60 genera and 500 species in this order.

Certain important characters of this order are given below :

1. The forms are either unicellular or colonial, *e.g., Chlamydomonas, Volvox, Eudorina, Pandorina, Pleodorina.*
2. The forms are either motile throughout life as in sub-order Chlamydomonadineae or forming sedentary colonies of palmelloid forms as in sub-order Tetrasporineae or dendroid forms as in sub-order Chlorodendrineae.
3. Almost all species are fresh water.
4. The chloroplasts are usually basin-shaped each with a single pyrenoid embedded in it.
5. Asexual reproduction is by means of biflagellated zoospores, palmella stages, aplanospores.
6. Sexual reproduction ranges from isogamy to oogamy.
7. The germination of zygote is followed by reduction division.

Classification. The order Volvocales is sub-divided into *three* sub-orders, *i.e.,* 1. Chlamydomonadineae, 2. Tetrasporineae and 3. Chlorodendrineae.

The sub-order Chlamydomonadineae (motile forms) has *four* families, *i.e.,* 1. Chlamydomonadaceae, *e.g., Chlamydomonas, Eudorina, Volvox,* etc., 2. Sphaerellaceae, 3. Polyblepharidaceae and 4. Phacotaceae.

The sub-order Tetrasporineae (palmelloid forms) is divided into *two* families, *i.e.,* 1. Tetrasporaceae and 2. Palmellaceae.

The sub-order Chlorodendrineae has only *one family, i.e.,* Chlorodendraceae.

Here genus *Volvox,* of order Volvocales has been described in detail.

COLONIAL VOLVOCALES

Family-Volvocaceae (10 genera; 30 species)

Characteristic features. 1. This family includes all motile colonial genera. The cells of these colonies either lie in a disk or in a hollow sphere. The number of cells in a colony is definite, *i.e.*, a multiple of two.

2. In sexual reproduction all or certain specific cells divide simultaneously forming daughter colonies.

3. Sexual reproduction ranges from isogamy to oogamy.

Genus VOLVOX (*Volvere*, to roll)

There are about 20 species.

Habitat or occurrence. All of them are fresh water forms. They. are found in puddles, ponds and pools. They are also found in the temporarily formed ponds during rainy season. Sometimes the colonies are found in such a great abundance that the whole water becomes green in colour. The colonies of *Volvox* are motile and met within rolling condition in the still fresh waters. Sometimes the colonies are intermingled with the tangled masses of other filamentous algae and small aquatic plants.

Structure. The colonies of *Volvox* are hollow. They are spherical or broadly elliptic. The cell number of each colony ranges from 500-50,000. In *Volvox globator,* there are 1500-20,000 cells in each coenobium, whereas 500-1000 in *V. aureus.* The inner hollow of the coenobium is filled up with mucilage. The whole coenobium is surrounded by a delicate mucilaginous lamella. The cells are chlamydomonad type. They possess cup-shaped chloroplasts. In each cell, there are 2-6. cotractile vacuoles, a single pyrenoid and a stigma. The cells are biflagellate and interconnected by

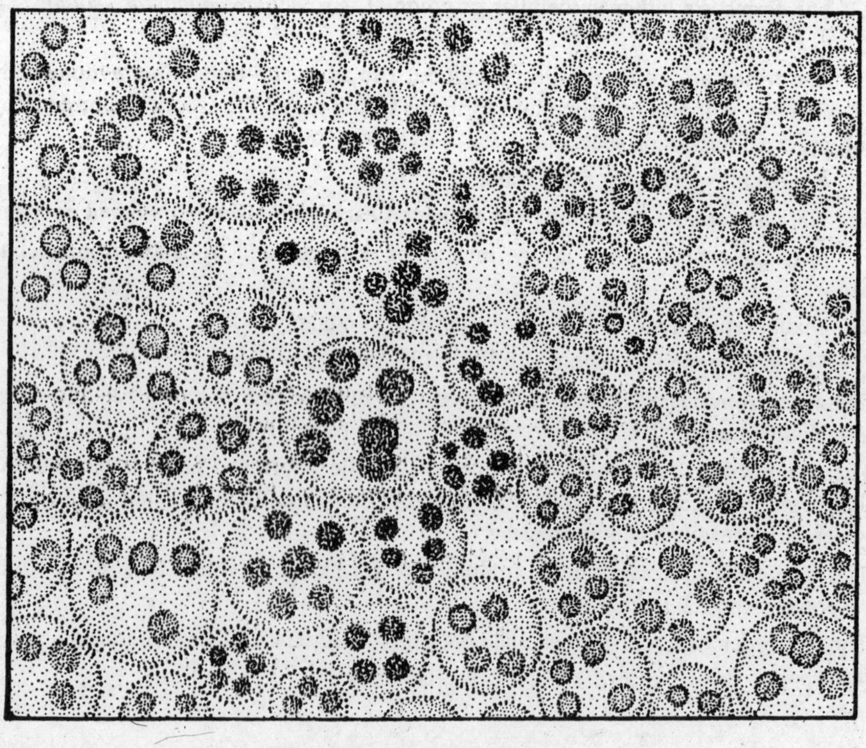

Fig. 4.6. *Volvox* sp. Colonies.

plasmodesmata. In *Volvox globator,* the cells look hexagonal from upper surface because of mutual compression. Each cell is individually ensheathed by a gelatinous sheath.

Mostly the cells of a mature colony are vegetative and only a few are reproductive. The reproductive cells are devoid of flagella, about ten times bigger than the vegetative cells, and possess many pyrenoids within their chloroplasts. These reproductive cells may be asexual or sexual. When the conditions are favourable asexual reproduction takes place. In the end of the season, when the conditions are comparatively drier, the colonies reproduce sexually.

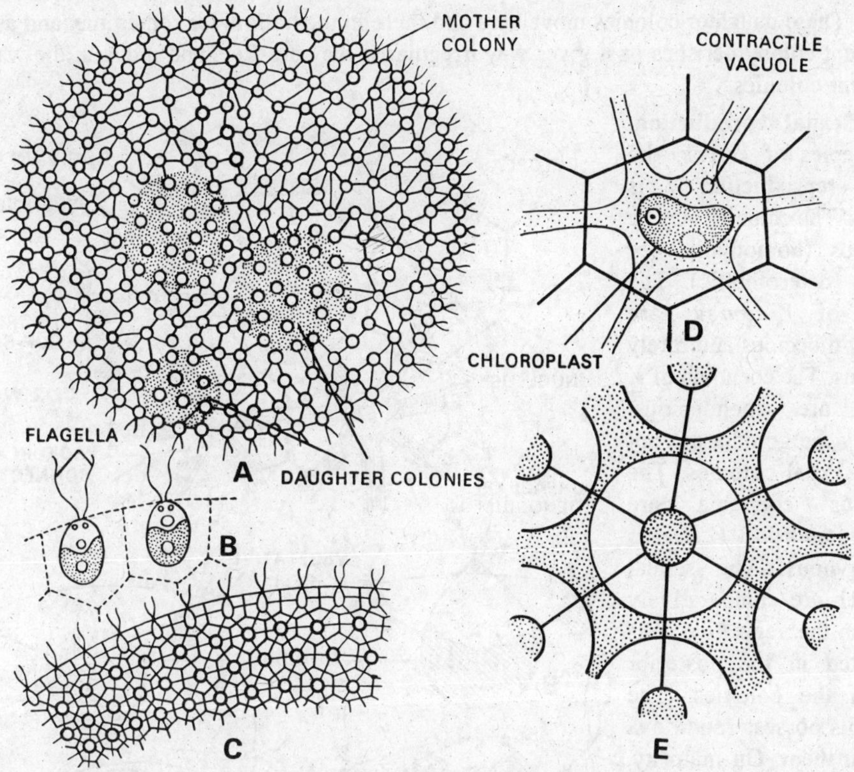

Fig. 4.7. *Volvox* sp. A, a colony of *V. mononae* with daughter colonies; B, enlarged peripheral cells; C, a part of the colony of *Volvox globator,* D, surface view of the cell of *V. globator,* E, surface view of the cell of *V. aureus.*

Reproduction. The reproduction takes place by means of asexual and sexual methods. Parthenogenesis has also been recorded in some species.

1. Asexual reproduction. Asexual reproduction takes place by means of special reproductive cells developed on the posterior end of the colony and called **'gonidia'**. They may be of 5 to 50 in number in different species. These gonidia undergo division and new colonies are produced. All the divisions are longitudinal. The gonidium divides vertically which is followed by second longitudinal division at right angles to the first. Thus four cells are formed. These four cells divide again longitudinally and the octant stage also known as **'plakea stage'** is attained. Here the cells are arranged in an incurved plate facing inward. The plakea stage is followed by another division and 16 cells are produced. They are arranged in a hollow sphere inside the large gonidial cell having a 'phialopore' at the anterior side. Such divisions are followed by several cell generations and a young colony of cells is developed. This young coenobium inverts and the phialopore changes its position during inversion and reaches the other side. Powers (1908) was the first man to investigate such inversion in two species of *Volvox* and since then this phenomenon has also been investigated in other species.

This invagination or inversion of the young coenobium takes place by a simple method. According to Pocock (1933), a slight constriction develops on the equatorial region of the young colony. Later on the portion of the coenobium opposite the phialopore infolds and gradually the complete inversion of the coenobium takes place. During this process, the phialopore reaches on the outer side of the sphere. Now anterior ends of the cells face outwards. Flagella are developed very shortly after invagination. Such daughter colonies are freed in the hollow of the parent colony of *Volvox*. These daughter colonies move here and there in the hollow for some time, and as soon as the parent colony perishes or it gives way irregularly, they are free, and move in the water as independent colonies.

2. Sexual reproduction. In the species of *Volvox*, the sexual reproduction is **oogamous**. The colonies may be monoecious (homothallic) or dioecious (heterothallic). The coenobia of *V. aureus* are commonly dioecious and rarely monoecious. The coenobia of *V. globator* are monoecious. Commonly the gonidia are not found in sexual colonies. The monoecious coenobia are usually protandrous. *V. aureus* is protogynous. The sexual cells which are comparatively bigger than the vegetative cells are situated in the posterior halves of the colonies. The female cells possess round ova or eggs in them. On maturity, such cells are devoid of flagella. The antherozoids enter the oogonia through neck at the time of fertilization. The oogonia may easily be distinguished from gonidia, that they do not divide like the latter.

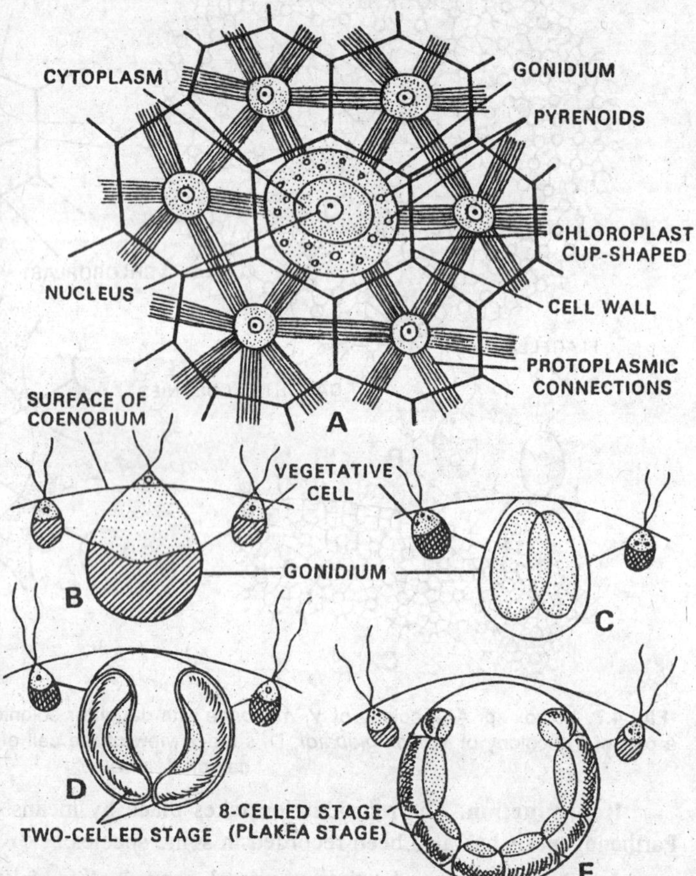

Fig. 4.8. *Volvox* sp. Asexual reproduction. A, a gonidium in the colony in surface view; B, side view of gonidium; C, first division

The antheridial cells resemble, gonidia, but they are fewer in number. The antheridial cell divides again and again and a plate of 16, 32, 64, 128, 256 or 512 antherozoids is formed. The antherozoids are arranged in a sphere. Usually the inversion of the coenobium takes place. Each antherozoid is spindle-like, biflagellate and with a small pale green chloroplast. The antherozoids are liberated from the cells in the form of bundles. These bundles swim here and there in the water and approaching the eggs. After approaching the female coenobium of oogonia the bundle breaks and the antherozoids are separated from each other.

At the time of fertilization, the antherozoids swim around the gelatinous lamella or the coenobium. One or more antherozoids enter through the gelatinous lamella and come in contact of the egg. Only one antherozoid penetrates the egg anteriorly. The flagella are retracted. The male nucleus fuses with the female nucleus, and the **oospore** is developed. The oospores of *Volvox* are commonly spiny. The oospores remain for some time in the hollow of the coenobium after fertilization is over.

Development and germination of oospores. After fertilization a thick wall develops around oospore which usually becomes ornamented, The oospores usually become orange red coloured,

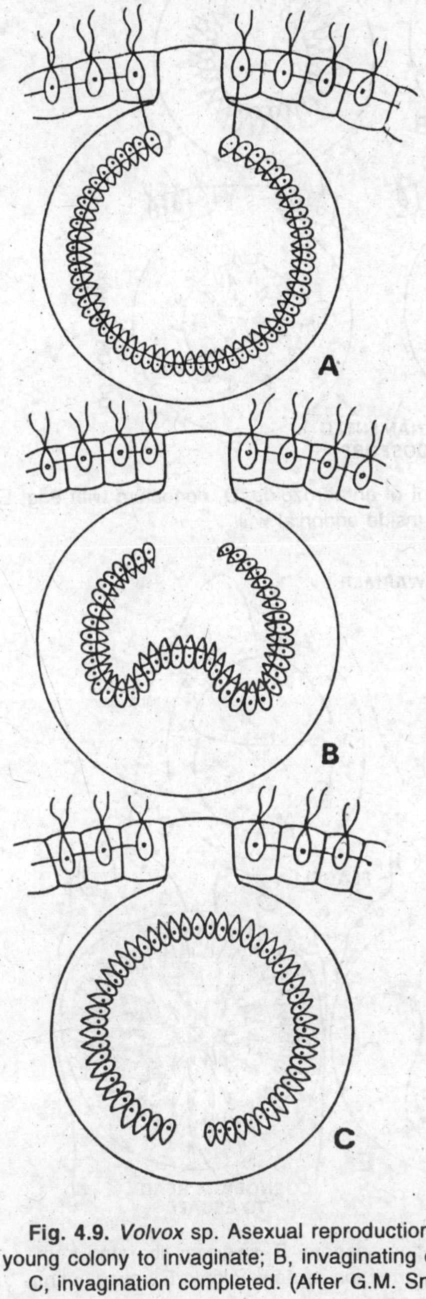

Fig. 4.9. *Volvox* sp. Asexual reproduction. A, young colony to invaginate; B, invaginating colony; C, invagination completed. (After G.M. Smith)

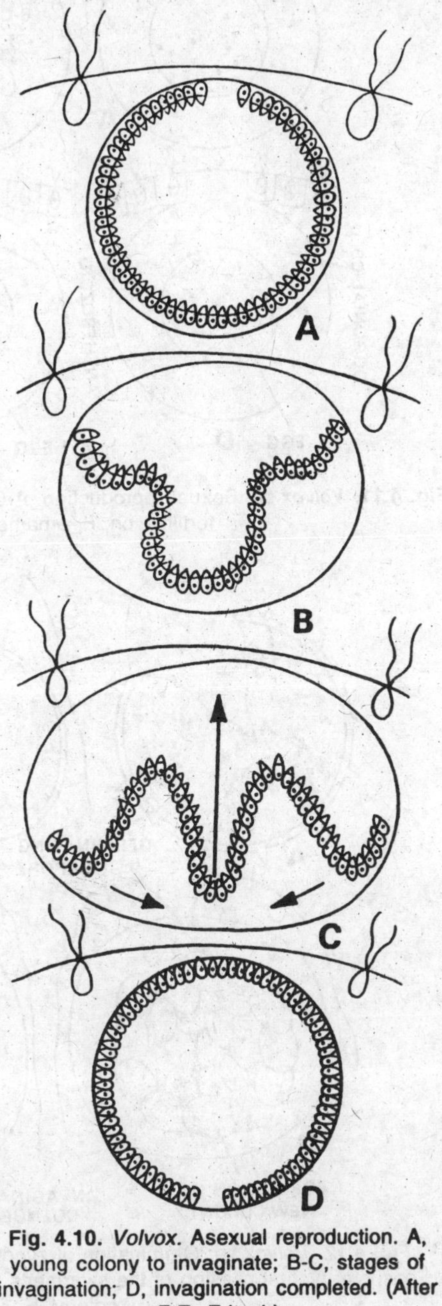

Fig. 4.10. *Volvox.* Asexual reproduction. A, young colony to invaginate; B-C, stages of invagination; D, invagination completed. (After F.E. Fritsch)

because of the presence of '**haematochrome**'. There is always a reduction division during germination. In *Volvox,* the contents of the oospore may develop into a large swarmer.

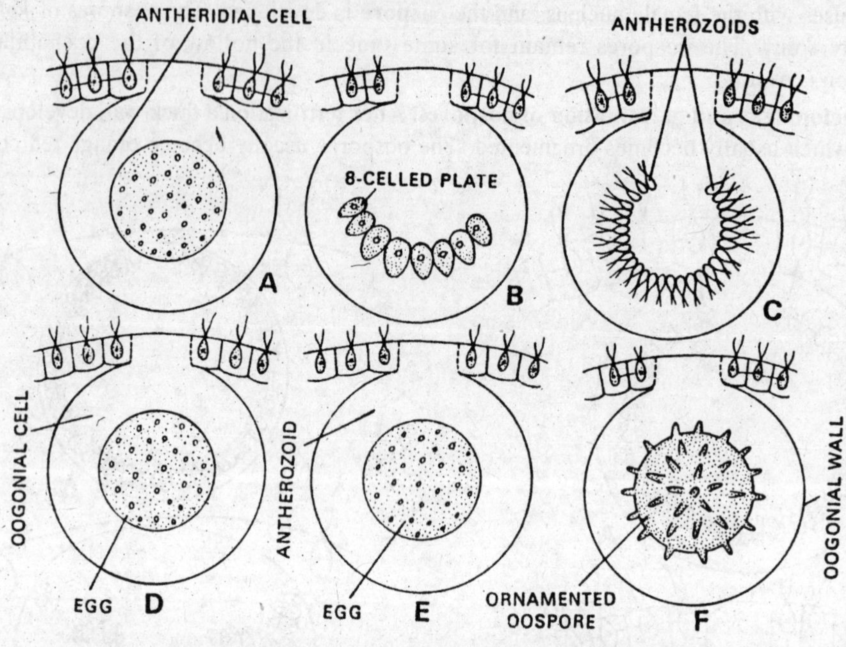

Fig. 4.11. *Volvox* sp. Sexual reproduction. A-C, Development of antherozoids; D, oogonium with egg; E, fertilization; F, ornamented oospore inside oogonial wall.

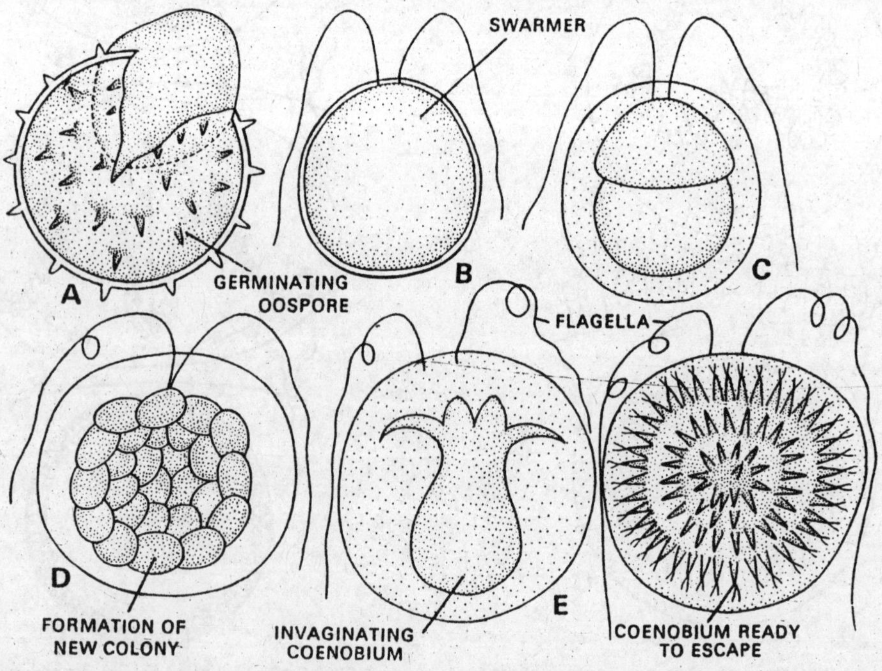

Fig. 4.12. *Volvox* sp. Germination of zygote. A, germinating oospore; B, swarmer liberated from oospore; C, first division of the swarmer's protoplast; D, formation of new colony; E, invaginating colony; F, colony ready to eacape.

This swarmer liberates and swims in the water. The protoplast of this swarmer divides again, and a hollow sphere of the cell is developed. Inversion of this hollow sphere also takes place and the newly developed coenobium liberates in the water. According to Miss Pocock, perhaps several swarmers are developed and only one survives from the lot.

Parthenogenesis. Parthenospores have also been recorded in *Volvox aureus.*

Phylogenetic relationships of Volvoclaes. The relationship of the members of the Volvocales is clearly shown by the similar type of vegetative cells found throughout the order. The daughter colony formation and manner of sexual reproduction, although it ranges from isogamy to oogamy are also similar in the order. In this order, the motility is preserved throughout. The complexity and differentiation of cell types within the colonies clearly show a line of evolution, known as the **Volvocine** line of which *Volvox* is the culmination. Botanists are of opinion that this line ended in *Volvox* and did not develop further.

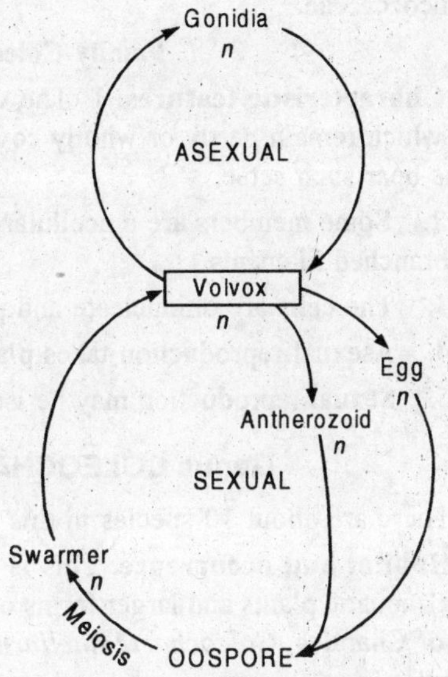

Fig. 4.13. *Volvox* sp. Graphic representation of life-cycle.

Systematic position. Division-Chlorophycophyta; Class-Chlorophyceae; Order-Volvocales; Family-Volvocaceae; Genus-*Volvox*.

Order CHAETOPHORALES

Characteristic features

1. Majority of them are fresh water species found in ponds, puddles and lakes, etc. They may grow in standing or flowing water. Most of them are attached to some substrata-like stones, grasses and other hydrophytes. Some are aerial and terrestrial, *e.g.*, *Trentepohlia.*Some are parasitic, *e.g.*, *Cephaleuros* sp.
2. The plant is fundamentally a heterotrichous filament and consists of prostrate and erect systems. In some cases, the prostrate system is well developed whereas in other cases the erect system is well developed.
3. The filaments are dichotomously branched and the terminal ends of the branches end in **'chaetae.'**
4. Usually the plant bodies are enveloped in thick mucilaginous envelope.
5. The cell wall consists of two layers, the outer layer is of pectose and the inner of cellulose.
6. Usually several pyrenoids are found in each chloroplast. The cells are uninucleate.
7. Reproduction takes place by vegetative, asexual and sexual methods. Vegetative reproduction takes place by means of cell division, fragmentation and akinetes. Asexual reproduction takes place by micro - and macrozoospores, aplanospores and palmelloid stages. Sexual reproduction ranges from isogamy to oogamy (*Coleochaete.*)

There are 90 genera and about 344 species in this order.

Classification. The order is sub-divided into five families.

1. Chaetophoraceae, 2. Trentepohliaceae, 3. Coleochaetaceae, 4. Chaetosphaeridiaceae and 5. Pleurococcaceae.

Family-Coleochaetaceae (10 genera; 25 species)

Characteristic features. 1. The vegetative cells of these algae possess long cytoplasmic setae which remain partly or wholly covered by a gelatinous sheath. All or only few cells of the thallus bear such setae.

2. Some members are unicellular, and others may be multicellular with the cells uniting to form branched filaments.

3. The cells are uninucleate and possess a single parietal laminate chloroplast.

4. Asexual reproduction takes place by means of zoospores.

5. Sexual reproduction may be isogamous or oogamous.

Genus COLEOCHAETE (*coleo,* sheath; *chaete,* hair)

There are about 10 species in this genus.

Habitat and occurrence. This is a fresh water genus, usually occurs epiphytically upon several aquatic plants and larger forms of algae. Some species grow endophytically inside the cell walls of Charales. *Coleochaete nitellarum,* is endophytic in *Nitella*. Majority of the species are epiphytic.

Structure. The thallus of *Coleochaete* is typically heterotrichous, *i.e.,* consists of prostrate and erect systems. The filaments of projecting system in many species, *e.g.*, *Coleochaete scutata,* are arranged so compactly that the whole system looks like a disc or cushion. The branching is dichotomous, which can be clearly seen in the species which possess free branches in projecting system, *e.g.,* C. *pulvinata.* The thalli of *Coleochaete scutata* are usually surrounded by a common gelatinous envelope. Almost in all thalli of the genus some of the cells possess pointed '**chaetae**'. The base of chaetae is usually surrounded by gelatinous material. This chaetae is the prolongation of the cell wall.

The cell wall consists of two layers. The outer layer is formed of pectose and the inner one of cellulose.

In *Coleochaete,* the cells are uninucleate. In each cell, there is girdle-like chloroplast which wholly or partially encircles the protoplast. There is a single large pyrenoid in each chloroplast.

The motile bodies are biflagellate and lack eye spots.

Reproduction. The reproduction takes place in *Coleochaete* by asexual and sexual methods.

1. Asexual reproduction. The asexual reproduction takes place by means of zoospores. The zoospores are produced in the ordinary cells. A single zoospore may be produced in each cell. The zoospores are large, ovoid and biflagellate. The zoospore liberates through a round pore formed at the apex of a short papilla of the parent cell. The zoospores swim for an hour approximately and then rest. The cilia are withdrawn and the protoplast of each zoospore becomes rounded and secretes a wall around it. The protoplast of this one-celled germling divides again and again and a multicellular thalloid structure is developed. Some of cells bear setae, and the shape of a cushion-like disc is attained.

According to Wesley (1928), in unfavourable conditions the aplanospores may also be developed. A single aplanospore develops in each cell, and germinates in the manner of zoospore on the approach of favourable conditions.

2. Sexual reproduction. The sexual reproduction in *Coleochaete* is oogamous. The oogamy is of highly specialized type among Chlorophyceae. The plants may be heterothallic (dioecious) or homothallic (monoecious) according to species. The oogonia develop on the terminal ends of the

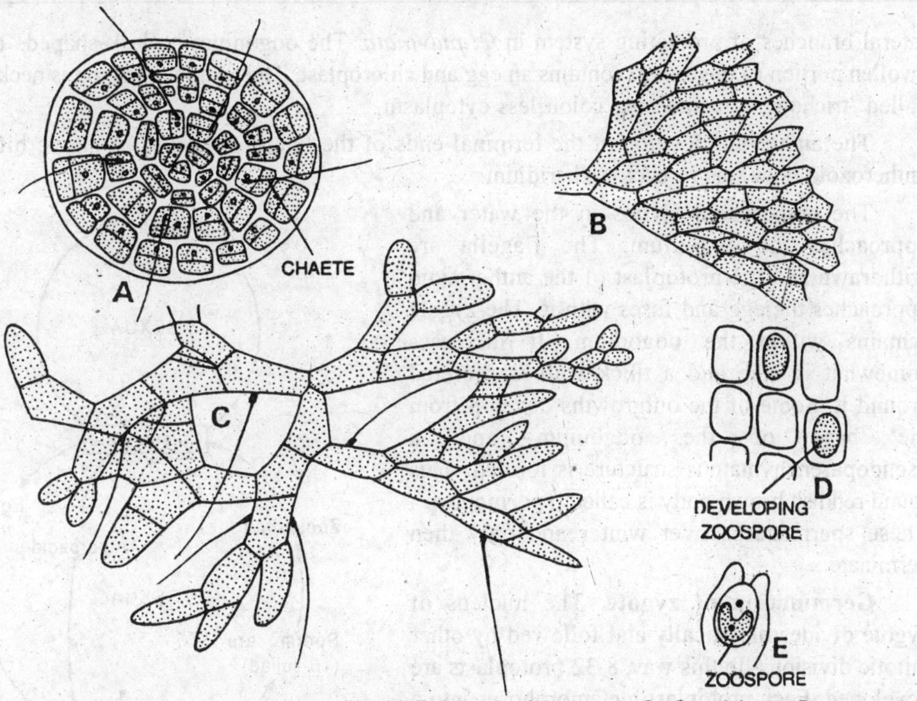

Fig. 4.14. *Coleochaete* sp. A, *C. scutata* discoid; B, branch of the same; C, *C. pulvinata; D,* zoospore formation.

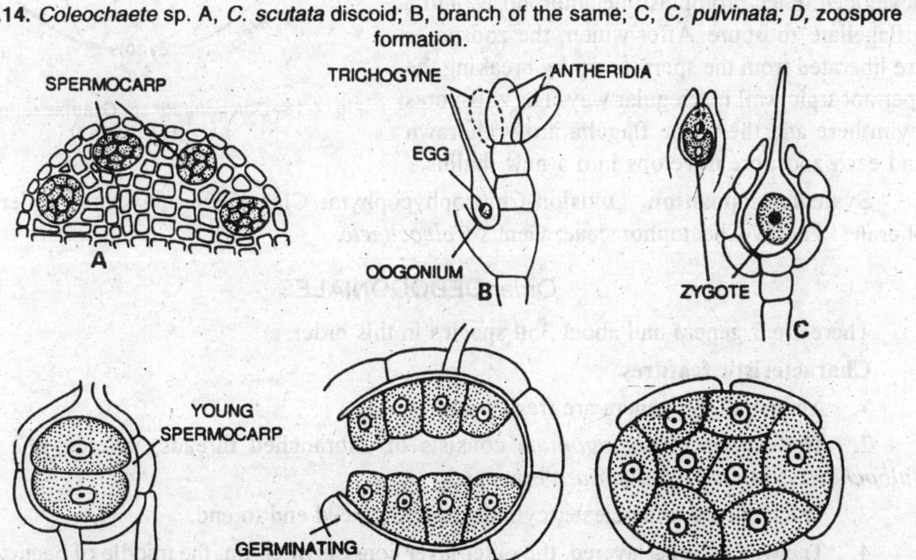

Fig. 4.15. *Coleochaete* sp. Sexual reproduction. A, a part of disc with spermocarp; B, antheridia and liberating antherozoid; C, oogonium; D, young spermocarp; E-G, germinating spermocarp; H. swarmers.

lateral branches of projecting system in C. *pulvinata.* The oogonium is flask-shaped. The basal swollen portion of oogonium contains an egg and chloroplast. The terminal portion is neck-like and called 'trichogyne' containing colourless cytoplasm.

The antheridia develop at the terminal ends of the lateral branches. A single biflagellate antherozoid develops in each antheridium.

The antherozoid swims in the water and approaches the oogonium. The flagella are withdrawn and the protoplast of the antherozoid approaches the egg and fuses with it. The zygote remains within the oogonium. It increases somewhat in size and a thick wall is secreted around it. Some of the outgrowths develop from the base of the oogonium and a pseudoparenchymatous structure is formed. This round reddish brown body is called **'spermocarp'**. These spermocarps over winter and only then germinate.

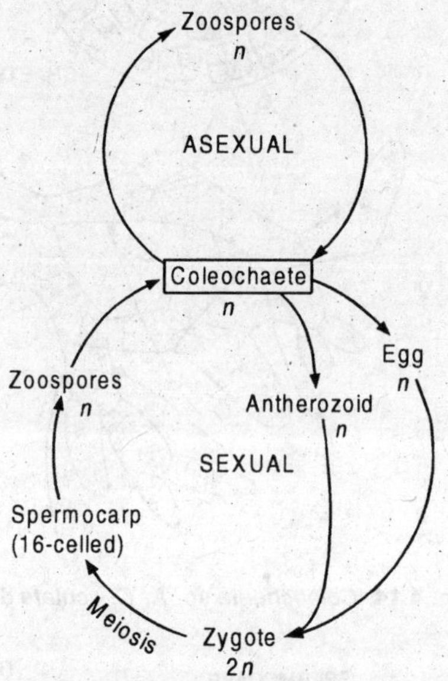

Fig. 4.16. *Coleochaete* sp. Graphic representation of life cycle.

Germination of zygote. The nucleus of zygote divides meiotically and followed by other mitotic divisions. In this way, 8-32 protoplasts are developed. Each protoplast metamorphoses into a biflagellate zoospore. After winter, the zoospores are liberated from the spermocarp by breaking the spermocarpic wall in irregular way. The zoospores swim here and there, the flagella are withdrawn and each zoospore develops into a new thallus.

Systematic position. Division-Chlorophycophyta; Class-Chlorophyceae; Order-Chaetophorales: Family-Chaetophoraceae; Genus-*Coleochaete.*

Order **OEDOGONIALES**

There are 3 genera and about 350 species in this order.

Characteristic features.

1. Almost all the genera are fresh water.
2. The thallus of *Oedogonium* consists of unbranched threads, whereas the thalli of *Bulbochaete* and *Oedocladium are* branched.
3. The cells are uninucleate, cylindrical and placed end to end.
4. The wall is three-layered, the outer layer consists of chitin, the middle of pectose and the innermost of cellulose.
5. There is special type of cell division and **'caps'** are formed.
6. The reproduction takes place by means of vegetative, asexual and sexual methods.
7. Stephanokontean type of motile bodies are found.
8. Vegetative reproduction takes place by fragmentation and akinete formation.
9. Asexual reproduction takes place by zoospores.
10. Sexual reproduction is oogamous.
11. Life-cycle is simple and haplontic.

Classification. A single family Oedogoniaceae is included in this order.

Family-Oedogoniaceae (3 genera; 350 species)

Characteristic features. 1. They possess simple or branched filaments.

2. The cylindrical uninucleate cells remain united end to end. Unique type of cell division. There is distinctive annular splitting of lateral cell wall.

3. Asexual reproduction takes place by zoospores formed singly within a cell. It also takes place by means of akinetes.

4. Sexual reproduction is always oogamous.

5. All species are fresh water in habit.

Genus OEDOGONIUM (*oedo*; swelling; *gonium*, vessel)

There are 285 species in this genus.

Habitat and occurrence. It is exclusively fresh water in habit. The young filaments are found to be attached to some substrate. On maturity, the filaments become free-floating. Sometimes they are epiphytic upon other large sized fresh water algae.

Structure. The filaments of *Oedogonium* are long and unbranched. The basal cell of the young filament is devoid of chlorophyll and attached to some substratum is called the holdfast. The terminal cell of the filament is broadly rounded at its apex. The intercalary cells are cylindrical, unifom in thickness and arranged end to end. These cells possess one or more striations at 'their apical ends called '**apical caps**'.

The cells are cylindrical and covered with well defined three-layered cell walls. The outer layer consists of chitin, the middle of pectose and the innermost of cellulose.

Each cell contains a large nucleus in it. The nucleus is usually situated in the peripheral cytoplasm of the cell. Each nucleus has well defined chromatin and one or two nucleoli. The chloroplast, is reticulate. At the anastamoses of the chloroplast pyrenoids are present. A starch sheath surrounds each pyrenoid.

Cell division. A special type of cell division is found. This is the peculiar feature of this genus. The following important points are to be considered in the cell division.

1. **Ring formation.** Just before the cell division a ring is developed immediately beneath the upper septum of the cell. According to Wisselingh, this ring consists of cellulose mixed with some starch substances.

2. **Nuclear division.** The nucleus of the parent cell divides into two and simultaneously a septum develops between them. This septum remains unconnected with longitudinal cell walls for some time.

3. **Formation of new cell.** At the level of the ring the cell membrane breaks transversely, the ring stretches gradually and a new cell develops between

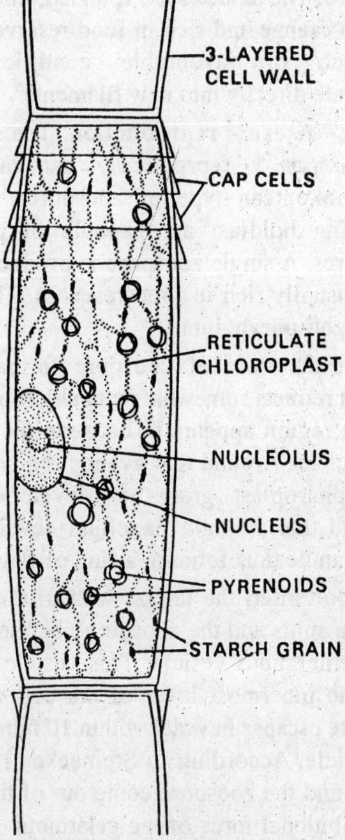

Fig. 4.17. *Oedogonium* sp. Detailed structure of a cell with conspicuous reticulate chloroplast.

sheath and cap of parent cell wall. Side by side the septum shifts upwards and acquires a permanent position at the edge of the lower end of the ruptured cell wall.

The wall of the upper daughter cell is mainly developed from the stretched ring. At the top of this, there is cap developed by original membrane. The cap cells usually divide again and again. Usually the number of cap cells is coincided with the number of cell divisions. The species of *Oedogonium* may easily be recognised because of the presence of these 'cap cells'.

Reproduction. The reproduction takes place by vegetative asexual and sexual methods.

1. Vegetative reproduction. ***(a)* By fragmentation.** Reproduction in *Oedogonium* takes place by means of fragmentation. These fragments develop into new individuals. This type of reproduction is especially found in free floating species and rarely in permanent attached forms.

***(b)* By akinetes.** In some of the species, the vegetative akinetes are developed. The akinetes are produced in the chains of 10-40 in small and inflated cells. They resemble oogonia in structure. The akinetes are rounded, thick-walled, reddish orange and rich in food reserves. On the approach of favourable conditions, they germinate directly into new filaments.

2. Asexual reproduction. The species of *Oedogonium* reproduce asexually by stephanokontean type of zoospores. All cells excluding holdfast are capable of producing zoospores. A single zoospore is produced within a cell usually rich in food reserves. This cell is called **'zoosporangium.'**

At the time of zoospore formation, the nucleus retracts somewhat from chloroplast and a hyaline region appears in between the cell wall and nucleus. Around this hyaline region, a circle of blepharoplast grains is developed. It is assumed that a single flagellum develops from each granule thus forming a ring of flagella.

Soon after, the lateral wall of the zoosporangium splits and the zoospore liberates out in a thin membranous vesicle. This vesicle develops from the innermost layer of the cell wall. The zoospore escapes in water within 10 minuets from the vesicle. According to Steinecke (1929), the vesicle and the zoospore come out of the cell by an imbibitional force of the gelatinous substance secreted by the protoplast. Very soon the delicate vesicle disappears and zoospore swims freely in the water. The zoospore swims for about an hour

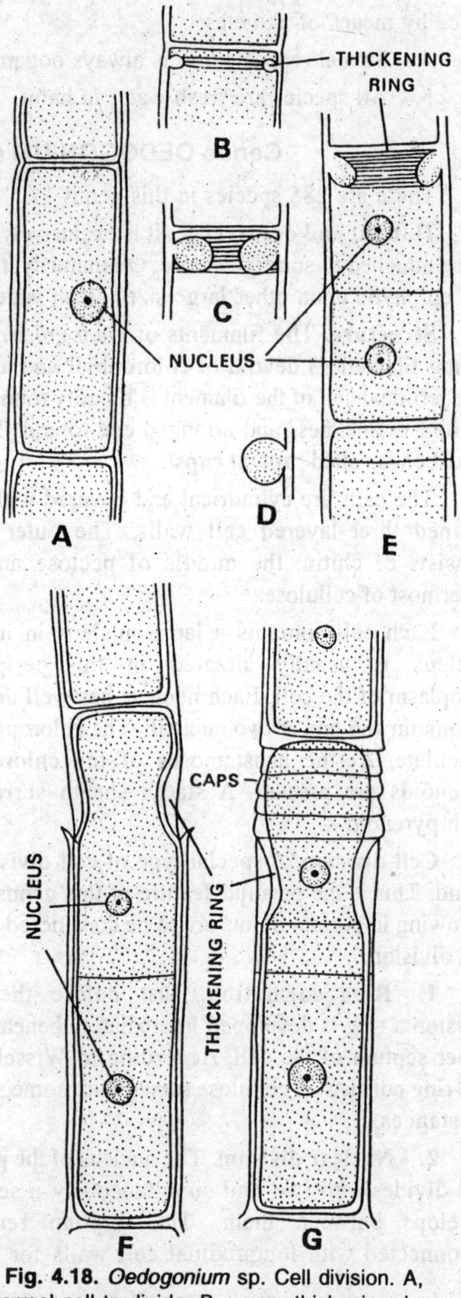

Fig. 4.18. *Oedogonium* sp. Cell division. A, normal cell to divide; B, young thickening ring; C, old thickening ring; D, side view of thickening ring; E, nuclear division and septum formation; F-G, stages of cell division.

and then settles down on some substratum from its anterior side, *i.e.*, hyaline spot. The flagella are withdrawn, the structure becomes elongated and the holdfast develops. The structure of the holdfast depends upon the nature of the substratum .and species. If the surface of substratum is rough the rhizoidal system is much branched and if it is smooth then the rhizoidal system is unbranched.

3. Sexual reproduction. It is always oogamous. The species may be homothallic or heterothallic. Sexual reproduction may be of **macrandrous or nannandrous** type. It frequently occurs in still water and rarely in flowing waters.

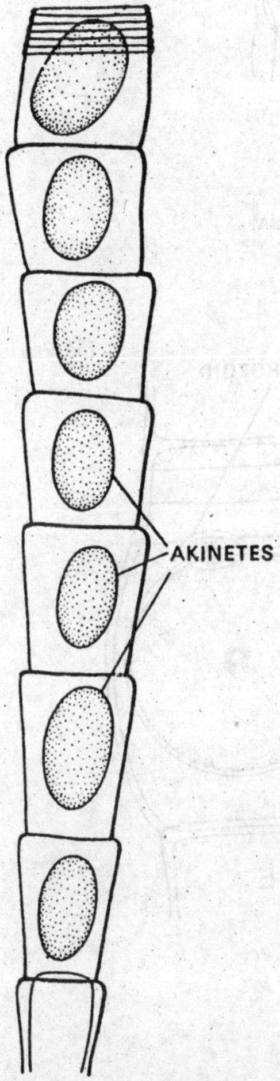

Fig. 4.19. *Oedogonium* sp. Vegetative reproduction-chain of akinetes.

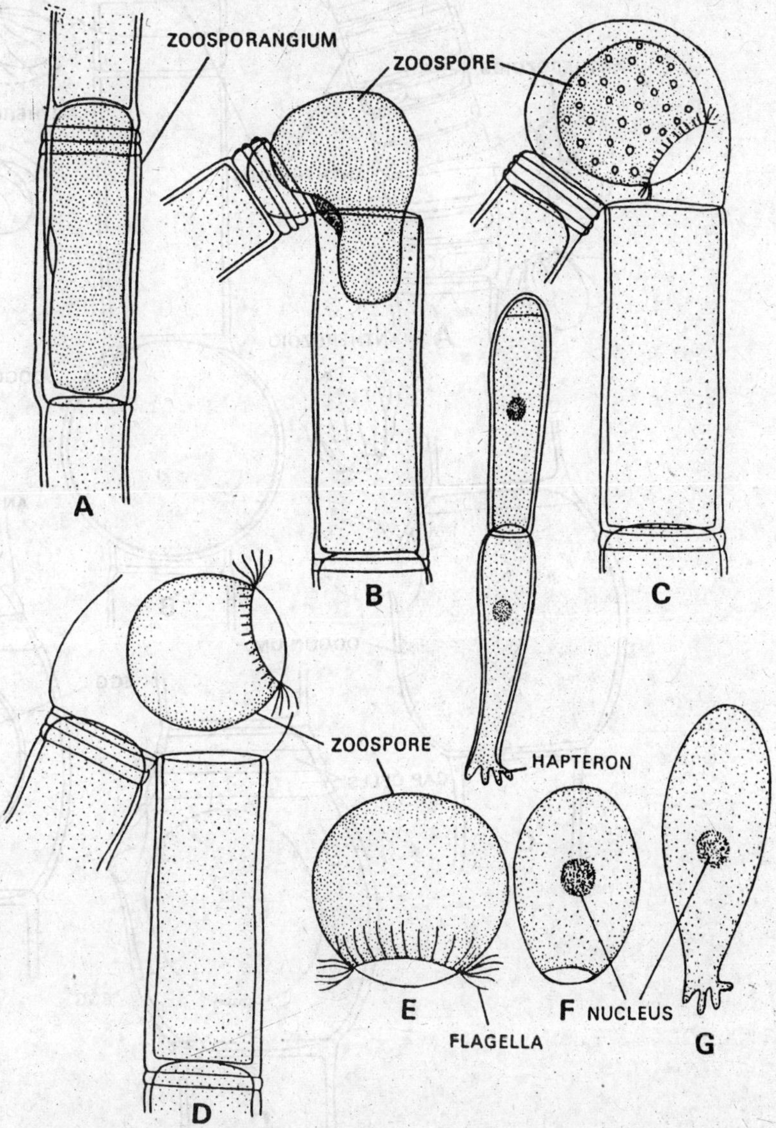

Fig. 4.20. *Oedogonium* sp. Asexual reproduction. By zoospores; A, cell acting as zoosporangium; B-C, formation and liberation of zoospore; D, zoospore liberating out of vesicle; E-G, germinating zoospore and germling;, H, two-celled germling.

Development of antheridia. In **macrandrous** species, the oogoma and antheridia may be developed on the same filament (homothallic or monoecious species) or on two different filaments (heterothallic or dioecious species). The antheridia are either terminal or intercalary. They are produced by the division of antheridial mother cell. Any vegetative cell may act as antheridial mother cell. This cell devides into two unequal cells at its upper end forming an antheridium. The

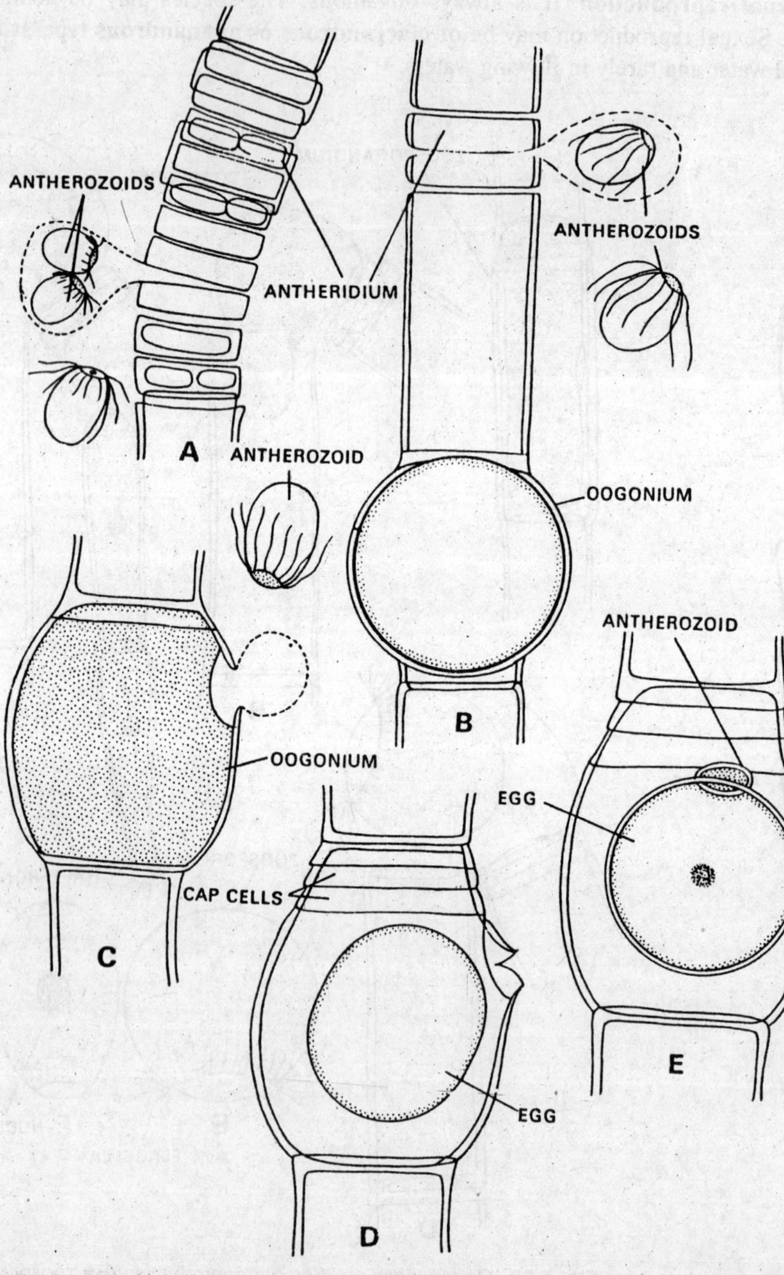

Fig. 4.21. *Oedogonium* sp. Sexual Reproduction-Macrandrous species A, antheridia and liberation of antherozoids; B, a monoecious species having antheridia and oogonium on the same thread; C, mature oogonium ready for fertilization; D, entrance of antherozoid in the oogonium through receptive spot; E, fertilization; F, single antherozoid.

lower sister cell divides repeatedly and a chain of 2-40 antheridia is produced. The protoplast of each antheridium metamorphoses in a single antherozoid. Sometimes two antherozoids are produced in one antheridium. The antherozoids resemble zoospores in structure and shape only with the difference that they are somewhat smaller in size and with fewer number of flagella. The antherozoids liberate in the same way as zoospores. They come out in thin membranous vesicles by the split of antheridial wall. Very soon the vesicle disappears and the antherozoid moves freely in the water in any direction.

In **nannandrous** species, the antheridia are produced on special dwarf males or nannandria. Majority of the nannandrous species are heterothallic (dioecious), *i.e.,* oogonia are produced on the ordinary filaments and antheridia on special filaments called 'dwarf males' or nannandria. These 'dwarf males' or nannandria originate by the germination of a special type of swarmers known as androspores produced singly in flat cells, *i.e.,* androsporangia formed by repeated transverse divisions of the ordinary vegetative cells. The androsporangia may occur on the same filaments on which the oogonia are produced or on different filaments. The androspores resemble in shape and structure to zoospores. They are somewhat smaller than zoospores and larger than antherozoids. After their small swarming period, they settle down either on the oogonium or on one of the

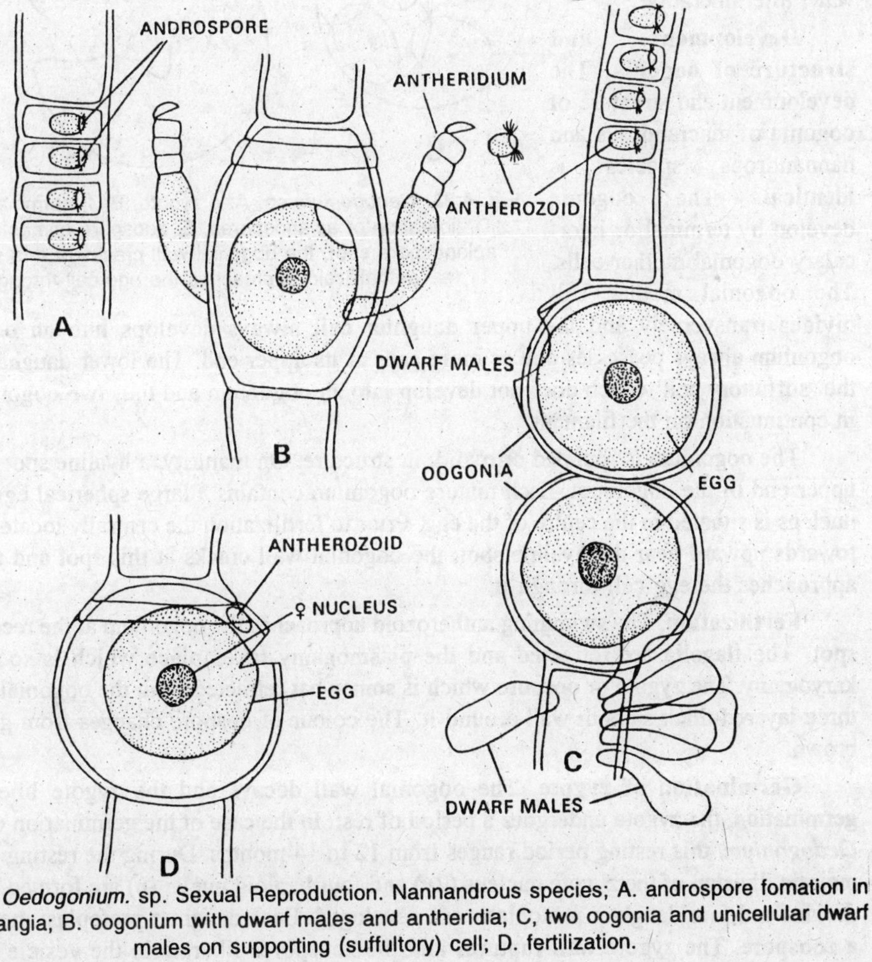

Fig. 4.22. *Oedogonium.* sp. Sexual Reproduction. Nannandrous species; A. androspore fomation in androsporangia; B. oogonium with dwarf males and antheridia; C, two oogonia and unicellular dwarf males on supporting (suffultory) cell; D. fertilization.

neighbouring cells. Here the androspore germinates and develops into a dwarf male. The nannandrium consists of one basal cell attached to the oogonium or adjacent cell, and one or more flat antheridia. In *Oedogonium,* the nannandrium consists of an attaching cell only producing two antherozoids develop in an antheridium. The structure of antherozoid is same as that of the antherozoids of macrandrous species. The antherozoids swim here and there in the water after liberation.

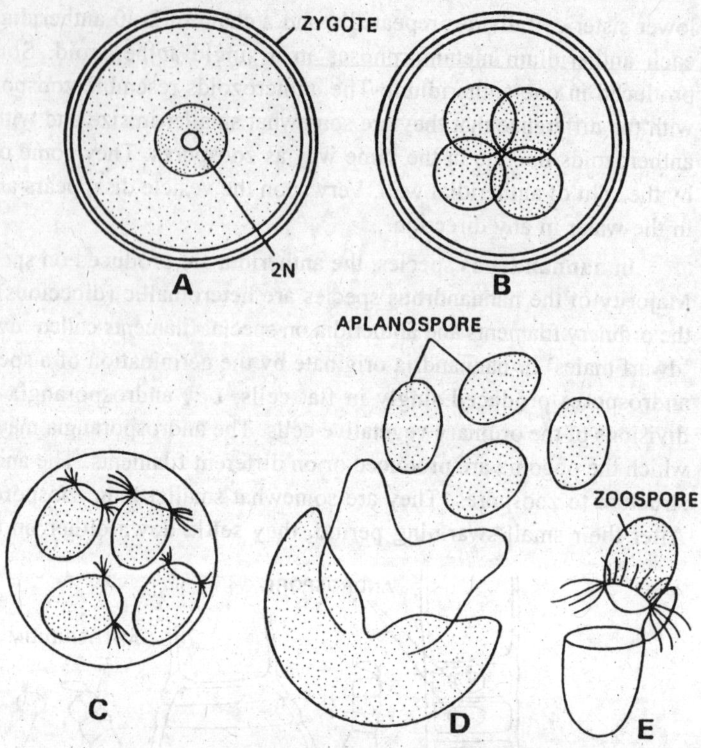

Fig. 4.23. *Oedogonium* sp. A, a zygote; B-C, germination of zygote; D, liberation of aplanospores; E. zoospore comes out from an aplanospore spot, the oogonial wall cracks at this spot and the antherozoid approaches the egg cell through it.

Development and structure of oogonia. The development and structure of oogonia of macrandrous and nannandrous species is identical. The oogonia develop by terminal or intercalary oogonial mother cells. The oogonial mother cell divides transversely and the upper daughter cell always develops into an oogonium. Each oogonium always possesses one or more caps at its upper end. The lower daughter cell is called the 'suffultory cell' which does not develop into an oogonium and thus two oogonia may be seen in continuation on the filament.

The oogonium is rounded or ovoid, in structure. On maturity, a hyaline spot develops at the upper end of the oogonium. Each mature oogonium contains a large spherical egg cell in it. The nueleus is situated in the centre of the egg. Prior to fertilization the centrally located nucleus shifts towards upward near the hyaline spot, the oogonial wall cracks at this spot and the antherozoid approaches the egg cell through it.

Fertilization. The swimming antherozoid approaches the oogonium at the receptive (hyaline) spot. The flagella are retracted and the plasmogamy takes place which is soon followed by karyogamy. The zygote or oospore which is somewhat retracted from the oogonial wall secretes a three layered-thick smooth wall around it. The colour of oospore changes from green to reddish brown.

Germination of zygote. The oogonial wall decays and the zygote liberates. Prior to germination, the zygote undergoes a period of rest. In the case of the germination of the zygote of *Oedogonium,* this resting period ranges from 12 to 14 months. During the resting period, there is meiotic division of the zygote nucleus (2*n*) and four haploid nuclei (*n*) are formed. The protoplast divides and four daughter protoplasts are developed. Each daughter protoplast metamorphoses in a zoospore. The zygote wall ruptures and the zoospores liberate in the vesicle which later on disappears and the zoospores swim freely in the water. Sometimes, instead of zoospores, four aplanospores are formed. Each aplanospore opens at its anterior end and a single zoospore liberates from it. Each zoospore is capable to give rise to a new filament.

Parthenogenesis. In some of the species, the unfertilized eggs develop into parthenospores. Pathenospores are zygote-like and germinate in the same way without meiosis.

The life cycle of *Oedogonium* is of haplontic type.

Systematic position. Division-Chlorophycophyta; Class-Chlorophyceae; Order-Oedogoniales; Family-Oedogoniaceae; Genus-*Oedogonium.*

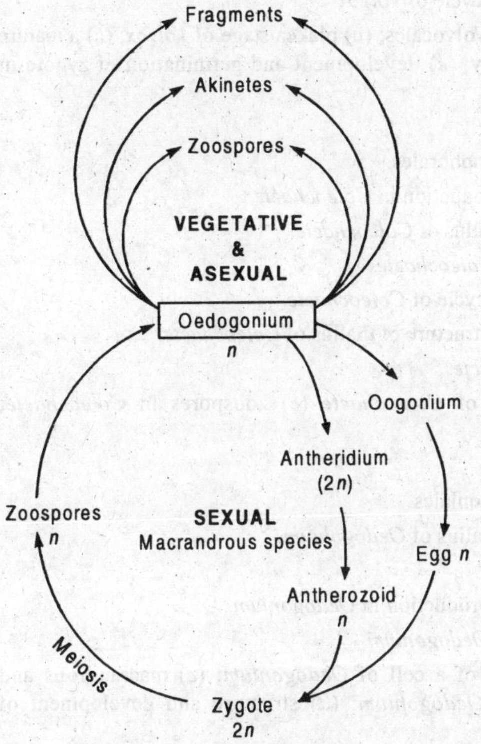

Fig. 4.24. *Oedogonium*. Macrandrous sp., graphic life-cycle.

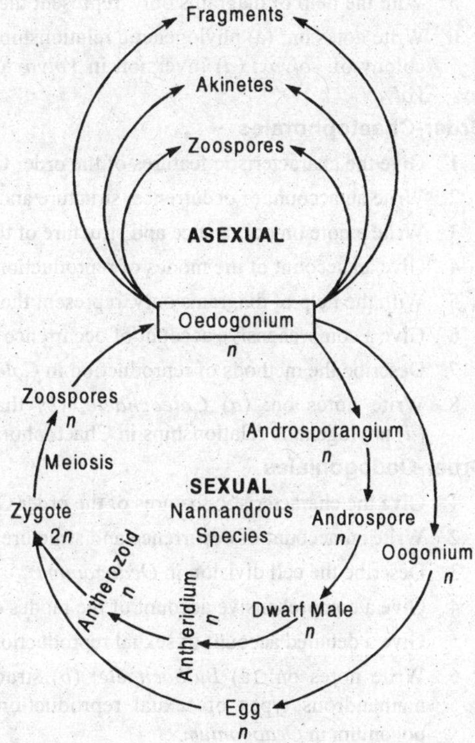

Fig. 4.25. *Oedogonium*. Nannandrous species, graphic life-cycle.

QUESTION BANK

N.B. *Undermentioned questions are asked repeatedly in various Indian Universities and competitive examinations.*

Class-Chlorophyceae

1. Give the occurrence and distribution of green algae.
2. Write a note on the range of structure in Chlorophyceae.
3. Describe the modes of reproduction found in Chlorophyceae.
4. What types of life-cycles are found in Chlorophyceae. Supplement your answer with suitable examples.
5. Write a brief account of phylogenetic relationships in the Chlorophyceae.
6. Give an outline of suitable classification of Chlorophyceae, also mention the criteria of classification.

7. Write notes on : (a) Neuromotor apparatus; (b) palmelloid forms in Chlorophyceae; (c) vegetative reproduction in Chlorophyceae; (d) evolution in Chlorophyceae; (e) economic importance of Chlorophyceae with suitable examples.

Order-Volvocales

1. Give the characteristic features of the order Volvocales.
2. Describe the habit and habitat and structure of *Volvox*.
3. Give the details of asexual reproduction found in *Volvox*.
4. Give an account of sexual reproduction and germination of oospore in *Volvox*.
5. With the help of diagrams only, represent the life-cycle of *Volvox*.
6. Write notes on: (a) phylogenetic relationships in Volvocales; (b) plakea stage of *Volvox*; (c) a mature colony of *Volvox*; (*d*) inversion in *Volvox* colony; (*e*) development and germination of zygote in *Volvox*.

Order-Chaetophorales

1. Give the characteristic features of the order Chaetophorales.
2. Write an account of occurrence, structure and reproduction in *Coleochaete*.
3. Write a note on occurrence and structure of the thallus of *Coleochaete*.
4. Give an account of the modes of reproduction in *Coleochaete*.
5. With the help of diagrams only, represent the life-cycle of *Coleochaete*.
6. Give a comprehensive account of occurrence and structure of thallus of *Coleochaete*.
7. Describe the methods of reproduction in *Coleochaete*.
8. Write notes on: (*a*) *Coleochaete*; (*b*) thallus of *Coleochaete* (*c*) zoospores in *Coleochaete*; (*d*) phylogenetic relationships in Chaetophorales.

Order-Oedogoniales

1. Give the characteristic features of the order Oedogoniales.
2. Write an account of occurrence and structure of thallus of *Oedogonium*.
3. Describe the cell division in *Oedogonium*.
4. Give a comprehensive account of the modes of reproduction in *Oedogonium*.
5. Give a detailed account of sexual reproduction in *Oedogonium*.
6. Write notes on: (a) *Bulbochaete*; (*b*) structure of a cell of *Oedogonium*; (*c*) macrandrous and nannandrous types of sexual reproduction in *Oedogonium*; (*d*) structure and development of oogonium in *Oedogonium*.
7. With the help of diagrams only, represent the life-cycle of *Oedogonium*.

Multiple Choice Type

1. Unicellular motile alga is:

   (i) *Pandorina,* (ii) *Eudorima,* (iii) *Volvox,* (iv) *Chlamydomonas.*
2. The multicellular filamentous alga is:

   (i) *Ulothrix,* (ii) *Vaucheria,* (iii) *Tetraspora,* (iv) *Pandorina.*
3. The star-shaped chloroplast is found in:

   (i) *Zygnema,* (ii) *Spirogyra,* (iii) *Ulothrix,* (iv) *Cladophora.*
4. The cup-shaped chloroplast is found in:

   (i) *Prasiola,* (ii) *Chlamydomonas,* (iii) *Spirogyra,* (iv) *Ulothrix.*
5. The girdle-shaped chloroplast is found in:

   (i) *Ulothrix,* (ii) *Spirogyra,* (iii) *Zygnema,* (iv) *Chlamydomonas.*
6. The rhizoplast is:

   (i) rhizoid, (ii) rhizomorph, (iii) fibre connecting blepharoplast with centrosome, (iv) fibre connecting two blepharoplasts.

7. The paradesmose is:
(i) rhizoid, (ii) rhizomorph, (iii) fibre connecting blepharoplasts wirh centrosome, (iv) fibre connecting two blepharoplasts.
8. Pigment cup is found in
(i) eye spot, (ii) pyrenoid, (iii) nucleus, (iv) nucleolus.
9. The sea lettuce is
(i) *Enteromorpha,* (ii) *Ulva,* (iii) *Spirogyra,* (iv) *Sargassum.*
10. The Conjugales are;
(i) exclusively fresh water, (ii) exclusively marine, (iii) 50 per cent fresh water, (iv) 90 per cent fresh water.
11. The cause of red snow is:
(i) *Volvox globator,* (ii) *Ulothrix zonata,* (iii) *Chlamydomonas nivalis,* (iv) none of above.
12. The life-cycle of *Ulothrix* is:
(i) haplontic, (ii) diplontic, (iii) haplobiontic, (iv) diplobiontic.
13. *Volvox* colonies are found in
(i) fresh water ponds, (ii) salt lakes, (iii) sea water, (iv) water falls.
14. The structure of *Volvox* is:
(i) Unicellular, (ii) filamentous, (iii) thalloid, (iv) colonial
15. The plakea stage of *Volvox* consists of
(i) 2 cells, (ii) 4 cells, (iii) 8 cells, (iv) 16 cells.
16. *Volvox* is a member of the group:
(i) Rhodophyceae, (ii) Phaeophyceae, (iii) Chlorophyceae, (iv) Myxophyceae.
17. The gonidia of *Volvox* are concerned with:
(i) sexual reproduction, (ii) asexual reproduction, (iii) vegetative reproduction, (iv) none of above.
18. The sexual reproduction in *Volvox* is:
(i) isogamous, (ii) anisogamous, (iii) oogamous, (iv) none of above.
19. The alga may be used in space ship as food:
(i) *Chlorella,* (ii) *Chlorococcum,* (iii) *Chlamydomonas,* (iv) *Haematococcus*
20. The attached form of alga is:
(i) *Spirogyra,* (ii) *Zygnema,* (iii) *Ulothrix,* (iv) *Mougeotia.*
21. The alga found on mollusc shells is
(i) *Ulothrix,* (ii) *Chara,* (iii) *Zygnema,* (iv) *Cladophora.*
22. The branched alga is:
(i) *Ulothrix,* (ii) *Cladophora,* (iii) *Spirogyra.* (iv) *Zygnema*
23. The terminal and intercalary akinetes are found in:
(i) *Cladophora* (ii) *Oedogonium,* (iii) *Ulothrix,* (iv) *Pithophora.*
24. The process of 'evection' is found in:
(i) *Cladophora,* (ii) *Oedogonium,* (iii) Vaucheria, (iv) *Spirogyra.*
25. Sort out the parasitic alga:
(i) *Cladophora,* (ii) *Pithophora.* (iii) *Cephaleuros,* (iv) *Coleochaete.*
26. *Bulbochaete* is member of:
(i) Zygnematales, (ii) Oedogoniales, (iii) Cladophorales, (iv) Chaetophorales.
27. Point out the members of order Oedogoniales:
(i) *Oedogonium,* (ii) *Oedocladium,* (iii) *Bulbochaete,* (iv) all the above.
28. The cap cells are found in:

(i) *Spirogyra.* (ii) *Vaucheria,* (iii) *Chara.* (iv) *Oedogonium.*

29. The 'pond silk' is:

    (i) *Spirogyra,* (ii) *Vaucheria,* (Hi) *Chara,* (iv) *Oedogonium*

30. The 'water net' is:

    (i) *Pediastrum.* (ii) *Hydrodictyon.* (iii) *Cladophora.* (iv) *Pithophora.*

31. Stephanokontean type of zoospores are found in:

    (i) *Vaucheria,* (ii) *Oedogonium,* (iii) *Chara,* (iv) *Ectocarpus.*

32. The nannandria produce:

    (i) oogonia, (ii) akinetes, (iii) zoospores, (iv) antheridia.

33. *Oedogonium* belongs to:

    (i) Chlorophyceae, (ii) Myxophyceae, (iii) Phaeophyceae, (iv) Rhodophyceae.

34. Chloroplast of *Oedogonium* is:

    (i) star-shaped, (ii) spiral, (iii) reticulate, (iv) cup-shaped.

35. The stellate chloroplasts are found in:

    (i) *Spirogyra,* (ii) *Cladophora,* (ii) *Zygnema,* (iv) *Pithophora.*

36. The semicells of the desmids remain joined together by:

    (i) raphe, (ii) sinus, (iii) isthmus, (iv) plasmodesma.

37. The desmids are:

    (i) Ch1orophyceae, (ii) Phaeophyceae, (iii) Rhodophyceae, (iv) Myxophyceae.

38. The species of *Caulerpa* are :

    (i) fresh water, (ii) marine, (iii) terrestrial, (iv) 50 per cent fresh water.

ANSWERS

1 (iv), 2 (i), 3 (i), 4 (ii), 5 (i), 6 (iii), 7 (iv). 8 (i), 9 (ii), 10 (i), 11 (iii), 12 (i), 13 (i), 14 (iv), 15 (iii), 16 (iii), 17 (i), 18 (iii), 19 (i), 20 (iii), 21 (iv), 22 (ii), 23 (iv), 24 (i), 25 (iii), 26 (ii), 27 (iv), 28 (iv), 29 (i), 30 (ii), 31 (ii), 32 (iv), 33 (i), 34 (iii), 35 (iii); 36 (iii), 37 (i) 38 (ii).

4A CHAPTER

Charophyta–Charophyceae

(Stoneworts)

Characteristic features.

Pigmentation. Chlorophyll *a* and *b,* β-carotene, γ-carotene, lycopene.

Storage products. Starch.

Flagellation. 2, subapical equal flagella (on sperm).

ORDER–CHARALES

There are 6 genera and 250 species which are living, rest are fossils. The living members of this order are known as **'stoneworts'**.

Characteristic features

1. The stoneworts usually occur in still and clear waters in attached condition to the mud of the bottom of the pools. They are found in less oxygenated water and best survive in clear and hard waters. Calcium carbonate is deposited on the plant body and the surface of the plant becomes rough.
2. The thallus is attached to the mud by a rhizoidal system. The plant body is erect and possesses nodes and internodes. Secondary laterals, also called **'leaves'** arise from the nodes which are of limited growth. The leaves may or may not be differentiated into nodes and internodes.
3. The reproduction takes place by vegetative and sexual methods. Asexual reproduction is altogether absent. Vegetative reproduction takes place by means of special vegetative bodies such as amylum stars, bulbils, secondary protonema, etc. Sexual reproduction is oogamous and takes place by oogonia (nucule) and antheridia (globule).
4. The zygote nucleus divides reductionally producing 4 haploid nuclei. Out of these 4 haploid nuclei one is functional and rest degenerate. The functional nucleus divides into two cells, the lower cell is rhizoidal and the upper one gives rise to main thallus.

Classification. According to Dr. F. E. Fritsch, this is last order of Chlorophyceae. There is a single family Characeae. G. M. Smith has divided his division Chlorophyta in two classes, *i. e.,* 1. Chlorophyceae and 2. Charophyceae. He includes his single order Charales and family Characeae along with other three families in class Charophyceae. He had included all the living genera in the family Characeae. There are about six genera and 250 species in this family.

There are also Characeae known only as fossils. The remaining three families are all fossil. According to Peck (1946) the Charales are known from as for back as the Devonian. The Characeae are known from as far back as the Upper Carboniferous. Another family the Trochilisceaceae, is known only from the Lower Carboniferous and from the Devonian.

Family–Characeae (6 genera; 250 living species)

Characteristic features. They possess an erect branched thallus with regular nodes and internodes. Each node bears a whorl of branches of limited growth, also known as 'leaves'. The branches of unlimited growth may also arise from the axils of the leaves.

1. Many members are calcified.
2. Vegetative reproduction takes place by means of amylum stars, bulbils and protonema like outgrowths.

3. Sexual reproduction is oogamous. The oogonia are one-celled and surrounded by a sheath of sterile cells. They are borne upon the 'leaves'. The antheridia are one celled and remain united in uniseriate branched filaments. They are surrounded by a common spherical envelope composed of eight cells. These are always borne on 'leaves'.
4. They are known as stoneworts. They grow submerged in fresh water upon muddy or sandy bottoms of pools and ponds. Few species are found in brackish water.

Genus-CHARA

There are about 90 species in this genus.

Occurrence: The species of *Chara* grow submerged attached to the muddy bottom of the pools and ponds of clear water. They prefer less oxygenated and hard water. The species of *Chara* are not found in the waters where mosquito larvae are present. They form sub-aquatic meadows extending to miles.

The plants of *Chara* are of great ecological value. As they are covered with calcium carbonate deposits, they deposit lot of calcium in the bottom of lake, etc., and after a considerable time the whole lake or pond is filled up with calcareous deposites.

Vegetative structure. The plant is always found attached to the substratum by a well developed rhizoidal system. The rhizoids are uniseriately branched and obliquely septate. The rhizoids may or may not be differentiated into nodes and internodes.

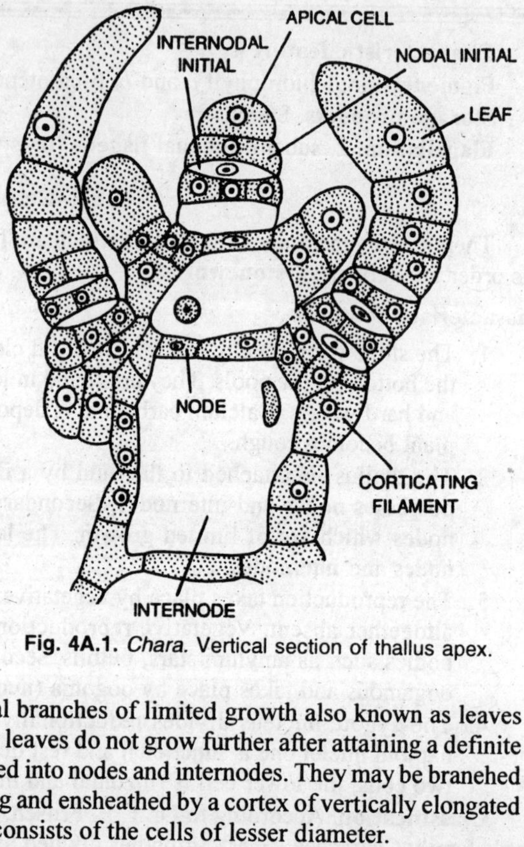

Fig. 4A.1. *Chara.* Vertical section of thallus apex.

The axis has nodes and internodes. Several branches of limited growth also known as leaves grow in whorls from the nodes of the axis. These leaves do not grow further after attaining a definite length. The leaves may or may not be differentiated into nodes and internodes. They may be branehed or simple. This ineternodal cell is sufficiently long and ensheathed by a cortex of vertically elongated cells. The cortex is one celled in thickness and consists of the cells of lesser diameter.

Apical growth. The apical growth of the axis takes place by the terminal dome-shaped cell which cuts other derivatives at its posterior face. The derivative of the apical cell divides by a transverse wall producing two daughter cells. The upper daughter cell is the **nodal initial** and the lower one the **internodal initial**. The lower cell (internodal initial) does not divide further and elongates several times of its original length to give rise to an internode. The upper cell (nodal initial) divides by a vertical wall producing two daughter cells. These daughter cells again divide vertically in such a way that two cells remain in the centre called central cells usually surrounded by 6 to 20 peripheral cells. The central cells may or may not divide further. The peripheral cells divide periclinally. In such a way each peripheral cell divides into two cells. The outer one acts as leaf initial and the inner as nodal which may or may not divide further, if divides the division is vertical.

Development of cortex. The peripheral cells of the nodal regions of the leaves never become apical cells but they may produce one-celled branch known as **stipules**, but this is not necessary for basal node of each leaf. On adaxial side, the upper and lower cells of the basal node protrude. After protrusion a transverse septum is formed at the base of these cells, thus initiating the cortical initials of corticating filaments. This way one of the cortical initials develops towards upper side and the other towards lower side. These cortical initials act as apical cells and give rise to ascending and descending cortical filaments respectively ensheathing the main axis. In young corticating filaments both the nodal and internodal cells are approximately of equal size, but later on the internodal cell and the two peripheral nodal cells elongate sufficiently. The central cell of the nodal region does not elongate. Sometimes one celled spines develop from this central cell . (See Figs. 4A.4 and 4A.7).

Development of leaf. As already mentioned the peripheral cells of the nodal region divide periclinally. The outer daughter cell acts as leaf initial and a leaf develops from it. The structure of the leaf resembles that of axis. The apical cell of the leaf cuts the derivatives in the same way as the main axis. 5-15 derivatives are cut from the apical cell. The first derivative becomes the leaf's basal cell. Rest of derivatives divide and give rise to an internodal and nodal initial. The internodal initial gives rise to a comparatively shorter internode than that of axis. The peripheral cells of the nodal region of the leaf do not give rise to the leaves as in the case of axis. Here the peripheral cell gives rise to a single celled structure called the **'stipule'.** The cortex initials found on the leaf give rise to corticating filaments in the same way as on the axis.

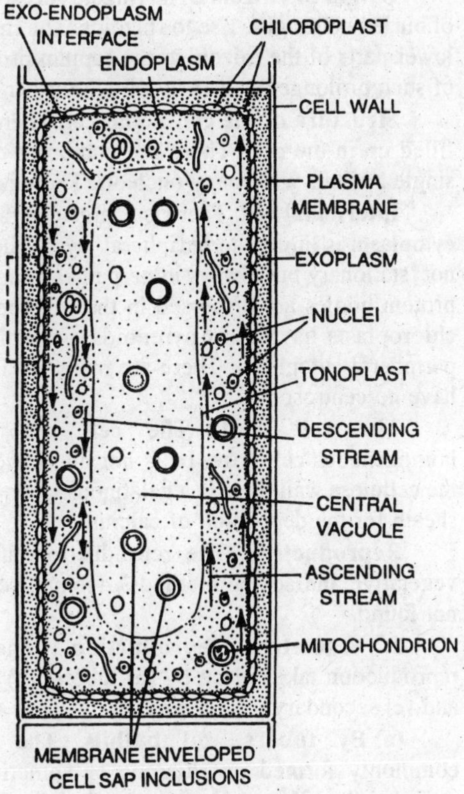

Fig. 4A.2. *Chara* spp. Structure of a mature internodal cell showing details.

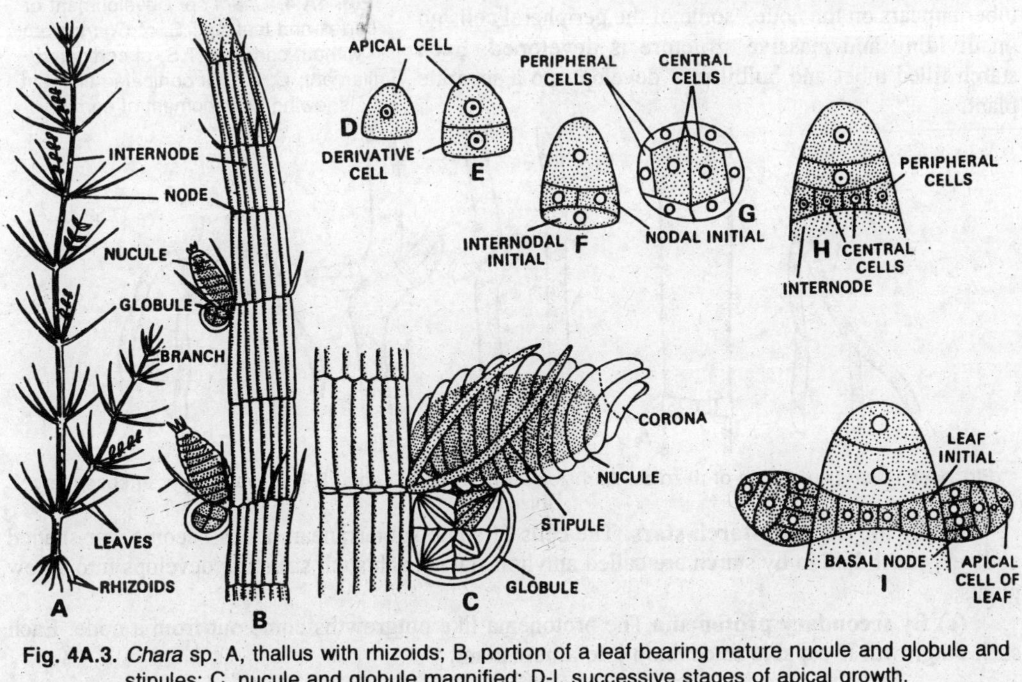

Fig. 4A.3. *Chara* sp. A, thallus with rhizoids; B, portion of a leaf bearing mature nucule and globule and stipules; C, nucule and globule magnified; D-I, successive stages of apical growth.

Origin of rhizoids. The rhizoids arise from the buried parts of the nodes. The peripheral cells of buried nodes give rise to rhizoids. The rhizoids possess oblique septa. Sometimes the upper and lower parts of the rhizoid at the septum protrude upward and downward in prolonged cells. Each of such prolonged cells may divide further and give rise to secondary rhizoids.

Structure of the cell-nodal cell. There is no large central vacuole. The dense cytoplasm is filled up in the cell. There are numerous discoid chloroplasts, devoid of pyrenoids. There is a single nucleus with many nucleoli. There is well defined chromatin within the nucleus.

Internodal cell. There is a large central vacuole. The cytoplasm is limited to peripheral layer. The cytoplams is not stationary but always in perpetual rotary motion. The protein bodies are dispersed in the cell sap. The discoid chloroplasts having no pyrenoids are embedded in the peripheral cytoplasm. There are many nuclei. The nuclei have no centrosomes.

The cell wall. The cell wall consists of homogeneous cellulose. It is not multilayered. Outer to the cellulose wall there is a gelatinous layer, and this is the sheath for the deposition of calcium.

Reproduction. The reproduction takes place by vegetative and sexual methods. Asexual reproduction is not found.

1. Vegetative reproduction. The vegetative reproduction takes place by (a) tubers; (b) amylum stars and (*c*) secondary protomema.

(a) By tubers and bulbils. The tubers are commonly formed on rhizoids or sometimes even on buried nodes. The whole structure is full of starch.

Sometimes the globule divides and becomes multicellular and known as **'simple tuber'.** When the tuber appears on the node, some of the peripheral cells go on dividing and massive structure is developed. Each starch filled tuber and bulbil may develop into a separate plant.

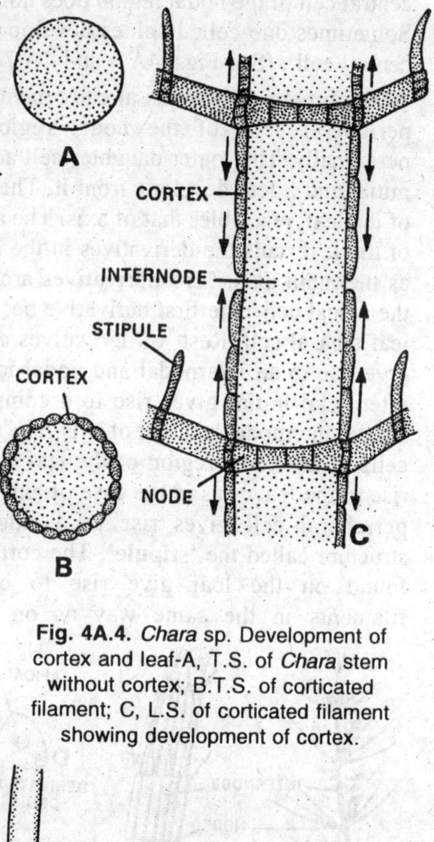

Fig. 4A.4. *Chara* sp. Development of cortex and leaf-A, T.S. of *Chara* stem without cortex; B.T.S. of corticated filament; C, L.S. of corticated filament showing development of cortex.

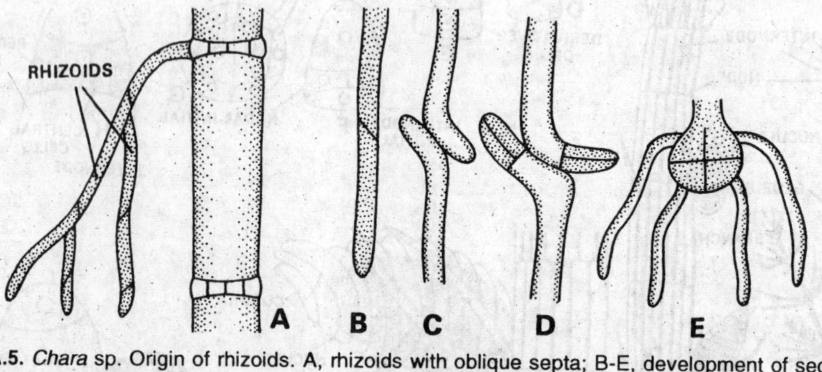

Fig. 4A.5. *Chara* sp. Origin of rhizoids. A, rhizoids with oblique septa; B-E, development of secondary rhizoids.

(b) By amylum or starch stars. The cells of some subterranean nodes become star-shaped and very much laid in by starch are called amylum stars. Each such structure develops into a new plants.

(c) By secondary protonema. The protonema like outgrowths come out from a node. Each such outgrowth is capable to develop into a new plant.

2. Sexual reproduction. The sexual reproudction is oogamous. A very advanced and specialized type of oogamy is found. There is a special terminology for the sex organs. The male fruiting body is called **globule** and the female **nucule.**

Most of the species are homothallic (monoecious) and few are heterothallic (dioecious). The globule is borne on secondary lateral of limited growth. One globule and one nucule borne on one node of the leaf. In homothallic species both the fructifications borne on the same node. In *Chara* the nucule is borne above the globule.

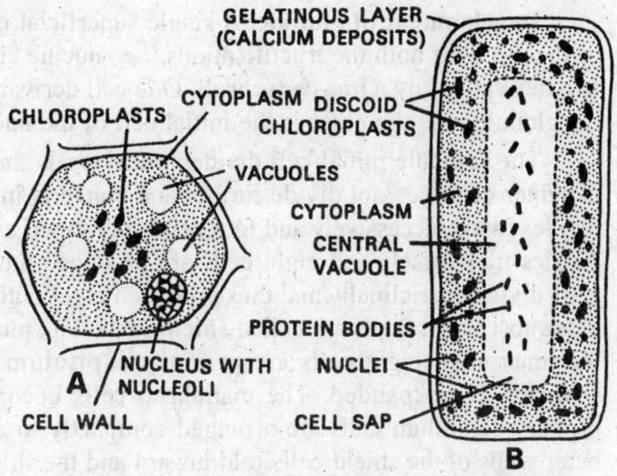

Fig. 4A.6. *Chara* sp, Cell structure. A, nodal cell; B, internodal cell.

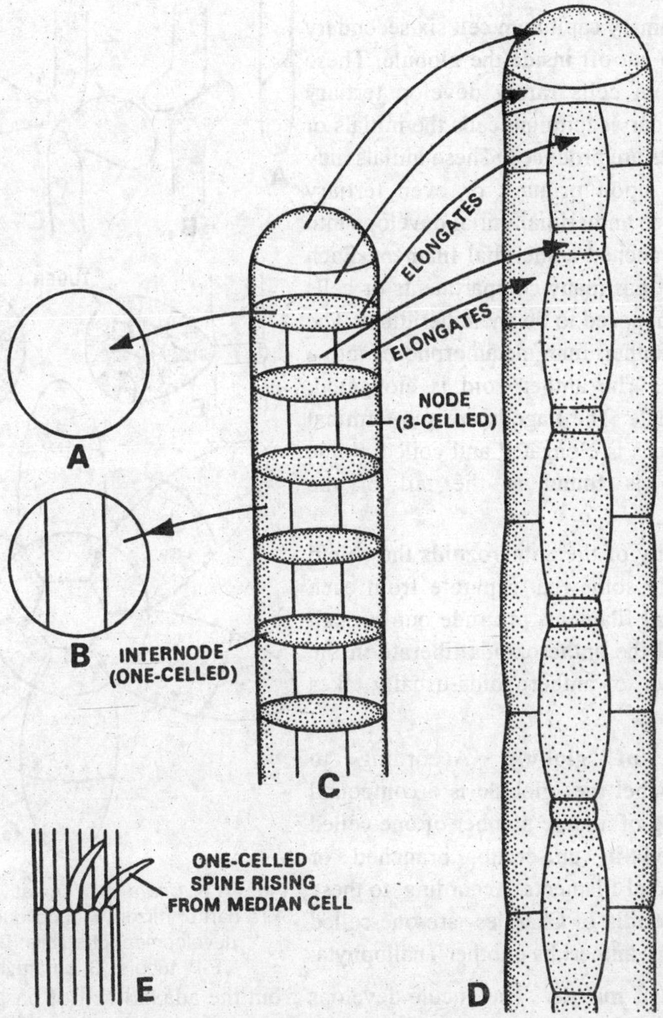

Fig. 4A.7. *Chara* sp. A, T.S. of internodal region of young corticating filaments; B, T.S. of nodal region of young corticating filament; C, young corticating filament; D, mature corticating filament; E, one celled spines rising from median cell.

Development of globule. A single superficial nodal cell of the adaxial side of the leaf acts as the initial of both the fructifications, *i.e.,* nucule and globule. This superficial cell divides into two derivatives by a transverse wall. One cell derivative of the superficial cell is the initial cell of the globule and the other is the initial cell of the nucule.

The globule initial cell divides transversely and two daughter cells are formed. The lower daughter cell does not divide further and converts into the pedicel cell. The upper daughter cell divides twice successively and four cells are formed arranged in quadrants. Each of these quadrants divides transversely and eight cells are produced thus attaining octant stage. Each of these eight cells divides periclinally and thus produced eight outer cells which divide further periclinally. The outermost eight cells are called **shield cells.** The middle cells are known as **manubrial cells** and the innermost eight cells are **primary capitulum cells.** The shield cells become very much enlarged and expanded. The manubrial cells become very much radially elongated, but the primary capitulum cells are arranged compactly to each other in the centre of the globule. The outer walls of the shield cells fold inward and the shield cells appear multicellular structures. The infoldings are incomplete. The shield cells develop red pigments in them and so the globules appear orange red in colour.

From each primary capitulum cell six secondary capitulum cells are cut off inside the globule. These secondary capitulum cells rarely develop tertiary cells. On the secondary capitulum cells the initials of antheridial filaments are produced. These initials may also be produced upon primary or even tertiary capitulum cells. Each antheridial initial develops into a branched or unbranched antheridial filament. Each antheridial filament has many compartments or cells in it. Each cell is supposed to be an antheridium. The protoplast of each antheridium metamorphoses into a single antherozoid. The antherozoid is elongated, coiled and biflagellate. The flagella are sub-terminal in origin. The nucleus is elongated and coiled. Some unused cytoplasm is found in the tail of the antherozoid.

On the maturity of the antherozoids the shield cells of the globule somewhat separate from each other, the antheridial filaments protrude out through these openings and the antherozoids liberate in the water. The liberation of antherozoids usually takes place in the morning.

Morphology of globule. According to Hofmeister and Goebel the globule is a compound structure comprising of a large number of one-celled antheridia arranged in uniseriate, branched or unbranched antheridial filaments. According to these biologists, the antheridia of Charales are one-celled and at the par with the antheridia of other Thallophyta.

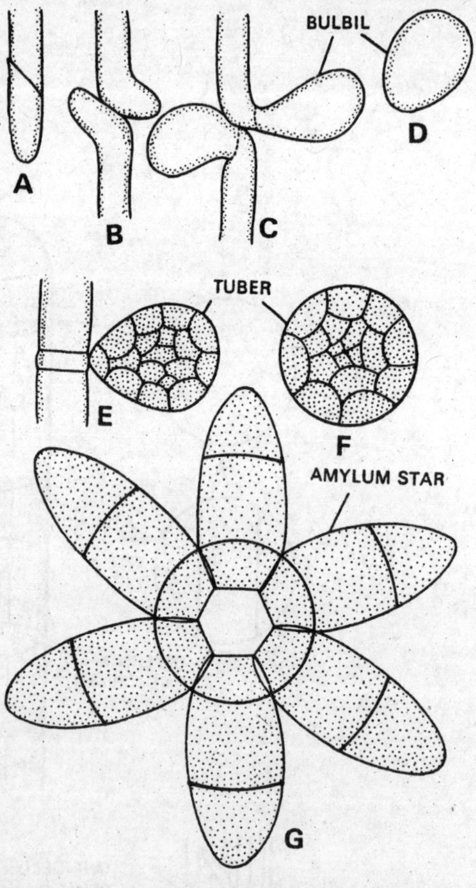

Fig. 4A.8. *Chara* sp. Vegetative reproduction; A, a part of rhizoid with oblique septum; B-C, development of bulbils; D, a bulbil; E-F, tubers; G, an amylum star.

Development of nucule. The nucule develops from the adaxial cell of basal node of the globule. The globule is homologous with the branch of limited growth and the nucule with the branch of unlimited growth.

The nucule initial divides twice and a row of three cells is formed. The terminal cell acts as oogonial mother cell which elongates sufficiently in vertical direction and transverse wall develops in the lower region of it dividing it into two cells. The lower small cell and the upper one is oogonium which contains an egg. The lowermost cell of the row of three cells does not divide and acts as a pedicel. The middle cell divides vertically in such a way so that a single central cell and five sheath initials are produced. The sheath initials surround the central cell. The sheath initials elongate vertically sometimes even before the vertical elongation of the oogonial mother cell and encircle it. Each of the sheath initials divides transversely forming the upper tier of coronary cells and lower tier of tube cells. The tube cells elongate several times to their original length and become spirally coiled around the oogonium. The coronary cells do not elongate much and act collectively as the **corona** of the nucule.

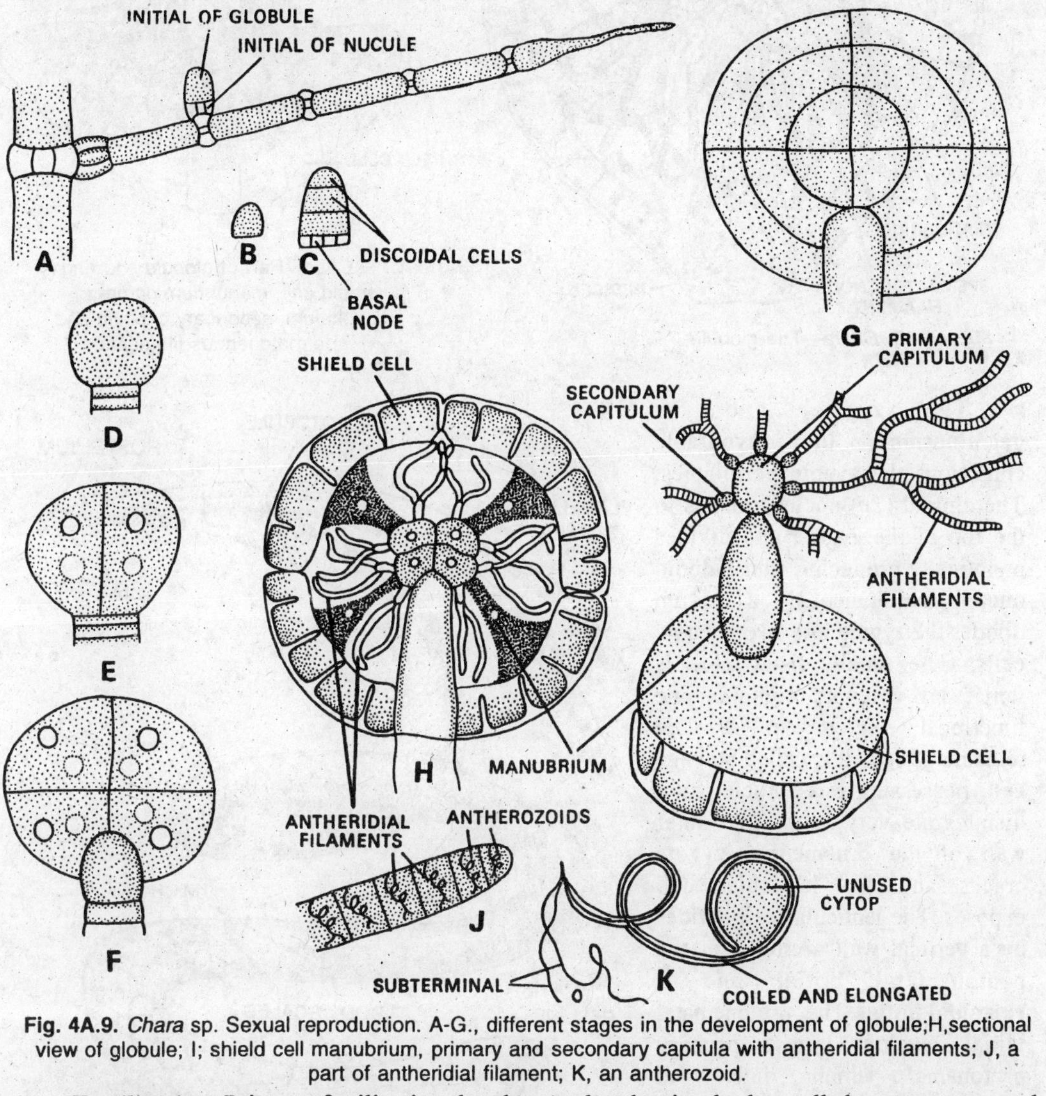

Fig. 4A.9. *Chara* sp. Sexual reproduction. A-G., different stages in the development of globule;H,sectional view of globule; I; shield cell manubrium, primary and secondary capitula with antheridial filaments; J, a part of antheridial filament; K, an antherozoid.

Fertilization. Prior to fertilization the elongated and twisted tube cells become separated from each other and five small slits are developed just below the corona. The swimming antherozoids around the nucule try to enter through these openings.

The flagella are withdrawn and one of the antherozoids penetrates the egg. The male nucleus travels downwards and fuses with the egg nucleus developing a diploid ($2n$) nucleus. This diploid nucleus situates in the bottom of the zygote. The zygote settles down in the mud, secretes a thick wall and germinates on the approach of favourable conditions.

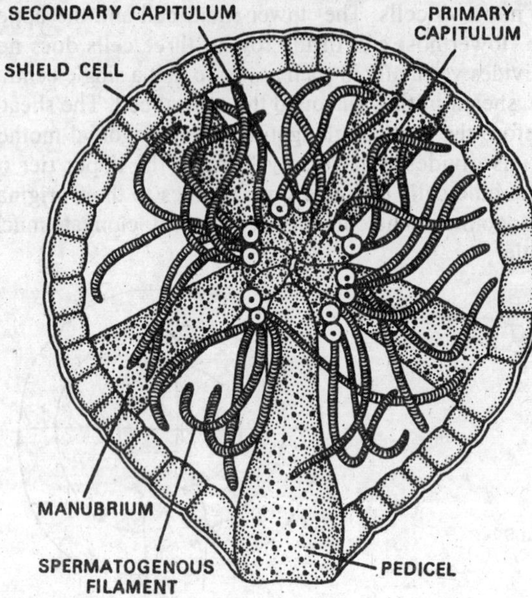

Fig. 4A.10. *Chara*—The globule.

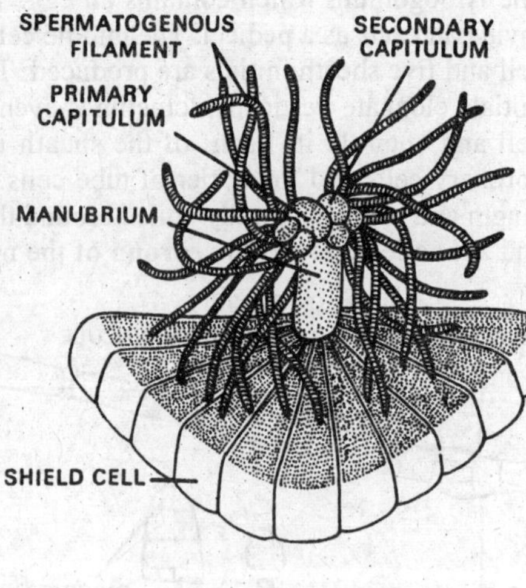

Fig. 4A.11. *Chara*. Part of globule showing shield cell, manubrium, primary capitulum, secondary capitula and spermatogenous filaments.

The zygote and its germination. In favourable conditions the zygote germinates. The diploid ($2n$) nucleus moves to the top of the zygote and divides meiotically producing four halpoid nuclei. Simultaneously a septum divides the zygote into two unequal cells. The small distal cell is **lenticular cell** and contains one functional **nucleus** in it. The remaining big cell is called **storage cell,** possessing three nuclei in it disintegrate very soon. The outer wall of the ornamented zygote cracks and the lenticular cell exposes. The lenticular cell divides by a vertical wall giving rise to a **protonematal initial and a rhizoidal initial.** The protonematal initial develops into a primary protonema which later on differentiates into nodes and internodes. The rhizoidal initial gives rise to a colourless rhizoid having nodes and internodes.

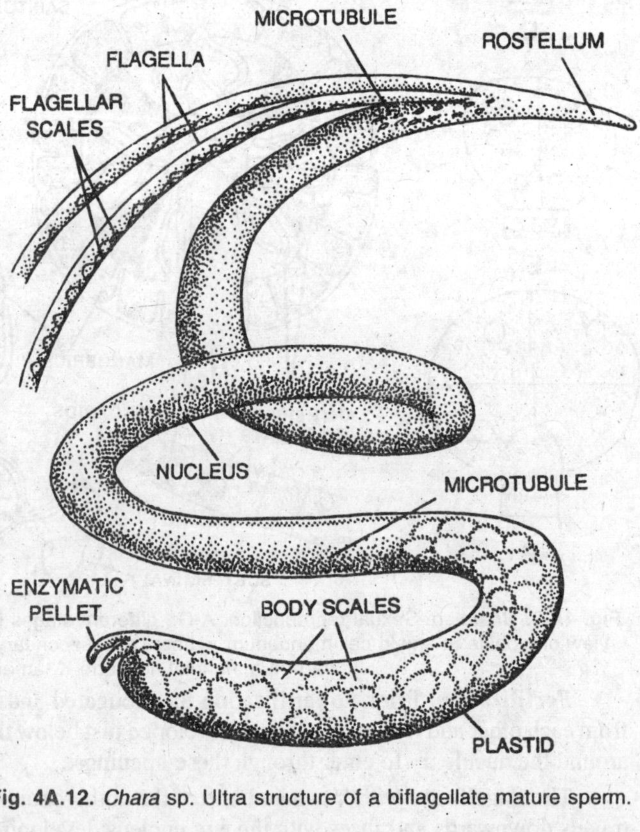

Fig. 4A.12. *Chara* sp. Ultra structure of a biflagellate mature sperm.

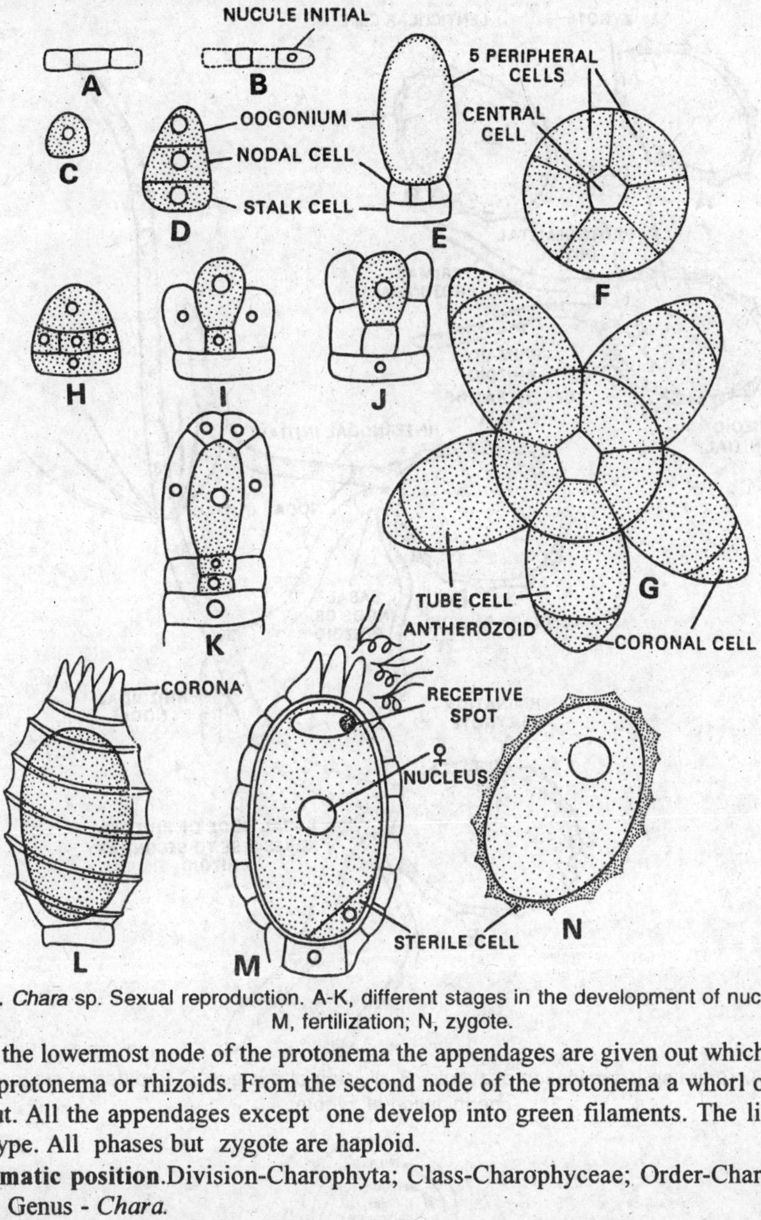

Fig. 4A.13. *Chara* sp. Sexual reproduction. A-K, different stages in the development of nucule; L nucule; M, fertilization; N, zygote.

From the lowermost node of the protonema the appendages are given out which develop into secondary protonema or rhizoids. From the second node of the protonema a whorl of appendages is given out. All the appendages except one develop into green filaments. The life cycle is of haplontic type. All phases but zygote are haploid.

Systematic position.Division-Charophyta; Class-Charophyceae; Order-Charales; Family-Characeae; Genus - *Chara.*

Advanced features of the Charales. The position of the Charales is controversial on account of the multicellular female organ, the complex antheridium, strong apical growth by vegetative shoots and a degree of specialization which is generally not found in other green algae. The female sex organ possesses a sterile jacket of cells, which is not found in the typical oogonium of other green algae. In 1875, Sachs referred the oogonium of *Chara* as a **nucule.** The presence of sterile jacket of cells is a new thing for the algae whereas on the other hand it is characteristic of the archegonium in the Embryophyta. But the jackets of these two groups are not supposed to be homologous. A parallel or analogous situation is found in the Rhodophyceae, where the female organ is a fairly complex structure.

In the same way the antheridium of the Charales is a much more complex structure than the typical antheridia and does not resemble the antheridia of either Bryophyta or Tracheophyta. Sachs (1875) called the antheridium as a **globule.**

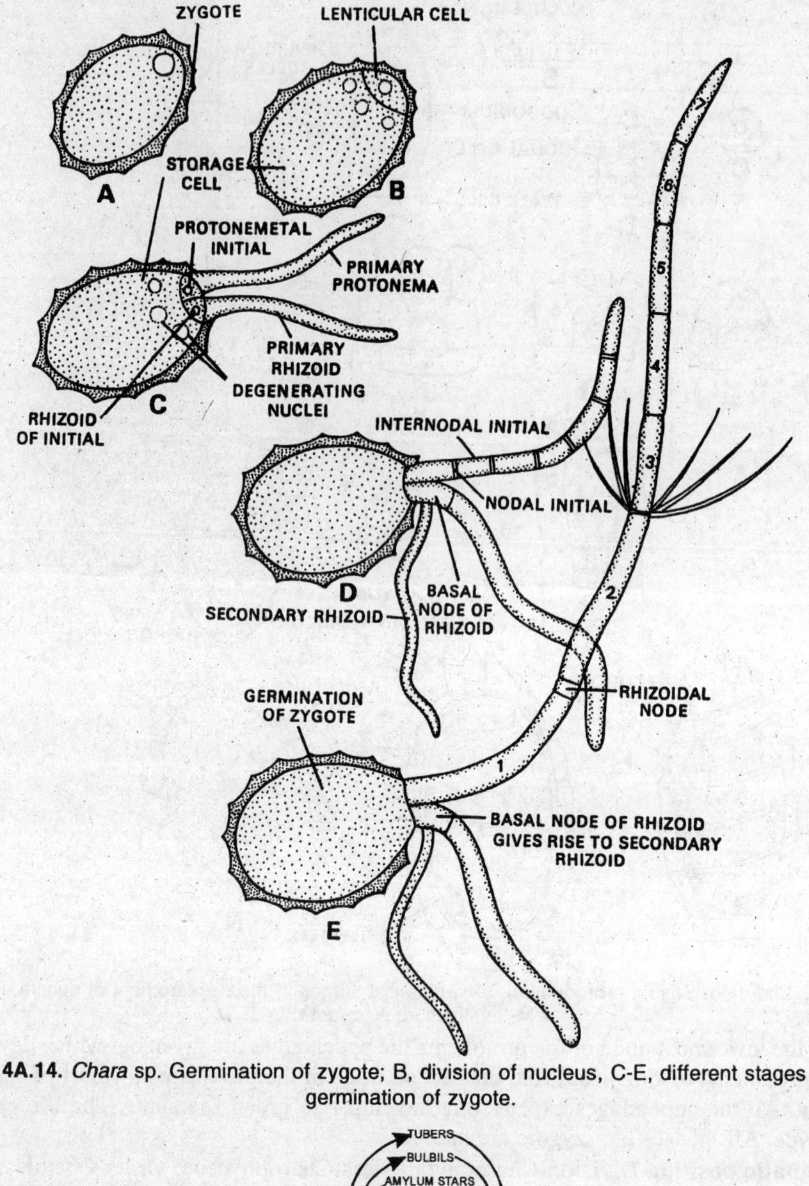

Fig. 4A.14. *Chara* sp. Germination of zygote; B, division of nucleus, C-E, different stages in the germination of zygote.

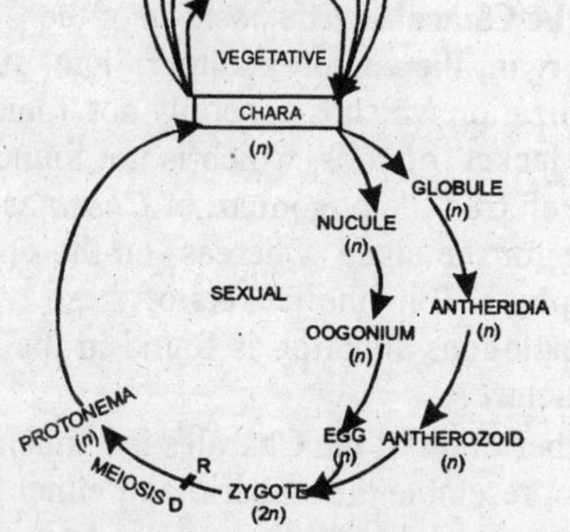

Fig. 4A.15. *Chara* sp. Graphic life-cycle.

The Charales have also achieved a degree of specialized apical growth which superficially resembles that of the Equisetales. However, the Charales are devoid of vascular system and they cannot be compared with Equisetales.

The Charales are a unique group so much so that Smith and others have placed this order in a separate class, the Charophyceae. The complex sex organs, *i.e.,* oogonia (nucule) and antheridia (globule) make a definite classification of the group difficult. However, the cellulose wall, pigmentation and food reserves are clearly of Chlorophycean type and according to Dr. Fritsch the order has been included is Chlorophyceae.

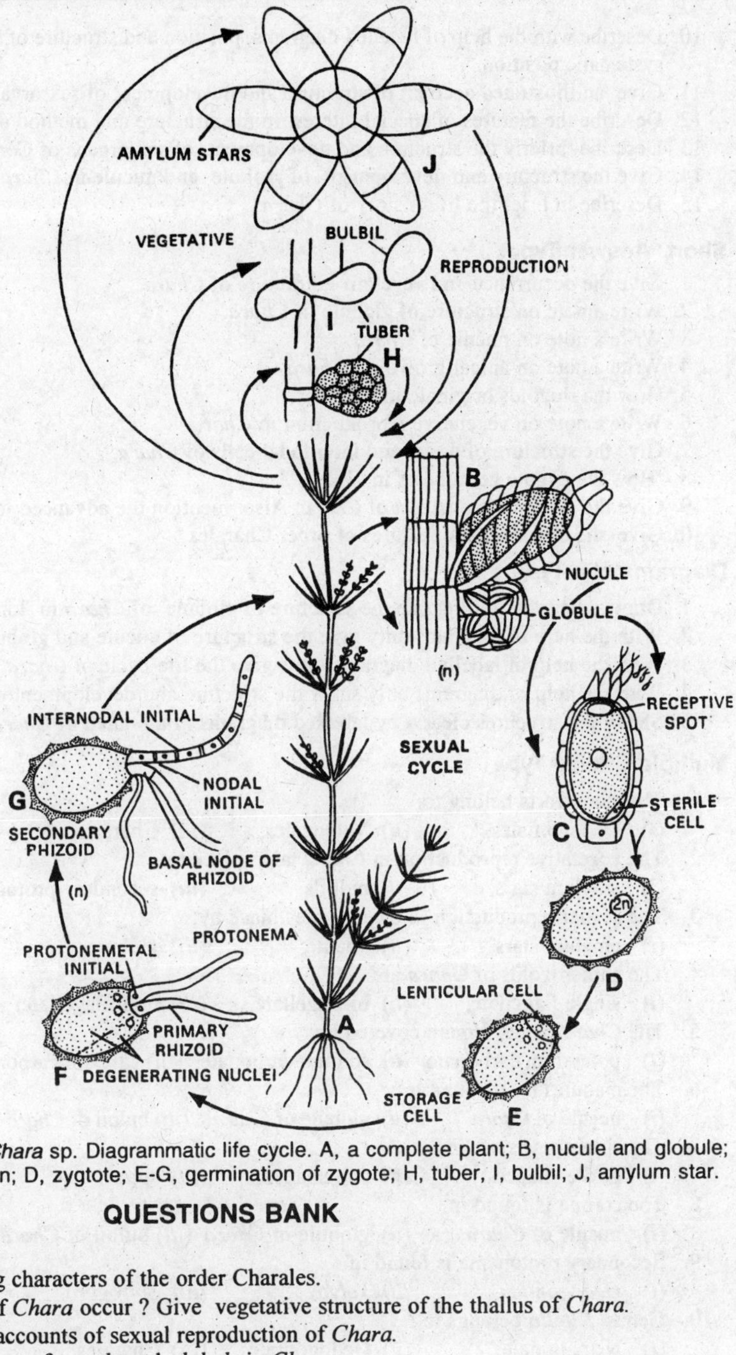

Fig. 4A.16. *Chara* sp. Diagrammatic life cycle. A, a complete plant; B, nucule and globule; C, fertilization; D, zygtote; E-G, germination of zygote; H, tuber, I, bulbil; J, amylum star.

QUESTIONS BANK

Long Answer Type

1. Give the distinguishing characters of the order Charales.
2. Where do the species of *Chara* occur ? Give vegetative structure of the thallus of *Chara.*
3. Give a comprehensive accounts of sexual reproduction of *Chara.*
4. Describe the development of *nucule* and *globule* in *Chara.*
5. Give the structure of nodal and internodel cells of *Chara.*
6. Give the advanced features of the order *Charales.*
7. Write note on (a) apical growth of axis in *Chara;* (b) development of cortex, leaf and rhizoids in *Chara;* (c) vegetative reproduction in *Chara;* (d) structure of nucule and globule in *Chara;* (e) the zygote and its germination in *Chara.*
8. Give the development of cortex, leaf and rhizoids in *Chara.* Also write briefly about its apical growth.
9. Describe the reproductive organs in *Chara.*

10. Describe with the help of labelled diagrams, position and structure of sex organs of *chara.* Discuss its systematic position.
11. Give an illustrated account of structure and development of sex organs in *Chara.*
12. Describe the features of special interest in the structure and method of reproduction in *Chara.*
13. Describe briefly the structure and development of sex organs of *Chara.*
14. Give the structure and development of globule and nucule in *Chara.*
15. Describe in brief the life-history of *Chara.*

Short Answer Type

1. Give the occurrence and vegetative structure of *Chara.*
2. Write a note on structure of globule in *Chara.*
3. Write a note on nucule of *Chara.*
4. Write a note on apical growth in *Chara.*
5. How the rhizoids originate in *Chara.*
6. Write a note on vegetative reproduction in *Chara.*
7. Give the structure of nodal and internodal cells of *Chara.*
8. How the zygote germinates in *Chara. ?*
9. Give the systematic position of *Chara.* Also mention the advanced features of the order *Charales.*
10. Give the characteristic features of order Charales.

Diagrammatic Type

1. Draw a diagram to illustrate the structure of globule of *Chara* in longitudinal section.
2. With the help of diagrams only give the structure of nucule and globule in *Chara.*
3. With the help of labelled diagrams only, give the life-cycle of *Chara.*
4. With the help of diagrams only show the structure and development of sex organs in *Chara.*
5. Show the strucures clearly by labelled diagrams of a branch of *Chara* with sex organs.

Multiple Choice Type

1. The stoneworts belong to:
   (*i*) Oedogoniales, (*ii*) Volvocales, (*iii*) Charales, (*iv*) Bryophytes
2. The vegetative reproduction in *Chara* takes place by :
   (*i*) amylum stars, (*ii*) bulbils, (*iii*) secondary protonema (*iv*) all the above
3. The sexual reproduction in *Chara* takes place by:
   (*i*) amylum stars, (*ii*) nucule, (*iii*) globule, (*iv*) nucule and globule
4. The antherozoids of *Chara* are:
   (*i*) single flagellate, (*ii*) bi-flagellate, (*iii*) multiflagellate, (*iv*) none of above
5. The *Chara* plants remain covered by:
   (*i*) potassium carbonate, (*ii*) sodium carbonate, (*iii*) calcium carbonate, (*iv*) none of above
6. The manubrium is found in :
   (*i*) nucule of *Chara* (*ii*) globule of *Chara,* (*iii*) bulbil of *Chara* (*iv*)amylum star of *Chara*
7. The chloroplasts of *Chara are :*
   (*i*) girdle-shaped, (*ii*) cup-shaped, (*iii*) spiral, (*iv*) discoid
8. The corona is found in:
   (*i*) nucule of *Chara,* (*ii*) globule of *Chara* (*iii*) bulbil of *Chara* (*iv*) protonema of *Chara*
9. Secondary protonema is found in :
   (*i*) *Oedogonium,* (*ii*) *Volvox* (*iii*) *Spirogyra,* (*iv*) *Chara*
10. Genus *Nitella* belongs to :
    (*i*) Nemalionales, (*ii*) Oedogoniales, (*iii*) Charales, (*iv*) Cladophorales
11. The number of coronal cells in nucule of *Chara :*
    (*i*) 2, (*ii*) 3, (*iii*) 5, (*iv*) 10.
12. The number of coronal cells in nucule of *Nitella:*
    (*i*) 2, (*ii*) 3, (*iii*) 5, (*iv*) 10

ANSWERS

1. (*iii*), 2 (*iv*), 3 (*iv*), 4 (*ii*), 5 (*iii*), 6 (*ii*), 7 (*iv*), 8 (*i*), 9 (*iv*), 10 (*iii*), 11 (*iii*), 12 (*iv*).

5

CHAPTER

Xanthophycophyta–Xanthophyceae

(Yellow-Green Algae)

Characteristic features.

Pigmentation. Chlorophyll *a*, β-carotene, Lutein, Neoxanthin.

Storage products. Chrysolaminarin, oils.

Flagellation. 2, unequal, apical flagella.

The members of this division were formerly included in the Chlorophycophyta until their distinctive pigmentation, storage products, and flagellation became apparent. The organization of the plant body in this group parallels in part that of the Chlorophycophyta in that both motile and nonmotile unicellular, colonial, filamentous and tubular genera are known.

There are 75 genera and about 375 species in this class.

Characteristic features.

1. **Pigmentation.** They have yellowish green chromatophores containing chlorophyll *a*, chlorophyll *b*, β-carotene and one xanthophyll.

2. **Food reserves.** In this class, the food reserves are found in the form of a carbohydrate of unknown combination known as leucosin. Sometimes oil is also found in small quantity. Starch is lacking altogether.

3. **Flagellation.** The motile cells possess two unequal flagella on their anterior ends. The shorter flagellum is of whiplash type and the longer one is of tinsel type. Due to the presence of unequal flagella on swarm cell they are also called **'heterokontae'**.

4. In certain genera, the wall consists of two equal or unequal halves overlapping each other.

5. Asexual reproduction takes place by means of zoospores, aplanospores and special type of statopores.

6. Sexual reproduction is known in only few genera. In majority of cases, it is isogamous. In one genus, oogamy is also found.

7. Majority of Xanthophyceae are fresh water in habit. They thrive best in soft waters. They may be aquatic, *e.g.*, *Tribonema*, or terrestrial, *e.g.*, *Botrydium*. Few genera are marine.

Classification. There are six orders in this class, *i.e.*, 1. Heterococcales and 2. Rhizochloridales, 3. Heterocapsales, 4. Heterotrichales, 5. Heterococcales and 6. Heterosiphonales.

1. Heterochloridales

(*i*) The plant body is flagellated, naked and unicellular.

(*ii*) Each cell contains two or more discoid or band-shaped chromatophores and one or more contractile vacuoles in the protoplast.

(*iii*) Reproduction takes place by cell division and in one genus statospores are formed.

(*iv*) There are 9 genera and about 15 species; 4 genera are marine and rest are fresh water in habit.

(*v*) Important genus–*Chloromeson.*

2. Rhizochloridales

(*i*) The plant body is amoeboid in pseudopodia. Sometimes several individuals may become jointed to one another by cytoplasmic bridges, mostly naked or partially surrounded by the **lorica** that may be attached to substratum by stalk.

(*ii*) Reproduction takes place by cell division, zoospores, aplanospores or statospores.

(*iii*) The protoplast contains one to many chromatophores. They may be uninucleate or multinucleate.

(*iv*) There are 7 genera and about 10 species. Majority of them are fresh water.

(*v*) Important genus–*Chlorachnion.*

3. Heterocapsales

(*i*) The plant body is palmelloid, *i.e.*, gelatinous colonies are found which may be amorphous or dendroid.

(*ii*) The cells have the capacity to return directly in the motile condition.

(*iii*) Reproduction takes place by means of fragments, akinetes and zoospores. The zoospores may divide into new zoospores.

(*iv*) There are 8 genera and about 90 species. All are fresh water in habit.

(*v*) Important genus–*Gloeochloris.*

4. Heterotrichales

(*i*) The plant body is simple or branched filament.

(*ii*) Reproduction takes place by means of fragementation, akinetes, zoospores, aplanospores and hypnospores.

(*iii*) Physiological isogamy is found in one genus, *i.e., Tribonema.*

(*iv*) There are 8 genera and about 35 species. All are fresh water in habit.

(*v*) Important genus–*Tribonema.*

5. Heterococcales

(*i*) They are nonfilamentous and incapable of returning directly to a motile form.

(*ii*) The cell may be uninucleate or multinucleate. In majority of cases, the wall consists of two halves.

(*iii*) The plant body consists of coccoid or colonial (gelatinous) forms.

(*iv*) Reproduction takes place by means of zoospores and aplanospores. The aplanospores are autospores. They have no capacity of vegetative cell division.

(*v*) There are 45 genera, and about 270 species. All are fresh water in habit.

(*vi*) Important genus–*Botrydiopsis.*

6. Heterosiphonales

(*i*) All the members are multinucleate, siphonaceous and unicellular in structure.

(*ii*) Asexual reproduction takes place by means of zoospores, aplanospores and hypnospores. These bodies are coenocytic.

(*iii*) Sexual reproduction is isogamous.

(*iv*) There are 3 or 4 fresh water genera.

(*v*) Important genera–*Botrydium* and *Vaucheria.*

Here the genus *Vaucheria* of order Heterosiphonales has been described in detail.

Phylogenetic Relationships in Xanthophyceae

Luther (1889) for the first time recognized Xanthophyceae as a separate class and named it Heterokontae. He placed the motile genera in one order and the remaining genera in another. Thereafter, Pascher (1913) pointed out that evolutionary lines found in the Chlorophyceae are also evident among Xanthophyceae. He also suggested the segregation of Xanthophyceae into orders homologous with those of Chlorophyceae. According to this classification, the Heterocapsales to the Tetrasporales, the Heterotrichales to the Ulotrichales, the Hetercoccales to the Chlorococcales and the Heterosiphonales to the Siphonales. The Rhizochloridales however, does not possess its counterpart among Chlorophyceae.

Genus VAUCHERIA
(Family-Vaucheriaceae)

There are about 40 species in this genus. Majority of them are fresh water in habit and about half a dozen are marine.

Occurrence. Majority of the species of *Vaucheria* are either aquatic or terrestrial. The terrestrial species are found upon damp soils and muds in yellowish green velvet-like carpets. In terrestrial species, the rhizoid-like branches are present at the base of the thallus which attach it to the substratum. The fresh water species are usually found in aerated water, *e.g.*, in water cataracts.

Structure. The thallus consists of branched, coenocytic unseptate and tubular filament. At the base of the thallus a branched rhizoidal system is present attaching thallus to the substratum is known as '**hapteron**'.The septation takes place at the time of the formation of reproductive bodies or injury to the plant.

The cell wall is thin and consists of pectose. In the centre of the tubular thallus, there is a big central vacuole. The cytoplasm is confined to the peripheral region. The numerous discoid chloroplasts are embedded in the cytoplasm towards outside and numerous minute nuclei are found towards the central vacuole. The chloroplasts are devoid of pyrenoids. The strach grains are lacking.

Reproduction. The reproduction takes place by vegetative, asexual and sexual methods.

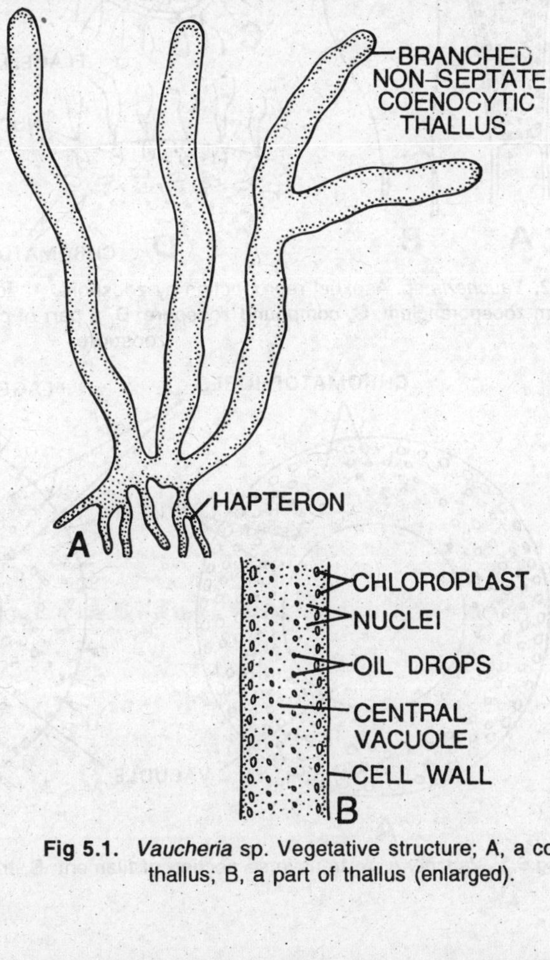

Fig 5.1. *Vaucheria* sp. Vegetative structure; A, a complete thallus; B, a part of thallus (enlarged).

1. Vegetative reproduction. This type of reproduction takes place by accidental breaking of thalli into small fragments. Each fragment of the filament is capable to give rise to a new plant.

2. Asexual reproduction. The asexual reproduction takes place by several means, *e.g.*, by zoospores, aplanospores, hypnospores, etc.

(*a*) By zoospores. The multiflagellate compound zoospores also known as **'synzoospores'** or **'coenozoospores'** are developed in this genus usually in aquatic forms. A single compound zoospore develops in a single zoosporangium. Any distal branch of the thallus may convert into a zoosporangium. Much of the food reserves, chloroplasts and nuclei accumulate in the distal end of the branch of the thallus. This distal end is comparatively swollen. Very soon a septum appears at the base of this swollen end. The central vacuole disappears, the nuclei and chloroplasts reverse their position, *i.e.*, the chloroplasts shift inwards and nuclei towards periphery just beneath the cell wall. The protoplast retracts from the zoosporangial wall and opposite to each nucleus two flagella are developed.

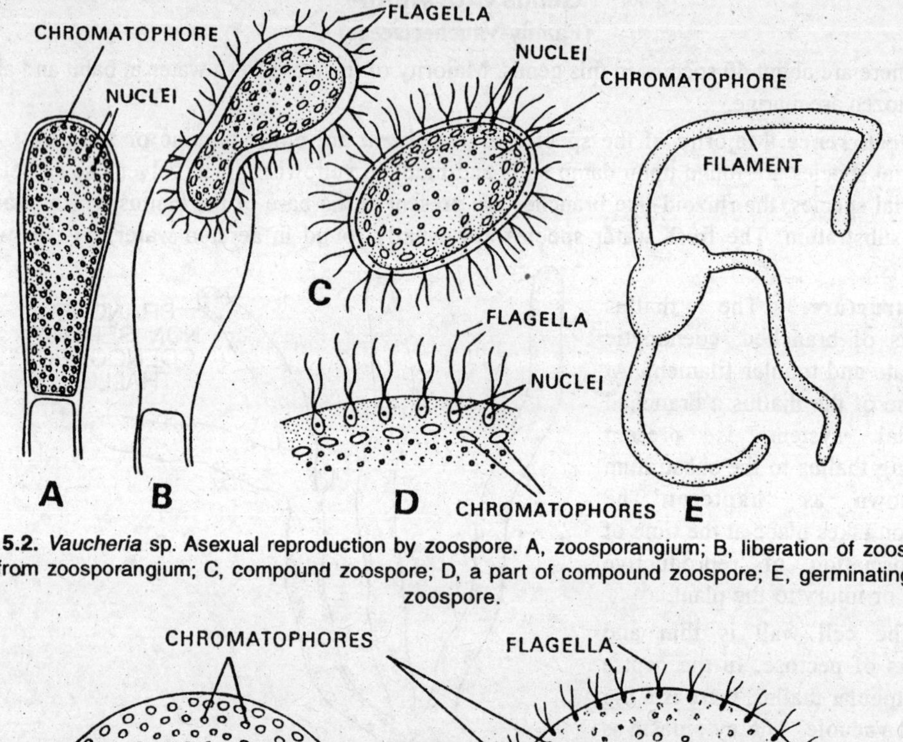

Fig. 5.2. *Vaucheria* sp. Asexual reproduction by zoospore. A, zoosporangium; B, liberation of zoospore from zoosporangium; C, compound zoospore; D, a part of compound zoospore; E, germinating zoospore.

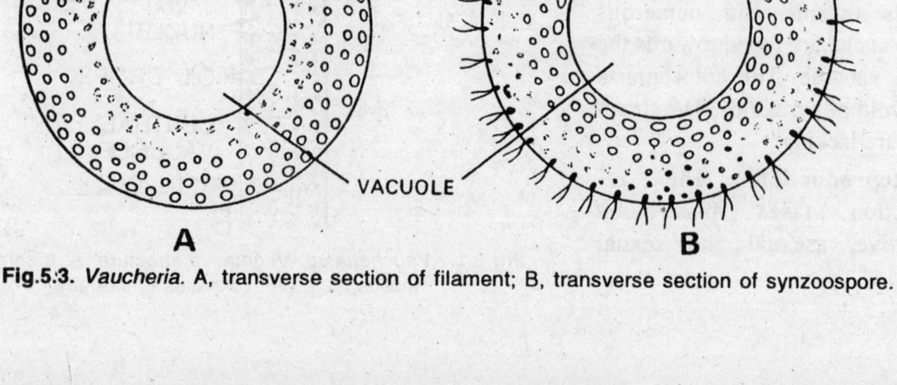

Fig.5.3. *Vaucheria.* A, transverse section of filament; B, transverse section of synzoospore.

The terminal portion of the zoosporangium softens and a small pore develops. Through this small terminal aperture the zoospore squeezes out and swims freely in the water. The zoospore is ovoid or elliptical. It is multiflagellate. From each peripheral nucleus, a pair of flagella is given out. There are several chloroplasts arranged inner to the nuclei around the large central vacoule. The nuclei and chloroplasts are embedded in cytoplasm. The zoospore swims for about 15 to 20 minutes and then settles down to some substratum, withdraws its flagella and secretes a wall around it. Very soon it germinates giving rise to 2 or 3 tubular outgrowths and a new branched thallus develops.

(*b*) By aplanospores. The aplanospores develop in dry conditions and especially in terrestrial forms. Here the contents of an aplanosporangium develop into a non-motile aplanospore. This aplanospore escapes by the irregular rupture of the aplanosporangium. On the approach of favourable conditions, the aplanospore germinates producing tubular outgrowths. They may germinate after liberation or within the aplanosporangium.

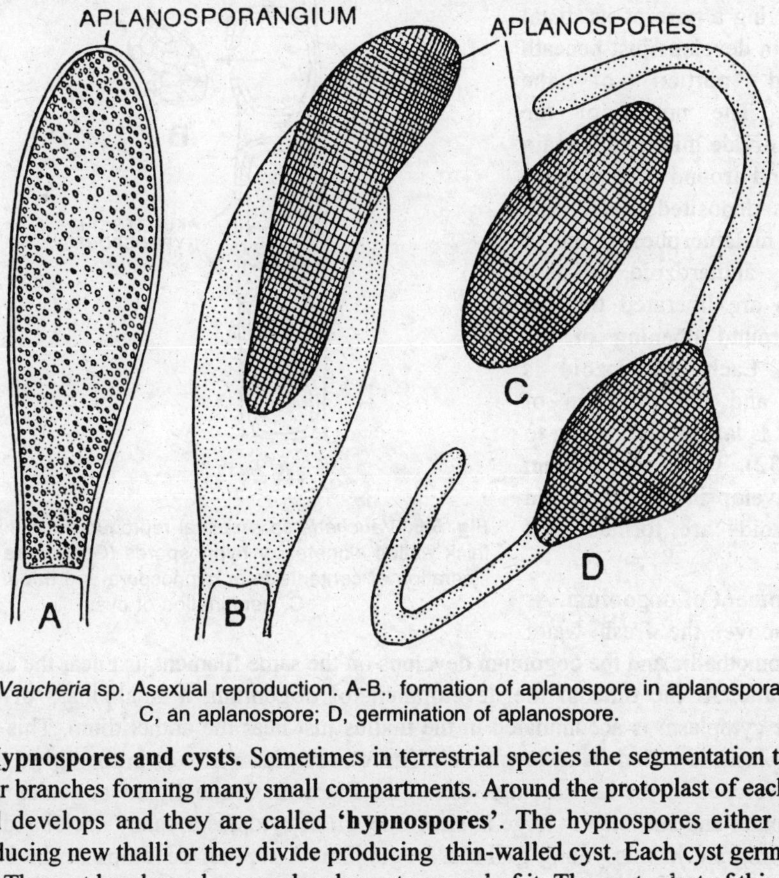

Fig 5.4. *Vaucheria* sp. Asexual reproduction. A-B., formation of aplanospore in aplanosporangium; C, an aplanospore; D, germination of aplanospore.

(*c*) Hypnospores and cysts. Sometimes in terrestrial species the segmentation takes place in the tubular branches forming many small compartments. Around the protoplast of each segment, a thick wall develops and they are called **'hypnospores'**. The hypnospores either germinate directly producing new thalli or they divide producing thin-walled cyst. Each cyst germinates in a special way. The cyst breaks and a pore develops at one end of it. The protoplast of this comes out in amoeboid fashion, becomes rounded and develops, into a new thallus. The segmented thallus looks like an alga *'Gongrosira'* and this stage is called **'Gongrosira stage'.**

3. Sexual reproduction. The sexual reproduction is oogamous. About all the species reproduce by this method. All the fresh water species are homothallic. The sexual reproduction takes place in the forms growing in damp soil or still waters. It does not take place in running water species. In the monoecious species the oogonia and antheridia develop on the same thallus and in dioecious species these organs develop on two different thalli. Here we will consider the sexual reproduction of fresh water forms (homothallic).

Development of antheridium. Mostly the species are protandrous, *i.e.*, antheridium develops first and oogonium afterwards.

The antheridia develop on the lateral branches at their ends shortly before the formation of oogonia. The part of the thallus giving rise to antheridium possesses abundance of cytoplasm, chloroplasts and nuclei. In most of the fresh water species, slender hook-like antheridium develops having a pore at its distal end. A septum develops just beneath the curved portion of the antheridium. The nuclei of the antheridium divide mitotically again and again, and around each nucleus cytoplasm is deposited. Each such small bit metamorphoses in a biflagellate antherozoid. These antherozoids are liberated through the apical round opening of the antheridium. Each antherozoid is spindle-like and the insertion of flagella on it is lateral. According to Couch (1932), the antheridium begins to develop in afternoon and the antherozoids are formed next morning.

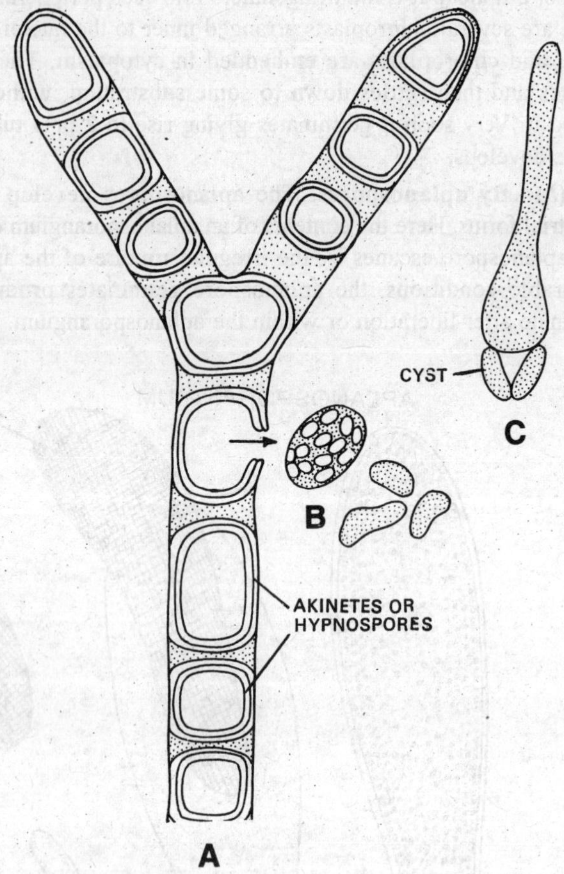

Fig. 5.5. *Vaucheria* sp. Asexual reproducion. A, formation of thick walled akinetes or hypnospores (Gongrosira stage); B, liberation of contents from hypnospore and three amoebae; C, germination of cyst.

Development of oogonium. As mentioned above, the fresh water species are homothallic and the oogonium develops on the same filament just near the antheridium after some time. At the time of the development of oogonium, a small mass of colourless multinucleate cytoplasm is accumulated in the thallus just near the antheridium. This colourless mass of cytoplasm is called **'wanderplasm'.** The wanderplasm moves in the apical part of the oogonium developed by the lateral outgrowth of the filament. Gradually the oogonium outgrowth becomes globular and sufficiently large in size. Enough of cytoplasm, many nuclei and numerous chloroplasts migrate in this young globular oogonium. A septum develops at the base of the oogonium which separates the oogonium from the rest of the plant body. A colourless beak-like structure develops at the upper end of the oogonium. All the nuclei but one disintegrate and a round uninucleate egg develops inside the oogonium. At maturity, the beak gelatinizes and an opening develops.

As soon as the antherozoids liberate from the apical opening of the antheridium in the morning, they approach the oogonium. Many antherozoids enter through the opening of the oogonium but only one lucky antherozoid penetrates the egg. The plasmogamy is followed by

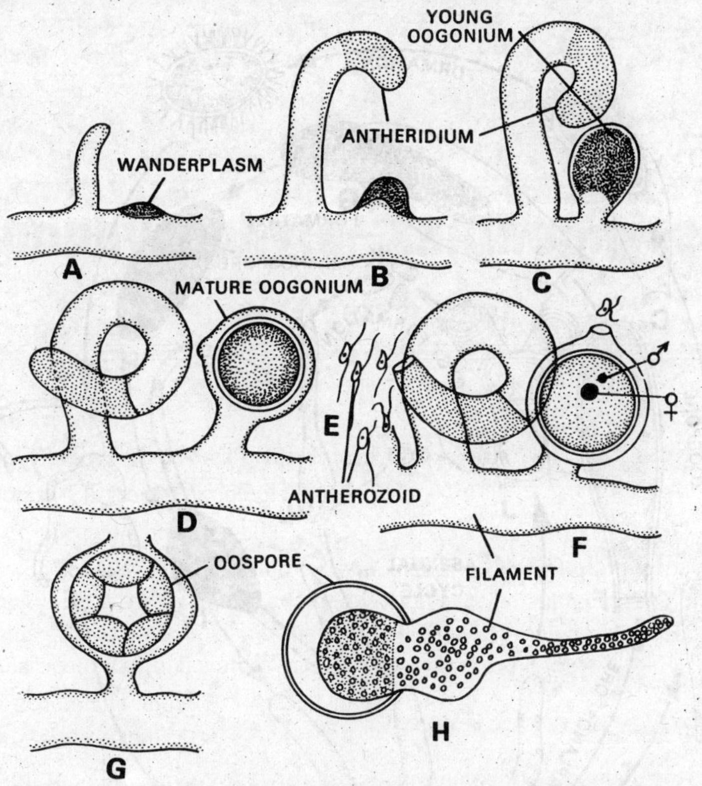

Fig. 5.6. *Vaucheria* sp. Sexual reproduction. A-C, simultaneous development of antheridium and oogonium; D, mature oogonium and coiled antheridium; E, antherozoids; F, fertilization; G, oospore; H, germinating oospore,

karygoamy. The small nucleus of antherozoid enlarges and becomes of the size of female nucleus and only then fuses with the large nucleus of the egg. Ofcourse, the fusion takes some time. At the time of the fertilization of the female nucleus shifts to the periphery of the egg to receive the male nucleus. After fusion, the large diploid nucleus again shifts in the centre of the zygote. The zygote secretes a multilayered wall around it and called the oospore. The oospore is sufficiently rich in oil reserves.

Germination of zygote. The thick walled zygote (oospore) undergoes a resting period. This resting period may be of several months. On the approach of favourable conditions, the zygote germintes producing tubular structure which develops into a new thallus. Gross (1937) has given an inconclusive data of the meiotic division of the zygote nucleus during germination.

Classification. The position of *Vaucheria* is not quite clear. F.E. Fritsch has placed this genus in the ninth and last family Vaucheriaceae of his order Siphonales. Gilbert M.Smith has placed this genus in his order, Heterosiphonales of division Chrysophyta. V.J. Chapman has placed this genus in his class Xanthophyceae of Phylum Chrysophycophyta. Modern algologists place this genus in the Class Xanthophyceae of Division Xanthophycophyta.

Many algologists for a long time questioned the practice of placing *Vaucheria* in the Siphonales of the Chlorophyceae. For the first time in 1945 Chadefaud formaly placed this genus among the Xanthophyceae. It was placed in this class because of the colour of chromatophores and lack of starch. Since then other evidences have also been collected which confirm its place among Xanthophyceae. The pigments of the chromatophores are those of typical Xanthophyceae. Koch

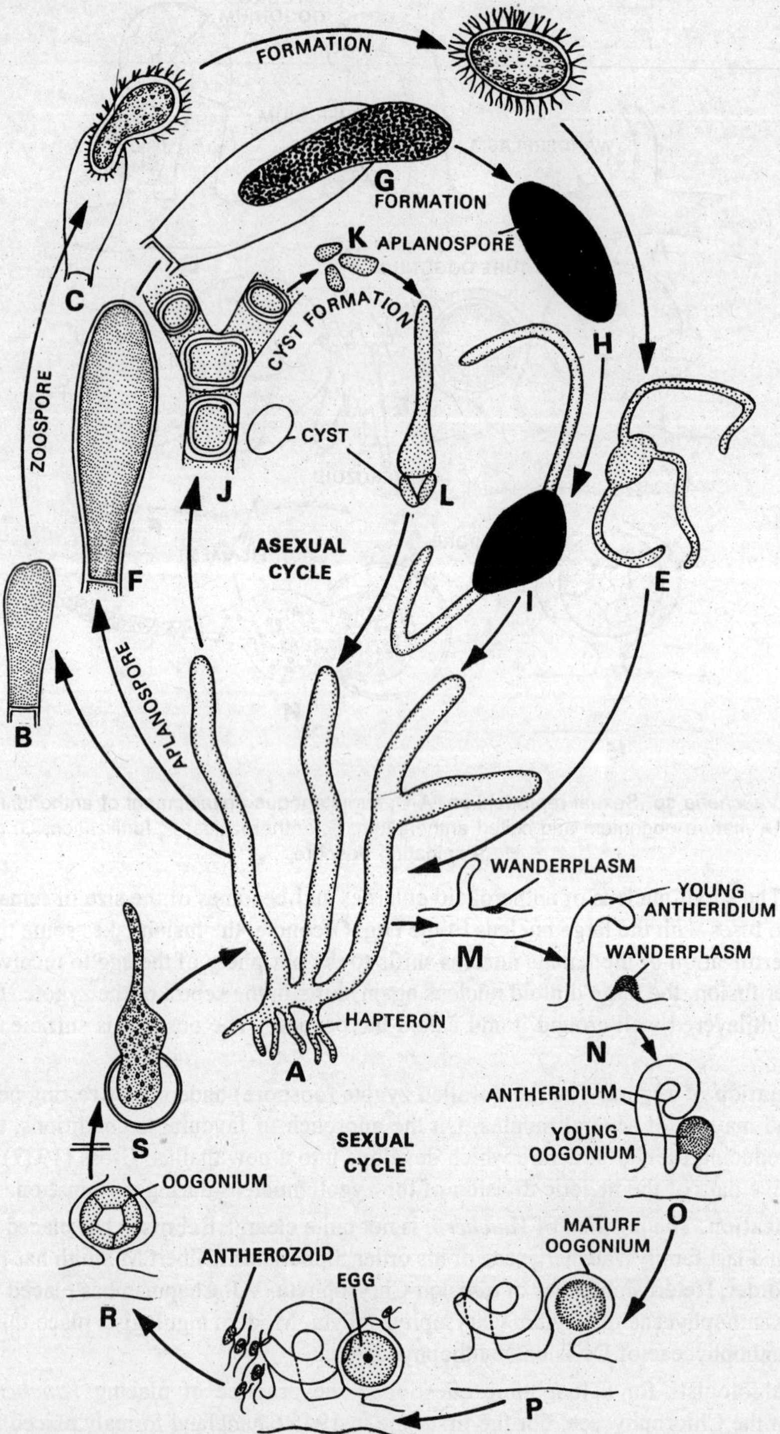

Fig. 5.7 *Vaucheria* sp. Diagrammatic life cycle. A, a complete plant; B-D, formation of synzoospore; E, germination of zoospore; F-G, aplanospore formation; H, aplanospore; I, germination of aplanospore; J-L, cyst formation and its germination; M-O, development of antheridium and oogonium; P, mature antheridium and oogonium; Q, fertilization; R, oospore; S, germination of oospore.

(1951) reported that of the two flagella borne by an antherozoid one is whiplash type and the other of the tinsel type.

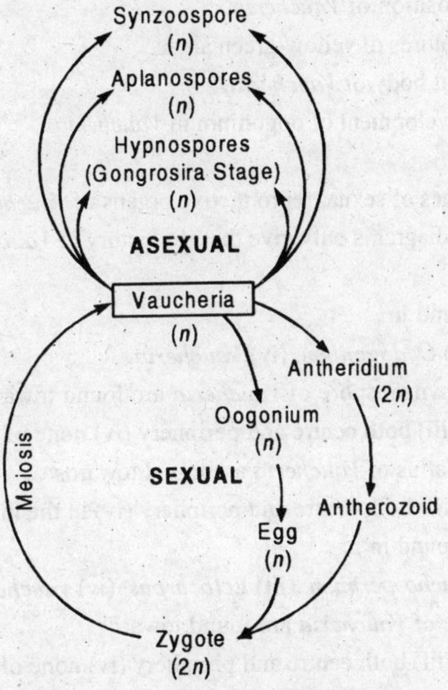

Fig. 5.8. *Vaucheria* sp. Graphic life cycle.

Systematic Position. Division–Xanthophycophyta; Class–Xanthophyceae; Order–Heterosiphonales; Family–Vaucheriaceae; Genus–*Vaucheria.*

QUESTION BANK

N.B. *Undermentioned questions are asked repeatedly in various Indian Universities and competitive examinations.*

Long Answer Type

1. Describe briefly the sexual life-cycle of *Vaucheria*
2. Describe the life-history of *Vaucheria.*
3. Discuss the characteristic features of *Vaucheria.* Explain its sexual reproduction.
4. Give the details of asexual and sexual reproduction in *Vaucheria.*
5. Give an illustrrated account of the vegetative structure and mode of reproduction in *Vaucheria.*
6. Give an account of the life-history of *Vaucheria* and discuss its relationship with other algae.
7. Discuss the systematic position of *Vaucheria.*Give a diagrammatic reproduction of its life-cycle illustrating the lenght of haploid and diploid phases.
8. Describe the structure and methods of reproduction in *Vaucheria.*
9. Write an illustrated account of the methods of reproduction in *Vaucheria.*
10. Give the development and structure of antheridium and oogonium in *Vaucheria.*

Short Answer Type

1. Give the characteristics of Xanthophyceae.
2. Write a note on phylogenetic relationships in Xanthophyceae.
3. Give the characteristic features of order Heterosiphonales.
4. Write a note on occurrence and structure of *Vaucheria.*
5. Write a note on: (a) synzoospore and (b) *Gongrosira* stage in *Vaucheria.*

6. Give the structure of antheridium and oogonium in *Vaucheria*.
7. Write a note on germination of zygote in *Vaucheria*.
8. Discuss the systematic position of *Vaucheria*.
9. Give the characteristic features of yellow-green algae.
10. Give the structure of plant body of *Vaucheria*.
11. Give the structure and development of oogonium in *Vaucheria*.

Diagrammatic Type

1. Draw well-labelled sketches of sexual reproductive organs of *Vaucheria*.
2. With the help of labelled diagrams only give the life-history of *Vaucheria*.

Multiple Choice Type

1. The synzoospores are found in:
   (i) *Chara,* (ii) *Volvox,* (iii) *Oedogonium,* (iv) *Vaucheria*.
2. The chromatophbores in synzoospore of *Vaucheria* are found towards:
   (i) centre, (ii) periphery, (iii) both centre and periphery (iv) none of above.
3. The chromatophores in thallus of *Vaucheria* are found towards:
   (i) centre, (ii) periphery, (iii) both centre and periphery (iv) in the middle
4. The *Gongrosira* stage is found in :
   (i) *Oedogonium,* (ii) *Batrachospermum,* (iii) *Ectocarpus,* (iv) *Vaucheria*.
5. The nuclei in synzoospore of *Vaucheria* are found towards:
   (i) centre, (ii) periphery, (iii) both centre and periphery (iv) none of above.
6. The nuclei in thallus of *Vaucheria* are found towards:
   (i) centre, (ii) periphery, (iii) both centre and periphery (iv) all of above.
7. Asexual reproduction in *Vaucheria* takes place by:
   (i) synzoospore, (ii) aplansospore, (iii) hypnospore, (iv) all of above.
8. The thallus of *Vaucheria* contains food material in the form of :
   (i) oil globules, (ii) starch grains, (iii) aleurone grains, (iv) none of above.
9. Each sporangium of *Vaucheria* contains:
   (i) single zoospore, (ii) two zoospores, (iii) four zoospores, (iv) sixteen zoospores
10. The thallus of *Vaucheria* is:
    (i) septate, (ii) aseptate (iii) none of above (iv) both of above
11. The synzoospore of *Vaucheria* is:
    (i) biflagellate, (ii) tetraflagellate, (iii) multiflagellate, (iv) all the above.
12. The synzoospore of *Vaucheria* is:
    (i) pectose, (ii) chitin, (iii) calcium carbonate, (iv) silica.
13. The 'wanderplasm' is found in:
    (i) gonidia of *Volvox*; (ii) nucule of *Chara,* (iii) young oogonium of *Vaucheria,* (iv) Oogonium of *Oedogonium.*
14. Formation of thick-walled akinetes or hypnospores in *Vaucheria* is called:
    (i) *Chantrantia* stage, (ii) *Gongrosira* stage, (iii) *Palmella* type; (iv) none of above.

ANSWERS

1 (iv), 2 (i), 3. (ii), 4. (ii), 5.(ii), 6 (i) ,7 (iv), 8 (i), 9 (i), 10 (ii), 11 (iii), 12 (i), 13 (iii), 14 (ii).

6

CHAPTER

Phaeophycophyta–Phaeophyceae

(The Brown Algae)

Characteristic features

Pigmentation. Chlorophylls *a* and *c*, α and β-carotenes, Diatoxanthin, Flavoxanthin, Fucoxanthin, Lutein, Violaxanthin.

Storage products. Laminarin, mannitol, oils.

Flagellation. 2, lateral

There are about 195 genera and 1,000 species in this class.

Occurrence. With the exception of three genera, *i.e., Heribendiella*, *Pleurocladia* and *Bodanella* which are fresh water in habit, rest of all Phaeophyceae are marine and commonly called brown sea weeds. The marine Phaeophyceae occur in littoral zones of arctic and antarctic oceans. As we pass on to the tropics the number of the members of brown algae gradually decreases. Certain forms, such as Dictyotales and *Sargassum* are predominant in warm waters. Almost all of Phaeophyceae are attached; *Sargassum natans* and *Sargassum fluitans* are pelagic, *i.e.*, they are secondary free floating in Sargasso sea of Atlantic ocean.

Most of the Fucaceae are found in the upper littoral zone of the sea. The Laminariales are restricted in the lower littoral zone, whereas the *Sphacelaria* and *Leathesia* are found in midlittoral zone. Certain species of *Sargassum* and *Desmarestia* are found in sub-littoral zones. They are not found below the depth of 25 metres. *Macrocystis* and *Nereocystis* occur in 13 to 20, metre deep waters. These giant kelps are attached to the rocks by their holdfasts. A stipe arises from the holdfast and the large leaf blades and bladders float on the surface of the water. The brown algae are not found below the depth of 110 metres.

Range of somatic structure of Phaeophyceae. No member is unicellular, colonial or unbranched filament. The simplest member is heterotrichous filament, which shows the highest organization in Chlorophyceae is the lowest organization in Phaeophyceae. The various vegetative organizations of the Phaeophyceae may be classified in the following groups.

Ectocarpoid. In *Ectocarpus* the thallus is profusely branched and the cells are joined end to end in a single series. The thallus consists of a prostrate branched portion, from which the erect branched system arises. This is heterotrichous type of organization.

Cable type. This type is represented by *Mesogloea.* Here the thallus consists of several axial strands like a bundle. Some radial branches come out at right angles from these threads, and the whole structure appears like a cable.

Multiseptate cable type. This type is represented by *Chorda filum.* The thallus is formed by many compact filaments. Actually there is no parenchyma but pseudoparenchyma.

Corticated type. *Desmarestia, Arthrocladia,* etc., represent this type. Here in the lower region of the plant body the lateral branches become rhizoidal and coil around the main axis to form a compact pseudoparenchymatous cortex around the main axis.

Truly parenchymatous forms. In such forms the thalli are leaf-like or laminate and truly parenchymatous, *e.g., Punctaria.* The members of Laminariales and Fucales possess improved type of parenchyma in them. The plant body of *Macrocystis* is 30 to 50 metres long. It has a dichotomously branched holdfast, a stipe and large leaf blades having gas bladders at their bases. This is the longest known giant kelp.

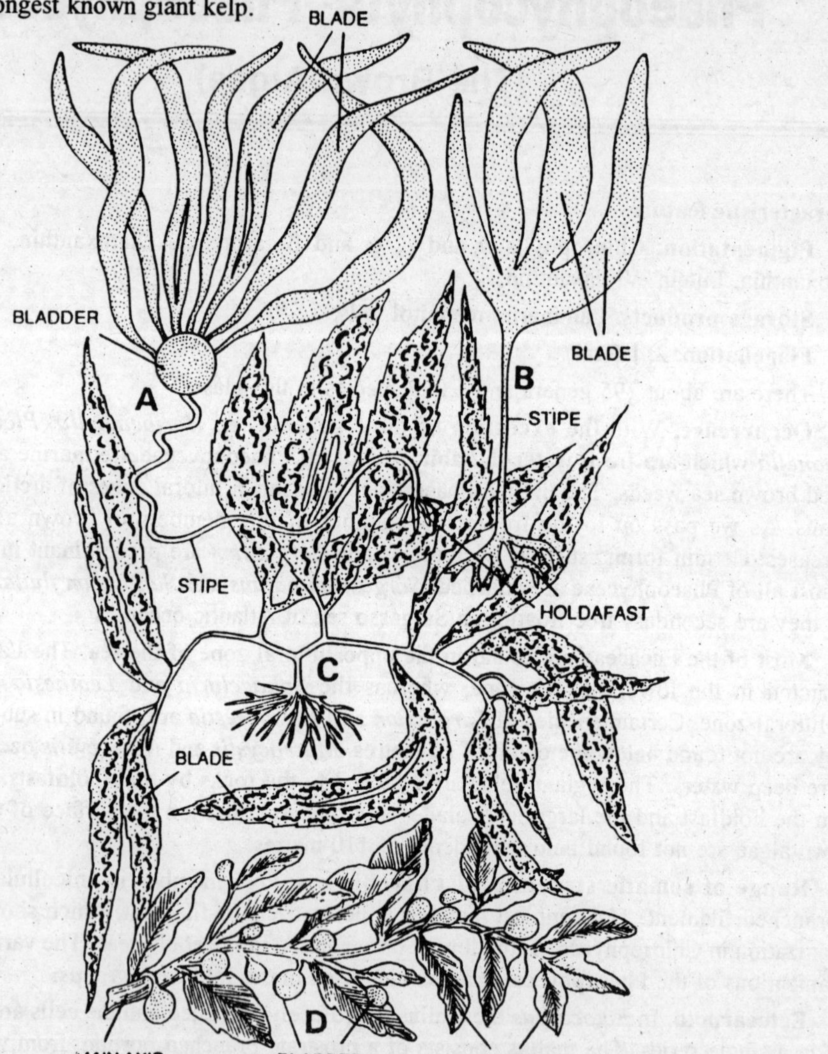

Fig. 6.1. Phaeophyceae. Range of structure. A, *Nereocystis luetkeana*; B. *Laminaria andersonii* (sporophyte); C, *Macrocystis pyrifera*; D, *Sargassum* sp.

The plant body of *Nereocystis* is 20 to 25 metres long. It consists of a haptera, a stipe and terminal large gas bladder having many blades on it.

Laminaria, Fucus, Sargassum, Postelsia, etc., all possess truly parenchymatous structures.

Internal structure. In *Laminaria,* the stipe consists of epidermis, cortex and medulla. There are 'trumpet hyphae'. At the junction of the mouths of two trumpets a callus pad develops in winter and dissolves in spring. This refreshes the memory of the callus pads of the sieve tubes of higher angiosperms.

Hairs. The hairs are present in all members of Phaophyceae. They are multicellular and generally ensheathed at their bases.

Cell structure. The cell wall is two layered. The inner layer outer to the protoplast consists of cellulose and the outer layer of mucilaginous substance. The mucilage contains gum like substances such as algin, fucoidin, etc. Callose is found in the cell membranes of *Laminaria digitata.* The callus pads consists of callose. In Laminariales special mucilage ducts are found, secreting mucilage.

The protoplast. It is less viscous. The cytoplasm contains mitochondria which are in contact of chromatophores. The cell contains numerous vacuoles. In *Dictyota* and *Fucus* each cell contains a large vacuole. In cell sap, the colloidal substances are found.

The special bodies called fucosan vesicles are abundantly found in meristematic, photosynthetic reproductive cells. They are highly refractive, colourless and rounded bodies.

Generally the chromatophores are discoid and parietal in position. In each cell, there may be one or more chloroplasts. The pyrenoids are usually attached to the inner margins of the chloroplasts. In many cases the pyrenoids are stalked and attached laterally to the chloroplasts.

The chromatophores contain chlorophyll *a,* chlorophyll *b,* xanthophyll and carotenoids as pigments. Fuco or phycoxanthin overmasks the other pigments, giving a tinge of brown colour to the plant body.

Photosynthetic products. According to Kylin, simple reducing sugar such as dextrose or pentose constitute the first products of photosynthesis. He also reports a disaccharide known as laminarose as first product of photosynthesis.

Food products. The important food products are (1) laminarin (2) certain hydrolases (3) mannitol and anhydride manniton (4) fats and (5) certain oxalates. Besides, in Laminariales and Fucales considerable quantity of vitamin C is found. Iodine is stored in many algae. In *Laminaria* (kelps) the percentage of iodine ranges from. 08 to 3.5.

Nucleus. Majority of Phaeophyceae have uninucleate cells. Multinucleate condition is rare. The nucleus is quite large in size. The chromatin material is found in less quantity. The nucleus contains one or two nucleoli. In Fucales the nucleoli are vacuolate. The extra or intranuclear centrosome is also present. In Phaeophyceae the number of chromosomes is from 8-32. The basic number is 8.

Swarmers. The swarmers may be zoospores or gametes. They are ovoid or pyriform. The swarmers are biflagellate and flagella are unequal in size. In majority of Phaeophyceae the flagellum directed forward is larger and directed backward is smaller in size. In order Fucales the position of flagella on the swarmer is reverse. Here the flagellum directed forward is smaller and directed backward is larger in size. In order Dictyotales the swarmers are single flagellate. The flagella are attached laterally to the swarmer. Each swarmer contains a large nucleus, 3 or 4 chromatophores, a stigma, a blepharoplast and a plastomere.

Reproduction. The reproduction takes place by 1. vegetative, 2. asexual and 3. sexual methods.

1. Vegetative reproduction. Vegetative propagation takes place by fragmentation of the thalli. In the case of *Sargassum* this type of reproduction is very prolific. In *S. natans* it is found abundantly. This is a free floating species.

In some of the cases special fragments known as **'propagula'** are developed. They occur in *Sphacelaria.*

The **'adventitious buds'** develop in many species of *Fucus.* They develop by the division of meristematic cells in young plants. Each bud develop into a new plant.

2. Asexual reproduction. Asexual reproduction takes place by means of zoospores and aplanospores formed inside the sporangia.

(*a*) By zoospores. The formation of zoospores is most common in all the members of Phaeophyceae except in *Dictyota* and *Fucus.* The zoospores are pyriform and biflagellate. The anterior flagellum is larger than the posterior one except in Fucales. In Dictyotales single flagellum is found on the zoospore. The zoospores are produced inside the zoosporangia, which may be of two types, *i.e.,* **unilocular sporangia** and **multilocular** or **plurilocular sporangia.**

The unilocular sporangium may be terminal or intercalary in position. In each sporangium 64 or 128 zoospores are produced. At the time of formation of zoospores reduction division takes place and zoospores are haploid. They germinate into haploid thalli (gametophytes). In some cases as Punctariales, Sphacelariales, Chordales, *Ectocarpus, Pyaeliella,* etc., the zoospores behave as gametes and unite in pairs. This is the example of origin of sex.

The plurilocular sporangia are always terminal in position. All the cells of sporangium are diploid and give rise to diploid zoospores. Such zoospores will give rise to sporophytes. The multilocular sporangia are unknown in Fucales and Laminariales.

(*b*) By aplanospores. Sometimes, in unilocular sporangia instead of producing zoospores the aplanospores may also be produced. They are non-motile and without flagella, *e.g.,* Dictyotales. The first division is always reductional. The aplanospores are always less in number. In *Dictyota* 4 aplanospores and in *Zonaria* 8 aplanospores are produced per sporangium.

3. Sexual reproduction. This ranges from isogamy to oogamy.

(*a*) Isogamy. This takes place by the fusion of two similar gametes. This is found in many Phaeophyceae such as Ectocarpales, Sphacelariales, Dictyosiphonales, etc. The gametophytes may be monoecious or dioecious.

(*b*) Aniogamy. Here the fusion of two dissimilar gametes takes place. The examples of this type are found in some members of Ectocarpales, *Cutleria, Soranthera,* etc. If fusion fails the female gametes may develop parthenogenetically.

(*c*) Oogamy. In majority of Phaeophyceae the sexual reproduction is oogamous. The species may be homo or heterothallic. The male sex organs are called the antheridia and the female sex organs are called the oogonia. In Dictyotales the antheridia are multicellular structures. Each cell of antheridium gives rise to a spermatozoid. In Desmarestiales and Laminariales the antheridia are unicellular and each antheridium produces a single spermatozoid.

Usually each oogonium produces a single ovum or oosphere, but in *Fucus* eight eggs are produced in the oogonium.

The eggs liberate in the water and only then the fertilization takes place so that the fertilization is external.

In Fucales the sex organs develop inside conceptacles which develop on special reproductive branches called the **'receptacles'**.

Life-cycle. There are three general types of life-cycle found in the class Phaeophyceae (brown algae).

1. An alternation of morphologically similar haploid and diploid generations, as found in *Ectocarpus.*

2. An alternation of morphologically dissimilar (heteromorphic) haploid and diploid generations, as found in Laminariales, *e.g., Laminaria, Nereocystis.*

3. The almost complete suppression of the haploid generation (*i.e.,* diplontic), as in *Fucus.* In this genus only haploid cells are gametes.

Origin of Phaeophyceae. It is very probable that Phaeophyceae are a series of considerable antiquity. The distinct features of Phaeophyceae such as pigmentation, the reserve food products and the type of flagellation are such unique characters that there appear to be no affinities with any other class of algae. The universal presence of a swarmer in the class supports the view of a flagellate ancestry for the class. According to Kylin, the Fucales arose directly unrelated with Phaeophyceae. But according to Gilbert M. Smith, the Fucales arose by dropping off of the gametophytic generation in a complex group of the brown algae with two alternating generations.

Classification. According to Dr. F.E. Fritsch, there are nine orders in this class-1. Ectocarpales, 2. Tilopteridales 3. Cutleriales, 4. Sporochnales, 5. Desmarestiales, 6. Laminariales 7. Sphacelariales, 8. Dictyotales, 9. Fucales.

The outline of Taylor's system is as follows. About 12 orders are included in this sytstem. The whole class is divided into three sub-classes.

Three sub-classes are:

I. Isogeneratae. Isomorphic alternation of generations.

II. Heterogeneratae. Heteromorphic alternation of genernaions.

III. Cylosporeae. Diplontic type of life cycle.

The sub-class **Isogeneratae** includes 5 orders.

1. Ectocarpales. Branched heterotrichous filamentous body; trichothallic growth; reproductive bodies terminal or intercalary, single or in chain; the sporophyte produces zoospores or neutral spors ($2n$); isogamy.

2. Sphacelariales. Growth by single large apical cell segments formed which frequently divide lengthwise in a regular polysiphonous manner; sporophyte may produce haploid or diploid zoospores; iso or anisogamy.

3. Tilopteridales. Thallus freely branched showing trichothallic growth, upper portion monosiphonous; lower portion polysiphonous; sporophytes produce unilocular sporangia only, each containing a single quadrinucleate aplanospore; gametangia are intercalary; doubtful oogamy.

4. Cutleriales. Thallus flattened, blade-like or disc-like, dichotomously branched, trichothallic growth; the sporophyte produces unilocular sporangia only; anisogamy.

5. Dictyotales. Erect flattened dichotomously branched parenchymatous thallus; growth by a single apical cell or a marginal row of special cells; the unicellular sporangia of sporophyte each produces 4 or 8 aplanospores; oogamy.

The sub-class **Heterogeneratae** includes two series and six orders.

Series **1. Haplostichineae.** Growth trichothallic; thallus consists of one or more filaments.

Series **2. Polystichineae.** Trichothallic growth absent; longitudinal and transverse intercalary cells form a parenchyma.

Three orders are included in Series **Haplostichineae.**

1. Chordariales. Branched filamentous sporophyte; isogamous.

2. Sporochanles. Each branch of the sporophyte terminates in a tuft of hairs; growth is trichothallic due to intercalary cell division at the base of each hair; the unicellular sporangia are borne terminally in dense clusters; plurilocular sporangia absent; oogamy.

3. Desmarestiales. The thallus has a single filament of each growing apex posterior to which there is pseudoparenchymatous cortication, the thallus is macroscopic; oogamy, gametophyte microscopic.

Three orders are included in Series **Polystichineae.**

1. Punctariales. Parenchymatous, sporophyte of medium size; reproductive organs may or may not be borne in sori; the sporophyte reproduces by zoospores; iso or anisogamy. gametophyte microscopic.

2. **Dictyosiphonales.** Profusely branched cylindrical thallus, growth by single apical cells; sporophyte produces unilocular sporangia only; isogamy, gametophyte microscopic.

3. **Laminariales.** Sporophyte differentiated into a holdfast stipe, and blades; growth by intercalary meristem; internal structure of thallus is differentiated into epidermis, cortex and medulla, the medulla shows 'trumpet hyphae' with callus pads; oogamy, gametophyte microscopic; the sporophyte bears only unicellular sporangia in sori.

The sub-class **Cyclosporeae** includes a single order Fucales.

Order-Fucales. Life cycle is diplontic; growth by single apical cell, thallus parenchymatous; unilocular sporangia borne in conceptacles; the egg is liberated in the water prior to fusion.

Economic importance. The people of orient use seaweeds as food than by people anywhere else. The Japanese people begin to collect kelp in July and continue through October. They use more than twenty species of brown algae as food. Many other brown algae are used as fertilizers. The acetic acid is extracted from the seaweed by means of fermentation. With one seaweed the Japanese make **kombu**, which is first dried and then cooked with meat. Sometimes it is powdered and added into soups and sauces. It may also be used as a vegetable, steeped as a beverage, or made into a confection by coating with sugar. In many places *Nereocystis* is used in the preparation of medicines, poultry feed and for the extraction of potash salts. About 30 percent of its dry weight is potassium chloride. The stipe of *Nereocystis* has been used in the preparation of **seatron**.

Macrocystis and *Laminaria* are the chief sources for the extraction of algin. About 44 pounds of algin is extracted from a ton of kelp. Algin is used in the preparation of paints and varnishes. It is also used in the preparation of ice cream. *Egregia*, a kelp, is used as fertilizer. It contains greater amount of potassium salt and about the same amount of nitrogen. Seaweeds also contain iodine and other salts.

Phylogenetic Relationships in Phaeophyceae

In a number of ways, the Phaeophyceae and Chlorophyceae have been evolved on parallel lines. However, the brown algae possess a more highly differentiated thallus and mostly multicellular reproductive organs. It is also thought that the Phaeophyceae have been evolved like Chlorophyceae from a flagellate ancestor. Although it is quite certain that the divergence from the green line was quite early and gave rise to laterally flagellated progenitors much like the simple forms of Chrysophyceae and Xanthophyceae of today. At present there are no known unicellular Phaeophyceae, but the reproductive cells of today's brown algae indicate that the original brown algae were unicells of this type. The pigments and reserve food substances of Phaeophyceae also resemble with those of Chrysophyceae and Xanthophyceae. Such characteristics suggest the common ancestry of these groups. The thalli of brown algae are highly specialized but the idea of the evolution of higher plants from Phaeophyceae has totally been given up.

Two divergent lines have been established very early in the evolution of Phaeophyceae. One of these lines has given rise to the groups having isomorphic alternation of generations, and the other to the groups having heteromorphic alternation of generations. In both series there was a progressive evolution towards the complexity of the thallus and from isogamous type of reproduction. It is thought that the Fucales are at the top of the heteromorphic series.

Order ECTOCARPALES

Characteristic Features

1. The plant body possesses a branched heterotrichous filamentous thallus in which the cell division is not localized.

2. The growth mostly intercalary.

3. The reproductive elements may be terminal or intercalary borne singly or in chains.

4. The plant itself is sporophyte bearing unilocular and neutral sporangia which produce zoospores or neutral spores.

5. The gametophyte produces iso or anisogamous gametes. There are about 50 genera in this order.

Genus **ECTOCARPUS**

Distribution. *Ectocarpus* is world-wide in distribution and possesses many species. This is most primitive of all the brown algae. The species are commonly found in the colder seas of the temperate and polar regions. They are very common along the Atlantic coast. The plants occur attached to the rocks in the littoral and sublittoral zones. Many species are epiphytic upon the members of Fucales and Laminariales. *E. tomentosus* is an epiphyte upon the species of *Fucus*. *E. fasciculatus* is found growing on the fins of certain fish in Swedish waters.

Structure. The plant body is filamentous and heterotrichous. It is differentiated into prostrate and erect thalli. The prostrate portion is found attached to the substratum and remains creeping on it. The branches of the erect system of the plant area given out from the prostrate system. Comparatively, the prostrate portion of the plant is much branched than that of the erect portion. The ultimate branches of the erect portion are generally thin and pointed. The cells of the branches of the erect portion are uniseriately arranged end to end. The branching is always lateral and the branches arise just beneath the septa. In certain species the older branches are ensheathed or corticalted by a layer of the branches of the rhizoids.

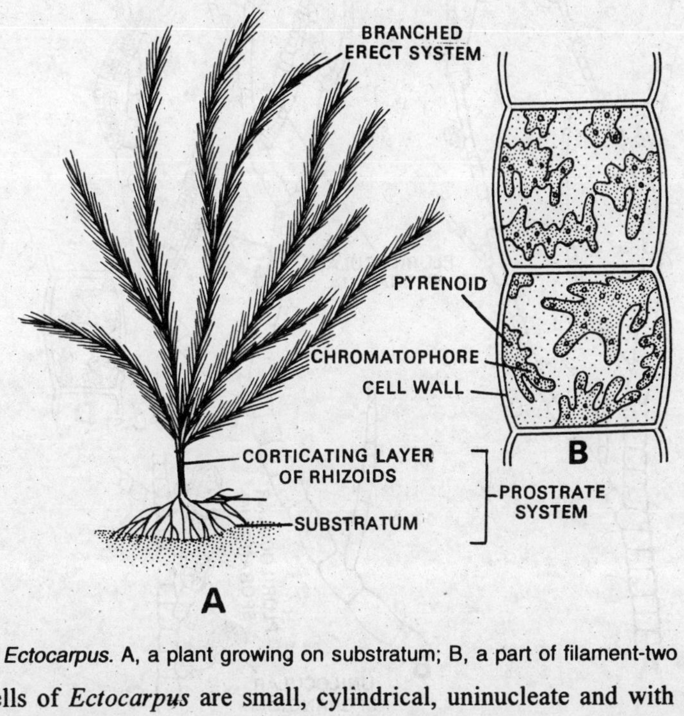

Fig. 6.2. *Ectocarpus.* A, a plant growing on substratum; B, a part of filament-two cells, (magnified)

The cells of *Ectocarpus* are small, cylindrical, uninucleate and with a few band-shaped chromatophores of irregular shape. Sometimes the cells contain many discoid chromatophores. The cell wall is thick and consists of three pectic cellulose layers. The cell wall is mucilaginous. The protoplast is differentiated into a single nucleus and cytoplasm. Naked pyrenoid-like structures are present in the golden-brown chromatophores. The plastids contain chlorohyll *a*, chlorophyll *b*, xanthophylls and a brown pigment **fucoxanthin**.

Growth. It is apical in prostrate system and trichothallic, *i.e.*, by means of intercalary meristem located at the base of multicellular hair. The meristematic cells cut off new cells towards the upper side transversely.

Reproduction. The reproduction takes place by means of asexual and sexual methods.

Asexual reproduction. The asexual reproduction takes place by means of biflagellate zoospores known as neutral spores produced within neutral and unilocular sporangia found upon asexual plants (sporophyte) also known as diploid thalli (2*n*).

Unilocular sporangia. The terminal cell of a short branchlet enlarges in size and develops into an unilocular sporangium. The number of chromatophores increases within the young sporangium, and the young sporangium becomes several fold in size. The single nucleus found within the sporangium divides and redivides and ultimately 32 or 64 nuclei are formed. The first division is reductional one. After the formation of the nuclei, there is cleavage, which results into uninucleate protoplasts. Each such protoplast possesses a chromatophore. Very soon each uninucleate protoplast metamorphoses into a biflagellate zoospore. The zoospores are pyriform in shape. The two unequal flagella are laterally inserted on each zoospore. The longer flagellum is directed forward and the shorter backward. The zoospores are extruded *en masse* through a small opening

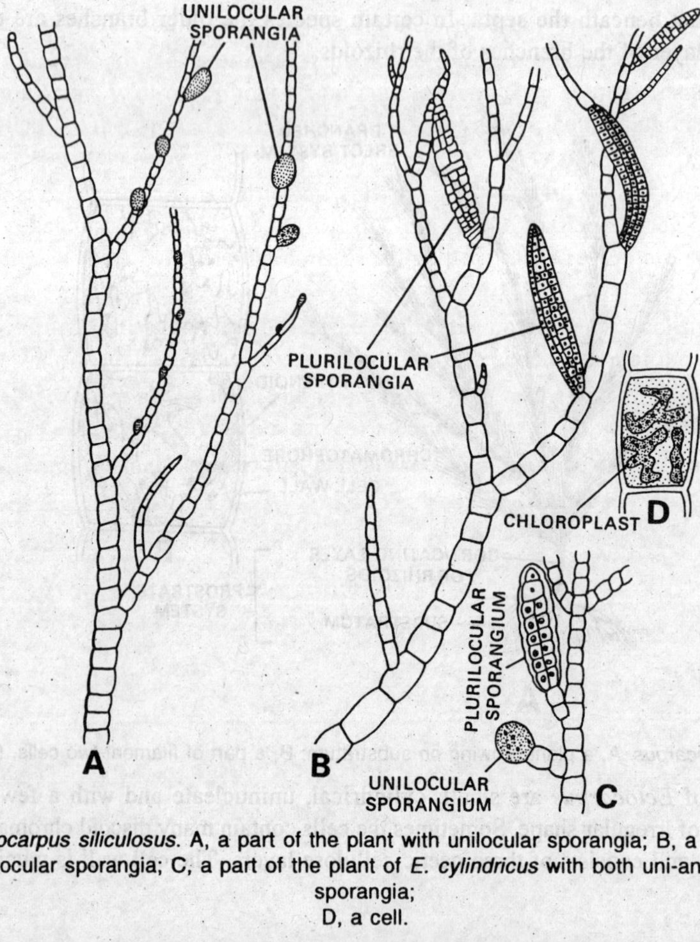

Fig. 6.3. *Ectocarpus siliculosus.* A, a part of the plant with unilocular sporangia; B, a part of the plant with plurilocular sporangia; C, a part of the plant of *E. cylindricus* with both uni-and plurilocular sporangia;
D, a cell.

found at the distal end of the sporangium. The zoospores of the mass remain inactive about for a minute and thereafter, they become motile swim here and there in the water. Ultimately the zoospores come to the rest on substratum and lose their flagella. The protoplasts become round and give rise to new haploid plants.

Plurilocular sporangia. The plurilocular sporangium develops from the terminal cell of a lateral branchlet. This cell divides several times repeatedly producing a vertical row of 6 to 12 cells. Several other vertical and transverse divisions take place and ultimately several hundred small cubical cells are formed which are arranged in 20 to 40 transverse tiers. There is no reduction division. The protoplast of each diploid cubical cell metamorphoses into a biflagellate neutral zoospore. The zoospores liberate outside to the sporangium by means of apical or lateral pore. The zoospores resemble in shape to the zoospores produced by unilocular sporangia.

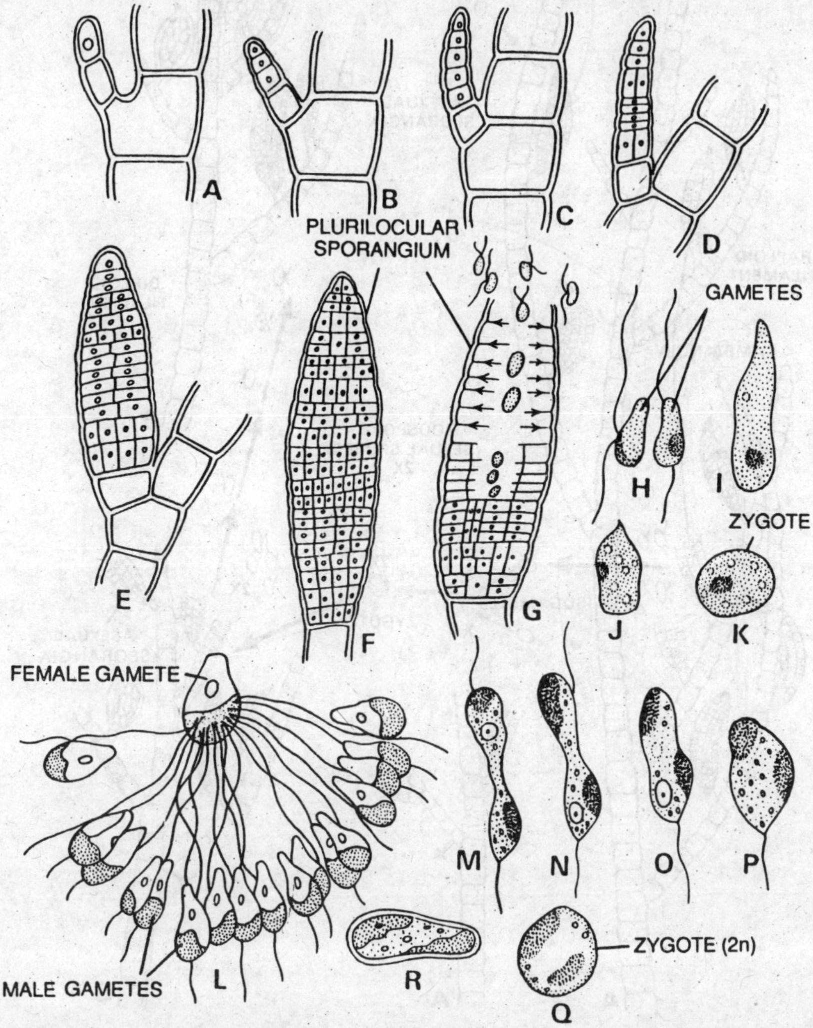

Fig. 6.4. *Ectocarpus* sp. A.E., stages in the development of plurilocular sporangium; F, plurilocular sporangium; G, liberation of neutral spores; H-K, gametes and their union resulting in the formation of zygote; L, clumping of gametes; M-Q, successive stages in the union of gametes and formation of zygote in *E. siliculosus*.

The zoospores swim for a short while and then settle down on some substratum, retract their flagella and directly give rise to new sporophytic (2*n*) plants. These plants once again bear unilocular and plurilocular sporangia. These sporophytic plants never produce plurilocular gametangia. The plurilocular gametangia are produced upon the plants which are developed from the zoospores of unilocular sporangia.

Sexual reproduction. The sexual reproduction ranges from isogamy to anisogamy. Majority of the species are monoecious. The most thoroughly worked out species *E. siliculosus* is dioecious.

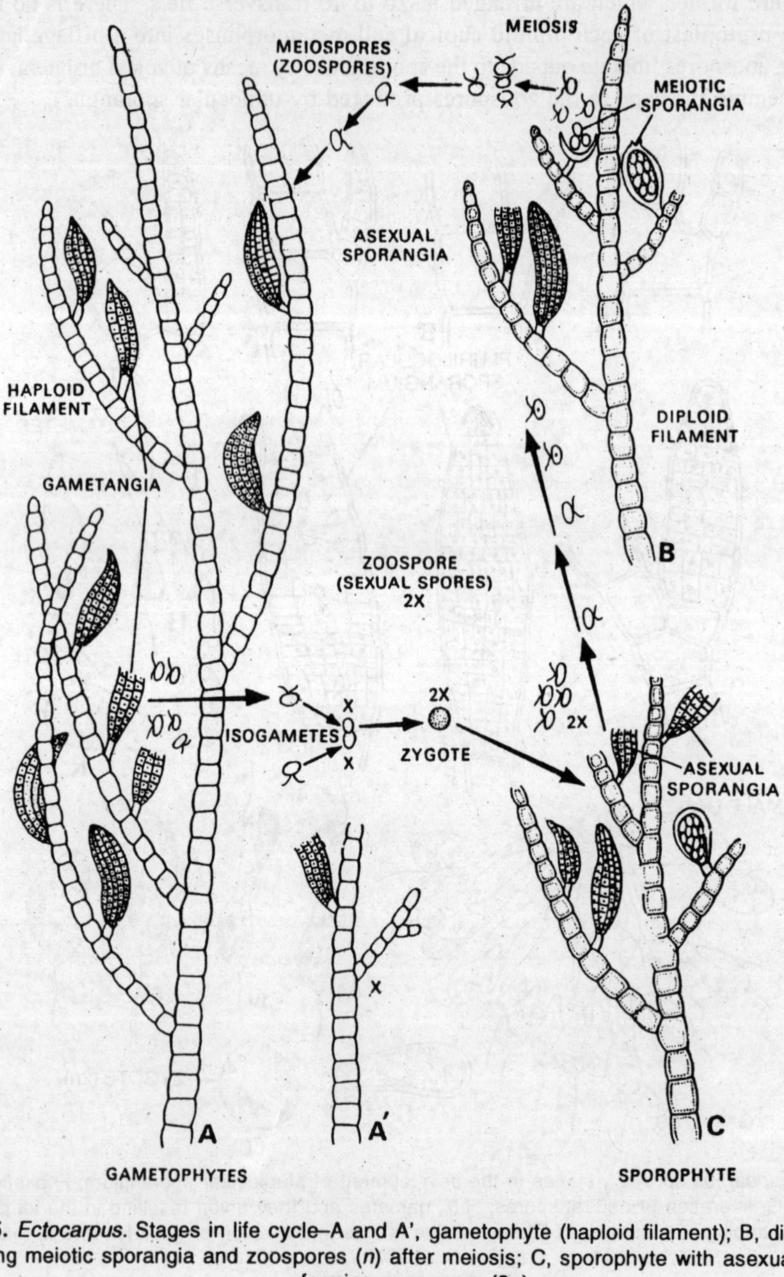

Fig. 6.5. *Ectocarpus.* Stages in life cycle–A and A', gametophyte (haploid filament); B, diploid filament showing meiotic sporangia and zoospores (*n*) after meiosis; C, sporophyte with asexual sporangia forming zoospores (2*n*).

The plurilocular gametangia develop from the terminal cell of a lateral branchlet. It divides several times repeatedly producing a vertical row of 6 to 12 cells. This is followed by several other vertical and transverse divisions and ultimately hundreds of small cubical cells are produced arranged in 20 to 40 transverse tiers. As already stated, the gametes are produced on haploid plants which develop from the zoospores produced in the unilocular sporangia. The gametophytic plants

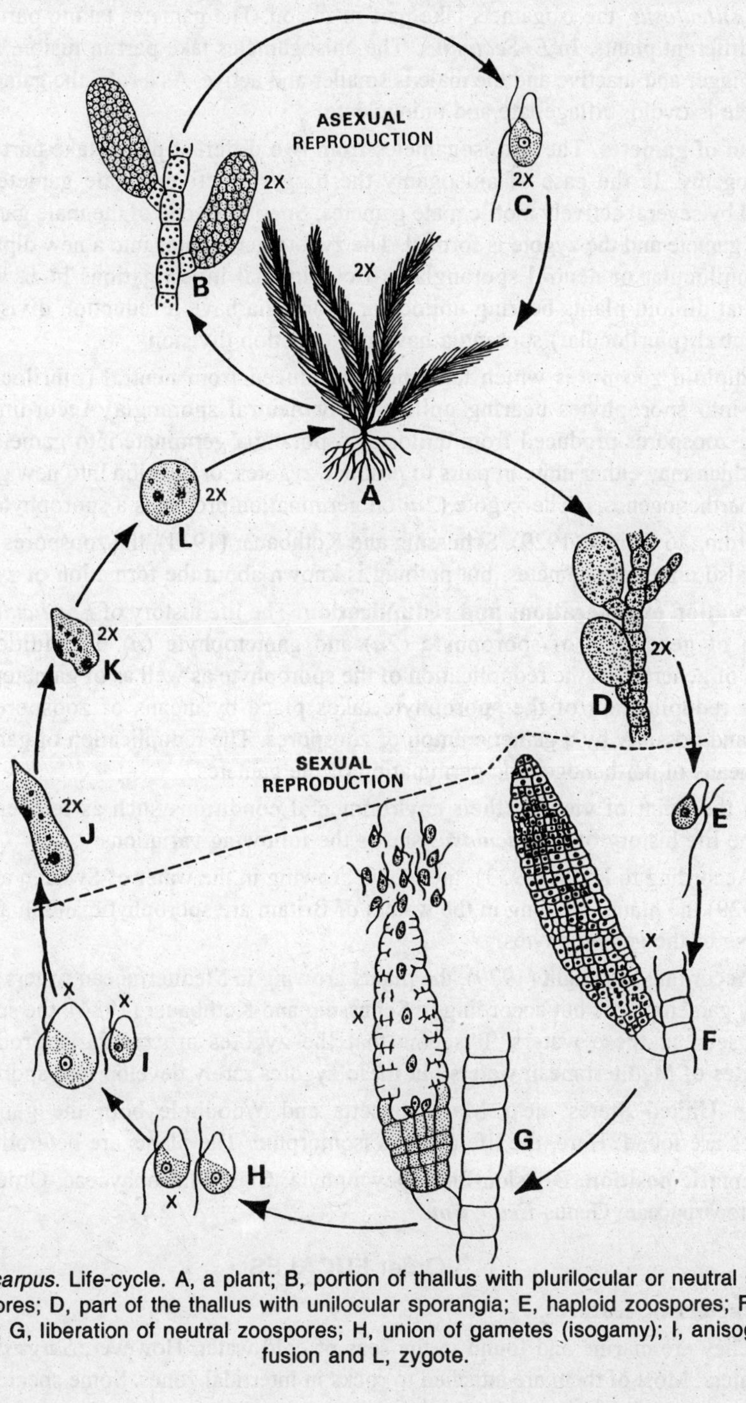

Fig. 6.6. *Ectocarpus*. Life-cycle. A, a plant; B, portion of thallus with plurilocular or neutral sporangia; C, neutral zoospores; D, part of the thallus with unilocular sporangia; E, haploid zoospores; F, plurilocular sporangium; G, liberation of neutral zoospores; H, union of gametes (isogamy); I, anisogamy; J-K, fusion and L, zygote.

resemble the sporophytic plants in shape but quite different in behaviour. These gametophytes bear plurilocular gametangia. The plurilocular gametangia are quite similar to the plurilocular or neutral sporangia in their shape.

The protoplasts of the cubical cells of the plurilocular gametangia metamorphose into biflagellate gametes which liberate through an apical pore on the gametangium gradually.

In *E. siliculosus*, the isogamets take part in fusion. The gametes taking part in fusion come from two different plants. In *E. Secundus,* The anisogametes take part in fusion. Here the female gamete is bigger and inactive and the male is smaller and active. As a rule, the gametes are haploid. Each gamete is ovoid, biflagellate and uninucleate.

Fusion of gametes. The two isogametes from two different plants take part in fusion in the case of isogamy. In the case of anisogamy the bigger inactive female gamete rests and gets surrounded by several actively motile male gametes. Soon after one of the male gametes fuses with the female gamete and the zygote is formed. The zygote germinates into a new diploid plant which produces unilocular or neutral sporangia. All cytological investigations in *E. siliculosus* have revealed that diploid plants bearing unilocular sporangia have a reduction division while those bearing neutral (plurilocular) sporangia have no reduction division.

The diploid zoospores which have been produced from neutral (plurilocular) sporangia germinate into sporophytes bearing unilocular or neutral sporangia. According to Papenfuss (1935), the zoospores produced from unilocular sporangia germinate into gametophytes bearing gametes which may either unite in pairs to produce zygotes, or develop into new gametophytes by means of parthenogenesis. The zygote ($2n$) on germination produces a sporophyte ($2n$).

According to Knight (1929), Schussnig and Kothbauer (1934), the zoospores from unilocular sporangia also unite like gametes, but nothing is known about the formation of zygotes.

Alternation of generations and reduplication. The life history of *E. siliculosus* exhibits an alternation of generatins of sporophyte ($2n$) and gametophyte (n). In addition to a regular alternation of generations the reduplication of the sporophyte as well as of gametophyte also takes place. The reduplication of the sporophyte takes place by means of zoospores from neutral sporangia and possibly by a gametic union of zoospores. The reduplication of gametophyte takes place by means of parthenogenetic germination of the gametes.

From the point of view of their environmental conditions such as temperature and light intensity the life history of *E. siliculosus* shows the following variations:

1. According to Kylin (1933), the plants growing in the water of Sweden and according to Knight (1929) the plants growing in the waters of Britain are sporophytic and in any case they do not give rise to the gametophytes.

2. According to Knight (1929), the plants growing in Mediterranean waters near Naples are exclusively gametophytic but according to Schussnig and Kothbauer (1934), the sporophytes have also been seen in these waters. It seems that the zygotes are regularly produced from the gametophytes of Mediterranean waters but these zygotes rarely develop into sporophytes.

3. In United States, near Massachusetts and Woodhole both the gametophytes and sporophytes are found. Here, the life-cycle is isomorphic. The plants are heterothallic.

Systematic position. Division-Phaeophycophyta; Class-Phaeophyceae; Order-Ectocarpales; Family-Ectocarpaceae; Genus-*Ectocarpus*.

Order **FUCALES**

Characteristic features.

1. They are marine and found in the seas of cold water. However, *Sargassum* is found in tropical waters. Most of them are attached to rocks in intertidal zones. Some species of *Sargassum* are floating.

2. The spores behave as gametes. The motile spores develop in microsporangia (antheridia) and called the antherozoids, and the non-motile aplanospores develop in macrosporangia. (oogonia), and called the eggs.

3. The anterior flagellum of the swarmer of Fucales is longer than the posterior one whereas in other orders their arrangement is reverse.

4. The plant itself is a sporophyte ($2n$). The reduction division takes place just prior to formation of gametes.

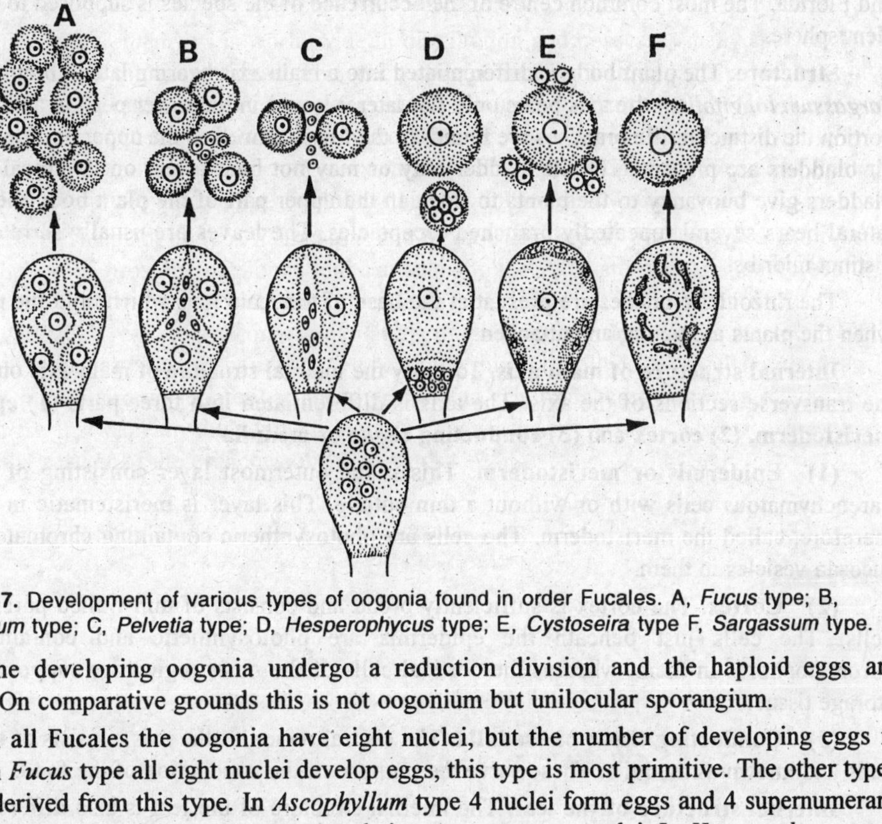

Fig. 6.7. Development of various types of oogonia found in order Fucales. A, *Fucus* type; B, *Ascophyllum* type; C, *Pelvetia* type; D, *Hesperophycus* type; E, *Cystoseira* type F, *Sargassum* type.

5. The developing oogonia undergo a reduction division and the haploid eggs are developed. On comparative grounds this is not oogonium but unilocular sporangium.

6. In all Fucales the oogonia have eight nuclei, but the number of developing eggs is different. In *Fucus* type all eight nuclei develop eggs, this type is most primitive. The other types have been derived from this type. In *Ascophyllum* type 4 nuclei form eggs and 4 supernumerary nuclei. In *Pelvetia* type there are two eggs and six supernumerary nuclei. In *Hesperophycus* type there is one functional large egg and other eggs with 7 degenerating nuclei. In *Cystoseira* type there is one egg and 7 supernumerary nuclei. In most advanced *Sargassum* type there is one egg and 7 degenerating nuclei.

There are about 40 genera and 350 species in this order.

Classification. Seven families are included in this order.

1. Fucaceae, *e.g., Fucus, Pelvetia.*
2. Himanthaliaceae, *e.g., Himanthalia.*
3. Cystoseiraceae, *e.g., Cystoseira.*
4. Sargassaceae, *e.g., Sargassum, Turbinaria.*
5. Hormosiraceae, *e.g., Hormosira.*
6. Durvilleaceae, *e.g., Durvillea.*
7. Ascoeiraceae, *e.g., Ascoseira.*

Genus **SARGASSUM**

There are about 150 species in this genus.

Occurrence. All species are marine. Majority of them are free floating on maturity but some are lithophytes when young. The species of *Sargassum* are found in tropical or warm waters. *Sargassum natans* and *S. fluitans* are abundantly found in the Sargasso sea of Atlantic ocean. The name of sea is given after the name of genus. The species of *Sargassum* are commonly met with along the coast of Australia, India and Japan. Some species are found in the waters of West Indies and Florida. The most common centre of the occurrence of the species is supposed to be Southern Hemisphere.

Structure. The plant body is differentiated into a main axis bearing lateral branches on it. In *Sargassum longifolium* the axis bears only flat lateral leaves in the lower portion and in the upper portion the distinct fertile branches are found. In the lower as well as the upper portions the stalked air bladders are present. The air bladder may or may not bear a leaf on its apical end. These bladders give buoyancy to the plants to float. In the upper part of the plant body the axis of the lateral bears several repeatedly branched receptacles. The leaves are usually serrated and with distinct midribs.

The rhizoidal branches are present at the base of the main axis which hold the mud or rock when the plants are young and attached.

Internal structure of main axis. To study the internal structure of main axis one has to cut the transverse sections of the axis. The axis is differentiated into three parts **(1) epidermis** or **meristoderm, (2) cortex** and **(3) conducting tissue** or **medulla**.

(1) Epidermis or meristoderm. This is the outermost layer consisting of thin-walled parenchymatous cells with or without a thin cuticle. This layer is meristematic in nature and, therefore, called the meristoderm. The cells are photosynthetic containing chromatophores and fucosan vesicles in them.

(2) Cortex. The cortex is sufficiently broad and consists of thin-walled parenchymatous cells. The cells just beneath the epidermis are photosynthetic and contain numerous chromatophores in them, whereas the cortical cells of the inner region are supposed to be the storage tissue.

(3) Conducting tissue or medulla. The central region of the axis consists of thick-walled cells and known as medulla. These serve the purpose of the conduction of the food.

Internal structure of the leaf. The internal structure of the leaf is also differentiated into three regions **(1) epidermis, (2) cortex** and **(3) medulla.**

The epidermis is single layeed. Just beneath the epidermis the assimilatory cells are found with numerous chromatophores in them. The outline of the leaf is flat. It may or may not be covered by a thin cuticle. The cryptoblasts opened outside by cryptostomata are also found on the epidermis.

The cortex is parenchymatous and serves as storage tissue.

The central region of the leaf is medulla consisting of thick-walled cells. They act collectively as conducting tissue.

Hairs. The multicellular hairs are present in the flask-like bodies called the **'cryptoblasts'** which are opened outside by stomata-like openings called the **'cryptostomata'**. The multicellular hairs arise from the base and extrude out through the openings. The function and nature of these hairs is a subject of controversy. According to Prof. Lloyd Williams, they are respiratory in function. They also absorb nutritive materials from the water. They also act as protective organs. According to Church, they are mucilage secreting hairs.

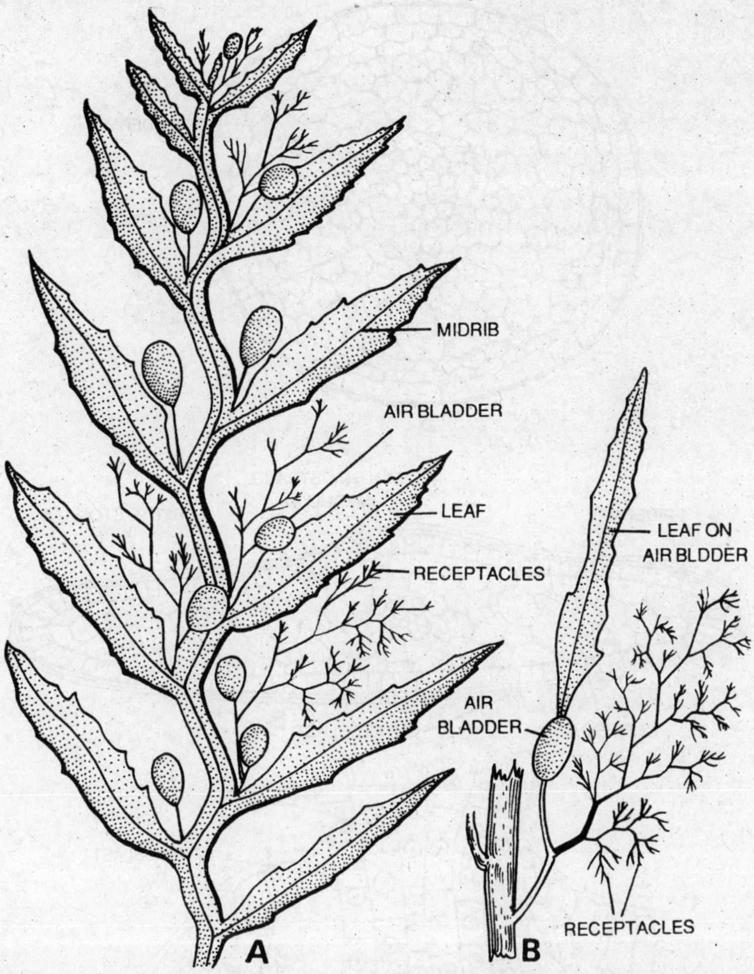

Fig. 6.8. *Sargassum* sp. A, thallus; B, a part of thallus with receptacle and a leaf.

Method of the growth of the thallus. The growth of the thallus is initiated by an apical cell found at the tip of the branch. The apical cell is four-sided, pyramid-like which cuts basal and lateral derivatives. The derivatives divide periclinally and anticlinally giving rise to the parenchymatous thallus.

Reproduction. The reproduction takes place by means of vegetative and sexual methods. Asexual method is altogether absent.

1. **Vegetative reproduction.** It takes place by means of fragmentation. In the case of *Sargassum natans* it is very prolific. This is a free floating species. Each fragment develops into a new plant.

2. **Sexual reproduction.** The sexual reproduction is **oogamous**. The male sex organs are called the **antheridia** and the female sex organs the **oogonia**. The species may be homo or heterothallic. In homothallic species the oogonia and antheridia develop in the same conceptacle whereas in heterothallic species in separate conceptacles situated on different plants. The conceptacles are found upon special repeatedly branched reproductive branches, the **receptacles**.

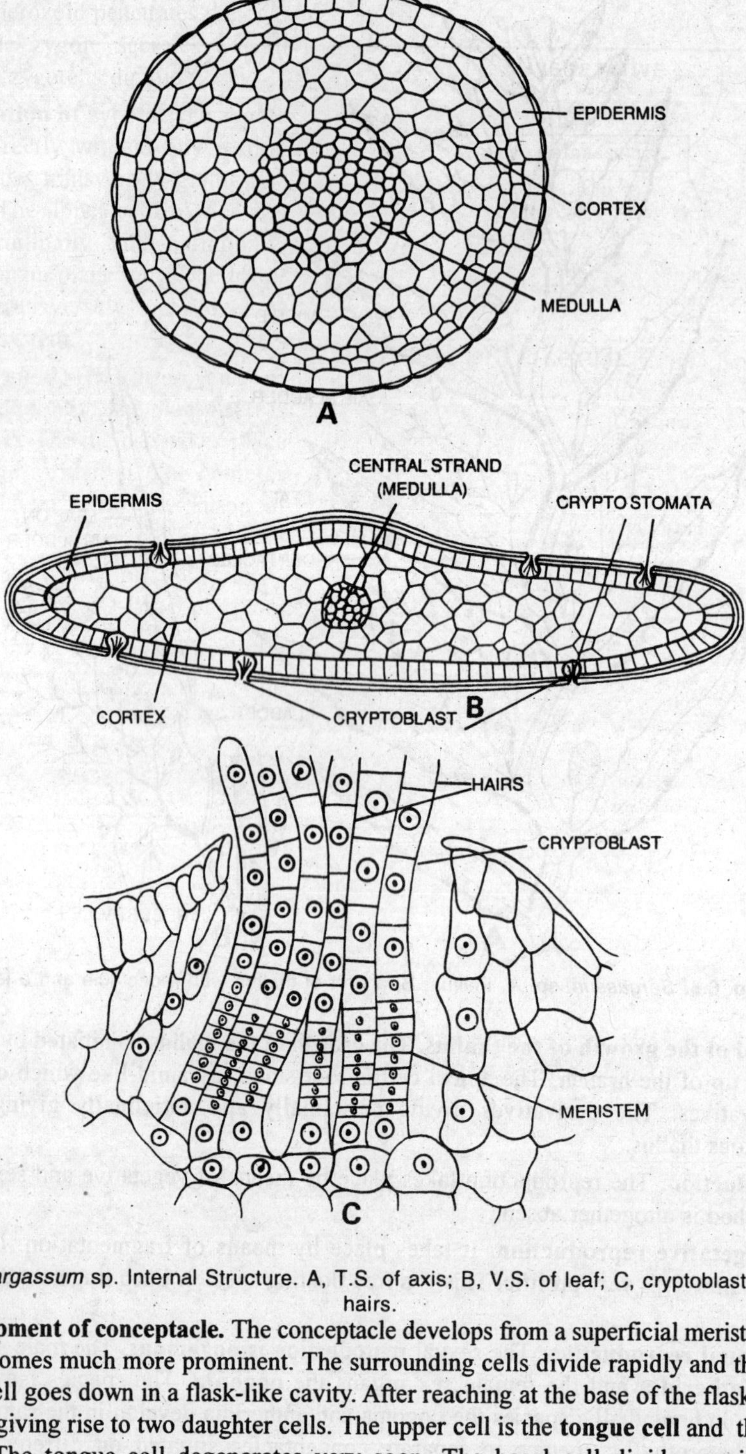

Fig. 6.9. *Sargassum* sp. Internal Structure. A, T.S. of axis; B, V.S. of leaf; C, cryptoblast with sterile hairs.

Development of conceptacle. The conceptacle develops from a superficial meristodermal cell. This cell becomes much more prominent. The surrounding cells divide rapidly and the prominent superficial cell goes down in a flask-like cavity. After reaching at the base of the flask this divides transversely giving rise to two daughter cells. The upper cell is the **tongue cell** and the lower one **basal cell**. The tongue cell degenerates very soon. The basal cell divides again and again anticlinally giving rise to a curved row of cells. These cells divide periclinally giving rise to two layers. These layers of the cells represent fertile sheet. The antheridia and oogonia develop on this fertile sheet from the cells of the upper tier.

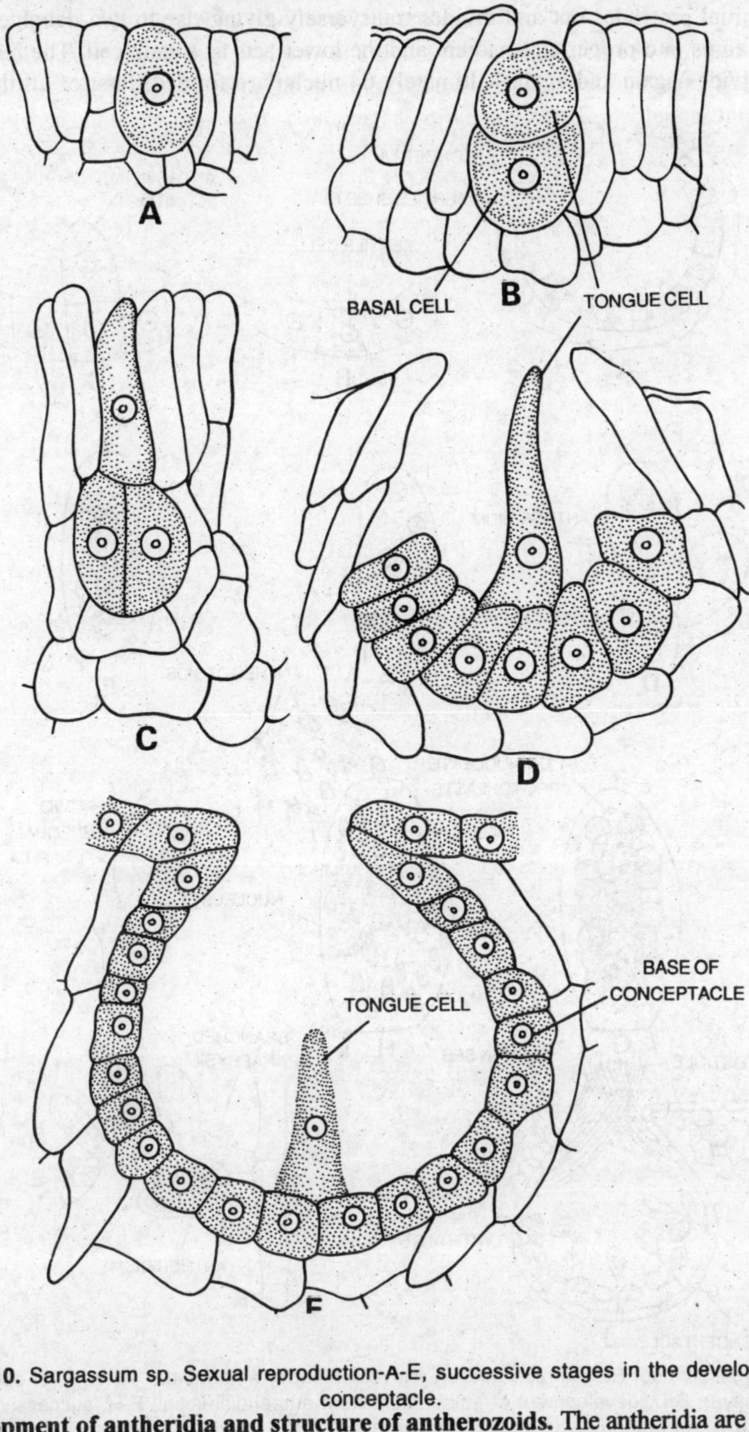

Fig. 6.10. Sargassum sp. Sexual reproduction-A-E, successive stages in the development of conceptacle.

Development of antheridia and structure of antherozoids. The antheridia are developed on the branched paraphyses situated in the conceptacle. The uniseriate unbranched paraphyses are also called the periphyses developing from the neck of the conceptacles and protruding out through the ostiole. They do not bear any sex organs.

The antheridia develop on the ultimate tips of the branches of branehed paraphyses. The antheridium initial protrudes out and divides transversely giving rise to two daughter cells. The upper cell develops into proper antheridium and the lower acts as a stalk cell. The nucleus of the antheridium divides again and again. Ultimately 64 nuclei are formed. First of all the reduction

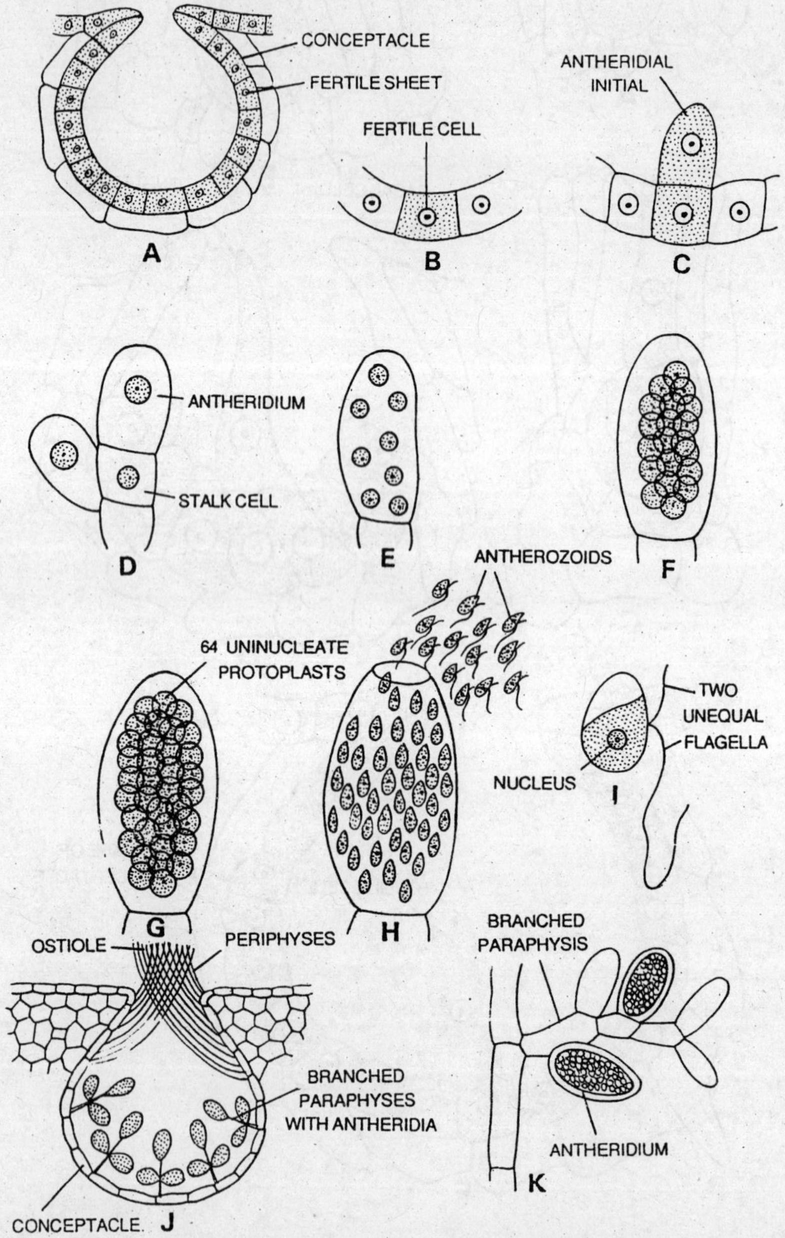

Fig. 6.11. *Sargassum* sp. Sexual reproduction. Development of antheridium. A, V. S. of conceptacle showing fertile layer; B-D, development of antheridium from antheridial initial; E-H, successive stages in the development of antherozoids; I, single antherozoid; J,V.S. of conceptacle showing branched paraphyses and antheridia; K, branched antheridium.

division takes place which is followed by several mitotic divisions. With the result of the cleavage of the contents of the antheridium 64 uninucleate protoplasts are developed. Each such uninucleate protoplast metamorphoses into an antherozoid. On maturity of the antheridium, the apical portion breaks and the antherozoids are liberated in the water.

Each antherozoid is ovoid or pyriform and bears two unequal flagella inserted on the lateral side. The anterior flagellum is shorter than the posterior one whereas in other Phaeophyceae the case is reverse. Each antherozoid contains a large nucleus, a blepharoplast and few chromatophores.

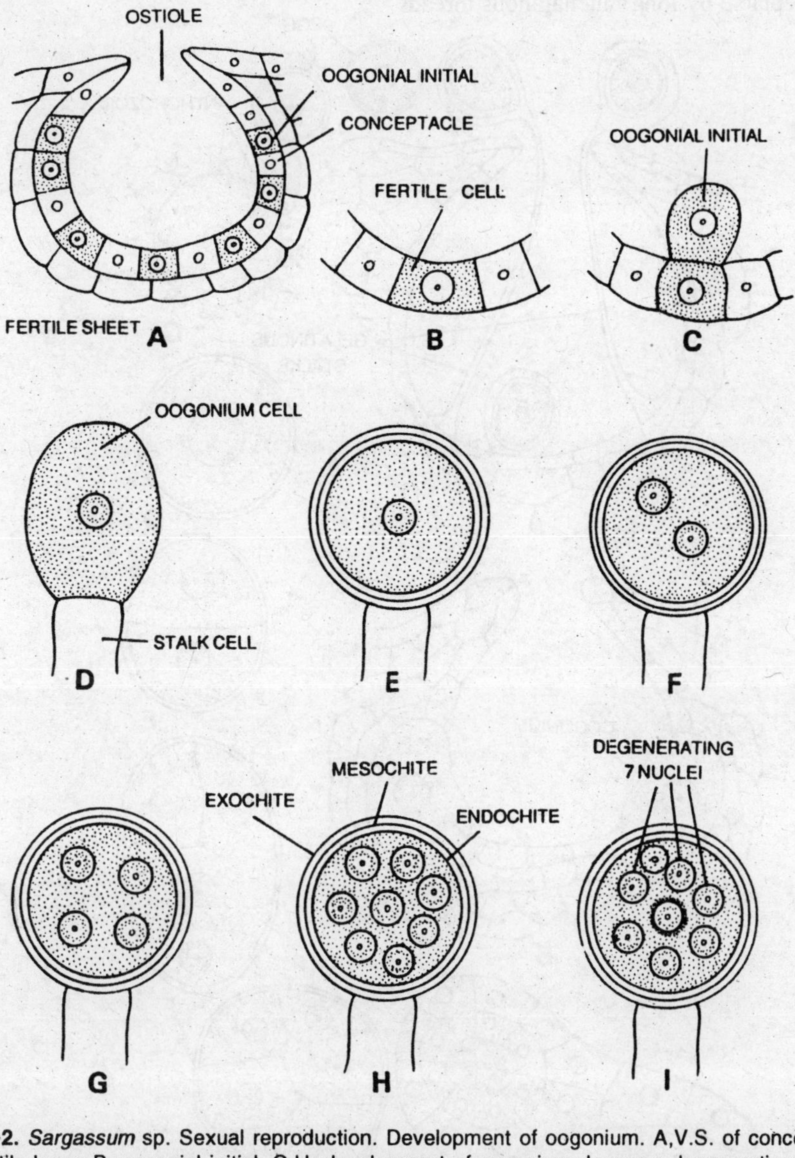

Fig. 6.12. *Sargassum* sp. Sexual reproduction. Development of oogonium. A,V.S. of conceptacle showing fertile layer; B, oogonial initial; C-H, development of oogonium; I, seven degenerating nuclei and one functional nucleus in the three layered oogonium.

Development and structure of oogonium. Some of the cells of the upper tier of the fertile sheet protrude out and act as oogonial initials. Each oogonial initial divides transversely and two daughter cells are formed. The upper daughter cell develops into oogonium proper and the lower acts as stalk cell.

The nucleus of the oogonium divides thrice, bringing the number of the nuclei to eight. The first division is reductional. One functional nucleus remains and the seven nuclei degenerate. This single egg develops in the oogonium.

The oogonium is three layered. The outer layer is **exochite,** the middle layer is **mesochite** and the inner layer is **endochite**. The mesochite is mucilaginous. On maturity the exochite breaks and the mucilaginous mesochite forms a long mucilaginous stalk. The oogonia are attached to the base of the conceptacle by long mucilaginous threads.

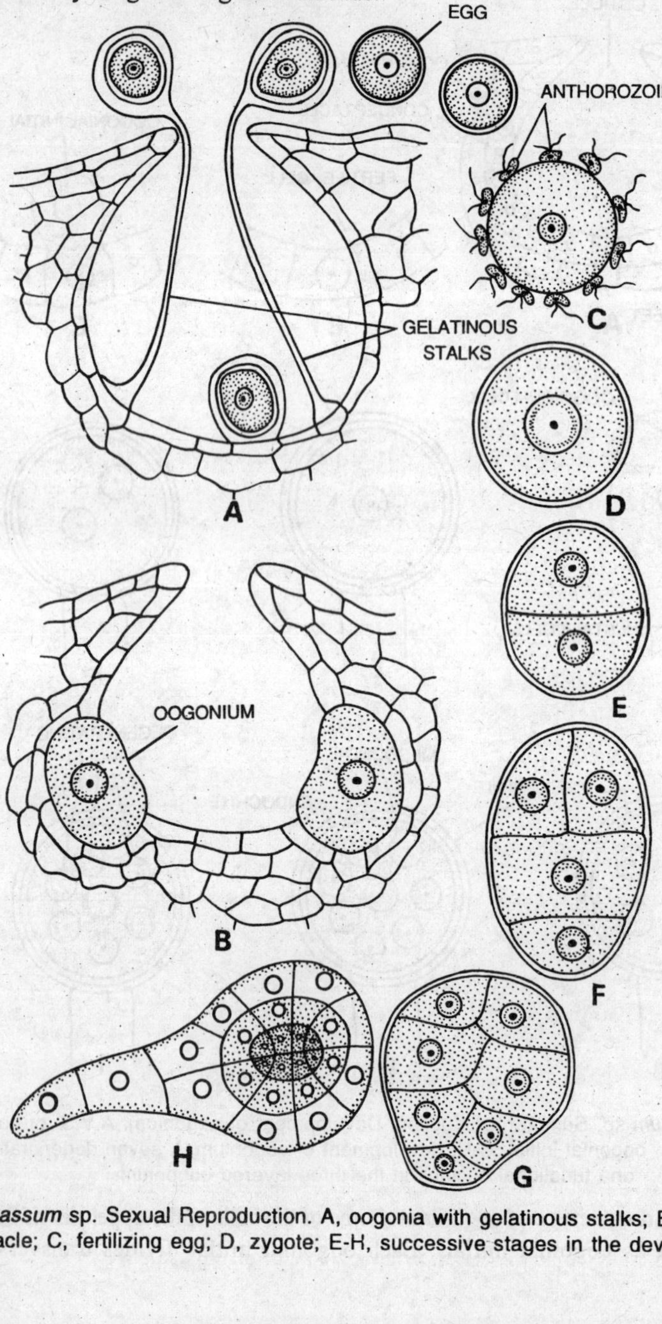

Fig. 6.13. *Sargassum* sp. Sexual Reproduction. A, oogonia with gelatinous stalks; B, two young oogonia in the conceptacle; C, fertilizing egg; D, zygote; E-H, successive stages in the development of embryo.

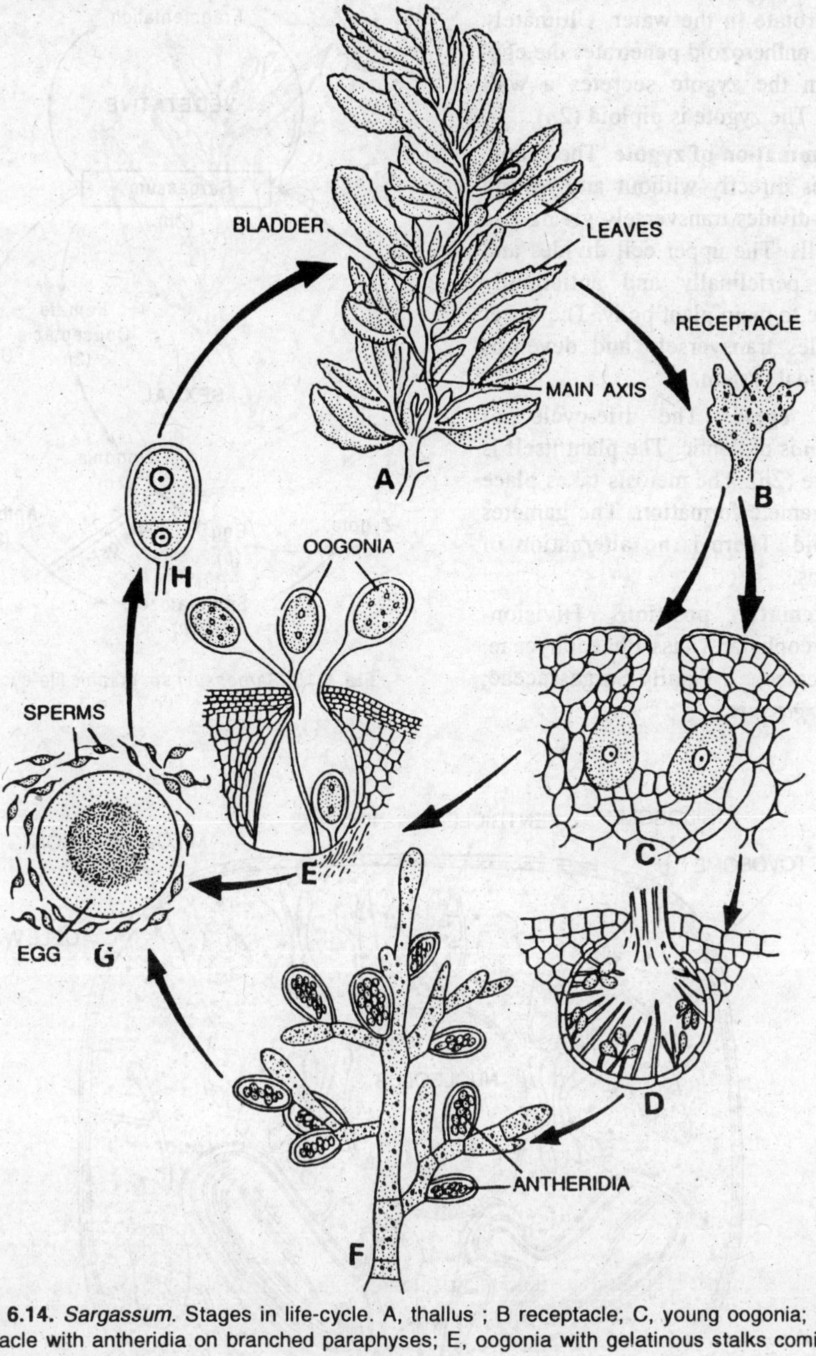

Fig 6.14. *Sargassum.* Stages in life-cycle. A, thallus ; B receptacle; C, young oogonia; D, conceptacle with antheridia on branched paraphyses; E, oogonia with gelatinous stalks coming out of the conceptacle through ostiole; F, antheridia on branched paraphyses; G, fertilizing egg; H, young embryo (reduction division in the life-cycle at gametogenesis).

Fertilization. Later on the endochite also ruptures and the egg is liberated in the water. Several antherozoids approach the egg. Sometimes, because of the presence of the many antherozoids, the egg appears to be a multiflagellate structure. With the help of these antherozoids the egg may rotate in the water. Ultimately one lucky antherozoid penetrates the egg. Very soon the zygote secretes a wall around it. The zygote is diploid (2*n*).

Germination of zygote. The zygote germinates directly without any resting period. It divides transversely giving rise to two cells. The upper cell divides and redivides periclinally and anticlinally giving rise to main plant body. The lower cell divides transversely and develops into rhizoidal system.

Life cycle. The life-cycle of *Sargassum* is diplontic. The plant itself is sporophyte (2*n*). The meiosis takes place prior to gamete formation. The gametes are haploid. There is no alternation of generations.

Systematic position. Division-Phaeophycophyta; Class-Phaeophyceae; Order-Fucales; Family-Sargassaceae; Genus-*Sargassum*.

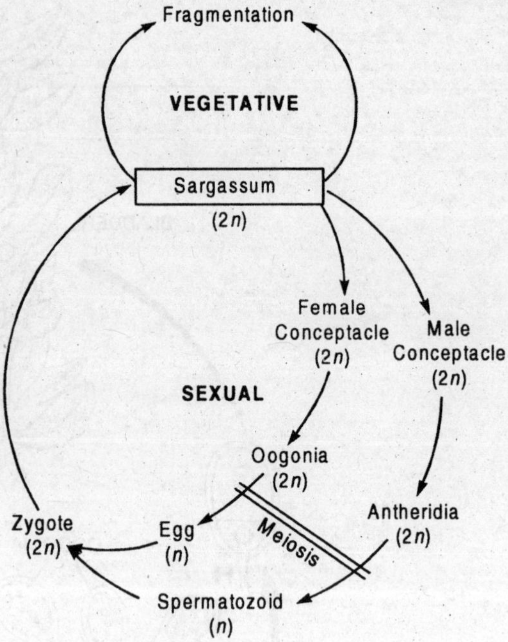

Fig. 6.15. *Sargassum* sp. graphic life-cycle.

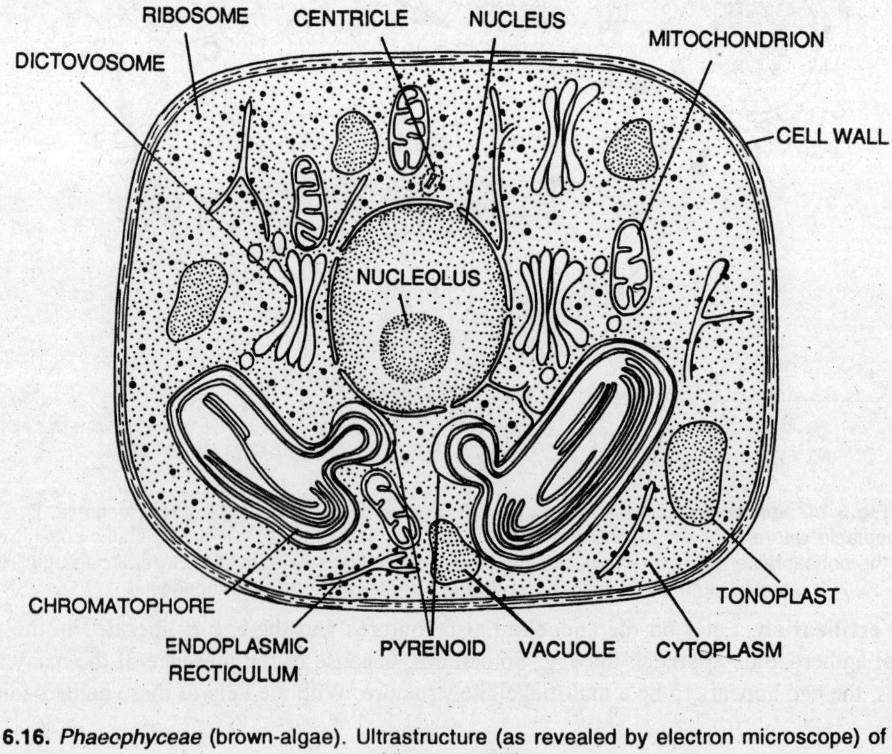

Fig. 6.16. *Phaecphyceae* (brown-algae). Ultrastructure (as revealed by electron microscope) of a cell (diagrammatic)

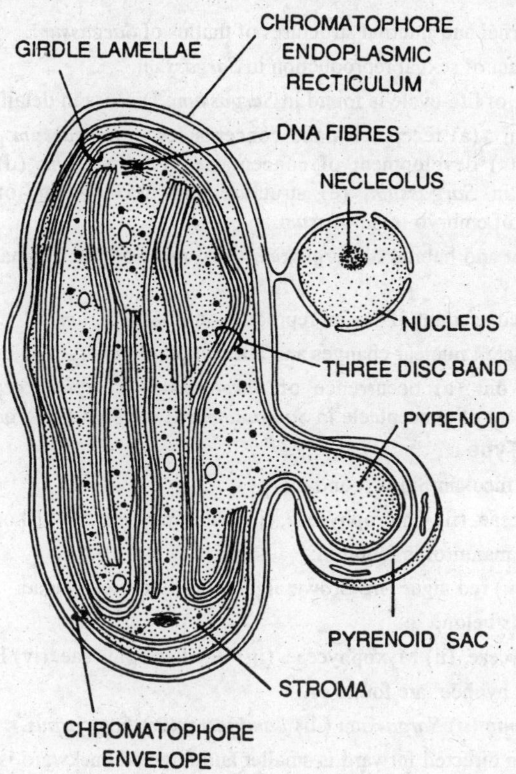

Fig. 6.17. *Phaeophyceae* (brown algae). Ultra structure of a chromatophore of a cell of brown alga.

QUESTION BANK

N.B. *Undermentioned questions are asked repeatedly in various Indian Universities and competitive examinations.*

Class. Phaeophyceae

1. Give the characteristic features of Phaeohyceae. Why they are called brown algae?
2. Where do the brown algae occur? Describe the range of somatic structure of Phaeophyceae.
3. Write a detailed note on pigmentation, food products and flagellation of Phaeophyceae.
4. Mention the modes of reproduction found in Phaeophyceae.
5. Describe in brief the classification of Phaeophyceae.
6. Write notes on: (a) economic importance of Phaeophyceae: (b) phylogenetic relationship in Phaeophyceae.

Order. Ectocarpales

1. Give the characteristic features of the order Ectocarpales.
2. Give the life-history of *Ectocarpus*. Also mention the phenomenon of alternation of generations and reduplication in the life-history of *Ectocarpus*.
3. Give a comprehensive account of the modes of reproduction in *Ectocarpus*.
4. Write notes on: (a) distribution and occurrence of *Ectocarpus*; (b) structure of plant body of *Ectocarpus;* (c) unilocular and plurilocular sporangia; (d) clump formation.

Order. Fucales

1. Give the characteristic features of the order Fucales.
2. With the help of diagrams only, show the stages in the development of various types of oogonia found in the order Fucales.

3. Give the external and internal structures of thallus of *Sargassum.*
4. Give an account of sexual reproduction in *Sargassum.*
5. What type of of life-cycle is found in *Sargassum.* Discuss in detail.
6. Write note on : (a) receptacles and conceptacles in *Sargassum*; (b) apical growth of thallus in *Sargassum;* (c) development of conceptacle in *Sargassum*; (d) structure and development of antheridium in *Sargassum*; (e) structure and development of oogonium in *Sargassum*; (f) development of embryo in *Sargassum.*
7. Give the habit and habitat of *Sargassum.* Also mention the internal structure of the axis and leaf of *Sargassum.*
8. Give a detailed account of sexual reproduction in *Sargassum.*
9. Describe in detail nuclear changes and alternation of generations in *Sargassum.*
10. Write notes on: (a) occurrence of *Sargassum*; (b) apical growth of thallus in *Sargassum*; (c) development of conceptacle in *Sargassum*; (d) oogonia of *Sargassum.*

Multiple Choice Type

1. The pigment fucoxanthin is found in:
   (i) Myxophyceae, (ii) Chlorophyceae, (iii) Phaeophyceae, (iv) Rhodophyceae.
2. Storage food mannitol is found in:
   (i) diatoms, (ii) red algae (iii) brown algae; (iv) blue-green algae.
3. The sea weeds belong to:
   (i) Chlorophyceae, (ii) Myxophyceae, (iii) Bacillariophyceae. (iv) Phaeophyceae.
4. The 'trumpet hyphee' are found in:
   (i) *Polysiphonia* (ii) *Sargassum* (iii) *Laminaria*, (iv) *Ectocarpus.*
5. The flagellum directed forward is smaller and directed backward is larger in size, in order:
   (i) Fucales, (ii) Ectocarpales, (iii) Laminariales (iv) Cutleriales.
6. The iodine is obtained from:
   (i) *Laminaria*, (ii) *Polysiphonia*, (iii) *Sargassum.* (iv) *Ectocarpus.*
7. An alternation of morphologically similar haploid and diploid generations is found in:
   (i) *Ulothrix*, (ii) *Chlamydomonas*, (iii) *Spirogyra*, (iv) *Ectocarpus*
8. Erect, flattened, dichotomously branched thalli are found in:
   (i) *Ectocarpus*, (ii) *Polysiphonia*, (iii) *Dictyota*, (iv) *Nemalion.*
9. Unilocular and plurilocular sporangia are found in:
   (i) *Polysiphonia*, (ii) *Batrachospermum*, (iii) *Ectocarpus.*
10. Sort out the branched alga:
    (i) *Ectocarpus*, (ii) *Oedogonium* (iii) *Spirogyra*, (iv) *Zygnema.*
11. Clumping of gametes is found in:
    (i) *Polysiphonia*, (ii) *Ectocarpus*, (iii) *Oedogonium*, (iv) *Volvox*
12. Motile zoospores are found in:
    (i) *Polysiphonia* (ii) *Ectocarpus*, (iii) *Spirogyra*, (iv) *Nemalion.*
13. The receptacles are found in:
    (i) *Ectocarpus*, (ii) *Sargassum*, (iii) *Polysiphonia*, (iv) *Nemalion.*
14. The life-cycle of *Sargassum* is :
    (i) haplontic, (ii) diplontic (iii) haplo-diplobiontic, (iv) none of above.

ANSWERS

1 (iii), 2 (iii), 3 (iv), 4 (iii), 5 (i), 6 (i), 7 (iv), 8 (iii), 9 (iii), 10 (i), 11 (ii), 12 (ii), 13 (ii), 14 (ii).

7

CHAPTER

Rhodophycophyta–Rhodophyceae

(The Red Algae)

Characteristic Features:

Pigmentation. Chlorophylls *a* and *b*; α and β-carotenes, Lutein, Taraxanthin, Allophycocyanin, Phycocyanin, Phycoerythrin.

Storage products. Floridean starch, oils.

Flagellation. Absent.

There are about 400 genera and 2500 species in this class.

Occurrence. Majority of Rhodophyceae are marine and mostly found in sub-littoral zones. When water is clear they can penetrate several metres of depth. Certain species of *Gigartina, Laurencia, Lomentaria* form extensive belts in littoral zones. *Catenella* and *Bostrychia* are found in saline waters. Several Rhodophyceae are found on the roots of mangrove vegetation. The red algae are commonly found in the warmer seas of Australia and Asia. Some species of red algae are found in the warmer seas in the depth of 30 to 92 metres. Some red algae have been reported even from the depth of 200 metres. Majority of red algae are lithophytes and some are epiphytes, *e.g., Polysiphonia violacea* is epiphytic on *Fucus vesiculosus*. True parasitic forms have also been reported from red algae. *Harveyella pachyderma* is parasitic upon *Gracilaria confervoides. Polysiphonia fastigiata* is parastic upon *Ascophyllum nodosum.* The photosynthetic tissue is greatly reduced in such forms and sometimes they are quite colourless. About 50 species of 19 or more genera are fresh water in habit.The fresh water species are found in well aerated waters such as water falls. *Batrachospermum* is found in the water falls of Dehradun. *Compsopogon* occurs in the plains of Uttar pradesh even in the rivers. The other well-known fresh water forms are *Hildenbrandia* and *Lemanea.*

Range of vegetative structure. There is great diversity in the vegetative structure of red algae. They may be of several types.

1. **Unicellular forms.** The species of *Porphyridium* and *Chroothece* are unicellular in structure, but the cells form globose mucilaginous colonies. True unicellular forms are unknown. This is the case of reduction.

2. **Heterotrichous filaments.** Such filaments are commonly found in the species of *Goniotrichum* and *Asterocystis*. In *Asterocystis* the thallus is dichotomously branched.

3. **True parenchymatous forms.** The species of *Porphyra* are *Ulva* like. The Thallus may be one-celled thick (monostromatic) or two-celled thick (distromatic).

4. **Branched filamentous thallus.** All florideae have branched filamentous thalli. In some genera the various branches are free from one another, *e.g., Chantransia,* in other genera they lie more or less intermingled with one another within a common gelatinohs matrix, *e.g., Batrachospermum;* in still other genera they are so closely applied to one another that the thallus

seems to be parenchtymatous. In the case of *Nemalion, Liagora,* etc., the filaments of unlimited growth are formed from protuberances of peripheral cells. The thalli of *Nemation, Cumagloia* are pseudoparenchymatous. These constructions are called multiaxial type or fountain type of construction. In *Nemalion* the apical cell forms a hair.

All Rhodymeniales are multiaxial. All Ceramiales are uniaxial. In other orders both types are represented in different families.

Methods of growth of thallus. The growth may be diffused or apical. In the order Bangiales, *Porphyra* and *Asterocystis* any cell of the thallus may divide in any direction. This is called diffused growth. The apical growth is strictly restricted to sub-class Florideae.

Structure of the cell. The cell wall is two layered. The outer layer consists of pectic substance and layer next to protoplast of cellulose. Sometimes the cell wall contains iodine in it. In majotity of forms a thick cuticle like structure of pectic substance is found. In Ceramiaceae the wall becomes stratified.

Pit connections. The members of order Bangiales lack connections. At the point of contact of two cells there is a pit membrane surrounded by a disc of pectic material. Schimitz reported that these are of cytoplasm.

The protoplasm. The protoplasm is highly viscous. In order Bangiales there is no central vacuole, but in all Florideae the cytoplasm possesses a conspicuous central vacuole. The cell sap is either alkaline or neutral and rarely acidic.

The nucleus. In lower Rhodophyceae the cells are uninucleate, *e.g.*, Cryptonemiales, Bangiales, etc. In other cases the cells are multinucleate, *e.g.*, Ceramiales, Rhodymeniales, etc. Some of the larger multinucleate cells possess 3000 to 4000 nuclei in each of them. The nuclei are with one or more prominent nucleoli. There is well developed chromatin network.

Chromatophores and pigments. In order Bangiales (lower Rhodophyceae) there is a single stellate chromatophore in each cell. The back cells of *Nemalion* have ridged chromatophores. In majority of Rhodophyceae the number of chromatophores per cell is more than one. They may be band-like, *e.g.*, in family Ceramiaceae, irregularly lobed or discoidal, *e.g.*, in *Polysiphonia.*

In each chromatophore there is a centrally placed naked (without starch sheath) pyrenoid.

The pigments of red algae are *r*-phycoerythrin (red-water soluble pigments), *r*-phycocyanin (blue water soluble pigment). Chlorophyll *a,* Chlorophyll *b* (in traces only), xanthophylls and carotenes (β-carotene).

Reserve food products. The most important food product is **floridean starch** (polysaccharide) which is found in the form of small grains usually around the nuclei. These grains are not found within the chromatophores. In many Rhodophyceae a soluble sugar **floridoside** is found. This is galactoside of glycerol. A trehalose polysaccharide water soluble sugar is reported from *Lemanea.* The fats are found in traces.

Reproduction. The reproduction takes place by means of vegetative, asexual and sexual methods.

1. **Vegetative reproduction.** In majority of cases it takes place by fragmentation.

2. **Asexual reproduction.** The asexual reproduction of the gametophytes takes place by **neutral spores, monospores,** and **polyspores.** The neutral spores develop in ordinary cells of thallus, *e.g., Asterocystis*. The monospores are developed in sporangia. A single spore develops in each sporangium, *e.g.*, *Scinaia.* The polyspores are formed in larger number in the sporangium.

The asexual reproduction of the sporophytes takes place by **tetraspores** formed in tetrads in the tetrasporangia, *e.g., Polysiphouia* and **paraspore** borne inside parasporangium in greater numbers. During the development of tetraspores reduction division takes place. In the development of paraspores there is no reduction division.

3. Sexual reproduction. The sexual reproduction is always oogamous. The oogamy is of special type. The sex organs of Rhodophyceae have a distinguished terminology. The male structures are called the **spermatangia** developing non-motile **spermatia** (male gametes) in them. The female sex organ is called the **procarp**. It has a **carpogonium** bearing a receptive structure **trichogyne.** The egg develops in the basal swollen part of carpogonium. The **auxiliary cell** may or may not be formed. In Ceramiales the auxiliary cell is formed after fertilization.

Zygote and its germination. In sub-class Bangioideae the zygote divides by vertical and transverse divisions. The number of cells may be 2, 4,or 16. The spores are known as **carpospores**. They are naked and liberated by disintegration of zygote wall and move about in amoeboid fashion. The first division of the nucleus of the zygote is reductional. In sub-class Florideae the carpospores develop indirectly from the zygote.

Life cycle. The life-cycle may be haplontic, haplobiontic or diplobiontic.

Classification. The class Rhodophyceae has been divided into two sub-classes (1) Bangioideae and (2) Florideae.

1. Sub-Class Bangioideae

(i) The growth of thallus is intercalary.

(ii) The zygote directly gives rise to carpospores.

(iii) The cytoplasmic connections between the cells are not obvious.

There is a single order Bangiales in this sub-class.

2. Sub-Class Florideae

(i) The growth of thallus is strictly apical.

(ii) The carpospores are formed indirectly from the zygote.

(iii) The plasmodesmata are very conspicuous.

There are six orders in this sub-class, *i.e.*, 1. Nemalionales, 2. Gelidiales, 3. Cryptonemiales, 4. Gigartinales, 5. Rhodymeniales and 6. Ceramiales.

1. Nemalionales

(i) They are non-tetrasporophytic Florideae.

(ii) Leaving aside a few where an auxiliary cell is found the carposporophyte develops directly from carpogonium.

(iii) The first division of zygote nucleus is reductional.

2. Gelidiales

(i) They are tetrasporophytic.

(ii) The carposporophyte develops directly from carpogonium. Auxiliary cell is not found.

3. Cryptonemiales

(i) They are tetrasporophytic Florideae.

(ii) The auxiliary cell develops on special filament.

4. Gigartinales

(i) They are tetrasporophytic Florideae.

(ii) The auxiliary cell is a vegetative cell of the gametophyte. Either it is borne upon the supporting cell or aloof from carpogonium.

5. Rhodymeniales

(i) They are tetrasporophytic Florideae.

(ii) The auxiliary cell is a special cell differentiated before fertilization.

6. Ceramiales

(i) They are tetrasporophytic Florideae.

(ii) The auxiliary cell is formed subsequently to fertilization directly borne upon supporting cell.

Economic importance of red algae. A few species of red algae also serve as food. They are processed in large quantities. The agar used in bacteriological and mycological media is extracted from red algae. An important genus known as *Porphyra* is used as an important ingredient in soups and is also cooked as a flavouring agent with meat. It is mostly used by Chinese people. This alga is used on large scale by Japanese. *Gelidium* is also a very important alga. Another most important edible alga is *Chondrus crispus* which is commonly known as Irish moss. This alga is utilized in the preparation of various pharmaceuticals including laxatives and cosmetics. This alga is used as stabilizer in prepared food such as chocolate milk, puddings, jellied fish and similar other products. Another important red alga is *Rhodymenia palmata*. This alga is used as a food and also in the preparation of medicines. This is commonly known as dulse. At the time of famine in Ireland dulse became a very important food there.

Phylogenetic Relationships in Rhodophyceae

The Rhodophyceae are very much obscure in their phylogenetic relationships. The Rhodophyceae are clearly an independent group, however, they possess several characteristics of resemblance with other groups. The Rhodophyceae and Myxophyceae both lack flagellated reproductive cells. They possess similar pigments although one is red and the other blue-green. Both red and blue green algae possess starch variants as reserve foods, *i.e.*, glycogen and floridean starch. Sometimes both the algae possess some food in the form of oil droplets and both may have protoplasmic connections between cells. The Rhodophyceae possess a much more highly specialized vegetative body and specialized reproductive structures quite different and advanced from other groups of algae.

Order **CERAMIALES**

Characteristic features.

1. They are tetrasporophytic Florideae.
2. The auxiliary cell is formed after fertilization.
3. The auxiliary cell is directly borne upon the supporting cell of the carpogonial filament.
4. Though the zygote nucleus divides, yet there is further postponement of meiosis to the formation of tetraspores in the tetrasporangia.
5. The carpospores are formed indirectly. They are diploid. They give rise to tetrasporophytes.
6. The life cycle is of diplobiontic type.
7. There is regular alternation of generations.

There are about 160 genera and 900 species in this order.

Classification. According to Fritsch, four families are included in this order:

1. Ceramiaceae.
2. Delesseriaceae.
3. Rhodomelaceae.
4. Dasyaceae.

Genus POLYSIPHONIA (Many siphons)

Occurrence. All species are marine. The species of *Polysiphonia* are abundantly found along the Atlantic coasts of America. Few species have reported from the coast of Karachi (Pakistan). The species have also been reported from Western Ghats of India. *Polysiphonia platycarpa* is a common Indian species and found in the Kumari harbour. Some species are

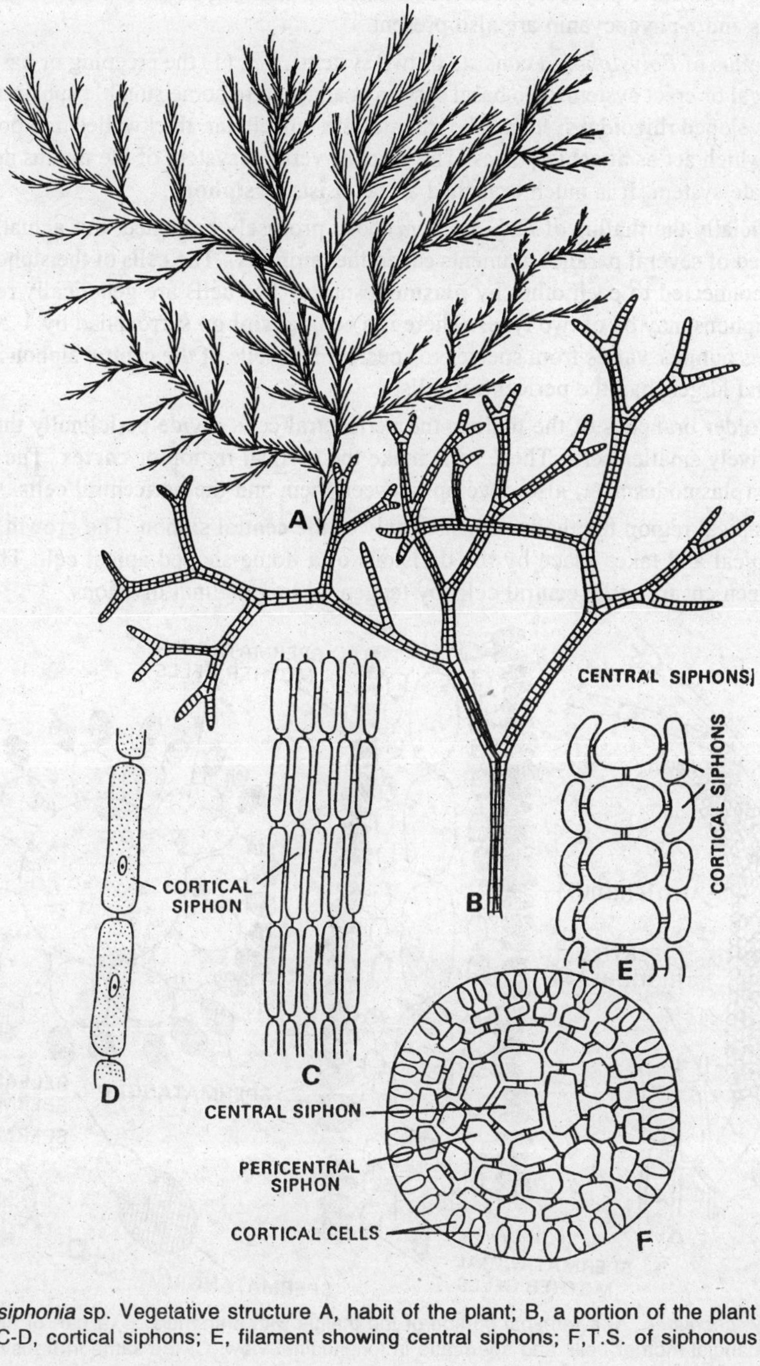

Fig 7.1. *Polysiphonia* sp. Vegetative structure A, habit of the plant; B, a portion of the plant (somewhat enlarged); C-D, cortical siphons; E, filament showing central siphons; F,T.S. of siphonous filament.

epiphytic. *Polysiphonia violacea* is epiphytic upon a brown alga *Fucus vesiculosus*. Some species are parasitic too. *Polysiphonia fastigiata* is parasitic upon *Ascophyllum nodosum*. *Polysiphonia variegata* is found in polluted waters near the estuaries of sea. Some species are very commonly found on the roots of mangrove vegetation.

Structure. *Polysiphonia* varies in colour from red to purple. This colour is because of the overmasking pigment r-phycoerythrin. In addition to this pigment chlorophyll *a*, β-carotene, xanthophylls and *r*-phycocyanin are also present.

The thallus of *Polysiphonia* consists of two systems, *i.e.*, (1) the creeping or basal system and (2) the vertical or erect system. The basal system is attached to some suitable substratum by means of a well developed rhizoidal system. The rhizoids are unicellular, thickwalled and possess swollen distal ends which act as attaching discs. The erect or vertical system of the thallus arises from the basal prostrate system. It is much branched and consists of **siphons**.

Superficially the thallus of *Polysiphonia* looks profusely branched, but actually it is not so and composed of several parallel filaments called the '**siphons**'. The cells of the siphons are placed end to end connected to each other by **plasmodesmata**. The cells are genetically related to each other. The siphons may be of two types. There is a **central siphon** surrounded by 4-24 **pericentral siphons**. This number varies from species to species. The cells of the central siphon are somewhat elongated and larger than the pericentral cells.

In the older branches of the thallus, the pericentral cells divide periclinally thus giving rise to comparatively smaller cells. These cells make the cortical region or **cortex**. The protoplasmic connections (plasmodesmata) also develop between them and the pericentral cells.

In the apical region the thallus consists only of the central siphon. The growth of the thallus is always apical and takes place by the divisions of a dome-shaped apical cell. The pericentral cells have been cut from the central cells by tangential longitudinal divisions.

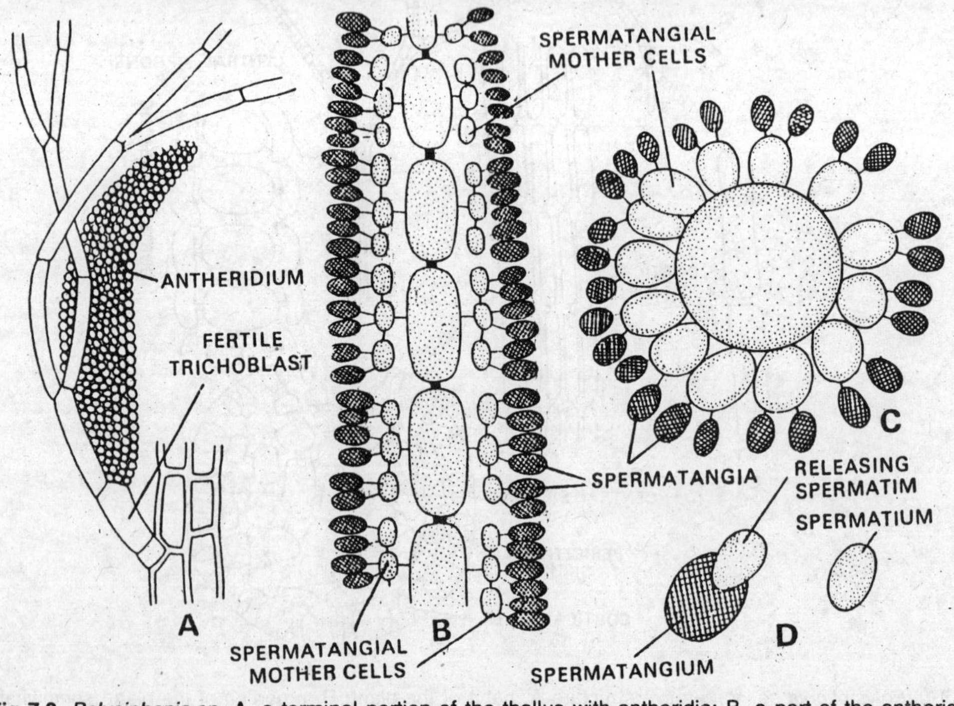

Fig 7.2. *Polysiphonia* sp. A, a terminal portion of the thallus with antheridia; B, a part of the antheridium with spermatangial mother cells and spermatia in longitudinal view; C, the same in transverse section; D, releasing spermatium from spermatangium.

Trichoblasts. The small monosiphonous branches of limited growth are called the **trichoblasts**. The trichoblast initial is initiated by the oblique division of the apical cell. The small apical trichoblast initial divides again and again giving rise to a multicelled uniseriate dichotomously forked filament, the trichoblast.

Structure of a cell. The cells are genetically related and connected to each other by cytoplasmic connections (pit connections) the plasmodesmata. The cell wall is two layered. The outer layer consists of pectic substances and the layer next to protoplast of cellulose. The protoplasm of the cell is highly viscous. There is a conspicuous central vacuole. Each vegetative cells is uninucleate. In each cell there are discoid chromatophores.

Reserve food product. The most important food product is floridean starch (a polysaccharide). This is found in the forms of small grains abundantly present around the nuclei but not formed inside the chromatophores. The other food product is floridoside. This is a galactoside of glycerol.

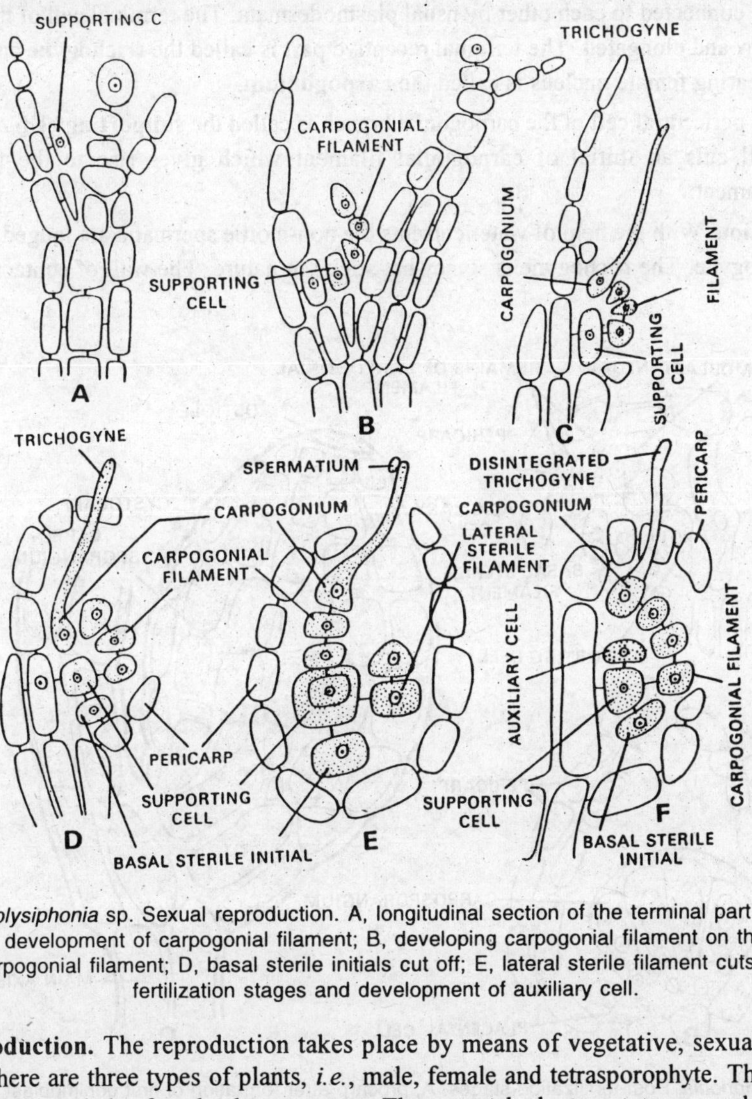

Fig. 7.3. *Polysiphonia* sp. Sexual reproduction. A, longitudinal section of the terminal part of the thallus before the development of carpogonial filament; B, developing carpogonial filament on the supporting cell; C, carpogonial filament; D, basal sterile initials cut off; E, lateral sterile filament cuts off; F, post-fertilization stages and development of auxiliary cell.

Reproduction. The reproduction takes place by means of vegetative, sexual and asexual methods. There are three types of plants, *i.e.*, male, female and tetrasporophyte. The male plants bear antheridia and the female the procarps. The asexual plant or tetrasporophyte produces tetraspores. All the three types are morphologically identical.

Sexual reproduction. The male plants bear the male sex organs the antheridia, and the female plants bear the female sex organs the carpogonia.

Antheridium and its development. The antheridia are borne upon short branches in clusters near the apical portion of the thallus. Each antheridial branch consists of a central trichoblast filament, which produces many lateral pericentral cells. These cells develop antheridial mother cells on them. On each antheridial mother cell two, three or four antheridial cells are developed. These antheridial cells are called the **spermatangia**. A single **spermatium** is liberated from each spermatangium. The new spermatangium may be proliferated from the wall of the old spermatangium. The spermatia are non-motile and uninucleate. They are dependent for their movement upon the mercy of water currents.

Carpogonium and its development. The **procarp** develops from a pericentral cell of greatly reduced trichoblast of female plant. It consists of a row of 3 or 4 celled the **carpogonial filament**. These cells are connected to each other by usual plasmodesmata. The terminal cell of this filament becomes narrow and elongated. The terminal receptive part is called the **trichogyne** and the basal swollen part bearing female nucleus is called the **carpogonium**.

The basal pericentral cell of the carpogonial branch is called the **supporting cell.** Actually the supporting cell cuts an initial of carpogonial filament which gives rise to the four celled carpogonial filament.

Fertilization. With the help of water currents the non-motile spermatia are lodged against the receptive trichogyne. The trichogyne is somewhat sticky in nature. The wall of contact dissolves,

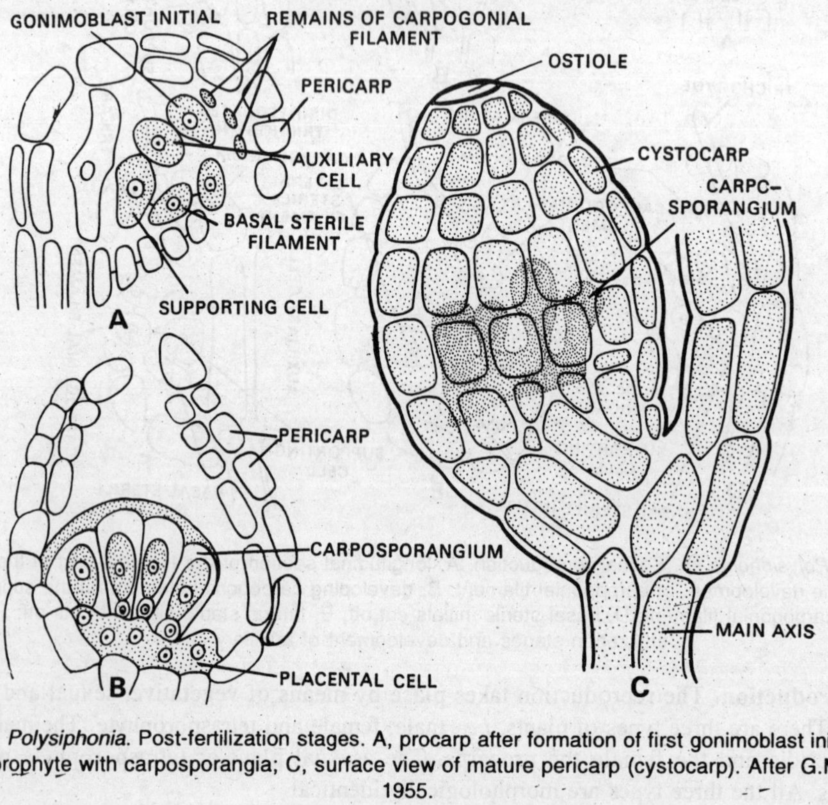

Fig 7.4. *Polysiphonia.* Post-fertilization stages. A, procarp after formation of first gonimoblast initial; B, Carposporophyte with carposporangia; C, surface view of mature pericarp (cystocarp). After G.M. Smith 1955.

the male nucleus passes through the trichogyne and ultimately reaches the female nucleus situated in the basal part of the carpogonium and fertilization takes place. Ultimately the trichogyne shrivels up and disintegrates.

Post-fertilization stages. The supporting cell cuts off another carpogonium and the supporting cell. The auxiliary cell becomes connected to the carpogonium by a somewhat wide tubular protoplasmic connection. By this time, the fertilized zygote nucleus (2*n*) divides mitotically giving to two nuclei, one óf the daughter nuclei passes into the auxiliary cell through the tubular protoplasmic connection. The haploid nucleus already present in the auxiliary cell disappears. The diploid nucleus received from the carpogonium divides mitotically in the auxiliary cell and gives rise to a cell called the **gonimoblast initial**. The gonimoblast rises from the upper face to the auxiliary cell. It consists of many compact gonimoblast filaments, each cell of the filament possesses a diploid nucleus in it.

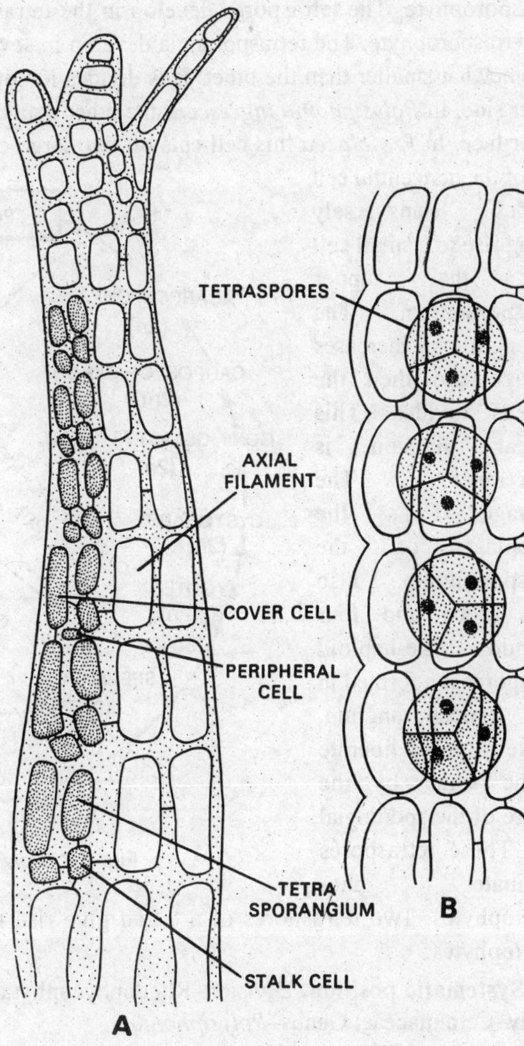

Fig. 7.5. *Polysiphonia* sp. A, development of tetrasporangia; B, tetraspores within tetrasporangia.

The terminal cells of the gonimoblast filaments develop into carposporangia. A single diploid carpospore develops in each carposporangium.

From the supporting cell basal and lateral sterile initials are given out. The basal sterile initial divides later on and the lateral sterile initial divides immediately giving rise to 4-10 celled sterile filament. The basal sterile initial gives rise to two cells only.

Ultimately, by the fusion of the supporting cell, auxiliary cell and sterile filaments, a large cell is developed called the **placental cell**. The placental cell is nutritional in function. The carpogonial cells are disintegrated by this time and they do not participate in the development of the placental cell.

A pseudoparenchymatous urn-shaped structure develops around the carposporangia, etc. This structure is called the **cystocarp**. It possesses a small terminal opening, the **ostiole**. Each cystocarp contains 50-60 carpospores in it, this structure is **carposporophyte**. The carpospores develop into tetrasporophytes.

Asexual reproduction. The carpospores liberate from the carposporangia and come out in the water through the apical pore of the urn-shaped cystocarp. Each carpospore (2*n*) develops into a tetrasporophyte. The tetraspores develop in the tetrasporangia situated on the apical branch of the tetrasporophyte. The tetrasporangia develop in several tiers. The fertile pericentral cell which is somewhat smaller than the other cells divides longitudinally and cuts off a daughter cell on its either side. In *Polysiphonia nigrescens* the other daughter cell cuts off two large cover cells on its upper face. In *P. violacea* this cell cuts off two large cover cells and a peripheral cell. The inner half of the pericentral cell divides transversely giving rise to a small cell and the upper tetrasporangium. The sporangial cell increases in size and then the nucleus divides. This nuclear division is reductional. The cleavage of the protoplast of the tetrasporangium also takes place and four pyramid like-haploid tetraspores are formed in each tetrasporangium. The tetraspores liberate in the water by the rupture of the sporangial wall. These tetraspores germinate into gametophytes. Two tetraspores of a tetrad give rise to male gametophytes and two to female gametophytes.

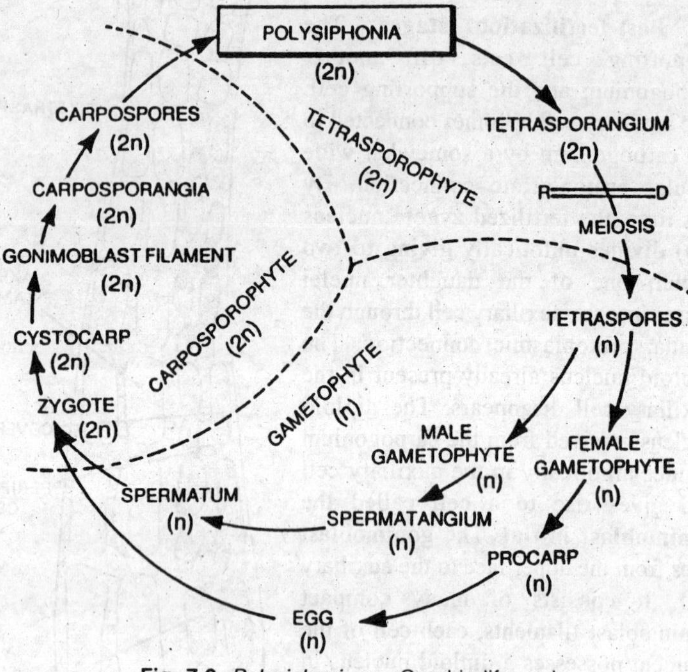

Fig. 7.6. *Polysiphonia* sp. Graphic life-cycle.

Systematic position. Division–Rhodophycophyta; Class-Rhodophyceae; Order-Ceramiales; Family–Ceramiaceae; Genus–*Polysiphonia.*

Order-NEMALIONALES

Characteristic features

1. Mostly they are marine, but few exceptions, ***e.g.***, *Batrachospermum.*
2. The vegetative structure is both uni and multiaxial.
3. There are no true auxilliary cells in many genera, *e.g., Batrachospermum, Nemalion;* while in others, *e.g., Galaxaura* the so-called auxiliary cell may be largely nutritive.
4. In the majority of the genera only haploid plants are represented in the life-cycle but in genus *Galaxaura* there is distinct alternation of generations. Commonly the life-cycle is haplobiontic.
5. The sporophytic phase is being represented by zygote only.
6. The carpospores are haploid and produced in haploid carposporangia.
7. On germination the carpospores produce a juvenile of *Chantransia* stage.
8. The zygote divides meiotically producing haploid carpospores by carposporophyte and have no free living tetrasporophytic generation.
9. Asexual reproduction takes place by means of monospores produced within monosporangia.

There are 35 genera and 250 species in this order. There are eight families in this order. The type genus *Batrachospermum* of Batrachospermaceae has been discussed here.

Family –Batrachospermaceae

Genus BATRACHOSPERMUM *(batracho,* frog; *spermum, seed).*

Occurrence. This is one of the fresh water forms of Rhodophyceae. This alga is found in slow running streams and on the banks of lakes and ponds. It is more commonly found in well aerated waters. The plants are blue-green, olive-green, violet and reddish in colour. The colour varies as a result of the differences in light intensity. The species which grow in deep water are reddish or violet in colour whereas the species growing in shallow water are olive-green in colour. The alga is also known as the **'frog spawn'.** The plants are mucilaginous, moniliform or beaded in appearance to the naked eye. The plants may reach a length of twenty centimetres and may easily be collected from the slow running streams around Dehradun especially in winter season.

Structure. The thallus is filamentous, profusely branched and with a mucilaginous feel. The filament consists of only one axial filament of main axis which has been produced by a single apical cell cutting off the segments parallel to the base. This way, the central filament or main axis consists of a uniseriate row of axial cells. The main axis bears the laterals at various points on its length. These laterals are very short in comparison to the main filament. They are called the branches or the laterals of limited growth.

At the nodal point just beneath the septum, usually four basal cells are formed each producing a whorl of the branches of limited growth. The laterals of limited growth possess constricted cells of moniliform or beaded appearance. The ultimate cells of the branches of limited growth usually terminate in unicellular colourless hairs. The clusters of the laterals at nodes are called the glomerules. In addition to the numerous branches of limited growth, resemble the main filament in structure. The basal cells from which the branches of unlimited growth arise also produce filaments growing downward and ensheathing the main axis. They are called corticating filaments and they form the pseudocortex.

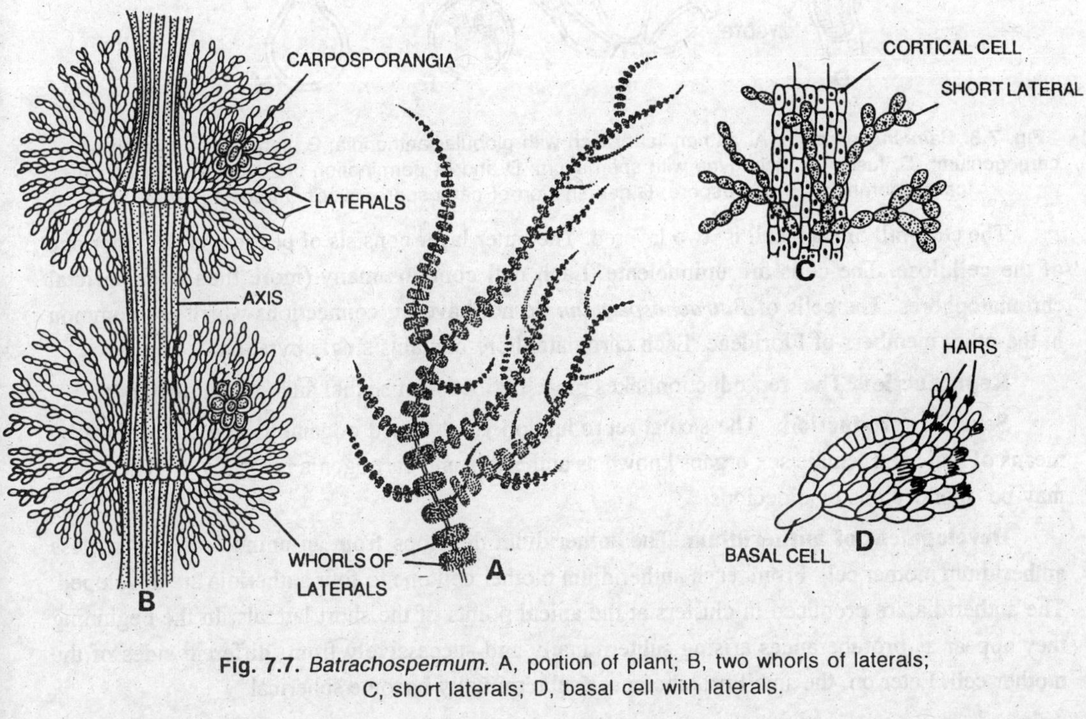

Fig. 7.7. *Batrachospermum.* A, portion of plant; B, two whorls of laterals; C, short laterals; D, basal cell with laterals.

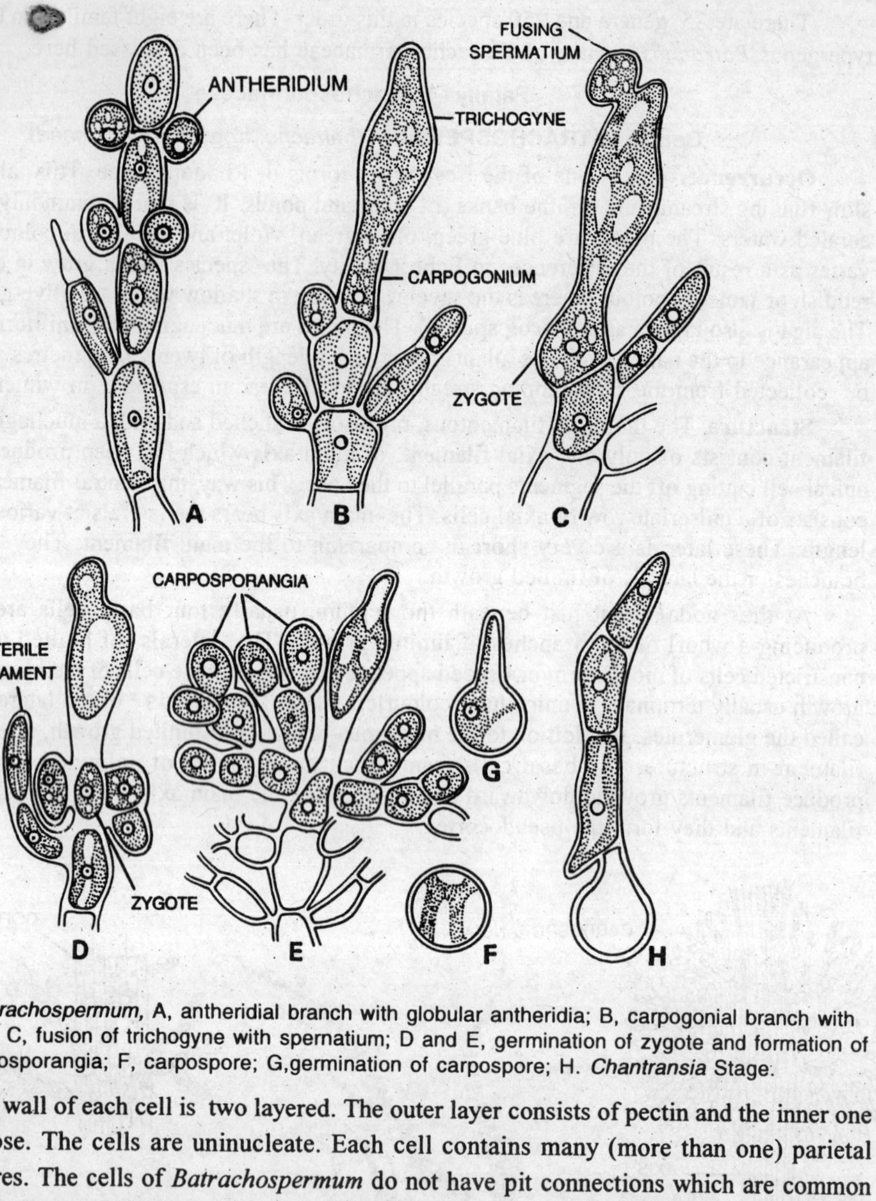

Fig. 7.8. *Batrachospermum*, A, antheridial branch with globular antheridia; B, carpogonial branch with carpogonium; C, fusion of trichogyne with spernatium; D and E, germination of zygote and formation of carposporangia; F, carpospore; G,germination of carpospore; H. *Chantransia* Stage.

The cell wall of each cell is two layered. The outer layer consists of pectin and the inner one of the cellulose. The cells are uninucleate. Each cell contains many (more than one) parietal chromatophores. The cells of *Batrachospermum* do not have pit connections which are common in the other members of Florideae. Each chromatophore contains single pyrenoid.

Reproduction. The reproduction takes place by means of sexual and asexual methods.

Sexual reproduction. The sexual reproduction is advanced oogamous and takes place by means of male and female sex organs known as antheridia and carpogonia respectively. The plants may be monoecious or dioecious.

Development of antheridium. The antheridium develops from an uninucleate, colourless antheridium mother cell. From each antheridium mother cell one to four antheridia are developed. The antheridia are produced in clusters at the apical points of the short laterals. In the beginning they appear as protuberances arising subterminally and successively from different sides of the mother cell. Later on, the small protuberances subterminally become spherical.

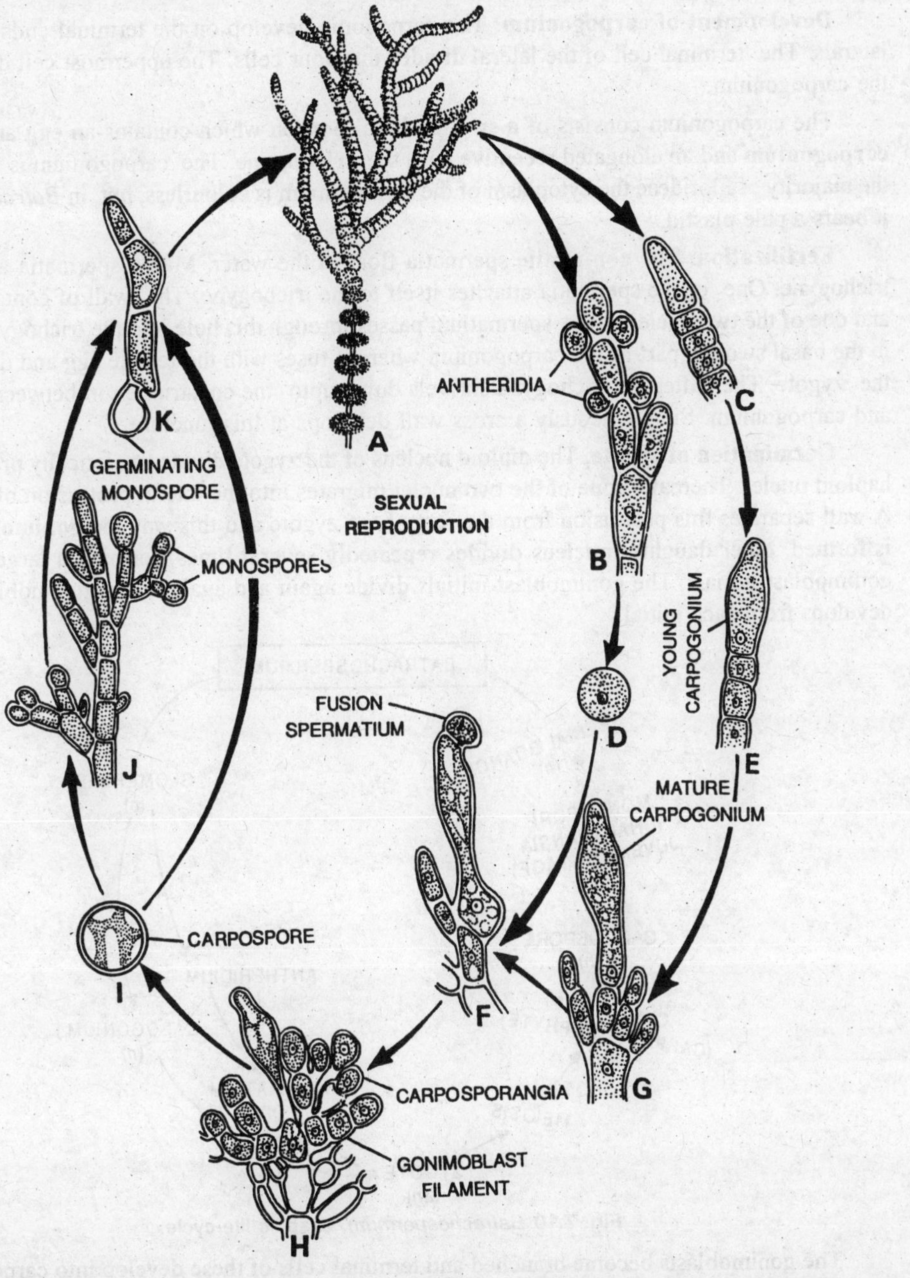

Fig. 7.9. *Batrachospermum*. Life-cycle, A. a portion of plant; B., antheridial branch with antheridia; C and E, formation of carpogonium; D, spermatium; E, young carpogonium; F, fusion of spermatiun; G, mature carpogonium; H. carposporophyte; I, carpospore; J, *Chantransia* stage producing monospores; K, germinating monospore.

Each antheridium contains a single spermatium on its maturity. The non-motile, spermatium liberates through a slit formed in the wall of antheridium. The spermatia remain floating in the water.

In *Batrachospermum*, the nucleus divides into two as soon as it contacts the trichogyne. So at the time of the fertilization spermatium contains two nuclei and sometimes known as spermatium complex.

Development of carpogonium. The carpogonia develop on the terminal ends of the short laterals. The terminal cell of the lateral divides into four cells. The uppermost cell develops into the carpogonium.

The carpogonium consists of a swollen basal portion which contains an egg and known as **carpogonium** and an elongated receptive part, the **trichogyne.** The carpogonium is flask-like. In the majority of Florideae the cytoplasm of the carpogonium is colourless, but in *Batrachospermum* it bears a pale plastid.

Fertilization. The non-motile spermatia float in the water. Many spermatia approach the trichogyne. One of the spermatia attaches itself to the trichogyne. The wall of contact dissolves and one of the two nuclei of the spermatium passes through this hole into the trichogyne, reaching in the basal swollen part of the carpogonium where it fuses with the female egg and develops into the zygote. Thereafter the trichogyne shrivels down upto the constriction in between trichogyne and carpogonium. Simultaneously a cross wall develops at this juncture.

Germination of zygote. The diploid nucleus of the zygote divides meiotically producing two haploid nuclei. Thereafter one of the two nuclei migrates into the lateral protrusion of the zygote. A wall separates this protrusion from the rest of the zygote and this way the **gonimoblast initial** is formed, other daughter nucleus divides repeatedly several times, forming a large number of gonimoblast initials. The gonimoblast initials divide again and again and a gonimoblast filament develops from each initial.

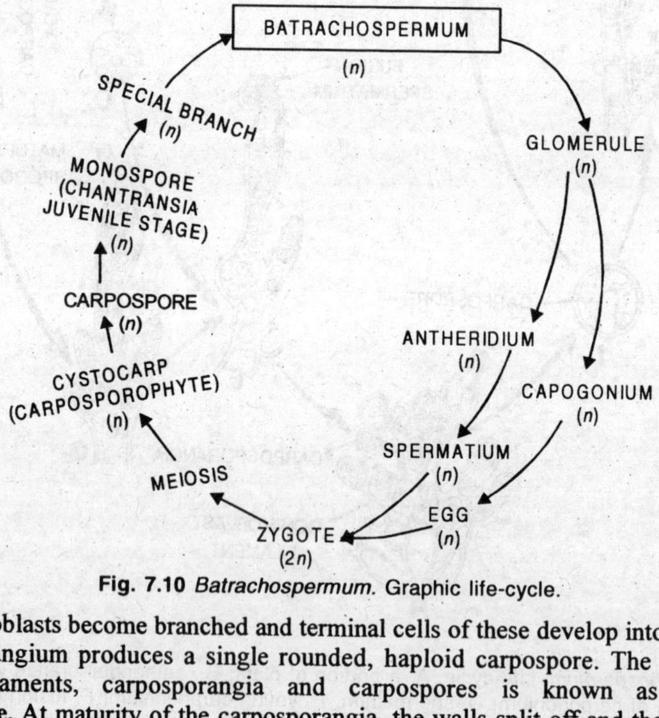

Fig. 7.10 *Batrachospermum.* Graphic life-cycle.

The gonimoblasts become branched and terminal cells of these develop into carposporangia. Each carposporangium produces a single rounded, haploid carpospore. The structure having gonimoblast filaments, carposporangia and carpospores is known as **cystocarp** or **carposporophyte.** At maturity of the carposporangia, the walls split off and the carpospores are liberated. Each carpospore gives rise to the juvenile branched filamentous body. The juvenile form of *Batrachospermum* resembles an alga known as *Chantransia*, and therefore, known as the *Chantransia* stage. The terminal cells of these plants act as apical cells and develop into new *Batrachospermum* plants.

Asexual reproduction. In several species of *Batrachospermum* the short branches of the filaments of *Chantransia* stage produce monospores. These monospores again produce *Chantransia* stage, and again the apical cells of this stage produce new plants

Systematic position. Division-Rhodophycophyta; Class-Rhodophyceae. Sub-class-Florideae; Order-Nemalionales; Family-Batrachospermaceae; Genus-*Batrachospermum.*

QUESTION BANK

N.B. *Undermentioned questions are asked repeatedly in various Indian Universities and competitive examinations.*

Class. Rhodophyceae

1. Give the characteristic features of red algae with special reference of pigmentation, food products and flagellation.
2. Write an account of distribution and occurrence of red algae. Also mention the range of vegetative structure in Rhodophyceae.
3. Give the apical growth of thallus and structure of typical cell in Rhodophyceae.
4. Give a comprehensive account of the modes of reproduction found in red algae.
5. Give the outline of Fritsch's system of classification of red algae.
6. Write notes on: (a) reserve food products in Rhodophyceae, (b) economic importance of red algae; (c) phylogenetic relationships in Rhodophyceae.
7. Describe briefly the general characteristic features and classification of Rhodophyceae.
8. Write notes on (a) reserve food products in Rhodophyceae, (b) asexual reproduction in Rhodophyceae (c) sexual reproduction in Rhodohpyceae, (d) germination of zygote in Rhodophyceae.

Order. Ceramiales

1. Give the characteristic features of the order Ceramiales.
2. Mention with the help of diagrams only, the important stages in the life-cycle of *Polysiphonia.*
3. Describe the sexual reproduction in brief, in *Polysiphonia.*
4. Give the occurrence and the structure of thallus of *Polysiphonia.*
5. Mention the post-fertilization stages of *Polysiphonia.*
6. Write notes on: (a) carposporophyte of *Polysiphonia*; (b) tetrasporophyte of *Polysiphonia*; (c) pigmentation and reserve food products of *Polysiphonia*; (d) structure and development of antheridium and carpogonium in *Polysiphonia.*
7. What type of life-cycle is found in *Polysiphonia*? Discuss its details.
8. Describe in detail the process of sexual reproduction in *Polysiphonia.*
9. Describe the structure and reproduction in *Polysiphonia.* Discuss alternation of generations in its life-cycle.
10. Describe the post-fertilization changes in the life-history of *Polysiphonia.*
11. Trace the development of cystocarp in *Polysiphonia.*
12. Write a detailed account of tetrasporic plant in *Polysiphonia.*
13. Write a detailed note on carposporophyte in *Polysiphonia.*
14. Write short notes on: (a) Gonimoblast filaments (b) Cystocarp .

Diagrammatic Type

1. Draw labelled diagram showing detailed structure of cystocarp.
2. Draw labelled diagram of the tetrasporohyte of *Polysiphonia.*

Multiple Choice Type

1. The pit-connections are found in:

   (i) *Chara*, (ii) *Ectocarpus*, (iii) *Polysiphonia*, (iv) *Oedogonium*,
2. Phycoerythrin is found in:

(i) Chlorophyceae, (ii) Xanthophyceae, (iii) Rhodophyceae, (iv) Bacillariophyceae.

3. *Polysiphonia* is member of:

   (i) Ceramiales, (ii) Gelidiales, (iii) Nemalionales, (iv) none of above

4. The carpospores are:

   (i) n, (ii) 2n, (iii) nn, (iv) none of above

5. On germination the carpospores give rise to:

   (i) tetrasporophyte, (ii) gametophyte, (iii) sporophyte, (iv) none of above

6. Floridean starch is food product of:

   (i) Chlorophyceae, (ii) Rhodophyceae, (iii) Myxophyceae, (iv) Phaeophyceae.

7. The floridoside is found in:

   (i) Chlorophyceae, (ii) Rhodophyceae, (iii) Myxophyceae, (iv) Phaeophyceae.

8. In *Polysiphonia* each spermatangium contains:

   (i) single spermatium, (ii) four spermatia, (iii) many spermatia (iv) none of above

9. The life-cycle of *Polysiphonia* is:

   (i) haplontic, (ii) diplontic, (iii) haplo-diplobiontic, (iv) none of above

10. The trichoblasts of *Polysiphonia* are:

    (i) polysiphonous (ii) monosiphonous, (iii) without siphons (iv) none of above

11. The gonimoblast initials give rise to:

    (i) gametophyte, (ii) tetrasporophyte, (iii) carposporangia, (iv) none of above

12. The gonimoblast initials give rise to:

    (i) male gametophyte, (ii) female gametophyte, (iii) tetrasporophytes, (iv) none of above

13. The non-motile male gametes are found in:

    (i) *Chlamydomonas*, (ii) *Oedogonium* (iii) *Ectocarpus*, (iv) *Polysiphonia.*

14. The urn-shaped structure in *Polysiphonia* is:

    (i) Procarp, (ii) Pericarp, (iii) Cystocarp, (iv) Archicarp.

15. The thalli of worm-like appearance are found in:

    (i) *Polysiphonia,* (ii) *Batrachospermum,* (iii) *Sargassum,* (iv) *Nemalion.*

16. Agar-agar is obtained from:

    (i) *Polysiphonia,* (ii) *Batrachospermum,* (iii) *Gracilaria,* (iv) *Nemalion.*

ANSWERS

1 (iii) 2. (iii) 3. (i) 4. (ii) 5. (i), 6. (ii), 7.(ii) 8.(i) 9. (iii) 10. (ii) 11. (iii) 12.(ii), 13. (iv) 14. (iii), 15. (iv), 16. (iii)

N.B. *Undermentioned questions are asked repeatedly in various Indian Universities and competitive examinations.*

Long Answer Type

1. Write short essay on economic importance of algae.
2. Describe the sexual reproduction in *Oedogonium.*
3. Describe the post-fertilization changes in *Polysiphonia.*
4. Give a brief account of the life-history of *Ectocarpus.*
5. Describe the structure and method of asexual reproduction in *Volvox*.
6. Discuss the phenomenon of alternation of generations in the members of Phaeophyceae you have studied.

7. Classify algae upto class level and give salient features of different classes.
8. What is coenobium? Describe asexual reproduction in any motile coenobial form prescribed in your syllabus.
9. Write an essay on the origin and evolution of sex in green algae.
10. Describe the habitat, structure and methods of reproduction in *Oedogonium* or *Chara*.
11. Give an account of vegetative multiplication and sexual reproduction in diatoms.
12. What are the methods of reproduction and perennation in Myxophyceae? Explain.
13. Describe the method of sexual reproduction in *Oedogonium*.
14. "*Chlamydomonas* throws some light on the origin and evolution of sex in algae." Comment upon the statement.
15. What are the prokaryotic and eukaryotic cells? Which of the two is more primitive type and why? Give suitable diagrames.
16. Describe the various methods of reproduction in *Chlamydomonas*.
17. Give an account of habit, structure and methods of reproduction of *Nostoc*.
18. List the major classes of algae whose representative genera have been prescribed in your course and describe their distinguishing characters in a tabular form.
19. What is diplobiontic type of life-cycle? Explain with the help of life-cycle of *Polysiphonia*.
20. Describe in brief the life-cycle of *Ulothrix*.
21. Describe in brief the sexual reproduction in *Vaucheria*.
22. Give an illustrated account of habit, structure and methods of reproduction of *Oscillatoria*.
23. Give an account of life-cycle of *Ectocarpus*, and mention how it is affected by environment
24. Describe the formation of cystocarp in red algae studied by you.
25. What are the main features on which the classification of algae is based? Discuss giving suitable examples.
26. What do you understand by alternation of generations? Explain this phenomenon with reference to the life-history of *Ecotcarpus*.
27. Dexcribe the structure of the cell, and asexual reproduction in *Chlamydomonas*.
28. Describe the characteristic features of blue-green algae. What is their economic importance?
29. Give an account of the external and internal features of *Sargassum* thallus.
30. Give an account of thallus structure and sexual reproduction of *Chlamydomonas*.
31. Write about sexual reproduction in *Vaucheria*. How does it resemble with that of *Cystopus* and mention its significance.
32. Describe thallus structure, and reproduction of *Sargassum*.
33. Describe thallus structure, and sexual reproduction of *Volvox*.
34. Give habit, occurrence and reproduction in *Cladophora*.
35. Give an account of structure, reproduction and systematic position of *Vaucheria*.
36. What are the common pigments present in algae? State the importance of these pigments in the classification of algae.
37. Describe the structure and method of sexual reproduction in *Volvox* or *Draparnaldiopsis*.
38. Describe distinguishing characters of Myxophyceae, Rhodohyceae and Phaeophyceae
39. Describe the various modes of perennation in the fresh water algae studied by you.
40. Describe the mode of sexual reproduction in *Coleochaete* giving suitable illustrations.

Short Answer Type

1. (a) Describe the structure and development of **globule** and **nucule**. In which alga they are found?

   (b) Give an account of the morphology of *Oscillatoria*.
2. Write short notes on any four: (a) palmella stage, (b) dwarf male, (c) heterocyst, (d) conceptacle, (e) lateral conjugation in *Spirogyra*.
3. Describe in detail the following: (a) structure of thallus in *Sargassum*, (b) neuromotor apparatus, in *Chlamydomonas*, (c) reproduction in *Nostoc*.
4. Write short notes on: (i) *Oscillatoria*, (ii) growth in *Chara* or conjugation in *Spirogyra*, (iii) germination of oospore of *Volvox*.
5. Write short notes on: (i) palmella stage, (ii) nannandrium, (iii) synzoospore, position and shape of chloroplasts in algae.
6. Write short notes on any four: (a) vegetative reproduction in *Chara*, (b) dwarf male, (c) gongrosira stage, (d) *Ectocarpus* cell structure, (e) alternation of generations.
7. Write short notes on: (i) structure of cell in Cyanophyceae, (ii) cystocarp of *Polysiphonia*, (iii) heterocyst, (iv) fertilization in *Vaucheria*.
8. Describe in brief the following: (a) sexual reproduction in *Oedogonium*, (b) vegetative thallus of *Polysiphonia*.
9. Write notes on: (a) *Sargassum* thallus, (b) *Stigeoclonium* thallus
10. Write short notes on any three: (i) zoospores of *Vaucheria*, (ii) lateral conjugation in *Spirogyra*, (iii) cell structure in Cyanophyceae, (iv) coenobium of *Volvox*.

Diagrammatic Type

1. Draw the life cycle of *Ectocarpus* with the help of well labelled diagrams alone.
2. Illustrate any two of the following with neat and labelled diagrams alone. (a) Structure of cell in Cyanophyceae as seen under electron microscope, (b) asexual reproduction in *Volvox*, (c) stages in scalariform conjugation in *Spirogyra*.
3. Illustrate the following with neat and labelled diagrams alone, (a) asexual reproduction in *Vaucheria*, (b) *Chlamydomonas* cell, as seen under electron microscope.
4. Illustrate the structure and development of sex organs of *Chara* with neat and labelled diagrams alone.
5. With the help of labelled diagrams alone show the structure of thallus and life-cycle of *Polysiphonia*.
6. What is heterotrichous form of plant body? With the help of labelled diagrams only show the life-cycle of a green heterotrichous alga.
7. What do you understand by cyclosporean kind of life-cycle? With the help of labelled diagrams only show the life-cycle of cyclosporean alga.
8. Describe thallus structure of *Polysiphonia*. Give its life-cycle with the help of figures only
9. Give well labelled diagrams of any four. (i) *Ectocarpus* cell, (ii) vegetative structure of thallus of *Spirogyra*, (iii) *Oedogonium* cell structure, (iv) *Volvox* cell structure, (v) structure of *Nostoc* thallus and cell.

Multiple Choice Type

1. Organisms which fix atmospheric nitrogen are: (i) mosses (ii) bacteriophages, (iii) liverworts, (iv) blue-green algae.
2. *Spirogyra* occurs in: (i) running salt water (ii) running fresh water, (iii) marine water, (iv) stagnant water.
3. *Ulothrix* is a green alga because: (i) it has a cell wall, (ii) each cell has a single nucleus, (iii) each cell has a single chloroplast, (iv) it produces gametes.

4. Blue-green algae are: (i) procaryotes, (ii) eucaryotes, (iii) acellular, (iv) actinomycetes.
5. The mode of sexual reproduction in *Spirogyra* is: (i) oogamy, (ii) anisogamy, (iii) isogamy, (iv) heterogamy.
6. Cell wall in algal cell consists of (i) cellulose, (ii) chitin, (iii) cutin, (iv) suberin.
7. *Spirogyra* resembles *Rhizopus* in having: (i) stored starch, (ii) cellulose cell wall, (iii) single celled gametangia, (iv) single nucleus in each cell.
8. Algae and other submerged green plants float in water during day time and sink at night because: (i) they come up to enjoy some time, (ii) they lose weight at night, (iii) they become buoyant due to accumulation of O_2 as a result of photosynthesis, (iv) they become light due to food material accumulation.
9. On germination each zygospore of *Spirogyra* gives rise to: (i) four plants, (ii) three plants, (iii) two plants (iv) one plant.
10. Name the organism that lacks archegonium: (i) *Pteris*, (ii) *Funaria*, (iii) *Cycas*, (iv) *Spirogyra*.
11. *Spirogyra*: (i) the filaments showing lateral conjugation are homothallic, (ii) the filaments showing lateral conjugation are heterothallic, (iii) the filaments showing scalariform conjugation are homothallic, (iv) asexual reproduction occurs by zoospores.
12. Ability to fix atmospheric nitrogen is found in: (i) leaves of some crop plants, (ii) some marine red algae, (iii) some blue-green algae, (iv) *Chlorella*.
13. Agar-agar is derived from: (i) fungi, (ii) algae, (iii) *Dryopteris*, (iv) gymnosperms.
14. In *Spirogyra* pyrenoid is found in: (i) nucleus, (ii) cell wall (iii) chloroplast, (iv) cytoplasm.
15. Which one of the following is rich in protein: (i) *Ulothrix* (ii) *Spirogyra*, (iii) *Oedogonium* (iv) *Chlorella*.
16. *Ulothrix* is considered more advanced than *Spirogyra* because in *Ulothrix*: (i) filament is differentiated into holdfast and apical cell, (ii) chloroplast is parietal and girdle-shaped, (iii) zoospores are formed, (iv) gametes are motile.
17. The pyrenoids are: (i) protein bodies, (ii) starch bodies, (iii) protein bodies with an envelope of starch, (iv) starch bodies with an envelope of protein.
18. Iodine is obtained from: (i) Bryophytes, (ii) Bacteria, (iii) fungi, (iv) sea weeds.
19. Parasitic alga is: (i) *Cephaleuros*, (ii) *Ulothrix*, (iii) *Oedogonium*, (iv) *Sargassum*.
20. The organ by which *Ulothrix* is attached to its substratum is called a: (i) rhizoid, (ii) holdfast, (iii) trichome, (iv) root.

ANSWERS

1 (iv) 2. (iv) 3. (iii) 4. (i) 5. (iii) 6. (i) 7. (iii) 8. (iii) 9. (iv) 10. (iv) 11. (i) 12. (iii) 13. (ii), 14. (iii) 15. (iv) 16. (i) 17. (iii) 18. (iv) 19. (i) 20. (ii)

UNIT - 3

Diversity of Cryptogams
Fungi

8
CHAPTER

Fungi: Introduction

The fungi constitute a large and diverse group of plant kingdom. They resemble algae in many respects and therefore, included in the large group Thallophyta. There are about 50,000 to 100,000 known species of fungi in the world. According to one estimate, there are 4,300 genera and 50,000 species of fungi (Ainsworth, 1961), but this number is constantly increasing because of the never ending search for these organisms, throughout the world. In the five kingdom system of classification, the fungi are treated as a separate kingdom.

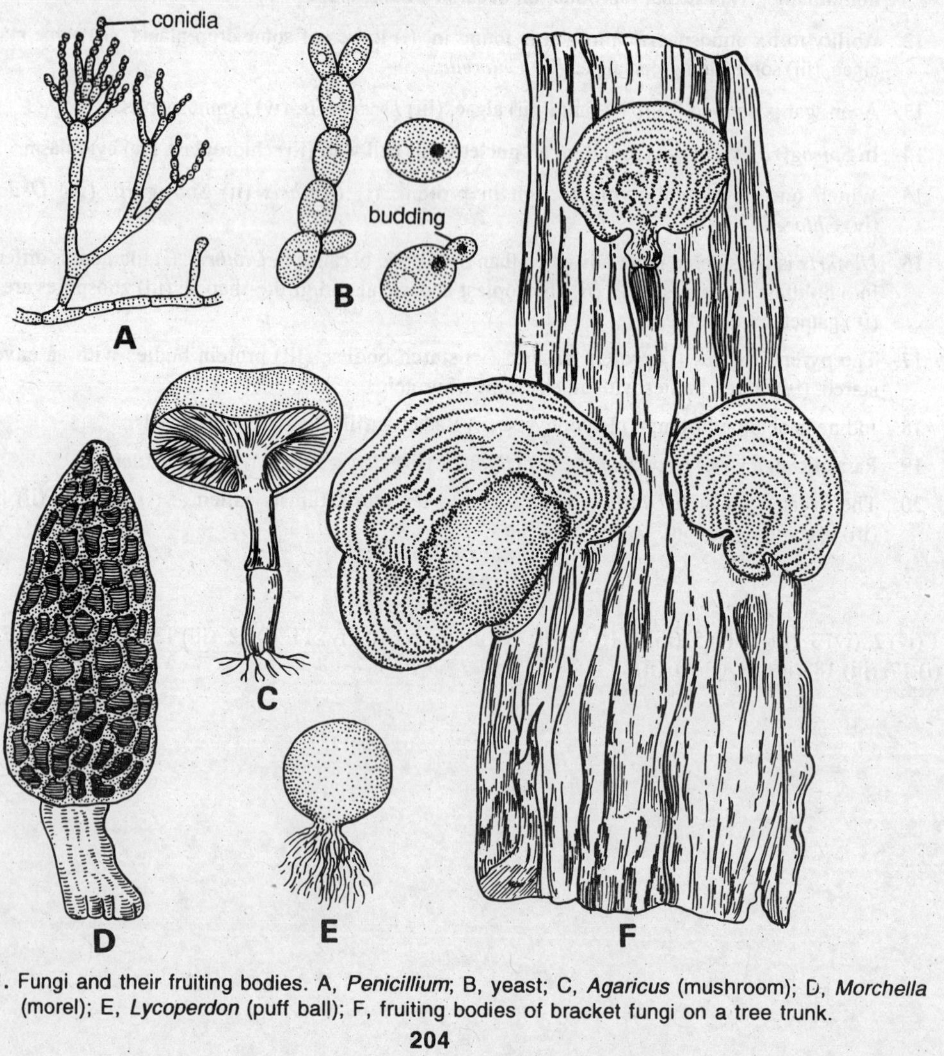

Fig. 8.1. Fungi and their fruiting bodies. A, *Penicillium*; B, yeast; C, *Agaricus* (mushroom); D, *Morchella* (morel); E, *Lycoperdon* (puff ball); F, fruiting bodies of bracket fungi on a tree trunk.

They lack chlorophyll and other photosynthetic pigments and cannot synthesize their food from carbondioxide and water in the presence of sunlight and their mode of nutrition is either saprophytic, parasitic or symbiotic. When fungi live as saprobes they bring about the decay of organic materials, and when they act as parasites they attack living protoplasm and cause disesases of plants, animals and human beings.

The body of fungi is very simple, and in majority of cases consists of a network of branched filaments called the **hyphae**. The tangled mass of hyphae is the **mycelium.** In few cases the mycelium is completely lacking, *e.g., Synchytrium* and in other cases the plant body may be unicellular, *e.g., Saccharomyces*.

The cell wall of the mycelium does not consist of true cellulose. It either consists of chitin or fungal cellulose along with other substances.

The chief food reserves are glycogen and oils. As they are devoid of chlorophyll the starch is not found.

The fields of the study of fungi. There are several idealistic points of view of the study of fungi such as 1. Plant Pathology; 2. Medical Mycology; 3. Industrial Mycology; 4. Antibiotics Production, *e.g.*, penicillin, streptomycin, etc.; 5. Microbiology (study of fungi in relation to chemistry or biochemistry and production of vitamins, etc.)

Origin of fungi. There are two important views about the origin of the fungi, *i.e.*, (1) Polyphyletic view and (2) Monophyletic view.

According to first view (polyphyletic) the fungi have originated from various groups of algae by attaining saprophytic or parasitic mode of life. There are several colourless and parasitic algae (*e.g.*, Volvocales and Chlorococcales) which give the idea of the fungi.

According to this view, it is assumed that in the very beginning the whole surface of the earth was covered with water and for the first time the algae appeared which thereafter transmigrated from water to land. During this transmigration many of them died and lost their green colour. A large quantity of dead organic material was available and many of algae converted into saprophytic fungi of today. It is also assumed that the parasitic fungi were the resultants of further transmigration of saprophytic fungi upon other living plants.

According to second view (monophyletic view), the fungi arose independently and not from algae. There are few points in support of this view.

That the sex organs are quite different in two groups.

That the two groups produce diverse forms.

That the fungi were present in the red sandstone of Devonian period which showed the earliest records of the plants.

That the algal and fungal structures are quite different, even on the loss of the chlorophyll of the algae.

That the fungi lack carbohydrates.

That the longitudinal division is absent even in the cell of the highest evolved fungi.

That the chemical composition of the cell wall of two groups is quite different.

However, in the present state of our knowledge, there is no valid reason to assume that all fungi originated from a common ancestor. Many mycologists of present day believe, that all fungi are not related. Some may have arisen from protozoan ancestor and others from a plant-like ancestor, perhaps some primitive alga, which later on assumed a saprobic or parasitic way of life and eventually lost its chlorophyll.

Differences between algae and fungi. The algae and fungi possess the following characteristic features of differentiation:

1. The algae are green thallophytes containing chlorophyll and other pigments; whereas the fungi are devoid of chlorophyll and other such pigments.

2. The algae are autotrophic whereas the fungi are heterotrophic in nutrition; they may be saprophytic, parasitic or symbiotic.

3. The body of algae may be truly parenchymatous whereas of fungi it is pseudoparenchymatous. The pseudoparenchyma consists of compactly interwoven hyphae.

4. The algal cell wall is composed of true cellulose along with other substances. The fungal walls lack cellulose. They consist of fungal cellulose and chitin.

5. The algae grow in water or on wet substrata whereas fungi are mostly parasites or saprophytes.

6. In algae there is regular increase in the complexity of reproduction from simpler forms to higher forms whereas in fungi there is definite and regular reduction in sexuality from lower to higher fungi.

Occurrence and distribution. The fungi are most diversified in their habitat. They are found in almost all possible types of habitats. They are devoid of chlorophyll and may be saprophytes, parasites or symbionts. Many species of Phycomycetes, are found in the water and are called the **aquatic fungi.** Some fungi are commonly found upon algae and other aquatic plants in epiphytic state. Some species are found on the dead organic materials present in the water. Some species are sub-terranean and found under the surface of the earth.

Many species of the fungi are recognizable only with the naked eye and even from the distance such as mushrooms, morels, puff balls, bracket fungi, cup fungi, etc. Whereas on the other hand they are microscopic and may be recognized with great care by the experts of the subject with the help of compound microscope.

The parasitic fungi are commonly found upon the hosts, *i.e.*, vascular plants where they cause various diseases and do great harm to the host plants.

Some fungi are found in the alimentary canals of mammals and human beings where they cause the stomach disorders. Some forms cause the skin diseases. In this way we see that fungi are much diversified and accustomed to unusual habitats.

Vegetative structure. Ainsworth (1971) has divided the fungal kingdom into two large divisions – 1. Myxomycota and 2. Eumycota. The Myxomycota includes the class Myxomycetes. In Myxomycetes (the slime molds) the thallus is a naked amoeboid mass of protoplasm. This may be a **plasmodium** (*i.e.*, single large multinucleate protoplast) or a pseudoplasmodium (*i.e.*, an aggregation of many small uninucleate protoplasts that retain their individuality).

In the division Eumycota (*i.e.*, *Saccharomyces, Blastocladiella*) and the entire surface of the cell is responsible for the absorption of food. The plant body which consists of a single cell, and doesn't bear any

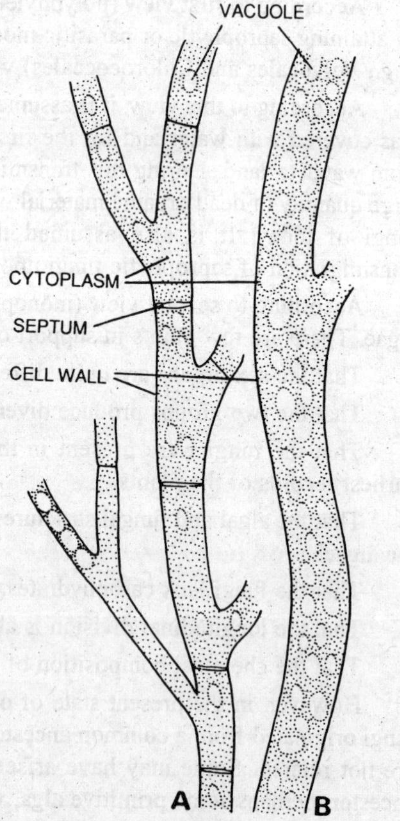

Fig. 8.2. The fungi. vegetative hyphae. A, portion of a septate hypha; B, portion of non-septate hypha.

additional structures for sucking. The food from the substrate or for reproduction, is called **holocarpic** when a part of the thallus is meant for reproduction, it is **eucarpic.** This eucarpic multicellular vegetative body of the fungus consists of a network of much branched thin filaments called the **hyphae** and the tangled mass of the hyphae is called **mycelium.** The hyphae grow by apical elongation. The hyphae may be segmented or non-segmented. The segmented hyphae possess cross walls in them at regular intervals called the **septa.** The mycelium having septa is called the **septate.** Usually each septum bears a small pore in its centre for the communication of the cytoplasm from one cell to another. The mycelium without septa is called the **aseptate.** The aseptate mycelium possesses many nuclei embedded in the cytoplasm and called the **coenocytic** mycelium. The cells of the septate mycelium may be uninucleate, binucleate, or multinucleate.

The septa in the higher fungi (*i.e.*, Ascomycotina, Basidio- mycotina and Deuteromycotina) possess one or more small perforations in the centre to maintain protoplasmic continuity between adjacent cells. This pore is a simple pore in most of higher fungi (all Ascomycotina) but in some (many Basidiomycotina excluding rusts) it is a complex structure called **dolipore.** The electron microscope reveals a curved double membrane on each side of the septum. Because this membrane structure looks like a parenthesis in sectional view, it is known as **parenthosome.** (Fig. 8.3.).

Some fungi lack mycelium altogether, *e.g.*, *Synchytrium*.

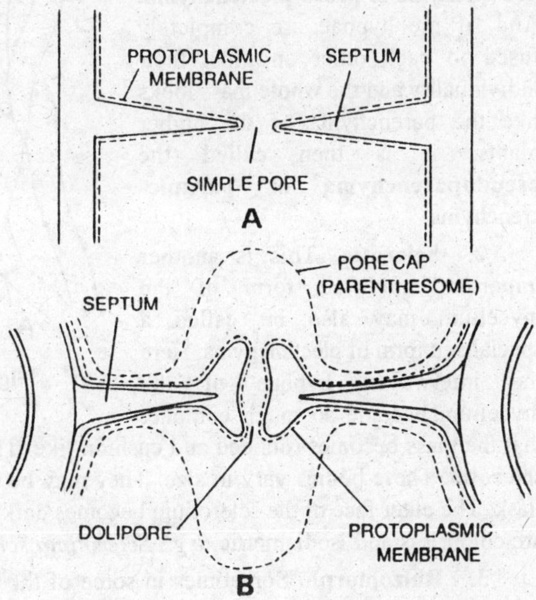

Fig. 8.3. Septal pores in fungi. A, simple pore; B dolipore in many Basidiomycetes.

The cell wall of fungi consists of a chemically different substance from normal cellulose called the **fungal cellulose.** The constituents of the fungal cellulose appear to be carbohydrates, cellulose, pectose, callose and related compounds mixed with other substances. In higher fungi the walls consist of chitin. The deposits of calcium carbonate and some other salts have also been reported from the cell wall of some fungi. The composition of the cell wall is not a constant feature and changes according to the age of mycelium, temperature, composition and the pH of the media.

The cells of a mycelium, as a whole are filled up with colourless cytoplasm. In coenocytic mycelium numerous nuclei are embedded. Many irregular vacuoles also found. The glycogen granules and oil droplets are present as food reserves in the cytoplasm. Most of the parasitic fungi on living plants secrete certain enzymes which dissolve the host cell walls.

Most of the fungi possess hyaline mycelium but some of them possess various coloured pigments. These may be yellow, green, smoky green, orange, brown and red. Certain fungi are named after the colour of their spores, *e.g.*, black molds, green molds and blue green molds. These pigments are nothing to do with the metabolic activity of the fungi.

The mycelium of the lower fungi is aseptate and coencocytic whereas of higher fungi it is always septate and the cells may be uni, bi or multinucleate. In lower fungi the septation takes place only at the time of the formation of sex organs.

There are certain modifications of the mycelium, which are as follows:

1. Plectenchyma. Sometimes the normal hyphae are so compactly interwoven that the whole mass becomes felt like and called the **plectenchyma.** If the hyphae of the mass retain their individuality and do not fuse, the mass is called the **prosenchyma** or **proso-plectenchyma.** And, if the hyphae are completely fused to each other and lost their individuality and the whole mass looks like the parenchyma of the higher plants, it is then called the **pseudoparenchyma** or **paraplectenchyma**.

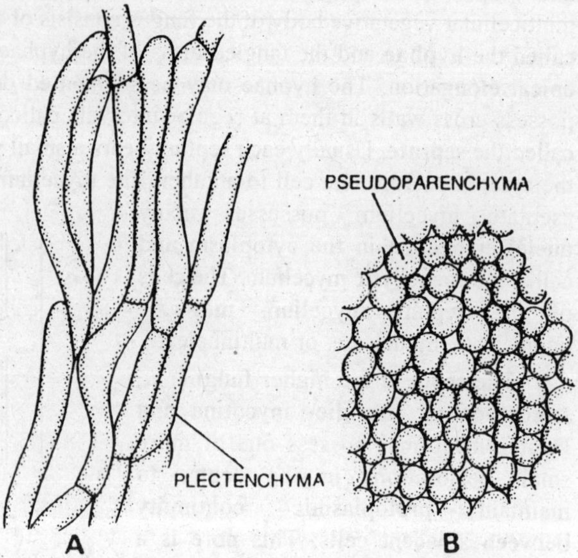

Fig. 8.4. Fungal tissues. A, plectenchyma; B, pseudoparenchyma.

2. Sclerotia. This is another important modified form of the mycelium may also be called a specialized form of plectenchyma. Here the interwoven hyphae of the mycelium become so much compact that the mass becomes rounded and cushion-like. These cushion-like compact masses are termed **sclerotia.** These bodies vary in size. They may be as small as a pin head and as large as a litre flask. The outer face of the sclerotium becomes dark brown to black and crust-like. The inner cells are colourless and isodiametric, *e.g.*, *Sclerotinia scleriotiorum*.

3. Rhizomorph. Sometimes in some of the fungi the hyphae fuse to each other forming rope-like structures running parallel to each other. These thick, gelatinous, dark brown and rope-like coiled structures are called the **rhizomorphs.** They resemble the finer roots of the trees. These structures are perennating and face the adverse conditions. They may survive even for several years. On the approach of favourable conditions these structures may give rise to new mycelia.

Fig. 8.5. Sclerotium. A, complete sclerotium; B cross section of sclerotium.

4. Sporophores. The sporophores bear spores on them. These are the modified forms of the mycelium. Usually the sporophores are erect and aerial. They arise from the prostrate hyphae. The sporophores may be branched (*e.g.*, *Peronospora*) or unbranched (*e.g.*, *Albugo*). They bear sporangia (*e.g.*, *Albugo*) or conidia (*e.g.*, *Peronospora*) on them. The sporophore bearing sporangia is called the **sporangiophore** and bearing conidia the **conidiophore**. Sometimes the sporophores are found in groups and form **pycnia, sporodochia, hymenia and acervuli** in various fungi, (Fig. 8.8)

5. **Stroma.** They are flat, cushion-like pesudoparenchymatous structures from which usually the sporophores arise.

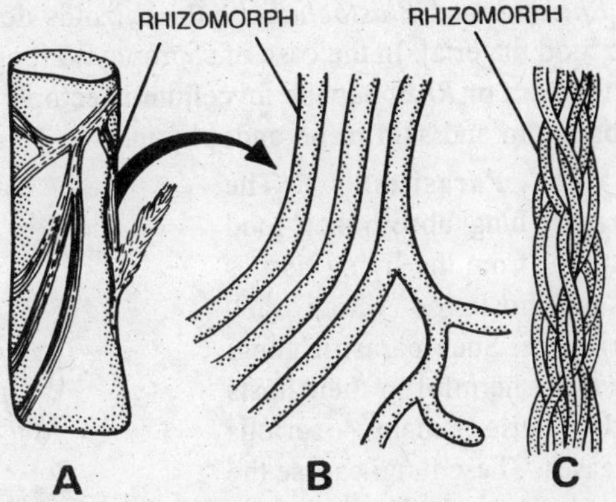

Fig. 8.6. Rhizomorph, A, as seen on root; B, parallel hyphae of rhizomorph; C, rope like structure.

Nutrition. The fungi are chlorophyllless plants and cannot synthesize their own food unlike green plants from carbondioxide and water in the presence of sunlight. They are so simple in structure that they cannot obtain inorganic food directly from the soil, and therefore they are always dependent for their food on some dead organic material or living beings. The fungi which obtain their food from dead organic materials are called the **saprophytes,** whereas the fungi obtaining their prepared food from living plants or animals are called the **parasites.** The living beings on which the fungi parasitize are called the **hosts**. Some grow in the association of other plants and are mutually beneficial. This association is called the **symbiosis** and the participants are **symbionts.** From the point of view of their nutrition the fungi may be classified as saprophytes, parasites, symbionts and predacious fungi. They are **heterotrophic** and never autotrophic.

(*a*) Saprophytes. The saprophytic fungi live on dead organic materials produced by the decay of animal and plant tissues. They grow upon dead organic matters such as rotten fruits, rotten vegetabes, moist wood, moist leather, jams, jellies, pickles, cheese, rotting leaves, plant debris, manures, horse dung, vinegar, moist bread and many other possible dead organic materials. *Saprolegnia, Mucor, Rhizopus, Penicillium, Morchella, Aspergillus, Agaricus* and many others are good examples of saprophytic fungi.

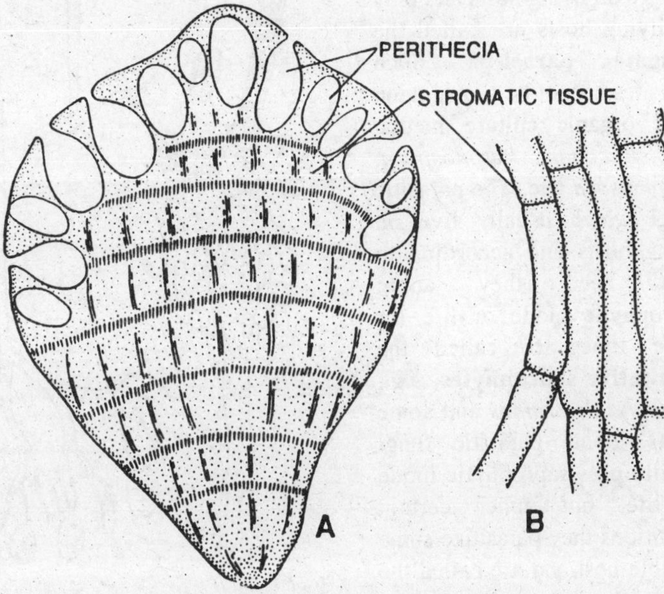

Fig. 8.7. Stroma. A, section through a stroma; B, structural detail of stroma.

The saprophytic fungi absorb their food from the substratum by ordinary vegetative hyphae which penetrate the substratum, *e.g., Mucor mucedo*. In other cases of the saprophytic fungi such as *Rhizopus* and *Blastocladiella* the rhizoids develop which penetrate the substratum and absorb the food material. In the case of saprophytic fungi the mycelium may be ectophytic or endophytic. In the case of *Rhizopus* the mycelium is ectophytic whereas the rhizoids remain embedded in the substratum and said to be endophytic.

(*b*) **Parasites.** The parasitic fungi absorb their food material from the living tissues of the hosts on which they parasitize. Such parasitic fungi are quite harmful to their hosts and cause many serious diseases. These fungi causse the great losses to the human beings or indirectly. Many diseases of the important crops are caused by parasitic fungi. The rusts, smuts, bunts, mildews and many other plant diseases are important examples of fungal diseases of crops. Their mode of life is parasitic and the relation of host and parasite is called the **parasitism.**

The parasites which survive on living hosts and only on living hosts are called the **obligate parasites.** Such parasites cannot be grown upon dead organic culture media, *e.g., Puccinia, Peronospora, Melampsora,* etc. The parasitic fungi which usually live on living hosts and according to their need they adopt saprophytic mode of life for some time are called the **facultative saprophytes**, *e.g., Taphrina deformans* and some smuts. Some parasitic fungi usually pass saprophytic mode of life, but under certain conditions they parasitize some suitable host and are called the **facultative parasites,** *e.g., Fusarium, Pythium,* etc.

The parasitic fungi absorb their food from the hosts in different ways. The fungus having the mycelium outside

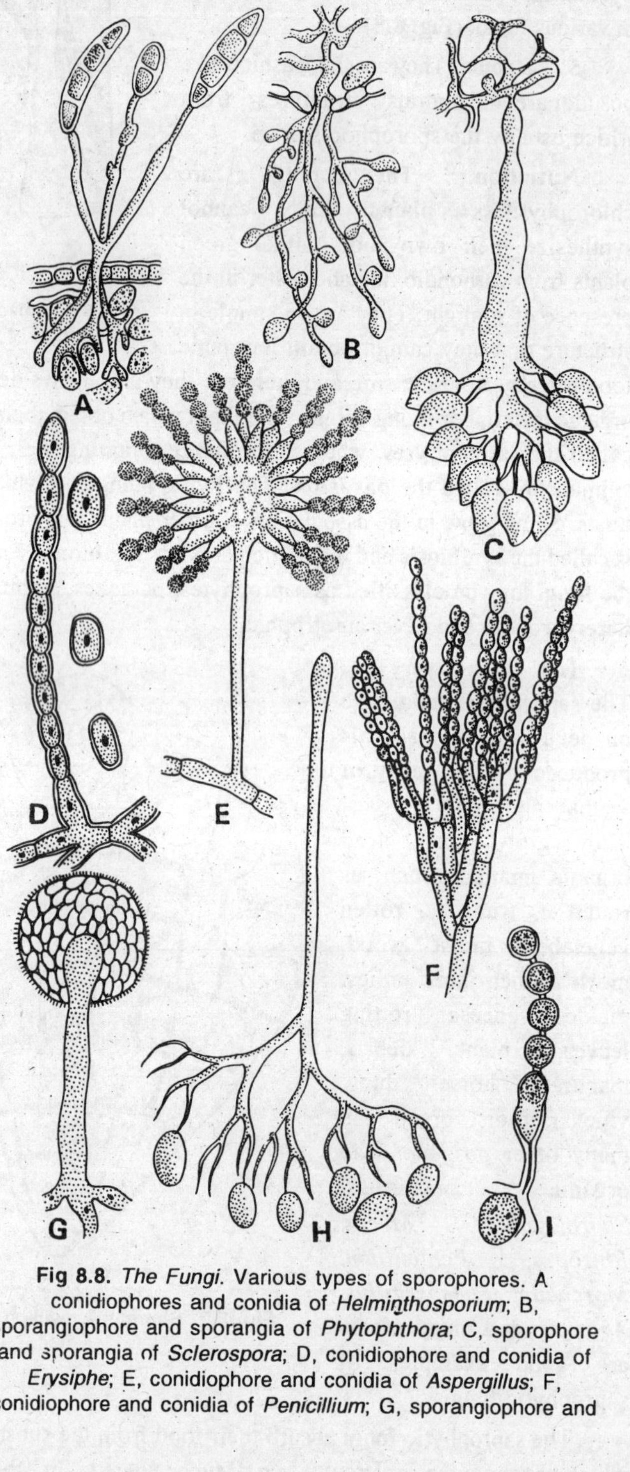

Fig 8.8. *The Fungi.* Various types of sporophores. A conidiophores and conidia of *Helminthosporium*; B, sporangiophore and sporangia of *Phytophthora*; C, sporophore and sporangia of *Sclerospora*; D, conidiophore and conidia of *Erysiphe*; E, conidiophore and conidia of *Aspergillus*; F, conidiophore and conidia of *Penicillium*; G, sporangiophore and

the host is called the ectoparasite, *e.g.*, *Erysiphe*, whereas the fungus having the mycelium embedded in the host tissue is called the **endoparasite**. In the former type certain cushion-like **appressoria** develop on the surface of the host and from each appressorium a peg-like structure develops which penetrates the host epidermal cell giving rise to a branched or unbranched absorbing organ called the **haustorium**. The haustoria may also develop from the mycelium of endoparasites. The haustoria vary in their shapes. They may be small, rounded, button-like as in *Albugo*, branched and convolute as in *Peronospora* and highly branched as in *Erysiphe*.

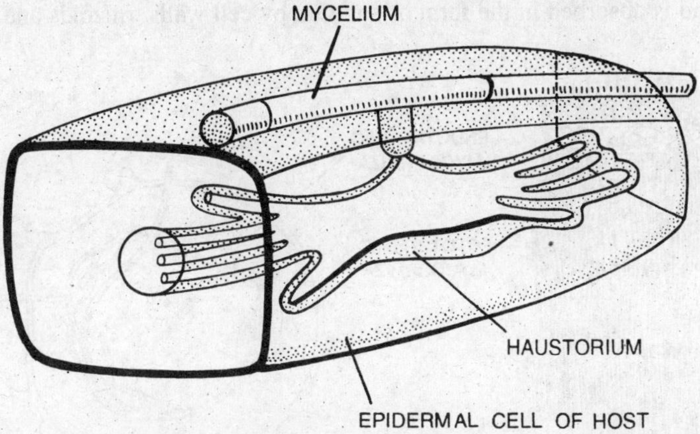

Fig. 8.9. Haustorium. Three-dimensional diagram of an infected epidermal cell with a branched haustorium of *Erysiphe* sp. (powdery mildew).

In the case of rusts and mildews the mycelium remains confined in the pustules and not in the whole body of the plant. This type of fungus is called the **localized fungus**. When the mycerlium prevails in the whole of the plant it is said to be **systemic fungus**, *e.g.*, smuts. When the mycelium is confined to the intercellular spaces it is called **intercellular mycelium** and in other cases the mycelium penetrates the host tissue and said to be **intracellular**. Usually the former bears haustoria and the latter does not.

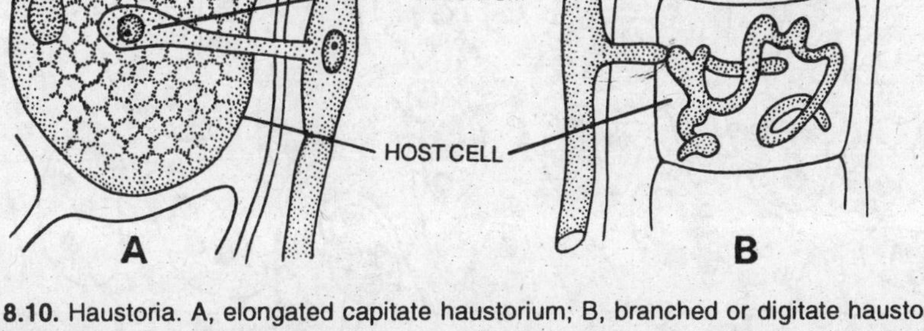

Fig. 8.10. Haustoria. A, elongated capitate haustorium; B, branched or digitate haustorium.

(c) Symbionts. Some fungi live in close association of other higher plants where they are mutually beneficial to each other. Such relationship is called the **'symbiosis'** and the participants the **'symbionts'**. The most striking examples are the **lichens** and **mycorrhiza**. The lichens are the resultants of the symbiotic association of algae and fungi. Here, both live together and are beneficial to each other. The algal partner synthesizes the organic food and the fungal partner is responsible for the absorption of inorganic nutrients and water. Certain fungi develop in the roots of higher plants and the **mycorrhiza** are developed. Here the fungi absorb their food from the roots and in response are beneficial to the plants. The mycorrhiza may be external or internal. The external mycorrhiza also called the ectophytic mycorrhiza are confined to the outer region of the roots whereas the internal mycorrhiza are found deeply in the root cells.

It is to be remembered that in all the cases whether they may be saprophytes, parasites or symbionts, the food is absorbed in the form of solution by cell walls, rhizoids and haustoria.

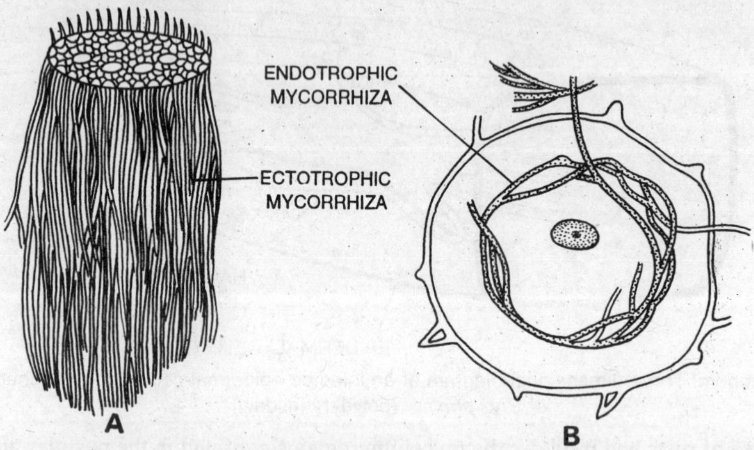

Fig. 8.11. Mycorrhiza. A, ectophytic mycorrhiza, B, endophytic mycorrhiza.

The hyphal cell walls are permeable and the plasma membranes lining to the cell walls are semipermeable. The osmotic pressure of the hyphal cells is higher than that of the host cells and thus the food materials are absorbed from the substratum and host cell. The fungi secrete some enzymes which dissolve the cellulose walls of the host, hydrolyse the starch and make it available to the fungus.

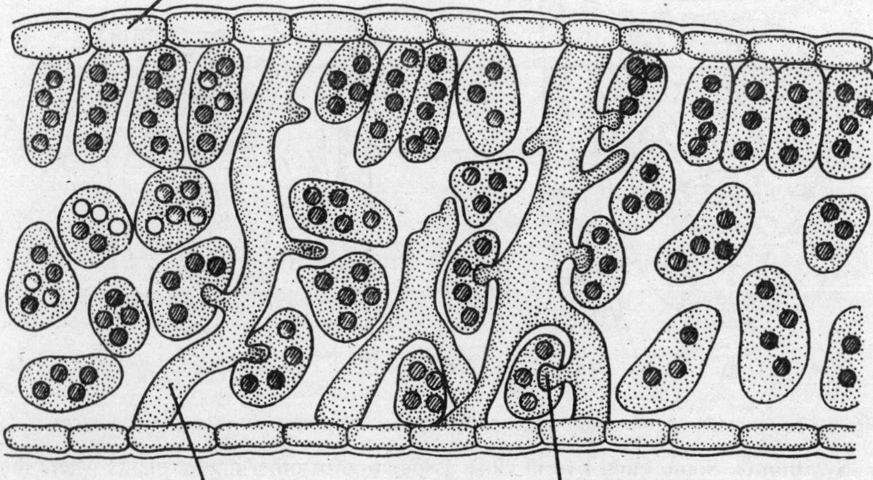

Fig. 8.12. Haustorium. Button-like or globular.

Many elements such as hydrogen, oxygen, nitrogen, small amounts of potassium, phosphorus and sulphur with the traces of magnesium and iron are required for the growth and other metabolic activities of the fungi. The carbon is always required in its organic form for growth.

When fungi are cultured in the laboratory on synthetic media, the necessary elements may be supplied in the following way: C is usually supplied in the form of a carbohydrate, such as glucose or maltose, sucrose and soluble starch are utilized by many fungi also. N may be supplied in the form of NH_4 salt or as amino acids. Many fungi can utilize NO_3 salts. Each fungus has its own specific requirements which may be known experimentally. Most fungi are able to synthesize the vitamins they need. However, several fungi may need thiamine or biotin or both these substances are generally added to synthetic media.

(*d*) Predacious fungi. There are many animal trapping fungi which have developed ingenious mechanisms for capturing small animals such as eellworms, rotifers or protozoa which they use for food. The most interesting of these mechanisms is that which utilizes a rapidly constricting ring around a nematode which holds it capitve while the hyphae sink haustoria into the body of the victim. Several species of fungi in the genera *Arthrobotrys, Dactylella* and *Dactylaria* employ this method. In the presence of an eelworm population, the hyphae of the fungi produce loops which are stimulated to swell rapidly and close the opening when an eelworm passing through the loop rules against its inner surface. It is assumed that the amount of osmotically active material in the ring cells increases greatly as a result of stimulation and causes water to enter the cells increasing their turgor pressure. The ring cells swell rapidly and the ring closes around the eelworm which is thus held tightly in the trap. Some predacious fungi secrete a sticky substance on the surface of their hyphae to which a passing small animal adheres. Haustorium like hyphae then grow into the body of the animal and absorb food. The animals ultimately die.

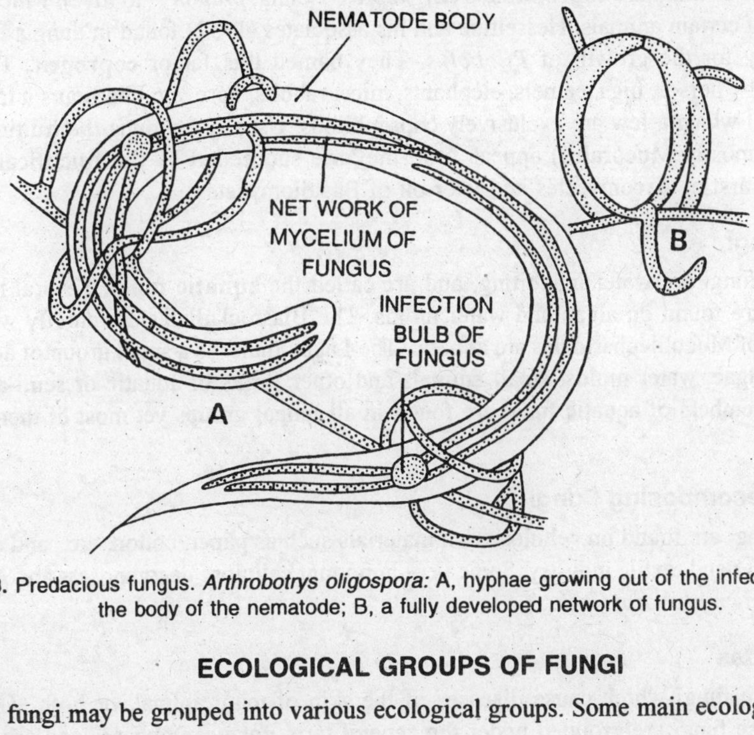

Fig. 8.13. Predacious fungus. *Arthrobotrys oligospora:* A, hyphae growing out of the infection bulb inside the body of the nematode; B, a fully developed network of fungus.

ECOLOGICAL GROUPS OF FUNGI

The fungi may be grouped into various ecological groups. Some main ecological groups are as follows:

Soil Fungi

Along with other microorganisms fungi are also associated with microbiological phenomena in the soil. When the plants grow in the soil their roots are added to the soil system. The greatest microbiological activity remains confined to the surface of the roots, called the **rhizoplane.** soil nearer the roots, under the influence of the exudations of the roots, is called the **rhizosphere.** The fungal flora of the rhizoplane and rhizosphere play an important role in the incidence of root-disease fungi. In many cases the roots of resistant varieties of plants have been found to produce hydrocyanic acid which is however, not produced in the susceptible varieties. It is presumed that this acid has an inverse effect on the pathogenic fungi by preventing them from flourishing near the roots.

Lignicolous Fungi

Lignin is quite resistant to the attack of fungi and other microorganisms. However, certain specific fungi (mostly higher fungi) are responsible for bringing about its distribution. The wood of trees is decayed by such fungi. For example *Polystictus sanguinens* causes the wood decay of the logs of *Shorea robusta* (sal tree).

Entomogenous Fungi

The fungi associated with insects are known as entomogenous fungi. The most important example of this type is *Entomophthora muscae* on house fly. Flies attached by this fungus are generally seen in moist weather attached to glass window panes surrounded by a white halo about 2 cm. in diameter formed by the discharged conidia.

Coprophilous Fungi

A number of fungi are **coprophilous** (*Gr. kopros* = dung; *philein* = to love), which grow only on the dung of certain animals. Hesseltine and his associates (1952) found in dung a factor which was necessary for the growth of *Pilobolus.* They named this factor **coprogen.** The dung of herbivores (*e.g.,* horses, pigs, camels, elephants, mice, rabbits, hare, etc.) harbours a large number of fungi out of which a few are exclusively coprophilous. Usually, on dung the fruiting bodies of lower fungi (mostly Mucorales) appear first, they are succeeded by the fructifications of the higher fungi (first of Ascomycetes and later on of Basidiomycetes).

Aquatic Fungi

Several fungi are water inhabiting, and are called the **aquatic fungi.** Several members of Chytridiales are found on algae and water molds. The Blastocladiales are chiefly water molds. Most species of Monoblepharidales are aquatic. The Lagenidiales is a small group of aquatic fungi parasitic on algae, water molds, small animals and other forms of aquatic or semi-aquatic life. Though the members of aquatic fungi are found in all fungal groups yet most of them belong to lower fungi.

Cellulose Decomposing Fungi

Some fungi are found on cellulose rich materials such as paper, cotton, etc., and cause much damage to paper and textile industry. Some very important cellulose decomposers are *Stachybotrys atra, Chaetomium globosum,* etc.

Dermatophytes

These are fungi which cause diseases of the skin of man, animal, or both. The diseases caused by these fungi are grouped under the general term **dermatomycoses** and are commonly known as ring worm, athlete's foot, etc. The important genera of these fungi belong to Fungi

Imperfecti (Deuteromycotina). Some important genera are–*Trichophyton, Microsporum, epidermophyton, Eeratinomyces,* etc.

Respiration. The fungi respire like higher plants. They intake the atmospheric oxygen and release the carbondioxide. The atmospheric oxygen is diffused through the hyphal cell walls. The fungi and bacteria which require free oxygen for their proper growth are called the **'aerobes'** whereas the micro-organisms which do not require free oxygen and may grow even in the absence of oxygen are called the **'anaerobes'.**

Reproduction. The fungi reproduce by means of vegetative, asexual and sexual methods.

1. Vegetative reproduction. The most common method of vegetative reproduction is **fragmentation.** The hypha breaks up into small fragments accidentally or otherwise. Each fragment develops into a new individual. In the laboratory the 'hyphal tip method' is commonly used for inoculation of saprophytic fungus.

In addition to above-mentioned common method of vegetative reproduction the fungi reproduced vegetatively by other means, such as **fission, budding, sclerotia, rhizomorphs, etc.** In fission, the cell constricts in the centre and divides into two giving rise to new individuals.

The budding is commonly found in *Saccharomyces.* The buds arise from the protoplasm of the parent cells and ultimately become new individuals.

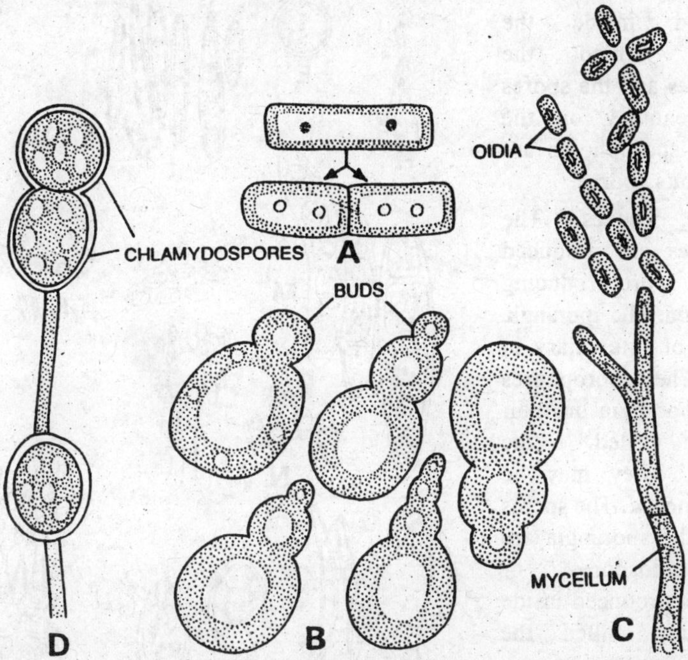

Fig. 8.14. Asexual reproduction, A, transverse cell division (fission); B, budding in yeast cell; C, hypha fragmenting into oidia or arthrospores in *Collybia conigena*; D, chlamydospore formation in *Fusarium.*

The sclerotia are resistant and perennating bodies. They survive for many years. Each sclerotium is cushion-like structure of compact mycelium. They give rise to new mycelia on the approach of favourable conditions.

As mentioned under the modified mycelium, the rope-like rhizomorphs are also resistant to unfavourable conditions and give rise to new mycelia even after several years on the approach of favourable conditions.

2. Asexual reproduction. The asexual reproduction takes place by means of spores. Each spore may develop into a new individual. The spores may be produced asexually or sexually and thus named (*a*) asexual spores and (*b*) sexual spores. Under asexual reproduction, only asexual spores will be considered.

Asexual spores. They are innumerable and produced on the diplont mycelium in Phycomycetes and Ascomycetes. In Basidiomycetes they are produced on the diplont mycelium. The spores are of diverse type and borne upon special structures called the **sporophores.** These spores are produced asexually and called the asexual spores. Usually the spores are uninucleate and nonmotile but multinucleate and motile spores are also found. The fungus producing more than one type of spores is called the **pleomorphic** or **polymorphic.** The spores produced inside the sporangia are termed the **endogenous spores** and the spores developing exogenously on the terminal ends of sporophores are called the **exogenous spores.**

Endogenous spores. The endogenous spores are produced within the special spore producing cell the **sporangium.** The sporangia may be terminal or intercalary in their position. The sporophores which bear the sporangia on their apices are called the **sporangiophores.** They may be branched or unbranched. The spores produced inside the sporangia are called the endospores or endogenous spores produced inside the sporangia are called the endospores or endogenous spores. They may be motile or non-motile. The motile spores are called the **zoospores** and the non-motile **aplanospores.** The zoospores are produced inside the zoosporangia. The protoplasm of the sporangium divides into uninucleate or multinucleate protoplasmic bits and each bit metamorphoses into a spore.

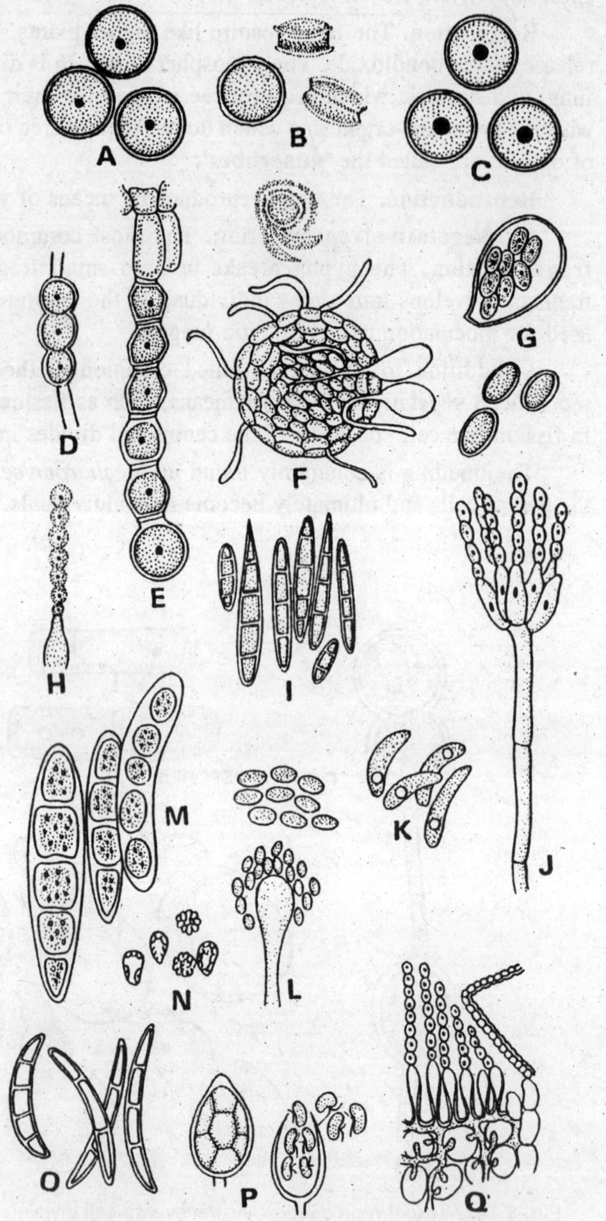

Fig. 8.15. The Fungi. Various types of spores. A, chlamydospores of *Ustilago hordei*; B, ascospores of *Aspergillus*; C, chlamydospores of *Ustilago tritici*; D, chlamydospores of *Fusarium*; E, ascospores of *Erysiphe*; F, cleistothecium of *Erysiphe*; G, ascus and ascospores of *Erysiphe*; H, conidia of *Aspergillus*; I, conidia of *Cercospora*; J, conidiophore and conidia of *Penicillium*; K, conidia of *Colletotrichum* sp; L, aplanospores of *Mucor*, M, conidia of *Helminthosporium*; N, aplanospores of *Rhizopus*; O, conidia of *Fusarium* sp; P, sporangium and zoospores of

The endogenously produced zoospores are uni or biflagellate. Each spore is without any cell wall, uninucleate and vacuolate. They can move with the help of their flagella. They are usually kidney-shaped or reniform and the flagella are inserted posteriorly or laterally on them. Such zoospores have been recorded from *Albugo, Pythium, Phytophthora* and many other lower fungi.

The aplanospores are non-motile, without flagella and formed inside the sporangia. They may be uni or multinucleate *(e. g., Mucor, Rhizopus)*. These spores lack vacuoles and possess two layered cell walls. The outer thick layer is **epispore** or **exospore** which may be ornamented in many cases. The inner thin layer is **endospore.**

Exogenous spores. The spores producing externally or exogenously are either called the exogenous spores or **conidia.** They are produced externally on the branched or unbranched **conidiophores.** The condiophores may be septate or aseptate. The conidia borne upon the terminal apices of the conidiophores or the ends of the branches of the condiophores. The conidia may be produced singly on each **sterigma** or in chains. The conidial chains may be **basipetal** to **acropetal** in succession. The conidia are diverse in their shape and size. They may be unicellular or multicellular, uninucleate or multinucleate. Different genera may be recognized only by the presence of various shaped and various coloured conidia. The conidia of Fungi Imperfecti are multicellular and variously shaped, whereas the conidia of *Aspergillus* and *Penicillium* are smoky green coloured and the fungi are called 'the blue-green molds'.

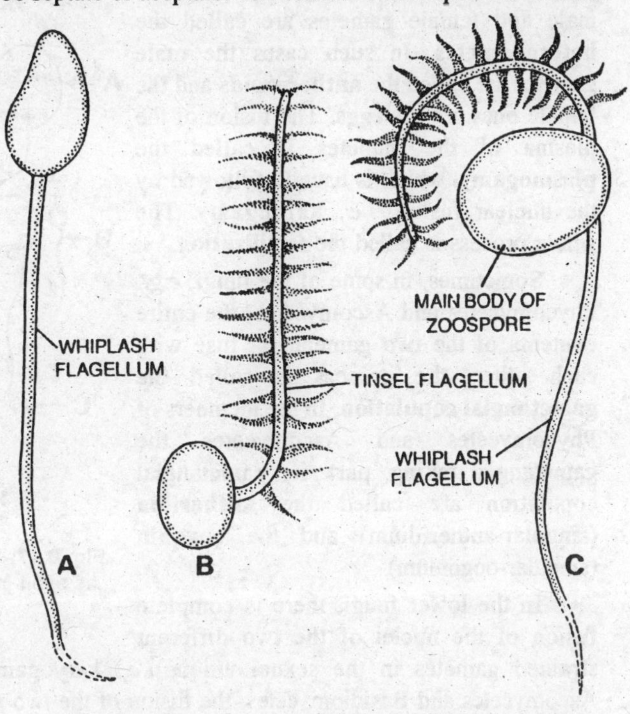

Fig. 8.16. Zoospores. A, posteriorly uniflagellate with a whiplash flagellum; B, anteriorly uniflagellate with a tinsel flagellum; C, biflagellate with an anterior tinsel and posterior whiplash flagellum.

In other type of exospores, the sporophores develop in groups and form the specialized structure called the **pustules, pycnia, aecidia, acervuli,** and **sporodochia.** The pycnia are flask-shaped producing **pycniospores** in them. The acervuli are saucer-shaped widely open bodies having developed conidia in them on small conidiophores. In mushrooms the sporophores are compactly arranged and form an umbrella-like fructification. The terminal expanded portion bears gills. In each gill there are hundreds of sporophores called the basidia bearing basidiospores. The sporophores (basidia) are arranged in hymenia.

3. **Sexual reproduction.** A large number of fungi reproduce sexually. However, the members of 'Fungi Imperfecti, or 'Deuteromycetes' lack sexual reproduction.

Usually two phases are found in the life cycle of the plants. These phases are called haploid and diploid phases respectively. The haploid phase possesses the (*n*) number of chromosomes in the nucleus, whereas this number becomes (2*n*) in the diploid phase. The gametes are always haploid (*n*) and by a sexual fusion they result in diploid (2*n*) sexual spores, such as **zygospores,**

oospores, etc. To bring haploid (n) phase once again in the life cycle the reduction division (meiosis) takes place and the number of chromosomes becomes half.

The gametes taking part in sexual fusion may be morphologically or physiologically different. Such two gametes taking part in fusion are of opposite sexes or strains, which may be called male and female sex organs or plus and minus strains. When both the sex organs or strains occur on the same mycelicum, the fungus is said to be **monoecious** or **homothallic,** and when the male and female sex organs or plus and minus strains occur separately on different mycelia the fungus is said to be **dioecious** or **heterothallic.**

The gametes taking part in fusion are usually formed in the cells of sacs called **gametangia** (singular-gametangium). The morphologically identical male and female gametes are called the **isogametes.** The morphologically dissimilar male and female gametes are called the **heterogametes.** In such cases the male gametes are called the **antherozoids** and the female ones are the **eggs.** The fusion of the plasma of the gametes is called the **plasmogamy,** which is usually followed by the nuclear fusion, *i.e.,* **karyogamy.** The whole process is called the **fertilization.**

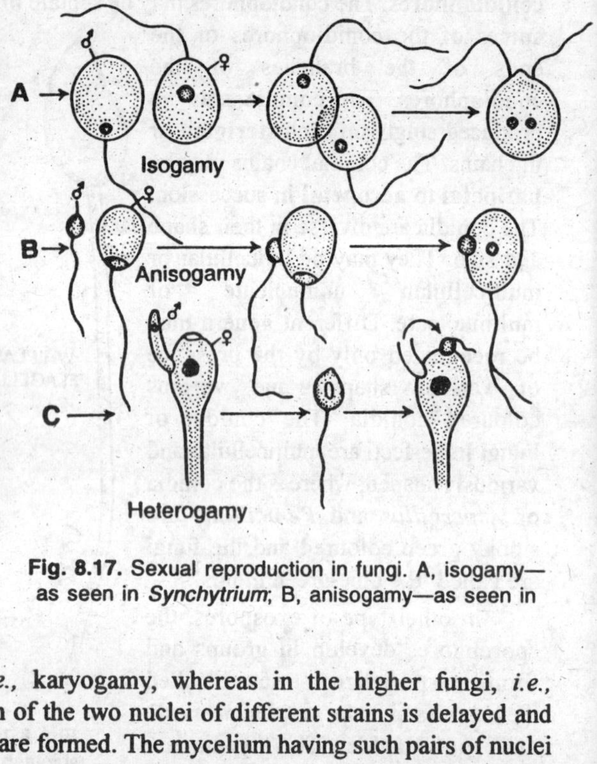

Fig. 8.17. Sexual reproduction in fungi. A, isogamy—as seen in *Synchytrium*; B, anisogamy—as seen in

Sometimes, in some of the fungi, *e.g.,* Phycomycetes and Ascomycetes, the entire contents of the two gametangia fuse with each other, the process is called the **gametangial copulation.** In the members of Phycomycetes and Ascomycetes the gametangia taking part in gametangial copulation are called the **antheridia** (singular-antheridium) and the **oogonia** (singular-oogonium).

In the lower fungi, there is complete fusion of the nuclei of the two different strained gametes in the sexual union, *i.e.,* karyogamy, whereas in the higher fungi, *i.e.,* Ascomycetes and Basidiomycetes, the fusion of the two nuclei of different strains is delayed and the pairs of the nuclei called the **'dicaryons'** are formed. The mycelium having such pairs of nuclei is called the **'dicaryotic mycelium'.** In the opposite cases where the mycelium possesses single haploid nucleus of either strain in each cell is called the **monocaryotic mycelium.** The most common methods of sexual reproduction are as follows:

1. Planogametic copulation. This type of sexual reproduction involves the fusion of two naked gametes one or both of them are motile. The motile gametes are known as planogametes. The most primitive fungi produce insogamous planogametes, *e.g., Synchytrium, Plasmodiophorq* etc. The anisogamous planogametes are only found in the genus *Allomyces* of order Blastocladiales. In *Monoblepharis* (order Monoblepharidales) the unique condition is present, here the female gamete is non-motile whereas the male gamete is motile. The male gamete enters the oogonium and fertilizes the egg.

2. Gametangial contact. This method of reproduction is found in many lower fungi (class Phycomycetes). In this method two gametangia of opposite sex (oogonium and antheridium) come in contact and one or more gamete nuclei migrate from the male gametangium (antheridium) to the female gametangium (oogonium). In no case the gametangia actually fuse. The male nuclei in some species enter the female gametangium through a pore developed by the dissolution of the wall of contact (*e.g.,* in *Aspergillus, Penicillium,* etc.).; in other species the male nuclei migrate through a fertilization tube (*e.g., Phythium, Albugo, Peronospora,* etc.). After the migration of the nuclei the antheridium eventually disintegrates but the oogonium continues its development in various ways.

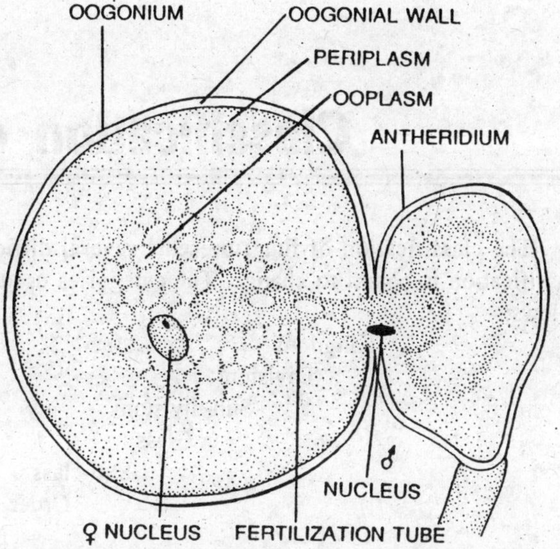

Fig. 8.18. Sexual reproduction. Gametangial contact by means of fertilization tube in *Pythium aphanidermatuam.*

3. Gametangial copulation. In this method of sexual reproduction the fusion of the entire contents of two contacting compatible gametangia takes place (*e.g., Mucor, Rhizopus, Entomophthora,* etc.)

4. **Spermatization**. The sexual reproduction in *Neurospora* (Class-Ascomycetes) and other fungi takes place by means of this method. The minute, uninucleate, spore-like male structures are known as spermatia. They are produced in several ways. The spermatia are carried by outer agencies to the receptive hyphae (trichogynes) of female gametangia, to which they become attached. A pore develops at the wall of contact and the contents of spermatium pass into the female gametangium through the receptive hypha.

5. Somatogamy. The sex organs are not produced. The somatic cells take part in sexual fusion, *e.g., Morchella,* many higher fungi.

Reduction in sexuality. In the series Flagellatae of Phycomycetes (sub-division Mastigomycotina, Ainsworth, 1971) the sex organs are quite conspicuous and called the antheridia (male) and the oogonia (female). The resultants of fusion are oospores. In the series Aplanatae (sub-division Zygomycotina, Ainsworth 1971) the gametangial copulation takes place. The gametangia are represented by + and – signs. The resultants of gametangial union are zygospores. In Ascomycetes (sub-division Ascomycotina, Ainswoth 1971), the sex organs are not so prominent and there is a regular phenomenon of reduction of sexuality. In Ascomycetes, the phenomenon of nuclear fusion is not completely understood as yet. In some of the Ascomycetes (sac fungi) (*e.g., Morchella*) no sex organs are produced and the somatic cells take part in sexual fusion. This process is called the **somatogamy.** In the Basidiomycetes (sub-division Basidiomycotina, Ainsworth 1971) there is further reduction of sexuality. In this class of fungi there are no sex organs and only the fusion of the nuclei of a dicaryon represents the process of sexuality. In the members of Fungi Imperfecti or Deuteromycetes (sub-division Deuteromycotina, Ainswoth 1971) the sexual process is altogether absent and they reproduce asexually by conidia.

9
CHAPTER

Classification of Fungi

The classification of fungi presents several difficulties. As the knowledge increases, the classification is bound to change. The categories used in the classification of the fungi are as follows:

Kingdom
  Division
    Sub-division
      Class
        Sub-class
          Order
            Family
              Genus
                Species

The kingdom is the largest of the categories and includes—divisions, sub-divisions, classes, sub-classes, orders, families and so on down to the species.

According to the recommendations of the Committee on International Rules of Botanical Nomenclature, the names of the divisions of fungi end in **-mycota**, of sub-divisions in **-mycotina.** of classes in **-mycetes**, sub-classes in **-mycetidae**, of orders in **-ales** and of families in **-aceae**. For example, *Puccinia graminis* may be classified as follows:

G. W. Martin (1961)

Kingdom	–	Plantae
Division	–	Mycota
Sub-division	–	Eumycotina
Class	–	Basidiomycetes
Sub-class	–	Heterobasidiomycetidae
Order	–	Uredinales
Family	–	Pucciniaceae
Genus	–	*Puccinia*
Species	–	*graminis*

C. J. Alexopoulos (1962)

Kingdom	–	Plantae
Division	–	Mycota
Sub-divison	–	Eumycotina
Class	–	Basidiomycetes
Sub-class	–	Heterobasidiomycetidae
Order	–	Uredinales
Family	–	Pucciniaceae
Genus	–	*Puccinia*
Species	–	*graminis*

G. C. Ainsworth (1971)

Kingdom	–	Fungi
Division	–	Eumycota
Sub-division	–	Basidiomycotina
Class	–	Teliomycetes
Order	–	Uredinales
Family	–	Pucciniaceae
Genus	–	*Puccinia*
Species	–	*graminis*

The binomial name of an organism is composed of two words. The first designates the genus and the second denotes the species. The genus name is always capitalized. The specific name starts with small letter.

Binomials when hand written should always be underlined, and when printed should be italicized.

Some mycologists are of opinion that the fungi have evolved from algae by loss of chlorophyll. If this is true, the fungi are plants and are properly placed in the plant 'kingdom'. Mycologists of other school of thought, however, believe that the fungi had a common ancestry, with the protozoa but split off at a very early stage of organic evolution. If this is true the fungi are neither plants nor animals—*they are fungi.* Ainsworth (1971) in his *dictionary of fungi* treated the fungi as a separate and independent kingdom.

The outline of *three* very important and modern systems of classification is being given.

G. W. Martin (1961)
Fungal Groups
(Division–MYCOTA)
Chlorophyll always lacking
Sub–divisions

MYXOMYCOTINA
(Assimilative phase a true plasmodium)

EUMYCOTINA
(Plasmodium lacking; thallus sometimes multinucleate or unicellular, but characteristically filamentous)

Sub-division MYXOMYCOTINA

Sub-Class–MYXOMYCETES

Sub–classes

Ceratiomyxomycetidae
(Spores borne externally on individual stalks each producing on germination a protoplasmic body which is transformed into eight swarm-cells; hypothallus massive, with erect and often branching extensions, the sporophores).

Myxogastromycetidae
(Spores borne internally in fructification of characteristic form, each spore producing on germination one or two, rarely more, swarm cells; hypothallus rarely prominent, usually inconspicuous, sometimes imperceptible.)

Orders:

1. Ceratiomyxales	1. Liceales
	2. Echinosteliales
	3. Trichiales
	4. Stemonitales
	5. Physarales.

Order-Stemonitales : Spores, black, rusty to deep violet in mass; neither peridium nor capillitium calcareous; lime rarely present and then restricted to hypothallus, stalk and columella; 2 families— 1. Collodermataceae; 2. Stemonitaceae.

Sub-division: EUMYCOTINA

PHYCOMYCETES	ASCOMYCETES	BASIDIOMYCETES	DEUTEROMYCETES
(Mycelium, when present, either continuous in active assimilative phase, or, if septate, the segments multinucleate, if lacking, reproduction not by budding; perfect stage represented by oospores or zygospores, or by zygotes resulting from union of differentiated gametes; imperfect stage by sporangiophores or modified sporangia or sporangial parts, or by conidia).	(Mycelium septate, the segments 1-2-or few-nucleate, rarely lacking and cell in such case reproducing by budding or very rarely by fission; perfect stage characterized by spores borne in asci).	(Mycelium septate the segments 1-2-nucleate; perfect stage characterized by spores borne on basidia).	(Mycelium septate with uni-or multinucleate cells: perfect stage not known or at least not normally seen).

Form-class-Lichens : Parasitic on algae, forming with them symbiotic subaerial structures often resembling green plants.

Class-PHYCOMYCETES

Sub-classes

Trichomycetidae	**Oomycetidae**	**Zygomycetidae**
(Thallus filamentous, single or branched, attached to digestive tract or external cuticle of living arthropods by a holdfast or basal cell; mycelium not immersed in tissues of hosts).	(Thallus mainly saprobic; asexual reproduction by zoospores or by detached zoosporangia; perfect stage represented by oospores produced by fusion of unlike gametangia without differentiated gametes or, less commonly, by fusion of differentiated gametes).	(Thallus mainly saprobic; asexual reproduction by aplanospores by modified sporangial segments functioning as conidia, or by true conidia; gametangia morphologically similar (sometimes differing in size; perfect stage represented by zygosores).
Orders:		
1. Harpellales	1. Hyphochytriales	1. Mucorales
2. Amoebidiales	2. Chytridiales	2. Entomophthorales.
3. Eccrianales	3. Blastocladiales	
4. Asellariales	4. Monoblepharidales	
5. Genistellales	5. Plasmodiophorales	

6. Saprolegniales
7. Leptomitales
8. Lagenidiales
9. Peronosporales

Order–Chytridiales : Zoospores posteriorly uniflagellate, usually formed inside the sporangium; thallus holocarpic or eucarpic, in the latter case with one or more reproductive structures attached to the substratum by basal vesicle or rhizoidal system; zoospore usually containing a single conspicuous oil globule. 9 families.

Order-**Blastocladiales** : Thallus always eucarpic; assimilatory portion hypha-like, scanty or profuse; zoospore without conspicuous oil globule; resistant sporangia always present; thin walled zoosporangia usually present; sexual reproduction when known, by isogamous or anisogamous planogametes; 3 families–1. Coelomomycetaceae; 2. Cateneriaceae and 3. Blastocladiaceae.

Order–**Monoblepharidales** : Thallus eucarpic; assimilative portion hypha-like; resistant sporangia lacking; thin-walled zoosporangia always present, sexual reproduction by union of free-swimming sperm and a non-motile egg, resulting in thickwalled oospore which germinates by a hypha; single family-Monoblepharidaceae.

Order–**Plasmodiophorales** : Always parasitic, assimilative phase a multinucleate thallus within host cells of plants, mostly vascular, sporulating in place and often causing hypertrophy, both flagella of zoospores of whiplash type; single family–Plasmodiophoraceae.

Order–**Saprolegniales** : Parasitic or saprobic; assimilative phase extensive mycelium; anterior flagellum zoospores always formed within the sporangium diplanetic, monoplanetic or rarely aplanetic; holocarpic or eucarpic; oogonium containing 1-several oospores; 3 families–1. Ectrogellaceae, 2. Thraustochytriaceae and 3. Saprolegniaceae.

Order–Peronosporales : Eucarpic; primarily terrestrial, living in soil or parasitic on vascular plants; zoosporangia in latter case, often functioning as conidia; 3 families–**1. Pythiaceae** (described genera: *Phythium* and *Phytophthora)*, 2. **Peronosporaceae** 3. **Albuginaceae** (described genus: *Albugo)*.

Order–Mucorales : Outer wall of zygospore formed by modification of walls of gametangia; imperfect spores typically sporangiospores; mostly saprobic, 9 families, described family-**Mucoraceae** (described genera *Mucor* and *Rhizopus.*)

Class-ASCOMYCETES

Sub Classes

Hemiascomycetidae

(Asci or ascus-like spore sacs formed singly, typically following karyogamy, no ascocarp formed).

Orders :

1. Protomycetales
2. **Endomycetales**
3. Taphrinales

Euascomycetidae

(Asci borne in ascocarps)

1. Myriangiales
2. Microthyriales
3. Dothideales
4. Pleosporales
5. Hysteriales
6. Ostropales
7. Coryneliales
8. Coronophorales
9. Laboulbeniales
10. Meliolales
11. Erysiphales
12. **Eurotiales**
13. Microascales
14. Onygenales
15. Hypocreales
16. Chaetomiales
17. Diaporthales
18. Xylariales
19. Phacidiales
20. Helotiales
21. **Pezizales**
22. Tuberales.

Order–**Endomycetales** : Individual asci formed; zygote or single cell transformed directly into an ascus; mycelium sometimes lacking, mostly saprobic; 4 families–1. Ascoideaceae. 2. **Saccharomycetaceae** (described genus–*Saccharomyces*), 3. Endomycetaceae, 4. Spermophthoraceae.

Order–**Taphrinales** : Hyphae bearing terminal chlamydospores or ascogenous cells, each of which produces a single ascus, usually forming a continuous hymenium-like layer on often modified tissues of host; parasitic on vascular plants; single family-Taphrinaceae.

Order–**Erysiphales** : Mycelium white, without appendages, usually with haustoria; perithecia without ostiole (cleistothecia), bearing appendages; the powdery mildews; single family–Erysiphaceae.

Order–**Eurotiales** : Ascogenous hyphae filling ascocarp and bearing asci at all levels, perithecia without ostioles (cleistothecia); 2 families—1. Gymnoascaceae, 2. **Eurotiaceae;** described genera: *Aspergillus* (*Eurotium*).

Order–**Hypocreales** : Ascocarp not a mazaedium; perithecia, and stromata if present, bright coloured, soft and fleshy; 2 families—1. Nectriaceae; 2. Clavicipitaceae.

Order–**Xylariales** : Perithecia dark coloured and usually possess firm walls; perithecia superficial or embedded in substratum or stroma; asci persistent, paraphyses present, persistent; centrum not pseudoparenchymatous; asci mostly in an hymenial layer lining wall sometimes in a basal cluster, their walls not notably gelatinous; 4 families — 1. Sordariaceae; 2. Phyllachoraceae; 3. Diatrypaceae and 4. Xylariaceae.

Order–**Pezizales** : Asci operculate, discharging by opening or separation of an apical lid; ascocarp an open apothecium or modified form of it, ascocarp epigeic at least at maturity, 2 families—**1. Pezizaceae** (described genus-*Peziza*) **2. Helvellaceae** (described genus-*Morchella.*)

Class-BASIDIOMYCETES

Sub-classes

Heterobasidiomycetidae

(Basidia septate or deeply divided, or arising from a teliospore or cyst; basidiospores often germinating by repetition, by budding, or by the production of conidia).

Orders :

1. Tremellales
2. **Uredinales**
3. **Ustilaginales**

Homobasidiomycetidae

(Basidia simple, cylindrical, uniform, or broadly clavate; cyst not as a rule differentiated; basidiospores on germination, usually producing a mycelium directly)

1. Exobasidiales
2. Polyporales
3. **Agaricales**
4. Lycoperdales
5. Phallales
6. Hymenogastrales
7. Nidulariales
8. Sclerodermatales

Order–**Uredinales** : Basidiocarps lacking, replaced by sori; accessory spore forms often present; always parasitic on vascular plants; sori, especially of accessory spores, usually yellow or orange, of teliospores dull orange brown to black; tube arising from teliospore cell becoming transversely septate into usually four cells, each cell producing a basidiospore; basidiospores often germinating by repetition but not budding; the rusts; 2 families — 1. Melampsoraceae 2. **Pucciniaceae** (described genus-*Puccinia*).

Order–**Ustilaginales** : Basidiocarps lacking, replaced by sori; sori usually dull black; promycelium (basidium) septate or not, bearing sessile basidiospores usually capable of germinating by budding, occasionally by repetition or teliospores rarely germinating to produce a mycelium directly; parasitic on vascular plants; the smuts; 3 families—1. Graphiolaceae; 2. **Ustilaginaceae** (described genus-*Ustilago*); 3. Tilletiaceae.

Order-**Polyporales** : Basidiocarp present, varying from an arachnoid subiculum bearing a loose hymenium to a complex and highly specialized sporocarp; hymenium amphigenous or unilateral, the surface smooth or ridged to warted, spiny or porose, or rarely lamellate, but if porose or lamellate, texture of basidiocarp not soft and putrescent; 6 families—1. Thelephoraceae; 2. Clavariaceae; 3. Cantharellaceae; 4. Hydnaceae; 5. Meruliaceae; 6. Polyporaceae.

Order–**Agaricales** : Basidiocarp present, hymemium inferior, borne on lamellae or lining the interior of tubes, in the latter case soft and putrescent; 5 families — 1. Boletaceae, 2. Paxillaceae, 3. Russulaceae, 4. Hygrophoraceae, 5. **Agaricaceae** (described genus-*Agaricus*).

Order–**Lycoperdales** : Hymenium present at least in early stages, lining chambers of gleba, gleba powdery and dry at maturity; glebal chambers usually breaking down completely; if persistent, as crumbling masses of spores; spores typically small, pale and accompanied by well-developed capillitium; 3 families—1. Arachniaceae; 2. Lycoperdaceae; 3. Geastracae.

Order–**Nidulariales** : Gleba waxy, the chamber or chambers thick-walled, each lined by an obscure hymenium and at maturity separating as peridioles which serve as disseminules; 2 families—1. Nidulariaceae; 2. Sphaerobolaceae.

Class-DEUTEROMYCETES
(Fungi Imperfecti)

Orders

Sphaeropsidales (Conidia borne in pycnidia or chambered cavities).	**Melanconiales** (Conidia borne in acervuli, definitely circumscribed, erumpent)	**Moniliales** (Conidiophores if present superficial, entirely free or bound in tufts or pulvinate masses)	**Mycelia sterilia** (No spores known, mycelium or masses of fungus cells).
Families : 1. Sphaeropsidaceae 2. Zythiaceae 3. Leptostromataceae 4. Excipulaceae	1. **Melanconiaceae** (described genus-*Colletotrichum*).	1. Cryptococcaceae 2. Sporobolomycetaceae 3. Moniliaceae 4. **Dematiaceae** (described genera-*Alternaria*, *Cercospora*.) 5. Stilbellaceae 6. Tuberculariaceae	

C. J. ALEXOPOULOS (1962)
Division–MYCOTA

Sub-divisions

MYXOMYCOTINA | **EUMYCOTINA**

Classes

MYXOMYCOTINA:
1. Myxomycetes

EUMYCOTINA:
1. Chytridiomycetes
2. Hyphochytridiomycetes
3. Plasmodiophoromycetes
4. Oomycetes
5. Trichomycetes
6. Zygomycetes

Form-class—Deuteromycetes

7. Ascomycetes
8. Basidiomycetes

DIVISION–MYCOTA (The Fungi) : Devoid of chlorophyll; the plant body varies from a microscopic unicell to an extensive mycelium; true nuclei with nuclear membranes, nucleoli present; cell walls contain chitin or cellulose, or a mixture of both, and other complex polysaccharides;

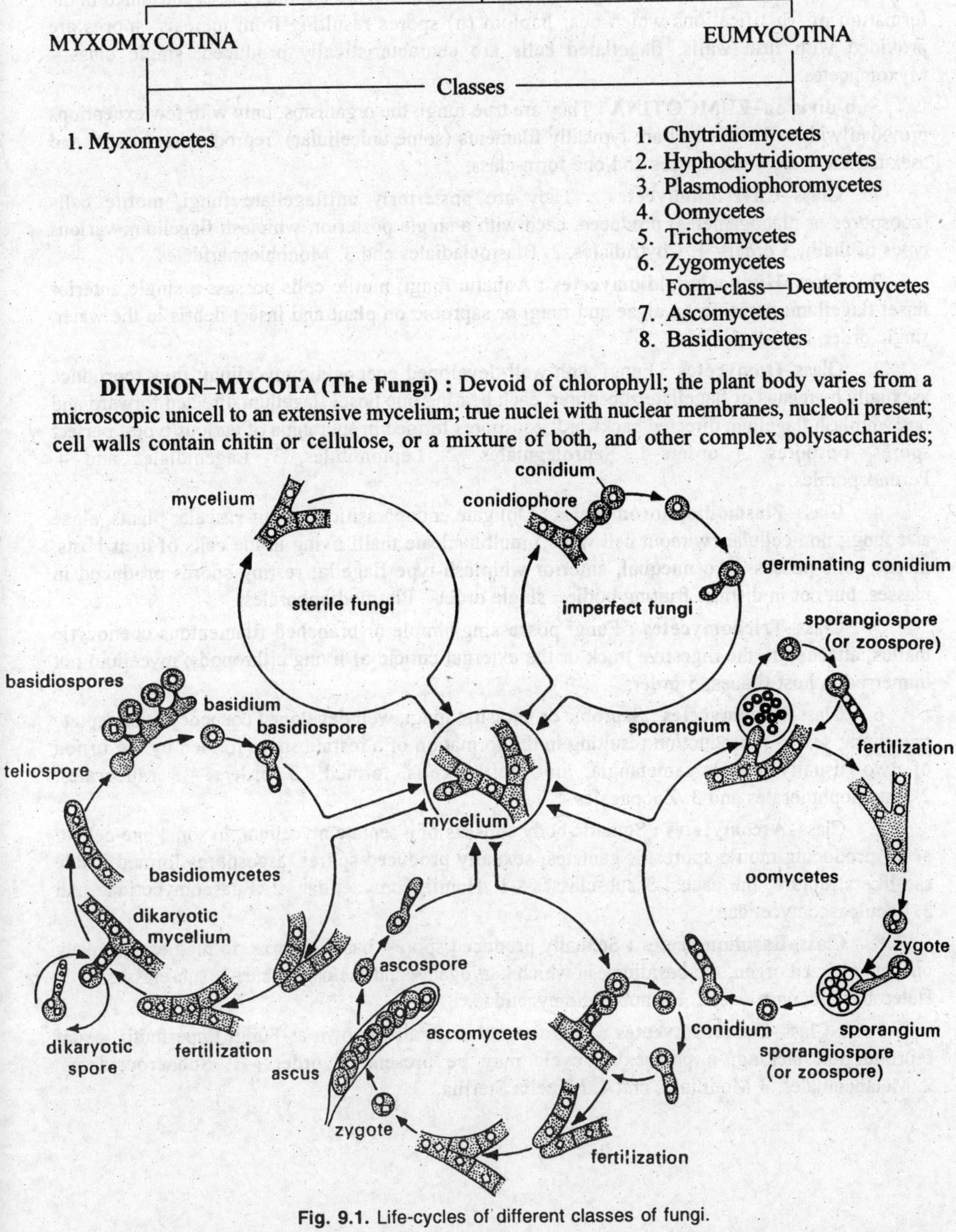

Fig. 9.1. Life-cycles of different classes of fungi.

reproduction asexual and sexual; propagative units—spores, two sub-divisions—1. Myxomycotina and 2. Eumycotina.

Sub-division–MYXOMYCOTINA : The definite cell walls are absent from their amoeba- like bodies; somatic structure, a free-living plasmodium, *i.e,* a multinucleate mass of protoplasm without definite cell walls, the entire plasmodium whose nuclei are diploid (2n), is consumed in the formation of fructifications which bear haploid (n) spores resulting from meiosis; spores are provided with firm walls, flagellated cells are characteristically produced; single class—Myxomycetes.

Sub-division–EUMCOTINA : They are true fungi, the organisms, only with few exceptions provided with cell walls and are typically filaments (some unicellular); reproduction–sexual and asexual; there are eight classes and one form-class.

1. Class–**Chytridiomycetes :** They are posteriorly uniflagellate fungi, motile cells (zoospores or planogametes) produced, each with a single posterior, whiplash flagellum; various types of thalli; 3 orders–1. Chytridiales, 2. Blastocladiales and 3. Monoblepharidales.

2. Class–**Hyphochytridiomycetes :** Aquatic fungi; motile cells possess a single anterior tinsel flagellum; parasitic on algae and fungi or saprobic on plant and insect debris in the water, single order–hypochytriales.

3. Class–**Oomycetes :** Fungi with well-developed coenocytic mycelium; they reproduce asexually by means of flagellate zoospores, each bearing one tinsel flagellum directed forward and one whiplash flagellum directed backward; zoospores formed in sporangia of various types; perfect spores—oospores; 4 orders–1. Saprolegniales, 2. Leptomitales, 3. Lagenidiales and 4. Peronosporales.

4. Class–**Plasmodiophoromycetes :** Obligate endoparasitic fungi of vascular plants, algae and fungi; non-cellular (without cell walls,) multinucleate thalli living in the cells of their hosts, motile cells possess two unequal, anterior whiplash-type flagella; resting spores produced in masses, but not in distinct fruiting bodies, single order—Plasmodiophorales.

5. Class–**Trichomycetes :** Fungi possessing simple or branched filamentous coenocytic thallus, attached to the digestive track or the external cuticle of living arthropods; mycelium not immersed in host tissues; 5 orders.

6. Class–**Zygomycetes :** Saprobic or parasitic fungi, well developed coencocytic or septate mycelium; sexual reproduction resulting in the formation of a resting spore formed by the fusion of two usually equal gametangia; no motile cells formed; 3 orders—1. Mucorales, 2. Entomophthorales and 3. Zoopagales.

7. Class–**Ascomycetes :** Somatic body consists of a septate mycelium, in some one-celled; never producing motile spores or gametes; sexually produced spores, ascospores formed inside sac-like structure, the ascus; 3 sub-classes—1. Hemiascomycetidae, 2. Euascomycetidae and 3. Loculoascomycetidae.

8. Class–**Basidiomycetes :** Sexually produced spores, basidiospores, formed exogenously on a specialized organ, the basidium, in which karyogamy and meiosis occur; 2. sub-classes — 1. Heterobasidiomycetidae. 2. Homobasidiomycetidae.

Form-Class–**Deuteromycetes :** This form-class is also known as Fungi Imperfecti; sexual reproduction lacking; a parasexual cycle may be present; 4 orders—1. Sphaeropsidales 2. Melanconiales, 3 Moniliales and 4. Mycelia Sterilia.

Class–MYXOMYCETES

Sub-classes

Ceratiomyxomycetidae (Spores borne externally on individual stalks)

Orders

1. Ceratiomyxales.

Myxogastromycetidae (Spores borne internally in fructifications)

Orders

1. Liceales
2. Echinosteliales.
3. Trichiales
4. Stemonitales
5. Physarales.

Class–CHYTRIDIOMYCETES

Orders

1. Chytridiales (True mycelium lacking, rhizomycelium present in some species).
2. Blastocladiales (True mycelium present, sexual reproduction of fungi by planogametes, thickwalled resistant sporangia characteristically formed.)
3. Monoblepharidales (True mycelium present, sexual reproduction by fusion of a male planogamete, with a female aplanogamete, no resistant sporangia).

Class–HYPHOCHYTRIDIOMYCETES

Single order

Hypochytriales

(Aquatic fungi, motile cells anteriorly uniflagellate possessing a flagellum of the tinsel type)

Class–OOMYCETES

Orders

1. Saprolegniales
2. Leptomitales
3. Lagenidiales
4. Peronosporales

Order–**Saprolegniales** : Holocarpic or eucarpic thallus, zoospores always formed within the sporgangium, diplanetic, monoplanetic or rarely aplanetic; oogonium not differentiated into ooplasm and periplasm; aquatic; saprophytic except few; 3 families—1. Ectrogelceae, 2. Thraustochytriaceae and 3. Saprolegniaceae.

Order–**Peronosporales** : Eucarpic thallus, monoplanetic; oogonium differentiated into ooplasm and periplasm, terrestrial, parasitic on higher plants; 3 families — 1. **Pythiaceae,** 2. **Albuginaceae,** 3. Peronosporaceae.

Class–PLASMODIOPHOROMYCETES

Single order

Plasmodiophorales

Always parasitic assimilative phase a multinucleate thallus within host cells of plants (mostly vascular); sporulating in place and often causing hypertrophy, both flagella of zoospores of whiplash type, single family Plasmodiophoraceae.

Class–ZYGOMYCETES

Orders

1. Mucorales
2. Entomophthorales (Two families : 1. Entomophthoraceae 2. Basidiobolaceae)
3. Zoopagales (Single family Zoopagaceae)

Order–**Mucorales** : Chiefly saprobic, some weakly parasitic on plants, a few endoparasitic in vertebrates; asexual reproduction by sporangia containing one to many aplanospores, sometimes by conidia; 9-11 families, **Mucoraceae** described in the text.

Class–TRICHOMYCETES

Class–ASCOMYCETES

Sub–classes

1. Hemiascomycetidae (Asci naked; no ascocarps; no ascogenous hyphae).
   - Orders :
     1. **Endomycetales**
     2. **Taphrinales**
2. Euascomycetidae (Asci usually unitunicate; borne in a cleistothecium, perithecium or apothecium).
   - Series
     1. Plectomycetes
        - Orders–
          1. **Eurotiales**
          2. Microascles
          3. Onygenales
     2. Pyrenomycetes
        1. Erysiphales
        2. Meliolales
        3. Chaetomiales
        4. Clavicipitales
        5. **Sphaeriales**
        6. Diaporthales
        7. Hypocreales
        8. Coryneliales
        9. Coronophorales
     3. Discomycetes
        1. Ostropales
        2. Helotiales
        3. **Pezizales**
        4. Tuberales
     4. Laboulobeniomycetes
        1. Laboulbeniales
3. Loculoascomycetidae (Asci bitunicate; ascocarp an ascostroma).
   - Orders :
     1. Myriangiales
     2. Microthyriales
     3. Hysteriales
     4. Pleosporales
     5. Dothideales

Order-Endomycetales : Asci arising directly from zygotes each derived from the copulation of two cells, or parthenogenetically from single cells; 4 families — 1. Ascoideaceae; 2. Endomycetaceae; 3. Spermophthoraceae and 4. **Saccharomycetaceae.**

Order-Taphrinales : Product of sexuality a dikaryotic thallus; asci arising directly from cells of this thallus; single many family—Taphrinaceae.

Order–Eurotiales : Ascocarp sessile and without an ostiole; 3 families — 1. Ascosphaeriaceae, 2. Gymnoascaceae 3. **Eurotiaceae.**

Order–Erysiphales : Ascocarp closed (cleistothecium), typically black or dark coloured, wall appendaged, mycelium largely superficial, single family — Erysiphaceae.

Order–Clavicipitales : Asci persistent; ascospores thread-like; ascocarp a perithecium with an ostiole; periphyses present; asci with enlarged thickened by apical cap penetrated by a narrow thread-like apical pore; single family — Clavicipitaceae.

Order-Sphaeriales : Ascocarps and stromata, if present, dark, membranous or carbonous; perithecia, typically white to bright coloured; periphyses and apical paraphyses present; mature asci attached to the inner perithecial wall; 4 families — 1. Sordariaceae, 2. Phyllachoraceae, 3. Diatrypaceae and 4. Xylariaceae.

Order–Pezizales : Ascocarp an open apothecium or a modified form of it; apothecia above ground (epigeous); asci operculate or sub-operculate; 3. families — 1. Sarcoscyphaceae, **2. Pezizaceae** and 3. **Helvellaceae.**

Class–BASIDIOMYCETES

Sub–classes

1. Heterobasidiomycetidae (Basidium septate)

Orders :

1. Tremellales
2. Uredinales
3. Ustilaginales

2. Homobasidiomycetidae (Basidium non-septate)

1. Exobasidiales

Series-Hymenomycetes

2. Polyporales
3. Agaricales

Series—Gasteromycetes

4. Hymenogastrales
5. Lycoperdales
6. Sclerodermatales
7. Phallales
8. Nidulariales

Order–Uredinales : Basidiocarp absent; basidium (promycelium) arising from a thick-walled probasidium, a teleutospore; plant parasites; basidiospores produced on sterigmata, forcibly discharged ; 3 families—1. **Pucciniaceae,** 2. Melampsoraceae and 3. Coleosporiaceae.

Order–**Ustilaginales**	:	Basidiocarp lacking; mostly parasitic on vascular plants; teleutospores formed in a manner similar to that of chlamydospores; basidiospores sessile, not forcibly discharged, 3 families—**1. Ustilaginaceae,** 2. Tilletiaceae and 3. Graphiolaceae.
Series–**Hymenomycetes**	:	Basidiocarp present; hymenium present and exposed before the spores are mature.
Order–**Polyporales**	:	Basidiocarp present; hymenium present; hymenium gymonocarpic texture of basidiocarp not soft and putrescent; 6 families—1. The lephoraceae, 2. Clavariaceae, 3. Cantharellaceae 4. Hydnaceae, 5. Meruliaceae and 6. Polyporaceae.
Order–**Agaricales**	:	Basidiocarp present; hymenium borne on lamellae (gills), or if lining the interior of pores then basidiocarp soft and putrescent; 5 families—1. Boletaceae, 2. Paxillaceae, 3. Russulaceae, 4. Hygrophoraceae and 5. **Agaricaceae.**
Series–**Gasteromycetes**	:	Hymenium present or absent, basidiocarps remaining closed at least until the spores have been released from the basidia (*i.e.,* angiocarpic).
Order–**Lycoperdales**	:	Gleba powdery; glebal chambers not separating from peridium; hymenium present in early stages; spores mostly light coloured, small; 3 families– 1. Arachniaceae, **2. Lycoperdaceae** and 3. Geastraceae.
Order–**Nidulariales**	:	Gleba waxy; glebal chambers forming waxy peridioles, or entire gleba separating as a unit from the peridium; 2 families — 1. Sphaerobolaceae and 2. Nidulariaceae.

Form–class-DEUTEROMYCETES
(The Imperfect Fungi)

Orders

1. Sphaeropsidales (Reproduction by means of conidia borne in pycnidia).	2. Melanconiales (Reproduction by) means of conidia borne in acervuli).	2. Moniliales (Reproduction by means of conidia borne otherwise, by oidia or by budding).	4. Mycelia Sterilia (No reproductive) structures known).
Form-Families 1. Sphaeropsidaceae 2. Zythiaceae 3. Leptostromataceae 4. Excipulaceae	**1. Melanconiaceae** Genus-*Colletorichum*	1. Moniliaceae **2. Dematiaceae** *Helminthosporium*, *Alternoria*, *Cerospora*, 3. Stilbellaceae 4. Tuberculariaceae Genus—*Fusarium*	Important form genera *Rhizoctonia*; and *Sclerotium*

G. C. AINSWORTH (1971)

G. C. Ainswoth (1971-Dictioary of Fungi) has upgraded the fungi upto **kingdom** level. He has treated the fungi in a separate kingdom, or a sub-kingdom of plant kingdom. The outline of Ainswoth's (1971) classification is given here:

Outline of Ainsworth's (1971) Classification

Kingdom–FUNGI

- Somatic body a mass of free-living plasmodium with no firm wall
  - Division—I MYXOMYCOTA
  - Single Class—Myxomycetes
- Somatic body unicellular or multicellular filamentous.
  - II EUMYCOTA

Division II–EUMYCOTA

- Motile cells (zoospores) present; perfect spore, oospore
- Motile cells absent
  - Perfect stage present
    - Perfect spore zygospore
    - Perfect spores, ascospores
    - Perfect spores, basidiospores
  - Perfect stage absent

Sub–divisions

III	IV	V	VI	VII
MASTIGO-MYCOTINA	ZYGOMYCOTINA	ASCOMYCOTINA	BASIDIO-MYCOTINA	DEUTERO-MYCOTINA

Sub-division–MASTIGOMYCOTINA

Posteriorly uniflagellate zoospore (flagellum whiplash type).	Anteriorly uniflagellate zoospore (flagellum tinsel type).	Biflagellate zoos pores (posterior flagellum whisplash type; anterior flagellum tinsel type) cellulose cell wall.	Biflagellate heterokontean swarmers (zoospore); flagella whiplash type.
Classes- CHYTRIDIOMYCETES	HYPHOCHYTRIDI-OMYCETES	OOMYCETES	PLASMODIOPHO-ROMYCETES
Orders- 1. Harpochytriales 2. Chytridiales 3. Blastocladiales 4. Monoblepharidales	1. Hyphochytriales	1. Saprolegniales 2. Leptomitales 3. Lagenidiales **4. Peronosporales**	1. Plasmodio-phorales

Sub-division–ZYGOMYCOTINA

Mostly saprophytic; sometimes weak parasites or mycoparasites some attacking insects but then developing mycelium inside instead of only attached to the inner lining of digestive tract; zygospores generally spherical in shape

Classes—

ZYGOMYCETES

Orders–

1. **Mucorales**
2. Entomophthorales
3. zoopagales

Mostly commensals with the guts of arthropods; hyphae attached to inner lining of digestive tract; rarely on external parts of aquatic living arthropods; zygospores where known biploar or biconical

TRICHOMYCETES

1. Harpellales
2. Asellariales
3. Eccrinales
4. Amoebidales

Sub-division–ASCOMYCOTINA

Ascocarp and ascogenous hyphae absent; thallus mycelial or yeastlike; asci naked

Ascocarp and ascogenous hyphae present; thallus mycelial.

Ascus bitunicate; ascocarp an ascostroma.

Ascus unitunicate; if bitunicate, then ascocarp and apothecium

Asci scattered at various levels within a cleistothecium; ascospores aseptate.

Asci regularly arranged forming a hymenium at the base or periphery of ascocarp.

Ectoparasitic on Arthropods; thallus reduced; ascocarp a perithecium; ascus inoperculate.

Not ectoparasitic on Arthropods.

Ascocarp a perithecium; ascus inoperculate.

Ascocarp an apothecium, often massive; hypogeal or epigeal, ascus inoperculate or operculate.

Classes: 1 2 3 4 5 6

Classes:

1. HEMIASCOMYCETES
2. LOCULOASCOMYCETES
3. PLECTOMYCETES
4. LABOULBENIOMYCETES
5. PYRENOMYCETES
6. DISCOMYCETES

Orders:

HEMIASCOMYCETES	LOCULOASCOMYCETES	PLECTOMYCETES	LABOULBENIOMYCETES	PYRENOMYCETES	DISCOMYCETES
1. **Endomycetales**	1. Myriangiales	1. **Eurotiales**	1. Laboulbeniales	1. Erysiphales	1. Helotiales
2. Protomycetales	2. Microthyriales	2. Microascales		2. Meliolales	2. **Pezizales**
3. Táphrinales	3. Hysteriales	3. Onygenales		3. Coroyophorales	3. Tuberales
	4. Pleosporales			4. **Sphaeriales**	
	5. Dothideales			5. Clavicipitales	

Sub-division–BASIDIOMYCOTINA

- Basidiocarps absent, basidium arising from thickwalled probasidium, a teleutospore, teleutosori on host tissue, parasitic on vascular plants.
  - Classes 1. **TELIOMYCETES**
    - Orders :
      - **1. Uredinales**
      - **2. Ustilaginales**
- Well-developed basidiocarp present basidia arranged in a hymenium saprophytic, rarely parasitic.
  - Basidiocarp gymnocarpous or angiocarpous, hymenium exposed throughout its development, *i.e.*, gymnocarpous, hymenium enclosed in the fruit body, *i.e.*, angiocarpous.
    - 2. **HYMENOMYCETES**
      - Basidium phragmobasidium
        - Sub-classes
        - PHRAGMOBASIDIOMYCETIDAE
          - 1. Tremellaels
          - 2. Auriculariales
          - 3. Septobasidiales
      - Basidium holobasidium
        - HOLOBASIDIOMYCETIDAE
          - 1. Exobasidiales
          - 2. Brachybasidiales
          - 3. Dacrymycetales
          - 4. Tulasnellales
          - 5. Aphyllophorales (Polyporales)
          - **6. Agaricales**
  - Basidiocarp angiocarpous, basidium holobasidium.
    - 3. **GASTROMYCETES**
      - 1. Podoxales
      - 2. Phyllales
      - 3. Hymenogastrales
      - 4. Lycoperdales
      - 5. Geitiriales
      - 6. Tulostematales
      - 7. Nidulariales
      - 8. Melanogastrales
      - 9. Sclerodermatales

Sub-division–DEUTEROMYCOTINA

- Without pseudomycelium or pseudomycelium with yeast-like budding cells; true mycelium absent or underdeveloped.
  - Classes: **BLASTOMYCETES**
    - Orders:
- Mycelium developed assimilatory budding cells absent.
  - Sterile mycelium, spores borne on sporophores, sporophores may be grouped together, but pycnia and acervuli are not formed
    - **HYPHOMYCETES**
      - **1. Moniliales**
      - 2. Mycelia sterilia
  - Spores or conidia formed in pycnidia or acervulii.
    - **COELOMYCETES**
      - **1. Melanconiales**
      - 2. Sphaeropsidales

10
CHAPTER

Sub-div. Mastigomycotina Class—Oomycetes

Kingdom–**PLANTAE**
Division–**MYCOTA**
Sub-division–**EUMYCOTINA**
Class–**OOMYCETES**
Water Molds, White Rusts, Downy Mildews
(Alexopoulos, 1962)

G. W. Martin (1961)
Kingdom–**PLANTAE**
Division–**MYCOTA**
Sub-div.–**EUMYCOTINA**
Class–**PHYCOMYCETES**

G. C. Ainsworth (1971)
Kingdom–**FUNGI**
Division–**EUMYCOTA**
Sub-division–**MASTIGOMYCOTINA**
Class–**OOMYCETES**

The fungi belonging to this class reproduce asexually by means of biflagellate zoospores. Each such zoospore bears one flagellum of tinsel-type directed forward and the other of whiplash type directed backward. The zoospores are borne endogenously in zoosporangia of various types.

The most advanced members of the Oomycetes are terrestrial and obligate parasites on vascular plants; several members are aquatic. The spores or spore-like sporangia of such parasites are wind disseminated.

The somatic structures of the Oomycetes range from a primitive unicellular thallus to a profusely branched filamentous mycelium, which grows abundantly in the substratum or in the host plant. Majority of the members are eucarpic. The characteristic components of the cell walls is cellulose.

The reproduction takes place by means of asexual methods. In all the members of this class the zoospores are produced except in the most advanced species (*e.g., Peronospora* sp.) where the sporangium behaves like a conidium or spore and germinates directly by a germ-tube which gives rise to the mycelium.

Sexual reproduction is always heterogamous. In the primitive Oomycetes, the entire thallus may act as a gametangium. With the result of the fusion of gametes the oospores are resulted except in the most primitive species. The oospores are developed within the oogonia and mature within them. The central part of the oogonium may be differentiated into one or more eggs (oospheres). The oospheres become uninucleate on maturation. In certain forms the oosphere becomes multinucleate and is known as **compound oosphere.**

Classification. There are *four* orders in this class. These orders are—1. Saprolegniales; 2. Leptomitales; 3. Lagenidiales and 4. Peronosporales. Of which the members of Peronosporales have been described in this text.

Order PERONOSPORALES

(15 genera; 550 species)

Characteristic Features

1. Majority of them are parasitic upon aerial portions of herbaceous plants. Besides several aquatic, amphibious and terrestrial species are found.

2. The mycelium is well developed, branched, aseptate and coenocytic. Majority of the species produce haustoria which obtain nourishment from the host cells. The haustoria are button-like or branched.

3. Asexual reproduction takes place by means of biflagellate zoospores produced within the zoosporangia. The sporangia are usually borne upon sporangiophores. Sometimes the sporangia germinate directly producing germ tubes and called the conidia. Here the sporophores are called the conidiophores.

4. The zoospores are kidnely-like, naked uninucleate and single vacuolate. Each zoospore after being liberated from the zoosporangium swims for some time, comes to rest, becomes encysted, and germinates producing a germ tube.

5. The sexual reproduction is oogamous and takes place by means of antheridia and oogonia. The antheridium develops by the side of oogonium. The wall of contact dissolves and a fertilization tube develops effecting the fertilization.

6. After fertilization the thick-walled diploid oospores are developed which may germinate directly producing a germ tube or indirectly producing the biflagellate zoospores. The outer wall of the oospore may be smooth or ornamented. The oospores are perennating bodies and germinate on the approach of favourable conditions.

Classification. This order has been sub-divided into three families: 1. Phythiaceae, 2. Albuginaceae and 3. Peronosporaceae.

Family–Pythiaceae

Characteristic Features

1. The family includes parasitic species with saprophytic tendencies.

2. Asexual reproduction takes place by means of sporangia. The sporangia may germinate directly producing germ tubes or liberate zoospores. These sporangia develop directly on the mycelium. The sporangiophores if at all present differ little from vegetative hyphae.

3. Sexual reproduction is oogamous. The sex organs develop on the apices of the hyphae.

The two genera *viz., Phytophthora* and *Pythium* are most important. These genera have been discussed here in detail.

Genus PHYTOPHTHORA (20 species)

(*Phytophthora infestans*)

The most important species *Phytophthora infestans* causes the 'late blight of potato'. This is a havoc for potato crop and causes sufficient damage. The symptoms of the disease appear both upon aerial and underground parts. The whole plant becomes blighted in severe conditions. Dry and wet rots damage the tubers.

The first sign of the disease is appearance of small brown patches on leaves which in cloudy and muggy weather rapidly increase to the whole leaf surface. In bad cases the crown also shows similar symptoms which fall over into a rotten pulp mass emitting foul odour. On the undersurface of such infected leaves a cottony growth of mycelium consisting of fructifications of the fungus is seen. This growth is absent in dry weather. The underground parts especially the tubers are also affected which often remain smaller in size and show dry rot with rusty brown markings in the flesh and brown depressions at certain places, in the skin.

Structure of mycelium. The mycelium is endophytic, branched, aseptate, coenocytic, hyaline, intercellular and nodulated. The rounded or branched haustoria are found which absorb food material from the host cells.

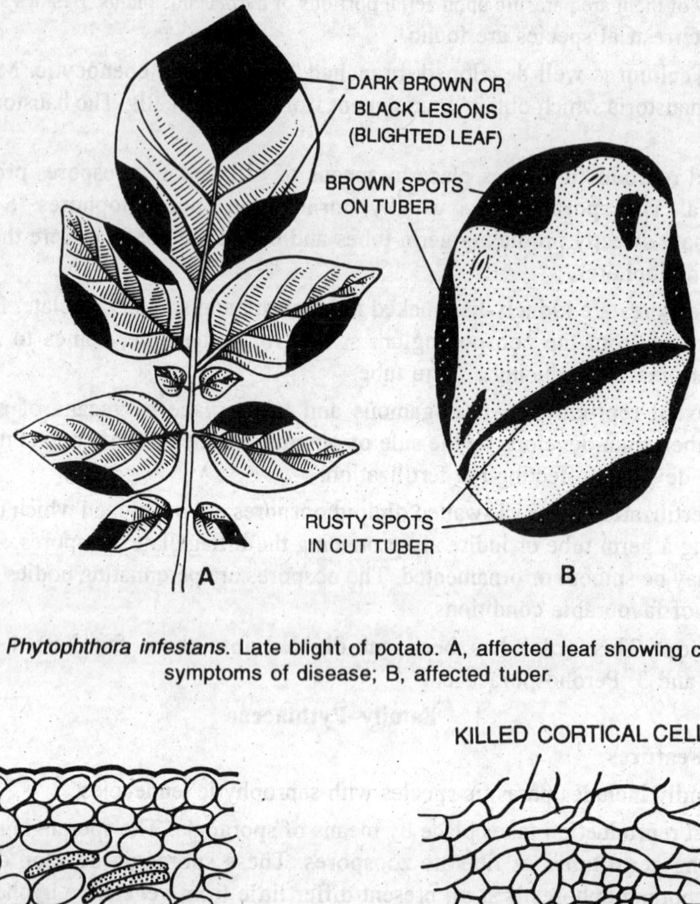

Fig. 10.1. *Phytophthora infestans*. Late blight of potato. A, affected leaf showing characteristic symptoms of disease; B, affected tuber.

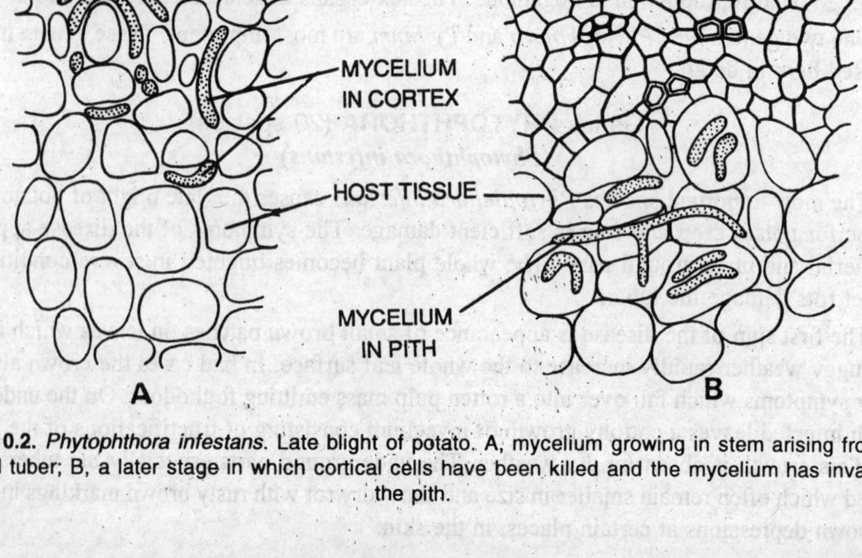

Fig. 10.2. *Phytophthora infestans*. Late blight of potato. A, mycelium growing in stem arising from infected tuber; B, a later stage in which cortical cells have been killed and the mycelium has invaded the pith.

Reproduction. The reproduction takes place by means of asexual and sexual methods.

Asexual reproduction. The asexual reproduction takes place by means of biflagellate zoospores produced inside the sporangia. In favourable conditions the sporangia are produced on the branched sporangiophores coming out through the stomata in groups on the lower surface of the infected leaves. The sporangia are formed on the branches of the sporangiophores. On maturation the sporangia are detached and leave the swellings at the points of contact on the sporangiophores.

The sporangia are rounded or lemon-shaped. At the anterior end of the sporangium there is a papilla. On maturation the protoplasm of the sporangium divides into several uninucleate, protoplasts. Each protoplast metamorphoses into a biflagellate, reniform, uninucleate, vacuolate and naked zoospore. The mature sporangium bursts at the papilla and the zoospores liberate in the film of water. The zoospores swim for some time with the help of their flagella, come to rest and become encysted. After some time in favourable conditions the encysted zoospore germinates producing a germ tube.

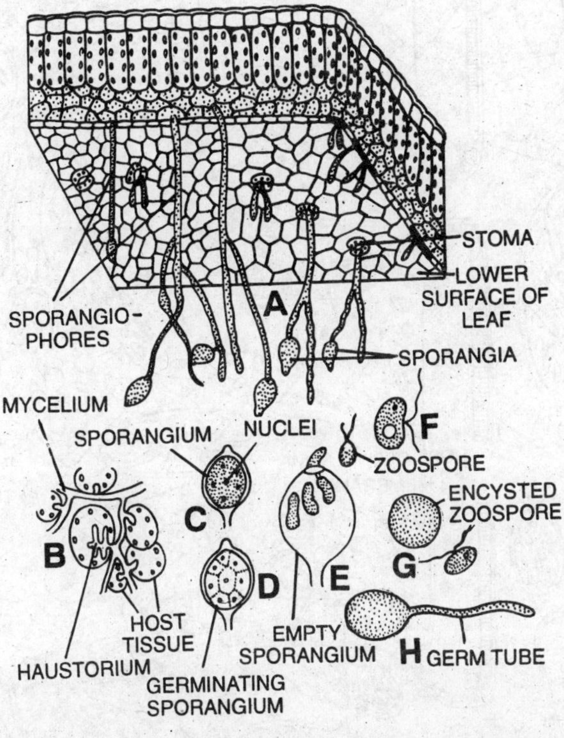

Fig. 10.3. *Phytophthora infestans.* Late blight of potato. A, sympodially branched sporangiophores coming out through stomata on the lower surface of potato leaf, swellings on sporangiophores are distinct, the sporangia are lemon-shaped; B, intercellular mycelium with haustoria; C, sporangium; D-E, germination of sporangium; F, biflagellate zoospores; G, encysted zoospore; H, germination of encysted zoospore.

Sometimes the phenomenon of **diplanetism*** is also observed. The optimum temperature for the germination of the sporangium producing zoospores is 12°C.

Conditions necessary for the production of sporangia. The sporangia are produced in abundnace in optimum conditions of weather. For the production of sporangia 100% relative humidity is required. If the relative humidity is below 90% the sporangia are not produced. The

**Diplanetism*—Condition of having two periods of motility in one life history as of zoospores in some Flagellatae.

sporangia are produced in the saturated atmosphere at 3°C-22°C temperature. The optimum temperature for the production of sporangia is 21°C.

Sexual reproduction. The sexual reproduction is oogamous. The female sex organ is oogonium and the male organ is antheridium. The antheridium develops first and the oogonium later. Both the sex organs may develop on the same hypha or on two adjacent hyphae lying side by side.

The antheridia are of two types. In some species the antheridium remains laterally attached to the oogonium. Such type of antheridium is known as **paragynous antheridium.** In other species

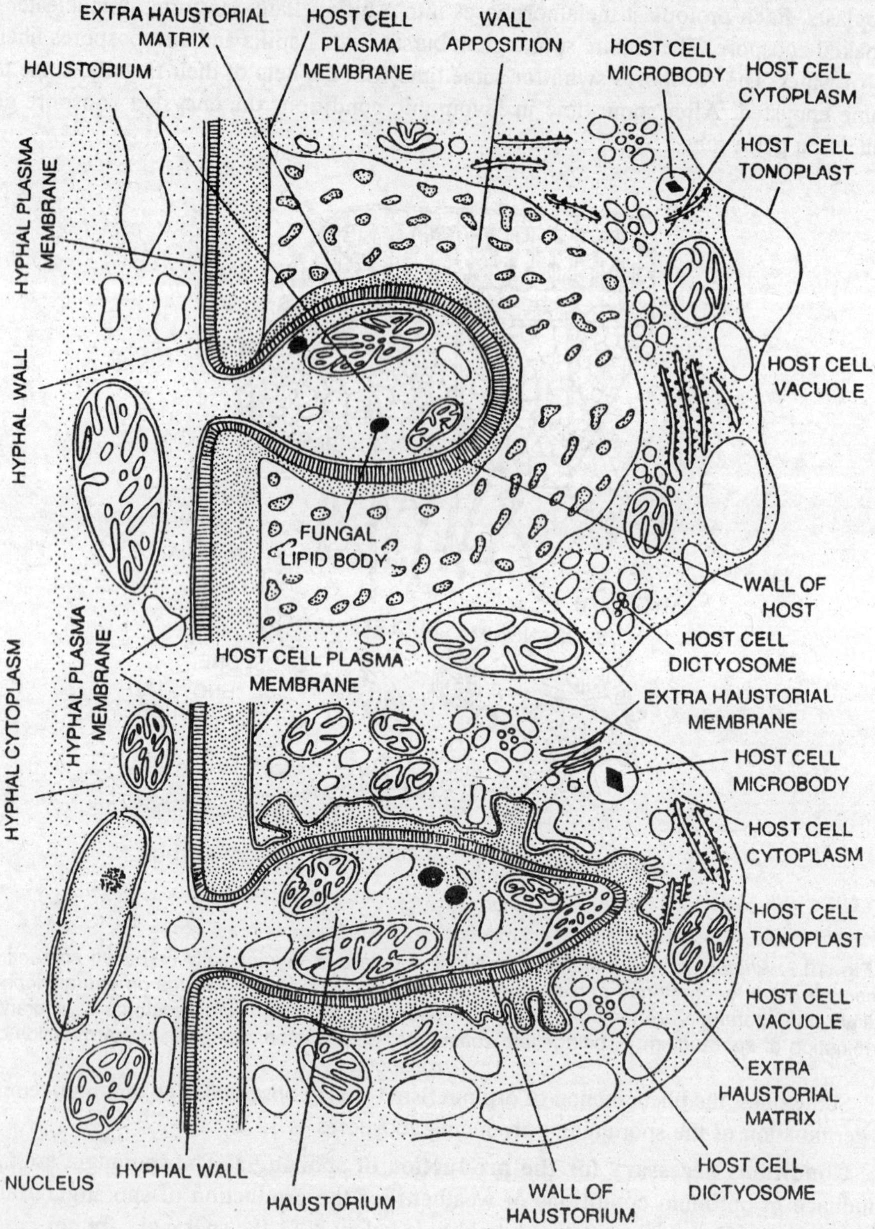

Fig. 10.4. *Phytophthora infestans.* Host parasite relationship. A, infection on partially resistant potato variety; B, infection on susceptible potato variety. Ultrastructure of mycelium and haustoria.

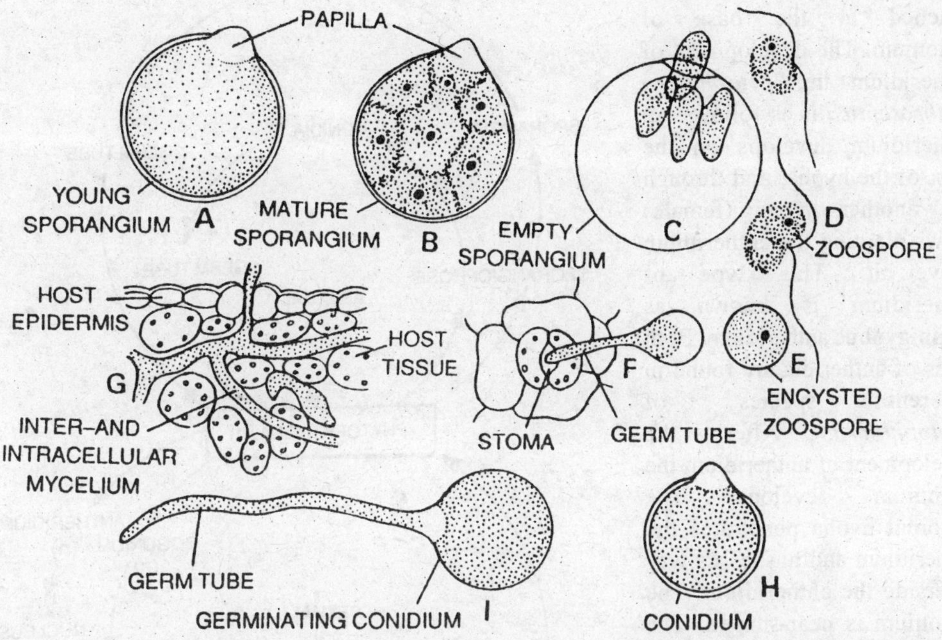

Fig. 10.5. *Phytophthora infestans.* Late blight of potato. A, young sporangium; B, mature sporangium; C, germination of sporangium and production of zoospores; D, biflagellate zoospores; E, encysted zoospore; F, germination of encysted zoospores; G, formation of inter-and intra-cellular mycelium within host tissue; H, conidium; I, germinating conidium.

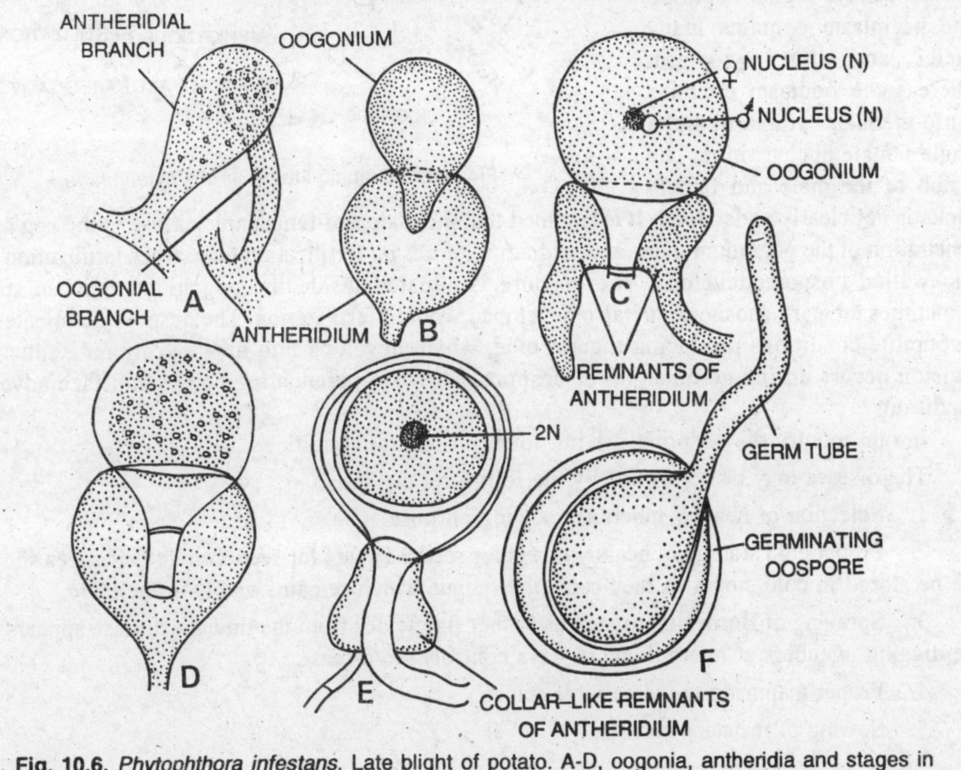

Fig. 10.6. *Phytophthora infestans.* Late blight of potato. A-D, oogonia, antheridia and stages in fertilization; E, oospore within oogonial wall, and remnants of antheridium; F, germination of oospore by germ tube.

the antheridium remains attached at the base of oogonium. The development of antheridium in *Phytophthora erythroseptica* is as follows: an antheridium develops on the apex of the hypha, and through this another hypha (female) grows piercing the antheridium above it. This type of antheridium is known as **amphigynous** antheridium. Both types of antheridia are found in different species of *Phytophthora*. After the development of antheridium the oogonium develops. The oogonial hypha penetrates the antheridium and pushes its way otherside the antheridium. The oogonium is pear-shaped. The protoplasm of the oogonium is demarcated into two regions. The outer region is periplasm and the central region ooplasm. The periplasm contains many nuclei and thin cytoplasm whereas the ooplasm contains dense granular cytoplasm and a single female nucleus in it. The fusion of the male and female nuclei is not clearly understood. It is assumed that the male and female nuclei fuse in the egg after penetration of the oogonium in the antheridium. There is no fertilization tube. After fertilization the thickwalled oospore develops. The oospore is situated inside the oogonium in loose state. Sometimes the parthenospores are also developed without fertilization. The oospore germinates in favourable conditions producing a germ tube, which develops into new mycelium. Reduction division occurs during germination of oospore. They are perennating bodies and face adverse conditions.

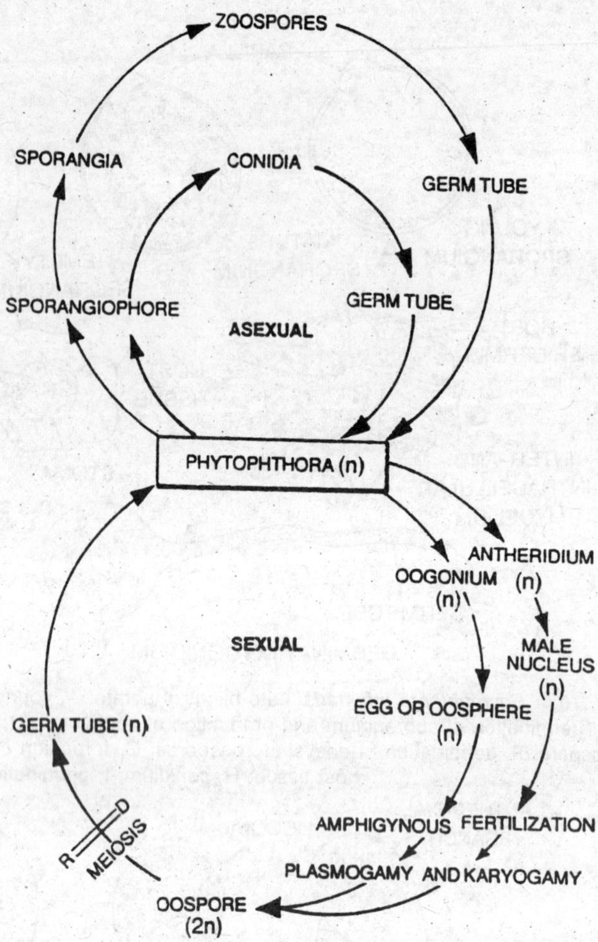

Fig. 10.7. Graphic life-cycle of *Phytophthora*.

In our country the oospores are not found in natural conditions.

The disease may be controlled by the following methods.

1. Selection of healthy tubers for sowing purpose.
2. Proper cold storage of healthy tubers for seeds. Tubers for seed from infected area should not be stored in cold stores as they carry the fungus which remains viable in the store.
3. Spraying of Bordeaux mixture and other fungicides from the time the disease appears till the digging of tubers at intervals of 15 days controls the disease.
4. Proper manuring to increase resistance.
5. Sowing of resistant varieties.

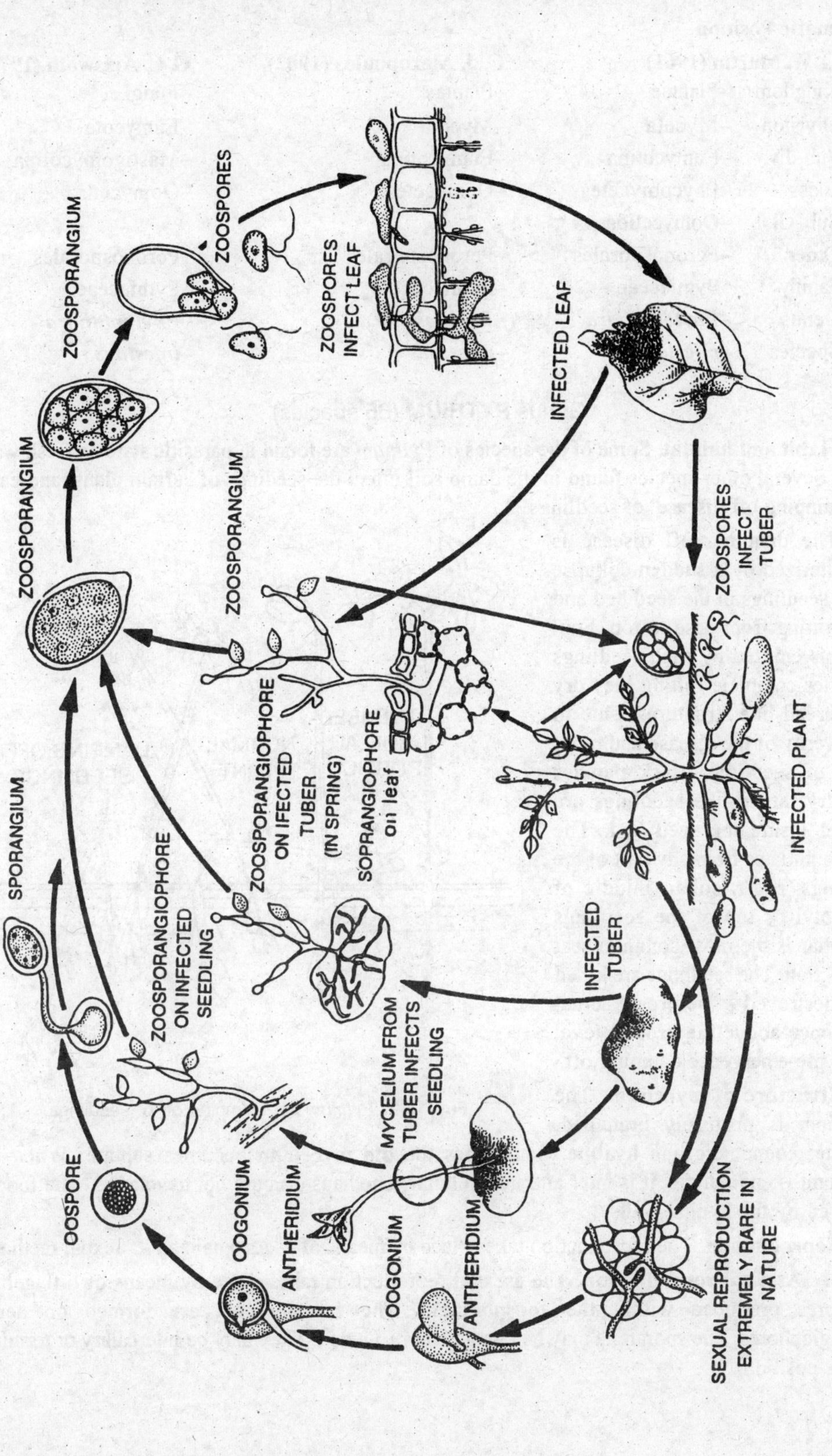

Fig. 10.8. Disease cycle of late blight of potato (*Phytophthora infestans*)

Systematic Position

	G. W. Martin (1961)	C. J. Alexopoulos (1962)	G. C. Ainswoth (1971)
Kingdom	–Plantae	–Plantae	–Fungi
Division	–Mycota	–Mycota	–Eumycota
Sub-div.	–Eumycotina	–Eumycotina	–Mastigomycotina
Class	–Phycomycetes	–Oomycetes	–Oomycetes
Sub–Cl.	–Oomycetidae		
Order	–Peronosporales	–Peronosporales	–Peronosporales
Family	–Pythiaceae	–Pythiaceae	–Pythiaceae
Genus	–*Phytophthora*	–*Phytophthora*	–*Phytophthora*
Species	–*infestans*	–*infestans*	–*infestans*

Genus **PYTHIUM** (65 species)

Habit and habitat. Some of the species of *Pythium* are found in parasitic state on fresh water algae. Several other species found in the damp soil infect the seedings of certain plants and cause the 'damping off disease' of seedlings.

The damping off disease is characterized by a sudden collapse of the seedlings in the seed bed and their rotting from base to top. Foul odour is emitted by such seedlings and after complete rotting they dry into thread-like structures. Due to the severity of the disease and rapid death of seedlings black patches with dry thread-like seedlings are marked inside the seed bed. The disease initiates from any part of the seedlings, roots, collar, middle of hypocotyl or top of the seedlings but collar is the most common seat of infection. The seedlings are killed or destroyed before actual emergence above the ground level due to pre-emergence damping off.

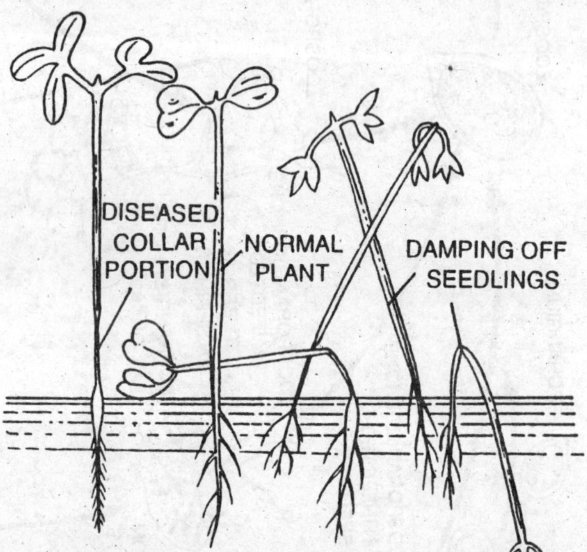

Fig. 10.9. *Pythium* sp. Damping off of seedlings.

Structure of mycelium. The mycelium is profusely branched, aseptate, coenocytic and hyaline. Sometimes the old mycelium becomes septate. When the mycelium is endophytic, it is inter and intracellular. The haustoria are not developed. The food is absorbed by the hyphal walls.

Reproduction. The reproduction takes place by means of 1. Asexual and 2. Sexual methods.

1. Asexual reproduction. The asexual reproduction takes place by means of biflagellate zoospores produced within the zoosporangia. The zoosporangia are formed on aerial sporangiophores. The sporangia may be elongated or rounded. They may be intercalary or terminal in their position.

possessing a nucleus in it. These bits are metamorphosed into zoospores. The vesicle bursts, the zoospores liberate and swim in the film of water.

Each zoospore is reniform, naked, uninucleate and biflagellate. The flagella are laterally inserted. There is a single vacuole in each zoospore. Very soon the flagella are retracted, the protoplasts become rounded and encysted. In the favourable conditions of moisture and temperature the encysted protoplast germinates producing a germ tube developing into a new mycelium.

Sometimes the sporangia behave like conidia and germinate directly producing germ tubes.

2. **Sexual reproduction.** The sexual reproduction is oogamous. The oogonia are developed on the hyphal ends. The hyphal end becomes somewhat swollen, and many nuclei and abundance of cytoplasm are accumulated in this portion. A septum develops at the base of this portion which separates this rounded oogonium from rest of mycelium. Majority of the species possess smooth walled oogonia. In some cases the oogonia are echinulate, *e.g., Pythium echinulatum.*

In the beginning of the development of the oogonium the nuclei are uniformly distributed in the protoplasm. On the maturation of the oogonium the protoplasm demarcates into outer **periplasm** and the central **ooplasm.** The periplasm is found just beneath the oogonial wall. In this region there are several nuclei and thin protoplasm. The nuclei of periplasm gradually degenerate. The ooplasm contains dense protoplasm and a single female nucleus. This is in the central portion of the oosphere or the egg.

By the side of the oogonium, the antheridium develops on the other hyphal end. The antheridium develops on the same hypha on which the oogonium is developed or on the adjacent hypha. In the development of the antheridium the hyphal end swells somewhat and becomes club-shaped. Simultaneously a septum at the base separates the antheridium from rest of the mycelium. Several nuclei and cytoplasm are found in the antheridium. All the nuclei degenerate but one which acts as male nucleus.

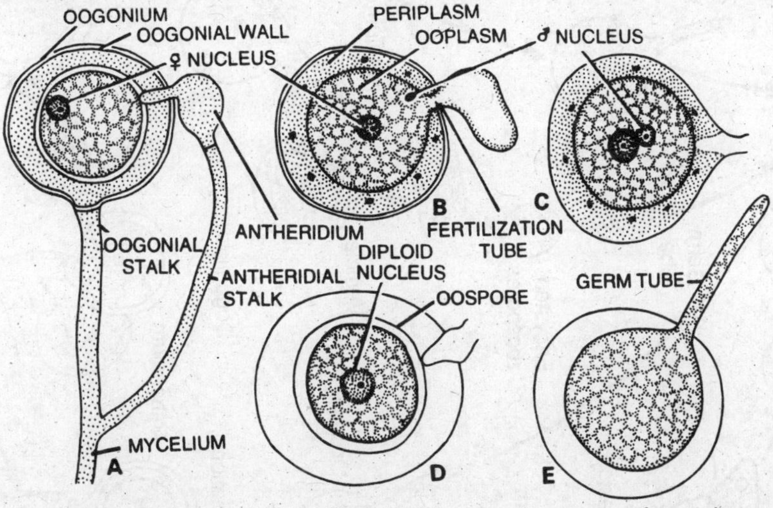

Fig. 10.14. *Pythium* sp. Damping-off of seedlings. A-C, successive stages of sexual reproduction of *Pythium*; D, oospore within oogonial wall; E, germination of oospore.

The antheridium attaches itself to the oogonium. The fertilization tube develops from the antheridium which penetrates the oogonium at the wall of contact. The fertilization tube reaches the egg through the periplasm. The male nucleus along with some protoplasm is transferred to the egg through this tube, where it fuses with the female nucleus. Very soon the process of karyogamy is completed and the oospore is formed. The oospore is thick-walled which germinates after going under a rest period.

To bring the haploid condition again, the reduction division (meiosis) takes place during the germination of the oospore.

Germination of oospore. The oospore germinates directly by producing germ tube at high temperatures (28°C). At low temperatures (10–17°C), they germinate indirectly by producing

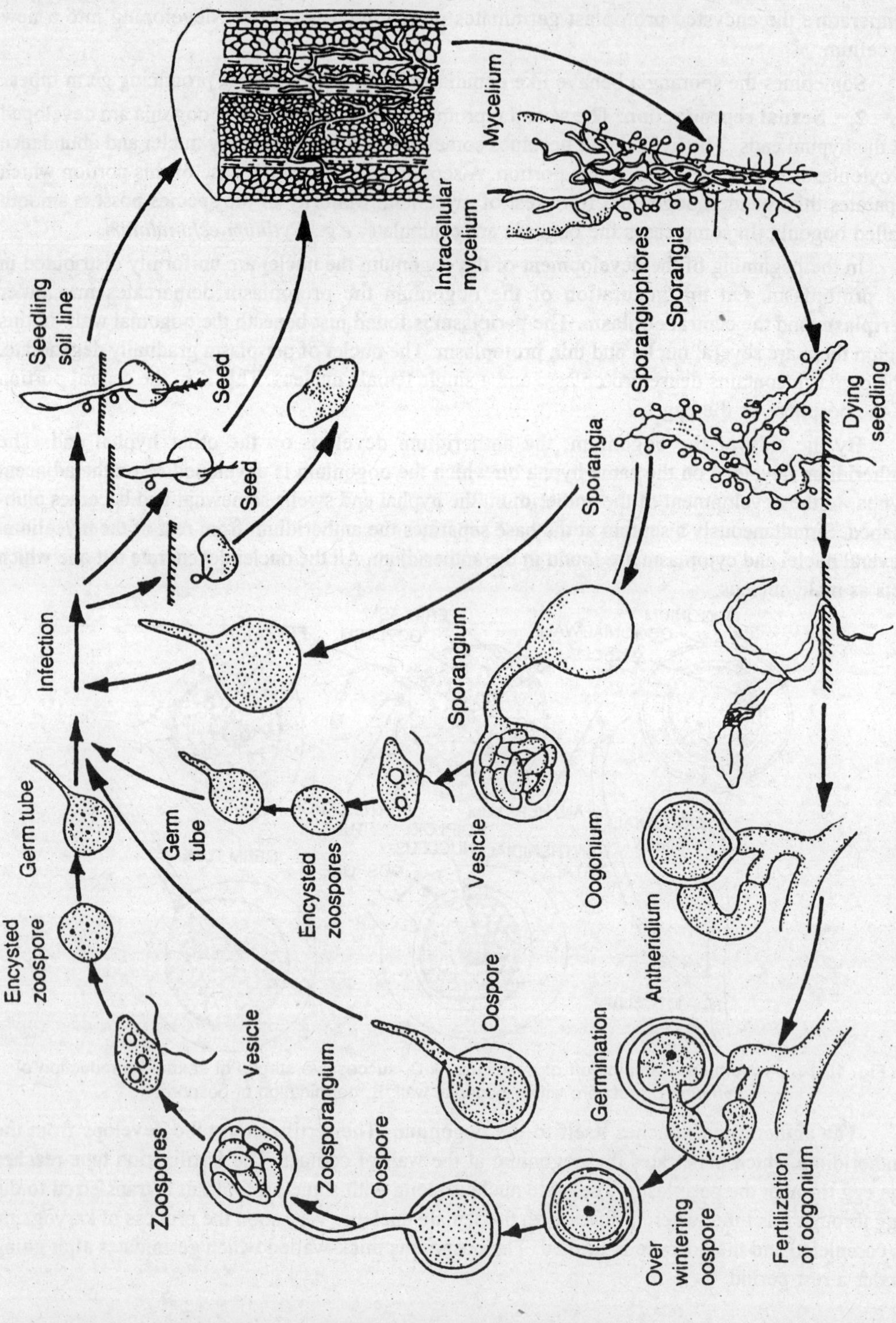

Fig. 10.15. *Pythium debaryanum*. Diagrammatic life-cycle and disease-cycle.

zoospores in the vesicle. The vesicle bursts, zoospores come out and germinate like that of asexual reproduction.

Economic importnace. The genus *Pythium* is commonly found in our country. It causes 'damping off of seedlings'. Some of the common diseases found in our cuontry are: 'Damping off of tobacco or chillies' caused by *P. debaryanum;* 'Soft rot of papaya' caused by *P. aphanidermatum;* 'Foot root of ginger' caused by *P. myriotylum*; 'Wheat rot', caused by *P. graminicolum*; ' Rhizome and root rot of turmeric' caused by *P. graminicolum;* 'Damping off of potato seedling' caused by *P. aphanidermatum*; 'Fruit rot of *Hibiscus esculentus* (*Abelmoschus esculentus*) caused by *P. indicum*.

Control. The disease is fundamentally soil borne. Attempts should therefore be made to kill the pathogen inside the nursery soil through soil treatments. Treating the seeds with various fungicides and chemicals would also protect from external attack. Water should not be allowed to stagnate in nursery beds which should be made about four inches raised from the soil level. Thin sowing of seeds would produce healthier and disease free seedlings which should always be a practice in badly affected areas.

Systematic Position

	G. W. Martin (1961)	C. J. Alexopoulos (1962)	G. C. Ainswoth (1971)
Kingdom	–Plantae	–Plantae	–Fungi
Division	–Mycota	–Mycota	–Eumycota
Sub–div.	–Eumycotina	–Eumycotina	–Mastigomycotina
Class	–Phycomycetes	–Oomycetes	–Oomycetes
Sub–Cl.	–Oomycetidae	—	—
Order	–Peronosporales	–Peronosporales	–Peronosporales
Family	–Pythiaceae	–Pythiaceae	–Pythiaceae
Genus	–*Pythium*	–*Pythium*	–*Pythium*

Family—Albuginaceae

Characteristic Features

1. The members of this family are obligate parasites. They cause diseases of vascular plants commonly known as 'white rusts'.

2. Asexual reproduction takes place by means of sporangia. The sporangiophores are short, clavate, usually unbranched, sharply differentiated from vegetative hyphae. The sporangia are found in chains.

3. Sexual reproduction is oogamous.

This family has a single genus *Albugo* or *Cystopus.* This genus has been discussed here in detail.

Genus **ALBUGO or CYSTOPUS** (30 species)
(*Albugo candida* or *Cystopus candidus*)

Habit and habitat. The genus *Albugo* or *Cystopus* is an obligate parasite and belongs to family Albuginaceae. *Albugo candida* causes the important disease of crucifers known as the 'white rust'. This species infects *Brassica campestris, B. oleracea, B. nigra, B. napus, Raphanus sativus* and *Eruca sativa.* Besides this genus also infects the memebers of Capparidaceae, Convolvulaceae and Amarantaceae.

Every part of the host plant is infected except the roots. White or cream coloured pustules develop on the surface of the leaf. Many pustules merge with each other and irregular patches are developed. These pustules are sub-epidermal. Due to the production of sporangia in abundance, the epidermis of the host bursts and powder like substance is exposed. In normal conditions the pustules appear on the lower surface of the leaf, but in severe conditions of the disease the symptoms appear on both the surfaces. The leaves of some crucifers become thick, fleshy and curled. In some conditions the leaves remain undeveloped and the plants become dwarf. The stem becomes hypertrophied and the floral parts become withered. The sepals and petals become fleshy and leaf-like. The ovary and stamens become flat, leaf like and sterile. The whole infected flower becomes hypertrophied and several times bigger than the normal size.

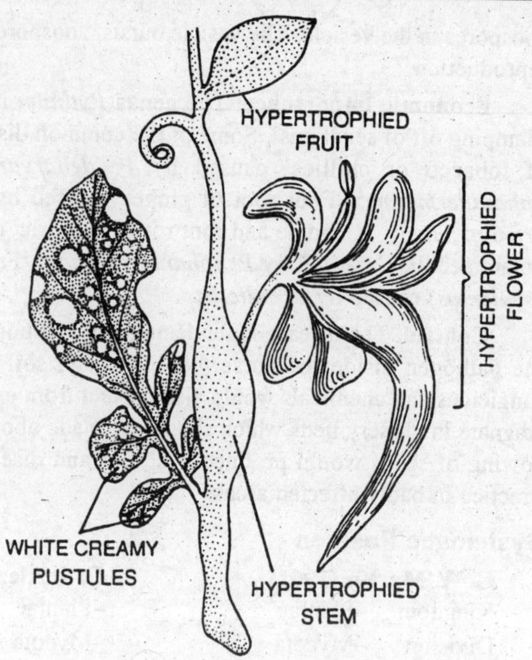

Fig. 10.16. *Albugo candida.* White rust of crucifers—infected twig, leaf with white creamy pustules, hypertrophied stem, hypertrophied flower and fruit

Structure of mycelium. The mycelium is endophytic, branched, aseptate, coenocytic, hyaline, intercellular and with knob-like haustoria for the absorption of food material from the host cells.

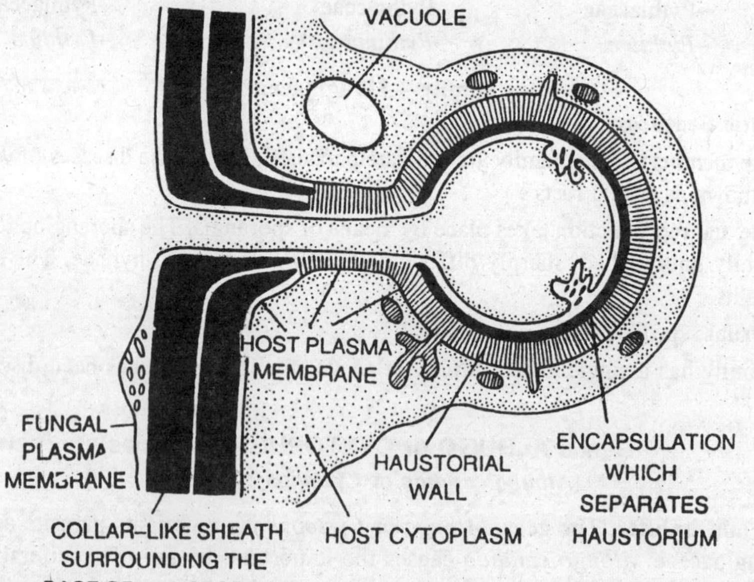

Fig. 10.17. *Albugo candida.* Knob-like haustorium as seen with electron microscope.

Reproduction. The reproduction takes place by means of asexual and sexual methods.

1. Asexual reproduction. The asexual reproduction takes place by means of biflagellate zoospores formed inside the sporangia. In the very beginning the hyphae accumulate just beneath the epidermis of the infected leaf. From these hyphae, certain thick-walled, clavate aerial

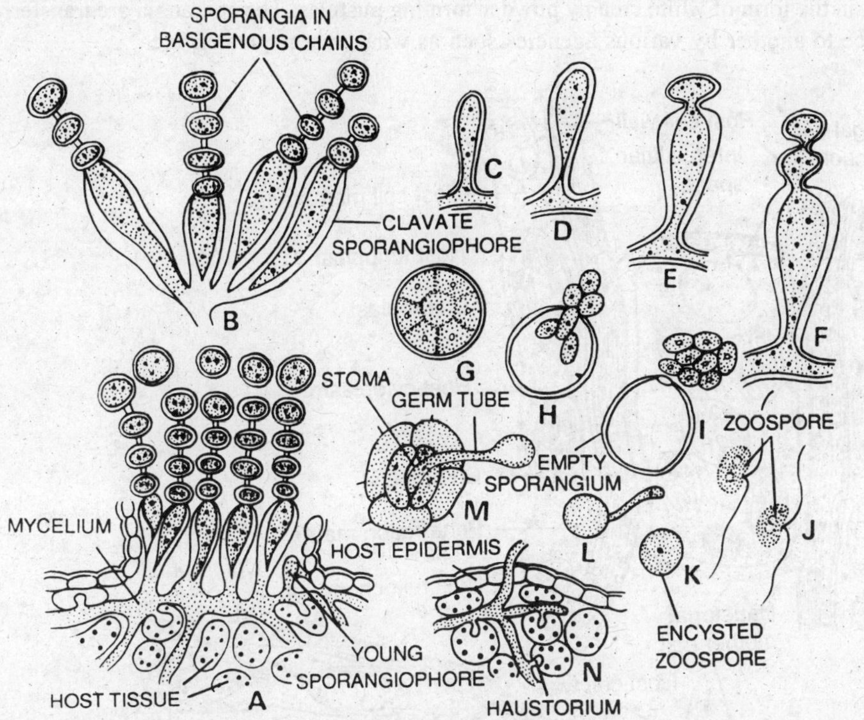

Fig. 10.18. *Albugo candida.* White rust of crucifers—asexual reproduction of *Albugo*—A, section of infected host leaf showing mycelium, haustoria, sporangiophores and sporangia; B, branching of sporangiophores with basigenous chains of sporangia; C-F, development of sporangiophore and sporangia; G-I, germination of sporangium and formation of zoospores; J, biflagellate zoospores; K, encysted zoospores; L, germinating encysted zoospores; M, infection; N, intercellular mycelium and rounded haustoria within host tissue.

sporangiophores come out. In each such sporangiophore there are about a dozen nuclei and sufficient cytoplasm. The terminal end of the sporangiophore becomes constricted and sporangium cuts off. This sporangium contains 5-8 nuclei and cytoplasm. Successively the sporangia develop by constriction method, in basigenous chains. In between each two sporangia a gelatinous pad develops acting as a separator of two sporangia from each other. The sporangium is smooth, double-walled and rounded. When the sporangia are formed in abundance on innumerable sporangiophores, the pressure is caused, the host epidermis ruptures and hundreds of sporangia are seen on the surface of

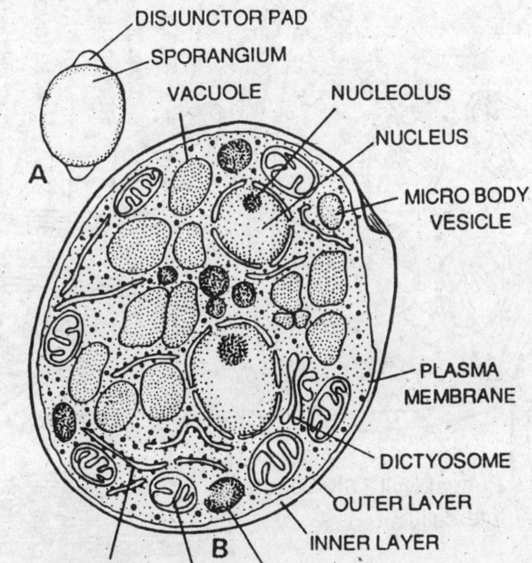

Fig. 10.19. *Albugo candida.* A, sporangium; B, ultrastructure of sporangium (diagrammatic)

the host in the form of white creamy powder forming pustules. The sporangia are transferred from one place to another by various agencies such as wind, insects, water, etc.

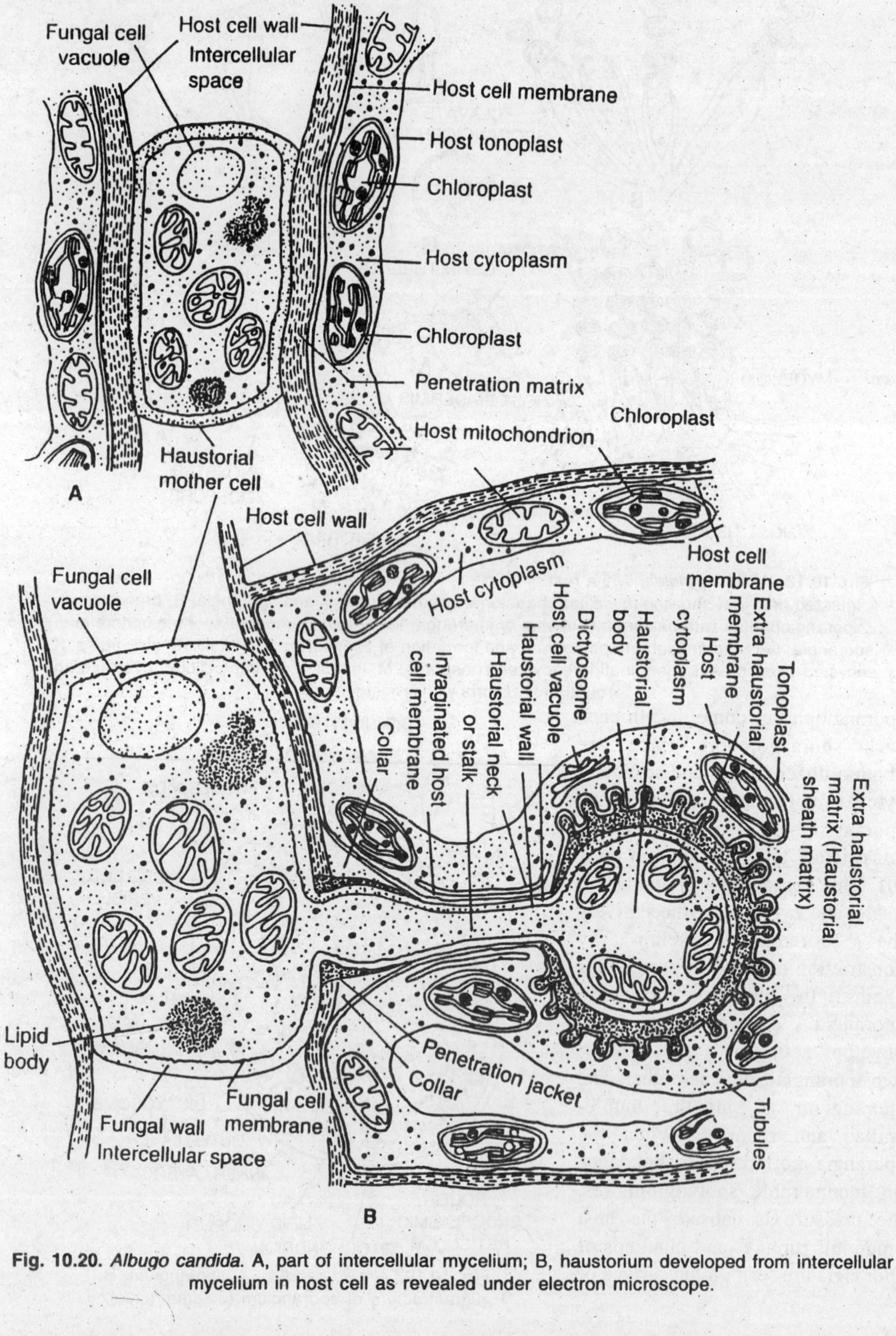

Fig. 10.20. *Albugo candida.* A, part of intercellular mycelium; B, haustorium developed from intercellular mycelium in host cell as revealed under electron microscope.

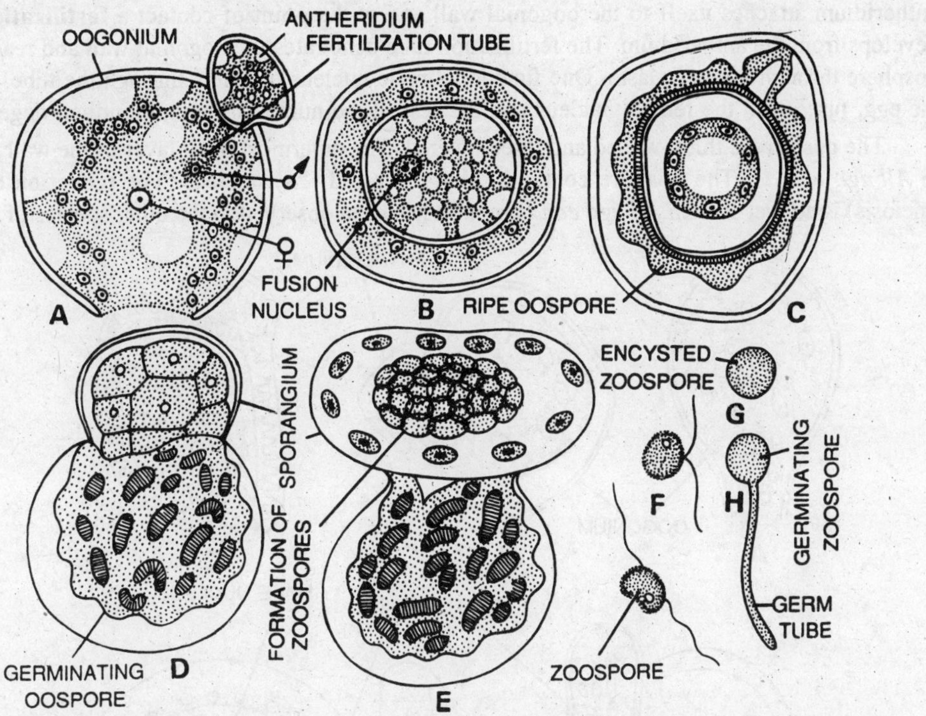

Fig. 10.21. *Albugo candida.* White rust of crucifers. A, oogonium and antheridium with fertilization tube; B, the oospore wall has begun to form, the fusion nucleus seen; C, the mature oospore in cross section; D-E, germination of oospore; F, biflagellate zoopsore; G, encysted zoospore; H, germinating zoospore.

On the maturation of the sporangium the protoplast is cleaved into uninucleate protoplasts. Each protoplast metamorphoses into a naked, biflagellate, uninucleate, reniform and vacuolate zoospore. The sporangium bursts anteriorly and the zoospores liberate in the film of water. The flagella are withdrawn and the zoospore becomes encysted. Each encysted protoplast germinates, producing a germ tube on the surface of the suitable host. The germ tube enters through stoma, develops into new mycelium and ramifies in the intercellular spaces of the host tissue.

Sometimes the sporangia behave as conidia and germinate directly producing germ tubes. The conidia may germinate from 3°C to 25°C temperature, but the optimum temperature is 10°C.

2. **Sexual reproduction.** The sexual reproduction is **oogamous.** The sex organs develop on the hyphal ends in the intercellular spaces of the deeper tissues of petioles and stems. The female sex organs are oogonia and the male sex organs are antheridia. The oogonium is rounded and the antheridium club-shaped. The developing oogonia and antheridia are separated from rest of the mycelium by septa. The cytoplasm, vacuoles and nuclei are uniformly distributed in the young oogonium. On the maturation of the oogonium the protoplasm of the oogonium differentiates into two regions. The outer region is called the **periplasm** containing thin cytoplasm, many nuclei and many vacuoles. The central protoplasm with denser consistency surrounded by periplasm is called the **oosphere** or the **egg.** The dense cytoplasm within the oosphere contains one female nucleus in it and called the **ooplasm.** In the beginning of the development of the oogonium there are many nuclei, which degenerate soon leaving one functional female nucleus.

The antheridium develops on the terminal end of another hypha lying very close to the oogonium. The hyphal end swells, becomes club-shaped and separates from rest of the mycelium by a septum. This swollen multinucleate club-shaped portion is called the antheridium. The antheridium attaches itself to the oogonial wall and at the point of contact a **fertilization tube** develops from the antheridium. The fertilization tube penetrates the oogonial wall and reaches the oosphere through the periplasm. One functional male nucleus transfers through the tube, reaches the egg, fuses with the female nucleus and the rest of the nuclei of the antheridium degenerate.

The **oospore** is thick-walled and three-layered. The outermost thick layer of the wall is warty in *Albugo cadida.* The oospore contains a large diploid ($2n$) nucleus. The reduction division (meiosis) is not yet seen in *Albugo candida* but it has been observed in the other species of *Albugo.*

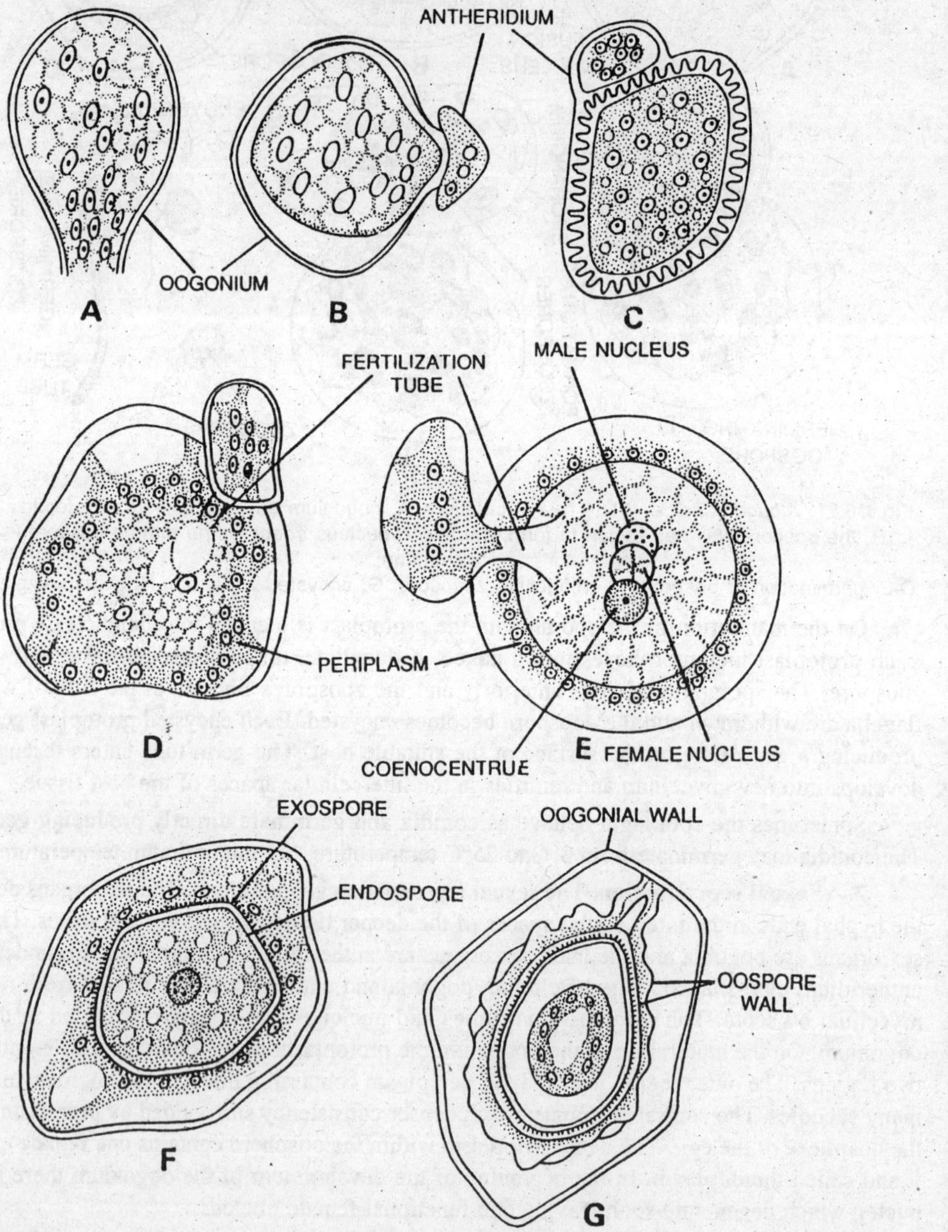

Fig. 10.22. *Albugo Candida.* A, oogonium with many nuclei; B, formation of the receptive spot towards the attached antheridium; C, nuclei in antheridium and oogonium; D, formation of fertilization tube and migration of the nuclei of oogonium in the peripheral region leaving single female nucleus in the centre of oogonium; E, fusion of male and female nuclei; F, oospore formation and karyogamy; G, mature oospore.

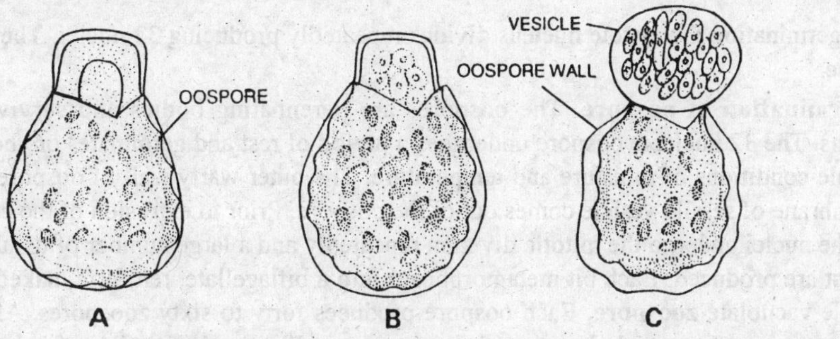

Fig. 10.23. *Albugo* sp. A-B, germination of oospore; C, the zoospores are being formed in the vesicle.

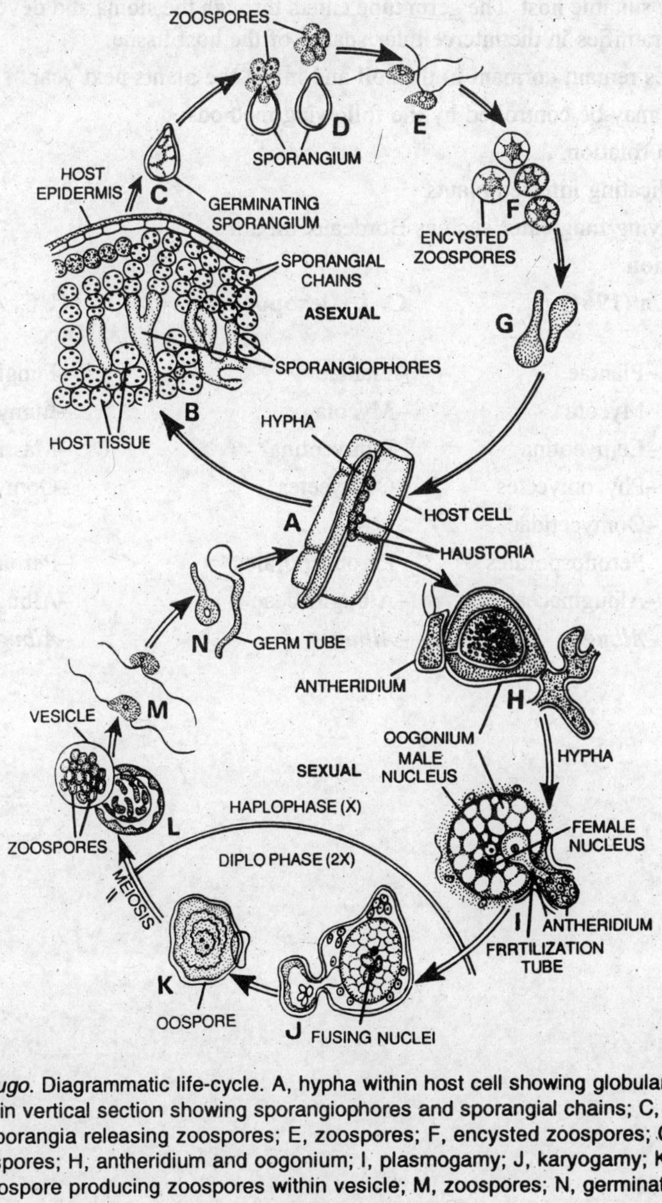

Fig. 10.24. *Albugo.* Diagrammatic life-cycle. A, hypha within host cell showing globular haustoria; B, infected leaf in vertical section showing sporangiophores and sporangial chains; C, germinating sporangium; D, sporangia releasing zoospores; E, zoospores; F, encysted zoospores; G, germination of encysted zoospores; H, antheridium and oogonium; I, plasmogamy; J, karyogamy; K, oospore; L, germination of oospore producing zoospores within vesicle; M, zoospores; N, germination of encysted zoospore.

Prior to germination the zygote nucleus divides repeatedly producing 32 nuclei. The first division is meiotic.

Germination of oospore. The oospores are perennating bodies and survive in adverse conditions. The 32 nucleate oospore undergoes a period of rest and germinates on the approach of favourable conditions of moisture and temperature. The outer warty wall of oospore bursts and a thin membrane of sessile vesicle comes out of the oospore. Prior to extrusion of the contents in the vesicle the nuclei undergo the mitotic division repeatedly and a large number of unnucleate bits of protoplast are produced. Each bit metamorphoses into a biflagellate, reniform, naked, uninucleate and single vacuolate **zoospore.** Each oospore produces forty to sixty zoospores. After extrusion from the oospore, the vesicle bursts and the zoospores liberate in the film of water where they move about with the help of their flagella. They swim about, encyst and germinate producing the germ tubes on the suitable host. The germ tube enters through the stoma and develops into the new mycelium which ramifies in the intercellular spaces of the host tissue.

The oospores remain dormant in the soil and infect the plants next year.

The disease may be controlled by the following methods:

1. By crop rotation.
2. By eradicating infected plants.
3. By spraying fungicides such as Bordeaux mixture.

Systematic Position

	G. W. Martin (1961)	C. J. Alexopoulos (1962)	G. C. Ainswoth (1971)
Kingdom	–Plantae	–Plantae	–Fungi
Division	–Mycota	–Mycota	–Eumycota
Sub–div.	–Eumycotina	–Eumycotina	–Mastigomycotina
Class	–Phycomycetes	–Oomycetes	–Oomycetes
Sub–Cl.	–Oomycetidae	—	—
Order	–Peronosporales	–Peronosporales	–Peronosporales
Family	–Albuginaceae	–Albuginaceae	–Albuginaceae
Genus	–*Albugo*	–*Albugo*	–*Albugo*

11
CHAPTER

Sub-div. Zygomycotina Class–Zygomycetes

Kingdom–**PLANTAE**
Division–**MYCOTA**
Sub-division–**EUMYCOTINA**
Class–**ZYGOMYCETES**
Bread Molds and Fly Fungi
(Alexopoulos, 1962)

G. W. Martin (1961)
Kingdom–**PLANTAE**
Division–**MYCOTA**
Sub-div.–**EUMCOTINA**
Class–**PHYCOMYCETES**

G. C. Ainsworth (1971)
Kingdom–**FUNGI**
Division–**EUMYCOTA**
Sub-div.–**ZYGOMCOTINA**
Class–**ZYCOMYCETES**

There are two main characteristics of the Zygomycetes.

1. They reproduce sexually by the fusion of two generally equal gametangia which results in the formation of a **zygospore.**

2. Asexual reproduction takes place by means of non-motile spores (aplanospores) in the form of sporangiospores or conidia.

They are terrestrial fungi. They may range all the way from saprobes, through facultative, weak parasite of plants,. to specialized parasites of animals and to obligate parasites of other Zygomycetes.

The thallus is filamentous. The hyphae are generally non-septate. Septa may be formed in old mycelium.

The characteristic component of the hyphal wall is *chitin,* but *cellulose* may also be present.

Economic importance. Some Zygomycetes attack the human body causing diseases known as Mucomycoses. They are employed in industry for the production of organic acids such as fumaric, oxalic and lactic and some are used in the production of cortisone. *Rhizopus oryzae* is employed in fermentation for making rice wine (sake) and other alcoholic preparations.

Classification. The Class Zygomycetes consists of *three* orders. They are—1. Mucorales; 2. Entomophthorales and 3. Zoopagales. Of which Mucorales has been discussed here.

Order MUCORALES
(150 genera; 250 species)

Characteristic Features

1. Majority of the species are saprophytic growing on dead organic materials. A few species are parasitic. They are called the 'black molds' because of the presence of black spores.

2. The mycelium is profusely branched, aseptate, coenocytic, vacuolate and with oil droplets.

3. Asexual reproduction takes place by means of aplanospores produced inside the globular sporangia borne terminally on aerial sporangiophores.

4. Sexual reproduction takes place by means of aplanogametes (coenogametes) produced singly in the gametangia. Most of the species are isogamous but some species are anisogamous.

5. The perfect or sexual spores are the zygospores resulted by the fusion of aplanogametes.

6. The species may be homothallic or heterothallic.

7. There are about 45 genera and 250 species in this order.

Classification. There are several families of which the family Mucoraceae has been discussed here. It is represented by *Mucor mucedo.*

Family–Mucoraceae

Characteristic Features

1. Most of the species are saprophytic.
2. The sporangiola and conidia are absent.
3. The sporangium possesses a columella. The sporangial wall is thin, fugacious, breaking or deliquescing.
4. Sporangia, chlmydospores or zygospores not enclosed in sporocarps.

The important genera are *Mucor* and *Rhizopus.* They have been discussed here in detail.

Genera **MUCOR** (60 species) and **RHIZOPUS** (5 species)

Habit and habitat. These are saprophytic fungi and grow on dead organic material. One can get these fungi in abundance by keeping moist bread under bell jar for two or three days at suitable temperature in the laboratory. The fungal mycelium looks like fine cottony threads on the surface of the bread. This is also called the 'black mold' because of its black spores.

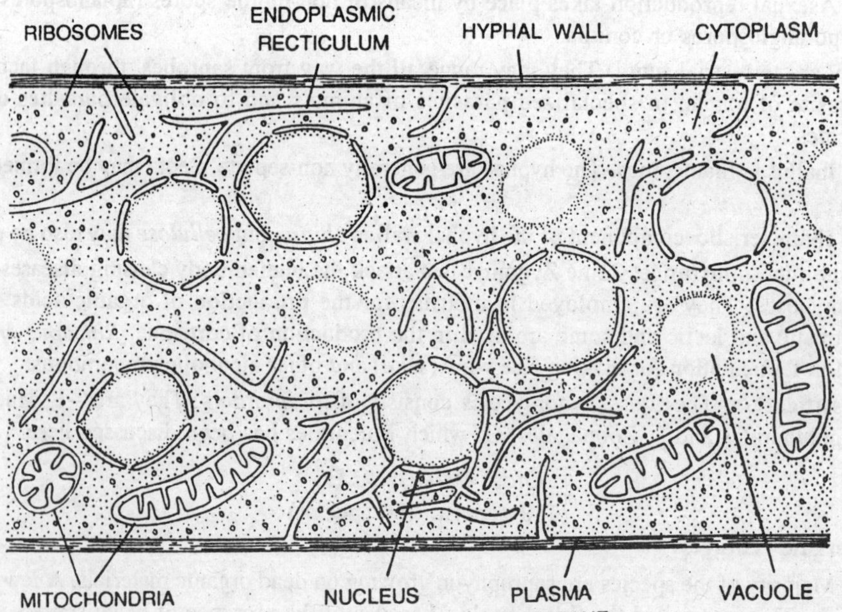

Fig. 11.1. *Mucor* sp. Part of hypha as seen under electron microscope.

Structure of mycelium. The mycelium is white cottony, branched, coenocytic, aseptate, hyline, vacuolate and with oil droplets.

Reproduction. The reproduction takes place by means of asexual and sexual methods.

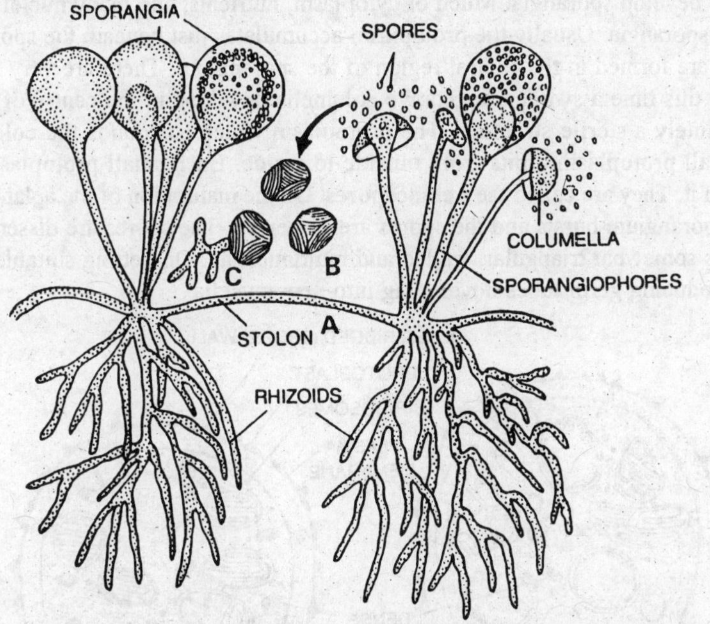

Fig. 11.2. *Rhizopus stolonifer.* A, fascicle of sporangiophores arising opposite the rhizoids, and connected by a stolon with another group; B, spores (aplanospores); C, germinating spore.

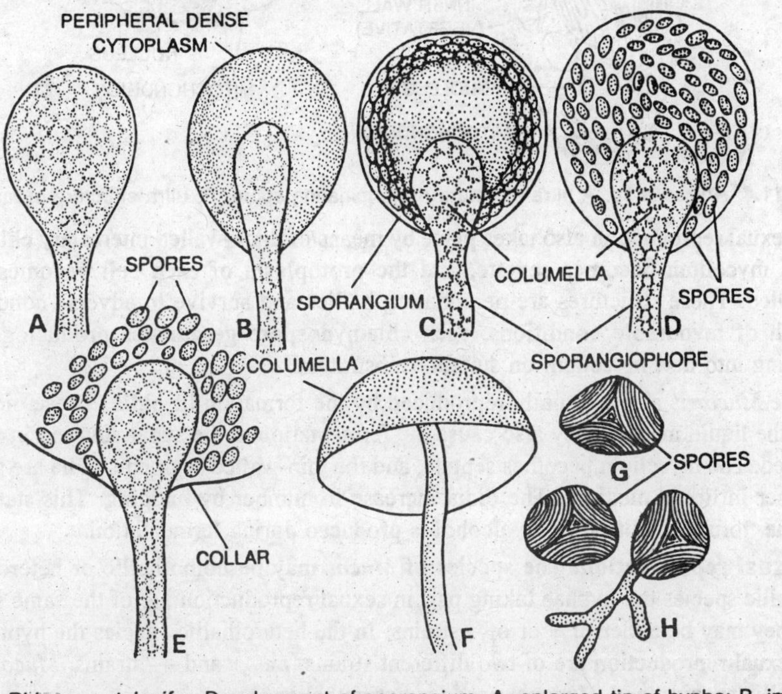

Fig. 11.3. *Rhizopus stolonifer.* Development of sporangium. A, enlarged tip of hypha; B, immature sporangial head with peripheral dense cytoplasm; C, formation of columella and spores; D, a fully developed sporangium with columella and aplanospores; E, dehiseed sporangium; F, Columella; G, aplanospore; H, germinating aplanospore.

Asexual reproduction. The asexual reproduction takes place by aplanospores formed within the sporangia. The sporangia are formed on the terminal ends of the aseptate, aerial sporangiophores. These sporangiophores directly come out from the mycelium, and there terminal ends swell to develop sporangia. Much of cytoplasm, nutrients, and many nuclei are accumulated in the young sporangia. Usually the protoplasm accumulates just beneath the sporangial wall, and the vaculoes are formed in the central region of the sporangium. There are very few nuclei in the vacuoles. By this time a swollen vegetative columella develops in the centre of this protoplasm. This is absolutely a sterile structure. The protoplasm which surrounds the columella begins to divide in small protoplasmic bits from outside to inside. Each small protoplasmic bit contains 5-10 nuclei in it. They are called the **aplanospores.** On the maturation of the aplanospores the outer wall of the sporangium bursts and the spores are dispersed. The spores are disseminated by wind. Each spore is somewhat triangular, walled and multinucleate. On getting suitable conditions they germinate producing germ tubes developing into new mycelia.

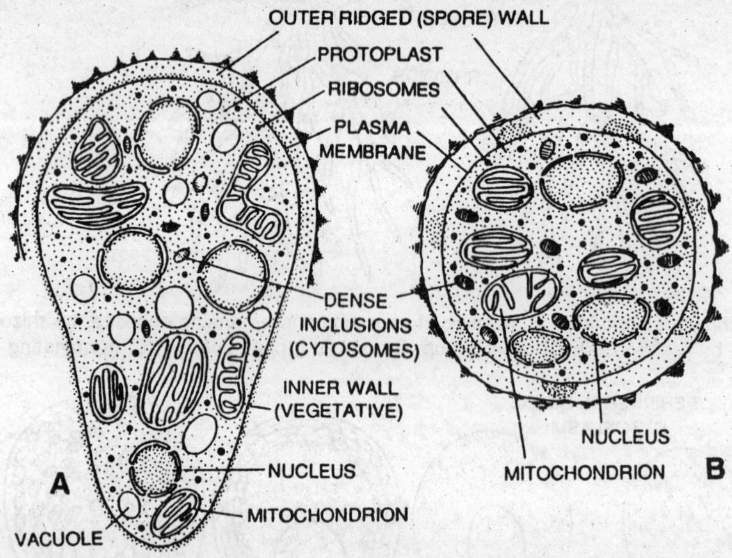

Fig. 11.4. *Rhizopus* sp. A ultra structure of germinating spore; B, ultra structure of mature spore.

Asexual reproduction also takes place by means of thick-walled intercalary **chlamydospores.** The old mycelium becomes septate, and the protoplasm of each cell becomes rounded and thickwalled. These structures are perennating bodies and survive in adverse conditions. On the approach of favourable conditions, each chlamydospore germinates producing a germ tube, developing into new mycelium on suitable substratum.

The *Mucor* is also asexually reproduced by the formation of **oidia.** These oidia are formed only in the liquid media. They also cause the fermentation in the sugar solution in which they are developed. The mycelium becomes septate, and the thin-walled segments oidia are separated from each other in liquid medium. The oidia increase in number by budding. This state of *Mucor* is called the 'torula condition'. The alcohol is produced during fermentation.

Sexual reproduction. The species of *Mucor* may be homothallic or heterothallic. In the homothallic species the hyphae taking part in sexual reproduction are of the same strain. In other words they may be either of + or of – strains. In the heterothallic species the hyphae taking part in the sexual reproduction are of two different strains, *i.e.*, + and — strains. *Mucor hiemalis* is a homothallic species, and *Mucor mucedo* is heterothallic.

In *Mucor mucedo* the sexual reproduction takes place as follows:

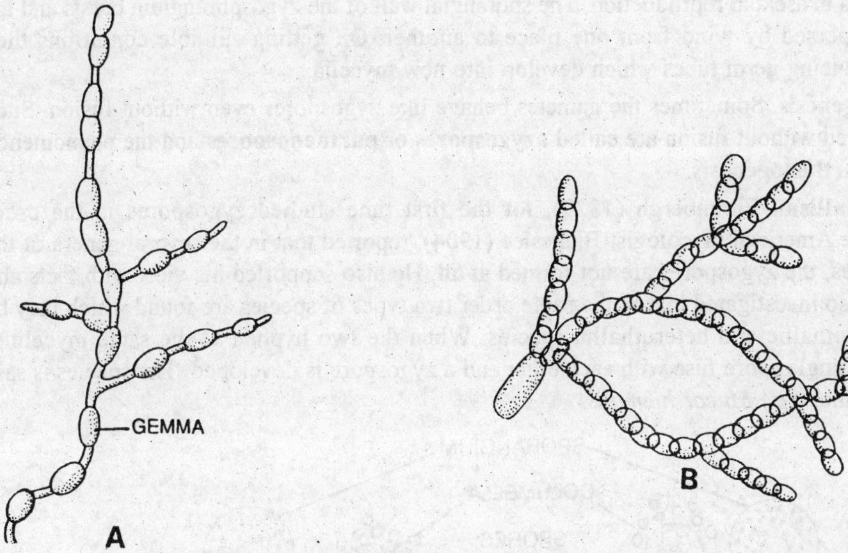

Fig. 11.5. *Mucor* sp. Asexual reproduction. A, gemma formation; B, oidia formation.

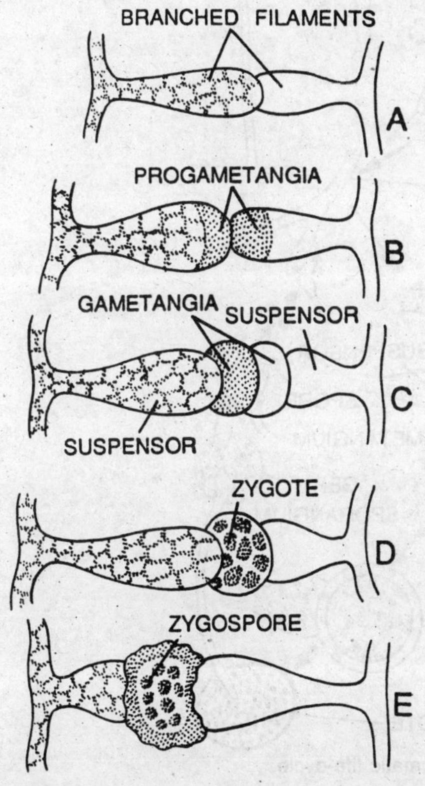

Fig. 11.6. *Rhizopus nigricans.* Sexual reproduction. A-E, successive stages in the formation of zygospore.

When two hyphae of different strains, *i. e.*, + and – come in contact, the outgrowths arise from these hyphae called the **progametangia.** Much of cytoplasm and nuclei are accumulated in terminal swollen ends of these **progametangia.** Very soon the septa are developed just behind the point of contact. The terminal multinucleate segments are called the **gametangia,** and the elongated cells behind the gametangia, the **suspensors.** The gametangia increase in size and the nuclei embedded in the cytoplasm increase in number by repeated mitotic divisions. At the point of contact the gametangial wall dissolves and the **coenogametes** found inside the gametangia are fused. With the result of fusion the **zygote** is formed. The zygospore is black, thick-walled and warty. At the time of the fusion the nuclei of different strained coenogametes form pairs and ultimately fuse forming diplid nuclei. The nuclei which do not form pairs remain haploid and disorganise very soon. The **coenozygospore** is a perennating body and faces adverse conditions. This undergoes the period of rest, which may be of several months. On the approach of favourable conditions the zygote spore germinates.

Germination of zygospore. In favourable conditions the outer warty wall of the zygospore bursts and a germ tube comes out known as **promycelium.** At the terminal end of this promycelium the **germsporangium** or the

zygosporangium develops. The apalanospores are produced inside the zygosporangium in the same manner as in asexual reproduction. The sporangial wall of the zygosporangium bursts and the spores are dispersed by wind from one place to another. On getting suitable conditions they germinate producing germ tubes which develop into new mycelia.

Pathenogenesis. Sometimes the gametes behave like zygospores even without fusion. Such spores developed without fusion are called **azygospores** or **parthenospores** and the phenomenon is known as **parthenogenesis.**

Heterothallism. Ehrenbergh (1829), for the first time studied zygospores in the order Mucorales. The American mycologist Blakeslee (1904), reported that in the several genera of the order Mucorales, the zygospores are not formed at all. He also supported his view with facts and reasons, and also investigated that in the same order two types of species are found which may be named as homothallic and heterothallic species. When the two hyphae of the same mycelium produced by a single spore fuse with each other and a zygospore is developed. The species is said to be homothallic, *e.g., Mucor hiemalis.*

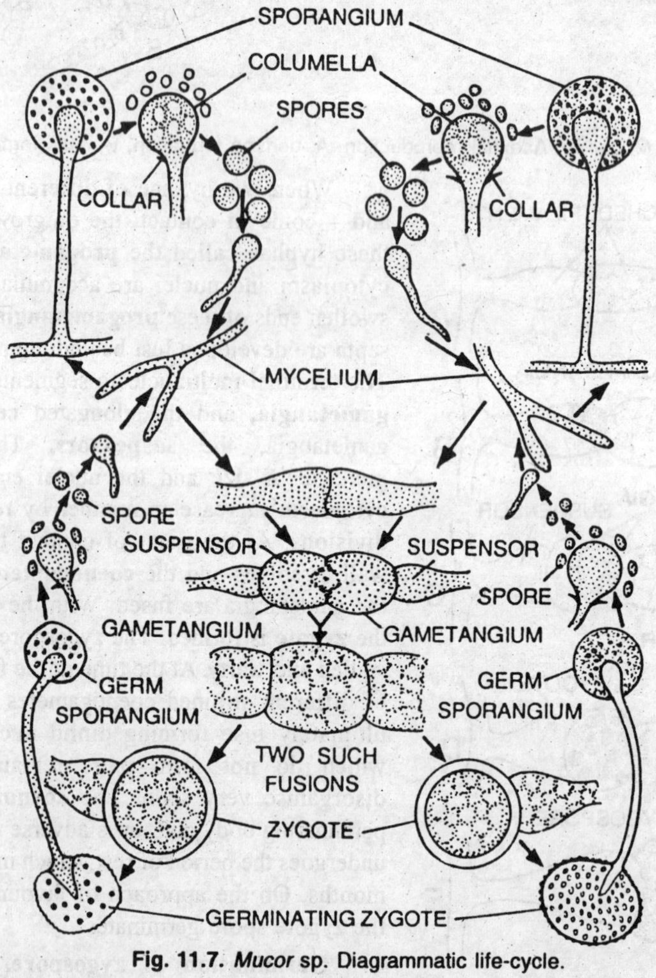

Fig. 11.7. *Mucor* sp. Diagrammatic life-cycle.

Mucor mucedo and *Mucor stolonifer* are the typical heterothalic species. In heterothallic species the fusion can take place only among the different strained hyphae, which develop on different mycelia of different (+ and –) strains. In these species the zygospores cannot be produced by the fusion of two hyphae of the same strain. In 1904 Blakeslee reported that in heterothallic

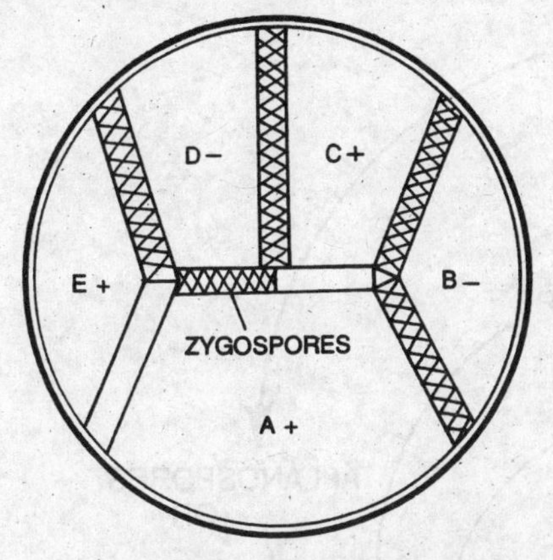

Fig. 11.8. Heterothallism in Mucorales. Blakeslee's experiment.

species whenever the mycelia of + and – strains remain apart from each other, the zygospores are not produced and only the sporangia are formed. On the other hand, when + and – mycelia grow together, the fusion takes place and the zygospores are produced. Morphologically the + and – strained mycelia are quite similar in structure, but different in physiological behaviour. In other words, they are morphologically indentical and physiologically different. Sometimes, it has also been observed that the growth of the + mycelium is comparatively faster, and the gametangia of + mycelium are bigger than that of the – mycelium, and they can be distinguished as female and male gametangia. Many mycologists do not support this view and advocate that this behaviour is only because of nutrition.

Blackeslee proved the phenomenon of heterothallism on the basis of experiments. He inoculated many + and – strained spores on the agar media in a petri dish and observed that the fusion occurs only at those points where the hyphae of different strains (*i.e.*, + and –) come in contact. The zygospores were only produced at the points of contact of different strained hyphae. Absolutely no zygospores were produced at the points of contact of same strained hyphae.

This experiment may be clearly understood by the figure given in the text. Five spores of different strains are incoculated in a sterilized petri dish on agar medium at different five points designated as A, B, C, D, and E. The zygospores are produced only at those points where the mycelia are opposite strained and produced from + and – spores, otherwise the zygospores are not formed. The spores designated as A, C and E are + strained, B and D are – strained. The zygospores are produced at the junctures of AB, BC, DE and AD where the opposite strained mycelia take part in fusion. The zygospores are not produced at the junctures of AC and AE because of the mycelia of same strains.

Blakeslee also investigated that if the + strained hyphae of one heterothallic species fuse with the — strained hyphae of another heterothallic species the **imperfect zygospores** are resulted. In such conditions the zygospores do not mature. He also investigated that in the heterothallic species *Mucor mucedo,* the zygospores produce germsporangia on germination, which contain single strained spores, and, therefore, the fusion is possible only when the mycelia are produced from the spores formed in different strained germsporangia. In the heterothallic species of *Phycomyces nitens* the + and – strained spores are produced in the same germsporangium.

The phenomenon of heterothallism was first reported in the order Mucorales, but now this has been established in rusts, smuts, Homobasidiomycetidae and other fungi.

Systematic Position

	G. W. Martin (1961)	**C. J. Alexopoulos (1962)**	**G. C. Ainsworth (1971)**
Kingdom	–Plantae	–Plantae	–Fungi
Division	–Mycota	–Mycota	–Eumycota
Sub-div.	–Eumycotina	–Eumycotina	–Zygomycotina

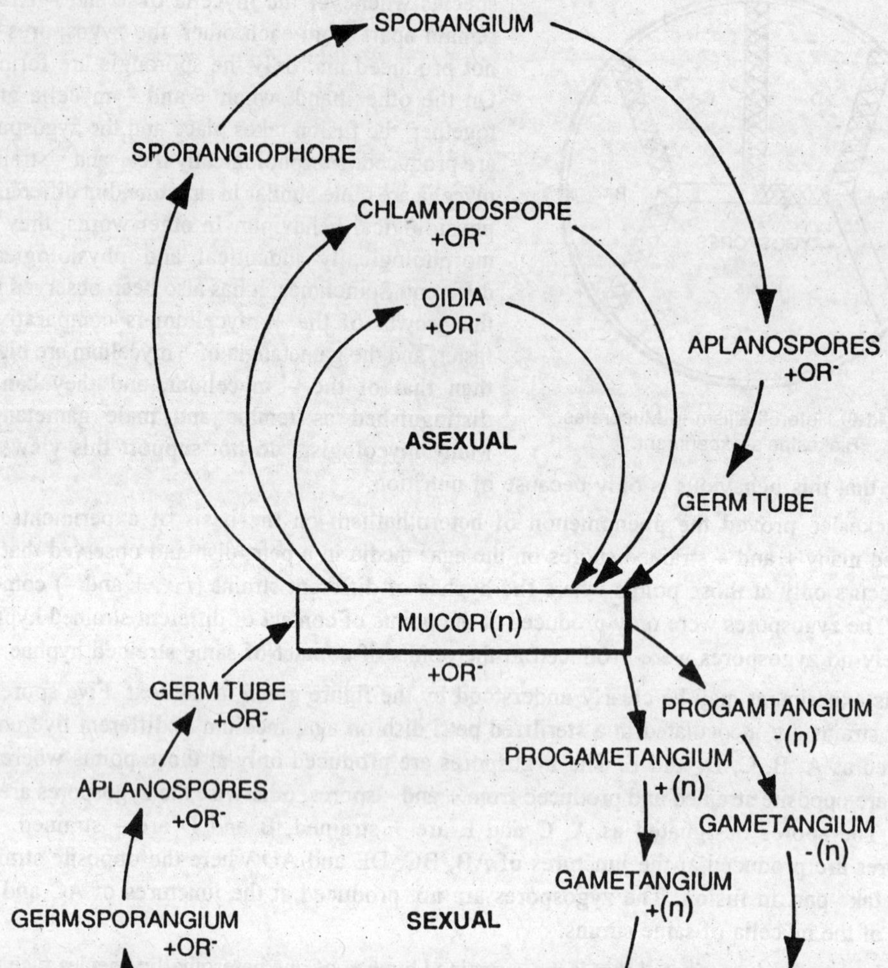

Fig. 11.9. Graphic life cycle of *Mucor*.

Class	–Phycomycetes	–Zygomycetes	–Zygomycetes
Sub-cl.	–Zygomycetidae		
Order	–Mucorales	–Mucorales	–Mucorales
Family	–Mucoraceae	–Mucoraceae	–Mucoraceae
Genera	–*Mucor* and *Rhizopus*	–*Mucor* and *Rhizopus*	–*Mucor* and *Rhizopus*

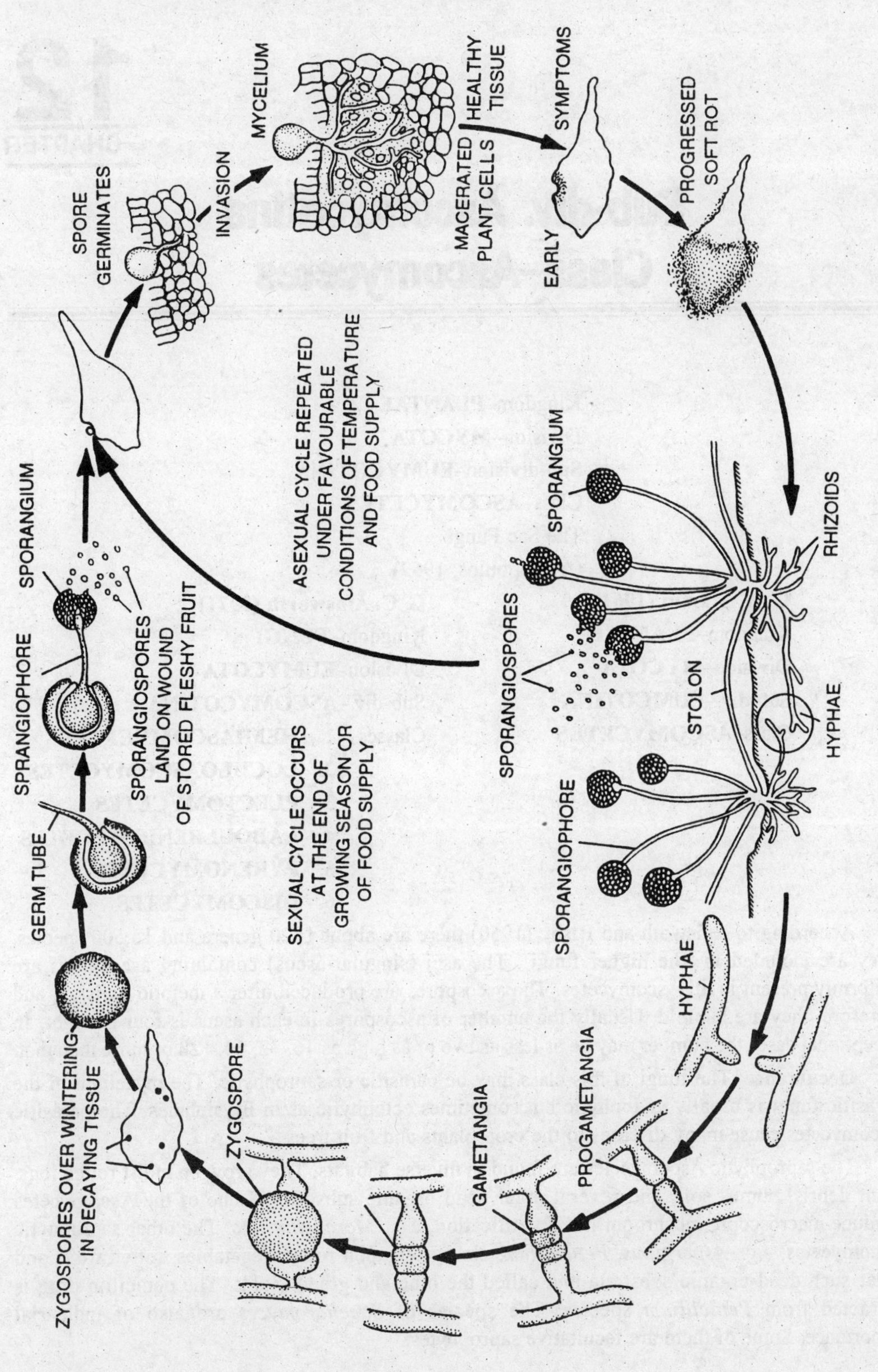

Fig. 11.10. Diagrammatic life-cycle of *Rhizopus*.

12

CHAPTER

Sub-div. Ascomycotina
Class–Ascomycetes

Kingdom–**PLANTAE**
Division–**MYCOTA**
Sub-division–**EUMYCOTINA**
Class–**ASCOMYCETES**
The Sac Fungi
(Alexopoulos, 1962)

G. W. Martin (1961)
Kingdom–**PLANTAE**
Division–**MYCOTA**
Sub-div.–**EUMCOTINA**
Class–**ASCOMYCETES**

G. C. Ainsworth (1971)
Kingdom–**FUNGI**
Division–**EUMYCOTA**
Sub-div.–**ASCOMYCOTINA**
Classes–
1. **HEMIASCOMYCETES**
2. **LOCULOASCOMYCETES**
3. **PLECTOMYCETES**
4. **LABOULBENIOMYCETES**
5. **PYRENOMYCETES**
6. **DISCOMYCETES**

According to Ainswoth and Bisby (1950) there are about 1820 genera and 15,500 species. They are included in 'the higher fungi'. The asci (singular-ascus) containing ascospores are uniformly present in all Ascomycetes. The ascospores are produced after a meiotic division, and therefore, they are haploid. Usually the number of ascospores in each ascus is four or eight. In exceptional cases the number may be as less as two or as high as 16, 32, 64, 128 or more than that.

Occurrence. The fungi of this class may be parasitic or saprophytic. The mycelium of the parasitic fungi is usually endophytic but sometimes ectophytic as in Erysiphales. The parasitic Ascomycetes cause many diseases to the crop plants and fruit trees.

The saprophytic Ascomycetes are found in diverse habitats. They grow on moist rotten logs, plant debris, humus soil, and several other dead organic substrata. Some of the Ascomycetes produce macroscopic mushroom like fructification, *e.g., Morchella,* etc. The other saprophytic Ascomycetes, *e.g., Aspergillus*, *Penicillium,* etc., grow upon rotten vegetables, rotten fruits and other such dead organic substrata and called the blue and green molds. The penicillin drug is extracted from *Penicillium* species. The species of *Saccharomyces* are also of industrial importance. Some of them are facultative saprophytes.

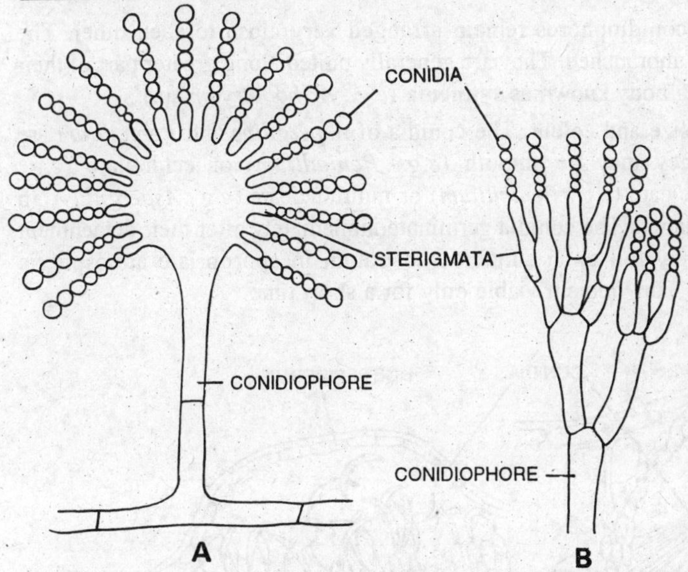

Fig. 12.1. Asexual reproduction by conidia in *Aspergillus* (A) and *Penicillium* (B).

Somatic structure. Except for some, such as yeasts, the vegetative body of an ascomycete consists of a well developed mycelium. The mycelium is branched and septate. The cells of the hyphae may be uninucleate, binucleate or multinucleate. There is a central pore in each septum through which the cytoplasm and nuclei are communicated from one cell to another. Sometimes the mycelium possesses densely interwoven hyphae and the pseudoparenchymatous sclerotia.

Asexual reproduction. The asexual reproduction in Ascomycetes takes place by several methods, *e.g.*, by fission, fragmentation, budding, chalamydospores and conidia. The methods of fission and budding are exclusively reserved for the yeast and some other Ascomycetes.

Chlamydospores. The chlamydospores have been reported from many Ascomycetes. They are commonly formed by rounding up of the protoplasm of the terminal and intercalary cells of the hyphae. These structures are thick-walled and act as perennating bodies. Much of the food reserves are accumulated in them. They undergo a period of rest prior to their germination.

Conidia. The most important method of asexual reproduction is by means of exogenously produced conidia. The conidia are produced on the conidiophores in basigenous or acrogenous chains. Sometimes the conidiophores are aerial and in other cases they are arranged in saucer-shaped acervuli. In certain cases the conidiophores are found within the flask-like structures the 'pycnia' and called pycniosporophores. In such cases the conidia are known as 'pycnidiospores'. The pycnia are open at their apical ends by ostioles.

Arrangement of conidiophores. The conidia are generally produced on conidiophores. The conidiophores may be produced free from each other, or they may be organized into definite fruiting bodies. The most common fruiting bodies are of following types.

1. Pycnidium. This type of fruiting body is a hollow, commonly globose or flask-shaped structure whose pseudoparenchymatous walls are lined with conidiophores. In such case the conidia are known as **pycniospores** or **Pycnidiospores.** The pycnidium opens by an opening called **ostiole.**

2. Acervulus. This type of fruiting body develops a mat of hyphae, below the epidermis of cuticle of the host plant and giving rise to short conidiophores closely packed together. On maturation such structure becomes saucer-shaped and looks like a flat open bed of conidiophores with conidia at their tips. The acervuli are usually formed by the parasitic fungi.

3. Sporodochium. The conidiophores may also be cemented together to form complex structures known as sporodochia. This is a hemispherical or barrel-like asexual fruiting body. The lower part of sporodochium consists of a cushion-like stromatic mass of hyphae. The conidiophores arise from the surface of the stroma. The conidiophores makes the upper part of the sporodochium.

4. Synnemata. Here the conidiophores remain arranged very close to each other. The conidiphores may be branched or unbranched. They are generally united along greater part of their length to form the asexual fruiting body known as **synnema** (*e.g.*, *Arthobotryum* sp.).

The conidia vary in shape, size and colour. The conidia of *Aspergillus* and *Penicillium* are spherical and smoky green. They may be smooth (*e.g.*, *Penicillium*) or echinulate (*e.g.*, *Aspergillus*). They may be uninucleate (*e.g.*, *Penicillium*) or multinucleate (*e.g.*, *Aspergillus*). In *Erysiphe* the conidia are barrel-shaped. The conidia germinate immediately after their detachment from the conidiophores when they fall upon suitable substratum in appropriate atmospheric conditions producing germ-tubes. They remain viable only for a short time.

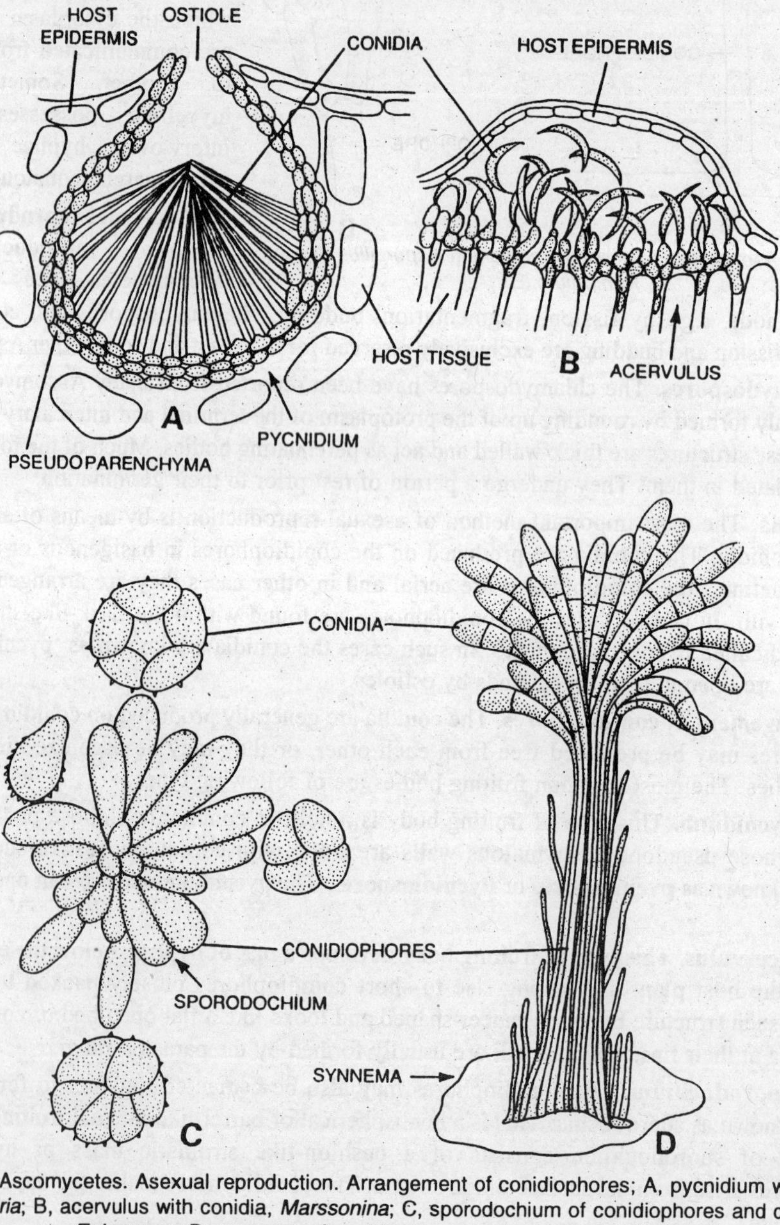

Fig. 12.2. Ascomycetes. Asexual reproduction. Arrangement of conidiophores; A, pycnidium with conidia, *Septoria*; B, acervulus with conidia, *Marssonina*; C, sporodochium of conidiophores and conidia, *Epicoccum*; D, synnema of conidiophores and conidia, *Arthbotryum*.

Oidia. In some ascomycetous fungi the hyphae break into small fragments. Such small, globose structures are known as **oidia.** They separate from each and act as spores. Such structures are found in *Oospora, Monilinia,* etc.

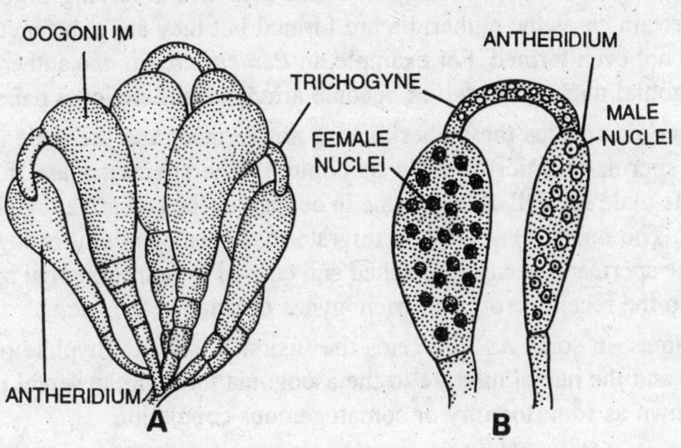

Fig. 12.3. Sexual reproduction in *Pyronema* of Ascomycetes showing sex organs and fertilization. (A and B).

Sexual reproduction. The sexual reproduction takes place by the union of two compatiable nuclei. The . species are either homothallic or heterothallic, and therefore, the fertilization is either isogamous or heterogamous. Commonly the male sex organs are called **antheridia** and female sex organs the **ascogonia.**

In the lower Ascomycetes, *e.g., Eremascus fertilis,* the sex branches are not distinguished into male and female sex organs. One cell with a nucleus arising from hypha acts as male cell and the other uninucleate cell arising from the same hypha acts as female cell. The two cells fuse and simultaneously the nuclei unite. The fusion nucleus enlarges in size and the cell is known as **ascus.** This nucleus divides thrice and eight nuclei are produced, each developing into an ascospore. The first division is meiotic one which is followed by ordinary mitotic division.

In the higher Ascomycetes, *e.g., Aspergillus,* the antheridia and ascogonia are developed. The ascogonium may be single celled or multicellular. In some Ascomycetes the receptive trichogyne is developed at the distal end of the ascogonium which may be septate or aseptate. The mature antheridia and oogonia come close to each other and the mature antheridium touches the trichogyne, the wall of contact dissolves and male nucleus passes into the ascogonium through trichogyne. The plasmogamy is not immediately followed by karyogamy as in Phycomycetes. The nuclei of both sexes arrange in pairs, each such pair is called a 'dikaryon'. Many **ascogenous hyphae** arise from the oogonium or ascogonium. The ascogenous hyphae are septate and the ultimate and penultimate cells develop into asci. According to one school of thought, the nuclei fuse in the ascogonium and the ascogenous hyphae contain diploid nuclei. The second fusion takes place within the ascus cell giving rise to a tetraploid nucleus. By the process of **brachymeiosis** the eight ascospores are formed in each ascus. The first division is reductional which is followed by brachymeiosis. In other cases the ascus cells contain diploid nuclei. This diploid nucleus divides thrice giving rise to eight ascospores. Here the first division is meiotic one.

The sexual reproduction in Ascomycetes shows great diversity. Generally it takes place by means of four important methods. They are as follows:

1. Gametangial copulation. This type of sexual reproduction is found in lower Ascomycetes where the male and female sex organs cannot be distinguished from each other. For example, in the homothallic species of *Eremascus* small vertical protuberances arise in pairs. Two such similar gametangia touch at their tips and fuse, the fusion cell developing into an ascus. There is no dikaryotic phase in such species. Karyogamy takes place immediately after plasmogamy. The somatic cells of yeasts act as gametangia; two of them fuse forming a zygote which later on directly converts into an asucs.

2. Gametangial contact. Some species of Ascomycetes produce morphologically differentiated gametangia known as antheridia (male) and ascogonia (female). The antheidium and ascogonium come in contact; the wall of contact dissolves and the male nucleus from antheridium passes through the lateral pore into the ascogonium. No fertillization tubes are formed but sometimes the ascogonium is provided with a trichogyne which acts as a receiving organ and receives the male nucleus. In certain cases the antheridia are formed but they are non functional; in other cases the antheridia are not even formed. For example, in *Penicillium* sp., the antheridium is non-functional, and the ascogonial nuclei themselves become arranged in functional pairs.

3. Spermatization. In certain species the antheridia are not formed, and the male nuclei reach ascogonia by means of spermatia, microconidia or conidia. The spermatia are minute, spherical, specialized unicucleate male sex cells. They come in contact of the receptive organs and empty their contents into them. The nuclei of spermatia migrate to the ascogonium through the pores of septa. On maturation the spermatia become detached and carried by some external agency such as insects, wind or water to the receptive organs (trichogynes or somatic hyphae).

4. Somatogamy. Sometimes, in some Ascomycetes the fusion of somatic hyphae of two compatible mycelia takes place and the nuclei migrate to the ascogonia through the septal pores. This type of sexual union is known as **somatogamy** or somatogamous copulation.

Ascus formation. As mentioned above the ultimate and penultimate cells develop into asci. The terminal cell of ascogenous hypha is binucleate. One nucleus of the pair is male and the other female. The terminal cell becomes curved on one side like a **crozier.** Simultaneously this enlarges in size. Both the nuclei of the pair (dikaryon) divide mitotically and four daughter nuclei are produced, the transverse septa develop in the crozier cell. This results in the formation of a uninucleate terminal cell, binucleate penultimate middle cell and uninucleate basal cell. The penultimate or middle cell is also known as **arch cell** or **crook cell.** The arch cell contains two nuclei of different parentage, one male and the other female. The arch cell acts as ascus mother cell and ultimately develops into an ascus. Both the nuclei of the dikaryon fuse and a diploid ($2n$) nucleus is formed. This is the process of karyogamy and represents fertilization. The fusion nucleus or synkaryon represents the diploid phase. This is a temporary phase. Very soon the cell containing

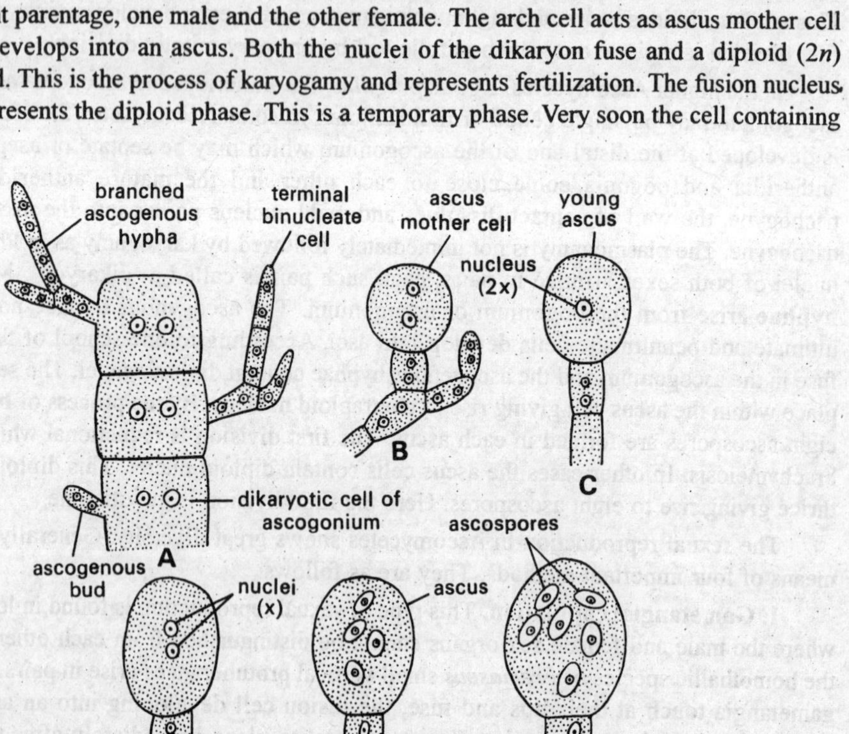

Fig. 12.4. Formation of ascus and ascospores in *Penicillium*.

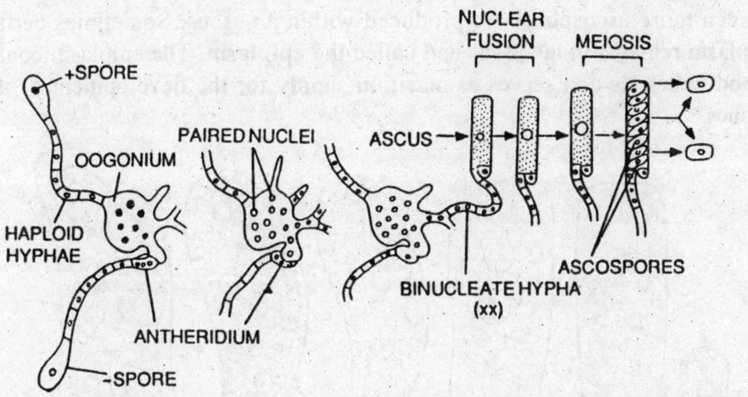

Fig. 12.5. Ascomycetes. Sexual reproduction: showing sexual fusion and ascus formation.

this nucleus elongates and the fusion nucleus divides. The first division is meiotic. This is followed by two mitotic divisions. In this way eight nuclei are produced. Around each nucleus the cytoplasm

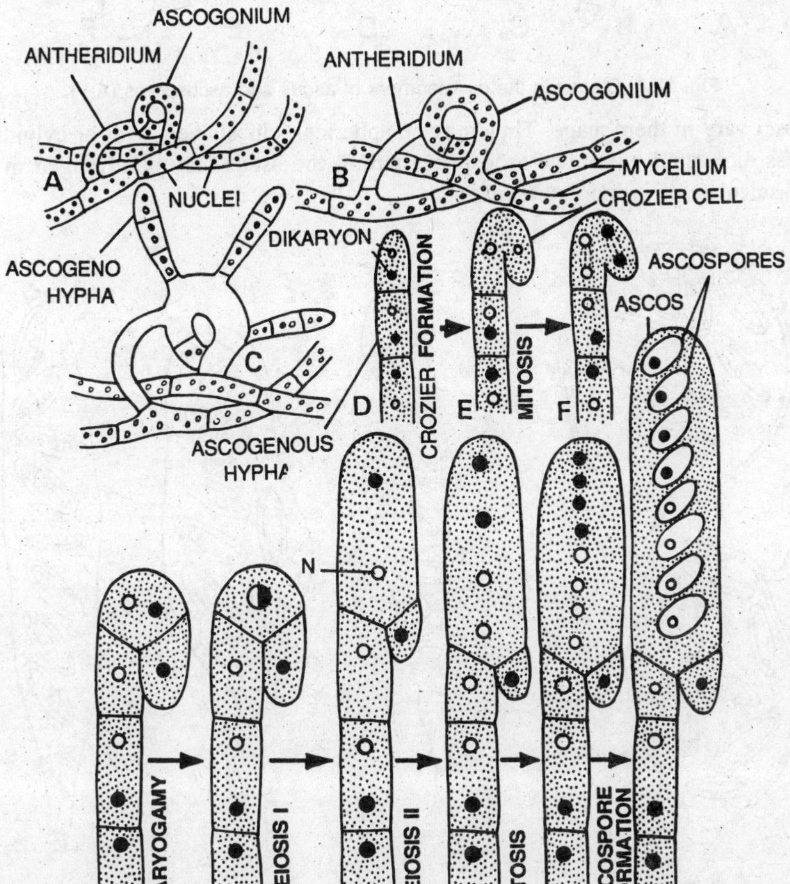

Fig. 12.6. Ascomycetes. A-L, diagrammatic representation of ascus and ascospore development in a heterothallic species. Black and white nuclei represent the two compatible mating types.

is deposited and eight ascospores are formed within the ascus. Sometimes in certain cases 16, 32, 64, 128 or even more ascospores are produced within an ascus. Sometimes certain amount of unused cytoplasm remains in the ascus and called the **epiplasm.** The epiplasm contains glycogen and other food materials and serves as nutrition supply for the development of the ascospores within the asucs.

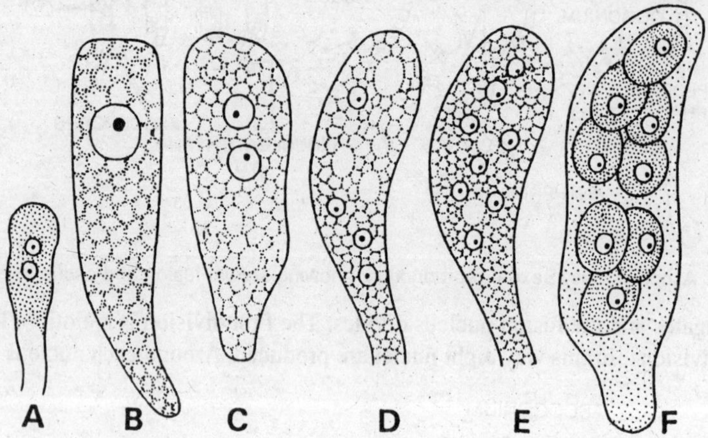

Fig. 12.7. Stages in the development of ascus and ascospores (A-F).

The asci vary in their shape. They may be spherical, clubshaped, oval or cylindrical with a more or less elongated base. The ascospores within the ascus may be arranged in uniseriate, biseriate, fasciculate or in inordinate manner.

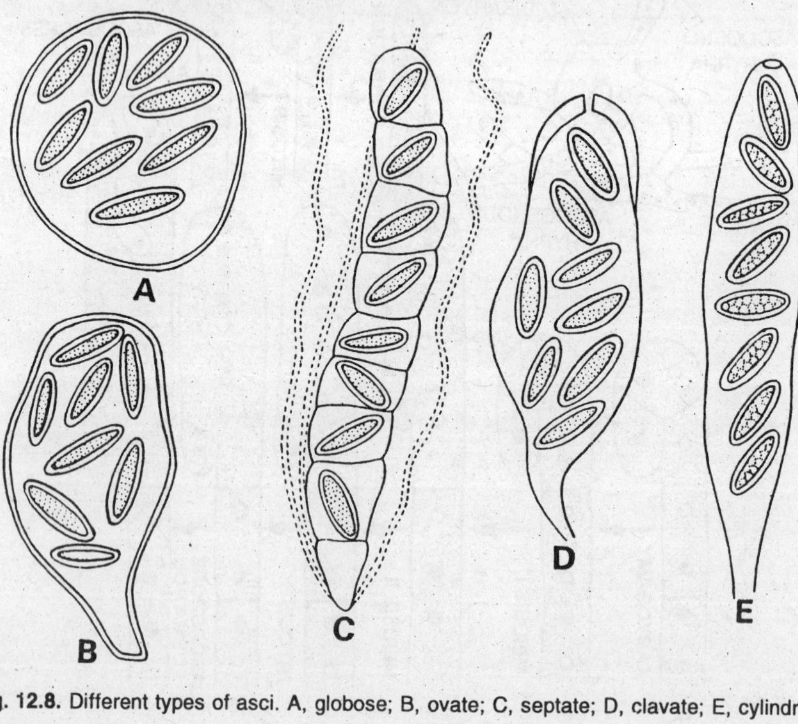

Fig. 12.8. Different types of asci. A, globose; B, ovate; C, septate; D, clavate; E, cylindrical.

The dehiscence of ascospores from asci differs from species to species. In majority of Ascomycetes the ascospores are released with a jerk from asci. The explosions take place successively. In certain cases the asci are provided with opercula. The operculum opens and the ascospores are released.

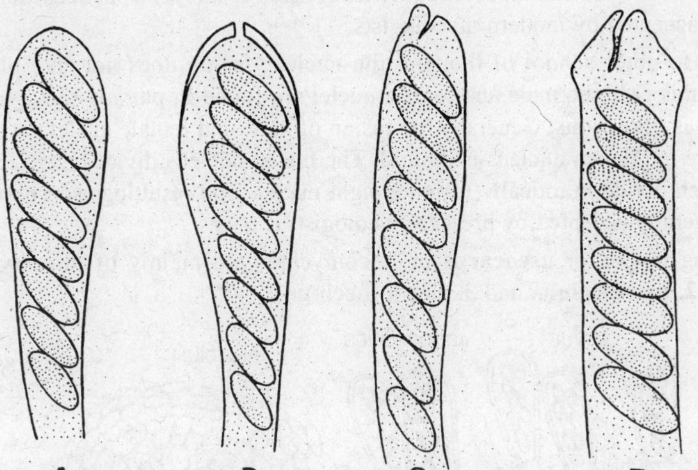

Fig. 12.9. Types of ascal openings. A, absent; B, porous; C, operculum; D, slit.

Cytology of ascus. Nuclear fusion, however, was first observed by Dangeard in 1894, not in the oogonium but in the ascus. As regards the cytology of ascus there are two schools of thought. According to one school of thought, the male and female nuclei fuse in the ascogonium and therefore each of the cells of the ascogenous hyphae contains a pair of diploid nuclei. These diploid nuclei once again fuse together within the ascus mother cell resulting in the formation of a tetraploid nucleus. The ascospores are always haploid. To bring about the haploid condition of

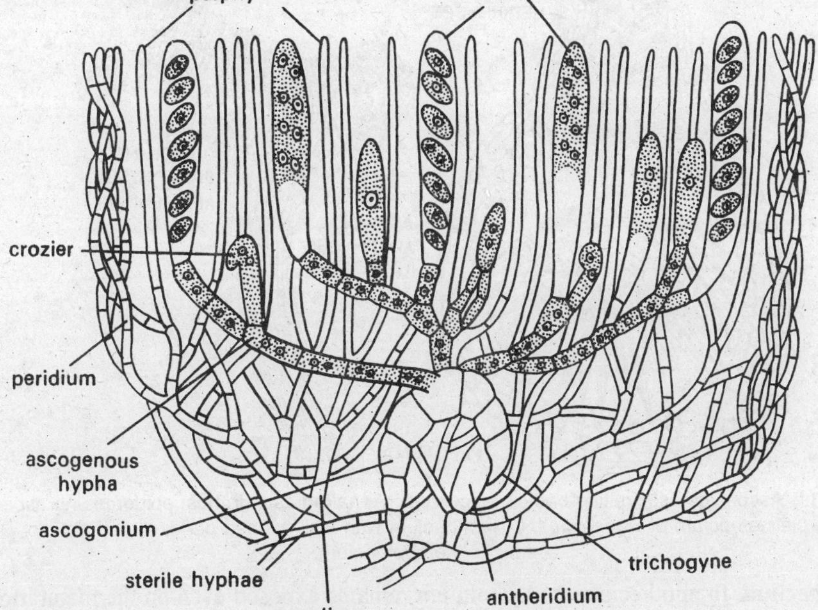

Fig. 12.10. Development of apothecium.

ascospores, there must be double reduction in the number of chromosomes during the three successive divisions of the tetraploid nucleus. It is held that meiosis takes place twice, *i.e.,* first and third divisions are meiotic. This double reduction division is known as **brachymeiosis.** However, this view is not accepted by modern mycologists.

According to other school of thought, the nuclear fusion does not take place within the ascogonium. In this case the male and female nuclei are found in pairs in the ascogonium. Such pairs are known as **dikaryons.** Generally the fusion of male and female nuclei occurs in the ascus mother cell. Here the fusion nucleus is diploid. The fusion nucleus divides meiotically producing four nuclei, which divide mitotically forming eight nuclei. The resulting ascospores are haploid. This view is generally accepted by present mycologists.

The fruiting bodies or **ascocarps** or Ascomycetes are mainly of three types. They are **1. apothecium, 2. perithecium,** and **3. cleistothecium.**

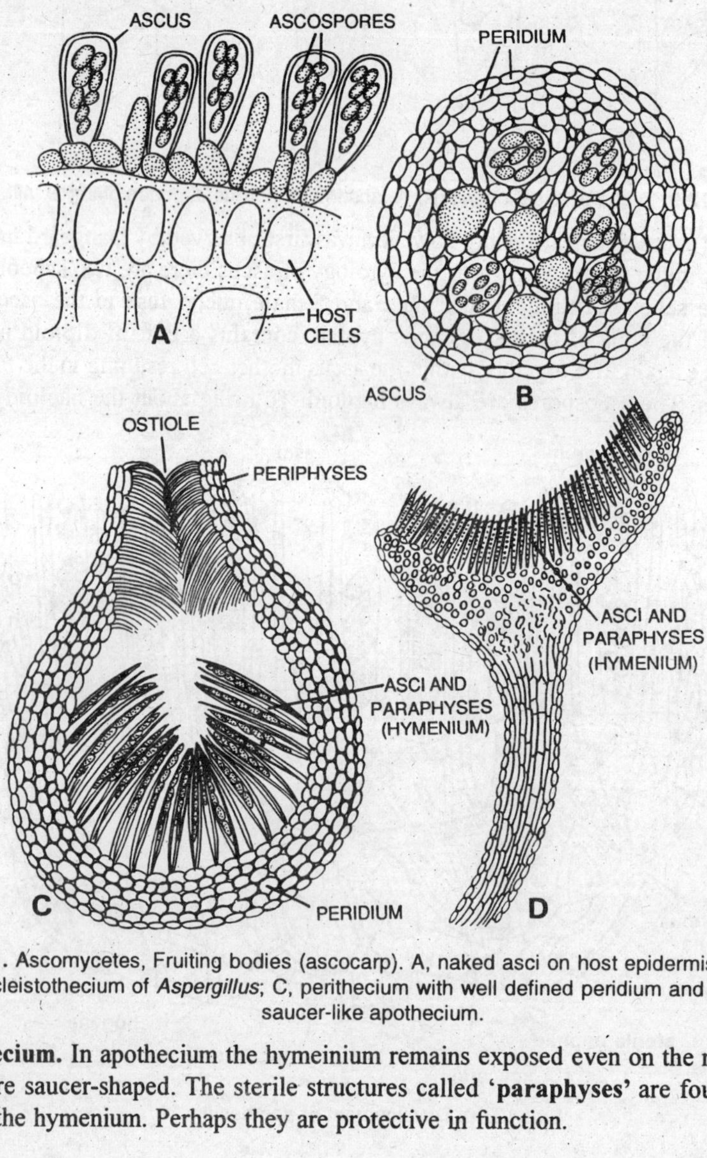

Fig. 12.11. Ascomycetes, Fruiting bodies (ascocarp). A, naked asci on host epidermis-no ascocarp, *Taphrina*; B, cleistothecium of *Aspergillus*; C, perithecium with well defined peridium and ostiole; D, stalked saucer-like apothecium.

Apothecium. In apothecium the hymeinium remains exposed even on the maturation of the asci. They are saucer-shaped. The sterile structures called **'paraphyses'** are found intermingled with asci in the hymenium. Perhaps they are protective in function.

Perithecium. The perithecia are spherical or flask-like. The perithecium is covered by a **peridium** except at ostiole. The discharge of ascospores takes place through this narrow opening called the **ostiole.** The asci are found on the ascogenous hyphae present in the round bottom of the pertithecium.

Cleistothecium. The cleistothecium is a closed body. Actually this is perithecium without an ostiole. The asci are filled up in these closed bodies. The cleistothecium is covered by a peridium layer. The appendages may or may not be present on the surface of cleistothecium.

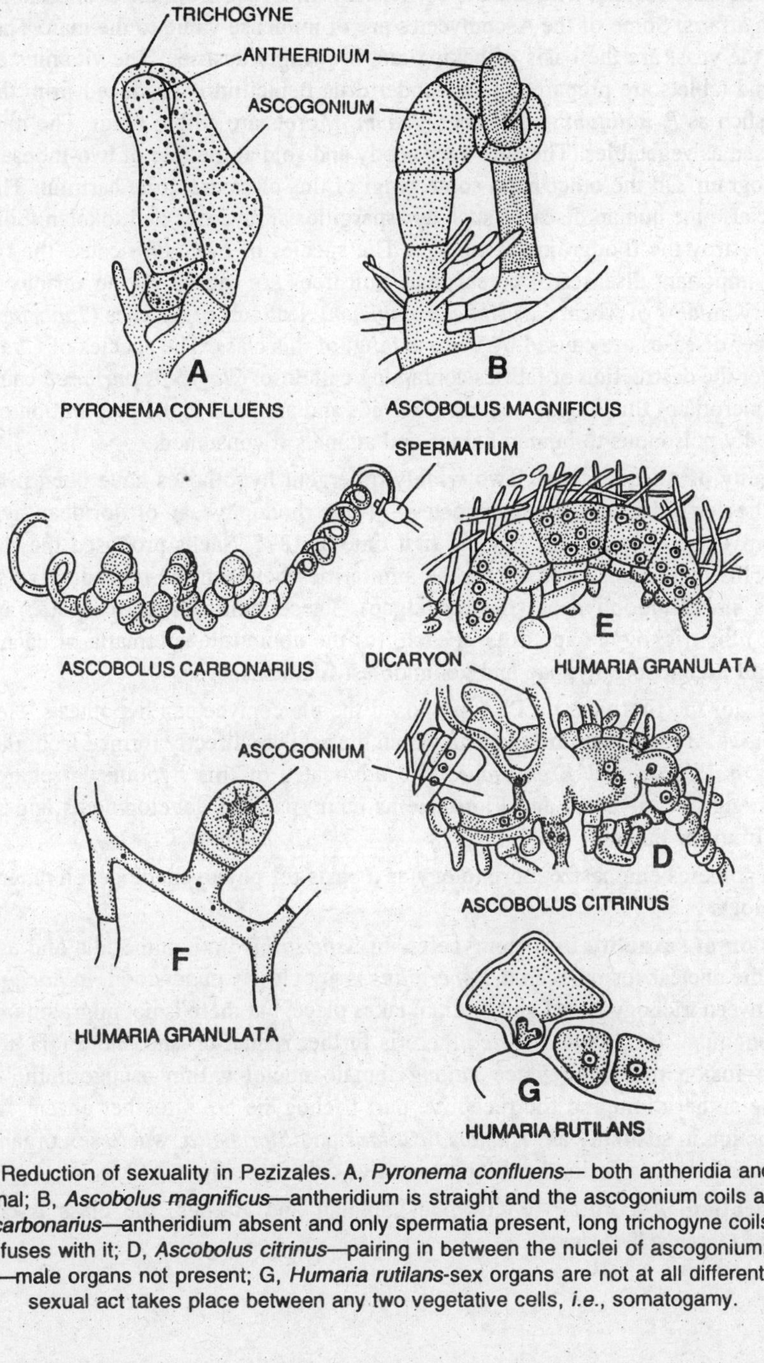

Fig. 12.12. Reduction of sexuality in Pezizales. A, *Pyronema confluens*— both antheridia and ascogonia are functional; B, *Ascobolus magnificus*—antheridium is straight and the ascogonium coils around it; C, *Ascobolus carbonarius*—antheridium absent and only spermatia present, long trichogyne coils around the conidia and fuses with it; D, *Ascobolus citrinus*—pairing in between the nuclei of ascogonium; E, *Humaria granulata*—male organs not present; G, *Humaria rutilans*-sex organs are not at all differentiated. The sexual act takes place between any two vegetative cells, *i.e.*, somatogamy.

The ascospores. They are usually elliptical, but in some species spherical and in some long and narrow. They contain densely granular cytoplasm and frequently oil drops. The epispore may be smooth or ornamented. They may be hyaline, opaque, colourless or variously coloured. The young spore is commonly unicellular and uninucleate, though in some species it contains two nuclei. Nuclear division and wall formation takes place during development. The mature spores are found to be arranged in a row or a mas of cells. Each spore germinates by putting out a germ tube. Most spores germinate immediately after their liberation from ascus.

Economic importance. The class is sufficiently important from the economic point of view to the human affairs. Some of the Ascomycetes are of immense value to the man. The fermenting activities of the yeast are the basis of baking and brewing industries. The vitamins are extracted from yeast and tablets are prepared. The wonder drug penicillin is produced from the species of *Penicillium* such as *P. notatum* and *P. crysogenum.* Morels are edible fungi. The mushroom-like morels are used as vegetables. They are very costly and sold at the rate of two-thousand rupees or more per kilogram. On the other hand some fungi of this class are quite harmful. The species of *Aspergillus* cause the human diseases such as aspergillosis, mycoses and tokelan. Some speies of *Aspergillus* destroy the foodgrains in storage. The species of *Penicillin* cause the rotting of the fruits. Many important diseases of cereals and fruit trees are caused by the various fungi of this class. Powdery mildew of wheat (*Erysiphe gramineum*), leaf curl of peaches (*Taphrina deformans*) and many other diseases are caused by various fungi of this class. The species of *Chaetomium* are responsible for the destruction of fabrics containing cellulose. *Claviceps purpurea* causes the ergot of rye. The sclerotia of this fungus contain alkaloids and are used in the preparation of medicines. They are deadly poisonous to human beings and animals if consumed.

Phylogeny of Ascomycetes. Two widely divergent hypotheses have been propounded to account for the derivation of the Ascomycetes— 1. the rhodophycean or floridean hypothesis and 2. the phycomycetean hypothesis. For the first time in 1875, Sachs proposed the rhodophycean hypothesis. This hypothesis is based on the similarities between the reproductive structures of Ascomycetes and Rhodophyceae (the red algae). These similarities are — (a) ascocarp and cystocarp, (b) the trichogyne appartus in each, (c) the nonmotile spermatia of each and (d) the resemblance of ascogenous hyphae and gonimoblast filaments.

In 1887, for the first time de Bary proposed the phycomycetean hypothesis. He maintained that the primitive Ascomycetes are those in which the asci are directly formed from the zygote, as they are in *Dipodascus* and *Eremascus.* The advocates of this hypothesis regard analogous structures among Ascomycetes and Florideae as being parallel developments and not as being derived one from the other.

Both hypotheses emphasize morphology as a basis for phylogenetic development and fail to stress physiology.

Reduction of sexuality in Ascomycetes. In *Aspergillus* both antheridia and ascogonia are present. But the nuclear fusion even in *Aspergillus* is not clearly understood. In *Lachnea stercorea* the fusion between trichogyne and antheridium takes place, but there is not migraton of male nuclei into the ascogonium. In *Lachnea cretea* there is further reduction and antheridia are altogether absent. Here fusion may take place among female nuclei within ascogonium. In *Humaria granulata* the antheridium and the receptive part trichogyne are altoether absent. The extreme cases of reduction in sexulaity are *Humaria rutilans* and *Morchella,* where sex organs are totally obliterated, and there is a fusion between vegetative cells. (see fig. 12.12).

Classification. According to Gwynne-Vaughan and Barnes, the class Ascomycetes is subdivided into three sub-classes.

1. **Plectomycetes.** Ascocarp, if present either with no definite ostiole, or shield-shaped, or with asci irregularly arranged.

2. **Plectomycetes.** Ascocarp flask-shaped when ripe; asci in parallel series.

3. **Discomycetes.** Ascocarp wide oepn when ripe; asci in parallel series.

According to latest trend, the Ascomycetes have been divided into two sub-classes — 1. Hemiascomycetidae (Hemiascomycetes) and 2. Euascomycetidae (Euascomycetes). (Also see chap. of classification.)

1. **Hemiascomycetidae.** Asci naked; ascocarp absent.

2. **Euascomycetidae.** Asci produced in an ascocarp,

Sub-class–HEMIASCOMYCETIDAE

Yeast and Leaf Curl Fungi
(Alexopoulos, 1962)

G.W. Martin (1961)	**G.C. Ainswoth (1971)**
Sub-class-**HEMIASCOMYCETIDAE**	Class-**HEMIASCOMYCETES**

The sub-class Hemiascomycetidae is also known as Protoascomycetidae. They include either very primitive Ascomycetes or degenerate forms. The characteristic features of these fungi are as follows:

1. The mycelium is poorly developed or altogether absent.
2. The ascogenous hyphae are not found.
3. The asci are naked and not enclosed inside an ascocarp.
4. The ascocarps are not formed.

The sub-class Hemiascomycetiadae includes three orders — 1. Protomycetales, 2. Endomycetales and 3. Taphrinales.

Protomycetales. The spore-sac is compound also known as synascus in which numerous asci are produced.

Endomycetales. The individual asci are formed. Plant body is greatly reduced often to a single cell. With the result of sexual fusion an ascus is produced. The asci are produced directly from zygotes.

Taphrinales. The individual asci are formed. The product of sexuality is a dikaryotic thallus. The asci are directly produced from the cells of this thallus.

Order **ENDOMYCETALES** (43 genera; 150 species)

Characteristic Features

1. The order includes both saprophytic and parasitic fungi.
2. The plant body is greatly reduced, often to a single cell, *e.g.*, in yeasts.
3. The zygote derived from the copulation of two cells becomes directly transformed into an ascus. In other forms a single cell becomes an ascus parthenogenetically. In still other forms the diploid zygote gives rise to a large number of diploid cells by budding, which, develop into asci.

Classification. According to G. W. Martin (1961), the order includes four families — 1. Ascoideaceae, 2. Endomycetaceae, 3. Spermophthoraceae and 4. Saccharomycetaceae.

Family–Saccharomycetaceae (10 genera; 100 species)

The asci are eight or few spored. The gametangia are not found. The copulation takes place by means of somatic cells.

The thoroughly worked out genus *Saccharomyces* of the family Saccharomycetaceae has been discussed here in detail.

Genus **SACCHAROMYCES** (30 species)

Occurrence and habit. Usually the yeasts grow in such organic materials where the sugar is found in abundance. They grow in toddy juice, grape juice and sugarcane juice very easily. The yeasts convert the sugars into alcohol by fermentation and only because of this quality they are used in making alcohols, wines and beer, etc. They are used in the bakeries to produce loafs. The spongy nature of the loaf is because of the presence of carbon dioxide. This carbon dioxide is released during fermentation.

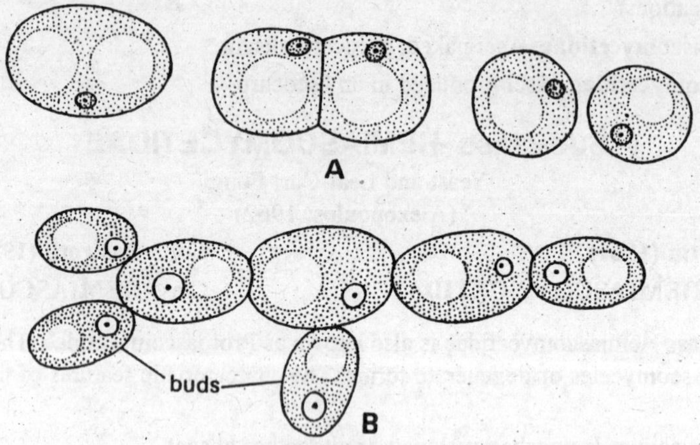

Fig. 12.13. Yeasts showing fission (A), and budding (B).

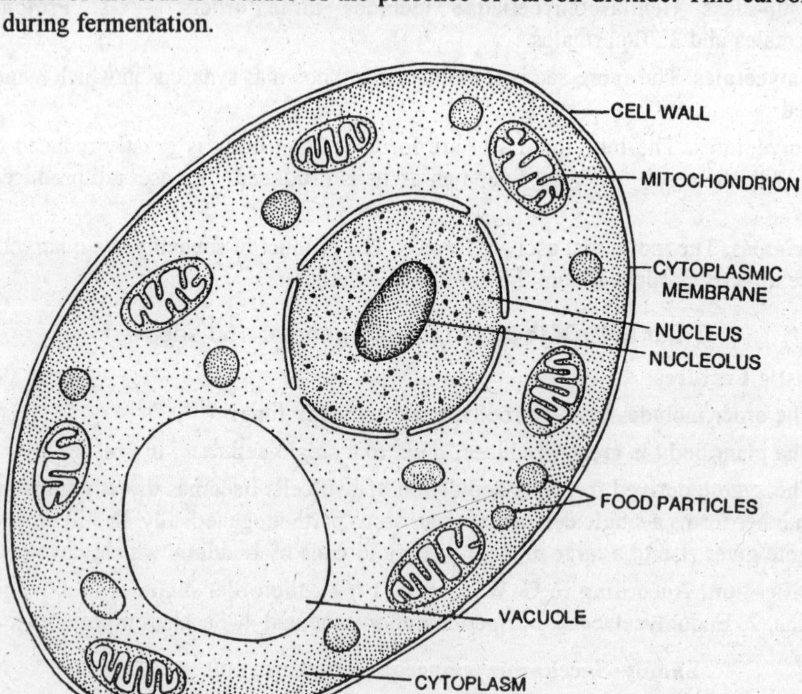

Fig. 12.14. Electron micrograph of single yeast cell.

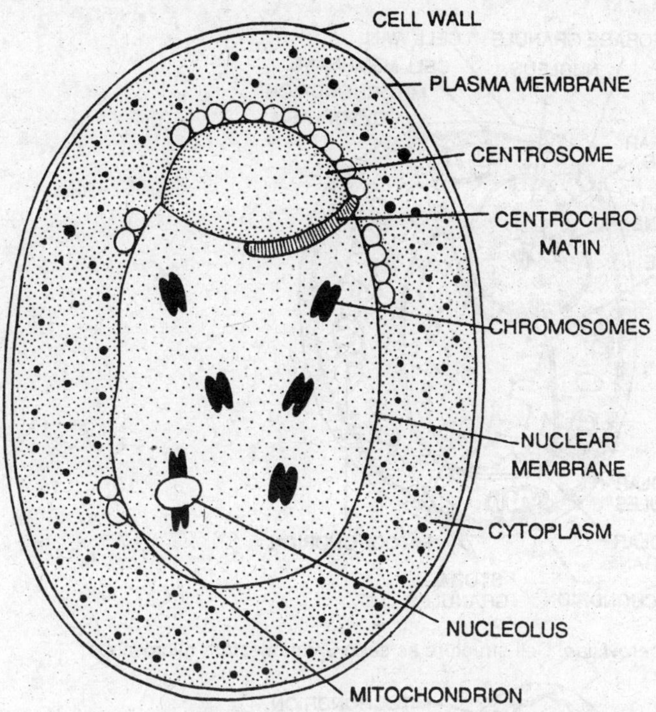

Fig. 12.15. A yeast cell. (After Lindgren)

Some yeasts occur as parasites, *e.g., Monosporella* is parasitic on the intestines of *Daphnia,* a crustacean; species of *Nematospora* are found to be parasitic on tomato, beans, etc. Certain yeasts are found in symbiotic relationship with certain bactria and moulds, *e.g.*, to prepare the sake wine in Japan, the fermentation is brought about by *Saccharomyces sake* (yeast) in association with *Aspergillus oryzae* (mould).

Structure. The yeast organisation are unicellular. All the metabolic activities are furnished by this single cell. These cells are microscopic and may be seen under the high power lens of the microscope as pin heads.

Each yeast cell is oval or spherical. The cell wall consists of a hard substance, the chitin. The cytoplasm is filled up in the cell. In the most common species *Saccharomyces cerevisiae* there is large rounded colourless area lying towards a pole of the cell. According to Lindgren (1949) and others this has been considered as a nucleus, but according to De Lamater (1950) this is the vacuole and the nucleus is the smaller body which lies on the side of the vacuole. If the former view is correct, the nucleus contains a big vacuole. The nuclear vacuole possesses a chromatin network and a small nucleus situated on one side. The glycogen granules, the oil droplets and the compounds of proteins are found embedded in the cytoplasm.

Historical. The cytology of yeast cells remained a topic of controversy for many years. Dangeard, Janssens, Le Blanc and others showed some nucleus like bodies to be present in the cell. This was assumed that a **diffused nucleus** of dispersed particles was present within the cells. Eischenschitz reported that these grains were abundantly found in the vacuole. This concept was recognized by Wagner (1898) and again by Wagner and Peniston (1910). These scientists described the necleus as **vacuole nucleare,** *i.e.,* a vacuole filled up with chromatin particles and carrying a distinct nucleolus towards its outside. Now this account is of historical value only.

ELECTRON MICROSCOPIC STRUCTURE

The electron microscope has now revealed astonishing details within the yeast cell : Yotsuyanagi, Y. (1959), Hashimoto, T. (1960), Naylor, H.B., Conti, S. F. and Thyangarajan, T. R. (1962).

The detailed electron microscopic structure of the yeast cell is as follows.

The cell remains surrounded by a definite two-layered cell wall. The cell wall consists of chitin and some other substances. Just benath the cell wall there is cytoplasmic membrane. There is a distinct nucleus within the cell. The nucleus is spherical to ovoid and remains surrounded by a double membrane which is interrupted by 'pores'. A distinct nuleolus is also present within the

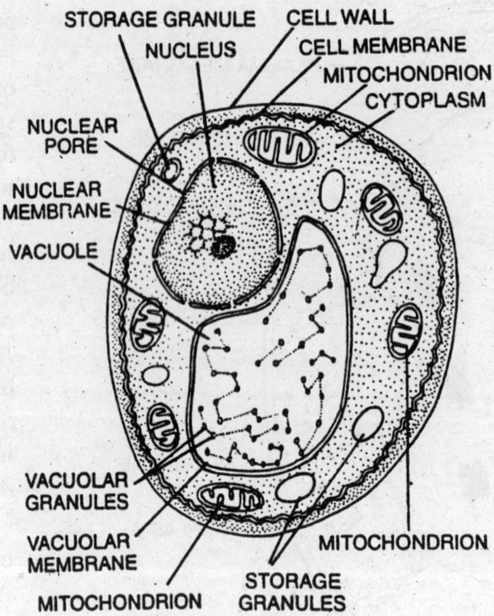

Fig. 12.16. *Saccharomyces cerevisiae.* Cell structure as seen under electron microscope.

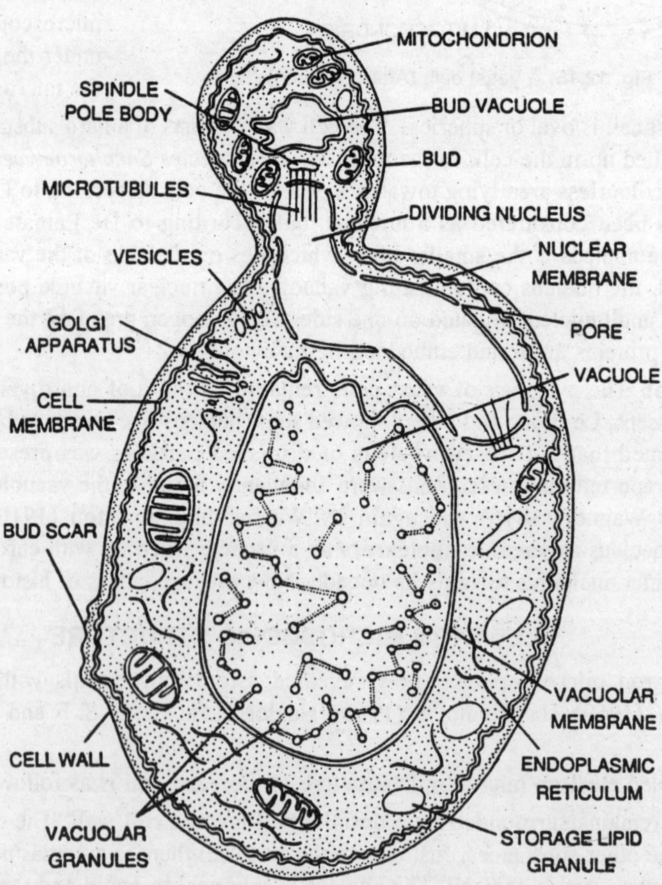

Fig. 12.17. Ultra structure of a budding yeast (*Saccharomyces cerevisiae*).

nucleus. The nucleolus is a spherical dense body and remains surrounded by its own membrane. Certain vacuole-like areas of lesser density are also found within the nucleolus. The granular and homogeneous nucleoplasm is filled up within the nucleus. There occur some irregular areas within the nucleus, which contain fibrils and correspond to the chromosomes of yeast cell. There is a distinct vacuole found within the cell that remains filled with a granular material, the volutin and is separated from the remaining cell by a membrane of its own. As assumed earlier the vacuole is not integral part of nuclear apparatus. The other cytoplasmic inclusions are mitochondria, endoplasmic reticulum and ribosomes. The number of mitochondria within a cell is 4-20. Certain other inclusions such as food particles of lipids, proteins, glycogen, etc., are also found.

Reproduction. The reproduction of yeasts takes place by means of 1. vegetative, 2. asexual and 3. sexual methods.

Vegetative reproduction. The vegetative reproduction takes place by means of 'budding'. This method of reproduction takes place in favourable conditions when the yeast cells grow in sugar solution. From each yeast cell one or more small outgrowths are given out, which gradually enlarge in size, detached from the mother cells and act as independent individuals. The nucleus of the mother cells divides amitotically and transfers to the daughter cell. Several other outgrowths develop from the newly formed outgrowths, and sometimes the chains of the cells are seen. Very soon the yeast cells are detached from each other and act as new independent individuals.

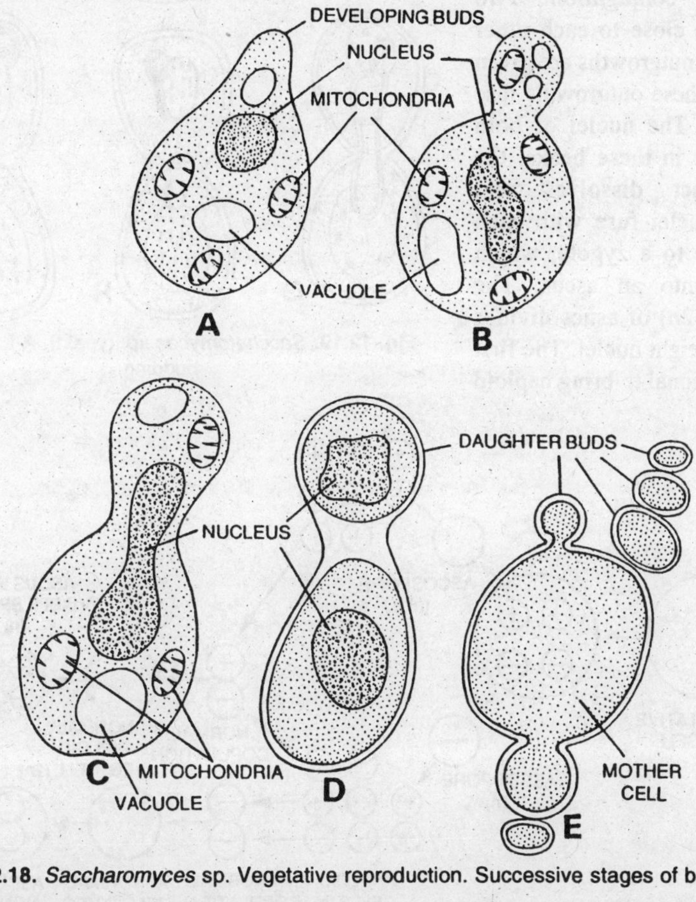

Fig. 12.18. *Saccharomyces* sp. Vegetative reproduction. Successive stages of budding (A-E).

Asexual reproduction. This type of reproduction probably takes place in adverse conditions, especially when there is scarcity of nutrients and abundance of oxygen. The yeast cell enlarges in size and called the 'ascus.' The nuclues of the ascus divides twice producing four nuclei. Now around each nucleus the cytoplasm deposits and the four ascospores are formed. Sometimes eight ascospores may also be produced. Each ascospore is surrounded by a thick wall. These spores are perennating bodies. They remain dormant in adverse conditions. On the approach of favourable conditions they germinate. The ascus wall bursts and the ascospores liberate in the atmosphere. They are dispersed by wind from one place to another. On getting suitable media and appropriate weather conditions the ascospores germinate and the new individuals are formed by budding.

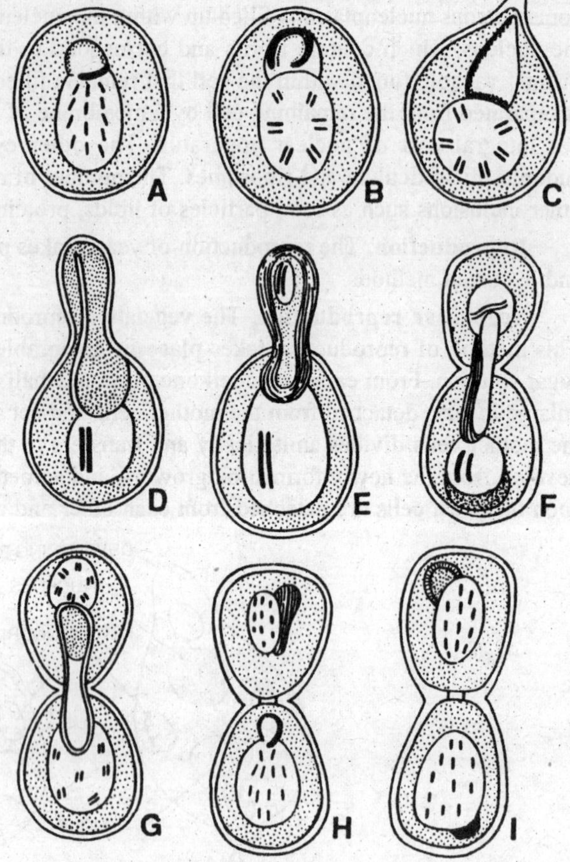

Fig. 12.19. *Saccharomyces* sp. (yeast). A-I, process of budding.

Sexual reproduction. The sexual reproduction takes place very rarely in some of the species of yeasts. This takes place by conjugation. Two individuals come close to each other and the beak-like outgrowths are given out from them. These outgrowths fuse with each other. The nuclei of both individuals come in these beaks, the wall of contact dissolves and ultimately the nuclei fuse with each other giving rise to a zygote, which soon converts into an ascus. The diploid nucleus ($2n$) of asucs divides thrice producing eight nuclei. The first division is reductional to bring haploid

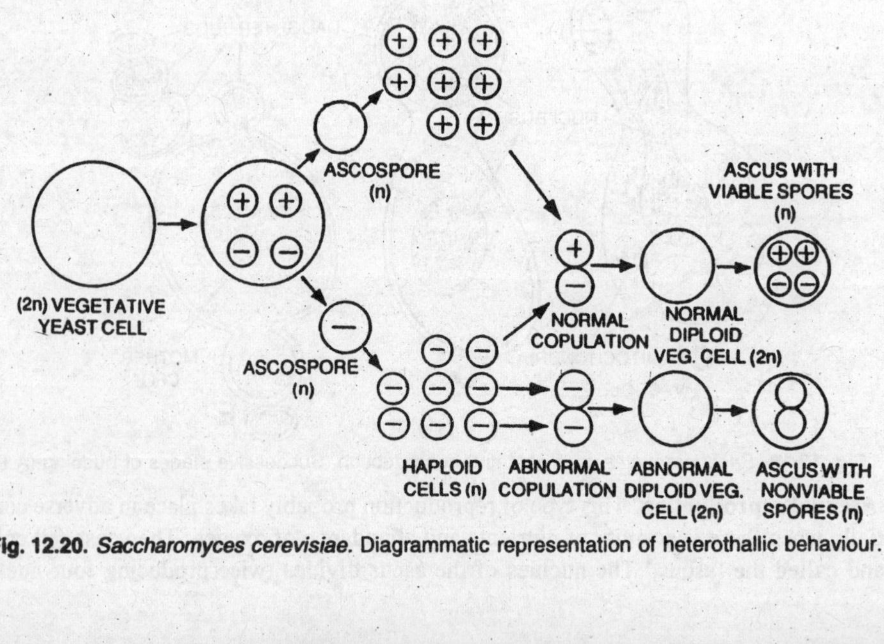

Fig. 12.20. *Saccharomyces cerevisiae.* Diagrammatic representation of heterothallic behaviour.

(n) condition again. Around each nucleus the cytoplasm is deposited, they become walled and called the ascospores. On bursting the wall of ascus the ascospores are liberated. On getting suitable conditions they germinate and the new individuals are produced by budding.

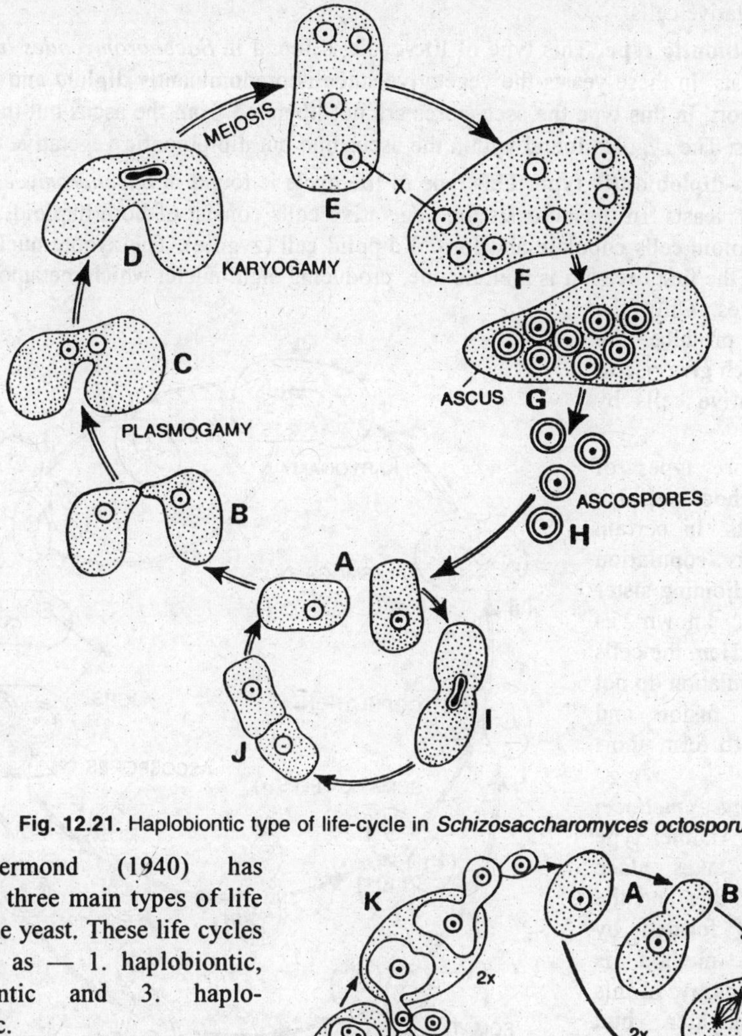

Fig. 12.21. Haplobiontic type of life-cycle in *Schizosaccharomyces octosporus*.

Guilliermond (1940) has recognized three main types of life cycles in the yeast. These life cycles are known as — 1. haplobiontic, 2. diplobiontic and 3. haplo-diplobiontic.

1. Haplobiontic type. This type of life cycle is found in *Schizosaccharomyces octosporus* and some other yeasts. In these yeasts the vegetaive stage is predominantly haploid and the diploid stage is very short. The diploid stage is represented by the zygote cell only which undergoes meiosis immediately after nuclear fusion. Here each somatic cell acts as a potential gametangium. During sexual union two cells fuse

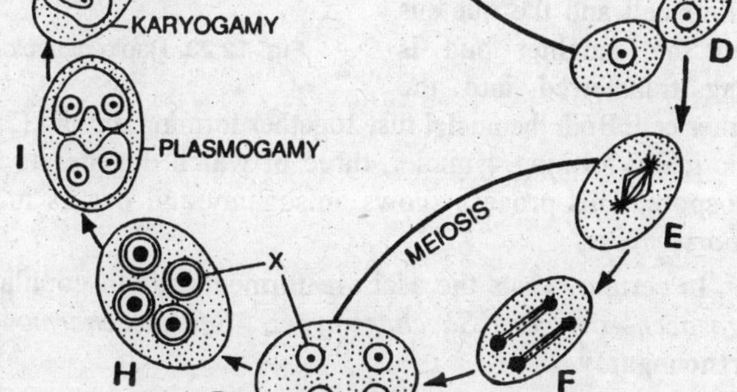

Fig. 12.22. Diplobiontic type of life-cycle in *Saccharomycodes ludwigii*.

(plasmogamy) and this is followed by the fusion of the two nuclei (karyogamy). The fusion or zygote nucleus divides thrice of which the first division is meiotic one. Now the zygote cell becomes ascus containing eight ascospores. After their liberation from the ascus the ascospores behave as vegetative cells.

2. Diplobiontic type. This type of life-cycle is found in *Saccharomycodes ludwigii* and some other yeasts. In these yeasts the vegetative stage is predominantly diploid and the haploid stage is very short. In this type the ascospores are not liberated from the ascus but they copulate within the ascus. The zygotes found within the ascus bud out diplobiontic vegetative cells.

3. Haplo-diplobiontic type. This type of life cycle is found in *Saccharomyces cerevisiae* and some other yeasts. In these yeasts the vegetative cells consist of both haploid and diploid phases. The haploid cells copulate and form a dipolid cell (zygote). The zygote nucleus divides thrice of which the first division is meiotic one, producing eight nuclei which metamorphose into eight ascospores. Only the diploid cells produce the ascospores which give rise to haploid vegetative cells by budding.

Some more types of copulation methods are also found in yeasts. In certain yeasts there is copulation between two adjoining sister cells; this is known as **adelphogamy**. Here the cells involved in copulation do not separate after fusion and remain united to form short chains, *e.g.*, *Schizosaccharmyces mellacei* and *S. pombe.* Another type of copulation takes place between the mother and the daughter cells formed by budding; this method is known as **pedogamy.** In this type the daughter bud remains continuous with the mother cell and the nucleus from the daughter bud is being transferred into the mother cell. Both the nuclei fuse together forming diploid (2*n*) nucleus. The fusion nucleus divides meiotically forming 4 nuclei, three of which degenerate and remaining one develops into an ascospore. This process shows anisogamy and occurs in a few yeasts such as *Nadsonia* and *Debaryomyces.*

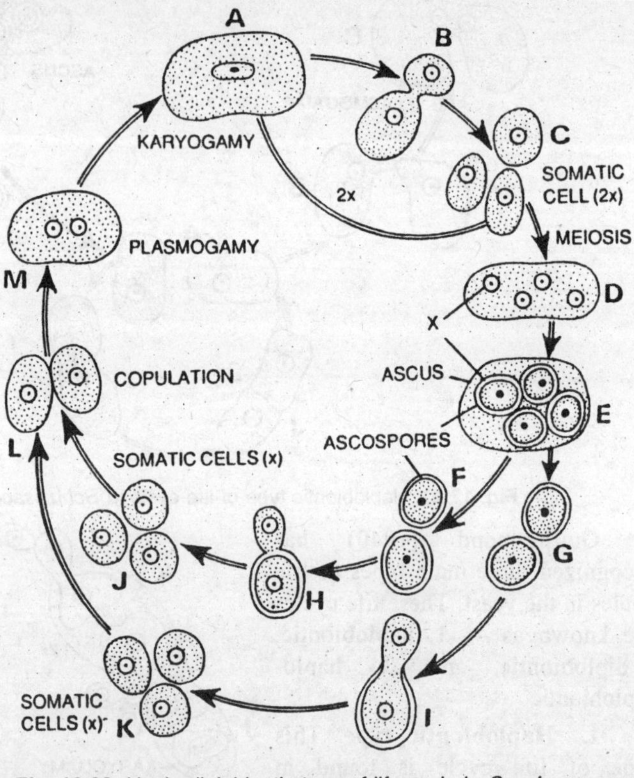

Fig. 12.23. Haplo-diplobiontic type of life-cycle in *Saccharomyces cerivisiae.*

In certain yeasts the asci are formed without copulation of two cells, *e.g.*, in species of *Zygosaccharomyces, Saccharomyces, Schizosccharmoyces.* This process is known as **parthenogamy.**

The number of ascospores per ascus varies from genus to genus, *e.g.*, the asci of *Monospora* and *Nadsonia* are monospores; 2-4 ascospres per ascus in *Debaryomyces* and *Hensenula;* 4 in *Saccharomyces* and 8 in *Schizosccharomyces.* The ascus in *Kluyveromyces polysporus* contains several spores.

The shape and structures of the ascospores is also variable, *e.g.*, the ascospores of *Saccharmyces* and *Schizosaccharomyces* are globose or ovoid; rough and warty of *Debaryomyces;* hat-shaped and flat on one side of *Hansenula;* needle-like and tapering at one end of *Nematospora.*

Fermentation. When the yeast cells grow in the sugarcane juice, toddy juice, or grape juice, then because of the activity of zymase enzyme the fermentation takes place. The sugar decomposes producing alcohol and carbon dioxide. In the abundance supply of oxygen the alcohol is formed in less quantity. On the other hand when there is scarcity of oxygen the large quantity of alcohol is produced.

The conversion of sugars into alcohol is termed **alcoholic fermentation.** The alcohol forming yeasts that grow at the top of the fermenting liquid are knwn as 'top yeasts'. The other yeasts which grow in the lower portions are known as 'bottom yeasts'. Though yeasts grow in less supply of oxygen yet fermentation does not take place in total absence of oxygen and most of the yeasts are intermediate between aerobic and anaerobic types. During fermentation besides the producion of alcohol and carbon dioxide other substances like glycerol, ethers, fatty acids, acetic acid and succinic acid are also produced in small quantities.

The important fermenting yeasts are the wine yeast (*Saccharomyces ellipsoides*); the beer and bread yeasts (*S. cerevisiae*); the ginger beer yeast (*S. piriformis*); the sake wine yeast of Japan (*S. sake*); African beer yeast (*S. pmbe*), etc.

The yeasts are rich in enzymes and vitamins and therefore, they make a valuable food and useful as medicine. The cubes or cakes with inactive starch as the base are made from the yeasts. Yeast tablets cure the stomach troubles. Certain pathogenic yeasts have also been reported.

$$C_6H_{12}O_6 + \text{Zymase} \rightarrow 2C_2H_5OH + 2CO_2 + \text{Zymase} + \text{Energy.}$$

Systematic Position

	G. W. Martin (1961)	**C. J. Alexopoulos (1962)**	**G. C. Ainsworth (1971)**
Kingdom	–Plantae	–Plantae	**–Fungi**
Division	–Mycota	–Mycota	**–Eumycota**
Sub-div.	–Eumycotina	–Eumycotina	**–Ascomycotina**
Class	–Ascomycetes	–Ascomycetes	**–Hemiascomycetes**
Sub-cl.	–Hemiascomycetidae	–Hemiascomycetidae	
Order	–Endomycetales	–Endomycetales	**–Endomycetales**
Family	–Saccharomycetaceae	–Saccharomycetaceae	**–Saccharomycetaceae**
Genus	–*Saccharomyces*	–*Saccharomyces*	**–*Saccharomyces***

Sub-class–EUASCOMYCETIDAE

Blue Molds, Perithecial Fungi, Cup Fungi and Morels
(Alexopoulos, 1962)

G. W. Martin (1961)	**G. C. Ainswoth (1971)**
Sub-class-**EUASCOMYCETIDAE**	Classes– **PLECTOMYCETES**
	PYRENOMYCETES
	DISCOMYCETES

Characteristic Features

1. The asci are usually unitunicate, *i. e.,* an ascus in which both the inner and outer walls are more on less rigid and do not separate during spore ejection.

2. The asci typically develop from special sexually produced ascogenous hyphae and in the majority of species are enclosed in a true ascocarp–the cleistothecium or apothecium.

3. They develop functional sex organs, but sexual degeneration is found at least in few groups.

Classification. This sub-class has been divided into three series — 1. Plectomycetes, 2. Pyrenomycetes and 3. Discomycetes.

Series–**Plectomycetes** or Class-**Plectomycetes** (Ainsworth, 1971)

Characteristic Features

1. The asci do not develop on any kind of special structures like hymenium or sporophores. They can develop from any part of mycelium.

2. The sexual reproduction results in the formation of asci and ascospores. The asci are globose or broadly club-shaped. They arise at different levels in the ascocarp. They are irregularly arranged.

3. The cleistothecia are formed which contain asci and ascospores. The asci release the ascospores by deliquescing in the ascocarp. Usually the cleistothecia do not possess definite pore.

Classification. The series has been divided into two orders– 1. Eurotiales and 2. Erysiphales.

Order **EUROTIALES** (60 genera; 120 species)

(Aspergillales)

Characteristic Features

1. Majority of them are saprophytic and known as green or blue green molds.

2. The hyphal ends penetrate the substrata to absorb the nutrition. The haustoria are not developed.

3. The conidia are formed in basigenous chains on conidiophores. The conidiphores may be septate or aseptate and branched or unbranched.

4. Sexual reproduction takes place by means of antheridia and ascogonia, resulting in the formation of asci and ascospores, within the cleistothecium in scattered condition. The outer wall of the cleistothecium is pseudoparenchymatous and called the 'peridium'. The asci are formed from ascogenous hyphae of various lengths and are found to be scattered at various levels within the ascocarp. There is no definite hymenial layer.

5. This order is sub-divided into two (Martin, 1961) to seven (Bessey, 1950) families.

Family–**EUROTIACEAE** (Aspergillaceae).

The family is characterized by having well-formed cleistothecia with definite peridia. The peridium consists of pseudoparenchyma formed by the compactly interwoven sterile hyphae. The conidial stages of the form genera *Aspergillus* and *Penicillium* are quite distinct and characteristic.

Genus **ASPERGILLUS** (20 sp.)

(*Eurotium*)

Occurrence. This is a saprophytic fungus. One can get the culture of *Aspergillus* by keeping a moist bread under the bell jar for few days at suitable temperature in the laboratory. In the beginning the white cottony cycelium appears, and afterwards on the same bread *Aspergillus* appears which can be very easily recognised by the presence of its smoky, green conidia. This fungus is found on moist leather, moist rotten wood, horse dung, rotten vegetables, rotten fruits and many other dead organic materials. In the rainy season one can very easily get the green coloured culture of *Penicillium* and *Asperfillus* on leather shoes and moist bread. This fungus is called 'blue green mould', because of the presence of its smoky green conidia in abundance.

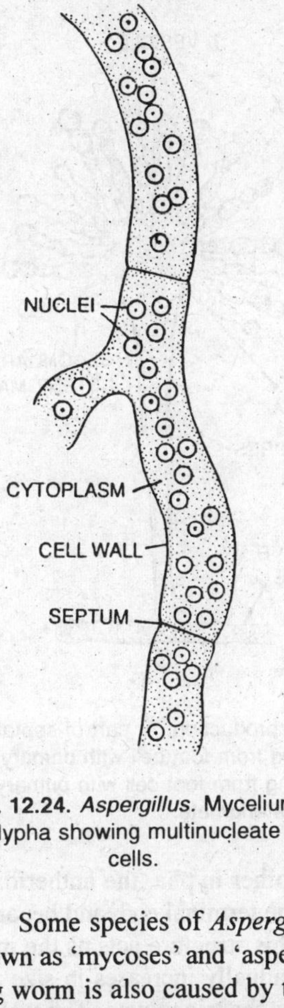

Fig. 12.24. *Aspergillus.* Mycelium. Hypha showing multinucleate cells.

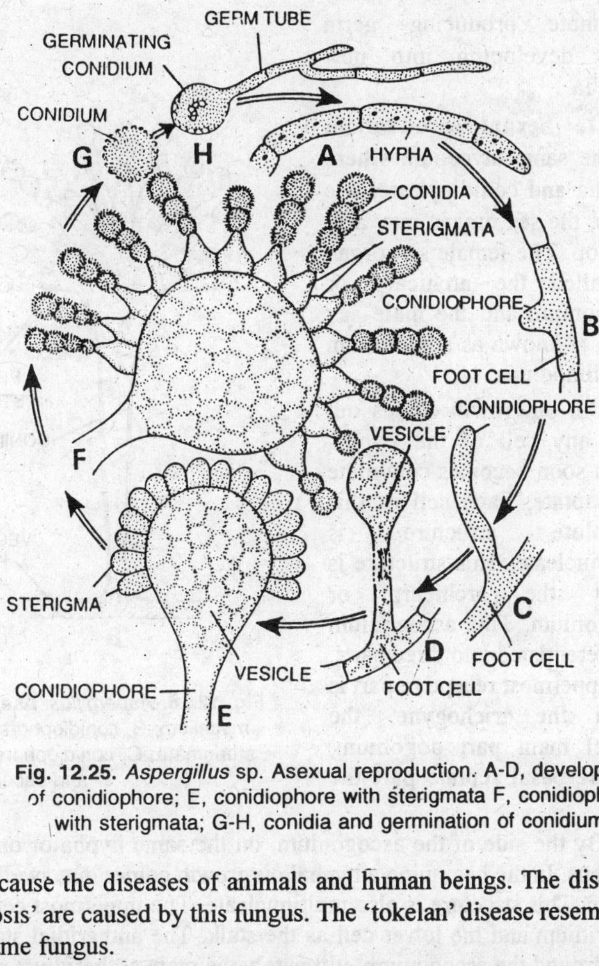

Fig. 12.25. *Aspergillus* sp. Asexual reproduction. A-D, development of conidiophore; E, conidiophore with sterigmata F, conidiophore with sterigmata; G-H, conidia and germination of conidium.

Some species of *Aspergillus* cause the diseases of animals and human beings. The diseases known as 'mycoses' and 'aspergillosis' are caused by this fungus. The 'tokelan' disease resembling ring worm is also caused by the same fungus.

Mycelium. The mycelium is found on the surface of the substratum. The hyphal ends penetrate the substratum and absorb food material. The mycelium is cottony, profusely branched, brownish and septate. Each cell is coenocytic having oil globules ad protoplasm in it. The cytoplasm of one cell to another is communicated through a central septal pore.

Reproduction. The reproduction takes place by means of asexual and sexual methods.

1. Asexual reproduction. The asexual reproduction takes place by means of conidia. The conidia are produced on the conidiophores. Each conidiophore is aseptate and globular at its terminal end. The swollen terminal portion is called the vesicle. Many bottle-shaped multinucleate primary sterigmata are produced on this vesicle. In certain species, the secondary sterigmata also develop on the primary sterigmata. Each conidium is spherical, echinulate, multinucleate and smoky green. Due to this green colour of conidia the colonies of *Aspergillus* look green and named as 'blue green moulds'. The conidiophore may develop from any cell of the mycelium. This is called the 'foot cell'. The conidia are dispersed by wind. Whenever they fall on the suitable media and get favourable conditions of moisture and temperature, they germinate producing germ tubes developing into new mycelia.

1. Sexual reproduction. On the same mycelium where conidia and conidiophores are borne, the sex organs may also develop. The female sex organ is called the archicarp or ascogonium and the male sex organ is known as antheridium or pollinodium.

An outgrowth comes out from any cell of the hypha, which soon becomes convolute and septate. Each cell of this convolute structure is multinucleate. This structure is called the archicarp or ascogonium. This ascogonium is differentiated into three parts. The uppermost receptive part is called the trichogyne, the central main part oogonium, and the basal septate part the stalk.

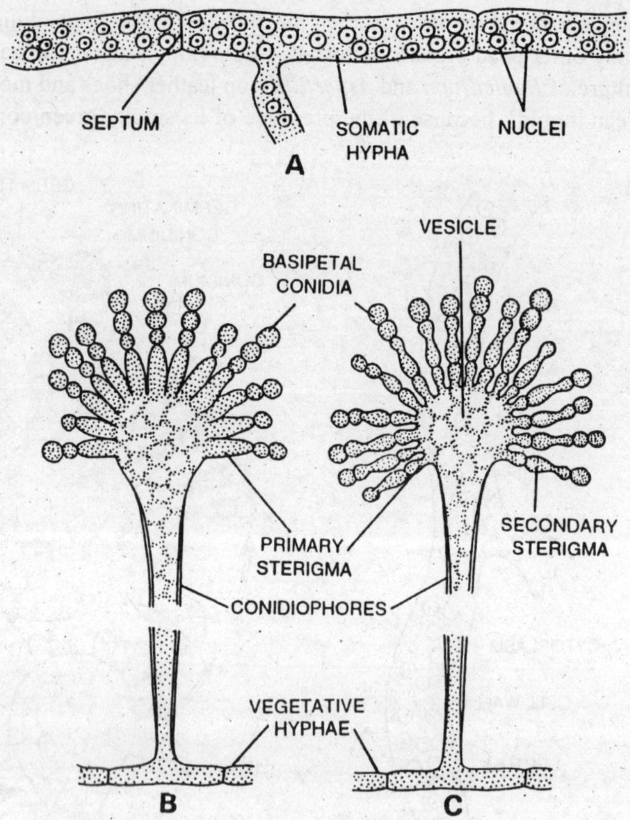

Fig. 12.26. *Aspergillus.* Asexual reproduction. A, part of septate mycelium; B, conidiophore arising from foot cell with primary sterigmata; C, conidiophore arising from foot cell with primary and secondary sterigmata.

By the side of the ascogonium, on the same hypha or on the other hypha, the antheridium develops. In the beginning a hyphal outgrowth comes out, swells at the terminal end, and becomes septate. This structure is also multinucleate. The uppermost cell of this structure acts as the main antheridium and the lower cell as the stalk. The antheridial stalk gradually increases in size and coils around the ascogonium. ultimately the main antheridium reaches the trichogyne. The wall of contact dissolves and the protoplasm of the antheridium along with nuclei passes to the oogonium through trichogyne, where each male nucleus makes a pair with female nucleus, but the paired nuclei do not fuse with each other. The fusion of the male and female nuclei in *Aspergillus* is not yet seen. In some species, the antheridium does not develop. According to Dangeard (1907), the antheridia are not found in *Aspergillus flavus* and *Aspergillus fischeri.* In such species the nuclei of the ascogonia make pairs. By now the main oogonial part of ascogonium becomes septate and the ascogenous hyphae develop on this part. The ascogenous hyphae are also septate. The asci develop from the ultimate and penultimate cells of the ascogenous hyphae. Each ascus contains 4-8 ascospores in it. By this time several sterile hyphae develop on septate stalk, which surround the ascogonium completely.

Development of ascus. The mycologists have different opinions about the development of the ascus.

According to one school of thought the male and female nuclei found in the pairs, in the ascogonium fuse with each other and the diploid ($2n$) nuclei are formed. By now the oogonium

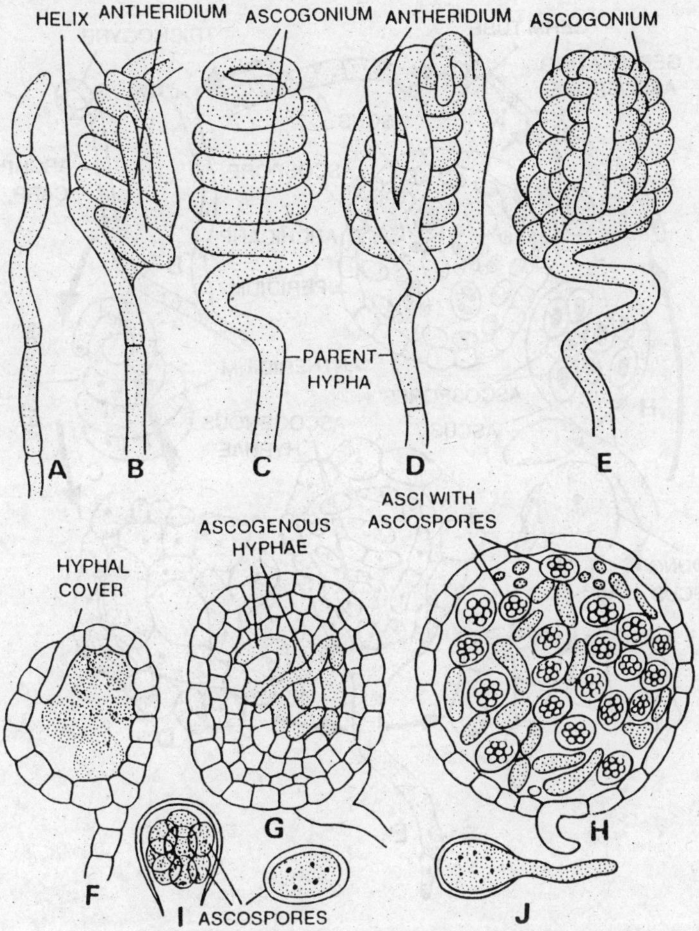

Fig. 12.27. Sexual reproduction in *Aspergillus repens.* Development of sex organs and cleistothecium (A-H); ascus, ascospores and germination of ascospore (I-J).

becomes septate and bears septate ascogenous hyphae on it. Each cell of the oogonium bears two diploid ($2n$) nuclei and each of the ascogenous hypha contains two diploid nuclei. The diploid nuclei of the ultimate and penultimate cells of the ascogenous hyphae fuse and give rise to tetraploid ($4n$) nuclei in these cells. This tetrapoid nucleus ($4n$) divides thrice. The first division is reductional (meiotic) which is followed by brachymeiosis. In this way eight nuclei are formed. The cytoplasm is deposited around each nucleus and eight ascospores are produced in each asucs. The ascospores are haploid. To bring the haploid phase from tetrapoid condition the meiosis is followed by **brachymeiosis.**

According to another school of thought, the oogonium becomes septate and each cell bears a pair of male and female nuclei. The septate ascogenous hyphae develop on the oogonium. The ascogenous hyphae are of unlimited growth and rarely branched. Each cell of the ascogenous hyphae contains a pair of male and female nuclei. The paired nuclei of the ultimate and penultimate cells become fused with each other, and a diploid ($2n$) nuclues develops in each cell. This diploid nucleus divides thrice and eight nuclei are formed. The first division is reductional. The nuclei, thus formed are haploid (n). The cytoplasm is deposited around each nucleus and eight ascospores are formed in each ascus.

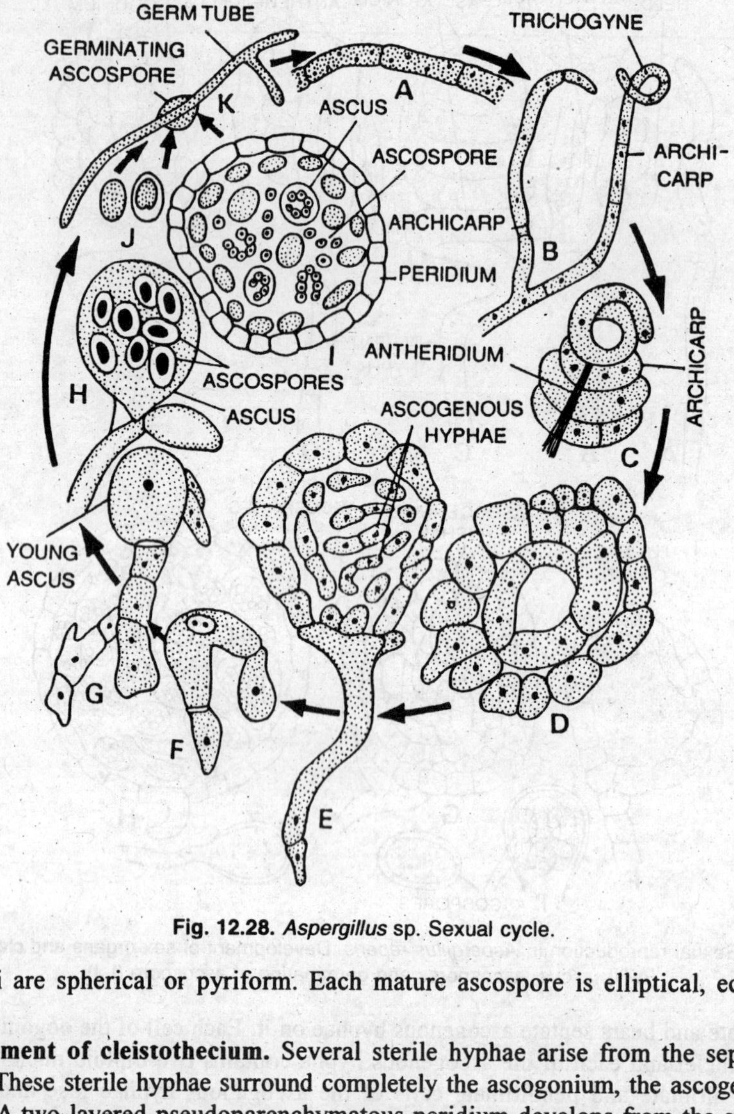

Fig. 12.28. *Aspergillus* sp. Sexual cycle.

The asci are spherical or pyriform. Each mature ascospore is elliptical, echinulate and biconvex.

Development of cleistothecium. Several sterile hyphae arise from the septate stalk of ascogonium. These sterile hyphae surround completely the ascogonium, the ascogenous hyphae and the asci. A two-layered pseudoparenchymatous peridium develops from the sterile hyphae which completely surround the ascogonium and its other parts. The inner layer of the peridium acts as nutritioal layer which provides food to ascospores whereas the outer layer is protective in nature. This rounded closed pseudoparenchymatous structre is known as **cleistothecium.** On the maturation of the cleistothecium, all parts of ascogonium, ascogenous hyphae and walls of asci disorganise and free ascospores remain sattered within the cleistothecium. The outer wall of the cleistothecium secretes some oily substance and appears yellowish in colour. The cleistothecia are perennating bodies. They survive in adverse conditions. They rupture by the pressure of outer agencies. The ascospores are liberated and dispersed by the wind. On getting the suitable media and optimum weather conditions, they germinate producing germ tubes developing new mycelia on the surface of the substratum.

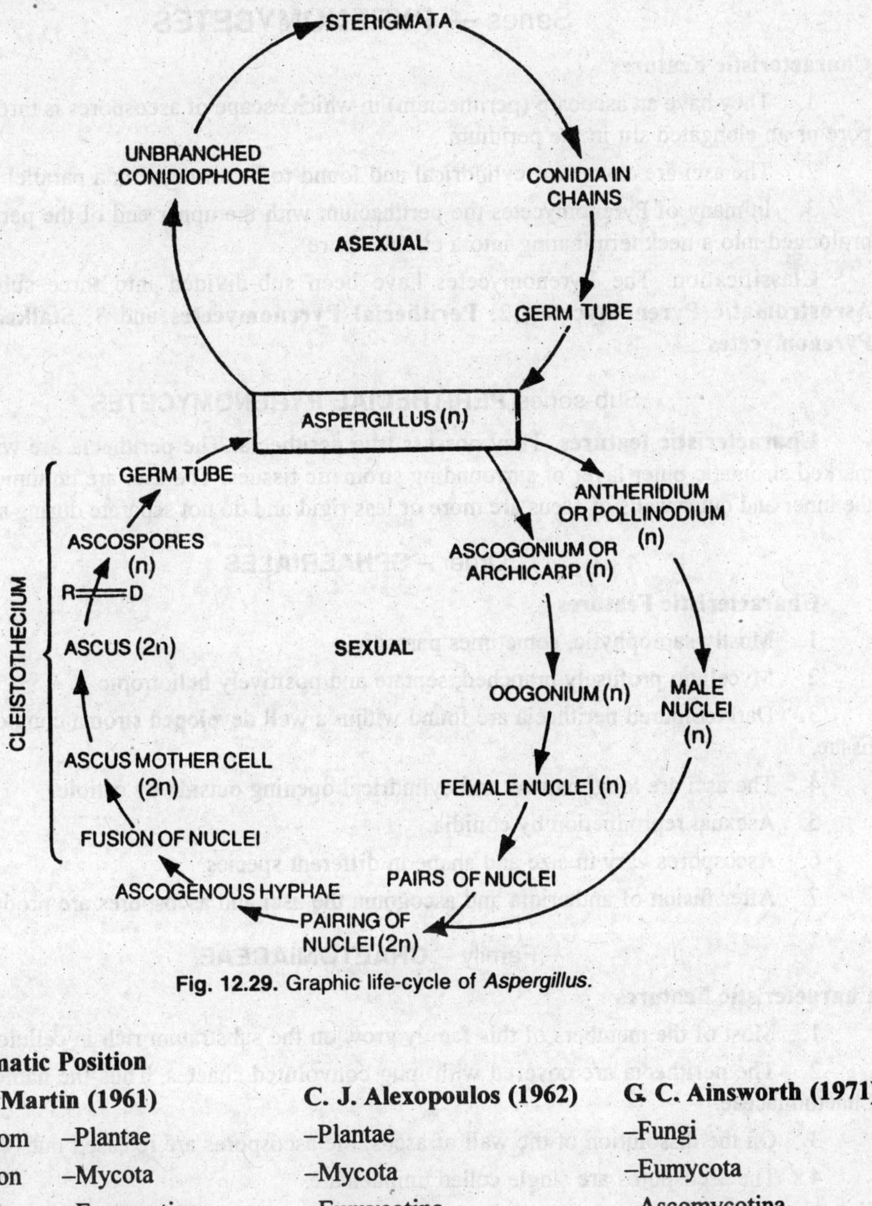

Fig. 12.29. Graphic life-cycle of *Aspergillus*.

Systematic Position

	G. W. Martin (1961)	**C. J. Alexopoulos (1962)**	**G. C. Ainsworth (1971)**
Kingdom	–Plantae	–Plantae	–Fungi
Division	–Mycota	–Mycota	–Eumycota
Sub-div.	–Eumycotina	–Eumycotina	–Ascomycotina
Class	–Ascomycetes	–Ascomycetes	–Plectomycetes
Sub-cl.	–Euascomycetidae	–Hemiascomycetidae	–
Series	–	–Plectomycetes	–
Order	–Eurotiales	–Eurotiales	–Eurotiales
Family	–Eurotiaceae	–Eurotiaceae	–Eurotiaceae
Genus	–*Aspergillus*	–*Aspergillus*	–*Aspergillus*

Series — PYRENOMYCETES

Characteristic Features

1. They have an ascocarp (perithecium) in which escape of ascospores is through a circular pore or an elongated slit in the peridium.

2. The asci are clavate or cylindrical and found to be arranged in a parallel series.

3. In many of Pyrenomycetes the perithecium with the upper end of the peridium is to be prolonged into a neck terminating into a circular pore.

Classification. The Pyrenomycetes have been sub-divided into three sub-series — 1. **Ascostromatic Pyrenomycetes**, 2. **Perithecial Pyrenomycetes** and 3. **Stalked Perithecial Pyrenomycetes**.

Sub-series PERITHECIAL PYRENOMYCETES

Characteristic features. They possess true perithecia. The perithecia are with or without marked stromatic outer layer or surrounding stromatic tissues. The asci are unitunicate, *i.e.*, both the inner and outer walls of ascus are more or less rigid and do not separate during spore ejection.

Order – SPHAERIALES

Characteristic Features

1. Mostly saprophytic, sometimes parasitic.
2. Mycelium profusely branched, septate and positively heliotropic.
3. Dark coloured perithecia are found within a well developed stroma composed of fungal tissue.
4. The asci are long, narrow and cylindrical opening outside by ostiole.
5. Asexual reproduction by conidia.
6. Ascospores vary in size and shape in different species.
7. After fusion of antheridia and ascogonia the asci and ascospores are produced.

Family – CHAETOMIACEAE

Characteristic Features

1. Most of the members of this family grow on the substratum rich in cellulose.
2. The perithecia are covered with long convoluted chaetes. Thus the name of family is Chaetomiaceae.
3. On the dissolution of the wall of ascus, the ascospores are released out.
4. The ascospores are single celled uninucleate.
5. Each ascospore contains a single pore.

Genus : CHAETOMIUM

There are about 200 species. Most of them are saprophytic which generally grow on dung, straw, paper, textiles that are rich in cellulose. *Chaetomium globosum* is capable to decompose cellulose. Some of the species cause 'soft rot' on wood. The enzymes C_1 and C_4 found in species of *Chaetomium* cause disintegration of cellulose in cotton, pulp and paper. They can also disintegrate cellulose found in substances present in acidic soils, where bacterial activity is low. Tevit (1956), obtained an antibiotic called chaetamin from *C. cochlioides.*

enerally the perithecia are elongated, flask-like with long neck. The perithecia remain covered with long hairs.

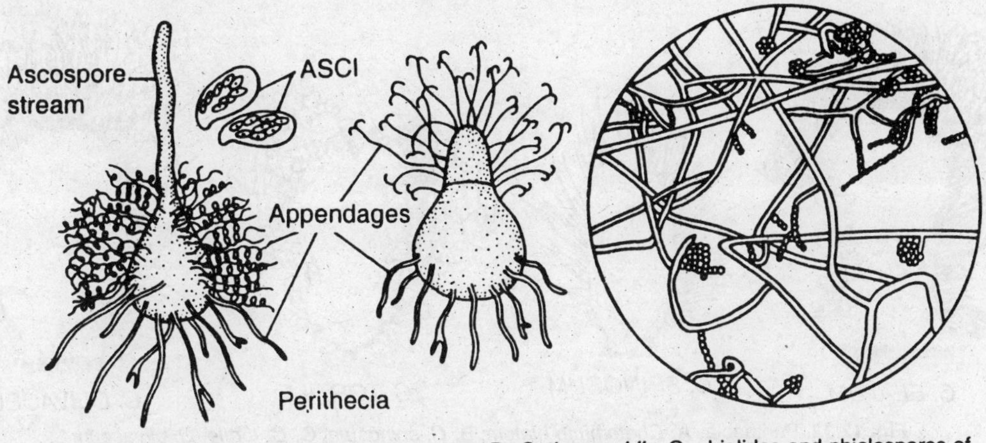

Fig. 12.30. Perithecia. A, *Chaetomium globosum*; B. *C. thermophila*; C, phialides and phialospores of *Chaetomium*.

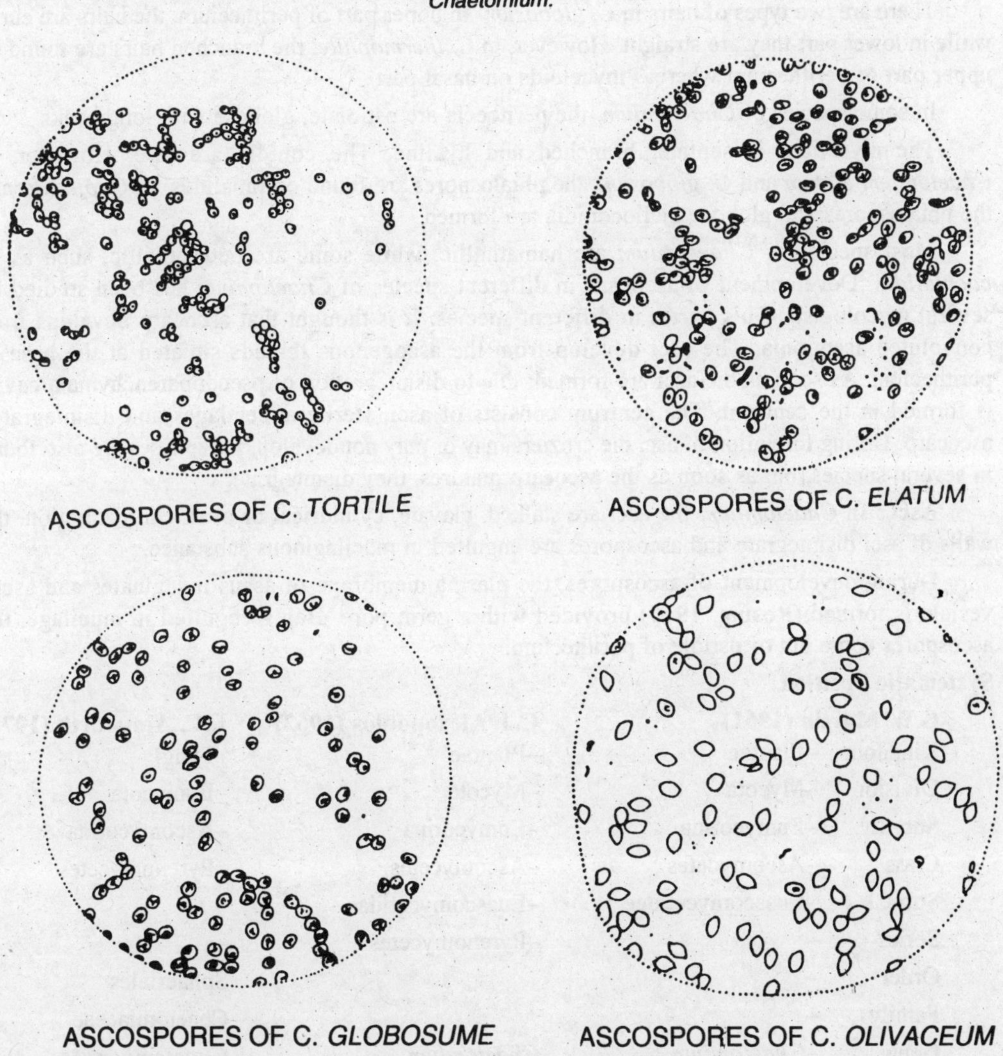

Fig. 12.31. Ascopores. A, *Chaetomium tortile*; B, *C. elatum*; C, *C. globosum*; D, *C. olivaceum*.

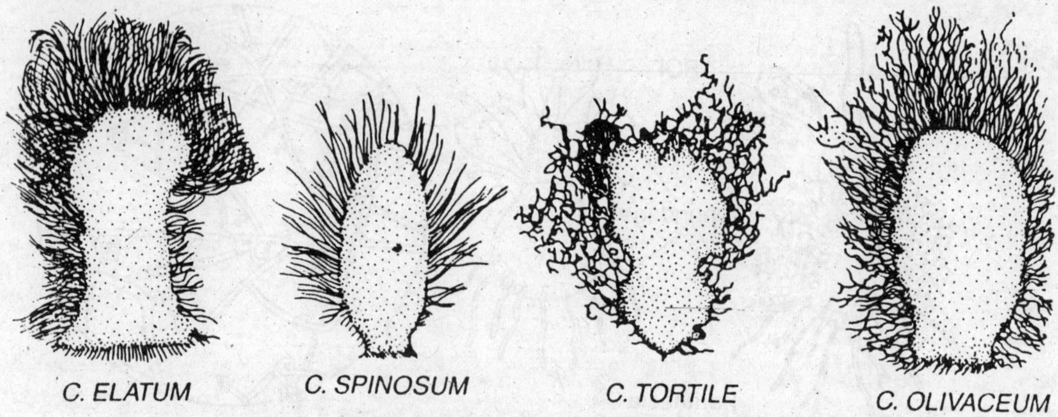

Fig. 12.32. Perithecia. A, *Chetomium elatum*; B, *C. spinosum*; C, *C. tortile*; *C. olivaceum*.

There are two types of hairs in *C. globosum*. In upper part of perithecium, the hairs are curly, while in lower part they are straight. However, in *C. thermophile,* the branched hairs are found on upper part of perithecium whereas myceloids on basal part.

In some species of *Chaetomium,* the perithecia are ostiolate, globose with long beaks.

The mycelium is septate, branched and hyaline. The conidia are rare. However, in *Chaetomium elatum* and *C. globosum,* the phialospores are found on phialides. In *C. piluliferum,* the phialospores and globose thalloconidia are formed.

Most species of *Chaetomium* are hamathallic, while some are heterothallic, such as *C. cochlioides*. Development of ascocarp in different species of *Chaetomium* has been studied by several mycologists. This varies in different species. It is thought that ascocarp develops from convoluted ascogonia. The asci develop from the ascogenons threads situated at the base of perithecium. As soon as the asci are formed; due to disintegration of pseudoparenchyma a cavity is formed in the centrum. The centrum consists of asci, sterile appendages and disintegrated ascocarp. During formation of asci the croziers may or may not develop. Paraphyses are also found in several species, but as soon as the ascocarp matures, they disintegrate.

Asci : In *Chaetomium,* the asci are stalked, clavate, cylindrical or oval. On maturation, the walls of asci disintegrate and ascospores are engulfed in mucilaginous substance.

During development of ascospores, the plasma membrane of ascus invaginates and ascus vesicle is formed (Rosing, 1975) provided with a germ pore. Being engulfed in mucilage, the ascospores come out of ostiole of perithecium.

Systematic Position

	G.W. Martin (1961)	**C.J. Alexopoulos (1962)**	**G.C. Ainsworth (1971)**
Kingdom	–Plantae	–Plantae	–Fungi
Division	–Mycota	–Mycota	–Eumycota
Sub-div.	–Eumycotina	–Eumycotina	–Ascomycotina
Class	–Ascomycetes	–Ascomycetes	–Pyrenomycetes
Sub-Cl.	–Euascomycetidae	–Euascomycetidae	–
Series	–	–Pyrenomycetes	–
Order	–		Sphaeriales
Family	–		Chaetomiaceae
Genus	–*Chaetomium*	*Chaetomium*	*Chaetomium*

Series **DISCOMYCETES** or

Class—**Discomycetes** (Ainswoth, 1971)

Characteristic Features

The apothecia of series Discomycetes are open and more or less cup-shaped at maturity. They possess incurved margins, which unfold on the maturation of the apothecium to expose the hymenial layer. They produce their apothecia on the ground, on buried sticks, rotten logs, overwintered leaves or fruits, or on animal dung. The apothecia of some species are variously coloured such as red, yellow or orange; some are brown and few are black. Besides the typical cup or disc-like forms of the apothecia there are sponges, bells, saddles, tongues and brain like fruiting bodies. However, all these apothecia are open and bear their asci on the surface or in large open cavities. The apothecium possesses a sheath of interwoven hyphae which is pushed open by the growth of the paraphyses so that it forms the outer wall of the cup. The lower part of the cup is filled by the hypothecium, a weft of some vegetative and some ascogenous hyphae. These give rise to the sub-hymenium. The later formed paraphyses and young asci arise from the sub-hymenial layer. The asci and paraphyses constitute the hymenium, which remains spread over the surface of the interior of the cup. The asci are club-shaped or cylindrical. They possess simple or branched paraphyses at the apices among them. The asci usually open by a lid (operculate type) or by the ejection of a plug (inoperculate type). The distinction between these modes of dehiscence has been used as a basis of classification.

Classification. Many authors divide the series Discomycetes into three orders 1. Pezizales, 2. Helotiales and 3. Tuberales.

Order **PEZIZALES** (65 genera; 500 species)

They are characterized by their fleshy or sometimes leathery ascocarp, bounded, except in the family Pyronemaceae, by a definite peridium which remains closed in the beginning but later on pushed open by the growth of a mass of paraphyes, giving the mature ascocarp its typical cup or saucer shape. The fungi of this order are known as the cup fungi. The apothecia are found above the ground (epigean) with or without the stalks.

The asci open by a lid (operculate). The ascospores are ellipsoidal and one-celled.

Most of them are saprophytic grow on dead and decaying wood, on soil rich in organic matter and on dung of animals.

There is a clear tendency towards the reduction of sexuality. In some cases both antheridia and ascogonia are present. In other cases the non-functional antheridia are found. In the extreme cases such as *Morchella* neither the antheridia and nor the ascogonia are present. In such cases the fusion of vegetative nuclei represents the sexual process and asci are produced.

Classification. According to Gaumann, the following families are included in the order Pezizales. 1. Pyronemaceae. 2. Pezizaceae and 3. Helvellaceae.

Family–Pezizaceae

They possess epigenous apothecia. The apothecia are mostly cup, disc or lentil shaped. They are sessile in some species while stalked in others. The apothecia are minute to very large. They are bright coloured to dark brown, smooth velvety, hairy or bristly. The peridium is well developed. It is either fleshy or waxy in texture. The ascospores are uniseriately arranged in asci.

Most of them are coprphilous. The species of many genera grow exclusively on dung, and some of them may even be specialized as to the animal on whose excreta they grow.

Most of the Pezizaceae do not possess any conidial stage. However, conidia have been found in *Peziza repanda* and *Peziza vesiculosa*.

Genus PEZIZA (50 species)

Peziza vesiculosa is one of the larger cup fungi. The cup-shaped, fleshy ascocarp with no hairs are found to be crowded together on mature piles, well fertilized gardens, green houses, etc. *Peziza badia* is commonly found on the ground in deciduous forests and open places. *P. repanda* is found on rotten logs in woods. The apothecia of the fungus are whitish or pale brown and exhibit minute pustules on the surface.

Somatic structure. The mycelium is well developed and consists of network of septate hyphae with uninucleate cells except in the fructification. The hyphae are found within the substratum and therefore, they are not visible from outside. They absorb the nutrition from the substratum.

Asexual reproduction. Most of the Pezizaceae are not known to reproduce asexually. However, conidia have been found in *Peziza repanda* and *Peziza vesiculosa*. In *Peziza pustulata* the conidiophores are found which bear heads of conidia developed from surface hyphae. The conidia are exogenously formed from three to four-celled conidiophores. The apices of them form narrow, clavate, oval or globose heads on which numerous conidia are attached. The conidia are abstricted from the apices of the conidiophores. Each conidium germinates forming a new mycelium.

Certain species produce thick-walled chalmydospores. The chlamydospores are intercalary in position. On the approach of suitable conditions of moisture and temperature they germinate and produce new mycelia.

Sexual reproduction. The sex organs are not found in the species of *Peziza*. The sexual process is extremely simple and consists in the association of two vegetative nuclei in a pair; the compatible nuclei are brought together by means of somatogamy. The vegetative cells are seen to

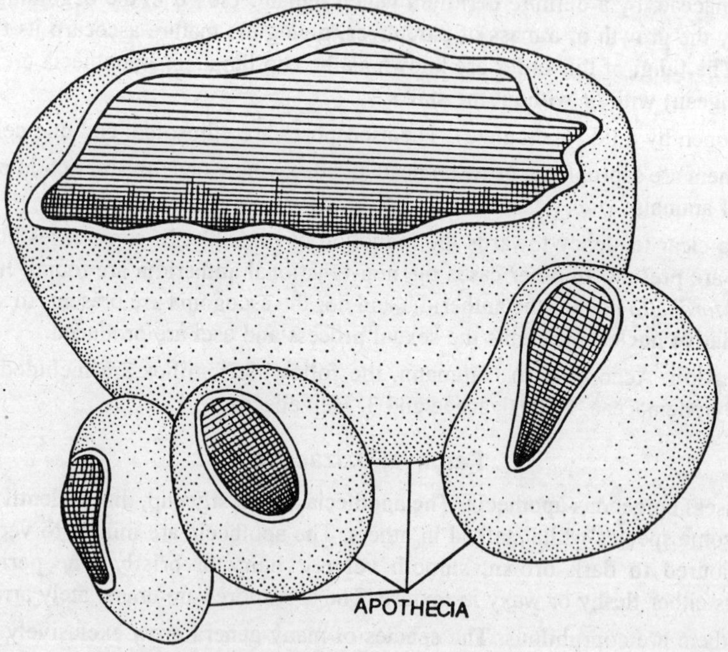

Fig. 12.33. *Peziza*. The apothecia.

possess the nuclei associated in pairs. These dikaryotic cells rise to ascogenous hyphae. The ascogenous hyphae become many celled by septa. The cells are binucleate. The terminal or sub-terminal cell of ascogenous hyphae acts as an ascus mother cell which develops into an ascus. The two nuclei of the ascus mother cell fuse together forming a synkaryon. This fusion nucleus divides meiotically and thereafter mitotically forming eight haploid nuclei which become ascospores. The asci are clavate, erect and lie side by side. They are intermingled with paraphyses.

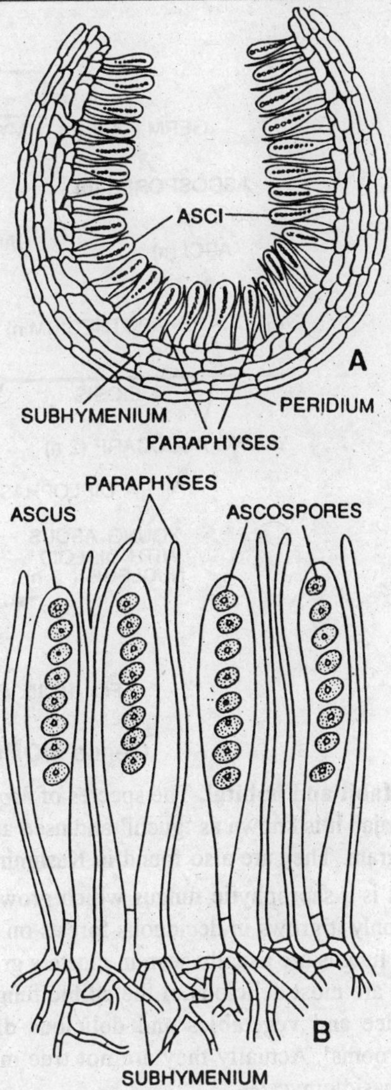

Fig. 12.34. *Peziza*. A, vertical section of mature apothecium; B, a portion of apothecium showing asci, ascospores and subhymenium.

Apothecium. The apothecial cups of *Peziza vesiculosa* are brown, 2 to 3 inches broad, and are commonly contorted and crimpled. The hymenial surface of *P. venosa* is convoluted. In sectional view the apothecium exhibits three layers. The uppermost layer is the hymenium; middle one the sub-hymenium and the outermost layer, the excipulum. The hymenium consists of erect asci and paraphyses; the sub-hymenium of the weft of hyphae and ascogneous hyphae; and the excipulum consists of pseudoparenchymatous tissue formed of compactly interwoven hyphae.

Asci and ascospores. The asci are clavate and form palisade-like layer lining the cavity of the ascocarp. Each ascus contains 8 ascospores. On maturation, the ascospores are shot off from asci and disseminated by wind. Each ascospore is haploid, smooth and elliptical. On germination it produces the haploid mycelium.

Systematic Position

	G.W. Martin (1961)	**C.J. Alexopoulos (1962)**	**G.C. Ainsworth (1971)**
Kingdom	–Plantae	–Plantae	–Fungi
Division	–Mycota	–Mycota	–Eumycota
Sub-div.	–Eumycotina	–Eumycotina	–Ascomycotina
Class	–Ascomycetes	–Ascomycetes	–Discomycetes
Sub-cl.	–Euascomycetidae	–Euascomycetidae	–
Series	–	–Discomycetes	–
Order	–Pezizales	–Pezizales	–Pezizales
Family	–Pezizaceae	–Pezizaceae	–Pezizaceae
Genus	–*Peziza*	–*Peziza*	–*Peziza*

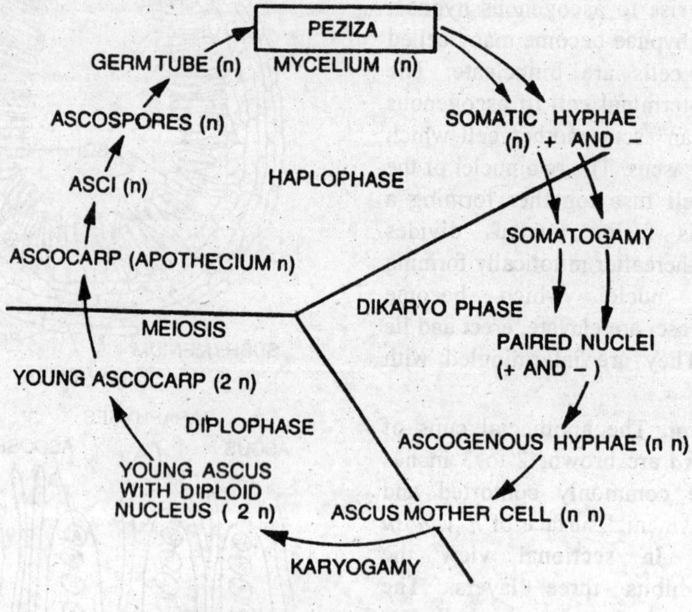

Fig. 12.35. *Peziza.* Graphic life-cycle

Genus **MORCHELLA** (15 species)

Habit and habital. The species of *Morchella* are commonly known as 'morels'. In H.P. and the Punjab it is known as 'guchi' and used as vegetable. It is sold as costly as two thousand rupees a kilogram. They are also found in Kashmir.

It is a saprophytic fungus which grows on humus and other dead organic substances. Very commonly it grows in deciduous forests on decaying leaves and wooden logs. The fructifications of this fungus are usually seen in clusters growing on organic substrata. The fructifications of this fungus are most delicious of the edible fungi. In Uttar Pradesh and Punjab this fungus is cooked with rice and vegetables and delicious dishes are prepared. Sometimes they are called the 'mushrooms'. Actually they are not true mushrooms. The true mushrooms are the members of Homobasidiomycetes.

Structure of mycelium. The mycelium of the fungus is found in masses in the humus soil. It is inconspicuous, septate and profusely branched. It penetrates the humus soil several inches deep. Prior to the development of fructifications the hyphal masses develop just below the surface of the soil. When abundance of food and moisture are found in the soil, the fructifications develop from the hyphal masses. The fruiting body is called the ascocarp. The ascocarp is composed of the compactly arranged interwoven hyphae. The fruiting body develops within a few hours.

The fructification. The fructifications of the different species of morels differ in size and external morphology. The fruiting body of *Morchella esculenta* is found upon a cream coloured, thick hollow and fleshy stalk about two or three inches in height and one half to one inch in diameter. The stalk bears at its terminal end more or less a conical hollow cap with ridges and depressions on it. It is as long as the stalk but sufficiently broader in diameter. In the beginning of the development of the fruiting body the cap is quite smooth but afterwards the ridges and depressions are brown.

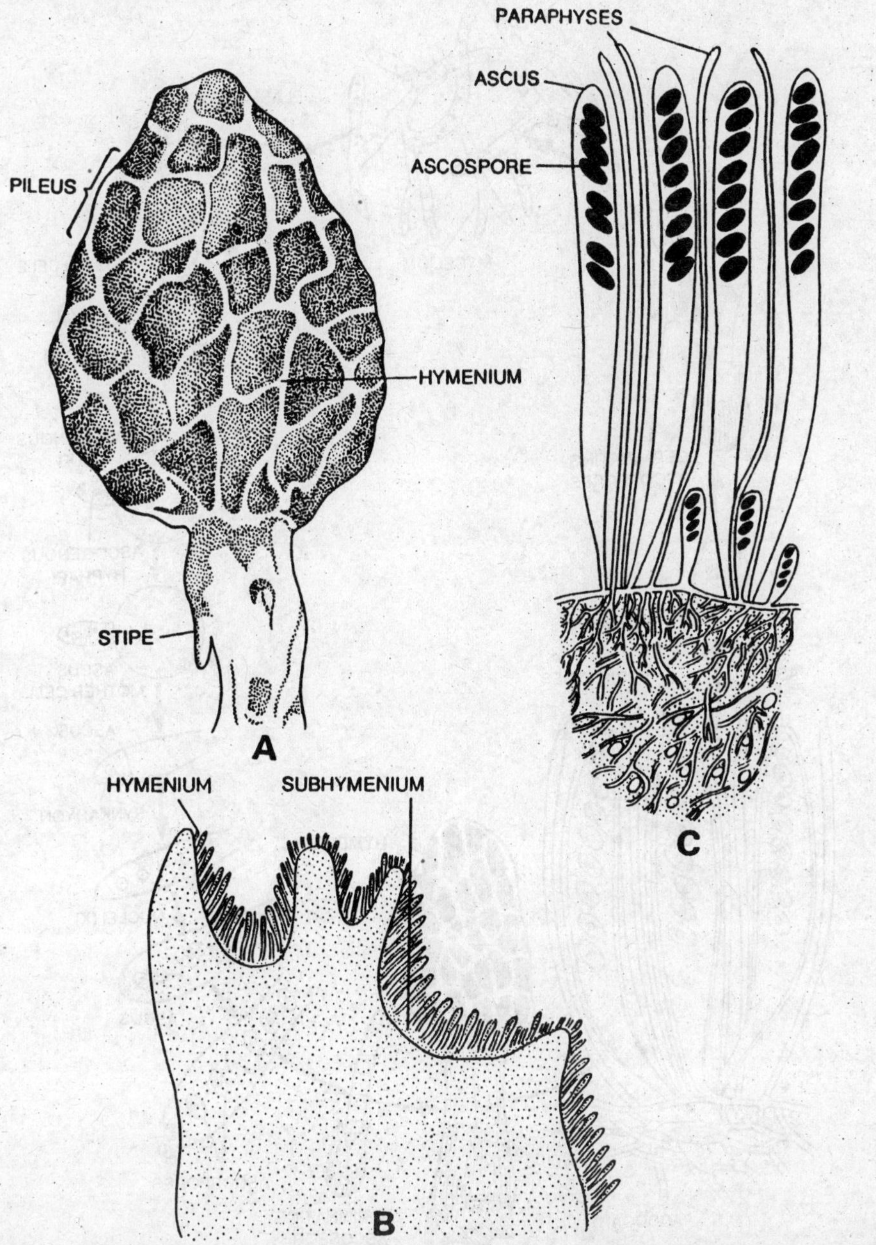

Fig. 12.36. *Morchella.* A, a fructification with stipe and pileus; B, sectional view of a part of pileus showing hymenium and subhymenium; C, paraphyses and asci with ascospores.

Internal structure of ascocarp. The stipe or stalk consists of compact interwoven hyphae and looks in sectional view as pseudoparenchymatous structure. In the depression of the ascocarp the hymenial layers are present which consist of innumerable elongated, cylindrical asci containing ascospores. The paraphyses remain intermingled with asci. On the maturation of the asci, they become elongated and break at the tips with the release of ascospores with a slight jerk. The ascospores are dispersed by wind. On the approach of suitable media and favourable conditions, they germinate producing germ tubes developing new mycelia.

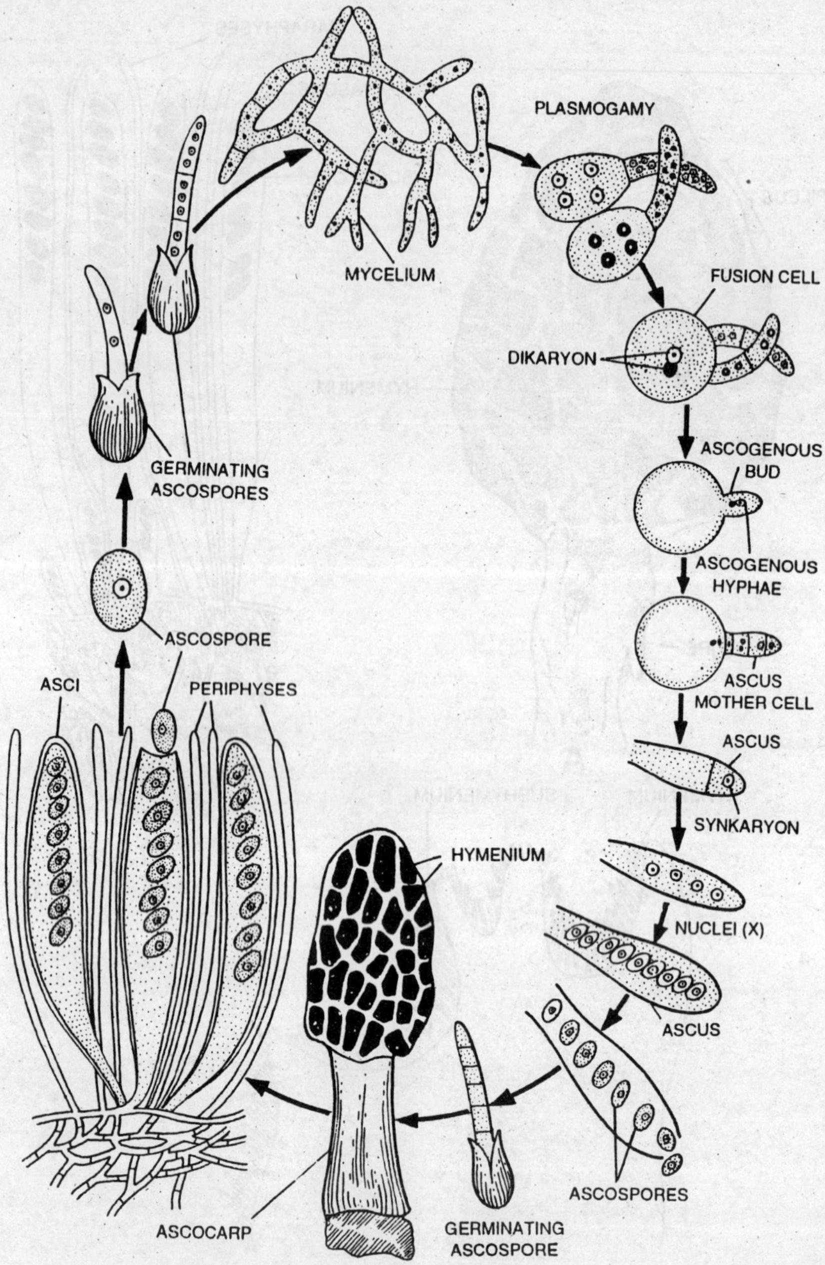

Fig. 12.37. *Morchella* sp. Diagrammatic life-cycle.

Sexual reproduction. The sexual reproduction in *Morchella* is very much reduced or altogether absent. The sex organs are not found. The nuclei found in the vegetative hyphae, however, fuse together. The two adjacent lying vegetative hyphae come together, the walls of contact dissolve and the nuclei of two different hyphae come in pairs and conjugate. This type of sexual reproduction is known as **somatogamy.** The ascogenous hyphae develop. The terminal cell of the ascogenous hypha behaves as ascus mother cell. The diploid nucleus divides thrice and 8 ascospores are developed in the ascus. The first division is reductional.

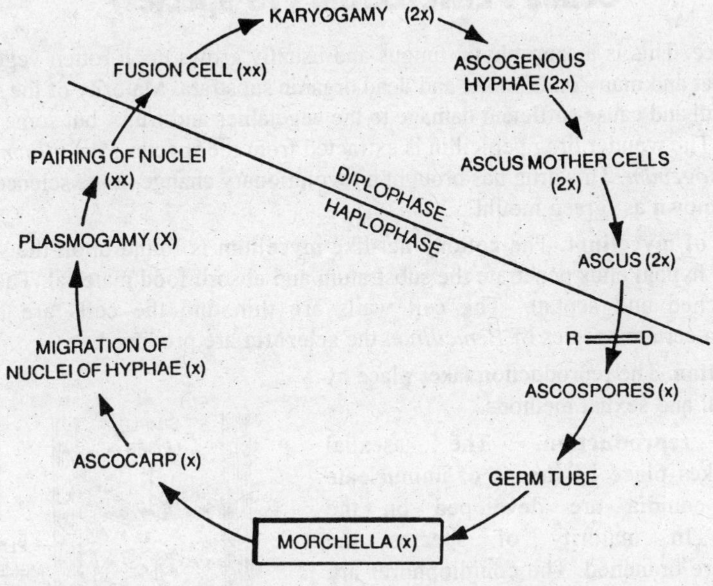

Fig. 12.38. Graphic life cycle of *Morchella.*

Systematic Position

	G.W. Martin (1961)	**C.J. Alexopoulos (1962)**	**G.C. Ainsworth (1971)**
Kingdom	–Plantae	–Plantae	–Fungi
Division	–Mycota	–Mycota	–Eumycota
Sub. div.	–Eumycotina	–Eumycotina	–Ascomycotina
Class	–Ascomycetes	–Ascomycetes	–Discomycetes
Sub-cl.	–Euascomycetidae	–Euascomycetidae	–
Series	–	–Discomycetes	
Order	–Pezizales	–Pezizales	–Pezizales
Family	–Pezizaceae	–Helvellaceae	–Helvellaceae
Genus	–*Morchella*	–*Morchella*	–*Morchella*

Genus PENICILLIUM (13 species)

Occurrence. This is a saprophytic fungus and usually grows upon rotten vegetables, rotten fruits, rotten meat and many other moist and dead organic substrata. Majority of the species of this genus are harmful and cause sufficient damage to the vegetables and fruits, but some of the species are very useful. The wonder drug penicillin is extracted from the species *Penicillium notatum* and *Penicillium crysogenum*. This drug has brought a revolutionary change in the science of medicine. This fungus is known as 'green mould'.

Structure of mycelium. The cottony net-like mycelium is found upon the surface of the substratum. The hyphal ends penetrate the substratum and absorb food material. The mycelium is profusely branched and septate. The cell walls are thin and the cells are binucleate or multinucleate. In certain species of *Penicillium* the **sclerotia** are produced.

Reproduction. The reproduction takes place by means of asexual and sexual methods.

Asexual reproduction. The asexual reproduction takes place by means of uninucleate conidia. The conidia are developed on the conidiophores. In majority of species the conidiophores are branched. The conidiophores are septate and the cells are multi-nucleate. On the terminal ends of the branches of the conidiophores the bottle-like sterigmata (phialides) are produced. The sterigmata are uninucleate and bear the uninucleate conidia in basigenous chains. When the conidiophores are unbranched, they directly bear the sterigmata on their ends producing the conidia in basigenous chains. Each conidial chain consists of hundred or more conidia. Each conidium is smooth uninucleate and green. Due to the presence of innumerable green conidia the fungus is called the 'green mould'. The conidia are dispersed by wind and carried from one place to another. Whenever they get suitable media and appropriate conditions for germination, they germinate producing germ tubes developing into new myclia.

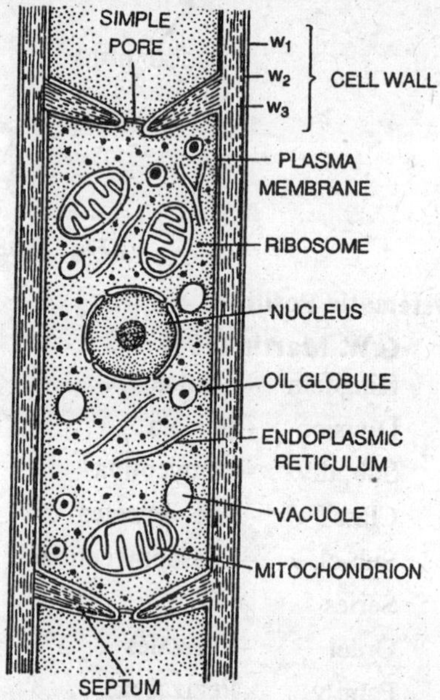

Fig. 12.39. *Penicillium* spp. Ultra structure of a cell (diagrammatic).

Sexual reproduction. Almost all the species of *Penicillium* are homothallic. Only one species is heterothallic. The structure of the sex organs and the production of asci on ascogenous hyphae differ from species to species. In the beginning of the development, the ascogonium is uninucleate, but gradually this nucleus divides mitotically and 2-64 nuclei are produced. A thin-walled uninucleate antheridial branch grows upward coiling around the ascogonium. This antheridial branch bears at its terminal end a small, swollen, uninucleate antheridium. The terminal end of the antheridium attaches itself to the ascogonium, the wall of contact dissolves and a pore develops. Actual process of fusion has not yet been seen and nothing has been confirmed about the sexual fusion. By this time the sterile hyphae surround the fused antheridium and ascogonium, and the central sterile portion of the cleistothecium develops. Now the ascogonium becomes septate and each cell contains a pair of male and female nuclei. In this way a series of binucleate cells is developed. From some of the cells of this row, the branched and septate ascogenous hyphae develop. The cells of ascogenous hyphae are binucleate. The ultimate and penultimate cells of the ascogenous hyphae develop into asci. Each ascus contains eight ascospores in it.

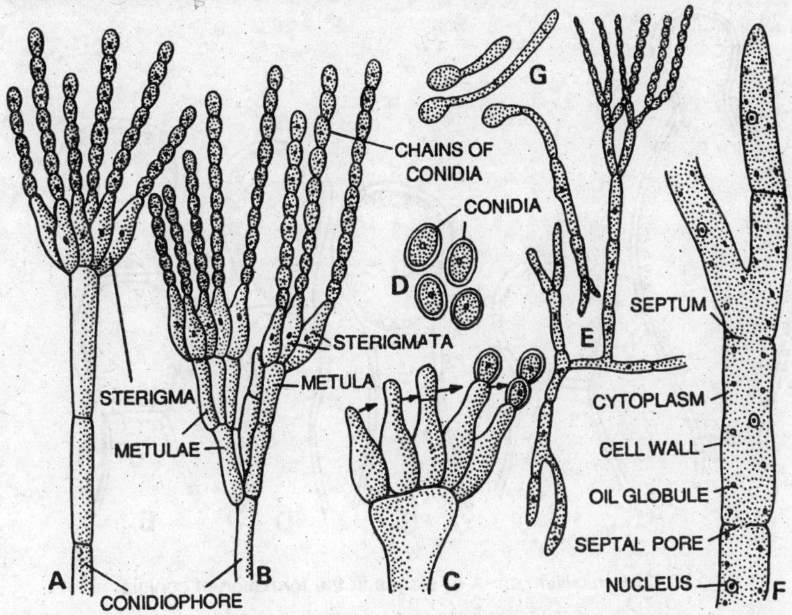

Fig. 12.40. *Penicillium* sp. Asexual reproduction. A, sterigmata and conidial chains arising on unbranched conidiophores; B, sterigmata and conidial chains developing on branched conidiophore; C, formation of conidia on sterigmata; D. conidia; E, conidiophore developed on mycelium; F, part of mycelium; G, germinating conidia.

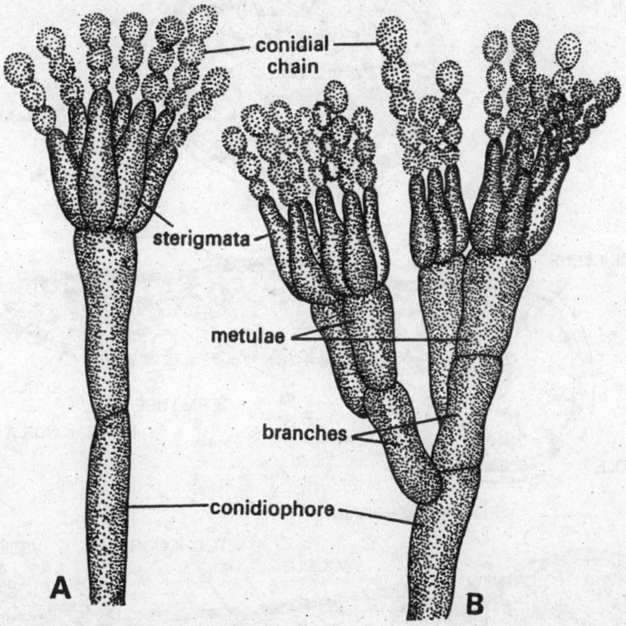

Fig. 12.41. Monoverticillate (A) and biverticillate (B) conidiophores of *Penicillium*.

The male and female nuclei found in the pairs in the ultimate and penultimate cells of the ascogenous hyphae fuse with each other and diploid nuclei are resulted. Now these cells with diploid nuclei are called the ascus mother cells. The diploid nucleus of each ascus mother cell divides thrice and the either nuclei are formed. To bring the haploid phase the first division is reductional. The cytoplasm deposits around each nucleus and eight ascospores are produced in the ascus. In each ascus there are 4 ascospores of + strain and 4 of — strain.

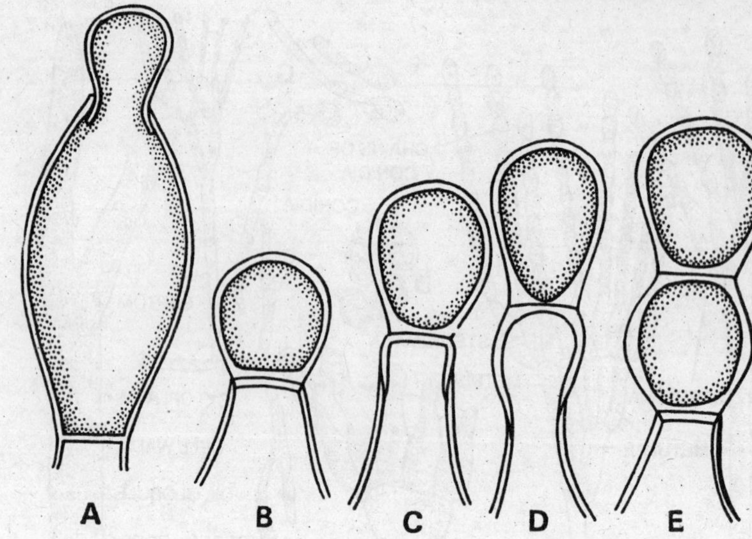

Fig. 12.42. *Penicillium* sp. A-F, stages in the formation of conidia.

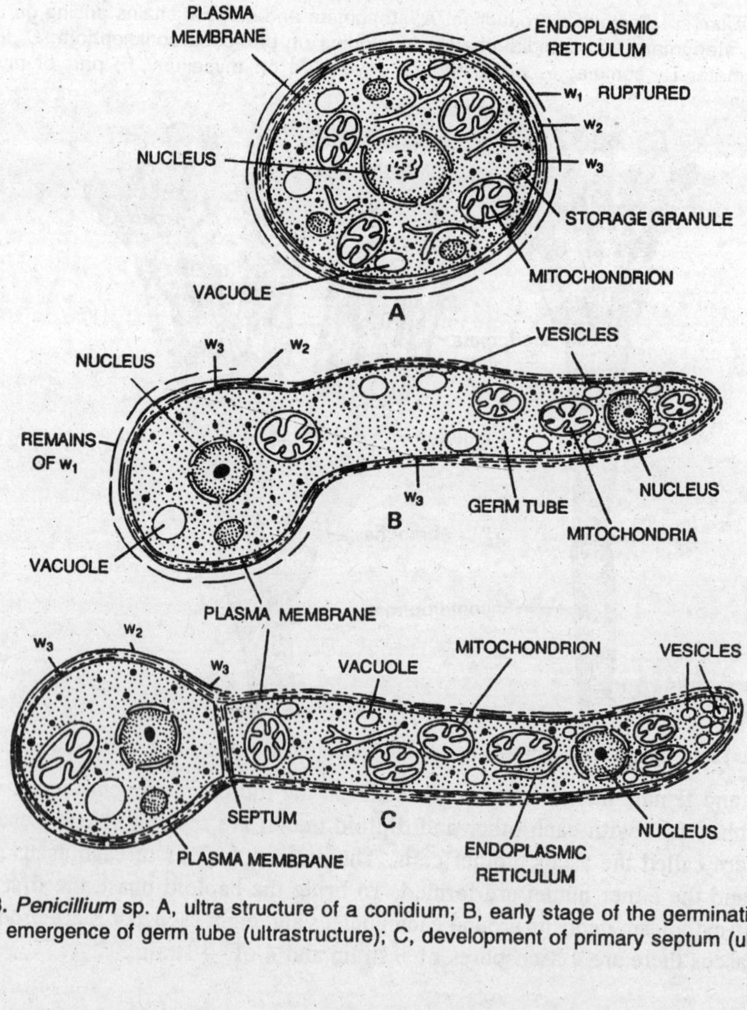

Fig. 12.43. *Penicillium* sp. A, ultra structure of a conidium; B, early stage of the germination of conidium showing emergence of germ tube (ultrastructure); C, development of primary septum (ultrastructure).

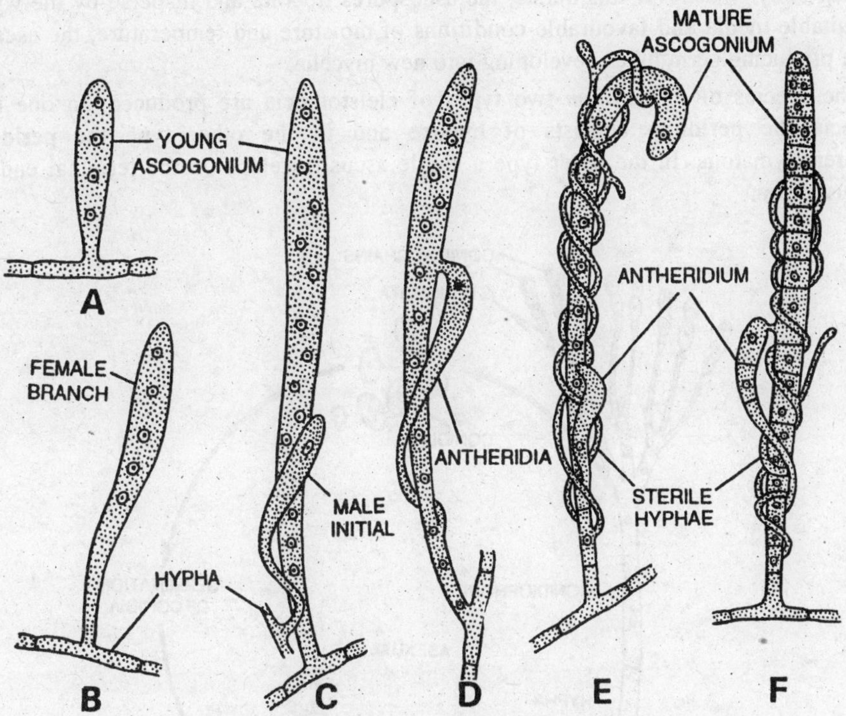

Fig. 12.44 *Penicillium.* Sexual reproduction — early stages.

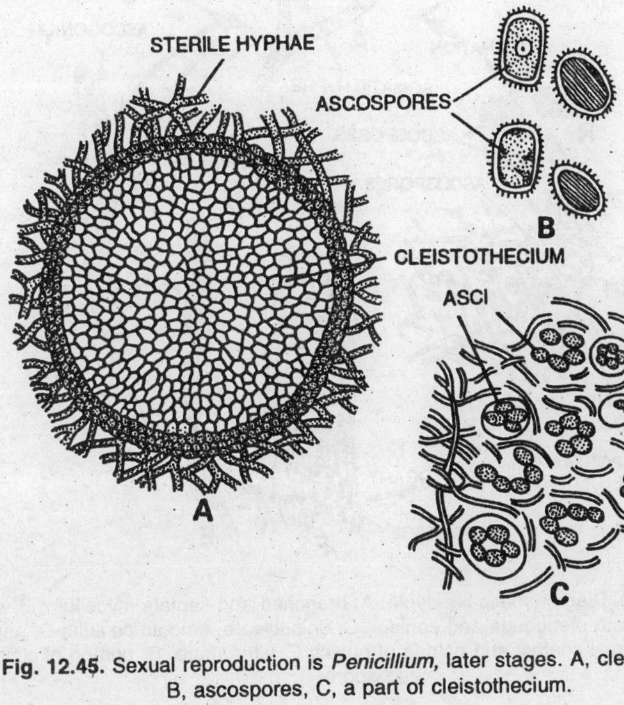

Fig. 12.45. Sexual reproduction is *Penicillium,* later stages. A, cleistothecium; B, ascospores, C, a part of cleistothecium.

The ascospores are uninucleate, ellipsoidal, biconvex and echinulate. The cleisothecium is a perennating body. Whenever this bursts, the ascospores liberate and disperse by the wind. On getting suitable media and favourable conditions of moisture and temperature, the ascospores germinate producing germ tubes developing into new mycelia.

In the species of *Penicillium* two types of cleistothecia are produced. In one type of cleistothecia the peridium consists of hyphae and in the other type the peridium is pseudoparenchymatous. In the latter type a single ascus develops on the terminal end of the ascogenous hypha.

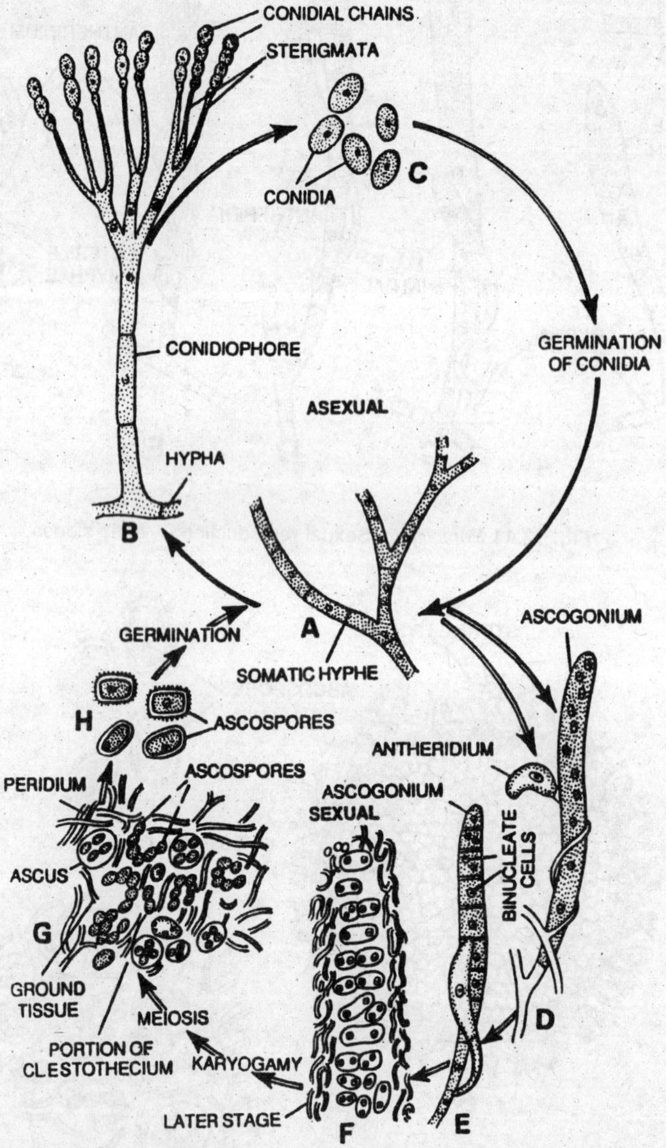

Fig. 12.46. *Penicillium.* Diagrammatic life-cycle. A, branched and septate mycelium; B, septate and branched conidiophore with sterigmata and conidia; C, uninucleate, smooth conidia; D, antheridium and ascogonium; E, gametangial contact and pairing of nuclei; F, later stage; G, portion of cleistothecium; H, ascospores.

Systematic Position

	G. W. Martin (1961)	**C. J. Alexopoulos (1962)**	**G. C. Ainswoth (1971)**
Kingdom	– Plantae	— Plantae	— Fungi
Division	— Mycota	1— Mycota	—Eumycota
Sub-div.	— Eumycotina	—Eumycotina	—Ascomycotina
Class	— Ascomycetes	— Ascomycetes	— Plectomycetes
Sub-cl.	— Euascomycetidae	— Euascomycetidae	
Series		— Plectomycetes	
Order	— Eurotiales	—Eurotiales	—Eurotiales
Family	—Eurotiaceae	—Eurotiaceae	—Eurotiaceae
Genus	— *Penicillium*	— *Penicillium*	—*Penicillium*

13
CHAPTER

Sub-div. Basidiomycotina Class–Basidiomycetes

Kingdom–**PLANTAE**
Division–**MYCOTA**
Sub–division–**EUMYCOTINA**
Class–**BASIDIOMYCETES**
Smuts, Rusts, Mushrooms, Puffballs, etc.
(Alexopoulos, 1962)

G. W. Martin (1961)
Kingdom–**PLANTAE**
Division–**MYCOTA**
Sub-div.–**EUMYCOTINA**
Class–**BASIDIOMYCETES**

G. C. ainsworth (1971)
Kingdom–**FINGI**
Division–**EUMYCOTA**
Sub-div.–**BASIDIOMYCOTINA**
Class– **1. TELIOMYCETES**
2. HYMENOMYCETES
3. GASTROMYCETES

This class consists of higher fungi. There are about 550 genera and 20,000 to 25,000 species.

Introduction. The Basidiomycetes include the fungi that bear their spores exogenously on **basidia.** The basidia may occur singly or be variously arranged on hymenia. On the basis of

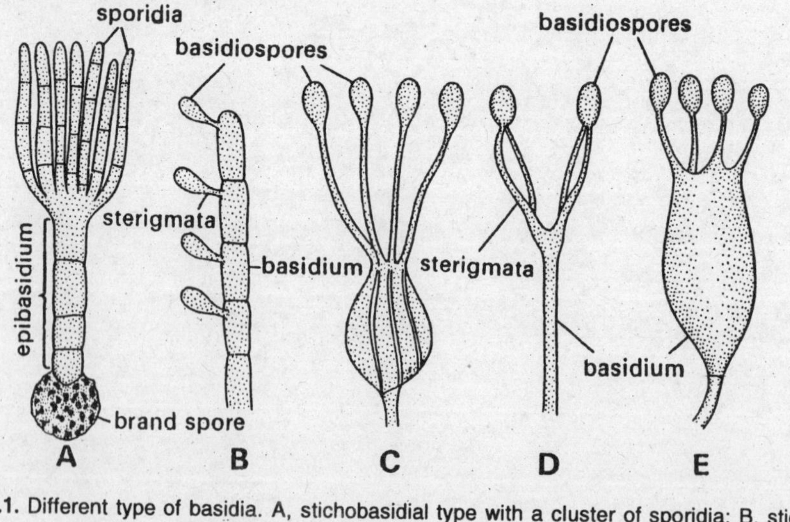

Fig. 13.1. Different type of basidia. A, stichobasidial type with a cluster of sporidia; B, stichobasidial type; C, chiastobasidial type; D, tuning fork type; E, holobasidial type.

morphology and arrangement of basidia, the Basidiomycetes are divided into two sub-classes, the Heterobasidiomycetidae (Heterobasidiomycetes) and Homobasidiomycetidae (Homobasidiomycetes). The Heterobasidiomycetidae include the fungi commonly known as smuts, rusts and jelly fungi, and the Homobasidiomycetidae include the gill, pore, leathery and coral fungi, puffballs, stinkhorns and bird's nest fungi. The fungi of sub-clas Heterobasidimycetidae are **gymnocarpous,** *i.e.*, their basidia arise on the free surface of the fructification. The Homobasidiomycetidae are further sub-divided into two series, the Hymenomycetes and the Gastromycetes. The Gastromycetes include those fungi whose hymenia formed within the basidiocarp (fruit body), and therefore, they are known as **angiocarpous.** The Hymenomycetes include those fungi which possess transitional conditions of gymnocarpy and hemiangiocarpy.

Somatic structure. The mycelium is well developed and filamentous. It consists of septate hyphae which penetrate into the substratum and absorb nourishment. The mycelium is usually white, yellow or orange and generally spreads out in a fan-shaped growth. In a few species a number of hyphae run parallel to each other and get joined together to form thick strands of mycelium. These structures are known as **rhizomorphs.** The rhizomorphs remain enveloped in a shealth of cortex and behave in the manner of a unit of tissue.

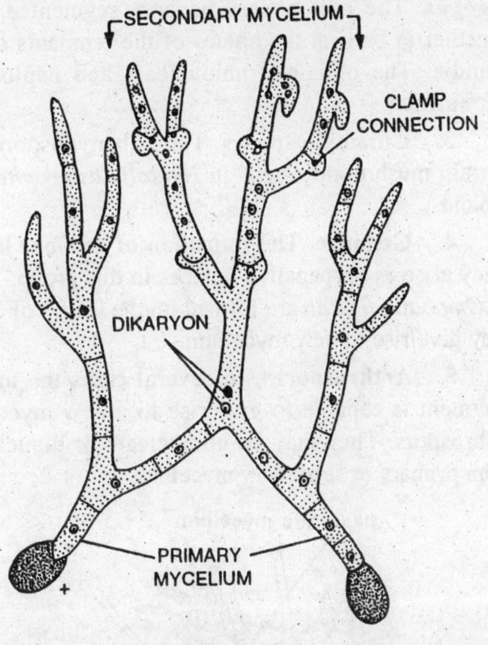

Fig. 13.2. Basidiomycetes. Primary and secondary mycelium.

The mycelium of the Basidiomycetes passes through three stages of development, the primary, secondary and teritiary.

The primary mycelium usualy develops from the germination of the basidiospore. It may be multinucleate at first, but later on, due to the formation of transverse walls, it divides into uninucleate cells. In certain species the primary mycelium is septate and uninucleate from the beginning. The primary mycelium may give rise to conidia or oidia and never forms basidia and basidiospores.

The secondary mycelium consists of binucleate (dikaryotic) cells. It originates from primary mycelium. The process is as follows.

(i) The protoplasts of the opposite strained (+&–) uninucleate cells of the primary mycelium fuse. The walls of contact between the two cells dissolve and the protoplasts from two opposite strained cells fuse together. Now the two (+&–) nuclei lie side by side. This process is **plasmogamy.** The binucleate cell thus formed may develop directly into a secondary hypha.

(ii) Two basidiospores or spermatia of opposite strains come together and conjugate. The binucleate (+ –) structre formed this way gives rise to secondary mycelium.

(iii) Sometimes the germinating oidium of one strain fuses with a cell of primary mycelium of opposite strain. The binucleate cell thus formed gives rise to secondary mycelium.

(iv) Sometimes a germinating basidiospore of one strain fuses with haploid cell of basidium of opposite strain. Thus formed the binucleate cell gives rise to secondary mycelium.

The secondary mycelium always gives rise to brand spores or teleutospores, which on germination produce basidia or promycelia.

Asexual reproduction. The asexual reproduction takes place by means of conidia, oidia, chlamydospores and gemmae. The methods are as follows:

1. Conodia. The basidiospores of many smuts bud indefinitely in yeast-like fashion, and these buds like conidia may eventually serve to initial infection. The uredospores of rusts are like conidia in their origin and since they function in dissemination may also be considered conidia. The basidiospores of many jelly fungi germinate by producing sprout cells which eventually develop mycelia. These conidia develop on secondary (dikaryotic) mycelium.

2. Oidia. Brodie (1931) showed lateral oidiophores arising from mycelium in *Coprinus logopus.* The oidiophores become segmented basipetally into ellipsoidal oidia which mass together in balls at the apices of the remnants of the oidiophores. These oidia may function as conidia. The oidia are uninucleate and haploid. They produce monokaryotic mycelium on germination.

3. Chlamydospores. The chlamydospores are formed as the upper surface of pileus of certain mushrooms, *e.g.,* in *Nyctalis asterophora* and *N. parasitica.* The chlamydospores are diploid.

4. Gemmae. The formation of gemmae has been reported from several Hymenomycetes. They arise as propagative devices in this group. According to Buller (1934), the stalked gemmae of *Omphalia flavida* are formed on the leaves of coffee and other tropical species. On germination they give rise to new mycelium.

5. Arthrospores. In several cases the mycelium breaks up into small fragments. Each fragment is capable to give rise to a new mycelium. These mycelial fragments are called the arthrospors. They may be uninucleate or binucleate, depending on whether they are produced from primary or secondary mycelium.

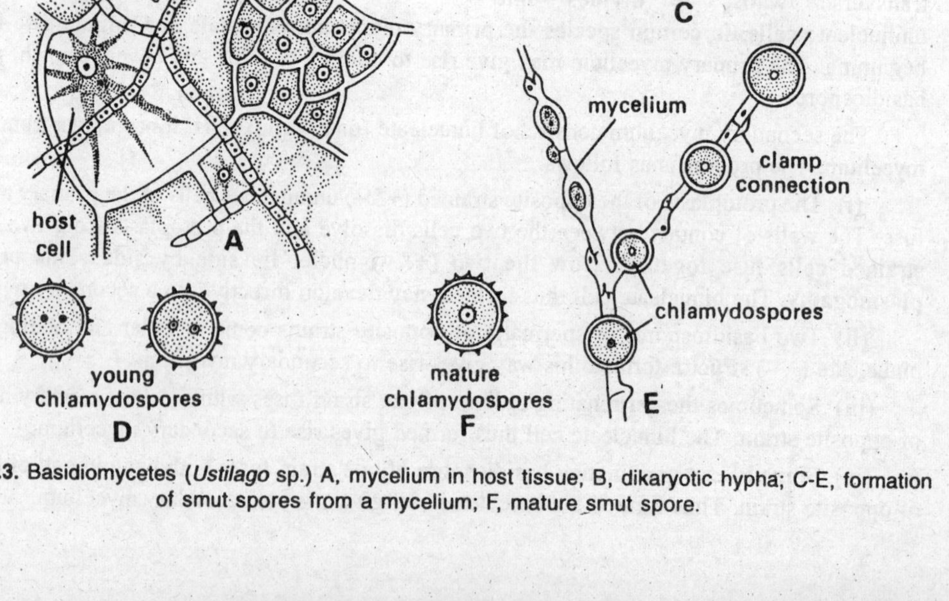

Fig. 13.3. Basidiomycetes (*Ustilago* sp.) A, mycelium in host tissue; B, dikaryotic hypha; C-E, formation of smut spores from a mycelium; F, mature smut spore.

Sexual reproductin. The sexuality in Basidiomycetes is very much reduced, and the sexual phenomenon is represented only by the fusion of the nuclei of oppositic strains (+ and –). Unlike, Phycomycetes and Ascomycetes the distinct male (antheridium) and female (oogonium or ascogonium) sex organs are never produced. The monokaryotic phase changes into a dikaryotic phase by means of **dikaryotization.** In the ultimate stages of the development of teleutospores and chlamydospores, the opposte strained nuclei get fused, resulting in the diploid nuclei. The diploid spores produce **promycelia** or **basidia** on germination. During this process the first division is reductional one, the four nuclei of the basidium are haploid. The basidiospores are haploid, uninucleate and single-celled structures. The basidiopore on germination produces haplont mycelium either of + or – strains. The phenomenon of heterothallism is quite conspicuous. The basidiospores produce monokaryotic hyphae, which dikaryotize by the process of dikaryotization and the dikaryotic hyphae are produced. Each pair of the opposite strained (+ and –) nuclei in each cell of the dikaryotic mycelium is known as a **dikaryon.** Ultimately the two nuclei of a dikaryon fuse and a diploid spore develops.

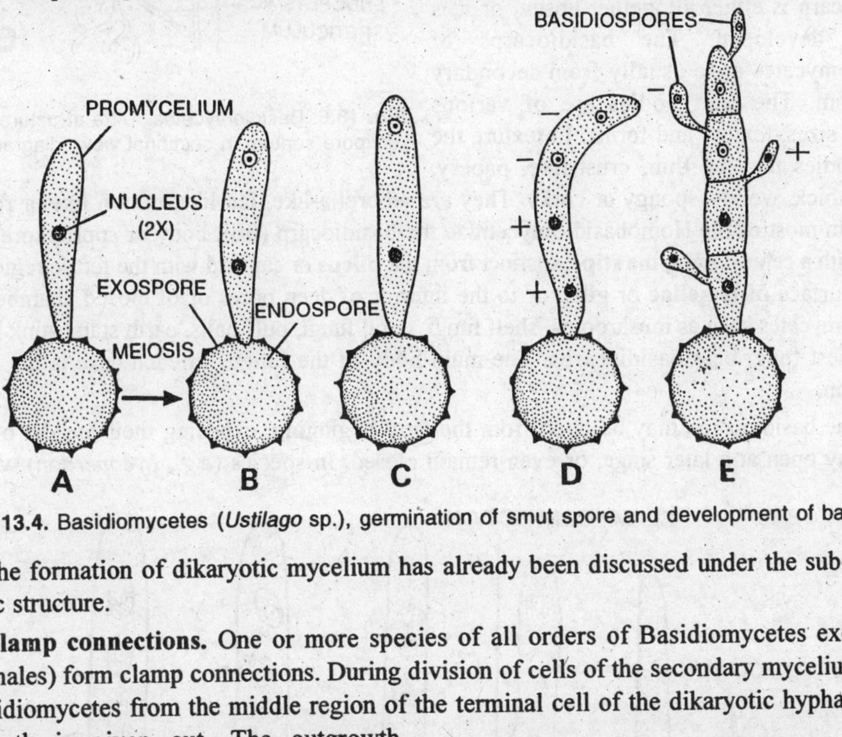

Fig. 13.4. Basidiomycetes (*Ustilago* sp.), germination of smut spore and development of basidium.

The formation of dikaryotic mycelium has already been discussed under the sub-heading–somatic structure.

Clamp connections. One or more species of all orders of Basidiomycetes except rusts (Uredinales) form clamp connections. During division of cells of the secondary mycelium in most of Basidiomycetes from the middle region of the terminal cell of the dikaryotic hypha a tubular outgrowth is given out. The outgrowth curves down ward making a hook-like structure, touches the cell wall and called the **clamp conection.** Both the opposite nuclei of the terminal cell divide producing four nuclei, two of + and two of – strain. Now one of the daughter nucleus of upper two nuclei passes through the clamp and moves in the basal part of the cell, and one of the daughter nucleus of the two lower nuclei moves upward in the cell. The wall of contact dissolves where the clamp touches the wall,

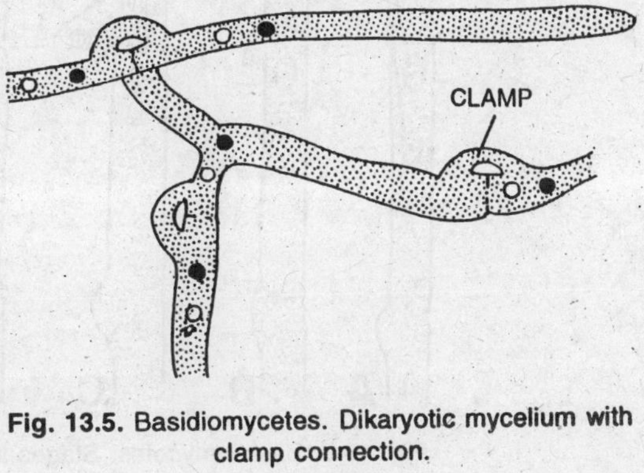

Fig. 13.5. Basidiomycetes. Dikaryotic mycelium with clamp connection.

and the nucleus comes through the clamp in the lower region of the cell. Now the clamp begins to degenerate and a septum appears at this point, which divides the mother cell into two cells, each possessing two nuclei of opposite strains, this way, one dikaryotic cell gives rise to two dikaryotic cells. The newly developed cells become elongated and the terminal cell is again ready for division and the hypha increases in length continuously.

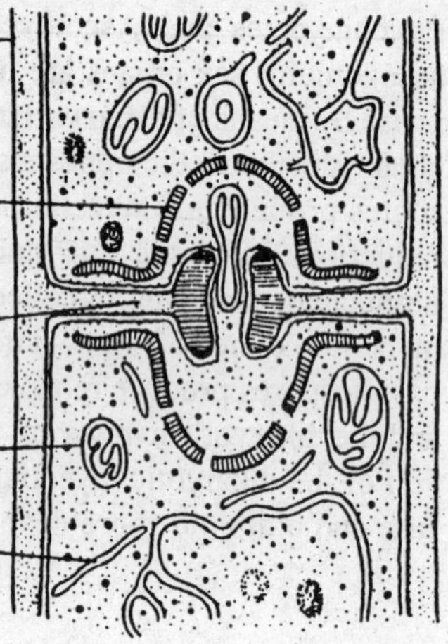

Fig. 13.6. Basidiomycetes. Ultra structure showing dolipore septum in sectional view (diagrammatic).

The basidiocarps. The fruit body of a basidiomycete is called a basidiocarp. In the orders Ustilaginales and Uredinales basidiocarp is either altogether absent or it is poorly developed. The basidiocarps of Basidiomycetes arise usually from secondary mycelium. The fruit bodies are of various shapes, sizes, texture and forms. In texture the fruit bodies may be thin, crust-like, papery, fleshy, thick, woody, spongy or corky. They are umbrella-like, fan-like, button like or rounded in shape. In most of the Homobasidiomycetidae the basidiocarp (fruit body or sporophore) is large, often with a central stalk, the **stipe,** distinct from the **pileus** or cap and with the fertile region limited to the surface of **lamellae** or **gills,** or to the interior of deep pores or of closed chambers. Many Basidiomycetes such as mushrooms, shelf fungi, coral fungi, puff balls, earth stars, stink horns and bird's nest fungi bear basidiocarps. The main body of the fungus in each case is the extensive mycelium.

The basidiocarps may be open from the very beginning, exposing their basidia outside, or they may open at a later stage, or even remain closed. In species (*e.g., Lycoperdon*) where fruit

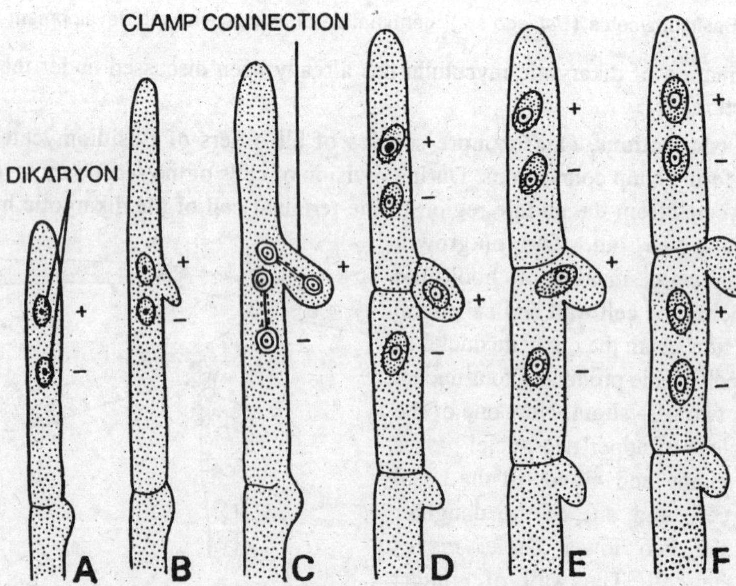

Fig. 13.7. Basidiomycetes. Stages in the formation of clamp connection.

bodies remain closed the spores liberate only when the basidiocarps are disintegrated or it breaks accidentally.

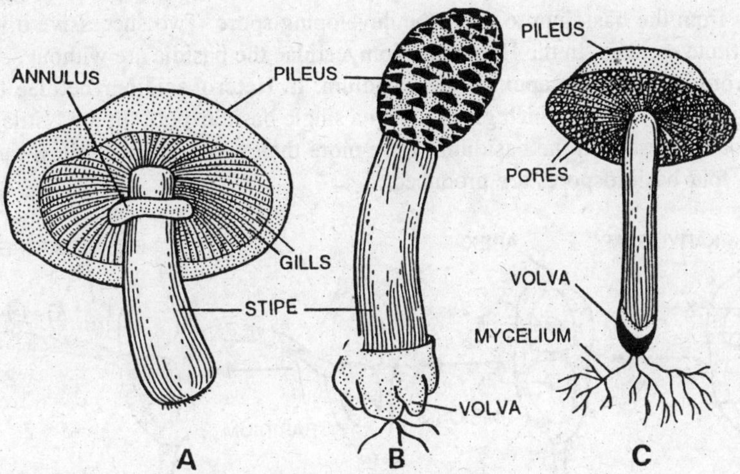

Fig. 13.8. Basidiomycetes. Basidiocarps of *Agaricus* (A), *Phallus* (B), and *Boletus* (C).

The basidia are formed in definite layers called hymenia. The hymenium consists of basidia and sterile paraphyses. In some species the sterile processes among the basidia become much larger than the basidia are called **cystidia.**

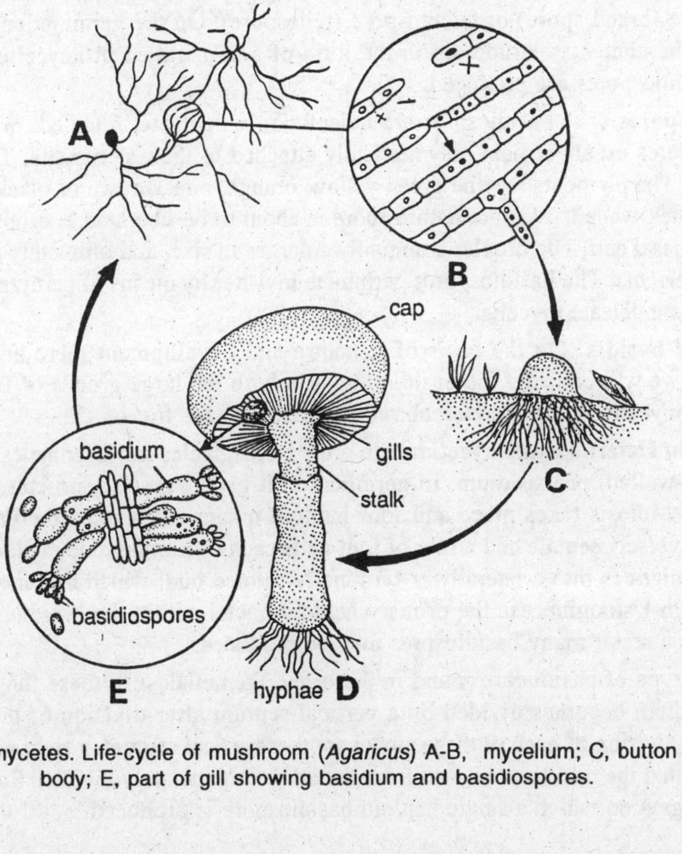

Fig. 13.9. Basidiomycetes. Life-cycle of mushroom (*Agaricus*) A-B, mycelium; C, button stage; D, fruit body; E, part of gill showing basidium and basidiospores.

Basidium. The basidiospores are borne externally on the mother cell, or **basidium.** The young basidium contains two nuclei, these fuse, the fusion nucleus divides forming the nuclei of the spores. The basidiospre develops at the end of the stalk known as **sterigma,** through which the nucleus passes from the basidium to enter the developing spore. Two successive divisions in the basidium constitute meiosis. In the Homobasidiomycetidae the basidia are without septa, and four spores arise from, or near, the apex of the basidium. In Heterobasidiomycetidae the basidium divides into four cells, each of which gives rise to a single basidiospore. In the Ustilaginales septa may or may not be present in the basidium. Here more than two divisions follow nuclear fusion, and more than four basidiospores are produced.

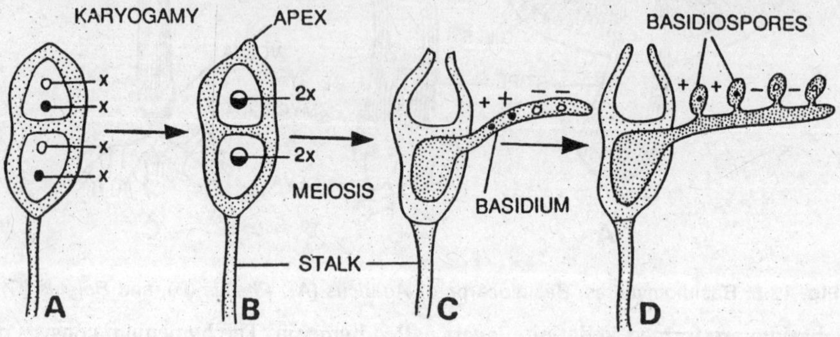

Fig. 13.10. *Puccinia* sp. A, young teleutospore; B, mature teleutospore; C, germinating teleutospore with young basidium; D, germinating teleutospore with basidium and basidiospores 2+ and 2–.

In the Ustilaginales and the Uredinales the basidium develops from a thick-walled chlamydospore (brand spore) or teleutospore (teliospore). On the germination of brand spore or teleutospore the contents extrude out in the form of a thin-walled promycelium or basidium on which the basidiospores are produced.

Basidiospores. The basidiospres are unicellular, uniculeate, rounded, or oval and haploid. The basidiospores usualy remain assymetrically attached to their sterigmata. They are colourless or pigmented. The pigments may be green yellow, orange, pink, brown or black. They are usually smooth and thick-walled. When a mature spore is about to be liberated, a droplet of water begins to form at its basal end. The droplet grandually enlarges in size, and ultimately the basidiospore is shot off the sterigma. The basidiospores germinate in water by means of germ tubes which later on develop into uninucleate mycelia.

Types of basidia. On the basis of strucutre and development there are several types of basidia. Here we will consider the basidia found in both the large groups of Basidiomycetes the Heterobasidiomycetidae and the Homobasidiomycetidae. (see fig. 13.1).

Basidia of Heterobasidiomycetidae. In order Ustilaginales and Uredinales, the teliospore is a diploid, thick-walled, probasidium. In germination it gives rise to a structure, the basidium or promycelium. Meiosis takes place and four haploid nuclei are formed thereafter the basidium becomes transversely septate and a row of four cells each containing one nucleus is formed. Each nucleus then migrates into a laterally or terminally formed bud, which becomes abstricted into a basidiospore. In Ustilaginaceae the primary haploid nuclei within the basidium may continue to divide, and as a result many basidiospres may be produced.

Another type of basidia are found in the order Tremellales. In these fungi the spherical or elongate basidium becomes divided by a vertical septum after division of the primary diploid nucleus. After division of each of these nuclei another vertical septum at right angles to the first is laid down, so that the basidium as seen from the above is cruciately divided. Each quadrant bears a slender sterigma on which a single haploid basidiospore is produced.

In the group Dacryomycetales of Heterobasidiomycetidae there are basidia of still another type. Here the primary diploid nucleus divides twice forming four haploid nuclei. A tuning fork shaped process arises at the apex of the basidium. Two of the nuclei migrate into basidiospores situalted at the apices of the tuning fork basidium and the remaining two nuclei, disintegrate. (see fig. 13.1 D).

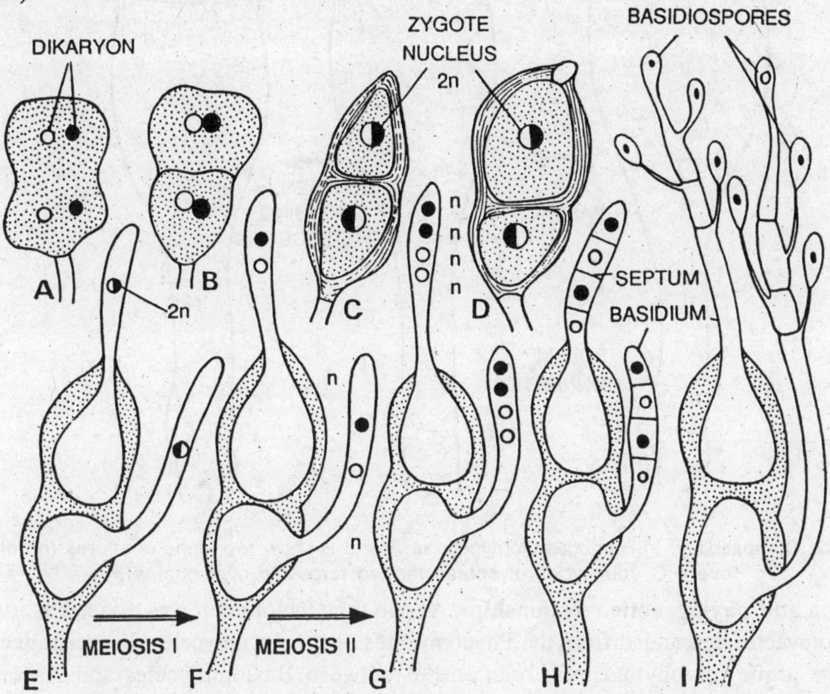

Fig. 13.11. *Puccinia graminis.* A-C, formation of teliospore; D, mature teliospore; E-I, germination of teliospore resulting in the formation of basidium and basidiospores.

Basidia of Homobasidiomycetidae. The basidia of Homobasidiomycetidae are of another type. In the series Hymenomycetes the basidia are slender clavate to braodly clavate whereas in the series Gastromycetes the basidia are globular. The diploid nucleus divides twice forming four haploid nuclei. Usually four sterigmata arise from the apex of the basidium and a uninucleate basidiospore is borne on each sterigma.

Compatibility. About 90% species of Basidiomycetes are heterothallic, *i.e.*, they require the union of two compatiable thalli for sexual reproduction. In about thirty-seven per cent of heterothallic species the sexual compatibility is governed by one pair of factors. As located on the same locus on different chromosomes and behave in the same manner as do the heterothallic, species in Mucorales. Such species are called **bipolar.** The remaining sixty-three per cent of heterothallic species are **tetrapolar,** *i.e.*, sexual compatibility is governed by two pairs of factors **Ab Bb** located on different chromosomes and segregating independently. This way, four types of basidiospores are produced by tetrapolar species, *i.e.*, **AB, Ab, aB, ab.** If crossing over does not take place between the mating type loci and their centromeres, then only two types of basidiospores will be formed:

AB AB ab ab OR **Ab Ab aB aB**

If crossing over takes place for one mating type locus, four types of spores will be formed: **AB Ab aB ab.** If crossing over takes place for both mating type loci, two or four types of spores may be formed on the basidium.

Fig. 13.12. Compatibility. In Homobasidiomycetidae A and B show two kinds of spores (no crossing over); C, four types of spores formed (crossing over occurs).

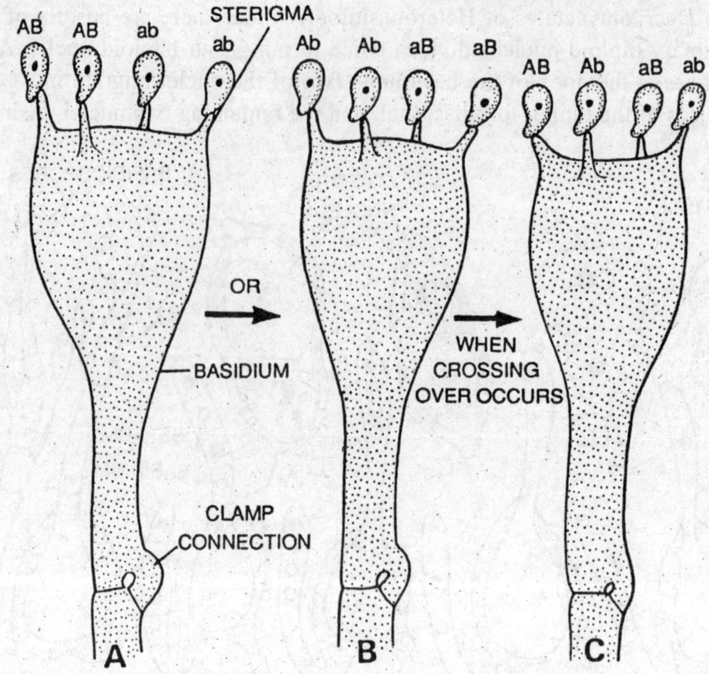

Fig. 13.12. Compatibility. In Homobasidiomycetidae A and B show two kinds of spores (no crossing over); C, four types of spores formed (crossing over occurs).

Origin and phylogenetic relationships. At one time the opinion was divided as to whether the Basidiomycetes descended from the Phycomycetes or the Ascomycetes. In recent decades the mycologists argue for phylogenetic relationship between Basidiomycetes and Ascomycetes. Kneip (1915) suggested the origin of Basidiomycetes from Ascomycetes on the basis that the hyphae of Basidiomycetes with clamp connections are homologous with the crozier formation in early development of asci. According to Kniep (1915), the apparently terminal binucleate cell which develops into a basidium is actually a penultimate cell which lies posterior to a terminal uninucleate cell. Here the fundamental change in evolution of an ascus into basidium would be a change from an endogenous to an exogenous method of spore formation.

Evolution within the Basidiomycetes. According to some mycologists (Bessey, 1950; Dietel, 1928; Jackson, 1944), the Basidiomycetes with phragmobasidia are most closely related to the ancestral Ascomycetes. These mycologists interpret that the Uredinales were the first Basidiomycetes to appear. According to other (Alexopoulos, 1952; Martin, 1945), the Auriculariales are the most primitive of phragmobasidial Basidiomycetes. According to another school of thought (Gaumann, 1949, 1953; Gaumann and Dodge, 1928), the Basidiomycetes with holobasidia and more primitive and closer to ancestral Ascomycetes. Many modern mycologists are of opinion that the Basidiomycetes with phragmobasidium are more primitive and called the Heterobasidiomycetidae. The Basidiomycetes with holobasidia are supposed to be advanced and are called the Homobasidiomycetidae.

Economic imortance. Majority of the species of this class are harmful. They cause various diseases to the plants of economic value. The most important diseases are smuts and rusts. Many bracket fungi do the great harm to the timber. Some species of Homobasidiomycetiadae are deadly poisonous (*e.g.*, *Amanita*). On the other hand some mushrooms are edible and sold very costly for eating purposes (*e.g.*, *Agaricus campestris*).

Classification. The outline of Gwynne-Vaughan and Barnes (1926) classification is as follows.

Class—BASIDIOMYCETES
(Sexually produced spores, basidiospores, formed exogenously on an organ, the basidium).

Sub-classes—	1. HEMIBASIDIOMYCETES (Number of basidiospores indefinite)	2. PROTOBASIDIOMYCETES (Number of basidiospores definite, usually four; basidium septate).	3. AUTOBASIDIOMYCETES (Number of basidiospores definite, usually four; basidium aseptate).
Orders—	1. Ustilaginales	1. Uredinales. 2. Auriculariales. 3. Tremellales.	1. Hymenomycetales. 2. Gasteromycetales.

Majority of modern mycologists believe that the phragmobasidial (basidium septate) forms are more primitive, being more closer to the ancestral Ascomycetes. The outline of the modern classification is as follows:

Class–BASIDIOMYCETES

Sub-classes–	1. HETEROBASIDIOMYCETIDAE (Basidium septate or deeply divided; or consisting of a teleutospore germinating into a promycelium; basidiospores usually capable of germinating by repetition).	2. HOMOBASIDIOMYCETIDAE (Basidium not septate or deeply divided; basidiospores usually germinating by germ tube).	
Series-		1. Hymenomycetes.	2. Gastromyctes.
Orders—	1. Ustilaginlaes. 2. Uredinales. 3. Tremellales.	1. Agaricales	1. Hymenogastrales. 2. Nidulariales. 3. Lycoperdales. 4. Sclerodermatales.

14
CHAPTER

Sub-class–Heterobasidiomycetidae

Rusts and Smuts (Alexopoulos, 1962)

G. W. Martin (1961)	**G. C. Ainsworth (1971)**
Sub-class–**HETEROBASIDIOMYCETIDAE**	Class–**TELIOMYCETES** and sub–class–**GMOBASIDIOMYCETIDAE** of Class–**HYMENOMYCETES**

Characteristic Features.

1. The basidium is transversely or vertically septate or incised at the apex.

2. The mature basidium typically consists of two portions; a basal hypobasidium and an elongated tube or inflated protuberance, the epibasidium, which typically bears basidiospores on sterigmata.

3. Many basidiospores show germination by repetition. Here the basidiospores on germination produce secondary basidispores which are shot off in the same manner as were the original spores.

4. In many of them a thick-walled probasidium also known as teleutospore or brand spore is found. The probasidium may undergo a number of morphological changes before it becomes differentiated into a mature basidium. Eventually the probasidium is transformed into hypobasidium which gives rise to the epibasidium.

5. Many members are without a fruit body or basidiocarp (*e.g.*, Ustilaginales and Uredinales) but some of them possess well developed fruit bodies (*e.g.*, Tremellales).

Classification. Martin (1961) has divided the sub-class into three oders—1. Ustilaginales (the smuts); 2. Uredinales (the rusts); and 3. Tremellales (gelatinous Basidiomycetes).

Order **USTILAGINALES** (The Smuts)

(42 genera; 700 species)

Characteristic Features

1. They cause the plant diseases known as 'smuts' and 'bunts'. The black powder of spores is known as 'smut' or 'soot' and because of this reason the diseases caused by Ustilaginales are called 'smuts'.

2. The smuts are parasitic on various plants, but they are not obligate parasites. They are facultative saprophytes.

3. The mycelium is inter and intracellular. The intercalary, thick-walled, round, dark brown chlamydospores or brand spores are produced on dikaryotic or secondary mycelium.

4. On germination, the chlamydospore produces a tube-like structure known as promycelium or basidium.

5. The basidium is either septate or aseptate, and from each cell of basidium, the basidiospores or sporidia are produced in basigenous chins. The basidiospores are not borne on sterigmata. They are not violently dischrged.

6. The smuts have a simple life-cycle in comparison to rusts.
7. They lack fruit bodies or basidiocarps.
8. Clamp-connections are more common in smuts.
9. There are about 40 genera and 700 species in this order.

Fig. 14.1. Smuts caused by *Ustilago* sp. A, covered smut of barley; B, loose smut of wheat; C, covered smut of oats.

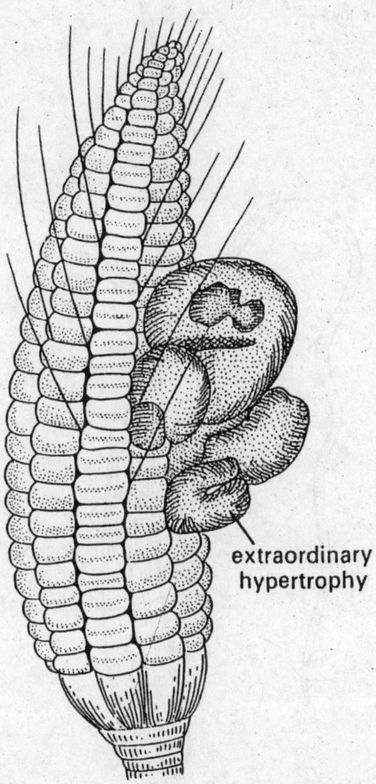

Fig. 14.2. Corn smut (*Ustilago maydis*).

Classification. The order Ustilaginales is sub-divided into three families: 1. Ustilaginaceae, 2. Tilletiaceae and 3. Graphiolaceae.

Family—Ustilaginaceae

The basidiocarps are absent. The promycelium or basidium is septate. The basidiospores are produced laterally from each cell of promycelium. The important genera — 1. *Ustilago* 2. *Sphacelotheca* and 3. *Tolyposproium.* Genus *Ustilago* has been discussed here in detail.

Genus **USTILAGO** (300 species)

Habit and habitat. There are about 300 species of this genus. Majority of them are parasitic upon vascular plants. Some species are parasitic upon cereals. They cause the 'smut diseases' of various crop plants. The diseases caused by this genus are of much economic value as they damage the various crops and cause sufficient losses to the country. In Uttar Pradesh and Punjab the diseases caused by this fungus are known as **'kangiari'**, **'karanjwa'**, **'kandua'**, etc.

Mycelium. The mycelium is branched and septate. There is a central septal pore in each septum through which the cytoplasm of one cell can move in other cell. The mycelium is endophytic, inconspicuous, hyaline and intercellular. Usually the haustoria are developed which absorb the food material from the host cells. However, the host cells are not destroyed. The mycelium is either localized or systemic.

According to nuclear behaviour of the cells of the fungus in the life cycle, there are two phases of the mycelium. They are primary and scondary mycelia.

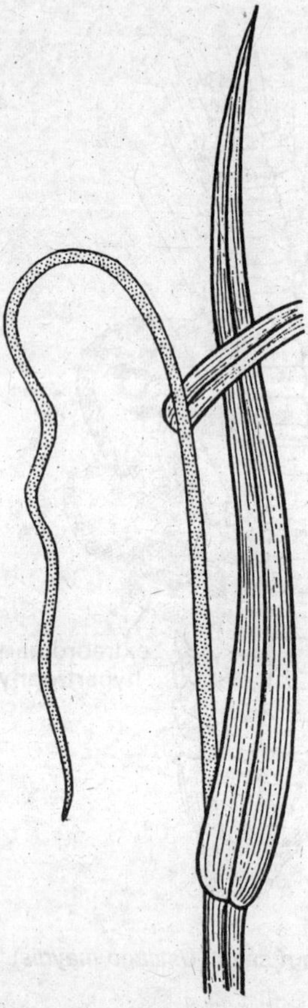

Fig. 14.3. Whip smut of sugarcane (*Ustilago scitaminea*)

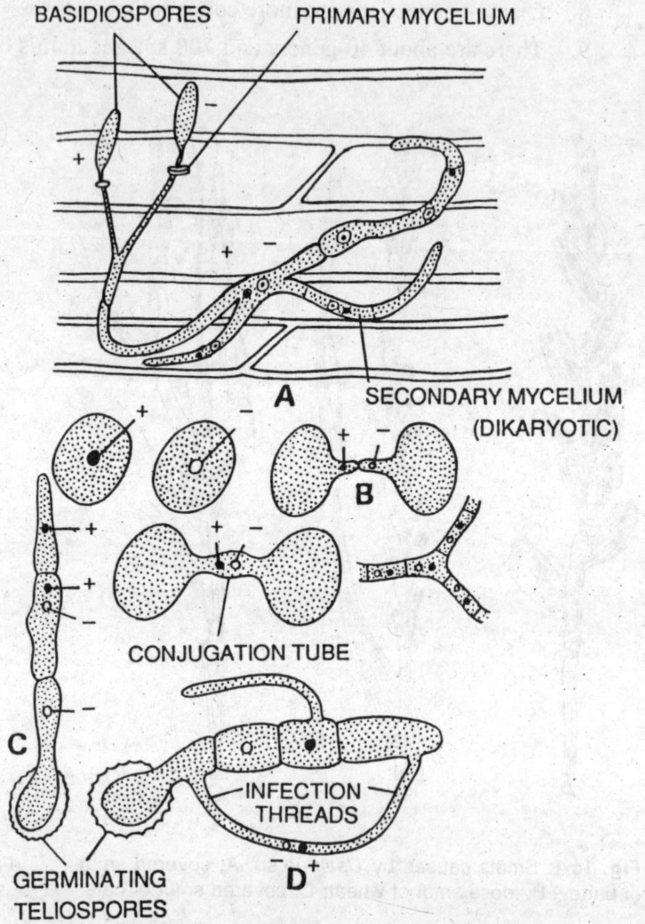

Fig. 14.4. *Ustilago.* Dikaryotization in different species. A, *Ustilago maydis;* B, *U. hordei*; C, *U. carbo*; D, *U. tritici.*

Primary mycelium. Each segment of the primary mycelium contains a single nucleus. The nuclei are haploid and the mycelium is **monokaryotic.** The monokaryotic mycelium develops from the basidiospore, and according to the sexual behaviour of the nucleus it is either of + or of – strain.

Secondary mycelium. Very soon the primary mycelium converts into the secondary mycelium. Each cell of the secondary mycelium is binucleate, but both the nuclei of the pair are haploid and of opposite strain, *i.e.*, + and –. This pair of opposite strained nuclei is called the **'dikaryon',** and the mycelium containing such pairs in its cells the **dikaryotic mycelium.** The phenomenon of the conversion of primary mycelium into secondary mycelium is called the **dikaryotization** or **diploidization.** The dikaryotization takes place by different methods. In *Ustilago maydis* the two hyphae of opposite strains come in contact, the walls of contact dissolve and the nuclei of a cell of monokaryotic hypha are transferred to the adjacent cells of the other monokaryotic hypha, giving rise to a dikaryotic cell, which develops into new dikaryotic mycelium. In *Ustilago tritici* the infection threads develop on the different strained cells of basidium which furnish the phenomenon of dikaryotization. In *Ustilago hordei* the dikaryotization takess place by the fusion of germ tubes developed from two basidiospores of oposite strains.

Spores. The dikaryotic phase of mycelium is found within the host tissue. Unless and until the infected stem is cut in sections, one cannot detect the presence of the fungus. The mycelium is found in endophytic state in the stem, leaves, etc. But it is found in abundance in the inflorescence, flowers and specially in the ovaries. In the ovaries the mycelium becomes closely septate and several small cells are produced. The small cells become swollen, rounded, thick walled and olive green or brownish in colour. These perennating structures are called either **chlamydospores, smut**

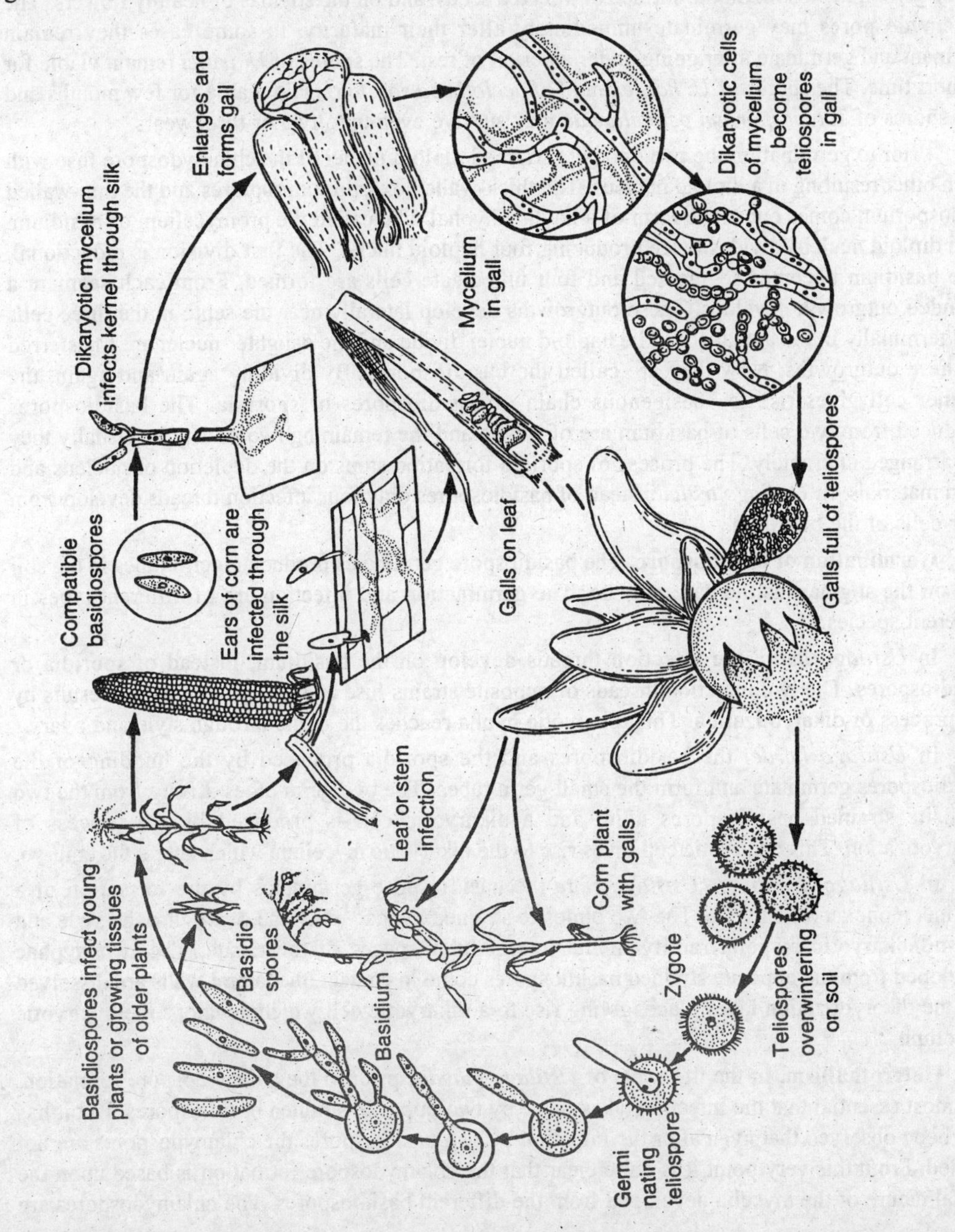

Fig. 14.5. Disease-cycle of corn smut (*Ustilago maydis*).

spores or **brand spores.** The outer thick wall of the chlamydospore, which may be smooth or echinulate, is exosporium and the inner thin wall the endosporium.

The chlamydospores are produced in abundance and look like black powder. The spores are found in sori. Leaving awns aside, the rest of the parts of the inflorescence of the cereals convert into smut sori.

Sometimes the smut sori are confined to ovaries only as in covered smut of oats caused by *Ustilago kolleri,* and sometimes the whole floral axis, converts into smut sori as in 'whip smut of sugarcane' caused by *Ustilago scitaminea.*

Germination of chlamydospore. The chlamydospores germinate in different situations. They germinate in soil, on the surface of infected seeds, and on the stigmas of healthy flowers. The chlamydospores may germinate immediately after their maturity. In some cases they remain dormant and germinate after going under a period of rest. The spores of *U. tritici* remain viable for a short time. The spores of *U. hordei* and *Sphacelotheca sorghi* remain viable for few months and the spores of *Tolyposporium penicillariae* may survive even upto two or three years.

Prior to germination the two opposite strained diploid nuclei of the chlamydospore fuse with each other resulting in a diploid nucleus. The thick-walled oxosporium ruptures and the thin-walled endosporium comes out in the form of a hayline hyphal tube called the promycelium or basidium. The diploid nucleus divides twice producing four haploid nuclei. The first division is reductional. The basidium becomes segmented and four uniculeate cells are formed. From each segment a rounded outgrowth develops. These outgrowths develop laterally near the septa in the three cells and terminally in the apical cell. The haploid nuclei divide and the daughter nuclei are trnasferred to these outgrowths. Now they are called the basidiospores. By dividing again and again, the mother cell gives rise to basigenous chain of basidiospores or sporidia. The basidiospores produced from two cells of basidium are of + type and the remaining two of – type. Usually they are arranged alternately. The process of sporidia formation stops on the depletion of nucleus and food materials. In *Ustilago tritici* instead of basidiospores only four infection threads develop from four cells of the basidium.

Germination of basidiospore. The basidiospore germinates producing germ tubes in the soil and on the stigmas, as the case may be. The germination and infection are of different types in different species.

In *Ustilago tritici* the infection threads develop on the basidium, instead of sporidia or basidiospores. The two infection threads of opposite strains fuse and a dkaryotic hypha results by the process of dikaryotization. This dikaryotic hypha reaches the ovule through style and ovary.

In *Ustilago hordei* the basidiospores and the sporidia produced by the budding of the basidiospores germinate and form the small germ tubes. The two germ tubes arising from the two opposite strained basidiospores unite and a dikaryotic cell is produced by the process of dikaryotization. This dikaryotic cell gives rise to the dikaryotic mycelium which enters the embryo.

In *Ustilago maydis* or *Ustilago zeae* the basidiospore germinates by the formation of a hayline, monokaryotic hypha. The two opposite strained hyphae reach the ovary through style and the monokaryotic mycelia ramify in the intercellular spaces of the tissue. The two hyphae developed from the opposite strained basidiospores come in contact, the lateral walls are dissolved and the dikaryotization takes place, giving rise to a dikaryotic cell which develops the dikaryotic mycelium.

Heterothallism. In the life cycle of *Ustilago maydis,* prior to the chlamydospore formation, it is most essential that the infection takes place by two opposite strained basidiospores. But it has also been observed that even after the infection by two basidiospores the chlamydospores are not formed. From this very point it is quite clear that the chlamydospore formation is based upon the sexual nature of the mycelia developed from the different basidiospores. The chlamydospores are produced only in those conditions when the monokaryotic hyphae giving rise to a dikaryotic cell by the process of dikaryotization are developed from the opposite strained (+ and –) basidiospores. The monokaryotic hyphae of similar strains cannot go under dikaryotization and the chlamydospores are not produced. It is very clear from this example that the monokaryotic primary mycelia, taking part in dikaryotization are always of different strains. Such fungus is called the heterothallic and the phenomenon heterothallism.

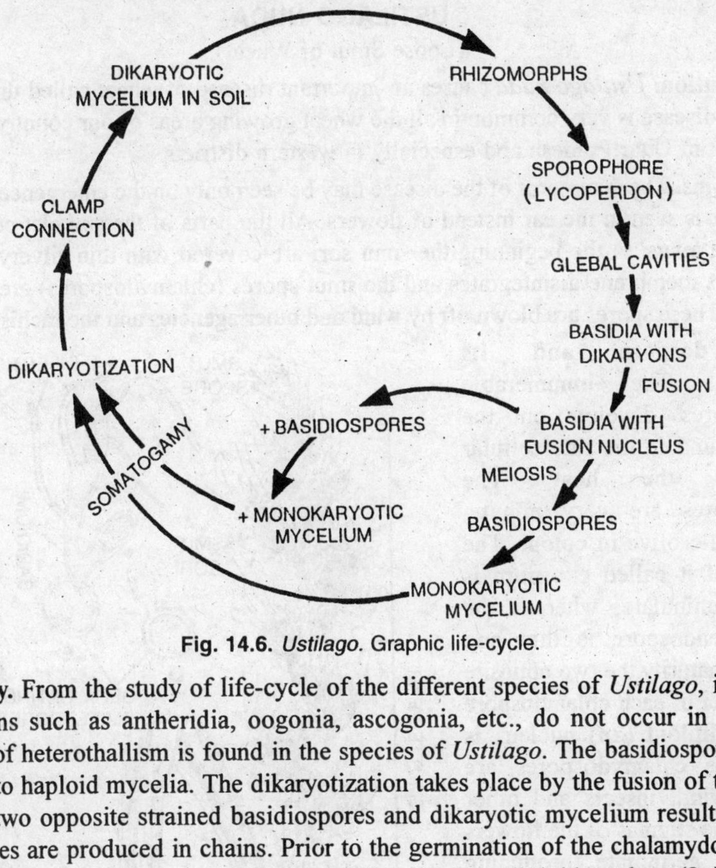

Fig. 14.6. *Ustilago.* Graphic life-cycle.

Sexuality. From the study of life-cycle of the different species of *Ustilago*, it is quite clear that sex organs such as antheridia, oogonia, ascogonia, etc., do not occur in any case. The phenomenon of heterothallism is found in the species of *Ustilago*. The basidiospores are haploid and give rise to haploid mycelia. The dikaryotization takes place by the fusion of the two hyphae produced by two opposite strained basidiospores and dikaryotic mycelium results on which the chlamydospores are produced in chains. Prior to the germination of the chalamydospore both the opposite strained (+ and –) nuclei of it fuse giving rise to a diploid nucleus (2*n*). This nucleus further divides twice producing 4 haploid nuclei (*n*) in the promycelium or basidium. The first division is reductional (meiosis). Morphologically, the basidiospores and the hyphae produced from them cannot be considered as sex organs. They are quite identical in outer appearance. The formation of diploid nucleus in the chlamydospore, and the production of 4 haploid nuclei in the basidium by meiosis, advocate the view that the two nuclei of a dikaryotic cell in *Ustilago* are of opposite sexual behavioour, *i.e.*, + and – strains. These nuclei behave as the sex organs of Phycomycetes and Ascomycetes. It is very obvious that the sexuality is very much reduced in *Ustilago* and can be represented only by the opposite strained nuclei present in dikryotic cell.

The different species of *Ustilago* cause various important crop diseases.

1.	Loose smut of wheat caused by	*Ustilago nuda*
2.	Covered smut of barley caused by	*U. hordei*
3.	Loose smut of barley caused by	*U. nuda*
4.	Covered smut of oats caused by	*U. kolleri*
5.	Loose smut of oats caused by	*U. avenae*

6. Smut of maize caused by *U. maydis or zeae*
7. Whip smut of sugarcane caused by *U. scitaminea*
8. Smut of doob grass caused by *U. cynodontis*

The characterstic features of an important disease caused by *Ustilago nuda i.e.*, loose smut of wheat is given here.

USTILAGO NUDA

(Loose Smut of Wheat)

Distribution. *Ustilago nuda* causes an important disease of wheat called the 'loose smut of wheat'. This disease is very common in all the wheat growing areas of our country. This disease is very common in Uttar Pradesh and especially in western districts.

Symptoms. The symptoms of the disease may be seen only on the emergence of the ears. The black powder is seen in the ear instead of flowers. All the parts of the spikelet convert into smut spores except awns. In the beginning the smut sori are covered with thin silvery membrane, but very soon this membrane disintegrates and the smut spores (chlamydospores) are liberated in the atmosphere. These spores are blown off by wind and other agencies and the rachis remains barren.

Chlamydospore and its germination. The innumerable chlamydospores develop on the mycelium found in the intercellular spaces of the host. The chlamydospores are very minute, round and pale olive in colour. The outer wall of it called exospore is thick and echinulate, whereas the inner wall endospore is thin and smooth. On maturity the two opposite strained nuclei of each chlamdospore fuse and a diploid ($2n$) nucleus is resulted. The chlamydospores are carried by wind, insects and other agencies to the stigmas of the flowers where they germinate producing basidia or promycelia. The exosporium ruptures and the basidium comes out in the form of a tube. The diploid nucleus divides twice giving rise to four haploid nuclei. The first division is reductional. The basidium becomes segmented and the four uninucleate cells are produced, two of + and two of – strains. From each segment single infection thread develops. The two opposite strained infection threads unite and a dikaryotic cell is produced. The dikaryotic hypha develops from this dikaryotic cell.

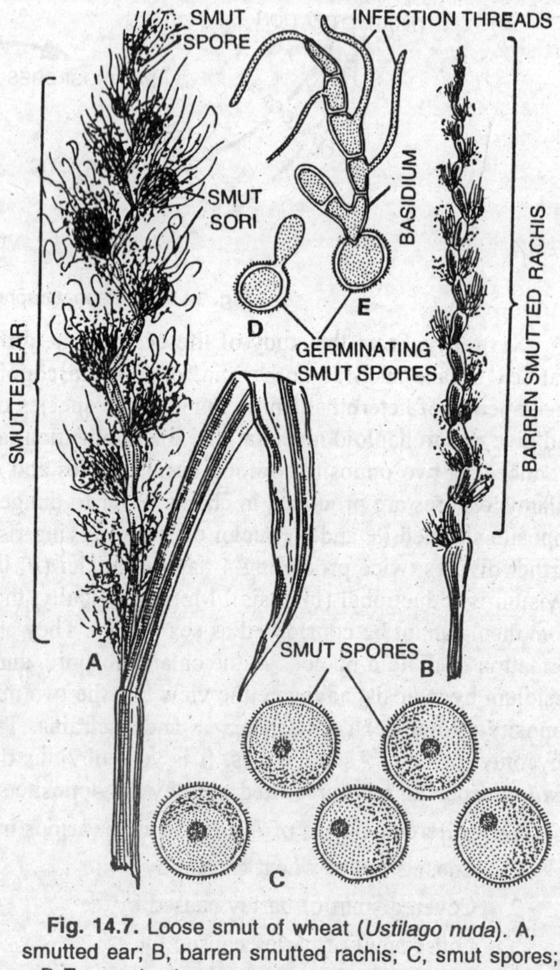

Fig. 14.7. Loose smut of wheat (*Ustilago nuda*). A, smutted ear; B, barren smutted rachis; C, smut spores; D-E, germinating smut spore showing basidium and infection threads.

This process is called dikaryotization. The dikaryotic hypha enters the style and reaches the embryo where it develops into thick, irregular and branched mycelium. The mycelium remains dormant in the embryo. The seeds with dormant mycelia in their embryos look quite healthy in outer appearance.

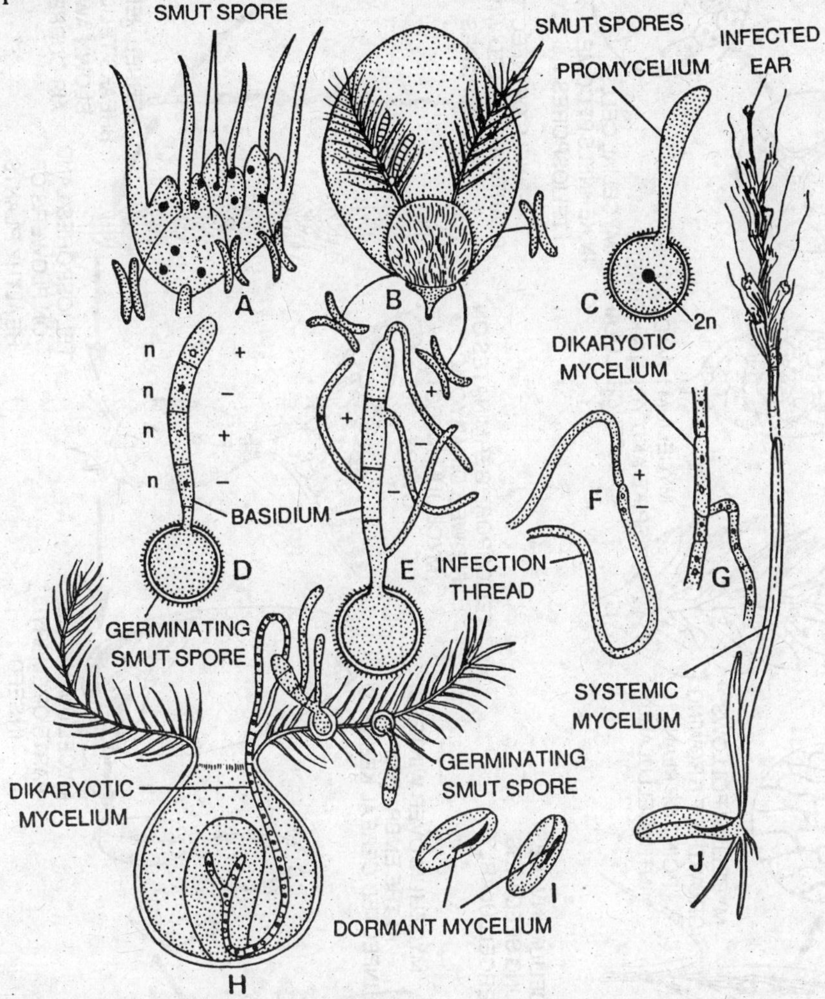

Fig. 14.8. *Ustilago nuda.* Loose smut of wheat. A, spikelet with smut spores; B, smut spores in feathery stigma of the flower; C-D, germination of smut spore; E, basidium and infection threads; F, fusion of compatible infection threads; G, dikaryotic mycelium; H, infection in embryo; I, infected grains; J, germination of infected grain and systemic infection.

Germination of infected seeds. The dormant mycelium situated in the embryo of the seed becomes active at the time of the germination of the infected seed and grows along with the apex of the coleoptile. The optimum temperature for the growth of the mycelium ranges from 20°C to 25°C. The fungus is systemic and the mycelium is found throughout in the stem. On the emergence of the ears the mycelium travels in the floral parts and becomes very active. The thousands of chlamydospores are produced in sori on the floral axis. All the floral parts become infected and turned into smut sori, except awns. The haustoria are not produced on the mycelium. The food is absorbed by diffusion method.

Recurrence of disease. In *U. nuda* the dormant mycelia perennate inside the embryos of infected seeds and whenever such seeds are sown the seedings become infected from the very beginning. This is an internally seed borne and systemic disease.

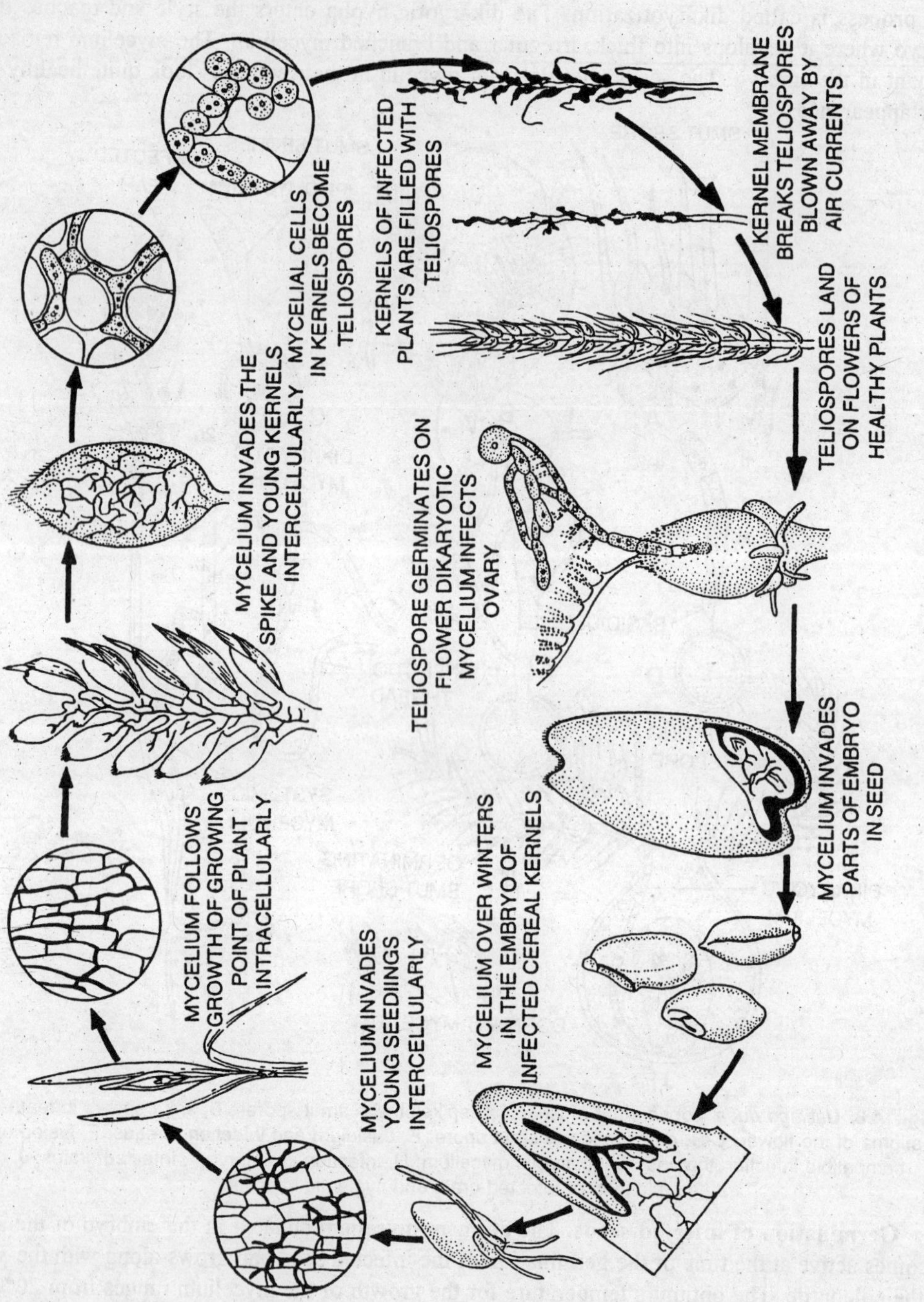

Fig. 14.9. Disease-cycle of loose smut of wheat (*Ustilago nuda*).

Control measures. Various methods have been proposed for the control of this internally seed borne disease.

1. Rogueing. The infected plants must be uprooted and burnt taking outside the fields. This method of control can easily be practised, because of the earlier emergence of the infected ears from the boot leaves.

2. Seed selection. As this is a seed borne disease, the seeds must be selected from healthy crops.

3. Hot water treatment. To activate the dormant mycelia, the seeds are soaked for four or five hours in the water at normal temperatures ranging fom 26°C to 30°C. Thereafter, the seeds are transferred in hot water at 54°C for ten minutes. The activated mycelium dies at this temperature. These grains may be used as seeds. The temperature of the water should not exceed in any case, because the embryo of the seed dies at 56°C. This method is not practicable in our country because of its complexity.

4. Solar treatment. This is the modification of hot water treatment. Luthra and Sattar proposed this treatment for Pubjab, Sind (Pakistan) and Uttar Pradesh. The seeds are soaked in water from 8 a.m. to 12 noon at normal termperatures in the months of May and June. Thereafter the seeds are dried up on the floors in the open sun in very thin layers up to 4 p.m. The dormant mycelium becomes active on soaking the seeds in the water for 4 hours, and dies when the seeds are dried in thin layers in the hot sun and the seeds become disinfected. The method of control may easily be practised in our country.

5. Resistant varieties. This is the simplest and most important method of the control of any disease. Every year many resistant varieties are developed at research centres and distributed to the farmers for sowing purposes.

Systematic Position

	G. W. Martin (1961)	C. J. Alexopoulos (1962)	G. C. Ainsworth (1971)
Kingdom	–Plantae	–Plantae	–Fungi
Division	–Mycota	–Mycota	–Eumycota
Sub-div.	–Eumycotina	–Eumycotina	–Basidiomycotina
Class	–Basidiomycetes	–Basidiomycetes	–Teliomycetes
Sub-cl.	–Heterobasidiomycetidae	–Heterobasidiomycetidae	–
Order	–Ustilaginales	–Ustilaginales	–Ustilaginales
Family	–Ustilaginaceae	–Ustilaginaceae	–Ustilaginaceae
Genus	–*Ustilago*	–*Ustilago*	–*Ustilago*
Species	–*nuda*	–*nuda*	–*nuda*

Order **UREDINALES** (The Rusts)

(130 genera; 3,600 species)

Characteristic Features

1. This order consists of obligate parasites. They are called the 'rust fungi'.
2. The mycelium is intercellular with round or branched haustoria.
3. Five types of spores are met with in the life histories of various genera of this order. They are pycniospores (o), aeciospores (i), uredospores (ii), teleutospores (iii), and basidiospores or sporidia (iv).
4. The rusts producing many types of spores are known as 'polymorphic' or 'pleomorphic'.
5. Leaving aside basidiospores the rest of the types of spores develop in definite sori, they are, pycnia, aecia, uredia and telia.
6. The rusts may be short-cycled (microcyclic) or long-cycled (macrocyclic)
7. They may be heteroecious (*e.g. Puccinia graminis*) or autoecious (*e.g., Melampsora lini*).
8. During dikaryotization the spermatium (pycniospore) of one strain fuses with the flexuous hypha of opposite strain, and dikaryotization takes place.

Classification. This order is divided into five families, (i) Pucciniaceae, (ii) Cronortiaceae, (iii) Melampsoraceae, (iv) Coleosporaceae, (v) Endophyllaceae.

There are about 130 genera and 4,600 species.

Family–**Pucciniaceae**

The teleutospore forms a septate promycelium (basidium) upon germination. The teleutospores are free or variously united, but never in the form of layer or crusts. The teleutospores are usually stalked. The promycelium is external. The important genus *Puccinia* has been discussed here in detail.

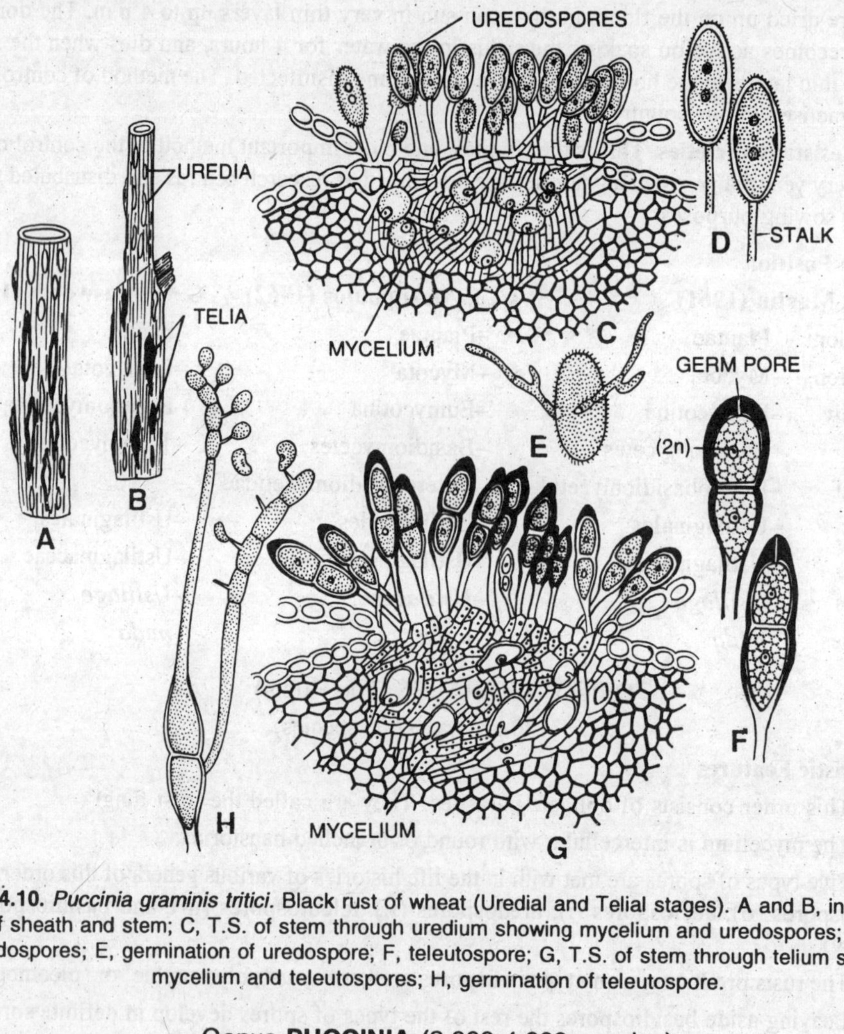

Fig. 14.10. *Puccinia graminis tritici.* Black rust of wheat (Uredial and Telial stages). A and B, infection on leaf sheath and stem; C, T.S. of stem through uredium showing mycelium and uredospores; D, two uredospores; E, germination of uredospore; F, teleutospore; G, T.S. of stem through telium showing mycelium and teleutospores; H, germination of teleutospore.

Genus **PUCCINIA** (3,000-4,000 species)

(*Puccinia graminis*)

The different species of *Puccinia* cause the diseases known as the rusts on various host plants. *Puccinia* is an obligate parasite which survives only on living host plants. The rusts are known as **rati, ratwa, gerua, gerui, gherir, tombora** in local languages in the different parts of our country. The characteristic feature of rusts is the presence of the pustules of rusty appearance on the surface of the host plants.

In our country the wheat crops are highly infected by the different species of *Puccinia.* The black or stem rust of wheat is caused by *Puccinia graminis tritici;* the yellow or stripe rust of wheat is caused by *Puccinia striiformis;* the brown, orange or leaf rust of wheat is caused by *Puccinia recondita.*

Symptoms. The symptoms of the black or stem rust of wheat (*Puccinia graminis tritici*) are commonly found on the stem but the symptoms also appear on the leaf sheath, leaf blade and even on the ears. The uredia developed by *P. graminis tritici* are oblong and reddish brown which unite to each other. These uredosori or uredia represent the uredial stage of the black rust. On maturity the uredia are ruptured and the uredospores are liberated in the form of brown powder. In the end of the season when it is somewhat hot the irregular elongated, dark brown or black, crusts like pustules develop on the stems. They are telia or teleutosori. On maturation the teleutosori rupture and the brown coloured teleutospores come out of them.

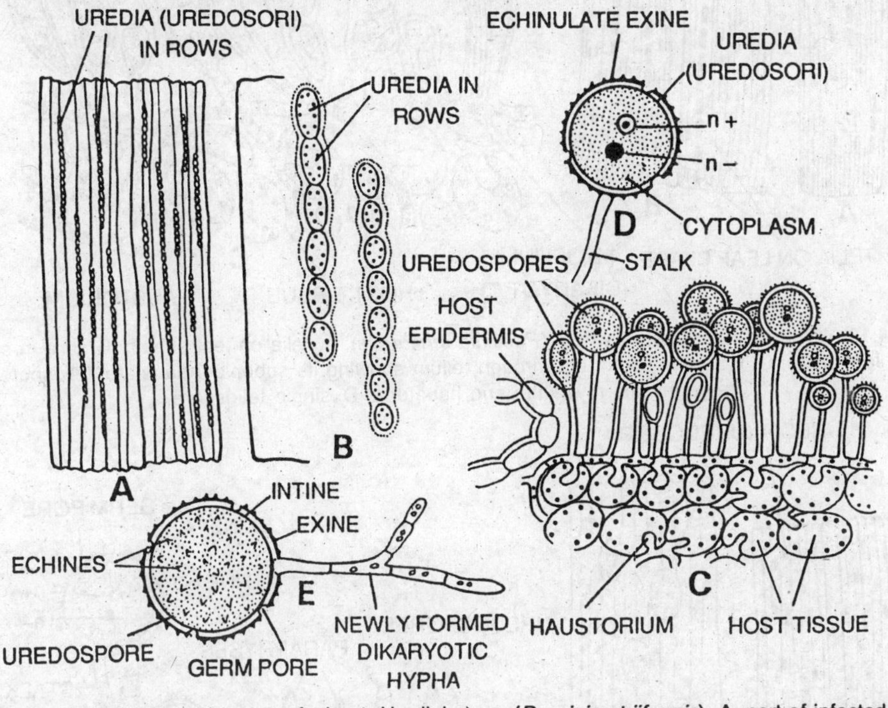

Fig. 14.11. Yellow or stripe rust of wheat. Uredial stage (*Puccinia striiformis*). A, part of infected leaf showing uredia arranged in rows; B, enlarged part of A; C, V.S. of infected leaf through uredium showing mycelium, haustoria and stalked uredospores; D, single uredospore; E, germinating uredospore.

The symptoms of black rust may also be accounted on its alternate host. The groups of pycnia are found on the upper surface of *Berberris* leaves, whereas on the lower surface the groups of aecidia are found. The stage represented by pycnia is called the 'pycnial stage' and by aecidia the 'aecidial stage'. The pycnia are flask-shaped and the aecidia cup-like.

The symptoms of yellow rust caused by *Puccinia striiformis* are somewhat different. Usually the symptoms appear on the leaves, but somewhat they appear upon leaf sheath, glumes, awns and even on the stem. The uredia are yellow coloured and arranged in parallel stripes, giving the name to the disease the 'yellow or stripe rust' of wheat. They are very close to each other but do not fuse. The green colour around the uredosori becomes fade, and yellow coloured parallel stripes are found on the leaf. In the end of the season dark brown and black telia appear on the leaf blade and leaf sheath. The telia are subepidermal and not exposed. The telia are elongated and arranged in parallel rows as in case of uredia.

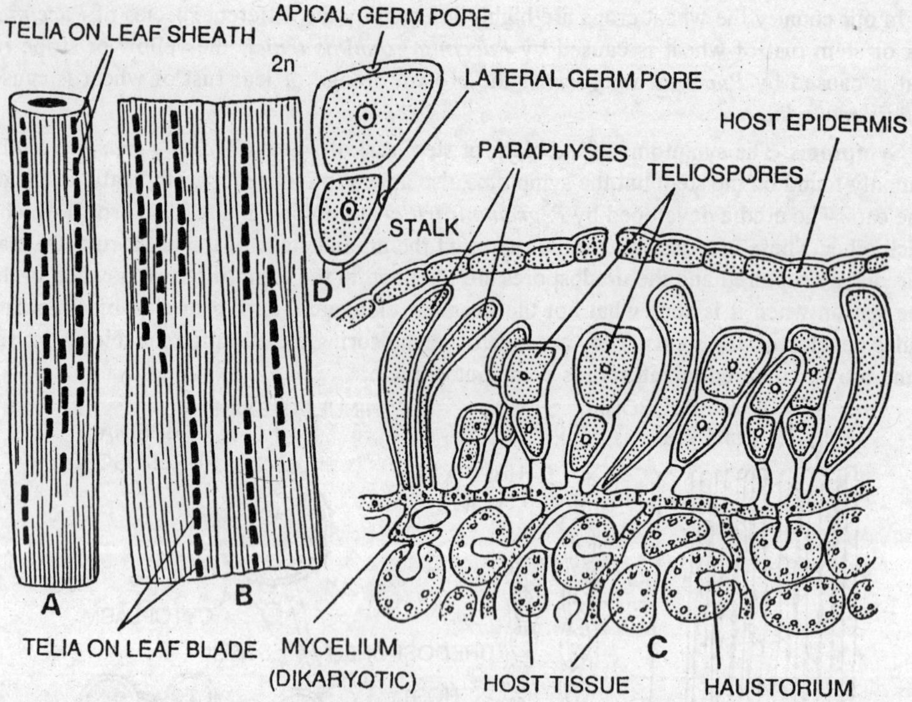

Fig. 14.12. Yellow or stripe rust of wheat (*Puccinia striiformis*). A, telia on leaf sheath; B, telia in stripes on leaf blade; C, V.S. of infected leaf through telium showing its subepidermal state, teliospores, paraphyses, mycelium and haustoria; D, single teliospore.

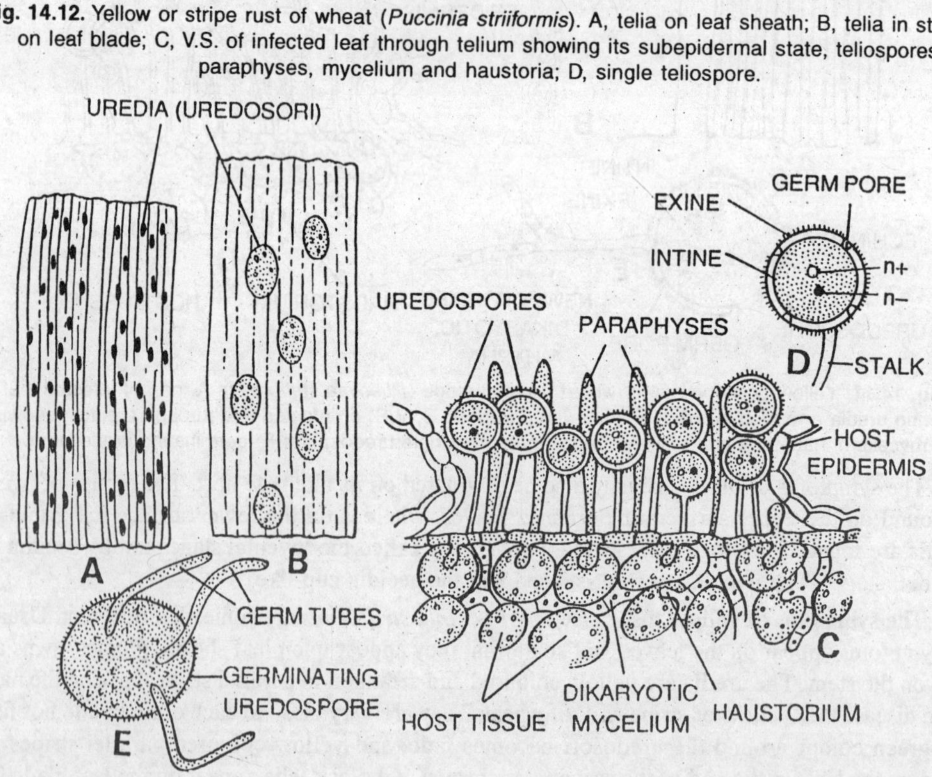

Fig. 14.13. Orange, brown or leaf rust of wheat (*Puccinia recondita*). A, a part of infected leaf with scattered uredia; B, magnified part of same; C, V.S. infected leaf through uredium showing stalked uredospores, paraphyses, mycelium and haustoria; D, single uredospore; E, germinating uredospore.

The symptoms of brown, orange or leaf rust caused by *P. recondita* appear exclusively on the leaf blade and the disease is called the 'leaf rust'. Rarely the uredia are found on the leaf sheath and stem. The uredia are rounded or oblong and brown or orange in colour. On maturation they rupture and the uredospores come out in the form of brown powder. Usually a fringe is seen around each ruptured uredium. The uredia are developed on both the surfaces of the leaf. The uredia represent the uredial stage. The telial stage is rarely found in the conditions of our country. Wherever the telia are present, they are small oval or somewhat elongated and sub-epidermal.

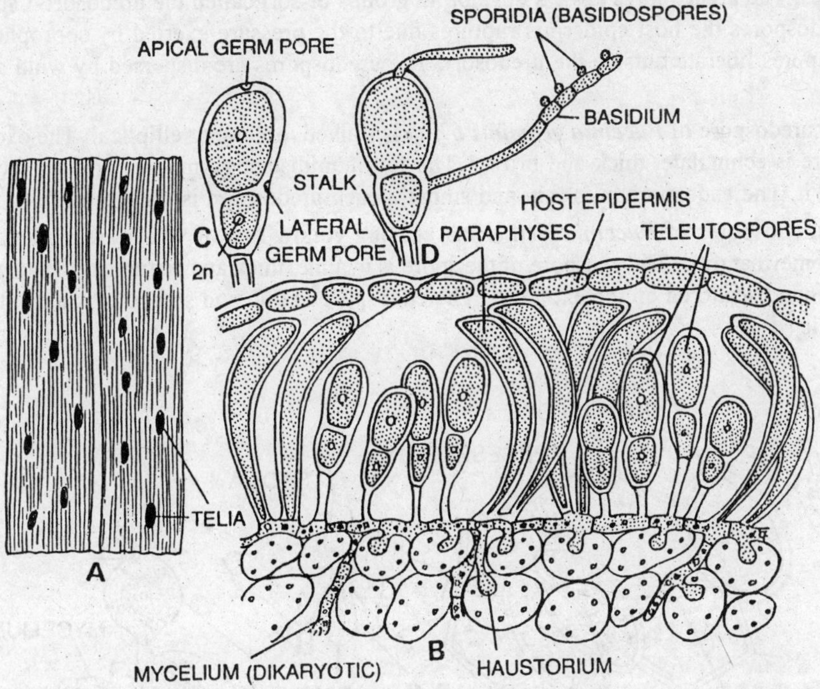

Fig. 14.14. Orange, brown or leaf rust of wheat-telial stage (*Puccinia recondita*). A, a part of infected leaf showing telia; B, V.S. of infected leaf through telium showing teliospores, paraphyses, mycelium and haustoria; C, single teliospore; D, germinating teliospore.

Structure of mycelium. The mycelium is localized, endophytic, intercellular, branched, septate and well developed. There is a single central septal pore in each septum for the communication of the cytoplasm from one cell to another. In certain phases of the fungus the mycelium is **monokaryotic** whereas in other phases it is **dikaryotic.** To absorb food material from host cells the haustoria are developed which may be knob-like, finger-like or convolute. In rare cases the hyphae penetrate the cell walls and the mycelium becomes intercellular.

The dikaryotic or secondary mycelium is confined to primary host, *i.e.,* wheat and the monokaryotic or primary mycelium is found on the secondary or alternate host, such as *Berberris* sp., in the case of black rust, and *Thalictrum* sp., in the case of brown rust. The alternate host of yellow rust has not been discovered so far. Such fungi which complete their life cycles on two different hosts, not related to each other in any way, called the **heteroecious** fungi and the phenomenon **heteroecism.** The black rust is heteroecious rust and completes its life-cycle on two different hosts called primary and alternate hosts such as wheat and barberry bushes respectively which are not at all related to each other in any way.

Spores. Several types of spores are found in the life history of *Puccinia graminis tritici.* They are pycniospores, pycnidiosores or spermatia, aeciospores or aecidiospores; uredospores, urediospores or uredeniospores; teliospores or teleutospores and; basidiospores. *Puccinia graminis* is said to be a **polymorphic** fungus, because of the presence of several types of spores and the phenomenon is called the "polymorphism".

Uredospores. The mycelium of *Puccinia* is dikaryotic in the primary host, *i.e.*, wheat. The hyphae accumulate in abudance just beneath the host epidermis. Many vertical hyphae arise from this accumulated mycelium. The terminal cell of each vertical hypha swels and becomes rounded or elliptical. These vertical hyphae are called the sporophores. The terminal cells develop into uredospores. The uredospores always develop in groups or sori called the uredosori. On maturity of the uredospores the host epidermis ruptures due to the pressure exerted by sporophores, and the uredospores liberate outside the uredosori. The uredospores are dispersed by wind and other agencies.

Each uredospore of *Puccinia graminis tritici* is stalked, oblong or elliptical. The exosporium of the spore is echinulate, thick and brown. The four equidistant germpores are arranged on the equator of it. The endosporium is thin and smooth. each uredospore is dikaryotic (*nn*).

The uredospores of *Puccinia striiformis* causing yellow rust of wheat are stalked, yellow, round or somewhat oval. The exospore of the spore is hyaline, thick and echinulate. Six to ten germ pores are found scattered on the exopores. The endospore is thin and smooth. As usual the spore is dikaryotic (*nn*).

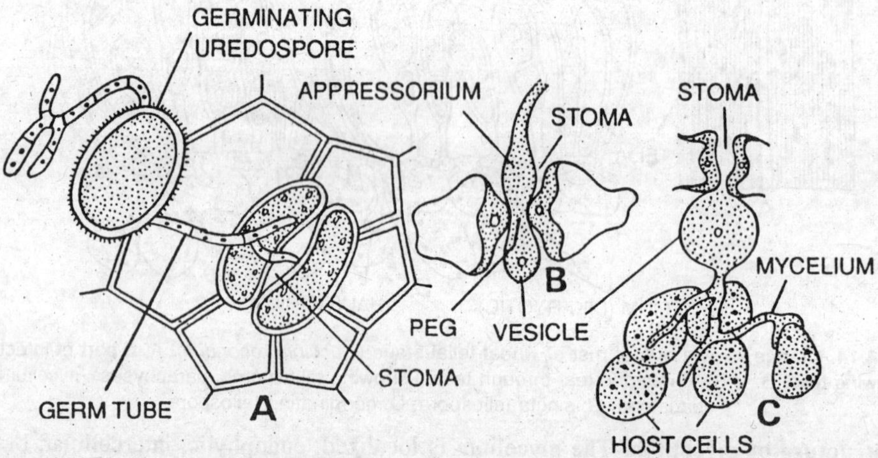

Fig. 14.15. Black or stem rust of wheat (*Puccinia graminis tritici*). A, germination of uredospore on host surface; B, formation of appressorium and vesicle; C, development of mycelium.

The uredospores of *Puccinia recondita* causing brown or leaf rust of wheat are stalked, brown and approximately rounded in structure. The exospore is thick, echinucleate and brown. About six germpores are scattered over it. The spore is diakaryotic (*nn*).

On getting suitable conditions of moisture and temperature for germination, they germinate producing germ tubes. The uredospore germinates on the epidermis of the host. More than one germ tubes come out of the germpores from a single spore. On approaching the stomata these germ tubes form appressoria. From the centre of the appressorium a peg-like structure develops and enters the stoma. The terminal portion of this peg swells into a vesicle. Many branched hyphae arise from this vesicle and ramify in the intercellular spaces of the host tissue, giving rise to normal knob-like or branched haustria for the absorption of the nutrition from the host cells. The uredial stage is several times repeated prior to the development of teliospores (telial stage) in the same season.

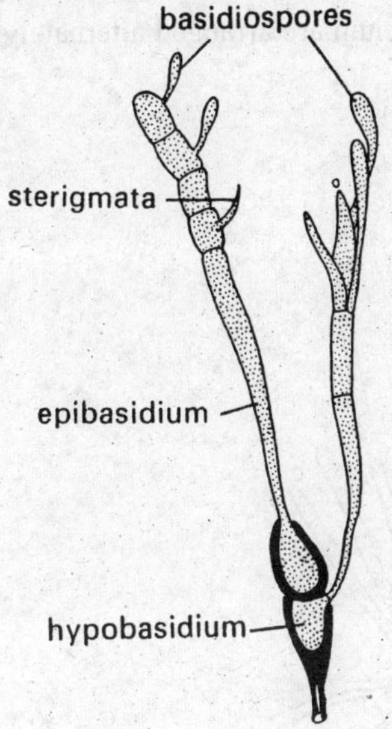

Fig. 14.16. *Puccinia graminis.* A germinating teliospore.

Teleutospores. After the formation of the uredospores for sufficiently a long period on the dikaryotic mycelium, the teleutospores develop in the same sori. These spores are called the **teleutospores,** the stage, **telial stage** and the sori in which the spores develop, the **teleutosori** or **teliosori.** In the transitional stages both the uredospores and teliospores are found in the same sori. But later on the telia develop independently having only teleutospores in them. The telia are dark brown or black coloured.

The telia of black rust (*P. graminis tritici*) are exposed, irregular, elongated, dark brown or black. The crust-like pustules develop on the stems. On maturation, the teleutosori rupture and brown coloured teleutospores are released. Each teleutospore is stalked, dark brown and double celled. The exospore is thick and dark brown in colour. The apex of the teleutospore is either rounded or pointed. There is one germpore at the apex in the upper cell. The lower cell of the spore possesses a lateral germpore. In each cell of a mature teleutospore, there is a conspicuous diploid nucleus (2*n*).

In *Puccinia striiformis* (yellow rust) the telia are sub-epidermal. They are somewhat elongated and arranged in parallel rows like uredia. The telium is divided into several chambers by the paraphyses attahced to the base of it. As usual each teleutospore is stalked, elongated, brown and bicelled. The apex of the spore is either flat or oblique. The wall of the apex is thinner than the wall of the apex of the spore of *P. graminis tritici.* There are two usual germpores, one apical and the other lateral. Each cell of the spore contains a diploid (2*n*) nucleus in it.

The teleutospore of *P. recondita* is stalked, oblong or cuneiform, bicelled, brown and somewhat flat or rounded at its apical end.

The fully matured teleutospores are thick-walled and dark brown in colour. The two nuclei of a dikaryon of each cell, fuse, and in each cell a diploid nucleus develops. Each cell of the mature teleutospore contains a dipoid nucleus. After the formation of the teleutospores they need a dormancy period prior to their germination. This is the resting period.

After the harvest of crop the teleustospores scattered in the soil undergo a period of rest. Prior to germination of the teleutospore the freezing temperature for a long period is essentially an important factor. The teleustospores germinate producing promycelia or basidia on which basidiospores are developed.

On the approach of the favourable conditios, the dormant protoplasm of the resting teleutospores becomes activated and the tubular promycelia or basidia come out of the germpores of each cell of the teliospore. The diploid nucleus is transferred in the basidium where it divides twice giving rise to four haploid nuclei. The first division is reductional. Each basidium becomes segmented forming four uninucleate cells. A sterigma comes out from each of the cells In the lower three cells, the sterigmata arise just beneath the septa, and in the upper cell at the apex. On the terminal end of each sterigma a small, rounded sporidium or basidiospore develops. A single basidiospore develops on each sterigma and four basidiospres are produced on the basidium, 2 of + strain and two of – strain. The promycelial cells become depleted after the formation of basidiospores. Usually the opposite strained nuclei of the basidium are arranged alternately in the segments. *Puccinia* is a heterothallic fungus.

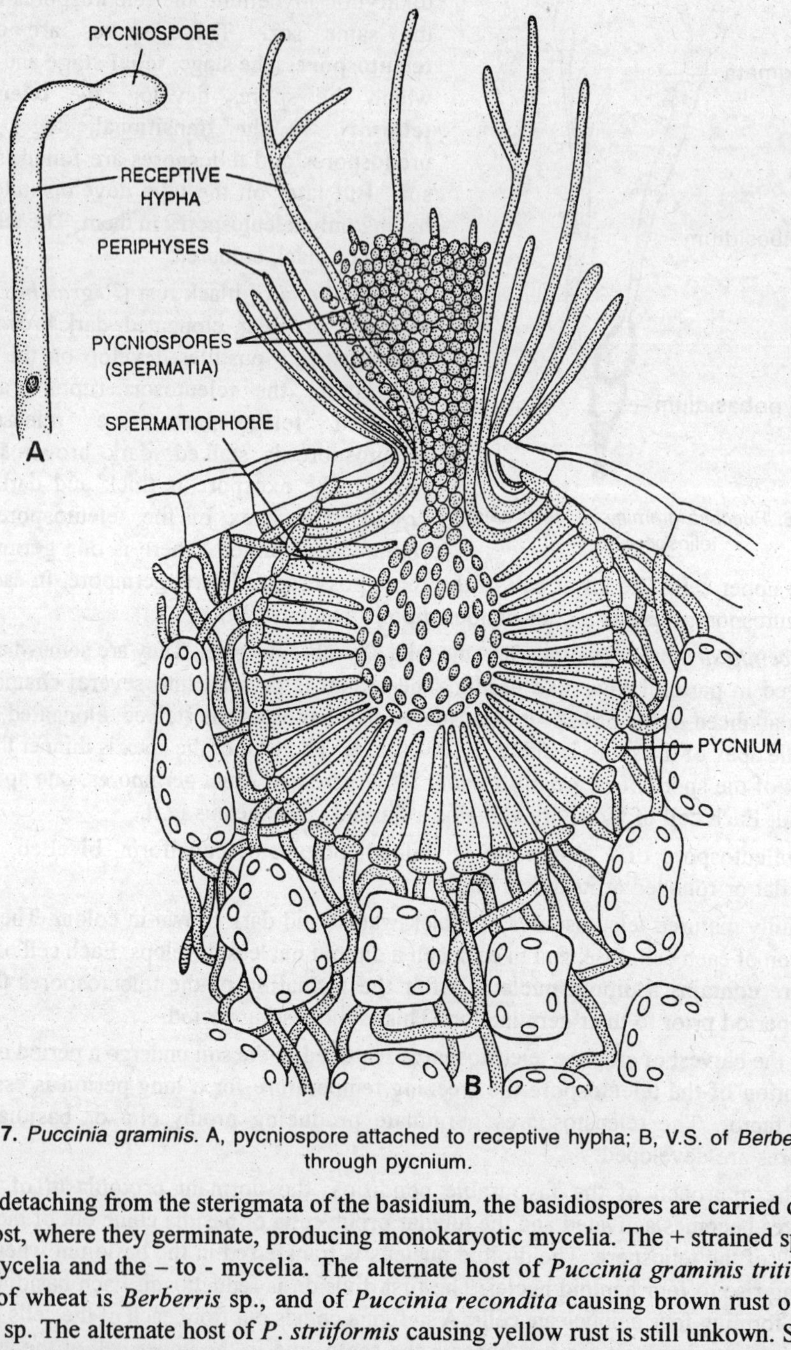

Fig. 14.17. *Puccinia graminis.* A, pycniospore attached to receptive hypha; B, V.S. of *Berberris* leaf through pycnium.

After detaching from the sterigmata of the basidium, the basidiospores are carried over to the alternate host, where they germinate, producing monokaryotic mycelia. The + strained spores give rise to + mycelia and the – to - mycelia. The alternate host of *Puccinia graminis tritici* causing black rust of wheat is *Berberris* sp., and of *Puccinia recondita* causing brown rust of wheat is *Thalictrum* sp. The alternate host of *P. striiformis* causing yellow rust is still unkown. Sometimes when suitable alternate host is not available, these spores give rise to secondary basidiospores which also germinate in the same way. A germ tube comes out from the basidiospore, swells at its terminal end, the nucleus along with cytoplasm transfers to this swelling and the secondary basidiospore develops.

Germination of basidiospore. After reaching the host surface, the basidiospores germinate producing germ tubes. The germ tube directly penetrates the epidermis, enters the host tissue, develops into monokaryotic mycelium which ramifies in the intercellular spaces of the alternate host. The mycelium is septate, branched and with haustoria for the absorption of food. Usually the opposite strained (+ and –) basidiospores germinate on the surface of the same host leaf and produce monokaryotic mycelia of both strains very close to each other.

The basidiospores are destroyed in the absence of suitable alternate host. About four days later after the germination of the basidiospore the pycniospores are produced on the monokaryotic mycelium.

Pycniospores, pycnidiospores or spermatia. The monokaryotic mycelium accumulates beneath the epidermis of the alternate host and develops into a flask-like structure called **pycnium.** A hymenial layer of vertical hyphae develops at the base of the pycnium. Each vertical hypha of this hymenium is called the pycniosporophore. From the terminal ends of the pycniosporophores, round, oval, haploid and uninucleate pycniospores develop in basigenous chains. The pycniospores detach from the pycniosporophores and accumulate in the middle of the pycnium. The pycnia open outside the host epidermis by small openings called the ostioles. Some of the sterile hyphae called the paraphyses are found in the neck region of the pycnium. Some vertical fertile hyphae called the flexuous hyphae also come out through the ostiole. They are of the same strain as that of the mycelium on which they are developed. Only one strained (+ or –) pycniospores and flexuous hyphae are produced in one pycnium.

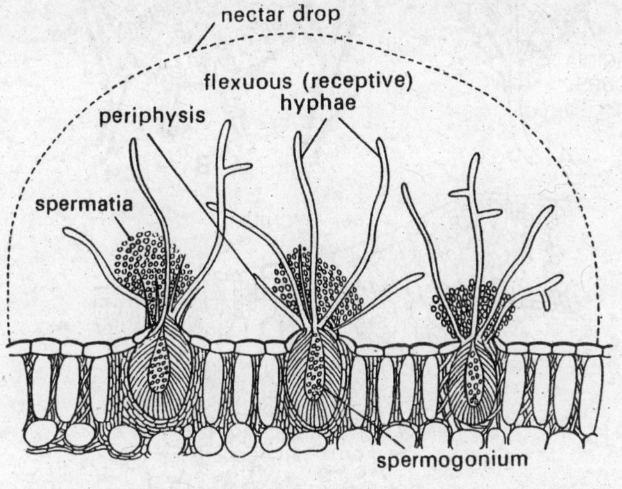

Fig. 14.18. *Puccinia graminis.* Pycnia in relation to nectar drop on the upper surface of barberry leaf.

On the maturity of the pycnium, the nectar oozes out through ostiole in the form of drops. Hundreds of pyoniospores come out along with the nectar. The pin head-like yellow pycnia are found on the upper surface of the infected leaf.

The pycniospores developed in the flask-like pycnia and the flexuous hyphae convert the monokaryotic phase of the mycelium of the fungus into dikaryotic phase by the process of dikaryotization. On this dikaryotic mycelium, the aeciospores are produced.

Dikaryotization or diploidization. The insects visit the mature pycnia in the search of nectar oozing out through the ostioles. The pycniospores found around the ostioles stick to their legs and proboscis and are trnasferred from one pycnium (+) to another (–). When the opposite strained pycniospores and the flexuous hyphae come in contact, the wall of contact dissolves and the nucleus of pycniospore moves downward in the flexuous hypha through septal pores.

Simultaneously, just beneath the pycnium, on the lower surface of the same leaf, the monokaryotic mycelium accumulates in the substomatal chamber and intercellular space. The aecidia are produced from this mycelium, after dikaryotization, and thus called the aecidial primoridia or protaecia. The nuclei of the pycniospores, after passing through the pores reach in the cells of the protaecia and diakaryotize all of them.

Besides this phenomenon, sometimes the pycniospores fuse with the sterile paraphyses. Sometimes, the opposite strained monokaryotic hyphae found in the intercellular spaces fuse with each other and the dikaryotization takes place. The nuclei of the cells of one hypha are transferred into the adjacent cells of opposite strained another hypha. The great mycologist, A.H.R. Buller observed in *Puccinia helianthi,* that the dikaryotization takes place by the fusion of monokaryotic hypha with the dikaryotic one. The walls of contact dissolve, and the nuclei lacking in the monokaryotic hypha are being transferred from the adjacent cells of the dikaryotic hypha, and dikaryotization takes place. This special process of dikaryotization is known as **Buller phenomenon.**

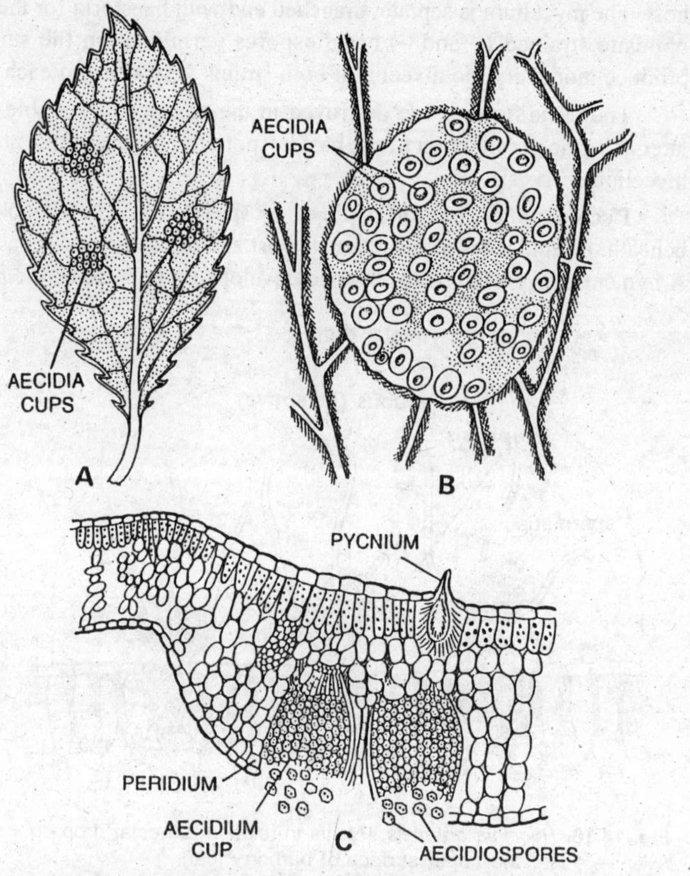

Fig. 14.19. *Puccinia graminis.* A, aecial cups on lower surface of barberry leaf; B, aecial cups; C, V.S. of barberry leaf through pycnium and aecial cups.

Aeciospores or aecidiospores. Just after dikaryotization the hyphae of protaecia form a stromatic layer. The cells of the stroma become dikaryotic. They are surrounded by the monokaryotic hyphae and pseudoparenchymatous structure is developed. The basal cells of this pseudosparenchymatous structure are somewhat longer than the lower cells. All the cells are monokaryotic.

The dikaryotic cells surrounded by pseudoparenchyma develop vertically elongated downwards in a thick stromatic layer, called the **hymenium.** The hymenium consists of aeciospore mother cells. The aeciospores develop first from the cells of the central region and afterwards from peripheral region of the hymenium. From the outermost mother cells of the hymenial periphery several cells are developed. The lateral walls of these cells fuse to each other and the peridium is develped. The cells of the peridium divide continuously in the transverse plane, increasing the height of the peridium. The aeciospore mother cells elongate and divide at their apical ends giving rise to a layer of the cells at their apices. The terminal cell separates from the aeciospore mother cell by a septum and two cells develop one larger in size and the other smaller. The terminal large cell develops into the aeciospore whereas the smaller cell acts as intercalary or disjunctor cell. The intercalary cells are sterile. The nucleus of aeciospore mother cell divides again and again giving rise to the basigenous chain of aeciospores and intercalry cells. On the maturation of the aeciospores, the intercalary or disjunctor cells degenerate and the aeciospores become free.

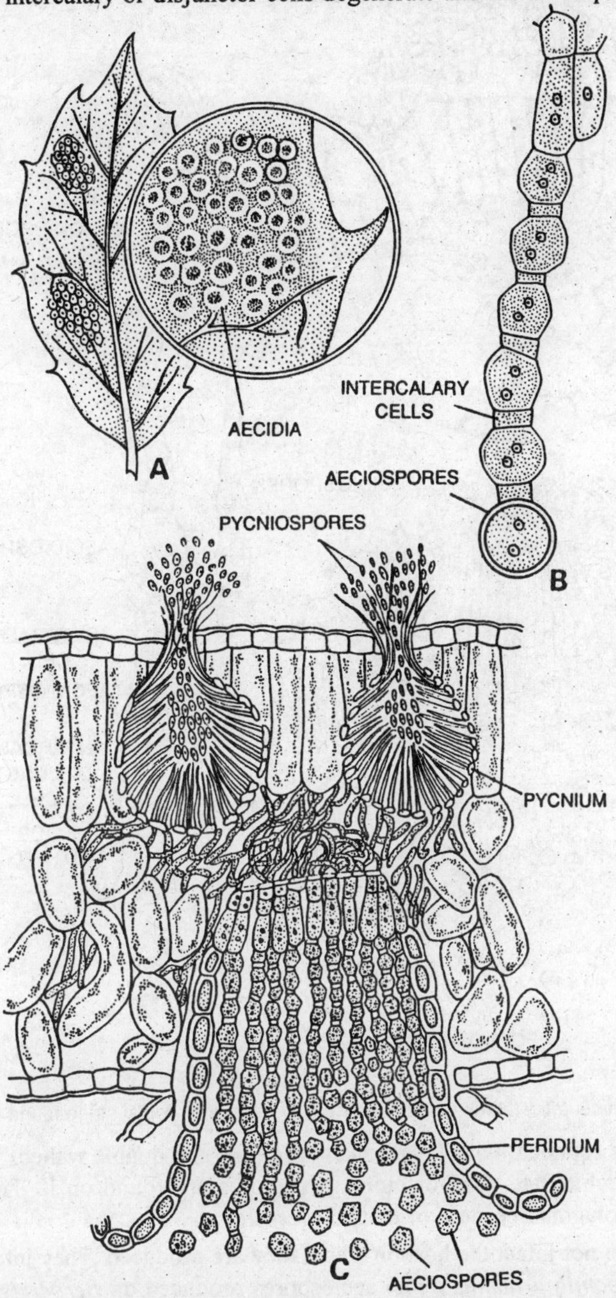

Fig. 14.20. *Puccinia graminis tritici* on *Berberris*. Black rust of wheat. A, infected leaf of *Berberris* showing aecidia; B, a chain of aeciospores; C, V.S. of infected *Berberris* leaf through pycnia and aecidium.

The aeciospores develop first in the centre and thereafter in the peripheral region, and therefore, the hymenial layer is thicker in the central region than the periphery and a domeshaped structure covered by peridium is formed. Very soon the peridium ruptures, due to the pressure exerted by the aeciospores and the whole structure called aecidium looks like a bell. The aeciospores are dispersed by wind, insects, water and other suitable agencies.

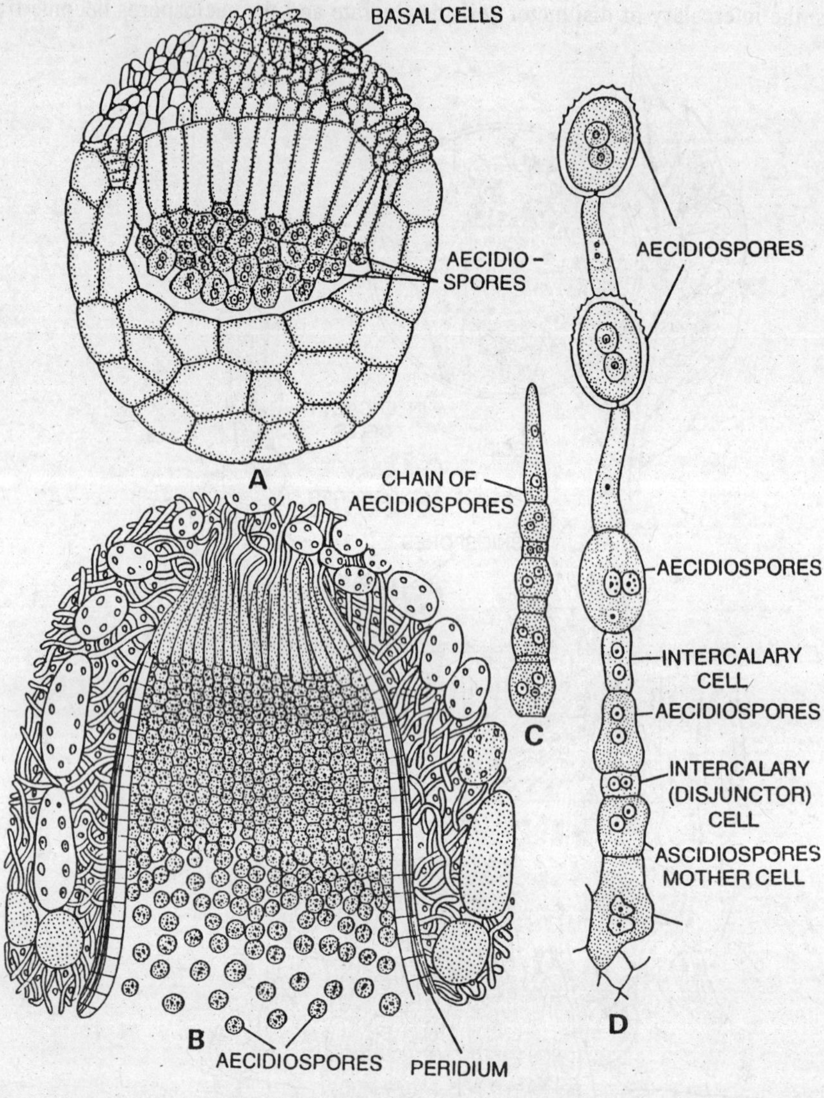

Fig. 14.21. *Puccinia graminis tritici* (black rust of wheat). A-D, details of aecial stage found in barberry leaf.

Each aeciospore is rounded, unicellular, dikaryotic (*nn*) and double walled. The exopore is smooth or sometimes echinulate. One or more germ pores are found on it. They are orange coloured and become polygonal because of mutual pressure.

The aeciospores do not infect the host on which they are produced. They infect the primary host. In the case of *Puccinia graminis tritici* aeciospores produced on *Berberris* sp., infect the wheat plants and not *Berberris* in any case. The aeciospores germinate producing a germ tube which comes out through the germ pore and enters the stoma of host leaf where it develops, into new mycelium. The mycelium ramifies in the intercellular spaces of the host tissue. It is dikaryotic and branched. This is seconadary mycelium, this dikaryotic mycelium produces uredospores and teleutospores in usual way.

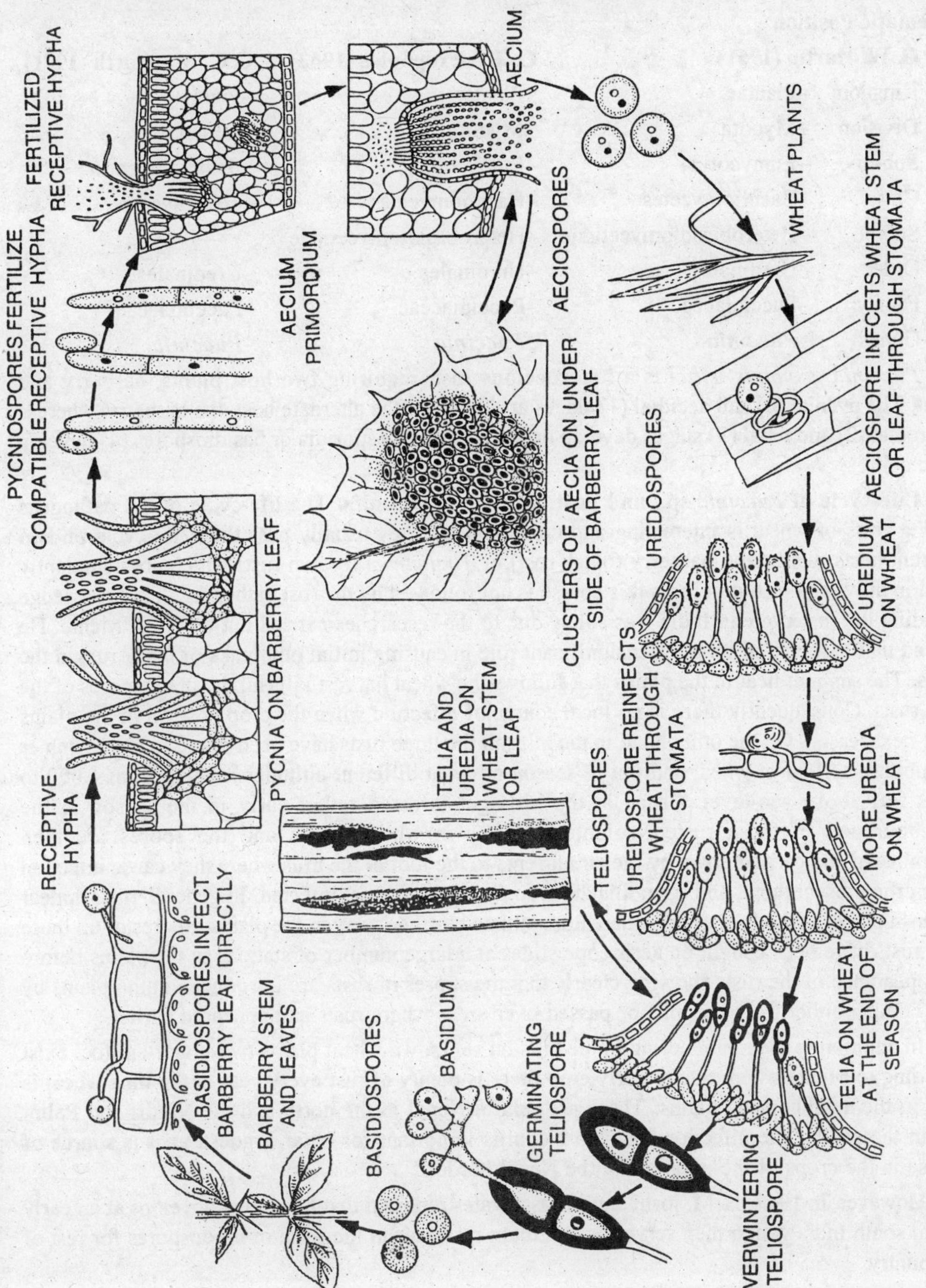

Fig. 14.22. Disease-cycle of black or stem rust of wheat caused by *Puccinia graminis tritici.*

The life-cycle of *Puccinia graminis tritici* completes on two different hosts called primary and secondary or alternate hosts and therefore, this is a heteroecious fungus. The uredial and telial stages are found on the primary host, the basidiospores are found in the soil, and the pycnial and aecidial stages are found upon the alternate host.

Systematic Position

	G. W. Martin (1961)	C. J. Alexopoulos (1962)	G. C. Ainsworth (1971)
Kingdom	–Plantae	–Plantae	–Fungi
Division	–Mycota	–Mycota	–Eumycota
Sub-div.	–Eumycotina	–Eumycotina	–Basidiomycotina
Class	–Basidiomycetes	–Basidiomycetes	–Teliomycetes
Sub-cl.	–Heterobasidiomycetidae	–Heterobasidiomycetidae	–
Order	–Uredinales	–Uredinales	–Uredinales
Family	–Pucciniaceae	–Pucciniaceae	–Pucciniaceae
Genus	–*Puccinia*	–*Puccinia*	–*Puccinia*

Puccinia graminis tritici is a **heteroecious** rust, requiring two host plants, barberry and wheat. The pycnial (0) and aecidial (1) stages are found on the alternate host, the barberry, whereas the uredial (2) and telial (3) stages develop on the wheat. The sporidia or basidiospores (4) develop in the soil.

Life-cycle of *Puccinia* sp., and rust recurrence in India. The life-cycle of the pathogens causing rusts of wheat is interesting and complicated. They usually pass their life-cycle on two different hosts, wheat and barberry (black rust) or *Thalictrum* (brown rust) which are abundantly found in hilly regions but in India this course is not followed by the rust pathogens. The knowledge regarding the rust cycle in India was solely due to the researches carried out by K. C. Mehta. He showed that barberry shrub plays insignificant role in causing initial outbreaks of black rust in the plains. The summer heat in the plains that follows the wheat harvest kills all the uredospores of the three rusts. Consequently there is no local source of infection when the crop is sown in the plains in the next season. On the other hand in the hills all the three rusts have been found to over summer on stubbles, self-sown wheat and out of season crops at different altitudes ranging from 4,000 to 8,000 feet above sea-level. Crops in the hills get infected rather early in the season in the neighbourhood of these rusted volunteer self-sown wheat plants and the spores are then disseminated and carried down by the air currents to the foot of the hills where they cause infection to the crop present there, and from this the crops in the plains get infected. In general, rusts appear earlier in places at the foot of hills than those situated farther away in the plains. Spores of all these three rusts have been caught on aeroscope slides at a large number of stations in the plains before the appearance of the rusts showing clearly that the spores of rusts are introduced in the plains by wind coming either from the hills or passed over areas where rusts had appeared earlier.

In addition to over summering in the hills on self-sown wheat plants two important foci exist according to Mehta where due to early crop there is plenty of rust every year at the time wheat is sown in the neighbouring plains. These are central Nepal in the north and the Nilgiri and Palini hills in the south. The infection focus in Nepal is important for Uttar Pradesh as it is source of disease in the crops of the plain along the Nepal border.

However, in 1947, L. M. joshi and his associates reported that black rust develops at an early date in south India which then serves as inoculum reservoir in the form of uredospores for rest of the country.

Brown rust of wheat has two foci of infection, the Himalayas in the North, and Nilgiri and Palini hills in the South. It is the most widely prevalent rust. The uredospores get transported by upper winds from Nilgiris and Palini hills and then are washed down over Central India (Madhya Pradesh) by rains. The northern infection moves slowly southwards until the two populations merge.

The air borne inoculum of brown rust as well as black rust and its deposition from southern hills (Nilgiri and Palini hills) has been traced by wind trajectories and weather satellite cloud images. From investigations of Nagarajan and Singh (1975), it appears that the Tropical cyclones have a great bearing on the deposition of spores in Central India (Madhya Pradesh) within a time span of about 70 hours. The cyclonic disturbances occurring in the Bay of Bengal/ Arbian Sea in late October or early November set a pattern of favourable winds which carry the inoculum (uredospores) of black and brown rusts from Southern foci to the target area in Madhya Pradesh along with rain clouds. It has been found that the invisible spore cloud movement can be successfully tagged with the visible cloud movement seen in satellite television cloud photographs. This method has shown that the uredospores are transported from Nilgiris by upper air wind and are deposited over Central India (Madhya Pradesh) along with rain. From these secondary foci, it spreads to other foci making a nation-wide appearance.

Yellow rust is restricted in its distribution and appearance in epidemic form only to northwestern egions. Normally the rust appears in the foot-hills and plains in December or first fortnight of January, when conditions in plains are favourable and rapid multiplication of inoculum (uredospores) takes place.

Spread of disease. The initial infection of few odd plants in any field in the plains is sufficient to spread the disease. The disease soon spreads from one field to another by the dissemination of spores by air currents. Every rust pustule contains 30,000 to 80, 000 spores and each such spore (uredospore) is capable of causing infection. They are easily carried away by wind

Table: 14.1. Comparison of the Symptoms of Three Wheat Rusts

Black rust *(Puccinia graminis tritici)*	**Yellow rust** (*P. striiformis*)	**Brown rust** (*P. recondita*)
Attacks the stem most severely, than the leaf sheaths leaf lamina and ears.	Attacks the leaf lamina most severely, stem and ears.	Attacks the leaf lamina almost exclusively, rarely the sheaths, very rarely the stems.
Uredosori large elongated running together and bursting early forming the large fragments of the epidermis in doing so. Colour dark brown, becoming darker gradually as teleutospores are formed in the same sori (pustules). Found on all green parts of the plants.	Uredosori (uredopustules) small, oval, do not usually run together, burst late and with little fringes of the epidermis. Almost always arranged in long stripes. Colour lemon yellow. Found on all green parts of the plant.	Uredosori (uredopustules) small but often larger than in yellow rust, oval or round do not usually run together, burst early with a fringe of broken epidermis around them. Always scattered the surfaces of leaf lamina. Colour bright orange when fresh. Found chiefly on upper surface of leaves.
Telia or teleutosori (teleutospustules) much elongated and black in colour. Burst rather early. Found on all green parts of the plants, but least on the leaf blades.	Telia or teleutosori (teleutopustules) like uredosori but more flattened dull black, and often run together. Donot burst through the epidermis (sub-epidermal). Arranged in rows. Found chiefly on under surface of leaves but also on all green parts of the plants.	Teleutosori (teleutopustules) often absent, when present resemble uredosori (uredopustules) but more flattened and dull black. Do not burst through the epidermis (sub-epidermal.) Found chiefly on under surface of leaves, very rare elsewhere.

leaf and other green surface of the plant. In presence of moisture, the spores germinate and cause infection. The new pustules are developed within 10-14 days of infection. Usually four or five short cycles are repeated during the season. The greater the number of these short cycles, the greater would be the damage to the crop.

The multiplication and spread of the rusts in the plains depend on the weather conditions specially the rain or heavy dew on the leaves of the wheat plants. Heavy dew for a long period or rain, moderately high temperatures and strong wind for dissemination of spores on the susceptible varieties are the important factors for rust epidemics.

Loss due to rusts. Rust pustules cover a greater part of photosynthetic area and thus hinder the plant in its food manufacturing process. When pustules burst through the epidermis, the water is lost by transpiration resulting in partial sterility, poorly filled ears and shrievelled grains. If rust becomes severe before milk stage of the crop, the losses are always great even resulting in crop failures.

Physiologic races. Stakman (1926) and his co-workers have defined physiologic specializations as "presence of entities within morphologic species, not readily distinguishable by structure, but differing from each other physiologically". These entities are—pathogenicity, host morphology, biochemical properties, cultural variability, spore germination and ecological relationships. These entities have been designated variously such as—physiologic races,biologic species, biologic forms, specialized races, parasitic strains, racial strains, physiologic forms, etc. Each physiologic race is denoted by a specific number such as—*Puccinia graminis tritici* 15, Eriksson termed these races as '*formae speciales*'. Specialization of parasitism of an extreme type is known as *biotype* and is denoted by an epithet such as 42A, 42B, etc.

Since 1930, twenty physiologic races and eight biotypes of black rust (*Puccinia graminis tritici*–11, 14, 17, 21, 24, 34, 40, 42, 72, 75, 117, 122, 184, 194, 196, 222, 295, X and Y, 15-C, 21-A, 21-A-1, 21-A-2, 34-A, 42-A, 42-B, 117-A), seventeen physiologic races and three biotypes of brown rust (*Puccinia recondita*–10, 11, 12, 16, 17, 20, 26, 61, 63, 70, 77, 106, 107, 108, 131, 162 and race D; 162-A, 162-B and 107-A), and fourteen physiologic races of yellow rust (*Puccinia striiformis*–13, 14, 19, 20, 24, 31, 38, 57, A, D, E, F, G, and H) have been identified. In recent years several new races have been identified. New races are known to arise through hybridization, mutation and introduction from neighbouring countries.

However, the barberries do not play any part in the annual recurrence of the rust in India, but they may be instrumental in creating new races by hybridization. Then it becomes necessary to eradicate the barberry bushes from pockets of infection in the hills.

Control measures. On the basis of the study of the life cycle of the rusts (*i.e.,* black, brown and yellow) and their annual recurrence in the Indian plains, a few important methods have been suggested. They are–A, clean-up campaign; B, cultural practices; C, chemical treatment and D, Resistant varieties.

A. *Clean-up-campaign.* In India, K. C. Mehta as a result of his extensive study of wheat rusts (*i.e.,* black, brown and yellow) recommended the vigorous destruction of self-sown plants and tillers which carry over the rusts in uredo-stage and which are responsible for infection of the crop in the plains. He had also recommended the suspension of the first crop (sown during April - June) in Nilgiris and Palini hills and postponement of sowing in the central Nepal to the normal period that is October.

A great doubt has been expressed on the above measures for their practicability and economic soundness. There are certain out of season and self-sown grasses (collateral hosts) which occur on the hills and are affected by the same rust as wheat and barley and they may serve as centres of foci of infection. There are places in the hills which are unaproachable where complete eradication

eradication of rusted plants is difficult. There is another factor that seems to mitigate against a thorough clean-up is the short period between the harvest of one season and the sowing of the next which is a monsoon period (July - September) during which it may not be possible to carry the eradication campaign effectively.

B. *Cultivation practices.* Certain modification in cultural practices, *viz.,* cultivation of early maturing varieties, early sowing, judicious manuring and irrigation help in minimising the incidence of rusts.

(a) *Cultivation of early maturing varieties.* Varieties of wheat which mature early should be sown so as to avoid the critical period that is milk stage of the crop from the attack of the rusts, the black rust assumes serious intensity during the month of February and if the crop has passed its milk stage at this time it will suffer negligible losses. There are certain varieties of wheat in Uttar Pradesh which come into full flowering as early as middle of January and thus are able to escape the attack of rust at their critical period. The varieties which come late into full flowering are the worst sufferers of the disease.

(b) *Early sowing.* Sowing of wheat varieties a bit earlier than normal is another good method to escape the severe rust incdidence at the milk stage of the crop. The early sown wheat flowers early and also matures early. Even the moderately late varieties when grown early would flower and mature early and would thus escape the severity of the rust at the critical stage.

The best time of wheat sowing in uttar pradesh is the later half of October. The wheat can, however, be grown profitably if sown a bit earlier. Early sowing can easily be accomplished where irrigational facilities are available. Late sowings in November and December are never recommended and often result in huge losses in yeild due to the severity of rust attcks. The practice of ealry sowing should, therefore, be adopted to save enormous losses of the valuable national food.

(c) *Proper manuring.* Manuring is an importnat factor governing the incidence of rusts. Nitrogenous manures in general are known to increase the susceptibility of the crop towards diseases by delaying maturity. The nitrogen requirement for wheat in Uttar Pradesh is about 50 lbs. per acre. Over dosages of nitrogenous manures are therefore, not only uneconomical but induce the crop being affected by rusts. The rust resistance is increased by the application of potassic manures to the crop. Phosphatic manures, *e.g.,* bone meal, superphosphate, etc., are good as a top dressing to the wheat crop.

(d) *Proper irrigation.* High humidity favours the rusts and, therefore, irrigation is an important factor which governs the rust appearance. Heavy irrigation at the time of rust apearance only favours the speedy development and spread of rusts. Irrigation therefore, should be given timely and properly. Light irrigation during January and February help in keeping the rusts under control.

C. *Chemical treatment.* The control of rusts by chemicals is not a new idea and dates back to at least 1900. Since then the efficacy of sulphur was proved in the field in Canada and it was recommended that 3 to 5 applications at the rate of 30 lbs. per acre per application would control stem rust and increase the yield from 25 to 100 per cent depending on the severity of the rust. Trials conducted at Kanpur have shown that increase in yield of wheat can be attained by four dustings of finely powdered sulphur (400 meshes per square inch) at the rate of 30 lbs. per acre at the interval of 10 days. Similar results were also obtained in Madhya Pradesh. The cost involved was, however, prohibitive and was not commensurate with the gain. The use of sulphur in the control of rusts on a wide scale, therefore, has never been recommended.

Search for chemicals that would be effective in small quantities has led to the formation of such fungicides as *Dithane, Zineb* and *Actidione.* Extensive tests in different countries indicate that rust can now be controlled economically with the help of these compounds provided that weather conditions do not become very adverse during these operations. Dithane has been used successfully and economically in the control of rusts in Japan. Four to five applications of *Nabam* and *Zinc sulphate* gave effective control of wheat rusts in Canada and the cost of the chemicals was about one-third of sulphur. *Actidione* is a promising antibiotic in the control of rusts as a systemic fungicide. It has been demonstrated that certain sulpha compounds including sodium salts of sulphadiazine, sulphapyrazine, sulphapyridine and gantrisin are effective against black rust even as post infection sprays. Certain nephthoquinones and phenols also exert a fungicidal action at low concentrations on cereal stem rust. Parzate liquid with zinc sulphate has been found to be effective in reducing rust infection appreciably under artificial conditions at I.A.R.I. New Delhi.

D. *Ressitant varieties.* The best control of wheat rusts lies in evolving resistant varieties. In India systematic work on the breeding of rust resistant wheats was started by B. P. Pal in 1935 in collaboration with K. C. Mehta and several rust tolerant varieties have by now been released for general cultivation. These varieties have done very well in different states and given much higher yield than the local varieties even under epidemic conditions.

The breeding for resistance and chemical protection must make parallel contribution in order to get the best out of improved varieties. It is much more economical and effective to use the fungicide on resistanct varieties than on local varieties. Since the annual rust recurrence in India takes place through the agency of uredospores in the hills, K. C. Mehta had emphasized the importance and urgency of breeding rust resistant varieties and gave priority to the hills which act as foci of infection.

15
CHAPTER

Sub-class–Homobasidiomycetidae

Mushrooms, Shelf Fungi, Puffballs, Bird's Nest fungi, etc.
(Alexopoulos, 1962)

G. W. Martin (1961)	**G. C. Ainsworth (1971)**
Sub-class–**HOMOBASIDIOMYCETIDAE**	Sub-class-**HOMOBASIDIOMYCETIDAE** of class–**HYMENOMYCETES** and class–**GASTEROMYCETES**

Characteristic Features.

1. This sub-class consists of mushrooms, toad stools, bracket fungi, puff balls, shelf fungi, earth stars, stink horns, bird's nest fungi etc.

2. The mycelium is found in the hidden substrata in tangled masses. The clamp connections and rhizomorphs are commonly present in the mycelium. The mycelium is dikaryotic.

3. The fructifications or sporophores are seen above the ground.

4. The basidia are unseptate and arranged in hymenia. Each basidium bears four basidiospores at terminal end on sterigmata. The basidia are palisade-like and intermingled with paraphyses.

5. This sub-class is divided into two groups or series. (*i*) Hymenomycetes and (*ii*) Gastromycetes.

6. According to Gaumann (1952), Wolf and Wolf (1947) and Alexopoulos (1956), the group Hymenomycetes consists of single order Agaricales.

Series-**Hymenomycetes**

The Basidiocarp is present. It varies from an arachnoid subiculum to a very complex sporocarp. The hymenium is present. It becomes exposed before the spores are mature. There is a single order-Agaricales.

Order **AGARICALES** (275 genera; 7,000 species)

1. The order consists of mushrooms, toad stools, puff-balls, etc.

2. The sporophores (fructifications) are macroscopic.

3. The basidia are arranged in hymenia intermingled with paraphyses. The basidiospores are disseminated with a jerk and alternately.

4. The hymenial layers may be protected in gills or exposed.

5. According to Wolf and Wolf (1947), there are six families in this order. (*i*) Exobasidiaceae, (*ii*) Theleophoraceae, (*iii*) Clavariaceae, (*iv*) Hydnaceae (*v*) Polyporaceae and (*vi*) Agaricaceae.

6. There are about 7,000 species in this order.

Family-Agaricaceae

They include the majority of mushrooms. They bear their basidia on gills or lamellae. The gills are neither waxy and nor may easily be separated from context of the pileus. The gills are narrow in section. The basidiocarp is fleshy. The hymenium lining both the sides of gills.

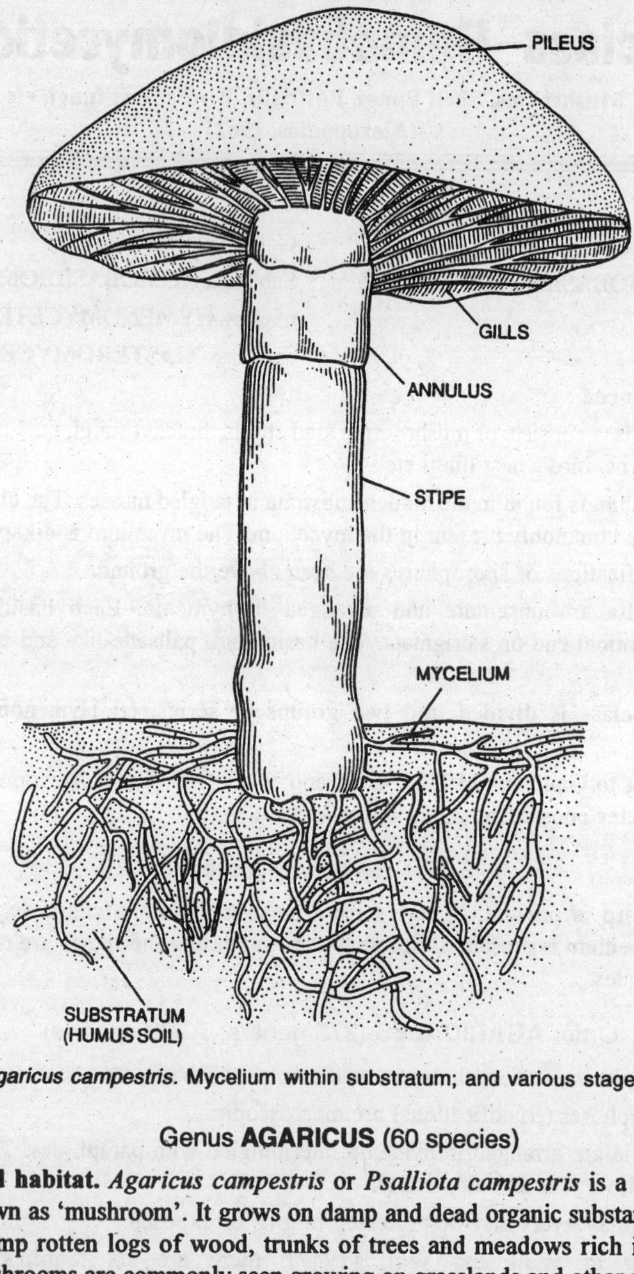

Fig. 15.1. *Agaricus campestris.* Mycelium within substratum; and various stages of fructification.

Genus **AGARICUS** (60 species)

Habit and habitat. *Agaricus campestris* or *Psalliota campestris* is a saprophytic fungus, commonly known as 'mushroom'. It grows on damp and dead organic substances such as humus, horse dung, damp rotten logs of wood, trunks of trees and meadows rich in decaying organic matter. The mushrooms are commonly seen growing on grasslands and other places rich in humus soil during rainy season. Some mushrooms are edible but some of them are deadly poisonous. So one must be very cautious about their selection for eating purposes. The common edible mushroom *Agaricus campestris* is cultivated in South India and other places. The group in which *Agaricus* is found includes many other forms commonly knows as mushroom and toadstools.

As regards its somatic structure the mushroom can be subdivided into two main parts, one part known as vegetative mycelium, is found living within the soil or humus and the other part, fructification or proper mushroom is aerial and above the soil. This portion is edible and bears the gills and spores.

Vegetative mycelium. The vegetative mycelium is found under the soil in thick, tangled wooly masses. Such masses are microscopic and very commonly known as **'spawns'.** On the maturation of such mass, the fructification commonly a 'mushroom' is developed on it. The hyphae are septate, hyaline, containing oil globules, vacuoles and thin protoplasm. The cells of the hyphae at first are uninucleate, but very soon dikaryotic stage is attained. The two adjacent hyphae of different strains, *i.e.*, + and – come in contact of each other fuse, and a nucleus from one hypha passes into the adjacent cell of the other hypha resulting in dikaryotization. Now the dikaryotic mycelium grows by the formation of clamp connections. Some of the hyphae are interwoven into thick rope-like strands known as **rhizomorphs,** but these strands or rhizomorphs and other separate hyphae can easily be broken if the soil is disturbed. Certain hyphae of rhizomorphs are of greater diameter than others, and in *Agaricus,* it is observed by Hein (1930) that the large hyphae can be produced by the lateral fusion of smaller ones. According to Jones and Buller (1949), the mycelium of mushrooms may remain alive for more than 400 years and every year the crop is produced.

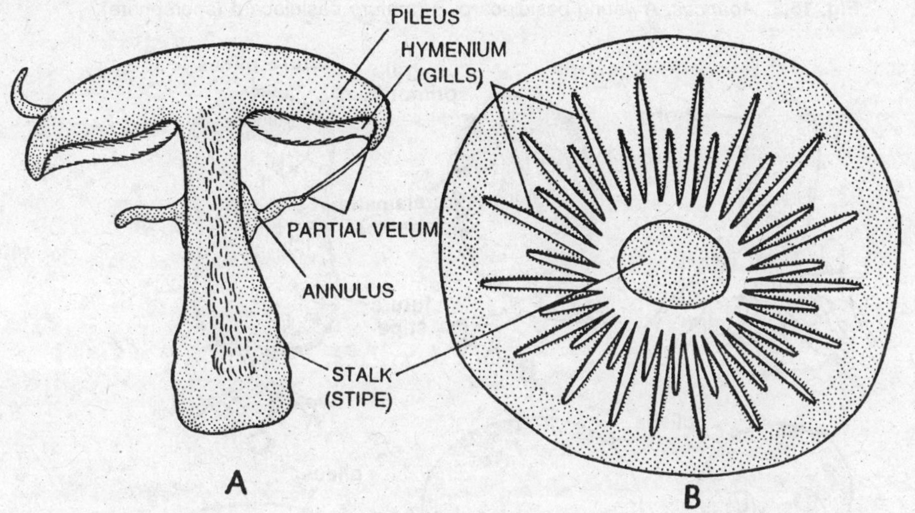

Fig. 15.2. *Agaricus* sp. A, L.S. of stipe and pileus; B, T.S. through pileus and stipe.

The fructification. The aerial portion of the fungus is fructification or fruiting body, and commonly known as mushroom. In the beginning, this is a very small tiny knot of compact hyphae developing from the rhizomorph and the subterranean mycelium. When young, the fructification is small, spherical or pear-shaped, and this stage is known as **'button stage'.** A constriction develops on this button-like structure and gradually it develops into a mushroom proper or sporophore. It is a conspicuous and well defined structure of six to nine centimetres when fully developed. From the study point of view the mushroom can be divided into two parts, (1) the stalk or stipe, which is a fleshy pinkish white portion, usually broader at the base, (2) the other part is somewhat more complex and termed as cap or pileus. The description is as follows.

The upper circular, umbrella-like expanded portion of mushroom is termed as cap or pileus. When young, it is spherical in shape and more or less button-like and is completely enclosed by a thin whitish membrane known as **velum.** As the pileus of mushroom grows in size; the velum ruptures and remnants of it remain surrounding the stipe are termed **annulus.** Ultimately the pileus develops in an umbrella-like expanded portion on the terminal end of the stipe. Under the pileus a

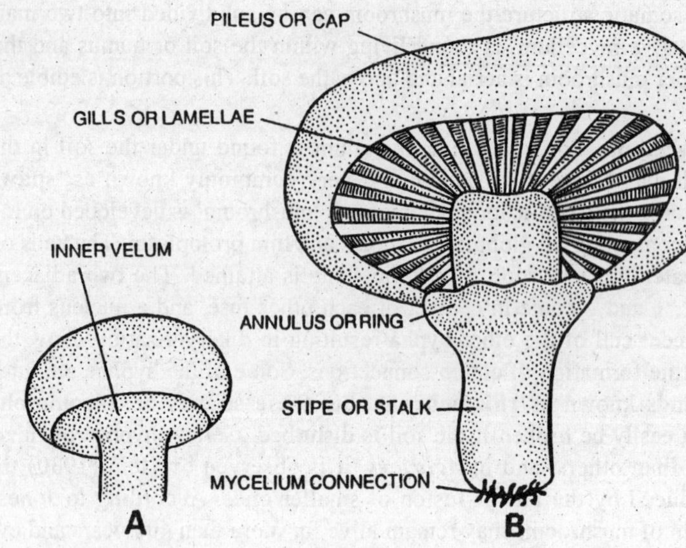

Fig. 15.3. *Agaricus*. A young basidiocarp; B, mature basidiocarp (sporophore).

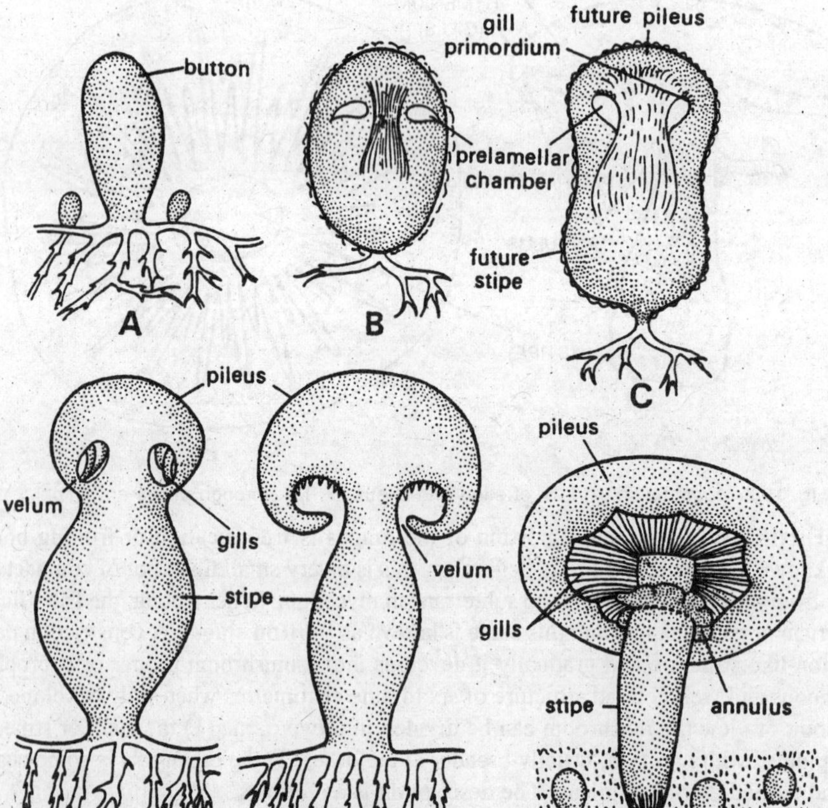

Fig. 15.4. *Agaricus campestris*. Development of fruiting body (basidiocarp) A-F.

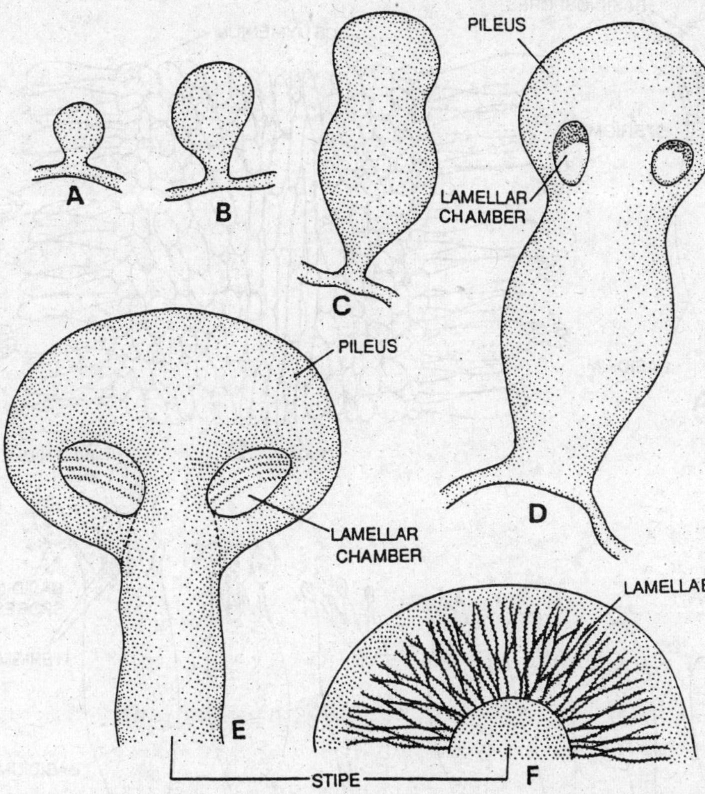

Fig. 15.5. *Agaricus campestris.* A-D, successive stages in the development of sporophore; E, L.S. of sporophore; F, V.S. of pileus Showing gills.

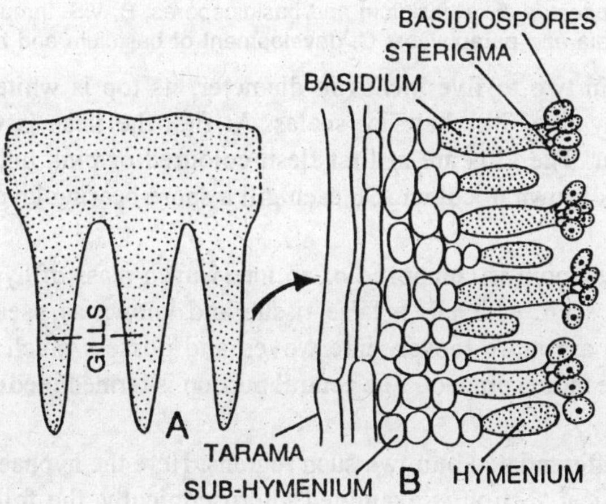

Fig. 15.6. *Agaricus* sp. Sporophore. A, tangential vertical section of pileus; B, details of a part of gill.

gill cavity is present and in this cavity a large number of thin, vertical, plate-like structures which run out towards the edge of the pileus like the radiating bars of a wheel are termed **gills** or **lamellae.** They vary in number from 300 to 600 in each pileus. On both the surfaces of each gill thousands of spores are produced.

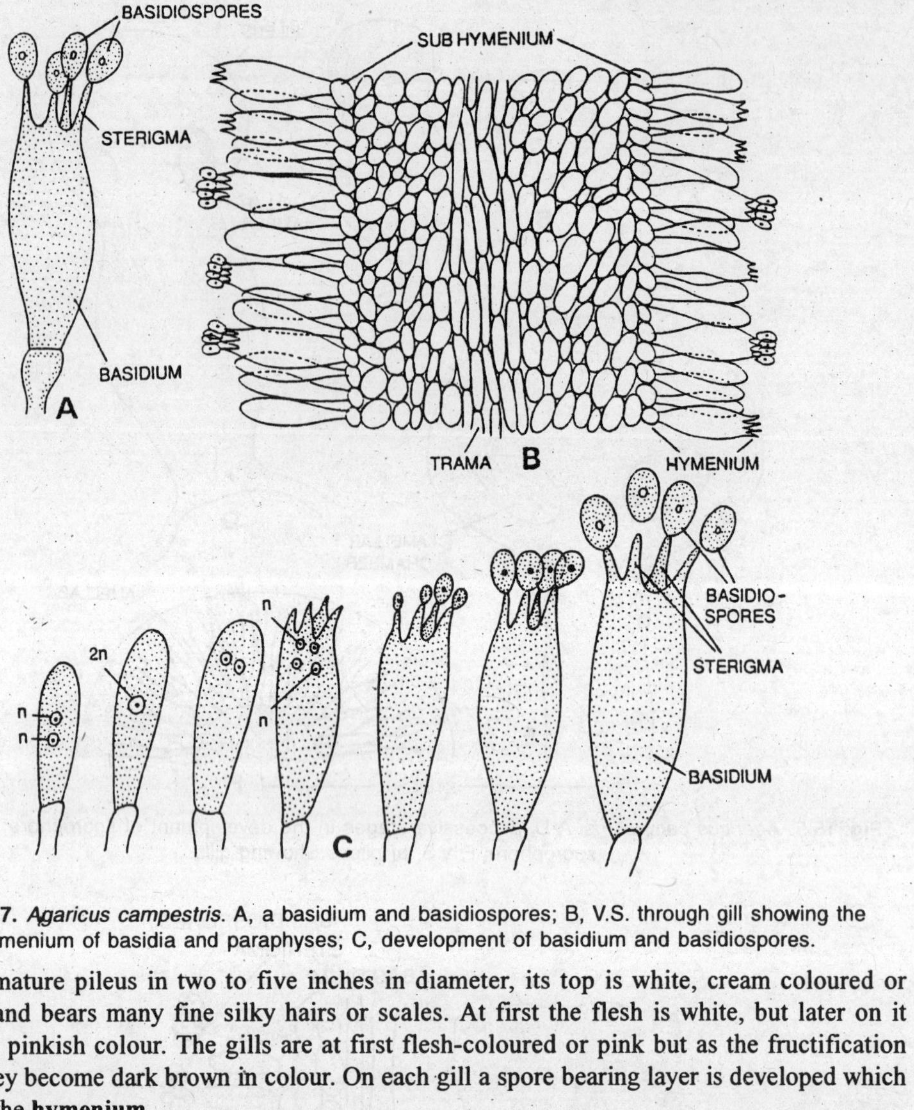

Fig. 15.7. *Agaricus campestris.* A, a basidium and basidiospores; B, V.S. through gill showing the hymenium of basidia and paraphyses; C, development of basidium and basidiospores.

The mature pileus in two to five inches in diameter, its top is white, cream coloured or brownish and bears many fine silky hairs or scales. At first the flesh is white, but later on it changes in pinkish colour. The gills are at first flesh-coloured or pink but as the fructification matures they become dark brown in colour. On each gill a spore bearing layer is developed which is termed, the **hymenium.**

The stipe of the mushroom is composed of an interwoven mass of hyphae, and in sections a tissue-like structure is seen, which is a false tissue and known as pseudoparenchyma. The hyphae of central region are more loosely interwoven and possess much air spaces while the hyphae of outer region are more compact. The central portion is termed **medulla** and the outer one **cortex.**

The pileus is also differentiated into two such regions. Here the hyphae of the gills are more compact. If a gill in vertical section is examined microscopically, the following three distinct portions will be seen: trama, sub-hymenium and hymenium. The central portion of the gill consisting of an interwoven mass of slender and long hyphae resulting in the formation of pseudoparenchyma is termed **trama.** The hyphae of the trama bend outwards on both the surfaces of the gill and terminate in a layer of small, rounded cells, called the **sub-hymenium.** Outside to it, on both the surface of the gill, a layer of clavate, palisade-like cells is developed, called the **hymenium.** The large cells of the hymenial layer are termed basidia, the shorter are known as paraphyses. At the terminal ends of the basidia, two or four sterigmata are developed. A single basidiospore develops on each sterigma. The basidiospores are haploid and uninucleate. Buller (1922), described the above mentioned paraphyses as immature basidia.

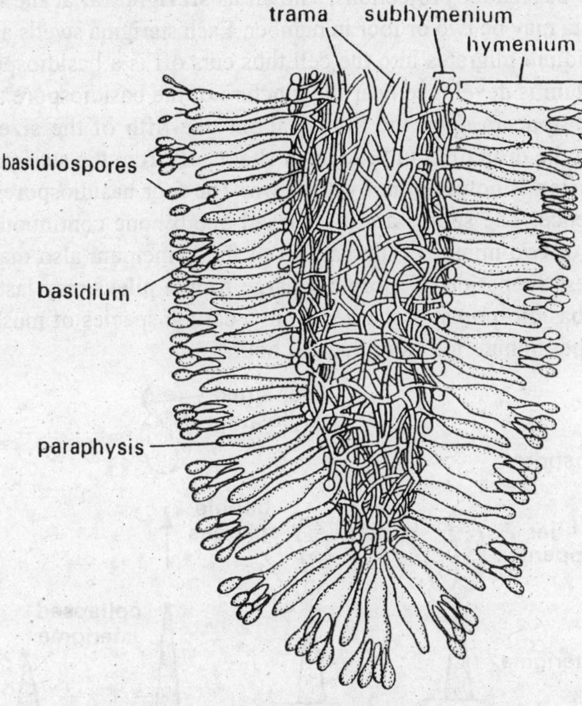

Fig. 15.8. *Agraicus.* L.S. through a gill.

Development of basidium and basidiospore. A young basidium is at first binucelate. As it increases in size there is a fusion of two haploid (n) nuclei and simultaneously the zygote nucleus

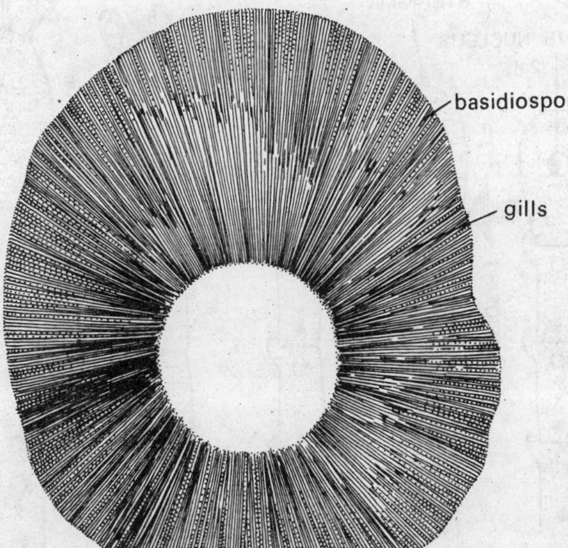

Fig. 15.9. *Agaricus campestris.* Spore print on paper from a mature pileus.

(2*n*) migrates towards the apex of the basidium. The zygote nucleus (2*n*) soon divides meiotically into four daughter nuclei, each with *n* chromosomes. After completion of the nuclear divisions there is a development of slender projections, known as **sterigmata,** at the terminal end of the basidium. The sterigmata may be two or four in number. Each sterigma swells at the distal end, and a nucleus from the basidium migrates into the cell thus cuts off is a basidiospore. A small lateral outgrowth known as hilum is developed near the juncture of the basidiospore and the sterigma. A small droplet of liquid appears upon the hilum, about one-fifth of the size of spore, and the basidiospore along with the drop of liquid suddenly shoots off from the sterigma. The mechanism of this explosive abscission is however, not yet known. The four basidiospores of a basidium are shed in a regular sequence. The spores are discharged one by one continuously at the intervals of few seconds or one or two minutes. The basidia of the hymenium also mature in succession. According to the species, the period of spore discharge from a pileus may last for hours, days or weeks. The number of basidiospores discharged from a certain species of mushroom is estimated at the rate of a million per minute for more than 50 hours.

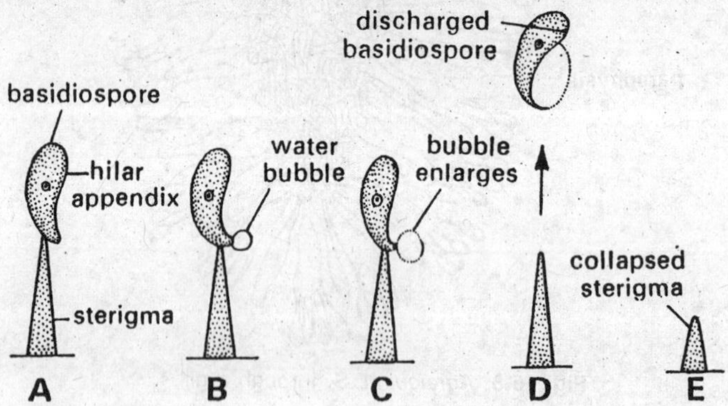

Fig. 15.10. *Agaricus.* A-E, water drop to release basidiospore.

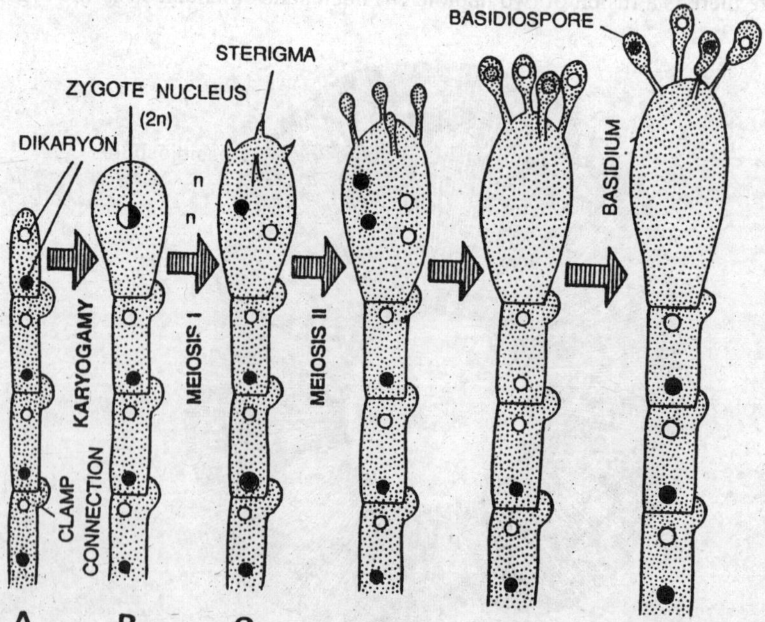

Fig. 15.11. *Agaricus* sp. A-F, diagrammatic representation of basidial and basidiospore development, without crossing over in a heterothallic species. Black and white nuclei represent two compatible mating types.

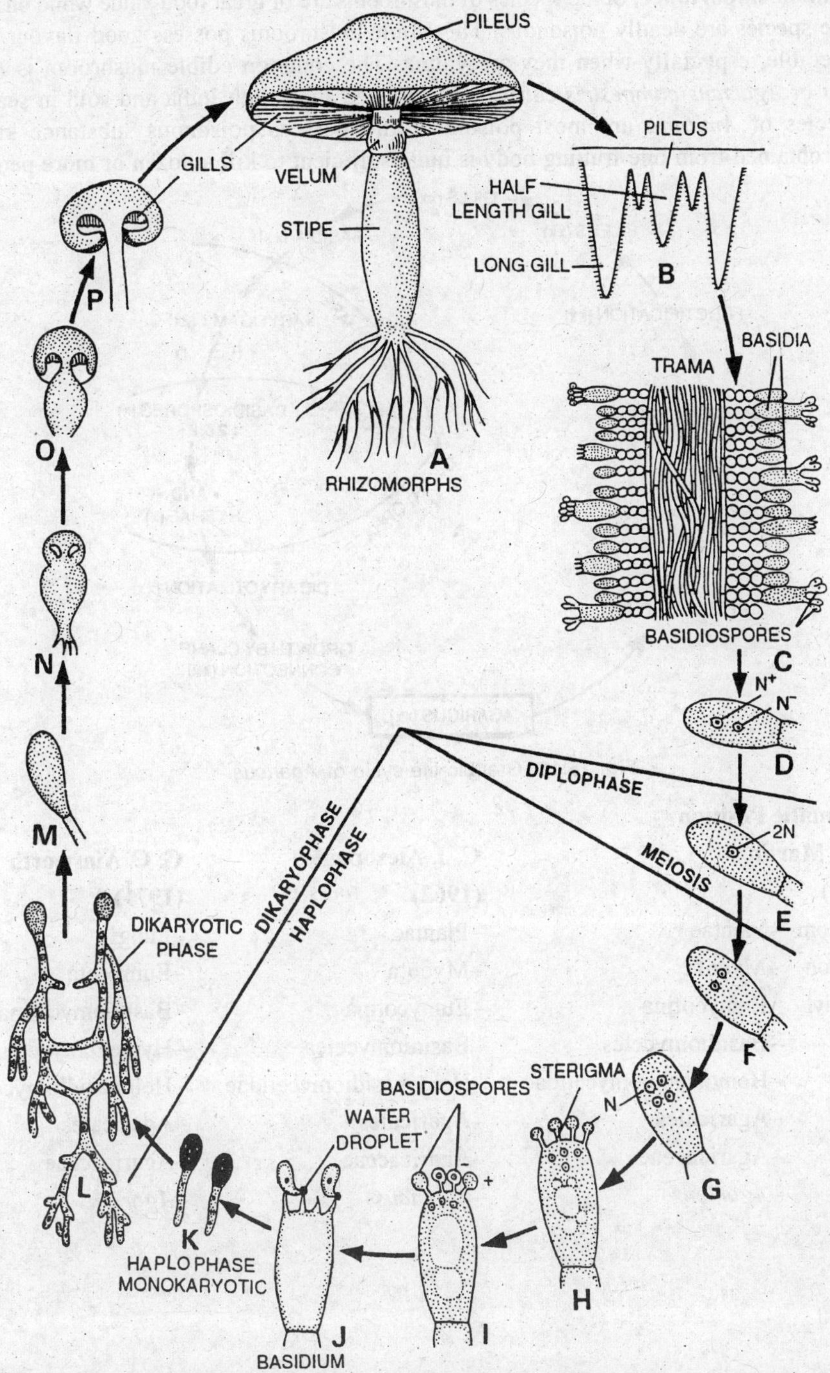

Fig. 15.12. Diagrammatic life-cycle of *Agaricus*, A-P.

As soon as the basidiospores are discharged from basidium they germinate on the advent of favourable conditions giving rise to primary monokaryotic, branched, hyaline and septate mycelium. Later on the hyphae of two different strains, *i. e.*, + and – anastamose with each other and the dikaryotic mycelium develops, which gives rise to spawn and fructification.

Economic importance. Some species of mushrooms are of great food value while on the other hand some species are deadly poisonous. The edible mushrooms possess good flavour. All puff balls are edible, especially when they are young. The common edible mushroom is *Psalliota campestris* or *Agaricus campestris* cultivated in France and South India and sold in sealed tins. Some species of *Amanita* are most poisonous and possess poisonous substance known as muscarine obtained from one fruiting body is quite sufficient to kill a dozen or more people.

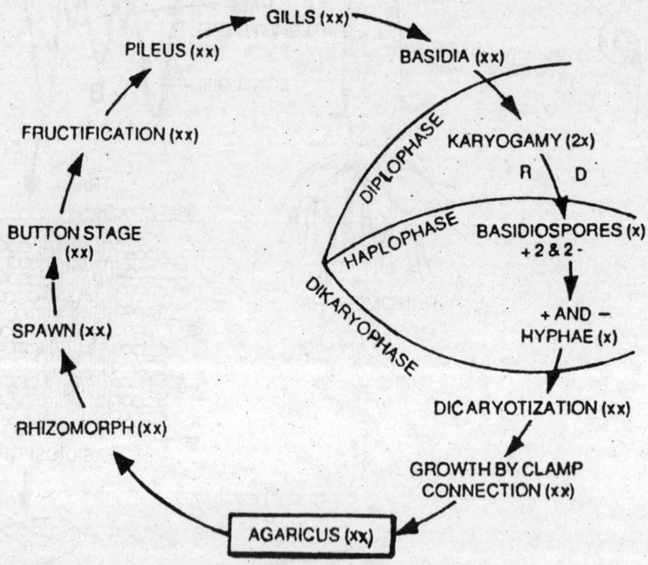

Fig. 15.13. Graphic life-cycle of *Agaricus*.

Systematic Position

	G. W. Martin (1961)	**C. J. Alexopoulos (1962)**	**G. C. Ainsworth (1971)**
Kingdom	–Plantae	–Plantae	–Fungi
Division	–Mycota	–Mycota	–Eumycota
Sub-div.	–Eumycotina	–Eumycotina	–Basidiomycotina
Class	–Basidiomycetes	–Basidiomycetes	–Hymenomycetes
Sub-cl.	–Homobasidiomycetidae	–Homobasidiomycetidae	–Holobasidiomycetidae
Order	–Agaricales	–Agaricales	–Agaricales
Family	–Agaricaceae	–Agaricaceae	–Agaricaceae
Genus	–*Agaricus*	–*Agaricus*	–*Agaricus*

16
CHAPTER

Sub-div. Deuteromycotina Class-Deuteromycetes

Kingdom–**PLANTAE**
Division–**MYCOTA**
Sub-division–**EUMYCOTINA**
From class–**DEUTEROMYCETES**
The fungi Imperfecti
(1425 genera; 15,000 species)
(Alexopoulos, 1962)

G. W. Martin (1961)
Kingdom–**PLANTAE**
Division–**MYCOTA**
Sub-division–**EUMYCOTINA**
Class–**DEUTEROMYCETES**

G. C. Ainsworth (1971)
Kingdom–**FUNGI**
Division–**EUMYCOTA**
Sub–div.–**DEUTEROMYCOTINA**
Classes – **1. BLASTOMYCETES?**
(Yeasts, etc.)
2. HYPHOMYCETES
3. COELOMYCETES

The members of this class lack perfect or sexual stages. Leaving aside the sexual structures, the fungi resemble to the members of Ascomycetes and Basidiomycetes. Majority of these fungi resemble to the members of Ascomycetes. The classification of the members of this class is based upon the structure of the conidia, conidiophores, mycelium and the formation of conidia. The perfect stages of many species have ben reported and they have been shifted to their class and genera. For example, the perfect stage of *Colletotrichum falcatum* is *Physalospora tucumanensis* of Ascomycetes. The fungi of this class cause many important plant diseases which cause sufficient damage to the crops. It is not at all necessary, that every member of this class possesses perfect stage. Majority of cases lack perfect stages, and are true representatives of this class.

Vegetative mycelium. The mycelium is endo or ectophytic, inter or intracellular, branched, hyaline or brown, septate and geniculate.

Asexual reproduction. It takes place by means of (a) conidia, (b) chlamydospores. The conidia are commonly found on conidiophores. The chlamydospores are perennating bodies and intercalary in origin.

Sexual reproduction. This is completely lacking.

Classification. This class is subdivided into four orders. 1. Sphaeropsidales; 2. Melanconiales; 3. Moniliales and 4. Mycelia sterilia.

Sphaeropsidales. The conidia are found within the flask-like pycnia. The pycnia are spherical hollow sporocarps. The conidia are borne on stalks, the conidiophores; 600 genera and 5,500 species.

Melanconiales. The conidia are found within acervuli. The acervuli may be regarded as pycnidia without their walls. It consists of an erumpent, cushion-like mass of hyphae possessing conidiophores, conidia, and sometimes setae are also present.

Moniliales. The conidia and conidiophores are not enclosed either in pycnidia or in acervuli.

Mycelia sterilia. The conidia are not formed. They produce sclerotia, rhizomorphs and various other forms of mycelia without spores.

Class Deuteromycetes

Order—	Moniliales				Melanconiales
Families—	Dematiaceae			Tuberculariaceae	Melanconiaceae
Genera—	*Alternaria*	*Cercospora*	*Helmintho-sporium*	*Fusarium*	*Colletotrichum*
Disease—	Early blight of potato	Tikka disease of groundnut	Stripe disease of barley	Wilt of arhar	Red rot of sugarcane
	A. solani	*C. personata*	*H. gramineum*	*F. udum*	*C. falcatum*

Order MELANCONIALES (105 genera; 1000 species)

Characteristic Features

1. There are about 105 genera and 1,000 species.
2. The mycelium is endophytic, branched, septate, brown or hyaline, geniculate, inter and intracellular.
3. The conidia and conidiophores are found in acervulus without peridium.
4. The acervuli are sub-epidermal and covered by the fringe of host epidermis.
5. The conidiophores are simple, hyaline or black.
6. The chlamydospores have also been reported.
7. The order consists of a single family, the Melanconiaceae.

Family–Melanconiaceae

Genus COLLETOTRICHUM (11 species)

(*Colletotrichum falcatum* Went)

Occurrence and symptoms. This fungus causes a disease called the 'red rot of sugarcane'. The symptoms of this disease appear on all parts of the plant, except roots. Particularly these symptoms appear on the stem and midrib of the leaves. The infected stem becomes somewhat greyish in colour and shrinks at nodes. The leaves become chlorotic, wither and droop downwards. The whole stem becomes rot. If, one examines the longitudinally cut stem, the long red stripes are quite visible, which are cut by transverse white patches, this is a characteristic feature of the disease. In the depressions, formed on the nodes of the infected stem, small black dots **acervuli** are seen. The red patches are developed on the midrib of the leaf. The straw coloured lesions develop in these red patches. The small black dots, called acervuli are found in these lesions. The presence of acervuli is also a characteristic feature of the disease. A foul odour emits from the infected plants, in the fields. The sucrose is converted in alcohol, because of this fungus.

Mycelium. It is endophytic and found within the host tissue. It is inter and intracellular. It consists of slender hyphae, profusely branched, hyaline, septate and with oil globules. The presence of oil globules in the mycelium is the peculiar feature of this fungus.

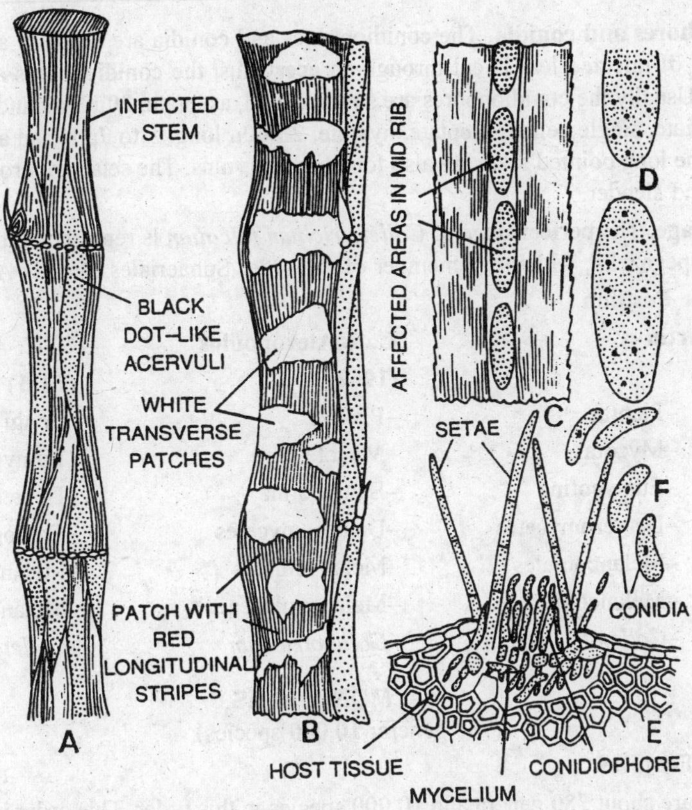

Fig. 16.1. *Collectotrichum falcatum.* Red rot of sugarcane. A, infected stem showing black dot-like acervuli; B, L. S. of infected stem showing red longitudinal stripes and white transverse patches; C, infected leaf showing black dot-like acervuli on straw coloured mid-rib patches; D, magnified straw coloured mid-rib patches acervuli; E, section of infected leaf through acervulus showing mycelium, conidiophores, conidia and setae.

Reproduction. The reproduction takes place by asexually developed conidia on conidiophores.

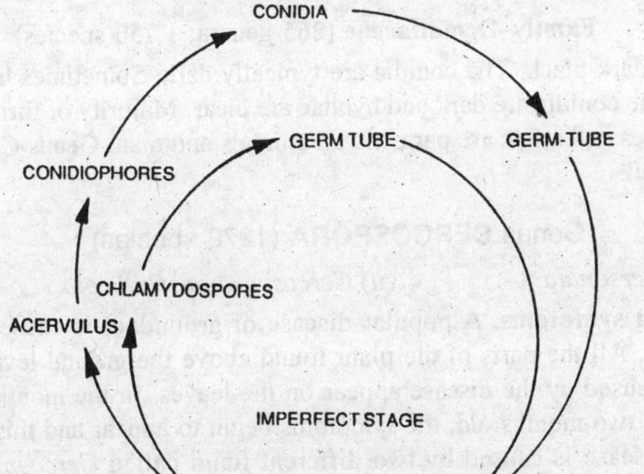

Fig. 16.2. Graphic life cycle of *Colletotrichum.*

Conidiophores and conidia. The conidiophores and conidia are found in acervuli. When a vertical section of infected leaf is cut through an acervulus, the conidiophores and conidia may easily be seen. Usually the conidiophores are single celled, aseptate, 20μ long and 8μ broad. Each conidium is falcate, single celled, aseptate, hyaline, 20-60μ long, 5 to 7μ broad and with a single oil droplet. Some long pointed setae are also found in acervulus. The setae are brown, septate, 100 to 200μ long and slender.

Perfect stage. The perfect stage of *Colletotrichum falcatum* is represented by *Physalospora tucumanensis* Spegazzini, which is a member of the order Sphaeriales of class Ascomycetes.

Systematic Position

	G. W. Martin (1961)	**C. J. Alexopoulos (1962)**	**G. C. Ainsworth (1971)**
Kingdom	–Plantae	–Plantae	–Fungi
Division	–Mycota	–Mycota	–Eumycota
Sub-div.	–Eumycotina	–Eumycotina	–Deuteromycotina
Class	–Deuteromycetes	–Deuteromycetes	–Coelomycetes
Order	–Melanconiales	–Melanconiales	–Melanconiales
Family	–Melanconiaceae	–Melanconiaceae	–Melanconiaceae
Genus	–*Colletotrichum*	–*Colletotrichum*	–*Colletotrichum*

Order **MONILIALES**

(780 genera; 10,000 species)

Characteristic Features

1. There are about 780 genera and 10,000 species in this order. This order is also known as Hyphomycetes.
2. The mycelium is branched, septate, geniculte, hyaline or brown, inter and intracellular.
3. The conidia develop on the apices and branches of conidiophores. The conidiophores are small, aerial, branched and sometimes arranged in stromatic structures.
4. The important families of this order are Dematiaceae, Moniliaceae, Stilbellaceae and Tuberculariaceae.

Family–Dematiaceae (265 genera; 1,750 species)

The hyphae are dark black. The conidia are typically dark. Sometimes hyphae are dark and conidia are clear, or the conidia are dark and hyphae are clear. Majority of forms are saprobic, but some are plant parasites and a few are parasitic on animals and man. Genus *Cercospora* has been discussed here in detail.

Genus **CERCOSPORA** (1270 species)

(*i*) *Cercospora personata* (*ii*) *Cercospora arachidicola*

Occurrence and symptoms. A popular disease of groundnut is caused by two different species of this fungus. All the parts of the plant found above the ground level are infected, but especially the spots caused by the disease appear on the leaves. In the month of July, when the plants become at least two months old, the symptoms begin to appear and this process continues upto maturity. The disease is caused by two different fungi called *Cercospora personata* and *Cercospora arachidicola* respectively. It is very clear that the symptoms caused by these two different fungi are different in nature and appearance. In our country, usually the disease is caused by *Cercospora personata.* The symptoms developed by *Cercospora arachidicola* are rarely seen.

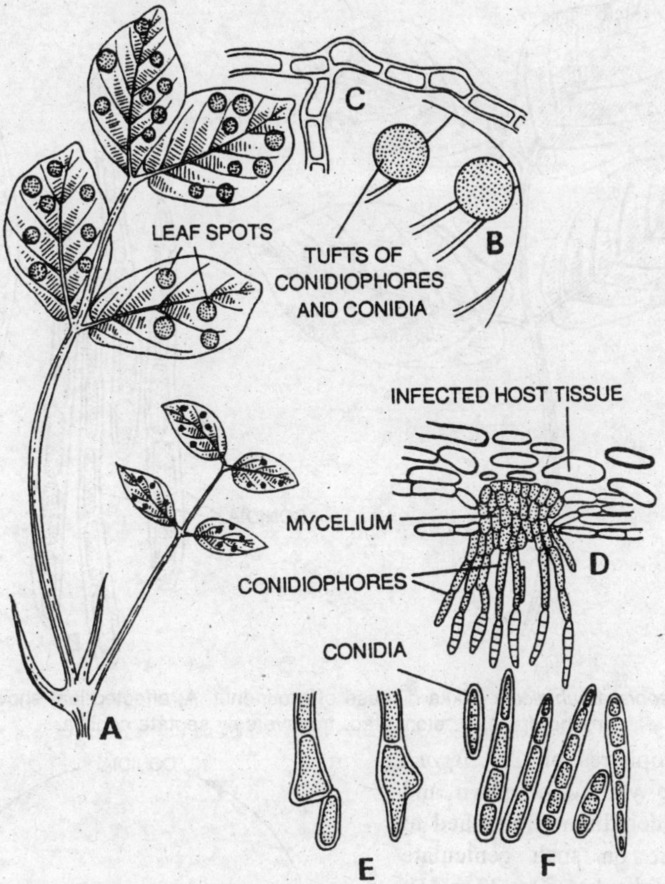

Fig. 16.3. *Cercospora personata.* Tikka disease of groundnut. A, infected leaf showing characteristic leaf spots; B, round spot surrounded by halo, tufts of conidiophores denoted by dots; C, geniculate mycelium; D, V.S. of infected leaf showing mycelium, conidiophores and conidia; E, apices of conidiophores; F, conidia.

The spots caused by *Cercospora personata* on the leaves are rounded and 1 to 6 mm. in diameter. The spots are dark brown or black and found on both the surfaces of the leaf. Afterwards, on maturity the yellow halo develops around each such spot. These yellow halos are confined to the upper surface of the leaf.

The spots caused by *Cercospora arachidicola* on the leaf are irregular and comparatively bigger in size. Each spot is 4 to 10 mm., in diameter. The yellow halos around the spots are found from the very beginning. The inconspicuous halos are found even on lower surface of the leaf.

Mycelium. The mycelium of *Cercospora personata* is intercellular, brown, septate, branched and slender. To absorb food material from the host tissue, the branched haustoria are developed.

The mycelium of *Cercospora arachidicola* is inter and intracellular on the host tissue. It is without any haustoria. It is branched, elongated, slender, septate, hyaline or somewhat brownish in colour.

Reproduction. The reproduction takes place by asexual method.

Conidiophores and conidia. The conidiophores of *Cercospora personata* are 24 to 54μ long and 5 to 8μ broad. They are aseptate of 1 to 2 times septate. The conidiophores are geniculate (knee like). The conidia are developed on such geniculate structures. The conidia are situated on the terminal ends of the conidiophores. Each conidiophore bears a single conidium. The conidiophores are arranged in stroma. The conidia are cylindrical, obclavate, 18 to 60μ long and 5 to 11μ broad. Each conidium is 1 to 7 times transversely septate.

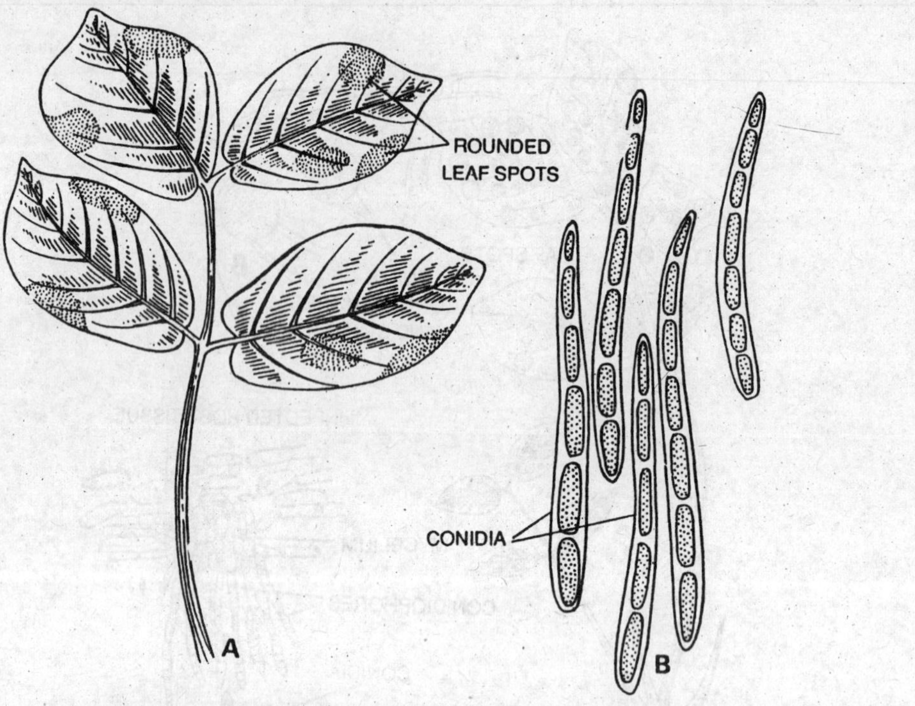

Fig. 16.4. *Cercospora arachidicola.* Tikka disease of groundnut. A, affected leaf showing characteristic round spots; B, elongated, transversely septate conidia.

The conidiophores of *Cercospora arachidicola* are yellowish brown and geniculate. The conidium is attached to the conidiophore on such geniculate structure. The conidiophores are 22 to 41μ long, 3 to 5μ broad, aseptate. The conidia are yellowish in colour, obclavate, 31 to 108μ long 3 to 6μ broad and curved. Each conidium is, 3 to 12 times septate.

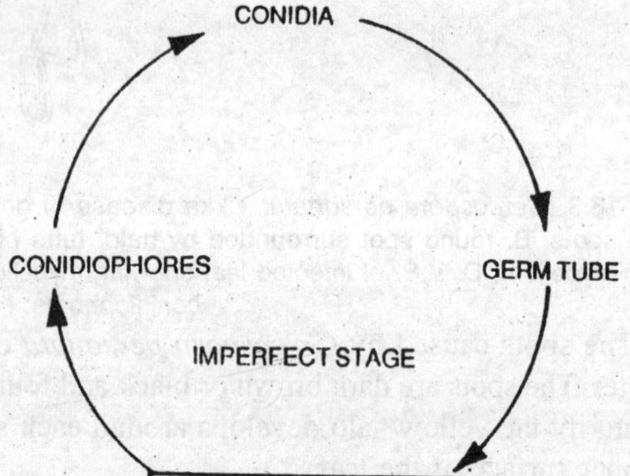

Fig. 16.5. Graphic life-cycle of *Cercospora.*

Perfect stages, (i) The perfect stage of *Cercospora personata* is *Mycosphaerella berkeleyji* W.A. Jenkins.

(ii) The perfect stage of *Cercospora arachidiola* is *Mycosphaerella arachidicola* W. A. Jenkins.

Systematic Position

	G. W. Martin (1961)	**C. J. Alxopoulos (1962)**	**G. C. Ainsworth (1971)**
Kingdom	–Plantae	–Plantae	–Fungi
Division	–Mycota	–Mycota	–Eumycota
Sub-div.	–Eumycotina	–Eumycotina	–Deuteromycotina
Class	–Deuteromycetes	–Deuteromycetes	–Hyphomycetes
Order	–Moniliales	–Moniliales	–Moniliales
Family	–Dematiaceae	–Dematiaceae	–Dematiaceae
Genus	–*Cercospora*	–*Cercospora*	–*Cercospora*

17 CHAPTER

Fungi: General Topics

ECONOMIC IMPORTANCE OF FUNGI

The fungi are related to both the useful and harmful activities to which the man is directly concerned. Several fungi are directly used as foodstuffs. Many mushrooms are edible and served as delicious dishes. The fungi, like yeasts are responsible for fermentation. The baking industry is dependent on the enormous budding of the yeast cells. Contrary to the useful activities the fungi act as great enemy to the human beings. Some of the mushroom-like fungi are highly poisonous and known as toadstools. If somehow, or other they are eaten as mushrooms, the death is sure. Many saprophytic fungi cause the sufficient damage and decay to the foodstuffs. Several fungi attack the paper wood pulp and give a death blow to the paper industry. Some fungi directly attack the book paper and cause sufficient damage. Many fungi cause the sufficient damage to the woollen and cotton textiles. Some fungi are the greatest enemy of the timbers, and cause the decay of the same. As we know, many of the parasitic fungi are responsible for hundreds of diseases of economic plants. Several fungi cause severe human and animal diseases. Conclusively, this can be said very safely that the fungi are our friends on one side and the greatest foes on the other side.

Useful Aspects

Direct utilization of fungi as food. Many Agaricales and Helvellales are directly used as food. There is a non-poisonous edible toadstool, *i.e., Coprinus* sp. found in lawns in the rainy season. *Agaricus campestris* is edible mushroom and cultivated for its fructifications. The fruiting bodies are quite fleshy and eaten directly as vegetable or with rice as **'pulao'**. These mushrooms are being successfully cultivated in South India. *Morchella esculenta* is another important edible fungus. It is found in Kashmir, Himachal and Punjab plains. Its local name is **'guchi'** and sold as costly as rupees two thousand or more per kilogram. *Torulopsis utilis*, is used for the large-scale production of yeast for food purposes. *Saccharomyces cerevisiae* is used in bread making industry.

Used as manures. The fungi with bacteria take part in the decay of organic plant and animal waste which is very essential for removing debris from our environments, for supplying carbon dioxide necessary for photosynthesis by green plants, and for providing humus, an important soil constituent for healthy plant growth.

Processing of food. A few species of *Penicillium* are being used in processing of food. *Penicillium camemberti* involved in ripening of Camembert cheese and *P. roqueforti* in ripening of Roquefort cheese. Danish blue cheese and the Italian Gorgonzola are also ripened with *Penicillium*. In Java, *Aspergillus wentii* is employed in processing soyabeans, because of its ability to loosen the hard tissues of the bean.

Production of antibiotics. *Penicillium* is best known to the non-botanist because it is the source from which the antibiotic penicillin is extracted. Penicillin was first discovered in *Penicillium notatum* Westling and for a time this was the species from which penicillin was extracted. Later investigation has shown *P. chrysogenum* Thom to be better for this purpose, and irradiation of it with X-rays and ultraviolet light has induced mutants with an even higher content of penicillin. In India at Pimpri and Rishikesh there are big factories of antibiotics.

Beer and wine. Strains of *Saccharomyces cerevisiae* are commonly used in the manufacture of beer. **Taette** is an alcoholic beverage prepared from milk. Yeasts cause the characteristic changes in flavour. **Sake** is widely used alcoholic beverage of Japan. It is a yellow rice wine containing 14 to 24 percent of alcohol. An alcoholic fermentation ensues in which several yeasts may be active. *Saccharomyces sake, S. tokyo* and *S. yeddo* are some of the yeasts characteristic of sake. In Japan *Aspergillus oryzae* is used to make sake, and to manufacture various fermented foods.

Preparation of medicines. The famous drug, ergotine, which has long been used as a drug for obsteric purposes to induce uterine contractions in cases of delayed childbirth; ergot is obtained from *Claviceps purpurea,* the causal organism of plant disease, ergot of rye. This is found in Nilgiris and South India. The other fungi, *Ashbya gosypii* and *Eremothecium ashbyii* are used in the synthesis of vitamin B-riboflavin. *Saccharomyces cerevisiae,* are employed in the synthesis of yeast tablets, rich in vitamins. *Saccharomyces cerevisiae*, are used in alcoholic fermentation.

Preparation of various acids. *Aspergillus niger* is however, employed in the production of gallic acid. The same fungus has also been employed in the production of citric and gluconic acids. Molliard (1922), found that gluconic acid was synthesized along with citric acid and oxalic acids by *Aspergillus niger*. *Aspergillus itaconicus* is used in the synthesis of itaconic acid. *Aspergillus terreus* also produces itaconic acid along with fumaric, succinic and oxalic acids. Kojic acid is obtained from the mycelium of *Aspergillus oryzae, A. glaucus, A. tamarii* and *A. flavus*. Many species of *Penicillium* are capable of producing organic acids, such as citric fumaric, oxalic, gluconic and gallic.

Synthesis of enzymes. Many important and useful enzymes have been synthesized from various fungi. The enzyme tyrosinase is obtained from *Neurospora crassa*. *Saccharomycees cerevisiae* synthesizes invertase enzyme. *Aspergillus oryzae* has been employed in active enzyme preparation. *A. niger* produces starch digesting enzymes. *Aspergillus versicolor* synthesizes an enzyme capable of destroying tartrates and recommended its use in the commercial preparation of grape juice.

Production of esters. According to Birkinshaw and other (1931), the ethyl acetate was synthesized by *Penicillium digitatum*.

Production of pigments. A maroon coloured pigment, **fumigatin** is produced by *Aspergillus fumigatus.* The red pigments, **catenarin** is produced by *Helminthosporium* sp. Another important blue-purple pigment, **spinulosin** is produced by *Penicillium spinulosum*. A number of pigments have been isolated from yeasts, particularly from species of *Rhodotorula* and *Cryptococcus.*

Harmful Aspects

Plant pathogens. Various parasitic fungi act as causal organisms and infect hundreds of species and varieties of plants of the economic value. A list of few important diseases of the plants of economic importance, with their causal organisms is given below.

Table 17.1

Name of Disease	Name of Crop	Causal organism
1. Club root	Cabbage	*Plasmodiophora brassicae*
2. Wart disease	Potato	*Synchytrium endobioticum*
3. Damping off	Various seedlings	*Pythium* sp.
4. Late blight	Potato	*Phytophthora infestans*
5. White rust	Crucifers	*Albugo candida* or *Cystopus candidus*
6. Downy mildew	Peas	*Peronospora pisi*
7. Downy mildew	Grapes	*Plasmopara viticola*
8. Powdery mildew	Grapes	*Uncinula necator*

9. Powdery mildew	Peas	*Erysiphe polygoni*
10. Powdery mildew	Wheat	*Erysiphe graminis*
11. Loose smut	Wheat	*Ustilago tritici*
12. Covered smut	Barley	*U. hordei*
13. Loose smut	Barley	*U. nuda*
14. Covered smut	Oats	*U. kolleri*
15. Loose smut	Oats	*U. avenae*
16. Whip smut	Sugarcane	*U. scitaminea*
17. Grain smut	Jowar	*Sphacelotheca sorghi*
18. Smut of	Bajra	*Tolyposporium penicillariae*
19. Black rust	Wheat	*Puccinia graminis tritici*
20. Brown rust	Wheat	*P. recondita*
21. Yellow rust	Wheat	*P. striiformis*
22. Rust of	Linseed	*Melampsora lini*
23. Early blight	Potato	*Alternaria solani*
24. Tikka disease	Groundnut	*Cercospora personata*
25. Red rot	Sugarcane	*Colletotrichum falcatum*
26. Flag smut	Wheat	*Urocystis tritici*
27. Leaf curl	Peach	*Taphrina deformans*

Besides, there are hundreds of other diseases which cause heavy losses to the pulses, vegetables, cereals, fruits and other important crops.

Decay of timber. Many species of *Polyporus* and other Basidiomycetes cause the damage and decay to the timber wood and sufficient losses are caused.

Destruction of textiles. According to Prindle (1935), the species of the genera *Alternaria, Penicillium,* etc., are especially destructive to woollen textiles. According to Duske (1943), the rayon is destroyed by the genera of *Aspergillus* and *Penicillium.* According to Hardy (1943), the chief destructing fungi to the textiles are *Mucor, Aspergillus, Fusarium* and *Trichoderma*. The deterioration of cotton in storage is caused chiefly by a species of *Stachybotrys*. He also investigated that the greatest amount of damage to textile materials is due to *Chaetomium globosum*.

Destruction of paper industry. Many fungi cause damage to the paper pulp-wood, *e.g., Polyporus adustus, Polysticus hirsutus,* etc. Many species of Basidiomycetes cause wood decay if suitable environmental conditions prevail. Many fungi destroy the paper of books and newspapers in moist condition. The fungi such as *Chaetomium, Aspergillus, Cephalothecium, Stachybotrys, Alternaria, Fusarium, Cladosporium, Stemphyllium, Dematium* cause sufficient damage to paper industry.

Spoilage of foodstuffs. *Penicillium digitatum* is responsible for the rotting of citrus fruits. According to Thom and Ayers (1916), many fungi, such as *Mucor, Aspergillus, Penicillium, Oidium* and *Fusarium,* cause damage to the milk and milk products. To avoid this damage to the milk normal pasteurization (*i.e.,* heating milk to 145°F) is necessary before distribution to the consumer. *Oidium lactis* develops the fishy odour of the butter and causes the damage to the butter. *Mucor mucedo* and *Aspergillus* are responsible for the bread spoilage.

Spoilage of meat. Many fungi are reported from the meat which cause sufficient spoilage to it, usually in tropical conditions. According to Yesair, *Mucor* sp., *Penicillium* sp., *Neurospora* sp., *Fusarium* sp., *Oidium sp.* and *Aspergillius sp.*, are commonly met with on meat and cause damage to it.

Medical mycology. Several important human diseases are caused by different species of fungi. *Aspergillus fumigatus* and *A. glaucus* are responsible for the lung disease 'aspergillosis.' The disease 'mycosis' and 'tokelan' are caused by other *species* of *Aspergillus.* Several other animal and human diseases are caused by various fungi.

The dermatophytes are fungi which cause diseases of the skin of man. The diseases they cause are grouped under the general term **dermatomycoses** and are commonly known as ringworm, athlete's foot, etc. *Epidermophyton flocossum* causes chronic infections commonly called athlete's foot. *Microsporum* generally infects the scalp of children, causing a condition called by medical personnel **tinea capitis**. The form genus *Trichophyton* also a cause of athlete's foot and other skin diseases is the largest of the form genera of dermatophytes. Several form-genera of the Moniliaceae which include serious human pathogens are *Blastomyces, Histoplasma, Geotrichum* and *Sporotrichum*. *Blastomyces dermatitidis* causes North America **blastomycosis.** South American **blastomycosis** caused by *Blastomyces braziliensis* is confined to South American and some parts of Central America. It is confined to the skin and mucous membranes around and in the mouth. Certain form species of *Geotrichum* are known to be pathogenic to man. Four forms of **geotrichosis** have been described–oral, intestinal, bronchial and pulmonary. *Sporotrichum schenkii* is the cause of **sporotrichosis** to man and animals.

MUSHROOMS AND THEIR CULTIVATION

Direct utilization of fungi as food. Many Agaricales and Helvellales are directly used as food. There is a non-poisonous edible toadstool, *i.e., Coprinus* sp. found in lawns in the rainy season. *Agaricus campestris* is edible mushroom and cultivated for its fructifications. The fruiting bodies are quite fleshy and eaten directly as vegetable or with rice as **'pulao'.** These mushrooms are being successfully cultivated in South India. *Morchela esculenta* is another important edible fungus. It is found in Kashmir and Punjab plains. Its local name is **'guchi'** and sold very costly. *Torulopsis utilis,* is used for the large-scale production of yeast for food purposes. *Saccharomyces cerevisiae* is used in bread making industry.

Of the many mushrooms that can be cultivated, only three kinds namely button mushroom (*Agaricus bisporus*), straw mushroom (*Volvariella volvacea*) and oyster mushroom (*Pleurotus sajorcaju*) are suitable for growing in India where suitable environmental conditions exist. The following account deals with cultivation of Button Mushroom (*Agaricus bisporus*).

Growing Season

Agaricus bisporus being a temperate mushroom grows best during winter throughout the plains of North India. It can however, be grown throughout the year in hills. The most suitable temperature for the spread of mycelium is 24-26°C. Temperature ranging from 16-18°C is essential for the formation of fruit bodies. Higher temperatures are harmful but the lower temperature retards the development of both mushroom mycelium and fruit bodies.

Mushroom House

The mushroom house can be any available room, shed, basement, garage, etc. The growing house should be well ventilated and not stuffy.

Compost

The cultivated mushroom is grown on a special compost. Two types of composts, natural and synthetic are used for growing this mushroom.

Composting Yard

The compost should be prepared on well cleaned concrete or *pucca* floor, which should be on a higher level so that the run-off water does not collect near the heap. Composting is usually done in the open, but it has to be protected from rain by covering it with polythene sheet. It can also be done in a shed with open sides or a large room to shelter it from rain.

Synthetic Compost

The following ingredients are required:

Wheat straw (chopped 8-20 cm. long)	250 kg.
Wheat/Rice bran	20 kg.
Ammonium sulphate/calcium ammonium nitrate	3 kg.
Urea	3 kg.
Gypsum	20 kg.

This will make compost sufficient for 15-16 trays of size 1m × 1/2m × 15 cm.

Ten kg molasses and/or 60 kg chicken manure can also be used if available.

Procedure for Long Method

Mixing the materials and making pile : Day-0.

The straw is uniformly spread over the composting yard in a thin layer and wetted (thoroughly by sprinkling water). All ingredients such as wheat bran, fertilizers, etc., except gypsum are mixed thoroughly in the wetted straw, which is finally heaped into a pile about one metre high, one metre wide and as long as to hold the entire quantity of straw mixture. The pile can be made with hand or stack mould. The straw should be firmly but not compactly compressed into the mould.

It is essential to open the entire pile and remake it a number of times according to the following schedule :

1st turning	4th day
2nd turning	8th day
3rd turning	12th day, add 10 kg. gypsum.
4rd turning	16th day, add 10 kg. gypsum.
final turning	20th day, spray 10 ml malathion in 5 litre water (any other available pesticide like, DDT, BHC, Lindane can also be used).

At each turning water should be sprinkled to make up the loss of water due to evaporation. If it is desired to add molasses, then 10 kg molasses diluted 20 times with water should be poured over the straw mixture during the first turning. Sixty kg chicken manure if available can also be added at the time of start of pile, *i.e.*, 0-day to improve the quality of the compost.

Natural Compost

It is prepared from horse dung which must be freshly collected and not have been exposed to rain. It should not contain admixture of dung of other animals. Chopped wheat straw (1/3, or more of weight of horse dung) is mixed with the horse dung to which 3 kg urea and/or 100-110 kg poultry manure per tonne is usually added. Horse dung and straw mixture is uniformly spread over the composting yard and sufficient water is sprinkled over it so that the straw becomes sufficiently wet so that it takes no more water. The manure is then heaped in a pile as for synthetic compost. After 3 days when the manure in the heap gets heated up due to fermentation and gives off an odour of ammonia, it is opened up and repeated 3 or 4 times after an interval of 3-4 days. Twenty five kg gypsum per tonne is added in two instalments at 3rd and 4th turning. At the final turning 20 ml malathion diluted in 10 litres of water is sprayed into the manure.

Filling

The compost when ready for filling and spawning has a dark brown colour and no trace of ammonia. There is no unpleasant odour but it smells like a fresh hay. The pH is neutral or near neutral. The compost should not be too dry or too wet at the time of filling in the tray which can be determined by the palm test. For this purpose a small quantity of compost is taken into the hand and pressed lightly, if a few drops of water ooze of the fingers then it is of right consistency. If relatively dry then the water should be added by sprinkling. If too wet, the excess water may be allowed to evaporate. The prepared compost is now filled in trays, which may be of any convenient size but its depth should be 15-18 cm. A standard size of tray 1 m × ½ m × 15 cm. The tray should be made of soft wood and provided with the pegs at the four corners so that they can be stacked one over the other leaving sufficient space between the two trays for various operations. The tray is completely filled with the compost, lightly compressed and the surface levelled.

Spawning

Spawning means sowing the beds with the mycelium (spawn) of the mushroom. Spawn can be had from mushroom Laboratory College of Agriculture, Chambaghat, Solan at a nominal cost. Small quantity of spawn is also available from the Division of Mycology, Indian Agricultural Research Institute, New Delhi. The grain spawn is scattered on the surface of the tray bed which is covered with the thin layer of compost. Spawning can also be done by mixing the spawn with compost before filling it in trays. Five hundred gram spawn (two ½ lit bottles/polypacks) is sufficient for five trays of stadard size. After spawning the compost surface is covered with old newspaper sheet, which is wetted by sprinkling water to provide the humidity but not water is directly added to the compost during spawn running.

The trays after spawning are stacked vertically one over the other in 4-5 tiers. One metre clear space may be left in between the top tray and ceiling. There should be about 15-20 cm space between the two trays.

The room should be maintained at 25°C or near about. The humidity should be built up by frequently watering the floor and walls. The room may be kept closed as no fresh air is needed during the spawn run. White cottony mycelium spreads and permeates through the compost.

Eventually the compost surface gets covered with the mycelium. It takes 12-15 days for complete spawn run. Low temperature prolongs the spread of the mycelium.

Casing

After the spawn run is complete as is evident by white mouldy growth, the surface of the compost is covered with 3 cm layer of casing soil. A suitable casing soil can be prepared by mixing equal part of well rotten cowdung (finely crushed and coarsely sieved) and garden soil. The casing material should possess high water holding capacity, good pore space and pH should not be lower than 7.4. The casing material is sterilized to kill insects, nematodes and molds. Sterilization can be accomplished either by steaming or by treating with formalin solution. For one cubic metre of casing soil, half litre of formalin diluted with 10 litre of water is sufficient. The casing soil is spread over a plastic sheet and treated with formalin by sprinkling. The treated soil is piled up in a heap and covered with another plastic sheet for 48 hours. The soil is turned frequently for about a week to remove all traces of formalin which can be tested by smelling it. After casing, the temperature of the room is maintained for further three days, after which it must be lowered to below 18°C. At this stage lot of fresh air is needed and therefore, the growing room should be ventilated by opening windows, etc.

Cropping and Harvesting

The first flush of the pin heads becomes visible 15-20 days after casing or say about 35-40 days after spawning. Small white buttons develop 5-6 days after pin head stage. The right stage of harvest is when the cap is still tight over the short stem. In case the buttons are allowed to mature, the membrane below the cap will rupture and the cap will open up in umbrella-like shape. Such mushrooms are considered to be the inferior. Harvesting is done by holding the cap with forefingers slightly pressed against the soil. The soil particles and mycelial threads clinging to the base of the stalk are chopped off. Mushroom can also be harvested by cutting off with a sharp knife at soil level.

Yield

The average yield of 3-4 kg per tray is considered normal. However, if compost is carefully prepared, spawn reliable and favourable termperature prevailing during the growing period, then a yield of 5-6 kg per tray is possible. Partial or complete failure may also happen due to negligence.

Storage :

The mushrooms are best consumed fresh. Storage in refrigerator for a few days is possible. The mushrooms should be placed between moist paper towel for storing in a refrigerator.

HETEROTHALLISM

Heterothallism. Ehrenbergh (1829), for the first time studied zygospores in the order Mucorales. The American mycologist Blakeslee (1904), reported that in the several genera of the order Mucorales, the zygcspores are not formed at all. He also supported his view with facts and reasons, and also investigated that in the same order two types of species are found which may be named as homothallic and heterothallic species. When the two hyphae of the same mycelium produced by a single spore fuse with each other and a zygospore is developed, the species is said to be homothallic, *e.g., Mucor hiemalis.*

Mucor mucedo and *Mucor stolonifer* are the typical heterothallic species. In heterothallic species the fusion can take place only among the different strained hyphae, which develop on different mycelia of different (+ and –) strains. In these species the zygospores cannot be produced by the fusion of two hyphae of the same strain. In 1904 Blakeslee reported that in heterothallic species whenever the mycelia of + and – strains remain apart from each other, the zygospores are not produced and only the sporangia are formed. On the other hand, when + and – mycelia grow together, the fusion takes place and the zygospores are produced. Morphologically the + and – strained mycelia are quite similar in structure, but different in physiological behaviour. In other words, they are morphologically identical and physiologically different. Sometimes, it has also been observed that the growth of the + mycelium is comparatively faster, and the gametangia of + mycelium are bigger than that of the – mycelium, and they can be distinguished as female and male gametangia. Many mycologists do not support this view and advocate that this behaviour is only because of nutrition.

Blakeslee proved the phenomenon of heterothallism on the basis of experiments. He inoculated many + and – strained spores on the agar medium in a petri dish and observed that the fusion occurs only at those points where the hyphae of different strains (*i.e.,* + and –) come in contact. The zygospores were only produced at the points of contact of different strained hyphae. Absolutely no zygospores were produced at the points of contact of same strained hyphae.

This experiment may be clearly understood by the figure given in the text. Five spores of different strains are inoculated in a sterilized petri dish on agar medium at different five points designated as A, B, C, D and E. The zygospores are produced only at those points where the mycelia are opposite strained and produced from + and – spores designated as A, C and E are + strained, B and D are – strained.

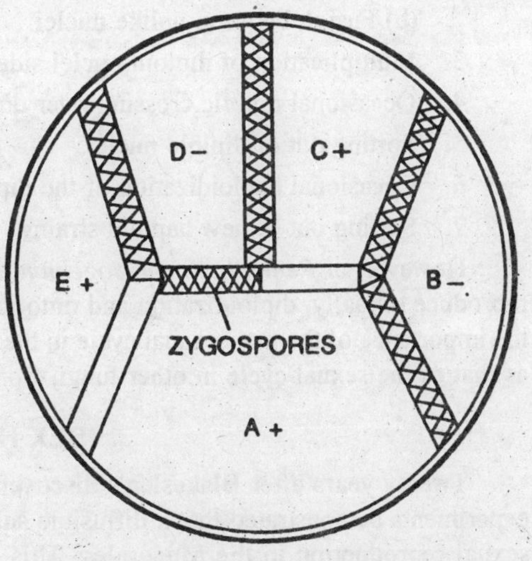

Fig. 17.1. Heterothallism in Mucorales-Blakeslee's experiment.

The zygospores are produced at the junctures of AB, BC, DE and AD where the opposite strained mycelia take part in fusion. The zygospores are not produced at the junctures of AC and AE because of the mycelia of same strains.

Blakeslee also investigated that if the + strained hyphae of one heterothallic species fuse with the – strained hyphae of another heterothallic species the **imperfect zygospores** are resulted. In such conditions the zygospores do not mature. He also investigated that in the heterothallic species *Mucor mucedo* the zygospores produced germsporangia on germination, which contain single strained spores, and therefore, the fusion is possible only when the mycelia are produced from the spores formed in different strained germsporangia. In the heterothallic species of *Phycomyces nitens* the + and – strained spores are produced in the same germsporangium.

The phenomenon of heterothallism was first reported in the order Mucorales, but now this has been established in rusts, smuts, Homobasidiomycetidae and other fungi.

PARASEXUALITY

Some fungi do not go through a true sexual cycle, but derive many of the benefits of sexuality through **parasexuality** (Gr. *para* = beside + sex). This is a process in which plasmogamy, karyogamy and haploidization take place, **but not at specified points in the thallus or the life cycle.** For example, the Deuteromycetes (Fungi Imperfecti) are fungi in which sexual reproduction does not take place and in which the parasexual cycle is of paramount importance. However, some fungi which reproduce sexually also exhibit parasexuality. The parasexuality has been reported also in *Aspergillus nidulans* (of Ascomycetes).

Parasexuality was first discovered in 1952 by Pontecorvo and Roper of the University of Glasgow in *Aspergillus nidulans,* the imperfect stage of *Emericella nidulans.* Since then parasexual phenomena have been indentified in several imperfect fungi (the Deuteromycetes) which possess no sexual stage, as well as in Basidiomycetes and in Ascomycetes other than *Emericella nidulans.*

According to G. Pontecorvo (1956, 1958), the sequence of events in a complete parasexual cycle is as follows:

1. Formation of heterokaryotic mycelium.
2. Fusion between two nuclei.
   (a) Fusion between like nuclei.
   (b) Fusion between unlike nuclei.
3. Multiplication of diploid nuclei side by side with the haploid nuclei.
4. Occasional mitotic crossing-over during the multiplication of the diploid nuclei.
5. Sorting out of diploid nuclei.
6. Occasional haploidization of the diploid nuclei.
7. Sorting out of new haploid strains.

However, in *Penicillium chrysogenum* and *Aspergillus niger,* neither of which is known to reproduce sexually, diploidization and mitotic crossing-over occur much more frequently, so that the importance of the parasexaual cycle in the Fungi Imperfecti (Deuteromycetes) may be as great as that of the sexual cycle in other fungi, from the stand point of the evolution of the species.

SEX HORMONES

Twenty years after Blakeslee's discovery of heterothallism, Burgeff (1924), in a series of experiments demonstrated that a diffusible substance is probably responsible for the initiation of sexual reproduction in the Mucorales. This was the first demonstration of a hormonal sexual mechanism in the fungi. Burgeff, working with a number of species, showed that, when + and – strains were separated by a collodion membrane, hyphae of opposite strains were still attracted towards each other. Subsequent work by other investigators (Plempel, 1957, 1960; Plempel and Braunitzer, 1958) has expanded upon Burgeff's discovery, and is giving us some knowledge as to the nature of hormones involved and their method of operation.

There are strong indications that a complicated hormonal mechanism operates in initiating plasmogamy, at least in some Ascomycetes. For example, the trichogynes are attracted to compatible male sex cells (spermatia) and will change their direction of growth in response to the proximity of sex cells.

In *Allomyees* (of order Blastocladiales and class Chytridiomycetes), attracted by the sexual hormone **sirenin** produced by the female gametes (Machlis, 1958), the male gametes copulate with the latter in parts very soon after their release from the gametangia.

18
CHAPTER

The Lichens

Definition. A lichen is a plant consisting of two separate plants, a fungus and an alga, so closely associated with each other as to appear a single plant. The algal cells of the association are always enveloped by the fungus. The combined growth of both parterns results in a constant definite form and internal structure of the lichen. This group has about 400 genera and 15,000 species and is treated separately instead of its connection with fungi or algae. The science of lichens is termed **lichenology** and one who studies this science is known as **lichenologist.**

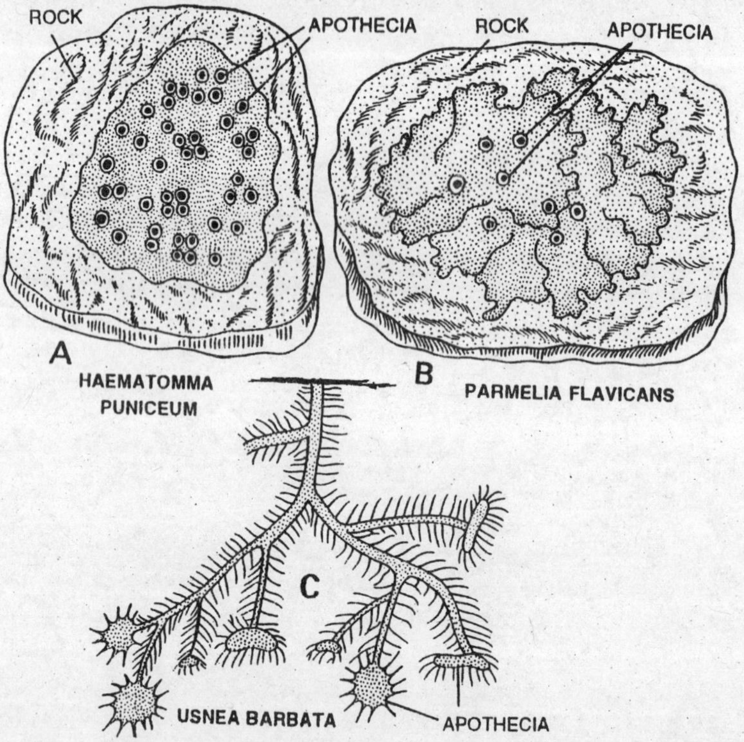

Fig. 18.1. Different types of lichens. A, crustose; B, foliose; C, fruticose.

Composition. The fungal component in four genera of lichens is a member of Basidiomycetes and in rest it is an Ascomycetous member as such the lichens are called **Basidiolichens** and **Ascolichens.** Basidiolichens have only four genera, all to restricted to tropical regions, while the Ascolichens are restricted to temperate regions. The algal components may belong to Myxophyceae (blue green algae) or Chlorophyceae (green algae). They may be filamentous or non-filamentous. Basidiolichens always possess a member of Myxophyceae. Some lichens have been observed, where fungi are associated with autotrophic bacteria. Each lichen is always formed of the same fungal and algal components and has constant internal structure and habit.

Nature. The nature of lichens is a subject of controversy. According to some authors, the fungal-algal partnership is of parasitic nature, while others believe, they are symbiotic in nature. In many lichens, the haustoria and appressoria from the fungus penetrate into the algal cells. This is a very strong point in support that the relationship between fungus and alga is of parasitic nature. They are of the view, "a lichen is a fungus which lives during all or a part or its life in a parasitic relation with an organic or inorganic substratum".

Others believe that the partnership (**consortium**) between a fungus and an alga is of symbiotic nature. Fungal hyphae absorb moisture from a spongy structure which retains it, and give support to the alga. The alga in return synthesizes the carbohydrates necessory for both the partners. This type of symbiosis is known as **heliotism,** (=master and slave).

The fungus separated from the lichen never grows to maturity. On the other hand the alga can live independently apart from the lichen but it grows better in the symbiotic association.

Habit and habitat. Lichens grow in the presence of sufficient moisture, cold or moderate temperature, direct sunlight and pure atmosphere. They grow on the leaves, tree trunks, old logs, floor of forests and on exposed rocks. Some lichens are cushion-like masses on bare rocks in extremely cold regions of arctic zones. The striking example of this type is *Cladonia rangifera* Web. Some lichens grow most luxuriently in tropical and subtropical rain forests, in abundance of moisture.

The rainfall in the Eastern Himalayas is much more than in the Western Himalayas. The lichens are more common in the Eastern Himalayas. Darjeeling, Gangtok, and places upto 10,000 feet en route to Jemu glacier in Sikkim are ideal for lichen collection. There, almost every rock, tree fence, road side wooden poles and house walls are seen covered with various forms of crustose, fruticose and foliose lichens. In the plains the lichens are almost absent as the conditions are entirely unfavourable for their growth.

Structure. Regarding their structure, the lichens are of three types–**Crustose, Foliose** and **Fruticose.**

1. Crustose. These lichens have flattened thalli, closely adherent to the substratum. Thallus of majority of such lichens has more or less leathery texture and is internally differentiated with the algal component always restricted to a definite portion of the thallus. Some lichens have gelatinous thallus, in which alga and fungus are uniformly distributed through a gelatinous matrix. Lichens vary in form, colour and thickness. Thalli of these lichens may remain partly or fully submerged in the substratum on which they grow. In the cases where body is fully embedded, only the fruiting bodies (ascocarps) of the fungus appear on the surface of the substrate.

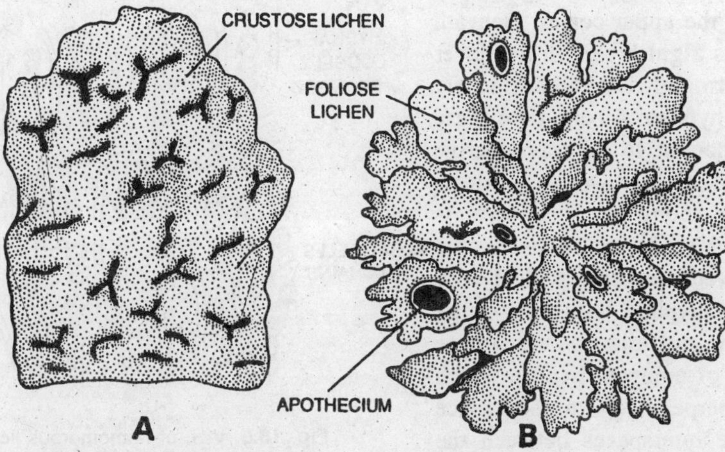

Fig. 18.2. A, Crustose lichen; Fig. 18.2. B, Foliose lichen.

2. **Foliose.** These lichens possess leaf–like thalli with lobed or irregularly folded margins. Some parts of thallus are more or less firmly attached to the substratum by means of hyphal outgrowths, **rhizines** from the lower surface. Rhizines may consist of separate single branched or unbranched hyphae, or of several parallel hyphae closely adhered to each other to form strands. A foliose lichen like *Gyrophora* may be attached to the substratum by a single rhizine growing from the centre of the thallus or a lichen may be attached by several rhizines.

3. **Fruticose.** These lichens have much branched, cylindrical, ribbon-like, flattened or sometimes filamentous thalli. They are much branched, and appear shrubby, and so the name **fruticose** (*frutex* = shrub) is given to them. They may be erect (*Cladonia*) or pendant (*Usnea*). Their thalli are attached to the substratum by the basal portion only which is composed of strands of densely packed hyphae.

Fig. 18.3. Fruticose lichen.

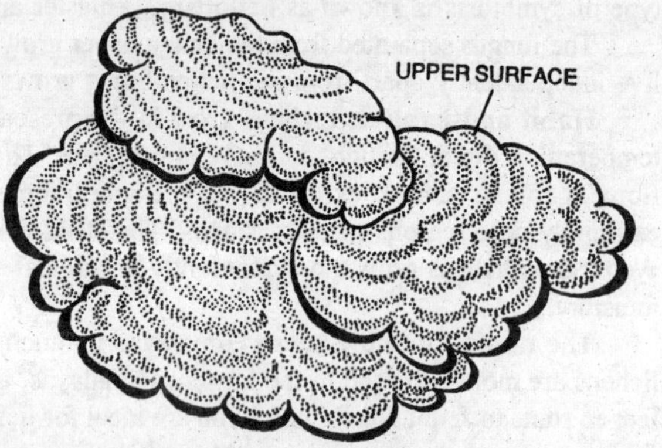

Fig. 18.4. Basidiolichen (*Cora puvina*)

The lichens are variously coloured commonly bluish-green or greyish-green. Many lichens are yellow, orange, reddish, brownish or black due to the presence of additional pigments.

Internal structure. The gelatinous thalli of crustose lichens have algal and fungal components uniformly distributed through a gelatinous matrix. Thalli of such lichens are not differentiated into layers of tissues and therefore, known as **homomerous.** Thalli or most foliose and fruticose lichens are differentiated into several layers of tissues, and therefore known a **heteromerous.** Thalli of foliose lichens have four layers of different tissues. The uppermost layer, **upper cortex,** is composed of more or less vertical hyphae. The intercellular spaces between the hyphae are either absent or filled with gelatinous substance. An epidermis-like layer formed of closely packed hyphae may or may not be present external to the upper cortex. Beneath this cortical layer is **algal layer.** In this layer algal cells lie intermingled with the loosely interwoven fungal hyphae. At one time these algal cells were thought to be reproductive cells of lichens and were known as **gonidia,** and therefore, even today the algal layer is often called **gonidial layer.** Beneath the gonidial layer is **medulla** composed of very loosely interwoven hyphae with very large interspaces. Below medulla is the **lower cortex.** It is made of compact hyphae which may be parallel to perpendicular to the surface of the thallus. The interspaces between the

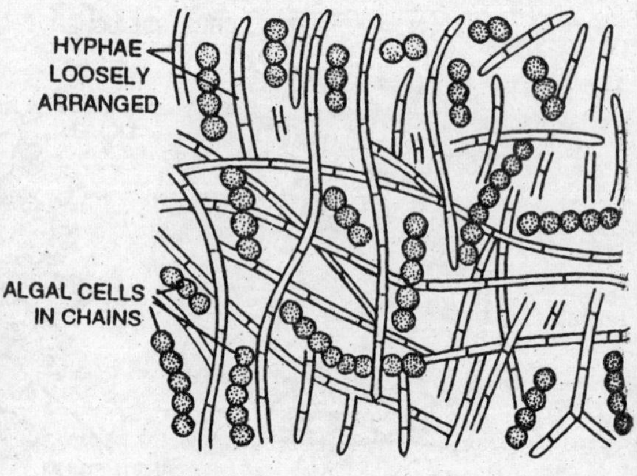

Fig. 18.5. V.S. of homomerous lichen.

hyphae are either absent or filled with gelatinous material. The rhizines grow out from the underside of the lower cortex and attach the thallus to the substratum. Basidiolichen has uppermost layer of loose perpendicular hyphae (**superior layer**), below it **algal layer** and the lowermost is the **inferior layer** made of dense mass of hyphae.

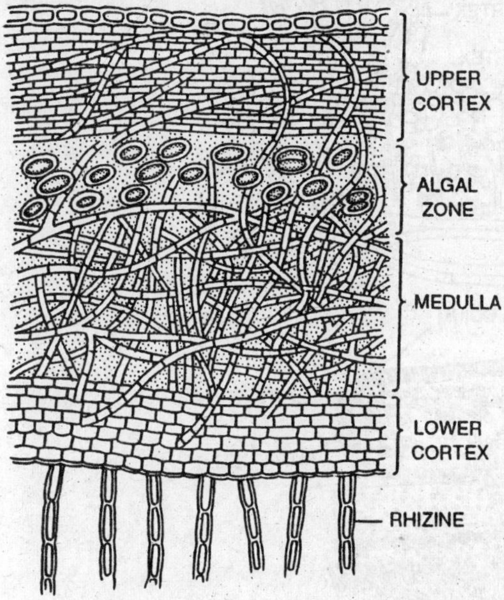

Fig. 18.6. V.S. of heteromerous lichen.

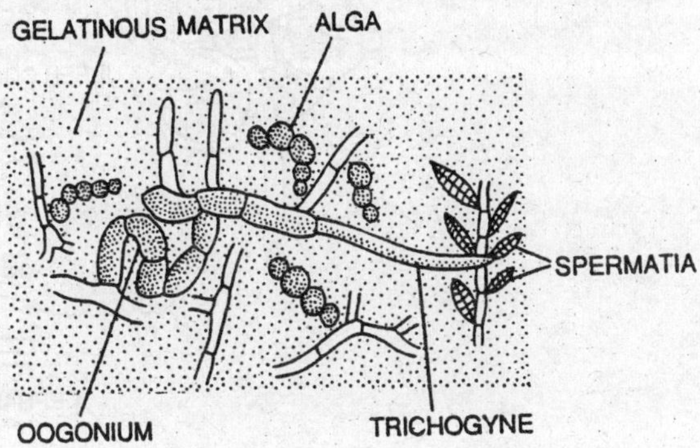

Fig. 18.7. V.S. thallus-crustose lichen.

In addition to the structure described above, lichens bear certain peculiar vegetative structures. These structures are:

(a) Breathing pores. In the upper cortex of some foliose and fruticose lichens, the hyphae are very loosely interwoven to facilitate the gaseous exchange between the thallus and the atmosphere. These localized areas may be depressions or cone like and are known as **breathing pores.**

(b) Cyphellae. They are small, concave circular depressions or cavities developed in the lower cortex of a few foliose lichens. Hyphae from the medulla grow out into these cavities, which function as aerating or respiratory organs.

(c) Soredia. They are small bud-like outgrowths, recognisable as greyish powder on the surface of the thallus. A soredium consists of one or more algal cells enclosed by a few hyphae. They develop in algal layers at places where the over lying upper cortex is absent. Soredia sometimes develop in pustule-like areas known as **sorelia.** Each soredium develops into a new thallus.

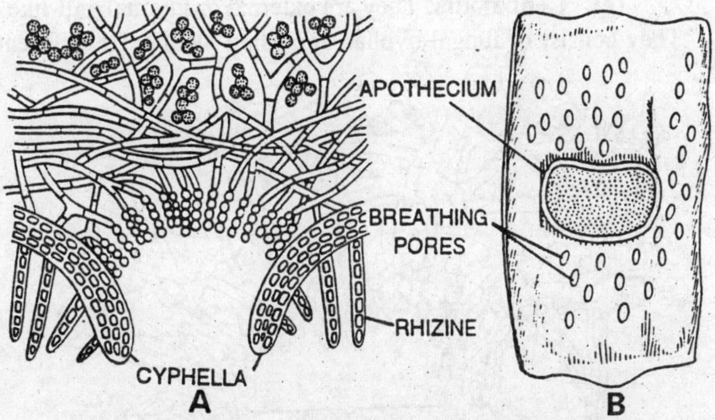

Fig. 18.8. Lichens. A, cyphellae; B, breathing pores.

(d) Isidia. They are the small papillate outgrowths from the upper surface of the thallus. They consist of an external cortical layer and an internal algal layer. Their primary function is to increase the photosynthetic surface of the thallus, but sometimes they get detached from the thallus and serve as vegetative reproductive bodies.

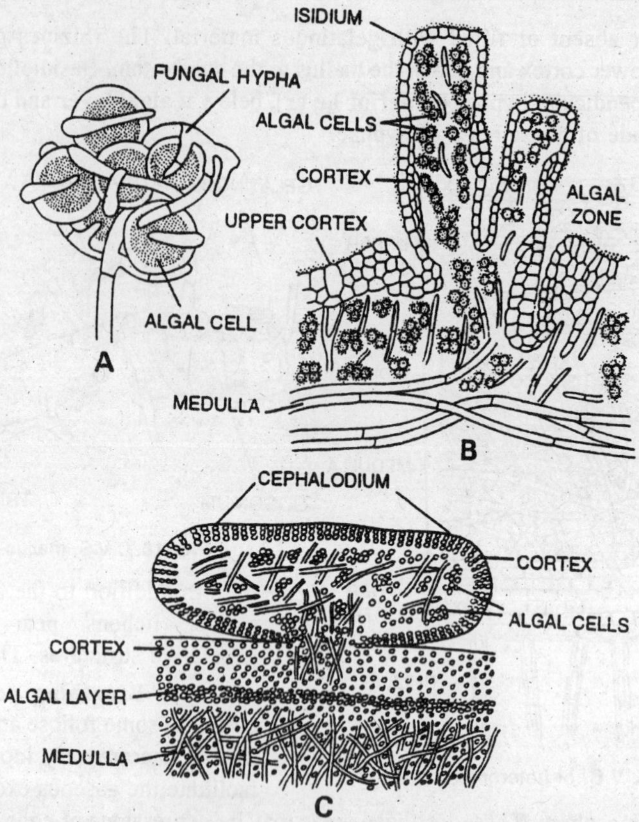

Fig. 18.9. Lichens. A, soredia; B, isidia; C, cephalodia.

(e) Cephalodia. They are external or internal gall-like outgrowths, generally of dark colour. They consist of fungal hyphae enclosing algal cells different from those of the thallus.

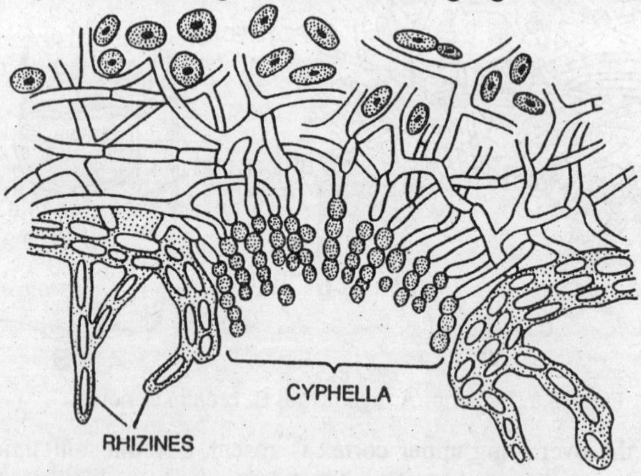

Fig. 18.10. V.S. thallus–Foliose lichen (*Sticta*). Cyphella.

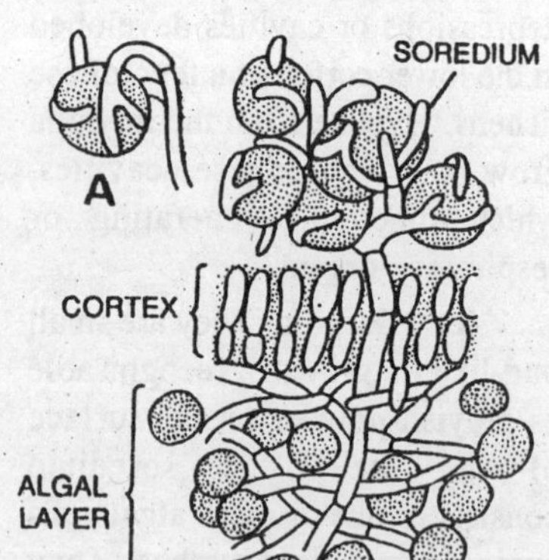

Fig. 18.11. Soredium (*Parmelia*). A, with single algal cell; B, with many algal cells

Reproduction. Commonly lichens reproduce vegetatively and sexually, but in some Ascolichens asexual spores are also formed.

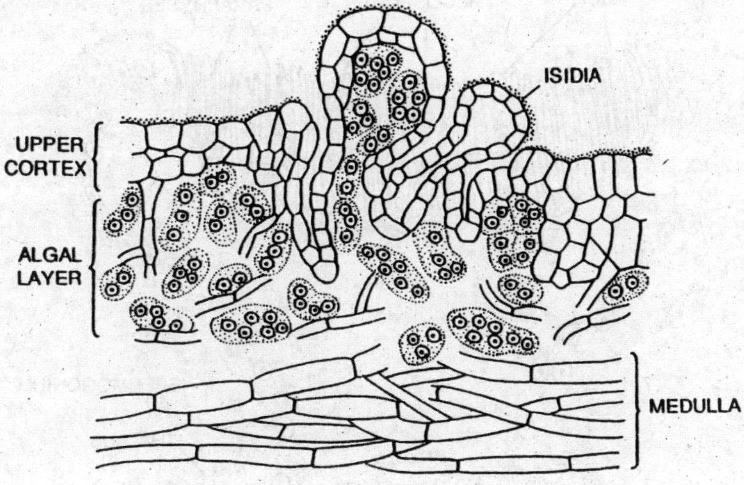

Fig. 18.12. Isidia (*Peltigera* sp.) V.S. thallus.

A. Vegetative reproduction. It takes place by following methods:

1. Fragmentation. On maturity the older portions of the thalli of lichens die and decay. The thallus breaks into pieces accidentally and each piece develops into a new plant. This occurs more frequently in pendant thallus, such as of *Ramalina reticulata.*

2. Isidia and soredia. As described above (c and d), the vegetative reproduction takes place by means of isidia and soredia as they detach from the mother thalli.

B. Asexual spores. Hyphae of few lichens break up into oidia, they germinate into new fungal hyphae and each oidium produces a lichen when comes in contact with suitable alga. Many lichens produce large number of small spore-like structures, **pycniospores,** within flask shaped **pycnia,** immersed within the thallus. These structures when act as male gametes are known as spermatia and spermagonia respectively.

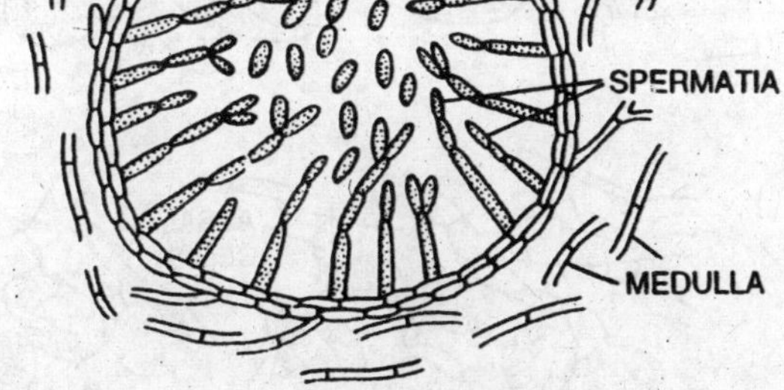

Fig. 18.13. Diagrammatic representation of spermagonium (pycnium) of *Physcia.*

C. Sexual reproduction. In Ascolichens the fungus belongs to Ascomycetes and the sexual reproduction results in the formation of **apothecia** or **perithecia.** These fruiting bodies are small cup-like or disc-like and may be embedded in, or raised above the surface of thallus by short or long stalks. The structure of the wall of an apothecium is similar to that of the thallus, it consists of an upper and a lower cortical layer with medulla, in between. Algal components may not be present in the vegetative part of the apothecium. The bottom of the cup, or the surface of the disc is the fertile part of the apothecium and is lined by the **hymenium.** Hymenium consists of asci and paraphyses growing vertically. Paraphyses contain a reddish oily substance in them and never project beyond asci. Each ascus contains eight ascospores, which become two celled prior to dissemination. Asci are the resultants of sexual union.

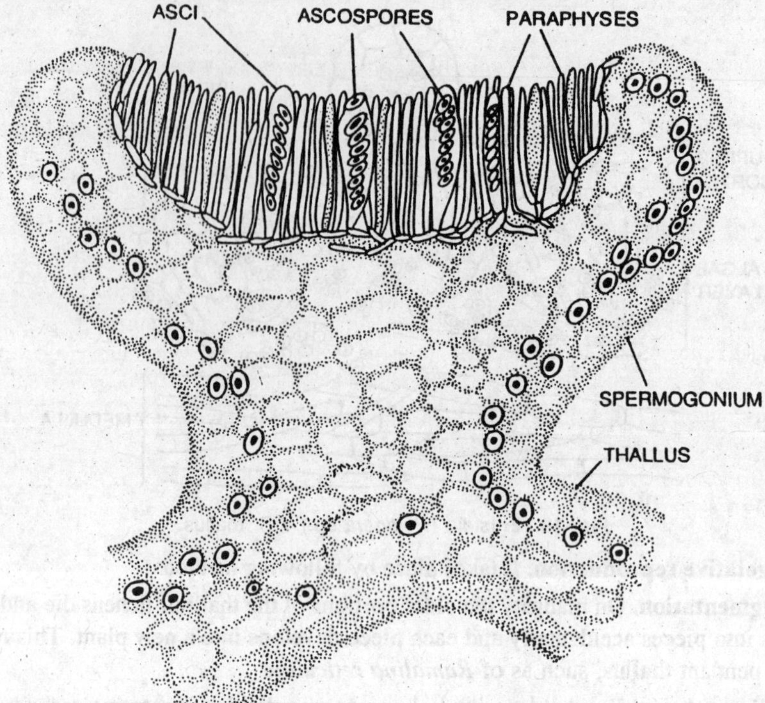

Fig. 18.14. V.S. apothecium (*Physcia*).

Sex organs. The female reproductive organ is an ascogonium (carpogonium) which develops from hypha deep in the algal layer. It is a long multicellular hypha, the coiled base of it is the oogonium and the straight portion above it the trichogyne. The trichogyne in some species projects beyond thallus. More than one ascogonia may develop at a point where an apothecium is later formed but only one becomes fertile.

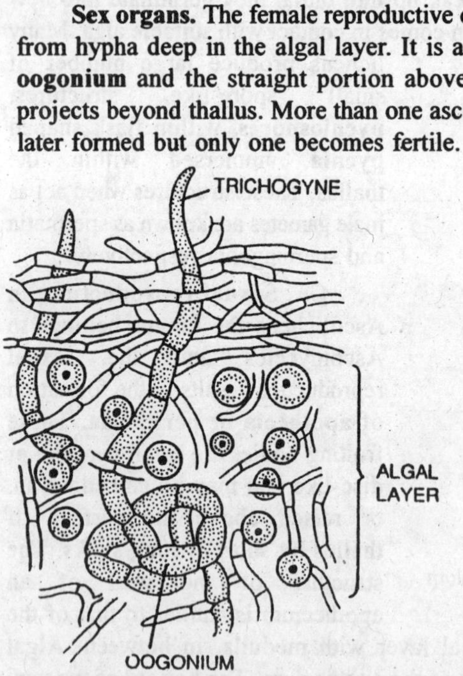

Fig. 18.15. V.S. thallus (*Physcia*).

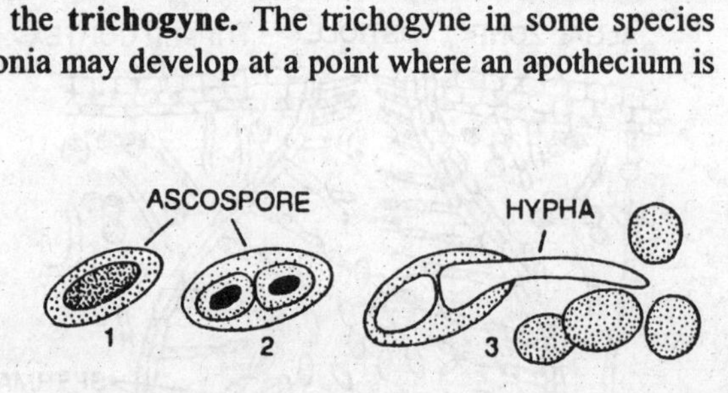

Fig. 18.16. Germination of ascospores and its association with algal cells to form lichen.

The male reproductive body is **spermagonium** (pycnium). It is flask-shaped cavity immersed in the thallus and opens to the exterior by small **ostiole.** The fertile hyphae lining the inner surface of the spermagonium produce large number of small non-motile gametes **spermatia.** The spermatia

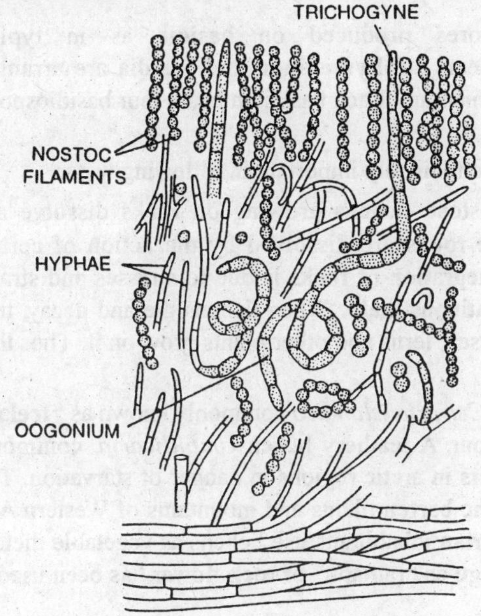

Fig. 18.17. V.S. of lichen thallus showing symbiotic components and trichogyne.

are functional male gametes. The spermatia are lodged against the sticky protruding tips of trichogynes, and the fact that ascogonia of thalli lacking spermagonia rarely produce ascocarps. In *Collemodes bachmannianum* a gelatinous lichen trichogyne does not protrude, but grows more or less horizontally in the thallus. Spermatia are borne laterally and terminally on the surface of the hyphae with the thallus.

The growing trichogyne comes in contact with spermatia. The walls of contact dissolve and the male nucleus gradually passes downward to the oogonium, where it fuses with the female nucleus of the egg and the fertilization is effected.

Numerous branched, septate ascogenous hyphae containing one, two or many nuclei developed from the oogonium. The ultimate or the penultimate cells of the ascogenous hyphae develop into asci. At the same time sterile hyphae develop from below the ascogonium and the wall of the ascocarp. As the ascocarp grows it breaks through the thallus and appears, above the surface as a cup or disc or remains embedded.

The development of asci and ascospores resembles to that of typical Ascomycetes. Spores are shed only during moist weather on germination, a spore produces a germ tube which grows in all directions, and as soon as it comes in contact with a suitable alga, additional branches are formed to engulf the alga. Combined growth of the fungus and the alga continues and results in a lichen. In absence of a suitable alga the germ tube dies.

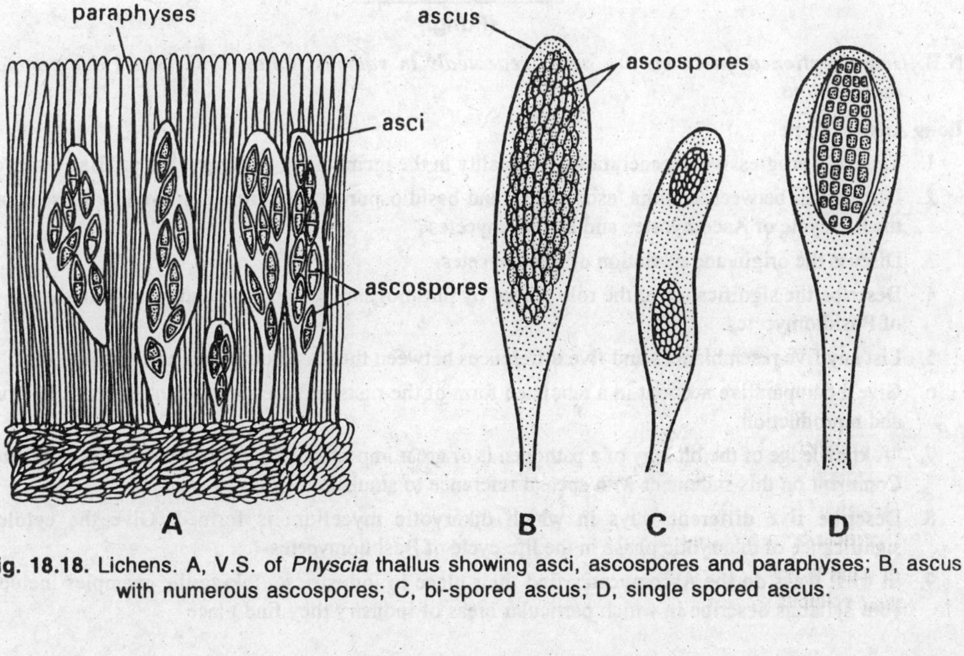

Fig. 18.18. Lichens. A, V.S. of *Physcia* thallus showing asci, ascospores and paraphyses; B, ascus with numerous ascospores; C, bi-spored ascus; D, single spored ascus.

Basidiolichens reproduce by basidiospores produced on basidia as in typical Basidiomycetes. Lower surface of the thallus bears **subhymenium,** and basidia are arranged palisade-like on the lowermost face of each subhymenium. Each basidium bears four basidiospores at the tips of sterigmata.

Economic importance. Many lichens are economically important in following ways:

1. **Ecological importance.** Many crustose lichens growing on rocks dissolve and disintegrate them into soil particles. Limestone rocks are dissolved by the action of certain chemicals, secreted by the lichens and the disintegration of rocks is due to stresses and strains induced by contraction and expansion of the gelatinous thalli. When lichens die and decay, they form humus soil together with rock particles, mosses, ferns and other plants grow on it. Thus they prepare ground for succession of vegetation.

2. **Food for man and animals.** In Iceland *Cetraria islandica* commonly known as **"Iceland moss"** is grounded up and mixed with wheat flour. A leathery lichen *Umbillicaria,* commonly known as **"root tripe"** has been eaten by travellers in arctic regions in danger of starvation. Two species of *Lecanora* have been used as food in the barren plains and mountains of Western Asia and Northern Africa. Certain classes of East Siberian inhabitants use lichens as vegetable diet. In Tamil Nadu a species of *Parmelia* known in Telugu as "rathapu" or rock flower has been used in curry preparation and is famous for its delicacy.

Cladonia rangifera, commonly known as **"reindeer moss"** grows in dense tufts of upto 12 inches height in extremely cold regions. It is a food for reindeers and cattle.

3. **Medicinal importance.** *Peltigera canina,* the dog lichen was used as medicine for hydrophobia in ancient days and *Lobaria pulmonaria,* **lungwort** was used for the diseases of lungs.

4. **Chemical uses.** The lungwort lichen was also used in tanning, in perfumery and as a substitute for hops in brewing. The cell walls of the fungi of certain lichens contain colouring matters. Species of *Rocella* and *Lecanora* yield a most important colouring matter known as Orchil or Cudbear. This is used in colouring woollen and silk fabrics.

QUESTION BANK

(Fungi)

N.B. *Undermentioned questions are asked repeatedly in various Indian Universities and competitive examinations.*

Long Answer Type

1. Trace the progressive degeneration of sexuality in the members of Ascomycetes and Basidiomycetes.
2. Distinguish between conidia, ascospores, and basidiospores. What role is played by these spores in the life cycle of Ascomycetes and Basidiomycetes?
3. Discuss the origin and evolution of Ascomycetes.
4. Describe the significance of the role played by plasmogamy, karyogamy and meiosis in the life-cycle of Basidiomycetes.
5. List any five resemblances and five differences between the Basidiomycetes and Ascomycetes.
6. Give a comparative account in a tabulated form of the rusts and the smuts with reference to structure and reproduction.
7. "A knowledge of the biology of a pathogen is of great importances in devising measures to control it". Comment on this statement with special reference to smuts and rusts.
8. Describe five different ways in which dikaryotic mycelium is formed. Give the cytological significance of dikaryotic phase in the life-cycle of Basidiomycetes.
9. In what ways do the Ascomycetes find their place in industries? Taking the examples included in your syllabus describe in which particular areas of industry they find place.

10. Give account of asexual reproduction in *Mucor.*
11. Describe briefly the structure and development of an ascus. How does an ascus differ from basidium?
12. Give a brief illustrated account of the life-history of a smut fungus. Discuss various kinds of its infection and possible methods of controlling it.
13. Give a comparative account of sexual reproduction in *Cystopus* and *Mucor.*
14. Give botanical names of two edible fungi along with their classification. Describe the structure and method of reproduction in any one of them.
15. Describe the habitat, structure and methods of reproduction in *Cystopus candida.* Give methods of its control.
16. Describe the structure and method of reproduction in *Ustilago.* Give methods of its control also.
17. Describe the method of asexual reproduction in *Albugo.* Name the disease caused by this fungus.
18. How does the reproduction take place in *Chaetomium.* What is its economic importance?
19. Describe the life-history of *Puccinia graminis tritici.*
20. Describe the life-history of *Puccinia* on alternate host (secondary host). Name three diseases caused by this fungus on wheat plant.
21. Describe the life-history of *Morchella.*
22. Describe the reproduction in yeasts.
23. Differentiate loose smut and covered smut. Give an account of life-cycle of *Ustilago.*
24. Describe life-cycle of gills bearing fungus of your course. Also describe structure of its basidiocarp.
25. Describe sexual reproduction of the fungus which causes, "white rust". Also explain symptoms of this disease.
26. Give an account of thallus, structure and sexual reproduction of *Aspergillus.* What light does it throw on the degeneration of sex?
27. Write an essay on annual recurrence of "wheat rust" in India. Also explain heteroecism.
28. Describe structure of the mycelium and reproduction of a 'smut fungus' studied by you.
29. Give the different biological categories of fungi based on their modes of nutrition. Write brief notes on each, mentioning at least one example of each.
30. Give the life-history and economic importance of *Saccharomyces* (yeast).
31. What material (host plant) will you collect for studying the following genera? (b) *Albugo,* (c) *Ustilago,* (d) *Puccinia.* Give the symptoms of the diseases caused by at least one species of the above genera.
32. Give a brief account of the different types of somatic structure in fungi.
33. Describe the methods of infection and control measures of the different species of *Ustilago.* Discuss how are they interlinked.
34. With the help of labelled diagrams only describe the life-cycle of *Puccinia graminis tritici.*
35. Give a comparative account of cleistothecium, perithecium, and apothecium with suitable examples.
36. On *Puccinia graminis tritici* both heterothallic and heteroecious conditions occur. Illustrate your answer with suitable sketches.
37. Write about the reproduction of a fleshy fungus of the class Basidiomycetes.
38. Describe the life-history of *Ustilago* or *Agaricus.*
39. Give systematic position, habit and habitat of *Chaetomium.* How does the reproduction take place in it and what is its economic importance?
40. Write an essay on annual recurrence of 'wheat rust' in India. Also explain heteroecism.
41. Give botanical names of two edible fungi along with their clasification. Describe the life-history of any one of them.
42. What is damping-off disease? Describe the life-cycle of the causative organism.
43. What are the different rusts of wheat? What are their distinguishing features? Explain the annual recurrence of rusts in the plain of Northern India.
44. Name any three important plant diseases occurring in India. Mention their symptoms, pathogenicity, and control measures.

45. What measures would you adopt to control the powdery mildew of peas and red rot of sugarcane and why?
46. Give a clear and illustrated account of the life-history of *Phytophthora* and explain its mode of perennation in the late blight disease of potato.

Short Answer Type

1. Give the structure of mycelium during somatic phase in the case of *Rhizopus.*
2. Draw the T. S. of a *Cystopus* infected leaf of *Brassica* showing the manner in which conidia are produced.
3. Give three modifications of fungal mycelium and mention their uses if any.
4. What type of ascocarp is produced in *Chaetomium*. Give at least two chief characters of such a structure?
5. Bring out the differences between oidia, chlamydospores, rhizomorphs and sclerotia.
6. Draw a neat labelled diagram of the L. S. of gill in *Agaricus.*
7. Very often people talk about the 'fairy rings' they most commonly see in the forests. What are they? How do you account for their appearance?
8. Illustrate the degeneration of sexuality in the fungi with examples.
9. Describe the external and internal structure of the basidiocarp of *Agaricus.*
10. Describe the sexual reproduction in *Mucor.*
11. Where are the yeasts found in nature? Mention their economic importance?
12. In what essential respects to the apothecium, perithecium and cleistothecium differ from each other?
13. Write illustrated notes on :
    (a) Clamp connection.
    (b) Buller's phenomenon
    (c) Mushroom's fairy rings
    (d) Heterothallism
    (e) Aecidial cup in *Puccinia*
14. Point out three differences between each of the following pairs.
    (i) Prosenchyma and pseudoparenchyma; (ii) Sporangiophore and conidiophore; (iii) Zygospore and oospore.
15. Explain the following terms with the help of suitable diagrams. (a) Symbiotism, (b) Parasitism, (c) Saprophytism.
16. Draw suitable diagrams of – (a) stage O of *Puccinia,* (b) V. S. of *Morchella* ascocarp, (c) sexual stage of 'damping off' discase.
17. How will you distinguish between? (a) rust and smut, (b) *Morchella* and *Agaricus* fructifications, (c) Cleistothecium, apothecium and perithecium.
18. Give the systematic position of *Cystopus, Saccharomyces, Aspergillus, Morchella, Puccinia. Agaricus, Mucor, Ustilago.*
19. How will you distinguish the following diseases in the field? (a) Smut disease, (b) white rust disease, (c) brown rust disease, (d) late blight of potato.
20. Write short notes on – (a) Mushroom cultivation, (b) algal origin of fungi.
21. Give the life-cycle of any holocarpic fungus.
22. Draw well-labelled diagrams of the sections passing through– (a) *Brassica* leaf infected with white rust disease, (b) potato leaf infected with late blight disease, (c) wheat leaf infected with brown rust.
23. Explain the following terms with the help of suitable examples–(a) Heterothallism, (b) autoecious and heteroecious forms.
24. Draw the symptoms and write the specific names of causal organisms of the following plant diseases– (i) loose smut of wheat, (ii) black rust of wheat, (iii) late blight of potato.
25. Distinguish between each of the following in pairs– (a) parasitism and saprophytism, (b) rust and smut, (c) annual recurrence of rust on wheat in India.

26. What are imperfect fungi? Describe the structure and reproduction of *Colletotrichum fatcatum*. Write the name and symptoms of the disease caused by it.
27. Distinguish between each of the following pairs–(a) parasitism and saprophytism, (b) rust and smut, (c) hypertrophy and hyperplasia.
28. Where do you expect to find the following in nature? (i) *Aspergillus,* (ii)*Saccharomyces,* (iii) *Cercospora.*
29. Write note on : Economic importance of fungi.
30. Write main identifying characters of the following giving their diagrams and importance– (a) *Saccharomyces,* (b) *Eurotium.*
31. Distinguish between any two of the following–(a) Heterothallism and heteroecism, (b) oospore and zygospore.
32. Write about the asexual reproductive structure of the following–(a) *Mucor,* (b) *Cercospora* (c) *Albugo.*
33. Describe the characteristic features of Ascomycotina.
34. Write notes on–(i) late blight of potato, (ii) fruit body of *Morchella.*
35. Why are Myxomycetes placed under fungi?
36. Write short notes on sex organs of *Pythium.*
37. Give a brief account of human diseases caused by fungi.
38. Comment on – gleba of *Agaricus.*
39. Comment on – asexual reproduction of *Chaetomium.*
40. Make labelled outline figures of the following, spore producing organs of fungi: (i) *Agaricus,* (ii) *Cercospora.*
41. By means of simple and labelled drawings, explain the structure of the reproductive organs in *Aspergillus.*
42. Describe the reproductive organs in *Pythium.*
43. Give the labelled diagram of teleutospores of *Puccinia graminis* as seen in a L. S. of stem of infected wheat.
44. Why do mushrooms frequently appear in circles are called fairy rings?
45. Write critical notes on –(a) *Colletotrichum,* (b) *Chaetomium,* (c) *Agaricus* (d) diplanetism.
46. What evidence is their to prove that Myxomycetes are fungi?
47. Trace the nuclear changes in the life-cycle of a macrocyclic rust.
48. During which process does the dikaryotization of the hyphae in *Puccinia* take place and mention the name of host in which it occurs.
49. Write short notes on the following : (i) smut disease, (ii) cleistothecium, (iii) symptoms of white rust, (iv) clamp connection.
50. Give well labelled diagrams of–(a) *Pythium* sexual stage, (b) *Aspergillus* –asexual reproduction, (c) a section through the teleutosorus, (d) ascocarp of *Morchella.*
51. Write short notes on the following– (a) smut disease, (b) economic importance of fungi, (c) ascocarp of *Morchella.*
52. Write short notes on the following– (a) Ascospore, (b) basidiospore, (c) oospore, (d) zygospore.
53. How will you distinguish the following ? (a) Zoospore and aplanospore, (b) oospore and zygospore, (c) ascospore and basidiospore, (d) uredospore and teliospore, (e) conidium and sporangiospore.
54. Discuss the mode of nutrition in any five of the following–(a) *Agaricus* (b) *Pythium* (c) *Albugo* (d) *Saccharomyces,* (e) *Chaetomium* (f) *Morchella,* (g) *Puccinia,* (h) *Agaricus,* (i) *Cercospora.*
55. Write short notes on the following : (a) Basidium (b) Ascus, (c) Sporangium, (d) Zoosporangium.
56. Compare and contrast sexual reproduction in the following pairs of genera– (a) *Rhizopus* and *Albugo,* (b) *Aspergillus* and *Chaetomium* (c) *Pythium* and *Albugo* (d) *Mucor and Phytophthora.*
57. Give names of the fungi which cause following : (a) Late blight disease of potato, (b) loose smut of wheat, (c) black rust of wheat, (d) white rust of crucifers.

Multiple Choice

1. *Erysiphe* causes the disease. (i) powdery mildews, (ii) downy mildews, (iii) covered smut, (iv) late blight of potato.
2. The coprophilic fungi inhabit. (i) dung substratum, (ii) dead wood, (iii) decaying leaves, (iv) food articles.
3. Mycotrophy is the symbiosis of a fungus with (i) bacteria (ii) algae, (iii) bryophytes, (iv) other fungi.
4. The process of self-fertilization in fungi is known as. (i) automixis, (ii) amphimixis, (iii) spermatization, (iv) somatogamy.
5. Diplanetism is exhibited by. (i) *Phytophthora,* (ii) *Saprolegnia,* (iii) *Mucor,* (iv) *Albugo.*
6. The position of the antheridium in connection with the oogonia in *Albugo* is called as paragynous when the antheridium. (i) grows beside the oogonium, (ii) encircles the oogonium, (iii) remains at the base of the oogonium, (iv) grows on the oogonial wall.
7. The fertile portion in the fuiting body of *Lycoperdon* is called (i) gleba, (ii) peridium, (iii) hymenium, (iv) gills.
8. Macrocyclic rust is the name given to some fungi, (i) which produces bigger spores, (ii) where all the five spore stages are produced, (iii) which completes its life-cycle on a single host, (iv) which selects many hosts to complete its life-cycle.
9. What is the fruit body of *Penicillium* called (i) perithecium, (ii) cleistothecium, (iii) apothecium, (iv) stroma.
10. *Rhizopus* multiplies by the production of. (i) zoospores, (ii) conidiospores, (iii) sporangiospores, (iv) chlamydospores.
11. Sclerotium is a perennating and vegetatively reproducing body, the structure of which has (i) hyphae aggregated to form strands, (ii) loosely interwoven hyphae which have not lost their identity, (iii) compactly interwoven hyphae which have lost their individuality, (iv) thick-walled hyphae giving a rounded or irregular configuration.
12. The obligate parasitic fungi absorb their nourishment from the host cells through: (i) the surface, (ii) haustoria, (iii) appressoria, (iv) rhizoids.
13. Biological specialization is a term used for fungus which : (i) can infect differential hosts, (ii) shows host specialization, (iii) can grow in a variety of substrata, (iv) are biologically useful.
14. Sporangial proliferation in *Saprolegnia* will be characterized by the : (i) development of secondary sporangium into the primary sporangium, (ii) the primary spornagiam cuts off spores from its apex, (iii) production of new sporangia from the vegetative hypha, (iv) germination of spore into a mycelium.
15. Which of the following depicts the position of antheridium in *Penicillium* in connection with the ascogonium would be: (i) coils loosely around the ascogonium, (ii) grows besides the ascogonium, (iii) remains at the base of the ascogonium, (iv) approaches the ascogonium only at its tip.
16. The fungus which is so important for its use in genetic studies is: (i) *Aspergillus,* (ii) *Rhizopus,* (iii) *Penicillium,* (iv) *Neurospora.*
17. White rust of crucifer is caused by : (i) *Puccinia,* (ii) *Ustilago,* (iii) *Cystopus,* (iv) *Peziza.*
18. Microconidia are found in : (i) *Claviceps,* (ii) *Neurospora,* (iii) *Rhizoctonia,* (iv) *Pyricularia.*
19. In *Agaricus,* the cell in which reduction division takes place is known as : (i) basidiospore, (ii)basidium, (ii) chlamydospore.
20. Coprophilous fungi are growing in : (i) grasses, (ii) dung, (iii) animals, (iv) wood.
21. Stroma is: (i) compact somatic hyphae with fruit bodies, (ii) loosely interwoven hyphae, (iii) a small hyphal branch, (iv) a group of spores.
22. Somatogamy is the : (i) fusion of gametes, (ii) fusion of vegetative cells, (iii) contact between two gametangia, (iv) copulation between two gametangia.
23. Fungus *Alternaria solani* belongs to class: (i) Ascomycetes, (ii) Deuteromycetes, (iii) Schizomycetes, (iv) Oomycetes.

24. The protective covering of sterile hyphae around an ascocarp is termed as : (i) periderm, (ii) in branched conidiophores, (iii) on unbranched conidiophores, (iv) on both branched or unbranched conidiophores.

25. In *Penicillium* conidia are produced : (i) in sori consisting of several conidiophores, (ii) in branched conidiophores, (iii) on unbranched conidiophores, (iv) on both branched or unbranched conidiophores.

26. A haustorium of a fungus is meant for: (i) fixing up to the mycelium to the host, (ii) increasing the spread of the disease, (iii) reproduction of the fungus, (iv) absorbing nourishment from the host.

27. The sexual reproduction of *Puccinia graminis* is of the type known as : (i) somatogamy, (ii) dikaryotization, (iii) sepermatisation, (iv) automixis.

28. In *Agaricus* the fruiting body is made up of : (i) tertiary mycelium, (ii) primary mycelium, (iii) secondary mycelium, (iv) diploid mycelium.

29. In the Ascomycetes karyogamy occurs within the : (i) ascogonium, (ii) antheridium, (iii) ascus, (iv) ascogenous hypha.

30. Haustoria are produced in the case of mycelium which is: (i) both intracellular and endoparasitic, (ii) ectoparasite, (iii) both intercellular and endoparasite, (iv) either ectoparasitic or intercellular.

31. Perfect stage of fungus means : (i) when the fungus is perfectly healthy, (ii) when it reproduces asexually, (iii) when it reproduces sexually, (iv) when it forms perfect resting spores.

32. Penicillin was discovered by : (i) Alexander Fleming, (ii) Edward Jenner, (iii) Louis Pasteur (iv) Ian Fleming.

33. In the fruit body of *Agaricus* basidia are produced on the : (i) gills, (ii) pileus, (iii stipe, (iv) rhizomorph.

34. A macrocyclic fungus is the one which (i) needs two different hosts to complete its life-cycle, (ii) produces many types of spores to complete the life-cycle, (iii) does not show any asexual reproduction, (iv) has a prolonged life cycle.

35. The gametes taking part in the sexual reproduction of *Rhizopus* are: (i) uninucleate, (ii) binucleate, (iii) multinucleate, (iv) dikaryotic.

36. The classification of the fungi is based mainly on : (i) The structure of vegetative mycelium, (ii) the asexual stage, (iii) the sexual reproductive stage, (iv) both the mycelial structure and sexual stages.

37. A phragmo-basidium means: (i) an entire basidium which is fully reproductive, (ii) a septate basidium which for all purposes is one structure, (iii) a septate basidium where only one cell is reproductive, (iv) a septate basidium where both cells are reproductive.

38. The phenomenon of heterothallism was first discovered in Mucorales by : (i) Charles Bessey, (ii) Gaumann, (iii) Blakeslee, (iv) Alexopoulos.

39. The dikaryotic mycelium of heterothallic forms is characterized by having in each of its cells: (i) single 2 n nucleus, (ii) two diploid nuclei belonging to opposite strains, (iii) two halpoid nuclei belonging to opposite strains, (iv) two haploid nuclei of similar strains.

40. A sclerotium refers to a modified mycelium which is: (i) an underground structure, (ii) a hard resting body, (iii) mainly a food storing organ, (iv) easily carried off by wind.

41. In *Phytophthora* the asexual reproductive bodies behave as : (i) conidia, (ii) sporangia, (iii) conidiosporangia, (iv) both conidia and conidiosporangia.

42. The azygospores produced in Mucorales are formed from : (i) zygotes, (ii) unfertilized gametangia (iii) vegetative myceluim, (iv) the female sex organ.

43. Early blight of potato is caused by : (i) *Albugo candida,* (ii) *Phytophthora infestans,* (iii) *Altenaria solani.*

44. The phenomenon of heterothallism was observed for the first time in the order: (i) Erysiphales, (ii) Mucorales, (iii) Ustilaginales, (iv) none of above.

45. Tikka disease of groundnut is caused by : (i) *Aspergillus,* (ii) *Puccinia,* (iii) *Cercospora,* (iv) *Fusarium.*

46. The name 'smut diseases' is given to those produced by *Ustilago* because : (i) its mycelium is black in colour, (ii) it parasitizes cereals, (iii) the host becomes completely black, (iv) the fungus produces black sooty spore masses.
47. White rust of crucifers is a pseudo-rust because : (i) the disease is not caused by basidiomycetous members, (ii) the colour of the pustule is not red, (iii) the disease is seen on crucifers, (iv) the disease is not seen on wheat.
48. Wilt of arhar is caused by : (i) *Pythium,* (ii) *Alternaria,* (iii) *Colletorichum,* (iv) *Fusarium.*
49. The whip smut of sugarcane is caused by: (i) *Ustilago maydis,* (ii) *Ustilago hordei,* (iii) *Ustilago scitaminea,* (iv) *Ustilago nuda.*
50. Downy mildews are caused by the members of : (i) Erysiphales, (ii) Taphrinales, (iii) Ustilaginales, (iv) Peronosporales.
51. The rusts are caused by : (i) Ustilaginales, (ii) Peronosporales, (iii) Uredinales, (iv) Erysiphales.
52. The wall of hyphae of *Rhizopus* is made up of : (i) cellulose, (ii) callose, (iii) pectin, (iv) chitin.
53. *Rhizopus* resembles a moss in the both develop: (i) mycelia, (ii) hyphae, (iii) archegonia, (iv) spore.
54. Penicillin was extracted by : (i) Flemming, (ii) Huxley, (iii) Lamarck, (iv) Brown.
55. Yeast is an important source of : (i) vitamin C, (ii) riboflavin, (iii) sugar, (iv) protein.
56. Fungi occurring on wood are : (i) epibiotic, (ii) eucarpic, (iii) epixylic, (iv) epigean.
57. Which is an ediable fungus : (i) *Rhizopus,* (ii) *Mucor,* (iii) *Agaricus,* (iv) *Polyporus.*
58. Gibberelin was first extracted from : (i) Bacteria, (ii) Fungi, (iii) virus, (iv) algae.
59. The mushroom is : (i) a plant consisting of fine green threads, (ii) an edible fungus, (iii) a bryophyte devoid of root, (iv) a flowering plant.
60. The brown gills of the mushroom : (i) have no function to perform, (ii) are meant for its respiration, (iii) help the plant to float in water after heavy rains, (iv) bears spores which help in reproduction.
61. Toadstools cannot manufacture their own food because : (i) they do not have roots, (ii) they do not have leaves, (iii) they do not have chlorophyll, (iv) they do not need food for their growth.
62. Reproduction in 'fairy rings' occurs by means of : (i) seeds, (ii) spores, (iii) flowers, (iv) gills.
63. Fungi are always : (i) parasitic, (ii) saprophytic, (iii) autotrophic, (iv) heterotrophic.
64. An organism which is normally a saprophyte, but can also become a parasite is called : (i) facultative saprophyte, (ii) partial saprophyte, (iii) faculltive parasite, (iv) partial parasite.
65. Which of the following is a good example of heterothallism : (i) *Pteris,* (ii) *Rhizopus,* (iii) *Cycas,* (iv) castor bean.
66. Why zygospores are not generally formed in a culture of *Rhizopus* developed from a single spore : (i) due to lack of light, (ii) due to shortage of oxygen, (iii) due to absence of plus (+) and (–) strains of mycelium.
67. Columella is present in the sporangium of : (i) *Spirogyra,* (ii) yeast, (iii) *Ulothrix,* (iv) *Rhizopus.*
68. A mushroom is: (i) saprophyte, (ii) photosynthetic organism, (iii) facultative parasite, (iv) obligate parasite.
69. What is the mode of nutrition in *Rhizopus,* (i) autotrophic, (ii) parasitic, (iii) symbiotic, (iv) saprophytic.
70. Which of the following diseases is caused by a fungus : (i) cholera, (ii) rust of wheat, (iii) T. B., (iv) tetanus.
71. Mycology is the study of : (i) Algae, (ii) Fungi, (iii) Bryophytes, (iv) Pteridophytes.
72. Which of the following is an edible fungus : (i) *Rhizopus,* (ii) *Penicillium,* (iii) *Mucor,* (iv) *Agaricus.*
73. Which of the following diseases is caused by a fungus : (i) small-pox, (ii) tuberculosis, (iii) cancer, (iv) black rust of wheat.
74. Which of the following is a good example of heterothallism : (i) *Spirogyra,* (ii) *Rhizopus,* (iii) *Pinus,* (iv) castor bean.

75. In which of the following, respiration in absence of oxygen too takes place : (i) man, (ii) yeast, (iii) potato, (iv) *Spirogyra.*

76. Gills are seen in ; (i) bacteria, (ii) *Osscillatoria,* (iii) *Ulothrix,* (iv) *Agaricus.*

77. The zygospore of *Mucor* is thick-walled and its colour is : (i) blue, (ii) white, (iii) green, (iv) black.

78. The vegetative cells of the *Saccharomyces* are recognised by the presence of : (i) chloroplasts, (ii) a large vacuolated nucleus, (iii) a small nucleus without a nuclear membrane, (iv) a distinct cell wall.

79. Heterothallism was discovered by : (i) Bessey, (ii) Blakeslee, (iii) Alexopoulos, (iv) Leuwenhoek.

80. The structure in which the ascospores are formed in : (i) basidium, (ii) sporangium, (iii) ascus, (iv) gametangium.

81. Fungal hyphae penetrate hard cell wall of their host with the help of : (i) enzymes, (ii) hormones, (iii) sharp tips, (iv) haustoria.

82. Loose smut of wheat is caused by : (i) *Ustilago tritici,* (ii) *Cystopus,* (iii) *Puccinia,* (iv) *Aspergillus.*

83. Obligate parasites are those organisms which : (i) are essentially saprophyte can become parasite, (ii) are essentially parasite but can also become saprophyte, (iii) live only in dead and decaying organic matter, (iv) live only in living hosts.

84. Fermentation of sugar occurs by : (i) *Mucor,* (ii) *Saccharomyces,* (iii) *Rhizopus,* (iv) *Penicillium.*

85. Fungal spores produced asexually at the tip of hypha are called: (i) conidia, (ii) sporangiophores, (iii) spores, (iv) arthrospores.

86. Mushroom is : (i) plant consisting of fine green threads, (ii) an edible fungus, (iii) bryophytes, (iv) flowering plants.

87. Same haploid structures of *Rhizopus* include the : (i) mycelia, sporangia and spores, (ii) hyphae, zygote and sporangia, (iii) mycelia, zygospore and spores, (iv) mycelia, zygospore and suspensor.

88. Eucarpic fungi are those: (i) in which the habitat is saprophytic, and mycelium coenocytic, (ii) in which reproductive organs arise from a part of thallus while rest carries out of somatic function, (iii) in which entire thallus may be converted into reproductive structure so that somatic and reproductive phases do not occur in same individual.

89. Fungal spores produced asexually at tips or side of hyphae are called: (i) Sporangiospores, (ii) Arthrospores, (iii) Conidia (iv) Spores.

90. When two host species are required for completion of parasitic fungi life-cycle, this condition is described as: (i) autoecism, (ii) autotrophic, (ii) heteroecism, (iv) heterokaryotic.

91. Mycorrhiza is a term to indicate: (i) fungus association with stem, (ii) bacteria association with root, (iii) fungi association with root, (iv) study of fungi.

92. After fusion of sexual gametes in *Mucor* the resultant structure is known as : (i) oospore, (ii) cleistothecium, (ii) zygospore, (iv) zygote.

93. Heterothallism was discovered by : (i) Blakeslee, (ii) Faraday, (iii) Alexopoulos, (iv) Gaumann.

94. Clamp connections are very common in : (i) Ascomycetes, (ii) Basidiomycetes, (iii) Phycomycetes, (iv) Deuteromycetes.

95. Haplo-diplo-biontic life-cycle is exhibited by (i) yeast, (ii) *Mucor,* (iii) *Penicillium,* (iv) *Aspergillus.*

96. The worker associated with mycology : (i) S. R. Kashyap, (ii) A. J. Eames, (iii) A. F. Blakeslec, (iv) Y. Bhardwaja.

97. Heterothallism in Mucorales was discovered in : (i) 1904, (ii) 1918, (iii) 1927, (iv) 1931.

98. Sort out the father of Indian mycology and plant pathology: (i) K. C. Mehta, (ii) B. P. Pal, (iii) E. J. Butler, (iv) B. B. Mundkur.

99. The aseptate mycelium is found in : (i) lower fungi, (ii) higher fungi, (iii) fungi imperfecti, (iv) none of above.

100. The mycelium of *Erysiphe* is : (i) aseptate, (ii) septate, (iii) both (i) and (ii), (iv) none of abvoe.

101. *Neurospora sitophila* is known as : (i) Black mold, (ii) red mold, (iii) blue mold, (iv) green mold.

102. The spermatization takes place in (i) *Neurospora,* (ii) *Penicillium,* (iii) *Peziza,* (iv) *Erysiphe.*

103. One of the following represents cup fungi : (i) *Peziza,* (ii) *Morchella,* (iii) *Agaricus,* (iv) *Amanita.*

104. *Peziza* is a member of : (i) Phycomycetes, (ii) Ascomycetes, (iii) Basidiomycetes, (iv) Deuteromycetes.

105. The basidiospores are : (i) exogenous spores, (i) endogenous spores, (iii) both (i) & (ii), (iv) none of above.

106. Black rust of wheat is caused by : (i) *Puccinia graminis*, (ii) *Puccinia recondita*, (iii) *Puccinia striiformis*, (iv) *Puccinia glumarum*.

107. The sexual reproduction lacks in : (i) Ascomycetes, (ii) Basidiomycetes, (iii) Phycomycetes, (iv) Deuteromycetes.

108. Only asexual reproduction is found in : (i) Ascomycetes, (ii) Basidiomycetes, (iii) Oomycetes, (iv) Deuteromycetes.

109. *Alternaria solani* causes : (i) late blight of potato, (ii) wart of potato, (iii) early blight of potato, (iv) leaf curl of potato.

110. 'Target board effect' is caused by : (i) *Alternaria*, (ii) *Colletotrichum*, (iii) *Pyricularia*, (iv) *Helminthosporium*.

ANSWERS

1 (i), 2 (ii), 3 (iv), 4 (iv), 5 (i) 6 (i), 7 (i), 8 (ii), 9 (ii), 10 (iii), 11(iv), 12 (ii), 13 (ii), 14 (i), 15 (iv), 16 (iv) 17 (iii), 18 (ii), 19 (ii), 20 (ii), 21 (i), 22 (ii), 23 (ii), 24 (ii), 25 (ii), 26 (iv), 27 (ii), 28 (iii), 29 (iv), 30 (iv), 31 (iv), 32 (i), 33 (i), 34 (ii), 35 (iii), 36 (iv), 37 (ii), 38 (iii), 39 (iii), 40 (ii), 41 (ii), 42 (ii), 43 (iii), 44 (ii), 45 (iii), 46 (iv), 47 (i), 48 (iv), 49 (iii), 50 (iv), 51 (iii), 52 (iv), 53 (iv), 54 (i), 55 (iv), 56 (iii), 57 (iii), 58 (ii), 59 (ii), 60 (iv), 61 (iii), 62 (ii), 63 (iv), 64 (iii), 65 (ii), 66 (iii), 67 (iv), 68 (i), 69 (iv), 70 (ii), 71 (ii), 72 (iv), 73 (iv), 74 (ii), 75 (ii), 76 (iv), 77 (iv), 78 (ii), 79 (ii), 80 (iii), 81 (iv), 82 (i), 83 (iv), 84 (ii), 85 (iv), 86 (ii), 87 (i), 88 (iii), 89 (ii), 90 (iii), 91 (iii), 92 (iii), 93 (i), 94 (ii), 95 (i), 96 (iii), 97 (i), 98 (iii), 99 (i), 100 (ii) 101 (ii), 102 (i), 103 (i), 104 (ii), 105 (i), 106 (i), 107 (iv), 108 (iv), 109 (iii), 110 (i).

UNIT - 4

Diversity of Cryptogams Bryophyta

19

CHAPTER

Bryophytes: Introduction

There are 960 genera and 24,000 species of bryophytes.

From the point of view of their evolution the bryophytes stand at level higher than that of thallophytes but lower than that of pteridophytes and phanerogams. The plants belonging to algae are commonly found in water and rarely on the land whereas the bryophytes are land inhabiting plants. The water is still needed for the movement of the gametes of bryophtes and in their vegetative structure they have adapted themselves to a terrestrial life. The bryophytes are known as **amphibians** of the plant kingdom. Some of the bryophytes are inhabitants of water, *e.g.*, *Riccia fluitans.* The other bryophytes like the swampy muddy shady and humid places for their survival. Most of the bryophytes are land inhabitants. In our country the bryophytes are very commonly found in the hilly regions of the Himalayas. They are also quite common in the hills of South India, whereas on the other hand certain species of *Riccia, Marchantia, Anthoceros* and *Funaria* are quite common in the plains.

Distribution. The bryophytes are cosmopolitan in distribution. They are found in all places where plants can live except in the sea. They are quite commonly found in the moist mountain forests of tropical regions as well as in the sub-tropical regions of Arctic Tundra. Some of the bryophytes have been recorded at altitudes as high as 18,000 to 20,000 feet a.s.l. in the Himalayas. A moss plant, *Aongstroemia julacea* has been recorded from an altitude of 19,800 feet a.s.l.

Many members of Jungermannineae (acrogynous Jungermanniales), Musci (mosses) and all species of *Dendroceros* are epiphytes and found in tropical rain forests. Certain small species are epiphyllous and grow on the surface of leaves (*e.g.*, *Radula protensa*).

Most of the bryophytes are autorophic but a few exceptions are there. For example, *Buxbaumia aphylla* and *Mnium hornum* are more or less saprophytic and grow upon organic matter such as rotten wood.

Characteristic features. The most characteristic features of the bryophytes are as follows:

The bryophytes occupy the position in between algae on one hand and the pteridophytes on the other.

With the exception of few aquatic forms they are truly land-inhabiting plants. They are found in humid and shady places. As water is indispensable for the act of fertilization they are treated to be the **amphibians** of the plant kingdom.

They are quite small and inconspicuous organisms. The gametophyte is highly developed and differentiated than that of a complex alga. It is independent plant at maturity. It consists of a flattened thallus (*e.g.*, *Riccia, Marchantia*) or an erect plant body (*e.g.*, *Funaria, Polytrichum,* etc.).

The true roots are always absent. Instead of true roots the unicellular or multicellular hair-like **rhizoids** develop from the thalli. The rhizoids absorb the nutrients from the moist soil.

They lack typical vascular tissue, *i.e.*, xylem and phloem from their gametophytes and sporophytes.

They are **homosporous.** It means, that the spores of a species are morphologically similar in size and form.

The spore germinates into a filamentous or thalloid green **protonema** which later on gives rise to the thallus (gametophyte).

The sexual reproduction is of oogamous type, *i.e.,* it takes place by means of gametes. The male gametes are motile and known as **antherozoids;** the female gametes are non-motile and known as **eggs** (oosphere). The gametes are produced within the sex organ known as **antheridium** (male) and **archegonium** (female).

The antheridia are usually club shaped. Each antheridium is surrounded by a single layer of protective sterile jacket cells. Within the jacket there are **antherozoid mother cells** or **androcytes**. Each androycyte metamorphoses into a motile biflagellate antherozoid.

The archegonia are usually flask-like. Each archegonium consists of a **venter** and a **neck.** The venter is basal swollen portion and the neck is elongated. Within the neck and the venter there is an axial row of cells surrounded by sterile jacket cells. This axial row consists of a few **neck canal cells, a ventral canal cell** and an **egg** or **oosphere.**

The water is essential for the act of fertilization. The motile ciliated antherozoids swim in the film of water and reach to the neck of an archegonium. The antherozoid enters the neck and ultimately approaches the egg. The antherozoid penetrates the egg and fertilization is effected. With the result of fertilization the zygote is formed. The zygote begins to develop into a multicellular embryo just after the act of fertilization without going under any resting period. The embryo remains within the venter of the archegonium and liberated as in the case of algae. The venter wall enlarges along with the developing embryo to form the protective envelope known as **calyptra.**

Ultimately the **sporogonium (sporophyte)** develops. The sporogonium consists of a **foot, seta** and **capsule.** In the case of bryophytes the sporogonium is completely dependent for its nourishment upon the gametophyte. The foot of the sporophyte remains embedded in the tissues of gametophyte and acts as haustorium.

The **spore mother cells** are produced within the sporogonium. The spore mother cells are diploid (2n) and they represent the last stage of the sporophyte generation. The spore mother cells undergo the reduction division (meiosis) and the tetrads of the haploid (n) spores are formed. The spores represent the beginning of the gametophyte generation.

SUMMARY OF A TYPICAL BRYOPHYTE

Habitat

1. The bryophytes grow frequently in humid and shady places. With the exception of few aquatic forms (*e.g., Riccia fluitans, Ricciocarpos natans* and *Riella* spp). They are truly land inhabiting plants. They may grow in bogs (*e.g., Sphagnum* spp.), and in xeric conditions (*e.g.,* many mosses); some are epiphyllous (*e.g., Frullania* spp., *Radula* spp., etc.), and some are saprophytes (*e.g., Buxbaumia* spp). *Porella plathypylloidea* grows on barks of trees and rocks and can face adverse desiccation.

Structure of Gametophyte

2. They have sharply defined heteromorphic alternation of generations in which the diploid (2n) asexual generation (*i.e.,* sporophyte), although morphologically distinct from the haploid (n) sexual generation (*i.e.,* gametophyte), is attached to it and never becomes an independent free living plant.

3. The gametophytic phase is the dominant one of the life cycle and generally the term 'plant body' is used to denote this stage.

4. The gametophyte is always an independent plant at maturity and is nutritionally self-sufficient because of the presence of chloroplasts within it.

5. In many cases the gametophyte is a thallus (*e.g., Riccia, Marchantia, Pellia,* etc.), that is, a plant body without differentiation into root, stem and leaf. In many mosses the plant body is externally differentiated into stem and leaf. In many mosses the plant body is externally differentiated into stem and leaves, but there are no roots and instead only one celled or multcellular rhizoids (absorptive in function) are present.

6. The thalli of primitive forms remain attached to the substratum, by unicellular unbranched rhizoids (*e.g.*, in *Riccia, Marchantia,* etc.), whereas in higher forms they remain attached by branched multicellular rhizoids (*e.g.*, in mosses).

7. In some bryophytes (*e.g.*, *Riccia* and *Marchantia)* the purple coloured multicellular scales are present.

8. As regards the internal structure, the plant body consists of simple parenchymatous cells. They do not possess xylem and phloem. The parenchynia may be differentiated into **chlorophyllous** and storage cells (*e.g.*, in *Riccia* and *Marchantia).*

Reproduction

In bryophytes, the reproduction takes place by means of vegetative (asexual) and sexual methods.

Vegetative Reproduction

9. Bryophytes largely multiply by means of **vegetative** reproduction. This takes place by death of older parts, *e.g.*, in *Riccia, Marchantia, Anthoceros, Notothylas*; by branch tips as in several species of *Riccia*; by adventitious branches, as in several species of *Riccia, Marchantia* and *Anthoceros;* by gemmae as in the species of *Marchantia, Lunularia, Funaria,* etc.; by tubers as in species of *Riccia, Anthoceros,* etc., by primary and secondary protonema as in species of *Funaria* and other mosses.

Sexual Reproduction

10. All bryophytes are oogamous. The spermatozoids (male gametes) are motile and small while the eggs (female gametes) are large and non motile.

11. The gametes (male and female) are produced within multicellular sex organs (antheridia and archegonia) in which there is an outer sterile layer of jacket cells.

12. The antheridium (male organ) consists of a central mass of androcytes enclosed by a single layer of sterile jacket cells. Each androcyte metamorphoses into a biflagellate spermatozoid.

13. The archegonium (female organ) is a multicellular flask-like structure. The basal swollen portion is venter and contains an egg (oosphere) in it. The elongated neck contains neck canal cells.

Fertilization

14. The water is essential for the act of fertilization. The motile flagellate spermatozoid swims to the neck of archegonium and ultimately approaches the egg effecting the fertilization. With the result the zygote (2n) is formed.

The Sporophyte

15. The zygote divides repeatedly immediately after fertilization. There is no resting period.

16. The first division of zygote is transverse and the embryo develops from the upper cell. Such emboryogeny is known as **exoscopic embryogeny.**

17. The embryo remains within the venter of the archegonium. The venter wall enlarges along with the developing embryo to form the protective envelope known as **calyptra.**

18. Generally, the sporongonium consists of **foot, seta** and **capsule.**

19. Here the sporogonium is completely dependent for its nourishment upon the gametophyte.

20. The spore mother cells are developed in the sporogonium. They are diploid (2n) and represent the last stage of the sporophytic generation. The spore mother cells (2n) undergo meiosis and the tetrads of haploid (n) spores are formed. All bryophytes are homosporous.

The Young Gametophyte

21. The spores (n) represent the beginning of the gametophyte generation.

22. The spores are cutinized and dispersed by means of wind.

23. The spores may germinate directly into the new gametophyte (*e.g.*, in *Riccia* and *Marchantia*) while in mosses they germinate to produce filamentous protonema from which buds are produced to develop into new plants (young gametophytes).

ALTERNATION OF GENERATIONS

The life-cycle of a bryophyte shows regular alternation of gametophytic and sporophytic generations. This process of alternation of generations was demonstrated for the first time in 1851 by Hofmeister. Thereafter in 1894 Strasburger could actually show the periodic doubling and halving of the number of chromosomes during the life-cycle.

The haploid phase (n) is the gametophyte or sexual generation. It bears the sexual reproductive organs which produce gametes, *i.e.*, **antherozoids** and **eggs.** With the result of gametic union a zygote is formed which develops into a sporophyte. This is the diploid phase (2n). The sporophyte produces spores which always germinate to form gemetophytes. During the formation of spores, the spore mother cells divide meiotically and haploid spores are produced. The production of the spores is the beginning of the gametophytic or haploid phase. The spores germinate and produce gametophytic or haploid phase. The spores germinate and produce

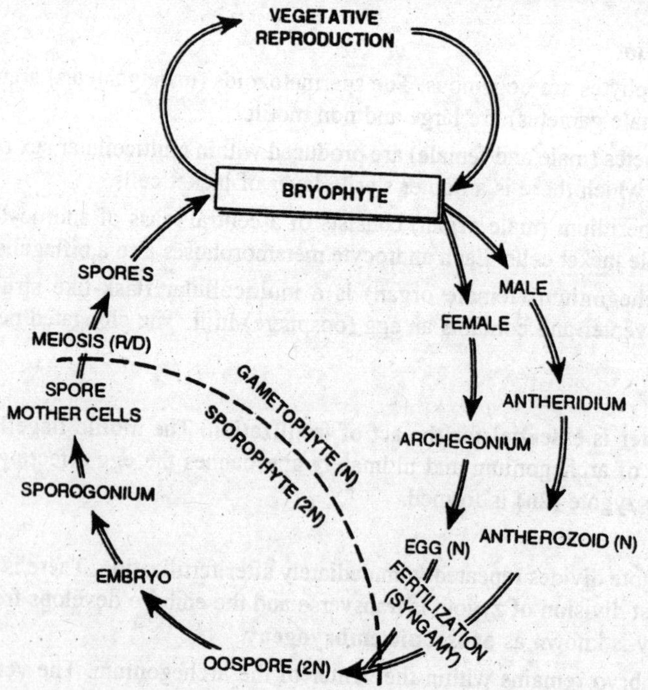

Fig. 19.1. Bryophyte. Typical graphic life-cycle.

gametophytes which bear sex organs. Ultimately the gametic union takes place and zygote is resulted. It is diploid (2n). This is the beginning of the sporophytic or diploid phase. This way, the sporophyte generation intervenes between fertilization (syngamy) and meiosis (reduction division); and gametophyte generation intervenes between meiosis and fertilization.

In bryophytes, where the two generations are morphologically different, the type of alternation of generations is known as **heteromorphic.**

In the case of bryophytes the gametophyte generation is conspicuous and longer-lived phase of the life-cycle in comparison to that of sporophyte generation. Here, the gametophyte is quite independent whereas the sporophyte is dependent somehow or other on the gametophyte for its nutritive supply. The gametophyte gives rise to sporophyte and sporophyte to the gametophyte and thus there is regular alternation of generations.

Resemblances of Bryophytes with Algae

In many ways the bryophytes resemble with algae in their morphology and physiology. More probably they are near to green algae. The main resemblances are as follows:

1. The plant body is thallus-like.
2. The plants are autotrophic, *i.e.,* similar in mode of nutrition.
3. The pigments (*i.e.,* chlorophyll, carotene, leutin, violaxathin and zeaxanthin) are similar in both.
4. Reserve food material is starch in both cases.
5. The vascular tissues are absent in both.
6. Roots are absent in bryophytes as well as in algae.
7. The cell wall is made up of cellulose in both.
8. The gametophyte is dominant both in bryophytes and algae.
9. Spermatozoids are motile and flagellated in bryophytes and many algae (mostly Chlorophyceae).
10. The spermatozoids are biflagellate and both the flagella are of whiplash type.
11. Water is required for fertilization.
12. The protonema, during early stage of development in many bryophytes resemble with the filamentous green algae.

The above mentioned similarities put the bryophytes akin to the green algae. This gives support to the view that bryophytes have evolved from algal ancestors.

Resemblances of Bryophytes with Pteridophytes

1. They are terrestrial in habit.
2. The primitive root-less and leafless sporophyte of Psilotales (pteridophytes) can be compared with sporophytes of the bryophytes.
3. In both cases the development of embryo stage after gametic union is similar.
4. The structure of archegonium is fundamentally similar in both cases and the oosphere (egg) remains protected by a layer of sterile cells.
5. In both cases the androcytes or antherozoid mother cells remain enclosed in a single layer of sterile jacket cells.
6. Antherozoids (spermatozoids) are flagellated.
7. Water is essential for fertilization.
8. Sexual reproduction is oogamous.
9. In bryophytes, as well as in pteridophytes a sharply defined heteromorphic type of alternation of generations is present.
10. Cuticle is present in both cases.
11. The stomata are present except in the liverworts.
12. The multicellular sporangia are present.

Points of Differentiation

1. In pteridophytes the plant body is sporophyte, while in bryophytes it is gametophyte.

2. In pteridophytes the sporophyte is independent and the dominant phase of life cycle whereas in bryophytes the sporophyte is dependent on gametophyte.

3. The plant body (*i.e.,* sporophyte) of a pteriodophyte consists of root, stem and leaves while in bryophytes the plant body (*i.e.,* gametophyte) is thalloid.

4. In pteriodophytes the vascular tissues are well developed whereas in bryophytes they are not developed.

Table 19.1. DIFFERENCES BETWEEN THALLOPHYTES AND BRYOPHYTES

Thallophyta (Algae)	Bryophyta
1. They are mostly aquatic.	1. Mostly terrestrial and prefer damp and shady places.
2. The thallus consists of a single cell to well developed uniseriate or branched filaments.	2. The thalles is not filamentous except in protonema stage. It is made up of parenchymatous cells.
3. There is no or ill defined tissue differentiation	3. There is well defined tissue differentiation.
4. In each cell one or few chloropasts are present. In other cases instead of chloroplasts, the chromatophores are present.	4. In each chlorophyllous cell many well developed chloroplasts are present.
5. The stomata or pores are not present	5. The pores or stomata are present.
6. Generally rhizoids are not present and if present they are of simple type	6. The rhizoids are present. They may be smooth walled, tuberculate or obliquely septate. The scales are also present.
7. Every cell is capable of growth and development.	7. Only the special cells are capable of growth and development.
8. Asexual reproduction takes place by means of zoospores or aplanospores. In many cases vegetative reproduction is commonly found.	8. Asexual reproduction is absent. Vegetative reproduction is common.
9. Sexual reproduction may be isogamous, anisogamous or oogamous.	9. Sexual reproduction is only oogamous.
10. The sex organs are unicellular, and when multicellular every cell forms a gamete. There is no jacket of sterile cells.	10. The sex organs are multicellular and remain protected by a jacket of sterile cells.
11. The female sex organ is single-celled oogonium. It is not covered by any sterile jacket layer.	11. The female sex organ is flask-shaped, multicellular archegonium which remains covered by a sterile jacket layer. Neck canal cells, ventral canal cell and egg are found within archegonium.
12. The zygote is liberated from the plant and passes into the resting stage.	12. The zygote is not liberated from the plant and does not undergo any resting period.
13. No embryo is formed after fertilization.	13. In all bryophytes the embryo is developed after fertilization.

14. They show ill defined homologous type of alternation of generations.	14. They show well defined heteromorphic type of alternation of generations.
15. The gametophyte and sporophyte are independent.	15. The sporophyte is dependent on gametophyte for its nutrition.

ORIGIN OF BRYOPHYTES

Origin of the Gametophyte

The bryophytes are quite soft and delicate and, therefore, they lack fossil records. There are no known fossil bryophytes more primitive than the forms of to-day. However, there are two schools of thought about their origin. According to one school of thought they are evolved from the green thallophyta the algae; and according to the other school they have been descended from the pteridophytes. Majority of the workers support their origin from the algal ancestors.

Origin from algae. This view of the origin of bryophytes has been supported by most of the bryologists. Though there is no fossil connection between algae and bryophytes yet there are so many points in support of this view, such as-the necessity of water for the act of fertilization; their amphibian nature and the presence of ciliated antherozoids. These points support the view that they have been originated from aquatic ancestors. Lignier in 1903, pointed out that the algae gave rise to a connecting link known as **'prohepatics'** and thereafter bryophytes originated from this connecting link on one hand and the pteridophytes on the other. Bower (1908) also supported this view and said that the Archegoniatae have been evolved from the aquatic ancestors, *i.e.,* the algae. The bryophytes resemble in many respects the green algae, *i.e.,* Chlorophyceae, and Fritsch (1916, 1945) has advocated that the Chaetophorales gave rise to the bryophytes.

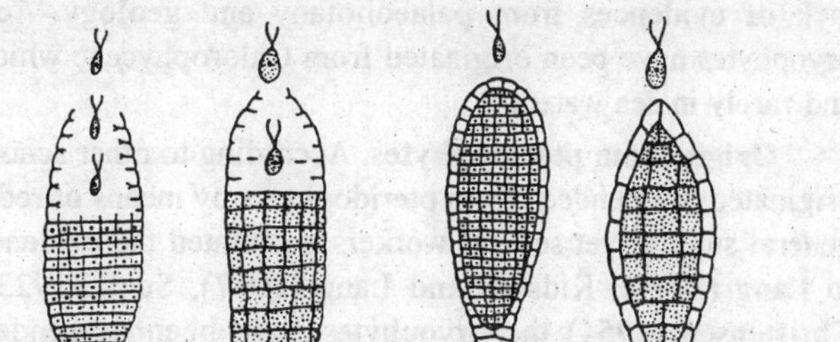

Fig. 19.2. Origin of bryophytes. Diagrams (A-D) illustrating the hypothetical origin of the antheridium and archegonium from the plurilocular sporangium of certain Phaeophyceae (*Ectocarpus*).

There seems no apparent relation between the antheridium and the archegonium of the bryophytes and the antheridium and the oogonium of the algae. In none of the algae the egg is surrounded by any cellular jacket as it is always enclosed within a protective layer (jacket layer) in the case of bryophytes.

According to many workers the sex organs of the bryophytes have been evolved from those of the algae as follows: According to this view the antheridium and archegonium of broyphytes originated from gametangia of a type similar to that of *Ectocarpus.* In *Ectocarpus* (Phaeophyceae) the gametangium consists of a number of cells, each of which gives rise to a gamete. As soon as the migration from the water to land took place, there arose the necessity for the protection of the gametes from desiccation. With the result the outer layer of the cells of the gametangium became sterile and functioned as a protective layer. This way, the antheridium has been derived from the algal gametangium. For the derivation of the archegonium from such structure , it has been suggested that after the formation of the protective wall, further sterilization took place, and in the centre an axial row of cells developed. According to this view the neck canal cells were originally female gametes, which later on lost their walls and cytoplasm. The ventral canal cell is the sister cell of the oosphere and very rarely it may be fertilized. However, *Ectocarpus* is not a member of Chlorophyceae, but it is presumed that bryophytes have been originated from green algae. According to Smith (1938) the reproductive cells of *Schizomeris* and antheridia of *Chaetonema* are quite alike to that of the gametangia of *Ectocarpus*.

According to Church (1919), the bryophytes have been originated from the marine ancestors and not from the fresh water ancestors. This theory could not get general support because of the lack of evidences from palaeobotany and geology. According to majority of workers the bryophytes have been originated from Chlorophyceae which are commonly found in fresh waters and rarely in sea waters.

Origin from pteridophytes. According to other school of thought the bryophytes have been originated (descended) from pteridophytes by means of reduction. Though this view could not get general support yet several workers postulated the evidences in support of this view. According to Lang (1917), Kidston and Lang (1917), Scott (1923), Halle (1936), Haskell (1949) and Christensen (1954) the bryophytes have been descended by the process of reduction from pteridophytes. Kashyap (1919) also supported the view, because of common resemblances of the two groups. Similarities between sporangia of some members of Psilophytales (*Rhynia, Horneophyton* and *Sporogonites*) with capsules of Anthocerotales, *Sphagnum* and *Andreaea* led to conclude this hypothesis. The Psilophytales are the oldest pteridophytes in which the sporophytes were rootless, leafless and dichotomously branched with terminal sporangia. Such sporophytes resemble the bryophytes, especially the members of Anthocerotales and are thought to have evolved by progressive reduction. Proskaeur (1960), thinks that if bryophytes are polyphyletic in origin, at least Anthocerotales originated from Psilophytales like *Horneophyton*. According to Kashyap (1919), "bryophytes represent a degenerate evolutionary line of pteridophytes or in more correct term, the bryophytes are descendents of pteridophytes."

Origin of the Sporophyte

It is an established fact now that an alternation of independent gametophytic and sporophytic generations has been evolved in several independent lines among aquatic algae. It is also thought that a migration from an aquatic to a land habit is not essential for the appearance of an alternation of generations. The algae have both **isomorphic** and **heteromorphic alternation of generations**.

Different views have been proposed concerning the origin of the sporophytes of bryophytes. There are *two* main theories — 1. **Modification (homologous) theory** and 2. **Intercalation (antithetic) theory.** According to modification or homologous theory the sporophyte is a direct modification of the gametophyte. The supporters of this theory advocate that the sporophyte is to be interpreted as a neutral generation and its primary function is the production of spores. The facts which support this theory are - the presence of isomorphic alternation in some algae; photosynthesis in sporophytes of bryophytes; presence of tracheids in gametophytes of pteridophytes; apogamy and apospory. Supporters of the modification theory (Church, 1919; Fritsch, 1945; Zimmermann, 1932) are of the opinion that the two generations were isomorphic, independent and of a complex external structure. They think that in time the sporophyte became permanently attahced and partially dependent upon the sporophyte and resulted in a reduction in the complex structure of the sporophyte.

According to intercalation or antithetic theory the sporophyte has been interpreted as a entirly new structure intercalated between two successive gametophytic generations. This theory further supports that the bryophytics sporophyte did not make its appearance until after a gametophyte had evolved an archegonium. The zygote found in the venter of archegonium divided mitotically forming a number of diploid (2n) cells, each of which divided meiotically to form four haploid (n) spores. Thus a rudimentary type of sporophyte was developed in which all cells were sporogenous. Further evolution of the sporophyte took place as a result of sterilization of sporogenous tissue and foot, seta and capsule were differentiated. According to this theory there was a progression from a simple to a more complex sporophyte by a progressive sterilization of sporogenous tissue.

Evolution Among Bryophyta

The major lines (*i.e.,* Hepaticopsida, Anthocerotopsida and Bryopsida) have been considered among bryophytes. According to G.M. Smith (1955) the primitive bryophytic gametophyte was a simple thallose plant, and that the primitive sporophyte was the simple globose type. According to this interpretation the bryophytes of primitive nature are found among the liverworts (Hepaticopsida-Hepaticae). Here the simplest gametophytes are found among Sphaerocarpaceae, and the simplest sporophytes are found among the Ricciaceae. For hypothetical bryophytes (*i.e.,* protobryophytes) more primitive than any known bryophyte, Lotsy (1909) has suggested a type combining the simple gametophyte of *Sphaerocarpos* with the simple sporophyte of Ricciaceae to form the imaginary genus *Sphaero-riccia.* The Marchantiales of Hapaticae is one characterized by gametophytes of external simplicity but internal complexity. Features in ontogeny of sex organs indicate that relationships of Jungermanniales and Calobryales with other Hepaticae are rather remote. According to Smith (1955) these orders are of an advanced type among the Hepaticae. No matter what the evolutionary order of their interrelationships the Hepaticae seem to constitute a blindly ending evolutionary side line. The mosses (Bryopsida) also make a blindly ending evolutionary side line. Their gametophytes are externally complex and most of them have sporophytes with greater internal differentiation of tissues. The mosses are sometimes thought to

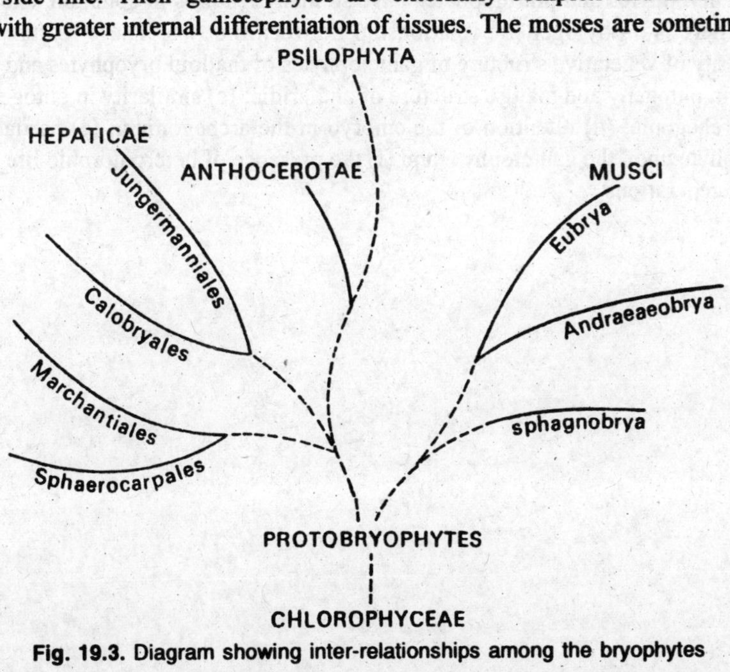

Fig. 19.3. Diagram showing inter-relationships among the bryophytes

have been evolved from leafy Hepaticae (*i.e.,* Jungermanniales and Calobryales). On the other hand the gametophytes of Anthocerotales, except for their embedded sex organs resemble those of thallose Hepaticae. However, the anthocerotean sporophyte is of a much more advanced type than is found in any Hepaticae or Bryopsida (moss). Because of their similarities in gametophytes, the Anthocerotae may be considered a series which departed from primative Hepaticae.

BRYOPHYTES AND PTERIDOPHYTES
(Evolutionary Tendencies)

1. The bryophytes and the pteridophytes remain at an evolutionary level higher than that of algae, but lower than that of spermatophytes (seed plants).

2. The features which keep the bryophytes and pteridophytes at a higher evolutionary level than algae are (a) multicellular sex organs with an outer jacket layer of sterile cells; (b) permanent retention of the zygote within the archegonium; (c) retention of the embryonic sporophyte within archegonium; and (d) parasitism of sporophyte upon gametophyte (as in bryophytes) for at least in early stages of development (as in pteridophytes).

3. The features responsible for the seed plants (spermatophytes) to be more advanced are: (a) migration of male gametes through a tubular structure between male and female gametophytes; (b) the permanent retention of female gametophyte within tissues of the sporophyte; (c) formation of seeds and (d) temporary cessation in growth of embryonic sporophyte at a certain stage in development.

4. The sporophytes of pteridophytes are more advanced than of bryophytes because of the presence of following features: (a) in bryophytes the sporophyte is never independent and takes its nutrition from the gametophyte throughtout its development; (b) the bryophytes have no conducting system differentiated into xylem and phloem; (d) whereas on maturity the sporophyte of pteridophyte is independent; (e) the sporophyte of a pteridophyte is potentially capable of unlimited growth and (f) the conducting system of the sporophyte of pteridophyte remains differentiated into xylem and phloem.

5. The bryophytes and pteridophytes have so many features in common, and therefore it is thought that there is a phylogenetic relationship between the two. These features include - (a) general similarity of vegetative structure of gametophytes of thalloid bryophytes and pteridophytes; (b) similarity in ontogeny and mature structure of antheridia; (c) similarity in ontogeny and mature structure of archegonia; (d) retention of the embryo in the archegonium; (e) partial parasitism of mature sporophyte upon the gametophyte and (f) the presence of heteromorphic life cycle showing alternation of generations.

20

CHAPTER

Classification of Bryophytes

For the first time Braun in 1864 gave the name Bryophyta of this group of plants. However, he included algae, fungi, lichens and mosses in his Bryophyta. Now, it is an established fact that algae, fungi and lichens comprise the group Thallophyta and the mosses are included in Bryophyta. Thereafter Schimper in 1879 gave the name Bryophyta which is used in full sense for the plants included in this group. Eichler in 1883 divided Bryophyta into two groups-Hepaticae and Musci. Engler (1892) divided the class Hepaticae into three orders-Marchantiales, Jungermanniales, and Anthocerotales. However, Bower (1935), Wettstein (1935), Bessey, Fritsch (1929), Evans (1938, 1939) and other bryologists still follow the same old system. Underwood (1894) and Gayeh (1897) however, withdrawn the order, Anthocerotales from class-Hepaticae.

According to Campbell, Smith, Takhtajan and others, the Bryophyta has been divided into three classes-Hepaticae, Anthocerotae and Musci.

In 1951, Rothmaler changed the class names. He recognized Hepaticae as Hepaticopsida; Anthocerotae as Anthoceropsida and Musci as Bryopsida. The above given classes have been also recognized by the international code of botanical nomenclature, 1956. Proskauer (1957) however changed the name Anthoceropsida to Anthoerotopsida and thus, the Bryophyta may be classified as follows:

Division	**Bryophyta**
Class 1.	Hepaticopsida (Hepaticae)
Class 2.	Anthocerotopsida (Anthocerotae)
Class 3.	Bryopsida (Musci)

Class 1. **Hepaticopsida.** There are 4 orders, 9 familles, 225 genera and 8,500 species.

1. The gametophytes are dorsiventrally differentiated. They may be thalloid (thallose) or differentiated into leaves and stem (foliose).
2. In foliose types the leaves are arranged in two or three rows on the axis and are always without mid-rib.
3. The sex organs develop from superficial cells on the dorsal side of the thallus, except when they are terminal in position.
4. The sporophyte may be simple, or differentiated into foot and capsule, or into a foot, seta and capsule.
5. The sporogenous cells develop from the endothecium of sporogonium.
6. The sporophyte is completely dependent on gametophytes for its nutritive supply.
7. The wall of sporogonium is one to several layered thick. The stomata are not present on the wall of sporogonium.
8. The dehiscence of sporogonium is irregular.

The class Hepaticopsida is further divided into several orders - (1) Sphaerocarpales; (2) Marchantiales; (3) Metzgeriales; (4) Jungermanniales; (5) Calobryales and (6) Takakiales.

Order-Sphaerocarpales (3 genera)-two families.

1. Family-Sphaerocarpaceae - *Sphaerocarpos* (seven species) and *Geothallus* (single species).

2. Family-Riellaceae-*Riella* (17 species).

The characteristic features of the order are as follows:

1. Vegetative structure of gametophyte is similar to that of order-Metzgeriales, but in which development and structure of sex organs, as well as the structure of sporophyte are similar to those of order-Marchantiales, and because of this the genera are placed in separate order Sphaerocarpales.

2. The main diagnostic feature by which the order is recognized is the presence of globose or a flask-like envelope or involucre around each of the sex organs (*i.e.,* antheridia and archegonia).

Order-Marchantiales (32 genera; 400 species)

The characteristic features are as follows:

1. The ribbon-like, dichotomously branched and dorsiventral thalli grow prostrate upon suitable substrata.

2. Excluding *Dumortiera, Monoselenium* and *Monoclea,* the rest of the genera possess internally differentiated air chambers on the dorsal side of the thallus, such chamber opens outside by an air pore of a particular design.

3. The ventral portion of the thallus consists of parenchyma which acts as storage tissue; oil and mucilage cells may be present in this region.

4. The scales and rhizoids are present on the ventral side of the thallus; the rhizoids are of two types (smooth-walled and tuberculate).

5. The antheridia and archegonia may be found directly on the dorsal surface of the thallus or they may be present on the special branches known as antheridiophores and archegoniophores respectively.

6. In most of cases the capsules of the sporophytes possess single layered jacket.

7. The capsule may be simple as in *Riccia* or it may be differentiated into foot, seta and capsule as in *Marchantia.*

8. The elaters may or may not be present. According to Campbell (1940), there are *five* families in this order. The characteristic features of *two* families are given here:

I. Family-Ricciaceae (3 genera; 140 species)

1. The gametophyte consists of a rosette-like dichotomously branched thallus.

2. In the thallus, the dorsal portion consists of chlorophyllous strips which may or may not have air canals among them; the ventral portion of the thallus is parenchymatous and acts as storage tissue.

3. The sex organs (antheridia and archegonia) are found in the longitudinal groove on the dorsal side from the growing apex to backward in basipetal succession.

4. The sporogonium consists of a simple capsule which is not differentiated into foot, seta and capsule.

5. Elaters not present.

6. The archesporium produces only the spores.

The important genera are-*Oxymitra, Ricciocarpus* and *Riccia.*

II. Family-Marchantiaceae (23 genera; 250 species)

1. The thallus is dorsiventral; it has distinct assimilatory and storage regions.

2. The assimilatory region remains divided into several chambers and each chamber contains branched assimilatory filaments.

3. The pores of the thallus may be simple or barrel-shaped.

4. The archegonia are borne upon special erect, stalked, vertical branches, the archegoniophores.

5. The antheridiophores may or may not present; however, in *Marchantia*, the antheridia are borne upon these erect, stalked antheridiophores.

6. The typical sterile elaters are found in the sporogonium mixed with the spores. The important genera are - *Conocephalum, Cryptometrium, Lunularia, Marchantia,* etc.

Order-**Metzgeriales** (23 genera; 550 species).

The characteristic features of this order are as follows:

1. The gametophyte may be thalloid or differentiated into stem and lateral leaves.
2. In most cases the gametophytes are without internal differentiation of tissues but certain genera have a central strand of thick-walled cells.
3. The ventral surface of a gametophyte bears smooth-walled rhizoids.
4. The sex-organs are found to be scattered on dorsal surface of thallus.
5. The archegonia arise from the young segments cut off by the apical cell.
6. The mature sporophytes lie some distance back from the growing apex of a gametophyte.
7. The sex organs (antheridia and archegonia) are produced on any branch of the gametophyte or only on special branches.

Family-Pelliaceae (Three genera- *Pellia, Noteroclada* and *Calycularia*).

1. The thallus is prostrate, dorsiventral and very often lobed by irregular incisions.
2. The rhizoids are simple, non-septate, smooth and thin-walled. The scales are absent.
3. The sex organs (antheridia and archegonia) remain scattered on the dorsal surface of the thallus.
4. The archegonial cluster always remains surrounded by an involucre which is an outgrowth of the thallus.
5. The capsule (sporogonium) is globose or oval in shape. It possesses a basal elaterophore.

Family-Riccardiaceae (two genera).

1. The gametophytes are wholly thallose or have thallose terminal branches.
2. The cells of thallus possess finely segmented oil bodies.
3. The sex organs (antheridia and archegonia) are borne on short lateral branches.
4. A well-developed calyptra is present but there is no involucre.
5. A distal elaterophore is present to which some elaters are attached.
6. The capsule is ovoid to cylindrical and dehisces longitudinally into four parts extending to the base.

Family-Fossombroniaceae (4 genera; *Fossombronia, Simodon, Petalophyllum* and *Sewardiella*).

1. Thallus is distinctly foliose.
2. The thallus is dorsiventral and prostrate.
3. The stem is branched. Growth of the main axis and of its branches is by means of an apical cell with two cutting faces.
4. The leaf is thin, one cell-thick except the basal portion which is 2 or 3 cells in thickness.
5. Antheridia develop in acropetal succession singly or in small groups.

6. Archegonia occur in small groups.

7. Young sporophyte remains surrounded by a calyptra and which is ensheathed by a cuplike involucre.

8. The mature sporophyte is surrounded and protected by a bell-shaped sheathing perianth.

9. Important genus-*Fossombronia*.

Order-**Jungermanniales** (220 genera; 8, 500 species).

The characteristic features of this order are as follows:

1. The gametophyte is differentiated into stem and leaves; the leaves are borne in a regular spiral succession along the stem.

2. The apical cell is pyramid-like with three cutting faces.

3. The stem generally bears three rows of leaves; two rows are lateral and consist of leaves of normal size; the third row consists of the under leaves which are generally smaller than the lateral leaves.

4. The archegonia are always restricted to the apices of the axis and its branches.

5. The sporophytes are always terminal in position.

6. The antheridia are borne singly or in groups in the axis of leaves.

Family-Porellaceae (single genus-*Porella*).

1. The leaves are arranged in three rows on the stem; ventral leaves are well developed and usually decurrent at the base; dorsal leaves are incubuous; postical lobe distinct.

2. The rhizoids are scarce and arise from the lower side of the stem in tufts generally near the base of underleaves (ventral leaves).

3. The archegonia are borne in terminal cluster on small lateral branches; the archegonia remain surrounded by a large inflated perianth.

4. The spherical capsule dehisces by four valves which split only to half way down.

Family-Frullaniaceae (three genera; important genus *Frullania*).

1. The thallus is pinnately branched and differentiated into stem and leaves.

2. The leaves are arranged in three rows two laterals unequally lobed and a ventral lobule.

3. The ventral leaves are bifid and trumpet-shaped.

4. The archegonia develop in a group.

Order-**Calobryales** (2 genera-*Calobryum* (8 spp.) and *Haplomitrium* (single spp.)

The characteristic features are as follows:

1. They possess erect leafy gametophytes with leaves in three vertical rows.

2. The leaves are dorsiventrally flattened.

3. They have a pale, substerranean, sparingly branched rhizome from which arise erect leafy branches.

4. Erect branches bearing sex organs have the uppermost leaves close together and in more than three rows.

5. They are devoid of rhizoids.

6. The antheridia are ovoid, stalked, and borne at the apex of the stem.

7. The jacket of the neck of archegonium has only four vertical rows of cells.

8. The sporophyte bears an elongate capsule whose jacket layer is only one cell in thickness except at the apex.

9. The number of chromosomes is-n=9.

Since there is single family Calobryaceae the characters are similar to that of the order. Two genera-*Calobryum* and *Haplomitrium.*

Order-**Takakiales** (1 genus; 2 species)

The characteristic features are as follows :

1. They possess cylindrical, rhizomatic and erect gametophores.
2. They are devoid of rhizoids.
3. They possess copious beaked or non-beaked mucilage hairs on them.
4. They possess terete bifid-trifid-quadrifid leaves or phyllids.
5. The gametophores are about 1 to 1.5 cm. in height.
6. The leafless branches facing downward known as 'flagella' or 'stolons' may be present.
7. Asexual reproduction is not known.
8. Only female (archegonial) shoots are known. They bear conspicuous pedestalled archegonia.
9. The male (antheridial) shoots and the sporophytes are not known.
10. They have lowest chromosome number (*i. e.,* n=4).
11. They are supposed to be most primitive and sometimes known as living fossil. There is one family Takakiaceae, and one genus *Takakia.*

Class II. **Anthocerotopsida.** There are 1 order, 1 or 2 families, 6 genera and 301 species.

1. The gametophyte is thalloid and dorsiventral, bearing simple and smooth-walled rhizoids; tuberculate rhizoids and ventral scales are altogether absent.
2. The tissues of the thallus are undifferentiated; air chambers and air pores are absent; each cell bears a large chloroplast and a conspicusous pyrenoid within it.
3. The sex organs are found to be embedded in the gametophytic tissue.
4. The antheridia arise from the hypodermal cells of the thallus on the dorsal side of it; they develop within the antheridial chambers, singly or in groups on the dorsal side of the thallus.
5. The archegonia are found in sunken conditions on the dorsal side of the thallus, they develop from superficial cells.
6. The elongated and cylindrical sporogonium arises from the dorsal side of the thallus.
7. The sporogonium consists of foot, meristematic region and capsule; the meristem is intercalary and continues its growth thoughout the growing season.
8. The wall of sporogonium contains chlorophyll.
9. The central sterile portion of sporogonium is columella, which remains surrounded by sporogenous tissue and spores; the elaters are also present.
10. The sporogenous mass develos from amphithecium and arches over the columella. The class Anthocerotopsida includes a single order, the Anthocerotales with the same characters. There are two families 1. Anthocerotaceae and 2. Notothylaceae.

Family-Anthocerotaceae (4 or 5 genera; important genus *Anthoceros*).

1. The capsule is linear and vertical.
2. The stomata are present on the wall of capsule.
3. The archesporium develops from amphithecium.
4. The eleaters are four-celled, smooth or thick-walled; thickening band may or may not be present.

Family-Notothylaceae (single genus-*Notothylas*)

1. The capsule is cylindrical and horizontal.
2. The stomata are not found on the wall of capsule.
3. Archesporium arises from endothecium and amphithecium.
4. Elaters are short and stumpy; they have irregular thickening bands.

Class III. **Bryopsida.** There are 3 orders, 28 families, 660 genera and 15,504 species.

1. The gametophyte consists of prostrate, thalloid, branched protonema and erect leafy gametophore.
2. The gametophytic plant body consists of the stem, spirally arranged leaves and the sex organs (antheridia and archegonia) at its apical portion.
3. The rhizoids are multicellular, branched and obliquely septate.
4. The sex organs (antheridia and archegonia) develop from the superficial cells of the gametophore.
5. The sporophyte is differentiated into foot, seta and capsule.
6. The capsular wall remains interrupted by stomata at several places.
7. The archesporium or sporogenous mass develops from outer layer of endothecium which in addition forms columella.
8. The elaters are not present in the sporogonium.

The class Bryopsida (Musci) has been divided into *three* sub-classes (1) Sphagnobrya (Sphagnidae); (2) Andreaeobrya (Andreaeidae) and (3) Eubrya (Bryidae).

I. Sub-class. Sphagnobrya. The sub-class has a single order, the Sphagnales and a single family, the Sphagnaceae. (single genus *Sphagnum* with 326 species). The characteristic features are as follows :

1. They are called 'bog mosses' or 'peat mosses'.
2. The protonema is broad and thallose; It produces one gametophore; the leaves or gametophores lack mid-rib and usually composed of two types of cells-(i) the narrow living green cells and (ii) large hyaline dead cells.
3. The branches arise in lateral clusters in the axis of the leaves borne on the stem.
4. The antheridia are borne in the axis of leaves on the antheridial branch.
5. The archegonia are terminal and formed acrogynously.
6. The sporogenous tissue of a sporophyte develops from the amphithecium.
7. The sporogonium remains elevated above the gametophyte due to elongation of a stalk of gametophytic tissue, the pseudopodium.

II. Sub-class. Andreaeobrya. This sub-class has a single order, the Andreaeales, and a singly family, the Andreaceceae. The important genus is *Andreaea.*

The characteristic features are as follows :

1. The gametophores are brittle, and can easily be broken.
2. There is practically no tissue differentiation in plant body.
3. The leaves are generally large, erect and convolute.
4. The archesporium and colonmella develop from the endothecium.

III. Sub-class Eubrya (650 genera; 14,000 species). This sub-class has been further divided into *three* cohorts and *fifteen* orders. The true mosses are included in this sub-class. The characteristic features are as follows:

1. The leaves of the gametophores are more than one cell in thickness and possess midtrib on them.

2. The protonema are filamentous.

3. The sporophyte bears a well differentiated, elongated seta which pushes out the capsule from the gametophore.

4. The sporogenous tissue is derived from the endothecium.

5. The archesporium does not overarch the columella; the columella continues upto the apex of the capsule; both columella and archesporium have been derived from the endothecium.

6. In between spore sac and columella, the partitioned air spaces are present.

7. The mature capsule possesses the complex structure made of many tissues.

8. The capsule opens at its apex by an operculum; the spore dispersal is regulated by a teeth like apparatus, the persistome.

Order-**Funariales** (26 genera; 356 species).

Characteristic features.

1. The plants are terrestrial; they are small in size and may be annual or biennial.

2. The leaves possess distinct mid-ribs and arranged in rosettes at the apex of the gametophyte.

3. The capsule is wide and provided with an unbeaked operculum.

4. The peristome of the capsule is double and consists of inner and outer peristomes called endostome and exostome respectively.

5. There are *five* families in this order, of which Funariaceae is most important.

Family-Funariaceae (9 genera; 200 species).

1. The leaves are one cell in thickness except at the mid-rib region.

2. The small mosses form the velvety appearance on the surface of the sustratum.

3. The calyptra are soon detached from the opercula of the capsules; the calyptra are provided with long beaks.

4. The capsules are pyriform and situated on the long, elongated setae.

Order-**Polytrichales**

Characteristic features.

1. The gametophyte is perennial and tall.

2. The leaves are narrow and possess longitudinal lamellae on the upper surface of the midrib.

3. The capsule is terminal.

4. The single annular series of cells gives rise to a peristome in the inner zone of the amphithecium.

5. There are 32 to 64 pyramidal teeth in peristome; the tips of the peristome teeth remain joined above to a thin membrance, the epiphragm covers the mouth of the capsule.

6. There is a single family, the Polytrichaceae in this order; the important genera of this family are — *Polytrichum* and *Pogonatum.*

21

CHAPTER

Hepaticopsida
Order-Marchantiales

This class includes 4 orders, 9 families, about 225 genera and 8,500 species (Smith, 1955). The Latin work *hepatica* means liver and there from the members of Hepatiocopsida are popularly known as liverworts. The most characteristic features of the class are as follows:

The gametophytes are dorsiventrally differentiated. The vegetative plant body (gametophyte) may be either a thallus (thallose) or a leafy axis, *i.e.,* differentiated into leaves and stem (foliose). In foliose types the leaves are arranged in two or three rows on the axis and are always without mid-rib.

The interior of the gametophyte is either simple or composed of various tissues. The assimilatory cells contain numerous chloroplasts which are devoid of pyrenoids.

The sex organs are always developed from superficial cells on the dorsal side of the thallus, except when they are terminal in position.

The sporophyte (sporogonium) may be simple, or differentiated into foot and capsule, or into a foot, seta and capsule. In each case the sporogenous cells develop from the endothecium of an embryo.

The sporophyte is completely dependent on gametophyte for its nutritive supply.

Classification. According to Engler (1892), the class Hepaticopsida had only two orders: Jugermanniales and Marchantiales. Cavers in 1910 isolated the family Sphaerocarpaceae from the order Jungermanniales and gave it the rank of order, *i.e.,* Sphaerocarpales. This way according to Cavers there are three orders in the class-Sphaerocarpales, Jungermanniales and Marchantiales. Campbell (1936) raised another family Calobryaceae of Jungermanniales to the rank of order, *i.e.,* Calobryales. Thus the classification of Hepaticopsida is as follows (Smith, 1955):

Class.**Hepaticopsida**

Order 1. Sphaerocarpales

Order 2. Marchantiales

Order 3. Jungermanniales

Order 4. Calobryales.

In the present text, however, *six* orders have been recognized in class-Hepaticopsida. They are as follows:

Order 1. Sphaerocarpales

Order 2. Marchantiales

Order 3. Metzgeriales

Order 4. Jungermanniales

Order 5. Calobryales

Order 6. Takakiales.

Order **MARCHANTIALES**

There are about 32 genera and 400 species in this order. The ribbon shaped, dichotomously branched and dorsiventral gametophytes of this order grow prostrate upon the suitable substrata. Excluding *Dumortiera, Monoselenium* and *Monoclea* the rest of the genera possess internally differentiated air chambers on the dorsal side of the thallus, such chamber opens outside by an air pore of a particular design. The ventral portion of the thallus consists of parenchyma which acts as storage tissue. Sometimes the oil and mucilage cells are also found in this region of parenchyma. Usually the scales and rhizoids are found on the ventral side of the thallus. Commonly the rhizoids are of two types. The antheridia and archegonia may be found directly on the dorsal surface of the thallus or they may be present on the special branches designated antheridiophores and archegoniophores respectively. The capsules of the sporophytes of most of the Marchantiales possess single layered jacket. The capsule may be simple as in *Riccia* or it may be differentiated into foot, seta and capsule as in *Marchantia.* The elaters may or may not be present.

Classification. Different bryologists have put forward the various systems of classification from time to time. The bryologists, such as Von Wettstein (1908), Kashyap (1919), Verdoorn (1932) and Evans (1939) supported the view of retrogression in Marchantiales from the complex type *Marchantia* to the simple type *Riccia.* This view was supported and put forward by Goebel (1910).

Verdoorn (1932) divided the order Marchantiales into six families. They are, (1) Marchantiaceae; (2) Operculatae; (3) Astroporae; (4) Targioniaceae; (5) Corsiniaceae and (6) Ricciaceae. Evans (1939), however, also recognized the six families. They are (1) Marchantiaceae; (2) Rebouliaceae; (3) Sauteriaceae; (4) Targioniaceae; (5) Corsiniaceae and (6) Ricciaceae.

Campbell (1940), has divided the order into five families. He followed the progression of the complexity from Ricciaceae to Marchantiaceae. The five families recognized by him are, (1) Ricciaceae; (2) Corsiniaceae; (3) Targioniaceae; (4) Monocleaceae and (5) Marchantiaceae.

The Campbell's classification has been followed by various authors of the different books. Gilbert M. Smith has followed the same classification in his text book Cryptogamic Botany Vol II. In the present text, the Campbell's classification has been followed.

Family-Marchantiaceae

There are 23 genera and about 250 species in this family. The archegonia of this family borne upon special erect, stalked, vertical brnaches, the **archegoniophores.**The male branches, the **antheridiophores** may or may not be present. In *Marchantia* the antheridia are borne upon these erect, stalked antheridiophores.

The typical sterile **elaters** are found in the sporogonium mixed with the spores.

The Marchantiaceae are separated into three sub-families, but according to Evans (1923, 1939) they are treated as separate families.

Genus MARCHANTIA

Occurrence and distribution. This genus comprises of about 65 species. They are cosmopolitan. *Marchantia polymorpha* the best known species, is widely distributed. According to R. S. Chopra (1943) there are about eleven Indian species. They are mostly confined in the Himalayas. Only one or two species have been recorded from the plains.

According to Udar (1970) six species are found in the Himalayas. The common Himalayan species are-*M. palmata, M. nepalensis, M. polymorpha, M. simalana* and *M. indica.* The common species of the Indian plains are *M. polymorpha* and *M. indica.* These two species are also found in the hills of South India. Brandis reported *M. simalana* from the hills of Simla. *M. polymorpha* grows luxuriantly at 8,000 feet on Himalayan and South Indian hills on moist rocks and banks of streams. The thalli with gemma cups are found throughout the year whereas the plants with sex organs occur abundantly during February-March in Himalayas and October-November in the hills of South India.

The thalli of *Marchantia* commonly thrive upon moist soil found on the rocks in shady places, in open wood lands or near the banks of streams. This genus grows best in the burnt soil. They commonly grow after the forest fires in the burnt soil. According to Richards (1958) this is perhaps because of the nitrification of the soil due to the fire.

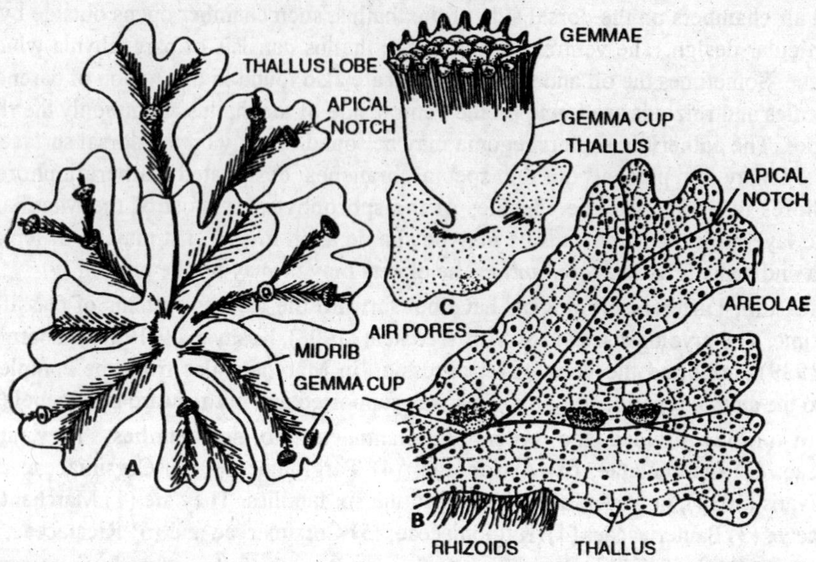

Fig. 21.1. *Marchantia* sp. A, thalli showing habit; B, part of thallus; C, gemma cup on thallus.

THE GAMETOPHYTE

External structure of the thallus (adult gametophyte). The thallus of *Marchantia* is prostrate, dichotomously branched and dorsiventral. Comparatively, the dorsal surface is deeper

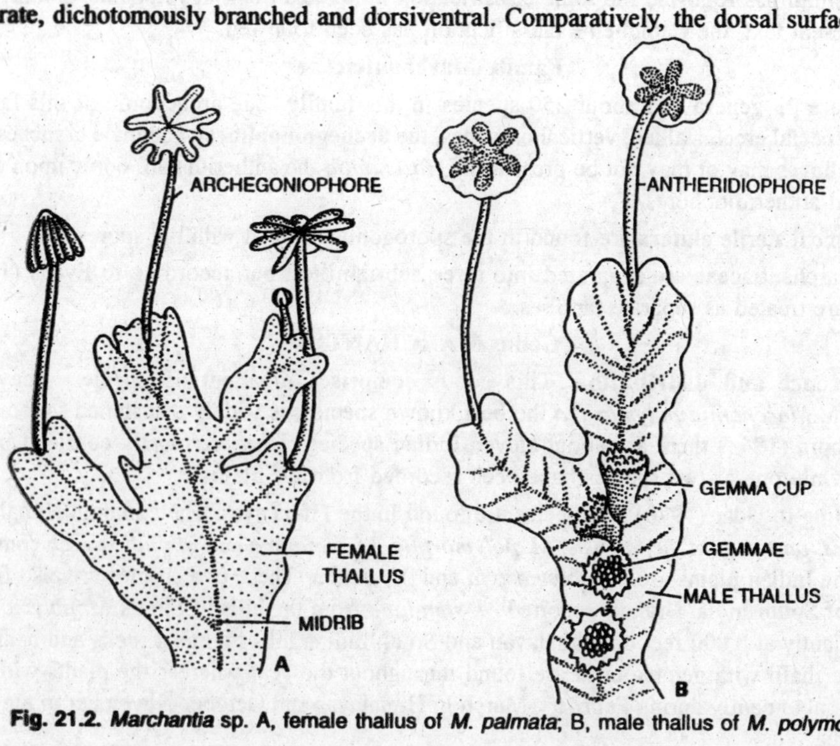

Fig. 21.2. *Marchantia* sp. A, female thallus of *M. palmata*; B, male thallus of *M. polymorpha*.

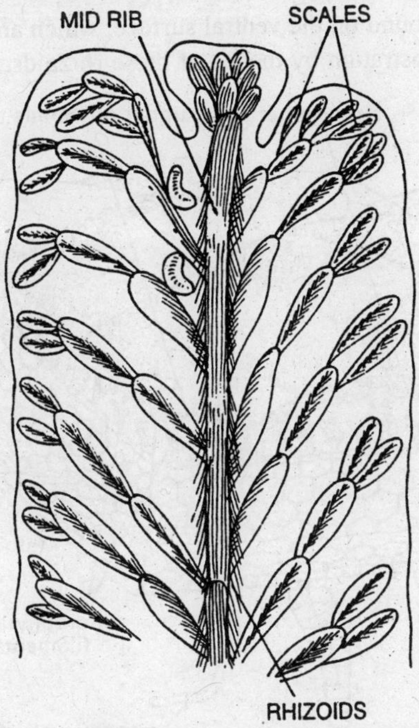

Fig. 21.3. *Marchantia* sp. Thallus seen from the underside showing scales and rhizoids.

the ventral one. The rhombiodal or polygonal small areas are found on the dorsal surface of thallus. Each such small area is provided with a small pore in its centre. The apex of each branch of the thallus bears a notch in which the growing point is situated. The thallus reaches to the length of 2 to 10 cm. On maturity the margin of the thallus is somewhat wavy. The thallus bears a conspicuous mid rib, which is well recognizable on the dorsal surface by the presence of a groove and on the ventral surface by presence of a ridge.

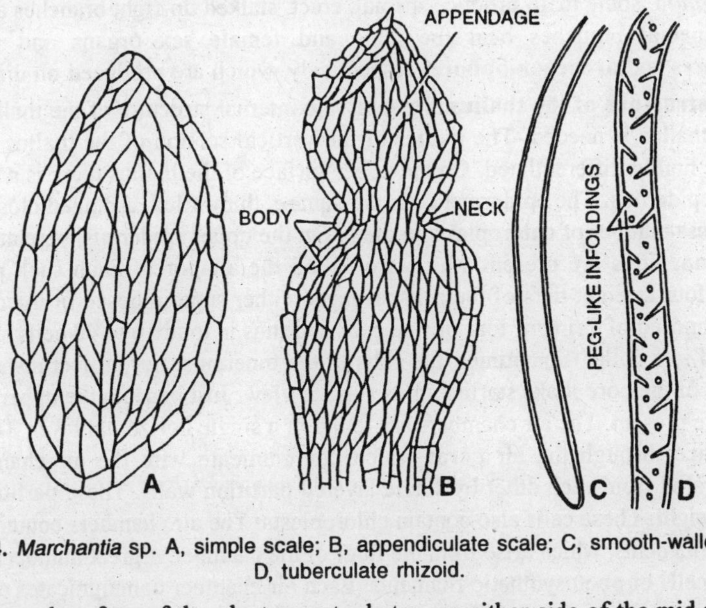

Fig. 21.4. *Marchantia* sp. A, simple scale; B, appendiculate scale; C, smooth-walled rhizoid; D, tuberculate rhizoid.

On the ventral surface of the plant gametophytes, on either side of the mid-rib two or more rows of the pinkish, multicellular scales are present, which protect the growing points. The simple and tuberculate rhizoids are also found on the ventral surface, which are absorptive in nature. The thallus remains attached to the substratum by means of these rhizoids.

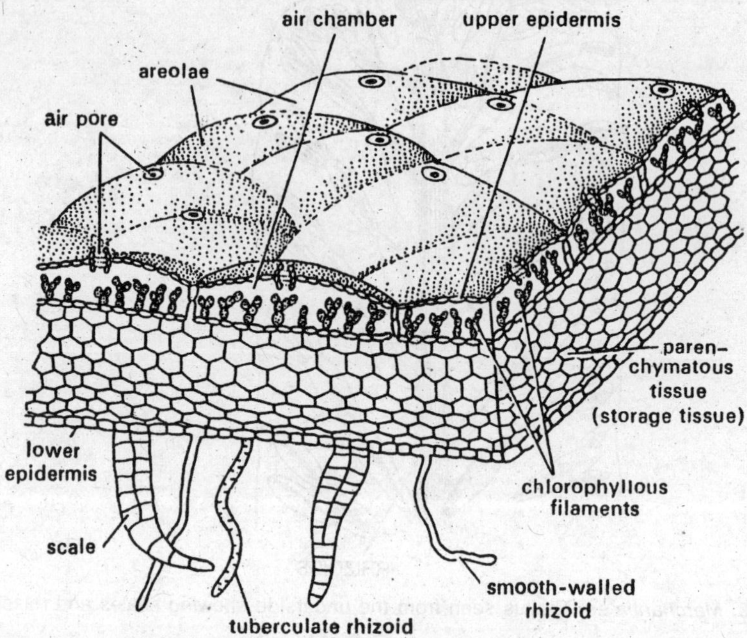

Fig. 21.5. *Marchantia* sp. Three dimensional diagram of a part of thallus.

On the dorsal surface of the plant gametophytes, the **gemma cups** are found. The gemma cups are usually found along the mid-ribs. The margins of the gemma cups are cut. The gemma cups contain the vegetative bodies, the **gemmae.** Such structures give rise to new plant gametophytes.

In *Marchantia,* some thalli produce special, erect, stalked up-right branches at their growing apices. The special branches bear the male and female sex organs and are called the **antheridiophores** and **archegoniophores** respectively, which are produced on different thalli.

Internal structure of the thallus. To study the internal structure of the thallus, the vertical section of the thallus is needed. The tissues of the vertical section of the thallus seen under the microscope are highly differentiated. On the upper surface of the thallus there is a clear cut single layered upper epidermis. The epidermal cells are square, thinwalled, arranged close to each other and contain a less number of chloroplasts in them. On the upper epidermis, here and there certain barrel-shaped openings are present which are called the air-pores. Each such pore is usually surrounded by four to eight tiers of the cells arranged either superimposed or in concentric rings. Each tier is composed of a ring of four or five cells and thus in total 16 to 40 cells are there around a pore arranged in a collar. Sometimes, the cells of the innermost tier project inward in the pore, and the openng of the pore looks starlike in its surface view. Just beneath the upper epidermis, the air chambers are present. The air chambers are found in a single horizontal layer. The exchange of gases takes place through the air pores which communicate with the air chambers. The air chambers separated from each other by single layered partition walls. These partitions are two to four celled in height. These cells also contain chloroplasts. The air chambers contain the branched photosynthetic filaments, which arise from the floor of the chamber. A great number of chloroplasts is found in the cells of photosynthetic filaments. Each air chamber communicates outside through a pore. The photosynthetic tissue and the air pores may easily be compared with the stomata and the spongy tissue present on the lower surface of an angiospermic leaf.

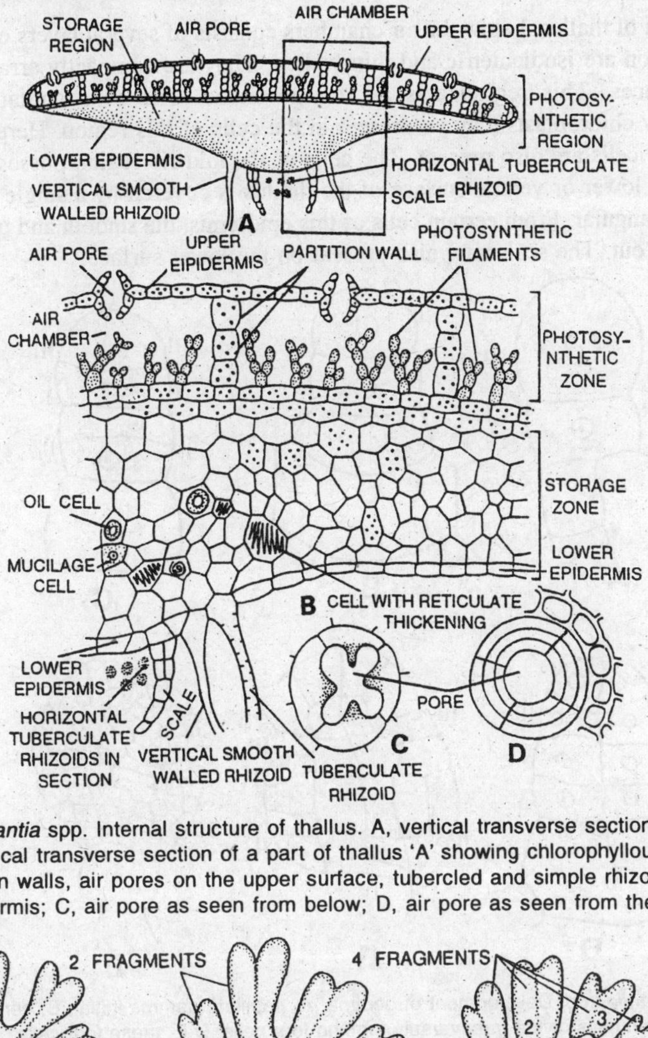

Fig. 21.6. *Marchantia* spp. Internal structure of thallus. A, vertical transverse section of thallus-outline sketch; B, vertical transverse section of a part of thallus 'A' showing chlorophyllous filaments, air chanbers, partition walls, air pores on the upper surface, tubercled and simple rhizoids and scales on the lower epidermis; C, air pore as seen from below; D, air pore as seen from the upper surface.

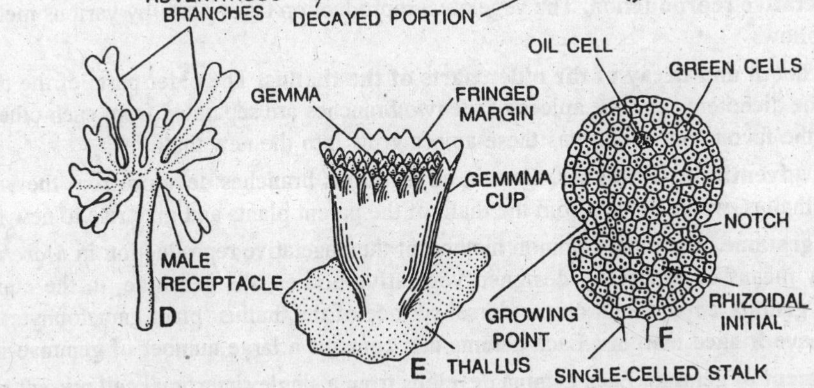

Fig. 21.7. *Marchantia* sp. Vegetative reproduction. A-C, by death and decay of the older parts of the thallus; D, by adventitious branches; E-F, by gemmae.

The portion of thallus, below the air chambers consists of several layers of parenchyma. The cells of this region are isodiametric and thin walled. They are compactly arranged, and, thus no intercellular spaces. This region acts as storage tissue. The cells contain starch in them. Sometimes a few chloroplasts may be present in the cells of this region. Here and there, certain oil and mucilage cells are also present. The cells of the mid-rib region possess certain reticulate thickenings. The lower or ventral surface of the thallus is covered by a single layered epidermis. The cels are rectangular. From certain cells of this epidremis, the smooth and tuberculate rhizoids have been given out. The scales are also present on the lower surface.

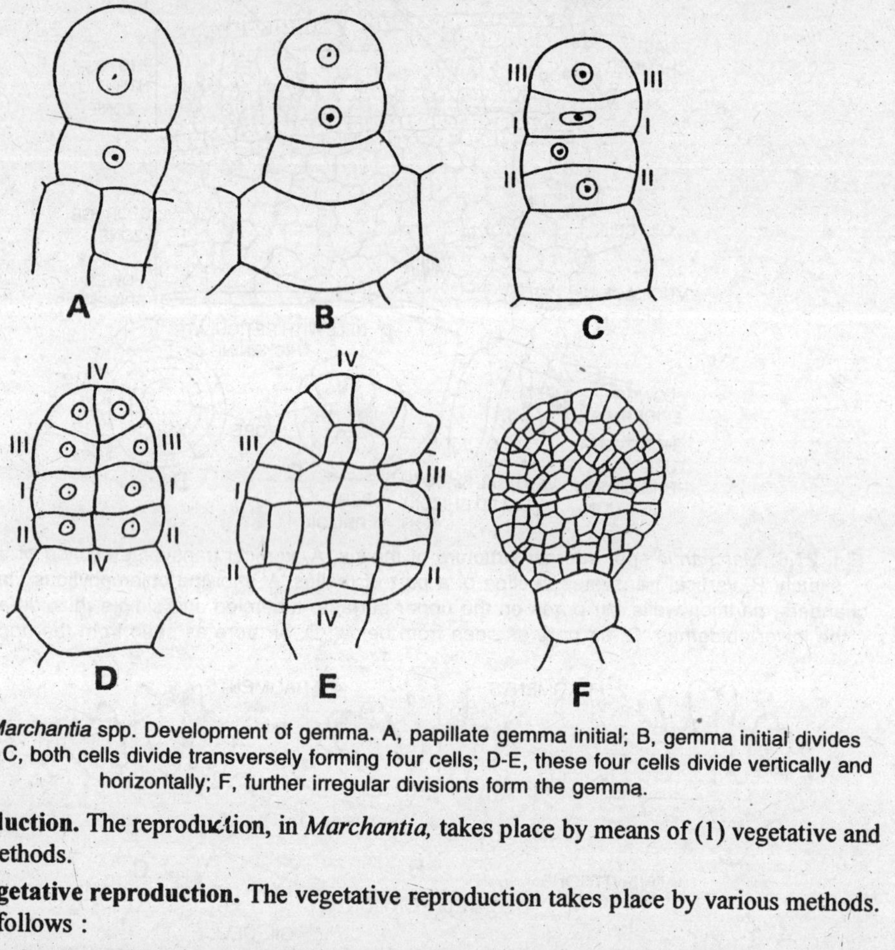

Fig. 21.8. *Marchantia* spp. Development of gemma. A, papillate gemma initial; B, gemma initial divides transversely; C, both cells divide transversely forming four cells; D-E, these four cells divide vertically and horizontally; F, further irregular divisions form the gemma.

Reproduction. The reproduction, in *Marchantia,* takes place by means of (1) vegetative and (2) sexual methods.

1. Vegetative reproduction. The vegetative reproduction takes place by various methods. They are as follows :

(a) By death and decay of the older parts of the thallus. The older parts of the thallus decay below the dichotomy and the apices of the two branches are separated from each other. On the advent of the favourable conditions these apices grow into the new thalli.

(b) By adventitious branches. Certain adventitious branches develop from the ventral surface of the thallus, which detach from the thalli of the parent plants and give rise to new thalli.

(c) By gemmae. The most common method of the vegetative reproduction in *Marchantia* takes place by means of especially designed vegetative bodies, the **gemmae,** in the cup like structures, the **gemma cups,** found on the dorsal surface of the thallus (plant gametophyte). The gemma cups have fringed margins. Each gemma cup contains a large number of gemmae in it.

Development of gemma. Each gemma develops from a single superficial cell present on the floor of the gema cup. Any of the superficial cells protrudes out from the floor and acts as a **gemma initial.** This initial divides twice transversely producing three cells. The basal cell, the

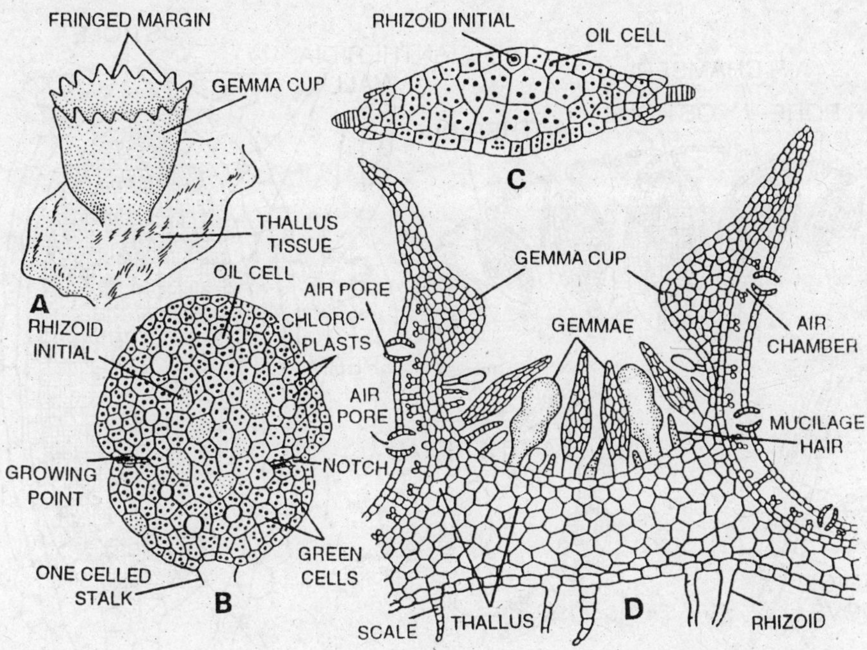

Fig. 21.9. *Marchantia* spp. Gemma cup and gemmae. A, gemma cup with fringed margin on upper surface of thallus; B, single gemma; C, transverse section of a mature gemma; D, vertical transverse section of thallus passing through a gemma cup.

stalk cell and the upper primary cell which changes into gemma proper. The stalk cell does not divide further and changes into a single celled stalk. The primary cell divides further by transverse, longitudinal and periclinal divisions giving rise to the gemma proper. In the beginning the gemma is flat and one cell in thickness, but in later stages, due to further periclinal divisions, it becomes three or four celled in thickness in the centre.

Structure of mature gemma. The mature gemma is found attached by a single celled stalk to the floor of the gemma cup (cupule). It is disciform and few celled thick in its centre, and quite thin at its margins. It is somewhat dump bell like. There are two depressions or notches situated on the lateral margins opposite to each other. The notches are growing points. A marginal row of apical cells is situated in each notch. About all the cells of the gemma contain chloroplasts in them. A few superficial cella are colourless which are somewhat larger in size than the other cells. There are certain isolated cells, here and there on the gemma, which contain oil in them and called the oil cells.

Mixed with the gemmae, certain mucilage cells are found on the floor of the cupule. These small mucilage cells discharge mucilage from them. This fluid absorbs water by imbibition, which helps, however, in detaching the gemmae from their stalks. The colourless cells, described in preceding paragraph are also called the **rhizoidal cells,** and the rhizoids come out from these cells at the time of germination.

Germination of gemma. As soon as the gemma falls on the soil, and if the suitable conditions for its germination prevail, it germinates. The rhizoids come out first from the rhizoidal cells and penetrate the soil. The marginal rows of the apical cells situated in the two notches of the gemma become activated and begin in function in normal way. From each growing point, a young thallus develops. This way, two young thalli develop in opposite directions from a single gemma. Eventaually, the central part of gemma dies and the two thalli are separated, which grow in opposite directions.

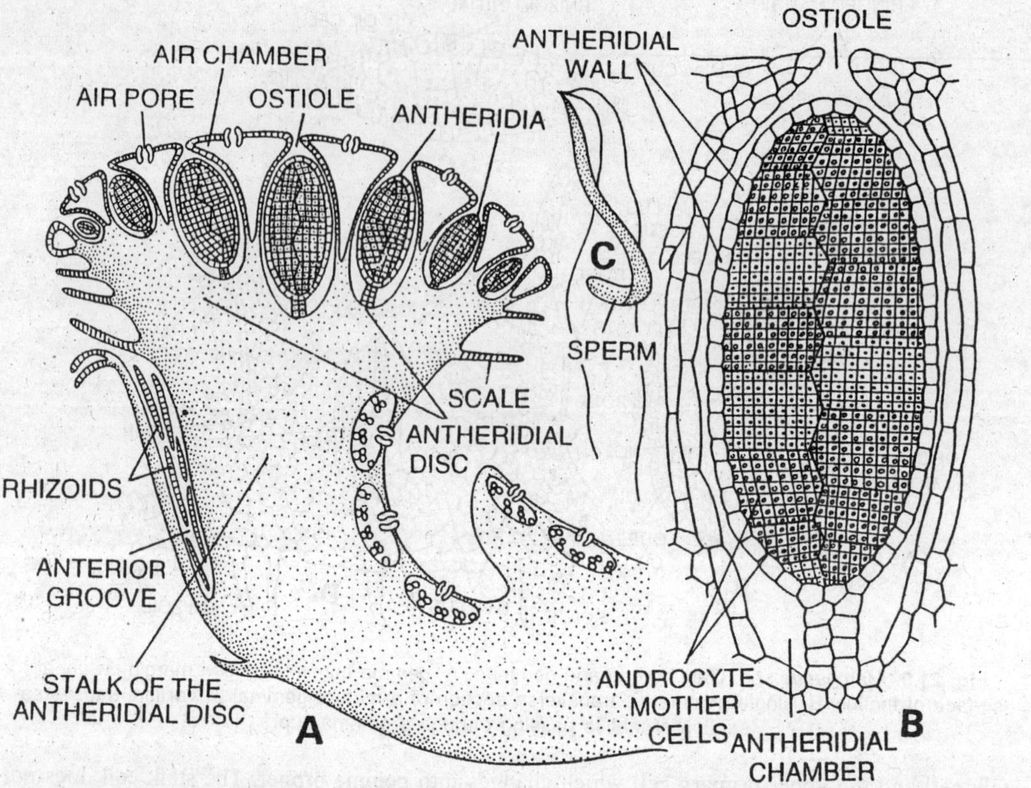

Fig. 21.10. *Marchantia* spp. Male sex organs. A, longitudinal section of antheridiophore with antheridia; B, a mature stalked antheridium covered by a jacket; C, a mature biflagellate sperm or antherozoid.

2. Sexual reproduction. In *Marchantia,* the sex organs are antheridia and archegonia, which are borne upon the special erect stalked branches, called the antheridiophores and archegoniophores, respectively. These branches are the modified prostrate branches. This fact may be proved by the morphology of the reproductive branch. The ventral side of an archegoniophore possesses rhizoids and scales situated in one or two longitudinal furrows. The dorsal side of it possesses the usual air chambers with photosynthetic filaments. The development of the reproductive branches depends upon several environmental and climatic factosrs. In *Marchantia,* the male (antheridiophores) and female (archegoniophores) reproductive branches develop on different thalli, *i.e.,* the genus is strictly heterothallic (dioecious). The upright sexual branches consist of a stalk and horizontal discs at their apical ends, in each case.

The structure of antheridiophore. The antheridiophore consists of a two to three centimentres long stalk, and a flattened, usually eight lobed disc at its apical end. As mentioned in the preceding paragraph the antheridiophore is a modified branch of the prostrate thallus. The terminal disc of the thallus shows a well developed branch system. Each lobe represents a branch. At the tip of each lobe a growing point is situated. Sometimes, in certain species, the number of lobes is reduced to four. The rhizoids and scales are present in the two grooves situated on the ventral surface of antheridiophore.

The disc of antheridiophore is somewhat convex at its upper face. The upper epidermis is interrupted by several usual barrelshaped air pores. These pores open in the air chambers, containing photosynthetic filaments in them. The flask like cavities are also found among and in the neighbourhood of these air chambers. Such flask-like cavities are developed by the rapid

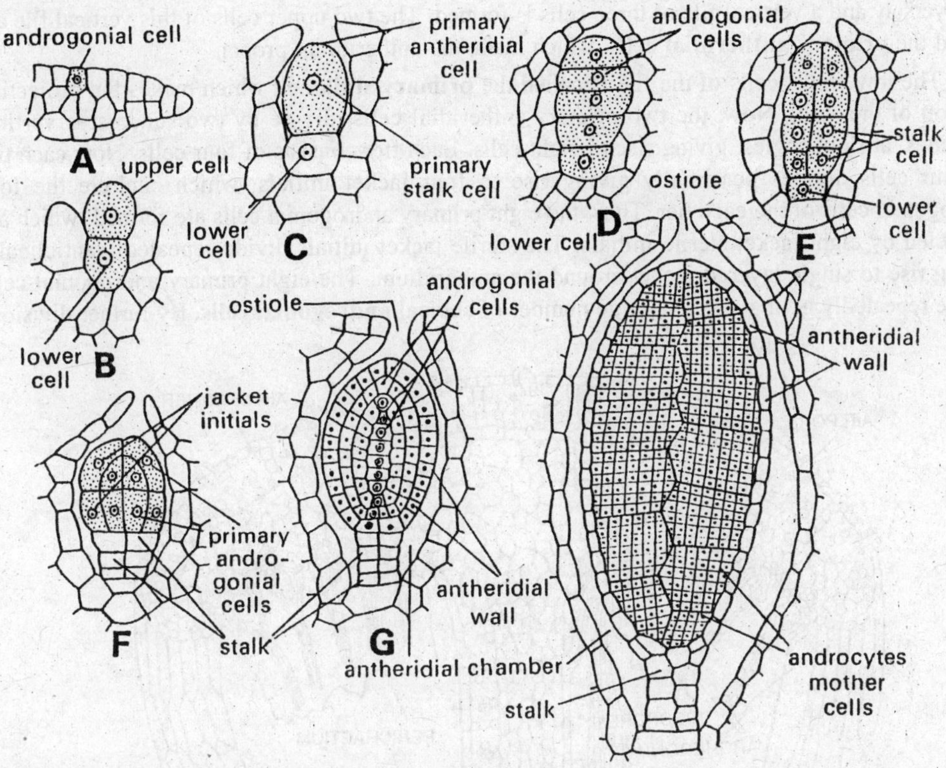

Fig. 21.11. *Marchantia* spp. Development of antheridium. A, antheridial initial or antheridium cell; B, transverse division of antheridial initial; C-G, successive stages in the development of antheridium; H, a stalked antheridium with androcytes.

growth of the tissues surrounding them. Each flask-like cavity contains a stalked antheridium within it, which partially or wholly fills up the cavity. The antheridial cavities open outside through small openings, the **ostioles.**

Development of antheridium. The antheridium develops from a superficial dosrsal cell called the **antheridial initial** which is found two or three cells back of the apical cell. The stages of the development of the antheridium in *Marchantia* are as follows :

The antheridial intial divides, and two cells are formed. The upper one is called the **outer cell** and the lower one the **basal cell.** The outer cell protrudes somewhat out of the thallus. The basal cell remains embedded in the thallus and develops into the embedded portion of the stalk of the

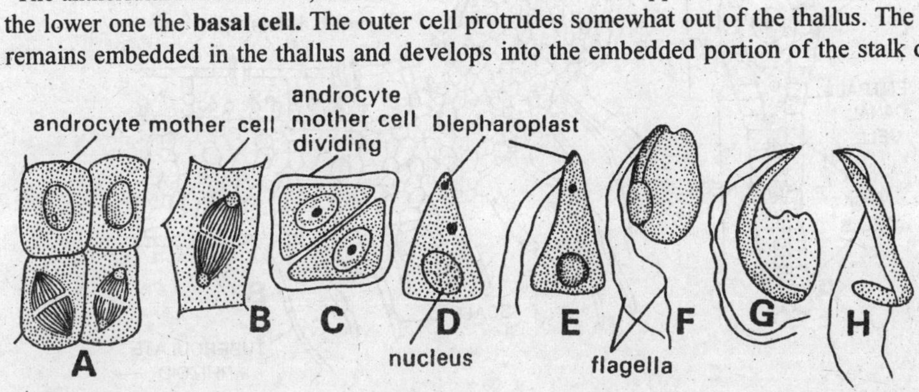

Fig. 21.12. *Marchantia* spp. Spermatogenesis. A, androcyte mother cell; B-C androcyte mother cell dividing diagonally into two androcytes; D-G, formation of antherozoid; H, single antherozoid.

antheridium. The outer cell develops into the main antheridium. Now the outer cell further divides transversely and a vertical file of three cells is formed. The two upper cells of this vertical file are called the **primary antheridial cells,** which make the antheridium proper.

The lowermost cells of the file is called the **primary stalk cell** which makes the projecting portion of the stalk. Now, the two primary antheridial cells, divide by two successive vertical divisions, at right angles, giving rise to eight cells. Each tier consists of four cells. Now each tier of four cells divides perclinally giving rise to four jacket initials, which encircle the four androgonial cells of the each tier. Thus, the eight primary androgonial cells are formed, which are encircled by eight jacket sterile initials. The sterile jacket initials divide repeatedly anticlinally giving rise to single layered jacket around the antheridium. The eight primary androgonial cells divide repeatedly giving rise to a large number of cubical **androgonial cells.** By further divisions

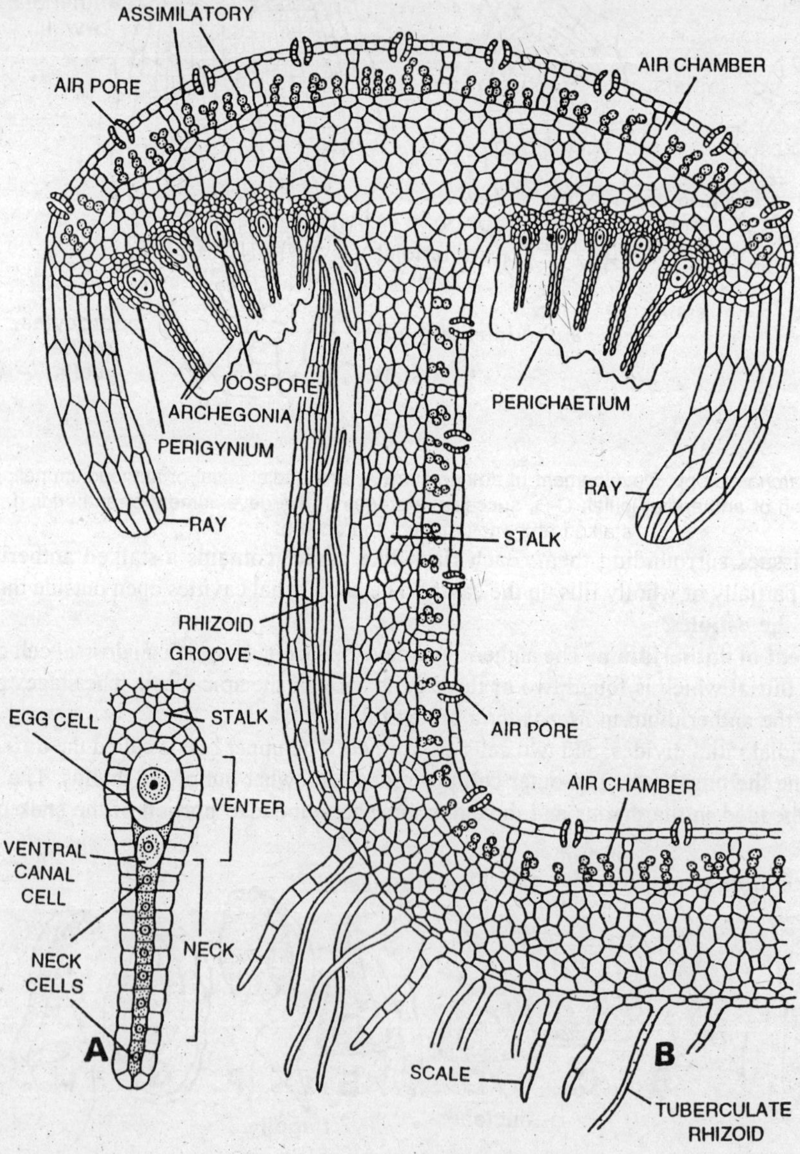

Fig. 21.13. *Marchantia* spp. Female sex organs. A, longitidunal section of archegoniophore; B, a mature archegonium.

these androgonial cells become smaller and smaller. The last cell generation of the androgonial mother cells consists of the **androcyte mother cells.** Each androcyte mother cell divides diagonally giving rise to two **androcytes.**

Soon after the formation of androcytes they are metamorphosed into **antherozoids.**

Structure of mature antheridium. The mature antheridium consists of a short stalk and a rounded structure above it. The antheridium proper is surrounded by a single layered jacket. Inside the jacket layer a large number of androcytes are filled up.

Discharge of antherozoids from antheridium. In usual cases, the antherozoids come out from the antheridia in the cavities, and thereafter they come outside of the cavity through the ostiole. However, the water enters the antheridial cavity through the ostiole, the jacket cells of the upper part of the antheridium become gelatinized, and the antheridium gives way. The androcytes are liberated from the antheridium in a long smoke-like column. As soon as, the mass of androcytes comes out, it breaks up into individual androcytes. Very soon, the antherozoids are liberated from the androcytes.

Structure of antherozoid. Each antherozoid is elongated, rod-like, uninucleate and biflagellate. The antherozoids move in the water with the help of their flagella.

Structure of archegoniophore. The archegoniophore as already stated is a modified branch of the prostrate thallus. This develops by an upward bending of a prostrate branch. According to Cavers (1904), there is a dichotomy of the young archegoniophore. The forked apex divides again giving rise to a four-lobed structure. This is followed by another dichotomy resulting in a eight lobed rosette-like disc. As soon as, the apical branching of archegoniophore stops, the archegonia are developed on dorsal side of it acrogenously. The fertilization takes place in this stage. As soon as the fertilization is over, the marginal portion of the disc becomes inverted and the archegonia become up side down, *i.e.,* their necks become downward. Now, the oldest archegonia are situated towards periphery of the disc and youngest towards the stalk. In many cases, the discs posses nine rays.

After being curved of the margin of the disc a single layered plate of tissue develops on either side of the groups of archegonia. In this way, each group of 12 to 15 archegonia is enclosed in a single layered fringed **perichaetium.** In many cases the archegonial lobes elongate further in the green cylindrical rays. In *M. polymorpha* there are nine rays coming out from the disc. The disc looks stellate in appearance from the upper side, because of the presence of these stout cylindrical rays.

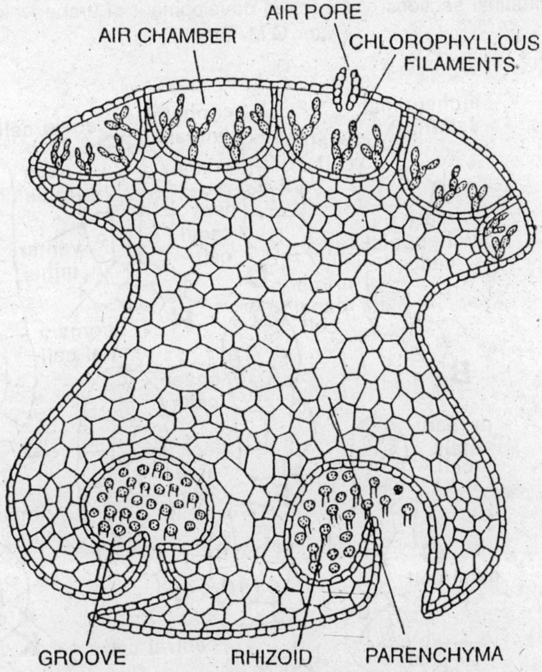

Fig. 21.14. *Marchantia* sp. T.S. of the stalk of archegoniophore.

Development of archegonium. The archegoniun develops from a superficial cell which lies two or three cells back from the apex of the thallus. This cell is called the **archegonial initial.** Now the archegonial initial divides by a transverse wall into two cells. The lower cell is the **basal cell** and the upper cell is the **outer cell.** From the basal cell, the lowermost portion of the archegonium is developed. Wheareas the outer cell produces the main body of the archegonium, the outer cell divides vertically producing a peripheral initial, and it is followed by two more such divisions, giving rise to three cells in total. These cells are peripheral initials. The three cells remain situated upon a **primary axial cell.** The three peripheral initials divide vertically giving rise to six **jacket initials.** All. the six cells encircle the central cell. Thereafter each jacket initial divides by a transverse wall producing six **neck initials** which are arranged in a tier. The tier of six neck cells is

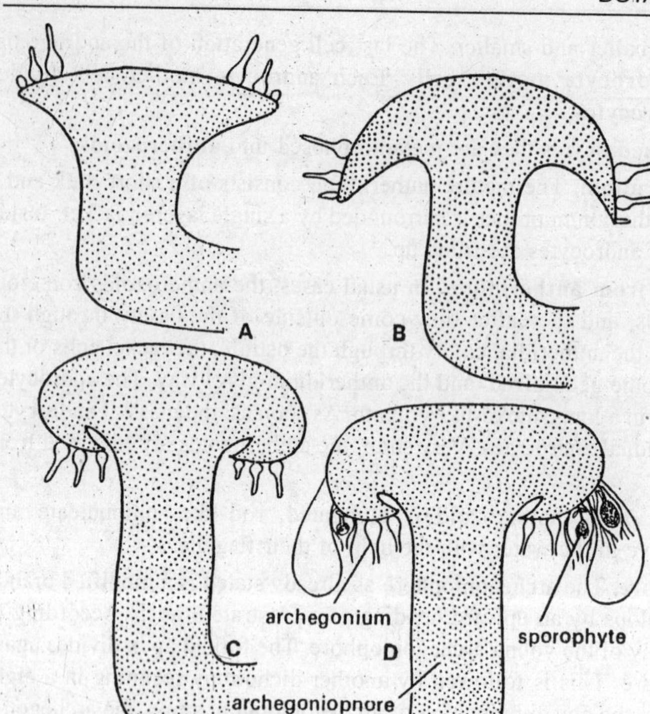

Fig. 21.15. *Marchantia polymorpha*–A–D, diagrams of vertical longitudinal sections of stages in development of archegoniophore (After G.M. Smith)

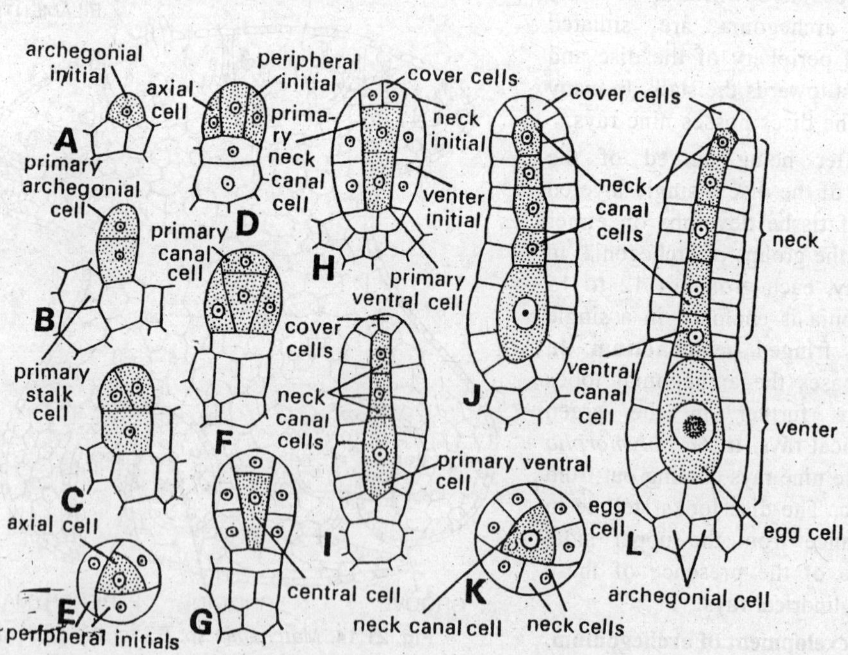

Fig. 21.16. *Marchantia* spp. Development of archegonium. A-K, early stages of development; E and K, transverse sections of archegonium; L, a nearly mature archegonium with venter and neck, the venter contains an egg and the neck contains canal cells.

found above the tier of six **venter initials.** By several transverse divisions, the neck initials give rise to a neck of archegonium 6 to 9 cells in length and six-celled in perimeter. The number of vertical rows of the cells of which the archegonium consists of is six. From the venter initials the jacket of the venter is developed. The initials of the jacket of venter are somewhat different from the neck cells in division. The neck initials divide only by transverse divisions producing daughter cells whereas, the venter initials divide both ways, transversely and vertically producing the bulbous venter jacket.

The primary axial cell divides by an unequal transverse division forming a **primary cover cell** and a **central cell.** The primary cover cell divides twice vertically producing four cover cells at the apex of jacket layer. Eventually, the central cell divides by a transverse wall producing an **uppre canal cell** and a **lower canal cell.** The upper canal cell divides twice forming four **neck canal cells,** which are found within the neck of the archegonium. The lower canal cell, which is somewhat longer and broader, divides by an unequal transverse division forming a **venter canal cell** (ventral canal cell) and a large **egg.** The ventral canal cell is situated above the large egg. On the maturation of the archegonium, the neck canal cells and ventral canal cell disintegrate, and produce a mucilaginous substance in the archegonium. When the archegonium attains maturity, the cover cells become spread and separated from each other.

Structure of archegonim. In *Marchantia,* the flask like archegonium is found attached to the lobe of the disc by a small stalk. The archegonium consists of an elongated neck and a bulbous venter. The neck consists of the jacket cells arranged in six vertical rows. Four or more neck canal cells are found

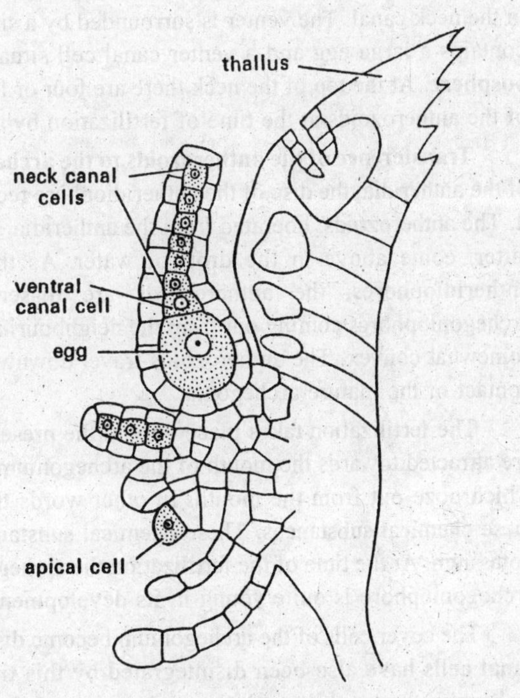

Fig. 21.17. *Marchantia polymorpha.* L.S. through a young archegoniophore showing archegonia originating on the upper surface of the disc. (After Goebel).

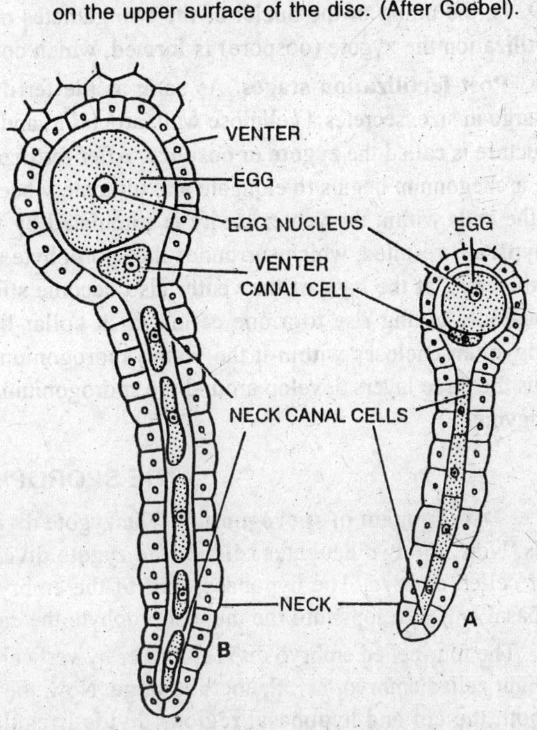

Fig. 21.18. *Marchantia polymorpha.* A, young archegonium; B, mature archegonium.

in the neck canal. The venter is surrounded by a single layered jacket. It is somewhat dialated and contains a large egg and a venter canal cell situated above the egg cell. The egg is also called **oosphere.** At the top of the neck there are four or five cover cells, which give way for the entrance of the antherozoids at the time of fertilization by their disintegration.

Transference of the antherozoids to the archegoniophore and fertilization. On the maturation of the antheridia, the disc of the antheridiophore receives a drop or two of water, which spread over it. The antherozoids, liberated from the antheridia, by the gelatinization of the upper portion of the latter, come above in the drop of water. As the rain drops fall on the concavities of the antheridiophores, the antherozoids are disseminated from these on the discs of the archegoniophores coming out from the neighbouring thalli. The disc of mature archegoniophore is somewhat convex. The antherozoids travel downward through the medium of water and come in contact of the mature archegonia.

The fertilization takes place only in the presence of water. The free-swimming antherozoids are attracted towards the mouth of the archegonium, because of the presence of certain chemicals, which ooze out from the mouth. In other words the antherozoids are positively chemotactic to these chemical substances. These chemical substances are certain proteins and inorganic salts of potassium. At the time of the fertilization the archegonia have their necks directing upward and the archegoniophore is quite young in its developmental stage.

The cover cells of the archegonium become disintegrated and an opening is created. The neck canal cells have also been disintegrated by this time. The antherozoids enter the mouth of the archegonium, travel through the neck and reach in the close vicinity of the egg (oosphere). One of the antherozoids penetrates the egg and fertilization is effected. As reported by Anderson (1929), the union of the nuclei of the two gametes may take a long time. With the result of the fertilization the zygote (oospore) is formed, which contains 2n number of chromosomes.

Post-fertilization stages. As soon as the fertilization is over the fertilized egg begins to enlarge in size, secretes a cellulose wall, around it, and ultimately fills the cavity of the venter. This structure is called the **zygote** or **oospore.** After fertilization many changes take place. The stalk of the archegonium begins to elongate and ultimately becomes 3 or 4 centimeters long. The division of the cells within the venter cavity is stimulated by fertilization and a two or more celled thick **calyptra** is resulted, which surrounds the sporophyte unless and until this becomes mature. Some of the cells, at the base of the venter also become stimulated by fertilization. These cells divide repeatedly giving rise to a one celled thick collar-like **perigynium** or **pseudoperianth.** The perigynium encloses within it the young sporogonium. The perigynium acts as protective layer. Thus the three layers develop around the sporogonium, *viz.,* the perichaetium, the calyptra and the perigynium.

THE SPOROPHYTE

Development of sporogonium. The zygote divides by a transverse wall, giving rise to two cells. Now, the two daughter cells of the zygote divide vertically giving rise to a quadrant type four celled embryo. The hypobasal half of the embryo gives rise to foot and seta, whereas the epibasal half develops into the main sporophyte the capsule.

The four-celled embryo divides further by vertical walls intersected at right angles, giving rise to eight celled embryo, *i.e.,* the **octant stage.** Now, the embryo elongates to some extent. The cell of both the epi and hypobasal regions divide irregularly giving rise to many cells in both the regions. The cells of the posterior region of the parenchyma, ultimately develop into bulbous mass of cells, the **foot.** The foot anchors the sporophyte to the gametophyte and absorbs food materials

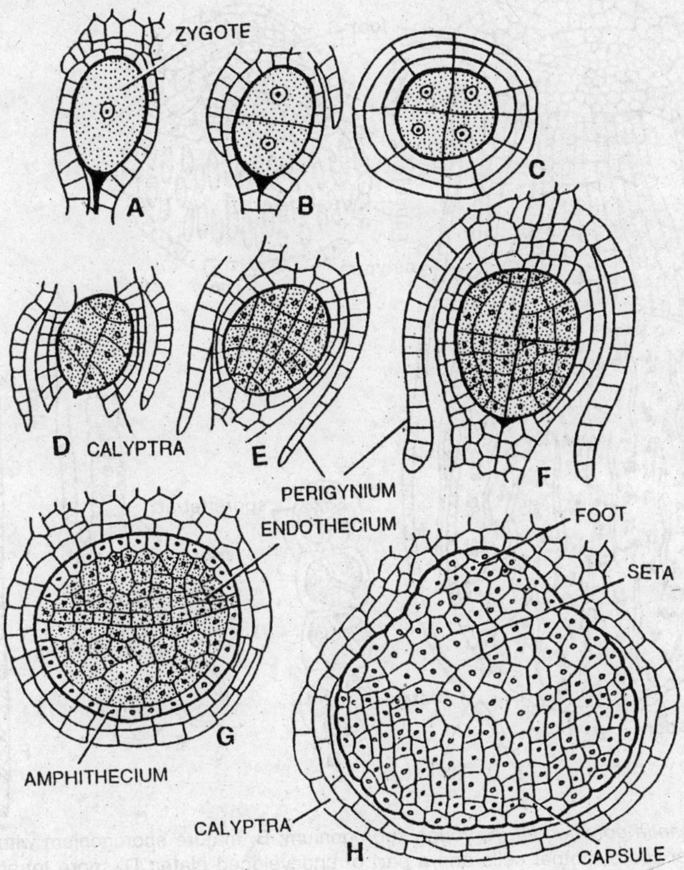

Fig. 21.19. *Marchantia polymorpha.* Development of sporogonium. A-H, successive stages in the development of sporogonium from a single-celled zygote (2n). (After Durand).

for the former from the latter. Next to the foot the **seta** is developed. The cells of the seta are arranged in vertical rows. The cells divide few times and thus the seta remains somewhat short in structure. But at the time of the sporogenesis, the cells of seta divide many times transversely. These cells elongate, thus an elongated seta is resulted which pushes the capsule, out of the three layers, *viz.,* the calyptra, perigynium and perichaetium.

The cells of the upper part of the sporogonium divide periclinally, giving rise to well differentiated regions the **amphithecium** and the **endothecium.** The amphithecium gives rise to a single jacket layer of the capsule. The endothecium develops into the **archesporium.** The cells of this region divide repeatedly giving rise to sporogenous tissue. Eventually, about half of the sporogenous cells divide many times transversely giving rise to the vertical rows of the **spore mother cells or sporocytes.** According to O'Hanlon (1926) each sporogenous cell divides five times resulting in a vertical file of 32 spore mother cells (sporocytes.) Half of the sporogenous cells, which do not divide become elongated and give rise to sterile **elaters.** The elaters are tapering at both the ends and possess spiral thickenings on their walls. Each spore mother cell divides meiotically producing 4 spores. In *Marchantia polymorpha* to each elater there are 32 sporocytes, which give rise to 128 spores. This way, there are 128 spores to each elater. Usually in most of the species the elaters and spores are evenly distributed in the capsule.

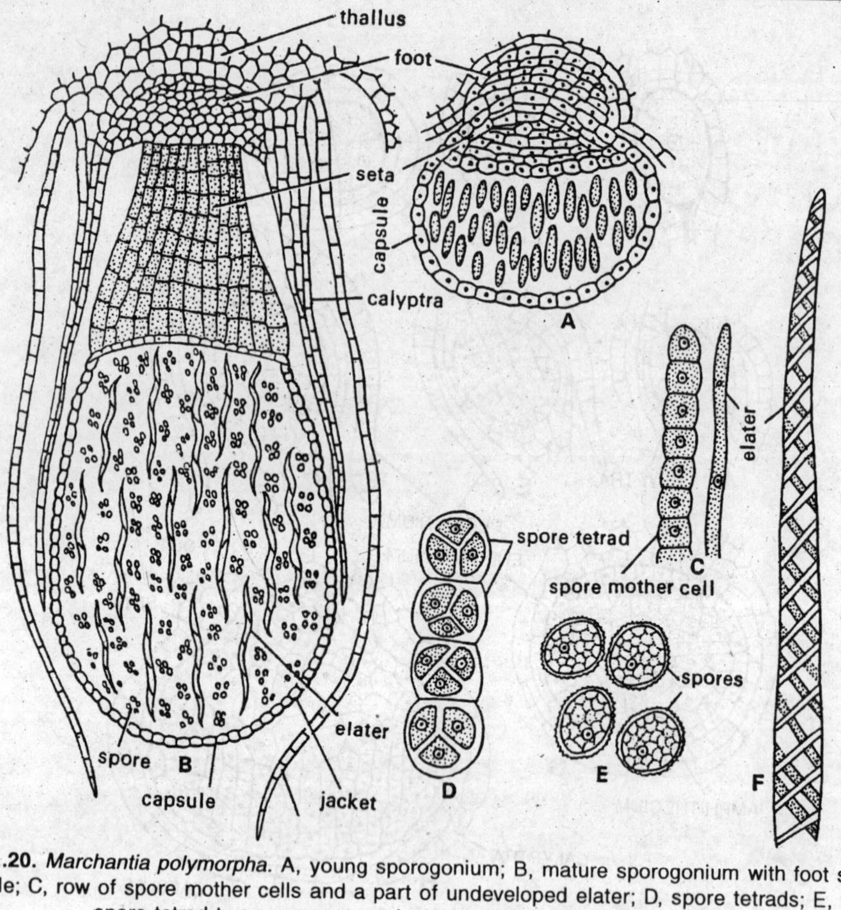

Fig. 21.20. *Marchantia polymorpha.* A, young sporogonium; B, mature sporogonium with foot seta and capsule; C, row of spore mother cells and a part of undeveloped elater; D, spore tetrads; E, in each spore tetrad two spores are male and two female; F, a part of elater.

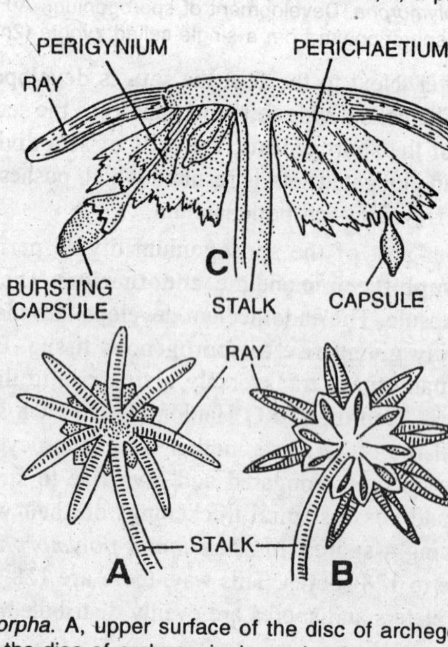

Fig. 21.21. *Marchantia polymorpha.* A, upper surface of the disc of archegoniophore showing rays and involucres; B, undersurface of the disc of archegoniophore showing sporogonia in involucres; C, V.S. of the disc of archegoniophore, (A, B after Kny).

According to O'Hanlon (1930), in *Marchantia* the output of spores from a single capsule is about 300,000 spores.

Structure of mature sporogonium. The mature sporogonium is differentiated into three regions, *viz.*, the foot, seta and capsule. Towards the base of the archegonium, there is a bulbous, absorptive structure, the foot. It absorbs nutrition and water for the sporogonium from the gametophyte. The seta is somewhat elongated and connects the foot and the capsule. As stated before, the seta pushes the capsule out of the three layers, the calyptra, perigynium and perichaetium. The capsule is somewhat oval shaped. It is covered by single layered jacket layer. Within the jacket layer, there are spores and sterile, hygroscopic, spindle-like elaters. The elaters are hygroscopic in nature and help in dehiscing the spores.

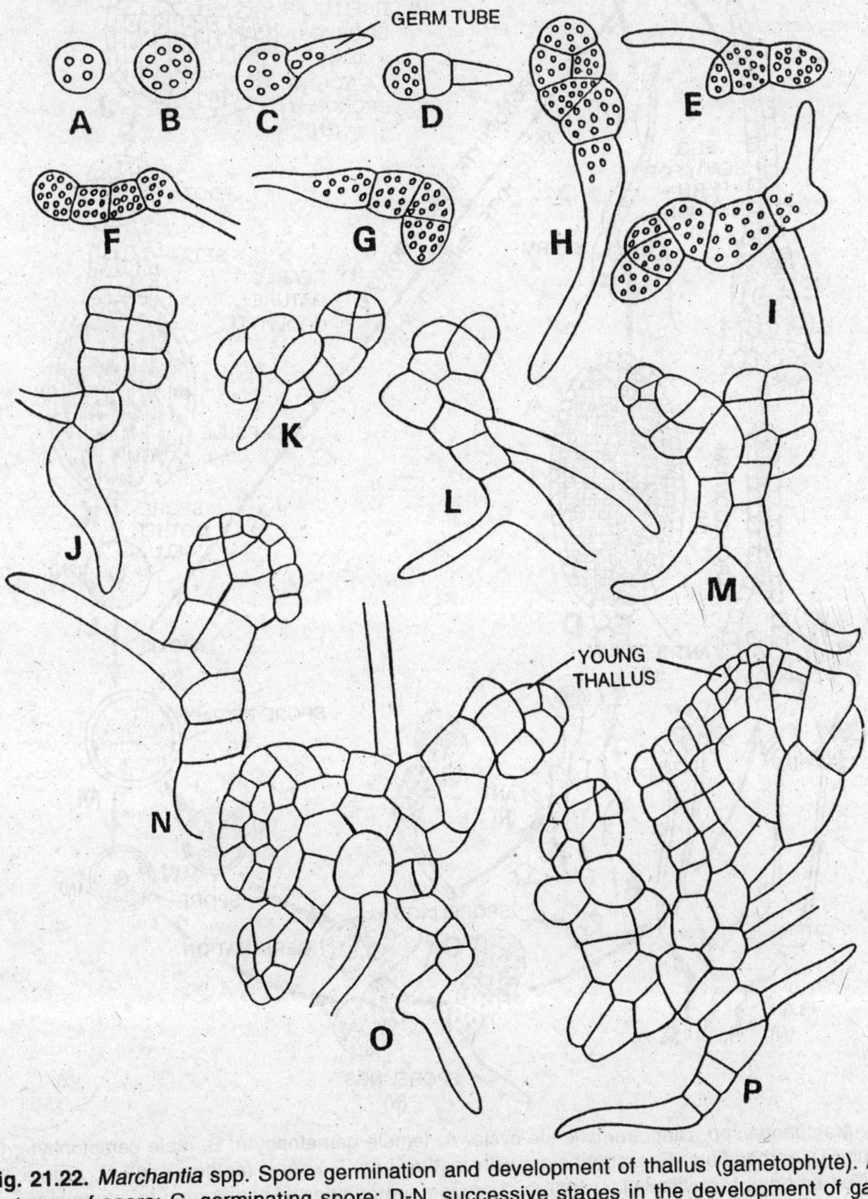

Fig. 21.22. *Marchantia* spp. Spore germination and development of thallus (gametophyte). A-B, different stages of spore; C, germinating spore; D-N, successive stages in the development of gametophyte (thallus); O-P, young gametophyte (thallus) with rhizoids.

Dispersal of the spores. On the maturation the capsule comes out of the three layers with the help of its seta. It hangs downward from the lower side of the disc of archegoniophore. The wall of the mature capsule ruptures and the longitudinal slits develop about half of the length of the capsule, giving rise to several teeth like lobes rolling backward. The eleaters are hygroscopic in nature. In dry atmosphere, they lose water to the atmosphere and become twisted. This action of

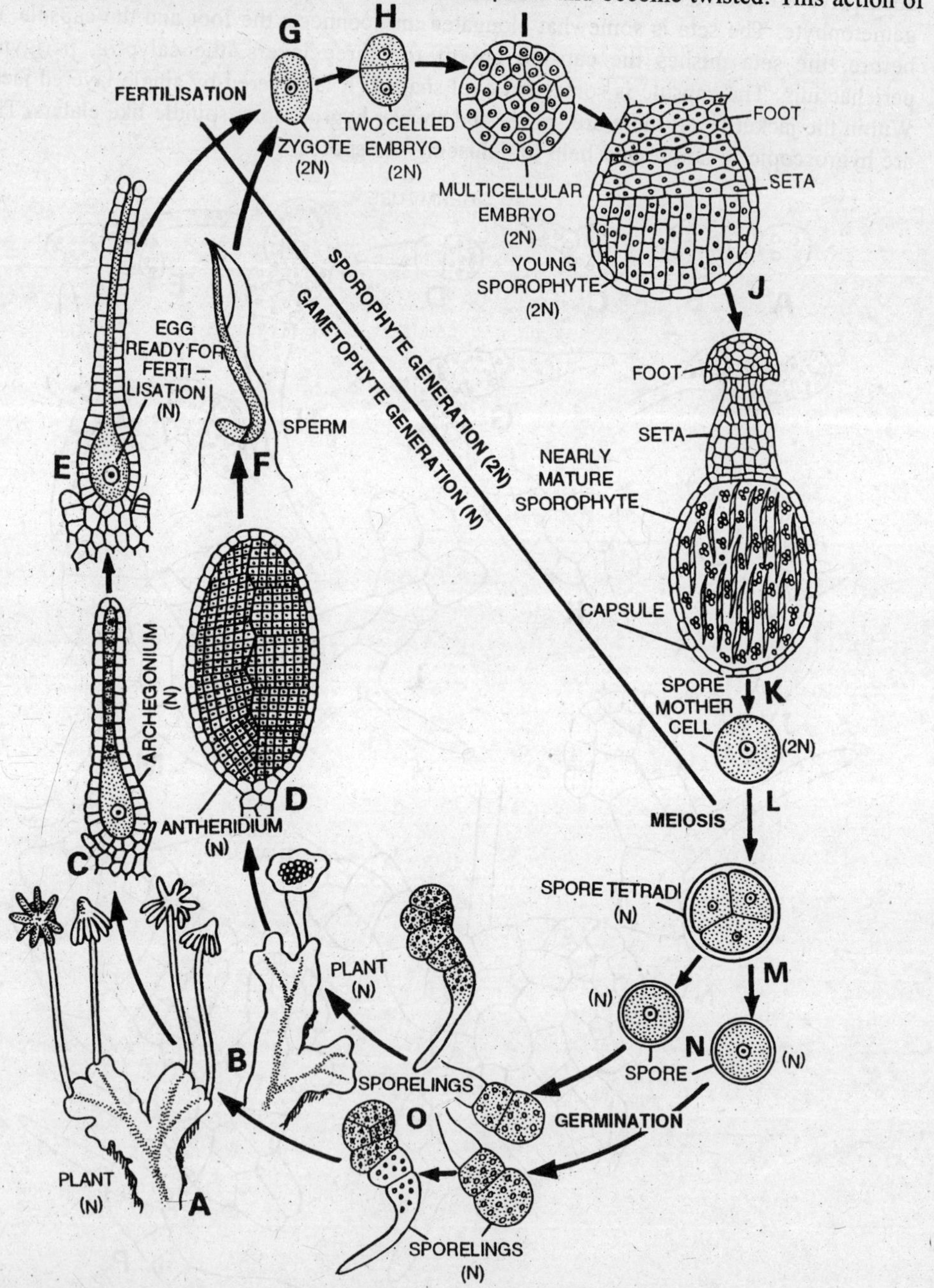

Fig. 21.23. *Marchantia* spp. Diagrammatic life-cycle. A, female gametophyte; B, male gametophyte; C, archegonium; D, antheridium; E, archegonium with mature egg; F, sperm (antherozoid); G, zygote; H, two-celled embryo; I, multicellular embryo; J, young sporophyte; K, nearly mature sporophyte (sporogonium); L, spore mother cell; M, spore tetrad; N, spores; O, germinating spores (sporelings).

the elaters help in the release of the spores from the capsule. As soon as, the spores are released and they come out of the capsule, they are dispersed by wind.

Structure of spore. The spores are small and spherical. They range from 12 μ to 30 μ in diameter. Each spore is covered by a thin coat, differentiated into two layers. The outer layer of the coat is **exospore** and the inner one is **endospore.** The exospore is somewhat thick and the endospore is thin and smooth. Each spore contains a small amount of granular cytoplasm and a nucleus within it.

Spore germination and development of young gametophyte. As soon as the spores fall on the soil, and if the other conditions are favourable, they germinate immediately. After coming in contact of moist soil, the spore enlerges at least double of the original and thereafter it germinates (Menge, 1930; O'Hanlon, 1926). The germinating spore immediately produces a long irregular filament of 6 to 8 cells. At the apex it is two cells wide. One of the cells acts as an apical cell with two cutting faces. This cell cuts alternately 5 to 7 segments both on right and left faces. This division continues. The derivatives of the first formed segments divide twice or thrice, but the derivative of the later-formed segments divide again and again producing a large number of cells. After cutting off a few cells, the two-faced apical cell itself divides repeatedly giving rise to a sheet of a large number of cells. In this sheet, a transverse row of the apical cells characteristic of adult gametophyte of *Marchantia* develops. When the young gametophyte becomes 30 to 40 celled, the notch appears in the apical portion of it.

Systematic position. Division. Bryophyta
Class. Hepaticospida (Hepaticae)
Order. Marchantiales
Family. Marchantiaceae

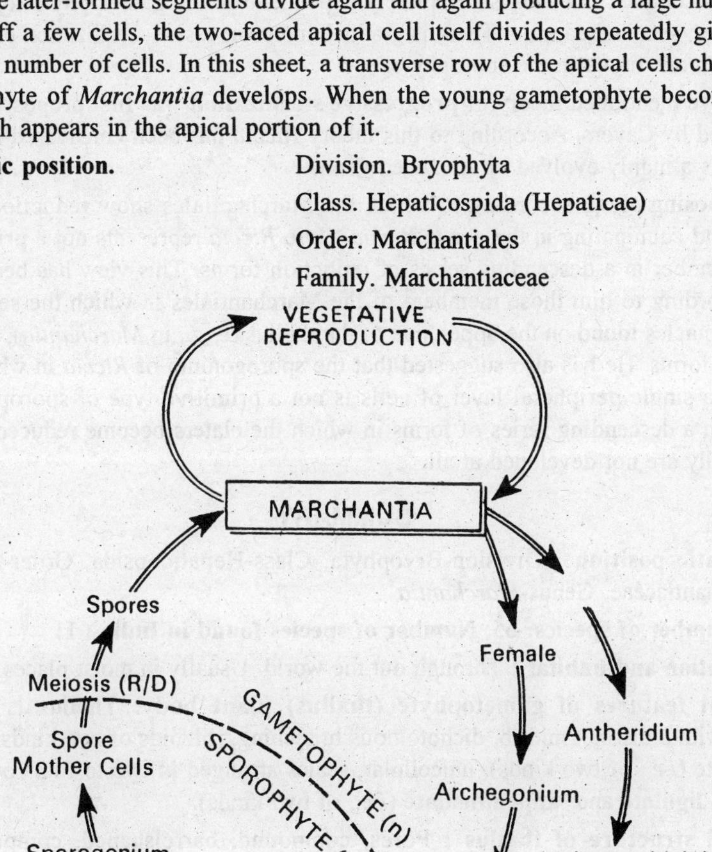

Fig. 21.24. *Marchantia.* Graphic life-cycle.

PHYLOGOENY OF MARCHANTIALES

In the order Marchantiales, starting from *Riccia,* there is a general advancement in the elaboration of vegetative organs, the aggregation of the sexual organs into complex gametophores, and a progressive differentiation of the sporogonium. This is known as **progressive evolution.** Schiffner, Cavers, Campbell, Bower, Fritsch, Smith are the chief supporters of this view.

The simplest thallus structure is seen in most species of *Riccia,* where the upper tissue is made up of the chlorophyllous filaments separated by narrow air spaces. In higher Marchantiales the structure of thallus is more advanced and differentiated into air chambers which contain the chlorophyllous filaments. The eleborated pores (*e. g.,* in *Marchantia*) remain surrounded by a single ring of cells, forming a barrel-like structure. The elaboration of the thallus took place independently in several distinct lines.

In *Riccia,* the sporophyte is the simplest known among the Hepaticopsida (Hepaticae). It is simply a sac of spores and incapable of self-nutrition. It does not possess foot and seta. The wall of capsule consists of one layer of cells in thickness. On the other hand the sporophyte of *Marchantia* is complex one and is completely dependent on the gametophyte for its nutrition. It consists of foot, seta and capsule. There is gradual and progressive sterilization of the fertile cells of the sporogenous tissue in the sporophyte.

The above mentioned theory of **progressive sterilization** was first proposed by Bower and later supported by Cavers. According to this theory *Riccia* has been considered as primitive and *Marchantia* as a highly evolved or advanced genus.

The opposing view, according to which the Marchantiales show reduction, starting from *Marchania* and culminating in the genus *Riccia.* Here *Riccia* represents not a primitive form but the lowest member in a descending series of reduction forms. This view has been supported by Goebel. According to him those members of the Marchantiales in which the sexual organs are borne in receptacles found on the upper side of the thallus (*e. g.,* in *Marchantia*), are reduced and not primitive forms. He has also suggested that the sporogonium of *Riccia* in which sterilization is limited to a single peripheral layer of cells is not a primitive type of sporophyte, but is the lowest term in a descending series of forms in which the elaters become reduced to short sterile cells and finally are not developed at all.

SUMMARY

Systematic position : Division-Bryophyta, Class-Hepaticopsida, Order- Marchantiales, Family-Marchantiaceae, Genus-*Marchantia.*

Total number of species: 65; **Number of species found in India** : 11

Distribution and habitat : Through out the world. Usually in moist places.

External features of gametophyte (thallus) plant body: Thallus is prostrate, flat, dorsiventral with a distinct midrib, dichotomous branching, rhizoids of two kinds, smooth-walled and tuberculate (*i.e.,* of two kinds), unicellular, scales arranged in two to four rows on each side of the midrib, **ligulate** and **appendiculate** (*i.e.,* of two kinds).

Internal structure of thallus : Pores, compound, barrelshaped, composed of 4 to 8 superimposed tier of cells, each tier consists of 4-5 cells, there is single horizontal layer of air chambers separated by partitions, the photosynthetic tissue consists of single or branched filaments of chlorophyllous cells arising from the floor of air chambers, the storage tissue is made up of the ventral region of the thallus which is composed of compact, colourless parenchyma.

Apical growth : By transverse row of 3-5 apical cells each with two cutting faces which cut off segments from dorsal and ventral faces.

Sex organs : *Marchantia* is dioecious as the antheridiophores and archegoniophores are borne on different thalli.

Antheridia : The antheridia are borne in groups on the upper surface of the disc of antheridiophore and remain immersed in antheridial chambers; the antheridial stalk, short and few celled; antheridial jacket, singlelayered; antherozoids biciliate.

Archegonia : The archegonia are situated on the disc of stalked archegoniophore; the archegonia are protected by perichaetium, rays and disc of the archegoniophore; archegonial stalk short; the neck consists of six vertical rows of cells; the venter is single-layered thick; there are 4 cover cells at the tip of neck; generally 4-6 neck canal cells are present.

Sporophyte : The **amphithecium** give rise to the jacket of capsule; the **endothecium** gives rise to archesporium; the archesporium gives rise to sporocyte and elaters; the foot is bulbous or broad-spreading; the long seta elongates rapidly after the maturity of spores; the capsule is nearly spherical structure at the distal end of the sporogonium; the wall of capsule is single-layered and persistent; long, narrow, spindle-shaped elaters with two spiral thickenings; the wall of capsule splits longitudinally into a number of lobes and the elaters become helpful in spore dispersel; sporophyte is dependent on gametophyte.

Young gametophyte : The spore is surrounded by 2 wall layers; the size of spore varies from 0.012 to 0.03 mm; on germination of spore, a cell filament is formed directly; thereafter a many celled tissue is developed, in which the apical cell is differentiated.

22

CHAPTER

Anthocerotopsida Order-Anthocerotales

The class Anthocerotopsida (Anthocerotae) consists of a single order, the Anthocerotales and a single family, the Anthocerotaceae, 6 genera and 301 species. According to Muller (1940), Reimers (1954) and Proskauer the order Anthocerotales includes two families, (i) Anthocerotaceae and (ii) Notothylaceae. The latter includes a single genus, *i.e.*, *Notothylas.* However, according to top bryologists, there is only one family, *i.e.*, Anthocerotaceae. About five or six genera are included in this family.

These genera are—*Anthoceros*, *Phaeoceros*, *Aspiromitus*, *Notothylas*, *Dendroceros* and *Megaceros*. Four genera are universally recognised, they are—*Anthoceros*, *Megaceros*, *Dendroceros* and *Notothylas*. This group differs in many respects from the other Bryophyta. However, the group is placed intermediate between Hepaticopsida (Hepaticae) and Bryopsida (Musci). The group is considered to be very important from the point of view of its morphology, because of its intermediate position between the two important groups, the Hepaticopsida and Bryopsida.

The most characteristic features of the group are as follows : The gametophytic plant body is thalloid and dorsiventral. The rhizoids are simple and smooth walled. Tuberculate rhizoids and ventral scales are altogether absent.

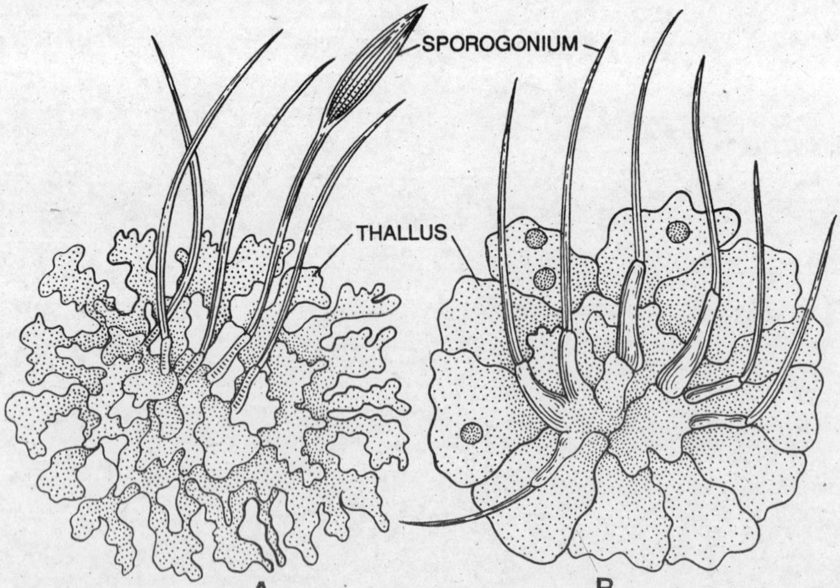

Fig. 22.1. *Anthoceros* sp. A, thallus with sporogonia of *A. punctatus*; B, thallus with sporogonia of A. *laevis*.

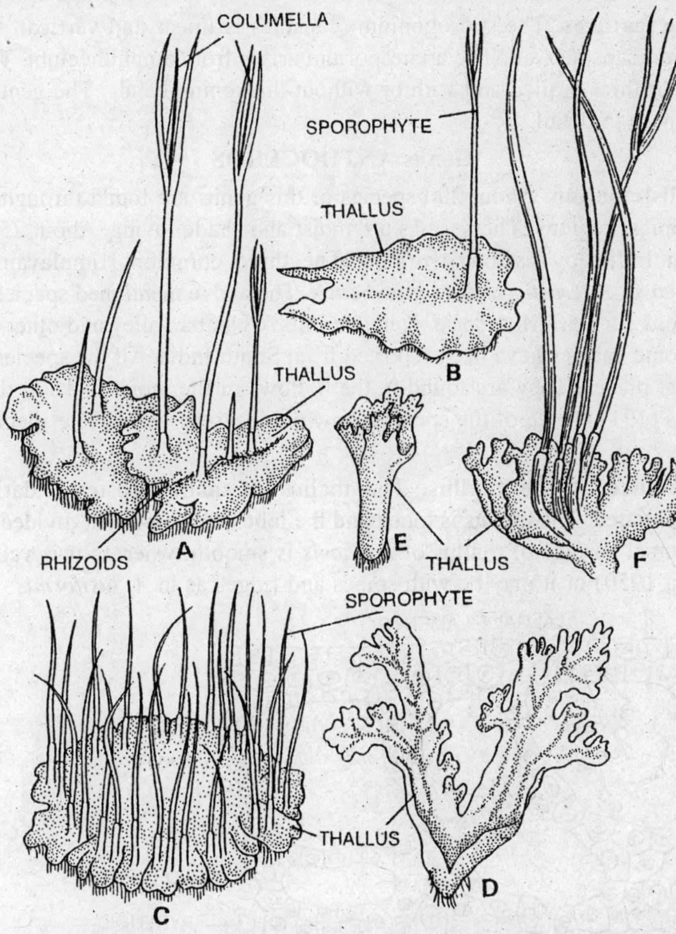

Fig. 22.2. *Anthoceros* sp. Gametophyte and sporophyte. A, gametophyte (thallus) bearing rhizoids and mature and immature sporophytes; B, part of a thallus with opened sporophyte; C, thallus of *Anthoceros crispulus* with several elongated horny sporophytes; D and E, thalli of (sterile) *A. erectus* without sporogonia; F, thallus (fertile) of *A. fusiformis* with one opened and three unopened sporophytes.

The tissues of the thallus are not differentiated. Air chambers and air-pores are absent. Each cell of the thallus possesses a large chloroplast and a conspicuous pyrenoid within it.

The antheridia are endogenous, *i.e.*, they arise from the hypodermal cells of the thallus on the dorsal side of it. The antheridia are developed within the antheridial chambers, singly or in groups on the dorsal side of the thallus.

The archegonia are found in sunken condition on the dorsal side of the thallus.

The sporogonium arises from the dorsal side of the thallus. It is elongated and cylindrical in structure. It consists of foot, meristematic region and capsule. It possesses intercalary meristem, and continues its growth throughout the growing season. The wall of sporogonium contains chlorophyll. The central sterile portion is columella, which is surrounded by sporogenous tissue and spores. The elaters are also present.

The sporogenous mass develops from amphithecium and arches over the columella.

Family-Anthocerotaceae.

There are *five* genera (*i.e.*, *Anthoceros*, *Phaeoceros*, *Aspiromitus*, *Dendroceros* and *Megaceros*) in this family.

Characteristic features. The sporogonium (capsule) is linear and vertical. Generally the stomata are present on capsule wall. The archesporium arises from amphithecium. The elaters are four-celled, smooth or thick-walled and with or without thickening bands. The genus *Anthoceros* has been discussed here in detail.

Genus ANTHOCEROS

Habitat and distribution. About 200 species of this genus are found throughout the world in temperate and tropical regions. The species are moist and shade loving. About 25 species have been reported from India, by various workers. The three common Himalayan species are, *Anthoceros himalayensis, A. erectus* and *A. chambensis.* The above mentioned species are common in various hilly regions such as, Mussoorie, Kumaon hills, Chamba valley and other places, 5,000 feet to 8,000 feet. Some species have been reported from South India. All the species are found in very shady and moist places. They are found in the hollows of the moist rocks in dense patches. According to Cavers (1911) some of the species grow on decaying wood. The species, are not at all drought resistant.

External structure of the thallus. The thallus is small, prostrate, dark green and dorsiventrally differentiated. The thallus is lobed and the lobes are somewhat divided. The mid-rib is not found. The dorsal surface of thallus of *A. laevis* is smooth, whereas it is velvet like in *A. crispulus* (Bhardwaj, 1950) or it may be with spines and ridges as in *A. fusiformis.* In every case

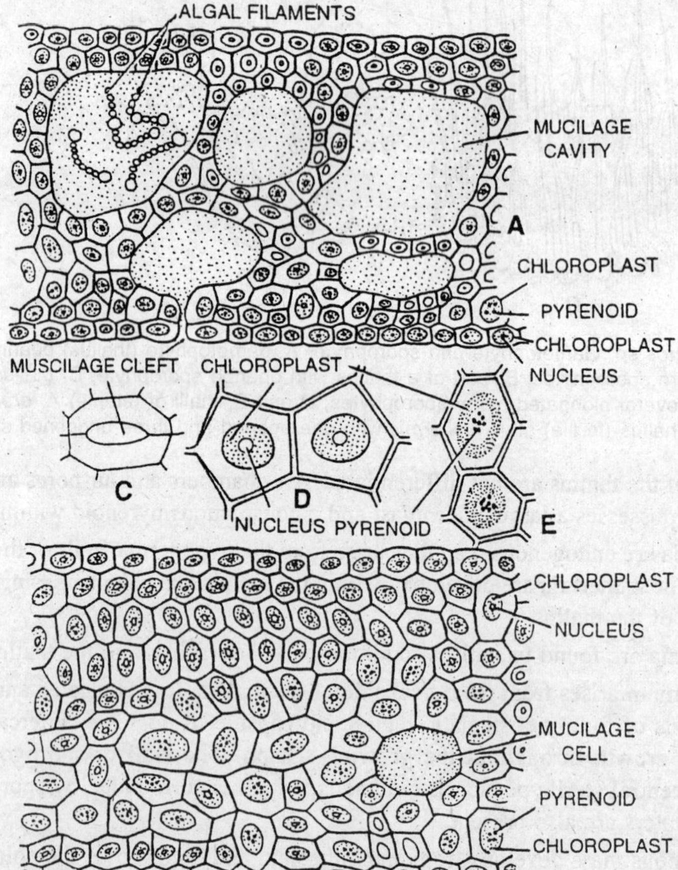

Fig. 22.3. *Anthoceros* sp. A, portion of a section of thallus of *A. punctatus*; B, portion of a section of thallus of *A. laevis*; C, mucilage cleft from the ventral surface of thallus; D-E, cells of thallus enlarged.

the smooth walled, simple rhizoids are found. The ventral scales and tuberculate rhizoids are altogether absent. The thalli are dark green, because of the presence of the *Nostoc* colonies, which may be easily seen with the help of lens from the underside of the thallus.

Internal structure of the thallus. The anatomy of the thallus is quite simple. Internal to the upper and lower epidermis there are simple, parenchymatous cells. The cells of parenchyma are isodiametric and uniform. Air chambers and air pores are absent. Each cell contains a big chloroplast which possesses a single pyrenoid in its centre. The chloroplasts are lens shaped. The chloroplasts of the superficial cells are longer than the chloroplasts of the other cells. According to McAllister (1914, 1927) the pyrenoids of *Anthoceros* are quite different in structure from those of Chlorophyceae. On the ventral side of the thallus certain intercellular **mucilage cavities** are found. These cavities open by small openings, the **slime pores** on the ventral surface of the thallus. Sometimes, the colonies of blue green algal form-*Nostoc*, are found in the mucilage cavities. There is no symbiotic relationship between these *Nostoc*, colonies, and thalli. According to Peirce (1960) these colonies, however, do harm to the thalli. The nucleus lies in the close vicinity of the chloroplast, near the pyrenoid. Sometimes the chloroplast enfolds the nucleus within it.

Apical growth. The apical growth of *Anthoceros* is a controversial topic, whether it takes place by a single apical cell or by a group of apical cells. According to Smith (1955) the apical growth of the thallus is initiated by a single apical cell. According to Leitgeb (1879) the apical growth in this genus takes place by several marginal apical cells. According to Mehra and Handoo (1953) the apical growth in *Anthoceros erectus* and *A. himalayensis* takes place by a group of cells situated in the depression at the apex of the thallus.

Reproduction. The reproduction takes place by means of (1) vegetative and (2) sexual methods.

1. **Vegetative reproduction.** The vegetative reproduction takes place by various ways.

(a) **By progressive growth and death of thallus.** The vegetative propagation takes place by progressive growth and death of the older part of the thallus reaching dichotomy. But this method is not so common in *Anthoceros* as in *Riccia* and *Marchantia.*

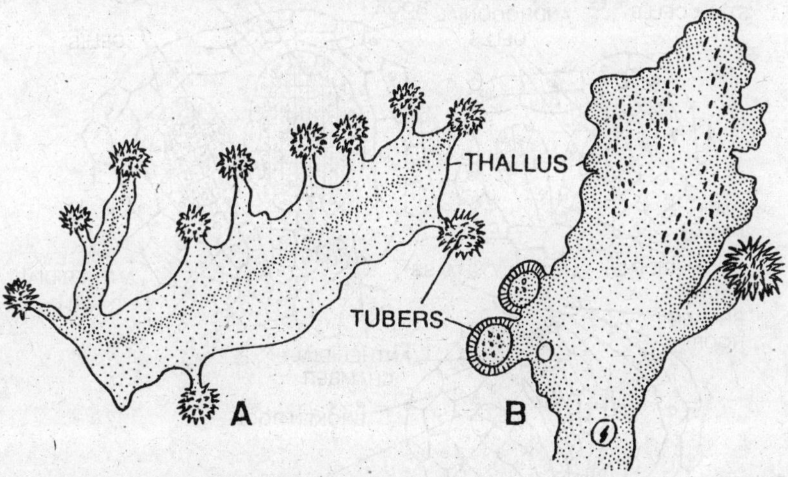

Fig. 22.4. *Anthoceros* sp. Vegetative reproduction. A, thallus of *A. himalayensis* with tubers; B, thallus of *A. laevis* with tubers.

(b) **By tubers.** In certain species of *Anthoceros,* the thallus becomes thickened at several places on the margins. Such marginal thickenings are called, the **tubers.** These tubers are perennating structures. They survive in the drought conditions. On the advent of the favourable conditions, they develop into new thalli. The tubers are formed in *A. laevis, A, tuberosus, A hallii, A. pearsoni* and *A. himalayensis.*

(c) **By gemmae.** In some of the species of *Anthoceros,* the gemmae have been found. The gemmae have been recorded from the species, *A. glandulosus, A. formosae,* etc. Each such gemma develops into a new thallus.

(d) **By persistent growing apices.** According to Campbell the thalli of the species *A. pearsoni* and *A. fusiformis* become completely dried up in summers, leaving growing apices with

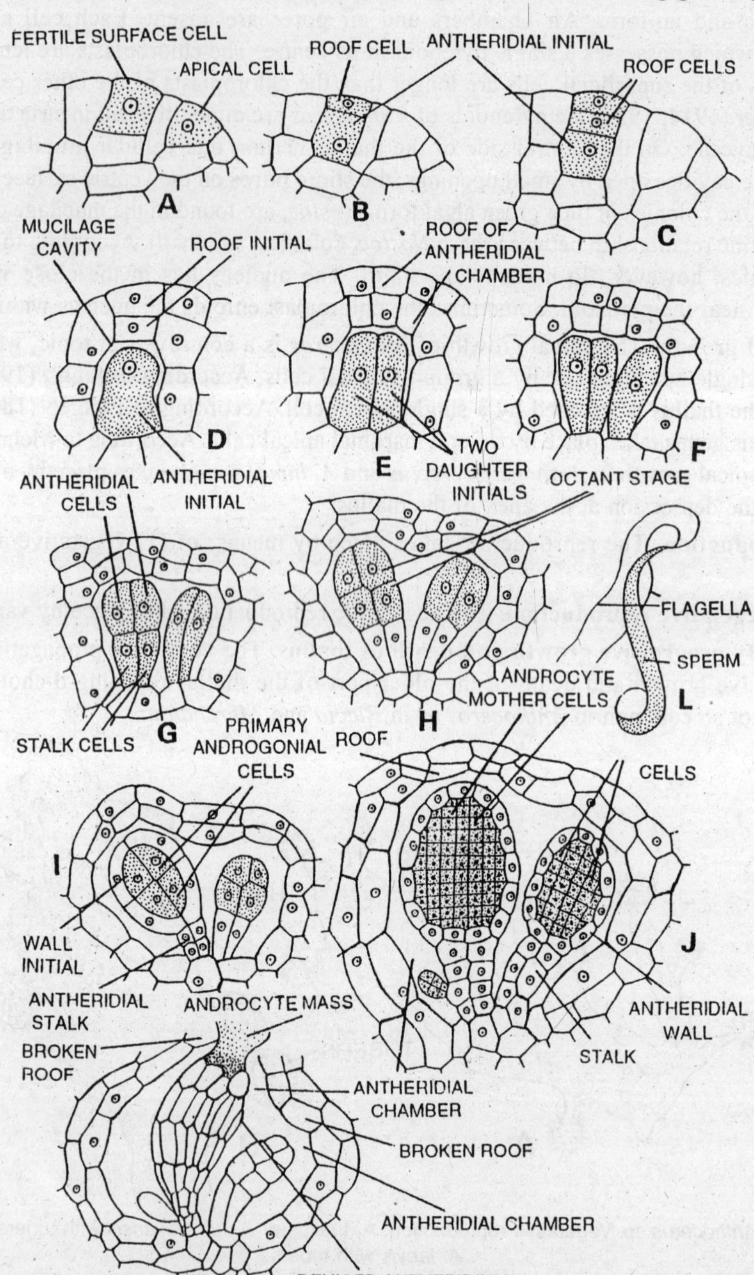

Fig. 22.5. *Anthoceros* sp. Development of antheridium. A, surface cell lies close to the apical cell; B, surface cell divides periclinally into two cells-roof cell and antheridial initial; C-J, further stages in the development of antheridium and roof; K, mature dehisced antheridium releasing out androcyte mass; L, biflagellate sperm or antherozoid with long anterior flagella.

adjacent tissues. These apices face the drought conditions. On the approach of favourable conditions, these apices develop into new thalli.

2. Sexual reproduction. The species of *Anthoceros* may be homothallic (monoecious) or heterothallic (dioecious). Some of the homothallic species are- *A. fusiformis, A. punctatus,* Kashyap (1915, 1929) has recorded, *A. himalayensis* as a dioecious (heterothallic) species, but Mehra and Handoo (1953) have reported the same as monoecious (homothallic) but protandrous. The heterothallic species are, *A. pearsoni, A. halli, A. erectus* and others. The sex organs, *i.e.*, antheridia and archegonia are found embedded in the tissues of the dorsal side of the thallus.

Development of antheridium. The antheridia are produced singly or in groups in the antheridial chambers. The development is endogenous. Though the antheridium develops from a superficial cell, yet it is enclosed within the antheridial chamber which does not open out by any opening.

A dorsal superficial cell of the thallus, situated near the growing apex divides periclinally giving rise to two daughter cells. According to Cavers, Campbell and Haupt the superficial cell divides transversely and not periclinally. The upper daughter cell acts as **roof initial** and the lower one as **antheridial initial.** Eventually, a mucilage filled space appears in between the roof initial and antheridial initial. This mucilage cavity enlarges in size and ultimately becomes the **antheridial chambers.** The roof initial is nothing to do with the development on the antheridium.

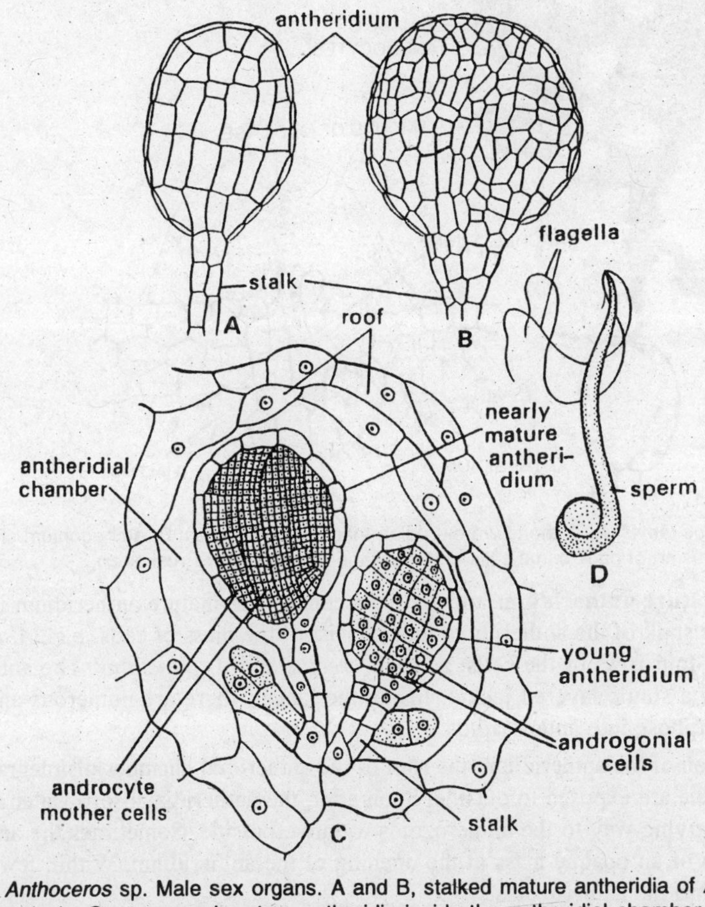

Fig. 22.6. *Anthoceros* sp. Male sex organs. A and B, stalked mature antheridia of *A. crispulus* and *A. laevis* respectively; C, young and mature antheridia inside the antheridial chamber, secondary budded antheridia are seen on the stalks of primary antheridia; D, biflagellate sperm or antherozoid.

It divides and redivides several times anticlinally and periclinally giving rise to a two layered roof of the antheridial chamber. Simultaneously, the antheridial initial develops into a single antheridium or in a group of antheridia. A single antheridium develops in *A. pearsoni* and sometimes in *A. himalayensis.* According to Mehra and Handoo (1953), in *A. erectus* a number of antheridia develop in an antheridial chamber. Here, the antheridial initial divides many times anticlinally producing many cells and each cell thus produced, develops into an antheridium. The further development is as follows:

The antheridial initial divides twice by vertical walls intersecting each other at right angles, giving rise to four cells. This is followed by another transverse division giving rise to two tiers of four cells each. The four cells of the upper tier divide transversely giving rise to eight cells, the octant stage. All of the cells of octant stage divide periclinally giving rise to eight outer **primary jacket cells** and eight inner **primary androgonial cells.** The four cells of the lower tier develop into a stalk of the antheridium consisting of the four rows of the cells. The so formed androgonial cells divide repeatedly. The last generation of these cells are androcyte mother cells. According to Bagchee (1924) each androcyte mother cell divides diagonally producing two **androcytes.** Each androcyte metamorphoses into a spindle like biciliate antherozoid.

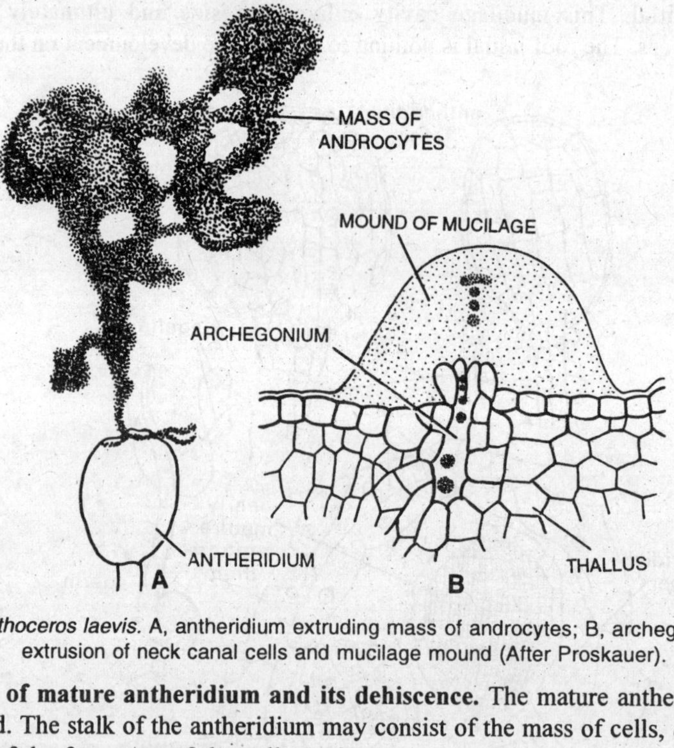

Fig. 22.7. *Anthoceros laevis.* A, antheridium extruding mass of androcytes; B, archegonium showing extrusion of neck canal cells and mucilage mound (After Proskauer).

Structure of mature antheridium and its dehiscence. The mature antheridium is stalked and club shaped. The stalk of the antheridium may consist of the mass of cells, *e.g., A. laevis,* or it may consist of the four rows of the cells, as in *A. erectus* and *A. punctatus.* The antheridium proper is covered by a single layered jacket. Inside the jacket, there are numerous androcytes which are to metamorphose into antherozoids.

On the maturation of the antheridium, the roof of the antheridial chamber disintegrates, with the result the antheridia are exposed to outside. Soon after, the antheridia absorb water and burst at their apical ends, giving way to the antherozoids to move outside. Sometimes the androcytes come out in the form of an opaque mass at the opening of the antheridium. Within few minutes they metamorphose into the antherozoids.

The antherozoid. The antherozoid is spindle like and biciliate. The cilia are attached to the anterior end of the body. Sometimes just near the attaching point of the flagella to the body, the blepharoplast is visible. The antherozoids swim in the water by the lashing moment of their flagella.

Development of archegonium. The archegonia are found embedded in the thallus. They remain in direct contact of the vegetative cells of the tissue of the thallus lateral to them. They do not possess jacket sterile cells around them. The development of archegonium begins from a single superficial cell. This cell becomes prominent and acts as **archegonial initial.** According to Mehra and Handoo (1953), this has been established that the archegonial initial functions directly as a **primary archegonial cell**. Formerly it was believed (Campbell) that it divides, and produces two cells, the primary archegonial cell and primary stalk cell. The archegonial initial first divides vertically, producing three **jacket initials** which surround an **axial cell.** The axial cell divides transversely, producing a **cover initial** and a **central cell**. Thereafter, the central cell divides by a transverse wall, giving rise to a **primary canal cell** and a **primary venter cell.** The primary canal cell divides repeatedly, producing a linear file of 4-6 neck canal cells. The primary venter cell divides once transversely, giving rise to two cells, a **ventral canal cell** and an **egg** (oosphere). Comparatively, the neck canal cells are narrower than the venter canal cell and the egg. Eventually, the three jacket initials also divide by transverse walls. The further development of the jacket layer

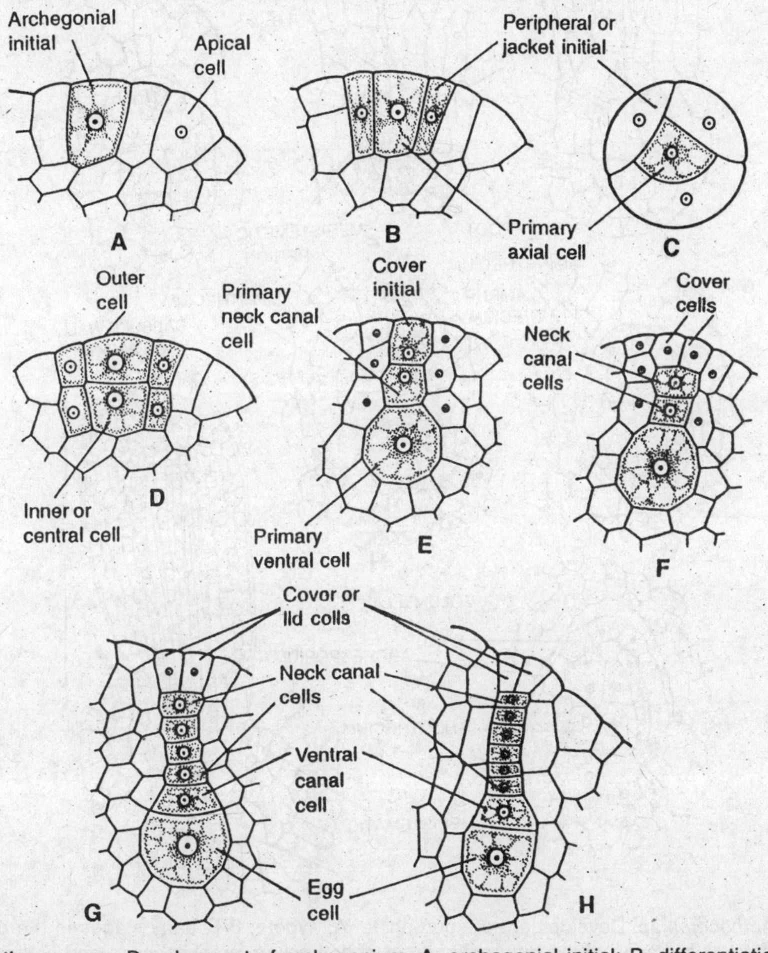

Fig. 22.8. *Anthoceros* sp. Development of archegonium. A, archegonial initial; B, differentiation of primary axial cell in L.S.; C, same in T.S.; D, division of axial cell into inner and outer cell; E, development of cover initial, primary neck canal cell and primary ventral cell; F, archegonium with egg, ventral canal cell, two neck canal cells and cover cells; G, nearly mature archegonium; H, mature archegonium (After Campbell).

is not at all clear. Since the cells on the lower face of the egg have been derived from the archegonial initial they cannot be treated as a part of archegonium.

Structure of archegonium. On the maturation of the archegonium, the venter canal cells and neck canal cells become gelatinized. Thus a mature archegonium is flask-like in shape, without neck canal cells and with an egg (oosphere) in its venter. At the top of the neck of the archegonium there are four cover cells, which become separated from the archegonium, as soon as the gelatinization of the venter and neck canal cells is over.

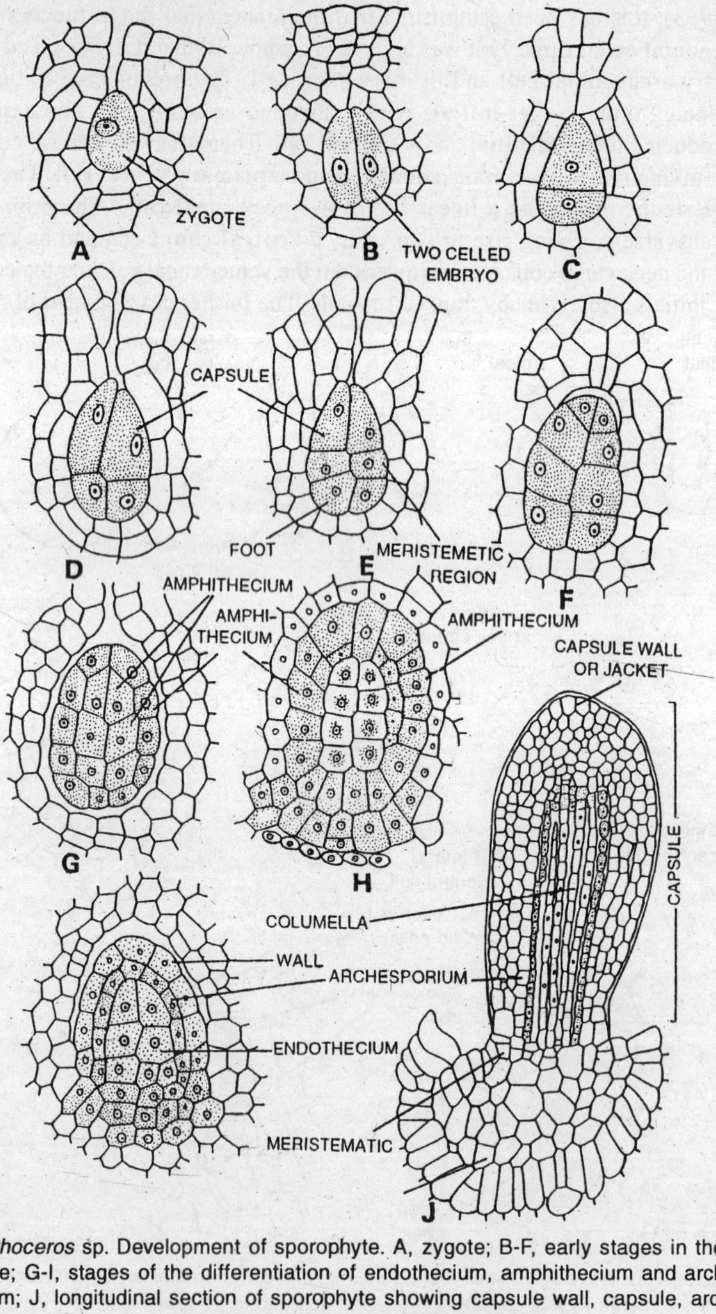

Fig. 22.9. *Anthoceros* sp. Development of sporophyte. A, zygote; B-F, early stages in the development of sporophyte; G-I, stages of the differentiation of endothecium, amphithecium and archesporium of sporogonium; J, longitudinal section of sporophyte showing capsule wall, capsule, archesporium, columella, meristematic region and foot.

Fertilization (syngamy). Prior to fertilization, the cover cells become detached from the archegonium, and the neck canal cells become gelatinized. Through the medium of water, the antherozoids enter the mouth of archegonium. The antherozoids are attracted chemotactically. Ultimately, one lucky antherozoid penetrates the egg, and the fertilization is effected. The male and female nuclei unite to each other, producing a zygote (oospore), zygote is 2n, and this is the beginning of the sporophytic stage.

Development of sporogonium. After fertilization, the zygote secretes a cellulose wall around it and enlarges in size still it completely fills the venter of the archegonium. The first division of the zygote is vertical. But according to Pande (1932) and Bhardwaj (1950), in certain cases this divides transversely. With the result of the first vertical division, the daughter cells are produced, which are subjected to a transverse division producing four cells of equal or unequal size. When the cells are unequal in size, definitely the cells towards the neck of the archegonium are larger cells. These cells again divide vertically, developing eightcelled embryo, four cells in each tier. The upper tier of four cells divides transversely. This way the three tiers of four cells each have been produced. The lowermost tier produces the **foot**, the middle tier produces partly the foot and mainly the **seta** and the upper-most tier produces the **capsule.**

The cells of the lowermost tier divide regularly or irregularly many times, producing the bulbous foot. The foot is haustorial in nature, which absorbs food from the tissue of the gametophyte.

The uppermost tier of four cells, divides once or twice transversely, producing two or three tiers of cells. Now the cells divide periclinally, giving rise to an outer layer the **amphithecium** and the central mass of cells the **endothecium.** In the young sporogonium, the columella consists of four vertical rows of the cells, the **endothecium.** In the young sporogonium, the columella consists of four vertical rows of the cells, but later on it is made up of sixteen rows of cells. According to Bhardwaj (1958) in *A. gemmulosus* the columella consists of 36 to 49 vertical rows of the cells. The amphithecium divides periclinally producing the outer sterile layer of the **jacket initials** and an inner sporogenous tissue, the **archesporium.** The jacket initials divide again and again periclinally producing the 4 to 6 layered wall of the capsule. The outermost layer develops into the single layered epidermis. The epidermal cells are cutinized. The stomata may or may not develop on the epidermal layer. The rest of the layers, beneath the epidermis develop in normal chlorenchyma. These chlorophyllous cells, however, help in synthesizing the food.

The layer of archesporium developed by periclinal division overarches the columella. The sporogenous layer may be one to four celled in thickness in its further development. In *Anthoceros hawaiensis* it remains one celled in thickness; in *A. pearsoni* and *A. himalayensis* it may become two, three or even four cells in thickness. In younger stages all the cells of sporogenous tissue remain somewhat rectangular in shape. Later on, the sporogenous tissue becomes differentiated into two types of cells, *i. e.,* (i) the **sporocytes** (spore mother cells) and (ii) the **sterile cells** (pseudoelaters).

The sporocytes or spore mother cells undergo the reduction division, each producing a tetrad of four spores. These spores are haploid. The spores, which develop in the upper part of the capsule, mature first. Each spore contains a nucleus and a chloroplast.

The sterile cells soon divide obliquely or transversely producing three to five celled pseudoelaters. On maturity, the pseudoelaters lose their protoplasm. The walls of the elaters may be smooth, irregularly thickened or with spiral thickenings. Such thickenings, vary from species to species. The pseudoelaters help in the dehiscence of the spores and behave like true elaters. In earlier stages their function seems to be nutritive.

The sporogonium and its dehiscence. The capsules arise from the thalli in the form of small horny structures. Usually they are two to three centimeters long. But in some species they range even from five to fifteen centimeters in their height and because of their horny appearance, the species are called 'hornworts'. The mature sporogonium consists of a bulbous foot and a projecting, slender and erect capsule. There is a meristematic zone above the foot, instead of seta. The bulbous foot consists of parenchyma. It remains penetrated in the thallus and acts as haustorium. The superficial cells of the foot are palisade like. The space in between the foot and the capsule is occupied by a meristematic zone. The cells of this zone divide throughout unless and until the food is completely exhausted. This way, the capsule increases in length. The capsule increases in height even after maturation. The cells developing from the meristem become

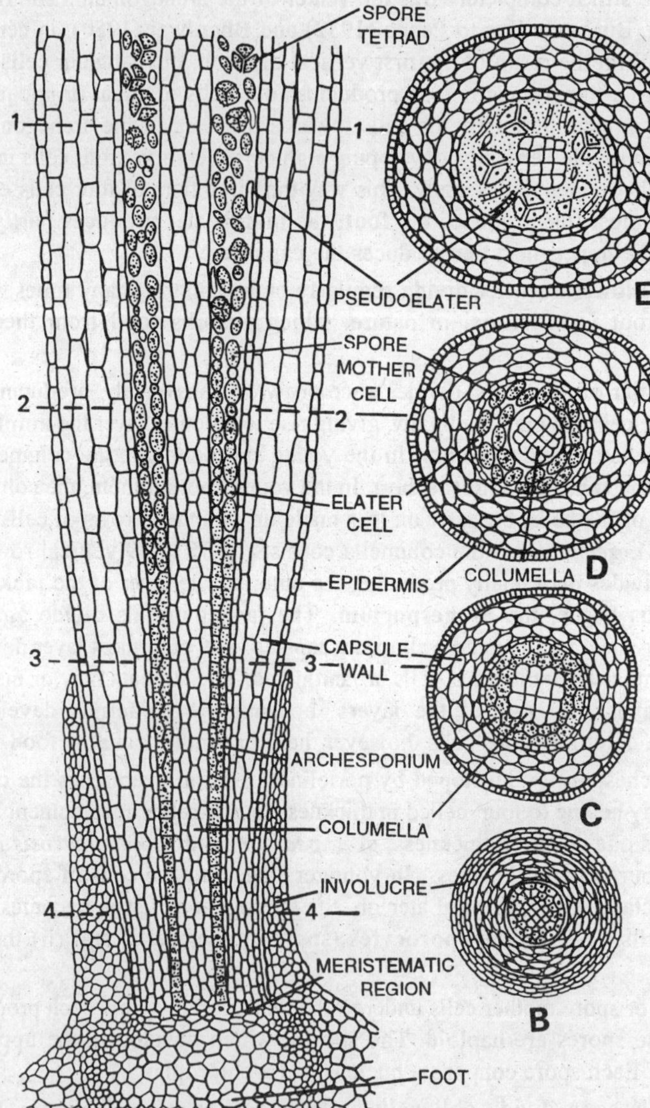

Fig. 22.10. *Anthoceros* sp. A, longitudinal section of sporogonium; B, cross section of sporophyte (A) at 4-4; C, cross section of sporophyte (A) at 3-3; D, cross section of sporophyte (A) at 2-2; E, cross section of sporophyte (A) at 1-1.

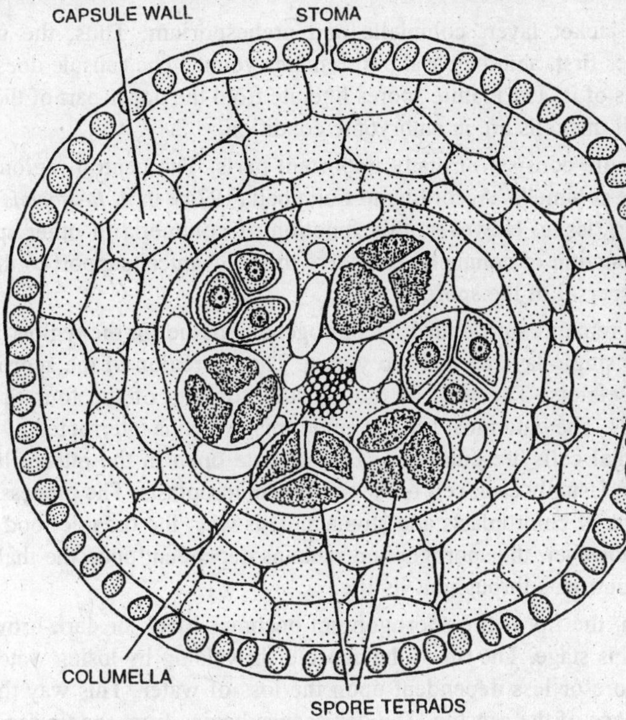

Fig. 22.11. *Anthoceros* sp. T.S. of sporogonium.

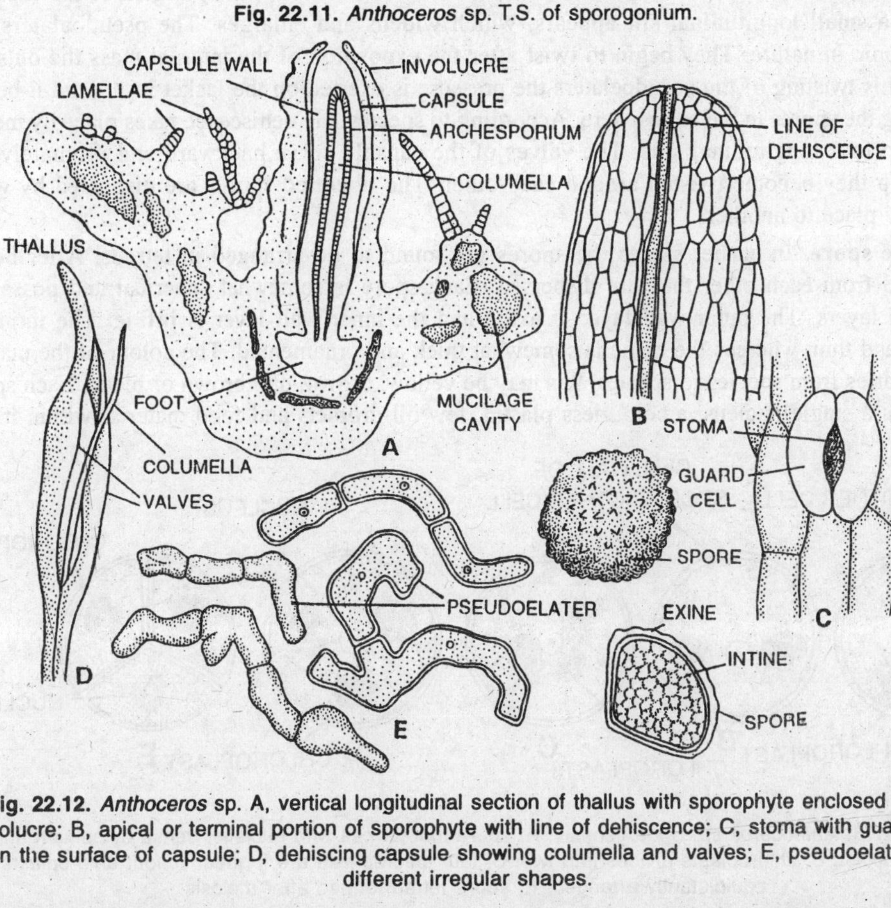

Fig. 22.12. *Anthoceros* sp. A, vertical longitudinal section of thallus with sporophyte enclosed within involucre; B, apical or terminal portion of sporophyte with line of dehiscence; C, stoma with guard cells on the surface of capsule; D, dehiscing capsule showing columella and valves; E, pseudoelaters of different irregular shapes.

differentiated into jacket layer, columella ana archesporium. Thus, the upper part of the sporogonium matures first, and the basal part remains young. The capsule does not mature at the same rate in all parts of it. The mature spores liberate from the upper part of the capsule whereas, in the basal part still the cells are in embryonic condition.

The main capsule consists of many important parts. The central region of the capsule is occupied by a sterile columella. According to Campbell (1924), in *A. fusiformis* the columella acts as water conducting tissue. However, this function of columella in other species is not very certain. The main function of columella is to give the mechanical support to the sporogonium. It also helps in the dispersal of the spores.

The columella remains surrounded by sporogenous tissue. In the region just above the foot the archesporium is single layered and too young. The tip region of the sporogonium possesses mature spores and elaters. The wall of the capsule consists of the four to six layers of the parenchymatous cells. The outermost layer is epidermis, which is interrupted by stomata at several places. The epidermal cells are cutinized. The stomata open in the intercellular spaces of the chlorophyllous cells. Usually each cell possesses two chloroplasts. The process of photosynthesis takes place by means of chloroplasts and stomata. This way, the organic food is synthesized for the sporogonium. However, the sporogonium remains dependent upon the thallus for the supply of water and other nutrients throughout its life.

On maturation, the tip of the sporogonium becomes black or dark-brown in colour. The capsule divides at this stage. The tip of the capsule shrivels up by losing water. The dehiscence of the capsule is more or less dependent upon the loss of water. This way the dry atmosphere helps in the dehiscence of the capsule. The dehiscence begins from the tip region of the capsule. At first a small longitudinal slit appears, which widens and enlarges. The pseudoelaters are hygroscopic in nature. They begin to twist after the exposition of the internal mass the outside. Due to this twisting of the pseudoelaters the pressure is exerted on the jacket layer, and it bursts liberating the spores in the atmosphere. According to species, the dehiscence takes place by means of one to four longitudinal slits. The valves of the capsule curve backward and ultimately, on drying up they become twisted around each other. The liberated spores are dispersed by wind from one place to another.

The spore. In earlier stages the spores are found to be arranged in tetrads. After being separated from each other they are dispersed. Each spore is somewhat spherical and possesses two wall layers. The outer wall layer is **exine** and the inner wall layer is **intine.** The intine is smooth and thin, whereas the exine is somewhat thick and ornamented. The colour of the mature spores varies from species to species, this may be yellow, brown, darkbrown or black. Each spore possesses a single nucleus, a colourless plastid, few oil droplets and food material within it.

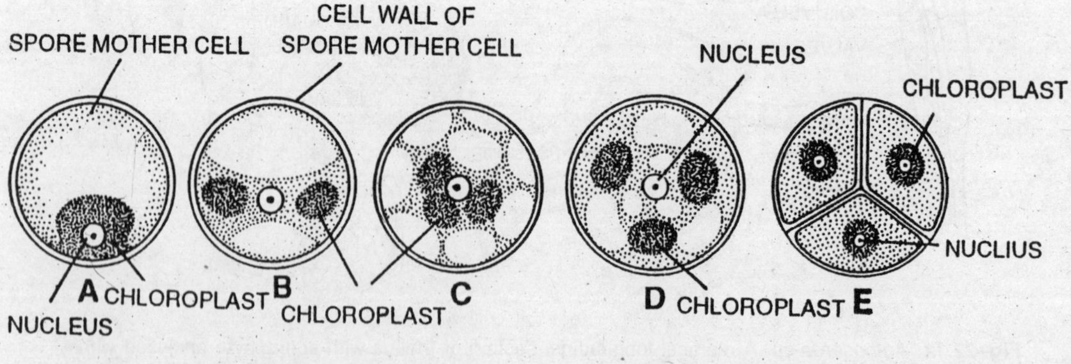

Fig. 22.13. *Anthoceros* sp. Sporogenesis. A, spore mother cell with nucleus, dense cytoplasm and chloroplast; B, chloroplasts divides into two; C, four chloroplasts are formed; D, four chloroplasts equidistantly arranged; E, spore tetrad formed after meiosis.

Germination of spore. After their liberation from the sporogonium the spores undergo a period of rest prior to germination which ranges from few weeks to few months. The exine of the spore ruptures, and the intine comes out in the form of a germinal tube or protonema of variable length.

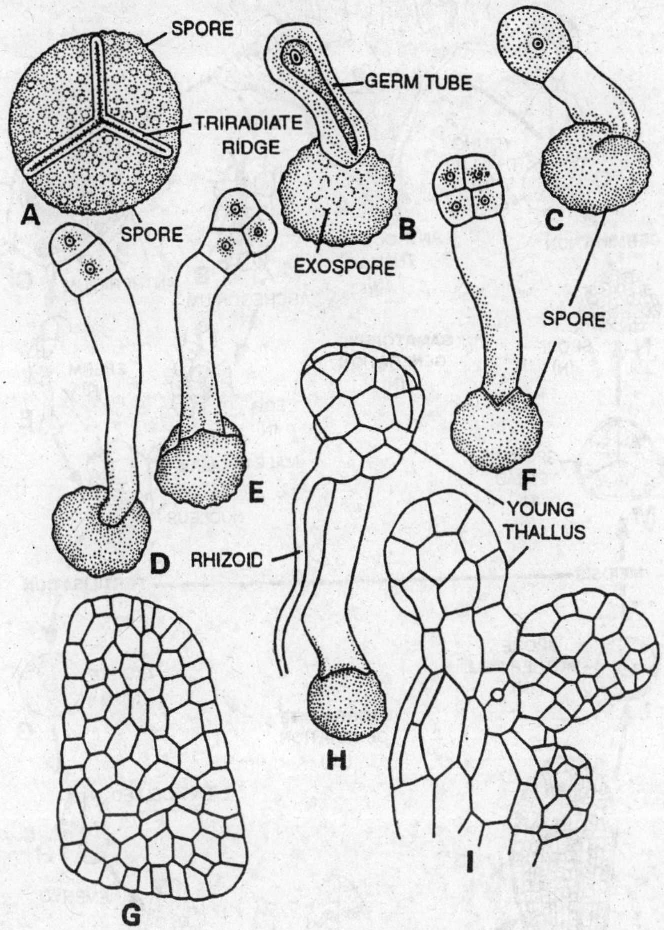

Fig. 22.14. *Anthoceros* sp. Spore and its germination. A, spore with triradiate ridge; B, germinating spore producing germ tube; C-F, further stages of spore germination; G-I, formation of young gametophyte (thallus) and rhizoids.

Development of young gametophyte. The chlorophyll present in the chloroplast of the spore passes in the germinal tube along with oil droplets and food material. Thereafter the germinal tube divides transversely at its apical end. This division is followed by another transverse division. Soon after, the so formed two cells divide by longitudinal intersecting walls and the quadrants are resulted. At the terminal end a growing point with an apical cell appears which by producing several segments develops into the young gametophyte. Soon after a mucilage slit develops on the ventral side of the thallus. Some of the marginal cells of the young thallus develop into smooth walled rhizoids. The *Nostoc* sp. penetrates the thallus through the mucilage slits, which later on form the colonies of the same.

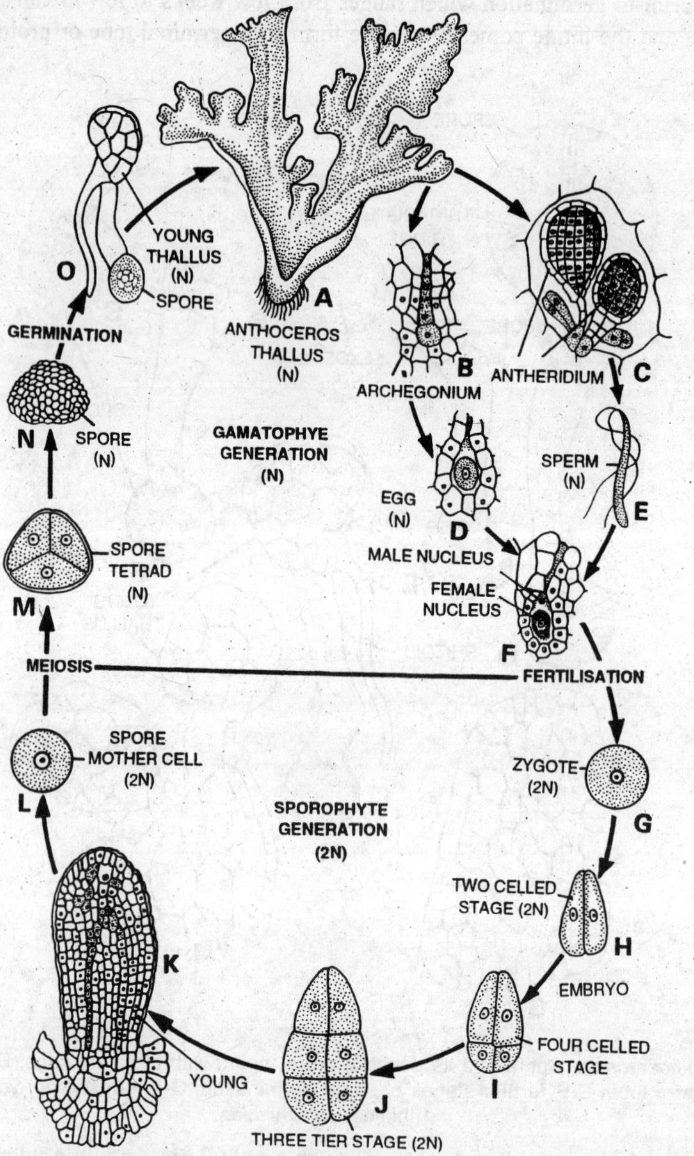

Fig. 22.15. *Anthoceros* sp. Diagrammatic life-cycle A, thallus; B, archegonium; C, antheridium; D, egg within venter; E, sperm or antherozoid; F, male and female nuclei are in close vicinity; G, zygote (2n); H, two-celled embryo; I, four-celled embryo; J, three tier stage; K, young sporophyte in L.S.; L, spore mother cell; M, spore tetrad; N, spore N, spore (n); O, germination of spore and development of young thallus.

Systematic position.

Division.	Bryophyta
Class.	Anthocerotopsida (Anthocerotae)
Order.	Anthocerotales
Family.	Anthocerotaceae
Genus.	*Anthoceros*.

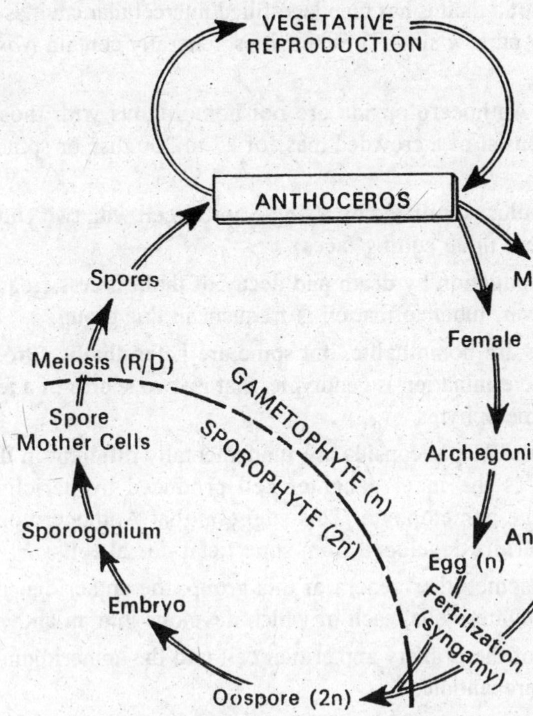

Fig. 22.16. *Anthoceros.* Graphic life-cycle.

Inter-relationships of Anthocerotopsida

The Anthocerotopsida differ from other bryophytes in a number of respects. These differences are :

(a) Among Anthocerotopsida the cells are with large chloroplasts and each chloroplast contains a pyrenoid.

(b) In this group the development of antheridia takes place from hypodermal cells on the dorsal side of a gametophyte.

(c) The archegonia of this group are almost completely embedded in the gametophyte.

(d) The growth of sporophyte is indeterminate because of a meristematic region continually adding to the base of capsule.

Because of the differences mentioned above the class has been placed in between Hepaticopsida on one hand and the Bryopsida on the other.

The other features of special interest of Anthocerotopsida are :

(a) The gametophytes are thallose, somewhat lobed or radially dissected, and sometimes show a tendency of dichotomous branching.

(b) The gametophytes are always dorsiventrally differentiated and possess numerous smooth-walled rhizoids on the ventral surface.

(c) The ventral surface is devoid of scales and mucilage hairs.

(d) Lateral margins of most of the genera of this group (except of *Dendroceros*) are more than one cell in thickness.

(e) There is no internal differentiation of tissues.

(f) The ventral portion of a thallus has mucilage-filled intercellular cavities on the dorsal side and opening to the surface by narrow slits. These cavities generally contain *Nostoc* (a blue green alga) colonies in them.

(g) The pyrenoids of Anthocerotopsida are not homologous with those of green algae (chlorophyceae) since they consist of a crowded mass of 25 to 300 disc or spindle shaped bodies (McAllister, 1914, 1927).

(h) The growth of a thallus is initiated by a single apical cell with two cutting faces (except in *Dendroceros* where there are three cutting faces).

(i) The vegetative reproduction by death and decay of thalli is less frequent in this group than in Hepaticopsida. However, tuber formation is frequent in this group.

(j) Most of the species are homothallic, but some are heterothallic (Proskauer, 1948). In heterothallic species the sex determination is genotypic, that is, two spores of a tetrad develop into male and two into female gametophytes.

(k) Anthocerotopsida and Hepaticopsida are fundamentally different in the fact that in the former an antheridial initial is the inner daughter cell produced by periclinal division of a superficial daughter cell of the gametophyte. This suggests that Anthocerotopsida are derived from ancestors in which antheridia developed from superficial dorsal cells.

(l) In *Anthoceros* and some other genera of this group, the antheridial initial may divide vertically into two or four daughter cells, each of which develops into an antheridium.

(m) The development of the primary antheridial cell into the antheridium proper is similar as in Sphaerocarpales and Marchantiales.

(n) In Anthocerotopsida, the spermatogenesis is much like that of other bryophytes and involves a metamorphosis of androcytes into biflagellate sperms on antherozoids.

(o) In Anthocerotopsida the archegonial initial functions directly as a primary archegonial cell instead of dividing into primary archegonial cell and primary stalk cell as found in other bryophytes.

(p) The first division of a zygote is vertical but cases have been found (Bhardwaj, 1950; Pande, 1932) where the transverse division occurs.

(q) The amphithecium divides periclinally, where the outer layer functions as the initial layer of the jacket and the inner layer as the archesporium.

(r) One unique feature of Anthocerotopsida is that cells of a capsule do not mature at the same rate and that cells in the basal portion of a capsule remain embryonic even after those in the apical portion are fully mature. This feature is not found in other bryophytes.

Affinities of Anthocerotopsida

This group exhibits similarities with green algae, Hepaticopsida (liver worts), Bryopsida (mosses) and Psilophytales of Pteridophyta.

(A) Features common with green algae

The green algae (Chlorophyceae) have been considered a group from which the bryophytes and pteridophytes are believed to have originated.

(i) Usually each cell of gametophyte has a single large chloroplast of a definite shape.

(ii) The pyrenoids are present in the chloroplasts of the cells of gametophyte. The presence of pyrenoids is the characteristic of cells of green algae only.

(iii) The pyrenoids of Anthocerotopsida and green algae are similar in function and form starch grains in their peripheral region.

(iv) The outline and branching of the gametophyte is similar in both cases.

(v) The presence of biciliate or biflagellate (both flagella of whiplash type) antherozoids.

(B) Features common with liverworts (Hepaticopsida)

Many taxonomists include Anthocerotales in Hepaticae.

(i) The gametophyte is thallus-like.

(ii) The smooth-walled rhizoids are found in the members of Anthocerotopsida as well as in the members of Jungermanniales of Hepaticopsida.

(iii) The apical growth of thallus is similar in both.

(iv) Archesporium giving rise to spores and sterile cells in both groups. The sterile cells with spiral bands are found in *Megaceros* of Anthocerotopsida which exhibit the similarity with many members of Hepaticopsida.

(v) Differentiation of the amphithecium and endothecium by periclinal walls is similar to that of many Hepaticae.

(C) Features common with the Bryopsida (mosses)

(i) Presence of central columella in both groups.

(ii) Large reduction of the sporogenous tissue in both Anthocerotopsida as well as Bryopsida groups.

(iii) Presence of functional stomata (*e.g.,* in *Funaria*).

(iv) Differentiation of archesporium from the inner amphithecium as in Sphagnales. This feature shows link between Anthocerotopsida and Bryopsida.

(v) The developmental stages of embryo are quite similar. The early divisions are very much alike.

(D) Features common with the Pteridophyta

(i) The presence of sunken sex organs is common in both groups.

(ii) The presence of similar vegetative structure of the gametophyte in *Anthoceros* and Fern.

(iii) The presence of highly developed sporogonium with photosynthetic tissue indeterminate growth and functional stomata.

The abovementioned facts support that the Anthocerotopsida are a distinct but **synthetic group** of plants. It forms a connecting link with liverworts and mosses on one hand and with pteridophytes on the other. There is also a remote connection with green algae (Chlorophyceae). Campbell (1928) suggested, "The fact that the primary sporogenous tissue in the Anthocerotales always arises from the amphithecium, while in all other liverworts it is developed from the endothecium, would seem to be a radical difference". Campbell is also of opinion that the sporophyte of *Anthoceros* with its assimilatory system with stomata and continued growth shows the close alliance to the independent, rootless dichotomously branched sporophyte of the primitive fossil group, Psilophytales. Mehra (1957) suggested that both Anthocerotopsida and Psilopsida arose from the common Anthorhyniaceae stock.

Biological significance of Anthocerotopsida

The sporophytes of Anthocerotopsida exhibit probable lines of biological progress as follows:

(i) The presence of elaborate ventilated assimilatory system suggests the beginning of the physiological independence of the sporophyte.

(ii) The discontinuation of continuous sporogenous tissue by the growth of sterile cells between the spore mother cells which suggests the beginning of the formation of sporangia.

(iii) The establishment of a well developed sterile columella from the central endothecium suggests the beginning of the formation of conducting system, and the initial stage in the formation of superficial sporangia.

(iv) The presence of intercalary meristematic zone suggests the beginning of the indeterminate and continuous growth of the sporophyte.

SUMMARY

Systematic position and distribution. Division-Bryophyta; class Anthocerotopsida (Anthocerotae); Order-Anthocerotales; Family- Anthocerotaceae; Genus-*Anthoceros.* 25 species are found in India; cosmopolitan in distribution.

Habit: external features. Moist shaded places. Thallus thin, dorsiventral with a broad indistinct mid rib; tendency towards dichotomous branching; rhizoids-unicellular, only smooth walled; unbranched scales are absent.

Internal structure. All cells photosynthetic; apical growth is either by a single apical cell or a group of apical cells.

Sexual organs. Monoecious and dioecious; antheridia endogenous, inside closed antheridial cavities on the dorsal surface of thallus, stalk is long, either slender or massive, body jacket is single layered, antherozoid is biciliate; archegonia-embedded in the dorsal surface of the thallus, neck is made up of 6 vertical rows of cells, venter is single layered thick, cover cells are 2 to 4, neck canal cells are usually 4.

Sporophyte. Amphithecium gives rise to jacket of capsule and archesporium; endothecium gives rise to columella; archesproium gives rise to sporocytes and pseudoelaters; foot here is bulbous; seta is absent; capsule is long and cylindrical and its wall is four to six-layered and persistent; pseudoelaters without thickenings; columella present, arched over by spore sac; dehiscence of spores occur when capsule wall splits into 1 to 4 valves, which remain united at the tip, pseudoelaters help in spore dispersal.

Young gametophyte. Two spore wall layers; size of spore is 0.025 to 0.05 mm; germinal tube forms at the time of spore germination.

23
CHAPTER

Bryopsida (Musci)
Order-Funariales

Sub-class EUBRYA

The Eubrya are called the true mosses. This group consists of a great majority of the mosses. There are about 650 genera and 14,000 species. The characteristic features of the true mosses are as follows:

The leaves of the gametophores are more than one cell in thickness and possess mid-rib on them.

The protonema are filamentous.

The sporophyte possesses a well differentiated, elongated seta which pushes out the capsule from the gametophore.

The sporogenous tissue is derived from the endothecium. The archesporium does not overarch the columella. The columella continues upto the apex of the capsule. Both columella and archesporium have been derived from the endothecium.

In between spore sac and columella, the partitioned air spaces are present.

The mature capsule possesses the complex structure. It is made of many tissues.

The capsule opens at its apex by an operculum. The spore dispersal is regulated by a teeth like apparatus, called the peristome.

Classification. According to Brotherus (1924-1925) there are 80 families in this sub-class. According to schaffner (1938) and Sharp (1939), the families have been grouped into seven orders. According to Brotherus (1924-1925) and Dixon (1932) there are fourteen orders in this sub-class. They have taken both gametophytic and sporophytic characters in grouping the families into orders. In the present day, the classification of Eubrya (Bryidae), proposed by Fleischer, Dixon and Wettstein has been followed. This is as follows:

Sub-class (*Unterclasse*)-**Bryidae**

I.	Cohort (*Reihengruppe*)	-	Eubryiidae
	1. " "	-	Fissidentales
	2. " "	-	Dicranales
	3. " "	-	Pottiales
	4. " "	-	Grimmiales
	5. " "	-	Funariales
	6. " "	-	Schistostegales
	7. " "	-	Tetraphidales
	8. " "	-	Eubryales
	9. " "	-	Isobryales
	10. " "	-	Hookeriales
	11. " "	-	Hypnobryales

II.	Cohort (*Reihengruppe*).		Buxbaumiidae
12.	Order (*Reihe*)	-	Buxbaumiales
13.	" "	-	Diphysciales
III.	Cohort (*Reihengruppe*).		Polytrichiidae
14.	Order (*Reihe*)	-	Polytrichales
15.	" "	-	Dawsoniales

Order FUNARIALES

The most characteristic features of this order are as follows: There are about 26 genera and 356 species. Usually the plants are terrestrial. They are quite small and may be annual or biennial. The leaves are with distinct mid-ribs and arranged in rosettes at the apex of the gametophyte. The capsule is wide. It is provided with an unbeaked operculum. The peristome of the capsule is double. It consists of inner and outer peristomes which are called **endostome** and **exostome** respectively.

The order **Funariales** has been further sub-divided into *five* families. They are as follows:

1. Gigaspermaceae
2. Funariaceae
3. Disceliaceae
4. Oedipodiaceae
5. Splachnaceae.

The genus *Funaria* is of family Funariaceae. In the present text, this genus has been discussed in detail.

Family-Funariaceae

The outstanding features of the family are as follows: This family includes about 9 genera and 200 species. The leaves are one cell in thickness except at the mid-rib region. This family includes the small mosses which form the velvety appearance on the surface of the substratum. The calyptra

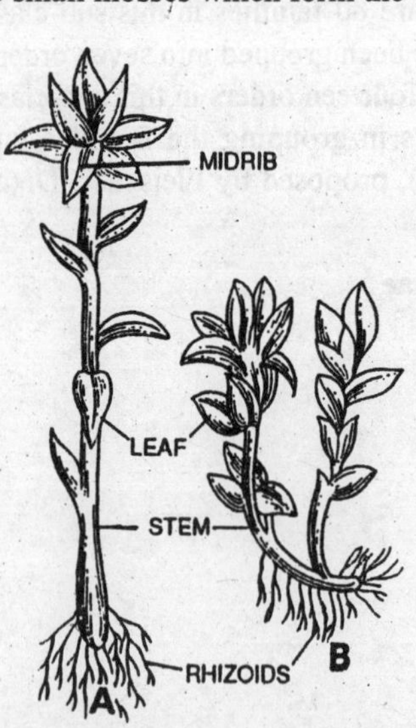

Fig. 23.1. *Funaria* sp. A, gametophytic plant with unbranched stem; B, a plant with branched stem.

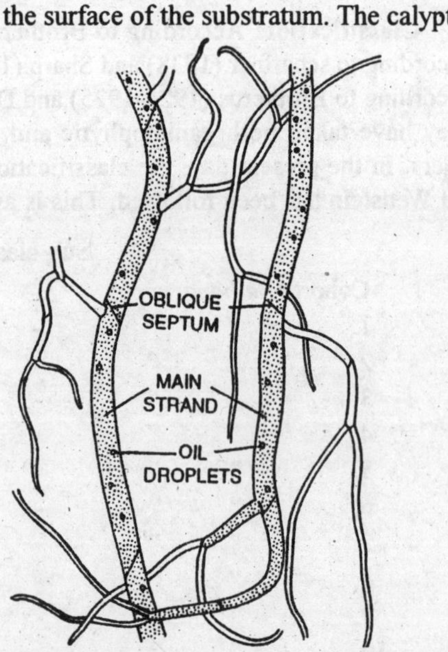

Fig. 23.2. *Funaria* sp. Branched septate (oblique septa) rhizoids containing oil droplets.

are soon detached from the opercula of the capsules. The calyptra are provided with long beaks. The capsules are pyriform in structure. They are situated on the long, elongated setae.

Genus FUNARIA

Habitat and distribution. The genus is cosmopolitan in its distribution. It includes about 117 species. Bruhl (1931) has recorded about 15 species of *Funaria* from our country. The species *Funaria hygrometrica* is best known among the mosses and found throughout the world. The mosses may grow luxuriantly in humus soil and on the soils burnt by fire. Sometimes they occur on the rocks and damp walls in the form of velvety mat. The green protonemata of the mosses appear in the recently ploughed fields. Some mosses are epiphytic in habitat and grow upon the trunks of the trees.

External structure. The small green moss plants may easily be collected from the damp walls after the rains, where they are found in velvety mats. The small, erect, green gametophytic plant arises from a prostrate alga-like filament the protonema. The gametophytic plant of *Funaria* may easily be differentiated into the rhizoids, stem and leaves. The gametophytic plant may also be called the gametophore, as it bears the sex organs at its apex.

The gametophytic plants are green, erect and one to three centimetres in height. As mentioned above, each plant consists of rhizoids, leaves and stem. Usually the stem is branched. The branching is of monopodial type.

At the base of the gametophytic plant the **rhizoids** are found. The rhizoids are branched, multicellular and thread-like. The septa of the rhizoids are oblique. In earlier stages the rhizoids are hyaline or whitish in colour whereas, in later stages they become brown or pink in colour. The rhizoids absorb the water and nutrients from the soil gametophytic plants as well as they help in anchoring the plants to their substrata. They usually contain oil in them, but if exposed to the light, develop chlorophyll in them and help in assimilation.

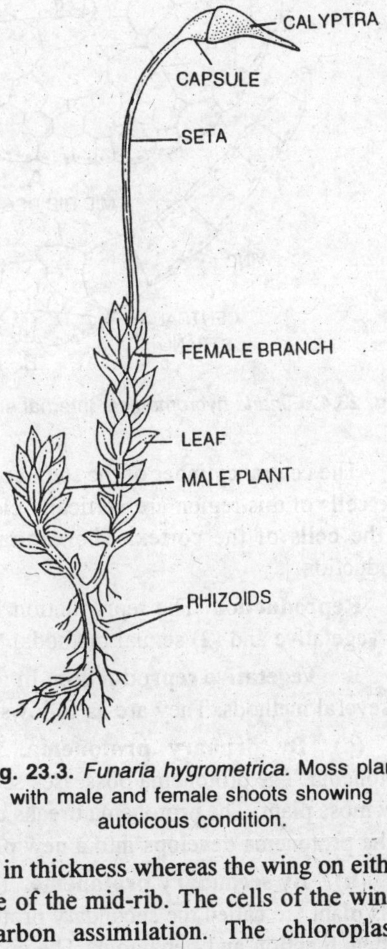

Fig. 23.3. *Funaria hygrometrica.* Moss plant with male and female shoots showing autoicous condition.

The **leaves** are small, ovate, bright green and spirally arranged on the stem. Each such leaf possesses a distinct mid-rib. The upper leaves are somewhat larger in size and crowded at the apex of the plant, whereas the lower leaves are smaller and scattered on the stem. The leaves are sessile, ovate and with broad bases at the point of attachment to the stem.

The **stem** is upright erect, green and monopodially branched. Sometimes, this may or may not be branched.

Internal structure-the leaf. To study the internal structure (anatomy) of the leaf, one has to cut the leaf in vertical sections. The leaf possesses a distinct mid-rib and wings. The mid-rib is several celled in thickness whereas the wing on either side is single layered. There is central strand in the centre of the mid-rib. The cells of the wings contain chloroplasts in them which facilitate the carbon assimilation. The chloroplasts continuously divide again and again. This process goes on even after the maturity of the leaves.

The stem. The internal structure of the stem may easily be understood by studying the transverse section under the microscope. It mainly consists of three parts, (1) the epidermis, (2) the cortex and (3) the central cylinder.

The epidermis is single layered. It is the outermost layer consisting of thin walled, compactly arranged cells with chloroplasts. This layer is assimilatory in function. It is devoid of cuticle and stomata. Sometimes, at certain places the epidermis becomes double layered.

The cortex is multilayered and consists of thin walled parenchymatous cells. In earlier stages the cortical cells contain chloroplasts but later on they disappear. The diagonal leaf traces may also be seen near the periphery.

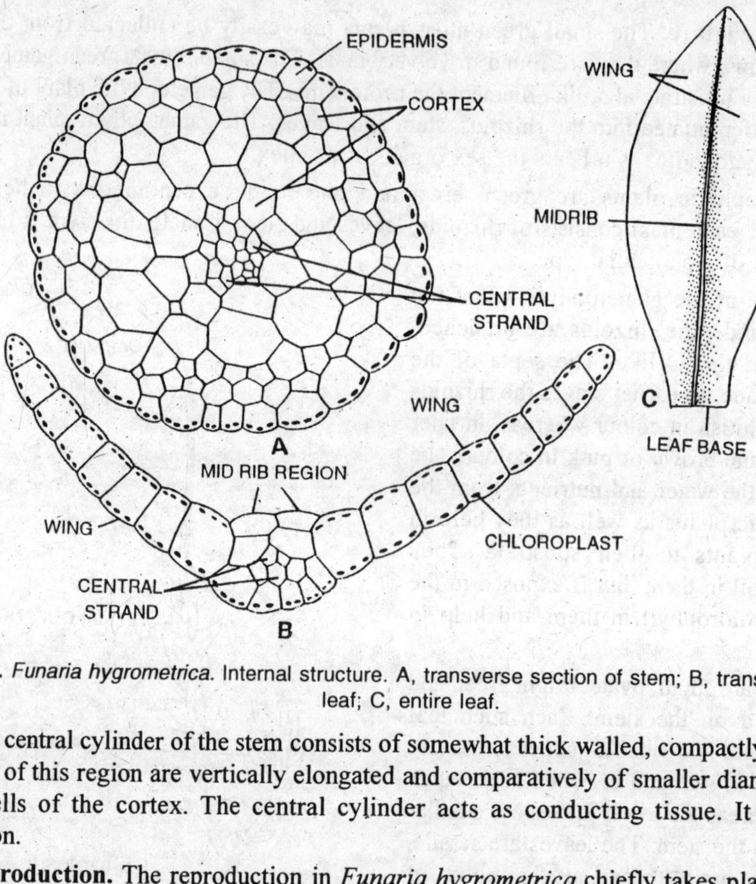

Fig. 23.4. *Funaria hygrometrica.* Internal structure. A, transverse section of stem; B, transverse section of leaf; C, entire leaf.

The central cylinder of the stem consists of somewhat thick walled, compactly arranged cells. The cells of this region are vertically elongated and comparatively of smaller diameter than those of the cells of the cortex. The central cylinder acts as conducting tissue. It helps in water conduction.

Reproduction. The reproduction in *Funaria hygrometrica* chiefly takes place by means of (1) vegetative and (2) sexual methods.

1. Vegetative reproduction. In mosses, the vegetative multiplication takes place by means of several methods. They are as follows-

(a) By primary protonema. After germination, the spore develops into primary protonema. The protonema possesses several small buds on it. Each such bud is to develop into a new moss plant. The protonema breaks up and propagates vegetatively. Each such small fragment of the protonema develops into a new plant.

(b) By secondary protonema. The protonema which develop from the other parts of the moss plant are called the secondary protonema. Such protonema resemble the primary protonema in their function and behaviour. The secondary protonema may develop from any detached part of the plant gametophyte, such as leaf, stem, antheridium, paraphyses and even from the exposed rhizoids. Such protonema may also occur from the wounded portions of the moss plants. Each protonema develops into a new plant.

(c) By gemmae. Many mosses produce small, multicellular gemmae, in groups at the apices of the leaves. Sometimes, solitary gemmae may be produced on the rhizoids. Subterranean gemmae, produced on the rhizoids are called the bulbils. On the advent of the favourable conditions, each such gemma or bulbil develops into a new moss plant.

(d) By death and decay of prostrate system. The prostrate branches of the

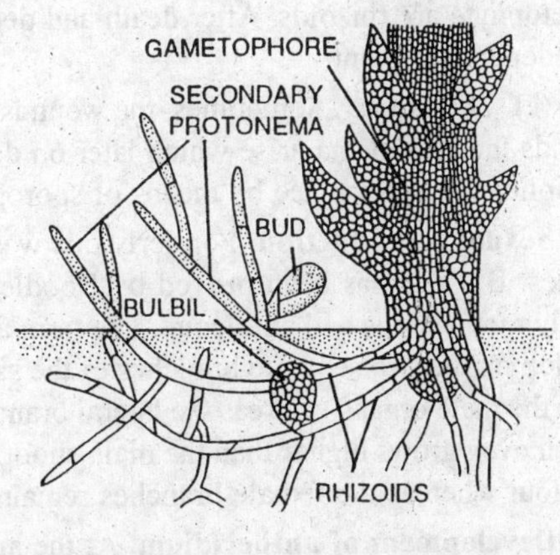

Fig. 23.5. *Funaria hygrometrica*. Vegetative reproduction. Rhizoids with oblique septa developing from lower part of gametophore; outside the substratum secondary protonema bears buds; the bulbil is seen on the rhizoids within substratum.

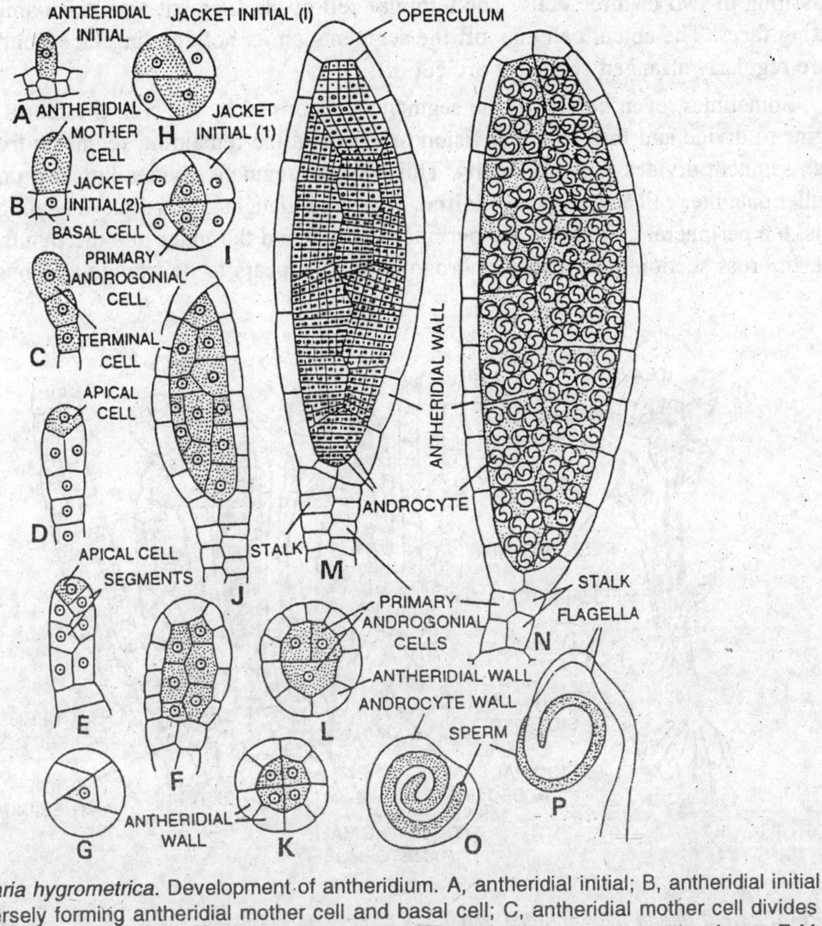

Fig. 23.6. *Funaria hygrometrica*. Development of antheridium. A, antheridial initial; B, antheridial initial divides transversely forming antheridial mother cell and basal cell; C, antheridial mother cell divides transversely forming a 2 or 3 celled filament; D, apical cell developed having two cutting faces; E-M, further successive stages in the development of antheridium; G,H,I and L, cross sections at different stages; N, antheridium containing androcytes; O, single androcyte; P, single biflagellate sperm or antherozoid.

gametophyte are rhizoids. After death and decay of such branches each erect branch behaves like an independent plant.

(e) Apospory. Sometimes, the wounds are developed on the sporophytic tissue. From these wounds the protonema arise which later on develop into new plants. So, here the reproduction or multiplication takes place by means of sporophytic tissue and not by the spores.

Sexual reproduction. Formerly this was believed that *Funaria hygrometrica* is a dioecious species. But this has been proved by Boodle (1906) and Miss Brown (1919) that the species is strictly monoecious and autoicous. The term autoicous means that the male and female sex organs develop on the two separate branches of the same plant. The main shoot of the gametophytic plant bears the male organs whereas the lateral branch bears the female sex organs. Sometimes the lateral branch even grows higher than the main shoot. On maturation the male branches become brownish in colour whereas the female branches remain green.

Development of antheridium. At the apex of the male branch a single superficial cell acts as an **antheridial initial.** This cell may be distinguished by its papillate and projecting nature. The antheridial initial divides first transversely, giving rise to two cells. The outer daughter cell gives rise to the complete main antheridium whereas the lower cell develops into the embedded portion of the stalk. The outer daughter cell divides twice or thrice transversely producing a filament consisting of two or three cells. The terminal cell of the filament acts as an apical cell with two cutting faces. The apical cell cuts off the segments on its both cutting faces, and this way 5-7 or more regularly arranged segments are cut off.

Sometimes, even before all the segments are formed by the cell, the fourth or fifth segment begins to divide and further this division proceeds in the remaining segments from base to apex. Each segment divides first diagonally. This way the segment divides into two unequal cells. The smaller daughter cell acts as **jacket initial.** The larger daughter cell divides again giving rise to two cells, the peripheral cell called the peripheral initial and the inner one the **primary androgonial cell.** In cross section the primary androgonial cell appears to be triangular. Thus, each segment,

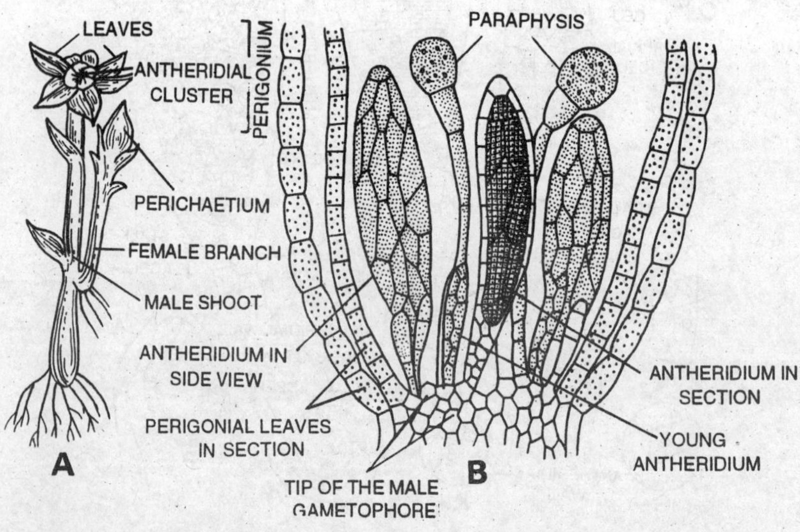

Fig. 23.7. *Funaria hygrometrica.* Male sex organs. A, plant with male and female branches; B, longitudinal section of the tip of the male gametophore showing antheridia, paraphyses and perigonial leaves.

gives rise to three cells, two jacket initials, and one primary androgonial cell. The primary androgonial cell of each segment divides further in all planes producing numerous **androcyte mother cells.** Simultaneously the jacket initials also divide repeatedly giving rise to a single-layered **jacket** of the antheridium. From each androcyte mother cell, two androcytes are produced. Later on each androcyte metamorphoses into a biflagellate antherozoid.

Structure of antheridium. The antheridium mainly consists of two parts, a short massive stalk and the main body. It is clavate in outline. The main body is surrounded by a single layered outer jacket. The jacket cells contain chloroplasts, they turn red or brown on maturity. The jacket layer surrounds, the central dense mass of androcytes. At the apical end of the male branch of the gametophore the antheridia are intermingled with several sterile **paraphyses.** They are hair-like in structure. Each paraphysis is multi-cellular and consists of 4 to 5 cells arranged in a uniseriate row. The lower cells of the paraphysis are elongated and the terminal cells are globular. The function of these paraphyses is not very definite. It is assumed that they hold water among them which is needed for the developing antheridia.

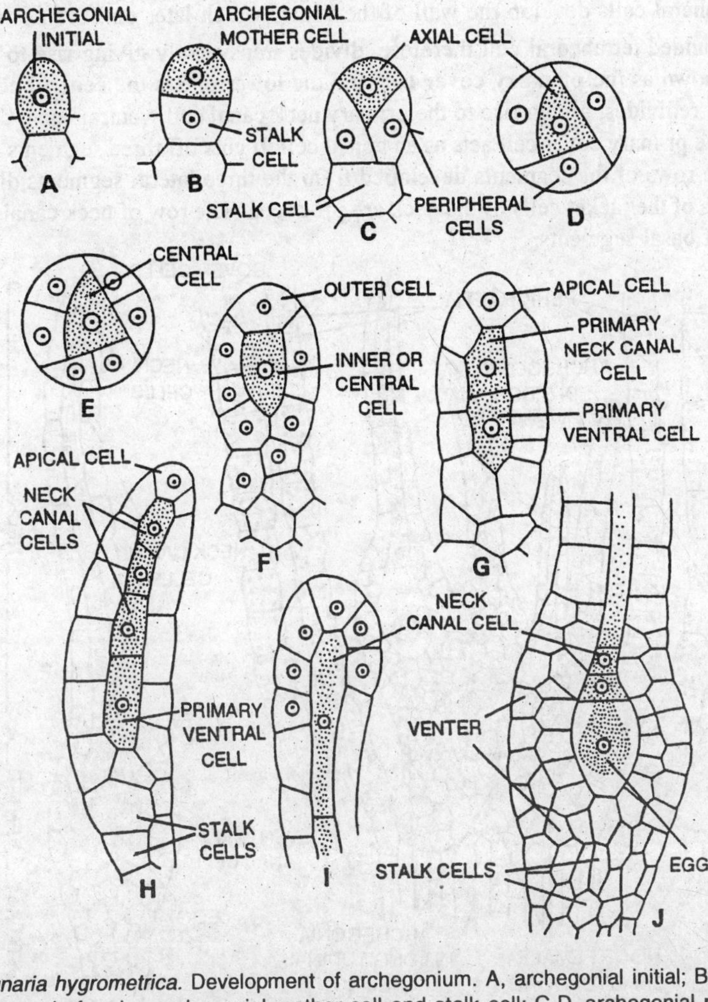

Fig. 23.8. *Funaria hygrometrica.* Development of archegonium. A, archegonial initial; B, archegonial initial divides transversely forming archegonial mother cell and stalk cell; C-D, archegonial mother cell divides by three oblique walls forming three peripheral cells surrounding the central axial cell; D and E, cross sections of the same; F-H, further successive stages in the development of archegonium; I, neck of nearly mature archegonium; J, venter containing egg of mature archegonium.

Dehiscence of antheridium. The water is essentially required for the dehiscence of the antheridium. At the time of the dehiscence the rain water or dew is collected at the apical end of the male branch of the gametophore in a cupular structure developed by perichaetial leaves. On the approach of water the opercular distinctive cell of the antheridium becomes mucilaginous and somewhat swollen. The inner wall of the opercullar cell disintegrates and it is soon followed by the disintegration of the upper wall of the cell, thus, giving rise to an apical pore, the mass of androcytes squeezes out through this pore. As soon as this mass comes in contact of the water collected in the cup of perichaetial leaves, the androcytes become separated and spread out in the film of water. They become motile, biflagellate antherozoids which swim here and there in the film of water.

Development of archegonium. The development of the archegonium in *Funaria hygrometrica* is as follows. According to Campbell the archegonial initial divides transversely giving rise to two cells, a basal cell and a terminal cell. The terminal cell, which is also designated as the mother cell of the archegonium divides by three oblique walls successively. These walls do not reach the basal cell. With the result, three peripheral cells surround a central polyhedral cell. The three peripheral cells develop the wall of the venter which later on becomes two layered.

The surrounded tetrahedral cell therefore, divides transversely giving rise to two cells. The upper cell is known as the **primary cover cell** and the lower one is the **central cell.** The central cell divides and redivides, giving rise to the primary neck canal cell, ventral canal cell and the egg or oosphere. The primary cover cell acts as an apical cell. It cuts off three segments from its lateral faces. The three rows of the segments developed from the three lateral segments divide vertically and the six rows of the jacket cells of the neck are produced. The row of neck canal cells develops from the row of basal segments.

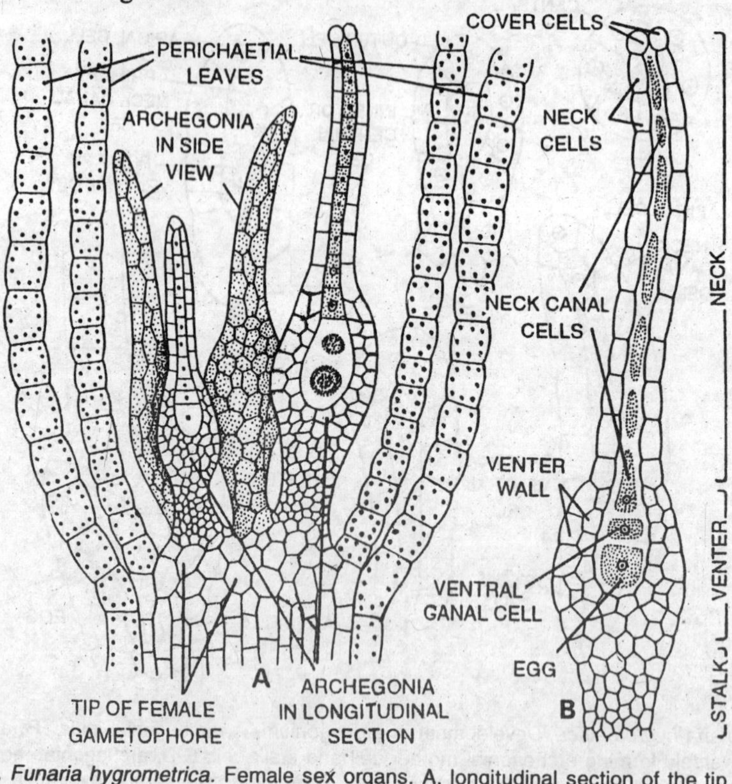

Fig. 23.9. *Funaria hygrometrica.* Female sex organs. A, longitudinal section of the tip of female gametophore showing archegonia and perichaetial leaves; B, a mature archegonium in sectional view containing egg, ventral canal cell and neck canal cells.

Structure of archegonium. Comparatively, the stalk of archegonium is longer than that of antheridium. The archegonium is flask-like. It possesses a massive stalk, the venter and the neck. The bulbous venter possesses a two-layered jacket, whereas the jacket of the neck is single-layered. The venter contains the egg or the oosphere and the ventral canal cell. The elongated neck contains six or more neck canal cells. On the maturation of the archegonium, the neck canal cells and the ventral canal cell disintegrate forming a mucilaginous substance. The terminal cells of the neck separate from each other forming a passage. The oosphere remains inside the venter.

Fertilization. The water is essentially required, for the process. On the maturity the neck canal cells are disintegrated forming mucilaginous mass. The opening is created by the separation of the apical cells of the neck. The antherozoids are attracted towards the mouth of the archegonium by its secretions. The antherozoids are chemotactic. Certain sugars ooze out through the opening which attract the antherozoids. Some of the antherozoids enter through the clear passage and reach to the oosphere. Ultimately the single antherozoid penetrates the egg and the

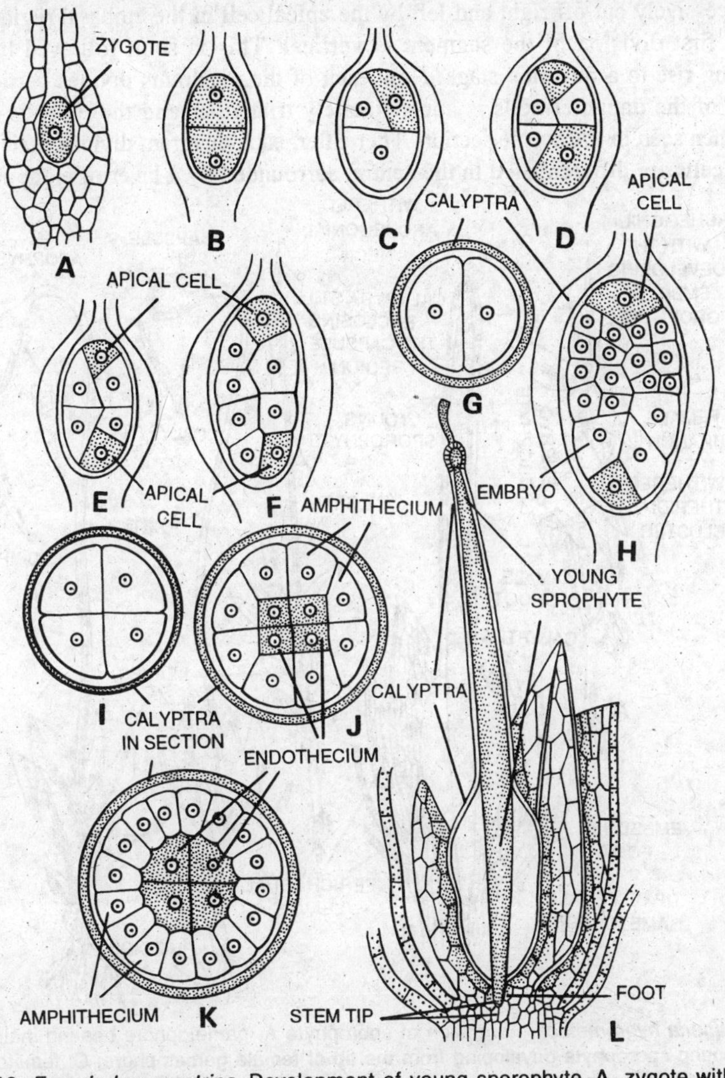

Fig. 23.10. *Funaria hygrometrica.* Development of young sporophyte. A, zygote within venter, B, transverse division of zygote; C-L, further successive stages in the development of young sporophyte; G, I, J and K are cross sections at different stages; rest are longitudinal sections.

fertilization is effected. The rest of the antherozoids within the venter, disintegrate. The male nucleus of the antherozoid fuses with female nucleus of the egg and the **oospore** is resulted.

The sporophyte. As soon as the fertilization is over, the zygote increases in size and begins to secrete a wall around it. There after the zygote divides transversely giving rise to two cells. The upper cell is epibasal cell, and the lower one is hypobasal cell. In the epibasal cell, two oblique intersecting walls are formed successively, giving rise to a two-faced apical cell. In the same manner another two-faced apical cell is formed in the hypobasal cell. Thus, the two growing points are established in the earlier stage of the development of the sporogonium, in its upper and lower cells respectively. The upper two-faced apical cell cuts off the segments to the right and left alternately and in the end the derivatives assemble in the form of a capsule and the upper part of the seta. The lower apical cell, also cuts off cells, but not so regularly as the upper one. The derivatives ultimately give rise to the lower part of the seta and the foot of the capsule.

Development of the sporogonium (capsule). The capsule of the sporophyte develops from the segments successively cut off right and left by the apical cell in the epibasal region of the young embryo. The first division of the segment is vertical. This is soon followed by a transverse division, giving rise to a quadrant stage. Now each of the quadrants divides vertically in such a way that one of the daughter cells is approximately triangular and the other is approximately rectangular when seen in transverse section. Thereafter, each quadrant divides periclinally and this way, the four cells are differentiated in the centre, surrounded by a layer of eight peripheral cells.

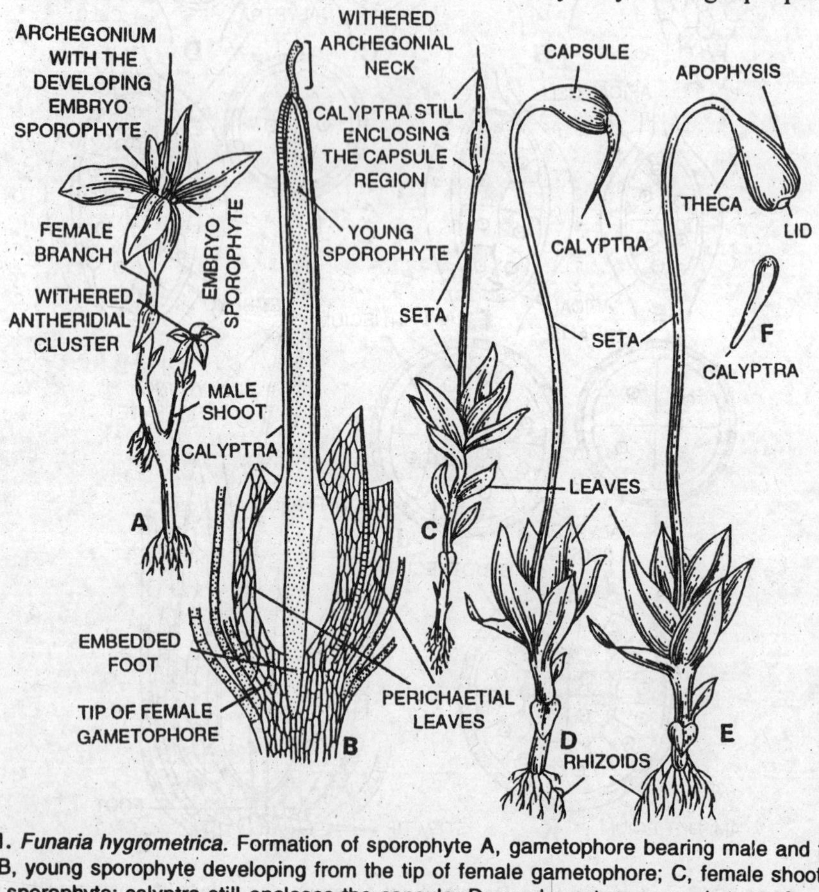

Fig. 23.11. *Funaria hygrometrica.* Formation of sporophyte A, gametophore bearing male and female shoots, B, young sporophyte developing from the tip of female gametophore; C, female shoot with young sporophyte; calyptra still encloses the capsule; D, nearly mature sporophyte on female gametophyte; E, mature sporophyte after the removal of calyptra; F, calyptra.

The centrally located four cells constitute the **endothecium** and the remaining eight, peripheral cells constitute the **amphithecium.** The endothecium and the ampithecium are the chief embryonic layers of the young sporogonium, which are to give rise to the different tissues of the sporogonium. For the sake of convenience the development of the whole capsule may be divided into three development zones. These zones are-(a) the fertile region (b) apical region of the capsule and (c) the lower region of the capsule. The development in the different regions of the capsule has been given below, under separate sub-headings.

(a) Development of the fertile region of the capsule. All the four cells of the endothecium divide diagonally or vertically, which is soon followed by periclinal division giving rise to four inner cells and eight outer cells. The four inner endothecial cells divide twice resulting in a group of sixteen cells which ultimately give rise to the **columella.** The outer eight endothecial cells divide by radial and periclinal divisions, and this way the two rings of the cells are resulted. The cells of the outer layer divide once again radially, and the **archesporium** is resulted. The cells of inner layer, adjacent to the columella develop into the **inner spore sac.** The archesporium may be one or two layered. The cells of the archesporium are spore mother cells. They divide by reduction division giving rise to haploid spores. In the earlier stage of the development, the spores are arranged in tetrads.

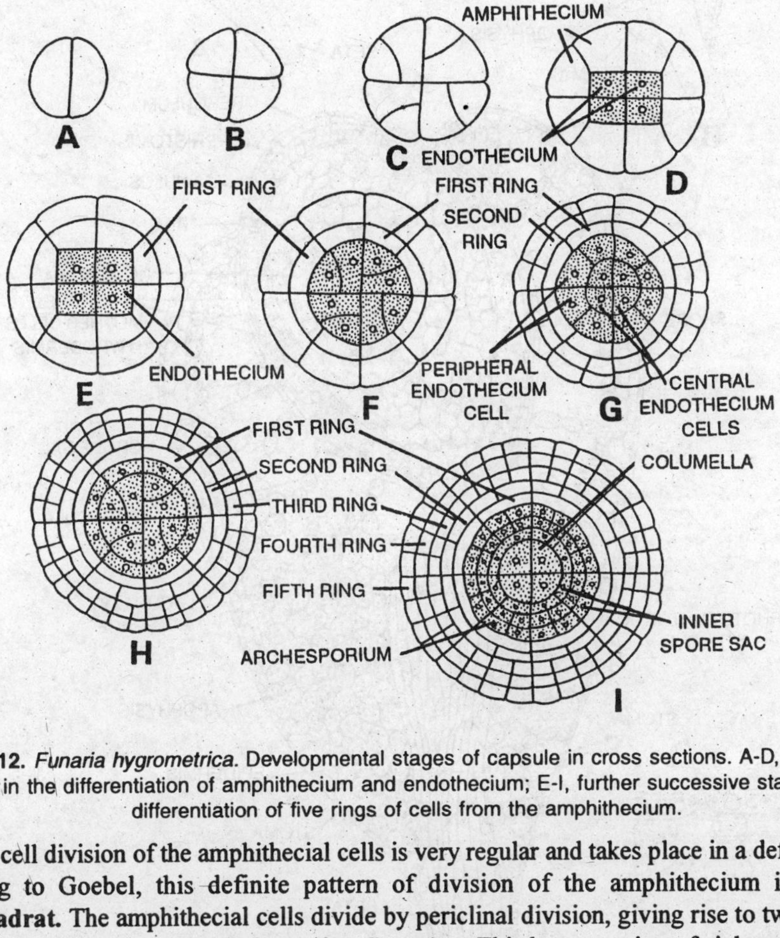

Fig. 23.12. *Funaria hygrometrica.* Developmental stages of capsule in cross sections. A-D, successive stages in the differentiation of amphithecium and endothecium; E-I, further successive stages in the differentiation of five rings of cells from the amphithecium.

The cell division of the amphithecial cells is very regular and takes place in a definite pattern. According to Goebel, this definite pattern of division of the amphithecium is called the **grundquadrat.** The amphithecial cells divide by periclinal division, giving rise to two concentric layers. The inner layer may be designated as **first ring.** This layer consists of eight cells. The cells of the outer concentric layer divide anticlinally giving rise to sixteen cells. This layer of sixteen cells divides periclinally forming two concentric layers. Each such layer consists of sixteen cells.

The inner layer of these two layers is designated as **second ring.** The outer layer of these two layers divides anticlinally giving rise to thirty two cells. These 32 cells divide periclinally giving rise to two layers of 32 cells each. The inner layer of these two layers may be referred as the **third ring.** The outer layer, consisting of 32 cells, divides periclinally, giving rise to two layers of 32 cells each. The inner layer of 32 cells may be referred as **fourth ring** and the outermost layer of

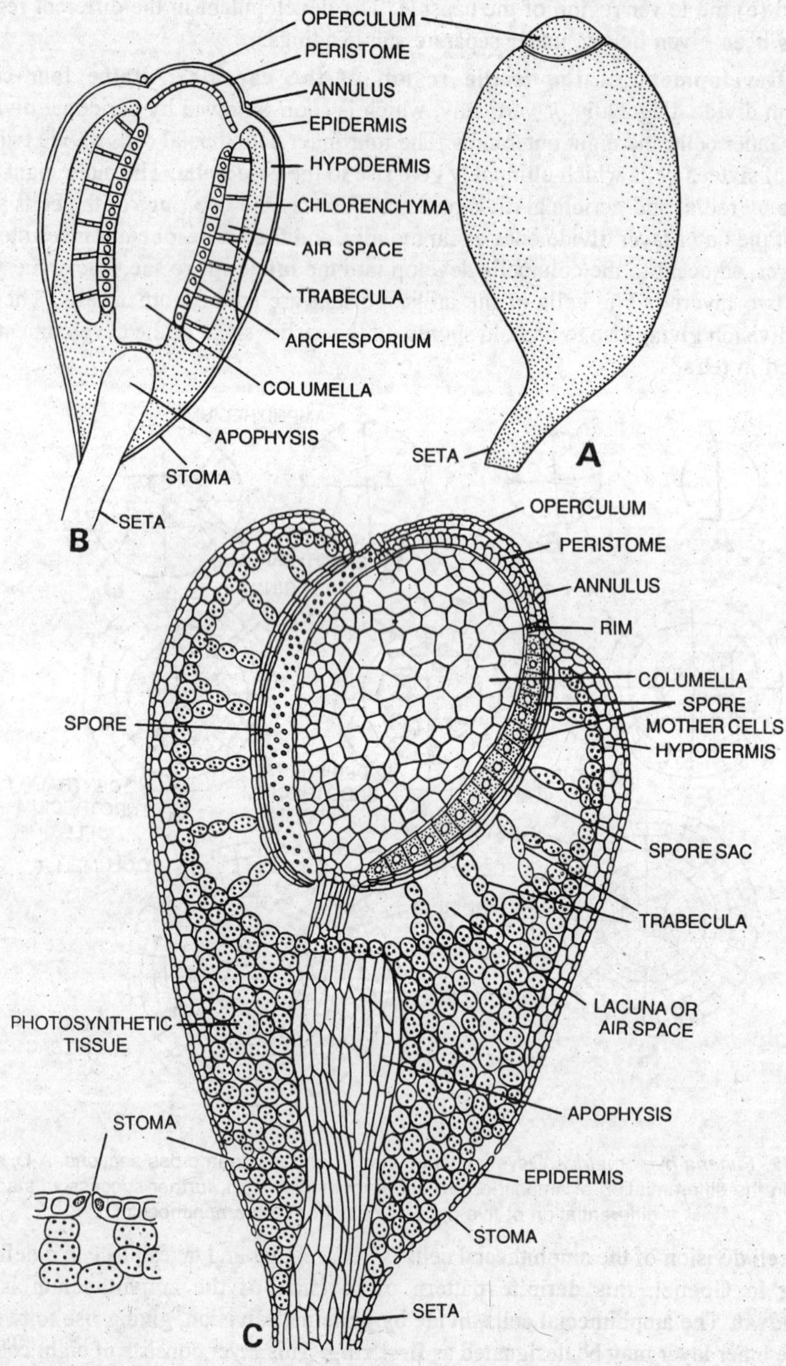

Fig. 23.13. *Funaria hygrometrica.* Capsule of sporophyte. A, capsule; B, longitudinal section of capsule (outline); C, longitudinal section of capsule showing details of B.

the same number of cells is the **fifth ring.** These five rings give rise to the different parts of the fertile region.

The cells of the first ring divide anticlinally giving rise to thirty-two cells. These cells again divide periclinally and a three-layered **outer spore sac** is formed. These cells lack chloroplasts.

The 16 celled second ring develops chloroplasts in them and elongate radially; thus giving rise to lacunae or air spaces. The cells become separated from each other. Each cell divides transversely twice or thrice giving rise to a filament of 3 or 4 cells. Such filaments connect the outer spore sac to the wall of the capsule.

The third ring which consists of 32 cells, remains one cell in thickness, and the chloroplasts develop in the cells on their maturation.

The fourth ring also consists of 32 cells. These cells divide periclinally giving rise to two or three concentric layers of cells without chloroplasts.

The fifth ring also possesses 32 cells. The cells of this ring divide anticlinally and develop into a distinct epidermis.

(b) Development of apical region of the capsule. The four endothecial and the eight amphithecial cells are also found in the apical region of the developing capsule. The endothecial cells divide again and again continuously giving rise to thin walled parenchyma in continuation of the columellà of the capsule. The eight cells of amphithecium divide alternately periclinally and anticlinally, giving rise to five usual concentric rings of the cells. Each concentric ring consists of twice as many cells as the next inner ring. The operculum and peristome, are, however, developed from the cells of these rings.

The peristome develops from the first and second rings. The first ring consists of 32 cells and forms the **inner peristomial layer,** whereas the second ring consists of 16 cells and forms the **outer peristomial layer.** The cells of these two inner and outer peristomial layers are so arranged that one cell of the inner layer is çovered by the two cells of the outer layer. On maturation, the wall material is added to the peristomial layers and a double set of the teeth of peristome is developed. In each series (set) of the peristome, there are sixteen teeth or clefts. The teeth are somewhat curved, narrow and triangular in outline. The teeth are attached to the rim of the mouth of the capsule. This complete structure, consisting of the two sets of the teeth is called the **peristome.**

External to the peristome, third to fifth rings of the amphithecium give rise to the operculum of the capsule. The third and fourth rings give rise to a three celled thick tissue. However, the fifth ring derived from the amphithecium, develops into the epidermis of the operculum. The epidermal cells at the base of the operculum enlarge radially and form a ring of **annulus.** The annulus is constituted of 5 to 6 layers of the cells of the epidermis. This is the broadest region of the operculum. The upper super-imposed layers of the epidermal cells of the annulus are narrow and thick-walled whereas the cells of the lowermost two layers of the annulus are thin walled.

(c) Development of the lower region of capsule. In the later stages of the development of sporophyte the seta and the capsule are differentiated. There is a conducting region of apophysis in between seta and capsule. The tissues in this region are developed in the same way as they are developed in the fertile region, only with the difference that they lack archesporium and spore sac layers. The endothecial cells in this region give rise to the conducting strand of the seta. The amphithecium forms the spongy chlorenchyma and the epidermis. The epidermis is interrupted by stomata here and there.

Structure of mature sporogonium. The sporophyte consists of a foot, seta and capsule.

The foot is a small, conical structure embedded in the apical portion of the female branch of the gametophore. This is a poorly developed structure. It absorbs water and nutrients from the gametophyte for the development of the sporogonium.

The seta is an elongated slender and thread like structure which bears the capsule at its apical end. The cells of the tissue of seta are elongated. The tissue helps somewhat in the conduction of water but mainly its function is mechanical.

The outline of a mature capsule is pear shaped. In the beginning, an irregular calyptra is found as a cap-like structure at the apical end of the capsule, but very soon this is blown off. The main parts of the capsule are as follows:

The apophysis. The elongated seta expands at its apical end to form the apophysis. This is assimilatory region. The stomata may be seen here and there in the epidermis. The stomata are meant for gaseous exchange. The cells with chloroplasts possess intercellular spaces among them. The central conducting strand consists of colourless cells. This strand is continuous with the conducting strand of seta. This region is assimilatory in function and makes the capsule self-sufficient in nutrition to some extent.

The theca. The central fertile region of the capsule is known as the theca proper. This region consists of many layers of the cells. The outermost layer is the epidermis, which lies in continuation of the epidermis of the apophysis of the capsule. Just below the epidermis, there is one or two layered hypodermis. The cells of the hypodermal region are colourless. Just next to the hypodermis there is one or two layered chlorenchyma tissue. The cells of this layer contain chloroplasts and act as photosynthetic tissue. This chlorenchymatous layer becomes several layered in the region of apophysis. Just beneath the chlorenchymatous layer there is a big air space, traversed by severs **trabeculae.** Each trabecula consists of 3 or 4 green cells. These filiform structures act as connections between chlorenchymatous layer and the outer spore sac of the capsule. The central region of the theca is occupied by a pith like solid cylinder or parenchyma, called the **columella.** The columella is narrow at its base and broad at its apical end. It is connected to the conducting strand of the apophysis of the capsule at its base. The columella is

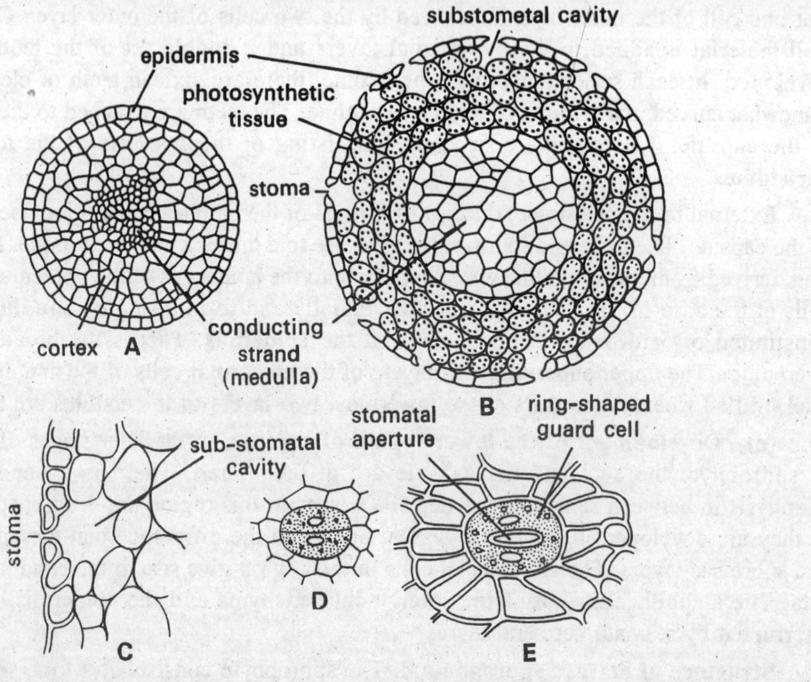

Fig. 23.14. *Funaria hygrometrica.* Anatomy of different parts of sporophyte. A, transverse section of seta showing cortex and conducting strand; B, transverse section of capsule through apophysis showing photosynthetic tissue, conducting, strand and epidermis with stomata; C, enlarged portion of B, showing epidermis, stoma and sub-stomatal cavity; D, young stoma in surface view; E, mature stoma in surface view; stomata confined to apophysis region only.

surrounded by the spore sac. The spore sac is barrel shaped. The spore sac has an inner single layered wall called the inner spore sac. The outer wall of the spore sac is 3 or 4 layered and known as the outer spore sac. In between these inner and outer spore sacs there are spore mother cells. As usual each spore mother cell develops into a spore tetrad of four spores. The first division is the reductional one.

The upper region. The upper region of the capsule consists of operculum and peristome. It is differentiated from the theca proper by a well marked constriction. The lid of the capsule is somewhat obliquely placed upon it. The lid or operculum is 4 or 5 layers in thickness. The outermost layer of it is epidermis. The epidermis consists of the cells of greatly thickened walls. The next 3 or 4 layers of the cells consist of thin walled parenchymatous cells. Just beneath the constriction there is a 2 or 3 layered rim of radially elongated cells. The rim is found in the form of a circular edge. It stretches inward from the epidermis and this way connects the peristome to the epidermis of the capsule.

The peristome lies just beneath the operculum. It consists of two sets of incurved teeth known as inner and outer peristome. Each set of teeth consists of sixteen incurved, triangular teeth. The teeth are spirally twisted towards left. The teeth of the outer peristome are red and of inner peristome are colourless. The inner teeth are shorter than the outer ones.

The annulus is found just above the rim. It consists of 5 to 6 layers of epidermal cells. The structure of annulus has already been described in foregoing pages.

The mouth of the capsule remains closed by means of lid or operculum.

Dehiscence of the capsule. The capsule begins to dry up on its maturation. The columella and the outer thin walled tissues of the capsule lose water and shrink up. This way, the spore containing space is teared off. The annulus breaks, and the operculum is thrown off. As soon as the operculum is detached from the capsule the peristome teeth move outward hygroscopically, and help in throwing away the operculum. The hygroscopic movements of the peristome teeth help in the gradual dispersal of the spores from the capsule. On becoming wet the outer layer of each tooth becomes long and on becoming dry each tooth becomes short. However, the inner layer of the teeth is not affected by the changing conditions of the moisture. The outer layers of the teeth of the outer peristome absorb water on the removal of the operculum and the teeth increase in length than the inner teeth which are not at all affected by the moisture contents. This way the two sets outer and inner of teeth of the peristome curve together inward and close the opening of the capsule caused by the removal of the lid or operculum, and check the dispersal of the spores. In dry weather outer layers of the teeth of the outer peristome lose water and become short. This

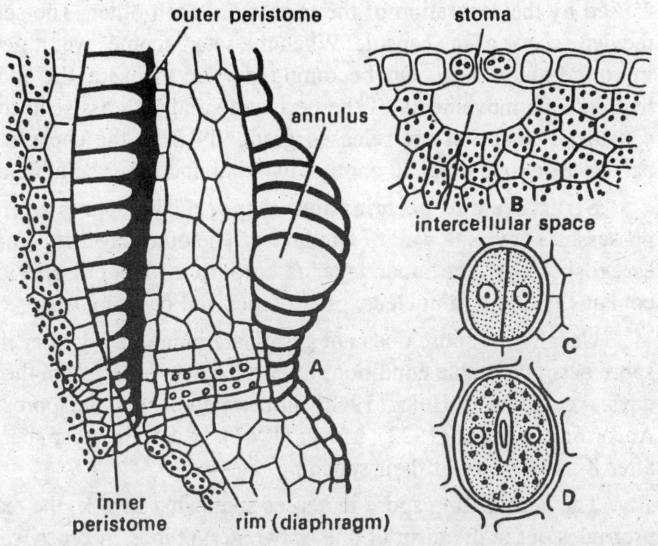

Fig. 23.15. *Funaria hygrometrica.* A, a portion of L.S. through operculum and upper part of theca exhibiting relationship of peristome and rim; B, T.S. through apophysis showing stoma; C, young stoma in surface view; D, mature stoma in surface view.

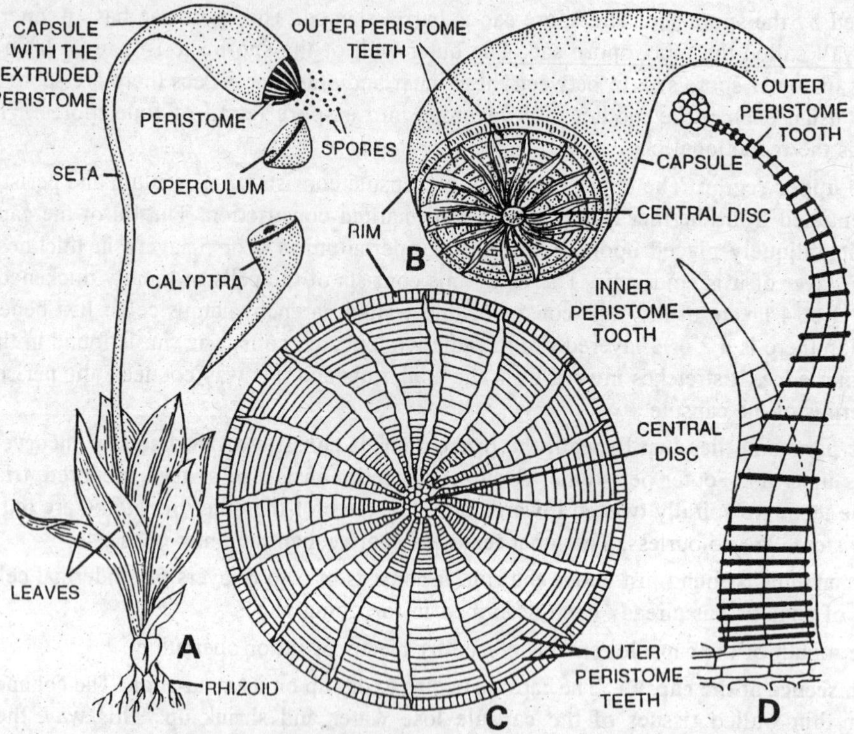

Fig. 23.16. *Funaria hygrometrica.* Dehiscence of capsule. A, female gametophore with mature sporophyte releasing spores; B, terminal portion of the capsule after the removal of operculum showing the peristomal teeth; C, outer peristomal teeth seen from above; D, outer and inner peristomal teeth.

way the teeth curve outwards with jerks and the spores are released gradually through the slits caused by the separation of the teeth from each other. The seta is also hygroscopic and helps in the dehiscence of the capsule. When the seta becomes wet it gets untwisted and swings the capsule in round-about way. On becoming dry of the seta the movements are reserved. Thus, the hygroscopic movements of the peristome and seta assist in the dispersal of the spores from the capsule. The spores are released gradually from the capsule for a long period. The spores are carried from one place to another by wind and other suitable agencies.

Structure and germination of spore. The spores are more or less spherical. Each spore possesses a covering wall of two layers. The outer layer is somewhat coloured, smooth and known as exosporium. The inner layer is colourless, smooth and known as endosporium. The spore contains within it, a nucleus, oil globules and chloroplasts.

Usually the spore does not germinate immediately after its shedding from the capsule. If the spore gets favourable conditions after it has been shed from the capsule, it germinates within a few days. According to Heitz (1942), in certain species the spores remain viable even for two years. According to Malta (1922), Meyer (1941), in certain other species, the spores may germinate even after 8 to 16 years of their shedding.

The germinating spore increases somewhat in size, the exospore ruptures and the endospore protrudes out in the form of one or two germ tubes. A cross wall appears in the germ tube near its point of emergence from the exosporium. The cell formed by this cross wall gives rise to a multicellular filament called the **primary protonema.** The protonema further grows strictly by the division of the apical cell. According to Correns (1899), Muller (1874) and Sironval (1947) the primary protonema is differentiated into two kinds of branches, known as **chloronema** or

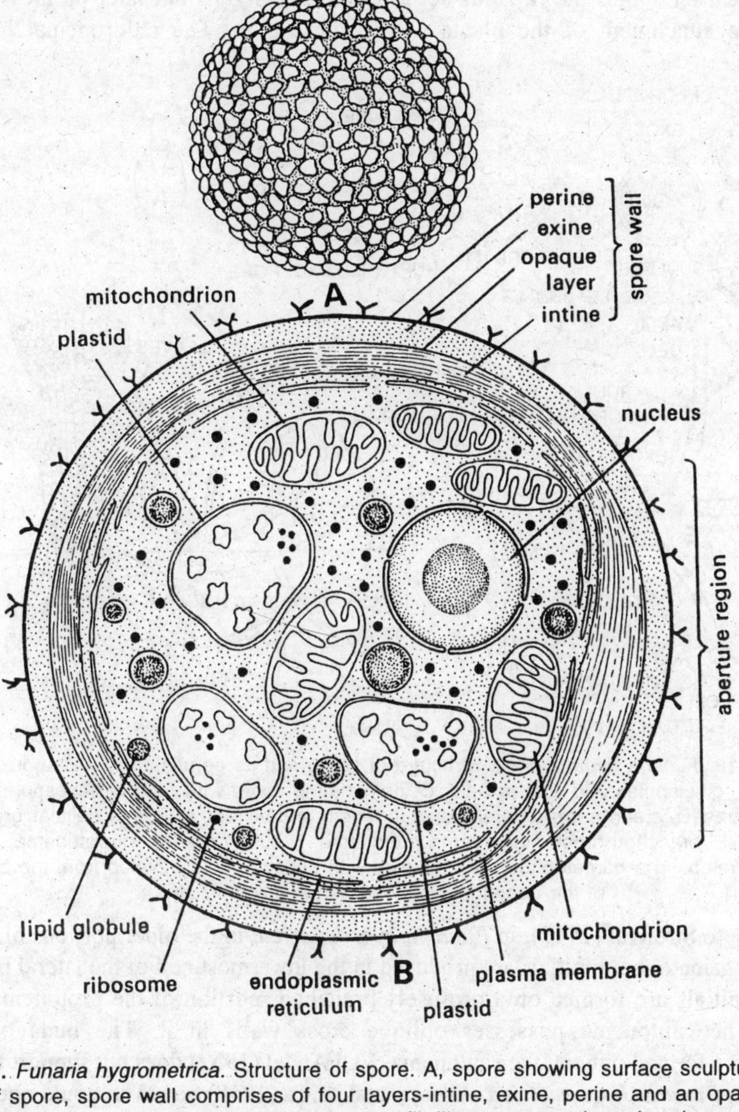

Fig. 23.17. *Funaria hygrometrica.* Structure of spore. A, spore showing surface sculpturing; B, ultra-structure of spore, spore wall comprises of four layers-intine, exine, perine and an opaque separating layer in between exine and intine; the intine shows a fibrillar structure; the exine does not show filbrillar structure; the perine is yellowish in colour; there is a plasma membrane within the spore wall, it makes a tight covering around cytoplasm; in the aperture region the exine is quite thin and a lamellate structure is present in the intine adjacent to the exine, the germ tube emerges through this region; besides the nucleus the cytoplasm contains plastids, mitochondria, endoplasmic reticulum, ribosomes and lipid globules; no vacuoles.

chloronemal branches and **rhizoids** or **rhizoidal branches.** The chloronemal branches are aerial and prostrately upon the surface of the substratum, whereas, the rhizoidal branches or rhizoids grow downward and penetrate the substratum. The chloronemal branches are thick, possessing hyaline cell walls, with transverse cross walls forming right angles to the lateral walls, and containing numerous chloroplasts in them. The rhizoidal branches are relatively thin, brown and with diagonal cross walls. They possess inconspicuous small chloroplasts or leucoplasts in them. Some of the rhizoidal branches which, however, grow in the light are provided with hyaline walls, conspicuous chloroplasts and cross walls at the right angles to the lateral walls. And, this way, the rhizoidal and chloronemal branches may mainly be differentiated on the basis of environmental conditions. In earlier stages the rhizoids act as absorbing organs but later on they perform the function of the attachment of the plants to the substratum. The chloronemal branches are photosynthetic.

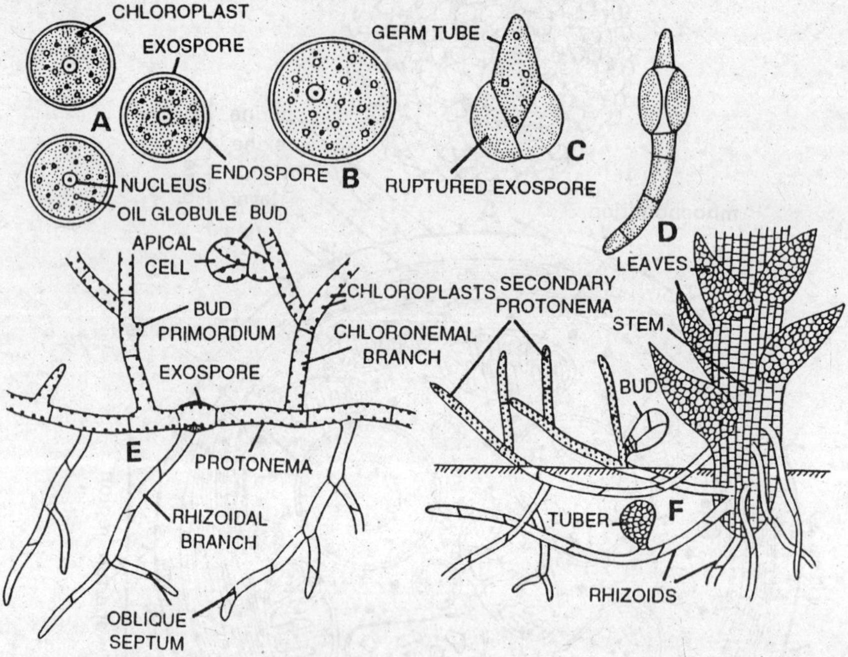

Fig. 23.18. *Funaria hygrometrica.* Structure of spore and its germination. A-B spore with nucleus, oil-globules and chloroplasts; C, germinating spore with ruptured exospore and germ tube; D, green germ tube enlarges in length and divides to form a cellular green filament; E, branched green filament the protonema; F, buds develop on protonema; each bud develops in a gametophore, the rhizoids with oblique septa develop from the base of the gametophore, tuber is seen on rhizoid.

According to Sironval (1947), in *Funaria hygrometrica,* in the older portions of the primary protonema, the gametophore initials are produced in the lowermost cell of the lateral branch. Such gametophore initials are formed on a profusely branched portion of the protonema called the **caulonema.** The caulonema possesses oblique cross walls in it. The buds of the leafy gametophores are formed only on the caulonema. Van Andel (1952) does not support the presence of caulonema in *Funaria hygrometrica.* Allsopp and Mitra (1958) however, could not see the two distinct stages the 'chloronema' and 'caulonema'. According to Allsopp and Mitra, the germinating spore produces a filamentous structure with cross walls at right angles to the long axis, hyaline cell walls and numerous discoid chloroplasts in it. This structure resembles to that of a typical chloronema and may be designated as chloronemal branches. These branches grow horizontally on the surface of the substratum. This structure develops the heterotrichous protonema with two types of branches. The upright and prostrate branches are provided with discoid conspicuous chloroplasts, hyaline cell walls, and cross walls at the right angles of the lateral walls. They are chloronemal branches. The branches which go downward (rhizoidal branches) possess incospicuous chloroplasts or leucoplasts, oblique cross walls and somewhat brownish walls. Finally the buds appear on the protonema. The buds usually appear just behind the cross walls. Each such bud develops into a new gametophyte. As soon as the leafy

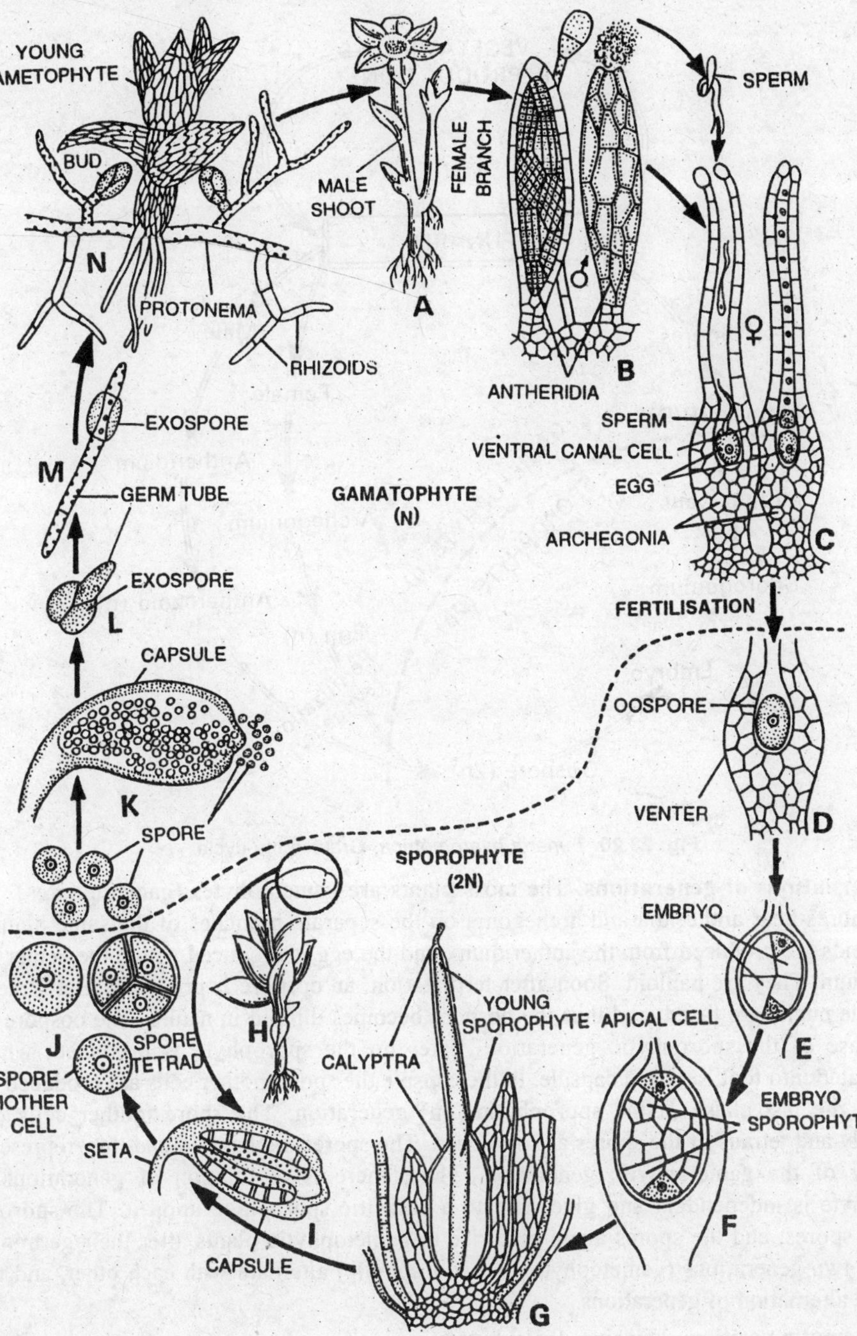

Fig. 23.19. *Funaria hygrometrica.* Diagrammatic life-cycle. A, gametophore with male and female shoots; B, antheridia; C, archegonia; D, venter containing oospore (2n) after fertilization, E-G, successive stages in the development of young sporophyte; H, gametophore bearing sporophyte; I, longitudinal section of sporogonium (outline sketch); J, spore mother cell (2n) and spore tetrad (n); K, spores (n); L-N, germination of spore and development of young gametophyte.

gametophores with few leaves and rhizoids have been established in the soil (substratum), the primary protonema disintegrates and disappears, and the gametophores are separated from each other and become independent.

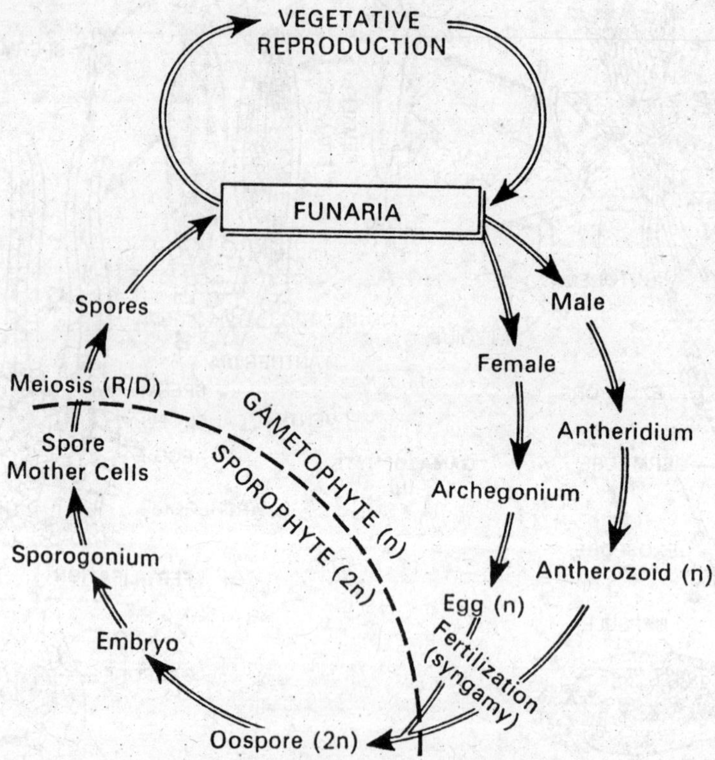

Fig. 23.20. *Funaria hygrometrica.* Graphic life-cycle.

Alternations of generations. The moss plants are gametophytes (gametophores). These gametophores bear antheridia and archegonia on the separate branches of the same plant. The antherozoids are produced from the antheridium, and the egg is produced within the venter of the archegonium. They are haploid. Soon after fertilization, an oospore is produced. Here, the male and female nuclei are fused, and thus the oospore becomes diploid in nature. The oospore is the first phase of the sporophytic generation. Later on the sporophyte is produced, which is differentiated into foot, seta and capsule. In the capsule the spore mother cells are produced. They represent the last phase of the sporophytic (2n) generation. The spore mother cells divide meiotically and tetrads of the spores are produced. The spores are haploid and they represent the beginning of the gametophytic generation. Thus, there is alternation of generations. The gametophyte is independent and gives rise to a parasitic sporophyte upon it. The sporophyte produces spores, and the spores again give rise to gametophytic plants after their germination. Here, the two generations (gametophytic and sporophytic) alternate with each other, and this is called the alternation of generations.

Systematic position. Division. Bryophyta
Class. Bryopsida (musci)
Sub-class. Bryidae
Cohort. Eubriidae
Order. Funariales
Family. Funariaceae
Genus. *Funaria.*

SUMMARY

Systematic position and distribution. Division-Bryophyta; Class-Bryopsida (Musci); Order-Funariales; Family-Funariaceae; Genus-*Funaria;* 15 species found in India; cosmopolitan in distribution.

Habitat. Lives on rocks, walls or in moist soil.

External features. Gametophore with an upright radial stem with leaves; branches lateral, below the leaf; rhizoids multi-cellular with oblique septa, usually branched rhizoids; scales absent; leaves spirally arranged, with a divergence of 3/8, a distinct midrib present, more than one-layered thick, wings single-layered.

Internal structure. Stem epidermis one or two-layered, without stomata; many layers usually thick walled; central cylinder consisting of long, narrow thin walled colourless cell without protoplasm; as in *Sphagnum.*

Sexual organs. Monoecious and dioicous; antheridia at the apex of an antheridial branch, stalk long, body jacket single layered; antherozoids biciliate; archegonia, at the tip of the archegonial branch, long stalk, neck made up of 6 vertical rows of cells venter 2-layered thick, single cover cell which acts as an apical cell with 4 cutting faces, neck canal cells usually 8 to 10.

Sporophyte. Amphithecium gives rise to the jacket of capsule; endothecium gives rise to columella and archesporium; archesporium gives rise to sporocytes; small, conical foot; seta long and elongates gradually; pyriform (pear-shaped) capsule; capsule wall more than four-layered, persistent; elaters absent; columella present, penetrates the spore sac; dehiscence takes place by separation of operculum; peristome present.

Young gametophyte. Spore with two-wall layers; spore size approximately 0.012 to 0.015 mm.; spore germination by formation of a branched filamentous protonema.

24
CHAPTER

Bryophytes: General Topics

VEGETATIVE PROPAGATION IN BRYOPHYTES

Most of the bryophytes propagate vegetatively. The vegetative reproduction takes place in favourable conditions for the vegetative growth. The vegetative reproduction takes place by various means and methods. Certain bryophytes, such as *Bryum* reproduces exclusively by vegetative methods, but most of the bryophytes reproduce both by sexual and vegetative methods. Some of the important methods of vegetative propagation are quoted below:

1. **By death of older parts.** This method of vegetative propagation is commonly found in many bryophytes. In *Riccia, Marchantia, Anthoceros, Notothylas* and others the progressive death and decay of older parts reach to the dichotomy and two branches are resulted, each of such branches may develop into a new plant.

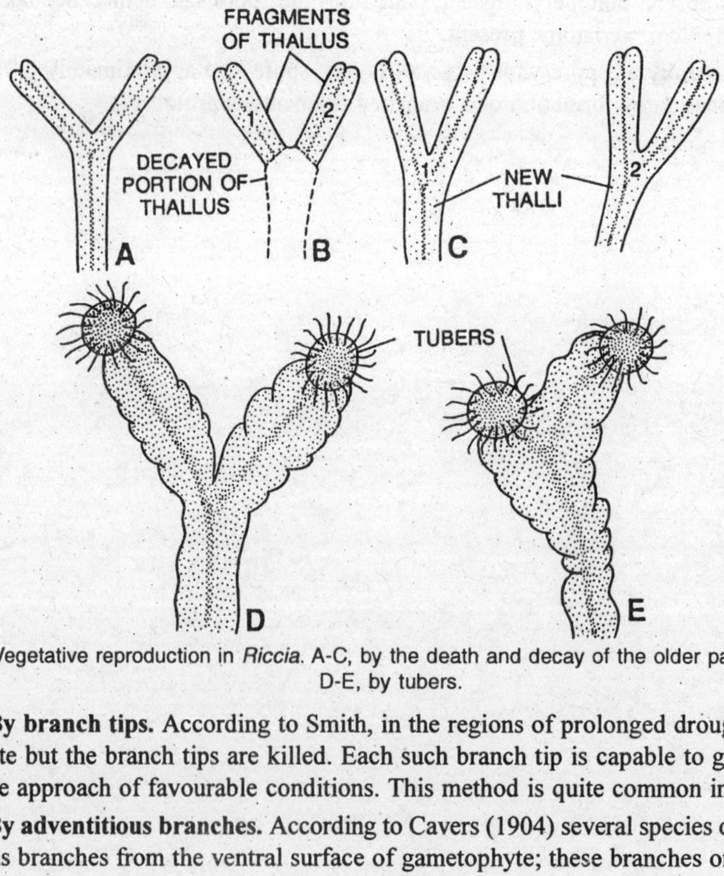

Fig. 24.1. Vegetative reproduction in *Riccia*. A-C, by the death and decay of the older parts of the thallus; D-E, by tubers.

2. **By branch tips.** According to Smith, in the regions of prolonged drought, all parts of a gametophyte but the branch tips are killed. Each such branch tip is capable to give rise to a new plant on the approach of favourable conditions. This method is quite common in *Riccia.*

3. **By adventitious branches.** According to Cavers (1904) several species of *Riccia* produce adventitious branches from the ventral surface of gametophyte; these branches on being detached

may result in the formation of new gametophytes. Such adventitious branches have also been recorded from other bryophytes, *e.g., Marchantia, Anthoceros* sp.

4. By Gemmae. According to Campbell (1918) even in *Riccia glauca,* the cell division takes place in the apices of young rhizoids resulting in gemma-like structures which later on develop into new thalli.

In *Marchantia* and *Lunularia* the gemmae are produced in gemma cups situated on the dorsal surface of the gametophyte. Many gemmae are found in each gemma cup. On being detached from the cup, each gemma gives rise to a plant. The gemma cups are circular in *Marchantia* and crescent-shaped in *Lunularia.*

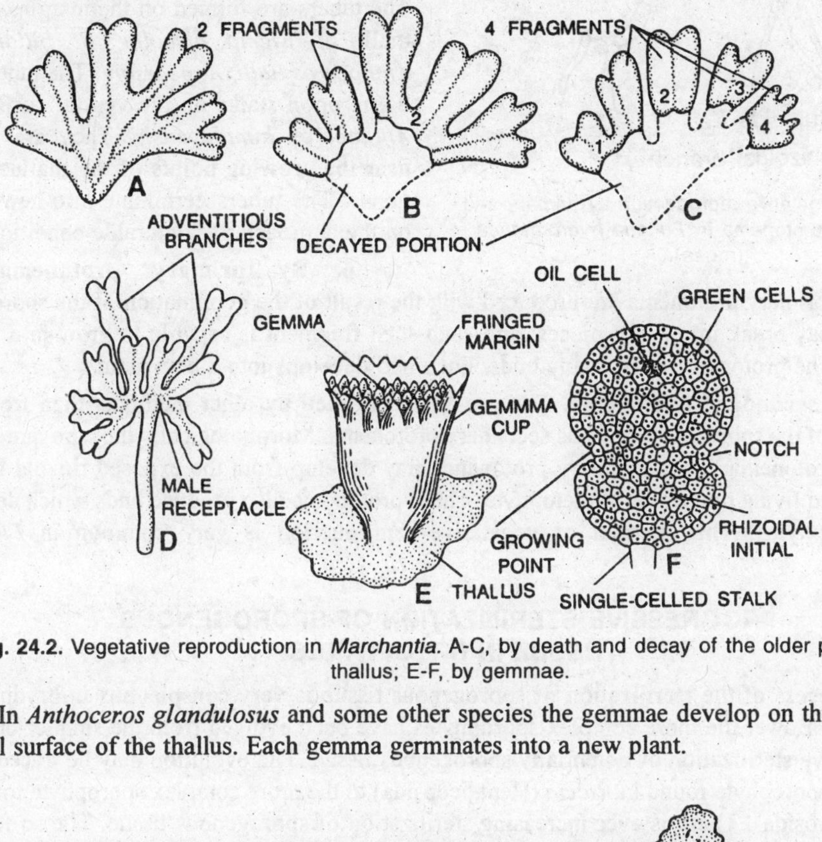

Fig. 24.2. Vegetative reproduction in *Marchantia.* A-C, by death and decay of the older parts of the thallus; E-F, by gemmae.

In *Anthoceros glandulosus* and some other species the gemmae develop on the margin and dorsal surface of the thallus. Each gemma germinates into a new plant.

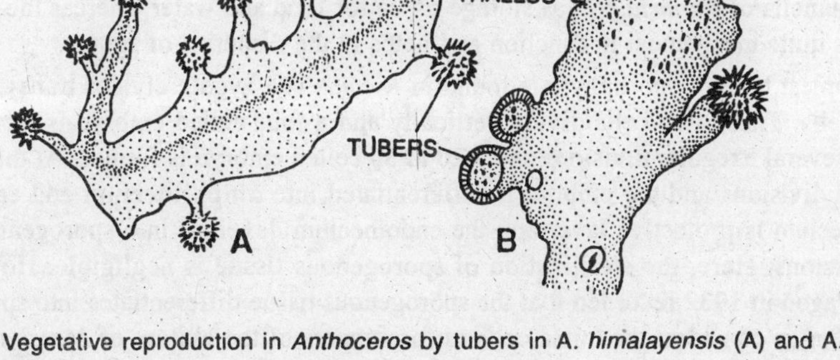

Fig. 24.3. Vegetative reproduction in *Anthoceros* by tubers in *A. himalayensis* (A) and *A. laevis* (B).

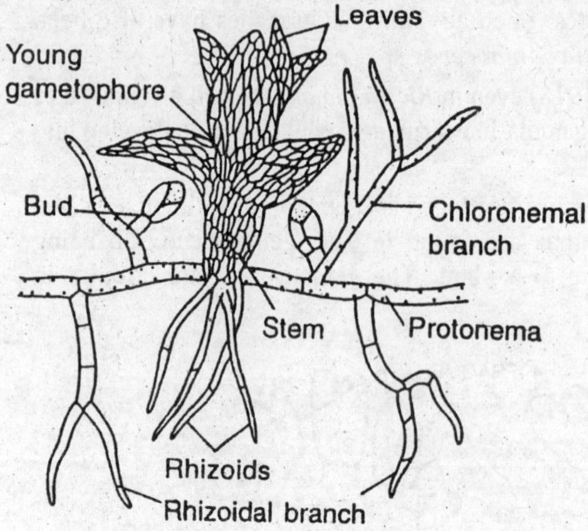

Fig. 24.4. Vegetative reproduction by primary and secondary protonema in *Funaria hygrometrica* (moss).

In *Funaria*, small, multi-cellular, linear green gemmae are borne at the leaf apices, shoot apex and on the rhizoids. Such gemmae also develop on the protonema of *Funaria hygrometrica*. Each gemma is capable to give rise to a new plant. The underground gemmae are known as bulbils.

5. By tubers. Many bryophytes reproduce vegetatively by means of tubers. The tubers are formed on the margins of the thalli of *Riccia discolor, R. billardieri, Anthoceros halli, A. pearsoni*. The tubers are borne upon stalks on the ventral surface of *Anthoceros himalayensis*. They are found near the growing points of the thallus in *A. laevis*. The tubers germinate into new thalli on the approach of favourable conditions.

6. By primary protonema. In *Funaria* the primary protonema are produced with the result of the germination of the spore. The protonema may break into small pieces; and each such fragment is capable to grow into a new protonema. The protonema bears many buds. Each bud develops into a new plant.

7. By secondary protonema. The protonema formed by other methods, than from the germination of the spore, are known as secondary protonema. Morphologically they are quite alike to primary protonema. The secondary protonema may develop from the exposed rhizoid or any other detached living part of the gametophyte. These protonema also produce buds which develop into gametophytes. This method of vegetative reproduction is very common in *Funaria hygrometrica*.

PROGRESSIVE STERILIZATION OF SPOROGENOUS TISSUE IN BRYOPHYTES

The process of the sterilization of sporogenous tissue is very conspicuous in bryophytes. According to Bower the more complex sporophytes have been evolved from the simpler ones by the progressive sterilization of potentially sporogenous tissue. The evolution may be traced from the simplest sporophyte found in *Riccia* (Hepaticopsida) to the more complex sporophyte found in *Funaria* (Bryosida). There is ever increasing sterilization of sporogenous tissue. The so formed sterile tissues function in different ways, *e.g.*, the foot may absorb the food and perform the function of anchorage; chlorophyllous tissue, stomata and inter-cellular spaces are meant for food manufacture; elaters, operculum and peristome help in the dehiscence of the spores somehow or the other; columella of *Funaria* acts as storage tissue for food and water whereas the columella of *Anthoceros* is quite mechanical in function and helps in the dispersal of spores.

The simplest type of sporophyte is found in *Riccia*. The zygote divides transversely giving rise to two cells. These two cells divide vertically and a four celled embryo is formed. This is followed by several irregular divisions and a 20 to 30 celled embryo is formed. At this stage there are periclinal divisions and the embryo is differentiated into **amphithecium** and **endothecium.** The amphithecium is protective layer and the endothecium develops into sporogenous tissue by repeated divisions. Here, the sterilization of sporogenous tissue is negligible. However, in *R. crystallina,* Pagon in 1932 recorded that the sporogenous tissue differentiates into sporocytes and **nurse cells**. Pagon considers the nurse cells as forerunners of the **elaters** of *Marchantia*.

In *Sphaerocarpos,* some cells of the sporogenous tissue become sporocytes which form spores by meiotic division. Other cells of the sporogenous tissue develop into sterile **nurse cells** which supply food to the spores. Here the sterilization of sporogenous tissue has progressed to some extent.

In *Corsinia,* the last generation of the sporogenous tissue differentiates into **soporocytes** and **sterile cells** which persist until the maturity of the spores.

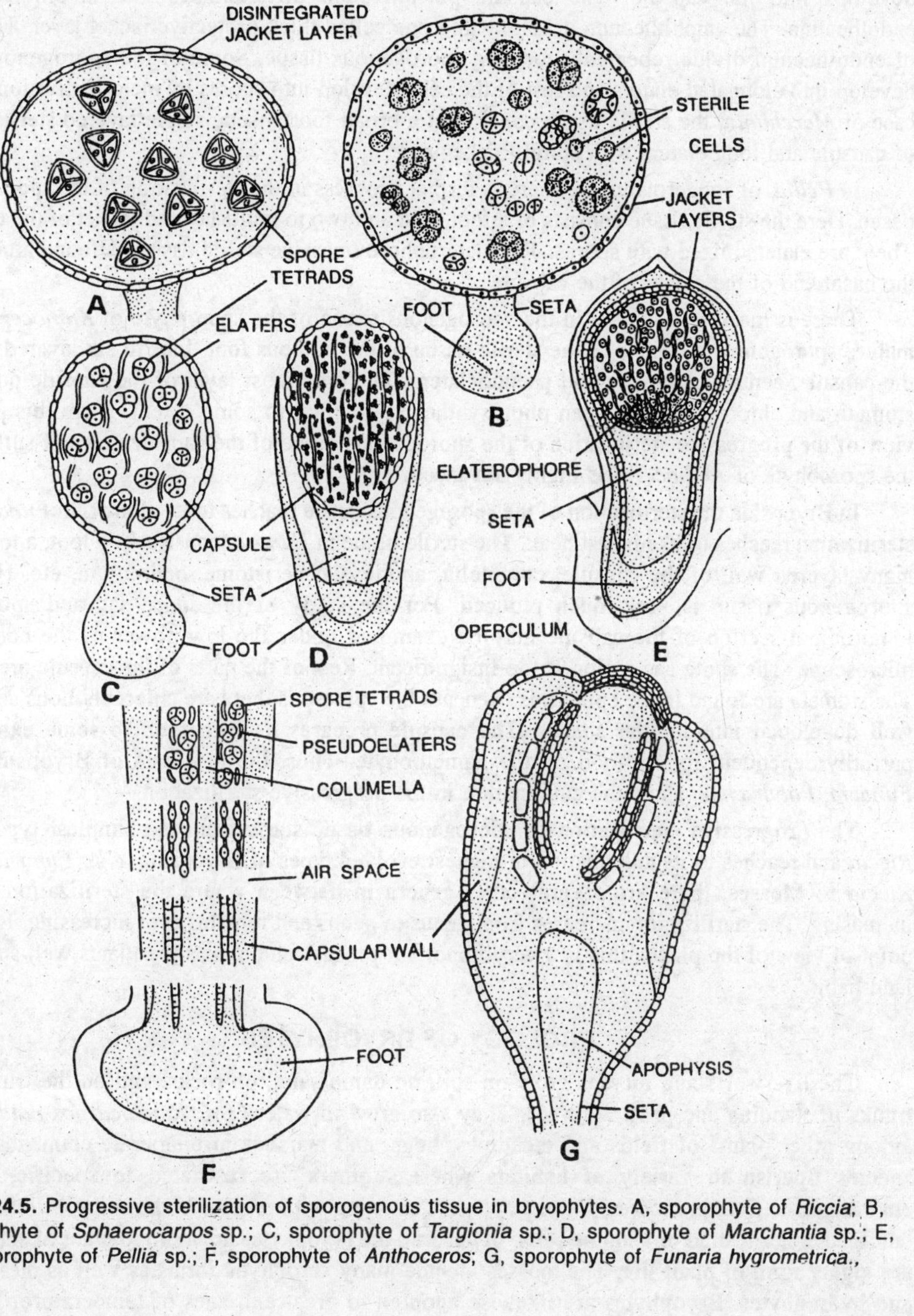

Fig. 24.5. Progressive sterilization of sporogenous tissue in bryophytes. A, sporophyte of *Riccia*; B, sporophyte of *Sphaerocarpos* sp.; C, sporophyte of *Targionia* sp.; D, sporophyte of *Marchantia* sp.; E, sporophyte of *Pellia* sp.; F, sporophyte of *Anthoceros*; G, sporophyte of *Funaria hygrometrica*.,

In *Targionia,* about half of the cells of the last generation in the sporogenous tissue become sporocytes while the rest half of the cells elongate greatly and mature into **elaters.** In *Targionia* the sporophyte consists of a broad foot, a seta and a single layered jacket of the capsule. Here, the sterilization of the sporogenous tissue shows an advance over that of *Riccia.*

In *Marchantia,* the zygote divides-transversely and thereafter vertically resulting into a four-celled embryo. The lower half of the embryo develops into foot and seta while the upper half develops into the capsule. The capsule portion soon differentiates into amphithecium a endothecium. The amphithecium develops into one celled thick protective jacket layer. The cells of endothecium divide repeatedly forming sporogenous tissue. Some of the sporogenous cells develop into elongated elaters and rest of the cells develop into spores by meiotic division. In the case of *Marchantia* the sterile tissue consists of a broad foot, a seta, single layered sterile jacket of capsule and long elaters with spiral thickenings.

In *Pellia,* of Jungermanniales there is further progress in the sterilization of the sporogenous tissue. Here the sterile tissue consists of a foot, seta and two to multi-layered jacket of the capsule. There are elaters mixed with spores. The elaterophore (massive sterile tissue) is found attached at the basal end of the cavity of the capsule.

There is marked reduction in the sporogenous tissue of the sporophyte of *Anthoceros.* The mature sporogonium consists of the sterile tissue of the bulbous foot, four to six layered wall of the capsule, central columella and pseudoelaters. The outermost layer of the capsule possesses stomata and chlorophyll. This can photosynthesize its food to some extent. From this point of view of the progressive sterilization of the sporogenous tissue of the capsule and self-sufficiency the sporophyte of *Anthoceros* is highly advanced.

In Bryopsida the sterilization of the sporogenous tissue reaches to its climax. In *Funaria,* the sterilization reaches to its highest peak. The sterile tissue in *Funaria* consists of a foot, a long seta, many-layered wall of the capsule, columella, apophysis, peristome, operculum, etc. Here the sporogenous tissue is very much reduced. For the study of the spore sac and spores the longitudinal section of the capsule may be examined under the low power of the compound microscope. The spore sac region is too insignificant. Rest of the parts of the capsule are sterile. The stomata are found in the epidermis. Beneath the epidermis there are chlorophyllous cells with well developed intercellular spaces. The capsule prepares its own food to some extent and partially dependent for its food on the gametophyte. Thus, the members of Bryopsida, *e.g., Funaria, Polytrichum* show the culmination in the progressive sterilization.

The progressive sterilization of sporogenous tissue starts from the simplest types, *e.g., Riccia* and reaches to its culmination in highest evolved types of Bryopsida, *e.g., Funaria.* From *Riccia* to Mosses, there are several other genera in between where the sterilization is ever increasing. The sterilization increases from genus to genus and remains ever increasing. From the point of view of the plants, this is a very important process and makes the plants well suited for land habit.

ECOLOGY OF BRYOPHYTES

The liverworts and mosses grow on soil, on damp sand, on rocks, and on the trunks and trunks of standing and prostrate trees. They also grow in water (*e.g., Ricciocarpos natans)* and among other plants of fields and meadows, bogs, and marshes in temperate countries. Some species flourish in variety of habitats whereas others are restricted to specific, limited environments. Certain mosses, for example, are exacting with respect to the soil reaction and are confined either to acid or alkaline soils. In general, the bryophytes are more tolerant of shade than are higher form of plant life. The mosses include many xerophytic forms as well as mesophytes and hydrophytes. Bryophytes are likewise adopted to great extremes of temperature, for they range from the arctic zone to the tropics and grow in the vicinity of hot springs. They reach their greatest development in cool, moist forests in temperate countries and the mountains of the tropical countries.

The bryophytes are an important component of the flora of the earth and play a significant role in the economy of nature. This is partly the result of the great number of individual plants produced by vegetative propagation. Mosses are so prolific that they form great masses or carpets covering the soil. Another characteristic of ecological importance is the ability of mosses to hold water, which is trapped among the leaves and stems. Many woodland mosses and species of *Sphagnum* absorb water through their leaves. By their structures and their mode of life, mosses contribute in many ways to the modification of their own environment.

The significance of mosses as solid formers following lichens or other lower forms of plant life on bare rock surfaces is of much importance in the succession of vegetation. Lichens are followed by mosses, which, like lichens, are able to survive in a dry environment. The mosses shade the lichens and successfully compete with them for water and nutrients. The death and decay of older mosses often produce a mat over the rock surface. As this mat becomes thicker and develops a water-holding capacity a new stage in the plant succession follows. This stage consists of annual and perennial herbs. The retention of water by mosses of leafy liverworts and mosses growing on fallen trees and other organic material hastens the processes of decomposition and hence the organic enrichment of the soil. Absorbing but little water from the substratum, they do not dry out the soil but protect it from desiccation. As a result of their ability to retain water, natural beds of mosses undoubtedly act as seed beds for herbaceous and woody flowering plants and for conifers.

One of the roles of the bryophytes is in the retardation of erosion. Carpets or felt like mosses possess a greater water retaining power than do layers of dead leaves. They therefore slow down the rapid run off rain water and melted snow. In addition to this, dense strands of moss collect and hold particles of soil. Insignificant as the individual plants of this group may appear, they play a part, together with other and more advanced forms of plant life, in making and changing man's environment.

ECONOMIC IMPORTANCE OF BRYOPHYTES

A few bryophytes are economically important.

Ecological importance. The liverworts, mosses and lichens are supposed to be the pioneers in establishing vegetation where other vegetation seems to be practically impossible. They colonize the barren rocks and exposed areas of hills, and make them suitable for growing angiospermic and other plants by depositing humus soil and plant debris. In the beginning the forms and grasses grow, and ultimately shrubs and trees also establish, and the whole area converts into dense wood.

However the *Sphagnum* plants are of great ecological importance. When these plants establish themselves in some lake or other areas full of water, sooner or later they cover the whole surface of the water. Due to deposition of plant debris the surface may be raised. The *Sphagnum* plants along with other hydrophytes form a dense surface covering over the water below. This covering gives the appearance of the soil from the surface. These areas are known as **quacking bogs.** Later on these bogs are converted into swamps. Ultimately these swamps are replaced by the forest growth of mesophytic type.

A few bryophyotes play an important role in checking the soil erosion. They are capable of holding the soil by their extensive carpets, and prevent the soil erosion to some extent. (Also see 'ecology of bryophytes').

Packing material. Most of the mosses are used as packing material after being dried. They make a fairly good packing material in the case of glass ware and other fragile goods. Especially the dried peat mosses (*Sphagnum* spp.) are used to pack bulbs, cuttings and seedlings for shipment.

Used in seed beds. Since the peat mosses have remarkable power to absorb and hold water like a sponge, they are extensively used in seed beds and green houses to root cutting. The peat mosses (Sphagna) are also used to maintain high soil acidity required by certain plants.

As a source of fuel. The peat is also a potential source of coal. Dried peat may be used as fuel. In Ireland, Scotland and other European countries the peat is used for fuel. In colder parts of the world where peat reaches its greatest development, the lower layers of peat become carbonized, and after the ages have passed, becomes available to human kind in the form of coal.

Absorbent bandages. The *Sphagnum* plants are slightly antiseptic and possess superior absorptive power. On account of these properties they may be used for filling absorbent bandages in place of cotton, in the hospitals.

PHYLOGENY OF BRYOPHYTES

There are *two* theories : (i) The first theory is known as **up-grade or progressive evolution theory.** (ii) and the second one is called **down-grade or regressive evolution theory.** According to these theories the evolution in bryophytes is as follows:

Progressive (Up-Grade) Evolution Theory

According to this theory there is progressive evolution which means that it starts with simple forms and terminates in complex forms. Cavers (1910), Campbell (1940), and Smith (1955) supported this view. According to them the primitive gametophytes possessed simple, dorsiventral, prostrate thallus showing simplicity in their external and internal structures. According to this theory the first evolved bryophyte was *Sphaero - Riccia*, a hypothetical type which combined the present day genus *Sphaerocarpos* (Proskauer, 1954). Contrary to it Campbell considered *Riccardia* and *Metzgeria* as primitive bryophytes. However, Cavers (1910) proposed progressive theory and showed a phylogenetic line of evolution in bryophytes.

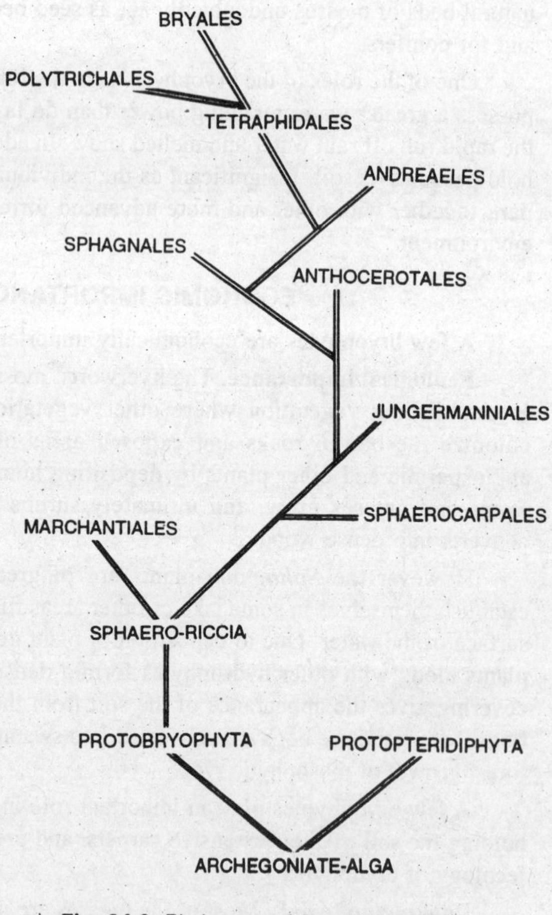

Fig. 24.6. Phylogeny of Bryophytes (Cavers)

According to the first line of evolution the primitive plants proceeded upto Marchantiales where the tissue differentiation in thallus took place in the form of epidermis, pores and air-chambers with chlorophyllous filaments in the sex-organs (antheridia and archegonia) in definite receptacles. The second line culminated in Jungermanniales and retained the simplicity in the thallus (gametophyte) without air pores but with lateral expansions known as wings or leaves. On the ventral surface of the thallus the multi-cellular filaments developed which later on converted into scales or under leaves with chlorophyll. On the other side, the thallus developed a distinct mid-rib and thin wings. Later on these wings divided up into definite segments and became leafy in structure. The Anthocerotales developed as one line from the Jungermanniales by elaboration of sporophyte while retaining the simple gametophyte. The Musci (*i.e.,* Sphagnales, Polytrichales, Bryales) arose from Anthocerotales.

Regressive (Down-Grade) Evolution Theory

According to this theory the primitive bryophytes were mass-like in appearance. They had erect leafy shoot with radial symmetry like that of mosses. From mosses the evolution proceeded through Acrogynous Jungermanniales, Anthocerotales and Marchantiales. This view has been supported by Wettstein (1908), Kashyap (1919), Church (1919), Evans (1930), Mehra (1957), Pande (1960), Praskauer (1960), Fulford (1965), Zimmermann (1966), Udar (1970), etc. Mehra (1957) proposed a common origin of Anthocerotales and Psilophytales through Antho-rhyniaceae stalk.

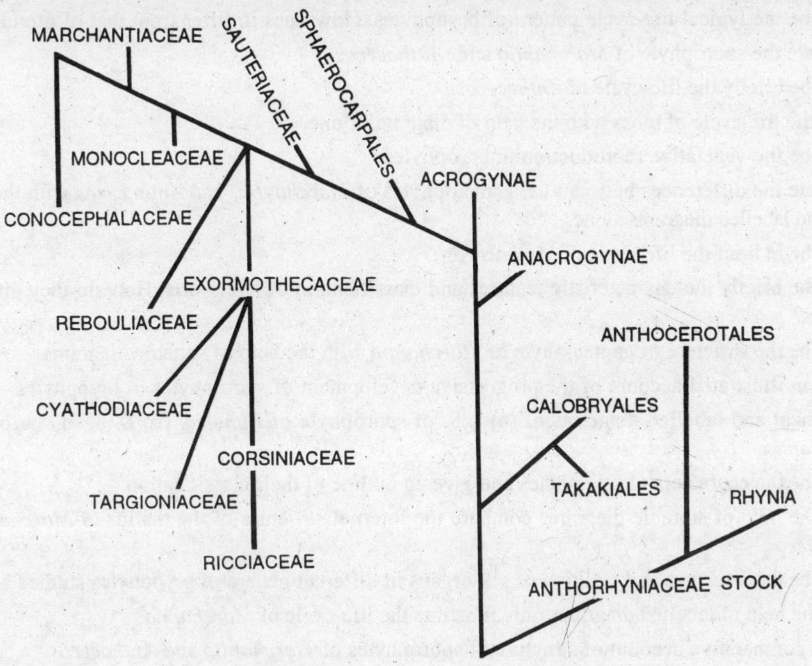

Fig. 24.7. Phylogeny of Bryophytes (Mehra's view)

On the basis of his **'condensation theory'** Mehra (1957) explained the course of evolution in Bryophytes. According to him it is after the regressive deviation of Marchantiales, a line of progressive evolution terminates in Marchantiaceae and both Marchantiaceae and Ricciaceae are the extreme forms in evolution in two different lines. Here the thallous forms originated from foliose forms in Metzgerineae. According to Mehra the Anthocerotales have not arisen from Hepaticeae but from the separate stock called Antho-rhyniaceae stock which splitted into Rhyniaceae and Anthocerotales.

QUESTION BANK

N.B. *Undermentioned questions are asked repeatedly in various Indian Universities and competitive examinations.*

Long Answer Questions

1. Discuss the evolutionary advancement of Musci over Liverworts.
2. Trace the evolution of the sporophyte in the various members of Bryophytes.

3. Describe the various views on the origin of Bryophytes in Plant Kingdom.
4. Explain the theory of "progressive sterilization of the potentially sporogenous tissue" in Bryophytes.
5. Give comparative account of the position and structure of the archegonia of *Marchantia, Anthoceros* and Moss.
6. By means of outline labelled diagrams only, illustrate the chief events in the life history of *Funaria.*
7. Describe the antheridiophore and archegoniophore of *Marchantia.*
8. Compare the vegetative thalli of *Marchantia* and *Anthoceros* and give the importance of the various structures found in them.
9. What are bryophytes ? What are their features of advancement over algae and how far they lag behind the pteridophytes ?
10. Compare the sporophytes of *Marchantia,* and *Anthoceros* on the basis of (a) origin of sporogenous tissue, (b) capsule wall and (c) extent of dependency on the gametophyte.
11. Give an account of the classification of the bryophytes and the criteria used.
12. Describe the typical life-cycle pattern of bryophytes. How does it differ from that of pteridophytes?
13. Compare the sporophyte of *Marchantia* and *Anthoceros*
14. Describe briefly the life-cycle of *Anthoceros.*
15. Draw the life-cycle of moss with the help of diagrams alone.
16. Describe the vegetative reproduction in bryophytes.
17. Illustrate the differences between the gametophytes of *Marchantia,* and *Anthoceros* with the help of neat and labelled diagrams alone.
18. Describe in brief the life-cycle of *Anthoceros.*
19. Describe briefly the characteristic features and classification of liverworts. How do they differ from mosses?
20. Describe the structure of gametophyte of *Marchantia* with the help of suitable diagrams.
21. Write an illustrated account of the progressive development of sporophytes in bryophytes.
22. Draw neat and labelled diagrams of (a) L.S. of sporophyte of *Funaria,* (b) L. S. of sporophyte of *Marchantia.*
23. Describe the characters of bryophytes and give an outline of their classification.
24. With the help of suitable diagrams compare the internal structure of the thallus of *Marchantia* and *Anthoceros.*
25. Describe the structure and position of sex organs in different genera of bryophytes studied by you.
26. With the help of labelled diagrams only illustrate the life-cycle of *Marchantia.*
27. Give a comparative account of structure of sporophytes of *Marchantia* and *Anthoceros.*

Short Answer Questions

1. Why are air pores in *Marchantia* not called stomata.
2. Is the sporophyte of *Funaria* wholly dependent for its nutrition on the gametophyte? Justify your answer.
3. Draw neat and labelled sketches of the following: T. S. of *Marchantia* thallus.
4. Write what you know about : Vegetative reproduction in the Hepaticopsida.
5. Show with the help of properly labelled diagrams only the microscopic structures of the following: Internal structure of *Marchantia* thallus.
6. Show with the help of properly labelled diagrams only, the microscopic structure of the following: V.S. mature capsule of *Anthoceros.*
7. Show with the help of properly labelled diagrams only, the microscopic structures of the following : V.S. antheridiophore of *Marchantia.*
8. Draw neat labelled diagram of the following : T.S. mature sporogonium of *Anthoceros.*
9. Describe the thallus of *Anthoceros.*
10. Draw the structure (L.S.) of sporogonium of *Marchantia* and label the parts.

11. Describe the structure of gemma-cup in *Marchantia.*
12. Draw the T. S. of *Marchantia* thallus and label the parts.
13. Describe the structure of mature archegonium of *Marchantia.*
14. Describe the leaves in *Funaria.*
15. Mention any two differences between *Anthoceros* and *Marchantia* with reference to sporophytes.
16. Describe the gametophytes of *Marchantia.*
17. Explain the post-fertilization changes in *Anthoceros.*
18. Describe the structure and development of antheridia in *Anthoceros.*
19. Describe the structure of sporogonium in *Funaria.*
20. Draw neat labelled diagrams representing the haploid phase (n) of *Anthoceros.*
21. Describe the female receptacle in *Marchantia* after the formation of mature archegonia.
22. Describe the internal structure of *Anthoceros* thallus, point out the significance of mucilage cavities.
23. Taking the sporogenous tissue into consideration do you consider sporophyte of *Anthoceros* to be more advanced than the sporophyte of *Marchantia.* Give at least two reasons.
24. Compare the gametophytes of *Anthoceros* with that of *Marchantia* taking the external morphology.
25. Describe the behaviour of gemmae when they get detached from the parent plant of *Marchantia.*
26. Differentiate between *Marchantia* and *Anthoceros..*
27. Draw a section cut parallel to the upper surface of the thallus of *Marchantia* and passing through the air chambers.
28. Comment briefly on the morphological nature of the male and female receptacles of *Marchantia.*
29. By mistake the materials of the following pair got mixed up. What are the criteria you will use to separate the components : Sterile thalli of *Anthoceros* and *Funaria.*
30. Among the sporogonia of *Marchantia* and *Anthoceros* which is more primitive and why ?
31. Write short note on : Experimental studies on the gemmae of *Marchantia.*
32. "In their general habits and adaptations the liverworts are comparable to the amphibians of animal kingdom". Comment upon this statement.
33. Describe the modes of perennation in the liverworts studied by you.
34. Give reasons for the inclusion of *Marchantia* under Bryophytes, Hepaticeae, Marchantiales and Marchantiaceae.
35. Discuss the evolutionary importance of the sporophytic generation of *Anthoceros.*
36. Draw labelled diagrams only : T.S. mature capsule of *Funaria.*
37. Draw labelled diagrams only : T.S. of the archegonial necks of *Marchantia*, *Anthoceros* and *Funaria.*
38. T.S. Capsule of *Marchantia* along with the calyptra and perigynium.
39. Draw labelled diagrams of the following : fertile thallus of *Marchantia* parallel to the longitudinal axis.
40. Draw labelled diagrams of the following : T.S. capsule of *Anthoceros* at the level of elaterophore.
41. Draw labelled diagrams of the following : Lower surface view of the vegetative thallus of *Marchantia.*
42. Briefly write about the following : Thallus of *Marchantia.*
43. Distinguish between : Rhizoid of a liverwort and root hair of an angiosperm.
44. In which habitat is *Anthoceros* found ?
45. Describe the T.S. of the thallus of *Marchantia.*
46. Describe the structure of mature sporophyte of *Anthoceros* which has started spore dispersal.
47. Write notes on : (a) peristome of *Funaria,* (b) rhizoids in bryophytes, (c) scales of *Marchantia.*
48. Give an account of vegetative reproduction in bryophytes.
49. Write notes on : (a) archegoniophore, (b) capsule of *funaria,* (c) pseudoelaters.
50. Write notes on : (a) archegoniophore, (b) rhizoids in bryophytes, (c) scales of *Marchantia*
51. Write notes on : (a) antheridiophore of *Marchantia,* (b) paraphysis, (c) dehiscence of moss capsule.

52. Write short notes on : (a) archesporium, (b) archegoniophore of *Marchantia,* (c) operculum.
53. Write short notes on : (i) rhizoids in bryophytes, (ii) liverworts, (iii) *Nostoc* cavity.
54. Write notes on (a) function of elaters in bryophytes, (b) differences between liverworts and moss, (c) function of peristome in moss.
55. Write short notes on any three of the following : (a) peristome, (b) gemmae, (c) alternation of generations, (d) elaters.
56. Describe in any three of the following : (a) structure and function of scales in *Marchantia,* (b) mechanism of spore dispersal in *Funaria,* (c) methods of vegetative reproduction in *Funaria,* (d) economic importance of bryophytes.
57. Draw neat and labelled diagrams of (a) V.S. of *Marchantia* thallus passing through dorsal furrow (b) *Marchantia* gemma structure and germination, (c) V.S. *Anthoceros* thallus passing through mucilage cavity.
58. Write notes on : (a) vegetative reproduction in *Marchantia,* (b) structure of sex organs of *Funaria.*
59. Write short notes on any four (a) calyptra, (b) protonema, (c) peristome, (d) paraphysis, (e) apophysis.
60. Discuss the following : (a) evolutionary significance of *Anthoceros,* (b) mechanism of spore dispersal in moss.
61. Draw neat and labelled diagrams of (i) V. S. of *Anthoceros* thallus passing through archegonium, (ii) L.S. of female branch of *Funaria,* (iii) V.S. of leaf of *Funaria.*
62. Draw neat and labelled diagrams of (a) V.S. of antheridiophore of *Marchantia,* (b) L.S. of moss capsule.
63. Name five important evolutionary features which enabled the plants to establish themselves successfully on land through the ages.
64. Write notes on the following : Elaters and Pseudoelaters.
65. Describe the external structure of the gametophyte of *Funaria.*
66. By means of outline labelled diagrams (cellular details are not required) only, illustrate the following: L.S. gametophyte of *Marchantia.*
67. Briefly write about the following : Biological importance of *Bryophytes.*
68. How do you distinguish a protonema of moss from a filamentous green alga ?
69. By mistake the materials of the following pair got mixed up. What are the criteria which you will use to separate the components : Gametophores of *Funaria* and *Marchantia.*
70. Write short note on : Mechanism of spore dispersal in mosses.
71. The stem of *Funaria* is said to be comparable to that of a vascular plant in function but not in structure. Why?
72. Draw labelled diagrams only : T.S. capsule of *Funaria* at the level of apophysis.
73. Draw lablled diagrams of the following : T. S. capsule of *Funaria* at the level of spore sac.
74. What are the taxonomic significance and function of peristomal teeth?
75. Does protonema suggest the ancestry of mosses? Comment.
76. Make well labelled sketches of the following : L.S. Moss capsule.
77. Write note on the following : Apophysis.
78. Write explanatory notes on the following : Habit and habitat of Moss.
79. Write brief note on : Bud formation in protonema of mosses.
80. Write short note on : Spore dispersal in *Funaria.*
81. Why the mosses are not considered as ancestors to the Pteridophyta in preference to the Anthocerotae.
82. Describe the stages of Development in the amphithecial and endothecial regions of the "theca zone" of Moss capsule.
83. Describe the mechanism by which spores are dispersed from the capsule of Moss.
84. Write short note on the following : Peristome.

85. **Describe the external structure of an adult 'Moss plant' bearing capsule.**

86. **What is the significance of protonemal stage in the life-cycle of a moss ?**

87. **How would you distinguish a true moss from a club moss.**

88. **List out the distinctive anatomical features of Moss stem.**

89. **Write short notes on any three : (a) economic importance of bryophytes (b) alternation of generations in bryophytes, (c) gemmae, (d) elaters, (e) peristome.**

90. **Write short notes on any three: (a) protonema, (b) economic importance of bryophytes, (c) peristome, (d) gemmae, (e) sporophyte of *Anthoceros*.**

91. **Describe the important features of *Funaria*.**

92. **Write short notes on any three (a) archegonophore, (b) structure of leaf of *Funaria*, (c) sporophyte of *Marchantia*, (d) characteristic features of Bryophyta, (e) elaters.**

93. **Write short notes on any three of the following : (a) sporophyte of *Anthoceros* (b) vegetative reproduction in Bryophyta, (c) economic importance of bryophytes, (d) elaters, (e) structure of leaf of *Funaria*.**

94. **Write short notes on : (a) sporophyte of *Anthoceros,* (b) protonema, (c) classification of bryophytes, (d) gemmae, (e) alternation of generations in bryophytes.**

95 **How would you distinguish with a hand lens : (a) *Anthoceros* plant from that of *Marchantia,* (b) antheridiophore from archegoniophore of *Marchantia,* (c) sporophyte of *Funaria* from that of *Anthoceros*.**

96. **How would you distinguish the follorring : (a) thallus of *Anthoceros* and *Marchantia* (b) thallus of *Riccia* and green leaves of *Funaria* (in a T.S.), (c) elaters and pseudoelaters.**

97. **If you are on a botanical tour, how would you differentiate *Funaria, Marchantia* and *Anthoceros* in their habitat.**

98. **Write short notes on : (a) gemma cup, (b) peristome, (c) rhizoids, (d) scales.**

99. **Write a brief account on any two : (a) elaters, their types and function in bryophytes, (b) vegetative propagation in bryophytes, (c) habit and habitat of bryophytes.**

Multiple Choice Questions

1. Development of a sporophyte directly from the sporophytic tissue is called (a) Apospory, (b) Parthenogenesis, (c) Apogamy, (d) Syngamy.
2. The simplest known sporophyte among the Bryophyta is seen in (a) *Riccia,* (b) *Funaria,* (c) *Marchantia,* (d) *Anthoceros.*
3. The theory of progressive sterilization of potentially sporogenous tissue was first proposed by (a) Campbell, (b) Bower, (c) Smith, (d) Cavers.
4. The advanced sporophyte is seen in (a) *Riccia,* (b) *Porella,* (c) *Sphagnum*, (d) *Anthoceros.*
5. Epiphragm is seen in (a) archegonium, (b) antheridium, (c) gametophyte, (d) sporophyte.
6. In archegonium the cell above the egg cell is (a) Neck-canal cell, (b) neck cell, (c) Cover cell, (d) Ventral-canal cell.
7. Calyptra is seen in (a) Gametophyte, (b) Sporophyte, (c) archegoniophore, (d) antheridiophore
8. Perichaetium is seen in (a) archegoniophore, (b) antheridiophore, (c) sporogonium, (d) none of the above.
9. Spore mother cell in Bryophytes is (a) haploid in nature, (b) diploid in nature, (c) triploid in nature, (d) tetraploid in nature.
10. Elaters are seen in (a) antheridium, (b) archegonium, (c) sporogonium, (d) none of the above.
11. Nourishing tissue surrounding the spore producing tissue in a young sporangium is (a) archesporium, (b) tapetum, (c) spore mother cells, (d) sporocytes.
12. In *Polytrichum* the spore liberation is regulated by (a) operculum, (b) columella, (c) peristome, (d) annulus.
13. Elongated cylindrical sporogonium is seen in (a) *Sphagnum,* (b) *Polytrichum,* (c) *Marchantia, Porella.*

14. Trabeculae are seen in the sporogonium of (a) *Sphagnum,* (b) *Polytrichum,* (c) *Anthoceros,* (d) *Riccia.*
15. In *Anthoceros* the meristematic tissue is present (a) at the apex of sporogonium, (b) at the base of sporogonium above the foot, (c) in the middle of sporogonium, (d) none of the above.
16. In *Marchantia* the chloroplast is (a) cup-shaped, (b) disc-shaped, (c) star-shaped, (d) ribbon-shaped.
17. In *Anthoceros* the sporogenous tissue is derived from (a) amphithecium tissue, (b) endothecium tissue, (c) gametophytic tissue, (d) columella tissue.
18. Peristomes are present in (a) *Marchantia,* (b) *Polytrichum,* (c) *Sphagnum,* (d) *Porella.*
19. In *Marchantia* the archegonia are situated (a) on the disc, (b) below the disc, (c) in the middle of the disc, (d) on the sides of the disc.
20. The most primitive type of sporogonium is seen is (a) *Anthoceros* (b) *Marchantia,* (c) *Porella,* (d) *Riccia.*
21. The nurse cells are present in the sporogonium of (a) *Riccia,* (b) *Marchantia,* (c) *Porella,* (d) *Anthoceros.*
22. Stomata are present on capsule wall of (a) *Marchantia,* (b) *Anthoceros,* (c) *Porella,* (d) *Riccia.*
23. In *Riccia* (a) smooth rhizoids are present, (b) pegged rhizoids are present, (c) multi-cellular scales are present, (d) both smooth and pegged rhizoids present.
24. In which of the following plants, sporophyte is completely dependent on gametophyte (a) *Riccia,* (b) *Funaria,* (c) Angiosperms, (d) None of above.
25. Obliquely septate rhizoids are found in (a) *Riccia,* (b) *Marchantia,* (c) *Funaria,* (d) *Anthoceros.*
26. Columella is absent in (a) *Anthoceros,* (b) *Funaria,* (c) *Riccia,* (d) *Anthoceros himalayensis.*
27. The gametophyte is leafy shoot in (a) *Riccia,* (b) *Marchantia,* (c) *Funaria,* (d) *Anthoceros.*
28. The sporophyte grows continuously for a long period due to basal meristem in (a) *Funaria,* (b) *Anthoceros,* (c) *Riccia,* (d) *Marchantia.*
29. Pseudoelaters are found in the sporophyte of (a) *Riccia,* (b) *Funaria,* (c) *Marchantia,* (d) *Anthoceros.*
30. Elevated areolae are present on the upper surface of the gametophytic thallus of (a) *Funaria,* (b) *Riccia,* (c) *Marchantia,* (d) *Anthoceros.*
31. In *Riccia* growth of the thallus takes place by the activity of a (a) group of intercalary initials, (b) group of apical initials, (c) single apical cell, (d) basal meristem.
32. Female sex organs in *Marchantia* are borne on (a) an elaterophore, (b) a rhizophore, (c) an antheridiophore, (d) an archegoniophore.
33. A dioecious gametophyte is seen in the case of (a) *Funaria,* (b) *Riccia,* (c) *Marchantia,* (d) *Anthoceros.*
34. A sporophyte of *Marchantia* is enveloped by (a) two protective sheaths, (b) one protective sheath, (c) three protective sheaths, (d) four protective sheaths.
35. An endophytic alga is present in the thallus of (a) *Anthoceros,* (b) *Riccia,* (c) *Funaria,* (d) *Marchantia.*
36. The true elaters are found in sporophyte of (a) *Riccia,* (b) *Anthoceros,* (c) *Marchantia,* (d) *Funaria.*
37. Spores are liberated from the sporogonium only by the decay of gametophyte in (a) *Marchantia,* (b) *Riccia,* (c) *Funaria,* (d) *Anthoceros.*
38. Leptoids are present in (a) *Sphagnum,* (b) *Polytrichum,* (c) *Anthoceros,* (d) *Marchantia.*
39. In *Sphagnum* the spore-sac is (a) spindle-shaped, (b) arc-shaped, (c) dome-shaped; (d) disc-shaped.
40. In *Sphagnum* the zygote first divides (a) Transversely, (b) Longitudinally, (c) Obliquely, (d) Irregularly.
41. Gametophyte is predominant in (a) *Cycas,* *(b)* pea, (c) moss, (d) fern.
42. In *Funaria* the stomata are present (a) in the epidermis of leaf, (b) in the epidermis of stem (c) in the epidermis of capsule wall, (d) in none of these.
43. In *Funaria* stomata are found on : (a) leaves, (b) stem, (c) archegonia, (d) capsule.
44. In *Funaria,* antheridia and archegonia are found on : (a) different plants, (b) different branch of the same plant, (c) different branches of plant but separately, (d) the same branches of a plant in inner middle condition.

45. The bryophyte of considerable economic importance is : (a) *Funarīa,* (b) *Sphagnum,* (c) *Marchantia.* (d) *Riccia.*
46. In *Funaria* stomata are present on the : (a) leaf, (b) stem, (c) upper part of capsule, (d) lower part of capsule.
47. When moss spores germinate, they form (a) leafy gametophyte directly, (b) capsule directly, (c) first protonema, then from bud a leafy gametophyte, (d) protonema which bears archegonia and antheridia.
48. The form differs from a moss in having : (a) an independent gametophyte, (b) an independent sporophyte, (c) swimming antherozoids (d) archegonia.
49. Multi-cellular and jacketed sex organs are present in : (a) *Funaria,* (b) *Spirogyra,* (c) *Saccharomyces,* (d) *Hibiscus.*
50. The antherozoids of moss are : (a) motile, coenocytic and multiciliate, (b) motile, uninucleate and uniciliate, (c) motile, uninucleate and biciliate, (d) non-motile and without cilia.
51. Xylem and phloem do not form the conducting strand of : (a) *Pinus,* (b) *Cycas,* (c) *Brassica,* (d) *Funaria.*
52. Bryophytes include : (a) liverworts and ferns, (b) mosses and ferns, (c) mosses and liverworts, (d) none of the above.
53. In *Riccia* the sporophyte has : (a) foot, seta and capsule, (b) seta only, (c) capsule only, (d) foot only.
54. The antheridium of *Riccia* is a: (a) unicellular structure, (b) multi-cellular structure, (c) structure having 20 cells, (d) structure having 60 cells.
55. Plants of *Funaria* grow in tufts and are : (a) thallose, (b) foliose, (c) thallose and foliose, (d) found in the hot springs of Rajgiri.
56. Rhizoids of *Funaria* are : (a) unicellular, (b) green coloured, (c) unicellular with septa, (d) multicellular with oblique septa.
57. Cortex cells of the young stem of *Funaria* are (a) chlorenchymatous, (b) collenchymatous, (c) parenchymatous, (d) sclerenchymatous.
58. The spores of moss on germination give rise to: (a) proteonema, (b) primary protonema, (c) secondary protonema, (d) operculum.
59. *Riccia* is a bryophyte because it : (a) occurs mostly on land and has motile sperms, (b) has heteromorphic alternation of generations, (c) has multicellular sex organs with a sterile jacket and lacks vascular tissues, (d) none.
60. Which one of the following statements is not true of Bryophyta : (a) they lack tracheids and sieve tubes, (b) they are photosynthetic, (c) their zygote undergoes meiosis and then produces the sporophyte, (d) their spores germinate, producing gametophytes.
61. Rhizoids of *Funaria* are : (a) unicellular, (b) green coloured, (c) unicellular with septa, (d) multicellular with oblique septa.

ANSWERS

1 (a), 2 (a), 3 (b), 4 (c), 5 (d), 6 (d), 7 (b), 8 (a), 9 (b), 10 (c), 11 (b), 12 (c), 13 (a), 14 (b), 15 (b), 16 (b), 17 (b), 18 (b), 19 (d), 20 (d), 21 (a) 22 (b), 23 (d), 24 (b), 25 (d), 26 (c), 27 (c), 28 (b), 29 (d), 30 (b), 31 (c), 32 (d), 33 (c), 34 (b), 35 (a), 36 (c), 37 (b), 38 (b), 39 (c), 40 (a), 41 (c), 42 (c), 43 (d), 44 (b), 45 (b), 46 (d), 47 (c), 48 (a), 49 (a), 50 (c), 51 (d), 52 (c), 53 (c), 54 (b), 55 (b), 56 (d), 57 (c), 58 (b), 59 (c), 60 (c), 61 (d).

UNIT - 5

Diversity of Cryptogams Pteridophyta

25

CHAPTER

Pteridophytes: Introduction

The pteridophytes occupy the intermediate position in between the bryophytes on one hand and the phanerogams on the other. They possess certain combinations of major characters which are not found in bryophytes or in phanerogams. The most characteristic feature of the pteridophytes is the presence of independent gametophytes and sporophytes at maturity. The pteridophytes are also known as 'vascular cryptogams'. The sporophytes of these plants possess a well developed system of conduction tissue which consists of xylem, phloem and other mechanical elements.

Origin of the pteridophytes. As regards their origin, the pteridophytes are known from as far back as the Silurian age of Palaeozoic era, *i.e.*, they are somewhat 380,000,000 years old.

According to one school of thought, the pteridophytes like bryophytes have been evolved from algae. According to this view both the groups are parallel to each other in their origin and they are not connected phylogenetically. Scott (1900), Eames (1936), however do not trace their origin from any particular group of algae. The workers (Church, 1919; Arnold, 1947) trace the origin of pteridophytes from a particular group of marine algae, *i.e.*, Phaeophyceae. According to Bohlin (1901), Fritsch (1916) and Lotsy (1909) they are thought to be originated from a filamentous Chaetophorales of Chlorophyceae. Later on in 1945, Fritsch suggested that the pteridophytes originated from an erect and parenchymatous green alga with an isomorphic life-cycle.

Some of the workers believe that the pteridophytes have been originated from bryophytes. The pteridophytes resemble the bryophytes in many respects. These resemblances are as follows: the general similarity in vegetative structure of gametophytes of thallose bryophytes and pteridophytes similarity in the ontogeny of antheridia and archegonia; the mature sporophyte is partially parasitic upon the gametophyte; in both the groups there is heteromorphic life-cycle.

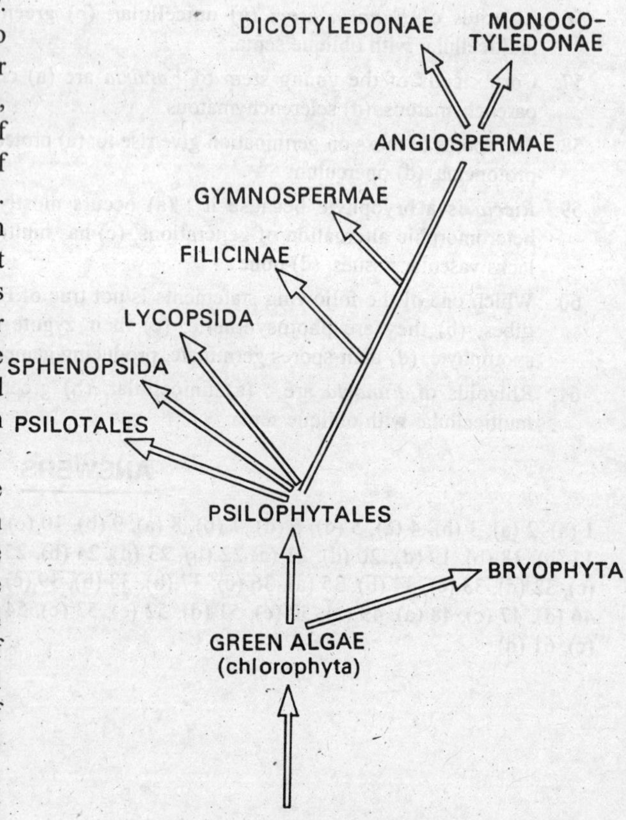

Fig. 25.1. Origin of Tracheophyta

According to Bower (1935) and Zimmermann (1938), the two groups (bryophytes and pteridophytes) are phylogenetically connected but they are divergent evolutionary lines evolved from hypothetical land plants of primitive archegonia type. According to Campbell (1899) they have arisen from an anthocerotean type of bryophyte.

This way, several hypotheses have been propounded by various workers from time to time.

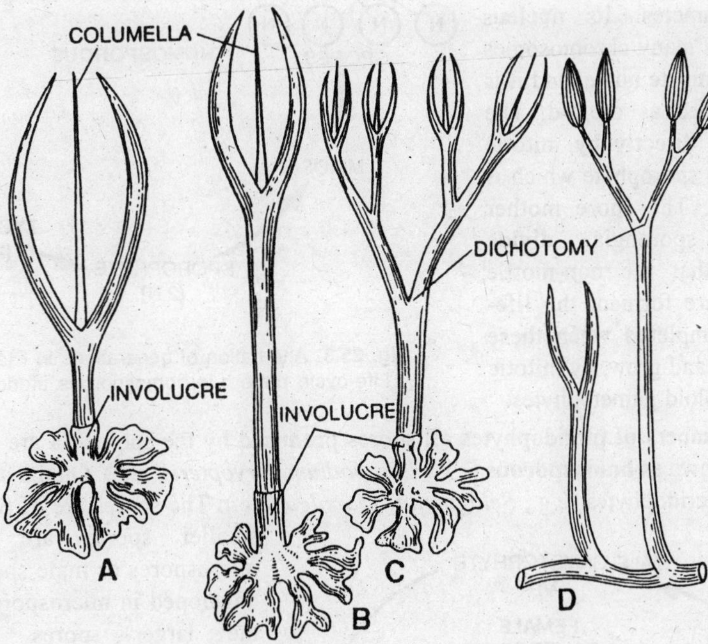

Fig. 25.2. Origin of the Pteridophytes. A-D, this scheme shows the origin of early land plants from *Anthoceros* (a bryophyte) like ancestors.

Characteristic features. The sporophyte is the conspicuous and familiar plant, while the gametophyte is always small and inconspicuous. In pteridophytes, the gametophyte is much less developed and simpler than that of the bryophytes. The sporophyte is a herbaceous plant differentiated into stems and leaves and nearly all cases have roots. In earlier stages of the development the sporophyte (embryo) is attached to the gametophyte by a foot and is dependent on the gametophyte for its nourishment, but later on, it becomes quite independent of the gametophyte, producing its own stems, leaves and roots. The sporophytes are long-lived plants which produce large number of sporangia and other vegetative organs.

Alternation of generations. There is a regular feature of alternation of generations. At the time of sporogenesis the reduction division takes place and the haploid spores are formed which represent the first stage of the gametophyte generation. Later on fertilization takes place and a zygote is produced which represents the first stage of sporophyte generation. This way the gametophyte gives rise to the sporophyte by the process of syngamy and the sporophyte gives rise to the gametophyte by the process of meiosis. The gametophyte of sexual generation intervenes between meiosis and syngamy while the sporophyte or asexual generation intervenes between syngamy and meiosis. The haploid phase (n) is the gametophyte and the diploid phase ($2n$) is the sporophyte. These two phases follow one another in the life-cycle, in regular succession.

Under normal circumstances there is a regular alternation between a gametophyte (sexual) phase and a sporophyte (asexual) phase. The male gametes, produced in numbers from antheridia, are known as antherozoids, since they are flagellated and are able to swim in water, while the female gametes (oospheres) are non-motile and are borne singly in flask shaped archegonia. Fusion between an egg cell and an antherozoid results in the formation of a zygote, which contains the combined nuclear material of the two gametes. Its nucleus contains twice as many chromosomes as either of the gamete nuclei and it is therefore described as diploid. The zygote develops directly by mitotic divisions into the sporophyte which is likewise diploid. The spore mother cells within sporangia divide meiotically so that the non-motile haploid spores are formed, the life-cycle is then completed when these spores germinate and grow, by mitotic division, into haploid gametophytes.

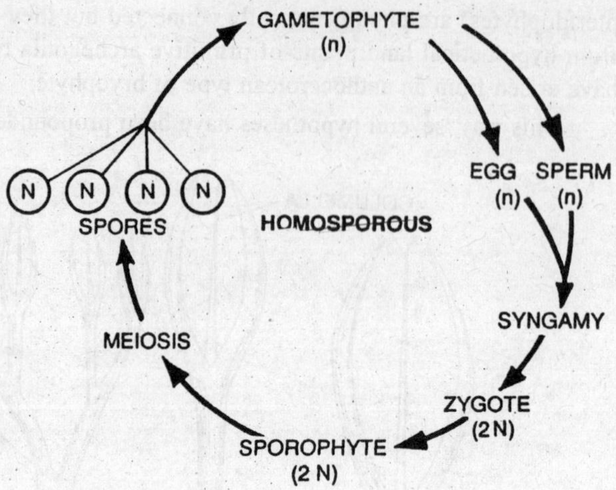

Fig. 25.3. Alternation of generations in Pteridophytes. Life-cycle pattern in homosporous pteridophytes.

In some members of pteridophytes all spores produced by the sporangia are of one type, such a plant is known as **homosporous** (*e.g.*, *Lycopodium, Dryopteris,* etc.). On the other hand, in **heterosporous** pteridophytes (*e.g.*, *Selaginella, Marsilea,* etc.). The spores are of two kinds-the smaller spores are known as microspores or male spores and are developed in microsporangia, while the larger spores are called megaspores or female spores and are formed in megasporangia. The microspores on germination produce the male gametophytes on which, the **sperms** are produced. The megaspores produce the female gametophytes on which, eggs are produced.

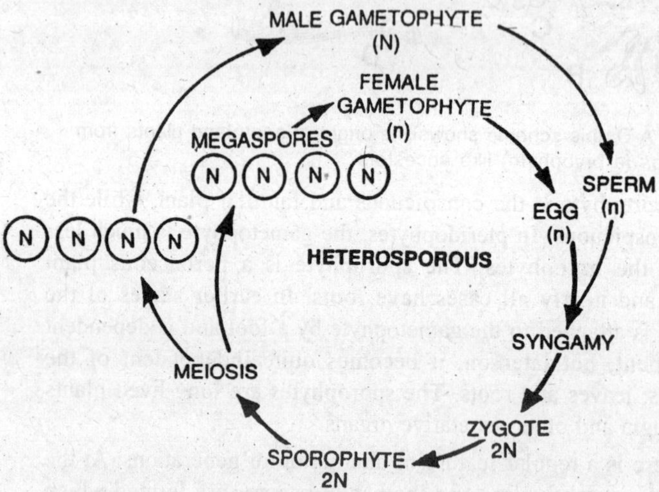

Fig. 25.4. Alternation of generations in Pteridophytes. Life cycle pattern in heterosporous pteridophytes.

In Bryophytes, the dominant phase in the life-cycle is the gametophyte, for the sporophyte is retained upon it throughout its life and is either partially or completely dependent on it for nutrition. Contrary to it in pteridophytes the sporophyte is the dominant generation, for it very soon becomes independent of the gametophyte (prothallus) and grows to a much greater size. The sporophytes are being differentiated into stems, leaves and roots. Only the sporophyte shows well developed conducting tissues (*i.e.*, xylem and phloem).

The Prothalli. The gametophytes, in some pteridophytes are subterranean and in others they are retained within the resistant wall of the spore and are thus able to survive in a much wider range of habitats. It is notable that wherever the gametophyte is retained within the spore the spores are of different sizes (heterosporous), the larger megaspores give rise to female prothalli which bear only archegonia, and the smaller microspores giving rise to male prothalli which bear only antheridia. The division between heterospory and homospory is one of the criteria used in the classification of pteridophytes.

The sporogonium. The sporangium, in all pteridophytes, is initiated by the laying down of cells. Since this wall is periclinal each initial cell is divided into an outer and inner daughter cell. If the sporogenous tissue is derived from the inner daughter cell, the sporangium is described as **'eusporangiate'** and if from the outer, as **'leptosporangiate'**. In leptosporangiate forms, the sporangium wall, stalk and the spores, are derived from the outer daughter cell, but in eusporangiate forms, the adjacent cells take part in the formation of part of sporangium wall and the stalk. In eusporangiate form the sporangium is large and massive, the wall is several cells in thickness and the spore content is high. In leptosporangiate forms, the sporangium is small, the wall is one cell in thickness and the spore content is low. Of these two types, the eusporangiate is primitive and the leptosporangiate advanced.

Classification. Firstly all the botanists divided the plant kingdom into two large groups-**Cryptogamia** and **Phanerogamia,** Linnaeus (1754) used the term **Cryptogams** for the first time for those plants which did not possess visible sex organs.

In 1880, the cryptogams of the plant kingdom were divided into three large divisions: Thallophyta, Bryophyta and Pteridophyta. The name-Thallophyta was first introduced by Endlicher in 1836, who called this division, a kingdom. Later on in 1866 Haeckel introduced the names Bryophyta and Pteridophyta.

For a long time the division Pteridophyta included-the ferns, lycopods and horsetails. On the basis of their anatomy Jeffrey in 1902 divided the vascular plants into two types or 'stocks'. He included ferns, gymnosperms and angiosperms in a 'stock' known as **Pteropsida,** and lycopods and horsetails in another 'stock' known as **Lycopsida.** Jeffrey, however, could not give the names divisions or sub-divisions to these stocks. Scott in 1923 ranked them as divisions. He seggregated the horsetails as a separate division, the **Sphenopsida**.

According to Eames (1936), Tippo (1942) and others there is a single division, the **Trachaeophyta** consisting of four subdivisions-**1. Psilopsida, 2. Lycopsida, 3. Sphenopsida** and **4. Pteropsida**.

In 1950, the International Code of Botanical Nomenclature amended and recommended that all names of divisions end in the suffix-phyta and that of sub-divisions end in the suffix opsida.

According to Smith (1955) the classification of the pteridophytes is as follows:

Division I	**Psilophyta**
Class1.	Psilophytinae
Division II	**Lepidophyta**
Class 2.	Lycopodinae
Division III	**Calamophyta**
Class 3.	Equisetinae
Division IV	**Pterophyta**
Class 4.	Filicinae.

The classification followed in the present text is as follows:

Division I	**Psilophyta**
Class 1.	Psilotopsida
Division II	**Lycophyta**
Class 1.	Ligulopsida
Class 2.	Eligulopsida

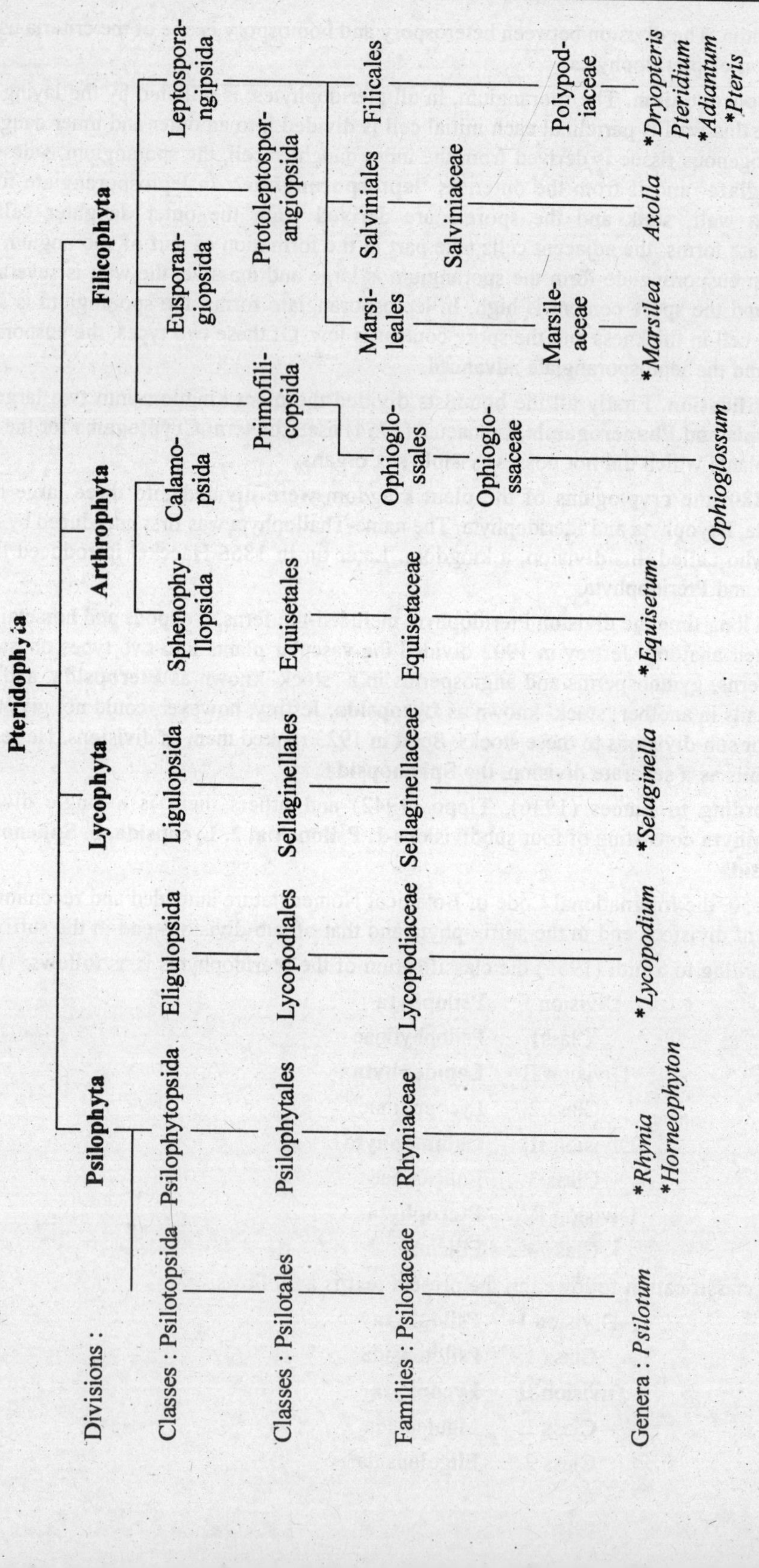

*Genera described in the text

Division III	**Arthrophyta**
Class 1.	Sphenophyllopsida
Class 2.	Calamopsida
Division IV	**Filicophyta**
Class 1.	Primofilicopsida
Class 2.	Eusporangiopsida
Class 3.	Protoleptosporangiopsida
Class 4.	Leptosporangiopsida

The characteristic features of these divisions are as follows:

Psilophyta. The sporophyte is differentiated into a rhizoid bearing subterranean rhizome and an aerial portion. The aerial portion is branched. The vascular system is of protostelic type. Leaf gaps are absent from vascular cylinder. The terminal sporangia are borne singly at the tips of short or long branches. The gametophyte is subterranean and colourless. They are homosporous. Antherozoids are multiciliate.

Lycophyta. The sporophyte is differentiated into stem, roots, and leaves. The leaves are microphyllous, and with a single vein. The vascular strands or steles may be protostelic, siphonostelic, or polystelic. The leaf-gaps are always absent, sporophylls produce a single sporangium on the adaxial side near its base. The sporophylls are borne in strobili. They are homosporous or heterosporous. The antherozoids are biflagellate or multiciliate.

Arthrophyta. The sporophyte is differentiated into stem, roots and leaves. The stem possesses distinct ridges and furrows. The foliage leaves are borne in transverse whorls upon stems and their branches. The vascular cylinder is protostelic or siphonostelic. The leaf-gaps are absent. The sporangia are produced upon a specialized structure, the sporangiophores present at the apex of the stem. The antherozoids are multiciliate. They are homosporous.

Filicophyta. The sporophyte is differentiated into stem, leaves and roots. With the exception of protostelic forms, the other siphonostelic forms possess leaf-gaps in their vascular cylinders. The leaves are macrophyllous. The leaf bears many sporangia on either the margin or the abaxial face of the leaf lamina. They are homosporous. The antherozoids are multiciliate. The sex organs are found on the ventral surface of the heart-shaped prothallus (gametophyte).

26
CHAPTER

Division-Psilophyta;
Class-Psilotopsida

Order-Psilophytales

There are about 25 known genera in this order. All genera are fossil. The fossil record indicates that the Psilophytales became extinct during the Upper Devonian. They however, lived from the Middle Silurian to the Upper Devonian. The fossil remains of the Psilophytales have been found in widely scattered regions, such as Canada, Scotland, France, United States of America, Australia, etc.

The characteristic features of the order are as follows:

The plant body of the sporophyte was rootless, dichotomously branched and was differentiated into a prostrate subterranean and an erect aerial portion. The aerial portion was leafless or with small simple leaves.

In majority of the genera, the sporangia were borne at the tips of the branches. However, a few genera had the sporangia in clusters, at the tips of their branches.

The stele was simple and protostelic type.

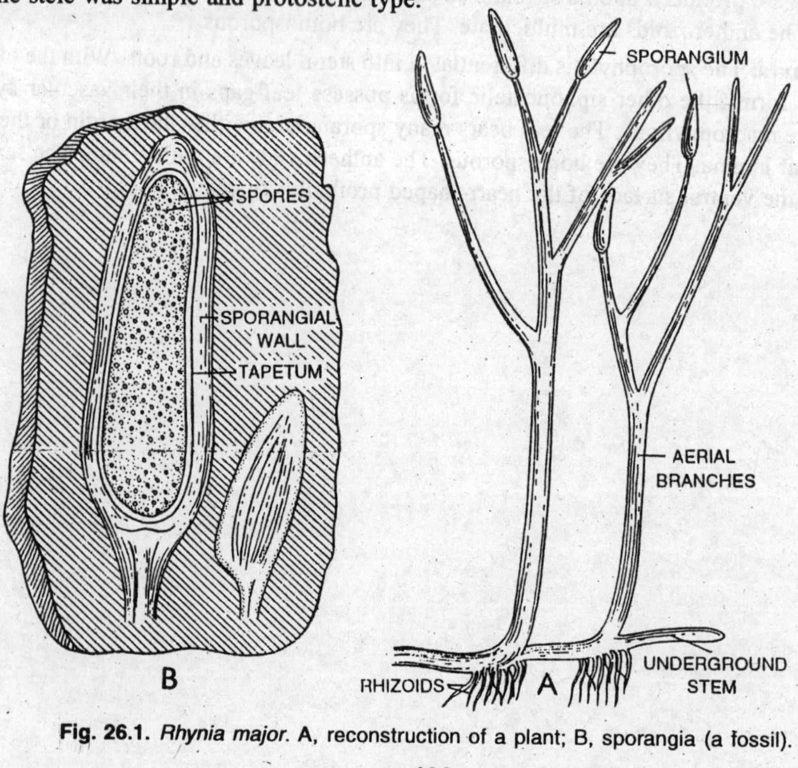

Fig. 26.1. *Rhynia major*. A, reconstruction of a plant; B, sporangia (a fossil).

Classification. The order Psilophytales has been divided into nine families. Some of the families consist of very imperfectly known genera. The important families are-Rhyniaceae, Psilophytaceae, Pseudoporochnaceae, Zosterophyllaceae, Asteroxylaceae, etc. Here the genera *Rhynia* and *Horneophyton* of Rhyniaceae have been described.

Genus RHYNIA

Systematic Position- Pteridophyta
Division-Psilophyta
Class-Psilophytopsida
Order-Psilophytales
Family-Rhyniaceae
Genus-*Rhynia*

Till now only two species of *Rhynia* named as *Rhynia major* and *R. gwynne-vaughani* have been described by Kidston and Lang (1917, 1921) from the Old Red Sandstone Beds of Middle Devonian. As the genus has been derived from the place Rhynie situated in Scotland and, therefore, it is named after the name of the place of its origin and known as *Rhynia*. A number of specimens of these two species have been found at Rhynie. The specimens are so well preserved that they give detailed information about the form and structure of this very primitive vascular plant.

Kidston and Lang (1921), thought and supported by evidences that in those times the plants grew in swampy marshes near the volcanoes.

Both the species of *Rhynia, i.e., R. major* and *R. gwynne-vaughani* were herbaceous plants. Comparatively the *Rhynia major* was larger in size. The plant body of both the species consisted of dichotomously branched horizontal rhizome and the erect, aerial dichotomously branched stem. The aerial stem of *R. major* were about 50 cms. high and 1.5 mm. to 6 mm. in diameter, whereas, the corresponding structures of *R. gwynne-vaughani* were 20 cms. in height and 1 to 3 mm. in diameter. The aerial branches were leafless shoots. The plants were rootless. Instead of the roots, the unicellular rhizoids were present in patches on the underside of the rhizome. The aerial branches were naked, leaf-less, cylindrical, dichotomously forked and tapering at their apices.

The terminal, elongated sporangia were found on the tapering vegetative apices.

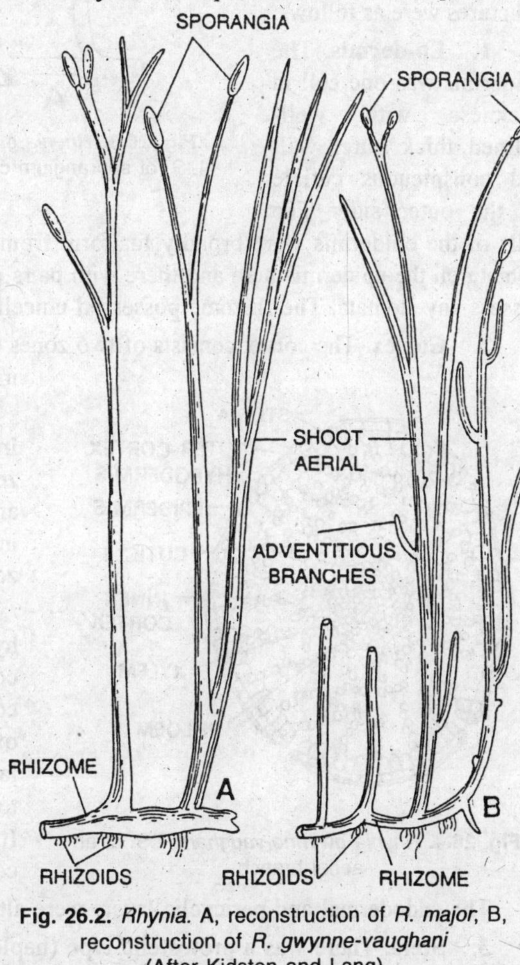

Fig. 26.2. *Rhynia*. A, reconstruction of *R. major*, B, reconstruction of *R. gwynne-vaughani* (After Kidston and Lang)

The stems of *R. gwynne-vaughani* bore Adventitious branches at the bases of aerial shoots and rhizome. It is thought that these branches got detached readily from the parent plant and helped in vegetative propagation.

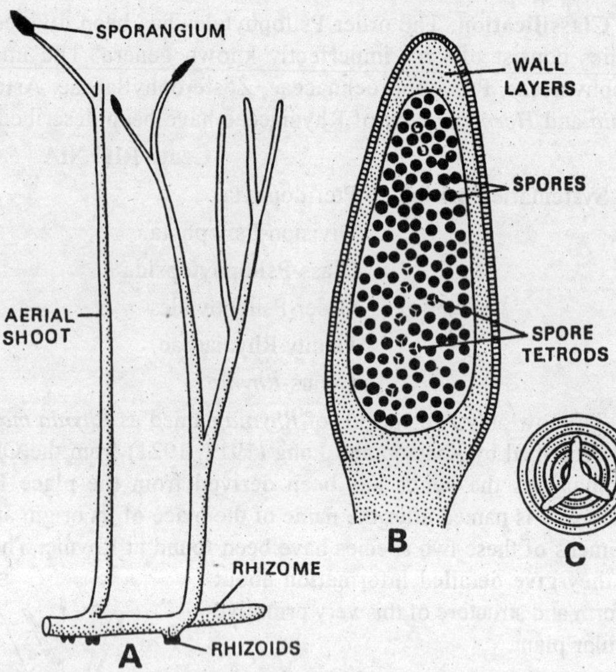

Fig. 26.3. *Rhynia gwynne-vaughani.*, A, reconstruction of plant; B, L.S. of sporangium containing spores and spore tetrads; C, a spore tetrad.

Anatomy. The internal structure of the rhizome and aerial stem was quite similar to each other. This was differentiated into epidermis, cortex and stele. The details of the structures were as follows:

1. Epidermis. The epidermis was one cell in thickness with well-defined thick outer wall and conspicuous cuticle on the outer side. The cells of the epidermis were broadly fusiform from the surface view. The aerial stem bore the stomata on the epidermis here and there with pairs of guard cells. However, the rhizome did not possess any stomata. The rhizome possessed unicellular rhizoids on its surface.

2. Cortex. The cortex consists of two zones (a) the outer cortex (hypodermis) and the (b) inner cortex.

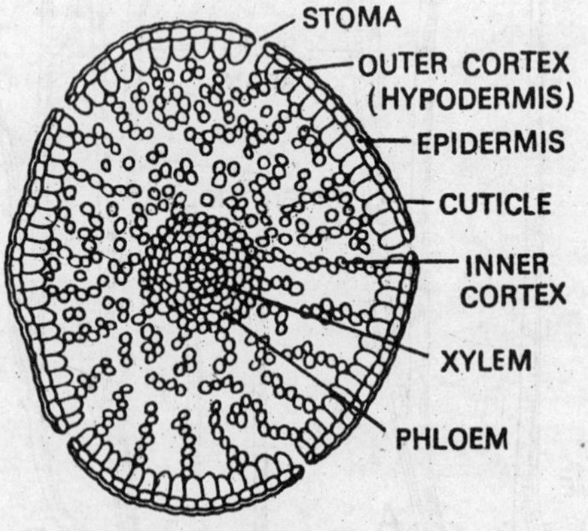

Fig. 26.4. *Rhynia gwynne-vaughani.* T.S. of an aerial branch.

(a) Outer cortex or hypodermis. Immediately beneath the epidermis there was a zone one to four cells in thickness consisted of angular parenchymatous cells having no intercellular spaces known as hypodermis. This zone was also known as the outer contex.

(b) Inner cortex. Inner to the hypodermis there was broad zone of inner cortex. In comparison of the central stele the cortex was ten fold. The inner cortex consists of rounded cells separated from each other by large intercellular spaces which communicate to the stomata through the gaps in hypodermis. It is also believed that this zone of the inner cortex acted as the chief photosynthetic tissue.

The endodermal and pericyclic layers were altogether absent.

3. Stele. There was a protostelic type (haplostele) of stele. The central xylem core was completely surrounded by phloem layer four or five cells in thickness. The xylem consists of tracheids only. In majority of the specimens the xylem had imperfectly preserved. According to

Kidston and Lang (1917), the tracheids had annular thickenings at their walls. Kidston and Lang (1920) have reported that in some of the specimens the tracheis were isodiametric, whereas in some other specimens the central tracheids were smaller in diameter and the peripheral ones larger in diameter.

The phloem consists of elongated thin walled cells with oblique end walls.

The reproductive structures (the sporangia). The elongated or oval sporangia were terminally developed at the apices of the branches of the aerial shoot. It is not clear whether the sporangia were borne on all the apices of the branches or on certain branches of the plant. Each apex bore a single sporangium.

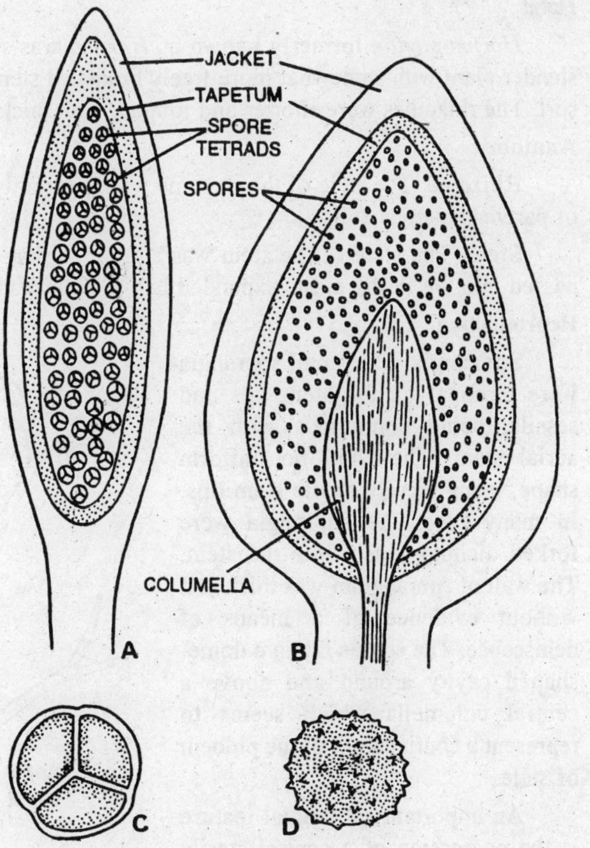

Fig. 26.5. *Rhynia gwynne-vaughani*. A, longitudinal section of sporangium; B, L.S. of sporangium of *Hornea* (*Horneophyton lignieri*); C, spore tetrad of *Rhynia*.

As regards the shape the sporangia were cylindrical and they possessed the greater diameter than those tips of the branches on which they were found. The jacket layer of the sporangium consists of several cells in thickness. The outermost cells of this layer were thick-walled and cutinized at their outer face. The middle layer or the sporangial wall possessed thin walled cells and inner-most layer consisted of thin walled rounded cells which probably acted as tapetal layer. The sporangium had no columella within it. The cavity contained many spores of the same size. The sporangia were indehiscent and it is thought that the spores were released only on the decay and disintegration of the thick sporangial wall.

The spores are spherical and cuticularized. They are found arranged in spore tetrads.

Genus HORNEOPHYTON

Systematic position. Pteridophyta
Division-Psilophyta
Class-Psilophytopsida
Order-Psilophytales
Family-Rhyniaceae
Genus-*Horneophyton*

Habit

Horneophyton formerly known as *Hornea* was similar to *Rhynia,* but was a smaller more slender plant with somewhat more freely branched stems, which were without appendages of any sort. The rhizomes were shorter and jointed, with thick tuberous segments.

Anatomy

Rhizome. The stele of the rhizome was apparently without vascular tissue consisting wholly of parenchymaıous cells.

Stem. The stele of the stem was similar to that of *Rhynia.* The steles of the aerial stems passed into the fleshy mass, expanded bell-like and faded out.

Reproduction

Spcrangia. The small sporangia were terminal on the stem tips and sessile, being continuous with the aerial stem. They had no uniform shape, being merely fertile stem tips. In many cases the sporangia were forked dichotomously like the stem. The wall of sporangium was thick and without evidence of a means of dehiscence. The spores lay in a dome-shaped cavity around and above a central columella which seems to represent a continuation of the phloem of stele.

An important sporangial feature is the possession of a central sterile column of cells called the **columella.** Thus, the spore bearing layer overarches the columella, as found in *Sphagnum* like mosses of today. This feature has strongly suggested that *Horneophyton* and its allies, though vascular in nature may have had affinities with the moss stock.

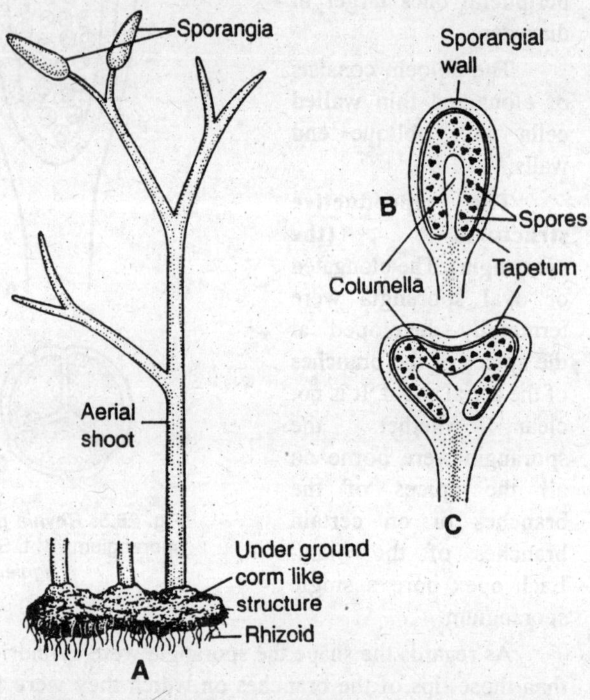

Fig. 26.6. *Horneophyton.* A, external structure; B and C, L.S. of sporangium.

In forked sporangia the columella too is forked, the spores are formed in tetrads and are 50μ in diameter. The cells of the columella are elongated and continuous with the phloem of the stem below. It served for the conduction of food as in the recent plants. Tapetum was a nourishing tissue for the developing spores.

Affinities. "The Rhyniaceae, simple as they are, show such a remarkable combination of characters and they have been placed by different botanists in three of the great sub-kingdoms of plants. Kidston and Lang, assign them to the pteriodophytes, because the spore bearing generation was a vasculer plant; Dr. Newell Arber, in whose time only *Rhynia* had been described referred the genus to Thallophytes, on the ground of its undifferentiated thalloid stem; the family has also been included under Bryophytes, on account of characters presented by *Horneophyton* and *Sporogonites.* This is a very extraordinary position and shows that Rhyniaceae are a unique race of plants," (D.H. Scott).

27
CHAPTER

Division-Lycophyta

Class-Eligulopsida

Order-Lycopodiales

Characteristic features. The living Lycopodiales are the representatives of a group which, during the Carboniferous period, formed the chief vegetation. Many of the types then growing, *e.g.*, *Lepidodendron*, were large trees.

The modern representatives are small and herbaceous sporophytes. The leaves are small and simple. Each leaf possesses an unbranched midrib. The leaves have no ligules. There are no leaf gaps in the stele of the stem. The sporophylls may or may not be restricted to the terminal portion of branches and organized into definite strobili. The sporophylls and simple vegetative leaves may be similar or dissimilar. They possess homosporous sporangia, *i.e.*, all the spores of one kind only. The gametophytes are wholly or partly subterranean. The antheridia remain embedded in the tissue of the prothallus. The antherozoids are biflagellate.

Classification. The class Eligulopsida includes the single order Lycopodiales. The living Lycopodiales contain a single family, the Lycopodiaceae. The Lycopodiaceae include two living genera *Lycopodium* and *Phylloglossum*. Genus *Lycopodium* has been described here in detail.

Genus LYCOPODIUM

Systematic Position. Pteridophyta
Division.Lycophyta
Class. Eligulopsida
Order. Lycopodiales
Family. Lycopodiaceae
Genus. *Lycopodium*

There are about 180 species in this genus.

Distribution and habit. The species of *Lycopodium* are world-wide in distribution. They are mainly found in tropical and sub-tropical forests. They are very commonly found on heaths and on humus soils in moist shady places. In India they are found in the hills of Eastern Himalayas. The plants are commonly known as **'ground pines'**, **'club mosses'** and **'trailing evergreens'** Many species occur in the tropics as hanging epiphytes (*e.g.*, *Lycopodium phlegmaria* and *L. squarrosum*). Chowdhury (1937) has reported eight species from our country. These species are: *Lycopodium clavatum; L. cernuum, L. heamiltonii; L. setaceum; L. phlegmaria; L. wightianum; L. serratum* and *L. phyllanthum*. The most common species is *L. clavatum*.

THE SPOROPHYTE

Habit. All species possess small, herbaceous or shrubby sporophytes. The stem in almost all the species is delicate and weak. Some species are epiphytic and with erect or pendant sporophytes while other species are terrestrial and have a trailing habit. The stem and its branches are densely covered with small leaves. *Lycopodium phlegmaria* is an epiphytic species.

According to Pritzel (1900) this genus is divided into two sub-genera which differ from each other in general organization of the sporophyte. These subgenera are-*Urostachya* and *Rhopalostachya*.

Stems. Species referred to the sub-genus *Urostachya* possess branched or unbranched stems that are erect or pendant but never creeping. This subgenus includes the species, *e.g., L. selago, L. lucidulum, L. phlegmaria* and others. If the stem is branched, the branching is always dichotomous. Usually the successive dichotomies are found at right angles to one another. The species belonging to this subgenus do not bear the adventitious roots along the stem.

The subgenus *Rhopalostachya* includes the species, *e.g., L. clavatum, L. inundulatum, L. complanatum, L. cernuum* and others. The species referred to this sub genus possess prostrate stems bearing upright branches. The branching of the stem in the first-formed portion may be dichotomous, but in later-developed portion it is always monopodial.

Leaves. The leaves are small, simple, sessile, numerous and cover the axis closely. Typically the leaves are 2 to 10 mm long. Usually the leaves are arranged in closed spirals (*e.g., in L. clavatum* and *L. annotinum*) while in other cases they are arranged in whorls (*e.g.,* in *L. verticillatum* and *L. cernuum*). In some species the leaves are found to be arranged in opposite pairs (*e.g., L. alpimum*); in others they are irregularly arranged. Usually the leaves are lanceolate

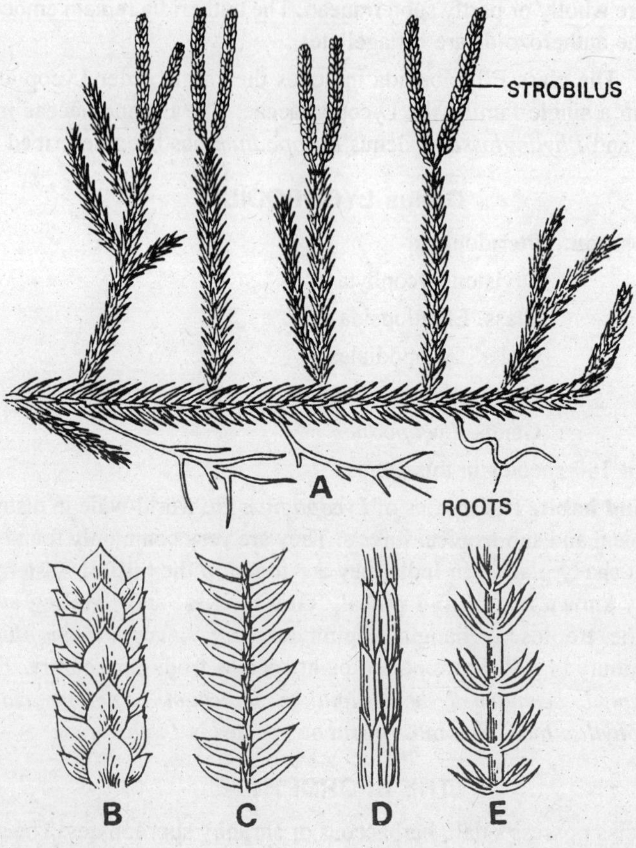

Fig. 27.1. *Lycopodium*. A, part of a plant of *L. clavatum* showing strobili; B-E, leaf form and arrangement in *Lycopodium*; B, *L. refescens*; C, *L. volubile*; D, *L. complanatum*; E, *L. cernuum*.

in outline possessing the broad bases. Usually the first-formed leaves on the stem do not bear the mid-rib, the later formed leaves usually possess a single unbranched vein that starts from the base but does not reach the apex. The leaves are all alike in size and shape, but in few species, such as *L. complanatum, L. volubile* and *L. chamaecyparissus* the leaves are **dimorphic** like that of *Selaginella* and found to be arranged in four vertical rows on the stem two being smaller and two larger in size.

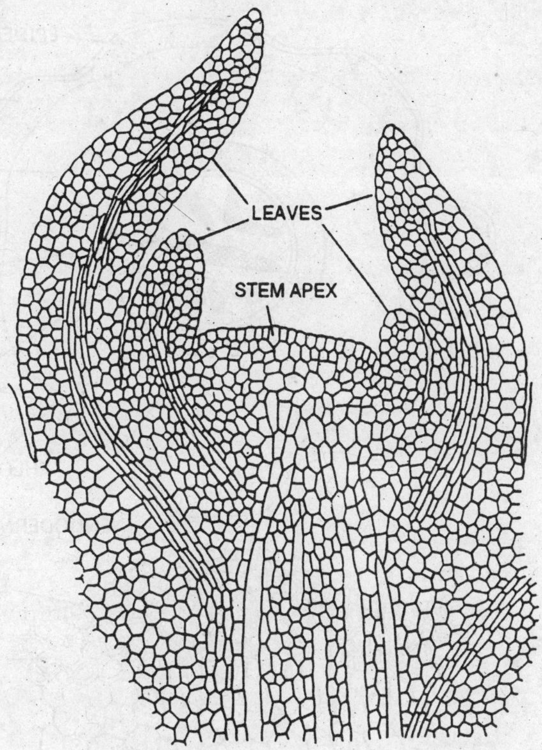

Fig 27.2. *Lycopodium* spp. L.S. of stem apex.

Apical growth. The apical growth of the shoot takes place by means of an apical meristem which consists of a group of apical cells.

Roots. The first root is short and atleast in some species does not live long. The older plant bears the adventitious roots that arise singly or in groups acropetally along the lower side of the stem. The species of subgenus *Urostachya* do not possess adventitious roots along the stem. The species of the subgenus *Rhopalostachya* with a creeping stem generally have their adventitious roots borne along the entire length of the prostrate portion.

In some species, *e.g., L. selago, L. phlegmaria* and others the roots that arise on the outside of the stele do not penetrate the cortical region of the stem at once. These roots turn downward and penetrate the soft middle cortex making canals through it, and ultimately they emerge only at the stem. Such roots are known as **'cortical roots'** or **'inner roots'**.

In some species, (*e.g., L. obscurum* and *L. lucidulum*) the branching of the roots is strikingly dichotomous. Here each successive forking is found at right angles to the preceding one. In many species the dichotomy is obscure. The endogenous lateral roots do not arise from the roots (as in seed plants). In terrestrial form the root hairs are found abundantly which persist over a long period.

Anatomy of the stem. A transverse section of the mature stem of *Lycopodium clavatum* shows the following structures:

The epidermis is one cell in thickness and has stomata that appear to be similar in structure to those of the leaf. The outer walls of the epidermal cells are thick and cutinized.

The cortex is quite broad and it varies greatly in relative thickness from species to species. In some species its radial thickness is several times that of the stele; in other species the two are approximately equal. The cortex has three zones. The peripheral and central zones consist of thickened sclerotic cells while the middle zone consists of larger and thin-walled cells containing a few chloroplasts.

The single stele occupies about half the area of the section. It consists of a number of irregularly shaped, parallel strands of xylem alternating with phloem bands. The xylem consists

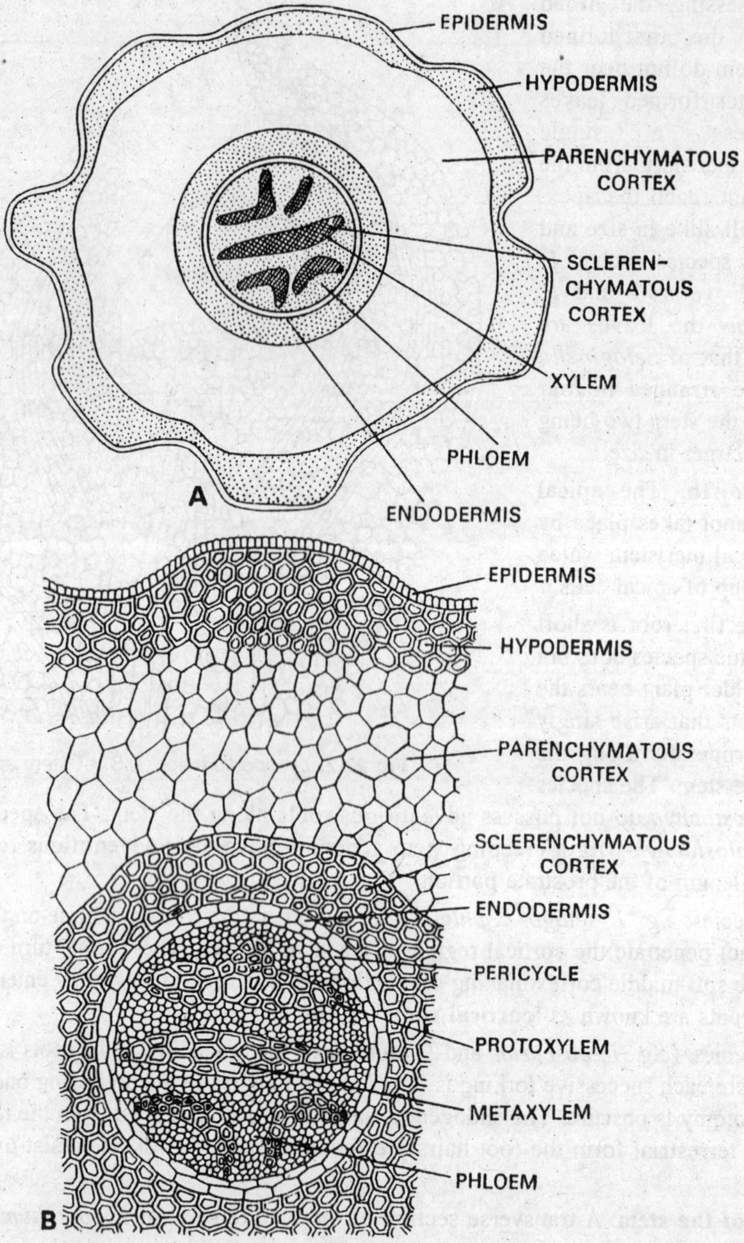

Fig. 27.3. *Lycopodium* sp. A, transverse section of stem (diagrammatic); B, transverse section of stem (detailed).

mainly of scalariform tracheids, with no vessels and no parenchymatous cells. Each Xylem plate remains surrounded by a single layer of parenchyma, with bordered pits on the inner walls, which lies between the xylem and the phloem. The phloem is composed of sieve tubes and parenchyma only. The development of both xylem and phloem strands is centripetal, that is, from the periphery towards the cente. The protoxylem consists of spiral and annular tracheids. The protoxylem is found to be situated at the outer edges of each xylem mass, that is to say, in the exarch position.

At the inner face of the cortex there may be an endodermal layer which, at least in younger portions of the stem has characteristic thickened radial walls.

The endodermis of *Lycopodium* is thought to be originated from the stele and not from the cortex. Internal to the endodermis there lies a multilayered pericycle. The pericycle layer is usually three to six cell in thickness.

There is a great variation of organization of the vascular elements, not only in different species but even within the same species may vary both from individual to individual and in different portions of the same individual. In stems the vascular elements are found to be arranged into an exarch actinostele resembling to the actinosteles characteristic of roots of vascular plants. In such steles the phloem lies in the space between the xylem plates. For example, in *L. phlegmaria,* the adult stem has the exarch actinostelic organization with rays of varying number. In *L. serratum* the xylem rays are expanded outwards like fan blades in appearance. According to J.B. Hill (1914), the species with actinostelic type of organization in the older portions of the stem are thought to be the most primitive of the genus as far as the vascular organization is concerned. Other species may have plectostelic organization of the vascular elements. In such stems the xylem core is found in the form of plate like lobes. Here the xylem and phloem bands are found to be arranged symmetrically in alternating transverse bands across the stele (*e.g.,* in *L. volubile* and *L. wightianum*). In still other species the stelic organization may have masses of xylem and phloem indiscriminately mixed with one another, *e.g.,* in *L. cernuum*. Holloway (1910), Jones (1905) hold the view that the plectostelic organization is found chiefly in horizontal portions of the plant and the actinostelic in erect portions. The stele lacks a cambium in all cases and hence there is no secondary growth.

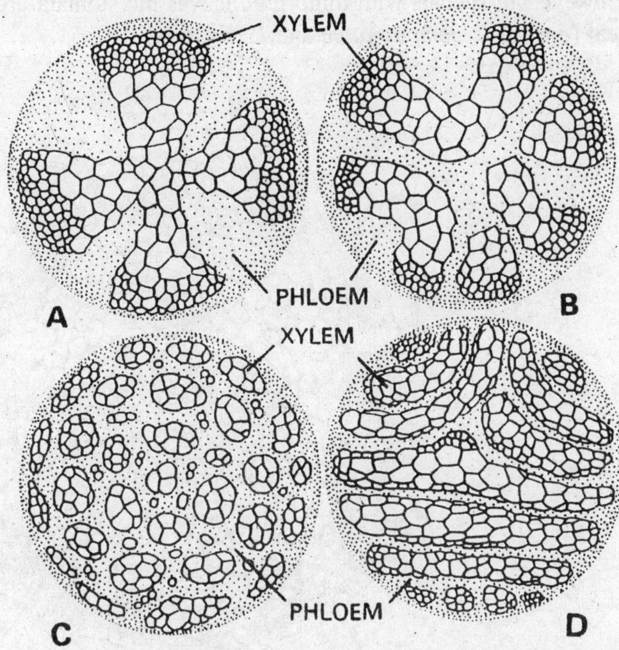

Fig. 27.4. *Lycopodium* sp. Transverse section of steles of various species of *Lycopodium*-A, *L. serratum*; B, *L. annotinum*; C, *L. cernuum*; D, *L. volubile.*

Anatomy of leaf. The transverse section of leaf shows following parts.

The leaf has a mid-rib which consists of a single concentric amphicribral bundle. The central xylem core is quite small. It is not always differentiated into meta-and protoxylem. The xylem is composed of thick walled annular and spiral tracheids. The phloem is composed

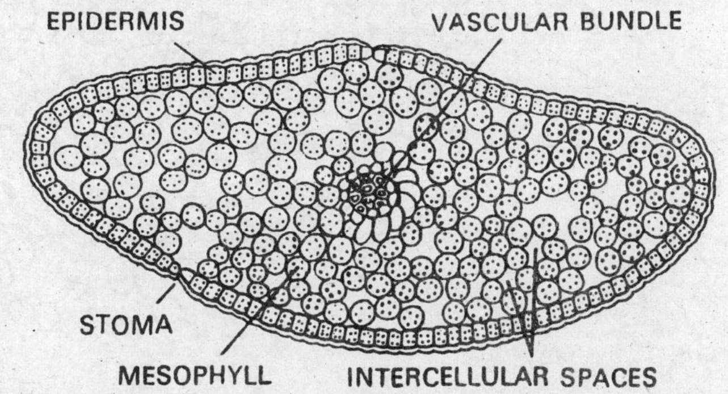

Fig. 27.5. *Lycopodium clavatum.* Transverse section of leaf.

of phloem parenchyma and narrow sieve-tubes. The endodermis is inconspicuous and may hardly be recognized. The vascular bundle remains surrounded by sclerenchymatous pericycle.

The mesophyll tissue lies between the vascular bundle and the epidermis. The epidermis of most species has stomata on both faces of a leaf (*e.g., in L. clavatum, L. selago* and others). However, in species with dimorphic leaves the stomata are frequently found on one face of the leaf (*e.g.,* in *L. complanatum* and *L. volubile*).

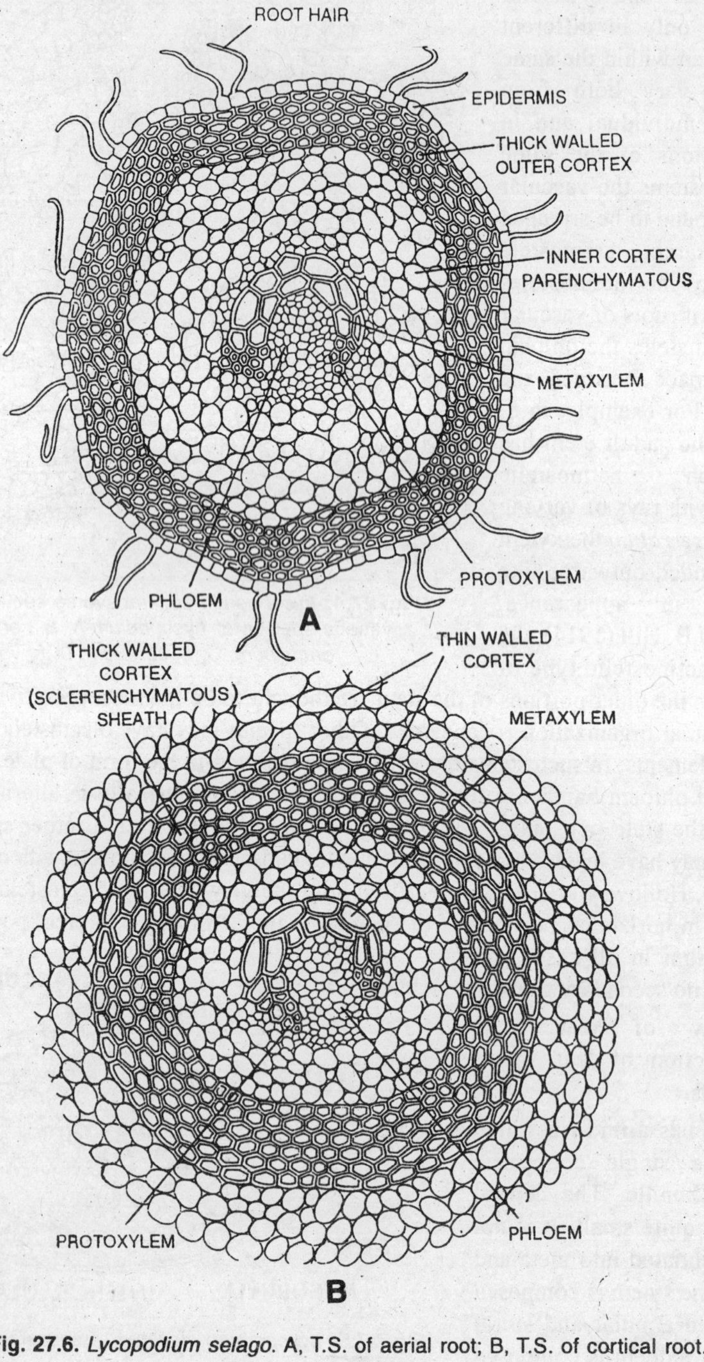

Fig. 27.6. *Lycopodium selago.* A, T.S. of aerial root; B, T.S. of cortical root.

Anatomy of root. The transverse section of the root shows the following parts:

The root is differentiated into epidermis, cortex and stele like other vascular plants. The single layered epidermis gives rise to numerous root hairs. The root hairs are formed to be arranged in pairs. According to Stokey (1907), the occurrence of root hairs in pairs results from the fact that the hair initials are found in twos by the oblique or anticlinal division of a young epidermal cell.

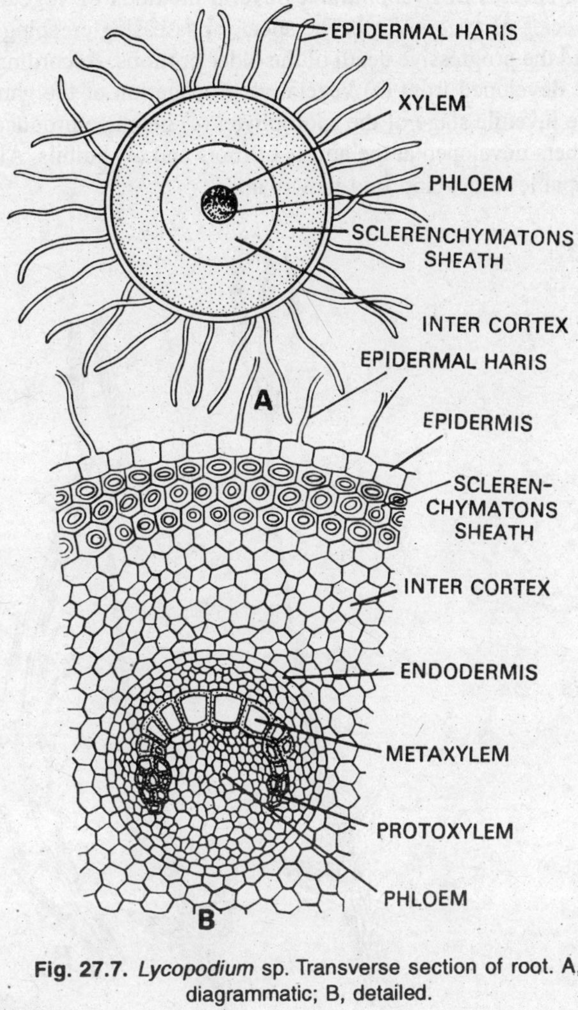

Fig. 27.7. *Lycopodium* sp. Transverse section of root. A, diagrammatic; B, detailed.

Just beneath the epidermis there lies the broad cortex. The cortex is several cells in thickness and often becomes greatly sclerified in the outer half as the root becomes older.

In some cases the stele is **monarch** with the protoxylem in one mass, *e.g.*, in *L. clavatum*. In other cases the stele is **diarch** with two protoxylem masses, *e.g.*, in *L. selago*, *L. reflexum* and other species. Here the two xylem masses are united into one continuous xylem group which is C-shaped or horseshoe shaped. In still other cases the stele is **triarch** with three groups of protoxylem. According to Saxelby (1908), sometimes the same root is diarch in one portion and tetrarch in another (*e.g.*, in *L. selago* and others). Most roots are diarch. These roots generally have but one mass of phloem and this lies between the protoxylem points of the C or U shaped xylem.

The spore bearing organs. In most species of the subgenus *Urostachya*, the foliage leaves and sporophylls are approximately of the same size and both are green. However, there is variation in the distribution of foliage leaves and sporophylls. In *Lycopodium selago* and many other species there are alternate sterile and fertile regions throughout length of stem and branches. These portions of the plant are called **strobili.** The strobili of different species are varied from each other. The strobili of *L. phlegmaria* and other related species are dichotomously branched and here the sporophylls, although green and of the same shape are comparatively much smaller than the foliage leaves.

All species of the subgenus *Rhopalostachya* have the sporophylls in well defined strobili which may be simple or branched. The sporophylls may be differentiated from foliage leaves in that they are smaller in size, paler in colour and have a dentate margin. In some species, *e.g., L. clavatum* and *L. complanatum* the strobili are usually borne on erect elongated stalks which remain covered with minute scaly leaves. In certain species the strobili are sessile at the apices of ordinary leafy shoots (*e.g.*, in *L. inundatum, L. alpinum,* etc.)

Vegetative propagation of the sporophyte. The vegetative propagation is frequently found in the species of *Lycopodium*. Several methods of vegetative propagation are found in different species of *Lycopodium*. The species possessing creeping stems have the apical growth branching and the progressive death of the older portions. According to Holloway (1917) the new plants may be developed from (a) vegetative propagation of the gametophyte, (b) vegetative propagation of the juvenile stage of the sporophyte, (c) gemmae produced from the cortical cells of the root, (d) tubers developed at the apices of roots and (e) bulbils. All above mentioned vegetative bodies are capable to develop into new plants.

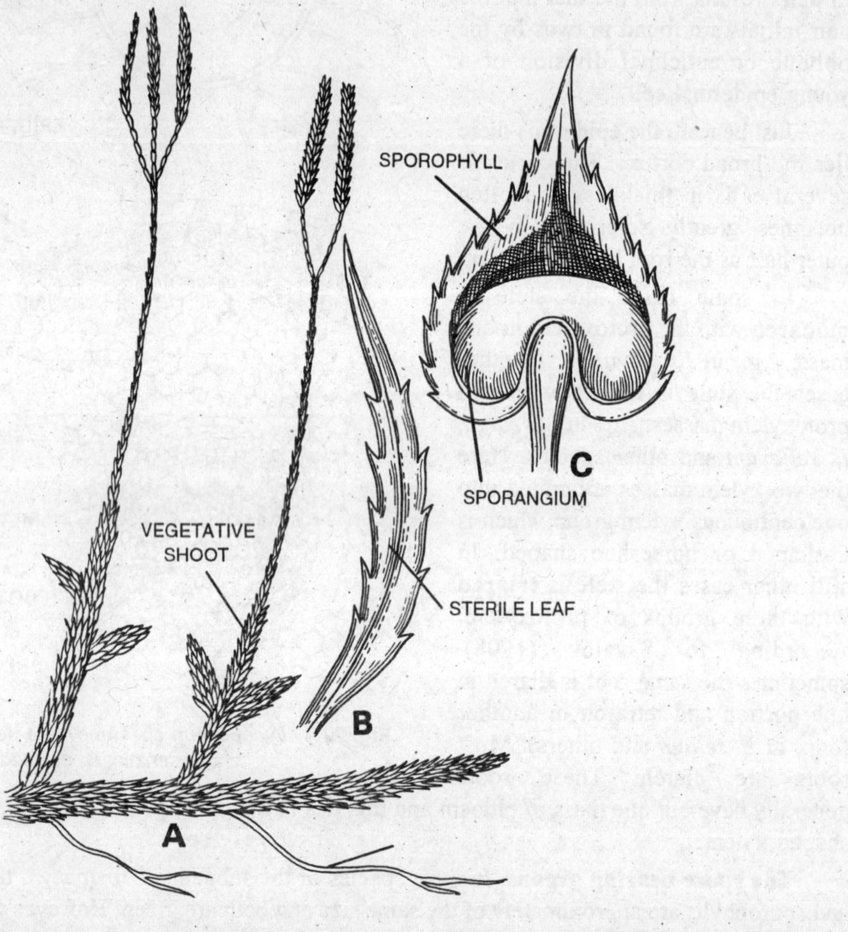

Fig. 27.8. *Lycopodium* spp. A, shoot bearing strobili; B, sterile leaf; C, sporophyll with sporangium.

The bulbils which are found only in species belonging to subgenus *Urostachya* resemble superficially to the protocormous stage (juvenile stage) of young sporophytes. The bulbils may develop any where on the stems. According to R.W. Smith (1920), these bulbils are interpreted as modified leaves while other workers interpret them as modified lateral branches or modified sporangia.

A sporophyll produces but one sporangium. The sporangium remains adaxial in its position and found near the base of the sporophyll. A mature sporangium is always kidney-shaped. Each sporangium possesses a long and slender or short massive stalk, according to species.

Development of sporangium. The sporangia begin to develop at a time when the sporophyll is composed of embryonic cells. First of all there is a periclinal division of a small group of epidermal cells on the adaxial side of a sporophyll and near its base (Bower, 1894). The inner daughter cells formed by this periclinal division give rise to the stalk and the basal portion of a sporangium. The outer daughter cells contribute to the formation to the bulk of the sporangium. The outer cells again divide periclinally forming an outer layer, the **jacket initials;** and an inner layer, the **archesporial cells**. The archesporial cells divide periclinally and anticlinally forming a massive **sporogenous tissue**. The cells of the last generation of the sporogenous tissue act as **spore mother cells**. They become rounded and are being separated from one another. Now these spore mother cells float about in a viscous liquid and divide meiotically into tetrads of spores.

The jacket initials, which are found external to the sporogenous tissue divide repeatedly forming a jacket layer of three or more cells, in thickness. Shortly before the development of the spore mother cells a nutritive **tapetal layer** is formed around the sporogenous tissue. This layer is partly formed from the inner-most layer of jacket cells and partly from sporangial cells found just beneath the sporogenous tissue. As found in most other pteridophytes, in *Lycopodium* there is no disintegration of the tapetum during spore formation.

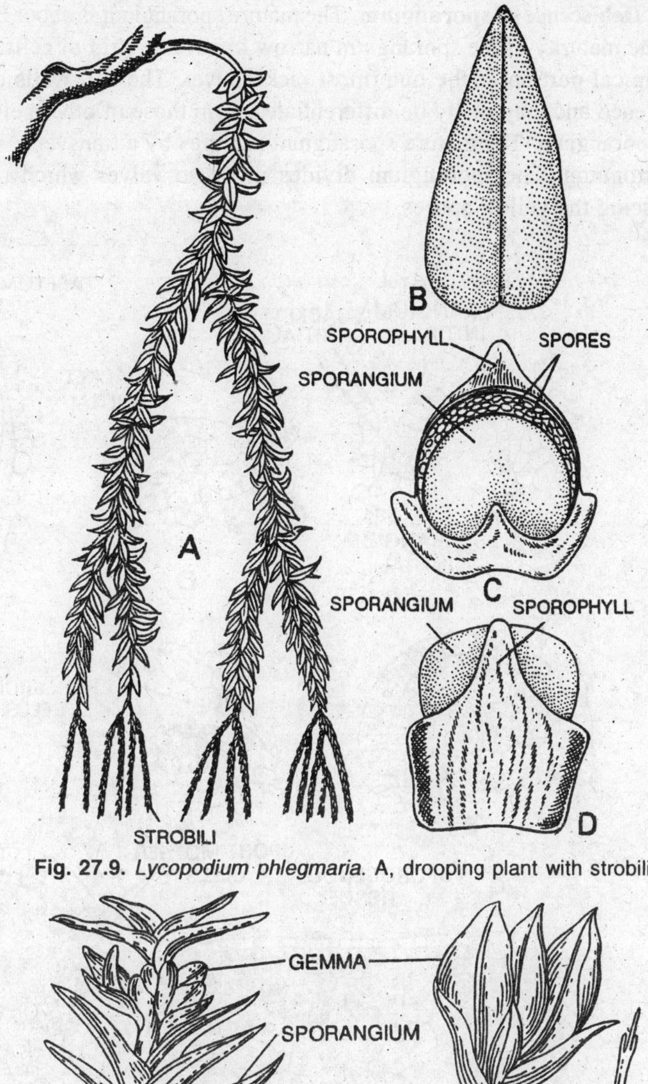

Fig. 27.9. *Lycopodium phlegmaria*. A, drooping plant with strobili;

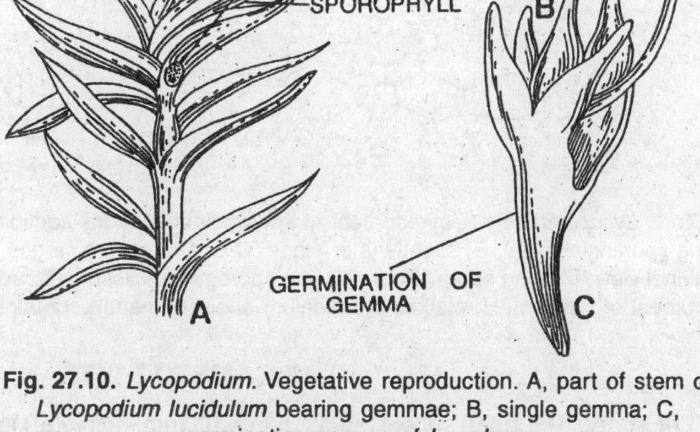

Fig. 27.10. *Lycopodium*. Vegetative reproduction. A, part of stem of *Lycopodium lucidulum* bearing gemmae; B, single gemma; C, germinating gemma of *L. selago*.

Dehiscence of sporangium. The mature sporangium is about 2 mm across and kidney-shaped. On the maturity of the sporangium narrow transverse strip of cells, the **stomium** is formed across the apical portion of the outermost jacket layer. The cell walls of the stomial portion become thickened and may easily be differentiated from those of other cells present in the jacket layer of the sporangium. The mature sporangium ruptures by a transverse slit at its apex along the line of the stomium. The sporangium divides into two valves which remain united at the base and dehiscing the yellow spores.

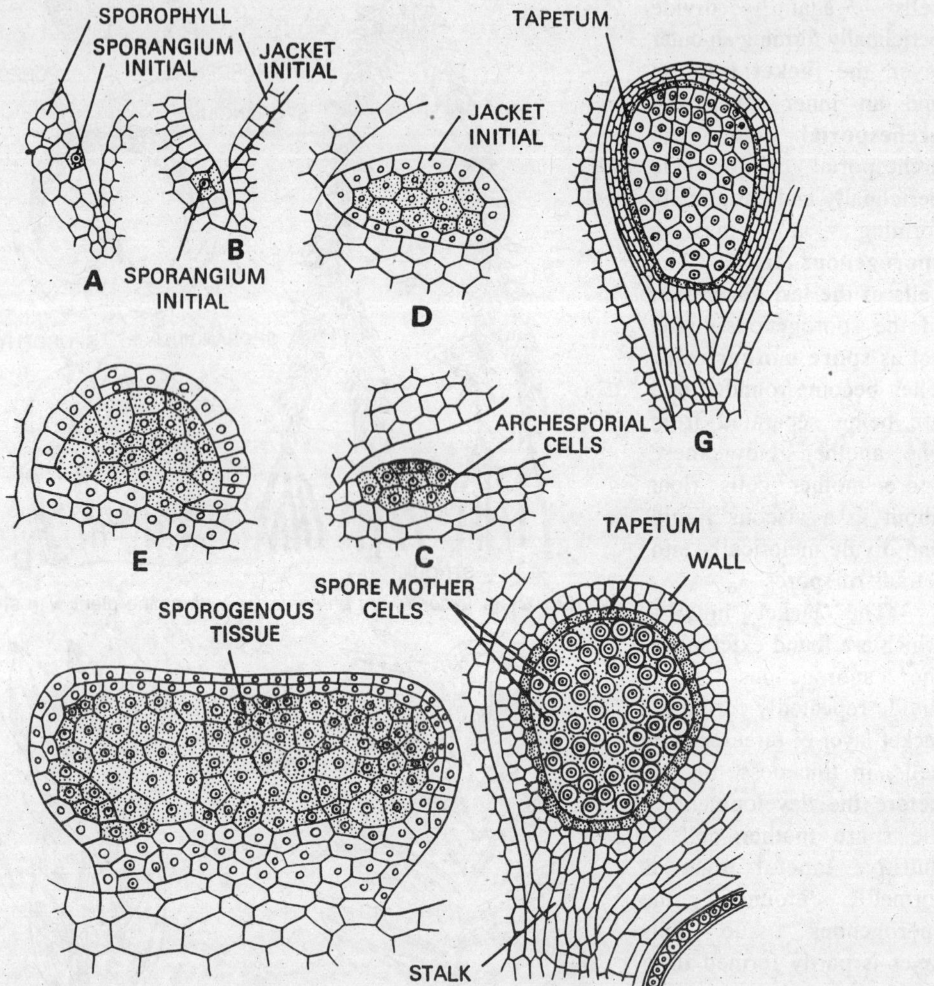

Fig. 27.11. *Lycopodium* sp. Development of sporangium-A, radial section through base of young sporophyll showing sporangial initial, B, slightly later stage; C-E, successive stages in the development of sporangium; F, sectional view of young sporangium showing sporogenous tissue; G, radial section of a later stage, showing development of tapetum; H, stalked sporangium about to mature containing spore mother cells.

THE GAMETOPHYTE

The spore. The small spores (about .03 to .05 mm. diameter) are uniform in size and shape, ***i.e.***, **homosporous.** This way *Lycopodium* is similar to most of the ferns and *Equisetum*. The spores are round or tetrahedral in shape. They are very light possessing thin exine formed of concentric lamellae, and a granular intine. Lustner (1898) classified the spores of *Lycopodium* into three groups. They are as follows:

1. Netzsporen. This type of spores possess an outer wall with reticulate ridges (*e.g.*, in *L. clavatum, L. complanatum* and *L. annotinum*). 2. **Tupfelsporen**. This type of spores possess knob-like outgrowths from the surface (*e.g.*, in *L. selago* and *L. phlegmaria*). 3. This is a transitional type possessing a fine ridged pattern (*e.g.*, in *L. cernuum* and *L. inundatum*).

Germination of spore and development of prothallus. The spores settle on the ground after their liberation from the sporangium and each germinates into a prothallus. Sometimes the germination of spore may, however, be delayed for years. According to Treub (1888), the spores germinate within a few days after they are shed (*e.g., in L. cernuum, L. inundatum* and *L. salakense*). In *L. clavatum* the spore germination begins after three to eight years of their shedding. Here the prothallus is a subterranean structure, solid and tuberous which takes from six to fifteen years to mature. The prothallus of *L. clavatum* is colourless and saprophytic and contains a mycorrhizal fungus.

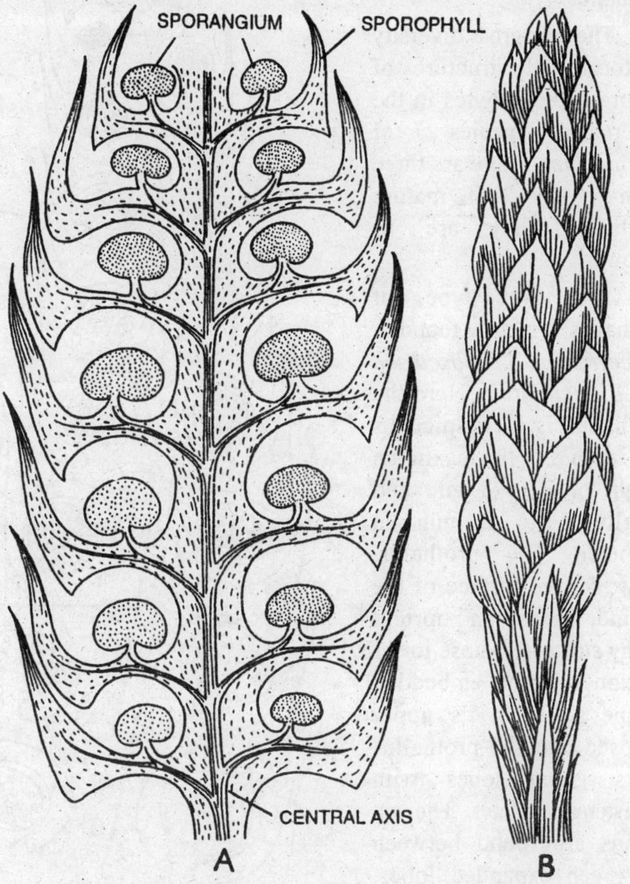

Fig. 27.12. *Lycopodium clavatum.* A, L.S. of strobilus bearing sporophylls and sporangia; B, a strobilus.

According to Bruchmann (1910), the first cell division of a germinating spore takes place before the outer spore wall (exine) is ruptured and results in two hemispherical daughter cells of approximately equal size. The lower or inner daughter cell also known as **basal cell** cuts off a small lens shaped rudimentary **rhizoidal cell** while the outer daughter cell divides twice successively producing an **apical cell** with two cutting faces. In the species with subterranean colourless prothalli a long period of rest, about one year is required between the five-celled stage and the mature prothallus. Upto four to five celled stage the early growth of the prothallus takes its nourishment from the reserve material contained in the mature spores. Thereafter 4-6 celled gametophytes may have an entrance of a symbiotic phycomycetous fungus into the basal cell. If there is no entrance of the fungus there is not further development of the gametophyte. After the establishment of the endophytic mycorrhiza within the cells of the young prothallus, the apical cell cuts of some half dozen segments and then becoming replaced by a group of meristematic cells. The segments thus cut off by the apical cell divide periclinally, and the outer cells thus formed become infected with the symbiotic fungus in the same way as does the basal cell. The apical group of meristematic cells, which are formed early in the development of a gametophyte produces the major portion of the adult prothallus.

There is great diversity in form and structure of mature gametophytes in the different species of *Lycopodium*. There are three main types of mature prothalli. They are as follows:

The first type of prothallus is mainly found in *L. cernuum, L. obscurum* and *L. inundatum*. Here the prothallus is inconspicuous and reaches to the maximum height of 2 to 3 mm and usually 1 to 2 mm in diameter. The prothallus grows on the surface of the ground. It is an upright fleshy structure whose lower portion remains embedded in the soil. On its upper exposed part the prothallus bears green lobes with meristematic plates. The sex organs are found between the green expanded lobes. The prothallus can manufacture its own food, though an endophytic mycorrhiza is always present. Rhizoids are also produced from the prothallus.

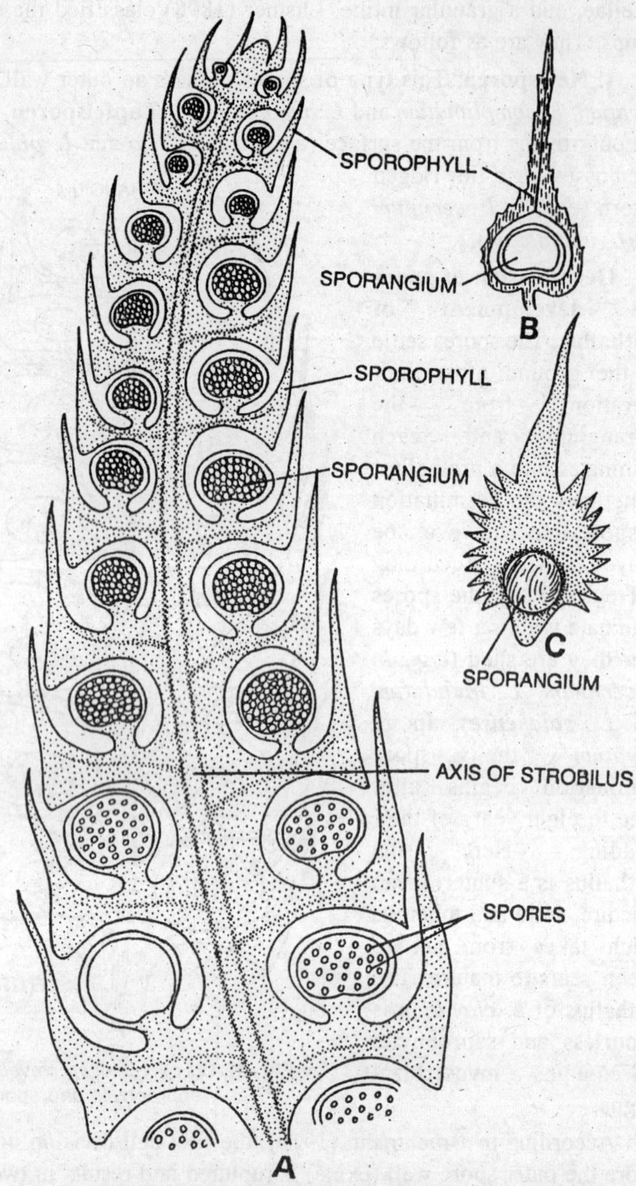

Fig. 27.13. *Lycopodium* spp. A, longitudinal section of the terminal portion of strobilus; B, sporophyll with sporangium of *L. clavatum*; C, sporophyll with sporangium of *L. cernuum*.

The second type of prothallus is found in the species, *e.g., L. clavatum, L. complanatum, L. obscurum* and *L. annotinum*. The prothalli are of various shades. They are brownish yellowish or colourless. They are saprophytic and found underground at various depths from one to eight centimetres. Much food material remains deposited in these tuberous prothalli. In *L. clavatum* the prothallus is much convoluted structure 1 to 2 cm. long with a pointed basal part. The surface is covered with an epidermis. Inside the epidermal layer there is a cortex of several layers. Just beneath the cortex there lies a palisade tissue composed of a single layer of elongated cells. These outer layers contain the endophytic fungus. This region is followed by a central tissue of hexagonal cells which store up food material. Long unicellular **rhizoids** are formed on the under surface of the prothallus. The sex organs are borne on the top surface of the prothallus. It is called the generative tissue. The cells of this tissue do not contain any food reserves and they remain meristematic for some time. The antheridia develop in the central portion and the archegonia towards the rim. The antheridia develop first and the archegonia appear later. The development is centripetal.

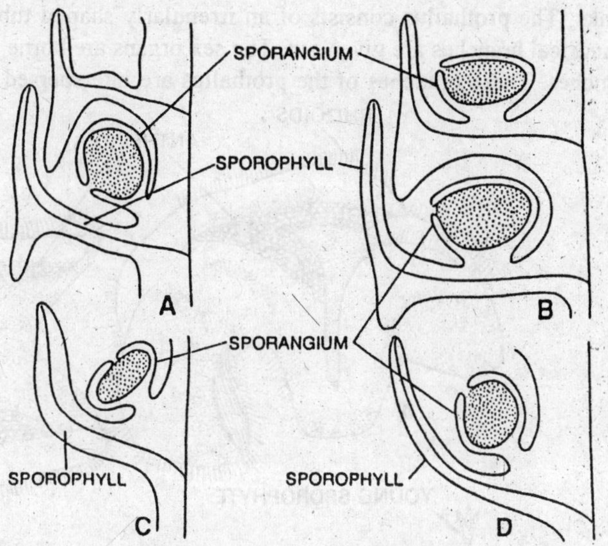

Fig. 27.14. *Lycopodium*. Position of sporangia in different species; A, *L. inundatum* (axillary); B, *L. cernuum* (foliar); C, *L. squarrosum* (sub-foliar); D, *L. lucidulum* (axillary).

The third type of prothallus is found in *L. phlegmaria* and other epiphytic species. These saprophytic colourless prothalli develop beneath the surface of the humus found on the tree

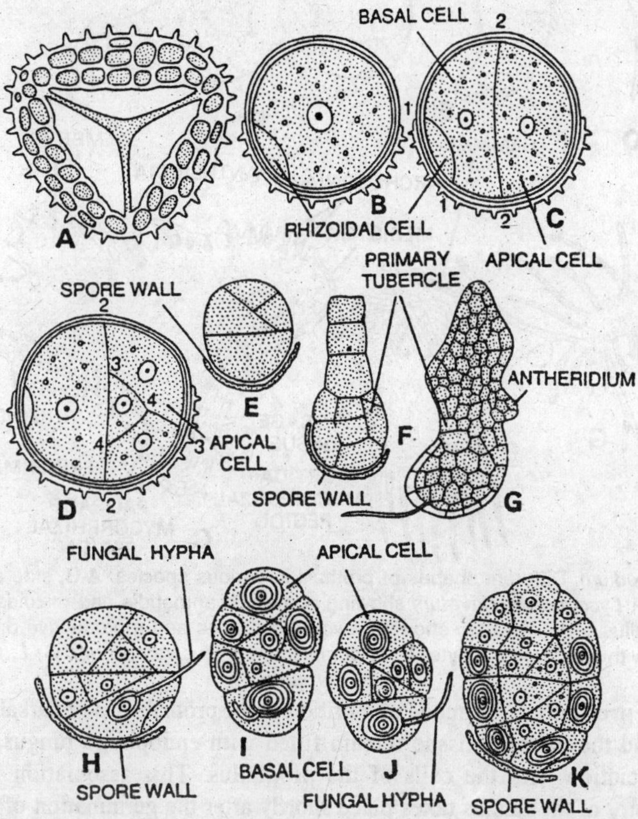

Fig. 27.15. *Lycopodium* spp. Spore germination and development of gametophyte. A, spore of *L. clavatum*; B-D, early stages in the development of gametophyte of *L. clavatum*; E-G, early stages in the development of gametophyte of *L. cernuum*; H-K, gametophytes of *Lycopodium* showing origin of mycorrhizal association.

trunks, The prothallus consists of an irregularly shaped tuberous body from which colourless, cylindrical branches are given out. The sex organs are borne on the upper surface of the enlarged branches. These branches of the prothallus are interspersed with slender paraphyses.

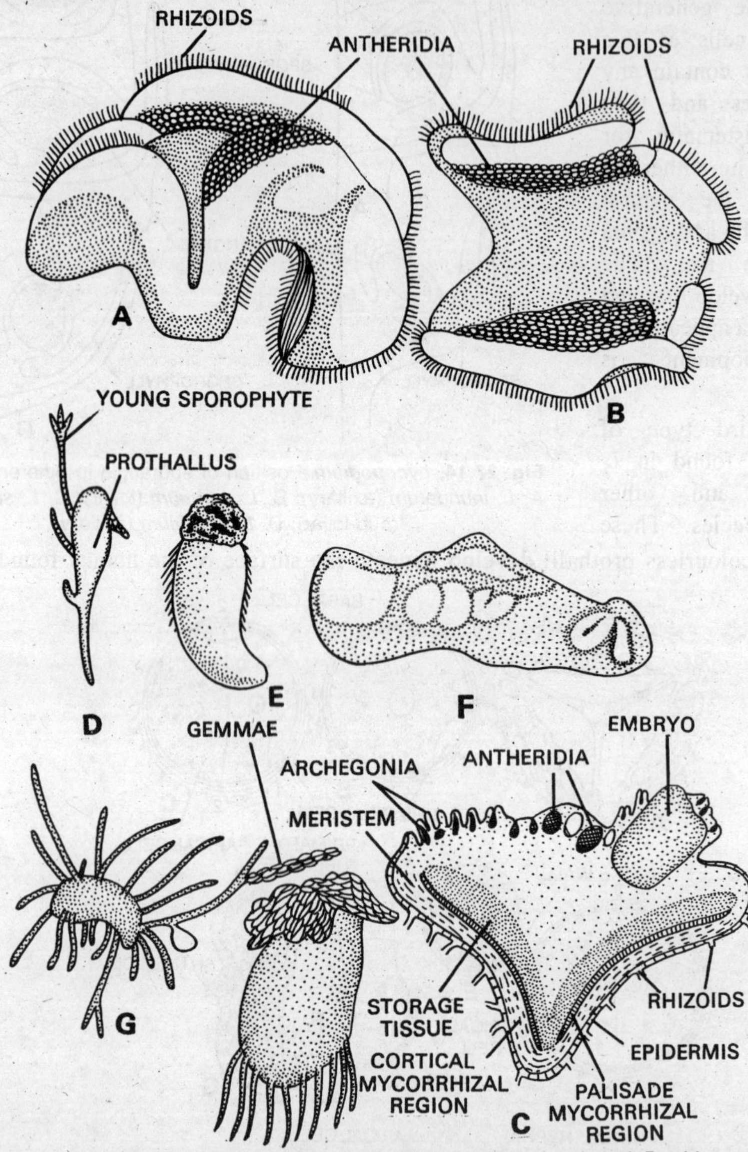

Fig. 27.16. *Lycopodium*. Different shapes of prothalli in various species. A-B, side and top aspects of mature prothallus of *Lycopodium clavatum* showing numerous antheridia and rhizoids; C, vertical section through the prothallus of *L. clavatum* showing mycorrhizal zone and reproductive organs; D, prothallus of *L. selago* with young sporophyte; E, *L. complanatum*; F, *L. annotinum*; G, *L. phlegmaria*; H, *L. cernuum*.

Nutrition of prothallus. The mode of nutrition of the prothallus is remarkable. The cells of the cortical region and the palisade tissue remain filled with endophytic fungus, and these form a mycorrhizal association with the cells of the prothallus. This association is supposed to be symbiotic. The entry of the fungus takes place shortly after the germination of the spore, and if it does not take place the gametophyte never develops more than five cells.

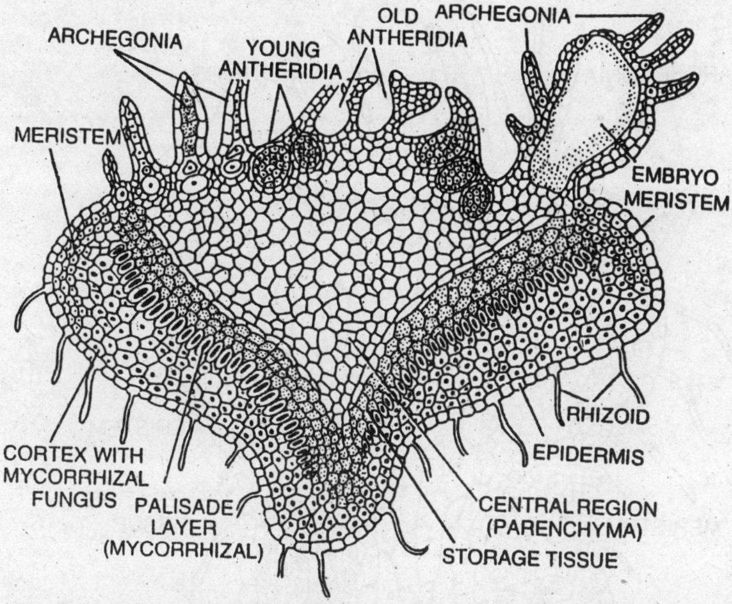

Fig. 27.16 (a). *Lycopodium clavatum.* Gametophyte; median vertical section through a mature prothallus showing mycorrhizal zone, archegonia, antheridia, embryo, etc.

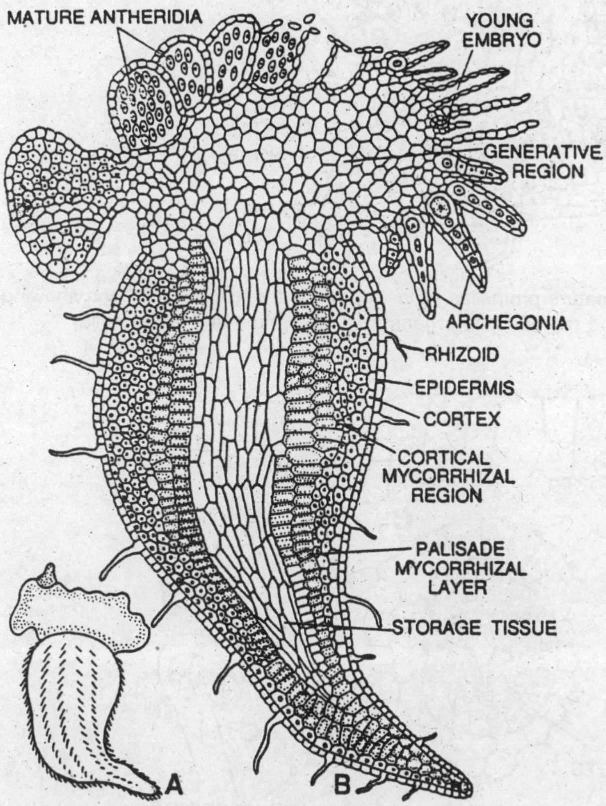

Fig. 27.17. *Lycopodium complanatum.* L.S. of a mature prothallus showing details of internal structure and position of male (antheridia) and female (archegonia) sex organs.

Development of antheridium. The antheridium develops from a superficial cell of prothallus found to be situated just back of the apical meristem. The antheridial initial divides periclinally into an outer cell, the **jacket initial** and an inner cell, the **primary androgonial initial.** The primary androgonial cell divides repeatedly forming a large mass of androgonial cells that lie embedded within the prothallus. Simultaneously the jacket initial divides and redivides anticlinally forming a jacket layer around the mass of androgonial cells, one cell in thickness. The jacket layer in *L. phlegmaria* is two layered towards the periphery. The androgonial cells of the last division being **antherozoid mother cells or androcytes.** Each

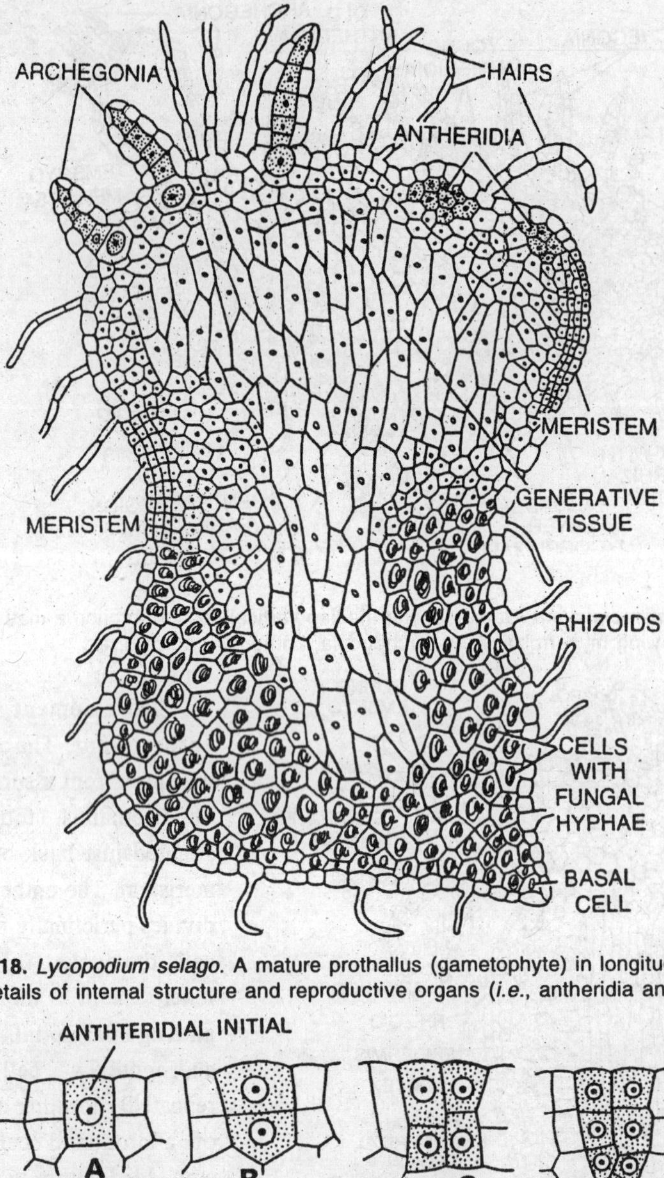

Fig. 27.18. *Lycopodium selago.* A mature prothallus (gametophyte) in longitudinal section showing details of internal structure and reproductive organs (*i.e.*, antheridia and archegonia).

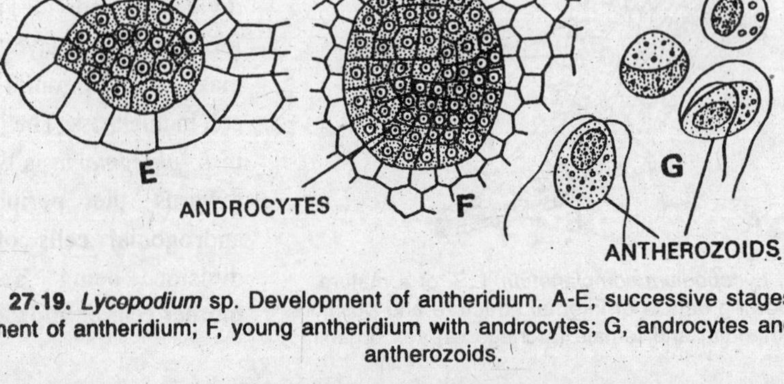

Fig. 27.19. *Lycopodium* sp. Development of antheridium. A-E, successive stages in the early development of antheridium; F, young antheridium with androcytes; G, androcytes and mature biciliate antherozoids.

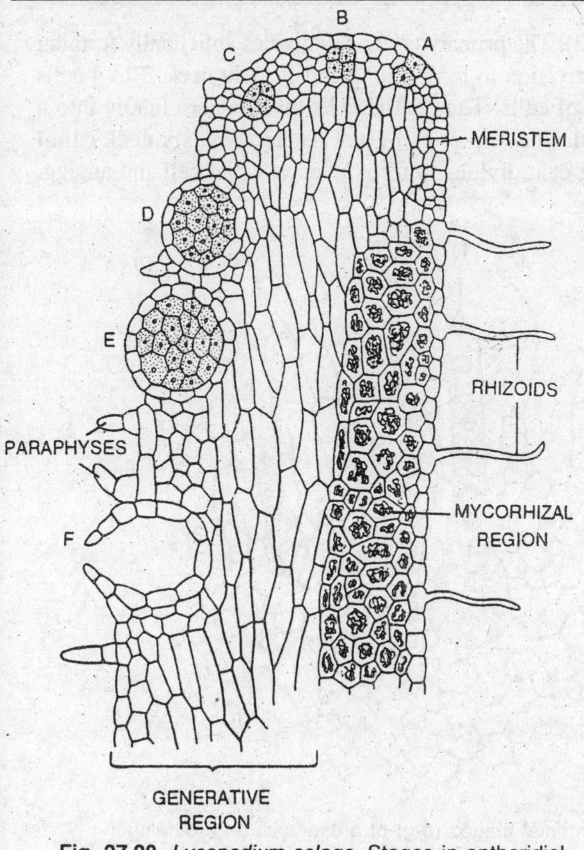

Fig. 27.20. *Lycopodium selago.* Stages in antheridial development on prothallus.

androcyte metamorphoses into an antherozoid.

The mature antheridium consists of an oval mass of antherozoid mother cells that projects out slightly from the prothallial tissue and surrounded by a wall, which is partly formed by the jacket layer and partly from prothallial cells.

The antherozoids are fusiform, broadly rounded at the posterior end with two flagella at the anterior end. The antherozoids closely resemble those of the Bryophyta. This suggests the closer linkage of Lycopodiales with the mosses than with the ferns.

Liberation of antherozoids takes place by a breaking down of the opercular cell in antheridium.

Development of archegonium. The archegonium develops from a single superficial cell of prothallus found just back of the apical meristem. This cell is known as archegonial initial. The arcehegonial initial divides periclinally into a **primary cover cell** and **central** cell

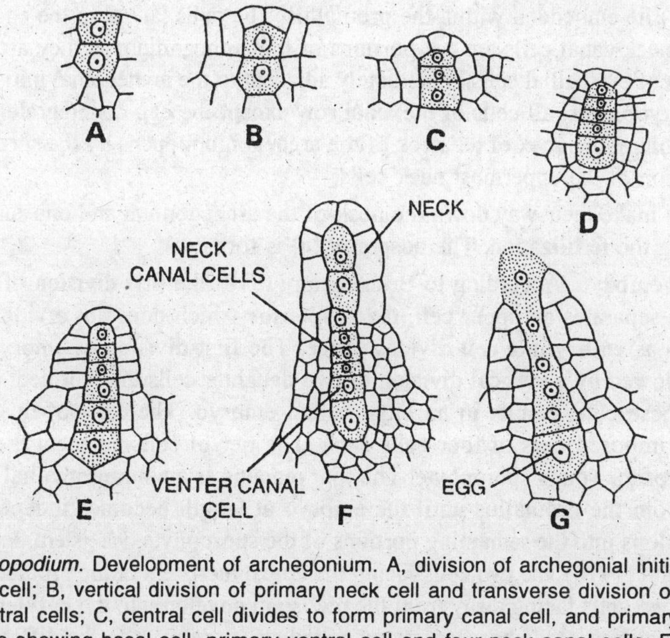

Fig. 27.21. *Lycopodium.* Development of archegonium. A, division of archegonial initial into primary neck cell and inner cell; B, vertical division of primary neck cell and transverse division of inner cell forming basal and central cells; C, central cell divides to form primary canal cell, and primary ventral cell; D-E, later stages showing basal cell, primary ventral cell and four neck canal cells; F, nearly mature archegonium with egg, ventral canal cell and neck canal cells; G, mature archegonium with egg and open neck canal.

(Lang, 1899; Bruchmann, 1910; Spessard, 1922). The primary cover cell divides anticlinally forming four **neck initials,** which undergo transverse division to form a long and straight neck, 3 to 4 cells in height and composed of four vertical rows of cells. The central cells divides periclinally into a **primary canal cell** and a **primary venter cell.** The former cell gives rise to about six **neck canal cells,** while the latter functions directly as egg or it divides to form a venter canal cell and an egg.

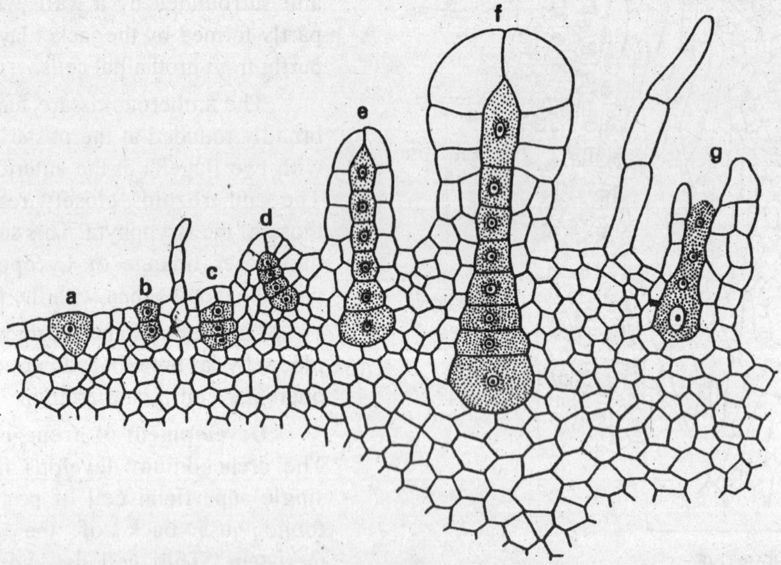

Fig. 27.22. *Lycopodium selago.* Developmental stages (a-g) of archegonia on prothallus.

The portion of the mature archegonium derived from the primary cover cell, together with the neck canal cells it ensheaths, projects vertically above the prothallus. The remaining canal cells and the oosphere (egg) lie embedded within the prothallus. The cells lateral to the egg and lower portion of the row of neck canal cells are not portions of the archegonium as they are formed by division and redivision of prothallial cells immediately adjacent to the archegonial initial. As found in the archegonia of bryophytes, all cells of the axial row except the egg disintegrate prior to the maturation of archegonium, the apex of the neck of the archegonium opens by the spreading apart and partial disintegration of the uppermost neck cells.

The antherozoids make their way down the neck of the archegonium and one fuses with the oosphere thus effecting the fertilization. The oospore (2*n*) is formed.

Development of embryo. According to Bruchmann (1910) the first division of a zygote is always transverse and separates an upper cell, the **suspensor** which does not divide again. The lower cell also known as **embryonic cell** divides thrice. The first division of embryonic cell is vertical and this is followed by a vertical division of two daughter cells thus formed. Transverse division of each of these cells results in an eight-celled embryo. The cells of this octant are arranged in two superimposed tiers of four cells each. The tier of cells adjacent the suspensor develops into the **foot** of the young sporophyte. The foot remains as an intraprothallial haustorium which derives food from the prothallus until the embryo at length becomes independent. The distal tier of cells develops into the remaining portions of the sporophyte, *viz.*, stem, leaf and root. According to Bruchmann (1910), the two cells facing the apical meristem of the prothallus develop into the **stem** and the two cells facing away from the meristem develop into the **cotyledon** and the primary root. The young sporophyte becomes many celled structure prior to the differentiation of the cotyledon and the embryonic stem. The primary root becomes evident even later and at one side of the region where cotyledon and foot adjoin each other. Until this time the embryo remains embedded within the prothallus and grows downward and forward toward the meristem of the prothallus. Thereafter the growing stem and cotyledon burst through the apical region of the gametophyte. Now the cotyledon and the stem start to grow upward into the air.

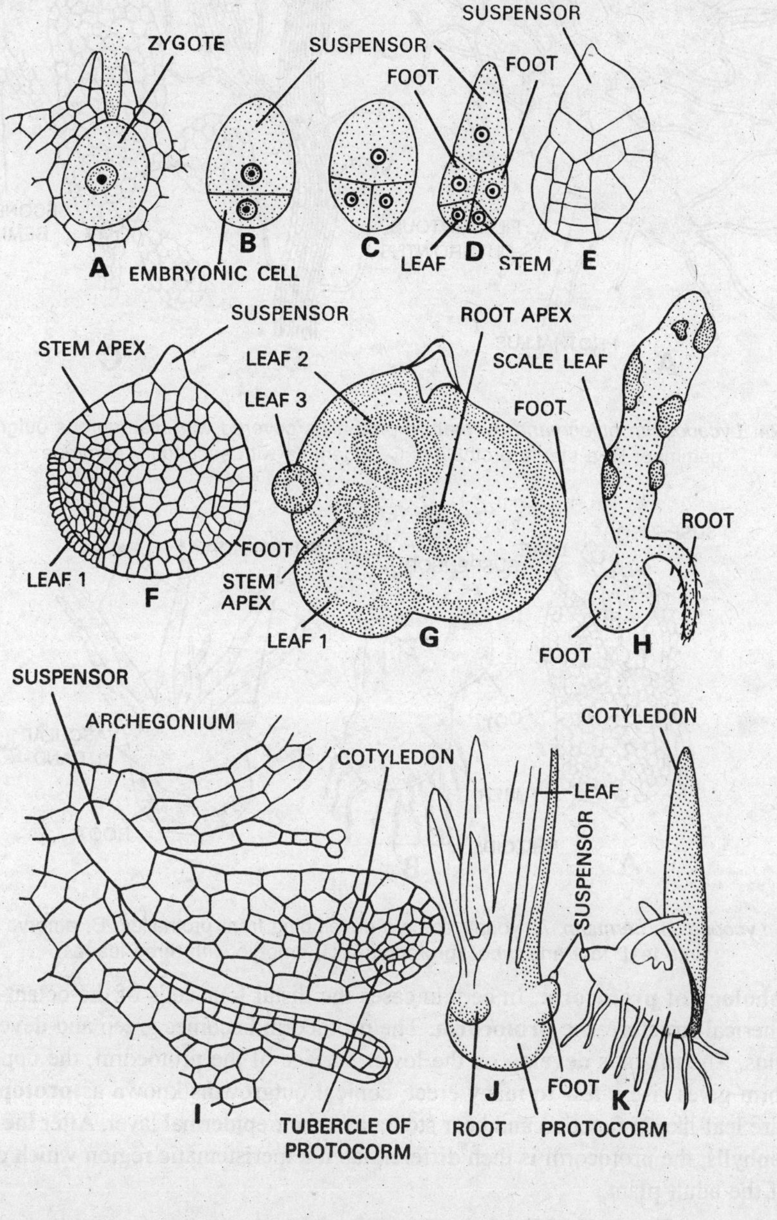

Fig. 27.23. *Lycopodium* sp. Development of embryo. A, zygote; B, first division of zygote; C, division of embryonic cell into two cells; D-F, further development in the embryo; G, embryo from above, showing origin of apex; H, young plant with foot and stem with reduced leaves; I, young embryo of *L. cernuum* emerging from the prothallus; J-K, embryo of *L. cernuum* showing protocorm.

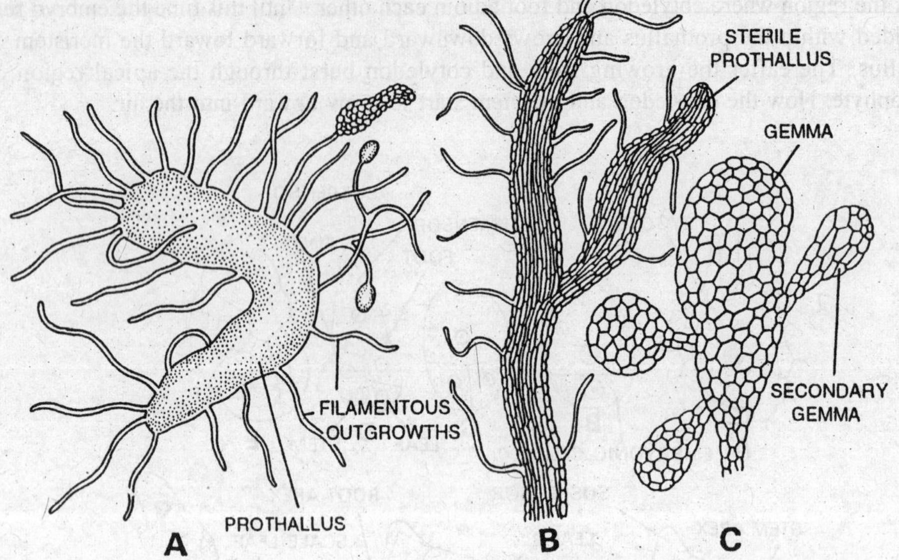

Fig. 27.24. *Lycopodium phlegmaria.* A, mature prothallus covered with filamentous outgrowths and gemmae; B, a sterile prothallus; C, a gemma with secondary gemmae.

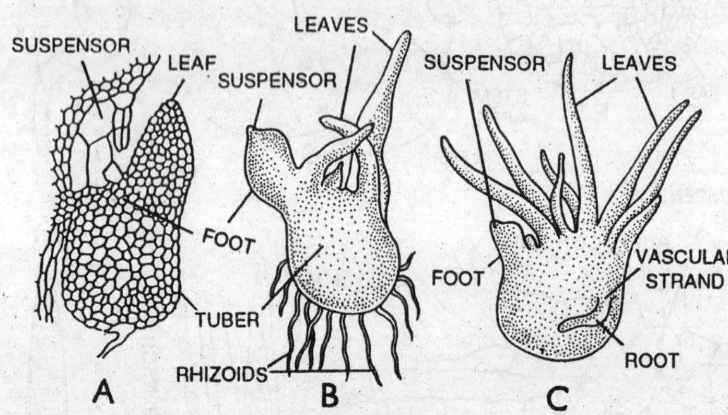

Fig. 27.25. *Lycopodium cernuum.* A, young embryo developing from prothallus; B, embryo differentiated into leaf, foot and tuber (protocorm); C, embryo with three leaves.

Morphology of protocorm. In certain cases the distal four cells of the octant-stage form a massive spherical structure, the **protocorm.** The protocorm becomes green and develops through the prothallus. The rhizoids develop on the lower surface of the protocorm, the upper surface of the protocorm gives rise a few to many erect, conical outgrowths known as **protophylls**. These structures are leaf-like in function and bear stomata in their epidermal layer. After the formation of many protophylls, the protocorm is then differentiated a meristematic region which develops into the stem of the adult plant.

According to Treub (1890), the protocorm is a structure of great antiquity and the phylogenetic precursor of the stem. However, Bower (1908), Holloway (1910) hold the opinion that is a late phylogenetic specialization and represents a 'gouty interlude' in the early ontogeny of the sporophyte.

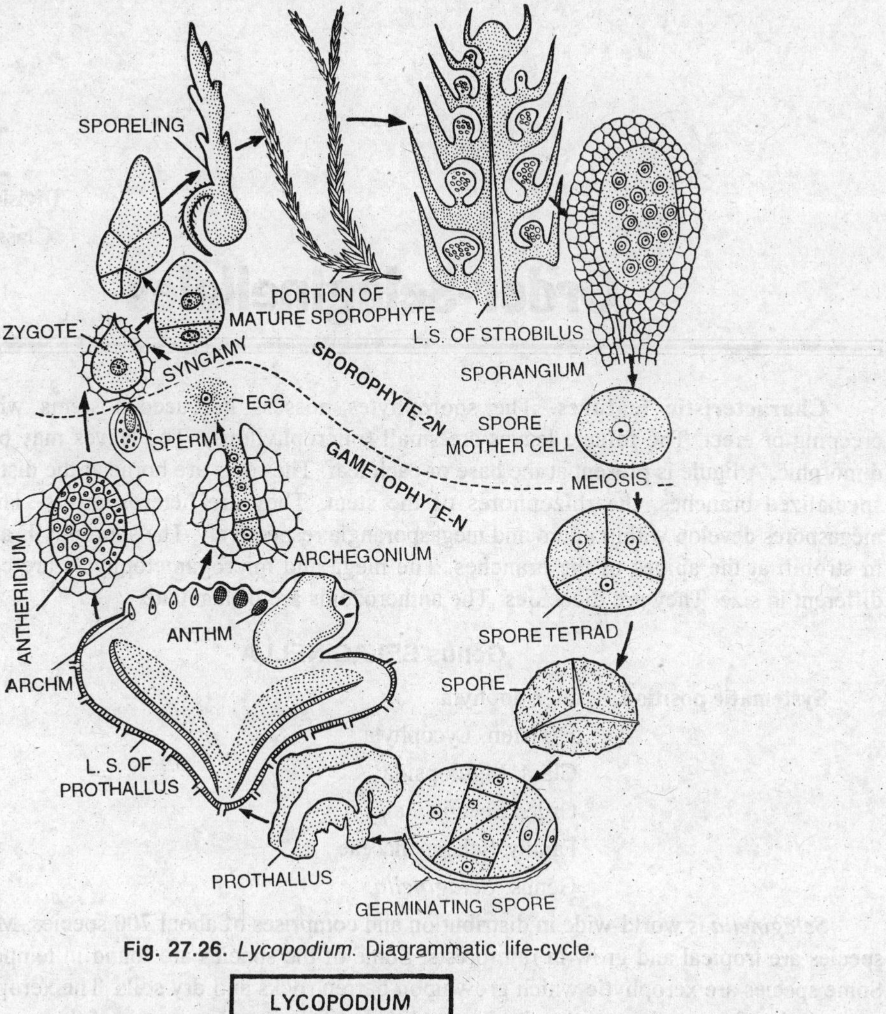

Fig. 27.26. *Lycopodium*. Diagrammatic life-cycle.

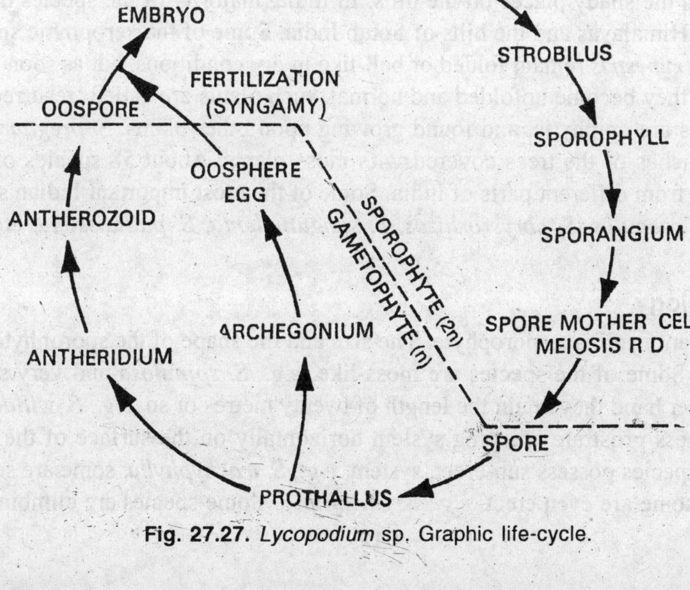

Fig. 27.27. *Lycopodium* sp. Graphic life-cycle.

28

CHAPTER

Division-Lycophyta

Class-Ligulopsida

Order-Selaginellales

Characteristic features. The sporophytes possess herbaceous stems which may be creeping or erect. The foliage leaves are small (microphyllous). The leaves may be all alike or dimorphic. A ligule is present at the base of each leaf. The roots are borne at the distal ends of the specialized branches, the rhizophores of the stem. They are heterosporous. The micro and megaspores develop within micro and megasporangia respectively. The sporophylls are aggregated in strobili at the apices of the branches. The mega and microgametophytes are conspicuously different in size. They are dioecious. The antherozoids are biflagellate.

Genus SELAGINELLA

Systematic position. Pteridophyta
Division. Lycophyta
Class. Ligulopsida
Order. Selaginellales
Family. Selaginellaceae
Genus. *Selaginella*.

Selaginella is world-wide in distribution and comprises of about 700 species. Majority of the species are tropical and grow in rainforests. Some of the species are found in temperate regions. Some species are xerophytic which grow upon barren rocks and dry soils. The xerophytic species may survive for months together in the conditions of desiccation. Some of the species are moist loving and found in the shady places on the hills. In India, majority of the species of *Selaginella* are common in the Himalayas and the hills of South India. Some of the xerophytic species, like *S. lepidophylla* and *S. rupestris* remain folded or ball-like in dry conditions, but as soon as they come in contact of water they become unfolded and normal, such plants are called 'resurrection plants'. Some of the species are epiphytic, and found growing upon other plants. *S. oregana* grows upon the trunks and branches of the trees covered with moss plants. About 58 species of *Selaginella* have been reported from different parts of India. Some of the most important Indian species are-*S. kraussiana, S. chrysocaulos, S. chrysrorhizos, S. sanguinolenta, S. pallidissima,* etc., and many others.

External Morphology

Habit. The plant itself is a sporophyte. The size and the shape of the sporophyte varies from species to species. Some of the species are moss-like, *e.g., S. spinulosa* and very small in size, whereas, on the other hand they attain the length of twenty metres or so, *e.g., S. willdenovii.* Some of the species possess prostrate creeping system horizontally on the surface of the soil, *e.g., S. kraussiana.* Somc spccics possess sub-erect, system, *e.g., S. trachyphylla,* some are scandent, *e.g., S. willdenovii* and some are even erect, *e.g., S. erythropus.* Some species are climbing, and climb with the help of rhizophores, *e.g., S. alligans*. A very few species of *Selaginella* are annual, *e.g., S. pygmaea* and *S. gracillima*. Most of the species are perennials.

On the basis of the general structure of the stem and arrangement of leaves upon it, the species have been divided into two groups. Hieronymus (1900) has placed these two groups of the species under the sub-genera, *Homoeophyllum* and *Heterophyllum*.

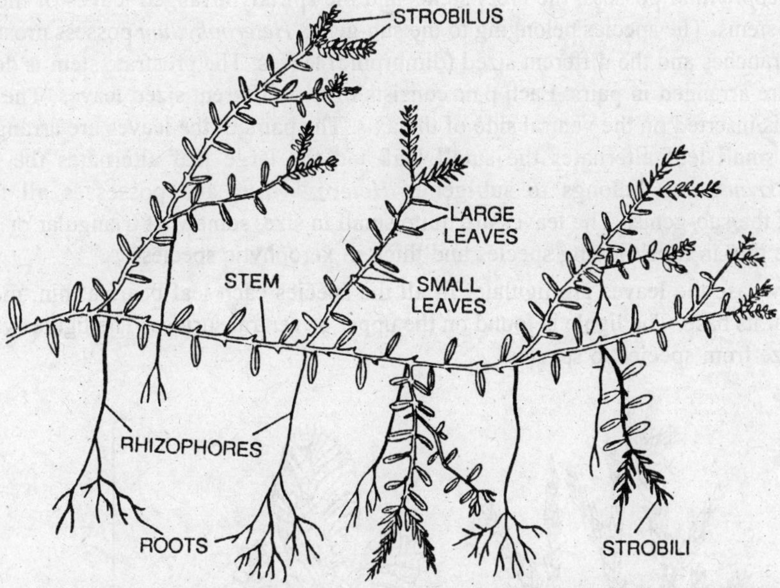

Fig. 28.1. *Selaginella kraussiana*. A, sporophyte with strobili, rhizophores roots, ligules, leaves, etc.

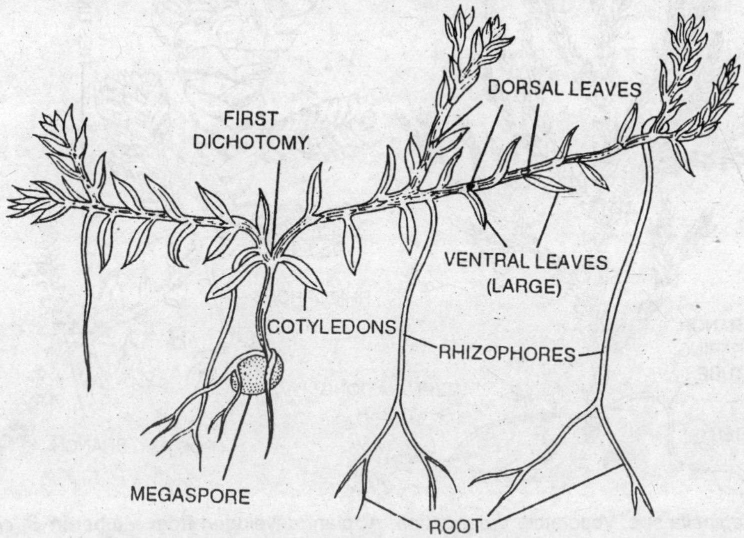

Fig. 28.2. *Selaginella kraussiana*. A, sporophyte developing from a megaspore with strobili, rhizophores, roots, leaves, etc.

The sub-genus *Homoeophyllum* comprises of about 50 or more species, *e.g., S. pygmaea, S. spinulosa* and *S. rupestris,* etc. These species possess erect stems and the spirally arranged leaves of the same size covering the stems.

The sub-genus *Heterophyllum* comprises of majority of the species, *e.g., S. kraussiana, S. lepidophylla, S. martensii* and many others. These species possess prostrate stems, short erect branches and the dimorphic leaves.

Stem and leaves. In each case the stem is covered over with four rows of leaves. The leaves may be of same size or of different size as the case may be. As mentioned above the species of sub-genus *Homoeophyllum* possess, the erect stems and the spirally arranged leaves of the same size covering the stems. The species belonging to the sub-genus *Heterophyllum* possess prostrate stems, short erect branches and the different sized (dimorphic) leaves. The prostrate stem is dorsiventral. The leaves are arranged in pairs. Each pair consists of two different sized leaves. The larger leaf of each pair is inserted on the ventral side of the axis. The pairs of the leaves are arranged in such a way, that small leaf alternates the small leaf and the large leaf alternates the large leaf, *Selaginella kraussiana* belongs to sub-genus *Heterophyllum* and possesses all the typical characters of the sub-genus. The leaves are quite small in size, somewhat triangular or lanceolate. They may be thin in moist loving species and thick in xerophytic species.

In every case the leaves are ligulate. In all the species each leaf bears a thin, membranous small ligule at its base. The ligule is found on the upper or ventral surface. The ligules vary in their shape and size from species to species.

Fig. 28.3. *Selaginella* spp. Vegetative propagation. A, plant developed from a tuber in *S. chrysorrhizos;* B, portion of *S. chrysocaulos* showing a resting bud.

The structure and the function of the ligule is as follows:

The mature ligule is thin, membranous, tongue like to the fan shaped. There is a cup-shaped sheath at its base. Just adjacent to the sheath, there is a hemispherical mass of thin-walled cells, known as **glossopodium.** The function of the ligule is controversial. It has been thought by many workers, that they are water absorbing or water secreting organs and they however, prevent the undue drying of the stem apex and the leaves by keeping them moist.

Rhizophores and roots. In many species, the prostrate axis bears an elongate, colourless, leafless, cylindrical, downwardly growing structure known as rhizophore. The rhizophores arise from the axis of the ramifications. Usually a single rhizophore comes out downward from the ramification of the stem, *e.g., S. kraussiana.* But in some cases two rhizophores are given out from each branching of the stem, *e.g., S. martensii.* In such cases the rhizophore coming from ventral side develops fully whereas from dorsal side develops only in the form of a small outgrowth. At the terminal end of the rhizophore, a tuft of adventitious roots develops, *e.g., S. kraussiana.* In *S. lepidophylla*, a knot like structure develops at the end of the rhizophore and from this knot like structure the adventitious roots are given out. The adventitious roots are thin, slender and delicate. They are dichotomously branched. The primary root perishes after a very short duration.

Strobilus or spike. In majority of the species, the reproductive structures develop at the apices of the branches. These reproductive structures are called **strobili** (singular = strobilus) or

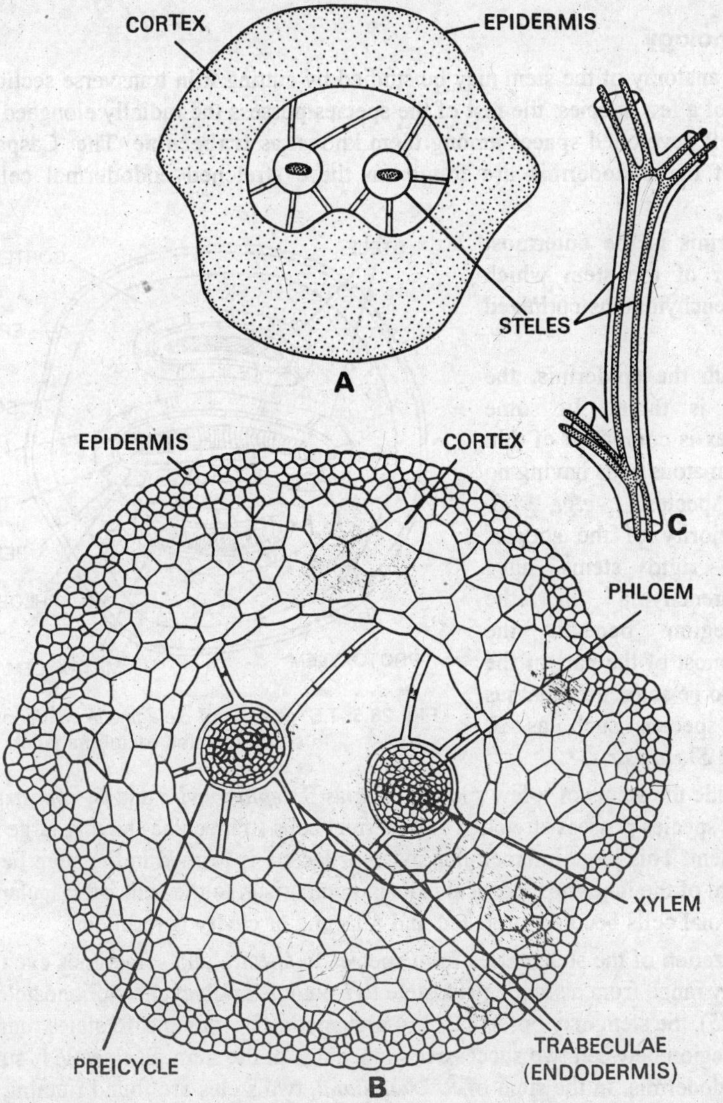

Fig. 28.4. *Selaginella kraussiana.* Anatomy of stem. A, transverse section of stem (diagrammatic); B, transverse section of stem (detailed); C, the stem in longitudinal sectional view to show the position of steles.

spikes. The strobilus is a compact structure of sporophylls. The sporophylls are quite similar in structure to that of ordinary vegetative leaves. Each sporophyll bears a ligule at its base. Each sporophyll bears a single stalked sporangium. The sporangia are of two types, *i.e.*, **microsporangia** and **megasporangia.** The sporophyll which bears microsporangium is microsporophyll, and which bears megasporangium is megasporophyll. Both micro and megasporophylls may be found in same strobilus or in different strobili. The sporophylls are always spirally arranged on the axis of the strobilus but in a very much condensed form. In majority of the species the micro and megasporophylls borne in the same spike, *e.g.*, *S. kraussiana.* According to Mitchell (1910), in certain species, *e.g.*, *S. gracilis* and *S. atroviridis* the strobili bear only one type of sporophylls. When the strobilus bears both kinds of sporophylls, the megasporophylls may be found in the lower portion of the strobilus and the microsporophylls in the upper portion of the strobilar axis, *e.g.*, *S. spinosa* and *S. rupestris*. In certain other species the mega and microsporophylls may be arranged in different vertical rows in the strobilus.

Internal Morphology

Stem. The anatomy of the stem may be studied by cutting it in transverse sections. Leaving aside the stems of a few species, the rest of the species possess the radially elongaed endodermal cells, having well developed spaces among them known as **trabeculae**. The 'Casparian strips', characteristic of the endodermis are found on these stretched endodermal cells so-called trabeculae.

The epidermis is the outermost superficial layer of the stem which consists of prosenchymatous cutinized cells.

Just beneath the epidermis, the cortical region is found. In some species the cortex is composed of thin walled parenchymatous cells having no intercellular species, *e.g.*, *S. kraussiana*. Majority of the species which possess rigid stems have compact sclerenchyma in the hypodermal region beneath the epidermis. The most of the part of the cortex is found to be sclerenchymatous in xerophytic species such as *S. lepidophylla* and *S. rupestris*.

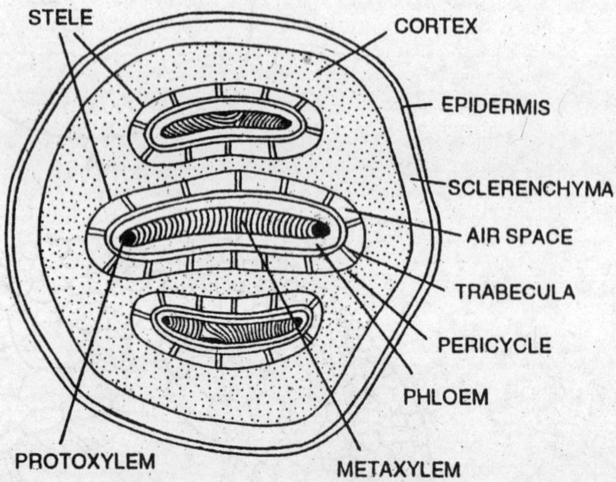

Fig. 28.5. T.S. of stem of *Selaginella willdenovii* showing three protosteles (diagrammatic).

Leaving aside the stems of a few species such as *S. lepidophylla* and *S. rupestris*, the stems of the rest of the species possess stretched endodermal cells (trabeculae) having large intercellular spaces among them. This large intercellular space or lacuna is formed in the stem because of the rapid enlargement of the innermost cortical cells in comparison to the cells of vascular tissue. This way the endodermal cells become stretched and a lacuna or cavity is resulted.

The organization of the stele varies from species to species and sometimes even in the same species. This may range from a simple protostele to a complex polycyclic siphonostele. According to Wardlaw (1925), the stem may possess 2, 3, 4 or sometimes even upto 16 steles running through the intervening region between two successive branchings of the stem. Each stele is surrounded by its trabecular endodermis. In the stem of *S. kraussiana*, two steles are found running through the region between two successive branchings. Both the steles are protostelic.

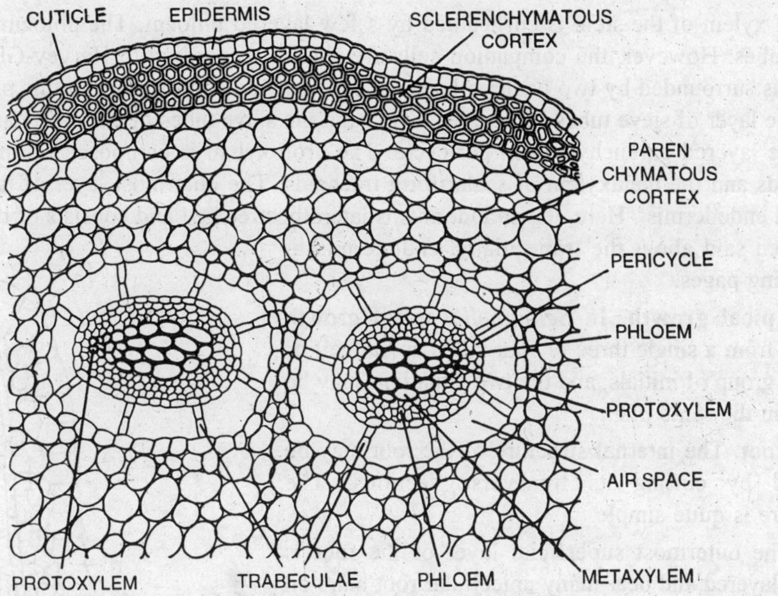

Fig. 28.6. *Selaginella* spp. T.S. of stem of *S. kraussiana* showing distelic condition. Note the trabeculae (endodermis).

Each stele possesses a single mass of exarch protoxylem in it. The masses of the exarch protoxylem of the two steles are so oriented that they face in opposite directions to each other. The metaxylem in each bundle is centrally placed and lies centripetal to the protoxylem. The centrally

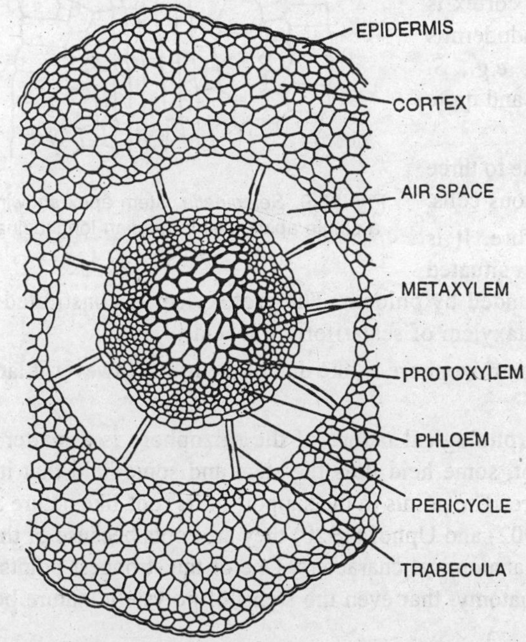

Fig. 28.7. *Selaginella kraussiana.* T.S. of stem showing monostelic condition.

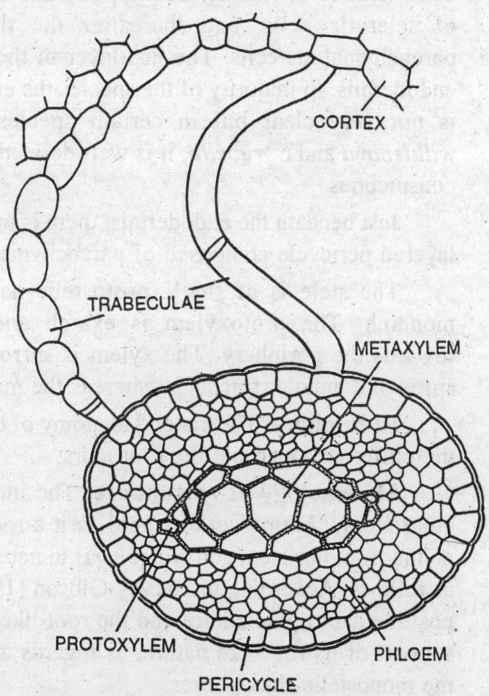

Fig. 28.8. *Selaginella kraussiana.* Part of a transverse section of stem to show inner cortex trabeculae and single stele.

located xylem of the stele is ensheathed by a few layered phloem. The phloem is composed of sieve tubes. However, the companion cells are absent. According to Harvey-Gibson (1894) the xylem is surrounded by two or three layers of parenchyma and external to this parenchyma there is single layer of sieve tubes. The parenchyma and the sieve tubes of each stele are surrounded as a single layered parenchymatous pericycle. The protoxylem is composed of annular and spiral tracheids and the metaxylem of scalariform tracheids. The outermost layer of the stele is single layered endodermis . Here, the endodermis is laterally stretched and consists of trabeculae. Much has been said about the trabeculated endodermis in preceding pages.

Apical growth. In *Selaginella* apical growth occurs from a single three or four-sided apical cell or from a group of initials, and the two situations may be found in the same plant.

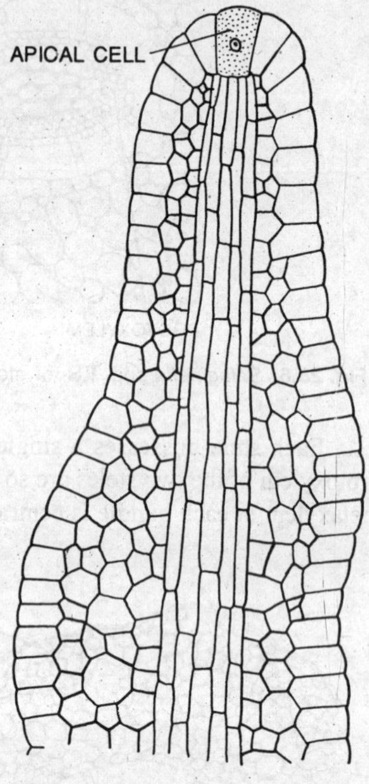

Fig. 28.9. *Selaginella.* Stem apex showing a single apical cell in median longitudinal section.

Root. The internal structure of the root may be studied by cutting its transverse sections. The structure is quite simple.

The outermost superficial layer of the root is single-layered and bear many unicellular root hairs on its surface and, therefore it is also known as piliferous layer. This layer consists of thin walled parenchymatous cells.

Inner to the piliferous layer, there is wide cortical region. In certain species, *e.g.*, *S. willdenovii* there is three to five layered hypodermis constituted of sclerotic cells, and thereafter the thin-walled parenchymatous cells. The last layer of the cortex is endodermis. In majority of the species the endodermis is not very clear but in certain species, *e.g.*, *S. willdenovii* and *S. rubella*, it is well-defined and quite conspicuous.

Just beneath the endodermis, there is one to three layered pericycle composed of parenchymatous cells.

The stele is of single protostelic nature. It is monarch. The protoxylem is exarch and situated towards the periphery. The xylem is surrounded by phloem. The protoxylem is constituted of spiral and annular tracheids whereas the metaxylem of scalariform tracheids.

Rhizophore. The general anatomy of the rhizophore is like that of the root. However it lacks the piliferous layer and the root hairs.

Morphology of rhizophore. The morphological nature of the rhizophore is a matter of controversy. Many workers held as it a root, some held it as the stem and some held that it is neither root nor stem and *sui generis* in nature. The points in the support of its root-like nature are as follows: According to Harvey-Gibson (1902) and Uphof (1920) they are roots because of their positive geotrophic nature and the root-like anatomical characters. One of the strongest points in support of its root-like nature, as regards anatomy, that even the stems of polystelic nature bear the monostelic rhizophores.

According to Wordell (1910) the rhizophore is the portion of the stem. This argument seems to be more strong and conclusive. The points in support of its stem-like nature are as follows-(a) Its origin is exogenous, (b) the cap is not found, (c) its branching or dichotomy is like that of the

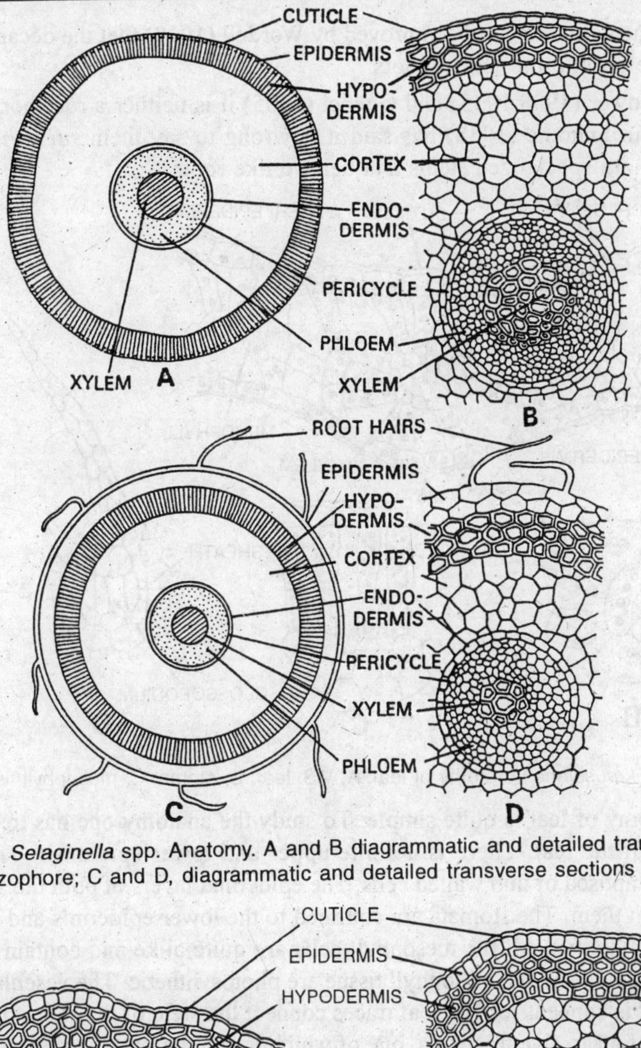

Fig. 28.10. *Selaginella* spp. Anatomy. A and B diagrammatic and detailed transverse sections of rhizophore; C and D, diagrammatic and detailed transverse sections of the root.

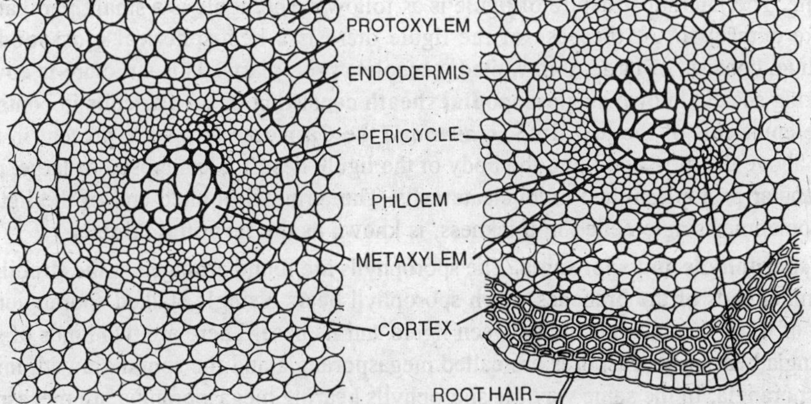

Fig. 28.11. *Selaginella.* A, T.S. of rhizophore, B, T.S. of root.

stem and, (d) experimentally this has been proved by Wordell (1910) that the decapitated apices of the rhizophores develop into the leafy shoots.

According to Bower (1908,1935) and Goebel (1905) it is neither a root nor a stem, it is an organ *sui generis*. But Schoute (1938) has said it is wrong to say them *sui generis* organs. He pointed out that they are specialized stems and behave like roots.

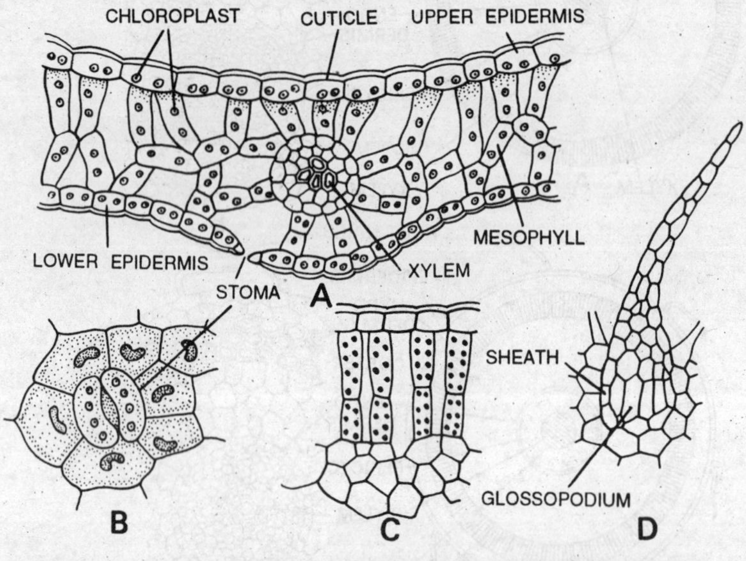

Fig. 28.12. *Selaginella kraussiana.* Anatomy of leaf. A, V.S. leaf; B, stoma; C, mesophyll tissue; D, a ligule.

Leaf. The anatomy of leaf is quite simple. To study the anatomy one has to cut the vertical microtome sections of the leaf. There is definite upper and lower epidermis. The epidermis is single-layered and composed of thin walled cells. The epidermal layers of both the surfaces contain discoid chloroplasts in them. The stomata are confined to the lower epidermis and sometimes also found on the upper epidermis. All the mesophyll cells are quite alike and contain chloroplasts in them. The epidermal layers and the mesophyll tissue are photosynthetic. The vascular bundle of the leaf is quite simple. It is concentric. The leaf traces connect the stele of the stem. In *S. kraussiana*, the xylem consists of four to five tracheids, one of which is annular and the rest spiral. The xylem is surrounded by phloem, composed of parenchyma cells and one or two sieve tubes.

Ligule. The internal structure of ligule is as follows: The ligule is a small, papillate, pointed tongue-like outgrowth. At the base of the ligule there is a hemispherical **glossopodium**. The glossopodium consists of large, thin walled, vacuolate protoplasts. This structure is covered over by a covering sheath, known as **glossopodial sheath** consisting of tubular cells in continuation of that of the epidermis of the leaf. In *S. kraussiana* the Casparian strips are present in the sheath cells. Just above the glossopodium, the body of the ligule lies. It is composed of large, polygonal, rich in granular protoplasm and uninucleate cells. The terminal pointed end of the ligule is very thin and sometimes single-celled in thickness, is known as the **terminal lamella**.

The sporophylls and sporangia. The sporophylls are found arranged in the strobili or spikes at the terminal ends of the branches. Each sporophyll bears a single stalked sporangium situated in the axil of it, on the adaxial side between ligule and its base. There are two kinds of sporangia. The sporangia bearing megaspores are called megasporangia and the sporangia with microspores are microsporangia. In the same way the sporophylls bearing megasporangia are megasporophylls and bearing microsporangia are microsporophylls.

Development of the sporangia. The development of mega and microsporangia is quite common upto the formation of spore mother cells. There are two schools of thought about the

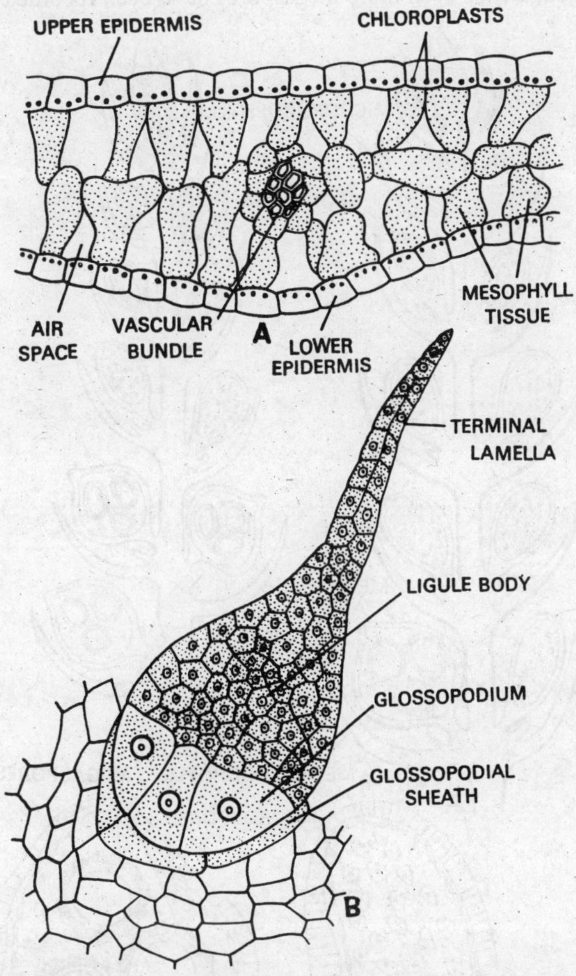

Fig. 28.13. *Selaginella* sp. A, vertical section of leaf showing chloroplasts in the epidermis; B, sectional view of a mature ligule.

nature of their development. According to one school of thought the sporangium develops from a single initial cell. According to other school, the sporangium arises from a transverse row of initials. In majority of the species the sporangia arise from the cells of the axis and said to be cauline whereas Bower holds the view that the sporangium arises from the base of the sporophyll and foliar in origin. Any how, the mature sporangium is a distinctly axillary structure at the base of the sporophyll.

Irrespective of whether there is single initial cell or a group of initial cells the further development proceeds on the same line.

The initial divides first periclinally giving rise to two cells. The outer cell (or cells) is **jacket initial** (or initials) and the inner cell (or cells) is **archesporial initial** (or initials). The jacket initial divides and redivides developing a two layered jacket. The archesporial initial divides vertically and transversely again and again and this way a mass of sporogenous tissue is resulted. The jacket is two layered, the outer layer is composed of thick-walled cells and the inner layer of thin-walled cells. The jacket completely covers the sporogenous tissue. The outermost cells of the sporogenous tissue develop into a nutritional layer the **tapetum**. The cells of the tapetum remain in close contact of the developing spores to nourish them. The tapetal cells degenerate after the maturity of the spores.

The sporangia which are to develop into microsporangia contain all the sporogenous cells. The sporogenous cells of the last generation are the sporocytes. During this process 10 to 20 percent of the sporogenous cells are disintegrated. Thus, the sporocytes become loosened within the sporangial chamber. The functional sporocytes become rounded and separated from each other. Each sporocyte (spore mother cell) divides meiotically giving rise to a tetrad of four spores. This way numerous haploid microspores are developed within the sporangium.

The sporangia which are to develop into megasporangia contain many sporocytes in them in the beginning. Later on all sporocytes but one are disintegrated. The functional megaspore mother cell undergoes the usual reduction division and tetrad of megaspores is developed. According to

Duerden (1929) in some species, 8, 12, 24, or even more megaspores have been recorded in one sporangium. In such cases more than one megaspore mother cells are functional. According to Duerden (1929) and Mitchell (1910) there are still other species in which all the four megaspores of the tetrad do not develop as usual. In these species the megasporangium contains one, two or three functional megaspores, rest of the spores are degenerated. As the megaspores develop less in number in each megasporangium they get much of nourishment from the degenerating cells and increase a lot in size and more or less fill up the sporangium. The megasporangium of *S. willdenovii* contains the highest recorded number of the megaspores, *i.e.,* 42 whereas *S. monospora* and *S. rupestris* possess monosporic megasporangia.

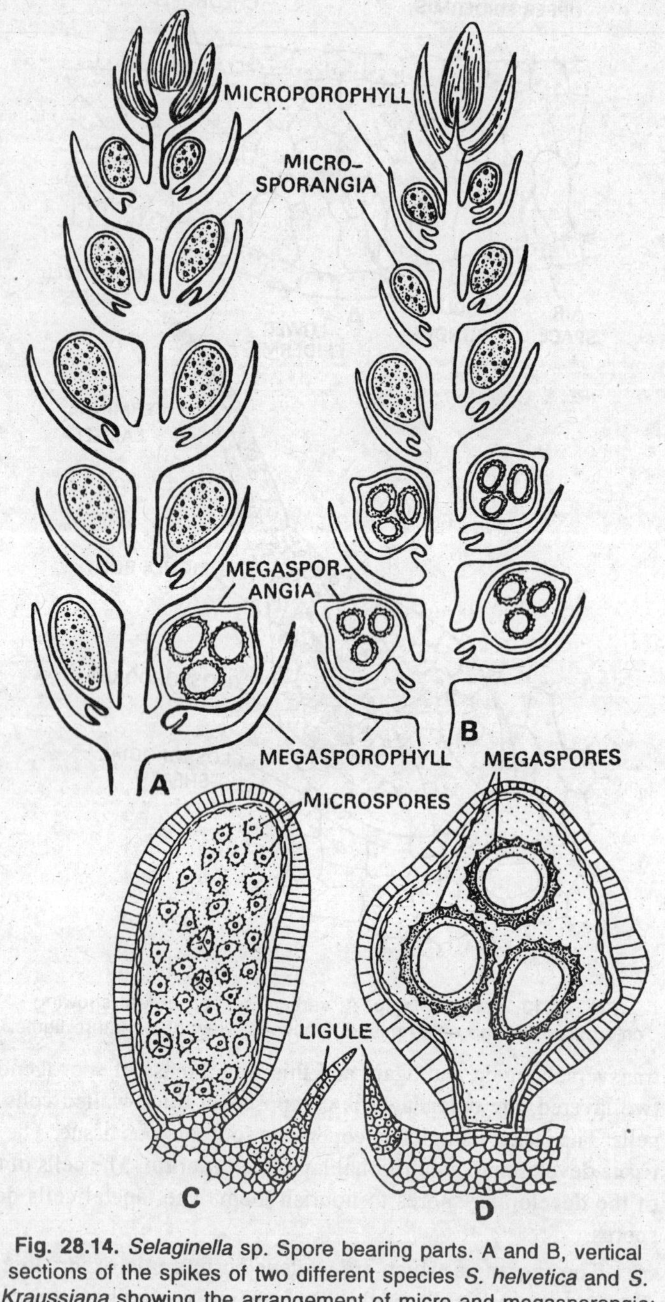

Fig. 28.14. *Selaginella* sp. Spore bearing parts. A and B, vertical sections of the spikes of two different species *S. helvetica* and *S. Kraussiana* showing the arrangement of micro and megasporangia; C, longitudinal section of microsporangia; D, longitudinal section of megasporangium.

Structure of sporangia. Both mega and microsporangia are stalked. In both the cases the wall of the mature sporangium is two layered. The outer layer of the wall is composed of thick walled columnar cells having chloroplasts in them. The inner layer of the wall is composed of thin walled cells. Just beneath the two layered sporangial wall there is a nutritional layer, the tapetum. This layer nourishes the developing spores and persists until the maturity of the spores and thereafter degenerates.

The ripe microsporangia are red yellow or brown in colour. This colour of the microsporangial wall is because of the coloured microspores within it. Comparatively the microsporangia are smaller in size. In *S. kraussiana*, each microsporangium contains about 600 microspores in it.

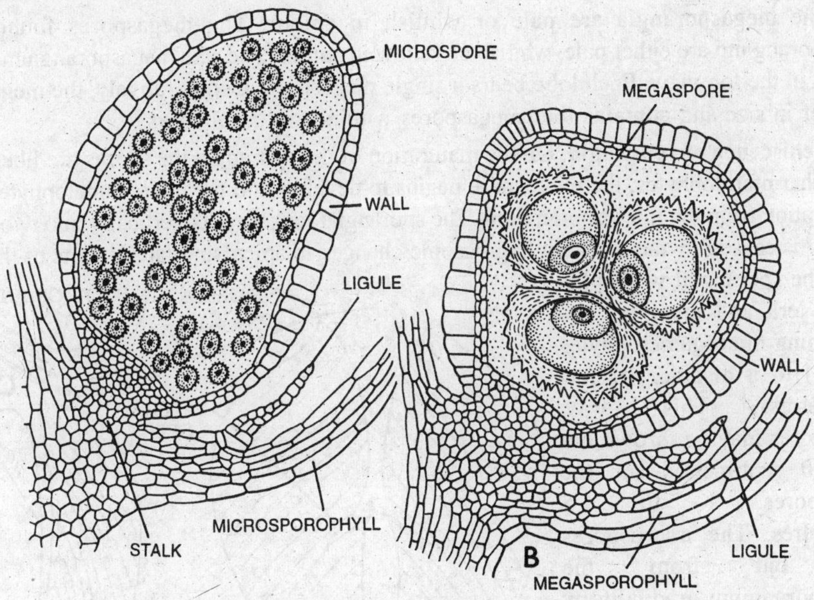

Fig. 28.15. *Selaginella*. Spore bearing organs. A, mature microsporangium containing microspores; B, mature megasporangium containing megaspores.

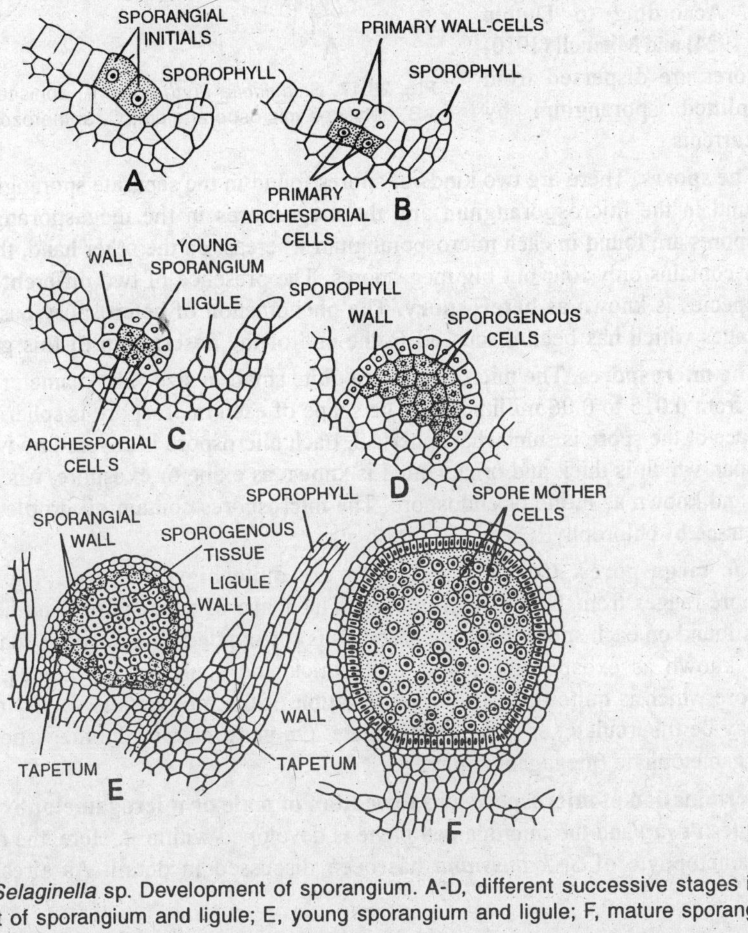

Fig. 28.16. *Selaginella* sp. Development of sporangium. A-D, different successive stages in the development of sporangium and ligule; E, young sporangium and ligule; F, mature sporangium.

The megasporangia are pale or whitish in colour. The megaspores found within the megasporangium are either pale, whitish or yellow in colour. The ripe megasporangium is four lobed as seen in the top view. Each lobe bears a single megaspore. Comparatively, the megasporangium is larger in size and contains four megaspores within it in usual cases.

Dehiscence of sporangia. On the maturation of the sporangia the spores are liberated. Unlike most other pteridophytes, the microspore begins to develop into the microgametophyte even before its liberation from the microsporangium. The sporangial jacket splits vertically into two valves. This splitting is caused by the different hygroscopic changes in apical and lateral parts of the sporangial wall. The sporangial wall splits with a jerk and eject out the developing microgametophytes (spores) to a distance of few centimetres. This way, megaspores may be thrown out 6 to 10 centimetres and the microspores 1 to 1.5 centimetres. The microspores eject out from the microsporangium gradually by subsequent openings and closings of the split of the jacket. According to Duthie (1923, 1924) and Mitchell (1910) the spores are dispersed from the splitted sporangium by wind currents.

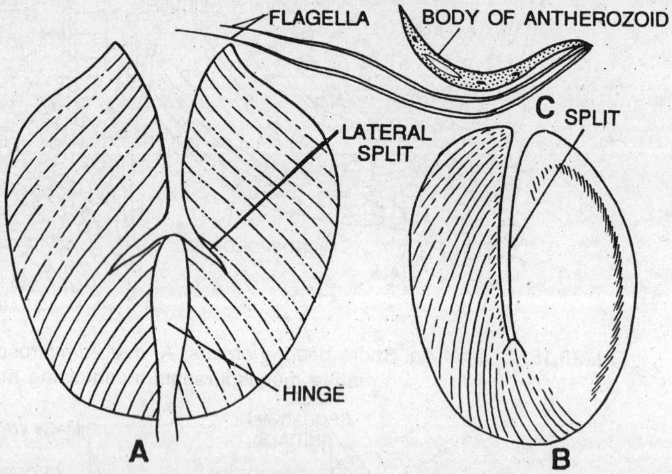

Fig. 28.17. *Selaginella erythropus.* A, dehisced megasporangium; B, dehisced microsporangium; C, antherozoid. (A and B, After

The spores. There are two kinds of spores found in the separate sporangia. The microspores are found in the microsporangium and the megaspores in the megasporangium. Hundreds of microspores are found in each microsporangium whereas, on the other hand, the megasporangium usually contains only four but big megaspores. The presence of two different sized spores in the same species is known as **heterospory**. The phenomenon of heterospory has its own biological advantages which has been discussed in the end of the description of this genus.

The microspores. The microspores are quite small in size. The diameter of the microspores ranges from 0.015 to 0.06 millimetres. The shape of each microspore is spherico-tetrahedral. The distal face of the spore is somewhat spherical. Each microspore is covered by two spore coats. The outer coat, which is thick and ornamented is known as exine or exospore, whereas, the inner coat is thin and known as intine or endospore. The microspores contain oil droplets. They are devoid of any trace of chlorophyll.

The megaspores. Comparatively they are quite bigger in size. The diameter of each megaspore ranges from 1.5 to 5 millimeters. The shape of each spore is tetrahedral. A triradiate ridge is found on each spore. Each megaspores is surrounded by two spore coats. The outer spore coat is known as exospore which is quite thick and sculptured, whereas, the inner coat is endospore which is quite thin. The ornamentation of the megaspores varies in different species. This may be tuberculate, spinose or reticulate. On germination, the megaspore develops into a female gametophyte (megagametophyte).

Germination of microspore and formation of male or microgametophyte. The microspore germinates *in situ* and the microgametophyte is developed within it. Here, the development of the microgametophyte of *S. kraussiana* has been discussed in detail. As already mentioned the

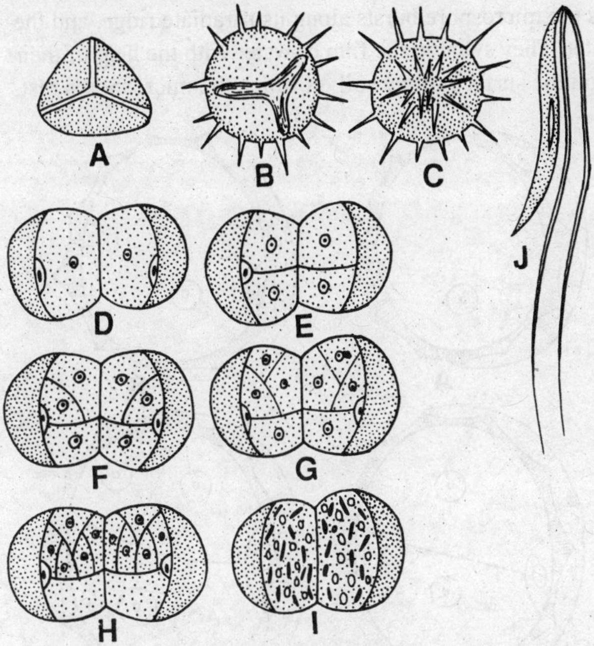

Fig. 28.18. *Selaginella kraussiana.* A, microspore just after the dissociation of a tetrad; B, complete microspore in apical view; C, complete microspore in basal view; D-I, schematic representation of the stages in the development of the microgametophyte; J, mature biflagellate antherozoid.

development of the microgametophyte begins even before the dispersal of the spore from the microsporangium. First of all an asymmetrical division takes place in the microspore, giving rise to two unequal cells. The small lens shaped cell is **prothallial cell** and the remaining large cell is **antheridial cell.** The antheridial cell further divides by a wall perpendicular to the prothallial cell, thus giving rise to two daughter cells. Both of these daughter cells divide at right angles to the first plane of division. One daughter cell of this pair does not divide further and becomes the first cell of the jacket of the antheridium. The other daughter cell of the pair divides by a plane parallel to the small prothallial cell. This way, two daughter cells are formed. The cell nearer to the prothallial cell divides periclinally giving rise to a **primary androgonial cell** and a **jacket cell.** The other cell of the two cells divides anticlinally and again one of the daughter cells divides periclinally resulting in the formation of two jacket cells and a primary androgonial cell. This way, a thirteen-celled microgametophyte is developed. It consists of one prothallial cell, four primary androgonial cells and eight jacket cells. In *S. kraussiana* only at this stage, the microgametophyte is shed from the microsporangium. After shedding of the thirteen-celled microgametophyte, further divisions take place. Each of the four androgonial cells divides successively six times thus producing the 256 androcytes. Each such androcyte metamorphoses into a spindle-like biflagellate antherozoid. Later on the jacket cells and the prothallial cell are disintegrated. This way the mature gametophyte possesses a mass of antherozoids swimming within the limits of the spore wall. **Duthie** (1923) supported in his statement that these later stages in the development of the microgametophyte take place in the soil. According to miss Lyon (1901), in certain other species this later development of the microgametophyte takes place on the partially opened megasporangium.

It is a notable point that the entire microgametophyte of *Selaginella* is very much reduced in size and developed within the limits of the sporangial wall. On account of its limited sphere and reduced size, there is difference of opinion regarding its homologies with the gametophytes of other homosporous pteridophytes such as *Equisetum* and *Lycopodium*. According to most of botanists, the prothallial cell is vegetative and homologous with the entire vegetative prothallus of the homosporous pteridophytes such as *Equisetum* and *Lycopodium.* The nature of the antheridial cell is also a matter of controversy. According to Campbell (1918) and Slagg (1932) it is the initial of a single antheridium. Bruchmann (1912) holds the opinion that antheridial cell divides and forms two or four antheridial initials according to the species.

On the maturation of the antherozoids the microspore bursts along its triradiate ridge, and the antherozoids are liberated in the film of water. They swim in the film of water with the help of their flagella. The water film is usually present on the surface of the soil on which the microspores rest.

Germination of megaspore and formation of female or megagametophyte. The megaspore also germinates *in situ*. The development of megagametophyte of *S. kraussiana* has been studied in detail. It has been discussed here in the following lines. The process of the development of the megagametophyte begins even before the maturity of the megaspore. The megaspore enlarges in size and a two-layered wall is secreted round it. When the megaspore is still enlarging, its nucleus divides, but there is no cell division. Their is a continuous nuclear division and ultimately a multinuclear protoplast with a large central vacuole is resulted. Just beneath the cell wall, the cytoplasm is found in a thin layer. After some time the enlargement of the multinucleate gametophyte is ceased, simultaneously the cytoplasmic layer underneath the cell wall becomes thick and thicker and this way the central vacuole is reduced in size. Ultimately the central vacuole disappears. Long before all this, the free cell formation takes place at the pyramid like end of the megagametophyte and a two to three layered thick cellular tissue is resulted. This cellular tissue remains separated from rest of the gametophyte by a diaphragm. The portion of the megagametophyte below this diaphragm remains multinucleate unless and until the development of the embryo has begun. Later on this region also becomes multicellular by free cell formation. All this region beneath the diaphragm is packed with starch.

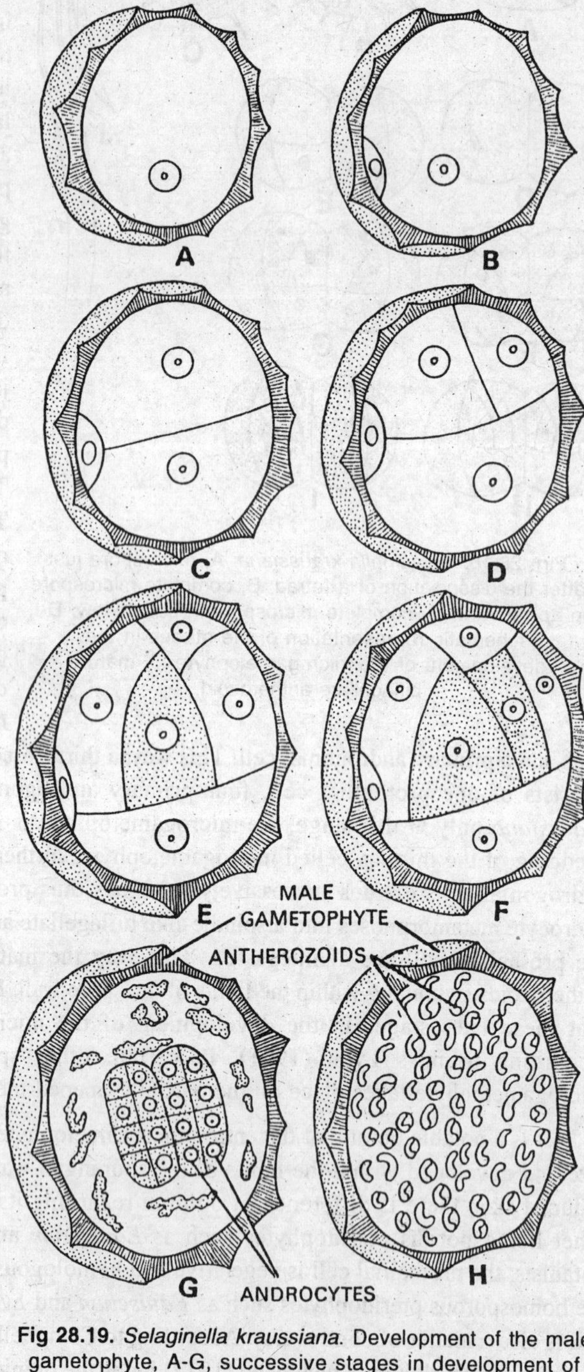

Fig 28.19. *Selaginella kraussiana*. Development of the male gametophyte, A-G, successive stages in development of microgametophyte; H, mature male gametophyte with antherozoids.

As soon as the apical tissue is formed, the wall of the megaspore cracks, and a portion of the tissue becomes exposed to the outside. Probably the exposed tissue becomes green. According to Bruchmann (1912) the rhizoids also develop from this portion. The food reserves for the development of the embryo have already been stored in the parent sporophyte.

Development of archegonium. Most of the superficial cells of the apical tissue of the megagametophyte are potential archegonial initials. Several of these archegonial initials develop into the archegonia. First of all the archegonial initial divides by a periclinal division so that a **primary cover cell** and a **central cell** are resulted. Again the central cell divides periclinally

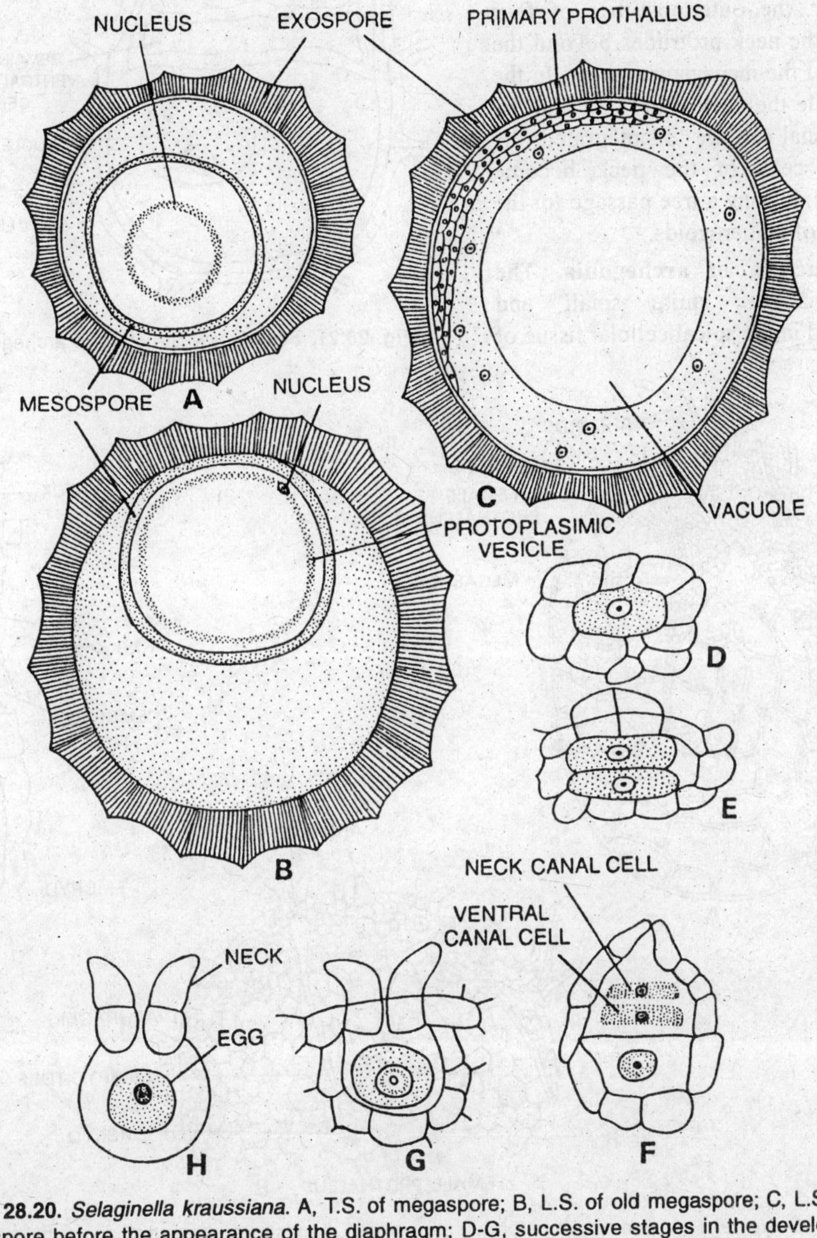

Fig. 28.20. *Selaginella kraussiana.* A, T.S. of megaspore; B, L.S. of old megaspore; C, L.S. of old megaspore before the appearance of the diaphragm; D-G, successive stages in the development of archegonium; H, mature archegonium.

resulting into a **primary canal cell** and a **primary venter cell**. The primary venter cell again divides periclinally producing a **venter canal cell** and an **egg**. This primary canal cell does not divide further and directly acts as **canal cell** of the archegonium. The primary cover cell formed in the beginning divides twice anticlinally at right angles to each other producing four **neck initials**. Each of the neck initials divides transversely, giving rise to neck two cells in height. However, the outermost tier of four cells of the neck protrudes beyond the surface of the megagametophyte. In the meanwhile the neck canal cell and the venter canal cell are disintegrated; the terminal cells of the neck become separated, leaving a free passage for the entrance of antherozoids.

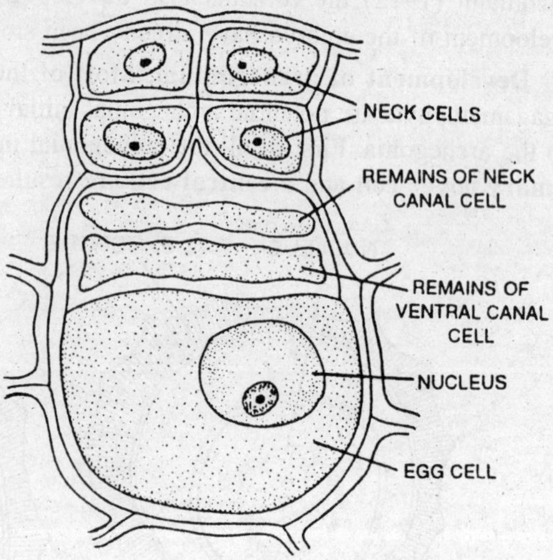

Fig. 28.21. *Selaginella*. Structure of archegonium.

Structure of archegonia. The archegonia are quite small and embedded in the apical cellular tissue of

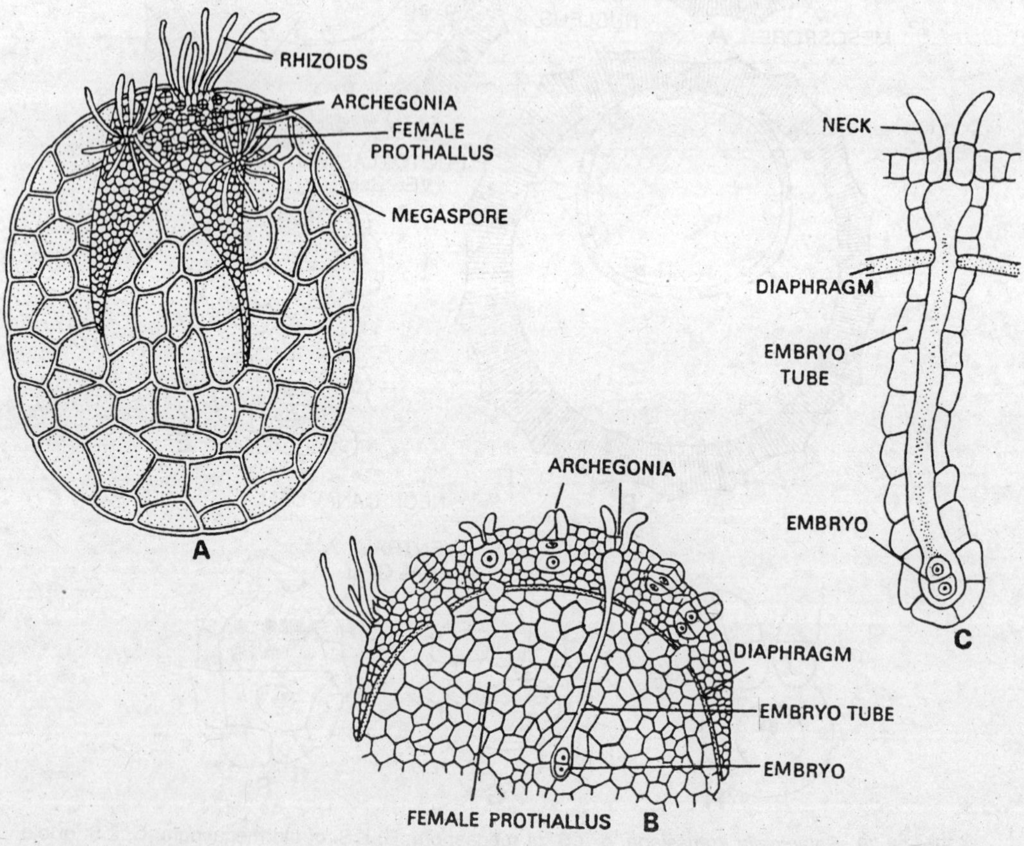

Fig. 28.22. *Selaginella* sp. Female gametophyte. A, splitted megaspore showing rhizoids, prothallus and archegonia; B, a part of the female gametophyte with diaphragm, rhizoid and young embryo; C, detailed structure of archegonium with young embryo.

megagametophyte. However, the outer tier of the neck of the archegonium, which is composed of four cells, projects out the gametophyte. The archegonium consists of a neck composed of two tiers of four cells each, a neck canal cell, a venter canal cell and an egg. On the maturity of the archegonium the neck and venter canal cells are disintegrated, the projecting cells become separated, laying a free passage for swimming the antherozoid to the egg.

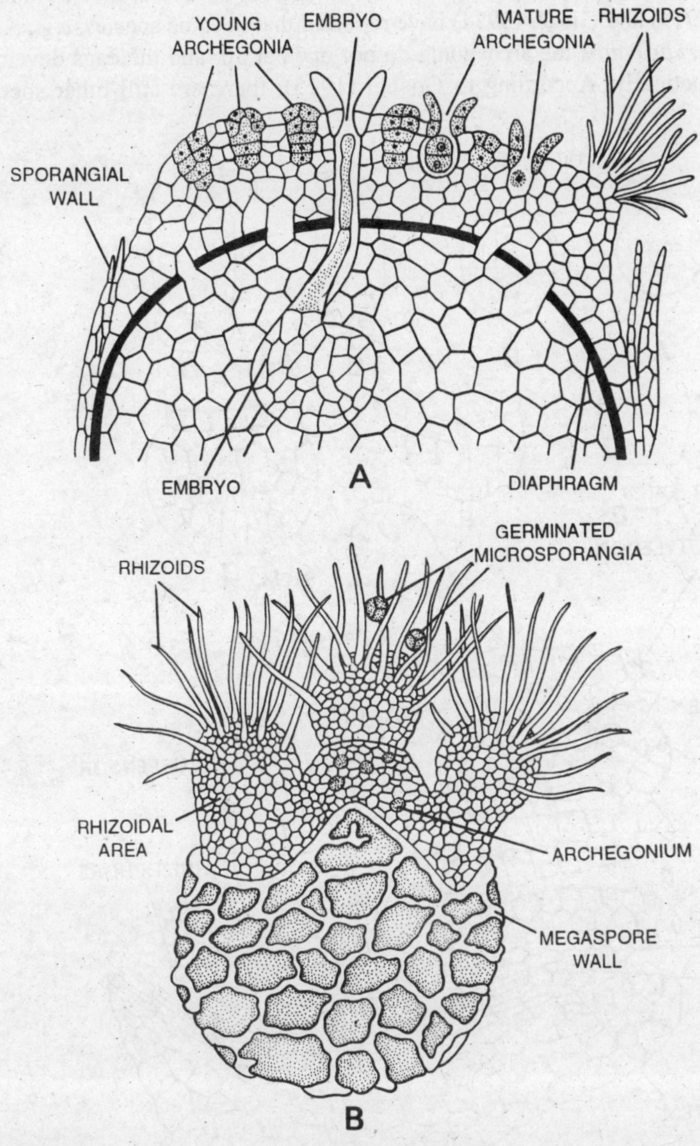

Fig. 28.23. *Selaginella*. A, median L.S. through mature female gemetophyte of *S. kraussiana* showing rhizoids, archegonia and embryo; B, megagametophyte (female gametophyte) of *S. galeotti* showing rhizoidal mounds.

Fertilization. According to Bruchmann (1912), the fertilization may take place while the megagametophytes are still in the megaspores situated in the megasporangia which may have fallen on the ground. The water is essentially required for the fertilization. After shedding from the microspore the antherozoids come in contact of the female gametophyte and approach the archegonia. The antherozoids swim down in the neck of the archegonium through the medium of water and approach the egg. Only one atherozoid penetrates the egg and the fertilization is effected. With the result the oospore is formed after fusion.

Apogamy. In some of the species the embryos develop apogamously. According to Miss Lyon (1904), the apogamy may take place due to the development of an archegonial initial into the embryo.

Parthenogenesis. This phenomenon has also been reported in several species of *Selaginella.* Bruchmann (1912, 1919) and Gieger (1934) have reported that in some species, *e.g.*, *S. helvetica*, *S. anocordia* and *S. rubricaulis* the archegonia do not open at all, and the eggs develop into the embryo parthenogenetically. According to Goebel (1915), there are still other species which

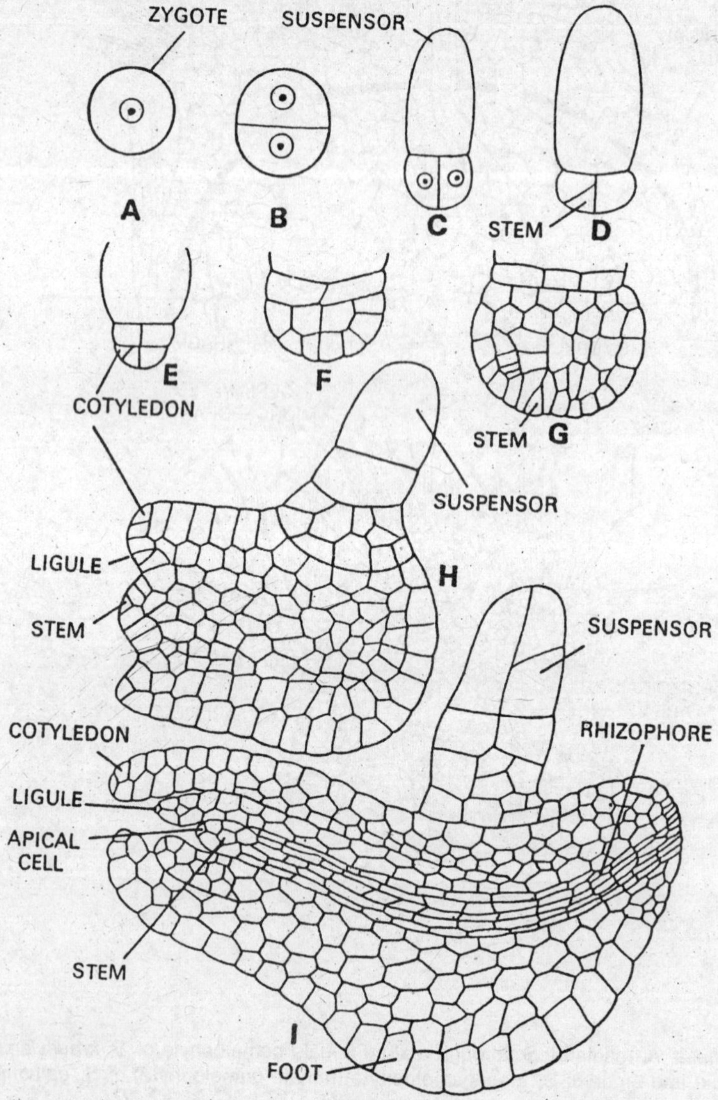

Fig. 28.24. *Selaginella* sp. Development of embryo. A, zygote; B, transverse division of zygote; C-I, different successive stages in the development of the embryo.

produce only megaspores, *e.g., S. intermedia*, and the eggs develop into the embryos parthenogenetically.

Development of embryo. As soon as the fertilization is over the zygote secretes a wall around it and develops into an embryo. First of all, generally, the zygote divides by a transverse wall giving rise to two cells. They are called the **epibasal** and **hypobasal** cells. Here, the further development of the embryo has been discussed of *Selaginella martensii*. The upper epibasal cell develops into a **suspensor**, whereas the hypobasal cell forms all the other parts of the embryo. The hypobasal cell divides vertically and this way two cells are formed. These two cells again divide by a vertical division and four cells are resulted. One of these four cells, divides diagonally by a vertical division forming the apical cell of the stem to be developed. Rest of three cells and the sister cell of the apical cell together divide by a transverse division, thus, two upper and lower tiers of four cells each are formed. Now the cells of both these tiers divide again and again vertically and transversely. There is more active cell division in one of the lower quardrants, and this leads to the formation of the **foot** of the embryo. The foot acts as haustorium and absorbs food for the embryo from the gametophyte. Simultaneously a superficial cell, in each of the two diagonally opposed quardrants in the upper tier begins to function as the apical cell of a **cotyledon**. Both these apical cells develop two identical cotyledons. These cotyledons possess ligules unlike the leaves near their bases. Along with the development of the cotyledons, the apical cell of the stem divides actively. The portion of the embryo found just behind the cotyledons develop into the hypocotyledonary portion of the stem.

As soon as the development of the cotyledon and the stem apex has begun, an apical cell differentiates from a superficial cell situated on one flank of the foot. The segments cut off from this cell develop into a root-like structure. According to Bruchmann (1909, 1912) this structure is considered to be a primary rhizophore. Later on from this primary rhizophore, the primary root comes out.

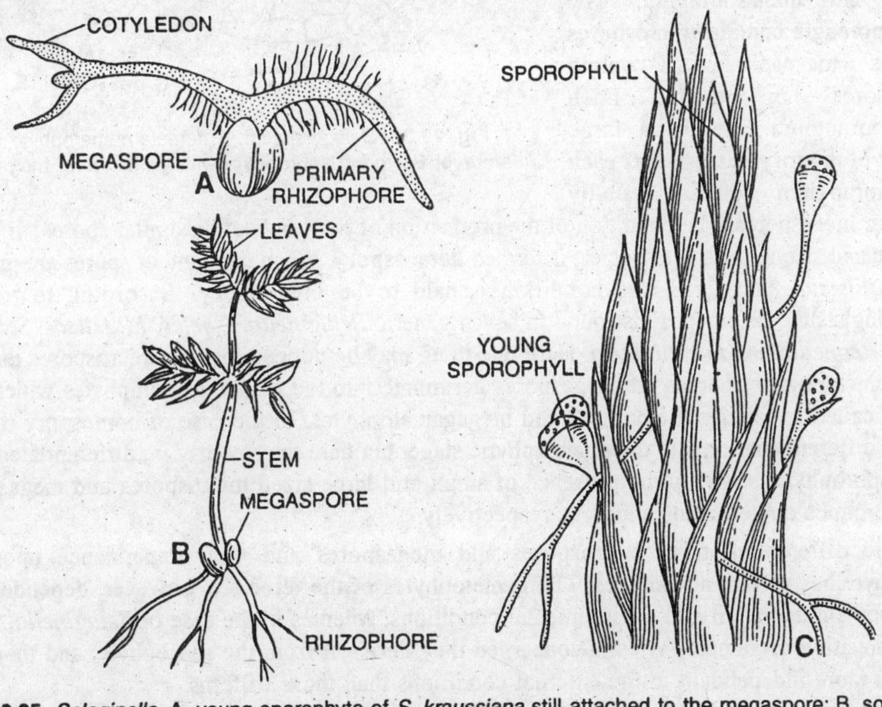

Fig. 28.25. *Selaginella*. A, young sporophyte of *S. kraussiana* still attached to the megaspore; B, somewhat older sporophyte showing the dichotomy of stem; C, portion of a strobilus of *S. rupestris* with young sporophytes.

The young sporophyte remains attached to the megagametophyte for sometime still situated within the megaspore wall. The young sporophyte grows further by sending its stem and cotyledons upward and the rhizophore and the roots downward in the soil. The most peculiar features of the juvenile sporophyte of *Selaginella* are as follows:

(a) The cotyledons are directly borne upon the stem (not found in other pteridophytes).

(b) Just posterior to the cotyledons a conspicuous hypocotyledonary part of the stem develops (not found in other pteridophytes). The apical cell of the stem is situated in between the cotyledons.

Later on, the gametophyte is finished, and the young sporophyte becomes independent by establishing its root system in the soil.

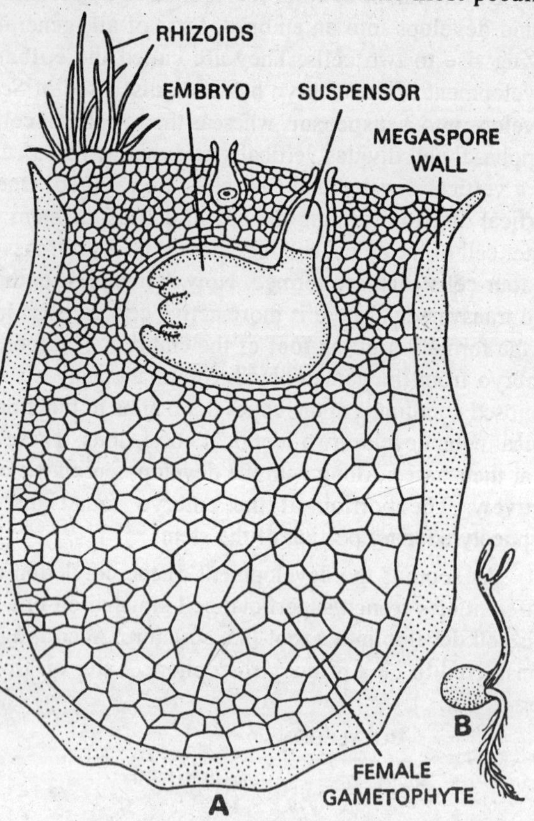

Fig. 28.26. *Selaginella* sp. A, female gametophyte with embryo; B, young sporophyte with germinating megaspore

Specialization of spores and heterospory. The sporangia show greater speciali...ion than those of ferns. They are differentiated into micro and megasporagnia. The microsporangia contain microspores whereas megasporagnia contain megaspores in them. Each microsporangium contains a large number of microspores whereas each megasporangium contains usually only four megaspores. The condition of the production of two kinds of dissimilar spores, differing in shape and size in the same species, is termed **heterospory.** When one kind of spores are present, *e.g., Equisetum* and ferns, the condition is said to be **homospory**. According to modern pteridologists the heterospory is found in seven genera: *Selaginella, Isoetes, Marsilea, Salvinia, Azolla, Regnellidium and Pilularia.* Here one thing may be noticed that the microspores develop male gametophytes whereas the megaspores germinate into the female gametophytes which may also be called as microgametophytes and megagametophytes. In the case of homospory the sex may be differentiated only at the gametophytic stage, but here the sex may be differentiated even at the sporophytic stage by the presence of small and large sized microspores and megaspores which produce antherozoids and eggs respectively.

The differentiation of microspores and megaspores and their dependence upon the sporophyte has certain advantages. The gametophytes of the ferns are, however, dependent for their nutrition upon soil and environmental conditions, whereas in the case of *Selaginella*, as far as the nutrition of gametophytes is concerned they derive it from the sporophyte, and therefore they are more independent to the external conditions than those of ferns.

Seed habit. The condition of heterospory in *Selaginella* constitutes one of the most important phases in the evolution of the plants. In certain species of *Selaginella* only one megaspore develops within the megasporangium. The highest evolved species, *i.e., S. apoda* has advanced almost upto the rank of a seed plant. Here, the megaspore remains within the sporangium; it itself fertilizes and this way for the first time the gametophyte of *Selaginella* shows complete dependence of the gametophyte upon sporophyte as in angiosperms.

Alternation of generations. Like other pteridophytes, there is an alternation of sporophytic and gametophytic generations. In *Selaginella* the sporophyte bears two kinds of sporophylls-sporangia-and spores, *i.e.*, microspores and megaspores giving rise to male and female gametophytes respectively. The antherozoid from the male gametophyte fertilizes an egg in the archegonium situated on the female gametophyte; the oospore is resulted which develops into a

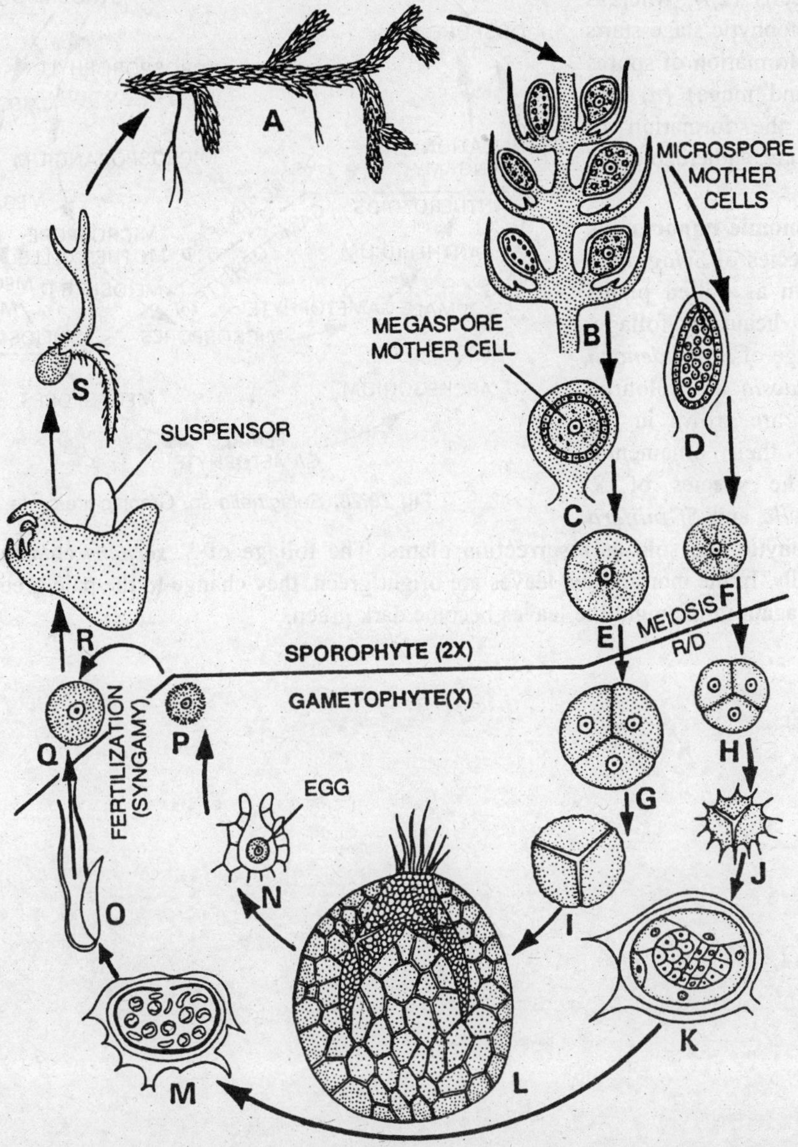

Fig. 28.27. *Selaginella* sp. Diagrammatic life-cycle. A, sporophyte with spikes and rhizophore; B, L.S. of spike; C, megasporangium with megaspore mother cell; D, microsporangium with microspore mother cells; E, megaspore mother cell; F, microspore mother cell; G, megaspore tetrad; H, microspore tetrad; I, megaspore; J, microspore; K, developing microgametophyte; L, megagametophyte; M, microgametophyte; N, archegonium with egg; O, antherozoid; P, egg; Q, oospore; R, embryo; S, young sporophyte.

sporophyte. Here, in comparison to the sporophyte, the gametophyte is very much reduced. On the other hand gametophytes are never independent for their nourishment. They take their nourishment from the sporophyte. Here in *Selaginella*, the gametophyte is greatly reduced in size, especially in the male, where it is represented by a single prothallial cell and a single antheridium. The sporophytic stage begins with the formation of oospore ($2n$) and ends with the formation of spore mother cells ($2n$), whereas the gametophytic stage starts with the formation of spores (micro and mega) (n) and ends in the formation of gametes, *i.e.*, antherozoid and egg (n).

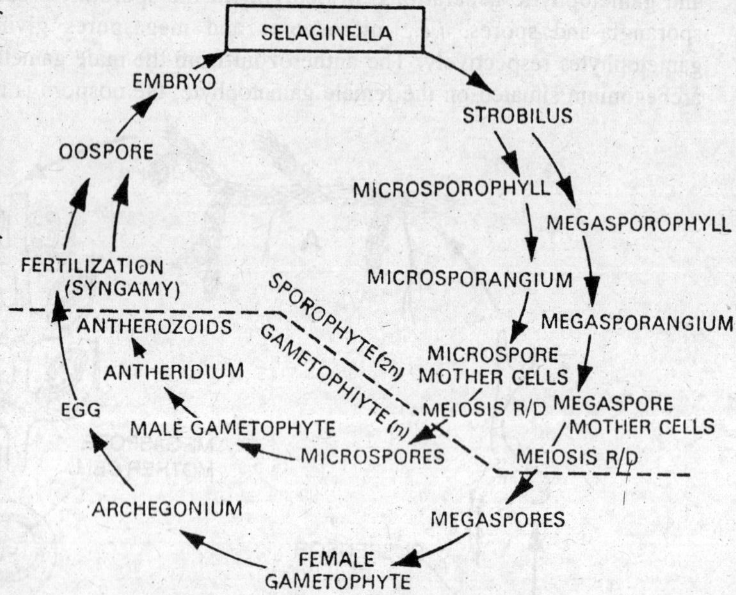

Fig 28.28. *Selaginella* sp. Graphic life-cycle.

Economic importance. Many species of *Selaginella* are grown as potted plants for their beautiful foliage. The foliage of *S. willdenovii* and *S. caesia* are coloured and they are grown in the pots for their ornamental value. The species of *S. lepidophylla* and *S. pilifera* are xerophytic and sold as resurrection plants. The foliage of *S. serpens* changes its colour periodically. In the morning the leaves are bright green, they change to the pale green in the day time and again in the night the leaves become dark green.

29
CHAPTER

Division-Arthrophyta
Class-Calamopsida

Order-Equisetales

Characteristic features. The sporophytes possess distinctly articulated stems. The ridges and furrows present on the internodes of the stems. The stems may or may not be branched at the nodes. If branched, the branches are present in the whorls alternating to the leaves. The leaves are scale-like. They form a toothed sheath around each node of the stem by their lateral fusion. The stele is siphonostelic. It consists of a number of collateral vascular bundles in the internode which, however, join to each other forming a continuous ring at each node. The sporangia are borne on the underside of peltate sporangiophores. The sporangiophores are arranged quite close to each other in terminal strobili or cone-like structures. The antherozoids are multiciliate.

Genus EQUISETUM

Systematic position - Pteridophyta
Division- Arthrophyta
Class. Calamopsida
Order. Equisetales
Family. Equisetaceae
Genus. *Equisetum*.

The genus comprises of about 25 species. They are world wide in distribution, however no species have been recorded from Australia and New Zealand. They grow in a variety of habitats. Majority of the species are found in north temperate zone. On the other hand many species are tropical and found in South America and West Indies. The common Indian species are: *E. debile, E. arvense* and *E. ramosissimum*. Certain species grow in ponds and marshy places, *e.g., E. palustre;* some species are shade loving and grow in damp shady places, *e.g., E. pratense* whereas some other species grow in exposed conditions. *E. arvense* is very widely distributed; *E. debile* is found along the banks of the rivers, canals and pools in the Indian plains. They are commonly known as horse tails.

External Characters

Habit. About all the species of *Equisetum* are perennial. They bear much branched, horizontal, creeping and subterranean rhizomes which sometimes penetrate 3 or 4 feet deep in the soil. The rhizomes bear the erect and aerial branches. Some of the aerial branches are sterile and some are fertile. The plants vary in their height from species to species. *E. scirpoides* is only few inches high, whereas, on the other extreme a species of American tropics, *i.e., E. giganteum* reaches upto 12 metres or 40 feet in height. However, its diameter is less than an inch. The plant of *E. debile* grows along the banks of the river, may attain the height of 10 to 15 feet when grows in dampy and shady places. Its diameter is about one and half centimetres.

Stem. The rhizome of *Equisetum* is horizontal, perennial, much-branched, subterranean and penetrates in the soil upto 3 to 4 feet and in many species it is spread over in the area of 10 to 15 feet. It consists of well differentiated nodes and internodes. At each node there is a whorl of small,

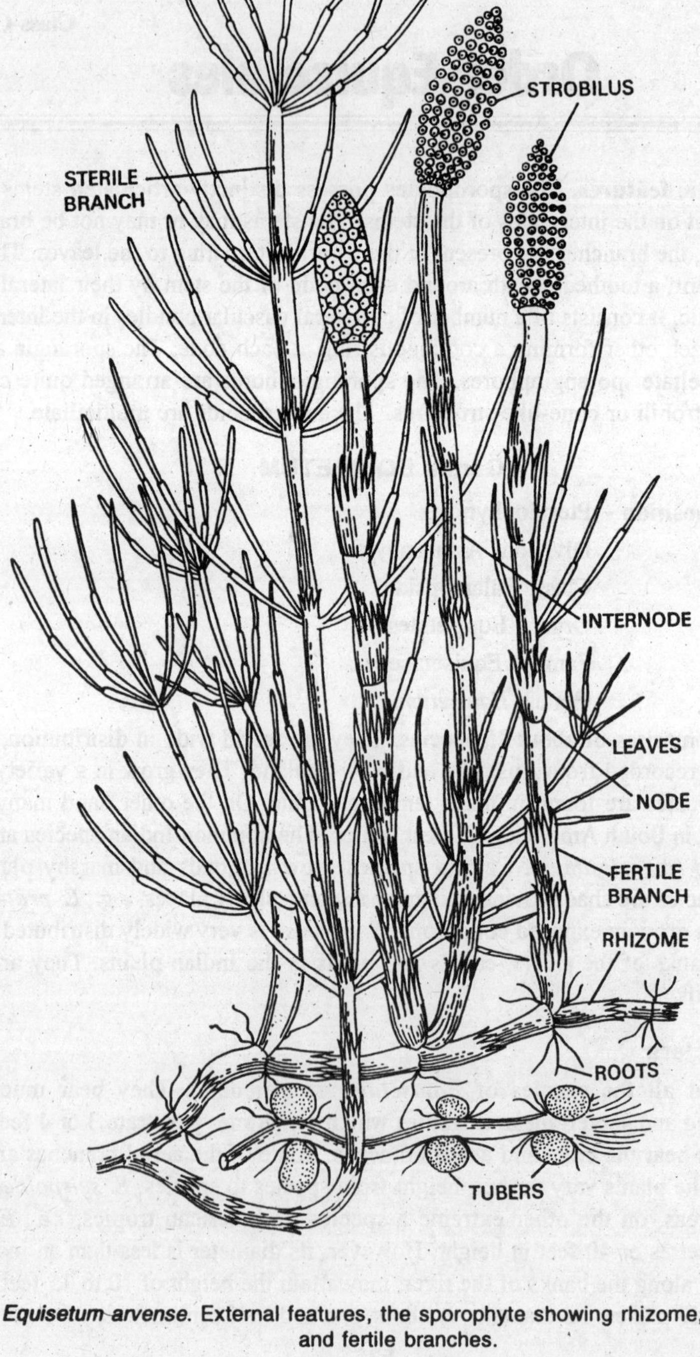

Fig. 29.1. *Equisetum arvense.* External features, the sporophyte showing rhizome, roots, tubers sterile and fertile branches.

scale-like leaves united laterally to each other forming a brown sheath. Alternating with each leaf of sheath at each node of the rhizome, there is a branch primordium or bud. These primordia may develop into aerial on subterranean branches immediately, or they may remain dormant for a considerable long period and develop into new branches on the approach of favourable conditions. Sometimes in certain species, *e.g.*, *E. telmateia* and *E. arvense* the primordia develop into short rounded branches composed of only one internode. These short rounded branches are known as tubers. After being separated from the parent rhizome, these tubers develop into new plants. This is a method of vegetative propagation of the sporophytic generation. The whorls of the much branched adventitious roots have been given from the successive nodes of the rhizome.

Usually the rhizomes bear two kinds of aerial branches-sterile and fertile, but there are still other species which bear either sterile or fertile branches.

The sterile branches are of green colour and bear a whorl of lateral branches at each node. These branches are assimilatory in function and practically synthesize all the food required by the plant. The lateral branches found on the primary branch also bear the whorl of branches which are comparatively smaller in size. Each whorl bears an equal number of branches to that of leaves, and alternate with the leaves. The sterile branches of a few species are unbranched, *e.g.*, *E. hyemale.*

The fertile branches are also found above the ground, they bear strobili at their apical ends. These branches arise first and shed their spores even before the appearance of the sterile branches. In *E. arvense* the fertile shoots are unbranched and devoid of chlorophyll, and they die as soon as the spores are shed. In certain other species, *e.g., E. cylvaticum* and *E. pratense* the strobili are thrown off after spore dissemination, and the fertile branch now acts as the sterile branch by developing lateral branches on it. Simultaneously, the chlorophyll is also developed in these branches. *E. palustre* bears three kinds of branches — (a) the sterile branches which are deep green and much branched, (b) the fertile shoots which are short lived and devoid of chlorophyll, (c) the intermediate type of branches which are fertile, devoid of chlorophyll and unbranched at first, but later on the strobili are thrown off and the branch becomes green, branched and persistent. This is a good example of division of labour in *Equisetum* plant.

As regards function, the rhizome acts as store house of the food material and helps in anchoring plant to the soil. The aerial sterile branches are green, and therefore, they synthesize all the food required by the plant. The fertile branches are without chlorophyll and bear the strobili at their apical ends.

The internodes of both the rhizome and aerial branches are ribbed longitudinally. These ribs alternate with the leaves in the subtending node. The ribs also alternate to the ribs of the successive node, in their position. The general structure of the stem is same both in the rhizome and the aerial shoot, but it may be more clearly understood in the aerial shoots.

Leaves. The leaves are arranged in whorls at each node of the aerial branches. They are simple, scaly, slender, uninerved, brownish and more or less fused at their bases laterally to form a sheath around the base of the internode. Their apices are, however, free and pointed. The number of the leaves in each whorl ranges 3 to 40 from species to species. At first, the leaves are green, but later on they become brownish, dry and scale-like. The mature leaves are protective in function. In certain other species they become brownish, dry and scale-like. The mature leaves are protective in function. In certain other species they become dead, dry, and scaly.

Roots. In *Equisetum* all the roots are adventitious except the primary root of the young sporophyte. The roots arise in whorls from each node of the creeping, sub-terranean rhizome. The roots may be many years together, but do not increase in their length. The roots do not arise directly, but from the lateral branch primordia situated on the nodes of the rhizomes and main aerial shoots.

Internal Structure

Stem. To study the internal structure of the stem, the transverse and longitudinal sections, passing through the nodes and internodes are required. Thus, the internodal and nodal anatomy of the aerial stem may be studied separately. It is as follows:

Internal anatomy. To study the anatomy of the node, one has to cut the aerial stem through internode in transverse sections. The outer surface of the stem looks wavy, because of the presence of ridges and furrows.

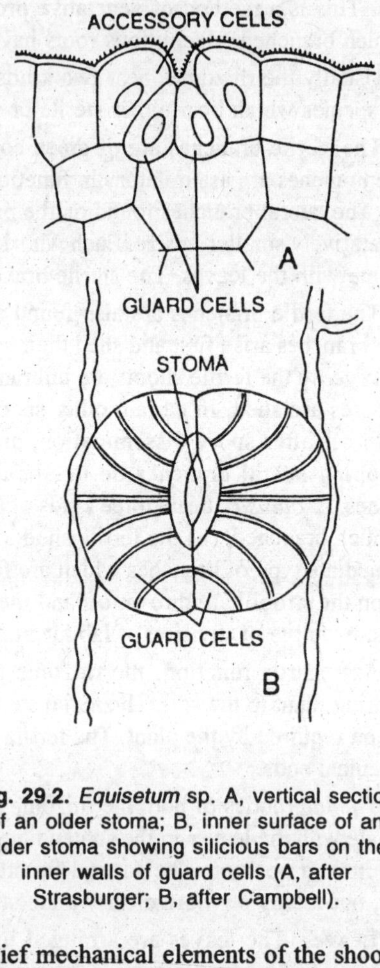

Fig. 29.2. *Equisetum* sp. A, vertical section of an older stoma; B, inner surface of an older stoma showing silicious bars on the inner walls of guard cells (A, after Strasburger; B, after Campbell).

The stem is covered over by a single layered epidermis interrupted by stomata situated in the grooves of furrows. The epidermis is always impregnated with a thick layer of silica. The deposition of the silica on the epidermal layer gives the rough appearance to the stem, and therefore, the *Equisetum* plants are also known as **'scouring rushes'**. The stomata are found in grooves of the aerial shoots. The development of the stomata is somewhat peculiar in the sense that the initial divides twice by successive longitudinal divisions and this way the two innermost cells develop into the guard cells, whereas, the two outermost cells develop into accessory cells, on being mature of the stomata, the two accessory cells completely overarch the guard cells and the stoma. In majority of the species, *e.g., E. hymeale, E.ramosissimum,* etc., the stomata are sunken, whereas in some other species, *e.g., E. palustre E. pratense,* etc., they are situated on the surface of the epidermis. The silica is deposited in the wall of guard cells in transverse radial bands.

Just underneath the epidermis, there is broad cortex. The cortex consists of mechanical and assimilatory tissue. In the outer cortex, just beneath each of the ridges, there is a strand of sclerenchyma. Usually the sclerenchyma is restricted to the periphery but in *E. giganteum* it extends inward to the endodermis. These columns of sclerenchyma are chief mechanical elements of the shoot. In addition to the abovementioned sclerenchymatous columns, the sclerenchyma strands are also found in the furrows. Here, each strand is situated in between the curved strands of chlorenchyma. The chlorenchyma possess well developed intercellular spaces. These spaces are very much distinct below the stomata. These chlorenchymatous bands are found beneath the sclerenchymatous strands situated underneath the ridges. The ends of the chlorenchymatous bands touch the epidermis in the grooves. Since the leaves are small, scaly and have less number of chloroplasts the chlorenchyma of the cortex of the shoot is the major photosynthetic tissue. The inner cortex is composed of thin walled parenchyma with well developed intercellular spaces. In this region of the cortex the air spaces known as **vallecular canals** are also present. Usually vallecular canals are found opposite the furrows in the deeper tissue of the cortex. They are alternate to the vascular bundles. These canals are filled up with air.

The endodermis is the last layer of the cortex. In *E. arvense*, *E. telmateia*, *E. palustre* and certain other species, the endodermis surrounds the entire stele. In *E. giganteum*, *E. limosum* and some other species, each vascular bundle is surrounded by separate endodermal layer. In still other cases, *e.g.*, *E. hyemale*, *E. ramosissimum,* etc., there is a common internal endodermis inside the ring of bundles delimiting the pith and common outer endodermis found outside the ring of bundles. Just beneath the endodermis a single layered pericycle is found. The pericycle is the outermost layer of the stele.

The vascular skeleton, *i.e.*, stele of *Equisetum* is siphonostelic. The separated vascular bundles are arranged in a ring. The vascular bundles are situated opposite the ridges and alternate

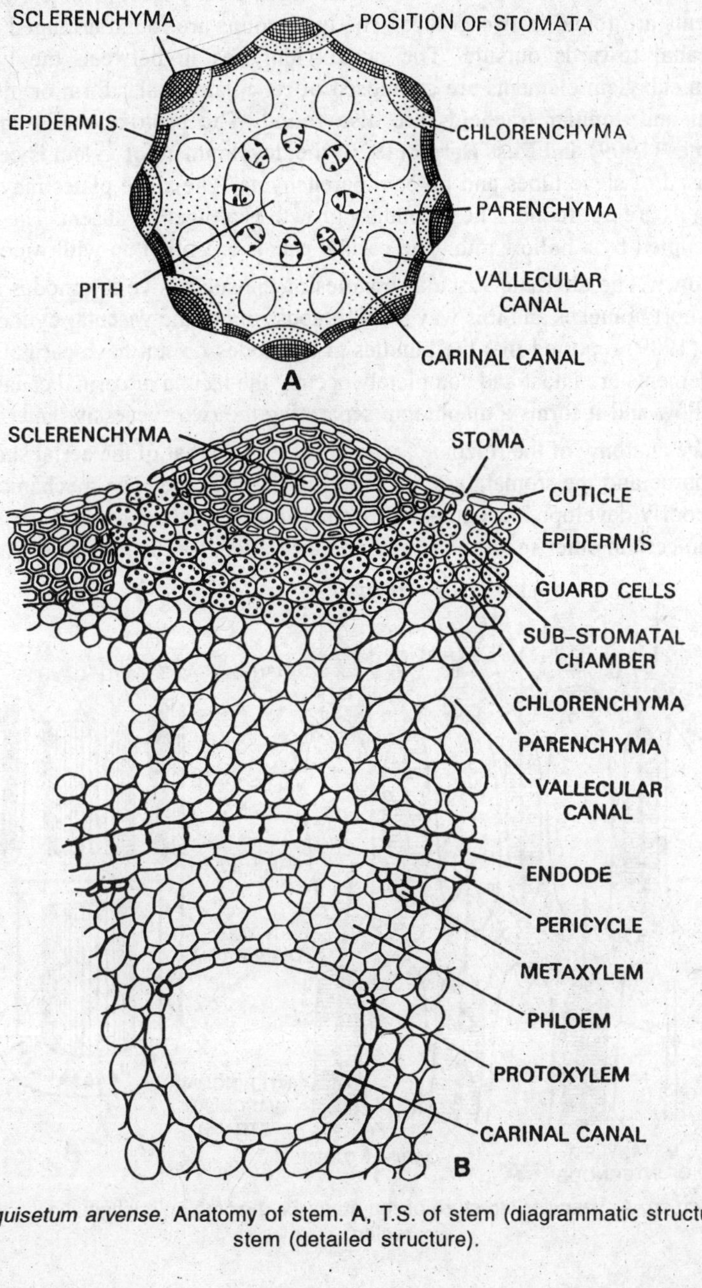

Fig. 29.3. *Equisetum arvense.* Anatomy of stem. A, T.S. of stem (diagrammatic structure); B, T.S. of stem (detailed structure).

to the vallecular canals. The number of bundles varies from species to species. In each internode, as many vascular bundles are found as there are leaves at the node. The vascular bundles alternate to each other in their position in the successive nodes. The internodal portions are longitudinally perforate, and each perforation runs the length of an internode. The perforations in the internode are situated above the traces. These perforations are not branch gaps. The vascular bundles are collateral type and resemble to some extent to that of monocotyledons. The vascular bundles contain both metaxylem and protoxylem. A **carinal canal** is developed in each bundle, because of the disintegration of the early-formed tracheids of the protoxylem during elongation of the surrounding cells of the internode. The remaining protoxylem elements are composed of few tracheids. These protoxylic elements are found arranged to the margin of carinal canal. The metaxylem elements are found in two groups. The two groups are found arranged on the margin of the carinal canal towards outside. The protoxylem lies in between the two groups of metaxylem. The metaxylem elements are composed of reticulate, scalariform or pitted tracheids. Sometimes spiral and annular tracheids are also found. The protoxylem is endarch and as supported by Eames (1909) and Miss Barrat (1920) the development of xylem is centrifugal. The phloem is composed of sieve tubes and phloem parenchyma. The sieve plates may also be seen. The companion cells are not found. The secondary growth is altogether absent. The central region of the stem is occupied by a hollow pith. The carinal canals are filled up with water.

Nodal anatomy. The alternate vascular bundles of the successive internodes are connected to each other by short branches and this way a continuous ring of the vascular cylinder is found in the node. Eames (1909) reported that the bundles at the nodes do not have carinal canals. Here, the protoxylem elements are intact and completely occupy the lacuna or carinal canal. At the node, the pith is not hollow and it forms a diaphragm separating the two successive internodes.

Rhizome. The anatomy of the rhizome is quite identical to that of the aerial shoot. However, the assimilatory tissue and the stomata are not found in the rhizome. The mechanical tissue, *i.e.*, sclerenchyma is poorly developed as compared to that of the aerial shoot. In *E. arvense,* the pith is solid, whereas in certain other species, the pith and the vallecular canals are very much reduced.

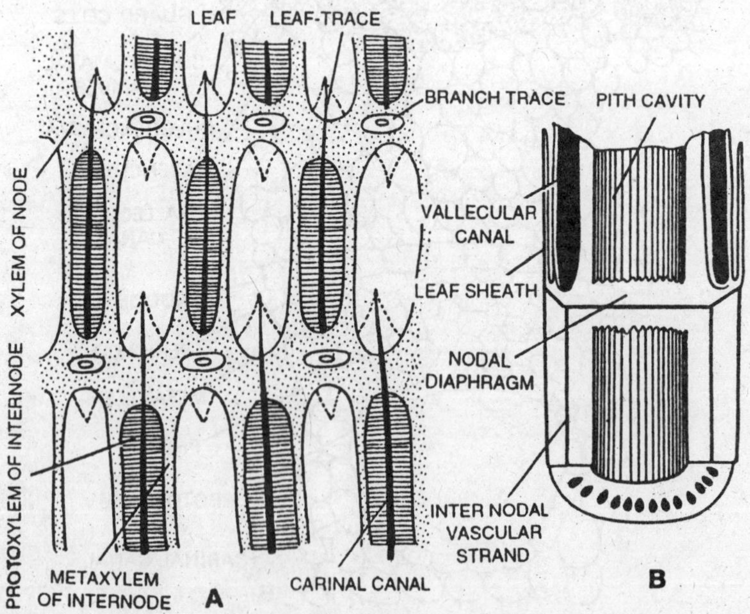

Fig. 29.4. *Equisetum* sp. A, vascular structure of the stem; B, rhizome split longitudinally through nodal region.

Leaf. The anatomy of the leaf is quite simple. The leaves are uninerved, this means that each leaf contains a single vascular bundle. The vascular bundles of the leaf sheath are simple and collateral. The carinal canals are not found. Individual bundles are surrounded by separate endodermal layers (endodermis). The outer tissues of the leaf sheath are composed of narrow sclerenchymatous bands. These bands of sclerenchyma pass up the leaf ridges and alternate with the strips of chlorophyllous tissue associated with stomata.

Root. The adventitious roots are borne at the nodes of the rhizome and aerial shoots. The anatomy of adventitious roots is quite simple. To study the anatomy the cross sections of the root are required. The outermost layer is known as piliferous layer which bears unicellular hairs upon it. Just beneath the piliferous layer there is a multilayered cortex. In small roots the cortex is divided into two zones. The outer zone is composed of three to four layered lignified exodermis. The inner zone of the cortex consists of thin walled parenchyma with well developed intercellular spaces.

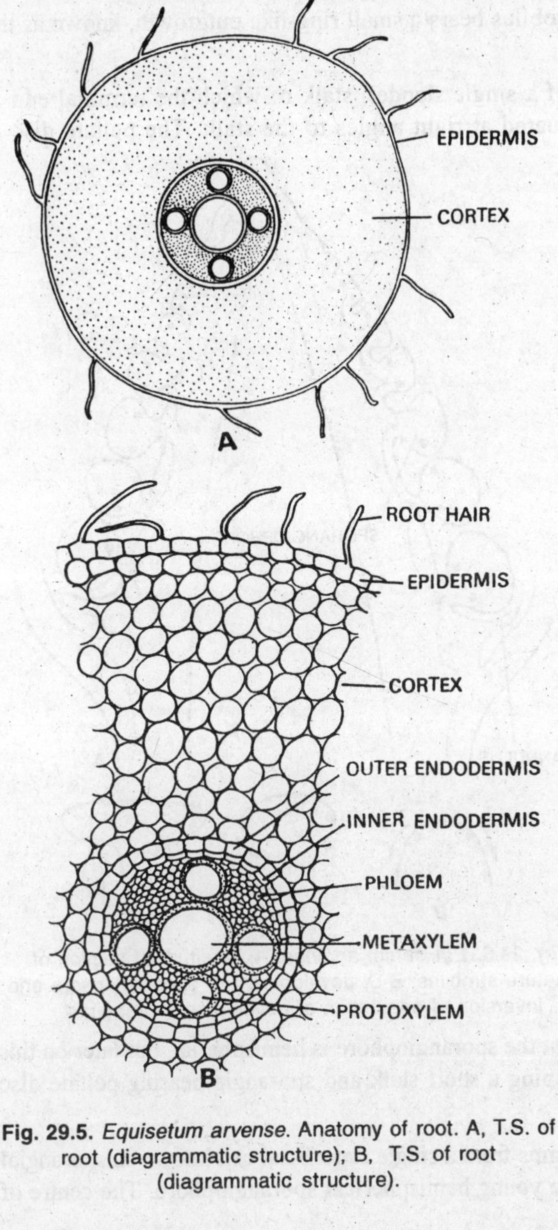

Fig. 29.5. *Equisetum arvense.* Anatomy of root. A, T.S. of root (diagrammatic structure); B, T.S. of root (diagrammatic structure).

The endodermis is two celled in thickness. The inner layer of the endodermis acts as pericycle. However, the pericycle is absent. The lateral roots originate from the inner layer of the endodermis. In between the two layers of the endodermis, intercellular spaces are found. The Casparian strips are strictly found in the outer layer of the endodermis.

The stele of the root varies in its nature from species to species. It is triarch to hexarch. There is a big central axial metaxylem elements surrounded by three to six narrow points of protoxylem. Each protoxylem point is represented by a single tracheid. The tracheids of xylem elements are spirally thickened. The angles between the protoxylem points are completely filled with phloem. The phloem is composed of phloem parenchyma and sieve tubes.

The apical growth. The apical growth of the main aerial shoot and its laterals takes place by means of a single pyramid like apical cell. This apical cell cuts the segments at its three faces. The segments are cut off regularly. Each segment divides anticlinally forming two upper and lower segments. Both these segments give rise to two tier of cells by successive divisions. The upper tier of the cells develops into the node and the lower into the internode.

In the roots, a young root primordium is differentiated into a group of initials which develop into a root cap and a pyramid-like apical cell, which develops into the root proper. Here, the behaviour of apical cell differs from that of the stem that it cuts off segments from all its four faces, whereas, in the stem only three faces of the tetrahedral apical cell cut off the segments. The lateral segments form the tissues of the root proper, and the terminal segment forms the root-cap.

The strobilus and sporangia. In *Equisetum,* each strobilus consists of a thick axis, having several whorls of densely crowded sporangia bearing peltate appendages situated on the sporangiophores. The sporangiophores are arranged in whorls. Usually, each such whorl is composed of twenty sporangiophores or so. In many cases, immediately below the sporangiophores, the central axis of the strobilus bears a small ring-like outgrowth, known as the **annulus.**

Each sporangiophore is composed of a single slender, stalk of which the terminal end is expanded into a flattened peltate disc situated at right angles to the stalk. The peltate disc is usually hexagonal in outline in its surface view. This hexagonal appearance of the discs is attained due to the mutual pressure of the discs of the crowded sporangiophores situated on the strobilus. Each peltate disc bears a ring of 5 to 10 isolated sporangia on its under side. Each sporangium is elongated and sac-like. On their maturity, the sporangia open by longitudinal slits for dispersal of the spores.

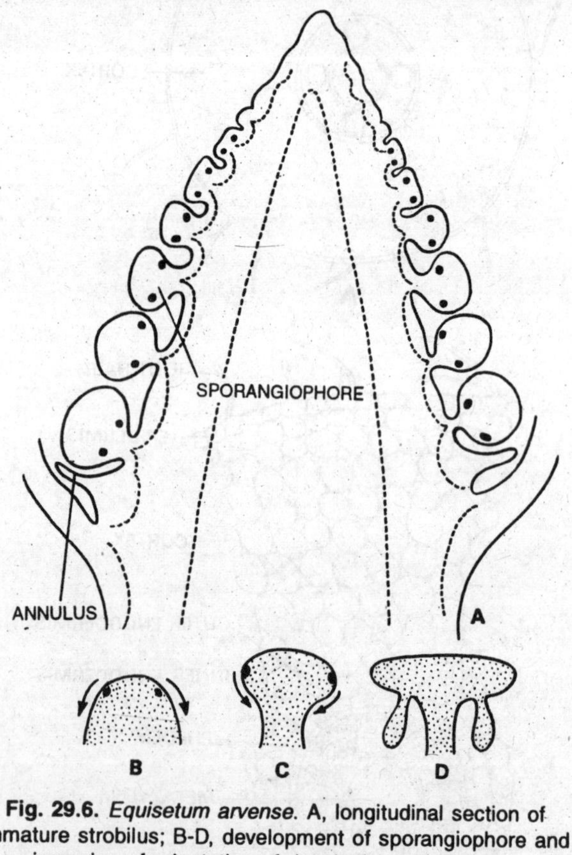

Fig. 29.6. *Equisetum arvense.* A, longitudinal section of immature strobilus; B-D, development of sporangiophore and inversion of orientation of developing sporangium.

Development of sporangium. The development of the sporangia of *Equisetum* is eusporangiate type. This means, that they are not entirely derived from a single sporangial initial. According to Bower (1894) all the sporogenous tissue in a sporangium are developed from a single superficial cell of the young sporangiophore.

The development of the sporangia begins as the strobilus ceases to grow in length. The young sporangiophores are found in the form of convex outgrowths in acropetal succession on the conical axis of the strobilus. In the beginning of the development the sporangiophore is hemispherial, but later on this soon becomes constricted at its base developing a short stalk and sporangia bearing peltate disc at its apical end.

The development of the sporangium begins from a single superficial cell known as sporangial initial situated in the peripheral region of the young hemispherical sporangiophore. The centre of the tip of young sporangiophore increases in size actively and the sporangial initials are shifted to its lateral sides. Sooner or later, because of further development, the outer edge of the sporangiophore becomes inverted and faces the axis, thus, the sporangial initials have also changed their position, and now facing the axis.

At first, the sporangial initial divides periclinally forming the two daughter cells. The inner daughter cell gives rise to the sporogenous tissue by repeated division, whereas the outer cell gives rise to both the sporogenous tissue and the portion of the jacket of the sporangium radial to the sporogenous tissue. The remaining portion of the jacket layer is developed from the cells laterally placed to the original initial. Ultimately the jacket layer becomes multilayered. The innermost layer is the tapetum and supplies nourishment to the developing spores. During the development of the spores, the tapetum and all but two layers of the jacket become disintegrated. Thus, the mature sporangium is surrounded by a two layered jacket. The cells of the outermost layer of the jacket are spirally thickened.

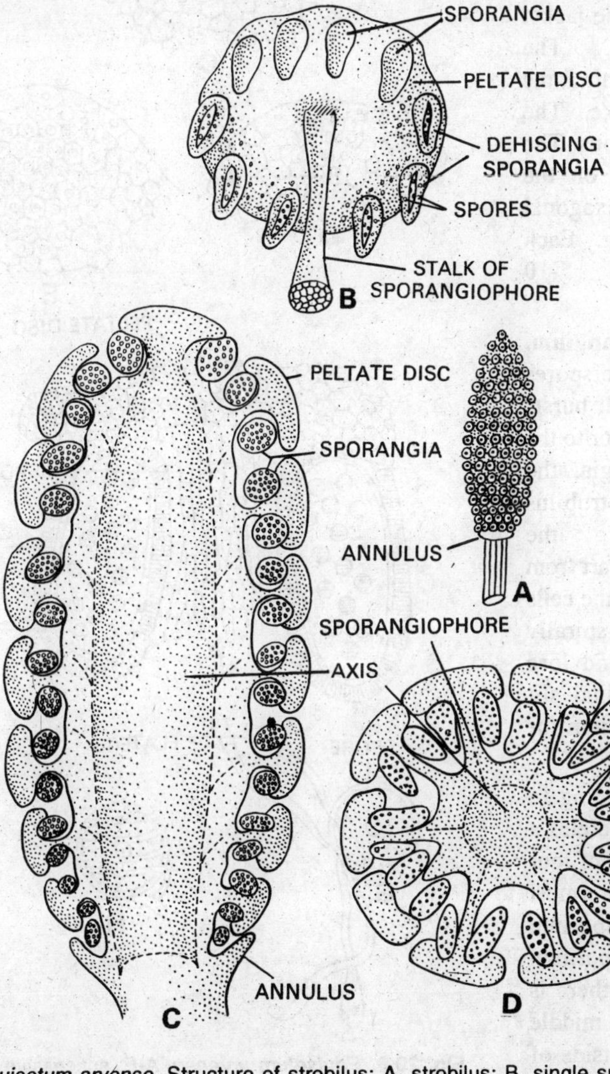

Fig. 29.7. *Equisetum arvense.* Structure of strobilus; A, strobilus; B, single sporangiophore with ten sporangia on the underside of the peltate disc; C, L.S. of strobilus showing sporangia; D, T.S. of strobilus showing sporangia.

During the development the cells of the sporogenous tissue become separated and rounded and act as spore-mother cells. About one third of the mother cells disintegrate. The disintegration of the mother cells, tapetum and the other jacket layers, together form a plasmodial fluid in which the remaining two third spore mother cells float about. Ultimately the spore mother cells undergo usual meiotic division developing the spore tetrads. Each tetrad is composed of four haploid spores.

Structure of sporangium. Each mature sporangium is elongated, sac-like and rounded at its apex. On maturity, it consists of single layered intact layer. By now, rest of the jacket layers are aborted. The sporangium contains within it the mature spores all alike. The *Equisetum* is homosporous. The sporangia are attached on the underside of the hexagonal peltate disc in a ring. Each sporangiophore bears 5-10 sporangia.

Dehiscenc of sporangium. On the maturation of the spores within the sporangium, it bursts by a longitudinal slit. Prior to the opening of the sporangia, the internodes of the strobilus elongate separating the sporangiophores quite apart from each other. The walls of the cells of the jacket layer are spirally thickened they shrink and lose water and the longitudinal slit develops on the sporangium for the release of the spores.

The spores. According to Beer (1909) the spore secretes a wall around it, composed of four concentric layers. The outermost layer is known as exosporium and the innermost endosporium. Outside the exosporium there is delicate cuticle called the middle layer. On the extreme outside of the spore there lies epispore which later on develops into four strips which get separated from

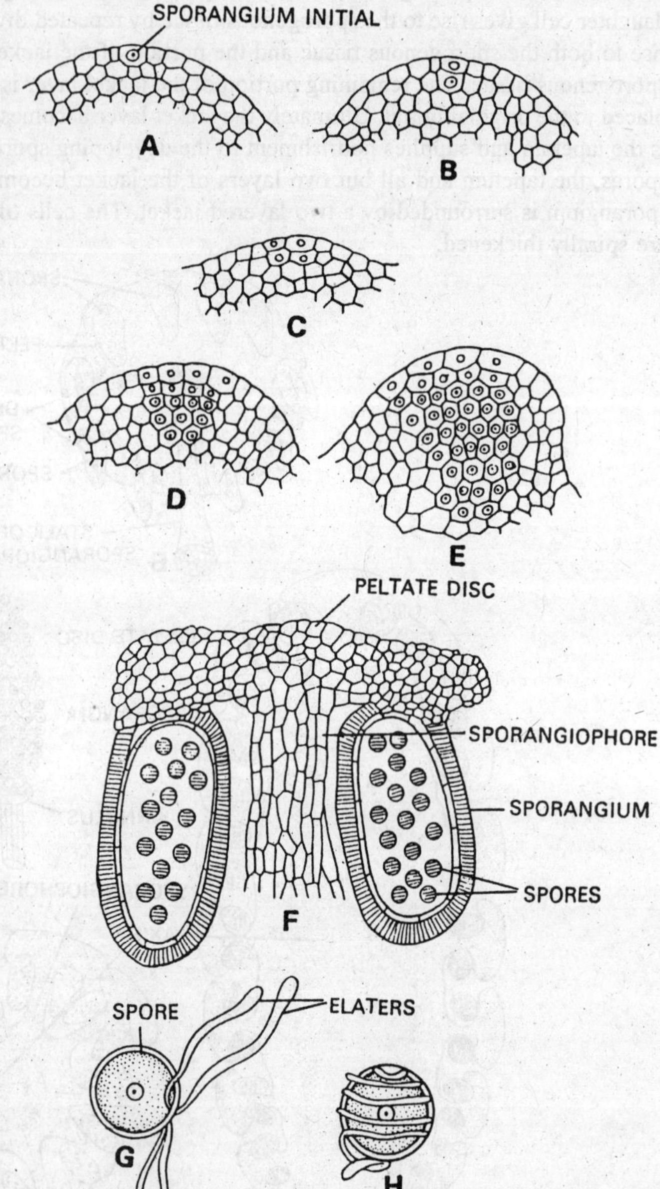

Fig. 29.8. *Equisetum arvense.* A-E, successive stages in the development of sporangium; F, sporangiophore with peltate disc and sporangia in longitudinal section; G, spore with uncoiled elaters; H, spore with coiled elaters.

each other from rest of the wall but for one common point of attachment. These strips are **elaters**. The strips or elaters remain coiled around the spore unless and until it is not disseminated from the sporangium. The elaters are hygroscopic. They remain coiled and uncoiled around the spore due to the changed conditions of the moisture of the air. When the air is moist they remain coiled around the spore and as soon as the air is dry the elaters are uncoiled. The elaters of *Equisetum* do not resemble in any respect to the elaters of Bryophyta. The ends of the spiral elaters are spoon-like. Each spore contains a centrally located nucleus and many small chloroplasts. The spores remain viable only for a few days after they are released from the sporangium.

Function of elaters. The function of elaters is not very certain, but it is assumed that their expansion may help in the dehiscence of sporangium. It is also thought that their hygroscopic movements may assist in the dispersal of the spores. According to Buchtien due to the presence of elaters, the spores are entangled together and thus dispersed in groups so the prothalli are found in close vicinity. According to Goebel the elaters help in fastening the spore to the substratum.

Germination of spore and development of prothallus (gametophyte). Usually the spores germinate in the clayey soil and mud along the banks of the rivers and streams. As already mentioned the spores contain chloroplasts and therefore, they can survive only for few days after their liberation from the sporangium. Prior to the germination the spore increases to some extent in its size. Within 10 to 12 hours, the first cell division in the development of the gametophyte takes place, after the spore is sown. This way, the two daughter cells are resulted one small and the other large in size. The small daughter cell develops into the first rhizoid and the large one gives rise to the remainder of the thallus (Campbell, 1918; Kashyap, 1914 and Walker, 1931). The large cell divides irregularly several times giving rise to one to several celled thick cushion-like tissue. This cushion-like structure bears numerous rhizoids on its lower surface. This cushioned structure is known as gametophyte or prothallus. Upon this cushion-like massive structure, the green lobes or branches are found. The prothallus is elongated and lobed. The cells of the lower portion of the cushion are colourless whereas those of upper portion bear chloroplasts and are green. The entire margin of the prothallus is meristematic. When the growth of the entire margin is uniform the prothallus becomes quite smooth and circular, whereas on the other hand when growth is more active on one side or the other, the irregularly lobed prothallus is resulted. The upright green lobes arising from the upper side of the prothallus are densely crowded and almost cover the whole upper side of the gametophyte. In certain species the gametophytes are quite big in size, *e.g.*, in *E. debile* Rox, the gametophyte is of 3 centimetres in diameter (Kashyap,1914), and two to three millimetres in thickness. In majority of the cases the gametophytes are less than a centimetre in size. The

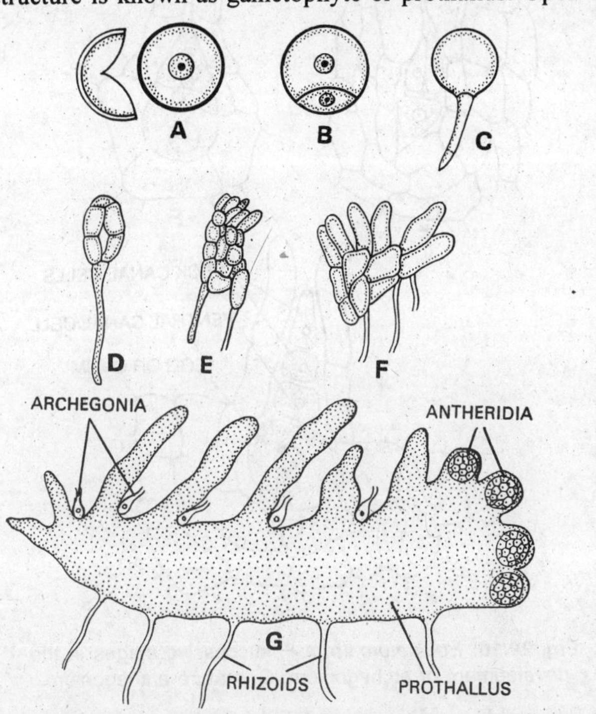

Fig. 29.9. *Equisetum arvense.* A-F, successive stages of the spore germination and formation of the gametophyte; G, mature prothallus with antheridia and archegonia.

gametophytes are long lived. According to Walker (1921), it has been reported on the basis of culture experiments that they can survive for more than two years. The large gametophytes survive for a long period, whereas small prothalli die soon.

Nature of prothalli. There are two types of prothalli-the small male prothalli bear only-antheridia whereas the large hermaphrodite prothalli bear both the antheridia and archegonia. Whether *Equisetum* is monoecious or dioecious is a matter of controversy. According to Kashyap (1917), in *E. debile* when spores are sown thickly, the prothalli, remain small in size and bear only one type of sex organs. On the other hand, if the spores are shown quite separate from each other the prothalli become large in size and bear both types of sex organs, *i.e.*, antheridia and archegonia. In such cases, however, the archegonia develop first and thereafter the antheridia. According to Walker (1921), *E. laevigatum* is a monoecious species. In *E. arvense*, under unfavourable conditions the prothalli remain small in size and only antheridia are developed, upon them. In favourable conditions, half of the prothalli are developed as males and the remaining half as females. According to Walker (1931) the normal prothalli are monoecious, which develop archegonia first and the antheridia later on. He also stated that the prothalli developing in crowded conditions remain small and dioecious. In *E. arvense*, the normal gametophytes produce archegonia at first and thereafter antheridia. In conclusion, the *Equisetum* is monoecious. This view has been supported by several eminent bryologists.

According to Joyet Lavergne (1931), morphologically the spores are alike, but physiologically they are different, and in the same lot some of the spores are male and some are female, so according to him, there is internal heterospory.

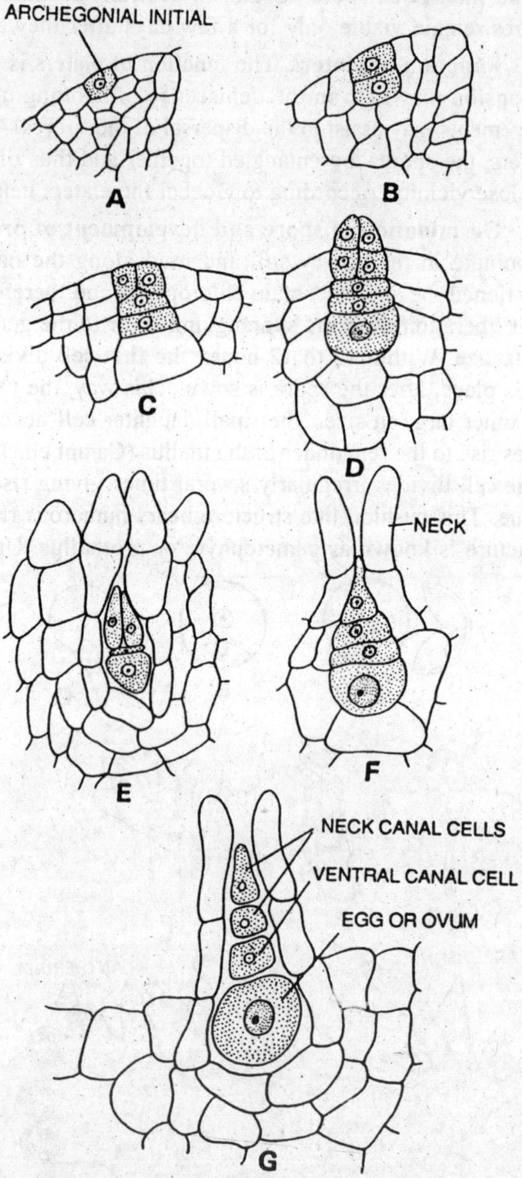

Fig. 29.10. *Equisetum* sp. A-F, successive stages in the development of archegonium; G, mature archegonium.

Development of archegonium. If the favourable conditions prevail, the first sex organs appear when the normal separated gametophytes are 30 to 40 days old. The sex organs appear only when the prothallus is differentiated into two regions-the upper green lobed region and the lower massive colourless tissue. Normally the archegonia develop first on the prothallus. The archegonia develop in the meristematic margin of the prothallus. They are situated among the up right green lobes.

The archegonial initial is a superficial cell situated in the meristematic margin of the prothallus. The archegonial initial divides first by a periclinal wall giving rise to two daughter cells. The outer daughter cell is known as the primary cover cell and the inner daughter cell is known as the mother cell of the central cell. The primary cover cell divides further producing four primary neck cells. The four primary neck cells divide transversely producing an archegonial neck three or four cells in height. The central cell divides by a transverse wall producing a primary neck canal cell and a primary centre cell. In some species, the primary neck canal cell rarely divides, whereas in other species it divides regularly producing two boot-shaped neck canal cells. In almost all the species, the primary venter canal cell divides by an asymmetrical transverse division producing a venter cell and an egg. Lastly, all the axial cells situated in the archegonium are disintegrated but the egg.

Structure of archegonium. An archegonium consists of a sunken base in the prothallus, and a projecting neck. The archegonial neck is composed of four vertical rows of cells. Each row is two to four cells in height. The terminal cells of the archegonial neck are quite long, which become separated and curved outward on their maturity. The axial row of the cells situated within the archegonium consists of three or four cells-the egg, the venter canal cell and one or two neck canal cells. When there are two neck canal cells, they are boot shaped, *e.g.*, *E. debile*. In the last stage of the development of archegonium the axial cells are disintegrated but the egg.

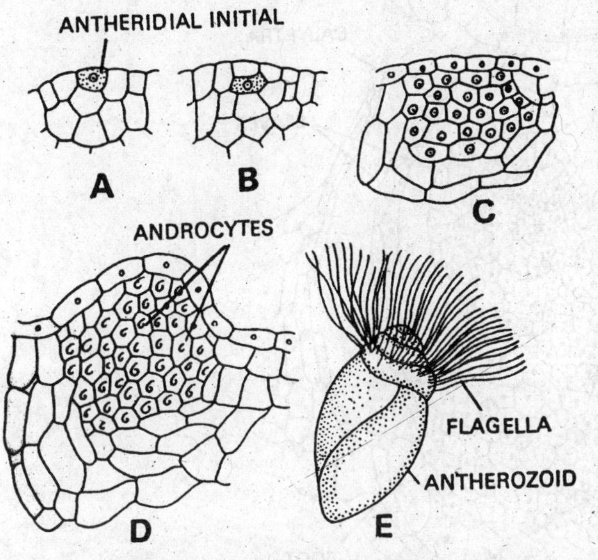

Fig. 29.11. *Equisetum* sp. A-D, successive stages in the development of antheridium; E, antherozoid.

Development of antheridium. In *E. arvense* as soon as the formation of archegonia is over the antheridia are formed on the same prothalli. At the time of the development of the antheridia the growth of the upright lobes on the prothallus is stopped, and the prothallus itself turns and expands upwards. The antheridia develop in these expanded and turned portions of the prothalli. On the other hand, in starved, crowded prothalli, only antheridia are developed. These prothalli may also be called the male prothalli. Usually, the antheridia are of two types- (a) the embedded type and (b) the projecting type. The former type of antheridia usually develop on the massive tissue of normal gametophytes and the later type of antheridia are developed in the meristematic margin of the starved crowded and dwarfed male prothalli. The development of the antheridium is as follows:

A superficial cell of the meristematic margin of the prothallus acts as an antheridial initial. The first division of the initial is periclinal. Thus, two daughter cells are formed. The outer cell is known as the jacket initial and the inner daughter cell is known as the primary androgonial cell. The jacket initial divides repeatedly giving rise to single layered jacket. Sometimes there lies a triangular opercular cell at the apex of the jacket layer. The primary androgonial cell divides repeatedly giving

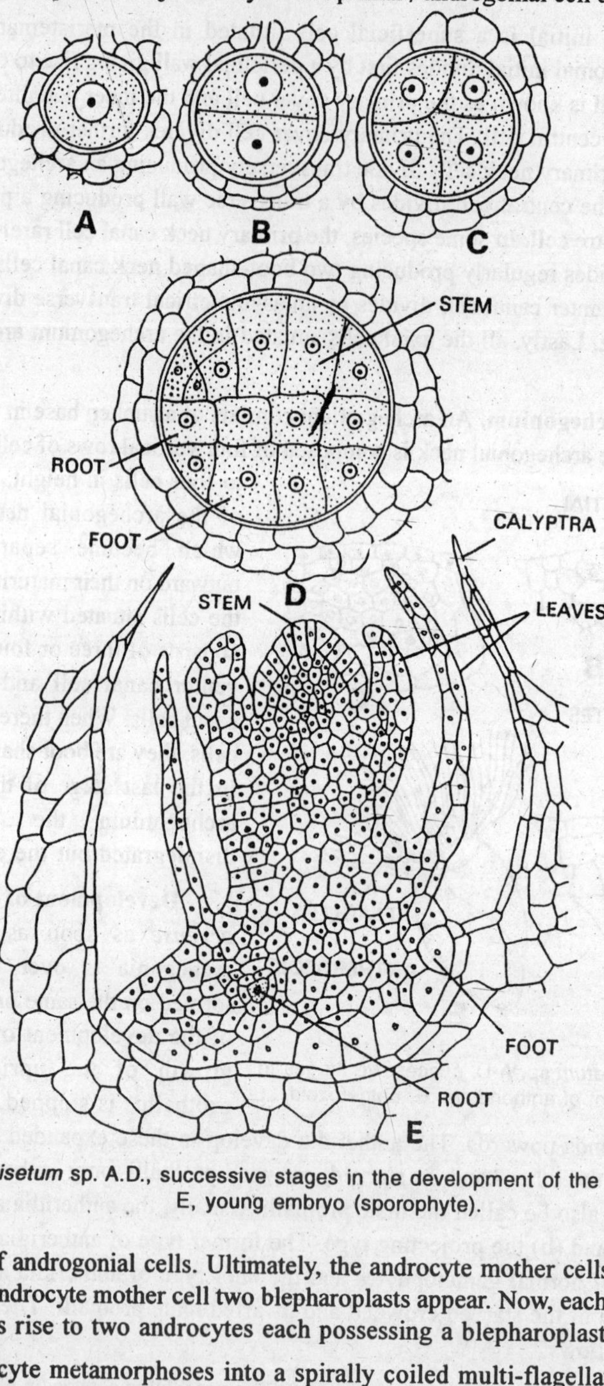

Fig. 29.12. *Equisetum* sp. A.D., successive stages in the development of the sporophyte (embryo); E, young embryo (sporophyte).

rise to a group of androgonial cells. Ultimately, the androcyte mother cells are formed. Near the nucleus of each androcyte mother cell two blepharoplasts appear. Now, each androcyte mother cell divides and gives rise to two androcytes each possessing a blepharoplast.

Each androcyte metamorphoses into a spirally coiled multi-flagellate antherozoid. In the beginning, the rounded nucleus and the blepharoplast of the androcyte become elongated. The blepharoplast is composed of many granules. Later on all the granules become united. The nucleus

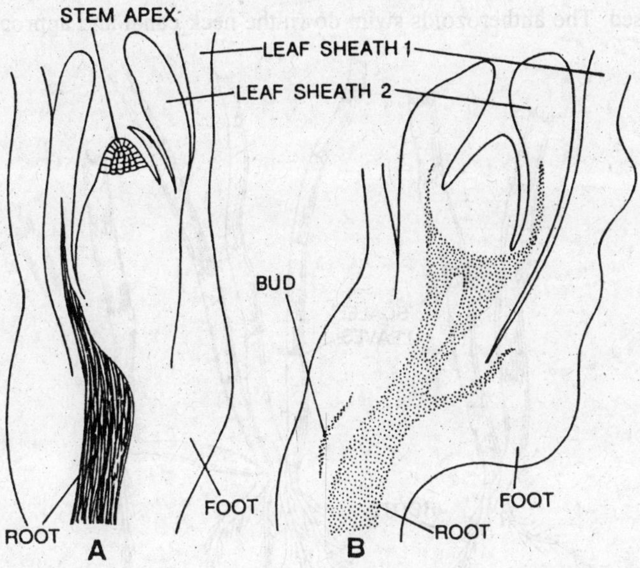

Fig. 29.13. *Equisetum* sp. A, L.S. of a young sporophyte showing the stem apex, the two first leaf sheaths, the foot and the base of root; B, as in A, but the bud for the secondary shoot is seen upon the primary root. (after Campbell).

elongates continuously and ultimately becomes crescent shaped. In the same way, the blepharoplast also becomes spirally elongated. Later on, the posterior end of the nucleus becomes expanded and flattened. A large vacuole is developed within androcyte. The cytoplasm of the antherozoid becomes foamy and coarse.

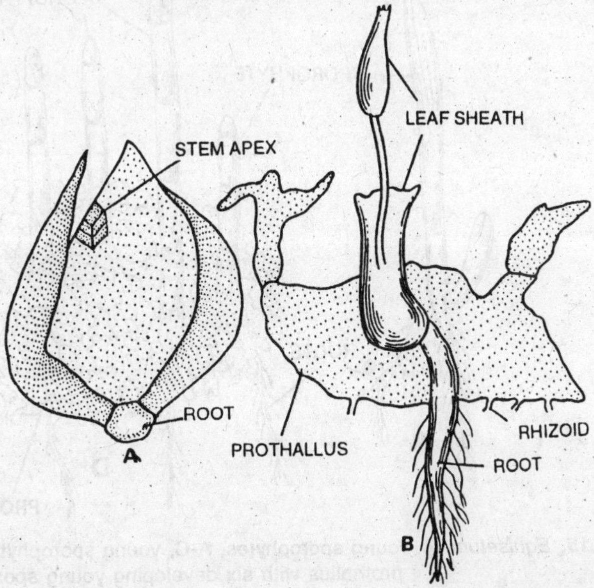

Fig. 29.14. *Equisetum arvense.* A, older embryo; B, V.S. of a lobe of prothallus with a young sporophyte (A, after Sadebeck; B, after Hofmeister).

Structure of antherozoid. The antherozoids are quite large in size. Each antherozoid is spirally coiled. The posterior part of the body of the antherozoid is thick and flat. The anterior part is conical and bears numerous flagella. The body is almost entirely derived from the nucleus of the androcyte. The numerous flagella at the anterior end of the antherozoid originate from the cytoplasm of the mother cell. The flagella are attached to the blepharoplast, which also lies at the anterior end.

Fertilization. On the maturity of the archegonium the neck canal cells and the venter canal cell are disorganised. The antherozoids swim down the neck canal and approach the egg. Several

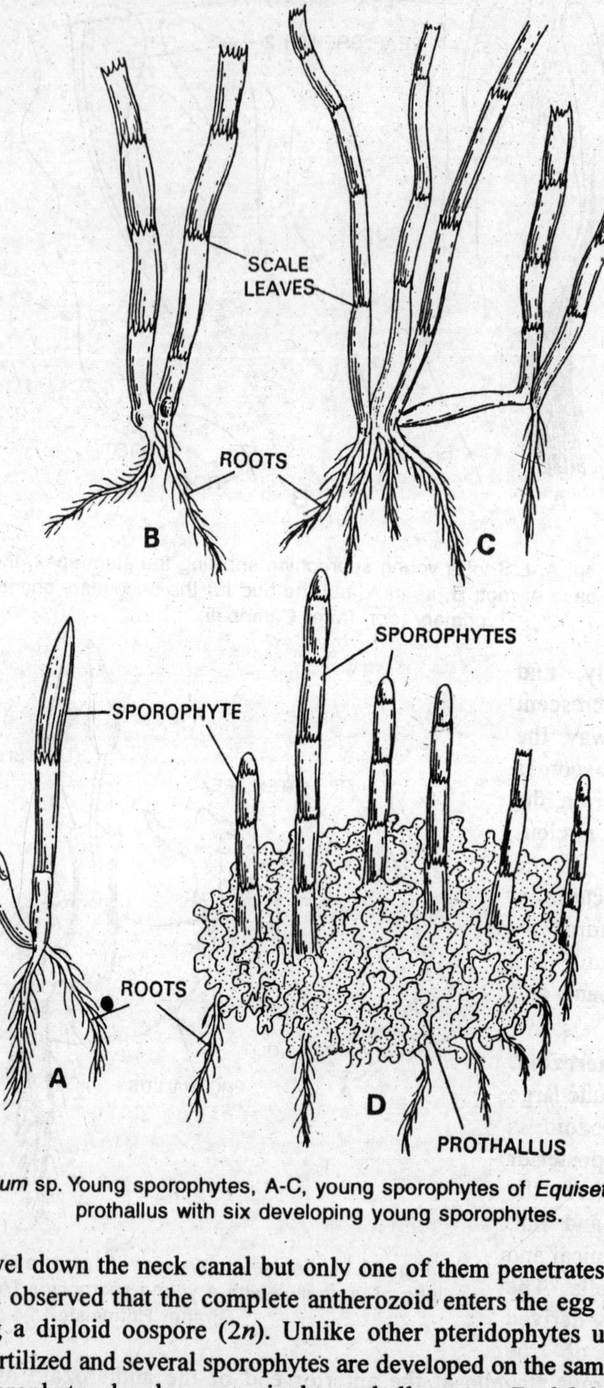

Fig. 29.15. *Equisetum* sp. Young sporophytes, A-C, young sporophytes of *Equisetum hiemale*; D, a large prothallus with six developing young sporophytes.

antherozoids travel down the neck canal but only one of them penetrates the egg. In *E. debile,* Sethi (1928) had observed that the complete antherozoid enters the egg and the fertilization is effected forming a diploid oospore (2*n*). Unlike other pteridophytes usually more than one archegonia are fertilized and several sporophytes are developed on the same prothallus. Usually 8 to 10 normal sporophytes develop on a single prothallus, *e.g., E. debile*. Kashyap (1914) has recorded as many as 15 sporophytes on a single gametophyte in *E. debile*.

Development of embryo (young sporophyte). As supported by Campbell (1918, 1928); Jeffery (1899) and Walker (1921), usually the first division of the oospore is transverse, giving rise to two cells-the upper one is epibasal and the lower one the hypobasal. Here in the case of *Equisetum* both the halves of the oospore take part in the development of the embryo proper and no suspensor is developed like that of *Selaginella*. The next division takes place at right angles to the first and the quadrant stage of the embryo is attained. According to Sadebeck (1878) in *Equisetum* the first division of the oospore is followed by two successive divisions at right angles to each other giving rise to eight cells. According to Campbell (1921), in *E. debile*, the entire hypobasal half develops into a foot, whereas in *E. arvense* (Sadebeck, 1878) the hypobasal half of the embryo gives rise to both foot and root. Early in the development, in *E. arvense* the epibasal cell divides by three intersecting walls in such a way that a tetrahedral apical cell of the stem is differentiated. The three cells, lateral to this apical cell are also differentiated, which develop in the three leaves of the first leaf sheath. Sometimes, there may develop two or four leaves instead of three. The so-developed leaves are quite small and scale-like. In *E. debile,* where the entire hypobasal half develops into foot, a superficial cell of the epibasal half lying near the hypobasal half develops into the apical cell of the root. The portions of the stem and root of the embryo grow and elongate quite rapidly. The stem portion of the embryo elongates and bursts through the neck of the archegonium, forming the calyptra and grows upright. On the other hand the root grows downward of the gametophyte and reaches the soil. Soon after, the stem differentiates into nodes and internodes. Generally a whorl of three leaves, or so is found at each node. The stem grows upward unless and until 10 to 15 nodes and internodes are developed. Thereafter a secondary branch arises from the base of the primary branch. This branch bears four or five leaves at each node. The growth of the secondary branch is also limited and reaches probably to the same height as that of primary branch. More vigorous branches arise successively from the base of the primary stem. The third or fourth branch turns downward into the soil and becomes rhizome of the *Equisetum* plant. According to Campbell (1928) the origin of secondary branch is endogenous and it arises from the tissue of the primary root. The secondary branch possesses adventitious roots. Several secondary branches of limited growth arise successively from the young sporophyte. All such branches are endogenous in their origin. From the point of view of its anatomy, the primary stem is solid and without carinal canals which develop later on in the adult stem. However, the xylem is better

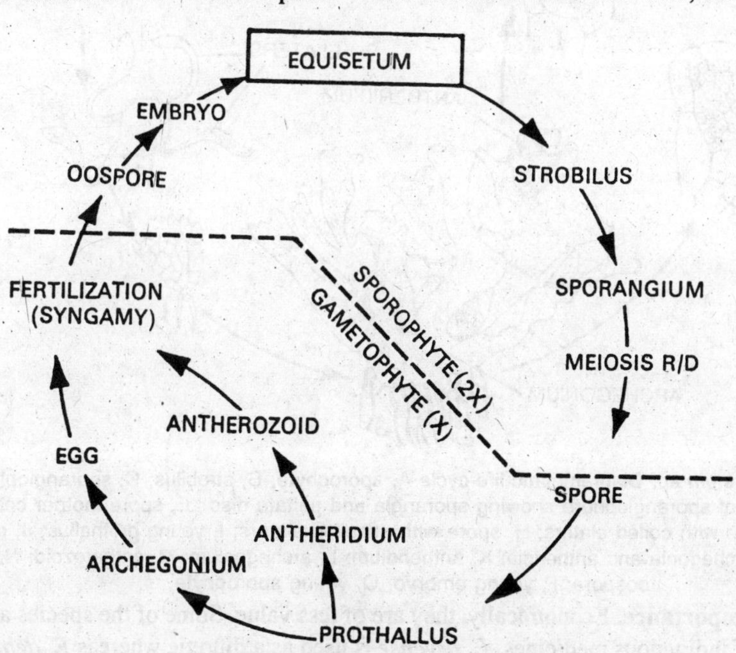

Fig. 29.16. Graphic life-cycle of *Equisetum*.

developed in the primary stem. The base of the seedling bears protostele, but later on the stem bears siphonostele at the level of the first branch.

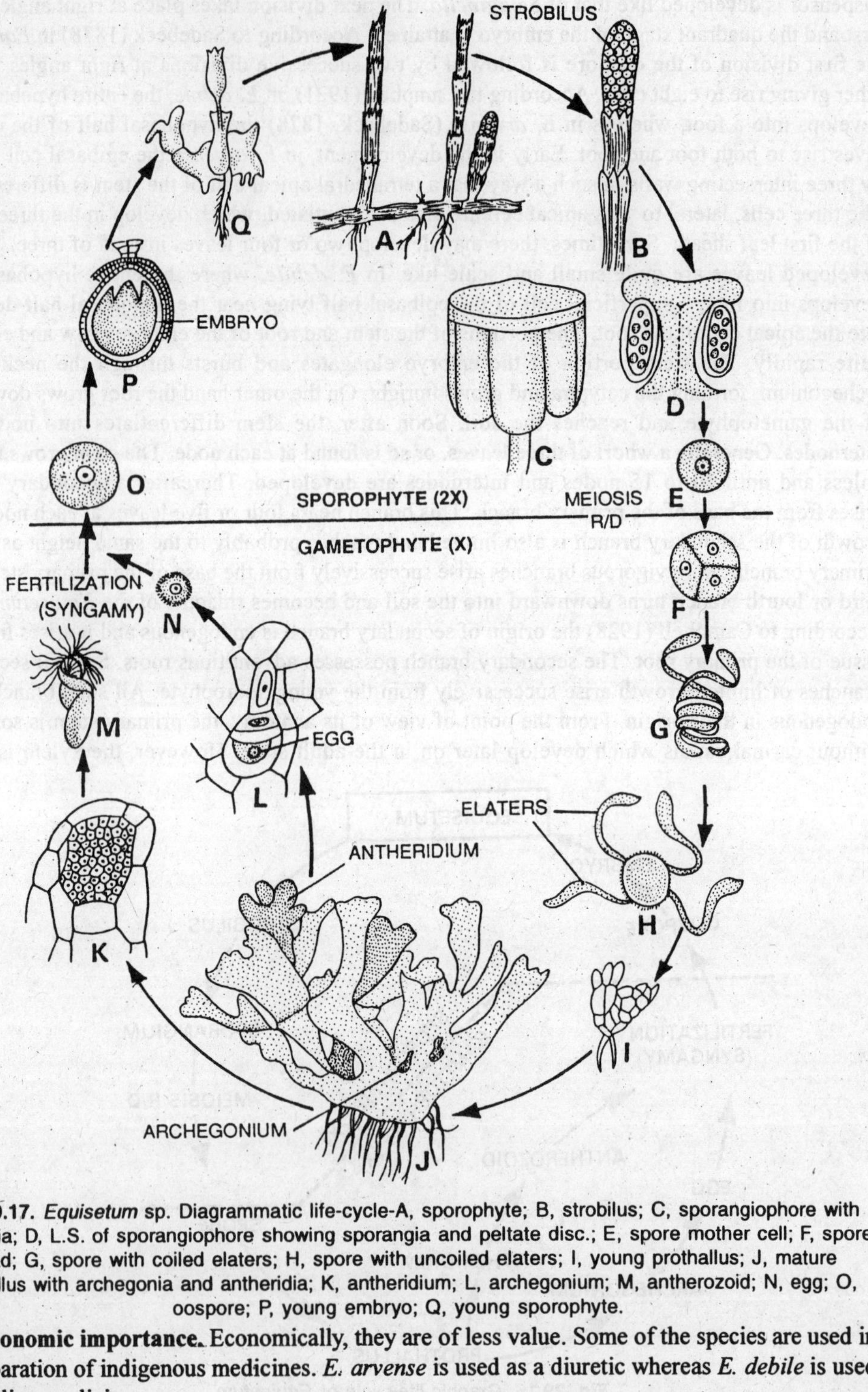

Fig. 29.17. *Equisetum* sp. Diagrammatic life-cycle-A, sporophyte; B, strobilus; C, sporangiophore with sporangia; D, L.S. of sporangiophore showing sporangia and peltate disc.; E, spore mother cell; F, spore tetrad; G, spore with coiled elaters; H, spore with uncoiled elaters; I, young prothallus; J, mature prothallus with archegonia and antheridia; K, antheridium; L, archegonium; M, antherozoid; N, egg; O, oospore; P, young embryo; Q, young sporophyte.

Economic importance. Economically, they are of less value. Some of the species are used in the preparation of indigenous medicines. *E. arvense* is used as a diuretic whereas *E. debile* is used as a cooling medicine.

30
CHAPTER

Division-Filicophyta

Class-Leptosporangiopsida

Order-Marsileales

Characteristic features. This order includes a single family, Marsileaceae. The family includes the living genera-*Marsilea*, *Pilularia* and *Regnellidium*. They are heterosporous aquatic ferns. The sporangia of these are produced within special structures known as **sporocarps.** Each sporocarp possesses many sori which bear microsporangia and megasporangia. The sori are **gradate,** *i.e.*, the oldest sori are found at their terminal end of the receptacle and youngest at the base.

We shall consider the life-history of *Marsilea* in detail.

Genus MARSILEA

Systematic position. Pteridophyta
Division. Filicophyta
Class. Leptosporangiopsida
Order. Marsileales
Family. Marsileaceae
Genus. *Marsilea.*

There are about 65 species of *Marsilea* distributed all over the world. They are more commonly found in tropical regions, such as, Africa and Australia. Gupta and Bhardwaja (1957) have recorded about ten species of *Marsilea* from our country. About six species of *Marsilea* are known to occur in the United States. They are hydrophytic or amphibious plants. They grow rooted in the mud of marshes and shallow pools. *Marsilea vestita* and some other species grow in shallow ponds. A single plant of *M. vestita* may grow in all directions until it covers an area twenty-five metres or more in diameter. In *M. vestita* and most of other species the sporocarps develop to maturity only on the plants that are not submerged. *M. minuta* and some other species produce sporocarps in the aquatic habitat, on the other hand *M. aegyptiaca* never produces the sporocarps in the aquatic habitat. An Australian species *M. hirsuta* can survive in dry conditions throughout the year.

The Indian species recorded by Gupta and Bhardwaj (1957) are as follows-

1. M. quadrifolia; 2. M, minuta; 3. M. aegyptiaca; 4. M. brachypus; 5. M. gracilenta; 6. M. poonensis; 7. M. rajasthanensis; 8. M. brachycarpa; 9. M. condensata and *10. M. coromandelica.*

External Morphology.

Stem. The species of *Marsilea* possess a rhizome which creeps on or just beneath the surface of the soil. The rhizome is slender, branched and possesses nodes and internodes. The leaves are borne alternately along the upper side of the rhizome usually at the nodes. One or more adventitious roots come out from each node on the underside of the rhizome. The rhizome is

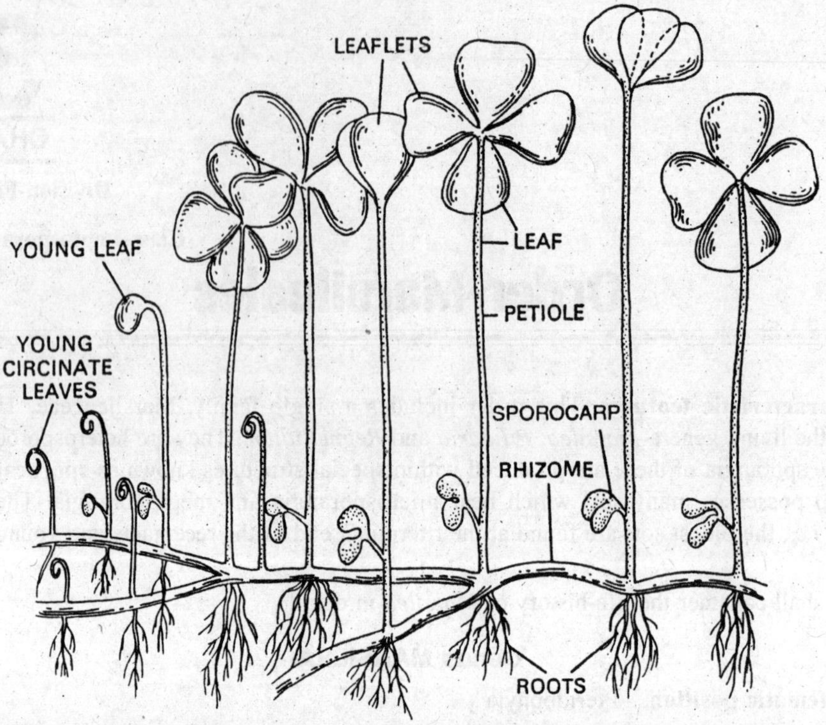

Fig. 30.1. *Marsilea quadrifolia.* A sporophyte with sporocarps, rhizome, leaves and roots.

dichotomously branched and is capable of indefinite growth in all directions and covers an area more than twenty-five metres in diameter.

Leaves. The leaves are borne alternately along the upper side of the rhizome at the nodes. The leaves possess circinate vernation typical of most Filicales (ferns). The young parts of the leaves of *M. minuta* and others are covered with numerous multi-cellular hairs. The leaves of submerged plants possess long flexible petioles and leaf lamina that float on the water surface. When the plants grow on mud or marshy places the leaves have got shorter and the petioles erect and stouter which spread the leaves in the air. The leaves are compound. The lamina of each leaf is divided into four leaflets pinnae arising from the apex of the petiole. Division of a leaf-blade into four leaflets results from two dichotomies arising in close succession to each other. According to Puri and Garg (1953) the leaf of *Marsilea* consists of single leaflet bearing four pinnules and not of four leaflets. Usually these pinnules have been referred as leaflets by many authors in their

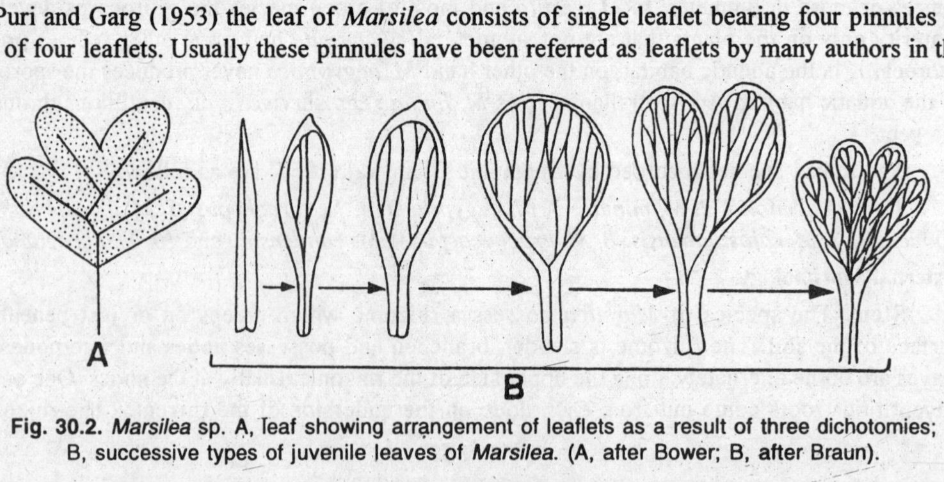

Fig. 30.2. *Marsilea* sp. A, leaf showing arrangement of leaflets as a result of three dichotomies; B, successive types of juvenile leaves of *Marsilea*. (A, after Bower; B, after Braun).

text-books. The veins of each pinna are dichotomously branched and have numerous cross connections resulting in close reticulum of veins. The shape of pinnae varies from obovate to obcuneate. The margin also varies from entire to crenate, *e.g.*, *M. minuta* or from crenate to lobed as in *M. aegyptiaca* a xerophytic form. The leaflets may be once or twice deeply lobed dichotomously, *e.g.*, in *M. macrocarpa* and *M. biloba.* In *M. minuta,* the pinnae become coarsely toothed. At night the pinnae of aerial leaves become folded upwards and exhibit the sleeping movement.

The **sporocarps** are borne on short peduncles near the base of the petiole. In majority of cases the peduncle or stalk of the sporocarp is unbranched and bears a single sporocarp at its apex. *Marsilea quadrifolia* possesses dichotomously branched peduncle, which bears two to five sporocarps. In *M. polycarpa* single dichotomously branched peduncle bears about 6 to 26 sporocarps.

Roots. One or more adventitious roots are borne at each node on the underside of the rhizome. The adventitious roots may arise even from internodes, *e.g.,* in *M. aegyptiaca.* In *M. minuta* the roots may develop laterally. The roots of *M. minuta* secrete carbonic acid and because of this secretion the roots could penetrate the shells of pond snails.

Internal Morphology (Anatomy)

Rhizome. The rhizome of *Marsilea* possesses an amphiphloic siphonostelic vascular cylinder. The outermost layer is the single-layered epidermis without any stomata. The cortex is differentiated into outer and inner cortical regions. The outer cortical region consists of compact parenchymatous tissue. This region may be one to several celled in thickness. Just beneath this the cortical region contains large lacunae or air spaces (aerenchyma). The lacunae are separated from each other by one layered parenchymatous septa.

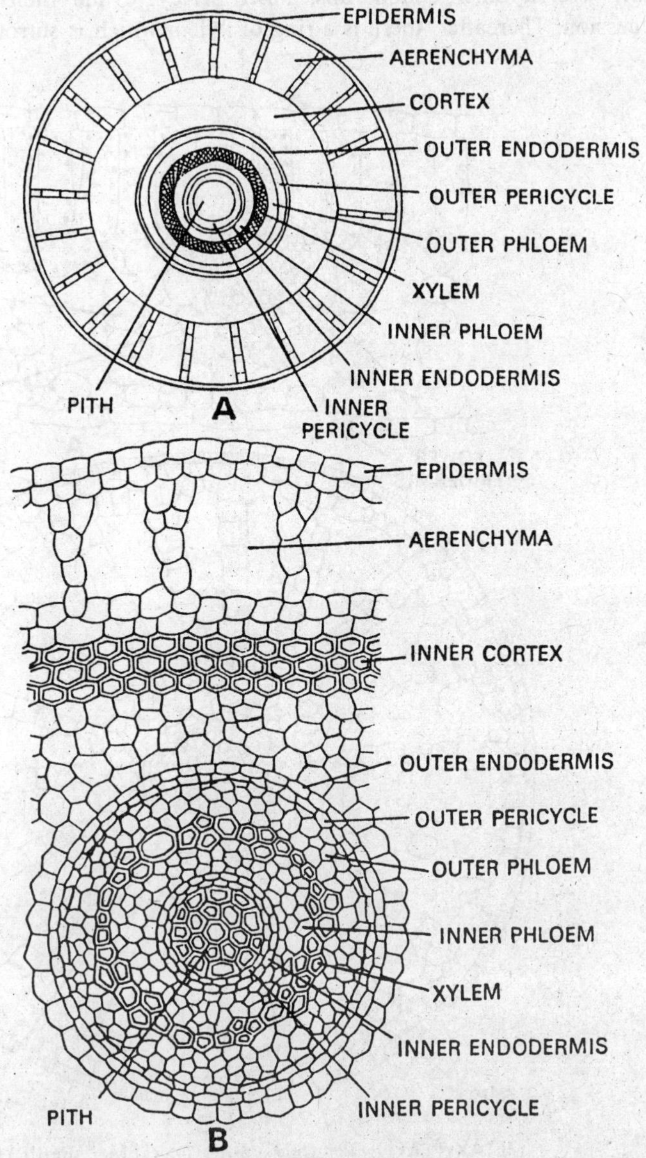

Fig. 30.3. *Marsilea quadrifolia.* Anatomy of rhizome. A, T.S. of rhizome (diagrammatic); B, T.S. of rhizome (detailed).

This region may be considered the middle cortex. Internal to this region the inner cortex is found. This consists of a sclerotic zone. The cells consisting this region are thick-walled sclerenchymatous cells (fibres). Beneath this region the cortex again consists of compact parenchymatous tissue acting as storage tissue containing starch in them. Some of the tannin cells may be found in this innermost region.

The stele is amphiphloic siphonostele. All species possess the siphonosteles externally and internally limited by a single layered endodermis. Just beneath the outer endodermis there is a single layered pericycle. In the centre is the pith whose structure depends upon environmental conditions. In submerged species, the rhizomes possess usually a parenchymatous pith, whereas the plants growing on mud have a more or less sclerenchymatous pith. Outside the pith there is a sindle-layered inner endodermis, inner pericycle and then inner phloem in a continuous succession. Thereafter, there is a ring of xylem which is surrounded by the successive rings of

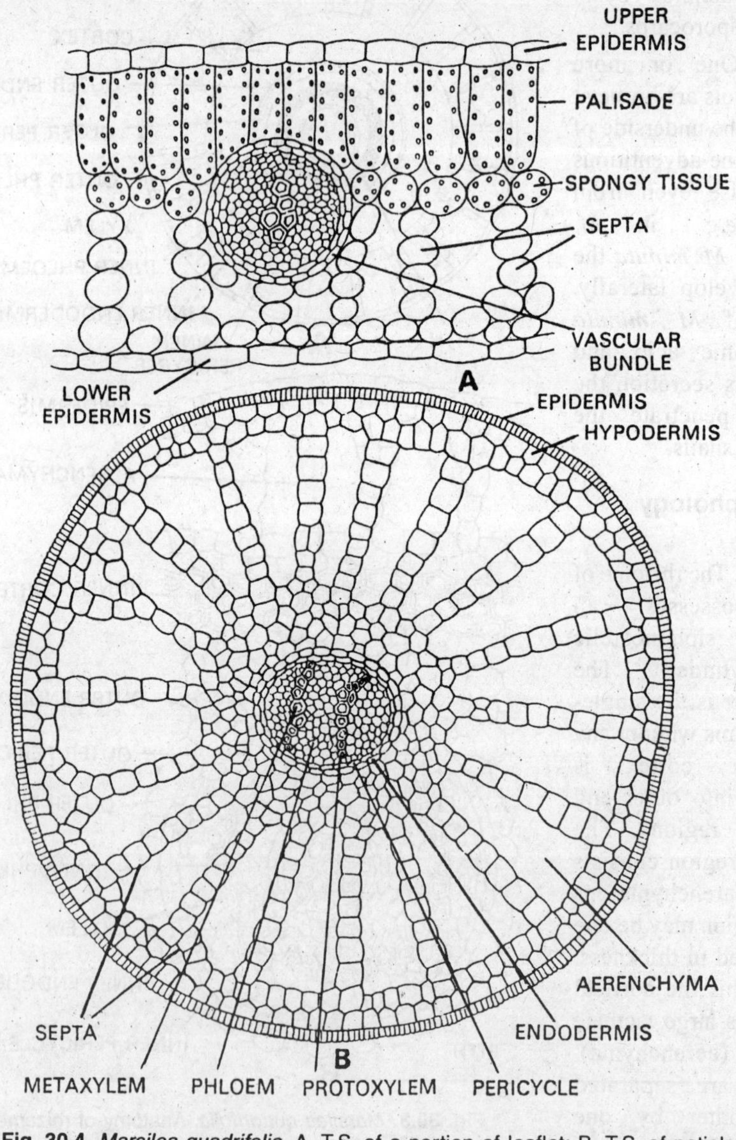

Fig. 30.4. *Marsilea quadrifolia*. A, T.S. of a portion of leaflet; B, T.S. of petiole.

outer phloem, outer pericycle and outer endodermis. In *M. quadrifolia* there is no definite protoxylem but in *M. vestita* there are well-defined protoxylem groups which are exarch. Gupta and Bhardwaja (1956) reported the mesarch protoxylem in *M. aegyptiaca*. This way, the continuation of different tissue in the amphiphloic siphonostele is as follows-outer endodermis, outer pericycle, outer phloem, xylem, inner phloem inner pericycle and then inner endodermis.

Petiole. The internal structure of a petiole as seen in transverse section is as follows— The outermost layer is single-layered epidermis. Just beneath the epidermis there is hypodermal layer. Then there is other cortex constituting of large air spaces or lacunae (aerenchyma) separated from each other by septa. The inner cortex is composed of compact parenchymatous cells. In this region some of the tannin cells may be present here and there. In the central region of the petiole there is stele which is more or less triangular in outline. The stele is surrounded by a single-layered endodermis. It possesses V-shaped mass of xylem with the exarch protoxylem. The two arms of V are quite separate and some what curved from each other. The V-shaped mass of xylem is surrounded by phloem and single-layered pericycle successively. The ends of each arm of V-shaped xylem are protoxylem groups whereas the middle regions of the arms of V consists of large tracheids representing the metaxylem groups. The opening of the V is always towards the axis, *i.e.*, adaxial side.

Leaflet. As shown in transverse section of the leaflet, it is bounded by two upper and lower epidermal layers. In the case of floating leaves the stomata are confined to the upper surface but in the case of the plants growing in mud and moist soil, the leaves are aerial, and the stomata are found on both the upper and lower surfaces. Just beneath the upper epidermis there is mesophyll differentiated into palisade and spongy tissues. The vascular bundles are concentric, *i.e.*, xylem elements are surrounded

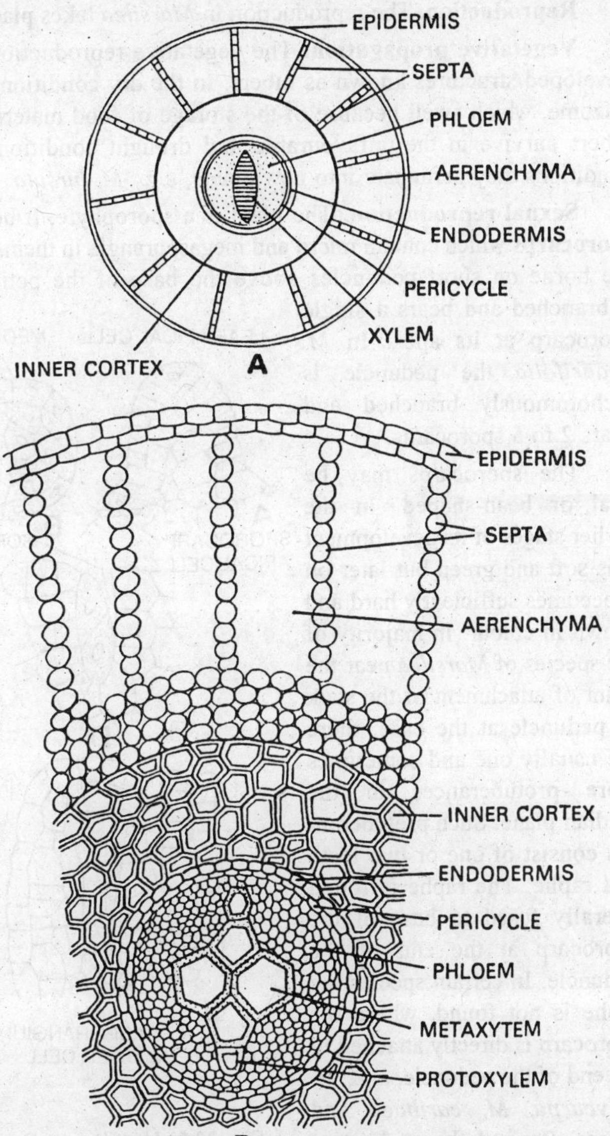

Fig. 30.5. *Marsilea quadrifolia*. Anatomy of root. A, T.S. of root (diagrammatic); B, T.S. of a portion of root (detailed structure).

by phloem. On the outside of the vascular bundle there is single layered endodermis. Towards lower side there are big air spaces separated by septa.

Root. In the transverse section of the root, the epidermis, cortex and stele may be seen. The epidermis is single layered and composed of thin-walled parenchymatous cells. The cortex of the root is differentiated into two regions. The outer region consists of large air spaces or lacunae separated by septa from each other, the inner zone of the cortex consists of a few layers of sclerotic cells. Inner to the cortex there is single layered endodermis. Just beneath the endodermis there is single-layered pericycle. In the centre of the stele there is typical diarch and exarch structure with a diametric plate like strand of xylem. The phloem bands are found on either side of the xylem plate.

Reproduction. The reproduction in *Marsilea* takes place by vegetative and sexual methods.

Vegetative propagation. The vegetative reproduction takes place by means of specially developed structures known as tubers. In the dry conditions certain branches develop from the rhizome, which swell because of the storage of food material, and are termed as **tubers**. These tubers survive in the unfavourable and drought conditions, and on the return of favourable conditions they germinate into new plants, *e.g., M. hirsuta.*

Sexual reproduction. The plant is a sporophyte. It bears the special structures known as **sporocarps** which contain micro and megasporangia in them. It is **heterosporous.** The sporocarps are borne on short peduncles above the base of the petiole. In most cases the peduncle is unbranched and bears a single sporocarp at its apex. In *M. quadrifolia* the peduncle is dichotomously branched and bears 2 to 5 sporocarps.

The sporocarps may be oval or bean-shaped. In the earlier stages of its development it is soft and green but later on it becomes sufficiently hard and brown in colour. In majority of the species of *Marsilea* near the point of attachment of the stalk or peduncle at the base, there are usually one and sometimes more protuberances in the median plane. Such protuberances consist of one or two teeth and raphe. The raphe is found laterally fused to base of the sporocarp at the end of the peduncle. In certain species the raphe is not found, where the sporocarp is directly attached to the end of the peduncle, *e.g., M. polycarpa, M. caribaea* and others. Beyond the raphe two teeth or projections are found. These teeth are unequal in size.

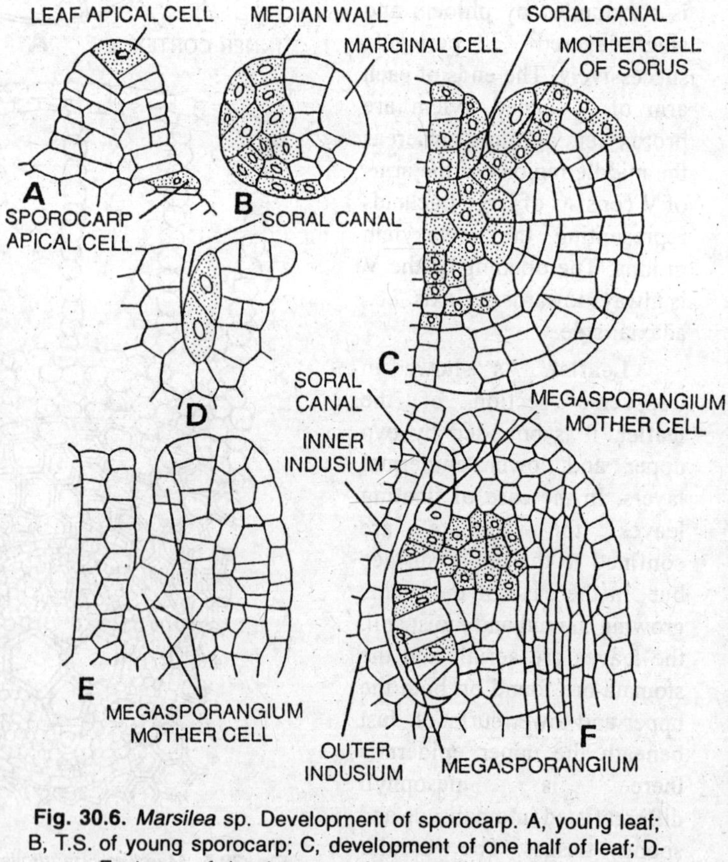

Fig. 30.6. *Marsilea* sp. Development of sporocarp. A, young leaf; B, T.S. of young sporocarp; C, development of one half of leaf; D-F, stages of development of sorus and soral canal.

The lower tooth is stouter than the upper one. In some species the teeth are very prominent, *e.g.*, *M. vestita* while in other species they are altogether absent.

The vascular supply of the sporocarp consists of a main vein along the narrow side facing the peduncle from which numerous lateral veins perpendicular to the main vein (dorsal bundle) are given off. These lateral veins are given off alternately right and left. Each lateral vein divides dichotomously about middle of its base and the apex. A receptacle develops at the region where each lateral vein forks dichotomously. The receptacles of one valve alternate with those on the opposite valve of the sporocarp. A flap-like outgrowth two-celled thick extends like a hood over the entire receptacle. Each receptacle projects inward from the wall of the sporocarp.

Development of sporocarp. The sporocarp of *Marsilea* develops from a marginal cell of a very young leaf. This marginal cell is situated on the inner face of the leaf and is being differentiated only after the apical cell of the leaf has cut off 6-8 segments. The marginal cells of the sporocarp initials divide and make a mass of undifferentiated cells which become curved so that the distal end gets directed horizontally. As the distal part enlarges to form the sporocarp two rows of soral mother cells appear on the ventral side, arising directly from the marginal cells of the earlier stage. Later the marginal cell and their derivatives are pushed downwards to the lower surface and slightly towards the median line due to unequal growth of the cells. As increase in diameter continues the cells lateral to the soral mother cells grow more rapidly than the developing sorus with the result that the sporangial mother cells become burried. Simultaneously each sorus towards the median line a small cresent-shaped depression appears on the side facing downwards which deepens quickly with the growth of tissue around the sorus (Fig. 30.6 C-F). These depressions become spaces above and around the sori opening to the outside by the small pores, which are called the soral canals (Figs. 30.6 D-E). The soral canals appear linear in the longitudinal section and crescent-shaped in cross section. Each soral canal is lined on its central face by an **indusium** of which two layers may be recognized, *e.g.*, outer indusium, and inner indusium (Fig. 30.6. F). In each soral canal the receptacle of the sorus faces the indusium, and sporangia arise on the receptacle in acropetal succession. The first sporangial initials appearing at the tip of the receptacle develop into megasporangia whereas the other initials appearing along the sides produce microsporangia. Lateral to the sori the massive tissue of the sporocarp develops and forms the massive outer wall which ultimately closes the soral canals, and the sori are completely enclosed within the sporocarp.

Internal Structure of the Sporocarp

To study the internal structure of the sporocarp the sections cut in longitudinal and transverse planes at right angles to the flat surface are required. The internal structure of the sporocarp in transverse, longitudinal and dorsiventral sections is as follows:

Transverse section. The following tissues are seen in the transverse section. Outside the sporocarp there is a very thick wall consisting of three layers. The outermost layer is known as epidermis which consists of broad and columnar cells. The epidermis is interrupted by a number of stomata. Below the epidermis there are two hypodermal layers. The outer hypodermal layer consists of elongated thick-walled cells, while the

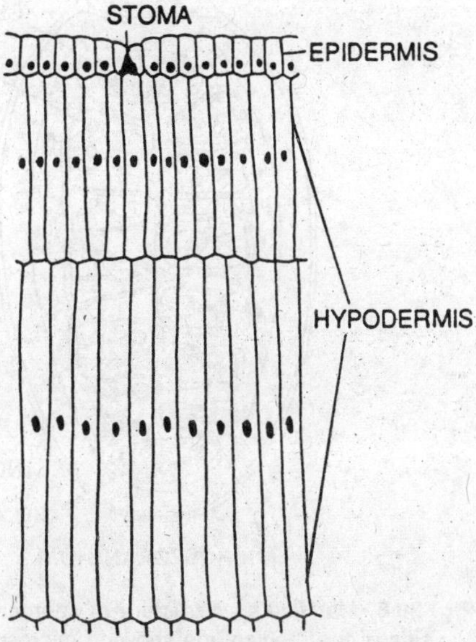

Fig. 30.7. *Marsilea*, portion of a transverse section of wall of sporocarp.

lower hypodermal layer consists of elongated and palisade like cells. In the inner region of the sporocarp only two chambers of sori are seen surrounded by separate indusia. The dorsal bundles are also seen. Each sorus possesses a receptacle which bears either micro or megasporangia.

Longitudinal section. The structure of the wall of sporocarp and the gelatinous ring is the same as in transverse section. In the interior of the sporocarp there are two alternating rows of sori each surrounded by a two layered indusium. A placental branch goes to each sorus from the lateral bundles. The receptacle bears a single megasporangium at the tip and the microsporangia on the lateral sides. Each megasporangium contains a large single megaspore while the microsporangium contains many small rounded microspores.

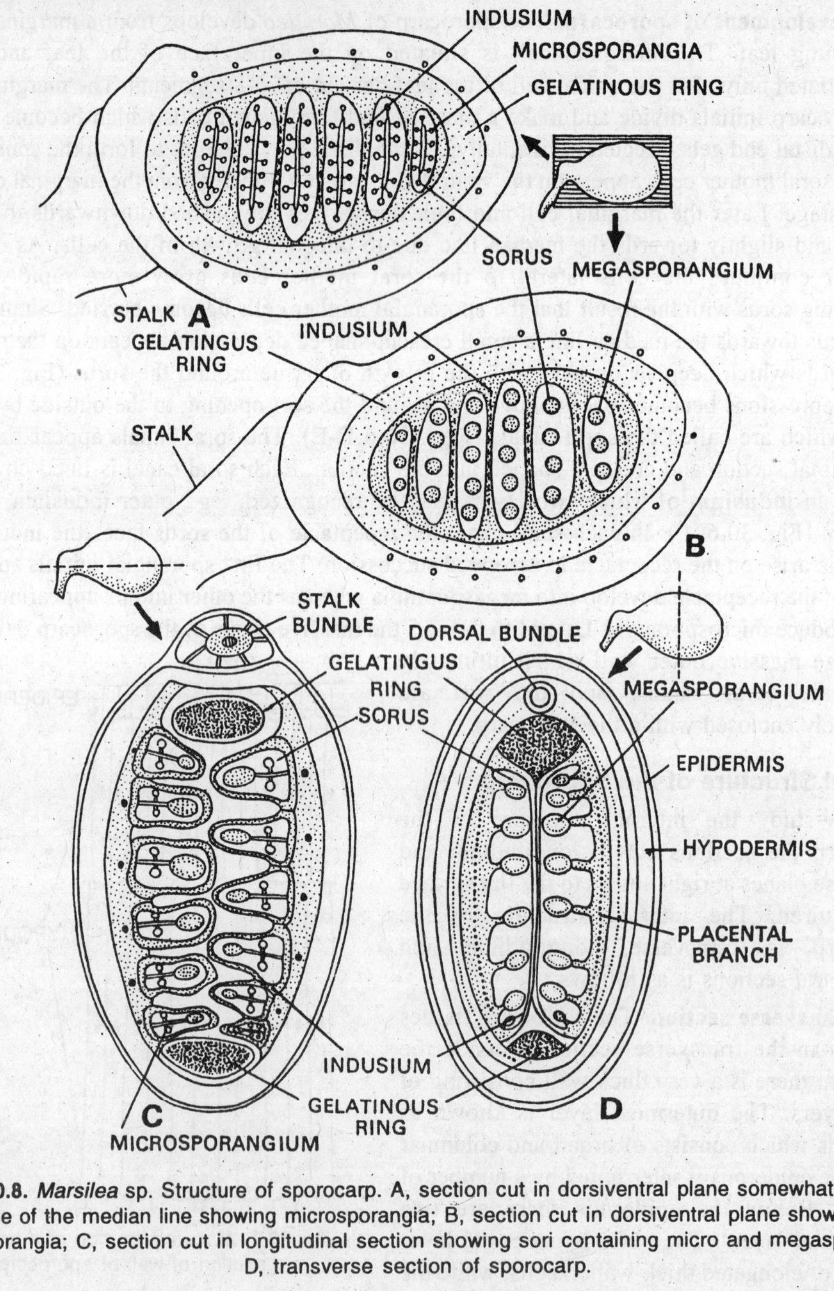

Fig. 30.8. *Marsilea* sp. Structure of sporocarp. A, section cut in dorsiventral plane somewhat to one side of the median line showing microsporangia; B, section cut in dorsiventral plane showing megasporangia; C, section cut in longitudinal section showing sori containing micro and megasporangia; D, transverse section of sporocarp.

Dorsiventral section. The structure of the outer wall of the sporocarp and gelatinous ring is the same as seen in transverse and longitudinal sections. Below the outer wall a continuous gelatinous ring is found. Within the sporocarp there is a group of sori. Each sorus appears to be formed in a cavity and is surrounded by a delicate, membranous indusium. The sorus contains micro and megasporangia. The megasporangia may easily be seen if the sporocarp is cut dorsiventrally somewhat to one side of the median line of the sporocarp, whereas, the microsporangia may be easily seen if the section of the sporocarp is cut in the same plane but slightly farther away than previous one.

Morphological nature of the sporocarp. As regards the morphological nature of the sporocarp, the interpretations have been given by various workers from time to time. There are two main hypotheses:

1. Leaf-segment or laminar hypothesis and 2. petiolar hypothesis.

The sporocarp of *Marsilea* has been interpreted as a lateral modified segment of the leaf. According to Bower (1926), Busgen (1890), Campbell (1905) **leaf segment or laminar hypothesis**—the sporocarp has been interpreted homologous with a modified fertile segment from the lower part of a leaf. **Petiolar or whole leaf hypothesis**-Johnson (1898, 1933) has interpreted the sporocarp to be homologous with an entire leaf. The argument for the latter interpretation by Johnson is that the apical growth of a sporocarp resembles to that of an entire leaf rather than to of a leaflet or leaf-segment. The leaf-segment or laminar hypothesis is supported by several workers and seems to be more correct. This hypothesis has been discussed in detail in the following lines:

Leaf segment or laminar hypothesis. However, the vascular supply to the peduncle of the sporocarp and the vascular supply to the interior of the sporocarp prove that the sporocarp is a modification of leaf-segment (pinna) rather than that of an entire leaf. The advocates of the leaf segment theory interpret it in various ways. Bugen (1890) interpreted that the sporocarp has been resulted from the opposition of two leaflets (pinnae), whereas Bower (1926) and Campbell (1905) interpreted that the sporocarp has been resulted from a single pinna by its enfolding. However,

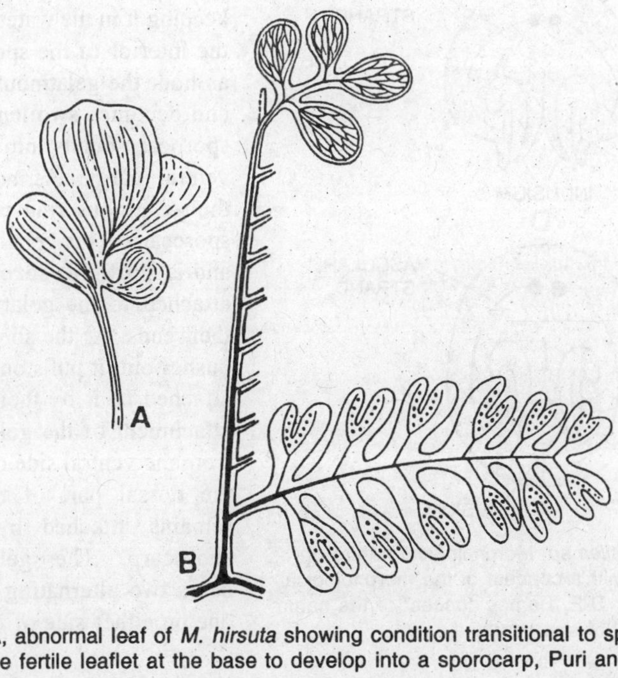

Fig. 30.9. *Marsilea.* A, abnormal leaf of *M. hirsuta* showing condition transitional to sporocarp; B, simple case with a single fertile leaflet at the base to develop into a sporocarp, Puri and Garg's view.

Puri and Garg (1953) interpreted that the sporocarp has been resulted from the enfolding of a pinna with several pinnules each with a marginal sorus. The presence of single dorsal main vein in the sporocarp and the single bundle in the peduncle indicate clearly that the sporocarp has been derived from a single pinna. It has been argued by Bower (1908, 1926) the ancestors of Marsileales were probably ferns with a gradate sorus surrounded by an involucroid indusium and not Schizaeaceae. In Schizaeaceae the sporangia are borne singly and develop simultaneously whereas in Marsileales the sporangia are borne in sori of a gradate type in which the sporangia at the apex of a receptacle develop earlier than those at the base of the receptacle. This way the relationships of the Marsileales appear to be more with the homosporous leptosporangiate ferns. There is no evidence that the heterospory of Marsileales arose earlier of the evolution of sporocarp. In all other ferns there is homospory and this point indicates that the origin of heterospory is somehow or other connected with evolution of the sporocarp.

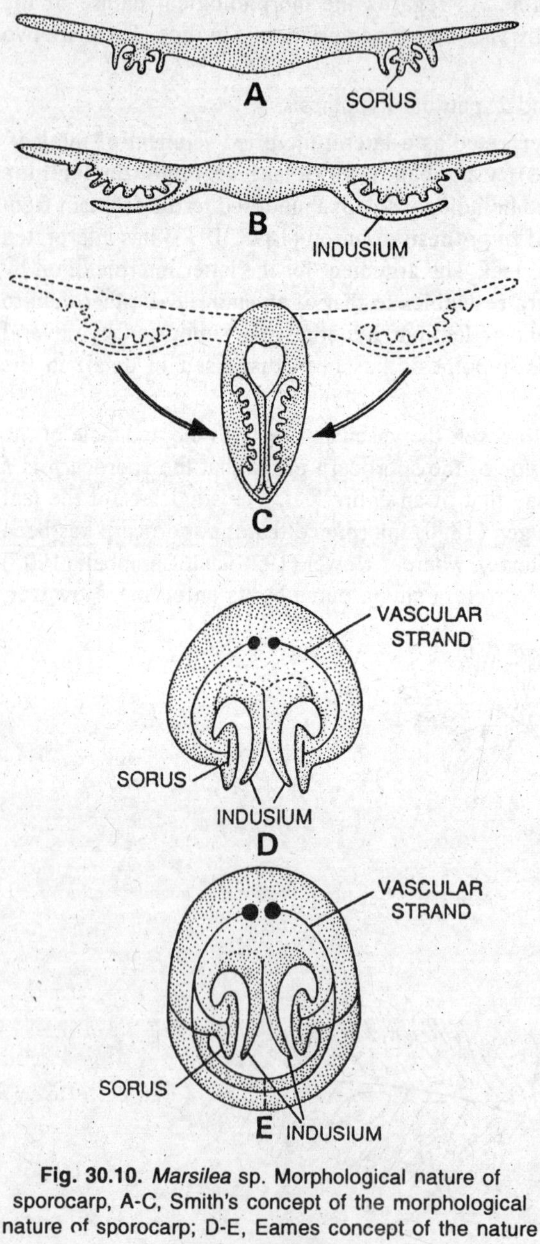

Fig. 30.10. *Marsilea* sp. Morphological nature of sporocarp, A-C, Smith's concept of the morphological nature of sporocarp; D-E, Eames concept of the nature of sporocarp.

Liberation of sorí from the sporocarp. The outer wall of the sporocarp is extremely hard and strongly resistant to mechanical injury and drying out. In natural conditions the sporocarp may burst even two or three years later its formation. The spores within the sporocarp may remain viable 20 to 30 years and sometimes upto 50 years. The sporocarp may easily be germinated by injuring it on the ventral median line and keeping it in the water. The water enters to the interior of the sporocarp. Within half an hour the gelatinous ring imbibes water and becomes swollen with the result the sporocarp bursts into two valves from its ventral margin. As more water is imbibed the gelatinous ring protrudes out of the sporocarp. This gelatinous ring bears sori and is known as **sorophore**. The sori are attached to the gelatinous sorophore by their ends. As the sorophore expands and pushes out, it pulls out the sori, which are attached to it by their ends. Usually the attachment of the gelatinous ring breaks from the ventral side of the sporocarp and the dorsal part of the gelatinous ring remains attached by one end to the sporocarp. The gelatinous sorophore bears two alternating rows of **soral sacs** one on either side of it. A fully elongated

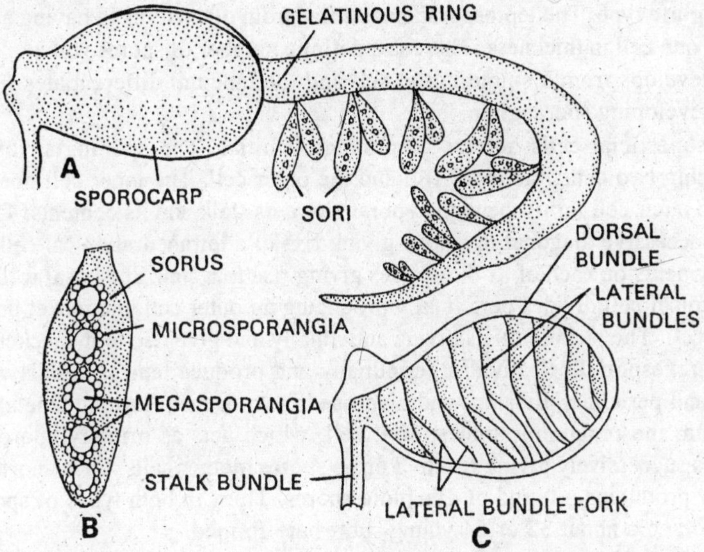

Fig. 30.11. *Marsilea quadrifolia*, A, extrusion of the gelatinous ring and sori; B, a sorus surrounded by indusium; C, vascular supply of the sporocarp in lateral view.

sorophore is a long gelatinous cylinder 15 to 20 times longer than the length of the sporocarp. At the tip of the sorophore there are certain small gelatinous projections, also alternate with one another. They are the rudimentary sori which could not develop. The gelatinous ring elongates and straightens and becomes a worm-like structure. At the time of their separation from the gelatinous ring the ventral ends of the sori are torn off and sporangia with spores escape from the ventral ends of the sori. However, the indusia and the jackets of sporangia become gelatinized and the spores are liberated. The germinating spores and the developing gametophytes remain embedded in the gelatinous matrix upto their maturity.

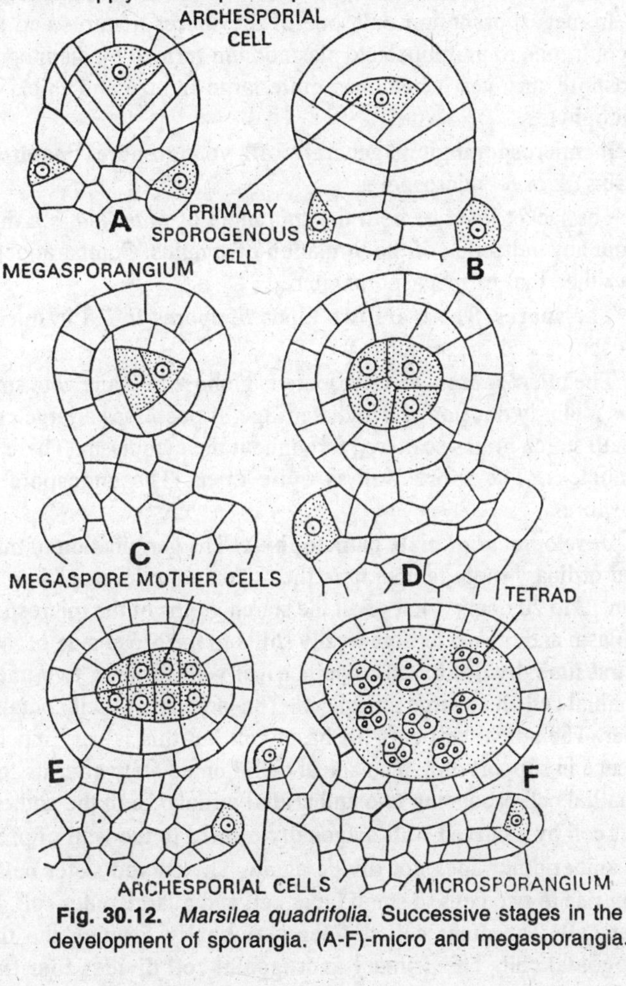

Fig. 30.12. *Marsilea quadrifolia*. Successive stages in the development of sporangia. (A-F)-micro and megasporangia.

Development of sporangia. The development of micro and megasporangia is of **lepto-sporangiate** type. The leptosporangiate differ from other ferns in having a sporangium with a jacket layer one cell in thickness. There is a definite number of spores within a sporangium. The sporangium develops from a single initial cell and the tapetum differentiates from a single internal cell of a developing sporangium.

One of the superficial cells acts as a sporangial initial. The sporangial initial divides periclinally producing two cells, the inner cell and the outer cell. The inner cell does not divide further whereas the outer cell forms the entire sporangium, its stalk and its contents. The outer cell divides thrice by successive diagonal divisions giving rise to a tetrahedral apical cell. The apical cell cuts of two segments on each of its three sides giving rise to a stalk and basal wall cells of the sporangium. The apical cell divides periclinally producing an outer cell, the jacket initials and an inner archesporial cell. The jacket initials divide anticlinally and give rise to the jacket layer of the sporangium. The archesporial cell divides periclinally and produce tapetal initials which divide both by anticlinal and periclinal divisions and give rise to two or three layered tapetal layer. After forming the tapetum the remaining archesporial cell, which acts as primary sporogenous cell divides 3 or 4 times successively giving rise to 8 or 16 spores mother cells. Each spore mother cell divides meiotically producing a tetrad of 4 haploid spores. Thus, in both types of sporangia, *i.e.*, micro and megasporangia about 32 or 64 young spores are formed.

The development in the micro and megasporangia is alike upto the formation of a one celled thick layer, two-celled thick tapetal layer and 32 to 64 young spores.

In megasporangium only one of the spores matures and the remaining spores disintegrate and contribute to multinucleate plasmodium formed by disintegration of the tapetum. As only one megaspore survives it becomes quite large in size and thick walled as in other heterosporous pteridophytes.

In microsporangium generally all young spores mature so that each microsporangium contains 32 or 64 microspores.

The jacket layer of both the sporangia of *Marsilea* is a thin walled homogeneous structure without any indication of the formation of annulus. Comparatively the stalk of microsporangium is longer than that of megasporangium.

The spores. There are two kinds of spores found in micro and megasporangia, known as microspores and megaspores.

The microspores. A microspore is globose in shape with small pyramid-like apex. Each spore shows a faintly developed triradiate ridge. It possesses a large central nucleus and a large number of small starch grains scattered throughout the cytoplasm. There may be one or two nucleoli in the big nucleus. The spore wall is quite thick. The endospore is surrounded by epispore and endospore.

Development of male gametophyte. The germination of microspore takes place quite rapidly and at ordinary room temperature the microspore develops into a fully mature microgametophyte within 12 to 20 hours. First of all the starch grains of the microspore migrate to the periphery of the cytoplasm and the nucleus generally shifts to the broad side of the microspore. The microspore for the first time divides by an asymmetrical wall forming two unequal cells—a small lenticular or prothallial cell and a large apical cell. The large apical cell contains granular contents and it divides further. The apical cell cuts off basal cell but this is very much doubted because of the lack of evidence in support of it. The apical cell, immediately after its formation divides diagonally to the prothallial cell producing two antheridial initials. Both the antheridial initials then cut off the first jacket cell by a curved wall diagonally formed to the wall of previous division. The jacket cell in each antheridium does not divide again. The curved sister cell divides by a periclinal division giving rise to two cells, a small inner cell and a large outer cell. The small inner cell is the second jacket cell. The outer cell divides periclinally forming the third jacket cell and the primary androgonial cell. The primary androgonial cell divides four times in a series giving rise to 16 androcytes. Simultaneously the prothallial cell distintegrates completely and the jacket cell disintegrate partially. The two androcyte masses lie within the old spore wall. In *Marsilea* two groups of androcytes have been produced in the two hemispheres.

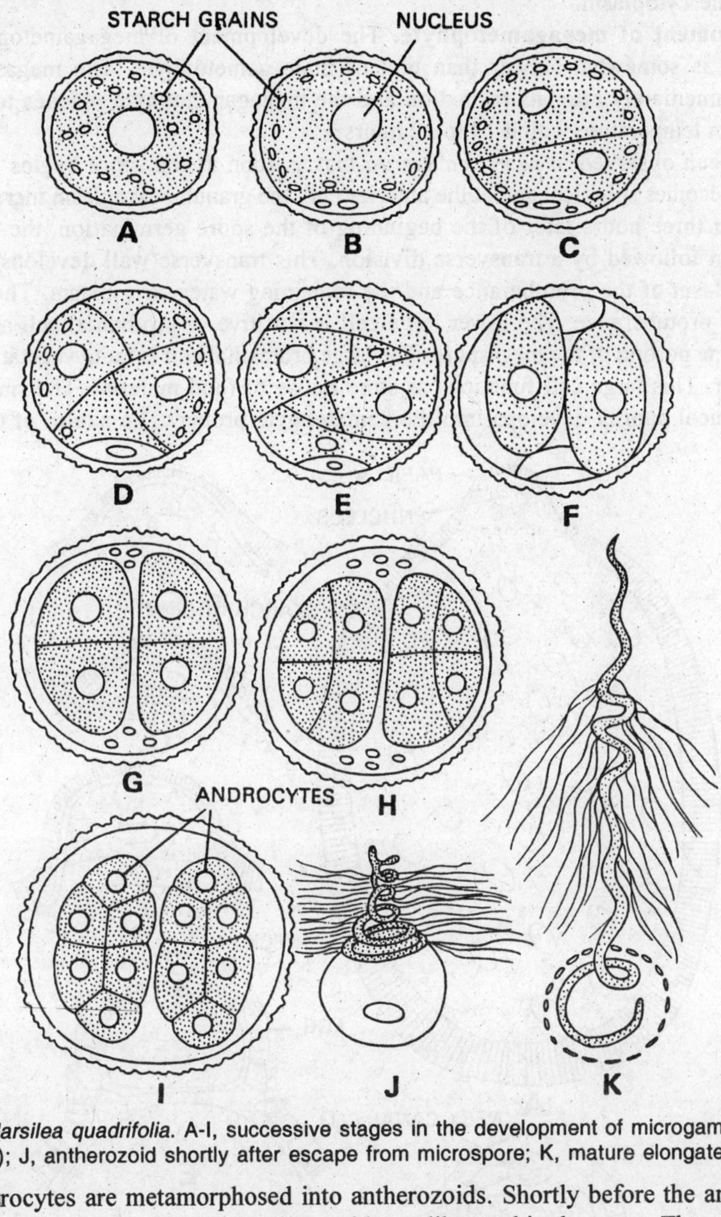

Fig. 30.13. *Marsilea quadrifolia.* A-I, successive stages in the development of microgametophyte (male gametophyte); J, antherozoid shortly after escape from microspore; K, mature elongated antherozoid.

The androcytes are metamorphosed into antherozoids. Shortly before the antherozoids are mature, the spore wall bursts and the antherozoids are liberated in the water. The antherozoids are liberated by the disintegration of the jacket cell of each antheridium.

Structure of antherozoid. The antherozoids of *Marsilea* have many coils, sometimes a dozen or more. They are cork-screw like in shape. Usually the upper three or four coils consist of blepharoplast and do not bear any flagella. The broad posterior coils bear 50 to 60 flagella. At the posterior end a globular vesicle is found which contains the cytoplasm.

The megaspore. The megaspore is ellipsoidal and possesses a small projection at the anterior end. The mucilaginous spore wall is quite thick around the megaspore but thin at its protuberant anterior end. In the remaining part of the megaspore the middle spore wall layer is much thicker and provided with radial fibrils. The anterior protuberant portion contains densely granular cytoplasm and nucleus. The remaining basal protoplast contains a large number of starch grains scattered throughout the cytoplasm.

Development of megagametophyte. The development of megagametophyte or female gametophyte is somewhat slower than that of male gametophyte. The megaspore begins to germinate immediately after its formation and the megagametophyte reaches to its maturity at ordinary room temperature within 14 to 22 hours.

It has been observed in *Marsilea vestita* that as soon as the spore begins its germination, the nucleus becomes spherical and at the anterior end, the granular cytoplasm increases in amount. About two or three hours later of the beginning of the spore germination, the nucleus divides which is soon followed by a transverse division. This transverse wall develops in between the denser cytoplasm of the protuberance and the remaining watery cytoplasm. The small cell lies within apical protuberance. The lower large cell is nutritive or prothallial cell and occupies the remaining large portion of the megaspore. The large prothallial cell is quite vegetative and does not divide further. This large cell functions as store house of food material. The small cell situated within the apical papilla develops into the gametophyte proper. The nature of the small cell is

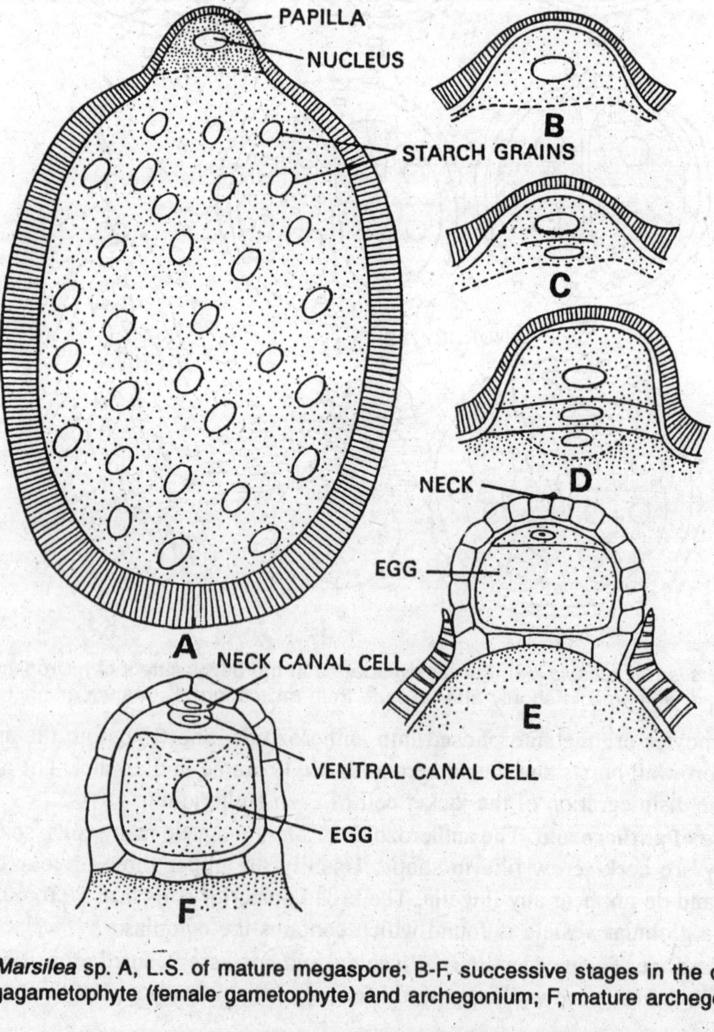

Fig. 30.14. *Marsilea* sp. A, L.S. of mature megaspore; B-F, successive stages in the development of megagametophyte (female gametophyte) and archegonium; F, mature archegonium.

somewhat doubtful. Sometimes this has been interpreted as an archegonial initial but it is much more probable that the apical cell bears four cutting faces, three lateral and one basal. Each of these cuts a single segment at each face before it acts as an archegonial initial. All the four segments divide by anticlinal divisions several times giving rise to a one-celled thick vegetative tissue. The archegonial initial divides periclinally forming a primary cover cell and central cell. The primary cover cell further divides by two successive anticlinal divisions forming four quadrately arranged neck initials. Each neck initial divides by an oblique wall forming a short neck comprised of two tiers of four cells each. Simultaneously, the central cell divides into two cells, a small primary canal cell and a large primary venter cell. In *M. drummondii* and some other species the primary

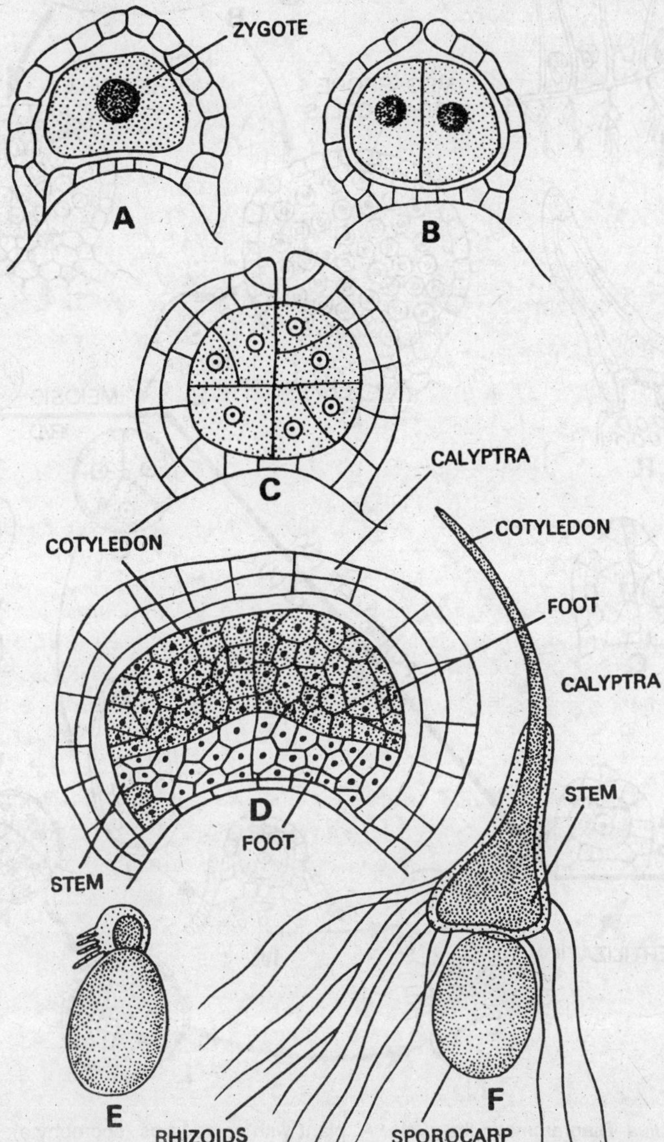

Fig. 30.15. *Marsilea* sp. A-D, successive stages in the development of the embryo (sporophyte); E, sporocarp with young embryo; F, more advanced stage of the embryo in median longitudinal section.

canal divides producing two small canal cells. In all species the **primary venter** cell divides into two cells, a small venter canal cell and a large spherical oosphere or egg. On the maturation of the archegonium the neck canal cell or cells and the venter canal cell disintegrate and the archegonial neck remains open.

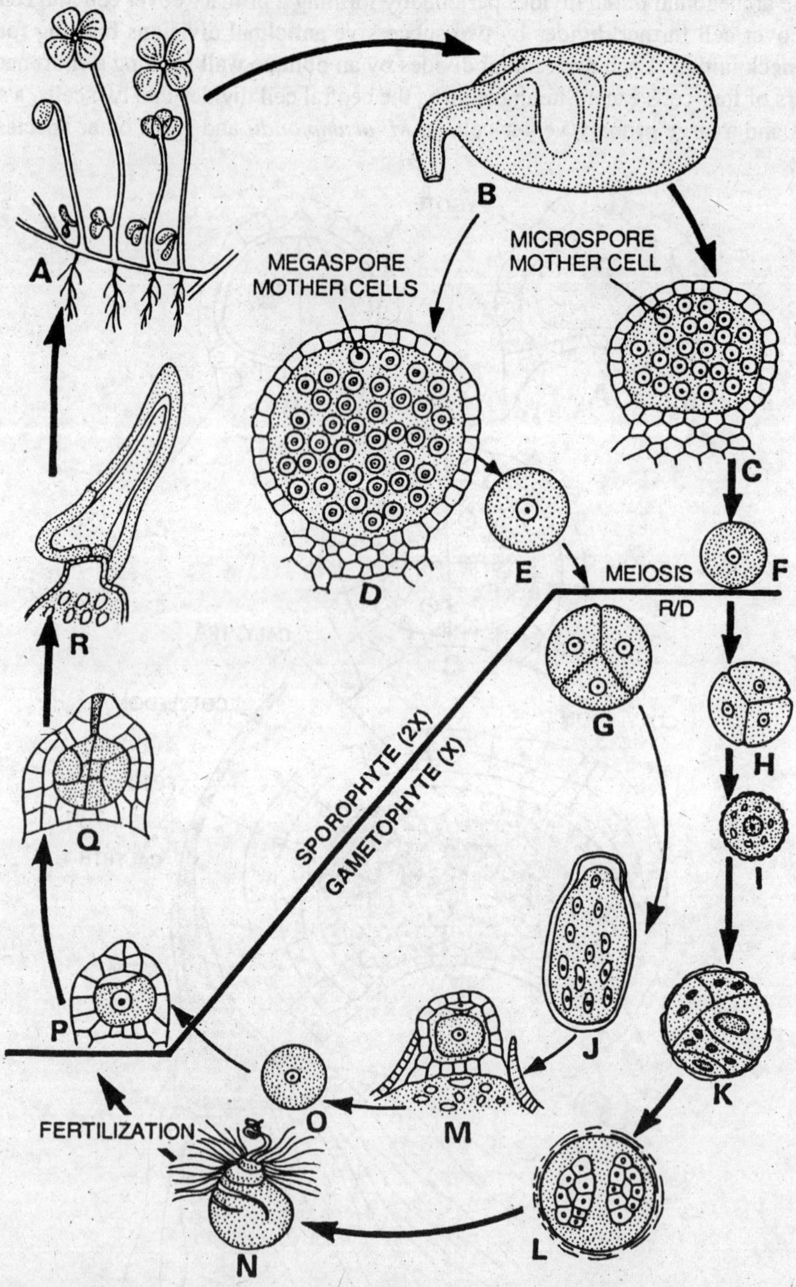

Fig. 30.16. *Marsilea*. Diagrammatic life-cycle. A, plant with sporocarps (sporophyte); B, a sporocarp; C, microspore mother cells within microsporangium; D, megaspore mother cells within megasporangium; E, megaspore mother cell; F, microspore mother cell; G, megaspore tetrad; H, microspore tetrad; I, microspore; J, megagametophyte; K-L, formation of microgametophyte; M, archegonium with egg; N, antherozoid; O, the egg; P, oospore; Q, early stage of embryo formation; R, young sporophyte.

A mature megagametophyte is surrounded by a broad, ovoid gelatinous zone which there in a watery funnel shaped opening directly upon the archegonium. A large number of antherozoids swim about the funnel shaped opening. Some antherozoids enter the gelatinous envelope, some reach the neck of the archegonium and others fail to do so. Many of the antherozoids die and their uncoiled elongated, non-motile bodies may be seen in the gelatinous envelope whereas the living actively swarming antherozoids may be seen in the gelatinous envelope coiled and moving. This way, the active and living antherozoids may be easily differentiated from the dead antherozoids which have become non-motile and inactive.

Fertilization. Many antherozoids are attracted towards the funnel shaped structure of megagametophyte. The antherozoids enter gelatinous envelope and reach to the archegonium. Only one antherozoid enters the egg's cytoplasm anteriorly, but later on its orientation becomes reversed and it penetrates the egg nucleus posterior end forward.

Marsilea drummondii shows parthenogenesis. Here the megaspore on germination produces a diploid ($2n$) megagametophyte which without fertilization develops into a sporophyte.

Development of embryo. In *Marsilea* the development of the embryo proceeds with great rapidity and only two to four days old sporophyte possesses the first leaf several millimetres long. In *Marsilea vestita* the zygote divides for the first time after fertilization only two or three hours later. The zygote divides first by a vertical division. Thereafter the two daughter cells divide vertically in a plane perpendicular to that of first division. Thus, a four-celled embryo is formed. Two sister cells of outer side develop into cotyledon and stem and the other two of inner side develop into the foot and root of the embryo. The adjoining cells of the megagametophyte divide periclinally forming a two to three-celled thick calyptra. Simultaneously, many superficial cells on the side of the calyptra facing downwards send out the rhizoids. For the first four or five days, the growth of the calyptra keeps pace with that of the embryo but later on the cotyledon and primary root penetrate through the calyptra are going upward and the other downward respectively.

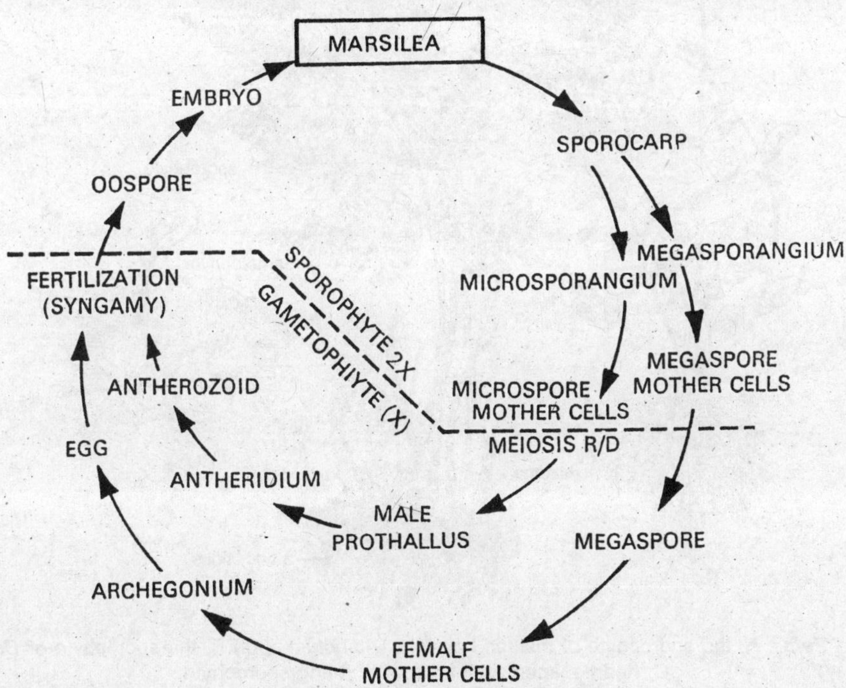

Fig. 30.17. *Marsilea*. Graphic life-cycle.

31

CHAPTER

Division-Filicophyta
Class-Leptosporangiopsida

Order-Filicales

Characteristic features. They are homosporous ferns. They produce usually heart-shaped prothalli (gametophytes) which bear sex organs (antheridia and archegonia) on their ventral surface. The prothalli are monoecious. The sporangia are produced in sori either on the margin or on the surface of the sporophyll. This order includes seven families, which are as follows:

1. Schizaeaceae
2. Gleicheniaceae
3. Matoniaceae
4. Hymenophyllaceae
5. Dicksoniaceae
6. Cyatheaeae
7. Polypodiaceae

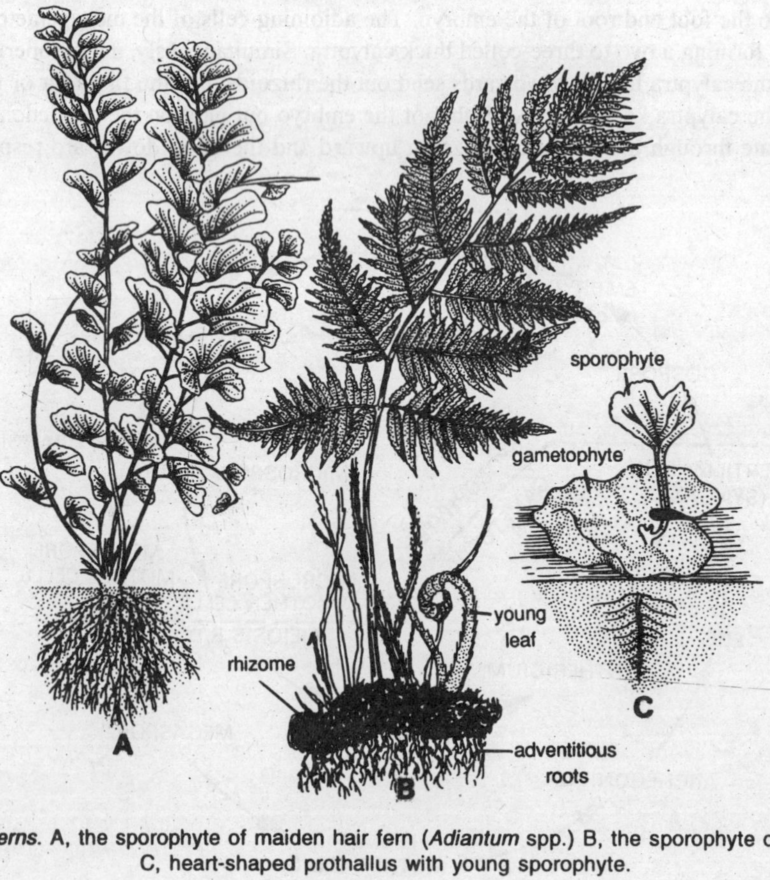

Fig. 31.1. *Ferns.* A, the sporophyte of maiden hair fern (*Adiantum* spp.) B, the sporophyte of *Dryopteris*; C, heart-shaped prothallus with young sporophyte.

Genus PTERIS Linn.

Systematic Position:

Division	:	Filicophyta
Class	:	Leptosporangiopsida
Order	:	Filicales
Family	:	Polypodiaceae
Sub-family	:	Pteridoideae
Genus	:	*Pteris*
Species	:	*vittata*

Habitat. This genus is widely distributed and represented by about 250 species that grow abundantly in the tropical and sub-tropical regions of the world. The common Indian species are: *Pteris vittata, P. cretica, P. biaurita, P. quadriaurita, P. stenophylla* and *P. wallichiana.* However, *Pteris vittata* is low level fern and brings out new leaves throughout the year. It is quite common along mountain walls and grows upto 1200 metres above sea level, while *P. quadriaurita* grows abundantly along road-sides and the valleys throughout the North-western and western Himalayas: *P. cretica* grows well from 1200-2400 metres a.s.l.

Habit. The plants have creeping rhizomes in *P. biaurita, P. vittata* and *P. grandiflora*. In *P. cretica* the rhizome is branched short, stumpy and semi-erect. The rhizome is covered with scales. Hair are absent. Roots arise from the lower surface of the rhizome or they may arise all over the surface, *e.g.*, in *P. biaurita* (Fig. 31.1).

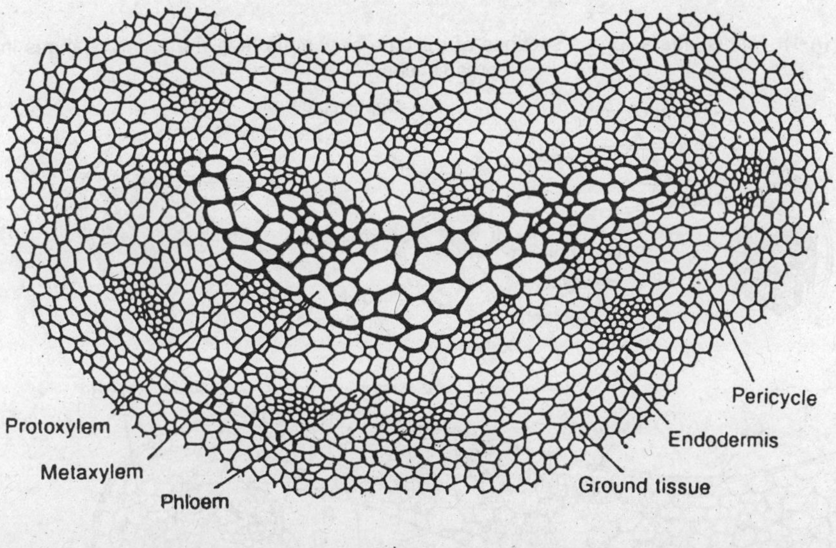

Fig. 31.2. A meristele from the rhizome of *Pteris vittata* showing detailed internal structure.

The leaves are pinnately compound. The petioles are covered with scales. Venation is of open furcate type. The leaves in *P. vittata* are once pinnate. The Pinnae are smaller near the base larger near the middle, and again smaller towards the apex of the leaf. Leaf apex is occupied by an old leaflet or pinna. Every pinna is traversed by a central midrib which gives off lateral veins which bifurcate near the tip. The pinnae are sessile and broad at the base, and gradually taper towards the apex, the leaves are bipinnate in *P. biaurita*. The pinnules are coriaceous or rough (Fig. 31.2).

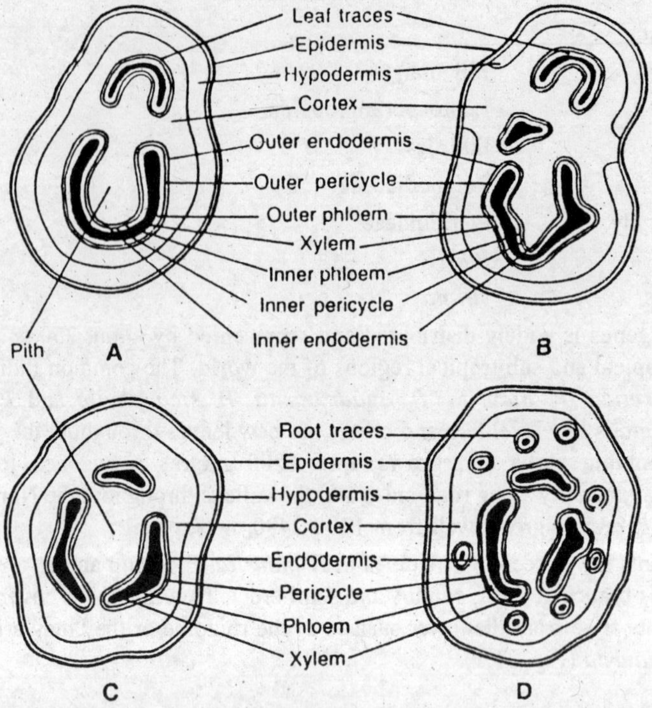

Fig. 31.3. (A-1). *Ptenis vittata.* Cross sections of rhizome at various levels showing changes in stelar organisation.

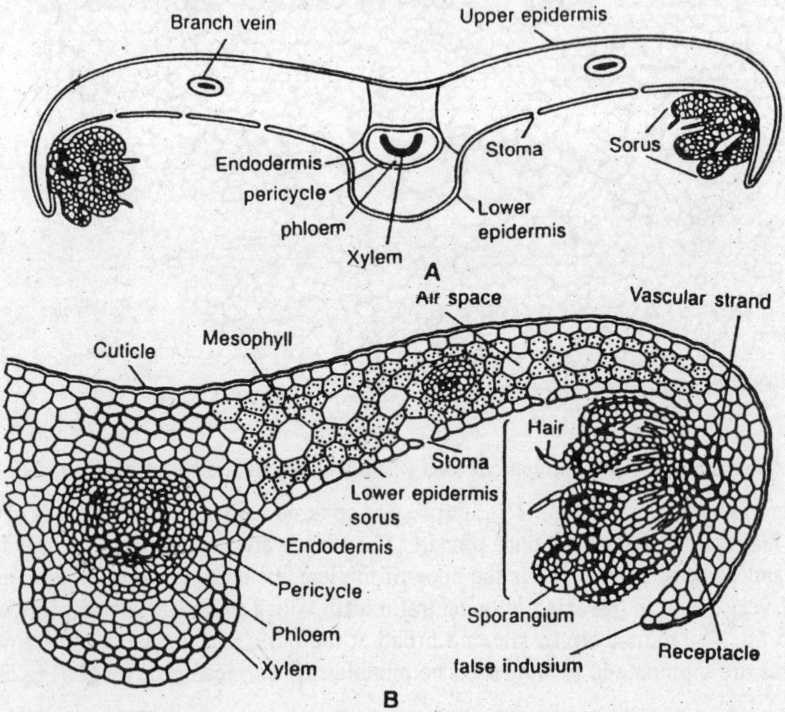

Fig. 31.4. (A–B). *Pteris vittata.* A. Outline figure of T.S. fertile leaflet showing internal structure; B, portion of 'A' in detail.

Anatomy. The stelar organization of rhizomes varies from species to species, and sometimes even in same species. It is solenostelic in *P. grandiflora* and *P. vittata*, a simple dictyostele in *P. cretica* and younger regions of *P. vittata*. In the younger branches of the rhizome in *P. biaurtia*, the stele is a mixed protostele in the lower region. It becomes siphonostelic a little higher up and solenostelic near the apex. In the main rhizome dictyostelic condition is also found. In *P. vittata* the stelar organization changes to a dicyclic dictyostele in the apical regions of the rhizome (Fig. 31.3). This condition is achieved by the separation of small, circular or oval strands or root traces from the outer surface of the parent meristeles that in turn were derived by the stelar perforation of the solenostelic arc. These strands lead to the formation of irregularly arranged root traces around 3-5 large meristeles (Fig. 31.4). However, there are no medullary strands.

The pinnule has upper and lower epidermal layers (Fig. 31.5). In *P. cretica* the upper epidermis has larger cells with less sinuous walls. The stomata are restricted to the lower epidermis which has smaller cells and the more sinuous walls. The mesophyll may not be differentiated into palisade and spongy parenchyma. The mid-rib region has single concentric type of vascular strand, *e.g.*, in *P. vittata*, with distinct endodermis. The bundle sheath extensions are prominent and occur as groups of thick-walled cells below the upper and above the lower epidermis. Palisade and spongy tissue is absent around the mid-rib.

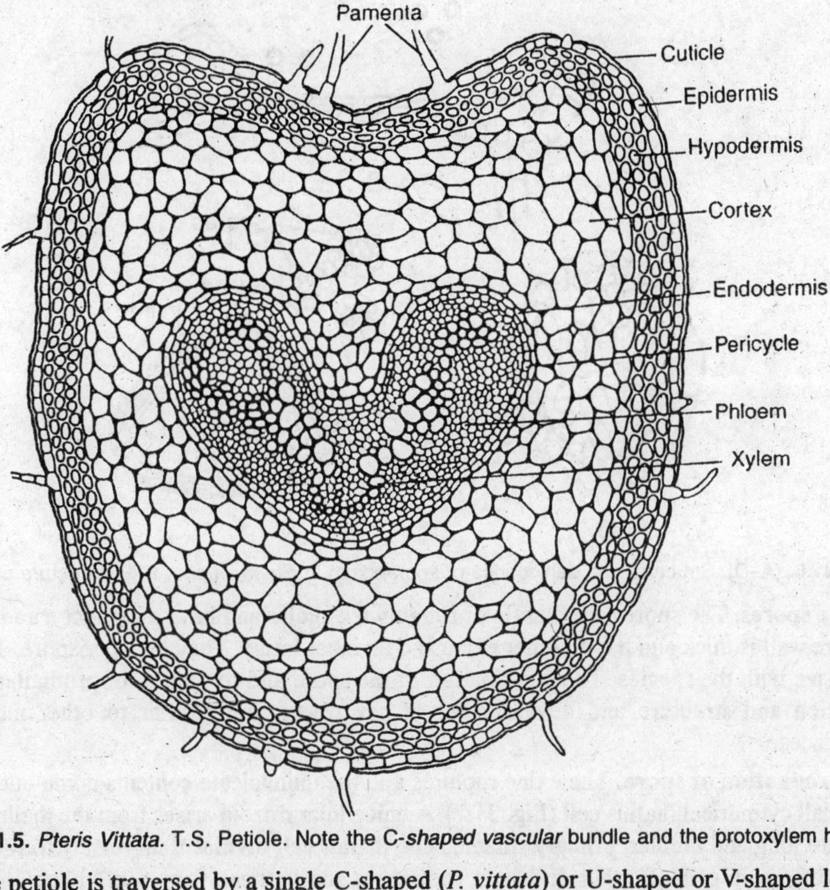

Fig. 31.5. *Pteris Vittata.* T.S. Petiole. Note the C-*shaped vascular* bundle and the protoxylem hooks.

The petiole is traversed by a single C-shaped (*P. vittata*) or U-shaped or V-shaped leaf trace (Fig. 31.6). In *P. biaurita* the U-shaped leaf trace enters the petiole and becomes V-shaped higher up. The xylem has two adaxial hooks. In the rachis the petiole trace gives off strands into its pinnae if the leaf is unipinnate or it divides to give rise to secondary and tertiary pinna traces in bipinnate leaves (*P. biaurita*). The rachis traces are marginal in origin and are usually flat U-shaped or shallow arc-like.

The sporophyll. The sorus is of continuous type. It is protected by the upper indusial flap that is formed by the margins of the pinnae (Fig. 31.7). The receptacle is intramarginal in origin. The lower or the true indusium is absent. The sporangia are unmixed and development of sporangia is similar to that of *Dryopteris*. Each sporangium produces 48 spores.

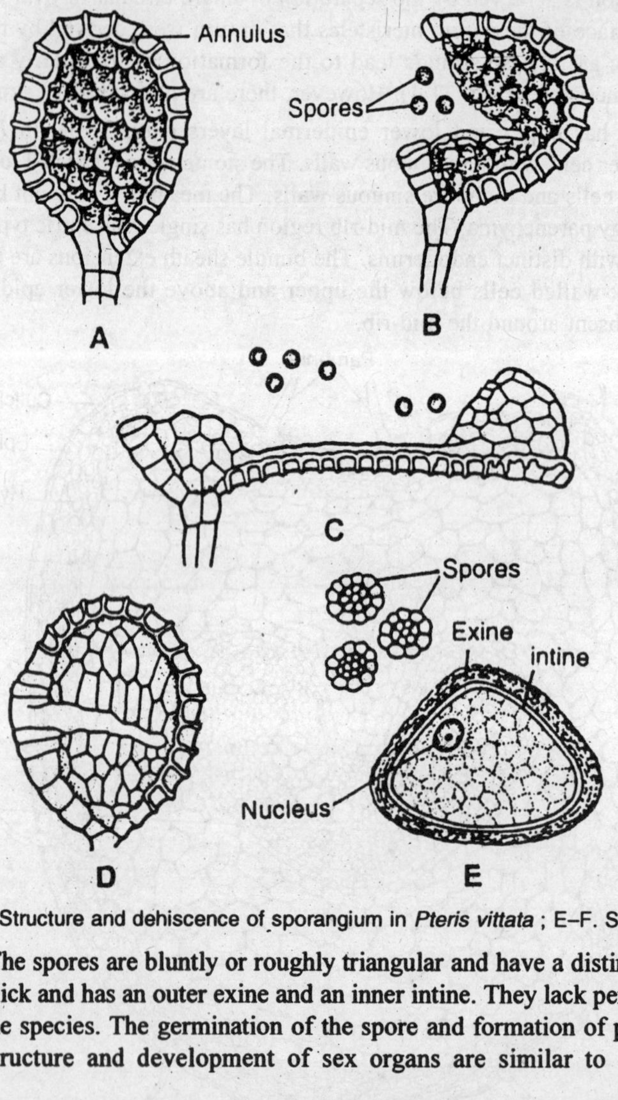

Fig. 31.6. (A–D). Structure and dehiscence of sporangium in *Pteris vittata* ; E–F. Structure of spore.

The spores. The spores are bluntly or roughly triangular and have a distinct triradiate mask. The spore wall is thick and has an outer exine and an inner intine. They lack perispore. The spores vary in size with the species. The germination of the spore and formation of prothallus position, distribution and structure and development of sex organs are similar to other members of polypodiaceae.

Germination of spore. The exine ruptures and the uninucleate contents come out and grow into a small cylindrical thallus cell (Fig. 31.7). A unicellular rhizoid arises from the thallus cell and grows into long and tubular, primary rhizoid. The thallus cell divides transversely to form a short filament of green cells. Further divisions, transverse and vertical, form a big plate one layer of cells in thickness. The cells behind the apical notch divide along two or more planes and form a thick central cushion. The secondary rhizoids arise from the posterior regions of the prothallus. Mature thallus cells act as rhizoidal mother cells. The rhizoidal cells differentiate and grow into secondary rhizoids.

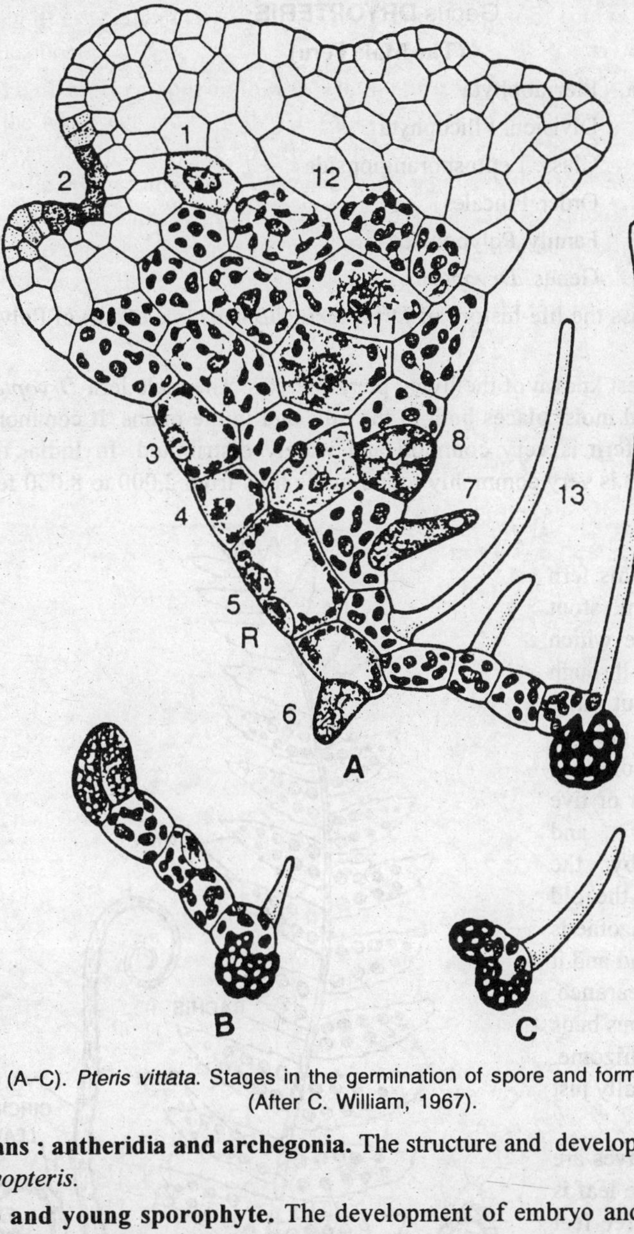

Fig. 31.7. (A–C). *Pteris vittata.* Stages in the germination of spore and formation of rhizoid cell (After C. William, 1967).

Sex organs : antheridia and archegonia. The structure and development of sex organs are similar to *Dryopteris.*

Embryo and young sporophyte. The development of embryo and young sporophyte is similar to that of *Dryopteris*:

Economic importance. The species of *Pteris* are of little economic value. They are noxious and toxic weeds, which may cause harm to livestock. However, some species of *Pteris* are useful. The plants have been used as a source of potash for soap and glass making industry. Fronds are used as fuel and thatching material. They are also used as medicine and anthelmintic. Plants are used for brewing purposes. They are used as tan or dye. Some plants are used for making paper and board.

Genus DRYOPTERIS

(The Male Fern)

Systematic position. Pteridophyta
Division. Filicophyta
Class. Leptosporangiopsida
Order. Filicales
Family. Polypodiaceae
Genus. *Dryopteris*

Here, we will discuss the life-history of *Dryopteris filix-mas* (male fern) of Polypodiaceae in detail.

The ferns are the best known of the living pteridophytes. The male fern-*Dryopteris filix-mas* is found in the shady and moist places both in the hills and in the plains. It commonly occurs in the thick forests. This fern is very common and widely distributed. In India, this is found throughout the country. It is very commonly found in the hills from 2,000 to 8,000 feet elevation.

External Morphology

Habit and stem. This fern possessess a short, blunt, stout and unbranched rhizome which grows obliquely upwards through the soil and appears but little above the surface of the ground. The rhizome is seldom more than a foot in length. It is four or five inches in diameter and completely covered by the persistent leaf bases of the old leaves. The top of the rhizome is broader than the lower end and it becomes conical in appearance. Sometimes, the adventitious buds also appear on the rhizome. These buds develop laterally just above the leaf bases.

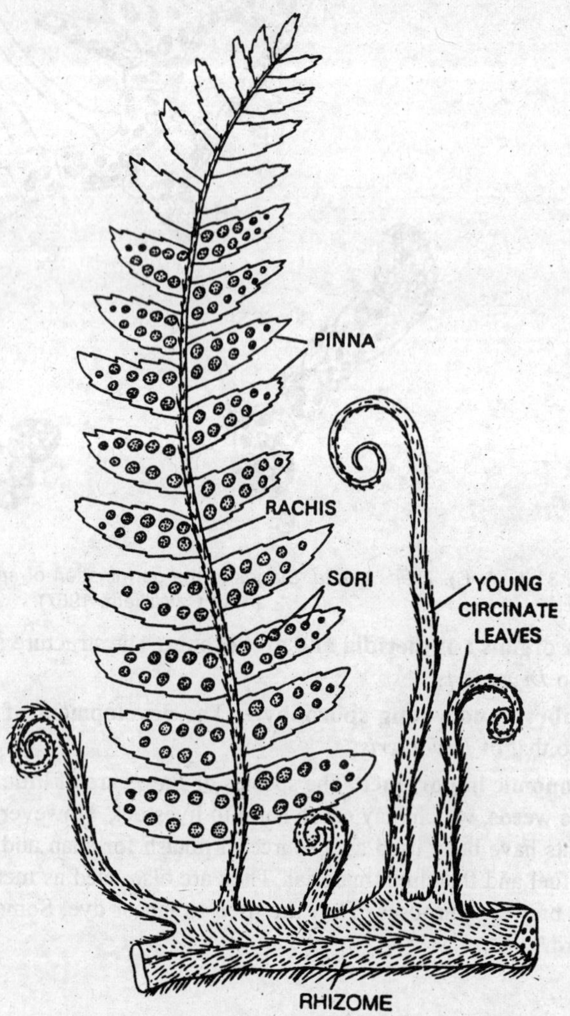

Fig. 31.8. *Dryopteris filix-mas* (Fern). Habit of fern showing rhizome, young circinate leaves and sporophyll.

The leaves. The leaves are also known as **frond.** The leaf is compound and one to three feet in length. Each leaf is composed of numerous leaflets or **pinnae.** The pinnae are again sub-divided into small **pinnules.** The leaf is petiolate. The petiole ends in a long rachis which bears pinnae. In young leaves, both the rachis and the petiole are covered with small, brown, hairy scales known as **ramenta.** The young leaves are **circinnate.** The leaf matures

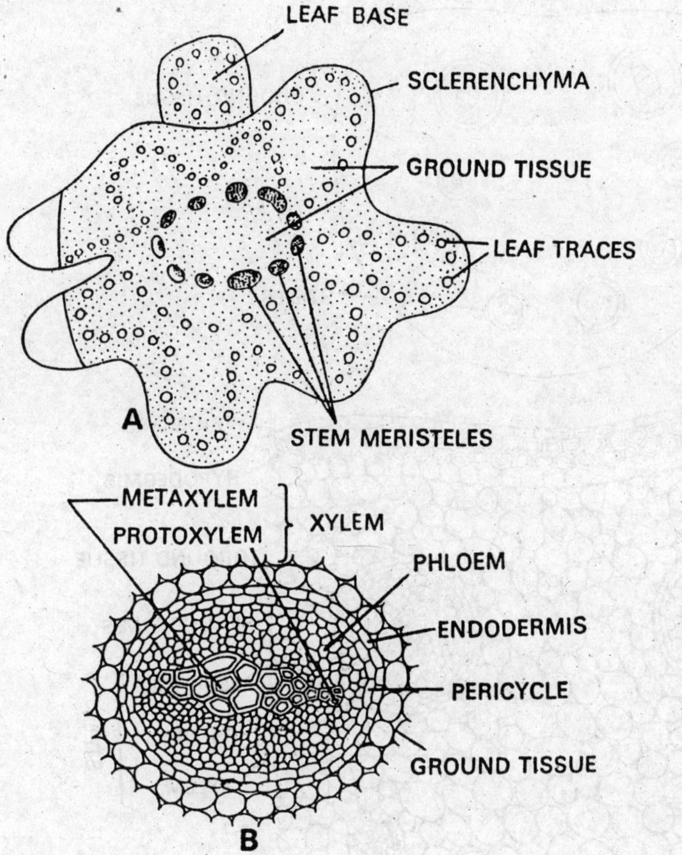

Fig. 31.9. *Dryopteris filix-mas.* A, T.S. of rhizome showing meristeles; B, detailed structure of a meristele.

in two years. In the first year, the petiole, rachis and the leaflets remain coiled. Such young leaves are called circinnate leaves. The mature leaves last only for one season and again ten to twenty leaves come out, which remain spirally arranged on the rhizome.

The root. The roots are of two kinds-primary and secondary. The primary root, produced by the embryo dies away earlier. The secondary roots remain, which are also called the adventitious roots. The adventitious roots are produced secondarily from the tissues of the stem. Usually three adventitious roots develop at the base of each petiole. The roots are slender and profusely branched, so that an old stem becomes densely covered by a mass of adventitious roots. The roots are monopodially branched and the branches arise acropetally.

Internal Morphology

Anatomy of rhizome. The vascular cylinder is a typical dictyostele. In many simple pteridophytes the stem possesses a single central stele surrounded by the cortex. In this fern, like that of many other ferns, the central stele breaks into a ring of steles. These steles are known as 'meristeles' and the stelar system is said to be dictyostelic. The meristeles are irregularly embedded in the ground tissue. In the case of *Dryopteris,* the stem is completely covered with leaf bases and there are no internodes. Any transverse section of the stem is surrounded by the leaf bases, showing the obliquely cut-leaf traces. The leaf-traces traverse the cortex and enter the petioles.

Each meristele is composed of a central mass of xylem surrounded by phloem. The scalariform tracheids and xylem parenchyma form the xylem. In the transection of the stem the tracheids appear polygonal and sufficiently wide. The vessels are altogether absent. In the centre of the meristele the tracheids bear spiral and reticulate thickenings. These tracheids represent the protoxylem. Rest of the wide tracheids of the xylem represent metaxylem. The phloem is composed of a single ring of large sieve tubes. The sieve tubes are quite long and tapering at their overlapping ends. At the junction of the two sieve tubes, the sieve plates are developed. The companion cells are not found. Just below the ring of sieve tubes, there is good deal of phloem parenchyma. The phloem is surrounded by a single-layered pericycle constituted of thin walled,

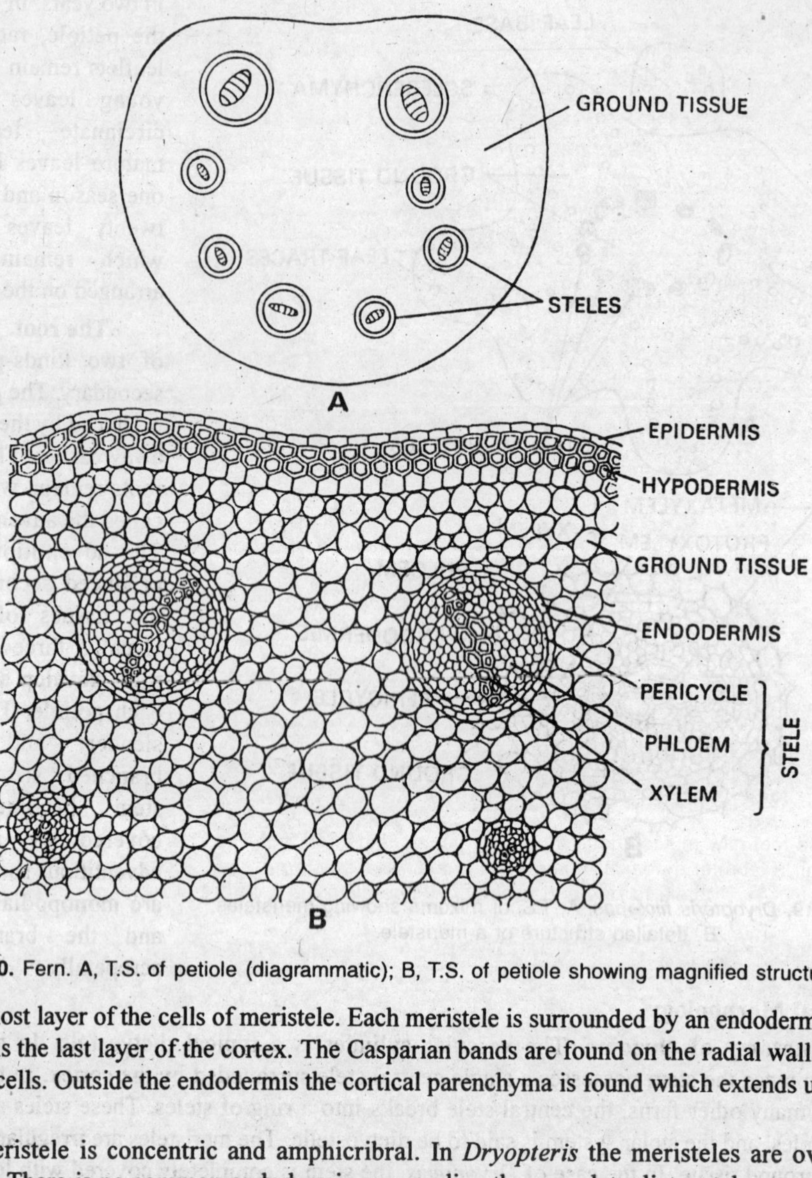

Fig. 31.10. Fern. A, T.S. of petiole (diagrammatic); B, T.S. of petiole showing magnified structure.

is the outermost layer of the cells of meristele. Each meristele is surrounded by an endodermis. The endodermis is the last layer of the cortex. The Casparian bands are found on the radial walls of the endodermal cells. Outside the endodermis the cortical parenchyma is found which extends upto the epidermis.

The meristele is concentric and amphicribral. In *Dryopteris* the meristeles are oval and independent. There is no common endodermis surrounding the complete dictyostele.

The cortex is composed of a mass of parenchyma (ground tissue) in which the meristeles are embedded. The outermost few layers of the cells of the cortex are sclerenchymatous. The outermost layer of the stem is epidermis, which is composed of the cells with thick outer walls. The epidermis bears one celled thick ramenta on it. These scale-like ramenta are considered to be modified hairs.

Anatomy of leaf. From the point of view of its anatomy, the leaf may be divided into two parts (a) the petiole and (b) the lamina.

(a) The petiole. To study the anatomy of the petiole, one has to cut the transverse sections of it. The petiole is approximately semicircular in outline. The upper (adaxial) side of the petiole is flat and remains towards the stem. The outermost layer is the epidermis. Just below the epidermis a few layers of sclerenchyma cells are found. The meristeles are embedded in the

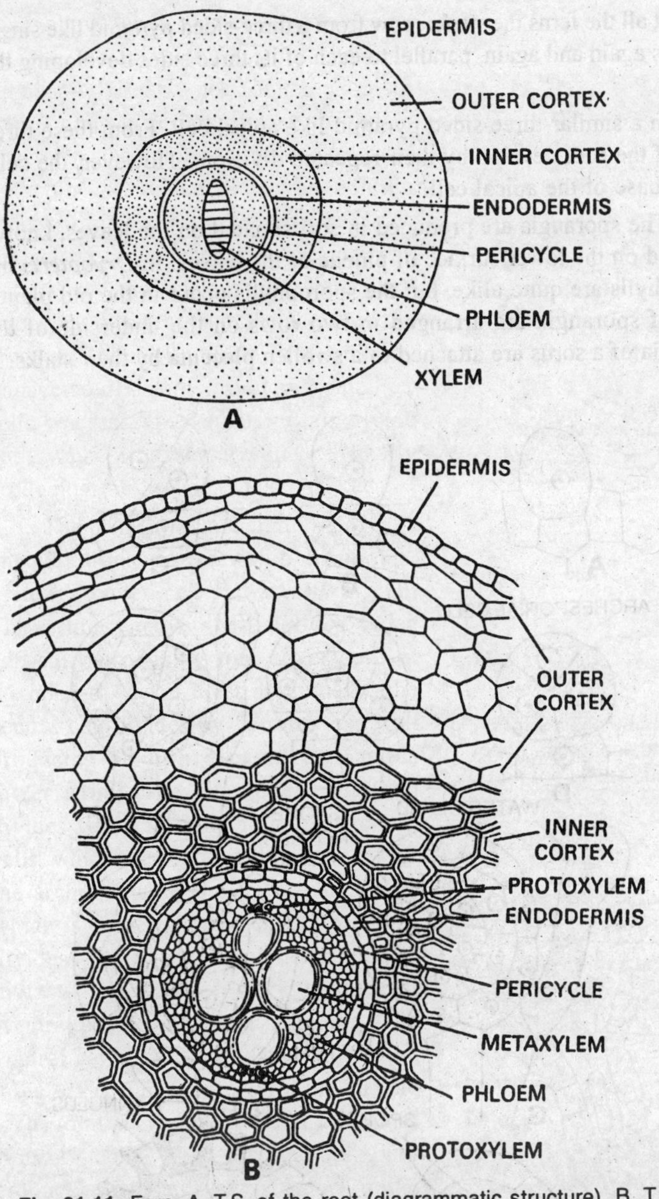

Fig. 31.11. Fern. A, T.S. of the root (diagrammatic structure), B, T.S. of a portion of the root (magnified structure)

parenchyma (ground tissue). In the arrangement the meristeles are horseshoe like. In the adaxial side of the rachis there are two meristeles, each giving off branches to each pinna.

(b) The lamina. The lamina has an upper and lower epidermis. Just beneath the upper epidermis there is mesophyll composed of irregular parenchymatous cells having well-developed intercellular spaces among them. Both the epidermal and mesophyll cells contain chloroplasts. The presence of the chloroplast in the epidermis is the characteristic of a shade loving plant. On the lower side of the lamina the lower epidermis is found. Just underneath the lower epidermis there is mesophyll, which is composed of irregularly branched and star-shaped cells having large intercellular spaces among them. The vascular strands are concentric.

Anatomy of root. To study the anatomy of the root, the transverse sections of the adventitious root are required. The internal structure of root is quite simple. There is a central core of xylem, surrounded by phloem. The xylem is with two protoxylem groups situated at its two ends. This way, the xylem is diarch. Outside the xylem there are phloem elements. The phloem is arranged laterally to the xylem in the form of phloem plates. A two-layered pericycle surrounds the phloem. The pericycle is the outermost part of the stele. Thereafter a single-layered endodermis is found around the stele. The Casparian strips are present on the radial walls of the cells of the endodermis. The outermost layer of the transverse section of the root is the piliferous layer. The piliferous layer is single-layered and bears root hairs here and there upon it. The cortical region is differentiated into two regions. The outer region consists of thin-walled cells whereas the inner region of thickwalled cells.

The apical growth. In about all the ferns the stems grow from a three-sided pyramid like single apical cell. The apical cell divides again and again, parallel to each of its three sides developing the cells of the mature tissue.

The roots also develop from a similar-three-sided pyramid like apical cell. From these three sides of the cell, all the organs of the root develop by their repeated divisions. However, the cells of the root cap are cut from the base of the apical cell.

Spore producing organs. The sporangia are produced in sori situated on the leaves, known as sporophylls. The sori are found on the lower surface of pinnae of the leaves. In *Dryopteris* the vegetative leaves and the sporophylls are quite alike, but the sporophylls occur in the old plants. The small kidney-shaped sori of sporangia are arranged in two rows on the underside of the pinnules of the leaf. The sporangia of a sorus are attached to a swollen placenta by their stalks. A delicate membranous structure, known as indusium also arises from the placenta and covers the sorus of sporangia like a hood. The mature sporangia are brown in colour.

Development of sporangium. The development of sporangium is of leptosporangiate type as in all other genera of Polypodiaceae. The sporangium develops from a single superficial cell of the placenta. This cell divides by a transverse wall forming two cells. The upper cell divides again by an oblique wall to the first wall. Another oblique division takes place, giving rise to a tetrahedral cell. This cell again divides by a transverse wall. The inner cell is known as archesporial cell. The lower cell of the pair formed by the first division divides several times and develops into the multicellular stalk. The cells surrounding the tetrahedral archesporial cell divide further giving rise to the sporangial wall. Thereafter, the archesporial cell divides to form a tapetal layer of the cells. At the same time it

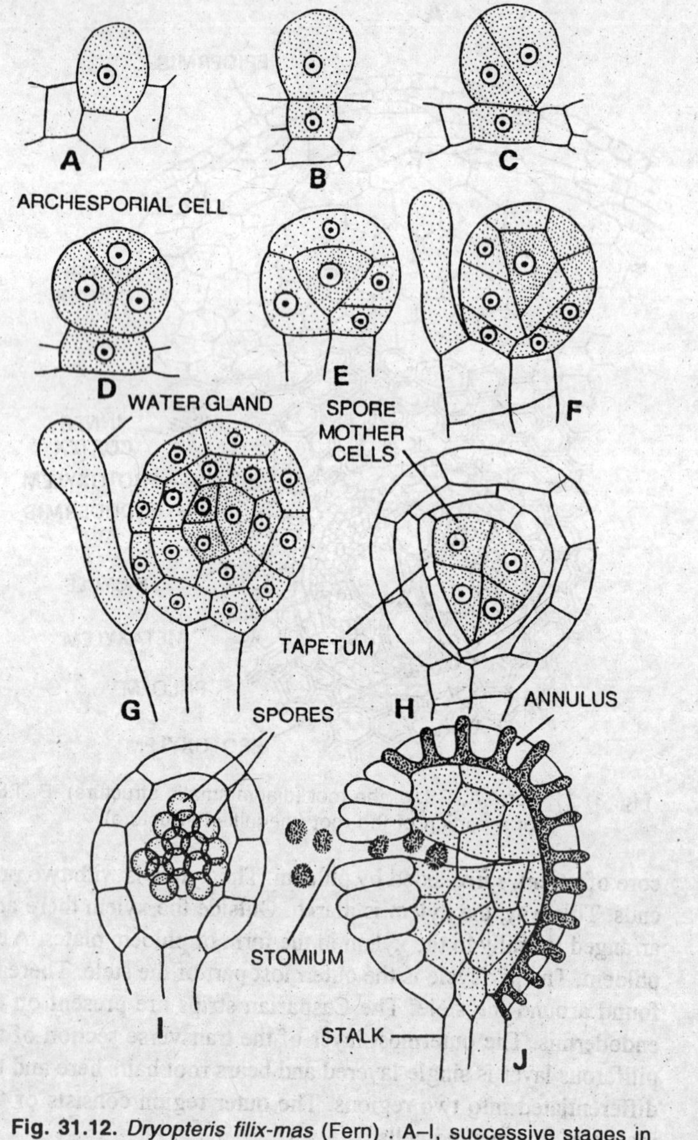

Fig. 31.12. *Dryopteris filix-mas* (Fern) . A–I, successive stages in the development of sporangium; J, mature sporangium.

also produces about a dozen spore mother cells. The spore mother cells give rise to spores. Each spore mother cell divides twice developing four spores. During division, there is meiosis and the spores are haploid. Usually each sporangium contains 48 haploid spores.

Structure of sporangium. Each sporangium is composed of a multicellular stalk and a capsule. The capsule of the sporangium is of the shape of a biconvex lens. The wall of the capsule is composed of a single layer of thin walled cells. A row of specially thickened cells, known as annulus partially surrounds the capsule. At one side of the annulus, there is a stomium composed of thin walled cells. On maturity, the sporangium contains about 32 to 64 dark, ragged walled spores in it.

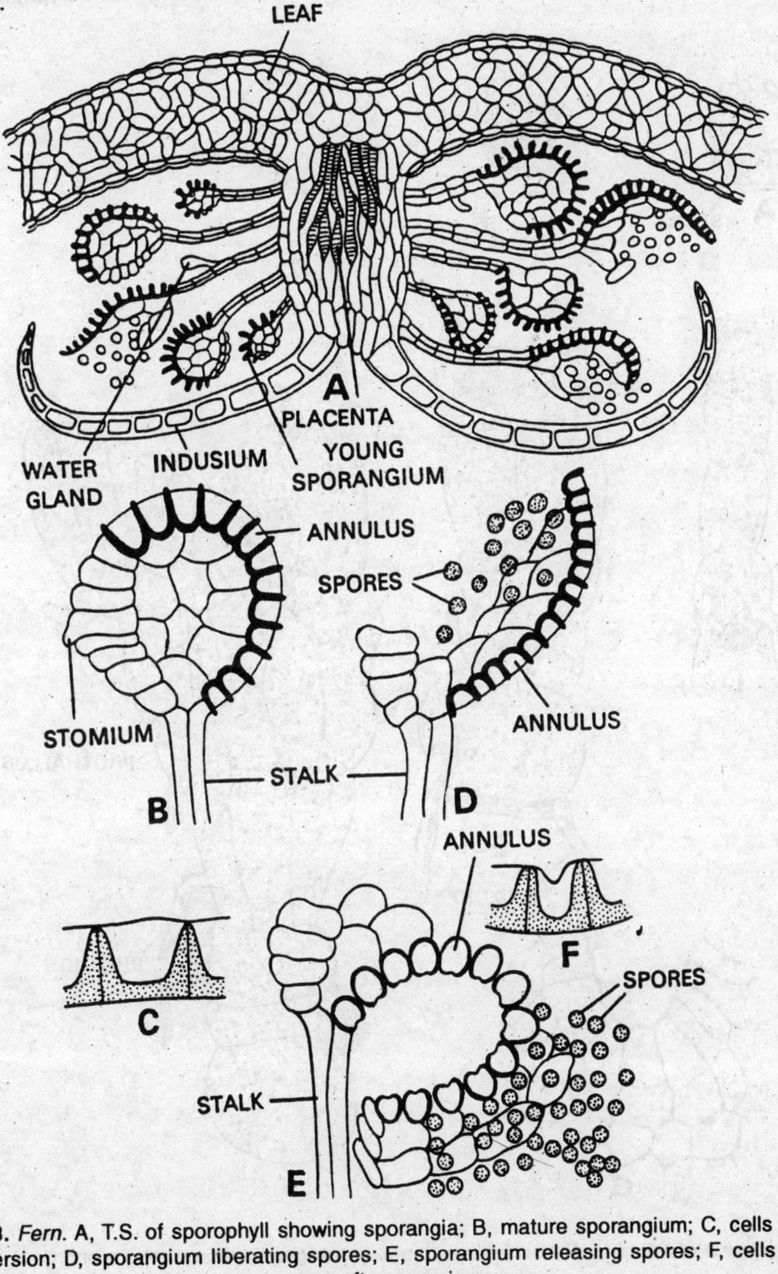

Fig. 31.13. *Fern.* A, T.S. of sporophyll showing sporangia; B, mature sporangium; C, cells of the annulus before eversion; D, sporangium liberating spores; E, sporangium releasing spores; F, cells of the annulus after eversion.

Dehiscence of sporangium. On the maturity of the sporangia, the indusium shrivels up and the sporangia are exposed to the outer atmosphere. The sporangium bursts by a special mechanism. The cells of the annulus are thickened in a peculiar way. The inner tangential and radial walls of the annulus are thick while the outer tangential walls are thin. The bursting of the sporangium is caused by the behaviour of the water in relation to the cells of the annulus. In young sporangia the water glands are present and they remain moist. On the maturity of the sporangia, the sporangia are exposed to dry air and the water is lost from the cells of the sporangia. This way, the thin outer walls of the annulus begin to contract, the stomium opens and the annulus along with some portion of the sporangium turns backward with a jerk, throwing the spores out of the sporangium.

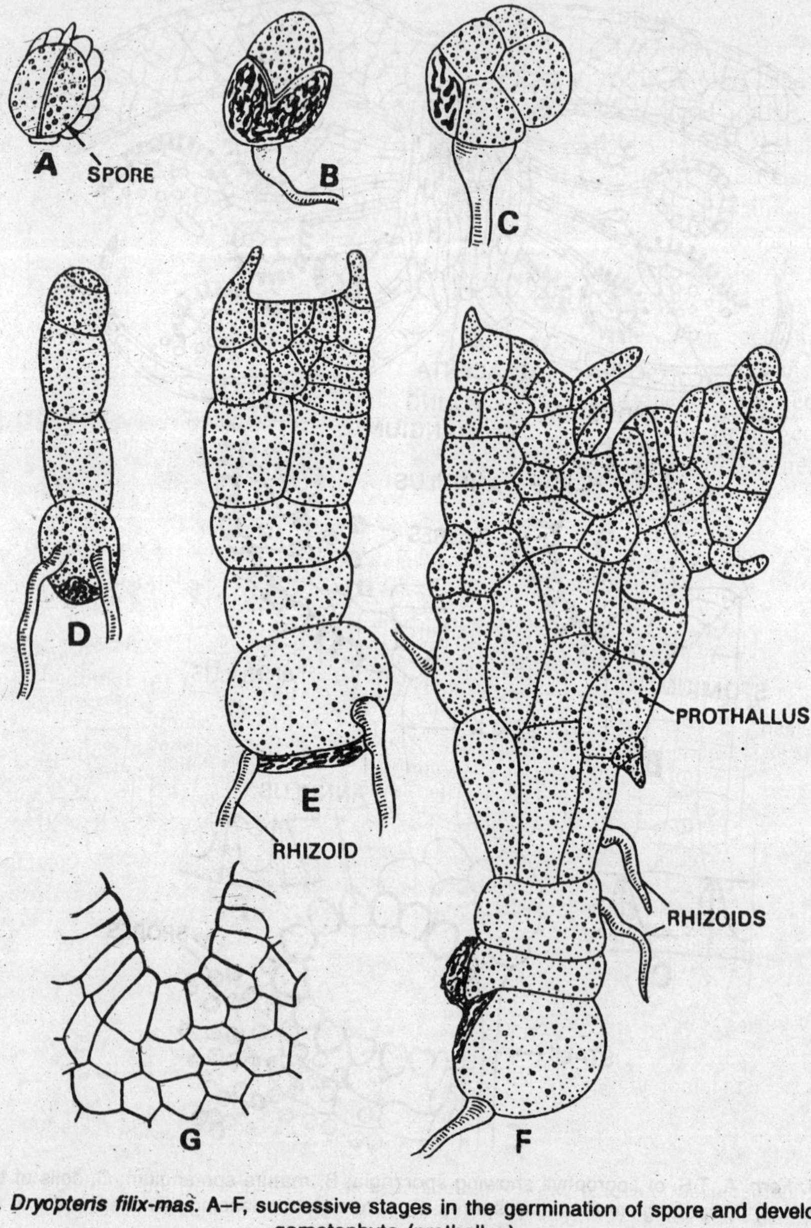

Fig. 31.14. *Dryopteris filix-mas.* A–F, successive stages in the germination of spore and development of gametophyte (prothallus).

The spore. The spores are double-walled. The outer wall is thick, ragged and brown in colour. It is known as exine. The inner wall is smooth and thin, and known as intine. Each spore contains a big central nucleus. The gametophytic stage of the fern begins from this spore.

Germination of the spore and development of the prothallus. After their dispersal the spores lie dormant for sometime and under the favourable conditions of moisture and temperature, they germinate. First of all the outer spore wall (exine) breaks and thereafter the intine protrudes out in the form of green germ tube. Very shortly this tube is cut off by a transverse wall and two unequal cells are resulted. The small colourless cell is known as rhizoidal cell. From this cell a colourless branch grows into the soil and develops a rhizoid. The large green cell elongates and divides

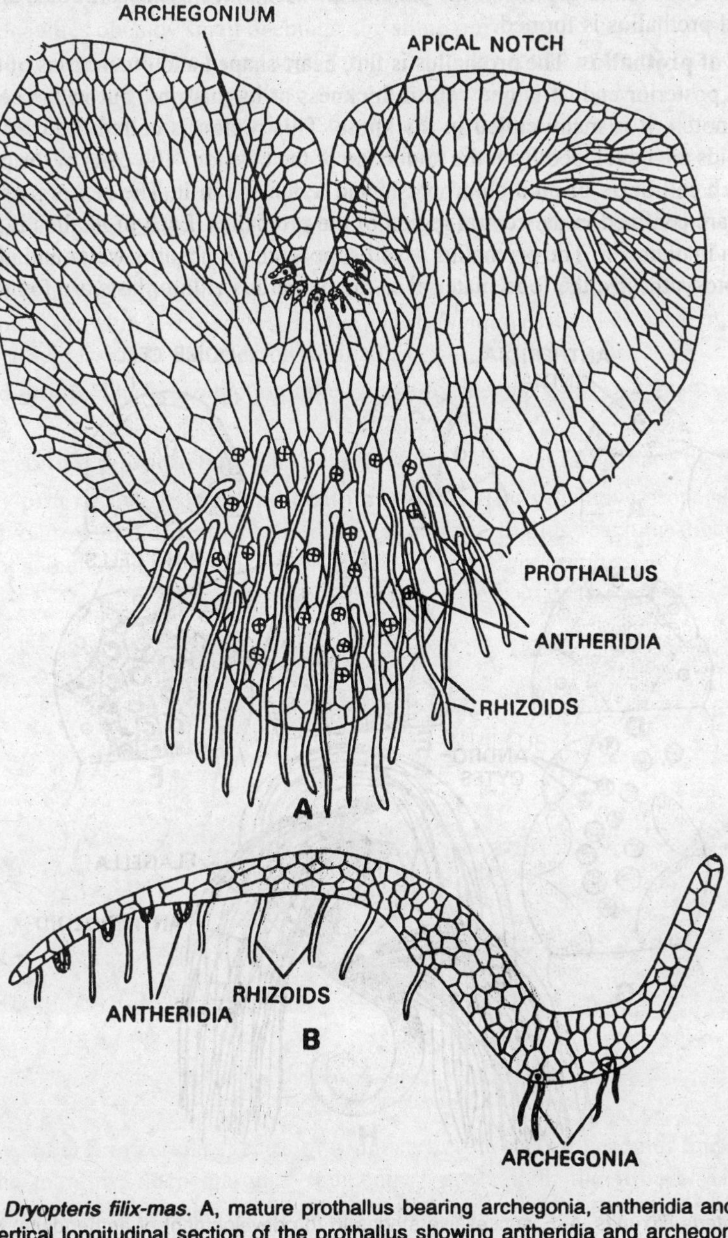

Fig. 31.15. *Dryopteris filix-mas.* A, mature prothallus bearing archegonia, antheridia and rhizoids; B, vertical longitudinal section of the prothallus showing antheridia and archegonia.

transversely several times developing into a filament 3 to 6 cells in length. The terminal cell of the filament known as an apical cell cuts off cells behind it and gives rise to this filamentous structure. The apical cell also divides laterally and flat prothallus is formed. The segments cut off from the apical cell divide again and again developing into the young prothallus. Usually the prothallus is heart-shaped. The prothallus is green and one celled thick at its margins. In well developed prothalli the central part of the prothallus is more than one cell in thickness. This region is called cushion. Many rhizoids develop on the underside of the cushion anchoring the prothallus to the soil. The rhizoids also serve as absorbing organs.

During the development of the gametophyte the original apical cell is replaced by a group of marginal initials. The further growth of the prothallus takes place by these initials and ultimately the heart-shaped prothallus is formed.

Structure of prothallus. The prothallus is flat, heart-shaped and green mass of tissue. There is a notch at its posterior end. It is one cell in thickness at its margins, but in the central region, near about the notch, it is many celled in thickness. This region is called the cushion. Several colourless rhizoids are developed from the underside of the cushion. The cells of the prothallus are thin walled. Each cell bears discoid chloroplasts and a nucleus in it. The sex organs (archegonia and antheridia) are confined in the central region of the prothallus. Each prothallus is probably 1/8 to 1/3 of an inch in diameter. The prothallus is self supporting. It absorbs water and nutrients from the soil and photosynthesize the food material as the numerous chloroplasts are found in its cells.

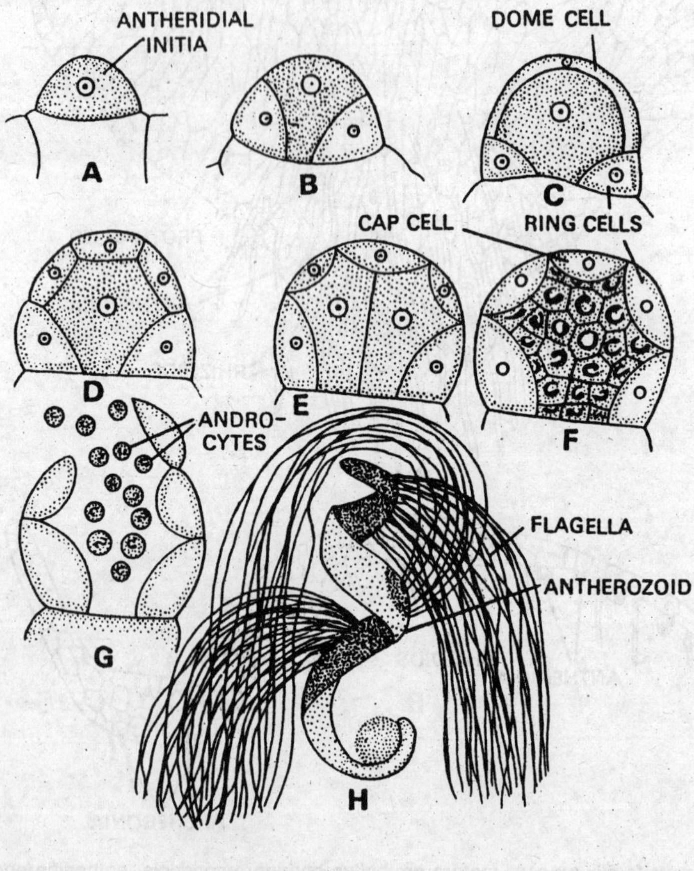

Fig. 31.16. *Dryopteris filix-mas*. A–F, successive stages in the development of antheridium; H, multiciliate antherozoid.

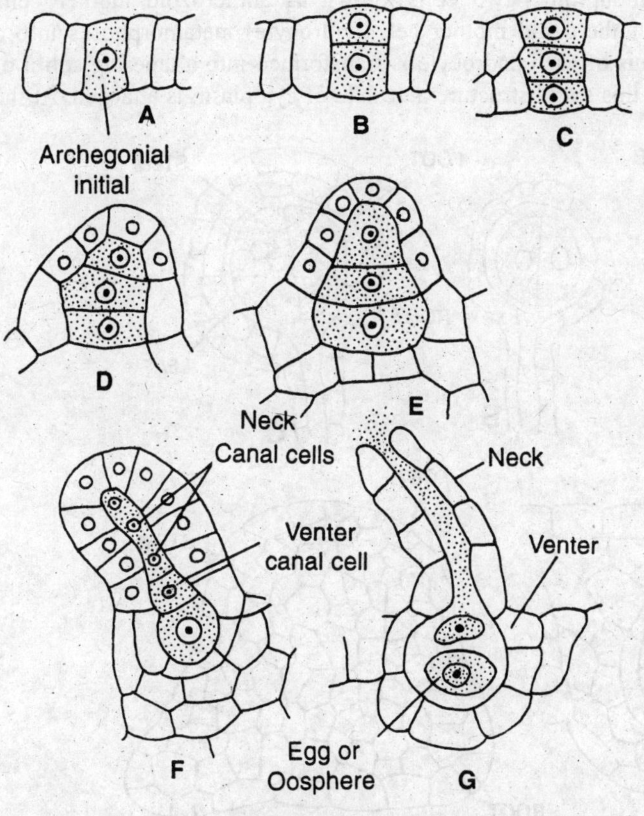

Fig 31.17. *Dryopteris filix-mas*. A–F, successive stages in the development of archegonium; G, mature archegonium.

The prothallus represents the gametophytic generation of the fern as it bears the sex organs upon it.

The sex organs. The sex organs develop on the underside of the prothallus (gametophyte). Usually the prothalli are monoecious and bear both male and female sex organs, *i.e.*, antheridia and archegonia. In the normal prothallus the archegonia develop around the notch upon the cushion and the antheridia in the basal central region of the prothallus among the rhizoids. The antheridia develop first and the archegonia later on.

Development of antheridium. One of the superficial vegetative cells of the prothallus protrudes out and divides by a transverse wall giving rise to two cells. The terminal cell acts as an antheridial initial. First of all the antheridial initial divides by an oblique funnel-shaped wall with the wide end upwards. By successive such oblique division a central cell is formed. This central cell is surrounded by a number of peripheral cells containing the chloroplasts in them of the original initial. The first division is followed by a dome shaped wall across the top of the funnel and thus a central cell is enclosed. Thereafter, the upper cell divides by a ring-shaped wall. This way, the semispherical antheridium is composed of two ring cells, a cap cell and a central cell. The central cell

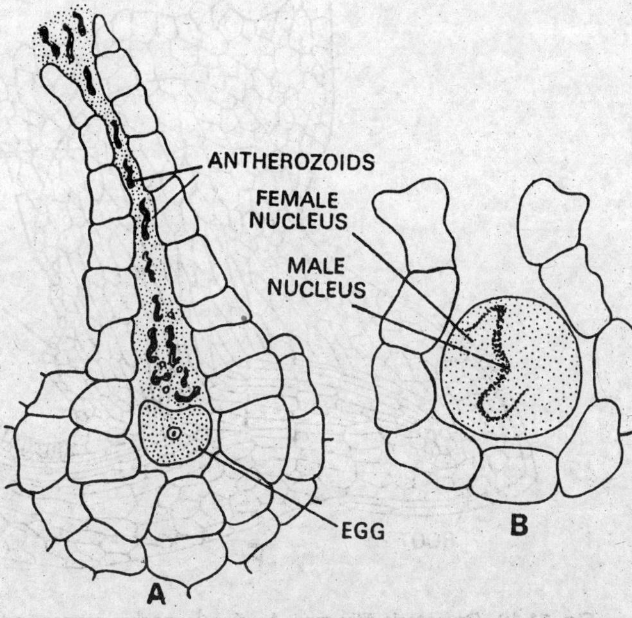

Fig. 31.18. *Dryopteris filix-mas*. A and B, fertilization.

divides further irregularly, and about thirty-two cells known as antherozoid mother cells (androcytes) are produced. Each antherozoid mother cell (androcyte) metamorphoses into a multiciliate antherozoid. The nucleus of each androcyte is transformed into a tapering spiral of three turns. At the posterior end of this spiral structure a vesicle of cytoplasm is attached. At the

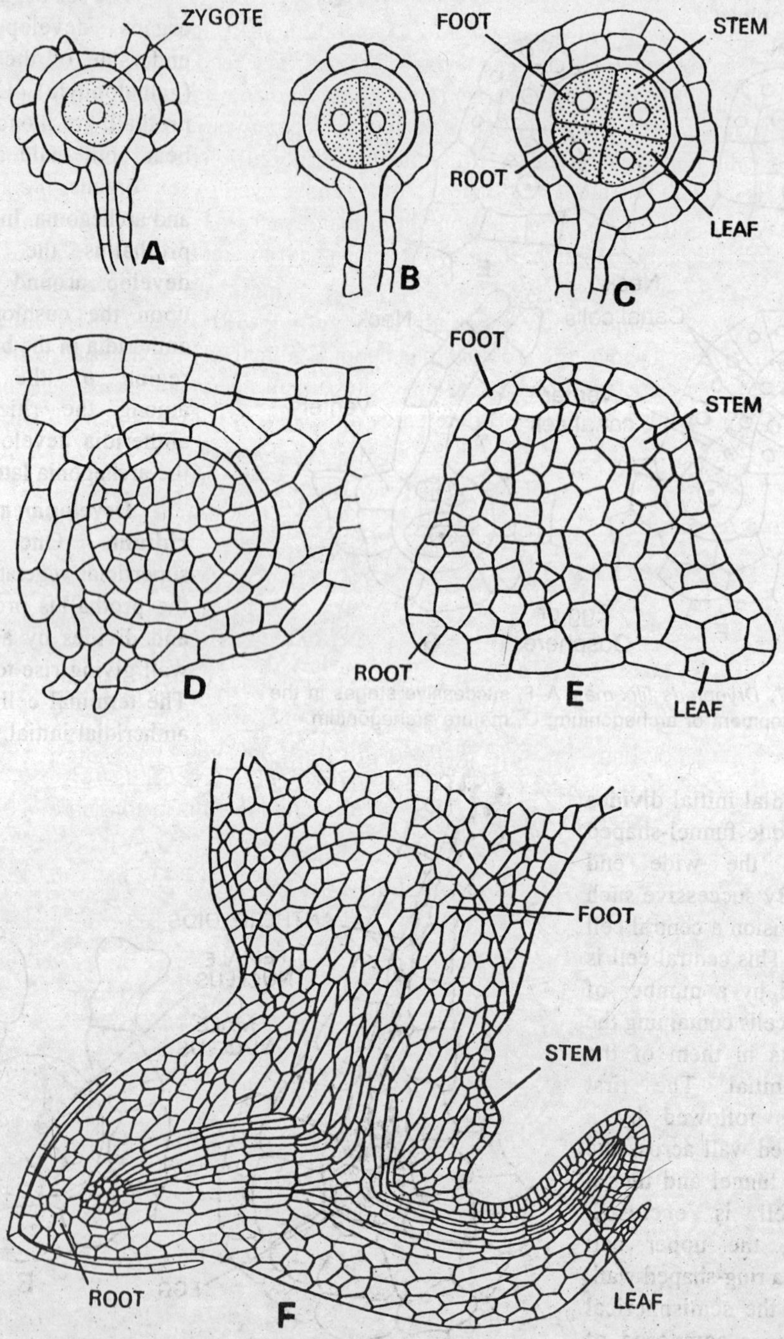

Fig. 31.19. *Dryopteris filix-mas.* A—F, successive stages in the development of the embryo (sporophyte); F, young sporophyte coming out of the prothallus.

anterior end there is an elongated blepharoplast bearing a group of very long flagella. By means of flagella the antherozoids can swim in the water film.

Dehiscence of antheridium. The water is essential for this process. When the mature antheridium comes in contact with the water, the walls of the androcytes situated within the mature antheridium become mucilaginous and swell and with the result the lid cell of the antheridium is thrown off. The antherozoids are ejected out. Each antherozoid is still within a thin membranous cell wall, which very soon dissolves in the water and the antherozoid is free in the water.

Development of archegonium. The archegonium develops from a single superficial cell of the prothallus, known as the archegonial initial. The initial divides into three cells by two transverse walls. Of these three cells the lower one is the basal cell which forms the wall around the venter cell. The remaining two cells are the central cell and the cover cell. The cover cell divides vertically twice forming four neck initials. Each neck initial divides transversely developing four vertical rows of the cells of the neck. The neck is 5 to 7 cells in height. The central cell grows up between the neck cells. This divides several times transversely producing a narrow neck canal cells a ventral canal cell and an egg or oosphere. The neck canal cell may or may not divide further into two cells. The neck is sharply curved. The two rows of the neck of the convex side contains 5 to 7 cells, whereas the rows of concave side have only four cells and thus the neck is strongly curved.

Structure of archegonium. The archegonium is flask-like, elongated structure consisting of a venter embedded in the tissues of the prothallus and a sharply curved neck, which projects outside the tissues of the prothallus. The curved neck consists of four vertical rows of cells. The archegonium within its neck and venter contains an axial row of cells. The lowermost cell found within the venter is the oosphere or ovum; above to it there is a ventral canal cell and thereafter a long neck canal which is generally binucleate. Prior to fertilization the neck canal cell and the ventral canal cell break down into a mucilaginous substance. This mucilaginous material swells and forces its way outside the neck. This way an opening is caused at the apex of the neck and the antherozoids enter through this opening to reach the oosphere.

Fertilization. The water is an essential factor for this process. Ordinarily the prothalli receive water by rain. The mature archeg-onium probably secretes malic acid, which attracts the swimming antherozoids towards it. The antheroz-oids swim down the neck canal through the mucilage and one of those penetrates the oosphere at a hyaline receptive spot. Soon after the entrance of the antherozoid in the egg the two male and female nuclei are fused together and the fertilization is effected. The oosphere secretes a wall around it and becomes oospore. Several archegonia of a single

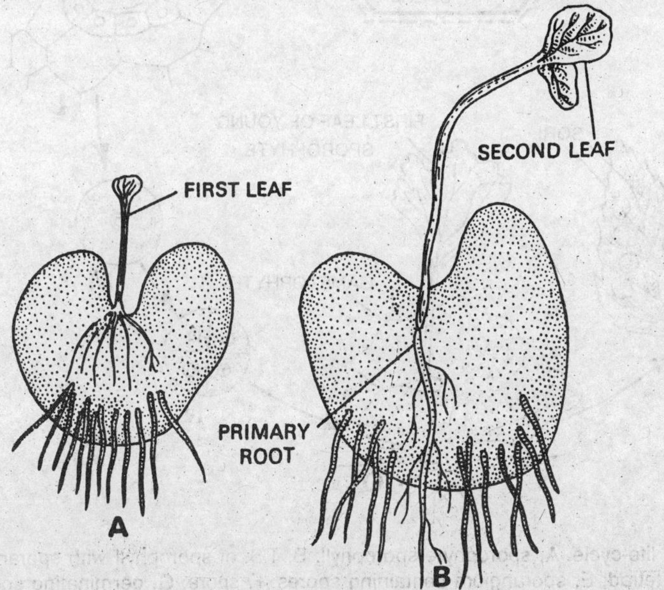

Fig. 31.20. *Dryopteris filix-mas.* (Fern). Young sporophytes still attached to prothalli.

prothallus may be fertilized but ultimately only one oospore develops into the complete sporophyte.

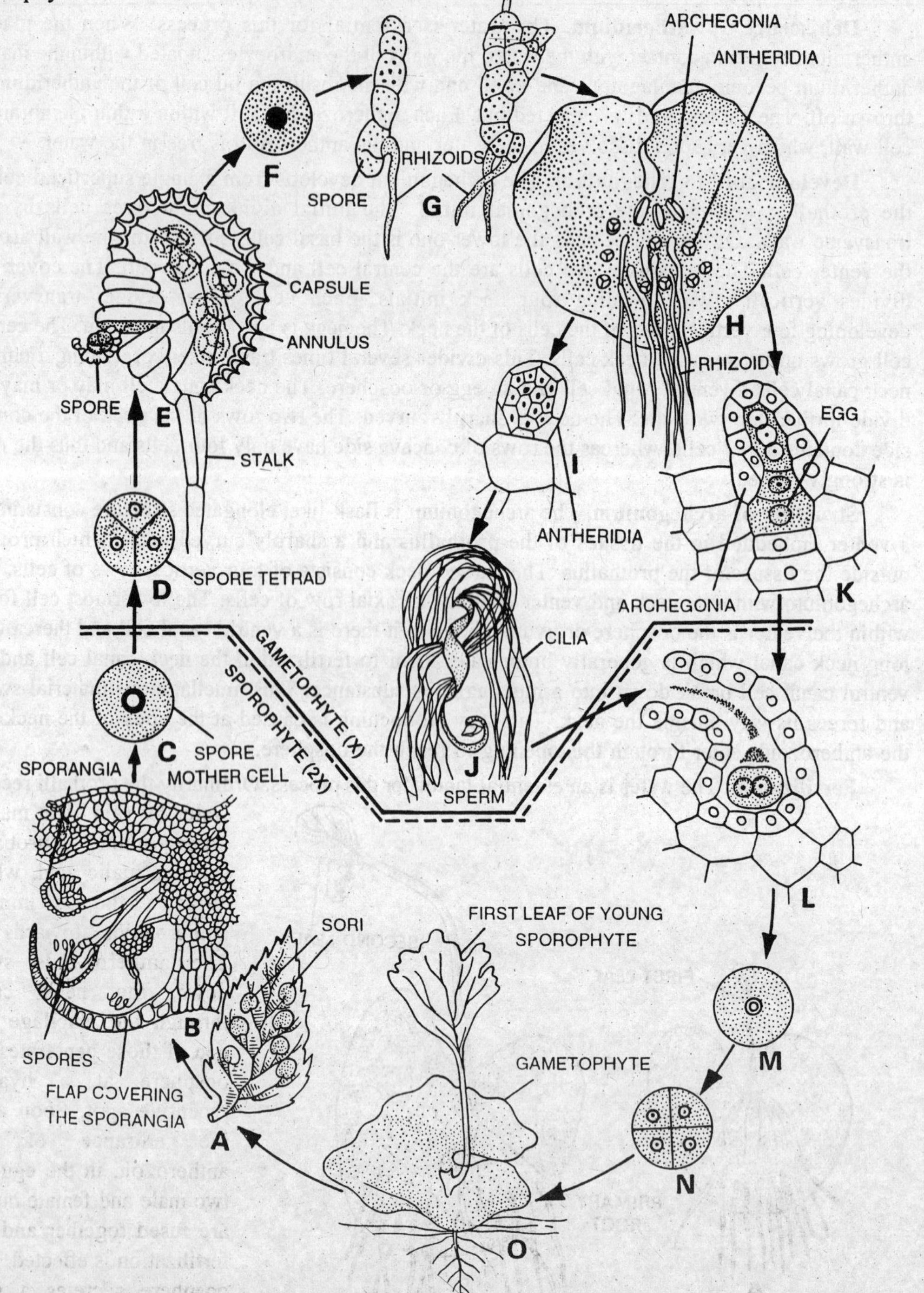

Fig. 31.21. Fern. Diagrammatic life-cycle. A, sporophyte-sporophyll; B, T.S. of sporophyll with sporangia; C, spore mother cell; D, spore tetrad; E, sporangium containing spores; F, spore; G, germinating spore; H, prothallus; I, antheridium; J, antherozoid; K, archegonium with egg; M, oospore; N—O, development of embryo and young sporophyte with prothallus.

The embryo and its development. Immediately after fertilization the zygote divides by a vertical wall, forming two cells. Soon after second and third divisions take place at right angles to each other resulting in an octant or a group of eight cells. Four of the cells of the octant found towards the apex of the parent prothallus form the epibasal half and the remaining four cells the hypobasal half. The epibasal cells produce the axis and the first leaf of the young sporophyte. The hypobasal half produces the primary root and the foot. The foot acts as haustorium and remains embedded in the venter. It absorbs nourishment for the young sporophyte from the prothallus and soil unless and until it becomes independent. The primary root dies soon and the adventitious roots develop. The prothallus disintegrates very soon. As soon as the first leaf becomes able to synthesize the food the prothallus disappears. The first leaf of the young sporophyte is generally small and two-lobed. Soon after other larger leaves develop which are dichotomously forked. Only after the formation of the fifth and sixth leaves, the pinnately compound leaves are developed.

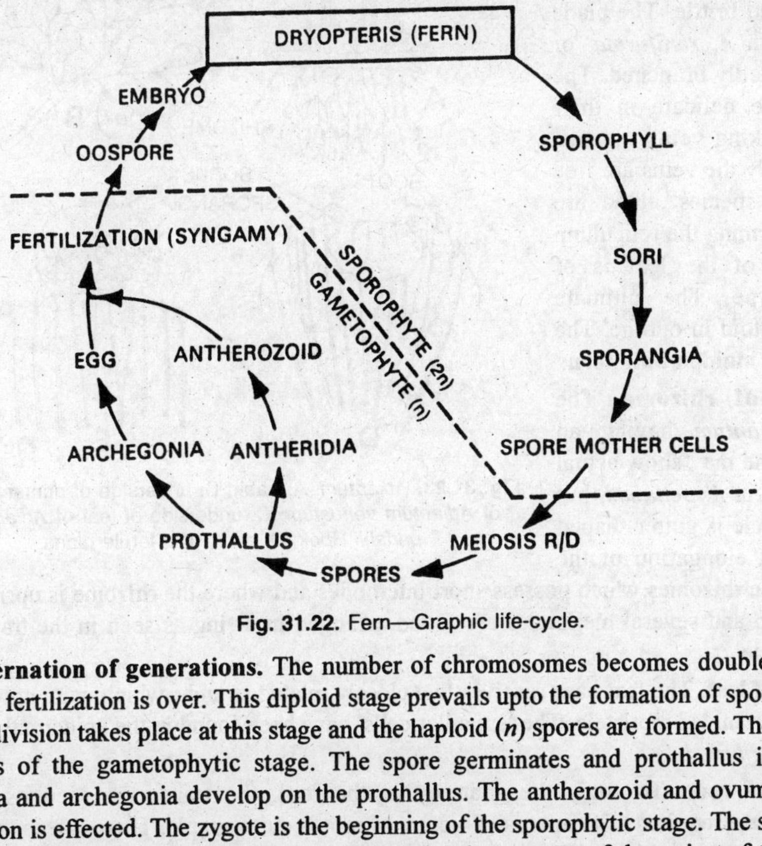

Fig. 31.22. Fern—Graphic life-cycle.

Alternation of generations. The number of chromosomes becomes doubled, *i.e.*, diploid ($2n$) after fertilization is over. This diploid stage prevails upto the formation of spore mother cells. Meiotic division takes place at this stage and the haploid (n) spores are formed. The spores are the beginners of the gametophytic stage. The spore germinates and prothallus is formed. The antheridia and archegonia develop on the prothallus. The antherozoid and ovum are fused and fertilization is effected. The zygote is the beginning of the sporophytic stage. The sporophyte ($2n$) gives rise to gametophyte (prothallus) and again by the process of the union of the gametes the sporophytic stage is attained. This way, it may be said that the fern exhibits an alternation of haploid (gametophytic) and diploid (sporophytic) generations.

Genus ADIANTUM

Systematic position. Pteridophyta
Division. Filicophyta
Class. Leptosporangiopsida
Order. Filicales
Family. Polypodiaceae
Genus. *Adiantum*

According to Bower there are about 184 species of ferns in this genus.

The species of this genus are widely distributed in the tropical regions of the world. They are found in shady and moist habitat. In our country the species are quite common in the hilly tracts of Himalayas as well as in the plains.

Habit. The species of *Adiantum* have a upright or creeping scaly rhizome. The leaves are arranged on the rhizome either spirally or alternately. The petiole is shining black and brittle. The blade may be entire in *A. reniforme,* or simple or repeatedly branched. The blades are quite delicate in their texture. The forking veins traverse the blade. Usually the veins are free but in some species they are anastamosing forming the reticulum. The branching of the leaf is of dichotomous type. The ultimate segments are deltoid in outline. The fern is known as **maiden hair fern**.

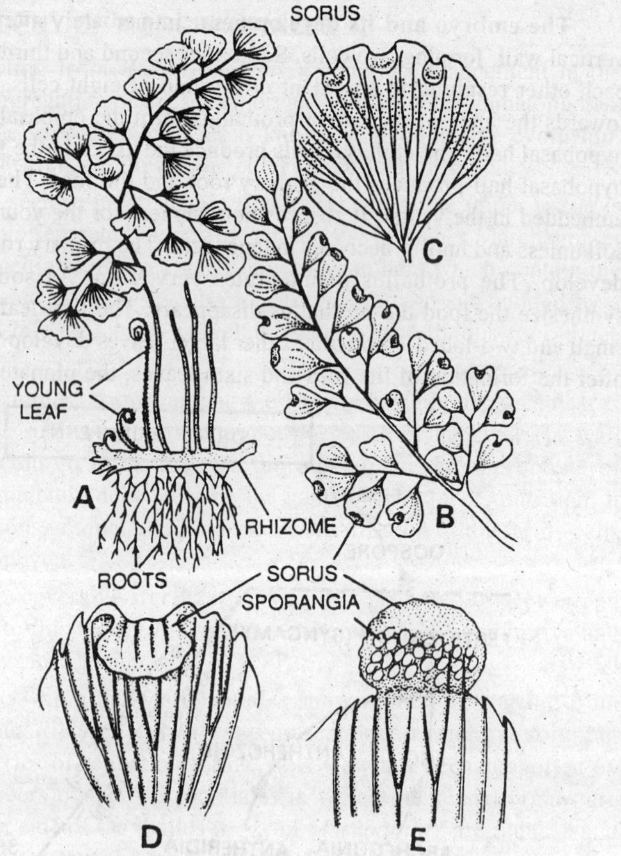

Fig. 31.23. *Adiantum.* A, habit; B, underside of primary pinna of *Adiantum venustum*; C, underside of leaf of *Adiantum parishii* Hook; D, parts of a fertile pinna.

Anatomy of rhizome. The species of *Adiantum* having an elongated rhizome may show actual solenostele, *e.g.*, in *A. pedatum*, but commonly the stele is gutter-shaped due to the great elongation of the leaf gaps. In these rhizomes which possess short internodes and where the rhizome is upright, the leaf gaps overlap and several meristeles are found arranged in a ring as seen in the transverse section of the stele.

Anatomy of leaf. The leaf base receives two closely related strands, which unite upwards to form a single four-angled meristele. The 'spicullar cells' have been found in the epidermal cells of the blade of *Adiantum*.

Anatomy of root. To study the anatomy of the root, the transverse sections of the adventitious root are required. The internal structure of root is quite simple. There is a central core of xylem, surrounded by phloem. The xylem is with two protoxylem groups situated at its two ends. This way, the xylem is diarch. Outside the xylem there are phloem elements. The phloem is arranged laterally to the xylem in the form of phloem plates. A two-layered pericycle surrounds the phloem. The pericycle is the outermost part of the stele. Thereafter a single-layered endodermis is found around the stele. The Casparian strips are present on the radial walls of the cells of the endodermis. The outermost layer of the transverse section of the root is the piliferous layer. The piliferous layer is single-layered and bears root hairs here and there upon it. The cortical region is differentiated into two regions. The outer region consists of thin-walled cells whereas the inner region of thick-walled cells.

The apical growth. In about all the ferns the stems grow from a three-sided pyramid like single apical cell. The apical cell divides again and again, parallel to each of its three sides developing the cells of the mature tissues.

The roots also develop from a similar three-sided pyramid like apical cell. From these three sides of the cell, all the organs of root develop by the repeated divisions. However, the cells of the root cap are cut from the base of the apical cell.

Spore producing organs. The characteristic of *Adiantum* is the presence of apparently marginal sori which are superficial in origin, and covered, by a sharply reflexed leaf margin which looks like an idusium because of its membranous nature and brown colour on its maturity. Actually there is no true indusium. The sporangia are inserted upon the distal region of the veins traversing the fertile lobe. The fertile region of the blade itself becomes sharply reflexed and serves as indusium. Each such fertile lobe bears a group of sori situated upon parallel veins. In certain cases the sporangia are found spread on the surface of the blade in between the veins.

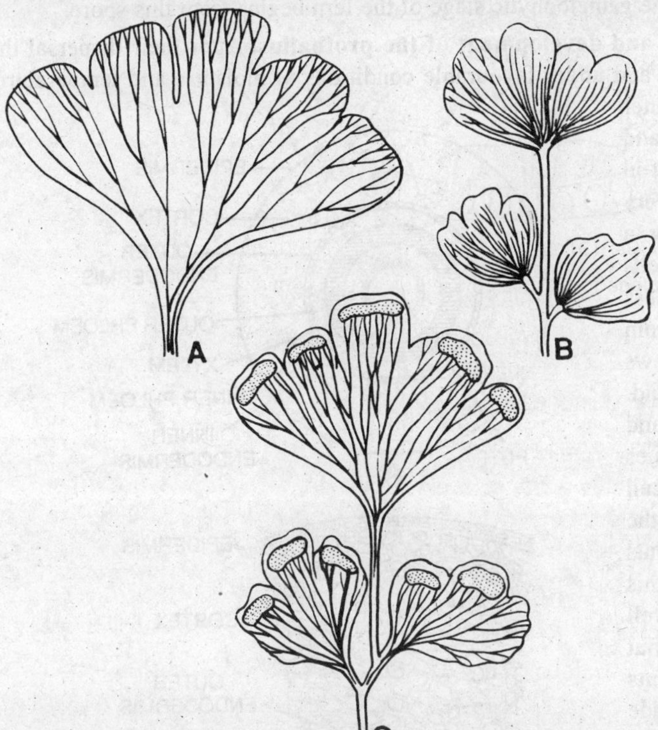

Fig. 31.24. *Adiantum* sp. A, a pinnule of *A. capillus-veneris* showing venation; B, sterile pinnules; C, fertile pinnules as seen from ventral side in *A. capillus-veneris*.

Development of sporangium. The development of sporangium is of leptosporangiate type as in all other genera of Polypodiaceae. The sporangium develops from a single superficial cell of the placenta. This cell divides by a transverse wall forming two cells. The upper cell divides again by an oblique wall to the first wall. Another oblique division takes place, giving rise to a tetrahedral cell. The cell again divides by a transverse wall. The inner cell is known as archesporial cell. The lower cell of the pair formed by the first division divides several times and develop into the multicellular stalk. The cells surrounding the tetrahedral archesporial cell divide further giving rise to the sporangial wall. Thereafter, the archesporial cell divides to form a tapetal layer of the cells. At the same time it also produces about a dozen spore mother cells. The spore mother cells give rise to spores. Each spore mother cell divides twice developing four spores. During division there is meiosis and the spores are haploid. Usually each sporangium contains 48 haploid spores. (Also see Fig. 31.6).

Structure of sporangium. Each sporangium is composed of a multicellular stalk and a capsule. The capsule of this sporangium is of the shape of biconvex lens. The wall of the capsule is composed of a single layer of thin walled cells. A row of specially thickened cells, known as annulus partially surrounds the capsule. At one side of the annulus, there is a stomium composed of thin walled cells. On maturity, the sporangium contains about 32 to 64 dark, ragged walled spores in it.

Dehiscence of sporangium. On the maturity of the sporangia, indusium shrivels up and the sporangia are exposed to the outer atmosphere. The sporangium bursts by a special mechanism. The cells of the annulus are thickened in a peculiar way. The inner tangential and radial walls of the annulus are thick while the outer tangential walls are thin. The bursting of the sporangium is caused by the behaviour of the water in relation to the cells of the annulus. In young sporangia the water glands are present and they remain moist. On the maturity of the sporangia, the sporangia, are exposed to dry air and the water is lost from the cells of the sporangia. This way, the thin outer walls of the annulus begin to contract, the stomium opens and the annulus along with some portion of the sporangium turns backward with a jerk throwing the spores out of the sporangium.

The spore. The spores are double walled. The outer wall is thick, ragged and brown in colour. It is known as exine. The inner wall is smooth and thin, and known as intine. Each spore contains a big central nucleus. The gametophytic stage of the fern begins from this spore.

Germination of the spore and development of the prothallus. After their dispersal the spores lie dormant for sometime and under favourable conditions of moisture and temperature, they germinate. First of all outer spore wall (exine) breaks and thereafter the intine protrudes out in the form of a green germ tube. Very shortly this tube is cut off by a transverse wall and two unequal cells are resulted. The small colourless cell is known as rhizoidal cell. From this cell a colourless branch grows into the soil and develops a rhizoid. The large green cell elongates and divides transversely several times developing into a filament 3 to 6 cell in length. The terminal cell of the filament known as an apical cell cuts of cells behind it and give rise to this filamentous structure. The apical cell also divides laterally and a flat prothallus is formed. The segments cut off from the apical cell divide again and again developing into the young prothallus. Usually the prothallus is heart-shaped. The prothallus is green and one-celled thick at its margins. In well developed prothalli the central part of the prothallus is more than one cell in thickness. This region is called cushion. Many rhizoids develop on the underside of the cushion anchoring the prothallus to the soil. The rhizoids also serve as absorbing organs.

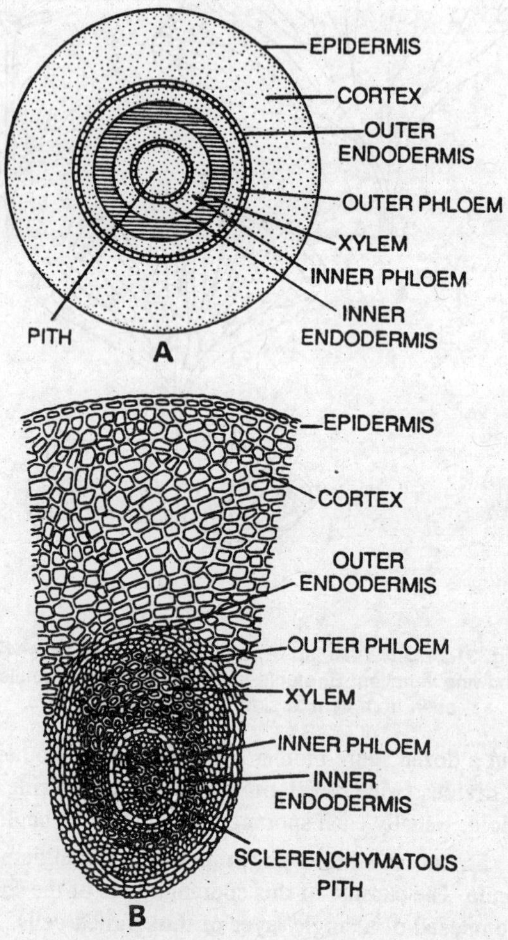

Fig. 31.25. *Adiantum* (Fern). T.S. of stem—A, diagrammatic; B, detailed diagram showing a siphonostele.

During the development of the gametophyte the original apical cell is replaced by a group of marginal initials. The further growth of the prothallus takes place by these initials and ultimately the heart-shaped prothallus is formed.

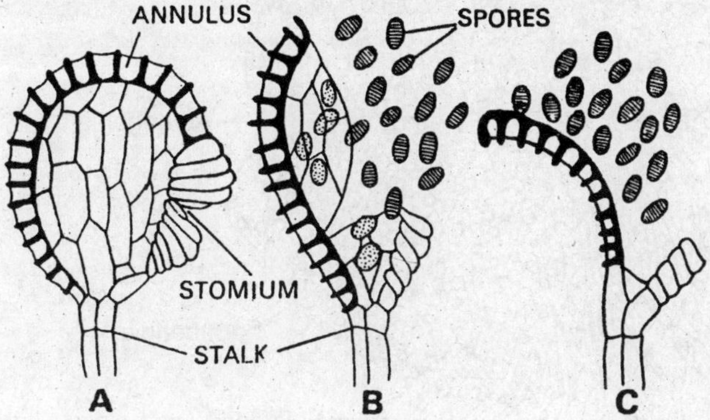

Fig. 31.26. *Adiantum* (Fern). A, sporangium; B—C, escaping spores from sporangium.

Structure of prothallus. The prothallus is flat, thin, heart-shaped and green mass of tissue. There is a notch at its posterior end. It is one cell in thickness at its margins, but in the central region, near about the notch, it is many celled in thickness. This region is called the cushion. Several colourless rhizoids are developed from the underside of the cushion. The cells of the prothallus are thin walled. Each cell bears discoid chloroplasts and a nucleus in it. The sex organs (archegonia and antheridia) are confined in the central region of the prothallus. The prothallus is probably 1/8 to 1/3 of an inch diameter. Each prothallus is self supporting. It absorbs water and nutrients from the soil and photosynthesize the food material as the numerous chloroplasts are found in its cells. The prothallus represents the gametophytic generation of the fern as it bears the sex organs upon it.

The sex organs. The sex organs develop on the underside of the prothallus (gametophyte). Usually the prothalli are monoecious and bear both male and female sex organs, *i.e.*, antheridia and archegonia. In the normal prothallus the archegonia develop around the notch upon the cushion and the antheridia in the basal central region of the prothallus among the rhizoids. The antheridia develop first and the archegonia later on.

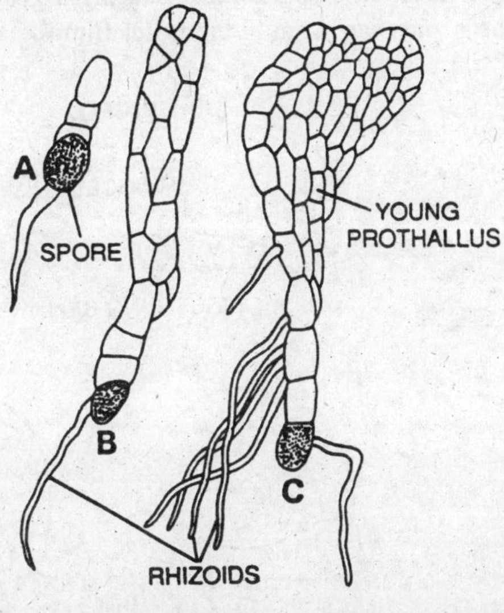

Fig. 31.27. *Adiantum* (Fern). A, spore germination and development of young prothallus.

Development of antheridium. One of the superficial vegetative cells of the prothallus protrudes out and divides by a transverse wall giving rise to two cells. The terminal cell acts as an antheridial initial. First of all the antheridial initial divides by an oblique funnel-shaped wall with the wide end upwards. By successive such oblique divisions a central cell is formed. This central cell is surrounded by a number of peripheral cells containing the chloroplasts in them of the original initial. The first division is followed by a dome shaped wall across the top of the funnel and thus a central cell is enclosed. Thereafter, the upper cell divides by a ring-shaped wall. This way, the semispherical antheridium is composed of two ring cells, a cap cell and a central cell. The central cell divides further irregularly, and about thirty two cells known as antherozoid mother cells (androcytes) are produced. Each antherozoid mother cell (androcyte) metamorphoses into a multiciliate

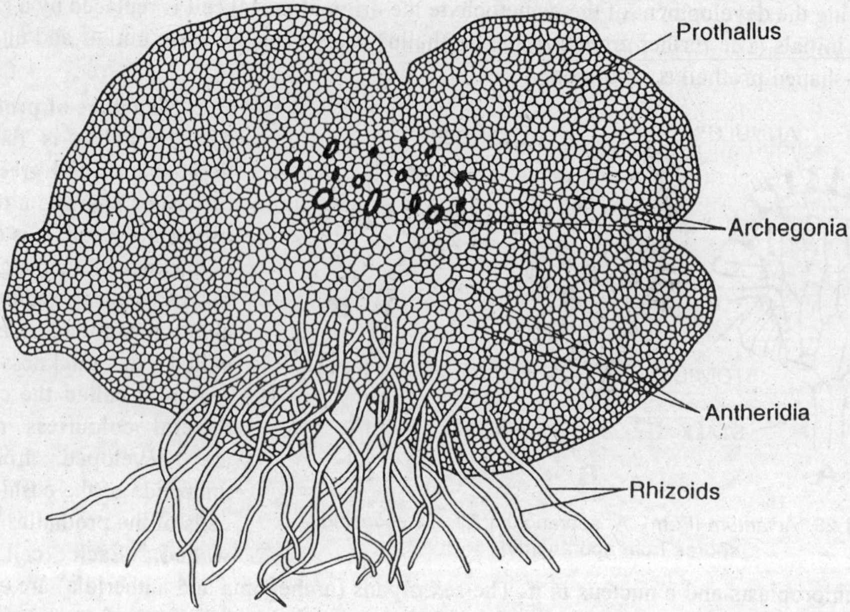

Fig. 31.28. *Adiantum* (Fern). Lower surface of prothallus showing antheridia, archegonia and rhizoids.

antherozoid. The nucleus of each androcyte is transformed into a tapering spiral of three turns. At the posterior end of this spiral structure a vesicle of cytoplasm is attached. At the anterior end there is an elongated blepharoplast bearing a group of very long flagella. By means of flagella the antherozoids can swim in the water film.

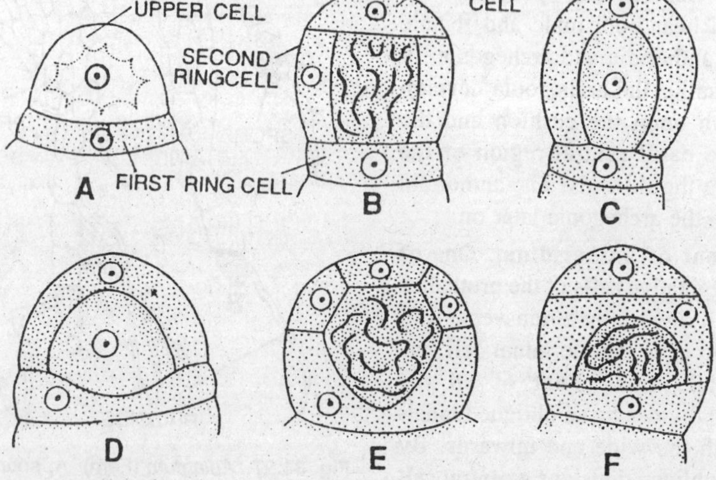

Fig. 31.29. *Adiantum* sp. A—F, stages in the development of antheridium.

Dehiscence of antheridium. The water is essential for this process. When the mature antheridium comes in contact with the water, the walls of the androcytes situated within the mature antheridium become mucilaginous and swell and with the result the lid cell of the antheridium is thrown off. The antherozoids are ejected out. Each antherozoid is still within a thin membranous cell wall, which very soon dissolves in the water and the antherozoid is free in the water.

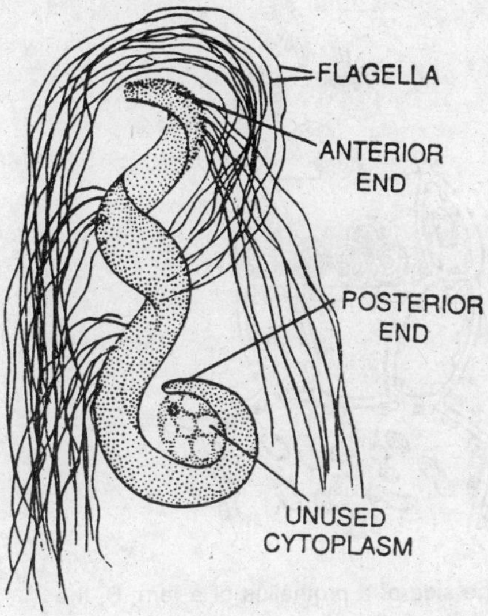

Fig. 31.30. *Adiantum* (fern). A single spermatozoid with flagella and unused cytoplasm.

Development of archegonium. The archegonium develops from a single superficial cell of the prothallus, known as the archegonial initial. The initial divides into three cells by two transverse walls. Of these three cells the lower one is the basal cell which forms the wall around the venter. The remaining two cells are the central cell and the cover cell. The cover cell divides vertically twice forming four neck initials. Each neck initial divides transversely developing four vertical rows of the cells of the neck. The neck is 5 to 7 cells in height. The central cell grows up between the neck cells. This divides several times transversely producing a narrow neck canal cell, a ventral canal cell and an egg or oosphere. The neck canal cell may or may not divide further into two cells. The neck is sharply curved. The two rows of the neck of the convex side contain 5 to 7 cells, whereas the rows of concave side have only four cells and thus the neck is strongly curved.

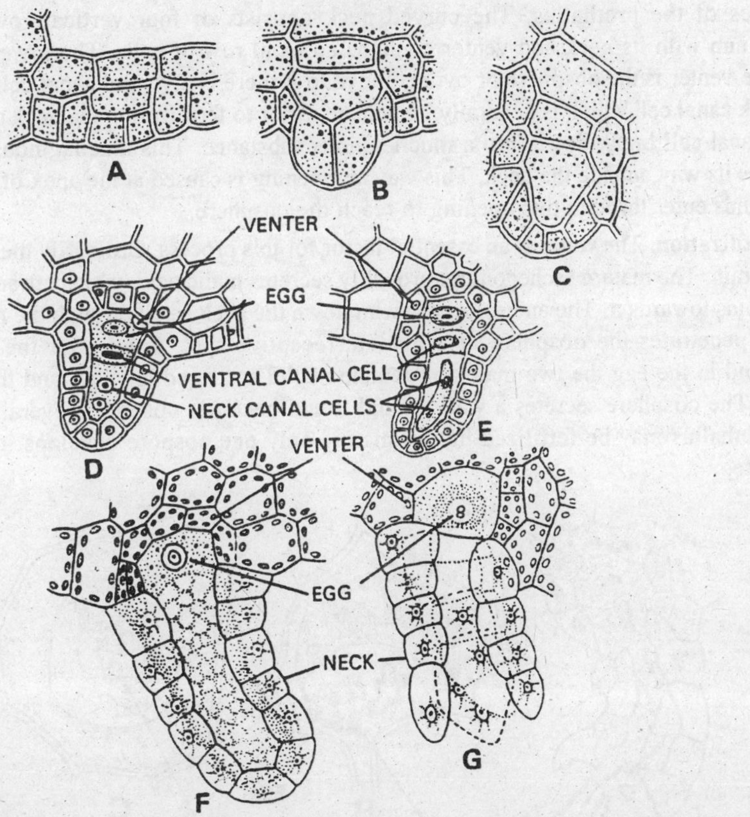

Fig. 31.31. *Adiantum* (fern). Development of archegonium—A-C, early stages of development; D, the venter contains an-egg, and exterior to it is the ventral canal cell, and in the neck two neck canal cells; E, the neck and venter canal cells have begun to disintegrate; F, the cells have been disorganized; G, archegonium open.

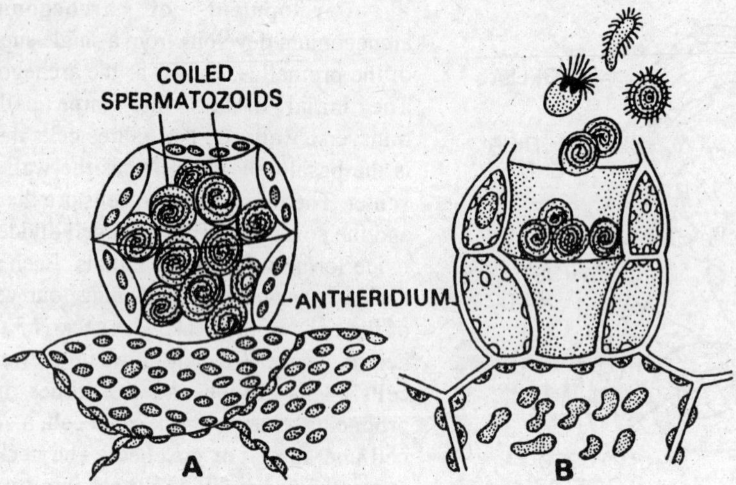

Fig. 31.32. *Adiantum* (fern). A, an antheridium on the side of a prothallus of a fern; B, the spermatozoids are escaping from the antheridium.

Structure of archegonium. The archegonium is flask-like, elongated structure consists of a venter embedded in the tissues of the prothallus and a sharply curved neck which projects outside the tissues of the prothallus. The curved neck consists of four vertical rows of cells. The archegonium with its neck and venter contains an axial row of cells. The lowermost cell found within the venter is the oosphere or ovum, above to it there is a ventral canal cell and thereafter a long neck canal cell which is generally binucleate. Prior to fertilization the neck canal cell and the ventral canal cell break down into a mucilaginous substance. This mucilaginous material swells and forces its way outside the neck. This way an opening is caused at the apex of the neck and the antherozoids enter through this opening to reach the oosphere.

Fertilization. The water is an essential factor for this process. Ordinarily the prothalli receive water by rain. The mature archegonium probably secretes malic acid, which attracts the swimming antherozoids towards it. The antherozoids swim down the neck canal through the mucilage and one of those penetrates the oosphere at a hyaline receptive spot. Soon after the entrance of the antherozoid in the egg the two male and female nuclei are fused together and the fertilization is effected. The oosphere secretes a wall around it and becomes oospore. Several archegonia of a single prothallus may be fertilized but ultimately only one oospore develops into the complete sporophyte.

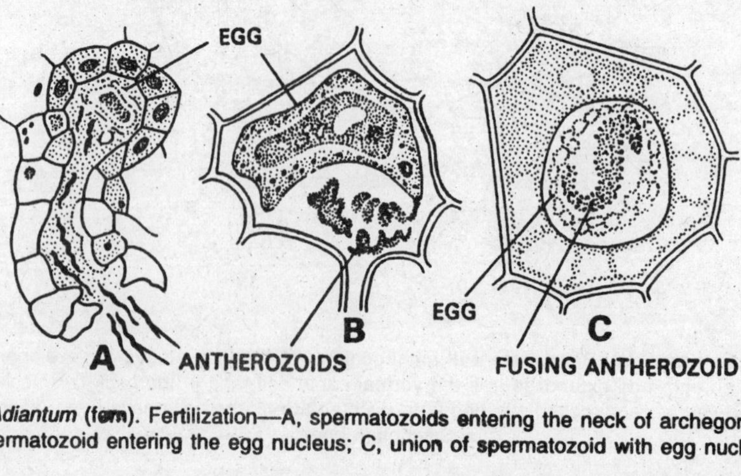

Fig. 31.33. *Adiantum* (fern). Fertilization—A, spermatozoids entering the neck of archegonium; B, single spermatozoid entering the egg nucleus; C, union of spermatozoid with egg nucleus.

The embryo and its development. Immediately after fertilization the zygote divides by a vertical wall, forming two cells. Soon after second and third divisions take place at right angles to each other resulting in an octant or a group of eight cells. Four of the cells of the octant found

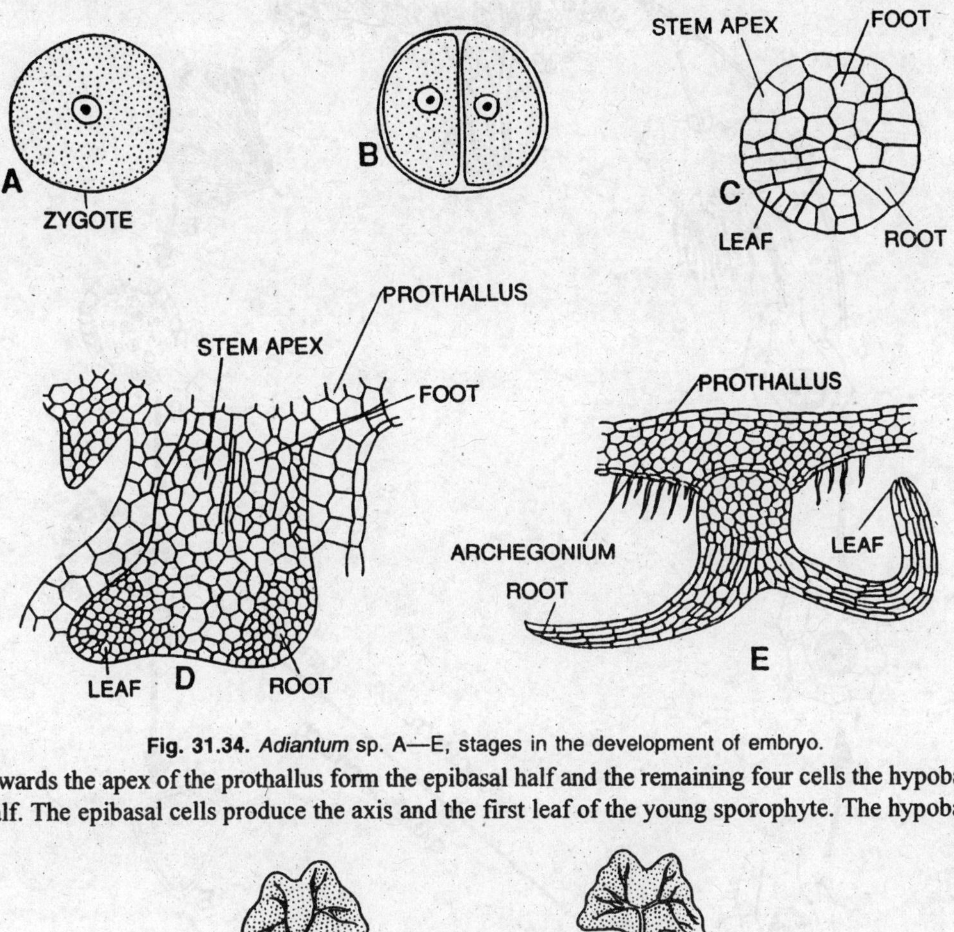

Fig. 31.34. *Adiantum* sp. A—E, stages in the development of embryo.

towards the apex of the prothallus form the epibasal half and the remaining four cells the hypobasal half. The epibasal cells produce the axis and the first leaf of the young sporophyte. The hypobasal

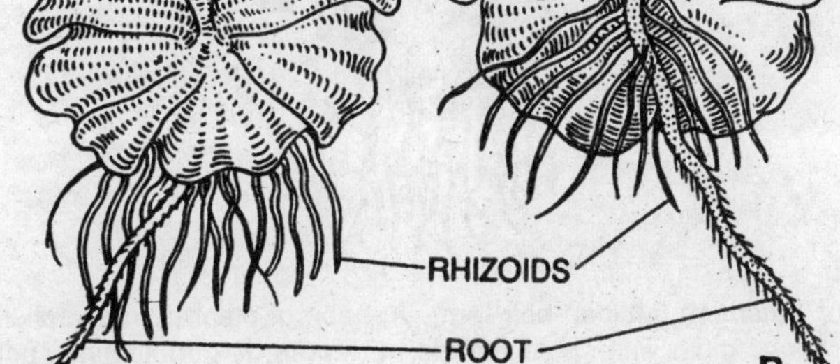

Fig. 31.35. *Adiantum* (fern), Prothallus. A, view from upper side; B, view from under side.

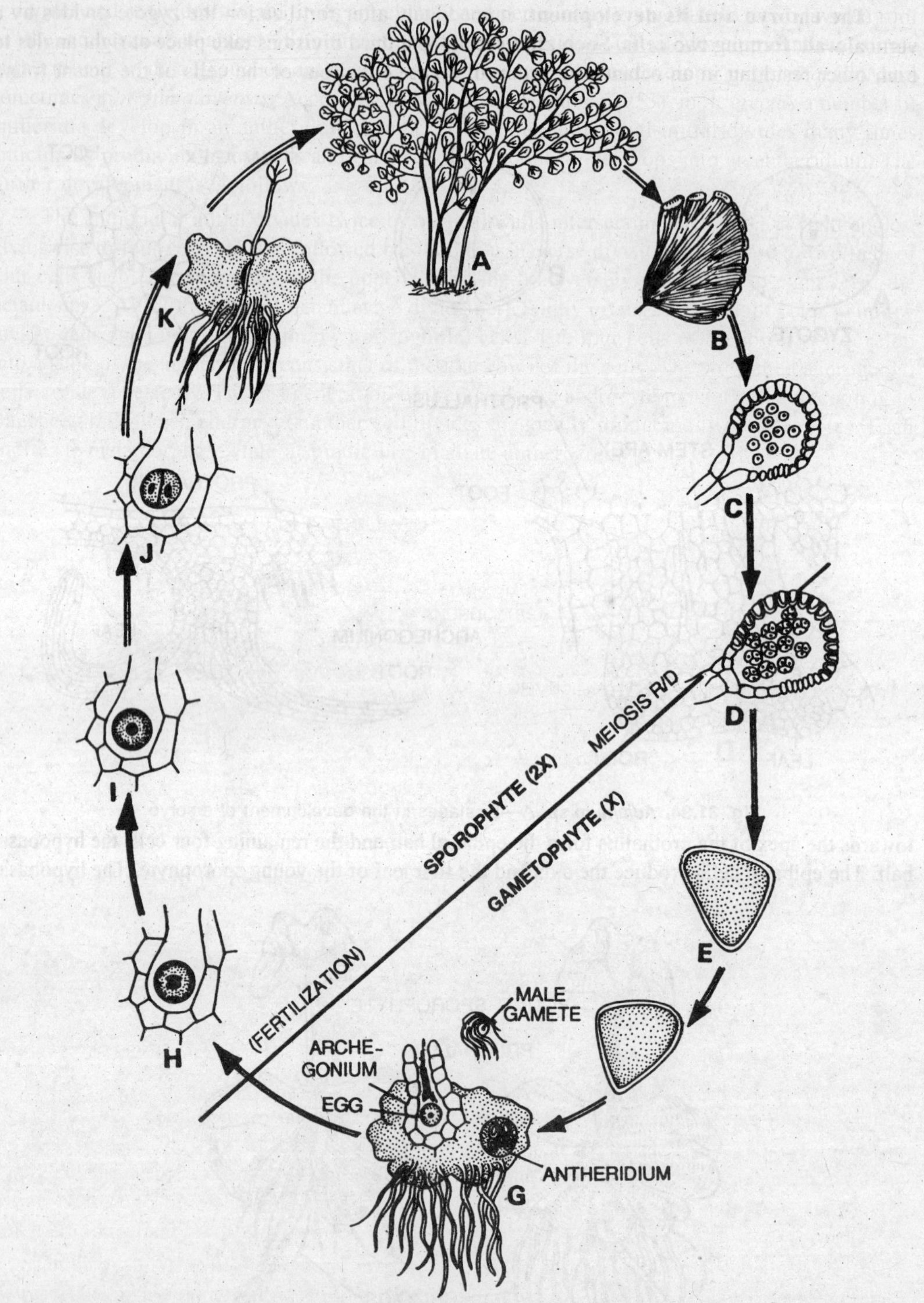

Fig. 31.36. Life cycle of *Adiantum* (Maiden hair fern). A, habit of plant B, sporophyll; C, sporangium with spore mother cells; D, sporangium with spore tetrads ; E, spore; F, germinating spore; G, prothallus with antheridia and archegonia; H, fusing male and female nuclei; I, oospore; J, embryo and K, prothallus with young sporophyte.

half produces the primary root and the foot. The foot acts as haustorium and remains embedded in the venter. It absorbs nourishment for the young sporophyte from the prothallus and soil unless and until it becomes independent. The primary root dies soon and the adventitious roots develop. The prothallus disintegrates very soon. As soon as the first leaf becomes able to synthesize the food, the prothallus disappears. The first leaf of the young sporophyte is generally small and two-lobed. Soon after other larger leaves develop which are dichotomously forked. Only after the formation of the fifth and sixth leaves, the pinnately compound leaves are developed.

Alternation of generations. The number of chromosomes becomes doubled, *i.e.*, diploid (2*n*) after fertilization is over. This diploid stage prevails upto the formation of spore mother cells. Meiotic division takes place at this stage and the haploid (*n*) spores are formed. The spores are the beginners of the gametophytic stage. The spore germinates and prothallus is formed. The antheridia and archegonia develop on the prothallus. The antherozoid and ovum fuse and fertilization is effected. The zygote is the beginning of the sporophytic stage. The sporophyte (2*n*) gives rise to gametophyte (prothallus) and again by the process of the union of the gametes the sporophytic stage is attained. This way, it may be said that the fern exhibits an alternation of haploid (gametophytic) and diploid (sporophytic) generations.

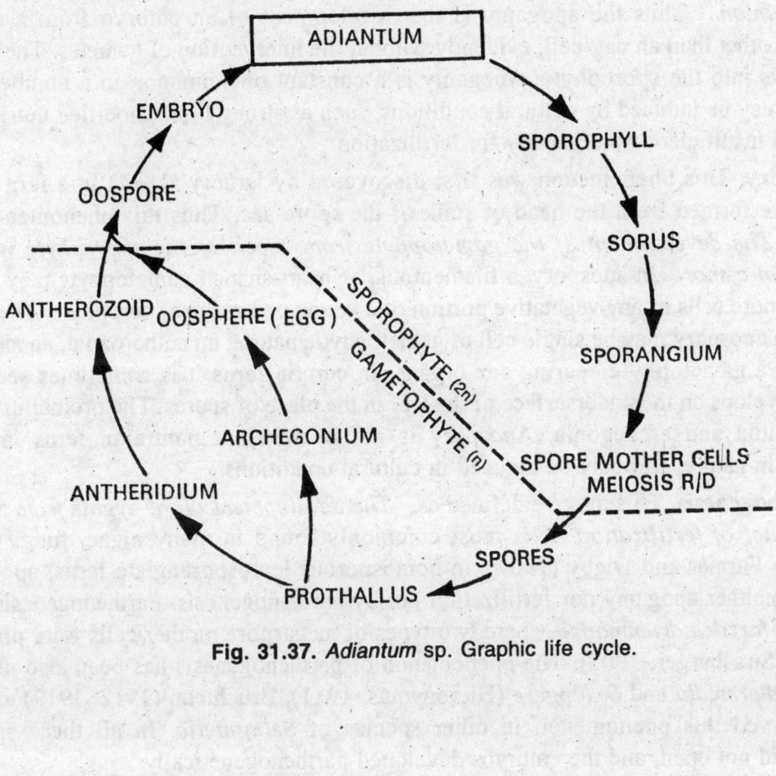

Fig. 31.37. *Adiantum* **sp. Graphic life cycle.**

32
CHAPTER

Pteridophytes: General Topics

APOGAMY, APOSPORY AND PARTHENOGENESIS

The normal life-cycle of bryophytes, pteridophytes and spermatophytes shows distinct alternation of generations. However, there are certain other modifications or abnormalities where the essential stages in the life-cycle are eliminated. These modifications may be called—**apogamy, apospory** and **parthenogenesis.**

Apogamy. According to Winkler (1908) the definition is as follows — *"It is the formation of a sporophyte directly, from the vegetative cells of the gametophyte without the act of syngamy or gametic union."* Thus the apogamy is the development of an embryo from a cell of the gametophyte other than an egg-cell, evidently without the intervention of gametes. The embryo so formed grows into the sporophyte. Apogamy is a constant phenomenon in a number of ferns. However, it may be induced by cultural conditions, such as strong light, modified nutrient culture solutions and insufficient water supply for fertilization.

Apospory. This phenomenon was first discovered by Druery (1884) in a fern where the prothalli were formed from the head or stalk of the spore sac. Thus this phenomenon may be defined as—*The development of the gametophyte from a cell of the sporophyte without the intervention of a spore.* In apospory, a filamentous, or heart-shaped gametophyte may be formed from one or more cells of any vegetative portion of a young or a mature sporophyte. The structures taking part in apospory may be single cell of gametophytic nature, an antherozoid, an antheridium, a rhizoid, or a gametophyte bearing sex organs. In certain ferns it is sometimes seen that the prothallus develops on the undersurface of the leaf in the place of spores. The prothallus so formed bears antheridia and archegonia. Apospory is not of constant nature in ferns, and occurs occasionally in nature, and may be induced in cultural conditions.

Parthenogenesis. This may be defined as—*The development of the zygote from the eggcell without the act of fertilization.* It is most commonly found in many algae, fungi and ferns. According to Farmer and Digby (1907), in homosporous leptosporangiate ferns, apospory was followed by neither apogamy nor fertilization but by parthenogenesis. Parthenogenesis has been observed in *Marsilea drummondii* where two types of megaspore mother cells were produced by this species (Strasburger, 1907). The phenomenon of parthenogenesis has been also observed in *Selaginella intermedia* and *S. langere* (Hieronymus, 1911). Bruchman (1912, 1919) and Goebel (1915) observed this phenomenon in other species of *Selaginelia.* In all these species the archegonia did not open, and the embryos developed parthenogenetically.

STELAR SYSTEM IN FERNS

According to the older botanists, the vascular bundle is the fundamental unit in the vascular system of pteridophytes and higher plants. Van Tieghem and Douliot (1886) interpreted the plant body of a vascular plant in the different way. According to them the fundamental parts of a shoot are the cortex and a central cylinder is known as **stele.** The name stele has been derived from a

Greek word meaning pillar. This way, the stele is defined as a central vascular cylinder, with or without pith and delimited the cortex by the endodermis. Van Tieghem and Douliot (1886) recognized only three types of steles. They also thought that the monostelic shoots were rare in comparison of polystelic shoots. It is an established fact that all shoots are monostelic and polystelic condition rarely occurs.

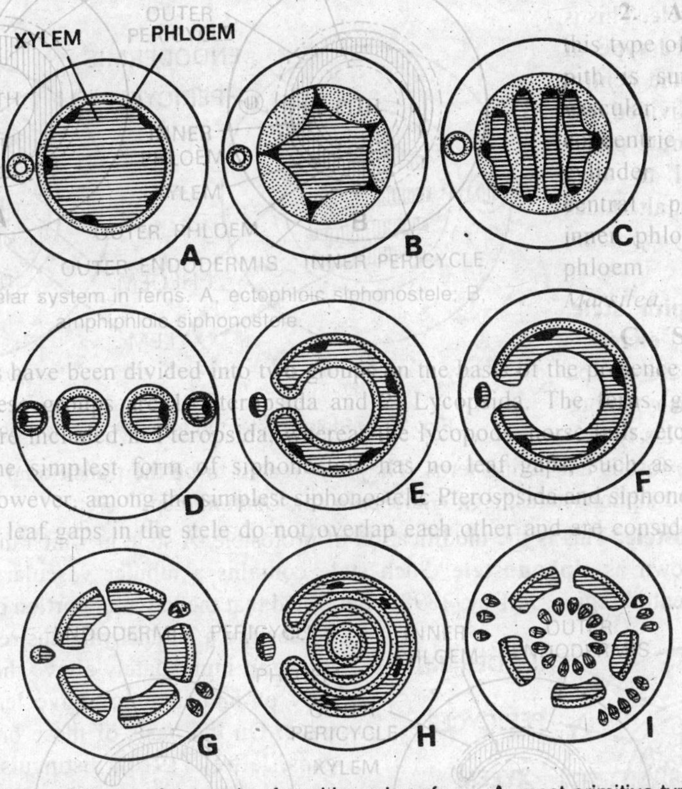

Fig. 32.1. Different types of the steles found in various ferns. A, most primitive type-haplostele; B, actinostele; C, plectostele; D, polystele; E, amphiphloic solenostele; F, ectophloic solenostele; G, dictyostele; H, polycyclic solenostele; I, polycyclic dictyostele.

The stele of the stem remains connected with that of leaf by a vascular connection known as the **leaf supply.**

The steles may be of the following types :

A. Protostele. Jeffrey (1898), for the first time pointed out the stelar theory from the point of view of the phylogeny. According to him the primitive type of stele is **protostele.** In protostele, the vascular tissue is a solid mass and the central core of the **xylem** is completely surrounded by a layer of phloem. This is the most primitive and simplest of steles.

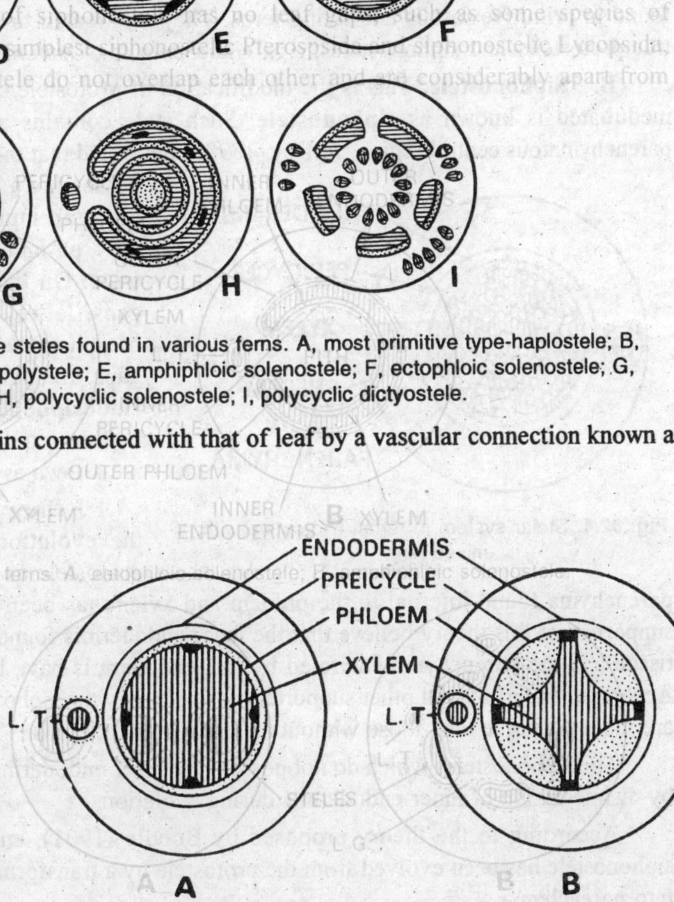

Fig. 32.2. Stelar system in ferns. A, haplostele; B, Actinostele.

There are several forms of the protostele which are as follows :

1. Haplostele. This is the most primitive type of protostele. Here the central solid smooth core of xylem is surrounded by a layer of phloem, *e.g., Salaginella* sp.

2. Actinostele. This is the modification of the haplostele and somewhat more advanced in having the central xylem core with radiating ribs, *e.g., Psilotum* sp.

3. Plectostele. This is the most advanced type of protostele. Here the central core of xylem is divided into a number of separate plates arranged parallel to each other. The phloem alternates the xylem, *e.g., Lycopodium* sp.

4. Mixed-pith stele. Here the xylem elements (*i.e.,* tracheids) are mixed with the parenchymatous cells of the pith. This type is found in primitive fossils and living ferns. They are treated to be the transitional types between true protosteles and siphonosteles, *e.g., Gleichenia* sp., *Osmunda* sp.

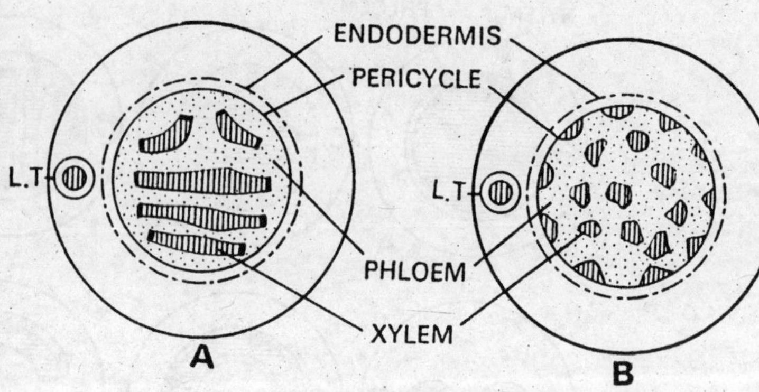

Fig. 32.3. Stelar system in ferns. A, plectostele; B, mixed

B. Siphonostele. This is the modification of protostele. A stele in which the protostele is medullated is known as **siphonostele**. Such stele contains a tubular vascular region and a parenchymatous central region. Jeffrey (1898) interpreted that the vascular portion of siphonostele possessing a parenchymatous region is known as a **gap** immediately above the branch trace only or immediately above leaf and branch trace. On the basis of these branch and leaf gaps, Jeffrey (1910), distinguished two types of siphonosteles. In one type however, the leaf gaps are not found and they are known as **cladosiphonic** siphonosteles. In the other type both leaf and branch gaps are present and they are known as **phyllosiphonic** siphonosteles.

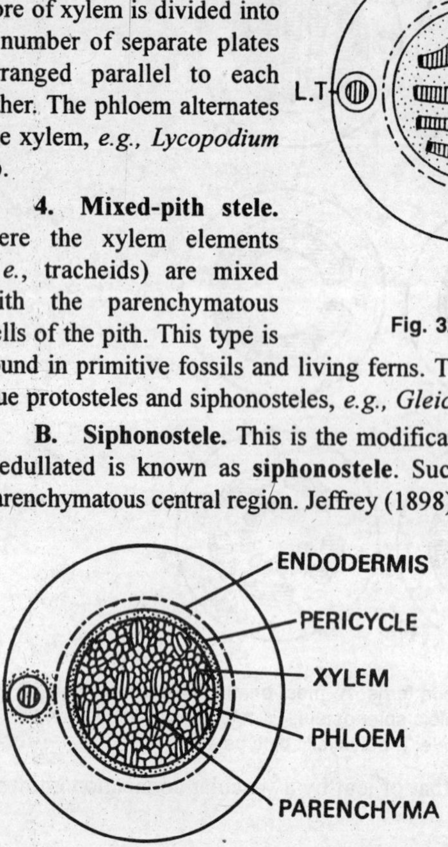

Fig. 32.4. Stelar system in ferns. Protostele with mixed pith.

Jeffrey (1902, 1910, 1917) interpreted the evolution of the siphonostele from the protostele as follows : He supported that the parenchyma found internal to the phloem and xylem has been originated from the cortex. The supporters of this theory believe that the inner endodermis found at the inner face of the vascular tissue and the parenchyma encircled by this endodermis have been originated from the cortex. According to Jeffrey and other supporters of this theory the siphonosteles with internal endodermis are more primitive than those without an internal endodermis.

The siphonosteles which do not possess the inner endodermis are believed to have originated by disintegration of inner endodermis during evolution.

According to the theory proposed by Boodle (1901), and Gwynne-vaughan (1903), the siphonostele has been evolved from the protostele by a transformation of the inner vascular tissue into parenchyma.

A siphonostele may be of the following types:

1. Ectophloic. In this type of siphonostele, the pith is surrounded by concentric xylem cylinder and next to xylem the concentric phloem cylinder.

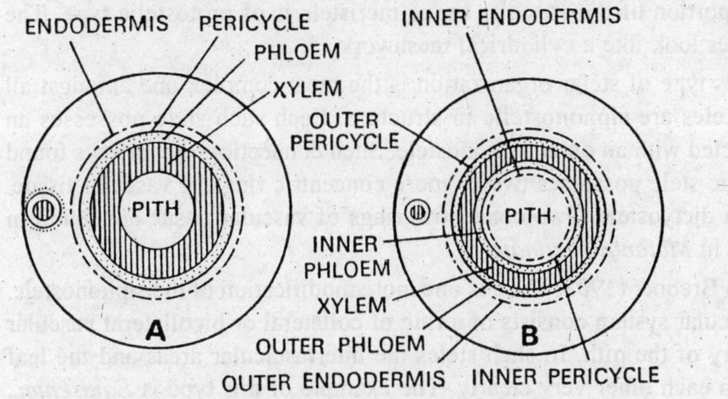

Fig. 32.5. Stelar system in ferns. A, ectophloic siphonostele; B, amphiphloic siphonostele.

2. Amphiphloic. In this type of siphonostele the pith is surrounded by the vascular tissue. The concentric inner phloem cylinder surrounds the central pith. Next to the inner phloem is the outer phloem cylinder, *e.g.*, *Marsilea*.

C. Solenostele. The vascular plants have been divided into two groups on the basis of the presence or absence of the leaf gaps. These groups are 1. Pteropsida and 2. Lycopsida. The ferns, gymnosperms and angiosperms are included in Pteropsida, whereas the lycopods, horse tails, etc., are included in Lycopsida. The simplest form of siphonostele has no leaf gaps, such as some species of *Selaginella*. However, among the simplest siphonostelic Pterospsida and siphonostelic Lycopsida, the successive leaf gaps in the stele do not overlap each other and are considerably apart from each other.

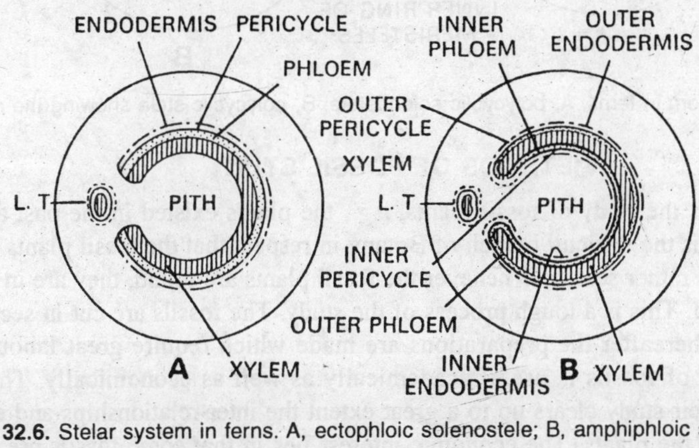

Fig 32.6. Stelar system in ferns. A, ectophloic solenostele; B, amphiphloic solenostele.

According to Brebner (1902), Gwynne-Vaughan (1901) such siphnosteles which lack overlapping of gaps are known as **solenosteles.** They may be ectophloic or amphiphloic in nature. Some authors (Bower, 1947; Wardlaw, 1952; Esau, 1953), however, interpreted the solenostele as an amphiphloic siphonostele.

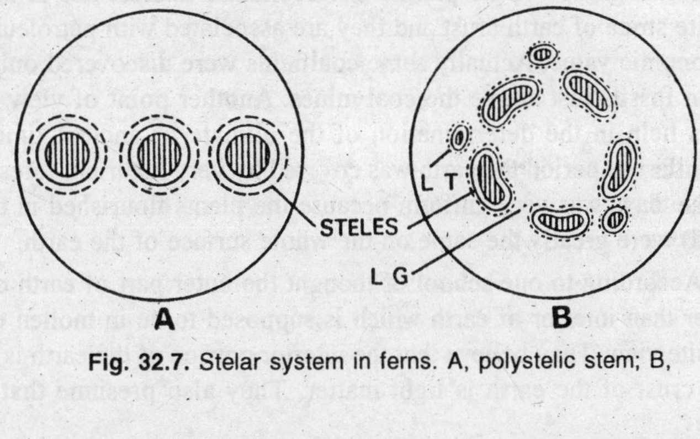

Fig. 32.7. Stelar system in ferns. A, polystelic stem; B,

D. Dictyostele. In the more advanced siphonosteles of Pteropsida, the successive gaps may overlap each other. Brebner (1902) called the siphonosteles with overlapping gaps as **dictyosteles**. In such cases the intervening portion of the vascular tissue meristele is of protostelic type. The dictyostele with many meristeles look like a cylindrical meshwork.

E. Polycyclic stele. This type of stelar organization is the most complex one amongst all pteridophytes. Such type of steles are siphonostelic in structure. Each such stele possesses an internal vascular system connected with an outer siphonostele. Such connections are always found at the node. A typical polycyclic stele possesses two or more concentric rings of vascular tissue. This may be a solenostele or a dictyostele. Two concentric rings of vascular tissue are found in *Pteridium aquilinum* and three in *Matonia pectinata*.

F. Eustele. According to Brebner (1902) there is one more modification of the siphonostele, known as eustele. Here the vascular system consists of a ring of collateral or bicollateral vascular bundles situated on the priphery of the pith. In such steles the interfascicular areas and the leaf gaps are not distinguished from each other very clearly. The example of this type is *Equisetum*.

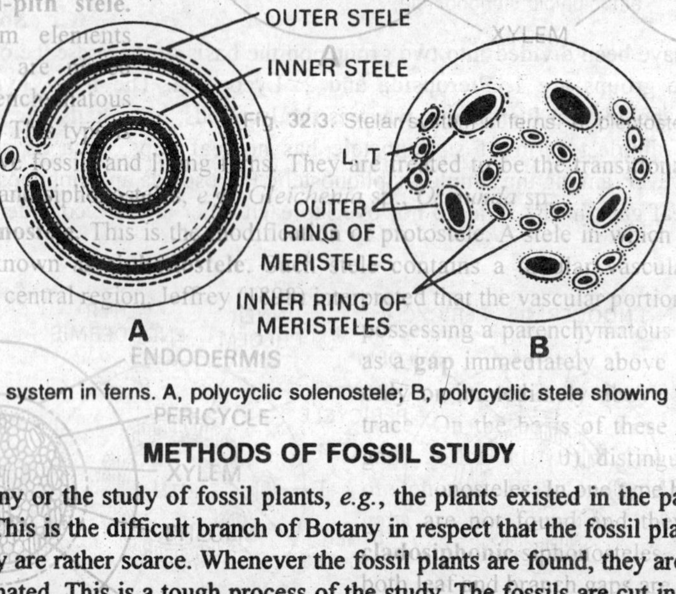

Fig. 32.8. Stelar system in ferns. A, polycyclic solenostele; B, polycyclic stele showing the meristeles.

METHODS OF FOSSIL STUDY

Palaeo-botany or the study of fossil plants, *e.g.*, the plants existed in the past and now are entirely extinct. This is the difficult branch of Botany in respect that the fossil plants are difficult to obtain and they are rather scarce. Whenever the fossil plants are found, they are in parts which are to be co-ordinated. This is a tough process of the study. The fossils are cut in sections with a great difficulty and thereafter the preparations are made which require great labour, time and technique. The study of fossils is useful academically as well as economically. The academic interest lies in that their study clears up to a great extent the inter-relationships and evolution of the ancient groups of the plants. The economic interest lies in that some fossils are confined to definite strata of earth crust and they are associated with petroleum, coal and similar other things of economic value. Actually some coalfields were discovered only on account of the presence of certain fossils just above the coal mines. Another point of view of academic interest is that the fossils help in the determination of the climate of ancient times in different regions. During carboniferous period the earth was covered by very luxuriant forests and it is assumed that climate in those days was very uniform because the plants flourished in those times (*e.g.*, Carboniferous period) were greatly the same on the whole surface of the earth.

According to one school of thought the outer part of earth crust is solid and comparatively thinner than interior of earth which is supposed to be in molten condition. Some workers are of opposite view. They believe that the interior portion of the earth is firmer and more solid while the outer crust of the earth is light matter. They also presume that the interior crust of the earth consists of many heavy metals like iron and lead. Whatsoever might have been the position but it is very clear that the first crust which formed the first surface must have been very uniform.

Later on the water and air appeared and with the appearance of these two destructive factors the surface of the earth began to change. Rain and air affected the surface of the earth to great extent and at certain places the surface rose up in high mountains by bursting of interior of matter and at other places it sank down and formed seas and oceans. The rocks withered off, the volcanic eruptions and other changes took place and the general topography of the earth was completely changed. Rivers coming down from mountains bringdown pieces of rocks and large quantity of sand when they flow in the plains. This sand settles down at the bottom of the water. Along with this sand the parts of the plants and animals which get into that sand would have a chance of being preserved. This preservation takes place in different forms.

The fossils are mainly found in the sand which has been brought down by the rivers from the mountains. This sand has become compressed and sedimentary rocks have been formed in the bottom water. They contain the pieces of plants or entire bodies or bodies of animals which have been surrounded by mud and salt and they got the chance of being preserved.

This preservation takes place in two forms :

1. **Impressions:** These impressions may be of the surface markings of the plants material or be the markings of the internal cavities in the plant material or gets into the cavity and their impressions are left. In some fine fossils even details of the venation, structure of stomata and epidermal hairs have been studied. In such impressions the plant material itself has decayed or turned into coal. However, the surface impressions cannot be relied upon as we know that the plants, *e.g., Equisetum* (a pteridophyte), *Ephedra* (a gymnosperm) and *Casuarina* (an angiosperm) gave same type of impressions, but all of them the widely differentiable and have no affinities and relationships with each other.

2. **Petrification.** Another method of preservation of great importance is that the plant material when lying in water has become infiltrated with mineral matter such as lime, silica, magnesium salts and other similar substances. These salts penetrate the minute cells and the organic matter is being replaced by mineral salts. The original organic matter may have completely or partially changed but its outline and general structure is left in the form of these infiltrated substances.

In coal mines generally the coal balls are found which are formed by the infiltration of calcium bicarbonate.

In another method of preservation in which rare and small parts of plant material such as seeds, pollen grains, spores, etc., have been preserved. In this method the plant material happens to fall in resin or amber extruded by some plants, such as pine trees and others.

Technique of Fossil Study

The different methods of the study of fossil plants are as follows :

It is laborious process and requires sufficiently great time. Usually the petrified specimens are cut in serial sections which give an idea of the actual structure of the fossil plant. These petrified pieces are cut into very fine slices by different methods. In one method each such piece is attached to glass plate and grounded to sufficient thinness and thereafter studied under the microscope.

Another improved method of the study of these petrified specimens is to prepare the films of the material by special techniques. The methods of preparing thin films is as follows: First of all the surface of the section of the petrified material is made smooth. If the material consists of calcium carbonate, then on the smooth surface of the slice a film of 5 percent hydrochloric acid is allowed to act for five minutes. If the slice of the petrified material is silica then the film of 10 percent hydroflouric acid (HF) is allowed to act on the smooth surface for ten minutes so that the silica is dissolved. The surface of the petrified section by the action of these acids becomes rough on account of the dissolution of the mineral matter. If any organic matter remains on the surface, now put hot gelatin on the surface. As soon as the things dry up, they are removed and studied under the microscope. This process may be successful only in the case when organic matter is left in petrified specimens. In case where organic matters are already decayed, such preparations are never good.

The fossils naturally would be pieces of plants. It is very rare that entire plant could have preserved. This way, only pieces can be studied. In such type of study the individual pieces are given botanical names, just as in living plants; the botanical names of the fossil plants are not so significant as those of living ones. As they are represented by the pieces of the plants and, therefore, their generic names would be according to stem, leaf and root or any reproductive structure. The stems are usually given the generic names which end with **'dendron'** (tree) or **'xylon'** such as, *Lygenodendron* or *Cladoxylon.* The leaves end with **'pteris'** or **'phyllum'** and reproductive parts end with **strobilus**. This way, the palaeobotany is the study of the parts of fossil plants and in certain case marvellous results have been obtained.

In the case of *Lyginopteris,* one of the Cycadofilicales of Carboniferous period which was found in pieces and later on the palaeobotanists supported that all the pieces belonged to the particular plant. Later on after few years the complete intact plant was found.

Important Strata

The Palaeozoic and Mesozoic strata are very important from the study points of view of the fossil plants. It is in the upper middle Palaeozoic, *i.e.*, the Devonian strata, we come across with the first land plants such as lycopods belonging to Lycopodiales, Equisetales, the seeded ferns, the primitive gymnosperms, the pteridophytes, etc. In the late Palaeozoic, *i.e.*, both in the upper and lower, Carboniferous strata the earth was covered up by the very luxuriant forests. These forests were formed by lycopods, horsetails, seeded ferns and later on with primitive gymnosperms. The Carboniferous strata is most important. The coal mines are situated in this strata. The coal mines are the result of dense forests having got submerged in those times. The Mesozoic is also very important from the point of view that the first angiosperms made their appearance, otherwise the higher gymnosperms formed luxuriant forests in those times. In the later Mesozoic some of the gymnosperms disappeared. Majority of the Cycads disappeared and only a few forms have been left upto the present day.

In India the most important strata is described technically the **'Godwana system'** named after Godwana Kingdom. For the first time the rocks of this period were discovered near Narmada river.

Work on Fossils in India

The first most work on Indian fossils was done by Fiestmental, the director of geological survey of India. He published a monograph of Indian fossil forms. Thereafter Seward and Bencroft made a number of additions. From 1895 to 1909 most of the work on Indian fossils was carried on in Europe. It has been since 1919 till the death of Dr. Birbal Sahni, *i.e.*, 1949, that work of Indian fossils had been added to a great deal. He first revised the known Indian fossils and then published a number of monographs describing numerous fossil forms. The most important Indian fossil of Cycads is *Williamsonia sewardiana.* In India it is only in few places that there is a good collection of fossil plants, *e.g.*, Museum of Calcutta, the Lucknow University Museum; Presidency College, Madras, Prince of Wales College, Jammu. The collections are made from Raniganj coal field, in Bengal, places near about Nagpur, Reeva state, Kotah state, in some parts of Kashmir near about the Satluj river, salt range in Punjab, and Rajmahal hills, Bihar.

THE TELOME THEORY

The interpretation of the plant body as a branched axis with its divisions specialized for various functions, but all like in fundamental structure, is called the **telome theory.** The telome

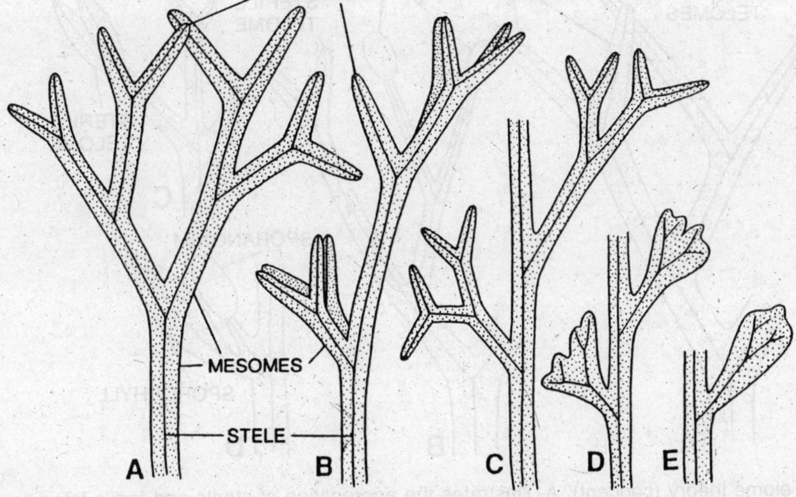

Fig. 32.9. Zimmermann's Telome concept. A, illustrates hypothesis of telomes and mesomes; B, overtopping-two sister branches of a dichotomy become unequal, the stronger one forms the axis, the weaker one is pushed aside and forms the leaf; C, *planation*-branching in more than one plane is being replaced by a dichotomy in a single plane, thus arranging the telomes and mesomes in a plane; D-E, *webbing* and *syngenesis*-the telomes and mesomes, become connected by the formation of parenchymatous tissue between them (parenchymatous webbing); this is accompanied by the fusion of their steles also.

theory is due to the work of Zimmermann. He goes back to the earliest and simplest type of vascular plant known, that in the Psilophytales in which the body consists of branching leafless axes, which are all substantially alike. In some cases these may be arranged dichotomously in others monopodially, and the difference may have arisen by 'overtopping', but in either case these naked axes as the primitive units of construction of the cormophytes, which is thus regarded as primarily a system of axes. Each of the terminal branches in such a system is called a **telome.** Each telome is an ultimate branch on an older axis or **mesome**. Some telomes are sterile, others bear terminal sporangia and are therefore fertile. Telomes also tend to unite into groups, called, **syntelomes,** which may be all sterile, all fertile or mixed. According to telome theory the earliest levees are flattened telomes or **phylloids,** and the sporangium in all cases, together with its stalk or sporangiophore, is a fertile telome. Compound sporangiophores, like those of *Equisetum*, are syntelomes. The phylloids are the prototypes of the microphyllous foliage leaves. In the most primitive

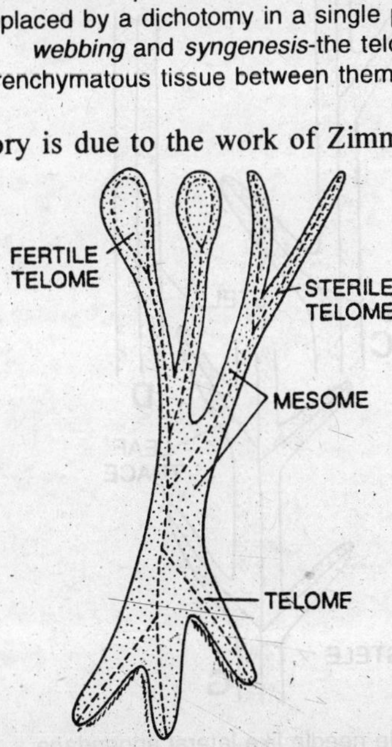

Fig. 32.10. Diagram of the organization in a theoretically primitive vascular plant showing a body built of telome and mesome limbs, the telomes being both fertile and sterile. (After Zimmermann).

conditions the sterile and fertile telomes were apparently quite independent of each other. To axillary position which the sporangial telomes occupy in Lycopsida has been arrived at by secondary shifting of the telomes and shows three ways in which the change may have taken place.

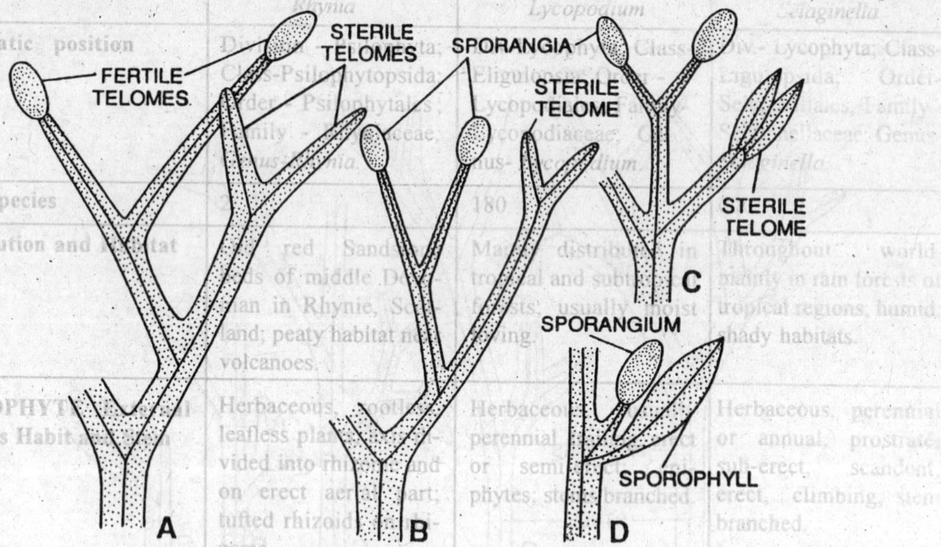

Fig. 32.11. Telome theory (concept). A, illustrates the aggregation of sterile and fertile telomes; B-C, shows reduction of supporting mesomes and reduction of sporangium number, D, shows one phylloid with axillary sporangium, (seen in origin of Lycophyta)

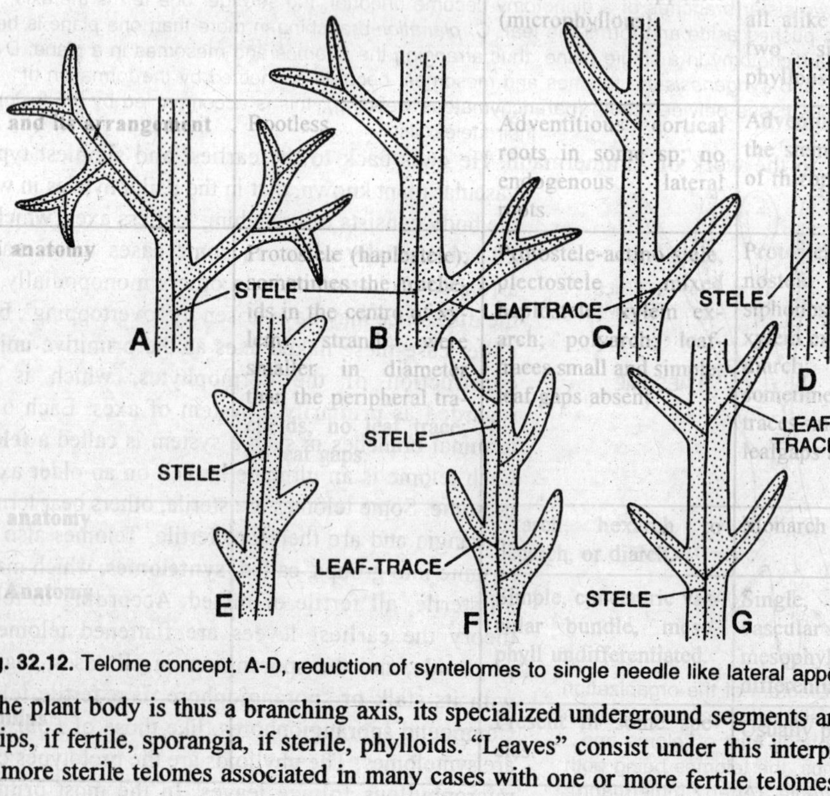

Fig. 32.12. Telome concept. A-D, reduction of syntelomes to single needle like lateral appendage.

The plant body is thus a branching axis, its specialized underground segments are roots, its aerial tips, if fertile, sporangia, if sterile, phylloids. "Leaves" consist under this interpretation, of one or more sterile telomes associated in many cases with one or more fertile telomes.

Leaf morphology thus supports the division of vascular plants into four major groups the Psilopsida are leafless or show leaves only partly differentiated; the leaf of the Lycopsida is an enation enlarged and elaborated; that of the Sphenopsida is a minor branch system, that of Pteropsida major one. The leaf is thus of different nature in vaseular plants; the evolutionary development of a photosynthetic organ has come about in more than one way.

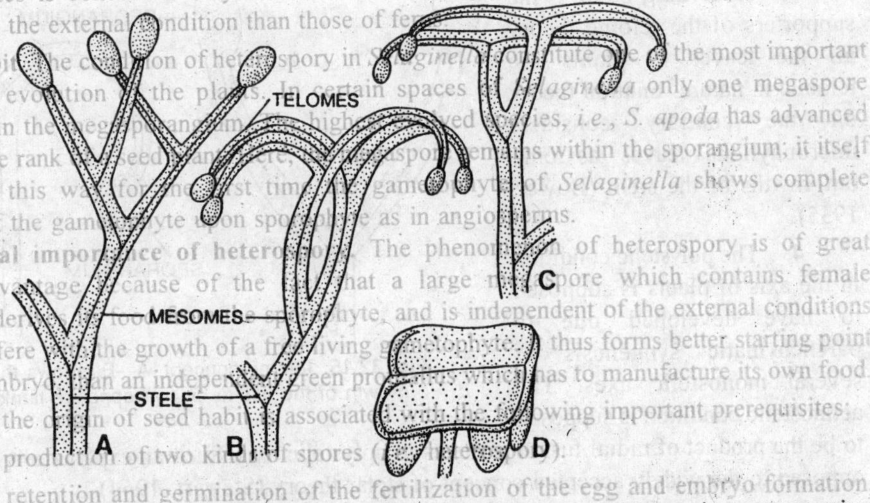

Fig. 32.13. Telome Concept. A-B, *incurvation*-downward bending of the sporangia toward a common axis; C-D, fusion and expansion of telomes and mesomes into a peltate structure (seen in Sphenopsida).

Merits. This is a simple concept and explains most of the morphological problems about different organs of a plant. According to Bierhorst (1971) this theory is too simple and too easily applicable but unfortunately its excessive use has greatly diminished its value.

Demerits. 1. Telome has been considered as a readymade unit. This difficulty was realized by Zimmermann (1949, 52) and subsequently he recognized several other elementary processes as (a) interconnection of cells; (b) rotation of cell axis; (c) differentiation of apical cell, etc., which have led to the formation of an apical cell with three cutting faces. However, these elementary processes do not satisfy the plant morphologists (Puri, 1956).

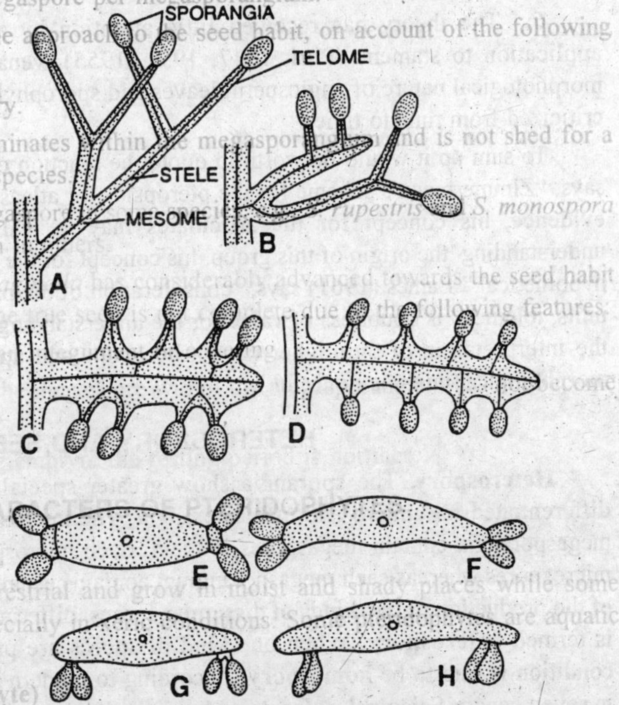

Fig. 32.14. Telome concept. Origin of Pteropsida. A-B, illustrates that overtopping leads to pinnate sporophyll; C, shows the fusion of mesomes laterally with each other; D, shows pinnately veined sporophyll with marginal sporangia; E-H, shows un-equal growth on the two opposite flanks of an organ (*i.e.*, incurvation).

2. Many other plants of much greater complexity, than *Rhynia* fossils have been discovered in beds of the same age or even earlier, *e.g., Zosterophyllum, Baragnnathia* (Leclercq, 1954) Lyon Hueber (1964), Hueber and Banks (1967) and Lyon (1964) observed lateral sporangia on short vascularized stalks in *Psilophyton* and *Asteroxylon* respectively instead of usual terminal sporangia.

3. According to the supporters of the telome theory all the leaves in plants are telomic in nature. Enation theory on the contrary considers microphyllous leaves as only outgrowths of the stem (Bower, 1935).

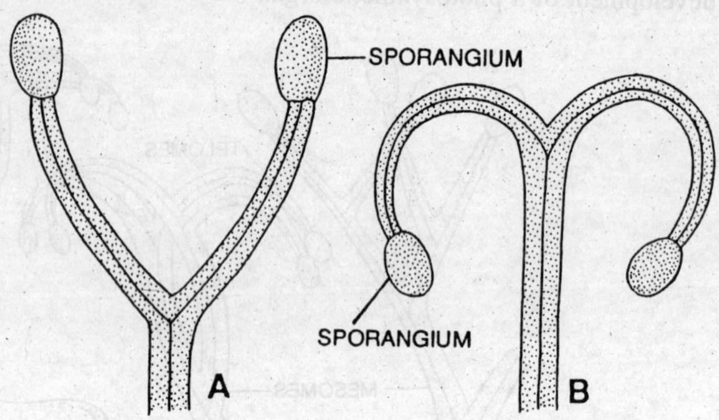

Fig. 32.15. Telome concept. A—B, shows the unequal growth of the tissue on two opposite flanks of the

4. The polystelic condition in the axis of plants is supposed to have developed due to parenchymatic syngenesis of several monostelic axes. The actinostelic condition is supposed to be the product of radial fusion of steles in polystelic axis. Such an explanation is diametrically opposed to the widely accepted concept of stelar theory (Stewart, 1964).

5. Andrews (1963) has given a series of diagrams of Palaeozoic seeds to explain the origin of cupule. Pettitt (1970) found them to be more or less of the same age.

6. The theory has received little attention by angiosperm centred morphologists. Its application to stamens (Puri, 1947, 1951, 1955), venation pattern of leaves (Foster, 1950), morphological nature of angiosperm leaves and sporophylls, and carpels (Eames 1961) have been criticized from time to time.

To sum up it would be worth to quote the reaction of Andrews and Eames. Andrews (1961) says, "Zimmermann's scheme for the pteropsids, or atleast some pteropsids, has much supporting evidence, his concept for the articulates may be valid, but we are only in the verge of understanding the origin of this group, his concept for the lycopsids is, so far as I am aware purely hypothetical." Eames (1961) says, "consideration of the primitive plant body as made up of basic units, telomes, is doubtless of value for the understanding of more primitive taxa, but its value in the interpretation of the higher taxa, where axis and appendages have become established as morphological units, is doubtful."

HETEROSPORY AND SEED HABIT

Heterospory. The sporangia show greater specialization than those of ferns. They are differentiated into micro and megasporangia. The microsporangia contain microspores whereas megasporangia contain megaspores in them. Each microsporangium contains a large number of microspores whereas each megasporangium contains usually only four megaspores. The condition of the production of two kinds of dissimilar spores, differing in shape and size in the same species, is termed **heterospory**. When one kind of spores are present, *e.g., Equisetum* and ferns, the condition is said to be **homospory**. According to modern pteridologists the heterospory is found in seven genera-*Selaginella, Isoetes, Marsilea, Salvinia, Azolla, Regnellidium* and *Pilularia.* Here one thing may be noticed that the microspores develop male gametophytes whereas the megaspores the megaspores germinate into the female gametophytes which may also be called as microgametophytes and megagametophytes. In the case of homospory the sex may be differentiated only at the gametophytic stage, but here the sex may be differentiated even at the sporophytic stage by the presence of small and large sized microspores which produce antherozoids and eggs respectively.

The differentiation of microspores and megaspores and their dependence upon the sporophyte has certain advantages. The gametophytes of the ferns are, however dependent for their nutrition upon soil and environmental condition, whereas in the case of *Selaginella*, as far as the nutrition of gametophytes is concerned they derive it from the sporophyte, and there they are more independent to the external condition than those of ferns.

Seed habit. The condition of heterospory in *Selaginella* constitute one of the most important phases in the evolution of the plants. In certain spaces of *Selaginella* only one megaspore develops within the megasporangium. The highest evolved species, *i.e.*, *S. apoda* has advanced almost upto the rank of a seed plant. Here, the megaspore remains within the sporangium; it itself fertilizes and this way for the first time the gametophyte of *Selaginella* shows complete dependence of the gametophyte upon sporophyte as in angiosperms.

Biological importance of heterospory. The phenomenon of heterospory is of great biological advantage because of the fact that a large megaspore which contains female gametophyte derives its food from the sporophyte, and is independent of the external conditions as might interfere with the growth of a free living gametophyte. It thus forms better starting point for the new embryo, than an independent green prothallus which has to manufacture its own food.

In brief, the origin of seed habit is associated with the following important prerequisites:

1. The production of two kinds of spores (*i.e.*, heterospory).
2. The retention and germination of the fertilization of the egg and embryo formation, which is still within germinating megaspore.
3. Development of only one megaspore per megasporangium.

In *Selaginella*, there is remarkable approach to the seed habit, on account of the following important features:

1. *Selaginella* shows heterospory.
2. The megaspore usually germinates within the megasporangium and is not shed for a time which may vary from species to species.
3. There is reduction to one megaspore in some species, *e.g.*, *S. rupestris* and *S. monospora* and a confirmed tendency to reduction in others.

It becomes quite evident that *Selaginella* has considerably advanced towards the seed habit in a few species, but its approach to the true seed is not complete due to the following features:

1. The megasporangium lacks an integument or covering.
2. The permanent retention of the megaspore within the megasporangium has not become established.
3. After the development of the embryo, the resting period is not there.

GENERAL CHARACTERS OF PTERIDOPHYTES

Occurrence

1. Most pteridophytes are terrestrial and grow in moist and shady places while some flourish well in open, dry places especially in xeric conditions. Some pteridophytes are aquatic and some are epiphytes.

The plant body (The adult sporophyte)

2. The sporophyte is the conspicuous and familiar plant body. It develops from the zygote, a diploid cell which results from the fertilization of the egg and antherozoid.

3. The sporophytic plant body remains differentiated into true roots, stem and leaves. Some primitive members lack true roots and well developed leaves (*e.g.*, in Psilophytales and Psilotales).

4. The branching of the stem may be of monopodial or dichotomous type.

5. There are two main categories of form and structure, one category comprises of megaphyllous types, in which the leaves are large in relation to the stem (*e.g.*, ferns); the second comprises of microphyllous types in which the leaves are quite small in relation to the stem (*e.g.*, Lycopods and Horse-tails).

6. All the vegetative parts of the sporophyte possess vascular supply.

Reproduction.

7. The sporophytes reproduce by spores which are produced within sporangia.

8. In some pteridophytes the sporangia develop on stems (*i.e.*, cauline in origin) while in other they are borne either on the leaves (foliar) or in the axils of the leaves. The leaves that bear sporangia are known as sporophylls.

9. The sporophylls may be widely scattered on a plant (*e.g.*, ferns) or may be clustered in definite areas and structures called cones or strobili (*Selaginella, Equisetum*).

10. In certain pteridophytes the sporangia are produced within specialized structure, the sporocarps (*e.g.*, *Marsilea, Salvinia, Azolla*).

11. The sporangium in all pteridophytes, is initiated by the laying down of a cross-wall in a superficial cell or a group of cells. Since this wall is periclinal each initial cell is divided into an outer and inner daughter cell. If the sporogenous tissue is derived from the inner daughter cell, the sporangium is described as **"eusporangiate"** (*e.g.*, in most of pteridophytes) and if from the outer as **"leptosporangiate"** (*e.g.*, in advanced pteridophytes).

12. The sporophyte plant may be homosporous (*e.g.*, *Lycopodium, Dryopteris*) or heterosporous (*e.g.*, *Selaginella, Isoetes, Marsilea*)

The gametophyte

13. The spores on germination give rise to the haploid gametophytes or prothalli which are small and inconspicuous. The gametophytes in some pteridophytes are subterranean and in others they are retained within the resistant wall of the spore. It is notable that wherever the gametophyte is retained within the spore the spores are of different sizes (heterosporous). The larger megaspores give rise to female prothalli which bear only archegonia, and the smaller microspores giving rise to male prothalli which bear only antheridia.

14. The gametophyte or prothallus bears the sex organs, antheridia and archegonia. Normally, the gametophytes formed from the homospores are monoecious, that is both antheridia and archegonia are borne on the same gametophyte or prothallus. The gametophytes formed from the heterospores are dioecious, *e.g.*, the antheridia and archegonia develop in separate male and female gametophytes.

The antheridia

15. The antheridia may be embedded in the gametophyte or they may project from it. The embedded antheridia are commonly found in eusporangiate pteridophytes while the projecting ones are usually found in the leptosporangiate ferns.

The mature antheridium is globular and consists of an outer sterile wall inside which are found a large number of androcytes. Each androcyte metamorphoses into a single motile antherozoid.

The archegonia

16. The archegonia are flask-shaped. Each archegonium consists of a basal swollen, embedded portion the venter and a short neck. The venter encloses the egg and ventral canal cell. The neck contains the neck canal cells.

At maturity the apical cells of archegonium separate, the neck canal cells disintegrate forming a passage for antherozoids to reach the egg.

Fertilization.

17. In all cases the fertilization is accomplished by the agency of water. With the result of the fusion of male gamete and female egg a diploid zygote (2*n*) is formed.

The embryo (The young sporophyte)

18. The zygote undergoes repeated divisions to form a new sporophyte. The young sporophyte remains attached to the gametophyte by means of a foot and draws nourishment from the prothallus until it develops its own stem, roots and leaves. The sporophyte is dependent on the gametophyte only during its early stages.

EUSPORANGIATE AND LEPTOSPORANGIATE SPORANGIA

The sporangium. The sporangium, in all pteridophytes is initiated by the laying down of a cross-wall in a superficial cell or group of cells. Since this wall is periclinal (*i.e.*, parallel to surface) each initial cell is divided into an outer and inner daughter cell. If the sporogenous tissue is derived from the inner daughter cell the sporangium is described as '**eusporangiate**' and if from the outer, as '**leptosporangiate**'. In leptosporangiate forms, the sporangium wall, stalk and the spores, are derived from the outer daughter cell, but in eusporangiate forms, the adjacent cells take part in the formation of part of sporangium wall and the stalk. In eusporangiate form the sporangium is large and massive, the wall is several cells in thickness and the spore content is high. In leptosporangiate forms, the sporangium is small, the wall is one cell in thickness and the spore content is low. Of these two types, the eusporangiate is primitive and the leptosporangiate advanced.

Table 32.1

Eusporangiate sporangium	Leptosporangiate sporangium
A superficial cell divides by a periclinal wall forming an outer and inner daughter cell. The sporogenous tissue is derived from the inner daughter cell.	A superficial cell divides by a periclinal wall forming an outer and inner daughter cell. The sporogenous tissue is derived from the outer daughter cell.
The adjacent cells are involved in the formation of part of the sporangium wall and the stalk.	The sporangium wall and the stalk, as well as the spores are derived from the outer cell.
The sporangium is large and massive; the wall is several cells thick and the spore content is high.	The sporangium is small, the wall is only one cell thick and the spore content is low.
Primitive.	Advanced.

Table 32.2. COMPARATIVE STUDY - I

	Rhynia	*Lycopodium*	*Selaginella*
Systematic position	Division - Psilophyta; Class-Psilophytopsida; Order - Psilophytales; Family - Rhyniaceae; *Genus-Rhynia.*	Div-Lycophyta; Class-Eligulopsia; Order - Lycopodiales; Family-Lycopodiaceae; Genus- *Lycopodium.*	Div.- Lycophyta; Class-Ligulopsida; Order-Selaginellales; Family.-Selaginellaceae; Genus-*Selaginella.*
No.of species	2	180	600
Distribution and Habitat	Old red Sandstone beds of middle Devonian in Rhynie, Scotland; peaty habitat near volcanoes.	Mainly distributed in tropical and subtropical forests; usually moist loving.	Throughout world, mainly in rain forests of tropical regions; humid, shady habitats.
SPOROPHYTE External Features Habit and Stem	Herbaceous, rootless, leafless plants; axis divided into rhizome and on erect aerial part; tufted rhizoids on rhizome.	Herbaceous, usually perennial trailing, erect or semi-erect; epiphytes; stems branched.	Herbaceous, perennial or annual, prostrate, sub-erect, scandent, erect, climbing; stem branched.
Leaves and their arrangement	leafless	Simple small, lanceolate; sessile, spiral, whorled or opposite (microphyllous).	Small simple ligulate, lanceolate to ovate; spiral or in four rows, all alike in size or of two sizes (microphyllous).
Root and its arrangement	Rootless	Adventitious cortical roots in some sp; no endogenous lateral roots.	Adventitious; borne on the stem or at the tips of rhizophores.
Stem anatomy	Protostele (haplostele); sometimes the tracheids in the centre of xylem strand were smaller in diameter than the peripheral tracheids; no leaf trace; no leaf gaps.	Protostele-actino stele, plectostele mixed haplostele; xylem exarch; polyarch; leaf-traces small and simple; leaf gaps absent.	Protostele; siphonostele to polycyclic siphonostele; polystelic; xylem exarch; mono or diarch; vessels sometimes present; leaf traces small and simple; leafgaps absent.
Root anatomy		Exarch, hexarch to decarch, or diarch.	Monarch and exarch.
Leaf Anatomy		Simple, concentric vascular bundle, mesophyll undifferentiated.	Single, concentric vascular bundle, mesophyll usually not differentiated.
Strobillus		Present in some species.	Usually present

	Rhynia	*Lycopodium*	*Selaginella*
Sporophylls		Resemble leaves in some species, or differ from them in size, form and colour.	Resemble leaves or differ from them slightly.
Sorus			
Sporangium	Cauline, terminal; homosporous; oval or slightly cylindrical; wall is massive; three layered; annulus absent; tapetum is probably onelayered; formed from the innermost layer of the wall; no special device for dehiscence.	Eusporangiate; foliar or cauline; homosporous; large, kidney shaped with a short massive stalk; wall is three or more layered; annulus absent; tapetum one layered; formed from the innermost layer of the wall; dehiscence by a transverse slit.	Eusporangiate foliar, axillary, or cauline; heterosporous; microsporangium-stalked, small with a smooth outline; megasporangium-stalked, large, four lobed, wall is two layered in both type of sporangia; annulus absent; tapetum is one layered, formed from the outermost layer of sporogenous tissue, persists until the spores are nearly mature and then degenerates; dehiscence by vertical apical splitting of the upper part of sporangial wall.
Average spore out-put from each sporangium.	many	many	Megasporangium-4 megaspores; microsporangium many microspores.
GAMETOPHYTE	One kind spores numerous; small tetrahedral in shape.	One kind spores numerous, small in size, tetrahedral in shape, (0.03 to 0.05 mm)	Two kinds-microspores and megaspores, microspores numerous, megaspores-four, microspores-small (0.015 to 0.06mm) tetrahedral, megaspores-large (1.5 to 5 mm), tetrahedral.
Prothallus		Monoecious; surface living or subterranean, or partially subterranean, short, 2 mm usually radial or dorsiventral, of various shapes, green, brownish, yellowish or nearly colourless.	Dioecious, microgameto-phyte wholly within micropore wall, megagametophyte also within megaspore wall with only the apical cushion protruding; both much reduced in sizes; non-green.

	Rhynia	*Lycopodium*	*Selaginella*
Sex organs: Antheridium		On crown or on flattened or hollowed upper surface, many in number, completely embedded or slightly projecting, antherozoids usually fusiform, biciliate.	Inside the microspore wall, single antheridium, spirally coiled, biciliate antherozoids.
Archegonium		On crown or on flattened or hollowed upper surface, many in number, sunken with neck projecting, neck is short 3 to 4 cells in height, neck canal cells 1, 2, 3, 4, 7, 14, to 16.	On the apical cushion, many in number, sunken with outer tier of neck slightly projecting, neck is short, 2 cells in height, neck canal cell-1.
Embryo		First division of zygote is transverse, suspensor present, weakly developed.	First division of zygote is transverse, suspensor usually well-developed.

Table 32.3. COMPARATIVE STUDY - II

	Equisetum	*Marsilea*	*Fern*
Systematic Position	Division- Arthrophyta; Class-Calamopsida; Order- Equisetales; Family- Equisetace; Genus- *Equisetum*.	Division- Filicophyta; Class Leptosporangiopsida; Order- Marsileales; Family- marsileaceae; Genus- *Marsilea*	Division - Filicophyta; Class- Leptosporangiopsida; Order- Filicales; Family- Polypodiaceae. Genus- *Dryopteris*; *Adiantum* etc.
No. of Species	25	65	
Distribution and Habitat	Throughout world except Australia and New Zealand; grow in exposed places, marshes meadows and along river and canal banks.	Cosmopolitan, mainly in warmer parts; aquatic or amphibious.	Cosmopolitan; mainly in humus rich places.
SPOROPHYTE			
External Features Habit and Stem	Herbaceous, perennial with creeping, horizontal, jointed ridged rhizomes	Herbaceous, rizomatous, creeping.	Underground creeping rhizome.
Leaves and their arrangement	Simple, small, scalelike uninerved fused laterally to form a sheath (microphyllous).	Compound, lamina divide into four leaflets, arranged alternately in two rows, dichotomous venation, circinate vernation (megaphyllous).	Pinnately compound; alternate, circinate vernation (megaphyllous).

	Equisetum	*Marsilea*	*Fern*
Root and its arrangement	Adventitious, developing from the base of lateral branch primordia.	Adventitious, borne at the nodes.	Adventitious, slender, little branched.
Stem anatomy	Perforated ectophloic siphonostele with nodal rings. In the internode separate endarch collateral bundles. Xylem endarch, leaf traces small and simple; leaf gaps absent.	Amphiphloic solenostele. Xylem exarch, leaf-traces large, leaf-gaps present, scattered.	Polycyclic dictyostele. Xylem mesarch, vessels present, leaf-traces large; inner and other series of vascular strands contribute to leaf traces; leaf-gaps and perforations present.
Root anatomy	Tri or tetrarch, exarch; two layers of endodermis.	Diarch, exarch.	Diarch, exarch.
Leaf anatomy	Single vascular bundle; collateral.	Many vascular bundles, concentric mesophyll differentiated.	Many vascular bundles, concentric or collateral, mesophyll differentiated.
Strobilus	Present.		
Sporophylls	Sporangiophores.	Sporocarp contains a large number of sori.	Resemble foliage leaves.
Sorus		Many marginal number of indusia equal soral sacs.	Superficial, coenosorus, surrounded by false and true indusium.
Sporangium	Eusporangiate; borne on peltate sporangiophores; homosporous; elongated sac like with a rounded apex, wall is single layerd or sometimes two layered in mature condition, annulus absent, tapetum one layered, irregular, ill defined, formed from the innermost layer of jacket, disorganises, dehisce along a longitudinal line facing inwards towards the stalk.	Leptosporangiate, borne in sporocarps, heterosporous, both micro and megasporangia stalked, wall is singlelayered, annulus absent, tapetum two to three-layered, formed from archesporial cell, disorganises, dehiscence by dissolution of the mucilaginous wall of the parietal cells of wall on coming in contact with water.	Leptosporangiate, on abaxial surface of sporophylls, homosporous, long stalked, capsule biconvex lens-shaped, wall is single layered, annulus vertical, incomplete, tapetum two layered, formed from the archesporial cell, disorganises, transverse dehiscence by means of annulus and stomium which form the opening mechanism.
Average spore output from sporangium	Many	Megasporangium one megaspore, microsporangium many microspores.	32 to 64.

	Equisetum	*Marsilea*	*Fern*
GAMETOPHYTE	One kind spores; numerous; small in size; globular or subspherical with no trace of triradiate ridge.	Two kinds-megaspores and microspores 64, megaspore-1, microspores-small 0.075 mm., globose with a pyramidate apex, megaspore ellipsoidal to oval, tetrahedral.	One kind spores, usually 64 in number, small in size, tetrahedral in shape, 0.03 mm. in diameter.
Prothallus	Usually monoecious or partially dioecious; surface living, usually less than 1 cm in diameter; usually dorsiventral consisting of massive cushion like base and upright green branched; green (autotrophic)	Dioecious, both megagametophyte and microgametophyte within spore wall, both much reduced, non-green.	Usually monoecious, surface living, about 6 mm. in diameter, dorsiventral, cordate, green (autotrophic).
Sex organs: antheridium	On upturned extensions in the margin of apex of delicate lobes; many in number; projecting, spirally coiled and flattened, multiciliate antherozoids.	Inside the microspore wall, two in number, cork screw like, spirally coiled, multiciliate, antherozoids.	On the basal or lateral regions of the ventral surface of the prothallus, many in number, slightly projecting, spirally coiled, multiciliate antherozoids.
Archegonium	In the meristematic rim where upright green lobes are being formed; many in number, base sunken in the prothallus with neck projecting, neck is short 3-4 cells in height, neck canal cells 1 or 2, in many species boot shaped.	Embedded in a cap or tissue fitted over the apex of the spore cavity, one in number, neck is short 2 cells in height, neck canal cell-I.	On the massive cushion just posterior to the apical notch on the ventral surface of the prothallus, many in number, sunken with neck projecting, neck-5 to 7 cells in height, neck recurved, neck canal cell 1, usually binucleate.
Embryo	First division of zygote is transverse, suspensor absent.	First division of zygote is vertical, merely coincident with the long axis of archegonium, suspensor absent.	First division of zygote is vertical nearly coincident with the long axis of archegonium, suspensor absent.

Question Bank

(Pteridophytes)

N.B. *Undermentioned questions are asked repeatedly in various Indian Universities and competitive examinations.*

Long Answer Type

1. Draw the life-cycles of a homosporous and a heterosporus pteridophyte. How do they differ? Which is more evolved and why?
2. Define the term stele' By means of labelled diagrammatic sketches only, depict the origin of leaf trace and structure of the various types of stele in pteridophytes. Arrange your illustrations in a evolutionary sequence.
3. Give in a tabular form, a comparative account of the spore producing organs of *Lycopodium, Selaginella* and *Equisetum.* Comment briefly upon features of special interest in them.
4. Give an illustrated account of the development of the male and the female gametophyte of *Selaginella* and state the relation between heterospory and the origin of seed habit in plants.
5. Describe briefly the various types of steles met within pteridophytes studied by you; add a note on their evolution.
6. With the help of a labelled diagram, describe the strobilus of *Equisetum* and how it differs from the cone of *Lycopodium*.
7. Draw a vertical section of the strobilus of *Selaginella* and name the parts. What are the main differences between the strobili of *Lycopodium* and *Selaginella.*
8. What is heterospory? Illustrate your answer with reference to life-history of *Selaginella.* Discuss the significance.
9. Where will you search for the plant of *Equisetum* in nature? Give a detailed account of the structure of its cone.
10. Describe the structure of sporocarp of *Marsilea* and write a short note on its morphological nature.
11. Describe the various types of steles found in pteridophytes you have studied.
12. Describe the male and female gametophytes of *Lycopodium.*
13. Write an essay on heterospory and seed habit in pteridophytes
14. Describe the internal structure of *Lycopodium* stem giving different types of steles found in it.
15. Give a brief account of the life-history of *Selaginella.*
16. Describe the male and female gametophytes of *Marsilea.*
17. Give an illustrated account of arrangement of various types of spores in sporocarp of *Marsilea.*
18. Describe the significant features in the life-history of *Lycopodium.*
19. What is alternation of generations? Explain it in the life-history of *Equisetum.*
20. Describe the significant features in the life-history of *Pteris* (fern).
21. Define the stele and describe the different types of steles in *Lycopodium* species with the help of neat and labelled diagrams only.
22. Describe the development of male gametophyte of *Selaginella* inside the sporangium.
23. Discuss the morphological nature of sporocarp of *Marsilea.*
24. Describe *Rhynia* on the basis of plant body and vascular structure. From where this genus discovered? How does it differ from *Horneophyton.*
25. What do you think were the essential features of first land plants? How far does *Rhynia* fit in your conception?
26. Give an account of the structure and position of sex-organs of *Lycopodium, Equisetum,* and *Marsilea.* In which genus do you find the most reduced type of antheridium? Give its significance, if any.
27. What is stele? Give an illustrated account of the various types of stelar structures seen in pteridophytes.

28. Describe the stelar system in pteridophytes.
29. Compare the gametophytes of *Selaginella* and *Marsilea.*
30. Give a comparative account of the strobili of *Lycopodium, Selaginella* and *Equisetum.*]
31. With the help of well-labelled diagrams, discuss the anatomy of stem and root of *Equisetum.*
32. What is stele? Give an account of the stelar organization in the pteridophytes studied by you.
33. Describe the male and female gametophytes of *Selaginella.*
34. Compare the spore producing organs of *Selaginella* and *Equisetum.*
35. Describe the morphology of *Marsilea* sporocarp. How does it differ from a true seed.
36. Compare the spore producing organs of *Selaginella* and *Lycopodium.*
37. Give a brief account of the salient features of Psilophytales and discuss the systematic position of *Rhynia.*

    OR

    Describe the stem structure and reproduction in *Rhynia.*
38. Compare structures of stem of *Equisetum*, *Selaginella* and *Marsilea* lying emphasis on their stelar system.
39. Describe the different types of steles present in various members of pteridophytes studied by you.
40. How many types of gametophytes found in *Lycopodium*? Illustrate your answer with suitable diagrams.

Diagrammatic Type

1. Give labelled diagrams only to show the main structural features of the following : (i) T.S. *Pteris* rhizome, (ii) T.S. *Marsilea* rhizome, (iii) T.S. *Equisetum* rhizome.
2. Draw well labelled diagrams of the following (description not needed): (i) T.S. *Marsilea* petiole, (ii) V.S. of fern leaf having sori.
3. Draw well labelled diagrams of the following: (i) T.S. *Marsilea* rhizome, (ii) T.S. *Selaginella* rhizophore.
4. Draw well-labelled diagrams of the following: (i) L.S. of *Equisetum* cone, (ii) V.S. of fern leaf with sori.
5. By means of labelled diagrams only illustrate the life-history of *Dryopteris* (fern)
6. By means of labelled diagrams only illustrate the life-history of *Marsilea.*
7. By means of labelled diagrams only illustrate the life-history of *Equisetum.*
8. With the help of neat and labelled diagrams only, show the internal structure of the *Marsilea* sporocarp.
9. Draw neat and labelled diagrams of the following : (i) T.S. of *Equisetum, internode stem,* (ii) L.S. of *Selaginella,* cone, (iii) V.S. of fern sporophyll.
10. With the help of neat and labelled diagrams only, show the different types of steles in pteridophytes.
11. Draw well labelled diagrams of the following: (i) T.S. of *Selaginella* stem, (ii) T.S. of *Marsilea* rhizome.
12. Differentiate the following with the help of neat and labelled diagrams only: (i) T.S. of stem of *Lycopodium clavatum* and *Lycopodium cernuum.* (ii) T.S. of microsporophyll of *Selaginella* and megasporophyll of *Selaginella,* (iii) T.S. of internode of *Equisetum* and node of *Equisetum.*
13. Draw well-labelled diagrams of longitudinal sections of strobili of *Equisetum, Lycopodium* and *Selaginella.* (Description not needed.)
14. Give well-labelled diagrams of any two of the following: (i) cone of *Equisetum,* (ii) rhizome of *Marsilea,* (iii) stem of *Lycopodium.*
15. Draw suitable diagrams of the following: (i) antherozoid of *Selaginella,* (ii) archegonia of fern, (iii) sporophyll of *Lycopodium.*
16. Illustrate the life-history of *Lycopodium* with the help of labelled diagrams only.
17. Draw well-labelled diagrams of the following: (i) T.S. *Selaginella* stem, (ii) T.S. *Marsilea rhizome,* (iii) T.S. fern petiole.

18. Draw labelled diagrams of the following: (i) T.S. *Equisetum* petiole, (ii) T.S. *Marsilea* petiole, (iii) V.S. leaf *pteris* passing through sori.
19. Draw labelled diagrams of T.S. stem of *Selaginella* and *Lycopodium.*
20. Draw only labelled diagrams of the following: (i) T.S. petiole of *Marsilea* (ii) T.S. stem of *Lycopodium.*
21. Illustrate the life-cycle of *Equisetum* with the help of necessary diagrams only.
22. Draw neat and labelled diagrams only of the following : (i) T.S.of *Equisetum* rhizome, (ii) T.S. of *Marsilea* petiole, (iii) T.S. of *Lycopodium* stem, (iv) T.S. of *Selaginella* stem.
23. Give labelled sketches of any three of the following : (i) L.S. gametophyte of *Lycopodium*, (ii) T.S. internode of *Equisetum* stem, (iii) L.S. strobilus of *Selaginella,* (iv) T.S. petiole of *Marsilea.*
24. With labelled diagrams only, illustrate the structure of spore bearing parts of *Lycopodium* and *Equisetum.*
25. With the help of labelled diagrams only, give the details of the following: (i) *Rhynia,* (ii) TLS and VLS of sporocarp of *Marsilea,* (iii) T.S. stem of *Selaginella.*

Short Answer Type

1. Give the classification, geological distribution and characteristic features of *Rhynia.*
2. Explain how the complex type of stele evolved from the simplest type. Give an example for each.
3. What is the essence of the Telome concept?
4. Enumerate the views on the morphology of the sporocarp of *Marsilea.*
5. What is the significance of heterospory?
6. Write note on 'morphological nature of the rhizophore in *Selaginella.*
7. Write short note on *Rhynia?*
8. What are the external features by which you will identify the sporophytes of *Selaginella, Equisetum,* and *Marsilea*?
9. Discuss in brief the morphology of the fertile spike of *Lycopodium.*
10. Describe how elaters are produced in *Equisetum.*
11. What are the xerophytic characters exhibited morphologically and anatomically by *Equisetum*?
12. Discuss in brief the morphology of the sporangiophore of *Equisetum.*
13. Write detailed note on stelar structure in *Lycopodium.*
14. Distinguish the rhizophores from that of roots in *Selaginella.*
15. Enumerate the root like characters of rhizophores of *Selaginella.*
16. How can the strobilus of *Lycopodium* be distinguished from that of *Selaginella*?
17. Explain the leaf arrangement in *Selaginella.*
18. Explain the stages of development of megasporangium which makes it different from microsporangium in *Selaginella.*
19. Mention any two root characteristics and any two stem characteristics of rhizophore of *Selaginella.*
20. Comment briefly on the morphological nature of the protocorm of *Lycopodium.*
21. Write notes on the gametophytes of *Lycopodium.*
22. Write short notes on : (i) amphiphloic siphonostele, (ii) false indusium, (iii) trabeculae.
23. Write short notes on : (i) heterospory, (ii) cone of *Equisetum.*
24. Write notes on : (i) *Equisetum* cone, (ii) *Marsilea* sporocarp.
25. Write short notes on : (i)cone of *Lycopodium,* (ii) rhizophore of *Selaginella.*
26. Write short notes on any three: (i) sorus, (ii) siphonostele, (iii) seed habit in *Selaginella,* (iv) trabeculae.
27. How would you distinguish between the following : (i) rhizophore and root, (ii) true indusium and false indusium (iii) *Equisetum* rhizome and *Marsilea* rhizome.
28. Write detailed notes on any two of the following : (i) prothallus of *Lycopodium,* (ii) protocorm of *Lycopodium,* (iii) strobilus of *Lycopodium.*

29. Write short notes on : (i) seed habit in *Selaginella,* (ii) *Equisetum* cone.
30. Write short notes on any four : (i) abaxial, (ii) basal, (iii) axillary, (iv) adaxial.
31. Write notes on: (i) telome, (ii) protostele (iii) heterospory, (iv) *Rhynia.*
32. Write notes on any four: (i) trabeculae (ii) sorus, (iii) spike of *Selaginella,* (iv) annulus, (v) fossil.
33. Write notes on : (i) meristele, (ii) heterospory, (iii) ligule, (iv) spore of *Equisetum.*
34. Write notes on : (i) *Rhynia,* (ii) cone of *Equisetum,* (iii) sporocarp.
35. Write short notes on: (i) *Rhynia,* (ii) rhizophore, (iii) rhizoids.
36. Distinguish between : (i) homospory and heterospory, (ii) apogamy and apospory, (iii) leaf traces and leaf gaps.
37. Distinguish between the following : (i) homospory and heterospory, (ii) sporophyll and sporangium, (iii) spore mother cell and spore as far as their chromosome number is concerned, (iv) strobilus and sori.
38. Explain the following : (i) carinal canal, (ii) vallecular canal, (iii) sporocarp, (iv) ligule, (v) annulus.
39. Distinguish between the following : (i) homospory and heterospory, (ii) apogamy and apospory, (iii) leaf trace and leaf gap, (iv) protophylls and protocorm. (v) ectophloic and amphiphloic siphonostele.
40. Write short notes on any three: (a) protostele, (b) dictyostele, (c) siphonostele, (d) heterospory.
41. Write brief and illustrated notes on: (i) protocorm of *Lycopodium*, (b) gametophyte of *Equisetum,* (c) ligule of *Selaginella* and its morphological nature, (d) sporocarp of *Marsilea.*
42. Write short notes on any three: (i) fossil types, (ii) rhizophore in *Selaginella*, (iii) nature of stele in *Marsilea* stem, (iv) spore of *Equisetum*, (v) leptosporangiate development.
43. Point out the differences in between: (i) T.S. of rhizome and petiole of *Marsilea*, (ii) T.S. of rhizome and stem of *Equisetum*, (iii) give the botanical names of atleast three species of each of the four representatives of pteridophytes studied by you.
44. Write short notes on any three : (a) heterospory, (b) sporangium of *Rhynia,* (c) rhizophore, (d) dehiscence of sporocarp of *Marsilea.*
45. Point out the differences in between : (a) T.S. of rhizome and petiole of *Marsilea,* (ii) T.S. of rhizome and stem of *Equisetum,* (c) T.S. of root and rhizophore of *Selaginella.*
46. Write short notes on any three : (a) heterospory; (b) ligule of *Selaginella;* (c) morphological nature of *Marsilea* sporocarp; (d) dehiscence of sporangium in *Equisetum;* (e) protocorm of *Lycopodium*.
47. Discuss any two of the following: (i) *Equisetum* has both hydrophytic and xerophytic characters; (ii) protocorm; (iii) morphological and anatomical differences occurring between the plants of *Marsilea* growing in water and land surface.

Multiple Choice Type

1. The living example of near telome is : (i) *Marsilea,* (ii) *Adiantum,* (iii) *Osmunda,* (iv) *Psilotum.*
2. Mixed protostele is found in : (i) *Selaginella,* (ii) *Lycopodium,* (iii) *Equisetum,* (iv) *Adiantum.*
3. Simple collateral vascular bundles are found in: (i) *Equisetum,* (ii) *Lycopodium,* (iii) *Pleopeltis,* (iv) *Selaginella.*
4. Common element of xylem in pteridophytes is : (i) tracheae, (ii) tracheids, (iii) sclereids, (iv) vessels.
5. Spores of pteridophytes are : (i) haploid, (ii) diploid, (iii) triploid, (iv) tetraploid.
6. Ligule and cone are present in : (i) *Selaginella*, (ii) *Lycopodium,* (iii) *Isoetes,* (iv) *Psilotum.*
7. The sporangia are developed in *Equisetum* on : (i) upper surface of sporophylls, (ii) sporangiophores, (iii) vegetative leaves, (iv) lower surface of sporophylls.
8. In the root of *Equisetum* the endodermis is : (i) absent, (ii) one layered, (iii) two layered (iv) three layered.
9. Vallecular canal in *Equisetum* is situated : (i) below the ridges, (ii) below the furrows, (iii) between ridges and furrows, (iv) between pith and endodermis.

10. A cavity formed by the disintegration of protoxylem tracheids in *Equisetum* stem is called : (i) vallecular canal, (ii) carinal canal, (iii) central cavity, (iv) air chamber.
11. In *Equisetum* sporangia are born in: (i) microsporophylls, (ii) megasporophylls, (iii) sporangiophore, (iv) fertile leaves.
12. Presence of air cavities in the stem of *Equisetum* indicate: (i) Xerophytic feature, (ii) hydrophytic feature, (iii) mesophytic feature, (iv) halophytic feature.
13. *Equisetum* is commonly called : (i) club moss, (ii) horse tail, (iii) quill wort, (iv) stone wort.
14. Intercalary meristem is seen in : (i) *Ophioglossum*, (ii) *Lycopodium*, (iii) *Psilotum*, (iv) *Equisetum*.
15. *Equisetum* is: (i) incipiently heterosporous, (ii) homosporous, (iii) heterosporous, (iv) aposporous
16. The stale of *Ophioglossum* is :(i) dictyostele, (ii) polystele, (iii) protostele, (iv) solenostele.
17. The sorus of *Pteris* is: (i) simple type (ii) mixed type, (iii) gradate type, (iv) none of above.
18. The sporangia of *Ophioglossum* are seen in the form of a : (i) strobilus, (ii) sporophyll, (iii) spike, (iv) tassel.
19. Leaf gaps and leaf traces are seen in the rhizome of : (i) *Psilotum*, (ii) *Ophioglossum*, (iii) *Pteris*, (iv) *Marsilea*.
20. Amphiphloic siphonostele is found in: (i) *Psilotum*, (ii) *Ophioglossum*, (iii) *Pteris*, (iv) *Marsilea*.
21. Stele in *Pteris* rhizome is : (i) dictyostele, (ii) polystelic, (iii) polycyclic, (iv) meristelic.
22. Gradate sorus is found in: (i) *Hymenophyllum*, (ii) *Marsilea*, (iii) *Pteris*, (iv) *Glychenia*.
23. Stele in the stem of *Salaginella* is: (i) endarch, (ii) exarch, (iii) mesarch.
24. Stele in *Lycopodium* is : (i) dictyostelic, (ii) protostelic, (iii) solenostelic (iv) siphonostelic.
25. *Rhynia* was discovered by : (i) Arnold, (ii) Kidston and Lang, (iii) Birbal Sahni, (iv) Campbell.
26. *Rhynia* belongs to a geological period called: (i) Carboniferous, (ii) Permean, (iii) Devonian, (iv) Cambrian.
27. In *Lycopodium* and *Selaginella* though the sporangia are developed on the adaxial surface of the sporophylls, they can be distinguished by: (i) different types of spores, (ii) nature of the sporophyll, (iii) pattern of vasculature, (iv) arrangement of sporangia.
28. The stele in the stem of *Selaginella* is: (i) protostele, (ii) solenostele, (iii) polystele, (iv) dictyostele.
29. The male gametes of *Selaginella* are: (i) biflagellate, (ii) multiflagellate, (iii) monoflagellate, (iv) aflagellate.
30. On which side is the sporangium seen on the sporophyll in *Lycopodium:* (i) abaxial, (ii) basal, (iii) axillary, (iv) adaxial.
31. Wall of sporangium in *Lycopodium* usually consists of : (i) single layer, (ii) two layers, (iii) three layers, (iv) many layers.
32. An adaxial outgrowth from the base of the leaves in *Selaginella* is called: (i) stipule, (ii) ligule, (iii) trabeculae, (iv) velum.
33. *Lycopodium* is commonly called as: (i) horse tail, (ii) club moss, (iii) quill wort, (iv) stone wort.
34. Trabeculae in *Selaginella* is modified: (i) pericycle cells, (ii) endodermal cells, (iii) pericycle and endodermal cells, (iv) cortical cells.
35. Trabeculated endodermis is seen in: (i) *Psilotum*, (ii) *Gleichenia*, (iii) *Ophioglossum*, (iv) *Selaginella*.
36. Double layer of endodermis is found in the root of : (i) *Pteris*, (ii) *Equisetum*, (iii) *Selaginella*, (iv) *Hymenophyllum*.
37. The rhizophore of *Selaginella* has : (i) triarch stele, (ii) monarch stele, (iii) diarch stele, (iv) tetrarch stele.
38. Rhizophore of *Selaginella* is like a : (i) root, (ii) organ suigeneris, (iii) stem, (iv) part of leaf.
39. A fern differs from a moss in having: (i) an independent gametophyte, (ii) an independent sporophyte, (iii) sperms, (iv) archegonia.
40. Fern gametophyte is : (i) homothallic, (ii) heterothallic, (iii) heteroecious, (iv) autoecious.
41. Formation of a sporophyte from a vegetative portion of prothallus is called: (i) apocarpy, (ii) apogamy, (iii) apomixis, (iv) apospory.

42. Young sporophyte of fern draws nourishment from the prothallus through: (i) haustoria, (ii) foot, (iii) root, (iv) rhizoids.
43. *Cycas* and *Pteris* resemble each other in the presence of: (i) cambium, (ii) ciliated sperm, (iii) vessels, (iv) seeds.
44. Pteridophytes differ from bryophytes in having: (i) motile sperm, (ii) alternation of generations, (iii) archegonia, (iv) vascular tissue.
45. *Cycas* differ from *Pteris* in having: (i) tracheids, (ii) motile sperms, (iii) pollen tubes, (iv) archegonia.
46. In *Dryopteris* the antherozoids are: (i) biciliate and coiled, (ii) biciliate and sickle shaped, (iii) multiciliate and coiled, (iv) multiciliate and sickle shaped.
47. A fern differs from a moss in having: (i) alternation of generations, (ii) swimming sperms, (iii) independent gametophyte, (iv) independent sporophyte.
48. Prof. Birbal Sahni is famous for his researches in the field of : (i) palaeobotany, (ii) physiology, (iii) cytology, (iv) ecology.
49. In ferns the xylem is: (i) exarch, (ii) mesarch, (iii) endarch, (iv) polyarch.
50. Which of the following conditions found in *Cycas* is also found in *Pteris*: (i) naked ovules, (ii) circinate vernation of leaves, (iii) presence of coralloid roots, (iv) presence of secondary growth.
51. The vascular bundle in which the xylem is surrounded by phloem is known as: (i) radial, (ii) collateral, (iii) amphicribral concentric, (iv) conjoint.
52. The spores of the fern plant germinate to form: (i) thick-walled zygospore, (ii) an embryo, (iii) the new fern plant, (iv) a prothallus.
53. The stele in the rachis of *Dryopteris* is : (i) protostelic, (ii) siphonostelic, (iii) dictyostelic, (iv) haplostelic.
54. The sori of *Pteris* are: (i) discontinous and circular, (ii) continuous and linear, (iii) discontinuous and vermiform, (iv) discontinuous and reniform.

ANSWERS

1. (iv), 2. (ii), 3. (i), 4. (ii), 5. (i), 6. (i), 7. (ii), 8. (iii), 9. (ii), 10. (ii), 11. (iii), 12. (ii), 13. (ii), 14. (iv), 15. (ii), 16. (i), 17. (i), 18. (iii), 19. (iii), 20. (iv), 21. (i), 22. (ii), 23. (iii), 24. (ii), 25. (ii), 26. (iii), 27. (i), 28. (i), 29.(i), 30. (iv), 31.(ii), 32. (ii), 33. (ii), 34. (ii), 35. (iv), 36. (ii), 37. (ii), 38. (i), 39. (ii), 40. (i), 41. (ii), 42. (ii), 43. (ii), 44. (iv), 45. (iii), 46. (iii), 47. (iv), 48. (i), 49. (ii), 50. (ii), 51. (iii), 52. (iv), 53. (iii), 54. (i),

UNIT - 6

Cell Biology

33

CHAPTER

The Cell: Structure and its Components

The plant body consists of a number of small, microscopic, box-like compartments called cells. In other words, **the cells are the universal elementary units of organic structure.** A plant cell may be defined as microcosm having a definite boundary or the cell-wall within which complicated chemical reactions are going on. A cell devoid of this chemical reaction is inert and is considered to be a dead cell.

THE CELL THEORY

Schleiden (1838) a German botanist and Schwann (1839) a German zoologist sponsored a theory that all animal and plant organisms are composed of cells.

COURSE - II

Cell Biology and Genetics

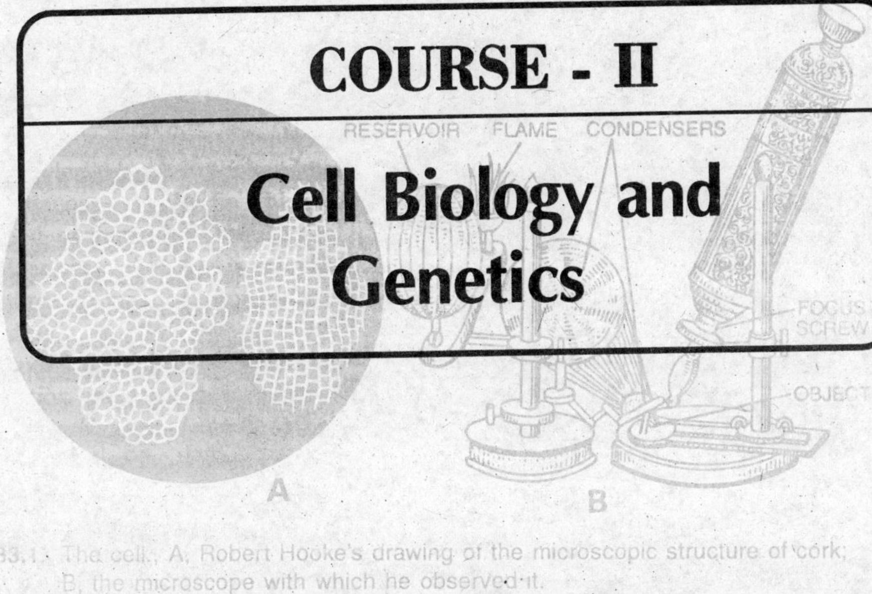

Fig. 33.1. The cell. A, Robert Hooke's drawing of the microscopic structure of cork; B, the microscope with which he observed it.

Schleiden (1838) and Schwann (1839) for the first time used the term **Cell Theory** and stated **"The cells are organisms, and animals as well as plants are aggregates of these organisms, arranged in accordance with definite law".** Schwann established the cell theory in a definite form. He had clear ideas both of morphological and physiological significance of the cell. According to him the cellular phenomena consists of two groups—(1) the plastic phenomena and (2) the **metabolic phenomena.** The former one corresponds to morphology of cell and also includes in it "the combination of molecules which form the cell". On the other hand the metabolic phenomena (physiologic phenomena) includes **"the chemical changes, whether in the particles constituting the cell itself, or in the surrounding "cytoblastema".** Thus, about a century back Schwann expressed the present view point of cytology and, therefore, he is generally considered as 'father of modern cytology'.

UNIT - 6
Cell Biology

33
CHAPTER

The Cell: Structure and its Components

The plant body consists of a number of small, microscopic, box-like compartments called **cells.** In other words, **the cells are the universal elementary units of organic structure.** A plant cell may be defined as microcosm having a definite boundary or the cell-wall within which complicated chemical reactions are going on. A cell devoid of this chemical reaction is inert and is considered to be a dead cell.

THE CELL THEORY

Schleiden (1838) a German botanist and Schwann (1839) a German zoologist sponsored a theory that all animal and plant organisms are composed of cells.

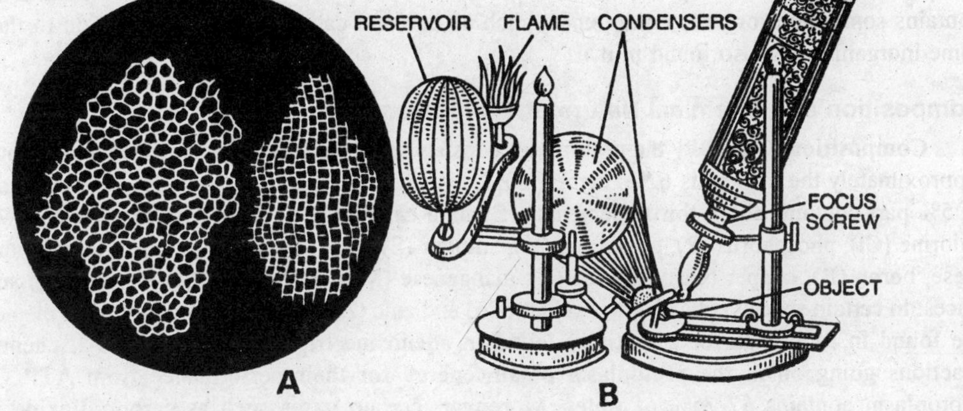

Fig. 33.1. The cell., A, Robert Hooke's drawing of the microscopic structure of cork; B, the microscope with which he observed it.

Schleiden (1838) and Schwann (1839) for the first time used the term **Cell Theory** and stated, **"The cells are organisms, and animals as well as plants are aggregates of these organisms, arranged in accordance with definite law".** Schwann established the cell theory in a definite form. He had clear ideas both of morphological and physiological significance of the cell. According to him the cellular phenomena consists of two groups—(1) the plastic phenomena and (2) the **metabolic phenomena.** The former one corresponds to morphology of cell and also includes in it "the combination of molecules which form the cell". On the other hand the metabolic phenomena (physiologic phenomena) includes **"the chemical changes, whether in the particles constituting the cell itself, or in the surrounding "cytoblastema".** Thus, about a century back Schwann expressed the present view point of cytology and, therefore, he is generally considered as 'father of modern cytology'.

The cell theory postulates, in brief that the cells from the structural units of living organisms and the latter show a characteristic cellular construction. In addition, they are also functional units in that all the vital functions of the living organism are traceable ultimately to the living substance, the protoplasm contained in them.

THE PROTOPLASM

According to Huxley **the protoplasm is the physical basis of life.** Inside the cell wall of living cell the living substance is known as *protoplasm*. The protoplasm is a thick fluid or jelly-like substance. These two forms are not much differentiable from each other and may easily be transformed from one form to another. When found fluid state it seems to be more active. Generally it is greyish or somewhat yellowish in colour. It is always transparent. Many small granules found in it are food granules. It is a very complex substance and found to be dispersed in the medium of water. It consists of ninety percent of water. The protoplasm found in the cells of the seeds contains less percentage of water, and therefore, it is of thick consistency. In such cases the protoplasm is somewhat inactive and becomes active only when sufficient amount of water is absorbed. In addition to water other substances are also found in the protoplasm. Among these substances the most important ones are proteins which are found in large quantity. The proteins remain dispersed in the water. The proteins are highly complex chemical compounds. The protein molecules are very large and of very high molecular weight. These complex compounds always contain carbon, hydrogen, oxygen and nitrogen. In addition to these elements sulphur and phosphorus are generally found. The proteins of protoplasm always contain sulphur and phosphorus. The protoplasm contains the proteins of various types. The protoplasm also contains some other organic compounds, such as fats and carbohydrates. In addition to these, some inorganic salts also found in it.

Composition and Chemical Nature of Protoplasm

Composition. Generally the protoplasm consists of oxygen, carbon, hydrogen and nitrogen. Approximately the oxygen is 62%, carbon 20%, hydrogen 10% and nitrogen 3%. The remainder of 5% part contains about thirty elements, of which calcium (Ca), iron (Fe), magnesium (Mg), chlorine (Cl), phosphorus (P), potassium (K), sulphur (S), etc., are important ones. In addition to these, boron (B), copper (Cu), fluorine (F), manganese (Mn) and silicon (Si) are found in small traces. In certain special cells alcohol, cobalt (Co) and zinc (Zn) are also found. All these elements are found in ionic state or essentially found in adenosine triphosphate (ATP). All chemical reactions going on in the protoplasm obtain energy for their performance from ATP. The protoplasm contains 67-75% of water. Moreover, certain gases such as carbon-dioxide and oxygen remain dissolved in it.

The protoplasm of each cell contains several organic substances of which **carbohydrates, fats, proteins,** and **nucleoproteins** are important ones. These organic substances make protoplasm by molecular combination.

Carbohydrates. About thirteen percent part of protoplasm consists of carbohydrates. The carbohydrates contain carbon, hydrogen and oxygen. The important carbohydrates are—glucose, sucrose, starch, cellulose, glycogen, etc. The granules of carbohydrates either remain suspended or dissolved in the cytoplasm. They are mainly responsible for the production of energy.

Fats. The fats or lipids consist of carbon, hydrogen and oxygen elements. They are formed by the combination of glycerol and fatty acids. They contain lesser amount of oxygen than carbohydrates. When they are chemically decomposed the energy is being liberated. They contain much more energy in comparison to carbohydrates. The cell membrane consists of fat.

Proteins. About fifteen per cent of the protoplasm consists of proteins. In addition to carbon, hydrogen and oxygen elements, the proteins essentially contain nitrogen. Usually they contain sulphur and sometimes phosphorus too. The proteins are formed by the combination of the molecules of amino-acids. About twenty amino-acids are found in the nature, that give rise to different kinds of proteins by their molecular combination. A protein molecule is made of hundreds or thousands of amino-acid molecules joined together by peptide links into one or more chains, which are variously folded. There are 20 different kinds of amino-acids commonly found in proteins, and most of these usually occur in any one protein molecule; they are arranged in the chain in a sequence which is exactly the same in all molecules of a given kind of protein. The possible different arrangement of the amino-acids are evidently practically infinite, and the diversity is fully exploited by living things, every species having kinds of protein molecule peculiar to itself. A protein molecule is very large (molecular weight from about 20,000 up to several millions), and dissolved proteins form therefore colloidal solutions. Proteins are not soluble in fat solvents. Many are soluble in water or dilute salt solutions (*e.g.*, globulins); others, with elongated (fibrous) molecules, are insoluble in these solvents (*e.g.*, scleroproteins, myosin). Proteins are synthesized from amino-acids by all living things; the precise sequence of the amino-acids being determined by the sequence of nucleotides in nucleic acids. The proteins are destroyed by proteolytic enzymes. They are frequently combined with other substances, especially nucleic acids (nucleo-proteins), carbohydrates (glycoproteins), fats (lipoproteins).

Nucleoproteins. They are most complex substances ever found. They are compounds of nucleic acid and protein. The protoplasm contains two types of nucleic acids—ribonucleic acid (RNA) and deoxyribonucleic acid (DNA). The RNA is found in the complete cell, whereas DNA remains confirmed in the nucleus. Ribonucleic acid (RNA), a molecule consisting of a large number of nucleotides attached together to form a long strand one nucleotide thick. Deoxyribonucleic acid (DNA) is mainly found in the genes of chromosomes. The DNA, a compound consisting of large number of nucleotides attached together in a single file to form a long strand. Usually two such strands are linked together parallel to each other by base pairing, and coiled into a helix. In each cell the DNA and RNA are mainly concerned with the metabolic activities. Numerous RNA and DNA are formed within the living cells. The nucleic acids of two different living beings are never identical. The RNA and DNA control all the metabolic activities going on within the living cells. They play an important role in the origin of life. DNA is the material of inheritance of almost all living beings.

Other chemical substances. Besides abovementioned organic substances several other inorganic substances are also found in the living cells in small quantities. These substances are specially concerned with the cell metabolism. Some of these substances are found in all cells whereas some are confined to certain special cells. Various pigments, latex, alkaloids, vitamins, hormones, antibiotics and some other substances are found within the plant cells.

Thus, the carbohydrates, fats, proteins, nucleoproteins and several other chemical substances make the protoplasm by their molecular combination. The molecules of all these substances are well organized and form protoplasm. In addition to these the molecules of living cells possess special characteristics by which various chemical reactions take place and with the result the energy liberates. This energy is utilized in the performance of the metabolic activities of the cells. Thus, this becomes an established fact that all constituents of the cell form a very powerful organization by their combination, which is living and quite active.

Physical Nature of Protoplasm

Colloidal state. The proteins of protoplasm are found in water in colloidal state. The proteins are formed in the form of very minute granules. Each granule consists of several molecules. Generally the protoplasm is an amulsoid, in which the proteins remain dispersed in the water. Many colloidal particles possess the power of holding and absorption of water. This is called **hydration** and the colloids which absorbed the water are said to be **hydrated.** The colloids of protoplasm remain hydrated, and the amount of hydration depends upon various conditions.

Distribution. Generally the protoplasm remains divided into two parts—**the nucleus** and the **cytoplasm.** Usually the young cells remain filled up with the protoplasm. In mature cells, the cytoplasm forms the inner layer of the cell wall and a layer around the nucleus. The nucleus remains embedded in this layer of cytoplasm. All the protoplasm of a cell excluding the nucleus is known as **cytoplasm.** The cytoplasm is usually a transparent slightly viscous fluid with inclusions of various sizes.

Permeability. Permeability of membrane, exhibits to which molecules of a given kind can pass through it. Substances differ greatly in the case with which they pass through a given membrane, *e.g.*, water or fatty substances pass easily, many ions and proteins with great difficulty, through plasma membrane. Permeability on any biological membrane is itself variable within wide limits. The very numerous membranes of animals and plants, *e.g.*, plasma membranes, produce many osmotic effects and are of great importance in determining local differences in composition of the organism. According to Virgin (1953), the permeability is the quality of a living membrane through which different ions and molecules may pass, and during this process the energy obtained from metabolic activities is not utilized. This is the permeability of the protoplasm which is dependable upon the physiochemical properties of the protoplasmic membrane of the cell. Virgin has also propounded that the permeability of the protoplasm is the control of active and passive forces on the particles passing through the living protoplasmic membrane. The cellulose cell wall is permeable for water and other solutes, whereas the protoplasmic membrane is permeable only to water and not to other solutes.

Properties of Protoplasm

The protoplasm is a living matter and makes the physical basis of life. On the basis of the properties given below it may be differentiated from non-living matter. The properties are as follows:

Absorption and excretion. The protoplasm takes essential substances for its growth by absorption and it gets rid of from non-required substances by excretion. The protoplasm of green plants absorbs water and minerals from the soil and carbon dioxide from the atmosphere. Thus, the substances quite different from the protoplasm are being absorbed by it.

Metabolism. In general the chemical processes occurring within an organism, or within part of one. They involve breaking down of organic compounds from complex to simple (catabolism) with liberation of energy available for the organism's many activities; and building up of organic compounds from simple to complex (anabolism) using energy liberated by catabolism and in the case of autotropic organisms, energy from external non-organic sources (particularly from sunlight). Catabolism is by no means confined to the breaking down of the 'food material' necessary for the organism's energy requirements. It involves, continuously, very many constituents of the organism, though they are broken down at various rates, and on the other hand anabolism involves, continuously, synthesis of many constituents. These processes go on even in an apparently static tissue. This metabolic whirlpool is a system of reactions predominantly controlled by enzymes. In most of the aspects of metabolism which have been sufficiently investigated, such as catabolism of carbohydrates, there is a fundamental similarity throughout a wide variety of organisms including plants, animals and bacteria.

Growth and reproduction. Growth means increase in size. It includes differentiation of cells, increase in plant size, even though such differentiation contributes nothing to size change. The growth of protoplasm is the result of the separation and growth of such a small part which has been derived from one or parental living beings. In other words, the reproduction is another form of growth.

Movement. The protoplasm has the power of movement and usually it remains in dynamic condition. The dynamic state of the protoplasm of the cell may be seen under the microscope. This phenomenon is known as Brownian movement. The protoplasm does not only possess the property of movement, but it also helps the living being or its organs to remain in dynamic state. Such movements are more clearly seen in animals than in plants, but in certain plants such as *Mimosa pudica*, it is quite obvious. Besides, the young parts of several plants bend towards the light; several unicellular and other types of water plants move like the animals from one place to another such as *Chlamydomonas, Volvox*, etc.

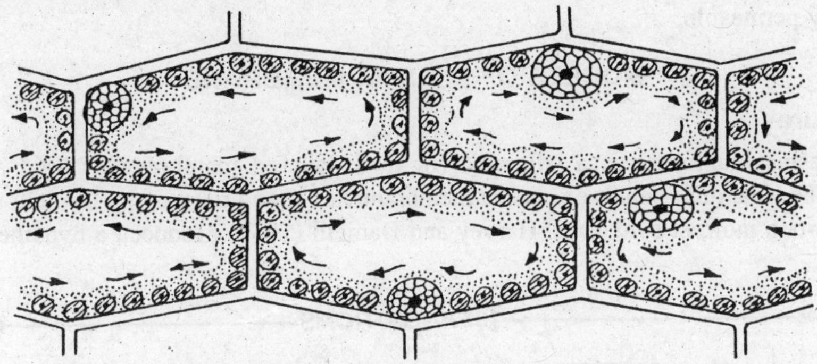

Fig. 33.2. Movement of protoplasm. Rotation in the leaf of *Vallisneria*.

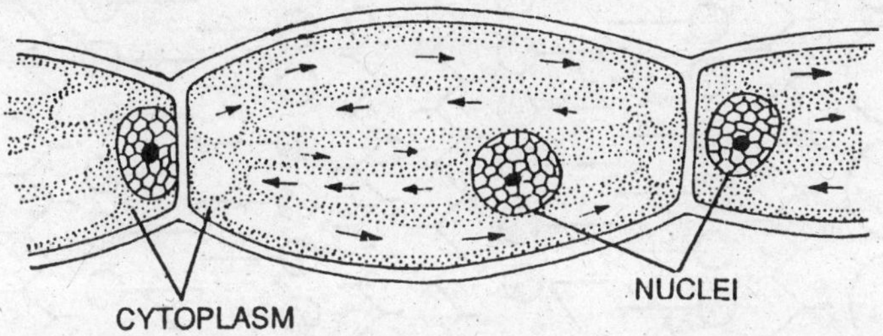

Fig. 33.3. Movement of protoplasm. Circulation in the staminal hair of *Commelina*.

Irritability. This is a universal property of living things (protoplasm). The responsiveness of protoplasm to outer stimulus is known as irritability. Temperature and several chemicals have definite effect upon the movement of the protoplasm. At ordinary temperature the rate of movement increases whereas at low temperature it decreases. The effect of certain chemicals increases the rate of movement while the effect of certain other chemicals decreases it. With the result of the stimulus upon the protoplasm, a complete organ of the plant may be affected, *e.g.*, *Mimosa pudica.*

THE PLASMA MEMBRANE
(Cell Membrane)

All cells remain covered by a **plasma membrane.** The plasma membrane is quite separate and should not be confused with the cell wall which lies outside the cytoplasm. Generally a cell wall is regarded as protoplasmic product and is not believed to be truly living. However, in certain cases the cell wall remains capable of growth.

The plasma membrane is generally considered to be the outer living limit of the cell, but not necessarily its outer boundary. Such outer boundaries are most commonly found in plant cells, many of which possess heavy walls of cellulose, but animal cells and unicellular organisms also exhibit a number of somewhat comparable external substances.

The cell membrane is double in nature and about 10 Å (Å = *Angstrom* unit) thick. This membrane remains connected with the nuclear membrane through the endoplasmic reticulum.

The double nature and molecular make-up of the plasma membrane are depicted in figure. The membrane has a lipoprotein (fat + protein) composition. The protein constituent of the membrane gives the cell wettability and flexibility. Since protein molecules, which are long and complex, can fold or unfold, the membrane can expand or contract, thus providing through molecular spacing a possible means of control over which molecules can enter the cell from the outside environment, or pass to the environment from inside the cell. Such a membrane is known as selectively permeable.

MODELS OF PLASMA MEMBRANE

Ultrastructure

The plasma membrane is a trilaminar or three-layered membrane of lipid and protein molecules. The trilaminar nature of plasma membrane has been proposed by Danielli and Davson (1935) by certain indirect evidences. Harvey and Danielli (1938) produced a hypothetical model

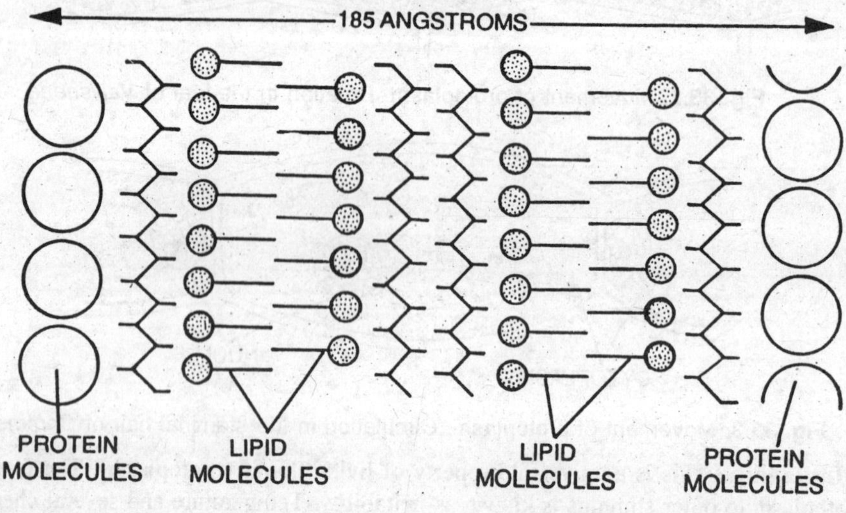

Fig. 33.4. The plasma membrane in molecular terms showing protein and lipid molecules.

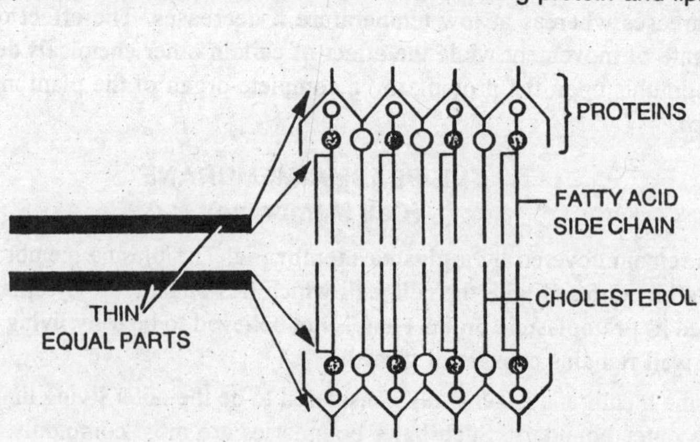

Fg. 33.5. Symmetrical pattern of plasma membrane.

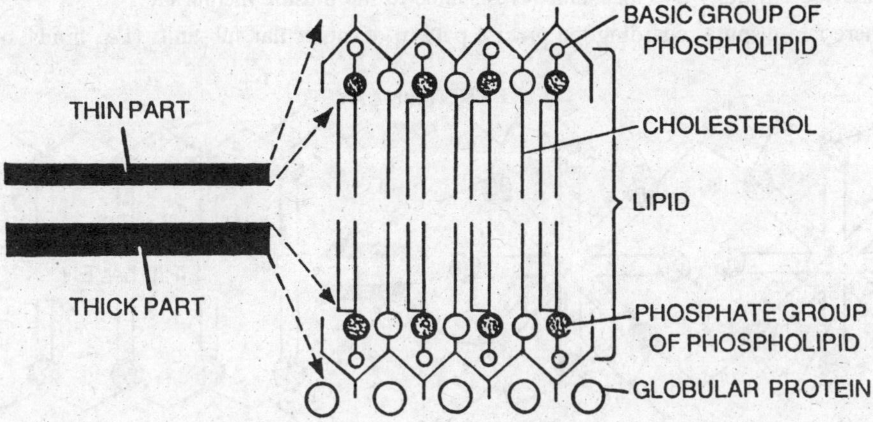

Fig. 33.6. Asymmetrical pattern of plasma membrane.

of the plasma membrane which has shown a bimolecular lipid structure in between two outer and inner layers of protein molecules. The studies by electron microscope have revealed the protein-lipid-protein arrangement in the plasma membrane. Robertson (1959) found wide occurrence of such trilaminar composition in most of the membranes of cellular organelles, and with the result the propounded the concept of the **unit membrane.** A unit membrane possesses the protein-lipid-protein arrangement. The unit membrane has been found in plasma membrane, endoplasmic reticulum, Golgi complex, lysosomes and plastids. It has been found, that both the outer and inner membranes of the mitochondria and the nucleus are the unit membranes.

The plasma membrane of most cells varies from 100 to 215 Å in thickness. The plasma membrane is not a continuous layer but contains certain minute pores of 8 to 15 Å diameter. The opening and closing of these pores depends on the metabolic phase of the cell.

Molecular Structure

The plasma membrane consists of two layers of protein molecules and two layers of lipid molecules. The lipid molecules occur in chains. The two molecular chains of lipids remain parallel to each other and form a bimolecular or double-layered structure. Both lipid layers remain linked with each other by the inner ends of lipid molecules which are non-polar and hydrophobic (Gr. *hydra* = water; *phobe* = hate) in nature. Both the layers of lipids are held together due to Vander waal's forces at these non-polar ends.

The lipid layers remain enclosed by outer and inner layers of proteins. The lipid molecules are linked with the molecules of protein layers by their outer, polar and hydrophilic (Gr. *hydra* = water; *phil* = loving) ends. In the hydrogen bonds, ionic linkages, or electrostatic forces bind the molecules of lipids and proteins together. The carbohydrate molecules are found in

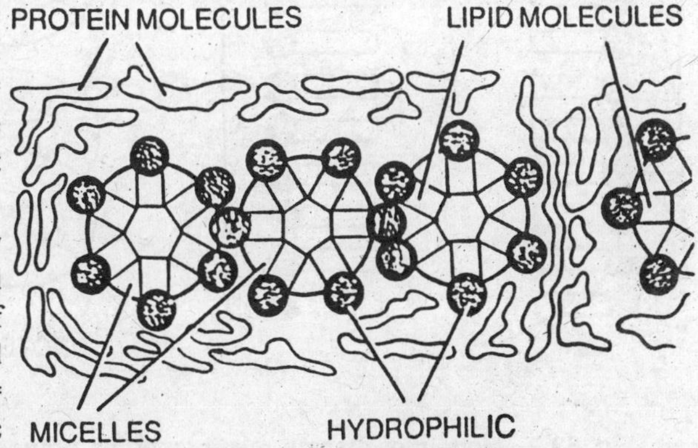

Fig. 33.7. Plasma membrane, based on Micellar theory (diagrammatic).

the association of protein molecules and provide stability to lipo-protein complex. The protein layers provide elasticity and mechanical resistance to the plasma membrane.

There is a dispute regarding the precise pattern of molecular sub-units (*i.e.*, lipids, proteins

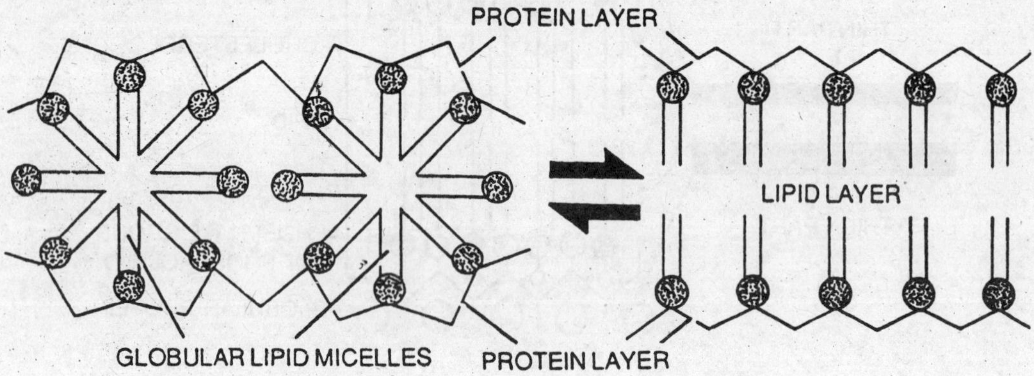

Fig. 33.8. Plasma membrane. Transformation between micellar and lamellar states (globular bilayertransformation).

and carbohydrates) in the plasma membrane. Following three kinds of theories and models, *viz*., **lamellar theory, micellar theory** and **fluid mosaic theory**, have been suggested for the possible molecular structure of plasma membrane.

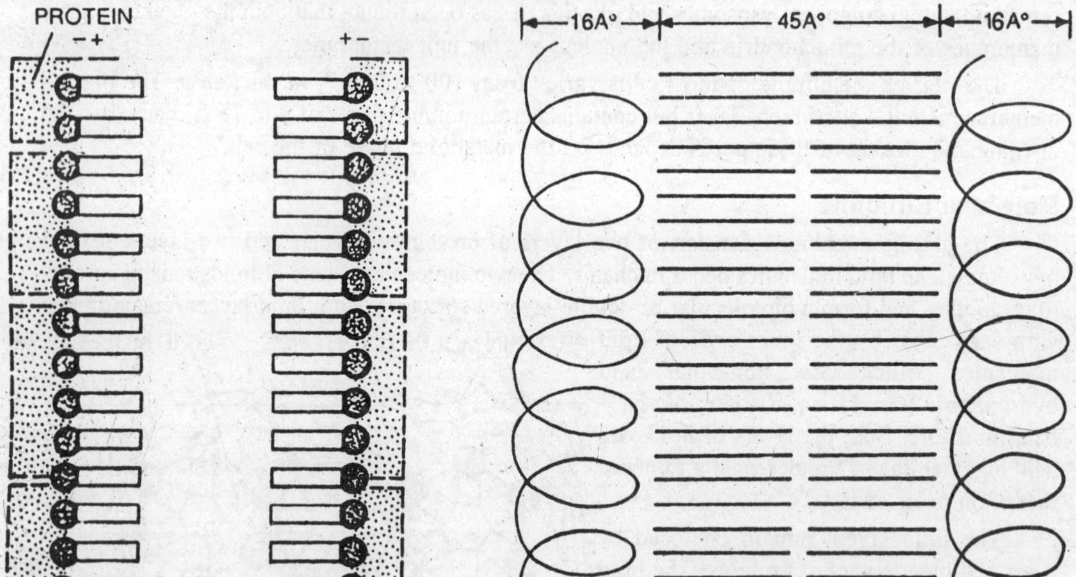

Fig. 33.9. Models of plasma membranes as proposed by (A) Davson and Danielli and (B) Robertson.

MEMBRANE MODELS

To explain the physical and biological features of plasma membranes two main categories have been proposed, *i.e.*, the **bilayer models** and the **micellar** or **sub-unit models**. In the bilayer models the protein and lipid that constitute the membrane are believed to occur in layers, while in the micellar model the membrane is thought to consist of a number of similar units.

Danielli-Davson Model

By studying the surface tension of cells the existence of protein was indicated (Harvey and Cole, 1931; Danielli and Harvey, 1935). This led Davson and Danielli (1935) to propose a lipoprotein model of the cell membrane. According to this model the bimolecular liquid layer consists of two layers of molecules with their polar regions on the outer side. Globular proteins are thought to be associated with the polar groups of the lipid.

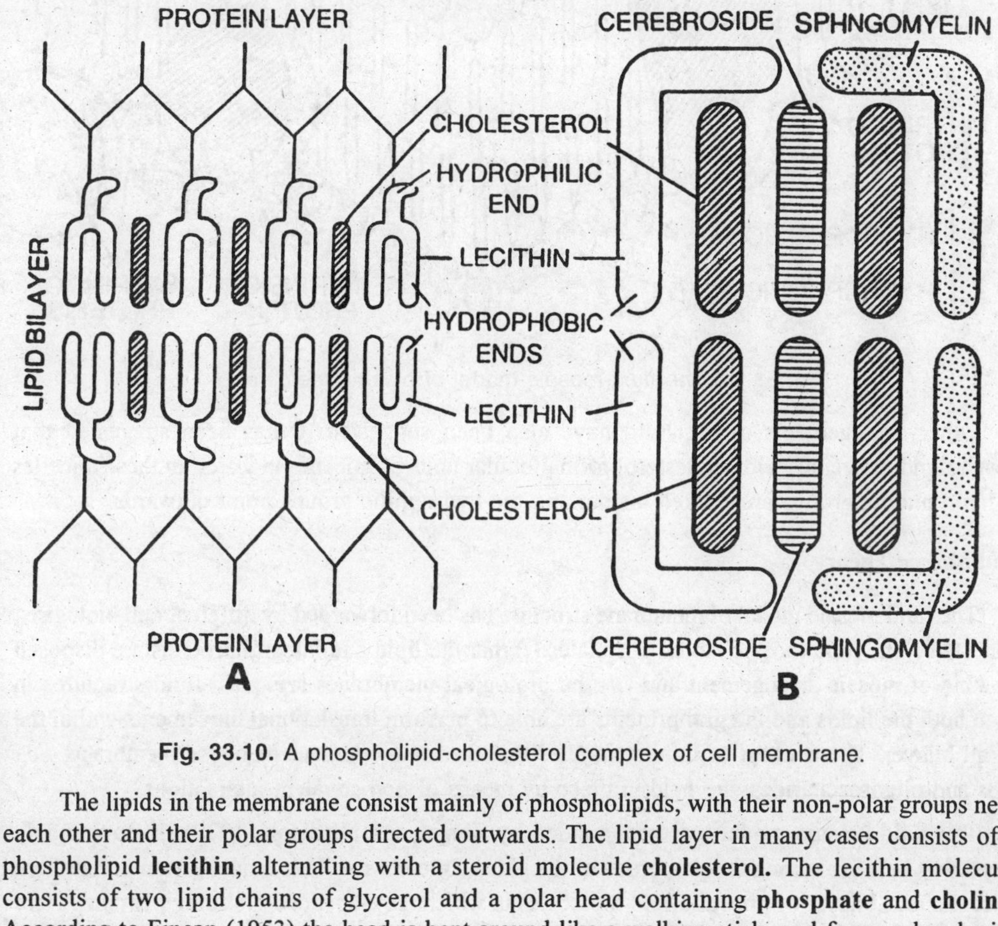

Fig. 33.10. A phospholipid-cholesterol complex of cell membrane.

The lipids in the membrane consist mainly of phospholipids, with their non-polar groups near each other and their polar groups directed outwards. The lipid layer in many cases consists of a phospholipid **lecithin**, alternating with a steroid molecule **cholesterol.** The lecithin molecule consists of two lipid chains of glycerol and a polar head containing **phosphate** and **choline.** According to Finean (1953) the head is bent around like a walking stick, and forms a bond with **cholesterol.**

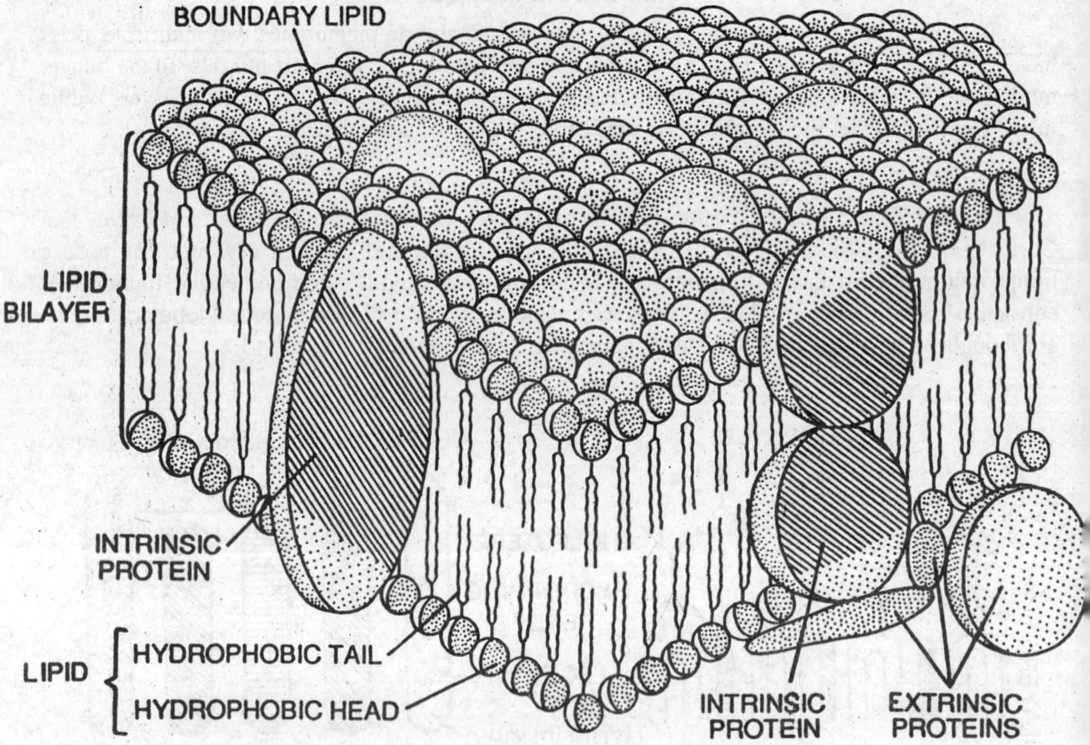

Fig. 33.11. The fluid mosaic model of plasma membrane

Other arrangements of the lipid have also been suggested. It has been suggested that phospholipid molecules with cholesterol form globular units or **micelles** in water. In these micelles the hydrophobic groups are packed inside, and the hydrophilic groups point outwards.

Fluid Mosaic Theory

The fluid mosaic model of membrane structure has been forwarded by different cell biologists, after, 1970. The fluid mosaic model postulates (*i*) that the lipid and integral proteins are disposed in a kind of mosaic arrangement and (*ii*) the biological membranes are quasi-fluid structures in which both the lipids and integral proteins are able to perform translational movements within the overall bilayer. The concept of fluidity implies that the main components of the membrane, *i.e.*, lipids and oligosaccharides, are held in place by means of non-covalent interactions.

The fluid mosaic model was propounded by Singer and Nicolson (1972), is now widely accepted as best explaining the properties of the plasma membrane. This model assumes that there is a continuous bilayer of **phospholipid** molecules in which are embedded **globular proteins**. Here the proteins have been compared to icebergs floating in the sea of the phospholipid bilayer. Thus plasma membranes are considered to be **quasifluid** structures in which lipids and integral proteins are arranged in a **mosaic** manner.

Chemical Nature of Plasma Membrane

The plasma membrane consists of a protein called **stromatin,** which is fibrous in nature and possesses a high molecular weight. These are also glycoproteins and mucoproteins. It also contains large amount of arginine and lysine. In addition to it, histidine, tyrosine, tryptophan, methionine and many other amino acids are also present. The protein is acidic in nature and forms greater percentage of plasma membrane as compared to lipid components. The lipids are mainly phospholipids and cholesterol. The phospholipids mainly consist of lecithin and cephalin. On an average the membrane possesses 75 to 90 molecules of lipid for every molecule of protein. Five phospholipids have been identified from the cell membrane. One is **phosphatidic acid** and the remaining four are more complex being comprising either cholein, inositol ethanolamine or serine attached to the phosphate groups in addition. The one phospholipid containing choline is called **lecithin.** The phospholipid molecules that are made up of glycerol and phosphate are water soluble, whereas the fatty acid molecules that constitute the tail of the whole structure are water insoluble. According to Parpart and Ballantine (1952) the structure of the plasma membrane is different in different cells. Salts are also present in the cell membrane. Lowell and Hokin (1965) demonstrated that the phospholipids are secretory in function and they increase the formation of the membrane phospholipids.

Functions

The main functions of plasma membrane are as follows:

Osmosis. The plasma membrane is semipermeable for differentially permeable membrane because water moves through it more easily than the large molecules. Osmosis accounts for the movement of water through the membrane with the implication that solute moves at slower rate. This means that the solute would remain isolated on one side of the cell membrane—this does not occur in living cell. Undoubtedly the smaller molecules like that of water move with greater facility through the membrane than the large molecules. Thus this membrane is **selective permeable.**

Active transport. Normally, the molecules of dissolved materials diffuse through the membrane towards a region of low concentration. But sometimes they move towards a region of higher concentration. For this the energy is needed to move the molecules in a direction opposite to that in which they would move by diffusion. This type of movement is called **active transport.** In brief, here the substance does not move by itself but is transported through the cell membrane. It can be defined as, **"the movement of molecule of a substance against a concentration gradient and (or) against an electrochemical."**

CELL STRUCTURE

The protoplast of each cell contains the protoplasmic and non-protoplasmic components. The protoplasmic or living components are—nucleus, mitochondria, plastids, endoplasmic reticulum, ribosomes, lysosomes, sphaerosomes, microtubules and Golgi bodies.

The chief non-protoplasmic or non-living components are—vacuoles, food products, secretory products and waste products. The protoplasmic and non-protoplasmic components of the cell are summarised on next page.

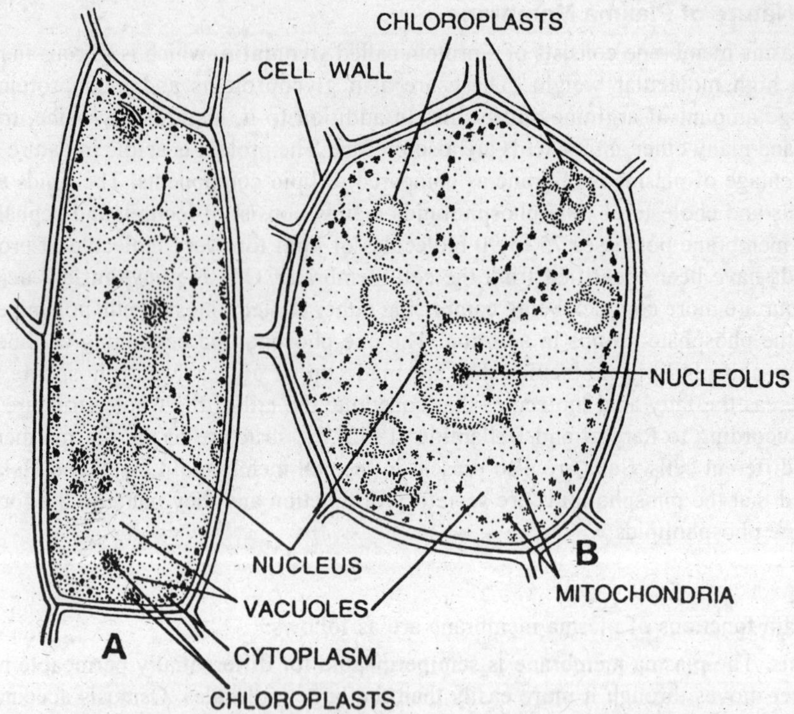

Fig. 33.12. Plant cells. A, cell from sugar-beet leaf with vacuolated cytoplasm, a nucleus and granular chloroplasts; B, a cell from starch-sheath of tobacco stem.

CELL

- **Protoplasmic Components**
  1. Nucleus
  2. Mitochondria
  3. Plastids
     - (*i*) Leucoplasts
     - (*ii*) Elaioplasts
     - (*iii*) Chromoplasts
     - (*iv*) Chloroplasts
  4. Endoplasmic reticulum
  5. Ribosomes
  6. Lysosomes
  7. Sphaerosomes
  8. Microtubules
  9. Golgi bodies
- **Non-Protoplasmic Components**
  - Vacuoles
  - Ergastic Substances
    - Food products
      - Non-nitrogenous (Carbohydrates)
        - (*i*) Starch
        - (*ii*) Inulin
        - (*iii*) Hemicellulose
        - (*iv*) Cellulose
        - (*v*) Sugars
      - Nitrogenous
        - (*i*) Proteins
        - (*ii*) Amino compounds
      - Fats and fatty oils
    - Secretory products
      - (*i*) Enzymes
      - (*ii*) Pigments
      - (*iii*) Nectar
    - Waste products
      - (*i*) Tannins
      - (*ii*) Mineral crystals
      - (*iii*) Latex
      - (*iv*) Essential oils

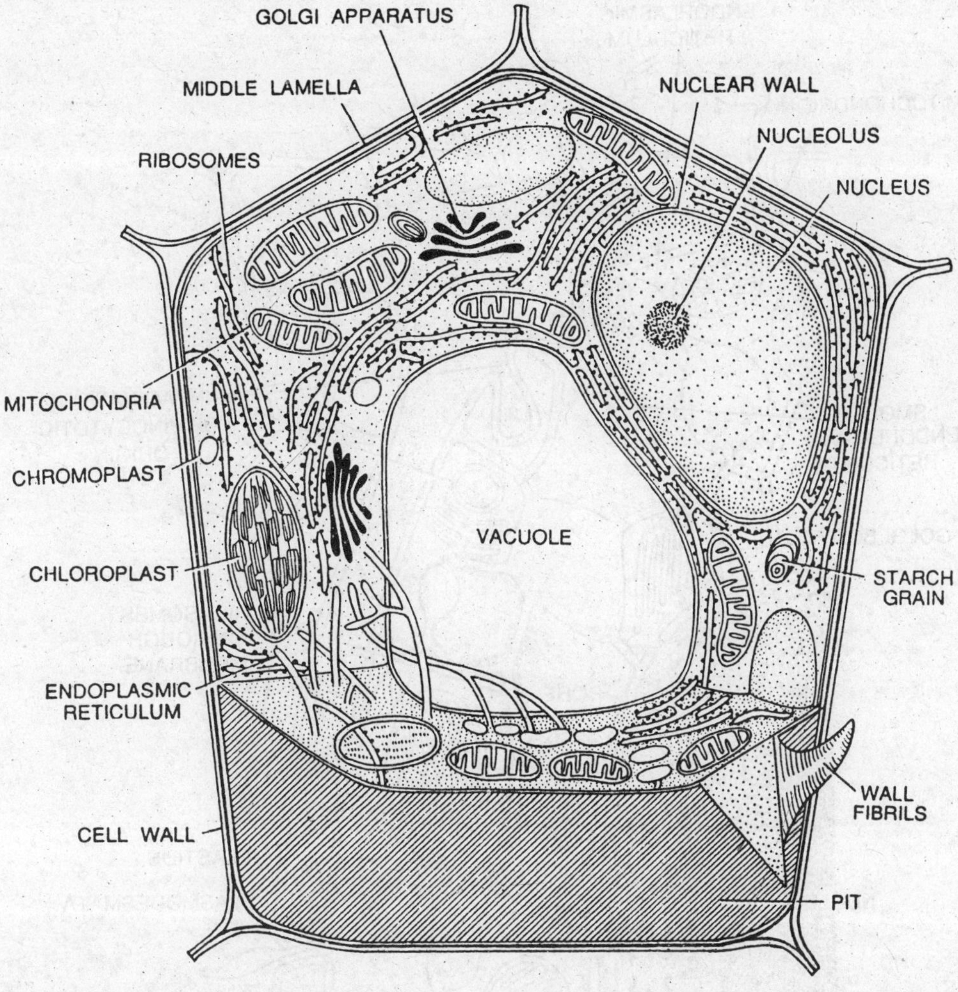

Fig. 33.13. The cell structure. Diagram of an electron micrograph of a cell from a leaf of *Phaseolus vulgaris.*

Cytoplasm. Kolliker (1862) gave the name '*cytoplasm*, to the substance found around the nucleus. According to Guilliermond (1941) and Sharp (1934) the word cytoplasm has been used to designate all the matter in the cell exclusive of the nucleus. The cytoplasm and various protoplasmic bodies (*e.g.*, nucleus, plastids, mitochondria, etc.,) have the same fundamental characteristics.

The cytoplasm is a transparent semifluid substance denser than water with granules, vacuoles and vesicles of various sizes embedded in it. It is highly complex, both physically and chemically. Usually it consists of water which may be 85–90% in it. Many organic and inorganic substances also occur in the cytoplasm either in true solution or in colloidal state. Salts, carbohydrates and other water soluble substances are found in dissolved state. Proteins and fats are also found in the form of very minute particles which are invisible in the microscopes of ordinary light. They are found in colloidal state. The studies of many workers such as Fry–Wyssling (1948); Seifriz (1935, 1936, 1945); Sharp (1943) and many others suggest the presence of a continuous framework of

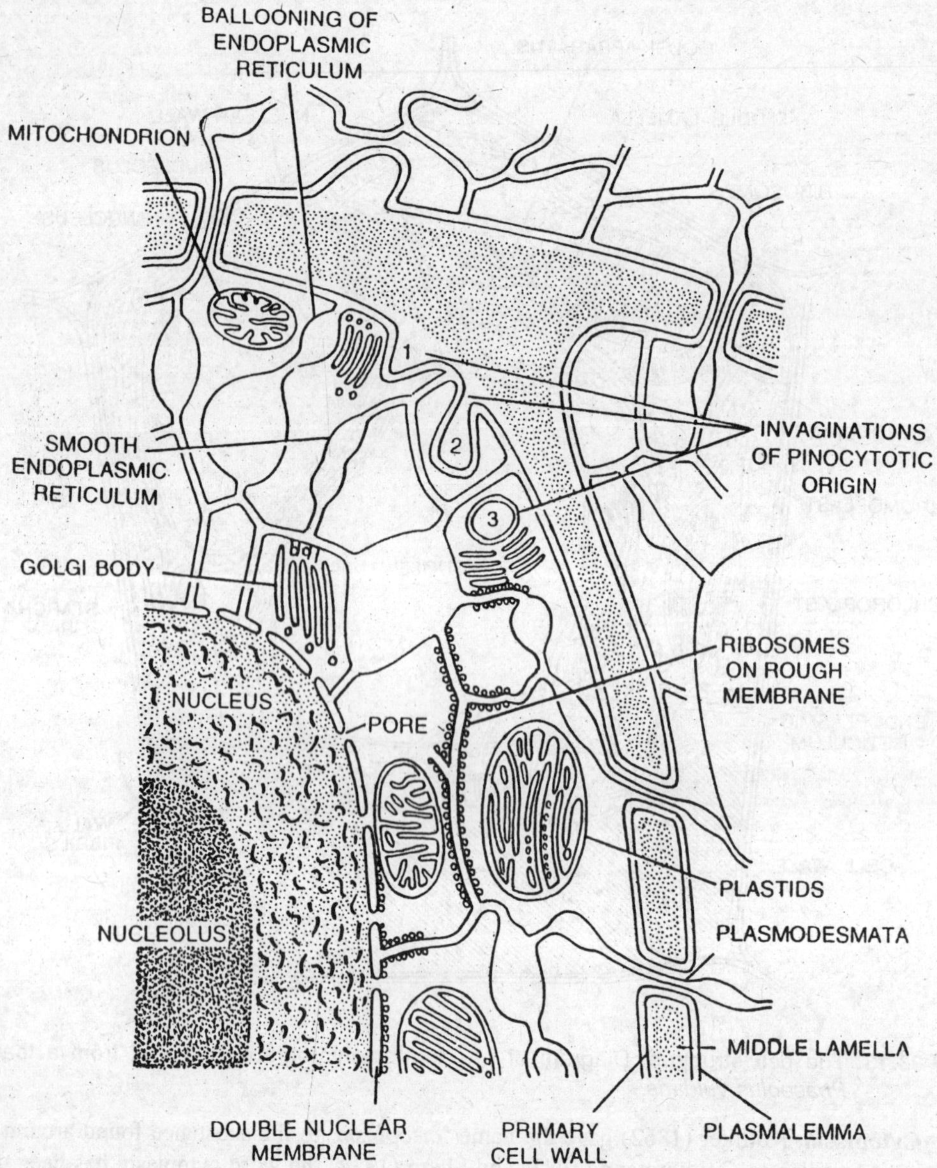

Fig. 33.14. Portion of a plant cell as seen under electron microscope. The invaginations of plasma membrane and ballooning of endoplasmic membrane are clearly seen.

proteins. It remains always in a dynamic state due to constant phase inversions. It is bounded by a monomolecular layer or a multimolecular layer on the outside called **ectoplast** (plasma membrane). Similar second layer on inside called **tonoplast** (vacuolar membrane). The protoplasmic layers are semipermeable in nature which involve in differential absorption. In the cytoplasm, fatty substances such as lipids and certain proteins take part in the formation of ectoplast (plasma membrane). The tonoplast (vacuolar membrane) also develops from the same substance and possesses similar structure to that of plasma membrane.

The electron microscope reveals membranous differentiations in the interior of the cytoplasm notably, the **endoplasmic reticulum** and the **dictyosomes**. The cytoplasm also includes granules of various sizes. Granules of 0.25 to 1 micron in diameter, containing lipids and proteins, constitute the **spherosomes**. These granules are highly mobile in living cells. At the submicroscopic level, a granule about 150 Å in diameter, the **ribosome** a globular macromolecule of ribonucleoprotein that takes part in protein synthesis. Ribosomes occur free in the cytoplasm or they remain associated with the endoplasmic reticulum.

Some of the resolvable entities in the protoplast, such as the **nucleus, plastids** and **mitochondria**, are referred to as **organelles**, whereas the endoplasmic reticulum and the dictyosomes are called sometimes membrane systems and sometimes organelles.

Cytoplasmic membranes. Among the cytoplasmic membranes, ectoplast and tonoplast, have long been associated with the important physiological characteristics of the protoplast. These membranes are difficult to recognize with the light microscope, but the electron microscopy has confirmed their morphologic identity (Mercer, 1960). The tonoplast is thinner than the ectoplast.

The endoplasmic reticulum is a system of membrane-bound cavities or **cisternae**. The cisternae are commonly much flattened and therefore, their sections appear as double lines. The endoplasmic reticulum is responsible for providing the cell with a large internal membrane surface along which enzymes are orderly distributed.

Dictyosomes are stacks of flattened sacks or cisternae, circular in outline, each surrounded by vesicles. The vesicles originate from the margins of the cisternae and pass into the cytoplasm. Dictyosomes are concerned with secretory activities and wall formation.

THE NUCLEUS
(The Controlling Centre)

The controlling centre of a cell is the ***nucleus.*** The ***chromosomes*** and ***genes*** are found within it which determine the character, activities and destiny of each individual cell.

Each nucleus is surrounded on the outside by a ***nuclear membrane.*** This membrane quite resembles the cell membrane. It is double-layered and consists of proteins and lipids. The main part of the nucleus consists of chromatin, which during cell division becomes more distinctly visible as a definite number of individual chromosomes. Certain coarser granules are also visible in the chromatin. They are known as **chromocentres** which stain darker than the rest of the net- work of the chromatin. Each nucleus consists of **nuclear membrane, nuclear sap (nucleoplasm), nucleolus or nucleoli and chromatin.**

Structure of the nucleus. Each nucleus remains surrounded by a limiting membrane as the

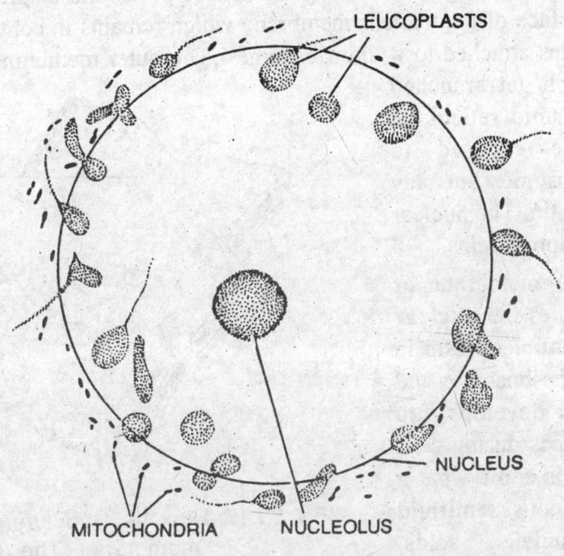

Fig. 33.15. The nucleus. Nucleus, amoeboid leucoplasts, and mitochondria from the cell of sugar-beet hypocotyl.

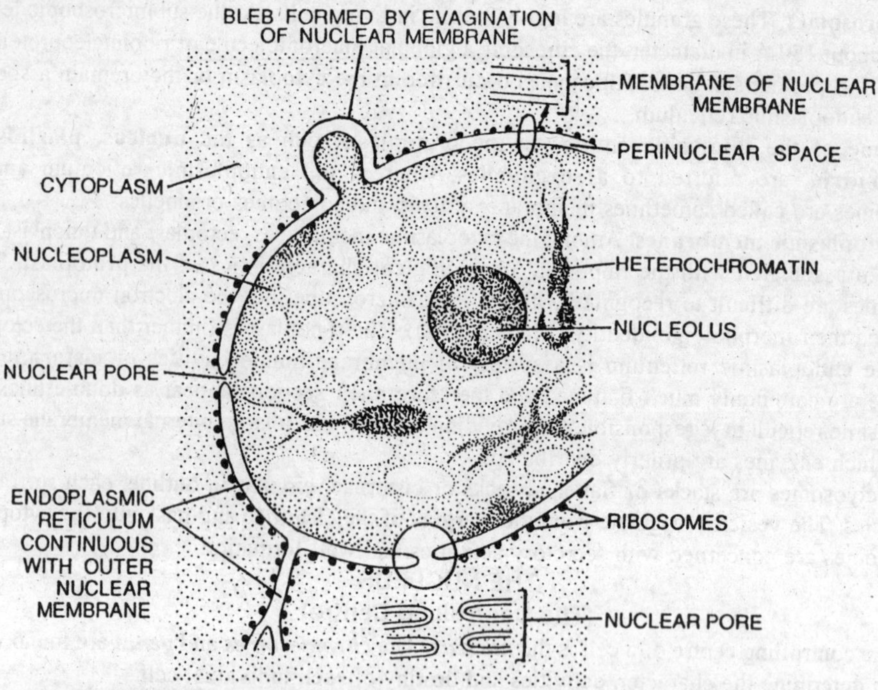

Fig. 33.16. The structure of the interphase nucleus (diagrammatic).

nuclear membrane. As electron microscopy reveals the nuclear membrane consists of two membranes each being 90Å thick and the space in between the two, the **perinuclear space** being 100-115Å wide. The nuclear membrane possesses pores here and there varying in diameter from 400-600Å. Through these pores the nucleoplasm communicates with cytoplasm. At the margins of the pores the two membranes are continuous. The number of pores varies from species to species and even from cell to cell. According to some workers, each pore remains encircled by a circular structure, the **annulus.** According to Watson (1959) these pores and annuli constitute the **pore complex.** The outer surface of the nucleus membrane which remains in contact of cytoplasm has RNA containing granules attached to it, the ribosomes. The outer membrane at places gives out tubular structures which get branched and form the endoplasmic reticulum. The inner membrane exposed to nucleoplasm lacks ribosomes but may have aggregated chromatin. The nuclear membrane consists of lipoproteins.

Inside the nuclear membrane or nuclear envelope their lies the **nuclear sap** or **karyolymph** or **nucleoplasm** in which are found nuclear ribosomes and chromatin network that develops into well defined chromosomes during cell-division. Nucleoplasm is some-what a granular and homogeneous semifluid which contains nucleic acids, nucleoprotein, etc.

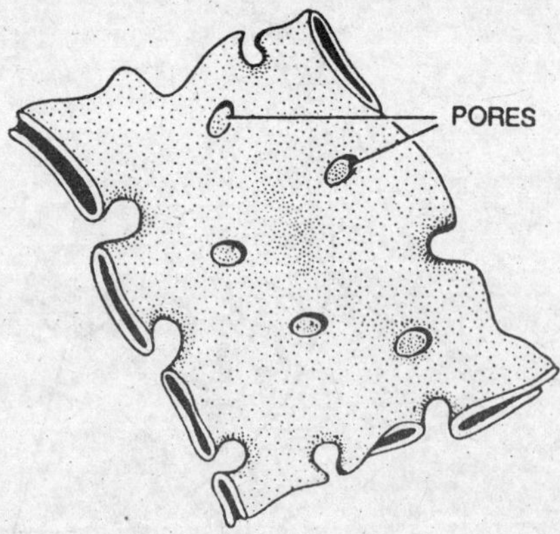

Fig. 33.17. Nucleus. Structure of a part of nuclear membrane. The membrane is two layered and the pores are clearly visible.

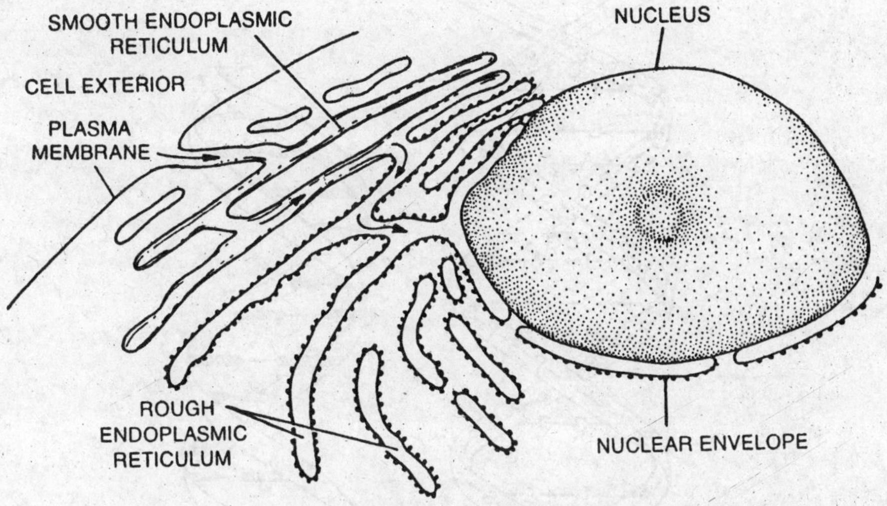

Fig. 33.18. The diagram depicts a eukaryotic cell that shows how the endoplasmic reticulum (ER) connects with both the plasma membrane and the double layered nuclear envelope. Connections of this type may form channels through which materials may pass to and from the cell's surroundings without penetrating the plasma membrane. However, such channels are formed only in cells with a highly developed system of endoplasmic reticulum (ER).

Certain distinct one or more rounded bodies known as **nucleoli (singular-nucleolus)** are also found within the nucleus. The number of nucleoli per nucleus of a particular species is definite. For example, nucleus of onion cell possesses four nucleoli, which sometimes fuse together.

The **nucleus** is a relatively large, generally spherical ball-like body found inside the nucleus. The nucleoli are made up of small granules containing proteins and RNA which has got some part to play in the activity of the nucleus. It is thought that nucleoli help in the synthesis of

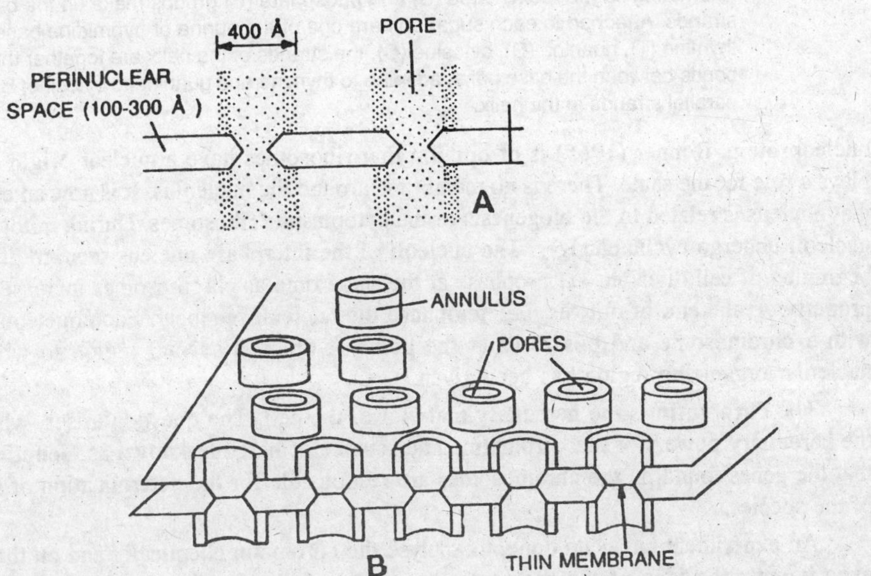

Fig. 33.19. The nucleus. Structure of the nuclear membrane. A, double membrane and pores; B, pore complex or annuli.

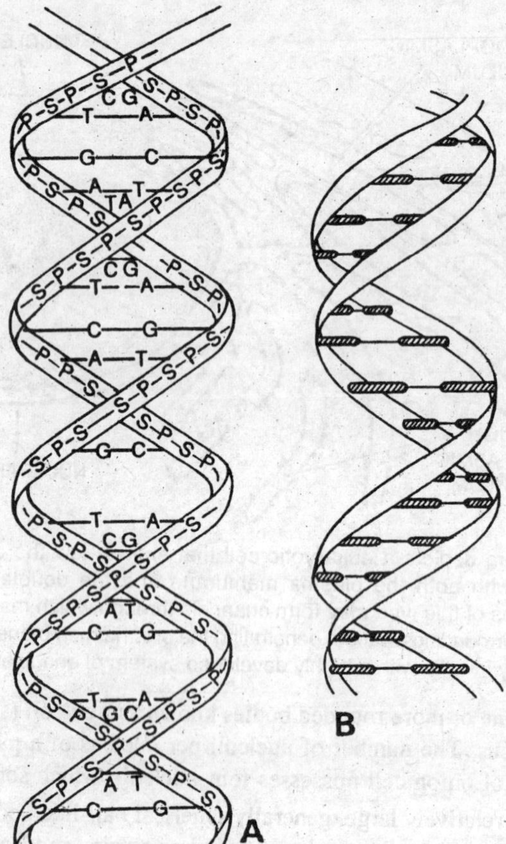

Fig. 33.20. DNA structure. A, schematic diagram of the double helix in a portion of a DNA molecule. Alternating sugar deoxyribose (S) and phosphate (P) groups make up the backbones of the strands. Attached to each sugar unit are one of the purine or pyrimidine bases, adenine (A), thymine (T), guanine (G), cytosine (C), the strands of the helix are together through hydrogen bonds between the base pairs, adenine to thymine and guanine to cytosine; B, diagram of the parallel strands in the helix.

nucleoprotein. Bonner (1965) is of opinion that ribosomes have a nuclear origin and nucleolus plays a role for the same. There is no membrane around the nucleolus. It is now an established fact that nucleus is related to the biogenesis of the cytoplasmic ribosomes. During mitotic division the nucleoli undergo cyclic changes. The nucleoli of the interphase nucleus seem to disappear at the beginning of cell division, *i.e.* prophase at the same time as chromosomes increase their staining property. At the end of mitosis, *i.e.,* telophase the nucleoli reappear. Each nucleolus has contact with a chromosome and possesses at the point of union a special region to which the name nucleolar organizing region has been given.

The **chromatin** is the hereditary material of the cell. The question arises, what determines the hereditary powers of the chromatin. The science of heredity known as **Genetics,** informs us that the genes found on the chromosomes are responsible for the determination of the characters of the species.

An experiment has been done, to analyse the chromatin chemically and on the basis of this **experiment :** the idea of the molecular basis of heredity has been gained. According to this **experiment** the chromatin has been resolved into four major molecules. They are—(*i*) **histone,** a low molecular weight protein (2000), (*ii*) a more complex protein than histone; (*iii*) deoxyribonucleic acid (DNA) and (*iv*) ribonucleic acid (RNA). These four molecules constitute the chromatin. But now, with the result of extensive studies this has been demonstrated that DNA is the molecule whose structure determines hereditary uniquences on the cell and the individual. In other words, this can be said that DNA is the hereditary material of the cell. The proteins and RNA are necessary for the functioning of the nucleus but not so important from the point of view of the heredity. In certain plant viruses only RNA is found in their structure and no DNA, in such cases the RNA acts as hereditary molecule.

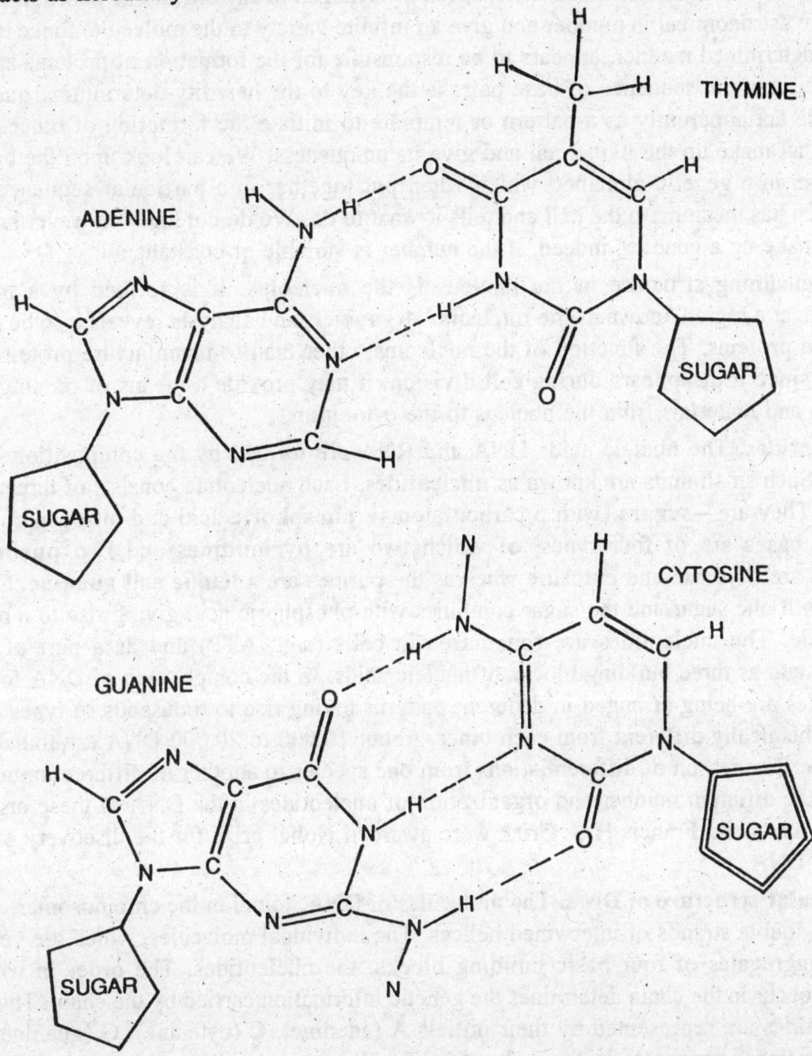

Fig. 33.21. Chemical configurations of the four bases found in the DNA molecule, arranged as base pairs. Thymine and cytosine are pyrimidines; adenine and guanine are purines.

DNA. The **deoxyribonucleic acid** (DNA) is the compound of high molecular wt. (over 1,000,000), *i.e.*, made up of a number of molecules linked together. These molecules include a sugar, **deoxyribose, phosphoric acid** and four bases of which two are **pyrimidines (thymine** and **cytosine)** and two are **purines (adenine** and **guanine)**.

Chemical structure of DNA. The alternating sugar and phosphate arrangement forms the outside boundaries of DNA while base pairs link the two sides together. The base pairs, however, are not at random, for adenine and thymine are always paired, as are guanine and cytosine. Hydrogen bonds tie the bases to each other. X-ray analysis of the molecular arrangements reveals that DNA is not a flat structure, as one might suspect, but a double helix, *i.e.*, a sort of spiral' stair case with alternating sugar-phosphate 'bannisters' and 'steps' of base pairs. This is the Watson-Crick model of DNA, so named after its discoverers. The DNA molecule may have thousand of turns in its spiral configuration and the steps, can be arranged in any order. The possible variations therefore are astronomical in number and give an infinite variety to the molecule. Since the DNA, in some undetermined manner, appears to be responsible for the formation of proteins and RNA, it is believed that the sequence of base pairs is the key to the heredity-determining qualities of DNA. These act apparently as a pattern or template to initiate the formation of other complex molecules that make up the living cell and give its uniqueness. We can look upon the base pairs as the letters in a genetic alphabet, which when put together in a particular sequence yield a 'word' which has meaning to the cell and tells it what to do. We do not know however how many base pairs make up a gene or, indeed, if the number is variable or constant.

The remaining structure in the nucleus is the nucleolus. It is formed by a particular chromosome at a region known as the **nucleolar organizer,** and analysis reveals it to be made up of RNA and proteins. The function of the nucleolus, other than to manufacture proteins, is not known but since it disappears during cell division, it may provide a means of passing genetic information and materials from the nucleus to the cytoplasm.

Nucleotides. The nucleic acids DNA and RNA are formed by the composition of small molecules. Such small units are known as **nucleotides.** Each nucleotide consists of three types of molecules. They are—**sugars** (with 5 carbon atoms), **phosphoric** acid and nitrogen containing bases. The bases are of four types, of which two are **pyrimidines** and two **purines.** The pyrimidines are **thymine** and **cytosine** whereas the purines are **adenine** and **guanine.** One base combines with one sugar and the sugar combines with phosphoric acid giving rise to a molecule of nucleotide. The nucleotides are found free in cells (*e.g.*, ATP) and as a part of various coenzymes, and as three building blocks of nucleic acids. In the composition of DNA four types of nucleotides are being arranged in different patterns giving rise to thousands of types of DNA which are chemically different from each other. About 10,000 to 20,000 DNA are found on one chromosome. The reason of differentiations from one species to another of different organisms is because of the different number and organization of nucleotides in the DNA of these organisms. James D. Watson and Francis H.C. Crick were awarded Nobel prize for the discovery of helical structure of DNA.

Molecular structure of DNA. The molecules of DNA, found in the chromosomes, occur in the form of double strands of interwined helices. The individual molecules, which are very long, are linear aggregates of four basic building blocks, the nucleotides. The order in which the nucleotides occur in the chain determines the genetic information carried by the chain. Thus, if the four nucleotides are represented by their initials A (adenine), C (cytosine), G (guanine) and T (thymine), a linear aggregate in the order ACGT will constitute different genetic information from—AGCT—or—ATCG—or ACCA–GTGT—chains. The inheritance or every organism consists in its essence of some repeated pattern of nucleotide units in the DNA molecules of the chromosomes contained in nuclei.

DNA replication

It is an established fact that the chemical nature and total amount of DNA remains constant in similar kind of cells from generation to generation. It means that quality and quantity of DNA remain same in similar cells, derived from the same parent cell. While explaining the DNA replication one has to keep in the view these both requirements. The weak hydrogen bond that holds together the two strands of the double helix works some-what like a zipper. If one starts at one end of the molecule, one can 'unzip' each purine from it pyrimidine partner one by one. This will leave unpaired purine and pyrimidine ends on both strands. It is well known that the cell is a

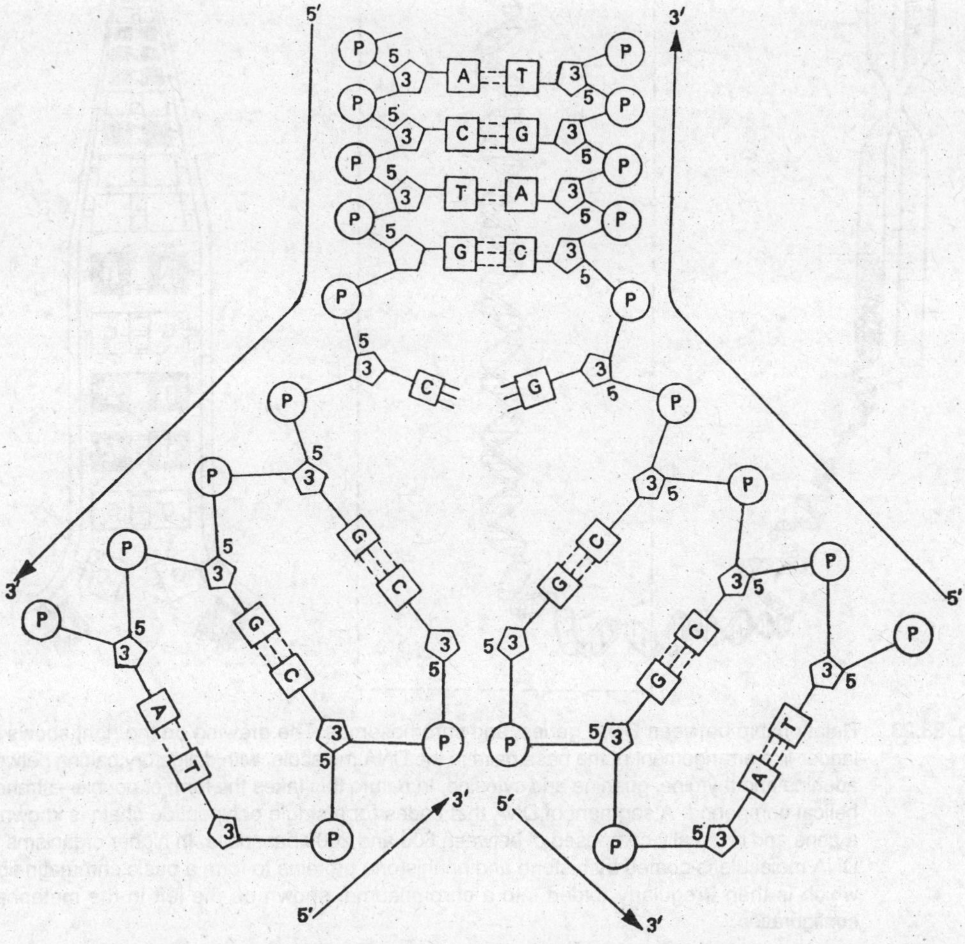

Fig. 33.22. DNA structure. The three components of DNA are the bases adenine, guanine, thymine and cytosine; a sugar molecule, deoxyribose; and phosphate groups. The phosphate groups are attached to the 3' and 5' carbons of the deoxyribose and serve to connect successive pairs of nucleotides. The replication process must one DNA strand by adding bases in the 5' → 3' direction and the other DNA strand, apparently by adding bases in the opposite direction.

storehouse of precursors of the four sorts of building blocks needed to make a new chain. These precursors are nucleotides. As such when the DNA helix 'unzips' new nucleotides from the cell of the proper kind fall into place, the improper ones are rejected as they will not fit. An adenine group will become bounded only with a new thymine group whereas the thymine group of the other strand will become bounded to a new adenine nucleotide thus completing the double chain in the two original strands. In the words of Crick the two strands of DNA 'unzip' as follows—'it is the growth of the two new chain that unwinds the original pair. This is likely in the terms of energy because, for every hydrogen bond that has to be broken, two new ones will be forming".

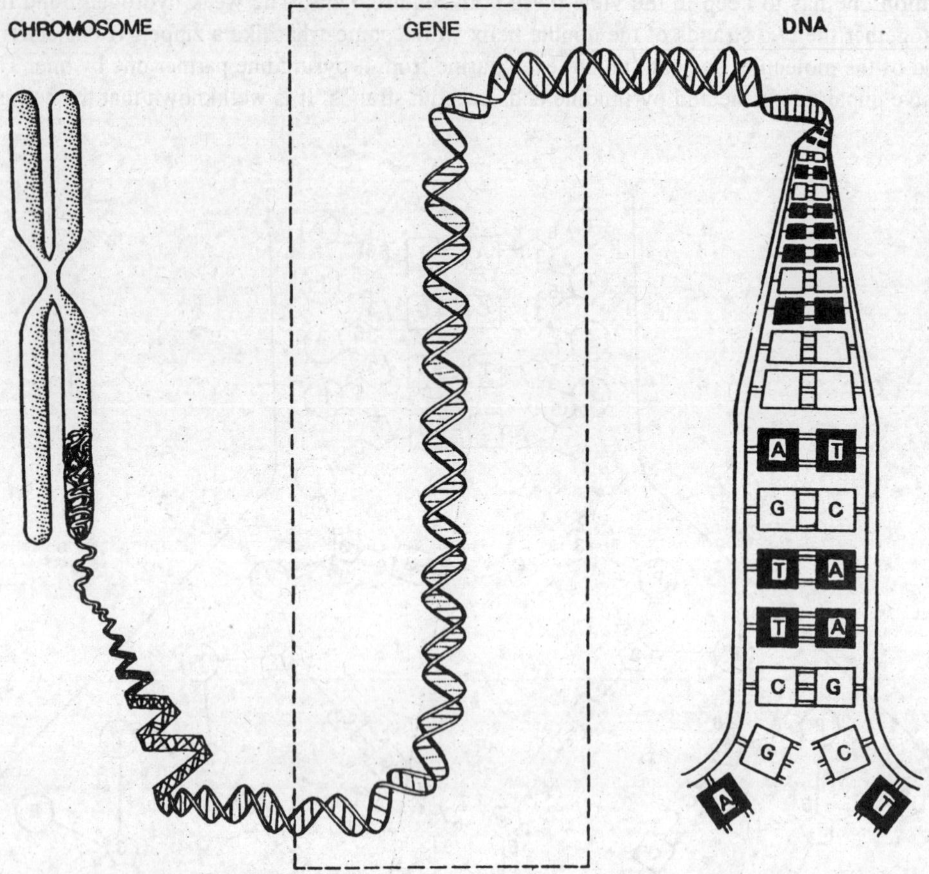

Fig. 33.23. Relationship between DNA, genes, and chromosomes. The drawing on the right shows the ladder-like arrangement of the base pairs in the DNA molecule, with deligatory pairing between adenine and thymine, guanine and cytosine. In nature this takes the form of double—stranded helical compound. A segment of DNA that codes for a single polypeptide chain is known as a gene and is usually composed of between 500 and 2000 basepairs. In higher organisms the DNA molecule is coated by histone and nonhistone proteins to form a basic chromatin fibre, which is then irregularly folded into a chromosome, shown on the left in the metaphase configuration.

The unzipping of DNA molecule at the weak forces (H-bonds) leaves the two strands apart, each strand with a purine or pyrimidine base attached to the sugar molecule. Each strand has not the capacity to replicate and exactly identical strand like the partner from which it has separated.

Classification of DNA

The DNA in the eucaryotic cells may be classified as follows:

(*i*) Nuclear DNA (*n*DNA)—It is found in the chromosomes of nucleus.

(*ii*) Mitochondrial DNA (*m*DNA)—It is found in the mitochondria.

(*iii*) Chloroplast DNA (*c*DNA)—It is found in the chloroplasts.

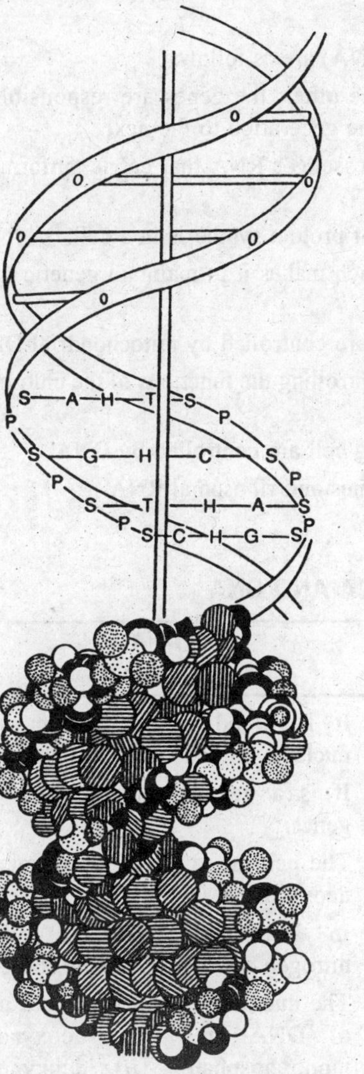

Fig. 33.24. The helix of DNA, with three different ways of representing the molecular arrangement. Top, general diagram of the double helix, with the phosphate sugar combinations making up the outside spirals and the base pairs the cross bars. Middle, a somewhat more detailed representation—phosphate (P), sugar deoxyribose (S), adenine (A), thymine (T), guanine (G), cytosine (C) and hydrogen (H). Bottom, detailed structure showing how the space is filled with atoms—carbon (C), oxygen (O), hydrogen (H), phosphorus (P) and base pairs.

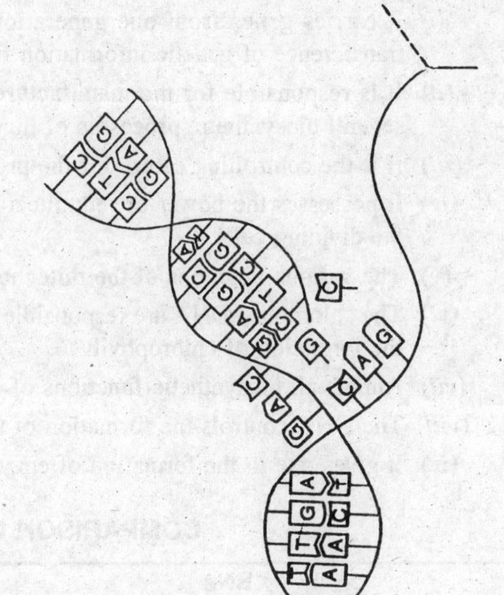

Fig. 33.25. Replication of DNA molecule. Diagrammatic representation.

Functions of DNA

The main functions of deoxyribonucleic acid (DNA) are as follows:

(*i*) It carries genes from one generation to the other; the genes are responsible for the transference of genetic information from one generation to the next.

(*ii*) It is responsible for the manufacture of messenger RNA that carries information for several biosynthetic processes of the cell.

(*iii*) It is the controlling centre for the process of protein synthesis in a cell.

(*iv*) It possesses the power of reduplication which makes it to maintain genetic balance in the dividing cells.

(*v*) The various functions of the mitochondria are controlled by mitochondrial DNA.

(*vi*) The chloroplast DNA are responsible for controlling the functions of the chloroplast and the formation of chlorophyll.

(*vii*) Almost all biosynthetic functions of a living cell are controlled by DNA.

(*viii*) The DNA controls the formation of ribosomes and ribosomal RNA.

(*ix*) It gives rise to the formation of enzymes.

COMPARISON OF RNA AND DNA

RNA **Ribonucleic acid**	**DNA** **Deoxyribonucleic acid**
1. It is found in nucleolus, karyolymph, cytoplasm and ribosomes.	1. It is found in the chromosomes of nucleus, mitochondria and chloroplasts.
2. It carries genetic information and helps in the protein synthesis.	2. It is a genetic material and contains genes.
3. The pentose sugar of RNA is known as ribose.	3. The pentose sugar of DNA is known as deoxyribose.
4. In RNA molecule, the pyrimidine nitrogen bases are Cytosine and Urasil.	4. In DNA molecule, the pyrimidine nitrogen bases are Cytosine and Thymine.
5. The nucleotides which make a molecule of RNA are – (*i*) adenosine monophosphate, (*ii*) guanosine monophosphate, (*iii*) cytidine monophosphate and (*iv*) Uridine monophosphate.	5. The nucleotides which make a molecule of DNA are – (*i*) deoxyadenosine monophosphate, (*ii*) deoxyguanosine monophosphate, (*iii*) deoxycytidine monophosphate and (*iv*) deoxythymidine monophosphate.
6. The nucleotides of a RNA molecule are arranged in a single chain.	6. The paired nucleotides of a DNA molecule are arranged in a double stranded helix.
7. Three main types of RNA are messenger (*m*RNA), transfer (*t*RNA) and ribosomal (*r*RNA).	7. Three main types of DNA are — nuclear (*n*DNA), mitochondrial (*m*DNA) and chloroplast (*c*DNA).

RNA

The molecule of RNA (ribonucleic acid) consists of a single strand, but it may be wound back upon itself in several places, producing helices in these places as found in DNA. In the region of helix formation, the bases are hydrogen bonded. The RNA differs chemically from the

DNA in that it contains a ribose sugar instead of deoxyribose sugar. The ribose sugar possesses an additional oxygen molecule which is not found in deoxyribose sugar. The base pairs in RNA are same, *i.e.*, three of the four bases—adenine, guanine and cytosine are the same as found in DNA however, the fourth base, thymine is replaced in RNA by **uracil** which possesses one methyl group less. There are three kinds of RNA—ribosomal RNA (rRNA), soluble or transfer RNA (tRNA) and messenger RNA (mRNA). The RNA contained in the ribosomes are known as rRNA. Electron microscopy reveals that ribosomes are composed of two sub-units—the larger and the smaller. The small subunits are attached to mRNA and the appropriate tRNA are specially bound to the large subunits.

Classification of RNA

The RNA in the eucaryotic cells may be classified as follows:

(*i*) The messenger RNA (mRNA)— This carries message from the DNA to the actual site of protein synthesis.

(*ii*) The transfer RNA (tRNA)— This carries a specific amino acid to the site of protein synthesis.

(*iii*) The ribosomal RNA (rRNA)— It is found in the ribosomes. The function of this is not very clear.

THE MITOCHONDRIA
(Power house of cell)

They are microscopic bodies occurring in cytoplasm of every cell in varying numbers (upto 2500 per cell in rat liver) except in bacteria, blue-green algae and human r.b.c. Most are granular, rod or thread shaped. They are generally known as 'powerhouses of cell'. They are the site of the chemical events that supply the energy to the cell. They are also held responsible for fat synthesis. They range in their size from 0.2μ to 3.0μ. They also vary in their shape from spheres to rods. They are always found in dynamic state.

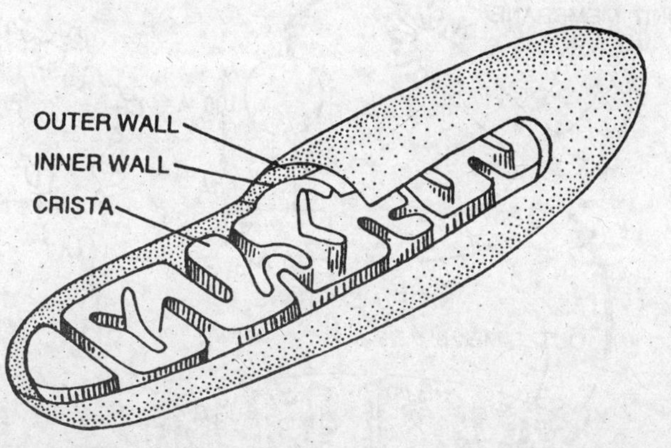

Fig. 33.26. Mitochondrion. Schematic diagram of a typical mitochondrion.

Each mitochondrion has a double-layered envelope with an outer and inner membranes. Each membrane is a typical unit membrane, being about 50–70 Å thick. The two membranes are generally 60–100 Å apart from each other. In between them remain filled the fluid rich in coenzymes. Extending from the inner membrane into the interior of the cavity, there are present a series of folds, called the **cristae.** The cristae are shorter in mitochondria of plant origin that in those of animal cells.In all the cases the purpose of the presence of cristae is to increase the surface area of the interior of the mitochondrion. The number of folds per unit of volume of a mitochondrion is variable. The cristae may be branched, tubular rather than lamellate, and may be

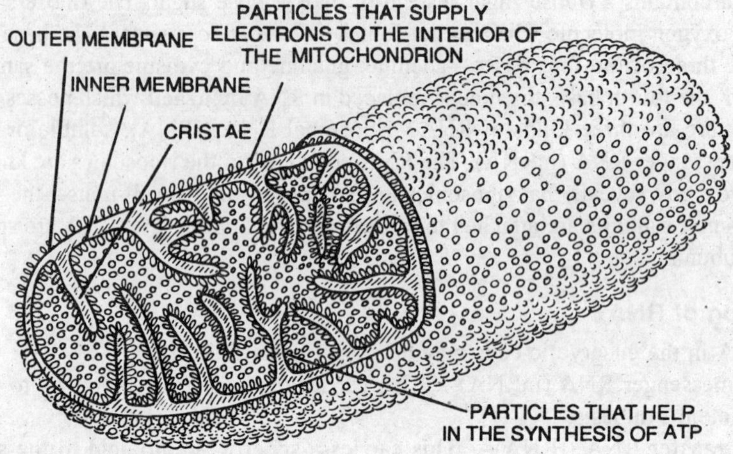

Fig. 33.27. Mitochondrion. Structure as viewed in electron microscope. The particles that supply electrons to the interior of the mitochondrion are clearly seen.

arranged parallel with long axis of mitochondrion, not, as is usual, at right angles. Thus each mitochondrion contains two cavities. The outer cavity found in between the two membranes is called **'outer compartment'** while the cavity limited by the folds of inner membrane is known as **'inner compartment** or **mitochondrial matrix'**. The matrix is normally homogeneous but it contains granules of 300–500 Å diameter. The outer covering membrane of the mitochondrion is quite elastic and may be stretched by swelling sometimes 200 times its normal dimensions. These facts advocate that the protein in molecules constituting the membrane can be greatly folded or stretched.

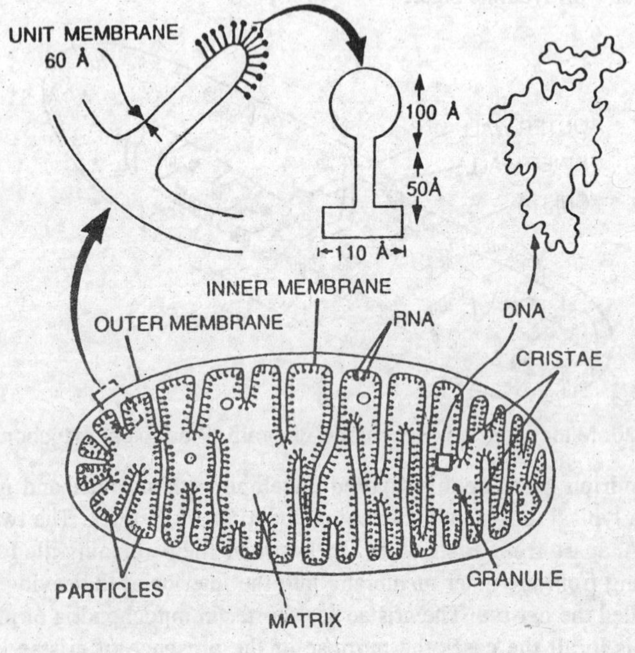

Fig. 33.28. Structure of mitochondrion as seen in electron microscope. Above, the various parts of mitochondrion magnified.

Electron microscopy has also revealed the presence of very small particles adhered to the outside of the outer membrane and the inside of the inner membrane. These particles were first described by Humberto Fernandez-Moran (1962). The particles of the two membranes differ considerably. Depending upon the size and type of the mitochondrion there are from 10,000 to 100,000 particles per mitochondrion. Each particle consists of a base, a stalk and a head. The base which remains attached to the inner membrane has about 80 Å diameter. The stalk is 50 Å long and 33 Å wide while the head has the same diameter as does the base. The length of the complete particle is about 160 Å.

The mitochondria contain DNA and many oxidative enzyme systems.

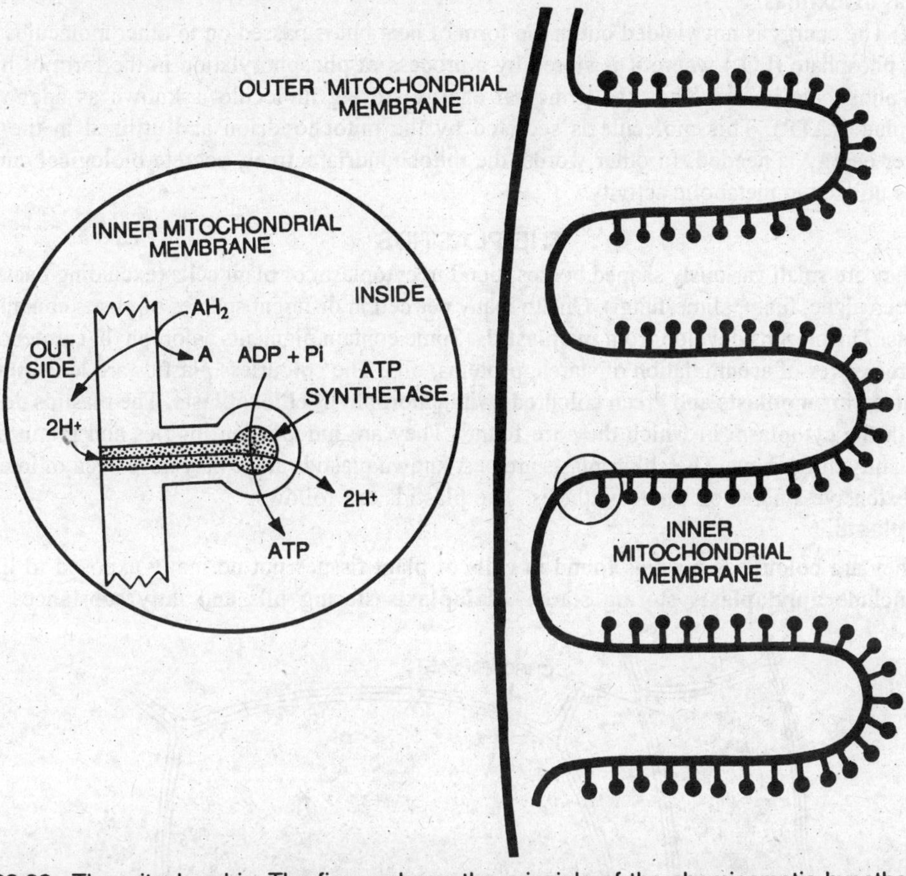

Fig. 33.29. The mitochondria. The figure shows the principle of the chemiosmotic hypothesis of energy conservation. The insert represents a magnified view of the encircled portion of the inner mitochondrial membrane.

Development. The mitochondria are essential structures where combustion of organic substances by oxygen takes place. Since this form of energy supply is essential for the life of most higher organisms the presence of functional mitochondria is essential for life. There are certain organisms, such as yeast, which can grow in the absence of oxygen. When such anaerobic cells are examined critically it is found that they contain small organelles having double membranes but the inner membrane does not have any of the invaginations (*i.e.*, cristae) typical of normal mitochondria. The characteristic enzymes and coenzymes which take part in aerobic respiration are also absent. Upon exposure to oxygen these structures soon develop to become normal respiring organelles. This observation shows that mitochondria can exist in an **embryonic** form, called a **promitochondrion,** and that upon stimulation by environmental factor (*i.e.*, oxygen) they undergo a process of development to become functional structures.

Functions

As regards the functions of the mitochondria they perform two important tasks. They are as follows.

(*i*) The carbohydrates, proteins and fats are broken down gradually into smaller and smaller molecules. During this process there is transference of energy. This process takes place through a series of steps, and each step is controlled by an enzyme. All these energy giving reactions are known as **oxidations.**

(*ii*) The energy is not yielded out in the form of heat but is passed on to other molecules that contain phosphate (PO_4) where it is stored by a process of phosphorylation in the form of high-energy phosphate bonds. Thus the principal energy storing molecule is known as adenosine triphosphate (ATP). This molecule is secreted by the mitochondrion and utilized in the cell wherever energy is needed. In other words, the mitochondria actively secrete biological energy which is utilized in metabolic activity.

THE PLASTIDS

They are small variously shaped bodies found in cytoplasm of plant cells (excluding bacteria, blue-green algae, fungi, slime-fungi), One to many per cell in different species of plants containing pigments. The plastids develop from **proplastids.** Some contain pigments chlorophyll, carotenoids, some are centres of accumulation of starch, proteins, oils. The colourless plastids are **leucoplasts,** pigmented **chromoplasts** and green coloured (with chlorophyll) **chloroplasts.** The plastids do not mix with the cytoplasm in which they are found. They are independent bodies and retain their individuality throughout. The chloroplasts are best known plastids since they have been of interest in the extensive studies on photosynthesis. The plastids are follows:

Leucoplasts

They are colourless plastids found in cells of plant tissues not normally exposed to light. They include **amyloplasts** storing starch, **elaioplasts** storing oils and fatty substances and

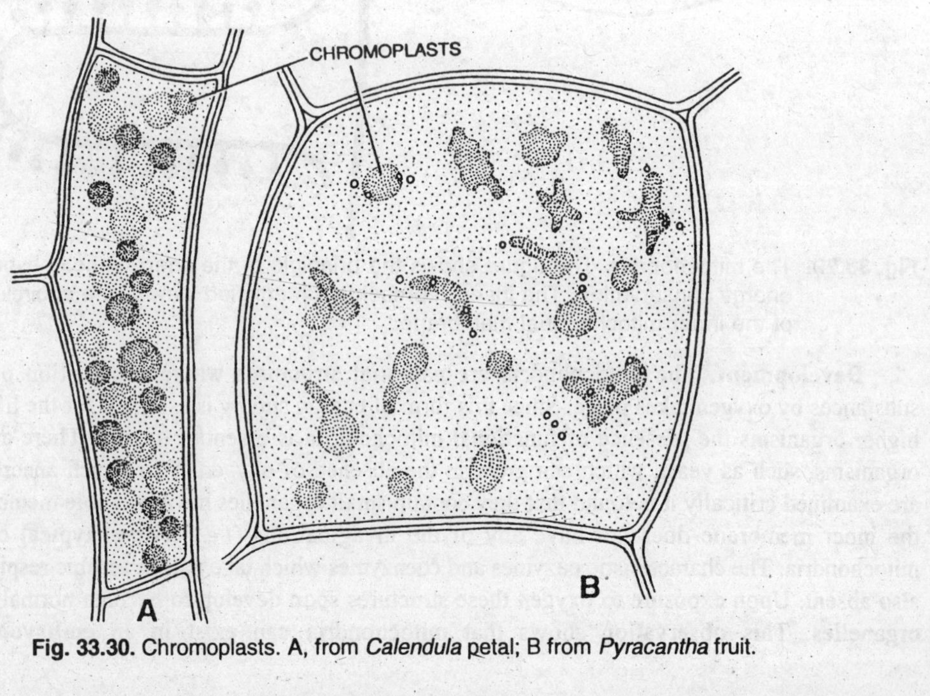

Fig. 33.30. Chromoplasts. A, from *Calendula* petal; B from *Pyracantha* fruit.

aleuroplasts, storing protein. The amyloplasts are found in cotyledons, endosperm and in storage organs such as potato tubers. The elaioplasts are commonly found in the tissues of liverworts and monocotyledons. They are refractive granules and also referred to as fatty or lipoid granules. The leucoplasts are found in many seeds. The leucoplasts often appear as small masses of protoplasm, variable and unstable in form. They commonly aggregate near the nucleus. The plastids of the epidermis frequently appear nonpigmented and are then classified as leucoplasts.

Chromoplasts

They are pigmented plastids of plant cells. They may be red, orange or yellow, *e.g.*, tomato fruits, carrot roots containing carotenoid pigments. They may appear rather granular with the pigments enclosed in it in a dispersed condition.

They show great variety in shape but are chiefly irregular; granular, angular, acicular and forked types occur. The irregular and sharp-pointed shapes are believed to be caused in part by the presence of the coloured substances, especially carotene and carotenoids, in crystalline form, as in the root of *Daucus*. They are associated with colour in flowers, fruits and roots. Usually they represent transformed chloroplasts, but may form directly from small leucoplasts. The development of chromoplasts with globular and fibrous inclusions from chloroplasts involves the destruction of the original grana system.

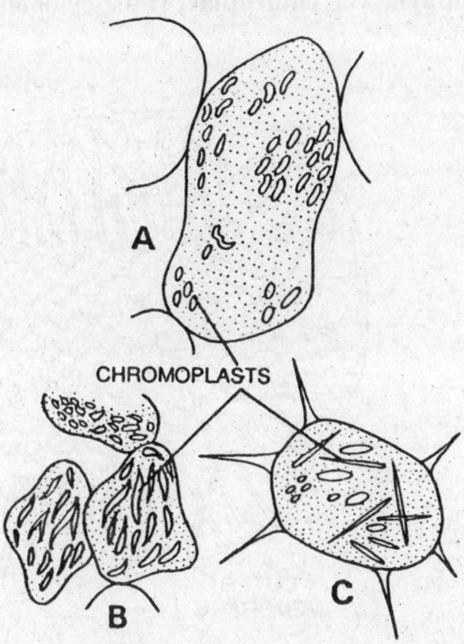

Fig. 33.31. Plastids. The chromoplasts. A, chromoplasts in pulp cells of tomato; B, chromoplasts in pulp cells of fruit of *Arisaema*; C, chromoplasts in the cells of *Daucus* root.

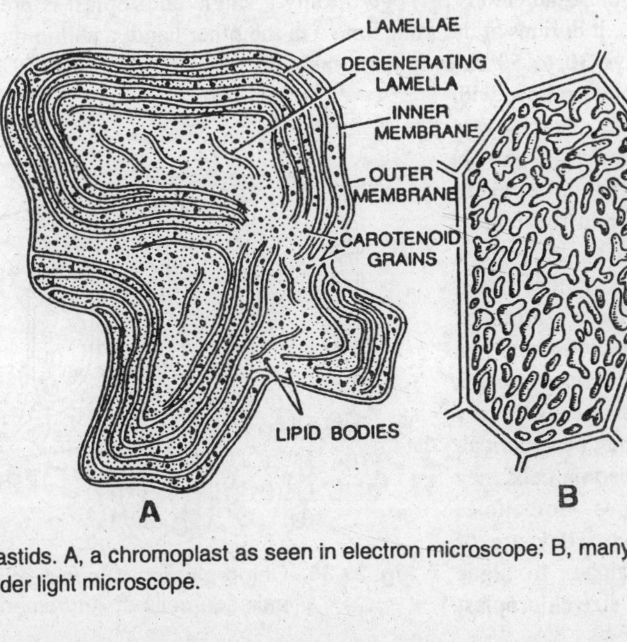

Fig. 33.32. Plastids. A, a chromoplast as seen in electron microscope; B, many chromoplasts as seen under light microscope.

Chloroplasts

The sequence of the trapping of light energy, its conversion into chemical energy, and its storage in molecules derived from CO_2 and water, is known as **photosynthesis.** The photosynthesis is initiated by the capture of light energy through absorption in the green pigment **chlorophyll.** The **chloroplast** is the cytoplasmic particle in which this takes place.

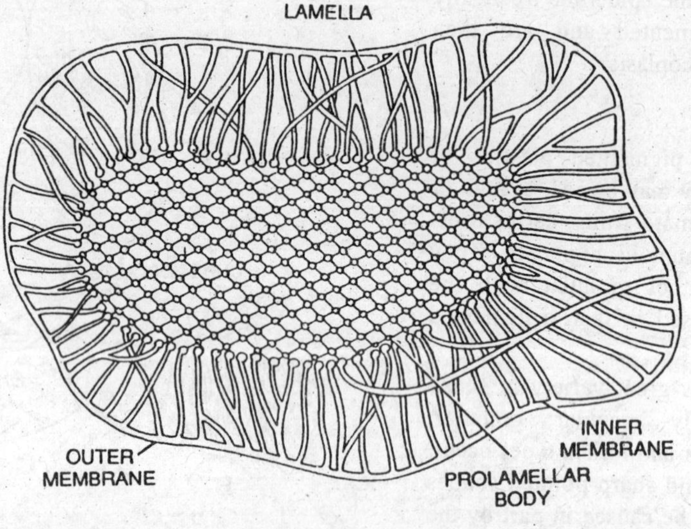

Fig. 33.33. Plastids. A, proplastid which gives rise to a plastid.

Morphology. The chloroplast is one of the largest cytoplasmic structures which can very well be seen under the low power of compound microscope. The size, shape and distribution of chloroplasts vary in different cells and species, but they remain relatively constant within the same tissue. The average size varies from 4 to 6 μ in diameter and 1 to 3 μ in thickness. The chloroplasts may assume many forms, and vary widely in number per cell in different plants. In some algae, such as the filamentous *Spirogyra*, only a single chloroplast is present in each cell; when the cell divides, it divides at the same time. On the other hand, a cell in the spongy tissue of a grass leaf may have 30 to 50 chloroplasts; their division which occurs in the immature, or proplastid state, is not correlated with cell division in any exact way. According to Haberlandt (1914) there are about 400,000 chloroplasts per square millimeter in the leaf of *Ricinus communis*. The blue-green algae lack definite chloroplasts; instead they possess loosely arranged membranes in the cytoplasm on which the photosynthetic pigments are layered. The shape too vary considerably. They may be spherical, ovoid or discoid. In certain cells they possess special shapes. Sometimes they are club-shaped. They are of various shapes in algae. In algae usually a single large size chloroplast

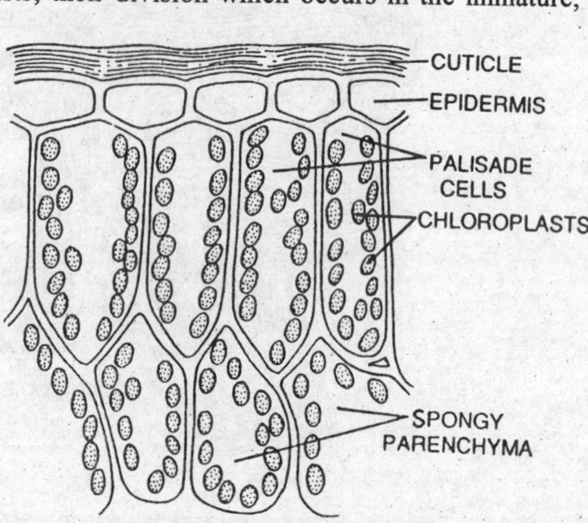

Fig. 33.34. Chloroplasts. Discoid chloroplasts in palisade cells of *Andromeda*.

is present which may be reticulate, spiral, band like or stellate. The chloroplass are sometimes homogeneously distributed within the cytoplasm, but are seldom packed near the nucleus or close to the cell wall. Their distribution depends largely on external conditions such as light intensity.

Structure. Electron microscopy reveals that the chloroplast is a structure of considerable complexity. A mature chloroplast remains surrounded by a semipermeable membrane. The membrane comprises of two separate layers, each being 40 to 60 Å thick and the space between

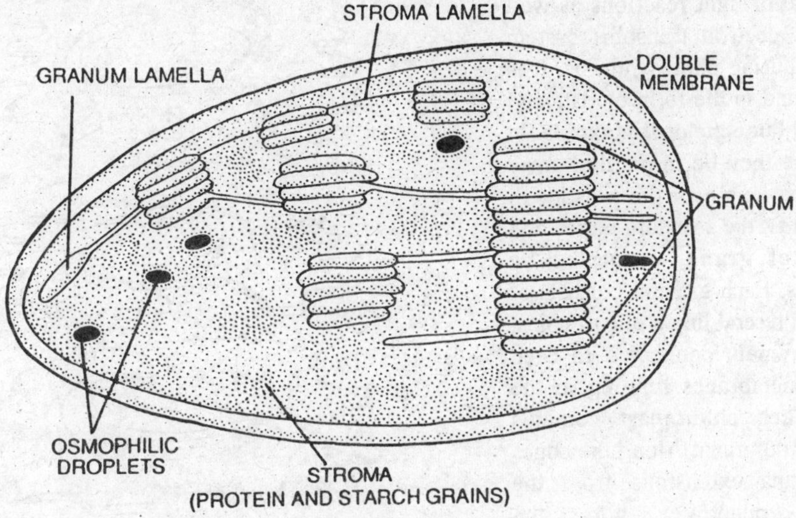

Fig. 33.35. A mature chloroplast (ultrastructure).

them vary from 25 to 75 Å. It is organized internally into series of lamellar areas **(grana)** and nonlamellar areas **(stroma).** Numerous small platelets, the **grana** remain embedded in stroma. The grana can be visualized as pieces of many-layered plywood lying in a less well-organized stroma. The number of grana is variable in different chloroplasts. The mesophyll cell of spinach has 40 to 60 grana per chloroplast whereas one granum per chloroplast is found in *Euglena*. Each granum consists of double membrane **discs** or **lamellae** which vary in thickness and are of two types, *i.e.,* **granum lamellae** and **stroma lamellae.**

Various particles and molecules may be found in the stroma; 175 Å diameter chloroplast **ribosomes;** the proteinaceous **stroma center,** starch grains **pyrenoids** in lower plants; osmophilic globule; and in some cases **phytoferritin** as well as fine fibrils of DNA—different from the nuclear DNA. The osmophilic globuli are

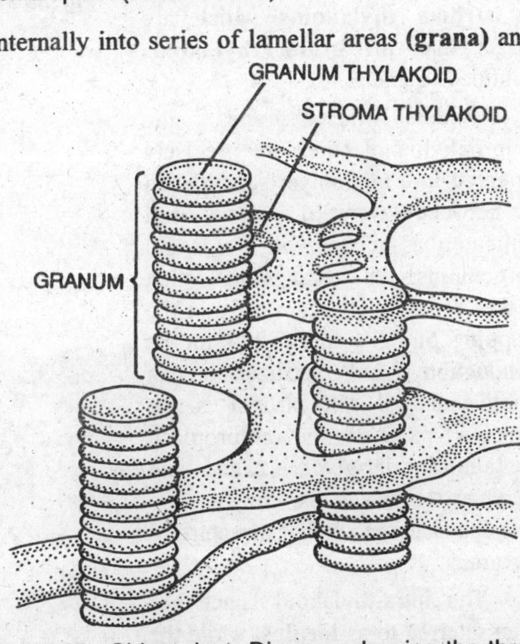

Fig. 33.36. Chloroplast. Diagram showing the three dimensional arrangement and inner relationships of grana (the scheme given in this diagram does not necessarily

generally referred to as **plastoglobuli.** They contain various lipid materials but no chlorophyll or carotenoid pigments.

In the stroma is also found suspended a chlorophyll containing lipoprotein membrane system. They are the site of light reactions as well as of the electron transport system operating during photosynthesis. It is usually found in the form of flattened sacs called **lamellae** or **thylakoids.** In many algae they lie in parallel array and run the length of the plastid. In higher plants, the structure varies and comprise of **grana,** connected by membranes. Each granum consists of **thylakoids** resembling a pile of coins, and they remain connected to each other by membranes running in the stroma. Each chloroplast contains about 20–100 grana. Along the edges of the grana extensions from the thylakoid penetrate into the intergranal regions. They are known as **stroma thylakoids** which are larger than the **grana thylakoids** confined to grana.

Within the grana, the chlorophyll molecules are precisely oriented in a monolayer sandwiched in between layers of proteins and intimately associated with lipids and carotenoids, an arrangement that makes for efficiency not only in the trapping of light energy but for its conduction and utilization in photosynthesis. The stroma is the aqueous part of the chloroplast, containing dissolved salts and enzymes. However, the enzymes are also found in the smaller structure of the grana.

The intra-thylakoid space has been referred to as **loculus,** while the lamellae between loculi as the **partition.** The connections between grana are termed as **frets** (Weier, 1966). Park and Pon (1961) reported

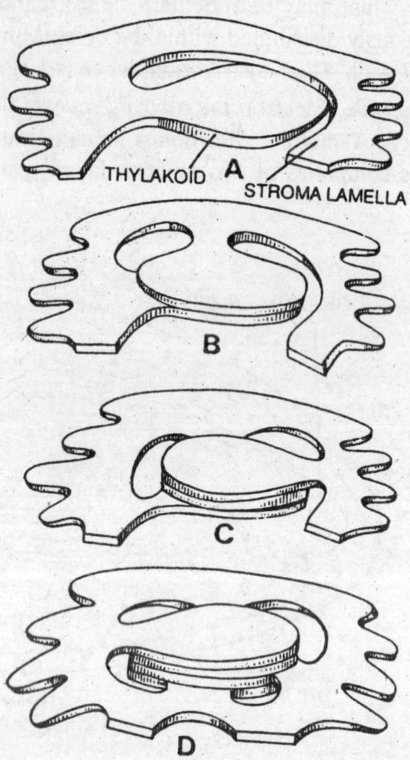

Fig. 33.37. Chloroplasts. A—D, stages showing the formation of thylakoids as outgrowths of stroma lamellae. The thylakoids are arranged one above the other forming the granum.

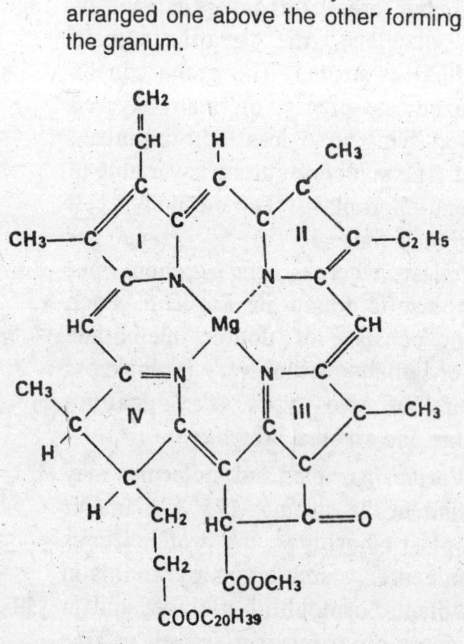

Fig. 33.38. Chlorophyll. Chemical structure of the chlorophyll— a molecule. The four pyrrole rings (1—4) attached to it.

the presence of **quantosome particles,** 100—200 Å in diameter arranged in rows in the chloroplast thylakoid membrane. They are the fundamental units responsible for the conversion of **quantum** of light energy into chemical energy.

Development. When a plant is germinated in the dark, its cells contain small double membraned structures. These structures appear colourless but they can be shown to contain a very low concentration of substances which are precursors of chlorophyll. Upon exposure to light these precursor substances are immediately converted to chlorophylls. At the same time a process of growth and development starts which results in the transformation of the small **prochloroplast** into a typical photosynthetic chloroplast. The entire process takes place in less than twenty four hours.

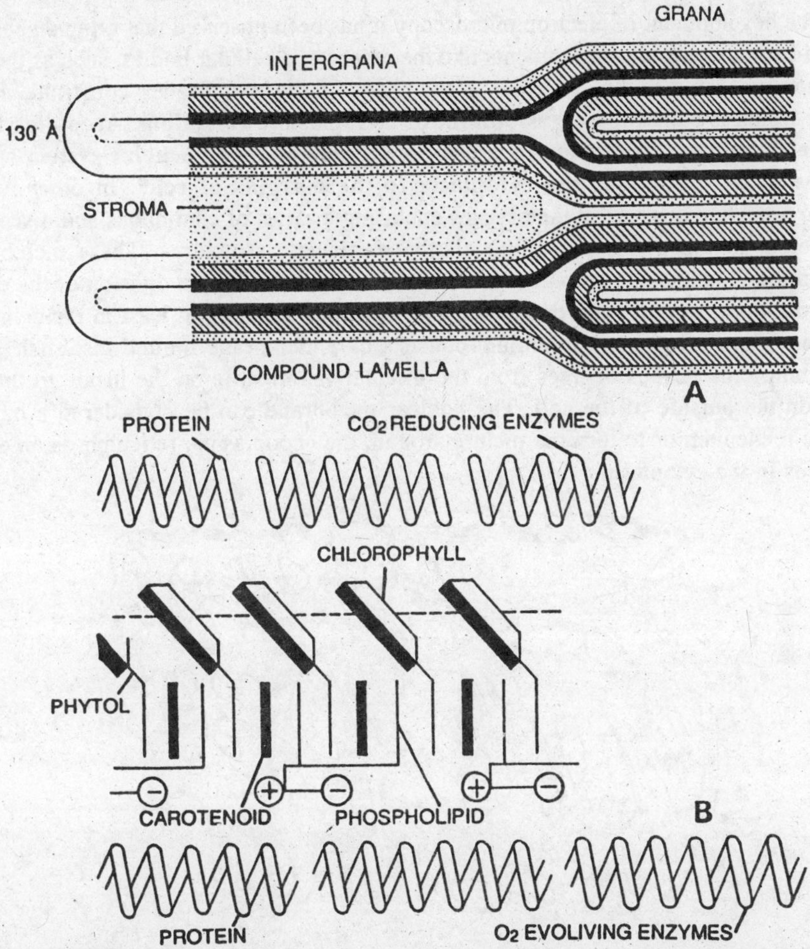

Fig. 33.39. Chloroplast. A, schematic representation of the layered arrangement of lamella, in the grana; B, possible arrangement of molecules in lamellae. The enzymes involved in photosynthesis are part of the protein layers, and the carotenoid and phospholipid molecules assist in the transfer of energy captured by the chlorophyll. (After A.J. Hodge and M. Calvin).

It seems quite obvious that light induces the activation of synthetic processes within the cell that result in the construction of the complex photosynthetic apparatus (*i.e.*, chloroplast). During the induced development process many new enzymes which were not present in the prochloroplasts are synthesized and organized. This phenomenon shows that control systems operate within the cells.

Functions

The main function of chloroplast is to take active part in photosynthesis. The photosynthesis is initiated by the capture of light energy through absorption in the green pigment chlorophyll inside the chloroplast.

The light energy trapped by chlorophyll can also be funneled through a series of enzymatically controlled reactions, into an energy compound called **adenosine triphosphate** (ATP). The chloroplast is, therefore, a dual energy converter, since the energy of sugars and ATP can be utilized by the cell in a variety of ways.

ENDOPLASMIC RETICULUM (ER)

With the development of electron microscopy it has been observed that ground substance of a cell is divided into several compartments like the other large cellular bodies, such as the nucleus and mitochondria. This highly ordered arrangement of the membranes constitute a cellular background which is called the **ergastoplasm** or **endoplasmic reticulum.** According to Porter (1961), **"the endoplasmic reticulum is a complex, finely divided vacuolar system extending from the nucleus throughout the cytoplasm to the margins of cell".** In other words the endoplasmic reticulum consists of membranes enclosing a series of continuous and discontinuous vacuoles, found in cytoplasm excluding all the particulate structures. These membranes of ergastoplasm may found to be arranged variously sometimes during the division of the cells they appear and disappear rapidly. But they form a conspicuous structure at least in the cells of liver and pancreas. The endoplasmic reticulum consists of a membrane-limited cisternal (sac-like) system extending, in various degrees from the **nuclear membrane** on the inside to the plasma membrane on the outside of the cell. The nuclear membrane can be considered a part of the endoplasmic recticulum or to turn the picture around, the endoplasmic reticulum is an extension of the nucleus in the cytoplasm.

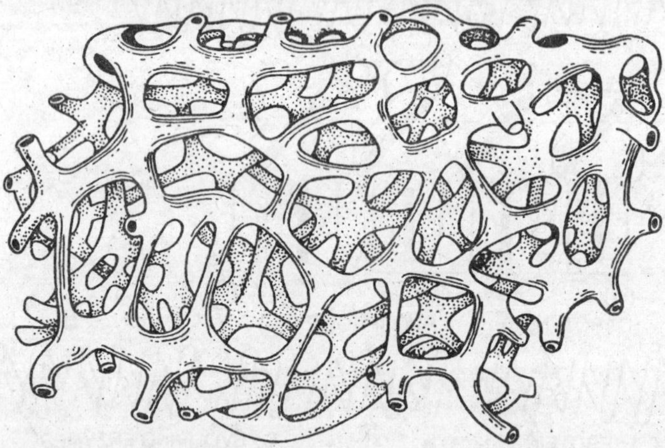

Fig.33.40. Endoplasmic reticulum (ergastoplasm). Three dimensional view.

Morphology. The endoplasmic reticulum is of a variable morphology, and each kind of cell has a characteristic ER. The ER can be loosely organized or its membranes can be tightly packed in the cytoplasm. Furthermore, the membranes can be either **rough or smooth.**

The rough, or **granular,** ER is found in great abundance in those cells engaged in protein synthesis. This chemical capacity resides in electron-dense particles (microsomes) that are rich in ribose nucleic acid (RNA). Since ribosomes live free in the cytoplasm, their association with ER is not necessary. On the other hand, the association of ribosomes with membranes, does not provide the intact cell with a means of compartmentalizing specific chemical reactions. Further the membranes of ER also provide a tremendous increase in surface area within the cell. If this is accepted that enzymes are part of membrane systems, then the cell can possess local patterns of synthesis.

The smooth, or **agranular,** ER, lacks the ribosomes. There is probably no sharp morphological discontinuity between smooth and rough kinds of ER, but the former is particularly prevalent in those cells engaged in the synthesis of fatty substances, *i.e.*, lipids in the cells of sebaceous glands, or steroid hormones in certain endocrine glands. The enzymes necessary for these syntheses appear to be a part of the membrane itself since they cannot be separated physically from the membrane fragments.

Function

Thus ER is a kind of cytoskeleton providing surfaces for chemical reactions, pathways for the transport of materials, and collection depots for synthesized materials. We also find the smooth ER in particular, in these cells that must maintain a particular shape; in these the ER is a structural component as well. The RNA-rich particles, called **microsomes,** are particularly active in the synthesis of proteins, while the membrane itself appears to be involved in the synthesis of **steroids,** a group of compounds which include certain of the hormones. We therefore look upon the endoplasmic reticulum as the principal manufacturing portion of our factory (cell), although we are far from a complete comprehension of its operation as whole.

RIBOSOMES

For the first time in 1955 Palade discovered the ribosomes in the cytoplasm. They are the intracellular sites of protein synthesis and are present in all the cells. They are very small granules (smaller than 1μ) and are visible only under the electron microscope. They are found frequently associated with membranes of endoplasmic reticulum. They are found attached to the outer surface of the double fold endoplasmic reticulum membrane. Sometimes this association is very tight. Their association with the endoplasmic reticulum has given rise to a number of different names, such as **'rough endoplasmic reticulum'** or **'granular endoplasmic reticulum'** as opposed to the **'smooth endoplasmic reticulum'** or **'agranular endoplasmic reticulum.'** In the smooth or agranular endoplasmic reticulum the membranes remain devoid of ribosomes. In tangential view the ribosomes seem to be arranged in definite patterns such as rosettes, spirals and circles. Though these two components (ER and ribosomes) are closely associated yet they are quite different in their functions and properties. They contain RNA (ribonucleic acid), nucleoproteins and enzymes needed in many synthetic reactions. The RNA is of very high molecular weight. The ribosome globules may be extracted out by centrifuging the cells. It is believed that they are mainly concerned with the synthesis of cellular proteins. These proteins control the chemical reactions going on within the cells. They are enzymic in nature and act as catalysts. The proteins of ribosomes appear to be basic in nature.

As electron microscopy reveals they appear as spherical to ellipsoidal bodies. The size varies from 150 Å in bacteria and chloroplast to 200 Å in the cytoplasm of plant and animal cells. Each ribosome consists of two distinctive sub-units, which differ in sizes. The two subunits are held together by bonds which appear to need magnesium ions for their continued stability.

Sub-units of ribosomes. The ribosome sizes are known in terms of speed with which they sediment in the centrifugal field. **Svedberg unit** (S) is the unit in measuring this speed. In this context, each plant cytoplasmic ribosome is typically a 80 S which possesses typically 60 S and 36 S particles or sub-units which in the natural conditions are probably held together by bonds which appear to need magnesium ions for their continued stability. The 60 S sub-unit (larger one) contains an RNA molecule whose molecular weight is about 1.5 million, *i.e.*, twice than those of

36 S sub-unit (smaller one) RNA. Further, the larger sub-unit contains about 30 proteins—all different from each other while the smaller sub-unit contains only 20.

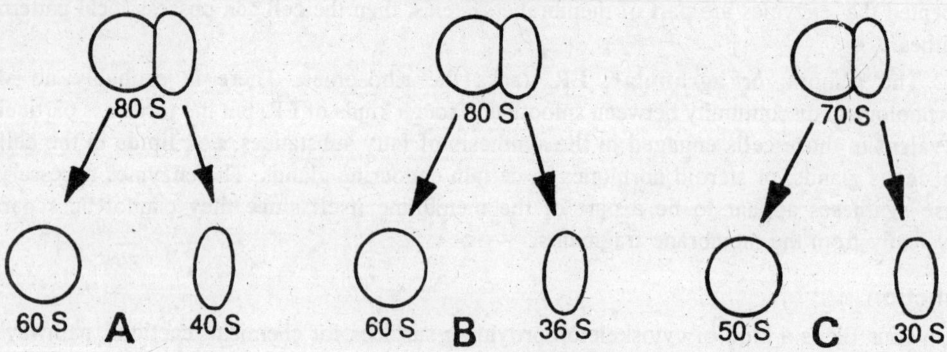

Fig. 33.41. Ribosomes. On the basis of sedimentation constant three classes of ribosomes can be distinguished—A, animal cytoplasmic ribosomes; B, plant cytoplasmic ribosomes; C, organelle bacterial ribosomes. The differences can be noted in the above-mentioned diagrams. (S—Svedberg unit).

The smaller particles are not active in protein synthesis. It is only the larger particles that are active. As regards protein synthesis, mRNA is specifically held by the smaller sub-units while tRNA seems to be specifically held by the larger sub-units. As a matter of fact, a portion of tRNA molecule is bound to the larger sub-unit and portion of it is bound to mRNA on the smaller sub-unit.

LYSOSOMES

They represent a new discovery among the cytoplasmic particles. They are found in most of the animal cells and in some meristematic cells of the plants. C. De Duve (1955) for the first time observed a fluffy layer of particles in the cytoplasm in which number of hydrolytic enzymes were present. The electron microscopy revealed that fluffy layer consists of membrane bounded structures whose appearance is markedly affected by the osmotic conditions present during preparation. De Duve called these structures, the lysosomes because of their contents of hydrolytic enzymes. As regards their origin it has been thought that they have been originated from the Golgi complex and the ribosomes. The lysosomes vary in size from 0.4μ to 0.8μ in diameter, but sometimes they are as large as 5μ in diameter, *e.g.*, in a mammalian kidney. They may be spherical,, rod like or irregular in shape, being covered by a single layered membrane composed of lipoprotein. They may be uniformly solid or some of them possess a very dense outer zone and airless dense core. The vacuoles or cavities are found within the granular material. They are distinguished from the mitochondria by their lack of internal cristae and by their characteristic enzyme content. The breakdown products diffuse out of the lysosomes and into the mitochondria, where they are further broken down in the process of respiration. They are also concerned with the digestion of proteins.

SPHAEROSOMES

The sphaerosomes are usually spherical bodies. They remain surrounded by membranes and range in diameter from 0.5 to 1.0μ. They are associated with the synthesis of fats and other such substances. They contain lipids and various types of lysosome-like enzymes. Sorokin (1966) reported the presence of a limiting membrane around each sphaerosome, however, James and Hilton (1972) do not agree with this view; they are in agreement with Yatsu that partial bounding membrane surrounds the sphaerosomes. Besides lipid storage and their transport, they are also thought to be intermediaries in the biosynthesis of waxes, cutin and other allied compounds, which result in cuticularization and suberization of cell walls.

LOMASOMES

They are generally small bodies found in algae, fungi and higher plants. Usually they are found in the periphery of cytoplasm. As regards their function they are supposed to be concerned with cell wall synthesis.

GLYOXYSOMES

They are small bodies and occur only in plant cells. They are supposed to contain enzymes which convert fatty acids into sugars

PEROXISOMES

These are microbodies and measure 0.2 to 2.0 mμ. They are found in all photosynthetic cells of higher plants in etiolated leaf tissue, in coleoptiles and hypocotyls, in tobacco stem and callus, in ripening pear fruits, and also in Euglenophyta, protozoa, brown algae, fungi, liverworts, mosses and ferns.

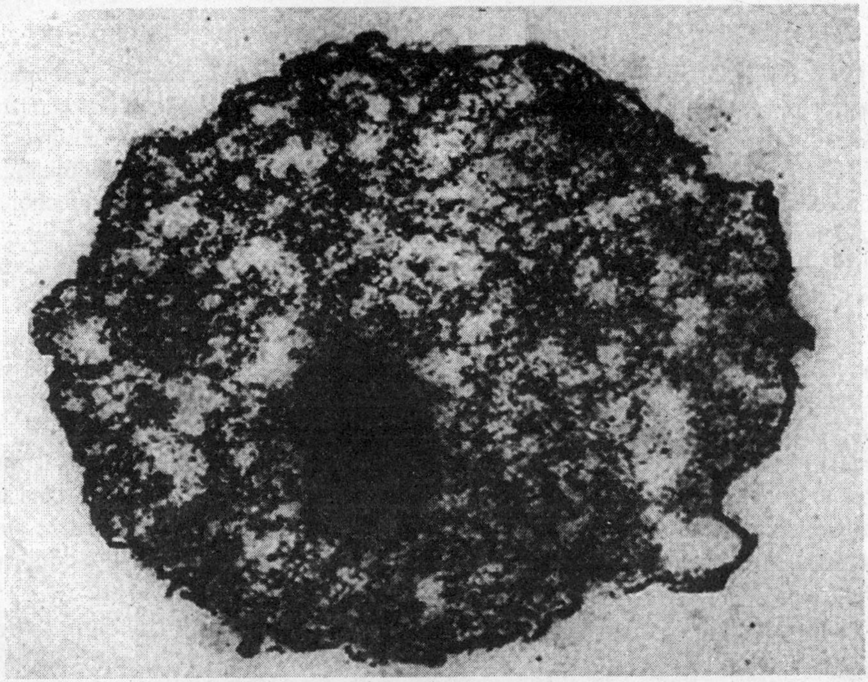

Fig. 33.42. Peroxisomes

Structure

Peroxisomes are variable in size and shape, but usually appear circular in cross section, having diameter between 0.2 and 2.0 mμ. They have a single limiting unit membrane of lipid and protein molecules, which encloses their granular matrix. In some cases (*e.g.*, *Festuca* grass) the matrix contains numerous threads or fibrils, while in others they are observed to contain either an amorphous nucleoid or a dense inner core which in many species shows a regular crystalloid structure (*e.g.*, tobacco leaf cell).

Functions

Peroxisomes are found to perform following two types of biochemical activities:

A. Hydrogen peroxide metabolism: Peroxisomes are so-called, because they usually contain one or more enzymes (*i.e.*), D-amino acid oxidase and urate oxidase) that molecular oxygen to remove hydrogen atoms from specific organic substances (R) in an oxidative reaction that produces hydrogen peroxide.

$$RH_2 + O_2 \longrightarrow R + H_2O_2$$

Catalase (which forms 40 percent of total peroxisome protein) utilises the H_2O_2 generated by other enzymes including alcohols, phenols, formic acid and formaldehyde by the peroxidative reaction.

$$H_2O_2 + R'H_2 \longrightarrow R' + 2H_2O$$

This type of oxidative reaction is particularly important in lirer and kidney cells, whose peroxisomes detoxify various toxic molecules that enter the blood stream. Almost half of alcohol one drinks is oxidised to acetaldehyde in this way. However, when excess H_2O_2 accumulates in the cell, catalase converts H_2O_2 to H_2O

$$2H_2O_2 \longrightarrow 2H_2O + O_2$$

H_2O_2 and aging. Most cytosolic H_2O_2 is produced by mitochondria and membranes of ER, alhough there are also H_2O_2– producing enzymes localised in the cytoplasm matrix. Catalase acts as a "safety valve" for dealing with the large amounts of H_2O_2 generated by peroxisomes; however, other enzymes such as glutathione peroxidase, are capable of metabolizing organic hydroperoxides and also H_2O_2, in the cytosol (cytoplasmic matrix) and mitochondria. The production of superoxide anion (O_2^-) in mitochondria and cytosol (cytoplasmic matrix) is regulated mainly by the enzyme superoxide dismutase. All of these protective enzymes are present in high levels in aerobic tissues.

B. Glycolate cycle. Peroxisomes of plant leaves contain catalase together with the enzymes of **glycolate pathway,** as glycolate oxidase, glutamate glyoxylate, serine-glyoxylate and asparate α-ketoglutarate aminotransferases, hydroxy pyruvate reductase and malic dehydrogenase. They also contain FAD, NAD and NADP coenzymes. The glycolate cycle is thought to bring about the formation of the amino acids - glycine and serene- from the non-phosphorylated intermediates of photosynthetic carbon reduction cycle, *i.e.*, glycerate to serene, or glycolate to glycine and serene in a sequence of reactions which involve chloroplasts, peroxisomes, mitochondria and cytosol. The **glycolate** pathway also generates C_1 compounds and serves as the generator of precursors for nucleic acid biosynthesis.

Photorespiration. In green leaves, these are peroxisomes that carry out a process called photorespiration which is a light stimulated production of CO_2 that is different from the generation of CO_2 by mitochondria in the dark. In photorespiration, glycolic acid (glycolate), a two-carbon product of photosynthesis is released from chloroplasts and oxidised into glyoxylate and H_2O_2 by a peroxisomal enzyme called glycolic acid oxidase. Later on, glyoxylate is oxidised into CO_2 and formate.

$$CH_2OH.COOH + O_2 \longrightarrow CHO—COOH + H_2O_2$$

$$CHO—COOH + H_2O_2 \longrightarrow HCOOH + CO_2 + H_2O$$

Photorespiration is so-called because light induces the synthesis of glycolic acid in chloroplasts. The entire process involves intervention of two basic organelles : chloroplasts and peroxisomes.

Lastly, photorespiration is driven by atmospheric conditions in which the O_2 tension is high and the CO_2 tension low. Apparently O_2 competes with CO_2 for the enzyme **ribulose diphosphate**

carboxylase which normally is the key enzyme in CO_2 fixation during photosynthesis. When O_2 is used by the enzyme, an unstable intermediate is formed which breaks down into 3-phosphoglycerate and phosphoglycolate. The latter tends to increase the glycolate concentration by removal of its phosphate group, and therefore, more glycolate is available for additional oxidation and CO_2 release.

C.–β– Oxidation. Peroxisomes of rat liver cells contain enzymes of β-oxidation for the metabolism of fatty acids. They are capable of oxidising palmitoyl-CoA (or fatty acyl-CoA) to acetyl–CoA, using molecular oxygen and NAD as electron acceptors. The acetyl-CoA formed by this process is, eventually, transported to the mitochondria where it enters into the citric acid cycle. If alternatively, acetyl-CoA remains in the cytosol, it is reconverted into fatty acids and ultimately to neutral fats. β-oxidation pathway of the peroxisomes is very similar to the one that occurs in mitochondria, with one very important exception. In mitochondria the flavin dehydrogenase donates its electrons to the respiratory chain. It doe not react with molecular oxygen. In peroxisomes, the dehydrogenase reacts directly with O_2 and in so doing generates H_2O_2. Mitochondria contain no catalase and therefore, cannot deal with the formation of toxic hydrogen peroxide. For peroxisomes, this is not a problem.

D. Other functions. Mammalian cells do not contain D-amino acids, but the peroxisomes of mammalian liver and kidney contain **D-amino acid oxidase.** It is suggested that this enzyme is meant for D-amino acids that are found in the cell wall of the bacteria. Thus, the presumed role of this enzyme is to initiate the degradation of D-amino acid that may arise from breakdown and absorption of peptidoglycan material of intestinal bacteria.

Uric acid oxidase (uricase) is important in the catabolic pathway that degrades purines. Thus, peroxisomes are unusually diverse organelles and even in different cells of a single organism may contain very different sets of enzymes. They can also adapt remarkably to changing conditions. For example, yeast cells grown on sugar have tiny peroxisomes. But when in some yeasts are grown on methanol, they develop large-sized peroxisomes that oxidize methanol; when grown on fatty acids, they develop large peroxisomes that break down fatty acids to acetyl-CoA.

Biogenesis of Peroxisomes

At one time it was thought that the membrane 'shell' of the peroxisomes is formed by budding of the endoplasmic reticulum (ER), while the 'content' or matrix is imported from the cytosol (cytoplasmic matrix) However, there is now evidence suggesting that now peroxisomes always arise from pre-existing ones being formed by growth and fission of old organelles similar to mitochondria and chloroplasts.

Thus peroxisomes are a collection of organelles with a constant membrane and a variable enzymatic content. All of their proteins are encoded by nuclear genes and are synthesised in the cytosol.

The proteins present in either lumen or membrane of the peroxisome are taken up post-translationally from the cytosol. For example, catalase enzyme is a tetrameric haeme-containing protein that is made in the cytosol (cytoplasmic matrix) as the haeme-free monomers; the monomers are imported into the lumen of peroxisomes, where they assemble into tetramers in the presence of haeme. Catalase and many peroxisomal proteins are found to have a **signal sequence** (comprising of three amino acids) which is located near their carboxyl ends and directs them to peroxisome. Peroxisomes contain **receptors** exposed on their cytosolic surface to recognise the signal on the imported proteins. All of the **membrane proteins** of the peroxisomes, including signal receptor proteins, are imported directly from the cytosol. The lipids required to make new peroxisomal membrane are also imported from the cytosol, possibly being carried by **phospholipid transfer proteins** from sites of this synthesis in the ER membrane.

MICROTUBULES

For the first time Ledbetter and Porter (1963) demonstrated the presence of fine elongated tubular structures in the peripheral cytoplasm of some plant cells. These tubules are usually found to be attached to the cell wall. They are thought to be responsible for the orientation of cellulose synthesis. These structures are very clearly visible during the process of mitosis as cytoplasmic fibrils. They are several microns in length and their diameter ranges from 150 to 250 Å with a wall of about 60 Å in thickness. According to some workers they are thought to be fibres of spindle apparatus.

GOLGI COMPLEX

Camillio Golgi (1898), an Italian neurologist, first discovered these organelles in animal cells. The **Golgi complex,** so named after its discoverer is characteristic system of cytoplasmic membranes. The term, **dictyosome,** is also used to describe this structure. It is comparable to the smooth endoplasmic reticulum, but discontinuous, smaller and more compact in nature. It is still debatable that there is a direct morphological relation between the endoplasmic reticulum and the Golgi complex. However, it was early recognized by its affinity for osmium of silver-containing stains. In modern days its existence has been clearly demonstrated with the electron microscope. They are found in living cells of both animals and plants, immersed in their cytoplasm.

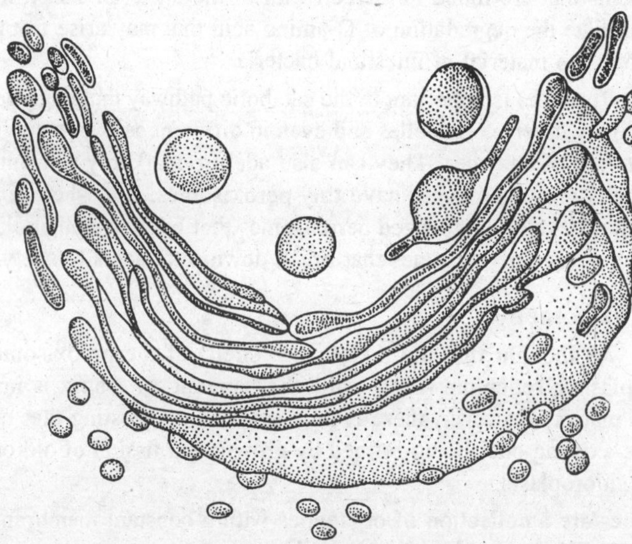

Fig. 33.43. The diagram indicates the organisation of the Golgi complex. At the convex "forming face" small vesicles containing the products of the synthetic activity of the cell are shown coalescing with the outermost saccule or cisterna. At the concave or "maturing face" of the complex, saccules containing-concentrated secretory product are rounding up to form secretory droplets or granules.

Various workers have given various names to this complex, such as **dictyosome,** idiosome, lipochondria, Golgi body, Golgi substance, Golgi apparatus and Golgi complex. Generally the name Golgi complex is used for the material invertebrates and dictyosome in invertebrates and plants. However, the Golgi complex is not found in bacteria and blue green algae.

Structure

The shape of Golgi complex is quite variable in somatic cells of plants and animals. Even in the same cell there are variations with functional stage. However, the shape is characteristic for each cell type. Their appearance depends partly on the position of cells within the plant or animal body and partly on the way in which cells are prepared for microscopic study. These bodies look like the complexes of the droplets and therefore, termed as Golgi complex. They appear thin plate like layers of irregular arrangement. Each complex consists of stacks of parallel lamellae. In some cases, it occurs as a dense reticulum of anastamosing trabeculae while in others as an irregular plaque, a ring, hollow spheres united together. In nerve cells it occurs as a reticulum of wide meshes around the nucleus. The size of Golgi complex is also variable. It is small in muscle cells but quite large in the nerve and gland cells. Generally the complex is well developed in an active cell while in old cells the complex diminishes in size and ultimately disappears.

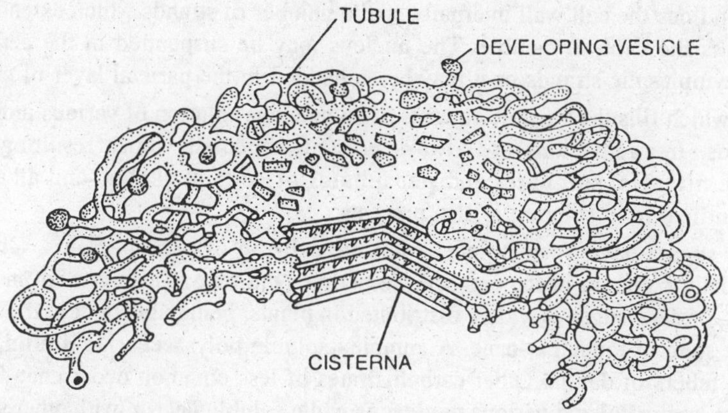

Fig. 33.44. Golgi complex. Diagrammatic representation of the morphology of Golgi body (dictyosome),

The electron microscopic observations reveal as follows. The Golgi complex comprises of 1. flattened sacs or cisternae, 2. large clear vacuoles and 3. clusters of dense vesicles.

1. Flattened sacs. These are also known as **cisternae.** These structures are similar to the smooth surface endoplasmic reticulum and appear in sectional view as dense parallel membranes. The packed cisternae are often arrayed concentrically, enclosing regions of the cytoplasm filled with numerous large vesicles.

2. Large vacuoles. These vacuoles are clear and commonly found at the edge of Complex. These structures represent the modified and expanded flattened sacs. In the cells of liver and pancreas these vacuoles of complex contain dense masses or granules.

3. Vesicles. These are the clusters of dense vesicles. They remain usually associated with cisternae and show continuity with them. The flattened sacs give rise to them by means of budding or pinching off. These structures remain dispersed in the surrounding hyaloplasm.

Functions

The Golgi bodies are believed to function in the manufacture of cellular secretion products. Since its membranes have little RNA it is not connected with protein synthesis. The general consensus of modern days is that it is somehow involved in the storage and possible modification of lipid substances, a point of view supported by observations that the appearance of the Golgi complex in animal cells can be modified greatly by changes in fat diet.

During the maturation of sperm, the Golgi complex plays a role in the formation of acrosome (Burgos and Fawcett, 1955).

The complex also activates mitochondria to produce the ATP which is utilized in respiratory cycle, nervous transmission, and nucleic acid and protein synthesis; the latter occurs in the vicinity of endoplasmic reticulum.

NON PROTOPLASMIC COMPONENTS: VACUOLES

The vaçuoles (latin, *vacuus* = empty) are non-living inclusions of cytoplasm. Some of these are found in the form of minute droplets in the cytoplasm of meristematic cells. Each droplet is a small vacuole filled up with cell-sap. These small vacuoles enlarge and finally coalesce to give rise to a large vacuole. The enlargement of the vacuoles is accompanied by the entry of water into the cell. It becomes quite evident that with the entry of water into the cell, the size of the cell will increase, and, in fact, the cell attains a volume many times greater than the volume of the original meristematic cell. During this process of cell enlargement by vacuolation the nucleus and cytoplasm do not greatly increase in size, and in the fully vucuolated cell the cytoplasm forms a parietal layer which lines the cell wall internally, and a number of strands which extend from this parietal layer across the central vacuole. The nucleus may be suspended in the centre of the vacuole by these cytoplasmic strands or it may be embedded in the parietal layer of cytoplasm.

The cell-sap which fills the vacuole consists of an aqueous solution of various inorganic and organic compounds. Inorganic ions are always present, especially those resulting from the dissociation of the **nitrates, sulphates** and **phosphates** of certain alkaline and alkaline-earth metals, and **carbondioxide** in solution is also present.

A number of sugars have been isolated from the cell-sap of various plants. **Glucose** and **fructose,** are specially common in the cell-sap. Sometimes **maltose** is also present in the sap of leaf cells. **Sucrose,** or cane sugar is widely distributed in plants, being abundant in the root of the sugar-beet and the stem of the sugar-cane. A complex soluble polysaccharide, **inulin,** occurs in the cell-sap of the tubers of dahlia. Other carbohydrates of less common occurrence in the cell-sap include **dextrin, mannitol** and various **pentosans** while soluble **pectin**, with other substances, sometimes occurs in appreciable amount. **Glycosides** are also of common occurrence in the cell-sap. **Mucilaginous compounds** also occur in cell-sap and they give the slimy character to the cytoplasm. **Mucilages** are found abundantly in the cell-sap of many bulbs, *e.g.,* onion, and in the leaves of many succulent plants. They are polysaccharides in nature. **Tannins** may also be present in cell-sap. They are complex substances and soluble in water and alcohol. **Anthocyanin pigments** are frequently present in cell-sap. These pigments give blue or red colours to the cell-sap. Malic, citric and tartaric acids are all of common occurrence, and the acidity of many fruits is due to the presence of these acids in the cell-sap. Oxalic acid is also found in the cell-sap, in the form of its insoluble calcium-salt, and as such is usually regarded as a waste product of cell metabolism. Calcium oxalate occurs in a crystalline form and the crystals may be of various types. **Calcium carbonate** occurs only infrequently in plants in the form of **cystoliths.** Soluble **nitrogenous compounds** may also be present in the cell-sap. **Ammonium compounds** may be present too.

Besides the proteins which form a part of living protoplasm, **non-living proteins** may occur either dissolved in the cell-sap or in the form of crystal-like bodies, **crystalloids.** In many seeds, especially those containing oil, the vacuoles contain large amounts of **dissolved proteins,** and as the seed dries out and ripens these proteins may be transformed into **aleurone grains.** Fats and oils are frequently present in plant cells, especially in certain seeds. They occur as droplets or globules, and may be present either in the cell-sap or protoplasm or both. Other substances sometimes present include basic substances such as the **purine bases** and the **alkaloids. Resins** as found also, but are of limited distribution, being confined to special cells or ducts in certain plants and essential oils are produced by some plants. These essential oils are usually mixtures of complex organic compounds such as **alcohols, esters, ketones, aldehydes** and **hydrocarbons. Resin** also is not itself a simple chemical substance, but a mixture of substances insoluble in water and containing **resin acids,** resin esters and **rosenes,** all of which are complex substances of high molecular weight.

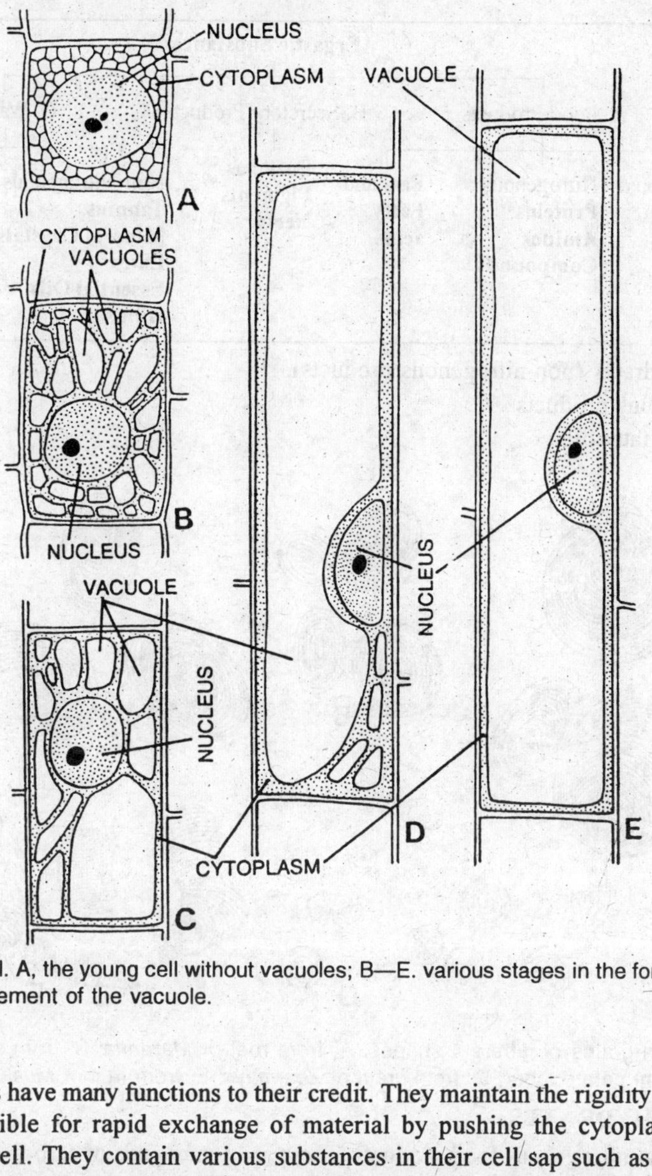

Fig. 33.45. The cell. A, the young cell without vacuoles; B—E. various stages in the formation and enlargement of the vacuole.

Functions

The vacuoles have many functions to their credit. They maintain the rigidity of the plant cell. They are responsible for rapid exchange of material by pushing the cytoplasm towards the periphery of the cell. They contain various substances in their cell sap such as enzymes, acids, salts, etc., which act in several ways. They store waste products. The vacuoles contain ions and molecules, and therefore, they maintain osmotic relationships of the cell. They help in the enlargement of the cell. Several metabolic activities take place in the cell sap of vacuoles.

ERGASTIC SUBSTANCES

Several chemical compounds occur as non-living inclusions in the cytoplasm. These inclusions remain dispersed either in cell-sap or in cytoplasm. Principally they are classified as follows:

A. FOOD PRODUCTS

These are the substances manufactured by the protoplasm of the cells from the simple inorganic substances like carbon dioxide and water and stored in the cells as food materials. The food thus manufactured is partly used up to make new protoplasm and partly it is broken down to provide necessary energy, and remainder of it is stored as reserve food material in protoplasm of cells. The reserve food materials may be classified as follows:

Ergastic Substances

- A. Food Products
  - Non-Nitrogenous: **Starch**, **Inulin**, **Hemicellulose**, **Cellulose**, **Sugars**
  - Nitrogenous: **Proteins**, **Amino-Compounds**
  - **Fats and Fatty acids**
- B. Secretory Products
  - Enzymes
  - **Pigments**
  - **Nectar**
- C. Waste Products
  - Non-Nitrogenous: **Tannins**, **Mineral Crystals**, **Latex**, **Essential Oils**, **Gums**
  - Nitrogenous: **Alkaloids**

1. Carbohydrates (non-nitrogenous products).
2. Nitrogenous products
3. Fats and fatty oils.

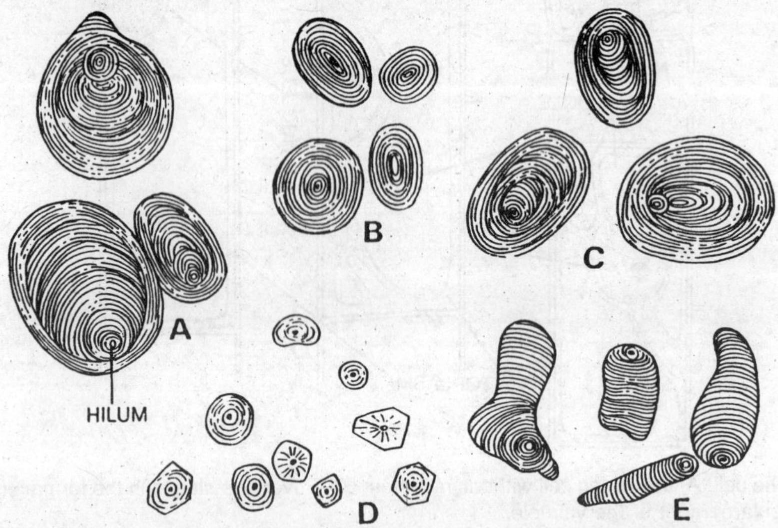

Fig. 33.46. Starch grains of different shapes. A, from root of *Marantia*; B, from seed of *Phaseolus*; C, from potato tuber; D, from grain of *Zea mays*; E, from fruit of *Musa*.

1. CARBOHYDRATES

They are non-nitrogenous food products. These are compounds of carbon (C), hydrogen (H), and oxygen (O). They are derived more or less directly from carbon dioxide (CO_2) and water (H_2O) during photosynthesis. Some carbohydrates are insoluble while some are soluble in water. The most important insoluble product is starch and the soluble products are inulin, sugar, etc.

***(i)* Starch.** It is an insoluble carbohydrate of polysaccharide type formed by the condensation of simple sugars like glucose. The starch is usually found in the form of starch grains of various shapes. The starch grains are abundantly found in the storage organs of plants, *e.g.*, tuberous roots, underground stems, cortex of stems, endodermis, grains of cereals, banana fruits, etc. The starch grains vary in their shape and may be used for the identification of plants. The starch grains are not found in fungi and certain algal groups. The starch grains have different shapes that are characteristic of the plant types, *e.g.*, they are oval-shaped in potato; flat in wheat; polygonal in maize; spherical in pulses and dumbell or rod-shaped in the latex cells of some *Euphorbias*. The starch grains of rice are smallest and those of *Canna* largest. The starch grains vary from 5–100μ in size. The starch is always derived either from chloroplasts of green cells or from leucoplasts (amyloplasts) of storage tissue.

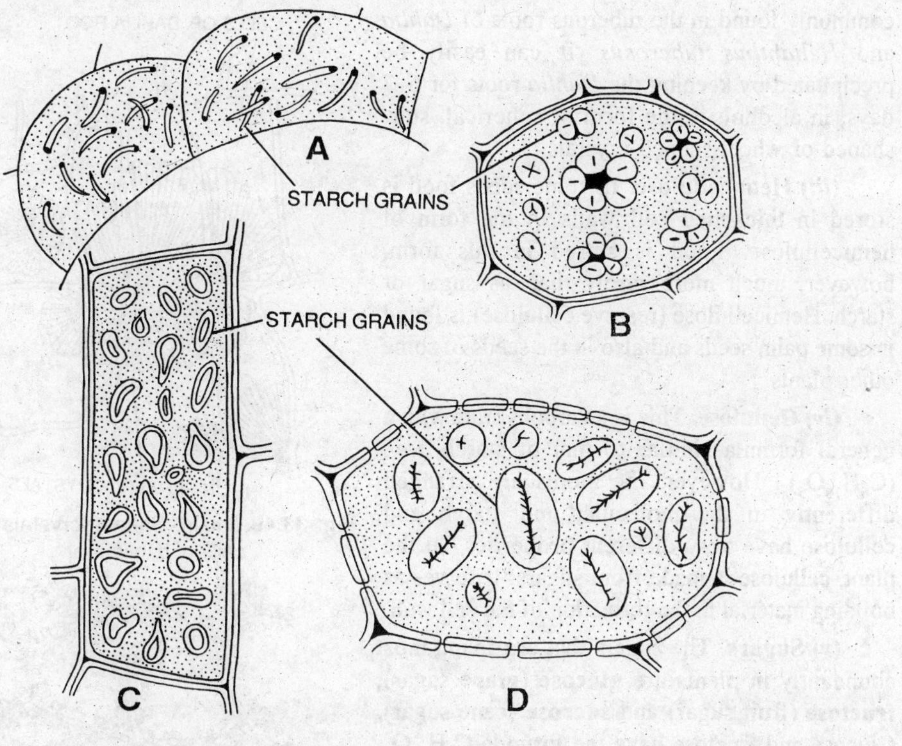

Fig. 33.47. Starch grains. A, starch grains in outer pericarp of *Musa*, B, starch grains in cotyledon of *Pisum*; C, starch grains in ray cell of phloem of *Ailanthus*; D, starch grains in cotyledon of *Phaseolus*.

The structure of the starch grain usually exhibits conspicuous concentric layers formed around a dark roundish spot, the **hilum.**

The layering may be conspicuous in some grains whereas inconspicuous in others. Most of the starch grains show this layering and are known as stratified starch grains. If the concentric layers of the starch are formed on one side of the hilum of starch grain, the grain is said to be **eccentric** (*e.g.,* potato) and when the layers are deposited concentrically around the hilum (*e.g.,* wheat) the grain is known as **concentric.** Concentric types of starch grains are quite common in most of the plants. If the starch grain possesses a single hilum, it is known as **simple.** Sometimes two, three or many grains, arranged in a group with as many hila as the starch grains, they are known as **compound** grains. Compound grains are commonly found in potato, sweet potato, rice and oats. The starch is turned blue or black in aqueous solution of iodine.

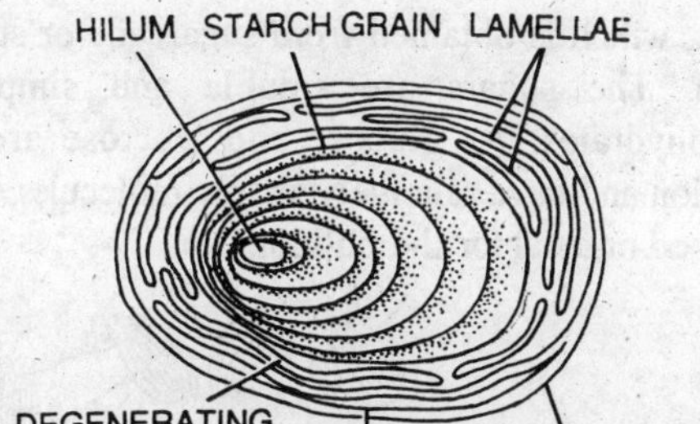

Fig. 33.48. Plastids. A, amyloplast with a starch grain; B, a leucoplast.

***(ii)* Inulin.** This is also a carbohydrate of polysaccharide type. It is soluble carbohydrate usually found in the cell-sap. Inulin has been reported from the roots of many Compositae. It is commonly found in the tuberous roots of *Dahlia* and *Helianthus tuberosus*. It can easily be precipitated by keeping the *Dahlia* roots for 6–7 days, in alcohol, in the form of spherical, star-shaped or wheel-shaped crystals.

***(iii)* Hemicellulose.** In some seeds food is stored in thickened cell walls in the form of hemicellulose. Food is stored in this form, however, much more rarely than as sugar or starch. Hemicellulose (reserve cellulose) is found in some palm seeds and also in the seeds of some other plants.

***(iv)* Cellulose.** This is a carbohydrate with a general formula similar to that of starch, *i.e.*, $(C_6H_{10}O_5)_n$. However, the atoms are arranged differently in the molecule, and starch and cellulose have very different properties. In the plant, cellulose is made from sugars. It serves as building material in the formation of the cell wall.

***(v)* Sugars.** The sugars that are found most abundantly in plants are **glucose** (grape sugar), **fructose** (fruit sugar) and **sucrose** (cane sugar). Glucose and fructose have the formula $C_6H_{12}O_6$. They are thus composed of the same kinds of atoms in the same proportion, but the arrangement in the molecule is different. The simplest natural sugars are monosaccharides. Sucrose $(C_{12}H_{22}O_{11})$ has twice as many carbon atoms and is a disaccharide. Sucrose is known as the ordinary sugar, which is obtained from sugarcane or sugar beets. The sugars are soluble and simplest carbohydrates. As glucose and fructose are in solution and have relatively simple molecules, they are good material for the building up

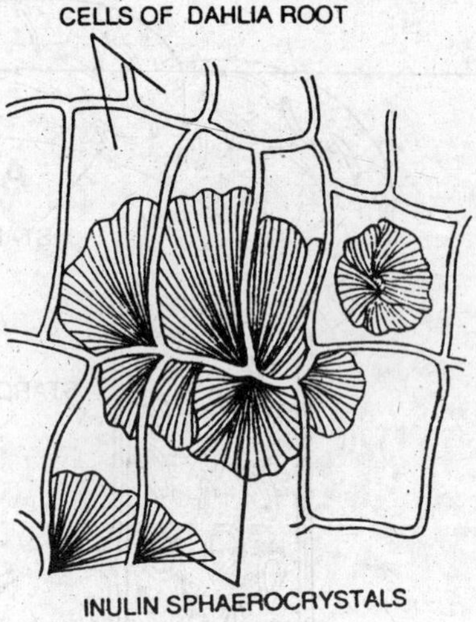

Fig. 33.49. Inulin. Sphaerocrystals of inulin in cells of dahlia root.

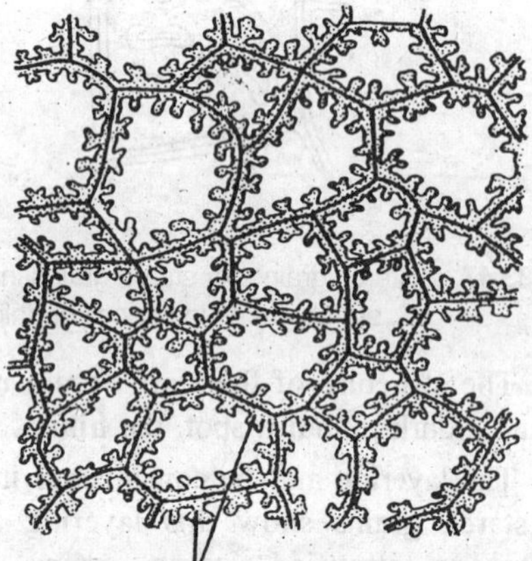

Fig. 33.50. Hemicellulose. Thickened walls of hemicellulose from *Areca* seed.

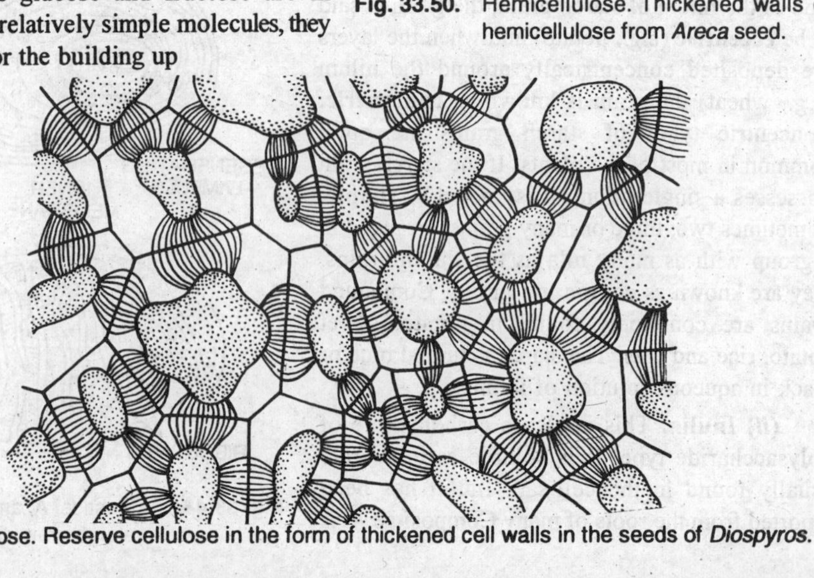

Fig. 33.51. Cellulose. Reserve cellulose in the form of thickened cell walls in the seeds of *Diospyros*.

of other substances or for the furnishing of energy. In sugarcane and sugarbeet sucrose is found as reserve food.

2. NITROGENOUS PRODUCTS

The important nitrogenous food materials are proteins and amino compounds.

***(i)* Proteins.** The proteins are the most important group of compounds found in plants, as they constitute the active matter of protoplasm, and the chemical phenomena of life processes are associated with them. The proteins are exceedingly complex organic nitrogenous substances constituting of carbon (C), hydrogen (H), oxygen (O) and nitrogen (N). In certain more complex proteins sulphur (S) and phosphorus (P) are also present. For example, representative proteins as zein $C_{736}H_{1161}N_{184}O_{208}S_3$ from Indian corn and gliadin $C_{685}H_{1068}N_{196}O_{211}S_5$ from wheat. Proteins are not only the main constituents of protoplasm but, in the form of solid granules, are frequently found in plants as reserve food material. Proteins are formed by a rearrangement of the atoms of carbohydrates with the addition of nitrogen, commonly sulphur and sometimes phosphorus. A protein molecule is made up of hundreds or thousands of amino acid molecules joined together by peptide links into one or more chains, which are variously folded. There are twenty different kinds of amino-acids commonly found in proteins, and most of these usually occur in any one protein molecule; they are arranged in the chain in a sequence which is exactly the same in all molecules of a given kind of protein. The possible different arrangements of the amino acids are evidently practically infinite, and the diversity is fully exploited by living things every species having kinds of protein molecule peculiar to itself.

Amino-acids are peculiar in that they have both basic and acid properties. The proportion of different amino-acids in different proteins varies; and some proteins lack amino-acid found in other proteins. Animal proteins are better for human food than plant protein because the amino-acid content of animal proteins is more like that of human proteins than is the amino acid content of plant proteins. Some plant proteins lack entirely some of the amino acids which are essential for the building of human proteins.

A common form of protein found in the endosperm of castor seed is known as **aleurone grains.** Each aleurone grain is a solid, ovate or rounded body which usually encloses a crystal like body in it,

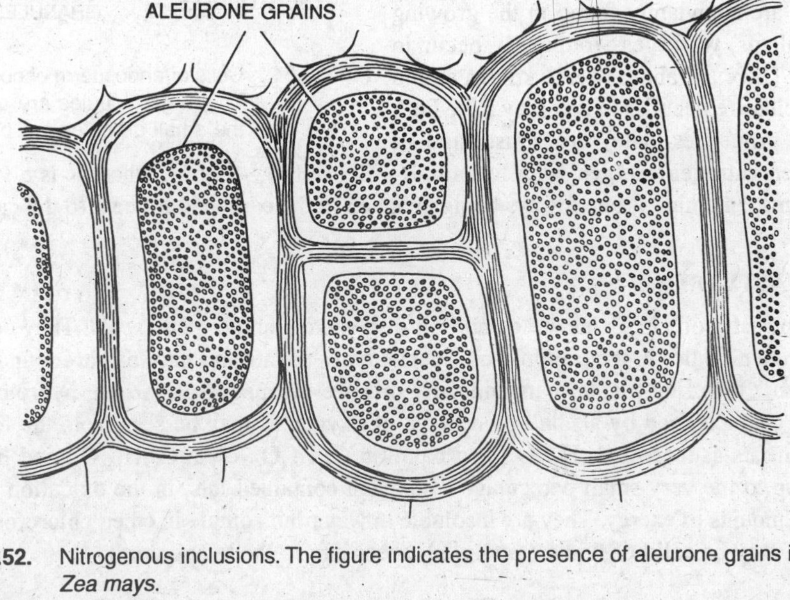

Fig. 33.52. Nitrogenous inclusions. The figure indicates the presence of aleurone grains in endosperm of *Zea mays.*

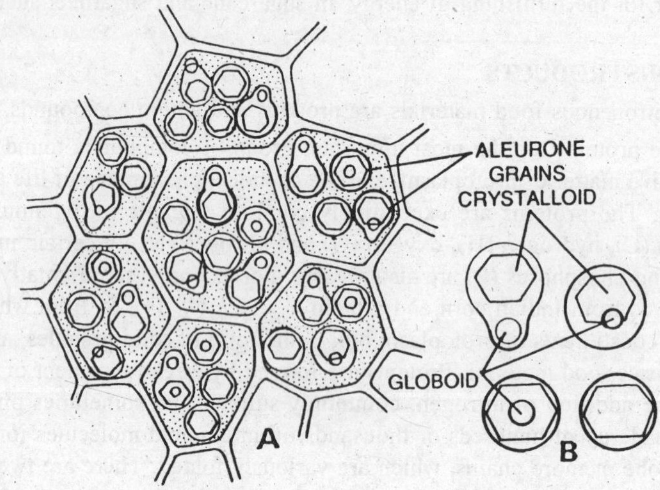

Fig. 33.53. Proteins. A, aleurone grains in the endosperm of castor seed; B, few grains magnified containing crystalloids and globoids in them.

known as **crystalloid** and a globule like body known as *globoid.* The crystalloid is proteinaceous in nature and occupies the major portion of aleurone grain, whereas the globoid is double phosphate of calcium (Ca) and magnesium (Mg) and occupies the narrower part of the grain. The aleurone grains vary in their shape and size in different plants. When they are found in oily seed of castor they are larger in size, while when found with starch they are very minute in size.

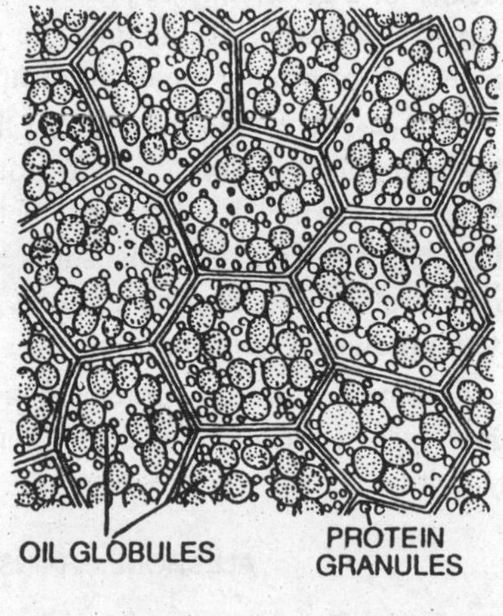

Fig. 33.54. Cells of endosperm of coconut. Thel arge globules are oil and the small granules are protein.

***(ii)* Amino compounds.** They are simple nitrogenous food materials. They are found in the form of amino-acids and amines which occur in the cell sap. These are abundantly found in the growing apices of the plants, while less frequently occur in storage tissues. There are about twenty known amino acids. Amino acids are peculiar in that they have both basic and acid properties. They are constituted of carbon, hydrogen, nitrogen, oxygen, with formula R—CH (NH_2)—COOH, where R is a variable grouping of atoms, an amino group always being attached to the carbon atom next to the carboxyl group.

3. FATS AND FATTY OILS

The fats and fatty oils in plants are composed of glycerine and organic acids. They occur in the form of minute globules in the protoplasm. Special types of fats and fatty oils are found in the seed and fruits of flowering plants. Fats and fatty oils are composed of carbon, hydrogen and oxygen and are characterized by a small percentage of oxygen as may be seen from the formula of such common fats as stearin $C_{57}H_{110}O_6$, and palmitin $C_{51}H_{98}O_6$, olein $C_{57}H_{104}O_6$, and linolein $C_{57}H_{98}O_6$. Owing to the very small percentage of oxygen contained fats, in the oxidation of fats produces large amounts of energy. They are insoluble in water but soluble in ether, chloroform and petroleum. Fats, may be solid or liquid (fatty oils), according to the temperature.

B. SECRETORY PRODUCTS

Besides food material several other products are also secreted by protoplasm which are not useful as nutritional products but they may help or accelerate the various reactions in the cells. These are as follows:

***(i)* Enzymes.** They are soluble nitrogenous substances secreted by the protoplasm. They are digestive in function and convert the insoluble substances into soluble and complex compounds into simple ones, *e.g.* **diastase** converts starch into sugar, so that because of the action of this enzyme an insoluble substance converts into soluble one. The **lipase** breaks up fats into their components, glycerine and fatty acids. **Papain** converts proteins into amino acids.

***(ii)* Colour in cell.** The substances that give colour to cells are usually found in the plastids. Chlorophyll is a green colouring matter secreted by the chloroplasts and performs the phenomenon of photosynthesis. Chlorophyll is not a single compound but it is mixture of two pigments known as chlorophyll *a* and chlorophyll *b.* The yellow pigments, carotenoids are also found in cell sap and give colours to the petals of the flowers. The anthocyanins are also secretary products of the protoplasm and are stored in the cell sap; they also impart colour to the petals of the flowers.

***(iii)* Nectar.** The nectar is another useful secretion of protoplasm. It is secreted by special glands or organs of the flower, called nectaries.

Nectaries. The nectaries occur on flowers (floral nectaries) and on vegetative parts (extrafloral nectaries). The floral nectaries are found in various positions on the flower, whereas the extrafloral nectaries occur on stems, leaves, stipules and pedicels of flowers.

The secretory tissue of the nectary is generally found in the epidermal layer. Usually the secretory epidermal cells have dense cytoplasm and may be papillate or elongated like palisade cells. In many nectaries, the cells found beneath the epidermis are also secretory. They have dense cytoplasm and thin walls. The nectary is covered by a cuticle.

The sugar of the nectaries (both floral and extrafloral), is derived from the phloem. Vascular tissue is found just near the nectary. In some nectaries the vascular tissue is merely that of the organ bearing the nectary, in others it is part of the nectary.

The nectar is excreted either through the cell wall and the ruptured cuticle or sometimes through stomata.

***(iv)* Osmophors.** The fragrance of flowers, is generally produced by volatile substances distributed throughout the epidermal region of perianth parts. However, in some plants the fragrance originates in special glands known as **osmophors** (Vogel, 1962). Examples of such special glands are commonly found in Asclepiadaceae, Aristolochiaceae, Araceae, Burmanniaceae and Orchidaceae. Various floral parts are differentiated as osmophors and they

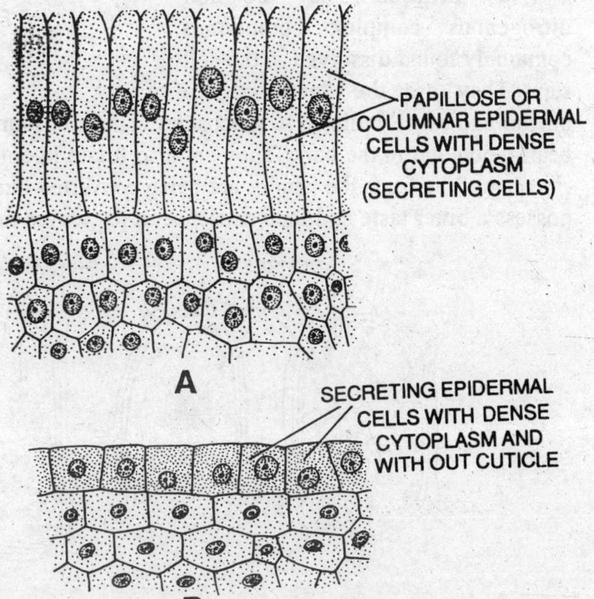

Fig. 33.55. Nectary. A, section of surface of nectary of *Euphorbia pulcherrima*; B, section of floral nectary of *Malus pumila.*

assume the form of flaps, cilia or brushes. The prolongation of the spadix of Araceae and the insect-attracting tissue in the flowers of Orchidaceae are osmophors. The osmophors have a secretory tissue usually several layers in depth.

C. WASTE PRODUCTS

They are excretions of the plant cells. Usually these products are stored in the dead cells. These products are formed as a result of metabolic activities of protoplasm, and therefore, they are known as metabolic wastes of the plants. Usually there are two types of metabolic wastes. They are 1. Non-nitrogenous waste products and 2. Nitrogenous waste products.

1. Non-nitrogenous waste products.

They are tannins, mineral crystals, latex, essential oils, gums, resins, and organic acids. They are either found in the cytoplasm or in the cell-sap.

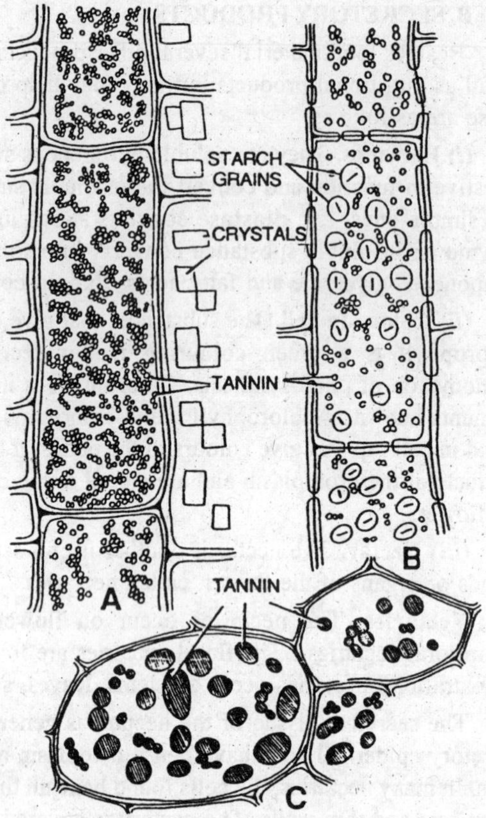

Fig. 33.56. Tannin, starch grains and crystals, A, Tannin and crystals in phloem parenchyma of *Pinus*; B, tannin and starch grains in ray cells of wood of *Malus pumila*; C, tannin in pith cells of *Fragaria*.

(*a*) Tannins. They are non-nitrogenous complex compounds, commonly found dissolved in the cell-sap. They are the derivatives of phenol and usually related to glucosides. They are found in the cell walls, in the dead cells, in the heart wood and in the bark. They are also found in abundance in leaves and unripe fruits. On the ripening of the fruits the tannins are being converted into glucose and other substances. The tannins possess a bitter taste and their presence in tea leaves make the tea decoction bitter. They are used

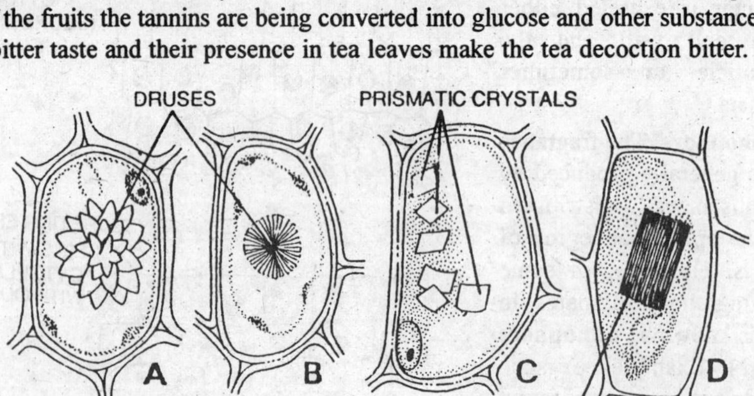

Fig. 33.57. Cells with different types of crystals. A, and B, druses from the cells of *Gnetum gnemon*; C, prismatic crystals from cell of *Gnetum indicum*; D, a bundle of raphides in leaf cell of *Vitis vinifera*.

in tanning industry. The commercial **katha** is prepared from the tannins of the heart wood of *Acacia catechu.* The tannins are poorly developed in monocotyledonous plants. They are either found in individual cells or in special organs, called the **tannin sacs.** The tannin-containing cells often form connected systems. In the individual cells the tannins occur in the protoplast and may also impregnate the walls, in cork tissue. Within the protoplast, tannins are a common ingredient of the vacuoles. As regards their function they act as protectants of protoplasm against injury, decay, termites and pests; as reserve substances related in some manner to the starch and metabolism; as substances associated with the formation and transport of sugars; as antioxidants; and as protective colloids maintaining the homogeneity of the cytoplasm.

(*b*) **Mineral crystals.** Various types of mineral crystals occur in the plant cells. They may occur either in the cell cavity or in the cell walls. The crystals usually lie loose in the cells but sometimes they are found to be suspended into the cell cavity from the cell wall.

The crystals vary in shape and size. Usually the crystals consist of calcium carbonate, calcium oxalate or silica.

(*i*) **Calcium carbonate crystals.** These crystals are commonly known as **cystoliths.** Usually they occur in the epidermal cells of leaves of many flowering plants. They are generally found in under mentioned angiospermic families—Moraceae, Urticaceae, Acanthaceae, Cucurbitaceae, etc. The main body of cystolith is a cellulose extension of the cell wall in which the calcium carbonate is deposited in the form of fine granules. In certain plants double cytoliths are also found.

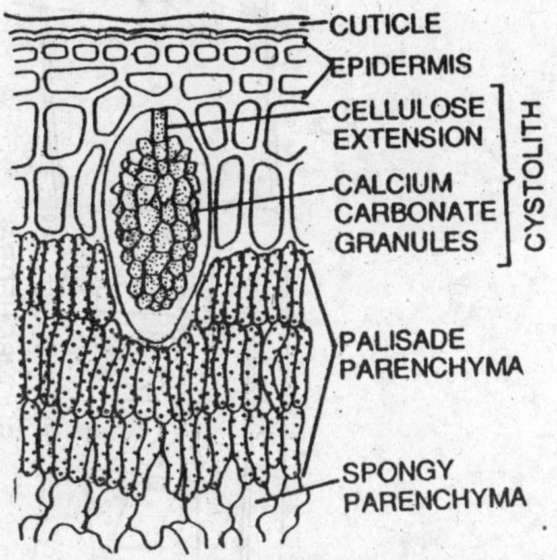

Fig. 33.58. Cystolith (calcium carbonate) in leaf of *Ficus elastica* (India rubber).

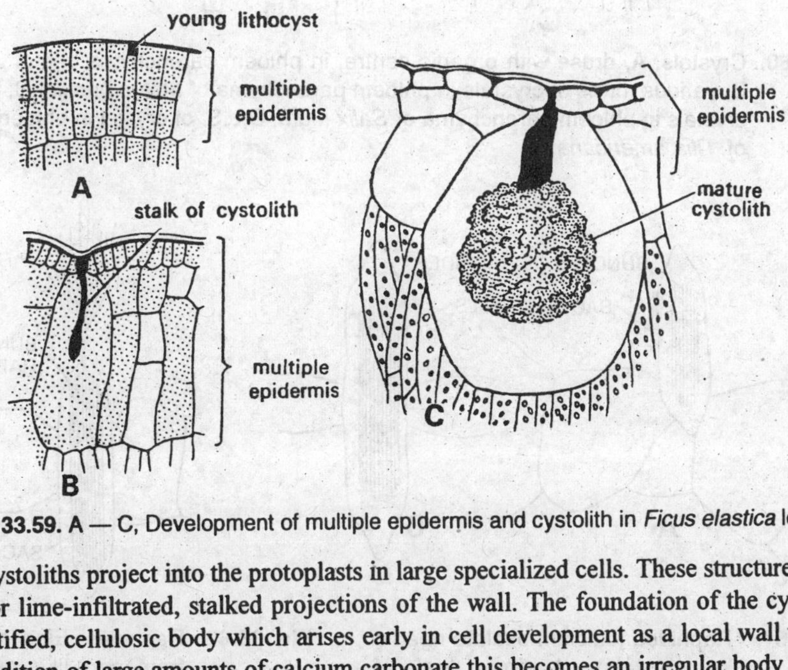

Fig. 33.59. A — C, Development of multiple epidermis and cystolith in *Ficus elastica* leaf.

The cystoliths project into the protoplasts in large specialized cells. These structures are lime-encrusted or lime-infiltrated, stalked projections of the wall. The foundation of the cystolith is a stalked stratified, cellulosic body which arises early in cell development as a local wall thickening. With the addition of large amounts of calcium carbonate this becomes an irregular body which may nearly fill the cell. In shape, cystoliths vary greatly in different genera and families.

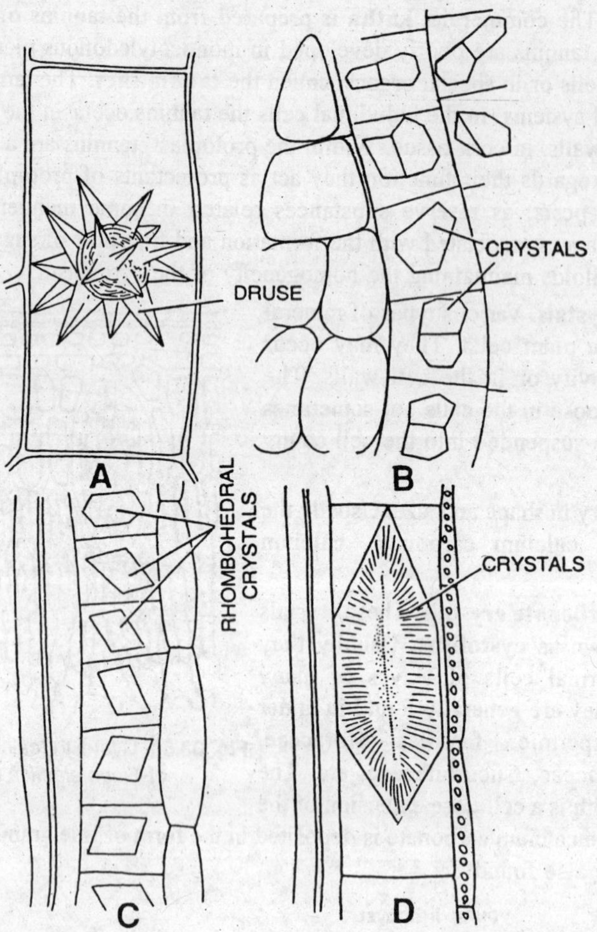

Fig. 33.60. Crystals. A, druse with organic centre, in phloem parenchyma cell of *Juglans nigra*; B, various forms of crystals in phloem parenchyma of *Malus pumila*; C, rhombohedral crystals in phloem parenchyma of *Salix nigra*; D,L.S. of crystal in phloem parenchyma of *Tilia americana*.

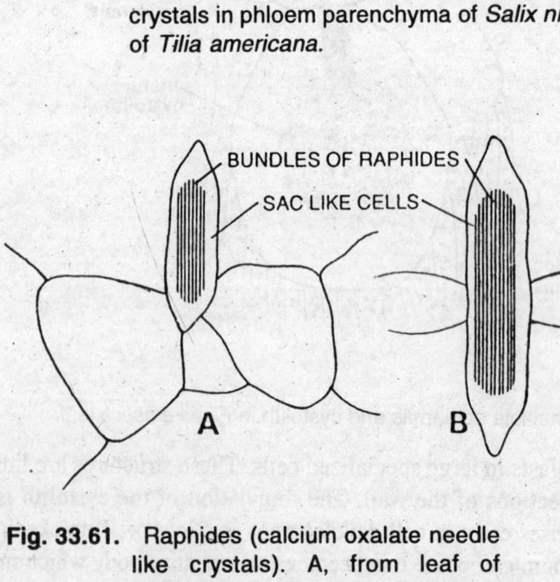

Fig. 33.61. Raphides (calcium oxalate needle like crystals). A, from leaf of *Colocasia*; B, from leaf of *Pistia*.

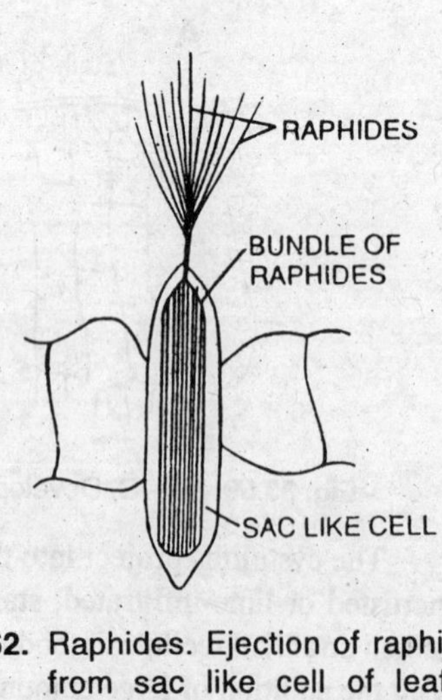

Fig. 33.62. Raphides. Ejection of raphides from sac like cell of leaf of *Colocasia*.

***(ii)* Calcium oxalate crystals.** The leaves and other organs of many plants contain conspicuous crystals of calcium oxalate. The forms of the crystals are very diverse. They are as follows:

Raphides. They are needle-like, long slender crystals usually lying parallel to each other in a bundle, which is sometimes found in a special sac like cell. When the sacs are injured mechanically, the raphides are released out with a jerk through a small hole. The raphides are commonly found in *Alocasia, Colocasia, Pistia,* etc. Certain raphides are very irritating and afford some protection to the plants from animals. Raphides are destroyed by boiling, and so food plants containing them are not irritating when cooked.

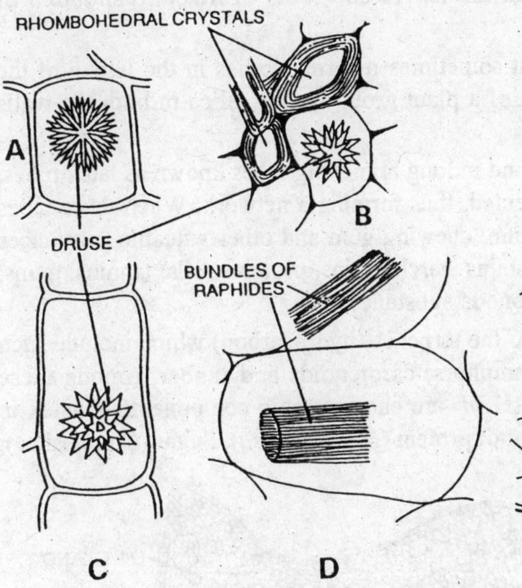

Fig. 33.63. Cell containing different types of crystals, A, druse in cortex cell of stem of Viburnum lentago; B, druse and rhombohedral crystals in stone cell of Carya glabra; C, druse in cortical cell of Carica papaya; D, bundles or raphides in cells of fruits of Smilacina racemosa.

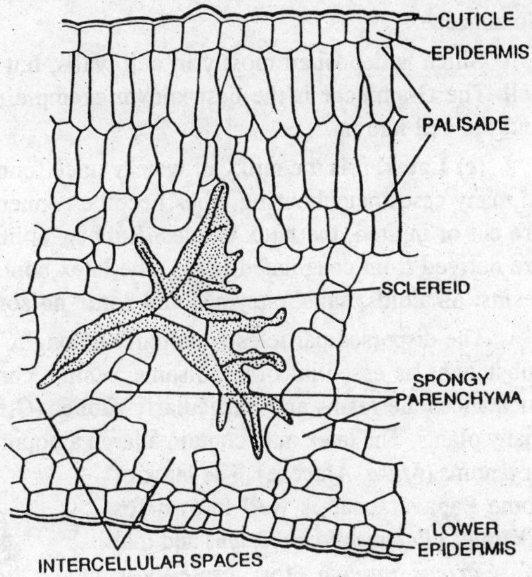

Fig. 33.64. Foliar sclereid. A branched foliar sclereid from the leaf blade of *Trochodendron.*

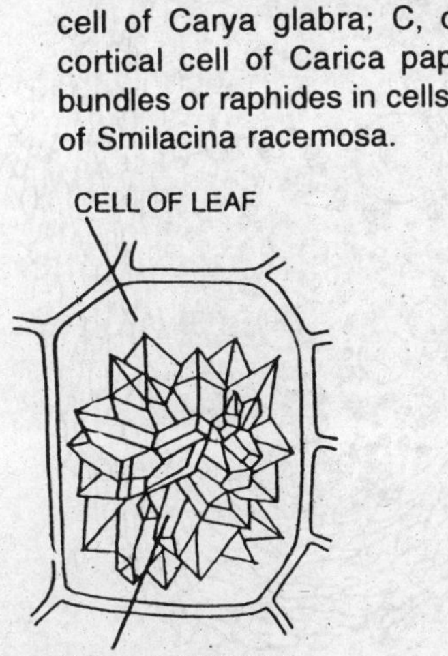

Fig. 33.65. Rosette crystal (calcium oxalate) from cell in a leaf.

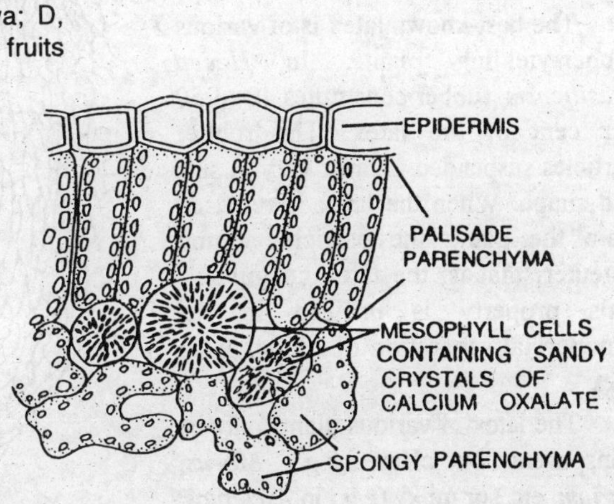

Fig. 33.66. Calcium oxalate crystals. Sandy crystals in leaf of *Atropa belladona.*

Idioblasts. They are stellate calcium oxalate crystals usually found in the aerenchyma of aquatic plants to give support to the tissue. They are quite common in *Limnanthemum, Nymphaea, Trapa,* etc.

Druses, rosette, crystals or cluster crystals. This is one of the commonest compound crystal having the appearance of a rosette and known as a **rosette crystal.** Such crystals are quite common in *Eucalyptus, Nerium, Ixora,* etc.

Prismatic crystals. They are single calcium oxalate crystals found in various plants. They may be square, prismatic, rhomboidal or pyramid-like in shape.

Sand crystals. They are commonly found in masses of micro sphenoidal crystals packed into a cell. Such crystals are generally found in the leaves and roots of *Atropa belladona* of Solanaceae.

Silica is deposited mostly in cell walls, but sometimes it forms bodies in the lumen of the cell. The Gramineae is the best-known example of a plant group having silica in both the walls and the cell lumina.

(*c*) Latex. It is the milky or watery juice found in long branching tubes known as latex tubes. In many cases neighbouring tubes become connected, thus forming a network. When these tubes are cut or injured, the latex exudes. Rubber, opium, chewing gum and other valuable substances are derived from coagulated latex. The latex contains starch grains, proteins, oils, tannins, gums, resins, alkaloids, salts, enzymes and some poisonous substances.

The dispersed particles commonly belong to the terpenes (hydrocarbon) which include such substances as essential oils, balsams, resins, camphors, carotenoids and rubber. Among these substances the resins and particularly rubber $(C_5H_8)n$ are characteristic components of latex in many plants. The latex may contain a large amount of protein (*Ficus callosa*), Sugar (Compositae), or tannins (*Musa,* Araceae). The latex of some Papaveraceae is well known for alkaloids (*Papaver somniferum*) and that of *Carica papaya* for proteolytic enzyme papain. The latex of *Euphorbia* species is rich in vitamin B_I. Crystals of oxalates and malates are abundantly found in latex. Certain plants contain starch grains in their latex.

The best-known latex is of various rubber-yielding plants. In *Hevea brasiliensis*, rubber constitutes 40 to 50 per cent of the latex. The rubber particles suspended in latex vary in size and shape. When the latex is released from the plant, the particles clump together; that is, the latex coagulates. This property is utilized in the commercial separation of rubber from latex

The latex of various plants may be transparent or clear (*e.g., Morus, Nerium,* etc.) or milky (*e.g.*, in *Asclepias,*

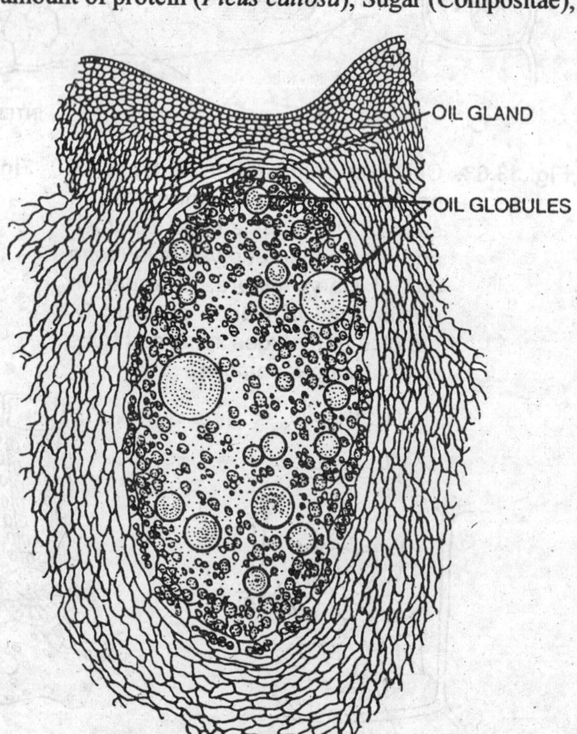

Fig. 33.67. L.S. of oil gland in orange peel.

Calotropis, Euphorbia, Ficus, etc.). It is yellow-brown in *Cannabis* and yellow or orange in several members of Papaveraceae.

(*d*) Essential or volatile oils. The essential or volatile oils frequently occur in oil glands. These oils are volatile and are usually very odoriferous. Well-known examples are eucalyptus oil and the oil from orange peel. The orange peel contains large oval glands. These glands originate in the splitting apart of certain cells, but are formed largely by the breaking down of the cells containing the oil. The disintegration of the cells brings the oil into the large cavity of the gland. The oil glands are also found in the petals of rose, jasmines and many other flowers.

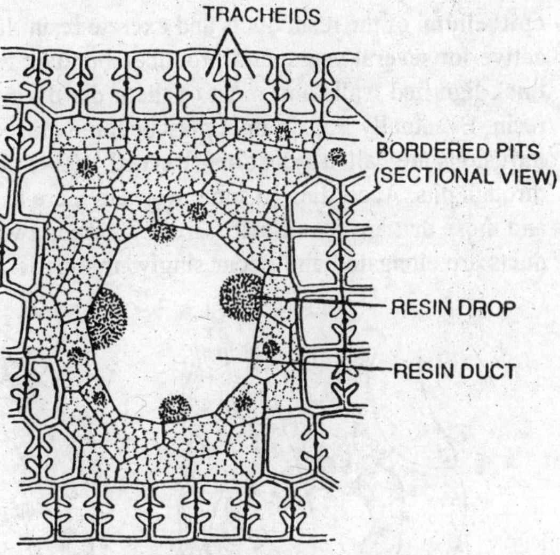

Fig. 33.68. Schizogenous cavities. Resin-duct of pine stem with resin.

(*e*) Gums. The gums are exuded out from the stems of many plants. The species of *Acacia* produce best gum. These spacies are *Acacia senegel, A modesta* and *A. arabica.* The gums are soluble in water and insoluble in alcohol. They swell up in water to produce a viscous mass.

(*f*) Resins. The resins are frequently found in the resin ducts of various conifers. The resin ducts are formed either by the separation of the neighbouring cells or by disintegration of cells. Turpentine obtained by cutting through the resin ducts of pine trees, after which the turpentine exudes and is collected. Various kinds of varnish and other resins are obtained by the same method from other conifers. The resins are insoluble in water but soluble in alcohol and turpentine. The well known mounting fluid, canada balsam is a resinous product of *Abies balsamea.*

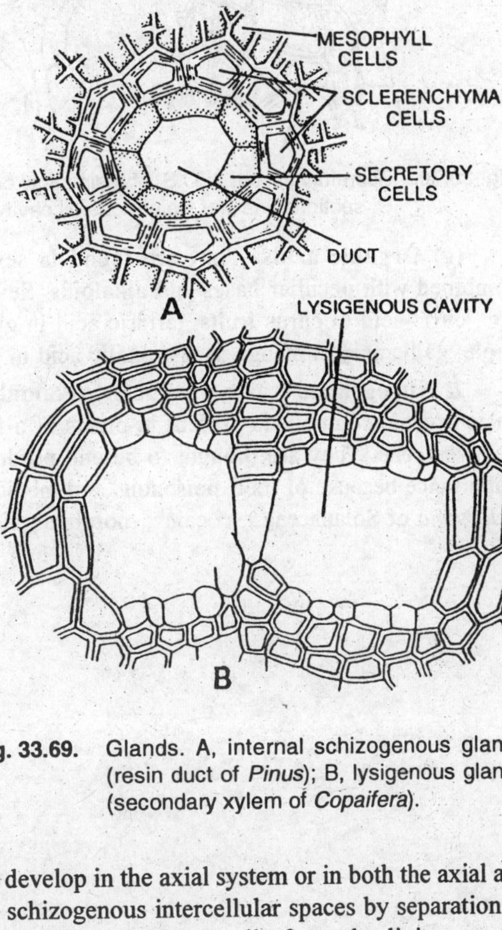

Fig. 33.69. Glands. A, internal schizogenous gland (resin duct of *Pinus*); B, lysigenous gland (secondary xylem of *Copaifera*).

Resin ducts. Resin ducts are long passages surrounded by glandular cells. They occur not only in stems but also in other parts of plants.

In *Pinus* and allied genera the resin ducts develop in the axial system or in both the axial and ray systems. Typically, resin ducts develop as schizogenous intercellular spaces by separation of parenchyma cells from each other. After some divisions these cells form the lining, or the

epithelium, of the resin ducts and excrete resin. In *Pinus* the epithelial cells are thin-walled, remain active for several years, and produce abundant resin. In *Abies* and *Tsuga* the epithelial cells have thick lignified walls and most of them die during the year of origin. These genera produce little resin. Eventually a resin duct may become closed by enlarging epithelial cells. These tylosis like intrusions are called **tylosoids** (Record, 1934). They differ from tyloses in that they do not grow through pits. According to some workers there is a distinction between resin ducts that are **normal** and those that are **traumatic** (Greek *trauma*; a wound), that is, arise in response to injury. Normal ducts are elongated and occur singly; traumatic ducts are cyst-like and occur in tangential series.

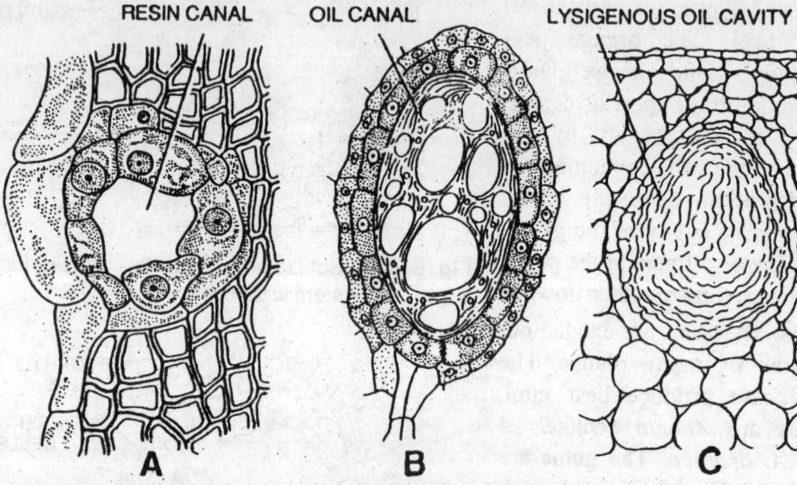

Fig. 33.70. Secretory tissue. A, C.S. of resin canal of *Pinus*; B, C.S. of oil canal of young fruit of *Angelica*; C, sectional view of lysigenous oil cavity of rind of *Citrus*.

(*g*) Organic acids. They are found in several vegetable and fruit juices and are often combined with peculiar bases and alkaloids. Several organic acids are found in various plants, *e.g.,* citric acid in citrus fruits; tartaric acid in grapes and tamarinds; malic acid in the fruits of apple; gallic acid in mango seeds; oxalic acid in *Rumex, Oxalis* and *Nepenthes.*

2. Nitrogenous waste products. Alkaloids. They make a group of nitrogen-containing, basic organic compounds present in plants of a few families of Dicotyledons, *e.g.,* Solanaceae, Papaveraceae. They are thought to be end products of nitrogen metabolism. They are of great importance because of their poisonous and medicinal properties, *e.g.*, atropine (found in *Atropa belladona* of Solanaceae.), cocaine, morphine, nicotine, quinine, strychnine, etc.

34

CHAPTER

The Cell Wall

The cells of the plant tissue possess protective cell wall in near about all flowering plants. The presence of non-protoplasmic walls is considered the outstanding characteristic differentiating the cell of plants from those of animals. Few plant cells lack cell walls. Examples of cells without walls in plants are the motile spores in algae and fungi and the sexual cells in lower and higher plants. As we know the cells are composed of very soft and delicate substance, called protoplasm and thus they need protection against other injurious agencies. The cell walls protect the protoplasts. The cell wall was discovered before the protoplast, and in the early history of botany has received more attention than the cell contents. Later the protoplast became the main object of study. In modern days the origin and the structure of cell wall have received critical attention and are now much better understood.

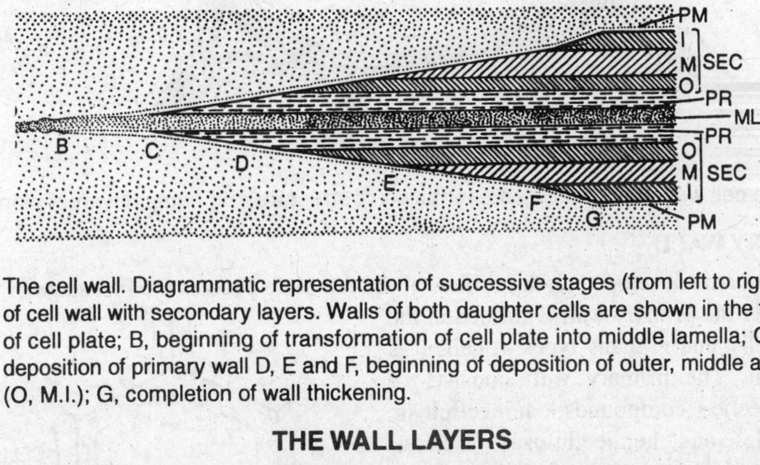

Fig. 34.1. The cell wall. Diagrammatic representation of successive stages (from left to right) in formation of cell wall with secondary layers. Walls of both daughter cells are shown in the figure. A, origin of cell plate; B, beginning of transformation of cell plate into middle lamella; C, beginning of deposition of primary wall D, E and F, beginning of deposition of outer, middle and inner layers (O, M.I.); G, completion of wall thickening.

THE WALL LAYERS

Each cell within a tissue has its own cell wall. The cell walls of plants vary much in thickness in relation to age and type of cells. Generally, young cells have thinner cell walls than the fully developed ones, but sometimes the wall does not thicken much after the cell ceases to grow. In mature thick-walled cells, a concentric layering is usually distinct in the cell wall. The layers differ from one another in physical and chemical nature. The cell wall is complex in its structure and usually consists of three layers, the **primary wall, the intercellular substance** or **middle lamella,** and the **secondary wall.** The intercellular substance cements together the primary walls of two contiguous cells, and the secondary wall is laid over the primary, that is, next to the lumen of the cell. The primary wall usually consists of a single layer, whereas the secondary wall is made up of one to many layers, most frequently of three.

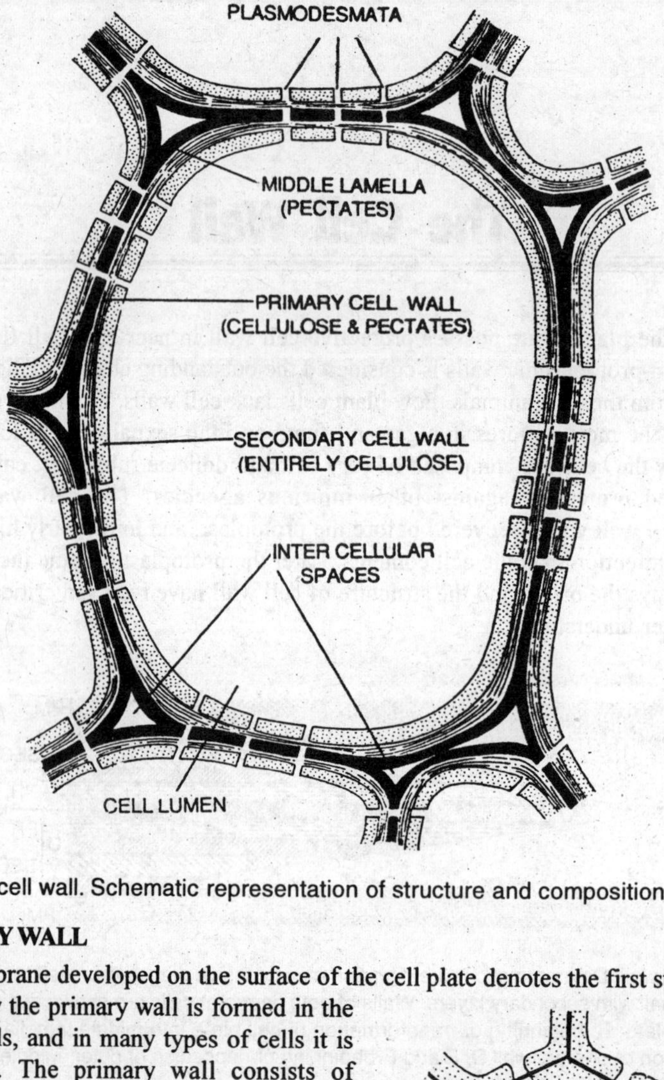

Fig. 34.2. The cell wall. Schematic representation of structure and composition of plant cell walls.

THE PRIMARY WALL

The membrane developed on the surface of the cell plate denotes the first stage of the primary wall. Generally the primary wall is formed in the developing cells, and in many types of cells it is the only wall. The primary wall consists of cellulose, pectic compounds, noncellulosic polysaccharides and hemicelluloses (Bonner, 1950). Sometimes it becomes lignified. The primary walls are usually associated with living protoplasm. The walls of dividing and growing meristematic cells are primary. The primary wall may change its shape and volume according to the growth of young protoplast. This wall is not uniformly thickened and may maintain its thickness and unevenness, or be alternately thickened and thinned.

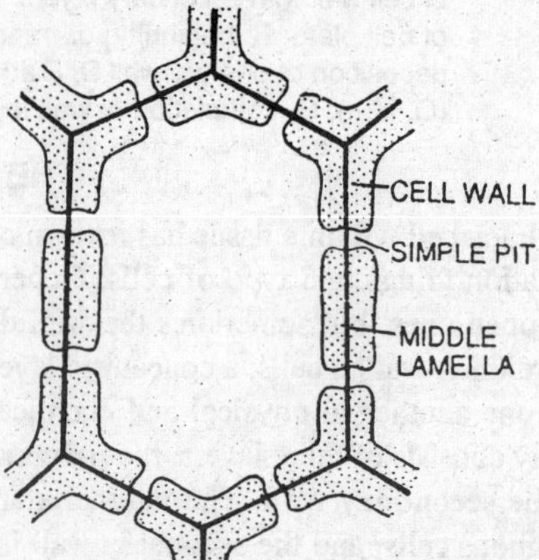

Fig. 34.3. The cell wall. Simple pits. A cell in sectional view showing simple pits in its wall.

As already mentioned, the walls of dividing and growing meristematic cells are primary, and so are those of most of the cells which retain

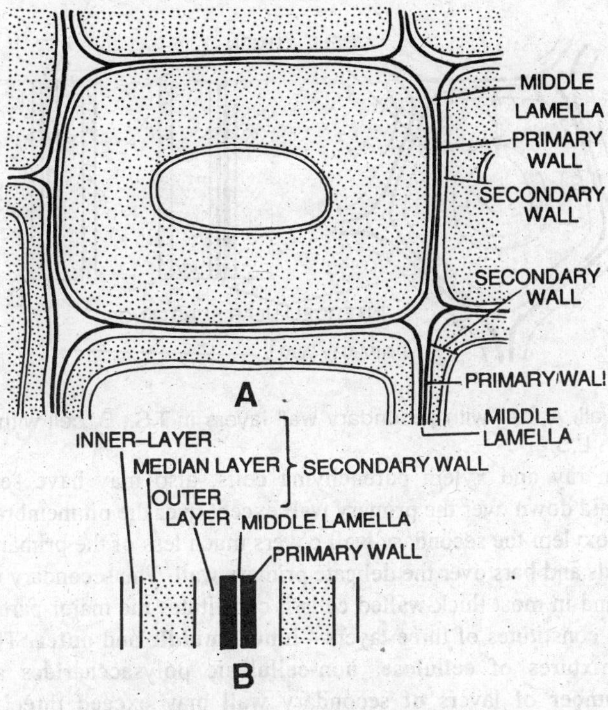

Fig. 34.4. The cell wall. The figure shows wall structure and middle lamella in wood fibres. A, T.S. of one wood fibre and parts of seven adjacent fibres; B, cross section of adjacent walls and middle lamella (highly magnified).

living protoplast. The changes that occur in primary walls are therefore reversible. The wall may lose a thickening previously acquired and chemical substances may be removed or replaced by others. For example, cambial walls show seasonal changes in thickness and in colloidal properties (Kerr and Bailey, 1934). In other cases the thick primary walls of the endosperm in many seeds are digested during germination.

THE MIDDLE LAMELLA

The intercellular substance which cements together the primary walls of two contiguous cells very firmly is called **middle lamella.** This is a complex layer in its structure and morphology. It is amorphous, colloidal, and optically inactive. It is composed mainly of a pectic compound which appears to be a mixture of calcium and magnesium pectate. In woody tissues the middle lamella is commonly lignified. The distinction between intercellular lamella and the primary wall is frequently obscured during the extension growth of the cell. In such cells as tracheids and fibres, which typically develop prominent secondary walls, the intercellular layer becomes extremely tenuous. As a result, the two primary walls of contiguous cells and the intervening middle lamella appear as a unit, particularly when all three become strongly impregnated with lignin. This triple structure is often known as middle lamella.

THE SECONDARY WALL

Usually in many fully developed cells further thickening of cell wall occurs. The wall then formed is the **secondary wall.** The secondary wall maybe considered a supplementary wall whose principal function is mechanical. Generally the cells with secondary walls are devoid of protoplasts at maturity. These walls are most characteristic of cells that are highly specialized and undergo irreversible changes in their development. However, cells with active, living protoplasts,

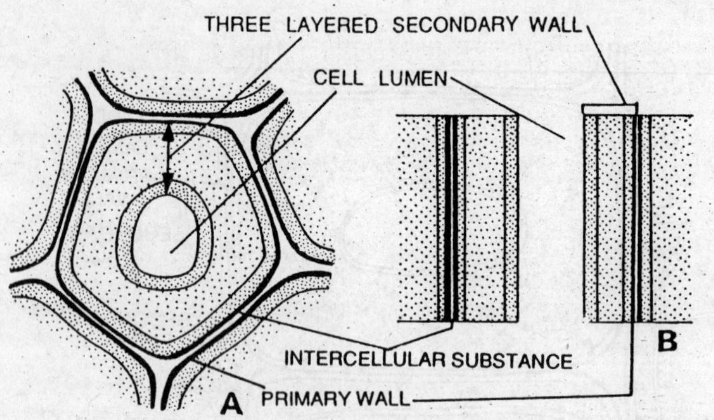

Fig. 34.5. A cell wall. A, cell with secondary wall layers in T.S.; B, cell with secondary wall layers in L.S.

such as the xylem ray and xylem parenchyma cells, also may have secondary walls. The secondary wall is laid down over the primary wall except over the pit membranes. In the tracheids and vessels of protoxylem the secondary wall covers much less of the primary wall; it forms only as rings, spiral bands and bars over the delicate primary wall. The secondary wall is more massive than the primary, and in most thick-walled cells it constitutes the major part of the wall. Usually the secondary wall constitutes of three layers—**inner, middle** and **outer.** This wall may consist of cellulose or mixtures of cellulose, non-cellulosic polysaccharides and hemicelluloses. Sometimes the number of layers of secondary wall may exceed three, and the innermost sometimes only of a helical band and such bands are called **tertiary spirals** or **spiral thickenings** (Eames and Mac Daniels, 1947).

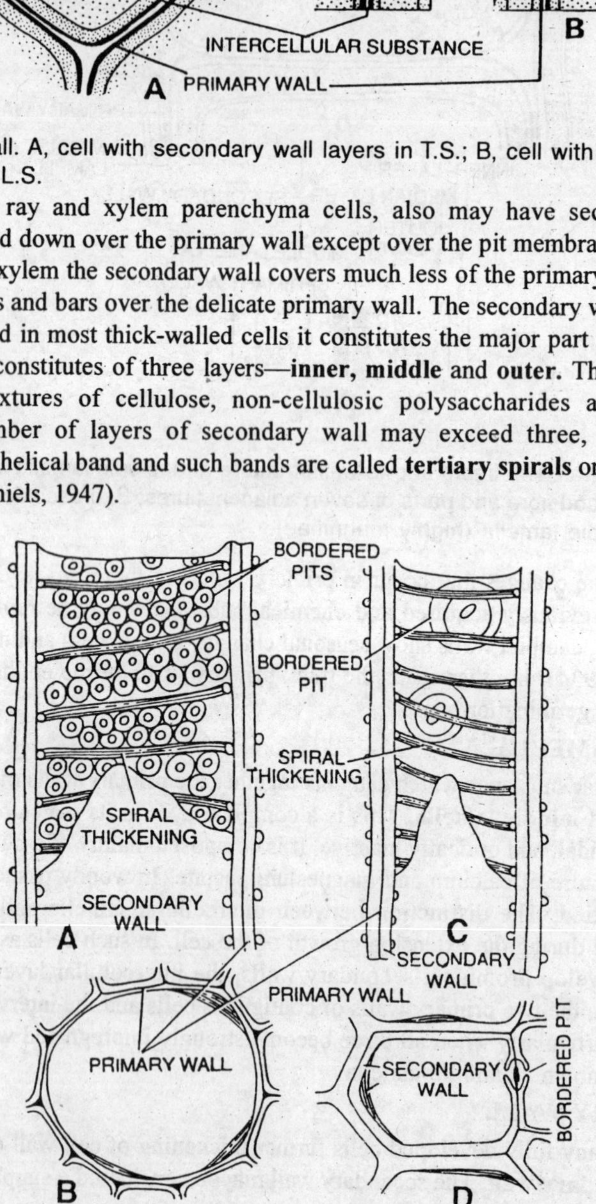

Fig. 34.6 The cell wall. This figure shows the primary wall and secondary wall with tertiary spirals. A and B, L.S. and T.S. of vessel of *Tilia;* C and D, L.S. and T.S. of tracheid of *Taxus* (a gymnosperm).

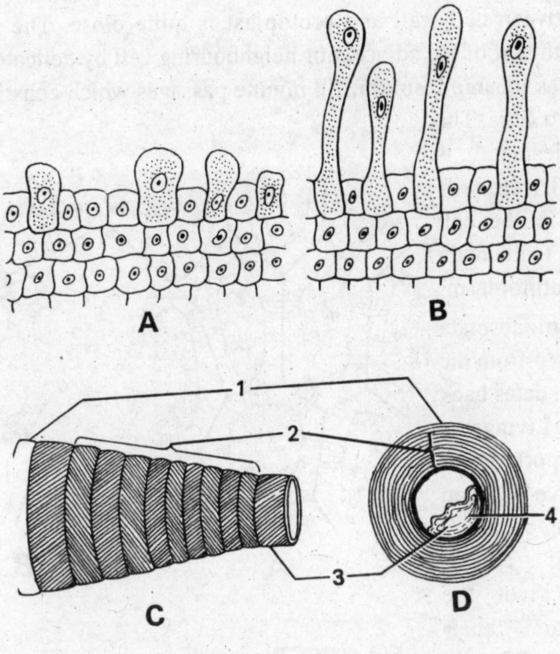

Fig. 34.7. Structure and growth of cotton fibre. A, outer layer of cells of young cotton seed showing the beginning enlargement of the fibre on the day of flowering; B, same after 24 hours, C, diagram of the various layers of the cellulose laid down in a mature cotton fibre (1) outer primary wall (2) concentric inner layers revealing the different orientation of the cellulose in the secondary thickenings, (3) last inner layer of the secondary wall, (4) remains of cell contents.

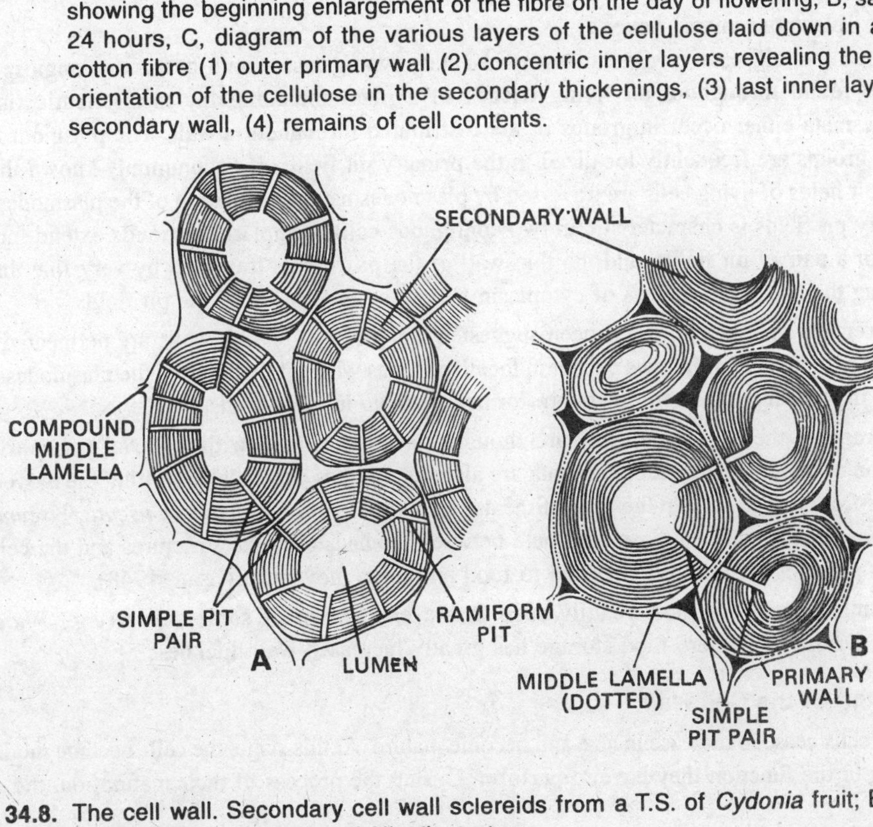

Fig. 34.8. The cell wall. Secondary cell wall sclereids from a T.S. of *Cydonia* fruit; B, phloem fibres from a transection of *Nicotiana* stem.

Plasmodesmata

The relation between cell wall and protoplast is quite close. The protoplast of one cell remains connected with that of the adjacent or neighbouring cell by delicate threads of cytoplasm. These threads known as **plasmodesmata,** fill minute passages which constitute the only breaks in the primary wall of two cells. The presence of plasmodesmata is the characteristic of living cells and they are found in the cell walls of all living cells to maintain the continuity of the protoplasm.

The name plasmodesmata (singular plasmodesma—from the Greek *desmos*, strand), dates back to Strasburger (1901). Livingston and Bailey (1946) reported that plasmodesmata have been recognized in the living cambium and ray cells of certain gymnosperms. This gives the best support to their interpretation as cytoplasmic structures.

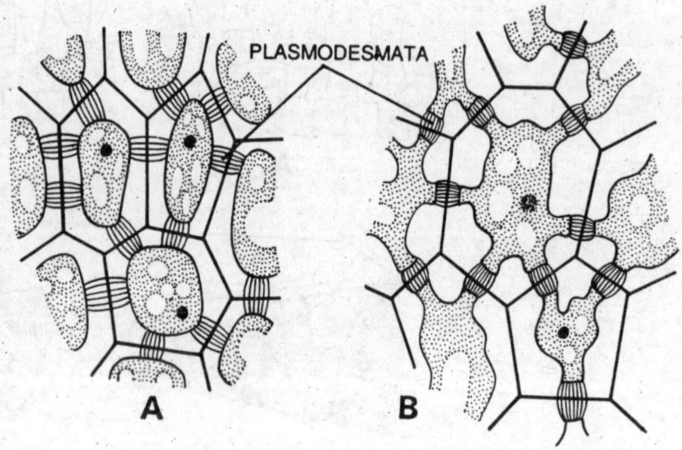

Fig. 34.9. Plasmodesmata. A, in endosperm of *Diospyros;* B, in endosperm of *Phoenix.*

The plasmodesmata have been seen in red algae, liverworts, mosses, vascular cryptogams, gymnosperms and angiosperms. They are found throughout all living tissues of a plant including the meristematic tissues. Plasmodesmata either occur in groups or are distributed throughout a wall. The plasmodesmata found in groups are frequently localized in the primary pit fields. It is commonly known that all primary pit fields of living cells are traversed by plasmodesmata. The relation of the plasmodesmata to primary pit fields is characteristic in two contiguous cells, cytoplasmic strands extend into the cavities of a pair of pit fields, and the thin wall of the pit field is traversed by very fine threads connecting the two small masses of cytoplasm filling the depression of the pit field.

As regards their origin, it has been suggested that young, growing walls are permeated with cytoplasm, which later withdraws except in localized areas where it then makes the plasmodesmata. It is also thought that plasmodesmata are formed *de novo* in older walls.

As regards their function, they are thought to be concerned with material transport and conduction of stimuli. The plasmodesmata are also regarded as channels permitting the movement of viruses from cell to cell. In the case of the angiospermic parasites such as *Cuscuta, Viscum* and *Orobanche*, the presence of plasmodesmata between the haustoria-like structures and the cells of their host plants is thought to be related to food and virus movement (Esau, 1948).

Plasmodesmata are most readily seen in the endosperm of some seeds (*e.g., Phoenix, Aesculus, Diospyros*) where food storage has greatly increased wall thickness.

Thickening of the Cell Wall

The cells cease to increase in size and become mature. At this stage, the cells become modified according to the function they have to perform. During the process of their maturation, the cells undergo additional or secondary thickening deposition of new materials such as cellulose, lignin, suberin, cutin, etc. Some of the cells (*e.g.*, parenchymatous cells) remain unthickened. The cells of certain parts (*e.g.*, pericycle, phloem, xylem, etc.) of the plants undergo heavy thickening of their walls. The thickening materials of the cells are secreted by the protoplasm. These materials are deposited in the cell walls in such manner that cell wall becomes stratified in appearance. The cells which ultimately develop into vessels, tracheids and fibres show the thickening of the cell wall in various ways. This thickening takes place due to the deposition of a hard substance, called **lignin,** on the inner surface of the cell wall. Usually the thickening material (*e.g.*, lignin) of the secondary wall is not laid down in uniform thickness but it may form the special patterns such as **annular, spiral, scalariform, reticulate** and **pitted.** In such patterns the whole of the wall is not thickened. Only those portions of the cell walls are thickened where the thickening material is laid down in the form of above mentioned patterns and the remaining portion of the wall remains thin. The details of these thickenings are as follows:

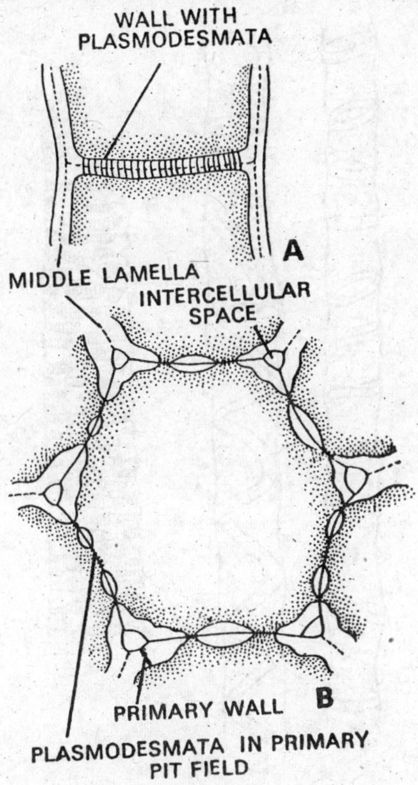

Fig. 34.10. Plasmodesmata. A and B, parenchyma cells without secondary walls from *Nicotiana* stem. Plasmodesmata dispersed throughout

Annular or ring-like thickenings: The deposition of lignin takes place in the form of rings on the inner surface of the cell wall. These lignified rings are placed one above the other like coins leaving sufficient space in between each other. The gaps of the walls remain unthickened. Such thickenings are commonly found in the vessels and tracheids.

Spiral thickenings. In such cases the deposition of thickening material (lignin) takes place in the form of complete spiral bands. The number of such bands may be one or more than one. This type of thickening is commonly found in the vessels or tracheae of angiosperms.

Scalariform or ladder-like thickenings. In such thickenings of the cell wall the lignin is being deposited in the form of the transverse rods of the ladder, and thus known as scalariform or ladder-like. The unthickened areas between the successive thickening layers appear as elongated transverse pits. This type of thickenings are common in xylem vessels and tracheids of protoxylem.

Reticulate or net-like thickenings. In such thickenings of the cell wall the thickening matter or lignin is being deposited in the form of a net or reticulum and thus known as reticulate or net like thickening of the cell wall. In such cases the unthickened areas of the cell wall are irregular in shape. These thickenings are commonly found in the vessels of the stems, roots and leaves of angiosperms and in the tracheids of protoxylem.

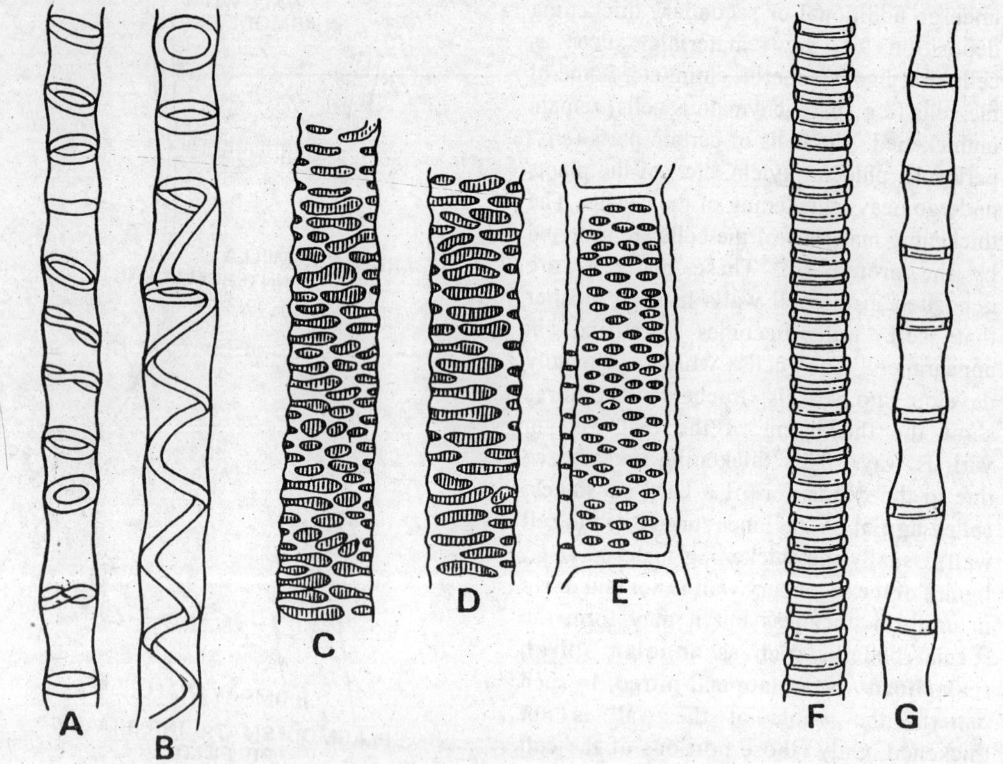

Fig. 34.11. Xylem-vessels. Secondary wall structure in primary tracheary (vessel) elements, the thickenings are visible. A, annular; B, partly annular, partly helical; C, reticulate; D, scalariform reticulate; E, pitted; F, annular in an unstretched vessel element; G, annular in a stretched vessel element.

Pitted thickenings. In such thickenings of the cell wall, the whole inner wall is more or less uniformly thickened, leaving here and there some small unthickened areas, the **pits.**

Pits

The pits are formed in pairs lying against each other on the opposite sides of the wall, and morphologically more correct they are called **'pit pairs'**. A pit pair is structural and directional unit constituted by two pits lying opposite to each other of contiguous cells.

Structure. The space found inside the pit is called the **pit cavity** or **pit chamber.** The separating membrane which separates the two chambers or cavities of a pit pair, is called the **pit membrane,** or **pit aperture.** The pit cavity opens internally in the lumen of the cell and is closed by the closing or pit membrane along the line of junction of two contiguous cells. A pit has two pit cavities, two pit apertures, and one pit or closing membrane. The pit membrane is common to both pits of a pit pair and consists of two primary walls and a middle lamella or intercellular substance. Usually two types of pits are met within the cells of various plants, *viz.*, **simple pits** and **bordered pits.** Two bordered pits make up a bordered pit pair, two simple pits form a simple pit pair. A bordered pit and a simple pit lying opposite to each other in contiguous cells, constitute a **half bordered pit pair.** A pit occurs opposite an intercellular space has no complementary pit and is known as **blind pit.** Fundamentally the bordered pit differentiates from a simple pit in having a secondary wall arching over the pit cavity, which constitutes the actual border and becomes narrow

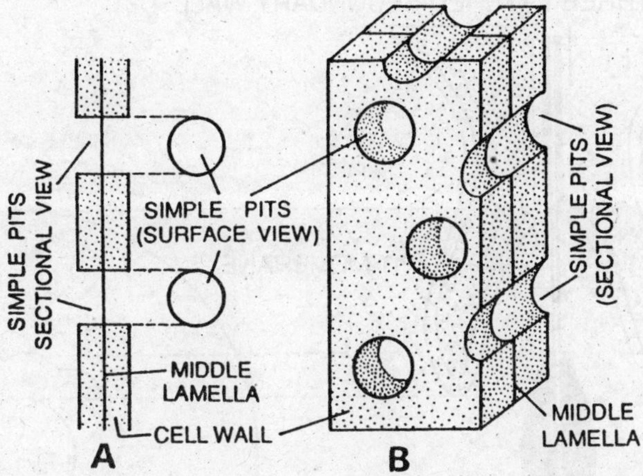

Fig. 34.12. Simple pits. A, cell-wall having two simple pits showing sectional and surface views; B, portion of a cell wall showing some simple pits—sectional view (right top) surface view (front).

like a funnel towards the lumen of the cell. In the simple pit, no such arching of the secondary wall and narrowing of the pit towards the lumen of the cell occurs.

Simple pits. Simple pit pairs occur in parenchyma cells, in medullary rays, in phloem fibres, companion cells, and in tracheids of several flowering plants. In the simple pits, the pit cavity remains of the same diameter and the pit or closing membrane also remains simple and uniform in its structure. The simple pit may be circular, oval, polygonal, elongated or somewhat irregular in its facial view. The simple pits occurring in the thin walls are shallow, whereas in thick wall the pit cavity may have the form of a canal passing from the lumen of the cell towards the closing or common pit membrane. The diffusion of protoplasm takes place through these pits.

Bordered pits. They are abundantly found in the vessels of many angiosperms and in the tracheids of many conifers. They are more complex and variable in their structure than simple pits. The overarching secondary wall which encloses a part of the pit cavity is called, the **pit border,** which opens outside by a small rounded mouth known as **pit aperture.** The overarching rim forms a border around the aperture and thus named 'bordered pits'. The pit aperture may be of various shapes in the facial view. It may be circular, lenticular, linear or oval. In the case of relatively thick secondary walls, the border divides the cavity into two parts. The space between the closing membrane and the pit aperture may be called the pit chamber and the canal leading from pit chamber to the lumen of the cell may be termed as **pit canal.** The pit canal opens in the pit chamber by an outer aperture and at the same time it opens in the lumen of the cell by an inner aperture. The closing membrane of a bordered pit pair which consists of the parts of two primary walls and the intercellular substance or middle lamella, is somewhat thickened in its central part. This thickening is called **torus** which remains surrounded by a delicate margin. In many bordered pits, the closing membrane may change its position within pit cavities. The torus may remain in central position or it may shift to the lateral position. As the torus is shifted to the lateral position the pit aperture closes, and the passage of the protoplasm may take place only by diffusion through torus.

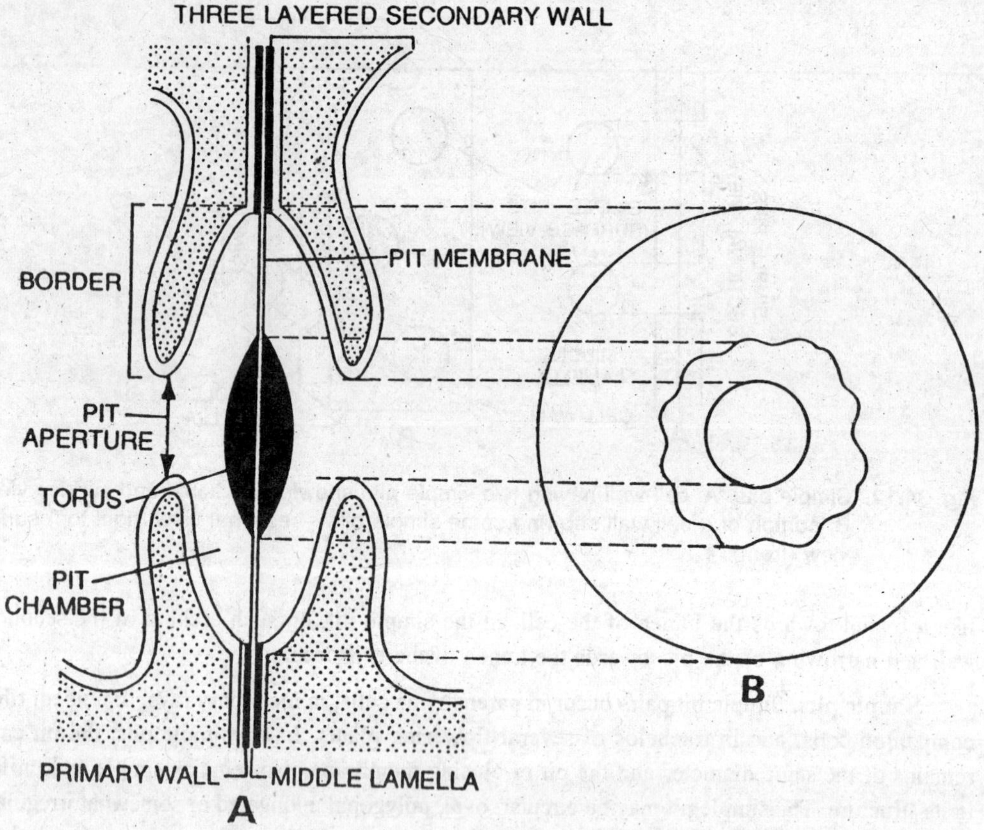

Fig. 34.13. Pits. Diagrams of bordered pit pairs. A, sectional view; B, front view. Pit membrane is made up of two primary walls and the intercellular lamella. The torus is formed by thickenings of primary wall. The front view shows that outline of torus is uneven.

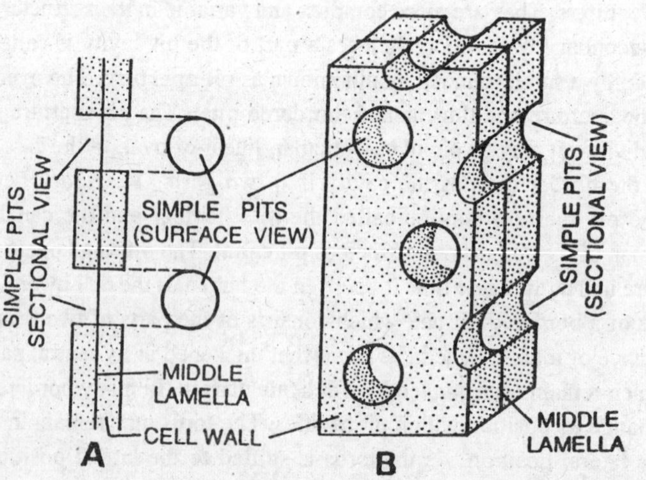

Fig. 34.14. Bordered pits. A, cell-wall with two bordered pits—sectional and surface views; B, a portion of the cell-wall with two bordered pits—sectional view (bottom) and surface view (top).

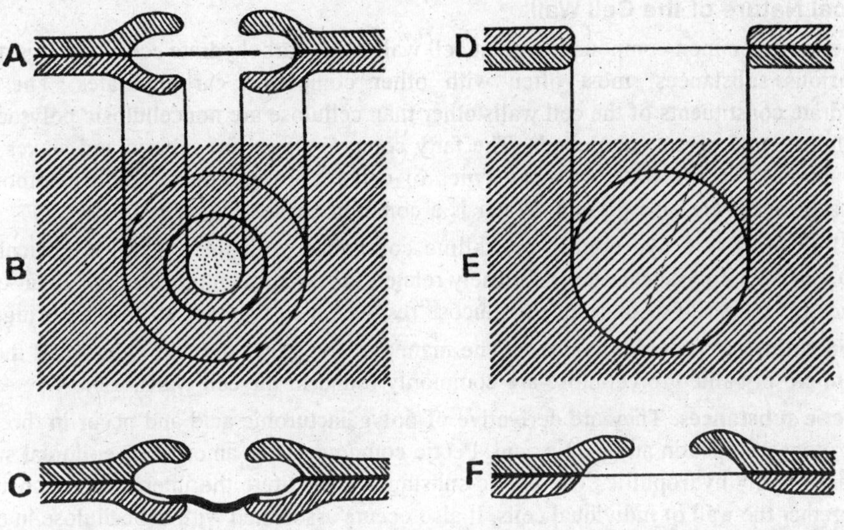

Fig. 34.15. The pits—diagrams of three kinds of pit-pairs. A, B, section and face view of bordered pit-pair showing overarching borders, pit cavities, closing membrane and torus; C, section of same showing closing membrane in lateral position, the torus closely appressed to the pit aperture; D, E, section and face view of a simple pit-pair, the closing membrane has no torus; F, section of a half bordered pit-pair, arches only on one pit cavity, no torus.

Patterns of bordered pits. The bordered pits in vessel walls of angiosperms show *three* main types of arrangement : (*i*) scalariform, (*ii*) opposite and (*iii*) alternate. If the pits are elongated or linear and make ladder-like arrangement, it is called **scalariform pitting** (Fig. 34.16 A). When the pits are arranged in horizontal pairs, it is called **opposite pitting** (Fig. 34.16 B). If such pits are crowded, their borders assume rectangular outlines in face view. When the pits occur in diagonal rows, the arrangement is called **alternate pitting** (Fig. 34.16C). Generally small simple pits are arranged in clusters, and such arrangement is called **sieve pitting.**

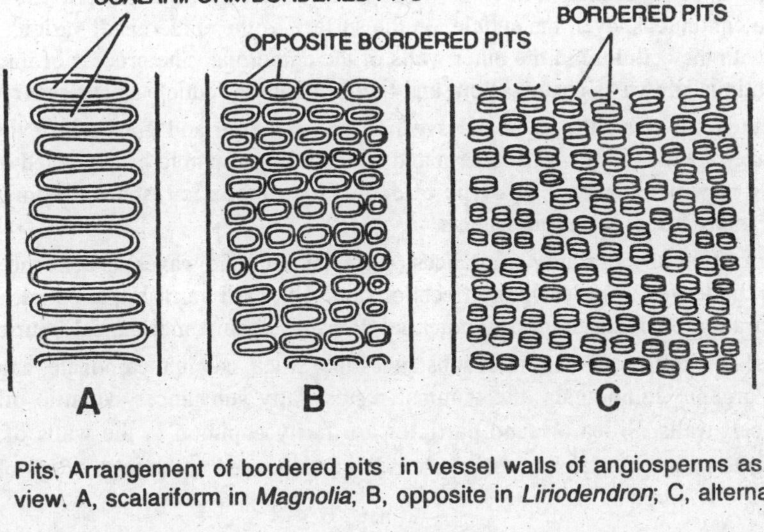

Fig. 34.16. Pits. Arrangement of bordered pits in vessel walls of angiosperms as seen in face view. A, scalariform in *Magnolia*; B, opposite in *Liriodendron*; C, alternate in *Salix*.

Chemical Nature of the Cell Wall

The most-common compound in plant cell wall is the carbohydrate cellulose. It is associated with various substances, most often with other compound carbohydrates. The common carbohydrate constituents of the cell walls other than cellulose are noncellulosic polysaccharides, hemicellulose, and pectic compounds. The fatty compounds, cutin, suberin and waxes, occur in varying amounts in the walls of many types of cells. Various other organic compounds and mineral substances are also present. Water is a common constituent of cell walls.

Cellulose. It is a hydrophilic crystalline compound with the general empirical formula $(C_6H_{10}O_5)_n$. It is a polysaccharide and closely related to starch. Its molecules are chain-or ribbon-like structures with 100 or more of the glucose residues held together by oxygen bridges.

Non-cellulosic polysaccharides. The mannans, the galactans, the xylans and the arabans which are closely allied to cellulose are commonly found in the cell walls.

Pectic substances. They are derivative of polygalacturonic acid and occur in three general types, protopectin, pectin and pectic acid. Pectic compounds are amorphous colloidal substances plastic and highly hydrophilic. The pectic substances constitute the intercellular substance that binds together the wall of individual cells. It also occurs associated with the cellulose in other wall layers, mainly the primary.

Gums and mucilages. They are also the compound carbohydrates of the cell walls. They are related to the pectic compounds, and both possess the property of swelling in water (Bonner, 1950). Gums exude as a result of physiological or pathological disturbances that induce a break down of walls and cell contents, whereas the mucilages occur in some gelatinous or mucilaginous type of cell walls.

Lignin. It is an organic compound of high carbon content, distinct from the carbohydrates. Lignin is found in all three wall layers—the primary wall, the middle lamella, and the secondary wall. The impregnation with lignin usually starts in the intercellular lamella and then spreads centripetally through the primary and secondary walls. The process of conversion of cellulose into lignin is called **lignification.**

Cutin, suberin and waxes. The cutin and suberin are fatty substances which are not meltable and exhibit considerable insolubility in fat solvents. Cutin and suberin are closely related, highly polymerized compounds consisting of fatty acids. Suberin occurs in association with the cellulose in cork cells of the periderm. Suberin also occurs in the endodermis and exodermis of many plants. The phenomenon of impregnation of walls with suberin is referred to as **suberization.** Usually the cutin forms a continuous layer, the cuticle, on the surface of the epidermis of various plants. Cutin also occurs with the cellulose in the outer walls of the epidermis. The process of impregnation of walls with cutin is known as **cutinization,** and the formation of cuticle as **cuticularization.**

The waxes are also fatty substances which are meltable and are easily extracted by fat solvents. Waxes are associated with suberin and cutin and are found to be deposited on the surface of the cuticle in various forms. This type of deposition of wax is responsible for the glaucous condition of many fruits, leaves and stems.

The deposition of these fatty substances (cutin, suberin and waxes) checks the transpiration and protects the leaves from leaching effects of rain. The hard varnish-like cuticle, may protect against penetration of living tissues by potential parasites against mechanical injuries.

Mineral substances. The mineral substances like silica, calcium carbonate, calcium oxalate and several organic compounds like tannins, resins, fatty substances, volatile oils, etc., may impregnate cell walls. Silica or sand particles are fairly common in the walls of grasses and horsetails (*Equisetum* sp.). Calcium carbonate is being deposited in the form of cystoliths in many species of *Ficus* and other plants. Organic compounds are frequently deposited in the xylem walls when this tissue changes from sapwood into heartwood.

INTERCELLULAR SPACES

Usually the meristematic or young cells do not possess any intercellular spaces among them but as they become older their walls split at certain places giving rise to small spaces called intercellular spaces, usually these are filled up with air or water. Generally there are two types of intercellular spaces—1. Schizogenous intercellular spaces and 2. Lysigenous intercellular spaces.

Schizogenous intercellular spaces. The most common intercellular spaces result from separation of cell walls from each other along more or less extended areas of their contact. In such cases, the intercellular substance dissolves partly and an intercellular space develops. Ultimately this becomes quite big in size and is known as **schizogenous cavity.** The ordinary intercellular spaces and schizogenous cavities form an intercommunicating system of long intercellular canals which facilitate diffusion of gases and liquids from one part of the plant body to the other. The resin ducts in the Coniferales, and the secretory ducts in the Compositae and Umbelliferae are the typical examples. The cells lining the cavity are secretory in nature and release their product in the intercellular canal.

Lysigenous intercellular spaces. This type of intercellular space arises through dissolution of entire cells, which are therefore called lysigenous intercellular spaces (*lysis*, loosening, Greek). These cavities of intercellular spaces store up water, gases and essential oils in them. The examples are commonly found in water plants and many monocotyledonous plants. The secretory cavities in *Eucalyptus*, *Citrus* and *Gossypium* are good examples.

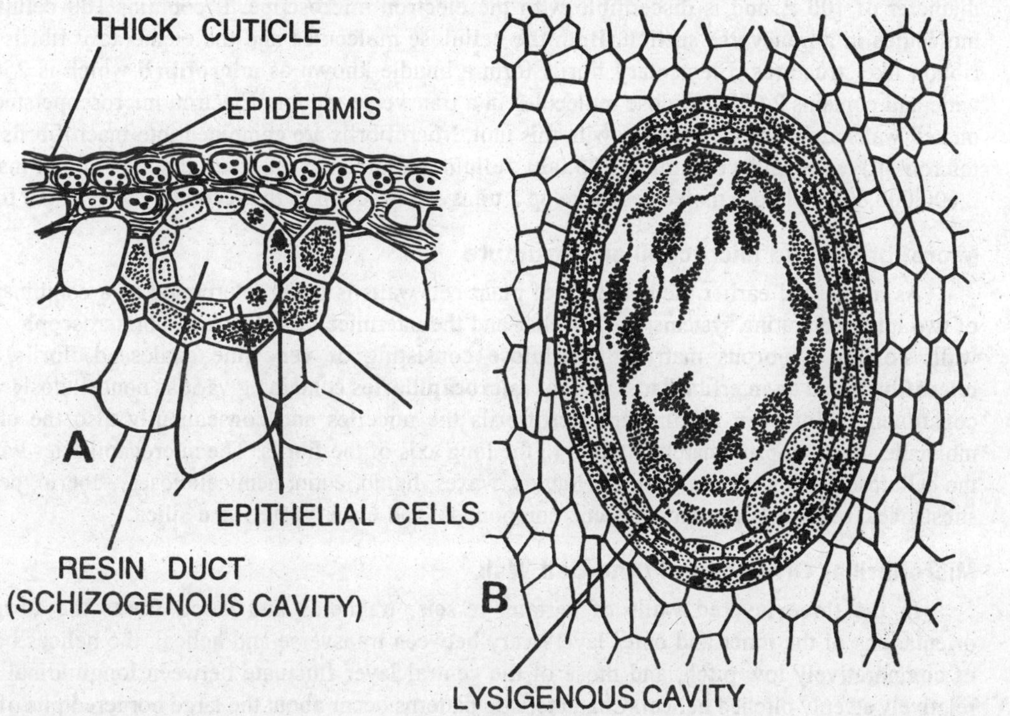

Fig. 34.17. Intercellular spaces. A, a schizogenous intercellular space in *Bryophyllum* stem; B, a schizogenous intercellular space in leaf of *Sequoia sempervirens*.

MICROSCOPIC AND SUBMICROSCOPIC STRUCTURE OF CELL WALLS

The chemical substances of cell walls remain combined physically and chemically with each other. Several physical and chemical methods have been employed for such investigations. At first the secondary wall was the main object of study, but with the refinement of methods the primary wall has been also investigated successfully. The significance of the study of primary walls is that it yields information on the methods of growth of cell walls in surface area. Investigators combine observations on differential staining; differential solubilities; coarse and fine structure variations; reaction to polarized and fluorescent light, to X-rays, and to dark-field illumination; refractive indices; and composition of ash. In modern days, the electron microscope is the main tool for the study of cell walls.

Micellar and Intermicellar Systems

The organization of the structure of the cell walls is based on cellulose. The fundamental units of the system are the chain-like cellulose molecules of different lengths. These chains are not dispersed at random but occur in aggregates and generally known as **micelles.** The chain molecules possess a parallel arrangement in a micelle and the glucose residues within a chain are spaced at uniform distances from each other. Thus, a bundle of cellulose molecules, the micelle, may be compared to a crystal in that its units are arranged symmetrically. This way, the bundles of cellulose molecules are interconnected by means of the lower chain molecules and from a porous coherent system, the **micellar system,** interpenetrated by an equally coherent **intermicellar system** in which various wall substances other than the cellulose are present.

Frey Wyssling (1959) graphically described these structural elements and their interrelations on the basis of the secondary wall of the fibre of *Boehmeria.* Here one cellulose molecule is 8 Å wide. Cellulose molecules are combined into an **elementary microfibril** which has a widest diameter of 100 Å and is discernible with the electron microscope. It contains 100 cellulose molecules in a transverse section. Both the cellulose molecules and the elementary fibrils are ribbon like structures. Elementary fibrils form a bundle known as **microfibril** which is 250 Å wide and contains 2,000 cellulose molecules in a transverse section. Electron microscope studies on cell walls are connected mainly with this unit. Microfibrils are combined into macrofibrils 0.4 micron (μ) wide and containing 500,000 cellulose molecules in transverse section. Finally, 2,000,000,000 cellulose molecules make up a transverse section of the secondary wall of the fibre.

Microfibrillar and Microcapillary Structure

As mentioned earlier the cellulose of plant cell walls is being interpreted as a combination of two interpenetrating systems, the micellar and the intermicellar. They are submicroscopic. The walls contain a porous matrix of cellulose consisting of very fine coalesced fibrils, the **microfibrils** and an interfibrillar system of **microcapillaries** containing various noncellulosic wall constituents. However, within the microfibrils the micelles and consequently also the chain molecules occur approximately parallel to the long axis of the fibrils. The microcapillaries within the cellulose framework may contain liquids, waxes, lignin, cutin, hemicelluloses, suberin, pectic substances, other less common organic compounds, and even crystals and silica.

Microfibrillar Orientation in the Cell Wall

In the three layered walls of certain vessels, tracheids and wood fibres the fibrillar orientations of the inner and outer layers vary between transverse and helical, the helices being of comparatively low pitch, and those of the central layer fluctuate between longitudinal and relatively steeply pitched helical. Characteristic patterns occur about the large bordered pits of the early wood tracheids.

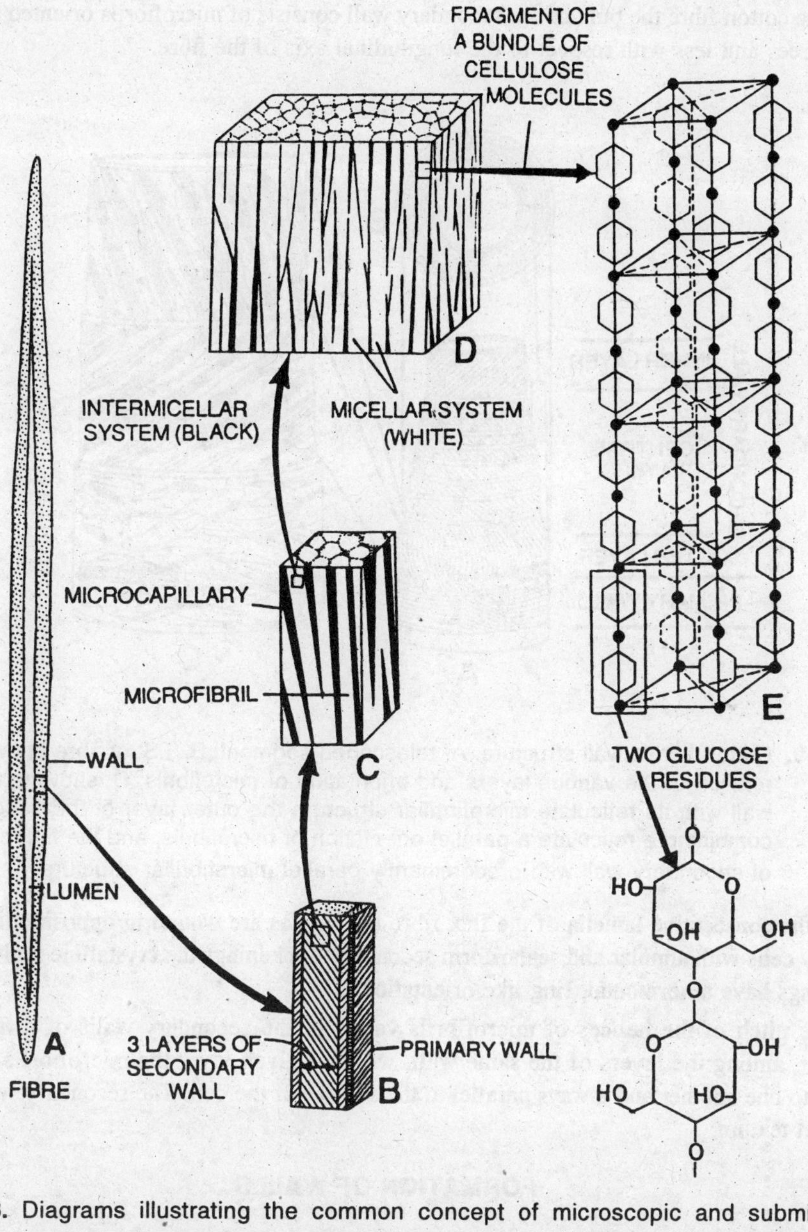

Fig. 34.18. Diagrams illustrating the common concept of microscopic and submicroscopic structure of secondary cellulose wall. Fibre (A) has a three layered secondary wall (B), In a fragment of the central layer of this wall (C), the microfibrils of cellulose (white), the other of micro capillaries (black) containing various noncellulosic wall materials. Microfibrils consist of inter micellar (black) and micellar (white) systems (D). The intermicellar system contains noncellulosic wall constituents, whereas the micellar system is composed of bundles of chain like cellulose molecules (E). The bundles of cellulose chains, the micelles, show an orderly arrangement of the glucose residues (F) that make the basic units of the cellulose molecules.

In the cotton fibre the bulk of the secondary wall consists of microfibrils oriented at an angle of 45 degrees and less with respect to the longitudinal axis of the fibre.

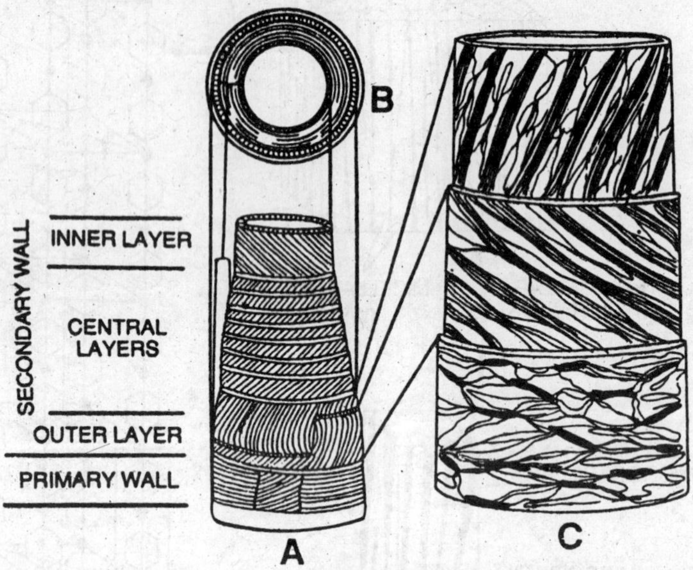

Fig. 34.19. Cotton fibre's wall structure. A, telescoped segment; B, T.S. of fibre showing spatial relation of the various layers and orientation of microfibrils; C, showing the primary wall with its reticulate microfibrillar structure the outer layer of the secondary wall combining a reticulate a parallel orientation of microfibrils, and the first central layer of secondary wall with predominantly parallel microfibrillar structure.

In the consecutive lamella of the flax fibre the helices are wound in opposite directions. In tracheary cells with annular and scalariform secondary thickenings the crystalline regions of these thickenings have a horizontal, ring-like orientation.

The pitch of the helices of microfibrils varies in the secondary walls of different cells. However, among the layers of the same wall, within a given layer the microfibrils are usually parallel to one another and always parallel to the surface of the cell. The secondary walls possess a parallel texture.

FORMATION OF WALLS

Initiation of Walls in Cell Division

The process of somatic division of a protoplast may be separated into two parts—(*i*) The division of the nucleus (*i.e.,* **mitosis** or **karyokinesis**) and (*ii*) The division of the part of the protoplast other than nucleus (*i.e.,* **cytokinesis**). In the cells which have cell walls the new wall develops during cytokinesis. The partition between the newly formed protoplasts, when first appears, is referred to as the **cell plate.** If cytokinesis takes place immediately after the nuclear division, the cell plate develops in the equatorial plane of a fibrous spindle, the **phragmoplast,** extending between the two groups of chromosomes that move apart during the anaphase of mitosis. As these two groups develop into the telophase nuclei, the phragmoplast becomes wider

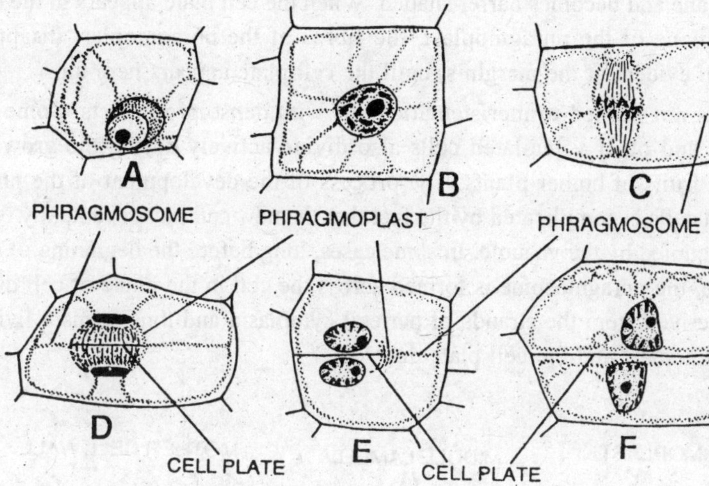

Fig. 34.20. Division of highly vacuolated cells. A, cell in a non-dividing state; B, nucleus in prophase and located in the middle of the cell; C, nucleus in early anaphase; laterally mitotic spindle is connected to parietal cytoplasm by a cytoplasmic layer the, phragmosome; D, daughter nuclei in telophase, the barrel-shaped spindle between the nuclei is the phragmoplast; cell plate appears in its equatorial plane; E, cell plate intersects one of the walls of mother cell; F, cell division is completed and the cell plate occupies the former position of phragmosome.

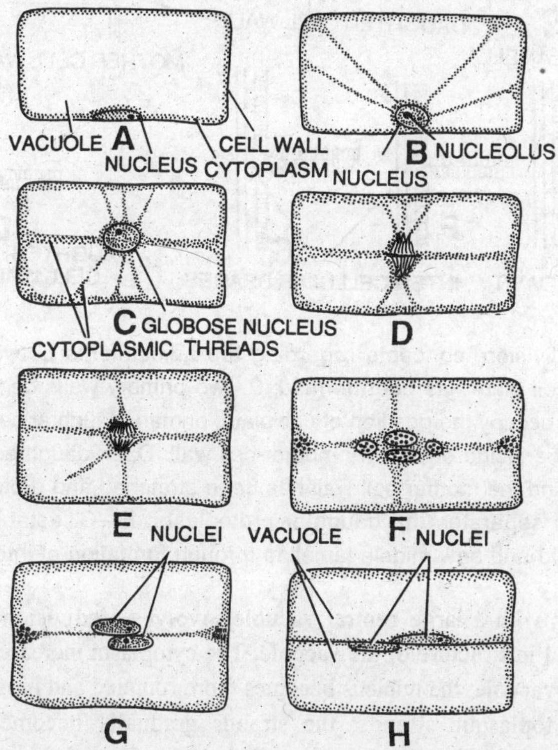

Fig. 34.21. Origin of the cell wall. Division in cell with large central vacuole. The cytoplasm increases in amount forming layer in position of new wall; nucleus becomes rounded and shifts to central position; normal nuclear division takes place with extending phragmoplast.

in the equatorial plane and becomes barrel-shaped. When the cell plate appears in the median part of the equatorial plane of the phragmoplast, the fibres of the phragmoplast disappear in this position but remain evident at the margin's, until the cell plate appears here too.

Cytokinesis is not limited to meristematic cells with dense protoplasts. Some vacuolated meristematic cells and other vacuolated cells also divide actively during the growth of roots, shoots, leaves and fruits of higher plants. The process of the development of the phragmoplast through vacuolated cells is complicated by the fact that the new cell plate eventually occurs in the region formerly occupied by the vacuole. In some cases, long before the beginning of cytokinesis, a cytoplasmic plate, the phragmosome is formed across the cell in the plane of cell division. The phragmosome is derived from the strands of parietal cytoplasm and thus forms a living medium in which the phragmoplast and the cell plate develop.

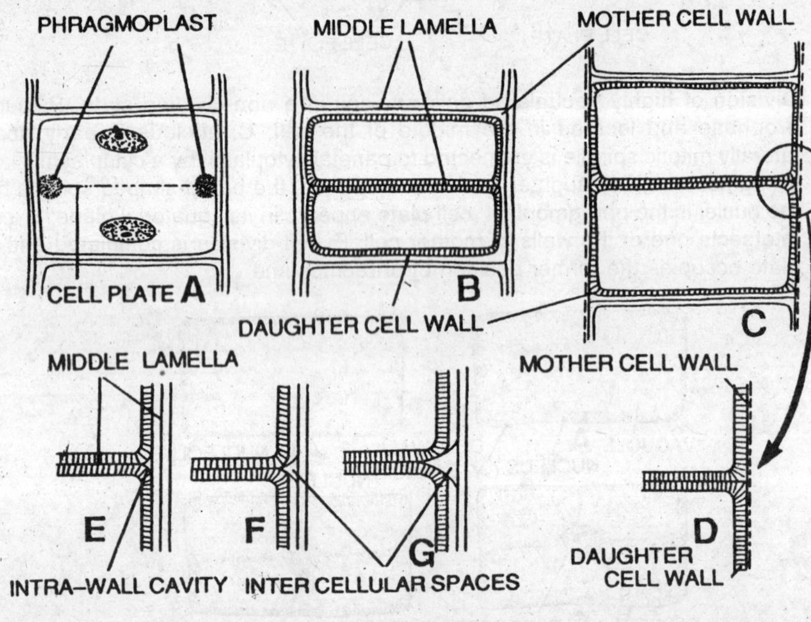

Fig. 34.22. After cell division, concepts regarding the adjustments between new and old cell walls. A, cell plate has been formed; B, two primary walls cemented by intercellular substance occupy the position of cell plate; primary daughter cell walls have been laid down on the inside of primary mother-cell wall; C, D, daughter cells have expanded vertically and the mother cell wall has been stretched and ruptured opposite the new wall which separates the daughter protoplasts; E—G, establishment of continuity between old and new middle lamellae through formation of intercellular space.

Division in cells with a large central vacuole involves predivision changes in form and position of nucleus and in structure of the vacuole. The cytoplasm increases in amount extending in strands through the vacuole; the nucleus becomes more rounded and moves away from the wall along one of the cytoplasmic strands; the strands gradually become concentrated in the **'phragmosome'** in the plane where the new wall is to be formed. Nuclear division and wall formation then take place as in other meristematic cells or cells with enough cytoplasm.

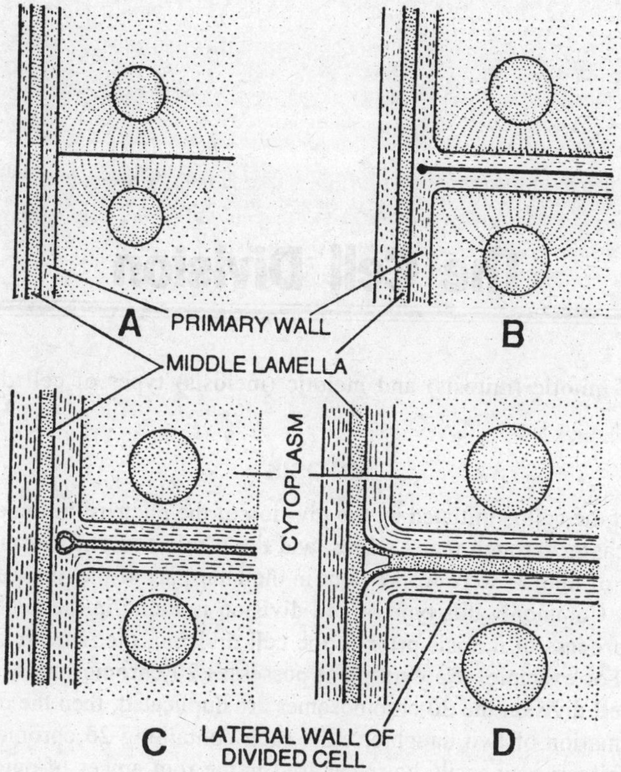

Fig. 34.23. The cell wall. This figure shows the stages in the establishment of the connection between the middle lamella of a newly formed wall with that of the lateral wall of the divided cell.

Growth of Walls

Growth in thickness of walls is evident in secondary walls as well as in primary walls. It occurs by a successive deposition of wall material, layer upon layer that is, by a process known as **apposition.** Growth of walls by apposition is usually centripetal, that is, it occurs from outside and toward the lumen of the cell. Centripetal growth is characteristic of cells forming tissues. Another important aspect of wall growth establishment of continuity between the new intercellular lamella and that located outside the primary wall of the mother cell. There is an extension and break down of the parent wall opposite the new middle lamella. The formation of intercellular spaces is thought to be associated with this phase of wall growth.

35

CHAPTER

The Cell Division

The details of mitotic (mitosis) and meiotic (meiosis) types of cell division have been discussed here.

MITOSIS

The division of the cell is initiated by the division of the nucleus. In the ordinary method of division a nucleus passes through many stages, and the whole complicated process is known as **mitosis.** The details of mitosis were worked out in the later part of the nineteenth century by W. Flaming and others. Commonly this type of cell division is found in the vegetative parts of the plant body. In the process of mitosis prior to the cell division, the number of chromosomes is always duplicated. For example, if a corn plant possesses 20 chromosomes in the somatic cell, then prior to each cell division the 20 chromosomes are duplicated, then the division takes place resulting in the formation of two daughter cells each containing 20 chromosomes again. The different stages of mitosis may easily be recognised in the root apices of onion. From the study point of view, the process of mitosis may be differentiated into two main phases — The **karyokinesis** and the **cytokinesis.** The actual division of the nucleus is known as karyokinesis whereas the division of the cytoplasm of the cell is called cytokinesis. Thus, during the process of mitosis the nucleus undergoes several changes which may easily be studied in the onion root apex by special cytological techniques. The chief function of mitosis seems to be to divide all parts of the chromatin equally between the two daughter nuclei. The important phases of mitosis are — **prophase, metaphase, anaphase** and **telophase.**

Prophase. In the resting nucleus the chromatin is spread out as a reticulum. It is actually composed of a number of separate units, the **chromosomes.** The number of chromosomes in the nuclei is definite in different species. Gradually the chromosomes become thick and condensed and each of them splits lengthwise forming two **chromatids.** The chromatids remain coiled around each other throughout their length. Gradually, they become much more thick and smooth. The chromatids coil around each other spirally and each chromosome itself remains surrounded by a membrane. In well fixed chromosomes some unstained gaps or constrictions are seen; they are the attachment regions, called **centromeres.** The nucleoli lose their staining power and disappear completely. The nucleus then rapidly passes into the next stage, the metaphase, through a complicated series of changes.

Metaphase. The nuclear membrane disappears and simultaneously a new structure, the **spindle,** appears in the cytoplasm, which chemically, consists of long chain protein molecules oriented longitudinally between two poles. The **fibres** of the spindle, however, are really fine tubules, not just protein threads. Chemical analysis of the cells has indicated that approximately 15 per cent of the cytoplasmic proteins go into its make-up. Once the spindle is formed, the chromosomes move through the cytoplasm to it, and become fastened by their **centromeres** to a region midway between the poles called the **equator** of the spindle, a position of apparent

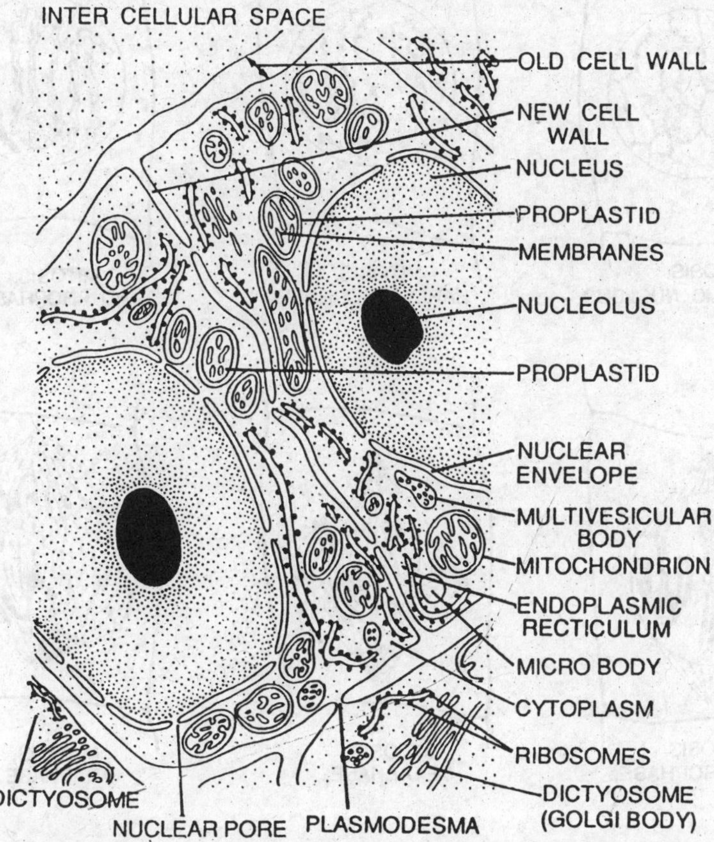

Fig. 35.1. The cell division. Diagram of an electron micrograph of meristematic cells in shoot apex of *Chenopodium album.*

equilibrium. The centromere of each chromosome always contacts the spindle at the equator; the arms of the chromosomes, not being so restricted, are randomly oriented.

The centromere is the organ of movement. Without it, a chromosome cannot orient on the spindle, and the chromatids cannot separate from each other later. The position of centromere is visible in a chromosome during metaphase by a constriction, and since the position of the constriction is characteristic for each chromosome, the centromere divides the chromosome into two arms of varying lengths. Very few chromosomes have strictly terminal centromeres.

Anaphase. Anaphase follows metaphase. At the end of the metaphase the centromeres of each pair of chromatids appear to repel each other. The centromeres now divide so that each chromatid has its own centromere; they then move apart from each other to initiate a slow movement that will take sister chromatids to opposite poles. Termination of anaphase movement occurs when the chromosomes form a densely packed group at the two poles.

Telophase. As soon as the chromosomes reach the poles, they collect into a more or less solid-appearing mass. This marks the beginning of telophase. The mass of chromosomes gradually converts into a nucleus. A new nuclear membrane forms. Spindle gradually disappears. The formation and enlargement of the spaces containing nucleoplasm continue until the

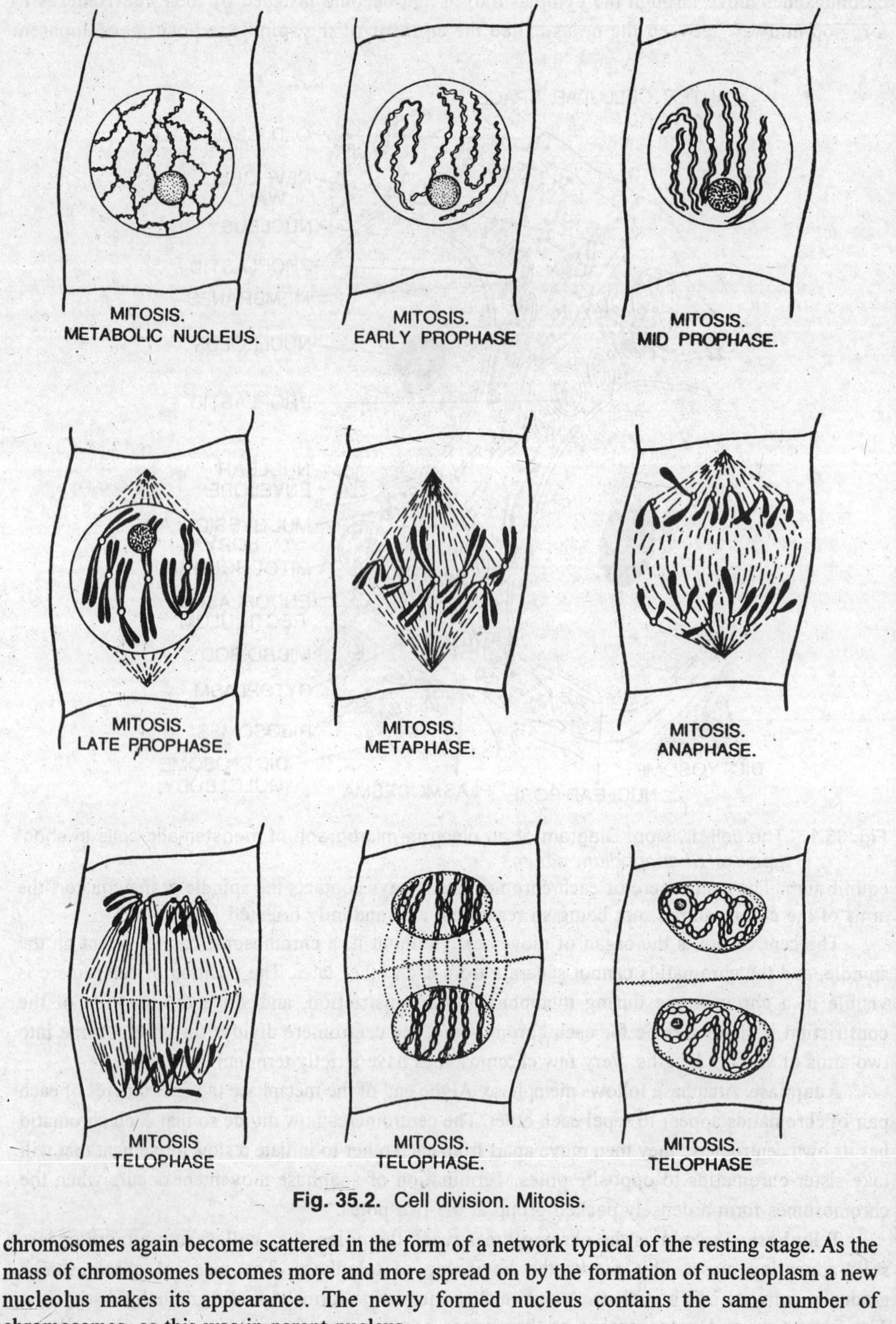

Fig. 35.2. Cell division. Mitosis.

chromosomes again become scattered in the form of a network typical of the resting stage. As the mass of chromosomes becomes more and more spread on by the formation of nucleoplasm a new nucleolus makes its appearance. The newly formed nucleus contains the same number of chromosomes, as this was in parent nucleus.

Cytokinesis. Just after the nuclear division, the division of cytoplasm takes place which is known as **cytokinesis.** The cytokinesis takes place in two ways. According to one method, much of the cellulose is being deposited in the centre of the cell, and the cell is resulted. This method is known as **cell plate method.** According to other method after the formation of young nuclei, a furrow develops in the cytoplasm and the cytoplasm is being divided into two equal parts, thus completing the cytokinesis.

Duplication of DNA and its transfer to daughter cells. With the result of mitotic cell division one parental cell gives rise to two daughter cells, and this process continues indefinitely. The newly formed daughter cells behave in the similar way as their parent cells. This shows that the daughter cells bearing the DNA molecule of one type, and they are also similar in quantity. As we know, the DNA molecules consist of two spirally coiled threads. This model of DNA is known as double helix DNA. During cell division, because of the presence of weak hydrogen bonds the threads of DNA helix separate from each other. In prophase stage of mitosis each chromosome splits into two chromatids. One of the DNA threads goes to one chromatid and the other to another chromatid. All the chemical substances, that give rise to the new thread of DNA are found in the protoplasm of the daughter cell. The newly formed thread coils around the old DNA thread and forms the double helix of DNA. The newly developed DNA is similar to that of original DNA of parental nucleus. By this process, the DNA molecules reach in the same quantity to each of the daughter cells.

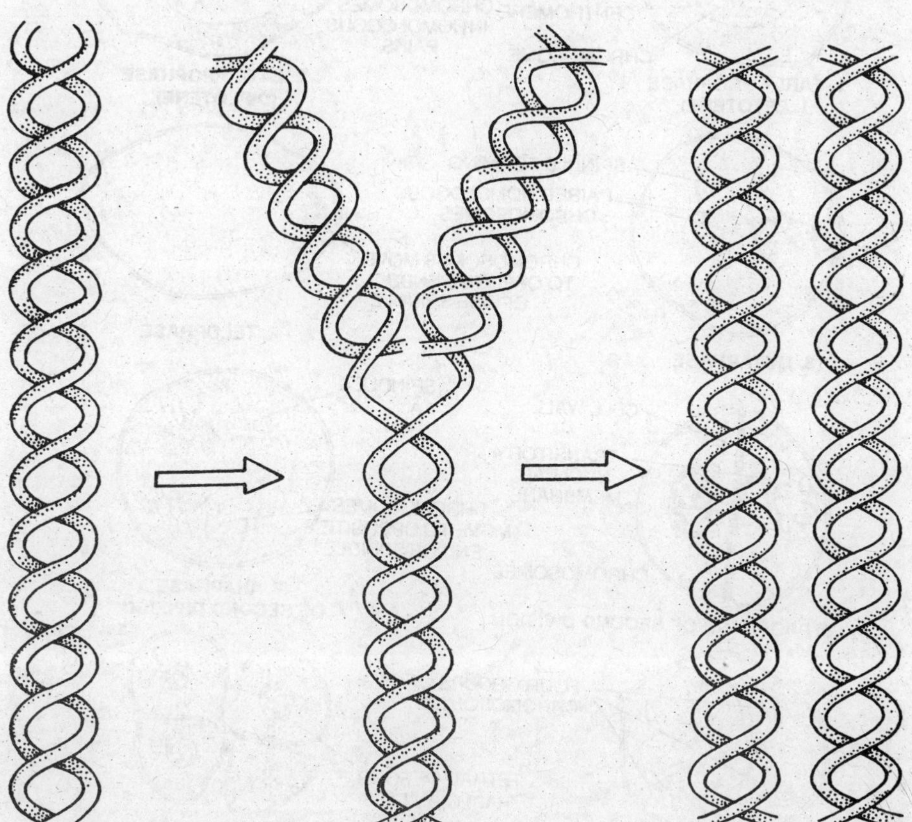

Fig. 35.3. Replication of the DNA molecule. The old helix unwinds (middle) and the two new helices are being formed.

Significance of mitosis

With the result of mitosis, the chromosomes split lengthwise into two chromatids. Each chromatid bears all those characteristics which were present in mother chromosome. In other words, with the result of mitosis, two identical cells have the same genetic constitution, qualitatively and quantitatively, as the parental cell from which they arose. Thus, the maintenance of the genetic integrity of the cell population and ultimately of the organism and its descendent depends upon the mechanism of mitosis. This process has been proved to be beneficial to vegetative reproduction. In the similar way, the characters of the plants grown by vegetative reproduction may be preserved for long time.

MEIOSIS

The division during which the number of chromosomes are reduced are known as the reducing or meiotic divisions, and the process of reduction division as **meiosis.** For the first time

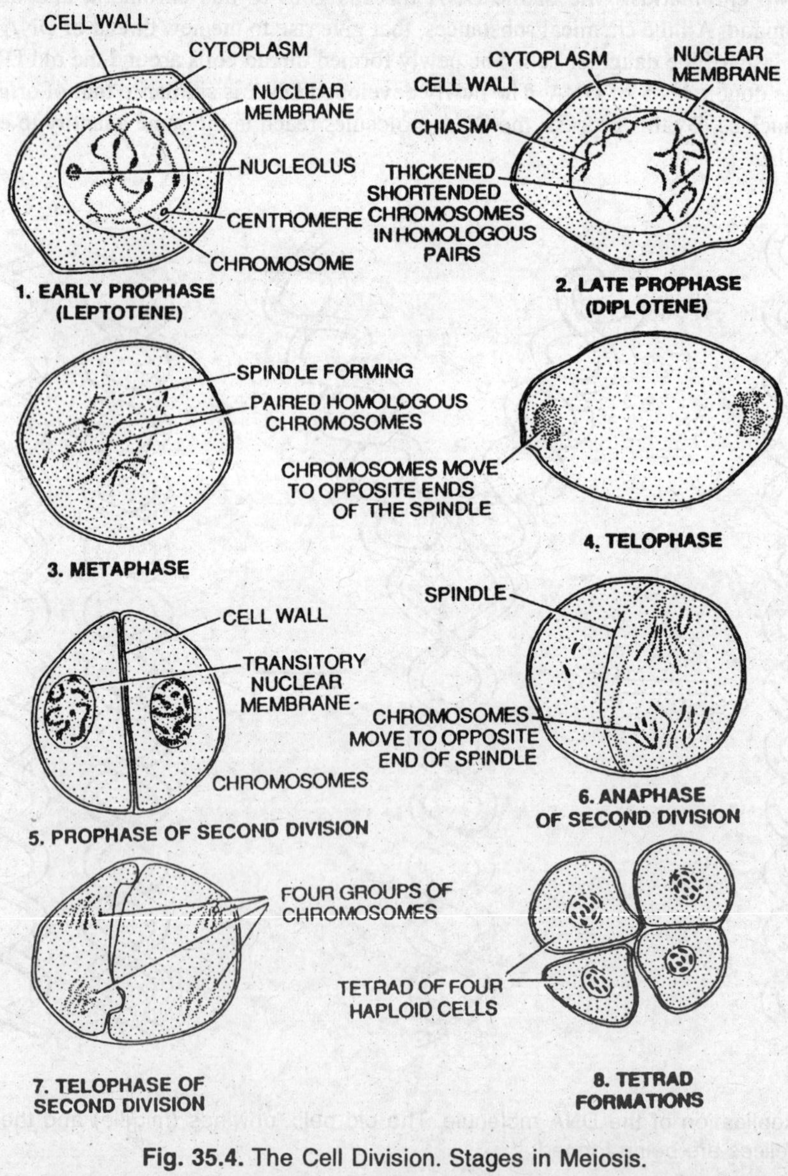

Fig. 35.4. The Cell Division. Stages in Meiosis.

in 1888, Strasburger discovered this phenomenon. There are two phases in the life-cycle of lower and higher plants — **sporophyte** and **gametophyte.** The sporophyte contains the 2n number of chromosomes, whereas the gametophyte possesses n number of chromosomes. If the nucleus of sporophyte contains 6 chromosomes, then its nucleus of gametophyte contains only 3 chromosomes, *i.e.*, half the number of the chromosomes of the nucleus of the sporophyte. With the result of meiotic division (meiosis), from one nucleus of the sporophyte four nuclei of the gametophyte are formed, and the number of chromosomes for each nucleus becomes half of the number of chromosomes of the nucleus of the sporophyte. In all the plants which are sexually reproduced, the meiosis takes place. In the sporophyte, the spore mother cells represent the last phase of the sporophytic status. Thus, each spore mother cell bears 2n number of chromosomes. Each spore mother cell eventually divides reductionally producing four haploid (n) spores. In such division, the first division is known as **heterotypic division.** The number of chromosomes is reduced to half during this division. The division is immediately followed by another ordinary division, which is known as **homotypic division.** The spore denotes the beginning of the gametophytic stage. The spore, when germinates gives rise to the gametophyte, that bears male and female gametes. The number of chromosomes in the gametes is always n. With the result of fertilization or conjugation the nuclei of male and female gametes fuse together forming **zygote** or **oospore.** The oospore (2n) is the beginning of the sporophytic stage. It bears 2n number of chromosomes. The oospore develops into the embryo and thereafter the young plant is formed. The mitotic division goes on continuously, and the plant remains diploid (2n). During the spore formation the meiosis takes place, thus the sporophytic and gametophytic stages of the plant bear 2n and n number of chromosomes in their nuclei respectively. This is only because of meiosis, the species of the plants bear definite number of chromosomes in their nuclei.

Mechanism of meiosis

It is already mentioned that during meiosis the mother nucleus divides twice, *i.e.*, heterotypic division is followed by homotypic division. The number of chromosomes is reduced to half in the heterotypic division whereas the homotypic division is mitotic one and here number of chromosomes remains same. The mechanism consists of four important phases — **prophase, metaphase, anaphase** and **telophase.**

Prophase. The prophase of meiosis is longer in duration, and consists of five separate states — **leptotene, zygotene, pachytene, diplotene** and **diakinesis**. When chromosomes first appear they are separate fine threads; this is the **leptotene** stage. The homologous members of each pair of chromosomes then associate closely side by side, corresponding loci (chromosomes) adhering together, a process called **pairing** or **synapsis;** each associated pair is a **bivalent;** this is the **zygotene** stage. The bivalents shorten and thicken. Each individual chromosome is now manifestly double; hence each bivalent consists of four **chromatids;** this is the **pachytene** stage. The two chromatids derived from one chromosome remain paired but they separate from the other two chromatids derived from the homologous chromosome; this is **diplotene** stage, followed by **diakinesis.** At certain places, however, they are held together by interchanges **(chiasmata)** occurring between chromatids derived from homologous chromosomes. Chiasmata are the visible expression of crossing-over of genes. There are usually one or more chiasmata per chromosome pair per meiosis. In a chiasma two chromatids, one from each of the original chromosomes, have apparently broken at corresponding places; and the broken ends of one chromatid have fused with the broken ends of the other. Where there is more than one chiasma in a single bivalent they may involve different pairs of chromatids, but always one from each of the original chromosomes.

As each of the two chromosomes which compose the bivalent chromosome is split into two chromatids, the bivalent chromosome is really a quadruple structure and is called a tetrad chromosome. After the formation of tetrad chromosomes, the nucleus has only half the number of

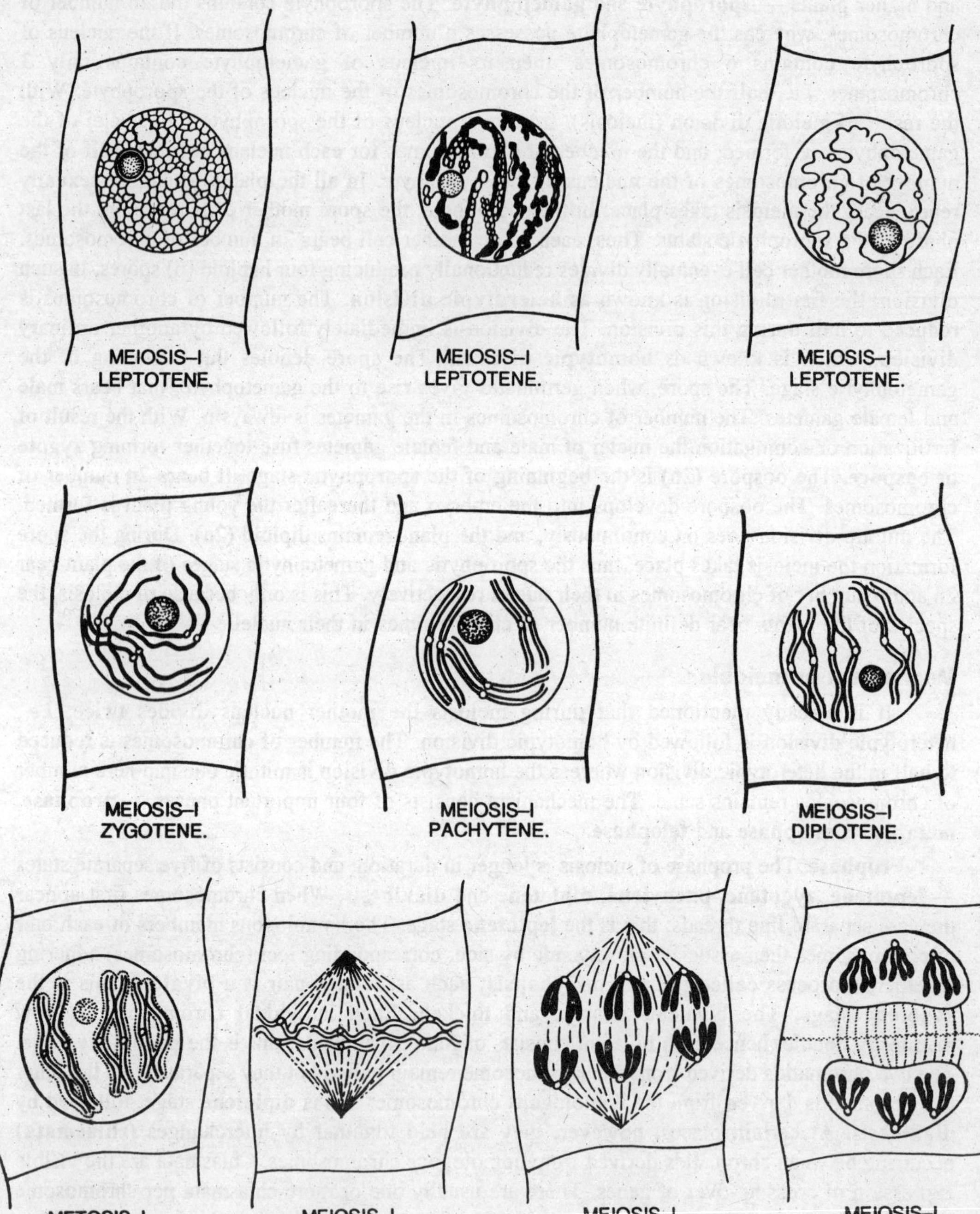

Fig. 35.5. Cell division. Meiosis I.

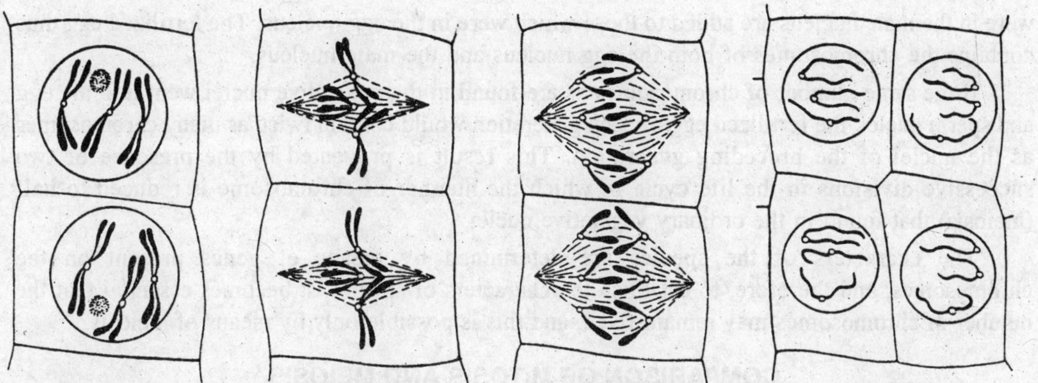

Fig. 35.6. Cell division. Meiosis II.

chromosomes that are found in ordinary vegetative cells. The subsequent events in this and the next division distribute one of the chromatids of the tetrad to each of the four cells.

Metaphase. At metaphase each of tetrad chromosomes is separated longitudinally into two daughter chromosomes which pass to opposite poles. Each of these chromosomes is, however, already split into two chromatids. At metaphase of the second of the reduction divisions these chromatids are separated from each other. We can now see why there are two reduction divisions. This is connected with the fact that the chromosomes which are formed in the prophase of the first meiotic division are quadruple and that it takes two divisions to divide them up into their constituent chromatids. At the end of the second reduction division not only has each chromosome been reduced to a single chromatid but the total number of chromosomes is also haploid (n).

Anaphase. At anaphase two of the four chromatids from each bivalent go to one pole of the spindle, and the other two to the other pole. The chromatids go in pairs because the spindle attachment of each original chromosome has not yet duplicated, so that two chromatids are united at that point. Owing to the effects of chiasmata, however, the united chromatids are not usually derived throughout their length from the same chromosome but are mixtures of one or more pieces from each of the original chromosomes. It is a matter of chance, and quite uninfluenced by behaviour of other bivalents, which spindle attachment with its chromatids, goes to which pole of the spindle.

Telophase. After anaphase there may be a short telophase and resting stage, or the second meiotic division may follow immediately, usually in both the daughter-cells formed by the first division. It is like a mitosis, except that it starts prophase with only half the normal number of chromosomes, each already divided into two chromatids before the previous prophase. The two chromatids separate at anaphase, one going to each daughter cell. Again, the distribution of the chromatids is a matter of chance, except that the two from each chromosome must go to opposite poles. On this chance distribution of the chromatids at both anaphase depends the law of independent assortment. Each daughter-cell, therefore, eventually receives only the haploid number of chromosomes. The reduction from the diploid number of the original cell to the haploid number of the four products is one of the bases of genetic segregation.

Significance of Meiosis

During the process of fertilization the male nucleus and the egg nucleus come together and the contents of the two nuclei come to be enclosed by a single nuclear membrane, a part of which comes from the male nucleus and a part from the egg nucleus. In this way, the chromosomes which were in the male nucleus are added to those which were in the egg nucleus. The fertilized egg thus contains the chromosomes of both the egg nucleus and the male nucleus.

If the same number of chromosomes as are found in the vegetative nuclei went into the egg and sperm nuclei, the fertilized egg of each generation would contain twice as many chromosomes as the nuclei of the preceding generation. This result is prevented by the presence of two successive divisions in the life cycle in which the number of chromosome is reduced to half (meiosis) that found in the ordinary vegetative nuclei.

The characters of the species are determined by means of genes present on the chromosomes, and therefore, to establish the characters of species, it becomes essential that the number of chromosomes may remain fixed, and this is possible only by means of meiosis.

COMPARISON OF MITOSIS AND MEIOSIS

Mitosis	Meiosis
1. It takes place in all the somatic cells, except those of spore mother cells, or germinal cells.	1. It always takes place either in the spore mother cells or in germinal cells.
2. The whole process is being completed in one sequence; two daughter cells or nuclei are produced from the mother cell.	2. The whole process is being completed in two sequences; with the result four daughter cells or nuclei are produced from one spore mother cell or germinal cell.
3. The daughter cells are identical to the parent cell as each cell possesses the similar number of chromosomes, *i.e.*, diploid (2n).	3. The daughter cells are dissimilar to that of parent cells as crossing over takes place and the number of chromosomes is reduced to half in each of daughter cells or nuclei, *i.e.*, haploid (n).
4. The prophase stage is of short duration and it does not bear any sub-stages. From the very beginning the chromosomes are double structures each formed of two chromatids. The split homologous chromosomes are not attracted towards each other. There is no chiasma formation and the crossing over does not take place. Synapsis does not occur.	4. The prophase is of long duration and it consists of several sub-stages. The chromosomes are granular and single structures in the beginning. In zygotene and later sub-stages, the homologous chromosomes attract towards each other forming bivalents. Each chromosome splits into two but the chromatids are not separated. The chromatids of one chromosome exchange with the chromatids of the second chromosome during the process of diplotene. The process of chiasma formation and crossing over takes place during diplotene. Synapsis takes place in the homologous chromosomes.
5. During metaphase the centromere of each chromosome divides, and the chromatids travel towards the opposite poles. No	5. During metaphase the centromeres of chromosome do not divide, and the homologous chromosomes travel towards

change takes place in the number of chromosomes and the resultant daughter cells remain similar to parent cells. In the last stage of metaphase, the chromatids are separated from each other in the divided parts of centromeres. This separation takes place because of repulsion.	the opposite poles themselves. The number of chromosomes travelling towards one of the poles remain half to the normal number. The resultant daughter cells are dissimilar to parent cell. The maternal and paternal chromosomes are separated because of repulsion of homologous chromosomes.
6. In anaphase the chromosomes travel upto poles.	6. As in mitosis.
7. As the result of telophase the two nuclei bear diploid (2n) number of chromosomes.	7. Two haploid (n) nuclei are formed during first telophase stages.
8. There is no second division in mitosis.	8. The second division occurs in meiosis. The reduction division takes place during the first telophase, whereas during second telophase ordinary mitotic division takes place.
9. As the result of mitosis, the number of chromosomes, their characters and constitution do not change.	9. As the result of meiosis, four daughter cells are formed. Each daughter cell or nucleus is haploid (n), *i.e.*, it contains half number of the chromosomes, thus the characters and the constitution of the chromosomes also change.
10. Cytokinesis always takes place after each division.	10. Cytokinesis does not necessarily occur in telophase one, normally it occurs in second telophase.

Unit 7: Genetics–
Continuity of Life

36
chapter

Genetic Basis of Inheritance

All the living organisms fundamentally have the capacity of reproduction. The offspring of all the organisms (plants and animals) resemble their parents in the several respects. Why is it so? The answer lies in the phenomenon of heredity. Heredity literally means "**tendency of like to beget like**", *i.e.*, all living organisms tend to produce young ones like themselves. It involves the transfer of biological information of parental generation to the new organism via the egg and sperm. For example, cow can produce cow and never cats or elephants. This tendency of individuals to resemble their progenitors is called heredity. Though the offspring of a species may resemble very closely to their parents, but they never resemble a particular set of parents differ from each other and from their parents in many respects and to different degrees. In other words we can say that each species has its **individuality**, *i.e.*, each species is recognisable by its certain specific morphological, physiological and behavioural characteristics. For example, each person has certain peculiar characteristics by which he is easily recognised by other persons. Similarly the dogs of same breed are distinguishable from each other by certain structural differences.

Thus the offspring are similar to their parents but not exactly. They differ not only one another but from their parents also in many characters. These differences are called **variations.** The variations may be of two types. 1. The variations which are inherited from one generation to other are called **hereditary variations**, and 2. The variations which are not inherited but are because of temperature, moisture, food, light or other environmental factors affecting the development of organism are called **environmental variations.** For example, the differences between a well-nourished and malnourished person are environmental because they are due to food factor.

The branch of biological science that deals with heredity and variations is termed **genetics.** Both heredity and variations play significant role in the organic evolution as well as in speciation, *i.e.*, formation of new species. The word **genetics** (Greek, *gen* to grow into or to generate) was proposed by William Bateson (1906). This branch of science seeks to understand the laws governing the transmission of hereditary potentialities from parents to offspring. The aim of the science of genetics was defined by **Bateson** in 1905 as "the elucidation of the phenomena of heredity and variation." The geneticist studies the causes both of the similarities and of the dissimilarities between the developmental patterns of different organisms.

Modern biologists know that the units of heredity are **genes** which are transmitted from generation to generation. The parallel behaviour of genes and chromosomes during inheritance demonstrates the location of genes on chromosomes. The genetic material which carries the information and the programme of biological potential of an organism is **deoxyribonucleic acid** (DNA), whereas in somes viruses this function is carried on by **ribonucleic acid** (RNA).

Several experiments have been performed which reveal the mechanisms how genes function in prokaryotic and eukaryotic cells. However, differential gene expression forms the molecular basis of normal processes of growth and differentiation as well as cancer. Advances in molecular biology have developed techniques to isolate individual genes and to transfer them from one kind of organism to another. Such researches have revolutionised the field of agriculture and medicine.

As scientists map **genome,** the total genetic content of organisms including of humans, the prospects of curing diseases, solving mysteries of forensics and biological evolution have become easy and within reach. The birth of the first human **clone** a baby girl called Eve by scientists, was announced by Brigite Boisselier, head of a company named **Clonaid.** This was announced in December, 2002.

There is an old proverb that "like begets like." Human babies always cast in the human mould and usually resemble in particular ways their parents. On the other hand puppies always grow to be dogs and usually have the body shape, size, colour and other characteristics of their breed. It involves the transfer of biological information of parental generation to the new organism *via* the egg and sperm. Thus, the offspring of the dogs always resemble with their parents and never with cats or tigers. This tendency of individuals to resemble their progenitors is called **heredity.** A scientific understanding of heredity could arise only out of knowledge of the basic facts of biology, that is, reproduction and the means by which living matter perpetuates, in particular.

It is obvious that a living individual always arises from another living individual of the same species and never from another species or from lifeless matter. It is a well known fact that individual living things grow old and die, and therefore, the continuity of life must be maintained by the transmission of heredity to new offspring. This process of reproduction occurs in different organisms in different ways. For example, in **asexual** or **vegetative** reproduction, the body of the parent is divided into two or more parts and each part grows into a new individual. In such cases only one "parent" inherits all its traits. It is really a carbon copy of parent and is called its **clone,** rather than offspring. In some plants a small portion of the body, when removed and placed under favourable conditions, may establish itself as a new individual. For example, potatoes are cultivated more easily from pieces of tubers than from seeds, and many plants of horticultural importance are propagated by the union of two sex cells, or **gametes** (*i.e.,* egg and sperm), which form a **zygote** from which the new individual develops. In both sexual and asexual reproduction there exists a physical, living link between the parents and the offspring/clone. A part, however small, of the parent grows and develops to become the body of the offspring/clone. Whatever is inherited must be contained in this part.

However, every living organism has a set of characters by which it is identified as a member of a particular species. Now question arises, what is that which makes a species having certain set of characters and how are they maintained generation after generation. The science which deals with the answers of these questions is **genetics.** Though genetics is a relatively young discipline, yet this has gained enormous importance.

Inheritance-Heredity and Variation

No two persons or individuals, even of the same sex, are never exactly alike. The characters that provide individuality to a species are said to cause **variation** among the species. The primary reason for this is that the environment of the organisms is never the same in different places and at different times. So long as an organism is alive, its heredity interacts with its environment; this interaction determines what an individual organism is like at a given moment and what it can develop into in the future. To facilitate this the Danish geneticist Johannsen (1911) proposed to distinguish the **genotype** of the organism from its **phenotype.** The genotype is the sum total of heredity, the genetic constitution that an organism receives from its parents. The observable characteristics, of an organism such as colour, form, size, behaviour, chemical composition, and structure, both external and internal gross and microscopic, constitute the phenotype. Two individuals with the same genotype may become different in phenotype when they come into contact with different conditions of food, temperature, light, humidity and other external factors. Such differences among organisms of similar heredity are known as **environmental variations.**

Among the sexually reproducing organisms no two individuals possess the same heredity, except for identical twins, which are likely to have identical genotypes. This is **hereditary** or genotypic variation. It arises from variations in the elements responsible for heredity, the genes, which may undergo changes called **mutations.**

However, hereditary component explains the variations which exist within a species. Another component of heredity and variation is associated with the mode of reproduction of a species.

Pre-Mendelian Ideas about Inheritance

In early nineteenth century, Schleiden and Schwann proposed their famous Cell Theory, and formerly proposed notion of spontaneous generation was rejected. Charles Darwin proposed the hypothesis of pangenesis. This hypothesis suggested that every part of the body produces a 'gemmule' which by the blood circulation, gets collected in the semen and thereafter these 'gemmules' transmit the characteristics to the offspring. Later on in 1892, this hypothesis was discarded by August Weismann. He proposed that living organisms consist of two kinds of materials, *i.e.,* (*i*) somatoplasm and (*ii*) germplasm. According to him the somatoplasm makes the whole body that undergoes growth, development and ultimately death. He believed that germplasm is responsible for continuity of subsequent generations.

Mendel provided the first proof for a theory that explained heredity by the transmission of units in the reproductive cell and thus put an end to vague notions such as those concerning 'bloods'.

Gregor Mendel and his Theory of Inheritance

Mendel was born in 1822 in a small village of Czech Republic. He made singular contributions in the field of genetics. He joined the Augustinian Monastery at Brno in 1843 and now he became Gregor Johann Mendel. He continued research throughout his life. In 1856, Mendel performed his first set of hybridization experiments with the garden pea. His interest in genetics was immense even when he did not get any recognition for his work in his life time. He died in 1884.

Mendel and his methods. Gregor Mendel (1822-1884) was a monk in the Augustinian Monastery at Brunn, Austria (now Brno, Czechoslovakia). He conducted his researches over a period of eight years on the common garden pea plant *Pisum sativum*. Thereafter he presented the results of his experiments, together with the generalization known as "Mendel's laws". The principle of segregation was formulated by Gregor Mendel in 1866 under peculiar circumstances that the scientific world failed to recognize or to appreciate it until after a lapse of 34 years. In 1900 A.D. when the law of segregation was rediscovered almost simultaneously by three different investigators, who had obtained results like Mendel's. These three scientists were — De Vries in Holland, Correns in Germany and Tschermark in Austria. They found Mendel's forgotten paper and proclaimed its importance. Immediately after his conclusions began to be confirmed and extended by experiments carried on in various parts of the world on many kinds of animals and plants. Mendel died in 1884, long before his scientific work came into its own.

About a century back Mendel did his experiments with garden peas which led to the discovery of the basic principles of inheritance. Subsequent work of several plant breeders showed three basic characteristics of inheritance. 1. Traits could be hidden or masked for one or more generations to reappear unchanged. 2. Traits may remain together in one generation and separate out in a later generation. 3. One form of a particular trait may be seen more often than its alternate form.

All sexually arising organisms owe their birth to the union of the male and female gametes and are the outcome of a single cell, the zygote. From this it follows that all resemblances between the offsprings and the parents must have come through the gametes. Thus the gametes are the carriers of all inheritance.

Why Mendel used garden pea for his experiments? Mendel's experimental use of the garden pea, *Pisum sativum* was evidently not an accident but the result of long careful thought. First, pollination could easily be controlled in this plant. Normally, the pea plant was self-fertilizing and, therefore, the use of Mendel's main techniques, 'selfing', presented no difficulties. When cross-fertilization between two pea plants was necessary, Mendel had merely to remove the stamens from one plant and transfer them to another plant without stamens. Second, the pea plant was easy to cultivate, and from one generation to the next took only a single growing season. Third, peas had many sharply defined inherited differences that had been long collected by seedsmen in the form of individual varieties. For his experiments. Mendel chose among these varieties seven different "**unit characters**" to follow in inheritance, ranging from stem size to shape of seed. Each character that he followed had two alternative appearances or "**traits**", *i.e.,* tall or short stems, round or wrinkled seeds, etc.

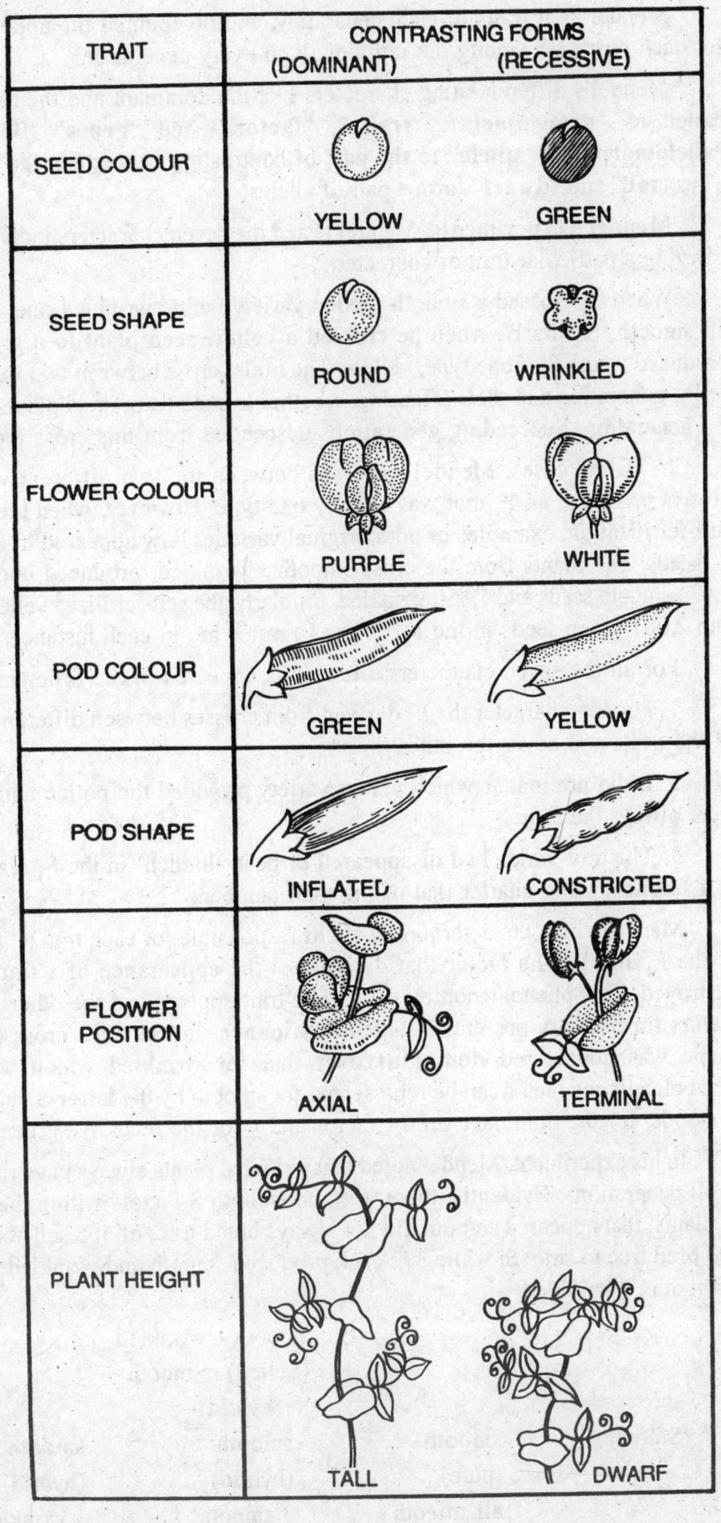

Fig. 36.1. Mendel's seven different unit characters.

Mendel kept traits of each separately, and he counted the appearance of the different traits for each character among the individuals in every generation.

Mendel's differentiating characters like the dominant and the recessive have been variously called as "**determiners**", "**traits**", "**factors**" and "**genes**". Bateson proposed the name "**allelomorph**" or "**allele**" to the pair of contrasting characters–the dominant and its recessive. Thus "**tall**" and "**dwarf**" form a pair of alleles.

Mendel's experiments. Mendel tested the seven characters individually by crossing a variety carrying a particular trait of character.

When he crossed a smooth-seeded variety with a wrinkled one, he obtained seeds that were all smooth. Similarly, when he crossed a yellow seed plant to a green seed variety, the seeds produced were all of one type, yellow. The initial cross between two varieties is called the parental, or P_1, generation and their offspring, whether in seed form or as plants are called the first filial, or F_1, generation. Succeeding generations descended from this cross are labelled F_2 and so on.

In all the cases, Mendel's crosses between the two different varieties for each character always produced an F_1 that was of only one type. However, when these F_1 plants reproduced by self-fertilization, examples of both original varieties now appeared in F_2. For example, the smooth F_1 seeds, *i.e.,* plants from the cross smooth *x* wrinkled, produced upon self-fertilization of F_2 of 5,474 smooth seeds and 1,850 wrinkled. Similarly, the self-fertilized yellow F_1 produced 6,022 yellow and 2,001 green seeds in the F_2. These F_2 ratios are in each instance very close to 3 : 1 ratio.

For all the seven characters, the results appeared in the following pattern:

1. For any character the F_1 derived from crosses between different varieties showed only one of the traits and never the other.

2. It did not matter which parent variety provided the pollen and which the ova; the results were always the same.

3. The trait which had disappeared or been 'hidden' in the F_1 reappeared in the F_2 but only in a frequency one quarter that of the total number.

Mendel called the determining **agent** responsible for each trait a "**factor**". From the evidence of the F_1 and F_2, the factor that determines the appearance of a trait could be hidden but not destroyed. This phenomenon by which one trait appears and the other does not, even through the factors for both are present, is called **dominance.** In Mendel's cross the factor for smooth seed shape was considered **dominant** over that for wrinkled which was considered **recessive.** Symbolically the factor can be represented for smooth by the **letter** S, and s for wrinkled. Similarly, Y stands, for the dominant yellow factor and *y* for the recessive green.

In his experiment, Mendel noted that wrinkled plants always gave rise ('bred true') to wrinkled in all generations. Evidently, there were no smooth S factors within them. On the other hand, the F_2 plants, that appeared smooth did not always breed true; of 565 self-fertilized smooth plants only 193 bred true to smooth while 372 each produced smooth and wrinkled plants in the proportion of 3 smooth : 1 wrinkled.

P_1 : Smooth × wrinkled

F_1 : (Selfed) : Smooth (hybrid)

F_2 (Selfed)	:	smooth (pure)	:	smooth (hybrid)	:	smooth (hybrid)	: wrinkled (pure)
F_3		all smooth		3 smooth		1 wrinkled	all wrinkled

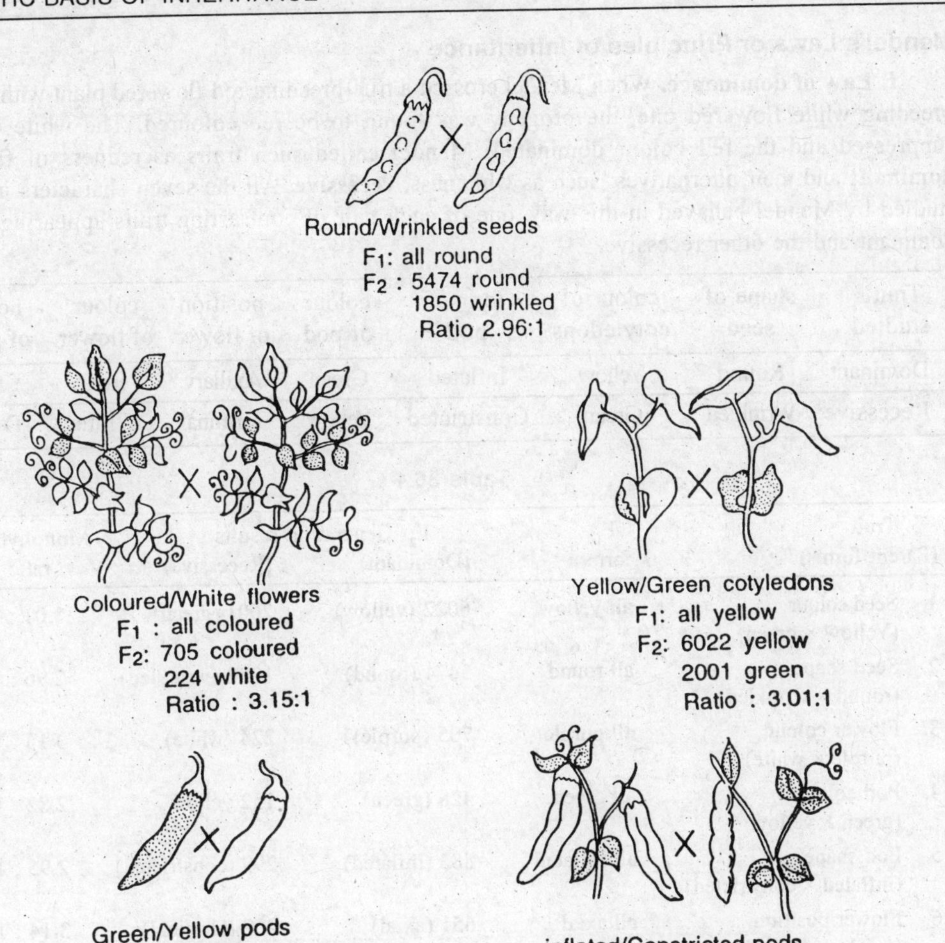

Fig. 36.2. Mendel's monohybrid crosses in garden pea plant for studying the transmission of seven different phenotypic traits of pea plants.

Mendel's Laws or Principles of Inheritance

1. Law of dominance. When Mendel crossed a true-breeding red flowered plant with a true breeding white flowered one, the progeny was found to be red coloured. The white colour suppressed and the red colour dominated. Mendel called such traits as redness of flowers **dominant** and their alternatives, such as whiteness, **recessive.** All the seven characters in peas studied by Mendel behaved in this way, one of each pair of contrasting traits appearing to be dominant and the other recessive.

Trait studied	shape of seed	colour of cotyledons	form of pod	colour of pod	position of flower	colour of flower	height of plant
Dominant	Round	Yellow	Inflated	Green	Axillary	Red	Tall
Recessive	Wrinkled	Green	Constricted	Yellow	Terminal	White	Dwarf

Table 36.1

Trait (Parent forms)	F_1 form	F_2 Generation results (Dominant)	(Recessive)	Monohybrid ratio
1. Seed colour (Yellow × green)	all yellow	6022 (yellow)	2001 (green)	3.01 : 1
2. Seed shape (round × wrinkled)	all round	5474 (round)	1850 (wrinkled)	2.96 : 1
3. Flower colour (purple × white)	all purple	705 (purple)	224 (white)	3.15 : 1
4. Pod colour (green × yellow)	all green	428 (green)	152 yellow	2.82 : 1
5. Pod shape (inflated × constricted)	all inflated	882 (inflated)	299 (constricted)	2.95 : 1
6. Flower position (axial × terminal)	all axial	651 (axial)	207 (terminal)	3.14 : 1
7. Plant height (tall × dwarf)	all tall	787 (tall)	277 (dwarf)	2.84 : 1

Later investigators have found many other characters that show similarly complete or almost complete dominance.

2. Law of segregation. Mendel demonstrated that a hybrid between two different varieties possesses both types of parental factors, which subsequently separate or segregate in the gametes. This is known as law of **segregation.** In contrast to the uniformity of the first generation hybrids, the second generation produced by self-fertilization of the F_1 red flowered plants which consist of two different kinds of plants—red ones like the red grand parent and white ones like the white grand parent. Mendel counted the numbers of individuals with each of the differentiating characters which reappeared by segregation in F_2. In the experiment for flower colour, for example, he raised 929 F_2 plants and found that 705 of them bore red flowers and 224 bore white flowers. The simple ratio was found that 3/4 of the F_2 resembled the dominant grand parent and 1/4 resembled the recessive one, *i.e.*, 3 : 1.

On the basis of his experiments, Mendel concluded as follows: contrasting characters, such as the red and white flower colours in peas, are determined by something that a transmitted from parents to offspring in sex cells, or gametes. This **something** of Mendel is now called a **gene.**

The different genes, for different flower colours or for round wrinkled seeds, do not blend, contaminate or affect each other while they are together in the hybrid. These different genes segregate, separate pure and uncontaminated, pass into different gametes formed by a hybrid and then go to different individuals in the offspring of the hybrid.

To understand how the genes are transmitted and distributed they are symbolized by letters. The dominant genes are provided with capital letters and their recessive alternatives by corresponding small letters. For example, in the cross of red and white peas *R* stand for the gene for red flowers and *r* for the alternative, or allelic, form of this gene, which gives white flowers *R* and *r* are allelic genes or **alleles.**

Since an individual develops from the union of two gametes, it receives a gene for flower colour from both parents. The true breeding red flowered parent is, therefore, represented by *RR* and its gametes are *R*; the true breeding white parent is *rr* and its gametes, *r*, or *vice versa*; and the resulting hybrid zygote will have both *R* and *r*; its genetic formula will be *Rr*.

When a gene pair in an organism contains two identical alleles, *e.g.*, *R* and *R*, the organism is considered **homozygous** for that gene pair and is called a **homozygote**. When two different alleles are present in a single gene pair, *e.g.*, *R* and *r*, the organism is **heterozygous** for that gene pair and is called a **heterozygote.** The red-flowered plants obtained by Mendel in the F_1 generation (*Rr*) were heterozygotes. They were red because the allele *R* is dominant over *r*.

3. Law of independent assortment. 'When two pairs of independent alleles enter into combination in the F_2, they exhibit independent dominant effects. In the formation of gametes the law of segregation operates but the factors assort themselves independently at random and freely.

Dihybrid and monohybrid crosses. Mendel studied seven pairs of characters in peas. These were—seed colour, seed surface, flower colour, plant height, colour of unripe pods, pod shape and position of flowers. One of his crosses was between a pea plant with round and yellow seed and one with wrinkled and green ones. This type of cross, which involves two character differences separable in inheritance is called a **dihybrid** cross and a cross which involves a single pair of alleles is **monohybrid.** The hybrids of F_1 possessed yellow and round seeds, since the yellow colour is dominant over the green and the round shape is dominant over the wrinkled. When F_1 hybrids were selfed, an F_2 generation was obtained consisting of 556 seeds of the following types:

315 round yellow
108 round green
101 round yellow
032 round green

In terms of proportion, these numbers are very close to a 9 : 3 : 3 : 1 ratio

P_1 : round × wrinkled	Yellow × green
F_1 : round	Yellow
F_2 : 423 round : 133 wrinkled	416 yellow : 140 green
or about	or about
3/4 round : 1/4 wrinkled	3/4 yellow : 1/4 green

The F_2 ratios therefore fit close to the ratios expected from crosses involving single gene pairs in which dominance occurs. When these two sets of results are combined in a single dihybrid cross, one finds that each gene pair acts independently of the other. This means that the chances for a plant to be round or wrinkled do not interfere with, or are independent of, its chances to be yellow or green. Therefore, if a seed has a 3/4 chance of being round and a 3/4 chance of being yellow, its chances to be round and yellow at the same time would be 3/4 × 3/4 = 9/16. The ratio

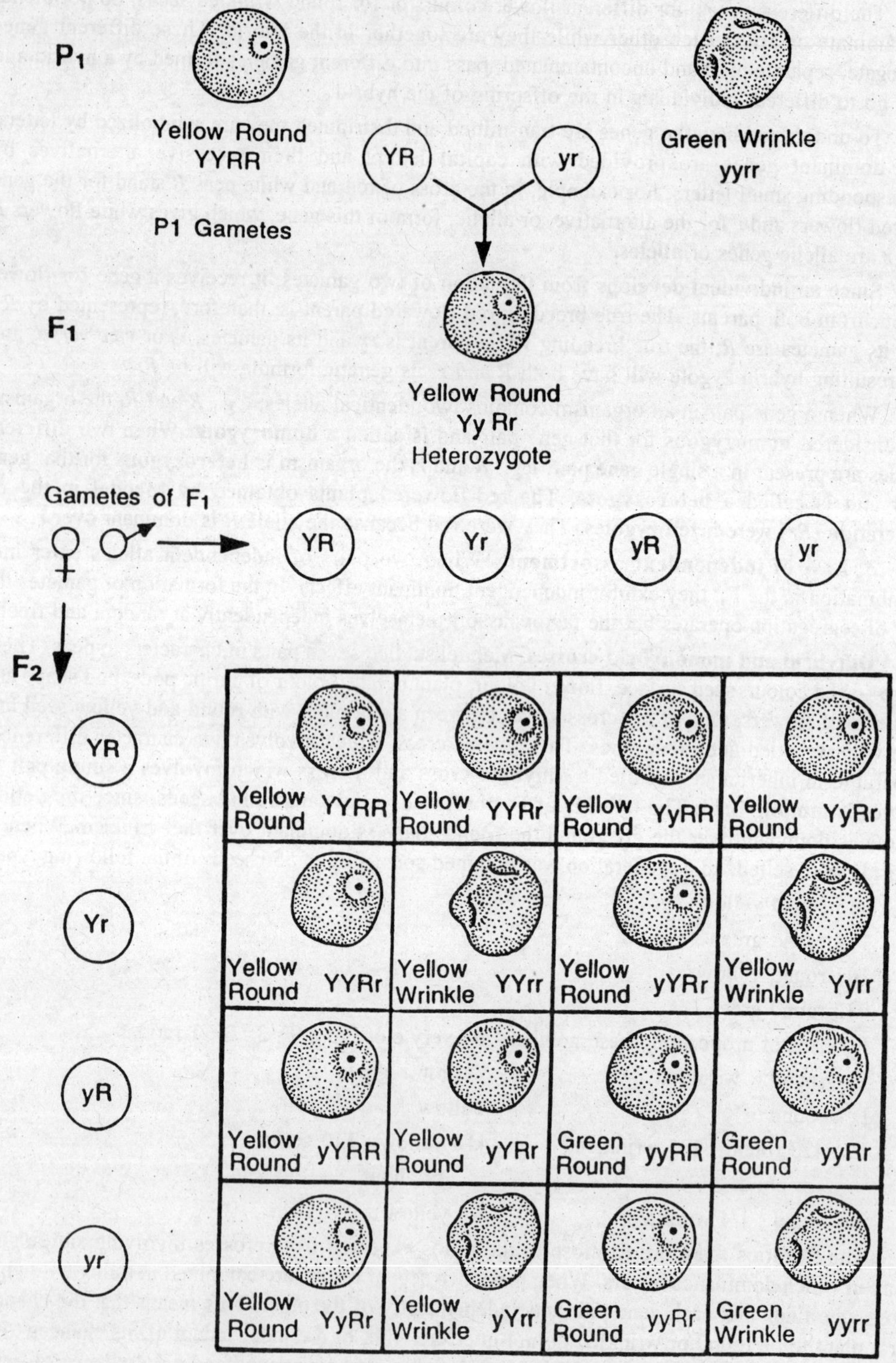

Fig. 36.3. A dihybrid cross between yellow round (YY RR) and a green wrinkled (yy rr) pea plant showing the formation of F_2 progeny in the ratio of 9 : 3 : 3: 1.

of each phenotypic combination may be obtained by multiplying the probabilities of the individual phenotypes:

3/4 round × 3/4 yellow	=	9/16 round yellow
3/4 round × 1/4 green	=	3/16 round green
1/4 wrinkled × 3/4 yellow	=	3/16 wrinkled yellow
1/4 wrinkled × 1/4 green	=	1/16 wrinkled green
		16/16 and a ratio of 9 : 3 : 3 : 1

Independent assortment. The gene for the round seeds is represented by *R* and its allele for wrinkled seeds by *r*, while the gene for yellow seeds by *Y* and for green seeds by *y*. Mendel's original yellow round seeded parent thus had the genes *RRYY*, and the green wrinkled parent *rryy*. The gametes produced by these parents were represented by *RY* and *ry*, and consequently the F_1 hybrid zygote formed by their union had the genotype *Rr Yy*. When the seed surface alone is considered, it is found that the F_1 *Rr* individuals produce gametes half of which carry *R* and half *r*. Every gamete must contain within itself not only a gene for seed surface but one for seed colour as well and indeed genes affecting every other character of the plant. Half these same gametes must, therefore, contain the allele *Y* and half the allele *y*; but in any given gamete it is a matter of chance as to whether the gene for round seeds is associated with that for yellow seeds or with that for green seeds. The assortment of the genes is independent. The F_1 hybrid produces four kinds of gametes in equal numbers *RY*, *Ry*, *rY* and *ry*. Now if two such F_1 plants are crossed, there will be 16 possible combinations among their gametes, because there are four kinds of pollen grains (♂) and four kinds of egg cells (♀). The 16 possible combinations that appear among the F_2 offspring will, therefore, equal in number. The parents, the F_1 and the F_2 of the cross, are represented diagrammatically in figure as to both their genotypes and their appearance, the 16 squares in F_2 represent the 16 possible combinations of gametes. Nine out of the 16 individuals are phenotypically round and yellow, 3 are wrinkled and yellow, 3 are round and green and only 1 is wrinkled and green. The ratio is 9 : 3 : 3 : 1.

Mendel's Postulates

The consistent pattern of results obtained from different monohybrid crosses led Mendel to propose four postulates which are also called **principles of inheritance.**

Postulate 1. Mendel proposed that each genetic character is controlled by a **pair of unit factors**, now known as **alleles** or **allelomorphic pair**

On the analysis of a monohybrid cross as per this postulate in the true bred parents, there are two unit factors for tallness and in the similar way two unit factors for dwarfness.

Thus, three combinations are possible: (*i*) either these are two factors for tall, or (*ii*) two factors for dwarf stems, or (*iii*) one of each factor in an individual.

Postulate 2. This postulate is based on the results obtained in F_1 generation. Mendel proposed that when two dissimilar unit factors are present in a single individual, only one is able to express and not the other.

The one which expresses itself is the **dominant unit factor** and the other which does not express is the **recessive unit factor**, and thus, tall stem is said to be dominant over the recessive dwarf stem. The dominant and recessive terms are used to denote a trait, *e.g.*, tall trait or recessive trait.

Now the unit factors of Mendel are called **genes.**

Postulate 3. This postulate is popularly known as the **principle of segregation** or **purity of gametes.** During gamete formation, the unit factors of a pair segregates randomly, so that, each gamete receives one or the other unit factor with equal probability. This ensures the purity of gametes.

During the formation of gametes in P generation (tall × dwarf) each gamete from the parent will receive one but the same unit factor either for tallness or for dwarfness.

After fertilization, F_1 plants will receive one unit factor for tallness and one for dwarfness. With the former (*i.e.,* factor for tallness) being dominant, all F_1 plants will be tall.

When F_1 tall plants will form gametes, the principle of segregation will demand that each gamete randomly received either the tall or the dwarf unit factor.

With subsequent fertilisation, which is a chance event, four combinations of these unit factors will result in equal frequency in F_2 generation:

(*i*) tall/tall (*ii*) tall/dwarf (*iii*) dwarf/tall (*iv*) dwar/dwarf.

When any individual produces gametes, the **alleles segregate**; so each gamete receives only one member of the pair of alleles and the paired condition is restored by random fusion of gametes during fertilisation.

Application of postulate 2. When postulate 2 is applied the combinations (*i*), (*ii*) and (*iii*) will result into tall plants, and only combination (*iv*) will result into dwarf plants. This means that the F_2 generation consists of three-fourths tall and one-fourth dwarf or a phenotypic ratio of 3 : 1. This is exactly what Mendel observed, and is found in all the monohybrid crosses.

Application of postulate 3. When postulate 3 is applied, the combinations (*i*) and (*iv*) will result into homozygous tall and homozygous dwarf respectively and combinations (*ii*) and (*iii*) will result in heterozygous tall and homozygous dwarf, respectively in F_2 generation.

Postulate 4. This postulate is popularly known as the **Principle of Independent Assessment of Factors.** As already described, Mendel extended his work by conducting dihybrid or two factor crosses, in which the parent plants differed in two pairs of contrasting traits.

For example, a cross in which plants producing round and yellow seeds were crossed with plants producing wrinkled and green seeds.

The F_1s were all yellow and round seeded, which suggests that yellow is dominant over green and round seed is dominant over wrinkled seed.

When F_1s were selfed, Mendel postulated that the segregation of one pair of unit factors will occur independently of the other pair or they will assort independently.

As per rule, the gametes must carry all possible combinations of the unit factors in equal frequency.

In such example, gametes with random distribution of unit factors will give following combinations: (*i*) yellow, round (*ii*) green, wrinkled (*iii*) green, round and (*iv*) green, wrinkled.

During fertilisation, each zygote will have equal probability of receiving two of these combinations, one from each parent. This will result into progeny as follows: 9/16 yellow round; 3/16 yellow, wrinkled; 3/16 green, round; 1/16 green, wrinkled, *i.e.,* 9 : 3 : 3 : 1. Mendel's dihybrid results.

On the other hand, if the two pairs of contrasting traits are inherited independently, the principle of independent assortment can also be stated as when two independent events occur simultaneously the combined probability of two outcomes is equal to the product of their individual probabilities of occurrence.

For example, the segregation of yellow × green and also of round seed × wrinkled seed is 3/4 yellow : 1/4 green, and 3/4 round seed : 1/4 wrinkled seed.

If these two events are combined:

(3/4) yellow × (3/4) round seed = 9/16 yellow, round;

(3/4) yellow × (1/4) wrinkled seeds = 3/16 yellow, wrinkled

(1/4) green × (3/4) round = 3/16 green, round seeds; and

(1/4) green × (1/4) wrinkled = 1/16 green, wrinkled seeds

This ratio is obtained in dihybrid cross.

Genetic Terms and Symbols

The pioneering study of Mendel led to a scientific study of heredity, which is known as **Genetics.** Central to this study is the **gene**, a unit of heredity and equivalent to the **factor** of Mendel. It is essentially a unit of information. The various forms of gene are called **alleles**, and therefore, the dominant and recessive factors of Mendel are alleles of a gene. A pair of alleles exists for each trait in the zygote of an organism, and in every cell of its body thereafter. However, the gametes it produces contain only one allele of a pair. Individual organisms in which the members of a pair of alleles are different, as in an Aa hybrid, are said to be **heterozygous.** Those in which they are alike are **homozygous** as in AA and aa forms. When selfed or inbred, homozygous organism will be truebreeding for the particular trait determined by the particular allele. However, the gene is chemically a stretch of DNA (or RNA in certain viruses).

The **genotype** is the genetic constitution of an organism. AA, Aa and aa are the genotypes of organisms with reference to these particular pair of alleles. The 1 : 2 : 1 ratio is a **genotypic ratio.** The observable characteristics of an organism constitutes its **phenotype.** The same dominant phenotype can be, as a result of two genotypes, homozygous dominant (AA) or heterozygous (Aa). The 3 : 1 monohybrid ratio is a **phenotypic ratio.** The phenotype is a result of how its genotype is expressed.

Symbols are assigned to the alleles of the genes determining a genetic trait. By the earlier convention of Mendel, genetic traits are usually assigned a letter symbol referring to their dominant state. Thus where yellow seed colour is dominant over green seed colour in peas, the dominant allele is symbolized by capital Y and the recessive allele with small y. Thus homozygous dominant will be genotype YY (phenotype yellow) and homozygous recessive genotype yy (phenotype green) and the heterozygous or hybrid form Yy (phenotype yellow).

Methods of Analysis

Punnett square. A monohybrid cross between yellow and green seeded forms is symbolized as follows. The visual diagram is called the **Punnett square.** The genotypes and phenotypes which result from various combinations of gametes can easily be determined by **Punnett squares.** Such squares were devised by Reginald C. Punnett (1875-1967). Here, each of the possible gamete is placed in an individual column or a row. The gametes are then arranged all possible combinations and the resulting genotypes are entered in the boxes along with phenotypes.

Testcross. A dominant phenotype (*e.g.*, yellow seed colour) may be exhibited by homozygous (*e.g.*, YY) or a heterozygous (*e.g.*, Yy) genotype, that is a **'pure'** or an **'impure'** form. For determining the genotype of such an offspring. Mendel devised a simple test, called **'testcross'**. The offspring with the unknown dominant phenotype is crossed with an individual homozygous recessive for the trait (green seed) and the phenotype of F_1 progeny is examined. This is illustrated as follows:

F_2 phenotypic ratio : 3 yellow (1 YY and 2Yy) : 1 green (yy)

genotypic ratio : 1 : 2 : 1 (1 YY : 2 Yy : 1 yy).

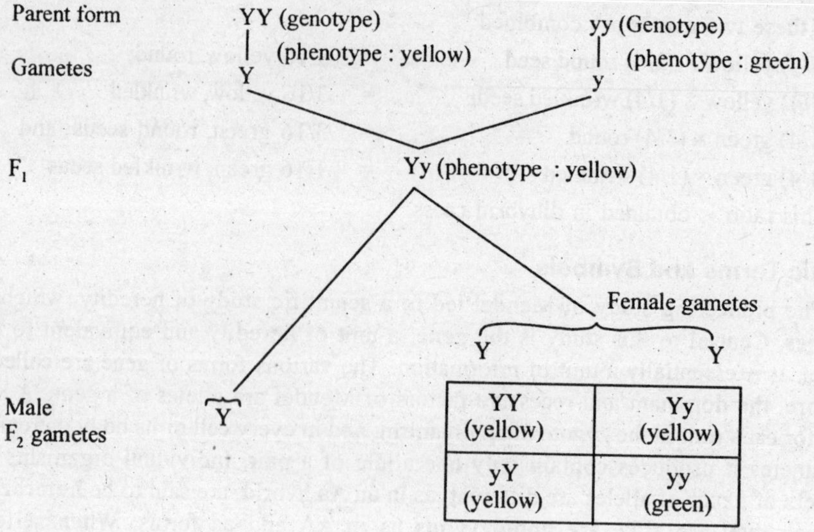

F_2 phenotypic ratio : 3 yellow (1 YY and 2 Yy) : 1 green (yy)
genotypic ratio : 1 : 2 : 1 (1 YY : 2 Yy : 1 yy).

Possibility 1. If the unknown homozygous (YY), then crossing with green recessive (yy) gives all yellow offspring, *i.e.,* all Yy. YY × yy: all offspring Yy, *i.e,* 100 per cent yellow.

Possibility 2. If the unknown heterozygous (Yy), then crossing with green recessive (yy) result in 50 per cent yellow (Yy) and 50 percent green (yy) progeny as undermentioned:

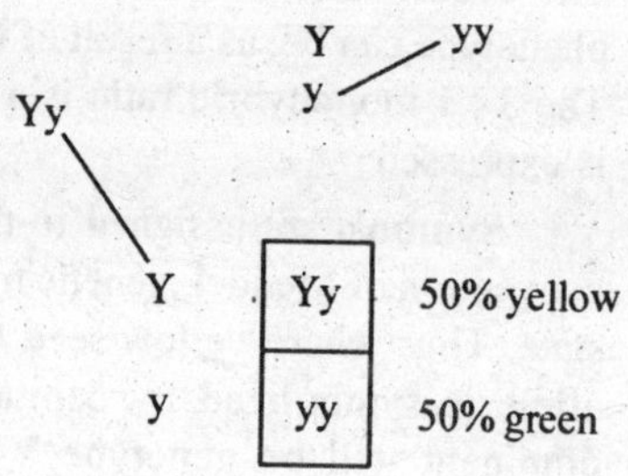

Backcross. This is another cross devised by Mendel where the F_1 is crossed with an individual with the same genotype as one or the other parent. Such crosses are used by animal and plant breeders to rapidly improve a breed or variety by making a useful trait (gene) homozygous, that is 'pure'.

Rediscovery of Mendel's Work

Mendel started his work on pea in 1856 but presented it for the first time in 1865. Mendel's work was unfortunately paid no attention from the biologists of his time. This is perhaps due to the controversies arose from the publication of Darwin's book 'Origin of Species'. Before the rediscovery of Mendel's paper in 1900, several investigators had almost deduced Mendelian principles, from their own experiments. But three biologists, who took pain and considered the conclusions of Mendel, were Hugo de Vries of Holland, Karl Correns of Germany and Erich Tschermark-Seysenegg of Austria. De vries had made many of the same observations in the course of his studies. In 1900, he reported that Mendelian principles apply to different groups of seed plants. De Vries also made an important contribution to genetics as a result of his investigations with the Evening Primrose. Such experiments of these three biologists led to the rediscovery of Mendelian work and birth of the science of Genetics.

Post-Mendelian Era-Other Patterns of Inheritance

After the re-establishment of Mendel's laws of inheritance in 1900, geneticists were used to repeat Mendel's experiments using other organisms and different phenotypes. Mendel explained inheritance in terms of separate hereditary determinants or factors now called **genes.** What Mendel discovered in peas for the first time that was later shown to be generally true for a variety of traits in many plants and animals. However, not all patterns of inheritance could be explained exclusively on the basis of Mendel's original principles alone and complexities were observed by later workers. However, many such complex situations could be resolved within the framework of his principles with suitable modifications and extensions.

1. Incomplete dominance. In many cases, however, dominance is absent, and the hybrid individuals resemble neither parent exactly but are more or less intermediate between the two. For example, in the snapdragon or four'o clock (*Mirabilis jalapa*) a red flowered plant crossed with a white one gives first generation hybrids that possess flowers of pink colour. In certain other cases the hybrid offspring resembles one parent much more closely than the other but does not resemble it exactly, so that dominance is incomplete. This is termed as **incomplete dominance.**

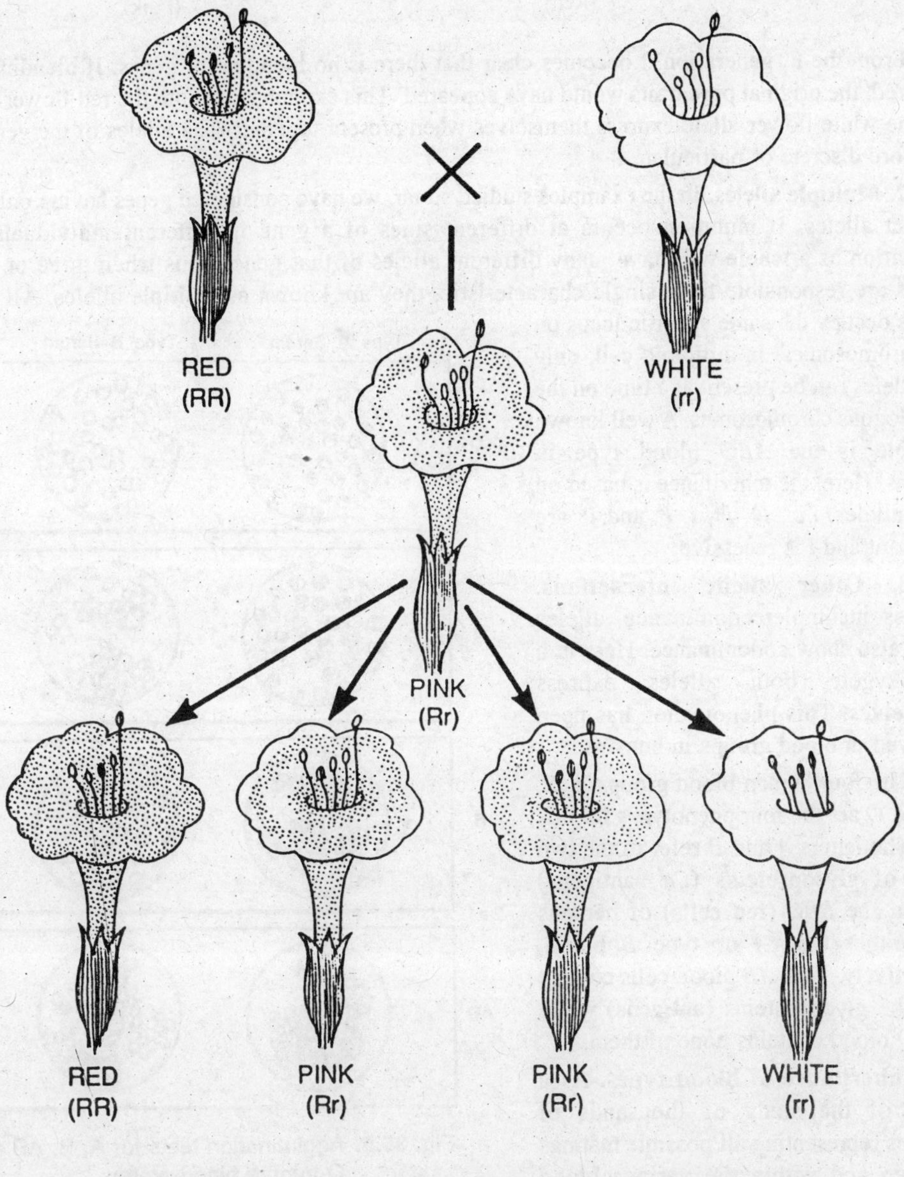

Fig. 36.4. Mendel's laws. Incomplete dominance in *Mirabilis jalapa*.

The F_1 when selfed gives F_2 progenies which are red, pink and white in the ratio of 1 : 2 : 1. Pink is therefore the heterozygous phenotype. The phenotypic and the genotypic ratio are the same in the case of incomplete dominance.

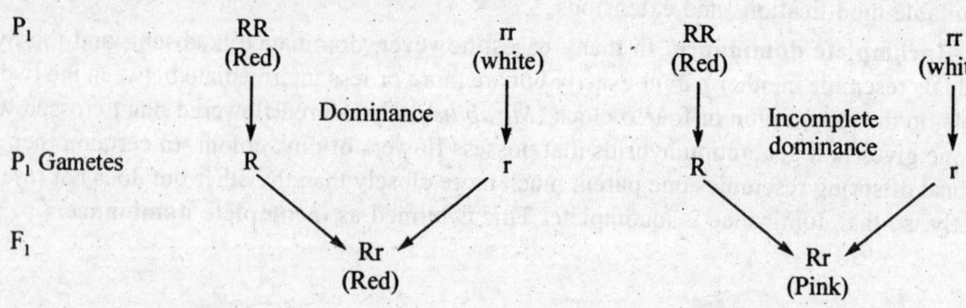

From the F_2 generation it becomes clear that there is no blending of genes. If blending had occurred, the original pure traits would have appeared. This explains that both the red flower allele and the white flower allele express themselves when present together. The alleles of the gene are therefore discrete or particulate.

2. Multiple alleles. In the examples studied so far, we have considered genes having only two distinct alleles. If mutation occurs at different sides of a gene in different individuals, the population as a whole will have many different alleles of that gene. Thus when three or more alleles are responsible for a single characteristic, they are known as multiple alleles. All these alleles occupy the same specific locus on the chromosomes. In a diploid cell, only two alleles can be present at a time on the homologous chromosomes. A well-known example is the *ABO* blood type in humans. Here, the inheritance is based on three alleles, *i.e.*, I^A, I^B, *i*, I^A and I^B are dominant and *i* is recessive.

3. Other allelic interactions. Besides incomplete dominance, alleles could also show codominance. Here in a heterozygote both alleles express themselves. This phenomenon has been observed in blood groups in humans.

The four human blood groups *A*, *B*, *AB* and *O* are the four phenotypes for this trait. The letters *A* and *B* refer to the two types of glycoproteins (*i.e.*, antigens) coating the *RBC* (red cells) of persons with either type *A* or type *B* blood, respectively. Type *AB* blood cells contain both the glycoproteins (antigens) while type *O* blood contains none of them.

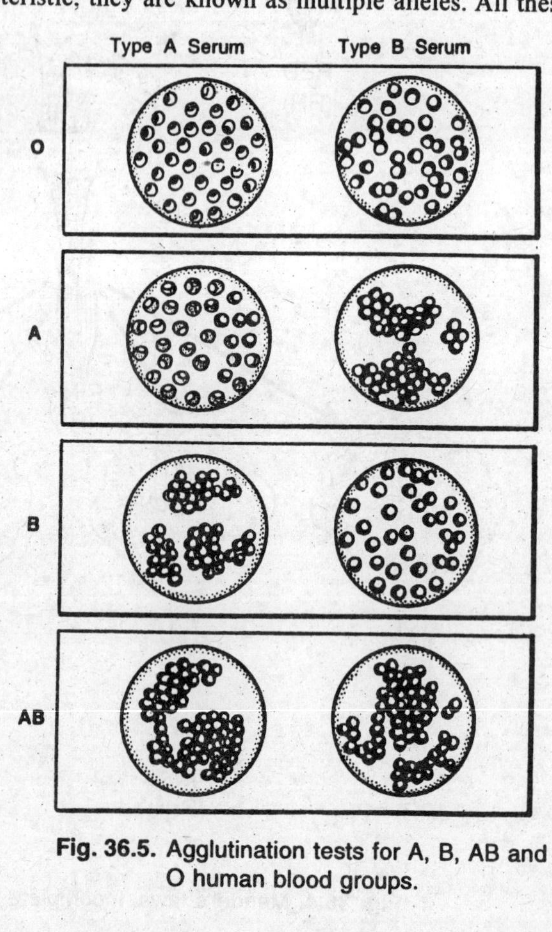

Fig. 36.5. Agglutination tests for A, B, AB and O human blood groups.

Inheritance of blood types. As a result of the study of thousands of families representing all possible matings between and within the various blood groups, the inheritance of the *A*, *B* and *O* alleles has been determined. The accumulated data support the hypothesis that there are three alleles concerned with the determination of the blood group of any given person. Each individual can have any two of these alleles, which are located in one of the 22 autosomes.

The method of inheritance of blood type is quite definite and well understood. We know that *A* antigen is produced by an autosomal gene. No person can carry this gene without having the *A* antigen, and, except for the very rare event of mutation, no child can have the *A* antigen unless at least one of his parents had it. In this respect the gene appears to act as a dominant, but it is not dominant in the same sense that the gene for normal pigmentation is dominant over the gene for albinism. When a person carries a gene for *A* antigen and the gene for *B* antigen, his blood will contain both antigens. Hence neither gene dominant over the other, a condition similar to intermediate inheritance, but which is more accurately described as a case of lack of dominance. Each of the blood group alleles seems to act independently and is not suppressed by the presence of any of the others.

Table 36.2. The genotypes which produce the four blood types

Genotype	*Blood type*
I^oI^o	O
I^AI^A & I^AI^o	A
I^BI^B & I^BI^o	B
I^AI^B	AB

Table 36.3. Blood groups of parents and the possible blood groups of their children.

Blood groups of parent	*Possible genotypes of parents*	*Possible genotypes of children*	*Possible blood groups of children*
A × O	$I^AI^A \times ii$	I^Ai	A
	$I_Ai \times ii$	I_A i and	A
		ii	O
B × O	$I_BI_B \times ii$	I_B i	B
	$I_B i \times ii$	I_B i and	B
		ii	O
A × B	$I_AI_A \times I_BI_B$	I_AI_B	AB
	$I_Ai \times I_BI_B$	I_AI_B and	AB
	I_Bi	I_bi	B
	$I_AI_A \times I_Bi$	I_AI_B and	AB
	I_Ai	I_Ai	A
	$I_Ai \times I_Bi$	I_B i and ii	A
			B or O
AB × O		I_A i and	A
	$I_AI_B \times ii$	I_B i	B

Because of this rather unusual method of inheritance, it has become customary to use the letter I (Isohemagglutinogen) as a basic symbol for the genes at this locus, with a second letter as an exponent to indicate which variation of the allele is represented. For instance, I^A represents the gene which produces the A antigen, and I^B represents the gene which produces the B antigen.

Persons heterozygous for these two genes will have both antigens in their blood, and it will therefore be classed as type *AB*. The third gene in the series is I^o. This gene is usually considered recessive to the other two, since a person heterozygous for the A and O genes to type A and a person heterozygous for B and O genes to type B. The type O individuals must therefore be homozygous, I^oI^o.

Table 36.4. Differences between co-dominance and incomplete dominance.

Co-dominance	Incomplete dominance
1. The heterozygotes show traits of both the alleles, *e.g.*, blood group *AB* has both the antigens *A* and *B*.	1. The heterozygote is of the intermediate type and shows none of the original parental forms, *e.g.*, pink flower colour in F_1 generation in Four 'O clock plant (*Mirabilis jalapa*).
2. Both the alleles are equally dominant, and therefore, both express themselves fully.	2. There is partial dominance of one allele, and therefore, the intermediate type is formed in F_1.
3. For example *ABO* blood group system in man.	For example colour of flowers in snapdragon and four 'o clock plants.

Two Gene Interactions

There are examples where a phenotype is not controlled by one gene but is an outcome of the contribution of more than one non-allelic genes. Such genes show an interaction at the phenotypic level depending upon the type of the alleles carried by an individual. Some cases of two gene interactions are as follows:

(*i*) In each case a distinct phenotype is produced.

(*ii*) Genes under consideration show normal dominance and assort independently.

(*iii*) On phenotypic level, homozygotes and heterozygotes, such as AA or Aa and BB or Bb, will show their respective phenotype governed by the dominant allele, such as A- or B- Here dash indicates the presence of either allele.

(*iv*) The F_1 generation must be heterozygous for the two gene pairs (Aa Bb) which will be selfed or intercrossed.

(*v*) In a normal dihybrid cross, the F_2 genotypes fall into four phenotypic classes, (*i*) 9/16 A-B-, (*ii*) 3/16 A-bb, (*iii*) 3/16 aa B-, and (*iv*) 1/16 aa bb.

Epistasis. This is a phenomenon that takes place when the presence of one gene suppresses the action of another gene at a different locus. In other words epistasis represents the marking of the effect of one gene locus by the effect of another gene locus. The epistatic genes are also called inhibiting genes. The gene that suppresses the effect is called **epistatic gene**, while the gene whose effect gets suppressed is called **hypostatic gene.**

In none of the combination of traits studied by Mendel did one gene affect the phenotypic expression of the other. The result was the typical phenotypic dihybrid ratio of 9 : 3 : 3 : 1. However, several types of mutual influences are now known, and termed as **epistasis**. This means *standing on*. Thus epistatic interactions between genes cause departures from the Mendelian phenotypic dihybrid ratio.

The suitable examples of epistasis is the expression of flower colour in sweet pea (*Lathyrus odoratus*):

Here the dominant allele *P* determines the formation of purple flower colour. However, neither *PP* nor *Pp* will show colour unless another gene *C* is also present in the dominant form (*CC* or *Cc*). Recessive (cc) will interfere with colour formation by *PP* or *Pp* thereby resulting in white flowers. A cross of purple *PPCC* and white ppcc gives all purple *PpCc* F_1 hybrid. Selfing the hybrid gives purple and white flowers progeny in the phenotypic ratio of 9 : 7 instead of normal 9 : 3 : 3 : 1.

All the genotypes recessive for *P* or *C*, or *P* and *C, i.e., ppcc, PPcc, PpCC* or *ppCc* express white phenotype.

Epistasis may be differentiated from dominance. The epistasis is due to the interaction of two separate genes, while the dominance is the interaction of the two alleles of the same gene.

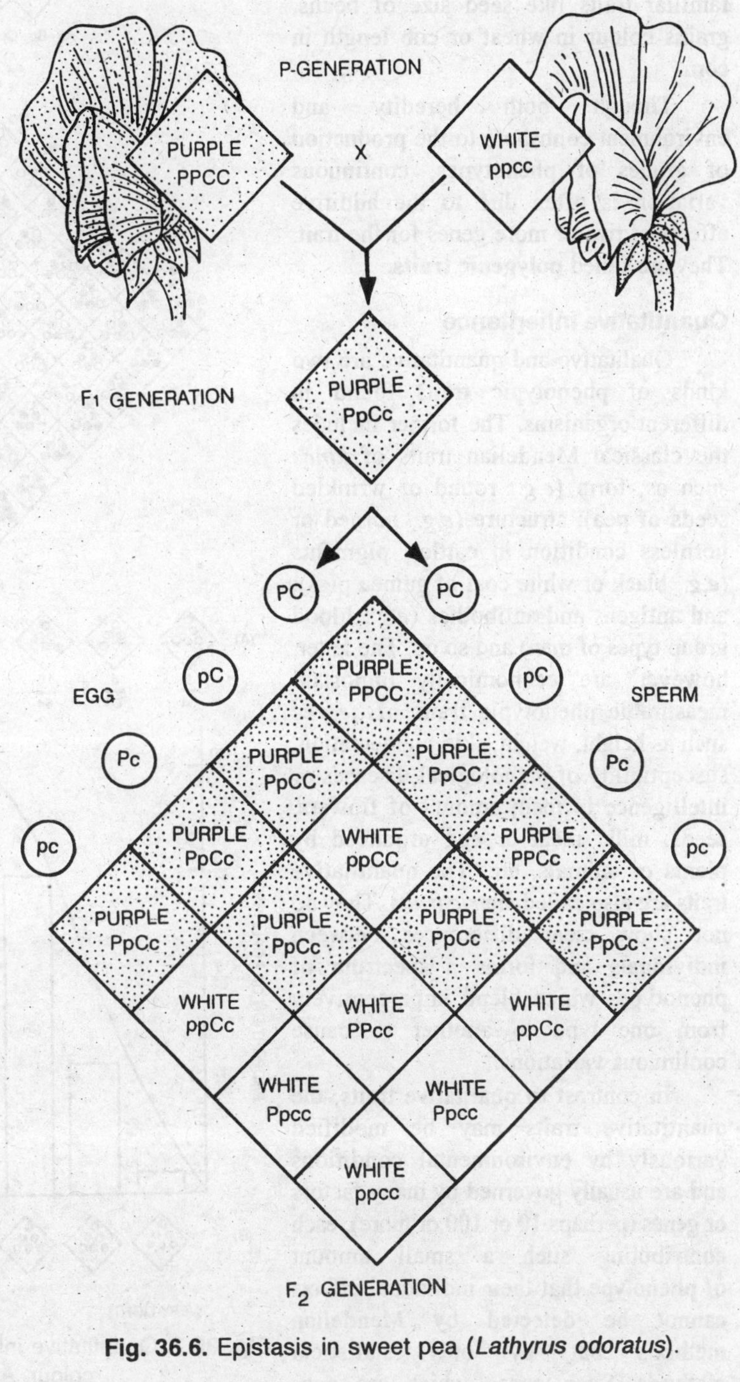

Fig. 36.6. Epistasis in sweet pea (*Lathyrus odoratus*).

Complementary Genes

A colour expression is observed in dramatic way when the two white genotypes ppCC and PPcc are crossed. All the progeny, *i.e.* PpCc are coloured. Here as one gene complements the expression of the other, they are called complementary genes.

Polygenic Traits

All the traits studied by Mendel has an either/or quality, and no intermediate forms. The seeds were either round or wrinkled, the plants tall or dwarf, the flowers purple or white. There were no intermediate or graded forms. However, several traits show gradation. For example, humans skin colour can grade from black to white with all intermediate shades, and so the human height also exhibits a gradation from short to tall forms. Gradations may also be seen in many less familiar traits like seed size of beans, grains colour in wheat or cob length in corn.

Though both heredity and environment contribute to the production of grades of phenotypes, continuous variation is often due to the additive effects of two or more genes for the trait. They are called **polygenic traits.**

Quantitative Inheritance

Qualitative and quantitative are two kinds of phenotypic traits found in different organisms. The former includes the classical Mendelian traits of *kinds* such as, form (*e.g.,* round or wrinkled seeds of pea); structure (*e.g.,* horned or hornless condition in cattle); pigments (*e.g.,* black or white coat of guinea pigs); and antigens and antibodies (*e.g.,* blood group types of man) and so on. The latter, however, are economically important measurable phenotypic traits of *degree* such as height, weight, skin pigmentation, susceptibility of pathological diseases or intelligence in man; amount of flowers, seeds, milk, meat or egg produced by plants or animals, etc. The quantitative traits are also called *metric traits*. They do not show clear differences between individuals and forms a spectrum of phenotypes which blend imperceptively from one type to another to cause continuous variations.

In contrast to qualitative traits, the quantitative traits may be modified variously by environmental conditions and are usually governed by many factors or genes (perhaps 10 or 100 or more), each contributing such a small amount of phenotype that their individual effects cannot be detected by Mendelian methods but by only statistical methods. Such genes which are non-

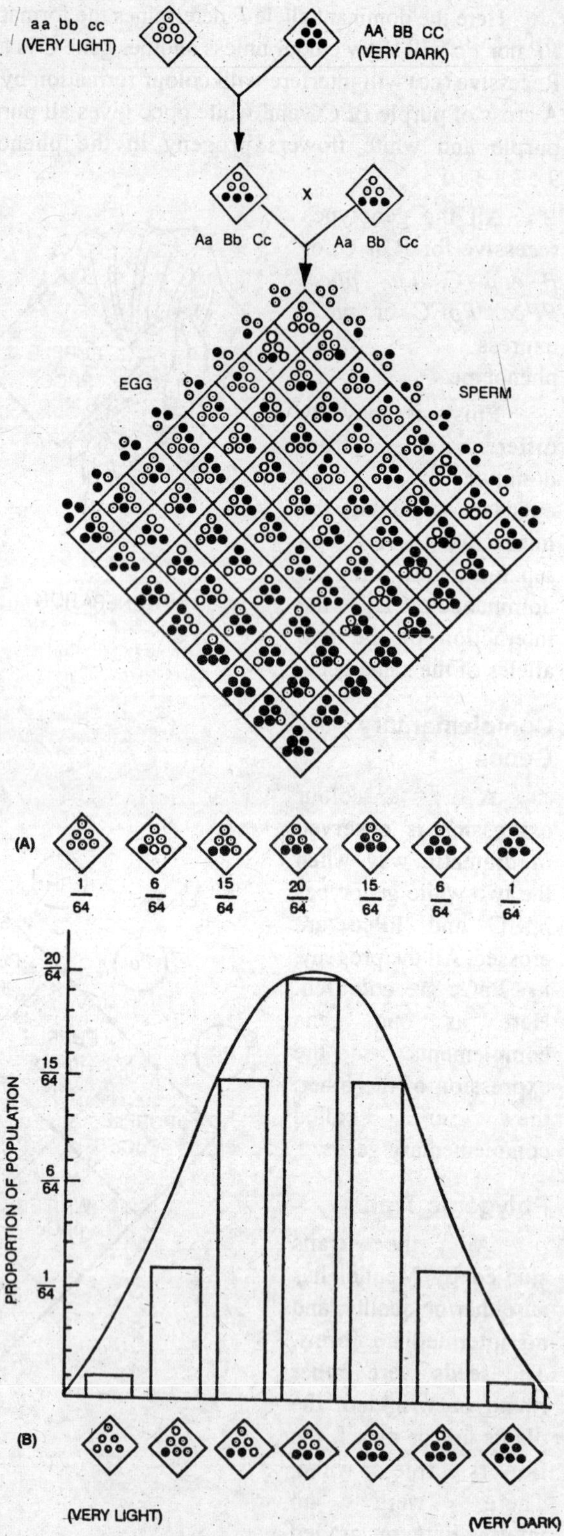

Fig. 36.7. Quantitative inheritance of human skin colour, A and B.

allelic and affect the phenotype of a single quantitative trait, are called *polygenes* or *cumulative genes*. The inheritance of polygenes or quantitative traits is called *quantitative inheritance.*

Human skin colour. Human skin colour is generally controlled by three separate genes. Each gene contributes to a unit of darkness due to incomplete dominance.

Here these three genes can be designated as A, B and C, and thus the skin shade has to vary from a very dark in a AABBCC individual to very light in aabbcc individual.

A person with AaBbCc, *i.e,* heterozygous for all three genes, will possess an intermediate colour.

The number of possible allele combinations in the gametes is eight for such a person, and therefore, a total of 64 phenotypic combinations are possible when two persons of similar genic constitution marry.

As shown in the figures (36.7 A-B), the phenotypic frequency for skin colour will gradually increase from 1/64 for the very dark coloured, *i.e.,* AABBCC to a maximum of 20/64 for the intermediate coloured AaBbCc, and again reduces to 1/64 for the very light coloured aabbcc.

The frequency distribution for skin colour can be shown by a bell-shaped normal distribution curve (Fig. 36.7.B).

Such inheritance is called **quantitative inheritance.**

Kernel colour in wheat. Nilsson-Ehle made a very interesting study of the colour of grain in wheat. He found that, when he crossed a variety of wheat bearing dark red kernels with a variety having white kernels, he obtained an intermediate shade of red in the first generation. This might suggest incomplete dominance, but when the offspring of this F_1 generation were crossed among themselves the zygotes in F_2 generation showed five different phenotypic classes in a ratio of 1 : 4 : 6 : 4 : 1. Noting that 1/16 of F_2 was an extreme in colour as either of the parental plants (red or white), they theorized that two pairs of genes controlling production of red pigment were operating in this cross. Each gene was supposed to contain two alleles. One allele produces a given quantity of the red pigment, while its counterpart did not produce any pigment. All alleles were equally potent in the production or lack or production of pigment.

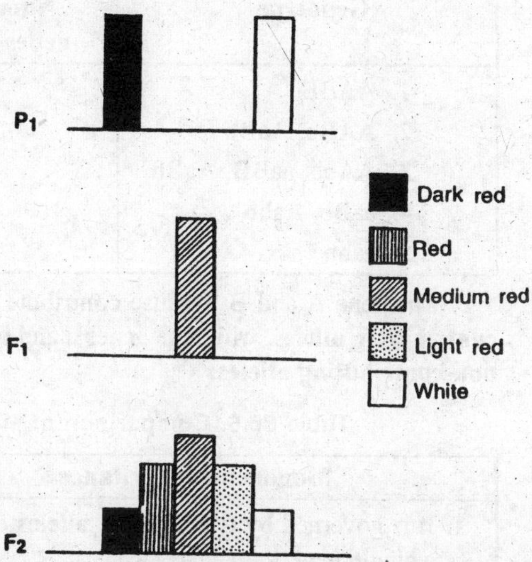

Fig. 36.8. Wheat colour as an example of quantitative trait.

If we symbolize the genes for red with the capital letters A and B and their alleles resulting in lack of pigment production by a and b. We can illustrate the results of this cross as follows:

P:	Red		White
	AABB	×	aabb
Gametes:	(AB) (AB)		(ab) (ab)
F_1: Mediun			Medium
Aa Bb		×	Aa Bb

F_1 gametes: Male ♂

Female ♀	AB	Ab	aB	ab
AB	AABB Red	AABb Dark	AaBB Dark	AaBb Medium
Ab	AABb Dark	AAbb Medium	AaBb Medium	Aabb Light
aB	AaBB Dark	AaBb Medium	aaBB Medium	aaBb Light
ab	AaBb Medium	Aabb Light	aaBb Light	aabb White

or 1/16 Red: 4/16 Dark: 6/16 Medium: 4/16 Light: 1/16 White.

If we suppose that each dose of a gene for pigment production increases the depth of colour, we can arrange the F_2 phenotypes according to the number of genes for red in the following table:

Table 36.5. Contributing and non-contributing alleles

Genotype	Number of genes for red	Phenotype	Fraction of F2
1. AABB	4	Dark red	1/16
2. AABb, AaBb	3	Medium red	4/16
3. AAbb, aaBB, AaBb	2	Intermediate	6/16
4. aaBb, Aabb	1	Light red	4/16
5. aabb	0	White	1/16

The gene A and B because contribute red colour in this case, therefore, are referred to as **contributing alleles**, while the genes a and b do not contribute red colour to kernel and are called **non-contributing alleles.**

Table 36.6. Comparison of Monogenic and Polygenic Inheritance.

Monogenic inheritance	Polygenic inheritance
1. It is governed by two or more alleles of a single gene.	**1.** It is governed by many non-allelic genes.
2. It gives a discontinuous variation.	**2.** It gives a continuous variation.
3. This is called qualitative inheritance.	**3.** This is called quantitative inheritance.
4. There are no intermediates.	**4.** Numerous intermediates may occur giving a complete gradation from one extreme to another.
5. Phenotype is not influenced by environment.	**5.** Phenotype is influenced by the environment.
6. Analysis of progeny is done by counting.	**6.** It is done by statistical method.
7. Example: All seven Mendelian traits.	**7.** Examples: Human skin colour, wheat kernel colour.

Pleiotropy

A single gene often influences more than one phenotypic traits. However, it may be that one gene may cause evidently well marked expression of some phenotypic trait (*i.e.,* **major effect**) than the others with less evident phenotype (*i.e,* **secondary effect**). Most genes have their multiple effects and are called pleiotropic genes. The ability of a gene to have many effects is known as **pleiotropy.** An important example of a pleiotropy is a hereditary disease called **sickle-cell anemia** or **cooley's anemia** found among certain African tribes. A recessive gene causes the disease. A codominant gene Hb^s in homozygous condition produces the disease sickle-cell anaemia which leads to death, generally at least by later adolescence. The gene causes the production of an abnormal hemoglobin. As a result the shape of the red blood cell containing it becomes sickle-shaped and distorted. However, heterozygous individuals which possess both normal and abnormal hemoglobin and have mild anaemia, are naturally protected against malaria as the malarial parasite cannot live in these distorted cells. The heterozygous individuals inherit the genes from generation to generation as shown here. Hb^A and Hb^s are the genes for the normal and abnormal hemoglobin.

SUMMARY

1. Heredity is the transmission of traits from one generation to following generations.
2. Variation in heredity is found in sexually produced offspring but not in asexually produced clones unless and until mutation occurs.
3. Traits combined in one generation may separate in later generations suggesting a particular nature for heredity.
4. Mendel selected seven distinct traits in garden pea (*Pisum sativum*) for his experiments.
5. Mendel hybridized alternate forms of single traits in the garden pea and studied their inheritance in subsequent generations following self-pollination.
6. Mendel selected garden pea as it normal self-pollinated but can readily be cross-pollinated.
7. Besides true breeding alternate forms for several traits such as yellow or green seeded forms were available.
8. For each of the seven traits hybridized. Mendel found that:
   (*i*) The F_1 hybrids always exhibited only one of the parental forms; (*ii*) both parental forms were present in F_2 and (*iii*) The form of the trait in F_1 called **dominant** form was present three times as frequently as its alternate **recessive** form, *e.g.,* a 3 : 1 ratio of dominant to recessive forms is termed Mendelian monohybrid ratio.
9. When F_3 generation was raised by selfing F_2 plants, all recessive F_2 plants produced only recessives. They were pure or true breeding.
10. On the other hand, only one-third of dominant F_2 were pure, the rest were impure and produced both dominant and recessive offspring like the F_1.
11. The phenotypic ratio of 3 : 1 was in reality a hidden 1 : 2 : 1 ratio, *i.e.,* of pure dominant: impure dominant or hybrid: pure recessive forms.
12. Mendel interpreted the 1 : 2 : 1 ratio as follows: pea plants possess two factors for each trait whereas each of its male or female gametes has only one factor (Mendel's law of segregation). If dominant factor is represented by symbol A and recessive factor by symbol a, the pure or homozygous dominant parent is AA and the pure recessive parent is aa. The gametes of dominant parent have only factor A and those of recessive parent have only a. These gametes unite to form F_1 hybrid offspring with factors Aa. The F_1 hybrid offspring form two types of male and female gametes having factors A and a. During random fertilisation the male and female gametes can unite in four possible ways, *i.e.,* AA, Aa, aA

and aa giving a 1AA (pure dominant): 2 Aa (impure or hybrid dominant): 1 aa (pure recessive) offspring.

13. Mendel's factor is now called a gene, and its alternative forms are termed alleles. The alleles lie in homologous chromosomes. The ratio 1 : 2 : 1 is known as Mendel's monohybrid genotypic ratio.
14. The separation of two alleles during gamete formation and the restoration of paired condition during random fertilization is called principle of segregation.
15. A cross between F_1 individual and a recessive parent is called test cross. A test cross can distinguish the pure dominant from the hybrid dominant.
16. When alternate forms of two traits controlled by two genes, *e.g.*, dominant round, yellow and recessive wrinkled green are hybridized, the F_1 generation shows the dominant trait of both genes, say round, yellow F_1 plants, on selfing, produce F_2 progeny, with two parental and two new recombinant phenotypes, *i.e.*, round yellow: round green: wrinkled yellow: wrinkled green in the dihybrid ratio of 9 : 3 : 3 : 1.
17. During the inheritance of two or more genes at a time, their distribution in the gametes and in the progeny is independent of each other. This is called Mendel's principle of independent assortment.
18. The results of genetic crosses can be predicted by Punnett squares, which use the principles of mathematical probability to follow the union of gametes.
19. Heterozygotes of incomplete dominance have phenotypes intermediate between the phenotypes of the two homozygotes. This is the law of incomplete dominance, or blended dominance.
20. Some genes have more than two alleles in a population and are called multiple alleles. For example the three I^A, I^B and i alleles of *ABO* blood groups. Here alleles I^A and I^B are codominant, both express themselves when in combination.
21. Epistasis is the influence of one gene on the expression of another gene.
22. Two independent pairs of genes, which interact to produce a trait together, but each dominant allele does not show its effect are called complementary genes. In the sweet pea complementary genes show the unusual 9 : 7 phenotypic ratio.
23. Polygenic traits are controlled by more than one gene such as human skin colour expression. They are called quantitative traits and show a gradation in their expression.
24. A gene affecting more than one trait is called a pleiotropic gene. For example, sickle cell gene.
25. Pedigree analysis is a method to analyse the distribution and movement of traits in human families.

TERMS TO REMEMBER

Allele. The various forms of a gene are called alleles. For example, blue colour and black colour of eye are two alleles of the eye colour gene.

Asexual reproduction. Any reproduction process that does not involve meiosis and fertilization.

Back cross. A cross between an individual offspring and one of its parents.

Clone. A carbon copy of the parent. A group of cells that are derived by mitotic divisions from a single ancestral cell.

Co-dominance. The condition in heterozygotes where both members of an allelic pair contribute to phenotype.

Complementary genes. As one gene complements the expression of the other, they are called complementary genes. The phenotypic ratio is 9 : 7.

Dominant allele. The allele that expresses itself in the homozygous and heterozygous condition. It masks the effect of recessive allele.

Dihybrid cross. A cross in which inheritance of two pairs of contrasting characters is studied simultaneously. The phenotypic ratio is 9 : 3 : 3 : 1.

Epistasis. The phenomenon where one gene suppresses the action of another gene at a different focus. The masked gene is called hypostatic.

Factor. The term used by Mendel. Now this is called gene. Each trait is controlled by a pair of factors.

F_1 generation. The first filial generation. The first generation resulting from a cross.

F_2 generation. The second filial generation. The generation that results from interbreeding members of F_1.

Gene. A basic unit of heredity located on a chromosome. It is equivalent to the 'factor' of Mendel. These words are units of information from one generation to another.

Genetics. The study of genes and how they produce characteristics and their inheritance.

Genotype. Genetic expression of an organism.

Heredity. The transmission of characteristics from one generation to successive generations of living things.

Heterozygous. A condition where a diploid organism has different allelic forms of a gene.

Homozygous. A condition where a diploid organism has same allelic forms of a gene.

Hybrid. An offspring resulting from a cross between two genotypically different parents.

Locus. The position on a chromosome occupied by a particular gene or one of its alleles.

Monohybrid. Individuals produced as a result of monohybrid cross. It contains genes of both the parents.

Monohybrid cross. A cross between two parents in which inheritance of only one pair of contrasting characters is studied.

Multiple alleles. More than two allelic forms exist for certain genes. A well known example is the *ABO* blood types in humans.

Offspring. Organisms produced as a result of sexual reproduction.

P. Parental generation in a cross.

Pedigree analysis. A system for following the movement and distribution of certain genetic traits.

Phenotype. The visible expression of the hereditary constitution possessed by an organism.

Pleiotropy. The ability of a gene to have many effects. Example: sickle-cell anaemia.

Polygenic traits. Very often the continuous variation is due to the additive effects of two or more genes for the traits. They are polygenic traits.

Polygenic inheritance. A particular trait is determined by interaction of more than one pair of non allelic genes. It is also known as quantitative inheritance.

Progeny. Offspring individuals.

Pure line. Pure breeding. A strain of individuals homozygous for all genes being considered.

Punnett square. A checker board that helps to study all possible results of crosses in genetics.

Quantitative inheritance. A particular trait is determined by interaction of more than one pair of non-allelic genes. Also known as polygenic inheritance.

Recessive allele. An allele that does not express itself when present along with the dominant cell.

Reciprocal cross. A second cross of the same genotype where gametes from the parental generation are reversed.

Supplementary genes. Where the obtained ratio is 9 : 3 : 4.

Siblings. Brother and sister relationship; where the individuals having the same material and paternal parents.

Test cross. A cross between an individual (F_1) with one having recessive phenotype.

Trait. It is any recognisable feature of an individual.

Zygote. A fertilized egg; the protoplast resulting from the fusion of two gametes (♂ and ♀).

QUESTION BANK

N.B. Undermentioned questions are asked repeatedly in various Indian Universities and competitive examinations.

Very Short Answer Questions

1. What is heredity?
2. Define variation.
3. What is 'genetics'?
4. Give name of father of genetics.
5. Name the plant on which Mendel worked.
6. How many different contrasting traits did Mendel notice in garden pea?
7. What term is used for a class of individuals which are morphologically similar?
8. What is a monohybrid cross?
9. What is a dihybrid cross?
10. What is the character that does not allow the expression of a contrasting character in a hybrid?
11. What are Mendel's hypothetical factors in modern terminology?
12. What is Mendel's monohybrid ratio?
13. What is dominant factor–round seeds or wrinkled seeds?
14. What is mendel's dihybrid ratio for phenotypes?
15. What do the letters P, F_1 and F_2 represent in heredity?
16. What is a recessive trait?
17. What is allele?
18. What does the phenomenon means where one gene suppresses the action of another gene at different loci?
19. What is the generic expression of an organism known as?
20. What is a locus?
21. What is hybrid?
22. What is a back cross?
23. When a hybrid of F_1 is crossed with the homozygous recessive parent, what is it called?
24. What is a gene pool?
25. What is principle of dominance?

26. Name three principles put forth by Mendel.
27. Blood groups in man is an example of.
28. What is a pleiotropic gene?
29. Which of the genes studied by Mendel in pea is now known as pleiotropic?
30. What is a clone?
31. Name the plant that shows incomplete dominance in respect to the colour of the flower?
32. Which type of inheritance deals with quantitative traits?
33. What is a Punnett square?
34. What are the phenotypes of TT and tt plants?
35. What is the allele of T?
36. What was the total number of true-breeding varieties of garden pea which Mendel had taken to start his experiments?

Answers

1. Study of transmission of characteristics from one generation to the next. **2.** When there is dissimilarities of characteristics in the individuals of a species or by the offspring of the same parents. **3.** The study of various traits and genes and how they are inherited from one generation to next. **4.** Gregor Johann Mendel. **5.** Garden pea (*Pisum sativum*). **6.** Seven. **7.** Phenotype. **8.** A cross between parents differing in a pair of contrasting characters. **9.** A cross of a dihybrid with a homozygous recessive individual. **10.** Dominant. **11.** Genes. **12.** 3 : 1 **13.** Round seeds. **14.** 9 : 3 : 3 : 1. **15.** Parental generation, first filial generation, second filial generation. **16.** It is a form of trait that has inability to express itself in the presence of the allele for other form of same trait. **17.** Alternative forms of a gene, *e.g.*, blue and black colour of eye are two alleles of the eye colour gene. **18.** Epistasis. **19.** Genotype. **20.** The point where an allele is located on chromosome. **21.** An individual resulting from a cross between two genetically unlike persons. **22.** The cross of a progeny individual with one of the parents. **23.** Test cross **24.** All genotypes of all the individuals in a population. **25.** Out of two contrasting alleles only one expresses itself in one individual. The factor which expresses itself if dominant and the other which has not shown its effect is recessive. **26.** (*i*) Principle of independent assortment, (*ii*) principle of segregation and (*iii*) principle of dominance. **27.** Co-dominance. **28.** A gene affecting more than one trait is called pleiotropic gene. **29.** Flower and seed coat colour. **30.** A group of cells derived by mitotic divisions from a single ancestral cell. **31.** *Mirabilis jalapa* (four 'o clock plant). **32.** Polygenic. **33.** The representation of a monohybrid or dihybrid cross by making squares. **34.** Tall and dwarf plants. **35.** t. **36.** Seven.

Short Answer Questions

1. What is Mendel's law of dominance?
2. Give a comprehensive account of law of segregation.
3. Describe Mendel's law of independent assortment.
4. What is epistasis? How does it differ from dominance?
5. What do you mean by a polygenic trait? Cite an example.
6. What heredity principle can be derived from a cross showing 3 : 1 ratio in the F_2 generation?
7. What heredity principle can be derived from a cross showing 9 : 3 : 3 : 1 ratio in the F_2 generation?
8. What do you mean by continuous variation?
9. Define clone. How does variation arise in a clone?
10. Make a list of the 7 pairs of contrasting traits selected by Mendel for breeding experiments.
11. A man with blue eyes and black hair (homozygous for both characters) marries a woman with black eyes and red hair. Black eyes (E) are dominant over black eyes (e) and black hair (H) is dominant over red hair (h). Work out the probabilities of phenotypes of the children of this couple using Punnett square method. Use only symbols for characters as given above.
12. How is skin colour inherited in man?
13. What kind of plants will be produced in F_1 and F_2 generations in the following crosses?
(*a*) Pure Tall × Pure Tall (*b*) Pure Tall × Pure Dwarf (*c*) Hybrid Tall × Pure Tall.

14. A pea plant with purple flowers was crossed with a plant with white flowers producing 40 plants with only purple flowers. On selfing, these plants produced 470 plants with purple flowers and 162 with white flowers. What genetic mechanisms account for these results?
15. The F_2 plants obtained in the Q. 4 were selfed. The white-flowered plants produced only white-flowered F_3 plants. Some of the purple flowered F_2 produced only purple-flowered F_3, while other produced both white and purple-flowered F_3 plants. Provide the genetic basis for this observation.
16. In dogs the barking trait is dominant over the silent one and crest ears are dominant over drooping ears. What is the expected phenotypic ratio of the offspring when dogs, heterozygous for both the traits are crossed? What would be the phenotypic ratio, if dogs, heterozygous for both traits are crossed with dogs homozygous recessive for both traits?
17. A man with type A blood has a wife with type B. They have a child with type O blood. Give the genotypes of all the three. What other blood groups can be expected in the future offspring of this couple?
18. What are the problems faced by a sickle cell anaemia carrier? Why is this defective gene still not eliminated by nature?
19. Taking the example of *ABO* blood types of humans, explain the phenomenon of multiple alleles and co-dominance.
20. What is a test cross? How does it differ from reciprocal cross.

Long Answer Questions

1. Give the scientific name of the plant selected by Mendel for his experiments. State three laws given by him. Briefly explain the law of independent assortment with the help of an example.
2. What are multiple allelic genes? Explain the blood groups in man as an example of multiple allelic genes.
3. Describe the phenomenon of epistasis as found in the flower colour of sweet pea and mention how epistasis is different from dominance.
4. What are complementary genes? Cite one example. Explain their interaction with the help of a cross.
5. What is pedigree analysis? What are the symbols used in such an analysis?
6. Why did Mendel choose garden pea for his experiments? How did he make sure that the plants were true-breeding?
7. Describe the mechanism of inheritance of the *ABO* system of blood groups, highlighting the principle of genetics involved in it.
8. State and explain Mendel's law of segregation.
9. State and explain Mendel's law of independent assortment.

Multiple Choice Questions (Objective type questions)

1. Father of genetics is
   (*a*) Morgan (*b*) Darwin
   (*c*) Mendel (*d*) Hutchinson.
2. Mendel worked on
   (*a*) *Lathyrus* (*b*) *Pisum*
   (*c*) *Dolichos* (*d*) *Vigna*
3. An organism with two unlike genes of a trait is called
   (*a*) heterozygous (*b*) homozygous
   (*c*) hermaphrodite (*d*) monozygous
4. When F_1 plants heterozygous for tallness are selfed, F_2 generation has both tall and dwarf plants. This depicts the principle of
   (*a*) dominance (*b*) law of segregation
   (*c*) law of independent assortment (*d*) blended inheritance

5. The principle of segregation was formulated by Mendel in
   (*a*) 1766 (*b*) 1866
   (*c*) 1900 (*d*) none of the above.
6. Mendel experimented on
   (*a*) wild pea (*b*) sweet pea
   (*c*) garden pea (*d*) lablab.
7. The law of segregation was rediscovered by
   (*a*) De Vries (*b*) Correns
   (*c*) Tschermark (*d*) all of the above.
8. Polygenic traits show
   (*a*) discontinuous variation (*b*) continuous variation
   (*c*) both (*d*) none of the above.
9. Inheritance of skin colour in man
   (*a*) monogenic (*b*) bigenic
   (*c*) trigenic (*d*) polygenic
10. A gamete contains
    (*a*) only one allele of a gene (*b*) two alleles of a gene
    (*c*) all alleles of a gene (*d*) none of the above
11. A breeding experiment dealing with a single trait is called
    (*a*) monohybrid (*b*) monozygous
    (*c*) dihybrid (*d*) heterozygous.
12. Mendel noted as many pairs of contrasting traits in pea plants
    (*a*) 2 (*b*) 5
    (*c*) 7 (*d*) 9
13. A cross that shows the principle of independent assortment is
    (*a*) monohybrid (*b*) dihybrid
    (*c*) both (*d*) none of the above
14. Human blood groups are determined by
    (*a*) single gene (*b*) two genes
    (*c*) alleles (*d*) multiple genes
15. An important example of pleiotropy is
    (*a*) bleeder's disease (*b*) smallpox
    (*c*) sickle-cell anaemia (*d*) none of the above

Answers

1. (*c*) **2.** (*b*) **3.** (*a*) **4.** (*b*) **5.** (*b*) **6.** (*c*) **7.** (*d*) **8.** (*b*) **9.** (*c*) **10.** (*a*)
11. (*a*) **12.** (*c*) **13.** (*b*) **14.** (*d*) **15.** (*c*)

37 chapter

Chromosomal Basis of Inheritance

As shown by Mendel the units of heredity, the factors or genes are stable and particulate. However, the location of these factors (now genes) in the cell was unknown to him. In those days neither the role of the nucleus in reproduction nor the existence of chromosomes in the nucleus was known. Mendel published his results in 1866 without the knowledge of genetic material. The processes of mitosis and meiosis were worked out only after Mendel had published his work in 1866. Mendel's observations remained unattended until 1900, less when De Vries, Corens and Tschermak realised the significance of his work.

(*i*) Mendel published his results in 1866 without the knowledge of genetic material.

(*ii*) Mendel's observations remained unattended until 1900, when De Vries, Corens and Tschermak realised the significance of his work.

(*iii*) By the time Mendel was finalising his work on pea, chromosomes were seen in the nuclei of Salamander cells by William Flemming in 1879. Flemming suggested the role of nucleus in the process of inheritance.

(*iv*) In 1902, Walter Sutton and Theodor Boveri came to the conclusion that the behaviour of chromosomes at meiosis can serve as the cellular basis of both segregation and independent assortment. They suggested that chromosomes can be equated with the Mendel's unit factors. Their work led to the chromosome theory of heredity.

(*v*) Work of W. Sutton and T. Boveri was further expanded by Thomas H. Morgan, Alfred H. Sturtevant and Calvin Bridges on fruit fly, *Drosophila.*

(*vi*) The term **factor** as the basic unit of inheritance was replaced by Johannsen, in 1909 with the term **gene.**

Now this is an established fact that in order to maintain the genetic continuity from cell to cell and from one generation to the next, in sexually reproducing organisms an orderly and coordinated process in required. During this process the haploid gametes are formed and there after by means of fertilisation diploid cells are produced. The structure which acts as the vehicle of transmission of genes in subsequent generations is **chromosome.** The manner in which the chromosomes and thus the genetic material is transmitted from one generation to another is accomplished by two cell divisional processes known as **mitosis** and **meiosis.** During mitosis the genetic material is duplicated and distributed equally to the two daughter cells, whereas in meiosis, the number of chromosomes is reduced to half and thus known as **reduction division.**

The principles of inheritance and the modifications in the phenotypic ratio by various types of interactions both at the allelic and non-allelic levels have already been discussed in previous chapter. The principles of inheritance are applicable to all living organisms including humans. In this chapter, the inheritance on the basis of chromosomes shall be dealt with. Though inheritance takes place by means of **genes,** yet it will be greatly influenced by their location on the chromosomes. However, several experiments have revealed that on many occasions genes do not follow the principle of independent assortment and they segregate en block. This type of segregation of genes was referred to as **linkage.** Later on this was also revealed that this characteristic could be assigned to genes located on the same chromosome. This was also demonstrated that linkage between two genes, located on the same chromosome can be broken by the process of **crossing over.** Crossing over takes place during prophase I of meiosis and results in the formation of new combination of alleles. This is called **recombination.**

Recombination is an important source of variation. This brings about allelic variation, which is called **mutation**. Alterations in chromosome number and structure may also bring about phenotypic changes.

Parallelism between Genes and Chromosomes

How genes control phenotypes, and how they are transmitted to the following generations have already been discussed. Now we shall have to learn how a parallelism can be drawn between genes and chromosome. The points in support are as follows :

(1) Each species has a specific and characteristic number of chromosomes in its somatic cells.

(2) In a diploid cell, these chromosomes are found in pairs called homologous pairs which resemble in their morphology and genetic content.

(3) One member of a homologous pair comes from maternal parent and the other comes from the paternal parent.

(4) In the similar way, each gene also exists in allelomorphic pair which are contributed equally by the maternal and paternal parents.

(5) Each chromosome has a double structure which consists of sister chromatids.

(6) During the process of mitosis, the two chromatids get separated into two daughter nuclei and then cells.

(7) Each member of an allele pair also goes in every daughter cell during mitosis, and thus a genetic composition is maintained in the cells of a multicellular organism.

(8) During meiosis the members of a chromosome pair are separated from each other and go to daughter nuclei.

(9) In the similar way, members of an allele pair also separate during meiosis, and therefore, each haploid cell or gamete contains only one allele.

(10) During process of fertilisation two haploid gametes unite and the chromosome number becomes diploid and also every allele pair is restored.

The Chromosome Theory of Inheritance

After the discovery of Mendel's laws of inheritance, biologists soon noticed the undermentioned similarities between the behaviour of genes (Mendel's factors) during inheritance and of chromosomes and genes during cell division.

1. Both (chromosomes and genes) were formed in pairs in diploid cell.
2. Homologous chromosomes separated and gene pairs (alleles) segregated during meiosis.
3. The paired condition of both chromosomes and genes was restored during fertilization.

Sutton and Boveri (1902) propounded **the chromosome theory of inheritance** on the basis of these similarities. According to this theory Mendelian genes are located on chromosomes, and the chromosomes are segregated and assorted independently. This takes place during meiotic cell division. Since Mendel's factors or genes, the units of heredity are present on the chromosomes, Mendel's principles of segregation and independent assortment can be explained better by the chromosome theory.

As the genes are located on the homologous pairs of chromosomes, both the principles of segregation and independent assortment can be explained in better form.

When the members of a homologous pair separate, alleles carried on them will also segregate and go to different gametes. This explains the segregation of alleles controlling the contrasting traits.

Contrary to it when two homologous pairs are considered, each carrying an allele pair, one chromosome pair will segregate independently of the other, and will give rise to four combinations which form the basis of independent assortment.

The number of genes are always much more than the number of chromosomes. This means that each chromosome must carry a large number of genes.

Whenever a gene is assigned a position on the chromosome, it is called a **locus.**

Each allele pair though specifies the same character, it must carry slightly different information so as to depict contrasting forms.

Table 37.1. Comparison of Behaviour of Chromosomes with Mendel's Factors/Genes.

Behaviour of factors at the time of Mendelian inheritance	Behaviour of chromosomes at the time of meiosis
1. The factors behave as independent units. In monohybrid cross in F_1, the factor that expresses is dominant, however, in F_2, recessive factor exhibits its unaltered expression.	1. The chromosomes behave as individual units during inheritance, and this can be studied directly under microscope during meiosis.
2. Mendel's factors maintain stability in the expression of characters.	2. The chromosomes also maintain their stability such as number, structure and functions.
3. The Mendel's factors are found in pairs, and each allelic pair segregates at the time of the formation of gametes.	3. Homologous chromosomes also segregate during meiosis at the time of formation of gametes.
4. Only one factor is present in a gamete.	4. Only one chromosome of a homologous pair passes into a gamete.
5. During fertilization the gametes unite to form a zygote. Each organism contains two factors, one from each parent.	5. During fertilization gametes unite to restore the chromosome number. Each homologous pair contains one paternal and one maternal chromosome.
6. Mendelian factors remain unchanged from generation to generation. Each factor segregates independently of every other pair.	6. Chromosomes also maintain their individuality during segregation. Each pair of homologous chromosomes segregates independently of every other pair.

Chromosomal Theory – Principle of Independent Assortment

As shown in the fig 37.1 it depicts the movement of the alleles of two genes (*i.e.,* seed shape and seed colour) located on two different chromosomes during a dihybrid cross of round-yellow (R–Y) and wrinkled-green (r-y) pea seed varieties which were studied earlier. There are two pairs of chromosomes, one pair carries the alleles for seed shape (R and r) and the other carries the alleles for seed colour (Y and y).

In the F_1 generation the genotype of the offsprings is YyRr. It can be depicted on the chromosomes as shown in fig. 37.2. The two duplicated and paired homologous chromosomes assembled at the metaphase stage of meiosis I can separate in two possible ways, and this leads to independent assortment of the chromosomes and the genes they carry in four possible ways. Four types of gametes are formed after meiosis II (fig. 37.2), each with half the number of chromosomes

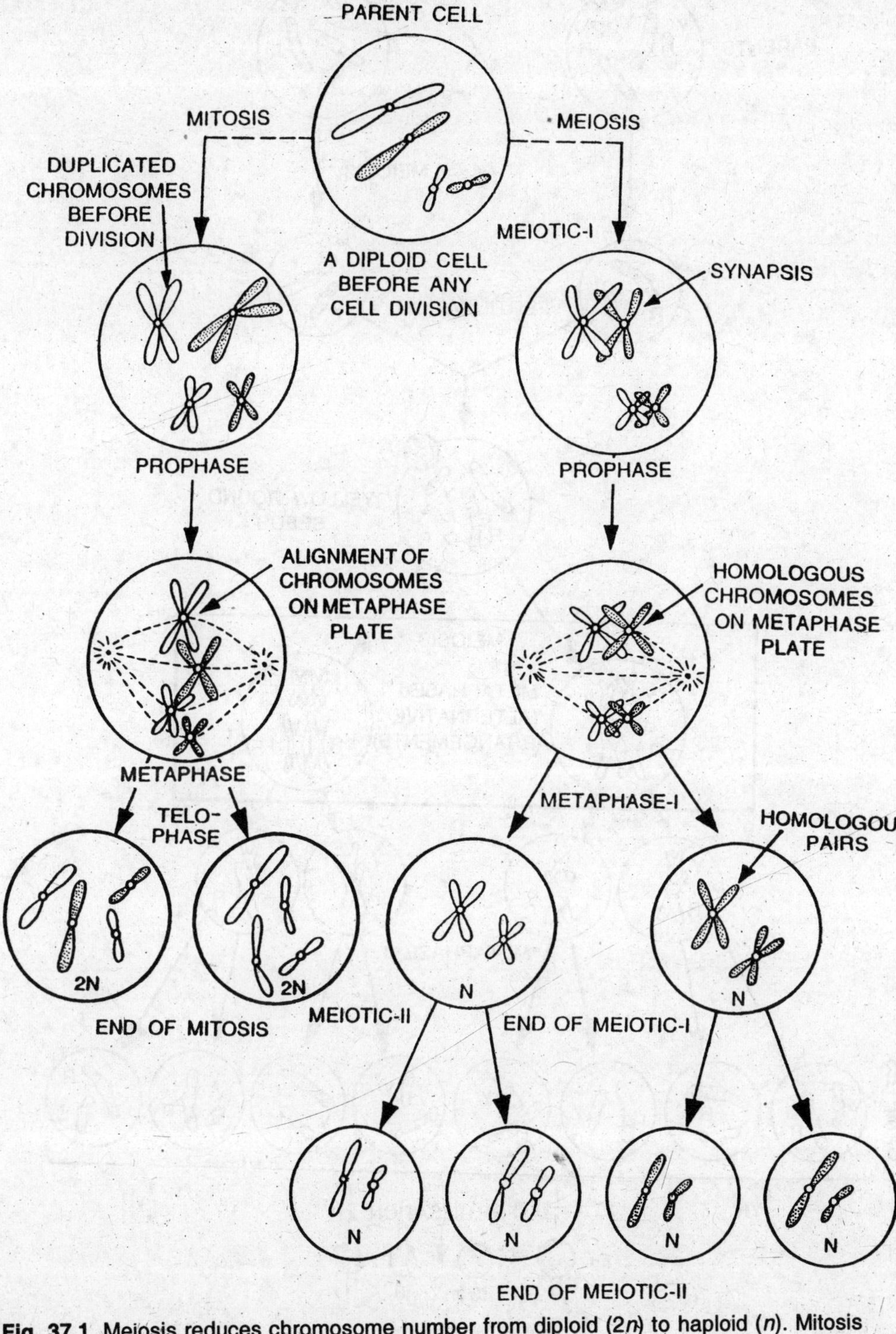

Fig. 37.1. Meiosis reduces chromosome number from diploid (2*n*) to haploid (*n*). Mitosis conserves chromosome number.

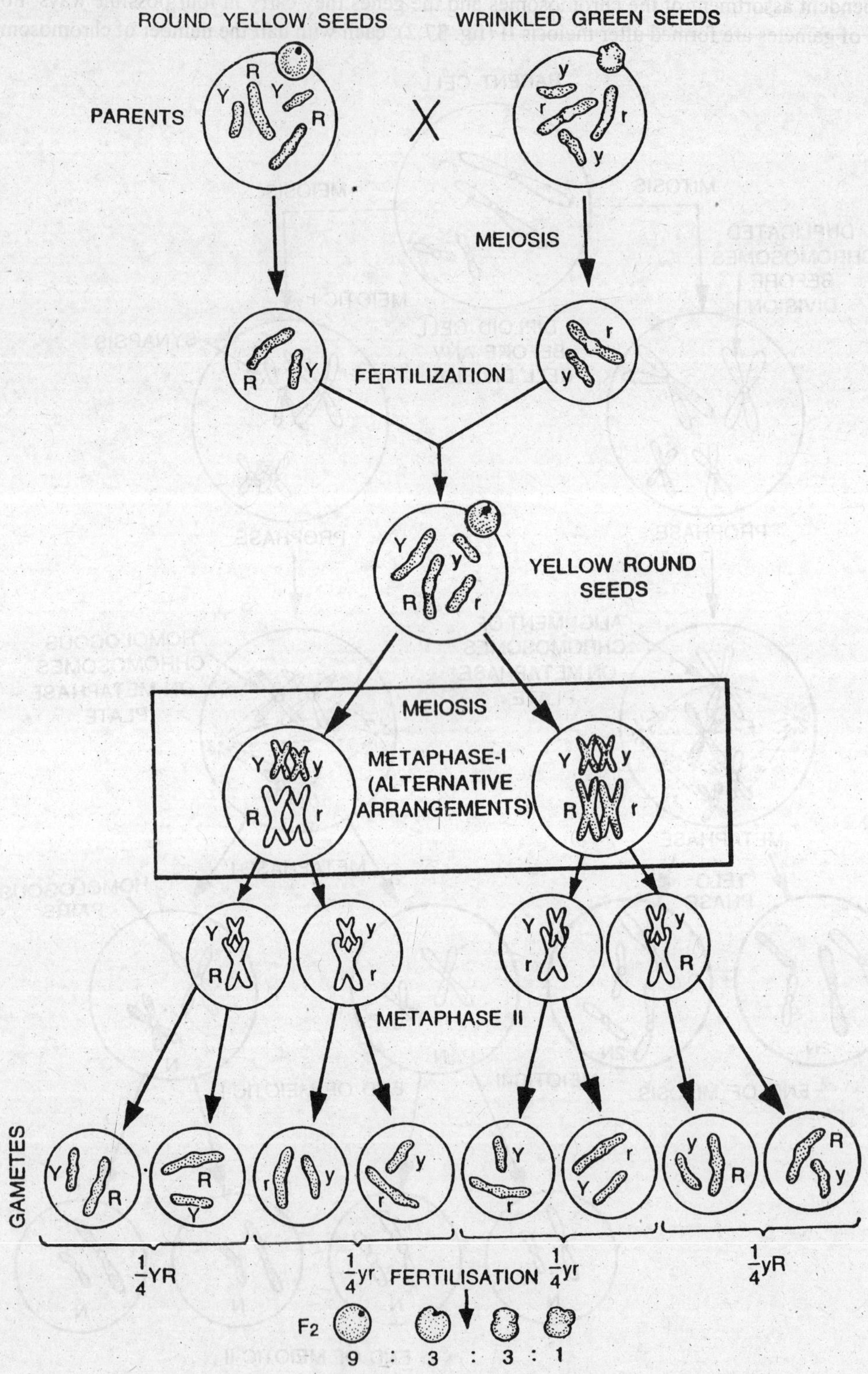

Fig. 37.2. Mendel's law of independent assortment and the behaviour of chromosomes.

and genes. The two parental types and the two new recombinants are formed in equal proportions, *i.e.,* 25 per cent each. Two genes located on separate chromosomes produce 50 per cent recombinants during meiosis. This results in the F_2 progeny showing a phenotypic dihybrid ratio of 9 : 3 : 3 : 1. There are millions of cells where meiosis occurs during gamete formation, and four types of possible gametes in equal numbers are produced from genotype RrYy when meiosis occurs in thousands of cells.

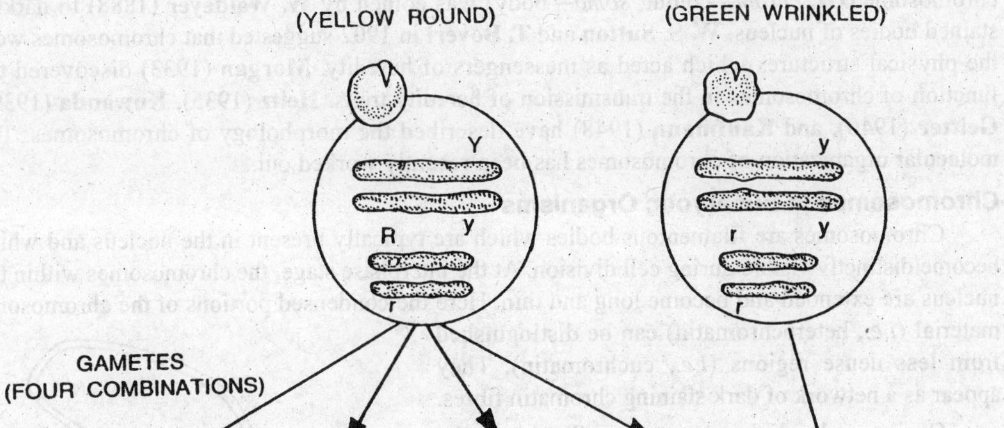

Fig. 37.3. Chromosomal explanation of the test cross. The genes located on separate chromosomes. Four types of progeny are obtained in equal proportions 50% parental and 50% recombinants.

Chromosomal Theory – Explanation of Test Cross

The progeny of a test cross between a dihybrid (heterozygous) round-yellow offspring and the double recessive wrinkled-green parent depicts as follows. With the genes located on separate chromosomes, four types of progeny are obtained in equal proportions. Two of these are of parental types and two are recombinants. Parental and recombinant frequencies are 50 per cent each. The two parental types (*i.e.*, round-yellow and wrinkled-green) and the two recombinants (*i.e.*, round-green and wrinkled-yellow) are formed in equal proportion, *i.e.*, 25 per cent each (fig. 37.3). This shows the confirmation of the proportion of each type of gamete formed by the dihybrid following independent assortment of chromosomes, that carry the alleles of two genes.

CHROMOSOMES

The chromosomes are the nuclear compounds of special organization and function. They are capable of self-reproduction and play a vital role in heredity, mutation, variation and evolutionary development of the species. **W. Fleming** (1879) first of all described the splitting of chromosomes and coined the term **chromatin** for stainable material of the nucleus. **W. Roux** (1883) suspected the involvement of the chromosomes in the mechanism of inheritance. **Benden** and **Bovery** (1887) reported that number of chromosomes for each species was constant. The present name chromosome (**Gr.** *chrom*—colour, *soma*—body) was coined by **W. Waldeyer** (1888) to darkly-stained bodies of nucleus. **W. S. Sutton** and **T. Boveri** in 1902 suggested that chromosomes were the physical structures which acted as messengers of heredity. **Morgan** (1933) discovered the function of chromosomes in the transmission of heredity traits. **Heitz** (1935), **Kuwanda** (1939), **Geitter** (1940), and **Kaufmann** (1948) have described the morphology of chromosomes. The molecular organization of chromosomes has been recently worked out.

Chromosomes in Eucaryotic Organisms

Chromosomes are filamentous bodies which are typically present in the nucleus and which become distinctly visible during cell division. At the interphase stage, the chromosomes within the nucleus are extended and become long and thin. Here the condensed portions of the chromosome material (*i.e.*, heterochromatin) can be distinguished from less dense regions (*i.e.*, euchromatin). They appear as a network of dark staining chromatin fibres.

During cell division, at the S phase the chromosomes replicate or make copies of themselves, with the two chromatids remaining attached. Each chromatid represents a strand of DNA. These two very long DNA molecules (chromatids) are attached to each other by a non-staining region called centromere.

During prophase of mitosis or meiosis I, the long replicated chromosomes condense. They get shortened and thickened by super coiling and condensation. Now two chromatids of a chromosome become visible.

At the metaphase state the replicated chromosomes are distinctly seen, and they remain connected together at the centromere. At this stage chromosomes may easily be counted and distinguished from each other. The chromosomes reveal a distinct pattern and possess specific size, shape and position of the centromere. At this stage the karyotyping may be done. A karyotype is systematic arrangement of chromosomes as per their size and shape.

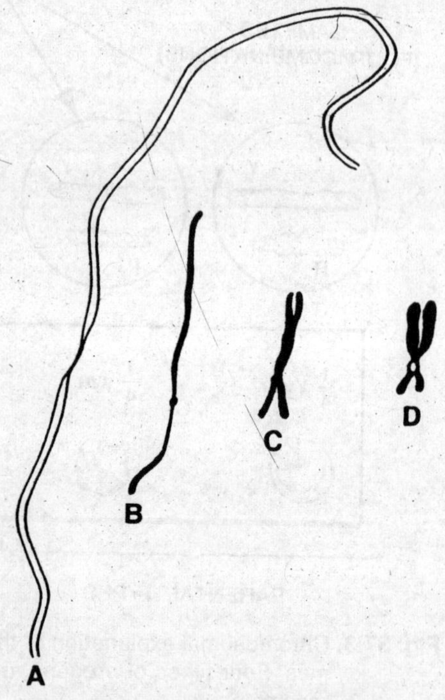

Fig. 37.4. Relative shape and size of a replicated chromosome at different stages of the cell cycle.

Chromosome Number in Eucaryotes

The number of chromosomes is constant for a particular species. Therefore, these are of great importance in the determination of the phylogeny and taxonomy of the species. The organism with the lowest number of the chromosome is the nematode, *Ascaris megalocephalus univalens* which has only two chromosomes in the somatic cells. Human body cells contain 46 chromosomes. This is the diploid (2n) number. Of these one set of 23 chromosomes (haploid, n) is derived from mother's egg cell and a set of 23 chromosomes from the father's sperm cell. The chromosome numbers vary a lot in different organisms.

The males of certain species of insects and round worms have one chromosome less than the females. Male honey bees are haploid, whereas females are diploid. Male honey bees are derived parthenogenetically from unfertilized egg cells. The chromosome numbers are not of much significance as the organisms such as *Amoeba*, adder's tongue fern, etc., are polyploids.

Chromosome numbers of some eucaryotes are tabulated here :

Table 37.2. Chromosome Number of Some Organisms.

Group	Common name	Scientific name	Chromosome number
Animals :			
Protozoa (Protist)	Paramecium	*Paramecium aurelia*	30-40
Cnidaria	Hydra	*Hydra vulgaris*	32
Nematoda	Round worm	*Ascaris lumbricoides*	24
Arthropoda	House fly	*Musca domestica*	12
	Mosquito	*Culex pipiens*	6
	Fruit fly	*Drosophila*	8
	Honey bee (male)	–	16 (n)
	Honey bee (female)	–	32 (2n)
	Cockroach (male)	*Blatta* sp.	23
	Cockroach (female)	*Blatta* sp.	24
Chordata	Frog	*Rana esculenta*	26
	Pigeon	*Columba livia*	80
	Rabbit	*Oryctolagus cuniculus*	44
	Gorilla	*Gorilla gorilla*	48
	Man	*Homo sapiens*	46
	Rat	–	42
	Chimpanzee	–	48
	Cow	–	60
	Horse	–	64
Plants :			
Algae	Chlamydomonas	*Chlamydomonas reinhardii*	10? 12? 16? (Haploid sets)
Fungi	Bread mold	*Mucor hiemalis*	2
		Neurospora	7
		Saccharomyces	17
Pteridophyte		*Ophioglossum*	1262 (largest number)
Gymnosperm	Yellow pine	*Pinus ponderosa*	24
Angiosperms	Cabbage	*Brassica oleracea*	18
	Coffee	*Coffea arabica*	44
	Potato	*Solanum tuberosum*	48
	Sugar cane	*Saccharum officinarum*	80
	Onion	*Allium cepa*	16
Protists (normally haploid)			
	Slime mold	*Dictyostelium*	7
		Amoeba proteus	250

Chromosomes in Procaryotic Organisms (Bacteria)

(1) It has been established now that a bacterial chromosome is always present in the centre of bacterial cell which is not capable of mitotic and meiotic divisions and is not enclosed within a definite nuclear membrane.

(2) The genetic material is found in a compact structure called **nucleoid.** The nucleoid consists of generally a large circular chromosome approximately 1.2 mm in length in the bacterium, *Escherichia coli.*

(3) An average bacterial cell contains one thousandth the DNA content of a typical eucaryotic cell.

(4) The bacterial chromosome is a thin elongated flexible and circular filament of DNA molecule associated with few proteins.

(5) These proteins have close similarities with histones found in eucaryotic cells. Also a large proportion of positively charged ions help them to counter balance the negative charge of phosphate groups in DNA.

(6) A chromosome of a bacterial cell (*Escherichia coli*) is a covalently closed circular structure consisting of several loops of supercoiled DNA.

Most fungi and protists are normally haploid (n) having one basic set of chromosomes. Wild plants may be diploid ($2n$) or polyploid having more than two basic sets of chromosomes. Bread wheat containing 42 chromosomes is a hexaploid, and many edible banana are triploid. Commonly the animals are diploid ($2n$), and some are polyploid. However, the organisms with large number of chromosomes as *Amoeba* and *Ophioglossum* (adder's tounge fern) are likely to be polyploids.

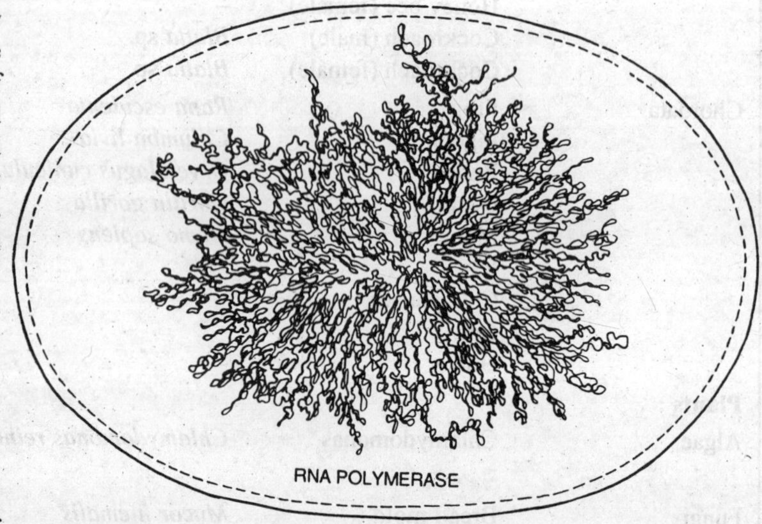

Fig. 37.5. Bacterial chromosome, chromosome of *Escherichia coli.*

Haploid and diploid. The number or set of the chromosomes of the gametic cells such as sperms and ova is known as the gametic, reduced or **haploid** sets of chromosomes. The haploid set of chromosomes is also known as the **genome.** The somatic or body cells of most organisms contain two haploid sets or genomes and are known as the **diploid cells.** The diploid cells achieve the diploid set of the chromosomes by the union of the haploid male and female gametes in the sexual reproduction.

Structure of Chromosome

Size. The size of the chromosomes varies from species to species and relatively remains constant for a particular species. The length of the chromosomes may vary from 0.2 to 50 μ. The diameter of the chromosome may become 0.2 to 20 . For instance, the human chromosomes are 6 in length. Moreover, the organisms with less number of chromosomes contain comparatively large sized chromosomes than the chromosomes of the organisms having many chromosomes. The monocotyledon plants contain large-sized chromosomes than the dicotyledon plants. The plants in general have large sized chromosomes in comparison to the animals. Further, the chromosome in a cell are never alike in size, some may be exceptionally large and others may be too small. The largest chromosomes are lampbrush chromosomes of certain vertebrate oocytes and polytene chromosomes of certain dipteran insects.

Shape. The shape of the chromosome is changeable from phase to phase in the continuous process of the cell growth and cell division. In the resting phase or interphase stage of the cell, the chromosomes occur in the form of thin, coiled elastic and contractile, thread-like stainable structures, the chromatin threads. In the metaphase and the anaphase, the chromosomes become thick and filamentous. Each chromosome contains a clear zone, known as **centromere** or **kinetocore**, along their length. The centromere divides the chromosomes into two parts, each part is called **chromosome arm.**

Chromonema. In less compact chromosomes, under the optical microscope, a coiled filament is seen in the chromosome. The coiled filament was called **chromonema** by **Vejdovsky** in 1912. The chromonema may be composed of two, four or more fibres according to the species. Further, the number of threads in the chromonema may depend on the different phases because at one phase the chromonema may contain one thread and at other phase it may contain three or four threads. The threads or fibrils of the chromonema remain coiled with each other.

Chromomeres. The chromonema of thin chromosomes of mitotic and meiotic prophases have been found to contain alternating thick and thin regions and thus giving the appearance of a necklace in which several beads occur on a string. The thick or bead-structures of the chromonema are known as the **chromomeres** and the thin region in between the chromomeres is termed as the **inter-chromomeres.** The position of the chromomeres in the chromonema is found to be constant for a given chromosome.

Centromere. The centromere consists of small granules or spherules. The centromere of the chromosome of the *Tilium* has the diameter of 3μ and the spherules have the diameter of 0.2μ. The chromonema remains connected with the spherules of the centromere. The chromosomes of most organisms contain only one centromere and are known as **monocentric** chromosomes. Certain chromosomes may have two or more centromeres and can be termed as **dicentric** and **polycentric** chromosomes respectively.

Material of the Chromosome

The composing material of the chromosomes is the chromatin substance which may be of the following two types :

1. Euchromatin. The euchromatin is the extended form chromatin and it forms the major portion of the chromosomes. It shows affinity to the basic nuclear stains. The euchromatin is rich in DNA. The euchromatin is considered as genetically active substance.

2. Heterochromatin. The highly stained region of the chromosome is said to contain heterochromatin. In the inter-phase the heterochromatin forms condensed chromosomal regions which are known as the **chromocenters** or false **nucleoli.** The heterochromatin remains closely associated with the nucleolus. The heterochromatin is considered as genetically inert substance but it controls the metabolism of the chromosome, biosynthesis of the nucleic acids and the energy metabolism. The sex chromosomes of the animals have been found rich in the heterochromatin.

Molecular compounds of chromosomes. The major molecular constituents of the eukaryotic chromosomes are DNA, RNA, histone and non-histone proteins and metallic ions. Recent studies have established that the chromosomes consist of a single, double stranded DNA molecule. The DNA is the most stable chemical molecule of biological world and carries genetic informations of one generation to next (Ris, 1969). The RNA is transcribed by the DNA. Most of the RNA is transported to the cytoplasm but some RNA remains associated with DNA along with proteins. The histone proteins are basic proteins and contain basic amino acids such as arginine and lysine in their molecules. They remain associated with DNA and serve as repressors of gene activity. The non-histone proteins are mostly acidic and most of them act as enzymes. The important non-histone proteins of chromosomes are phosphoproteins, DNA polymerase, RNA polymerase, DPN pyrophosphorylase and nucleoside triphosphatase. The chromosomes contain many metallic ions, *e.g.*, Mg^{++}, Ca^{++}, etc., which make the organization of chromosomes intact.

Functions of the Chromosomes

The chromosomes are the most significant components of the cell. They control most of the cell biological and genetical activities of a species. They contain the genetical material, the DNA which ultimately influences all the biological phenomena at molecular, physiological and gross morphological levels.

The chromosomes are considered as the organs of heredity because of the following reasons :

(*i*) They form the only link between two generations.

(*ii*) A diploid chromosome's set consists of two morphologically similar (except the X and Y sex chromosomes) sets, one is derived from the mother and another from the father at fertilization.

(*iii*) The genetic material, DNA or RNA is localized in the chromosome and its contents are relatively constant from one generation to the next.

(*iv*) The chromosomes maintain and replicate the genetic informations contained in their DNA molecule and this information is transcribed at the right time in proper sequence into the specific type of RNA molecules which directs the synthesis of different types of proteins to form a body form like the parents.

LINKAGE AND CROSSING OVER

Mendel's principle of independent assortment applies both to genes and chromosomes. It is very obvious that larger the number of chromosomes, the greater will be the number of combinations of the genes and trait resulting from independent assortment. For example, in *Drosophila,* hundreds of genes have been studied, yet there are only four pairs of chromosomes. Each chromosome contains many genes. Thus genes in the same chromosome will not be assorted independently. Mendel's principle of independent assortment is not universal but is limited to genes in different chromosomes.

An exception of law of independent assortment was discovered by Bateson and Punnett in 1906 in which they found two pairs of alleles in sweet peas which did not assort independently. They noted, when two alleles such as A and B, came from the same parent (AA BB × aa bb) they tended to enter the same gamete and to be transmitted together. This was called **coupling**. On the other hand, when the same alleles came from different parents (AA bb × aa BB), they tended to enter different gametes and to remain apart this phenomenon was called **repulsion**. No satisfactory reply of coupling and repulsion was developed until Morgan (1910) found similar situation in *Drosophila* and noted that coupling and repulsion are two aspects of a single phenomenon known as linkage.

Linkage

General definition. "The phenomenon in which linked genes tend to go together in the same gamete as a result of which the parental combinations are obtained in greater frequency than expected in the progeny, is known as **linkage** or the co-existence of two or more genes or factors in the same chromosome as a result of which they are transmitted from one generation to another in group is known as **linkage**".

In other words genes on same chromosome are called **linked genes,** and the phenomenon is termed as **linkage.**

According to Morgan the tendency of two genes to remain together in the same chromosome during hereditary transmission is called linkage. Morgan supposed that this tendency of linked genes to remain in their original combinations was due to their location in the same chromosome. Further, he advanced the basic idea that the degree or strength of linkage depends upon the distance between the linked genes in the chromosome.

Linkage groups. All the genes that are linked together form the linkage group. If a certain gene A is linked with B and B with C and C with D and so on then all the genes AB, CD, etc., are linked together and all the genes that exhibit linkage with one another form a linkage group. Extensive work on *Drosophila melanogaster* has shown that the known genes fall into four linkage groups. The haploid number or the number of pairs of chromosomes which this species possesses is also four. The number of genes which form into linkage groups are also somewhat relative to the lengths of the chromosomes. The smallest linkage group corresponds to the shortest chromosome pair. In maize there are ten linkage groups and ten pairs of chromosomes; in *Pisum sativum* seven linkage groups and seven pairs of chromosomes and in tomato 10 linkage groups for 12 chromosomes have been identified. Thus it can be said that the number of linkage groups is restricted to the haploid number of chromosomes. This principle is known as the **limitation of linkage groups.**

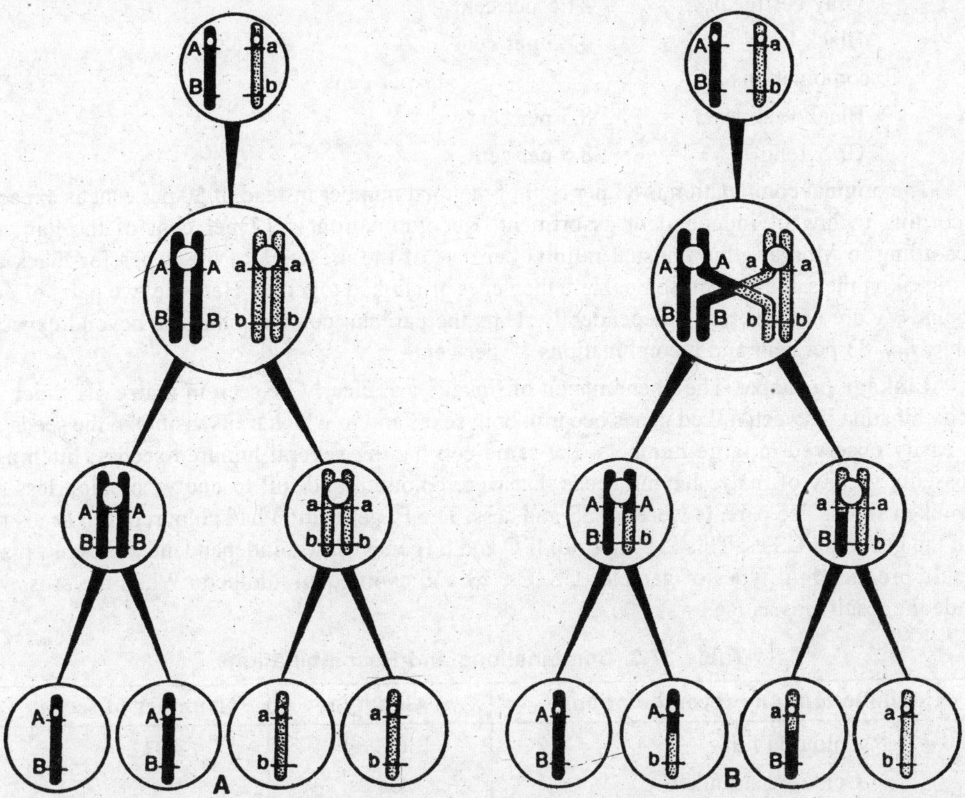

Fig. 37.6. Linkage. A, diagram of the segregation of two pairs of allelomorphic genes localized on the same pair of chromosomes without crossing over. The result is two types of gametes, AB and ab. A case of complete linkage. B, diagram of the same chromosome between which crossing over takes place during meiosis. Four types of gametes result AB, ab, aB and ab. A case of incomplete linkage.

Types of linkage. There are two main types of linkage : (1) complete and (2) incomplete. In complete linkage two or more characters are inherited together and these characters appear in two or more generations. This type of linkage is rarely found in nature. This takes place due to non-break in the gene combinations, situated on any chromosome. It is found in *Drosophila* and certain other insects.

The incomplete type of linkage has been found in many plants and animals. Here, the chromosomes break accidentally either in one or several places. Such type of linkage is found in maize.

Linkage in Drosophila. Morgan (1920) discovered two recessive mutants in *Drosophila* one gray-bodied vestigeal-winged and the other black-bodied long-winged. If a gray-bodied vestigeal-winged fly is crossed to a black-bodied long-winged one, the F_1 generation consists of gray long-winged (normal) flies. Thus the gene for gray body colour, B, is dominant over its allele, which is responsible for black body colour, b, : and the gene for long wings, V, is dominant over its vestigeal allele, v. When the F_1 male hybrids are crossed to double recessive females (*i.e.*, black-bodied vestigeal-winged females), only two kinds of offspring are produced, *i.e.*, gray vestigeal and black long. The expected types, *i.e.*, gray long and black vestigeal do not appear at all. If F_1 female fly is crossed with a black vestigeal male the four expected types are produced as follows :

Parental combinations			83 per cent
Gray vestigeal	=	41.5 per cent	
Black long	=	41.5 per cent	
Recombinations			17 per cent
Black vestigeal	=	8.5 per cent	
Gray long	=	8.5 per cent	

The original combination is 83 per cent of the total number instead of 50 per cent as expected according to law of independent assortment. Recombination is 17 per cent of the gametes. According to Morgan, this unusual ratio is because of the presence of the genes for black and vestigeal on the same chromosome. Here the genes are linked together. Here the two pairs of gene Bb and Vv are not assorted independently. Thus the parental combinations are beyond expected value, *i.e.*, 83 per cent and recombinations 17 per cent.

Linkage in maize. The phenomenon of linkage can clearly be seen in maize, in which the recombination between linked genes occur in both sexes and in which traits visible in the seeds can be easily observed in large numbers, the same cob having several hundred seeds. Hutchinson crossed a variety of maize having seeds that were coloured and full to one with colourless and shrunken seeds. The parents were CCSS and ccss. The F_1 generation had coloured, full seeds that had the genotype CS/cs. This is F_1 hybrid. If C and S (genes) assort independently, these F_1 plants should produce four types of gametes CS, Cs, cS and cs in equal numbers. When the cross was made the result was as follows :

Table 37.3. Combinations and Recombinations

Combinations and recombinations	Genotype	Number of seeds
Coloured full	CS/cs	4032
Coloured shrunken	Cs/cs	149
Colourless full	cS/cs	152
Colourless shrunken	cs/cs	4035

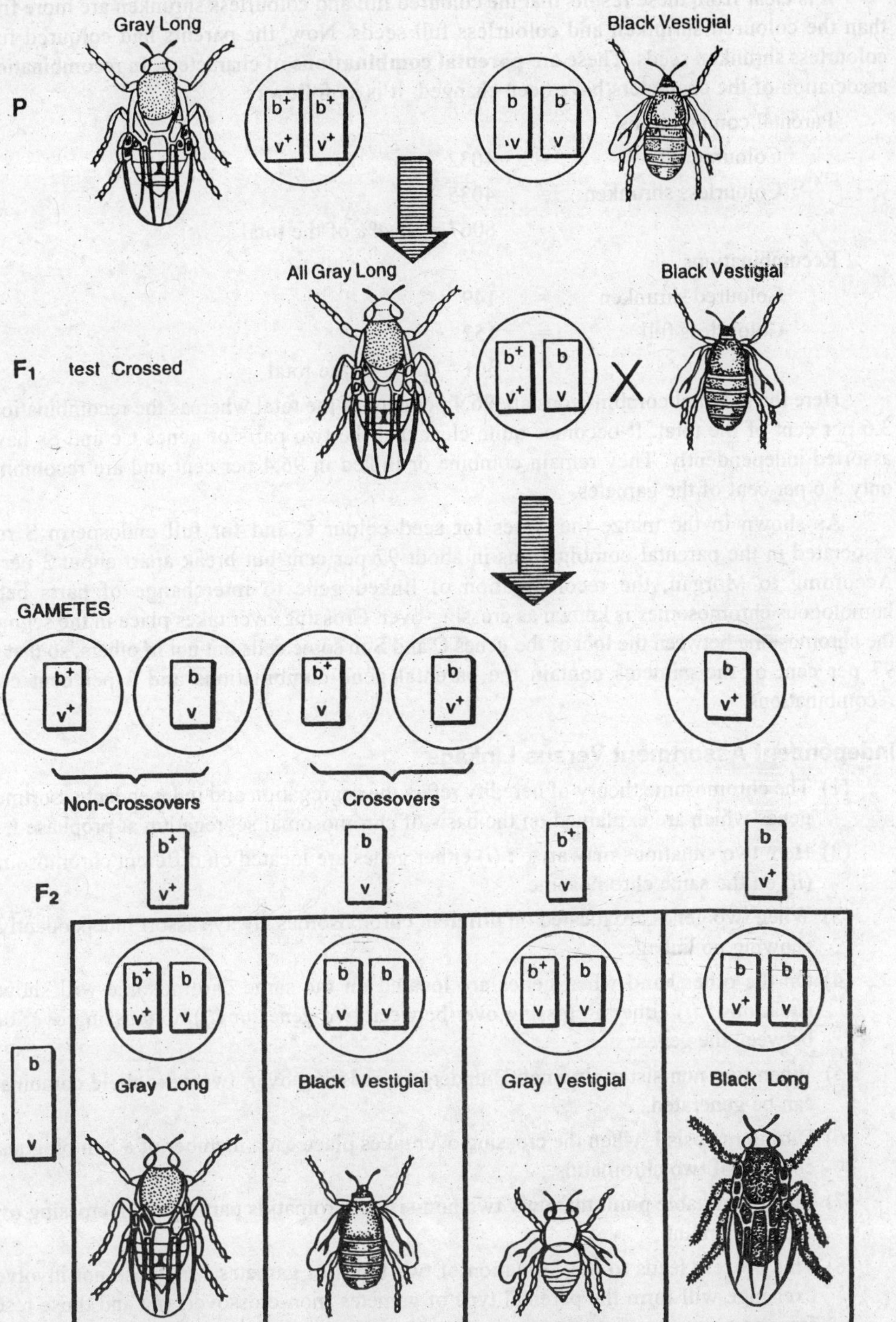

Fig. 37.7. Diagram of cross involving linkage and crossing over. The genes for vestigeal versus normal (long) and black versus gray body in *Drosophila* are linked. They are located in the same chromosome.

It is clear from these results that the coloured full and colourless shrunken are more frequent than the coloured shrunken and colourless full seeds. Now, the parents had coloured full and colourless shrunken seeds. These are **parental combinations** of characters. In recombinations the association of the characters have been changed. It is as follows :

Parental combinations

Coloured full	=	4032
Colourless shrunken	=	4035
		8067 = 96.4% of the total

Recombinations

Coloured shrunken	=	149
Colourless full	=	152
		301 = 3.6% of the total

Here the parental combinations are 96.4 per cent of the total whereas the recombinations are 3.6 per cent of the total. It becomes quite clear that the two pairs of genes Cc and Ss have not assorted independently. They remain combine or linked in 96.4 per cent and are recombined in only 3.6 per cent of the gametes.

As shown in the maize the genes for seed colour C and for full endosperm S remain associated in the parental combinations in about 97 per cent but break apart about 3 per cent. According to Morgan, the recombination of linked gene to interchange of parts between homologous chromosomes is known as crossing-over. Crossing-over takes place in the segment of the chromosome between the loci of the genes C and S in some cells but not in others, so that about 97 per cent of the gametes contain the parental gene combinations and 3 per cent contain recombinations.

Independent Assortment Versus Linkage

(1) The chromosome theory of heredity refers the segregation and independent assortment of genes which are explained on the basis of chromosomal segregation at prophase I.

(2) Here two situations may arise : (*i*) either genes are located on different chromosomes or (*ii*) on the same chromosome.

(3) When two genes are located on different chromosomes always assort independently thus showing no linkage.

(4) On the other hand when genes are located on the same chromosome will show two situations : (*i*) either a crossing over between two genes or (*ii*) no crossing over occurs between the genes.

(5) When two non-sister chromatids undergo a crossing over, two new allele combinations can be generated.

(6) During meiosis I, when the crossing over takes place each member of a homologous pair consists of two chromatids.

(7) This is a notable point that only two non-sister chromatids participate in crossing over at any given point.

(8) This process leads to the formation of two types of gametes : (*i*) those not involved in exchange will form the parental type of gametes (non-crossover) (*ii*) and those resulted from crossing over are the recombinant types.

(9) When there is no crossing over between the genes, they will be transmitted intact. Such behaviour leads to the formation of only parental (non-crossover) gametes and the phenomenon is called **complete linkage.**

(10) The frequency of crossing over between any two genes is generally proportional to the distance between them on the chromosome.

Complete and Incomplete linkage

When two genes located on the same chromosome are considered, the situations may be explained as follows :

If the two genes are very far from each other, crossing over is possible in all the meiotic cells. In such a case 50 per cent of parental and 50 per cent of recombinants will be produced and the gametes will appear in a genotypic ratio of 1 : 1 : 1 : 1. In such a case, transmission of two genes can not be distinguished from that of two genes located on different chromosomes and thus assorting independently.

Contrary to above situation, when the two genes are close to each other there can be two situations. In situation (*i*) because of it, no crossing over will result into only parental or non-cross over types of gametes (*i.e.*, complete linkage). Actually, such a situation does not exist and crossing over takes place even between closely situated genes. This situation is referred to as incomplete linkage. Here, there is a clear predominance of parentals or non-cross over types of gametes. This relationship was demonstrated for the first time in 1911 in *Drosophila* (fruit fly) by Morgan and Sturtevant.

This shows that all the genes located on the same chromosome will show some linkage between them.

The strength of linkage shall be determined by the distances between them.

When the genes are farther, linkage can be easily broken by crossing over.

Thus, genes located on the same chromosome constitute a linkage group.

The number of linkage groups in any organism is proportional to its haploid chromosome number.

Crossing Over

According to Morgan the linkage is caused due to linked genes being carried out by the same chromosome. He also pointed out that the phenomenon of complete linkage occurs rarely because sometimes the linked genes show the tendency to separate during meiosis. This way, the crossing over may be defined as "a mechanism of the recombination of the genes due to interchange of chromosomal segments at the time of pairing."

The cross-over between linked genes allows their recombination during meiosis. The mechanism of cross-over is as follows :

Mechanism of crossing over. The figure shows that crossing over, leading to recombination of linked genes, is due to interchange of sections of homologous chromosomes. At meiosis, the homologous (maternal and paternal) chromosomes come together and pair, or synapse, during prophase. The pairing is remarkably precise and is evidently brought about by mutual attraction of the parts of the chromosomes that are similar or homologous because they contain allellic genes. Further, during the transition between the pachytene and the diplotene of the prophase of meiosis, the paired chromosomes divide each into two chromatids, so that the bivalent is now composed of four chromatids. At about the time when the chromosomes are first seen to be divided, the chromatids establish one or more exchange, or chiasmata, per bivalent. At each chiasma, two of the chromatids have become broken and then rejoined, so that the new chromatids are now formed of sections of the original ones. Owing to chiasma formation, maternal and paternal chromosomes are not transmitted as indivisible units; they exchange sections, and the new chromosomes that arise as a result of meiosis carry genes that before meiosis were located in different members of the pair of homologous chromosomes. Janssens (1909) for the first time discovered the process of chiasma formation. This is a highly precise process since the two chromatids at a chiasma exchange exactly equivalent segments, so that, with very rare exception, neither chromatid gains or loses any genes.

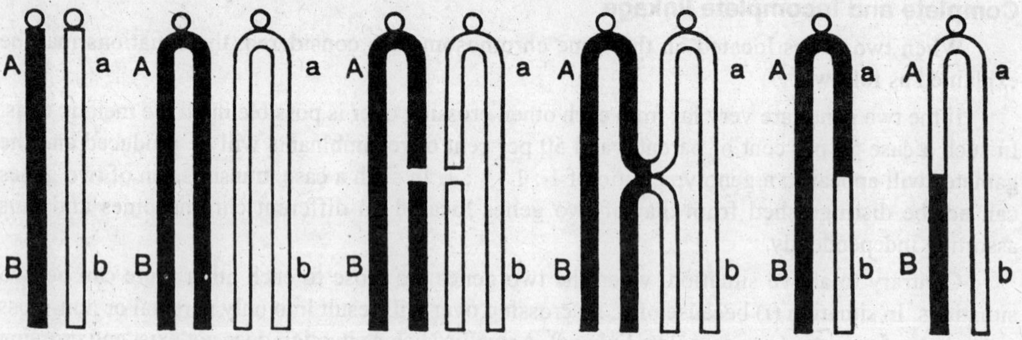

Fig 37.8. Diagram showing the mechanism of crossing over.

Absence of crossing over in Drosophila males. In most of the organisms, recombination of linked genes takes place both in females and males. Similarly, in most organisms, chiasmata formation takes place in both female and male meiosis. But in *Drosophila* the situation is very different. In *Drosophila* crossing over rarely or never takes place in the male. However in the silkworm moth it seems to be absent in the females.

Recombination and crossing over. The cytological observations have established one important fact about crossing over that the exchanges of chromosome segments which cause recombination of linked genes occur at the four-strand, or tetrad stage, in the prophase of meiotic division. Here the homologous chromosomes have paired, each homologue has split into two chromatids held together by undivided centromere, and chiasmata are observed in the tetrad. It is revealed by careful observation that at each chiasma only two of the four chromatids cross over whereas the two other chromatids preserve their original continuities. Genetically it means that of the four chromosomes which emerge from meiosis, only two can show crossing over at any one level.

Recombination is a process which can occur in a variety of situations but the most popular one is in relation to meiosis. Thus, this is a meiotic process that generates haploid product whose genotype is different from the two haploid genotypes constituting the diploid. The product of recombination is a **recombinant.**

Chiasma – type hypothesis. Morgan suggested that the formation of recombinants can be explained by the occurrence of crossing over (*i.e.,* chiasma formation) during meiosis I. This led to chiasma-type hypothesis. The points in support of this hypothesis are as follows :

1. During prophase I (*i.e.,* zygotene), homologous chromosomes pair (*i.e.,* synapsis), with each chromosome consisting of two chromatids.

2. During pachytene of prophase I two non-sister chromatids break at homologous sites and then rejoin.

3. Such rejoining takes place between non-sister chromatid and leads to physical exchange.

4. This physical exchange leads to reshuffling of genes and thus recombination takes place.

Synaptonemal complex and the process of crossing over. Since it is an established fact that DNA is the material of the gene it becomes clear that crossing over involves break and rejoining of DNA. During synapsis, a molecular scaffold known as **synaptonemal complex** aligns the DNA molecules of the two homologous chromosomes side by side. With the result the DNA strand of one chromosome, *i.e.,* a chromatid becomes able to pair with the corresponding DNA strand of the other where their base sequence match. Break and new synthesis of one DNA strand using a strand of other chromatid DNA as the template results in crossing-over.

Crossing-over involves the breaking and rejoining of chromosomes in the synaptonemal complex, which can be observed by cytological methods. However, no crossing-over has been observed in male *Drosophila* during sperm formation. Likewise no recombination of genes occur during meiosis.

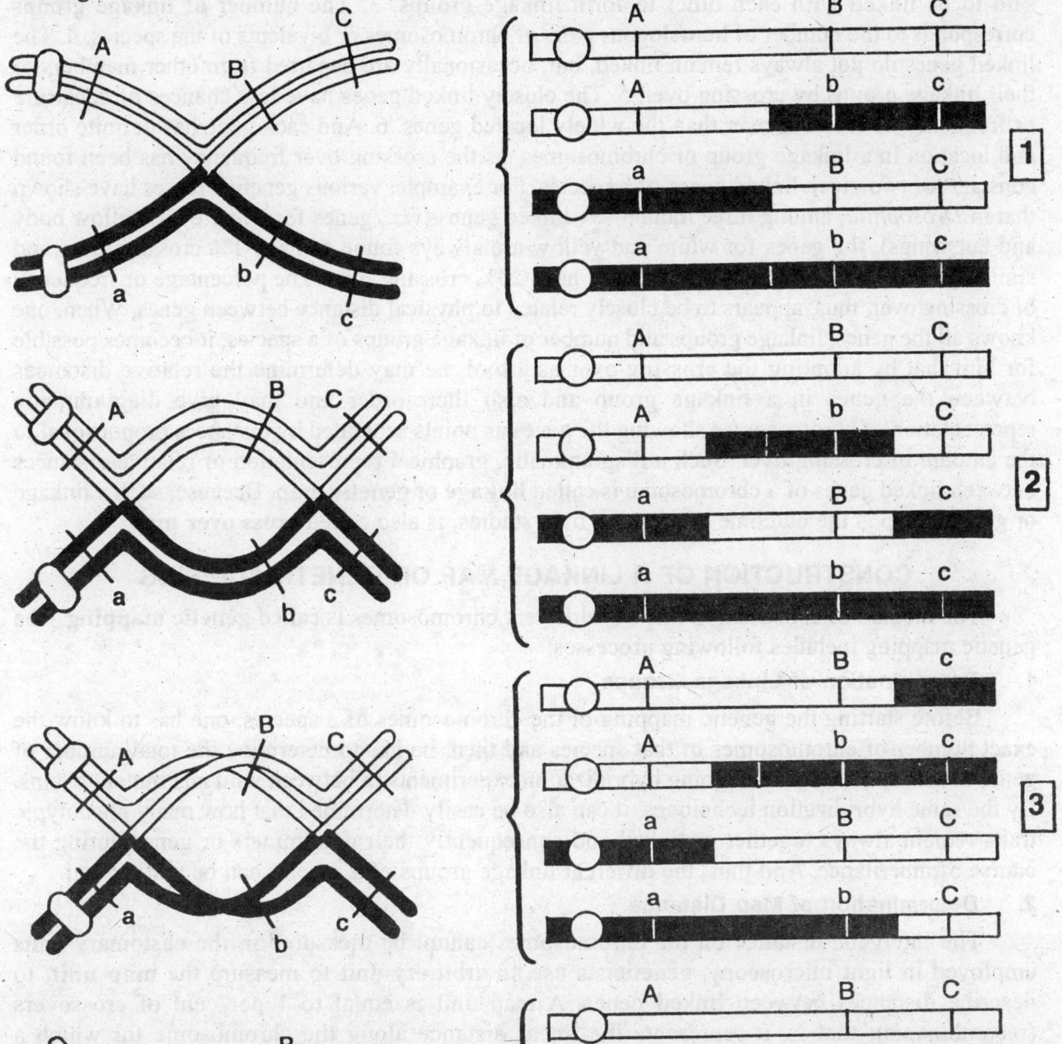

Fig. 37.9. Crossing over. Diagram showing the results of single and double chiasma. 1. Single chiasma. 2. Two strand double cross over. 3. Four strand double cross over. 4. Three strand double cross over.

MAPPING OF CHROMOSOMES

Study of linkage and crossing over has made clear so far, that 1. Because the number of genes usually exceeds the number of chromosomes in different species, therefore, many genes have to be located on same chromosomes. 2. The genes remain arranged linearly on the chromosomes and they have no option before them, except to behave according to the chromosomes during gametogenesis and inheritance, therefore, all the genes on a chromosome inherit together and are said to be linked with each other to form linkage groups. 3. The number of linkage groups corresponds to the number of homologous pairs of chromosomes or bivalents of the species. 4. The linked genes do not always remain linked, but, occasionally are departed from other members of their linkage groups by crossing over. 5. The closely linked genes have less chances of departure or frequency of crossing over than the widely located genes. 6. And each gene has definite order and location in a linkage group or chromosomes, as the crossing over frequency has been found constant for two given linked genes of a species. For example, various genetic crosses have shown that in *Drosophila*, among three mutant sex-linked genes (*viz*., genes for white eyes, yellow body and cut wings), the genes for white and yellow are always found to have 1% crossing over and similarly, white and cut genes are found to have 20% crossing over. The percentage or frequency of crossing over, thus, appears to be closely related to physical distance between genes, When, one knows all the genes, linkage groups and number of linkage groups of a species, it becomes possible for him that by adopting the crossing over as a tool, he may determine the relative distances between the genes in a linkage group and also their order and may give diagrammatic representation of chromosomes showing the genes as points separated by distances proportional to the amount of crossing over. Such a diagrammatic, graphical representation of relative distances between linked genes of a chromosome is called **linkage** or **genetic map.** Because, such a linkage or genetic map is the outcome of crossing over studies, is also called **cross over map.**

CONSTRUCTION OF A LINKAGE MAP OR GENETIC MAPPING

The method of construction maps of different chromosomes is called **genetic mapping**. The genetic mapping includes following processes:

1. Determination of Linkage Groups

Before starting the genetic mapping of the chromosomes of a species, one has to know the exact number of chromosomes of that species and then, he has to determine the total number of genes of that species by undergoing hybridization experiments in between wild and mutant strains. By the same hybridization techniques, it can also be easily determined that how many phenotypic traits remain always together or linked and consequently their determiners or genes during the course of inheritance. And thus, the different linkage groups of a species can be worked out.

2. Determination of Map Distance

The intergene distance on the chromosomes cannot be measured in the customary units employed in light microscopy, geneticists use an arbitrary unit to measure the **map unit**, to describe distances between linked genes. A map unit is equal to 1 per cent of crossovers (recombinants); that is, it represents the linear distance along the chromosome for which a recombination frequency of 1 per cent is observed. These distances can also be expressed in **morgan units;** one morgan unit represents 100 per cent crossing over. Thus, 1 per cent crossing over can also be expressed as **1 centimorgan (1 cM),** 10 per cent crossing-over as **1 decimorgan** and so on. The Morgan unit is named in honour of **T.H. Morgan**; however, most geneticists prefer map units.

Examples

1. If a F_1 hybrid having the genotypes Ab/aB produces 8% of cross over gametes AB and ab, then the distance between A and B is estimated to the 16 map units or centimorgan.

2. If the map distance between the gene loci B and C is 12 centimorgan, then 12% of gametes of genotypes BC/bc should be crossover types, *i.e.,* 6% bC.

Because, each chiasma produces 50% crossover products, 50 per cent crossing over is equivalent to 50 map units or centimorgans. If the mean number of chiasmata is known for a chromosome pair, the total length of the map for that linkage group may be predicted :

Total length = mean number of chiasmata × 50

Two Point Test Cross

The percentage of crossing over between two linked genes is calculated by test crosses in which a F_1 dihybrid is crossed with a double recessive parent. Such crosses because involved crossing over at two points, so called **two point test crosses.** For example, a dihybrid having the genotype Ac/ac is test crossed with a double recessive parent (ac/ac), then among F_2 test cross hybrids we may get 37% dominant genes at both gene loci (AC/ac), 37% recessive genes at both gene loci (ac/ac), 13% dominant gene at first gene locus and recessive gene at the second gene locus (Ac/ac). The last two groups (*i. e.,* 13% Ac/ac) and 13% aC/ac) were produced by crossover gametes (13 + 13) from the dihybrid parent. Thus 26% of all gametes (13 + 13) were of cross over types and the distance between the loci A and C is estimated to be 26 centimorgans. Because, double crossovers usually do not occur between genes less than 5 centimorgans apart, so for genes further apart, the **three point test crosses** are used.

Three Point Test Cross

A three point test cross or trihybrid test cross (involving three genes) gives us information regarding relative distances between these genes, and also shows us the linear order in which these genes should be present on chromosome. Such a three point test cross may be carried out if three points or gene loci on a chromosome pair can be identified by marker genes. If, in addition to genes A and C indicated above, a third marker gene B is located in fairly close proximity in the same linkage group, all three markers may be used together in conducting a more precise analysis of the map distance and the relative position of the three points.

Suppose that we testcross trihybrid individuals of genotype ABC/abc and find in the progeny the following:

36% ABC/abc	9% Abc/abc	4% ABc/abc	1% AbC/abc
36% abc/abc	9% aBC/abc	4% abC/abc	1% aBc/abc
72% Parental type	18% Single crossovers between A and B (region I)	8% Single crossovers between B and C (region II)	2% Double crossovers

To find the distance A–B we must count all crossovers (both singles and doubles) that occurred in region I = 18% + 2% or = 20% or 20 map units between the loci A and B. To find the distance B–C we must again count all crossovers (both singles and doubles) that occurred in region II = 8%+2% = 10% or 10 map units between the loci B and C. The A–C distance is, therefore, 30 map units when double crossovers are detected in a three point linkage experiment and 26 map units when double crossovers are undetected in the two-point linkage experiment above.

Without the middle marker (B), double crossovers would appear as parental types and hence we underestimate the true map distance (crossover percentage). In this case the 2% double crossovers would appear with the 72% parental types, making a total 74% parental types and 26% recombinant types. Therefore, for any three linked genes whose distances are known, the amount of detectable crossovers between the two outer markers A and C when the middle marker B is missing is; (A–B crossover percentage) plus (B–C crossovers percentage) minus (2 × double crossover percentage).

3. Determination of Gene Order

After determining the relative distances between the genes of a linkage group, it becomes easy to place genes in there proper linear order. For example, if the linear order of three genes

ABC is to be determined, then these three genes may be in any one of three different orders depending upon that which gene is in the middle. For the time being we may ignore left and right end alternatives. If double crossovers do not occur, map distances may be treated as completely additive units. Now, if we suppose that the distance between the genes A–B = 12, B–C = 7, A–C = 5, we can determine the order of genes correctly in the following manner :

Case I. Let us assume that gene A is in the middle (*e. g.*, B–A–C) :

B 12 A	A 5 C
B 7 C	

In this case because, the distances between B–C are not equitable, genes A cannot be in the middle.

Case II. Let us assume that gene B is in the middle (*e. g.*, A–B–C) :

A 12 B	B 7 C
A 5 C	

In this case, because the distance between A–C are not equitable, therefore, gene B cannot be in the middle.

Case III. Let us assume that gene C is in the middle (*e. g.*, A–C–B) .

A 5 C	C 7 B
A 12 B	

In this case, because the distances between A–B are equitable, therefore, gene C must be in the middle.

Thus, the relative distances and ordering of genes in a linkage group are determined in separate segments by two point test crosses or three point crosses, as the case may be.

4. Combining Map Segments

Finally, the different segments of maps of a complete chromosome are combined to form a complete genetic map of 100 centimorgans long for a chromosome.

Example: For example, suppose we have to combine following three map segments.

(I) a 8 b 10 c

(Ii) c 10 b 22 d

(III) c 30 e 2 d

We can superimpose each of these segments by aligning the genes shared in common.

(I) a 8 b 10 c

(II) d 22 b 10 c

(Iii) d 2 e 30 c

Then finally we may combine the three segments into one map:

The a to d distance = (d to b) – (a to b) = 22 – 8 = 14

The a to e distance = (a to d) – (d to e) = 14 – 2 = 12

d 2 e 12 a 8 b 10 c

Interference and Coincidence

In most higher organisms it has been found that one chiasma formation reduces the probability of another chiasma formation in an immediately adjacent region of the chromosome, probably because of physical inability of the chromatids to bend back upon themselves within certain minimum distances. The tendency of one crossover to interfere with the other crossover is called **interference.** Thus, the proximity of one crossover to another decreases the probability of another very close by. The centromere has a similar interference effect; frequency of crossing over is also reduced near the ends of the chromosome arms. The net result of this interference in the observation of fewer double crossover types than would be expected according to map distances. The strength of interference varies in different segments of the chromosome and is usually expressed in terms of a **coefficient of coincidence,** or the ratio between the observed and the expected double crossovers.

$$\text{Coefficient of coincidence} = \frac{\text{\% observed double crossovers}}{\text{\% expected double crossovers}}$$

The coincidence is the complement of interference, so:

Coincidence + Interference = 1.0

When interference is complete (1.0), no double crossovers will be observed and coincidence becomes zero. When, interference decreases, coincidence increases. Coincidence values ordinarily vary between 0 and 1. Coincidence is generally quite small for short map distance. There is no interference across centromere.

Example:

For explaining interference and coincidence, we can consider the results of one of the experiment of **Hutchison** (1922). He reported the map distance for three genes, c(colourless aleurone), Sh (shrunken grains), and wx (waxy endosperm) of corn and observed following crossing over frequencies between these genes:

Table 37.4. Crossing over frequencies between genes c, Sh and wx of corn.

Regions	Genes	Percentage crossovers	Map-distances (in map units)
I	c-Sh	3.4	3.4 + 0.1 = 3.5
II	Sh-wx	18.3	18.3 + 0.1 = 18.4
Double corssover	c-Sh-wx	0.1	

If crossing over in region I and II were independent, we should predict 0.035 ×0.184 = 0.6 per cent double crossovers, where as only 0.1 per cent was observed.

$$\text{So, coincidence} = \frac{0.1}{0.6} = 0.167$$

Linkage Maps of different Organisms

By adopting the above mentioned techniques, geneticists have constructed the linkage or genetic maps of various organisms, such as, viruses, bacteria, fungi, tomato, barley, wheat, rice, sorghum, morning glory, garden pea, maize, *Drosophila,* chickens, mice man, etc. The first linkage map has been constructed for two chromosomes of *Drosophila* by **Strutevant** in 1911. The linkage or genetic mapping in maize has been done by **McClintock** under the leadership of **R. A. Emerson.**

Syntenic Genes

If two or more specific human gene products and a given human chromosome are both present in the same hybrid cells, then those genes are located in the same chromosome ; that is, they are **syntenic.** The term **synteny** refers to genes that are located on the same chromosome, whether or not they show recombination; **linkage** refers only to genetic loci that have been shown by recombination studies to be in the same chromosome. Syntenic genes may be so far apart in their chromosome that they seem to segregate independently; that is, they may show as much as 50 per cent recombination as would be exhibited by non-syntenic genes.

Linkage Map of Drosophila melanogaster

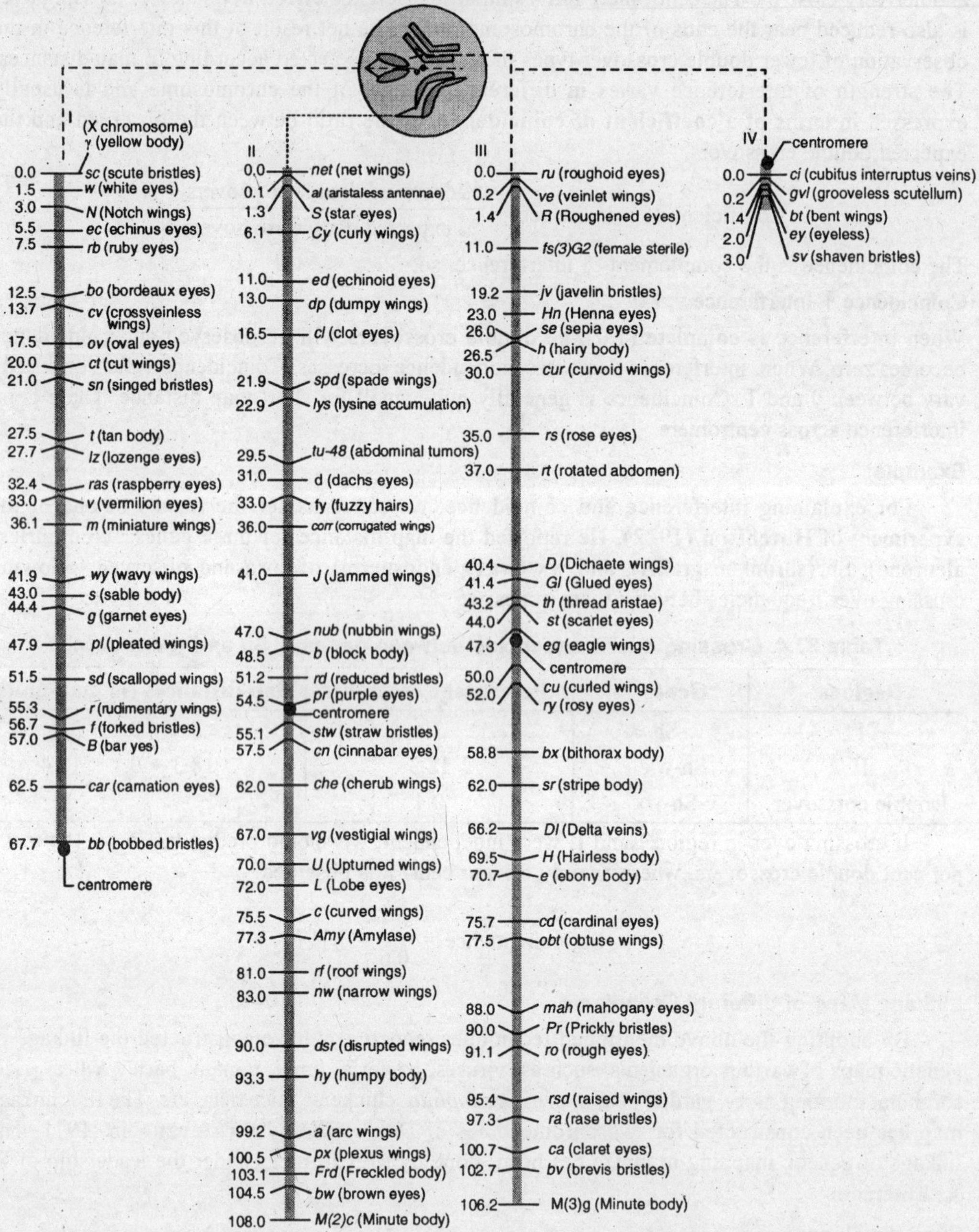

Fig. 37.10. Abbreviated linkage map of *Drosophila melanogaster*. The correlation of each linkage group with one of the pairs of chromosomes (circular inset) is shown by the dashed lines. Map positions are given on the left of each linkage map; the genetic symbols used to represent the mutant allele(s) of each gene are shown on the right of each linkage map. The phenotypic description commonly used for each mutant allele is given in parentheses after the gene symbol.

Linkage Map of Maize (*Zea Mays*)

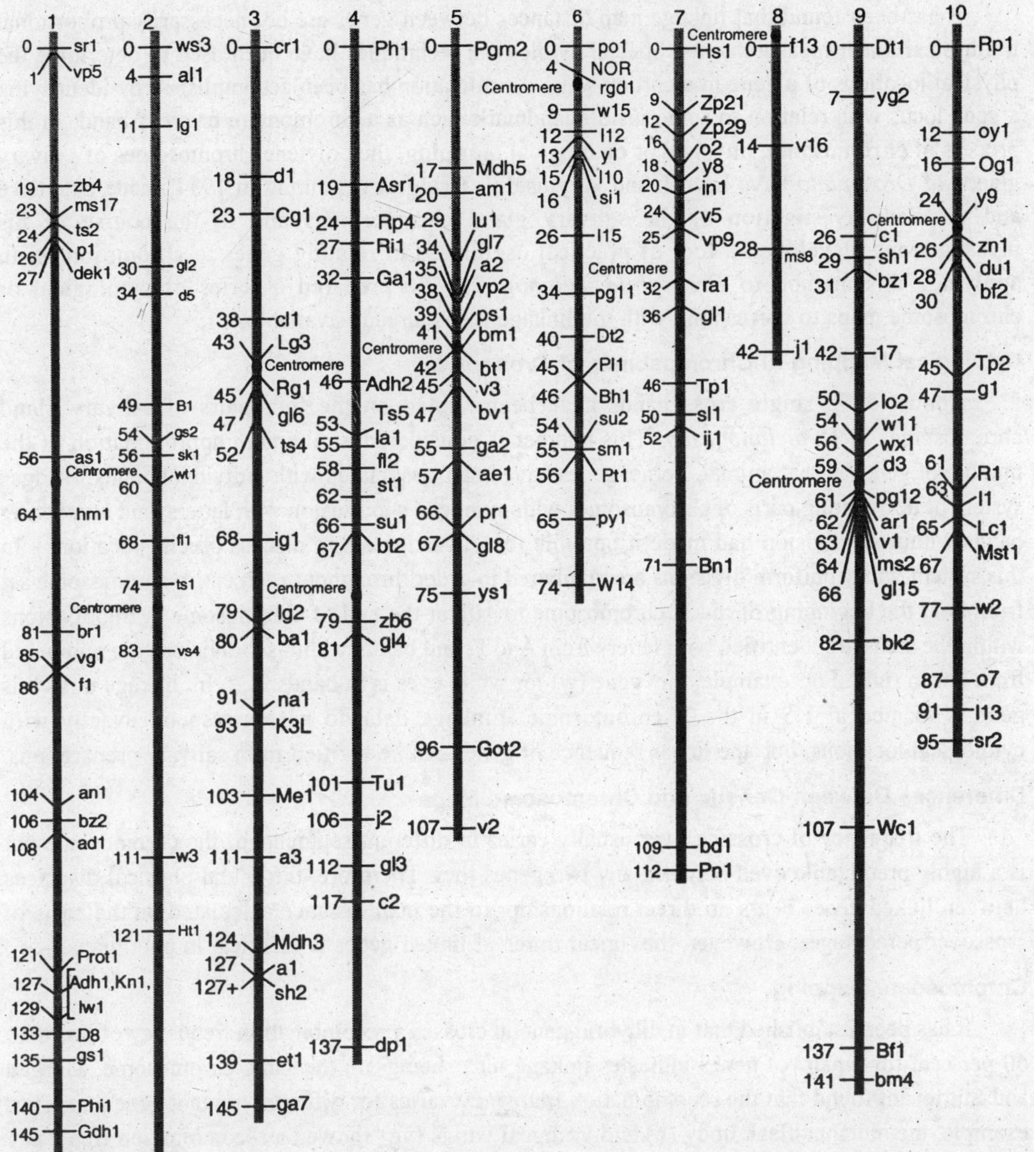

Fig. 37.11 Linkage Map of maize. The following are descriptions of phenotypes caused by mutant alleles of a few genes. Designations are in the form: chromosome no-map position, gene symbol, name of mutant allele mutant phenotype. 1-0 sr1, striate-1, white striations on leaves;; 1-56, as1, asynaptic-1 failure of the meiotic prophase chromosomes to synapse; *2-30, g12,* glossy-2, cuticle wax altered such that leaf surface appears glossy; 2-121, *Ht1, Helminthosporium turcicum* resistance, resistance to the northern leaf blight caused by this fungus; *3-0, cr1,* crinkly-1, leaves broad and crinkled; *3-139, et1,* etched-1, pitted or scarred endosperm; *4-24, Rp4, rust resistance-4, resistance to the fungus Puccinia sorghi;* 4-101, *Tu1,* tunicate-1, kernels enclosed in long glumes; *5* -20, *am1, ameiotic-1, meiosis fails; 6*-3 NOR, *nucleolus* organizer tandem repeats of genes coding for rRNA; 6-74, *w14,* white-14, white seedling; 7-16, *o2,* opaque-2, endosperm starch soft and opaque, has high lysine content; 8-28, *ms8,* male sterile-8, anthers not exserted; 9-26, c1, *colorless-1,* nonpigmented aleurone; 9-56, *wx1,* waxy-1, no amylose in endosperm or pollen; 10-16, *Og1, old gold1,* old gold-1, variable bright yellow stripes on leaf blades.

B. CHROMOSOME, PHYSICAL OR CYTOLOGICAL MAPPING

It has been found that linkage map distances between genes are not necessarily proportional to physical linear measurements. Special cytological techniques have been used to determine the physical locations of a gene in a chromosome. Localization has been accomplished by identifying a gene locus with relation to some visible landmark such as a chromomere or cross band. In this process of **chromosome, physical** or **cytological mapping,** the polytene chromosomes of salivary glands of *Drosophila* have been found very useful. **Bridges,** beginning in 1934, made extensive and detailed investigation of the salivary gland chromosomes and in the course of his investigations, developed a tool of practical usefulness in relating genes to chromosomes. In applying this method to *Drosophila melanogaster,* he prepared a series of cytological or chromosome maps to correspond with the linkage maps already available.

Cytological Mapping of Chromosomes of *Drosophila*

About 5000 single cross bands have been noted on the four pairs of salivary gland chromosomes in *D. melanogaster.* This number is considered a minimum approximation of the number of genes in that animal. Some genes have been associated with individual bands. Bridges system of designating parts of chromosomes with numbers, subdivision with letters, and numbering bands within subdivision had made it possible for investigators to discuss precise locations . In this system, fairly uniform divisions are numbered in order throughout each entire chromosome set from O at the beginning of the X chromosome to 102 at the end of chromosome 4. Subdivisions within the areas are identified with letters from A to F, and bands within subdivisions are numbered from left to right. For example, the gene (w) for white eyes is in bands 3C2. In linkage units this gene is located at 1.5 in the X chromosome. Linkage data do not correspond exactly with cytological locations, but the linear sequence of genes can be verified from salivary preparations.

Differences Between Genetic and Chromosome Maps

The frequency of crossing over usually varies in different segments of the chromosome but is a highly predictable even between any two genes loci. Therefore, the actual physical distances between linked genes bears no direct relationship to the map distance calculated on the basis of crossover percentages. However, the linear order of linked genes is identical in both maps.

Chromosome Mapping

It has been established that in dihybrid genetic crosses a recombination frequency of less than 50 per cent for a pair of genes indicates linkage, *i.e.,* being on the same chromosome, Morgan and Sturtevant found that the recombination frequency varies for different mutant gene pairs. For example, the mutants black body (b) and vestigeal wings (vg) showed a recombination frequency of 18 per cent while black body (b) and cinnabar eye (cn) showed 9 per cent. Thus, cross-over between b and vg are twice as frequent as those between b and cn. The difference in the recombination frequencies was found to be associated with the distance between the genes. The genes located farther apart in a chromosome undergo crossing-over more frequently than the genes laying nearby. This shows that the distance between b and vg should be about twice as that between b and cn. Hence, the recombination frequencies can help in the determination of the positions called **loci,** of the genes in the chromosomes and in preparing chromosome maps.

Thus, a **chromosome map** is a graphic representation of a linkage group in the form of a line which shows by points the sequence of the genes and the relative distances between the genes it contains. A chromosome map is also called a **linkage map** or **genetic map.** The method of preparing chromosome map of a species is known as **chromosome mapping.**

Chromosome mapping is based on two undermentioned genetic principles. They are –

1. That the genes are arranged in a linear order in the chromosome.

2. That the frequency of crossing over between two genes is directly proportional to the distance between them in the chromosome. Since the frequency of crossing over is used in preparing chromosome maps, the latter are also called **cross over maps.**

Chromosome map unit. A frequency of one per cent crossing over between two genes is known as one unit of map distance between these genes and it is termed a **morgan** after the name of T. H. Morgan.

How to prepare chromosome maps. Certain examples are quoted here to explain the technique of preparing a chromosome map. Suppose the five genes A, B, C, D and E are located in a single chromosome. If the crossing over percentage between A and B is 10 morgans, these genes are plotted 10 arbitrary units apart in a linear scale (Fig. 37.12 A).

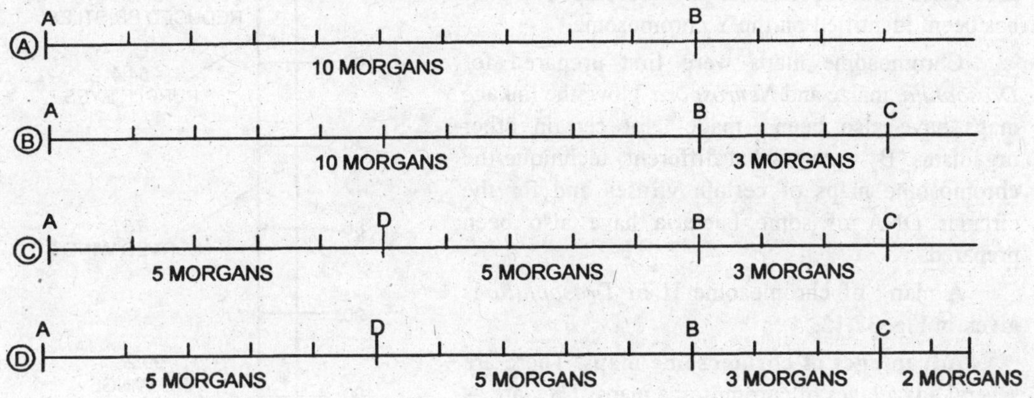

Fig. 37. 12. The technique of plotting the gene loci in a chromosome.

If the crossing over percentage between B and C is three, the crossing over percentage between A and C will determine the position of C, since it is three units from B towards A or away from A. It is assumed that A-C cross over rate is 13 per cent, and therefore, C will be plotted 3 morgans on the right side of B (Fig. 37. 12 B).

As per above mentioned examples, the location of the genes D and E can be plotted. Let A-D cross over rate be 5 per cent and B-D cross over rate also 5 per cent. D will then be plotted between A and B, *i.e.*, 5 morgans from each (Fig. 37. 12 C). If cross over rate between C and E happens to be two per cent and between B and E 5 per cent. E will be plotted two morgans on the right side of C (Fig. 37. 12D).

Fig. 37.13 (*a*). Morgan and Sturtevant. Mapping genes.

Morgan and Sturtevant got recombination frequencies of (*i*) 18 between the genes B for black body and for vestigial wings, (*ii*) 9 between the genes b for black body and cn cinnabar eye and (*iii*) 9.5 between the genes cn for cinnabar eye and vg for vestigial wings. If the gene b is plotted at one end of the chromosome, the gene vg will be 18 morgans away from it (Fig. 37. 13 (a). The gene *cn* being 9 morgans from *b* and 9.5 morgans from *vg* will be plotted approximately in the middle between the genes *b* and *vg*. Thus, the sequence of genes on the chromosome is *b, cn* and *vg*.

The recombination frequency of 9.5 was obtained between *cn* and *vg*. This revealed the sequence to be *b* – *cn* – *vg* with a total distance of 9 + 9.5 = 18.5 map units which is close to the observed 18 map units for *b* – *vg*. If it were *cn* – *b* – *vg, c* – *vg*, distance should be 9 + 18 = 27 map units apart. Using this method the relative distances and position of hundreds of mutant genes of *Drosophila* have been mapped.

All these mutants cluster together to form four groups of linked genes called **linkage groups.** *Drosophila* possesses four types of chromosomes, *i.e.*, I, II, III and IV. Each linkage group remains associated with a particular chromosome. No gene has been identified on the Y chromosome.

Chromosome maps were first prepared for *Drosophila,* maize and *Neurospora.* Now the linkage maps have also been made for certain other organisms. By a somewhat different technique the chromosome maps of certain viruses and for the circular DNA of some bacteria have also been prepared.

A map of chromosome II of *Drosophila* is given in Fig. 37.13.

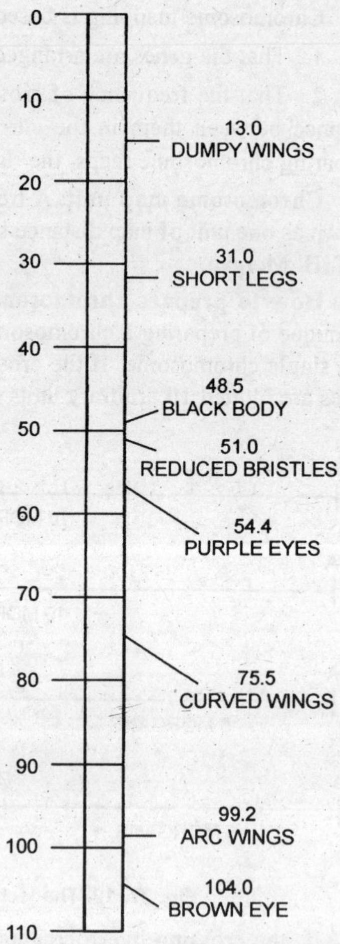

Fig. 37.13. A linkage map of chromosome II of *Drosophila*.

Advantages of chromosome maps. There are several advantages of chromosome maps. They are as follows:

1. Such maps give exact location of genes in a chromosome.
2. They show that which genes lie in which chromosome.
3. They show the strength of linkage between two genes and chances of crossing over between them.
4. They help in predicting the results of crosses.
5. According to these, the genes have linear arrangement in chromosomes; they have definite positions, the **loci,** in the chromosomes, they indicate specific genes in specific chromosomes.
6. They also show which gene/genes are missing.
7. In recent years, the most detailed genetic maps have been generated by determining the nuculeotide sequence of the DNA that makes up each chromosome.
8. Such mapping is not only important in genetic engineering, but is also of fundamental importance in understanding the overall architecture of the chromosome and in fact of the entire genome.

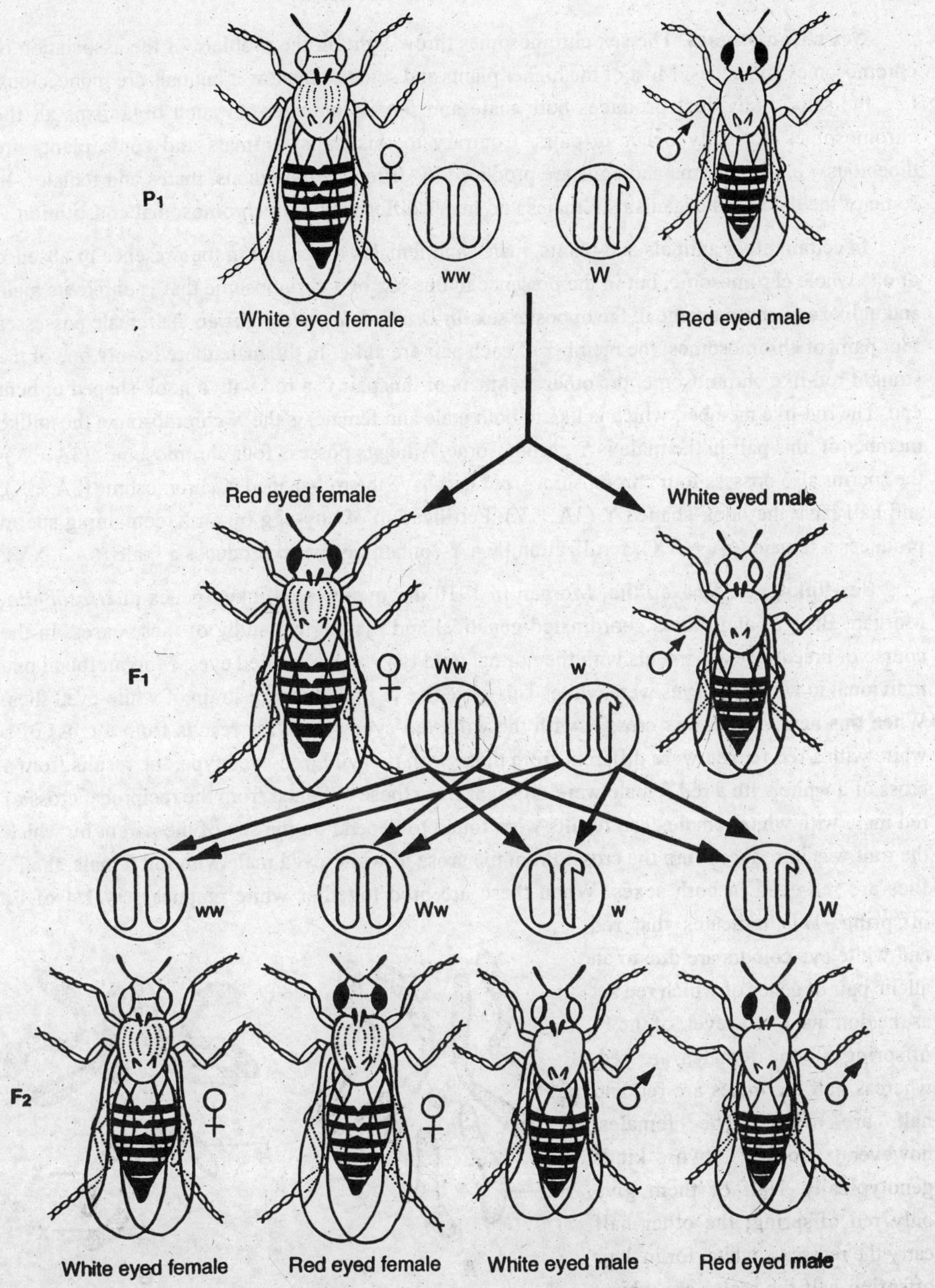

Fig. 37.14. A cross between red-eyed female and white-eyed male showing sex-linked inheritance in *Drosophila*.

SEX LINKED INHERITANCE

Sex chromosomes. The sex chromosomes throw light on the problem of the association of chromosomes and genes. Most of the higher plants and some of the lower animals are monoecious, *i.e.*, the same individual produces both male and female gametes. In such organisms all the chromosomes normally occur in pairs. Contrary to this, most animals and some plants are dioecious, *i.e.*, the sperms and eggs are produced by different individuals, males and females. In some of the dioecious organisms females and males differ visibly in chromosomal constitution.

In certain other animals and plants, males and females differ, not in the presence or absence of one whole chromosome, but in the presence in one sex of a chromosome that is unlike its male and unlike any chromosome in the opposite sex. In *Drosophila melanogaster,* the female possesses four pairs of chromosomes, the member of each pair are alike. In the male there is only one of the straight rod-like chromosome, the other members of this pair is a rod with a hook-shaped or bent end. The rod-like member, which is like in both male and female, is the X chromosome; the unlike member of this pair in the male is Y chromosome. All eggs possess four chromosomes (3A + X); the sperms also possess four chromosomes, but half have the rod-shaped X chromosome (3A + X), and half have the hook-shaped Y (3A + Y). Fertilization of any egg by an X-containing sperm produces a female (6A + XX); fertilization by a Y-containing sperm produces a male (6A + XY).

Sex-linkage in Drosophila. Morgan in 1910 discovered sex-linked genes in *Drosophila*. Morgan, Bridges and others coordinated genetical and cytological study of these cases. In the course of breeding experiments with the normal wild type which has red eyes, Morgan found one individual in which the eyes were white. This gave rise to true breeding strain of white-eyed flies. When this new variety was crossed with the wild, red-eyed typed, the results from a cross of a white with a red female were different from those with the wild, red-eyed type, the results from a cross of a white with a red female were different from those obtained from the reciprocal cross of red male with white female. The results were found to depend on the sex of the parent by which the trait was introduced into the cross. From the cross of white-eyed male with red female, the F_1 flies are red-eyed in both sexes. When these are bred together white reappears in 1/4 of F_2 offspring. This indicates that red and white eye colours are due to an allelic pair of genes of which red act as the dominant. However, of the F_2 offspring all the females are red, whereas half the males are red and half are white. The females, however, are of two kinds genotypically. Half of them give only red offspring; the other half carry the recessive white, for in their offspring half the males are white-eyed.

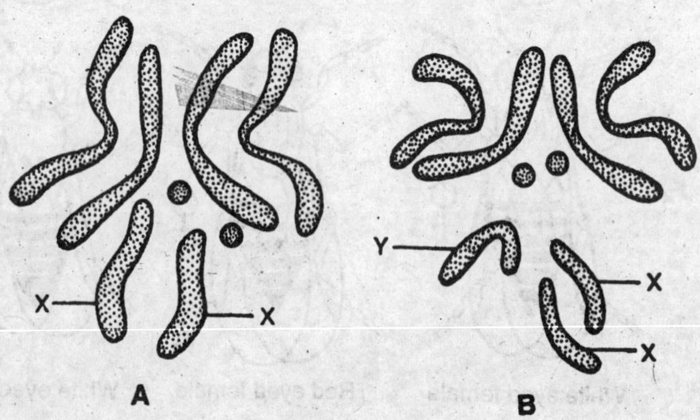

Fig. 37.15. *Drosophila melanogaster*. A, normal *XX* (female) chromosome; B, XXY chromosome in rare white-eyed phenotypically female.

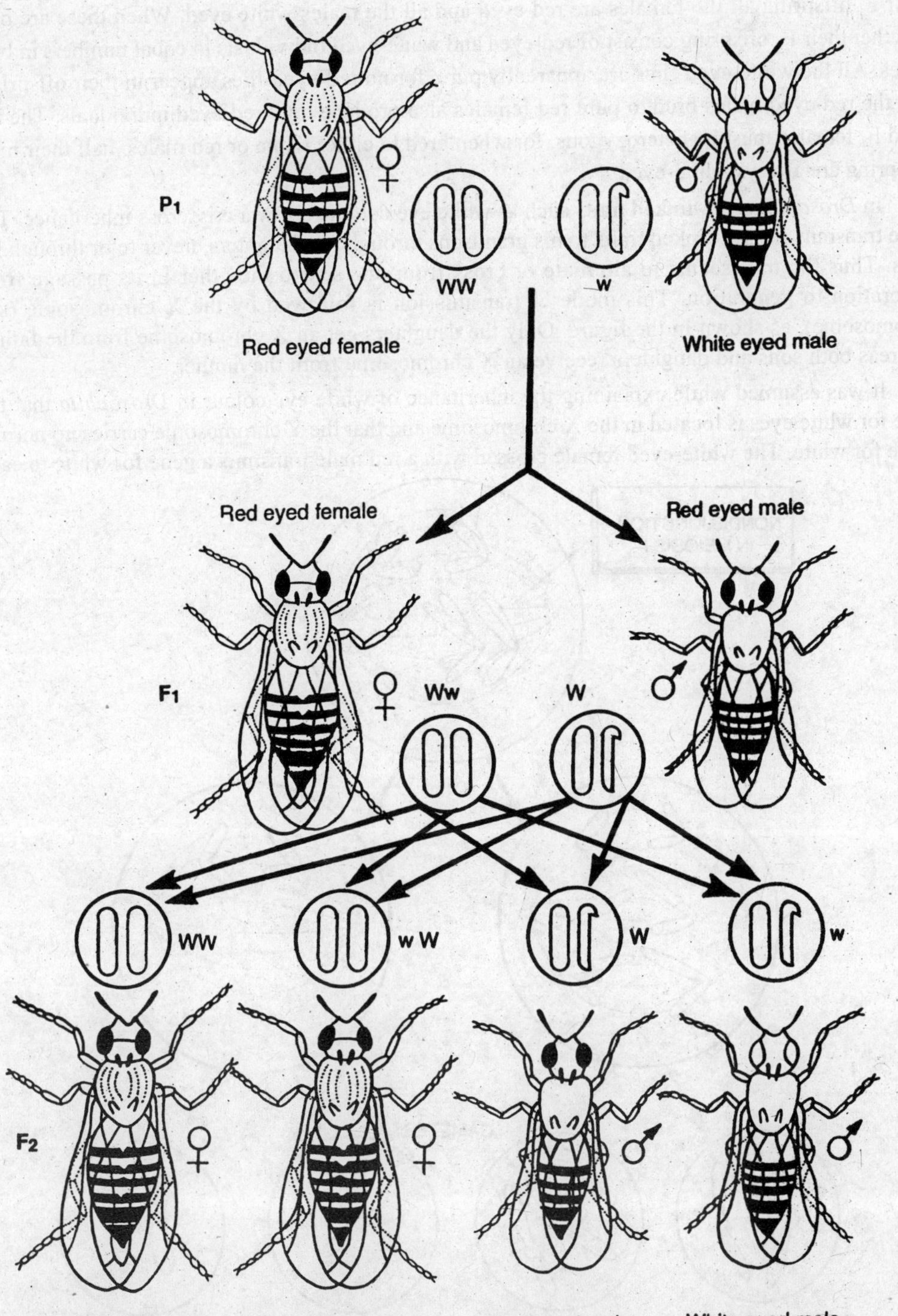

Fig. 37.16. A cross between white eyed-female and white-eyed male snowing sex-linked inheritance in *Drosophila.*

When a red male bred to a white female, however, quite a different result is obtained. Among their F_1 offspring all the females are red-eyed and all the males white eyed. When these are bred together their F_2 offspring consist of red-eyed and white-eyed individuals in equal numbers in both sexes. All the white-eyed flies are apparently pure, for no red-eyed flies appear in their offspring; and the red-eyed males bred to pure red females also produce only red-eyed individuals. The red eyed F_1 females must be heterozygous, for when bred to either white or red males, half their male offspring are always white-eyed.

In *Drosophila,* sex-linked traits such as white eye colour follow a crisscross inheritance. The male transmits his sex-linked traits to his grandsons through his daughters, never to or through his sons. Thus the trait seems to alternate or cross from one sex to the other in its passage from generation to generation. This mode of transmission is followed by the X chromosome (sex chromosome), as shown in the figure. Only the daughters get an X chromosome from the father, whereas both sons and daughters receive an X chromosome from the mother.

It was assumed while explaining the inheritance of white eye colour in *Drosophila* that the gene for white eyes is located in the X chromosome and that the Y chromosome carries no normal allele for white. The white-eyed female crossed with a red male transmits a gene for white to each

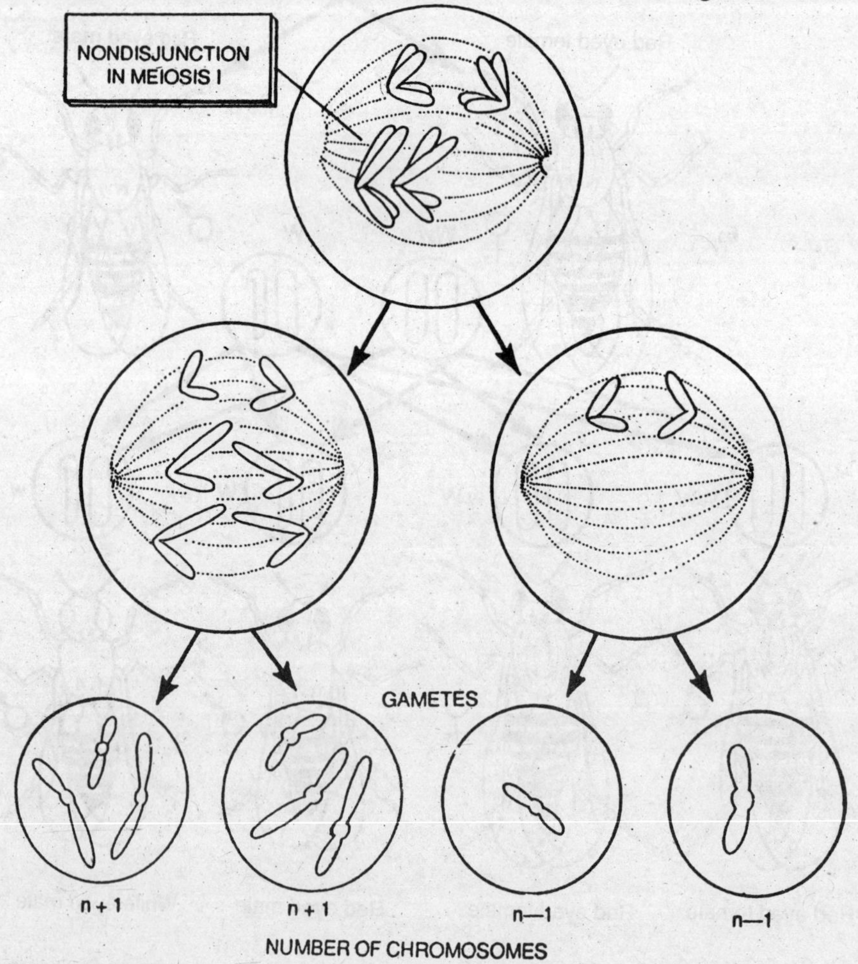

Fig. 37.17. Non-disjunction of a pair of homologous chromosomes during meiosis I produces abnormal gametes with (n + 1) and (n – 1) chromosomes in the gametes.

offspring and an X chromosome to each offspring. The daughters receive also an X chromosome from the father carrying the dominant allele of white and therefore, are red-eyed. The sons, however, receive a Y chromosome from the father and hence no allele of white and they are thus white eyed. The gene for white follows exactly the transmission of the X chromosome. According to Morgan this is actually carried in the X chromosome.

The evidence that the white-eye gene allele is indeed located on the X chromosome was obtained by Bridges, an associate of Morgan. In crosses of white-eyed female and red-eyed male, an exceptional white-eyed female was sometimes found among the otherwise all red-eyed female offspring. Bridges proved by genetic and cytological experiments that such white-eyed females had an unusual XXY rather than the normal XX sex-chromosome complement in its cells (Fig. 37.16)

The XXY condition arises as follows : Sometimes homologous chromosomes do not separate during meiosis. This phenomenon is termed **non-disjunction**, as shown in figure (Fig. 37.17). Non-disjunction of X chromosomes in the white-eyed female parent results in an egg with two XX chromosomes having only the white eye alleles. When such an XX egg is fertilized, by a Y chromosome containing sperm which has no allele for eye colour, the resulting offspring is a white-eyed XXY individual that shows in *Drosophila* a female phenotype.

A mutant trait located on the X chromosome exhibits the same pattern of inheritance as the white-eye allele in *Drosophila*. Fathers pass the X-linked allele trait to their daughters but not to their sons. Mothers pass the allele of the trait to both sons and daughters. If the trait is recessive, a female has to be homozygous to express it, and she should receive the mutant allele from both her parents. On the other hand, any male receiving the mutant allele from his mother will express the trait as he has only one X chromosome.

BASIS OF SEX DETERMINATION

Sex Chromosomes and Autosomes

In diploid ($2n$) organisms, where the sexes are separate, a particular chromosome or a pair of chromosomes are concerned with sex determination. They are designated as **sex chromosomes.** The other chromosome pairs are called **autosomes.** For example, *Drosophila* (fruit fly) has four pairs of chromosomes. Three pairs are identical in both the sexes while the fourth pair differs. The pair that differs is known as sex chromosome pair, while the rest are called autosomes. Also in humans and other mammals the similar situation is found. Here females have two X chromosomes (*i.e.*, XX) whereas males have one X and a morphologically distinct Y chromosome (*i.e.*, XY). Y chromosome, though different in size and shape, pairs with X during meiosis. Thus females are XX and the males are XY.

Since males produce two types of gametes either X or O in XO type, and X or Y in XY type.

They are known as **heterogametic** whereas females are **homogametic**, and produce only one type of gamete with an X chromosome.

Table. 37.4. Sex Chromosomes in Different Organisms.

	Genotype of female	Genotype of male
Insects (*Drosophila*)	XX	XO
Round worms	XX	XO
Humans and other mammals	XX	XY

The females having XX genotype are termed as **homogametic**, and they produce only one type of gametes. All gametes contain X chromosome.

The males having XY genotype are termed as **heterogametic**, and they produce two kinds of gametes. 50 per cent of gametes contain X chromosomes and 50 per cent Y chromosomes.

The chromosomal determination of sex in XY organisms is as follows. Here the sex is determined at the time of fertilization. Since the two sexes are produced in equal numbers, there is equal probability for an egg to be fertilized by a sperm having an X or a Y chromosome.

Parent :	Female ×	Male
	XX	XY
Gametes :	Egg	Sperm
	X	X or Y
Offspring :	Female	Male
	XX	XY
Probability	50 per cent	50 per cent.

In birds and some reptiles, females are heterogametic while males are homogametic. To avoid confusion with the male heterogametic X-Y system, this is referred to as W-Z system. Here males are WW and females WZ respectively.

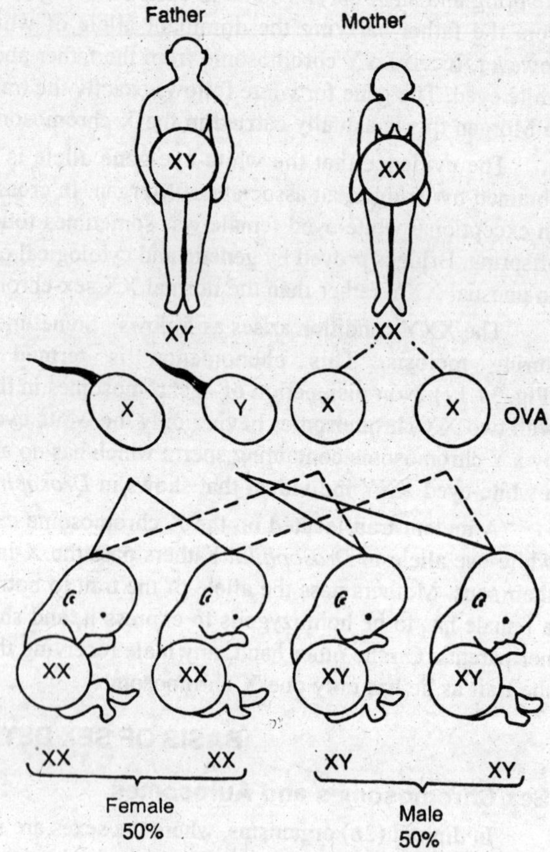

Fig. 37.18. Sex determination in man. Here all the ova carry X chromosome, but one-half of the sperm carry an X chromosome, and one-half carry a Y chromosome.

Sex determination in XX-XY organism. The sex determination in the offspring is being done by the chance union of an egg and a sperm. All eggs have a single X chromosome, whereas half of the sperms contain the X chromosome, and half the Y chromosome. The results are as follows :

(*i*) If the sperm containing X chromosome fertilises with the egg, the offspring is female.

(*ii*) If the sperm containing Y chromosome fertilises with the egg, the offspring is male.

Morgan and *Drosophila*

Morgan (1910) began his genetical work with the fruit fly, *Drosophila melanogaster.* Thomas H. Morgan was an American Zoologist, and he was awarded with Nobel Prize in 1933 (in medicine or physiology) for his research in genetics. T. H. Morgan's study to work on *Drosophila* proved most rewarding for investigations in genetics. The fruit fly, *Drosophila melanogaster*, a small fly about 2 mm. in size seen hovering around over ripe fruits such as banana, papaya, mango, etc. The fly is attracted due to yeast cells that grow on the fruit surface.

Drosophila is a suitable material for the genetical research experiments due to the undermentioned reasons :

(*i*) The fruit flies breed quickly and prolifically.

(*ii*) A large number of flies are produced to each progeny. A pair of flies in a small milk bottle can produce in a single mating hundreds of progeny.

(*iii*) A new generation can be bred every two weeks.

(*iv*) Fly breeding could be done throughout the year inside laboratory with inexpensive material.

(*v*) Each cell of *Drosophila melanogaster* has four pairs of chromosomes, *i.e.*, I, II, III and IV. All different in size and easily distinguishable. Three pairs, *i.e.*, II, III and IV are autosomes and the fourth pair, *i.e.*, I, sex chromosomes; XX in females and XY in males. The Y chromosome is hook shaped.

Morgan and his associates with their breeding experiments on *Drosophila* concluded that genes are located on chromosomes in a linear fashion. Some of them lie close to each other and tendency to have linked together. They concluded that the genes are physically associated with chromosomes. Morgan's discovery of sex linked traits in *Drosophila* led to the identification of genes carried by each chromosome and the approximate location of each gene on a chromosome (chromosome maps).

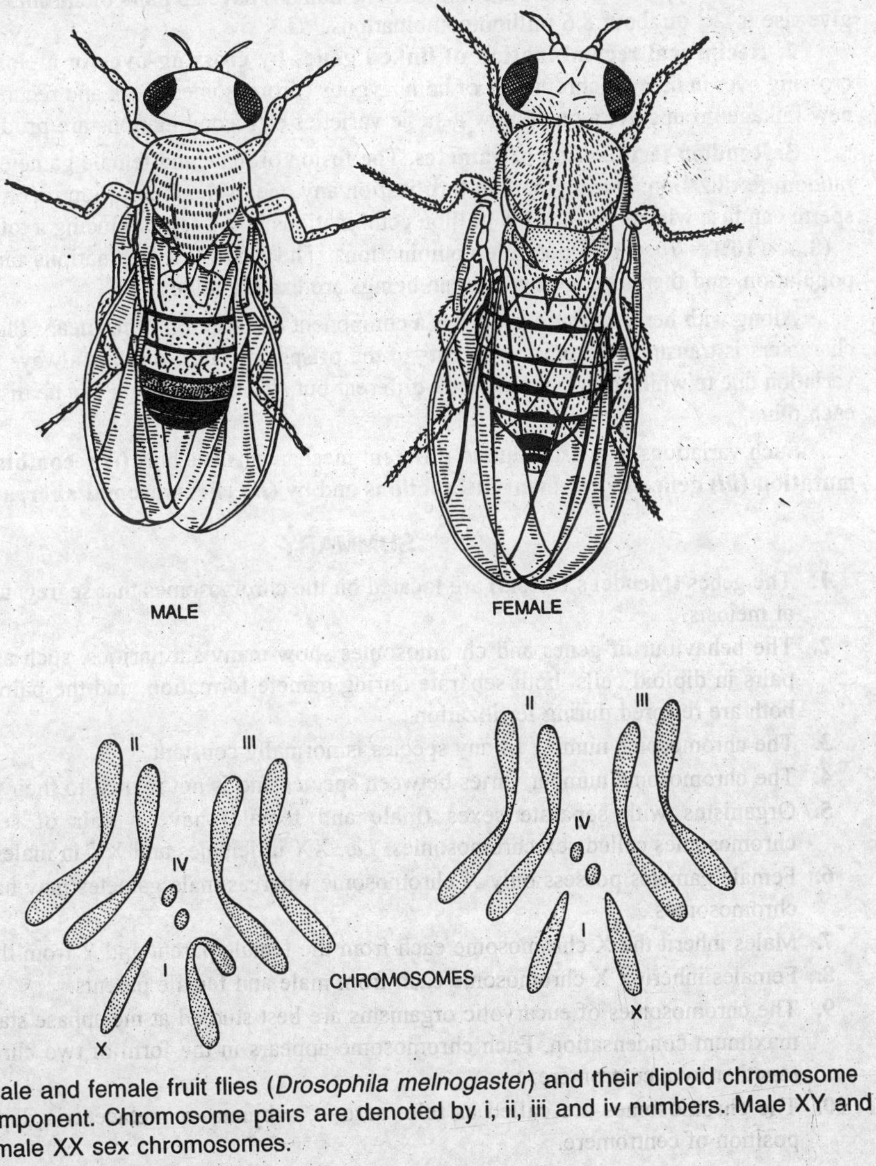

Fig. 37.19. Male and female fruit flies (*Drosophila melnogaster*) and their diploid chromosome component. Chromosome pairs are denoted by i, ii, iii and iv numbers. Male XY and female XX sex chromosomes.

GENETIC VARIATION

This is not always that offspring only resemble partners but they also vary from them. Of course, the resemblances are due to the units of information, the **genes,** they inherit. However, not all the inherited genes express themselves, *i.e.*, exhibit their phenotype. Some may be hidden or be recessive in the presence of dominant counter-parts, during expression. Genetic variation in sexually reproducing species develops primarily because parental genes are shuffled into new combinations, *i.e.*, new genotypes, in the offspring. The recombinations are produced by the routine reshuffling of genes as follows :

1. Independent assortment of chromosomes during meiosis. During the formation of gametes in the metaphase I of meiosis homologous chromosomes come together in pairs and then segregate into daughter cells independent of each other.

For example the maternal and paternal chromosomes of an organism with two pairs of chromosomes can recombine in $2^2 = 4$ possibilities in the gametes. With 3 pairs of chromosomes $2^3 = 8$ possible types of gametes are formed. The humans have 23 pairs of chromosomes which can give rise to 2^{23} or about 8.6 million combinations.

2. Reciprocal recombination of linked genes by crossing-over in meiosis. Here during crossing over in meiosis chromatids of homozygous chromosomes break and recombine to produce new linkage groups. As a result new genetic varieties or recombinations are produced.

3. Random fertilization of gametes. The fusion of male and female gametes takes place by random fertilization. During random fertilization any one of the 8.6 million recombinations in the sperm can fuse with any of the 8.6 million combinations in the egg producing a total of $(8.6 \times 10^6) \times (8.6 \times 10^6) = 70 \times 10^{12}$ possible combinations. The possible combinations far exceed human population, and therefore, no two human beings are exactly alike.

Along with heredity there is always a component of variation. This means that a whole set of characters is transmitted from the parents to the offspring. Thus, there is always a component of variation due to which they not only look different but also the offspring can be distinguished from each other.

Such variations may arise due to different mechanisms, such as ***(i)* recombination *(ii)* gene mutation *(iii)* gene environment interactions** and by ***(iv)* chromosomal aberrations.**

SUMMARY

1. The genes (Mendel's factors) are located on the chromosomes that segregate independently at meiosis.
2. The behaviour of genes and chromosomes show many similarities, such as both occur in pairs in diploid cells, both separate during gamete formation, and the paired condition of both are restored during fertilization.
3. The chromosome number of any species is normally constant.
4. The chromosome number varies between species and is not related to their complexity.
5. Organisms with separate sexes (male and female) have a pair of sex determining chromosomes called sex chromosomes, *i.e.*, XX in females and XY in males.
6. Female gametes possess only X chromosome whereas male gametes may have X or Y sex chromosomes.
7. Males inherit the X chromosome each from the female parent and Y from the male parent.
8. Females inherit a X chromosome each from male and female parents.
9. The chromosomes of eucaryotic organisms are best studied at metaphase stage due to their maximum condensation. Each chromosome appears in the form of two chromatids linked together by a centromere.
10. The chromosomes are called as telo, acro—, submeta or metacentric depending on the position of centromere.

11. Morgan carried out his experiments on the fruit fly (*Drosophila melanogaster*) and brought out the conclusions that eye colour is present on the X-chromosome, and follows a different pattern called sex-linked inheritance.
12. In all sex-linked or X-linked traits, males express them more often as they have only X chromosome.
    (*i*) The heterozygous female is normal but a carrier of the trait and can pass it on to both male and female offspring.
    (*ii*) For a female to have a disease, it has to be homozygous for the trait.
13. Non-disjunction at meiosis may give rise to abnormal individuals with monosomic (XO) or trisomic (XXY) condition giving rise to various disorders.
14. The genes located on the same chromosome are called linked genes and the phenomenon is known as 'linkage'. The linked genes move together and do not independently assort completely.
15. The linked genes show a recombination frequency of less than 50 per cent.
16. The linked genes can show recombination by crossing-over, *i.e.*, the physical change of parts of non-sister chromatids of homologous chromosomes just after synapsis at meiosis.
17. Even linked genes do not remain together due to crossing over. Cytologically crossing-over is observed as a chiasma.
18. The crossing-over occurs at the equivalent locus of the non-sister chromatids.
19. When the genes are closer, there are less changes of crossing-over that occurs between them. The crossing-over between two genes near each other is directly proportional to the distance between them.
20. Genes on chromosome can be mapped by determining the recombination frequency between pairs of genes and arranging them in the order of their relative distance.
21. A linkage map shows a linear arrangement of linked genes on a chromosome. Distances between genes can be expressed in map unit, where one map unit is defined as one per cent recombination.
22. All genes on a chromosome form a linkage group. The number of linkage groups is equivalent to the haploid number of chromosomes in an organism. For example, in *Drosophila* there are four linkage groups.
23. Genetic variations occur due to recombinations and mutations. Recombinations take place due to independent assortment of chromosomes, crossing-over of genes during meiosis and by random fertilization of gametes.
24. Gametic variation also takes place by mutation. Mutation is a heritable change in the structure of a gene or chromosome, or change in chromosome number.
25. Gene mutations alter the information conveyed by a gene. Chromosomal mutations may be deletions, duplications, inversion or tanslocation. Mutations are random spontaneous changes but can also be induced by mutagens like X-ray, UV-rays and several chemicals.
26. Sex-linked or X-linked genes are transmitted with a definite regular pattern. Fathers pass the X-linked trait to daughters but not to sons, while the mothers pass the trait to both sons and daughters.
27. Certain extranuclear genes are present in mitochondria and plastids, found in the cytoplasm and are inherited by the offspring through egg cytoplasm only. This type of cytoplasmic inheritance is also known as maternal inheritance and does not exhibit Mendelian inheritance pattern.

28. Haploid organisms have only allele of a gene, and so their genotype directly expresses the phenotype. Phenotypic ratio of haploid offspring after meiosis give a ratio of 1 : 1 instead of the 3: 1 ratio in diploid organisms.

TERMS TO REMEMBER

Acentric chromosome. A chromosome having no centromere.

Allopolyploid. A polyploid having chromosome sets from different sources.

Amphidiploid. Having two different diploid (2*n*) chromosome sets.

Aneuploid. Having non-balanced number of chromosomes, and not an exact multiple of haploid set.

Aneuploidy. A condition where the cells have either one less or one more chromosome than the usual normal number.

Ascus. A sac that contains meiotic ascospores.

Autopolyploid. A polyploid where all chromosomes come from the same species.

Autosomes. Chromosomes other than the sex chromosomes. They are similar in both males and females.

Centromere. The region of a chromosome to which spindle fibre attaches.

Chiasma. The X-shaped structure found in diplotene following a cross-over.

Chromosomal mutation. The mutation that involves a change in the number or structure of a chromosome.

Colchicine. A chemical that inhibits the mitotic division.

Cytoplasmic inheritance. Inheritance that takes place the genes present in cell organelles or the particles present in the cytoplasm.

Deletion. The absence of chromosome segment.

Dicentric. Having two centromeres.

Disjunction. Regular separation and movement of chromosomes to opposite poles during cell division.

Euploid. Having a balanced number of chromosomes, *i.e.*, an exact multiple of the haploid chromosome set.

Extranuclear genes. The genes that present outside the nucleus, *e.g.*, in mitochondria and plastids.

Gene mutation. Mutation that involves a minor change in a DNA sequence, such as deletion, substitution or insertion of a base.

Heterogametic. Organisms with XY genotype that produce two different kinds of gametes, half containing X and half containing Y chromosomes.

Homogametic. Organisms with XX genotype producing only one kind of gametes, *i.e.*, containing only X chromosome.

Isogenic. Having identical genotypes.

Linked gene. Genes present in the same chromosome.

Linkage group. Genes that are present on the same chromosome make one linkage group. They inherit together.

Monosomy. The occurrence of a single chromosome instead of a pair of homologous chromosomes in a diploid cell.

Mutagens. A chemical or some other agent that causes mutations.

Mutations. A permanent heritable change in the DNA molecule of an organism.

Non-disjunction. The chromosomes that fail to go to opposite poles during cell division and lead to unequal number of chromosomes in the daughter cells.

Polyploidy. The occurrence of more than two haploid sets of chromosomes in a cell.

Sex chromosomes. The X or Y chromosome, or the chromosomes that determine sex in a species.

Sex-linked genes. Genes occurring on the X chromosomes. They are inherited in sex-linked patterns, and also known as X-linked genes.

Transition. Mutation of a purine to another purine, or pyrimidine to another pyrimidine.

Transversion. Mutation of a purine to a pyrimidine or pyrimidine to a purine.

Trisomy. Having three representatives of chromosome instead of two homologous of a chromosome in the cells of a normal diploid organism.

QUESTION BANK

N.B. Undermentioned questions are asked repeatedly in various Indian Universities and competitive examinations.

Very Short Answer Questions

1. Who discovered the chromosomes ?
2. Who had proposed the chromosomes theory of inheritance ?
3. Name the scientist, who had proposed the mutation theory.
4. What is the least number of chromosomes a eukaryotic organism can have?
5. Who was the scientist to associate a specific gene with a specific chromosome ?
6. What is the number of chromosomes found in human cells ?
7. You know that the number of chromosomes is specific to a species. Give the number of chromosomes in *Drosophila melanogaster*?
8. Define autosomes.
9. What are sex chromosomes ?
10. How many sets of chromosomes does a diploid cell have ?
11. What do you understand by the term aneuploidy ?
12. When a cell has more than two cells of chromosomes what is it called ?
13. What is the characteristic shape of Y chromosomes of *Drosophila* ?
14. Which genes are known as linked ?
15. What is the tendency of genes located in the same chromosome to stay together in heredity called ?
16. What are the two types of linkages ?
17. Which scientists had shown that genes lie in chromosomes ?
18. How many homologous chromosomes are present in male *Drosophila* ?
19. The gametes of which parent in human determine the sex of the offspring during fertilization ?
20. What is a locus ?
21. What is known as pea of animal kingdom?
22. What is a karyotype ?
23. Who for the first time suggested a close relation between genes and chromosomes?
24. What is a linkage map ?
25. Name the process which is antagonistic to linkage.
26. What is crossing over ?

27. What is variation ?
28. What organisms are called recombinations?
29. What term Hugo de Vries used to describe phenotypic changes that are heritable ?
30. On which plant Hugo de Vries did his work ?
31. Why is gene mutation called point mutation?
32. What are mutagens ?
33. What is an exchange of segments between two non homologous chromosomes called?
34. Name two sex-linked diseases.
35. What are homologous chromosomes ?
36. What is the primary constriction of a chromosome ?
37. How many chromatids are involved in crossing over at one chiasma ?
38. How many linkage groups are found in man ?
39. Name the two main types of mutations.
40. Name three kinds of mutagens.
41. When was the foundation of human genetics laid ?

ANSWERS

1. Hofmeister; **2.** Sutton and Boveri; **3.** Hugo de Vries; **4.** Two; **5.** T. H. Morgan; **6.** 46; **7.** Four; **8.** The chromosomes other than the sex chromosomes; **9.** The pair of chromosomes associated with sex determination; **10.** Two; **11.** Not having an exact number of chromosomes in a multiple of haploid set, either one less or one more chromosome than the usual number; **12.** Polyploid; **13.** J-shaped; **14.** The genes present on the same chromosome; **15.** Linkage; **16.** Complete and incomplete; **17.** Morgan and Bridges; **18.** Three pairs; **19.** Male parent; **20.** Site of a gene in a chromosome; **21.** *Drosophila*; **22.** When chromosomes are systematically arranged and grouped according to length and position of the centromere, it is called karyotype; **23.** Beadle and Tatum.; **24.** Diagrammatic representation of relative distance between the genes in the chromosomes in linear fashion; **25.** Crossing over; **26.** Mutual exchange of blocks of homologous genes between a pair of homologous chromosomes; **27.** The process by which closely related organisms come to differ amongst themselves; **28.** The organisms which express characters of both the parents; **29.** Mutation; **30.** *Oenothera lamarckiana* (Evening primrose); **31.** Since a gene is located at a fixed point on the chromosome; **32.** The agents which may be physical, chemical or biological that bring about attraction in the chemical structure of genes; **33.** Translocation; **34.** Haemophilia and colour blindness; **35.** Chromosomes similar in form, size and traits; **36.** Where a kinetochore lies to which spindle fibres join in cell division; **37.** Two; **38.** 23; **39.** Chromosomal and genic; **40.** Radiation, temperature and chemicals; **41.** 1901.

Short Answer Questions

1. Why *Drosophila* considered suitable for genetic experiments ?
2. How is sex determined in *Drosophila* or human beings ?
3. What are the types of chromosomes present in *Drosophila* ?
4. Why is man unable to pass a sex linked gene to his son ?
5. Why is *Neurospora* selected as a material for genetic studies ?
6. What is aneuploidy and polyploidy ? Give an example of each.
7. What are lampbrush chromosomes ?
8. Briefly explain the structure of a metaphase chromosome ?
9. Explain the terms (*a*) crossing-over (*b*) linkage group ?
10. How does recombination occur in linked-genes ?
11. Why is karyotype done at metaphase stage ? How has karyotype analysis helped in solving some of the human genetic disorders ?
12. How does banding pattern help in studying human chromosomes ?
13. The diploid chromosome number of garden pea is 14. How many linkage groups does this represent ?

14. The map distance between genes A and B is three units, between B and C 10 units, and between C and A 7 units. What is the order of genes on a linkage map ?
15. How is sex determined in human beings?
16. A normal man marries a woman who is a carrier for colour blindness. What will be the phenotypes of the children born to them.
17. If marriage has taken place between a colour blind female and a normal male what will the offspring be ?
18. In humans the sex of the child is determined by the father and not by the mother. How ?

Long Answer Questions

1. What is linkage ? What is its relationship with independent assortment and crossing-over ?
2. Give a comprehensive account of the types of crossing-over.
3. Discuss the salient features of the mechanism of crossing-over.
4. Enumerate the modern approaches to human genetics.
5. Give an account of the types of mutation.
6. What are mutagens ? Give the nature of chemical and physical mutagens.
7. Give a comprehensive account of the role of mutations in evolution.
8. What are continuous and discontinuous variations ? Give details.
9. Give a brief and complete account of sex-linked inheritance.
10. Give a comprehensive account of sex-linkage of *Drosophila*.
11. Give a comprehensive account of cytoplasmic inheritance.
12. Describe the structure of eucaryotic chromosome.
13. Discuss the functions of the chromosome.
14. What is the significance of chromosome number ? Write down the chromosome number of four animals and four plants.
15. Point out the factors affecting the crossing-over.
16. What are linkage groups ?
17. Give the structure of chromosomes of man.
18. Write short notes on : (*a*) polyploidy in man, (*b*) Down syndrome, (*c*) Turner's syndrome, (*d*) chimera, (*e*) gynandromorphism in man.
19. What is mutation ? How this phenomenon was discovered ?
20. What is the role of mutation in agriculture and animal husbandry ?
21. How cytoplasm is held responsible for inheritance of characters ?
22. With the help of diagram only, give the morphological types of the chromosomes according to the position of centromere.
23. Give the diagram of a cross involving linkage and crossing-over in *Drosophila*.
24. With the help of diagram only, show the human chromosomes.
25. Give a cross between red-eyed female and white-eyed male showing sex-linked inheritance in *Drosophila*.
26. Give a cross between white-eyed female and red-eyed male showing sex-linked inheritance in *Drosophila*.
27. What is a chromosome map ? What is its basis ? How can it be prepared ?
28. Describe various kinds of gene mutations.
29. Discuss the modes of inducing mutations.
30. A heterozygous red-eyed *Drosophila melanogaster* female fly is crossed with a white-eyed male.
    (*a*) What type of cross is this ?
    (*b*) Represent the cross schematically.
    (*c*) Give the genotypes and phenotypes of the progeny.
    (*d*) What special type of inheritance does this trait show in this fly.

Multiple Choice Questions (Objective Type Questions)

1. The number of chromosomes in *Drosophila melanogaster* is
(*a*) six (*b*) eight
(*c*) ten (*d*) twelve.

2. *Homo sapiens* has
(*a*) 23 pairs of chromosomes (*b*) 24 pairs of chromosomes
(*c*) 25 pairs of chromosomes (*d*) 26 pairs of chromosomes

3. Sex chromosomes in human male are known as
(*a*) XX (*b*) XY
(*c*) XO (*d*) XZ

4. The prokaryotic chromosomes are found in
(*a*) *Bacillus* (*b*) *Nostoc*
(*c*) *Escherichia* (*d*) all of above.

5. Extrachromosomal hereditary determinants are known as
(*a*) plastids (*b*) ribosomes
(*c*) plasmids (*d*) none of above.

6. Eucaryotic chromosomes are found in
(*a*) cyanobacteria (*b*) blue-green algae
(*c*) fungi (*d*) none of above.

7. Chromosomes are best seen in
(*a*) interphase (*b*) prophase
(*c*) metaphase (*d*) telophase.

8. The components of the chromosome that controls heredity is
(*a*) proteins (*b*) DNA
(*c*) RNA (*d*) none of above.

9. A chromatid has
(*a*) one chromosome (*b*) two chromonemata
(*c*) four chromonemata (*d*) many chromonemata

10. Homologous chromosomes similar in both sexes are called
(*a*) androsomes (*b*) autosomes
(*c*) heterosomes (*d*) none of above

11. The term mutation was introduced by
(*a*) Mendel (*b*) Hugo de Vries
(*c*) Darwin (*d*) Lamarck.

12. Point mutations are caused by
(*a*) substitution (*b*) deletion
(*c*) insertion (*d*) all of above.

13. Mutation can be induced by
(*a*) radiation (*b*) temperature
(*c*) chemicals (*d*) all of above.

14. Crossing over during meiosis occurs in
(*a*) leptotene (*b*) zygotene
(*c*) pachytene (*d*) diplotene.

Answers

1. (*b*) **2.** (*a*) **3.** (*b*) **4.** (*d*) **5.** (*c*) **6.** (*c*) **7.** (*c*) **8.** (*b*) **9.** (*a*) **10.** (*b*)
11. (*b*) **12.** (*d*) **13.** (*d*) **14.** (*d*)

38
chapter

Nature of Gene: Its Expression and Regulation

As we know, the children acquire the characteristic physical features of their parents or grand-parents. The inheritance of these hereditary characters can be explained on the basis of ancestral genes that are found to be arranged on chromosomes, and they make the total genomic make-up of the organism. The specific questions may be asked about these genes such as what are their functions and what is their chemical nature, and how the characters are transmitted from the parent to the progeny ?

Now this is an established fact that parents transmit the heritable characters to the offspring.

Such transmission forms the basis of maintaining continuity among the numbers of a species.

Factors or genes responsible for this purpose are located on the chromosomes and through the gametes enter a diploid zygote cell.

This process is found in all sexually reproducing organisms.

Thus each offspring receives 50 per cent component of chromosomes and the genes from male parent and 50 per cent from female parent.

On the basis of modern genetic knowledge a gene is expected to have following important characteristics :

1. A gene should be able to store and express information of the heritable characters of living organisms.

2. A gene should be capable to make its own replica so that a copy of it may be transmitted to the offspring.

3. A gene should have a mechanism so that it can undergo mutations which will generate the biological variation essential for evolution.

DNA (deoxyribonucleic acid) and RNA (ribonucleic acid) are the principal genetic materials of living organisms. They are chemically called **nucleic acids** and are complex molecules larger than most proteins and contain carbon, oxygen, hydrogen, nitrogen and phosphorus. The nucleic acids were first isolated by **F. Miescher** in 1868 from the nuclei of pus cells and gained their name from the fact that they are acidic and were first identified in nuclei.

Genetic Material and its Nature

Highly purified DNA extracted from a wide variety of plants, animals, bacteria and viruses has been found to be complex macromolecular or polymeric chemical compound which contains, four kinds of similar building blocks (monomers) called **deoxyribotids or deoxyribonucleotides.** Each deoxyribonucleotide is made up of three molecules : **a phosphoric acid molecule** (biologically called phosphate and discovered by **Levene** in 1910's); a pentose sugar called 2 **deoxyribose** (discovered by **Levene** in 1910's); and **pyrimidine** and **purine** nitrogenous bases (discovered by **Fischer** in 1880's). Four major kinds of nitrogenous bases have been found in four kinds of deoxyribonucleotides of DNA : two are heterocyclic and two-ringed purines, **adenine** (A) and **guanine** (G), and two one-ringed pyrimidines, **cytosine** (C) and **thymine** (T).

It is now known that the material in the genes responsible for heredity is capable of performing *three* main functions. They are as follows :

1. The genes should be able to replicate and be inherited by the progeny.

2. The genes should be susceptible to an occasional change by mutation and such a change should be inheritable.

3. The genes should be able to carry all the information necessary for the functions of a cell.

Formerly this was thought that proteins could make an ideal genetic material. As we know, proteins are made up of twenty amino acids and thus the diversity that could be generated on the basis of the number of residues of each amino acid in a given protein and the sequence of amino acids would be enormous. However, this has been proved on the basis of experimental evidence that deoxyribonucleic acid (DNA) is the genetic material.

Bacterial Transformation : Griffith's Experiment

The lines of inquiry that led to an understanding of the chemical nature of genetic material arose from a study of the pestilent organism *Diplococcus pneumoniae.* This bacterium causes pneumonia in mammals. In 1928, Frederick Griffith, found that there are two strains of *D. pneumoniae*, one that forms smooth colonies protected by a capsule, and the other one that formed irregular or rough colonies without a capsule when grown on a suitable medium in petri dishes. When injected into mice (A) only capsulated smooth cells (virulent) produced the disease, but not the non-virulent rough cells (B). On the other hand when the heat killed capsulated (virulent) smooth cells were mixed with non-virulent rough cells (D) and then were injected in the mice the disease was produced. This shows that some factor from the dead capsulated smooth cells, converted the living non-virulent rough cells into living smooth capsulated (virulent) cells.

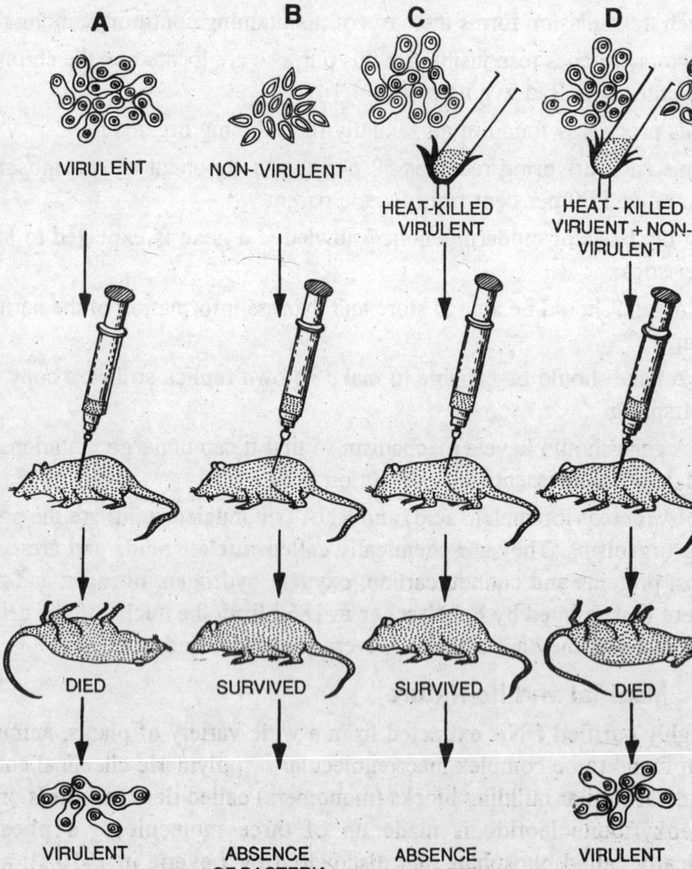

Fig. 38.1. Griffith's experiments with *Diplococcus pneumoniae* and mice.

In 1944, Avery, McCarty and Macleod supported the Griffith's experiment by molecular explanations. They found that the DNA isolated from the heat killed smooth (S) cells, when added to rough (R) cells changed their surface character from rough (R) to smooth (S), and also made them virulent. By this experiment, this was shown that DNA was the genetic material responsible for inducing the smooth (S) character of the cells and their property of virulence in mice. Their experiment proved that bacterial transformation involves transfer of a part of DNA from the dead bacterium (*i.e.*, donor) to the living bacterium (*i.e.*, recipient), that expresses the character of dead cell, and so is known as a **recombinant.**

The phenomenon by which the DNA isolated from one type of cell, when introduced into another type, is able to bestow some of the properties of former to the latter. This phenomenon is known as **transformation.**

Live S + mice ⟶ mice died

Live R + mice ⟶ mice survived

Heat killed S + mice ⟶ mice survived

Heat killed S + live R + mice ⟶ mice died.

Blood of the dead mice showed the presence of live S bacteria.

Another definitive evidence came from experiment on T_2 bacteriophage (virus). This is as follows :

Viral-infecting Agent is DNA : Hershey and Chase Experiment

A bacteriophage (T_2 virus) infects the bacterium *Escherichia coli.* After infection, the virus multiplies and the T_2 phages are released with the lysis of the bacterial cells. As we know, the T_2 phage contains both DNA and proteins. Now the question arises, which of the two components has the information to programme for the multiplication of more viral particles.

To solve this problem, Hershey and Chase (1952) devised an experiment with two different preparations of T_2 phage. In one preparation they made protein part radioactive and in the other preparation the DNA was made radioactive. Thereafter a culture of *E. coli* was made infected by these two phage preparations. Immediately after infection, and before lysis of bacteria the *E. coli* cells were gently agitated in a mixer so that the adhering phage particles were loosened and then the culture was centrifuged. With the result the heavier pellets of infected bacterial cells were settled down in the bottom of the tube. The lighter viral particles and those particles which did not enter the bacterial cells were found in the supernatant. It was found that when T_2 phage with radioactive DNA was used to infect *E. coli* in the experiment, the heavier bacterial pellet was also radioactive. On the other hand when T_2 phage with radioactive protein was used, the bacterial pellet had very little radioactivity and most of the radioactivity was found in supernatant. This experiment has proved, that during infection with T_2 phage, it is DNA that actually entered the bacteria when such infected bacterial cells were developed further they lysed and new phage particles were formed. This famous experiment conducted by Hershey and Chase (1952) has proved that it is the viral DNA and not protein that contains information for the production of more T_2 phage particles hence DNA is genetic material. However, in some viruses (*e.g.*, TMV, influenza virus and polio virus) RNA serves as genetic material.

Hershey and Chase conducted two experiments. In one experiment *E. coli* was grown in a medium containing the radio-isotope S^{35} and in the other experiments *E. coli* was grown in a medium containing the radio-isotope P^{32}. In these experiments *E. coli* cells were made infected with T_2 phage released from *E. coli* cells grown in S^{35} medium have S^{35} in their protein capsid, and those from P^{32} medium had P^{32} in their DNA. When these phages were used to infect new *E.coli* cells in normal medium, the bacterial cells which had infection by S^{35} labelled phages showed the radioactivity in their cell wall and not in cytoplasm. Whereas the bacteria infected with P^{32} labelled phages had shown the reverse condition. Thus it can be said that when T_2 phage infects the bacterial cell, its protein capsid remains outside of the bacterial cell but its DNA enters the cytoplasm of the bacterium. When the infected cells of bacteria gets lysed, new complete viral particles (T_2 phages) are formed. This proves, that viral DNA carries the information for synthesis of more copies of DNA and protein capsids. This shows that DNA is genetic material.

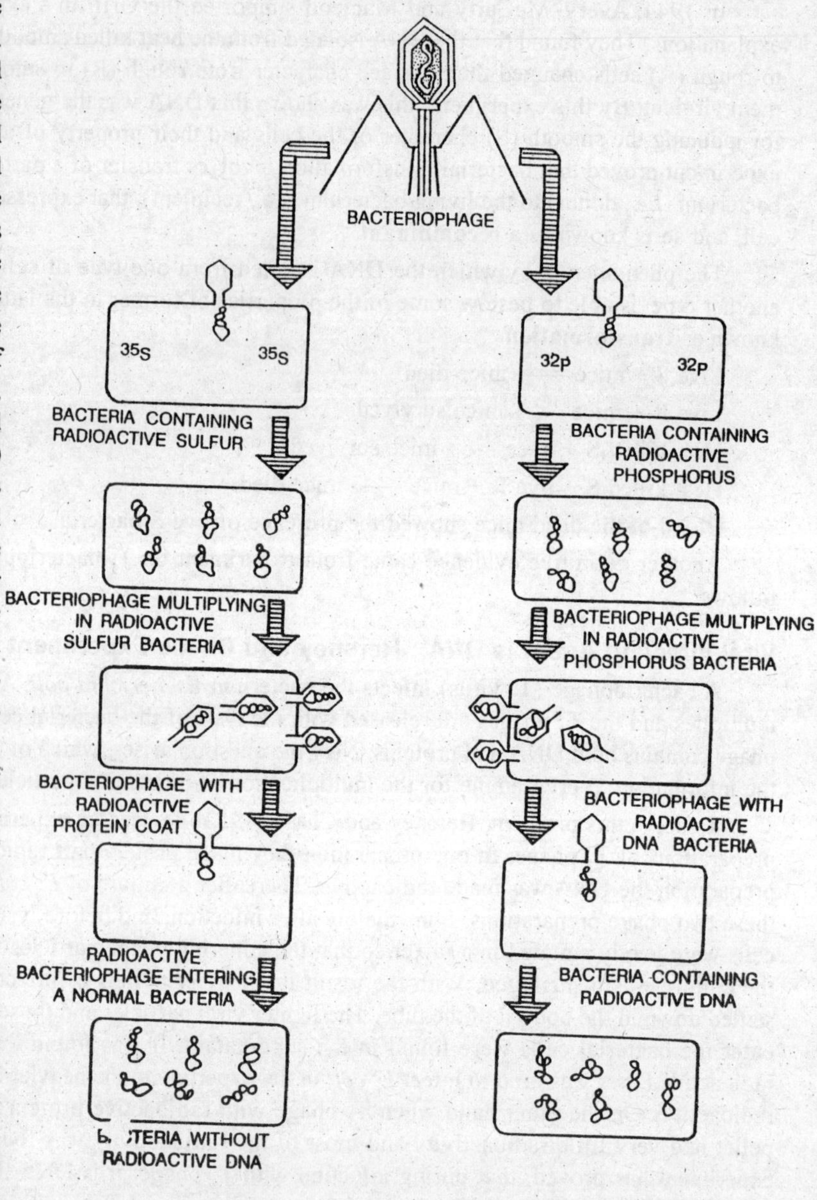

Fig. 38.2. Hershey and Chase's experiments show that DNA is hereditary material in bacteriophages.

That part of each nucleotide which contains a nitrogenous base and deoxyribose is called **deoxyribonucleoside**.

The four deoxyribonucleotides besides occurring in DNA molecule, occur also in nucleoplasm and cytoplasm, but in their triphosphate forms such as deoxyadenosine triphosphate (dATP), deoxyguanosine triphosphate (dGTP), deoxycytidine triphosphate (dCTP) and thymidine triphosphate (TTP). The significance of occurrence of deoxyribonucleotides in triphosphate forms lies in fact that during DNA polymerase enzyme can act only on triphosphates of deoxyribonucleotides.

This experiment shows that only the phage DNA enters the bacterial cell and not the protein coat. It depicts clearly that DNA alone is responsible for the synthesis of new viruses inside the bacterial cells. This shows that DNA is genetic material and not protein. In tobacco mosaic virus

(TMV) the inner core is made up of RNA instead of DNA, hence in this case RNA is genetic material.

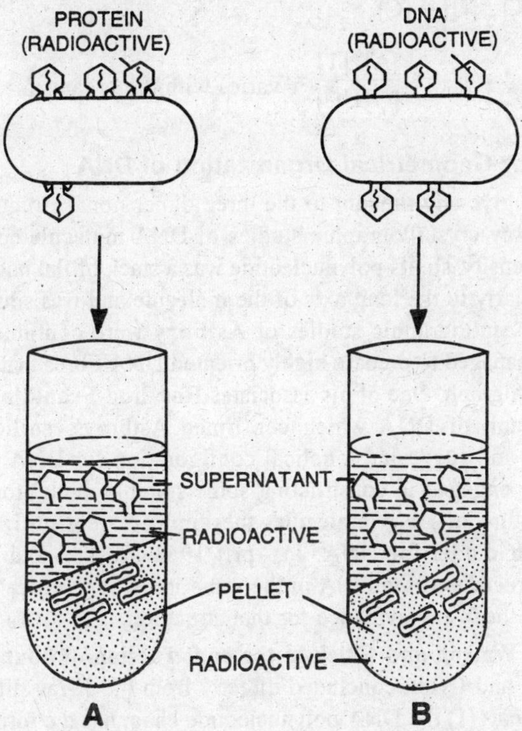

Fig. 38.3. The viral infecting agent is DNA. A, the viral protein coat is radioactive; B, viral DNA is radioactive. The infected bacterial pellet picks up radioactivity only when the viral DNA is radioactive.

DNA AND STRUCTURE OF DNA

Molar Ratios of Nitrogen Bases in DNA Molecule

When many samples of DNA were isolated, purified and analysed by various techniques, such as paper chromatography, etc., it was found by **Hotchkiss** and **Chargaff** in 1948 that contrary to **Levene's** (1920's) tetranucleotide theory which considered DNA as a monotonous polymer having four DNA bases in approximately equal molar proportions, the four nucleotide bases are not necessarily present in DNA in exactly equal proportions. Thus **Chargaff** reported that the DNA extracted from calf-thymus nuclei contains the four bases in following molar proportion : 28% adenine, 24% guanine, 20% cytosine and 28% thymine. When he analyzed the DNA samples of varied animals, he found that the exact base composition of DNA differs according to its biological source.

These different values of A, G, C, and T in different samples of DNA are suggesting that DNA rather than being a monotonous polymer, carries genetic information in the form of specific nucleotide base sequences.

Erwin Chargaff in 1949 proposed Chargaff's rules regarding the composition of bases in DNA are as follows :

(*i*) Total amount of purine nucleotides always equals the total amount of pyrimidine nucleotides, *i.e.*, [A] + [G] = [T] + [C] (*ii*) The proportion of A is equal to T and so also of G is equal to C, but amount of [A] + [T] is not necessarily equal to [G] + [C].

Therefore [A] = [T] : [G] = [C]

But $$\frac{[A]+[T]}{[G]+[C]} = \text{varies with the organism}$$

Physical, Molecular or Geometrical Organization of DNA

The first person to give any thought to the three dimensional structure of DNA was **W. T. Astbury** who by his X-ray crystallographic studies of DNA molecule concluded in 1940's that because DNA has high density, so, its polynucleotide was a stack of flat nucleotides, each of which was oriented perpendicularly to the long axis of the molecule and was situated every 3.4 Å along the stack. The X-ray crystallographic studies of **Astbury** were continued by **Wilkins** and his associates (1953), who managed to prepare highly oriented DNA fibres that allowed them to obtain an X-ray diffraction photograph. One of his associates **Rosalind Franklin** obtained a superior X-ray diffraction photograph of DNA which confirmed Astbury's earlier inference of 3.4 Å internucleotide distance and suggested a helical configuration for DNA molecule. **Watson** and **Crick** who were already engaged in constructing some suitable model for DNA structure, when observed Franklin's picture of DNA molecule, they immediately utilized that information in constructing a molecular model for DNA. In April 1953 **Watson** and **Crick** published their conclusions about the structure of the DNA in the same issue of **'Nature'** in which **Wilkins** and his colleagues presented the X-ray evidence for that structure.

Considerations of Watson and Crick in the construction of double helical structure of DNA molecule. Watson and **Crick** concluded directly from the X-ray diffraction photograph of DNA taken by Franklin that **(1)** the DNA polynucleotide chain has the form of a regular helix, **(2)** the helix has a diameter of about 20 Å and, **(3)** the helix makes one complete turn every 3.4 Å along its length, and hence, since the internucleotide distance is 3.4 Å, consists a stack of ten nucleotides per turn. Considering the known density of the DNA molecule, **Watson** and **Crick** next concluded that the helix must contain two polynucleotide chains, or two stacks of ten nucleotides each per turn, since the density of a cylinder 20 Å in diameter and 34 Å long would be too low if it contained but a single stack of ten, and too high if it contained three or more stacks of ten nucleotides each. Before trying to arrange these two polynucleotide chains into a regular helix of the required dimensions, however, **Watson** and **Crick** placed a further restriction on their model — a restriction that derived from their knowledge that DNA, is after all, the genetic material. If DNA is to contain heredity information so they reasoned, and if that information is inscribed as a specific sequence of the four bases along the polynucleotide chain, then the molecular structure of DNA must be able to accommodate any arbitrary sequence of bases along its polynucleotide chains. Otherwise, the capacity of DNA as an information carrier would be too severely limited. Hence, they felt the need of construction of such a regular helix that though composed of two polynucleotide chains containing an arbitrary sequence of nucleotide bases every 3.4 Å along their length, would nevertheless has a constant diameter of 20 Å. Since the dimension of the purine ring is greater than that of the pyrimidine ring **Watson** and **Crick** hit upon the idea that the two chain helix could have a constant diameter if there existed a complementary relation between the two nucleotide stacks, so that at every level one stack harbours a purine base and the other a pyrimidine base. Finally, to endow the helix with thermodynamic stability, the structure would have ample opportunities for the formation of hydrogen bonds between amino-or-hydroxyl hydrogen and keto-oxygens or immuno-nitrogens of the purines and pyrimidine bases. These considerations led them to construct a double helix model for the molecular structure of DNA molecule.

Watson and Crick's Structural Model of DNA

In DNA molecule the adjacent deoxyribonucleotides are joined in a chain by phosphodiester bridges or bonds which link the 5 carbon of the deoxyribose of one mononucleotide unit with the 3 carbon of the deoxyribose of the next mononucleotide unit. According to **Watson** and **Crick** DNA molecule consists of two such polynucleotide chains wrapped helically around each other, with the sugar-phosphate chain on the outside (forming ribbon like back bone of double helix) and purines and pyrimidines on the inside of the helix (projecting between two sugar-phosphate backbones as transverse bars). The two polynucleotide strands are held together by hydrogen bonds between specific pairs of purines and pyrimidines.

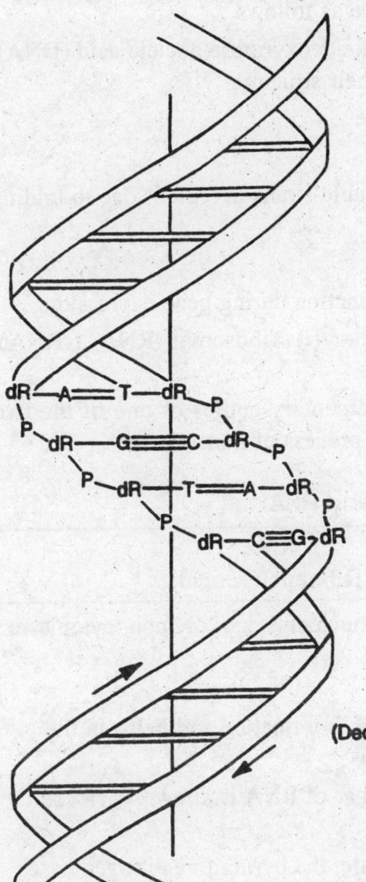

Fig. 38.4. Watson and Crick's double helical DNA molecule.

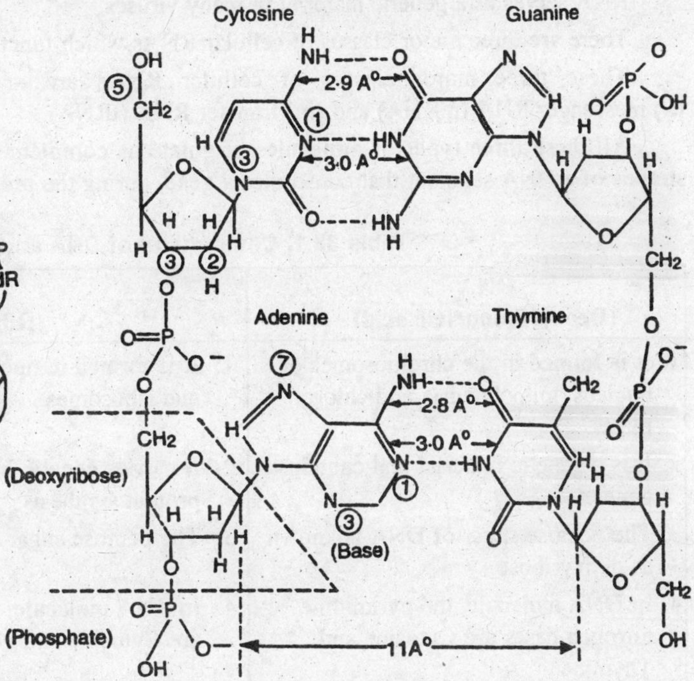

Fig. 38.5. Segment of a DNA molecule.

The hydrogen bonds between purines and pyrimidines are such that adenine can bond only to thymine by two hydrogen bonds, and guanine can bond only to cytosine by three hydrogen bonds and no other alternative is possible between them. The specificity of the kind of hydrogen bonds that can be formed assures that for every adenine in one chain there will be thymine in the other. For every guanine in first chain there will be a cytosine in the other and so on. Thus, the two chains are complementary to each other; that is sequence of necleotides in one chain dictates the sequence of nucleotides in the other. The two strands run antiparallely – that is, have opposite directions. One strand has phosphodiester linkage in $3' - 5'$ direction, while other strand has phosphodiester linkage in just reverse or $5' - 3'$ direction. Further, both polynucleotides strands remain separated by 20 Å distance. The coiling of double helix is right handed and a complete turn occurs every 34 Å. Since each nucleotide occupies 3.4 Å distance along the length of a polynucleotide strand, ten mononucleotides occur per complete turn.

The most crucial outcome of Watson and Crick's double helix model of DNA was the implicit suggestion for a mechanism by which the genetic material (DNA) can be copied and transmitted to progeny. This is known as **DNA replication**.

RNA AND STRUCTURE OF RNA

The characteristic features of ribonucleic acid (RNA) are as follows :

Ribonucleic acid (RNA) is another nucleic acid type. Like deoxyribose nucleic acid (DNA), this is a polynucleotide but several differences are found in their structure.

In RNA, the pentose sugar is ribose and not deoxyribose

RNA contains uracil in place of thymine.

Most RNAs are generally single stranded with partial double-stranded regions due to folding back of its single chain.

RNA serves as a genetic material in many viruses.

There are three major classes of cellular RNAs which function during gene expression.

These three major classes of cellular RNAs are – ***(i)*** ribosomal RNA (rRNA), ***(ii)*** messenger RNA (mRNA) and ***(iii)*** transfer RNA (tRNA).

All these three types of molecules originate as complementary copies of one of the two strands of a DNA segment that constitutes a gene, during the process of transcription.

Table 38.1. Comparison of DNA and RNA

DNA (Deoxyribonucleic acid)	**RNA (Ribonucleic acid)**
1. It is formed in the chromosomes of nucleus, mitochondria and chloroplasts.	1. It is formed in nucleolus, karyolymph, cytoplasm and ribosomes.
2. It is a genetic material and contains genes.	2. It carries genetic information and helps in the protein synthesis.
3. The pentose sugar of DNA is known as deoxyribose.	3. The pentose sugar of RNA is known as ribose.
4. In DNA molecule, the pyrimidine nitrogen bases are Cytosine and Thymine.	4. In RNA molecule, the pyrimidine nitrogen bases are Cytosine and Uracil.
5. The nucleotides which make a molecule of DNA are (*i*) deoxyadenosine monophosphate (*ii*) deoxyguanosine monophosphate, (*iii*) deoxycytidine monophosphate and (*iv*) deoxythymidine monophosphate.	5. The nucleotides which make a molecule of RNA are (*i*) adenosine monophosphate, (*ii*) guanosine monophosphate, (*iii*) cytidine monophosphate and (*iv*) Uridine monophosphate.
6. The paired nucleotides of a DNA molecule are arranged in a double stranded helix.	6. The nucleotides of a RNA molecule are arranged in a single chain.
7. Three main types of DNA are – nuclear (n DNA), mitochondrial (m DNA). and chloroplast (c DNA).	7. Three main types of RNA are – messenger (m RNA), transfer (t RNA) and ribosomal (r RNA).

This shows that RNA will have the same sequence as the other strand of DNA, except that uracil will replace thymine against adenine.

The abovementioned RNA types can be differentiated on the basis of their size, sedimentation behaviour and genetic functions.

Ribosomal RNA (rRNA) is generally the largest and most prevalent of the cellular RNA species. This is an important structural component of ribosomes, the sites where translation occurs during protein synthesis.

Messenger RNA (mRNA) is responsible to carry the genetic message from DNA to the ribosome. Their length and sequence vary depending upon the gene which is being transcribed into m RNA.

Transfer RNA (tRNA) is the smallest of the three types. It carries amino acids to the ribosomes during translation. This species of RNA contains number of modified bases.

REPLICATION OF DNA

As a carrier of genetic informations DNA has following two important functions :

1. Heterocatalytic function. When DNA directs the synthesis of chemical molecules other than itself (*e.g.*, synthesis of RNA, proteins, etc.), then such functions of DNA are called heterocatalytic functions. **2. Autocatalytic function.** The functions of DNA which direct the synthesis of DNA itself, are called autocatalytic functions. Here, we are concerned only with autocatalytic function of DNA.

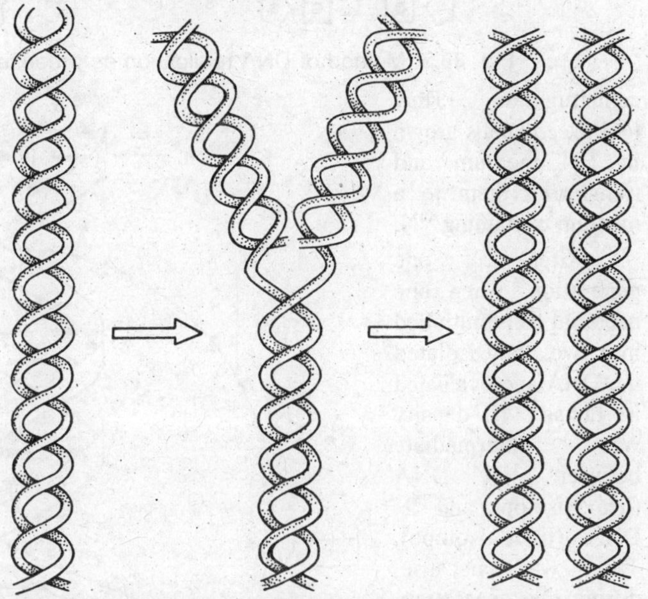

Fig. 38.6. Replication of the DNA molecule. The old helix unwinds (middle) and the new helices are being formed.

The copying mechanism proposed by Watson and Crick is quite simple. When the two strands of DNA separate and each strand serves as a template to form a new strand, one can easily predict how the new strand would look. Since the bases in the old strand would automatically decide the complementary bases that would appear in the new strand on the basis that A will pair with T and G will pair with C. Thus, one DNA molecule gives rise to two DNA molecules identical to that of the parent. However, in the newly formed daughter DNA molecule, one strand will be old (that comes from the parent) and another will be new. This is known as **a semi-conservative model of replication.**

Meselson and Stahl (1958) Experiment. This experiment showed that DNA replication is **semi-conservative.** In their experiment these scientists grew the bacterium *Escherichia coli* in a medium containing nitrogen salts labelled with the heavy isotope, *i.e.*, ^{15}N. ^{15}N was incorporated into both strands of DNA and this DNA was heavier than the DNA obtained from *E. coli* cells grown in ^{14}N medium (*i.e.*, common isotope of N). The heavier DNA was separated by

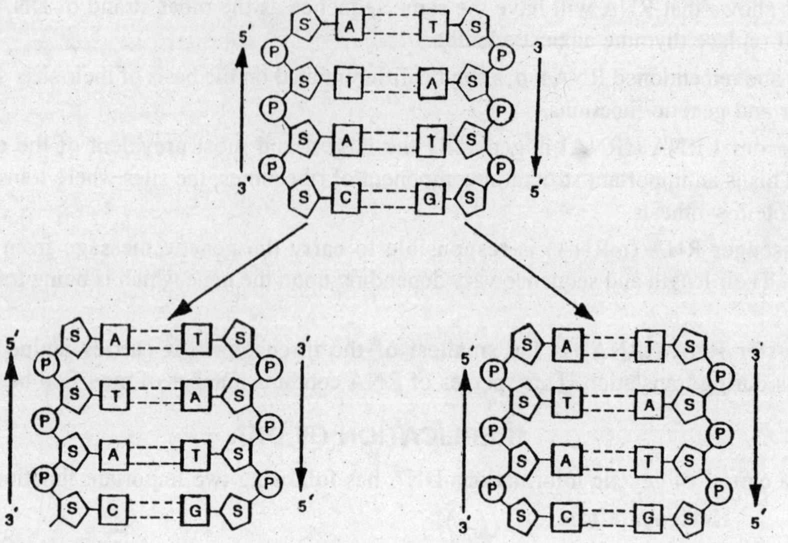

Fig. 38.7. Method of DNA replication as proposed by Watson and Crick.

centrifugation. They took *E. coli* cells grown in ^{15}N medium and transferred them to a medium containing ^{14}N.

After one generation, when one bacterial cell multiplied into two, they isolated the DNA and evaluated its density. Its density was intermediate between ^{15}N DNA (heavy isotope) and ^{14}N DNA (light isotope). This was because, during one generation, new DNA molecules with one ^{15}N old strand and a complementary ^{14}N new strand were formed and this had an intermediate density between the ^{15}N DNA in both the strands and ^{14}N DNA in both the strands. This experiment showed the semi-conservative nature of DNA replication.

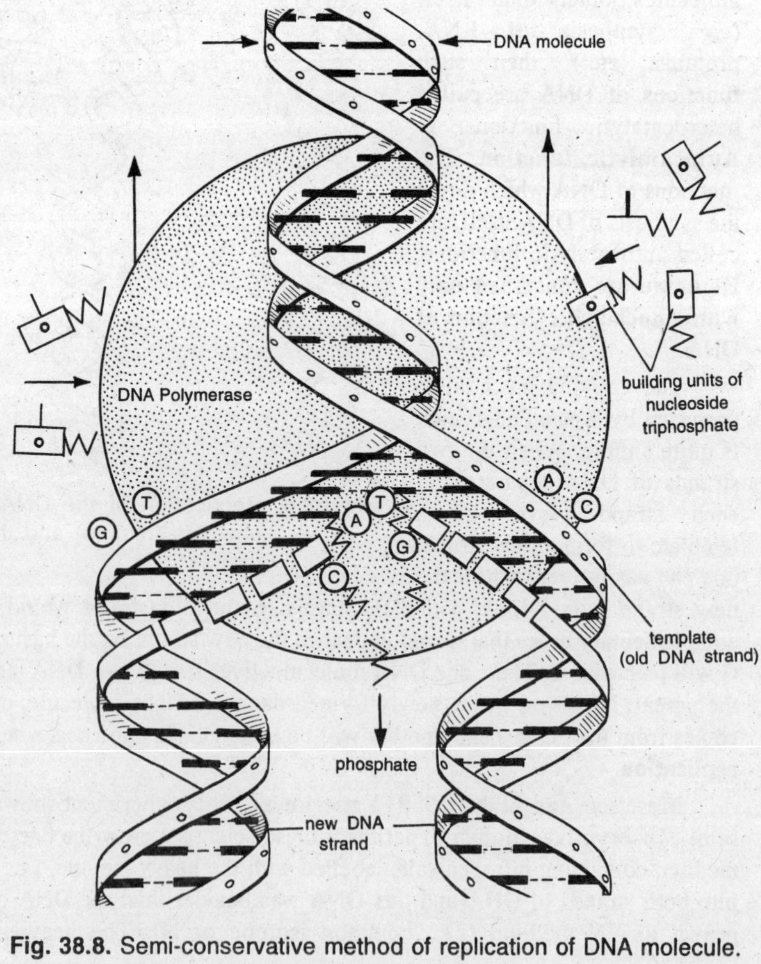

Fig. 38.8. Semi-conservative method of replication of DNA molecule.

Details of DNA Replication in Procaryotes and Eucaryotes

DNA replication in procaryotes and viruses. The procaryotes, such as bacteria possess a single circular molecule of DNA. This type of DNA molecule is much smaller in comparison to a single chromosome of a eucaryote. In this circular DNA molecule there is only one origin of replication. From this point of origin, two replication forks move in opposite direction and ultimately meet at half way round the circle at the termination points. Since each replication fork makes a replica of the original chromosome and therefore in the end the identical daughter DNA circles are formed. In viruses also DNA is in the form of single strand and there is only one origin of replication.

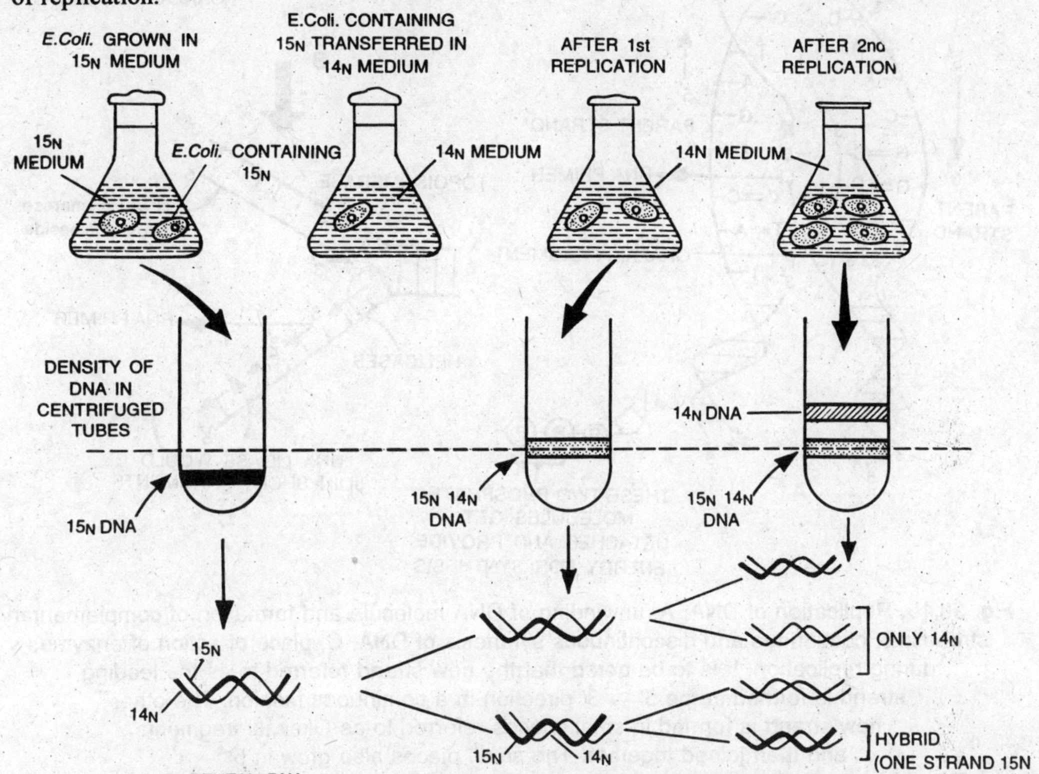

Fig. 38.9. Meselson and Stahl's experiment to demonstrate the semi-conservative replication of DNA.

DNA replication in eucaryotes. In eucaryotes with giant DNA molecules, there are several origins of replication and they can merge with each other while replication is under progress.

The second point is that the two strands of DNA should separate, before each acts as a template for the synthesis of a new strand. For this purpose of unwinding there are enzymes called **helicases** that unwind the helix. There are other enzymes known as **topoisomerases** which are responsible for breaking and resealing of one strand of DNA.

A **primer** is also needed for replication of DNA, that is also formed on the template. This primer is actually a short stretch of RNA formed on the DNA template and the enzyme that polymerases the RNA building blocks, *i.e.*, A, U, G, C into the primer is known as **primase.** The primers are later removed, and the gaps so left are filled up with deoxyribonucleotides make DNA strand continuous.

Synthesis of new strand. The synthesis of new strand of DNA takes place as follows : Here the enzyme DNA polymerase plays an important role in adding the building blocks to the primer in a sequence as influenced by the template. Formation of the complementary DNA strand cannot begin without the formation of an RNA primer.

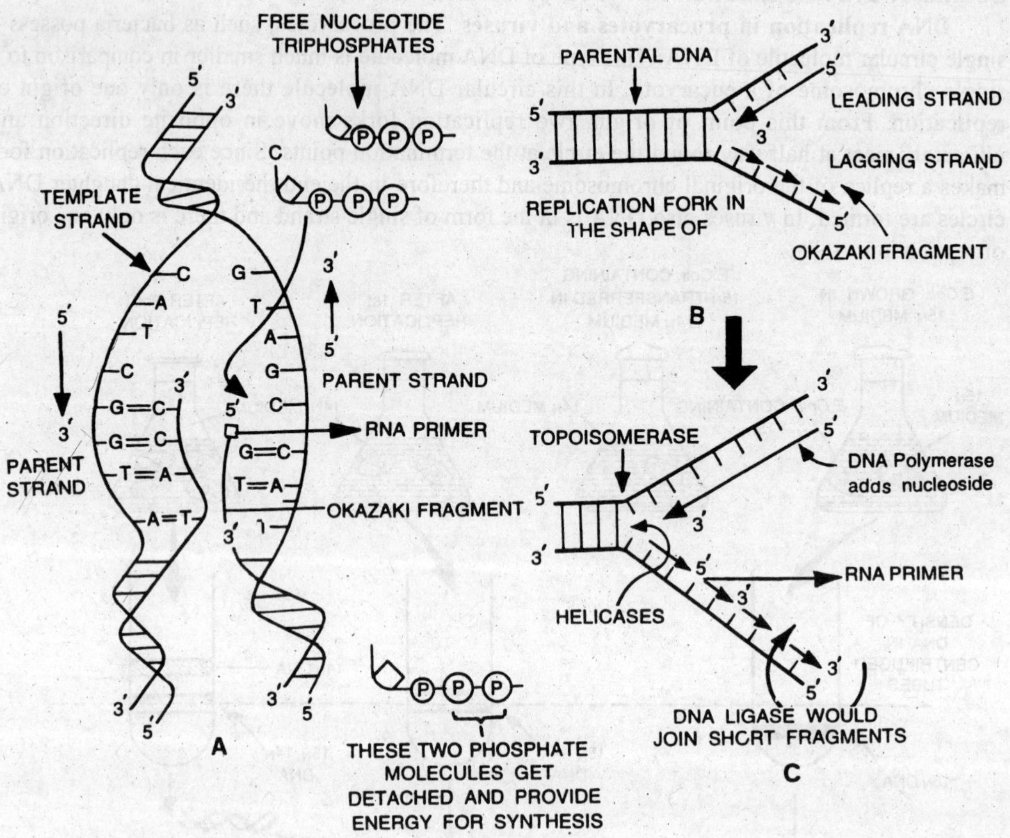

Fig. 38.10. Replication of DNA. A, unwinding of DNA molecule and formation of complementary strands; B, continuous and discontinuous synthesis of DNA; C, place of action of enzymes during replication. It is to be noted that the new strand referred to as the leading strand is formed in the $5' \rightarrow 3'$ direction in a continuous fashion. The other new strand is formed in small pieces referred to as Okazaki fragments and then joined together. The small pieces also grow in $5' \rightarrow 3'$ direction, but the overall direction of the lagging strand is $3' \rightarrow 5'$.

When the double stranded DNA is unwound upto a point, a Y-shaped structure is developed, which is referred as **replication fork.** New strands develop from the fork and as replication progresses this appears as if the point of divergence at fork is moving. The enzyme DNA polymerase can polymerise the nucleotides only in the $5' \rightarrow 3'$ direction. Since the two strands of the DNA are formed in antiparallel orientation, the two new strands will form by the growth taking place in opposite directions. Here the enzyme forms one new strand in a continuous stretch in the $5' \rightarrow 3'$ direction and this is referred to as the **leading strand.**

On the other parent strand, the enzyme forms short stretches of DNA once again in the $5' \rightarrow 3'$ direction starting from an RNA primer. The primer is synthesized by the enzyme primase. These DNA short fragments are known as **Okazaki fragments**, and the strand is called as **lagging strand,** because it is synthesized in small pieces and then joined to each other. These short fragments of DNA are joined together by the enzyme DNA ligase after replacing the RNA primer with DNA.

Proof reading and DNA repair. During DNA replication the specificity of base-pairing introduces a high degree of accuracy. Any error that may be one in 10,000, is corrected by removing the wrong base and replacing the correct one by repair enzymes.

However, bacterial DNA polymerase can do proof reading where it goes back and removing the wrong before it proceeds to add new bases in the $5' \rightarrow 3'$ direction. This type of proof reading ensures formation of identical DNA strands during DNA replication.

Sometimes abnormal base pairs are formed in the DNA due to mutation, that escape the proof reading mechanism of DNA polymerase can still be rectified by repair enzymes, which excise the damaged region and replace it with normal segment.

Enzymes involved in DNA replication. Several enzymes are involved in DNA replication. They are as follows :

(*i*) **Helicases.** They are unwinding enzymes, and help in breaking the weak hydrogen bonds that hold the two strands of DNA together.

(*ii*) **Topoisomerases.** These enzymes are responsible for breaking and resealing of one strand of DNA.

(*iii*) **Primase.** This helps in the formation of primer. Primer is a short RNA segment that is formed on the DNA template. It is absolutely essential.

(*iv*) **DNA polymerase.** This enzyme can link free DNA nucleotides to form the complementary strand of DNA. It polymerases nucleotides in $5' \rightarrow 3'$ direction only.

(*v*) **DNA ligase.** This enzyme can join together the short fragments after replacing RNA primer with DNA.

(*vi*) **Repair enzymes.** There may be some errors during DNA replication. Repair enzymes can excise the wrong bases and replace them with correct ones.

Functions of DNA

The DNA has a key role in all biosynthetic and hereditary functions of the living organisms. Its few important functions are as follows :

1. The most important function of the DNA is that it acts as a carrier of genetic informations from generation to generation. It is most stable compound of the biological world which can be considered immortal.

2. The DNA controls all the life activities of the cell directly or indirectly.

3. The DNA synthesizes the RNA.

4. The DNA guides the process of protein synthesis.

Mechanism of DNA Replication

The process of DNA replication is much complicated. The major steps of the process are as follows : (fig. 38.10.)

During this process a whole range of enzymes is required to take care of various steps.

The DNA replication in procaryotic cells (bacteria) starts at a single point known as **origin of replication**, and moves bidirectionally.

On the other hand, in eucaryotic cells there are several points of origin on the length of DNA per chromosome.

The first requirement before any type of synthesis is to unwind the double helix of DNA, so that the two strands become free to act as templates.

This function of unwinding of double helix is carried out by the enzyme **helicase**, which unzips the two strands beginning at the origin site.

As soon as unwinding takes place, other proteins called single-stranded binding proteins, associate with single strands and make this condition stable.

Unwinding also creates a coiling tension ahead of the moving **replication fork**, a structure that will be formed when DNA replication begins.

The coiling tension created by unwinding of double helix is reduced by the enzymes known as **topoisomerases.**

One of the most important DNA synthesizing enzyme is DNA polymerase III. This enzyme along with other DNA polymerases (*i.e.*, I and II) can elongate an existing DNA strand but cannot initiate the synthesis of DNA.

All the above mentioned three DNA polymerases (*i.e.*, I, II, and III) function in $5'$ to $3'$ direction only for DNA polymerisation and have $5'$ to $3'$ direction for exonuclease activity.

Now, to initiate DNA synthesis, a small segment of RNA, known as RNA primer complementary to the template DNA is synthesized by an unique RNA polymerase known as **primase.**

It is to this RNA primer, that DNA polymerase III adds $5'$ deoxyribonucleotides and extends the DNA.

A problem arises, when two strands of DNA run antiparallel to each other and DNA polymerase III can act only in $5' \rightarrow 3'$ direction. This problem is solved as follows :

While on the one strand, the DNA synthesis goes on continuously in $5' \rightarrow 3'$ direction, on the other strand, DNA is synthesised in small stretches resulting in discontinuous DNA synthesis.

This process takes place in the opposite direction to the first strand but maintains the over all $5' \rightarrow 3'$ direction as required and such a process is sometimes called semi-discontinuous replication.

The short stretches of DNA, each primed by RNA are called **Okazaki fragments**, named after Japanese scientist who discovered them.

Thereafter RNA primers are removed, and the gap is filled by DNA synthesis. Both these steps are performed by DNA polymerase I.

Now the Okazaki fragments are sealed by the enzyme **ligase**.

The strand which supports the continuous DNA synthesis is the **leading strand** and one which is replicated in short stretches is called the **lagging strand**.

The process of DNA replication ensures the accuracy to maintain the nucleotide sequence of the original DNA.

DNA synthesis is slower in eucaryotes, as larger DNA's need to be replicated.

The general steps of DNA replication are similar both in eucaryotes and procaryotes.

GENE EXPRESSION

The another important characteristic of the gene is to store and express the genetic information which will contribute towards the phenotype and will be passed on to successive generations.

In 1902, an English physician Archibald Garrod, advocated that the diseases which inherit from the parent by the progeny are **inborn errors of metabolism** and they are due to a defect in an enzyme that catalyses the conversion of one metabolic substance to another. Hence this inherited genetic defect is due to the deficiency of an enzyme. This was believed that genes directly influenced the biochemical metabolism by determining the pressure or absence of a specific enzyme and that the genes have the information to produce enzymes.

For example, in one disorder alkaptonuria, afflicted individuals cannot metabolise homogentisic acid, as a result it gets accumulated and is excreted in the urine.

Garrod found that this disorder is inherited as a recessive trait and reasoned that normal individuals must be metabolising homogentisic acid. On the other hand, in affected individuals, the metabolic pathway is blocked resulting into disorder alkaptonuria.

Later on, several other human diseases were identified and described which fell into the category of inherited metabolic defects. This led to idea that many disorders may be inherited as any other gene and that genes function by controlling the metabolism.

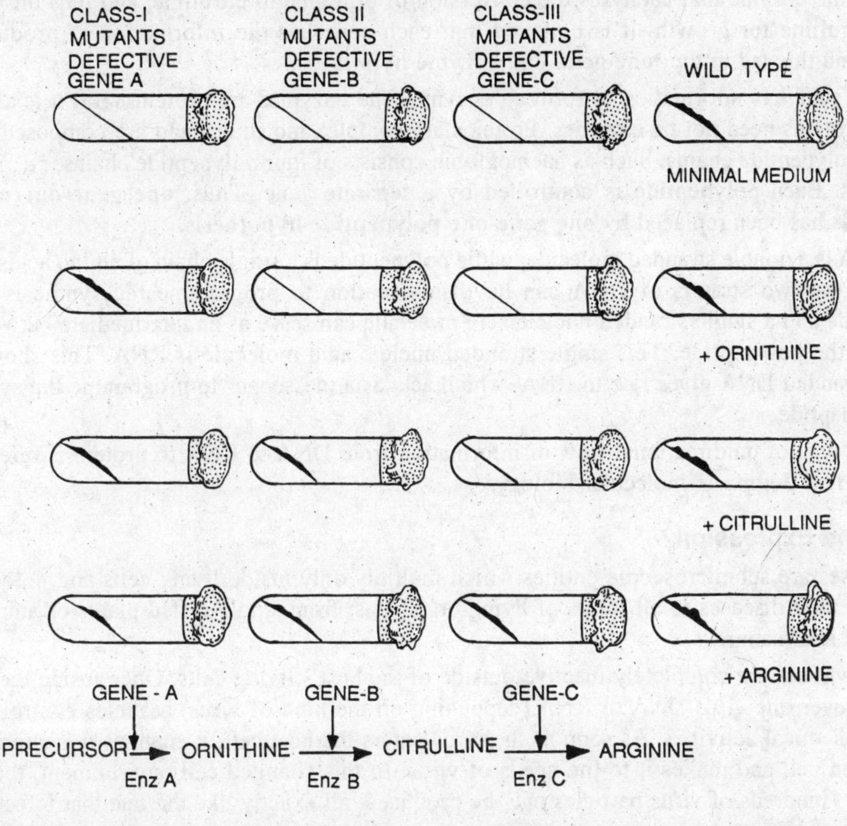

Fig. 38.11. One gene-one enzyme hypothesis—Experiments with *Neurospora crassa.* Wild type of *N. crassa* will grow in minimal condition and it does not require any supplement for growth. Class I mutants are defective in gene A (enzyme A) and thus cannot grow on minimal medium unless supplemented with ornithine or citrulline or arginine. Class II mutants are defective in gene B (enzyme B) and will grow in medium supplemented with citrulline or arginine. Class III mutants are defective in gene C (enzyme C) and will grow in medium supplemented with arginine.

The concept that genes have to information to produce enzymes which are proteinaceous in nature was put on a firm footing by the geneticist George W. Beadle and the biochemist Edward L. Tatum, for which they were awarded the Nobel Prize in medicine in 1959. It is known that UV rays, X-rays and environmental chemicals can bring about a change in the nucleotide sequence of DNA and thus cause mutations.

In 1941, Beadle and Tatum proved the **one gene one enzyme hypothesis**, for which they used X-rays to cause mutations in the mould *Neurospora crassa.* This mould can normally grow in a simple minimal medium containing salts and sugar, making all other chemicals, *i.e.*, amino acids, purines, pyrimidines, etc., through enzyme-catalysed reactions.

However, the mutants cannot grow on a minimal medium. Each type of mutant needs some extra nutrients, and such nutritional mutants are called **auxotrophs.** In their experiments, Beadle and Tatum obtained different mutants that require the amino acids ornithine or citrulline or arginine for growth. It is established that in the cell ornithine converts into citrulline which in turn converts into arginine, here each step is being catalysed by a specific enzyme. Beadle and Tatum proved that the mutant requiring ornithine for growth is defective in the enzyme producing ornithine as a result of gene mutation. In the similar way, a defect in another gene may lead to a defect in the enzyme that catalyses the conversion of ornithine to citrulline, and thus this mutant needs citrulline for growth. It is believed that each gene has the information to produce one enzyme and this led to the **'one gene-one enzyme hypothesis.**

The gene has information to form an enzyme. The enzymes are proteins. But it is also true that all proteins need not be enzymes. Proteins are complex molecules and are composed of one or more polypeptide chains. Such as haemoglobin consists of four polypeptide chains, *i.e.*, 2α and 2β chains. Each polypeptide is controlled by a separate gene. Thus, **one gene-one enzyme hypothesis** has been replaced by **one gene-one polypeptide hypothesis.**

DNA is a double stranded molecule, while polypeptide is a single chain of amino acids. Thus, either of the two strands of DNA can have information to programme the synthesis of the polypeptide and a single stranded nucleic acid molecule can serve as an intermediate between the gene and the polypeptide. This single stranded nucleic acid molecule is RNA. This shows that double stranded DNA gives rise to RNA, which acts as a messenger to programme the synthesis of a polypeptide.

This type of unidirectional flow of information from DNA to RNA to protein is referred to as the **central dogma** in molecular biology.

Viral Gene Expression

Viruses are submicroscopic entities which multiply only inside living cells and possess the ability to cause diseases to all forms of living organisms, from single-celled plants or animals to large trees and mammals.

The viruses are completely inactive outside of the host's living cells. Once inside the host's cells, however, the virus DNA or RNA (depending on the kind of virus) assumes control of the cell's biochemical activities. As soon as the virus enters the host cell, it changes the activities of the infected cell and makes it to the needs of virus. In this changed cell environment, the virus replicates. Hundreds of virus particles may be produced, all exactly like the one that infected the host cell.

Tobacco mosaic virus (TMV) contains RNA as the genetic material. Other viruses may contain DNA as the genetic material. There are also viruses that contain single stranded DNA as the genetic material. Viruses are diverse in size and shape but basically contain the genetic material wrapped around by a protein coat, generally known as the **capsid.**

Viral genomes are small and possess the information to code for a few proteins only. Thus, viruses use the host machinery such as enzymes, ribosomes and other components to replicate, to form the coat and to form more virus particles. The viruses infecting bacteria are known as **bacteriophages.** The expression of viral genes is regulated in such a way that the lysis of the host cells is caused. Thereafter many virus particles **(lytic)** are liberated which can infect new host cells to repeat the **lytic-cycle.** In some viruses after infection of the host cells viral particles get integrated with the host chromosome (DNA) and replicate along with it **(lysogeny).**

Lytic-cycle

Bacteriophages are the best understood viruses in terms of their gene structure and expression. For example T_2-bacteriophage infects *Escherichia coli* and causes lytic cycle. This phage has a double stranded circular DNA enclosed in a hexagonal proteinaceous head, cylindrical hollow tail and six tentacles for attachment to the host. The phage gets attached to bacterial wall by its tentacles. The lysozyme found at the tip of the tail dissolves the host cell wall to form a pore. The tail now contracts and injects the viral DNA into the host cell. Thereafter the viral DNA synthesises an enzyme known as nuclease which degrades the host DNA. Only after few minutes the host DNA is destroyed by nuclease. On the other hand, the viral DNA is resistant to nuclease because its cytosine residues are methylated. Now the viral genes make use of host ribosomes for multiplication. First of all the replication of viral DNA takes place. Thereafter, the proteins of head, tail and tentacles are synthesized. An independent synthesis of viral components occur within the cell followed by an assembly of those components into complete virus particles. After a well defined period the bacterial cell bursts and releases the newly formed virus particles. The liberated virus particles again infect the bacterial cells and the lytic cycle is repeated.

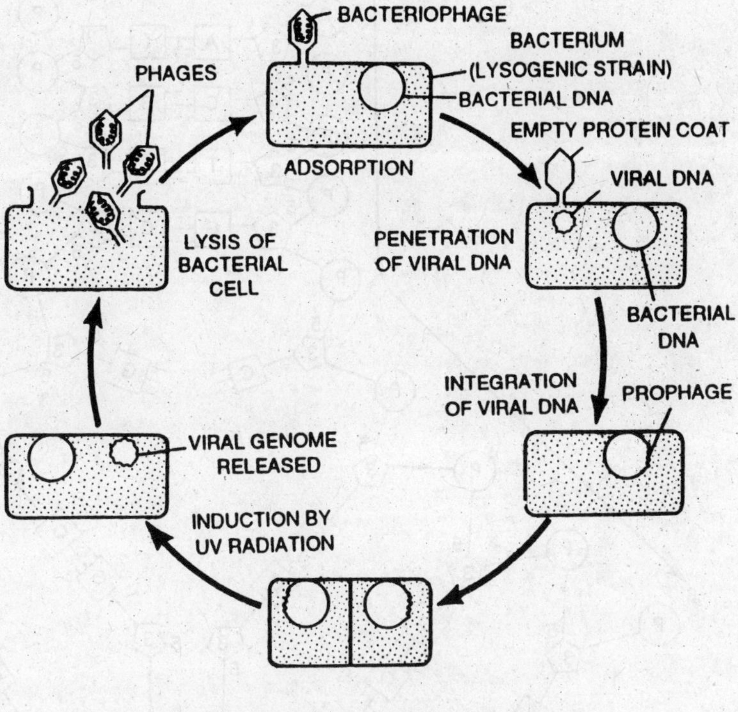

Fig. 38.12. Bacteriophages-life cycle.

Gene Expression in Prokaryotes

Bacterial chromosomes. It has been established now that a bacterial chromosome is always present in the centre of bacterial cell which is not capable of mitotic and meiotic divisions and is not enclosed within a definite nuclear membrane. An average bacterial cell contains one thousandth the DNA content of a typical eucaryotic cell. The bacterial chromosome is a thin elongated flexible and circular filament of DNA molecule associated with few proteins. The bacterial cells can divide very rapidly. The cell division is completed by the doubling of all the cell constituents followed by partition of the cell into two daughter cells. The first step in cell division is DNA duplication. The duplication of bacterial chromosome is semiconservative. Two chains of parental double helix of DNA unwind by breaking of hydrogen bonds between the bases in the presence of the **unwinding enzyme.** The separation of the parental chains results in the formation of Y-shaped growing points called replication forks. The free single chain is used as template and DNA synthesized is started. The chain with $3'$ end has continuous strand called **leading strand** is synthesised by DNA **polymerase** in $5' \rightarrow 3'$ direction only. The second complementary free parental chain with $5'$ end cannot be copied by DNA polymerase in the same direction, and so is copied in the opposite direction but in discontinuous fashion strand with the initial action of an enzyme called DNA

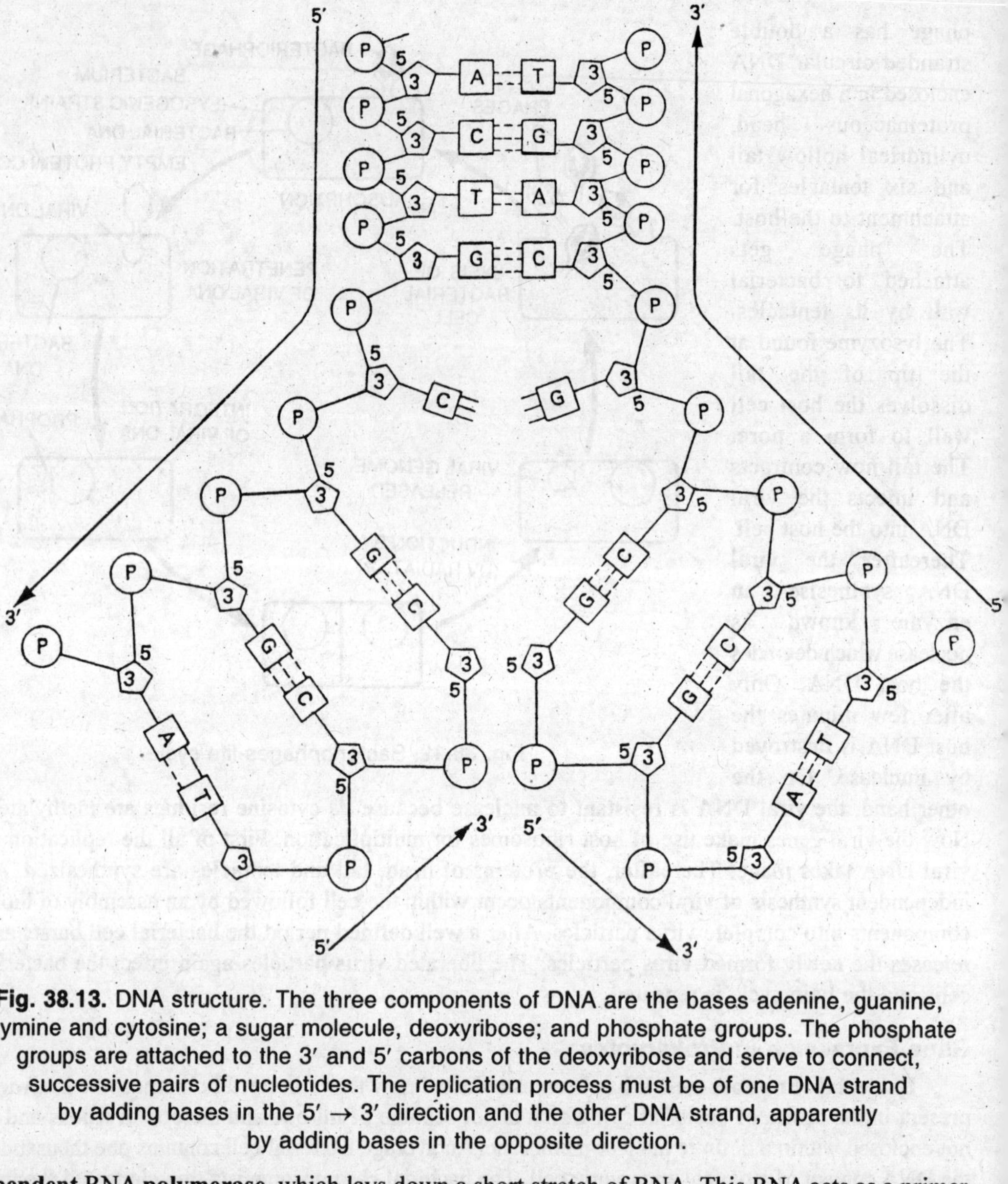

Fig. 38.13. DNA structure. The three components of DNA are the bases adenine, guanine, thymine and cytosine; a sugar molecule, deoxyribose; and phosphate groups. The phosphate groups are attached to the 3′ and 5′ carbons of the deoxyribose and serve to connect, successive pairs of nucleotides. The replication process must be of one DNA strand by adding bases in the 5′ → 3′ direction and the other DNA strand, apparently by adding bases in the opposite direction.

dependent RNA **polymerase**, which lays down a short stretch of RNA. This RNA acts as a primer at its 3—OH end for the enzyme DNA polymerase which then results the synthesis of lagging strand. Finally the enzyme DNA **ligase**, bonds the strands together and closes DNA 'nicks' with the result a continuous strand is formed like the leading strand. After the synthesis of new strands two new double helical DNA molecules are formed each with an old strand from the mother molecule. Now, each of the DNA strands still attached to the cell membrane, are distributed in two daughter cells by a simple pinching off process.

Genetic Recombination in Bacteria

This is a process where genetic materials, contained in two separate genomes, are brought together within one unit. In bacteria the recombination takes place by **(1) transformation, (2) transduction** and **(3) conjugation.**

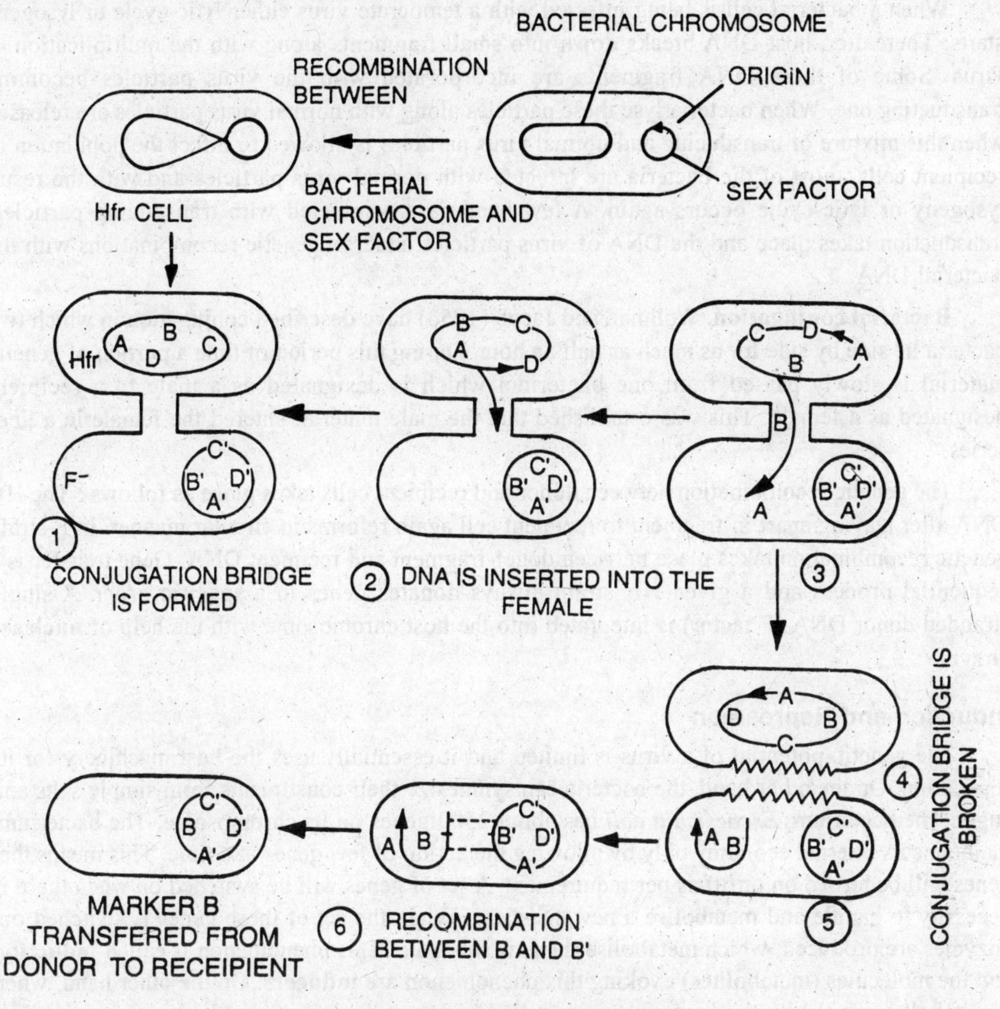

Fig. 38.14. Bacterial conjugation.

Transformation. The genetic transfer in bacteria also occurs by **transformation** in which the DNA molecule of the donor cell, when liberated by its disintegration, is taken up by another recipient cell and its offspring inherit some characters of the donor cell. When different strains of bacteria are found in a mixed stage either in culture or in nature, some of the resultant offspring possess a combination of characters of the parent strains. This phenomenon is known as **recombination**.

The phenomenon of transformation was first recorded by Griffith (1928). Avery, Macleod and McCarty (1944) demonstrated that the transforming principle being DNA in the sequence of events in bacterial transformation.

Transduction. The genetic transfer in bacteria is achieved by a process known as **transduction**. Laderberg and Zinder's (1952) experiment in U-tube *Salmonella typhimurium* indicated that bacterial viruses or phages are responsible for the transfer of genetic material from one to the other lysogenic and lytic phages. Thus the host acquires a new genotype. Transduction has been demonstrated in many bacteria.

In this process, the DNA molecule that carries the hereditary characters of the donor bacterium, is being transferred to the recipient cell through the agency of the phage particle. In this

When a bacterial cell is being infected with a temperate virus either lytic-cycle or lysogeny starts. Thereafter, host DNA breaks down into small fragments along with the multiplication of virus. Some of these DNA fragments are incorporated with the virus particles becoming transducting one. When bacteria lyse these particles along with normal virus particles are released when this mixture of transducing and normal virus particles is allowed to infect the population of recipient cells, most of the bacteria are infected with normal virus particles and with the result lysogeny or lytic-cycle occurs again. A few bacteria are infected with transducing particles, transduction takes place and the DNA of virus particles undergo genetic recombinations with the bacterial DNA.

Bacterial conjugation. Wollman and Jacob (1956) have described conjugation in which two bacteria lie side by side for as much as half an hour. During this period of time a portion of genetic material is slowly passed from one bacterium which is designated as a male to a recipient designated as a female. This was established that the male material entered the female in a liner series.

The genetic recombination between donor and recipient cells takes place as follows : The Hfr DNA after leaving apart in fragment to recipient cell again reforms in circular manner. In F-strain genetic recombination takes place between donor fragment and recipient DNA. Gene transfer is a sequential process and a given Hfr strain always donates genes in a specific order. A single stranded donor DNA (F factor) is integrated into the host chromosome with the help of **nuclease enzyme**.

Induction and Repression

The genetic potential of a virus is limited and it essentially uses the host machinery for its replication. On the other hand, the bacteria can synthesize their constituents from simple salts and sugar. The bacterium, *Escherichia coli* has about 2500 genes on its chromosome. The bacterium, maintains its internal economy only by allowing the action of few genes at a time. This means that genes will be turned on or off as per requirement. A set of genes will be switched on when there is necessity to handle and metabolise a new substrate. When the set of these genes is switched on, enzymes are produced, which metabolise the new substrate. This phenomenon is called **induction** and the molecules (metabolites) evoking this phenomenon are **inducers**. On the other hand, when a metabolite needed by the bacterium is supplied in excess from outside, the bacterium inhibits making it and this stops the further production of the metabolite by the bacterium. These inactivated genes are thus named **repressible**, and the phenomenon is known as **repression** or **feed-back repression**. However, some of the genes are constantly expressed to take care of normal

Table 38.2. Comparison Between Induction and Repression.

Induction	Repression
1. It is responsible for starting transcription and translation.	1. It stops transcription and translation.
2. It turns the operon on.	2. It turns the operon off.
3. It acts in a catabolic pathway.	3. It acts in an anabolic path way.
4. It is caused by new metabolite that requires enzymes to get metabolised.	4. It is due to excess of existing metabolite.
5. Repressor is checked by the inducer from combining with operator gene.	5. Repressor is helped by a co-repressor to combine with operator gene.

cellular activity such as glycolysis. These genes are known as **constitutive**, and the enzymes produced by them are known as **constitutive enzymes**, such as dehydrogenases.

Gene Expression in Eucaryotes

Higher eucaryotes have thousands of genes, in comparison to procaryotes. They are multicellular organisms which can also undergo differentiation. Thus, the cells in the undifferentiated stage not only grow and divide, but they also become part of specialised tissues such as the liver, spleen, heart, etc., in an animal and the leaf, root, stem, flower, etc., in a flowering plant (angiosperm). Hence, regulation gene expression in the eucaryotic cell is very complex.

The structure of the eucaryotic gene is also complex (fig. 38.15).

For example, in bacteria, if the base sequence of the DNA is known, the amino acid sequence of the protein can be predicted. Here the gene and the protein are collinear indicating that the linear arrangement of nucleotides (bases) in the gene correlates with the linear arrangement of amino acids in proteins. In most eucaryotic genes the match between the bases in the gene and the amino acids in the protein will only hold in parts. Here the gene will have stretches of bases, which will code for amino acids, that are interspersed with stretches of bases, which do not code for any amino acid. Thus, the information on the eucaryotic gene for assembling a protein is not continuous, but split.

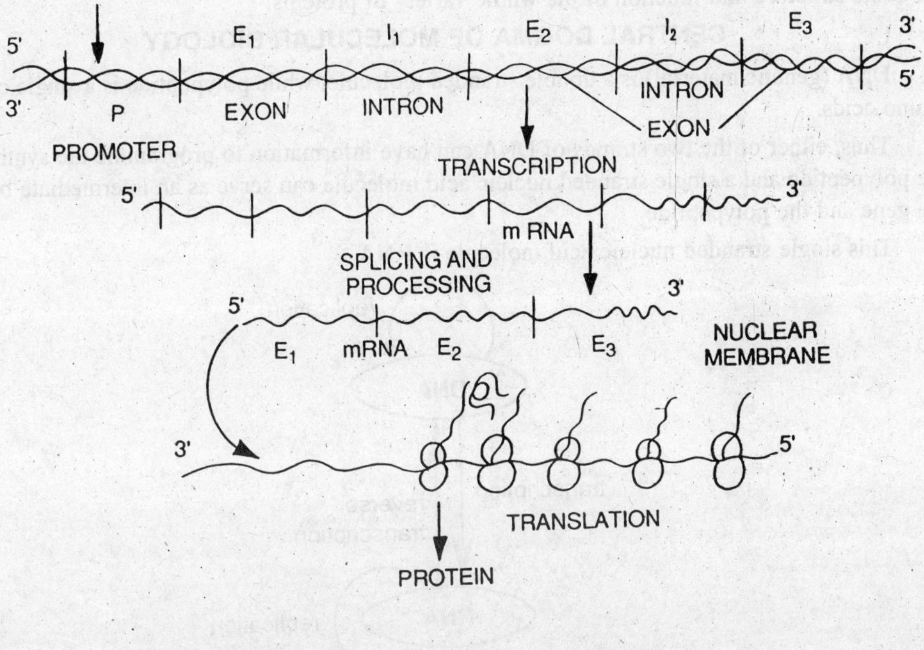

Fig. 38.15. Eucaryotic gene structure. Details in the text.

However, when mRNA is formed from such eucaryotic genes, the unwanted RNA regions are removed and the regions coding for amino acids are joined together, and this process is called **splicing.**

Thus, bases in mRNA and amino acids in proteins are collinear in eucaryotic cells also, though the genes are split. The regions of a gene that become part of mRNA and code for different regions of the protein are known as **exons.** The regions which do not form part of mRNA and are removed during RNA processing before mRNA formation are known as **introns.**

In eucaryotes, genes involved in coding for the enzymes of a definite metabolic pathway need not be linked. Sometimes they are found even on different chromosomes. However, such genes are regulated together as in bacterial operons.

The basic processes of induction and repression operate in eucaryotic cells, but through a complex network of regulatory genes. The gene expression is constantly regulated by the changing environment in the cell. Thus during growth and development, small molecules such as hormones, vitamins, metal ions, chemicals and invading pathogens can induce or repress certain genes, and this will result in the production or absence of certain proteins. This ultimately leads to the operation or non-operation of metabolic pathways leading to altered cell function. This is how the environment influences the molecular basis of the gene action.

Gene and Protein

Now this has been established that DNA is the genetic material, and the genes function by exerting their control over proteins or enzymes.

Now, we shall study, how is the genetic information stored in DNA and then how this information coded in DNA is transmitted in the process of protein synthesis.

The work done in the late 1950s and early 1960s proved that a specific sequence of four bases in the DNA serves as the store house of all genetic information and makes the basis of all life on earth.

The unique way in which these bases are arranged serve as genetic code, which determines the basic structure and function of the whole variety of proteins.

CENTRAL DOGMA OF MOLECULAR BIOLOGY

DNA (genetic material) is a double stranded molecule, while polypeptide is a single chain of amino acids.

Thus, either of the two strands of DNA can have information to programme the synthesis of the polypeptide and a single stranded nucleic acid molecule can serve as an intermediate between the gene and the polypeptide.

This single stranded nucleic acid molecule is RNA.

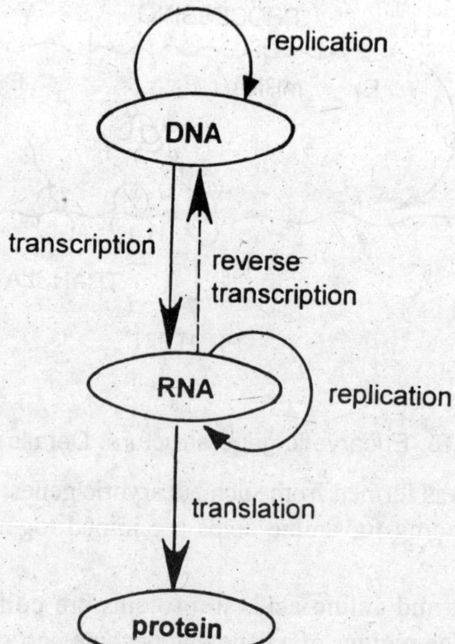

Fig. 38.16. The flow of genetic information.

This shows that double stranded DNA gives rise to RNA, which acts as a messenger to programme the synthesis of a polypeptide.

This type of unidirectional flow of information from DNA to RNA to protein is referred to as the **central dogma of molecular biology.**

The expression of the genetic material (DNA) which occurs generally through the production of protein involves two consecutive steps.

These two steps are known as ***(i)*** transcription and ***(ii)*** translation.

In transcription, the genetic information, stored in DNA, is transferred to an RNA intermediate, which in turn uses this information to direct the synthesis of proteins during translation.

This unidirectional flow of information was described by F.H.C. Crick in 1958 as the **central dogma of molecular biology.**

However, in 1970, an important modification of this information was brought to light by H. M. Temin and D. Baltimore.

According to them many tumor viruses contain RNA as genetic material and replicate by first synthesising a complementary DNA. This process is called **reverse transcription.**

Such process is carried out by an RNA- dependent DNA polymerase called reverse transcriptase.

These viruses are known as retroviruses and include **Human Immunodeficiency Virus** (HIV) which is responsible for causing AIDS.

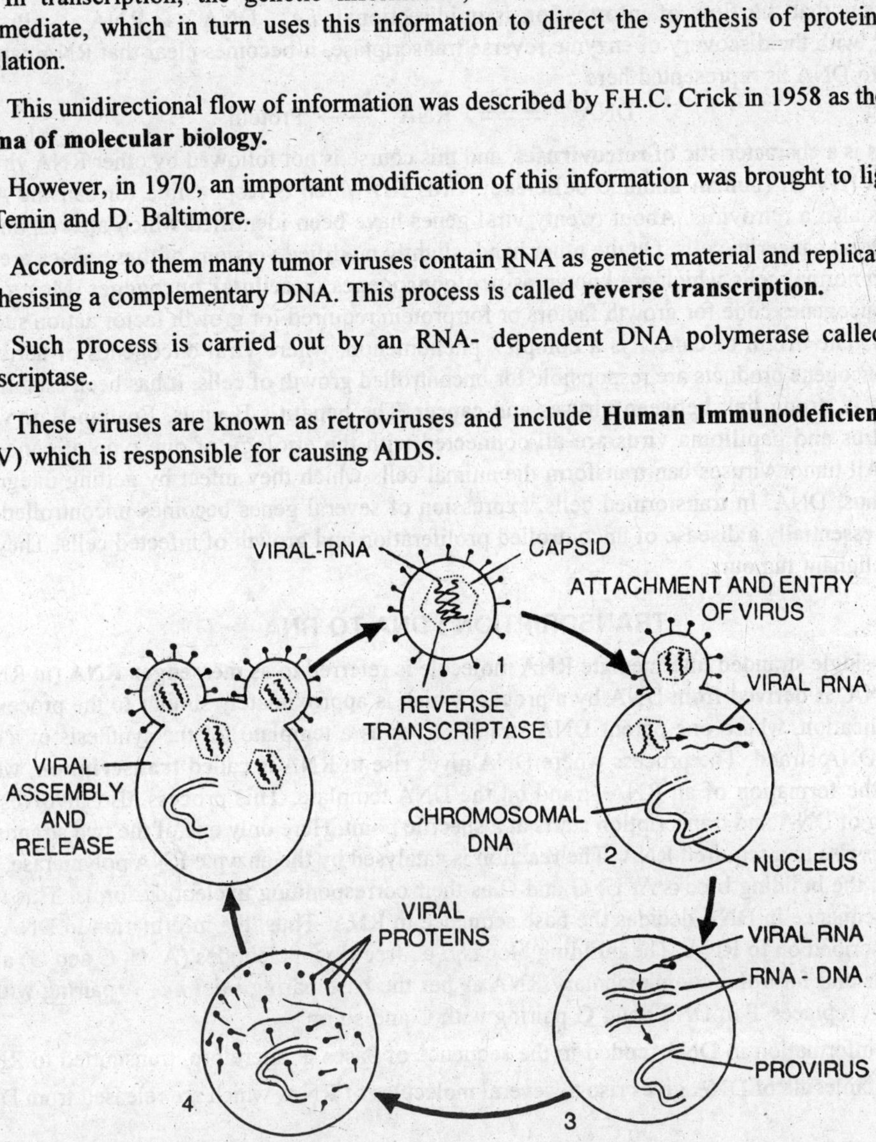

Fig. 38.17. Life-cycle of retrovirus. Details in the text.

Reverse Transcription

This phenomenon was studied by Temin and Baltimore (1972), and thus known as **Teminism.** While studying the activity of double stranded RNA containing Rous sarcoma virus (RSV) in the cancerous tissue, they found that viral RNA synthesized DNA complementary to its strands. As soon as the RNA of RSV comes in contact of the cytoplasm of host cell, it synthesizes the enzyme **reverse transcriptase** that catalyses the conversion of RNA to DNA copy and exhibits a reverse phenomenon of the normal process. Such phenomenon is known as **reverse transcription** or **Teminism**. Once the DNA complementary to viral RNA is synthesized it follows the normal rule and synthesizes viral RNA and viral proteins as depicted in the figure (fig. 38.16). The central dogma says that the flow of information is unidirectional, *i.e.*, DNA → RNA → Protein. However, with the discovery of enzyme reverse transcriptase, it becomes clear that RNA can also go back to DNA as represented here :

$$\text{DNA} \rightleftharpoons \text{RNA} \longrightarrow \text{Protein}$$

This is a characteristic of **retroviruses**, and this course is not followed by other RNA viruses. The virus HIV-III (human immuno deficiency virus III) which is responsible for causing AIDS disease is also a retrovirus. About twenty viral genes have been identified which are responsible for triggering cancer in cells. On the other hand, slightly modified versions of these genes are also present in normal cells which are known as **protooncogenes** or **cellular oncogenes**. Most of the cellular oncogenes code for growth factors or for protein required for growth factor action such as receptors. The origin of cancer is a complex phenomenon, where viral oncogenes or activated cellular oncogene products are responsible for uncontrolled growth of cells. It has been established that there is strong link between viruses and cancer. The hepatitis B virus, Epstein-Barr virus, herpes virus and papilloma virus are all connected with the etiology of one type of cancer or another. All tumor viruses can transform the animal cells which they infect by getting integrated with the host DNA. In transformed cells, expression of several genes becomes uncontrolled and cancer is essentially a disease of uncontrolled proliferation and growth of infected cells. They are called malignant tumours.

TRANSCRIPTION : DNA TO RNA

The single stranded intermediate RNA molecule is referred to as **messenger RNA** (m RNA). The m RNA, is derived from DNA by a process which is approximately similar to the process of DNA replication, where one parent DNA strand serves as a template for the synthesis of a new daughter DNA strand. The process where DNA gives rise to RNA is called **transcription**, which involves the formation of an RNA strand on the DNA template. This process also involves the unwinding of DNA and transcription starts at a specific point. Here only one of the two strands, of DNA gives rise to a required RNA. The reaction is catalysed by the enzyme RNA polymerase, and it requires the building blocks A, U, G and C as their corresponding nucleotide forms. This way, the base sequence in DNA decides the base sequence in RNA. Thus, the information in DNA has been transcribed on to RNA. The building blocks, *i.e.*, free ribonucleotides (A, U, C and G) align themselves and form the complementary RNA as per the base paring rule, *i.e.*, A pairing with U (U in RNA replaces T in DNA) and G pairing with C and so on.

The information in DNA, coded in the sequence of bases is, therefore, transmitted to RNA.

One molecule of DNA gives rise to several molecules of RNA, which are released from DNA template.

In eucaryotes, three different RNA polymerases (enzymes), I, II, and III, catalyse the synthesis of **ribosomal RNA** (r RNA), **messenger RNA** (m RNA), and **transfer RNA** (t RNA) respectively.

However, in procaryotes, a single RNA polymerase (enzyme) that has different subunits performs this function.

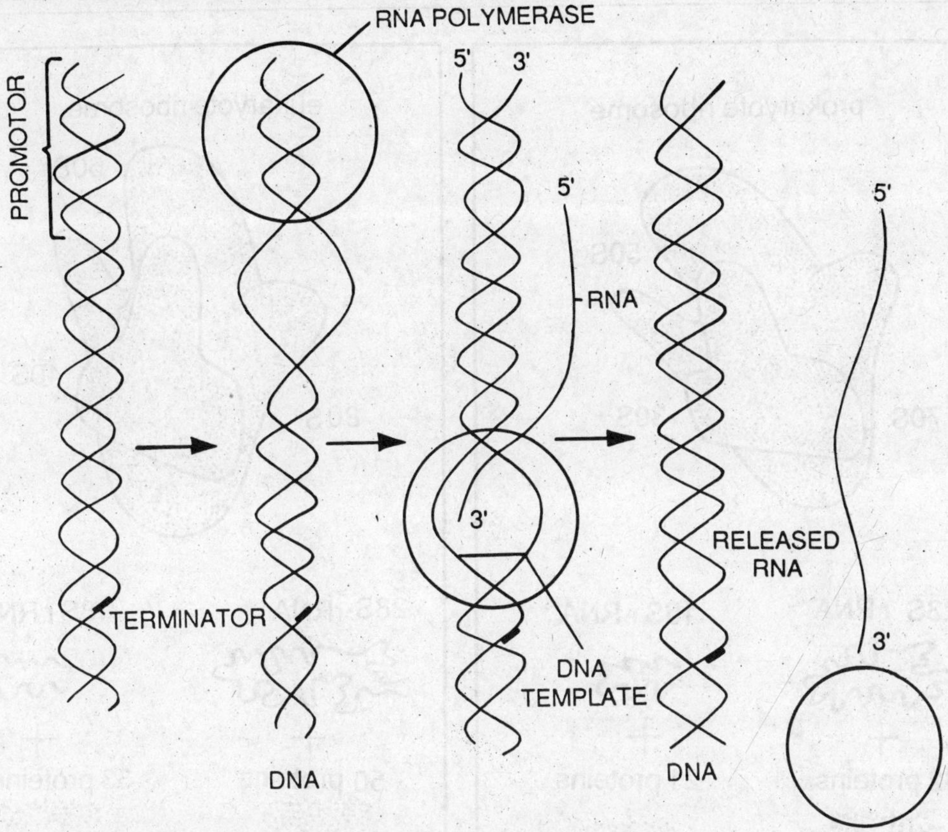

Fig. 38.18. Process of transcription. RNA polymerase binds to the promoter region of the DNA and starts the transcription process. The DNA unwinds and one of the two strands serves as a template for RNA synthesis, and a complementary copy of RNA is formed on the template strand.

TRANSLATION : BIOSYNTHESIS OF PROTEINS

Translation is the process in which genetic message carried by messenger RNA (m RNA) from the DNA is converted in the form of a polypeptide chain having a specific sequence of amino acids, *i.e.*, RNA to protein.

Besides m RNA there are three other key components which take part in this process. They are **(*i*)** ribosomes, **(*ii*)** transfer RNA (t RNA) and **(*iii*)** amino acids.

(*i*) Ribosomes. They are ribonucleoprotein particles which provide the sites for protein synthesis. On the basis of their sedimentation coefficients, procaryotic and eucaryotic ribosomes are characterised, *i.e.*, 70 S and 80 S respectively. Both of them consist of a small and a large subunit.

In procaryotes, the large or 50 S subunit consists of 23 S r RNA and 5 S r RNA + 32 different proteins and the 30 S small subunit consists of 16 S r RNA + 21 different proteins. (see fig. 38.19).

In eucaryotes, the large or 60 S subunit consists of 28 S r RNA and a 5 S r RNA + 50 different proteins and the 40 S small subunit consists of 18 S r RNA and a 5.8 S r RNA + 33 different proteins.

(*ii*) Transfer RNA (t RNA). The t RNA is the smallest of the known RNA species. This displays somewhat identical structure in procaryotes and eucaryotes.

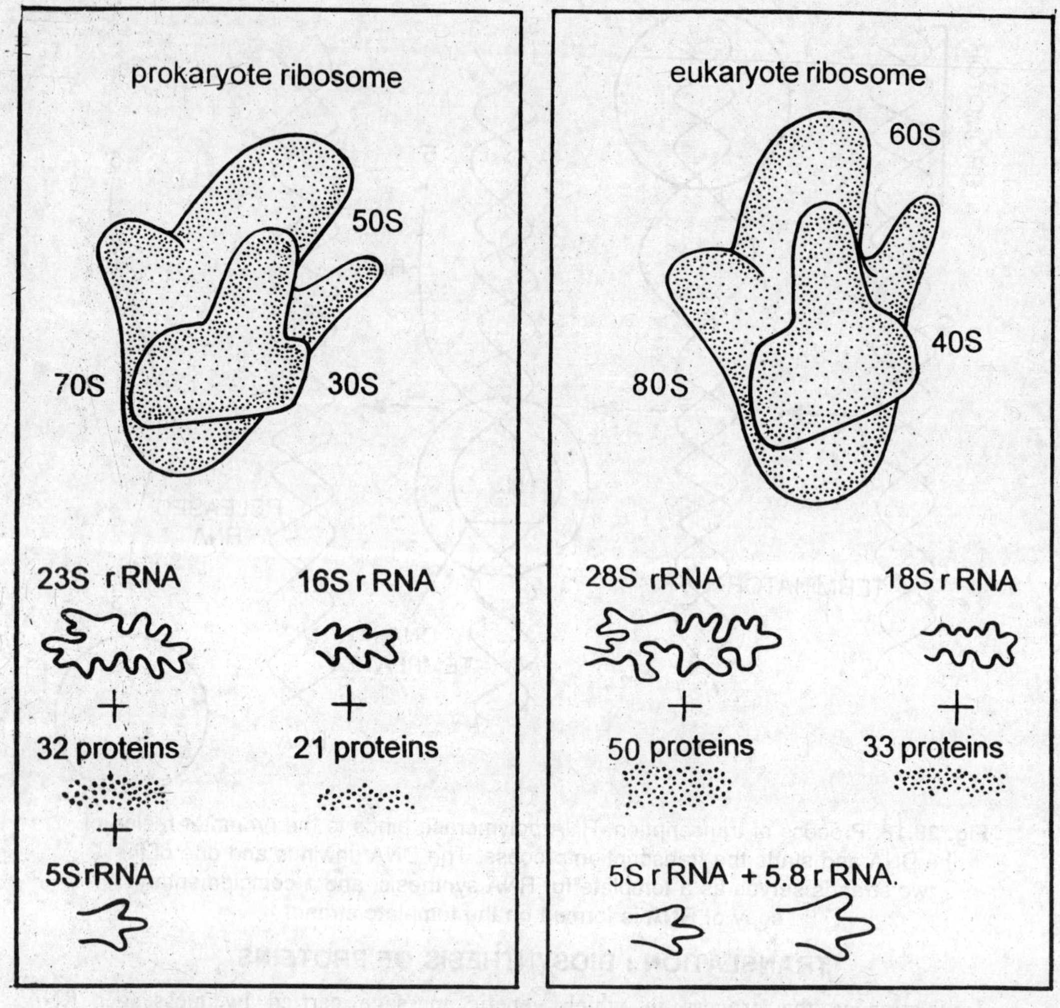

Fig. 38.19. Generalised structure of ribosome in prokaryotes and eukaryotes.

In 1965, R. W. Holley proposed a clover-leaf model of t RNA which results from intra-strand base pairing creating paired stems and unpaired loops. In three dimensions, t RNA looks like an L-shaped molecule. (see fig. 38.20)

***(iii)* Amino acids.** The 3′ end of t RNA contains the sequence for amino acid attachment. The anticodon loop carries an anticodon complementary to the known amino acid for which it is specific.

Steps in Translation

Twenty naturally occurring amino acids take part in protein synthesis.

Translation exhibits that a t RNA is specifically linked to an amino acid and this process is known as charging. This process is brought about under the direction of an enzyme called aminoacyl t RNA synthetase which is highly specific, *i.e.*, each recognizes only one amino acid.

Charging takes place in two steps. They are as follows :

$$\textbf{1.}\ \text{Amino acid} + \text{ATP} \xrightarrow[\text{t RNA synthetase}]{\text{Aminoacyl}} \text{Aminoacyl adenylic acid} + \text{P} \sim \text{P}$$

2. Aminoacyl adenylic acid + t RNA → Charged t RNA (t RNA-amino acid)
= P – A + aminoacyl t RNA synthetase

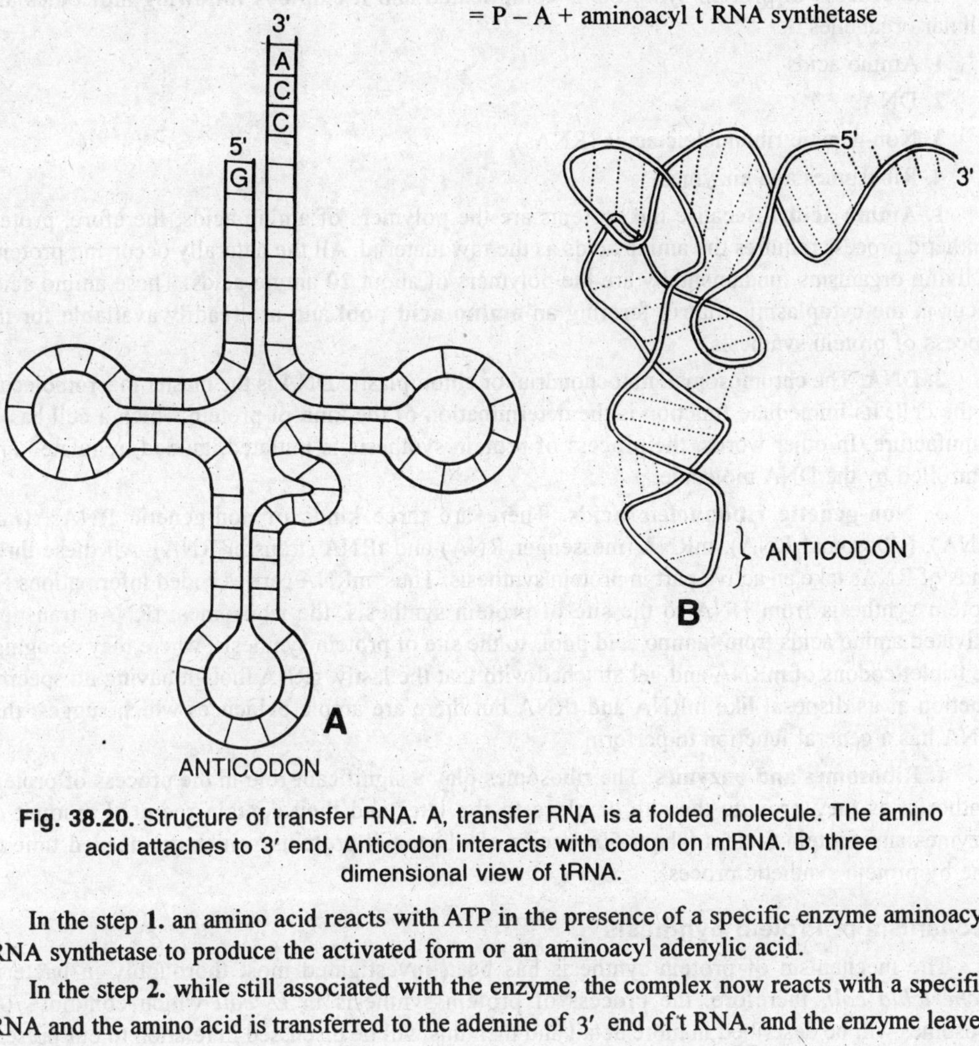

Fig. 38.20. Structure of transfer RNA. A, transfer RNA is a folded molecule. The amino acid attaches to 3′ end. Anticodon interacts with codon on mRNA. B, three dimensional view of tRNA.

In the step 1. an amino acid reacts with ATP in the presence of a specific enzyme aminoacyl t RNA synthetase to produce the activated form or an aminoacyl adenylic acid.

In the step 2. while still associated with the enzyme, the complex now reacts with a specific t RNA and the amino acid is transferred to the adenine of 3′ end of t RNA, and the enzyme leaves the complex.

The more important steps in the process of protein synthesis (*i.e.*, translation) are as follows :

PROTEIN SYNTHESIS

The proteins have significant role in the structural and functional organisation of the cell. The structural proteins constitute various cellular components and some extracellular parts such as cuticle and fibres, etc. The functional (*e.g.*, enzymatic and hormonal) proteins control almost all metabolic, biosynthetic, bioenergetic, growth regulating sensory and reproductive activities of the cell. All the proteins which are required by the cell for its different purposes are synthesized by the cell itself intracellularly.

A cell, however, cannot synthesize all the required proteins of multicellular individual, therefore, different kinds of proteins are synthesized by different types of cells. Thus, the liver cells are specialized for the synthesis of liver and plasma proteins, the β-cells of pancreatic islets are specialized for synthesis of proteins of insulin hormone and cells of pituitary are specialized of synthesis of protein component of pituitary hormone and so on.

Components of Protein Synthetic Machinery

The process of protein synthesis is complicated and it employs following molecules and cellular organelles :

1. Amino acids

2. DNA

3. Non-genetic ribonucleic acids (RNAs),

4. Ribosomes and enzymes.

1. Amino acids. Because the proteins are the polymers of amino acids, therefore, protein synthetic process requires the amino acids as the raw material. All the naturally occurring proteins of living organisms fundamentally are the polymers of about 20 amino acids. These amino acids occur in the cytoplasmic matrix forming an **amino acid pool** and are readily available for the process of protein synthesis.

2. DNA. The chromosomal, mitochondrial or chloroplastic DNA is the master macromolecule of the cell. Its immediate function is the determination of the kind of protein which a cell has to manufacture. In other words, the process of protein synthesis is initiated, guided, regulated and controlled by the DNA molecule.

3. Non-genetic ribonucleic acids. There are three kinds of non-genetic RNAs, (*i.e.*, rRNA), (ribosomal RNA), mRNA (messenger RNA) and tRNA (transfer RNA). All these three kinds of RNAs take an active part in protein synthesis. Thus, mRNA carries coded informations for protein synthesis from DNA to the site of protein synthesis, the ribosomes; tRNAs transport activated amino acids from 'amino acid pool' to the site of protein synthesis, where they recognise the triplet codons of mRNA and get attached with that the lastly, rRNA though having no specific function at its disposal like mRNA and tRNA but there are ample evidences which suggest that rRNA has a general function to perform.

4. Ribosomes and enzymes. The ribosomes play a significant role in the process of protein synthesis, as they provide the site or place to the later and they contain required amount of enzymes and protein factors (about 50 species of ribosomal proteins) which are needed time to time by protein synthetic process.

Mechanism of Protein Synthesis

The mechanism of protein synthesis has been investigated most thoroughly in bacteria *Escherichia coli*, therefore, the process of protein synthesis in *E. coli* which contains 70S ribosomes will be described in more detail and then this will be discussed in relation to our present knowledge of protein synthesis of 80S ribosomes of plants and animals.

A. Protein Synthesis on 70S Ribosomes

The mechanism of protein synthesis on 70S ribosomes includes the following stages :

1. The transcription. The process of protein synthesis is started by the uncoiling of strands of DNA molecule. One strand of DNA molecule acts as a template for the formation of mRNA. The mRNA is formed according to the triplet codes of DNA by the copying or transcription process. As soon as the mRNA is formed it leaves the nucleus and reaches in the cytoplasm where it attaches with the 30S subunit of the ribosomes. The mRNA carries the triplet codons for the synthesis of the proteins.

2. The attachment of mRNA with 30S ribosome and formation of polyribosome. In prokaryotic cells it has been observed that before the process of protein synthesis the ribosomes occur in dissociated and inactive state. The mRNA binds with the 30S ribosomal subunit in the presence of F_2 protein factor. Soon the N-formylmethionyl-tRNA (F-met-tRNAf) comes from the cytoplasm and binds with the first triplet codon (AUG or AUA) of the mRNA to initiate the process of protein synthesis and to form the **initiation complex.** The formation of initiation complex is aided by the GTP (Guanosine triphosphate) and 3 protein factors (F_1, F_2 and F_3). After the formation of initiation complex the 30S ribosomal subunit unites with 50S ribosomal subunit to form the 70S ribosome. The union of ribosomal subunits occurs in the presence of Mg^{++} ions and F_1 and F_2 factors. The codes of mRNA are not read only by one ribosome but many ribosomes move and read the codes of the mRNA. When many ribosomes bind with the mRNA the formation of polysome or polyribosome occurs.

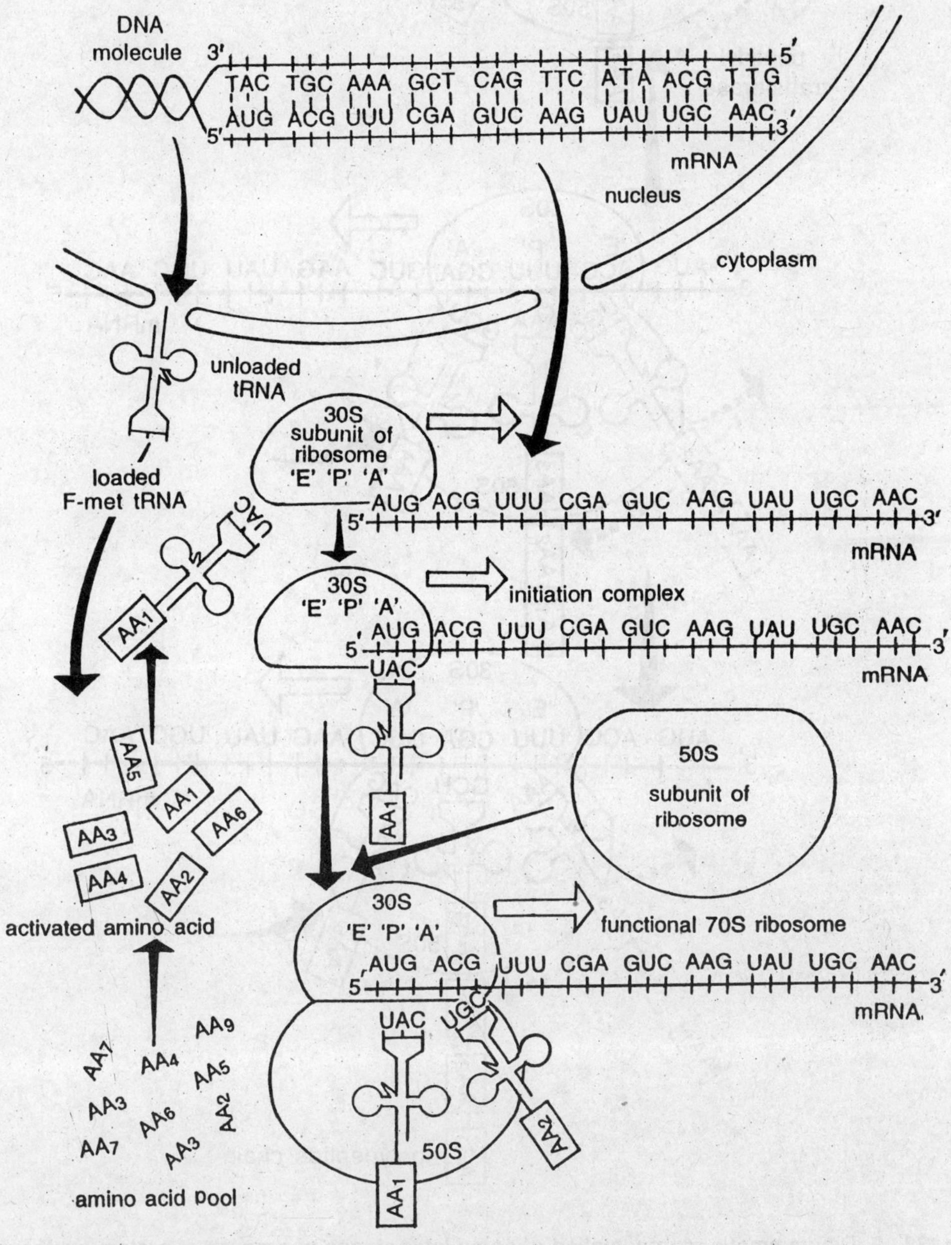

Fig 38.21. Diagrammatic representation of the some early steps of the process of protein synthesis of *E. coli.*

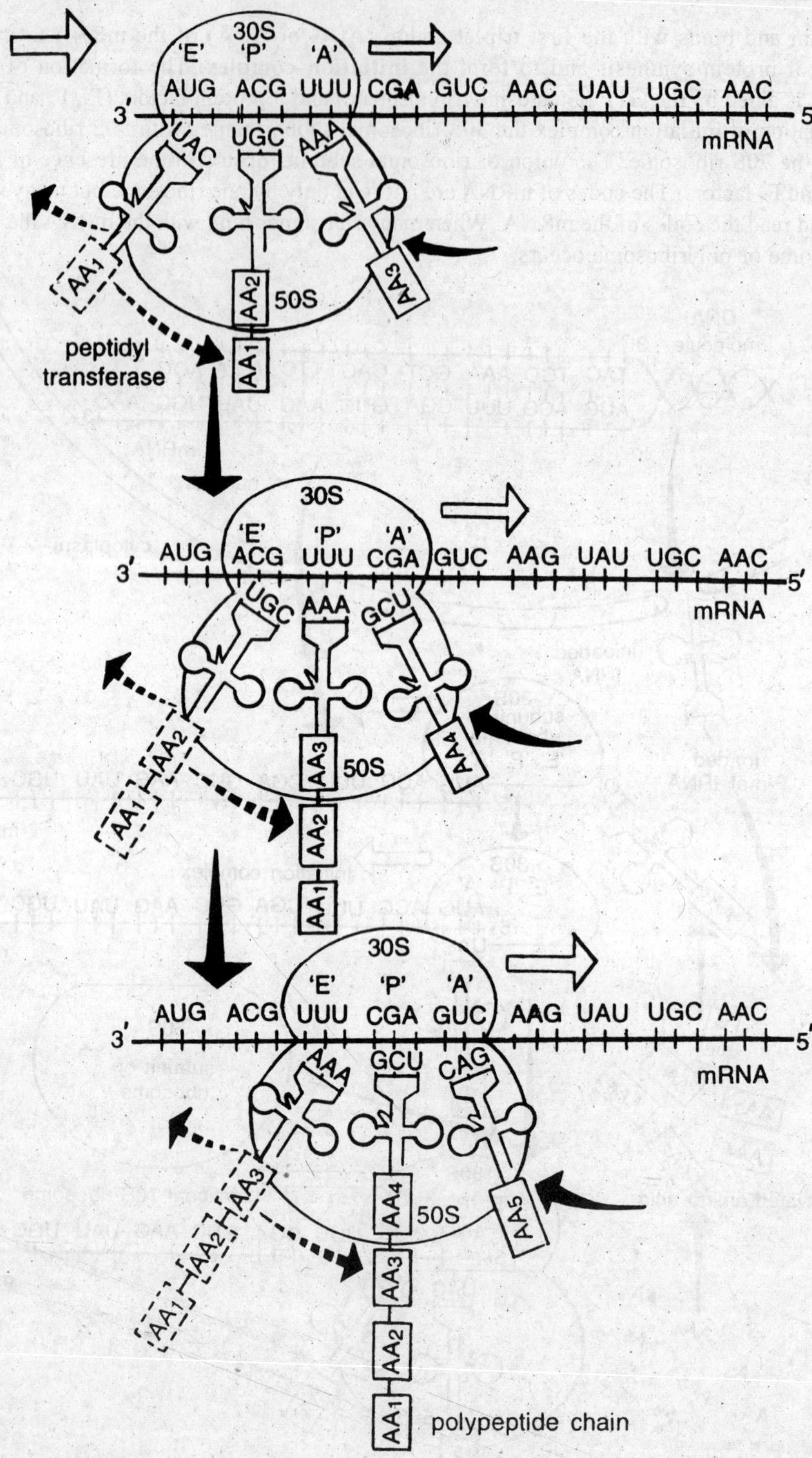

Fig. 38.22. Diagrammatic representation of some later stages of process of protein synthesis of *E. coli.* AA_3 = phenylalanine, AA_4 = arginine, AA_5 = valine, AA_6 = lysine.

3. The transfer of amino acids to the site of protein synthesis. The amino acids are transferred from the intra-cellular amino acid pool to the ribosomes by the tRNA. The transfer processes occur in various stages which are as follows.

(*a*) Activation of amino acids. Each of the 20 amino acids occurs in the cytoplasm in an inactive state. Each amino acid before its attachment with its specific tRNA is activated by a specific activating enzyme known as the **aminoacyl synthetase** and ATP. The free amino acids react with ATP, resulting the production of aminoacyl adenylate and pyrophosphate :

$$\underset{\text{Amino acid}}{AA} + ATP + \underset{\text{Aminoacyl synthetase}}{\text{enzyme}} \longrightarrow \underset{\text{Aminoacyl adenylate synthetase complex}}{AA \sim AMP - \text{enzyme}} + \underset{\text{Pyrophosphate}}{PP}$$

The reaction product aminoacyl adenylate is bound to the enzyme in the form of a monocovalent complex. This aminoacyl adenylate enzyme complex then esterifies to specific tRNA molecule. The cell has at least 20 aminoacyl synthetase enzymes for the 20 amino acids. Each enzyme is specific and it attaches with the specific amino acid without any error.

(*b*) Attachment of activated amino acid to tRNA. The aminoacyl adenylate remains bounded with enzyme until it collides with the specific tRNA molecule. Then, the carboxyl group of acid of amino acid residue of aminoacyl adenylate is transferred to the 3 OH group of the ribose of the terminal adenosine at CCA end of tRNA. As a result AMP and enzyme are released and a final product **aminoacyl tRNA** is formed by the following method :

'A' = aminoacyl site or decoding site, 'P' = peptidyl site or condensing site, 'E' = exist site,

AA_1 = N-formyl methionine.

AA_2 = threonine.

$$\underset{\text{Aminoacyl adenylate and enzyme}}{AA\text{—}AMP\text{—}Enzyme} + tRNA \longrightarrow \underset{\text{Aminoacyl—tRNA}}{AA\text{—}tRNA} + AMP + Enzyme$$

The aminoacyl-tRNA moves towards the site of protein synthesis, *i.e.*, ribosomes with mRNA.

4. Initiation of protein synthesis. As we have already mentioned that the initiation of protein synthesis in the bacterium *Escherichia coli* (*E. coli*) involves the formation of 70S complex. In it the mRNA always has first triplet codon as AUG at its beginning (*i.e.*, 5′ end). The AUG codons are the codes for the amino acid methionine. The methionine remains formulated and it has a very important role in initiating the process of protein synthesis. In every type of protein the formyl methionine occupies the first place in the molecule and when the protein molecule is completely synthesized then the formyl methionine often detaches from the newly synthesized protein molecule by the activity of a hydrolytic enzyme.

Since in protein synthesis the peptide chain always grows in the sequence from the free terminal amino (—NH_2) group towards the carboxyl (—COOH) end, the function of formyl methionine —tRNA (AA_1—tRNA in figs. 38.22 and 38.23) is to ensure that proteins are synthesized in that direction. In the formyl methionine—tRNA, the amino (—NH_2) group is blocked by the formyl group leaving only the —COOH group available to react with the—NH_2 group of the second amino acid (AA_2). In this way the synthesis of protein chain follows in the correct sequence.

5. Elongation of polypeptide chain. With the formation of functional 70S ribosome (*i.e.*, 70S—mRNA—F met tRNA), the elongation of polypeptide chain is brought about by the regular addition of amino acids and relative movements of ribosome and mRNA in the presence of GTP molecules, so that a new triplet codon remains available for new aminoacyl tRNA at the decoding or 'A' site of ribosome in each step. Thus, F-met tRNA, must move from decoding site ('A' site) to peptidyl site or 'P' site before the second aminoacyl—tRNA (*i.e.*, AA_2 tRNA) can bind to the next triplet codon occurring at decoding or 'A' site of ribosome. The aminoacyl-tRNA (AA_2 tRNA) binds with the codon of 'A' site in the presence of GTP and two proteins, called **transfer factors** (designated Tn and Ts) which remain associated with ribosomes. During this binding process, a complex is formed between GTP, the transfer factors and the aminoacyl tRNA (*viz.*, AA_2 -tRNA), which ultimately deposits aminoacyl tRNA at the 'A' site of ribosome with the release of transfer factors-GDP complex and inorganic phosphate.

In the next step, due to relative movement of ribosome and mRNA in the presence of single GTP molecule, the next triplet codon (*i.e.*, UUU codon in figs. 38.21 and 38.22) becomes available for next aminoacyl tRNA (*viz.* AA_3-tRNA) at the 'A' site of ribosome. At this stage, f met-tRNA occurs at exit of 'E' site, while, AA_2-tRNA occurs at peptidyl 'P' site. Now, an enzyme known as **transferase** I kicks off tRNA from formyl-methionine (f-met or AA_1) and flips the formyl-methionine to the aminoacyl-tRNA (AA_2-tRNA) bound at the peptidyl or 'P' site. The **'G' factor** is supposed to release the discharged or deacylated tRNA from the 'E' site of the ribosome.

It follows the next stage of elongation process which involves the synthesis of a peptide bond by a reaction between the free amino group of the incoming amino acid (*i.e.*, AA_2) and the carboxyl group of the first amino acid (AA_1) which is esterified to tRNA. The enzyme which catalyses this reaction is called **peptidyl transferase** (or peptide synthetase) and is an integral part of the 50S subunit. The energy for peptide bond synthesis is derived from cleavage of the ester link between an amino acid and its tRNA.

Thus, during the elongation of polypeptide chain each charged or loaded tRNA (Aminoacyl tRNA) enters the decoding or 'A' site, moves to the condensing or 'P' site, transfers its amino acid to the carboxyl end of the polypeptide, moves to the exit site, where the polypeptide chain is transferred to the adjacent tRNA on the condensing site, and tRNA is then released from the ribosome. This process is known as **translocation.** This results in the third codon coming into the A site and an appropriate tRNA charged with a third amino acid would bind at the A site. The process of peptide bond formation and translocation would be repeated. Thus, as the mRNA moves with respect to the ribosome, all the codons would be exposed at A site and the peptide chain grows. This sequence of events involved in elongation must take place very rapidly since it has been calculated that, *E. coli* growing under optimal conditions, a polypeptide chain of about 40 amino acids can be produced in 20 seconds.

6. Termination and release of polypeptide chain. When the synthesis of polypeptide chain is completed according to the codons of mRNA, then process of termination and release of polypeptide chain takes place. The termination of the polypeptide chain and the release of the completed chain from the ribosome is controlled by two factors. The termination of the chain is indicated by three special termination triplets in the mRNA which, in bacteria, are UAG, UAA and UGA. These are the so called **non-sense codons** which do not encode for any amino acid. The polypeptide chain, however, still bound to the tRNA which in turn is attached to the mRNA. The chain is released from the ribosome under the direction of three distinct proteins which are called **release factors** and are designated R_1 R_2 and S. These are bound to the ribosome and control the hydrolysis of the ester link between tRNA and the polypeptide chain. Once the chain has been terminated and released, the ribosome separates from mRNA and dissociates into its two subunits due to the action of F_3 factor. It is ready to enter a new cycle of polypeptide synthesis.

7. Modification of released polypeptide. The released polypeptide chain contains the formulated methionine at its one end. An enzyme **deformylase** removes the formyl group of methionine. The **exopeptidase** enzyme may remove some amino acids from N-terminal end or the C-terminal end of polypeptide chain. At this stage the polypeptide (protein) possesses its primary structure; at least some portion of many proteins have a secondary structure in the form of an alpha-helix. The protein chain may then fold back up itself, forming internal bonds (including storing disulphide bonds) which stabilize its tertiary structure into a precisely and often intricately folded pattern. Two or more tertiary structures may unite into a functional quarternary structure. For example, haemoglobin consists of four polypeptide chains, two identical α-chains and two identical β-chains. A protein does not become an active enzyme until it has assumed its tertiary or quaternary pattern.

The process of initiation, elongation and termination involve the mediation of several protein factors and hydrolysis of GTP (guanosine triphosphate) to provide energy. The detailed steps discussed are applicable to procaryotes and with some modifications also to eucaryotes.

B. Protein Synthesis of 80S Ribosomes

The process of protein synthesis in eukaryotes is found to be similar to the mechanism of protein synthesis of 70S ribosomes of prokaryotes. However, the process of initiation of polypeptide of 80S ribosomes of eukaryotes, differs with that of prokaryotes, in following two aspects :

1. In eukaryotes, formulation of methionine does not take place.

2. In eukaryotes, the smaller subunit associates with met $t\text{RNA}_f^{met}$ without the help of *m*RNA.

The initiation process of eukaryotes (*e.g.,* mammals, wheat and bean) is found to include following stages :

(*i*) $40S + \text{met} - t\text{RNA}_f^{met} \longrightarrow 40S - \text{met} - t\text{RNA}_f^{met}$

(*ii*) $40S\text{—met} - t\text{RNA}_f^{met} + m\text{RNA} \longrightarrow 40S\text{-}m\text{RNA-met-}t\text{RNA}_f^{met}$

(*iii*) $40S\text{—}m\text{RNA-met-}t\text{RNA}_f^{met} + 60S \longrightarrow 80S\text{-}m\text{RNA-met—}t\text{RNA}_f^{met}$

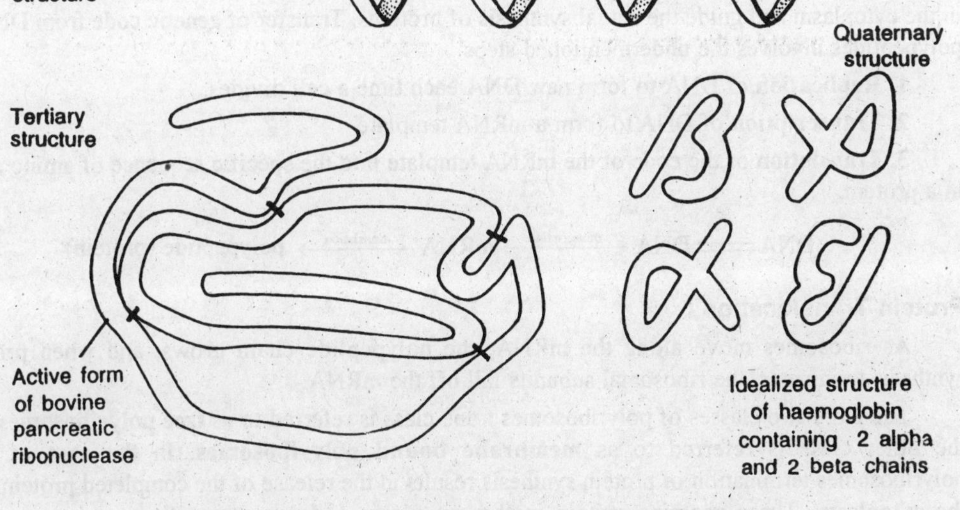

Fig. 38.23. Stages in the development of a functional protein.

Role of Polyribosomes

Since each ribosome moves along the mRNA that synthesises a polypeptide chain by translating the base sequence into the amino acid sequence, more ribosomes get attached one after another. Each ribosome takes the course from the initiation codon the termination codon. The

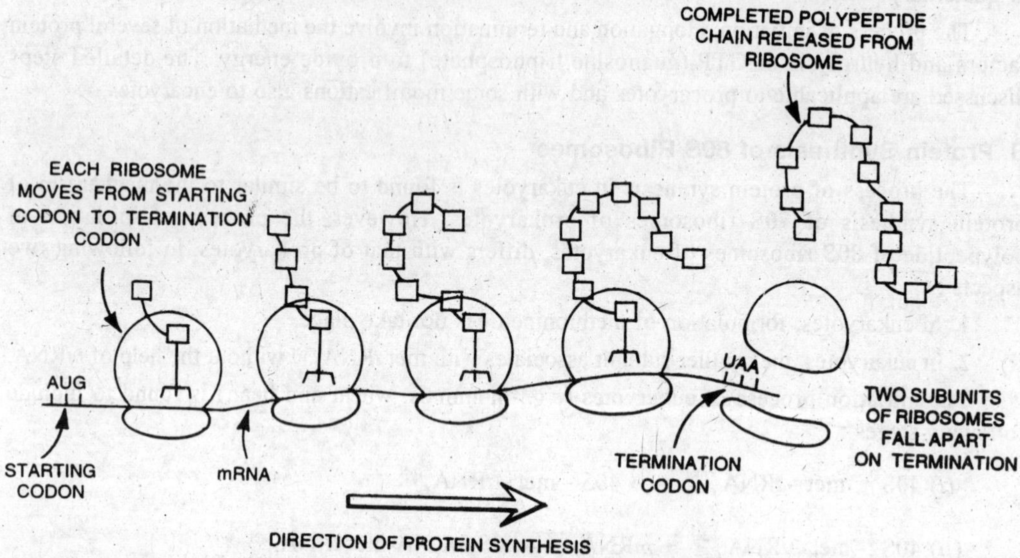

Fig. 38.24. A polyribosome (polysome).

whole complex with several ribosomes attached to a single mRNA looks like beads on a string and is known as **polyribosome** or **polysome.** This ensures the synthesis of several polypeptides simultaneously.

DNA, the Master Copy

DNA does not directly control the formation of a polypeptide. However, it forms an intermediate template mRNA molecule, which directs the synthesis of polypeptide. The DNA can be compared to a **master copy** that is carefully preserved in the nucleus and used for preparing **working copies** in the form of mRNA molecules. The mRNA molecules pass out to the ribosomes in the cytoplasm and guide the actual synthesis of proteins. Transfer of genetic code from DNA to polypeptides involves the undermentioned steps.

1. Replication of DNA to form new DNA each time a cell divides.

2. Transcription of DNA to form a mRNA template

3. Translation of the code of the mRNA template into the specific sequence of amino acids in a protein.

$$\text{DNA} \rightleftharpoons \text{DNA} \xrightarrow{\text{transcription}} \text{mRNA} \xrightarrow{\text{translation}} \text{polypeptide (protein)}$$

Protein Translocation

As ribosomes move along the mRNA, the polypeptide chain grows, and when protein synthesis terminates, the ribosomal subunits fall off the mRNA.

There are two classes of polyribosomes : one class is referred to as **free polyribosomes** and the other class is referred to as **membrane bound polyribosomes**. In the case of free polyribosomes termination of protein synthesis results in the release of the completed protein into the cytoplasm. These proteins remain in the cytoplasm and some specific proteins are then

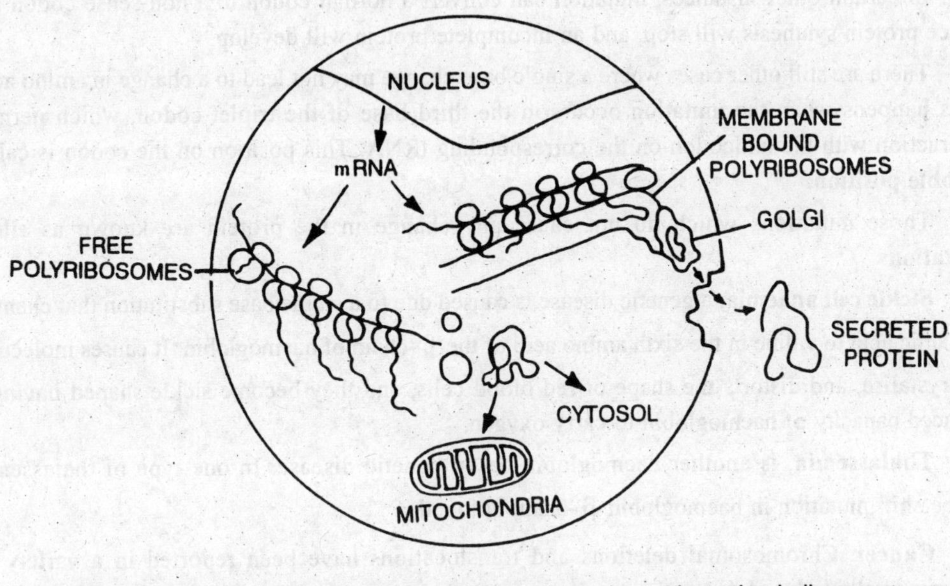

Fig. 38.25. Synthesis of proteins on free and membrane bound polyribosomes.

translocated into mitochondria and nucleus through special mechanisms. While in the case of membrane-bound polyribosomes, the polypeptide chain as it grows on the mRNA is inserted into the lumen of the ER (endoplasmic reticulum) membrane. Some of the proteins are integrated into the membrane and become integral membrane proteins. Some others are released into the lumen and are incorporated in Golgi vesicles. Now by the addition of certain sugar residues, these proteins undergo **glycosylation.** The Golgi vesicles fuse with the plasma membrane and the proteins are secreted out.

Effect of Mutations on Protein Structure and Function

As already stated that during DNA replication, the positioning of the correct base is maintained; and if any error persists, that is being corrected by proof reading and repair mechanisms. However, base changes are possible in DNA due to environmental mutagens such as UV rays, X-rays and certain chemicals. Hence, an A-T base pair may be changed into a G-C pair and this point mutation may be inherited by the progeny. In the similar way, insertion of an additional base may also cause mutation.

Under certain special condition cross-over and recombination between DNA strands can also lead to chunks of DNA being deleted or added. Thus mutations in genes lead to changes in mRNA and hence changes in the protein structure and function.

As already stated during cross-over and recombination between DNA strands may lead to chunks of DNA being deleted or added. Here shuffling of genes from one location to another is possible based on similarity of DNA sequences on the flanks of these genes. Such genes are popularly known as **'jumping genes'**.

The effect of such mutations will be reflected in the proteins. This is because the mRNA derived from the gene will lack a base, or will have an altered base. If the mutation involves loss or addition of a single base or a segment of DNA, then the enter reading frame will change from the site of mutation and one may end up in a protein with a new set of amino acids, and this is known as **frameshift mutation**.

In certain other instances, mutation can convert a normal codon to a non-sense codon and hence protein synthesis will stop, and an incomplete protein will develop.

There are still other cases where a single base change may not lead to a change in amino acid. This happens when the mutation occurs in the third base of the triplet codon, which permits interaction with the anticodon on the corresponding tRNA. This position on the codon is called **wobble position.**

Those mutations which do not cause any change in the protein are known as **silent mutations.**

Sickle cell anaemia, a genetic disease is caused due to a single base substitution that changes glutamic acid to valine in the sixth amino acid of the β-chain of haemoglobin. It causes molecules to crystalise, and distorts the shape of red blood cells, and they become sickle shaped having a reduced capacity of haemoglobin to carry oxygen.

Thalassemia, is another haemoglobin based genetic disease. In one type of thalassemia frame-shift mutation in haemoglobin β-chain is found.

Cancer. Chromosomal deletions and translocations have been reported in a variety of experimentally induced cancers.

Molecular Mechanism of Mutation

With the identification of biochemical nature of gene as a defined segment of DNA (deoxyribose nucleic acid), mutation mechanisms were worked out at molecular level.

This is now known that in DNA specific sequence of bases forms the reading frame in the form of genetic code for a gene.

Mutations are brought about by two main mechanisms which, alter the reading frame. (*i*) substitution and (*ii*) frame-shift mutations.

(*i*) **Substitution**. Here, one base is replaced by another base. For example, substitution of purine to purine or pyrimidine to pyrimidine is called **transition**, and purine to pyrimidine and *vice-versa* is known as **transversion**.

(*ii*) **Frame-shift mutation**. When the mutation involves loss or addition of a single base or a segment of DNA, then the entire reading frame will change from the site of mutation and one may end up in a protein with a new set of amino acids.

A specific base sequence within gene is crucial for it to express a particular phenotype.

Any change in the base sequence may change the codon and leads to mutation.

Each living cell has repair mechanisms to rectify such alterations. Only when these mechanisms are defective, the mutations can be induced.

GENETIC CODE

As DNA is a genetic material, it carries genetic informations from cell to cell and from generation to generation. At this stage, an attempt will be made to determine that in what manner the genetic informations are existed in DNA molecule? Are they written in articulated or coded language on DNA molecule? If in the language of codes what is the nature of genetic code?

A DNA molecule is composed of three kinds of moieties: (i) phosphoric acid,(ii) deoxyribose sugar, and (iii) nitrogen bases. The genetic informations may be written in any one of the three moieties of DNA. But the poly-sugar-phosphate backbone is always the same, and it is, therefore, unlikely that these moieties of DNA molecule carry the genetic informations. The nitrogen bases, however, vary from one segment of DNA to another, so the informations might well depend on their sequences. The sequences of nitrogen bases of a given segment of DNA molecules, actually, has been found to be identical to linear sequence of amino acids in a protein molecule. The proof of such a **colinearity** between DNA nitrogen base sequence and amino acid sequence in protein molecules has first obtained from an analysis of mutants of head protein of bacteriophage T_4 and the A protein of tryptophan synthetase of *Escherichia coli*. The colinearity of protein molecules and DNA polynucleotides has given the clue that the specific arrangement of four nitrogen bases (*e.g.*, A, T, C and G) in DNA polynucleotide chains, somehow, determines the sequence of amino acids in protein molecules. Therefore, these four DNA bases can be considered as four alphabets of DNA molecule. All the genetic informations, therefore, should be written by these four alphabets of DNA. Now the question arises that whether the genetic informations are written in articulated language or coded language? If genetic informations might have occurred in an articulated language, the DNA molecule might require various alphabets, a complex system of grammar and ample amount of space on it. All of which might be practically impossible and troublesome too for the DNA. Therefore, it was safe to conclude for molecular biologists that genetic informations were existed in DNA molecule in the form of certain special language of code words which might utilize the four nitrogen bases of DNA for its symbols. Any coded message is commonly called **cryptogram.**

BASIS OF CRYPTOANALYSIS

The basic problem of such a genetic code is to indicate how information written in a four letter language (four nucleotides or nitrogen bases of DNA) can be translated into a twenty letter language (twenty amino acids of proteins). The group of nucleotides that specifies one amino acid is a **code word** or **codon**. By the **genetic code** one means the collection of base sequences (codons) that correspond to each amino acid and to translation signals. The simplest possible code is a **singlet code** (a code of single letter) in which one nucleotide amino acid could be specified. A **doublet code** (a code of two letters)is also inadequate, because it could be specify only sixteen (4×4) amino acids, whereas a **triplet code** (a code of three letters) could specify sixty four ($4 \times 4 \times 4$) amino acids. Therefore, it is likely that there may be 64 triplet codes for 20 amino acids. The possible singlet, doublet and triplet codes, which are customarily represented in terms of **"mRNA language"** [mRNA is a complementary molecule which copies the genetic informations (crytogram of DNA) during its transcription], have been illustrated in Table 38.3

The first experimental evidence in support to the concept of triplet code is provided by **Crick** and coworkers in 1961. During their experiment, when they added or deleted single, or double base pairs in a particular region of DNA of T_4 bacteriophages of *E. coli,* they found that such bacteriophages ceased to perform their normal functions. However, bacteriophages with addition or deletion of three base pairs in DNA molecule, had performed normal functions. From this experiment, they concluded that a genetic code is in triplet form, because the addition of one or two nucleotides has put the reading of the code out of order, while the addition of third nucleotide resulted in a return to the proper reading of the message.

Table 38.3. Possible singlet, doublet and triplet codes of mRNA.

Single code (4 words)	Doublet code (16 words)				Triplet code (64 words)			
					AAA	AAG	AAC	AAU
					AGA	AGG	AGC	AGU
					ACA	ACG	ACC	ACU
					AUA	AUG	AUC	AUU
					GAA	GAG	GAC	GAU
					GGA	GGG	GGC	GGU
A	AA	AG	AC	AU	GCA	GCG	GCC	GCU
G	GA	GG	GC	GU	GUA	GUG	GUC	GUU
C	CA	CG	CC	CU	CAA	CAG	CAC	CAU
U	UA	UG	UC	UU	CGA	CGG	CGC	CGU
					CCA	CCG	CCC	CCU
					CUA	CUG	CUC	CUU
					UAA	UAG	UAC	UAU
					UGA	UGG	UGC	UGU
					UCA	UCG	UCC	UCU
					UUA	UUG	UUC	UUU

CODON ASSIGNMENT

(Cracking the Code or Deciphering the Code)

The genetic code has been cracked or deciphered by the following kinds of approaches :

A. Theoretical Approach

George Gamow proposed the **diamond code** (1954) and the **triangle code** (1955) and suggested the following properties of the genetic code :

(*i*) **A triplet codon** corresponding to one amino acid of the polypeptide chain.

(*ii*) **Direct template translation** by codon-amino acid pairing.

(*iii*) Translation of the code in an **overlapping** manner.

(*iv*) **Degeneracy** of the code, *i.e.,* an amino acid being coded by more than one codon.

(*v*) **Colinearity** of nucleic acid and the primary protein synthesized.

(*vi*) **Universality** of the code, *i.e.,* the code being essentially the same for different organisms.

B. The *in vitro* codon Assignment

1. Discovery and use of polynucleotide phosphorylase enzyme. Manago and **Ochoa** isolated an enzyme from the bacteria (*e. g., Azobacter vinelandii*) that catalyzes the breakdown of RNA in bacterial cells. This enzyme is called **polynucleotide phosphorylase.** They found that outside of the cell (*in vitro*), with high concentrations of ribonucleotides, the reaction could be driven in reverse and an RNA molecule could be made. Incorporation of bases into the molecule is random and does not require a DNA template. Thus, in 1955 **Manago** and **Ochoa** made possible the artificial synthesis of polynucleotides (=mRNA) containing only a single type of nucleotides (U, A, C, or G respectively) repeated many times.

Polynucleotide	Configuration
1. Polyuridylic acid or poly (U)	UUUUUU
2. Polyadenylic acid or poly (A)	AAAAAA
3. Polycytilic acid or poly (C)	CCCCCC
4. Polyguanidylic acid or poly (G)	GGGGGG

Thus, the action of polynucleotide phosphorylase can be represented in the following way:

$$(RNA)_n + \text{ribonucleoside diphosphate} \xrightleftharpoons[]{\text{polynucleotide phosphorylase}} (RNA)_{n+1} + Pi$$

The polynucleotide phosphorylase enzyme differs from RNA polymerase used to transcribe mRNA from DNA polymerase used to transcribe mRNA from DNA in that : (i) it does not require a template or primer; (ii) the activated substrates are ribonucleoside diphosphates (*e. g.*, UDP, ADP, CDP and GDP) and not triphosphates; and (iii) orthophosphate (Pi) is produced instead of pyrophosphates (PPi).

The deciphering of the genetic code was made possible by the use of synthetic (or artificial) polynucleotides and trinucleotides. The different types of techniques used include the use of polymers containing a single type of nucleotide (called **homopolymers),** the use of mixed polymers **(copolymers)** containing more than one type of nucleotides **(heteropolymers)** in random or defined sequences and the use trinucleotides (or **"minimessengers")** in ribosome-binding or filter-binding.

2. Codon assignment with unknown sequence. (i) Codon assignment by homopolymer. The first clue to codon assignment was provided by **Nirenberg and Matthaei** (1961) when they used in *vitro* system for the synthesis of a polypeptide using an artificially synthesized mRNA molecule containing only one type of nucleotide (*i.e.,* homopolymer). Prior to performing the actual experiments, they tested the ability of a cell-free protein synthesizing system to incorporate radioactive amino acids into newly synthesized proteins. Their cell-free extracts of *E.coli* contained ribosomes, tRNAs, aminoacyl-tRNA synthetase enzymes, DNA and mRNA. The DNA of this extract was eradicated by the help of **deoxyribonuclease** enzyme, thus, the template which might synthesize new mRNA was destroyed. When twenty amino acids were added to this mixture along with ATP, GTP, K^+ and MG^{2+}, they were incorporated into proteins. This incorporation continued so long as mRNA was present in such a cell-free suspension. It also continued in the presence of synthetic polynucleotides (mRNAs) which could be made with the help of polynucleotide phosphorylase enzyme.

The first successful use of this technique was made by **Nirenberg and Matthaei** who synthesized a chain of uracil molecules (poly U) as their synthetic mRNA (homopolymer). Poly (U) seemed a good choice, because there could be no ambiguity in a message consisting of only one base. Poly (U) was good choice for other reasons: it binds well to ribosomes and, as it turned out, the product protein was insoluble and easy to isolate. When poly(U) was presented as the message to the cell-free system containing all the amino acids, one amino acid was exclusively selected from the mixture for incorporation into the polypeptide, called **polyphenylalanine.** This amino acid was phenylalanine and it could be concluded that some sequence of UUU coded for phenylalanine. Other homogeneous chains of nucleotides (Poly A, Poly C and Poly G) were inactive for phenylalanine incorporation. The mRNA code word of phenylalanine was, therefore, shown to be UUU. The corresponding DNA code word for phenylalanine can be deduced to be AAA. Thus, the first code word to be deciphered was UUU.

This discovery was extended in the laboratories of **Nirenberg and Ochoa.** The experiment was repeated using synthetic **poly (A)** and **poly (C)** chains, which gave **polylysine** and **polyproline** respectively. Thus, AAA was identified as the code for lysine and CCC as code of proline. A poly

(G) message was found non-functional in *vitro,* since it attains secondary structure and, thus, could not attach to the ribosomes. In this way three of 64 codons were easily accounted for.

(ii) Codon assignment by heteropolymers (Copolymers with random sequences). Further exposition of the genetic code took place by using synthetic messenger RNAs containing two kinds of bases. This technique was used in the laboratories of **Ochoa** and **Nirenberg** and led the deduction of the composition of codons for the 20 amino acids. The synthetic messengers contained the bases at random (called **random copolymers).** For example, in a random copolymer using U and A nucleotides eight triplets are possible, such as UUU, UUA, UAA,UAU, AAA, AAU, AUU and AUA. Theoretically, eight amino acids could be coded by these eight codons. Actual experiments, however, yielded only six phenylalanine, leucine, tryosine, lysine asparagine and isoleucine. By varying the relative compositions of U and A in the random copolymer and determining the percentage of the different amino acids in the proteins formed, it was possible to deduce the composition of the code for different amino acids.

3. Assignment of codons with known sequences. (i) Use of trinucleotides or minimessengers in filter binding (Ribosome-binding technique). Ribosome binding technique of **Nirenberg** and **Leder** (1964) made use of the finding that aminoacyl-tRNA molecules specifically bind to ribosome-mRNA complex. This binding does not require the presence of a long mRNA molecule; in fact, the association of a **trinucleotide** or **minimessenger** with the ribosome is sufficient to cause aminoacyl-tRNA binding. When a mixture of such small mRNA molecules-ribosomes and amino acid-tRNA complexes are incubated for a short time and then filtered through a **nitrocellulose membrane,** then the mRNA-ribosome-tRNA-amino acid complex is retained back and rest of the mixture passes through the filter. By using a series of 20 different amino acid mixtures, each containing one radioactive amino acid at a time, it is possible to find out the amino acid corresponding to each triplet by analysing the radioactivity absorbed by the membrane, *e.g.*, the triplet GCC and GUU retain only alanyl-tRNA and valyl-tRNA respectively. All 64 possible triplets have been synthesized and tested in this way. Forty five of them have given clear-cut results. Later on, with the help of longer synthetic messages it has been possible to decipher 61 out of the possible 64 codons.

C. The *in vivo* Codon Assignment

The cell free protein synthetic systems, though have proved of great significance in decipherment of the genetic code, but they could not tell us whether the genetic code so deciphered is used in the living systems of all organisms also. Three kinds of techniques are used by different molecular biologists to determine whether the same code is also used in *vivo* (a) amino acid replacement studies (*e.g.*, tryptophan synthetase synthesis in *E. coli* (**Yanofsky** *et al.* 1963) and haemoglobin synthesis in man), (b) frameshift mutations (*e.g.*, investigations of **Terzaghi** *et. al.* 1966, on lysozyme enzyme of T_4 bacteriophages, and (c) comparison of a DNA or mRNA polynucleotide cryptogram with its corresponding polypeptide clear text (*e.g.*, comparison of amino acid sequence of the R_{17} bacteriophage coat protein with the nucleotide sequence of the R_{17} mRNA in the region of the molecule that dictates coat-protein synthesis by **S. Cory** *et. al.*, 1970).

Thus, *in vitro* and *in vivo* studies, so far described, gave the way to formulate a code table for twenty amino acids (see Table 38.4 and Table 38.5).

CHARACTERISTICS OF GENETIC CODE

The genetic code has the following general properties:

1. The code is a triplet codon. The nucleotides of mRNA are arranged as a linear sequence of codons, each codon consisting of three successive nitrogenous bases, *i.e.*, the code is a triplet codon. The concept of triplet codon has been supported by two types of point mutations: frameshift mutations and base substitutions.

Table 38.4. The genetic dictionary. The trinucleotide codons are written in the 5′→3′ direction.

First base	Second base U		Second base C		Second base A		Second base G		Third base
	U		C		A		G		
	UUA	Phe	UCU		UAU	Tyr	UGU	Cys	U
	UUC		UCC	Ser	UAC		UGC		C
U	UUA	Leu	UCA		UAA	Non-sense	UGA}	Non-sense codon	A
	UUG		UCG		UAG	codon	UGG}	Trp	G
	CUU		CCU		CAU	His	CGU		U
	CUC	Leu	CCC	Pro	CAC		CGC	Arg	C
C	CUA		CCA		CAA	Gln	CGA		A
	CUG		CCG		CAG		CGG		G
	AUU		ACU		AAU	Asn	AGU	Ser	U
	AUC	Ileu	ACC	Thr	AAC		AGC		C
A	AUA		ACA		AAA	Lys	AGA	Arg	A
	AUG*	Met	ACG		AAG		AGG		G
	GUU		GUC		GAU	Asp	GGU		U
	GUC	Val	GCC	Ala	GAC		GGC	Gly	C
G	GUA		GCA		GAA	Glu	GGA		A
	GUG		GCG		GAG		GGG		G

***AUG – Met or chain initiation codon.**

Table 38.5. Amino acids and their messenger RNA codons.

Amino acid	Codon	Amino acid	Codon	Amino acid	Codon	Amino acid	Codon
Alanine –	GCA	Glycine –	GGA	Leucine –	CUA	Threonine –	ACA
(Ala)	GCG	(Gly)	GGC	(Leu)	CUG	(Thr)	ACG
	GCC		GGC		CUC		ACC
	GCU		GGU		CUU		ACU
Arginine –	AGA	Glutamine –	CAA		UUA		
(Arg)	AGG	(Gln; Glun)	CAG		UUG	Tryptophan –	UGG
	CGA	Glutamic		Methionine –	AUG	(Trp)	
	CGG	acid –	GAA	(Met)		Tyrosine –	UAC
	CGC	(Glu)	GAG	(Starting		(Tyr)	UAU
	CGU	Histidine –	CAC	codon)		Valine –	GUA
Aspartic		(His)	CAU	Phenylalanine	UUC	(Val)	GUG*
acid –	GAC	(Isoleucine –	AUC	(Phe)	UUU		GUC
(Asp)	GAU	(Ile)	AUU				GUU
Asparagine–	AAC		AUA	Proline –	CCA		
(Asn; Aspn)	AAU			(Pro)	CCG	Terminator –	UAA
Cysteine –	UGC	Lysine –	AAA		CCC	(Nonsense	(ocher)
(Cys)	UGU	(Lys)	AAG		CCU	codons)	UAG
				Serine –	AGC		(amber)
				(Ser)	AGU		UGA
					UCA		(opal)
					UCG		
					UCC		
					ICU		

*GUC is also used as a start codon for some proteins.

(i) Frameshift mutations. Evidently, the genetic message once initiated at a fixed point is read in a definite frame in a series of three letter words. The framework would be disturbed as soon as there is a deletion or addition of one or more bases. When such frameshift mutations were intercrossed, then in certain combinations they produce wild type normal gene. It was concluded that one of them was deletion and the other an addition, so that the disturbed order of the frame due to mutation will be restored by the other (Fig. 38.26).

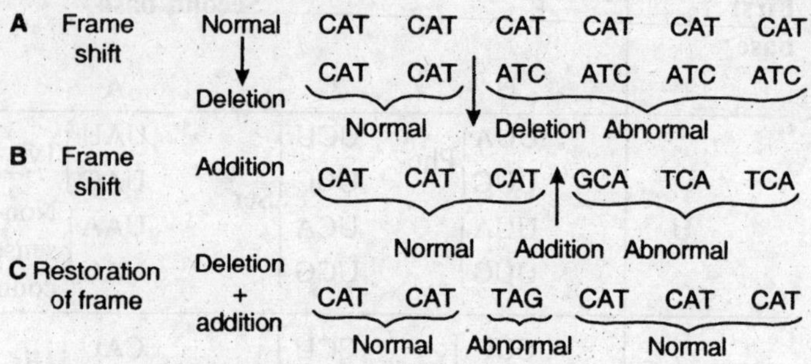

Fig. 38.26. Frame-shift mutations: A — Delection; B— Addition; C—Restoration of frame.

Fig. 38.27. Basses in mRNA are read sequentially in the 5′ to 3′ direction, in group of three bases.

(ii) Base substitution. If in a mRNA molecule at a particular point, one base pair is replaced by another without any deletion or addition, the meaning of one codon containing such an altered base will be changed. In consequence, in place of a particular amino acid at a particular position in a polypeptide, another amino acid will be incorporated. For example, due to substitution mutation, in the gene for tryptophan synthetase enzyme in *E.coli*, the GGA codon for glycine becomes a missence codon AGA which codes for arginine. **Missence codon** is a codon which undergoes an alteration to specify another amino acid.

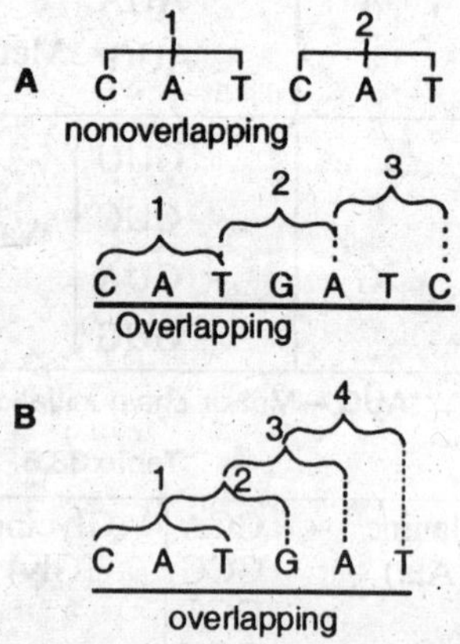

Fig. 38.28. A — Non-overlapping codons; B— Overlapping of codon due to one base; C—Overlapping of codon due to two bases.

A more direct evidence for a triplet code came from the finding that a piece of mRNA containing 90 nucleotides, corresponded to a polypeptide chain of 30 amino acids of a growing haemoglobin molecule. Similarly, 1200 nucleotides of "satellite" tobacco necrosis virus direct the synthesis of coat protein molecules which have 372 amino acids.

2. The code is non-overlapping. In translating mRNA molecules the codons do not *overlap* but are "read" sequentially (Fig. 38.27). Thus, a **non-overlapping code** means that a base in a mRNA is not used for different codons. In Figure 38.28, it has been shown that an overlapping code can mean coding for four amino acids from six bases. However, in actual practice six bases code for not more than two amino acids. For example, in case of an overlapping code, a single change (of substitution type) in the base sequence will be reflected in substitutions of more than one amino acid in corresponding protein. Many examples have accumulated since 1956 in which a single base substitution results into a single amino acid change in insulin, tryptophan synthelase, TMV coat protein, alkaline phosphatase, haemoglobin, etc.

However, it has been shown that in the bacteriophage ϕ ×174 there is a possibility of overlapping the genes and codons (**Barrel** and coworkers, 1976; **Sanger**, et al., 1977).

3. The code is commaless. The genetic code is commaless, which means that no codon is reserved for punctuations. It means that after one amino acid is coded, the second amino acid will be automatically, coded by the next three letters and that no letters are wasted as the punctuation marks (Fig. 38.29).

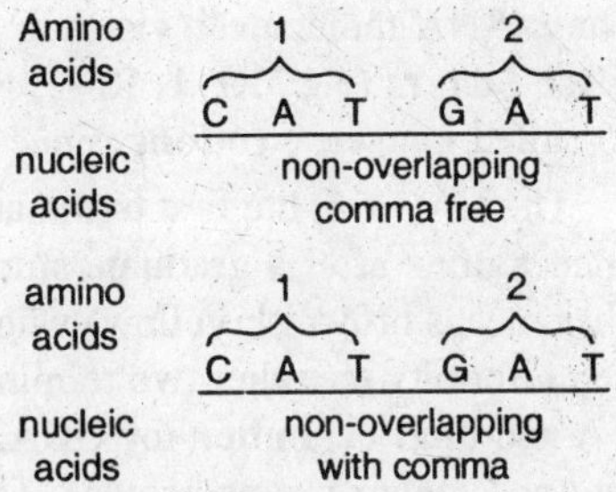

Fig. 38.29. Two forms of genetic code. A—Genetic code without comma; B—Genetic code with comma.

4. The code is non-ambiguous. Non-ambiguous code means that a particular codon will always code for the same amino acid. In case of ambiguous code, the same codon could have different meanings or in other words, the same codon could code two or more than two different amino acids. Generally, as a rule, the same codon shall never code for two different amino acids. However, there are some reported exceptions to this rule: the codons AUG and GUG both may code for *methionine* as initiating or starting codon, although GUG is meant for *valine*. Likewise, GGA codon codes for two amino acids *glycine* and *glutamic acid*.

5. The code has polarity. The code is always read in a fixed direction, *i.e.,* in the 5′→3′ direction. In other words, the codon has a **polarity**. It is apparent that if the code is read in opposite directions, it would specify two different proteins, since the codon would have reversed base sequence :

Codon :	UUG	AUC	GUC	UCG	CCA	ACA	AGG
Polypeptide:→	Leu	Ile	Val	Ser	Prof	Thr	Arg
	Val	Leu	Leu	Ala	Thr	Thr	Gly←

6. The code is degenerate. More than one codon may specify the same amino acid; this is called **degeneracy** of the code. For example, except for *tryptophan* and *methionine*, which have a single codon each, all other 18 amino acids have more than one codon. Thus, nine amino acids, namely *phenylalanine, tyrosine, histidine, glutamine, asparagine, lysine, aspartic acid, glutamic acid* and *cysteine*, have two codons each. Isoleucine has three codons. Five amino acids, namely *valine, proline, threonine, alanine* and *glycine*, have four codons each. Three amino acids, namely *leucine, arginine* and *serine*, have six codons each (see Table 38.5).

The code degeneracy is basically of two types : partial and complete. **Partial degeneracy** occurs when first two nucleotides are identical but the third (*i.e.,* 3′ base) nucleotide of the degenerate codons differs, *e.g.,* CUU and CUC code for leucine, **Complete degeneracy** occurs when any of the four bases can take third position and still code for the same amino acid (e.g., UCU, UCC, UCA and UCG code for serine).

Degeneracy of genetic code has certain biological advantages. For example, it permits essentially the same complement of enzymes and other proteins to be specified by microorganisms varying widely in their DNA base composition. Degeneracy also provides a mechanism of minimising mutational lethality.

7. Some codes act as start codons. In most organisms, AUG codon is the **start** or **initiation codon**, *i.e.,* the polypeptide chain starts either with **methionine** (eukaryotes) or **N-formylmethionine** (prokaryotes). Methionyl or N-formylmethionyl-tRNA specifically binds to the **initiation site** of mRNA containing the AUG initiation codon. In rare cases, GUG also serves as the initiation codon, *e.g.,* bacterial protein synthesis. Normally, GUG codes for valine, but when normal AUG codon is lost by deletion, only then GUG is used as initiation codon.

8. Some codes act as stop codons. Three codons UAG, UAA and UGA are the chain **stop or termination** codons. They do not code for any of the amino acids. These codons are not read by any tRNA molecules (via their anticodons), but are read by some specific proteins, called **release factors** (e.g., RF-1, RF-2, RF-3 in prokaryotes and RF in eukaryotes). These codons are also called **nonsense codons,** since they **do not** specify any amino acid.

The UAG was the first termination codon to be discovered by **Sidney Brenner** (1965). It was named **amber** after a graduate student named **Bernstein** (= the German word for 'amber' and amber means brownish yellow) who help in the discovery of a class of mutations. Apparently, to give uniformity the other two termination codons were also named after colours such as **ochre** for UAA and **opal** or **umber** for UGA. (*ochre* means yellow red or pale yellow; *opal* means milky white and *umber* means brown). The existence of more than one stop codon might be a safety measure, in case the first codon fails to function.

9. The code is universal. Same genetic code is found valid for all organisms ranging from bacteria to man. Such universality of the code was demonstrated by **Marshall, Caskey** and **Nirenberg** (1967) who found that *E.coli* (Bacterium), *Xenopus laevis* (Amphibian) and guinea pig (mammal) amino acyl-tRNA use almost the same code. **Nirenberg** has also stated that the genetic code may have developed 3 billion years ago with the first bacteria, and it has changed very little throughout the evolution of living organisms.

Recently, some differences have been discovered between the universal genetic code and mitochondrial genetic code (Table 38.6).

Table 38.6. Differences between the 'universal genetic code' and two mitochondrial genetic codes.

Codon	Mammalian mitocondrial code	Yeast mitochondrial code"	"Universal Code
1. UGA	*Trp* *	*Trp*	Stop
2. AUA	*Met*	*Met*	Ile
3. CUA	Leu	*Thr*	Leu
4. AGA 5. AGG	*Stop*	Arg	Arg

* Italic type indicates that the code differs from the 'universal' code.

Codon and Anticodon

The codon words of DNA would be complementary to the mRNA code words (*i.e.*, DNA codes run in 3′→5′ direction and mRNA code words run in 5′→3′ direction) and so thereby the three bases forming the **anticodon** of tRNA (*i.e.*, bases of anticodon run in 3′→5′ direction). Three bases of anticodon pair with the mRNA on the ribosomes at the time of aligning the amino acids during protein synthesis (*i.e.*, translation of mRNA into proteins which proceeds in mRNA → COOH direction). For example, one of two mRNA and DNA code words for the amino acid phenylalanine is UUC and AAG respectively, and the corresponding anticodon of tRNA is CAA. This indicates that codon and anticodon pairing is antiparallel. In this case, C pairs with G and U pairs with A.

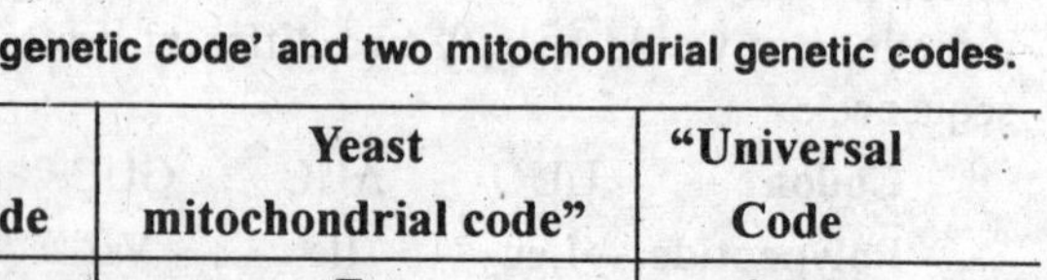

Fig. 38.30. Directions of synthesis of RNA and protein with respect to the coding strand of DNA.

WOBBLE HYPOTHESIS

To explain the possible cause of degeneracy of codons, **Crick** (1966) proposed the **wobble hypothesis** (wobble means to sway or move unsteadily). Since there are 61 codons specifying amino acids, the cell should contain 61 tRNA molecules, each with a different anticodon. Actually, however, the number of tRNA molecule types discovered is much less than 61. This implies that the anticodons of tRNAs read more than one codon on mRNA. For example, yeast $tRNA^{ala}$ with anticodon bases 5′ IGC 3′ (where I stands for inosine, a derivative of adenine or A) could bind to three codons in mRNA such as 5′ GCU 3′, 5′GCC3′ and 5′GCA3′. Inosine is frequently found as the 5′ base of the anticodon; at the time of pairing with the base of the codons it wobbles and can pair either with U, C or A of three codons (see Table 38.7). Thus, Crick's wobble hypothesis states that the *base at 5′ end of the anticodon is not spatially confined as the other two bases allowing it to form hydrogen bonds with any of several bases located at the 3′ end of a codon* (Fig. 38.31).

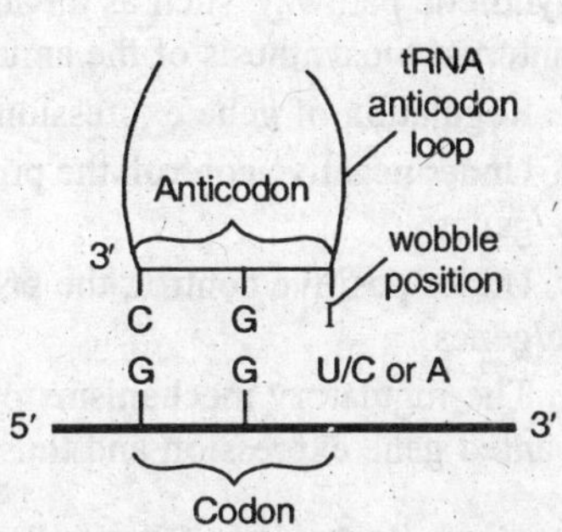

Fig. 38.31. Wobble hypothesis. In the third site (5′ end)) of the anticodon I can take either of three wobble positions, thus, being able to pair with either U, C or A. This means that a single tRNA species carrying an amino acid can recognize three codons in mRNA: GCU, GCC and GCA.

Table 38.7. Allowed base-pairing combinations according to the wobble hypothesis.

5′ base of codon	3′ base of anticodon
C	G
A	U
U	A or G
G	U, C or A
I	U, C or A

REGULATION OF GENE EXPRESSION

Now this is known to us that how the genetic information stored in DNA or gene is expressed through the process of protein synthesis.

The analysis reveals, there are vast differences in the concentration of different proteins, even in a simple bacterium *Escherichia coli*, where some proteins may be synthesised in low concentration as 5 to 10 molecules, while in others may be produced as 100,000 molecules per cell.

This phenomenon shows the regulation of gene expression.

Since 1900, this was known that lactose-metabolising enzymes are synthesised by yeast only when it is grown in the presence of lactose sugar.

Thereafter, bacteria were shown to adapt to their chemical environment by synthesising certain enzymes depending upon the substrate present.

Such adaptive enzymes are called **inducible enzymes.**

Contrary to it, some enzyme types were synthesised all the times irrespective of the chemical constitution of the environment, and are called **constitutive enzymes.**

Another type of regulation of gene expression was revealed when the end product of a biosynthetic pathway, such as an amino acid is provided in the medium. Under such a condition, the internal biosynthesis of the amino acid stops. Such regulation is called **repressible** regulation.

Regulation of gene expression can also be described as **negative** or **positive** control.

Under negative control, the product of the regulatory gene shuts off the expression of the said gene/genes.

Under positive control, the product of the regulatory gene activates transcription of the said gene/genes.

The regulatory mechanisms play an important role in the life of a cell as they prevent any unwanted gene expression and thus efficient energy management is ensured.

Induction (Inducible Control) and Repression (Repressible Control)

Induction. As we know, the genetic potential of a virus is limited, as it uses the host machinery for its replication.

On the other hand, the bacterial cell can synthesise its constituents from simple salts and sugar.

The bacterium *Escherichia coli* has about 2500 genes. The expression of these genes is usually controlled to achieve maximum cellular economy. This shows that genes will be turned on or off as per requirement.

A set of genes will be switched on when there is necessity to metabolise a new substrate.

This phenomenon is called **induction** and the small molecules responsible for this are known as **inducers.**

Repression. Similarly, when a metabolite needed by the bacterium (*E. coli*) is provided in excess from out side, the bacterium stops making it and thus conserves its reserves.

This condition is achieved by the added metabolite turning off a set of genes involved in producing that metabolite in the bacterial cell. This phenomenon is called **feed-back repression.**

The phenomena of induction and repression can be described on the basis of **operon concept.** This is as follows :

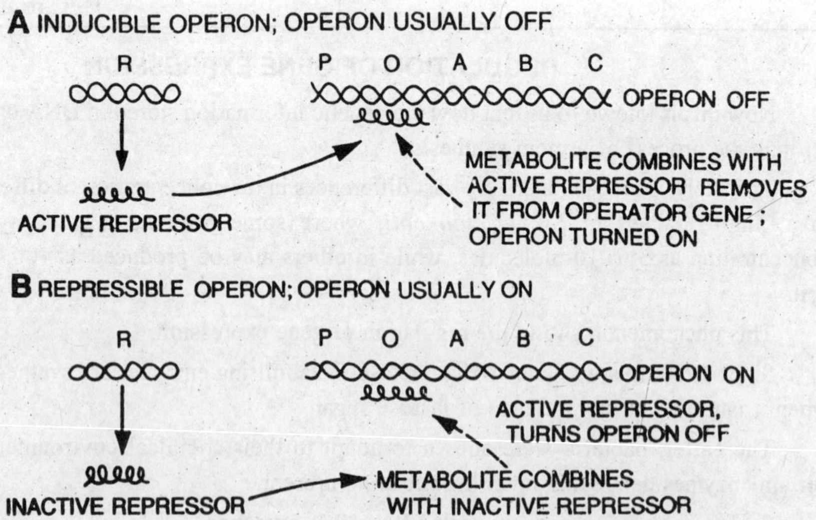

Fig. 38.32. Operons. A, inducible; B, repressible. Both the operons become off when operator site is bound by the repressor formed by regulator gene R.

Operon Concept

Francois Jacob and Jacques Monod (1961), while working at the Pasteur Institute, Paris, proposed that metabolic pathways are regulated as a unit. They studied the change in the nature of enzymes of *E. coli* when provided with different kinds of sugars. They found, when the sugar lactose is added to the cultures of *E. coli*, it induces three enzymes necessary to break down the lactose into galactose. These newly synthesized enzymes are – **(*i*)** β -galactosidase, **(*ii*)** permease and (*iii*) transacetylase. According to them, the synthesis of these three enzymes is controlled by a long DNA segment known as **operon** which is divisible into an operator site O and three structural genes Z, Y and A. The action of structural genes is regulated by operator site with the help of a **repressor** protein produced by the action of gene 'i' known as the **regulator gene.** The genes are expressed or not expressed that depends on whether the operator switch is on or off. When the switch is on, the three genes are transcribed by RNA polymerase into a single stretch of messenger RNA covering all the three genes. Each structural gene is generally known as **cistron** and the transcribed long m RNA covering all the cistrons is known as **polycistronic.**

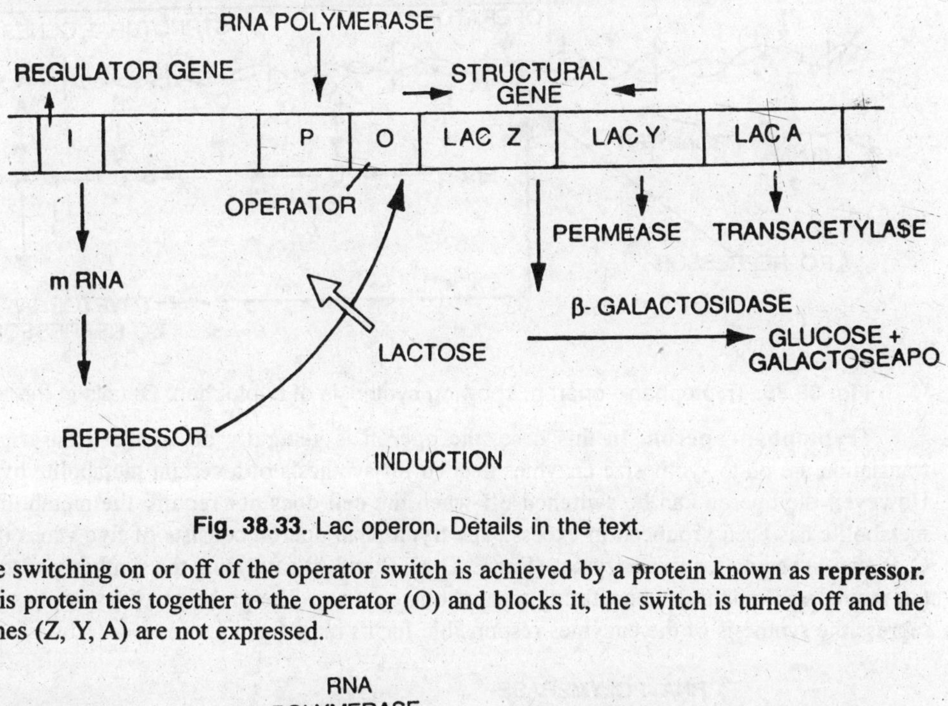

Fig. 38.33. Lac operon. Details in the text.

The switching on or off of the operator switch is achieved by a protein known as **repressor.** When this protein ties together to the operator (O) and blocks it, the switch is turned off and the three genes (Z, Y, A) are not expressed.

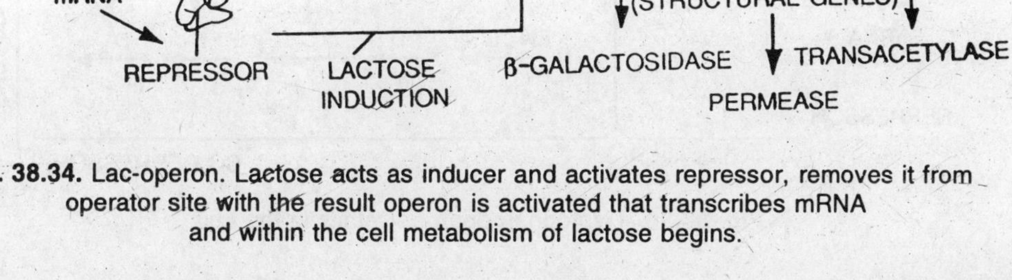

Fig. 38.34. Lac-operon. Lactose acts as inducer and activates repressor, removes it from operator site with the result operon is activated that transcribes mRNA and within the cell metabolism of lactose begins.

Lac-operon. This is an example of **inducible operon**. A few molecules of lactose are added into the cell by the action of the enzyme permease, a small quantity of which is found even under repressed conditions. These few molecules are then converted into an active form of lactose which binds to the repressor (i gene product) and thus, the repressor cannot bind to the operator switch any more. When the operator is free of repressor RNA polymerase starts the transcription of the operon by binding the promotor site P. Messenger RNA corresponding to all the three enzymes is synthesized which gets translated to produce the three enzymes— β -galactosidase, permease and transacetylase. With the production of these three enzymes metabolism of lactose begins. The synthesis of enzymes is continued unless and until all lactose molecules are consumed by the cell. When the last molecules of lactose bound to repressor are also consumed, the inactive repressor becomes active and binds to operator site (O) to switch off the operon as normal.

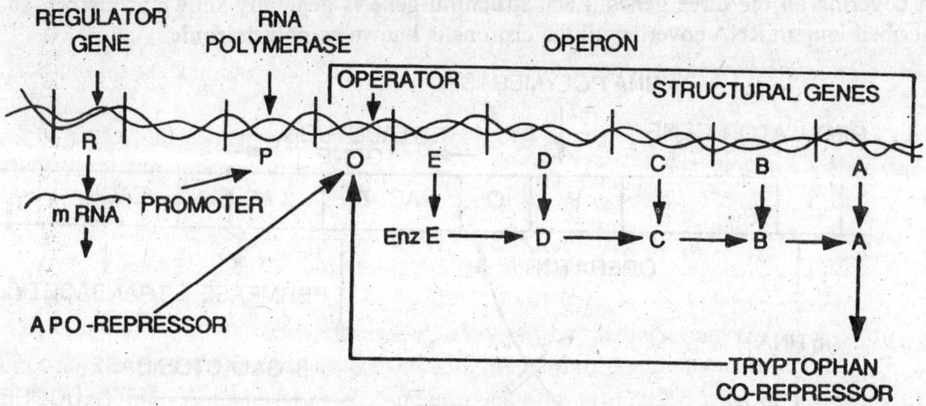

Fig. 38.35. Tryptophan—operon, showing synthesis of tryptophan. Details in the text.

Tryptophan–operon. In this case, the operon is generally on so that transcription and translation are on to synthesize enzymes needed for synthesis of a certain metabolite by the cell. However, the operon can be switched off when the cell does not require the metabolite or the metabolite has been produced in excess. The tryptophan operon consists of five genes (trp E, D, C, B and A) coding for five enzymes (Enz E, D, C, B and A) catalysing the synthesis of tryptophan (an amino acid) and thus constituting an anabolic pathway. The presence of tryptophan serves to repress the synthesis of the enzymes responsible for its manufacture.

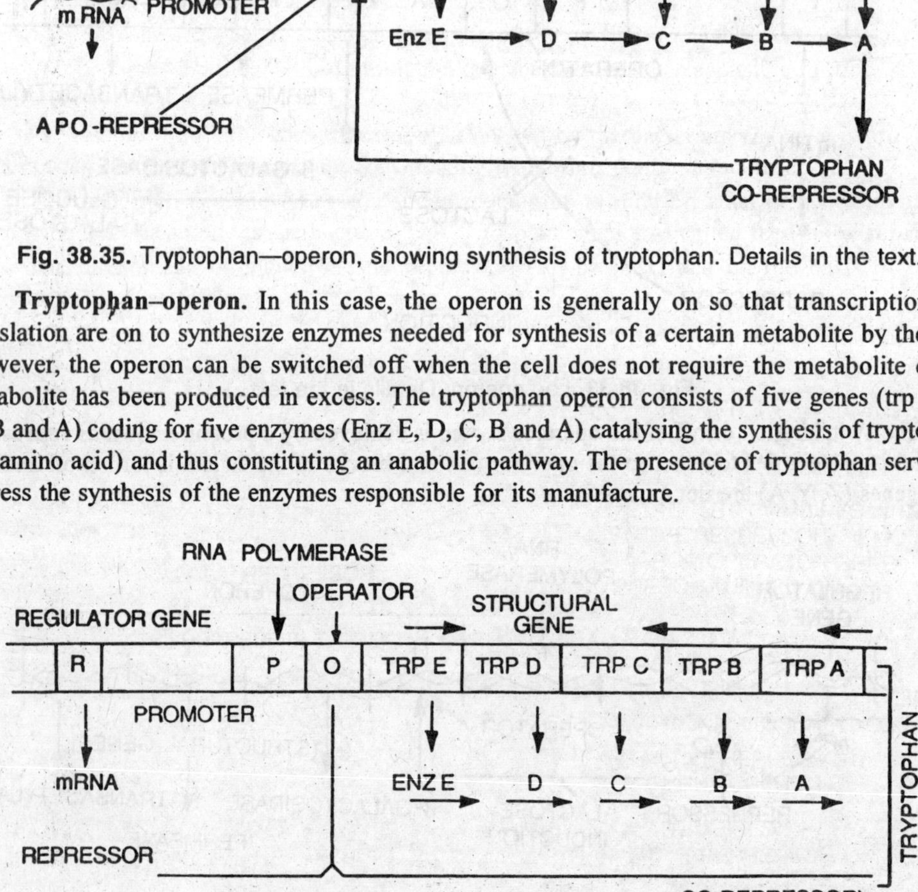

Fig. 38.36. Tryptophan operon. Details in the text.

The five structural genes coding for these five enzymes are regulated by a single operator switch. Here, the repressor protein produced by the regulatory genes does not bind to the operator by itself. However, when tryptophan is added, it binds to the repressor and is known as the **co-repressor.** The repressor-corepressor complex binds to the operator and blocks the expression of this operon.

The structure of the tryptophan operon is more or less similar to that of the lac operon, but has a functional variation. In this case the R gene product is equivalent to the i gene of lac operon. The R gene product produces protein which by itself is unable to bind to operator. This is referred to as **apo-repressor.** In the presence of tryptophan a co-repressor, the functional repressor is formed which now binds to the operator and prevents the transcription of the operon and production of tryptophan. Here the operon model explains both the induction and repression phenomenon in procaryotes.

House-keeping Genes

However, all the information present in the gene is not always used. Many genes remain silent, and expressed only when the gene product is needed. There are certain genes whose products are constantly needed during various metabolic activities of a cell, and therefore, they are known as **house-keeping genes**, and their function is referred to as house-keeping or constitutive activity.

DIFFERENTIATION AND DEVELOPMENT

The most complex of all the physiological processes is **growth**. The process of growth results from an increase in protoplasm, leading to an irreversible increase in size and weight. The growth involves three stages-stage of **cell division**, stage of **cell elongation** and stage of **cell differentiation.** Growth is associated with differentiation of cells which is a change in form and physiological activity. Development refers to the sequence of processes in the overall life history of a cell or an organism including growth, differentiation, maturation and senescence. In unicellular organisms growth involves increase in their volume and numbers. Cell division serves as a means of reproduction in these organisms, and leads to increase in population. In the multicellular organisms, however, growth is much more complex.

The life of flowering plant begins with the fertilized egg. Through a series of divisions an embryo is formed which is enclosed in a seed. Germination then takes place leading to the formation of a seedling. After passing through the juvenile phases a mature plant is finally formed.

When we observe the changes in the zygote as if gradually develops into an embryo and later into an adult organism (plant or animal), we find that physiological behaviour of cells changes with time. Generally the embryonic cells remain totipotent, or in other words each embryonic cell is capable of giving rise to an embryo and give rise to all tissues of the adult organism. However, as cell division proceeds, the cells, of the adult organism lose the property of totipotency. Now the group of cells give rise to specialized tissues such as leaf and stem in a plant and brain and liver in an animal. This process, where the totipotent embryonic cells are being converted into specialized cells forming specific tissues is known as **differentiation**. As embryonic cells divide, and the function of these specialized cells becomes restricted, to an organ of an adult organism, it is known as **determination.** This way, development contains both growth and differentiation. It has been revealed that these developmental changes are preceded by many molecular changes. Some of the developing cells start producing a new set of proteins and at the same time, they stop producing those proteins which were synthesized earlier. For example, muscle cells synthesize myosin; erythroid cells synthesize globin; brain cells synthesize myelin and leaf cells synthesize Ru BP carboxylase. The synthesis of these proteins is due to differentiation.

Robert Briggs and Thomas King have brought out some experiments on nuclear transplantation to prove that how these primary events trigger differentiation along a particular pathway. They isolated some mature cells from the body of frog and toad. They destroyed the nuclei of these cells and transplanted fresh nuclei from embryonic and tadpole cells into the enucleated eggs. Many of the embryos having transplanted nuclei developed into normal tadpoles, on the other hand, if the cells were transplanted with nuclei from differentiated organs such as intestine, tadpoles were not developed. Thus it becomes clear that the nuclei undergo some change during differentiation. Further investigations revealed that totipotency in animal systems was retained by embryonic cells upto a specific stage, and beyond that the cells acquired the function of the developed organ. However, in plants even a mature differential cell of leaf or stem can be dedifferentiated to form a **callus** which can give rise to a whole plant.

The main process of differentiation involves altered gene activity. Thus, different genes act at different times to suit the requirement of the organism. It is believed that the change in gene activity is brought about by its interaction with the cytoplasm. The components of the cytoplasm are further determined by a wide variety of environmental factors such as nutrition, light, temperature, cell to cell interaction and the pH. In other words the modulation of the gene and its cytoplasm is interdependent.

A lot of information has been collected by using *Drosophila melanogaster* (fruit fly), as a working model and the developmental mutants that have been isolated.

It is quite clear that during development, a single fertilized egg gives rise to cells that have different developmental fates.

This asymmetry may be a part of the egg itself in terms of distribution of its cytoplasmic components as in *Drosophila* or in the initial division cycles of mammals.

It is supposed that each cell type is characterised by its pattern of gene expression or by particular gene products which it produces.

In fruit fly, many types of genes have been identified by mutation and analysing their functions.

These genes not only regulate on another but perhaps also code for structural proteins.

Many of these genes have conserved motifs, the most common of which is known as **homebox**. Such homeboxes are distributed in other eukaryotes also, such as worms, frogs and mammals. It is thought that they may be characteristics of all animals.

Cancer and Oncogenes

Cancer can be cited as a problem associated with differentiation and development. The process of growth and differentiation are regulated by various growth controlling mechanisms, which operate to maintain the structure and coordinated function of the various parts of the organism. In malignant disease (cancer) there is a breakdown in this process-a cell/cells escapes from the usual growth control restrictions and is able to grow and divide, and to move more freely than normal cells. This results frequently in the ability to invade and damage normal tissues, and to spread by way of blood or lymphatic systems to other regions of the organism. Malignant cells may arise in any tissue in which cell proliferation takes place.

Thus, cancer may be defined as unorganized growth in which the controlling and regulating mechanisms have disappeared or have been ineffective.

The cancer cells continue to divide unchecked, resulting in an ever-increasing number of cancerous cells and, therefore, the size of the tumour keeps growing. Two hypotheses have been proposed to explain how a normal cell changes into a cancerous cell — (*i*) According to one hypothesis cancer may be essentially due to alterations in one or more chromosomes or gene mutations in the nuclei of the somatic cells. **(*ii*)** According to the other hypothesis, a genetic change may not be an essential factor. Instead, some change occurs in the cytoplasm, resulting in the loss of control on the nuclear and cytoplasmic divisions.

There are three main differences between the cancer cells and the normal cells. They are as follows.

(*i*) Unlike the normal cells, cancer cells show indefinite growth.

(*ii*) As the cancer cells are derived from normal cells they seem to arise from **transformation** where normal constraints of growth are bypassed.

(*iii*) At some stage cancer cells acquire the ability to **metastasize**, that is, they become capable of invading normal tissue and start a new chain of transformation away from the original site.

Oncogenesis. Multiple genetic changes also occur during the transformation of a normal cell to cancer cell. This process is called **oncogenesis**. Such a transformation may be caused spontaneously or by the agents known as **carcinogens.** They include agents such as radiations, chemical compounds and viruses.

Viruses and cancer. Rous (1910) observed that a particular type of virus causes tumours to develop in chickens. The virus is an RNA virus now known as the Rous sarcoma virus. Recently, viruses have been associated with several kinds of tumours in mammals, especially in mice. The virus transforms the cells from their normal state to a cancerous state; the transformation is accompanied by changes in the DNA of the infected cell and profound functional disturbances, including changes in cell surface, properties and behaviour. Although viruses have been isolated from human tumour material, in no case has any human cancer yet been conclusively shown to have been caused by a virus.

Now, two types of genes, *i.e.*, **(*i*) oncogenes** and **(*ii*) tumour suppressor genes** have been identified, whose mutations result in cancerous transformations.

Oncogenes. Oncogenes stimulate cell division. However, these genes have normal cellular counterparts known as proto-oncogenes (c- onc) which are involved in normal cell functions. Mutations of proto-oncogenes lead to the formation of oncogenes.

Many tumour causing viruses may carry similar c-onc gene (v-onc) which induce carcinogenesis.

Tumour suppressor genes. Such genes normally suppress the tumours and the disease appears only when the appropriate active gene is lacking or both the alleles are lost. They impose some constraints on cell cycle. Their absence leads to the release of these constraints and the tumours are initiated.

Regeneration

As already mentioned that development does not end with the completion of embryogenesis. Development potentialities persist into the post-embryonic life more or less in all organisms. These enable them to replace or repair the lost or damaged structures during their larval and adult life. For example, when you cut off part of the tail of a frog tadpole or a house wall-lizard, the missing part is developed again from the remaining portion of the tail. This restoration of the removed part of the tail is called **regeneration.** Thus, regeneration may be defined as replacement, repair or restoration of the lost or damaged structures or reconstitution of a whole body from a small fragment of it during the post-embryonic life of an organism.

The mammals have great capacity to regenerate the liver. When a part of the liver is removed the remaining portion grows by repeated division of its cells until the original volume of the liver tissue is attained.

In many respects, the process of regeneration is similar to the process of the development of embryo. As in the embryonic development, regeneration involves cell division, cell movements, cell and tissue differentiation and morphogenesis. However, for the process of regeneration cells arise through a process of de-differentiation from the already differentiated and specialized cells of well-developed tissues.

Aging

Failing disease or injury, a multicellular organism dies from the process of **aging**, which takes different forms in different types of cell. The average lifespan differs very widely for different organisms. While some live for only short periods, other may have a life of several decades or even centuries. For example, in plants there are giant sequoia and redwood trees at one end of the longevity scale, living to thousands, of years and always continuing to grow, whereas annual plants largely cease to grow soon after the flowering period when the fruit begins to form, after which they soon age and die. On the other hand among animals, the process of aging appears to be accompanied by obvious external changes in warm blooded animals, whereas in some long-living animals such as the sturgeon (with a lifespan of more than 100 years) and the tortoise (a life-span of more than 150 years), in which continuous growth during life occurs, the process of aging is so slow as to almost undetectable. However, none lives for ever. Even if an individual meets no total accident, is not eaten up by a predator or does not suffer a killing disease, death still comes as the natural final result of old age.

Thus *aging is defined as the process of progressive deterioration in the structure and function of the cells, tissues and organs of the organism as it grows older.* The area of developmental biology that is concerned with the study of the process of aging is known as **gerontology.**

Hormonal changes affecting adversely the functioning of various organs, the accumulation of waste material from various metabolic processes inside certain cells, loss of elasticity in elastic fibres and an increase in the amount of collagen in connective tissue—all play a part in aging, resulting in such well known changes as stiffening of the joints and hardening of the arteries. No new nerve cells are formed, but there is a replacement of intracellular material at the molecular level, and this process also becomes slower with advancing age. In the absence of death from injury or disease, the changes brought about by aging eventually results in death due to the breakdown of some vital structure such as the rupture of a blood vessel.

The reason why the various processes of aging take place are not yet clear. However, some biologists suggest that adverse changes in the environment are the causes of aging in the organism. Others believe that aging is an intrinsic genetic property of the cells of an organism. According to another theory aging is due to an interaction between hereditary factors (*i.e.*, genes) and the environment. Another theory proposes that the cells and the organisms with a high rate of metabolism are more rapidly and die sooner than those with a relatively lower rate of metabolic activity. The theory of immunity suggests that the decline and disappearance of the thymus gland by late middle age in man is the primary cause of aging. However, the cell death is programmed by the cell itself. As the cell ages the gene activity of it is altered to the extent that it gradually shows impaired metabolic reaction. It may involve the stop of synthesis of an essential enzyme (*e.g.*, dehydrogenase) or production of a toxin which stops electron transport system in the cell leading to its death.

Thus, it becomes clear that every development change in a biological cell takes place at the gene level that is DNA, which is directly exposed to cytoplasm in procaryotes and located in the nucleus of eucaryotes. The constant interaction of the genes (*i.e.*, DNA) with the cytoplasm and environment leads to the end result.

The problems of cell death, cell growth, malignancy and aging are all different aspects of cell development.

SUMMARY

1. Deoxyribonucleic acid (DNA) is the genetic material.
2. The identification of DNA is based on bacterial transformation and viral infection experiments.
3. In the bacterial transformation experiments it has been shown that DNA is responsible for inducing a smooth surface character and the property of virulence, when introduced into cells do not have these properties.
4. In the case of viral infection experiments it has been shown by radioactive labelling that the ability to infect and to give rise to more viral particles rests with the DNA and not with the proteins.
5. In certain viruses RNA also serves as genetic material.
6. The semi-conservative mode of replication is easily explained by the double stranded structure of DNA.
7. In this semi-conservative mode, the daughter cells derive from the parent one old strand of DNA and new strand is formed using the former as template.
8. The DNA replication involves unwinding of DNA and formation of new strands on the separated old strands. Third process is catalysed by the enzyme DNA polymerase.
9. During DNA replication, accuracy of replication and insertion of correct bases are preserved by proof-reading mechanisms.
10. In molecular biology the central dogma involves-DNA ⟶ RNA ⟶ Proteins.
11. The process by which DNA gives rise to RNA is called transcription.
12. The transcription is involved by unwinding of DNA, where one DNA strand gives rise to complementary RNA. This reaction is catalysed by the enzyme RNA polymerase.
13. The process by which proteins are derived from RNA is known as translation.
14. The sequence of bases in mRNA decides the sequence of amino-acids in proteins. Each amino acid is formed by a triplet code, which consists, of a sequence of three bases in the DNA and the complementary bases in mRNA.
15. Protein synthesis involves the step of —(*i*) initiation, (*ii*) elongation and (*iii*) termination. After termination the proteins are translocated to different parts of the cell or transported out of the cell.
16. The mutations cause changes in the genes, and thus the mutations in the genetic material are reflected in the proteins. Hence, a single base change in the DNA can affect the sequence of amino acids in the protein and its function too.
17. All processes of life are controlled by genes.
18. The genes have the information to code for proteins.
19. The gene expression that leads to the protein synthesis is regulated. Some genes are expressed only when their products are required.
20. The regulation of the expression of genes operates in all forms of life ranging from viruses to eucaryotic organisms.

21. In viruses, regulation of gene expression decides whether it follows the lytic or lysogenic cycles.
22. In retroviruses, RNA is the genetic material. Here, DNA is formed from RNA by the catalysis of enzyme reverse transcriptase. With the invention of this enzyme, the central dogma has been modified to read as —

    $DNA \rightleftharpoons RNA \rightarrow$ Proteins. Hence this type of conversion is bidirectional.
23. The expression of retroviral genes, referred to as oncogenes, is responsible for the onset of cancer in various living cells.
24. The regulation of gene expression in bacteria is generally effected through operons. Each operon is made up of a group of genes to code for enzymes involved in a specific metabolic pathway.
25. The operon concept can explain the phenomenon of both induction and repression.
26. During induction some enzymes are induced by an inducer, so that, it (inducer) could be metabolized.
27. Repression is the phenomenon by which the synthesis of a set of enzymes takes place that leads to a product is shut down when the product is provided from outside.
28. The regulation of gene expression in eucaryotic cells is more complex than in procaryotic cells.
29. The eucaryotic gene is splitted into exons and introns whereas procaryotic gene does not.
30. When the DNA is transcribed, the exonic portions of DNA are spliced together to give rise to functional mRNA, whereas the intronic portions are removed and degraded.
31. Cancer and ageing are basically due to differential gene expression.

TERMS TO REMEMBER

Aging. A progressive deterioration in structure and function of cells as the age advances. It is found that each type of animal cells starts dying after a certain number of mitotic divisions.

A-site. (Amino acyl site). A functional site in larger subunit of ribosome where the second and next amino acyl tRNA gets attached.

Anticodon. A triplet of bases present on tRNA complementary to the triplet code of mRNA.

Auxotrophs. The term refers to nutritional mutants of *Neurospora* that need one extra nutrient to grow on minimal medium.

Bacteriophages. The viruses that infect the bacterium, also termed bacterial viruses.

Cancer. The cells continue to divide and give rise to a malignant tumour. These cells can also invade other tissues and tumours can arise in new locations.

Capsid. In viruses, the genetic material is wrapped around by a protein coat called capsid.

Cistron. A unit of function, *i.e.,* segment of DNA that determines a single polypeptide chain.

Costitutive genes. The genes that are constantly expressed.

Co-repressor. A molecule that binds with the repressor protein to form a functional repressor complex.

Conjugation. The genetic transfer from one bacterial cell to another.

Central dogma. The unidirectional flow of information from DNA ⟶ RNA ⟶ proteins.

Core enzyme. The enzyme without the subunit, and capable of transcribing bacterial DNA into RNA.

Codon. Triplet (group of three) of bases on mRNA that specify a particular amino acid in protein synthesis.

DNA. Deoxyribonucleic acid.

DNA ligase. An enzyme that joins newly synthesized short fragments of DNA.

DNA polymerase. An enzyme responsible for linking free DNA nucleotides to form the complementary strand. It can polymerise the nucleotides only in 5′ – 3′ direction.

Exons. The regions of a gene which contain the information necessary to form a protein.

Feed-back repression. When a metabolite needed by the bacterium is provided in excess from outside, the bacterium stops making it and thus conserves its resources.

F factor. Fertility factor on sex faction present in the bacterium. F^+ is male bacterium while F^- is female bacterium.

Frameshift mutation. Mutation caused due to shifting of entire reading frame by loss or addition of single base or a segment of DNA.

Genetic code. The complex process by which the information in RNA is decoded into a polypeptide chain.

Glycosylation. The addition of sugar residues by modification of some proteins, that are released in the lumen and are incorporated in Golgi vesicles.

Helicases. The enzymes which help in unwinding of two strands of DNA.

Hemizygous. The genotype of a male with regard to an X-trait, since males have only one set of X-linked genes.

Hydrogen bond. The bond found in between nitrogen bases of DNA forming two nucleotide chains.

Hfr strain. High frequency strain of bacteria. This donates all or large portion of their DNA into another bacterial cell.

Inducer. The small molecules eliciting the phenomenon of induction are referred to as inducers.

Induction. When the genes are turned on enzymes are produced, which metabolize the new substrata, this is called induction.

Inducible enzymes. The enzymes whose production is induced by a compound called inducer, such as β-galactosidase.

Intron. Those regions of a gene which do not have the information for a protein.

Jumping genes. The genes which are reshuffled from one place to another.

Lysogenic cell. It refers to the bacterial cell that carries the provirus or prophage.

Lysogeny. The phenomenon by which the phage DNA integrate with the host chromosome and replicate along with it.

Oncogene. It involves the origin of cancer.

Operator. Gene that regulates the activity of structural genes.

Operon. The structural genes and an operator made an operon.

Okazaki fragments. Newly synthesized short fragments of DNA that are formed on the lagging strand of DNA during DNA replication.

P-site or peptidyl site. A functional site in ribosomes, where tRNA from A-site are translocated to P-site.

Phenocopy. When a normal gene under different environmental conditions copies down the phenotypic characters of a mutant.

Phosphodiester bond. The bond between two adjacent nucleotides of two adjacent sugar molecules at 3′ and 5′ positions with phosphoric acid.

Polyribosome or polysome. The series of ribosomes translating the same single strand of mRNA.

Primer. This is a short RNA segment formed on DNA template prior to the beginning of replication.

Primase. It polymerises ribonucleotide bases in the formation of a primer.

Plasmids. They are extrachromosomal DNA elements found mostly in bacteria.

Polycistronic mRNA. The long mRNA covering all the cistrons is known as polycistronic mRNA.

Promoter gene. Gene that forms the binding site of RNA polymerase.

Protooncogenes. Oncogenes present in the normal cell before triggering cancer in cells.

Prophage or provirus. A stage, when the DNA of virus gets incorporated in the host DNA.

Regulatory gene. A gene that regulates the function of the operator gene.

Repressor. The functioning of the operator by a protein is referred to as repressor.

Repression. A phenomenon by which the synthesis of a set of enzymes leading to a product is switched off when the product is provided from outside.

Repressible enzymes. Enzymes whose production is stopped or repressed by the presence of certain metabolite.

Repressor protein. Protein formed by regulatory gene.

Restriction enzyme. An enzyme that recognises specific nucleotide sequences in DNA and then makes a double strand cleavage of DNA molecule.

Retroviruses. A group of viruses possessing RNA as a genetic material. These viruses carry the gene for reverse transcription, *i.e.*, RNA → DNA.

Reverse transcriptase. The enzyme responsible for catalysing the conversion of RNA to DNA.

Replication. A phenomenon that refers to DNA synthesis or duplication.

RNA. Ribonucleic acid.

RNA polymerase. An enzyme that acts as a catalyst in the formation of RNA from one strand of DNA.

Silent mutation. Mutation that does not cause any change in the protein.

Semi-conservative. That conserves half of the original self as in the replication of DNA.

Structural genes. The genes that contain the information to determine the sequence of amino acids present in the proteins. They have the information to code for the amino acid sequence and decide the structure and function of the individual proteins.

Template. That refers to a DNA strand from which complementary or RNA strand is formed.

Transcription. Where a single strand of RNA is transcribed from one of the strands of DNA.

Transformation. The process by which DNA isolated from one type of cells is introduced into another type and that inherits some of the properties from former to the latter.

Translation. A phenomenon that changes the coded message of mRNA into the sequence of amino acids.

Translocation. In protein synthesis, where a tRNA molecule moves from A site to P site because of relative movement of ribosomes on mRNA.

Viral oncogenes. The origin of cancer is a complex phenomenon where viral oncogenes lead to uncontrolled growth of cells.

Wobble position. The position on the codons of the third base of the same amino acid.

QUESTION BANK

N.B. Undermentioned questions are asked repeatedly in various Indian Universities and competitive examinations.

Very Short Answer Questions

1. Who had first proposed the double helix structure of DNA ?
2. Which British medical officer had put forth the phenomenon of bacterial transformation ?
3. Who had coined the term gene ?
4. Who proposed the theory of continuity of germplasm ?
5. Which bonds hold the two polynucleotides chains of DNA molecule ?
6. Who had deciphered the genetic code?
7. Who had used X-rays to cause mutations in *Neurospora* ?
8. Who proved experimentally that DNA replication is semi-conservative ?
9. What does a nucleotide comprise of ?
10. Name the enzyme which can break and reseal the strand of DNA ?
11. In which direction $5'-3'$ or $3'-5'$ are the new strands of DNA formed during replication ?
12. Name the enzymes that catalyse (*a*) replication of DNA, (*b*) formation of RNA.
13. What is the linkage between two adjacent nucleotides called ?
14. What is a codon ?
15. Who performed the experiment that showed that DNA is the genetic material in bacteriophages ?
16. What is considered to be the central dogma in molecular biology ?
17. What is the sugar in DNA and RNA called?
18. Name the two types of nitrogenous bases.
19. Name the purines and pyrimidines.
20. Name the enzymes that unwind DNA helices.
21. What is transformation ?
22. What is an anti-codon?
23. What is transcription ?
24. Name the process that blocks the amino group of the first amino acid and thereby preventing the formation of dipeptide bond.
25. What are non-sense codons ?
26. Give two examples of amino acids that have one codon ?
27. Which triplet is known as terminating codon ?
28. What is wobble position ?
29. What is a genetic code ?
30. How many bases code for one amino acid?
31. Give two significant functions of DNA.
32. Which base triple are to code for the amino acid, phenylalanine ?
33. What does RNA control ?
34. What are the initiation codons for protein synthesis ?
35. Which is the first amino acid in any polypeptide chain ?

36. Who had proposed the one gene, one enzyme hypothesis ?

37. What is a polyribosome ?

38. What are Okazaki fragments ?

39. What are auxotrophs ?

40. Who had discovered jumping genes ?

41. Of the 64 possible code triplets how many code for amino acids and how many for stop signals ?

42. The amino acid arginine has 6 mRNA codons : CGU, CGC, CGA, CGG, AGA and AGG. Give the DNA codon for it.

43. What are virus infecting bacteria called?

44. Name the virus which is composed entirely of protein and RNA ?

45. What is the protein coat of virus cells ?

46. Name the group of virus to which the human immuno deficiency virus belong?

47. What is the phenomenon of destruction of a cell following infection by virus ?

48. Which is the enzyme that catalyses synthesis of DNA from RNA ?

49. Name the scientists who proposed the operon model ?

50. What are oncogenes ?

51. Which are the three modes of genetic transfer between bacterial cells ?

52. What is transduction ?

53. What are plasmids ?

54. What are inducers ?

55. Which is the strain of bacterial cells that have the chromosomal F-factor ?

56. Name the gene that controls the functioning of structural genes ?

57. Which gene is responsible for the switching on and off a structural gene?

58. What is the entire unit of regulator gene, operator gene and structural gene called?

59. What is splicing ?

60. Give the name of the science of ageing.

61. What is recombinant DNA ?

62. What is the enzyme that cuts DNA at specific sites ?

63. What is the process that synthesises protein on mRNA ?

64. Name the virus that is responsible for causing AIDS in humans.

65. What is the phenomenon that controls the expression of genes to achieve maximum cellular economy?

66. What is genetic engineering ?

67. Name a method for producing new combination of genetic material in bacterial cells.

68. Expand HIV.

69. Who synthesised the first artificial gene without a primer ?

70. Name the enzymes which cut the DNA molecule into fragments with sticky ends.

ANSWERS

1. Watson and Crick; **2.** Griffith; **3.** Johannsen; **4.** Weismann; 5. Hydrogen bond between the base pairs; **6.** Nirenberg, Ochoa, Hargobind Khorana and Francis Crick; **7.** Beadle and Tatum; **8.** Messelson and Stahl; **9.** It consists of a sugar, phosphate and nitrogen base; **10.** Topoisomerase; **11.** $5' - 3'$ direction; **12.** (*a*) DNA polymerase, (*b*) RNA polymerase; **13.** Phosphodiester bonds; **14.** Triplet of bases on mRNA coding for one amino acid; **15.** Hershey and Chase; **16.** The unidirectional flow of genetic information for DNA ⟶ RNA ⟶ Protein ; **17.** Deoxyribose and Ribose; **18.** Purines and Pyrimidines; **19.** Purines are — adenine (A) and guanine (G); Pyrimidines are-cytosine (C), thymine and uracil (U); **20.** Helicases;

21. The phenomenon where DNA is isolated from one type of cell, and is able to bestow some of its properties to another type of cell where the former is introduced; **22.** A triplet of bases present on tRNA, complementing the mRNA codon; **23.** The process of formation of a RNA strand on the DNA template; **24.** Formylation; **25.** The codon that do not code for any amino acid; **26.** Methionine and Tryptophan; **27.** The triplets UUA, and UAG, UGA are terminating codons; **28.** The third position in the codon; **29.** Sequence of base triplets in DNA molecule; **30.** Three; **31.** Replication and expression of genetic information; **32.** UUU, UUC; **33.** The synthesis of enzymes; **34.** AUG, GUC; **35.** Formylated methionine; **36.** Beadle and Tatum; **37.** During protein synthesis a group of ribosomes line up on the mRNA chain are called polyribosomes; **38.** Short segments of DNA formed in $5' - 3'$ direction starting from a RNA primer are called Okazaki fragments; **39.** The nutritional mutants; **40.** Barbara McClintock; **41.** 61,3; **42.** GCA, GCG, GCC, TCT and TCC. **43.** Bacteriophage; **44.** Tobacco Mosaic Virus (TMV); **45.** Capsid; **46.** Retrovirus; **47.** Lysis; **48.** Reverse transcriptase; **49.** Francis Jacob and Jacques Monod; **50.** They are genes that trigger the normal cells to become cancerous; **51.** Transformation, transduction and conjugation; **52.** It is the transfer of DNA from one bacterial cell to another with the help of bacteriophage; **53.** In addition to chromosome there are circular DNA molecules in bacteria called plasmids; **54.** Metabolites that activate the inactive genes in the cells; **55.** H fr (High frequency recombination); **56.** Operator gene; **57.** Operator gene; **58.** Operon; **59.** The process of removing unwanted regions of mRNA and joining the wanted regions again; **60.** Gerontology; **61.** DNA formed as a result of combination of DNA's from two organisms; **62.** Restriction endonuclease; **63.** Translation; **64.** HIV-III; **65.** The phenomenon of induction and repression; **66.** The science that deals with the synthesis of artificial genes, repair of genes, combining of genes from two organisms and manipulation of artificial genes for improvement of organisms; **67.** Gene cloning; **68.** Human immuno-deficiency virus; **69.** Hargobind Khorana; **70.** Restriction endonucleases.

Short Answer Questions

1. What are the essential requirements of the genetic material ?
2. Draw a neat, labelled diagram of the DNA molecules as suggested by Watson and Crick.
3. Do the two complementary strands of DNA carry the same or different biological message ?
4. What do you mean by proof-reading in DNA synthesis ?
5. What is DNA repair ?
6. What is genetic code ?
7. What is the role of non-sense codons in protein synthesis ?
8. DNA polymerase and RNA polymerase differ in their requirements while functioning. Explain.
9. What do you mean by frameshift mutation ? Give an example.
10. Write the transcribed m-RNA from DNA strand with the base sequence TAG TAC ACT. What is the specific term used for the last codon of the transcribed m-RNA in this case ?
11. Why are DNA and ribosomes are located at different sites ?
12. How is protein synthesis initiated in a cell?
13. Draw a neat and labelled diagram of T_4 bacteriophage.
14. What are oncogenes ?
15. Define ageing. What are the genetic causes of ageing ?
16. Differentiate between :
    (*a*) Lysis and lysogeny; (*b*) Induction and repression.
17. Explain the concept of colinearity of gene and protein structure.
18. What is contact inhibition ? What is its significance in cell division ?
19. What is an operon ? Name the three types of genes which make up an operon.
20. Viruses are intracellular obligately parasitic organisms. Discuss.
21. How does a bacterial chromosome differ from an eukaryotic chromosome.
22. How does an excess of tryptophan cause a switch off of the tryptophan-operon ?
23. How would you differentiate the following :
    (*a*) DNA and RNA; (*b*) initiation codon and termination codon.

24. What was Avery, Macleod and McCarty's contribution to confirm the transformation principles in *Diplococcus* ?
25. What did Hershey and Chase prove by their blender experiment ? Explain how they did the experiment to prove it.
26. Why is genetic code a triplet one ?
27. Explain the following :
    Replication, Transcription, Translation, Codon, Anticodon.
28. Which molecule bears codons and which molecules anticodons ?
29. How is the DNA content in our diet digested in the body?
30. What are the functions of m-RNA and t-RNA ? What anticodons will be required to recognize the following codons :
    (*i*) AAU (*ii*) CGA (*iii*) UAU (*iv*) GCA.
31. Give the structure of a t-RNA molecule.
32. Write a note on central dogma of molecular biology.
33. A tRNA is charged with the amino acid phenylalanine.
    (*i*) At what end of tRNA is this amino acid attached.
    (*ii*) What is the mRNA codon that coded for phenylalanine.
    (*iii*) What is its anticodon ?
    (*iv*) Name the enzyme responsible for this attachment.
34. How did Hershey and Chase prove that DNA is the genetic material ?
35. What are genomic and cDNA libraries ? How can they be prepared ? What are its advantages ?
36. What are plasmids ? How does transfer of genetic material take place by employing plasmids ?
37. All cells in a multicellular organism have same genetic constitution, yet they function differently. How would you account for it ?
38. What are introns and exons ? What process ensures a linear arrangement of amino acids although the genes are discontinuous.
39. How is functioning of lac operon different from that of tryptophan operon ?

Long Answer Questions

1. Provide experimental evidence for the semi-conservative mode of replication of DNA.
2. Genetic material is DNA and not protein. How did Griffith prove this experiment ?
3. How did Hershey and Chase prove that DNA is the genetic material in bacteriophage ?
4. Discuss the mechanism of DNA replication.
5. Explain briefly the steps of protein synthesis.
6. Explain one gene-one enzyme concept with experiments on *Neurospora* by Beadle and Tatum.
7. Describe the transcription of RNA from DNA.
8. Discuss the characteristics of genetic code.
9. Discuss the effect of mutations on the structure and function of proteins.
10. Why is the DNA molecule compared to spiralling stair-case ? Which three components made up the nucleotide ?
11. Give the structure of DNA molecule.
12. Define bacterial transformation. Who demonstrated it experimentally and how?
13. Describe bacterial conjugation as a means of genetic transfer. What is the significance of this process?
14. Explain the steps involved in a polypeptide synthesis. How are the amino acids activated during polypeptide synthesis ?
15. What is reverse transcription ? Explain how single-stranded RNA of viruses give rise to double stranded RNA.
16. What is conjugation in bacteria ? With the help of labelled diagrams explain the process.

17. Explain the operon concept by Jacob and Monod with the help of a suitable example.

18. Describe the modes of genetic transfer between the bacterial cells.

19. What is the inducer in a lac operon ? How does it ensure the switching off of genes ?

20. How is gene expressed in eukaryotes ?

21. What is genetic engineering ? Briefly describe the benefits of genetic engineering to human welfare.

22. What does lac operon consist of ? How is the operon switch turned on and off in the expression of genes in this operon? Explain.

Multiple Choice Questions (Objective Type Questions)

1. The principal genetic material of living beings are
(*a*) DNA (*b*) RNA
(*c*) both a and b (*d*) none of above.

2. Synthesis of DNA is known as
(*a*) replication (*b*) transcription
(*c*) translation (*d*) migration.

3. Transcription means the synthesis of
(*a*) RNA (*b*) DNA
(*c*) Proteins (*d*) Lipids.

4. Transfer of information from m RNA to a polypeptide chain is called
(*a*) amination (*b*) deamination
(*c*) translation (*d*) transamination.

5. Genetic code is
(*a*) singlet (*b*) doublet
(*c*) triplet (*d*) none of above.

6. The initiation codons are
(*a*) AUG (*b*) GUG
(*c*) both (*a*) and (*b*) (*d*) none of above.

7. Autocatalytic function is depicted by
(*a*) hormones (*b*) enzymes
(*c*) RNA (*d*) DNA.

8. Transfer of information from DNA to mRNA is called
(*a*) transcription (*b*) translation
(*c*) transduction (*d*) migration.

9. RNA is genetic material in
(*a*) bacteria (*b*) mycoplasma
(*c*) plant viruses (*d*) fungi.

10. DNA molecule has a uniform diameter of
(*a*) 20 Å (*b*) 60 Å
(*c*) 80 Å (*d*) 27 Å.

11. Griffith's experiment was with
(*a*) *Escherichia coli*
(*b*) *Diplococcus pneumoniae*
(*c*) *Rhizobium* (*d*) *Mycorrhiza*.

12. New strands of DNA are formed in the direction
(*a*) 5′ to 3′ (*b*) 4′ to 3′
(*c*) 3′ to 3′ (*d*) 2′ to 3′.

13. The segments of DNA strand on which a new strand is produced is known as
(*a*) codon (*b*) template
(*c*) nucleotide (*d*) none of above.

14. The genes that shuffle from one location to another are called
(*a*) walking genes (*b*) running genes
(*c*) jumping genes (*d*) none of above.

15. The abnormal ratio 2 : 1 brought about due
(*a*) normal genes (*b*) lethal genes
(*c*) jumping genes (*d*) none of above.

16. One gene-one enzyme hypothesis was proposed by
(*a*) Mendel (*b*) Morgan
(*c*) Punnett (*d*) Beadle and Tatum.

17. Synthesis of a gene in a test tube and its transfer to a cell is known as
(*a*) conjugation (*b*) hybridization
(*c*) genetic engineering (*d*) transduction.

18. The more correct statement is
(*a*) one gene-one enzyme (*b*) one gene-one protein
(*c*) one gene-one polypeptide (*d*) one gene-one nucleotide.

19. The transfer of genetic material from one cell to another by a phage is
(*a*) transduction (*b*) transcription
(*c*) transformation (*d*) conjugation.

20. The enzyme that cleaves DNA at specific sites producing sticky ends is known as
(*a*) cleaving enzyme (*b*) lysing enzyme
(*c*) endonuclease (*d*) restriction endonuclease.

21. HIV is responsible for
(*a*) cancer (*b*) AIDS
(*c*) pneumonia (*d*) tuberculosis.

22. Phages infect the
(*a*) algal cells (*b*) bacterial cells
(*c*) fungal cells (*d*) none of above.

23. TMV contains a genetic material
(*a*) chromosome (*b*) DNA
(*c*) RNA (*d*) none of above.

24. Reverse transcription was studied by
(*a*) Temin and Baltimore (*b*) Smith
(*c*) Jacob (*d*) Watson and Crick.

25. *Escherichia coli* hydrolyses lactose to glucose by
(*a*) β-galactosidase (*b*) permease
(*c*) transacetylase (*d*) all of the above.

ANSWERS

1. (*c*) **2.** (*a*) **3.** (*a*) **4.** (*c*) **5.** (*c*) **6.** (*c*) **7.** (*b*) **8.** (*a*) **9.** (*c*) **10.** (*a*)
11. (*b*) **12.** (*a*) **13.** (*b*) **14.** (*c*) **15.** (*b*) **16.** (*d*) **17.** (*c*) **18.** (*a*) **19.** (*a*) **20.** (*d*)
21. (*b*) **22.** (*b*) **23.** (*c*) **24.** (*a*) **25.** (*d*)

39

CHAPTER

Allelic and Non - allelic Interactions

We know that for determination of single phenotypic trait of an organism, two alleles or allelomorphs of a single gene interact in various ways. Out of the two alleles one may show dominance over the other (recessive) or both have partial or incomplete dominant relationship or both may be codominant.

Different kinds of genic or genetic interactions occur between the two allelomorphs of a single gene. These interactions are called **intraallelic** or **allelic** or **interallelic genetic** or **genic interactions** also. In such case the independent or non homologous genes located on the same or on different chromosomes interact with one another for expression of single phenotype.

According to **Bateson** and **Punnett,** the genes were not merely separate elements producing distinct individual effects, but they could interact with each other to give completely novel phenotypes.

For example, If 'A' allele is characterised by 'A' phenotype and 'B' allele is characterised by 'B' phenotype. But an organism with 'AB' both, is characterised by 'C' phenotype. It suggests that phenotype is not present in a gene but arises from a complicated developmental process. They are no genes which individually create specific organ structure or organisms. This concept is called **epigenetic concept,** where development occurs through a complex network of reactions and interactions influenced by genes.

When two gene pairs, each having two alleles, *i.e.,* A, a and B, b segregate independently in a diploid organism, the genetic consequences are easily predictable. A cross between heterozygotes for both pairs Aa Bb × Aa Bb will give offspring with following genotypes.

1/16	AABB
2/16	AABb
2/16	AaBB
4/16	AaBb
1/16	AAbb
2/16	Aabb
1/16	aaBB
2/16	aaBb
1/16	aabb

As long it is independently associated the above genotype exists. Phenotypic ratio is directly affected in changes as in the phenotype which can be observed and not in the genotype.

Time of genetic Interaction

Gene is a chemical determiner. Phenotypic trait results from the combined action of many genes and their products. Environment and ecological factors, *e. g.,* temperature, quality of light and internal factors, *e. g.,* hormones and enzymes also play an important role. Enzymes are proteins and organization of proteins is determined by genes. Each biosynthetic pathway includes various steps catalysed by specific enzyme which in turn depends on a specific gene. The reaction starts with a **precursor** and ends at the **product.** In between, it includes many metabolities which are produced by various genes. Mutant genes may cause **metabolic blocks** in the pathway.

INTRA ALLELIC GENIC INTERACTION

The expression of gene in the form of phenotype is dependent on product. The genotype can hide or mask the phenotype expression of alleles at one locus. The gene which suppresses or masks the other gene is called **epistatic gene** and the gene which is suppressed is called **hypostatic gene.**

Difference between epistasis and dominance

The phenomenon of dominance involves intraallelic gene suppression on the same locus while the epistasis involves intraallelic gene suppression or masking of other gene locus.

Kinds of epistasis

The epistasis is in between two genes, that is atleast a dihybrid and the phenotypes are less than 4.

(a) Dominant epistasis (12 : 3 : 1)

When dominant allele 'A' masks the expression of 'B' 'A' is epistatic gene of 'B'. A can express itself only in the presence of 'B' or b allele. Therefore it is called **dominant epistasis. B** expresses only when **'aa'** is present. Therefore, in 9 : 3 : 3 : 1 ratio both **9 and 3** are expressing the **'A'** gene the ratio is now 12 : 3 : 1.

Example : Complete dominance at both gene pairs, but one gene, when dominant, epistatic to the other.

Fruit colour in summer squash

Gene pair **A :** White dominant to colour 12/16 white

Gene pair **'B'** – Yellow dominant to green 3/16 yellow

Green – 1/16 both recessive aabb

Dominant white hides the effect of yellow or green.

AaBb × AaBb
white | white

	AB	Ab	aB	ab
AB	AA BB White	AA Bb White	Aa BB White	Aa Bb White
Ab	AA Bb White	AA bb White	AA Bb White	Aa bb White
aB	Aa BB White	Aa Bb White	aa BB Yellow	aa Bb Yellow
ab	Aa Bb White	Aa bb White	aa Bb Yellow	aa bb Green

AB – 9 White
Ab – 3 White
aB – 3 Yellow
ab – 1 Green

(b) Recessive epistasis (9 : 3 : 4)

Recessive alleles at one locus (aa) mask the phenotypic expression of other gene locus (BB, Bb or bb) such epistasis is called **recessive epistasis.** The alleles of 'B' locus express themselves only when epistatic locus 'A' has dominant allele like AA or Aa. The phenotypic ratio is 9 : 3 : 4.

Example : Complete dominance at both gene pairs, but one gene, when homozygous recessive, is epistatic to the other.

In Mouse coat colour.

Gene pair **A** : colour dominant over albino.

Gene pair **'B'** agouti colour dominant over black.

Interaction : homozygous albino is epistatic to aguoti and black.

Agouti 9/16
black 3/16
albino 4/16

Agouti (gray) × white albino
AABB ↓ aabb

Aa Bb × Aa Bb
Agouti Agouti

	AB	Ab	aB	ab
AB	AABB Agouti	AABb Agouti	AaBB Agouti	Aa Bb Agouti
Ab	AABb Agouti	AAbb Albino	AaBb Agouti	Aabb Albino
aB	AaBB Agouti	AaBb Agouti	aaBB Black	aaBb Black
ab	AaBb Agouti	Aabb Albino	aa Bb Black	aabb Albino

AB = 9 Agouti
Ab = 3 Albino
Ba = 3 Black
ab = 1 Albino

9/16 : 3/16 : 4/16

Agouti : Black : Albino

(c) Duplicate recessive gene (9 : 7)

If both gene loci have homozygous recessive alleles and both of them produce identical phenotype the F_2 ratio 9:3:3:1 would be 9:7. The genotype aaBB, aaBb, AAbb, Aabb and aabb produce same phenotype. Both dominant alleles when are present together only then they can complement each other. This is known as **complementary gene.**

Examples : Complete dominance at both gene pairs, but either recessive homozygote is epistatic to the effect of the other gene.

In sweet pea flower colour.

Gene pair **'A'** - Purple dominant over white.

Gene pair **'B'** - Colour dominant over colourless (white).

Interaction: Homozygous recessive of either gene A or B produce white

9/16 Purple
7/16 White

White flowers White flower
CCpp × PPcc
CcPp × CcPp
Purple Purple

	CP	Cp	cP	cp
CP	CCPP Purple	CCPp Purple	CcPP Purple	CcPp Purple
Cp	CCPp Purple	CCpp White	CcPp Purple	Ccpp White
cP	CcPP Purple	CcPp Purple	ccPP White	ccPp White
cp	CcPp Purple	Ccpp White	ccPp White	ccpp White

CP = 9 Purple
Cp = 3 White
cP = 3 White
cp = 1 White

9 : 7
Purple White

(d) Duplicate genes with cumulative effect (9 : 6 :1)

Certain phenotypic traits depend on the dominant alleles of two gene loci. When dominant is present it will show its phenotype. The ratio will be 9 : 6 : 1.

Example: Complete dominance at both gene pairs, interaction between both dominance to give new phenotypes.

Fruit shape in summer squash

Gene pair **'A'** sphere shape dominant over long.

Gene pair **'B'** sphere shape dominant overlong.

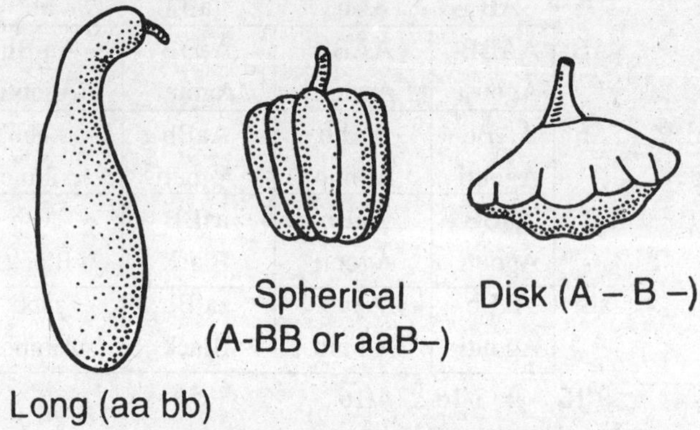

Fig. 39.1. Shape of summer squash.

Interaction at 'AB' when present together, form disc-shaped fruit (Fig .39.1)

Disc shaped fruits 9/16
Sphere shaped fruits 6/16
Long shaped fruit 1/16

Sphere shape × Sphere shape
AAbb aaBB
AaBb × AaBb
Disc shape ↓ disc shape

AB = 9 Disc shape
Ab = 3 = Sphere shape
aB = 3 = Sphere shape
ab = 1 = Long shape
= 9 : 6 : 1
Disc Sphere Long

	AB	Ab	aB	ab
AB	AABB Disc	AABb Disc	AaBB Disc	AaBb Disc
Ab	AABb Disc	AAbb Sphere	AaBb Disc	Aabb Sphere
AB	AaBB Disc	AaBb Disc	aaBB Sphere	aaBb Sphere
ab	AaBb Disc	Aabb Sphere	aaBb Sphere	aabb Long

(e) Duplicate dominant genes (15 :1)

If a dominant allele of both gene loci produces the same phenotype without cumulative effect, *i.e.*, independently the ratio will be 15 :1.

Example : Complete dominance at both gene pairs, but either gene when dominant, epistatic to the other.

Seed capsule of **shepherd's** purse (*Capsella bursa-pestoris)*

Gene pair **'A'** : Triangular shape dominant over ovoid

Gene pair **'B'** : Triangular shape dominant over ovoid (double recessive)

15/16 = Triangular

1/16 = ovoid (topshaped)

AABB × aabb

Triangular Ovoid

AaBb × AaBb

Triangular ↓

	AB	Ab	aB	ab
AB	AABB Triangular	AABb Triangular	AaBB Triangular	AaBb Triangular
Ab	AABb Triangular	AAbb Triangular	AaBb Triangular	Aabb Triangular
aB	AaBb Triangular	AaBb Triangular	aaBB Triangular	aaBb Triangular
ab	AaBb Triangular	Aabb Triangular	aaBb Triangular	aabb ovoid

AB = 9 = Triangular

Ab = 3 Triangular

aB = 3 Triangular

ab = 1 Ovoid (top shaped)

15 : 1

Triangular : Ovoid

(f) Dominant and recessive interaction (13 : 3)

Sometimes the dominant alleles of one gene locus (A) in homozygous and heterogygous (AA, Aa) condition and homozygous recessive alleles bb of another gene locus (B) produce the same phenotype. The F_2 ratio will become **13 : 3.** The genotype AABB, AaBB, AAbb, Aabb and aabb produce one type of phenotype and genotype aaBb, aaBB will produce another type of phenotype.

Example: Complete dominance at both gene pairs, but one gene when dominant epistatic to the other, and the second gene when homozygous recessive, epistatic to the first.

Feather Colour of Fowl

Gene pair **'A'** : colour inhibition is dominant to colour appearance.

Gene pair **'B'** : colour is dominant to white.

Interaction. Dominant colour inhibition prevents colour even when colour is present, colour gene, when homozygous recessive prevents colour even when dominant inhibitor is absent.

13/16 = White

3/16 = coloured

White leg horn × White plymouth Rock

AABB ↓ aabb

AaBb × AaBb

White White

	AB	Ab	aB	ab
AB	AABB White	AABb White	AaBB White	AaBb White
Ab	AABb White	AAbb• Coloured	AaBb White	Aabb• Coloured
aB	AaBB White	AaBb White	aaBB White	aaBb White
ab	AaBb White	Aabb• Coloured	aaBb White	aabb White

AB = 9 white

Ab = 3 coloured

aB = 3 white

ab = 1 white

13 : 3

White : Coloured

Non epistatic Inter allelic genetic interaction

In some cases two pairs of genes determine same phenotype but assorted independently, produce new phenotypes by mutual epistatic interaction. The F_2 ratio remains the same as 9 : 3 : 3 : 1.

Example : Each gene pair affecting the same character complete dominance at both gene pairs, new phenotypes resulting from interaction between dominants, and also from interaction between both homozygous recessives.

Comb Shape in Poultry (Fig. 39.2)

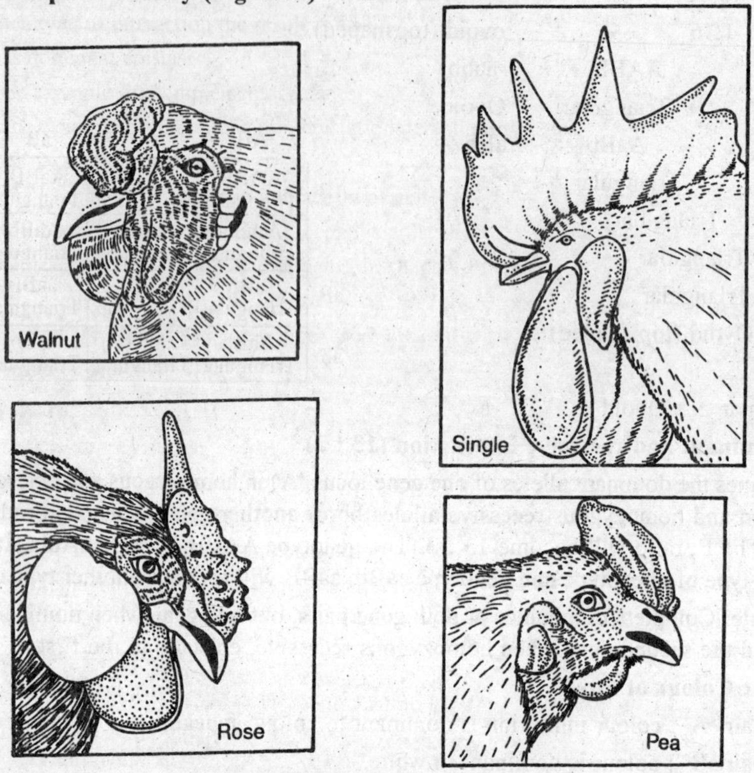

Fig. 39.2. Various types of combs in cock.

Gene Pair **A** : **Rose comb** dominant over non rose.

Gene Pair **B** : **Pea comb** dominant over non pea.

Interaction : Dominant of rose and pea produce **walnut comb.** Homozygous recessive for rose and pea produce **single comb.**

Rose comb × Pea comb

AAbb (Rrpp) ↓ aaBB (rrPP)

× AaBb (RrPp AaBb(RrPp)

Walnut Walnut

9/16 = Walnut
3/16 = rose
3/16 = Pea
1/16 = single
AB = 9 = Walnut
Ab = 3 = Rose
aB = 3 = Pea
ab = 1 = Single

	AB	Ab	aB	ab
AB	AABB Walnut	AABb Walnut	AaBB Walnut	AaBb Walnut
Ab	AABb Walnut	AAbb* Rose	AaBb Walnut	Aabb* Rose
aB	AaBB Walnut	AaBb Walnut	aaBb⁺ Pea	aaBb⁺ Pea
ab	AaBb Walnut	Aabb* Rose	aaBb⁺ Pea	aabb* Single

9 : 3 : 3 : 1
Walnut Rose Pea Single

According to **Bateson** and **Punnett** such result is obtained because by combination of homozygous recessive genes (bb) and a homozygous or heterozygous dominant AA or Aa the Rose comb is produced ; and by combination of homozygous recessive (aa) and homozygous or heterozygous dominant condition **BB** or Bb produce pea comb while the single type comb is produced by double recessive **aabb** genes. '**A**' gene determines the shape of rose comb and 'B' gene determines the shape of Pea comb, but when both genes combine a new shape Walnut appears. In the cross between two types of Walnut chickens 4 phenotypes appear. The genes here do not determine themselves in the development of character but **modify** the character determined by a basic gene. These genes are called **supplementary** or **modifying** genes.

LETHALITY

In 1905 a French geneticist **Cuenot** reported that on the inheritance of a mouse body colour gene, **yellow**, that did not appear to fit the **Mendelian** segregation pattern. **Yellow** has a dominant phenotypic effect on coat colour of mouse whenever yellow x yellow the progeny is yellow and agouti but when yellow is test crossed it is always found **heterozygotic. Castle and Little** explained it as some effect where homozygotes for yellow die. It has a lethal effect. **Robertson, Eaton, Green,** *et al* also have shown that approximately 25% inviable embryos can be obtained. **Xeroderma pigmentosum** of human is also a heterozygotic lethal form. The gene is called **recessive** lethal as long as it is lethal in homozygous condition.

Dominant lethal genes are having the effect on heterozygote individuals, *e.g.*, *Epiloia* gene causes abnormal skin growth, severe mental defect and multiple tumors in human. Heterozygotes die in young age. In *Drosophila* X-ray treatment of male spermatozoa may cause the death of fertilized offspring through the entire range of developmental stages, *i.e.*, from egg of adult. In *Drosophila* curly wings (Cy), plum eyes (Pm) and stubble bristles (Sb) influence the viability of flies in homozygous conditions.

Penetrance of lethals, recessive or dominant, may also vary, so that not all genotypically affected individuals are phenotypically lethal. Some genes have high degree of penetrance and expression so that no survival beyond embronic or infant stage is found. Some are called **Semilethal** or **subvitals** which permit larger proposition of genotype to survive. Lethal genes in general are regarded as those whose effects cause death to the organism, usually in the early growth stage.

Balanced lethal system

Muller in 1918 reported an unusual stock of *Drosophila* heterozygous for a particular lethal gene **beaded**. The beaded gene has dominant effect on the shape of the wings but has been shown to be lethal in homozygous conditions. A cross between **beaded x beaded** always shows 2:1 ratio (Fig 39.3).

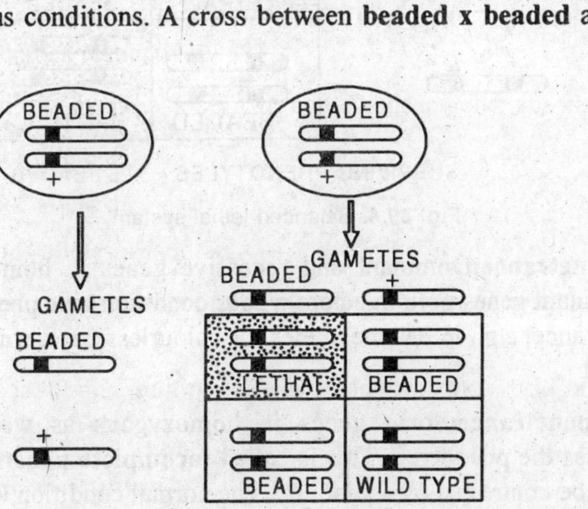

Fig. 39.3. Lethal system.

Muller analysed that the homologous chromosomes carry a normal allele for **beaded** gene also carry a gene which was lethal in homozygous condition, *i.e.*, **le.** But this was not an allele of beaded gene. In addition to beaded heterozygotes **the beaded x beaded** produce two types of homozgyous lethal **Beaded/Beaded, le** (Fig. 39.4). Since **Beaded** and **le** are recessive lethals and are not allelic, it is only the Beaded +/+ le individuals that survive because each of these homologous chromosomes carries the wild-type allele for the recessive lethal factor. Balancing effect between two different lethals in a self perpetuating stock was called **balanced lethal system by Muller.**

Changes in lethality

Lethality is also influenced by the environment many times. **Dobzhansky** in 1946 in certain genotypes of *Drosophila pseudoscura* showed that the temperature upto 16.5°C is optimum for growth but is lethal at 25.5°C. Such lethals are called **conditional lethals.**

Suzuki, *et al,* suggested that the conditions of lethality may be restricted to certain phase of life *e.g.*, ***D. melanogaster*** at high temperature, during post larval stage is lethal.

Some strains of yeast are unable to ferment the sugar galactose which may be lethal to them on galactose medium but no effect in glucose medium.

PENETRANCE

The ability of a given gene or gene combination to be expressed phenotypically to any degree is called **penetrance**. It may be of 2 types.

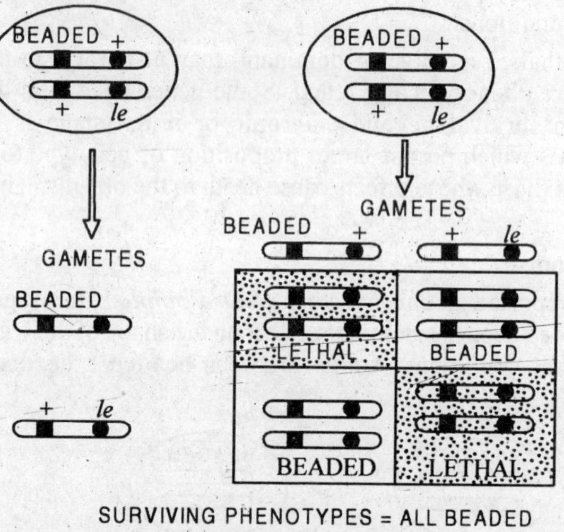

Fig. 39.4. Balanced lethal system.

(a) Complete penetrance: Dominant and recessive genes in homozygous condition and many completely dominant genes even in heterozygous condition give phenotypic expression. It is called complete penetrance, *e.g.*, in Pea the alleles for colourless (rr) are incomplete penetrance in homozygous condition.

(b) Incomplete penetrance: Some genes in homozygous as well as in heterozygous condition fail to express the phenotype. This is called **incomplete penetrance,** *e.g.*, polydactyly in man was thought to be controlled by a gene *'P'*. The normal condition with 5 digits is produced by recessive (pp). Some heterozygous individuals (Pp) are not polydactylus. The penetrance is less than 7%.

EXPRESSIVITY

The expression of different traits is different. The degree of effect produced by a penetrant genotype is called **expressivity,** *e.g.*, if the penetrance of gene is in left side, the man will have 6 digits in left hand and 5 in right.

PLEIOTROPISM

If a single gene influences more than one phenotypic trait, it is called **pleiotropic gene**. If it has more expressive effect on the phenotype, called as a **major effect** and if less effective it is **secondary effect**. The phenomenon of multiple effect of a single gene is known as **pleiotropism,** *e.g.*, in *Drosophila* the recessive gene for vestigial wings cause vestigial wings in homozygous condition. It also affects tiny wing like balancer behind the wings, bristles, structure of reproductive organs, egg production and longevity, etc.

In human the gene for **Phenylketonuria** has pleiotropic effect as it produces many effects called **syndrome**. The diseased individual secretes excessive quantity of amino acid phenylalanine in urine, cerebrospinal fluid and blood. The major characteristics are short stature, mentally retarded, pigmented patches on skin, non pigment hair and eyes, excessive sweating and wide spaced incisors, etc.

MULTIPLE ALLELES

An allele is specific form of a given gene. An allele may be a mutant allele, resulting, in an altered phenotype, or a wild type allele producing an active gene product and a 'normal' phenotype. Different wild type alleles are found in populations producing phenotypes within the 'wild type' range. When more than two different forms of a given gene exist in a species, they are called **multiple alleles.**

In *Drosophila melanogaster,* wild type *Drosophila* have red eyes, but a vast array of eye colour mutants are present. The colour ranges from pure white to intermediate types to red colour. They are named as w^a (white apricot). w^e (white eosin), w^{ch} (white cherry). w^{co} (white coral), w^{col} (white coloured), w^w (white wine), w^{bl} (white blood), w^{cre} (white carrot), w^{cf} (white coffee), etc. Different colours are the result of relative concentration of total pigment.

The number of different genotypes possible among diploid organisms is, a function of the number of alleles that exist for any given gene. If n is the number of alleles of a given gene, the number of different genotypes possible is $n(n + 1)/2$. Thus with 2, 3, 4 or 5 alleles there are 3, 6, 10 and 15 possible genotypes respectively.

Since each diploid organism has only two copies of the gene, an individual must be either

(a) homozygous for allele a_1.

(b) homozygous for allele a_2.

(c) heterozygous for alleles a_1 and a_2.

This individual organism can not have a_3 and a_4 alleles. These may be present in some other organism in the population.

In case of crosses involving multiple alleles standard monohybrid ratio will be observed as $1/4\ a_1 a_1 : 2/4\ a_1 a_2 : 1/4\ a_2 a_2$, etc.

A, B, O Blood groups in Human

Important multiple allele in human involves genetic locus controlling blood types A, B, AB and O. The locus has three common alleles I^A I^B, I^O, I^A and I^B are codominant as I^A I^B (in heterozygotes with both A and B antigens in their red blood cells). I^O is recessive (I^O I^O homozygous has no A, B, O, antigen)while I^A I^O has A antigen and I^B I^O has B antigen in RBC. ABO locus controls the types of glycolipids found on the surface of RBC by specifying the type of glycosyl transferase (enzyme catalysing systems of polysaccharide) synthesized in the RBC. It determines the antigenic reaction with specific antibodies present in the blood serum. When 'A' blood is given to 'B' type, A antibodies will react with 'B' antibodies in the serum or vice versa; producing agglutination or dumping of cells (Fig. 39.5). And therefore before transfusion of blood, the blood groups are matched. Type O is **universal donor** and type 'AB' is **universal recipient** of blood (Table 39.1).

Table 39.1. Blood transfusion capabilities for AB blood group.

Blood Group	Terminal sugars of antigens present	Antibodies present	Red cell types agglutinated	Transfusion accepted from
A	A (galactosamine)	Anti – B	B, AB	A or O
B	B (galactose)	Anti A	A, AB	B or O
AB	A (galactosamine) and B (galactose)	None	None	A, B, AB or O.
O	None	Anti A and Anti B	A, B and AB	O

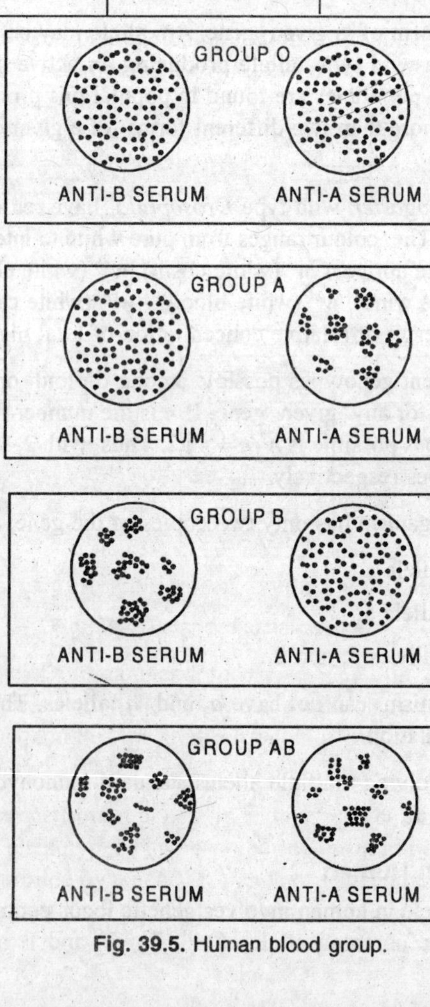

Fig. 39.5. Human blood group.

Rh factor allele in human

The **Rh** factor was discovered in 1940 by **K. Landsteiner** and **A. S. Wiener** from rabbits immunized with the blood of *Macaca rhesus* monkey. The resulting antibodies were found to agluttinate not only the RBC of monkey but those of a high percentage of **Caucasian** people of New York. Individuals whose RBC react with **Rh** antibody are called **Rh positive**, those who do not react are **Rh negative.** If a drop of blood from the individual is mixed with anti **Rh serum** agglutination of erythrocytes takes place, it indicates incompatibility Rh negative mothers are often immunised by blood from Rh-positive foetus. (When father is Rh positive) to which they give birth (Rh positive is dominant over Rh negative). If the mother is not treated properly for anti Rh serum the subsequent child may show haemolytic jaundice, anemia, etc., called **erythroblastosis fatalis** which may result even in death of the foetus or newborn infant.

SUMMARY

1. Influence of alleles and nonalleles on the normal phenotype expression of genes is callled **gene interaction.**
2. The interaction of gene may be **intergenic** (non allelic) or **intragenic** (allelic).
3. When the interference of genes cause death of individual such genes are called **lethal genes.**
4. Alleles expressing themselves within the same phenotypic range are called **isoalleles.**
5. When one gene hides the effect of the other, it is called **epistasis.**
6. If the epistasis is due to recessive allele ('a' hides the effect of 'B') the result will be 9 : 3 : 4.
7. If the epistasis is due to dominant allele ('A' hides the effect of 'B' or vice versa) the result will be 13 : 3.
8. The result of double recessive epistasis is 9 : 7.
9. When a character is controlled by two different genes, it is called complementary gene.
10. **Shull** studied in shepherd's purse (*Capsella bursa-pestoris)* the fruit shape is either triangluar or top shaped. The triangular shape is controlled by two dominant genes. It is called **duplicating** genes, the result will be 15 : 1.
11. The genes modifying the effect of other genes is called **modifier gene.**
12. The genes with more than one effect are called **pleiotropic genes.**
13. The effect of pleiotropic gene may be more for one trait (major effect) and less evident for other trait (secondary effect).
14. The characters which were expressed as intermediate in F_1 generation, *i.e.,* **blending inheritance** and incomplete dominance are controlled by many genes called **multiple factors** and polygenes.
15. **Penetrance** is the percentage of individuals with a given genotype exhibiting the phenotype associated genotype.
16. The degree or extent to which a given genotype is expressed phenotypically in a particular individual is called **expressivity.**
17. Mendelian genetics is also known as **forward genetics** where the study is **phenotype** to **DNA sequence in gene.**
18. Mapping and isolation of genes with unknown protein products is **reverse genetics.**
19. **Paul Berg** (1991) suggested the use of term **reverse genetics** to the studies starting with **DNA sequences with unknown phenotypic effect.**
20. Reverse genetics is used in production of transgenic plants and animals.

QUESTION BANK

Very Short Answer Questions

1. Give one example of pleiotropism.
2. Define penetrance.
3. Define expressivity.

4. In which type of interaction the result is 9 : 3 : 4.
5. In which type of interaction the result is 13 : 3.
6. What is dominant epistasis.
7. Give one example of complementary genes.
8. Give one example of intraallelic non epistatic interaction of genes.
9. Give one example of lethality.
10. Give one example of interallelic interaction of gene.
11. What will be the results of dominant recessive epistasis.
12. Who studied the effect of duplicating genes in shepherd's purse.
13. What are isoalleles.
14. In chiken when rose and pea comb types were crossed what will be the result of F_1.
15. What is major effect.

Short Answer Questions

1. Explain recessive epistasis.
2. Explain duplicating genes with the help of one example.
3. Describe complementary genes.
4. What is dominant and recessive interaction.
5. Explain non epistatic interallelic genetic interaction.
6. Write short notes on lethality.
7. Explain balanced lethal system.
8. Write short note on penetrance.
9. Write short note on expressivity.
10. Write short note on pleiotropism.

Long Answer Questions

1. Describe allelic and non allelic interaction with the help of suitable examples.
2. Explain different types of epistasis with the help of examples.

Multiple Choice Questions (Objective Type Questions)

1. Inherited traits tend to remain discrete was shown by a german botanist:
   (a) Ubish (b) Nageli
   (c) Kolreuter (d) Von Mohl
2. The trait which does not blend at all is:
   (a) Colour (b) Shape
   (c) Size (d) Sex
3. If pure form black and white andulasian fowls were crossed together the F_1 generation will show
   (a) Black fowls (b) White fowls
   (c) Blue fowls (d) Grey fowls
4. Influence of alleles and non alleles on the normal phenotype expression of genes is called:
   (a) Gene expression (b) Gene interaction
   (c) Gene mixing (d) Gene overlapping
5. If the interference of a particular gene is fatal, it is called:
   (a) Epistatsis (b) Lethal
   (c) Mixed (d) Duplicating
6. When more than two alternate forms of a gene are present on the same locus; they are called:
   (a) Intergenic alleles, (b) Intragenic alleles
   (c) Multiple alleles (d) Lethal genes

7. Human blood groups were established by:
(a) Wakesman (b) Landsteiner
(c) Kolreuter (d) Weisman
8. Universal recipient of Blood is :
(a) A (b) B
(c) AB (d) O
9. Universal donor of Blood is :
(a) A (b) B
(c) AB (d) O
10. Antigens are also called as :
(a) Agglutinogen (b) Agglutinin
(c) Thrombin (d) None of the above
11. Antigens are present in :
(a) Serum (b) RBC
(c) WBC (d) Cytoplasm
12. Antibodies are present in :
(a) Serum (b) RBC
(c) WBC (d) Cytoplasm
13. When two gene pairs affect the same character it is called :
(a) Supplementary gene (b) Duplicating genes
(c) Complementary gene (d) Lethal gene
14. Agglutinins are also called as :
(a) Antigens (b) Antibodies
(c) Agglutinogen (d) All the above
15. When a gene hides the effect of other, it is called:
(a) Epistasis (b) Complementary gene
(c) Incomplete dominance (d) Codominance
16. The result of F_2 generation in recessive epistasis is :
(a) 9 : 3 : 3 : 1 (b) 9 : 3 : 4
(c) 15 : 1 (d) 12 : 3 : 1
17. When one character is controlled by two different genes, it is called :
(a) Duplicating gene (b) Complementary gene
(c) Lethal gene (d) Supplementary gene
18. The F_2 ratio of complementary gene :
(a) 9 : 3 : 4 (b) 9 : 3 : 3 : 1
(c) 9 : 7 (d) 15 : 1
19. 15 : 1 is the result of F_2 generation is :
(a) Duplicating gene (b) Complementary gene
(c) Lethal gene (d) Supplementary gene
20. When a character is controlled by two different genes independently or together the effect is called :
(a) Duplicating genes (b) Supplementary gene
(c) Epistasis (d) Complementary gene
21. The result of F_2 generation in dominant epistasis is :
(a) 9 : 7 (b) 12 : 3 : 1
(c) 3 : 1 (d) 9 : 3 : 4

22. Genes modified the effect of other gene is called :
    (a) Modifier (b) Suporter
    (c) Suppressor (d) Multiplier
23. The genes with more than one effect are known as :
    (a) Pleiotropic genes (b) Polygene
    (c) Complementary gene (d) Multiple allele
24. The degree or extent to which a genotype is expressed phenotypically in an individual is called :
    (a) Polygene (b) Pentrance
    (c) Expressivity (d) Aclamatization
25. The percentage of individuals with a given genotype exhibiting the phenotype associated of genotype is known as :
    (a) Expressivity (b) Penetrance
    (c) Both 'a' and 'b' (d) None of the above
26. In *Lathyrus odoratus,* cross between two purple flowered plants give a pink/white progeny, it is due to :
    (a) Incomplete dominance (b) Codominane
    (c) Epistasis (d) Segregation
27. Gene on one locus influence the gene on other locus, it is called :
    (a) Epistatic (b) Complementary
    (c) Duplicating (d) None

ANSWERS

1. (c)	2. (d)	3. (c)	4. (b)	5. (b)	6. (c)
7. (b)	8. (c)	9. (d)	10. (a)	11. (b)	12. (a)
13. (a)	14. (b)	15. (a)	16. (b)	17. (b)	18. (a)
19. (a)	20. (a)	21. (b)	22. (a)	23. (a)	24. (c)
25. (b)	26. (d)	27. (b).			

40 CHAPTER

Genetic Variations: Mutations

Genetic variations are brought about by crossing over in meiosis and mutations. **Mutations** are inheritable changes. **Hugo de Vries** was first to use the term **mutation.**

GENETIC VARIATION

This is not always that offspring only resemble partners but they also vary from them. Of course, the resemblances are due to the units of information, the **genes,** they inherit. However, not all the inherited genes express themselves, *i.e.,* exhibit their phenotype. Some may be hidden or be recessive in the presence of dominant counter-parts, during expression. Genetic variation in sexually reproducing species develops primarily because parental genes are shuffled into new combinations, *i.e.,* new genotypes, in the offspring. The recombinations are produced by the routine reshuffling of genes as follows :

1. Independent assortment of chromosomes during meiosis. During the formation of gametes in the metaphase I of meiosis homologous chromosomes come together in pairs and then segregate into daughter cells independent of each other.

For example, the maternal and paternal chromosomes of an organism with two pairs of chromosomes can recombine in $2^2 = 4$ possibilities in the gametes. With 3 pairs of chromosomes $2^3 = 8$ possible types of gametes are formed. The humans have 23 pairs of chromosomes which can give rise to 2^{23} or about 8.6 million combinations.

2. Reciprocal recombination of linked genes by crossing-over in meiosis. Here during crossing over in meiosis chromatids of homozygous chromosomes break and recombine to produce new linkage groups. As a result new genetic varieties or recombinations are produced.

3. Random fertilization of gametes. The fusion of male and female gametes takes place by random fertilization. During random fertilization any one of the 8.6 million recombinations in the sperm can fuse with any of the 8.6 million combination in the egg producing a total of $(8.6 \times 10^6) \times (8.6 \times 10^6) = 70 \times 10^{12}$ possible combinations. The possible combinations far exceed human population, and therefore, no two human beings are exactly alike.

Along with heredity there is always a component of variation. This means that a whole set of characters is transmitted from the parents to the offspring. Thus, there is always a component of variation due to which they not only look differnt but also the offspring can be distinguished from each other.

Such variations may arise due to different mechanisms, such as ***(i)* recombination *(ii)* gene mutation *(iii)* gene environment interactions** and by ***(iv)* chromosomal aberrations.**

MUTATIONS

It is sudden and heritable change independent of the environment and liable to provide a permanent step of which evolutionary changes may be built. Very precisely, mutation is a sudden change in chromosomal DNA, *i.e.,* in a strict sense it covers ony those changes which alter the chemical structure of the gene at the molecular level. These are commonly called **gene mutations or point mutations.**

From where the new genetic traits come from ? For example, how can a white-eyed fruit fly suddenly appears in a pure culture of red-eyed flies ? Or how can a short-legged sheep appears in a population of sheep with legs of normal length ? In the early 1900s, Hugo de Vries attempted to answer this question. At the time, he was working on the genetics of the primrose plant (*Oenothera lamarckiana).* He noticed that several new traits appeared in the offpsring of the plants. He guessed the cause of the new traits to be permanent changes in the genetic material. Such changes are now called **mutations.** Thus, (*i*) a mutation is a permanent change in the DNA molecule of an organism (*ii*) Mutations are heritable changes, that is, if they appear in somatic cells they are inherited to daughter cells by mitosis, and if they appear in gametes, they are inherited to the offspring. The former are known as **somatic mutations** and latter as **germ mutations,** (*iii*) Mutations bring about a change in the genetic message. (*iv*) Mutations occur randomly and spontaneously.

Types of Mutations

There are two types of mutations : (*i*) **Gene mutations** or **point mutations,** and (*ii*) **Chromosomal mutations.**

1. Gene mutations. A chemical change that occurs in the DNA of a cell is called a **gene mutations** or **point mutations**. Such a mutation may alter the sequence of the nucleotides within a part of the DNA molecule. This alternation changes the information on the DNA chain and results in differences in the proteins being produced. For example, in sickle cell anemia, the mutation of a single gene causes a slight change in the structue of the protein molecule of hemoglobin, and because of that slight change, the blood cell that carries the hemoglobin takes a sickle shape.

According to many scientists, the mutations may be caused naturally by the radiation that constantly enters the earth's atmosphere from the cosmos. For example, gene mutations are probably caused when the sex cells (gametes) of an organism are exposed to X-rays, gamma rays, cosmic rays and ultraviolet rays. In addition to this radiation, certain chemicals called **mutagens** can change nucleotides within DNA molecules. The details of mutagens are as follows :

Physical mutagens. Energetic, penetrating radiations such as X-rays and ultraviolet light are powerful physical mutagens. Very high energy radiations can penetrate all organisms and produce large number of a great variety of mutations. For example,X-radiation is a powerful physical mutagen which can break chromosomes, oxidise deoxyribose, deaminate and dehydroxylate bases and form peroxides. Although less energetic than X-rays, ultraviolet light is also a powerful mutagen, largely because the base of nucleic acids absorb energy of its wavelengths. While UV's energy can also break chromosomes, its chemical effect on bases is particularly significant mutagenesis.

Chemical mutagens. Chemical mutagens such as nitrous acid, formaldehyde, ethylethane, sulphonate, etc., produce mutations by the addition or subtraction of whole nucleotides. C. Auerbach was the first to find mutations can also be induced due to certain chemicals . She made this important discovery during World War II.

The chance for a particular human gene to mutate in one generation is between 1 in 10,000 to 1 in 1,000,000. Since humans have atleast 30,000 genes, it is likely that each person carries atleast one mutation.

2. Chromosomal Abberrations (Chromosome mutations). Another way for the genetic traits of an organism to be altered is through changes involving whole chromosomes or parts of chromosomes. Structural changes in chromosomes are also caused by radiation, chemicals, and even by some virus infections.

Changes occur in the structure of chromosomes, during cell division. When homologous chromosomes pair up, linked genes on the chromosomes may break apart. The genes may join another chromosome, or they may be lost.

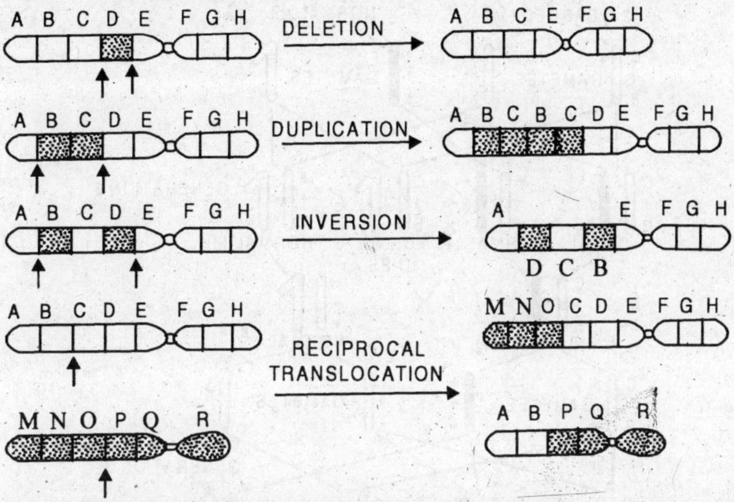

Fig. 40.1. Forms of chromosomal mutations showing deletion, duplication, inversion and receiprocal translocation. Apart of chromosome may be deleted, duplicated or inverted. It brings about the loss of genes, duplication of genes or inversion of gene order in a chromosome. A segment of one chromosome may be reciprocally exchanged with a pair of another chromosome.

Thus, a **deletion** involves the loss of a piece of chromosome. If a chromosome breaks and the parts do not reattach, the pieces may be lost. This is the most serious kind of chromosome mutation. Here, bits of genetic information are not available to the offspring.

A **duplication** occurs when one extra, but identical piece of chromosome is added to the normal chromosome.

When an **inversion** occurs, the pieces of chromosome break apart and pieces rejoin the same chromosome in a different order. Usually inversions have no harmful effects on the offspring.

Due to the mutagens two types of mutation **visible** and **lethal** are found. Mutations are located on either sex chromosomes or autosomes.

Detection of Sex linked lethals

CIB Method : The method involves use of CIB stock carried inversion of heterozygous state to work as **crossover supressor (C), a recessive lethal (i) X chromosome in heterozygote state and a dominant marker for barred eye (B)** or narrow eye.

Female heterozygote fly carries all the three features on one X chromosome (Fig. 40.2) and termed as **CIB X** flies. Male with CIB X chromosome will die. When male normal is irradiated for induction of mutation and crossed to CIB barred female in F_1 progeny male receiving **CIB X** chromosome will die. The **CIB** female will be barred type while the other female though normal but has one irradiated X chromosomes. When barred CIB female is crossed with normal male again the CIB male will die but 50% male will survive with irradiation X chromosome.

Muller-5-method : *Muller-5-Drosophila* with 2 marker genes, dominnant **Bar** (barred eye) and recessive **apricot.** It does not carry a lethal gene (Fig. 40.3) When **Muller**-5 ♀ barred apricot is crossed with irradiated normal male the progeny will have heterozygote barred ♀ (female) with one X chromosome irradiated and **Muller 5 barred aprict ♂** (male) type. If a lethal mutation is induced in X chromosome of irradiated male, no wild type male would apper in F_2 generation. The absence of wild type male in F_2 indicates induction of lethal mutation.

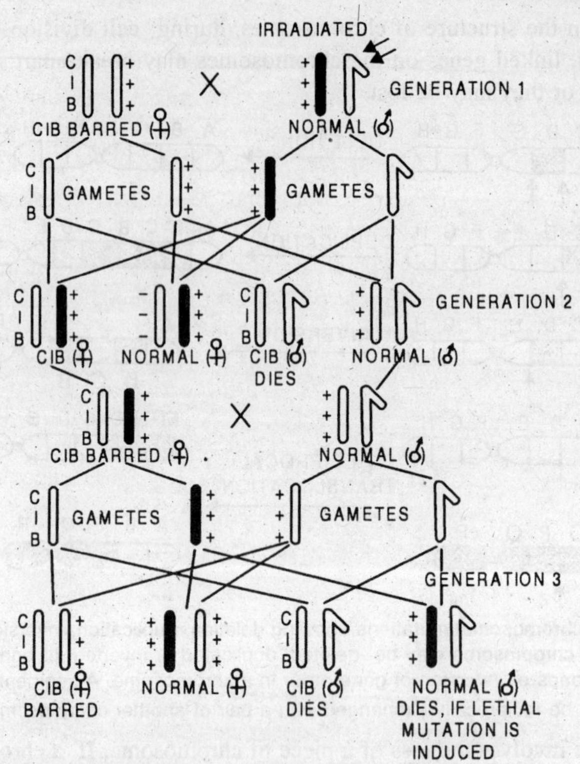

Fig. 40. 2. CIB method for detection of sex linked lethilnulat in *Drosophila.*

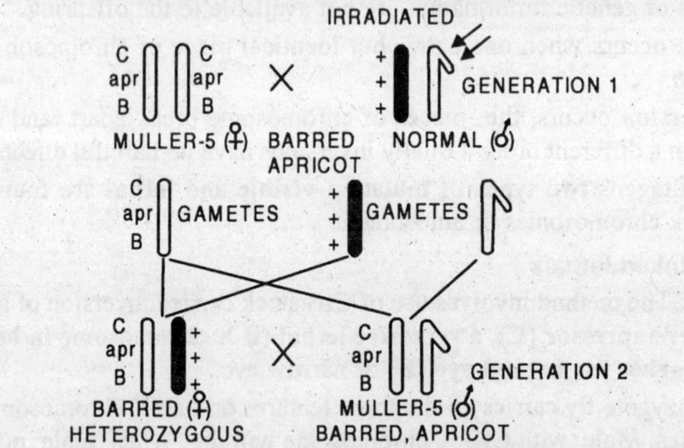

Fig. 40. 3. Muller-5-method in *Drosophila.*

Detection of sex linked visible mutation attached 'X' method : Muller-5 and attached **X** chromosomes were used to detect visible mutation. The attached X female (XXY) is crossed with irradiated male. Irradiated 'X' chromosome of male will form a superfemale (XXX) with visible induced mutation (Fig. 40.4).

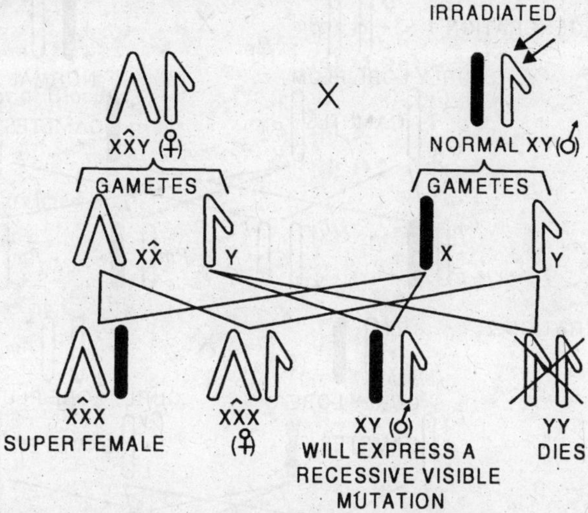

Fig. 40.4. Detection of sex linked visible mutator attached X method.

Balanced lethal system and detection of autosomal mutation

In *Drosophila* the female carrying dominant **Cy** gene (curly wing) and **L** (lobed eye) and the other homologous chromosome carrying gene of plum eye (**pm**) is crossed with same type of male, *i.e.,* curly, lobe, plum type (cyL/pm). the progeny is always heterozygous. Because homozygous with same genes on both the homologous chromosomes Cy L/Cy L or pm/pm will die. (Fig. 40.5).

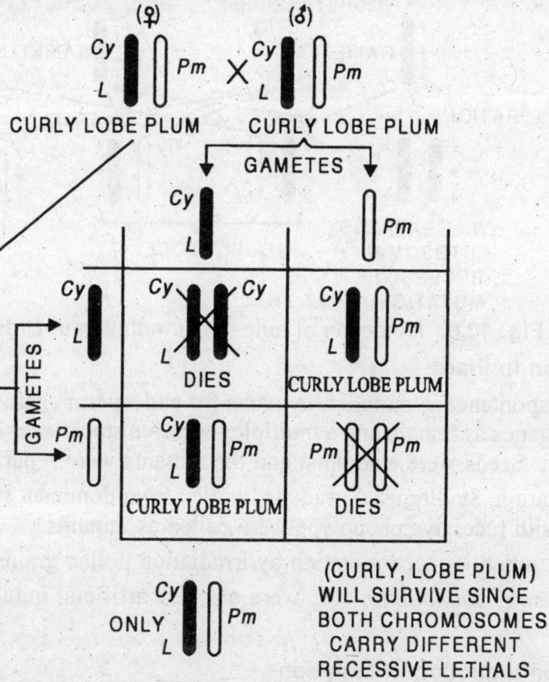

Fig. 40.5. Balanced lethal system in Cy L/pm *Drosophila*.

If Cy L/pm female is crossed with normal irradiated male, recessive mutated forms in females will survive but homozygous Cy L/cy L type will die (Fig. 40.6).

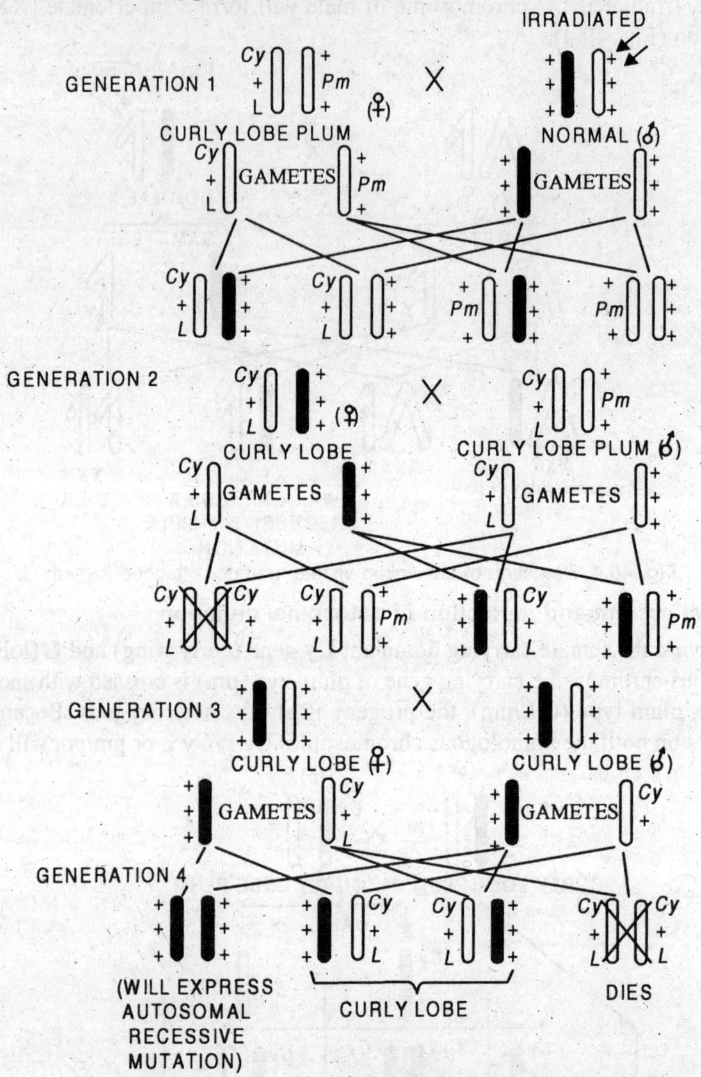

Fig. 40.6. Detection of autosomal mutation in *Drosophila.*

Detection of mutation in Plant

Stadler studied spontaneous mutation in maize for endosperm characters by growing a stock dominant for seve. . genes as female and a multiple recessive stock was sown on every 5th row as male for pollens grain. Seeds were examined and the mutants were separated.

For induced mutation, Stadler used irradiated pollen from dominant stock and a recessive was pollinated. Progeny with recessive phenotype were called as mutants.

Singleton also studied induced mutation by irradiation pollen grains.

Neurospora, E. coli, Salmonella, etc., were also for artificial induction of mutation using various mutagens.

Effect of genotype on induction of Mutation

Stadler showed that freqency of spontaneous and induced mutations vary from locus to locus in maize. The genetic differences influence induced mutation at **mutator** gene and **paramutation.**

Mutator gene : In maize a gene 'a' is located on chromosome 3 causing absence of anthocyanin leading to loss of purple colour in kernel endosperm and other parts. If chromosome 9 has a dominant gene '**Dt**', 'a' gets mutated to 'A' and allows anthocyanin product in heterozygote (Aa). The kernel will have spots in aleurone layer (Fig. 40.7) or stipes in leaves. Such gene is called mutator gene.

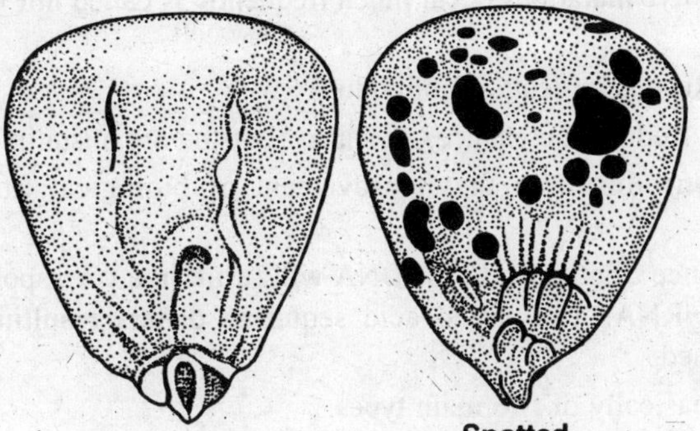

Colourless (aa)

Spotted
(aa, with mutator gene Dt)

Fig. 40. 7. Effect of mutator gene (Dt) in corn.

Ds-Ac system in cron was studied by **Barbara Mc Clintock. Ds** is dissociation gene on chromosome causing breakage but it is only active in the presence of **Ac (activator)** gene located on other chromosome.

Paramutation:

Brink (1956) discovered the phenomenon of paramutation in corn. He showed that the gene 'R' is for pigment with various phenotype effect R^r (full colour), R^{st} (stippened) and R^{mb} (marble) alleles is called paramutation. It is a rare phenomenon. The examples are found in *Oenothera*, tomato and fern, etc.

Practical applications of Mutation

According to **Gustafsson** less than one in 1,000 mutations produced may be useful in plant breeding.

Mutant varieties in wheat are produced for branched ears, high protein lysine control, amber seed colour, resistance and awned spikelets, etc. **M.S. Swaminathan** in IARI (Indian Agricultural Research Institute) New Delhi has produced a very important mutant variety of wheat, *i.e.*, **Sarbati sonara.**

Reimei is mutant variety of Rice produced by gamma radiation. Mutant varities of rice for increased protein, lysine content, etc., are produced. Mutations in *japonica* variety are created for *indica* type grains.

Erectoides and *Eceriferum* are the mutants of Barley for high yield. **Todd's Mitchem** variety of peppermint which is disease resistant and with improved quality of oil. **Aruna** variety of castor plant is an early maturing variety (110 days instead of 270). It can be harvested before the spread of disease and insects.

Penicillium is irradiated for increased yield of **penicillin.**

GENE MUTATION

Genes are segments of DNA molecules. Mutation is an abrupt qualitative or quantitative change in the genetic material of an organism. Mutations are **intragenic or intergenic. Intragenic mutations** are called **point mutations**. It includes change or alteration in DNA molecule in a gene. Change in base sequence of DNA molecule may bring about point mutation. The unit of gene bringing about mutation is called **'muton'. Intergenic mutations** involve changes in many genes (long chain of DNA). It includes changes in chromosomes.

Mutations occur in random manner. The frequency of spontaneous mutations is usually low, ranging from 10^{-7} to 10^{-12} per organism. The rate of detectable mutations in average gene is 1 in 10^6. the regions of gene where mutations occur much frequently is called **hot spots.** Mutations are classified as :

Lethal mutation : Results in death of individual.

Subvital mutation : Reduce the chances of survival of the organism in which they occur.

Supervital mutation: Result in the improvement of biological fitness under certain conditions.

Any change in sequence of nucleotides of DNA will result in a corresponding change in the nucleotide sequence of *m*RNA. The amino acid sequence differs resulting in structure and properties of enzyme formed.

Gene mutations are basically of two main types.

(a) Base pair mutation and (b) Frame shift mutations

(a) Base Pair Substitutions

This is the commonest type of mutation. The base pair switches are A-T, G-C, C-G or T-A. One base of a triplet codon is substituted by another, resulting is a changed codon.

(1) Original message reading frame CAT GAT CAT GAT CAT

(2) **Substitution or replacement A** is replaced by **'G'** CAT GAT **CGT** GAT CAT, message out of frame.

The mutated codon specifies another amino acid it will result in amino acid substitution in the polypeptide chain during translation (Fig. 40.8). Such changes are called *missense mutations*. If a single base substitution modifies a codon to a termination codon (CAG → UAG) the chain terminates at the place of mutation. This mutation is called **nonsense mutation**. Base pair substitution is of two types (Fig. 40.9).

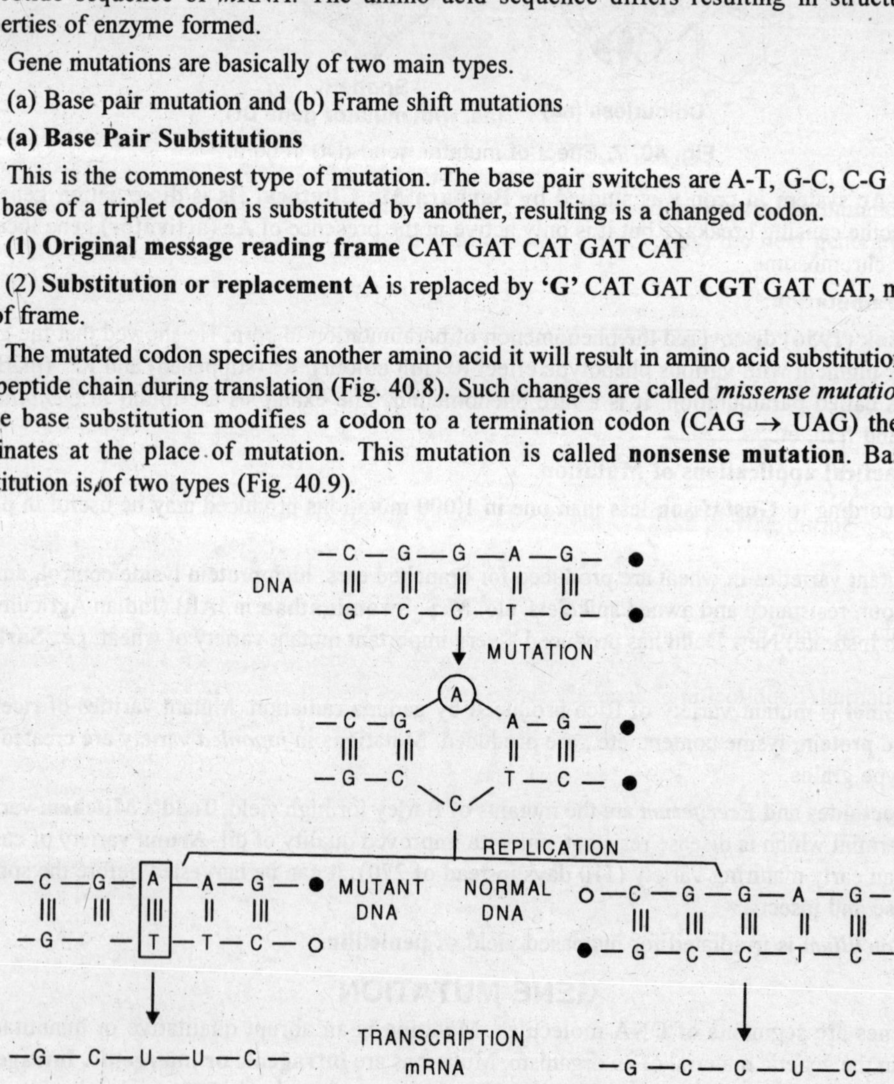

Fig. 40.8. Base pair substitution.

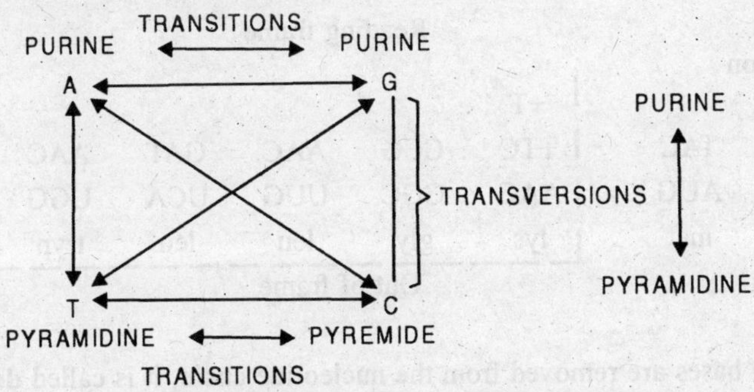

Fig. 40.9. Transition and Transversion.

(i) Transititon : Purine base is replaced by another purine, *i.e.*, 'A' by 'G' or 'G' by 'A', and pyramidine base is replaced by another pyramidine, *i.e.*, 'T' by 'C' or 'C' by 'T'.

(ii) Transversion: Purine base is substituted by pyramidine base is called transversion, *e.g.*, 'A' or 'C' is replaced by 'T' or 'G'.

Transition mutation generally code for chemically similar amino acids. Transversion shows change in amino acid (Fig. 40.10).

(b) Frame Shift mutation

A mutation in which there is deletion or insertion of one or a few nucleotides is called a **frameshift mutation.** Addition or deletion of one or two bases results in a new sequence of codons coding for entirely different amino acids. The protein synthesized will be entirely different.

Insertion

If one or more bases added to the message, it is called **insertion,** *e.g.*,

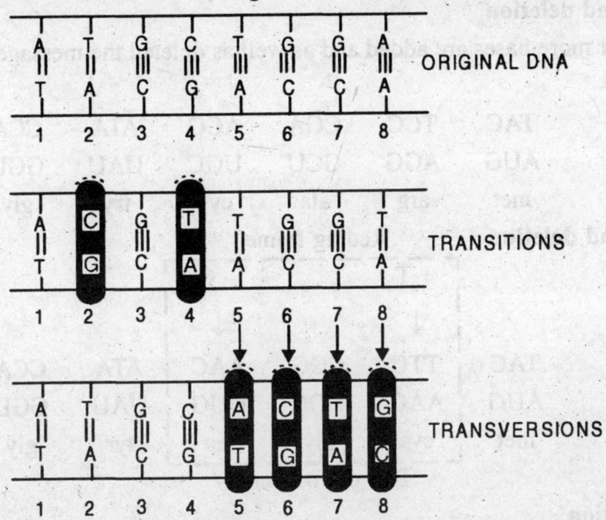

Fig. 40.10. Example of specific base pair change in transition and transversion.

Wild type

DNA	TAC	TCC	CGA	ACG	ATA	CCA	GAG
mRNA	AUG	AGG	GCU	UGC	UAU	GGU	CUC
Protein	met	arg	ala	cys	tyr	gly	leu

After Insertion

Reading frame

+T

DNA	TAC	TTC	CCG	AAC	GAT	AAC	AGA
mRNA	AUG	AAG	GGC	UUG	UCA	UGG	UCU
Protein	met	lys	gly	leu	leu	tryp	Ser

Out of frame

Deletion

If one or few bases are removed from the nucleotide chain, it is called **deletion,** *e.g.*,

Wild type

DNA	TAC	TCC	CGA	ACG	ATA
mRNA	AUG	AGG	GCU	UGC	UAU
Protein	met	arg	ala	cys	tyr

Reading frame

After deletion

–G

↑

DNA	TAC	TCC	CGA	ACA	TAC
mRNA	AUG	ACG	GCU	UGU	AUG
Protein	met	arg	ala	cys	met

Frame out

Insertion and deletion

When one or more bases are added and as well as deleted the message will be changed, *e.g.*,

Wild type

DNA	TAC	TCC	CGA	ACG	ATA	CCA
mRNA	AUG	AGG	GCU	UGC	UAU	GGU
Protein	met	arg	ala	cys	try	gly

Insertion and deletion Reding frame

+ T ↓   – G ↓

DNA	TAC	TTC	CCG	AAC	ATA	CCA
mRNA	AUG	AAG	GGC	UUG	UAU	GGU
Protein	met	cys	gly	leu	tyr	gly

Part out of frame

Triple insertion

+ T ↓   + T ↓   + G ↓

DNA	TAC	TTC	CCT	GAA	GCG	ATA	CCA
mRNA	AUG	AAG	GGA	CUU	CGC	UAU	GGU
Protein	met	lys	gly	leu	arg	tyr	gly

Part out of phase

Triple deletion

		–T	–A	–T		
		↑	↑	↑		
DNA	TAC	CCC	GAC	GAA	CCA	GAG
mRNA	AUG	GGG	CUG	CUU	GGU	CUC
Protein	met	gly	leu	leu	gly	leu
		part out of frame				

Example of **sickle cell anaemia** is of point mutation. The haemoglobin protein in RBC molecule consists of 4 polypeptide chains among which 2 are α chains and 2 are β chains, changes in haemoglobin structure results in certain types of anaemia. In sickle cell anaemia RBC become sickle shaped. Normal haemoglobin has amino acid **glutamic acid** in 6th position. In sickle cell gutamic acid substituted by **valine** and in haemoglobin C it is substituted by **Iysine**.

	1	2	3	4	5	6	7	8
A.	Val	his	leu	thr	Pro	**glu acid**	glu acid	lys
B.	Val	his	leu	thr	pro	**val**	gly acid	lys
C.	Val	his	leu	thr	pro	**lys**	glu acid	lys

Where 'A'is normal haemoglobin 'B' is sickle cell haemoglobin and 'C' is haemoglobin 'C'.

Missense mutation

Due to change in one base the amino acid in a polypeptide chain is replaced by other amino acid, the protein has only one different amino acid, *e.g.*, **UUU** codes for phenyl alanine and the base **'U'** is substituted by **'G'** the code will be **UGU** coding for **cysteine** (Fig. 40.11)

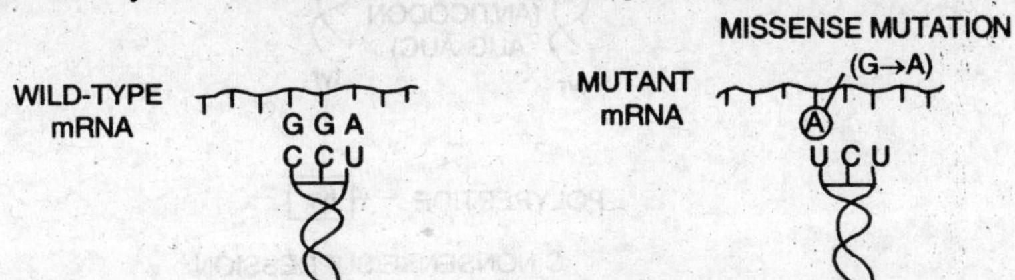

Fig. 40.11. Missense mutation and suppression.

Nonsense mutation

If the change in base sequence codes for termination codon UAA, UAG and UGA the mutation is called nonsequence mutation, *e.g.*, in UAC if C is substituted to G it will become UAG which is a termination codon (Fig. 40.12).

In the same way a termination codon may also be converted into nonsense codon, *e.g.*, in UAA, U is substituted by 'C' then it will become CAA coding glutamine.

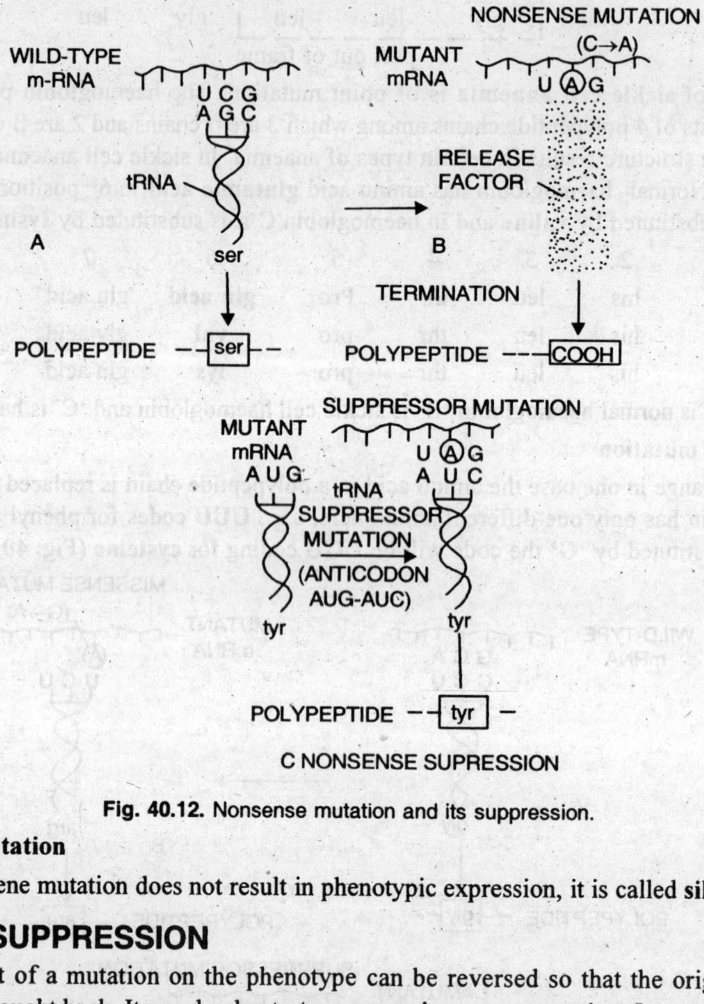

Fig. 40.12. Nonsense mutation and its suppression.

Silent mutation

When a gene mutation does not result in phenotypic expression, it is called **silent mutation.**

GENETIC SUPPRESSION

The effect of a mutation on the phenotype can be reversed so that the original wildtype phenotype is brought back. It may be due to **true reversion** or **suppression.** In **true reversion** the original genetic change is reversed, *e.g.*, if **A** is changed to C in UAV to UCU the true reversion is change to 'C' to 'A' in the same UCU to UAU. Such mutation is called **back mutation**.

When correction of the mutation is on a **different site** it is called **suppression.** Suppression mutation can yield recombinants. Suppression mutation is of 2 types.

(a) Intragenic suppression: Where a gene is suppression by another mutation in the same gene. It may be due to deletion or insertion or both.

(b) Extragenic or intergenic suppression: If mutation in a gene is overcome by a mutation in another gene, the process is called **intergenic suppression**, the effect of nonsense mutation can be suppressed by mutation in other gene.

Suppressor genes: Genes whose activity results in suppression of mutation in other genes are called **suppressor genes.** Suppressor genes are present for each termination codon. The **amber** (UAG) mutants of T_4 bacteriophage are able to grow on only **permissive** strains (mutant strain containing suppressor genes) of *E. coli*. The suppressor genes restore the function of mutant genes in T_4 DNA enabling them to grow on host cells. 3 **amber** suppressor genes **Su^1, Su^2** and **Su^3** or **sup D**, **sup E**, and **sup F** are known for amber termination codon. These genes produce changes in the anticodons of *t*RNA so that they can read the termination codon UAG. Among the three suppressor genes are, inserts **serine,** other **glutamine** and the third **tyrosine** in the termination position.

Ochre (UAA) mutant suppresses the mutation of **ochre** termination codon and **opal** (UGA) mutant suppresses opal termination codon.

Frameshift suppression

Certain suppressor genes can mask the frameshift mutations resulting from the insertion of nucleotides, *e.g.*, anticodon for phenyl alanine *t*RNA reads quadruplet UUUC instead of UUU as codon for phenylalanine. The extra nucleotide permits the reading of form nucleotides in *m*RNA at a time and thus restores the reading frame to the original position. In *Salmonella typhimerinum* glycine *t*RNA contains the nucleotide quadruplet CCCC instead of triplet (CCC) in the anticodon position. The extra nucleotide permits the reading of 4 nucleotides in *m*RNA. (Fig. 40.13)

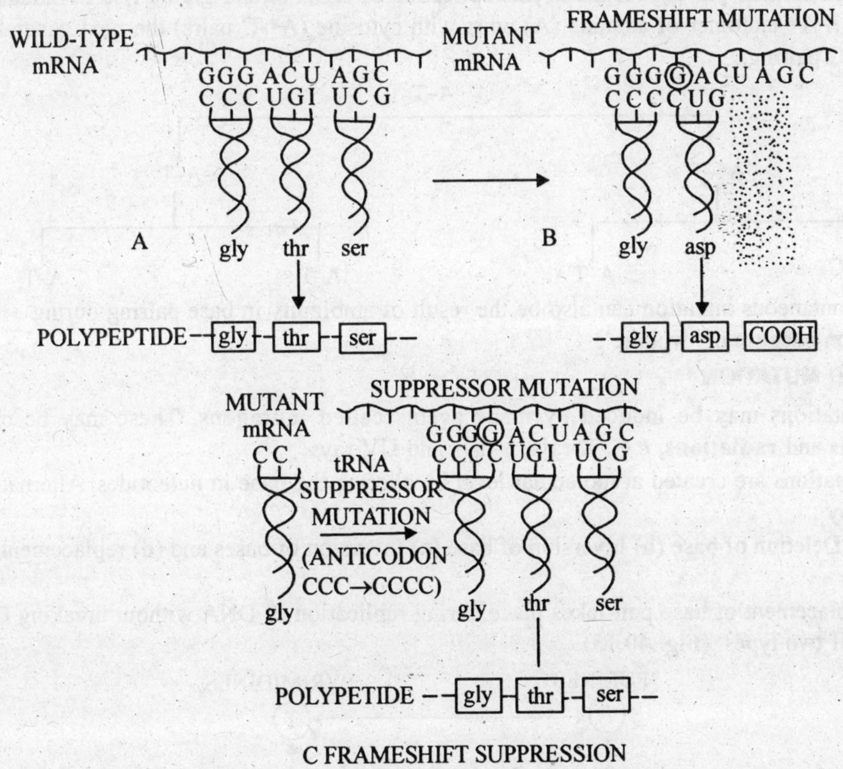

Fig. 40.13. Frameshift mutation and suppression.

SPONTANEOUS MUTATION

Mutation occurring under natural conditions are called **spontaneous mutations**. Such mutations arise by action of mutagens, *e.g.*, cosmic radiation, radioactive compounds, heat, etc.

Tautomerism

The ability of a molecule to exist in more than one form is called **tautomerism.** Normal base pairing in DNA is A-T and G-C. The tautomerism forms are, however capable of unusual base pairing like G-T,T-G,C-A and A-C (Fig. 40.14).

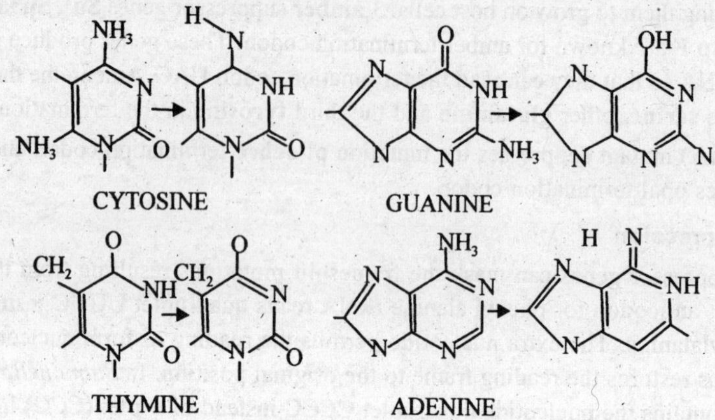

Fig. 40.14. Tautomeric forms.

Unusual base pairing results in misreplication of DNA strand giving rise to mutants in the progeny if A* tautomer of adenine (A) pairs with cytosine (A*-C pairs) the next generation will show G-C pairing.

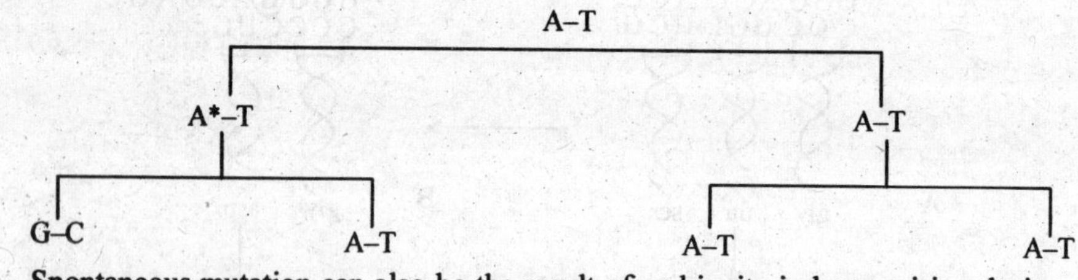

Spontaneous mutation can also be the result of ambiguity in base pairing during replication because of 'wobble'.

INDUCED MUTATION

Mutations may be induced by many agents called **mutagens.** These may be chemical **mutagens and radiations,** *e.g.*, X-rays, γ-rays and UV rays.

Mutations are created at molecular level by altering the **base** in nuleotides. Alternations are created by :

(a) Deletion of base (b) Inversion of base (c) Inversion of bases and (d) replacement of base pairs.

Replacement of base pair takes place during replication of DNA without breaking DNA. It may be of two types. (Fig. 40.15)

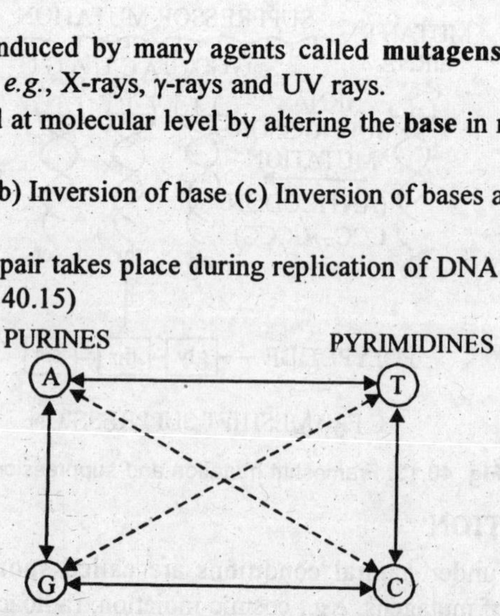

Fig. 40.15. Induced transition and transversion.

(a) **Transition** :Purine is replaced by another purine or pyrimidine is replaced by another pyrimidine.

(b) **Transversion** : Purine is replaced by pyrimidine.

The discovery of mutagenic effects produced by various types of radiations was shown for the first time as an experimental probe for changine gene structure and function. It was very difficult to distinguish between direct or indirect effects of irradiation and to analyse the exact nature of biochemical compounds produced. Chemical mutagens are more effective and their results are characterized.

Thomas and **Steinberg** found nitrous acid effective causing mutation in *Aspergillus.* **Auerbach** and **Robson** found that mutations can be induced by nitrogen and sulphur gas in *Drosophila*. Mutagenic activity in formaldehyde, diethylsulfate, diazomethane, etc., are discovered by **Rapoport.** Chemical mutagens cause severe skin irritations in mammals and also can produce cancer. There are some chemicals which affect some organisms but not other. **Watson** and **Crick** were first to suggest that mutation could occur as a result of occasional changes in the hydrogen bonding of nucletodie bases, *e.g.,* **adenine** normally bears an NH_2 (amino) group providing hydrogen atom for bonding with the complementary **keto** (C = O) group of thymine. In a **Tautomeric shift** the amino group is changed to amino group (NH). This base now bonds with **cytosine** (instead of thymine). In **thymine** the tautomeric shift from keto to **enol** (COH) form allows it to bond with **guanine** (instead of adenine) (Fig. 40.16).

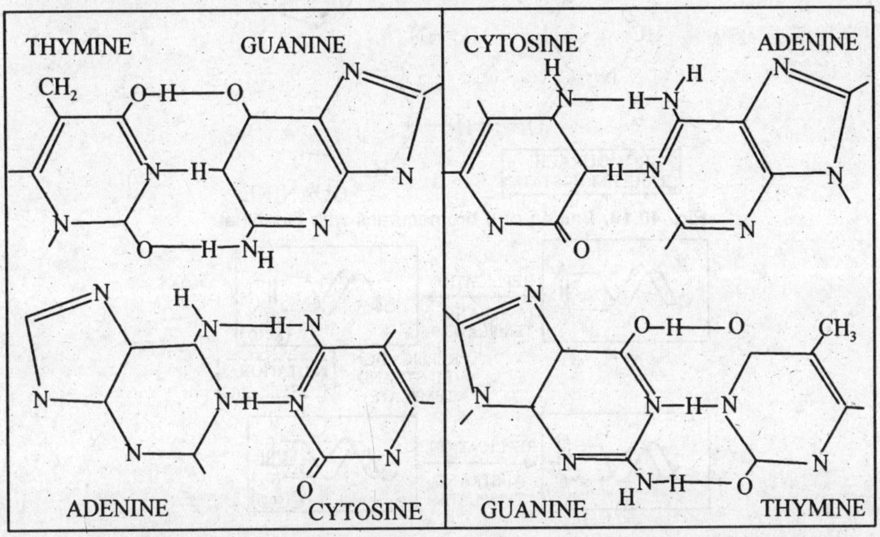

Fig. 40.16. Tautomeric shift : T with G instead of 'A', 'C with 'A' instead of 'G', A with 'C' instead of 'T', G with 'T' instead of 'C'.

If the tautomeric shift produces the error, it is necessary that DNA replication should take place.

Base analogues

A chemical substance resembling a base is called **base analogue**. It may be incorporated into newly synthesized DNA instead of a normal base. Pyrimidine analogue 5-bromouracil (5-BU) is structurally similar to thymine. 5-chlorouracil (5 CU) and 5 iodouracil (5 IU)can also replace thymine in DNA. 2-amino purine (2 AP) is incorporated in very small amount that it could not be possible to find out which base it replaces. 2,6 diamino purine is highly mutagenic. **5-bromouracil** can pair with adenine just like **thymine** (Fig. 40.17).

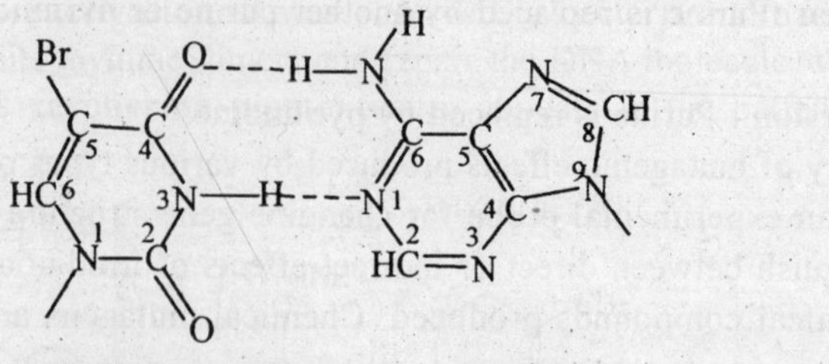

Fig. 40.17. Pairing of "5" bromouracil (keto form) with adenine.

5-bromouracil (5-BU)and bromodeoxyuridine (BUdR) are analogues of thymine which are **keto** forms but can undergo tautomeric shifts, they are **enol** form and pairs with Guanine (G) instead of Adenine (A) (Fig. 40.18). 5-BU produces G-C substitution for the original A-T,or it may occasionally be incorporated in the **enol** form as a pairing mate with guanine and then revert to its keto form to produce A-T substituion for the original G-C. **Lawley** and **Brookes** suggested that mispairing may be caused by ionization of bases rather than by tautomeric shifts. In this mechanism a base, *e.g.*, 5-BU loses the hydrogen normally associated with its 3 nitrogen atom (Fig. 40.19 A,B).and may now pair with guanine (G).

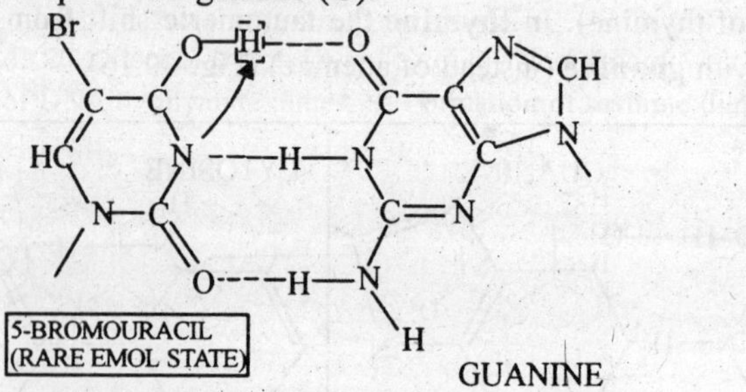

Fig. 40.18. Pairing of 5 bromocuracil with Guanine.

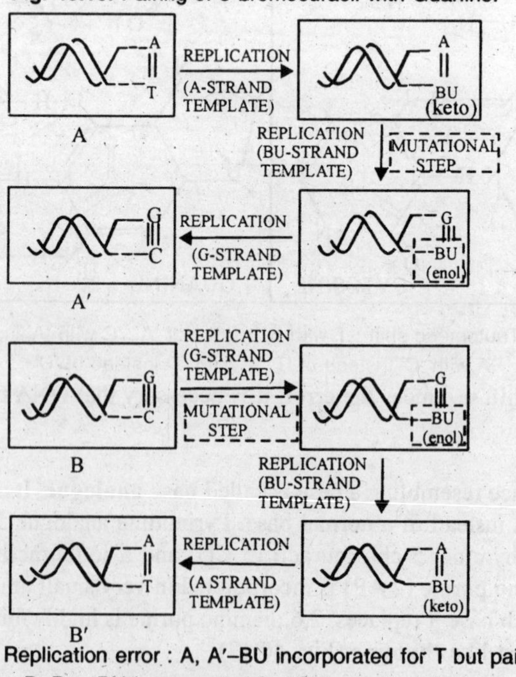

Fig. 40.19. Replication error : A, A′–BU incorporated for T but pairs with G,
B, B′ – BU incorporated for 'C' but pairs with A.

The base analogue 2 amino purine (2 AP) shows mutational properties enabling it to be incorporated as a substitute of adenine but to pair subsequently with cytosine, or to pair initially with cytosine and subsequently with thymine. Incorporation a of AP at place of guanine (G) to give AP–C base pair will cause mutation in subsequent generation.

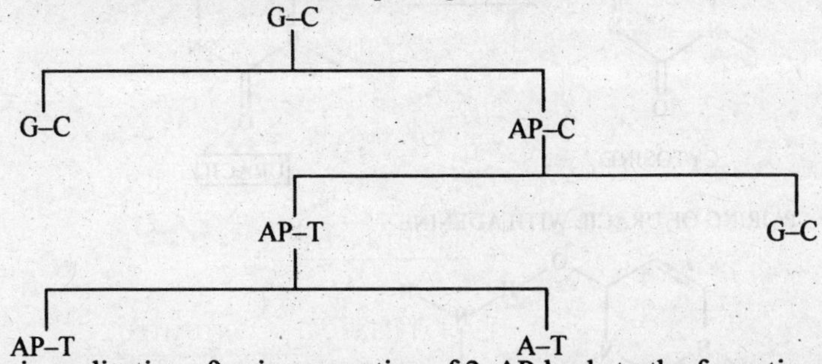

A mistake in replication after incorporation of 2–AP leads to the formation of AP–T base pair inducing transition.

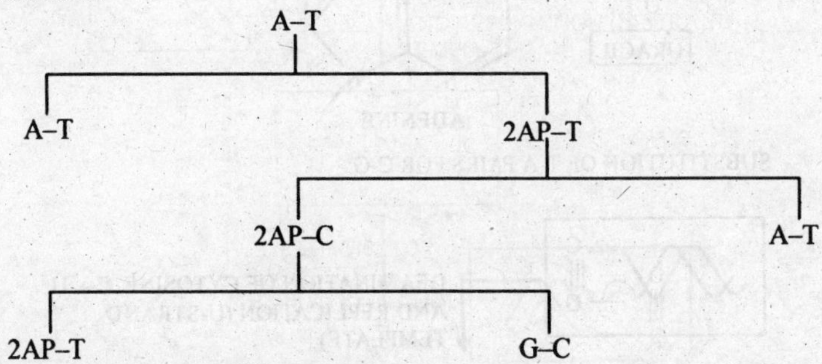

Agents modifying purines and pyrmidines

The agents which modify purines and pyrimidines or agents which stabilize the bases include **nitrous oxide** (HNO_2), hydroxylamine and alkylating agents.

Nitrous oxide (HNO_2): It reacts with bases containing amino groups. It changes the structure by deamination (removal of the amino group). The **amino group** (NH_2) is replaced by **hydroxyl group** (OH^-). Nitrous acid deaminates, the bases, G, C and A with decreasing frequency. Deamination of **adenine** results in the formation of **hypoxanthine** (Fig. 40.20). Hypoxanthine pairs with **cytosine** instead of thymine. Thus **A–T** pairing is replaced by G–C pairs. Deamination of **cytosine** at 6 position results in the formation of uracil (U) (Fig. 40.21) and the pairing of C–G instead of U–A is formed. **Guanine** deaminates to **xanthine. Xanthine** behaves like Guanine and pairs with cytosine, the pairing is **X–C** instead of G–C. Deamination of Guanine has no mutagenic affect (Fig. 40.22). The change in base pairing result in change in DNA in the 50% progeny. Deamination of gaunine, does not show any heritable mutation.

Table: 40.1. The structural and pairing behaviour change of DNA due to deamination by nitrous oxide.

Normal Base	Normal pairing	Deaminated Base	New pair
Adenine	A–T	Hypoxanthine	G–C
Cytosine	C–G	Uracil	U–A
Guanine	G–C	Xanthine	X–C

Hydroxylamine (NH_2OH): It reacts with **cytosine** and **guanine**, the hydroxylation of cytosine at amino group forms **hydroxylcytosine** which pairs with **adenine** because the hydroxyl amino group should be more electronegative than amino group. The hydroxylated molecule is in tautomeric form having a hydrogen atom in place of nitrogen at position 3. The effect of hydroxylamine on 'C' produces a transition in base pairing (Fig. 40.23)

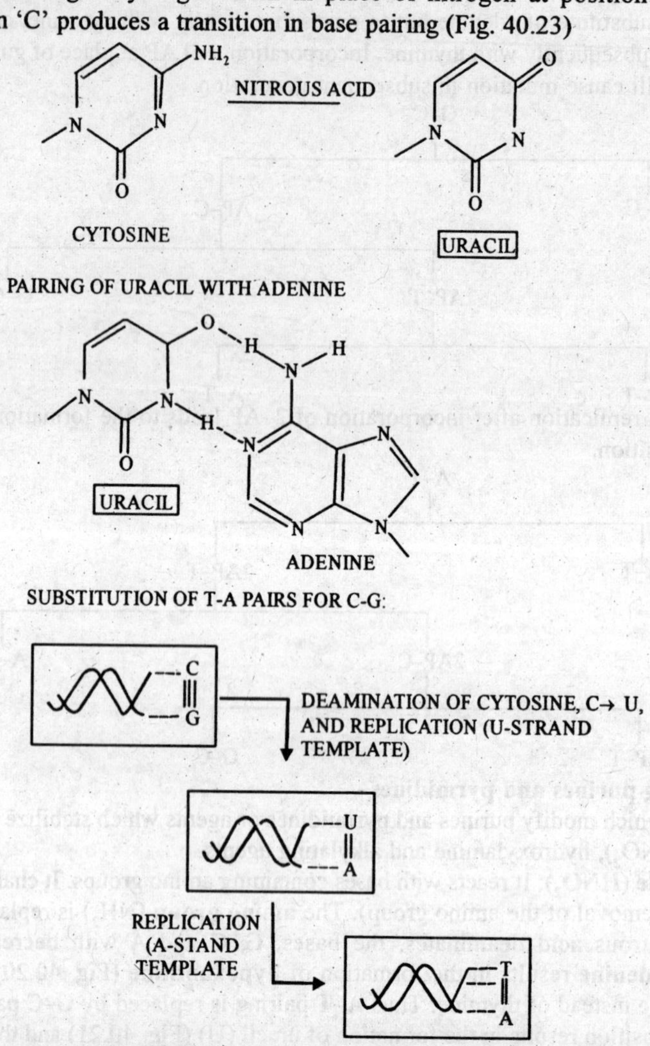

Fig. 40.20. Action of nitrous acid on cytosine.

Hydrazine (NH_2NH_2) breaks the rings of **uracil** and **cytosine** forms **pyrazolone** and **3-aminopyrasole.** When DNA is treated with hydrazine it produces **"apyrimidinic acid".** While when RNA is treated with hydrazine it produces **"ribo-apyrimidinic acid".**

Alkylating agents : Many mutagenic agents carry one or more **alkyl** groups. These are called **mono- , bi-** or **poly** functional alkylating agents, *e.g.*, **dimethyl sulphate** (DES), **dimethyl sulphate** (MMS), ethyl ethane sulphonate (DMS), **methyl methane sulphonate** (EES) and **ehtyl methane sulphonate (EMS),** etc. All of them act as mono-functional groups.

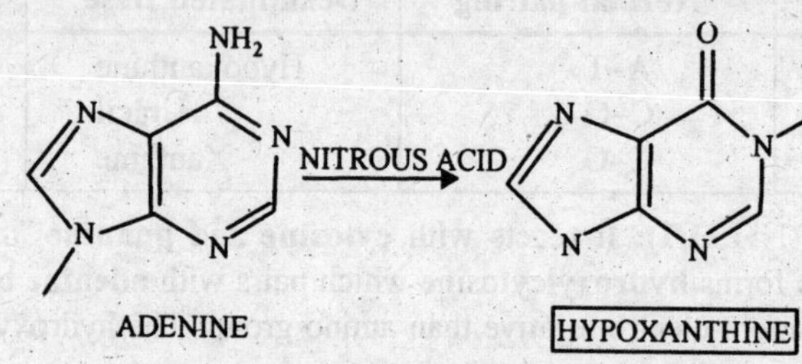

PAIRING OF HYPOXANTHYINE WITH CYTOSINE:

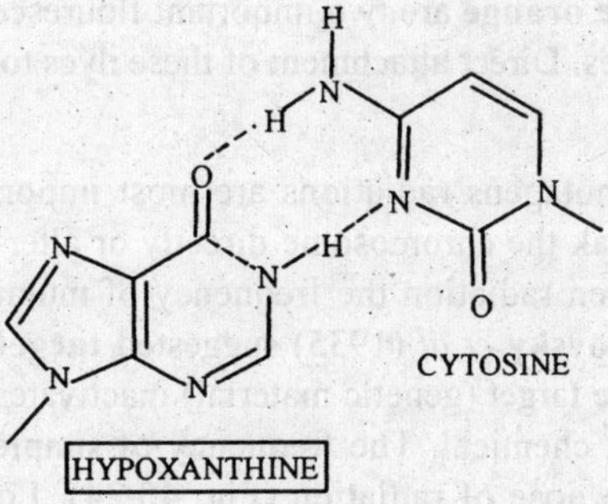

SUBTITUTION OF G-C PAIRS FOR A-T:

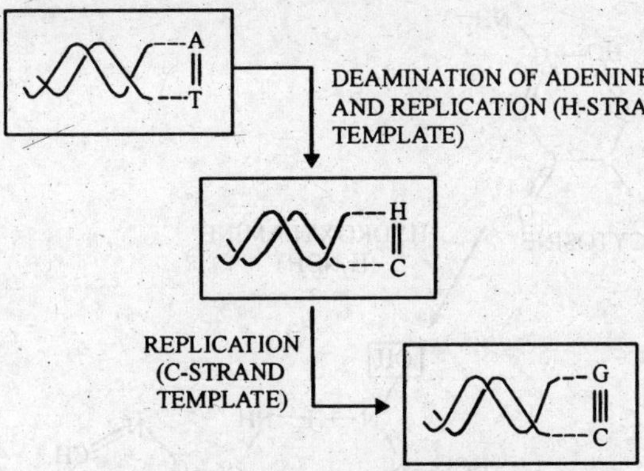

Fig. 40.21. Action of nitrous acid and adenine.

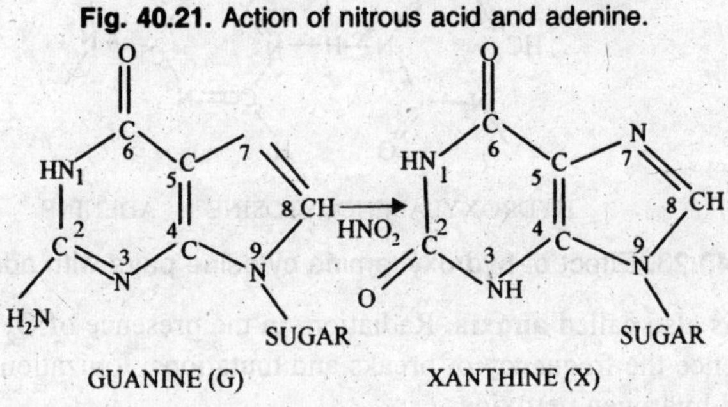

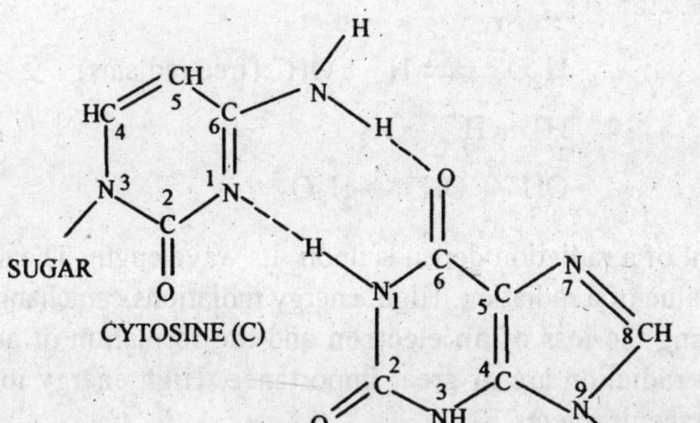

Fig. 40.22. Deamination by Nitrous acid : A, guanine converted to xanthine, B.Base pairing in cytosine and xanthine.

Agents producing distortion in DNA

Proflavin and **acridine orange** are two important flourescent dyes which cause mutation by **insertion** or **deletion** of bases. Direct attachment of these dyes to the nucleic acid causes mutation.

RADIATIONS

Among the physical mutagens radiations are most important. They have direct effect on chromosome. They may break the chromosome directly or alter the DNA bases. If chromosomes at meiotic prophase are given radiation the frequency of mutant per viable organism increases linearly with the dose. **Ressovsky** *et al* (1935) suggested **target theory** stating that single hit of the particle (radiation) on the target (genetic material) inactivates or mutates it. The radiation may act through production of a chemical. The frequency of simple chromosomal aberrations, *e.g.*, deletion, is proposed to the dose of radiation (Fig. 40.24). Low O_2 concentration reduces the frequency of chromosome breaks induced by radiations.

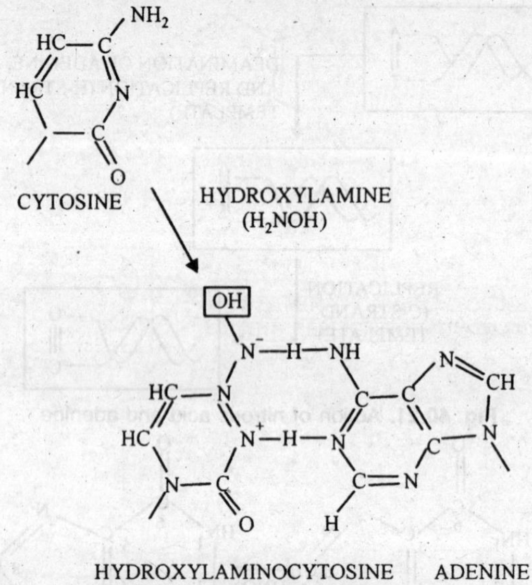

Fig. 40.23. Effect of hydroxyl amine cytosine pairs with adenine.

Oxygen effect is also called **anoxia**. Radiation in the presence of O_2 forms some **peroxide radicals** which influence the frequency of breaks and mutations. Ionization of water in cells may give free radicals and hydrogen peroxide

$$H_2O \rightleftharpoons H^+ + OH^- \text{ (free radicals)}$$

$$H^+ + H^+ \rightarrow H_2$$

$$OH^- + OH^- \rightarrow H_2O_2$$

The energy content of a radiation depends upon its wavelength. The shorter the wavelength the greater the energy value of a radiation. High energy radiations can change the atomic structure of a substance by causing the **loss of an electron** and the formation of an **ion.** Alternations in nucleic acid caused by radiation are of great importance. High energy ionizing radiations and ultraviolet light are mutagenic agents.

DNA and RNA absorbs UV light resulting in highly reactive free radicals in nitrogen containing bases. The unstability causes transition. If such changes occur in *m*-RNA only few inactive proteins are formed in substitution in DNA have lasting effect producing defective protein. UV light produces **thymine dimers** (Fig. 40.25). 5, 6 unsaturated bonds of adjacent pyrimidines become covalently linked and form **cyclobutance** ring. Three possible types of pyrimidine dimers in DNA are found in irradiated bacterial culture.

Thymine–thymine–50%
Thymine–cytosine–40%
Cytosine–cytosine–10%

In RNA pyrimidine dimers are formed between adjacent **uracil** and cytosine ring. These dimers cannot fit into the DNA double helix causing distortion of DNA molecules. If this damage is not repaired the replication is blocked and it is lethal. **Exonuclease** recognizes the distorted region and corrects it. **DNA polymerase** inserts correct bases in the gap and **DNA ligases** joins the inserted base.

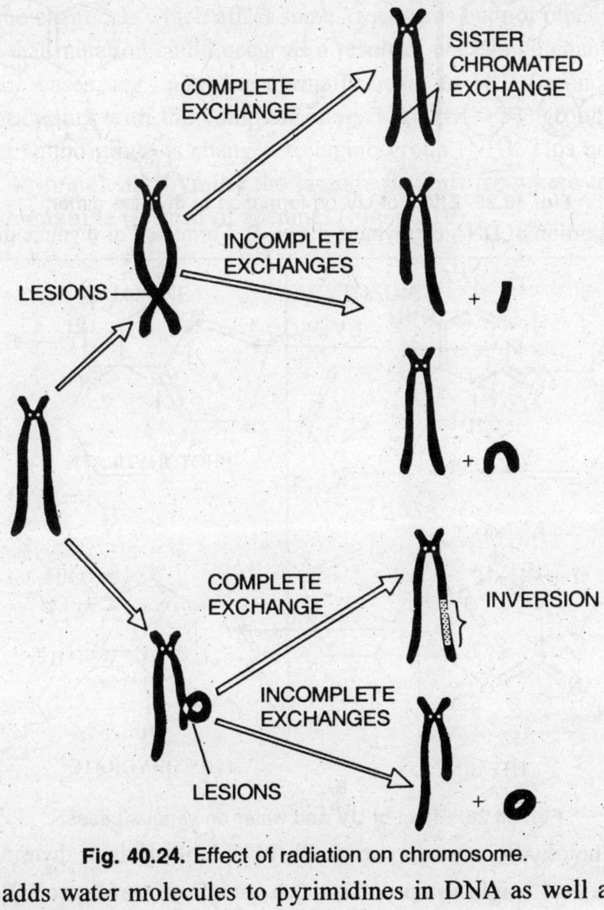

Fig. 40.24. Effect of radiation on chromosome.

UV radiation adds water molecules to pyrimidines in DNA as well as RNA resulting in photohydrates (Fig. 40.26).

X-ray causes mutation by breaking the phosphates ester linkage in DNA at one or more points causing a large number of deletion of bases or rearrangement. In double stranded DNA breaks may occur in one or both the strands. If it is found in both the strands it is lethal. Sometimes two double stranded breaks may occur in the same molecule and the two broken ends may rejoin. The part of DNA between the two breaks is eliminated resulting in deletion.

Photoreactivation: UV induced mutation discovered by **Kelner** *et al* shows that UV effect can be reversed by exposing cells to visible light containing wave length in the blue region of spectrum. It is called **photoreactivation.** It was observed in bacteria and bacteriophages. It is caused by enzyme which splits thymine dimers and repair the DNA molecule. When DNA repair system is absent in humans **xeroderma pigmentosum** appears in the patients susceptible to sunlight.

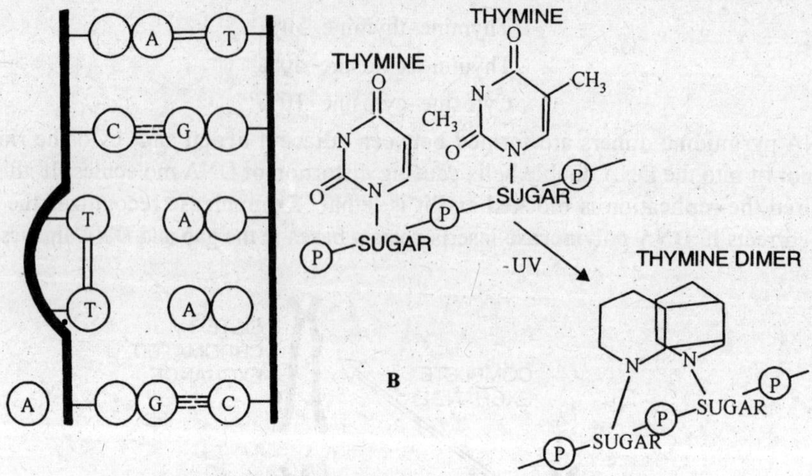

Fig. 40.25. Effect of UV on formation of thymine dimer:
A. Distortion of DNA by thymine dimer; B. Formation of thymine dimer.

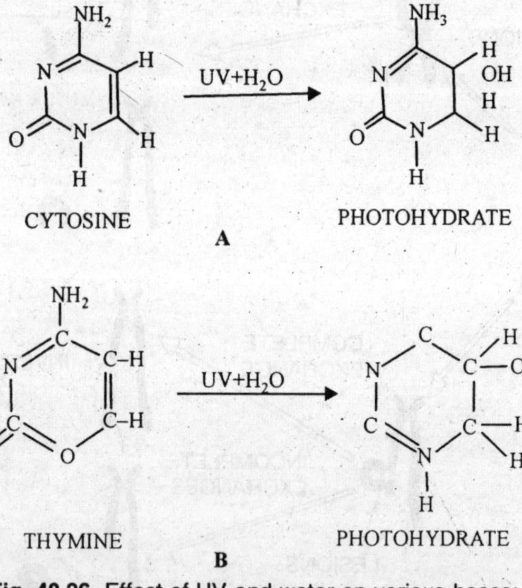

Fig. 40.26. Effect of UV and water on various bases:
A Photohydrolysis of cytosine; B. Photohydrolysis of thymine.

Mutation's role and origin in speciation (*i.e.*, origin of new species). It is learnt that all mutations are not bad. In fact mutations are the source of the great variety that we see among members of a population. Such variations often help some individuals to fit into the environment better than other individuals. However, in general, random mutations are of no help. In fact, most cause the death of the organism. Mutations that survive are usually recessive and remain hidden for many generations.

A species is one or more populations of similar organisms. They are similar because they share a common **gene pool.** A gene pool of a population is the total number of different kinds of genes (alleles) pooled by all its members. Thus, a gene pool is available to a population for adapting itself to the environment. Populations of a species that live in different geographical areas and in different environment conditions, pass through the elemental forces of mutation and natural selection. Thus, the population get adapted to their environment.

Selection of mutations. When a beneficial mutation occurs, this will probably appear with increasing frequency in subsequent generations. The rate at which the new gene replaces the original one in the population will be determined by many factors, such as (i) whether the new mutation is dominant or recessive, (ii) the intensity of the selection, (iii) the rate at which the mutation occurs in the population, and (iv) the size of the population. For this process, many generations are needed for the incorporation of an ordinary mutation into the majority of the individuals of a large population.

Mutations and their origin in organisms. All organisms living in the world are the products of hundreds of thousands of years of this process of selecting favourable mutations, and they are well adapted for the particular environment in which they live. During this process of evolving, when the environmental conditions have remained relatively constant for a long period of time, most of the mutations that could give survival value have occurred and have been selected. Thus, the beneficial mutations occurred with great frequency. George Simpson, in his study of horse evolution, has established that something over five million years was required for the formation of each of the genera in the evolution of the horse.

Under changed environmental conditions, populations either become adapted or die out. As a result of the adaptive changes, originally identical populations become more and more dissimilar. Over a period of time, a population may become so different, from the ancestral population that they are no longer able to interbreed with one another. They then become different species.

Mutations in haploid organisms. In all haploid organisms every mutation behaves as if it were a dominant. Because of this valuable characteristic, many important genetic discoveries have been made on such organisms during recent years. An example of how *Neurospora* can be used will illustrate how mutations with a biochemical effect can be detected. We start with a wild-type mold growing on a minimal medium. Asexual spores from this culture are transferred to a complete medium containing vitamins and amino acids. These spores may be treated with X-rays or in other way to increase the mutation rate. From this tube single spores of sexual fruiting bodies are transferred to tubes containing complete media. Then from these tubes transfers are made to tubes which are deficient vitamins and amino acids. Whenever a tube is found with no growth, that is a sign that a mutation has occurred in some gene which is necessary to synthesize the specific vitamin or amino acid which is lacking in this particular tube.

Among animals there are certain insects, such as male honey bees, which have only one of each type chromosome in their bodies. Unfortunately for genetics, the breeding habits of the bees and the problems of studying them do not make them favourable subjects for investigation. However, other members of the order Hymenoptera make very satisfactory forms for genetic breeding and study. One is the small parasitic wasp, *Habrobracon juglandis,* which has been investigated extensively by P. W. Whiting who has found that females carrying recessive genes yield male offspring which show these genes regardless of what type of male the female is mated to. This is due to the fact that the males develop from the unfertilized eggs. These tiny wasps are parasitic on the meal worm, *Tenebrio,* and may be raised very conveniently in the laboratory with a little corn meal and the *Tenebrio* worms.

Thus, haploids are better for mutation work becaue in them all the mutations, whether recessive or dominant, are expressed, as there is only one allele of each gene present in each cell.

Role of mutations in evolution. On the basis of the sudden changes in organisms. Hugo de Vries (1840-1935) proposed **mutation theory** for evolution. According to it mutation was the whole truth of evolution. Since mutations were known to be random, evolution was also considered to proceed at random. Many geneticists supported such a theory and they did not attach any importance, whatsoever, to the natural selection. They also thought that there are no adaptations, but only chance preadaptions.

Mutation theory of de Vries could explain some of the objections raised against Darwin's natural selection theory. As opposed to Darwin's hypothesis, de Vries attached importance to sudden changes or discontinuous variations, called "sports" by the former and "mutations" by the latter. On the basis of this, cases of overspecialization can be easily explained. One can also explain how the hereditary variations could originate and distinguish them from other environmental variations.

Considerable amount of work on frequency, nature, spectrum and size of mutations has been done in the modern times. It should be realized that hereditary changes can be brought about due to structural changes in chromosomes or due to changes at the genic level. These are called chromosomal mutations and gene mutations, respectively. Both kinds of changes play their roles in evolution. There are a large number of cases like *Drosophila, Crepis,* etc., where speciation has proceeded due to structural changes in chromosomes.

Hugo de Vries thought that mutations would always involve bigger and sudden changes. We now definitely know that the changes due to single mutation may vary from a major change, used for distinguishing species, genera or families to the changes which cannot be easily detected. The former are called **systematic macromutations,** while the latter are **micromutations.** Here, the cases are known where bigger mutations could be important, but in majority of the cases the smaller changes are more important. It is because if a genotype which is well adjusted into an environment, is too drastically altered, it will not be able to adjust. Minute changes, on the other hand, can be useful and help the organism to adjust in the new environmental conditions. Thus, mutation theory in its original form, as proposed by de Vries, does not hold good. However, evolution is a complex process and involves a complex set of factors. Mutation does play a very important role in evolution, but by no means is the only force working for organic evolution.

SUMMARY

1. Mutations can be artificially induced by **X-rays, gamma rays, mustard** gas, etc.
2. Mutation from the wild type to a new type is called **forward mutation.**
3. When mutant type is changed to the wild type is called **reverse mutation.**
4. Mutation involves deletion and insertion of base (s), the condition is known as **frame shift** mutations.
5. Mutation takes place at a definite locus in **point mutation**.
6. In **transition** purine base is replaced by other purine base or a pyrimidine is replaced by another pyrimidine base.
7. In **transversion** purine is replaced by pyrimidine base or vice-versa.
8. *Neurospora* or red bread mold is haploid in its vegetative phase and has a very short life cycle.
9. **Beadle and Tatum** (1941) used *Neuropora* to show biochemical mutations.
10. The whole process in *Neurospora* is controlled by a specific enzyme under control of specific gene.
11. One gene-one enzyme hypothesis in *Neurospora* was suggested by **Beadle** and **Tatum.**
12. Haemoglobin consists of 4 polypeptide chains (2α and 2β). Hence, one gene one enzyme hypothesis has become one gene-one polypeptide hypothesis.
13. **C. Auerback** first of all explained that certain chemicals can induce mutations. These are called as **chemical mutagens,** ***e.g.*****, phenol, mustard gas, formaldehyde, maleic hydrazide,** etc.

14. A **gene pool** is the total number of genes in reproductive gametes of a population.
15. **Gene frequency** is the proportion of different alleles of a gene in a population.
16. Changes in genetic frequencies are due to mutations, selection, genetic drift and migration.
17. Ionizing radiations X-rays, γ-rays, β-rays, neutrons and UV light can produce point mutations or chromosomal aberrations. The number of mutations increases proportionally with the dose of the effect of radiations cumulative.
18. Somatic mutations are not transmitted from one generation to another. But germ cell mutations may be passed to offsprings.
19. Mendelian recombinants are due to crossing over.
20. Mutant strains are called **auxotrophs.**
21. Wild strain capable of growing in minimal media is called **prototroph.**
22. All mutations are easily and immediately expressed because allele of each gene is present.
23. Mutation theory was proposed by **Hugo de Vries.**
24. Lethal mutations are mostly **recessive.**
25. **Hereditary variations** in plant can be produced by **X-rays.**
26. No two individuals exhibit exactly similar characters (except identical twins). This is known as **variation.**
27. Variations are not necessarily heritable.
28. **Variations** are more prominent in sexually reproducing organisms.
29. Sexual reproduction results in the formation of monoparental offspring, a carbon copy of parent called **clone.**
30. The progeny formed after sexual reproduction is called as **offspring.**
31. Huntigatons disease is due to insertion of copier of a repeat of CAG triplet in the working gene itself.
32. Cancers are due to mutation contains genes, *e.g.*, oncogenes and Tumor suppressor gene.
33. In colorectal cancers there is genetic instability in cancer cells with the expansion of trinucleotide repeats.
34. Certain carcinogens are also bacterial mutagens.
35. Superoxide dismutase (enzyme) destroys mutagen.
36. Sickle cell anaemia is due to change in single gene (point mutation).

QUESTION BANK

Very Short Answer Questions

1. Name various types of mutations.
2. What is mutation?
3. Name 3 chemical mutagens.
4. Name 2 physical mutagens.
5. Who noticed the **ancon** sheeps for the first tisue?
6. Who was the discover of mutations?
7. Who worked on ***Oenothera lmarckiana*?.**
8. Who worked on mutagenic effect on *Drosophila*?
9. What is micromutations?
10. What is Muller-5-method?
11. Who worked on detection of mutation in plant?
12. Who worked in induced irradaition effects on *Neurospora*?.
13. What is 5 bromouracil?
14. What is the effect of HNO_2?

15. What happens when DNA is exposed to low pH?

Short Answer Questions

1. What are the practical applications of mutation?
2. What are the agents producing distortion in DNA?
3. Write short notes on:
   (a) Alkylation agents (b) Base analogue
4. Write notes on effect of radiation on mutation.
5. Write short note on photoreactivation.
6. Write short note on spontaneous mutations.
7. Write short note on silent mutations.
8. Write short note on base pair substitution.
9. Write short note on frame shift mutation.
10. Write short note on mutator gene.
11. Write note on detection of sex linked lethals by CIB method.

Long Answer Questions

1. Describe genetic induced variation.
2. Describe gene mutation.
3. Describe induced gene mutation.
4. Describe missense and nonsense mutations.
5. Write an essay on mutations.

Multiple Choice Questions (Objective Type Questions)

1. Replacement of a purine nucleotide with a pyrimidine nucleotides:
   (a) Transversion (b) Transition
   (c) Frame shift (d) Inversion
2. De Vries attributed evolution to:
   (a) Slow differentiation (b) Polygenes
   (c) Continuous variations (d) Discontinuous variation
3. The major source of variations are:
   (a) Mutations (b) Polyploidy
   (c) Segregation (d) Chromosome abberation
4. Superoxide dismutase enzyme helps in destroying:
   (a) Mutagen (b) Carcinogen
   (c) Enzyme receptor (d) None of the above
5. Mutations used in agriculture are commonly:
   (a) Recessive and lethal (b) Lethal
   (c) Spontaneous (d) Induced
6. Gene mutation is due to:
   (a) Linkage and crossing over (b) Change in base sequence
   (c) Change is sequence of cistrons in DNA (d) Randomsegregation
7. Point mutation results in:
   (a) Free martin (b) Intersex
   (c) Sickle cell anemia (d) Down's syndrome
8. When mutation is from wild type to new type it is called as:
   (a) Forward mutation (b) Backward mutation
   (c) Reverse mutation (d) Point mutation

9. Lethal mutations are:
(a) Recessive (b) Dominant
(c) Homozygous (d) Fragile
10. Lethal gene of *Drosophila* is:
(a) Curly wings (Cy) (b) Plum eyes (Pm)
(c) Stubbles (Sub) (d) All the above
11. Gene mutation is produced in:
(a) Nucleus (b) Chromosome
(c) DNA (d) RNA
12. 5' bromouracil is a :
(a) Nucleotide (b) Base analogue
(c) Purine (d) All the above
13. Which of the following is the genetical induced (irradiated) variety:
(a) Sharbati sonara (b) Pusa Parwati
(c) Co 997 (d) All the above
14. Prof. M.S. Swaminathan has produced a mutant variety (irradiation) of wheat called:
(a) Pusa larma (b) Pusa 96
(c) Sharbati sonara (d) Pusa Parwati
15. Sudden change in structure and activity of gene is called:
(a) Variation (b) Evolution
(c) Mutation (d) Hybridization
16. Lethal mutants were first time observed by:
(a) Mischer (b) Muller
(c) Morgan (d) Mc Clincook
17. The rays absorbed by chromosome is:
(a) X rays (b) Gamma rays
(c) UV rays (d) β rays
18. Superoxide dismutase destroy:
(a) Mutagens (b) Transposon
(c) Virus (d) Fungus
19. Acridine mutagens, *e.g.*, acridine orange and proflavin, etc., are discovered by:
(a) Watson (b) Crick
(c) Wobble (d) Nirenberg
20. Ethyl ethane sulphonate, ethyl methane sulphonate, mustard gas, etc., are:
(a) Base analogues (b) Chemical mutagens
(c) Radiation (d) Physical Mutagens

ANSWERS

1. (a)	2. (d)	3. (a)	4. (a)	5. (d)	6. (b)
7. (c)	8. (a)	9. (c)	10. (d)	11. (c)	12. (b)
13. (d)	14. (c)	15. (c)	16. (b)	17. (a)	18. (a)
19. (b)	20. (b).				

41
CHAPTER

Transposable Genetic Elements

In fact, most genes occupy fixed sites on the chromosomes, and the overall structure of the genetic map is practically invariant. However, in early 1940's researchers have found that some DNA sequences (*viz.*, genes) can actually change their position.

These mobile elements have been variously called **'jumping genes'**, **'mobile elements'**, **'cassettes'**, **'insertion sequences'** and **'transposons'**. The formal name for this family of mobile genes is **transposable elements**, and their movement is called **transposition**. The term transposon was coined by **Hedges** and **Jacob** in 1974 for a DNA segment which could move from one DNA molecule (or chromosome) to other and carried resistance for antibiotic ampicilin. Transposable elements can be defined as small, mobile DNA sequences that move around chromosomes with no regard for homology and insertion of these elements may produce deletions, inversions, chromosomal fusions and even more complicated rearrangements.

MODE OF DISCOVERY OF TRANSPOSABLE ELEMENTS

Transposable elements were discovered by **Barbara McClintock** (1965) through an analysis of genetic instability in maize (corn). The instability involved chromosome breakage and was found to occur at sites where transposable elements were located. In McClintock's analysis, the events of breakage were detected by following the loss of certain genetic markers. In some experiments, McClintock used a marker (gene) that controlled the deposition of pigmentation in the aleurone which is the outer most layer of endosperm of maize kernels. Recall that the endosperm is **triploid**, being produced by the union of two maternal nuclei and one paternal nucleus. McClintock's marker was an allele of the *C* locus on the short arm of chromosome 9. Since this allele, called C^I, is a dominant inhibitor of aleurone colouration, any kernel possessing it should be colourless. McClintock fertilized *CC* ears with pollen from C^IC^I tassels, producing kernels in which the endosperm was C^ICC. Although many of these kernels were colourless, as expected, some showed patches of brownish-purple pigment. **McClintock** (1951) guessed that in such mosaics, the inhibitory C^I allele had been lost sometime during endosperm development, leading to a clone of tissue that was capable of producing pigment. Thc genotype in such a clone would be *–CC*, where the dash indicates loss of the C^I allele. Further study demonstrated that C^I allele had been lost through chromosome breakage.

As shown in Fig. 41.1 a break at the site labelled by the arrow would detach a segment of the chromosome from its centromere, creating an **accentric fragment**. Such a chromosomal fragment tends to be lost during cell division, so all the descendants of this cell would lack part of the paternal chromosome. Since the lost fragment carried the C^I allele, none of the cells in this clone would be inhibited from forming pigment, and if any of them produced a part of the aleurone, a patch of colour would appear.

McClintock established that such kernel mosaics frequently resulted from breaks at a particular site on chromosome 9. She named the factor that produced these breaks, ***Ds***, for **"dissociation"**. In her experiments, the chromosome that carried the C^I allele also carried the *Ds* factor. However, by itself, the *Ds* factor was not capable of inducing chromosome breakage. She found that *Ds* had to be activated by another factor, called *Ac*, for **"activator"**. The *Ac* factor was present in some maize stocks, but absent in others. By crossing different stocks, *Ac* could be combined with *Ds*, creating the condition that led to chromosome breakage.

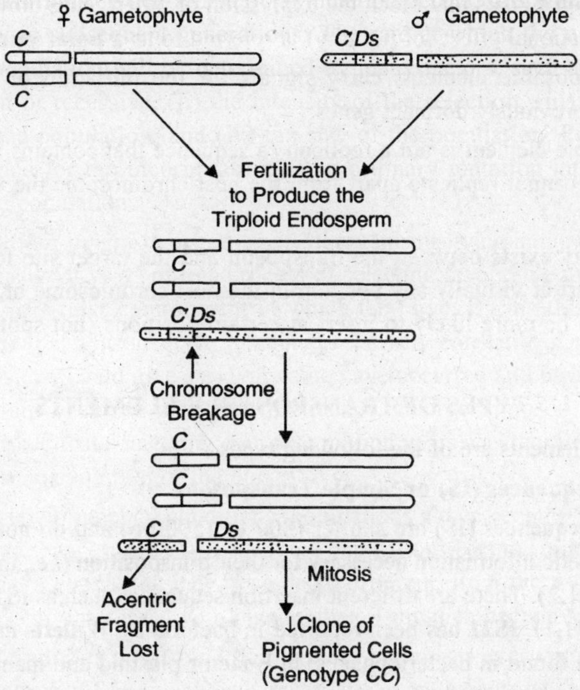

Fig. 41.1. Chromosome breakage caused by the transposable element *Ds* in maize. The allele *C* on the short arm of chromosome 9 produces normal pigmentation in the aleurone (of maize kernel); the allele *C* inhibits this pigmentation.

This two-factor system (*i.e.*, *Ac-Ds* family) provided an explanation for the genetic instability that McClintock had observed on chromosome 9 of maize. Her further studies have established that both *Ac* and *Ds* are members of a family of transposable elements. These elements are structurally related to each other and can insert at many different sites on the chromosomes. In fact, there often are multiple copies of the *Ac* and *Ds* elements present in the maize genome. Through genetic analysis, McClintock demonstrated that both *Ac* and *Ds* elements could move. When one of these elemants was inserted in or near a gene, McClintock sometimes found that the gene's function was altered. In extreme cases, the function was completely suppressed. Because of this effect on gene expression; **McClintock** (1956) referred to *Ac* and *Ds* as **controlling elements**.

CHARACTERISTICS OF TRANSPOSABLE ELEMENTS

Some salient features of the transposable elements are the following :

1. They were found to be DNA sequences that code for enzymes which bring about the insertion of an identical copy of themselves into a new DNA site.

2. Transposition events involve both recombination and replication processes which frequently generate two daughter copies of the original transposable elements. One copy remains at the **parent site** while the other appears at the **target site** (on the host chromosome).

3. The insertion of transposable elements invariably disrupts the integrity of their target genes. For example, if an IS (= insertion sequence) becomes inserted into an operon, it interrupts the coding sequence and inactivates the expression of the target gene into which it inserts as well as any gene downstream in that same operon. This is because an IS contains transcription and/or translation termination signals that block the expression of other genes downstream in an operon. This "one-way" mutational effect (or polarity) is referred to as a **polar mutation**.

4. Since transposable elements carry signals for the initiation of RNA synthesis, they sometimes activate previously dormant genes.

5. A transposable element is not a replicon (a sequence that contains a site for the origin of replication), thus, it cannot replicate apart from the host chromosome the way that plasmids and phage can.

6. No homology exists between the transposon and the target site for its insertion. Many transposons can insert at virtually any position in the host chromosome or into a plasmid. Some transposons seem to be more likely to insert at certain positions (hot spots), but rarely at base-specific target sites.

TYPES OF TRANSPOSABLE ELEMENTS

Transposable elements are of the following types :

1. Insertion Sequences (IS) or Simple Transposons

The insertion sequences (IS) are shorter (800 to 1500 bp) and do not code for proteins. In fact, IS carry the genetic information necessary for their transposition (*i.e.*, the gene for the enzyme **transposase**) (Fig. 41.2). There are different insertion sequences such as IS1, IS2, IS3, IS4 and so on in *E.coli* (Table 41.1). **IS21** has been reported in bacteria by **Willetts** *et al.*, (1981). Insertion sequences have been found in bacteriophages, in F factor plasmid and many bacteria.

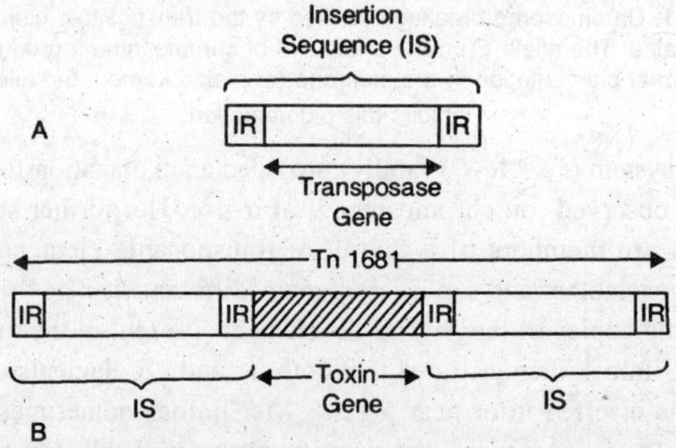

Fig. 41.2. A–A simple transposon (IS) carries only the genetic information necessary for transposition which is often enclosed by short regions of inverted repeat (IR) sequences; B– The complex transposon (Tn) is composed of two insertion sequences (IS) and contains genetic information in addition to that needed for Transposition. In Tn 1681 transposon carries the gene for a heat-labile toxin.

2. Transposons (Tn) or Complex Transposons

Transposons (Tn) are several thousand base pair long, and have genes coding for one or more proteins (including resistance factors in bacteria which act against antibiotics). The hallmark of a transposon is the presence of identical, **inverted terminal repeat (IR) sequences** of 8 to 38 base pairs (b.p.) (Fig. 41.2B). Each type of transposon has its own unique inverted repeat. On either side of a transposon is a short (less than, 10 b.p,) **direct repeat**. If a transposon exists in multiple copies, these direct repeats are of different base composition at each site where the transposon exists in the chromosome; the inverted terminal repeats, however, remain the same for a given transposon. The sequence into which a transposable element inserts is called the target sequence. During insertion of a transposon, the singular target sequence becomes duplicated and, thus, appears as direct repeats flanking the inserted transposable element. The direct repeats are not considered part of the transposon. These repeat sequences themselves act like IS or IS-like segments.

Table 41.1. Some examples of insertion sequences (IS) and complex transposons of *E.coli* (bacterium).

Type of transposable element	Size base pairs (bp)	Target DNA repeat (bp)	Known functions or proteins encoded (besides transposase)
A. Insertion sequences (IS)			
IS 1	768	9	–
IS 2	1327	5	–
IS10-R	1329	9	–
B. Complex Transposons (Tn)			
Tn 3	4957	5	Ampicillin resistance
Tn 5	5700	9 (IS 50 at end)	Kanamycin resistance
Tn 10	9300	9 (IS 10 at end)	Tetracycline resistance
Tn 1681	2100	9 (IS 1 at end)	Heat stable enterotoxin
Tn 2571	23000	9 (IS 1 at end)	Resistance to chloramphenicol, fusidic acid, streptomycin, sulphoamides and mercury.

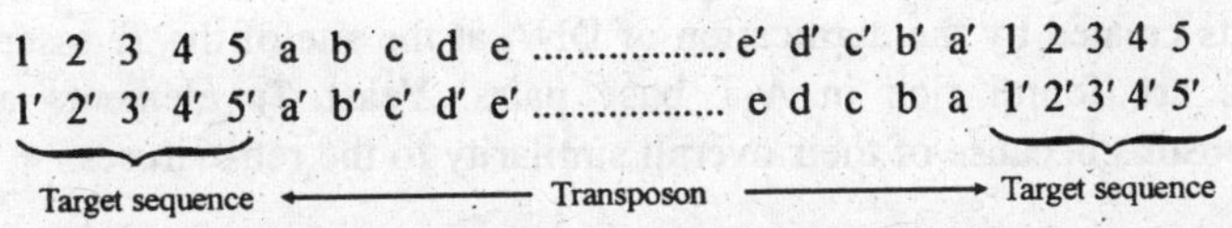

Fig. 41.3. Diagram of a simple transposon. In the diagram, letters and numbers have been used instead of nucleotides to make repeats easier to read. Primed and unprimed symbols of the same kind (*e.g.*, 3 and 3¢, d and d¢) represent complementary base pairs. The ends of a transposon consist of inverted repeats (represented by letters). Direct repeats of the target gene (represented by numbers) flank the transposon. The dotted central region of simple transposons contains only genes necessary for translocation, but in complex transposons it may contain one or more additional genes (consult Fig. 41.2).

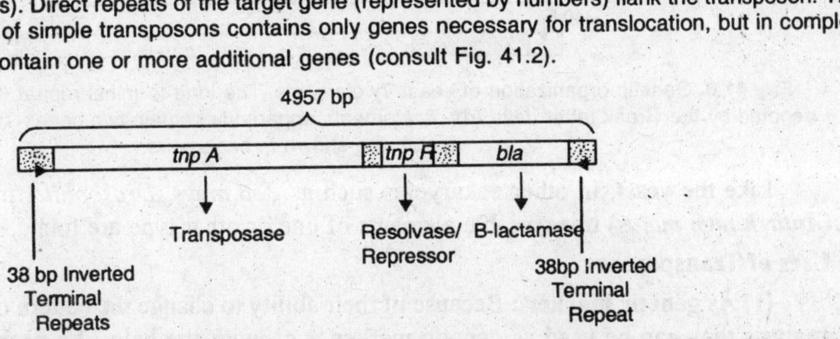

Fig. 41.4. Genetic organisation of Tn 3 transposon of *E.coli*. This complex transposon contains three genes that produce the protein indicated. Lengths of DNA squences are given in base pairs (bp).

Examples of Transposons

1. Tn 3 transposon of *E.coli*. The molecular structure of transposon Tn 3 of *E.coli* has been worked out (Fig. 41.4). Tn 3 has 4957 bp and contains three genes such as *tnp A, tnp R* and *bla,* coding respectively for the following proteins : 1. **transposase** having 1015 amino acids and required for transposition; 2. a **repressor** (also called **resolvase**) containing 185 amino acids which regulates the transposase; and 3. **β-lactamase** enzyme which confers resistance to the antibiotic ampicillin. On both sides of the Tn 3 is an inverted repeat of about 38 bp.

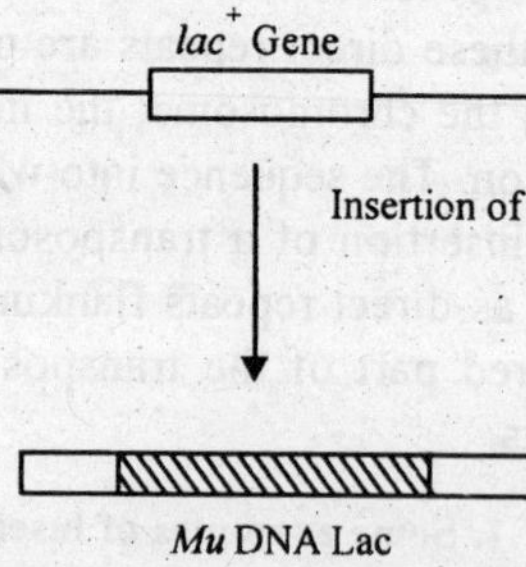

Fig. 41.5. Insertion of phase *Mu* into a *loc* gene renders the gene inactive (lac).

2. Bacteriophage Mu. The bacteriophage *Mu* (*Mu* = mutator) is a temperate bacteriophage having typical phage properties and could be regarded as a giant transposon. Phage *Mu* was thought to behave in a unique manner since it does not multiply in a lytic way. It always starts its life-cycle by inserting itself into the *E.coli* chromosome at random locations. The resulting *prophage* often mutates the genes into which they become inserted (Fig. 41.5). Thus, phage *Mu* was originally known for its mutagenic properties.

During later part of phage life cycle, upon receipt of an appropriate signal, the *Mu* prophage somehow generates its transposition, but also activates the genes encoding for structural proteins which package its DNA. *Mu* phage is unique also in transposing much often than other transposons, providing nearly a hundred new copies of *Mu* during its hour-long life-cycle.

Like other transposons, *Mu* phage has inverted repeats (IR) at or near the ends of its DNA and it makes a short duplication with 5 base pairs of the bacterial DNA intc which it is inserted.

3. Yeast *Ty* elements. The yeast *Saccharomyces cerevisiae* carries about 35 copies of a transposable element called *Ty* in its haploid genome. These transposons are about 5900 base pairs long and are bounded at each end by a DNA segment called the **delta (δ) sequence**, which is approximately 340 base-pairs long. Each δ sequence is oriented in the same direction, forming what are known as direct **long terminal repeats** or **LTRs**. *Ty* elements are flanked by five base-pair direct repeats created by the duplication of DNA at the site of the *Ty* insertion. These target site duplications are found rich in A-T base pairs. Yeast ***Ty*** elements are sometimes called **retrotransposons** because of their overall similarity to the retroviruses.

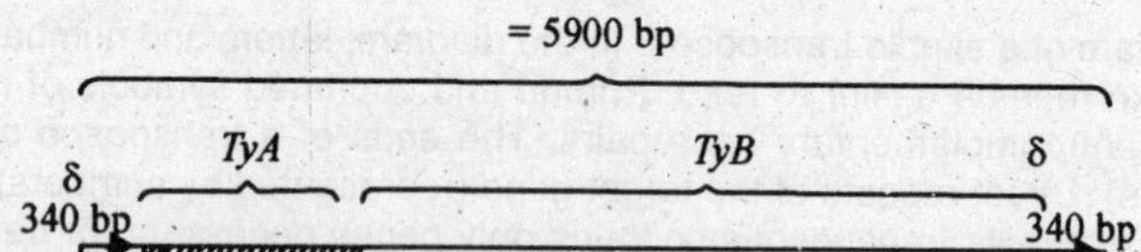

Fig. 41.6. Genetic organization of yeast *Ty* elements. The long terminal repeat (LTR) sequences are denoted by the Greek letter delta (d). *Ty* elements apparently contain two genes *TyA* and *TyB*. Sequence length are shown in base pairs.

Like the yeast, in other eukaryotes such as *Zea mays, Drosophila,* mice and snapdragons (*Antirrhinum majus*) transposable elements of one or other type are found.

Uses of Transposons

(1) As genetic marker : Because of their ability to change the pattern of restriction fragment analysis they can be used as genetic marker, *e.g.*, with the help of a probe, it was possible to characterise different strains of *Plasmodium falciparum*. The strains may be identified as they differ in the distribution of transposons.

Transposons are used as genetic markers to construct linkage maps.

Transposons are used in human for distinguishing carriers or non-carriers of sickle cell trait.

(2) As mutagens : Spontaneous mutation can be created by inserting transposons. Transposons cause mutation due to insertion in structural and regulatory region.

(3) For isolation of genes : Transposons are used to search the restriction fragments containing mutant genes. It is used for isolation of genes through **transposon tagging.**

(4) As transformation Vector : Transposons are used in gene transfer artificially if carried in a plasmid.

SUMMARY

1. Mutator genes are supposed to hop around the genome appearing at one place sometimes or on the other place and are known as **transposable elements.**
2. Some genes present in a genome which increase the mutation rate at several loci simultaneously are called **mutator genes.**
3. Transposable element is very typical about 4,500 base pairs long and code for an enzyme called **transposons** that copies the DNA element to another part of chromosome.
4. The effect of transposable elements were found to be due to mobilization of a **'P'-element** which is a piece of transposable DNA.
5. Unstable or hoping DNA is responsible for human mutations also. Parts of human genes are more fluid and open to change the classical geneticist opinion.
6. Inherited mental illness is due to **fragile 'X' syndrome.** The children usually have a distortion of X-chromosome, two small pieces break away from the end of long arm of X-chromosome.
7. **Fragile X-mutation** is unpredictable because of mobile DNA present next to actual locus with repeat triplet of **CCG.**
8. Normally there are 6-60 repeats of sequence but those with fragile X-syndrome have several thousand copies of CCG repeat.

QUESTION BANK

Very Short Answer Questions

1. The first transposable element were observed in which plant ?
2. What is IS element ?
3. T_3 transposon consists of how many base pairs ?
4. What is target sequence ?
5. What is replicon fusion ?
6. Who observed transposable element in Maize ?
7. How many types of transposons are found in eukaryote ?
8. How many copies of Ty elements are found in Yeast ?
9. *Tam* I element in *Snapdragon* has how many base pairs.
10. Name one disease in human due to transposable elements.

Short Answer Questions

1. Write the short note on structure of transposable element.
2. Write short note on IS element.
3. Write short note on T_3 transposon family.
4. Write short note on transposition.
5. Write short note on transposable elements in Eukaryotes.
6. Write short note on transposable elements in Human.
7. Write short note on transposable elements of *Drosophila.*
8. What is Ty element in Yeast ?
9. Write short note on Ac-Ds system.

10. Write short note on suppressor mutation system.
11. Write short note on uses of transposon.

Long Answer Questions

1. What is transposable element. Describe its structure or function.
2. Comment on Ms. Barbara McClintock's investigations regarding the controlling elements of maize.
3. Define the transposable elements. Describe different types of complex transposons of the eukaryotes.
4. Describe the transposable elements of *E.coli* and bacteriophage *Mu*.

Multiple Choice Questions (Objective Type Questions)

1. Mutator gene which hop from one place to the other, 4,500 base pairs long coding for an enzyme that copies that DNA element to another part of the chromosome is called :
   (a) Transposable element (b) Transferable element
   (c) Transmittable element (d) Transpogene
2. Part of DNA which switch their position are called as :
   (a) Exon (b) Transposon
   (c) Intron (d) Cistron
3. Transposable element is made up of :
   (a) 2,000 bp (b) 4,500 bp
   (c) 5,000 bp (d) 1,500 bp
4. Transposable elements were first detected in :
   (a) *Drosophila* (b) *Acetabularia*
   (c) *E. coli* (d) Human
5. Inverted repeat sequences in the terminus are present in :
   (a) Alleles (b) Plasmagens
   (c) IS element (d) Tn 3 element
6. R-plasmids are for types containing gene for :
   (a) Revolution (b) Resistence
   (c) Retranscription (d) All the above
7. The first record of transposable elements from Maize were given by :
   (a) Morgon (b) Sutton
   (c) Muller (d) Mc Clintock
8. Dotted elements were discovered by Marcus Rhoades in :
   (a) 1938 (b) 1928
   (c) 1908 (d) 1918
9. Transposable element involving RNA phase is called :
   (a) RNA transpose (b) RNA some
   (c) Retroelement (d) Retrovirus
10. Transposon caused mutation in human is :
    (a) Dowan's syndrome (b) Fragile 'X'-syndrome
    (c) Turner's syndrome (d) Edward's syndrome
11. The effect of Transposon is due to mobilization of :
    (a) DNA fragment (b) IS element
    (c) P-element (d) Rotroelement
12. The fragile X syndrome has several thousands copies of CCG repeats :
    (a) At the centromere (b) At the locus
    (c) Next to the actual locus (d) At the end of chromosome

ANSWERS

1. (a)	2. (b)	3. (b)	4. (a)	5. (c)	6. (b)
7. (d)	8. (a)	9. (c)	10. (b)	11. (c)	12. (c).

42

CHAPTER

Plasmids

Plasmids are extrachromosomal circular DNA molecules found in most bacterial species and in some species of eukaryotes. Under normal conditions, a particular plasmid is dispensable to its host cell. Many plasmids contain plasmid genes that may be essential in certain environments, *e.g.*, **'R'** plasmid carry genes providing resistance to numerous antibiotics so that the cell containing this plasmid can resist antibiotic action.

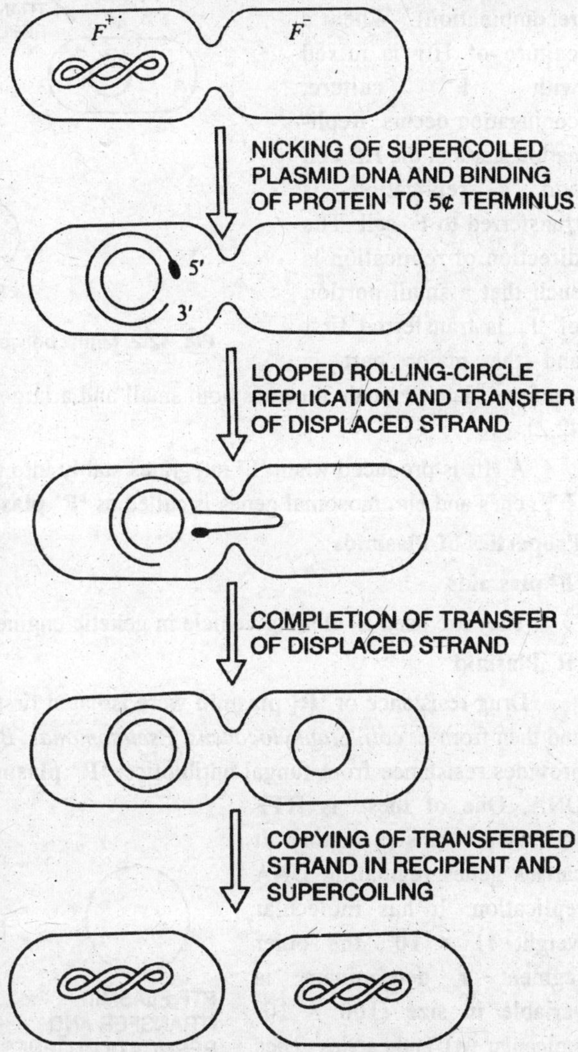

Fig. 42.1. Transfer of 'F' plasmid DNA from F⁺ cell by looped roling circle mechanism.

With only one exception (killer plasmid of Yeast, a RNA molecule) all known plasmids are supercoiled circular DNA. The molecular weight ranges from 10^6 – 10^8. The number of copies of a plasmid/cell ranges from 1–2 to 10—60.

Plasmid DNA can be isolated from bacteria by culturing plasmid containing bacteria. Detergent is added to it to get lysed bacteria. The lysate is centrifuged. The smaller plasmid DNA remains in the supernatant. **CsCI and ethidium bromide** is added to the supernatant CsCI forms a density gradient. The supercoiled DNA with high density can be identified by ethidium bromide and easily removed.

Transfer of Plasmid DNA

The bacterium *E. coli* possesses 2 mating types, *i.e.*, **donors** or male and **recipients** or female. The determinant of maleness is the 'F' or **sex plasmid** or donor, a male cell is designated as $\mathbf{F^+}$. Female lacks 'F' plasmid and designated as $\mathbf{F^-}$. When the culture of male and female is mixed, **conjugation** takes place between male-female pairs. Conjugation bridge is formed (Fig. 42.1). Pairing induces looped rolling circle replication of 'F'. One copy of 'F' is transferred to the female in a minute. The female is converted to a male often mating, the recipient cell contain the 'F' plasmid. During transfer both Donor and recipient cells show DNA synthesis. In donor, the synthesis replaces single strand to be transferred, synthesis in the recipient converts the transferred single strand to double stranded DNA.

'F' plasmid has an ability to integrate into the bacterial chromosome. 'F' fuses with chromosome increasing its size. The cell is now called **Hfr** cell (high frequency recombination). When a culture of Hfr is mixed with F^- culture, conjugation occurs. Replication starts in the Hfr cell and a replication is transferred to F^- cell. The direction of replication is such that a small portion of 'F' is transferred first and the major part is tranferred later. Female receives both small and a large fragement of the male chromosome (Fig. 42.2).

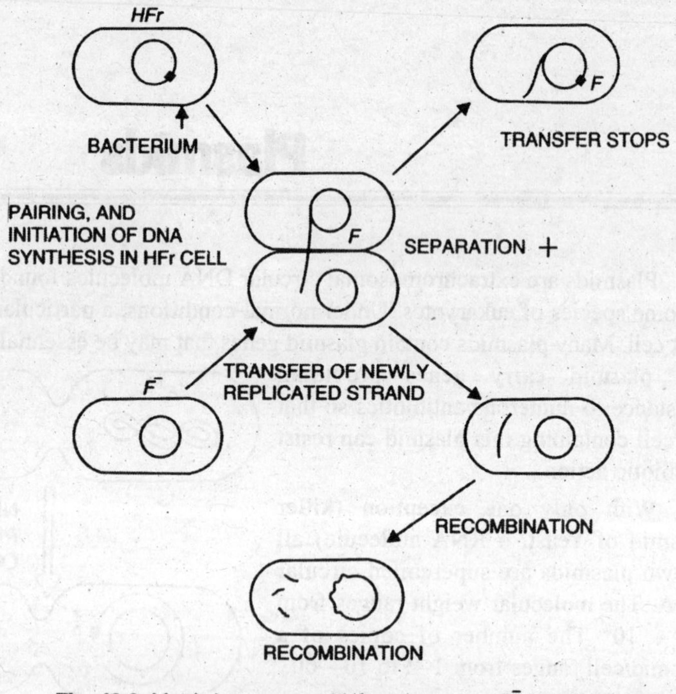

Fig. 42.2. Matrix between and Hfr male and an F^- female bacterium.

A Hfr is produced when 'F' integrates stably into the chromsome A plasmid containing both 'F', genes and chromosomal genes is called as **'F' plasm.**

Properties of Plasmids

'F' plasmids

They are used as cloning vehicle in genetic engineering.

'R' Plasmid

Drug resistance or **'R' plasmid** were isolated first from the bacterium *Shigella dysenteriae* and then from *E. coli*, *Staphylococcus*, *Pseudomonas*, *Bacillus*, *Vibrio*, *Salmonella*, etc. 'R' factor provides resistance from fungal antibiotics. **'R' plasmid** consists of 2 contiguous segments of DNA. One of these is **RTF** (resistance transfer factor). It carries genes regulating DNA replication. It has molecular weight **11 × 10⁶**, the other segment **r** determinant is variable in size (100 × 10^6 molecular wt), and carries genes for antibiotic resistance. Resistance to the drug

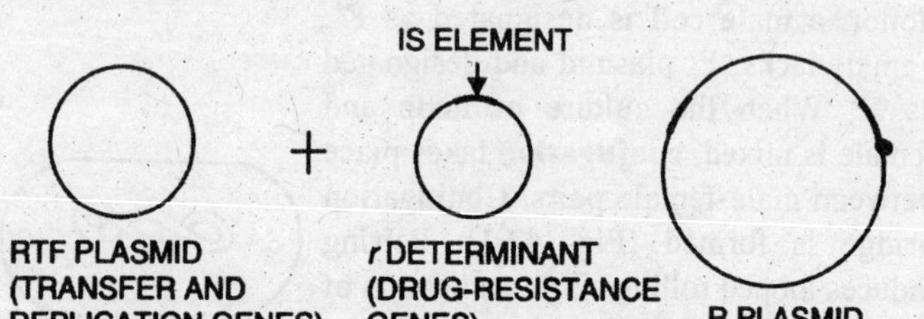

Fig. 42.3. Components of an infectious 'R' plasmid.

penicillin (pen), **ampicillin** (amp), **chloramphanicol** (cam), **streptomycin** (str), kanamycin (kan), and sulphonamide (sul), in combination is common (Fig 42.3).

Col Plasmid

Col plasmid is *E. coli* plasmid encoding colcins, proteins capable of preventing growth of a bacterial strain that does not contain **col** plasmid. Col plasmids are a class of general type of plasmid called **bacteriocinogenic** plasmid, which produce bacteriocins in many bacterial species. Bacteriocins, of which colicins are one example which are proteins binding to the cell wall of sensitive bacterium and inhibit essential processes, *e.g.*, replication, transcription, translation or energy metabolism, etc. The best studied **col** plasmid is col E_1, with a mol wt. 4.2×10^6. It is used in recombination DNA research and in an *in vitro* DNA. replication system.

Ti plasmid

Crown gall tumor is caused by *Agrobacterium tumefaciens.* Ti plasmid is the most important tool in the genetic engineering of plants. The wounds are infected by *A. tumefaciens*

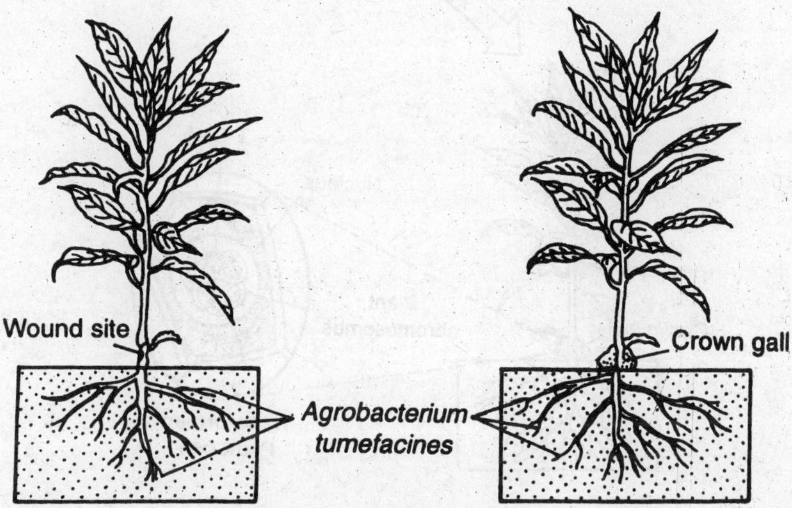

Fig. 42.4. Crown gall formation.

and induce tumour at the site (Fig. 42.4). At the onset of infection they begin to synthesize an arginine derivative called **opine,** *e.g.*, **nopaline** or **octopine** and the cells proliferate to form tumor. The induction of tumor is controlled by genetic information carried on **plasmid** called **Ti** plasmid. **T-DNA** and **vir** region of the Ti plasmid is essential for transformation of plant cells. The plasmid causing tumour is **Ti** plasmid. In an infected plant some of the bacteria enter and grow within the plant cell and lyse there releasing their DNA in the cell. The small fragment of **Ti** plasmid containing genes for replication, becomes integrated into the plant cell chromosomes. The integrated fragment breaks down the hormonally regulated system and control cell divison. The cell turns to tumour cell. **Ti** plasmid of *Agrobacterium tumefaciens* provides an excellent gene-transfer system for use in dicot plants. The host range is limited to dicots only. Fig. 42.5 **Ti** plasmids are very important as specific genes can be inserted into the **Ti** plasmid by recombinant DNA techniques. These genes can be inserted in plant chromosome changing the genotype or phenotype of plant. New varieties of value are produced.

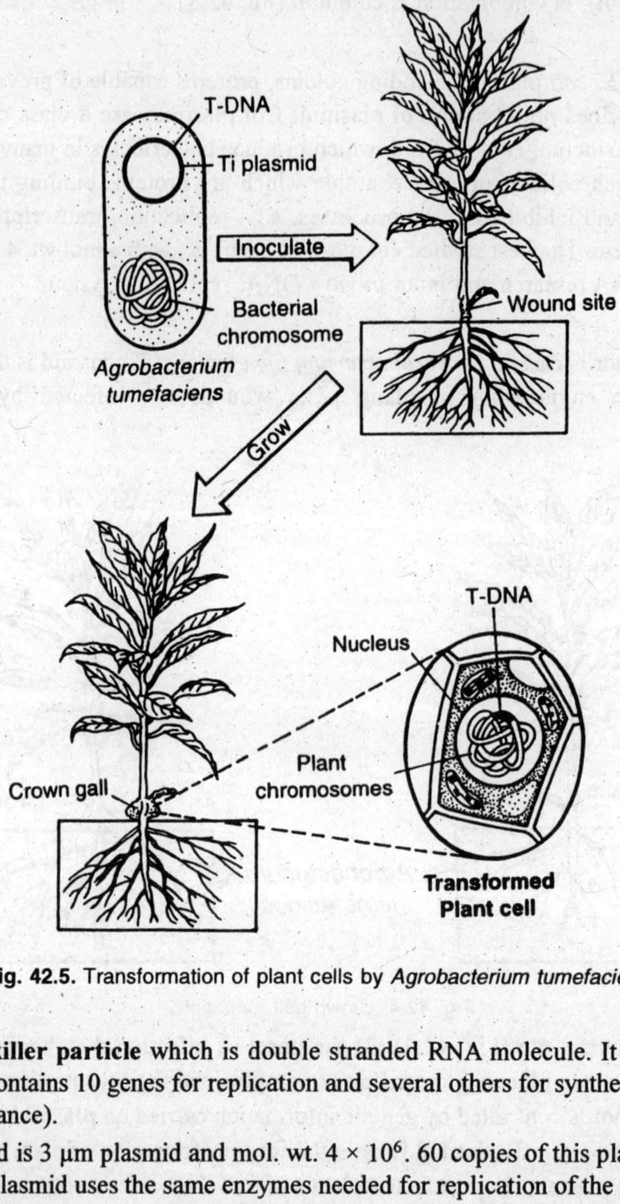

Fig. 42.5. Transformation of plant cells by *Agrobacterium tumefaciens*.

Yeast Plasmid

Yeast have **killer particle** which is double stranded RNA molecule. It is the only plasmid without DNA. It contains 10 genes for replication and several others for synthesis of killer particles (colicin like substance).

Yeast plasmid is 3 μm plasmid and mol. wt. 4×10^6. 60 copies of this plasmid/cell are found in yeast nucleus. Plasmid uses the same enzymes needed for replication of the yeast chromosomes. The plasmid apparently does not integrate into the host DNA. This plasmid is useful system for studying DNA replication in yeast. It acts as a vector for cloning foreign genes in yeast by the recombination DNA techniques.

PLASMID REPLICATION

Plasmid can only replicate within a host cell. The difference of plasmid and phage is :

(i) Reliance on host enzyme : Some plasmids use host enzymes exclusively, other encode some of their own enzymes.

(ii) Identity of polymerase : Most *E. coli* plasmids use Pol III for chain growth and Pol I for synthesis of precussor fragments, as in replication of *E. coli* chromosomes. Where as col E_1 uses Pol I for chain growth.

(iii) Direction : Both purely unidirectional and bidirectional replication is found. Plasmid RK_6 replicates first in one direction and then later in the opposite direction.

(iv) Termination : In unidirectionally replicating plasmids termination necessarily occur at the origin after one cycle of replication.

(v) Replicating form : Plasmids show replication by so called **butterfly** mode (Fig. 42.6). In a partially replicated molecules the replicated portions are untrusted as in 'θ' replication, but the unreplicated part is supercoiled. When the replication completes, one of the circles must be cleaned by DNA gyrase enzymes.

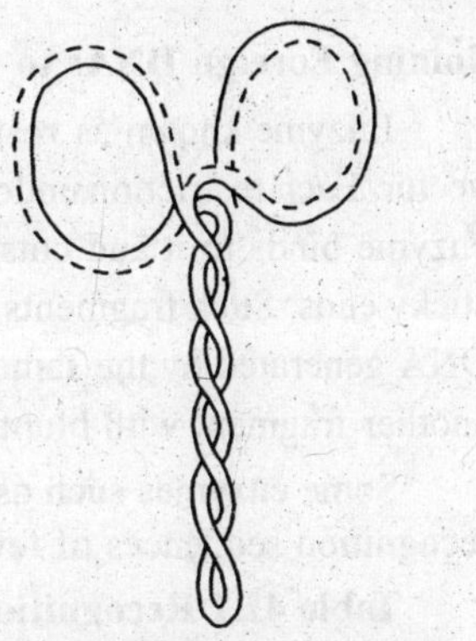

Fig. 42.6. Col E1 DNA replication butterfly molecule.

As plasmids are small circular DNA molecules found in bacteria, containing an origin of replication thus can replicate independently of bacterial chromosome and can produce many copies of itself. Plasmids often carry genes that confer antibiotic resistance on the host bacterium. So it is very easy to select the plasmid containing bacteria from a population of bacteria by applying the antibiotic. Only those bacteria with antibiotic resistance gene will survive (Fig. 42.7). Bacteriophages are viruses that infect bacteria and utilize the host cell's component for their own replication.

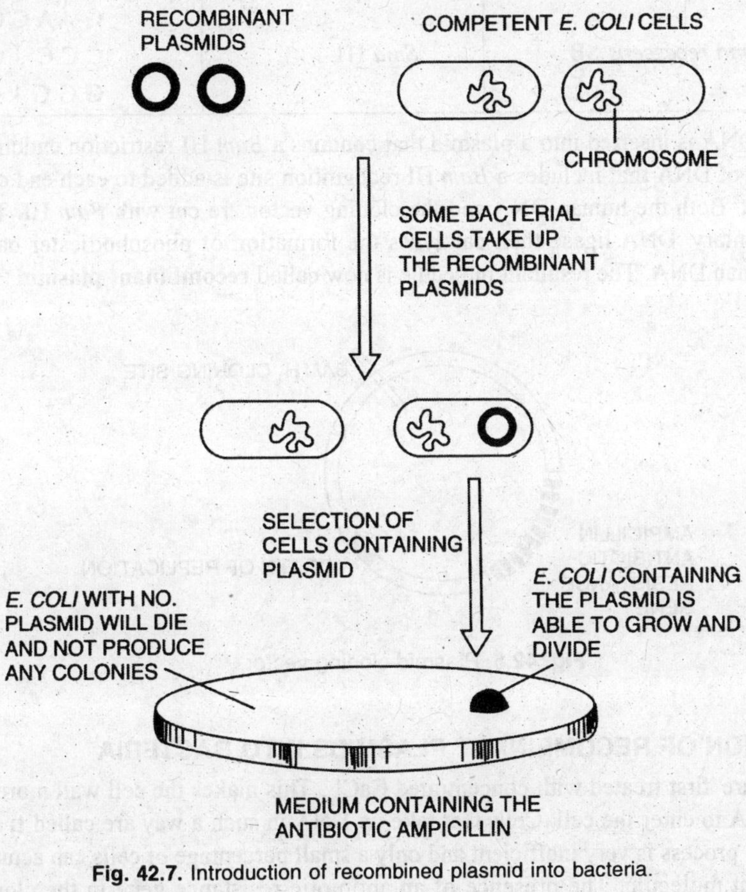

Fig. 42.7. Introduction of recombined plasmid into bacteria.

Joining Foreign DNAs to a Cloning Vector

Enzyme known as **restriction endonucleases** are used to insert foreign DNA into a cloning vector. Each restriction endonuclease recognizes a particular DNA sequence of 4–6 base pairs. The enzyme binds to it and cuts both strands of the double helix. The resultant DNA molecules have sticky ends. Such fragments can associate by complementary base pairing to any other fragment of DNA generated by the same enzyme. DNA fragments produced in this way can only be joined to another fragment with blunt ends.

Some enzymes such as *Bam* HI, *Eco* RI and *Pst* I make cuts one each strand. The names of recognition sequences of few of the common enzymes are shown in Table 42.1.

Table 42.1 Recognition sites of some common restriction enzymes.

Bacterial strain	Enzyme name	Recognition sequences and cleavage sites
1. *Bacillus amyloliquefaciens* **H**	*Bam* H1	G↓GATCC CCTAG↑G
2. *Escherichia coli* Ry 13	*Eco* R1	G↓AATTC CTTAA↑G
3. *Provedencia stuartii* 164	*Pst* 1	CTGCA↓G G↑ACGTC
4. *Serratia ma recescens* SB	*Sma* H1	CCC↓GGG GGG↑CCC

Human DNA is inserted into a plasmid that contains a *Bam* HI restriction endonuclease site. A short length of DNA that includes a *Bam* HI recognition site is added to each end of the human DNA fragment. Both the human DNA and the cloning vector are cut with *Bam* HI. The cut ends are complementary. DNA ligase then catalyzes the formation of phosphodiester band between vector and human DNA. The resultant molecule is now called **recombinant plasmid** (Fig. 42.8, 9)

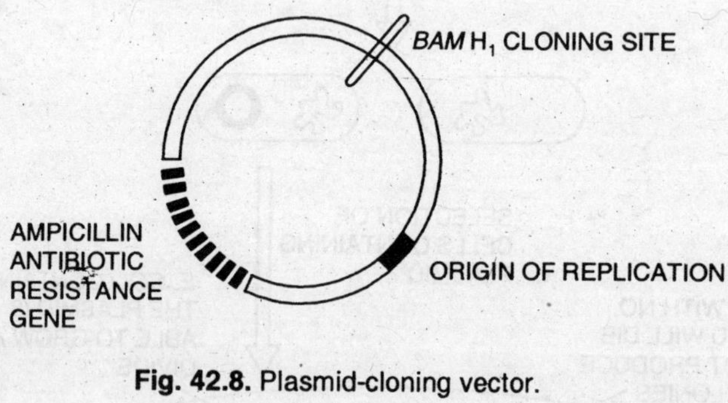

Fig. 42.8. Plasmid-cloning vector.

INTRODUCTION OF RECOMBINANT PLASMIDS INTO BACTERIA

Bacteria are first treated with concentrated $CaCl_2$. This makes the cell wall more permeable and allows DNA to enter the cell. Cells that take up DNA in such a way are called **transformed.** Transformation process is very inefficient and only a small percentage of cells can actually take up the recombinant molecule. The presence of an antibiotic resistance gene in the cloning vector makes it possible to select those bacteria that have taken up a molecule of foreign DNA with its resistance gene. Only transformed cells can survive in the presence of antibiotic. The collection of bacterial colonies produced after this selection process is a **clone library**. Each cell in a colony contain an identical recombinant molecule. Other colonies may contain plasmids carrying different DNA segments. Isolation of these colonies produce different clones of foreign DNA. The collection of such clones is called **c-DNA** library.

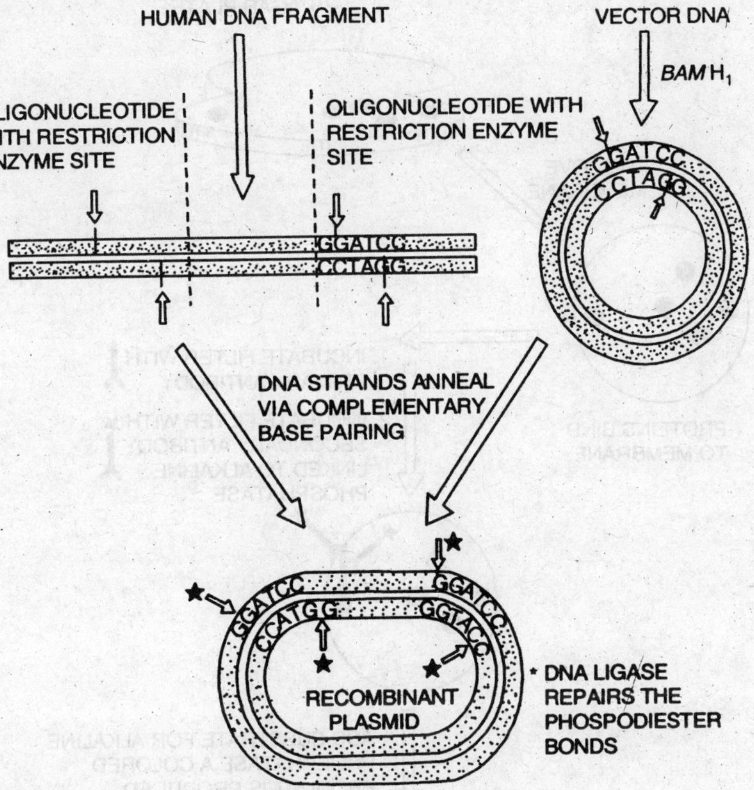

Fig. 42.9. DNA molecule cut at various sites.

SELECTION OF cDNA CLONES

In a library many thousands of different clones are available. The bacteria containing cDNA of particular type is identified. It may be done by various ways, *e.g.*, making use of specific antibodies to detect the bacteria expressing the corresponding antigen. The foreign DNA should be expressed is bacterial cell, its information should be copied first into *m*-RNA and then into protein. To ensure efficient expression the vector with bacterial promoter sequence is used to control transcription of foreign DNA. Such cloning vectors are called **expression vectors.**

To see that bacteria without plasmid will die in antibiotic, a membrane is placed over the bacterial colonies and then lifted off. Bacterial protein stick to the membrane, which is then incubated with appropriate antibody. Bacteria which synthesized corresponding antigen are identified by carrying out a simple colour test for enzyme action (Fig. 42.10).

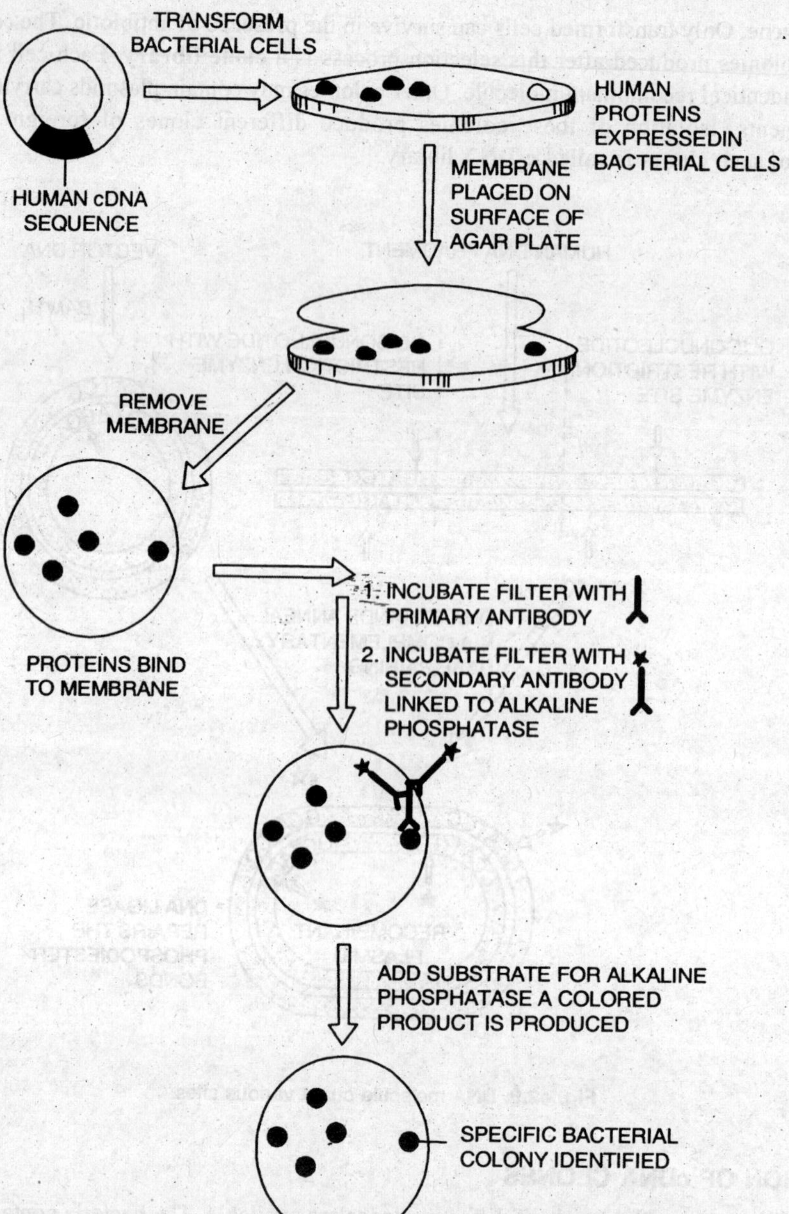

Fig. 42.10. Selection of cDNA clone.

GENOMIC DNA CLONES

The genomic DNA clones containing fragments of chromosomal DNA with promoter gene are developed to study the regulation of transcription of gene. Genomic DNA clones are much larger than cDNA clones as they contain both exon and intron sequences. The cloning vectors for isolation of genomic sequences can accomodate large piece of DNA. Generally vectors based on the bacteriophage lambda (λ) are used as they accept large DNA fragments of about 20,000 base pairs. Large DNA fragments are developed by digesting genomic DNA with a restriction enzyme for a short time. The large fragments are produced by partial digestion. The genomic DNA fragments are joined to bacteriophage vectors in the normal way. Bacteria are infected by bacteriophage with low efficiency of infection so that each bacterium is infected by one phage particle. The collection of bacteria produced in this way is called a **genomic DNA library** (Fig. 42.11).

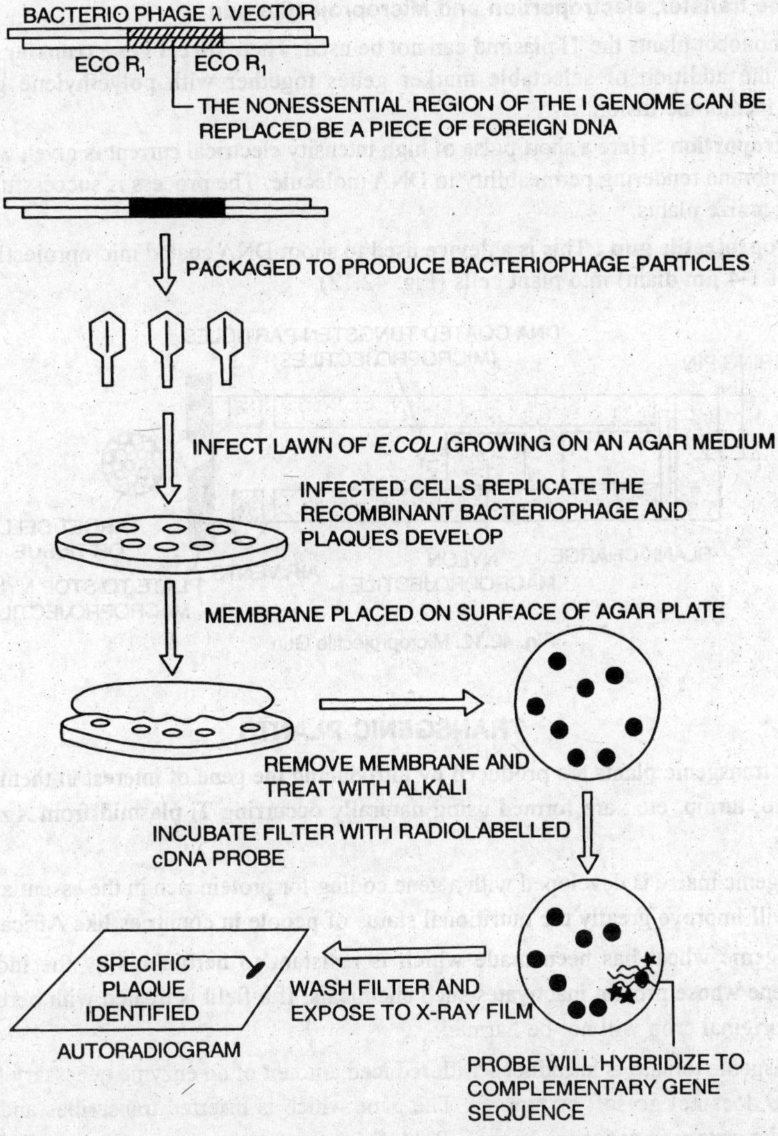

Fig. 42.11. Selection of genomic DNA clone.

To select the genomic DNA sequence of interest, the library is plated onto a layer of cultured bacteria so that many copies of recombinant phages are produced. The recombinant phage multiply inside host bacterial cell. The cell lyses and bacteriophages are spread to surrounded layer of bacteria and infect them. The dead cells give rise to a clear area on the culture plate called **plaque.** Each plaque contains many copies of a recombinant bacteriophage that can be transferred to a membrane. Specific DNA clones are selected by incubating the membrane with a radiolabelled *c*DNA probe complementary to the genomic sequence. This produces a radioactive area on the membrane which is identified by autoradiographic technique. If a protein product of a particular gene is unknown, it is impossible to isolate an appropriate *c*DNA clone using conventional techniques. Reverse genetics is applied with a genome and a protein product is identified. It is very time consuming and difficult process.

Direct gene transfer, electroportion and Microprojectile gun

For monocot plants the Ti plasmid can not be used. There **direct gene transfer** is attempted; involving the addition of selectable marker genes together with polyethylene glocyl which stimulates membrane fusion.

Electroportion : Here a short pulse of high intensity electrical current is given which disrupts the all membrane rendering permeability to DNA molecule. The process is successfully employed in 1990 in **maize** plants.

Microprojectile gun : This is a device used to shoot DNA coated microprojectiles (tungsten particles of 1-4 μm diam) into plant cells (Fig. 42.12).

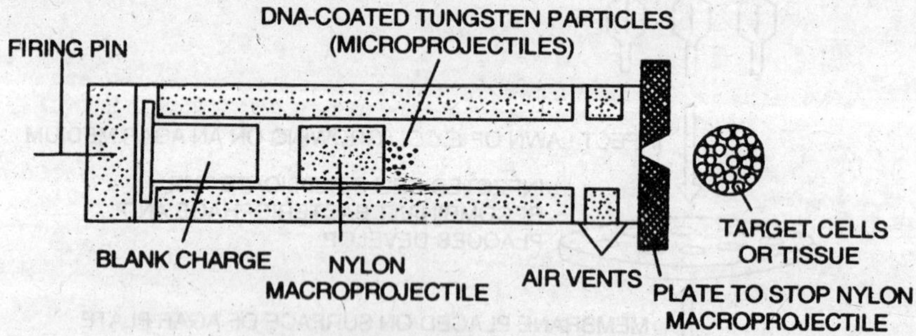

Fig. 42.12. Microprojectile Gun.

TRANSGENIC PLANTS

Many transgenic plants are produced by introducing the gene of interest in them. Transgenic trees, potato, turnip, etc., are formed using naturally occurring Ti plasmid from *Agrobacterium tumifaciens.*

Transgenic maize is developed with a gene coding for protein rich in the essential amino acid **lysine.** It will improve greatly the nutritional status of people in countries like Africa.

Transgenic wheat has been made which is resistant to herbicides by the induction of a bacterial gene whose protein inactivates such chemicals. If a field is treated with herbicide to kill weeds the original crop will not be harmed.

A transgenie tomato is introduced with reduced amount of an enzyme necessary for ripening. This tomato does not go soft on storage. The gene which is inserted transcribes and produces a RNA complementary in sequence to the *m*RNA for the ripening protein. The two RNAs bond by complementary base-pairing, and so the translation of the normal *m*-RNA is inhibited. Genetically engineered **"Flavr Savr"** tomatoes are produced in 1995 in USA. These have increased yield and full development of flavour.

Virus resistant plants of potato, tomato and alfa-alfa are also produced.

Genetically engineered herbicide resistance could double or even quadruple yields. Parasitic weeds such as *Orobanche* and *Striga*, etc., affect the crops a lot.

Herbicide resistent corn, sugarbeet and oil seeds are produced.

Gene for nitrogen fixation into crop plants in introduced. It involves 15 genes called *'Nif'* genes.

Genetic engineering produces transgenic plants for new colour, pattern, shapes of flower. Such plants in **blue roses** are produced.

Medical drug crops are produced which are cheaper, *e.g.*, human **enkephalin** gene has been expressed in plants.

TECHNIQUE OF INTRODUCING Ti PLASMID FOR PRODUCING TRANSGENIC PLANTS

The foreign genes are incorporated in 'Ti' plasmid of *Agrobacterium tumefaciens* (Fig.42.13). It causes a tumorous lump or crown gall. During infection the Ti plasmid is passed from bacterium into the plant cell. It becomes incorporated with plant cell chromosome and induces the cell to proliferate and provide nutrients for the bacterium. The discs cut from leaves or cultured in a culture dish. The transfected cells from leaf cells (discs) are allowed to grow into whole plants. These plants carry transgenes.

Microinjection

Microinjection is the device used for producing transgenic animals. The plasmid is microinjected into the male nucleus (sperm haploid nucleus) of the fertilized egg through a very fine tipped glass needle (Fig. 42.14). The injected DNA can be linear or circular.

TRANSGENIC ANIMAL

The transgenic animal is produced by injecting a foreign gene into the nucleus of a fertilized egg. The egg is then implanted into a foster mother, and the offspring are tested to determine whether they carry the foreign gene. Transgenic mouse are commonly used to investigate the regulation of gene expression in mammals. The production of the protein metallothionein is known to be increased by exposure to heavy metal ions.

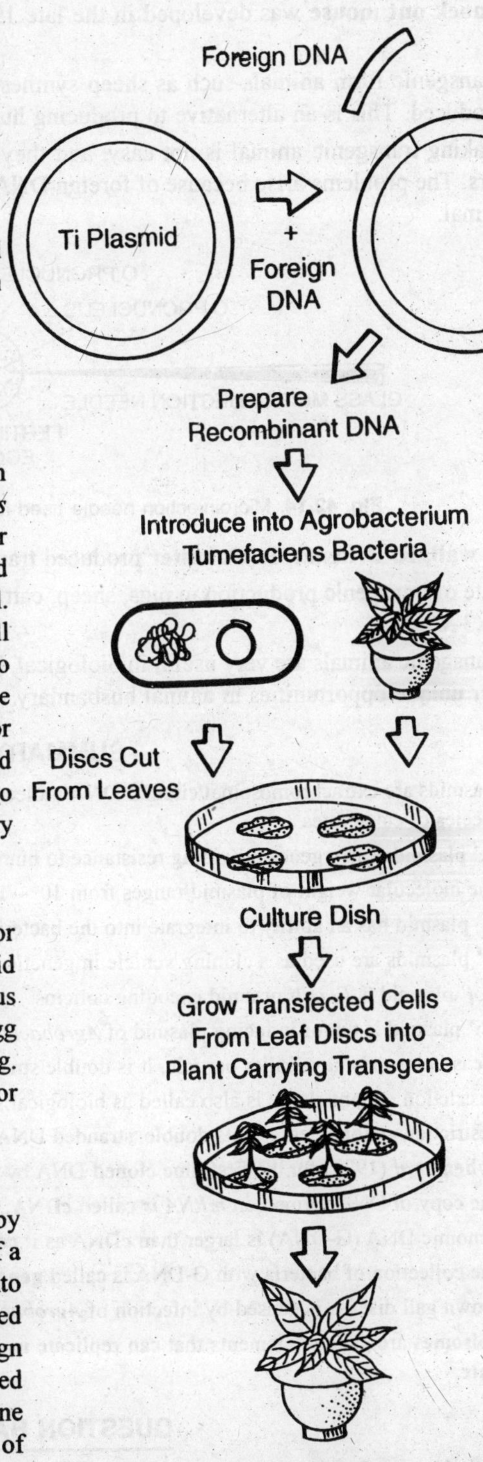

Fig. 42.13. Formation of transgenic plant with the help of Ti plasmid.

Knock out mouse was developed in the late 1980's by **Mario Capecchi** at University of Utah.

Transgenic form animals-such as sheep synthesizing human factor VIII in their milk-have been produced. This is an alternative to producing human proteins in bacteria.

Making transgenic animal is not easy, and they are sometimes sterile or suffer from other disorders. The problems arise because of foreign DNA is integrated at random into the genome of host animal.

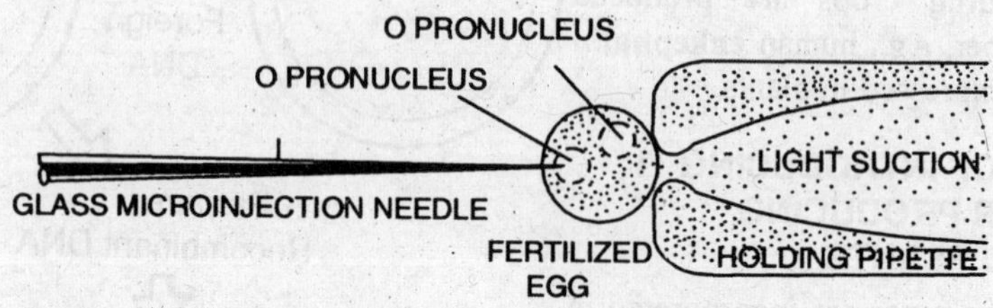

Fig. 42.14. Microinjection needle used for making transgenic animal.

R. wall, R. Brinstar, R. Palmiter produced transgenic pigs.

Rate of transgenic production in pigs, sheep, cattle and goats is much lower ($< 1\%$) than that is mice (3–6%).

Transgenic animals are very useful in biological, biomedical, and biotechnological researches and offer unique opportunities in animal husbandary.

SUMMARY

1. Plasmids are extrachromosomal circular DNA molecules found in most bacterial species and in some species of eukaryotes.
2. 'R' plasmids carry genes providing resistance to numerous antibiotics.
3. The molecular weight of plasmid ranges from $10^6 - 10^8$.
4. 'F' plasmid has an ability to integrate into the bacterial chromosome.
5. 'F' plasmids are used as a cloning vehicle in genetic engineering.
6. '*col*' plasmid is *E. coli* plasmid encoding colicins.
7. **'Ti'** plasmid is tumor inducing plasmid of *Agrobacterium tumefaciens.*
8. 'Yeast' plasmids have killer particle, it is double stranded RNA molecule.
9. Restriction endonuclease is also called as biological scissors.
10. Restriction endonucleases cuts double stranded DNA at a particular place.
11. **Cohen** *et al* (1973) for,the first time cloned DNA by using plasmid as vector.
12. The copy of DNA formed on *mRNA is* called **cDNA.**
13. Genomic DNA (G-DNA) is larger than *c*DNA as it contains both exon and intron sequences.
14. The collection of bacteria with G-DNA is called **genomic DNA library.**
15. Crown gall disease is caused by infection of *Agrobacterium tumefaciens.*
16. **Episomes** are genetic elements that can replicate in an **integrated state**, or in **extra chromosomal state.**

QUESTION BANK

Very Short Answer Questions

1. Ti plasmid is found in natural form in which plant ?
2. Name two transgenic plants.

3. What is 'R' plasmid ?
4. What is 'col' plasmid ?
5. What is the molecular weight of RTF ?
6. Yeast plasmid is made up of which genetic material.
7. Name 2 restriction endonucleases.
8. What is microprojectile gun ?
9. Name one transgenic animal.
10. What is "Flavr savr" Tomato ?

Short Answer Questions

1. Write note on Ti plasmid.
2. What do you understand by Genomic DNA clones ?
3. Write short notes on Transgenic plants.
4. Make only diagrammatic presentation of making transgenic plant by Ti plasmid introduction.
5. Make a diagram of microprojectile gun.
6. Make a diagram of microinjection.
7. Write short note on restriction endnoucleases.

Long Answer Questions

1. What do you mean by plasmids? Describe different types of plasmids.
2. Describe properties of different plasmids.

Multiple Choice Questions (Objective Type Questions)

1. DNA cloning is a part of :
   (a) DNA sequence (b) Recombinant DNA technology
   (c) DNA structure (d) All the above
2. The biological scissors used to cut DNA segment is :
   (a) Protein (b) Restriction endonucleases
   (c) Restriction exonucleases (d) Restriction DNA
3. A number of bacteria with recombinant DNA of same type form :
   (a) Clone library (b) Gene library
   (c) Gene pool (d) Gene frequency
4. *Bam* HI, *Eco* RI, are the types of :
   (a) Restriction endooxidases (b) Restriction endonucleases
   (c) Restriction exonucleases (d) Restriction polymerases
5. Transgenic animals are produced by injecting foreign gene into the :
   (a) Egg (b) Nucleus of unfertilized egg
   (c) Nucleus of fertilized egg (d) Nucleus of sperm
6. Transgenic plants are produced by using Ti plasmid from the :
   (a) *Agrobacterium tumefaciens* (b) *E. coli*
   (c) Bacteriophage (d) *A. varians*
7. The plasmid which carries a foreign DNA molecule of known sequence is called :
   (a) Bacteria (b) Vector
   (c) Viron (d) Virus
8. Plasmid is :
   (a) Linear DNA with several genes (b) A double stranded circular DNA
   (c) A single stranded linear DNA (d) Double stranded RNA

9. Plasmids that can integrate into the bacterial DNA is called : ___
   (a) Episome (b) Endosome
   (c) Exozyme (d) Hypnosome
10. Transfer of *nif* genes is carried out by :
   (a) Recombinant DNA technology (b) Genetic engineering
   (c) Interferon (d) Both 'a' and 'b'
11. Endonucleases act upon genetic material and :
   (a) Cleave the single strand of DNA
   (b) Cleave the specific point of single stranded DNA
   (c) Cleave the double stranded DNA at any point except the ends
   (d) Cleave the double stranded DNA at the ends only
12. Extrachromosomal self replicating double stranded circular DNA in bacterial cell is called:
   (a) Plasmid (b) Cosmid
   (c) Phasmid (d) All the above
13. Who cloned DNA by, using plasmid as a vector for the first time :
   (a) Robertis (b) Dahl
   (c) Cohen (d) Kornberg
14. Plasmids are :
   (a) Host specific (b) Resistant to antibiotic and heavy metals
   (c) Bacteriocin and toxic materials (d) All the above
15. The disease resistant gene of plasmid is called
   (a) D gene (b) DR gene
   (c) R gene (d) dr gene

ANSWERS

1. (b)	2. (b)	3. (b)	4. (b)	5. (d)	6. (a)
7. (b)	8. (b)	9. (a)	10. (d)	11. (c)	12. (a)
13. (c)	14. (c)	15. (c).			

43 CHAPTER

Chromosomal Aberrations : Structural Changes

The arrangement and presence of many genes on a single chromosome provides a change in genetic information not only through change in chromosome number but also by a change in chromosome structure. The change in chromosome is due to alteration in genetic material through loss, gain or rearrangement of a particular segment. Such changes are called **chromosomal aberrations.** The modification brings about chromosomal mutations. **Chromosomal mutations** are very rare in nature but can be created artificially by 'X' rays, atomic radiation and chemicals, etc.

The structural changes in chromosomes are due to breaks in chromosome, or in its cell division subunit, *i.e.*, chromatid. Each break produces 2 ends which may then follow three different paths. (Fig.43.1).

(a) They may **reunite**, leading to eventual loss of that chromosomal segment which does not contain the centromere.

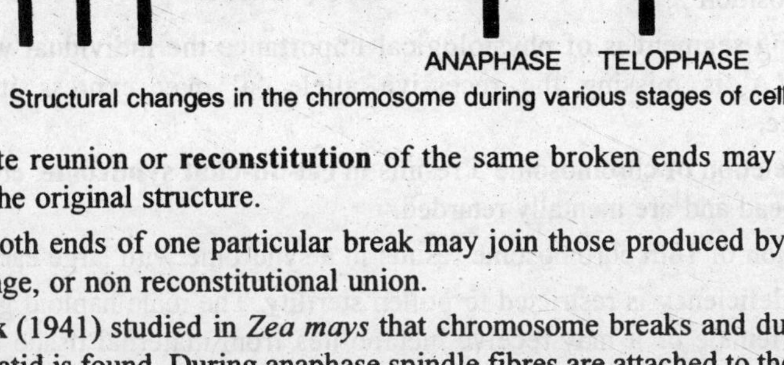

Fig. 43.1. Structural changes in the chromosome during various stages of cell division.

(b) Immediate reunion or **reconstitution** of the same broken ends may occur, leading to reconstitution of the original structure.

(c) One or both ends of one particular break may join those produced by a different break causing an exchange, or non reconstitutional union.

Mc Clintock (1941) studied in *Zea mays* that chromosome breaks and duplication follows. A dicentric chromatid is found. During anaphase spindle fibres are attached to the two centromeres resulting in the formation of bridge from one pole to other. The bridge breaks causing **deficiency** or **duplication.**

Chromosomal aberrations are of 4 major types :

(a) deletion (b) duplication (c) inversion and (d) translocation. (Fig. 43.2).

(A) DELETION OR DEFICIENCY

Deletion or deficiency as the name suggests there is a loss of segment of chromosome. After break the part without centromere is lost. On the other hand the part attached to the centromere acts as deficient chromosome. **Bridges** (1917) for the first time observed deficiency in the *Bar* locus of *Drosophila.*

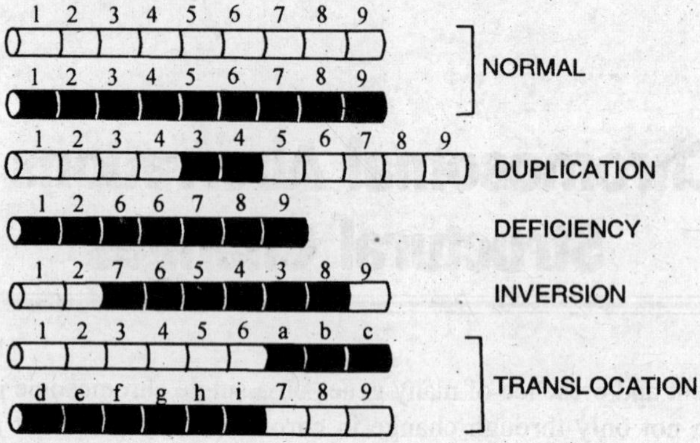

Fig. 43.2. Types of Chromosomal aberrations.

Two types of deletions are found :

Terminal deletion : A single break near the end of the chromosome. Described in maize but otherwise not common.

Interstitial deletion : Chromosome breaks and reunites but the part is lost from in between. (Fig. 43.3). Deletions are detected at the time of homologous pairing. If a part of chromosome is missing then the other chromosome also has to omit it in the form of bulging in order to make synapse. *e.g.*, if a chromosome has 1, 2, 3, 4, genes. The part 2 is missing from one chromosome leaving, 1, 3, 4. The other homologous chromosome at the time of synapse bulge out or form loop at position 2.

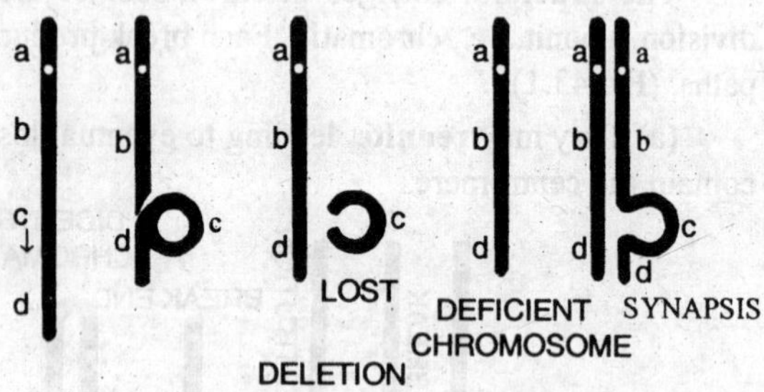

Fig. 43.3. Various steps involved in Deletion

If the missing segment is of physiological importance the individual will not survive. If dominant gene 'A' is missing the recessive allele 'a' may express itself. It is called **pseudodominance.**

In human, deletion of chromosome 5 results in **cri-du-chat syndrome**, children cry like cat, they have small head and are mentally retarded.

Partial deletion of 18th chromosome results in a syndrome with large ears and long fingers.

In corn the deficiency is restricted to pollen sterility. The male haploid gemetophyte shows deficiency while female of it may receive metabolites from maternal tissue supplementing the deficiency. The omitted segment forms buckles. (Fig. 43.4)

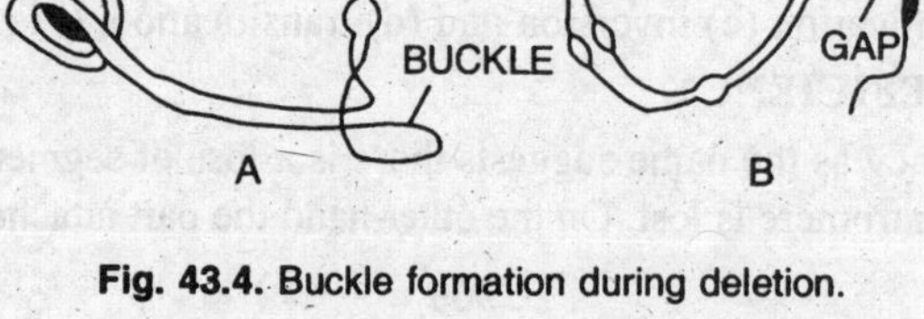

Fig. 43.4. Buckle formation during deletion.

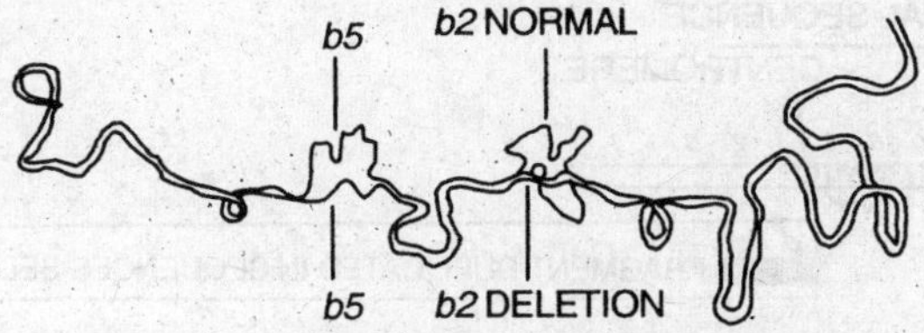

Fig. 43.5. Collapsed loop or brush in single strand DNA due deletion.

Deficiency in *E. coli* is also noted. The deletion points that the DNA is single stranded and looks like collapsed loop or brush. (Fig. 43.5).

(B) DUPLICATION

Here a segment of chromosome is repeated twice, *i.e.*, duplicated. Duplication was discovered in *Drosophila* 'X' chromosome for the first time carrying wild type allele for **vermilion** (v^+) and has been transposed to an 'X' chromosome carrying the mutant **vermilion** allele (v), **Bridges** found that due to the fact that 'X' chromosome was carrying allele v and v^+ both it was wild type instead of vermilion. Equal properties of v and v^+ produced wild type effect. Such 'duplication females' when crossed with nonduplicated vermilion males all female progeny was **vermilion** and all male progeny, *i.e.*, y was **wild type.** (Fig.43.6.)

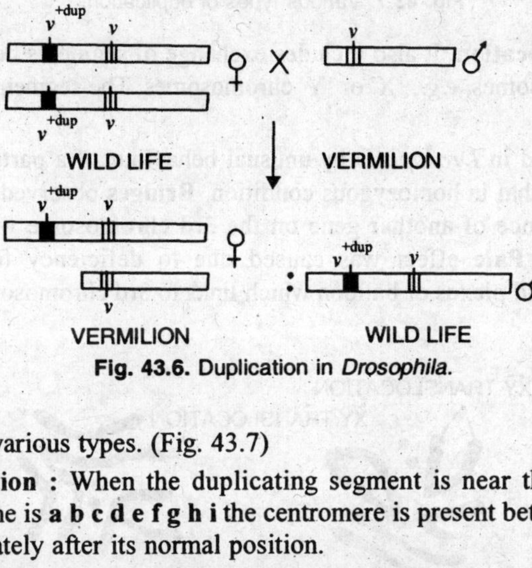

Fig. 43.6. Duplication in *Drosophila*.

Types of duplication

Duplication is of various types. (Fig. 43.7)

Tandem duplication : When the duplicating segment is near the centromeres *e.g.*, the sequence on chromosome is **a b c d e f g h i** the centromere is present between **e** and **f** the segment **d e** is repeated immediately after its normal position.

Reverse tandem : When the segment is reversed in duplication, *e.g.*, it is **d e** segment that is duplicated it will be duplicated as **d e e d** instead of **d e d e.**

Displaced tandem : The segment is repeated some where away from its original location but on the same arm (homobrachial displacement) or on the other arm (heterobrachial displacement).

Transposition : When the segment is duplicated on the non homologous chromosome it is called **transposition.**

Extrachromosomal : Duplication involves centromere it is called extra chromosomal. In salivary gland chromosome duplications are common either as buckling in the duplication heterozygote or as cross pairing between section of different chromosomes.

(C) TRANSLOCATION

Transfer of a section of one chromosome to non homologous chromosome is known as **translocation.** When there is exchange of segments on two **non homologous** chromosomes it is

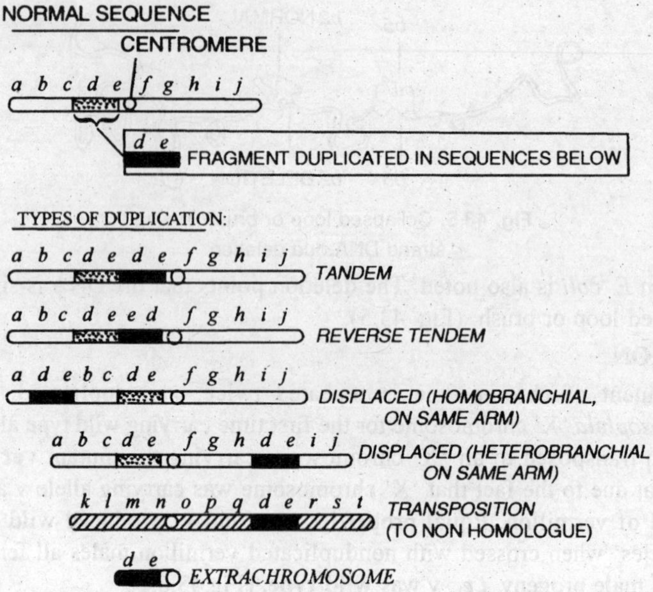

Fig. 43.7. Various types of duplication.

called **reciprocal translocation.** It also includes exchange of segments between non homologous parts of a pair of chromosomes, *e.g.*, 'X' or 'Y' chromosomes. The segment is neither lost or added it is just exchanged.

It was first observed in *Drosophila* by unusual behaviour of a particular 2nd chromosome gene called **Pale.** It is lethal in homozygous condition. **Bridges** observed that its lethality can be suppressed by the presence of another gene on the 3rd chromosome which was also lethal in homozygous condition. **Pale** effect was caused due to deficiency for a small tip of 2nd chromosome including and plexus or balloon which links to 3rd chromosome gene between ebony or rough.

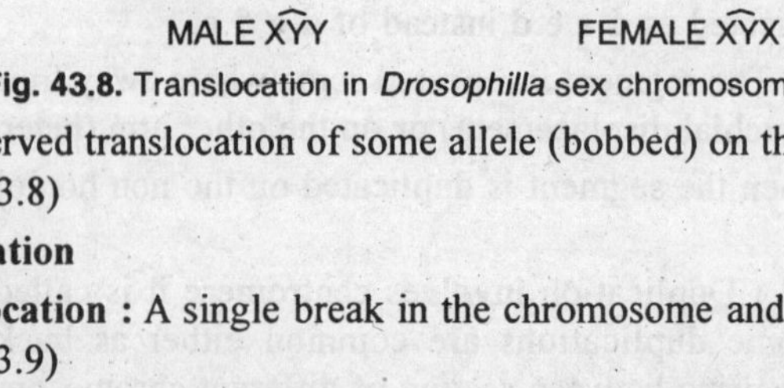

Fig. 43.8. Translocation in *Drosophilla* sex chromosome.

Stern in 1926 observed translocation of some allele (bobbed) on the 'Y' chromosome to the 'X' chromosome. (Fig.43.8)

Types of translocation

(a) Simple translocation : A single break in the chromosome and it is transferred onto the end of the other. (Fig. 43.9)

(b) Shift or intercalary translocation : Common type of translocation involving 3 breaks so that a two break section of one chromosome (*e.g., Pale)* is inserted within the break produced in a non homologous chromosome. (Fig. 43.9B)

(c) Reciprocal translocation or Interchange : Frequently observed translocation where single break in two homologous chromosomes produces an exchange of chromosome segment between them. (Fig. 43.9c)

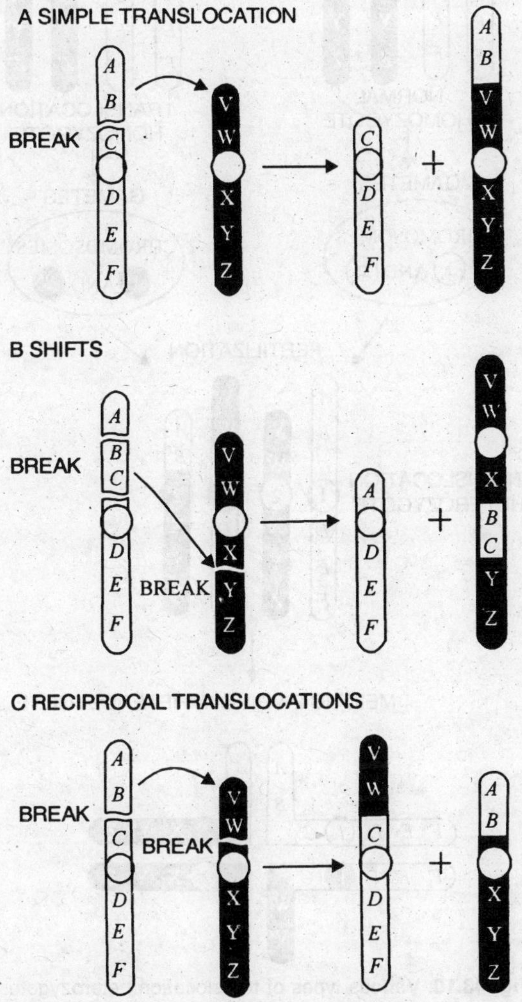

Fig. 43.9. Various types of translocations.

Translocation homozygote forms the same number of homologous pairs as the normal homozygote as long as centromere is not lost.

The result of pairing and meiosis are different in translocation heterozygote bearing two translocated segments and their normal counterpart. (Fig. 43.10). A reciprocal translocation forms a 4-chromosome complex at the pachytene stage. The chiasmata between such chromosomes may form a quadrivalent which can then disjoin in 3 different segregation patterns in the first meiotic division. (Fig. 43.11).

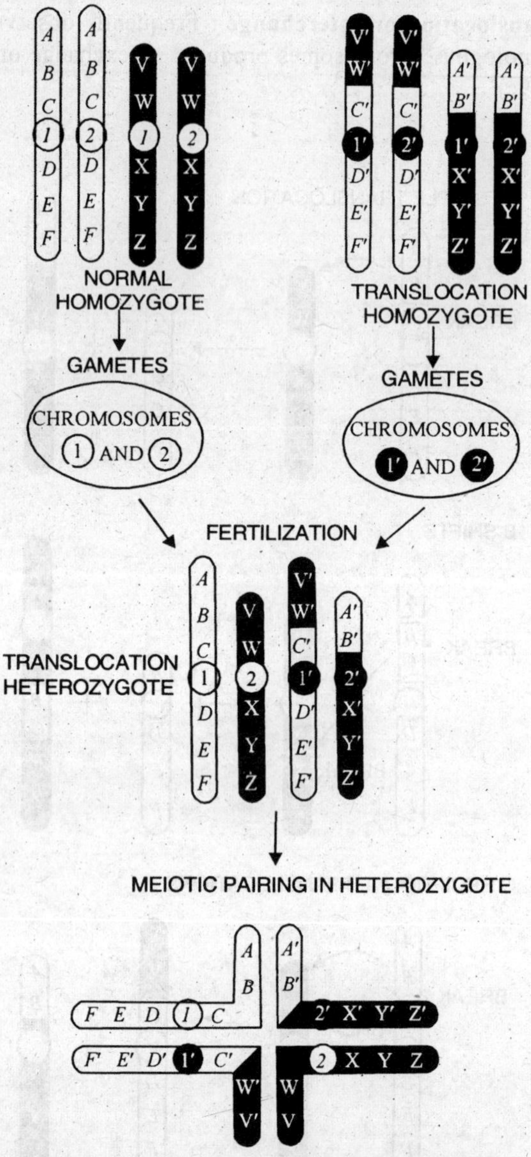

Fig. 43.10. Various types of translocation heterozygote.

(i) Alternate segregation : Opposite or alternate nonhomologous centromeres go to the same pole in a zigzag fashion, so that the nontranslocated (1, 2) and translocated (1′, 2′) chromosomes are in separate gametes. Gametes have complete balanced complement of genes without duplication or deficiency (Fig. 43.11).

(ii) Adjacent-1 segregation : Non homologous adjacent chromosomes go to the same pole but each gamete contains both translocated and non translocated chromosome (1 2′, 1′2) both duplication deficiencies in each gametes are present (Fig. 43.11).

(iii) Adjacent-2 segregation : Adjacent centromeres again go to the same pole but these are now homologous as well as containing both translocated and non translocated chromosomes (1, 1′; 2, 2′). Duplication and deficiencies produce unbalanced components of gene (Fig. 43.11C).

Adjacent-1 and adjacent-2 segregation produce unbalanced gametes. Fertile gametes of translocation heterozygotes will be mostly restricted to alternate segregation.

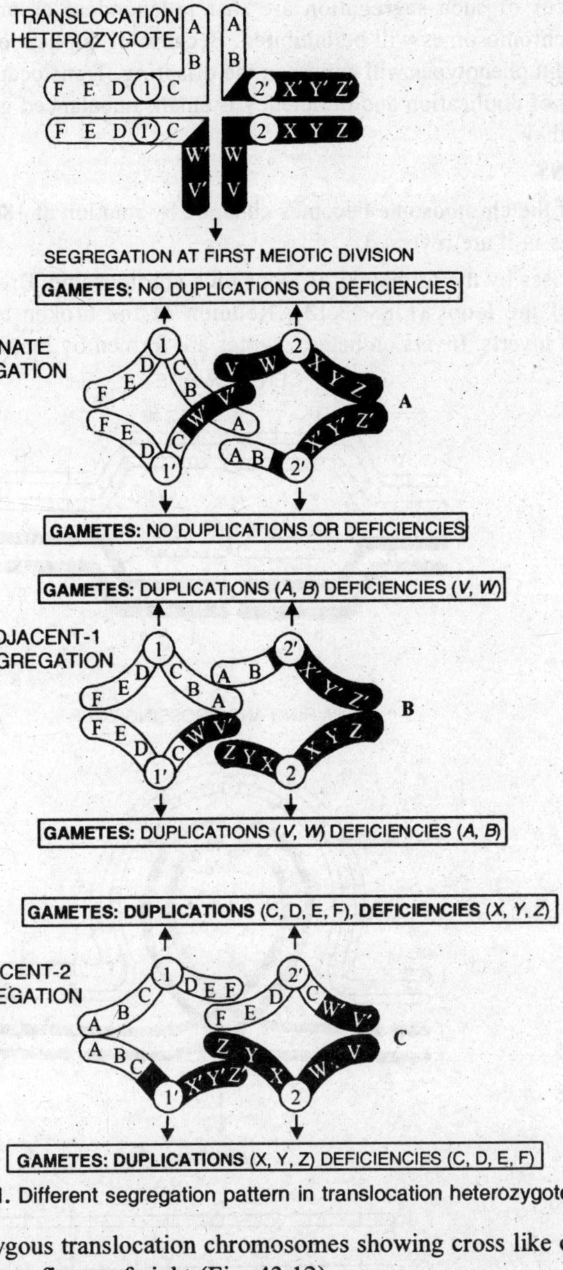

Fig. 43.11. Different segregation pattern in translocation heterozygote.

Synapsis of heterozygous translocation chromosomes showing cross like configuration later on opens out into a ring or a figure of eight (Fig. 43.12).

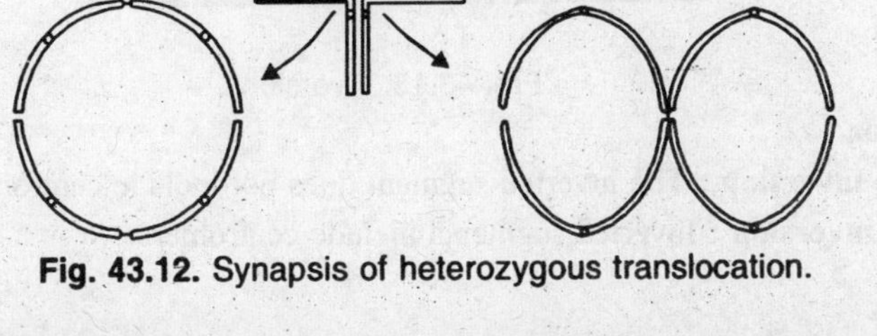

Fig. 43.12. Synapsis of heterozygous translocation.

Consequences of such segregation are that independent assortment between genes and nonhomologous chromosomes will be inhibited. Because of duplications and deficiencies neither of the single mutant phenotypes will appear in the offspring. Translocation heterozygotes have low fertility. If extent of duplication and deficiency is small, unbalanced gametes or zygotes may not necessarily be lethal.

(D) INVERSIONS

A section of the chromosome becomes changed by rotation at 180° is called **inversion.** The order of the genes in it are reversed.

Inversion arises by the formation of loops on a chromosome. Breaks may occur at the point of intersection of the loops (Fig. 43.13). Reunion of the broken ends takes place in a new combination, and inverts. Inversion heterozygotes are formed by loops and bulges in pairs.

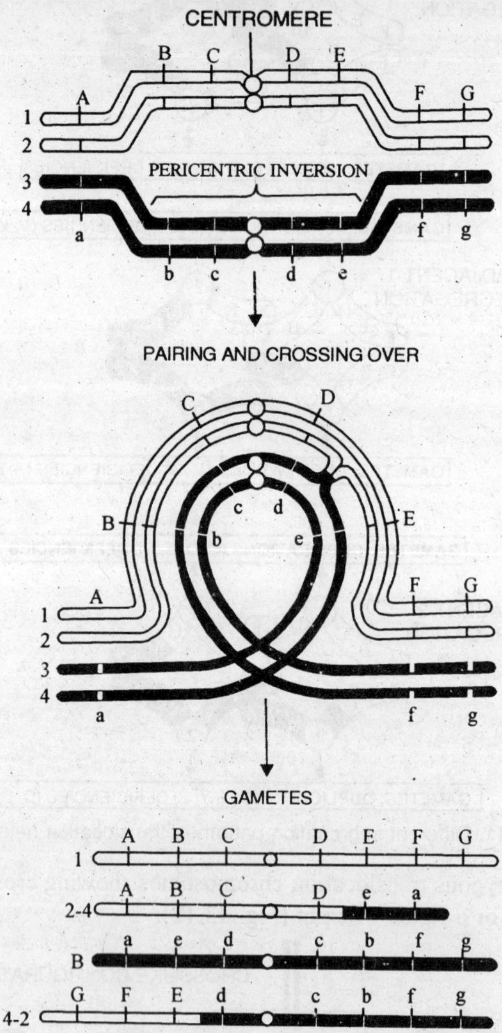

Fig. 43.13. Inversions.

Types of Inversion

Paracentric inversion : The inverted segment does not include centromere.

Pericentric inversion : Inverted segments include centromere.

Paracentric inversion

A single crossing over inverted region will result into formation of a **dicentric** chromosome (with 2 centromeres) and an **acentric** chromosome (with no centromere). Of the remaining 2 chromatids one will be normal and the other will carry inversion. The **dicentric chromatid** and **acentric chromatid** will be observed at anaphase I in the form of a bridge and a fragment (Fig. 43.14). Double crossover shows deficiencies and duplication (Fig. 43.15) giving rise to variations in anaphase I configurations.

Pericentric inversion

In pericentric inversion centromere is in the inverted segments. In pachytene stage 2 of the 4 chromatids resulting after meiosis will have deficiencies and duplications. No dicentric bridge or acentric fragment will be observed (Fig.43.16). In pericentric inversion, if two breaks are not

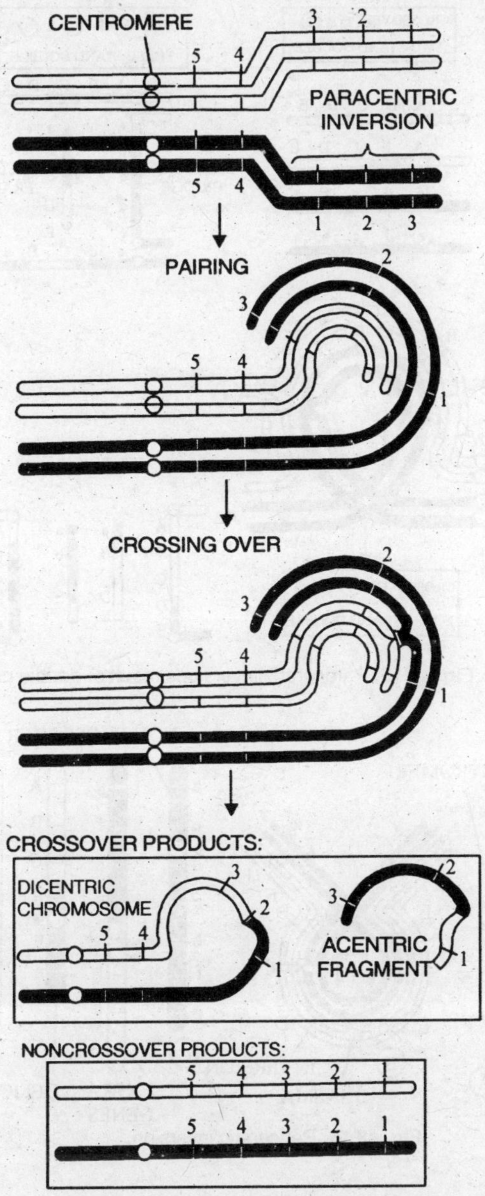

Fig. 43.14. Paracentric inversions.

situated equidistant from the centromere, a change in shape of chromosome results. A metacentric chromosome may become submetacentric and vice versa (Fig. 43.17).

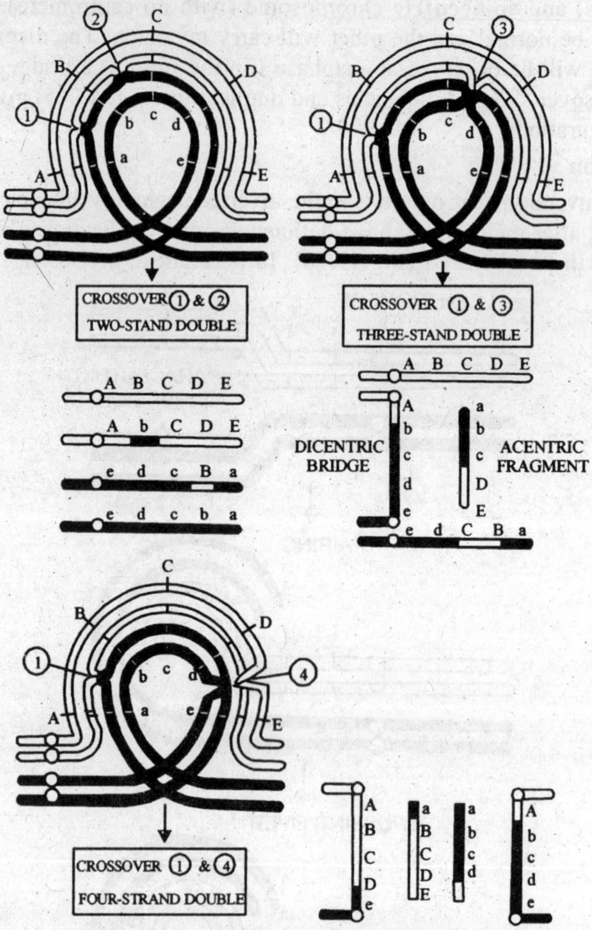

Fig. 43.15. Paracentric inversions result of double cross.

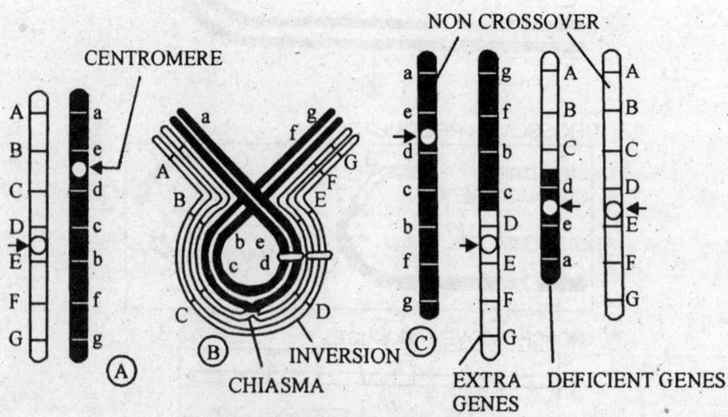

Fig. 43.16. Pericentric inversion.

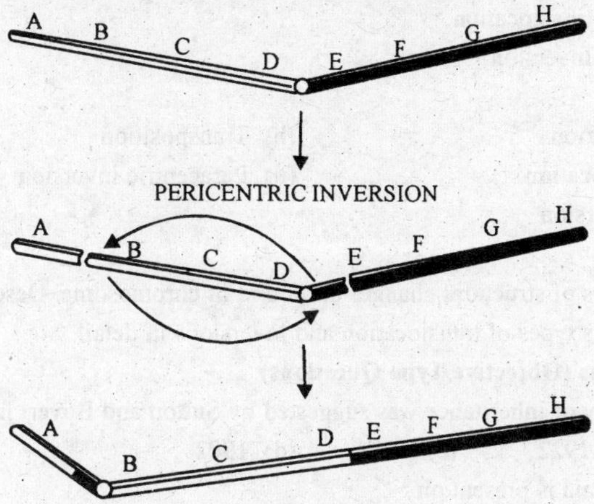

Fig. 43.17. Pericentric inversion and metaphase.

SUMMARY

1. Chromosomal mutations are due to
   (a) **Deletion :** Fragment of chromosome is lost.
   (b) **Duplication :** When the segment is duplicated by joining a homogolous chromosome.
   (c) **Inversion :** Reverse orientation of a segment.
   (d) **Translocation :** Joining of segment of non homologous chromosome.
2. **Cri-du-chat** syndrome is due to deletion of a segment of 5th chromosome.
3. **Myeloid Leukemia** is due to exchange of part of 22nd chromosome with the long arm of 9th chromosome. Such 22nd chromosome is called **philadelphia** chromosome (pH chromosome).
4. Ear and Finger are enlarged when a segment of 18th chromosome is lost.
5. *Drosophila* mutant bar (shape of eye) is the result of doubling up of short piece X chromosome.
6. Loss of a segment of 22nd chromosome (containing gene for **Schizophrenia**) or translocation of the segment cause **Schizophrenia.**

QUESTION BANK

Very Short Answer Questions

1. Define deletion.
2. Give one example of translocation in human.
3. Give one example of deletion in human.
4. In which type of alteration loops are formed.
5. What is reciprocal translocation?
6. What is chromosomal aberration?
7. What is pseudodominance?
8. What is cri du chat syndrome?
9. What is buckle?
10. Give one example of deficiency in *Drosophila.*
11. What is adjacent-1 segregation?
12. What is paracentric inversion?

Short Answer Questions

1. Write short note on deletion.
2. Write short note on insertion or duplication.

3. Write short note on translocation.
4. Write short note on Inversion.
5. Write short note on
   (a) Tandem duplication
   (b) Transposition
   (c) Types of translocation
   (d) Paracentric inversion
   (e) Pericentric inversion

Long Answer Questions

1. How many types of structural changes are found in chromosome. Describe with diagram.
2. Describe various types of translocation and inversions in detail.

Multiple Choice Questions (Objective Type Questions)

1. Chromosomal theory of inheritance was suggested by Sutton and Boveri in :
   (a) 1932 (b) 1922 (c) 1912 (d) 1902
2. Gene of Schizophrenia is present on :
   (a) 18th chromosome
   (b) 22nd chromosome
   (c) 25th chromosome
   (d) 17th chromosome
3. The change in the chromosome due to deletion, duplication, etc., is called :
   (a) chromosomal aberration
   (b) chromosomal deterioration
   (c) chromosomal arrangement
   (d) None of the above
4. Loss of a segment of chromosome is :
   (a) Deletion
   (b) Duplication
   (c) Translocation
   (d) Inversion
5. When the duplication segment is near the centromere it is called :
   (a) Translocation
   (b) Tandem duplication
   (c) Reverse duplication
   (d) Reverse tandem
6. When the chromosomal segment is repeated away from the original location the duplication is called :
   (a) Tandem duplication
   (b) Reverse tandem
   (c) Displaced tandem
   (d) Displaced duplication
7. In transposition the segment duplicates on :
   (a) Same chromosome
   (b) Homologous chromosome
   (c) Non-homologous chromosomes
   (d) It does not duplicate
8. The second chromosomal gene in *Drosophila* is called :
   (a) Amber
   (b) Pale
   (c) Ochre
   (d) All the above
9. When the inverted segment do not include centromere the inversion is known as :
   (a) Paraçentric inversion
   (b) Pericentric inversion
   (c) Reciprocal inversion
   (d) Accentric inversion
10. Which statement is correct for pericentric inversion :
   (a) The inversion involves, include centromere
   (b) No dicentric bridge is formed
   (c) No accentric fragement is formed
   (d) All the above.

ANSWERS

1. (d) 2. (b) 3. (a) 4. (a) 5. (b) 6. (c)
7. (c) 8. (b) 9. (a) 10. (d).

44

CHAPTER

Variation in Chromosome Number

Generally a cell possesses complete set of chromosomes. The condition is called **euploidy.** When the number of chromosomes differs from entire or complete set, *i.e.*, incomplete set of chromosomes it is called **aneuploidy.**

EUPLOIDY

When a cell bears one complete set of chromosomes, it is called **haploid.** Normally individuals bear two similar sets of chromosomes in a cell, the condition is called **diploid.** If an organism bears more than 2 sets of chromosomes in a cell, it is called **polyploid.**

HAPLOID (MONOPLOID)

Haploid organisms possess one complete set of chromosomes. In lower plants, the main plant is a gametophyte and bears haploid set of chromosomes. It is also called **hemizygous.** Among higher organisms, haploidy is found in gametes (sperms, egg). Some insects are also haploids (honey bee). Such haploids reproduce parthenogenetically.

Monoploids may be the result of defective meiosis, *e.g.*, *Datura.* Monoploids are usually sterile. Monoploids are usually smaller in size than the corresponding diploids with less vigour. They show sterility and have small leaves. Monoploids are very important in plants. Doubling of chromosomes in monoploids results in diploids (homozygous).

The chromosome behaviour in haploid differs from normal condition. Haploids of pepper are of same size as diploids.

Haploid generally is half the number of chromosomes. If diploid has 2n then haploid will be having n, *e.g.*, wheat is hexaploid, *i.e.*, 2n=6 X then the haploid will be n = 3 X.

Monoploid means only one set of chromosomes, *e.g.*, n = X *e.g.*, in pea, n = X = 7 or in maize, it is n = X = 10, etc.

POLYPLOIDY

When an organism has more than 2 sets of genomes, *i.e.*, if 3 sets then **triploid,** if 4 sets **tetraploids,** it five sets = **pentaploids,** if 6 sets then **hexaploids,** etc. When the polyploids are even number they are called **primary polyploid.** *e.g.*, 2, 4, 6n, etc.

Secondary polyploids usually arises by crossing between organisms of different ploidy levels, *e.g.*, 4n × 2n → 3n.

Autopolyploids

The chromosome sets are homologous (ABC, ABC, ABC), etc. Since it is a 'self' (auto) increase in ploidy level. Among the ways in which polyploidy can occur in sexual organisms are (Fig. 44.1) :

(a) The fertilization of egg by more than one sperm then zygote nucleus will have 3 or more sets of chromosomes (Fig. 44.1A).

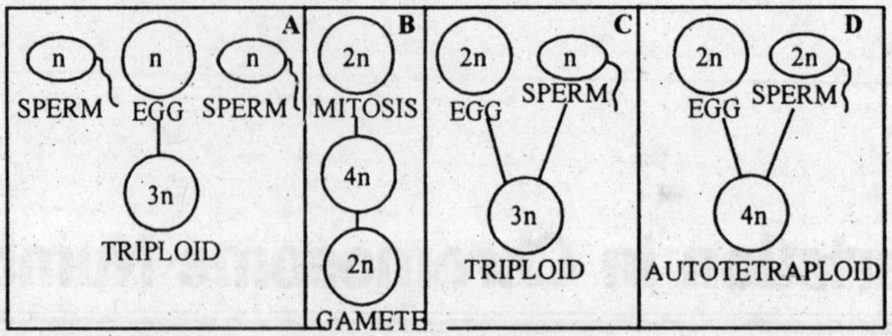

Fig. 44.1. Autopolyploids

(b) Failure of mitosis, multiplying the number of chromosomes there by increasing the number of gametic chromosomes (Fig. 44.1B).

(c) Failure of meiosis producing a diploid gamete instead of a haploid (Fig. 44.1 C, D).

The failure of normal mitotic division in asexual organisms may also produce nuclei with double set of chromosomes. Natural autopolyploids are rare because of the resulting anomalies in cell division.

Triploids may arise by many ways. If normal meiosis fails, a diploid (2n) gamete will be formed. If such gamete (2n) is fertilized by normal haploid gamete (n) an **autotriploid** (3n) is formed (Fig. 44.2A). Autotriploids are generally sterile; because of defective gamete formation. Because of 3 chromosome 2 form a pair **(bivalent)** but one remains **univalent.** If the third also pairs, it forms a **trivalent** (Fig. 44.2C). The gametes will have one and two chromosomes. In the third condition if there are 3 univalents formed, they are randomly distributed in gametes (Fig. 44.2C), *e.g.*, ***Hyacinth*** having triploid constitution 3 n = 8 + 8 + 8 = 24. The gametes may have 8 or 16 chromosomes which is unbalanced.

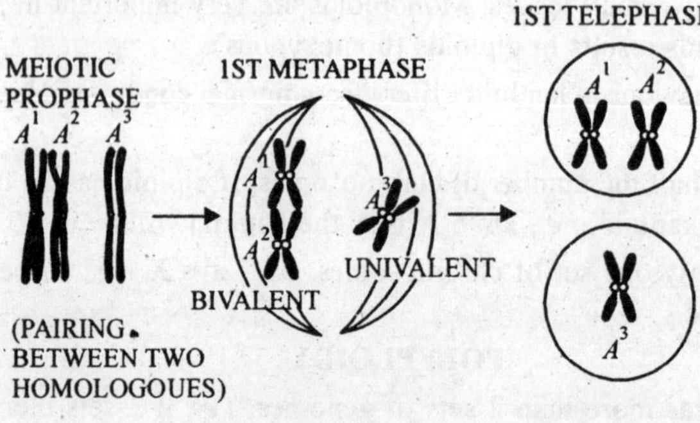

Fig. 44.2 A. Normal Triploid.

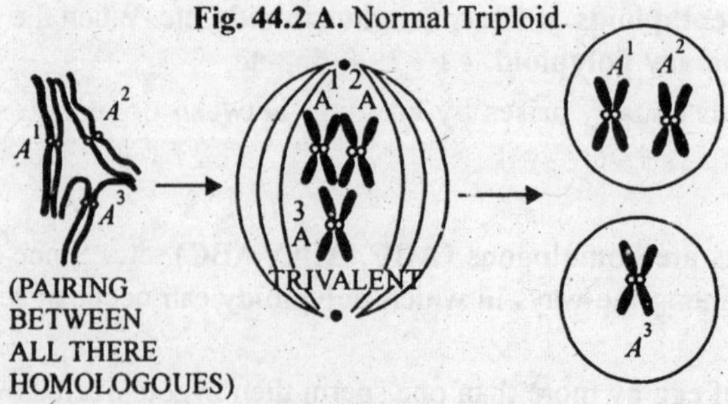

Fig. 44.2 B. In triploid 2 chromosomes forming bivalent and one remains univalent.

Similarly all odd numbered polyploids pentaploids (5n), heptaploids (7n) etc., produce sterile individuals.

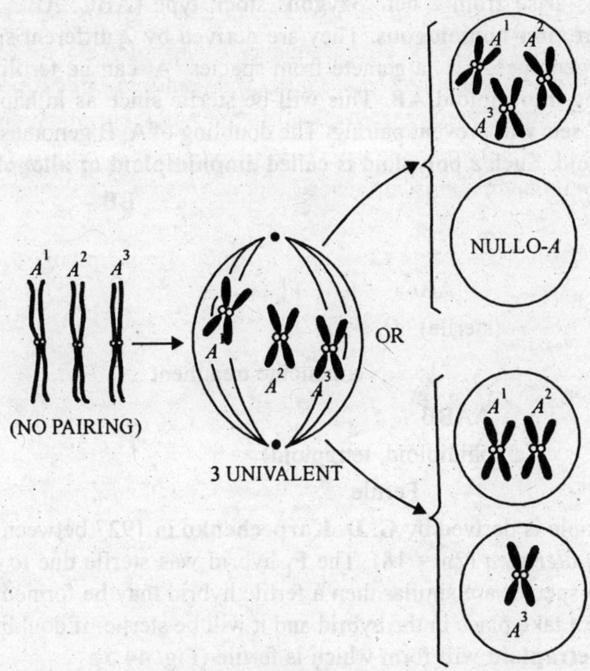

Fig. 44.2 C. Univalents in triploid.

In **autotetraploids** (4n) of *Datura,* there is an equal separation of chromosomes during meiosis. Autopolyploidy was observed by **de Vries** in *Oenothera lamarkiana* (autotetraploid). Autopolyploids are found in plum, moss, tomato, corn, rye, *Datura*, etc.

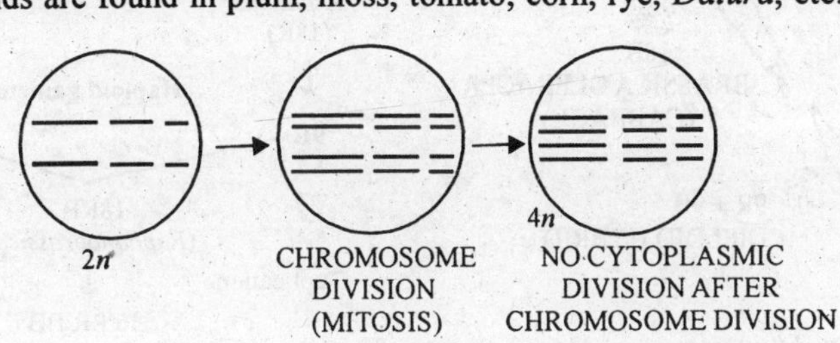

Fig. 44.3. Tetraploids.

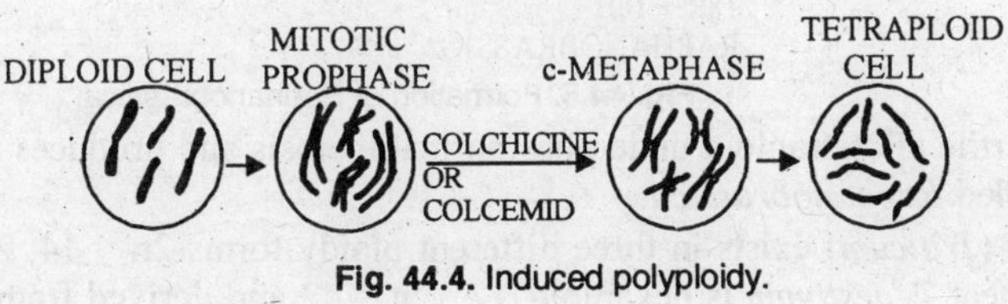

Fig. 44.4. Induced polyploidy.

Normal seedless banana is triploid. Alfa alfa, potato coffee, peanuts, etc., are tetraploids (Fig. 44.3). Polyploidy is induced by inhibiting normal mitosis in some plants. (Fig. 44.4).

The segregation pattern of polyploids is different from normal diploidy. The ratio is based on chromosome segregation or distance of gene form centromere (chromatid segregation). The main effect of autopolyploidy on segregation reveals different kind of inheritance. Tetrasomic inheritance and trisomic inheritance are useful to differentiate between auto and allopolyploids.

Allopolyploidy :

Allopolyploids arise from a heterozygous stock type (ABC, ABC A′B′C′, A′B′C′). The chromosome sets are **non-homologous.** They are derived by 2 different species. Allopolyploids contain 2 different genomes, *e.g.*, a gamete from species **'A'** can be fertilized by a gamete from species **'B'** resulting into diploid **AB.** This will be sterile since as in haploids the homologous between 'A' and 'B' sets will prevent pairing. The doubling of A, B genomes due to (AB) F_1 hybrid gives rise to tetraploid. Such a polyploid is called **amphidiploid** or **allopolyploid.**

AA × **BB**

↓

AB Fi

(sterile)

↓ colchicine treatment

AABB

(amphidiploid, tetraploid)

Fertile

Classical example is derived by **G. D. Karpechenko** in 1927 between *Raphanus sativus* (2n = 18) and *Brassica oleracea* (2n = 18). The F_1 hybrid was sterile due to lack of pairing. If the genomes of the two species are similar then a fertile hybrid may be formed. If they are dissimilar then synapsis will not take place in the hybrid and it will be sterile. If doubling of the chromosome is carried out, **allotetraploid** will form which is fertile (Fig. 44.5).

Fig. 44.5. Formation of Raphanobrassica.

The fertile allotetraploid undergoes normal meiosis and produces normal gametes. The new genus is called *Raphanobrassica.*

Wheat (*Triticum*) exists in three different ploidy forms 2n = 14, 2n = 18 and 2n = 42. The common wheat *T. aestivum* is hexaploid, *i.e.*, 2n = 42 and derived from 3 diploid species.

(a) AA = *T. aegilopoides* (2n = 14)

(b) BB = *Aegilops speltoides* (2n = 14)

(c) DD = *Aegilops squarrosa* (2n = 14)

The hexaploid wheat is AABBDD where tetraploid is 2n = 28 and designated as AABB, and diploid is 2n = 14 and designated as AA.

The common hexaploid wheat is **segmental allopolyploid** as the three ancestors are not much different.

Triticum spelta is a hexaploid which was artificially synthesized in 1946 by **E. S. Mc Fadden, E. R. Sears** and **H. Kihara.** It was obtained by crossing *Aegilops squarrosa* (diploid 2n = 14) and *T. dicoccoides* (tetraploid 2n = 28, commonly known as emmer wheat).

T. dicoccoides		*Aegilops squarrosa*
AA BB		DD
2n = 28	×	2n = 14

ABD
2n = 21
(Triploid hybrid, sterile)
↓ Colchicine treatment
T. spelta
AA BB DD
2n = 42
(*hexaploid wheat*)
Fertile

There are several other natural and synthesized polyploids, *e.g.*, *Gossypium hirsutum* (upland cotton with 2n = 52 is new world cotton raised by crossing *G. herbaceum* (old world cotton 2n = 26) and *G. raimondii* (American cotton 2n = 26).

Triticale is a new man made cereal. It is obtained by crossing *Triticum durum* (2n = 28) and *Secale cereale* (2n = 14). The F_1 hybrids which were sterile (2n = 21) were doubled to get a **hexaploid** *Triticale* (2n = 6X = 42) and **octaploid** *Triticale* (2n = 8X = 56). It was first reviewed by **Muntzing** (1979), **Gupta** and **Priyadarshan** (1982) and **Gupta** and **Reddy** (1991).

(*i*) *Triticum durum* × *Secale cereale*
↓
2n = 28 2n = 14
F_1 hybrid (sterile)
2n = 21
↓ colchicine treatment
2n = 42
Hexaploid *Triticale*

(*ii*) *Triticum aestivum* × *Secale cereale*
2n = 42 ↓ 2n = 14
F_1 hybrid (sterile)
2n= 28
↓ colchicine treatment
2n = 56
Octaploid *Triticale*

Clausen and Goodsped synthesized a new species of *Tobacco*. They crossed *Nicotiana tabacum* (48 chromosomes, gamete n = 24) with *N. glutinosa* (24 chromosomes, gamete n = 12) and obtained F_1 plants where most of them are sterile but some are viable. Fertile plant was having 72 chromosomes. The new species is called *N. digluta.*

Allopolyploids were also obtained by fusing the protoplast of somatic cells.

Relationship of different kinds of polyploids

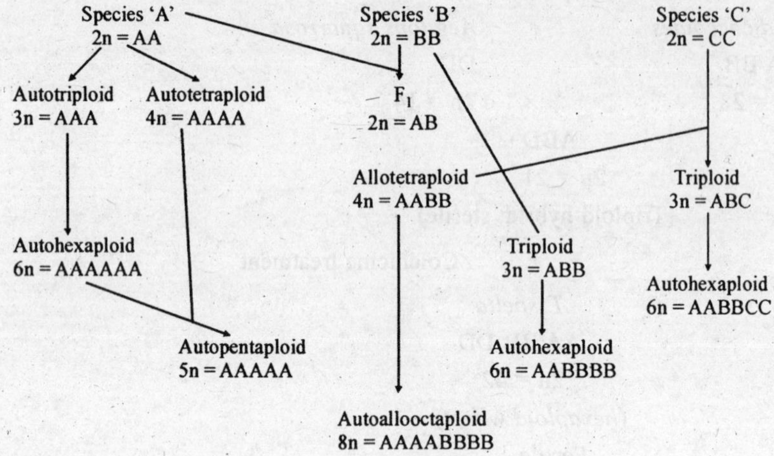

In animal polyploidy is greatly restricted, since the XY mechanism of sex determination is disturbed with duplication of sex chromosome numbers, *e.g.*, male tetraploid with XXYY would produce 'XY' gamete and female tetraploid XXXX will produce XX gamete. The union between XX and XY will form XXXY which is neither male nor female.

Segmental allopolyploids

In some allopolyploids the different genomes which are present are not quite different from one another. In the polyploids chromosomes from different genomes pair together to some extent and multivalents are found. The segments of chromosomes and not the whole chromosomes are homologous. Therefore, such allopolyploids are known as **segmental allopolyploids.** It was suggested by **Stebbins** in 1950. These segmental allopolyploids are intermediate between autopolyploids and allopolyploids. Generally the naturally occurring polyploids are segmental allopolyploids.

APPLICATIONS OF POLYPLOIDY

Tetraploids are more vigourous than diploids bearing larger fruits and seeds. Polyploids are artificially raised for disease resistance, large fruits and high nutritive contents. Disease resistant varieties of tobacco, large fruits of apples and watermelon, large grains in oats and wheat, and disease resistance in wheat, etc., are successfully obtained as polyploids (Fig. 44.6), tetraploid tomato has high vitamin content, tetraploid maize contains 2% more vitamins, etc.

New varieties of wheat, cotton, tobacco, etc., are the examples of polyploids.

INDUCTION OF POLYPLOIDY

Polyploidy may be induced by temperature treatment, decapitation and chemical treatment.

(a) High temperature : High temperatures are helpful in getting polyploids.

(b) Decapitation : The method is used by removing, buds, etc., in tomatoes.

(c) Chemical treatment : Polyploids may be obtained by using colchicine solution. Colchicine inhibits spindle formation. Other chemicals are **acenapthene** and **indole acetic acid** (IAA).

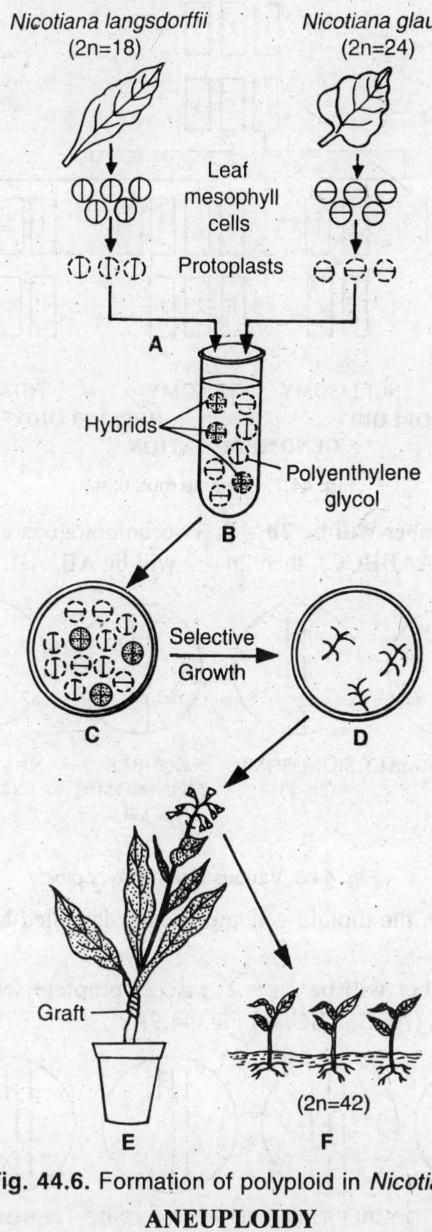

Fig. 44.6. Formation of polyploid in *Nicotiana.*

ANEUPLOIDY

Variation in chromosome number involving individual chromosome is called **aneuploid.** The genome in this is incomplete. It is either less or more than the diploid number. Aneuploidy is also called as **genome mutation** (Fig. 44.7).

Hypoploidy : Less than the diploid genome number (2n) is hypoploidy, *e.g.*,

(a) Monosomic : The number is **2n** – 1. One chromosome less than the diploid complement of 2n = 6 CAABBCC, 2n – 1 will be ABC, AB Fig. 44 .8A.

(b) Double monosomics : The number is **2n – 1 –1,** *i.e.*, two chromosomes less than the whole diploid genome, *e.g.*, 2n = 6. (AABBCC), then 2n – 1 – 1 will be ABC, AB Fig. 44.8B.

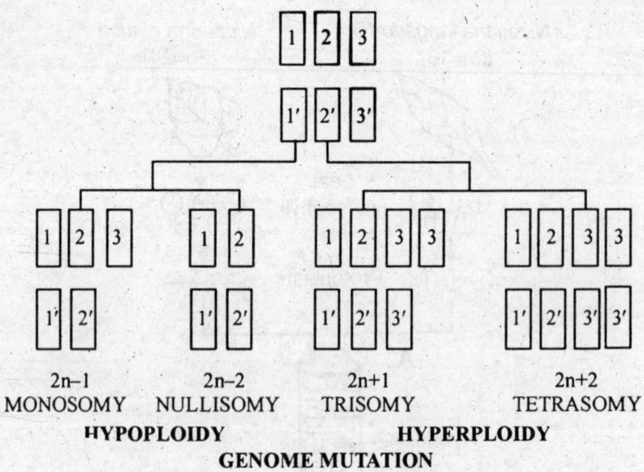

Fig. 44.7. Genome mutation.

(c) Nullisomic : The number will be **2n – 2.** Two homologous chromosomes less than the diploid genome, *e.g.*, 2n = 6 (AABBCC), then 2n – 2 will be AB, AB. Fig. 44.8C.

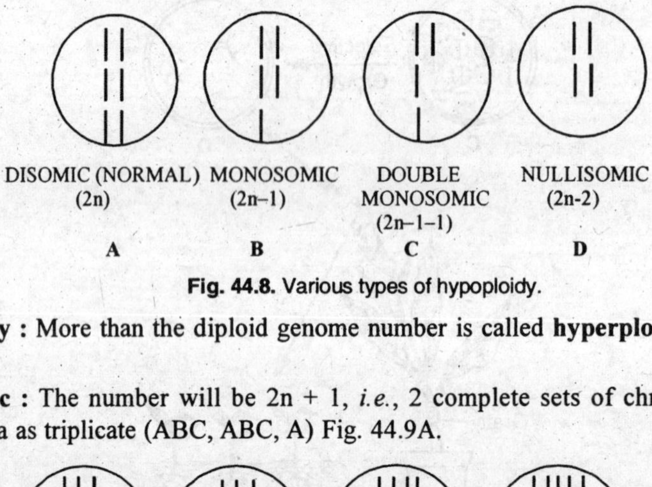

Fig. 44.8. Various types of hypoploidy.

Hyperploidy : More than the diploid genome number is called **hyperploidy** or **polysomy,** *e.g.*,

(a) Trisomic : The number will be 2n + 1, *i.e.*, 2 complete sets of chromosomes but 1 chromosome extra as triplicate (ABC, ABC, A) Fig. 44.9A.

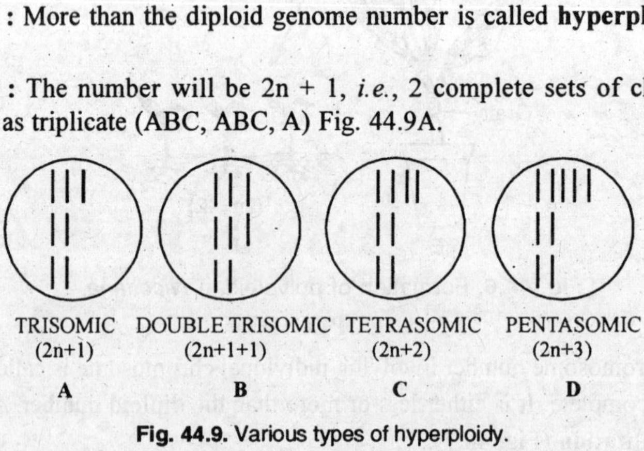

Fig. 44.9. Various types of hyperploidy.

(b) Double trisomics : The number of chromosome will be 2n + 1 + I, *i.e.*, two extra chromosomes than the diploid genome (ABC, ABC, AB) Fig. 44.9B.

(c) Tetrasomics : The number of chromosomes is **2n + 2,** *i.e.*, one chromosome in **quadriplicate** while others are in duplicate number (ABC, ABC, AA) Fig. 44.9C.

(d) Pentasomics : The number is **2n + 3.** One chromosome is **pentaplicate** while others in duplicate (ABC, ABC, AAA) Fig. 44.9D.

Aneuploids are the result of

(a) Union of gametes which contains an unbalanced number of chromosome.

(b) Perpetuation of a cell in which through a failure of mitosis, one or more chromosomes may be added or lost.

Aneuploids generally do not survive. Aneuploids in polyploids are more viable than diploids. Aneuploids in animals are much less viable than in plants.

True aneuploids have one complete set of genome + one fraction of genome. **False aneuploids** contain supernumerary chromosomes or may contain 2 or more complements derived from parents with different chromosome number.

Monosomic and Trisomic in *Drosophila*

Drosophila melanogaster having 2 pairs of chromosomes, among which 3 pairs are of autosomes and one of heterosome or sex chromosomes. In females it is XX and in male it is XY. **First aneuploid** was described by **Bridges** in 1916. Where **non-disjunction** in meiosis produced ***XO*** and XXX or XXY. After meiosis 2 types of eggs are produced one **XX** and one with no sex chromosome. If these eggs mate normal X and Y sperms four types of zygotes XXX, XXY, X and Y are produced. XXX and XXY are **trisomics** while X and Y are **monosomic.**

XX female	XX female
↓	↓
gamete	gamete
X X	X X
normal egg (n)	n + 1 n - 1
	egg egg
	Aneuploids

'XX' x 'X' or 'Y' = XXX – super female (die)

XXY = normal female

'O' x 'X' and 'Y' = XO and normal male

YO = super male (die)

'XXX' will be super female and will die, 'XXY' will act as normal female while 'XO' is normal male and 'YO' is super male and will die without 'X' chromosome.

E.R. Sears produced 2 1 monosomics in wheat variety **chinese spring. J. E. Endrizzi,** etc. have produced monosomics in **cotton** (2n = 52) and **E. R. Clausen** and **D. R. Cameron** produced. In tobacco (2n = 48). Double monosomics or triple monosomics in wheat (2n – 1 – I and 2n – 1 – 1 – 1) are also produced. A progeny of monosomic produces a mixture of disomic (2n), monosomic (2n – 1) and nullisomic (2n – 2) but a progeny of nullisomy can not possess a gene on specific chromosome.

Sears isolated all 21 **nullisomic** in wheat.

Trisomic in *Datura* :

Belling and **Blakeslee** obtained trisomics in *Datura stramonium* (Jinsson weed). The haploid chromosome number of it is n = 12. 12 primary trisomics, 24 secondary trisomics are identified by size, shape and other morphological features of fruit (Fig. 44.10).

The trisomics has 25 chromosomes instead of 24 Table 44.1 shows different types of trisomy in *Datura* as 2n + A, 2n + B, etc., and their phenotypes.

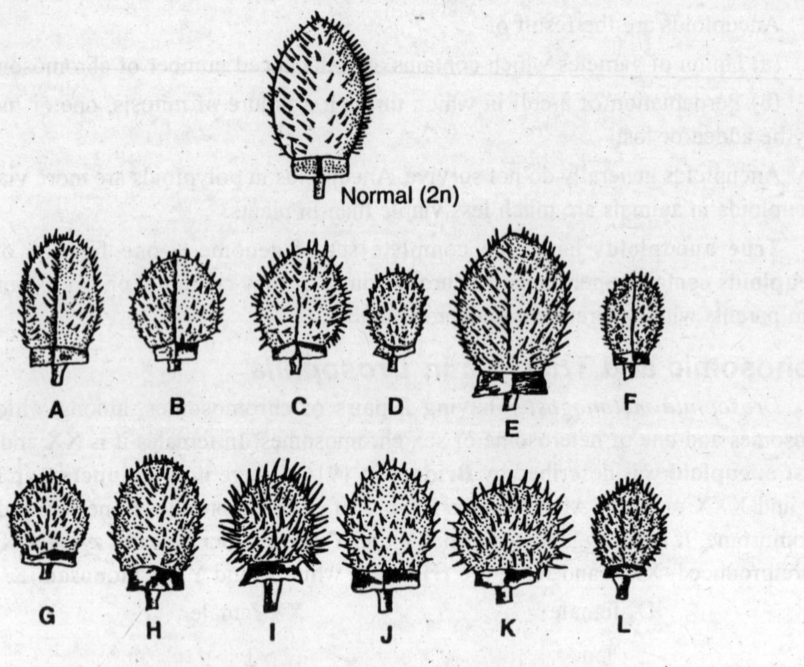

Fig. 44.10. Various shapes of fruits in *Datura* (trisomica).

Table 44.1. Trisomy in *Datura.*

Chromosome no.	Type of Trisomy	Trisomic name	Phenotype
A	2n + A	Rolled	Inrolled leaves
B	2n + B	Glossy	Glossy leaves
C	2n + C	Buckling	Leaves with distorted surface
D	2n + D	Elongate	Elongated leaves
E	2n + E	Echinus	Large capsules very long spines
F	2n + F	Cocklebur	Small, narrow capsule
G	2n + G	Microcarpic	Small, flat capsule
H	2n + H	Reduced	Short, blunt spines
I	2n + I	Poinsettia	Leaves long, clustered
J	2n + J	Spinach	Dark, leaves
K	2n + K	Globe	Globose capsule
L	2n + L	Ilex	Small capsule, sharply toothed leaves

One of the most extensively studied trisomic series is that produced and studied by **T. Tsuchiya** in Barley.

Aneuploidy in Sex chromosome

In *Homosapiens* (man) both monosomics and trisomics of sex chromosomes are found. In monosomic (Fig. 44.11) only one 'X' chromosome is found, *i.e.*, 45 chromosomes. The individuals show **Turner syndrome** (one in 5000). The syndromic individuals are phenotypically female, short statured, webbed neck, under developed breasts, small uterus and reduced overies. Such a condition arises when abnormal eggs without X chromosome fertilizes with 'X' sperm or a normal egg with 'X' chromosome fertilized with sperm without sex chromosome (Fig. 44.12).

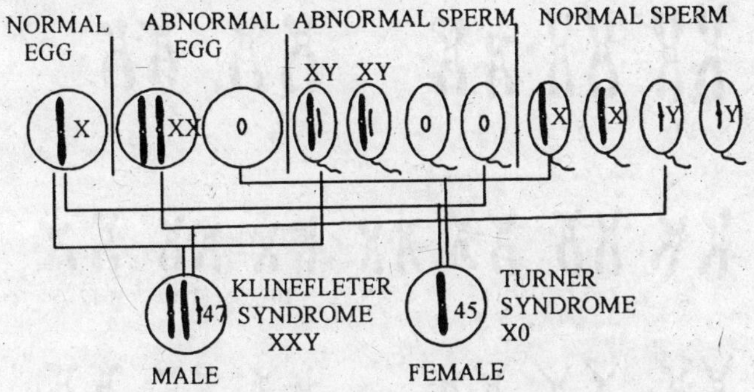

Fig. 44.11. Aneuploidy in sex chromosome of human.

Fig. 44.12. Turner syndrome in human.

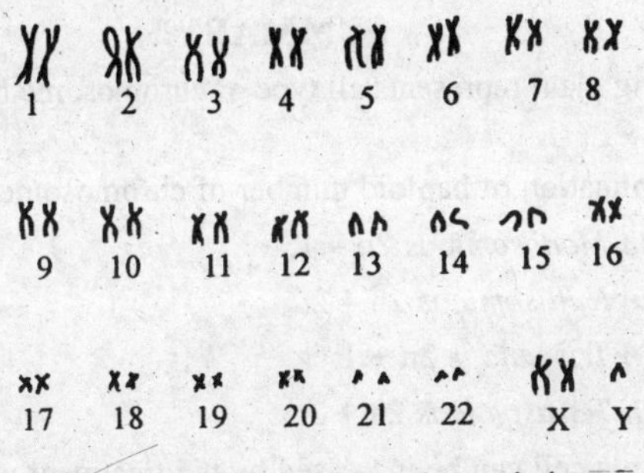

Fig. 44.13. Klinfelter syndrome in human.

Females with XXX (trisomics) XXXX (tetrasomics) and XXXXX (pentasomics) are superfemales with mental abnormality.

Trisomic in human sex chromosome with **XXY** are **Klinefelter syndrome.** They have 47 chromosomes. Such individuals develop small testes, external genitalia of male, female like breasts, sparse body hairs. They develop from abnormal egg (XX) fertilized with normal sperm (Y) or normal egg (X) fertilized with abnormal sperm (XY). XYY males are also found with high criminal tendency (Fig. 44.13).

Aneuploidy in autosome

In human being trisomic of autosome 21st is found. It is called **Down's syndrome** or **Mongolism.** It is characterised by mental retardation, short body, swollen tongue, folded eyelids, etc. (Fig. 44.14). Monosomy of 21st chromosome is fatal. Trisomy of 13, 14, 15, 17, 18 chromosomes results in mental retardation. Trisomy of 18th chromosome results in death within few months after birth. The individuals have laterally flat head, smaller short hands, etc.

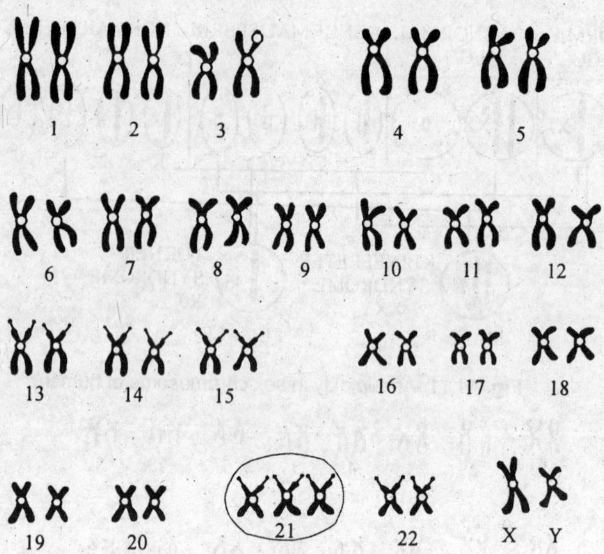

Fig. 44.14. Trisomy of 21st chromosomes.

Production of trisomics

It may be due to production of n + 1 type of gametes because of non disjunction of bivalents. They may be produced by crossing, diploids, autotetraploids or triploid females with diploid male.

SUMMARY

1. Life cycle of a flowering plant represents all type of chromosome numbers, *e.g.*, **haploid**, **diploid** and **polyploid.**
2. Euploidy means multiplication of haploid number of chromosomes.
3. *Chromosome number in Monosomic is* 2n – 1.
4. *Chromosome number in Nullisomic is* 2n – 2.
5. *Chromosome number in Trisomic is* 2n + 1.
6. *Chromosome number in Tetrasomic is* 2n + 2.
7. Chromosome number in a cell can be increased by the treatment with **colchicine.**
8. If a nucleus contains 3 sets of genome the organism is known as triploid and if more it is polyploid
9. In **aneuploid** the number of chromosome is less than the genome, *i.e.*, 2n – 1. **(hypoploidy)**
10. If the chromosome number is more than the genome, *i.e.*, 2n + 1 **(hyperploidy)**.
11. Aneuploids in mammals are lost in foetus generally.
12. Trisomy of 21st chromosome in human is called **Down's** syndrome.
13. Trisomy of 13th chromosome is called **Patau's** syndrome.
14. Edward's syndrome is due to extra copy of 18th chromosome.
15. **'Klienfelter'** syndrome is due to extra copy of '*X*', *i.e.*, XXY (superfemale) if it is extra copy of 'Y' it will be super male.
16. **Blackslee** discovered induction of polyploidy by colchicine in 1937.
17. **Stebbins** (1950) distinguished 3 types of polyploids : autopolyploids, allopolyploids and autoallopolyploids.
18. **Winklar** (1916) discovered heteroploidy or aneuploidy.
19. **Turners syndrome** is monosomy of sex chromosome with 44A + X.

QUESTION BANK

Very Short Answer Questions

1. What is Edward's syndrome?
2. What is Patau's syndrome?
3. What is Kleinefelters syndrome?
4. Give an example of trisomy in plants.
5. Give an example of trisomy in human.
6. Give an example of monosomy in human.
7. Give an example of allopolyploid.
8. Give the name of one allopolyploid in plants raised by man.
9. Give two examples of sterile hybrids.
10. What is polyploidy ? Define.
11. What is aneuploidy? Define.
12. Name one example of polyploid in plants which is used daily.
13. Name one example of aneuploidy in sex chromosome.
14. Name one example of aneuploidy in autosome.
15. Give two examples of induced polyploidy.

Short Answer Questions

1. Write short notes on autopolyploidy.
2. Write short notes an allopolyploidy.
3. Write short notes on hypoploidy.
4. Write short notes on hyperploidy.
5. Write short notes on.
   (*a*) Monosomy (*b*) *Triticale*
   (*c*) Nullisomy (*d*) Segmental polyploids
6. Write shorts notes on:
   (*a*) *Raphanobrassica* (*b*) Trisomy
   (*c*) Tetrasomy
7. Write short notes on hyperploidy giving suitable examples from human being.
8. Write notes on application of polyploidy.
9. Write short notes on induced polyploidy.
10. Make a line diagram showing relationship between various types of polyploids.

Long Answer Questions

1. What do you mean by polyploidy. Explain polyploidy and give suitable examples.
2. What is aneuploidy? How many types of aneuploidy are found. Explain.

Multiple Choice Questions (Objective Type Questions)

1. If an organism is tetraploid, its haploid will be :
   (a) n (b) 2n
   (c) 3n (d) 4n
2. Banana is generally :
   (a) Triploid (b) Diploid
   (c) Hexaploid (d) Tetraploid
3. Haploid set of chromosome is called :
   (a) Individual (b) Genome
   (c) Nucleus (d) Gene

4. Man made allopolyploid is :
   (a) Water melon (b) *Gossypium*
   (c) *Triticale* (d) *Triticum*
5. Colchicin brings about :
   (a) Polyploidy (b) Cell division
   (c) Cell elongation (d) Cell differentiation
6. Monosomic are :
   (a) n (b) 2n + 1
   (c) 2n – 1 (d) 2n + 2
7. Wheat plant is 6n = 42, what will be the number of chromosomes in its monosomics, haploid and monoploid :
   (a) 43, 21, 7 (b) 15, 7, 7
   (c) 41, 21. 7 (d) 14, 7, 7
8. Chromosome aberration occurs due to :
   (a) Aneuploidy (b) Polyploidy
   (c) Physical effect (d) All the above
9. A material which arrests cell division is obtained from :
   (a) *Chrysanthemum* (b) *Crocus*
   (c) *Cycas* (d) *Colchicum*
10. In Mongolism, the patient has :
   (a) Barr body (b) Trisomy
   (b) Nullisomy (d) Monosomy
11. Extra 18th chromosome in human results in :
   (a) Edwards' syndrome (b) Patau's syndrome
   (c) Down's syndrome (d) None of the above
12. Hexaploid wheat is :
   (a) *Triticum durum* (b) *T. monococcum*
   (c) *T. aestivum* (d) *T turgidum*
13. If haploid number is 10 then what will be the tetrasomic number :
   (a) 19 (b) 40
   (c) 20 (d) 22
14. A polyploid with genome derived from same original species is :
   (a) Autopolyploid (b) Allopolyploid
   (c) Autoallopolyploid (d) Amphiploid
15. A polyploid derived from F_1 hybrid between two species :
   (a) Autopolyploid (b) Autodiploid
   (c) Amphidipioid (d) Hexaploid
16. *Datura* is a classical example of :
   (a) Monoploidy (b) Triploidy
   (c) Trisomy (d) Monosomy
17. A trisomic individual possesses extra chromosomes :
   (a) Three (b) One
   (c) Two (d) Four
18. *Triticale* has been developed from intergeneric hybridization between :
   (a) Wheat and *Aegilops* (b) Rice and Maize
   (c) Wheat and Rice (d) Wheat and Rye

19. Exact multiple of haploid number is :
    (a) Heteroploid (b) Euploid
    (c) Aneuploid (d) Hypoploid
20. The mule is :
    (a) Hybrid (b) Sterile hybrid
    (c) Interspecific hybrid (d) All the above
21. Polyploidy means occurrence of :
    (a) Diploid set of chromosomes (b) Haploid set of chromosomes
    (c) More than diploid set of chromosomes (d) Both 'a' and 'b'
22. The hybrid between *Dianthus* and *Centauria is* called :
    (a) Fairchild's mule (b) Hybrid flower
    (c) Fairchild's centromic (d) All the above
23. Induction of polyploidy by colchicine treatment was suggested by Blakslee in :
    (a) 1930 (b) 1937
    (c) 1913 (d) 1952
24. Stebbins in 1950 distinguished :
    (a) Aneuploids (b) Barrbodies
    (c) Y spot (d) Polyploids
25. Hexaploid wheat is developed through :
    (a) Hybridization and chromosome doubling (b) Hybridization
    (c) Chromosome doubling (d) Hybridoma
26. Monosomy in sex chromosome in human results in
    (a) Klinefelter's syndrome (b) Turner's syndrome
    (c) Edward's syndrome (d) Patau's syndrome
27. In Klinefelters syndrome is extra copy of .... is found :
    (a) Autosome (b) Sex chromosome
    (c) Nucleotide (d) DNA
28. If aneuploid number is less than 2n it is called :
    (a) Hyperploid (b) Hypoploid
    (c) Polyploid (d) Euploid
29. In trisomic the chromosome number will be :
    (a) $2n + 1 + 1$ (b) $2n - 1 - 1$
    (c) $2n + 1$ (d) $2n - 1$
30. Non disjunction of a chromosome results in :
    (a) Polyploidy (b) Aneuploidy
    (c) Transversion (d) Inversion.

ANSWERS

1. (b)	2. (a)	3. (b)	4. (c)	5. (a)	6. (c)
7. (a)	8. (d)	9. (d)	10. (b)	11. (a)	12. (c)
13. (d)	14. (a)	15. (c)	16. (c)	17. (b)	18. (d)
19. (b)	20. (b)	21. (c)	22. (a)	23. (b)	24. (d)
25. (a)	26. (b)	27. (b)	28. (b)	29. (c)	30. (b)

45
CHAPTER

Chromosome Organization

Chromosomes are filamentous bodies present in nucleus. They are the carriers of genes (unit of heredity). These are the structure of special organisation and function; playing a vital role in heredity, variation, mutation and evolution. Chromosomes have an ability of self-replication or reproduction.

The chromatin condenses at the time of cell division and becomes clearly visible in the form of **chromosome.** Chromosomes are not visible in the active nucleus due to their high water content. The chromosomes in a cell are of two types (a) **autosomes** and (b) **sex chromosomes** or **heterosomes.**

HISTORY

Chromosomes were first described by **W. Fleming** (1879). He also coined the term **chromatin** to the material that takes stain in nucleus. **W. Roux** (1883) reported the involvement of chromosomes in inheritance. **Bovery** (1887) suggested that the number of chromosomes remain constant in each species. **Waldeyer** in **(1888)** for the first time named **chromosome** for dark stained bodies of nucleus. **Sutton** and **Boveri** 1902 explained chromosomes as **physical basis** of **heredity. Morgan** (1933) discovered that chromosomes are involved in transmission of hereditary traits. The structure of chromosome was described by **Heitz** (1935), **Kuwanda** (1939), and **Geitter** (1940), etc.

NUMBER OF CHROMOSOME

The number of chromosome of a species is constant. Closely related species may also have same number of chromosome. The chromosome number plays an important role in determination of phylogeny and taxonomy of species.

The lowest haploid number recorded in eukaryotes is **'2'** *(Mesotoma,* flat worm and *Ophryotrocha puerilis,* polychaete). In *Ascaris megalocephala,* however, the chromosome number is **'1'** but it a compound chromosome which divides into about 190 chromosomes in somatic cells. In other animals, the number ranges upto 50 (Table 45.1). The highest chromosome number in animals is upto 127 found in *Eupagurus schotensis* (hermit crab). In plants, the chromosome number ranges from '2' (*Mucor*) to approximately 630 (*Ophioglossum*) (Table 45.1).

Ploidy

Presence of whole set of chromosome is called **euploidy,** *e.g.*, haploid, diploid, triploid, tetraploid, etc. One set of chromosome is called **haploid,** *e.g.*, gamete (n). Somatic cells usually contain 2 sets of chromosomes, *i.e.*, **diploid** (2n). If 3 sets are present it is called **triploid** and so on. Many sets of chromosome in a cell it is called **polyploid.** A change in chromosome number in a set is called **aneuploidy.** If less then it is known as **hypoploidy,** *e.g.*, 2n – 1 is **monosomic,** 2n – n – 1 is nullisomic and if more it is called **hyperploidy,** *e.g.*, 2n + 1 is **trisomic,** 2n + 1 + 1 is **tetrasomic,** etc. Haploid set of chromosome is also called **genome.**

Table 45.1 : Haploid chromosome number of some animals and plants.

S. No.	Name	Scientific name	Chromosome number (haploid)
1.	Flat worm	*Mesotoma*	2
2.	Polychaete	*Ophryotrocha puerilis*	2
3.	Mosquito	*Culex piplesis*	3
4.	Fruit fly	*Drosophila melanogaster*	4
5.	Housefly	*Musca domestica*	6
6.	Frog	*Rana tigrina*	13
7.	Rat	*Rattus rattus*	21
8.	Rabbit	*Oryctogalus cuniculus*	22
9.	Man	*Homosapiens*	23
10.	Gorilla	*Gorilla gorilla*	24
11.	Pigeon	*Columba livia*	40
12.	Hermit crab	*Eupagurus schotensis*	127
13.	Bread mold	*Mucor*	2
14.	Haplopappus	*Haplopappus gracilis*	2
15.	Penicillium	*Penicillium*	2
16.	Neurospora	*Neurospora crassa*	7
17.	Pea	*Pisum sativum*	7
18.	Barley	*Hordeum vulgare*	7
19.	Onion	*Allium cepa*	8
20.	Cabbage	*Brassica oleracea*	9
21.	Mays	*Zea mays*	10
22.	Rice	*Oryza sativa*	12
23.	Yellow Pine	*Pinus* sp	12
24.	Yeast	*Saccharomyces*	17
25.	Wheat	*Triticum aestivium*	21
26.	Oat	*Avena sativa*	21
27.	Coffee	*Coffea arabica*	22
28.	Potato	*Solanum tuberosum*	24
29.	Sugar cane	*Saccharum officinarum*	40
30.	Ophioglossum	*Ophioglossum*	631

SIZE OF CHROMOSOME

The size of chromosome varies from species to species but remains constant in one species. The length varies from 1 μ to 30 μ. The diameter may vary from 0.2 – 20μ. If all the chromosomes of a species are of similar size they are termed **symmetrical karyotype** and if they are of different size they are called as **asymmetrical karyotype.** In man and *Yucca*, etc., there are distinct groups of different sizes. The chromosomes of *Trillium* are hundred times more in length as compared to the related genera *Medeola.* The chromosomes of *Allium porrum* are half the size of chromosomes in *Allium sativum*. Size variations may also take place in different regions or in different stages of the same organism. In *Medeola*, the root tip chromosomes are 50% longer than shoot tip chromosome. The chromosomes of a cell may be very differnt in size, some may be exceptionally large and some are small (human).

SHAPE OF CHROMOSOME

The shape of the chromosome depends on the position of centromere on the chromosome. It may be 'I' shape, 'J' shape, 'V' shape 'L' shape, etc.

STRUCTURE OF CHROMOSOME

Earlier the light microscopic structure of chromosome revealed that it consists of coiled thread called *charomonema* lying in a matrix and covered by membranous **pellicle.** But the electron microscopic studies revealed that there is no definite membranous pellicle present on the chromosome. Chromosome includes centromere, secondary constriction, nucleolar organiser, telomeres and satellite, etc. (Fig. 45.1)

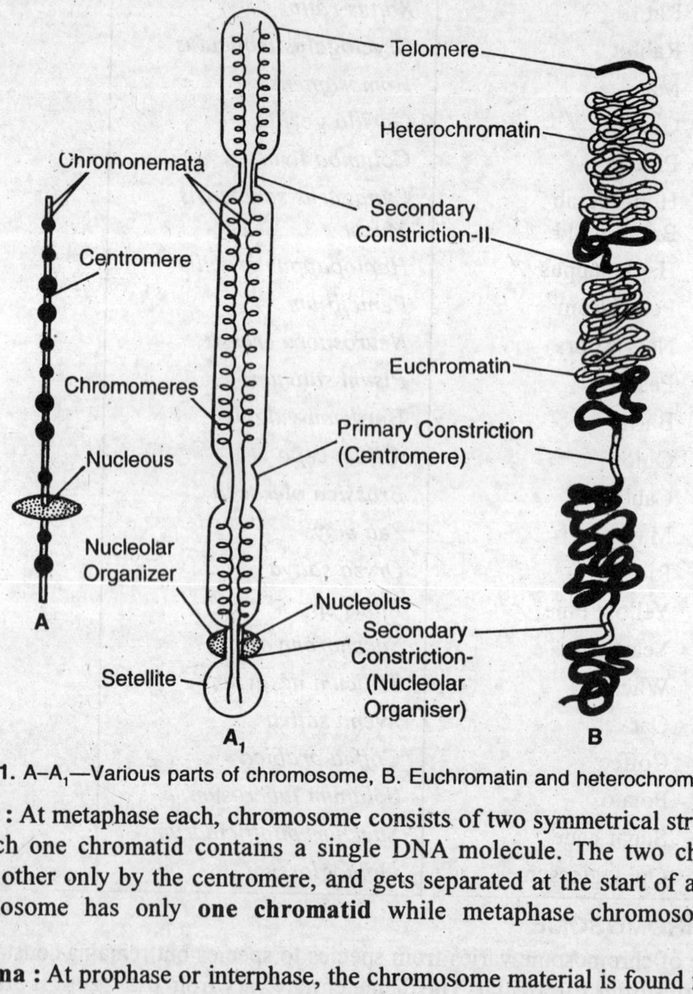

Fig. 45.1. A–A_1—Various parts of chromosome, B. Euchromatin and heterochromatin.

Chromatid : At metaphase each, chromosome consists of two symmetrical structures called **chromatids.** Each one chromatid contains a single DNA molecule. The two chromatids are attached to each other only by the centromere, and gets separated at the start of anaphase. The anaphase chromosome has only **one chromatid** while metaphase chromosome has **two chromatids.**

Chromonema : At prophase or interphase, the chromosome material is found in the form of thread called chromonemata, representing the chromatids in early stages of condensation. Therefore chromanemata is the chromatid as later stage, *i.e.*, a single linear DNA molecule with associated proteins.

(B) Chromomere : The beaded structures present on the chromatid are called **chromomeres.** Chromomeres are visible on the interphase chromosomes also. On polytene chromosomes they are aligned side by side, constituting the chromosome bands. These are the tightly folded regions of DNA. They correspond to the units of genetic function of chromosome. At metaphase, the chromosome is tightly coiled and the chromomeres are not visible.

(C) Centromere or kinetochore : The centromere lies within a linear segment of the chromosome, the **primary constriction.** It is the region of the chromosome that becomes attached to the mitotic spindle. The region adjacent to the centromere frequently contains highly repetitive DNA; and may stain more intensely with basic dyes (heterochromatin). It contains specific DNA sequences with special proteins, forming a disc shape structure. This disc is called **kinetochore.** Electron microscopic picture reveals that the kinetochore disc is **trilaminar structure** with a dense outer proteinaceous, middle low density and inner dense layer tightly bound to centromeric DNA 4–40 microtubules are attached to the kinetochore providing force for chromosomal movement during mitosis.

Depending upon the kinetochores the chromosomes are:

(a) Monocentric chromosomes : Chromosomes have only one kinetochore.

(b) Holocentric chromosomes : The kinetochores are diffuse with microtubules attached along the length of chromosome. In some animals like hemipteran and homopteran insects, *Ascaris*, etc., in some plants also such chromosomes are found. These are also called polycentric chromosomes of diffuse centromere by some cytologits.

(c) Acentric chromosomes : The chromosomes are without any kinetochore. Such type of abnormality is produced by X rays.

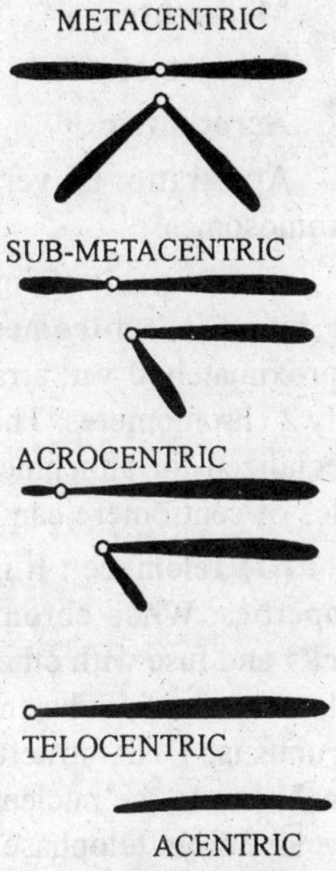

Fig. 45.2. Chromosome : various types of chromosomes based on centromere position.

(d) Dicentric chromosomes : The chromosome may have two kinetochores. This condition is also an abnormality where a part of other chromosome with centromere gets attached to the chromosome. The condition may not be stable.

According to the position of centromere (kinetochore), the chromosomes are of many types, *e.g.*, (Fig. 45.2).

(a) Metacentric : The centromere is present nearly at the middle of the chromosome (median). The two arms of the chromosome are approxiamtely equal. The chromosome appears 'V' shaped at anaphase. The chromosome is called **metacentric chromosome.**

(b) Submetacentric : The centromere is present away from the middle point on the chromosome. One arm of the chromosome is long and the other is small. The chromosome appears 'J' or `L' shaped. The chromosome is called **submetacentric chromosome.**

(c) Acrocentric : The centromere is present near the end or **subterminus.** It is rod shaped, the one arm is very long and the other arm is very small.

(d) Telocentric : The centromere is **terminal,** present at the tip of the chromosome. The chromosome lies only one arm. Such chromosomes are very rare. The IV chromsome of *Drosophila* was thought to be **telocentric** but now a very small arm has been identified so it is called **acrocentric.** Truely telocentric chromosomes are found in plants, mammals, etc.

In **grasshopper**, all chromosomes are acrocentric and in *Trillium* and *Tradescantia* all the chromosomes are metacentric. In human, the chromosomes are metacentric, submetacentric and acrocentric type.

Arm ratio : The length of the two arms of the chromosome are found in different ratio. It is called **arm ratio.** The arm ratio is characteristic for each single chromosome. It is important and useful in distinguishing chromosomes with similar total length. In human beings **Vogel** in 1961 suggested the **arm ratio** as

Metacentric — from 1·07 - 1.95
Submetacentric — from 2·05 - 3.75
Acrocentric — from 5·00- 11·94.

Arm ratio is very high in acrocentric chromosomes and very low in metacentric chromosomes.

Centromeric chromomeres : In metaphasic chromosomes with two chromatids, 4 granules are found in **centromere** which are called **centromeric chromomeres.** The size of them is approximately 0·5 μ; arrange in square. In anaphase due to separation of each chromatid, it shows only 2 chromomeres. The centromere is also considered to be an extended region of chromonema; specialized for attachment and formation of spindle fibres. Spindle fibres extending from either sides of centromere can be seen during cell division.

(D) Telomere : It is the tip of the chromosome which is called **telomere.** It has very special properties. When chromosomes are broken by **X-rays** the free ends without telomere become **sticky** and fuse with other broken chromosome. They however do not fuse with a normal telomere. Telomeres contain the ends of the long linear DNA molecule contained in each chromatid, so have an unusual DNA structure. Telomeres are specially modified regions of the chromosome for attachment to the nuclear envelope. The ends of the chromosome are associated with the nuclear envelope from telophase to prophase through interphase. Two telomeres never unite. They do not unite with the broken end also.

(E) Secondary constriction : The chromosome has another thin area called **secondary constriction.** The secondary constriction II is different from secondary constriction I which is also called as nucleolar organizer. The location of secondary constriction is constant for a particular chromosome, and is useful for identification of chromosomes. It represents the site of breakage and subsequent fusion. In human beings, the secondary constriction is found on the long arms of chromosome number 1, 10, 13, 16 and Y'chromosome. Secondary constrictions are distinguished from the primary constriction by the absence of marked angular deviations of the chromosomal segment during anaphase.

(F) Nucleolar organizer : This area on the chromosome is also called **secondary constriction I.** It contains the genes coding for **18 S** and **28 S** ribosomal RNA and genes for inducing formation of **nucleolus.** The constriction may arise because the *r*RNA genes are transcribed very actively, interfering with chromosomal condensation. In man, the nucleolar organisers are located in the secondary constrictions of 13, 14, 15, 21, 22 chromosomes (acrocentric with satellites) and 'Y' chromosomes.

Nucleolar organisers have been reported in several animals and plants, and are probably of universal occurrence. Each nucleolar organiser may be associated with the nucleolus or both nucleolar organiser (diploid) may be associated with nucleolus. In polyploids and some diploids, the number of nucleoli and nucleolar organisers may be more than two.

(G) Satellites : In some chromosomes, **satellite** is found. It is a round body separated from the rest of the chromosome by a secondary constriction. It is consant in shape and size for each particular chromosome.

KARYOTYPE AND IDIOGRAM

The whole group of characteristics allowing to identify a particular chromosomal set, *i.e.*, number, relative size, position of cetromere, length of arms, secondary constructions and satellites, etc., are called karyotype. The **karyotype** is characteristic of an individual, species, genus or group. It may be represented by a diagram in which pairs of homologous chromosomes are ordered in a series of decreasing size is called **idiogram.**

Some species may have characteristic features, *e.g.*, plants frequently have heterochromatic regions at the telomeres. Broad beans *(Vicia faba)* or onion (*Allium cepa)* have large chromosomes producing nice pattern in mitotic cells. In mouse, the chromosomes are acrocentric. In amphibians, they are metacentric, etc.

A set of chromosome of an individual or species is called **karyotype.** In man, 23 pairs of chromosomes are found in a somatic cell, forming its karyotype. The chromosomes are identified on the basis of total length, arm ratio, position of secondary constriction and nucleolar organiser and euchromatic and heterochromatic regions. They are grouped in 7 groups (Fig.45.3).

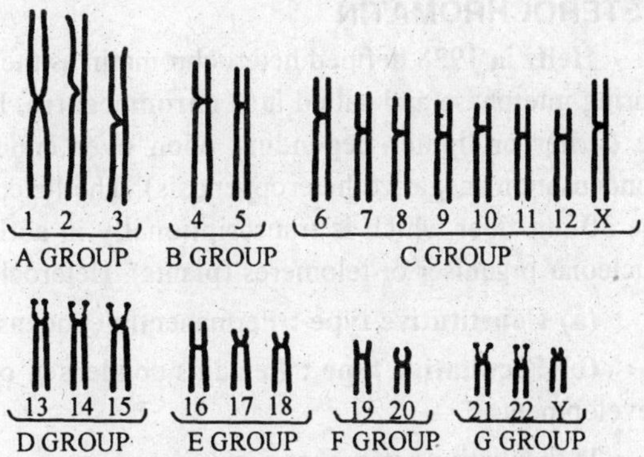

Fig. 45.3. Idiogram of normal human male karyotype.

Group I = chromosome no. 1, 2, 3.
Group II = chromosome no. 4, 5
Group III = chromosome no. 6–12 and 'X' chromosome.
Group IV = chromosome no. 13–15
Group V = chromosome no. 16–18
Group VI = chromosome no. 19, 20
Group VII = chromosome no. 21, 22, and 'Y' chromosome

Normally a single chromatid has single DNA molecule to whole of its length which replicates by semiconservative method. The fact is established by using H-thymidine as labelled DNA. The metaphase, chromosomes are symmetrical. The two sister chromatids are mirror images of each other. The morphological characters of each chromatid are same. The sister chromatids contain identical DNA molecules.

COILING OF CHROMOSOME

The coiled filaments are for the first time called **chromosome** by **Velovsky** in 1912.

The chromosome may have one to four threads or fibrils coiled with each other. The coiling may be of following types (Fig. 45.4).

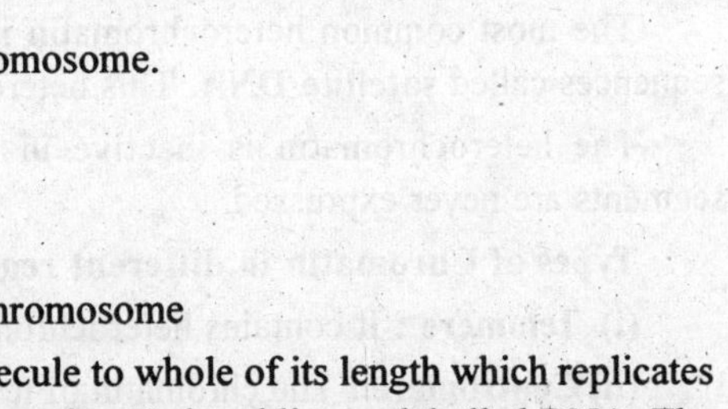

Fig. 45.4. Various type of coiling.

(a) Paranemic coiling : When the fibrils are easily separable from the coils, the coiling is called paranemic.

(b) Plectonemic coiling : When the fibrils are intertwined in such a way that they cannot be separated at all, the coiling is called plectonemic coiling.

The degree of coiling of fibrils may be different depending upon the length of chromosome. It may be major, minor or standard type.

EUCHROMATIN

It is the segment of chromosome that takes light stain and takes part in cell cycle. This part is non-condensed and extend during interphase, also called as euchromosome. Euchromatin never shows heteropycnosis.

HETEROCHROMATIN

Heitz in 1928 defined heterochromatin as the region of chromosome that remains condensed during interphase and called it as **chromocentre.** It takes dark stain upon staining. The stain may be darker or lighter depending upon over condensation (positive heteropycnosis) or under condensation (negative heteropycnosis). The heterochromatin of DNA remains tightly packed in the 30 nm fiber which is transcriptionally in active chromatin. It is generally present near the nucleolar organiser or telomeres (plants). Heterochromatin is of two types:

(a) Constitutive type : Permanently condensed in all types of cells.

(b) Facultative type : Remains condensed only in certain cell types or at special stages of development.

In facultative heterochromatin, one chromosome of the pair becomes either totally or partially heterochromatic, *e.g.*, 'X' chromosome of mammalian female where one is active and remains euchromatic and the other is inactive and forms sex chromatin or **barr body** at interphase.

The most common heterochromatin is constitutive type. It contains highly repeated DNA sequences called **satellite DNA.** This heterochromatin shows late replication.

The heterochromatin is inactive in RNA synthesis. Genes present on heterochromatic segments are never expressed.

Types of Chromatin in different regions of the chromosome

(i) Telomere : It contains heterochromatin of repetitive DNA sequences.

(ii) Centromere : The chromatin of it is continuous on either sides with the heterochromatin (Rae, 1970). The heterochromatin present is of two types (Heitz, 1934), *i.e.*, α-present around the centromere (compact and resistant to destaining); and β-present in rest of the centromeric region (diffused and less resistant to destaining).

(iii) Nucleolar organiser : Euchromatin is predominantly present. The genes for 18 S and 28 S *r*-RNA are found. RNA synthesis takes place throughout the cell cycle except at mid-mitosis.

(iv) Secondary constriction : It consists of euchromatin taking part in *m*-RNA synthesis throughout the cell cycle. According to some cytologists, it is also the place of 5S *r*-RNA synthesis.

(v) Satellite : The satellite is the mass of heterochromatin.

(vi) Euchromatin : It is found in bulk of the chromosome. It is partly composed of non-repeated DNA sequences taking part in *m*-RNA synthesis during interphase.

(vii) Intercalary heterochromatin : It is found within the euchromatin here and there.

MOLECULAR COMPONENTS OF CHROMOSOME

The components of eukaryotic chromosome are DNA, RNA, histone, and non-histone protein, metallic ions, etc. The DNA is the most stable molecule carrying out transfer of genetic information from one generation to the other. RNA is transcribed over it. Most of *m*-RNA is transported to the cytoplasm where it is translated in terms of proteins. Some *m*RNA remains in the coil associated with the protein.

Proteins : Two types of proteins, *i.e.*, acidic and basic proteins are found associated with DNA in chromosome.

Protamines : These are simple proteins of molecular weight less than 4000 daltons. Rich in basic amino acid **arginine,** found in spermatozoa of some fish (salmon), snail, fowl, etc. They are helically rapped around DNA. Polypeptides of protamine consists of 28 amino acid residue (19 arginines, 8-9 non-basic amino acids).

Histones : Small basic proteins associated with DNA of eukaryotic cells. They are structural proteins of chromatin and act as **gene repressors** also. In a wide variety of plants and animals, the ratio of DNA : Histone = 1 : 1 . There are 5 types of histones *i.e.*, Hl (H5), H2A, H2B, H3 and H4. Histones are rich in **arginine** and **lysine** (basic amino acids). They lack **tryptophan.** Histones are highly modified as well as conserved proteins; with very little difference in amino acid sequences. On the basis of arginine and lysine contents, they are divided into 3 groups.

(i) High lysine rich — Hl

(ii) Lysine rich — H_2A, H_2B

(iii) Arginine rich - H_3, H_4

Histones tend to **depress genetic acitivity.** They make DNA. They are structurally important in packaging DNA molecules.

Non-Histone proteins (NHP) : They have structural, enzymatic and regulatory function in chromatin. They are generally acidic proteins. The acidic to basic residues ratio in these proteins is 1.2 — 1.6.

The molecular weight ranges from 11000-21,500 daltons. They may show structural variation in different species and even tissues. These proteins are synthesised throughout the cell cycle. These proteins stimulate genetic activity.

Enzymes of chromosomal metabolism like nucleic acid polymerase, nucleases, DNA pyrophosphorylases and nucleoside triphosphatases, etc., are non-histone proteins. Such proteins play important role in the interaction of **steroid hormones** with target cell nuclei.

The chromosomes also contain metallic ions like Mg^{++} and Ca^{++}, etc.

ULTRASTRUCTURE OF CHROMOSOME

Electron microscopic studies revealed that chromosomes have very fine fibrils with a thickness of 2nm—4 nm. Since DNA is 2 nm in width a single fibril may correspond to a single DNA molecule. **Dupraw** (1965, 66) suggested that interphase chromatin fibres consist of fibrils of 230 Å and metaphase chromosomal fibrils have 300 Å diam. There are different views about chromosomal strands.

(A) Multistranded view : According to this view, the smallest unit of chromosome is 100 Å thick fibril. Each fibril is made up of 2 DNA molecules at 25 Å distances. A chromosome has 2 chromatids, each chromatid consists of 4 fibrils. According to this view a chromosome has a thickness of 1600 Å suggesting that 16 chromonemal fibrils and 32 DNA molecules. This does not hold the electron microscope observation, so was not supported by cytologists.

(B) Unistrand view : The model was proposed by **Dupraw** (1965) as **folded fibre model** of chromatid. According to this view, the DNA molecule is wrapped in a nucleoprotein and this DNA-protein fibre is coiled tightly to form 250–300 Å fibre (Fig. 45.5). The DNA-protein molecule of both the chromatids remain held at the centromere by a unreplicated fibre segment of DNA until anaphase.

This model was accepted by many workers. But this was discovered after the discovery of **nucleosome.**

STRUCTURE OF CHROMOSOME AT MOLECULAR LEVEL

(A) Nucleosome subunit of chromatin : In 1974, **R. D. Kornberg** and **J. O. Thomas** proposed an active model of DNA and histones. They suggested that DNA interacts with a **tetramer** ($H3_2$ – $H4_2$) and two molecules of an oligomer (H2A-H2B). Tetramer involves 2 molecules each of H3 and H4 histogen is associated with 200 base pairs of DNA. Each repeat unit has one molecule of Hl histone. **P Oudet** *et al* 1975 proposed the term **nucleosome** (Fig. 45.6). Nucleosome contains a **nucleosome core particle** consisting of 146 base pairs of supercoiled

DNA wrapped almost twist around a disc-shaped complex of 8 histone molecules consisting of 2 copies each of histone H2A, H2B, H3B and H4 assembled into an octamer. These nucleosomes are connected by **linker** DNA of about 60 base pairs. Together a nucleosome (12.5 nm in diam) = 200 base pairs + 2 mole each of H2A, H2B, H3 and H4.

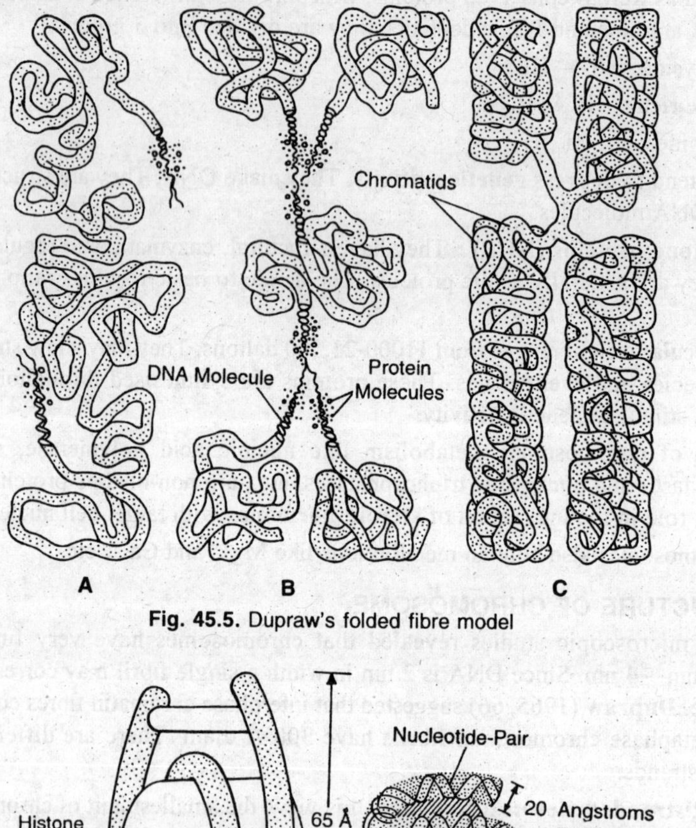

Fig. 45.5. Dupraw's folded fibre model

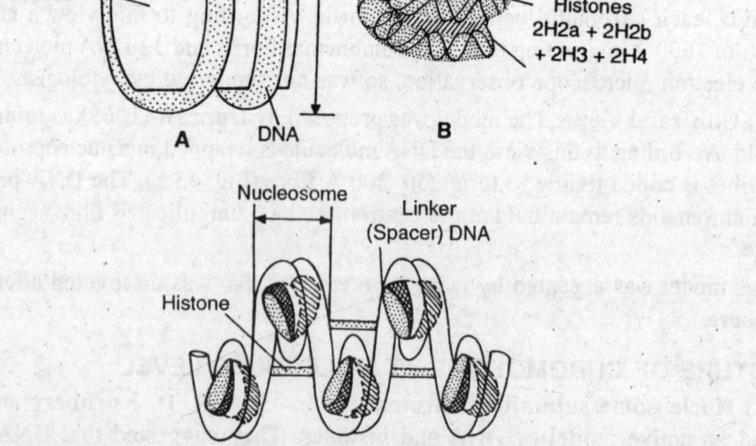

Fig. 45.6. A. Single nucleosome, B. Details of nucleosome, C. Nucleosome with linker DNA.

The model was proposed on the basis of X-ray diffraction, electron microscopy, nuclease digestion and chemical cross-linking (Fig. 45.7).

H1 histone is located just outside the nucleosome core particle and is associated with both ends of the DNA as it enters and exits from the core particle. The nucleosome core particle is a flat structure having dimensions of about 57 × 110 × 110 Å.

The unit fibre according to **Zeuthen** (1978) shows three levels of coiling.

(i) First level of coiling of DNA is in the string of nucleosome.

(ii) The string of nucleosome coils with 300 Å diameter **solenoid** (6–8 nucleosomes).

(iii)Solenoid further coils into **supersolenoid** structure **(Finch and Clug, 1976)** with a diameter of 4000 Å and a wall 300 Å thick. The supersolenoid structure is a **unit fibre.**

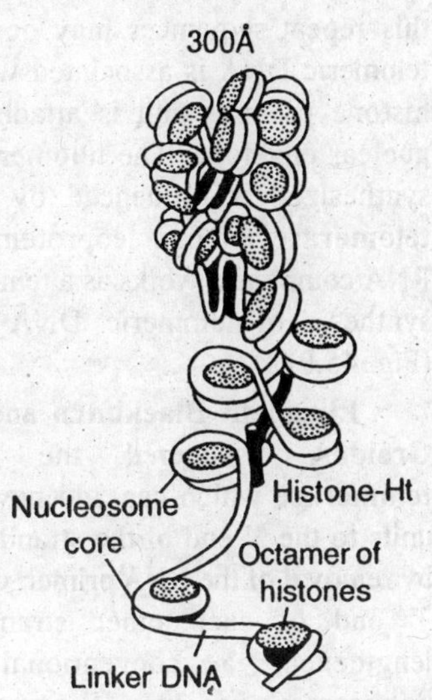

Fig. 45.7. Solenoid structure

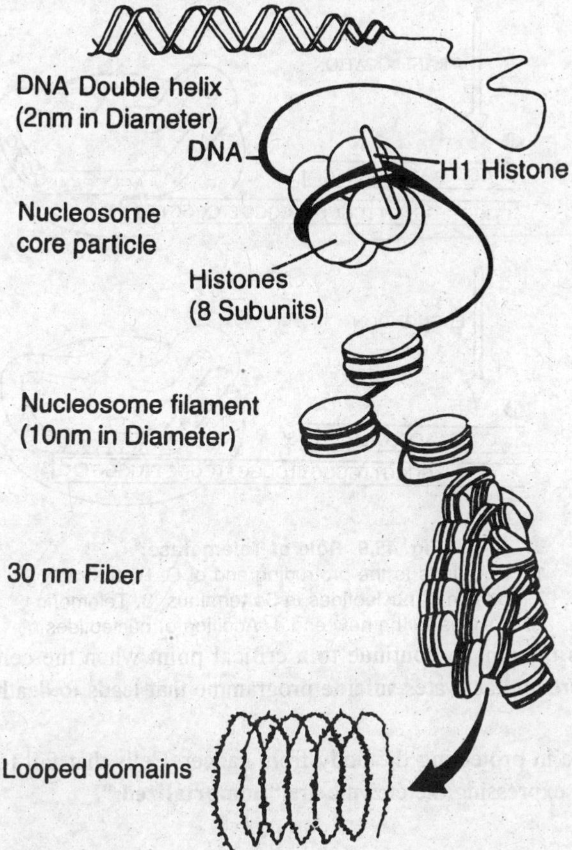

Fig. 45.8. Various levels of organization of chromosome.

The coiling ratio at each level is 7 (nucleosome), 6 (solenoid) and 30–40 (supersolenoid). Overall contraction of DNA within the unit fibre would be 1,300 to 1,500 folds (Fig. 45.8). Unit fibre corresponding **chromonema** which further coils in chromosome.

(B) Telomeres : Telomeres are most conservative elements both in structure and function. Telomeric DNA consists of randomly, repeated sequence of clusters of 'C' on one strand and 'G', on the other. There are 12 + 6 nucleotides in length of G rich strand, *e.g.*, the telomeric repeat sequence in *Tetrahymena* is **GGGGTT,** in *Homosapiens, Neurospora,* slime molds, *Trypanosomas,* etc., it is **AGGGTT,** In *Plasmodium,* it is **AGGG (T/C) TT.** In *Chlamydomonas* it is **AGGGTTTT,** etc.

The same repeated sequence is found at the end of all chromosomes in a species. The same telomeric sequence may occur in widely divergent species. At every telomere at least 10 kilobases of this repeat sequences may occur. The telomeric DNA is associated with non-histone proteins and is attached with nuclear envelope. The telomeric DNA synthesized is influenced by enzyme **telomerase** (ribonucleoprotein whose RNA component works as a template for synthesis of telomeric DNA repeats (Fig. 45.9).

Elizabeth Blackburn and **Carol Greider** discovered the enzyme **telomerase** which can add new repeat units to the 3′ end of the strand created by removal of the RNA primer. Once the 3′ end of each other strand been lengthened, a conventional RNA polymerase can use the newly 3′ segment as a template to return the 5′ ends of the complementary strands to their previous length. Telomerase is a reverse transcriptase that synthesizes DNA using an RNA temptale. The enzyme itself contains the RNA that serves as its template.

Telomere is very important part of a chromosome. It is required for the complete replication of the chromosome. It forms caps that protect the chromosome from nuclease. It prevents the ends of chromosomes from fusing with one another. It facilitates interaction between the ends of the chromosome and the nuclear envelope in some types of cells.

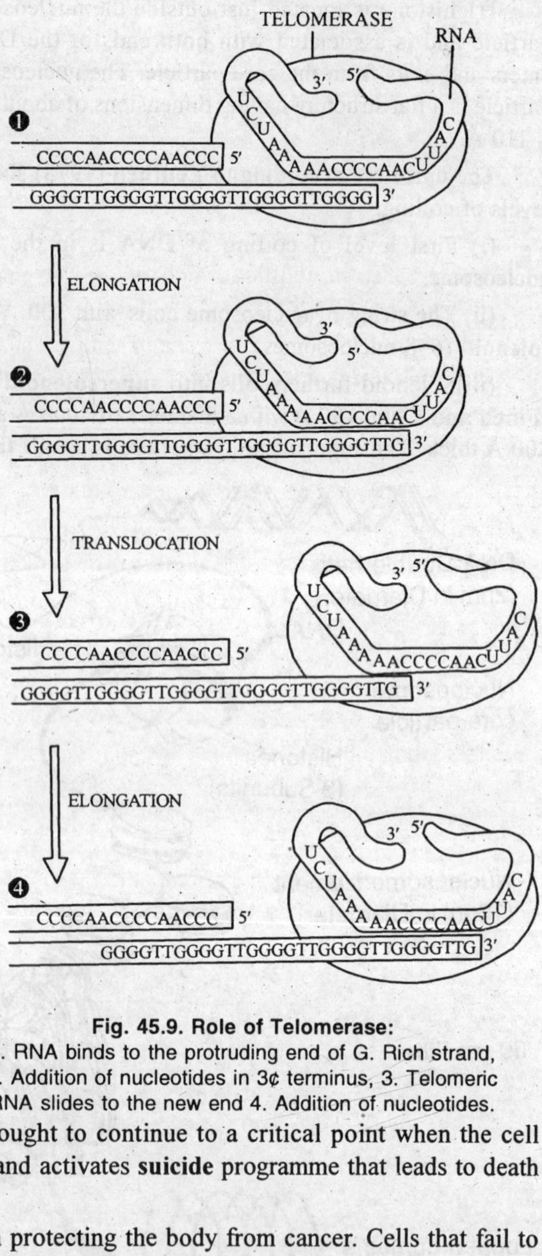

Fig. 45.9. Role of Telomerase:
1. RNA binds to the protruding end of G. Rich strand, 2. Addition of nucleotides in 3¢ terminus, 3. Telomeric RNA slides to the new end 4. Addition of nucleotides.

Telomere shrinking or shortening is thought to continue to a critical point when the cell recognizes the impending loss of a telomere and activates **suicide** programme that leads to death of a cell.

Telomere shortening plays a key role in protecting the body from cancer. Cells that fail to express telomerase die out, where as cells expressing the enzyme are **"immortalized."**

FUNCTIONS OF CHROMOSOME

(i) It carries genetic information from one generation to the next.

(ii) DNA of the chromosome is the genetic material.

(iii) Chromosome replication is the replication of DNA.

(iv) The number of chromosome and structure are important for identification of characters of an organism as they can cause evolution, mutation, variations, etc.

(v) Nucleolar organiser takes part in the formation of nucleolus.

SPECIAL TYPES OF CHROMOSOMES

In certain organisms special types of chromosomes are found, *e.g.*, **lampbrush chromosomes** of vertebrate and **giant chromosmes** of salivary glands of certain insects.

Lampbrush Chromosome

Such type of chromosomes are found in primary oocyte nuclei in amphibians and in some vertebrates. These are found during the prolonged diplotene stage of first meiotic division in primary oocytes of amphibians, and in spermatocyte nuclei of *Drosophila*. The chromosome is very lengthy. Upto 1 mm long chromosomes are observed in urodeles, amphibians.

Lampbrush chromosomes are so named because of their appearance like glass chimneys of kerosene lamps. The first description of lampbrush chromosomes was given by **Flemming** in 1882. The details were provided by **J. Ruckert** in 1982 in oocytes of Shark. Lampbrush chromosomes are also observed in *Sagitta, Sepia, Echinaster,* Shark, reptiles and birds beside amphibians and insects.

Lampbrush chromosomes of vcrtebrates are larger than invertebrates. **Grun (1958)** reported them in plant cells.

The size of lampbrush chromosomes may be more than 1,000 μ in length and 20 μ in width. At dikinesis, the chromosomes contract and reduce in size. The lampbrush chromosome of salamander oocytes is of 5900 μ in length.

Structure of Lampbrush Chromosome

During early prophase, they are in the form of homologous chromosomes with some **chiasmata.** It has an **axis** along which a row of dense granules or **chromomeres** are found. From each chromomere a pair of lateral loop arises (Fig.45.10).

Chromosomal axis : The two chromatids of the chromosome are made up of **axial filaments.** The pair of homologous chromosomes has **four** filametns in all. The axial filaments and chromomeres consist of DNA. The loops are extensions of axial filaments. The ends of axial filaments without loops is the **telomere.** Centromere is also present which is loop free.

Chromomere : The axial filaments at some points are tightly coiled. These points are the **chromomeres.** According to **Ris,** (1957) they are found in pairs, *i.e.*, one chroromomere on each filament; corresponding to heterochromatin.

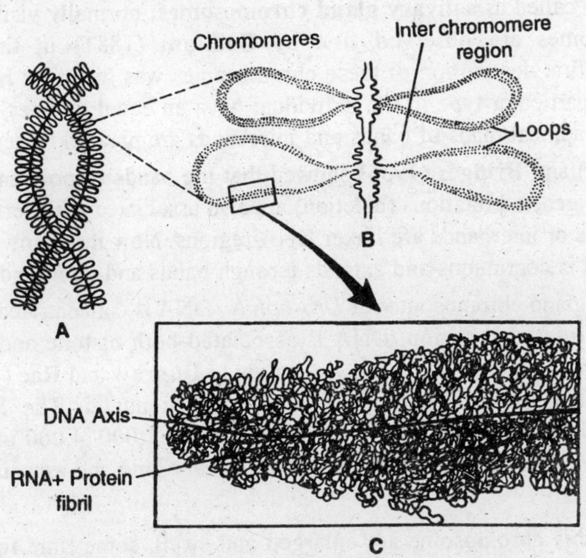

Fig. 45.10. Structure of Lampbrush chromosome.

Loops : Loops are **typical** and **special** type. **Typical loops** consist of a central axis from where RNA fibrils of progressively increasing length come out, making the loop thick. **Special loops** show asymmetry and have granules at the fibrils. Length of each loop may vary between 5-100 μm.

Each loop consists of an **axial fibre** covered with matrix (Fig. 45.11). The loop is made up of DNA and the matrix is made up of RNA. Because when the loop is treated with **deoxyribonuclease** enzyme the loop breaks down and when treated with **ribonuclease** the matrix breaksdown. The loop is believed to represent one long operon consisting of a series of identical copies of structural genes or cistron. Each gene codes for 45 S *r*-RNA and produces a very long RNA molecule interacting with protein to form ribonuceloprotein. If the activity of genes is stopped by **actinomycin** D, the loops will collapse. The matrix free regions between genes loci is the region of 'spacer' DNA which do not produce RNA. At one end of spacer DNA region, dense granules of RNA-polymerase are attached to the **promoter region** the other end is the termination region. Many of the loops correspond to a single transcriptional unit and show a continuous thin thick matrix. Other loops contain several units of transcription. Some transcriptional units in lampbrush chromosomes are extremely long even upto 100 μm in length (1 μm of DNA = 3000 base pairs). The long transcriptional units of lampbrush chromosomes may be due to a failure of transcription termination.

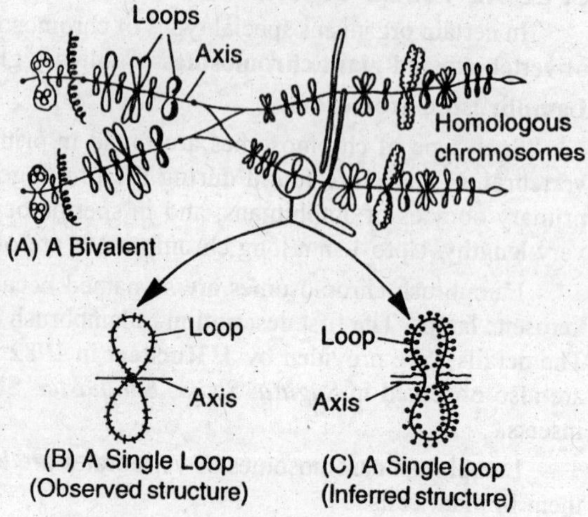

Fig. 45.11. Detailed structure of lampbrush chromosome.

Some *m*RNA produced by lampbrush chromosome is also stored as **informosome,** *i.e., m*-RNA coated by protein for producing biochemicals during the early development of embryo.

Polytene chromosome

These are also called as **salivary gland chromosomes;** normally visible during interphase. Polytene chromosomes are observed first by **Balbiani** (1881) in the salivary gland of *Chironomous*. The first description of these chromosomes was given by **Kollar** (1882). All the chromosomes of a particular type in the individual have an equal number of bands with similar distribution. Topographical maps of bands and interbands are parallel to genetic maps.

Painter (1933) and **Bridges** (1936) showed that the bands of polytene chromosomes have close relationship to genes. Mutations (deletion) are also associated with certain bands. The bands are the sites of genes or interbands are linker DNA regions. Now it is known that interbands also contain genes. DNA is continuous and extends through bands and interbands (Fig. 45.12).

In the salivary gland chromosomes of *Drosophila*, DNA is concentrated in the bands while it is very little in the interband region. DNA is associated both histone and non-histone protein forming nucleoprotein fibres of 100-500 Å. According to **Dupraw** and **Rae** (1966) these fibres are comparable to nucleoprotein fibres found in interphasic nuclei. The fibres correspond to chromatids. The polytene chromosome is formed by about 1,000–4,000 unit chromatids. Each chromatid consists of a single DNA molecule which is very long and associated with protein.

Chromosome puffs

The bands of this chromosome are enlarged and swell, some time forming **chromosome puffs** or **Balbiani** rings. According to **Beerman** and **Bahr** (1954) the puff represent the regions of tightly coiled chromosomal fibres to form many loops. Puffing is due to unfolding or uncoiling of individual chromomere in a band. Puffs are active genes and sites of RNA synthesis. This can be proved as :

(a) The puffs contain large amount of RNA while puffless area contain DNA and histones. This can be seen by staining the area with toludine blue; RNA takes violet strain and DNA blue.

(b) Radioactive **uridine** when injected, the chromosome shows black spots indicating RNA.

(c) Actinomycin D inhibits RNA formation.

(d) Radioactive leucine is injected to see that protein is labelled but the protein in puff is not labelled. It shows that the proteins in the puff are not made in chromosome. **Chloramphenicol** and **puromycin** (protein synthesis inhibitor) has no effect on puffs.

***m*-RNA synthesis**

Puffs actively synthesise ***m*-RNA** but no protein synthesises the *m*-RNA of one puff varies from the *m*-RNA of other. The *m*RNA formed in puffs moves out in the cytoplasm where it synthesises special type of protein.

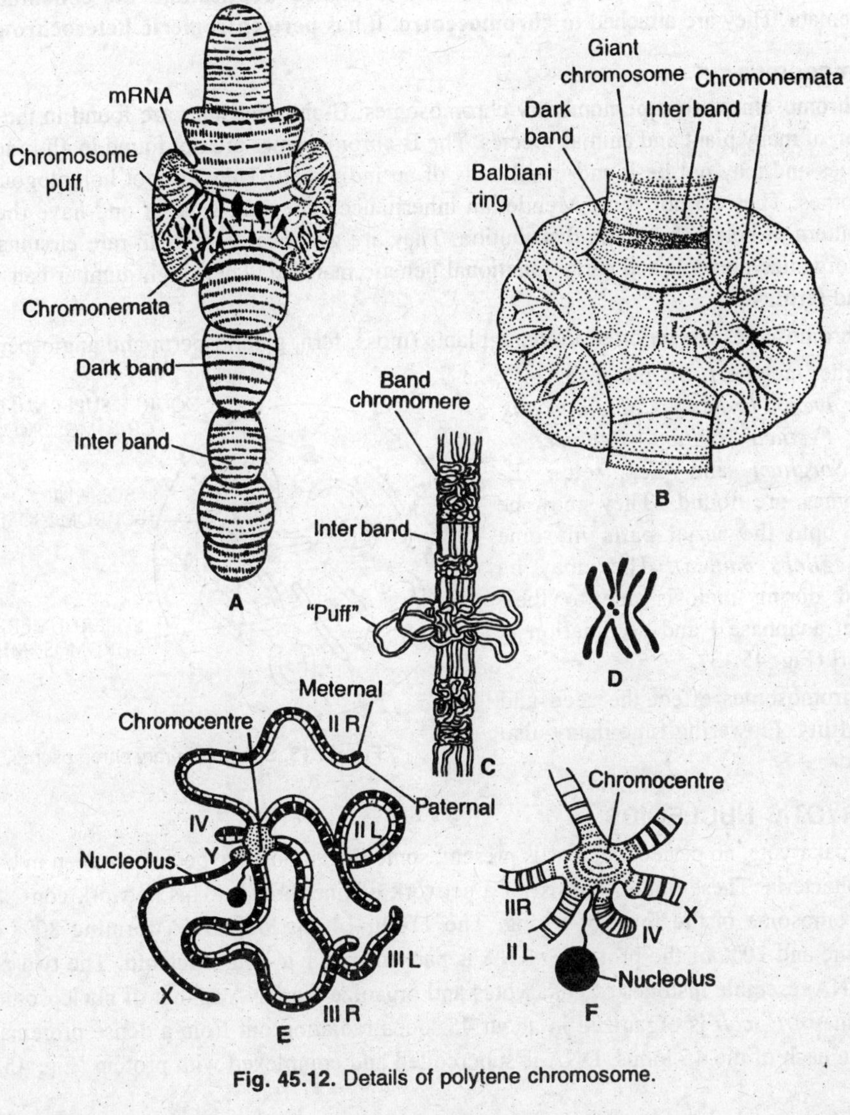

Fig. 45.12. Details of polytene chromosome.

The giant chromosomes represent a bundle of fibrils which arise by repeated cycles of endoreduplication of single chromatids. The chromatin replicates without cell division resulting into a number of chromonemata. Due to many chromosomes, it is also called as **polytene chromosome** and the process as **polyteny.** The number of chromonemata/chromosome may reach upto 2000 to 16000.

In *Drosophila melanogaster,* the giant chromosome is in the form of 5 long and 1 short strand radiating from a single chromocentre (Fig. 45.12 E, F). Qne long strand represents the 'X' chromosome and the other four long strands are the arms of II and III chromosomes. The centromeres of these chromosomes fuse to form **chromocentre.** In male flies, Y chromosome also fuses with chromocentre.

Polytene chromosomes are found in malpighian tubules, fat bodies, ovarian nurse cells, gut epithelium, etc., apart of salivary gland.

Polytene chromosomes are reported in endosperm cells of embryosac by **Malik** and **Singh** (1979). Length of this chromosome may be upto 2000 μm. The chromosome is formed by somatic pairs between homologous chromosomes and repeated replication or endomitosis of chromonemata. They are attached to **chromocentre.** It has **pericentromeric heterochromatin.**

B-Chromosomes

B-chromosomes are supernumerary chromosomes. B-chromosomes are found in the natural population of many plant and animal species. The B-chromosomes are not found in all individuals of a species and may not be found in all cells of an individual. They are not homologous to A-chromosomes. They do not show Mendelian inheritance. They are smaller and have their own unique pattern of heterochromatin distribution. They are genetically inert. In rare circumstances, they can organize nucleoli and carry functional genetic material. Their high number can supress vigour and fertility.

B-chromosomes are reported in many plants (moss, fern, gymnosperm and angiosperm).

In the members of poaceae, *e.g.*, *Agrostis, Avena, Bromus, Calamogrostis, Dactylis, Festuca, Lolium, Pennisetum, Secale, Sorghum* and *Zea*, etc., B-chromosomes are found. They may be restricted upto the aerial parts in some plants *(Aegilops mutica).* They may be eliminated during meiosis due to their division at anaphase I and elimination at anaphase II (Fig. 45.13).

B chromosomes effect the seed and pollen fertility. Flowering time many also be delayed.

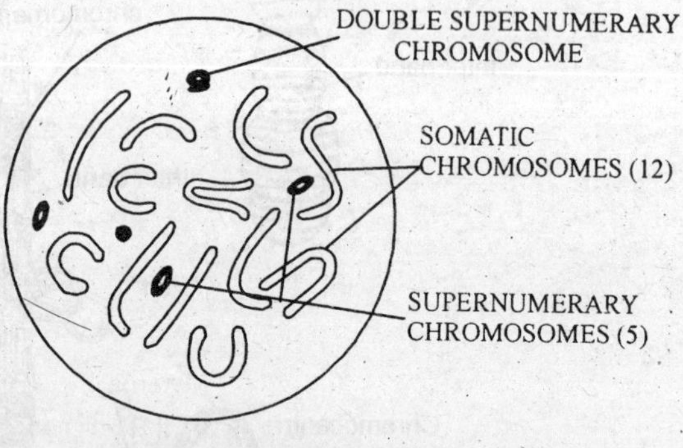

Fig. 45.13. Supernumerary chromosome.

PROKARYOTIC NUCLEOIDS

In proakaryote, no perfect nucleus is present, some fulgen positive bodies are seen in bacteria and cyanobacteria. These bodies consist of a network of fine threads. This network consists of a single chromosome in the form of a ring. The 1100μ-1400μ long DNA forming 80% of the chromosome and 20% of the protein + RNA is packed in a 1 μ long **nucleoid.** The two protein binding DNA resemble histones of eukaryotes and organize the DNA as that of nucleosome. The chromosome of *E. coli* is organized in about 45 loops radiating out from a dense proteinaceous scaffold. In each of the 45 loops, DNA is supercoiled and complexed with protein (Fig. 45.14).

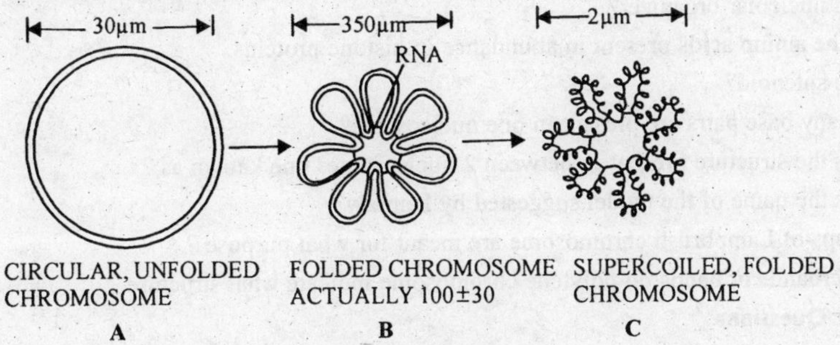

Fig. 45.14. Structure of nucleoid.

SUMMARY

1. Chromosomes were first observed by **Strasburger** (1875) who gave the law of constant number of chromosomes.
2. **Waldeyer** (1888) coined the term chromosome.
3. **Sutton** and **Boveri** (1902) proved that chromosome is physical basis of heredity.
4. Chromosome is so called because it is stained by basic stains.
5. **Morgan** (1933) showed the importance of chromosome in heredity.
6. Chromosome number for a species remain always the same. It is diploid in somatic cell and haploid in gametes.
7. Haploid number is shown in 'n' and diploid by '2n'.
8. Length of chromosome is 0.1μ – 30μ and diameter 0.2μ–2μ.
9. Shape of chromosome is clear in anaphase of cell division. On the basis of centromere position, they are I, J, L and V shaped.
10. The secondary constriction possessing chromosomes are called SAT chromosome.
11. Beaded structures present on chromosome are called chromomeres.
12. Thread-like structures at the metaphase stage of cell division in each chromosome are called **chromatid.** Each chromatid is made up of one molecule of DNA.
13. Barenetzky (1888) first observed chromatid in *Tradescantia* pollen grains.
14. The term **chromonema** is coined by **Vegdovsky** (1912).
15. **Balbiani** (1876) discovered chromomeres.
16. **Belling** (1928) shown 1500-2500 chromomere on a haploid chromosome in *Aloe* and *Lilium*.
17. **Johanssen** (1909) coined the term gene.
18. **Callened Lloyd** (1960) suggested that one loop of lampbrush chromosome is equal to a gene.
19. **Berman** (1967) suggested that one black band of polytene chromosome is equal to a gene.
20. **Woodcock** (1973) termed the beaded structure on chromosome as **nucleosome.**
21. Nucleosome is an octomer of 4 histone proteins complexed with DNA (Oudet *et al*).
22. Every 3/4 turns of DNA have 166 base pairs. When H_1 histone is added the nucleotide number becomes 200.
23. 6 nucleosomes coil to form a solenoid.
24. Many solenoids coil to form a chromatid.
25. **Klug** (1982) discovered **solenoid.**

QUESTION BANK

Very Short Answer Questions

1. Write the names of chromosomes according to the position of centromere.
2. What is primary construction known as?

3. What is nucleolar organizer?
4. Name the amino acids present in abundance in histone proteins.
5. What is solenoid?
6. How many base pairs are present in one nucleosome?
7. What is the structure present in between 2 nucleosomes one known as ?
8. What is the name of the model suggested by Dupraw?
9. The loops of Lampbrush chromosome are meant for what purpose ?
10. The interbands or bands on polytene chromosome indicate what structure ?

Short Answer Questions

1. What is nucleosome; describe and make well labelled diagram.
2. Draw the diagram of chromosome structure.
3. Write short note on giant chromosome.
4. Write in brief the strucutre of polytene chromosome.
5. Write about the structure of chromosome at molecular level.
6. Write briefly about folded fibre model of Dupraw.
7. Write short notes on (a) Telomere (b) Centromere.
8. Write notes on :
   (*a*) Histone protein (b) Non-histone protein
9. Write briefly about the different types of chromatin present at different parts of chromosome.
10. Write short note on kinetochore.

Long Answer Questions

1. Write in detail the structure of chromosome. Describe its various functions with suitable diagrams.
2. Describe the structure and functions of Lampbrush chromosome.
3. Describe the structure and functions of Polytene chromosome.

Multiple Choice Questions (Objective Type Questions)

1. One genome is equal to :
   (a) A set of chromosome (b) A name of animal
   (c) A name of plant (d) A set of gene
2. Centromere is found in :
   (a) ER (b) Ribosome
   (c) Golgi body (d) Chromosome
3. Kinetochore is related to :
   (a) Centrosome (b) Centriole
   (c) Centromere (d) Chromomere
4. The basic unit of eukaryotic chromosome is :
   (a) DNA (b) Chromonema
   (c) Nucleosome (d) Nucleoid
5. Salivary gland chromosomes or polytene chromosomes are found in salivary glands of :
   (a) Rabbit (b) Mouse
   (c) Dog (d) Insect
6. One chromatid has :
   (a) One DNA molecule (b) Two DNA molecules
   (c) Three DNA molecules (d) One DNA and one RNA molecule
7. Nucleolar organizer is :
   (a) Primary constriction (b) Secondary constriction
   (c) Centromere (d) Chromatid

8. Chromosomes are chemically made up of :
(a) DNA only (b) DNA + RNA
(c) RNA only (d) DNA + protein

9. The term gene was given by :
(a) Johanssen (b) Waldeyer
(c) de Duve (d) Virchow

10. Centromere is related with :
(a) DNA replication (b) Breaking of chromosomes
(c) Polar movement of chromosome (d) Spindle formation

11. Balbiani rings contain high amount of :
(a) DNA and *m*-RNA (b) Only DNA
(c) Only *m*-RNA (d) Heterochromatin

12. Chromosome number of hexaploid is 48, then its basic chromosome number would be :
(a) 8 (b) 16
(c) 32 (d) 24

13. Prokaryotic organism lack :
(a) True nucleus (b) Nuclear membrane
(c) Histone protein (d) All the above

14. Histone proteins have large amount of :
(a) Glycine and glutamic acid (b) Prolein and phenylalanine
(c) Lysine and argenine (d) Methionine and argenine

15. Chromosomes with equal arms are called:
(a) Telocentric (b) Acentric
(c) Metacentric (d) Acrocentric

16. Allosomes are:
(a) Sex chromosomes (b) Autosomes
(c) Somatic chromosome (d) Supernumerary chromosome

17. Histone rich in arginine is :
(a) H_3 (b) H_1
(c) H_2A and H_2B (d) H_3 and H_4

18. Nucleosome contains :
(a) 8 carbohydrate molecules (b) 8 histone molecules
(c) 8 lipid molecules (d) 8 nucleotide molecules

19. Part of chromosome staining dark is :
(a) Euchromatin (b) Heterochromatin
(c) Histone protein (d) All the above

20. Non-histone chromosomal protein help in :
(a) Transcription (b) Translation
(c) Both 'a' and 'b' (d) Transduction

21. Physical basis of heredity is :
(a) Protoplasm (b) Nucleus
(c) Chromosome (d) DNA

22. Paranemic coiling is found in .... stage of cell division.
(a) Interphase (b) Prophase
(c) Metaphase (d) G_o phase

23. Nucleosomes are also called as :
    (a) Histone core (b) nu particles
    (c) Gamma particles (d) All the above
24. Nucleosomes do not contain :
    (a) H_1 histone (b) H_2A histone
    (c) H_2B histone (d) H_4 histone
25. 'L' shaped chromosomes are :
    (a) Acrocentric (b) Submetacentric
    (c) Telocentric (d) Subcentric
26. A genophore or nucleoid consists of :
    (a) A single stranded DNA (b) Histones only
    (c) Double stranded DNA (d) Protein and non-histones
27. Spring *et al* in 1975 reported lampbrush chromosomes in :
    (a) *Drosophila* (b) *Lychnis*
    (c) *Acetabularia* (d) *Coccinia*
28. B-chromosomes are also called as :
    (a) Autosome (b) Allosome
    (c) Holandric chromosome (d) Supernumerary chromosome
29. Malik and Singh in 1979 reported presence of Polytene chromosomes in :
    (a) *Drosophila* (b) Salivary glands
    (c) Endosperm cells of embryosac (d) Oocytes
30. **Lampbrush** chromosomes are also called diplotene chromosomes because they :
    (a) Resemble-diplotene chromosomes (b) Show chiasmata
    (c) Are in permanent diplotene stage (d) All the above

ANSWERS

1. (a)	2. (d)	3. (c)	4. (c)	5. (d)	6. (a)
7. (b)	8. (d)	9. (a)	10. (c)	11. (b)	12. (a)
13. (d)	14. (c)	15. (c)	16. (a)	17. (d)	18. (b)
19. (b)	20. (a)	21. (c)	22. (b)	23. (b)	24. (a)
25. (b)	26. (c)	27. (c)	28. (d)	29. (c)	30. (c)

46
CHAPTER

Sex Chromosomes

Every cell of the organism bears two types of chromosomes.

(a) Autosomones

(b) Allosomes

Allosomes are the **sex chromosomes** which are very much necessary. One set of sex chromosomes is necessary to determine the sex of the organism. **Sex linkage** occurs when the genes controlling various traits are present on the **sex chromosome.** The sex chromosomes are not necessarily present in morphologically identical pairs in both sexes. The traits present on them exhibit unique **sex-linked patterns of inheritance.**

The sexually reproducing organisms are classified into 3 types:

(a) Hermaphrodite : In plants, generally both male and female reproductive parts are present in the single individual.

(b) Monoecious : When the flowers are unisexual (male or female) but present on the same plant.

(c) Dioecious : When both male or female individiuals are different.

Sexus means section or separation. The male has **male gonad** and the female has ovaries or **female gonad.** Such kind of separation is called **gonochorism.** The difference in sex cells and reproductive organs is called **primary sexual characters.** The somatic or morphological differences between male and female is called **secondary sexual character.**

The phenomenon of molecular, morphological, physiological or behavioural differentiation between male and female sexes is known as **sexual dimorphism**

Genetically controlled sex determining mechanism is classified into following categories:

(a) Sex chromosome mechanism or heterogamesis

(b) Genic balance mechanism

(c) Male haploidy or haplo-diploidy mechanism

(d) Single gene effect

(a) Sex chromosome mechanism (Heterogamesis)

Besides sexual dimorphism, the chromosomes also differ in the male and female organisms. **Henking** (1891) a German biologist studied the spermatogenesis of **squash bug,** *Pyrrhocoris*, and observed that half of the spermatozoa contained an extra chromosome and called them as **'X'** body. In 1902 **McClung** observed gametogenesis in grasshoppers and suggested the importance of 'X' body in determination of sex. He reported that the somatic cells of the female grasshopper (*Xiphidium fasciatum*) contained 24 chromosomes where as the male grasshopper has only 23 chromosomes. After intensive studies by **Stevens** and **Wilson** it was confirmed that 'X' body is nothing but the **'X' chromosome.**

The 'X' chromosome is also called sex chromosome because it is important in determining the sex. In *Lygaeus turicicus* (milkweed bug) both male and female carry 14 chromosomes. In female, all chromosomes are paired and the homologues are of equal size, while in male all chromosomes are paired but the homologue of 'X' chromosome is not identical it is called 'Y' chromosome.

STRUCTURE OF SEX CHROMOSOME

There is definite difference between 'X' and 'Y' chromosome. 'X' chromosomes are straight rod like and larger than 'Y' chromosome. However, the 'Y' chromosome is smaller and **curved** in *Drosophila* (Fig 46.1). In man and *Melandrium* the 'Y' chromosome is also straight (Fig. AB). The 'X' chromosomes have large amount of euchromatin. The 'Y' chromosomes have little amount of euchromatin. The 'X' chromosome contains many genes and 'Y' a few genes. 'Y' chromosome is also called genetically **inert** chromosome.

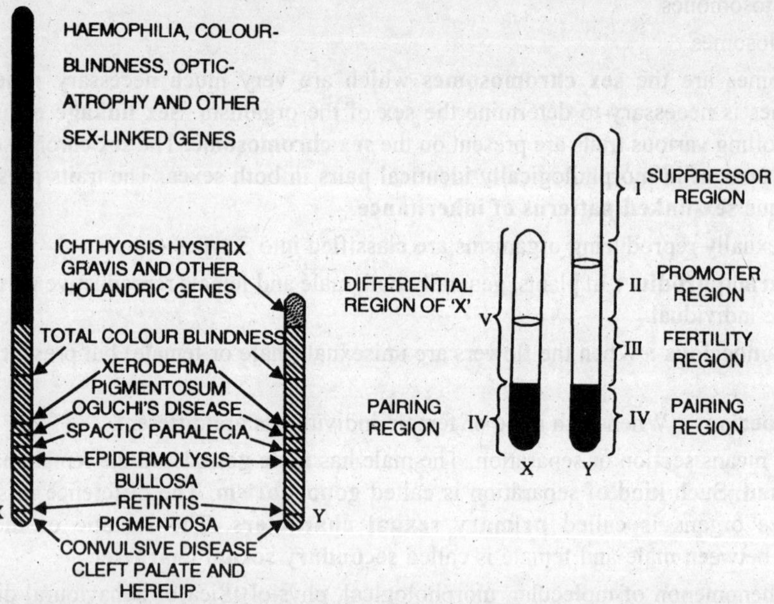

Fig. 46.1. Sex chromosome showing various genes: A. Man, B. *Melandrium.*

COMPOUND SEX CHROMOSOME

'X' chromosome is most commonly found as a single chromosome on single homologous pair of chromosomes. In some species, it is compound or more than one kind of 'X' chromosome that act together as a **sex-determining** group. In such species, there is a large difference in the number of chromosomes between male and females, *e.g.*, in the nematode *Ascaris incurva* the number of autosome pair is **13** but the number of compound **X** chromosome is **8.** A male therefore forms two types of gametes 8 X + A (21 chromosome) and Y +A (14 chromosomes). While the female forms only 8X+A (21 chromosome type), upon fertilization, the diploid number of chromosomes is therefore **35** in males and **42** in females

$$\underset{\text{sperm}}{(8X + A)} + \underset{\text{egg}}{(8X + A)} = 16X + 2A = \text{female}$$

$$\underset{\text{sperm}}{(Y + A)} + \underset{\text{egg}}{(8X + A)} = 8X + Y + 2A = \text{male}$$

The 'Y' chromosome may also exist as a compound group, and in some cases both compound Xs and compound Ys may be found together in the same species. *Blaps polychresta* (beetle) also has compound sex chromosome. The male has 12 Xs and 6 Ys in addition to 18 autosomes.

SEX DETERMINATION

Types of sex chromosomal mechanism in sex determination

In dioecious organism two system of sex chromosomal determination of sex is observed.

(A) Heterogametic male

The female sex has 2 'X' chromosomes (XX) while male has only one 'X' chromosome and at the time of gametogenesis produces 2 types of gametes. 50% with 'X' chromosome while other 50% without 'X' chromosome. Because of 2 types of gametes produced by males it is called **heterogametic sex** (male). While the females are **homogametic.**

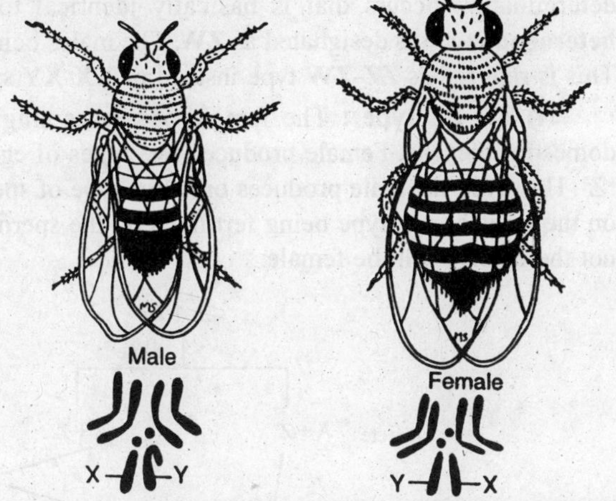

Fig. 46.2. Sex chromosomes in *Drosophila:* A. Male and female. B. Shape of chromosomes in male and female

Heterogametic males are of two types :

(i) XX-XY type : In man, other mammals, plants and many insects like *Drosophila,* etc., the female has homogametic XX type while male has X and Y chromosomes. The female produces only one type of gametes while male produces two types of gametes 'X' and 'Y'. The sex of the embryo depends on the type of sperm or male gamete (X or Y type). If the female gamete is fertilized by 'X' sperm the embryo will be carrying female sex while if it is fertilized by 'Y' sperm the embryo will be male (Fig. 46.2).

(ii) XX-XO type : In some insects like grasshopper, bugs, etc., females are homogametic XX type but the male has only one sex chromosome, *i.e.*, 'X' only, there is no 'Y' chromosome, hence the chromosome number of the male and female will be different. The male has chromosome number one less than that of female. The female will produce only one type of eggs ('X') but male produces 50% sperm with 'X' and other 50% without 'X'. The sex of offspring depends upon the sperm that fertilizes the egg (Fig. 46.3).

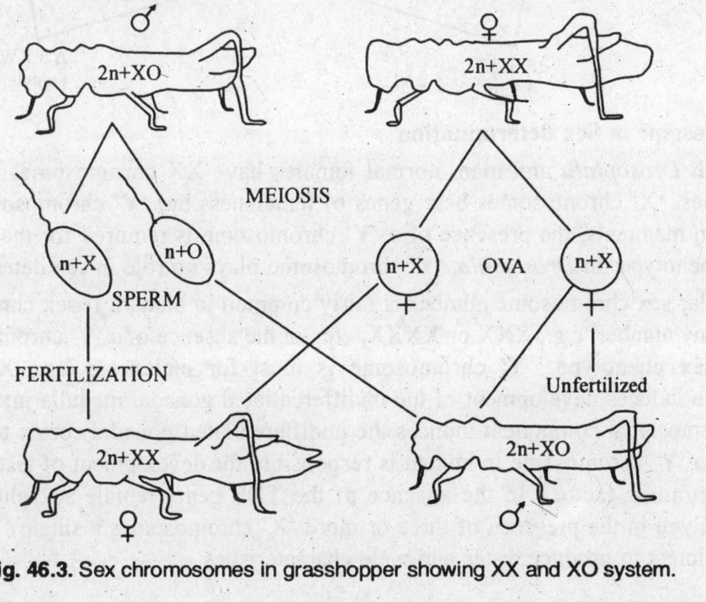

Fig. 46.3. Sex chromosomes in grasshopper showing XX and XO system.

(B) Heterogametic females

In most of species like birds, moths and some fish a chromosomal mechanism of sex determination occurs that is basically identical to the **XX-XY** mechanism. As females are **heterogametic** it is designated as **ZW.** The males being **homogametic,** they are designated as **ZZ.** This is referred as **ZZ-ZW** type instead of **XX-XY** system.

(i) ZO-ZZ type : The female possesses single Z chromosome in moth, butterflies and domestic chickens. Female produces two types of eggs one with 'Z' chromosomes other without 'Z'. However, the male produces only one type of sperm carrying 'Z' only. The sex is determined on the basis of egg type being fertilized by the sperm if it contains Z the embryo will be male if not the embryo will be female.

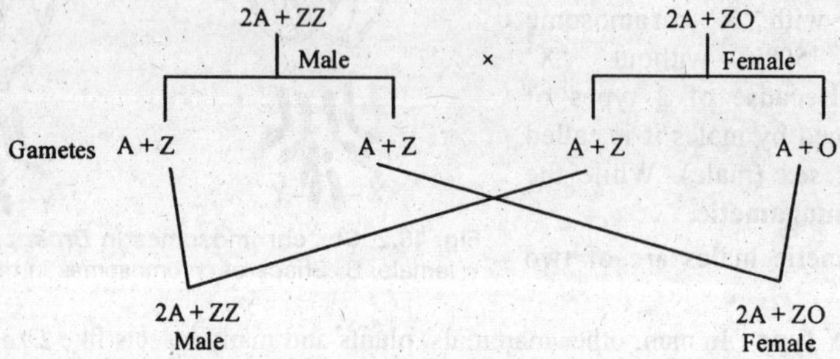

(ii) ZW-ZZ type : It is common in insects, vertebrates like fish reptiles, birds, etc. Females are heterogametic with ZW and males are homogametic with ZZ. Female produces two types of eggs, *i.e.*, 50% with 'Z' or 50% with 'W' while male produces only one type of sperms, *i.e.*, with 'Z'. The sex of the offspring depends on the type of egg it was fertilized.

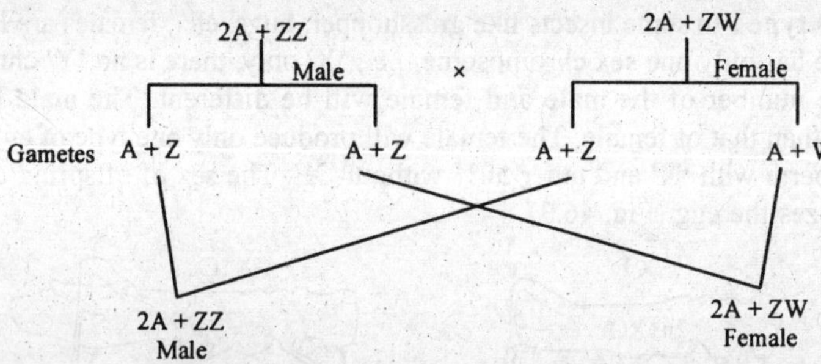

'Y' Chromosome in Sex determination

In both *Drosophila* and man, normal females have XX chromosomes and male has XY chromosomes. 'X' chromosomes bear genes of femaleness but 'Y' chromosome bears genes of maleness. In mammals, the presence of a 'Y' chromosome is required for the development of a male sex phenotype. In *Drosophila,* 'Y' chromosome plays no role in sex determination.

Irregular sex chromosome number is fairly common in human. If sex chromosomes 'X' are present in any number, *e.g.*, XXX or XXXX, etc., in the absence of a 'Y' chromosome give rise to a female sex phenotype. 'Y' chromosome is must for maleness, *e.g.*, XXXXY. The 'Y' chromosome induces development of the undifferentiated gonadal medulla into a **testis,** whereas an XX chromosomal component induces the undifferentiated gonadal cortex to develop ovaries. The gene on 'Y' chromosome in human is responsible for development of testis is called **'TDF'** (testis determining factor). In the absence of the TDF gene, female sex phenotype would be expressed. Even in the presence of three or more 'X' chromosomes a single 'Y' chromosome is usually sufficient to produce testes and male characteristics.

In *Melandrium album* (XY type) **Warmke** and **Westergaard** and others have shown that sex is determined by a balance between male-determining genes on the 'Y' chromosome and female determining genes on the 'X' and autosomes. In this plant, which is dioecious 'XY' individuals are staminate and 'XX' plants are pistillate. The 'Y' chromosome is the largest and most conspicuous (Fig. 46.1B). Three distinct regions of this chromosome influence in sex determination and male fertility.

Region I suppresses femaleness, in the absence of it plants are bisexual, *i.e.*, they express both male and femaleness.

Region II promotes male development when this region is missing with or without I region a female plant is produced.

Region III carries male fertility genes loss of this region results in male sterility. Only a part of 'Y' chromosome (Region IV) is homologous to 'X' but the major part of 'X' is differentiated with no structural counterpart on the 'Y'. **Westergaard** found that autosomes were also involved in female determination.

Genic Balance Concept of Sex determination In *Drosophila*

C. B. Bridges showed that female determiners were located on the 'X' chromosomes and male determiners were on autosomes. No specific loci were identified at that time. But recent evidences demonstrate that many chromosome segments are involved specifically, female determining genes were shown to be carried on the 'X' chromosomes and male determiner genes were shown to be located on the three autosomal chromosomes of *Drosophila.* The **genic balance theory** of sex determination was devised to explain the mechanics of sex determination in *D. melanogaster.* (Fig. 46.4)

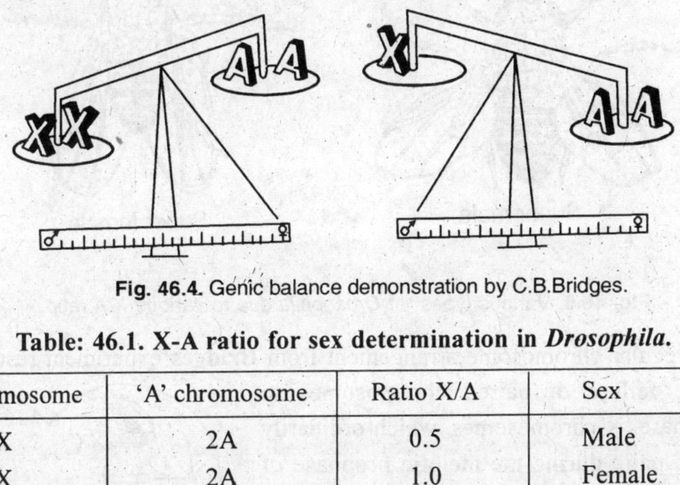

Fig. 46.4. Genic balance demonstration by C.B.Bridges.

Table: 46.1. X-A ratio for sex determination in *Drosophila.*

'X' chromosome	'A' chromosome	Ratio X/A	Sex
1X	2A	0.5	Male
2X	2A	1.0	Female
3X	2A	1.5	Metafemale
4X	3A	1.33	Metafemale
4X	4A	1.0	Tetraploid female
3X	3A	1.0	Triploid female
3X	4A	0.75	Intersex
2X	3A	0.67	Intersex
2X	4A	0.5	Tetraploid male
1X	3A	0.33	Meta male

Bridges experimentally produced various combinations of X chromosomes and autosomes (A) in *Drosophila* and deduced from comparisons that one 'X' chromosome and two sets of autosomes (A) produce a normal male. Normal males had a ratio of 'X' chromosomes to sets of autosomes of 0.5 (Table 46.1) (Fig. 46.5).

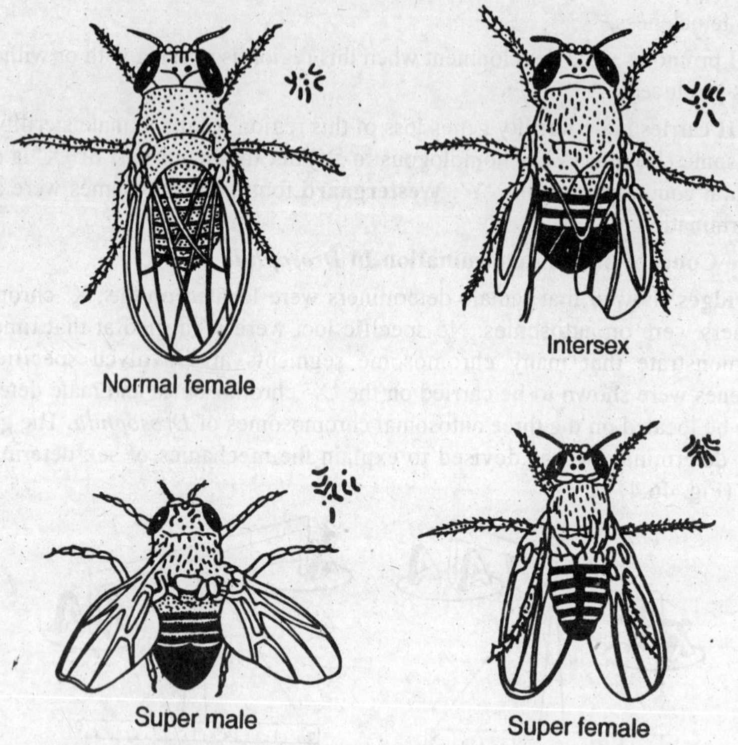

Fig. 46.5. Various types of *Drosophila* due to various *X/A* ratio.

The first irregular chromosome arrangement from **Bridges** experiment resulted from **non-disjunction**, the failure of paired chromosomes to separate in anaphase. X chromosomes, which ordinarily come together in pairs during the meiotic prophase of oogenesis and separate to opposite poles during anaphase, remain together and migrate to the same pole. As a result some female gametes receive 2X chromosomes and the other no X chromosome (Fig. 46.6). After fertilization by sperm from wild type males (2A + XY), all zygotes had 2 sets of autosomes (2A) but some received 2X (XX) from mother and one X from father and became 3X (XXX). The ratio of A/X chromosome was 3 : 2 and the flies were metafemales. The XO males were sterile and those with aY chromosomes and no X chromosome did not survive (Fig. 46.7).

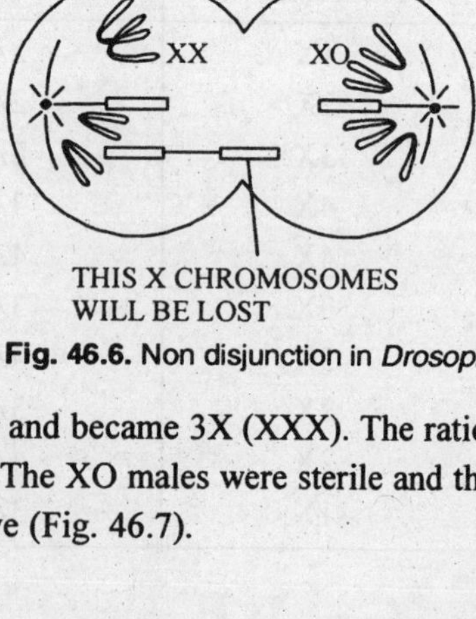

Fig. 46.6. Non disjunction in *Drosophila*.

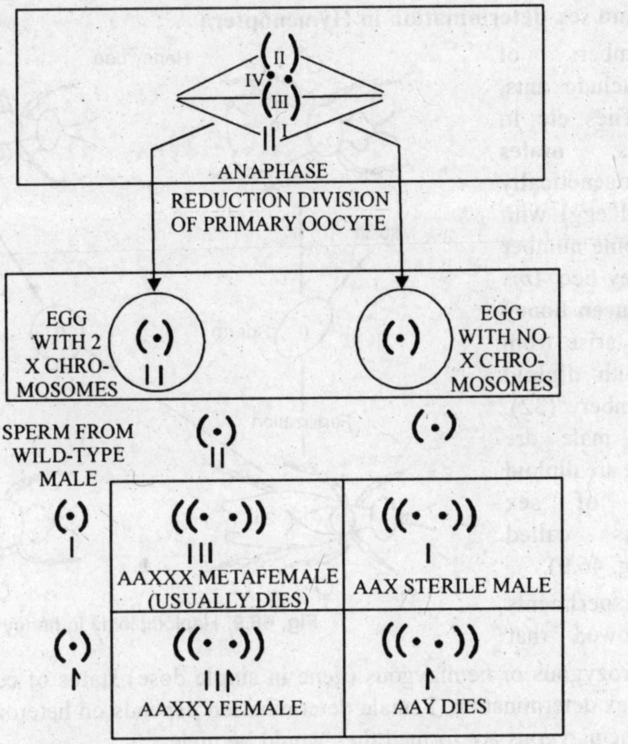

Fig. 46.7. Results of anaphase non-disjunction in *Drosophila*.

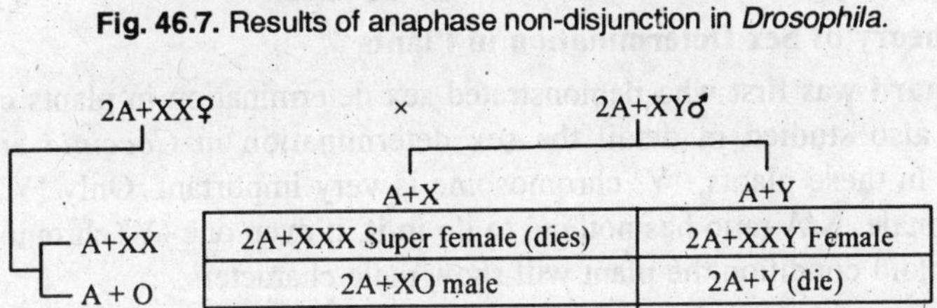

The presence of triploid intersexes in the experiment conducted by Bridges (Fig. 46.8) is a definite proof of autosome carrying factors of sex determination. According to it (genic balance theory) ratio of number of 'X' chromosome and number of complete sets of autosomes determine the sex. The female determiners were located on X chromosome and male determiners on **autosomes.** Recent evidence has demonstrated that many chromosome segments are involved in this process. Female determining genes are carried on 'X' and male determiner genes were located on the three autosomes of *Drosophila.*

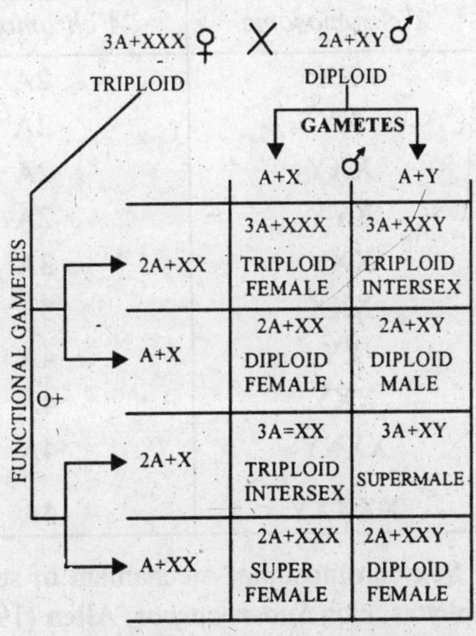

Fig. 46.8. Triploid intersex formation.

Haploidiploidy and sex determination in Hymenoptera

The members of **Hymenoptera** include ants, bees, wasps, sawflies, etc. In several species, **males develop** parthenogenetically (from unfertilized egg) with haploid chromosome number (16 in drone honey bee *Apis mellifera).* The **queen** honey bee and **workers** arise from **fertilized** egg with diploid chromosome number (32). Because normal male are haploid and female are diploid the mechanism of sex determination is called **haplodiploidy** (Fig. 46.9).

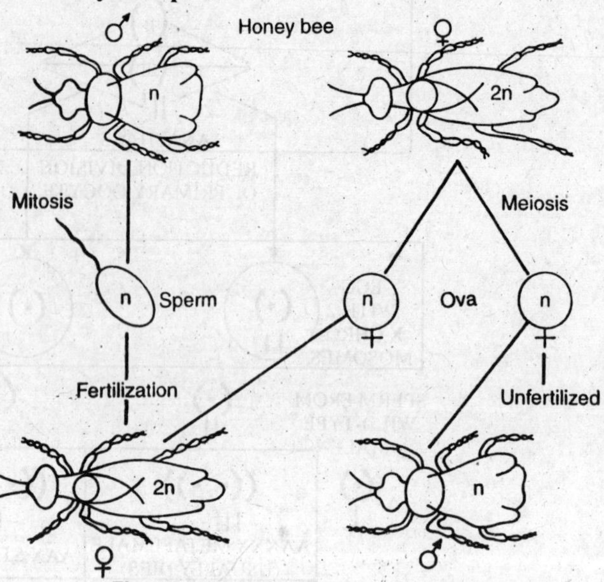

Fig. 46.9. Haplodiploidy in honey bee.

Results of experiments by **Whiting** showed that homozygous, heterozygous or hemizygous (gene in single dose) status of certain chromosome segments control sex determination. Female determination depends on heterozygosity for part of a chromosome. If hemizygous are formed they would be male.

Genic Balance Theory of Sex Determination in Plants

M. Westergaard was first who demonstrated sex determination in plants considering X/A ratio. **R. P. Roy** also studied in detail the sex determination in *Coccinia* and *Melandrium* (unisexual plant). In these plants, 'Y' chromosome is very important. Only 'Y' chromosome's presence makes it male. X/A ratio has nothing to do in it, if even one 'Y' chromosome is present in diploid or polyploid condition the plant will show male characters.

Table 46.2. Sex determination in *Coccinia* and *Melandrium*

X chromosome	*A chromosome*	*X/A ratio*	*Sex*
XX	2A	1.00	Female
XY	2A	0.50	Male
XXY	2A	1.00	Male
XYY	2A	0.50	Male
XXY	3A	0.67	Male
XXX	3A	1.00	Female
XY	4A	0.25	Male
XXY	4A	0.50	Male
XXXY	4A	0.75	Male
XXXXY	4A	1.00	Hermaphrodite

Sex chromosomal mechanism of sex determination has also been observed in monoploid bryophytes, *e.g., Spaerocarpos*. **Allen** (1919) found that the sporophyte of *Spaerocarpos* contains two sex chromosomes (XY) and it produces two kinds of meiospores (X and Y type) 'X' meiospores germinate into female gametophyte and 'Y' meiospores germinate into male gametophyte.

Mosaic and Gynandromorphs

Abnormal chromosomal behaviour in insect can result in the formation of **'gynandromorph'** or **sexual mosaic,** in which half part of the animal is male and the other half is female. Gynandromorphs in ***Drosophila*** (Fig.46.10) are bilateral intersexes, with male colour pattern, body shape and sexcomb on one half of the body and female charactersitics on the other half. Both male and female gonads and genitalia are present.

Such gynandromorphs are the result of irregularity in mitosis at the first cleavage of zygote. A chromosome lags in division and does not arrive at the pole in time to be included in the reconstructed nucleus. When one of the 'X' chromosomes of an XX (female) zygote lags in the spindle, one daughter nucleus receives only one 'X' chromosome. A mosaic body pattern is established one nucleus in the two nuclei stage would be **XX** (female) and the other would be **XO** (male). It develops into one-half as female and the other as male. In *Bracon bebetor,* gynandromorphs may occur in the anterior posterior plane giving rise to wasp with such peculiar arrangement as **male head** with **female abdomen** or **female head** with **male abdomen.**

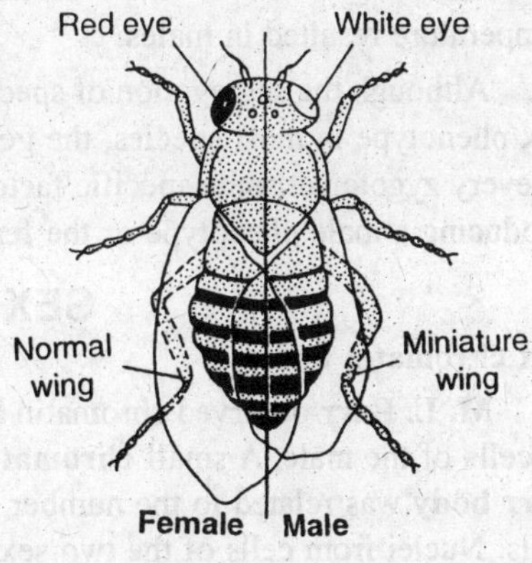

Fig. 46.10. Gynandromorph in *Drosophila.*

Environment factor and sex determination

In some lower animals, the sex determination is non-genetic and depends on factors in the external environment. The genotype of male and female is the same. Stimuli from environment help initiate development toward one sex or the other, *e.g.*, male of marine worm *Bonellia* are small and degenerate and live within the reproductive tract of the larger female (Fig. 46.11). All

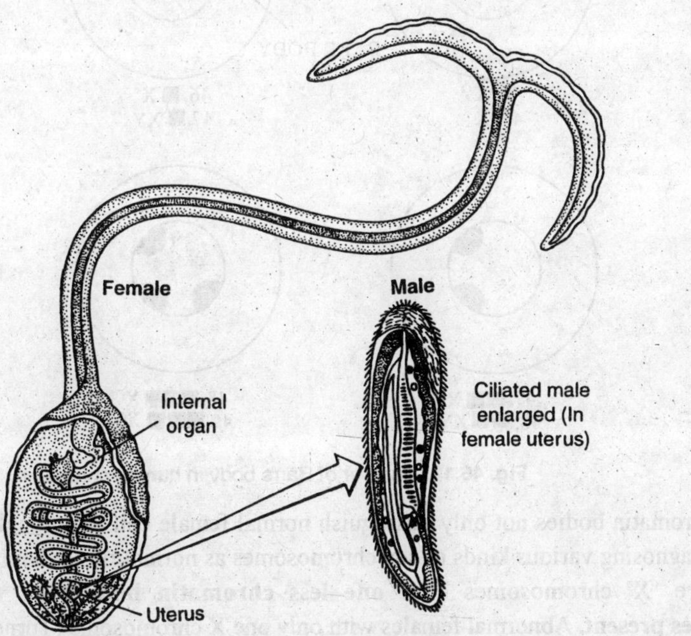

Fig. 46.11. Environmental determination in *Bonellia.*

organs of its body degenerate except the reproductive system. **F Baltzer** found that young worm reared from a single isolated egg became a female. Newly hatched worms in water containing mature females attach to the female proboscis transform into males, and eventually migrate into the female reproductive tracts as parasite. It shows that extract of female proboscis influence the young worm to become male.

In some reptiles, the temperature at the time of incubation prior to hatching plays a major role in determinting sex of offspring. The eggs hatched at 26-27°C developed into female and at 29°C they became male. In turtles (*Chrysema picta*) incubation high temperature (more than 30°C) produced female and at low temperature as male. In lizard (*Agama agama*) high incubation temperature resulted in males.

Although the segregation of specific sex determiner gene and chromosome is responsible for sex phenotype in most species, the genetic potential for both maleness and femaleness is present in every zygote yet some specific factor in the environment triggers the expression of either genes producing a male phenotype or the female phenotype.

SEX DIFFERENTIATION

Sex chromatin bodies

M. L. Barr observed chromatin bodies in the nerve cells of female cats that were not present in cells of the male. A small **chromatin** body was identified in the nucleus of female cells. This **barr body** was related to the number of 'X' chromosomes and was not observed in normal male cells. Nuclei from cells of the two sexes are shown in (Fig.46.12).

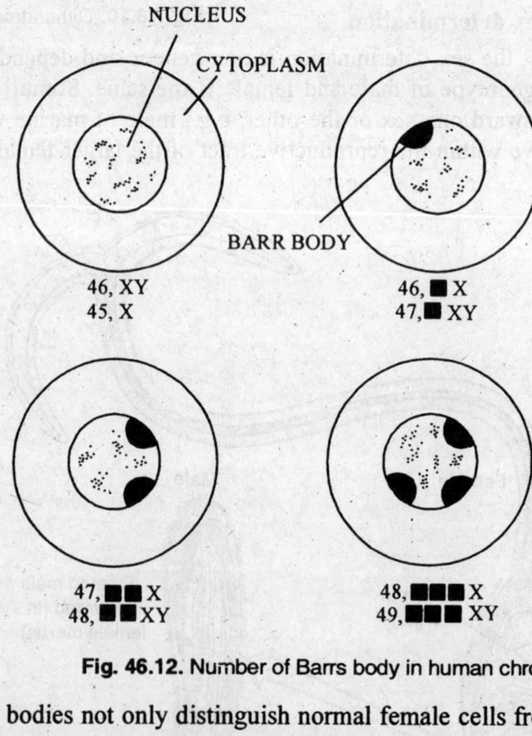

Fig. 46.12. Number of Barrs body in human chromatin.

Sex chromatin bodies not only distinguish normal female cells from male cells, but are also useful for diagnosing various kinds of sex chromosomes as normalities in human. Those who have two or more 'X' chromosomes have **one less chromatin body than the number of X chromosomes present.** Abnormal females with only one X chromosome (Turner's syndrome) have no sex chromatin body and cells of male with **two X** chromosomes and one **Y** chromosome

(Klinefelter's syndrome) have one sex chromatin body. Abnormal females with three **X** chromosomes have two sex chromatin bodies in their nuclei.

The sex chromatin appears in the interphase nucleus as a small chromocentre heavily stained with basic dyes (feulgen). It can be found in four positions

(a) Attached to the nucleus of nerve cells,

(b) Attached to the nuclear membrane (epidermis or oral mucosa),

(c) Free in the nucleoplasm (neurons)

(d) As nuclear expansion (neutrophil leucocyte of female where the barr body appears as a small rod called **drum stick**)

The relationship between sex chromatin and sex chromosome is elucidated by **Ohno, Kapan** and **Kinosita** in 1959. The sex chromatin is derived only from one of the 2 **'X'** chromosomes. The other **X** behaves like autosome and is not **heteropycnotic** at interphase.

Fragile 'X' chromosome

The inactivation of one of the two chromosomes in female mammals must be reversible since female transmit both 'X' chromosomes to their progeny, in functional state 'X' chromosome contains many genes that are vital to growth and development or to survival. The reactivation of the inactive or heterochromatic `X' chromosomes of mammalian females occur in germ cell prior to oogenesis. The abnormal reactivation of heterochromatic X chromosome is partly responsible for inherited mental retardation disease called **'fragile X'syndrome.**

This **'fragile X' chromosome** of humans contains a **fragile site** near the end of long arm (Fig. 46.13). It is the specific region that exhibits constriction gap or break in cells growing in culture. This particular 'X' chromosome contains a fragile site at position X_q 27 (Fig. 46.13B). The fragile site is inherited as a codominant with both normal and gapped X chromosomes in cells from heterozygotes. The affected individuals show a wide range in severity of mental retardation and other phenotypic manifestation of the syndrome, *i.e.*, hyperactive behaviour, prominant jaws and foreheads, longer than normal ears.

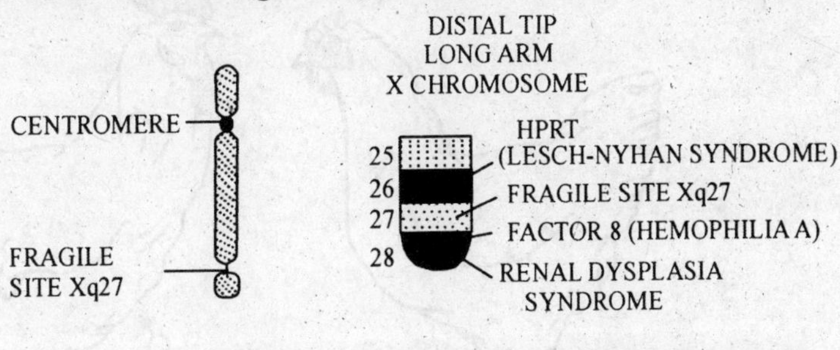

Fig. 46.13 Fragile X chromosome : A. Fragile site at the end of long arm B. Fragile site at position Xq^{27}

The daughter of male who carried the "fragile X" chromosome, did not exhibit phenotypic characteristics of the syndrome, and did not show any manifestation of the syndrome. Fathers always transmit the "fragile X" to their daughters in the same state as they received it from their mothers.

Hormone and Sex differentiation

The hromonal system that regulates the internal or physiological environment of the organism does not directly influence the fundamental process of sex determination. Hormones are necessary to develop the **secondary sex characters.** Sex hormones of higher animals are secreted by ovaries, testes and adrenal glands with stimulation from pituitary gland hormones. Ovaries and testes have dual functions, *i.e.*, production of gamete (sperms or egg) and sex hormones. Hormones influence the development of sexual characters, *e.g.*, physiological difference, bone structure, voice, breast development and hairs, etc. Adrenal gland produces steroids.

Sex Influenced Dominance

The dominance of alleles may differ in heterozygote of the two sexes. This phenomenon is called "sex influenced dominance", *e.g.*, autosomal genes responsible for horns in some breeds of sheep may behave differently in the presence of the **male and female** sex **hormones.**

Some human traits like enlargement of the terminal joints of the fingers, premature patterns of baldness, etc., are types of **sex influenced inheritance. The traits that show the pattern of inheritance characteristic of sex influenced dominance are controlled by genes located on autosomal chromosomes, not on sex chromosomes.** Traits that are governed by genes located on the sex chromosomes exhibit a very distinctive pattern of transmission called sex-linked inheritance.

Sex Limited Gene Expression

One sex may be uniform in expression of a particular trait and yet transfer genes that produce a different phenotype in offspring of the other sex. It is called **sex limited gene expression. Sex hormones** are apparently the **limiting factors** in the expression of some genes, *e.g.*, milk production in mammals is limited to females, but certain bulls are in great demand among dairy breeders and artificial insemination association because their mothers and daughters have increased milk production records. In chickens, redostem generally have long, pointed tail feathers, whereas hens have rather short, blunt tail feather. Same type of feathers are found in many domestic birds (Fig 46.14).

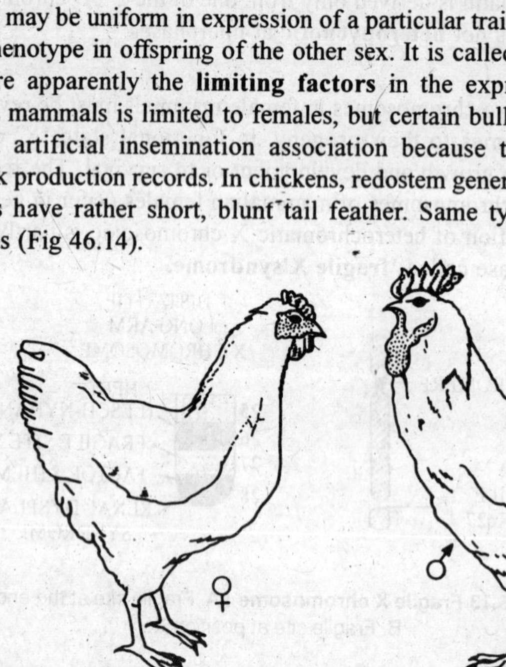

Fig. 46.14. Feathers in hen.

Single gene control of sex

In some organisms like *Chlamydomonas, Neurospora,* yeast, *Asparagus*, maize, *Drosophila*, etc., individual single gene is responsible for the determination or expression of sex.

(a) Sex determination in *Asparagus* : It is a dioecious plant. The female flowers may bear rudimentary anthers and male flowers can bear rudimentary pistils. Rarely a male flower can

Rare male plant

P_P

↓

selfing

P_P 25%	P_P 50%	pp 25% ♂

⚥ 75%

develop pistil and may set seeds. Plants raised from such seeds show male and female plants in 3 : 1 ratio. When male plants are used to pollinate female flowers of female plant only two-third of them segregate. It shows that the sex is controlled by single gene.

(b) Monogenic sex determination in maize : It is a monoecious plant with male inflorescence called **tassel** and female inflorescence as **silk.**

A gene **tassel seed** converts the tassel into seed bearing inflorscence. While gene **silkness** (Sk) is responsible for the absence of silks. A plant with genotype Sk/Sk will be a male plant. Plant with ts/ts will be female. But individual with Sk ts can impose bisexuality in maize.

Metabolically controlled sex determination

Crew suggested that sex is a physiological equitable division between catabolic and anabolic types. **Shull** and **Whitney** showed that by increasing metabolic rate in rotifers the occurrence of male individuals increases than females. **Riddle** suggested that metabolism in pigeon and doves plays an important role as increased rate of metabolism develop maleness.

Hormonally controlled sex determination

Sexual differentiation is controlled by hormones controlling sex of an individual.

Sex reversal in hen : In birds only one gonad of a normal female develops into a functional ovary. If this is destroyed the rudimentary other gonad develops into testes (**Crew** 1925). The female sex is reversed to male. It is called **sex reversal.** Birds have heterogametic (ZW) female. The ovary cells produce female hormones in such an amount that it supresses the male hormone producing cells. If ovary is removed the adrenal gland produces steroid which develops the rudimentary gonad into testis.

Witschi reported sex reversal by hormones in **salamender** also.

Hormones and free martin : In cattle when twin calves of different sexes occur, the female member is usually a sterile intersex called a **free martin.** The freemartin has external female genitalia but internal sex organs are more or less like those of male. The male twin is usually normal. Freemartin is due to a fusion of the foetal membranes of the twin calves, while they are in uterus of the mother. The fusion of the foetal membranes permits blood of each twin to circulate in the blood vessel of the other. The male hormones produced by the male twin are presumed to suppress the differentiation of the female internal sex organs of the co-twin.

XX males and XY females

Sex reversal in mice

Some mice are identified as XX male and XY females. It is due to the transfer of a sex reversing factor (sxr) from Y chromosome (translocation of genes). The Y chromosome is called Y^{sxr}. The duplicated segment of $\mathbf{Y^{sxr}}$ is transferred to X-chromosomes due to crossing over during male meiosis; and gets attached to long arm of 'X' chromosome, which pairs with a segment of 'Y' chromosome (Fig. 46.15). When X derived from male parent sxrXX was inactivated. The sxrXX develops into hermaphrodites or female instead of male.

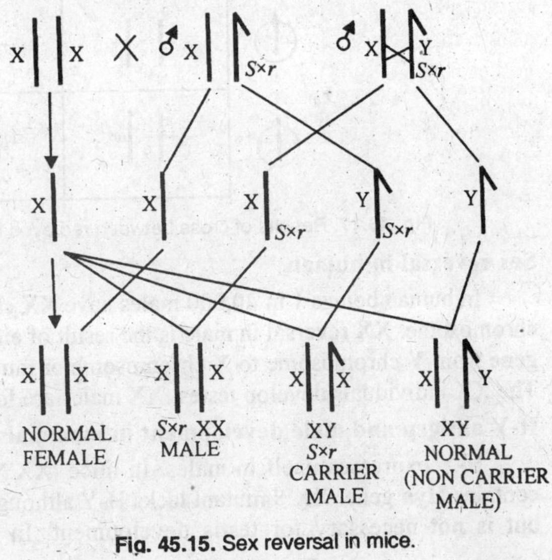

Fig. 45.15. Sex reversal in mice.

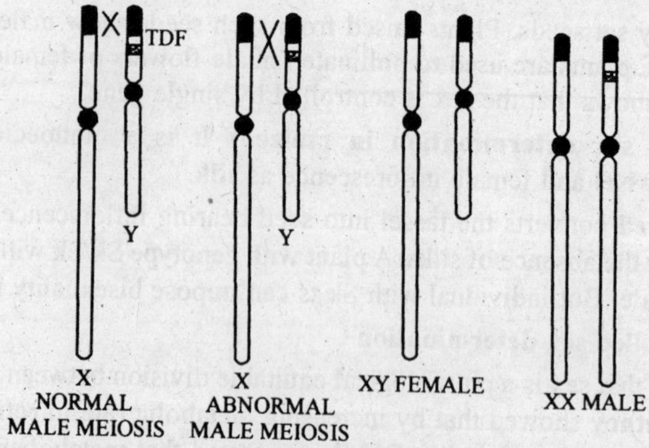

Fig. 46.16. Sex reversal in man.

When a chromosome specific DNA sequence like **Bkm** (present on Y chromosome) was used for *in situ* hybridization could be located on X chromosome of **sxrXX** males.

If a normal female is crossed with an XY^{sxr} carrier male, the normal female, sxrXX males, XYsxr carrier males and non-carrier XY males are obtained (Fig. 46.17).

XY females or hermaphrodite are due to the involvement of **Tdy** (testis determining Y) gene lying close to **Bkm** gene in centromeric region of 'Y" and **Tda-1** (Testis determining autosomal) gene which is autosomal.

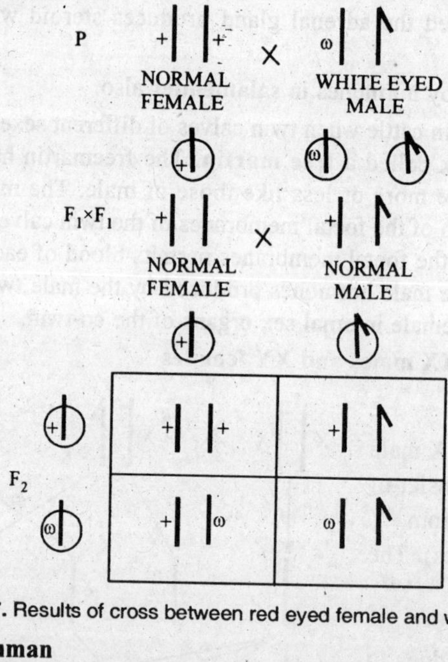

Fig. 46.17. Results of cross between red eyed female and white eyed male in *Drosophila*.

Sex reversal in human

In human beings 1 in 20,000 males have **XX** chromosome and 1 is 1,00,000 females have **XY** chromosome. **XX** reversal in male is the result of either transfer of **TDF** (testes determining factor) gene from **Y** chromosome to **X** chromosome or mutation in a locus which represses development. The XX individuals develop testes. XX males are found to possess 'Y' derived genes (Fig. 46.18).

H-Y antigen and male development in mammals

H-Y expresses itself in males. In mice, XX^{sxr} males also express H-Y because **sxr** region contains **Hya** gene. XX^{sxr} mutant lacks H-Y although it was male. **Hya** is closely linked with **Tdy** but is not necessary for testis development. In human, **Hya** is present on short arm of Y chromosome and TDF is present on long arm. XX males are H-Y negative suggesting that HY antigen does not play any role in male sex development.

SEX-LINKED INHERITANCE

Genes other than sex determiners are also located on the **'X'** chromosomes. Many mutant genes in human have been shown to be located on **X** chromosome. The inheritance of such genes are called **sex-linked inheritance.**

The first evidence for sex linkage in a particular species was given by T.H. Morgan in 1910 in white-eyed mutant in *Drosophila.* When white-eyed male was mated with a red-eyed female the F_1 flies were all red-eyed. F_2 generation of it included 3 : 1 ratio of red and white-eyed flies. But all white eyed flies of F_2 generation were males only. When normal female of F1 is crossed with normal male 50% of males were white-eyed and 50% were red-eyed (Fig. 46.17). It shows that the recessive allele is expressed in male only. **Morgan** explained it as the association of the gene on 'X' chromosome. Because male has only one X chromosome all characters present on it will be expressed. The mutant allele (white eye) of the male is transferred to the female (heterozygous). The hemizygous males obtain their X chromosome from their heterozygous mother only then they will be 50% red-eyed (w^+) and 50% white eyed (w).

The female can be white-eyed (ww) only when both 'X' carry gene for it. When heterozygous female (w^+) is crossed with white-eyed male (w) then white-eyed females can be obtained (Fig. 46.18).

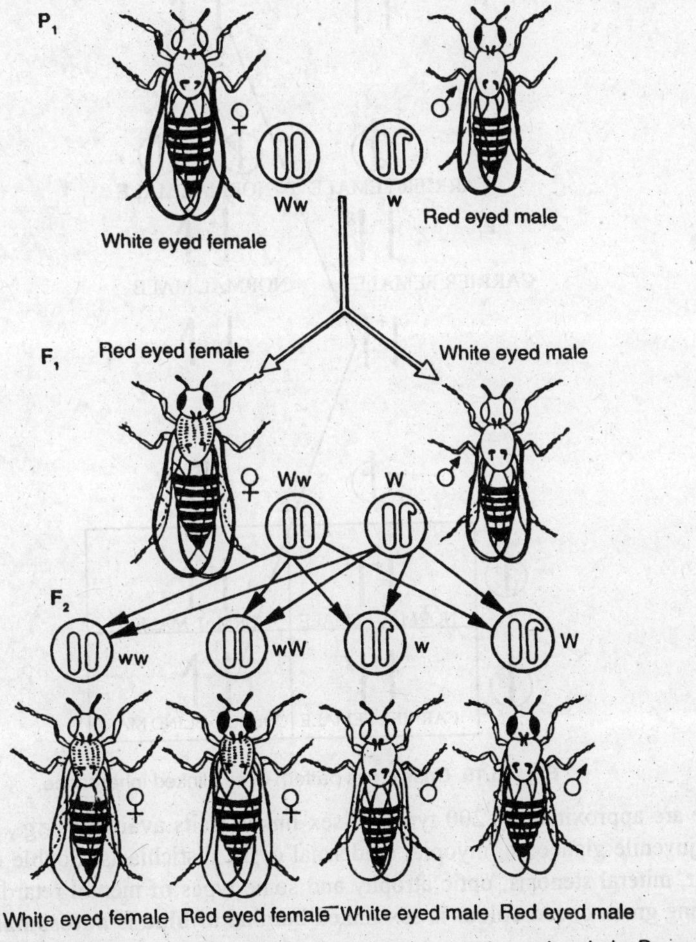

Fig. 46.18. Results of cross between white, eyed female and red, eyed male in *Drosophila.*

'Y' linked inheritance

Most sex-linked genes in male (heterogametic) are on 'X' chromosome. But a few genes are present on 'Y' chromosome also. These are called **'Y' linked** or **holandric** genes. They are transmitted from father to son directly.

'Y' chromosome in *Drosophila* is mostly heterochromatic. The X chromosomes with many genes and 'Y' with virtually none establish the characteristic pattern of inheritance for sex-linked traits. The mother with two 'X' chromosomes transmit one X to each gamete. Zygotes that receive a Y chromosome from father, develops into male. The hemizygous male progeny expresses the sex-linked genes received from the mother. The expression of sex-linked traits in females follow the same pattern as autosomal traits with the recessive phenotype appearing only in homozygous condition.

'X' linked traits in human

The inheritance pattern associated with sex linkage was recorded by Greek philosopher for the first time. An inherited character was observed in a father but not in any of his children (either male or female) and then it reappears in the males of next generation. It is the **criss cross** pattern, from father to daughter to grandson. It is the 'X' linked inheritance. (Fig. 46.19).

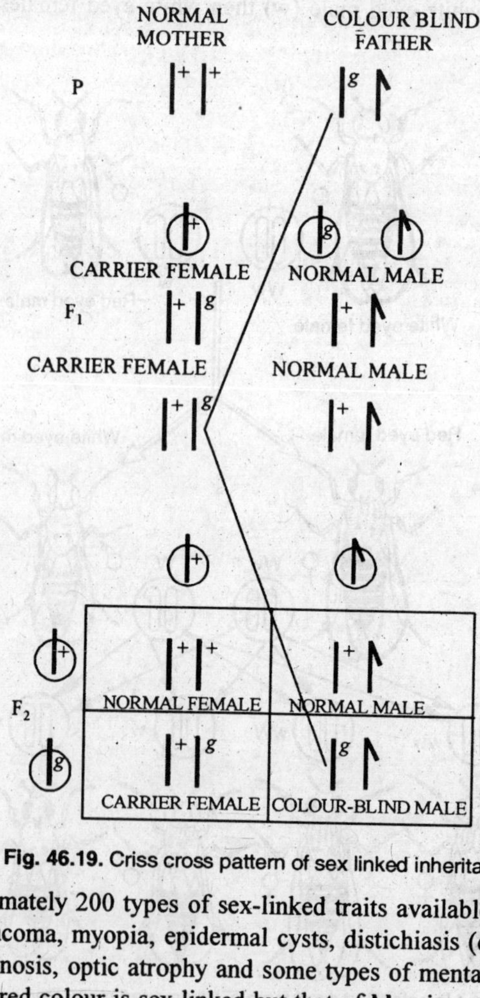

Fig. 46.19. Criss cross pattern of sex linked inheritance.

There are approximately 200 types of sex-linked traits available, *e.g.*, haemophilia, colour blindness, juvenile glaucoma, myopia, epidermal cysts, distichiasis (double eyelashes), occipital lock of hair, miteral stenosis, optic atrophy and some types of mental retardation, etc. Defect in distinguishing green or red colour is sex linked but that of **blue** is **autosomal.**

Sex-linked (X-linked) recessive traits are transmitted as:

(a) The traits occur much more frequently in males than in females.

(b) Traits are transmitted from an affected man through his daughter to half of his grandsons.

(c) An X-linked allele is never transmitted directly from father to son.

(d) All affected females have an affected father and a carrier or affected mother.

Dominant X-linked traits : Males expressing the trait would be expected to transmit it to all of their daughters but none of their sons. Heterozygous females would transmit the trait to half of their children of either sex. If a female expressing the trait is homozygous, all her children will inherit the trait. Sex linked dominant inheritance cannot be distinguished from autosomal inheritance in the progeny of females expressing the trait, but only in the progeny of affected males.

Sex-linked disease **hemophilia** and pedigree analysis in European royalty which generally showed transmission through females who were unaffected to some of their male offspring who were hemophilic. The disease starts with **Queen Victoria of England.** Who passed the hemophilic gene to 3 of 9 children half of her son were hemophilic. Queen was heterozygous for X chromosome. The females in the pedigree have no phenotypic effect. They may be the carriers and their mating partners are normal and the male who are free of the disease do not pass it on. Male shows hemophilia. The homozygotic diseased female will not survive.

Three general types of hemophilia are known, each affecting the production of a **prothrombin proteinase** (factor X), an enzyme necessary for blood clotting.

(a) The 'classic' sex linked type, or hemophilia 'A' which can be detected by a reduction in the amount of a substance called **antihemophilic factor** (AHF or factor VIII) accounts for approximately 80 per cent of hemophilia.

(b) **Hemophilia 'B'** or **christmas disease** is caused by reduction in the amount of **plasma thromboplastin component** (PTC or Factor IX) while accounts for 20 per cent of **hemophiliacs.** The gene for hemophilia 'B' is not allelic to that of hemophilia 'A'.

(c) **Hemophilia** 'C' is due to the interaction of rare autosomal gene with the production of **plasma thromboplastin antecedent** (PTA or factor XI) is responsible for less than 1 per cent hemophiliacs.

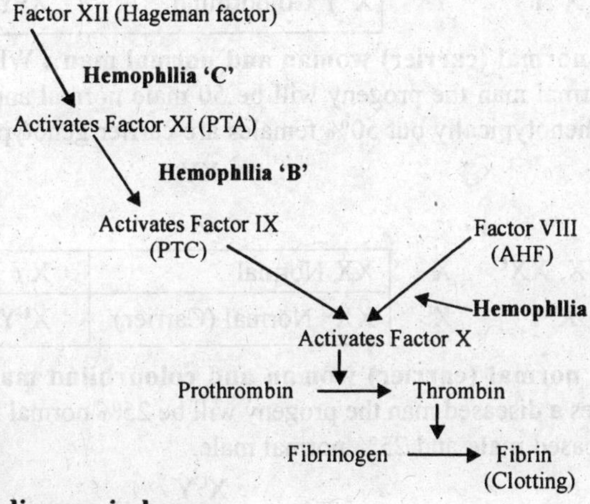

Some X-linked diseases in human

1. Glucose-6-phosphate dehydrogenase deficiency.
2. α-galactosidase deficiency.
3. Hemophilia A (factor VIII deficiency)
4. Hemophilia B (factor IX deficiency)
5. Phosphoglycerate kinase (Glycolysis)
6. Hypoxanthin-guanine-phosphoriboryl transferase (HGPRT or Lesch Nyhan syndrome).

7. Duchenne-Becker muscular dystrophy.
8. Vitamin D resistant ricketts (hypophosphatemia).
9. Ornithine transcarbamylase deficiency.
10. Anhydrotic ectodermal dysplasia.
11. Xg blood cell antigen.
12. Copper transport deficiency (Menkes Kinky-hair syndrome).
13. Bruton type agammaglobulinemia.
14. Testicular feminization syndrome (Tfm).

Various results of crosses in sex-linked inheritance (*e.g.*, colourblindness (X^c) or hemophilia (X^h).

(i) Cross between colourblind man X^CY and normal woman (XX) : When colourblind man marries normal woman the progeny will be, female normal visioned and male 50% colourblind and 50% normal. Though female will be carrier genotypically but normal phenotypically (carrier).

XX × X^CY

↓

		X^C	Y
Female = XX^C, XX^C	X	XX^C Normal carrier	X^CY Colourblind
Male = X^CY, XY	X	XX^C Normal (Carrier)	XY Normal

(ii) Cross between colourblind woman and normal man : When colourblind woman marries a normal man the progeny will be all male colourblind and all female normal but carriers.

XY × X^CX^C

↓

		X^C	X^C
Female —X^CX, X^CX	X	X^CX Normal (Carrier)	X^CX Normal (Carrier)
Male—X^CY, X^CY	Y	X^CY Colourblind	X^CY Colourblind

(iii) Cross between normal (carrier) woman and normal man : When woman is normal (carrier) and marries a normal man the progeny will be 50 male normal and 50% male diseased while all female normal phenotypically but 50% females are carrier genotypically

XX^C × XY

↓

		X	Y
Female — XX, XX^C	X	XX Normal	XY Normal
Male — XY, X^CY	X^C	XX^C Normal (Carrier)	X^CY Colourblind

(iv) Cross between normal (carrier) woman and colourblind man : When woman is normal (carrier) and marries a diseased man the progeny will be 25% normal female (carrier), 25% diseased female, 25% diseased male and 25% normal male.

XX^C × X^CY

↓

		X^C	Y
Female — X^CX^C, XX^C	X	XX^C Normal (Carrier)	XY Normal
Male — XY, X^CY	X^C	X^CX^C Colourblind	X^CY Colourblind

Sex-linkage In *Drosophila*

Example in *Drosophila* where the red eye is dominant over white eye. The genes are present on 'X' chromosome.

(i) **White eyed female crossed with red eyed male** the progeny will be all red eyed female and all white eyed male.

$X^W X^W$ × X^+Y

↓

		X^+	Y
Female = X^WX^+, X^WX^+	X^W	X^WX^+ Red eye	X^WY White eye
Male = X^WY, X^WY	X^W	X^WX^+ Red eye	X^WY White eye

(ii) **When red eyed-female is crossed with white eyed male** the progeny will be all female and male red eyed.

X^+X^+ × X^WY

↓

		X^W	Y
Female = X^+X^W, X^+X^W	X^+	X^+X^W Red eye	X^+Y Red eye
Male = X^+Y, X^+Y	X^+	X^+X^W Red eye	X^+Y Red eye

(iii) **When red eyed female (carrier) is crossed with red eyed male** the progeny will be all female red eyed (50% female are red eyed carrier) and 50% male red eyed and 50% male white eyed.

X^+X^W × X^+Y

↓

		X^+	Y
Female = X^+X^+, X^+X^W	X^+	X^+X^+ Red eye	X^+Y Red eyed
Male = X^+Y, X^WY	X^W	X^+X^W White eyed	X^WY White eyed

The reciprocal cross in sex linked inheritance are always different.

Inheritance of X-Y linked genes

The genes present in the homologous part of X-Y chromosome show inheritance same as autosomal genes. The X-Y linked genes are partially sex linked because some times crossing over may occur in the homologous part of X and Y chromosomes (Fig. 46.20). In human beings several

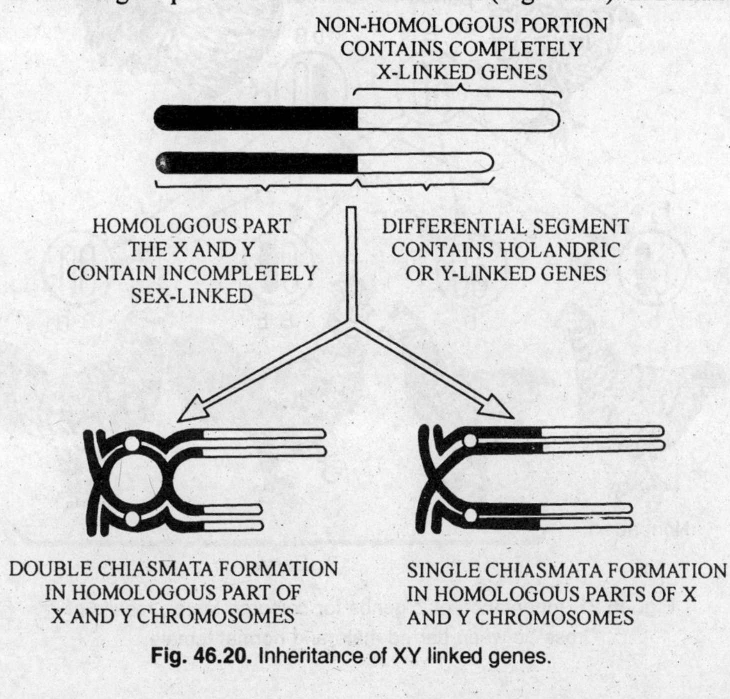

Fig. 46.20. Inheritance of XY linked genes.

diseases are XY linked, *e.g.*, **xeroderma pigmentosum and epidermolysis bulloso**, (skin disease), **retinitis pigmentosa** and **spastic paraplegia**, etc.

Inheritance of Z-linked gene for colured feathers in chicken.

In birds, the males are homogametic (ZZ) and females are heterogametic (ZW). The sex linked genes follow Criss cross pattern.

In plymoth rock chicken, the gene for barred feather is dominant (B) and for red or black unbarred feathers is recessive (b). Both the genes are Z linked. When barred male chicken is crossed with non barred female, F_1 produces only barred males and females. F_1 barred males when crossed with hemizygous barred female the F_2 shows 50% non-barred and 50% barred females but 100% barred males (Fig. 46.21).

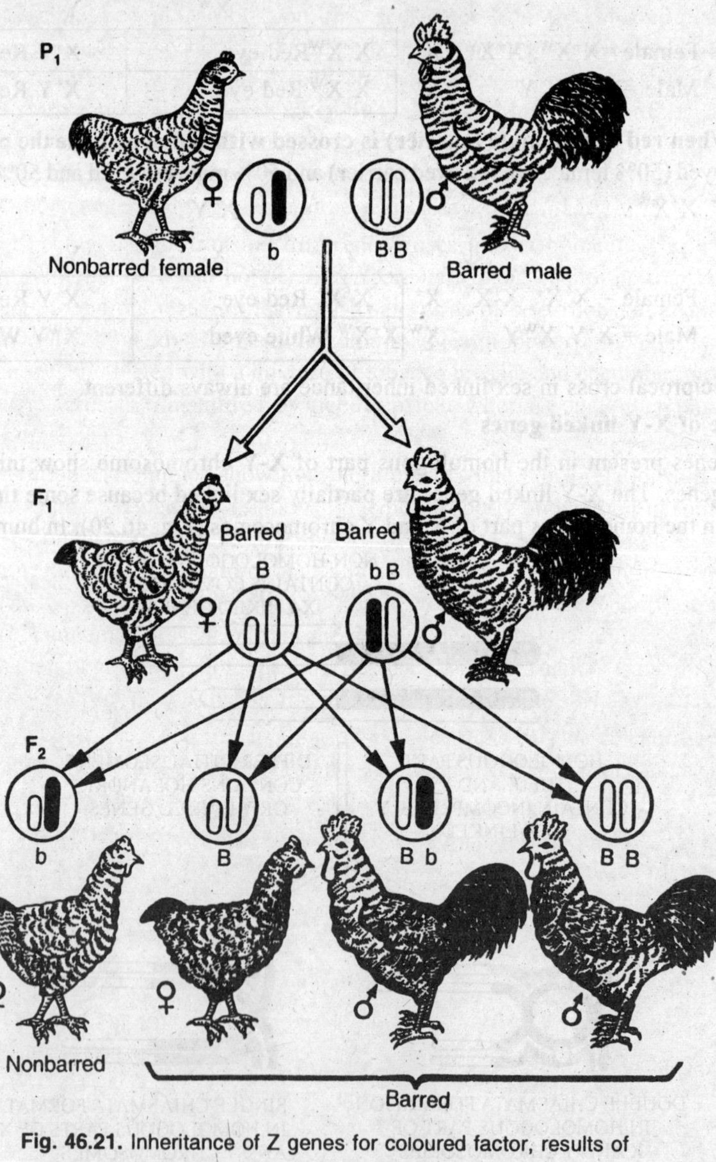

Fig. 46.21. Inheritance of Z genes for coloured factor, results of Cross between barred male and normal female.

When barred female is crossed with non-barred male the F_1 produces 50% non-barred female and 50% barred male when F_1 is self crossed in F_2 generation among the females (hen) 50% are barred and 50% non-barred while among the males (cock) 50% are barred and 50% are non-barred (Fig.46.22)

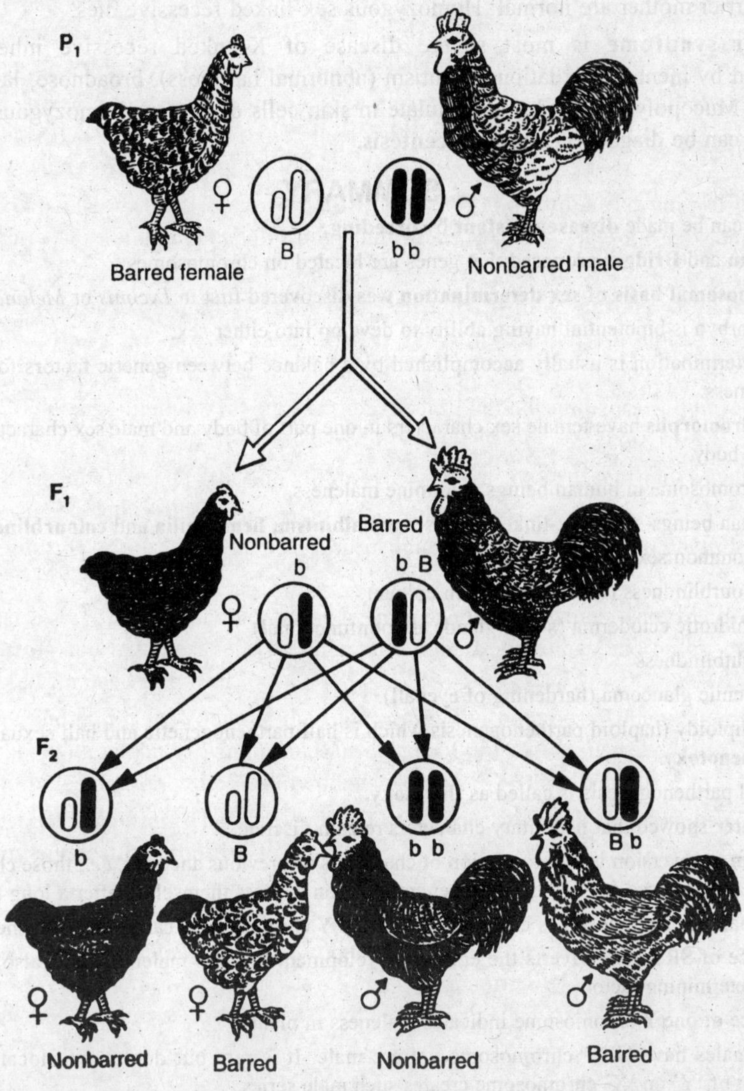

Fig. 46.22. A cross between a barred hen and a non-barred cock showing sex linked (z-linked) inheritance.

Deleterious Recessive sex-linked genes is human

Congenital hyperuricemia (Lesch-Nyhan syndrome), is characterized by excess production of uric acid, inherited through a **sex-linked recessive** gene. The gene for this is transferred from mother's 'X' chromosome to the zygote. The individual is deficient for enzyme **hypoxanthine guanine phosphoribosyltransferase** (HGPRT) which is involved in nucleotide synthesis. The symptoms start at the age of 10 months with orange uric acid crystal in urine; weak flabby muscles the child is unable to sit, walk, speak normally. Death is due to severe renal and neurological damage at the age of twenties.

Duchene type muscular distrophy

It is another sex-linked recessive gene disease. If mother is carrier of this gene half of her male children are expected to be affected. The disease affects males usually before they reach their teens with muscular deterioration. The individual dies before the age of 21. All female children born to a carrier mother are normal. Homozygous sex-linked recessive dies.

Hunter syndrome is most severe disease of X-linked recessive inheritance. It is characterized by mental retardation, hirsutism (abnormal hairiness), broadnose, large protruding tongue, etc. Mucopolysaccharides accumulate in skin cells of persons homozygous for the gene. The disease can be diagnosed by **amniocentesis.**

SUMMARY

1. Plants can be made **disease resistant** by **breeding.**
2. **Morgan** and **Bridges** proposed that genes are located on chromosomes.
3. **Chromosomal basis** of **sex determination** was discovered first in *Lychnis* or *Melandrium.*
4. The emrbyo is bipotential having ability to develop into either sex.
5. Sex determination is usually accomplished by a balance between genetic factors for maleness and femaleness.
6. **Gynadromorphs** have female sex characters in one part of body and male sex characters on the other part of body.
7. 'Y' chromosome in human beings determine maleness.
8. In human beings-some sex-linked diseases are **albinism, hemophilia** and **colourblindness,** etc.
9. Most common sex-linked diseases
   (a) Colourblindness (for red and green colour)
   (b) Anhidrotic ectoderma (sweat glands are nonfunctional)
   (c) Nightblindness
   (d) Juvenile glaucoma (hardening of eyeball).
10. Haplodiploidy (haploid parthenogenesis which is half parthenogenetic and half sexual) is also called as **arrhenotoky**.
11. Diploid parthenogenesis is called as **thelytoky.**
12. **Kolreuter** showed that hereditary characters remain discrete.
13. **Atavism** or reversion is the expression of characters of previous ancestor, *i.e.*, those characters which remain unexpressed for two or more generations can express themselves after a long time also.
14. In Human, 'X' chromosomes carry 102 genes and Y chromosomes carry only 10 genes.
15. Presence of SRY gene diverts the embryo development towards maleness. It is also called TDF or testis determining factor.
16. Presnece of one Y chromosome indicates maleness in plants.
17. Some males have 'XX' chromosome yet are male. It is rare but due to translocation of a small segment of 'Y' on 'X' chromosome creates such male series.
18. In alligators, sex is determined by temperature at which the egg is incubated.
19. Pseudogenes have poly 'A' tail.
20. Male worm *Ophryotrocha* produces sperm when young (male) but when it grows older it elongates upto 20 segments and become female and start producing eggs. If the size of it is reduced by cutting it again becomes male.
21. The discovery of sex-linked inheritance showed that genes were on sex chromosomes.
22. Sex-linked diseases inherited from X chromosomes in human beings are: colourblindness, hemophilia, opticatrophy, hairlessness, myopia, juvenile glaucoma, juvenile muscular dstrophy. distichiasis (double eye lashes), mitrasterosis (abnormal oscillation of eyeball).
23. XY-linked inheritance is due to genes located on homologous parts of X and Y chromosomes.

24. In X-linked inheritance, the characters are controlled by genes located on X chromosomes.
25. In Y linked inheritance, the characters are controlled by genes located on non-homologous points of 'Y' chromosome. These genes are called **holandric** genes.
26. In poultry, the sex linkage is **abraxis** type (ZZ, ZW).
27. **R. P. Roy** from Patna University studied the sex in diploid, triploid and tetraploid plants with and without 'Y' chromosome.
28. **C. B. Bridges** (1922) studied **genic balance** mechanism of sex determination in *Drosophila.*
29. **Allen** for the first time studied sex determination in liverworts (male gametophyte has 7A + Y, female gametophyte contains 7 A + X and the sporophyte has 14A + XY).
30. In human beings, 'Y' chromosome plays a definie role in sex determination (XO=Turner syndrome with sterile female; XXY = Klinefelter's syndrome with superfemales).
31. XY females are due to deletion of IA segment of short arm of 'Y' chromosome.
32. XY linked diseases are **epidermolysis bullosa** and **xeroderma pigmentosum.**
33. The genes on 'Y' chromosome are SRY (sex determing) or TDF (testis determins factor); ZFY (sperm formation), P ϕ 2V, hypertrichosis, keratoderma dissiapalum, etc.
34. *Asparagus* and maize show single gene cotnrol of sex.
35. Hormonal control of sex is shown by *Bonellia viridis*, some female owl, and cattles, etc.

QUESTION BANK

Very Short Answer Questions

1. Give an example of sex reversal.
2. Give an example of environment factor affecting sex determination.
3. Give an example of harmonal factor controlling sex determination.
4. What are holandric genes?
5. What is criss cross inheritance?
6. The character present on which chromosome will be transferred from father to son.
7. Give an example of sex linked inheritance.
8. Give an example of sex influenced expression.
9. In which plant single gene control of sex detrmination is studied.
10. What is X-Y linked gene.
11. Give two name of sex-linked diseases present on 'X'-chromosome.
12. Give two traits present on 'Y' chromosomes.
13. Is 'Y' chromosome important in *Drosophila* for sex determination.
14. Is 'Y' chromosome important is *Melandrium* for sex determination.
15. Name two sex-linked syndromes in man.

Short Answer Questions

1. Write notes on types of sex determining mechanism.
2. Write various types of combinations in sex linked inheritance.
3. Write short notes on 'Y' chromosome and sex determination in human.
4. Describe various regions on X and Y chromosomes of *Melandrium.*
5. Write notes on genic balance concept of sex determination in *Drosophila.*
6. Write short notes on hapdiploidy in insect.
7. Write short notes on gynandromorph.
8. Write short notes on genetic balance theory of sex determination in plants.
9. Write notes on environmental factors of sex determination.

10. Write short notes on : (a) Fragile X chrorrosome (b) Sex differentiation.
11. Write short notes on :
    (a) Sex linked traits (b) Sex influenced traits
12. Write short notes on single gene control of sex.
13. Write short notes on sex reversal.
14. Write short notes on compound sex chromosome.
15. Write in brief sex linked inheritance.

Long Answer Questions

1. What is sex chromosome ? What is its role in sex determination ?
2. How does a sex chromosome take part in sex-linked inheritance.
3. Describe the structure of sex chromosome in *Drosophila*, man and *Lychnis* or *Melandrium*.
4. Describe by giving various example environment factors and hormonal factors or sex determination.

Multiple Choice Questions (Objective Type Questions)

1. Chromosomes responsible for sex differentiation is called :
   (a) Sex chromosome (b) Somatic chromosome
   (c) Genome (d) Autosome
2. Males are generally :
   (a) Heterogametic (b) Homogametic
   (c) Homozygotic (d) Heterozygotic
3. Heterogametic females are found in :
   (a) Birds (b) Human
   (c) Both 'a' and 'b' (d) Insects
4. 'Y' chromosome is also known as :
   (a) Autosome (b) Androsome
   (c) Sex chromosome (d) Allosome
5. Morgan in *Drosophila* suggested that 'X' chromosome is rod-like and 'Y' chromosome is:
   (a) Rod-like (b) Hook-like
   (c) Cross-like (d) Dot-like
6. 'XO' method was proposed in 1902 while studying grasshopper by :
   (a) Mcclung (b) Thury
   (c) Morgan (d) Corrus
7. Haplodiploidy is also called as :
   (a) Arthotoky (b) Arrhenotoky
   (c) Thalytoky (d) Epigamic
8. Sex determination in dioecious plants was studied by:
   (a) Morgan (b) Wastergared and Warne
   (c) Mcclung (d) Avery and Mclead
9. Gynandromorphs are the result of first division in zygote as :
   (a) 2A + XX, 2A + X (b) 2A + XXX, 2A + Y
   (c) 2A + 2X, 2A + 2Y (d) 2X + 2Y, 2A + XX
10. Gynandromorph suggests that in sex determination in *Drosophila* there is no role of :
    (a) A-chromosome (b) B-chromosome
    (c) X-chromosome (d) Y-chromosome
11. In human sex determination major role is played by:
    (a) X-chromosome (b) Y-chromosome
    (c) A-chromosome (d) B-chromosome

12. Sterile female with abnormalities in XO condition is found in :
(a) Turner's syndrome (b) Down's syndrome
(c) Klinefelter's syndrome (d) Patau's syndrome

13. Despite the presence of **2X** chromosome and additional **Y** chromosome giving maleness to the individual in :
(a) Down's syndrome (b) Klinefelter's syndrome
(c) Turner's syndrome (d) Super male

14. In human being, rare XY females and XX males are found respectively due to:
(a) Deletion and translocation (b) Translocation and deletion
(c) Additon and addition (d) Translocation and transversion

15. IA region on segment of Y chromosome is responsible for
(a) Maleness (b) Femaleness
(c) Intersex (d) Sexual dimorphism

16. Single gene control of sex is found in :
(a) Maize (b) *Asparagus*
(c) Mango (d) Both 'a' and 'b'

17. In human female XX can be converted into XY male by the presence of :
(a) Sex linked genes (b) Sex reversal gene sxr
(c) Sex determination gene (SRY) (d) All the above

18. In *Bonellia viridis* (a marine worm), if the larva settles near the proboscis of adult female it will develop into :
(a) Male (b) Female
(c) Superfemale (d) Intersex

19. The studies of hormonal control of sex in fowl where fertile fowl changes to fertile male due to damage of ovary in female were done by :
(a) Puri and Garg (b) Hutington
(c) Crew (d) Grahm

20. 22 AA + XO and 22 AA + XXY show no bars body or one barr body respectively in :
(a) Turner's syndrome, Klinefelter's syndrome (b) Klinefelter's syndrome, Turner's syndrome
(c) Down's syndrome, Turner's syndrome (d) Turner's syndrome, Down's syndrome

21. Holandric inheritance is form :
(a) Female to son (b) Male to son
(c) Male to daughter (d) Female to daughter

22. Sex limited traits are those for which the genes are present but expressed due to some hormones :
(a) In males only (b) In females only
(c) Both male and female (d) Either male or female only

23. Sex influenced traits are those :
(a) Influenced by sex of bearer (b) Not influenced by sex of bearer
(c) Both 'a' and 'b' (d) Either 'a' or 'b'

24. Gene transfer froms parent to grand children by opposite sex children is called as :
(a) Sex linked inheritance (b) Autosomal inheritance
(c) Cris cross inheritance (d) Mitotic inheritance

25. Heteropyenotic X chromosome occurs as a round pedunculated body remains attached with nucleus is called as :
(a) Drumstick (b) Red spot
(c) Round body (d) Heteropycnotic body

26. A colourblind girl will only be born when her :
   (a) Parents are normal but grand parents are colourblind
   (b) Father and maternal grand father colourblind
   (c) Mother and maternal grand father are colourblind
   (d) Mother and paternal grand father are colourblind
27. If a physically female develops secondary male character the syndrome is called as :
   (a) Down's syndrome (b) Turner's syndrome
   (c) Klinefelter's syndrome (d) Patau's syndrome
28. Gynandromesph in *Drosophila* represent :
   (a) Head of female and abdomen of male (b) Half side (lateral) female and half male
   (c) Young male and old female (d) Old male and young female
29. Homogenetic member with ZZ sex chromosome was found in :
   (a) Hen (b) Cock (fowl)
   (c) *Drosophila* (d) All the above
30. What is the name of Indian scientist from Patna University who studied sex is diploid, triploid and tetraploid plants with and without 'Y' chromosomes :
   (a) C. B. Gupta (b) R. P. Singh
   (c) R. P. Roy (d) S. P. Roychaudhry

ANSWERS

1. (a)	2. (a)	3. (a)	4. (b)	5. (b)	6. (a)
7. (b)	8. (b)	9. (a)	10. (d)	11. (b)	12. (a)
13. (b)	14. (a)	15. (a)	16. (d)	17. (b)	18. (a)
19. (c)	20. (a)	21. (b)	22. (d)	23. (a)	24. (c)
25. (a)	26. (b)	27. (c)	28. (b)	29. (b)	30. (c)

47

CHAPTER

Extra-Nuclear or Cytoplasmic Inheritance

Although, the genes of nuclear chromosomes have a significant and key role in the inheritance of almost all traits from generations to generations, yet they altogether cannot be considered as the sole vehicles of inheritance, because certain experimental evidences suggest the occurrence of certain extranuclear genes or DNA molecules in the cytoplasm of many prokaryotic and eukaryotic cells. For example, bacterial cells such as *Escherichia coli* possesses a single main chromosome in the nucloid and often extra DNA elements called **plasmids** in the cytoplasm; the eukaryotic cells possess a main complement or chromosomes in the nucleus and extra-DNA molecules or chromosomes in their mitochondria and chloroplasts. Qualifying also as extra hereditary elements are certain viruses, bacteria, and algae (which live as endosymbionts inside the eukaryotic cells). These cytoplasmic extra-nuclear genes or DNA molecules of plasmids, mitochondria, chloroplasts, endosymbionts and cellular surfaces have a characteristic pattern of inheritance which does not resemble with that of genes of nuclear chromosomes and is known by different terms such as **non-Mendelian, non-chromosomal, uniparental, maternal, extra-chromosomal, cytoplasmic** and **extra-nuclear** inheritance.

EVIDENCES FOR CYTOPLASMIC FACTORS

Traits with extranuclear basis are identified by the accumulated evidence from a number of diagnostic criteria such as follows :

1. Because the female gamete contributes almost all of the cytoplasm to the zygote and male gamete (sperm or pollen) contributes only a nucleus, an inheritance pattern that differs between reciprocal crosses suggests a cytoplasmic involvement. This is clearly the basis for **uniparental** or **maternal inheritance** where the progeny always resemble one parent, most commonly the female parent (*e.g.*, shell coiling in *Limnaea peregra*).

2. Differences in reciprocal crosses which cannot be attributed to sex-linkage or some other chromosomal basis tend to implicate extranuclear factors (*e.g.*, chloroplast inheritance in *Mirabilis jalapa*).

3. The uniparental inheritance of a trait which cannot be attributed to unequal cytoplasmic contributions from parental gametes may, however, involve cytoplasmic factors (*e.g.*, streptomycin resistance in *Chlamydomonas*).

4. Whenever trait fails to demonstrate the classical segregation patterns and deviates from standard ratios, the conclusion is again a cytoplasm based type of inheritance (*e.g.*, mitochondrial inheritance in yeast).

5. When the trait fails to show linkage to any known nuclear linkage groups and assort independently from nuclear genes, a cytoplasmic mode of inheritance is suggested.

6. Many types of mutants that fit the above criteria will show segregation during mitotic division. This is very common in variegated plants that carry more than one type of plastid (chloroplast) per cell.

The cytoplasmic inheritance, therefore, will be, understood to be based on cytoplasmically located, independent, self-replicating nucleic acids, which differ from chromosomal genes by their location within the cell, and have their own nucleotide sequences. The smallest heritable extra chromosomal unit is called a **plasmagene.** All of the plasmagenes of a cell constitute the **plasmon.**

A. EXTRA-NUCLEAR INHERITANCE IN EUKARYOTES

Many geneticists have studied various cases of extra-nuclear inheritance in different eukaryotes. Certain most important examples of extra-nuclear inheritance in eukaryotes are the following:

1. Maternal Inheritance

In certain cases, it has been observed that certain characteristic phenotypic traits of F_1, F_2 or F_3 progeny are not the expression of their own genes, but rather those of the maternal parents. Such phenotypic expressions of maternal genes (genotype) may be short-lived or may persist throughout the life-span of the individual. The substances which produce the maternal effects in the progeny are found to be transcriptional products (*i.e.*, mRNA, rRNA and tRNA) of maternal genes which have been manufactured during oogenesis and which exist in the ooplasm of unfertilized eggs in the form of inactive protein coated and late translating mRNA molecules **(informosomes)** or inactivated rRNA and tRNA. These transcriptional products of maternal genes produce their phenotypic effects during early cleavage and blastulation when there occur little or no transcription since, maternal and paternal genes of zygote remain engaged in mitotic replication or duplication of DNA. There may be other reasons of maternal affect which are still little understood. The maternal inheritance has been studied in ***Limnaea*** (a snail).

Shell coiling in *Limnaea*. In the snails (gastropods), the shell is spirally coiled. In most cases the direction of coiling of the shell is clockwise, if viewed from apex of the shell. This type of coiling is called **dextral.** However, in some snails the coiling of shell may be counter clockwise or **sinistral.** Both types of coilings are produced by two different types of genetically controlled cleavages, one being **dextral cleavage,** another being **sinistral cleavage** (Fig. 47.1).

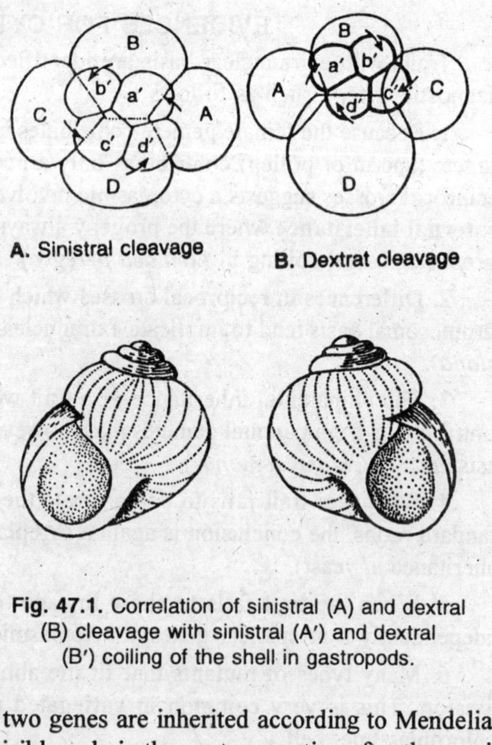

Fig. 47.1. Correlation of sinistral (A) and dextral (B) cleavage with sinistral (A′) and dextral (B′) coiling of the shell in gastropods.

There are some species of gastropods in which all the individuals are sinistral but the main interest attaches to a species in which sinistral individuals occur as a mutation among a population of normal dextral animals. Such a mutant was discovered in the freshwater snail *Limnaea peregra* (**A. Sturtevant,**1923). Breeding and cross breeding of dextral and sinistral snails showed that the difference between the two forms is dependent on a pair of allelomorphic genes, the gene for sinistrality being recessive (S), and the gene for the normal dextral coiling being dominant (S^+). The two genes are inherited according to Mendelian laws, but the action of any genic combination is visible only in the next generation after the one in which a given genotype is found.

The eggs of a homozygous sinistral individual (SS) are fertilized by the sperm of a dextral individual (S^+S^+), the eggs cleave sinistrally and all the snails of this F_1 generation show a sinistral coiling of the shell. Thus, the gene of sperm (S^+) do not manifest themselve, although the genotype of the F_1 generation is S^+S. If a second generation (F_2) is bred from such F_1 sinistral individuals, it is all dextral, instead of showing segregation as would be expected in normal Mendelian inheritance. In fact, segregation does take place in the F_2 generation so far as the genes are concerned, but the new genic combinations fails to manifest themselves, since the coiling is determined by the genotype of the mother.

The genotype of F_1 mother being S^+S, the gene for dextrality dominates and is responsible for the exclusively dextral coiling of the second generation. Only in the F_3 generation does segregation in the ratio of 3 : 1 becomes apparent, since the individuals of the F_2 generation had the genotypes —$1S^+S^+$; 2 S^+S, 1 SS, 1/4 of them, on the average, produce eggs developing into sinistral individuals (Fig. 47.2).

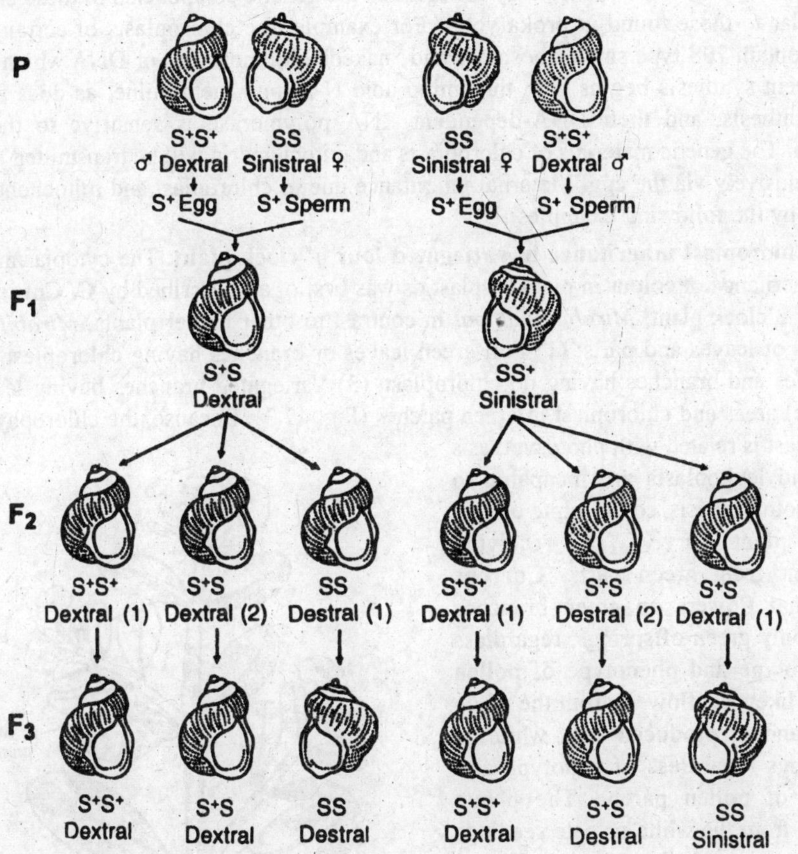

Fig. 47.2. Maternal effect in direction of coiling of the shell in *Limnaea*.

It is easy to understand that the results of a reciprocal cross that is, of the fertilization of the eggs of a homozygous dextral individual (S^+S^+) by the sperm of a sinistral individual (SS)—will lead to a somewhat different type of pedigree: the F_1 generation will be dextral (with genotype S^+S) and the F_2 generation again all dextral (with genotypic ratio of $1S^+S^+$:$2S^+S$: ISS). The F_3 generation will show segregation among broods, just as in the cross examined first.

The whole case becomes clear if it is realized that the type of cleavage (sinistral or dextral) depends on the organization of the egg which is established before the maturation division of the oocyte nucleus. The type of cleavage is, therefore, under the influence of the genotype of the maternal parent. The sperm enters the egg after this organization is already established. Lastly, the direction of coiling of shell depends upon the orientation of the mitotic spindle of first cleavage of the zygote. If the spindle is tipped toward the left of the median line of the egg cell, the sinistral pattern will develop; conversely if the mitotic spindle is tipped toward the right of the median line of the cell, the dextral pattern will develop. The spindle orientation is, thus, controlled by the organization of ooplasm which becomes established during oogenesis and before fertilization.

2. Extra-nuclear Inheritance by Cellular Organelles

Chloroplasts and mitochondria and organelles that contain their own DNA and protein-synthesizing apparatus. A widely held theory concerning their origin proposes that they were once infectious endosymbiotic prokaryotes that evolved such a dependence on the gene products of the host that they are no longer able to function autonomously.

This theory has been supported by the fact that the genetic components of these organelle are often similar to those found in prokaryotes. For example, the chloroplasts of certain algae and *Euglena* contain 70S type small ribosomes and "naked" chromosomes or DNA which is circular. Their protein synthesis begins with the amino acid N-formyl methionine, as does prokaryotic protein synthesis, and their DNA-dependent RNA polymerase is sensitive to the inhibitor **rifampicin.** The genetic materials of chloroplasts and mitochondria will be transmitted to offspring almost exclusively via the egg. Maternal inheritance due to chloroplast and mitochondria is well illustrated by the following examples :

(a) Chloroplast inheritance in variegated four o' clock plant. The cytoplasmic or extra-nuclear inheritance of colour in plant by plastids was first of all described by **C. Correns** in 1908 in the four o'clock plant, *Mirabilis jalapa.* In contrast to other higher plants, *Mirabilis* contains three types of leaves and parts: (1) Full green leaves or branches having chloroplast, (2) White (pale) leaves and branches having no chloroplast, (3) Variegated branches having leukoplast in white (pale) areas and chloroplast in green patches (Fig. 47.3). Because, the chlorophyll pigment of chloroplast is related with photosynthesis of food and leukoplasts are incapable to perform photosynthesis, so the white or pale parts of plant survive by receiving nourishment from green parts. **Correns** reported that flowers on green branches produced only green offsprings, regardless of the genotype and phenotype of pollen parent and likewise, flowers from the white or pale branches produced only white or pale seedings regardless of genotype and phenotype of pollen parent. The plants developing from the white or pale seedings die because they lack chlorophyll and cannot carry on photosynthesis. **Correns** further reported that flowers from the variegated branches yielded mixed progeny of green, white (pale) and variegated plants in widely varying ratios (Fig. 47.4). These results are summarized in Table 47.1.

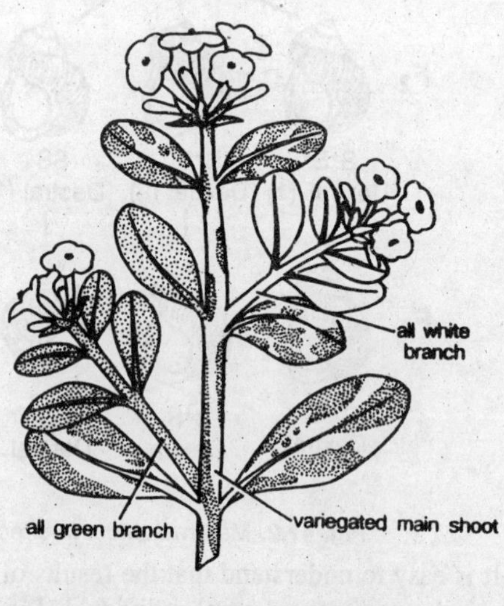

Fig. 47.3. Leaf variegation in *Mirabilis jalapa*, the four-o' clock plant. Flowers may form on any branch (variegated, green, or white), and these flowers may be used in crosses.

The irregularity of transmission from variegated branches could be understood by considering cytoplasmic genes (plasmagenes) of plastids. A study of the egg during oogenesis in *Mirabilis* reveals that the ooplasm contains plastids like cytoplasm of other plant cells. If the egg cell is derived from green plant tissues, its ooplasm will contain coloured plastids; if derived from white plant tissues, its ooplasm will contain white plastids; if derived from variegated tissues, its cytoplasm may contain coloured plastids only, white plastids only or a mixture of coloured and white plastids. A study of the pollenogenesis, however, reveals that pollen contains very little cytoplasm which in most cases is devoid of plastids. Without the plastids, the pollen cannot affect this aspect of the offspring's phenotype.

Mitotic segregation. Variegated branches of *Mirabilis jalapa* produce three kinds of eggs : some contain only white chloroplasts, some contain only green chloroplasts and some contain both types of chloroplasts. In the subsequent mitotic divisions, some form of cytoplasmic segregation occurs that segregate the chloroplast types into pure cell lines, thus, producing the variegated phenotype in the progeny individual. This process of sorting might be described as **"mitotic segregation"**of this is a pure extranuclear phenomenon. In mitotic segregation since both segregation and recombination of organelle genotype takes place, so it is called **cytoplasmic segregation and recombination** (its acronym is **CSAR**).

(b) Maternal inheritance by iojap gene of corn: Another example from higher plants also suggests the existence of plastid genes controlling plastid integrity. A gene in corn plant called **iojap (ij)** has been mapped by **M. Rhoades** (1946) to nuclear chromosomes 7. Plants homozygous for ij are either inviable white seedings or variegated with a characteristic white striping, the phenotype being known as **striped.** When the variegated plants serve as females in a cross, they give rise to green, white, and striped progeny, regardless of the nuclear genotype of the paternal parent. Thus, if the pollen derives from a normal green Ij/Ij plant as in Figure 47.5 b, the resulting progeny will be Ij/ij heterozygotes, but many will exhibit abnormal plastid pigmentation: the presence of the "normal" Ij gene has no curative effect. In the reciprocal Ij/Ij female X ij/ij male cross (Fig. 47.5). On the other hand, the Ij/ij progeny are all normally pigmented.

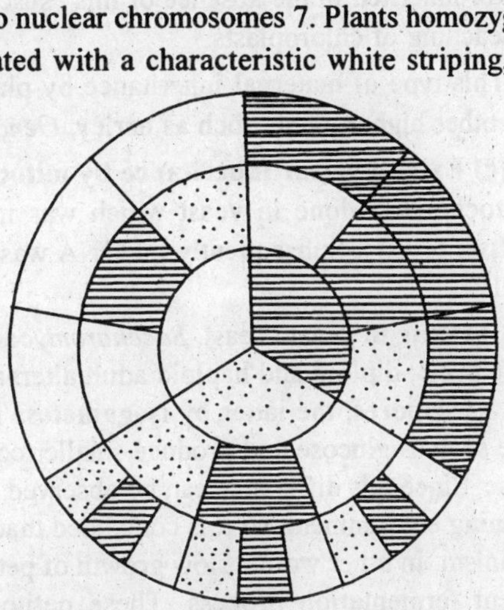

Fig. 47.4. Plastid inheritance of *Mirabilis jalapa*. The central circle represents the type of branch that produces flowers which are pollinated. Intermediate circle represents branch from which pollen is used and outer circle shows the progeny.

The iojap trait, thus, exhibits classical maternal inheritance once it has become established in an ij/ij plant. Moreover, once established, it becomes independent of the ij gene, as can be demonstrated by crossing F_1 Ij/ij variegated females to Ij/Ij normal males. As shown in Figure (47.5c), a mixture of green, striped and white progeny again results, even though some of the striped and white plants now have an Ij/Ij genotype. Thus, the iojap trait, once established, is permanent.

Table 47.1. Chloroplast inheritance in variegated four o'clock plants.

Branch of origin of the male parent	Branch of origin of the female parent	Progeny
Green	Green	Green
	Pale or white	Pale or white
	Variegated	Green, pale or white, variegated
Pale or white	Green	Green
	Pale or white	Pale or white
	Variegated	Green, pale or white, variegated
Variegated	Green	Green
	Pale or white	Pale or white
	Variegated	Green, pale or white, variegated

The iojap phenomenon has been explained by two hypotheses. One hypothesis holds that the ij/ij genetic constitution could bring about or permit, frequent mutations in the chloroplast genome that result the production of lines of abnormal plastids. Another hypothesis suggests that certain cytoplasmic elements other than chloroplast mutations come into being or residence in ij/ij cells, are later inherited in the absence of this "susceptible" or "permissive" genotype, and bring about the bleaching of chloroplasts.

This type of maternal inheritance by plasmagenes of chloroplasts has been also studied in many other higher plants such as barley, *Oenothera* sp., rice, etc.

(c) Extra-nuclear inheritance by mitochondria. The most important work on the genetics of mitochondria done in yeast which was initiated by the discovery of petite mutants by **B. Ephrussi** (1953). Subsequently mt DNA was studied in several organisms including plants and animals.

(i) Petite in yeast. Yeast, *Saccharomyces cerevisiae,* are single-celled ascomycetes fungi. In the life cycle, diploid and haploid adult alternates, the former reproducing by asexual meiospores called **ascospores**, the latter by **isogametes.** The **petite** mutants in yeast fail to grow on carbon source such as glucose and produce smaller colonies (the "littles") when grown on sugars such as glucose. Since this difference can be observed only when such yeast cultures are kept in a oxygen-containing environment; so it is concluded that petite mutants have a defective aerobic respiratory mechanism. In other words, slow growth of petite can be attributed to yeast cells utilization of less efficient fermentation process. These petites differ from wild type, called **grande** and are characterized by (i) their insensitivity to inhibitors of aerobic pathways (such as cyanide), (ii) absence of cytochromes a, a_3, b and a number of other changes in mitochondrial respiratory enzymes; (iii) incomplete development of mitochondria; and (iv) lack of stainability of petite mitochondria.

The petite mutants can be **segregational,** *i.e.*, they follow mendelian segregation and, therefore, presumably controlled by chromosomal genes. They may also be **vegetative,** *i.e.*, non-segregational or extra-chromosomal. The genetic basis of petite character is a cytoplasmic factor ρ^+ (rho) which may be absent or defective in petites. Thus, a vegetative petite can be **neutral** (ρ^0) which completely lack ρ^+ or it may be **suppressive** (ρ^-) having a defective ρ^+. The neutral petites are not transmitted while suppressive petites are transmitted to a fraction of vegetative diploid progeny. In various strains of yeast, the suppressiveness varies from 1–99 per cent petites. The

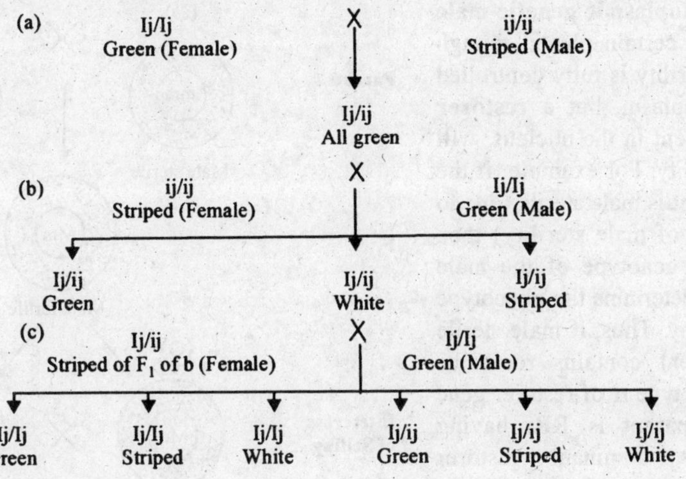

Fig. 47.5. (a) Cross between green (normal) and striped (iojap) plants. (b) Reciprocal of cross (a) (c) Cross of F_1 striped females (of cross b) to normal (green) males.

following two lines of evidences have suggested the association of ρ^+ with mitochondrial DNA (mt DNA); (1) **Ethidium bromide,** which induces pétite mutations with 100 per cent efficiency, causes degradation of mt DNA after prolonged exposure of cells. In fact, neutral petites have been found lacking in mt DNA. (2) Supressive petites contain mt DNA which is greatly altered in base composition with respect to wild mt DNA.

(ii) Poky strain of *Neurospora.* In fungi, *Neurospora crassa* a number of mutations of mitochondria are inherited via the female parent. The best studied of these is the **poky strain** of *N. crassa,* first isolated by **Mitchell** and **Mitchell** (1952). A poky mutant differs from wild type strain of *Neurospora* in the following aspects: (1) it is slow growing; (2) it shows maternal inheritance, and (3) it has abnormal cytochromes. Of the three cytochromes—cyt a, b and c found in wild type, cyt a and cyt b are absent, and cyt c is in excess in poky mutant. In reciprocal crosses, poky character shows maternal inheritance:

poky (female) × wild type (male) → all poky

wild type (female) × poky (male) → all wild type

However, there are other marker nuclear genes (ad^+/ad^-) which show 1:1 mendelian segregation. The following evidences suggested that poky trait may be located in mitochondrial DNA: (i) slow growth may be due to lack of ATP energy and source of this energy is mitochondria; (ii) cytochromes in poky strain differ from those in wild type in quality and quantity and these cytochromes are found in mitochondria.

(iii) Male sterility in plants. In plants, the phenotype of male sterility is found to be controlled either by nuclear genes or plasmagenes (cytoplasm) or by both. Therefore, the trait of male sterility of plants is controlled by the following three methods :

(a) Genetic male sterility. In this type of male sterility the sterility is controlled by a single nuclear gene which is recessive to fertility, so that the F_1 progeny would be fertile and in F_2 generation, the fertile and sterile individuals will be segregated in the typical 3:1 ratio (Fig 47.6).

(b) Cytoplasmic male sterility (CMS). In maize and many other plants, cytoplasmic control of male sterility is known. In such cases, if the female parent is male sterile (having plasmagene for male sterility), the F_1 progeny would always be male sterile, because the cytoplasm is mainly derived from the egg which is obtained from the male sterile female parent (Fig 47.7).

(c) Cytoplasmic genetic male sterility. In certain plants, though the male sterility is fully controlled by the cytoplasm, but a **restorer gene** if present in the nucleus, will restore fertility. For example, if the female parent is male sterile (due to plasmagene of male sterility) then the nuclear genotype of the male parent will determine the phenotype of F_1 progeny. Thus, if male sterile female parent contains recessive nuclear genotype rr of restorer gene and male parent is RR, having homozygous dominant restorer genes. Their F_1 progeny would be male fertile Rr. However, if the male parent is male fertile rr, the F_1 progeny would be male sterile rr. If the F_1 male fertile heterozygote (Rr) is test crossed with male fertile rr male, a progeny with 50 per cent male fertile and 50 per cent male sterile will be obtained (Fig. 47.8).

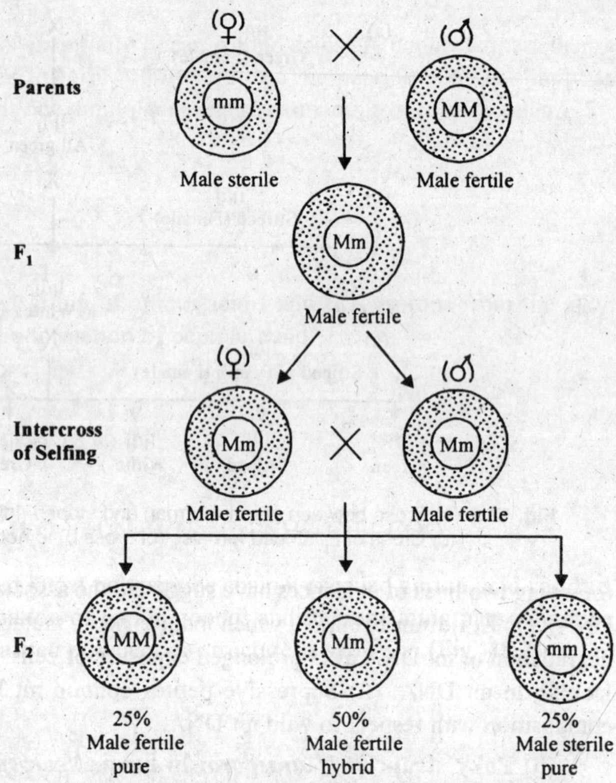

Fig. 47.6. Inheritance pattern of genetic male sterility.

Since, in maize expression of male sterility depends on an interaction between nuclear and extrachromosomal genes. Male sterile lines can bear seeds only after cross-pollination. For this reason they are useful in raising hybrid seeds, especially on large scale.

Later on, in maize the following four types of cytoplasms have been recognized : the normal (N) cytoplasm and three types of male sterile cytoplasms (T, C and S). The recent studies of mitochondria in these cytoplasm revealed that the factors responsible for cytoplasmic male sterility are located in mitochondrial DNA (mt DNA) and mt DNA of N, T, C and S cytoplasms are found to be different. The cytoplasmic male sterility (CMS) of C and S type can be reversed by nuclear storer genes, however, the CMS-T cannot.

3. Extra-Nuclear Inheritance by Endosymbionts

Certain intra-cellular parasites such as bacteria and virus particles maintain symbiotic relationship with host cells. They are self-reproducing and look like the cytoplasmic inclusions. Sometimes they exhibit an infection like transmission with a hereditary continuity of their own. Generally such symbionts are coined by letters of the Greek alphabets (sigma, kappa, mμ, etc.). The various types of infective symbionts are as follows :

(i) Sigma virus in *Drosophila* L. Heritier and **Teissier** found that a certain strain of *Drosophila melanogaster* shows a high degree of sensitivity to carbon dioxide, where as the wild type strain can be exposed for long periods to pure CO_2 without permanent damage, the sensitive strain quickly becomes uncoordinated in even brief exposure to low concentrations. This trait (extra-sensitivity) is transmitted primarily, but not exclusively, through the maternal parent. Tests have disclosed that CO_2 sensitivity is dependent upon an infectious DNA virus called **sigma**, found in the cytoplasm of **CO_2 sensitives** *Drosophila*. These infective particles are transmitted normally via the egg's larger amount of cytoplasm but occasionally through the sperms as well. Carbon dioxide sensitivity may even be induced in normal flies by injections of cell free extracts of sigma particles from CO_2 sensitive flies.

(ii) Spirochaetes and maternal sex ratio in *Drosophila*. Females of many *Drosophila* species can harbour a population of **spirochaete bacteria** known generally as **SR.** When **SR** spirochaetes infect the eggs of the host and when these eggs are fertilized, virtually all XY zygotes are killed early in embroyonic development and XX zygotes survive. Thus, the spirochaete can be considered as an endosymbiont of female but not of male *Drosophila*, and its presence in the female gives rise to the condition called **maternal sex ratio**, in which the progeny are exclusively or almost entirely female.

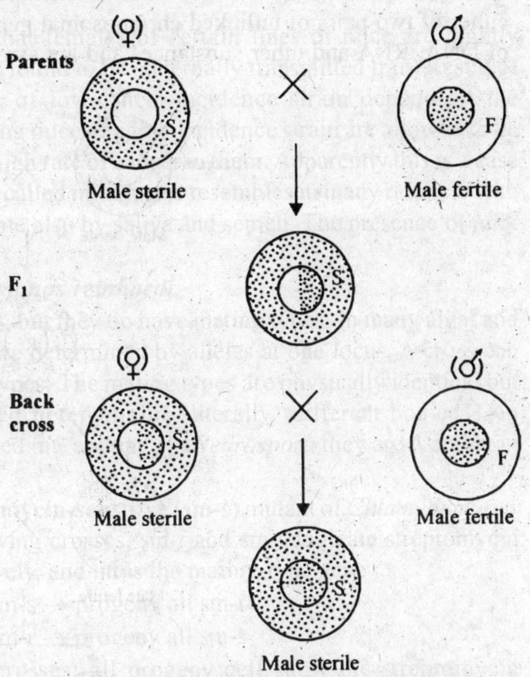

Fig. 47.7. Maternal inheritance of cytoplasmic male sterility.

The SR spirochaete is infectious, for when isolated from the haemolymph of female carriers and introduced into normal females the latter become carriers. Why the female genotype permits their retention and conversely, why XY cells are sensitive to their presence is not yet known. **K. Oishi** and **D. Poulson** (1970) have reported DNA-containing viruses in these endosymbiont spirochaetes of female *Drosophila*.

(iii) Kappa particles. In 1938, **T.M. Sonneborn** reported that some races (known as **"killers"** or **killer strain**) of the common ciliate protozoan, *Paramecium aurelia* produce a poisonous substance, called **paramecin** which is lethal to other individuals called **"sensitives"**. The paramecin is water soluble, diffusible and depends for its production upon cytoplasmically located particles called **kappa.** Electron microscopic observations have shown that kappa particles are about 0.4µ long symbiotic bacteria, *Caedobacter taeniospiralis*; 20 per cent of kappa bacteria of the killer strain contain a refractile protein containing **"R body"** and are called **"brights"**. They are infected with a virus that controls the synthesis of toxic viral protein, the paramecin. A killer *Paramecium* may contain hundreds (*e.g.*, 400) of kappa particles. The presence of kappa particles in the killer *Paramecium* is dependent for their maintenance and replication on the chromosomal dominant gene K. Paramecia with nuclear genotype kk are unable to harbour kappa particles.

When a *Paramecium* of killer strain having the genotype KK or (K^+) conjugates with the *Paramecium* of non-killer strain having , the genotype **kk**, the exconjugants are all heterozygous for Kk gents (Fig. 47.9). The Kk genotype suggests that both exconjugant should be killers. But this is not the case. If conjugation is normal, *i.e.*, lasts only for a short time, and no exchange of cytoplasm takes place between the two, both killers and non-killers (sensitive) are produced. However, rare or prolonged conjugation (*i.e.*, lasting for long time) permits mixing of cytoplasm of both conjugants and results killers only. The killer trait is stable only in killer strain with KK genotype and is suitable in sensitive strain with kk genotype.

(iv) mµ particle. Another type of killer trait known as **mate killer** has been reported in *Paramecium* by **R.W. Siegel** in 1952. The mate killer trait is imparted by a cytoplasmic mµ particle and a *Paramecium* with a mµ particle is called **mate killer** because when it conjugates with a *Paramecium* without any mµ particle is called **mate sensitive**, then it kills the latter. The mµ particles exist only in those cells whose micronucleus contains at least one dominant gene of either of two pairs of unlinked chromosomal genes (M_1 and M_2). The mμ particles are composed of DNA, RNA and other substances and are symbionts.

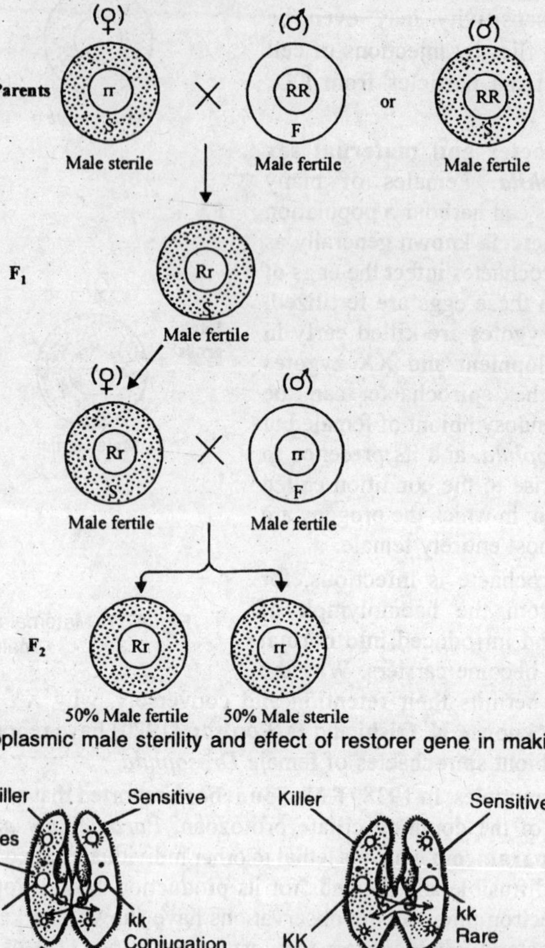

Fig. 47.8. Inheritance of cytoplasmic male sterility and effect of restorer gene in making male steriles into male fertiles.

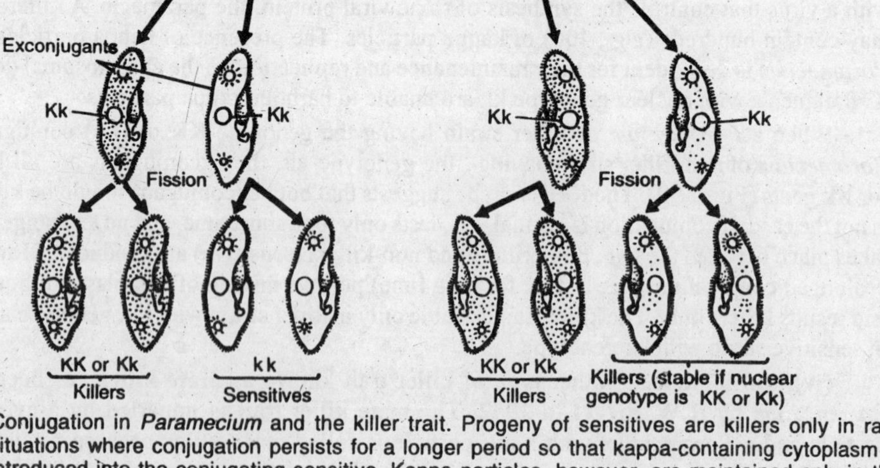

Fig. 47.9. Conjugation in *Paramecium* and the killer trait. Progeny of sensitives are killers only in rare situations where conjugation persists for a longer period so that kappa-containing cytoplasm is introduced into the conjugating sensitive. Kappa particles, however, are maintained only in the presence of K-nuclear genotype.

(v) Milk factor in mice. Bittner found that females of certain lines of mice are hightly susceptible to mammary cancer and this trait was found to be maternally transmitted trait. Result of reciprocal crosses between these and animals of low-cancer-incidence strain depend on the characteristic of the female parent. When the young mice of a low-incidence strain are allowed to be nursed by susceptible foster mothers produces a high rate of cancer in them. Apparently this is a case of infective agent transmitted in the milk. This so called milk factor resembles in many respects with a virus and has been discovered to be transmissible also by saliva and semen. The presence of milk factor also depends on nuclear genes.

4. Uniparental Inheritance *in Chlamydomonas reinhardi*

Like fungi, algae rarely have different sexes, but they do have mating types. In many algal and fungal species, there are two mating types that are determined by alleles at one locus. A cross can occur only if the parents are of different mating types. The mating types are physically identical but physiologically different. Such species are called **heterothallic** (literally "different bodied"). In *Chlamydomonas,* the mating type alleles are called mt^+ and mt^- (in *Neurospora* they are A and a; in yeast a and α).

In 1954, **Ms. Ruth Sagar** isolated a **streptomycin-sensitive (sm-s)** mutant of *Chlamydomonas* with a peculiar inheritance pattern. In the following crosses, sm-r and sm-s indicate streptomycin resistance and streptomycin sensitivity, respectively, and mt is the mating-type gene :

$$mt^+ \text{ sm-r} \times mt^- \text{ sm-s} \rightarrow \text{progeny all sm-r}$$
$$mt^+ \text{ sm-s} \times mt^- \text{ sm-r} \rightarrow \text{progeny all sm-s}$$

Here, occurs a difference in reciprocal crosses; all progeny cell show the streptomycin phenotype of the mt^+ parent. Like the maternal inheritance this is a case of **uniparental inheritance.** In fact, **Sagar** now refers to the mt^+ mating type as the female, using this analogy.

Uses of Extra-nuclear Genome

1. It prevents total loss of organelles due to a single mutation in nuclear gene.
2. It provides a reservoir of cytoplasmic mutations.
3. It is useful under adverse environmental condition.

SUMMARY

1. Cytoplasmic inheritance due to chloroplastic genome is called **plastidome**.
2. Cytoplasmic inheritance includes the transfer of characters from mother (egg) only.
3. Cytoplasmic inheritance never shows mendalian ratio.
4. Cytoplasmic inheritance is carried out by **plasmagenes.**
5. Plastid transmission in plants is reported by **Correns** in *Mirabilis jalapa.*
6. **T. M. Sonneborn** studied the transmission of certain cytoplasmic particles in *Paramecium aurelia.* These particles are called **kappa** paticles.
7. **L. Heritier** and **Teissier** studied *Drosophila melanogaster* sensitive to CO_2 due to transmission of **sigma factor** to the zygote cells through ova.
8. **J. J. Bittner** studied a strain of mice developing breast cancer which are transmitted to offspring through milk of mother. This is called **'milk factor'.**
9. **B. Ephrussi** investigated that yeast red strain is due to the presence of some cytoplasmic particles.
10. **M. B, Mitchell** and **H. K. Mitchell** reported some poky (p^0) strains in *Neurospora.*
11. **Episomes in bacteria** are the best examples of cytoplasmically located genetic determiner called bacterial episome 'F'.
12. **Shell coiling of *Limnaea*** (snail) is determined by a pair of genes.The coiling may be sinistral (left) or dextral (right). The coiling is determined by maternal snail. It was studied by **Conklin** et al.
13. Cytoplasmic inheritance is significant because :
    (a) It prevents the total loss of organelles due to a single mutation in nuclear genes.
    (b) It enhances the chances of cytoplasmic mutation useful under favourable conditions.
14. **Cytoplasmic** inheritance is usually **uniparental.**
15. When maternal genes have effect on the offspring through cytoplasm of egg but not involving plasmagenes, it is called **predetermination effect.**

QUESTION BANK

Very Short Answer Questions

1. Cytoplasmic inheritance in maize is studied by whom ?
2. Name one example of plastidal inheritance.
3. Who explained the streptomycin resistance gene in *Chlamydomonas* ?
4. Name two examples studied by you for haploid cytoplasmic inheritance.
5. Who discovered cytoplasmic male sterility in maize.
6. Name the types of plasmon.
7. In which plant **poky** mutants are found.
8. Write 2 uses of extra nuclear genome.

Short Answer Questions

1. Write the characters of extrachromosomal inheritance.
2. Explain the type of inheritance giving example of snail coiling.
3. Describe streptomycin resistance in *Chlamydomonas.*
4. Describe in short one example of plastogene inheritance.
5. Give one example of chondriogene inheritance.
6. Write short notes on :
   (a) Petite mutants (b) Sigma factor (c) rho factor
7. What are the resemblances between chloroplast and prokaryotes.

Long Answer Questions

1. What is cytoplasmic inheritance. Explain it with two suitable examples.
2. Explain the inheritance of kappa particles in *Paramecium*.
3. Give an illustrated account of the extra chromosomal inheritance in maize.
4. Describe in brief the cytoplasmic inheritance in yeast cells.

Multiple Choice Questions (Objective Type Questions)

1. Extranuclear chromosome is :
   (a) Circular and without protein coat (b) Circular and with protein coat
   (c) Linear and without protein coat (d) Linear and with protein coat
2. Cytoplasmic inheritance is also known as :
   (a) Paternal inheritance (b) Cris cross inheritance
   (c) Maternal inheritance (d) Sexlinked inheritance
3. Cytoplasmic inheritance in *Mirabilis jalapa* was reported by :
   (a) Morgan (b) Correns (c) Mc clincook (d) De Veries
4. Cytoplasmic inheritance of kappa particles was reported by T.M. Sonneborn in :
   (a) *Paramecium aureus* (b) *Paramecium aurelia*
   (c) *Entamoeba histolytica* (d) *Ascaris lumbricoides*
5. Studies on *Drosophila melanogaster* that is sensitive to CO_2 due to **sigma** particles transmission from ova to zygote was reported by :
   (a) Korper and Kappa (b) Bittner (c) Morgan (d) Heritier and Teisier
6. Bacterial episome 'F' is cytoplasmically located :
   (a) Genetic factor (b) Milk factor (c) Genetic determiner (d) Environmental factor
7. If dextral female snail is crossed with sinistral male, the resultant will be :
   (a) All female snails dextral (b) All male snails dextral
   (c) All snails dextral (d) 3 snails dextral and one sinistral
8. *Drosophila is* CO_2 sensitive due to the presence of :
   (a) *rho* factor (b) *sigma* factor (c) *milk* factor (d) none of the above
9. Iojap inheritance in corn is due to the presence of gene :
   (a) IJ (b) Ij (c) iJ (d) ij
10. Non chromosomal genes in *Chlamydomonas* is studied by :
    (a) Sagar (b) Strutvent (c) Correns (d) Morgan
11. Petite mutants of yeast are due to :
    (a) Genome (b) Chondriogenes (c) Plastogenes (d) Kinetogenes

ANSWERS

1. (a) 2. (c) 3. (b) 4. (b) 5. (d) 6. (c)
7. (c) 8. (b) 9. (d) 10. (a) 11. (b).

48
CHAPTER

Protein Structure

Proteins are most important component of cell. Proteins are the work horses of the cell. **Berzelius** (Swedish chemist) coined the term **protein** (*Protios* = holding the first place; Greek word). **Mulder** (Dutch chemist) extracted very similar nitrogen rich substances from animal and plant sources and depending upon the properties called them proteins. More than half dry weight of a cell is of protein.

Proteins are polymer of Amino acids

The unit constituent of protein in **amino acid.** So they are called as **polymers of amino acids.** Protein amino acids have a central **carbon atom** called **α-carbon;** to which a carboxyl group (—COOH) and amino group (—NH_2) are attached. The general formula may be represented as **ball and stick formula** and **Fischer projection formula** (Fig. 48.1). As the amino group is attached to the α-carbon, the amino acids are called α-amino acids. If R-group is not equivalent to 'H' the α-carbon atom is known as asymmetric; α-amino acids are **chiral** or optically active compounds.

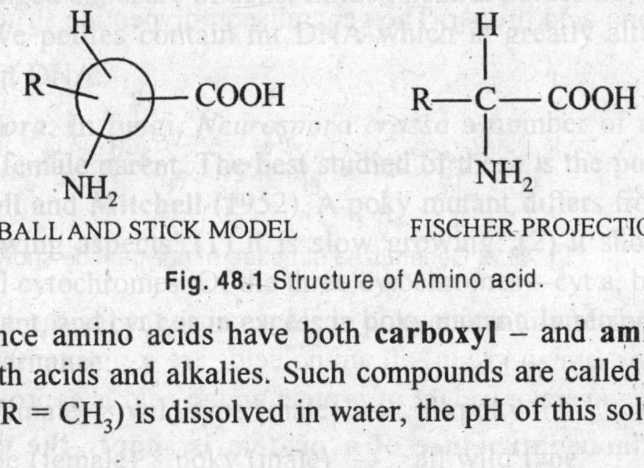

BALL AND STICK MODEL FISCHER PROJECTION FORMULA

Fig. 48.1 Structure of Amino acid.

Amino acids are also known as other, **zwitter ions,** since amino acids have both **carboxyl** – and **amino** + group balancing each. They can react with acids and alkalies. Such compounds are called **amphoteric substances,** *e.g.*, if solid **alanine** (R = CH_3) is dissolved in water, the pH of this solution will be approximately neutral.

Bjerrum in 1923 proposed the term zwitter ion, derived from German word **'zwitter'** meaning hybrid :

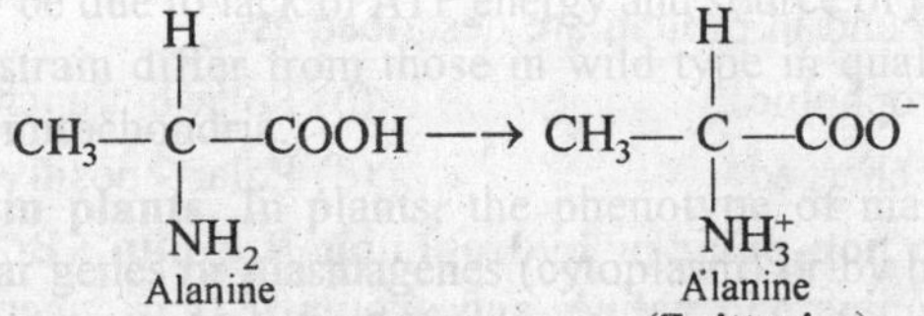

Alanine Alanine **(Zwitter ion)**

The fundamental structural unit of protein is α-amino acid. The proteins are high molecular weight compounds. The atomic structure of protein molecule generally has 30-33% 'C', 50% H, 9-11% 'O' and 7 to 9% N with small amount of 'S'. The atomic structure of hemoglobin is

$$C_{2796}H_{4592}O_{832}N_{812}S_8$$

PRIMARY STRUCTURE OF PROTEIN

The covalent structure of protein is essentially linear. Amino acids are connected together in the form of a chain with **peptide** bonds to form a **polypeptide.** The peptide bond is simply an amide bond between carboxyl carbon of one amino acid and the amino nitrogen of another

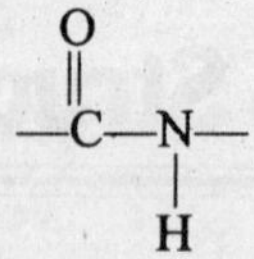

Oxygen is removed from carboxyl group of amino acid I and two protons from the alkyl ammonium group of amino acid II in the formation of peptide bond.

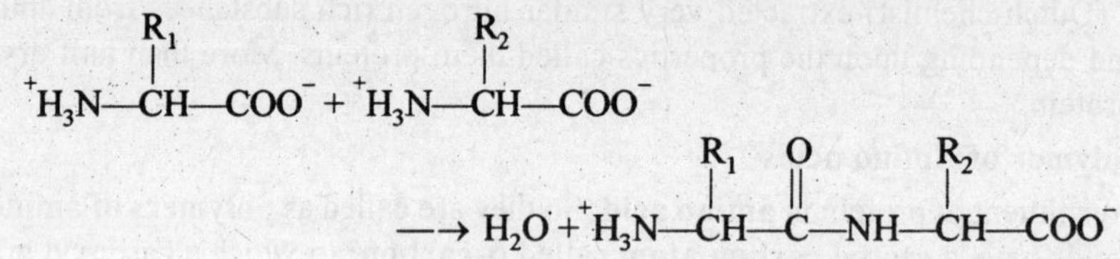

When an amino acid is incorporated into a polypeptide chain it is called **amino acid residue,** because water is 'split out' in the formation of peptide bond; and the remaining part cannot be called as amino acid proper. The generalised polypeptide structure is shown in (Fig. 48.2).

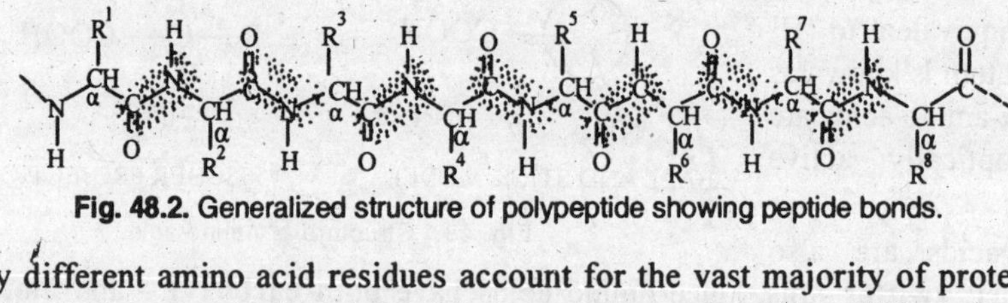

Fig. 48.2. Generalized structure of polypeptide showing peptide bonds.

Twenty different amino acid residues account for the vast majority of protein structures. Except **proline** (imino group) all amino acids are **α-amino acids.** They differ only in the type of 'R' group. The average weight of protein amino acid is approximately **110 kD.**

The primary structure of a protein is simply the order of amino acid residues in polypeptide chain. 20 types of amino acids are found in live forms. According to their chemical nature they are classified as **aliphatic, aromatic** and **heterocyclic** types depending upon 'R' group. On the bases of polarity of 'R' group, functional roles and contribution to the folding of the peptide chain the 20 amino acids found in protein are described as

(a) Non-polar or hydrophobic (b) Polar but uncharged

(c) Polar - negatively charged (d) Polar - positively charged

(a) **Amino acids with non-polar or hydrophobic R group :** Both **aliphatic,** *e.g.*, alanine; leucine, isoleucine, methionine and valine; and **aromatic,** *e.g.*, phenylalanine and tryptophan residues are included in it. The R group in this is hydrophobic (water healing). **Proline** amino acid is unusual in that its nitrogen atom present is a **secondary** rather than a primary amine. It is called **imino** acid rather than an **amino** acid.

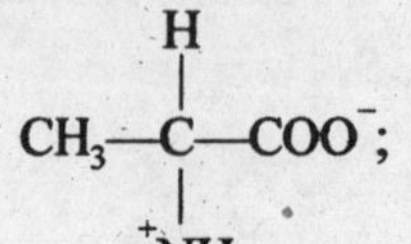

L$_8$-Alanine

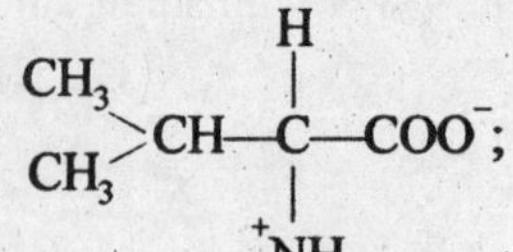

L$_8$-Valine

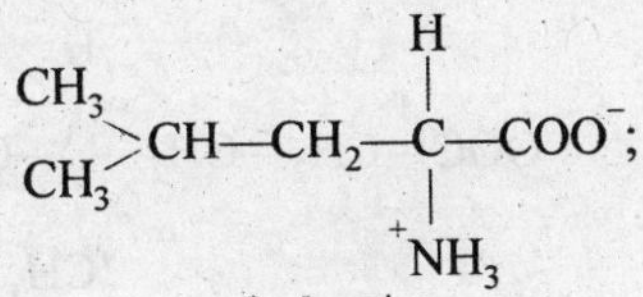

L$_{\delta}$-Leucine

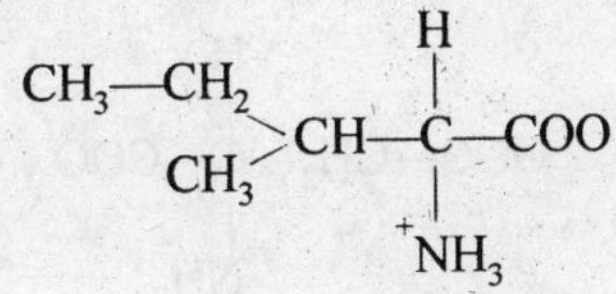

L$_{\delta}$-Isoleucine

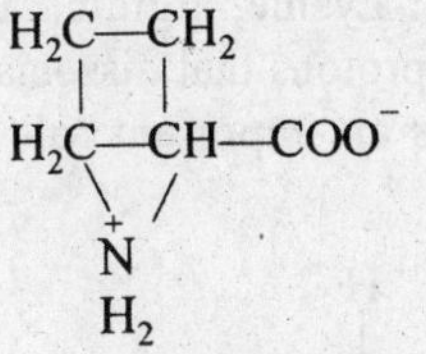

L$_{\delta}$-Proline

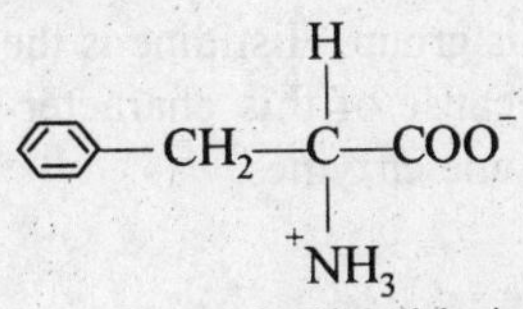

L$_{\delta}$-Phenylalanine

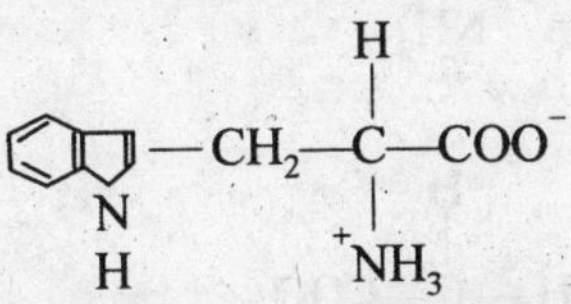

L$_{\delta}$-Tryptophan

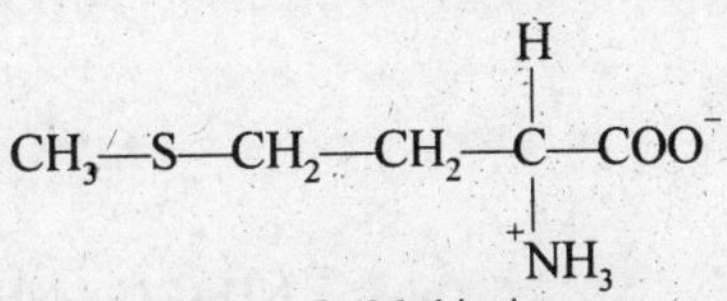

L$_{\delta}$-Methionine

(b) **Amino acids with polar but uncharged 'R' group :** These amino acids contain polar **'R'** residues participating in hydrogen bond formation. Several of the amino acids in this group possess a **hydroxyl group,** *i.e.*, serine, threonine and tyrosine, or **sulfhydryl group,** *e.g.*, cysteine and **amide group,** *e.g.*, asparagine and glutamine etc. **Glycine** lacks 'R' group. Both aliphatic and aromatic compounds are included in this group.

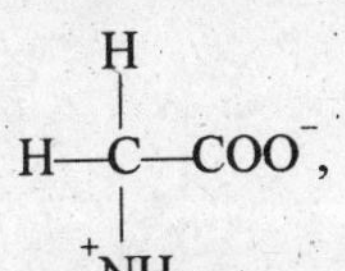

L$_{\delta}$-Glycine

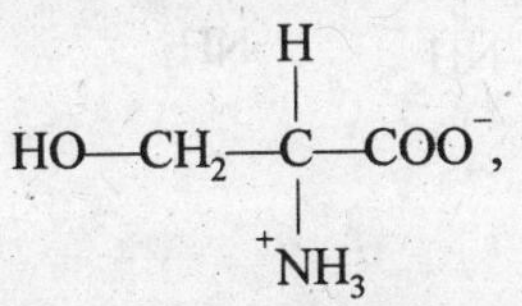

L$_{\delta}$-Serine

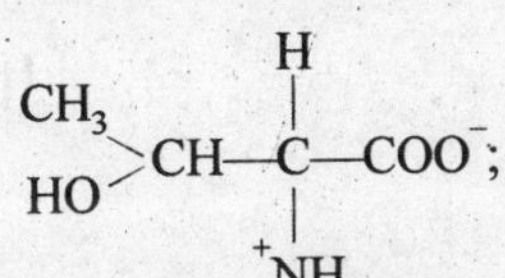

L$_{\delta}$-Threonine

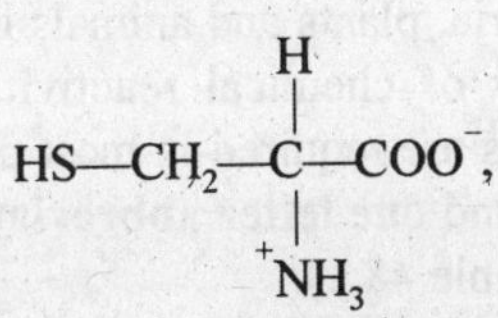

L$_{\delta}$-Cysteine

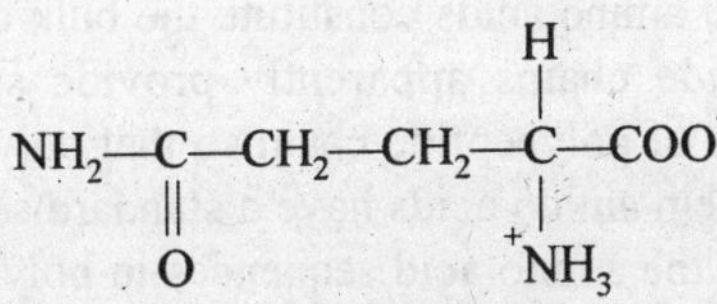

L$_{\delta}$-Glutamine

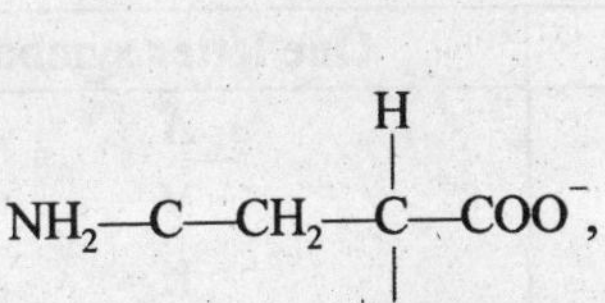

L$_{\delta}$-Asparagine

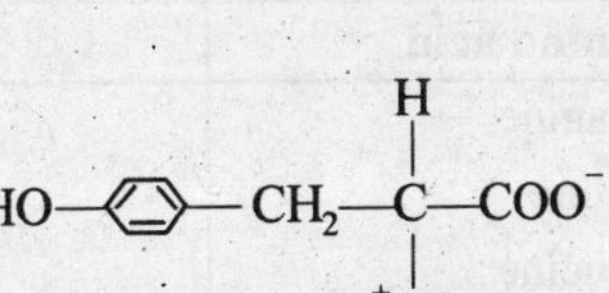

L$_{\delta}$-Tyrosine

(c) **Amino acids with negatively charged 'R' groups :** It includes the two dicarboxylic amino acids aspartic and glutamic acid.

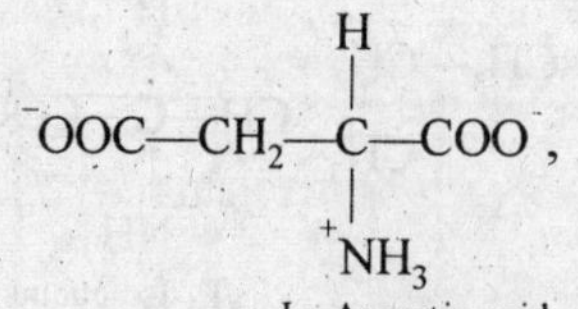

L_δ-Aspartic acid

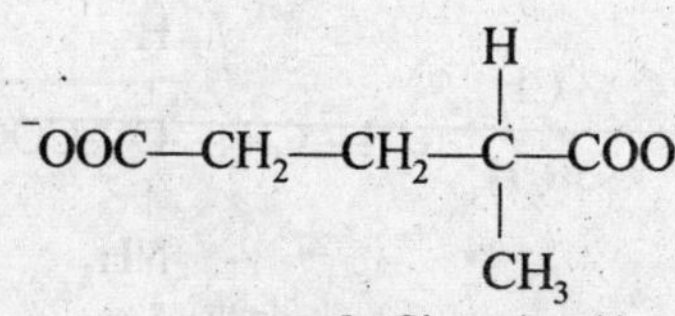

L_δ-Glutamic acid

(d) **Amino acids with positively charged 'R' group :** Lysine, arginine and histidine are included in this group. Histidine is the only amino acid with protons that dissociate in the neutral pH range. Because of this character histidine residue plays an important role in the catalytic activities of some enzymes.

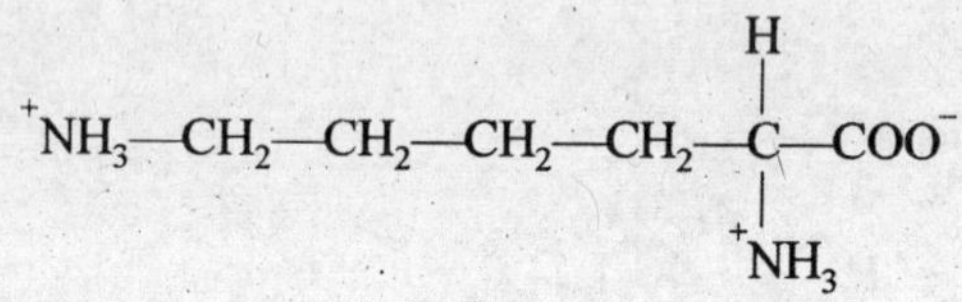

L_δ-Lysine

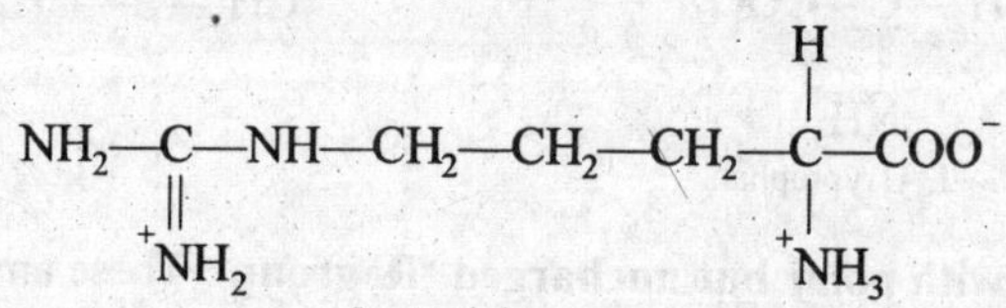

L_δ-Arginine

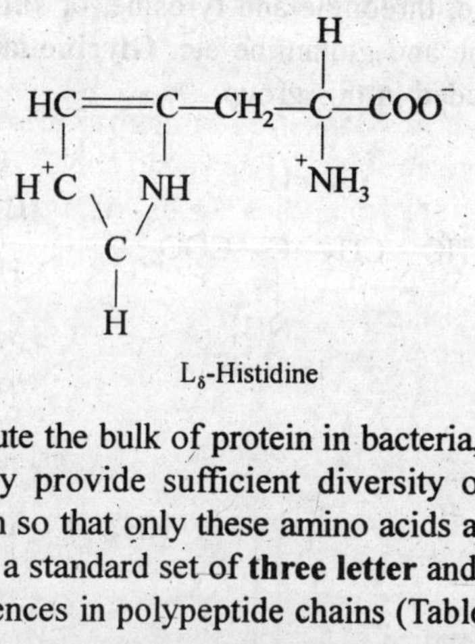

L_δ-Histidine

These 20 amino acids constitute the bulk of protein in bacteria, plants and animals in living system. 20 side chains apparently provide sufficient diversity of chemical reactivities and conformations of polypeptide chain so that only these amino acids are required in most proteins. Common protein amino acids have a standard set of **three letter** and **one letter abbreviation** for use in writing the amino acid sequences in polypeptide chains (Table 48.1).

Table 48.1. Amino acids and their abbreviations.

Amino acid	3 letter symbol	One letter symbol
Alanine	Ala	A
Valine	Val	V
Leucine	Leu	L
Isoleucine	Ile	I
Proline	Pro	P
Phenylalanine	Phe	F
Tryptophan	Trp	W

Methionine	Met	M
Glycine	Gly	G
Serine	Ser	S
Threonine	Thr	T
Cysteine	Cys	C
Glutamine	Gln	Q
Asparagine	Asn	N
Tyrosine	Tyr	Y
Lysine	Lys	K
Arginine	Arg	R
Histidine	His	H
Aspartic acid	Asp	D
Glutamic acid	Glu	E

Biologically Active Oligopeptides

An oligopeptide is simply a short polypeptide. It is composed of a smaller number of amino acids (10 or less amino acids.) A peptide has an α-amino and an α-carboxyl group. Peptides have roles in horomones, antibiotics, toxins and metabolic intermediates. Metabolically important peptide is **glutathione.**

It is a tripeptide (3 aminoacids). It is required in mammals to prevent oxidative damage to RBC. Glutathione is also called as V glutaminyl cysteinyl glycine.

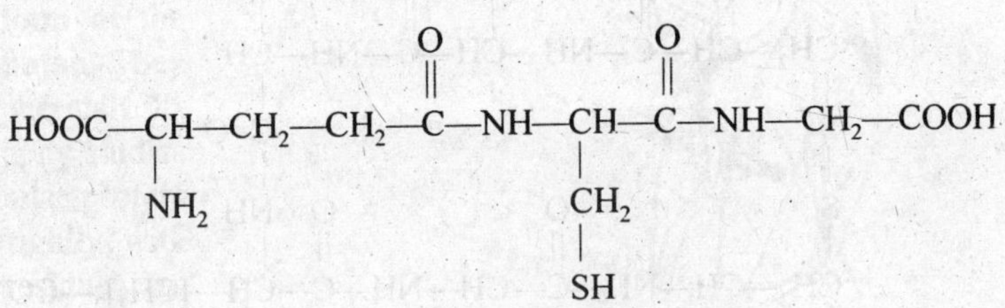

V Glutaminyl cysteinyl glycine
(Glutathione)

Peptide hormones responsible for intercellular and interorgan communication are important group of naturally occurring peptides. Vasopressin and oxytocin (nonapeptides) are formed in pituitary gland. 7 of the 9 peptides are similar in both but the physiological activities are very different. Oxytocin causes contraction of smooth muscles and vasopressin causes a rise in high blood pressure constricting peripheral blood vessels. (Fig. 48.3).

Several antibiotics are peptides of comparatively simple structure. Penicillin contains the valine and cysteine residues but are not linked by peptide bonds.

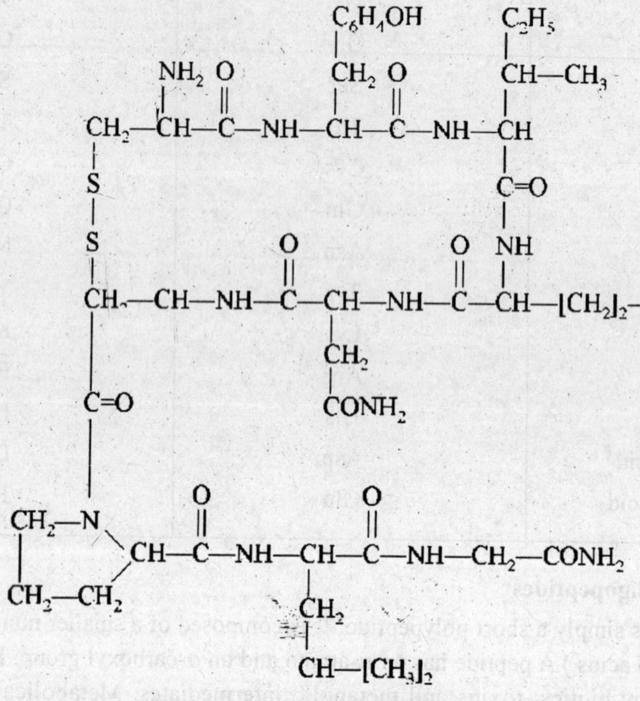

OXYTOCIN

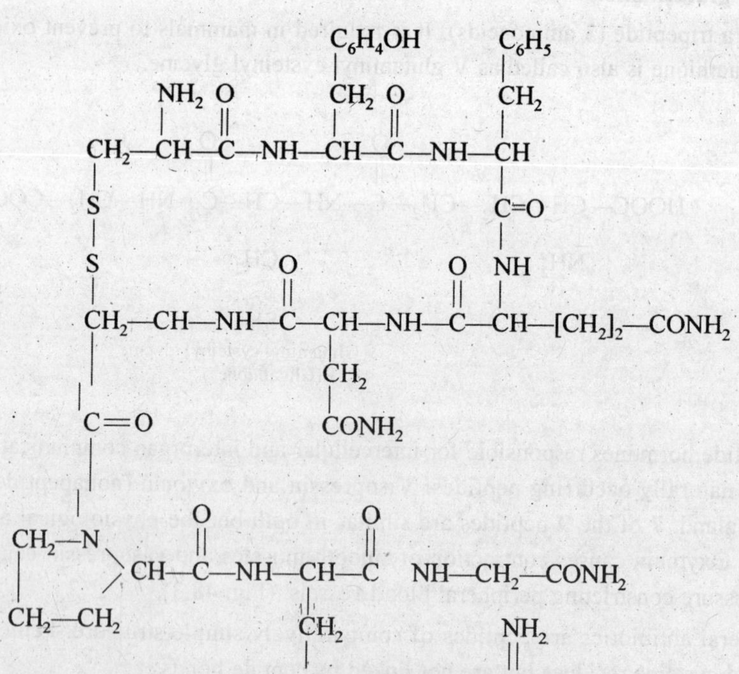

VASOPRESSIN

Fig. 48.3. Structure of Oxytocin and Vasopressin.

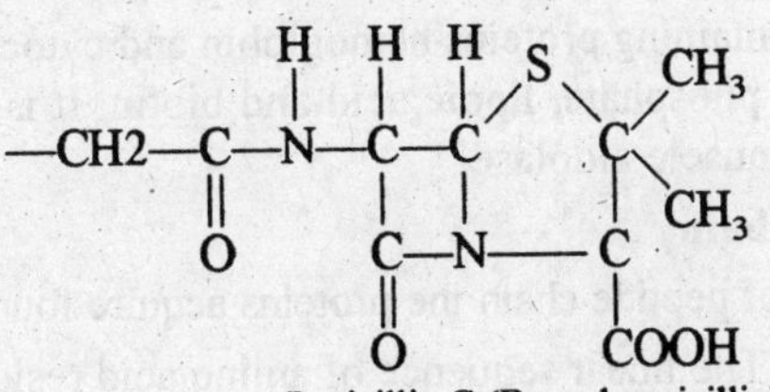

Penicillin G (Benzyl penicillin)

Artificial sweetner **aspartame** is L-aspartyl-L-phenyl alanyl-methyl ester. It is 200 times as sweet as sucrose and is considered to lack some of the unpleasant after taste associate with other synthetic sweetner.

FORMS OF PROTEIN

Proteins have not only a definite covalent structure but also a definite **3-dimensional** structure.

3D structure is controlled primarily by the order of amino acid residues. The stability of 3D structure is due to the non-covalent bonds. The position of reactive functional groups relative to one another in an environment of controlled polarity allows proteins to bind with other molecules.

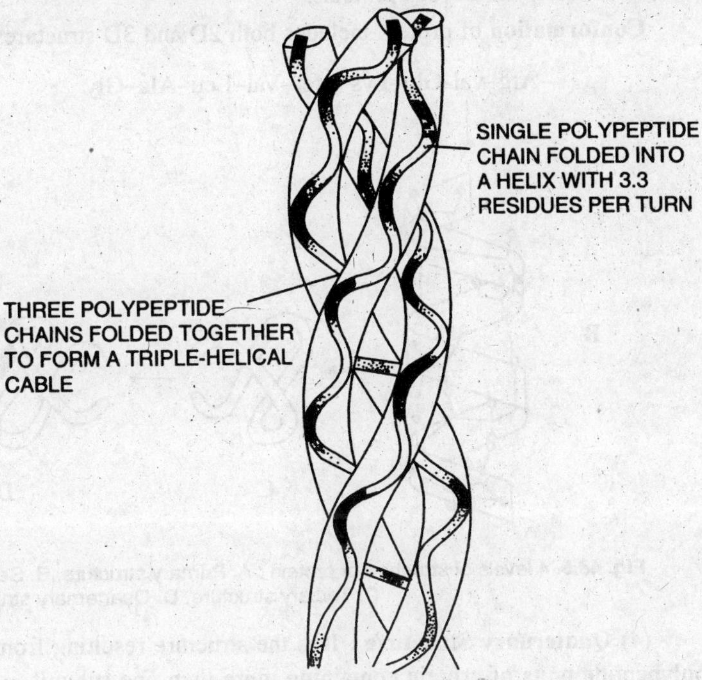

Fig. 48.4. Triple helix structure of protein.

Globular Protein

General form of the protein is **globular.** They are roughly spherical. X-ray crystallography studies show that globular protein is packed as tightly with amino acid residues as amino acids are packed in amino acid crystals.

Most of the side chains of amino acids are non-polar, the interior of the protein molecule is predominantly a polar. Polar amino acids are confined mainly to the exterior side of the protein.

Fibrous protein

The major fibrous proteins are **keratins**, the main constituents of hair, beaks, nail and claws, scales, horn, hoove and wool, etc. Collagen of skin, cartilage and bone also are fibrous proteins. Collagen are the glycoproteins composed of glycine, proline hydroxyproline and alanine. Collagen forms a **triple** helix (Fig. 48.4), cross-linked to give a rigid and inextensible material. Silk and the contractile proteins **tropomyosin** and paramyosin are also fibrous protein.

Some proteins are neither globular nor fibrous. **Myosin,** a muscle protein which non-contractile system as well, is one of the largest protein known with more than 1700 amino acid residues in its polypeptide chain. It is neither fibrous nor globular structure. Histidine with its lone electron pair in the ring nitrogen often acts as an efficient acid-base catalyst. Histidine also serves as a metal ligand in the iron containing proteins hemoglobin and cytochrome-c. Lysine is intimately involved in binding pyridoxal phosphate, lipoic acid and biotin. It is the component of the active site of certain enzymes, *e.g.*, muscle aldolase.

The folding of polypeptide chain

According to the folding of peptide chain the proteins acquire four structural levels (Fig. 48.5).

(1) Primary Structure : The linear sequence of amino acid residues in the polypeptide chain is called **primary** structure or 1-D structure.

(2) Secondary Structure : Regular folding pattern of contiguous parts of the polypeptide chain give a protein the secondary structure or 2D structures, *e.g.*, α-helix, β-pleated sheet form.

(3) Tertiary Structure : It is the 3D structure formed by the bonds between amino acid residues which are distant from each other in the polypeptide chain and the arrangement of secondary structure elements relative to one another. The term is applied generally to globular protein and rarely to fibrous proteins.

Conformation of protein includes both 2D and 3D structures.

A –Arg–val-Glu–Lys–Met–Val–Leu–Ala–Gly

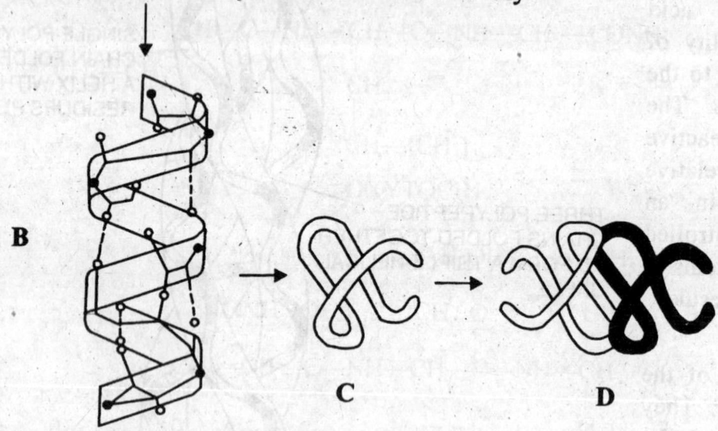

Fig. 48.5. 4 levels of structure of protein : A. Primary structure, B. Secondary structure (a helix). C. Tertiary structure, D. Quarternary structure.

(4) Quaternary Structure : It is the structure resulting from the chains between separate polypeptide units of protein containing more than one subunit, *e.g.*, enzyme **phosphorylase** a contains two identical subunits which alone are catalytically inactive but in a dimer form an active enzyme. Such structure is called **homogeneous quaternary structure.** If the two subunits are dissimilar it is **heterogeneous quaternary structure.** The subunits are also called as **protomers.** Protein made up of more than one protomer is called **oligomeric** protein, *e.g.*, **hemoglobin** is an oligomeric protein with heterogeneous quaternary structure consisting of 2 identical α-chain protomers and two identical β-chain protomers (*i.e.*, $\alpha_2\ \beta_2$).

Secondary Structure of Protein (2D)

The secondary structure of protein molecule generally is accomplished by consideration of polypeptide back bone alone, without analysis of amino acid side chain. The secondary structures are stabilized by **hydrogen bonds** between peptide imide and carboxyl groups of polypeptide back bone and not by bonds between -side chains. 3 classes of secondary structure are recognised.

(a) Helices, (b) Sheets and (c) Bonds.

The peptide bond, which is an imide (substituted amide) bond, has a planner structure. The electrons are delocalized in amide linkage, giving the C—N bond show considerable double bond characters.

Planner peptide bonds can be shown as

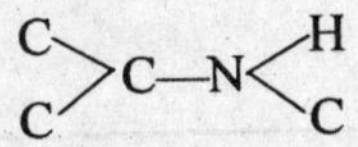

3 bonds make peptide chain, *i.e.*, α-carbon to carboxyl carbon bond, imide nitrogen to α-carbon bond and C—N bond. Double bond of C—N bond limits rotation about it (Fig. 48.6). Due to rotations of groups around α-carbon (ψ and ϕ value) the chain assumes a helical configurations or a circle or a zig-zag structure (Fig. 48.6B).

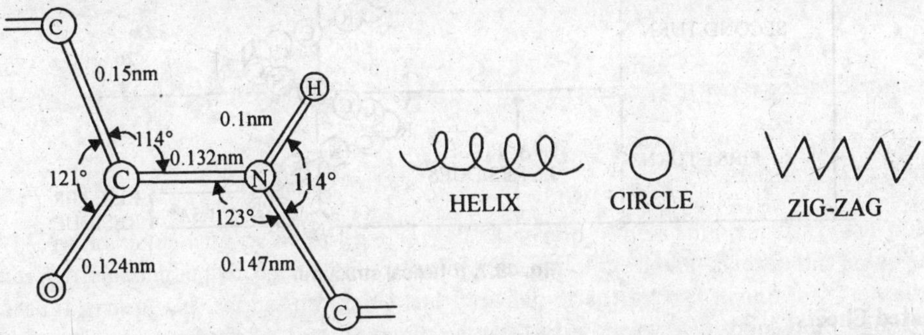

Fig. 48.6. A. Various angles between elements of amino acid, B. Various shapes of secondary structure of protein.

All combinations of ϕ a ψ angles are not possible, as many lead to clashes between atoms in adjacent residues. **Ramachandran** worked on possible combinations of ϕ of ψ angles that do not clash and plotted on a conformation map. It is called **Ramachandran plot.**

α-Helix

The protein α-helix is a right handed helix. It was demonstrated by **L. Pauling** and **R.B. Torey** in 1955. The chain rotates clockwise as one views down the helix axis at the polypeptide chain proceeding into the distance. The α-helix has 3.6 amino acid residues per turn and is stabilized by nearly straight hydrogen bonding between an imide group (—NH—) in the polypeptide chain and a carboxyl group (—CO—) at a position four residue away in the same chain (Fig. 48.7). Every —NH— and —CO— group can form a straight hydrogen bond, in this manner the α-helix is especially well stabilized.

The α-helix is characterized by the location of the atoms of the polypeptide chain, the amino acid side chains do influence the probability that a given sequence of amino acid residues will be four in an α-helix. Glycine residues do not foster α-helix as in the side chain only an 'H' atom is present. The polypeptide chain is too flexible at a glycine residue. Protein can also assume value necessary for the α-helix. Both glycine and protein residue are considered to be **helix breakers.** Glutamic acid, leucine, methionine, phenylalanine, etc., occur very frequently in α-helices and are considered to be α-helix promoting residues. Other features of α-helix are :

(a) Overlapping of repeating units after 18 amino acid residues, *i.e.*, after, 5 turns.

(b) Pitch (centre to centre) per turn = 5.4 Å.

(c) Pitch (rim to rim) of each turn = 5.1 Å.

(d) Rise per residue = 1.5 Å.

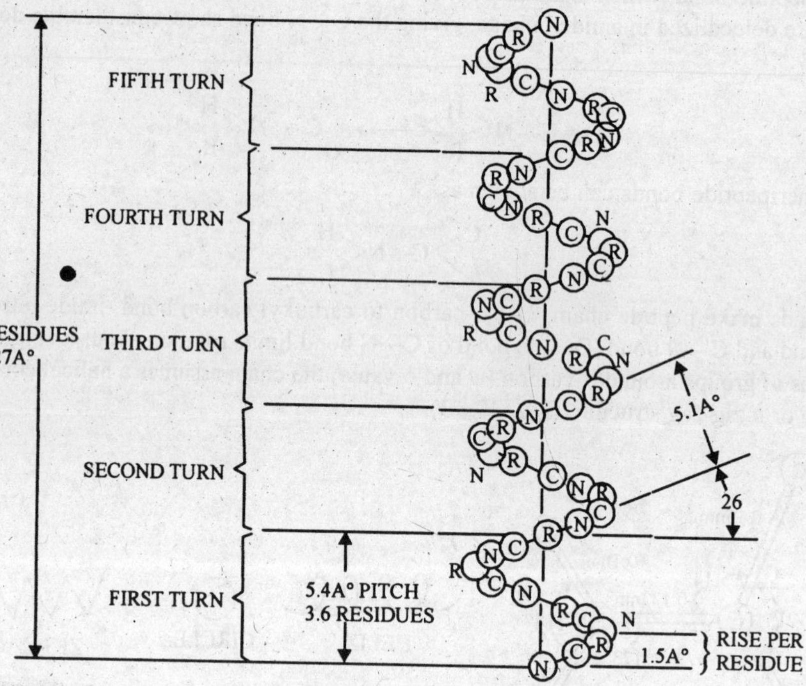

Fig. 48.7. α helical strucutre.

β-pleated Sheets

The β-pleated sheet structure in protein is not exactly in the same plane. The sheets are pleated (Fig. 48.8). The atoms lie in folded plane, and the side chains of succeeding residues along the polypeptide chain protrude alternately above and below the general plane of the structure.

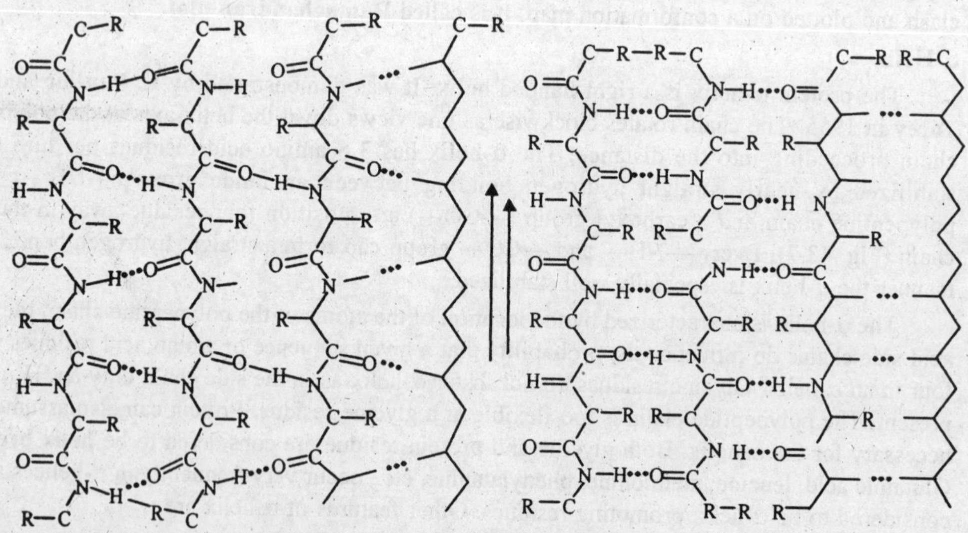

Fig. 48.8. β-Pleated structure.

The β-pleated sheet structures, like the α-helix, allow the maximum amount of hydrogen bonding between polypeptide back bone inside and carboxyl groups, and the H-bonds formed are straight enough to be stable. The β-pleated sheets seen in protein structures are not only pleated but also slightly twisted.

α-helix and β-pleated sheet structures are found in both globular and fibrous proteins. The wide and frequent occurrence of the α-helix and β-pleated sheet structures appears to be due to the combination of sterically favoured disposition of the polypeptide chain and favourable hydrogen bonding.

α-keratin and paramyosin are examples of fibrous proteins that have α-helix structures. Parallel β-pleated sheets and antiparallel β-pleated sheets form the structures of the fibrous proteins stretches of polypeptide chain entirely folded into the secondary structures. In globular proteins the α-helical and β-pleated structures are smaller than those of fibrous proteins.

Reverse Bending

Reverse turn or reverse bending occurs at the surface of protein in globular protein.

Super secondary structure is presented by some fibrous proteins like collagen. The amino acid composition of callagen is unusual being composed of 25% glycine and 25% proline and hydroxyproline. Because of high **glycine and proline** contain no α-helic occurs.

Tertiary Structures of Protein (3D Structure)

The next level after secondary structure is tertiary structure, which describes the conformation of the entire protein. The secondary structure is stabilized primarily by hydrogen bonds between atoms that form the peptide bonds of the backbone. The tertiary structure is stabilized by an array of non-covalent bonds between the diverse side chains of the protein. The secondary structure is limited to the small number of conformations but tertiary structure is virtually unlimited.

The detailed tertiary structure of a protein is usually determined using X-ray crystallography technique. Most proteins can be categorized on the basis of their overall conformations being either **fibrous** (elongated shape) or **globular** (compact shape).

Myoglobin : The first globular protein whose tertiary structure was determined. The polypeptide chains of globular proteins are folded and twisted into complex shapes. Distant points on the linear sequence of amino acids are brought next to each other and linked by various types of bonds. **John Kendrew** *et al* in 1957 produced the 3D structure of myoglobin. Myoglobin is present in muscle tissue as a storage site for oxygen. The oxygen molecule is bound to an iron atom in the centre of a heme group. Heme is the prosthetic group, *i.e.*, a part of protein not composed of amino acids. The heme of myoglobin gives most muscle tissue a red colour. Myoglobin is a globular protein the polypeptide chain was folded back on itself in a complex arrangement. Myoglobin is composed of δ-rodlike stretches of a α-helix ranging from

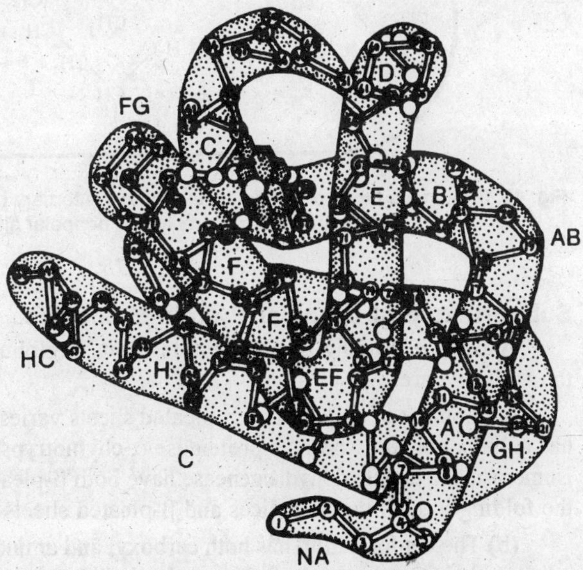

Fig. 48.9. 3D structure of myoglobin.

7-24 amino acids. 75% of the 153 amino acids in the polypeptide chain are in the α-helical conformation. No, β-pleated sheet is present. (Fig. 48.9). The 3D structure reveals that heme is present within a pocket of hydrophobic side chains that promotes the binding of oxygen without the oxidation of iron atom. No disulphide bond is present in myoglobin. Tertiary structure is held together purely by non-covalent bonds between side chains within proteins, *i.e.*, hydrogen bonds, ionic bonds and hydrophobic interactions (Fig. 48.10).

Triosephosphate isomerase (enzyme) consists of β-pleated sheets largely and some α-helices.

There are various types of non-covalent bonds stabilizing tertiary protein structure, *e.g.*, ionic bonds due to electrostatic interaction, hydrogen bonds between heteroatoms of side chains and between side chain and polypeptide backbone, hydrophobic interaction between non-polar side chains and disulphide bonds (Fig. 48.10).

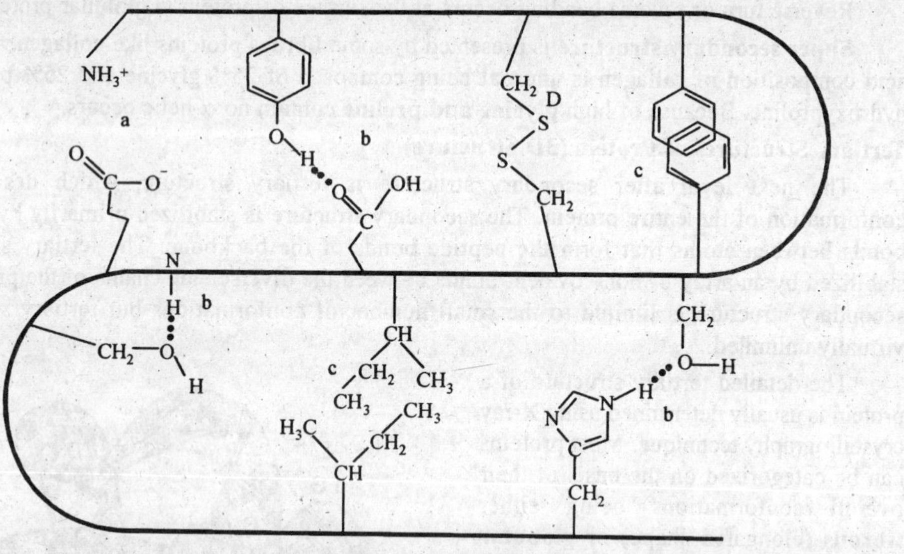

Fig. 48.10. Various types of bonds in stabilizing the tertiary protein structure : A. ionic bonds, B. H-bonds, C. hydrophobic interaction in nonpolar side chain, D. disulfide bonds.

Sub-structure of Globular Protein Molecules

Aspects of folding of polypeptide chain in globular protein other than non-polar interior of the molecule are:

(a) Content of α-helix and β-pleated sheets varies widely from protein to protein. Myoglobin has α-helix content where as proteinase α-chymotrypsin has β-sheet structures. Triose phosphate isomerase and lactate dehydrogenease have both β-pleated sheets and α-helices. In these molecules the folding is such that α-helices and β-pleated sheets alternate along the chain.

(b) The conformation has both carboxyl and amino terminus free, *i.e.*, if they are stretched out a linear chain would result. Polypeptide chain is not rolled up like a ball or string; but the course is from one surface of the molecule to another where it reverses its course.

(c) Useful general concept of globular protein structure is that of **domain.** Statistical tests that analyse the density of atom population around each amino residue in the 3D structure sometimes detect domains not apparent by viewing the protein structure model. Often, the course of the, polypeptide chain leaves one domain for a second and, after forming the second domain, returns to the first, making two polypeptide connections between the two domains. Most recognized domains have 100-200 amino acid residues, though both larger and smaller domains have been identified.

Domains have structural and evolutionary significance. An active site of an enzyme is the localized region of the molecule at which the enzyme bonds and acts on its substrate. Active sites are formed on the junction of 2 or more domains of the enzyme protein.

Domains have been postulated to facilitate the evolution of protein by providing **preformed** sequences of amino acids that will assume a stable conformation and provide a specific function.

The enzyme phosphofructokinase is an example of gene duplication. The phosphofructokinase of some bacteria and mammals are homogeneous tetramers, but mammalian enzyme is twice the size of bacterial enzyme. The two halves of the mammalian enzyme appear to be related **super domains.**

Bacterial and mammalian enzymes have the same number of active sites, *i.e.*, four per tetramer.

STRUCTURE OF SOME PROTEINS

(1) Wool α-Keratin

(i) It is a fibrous protein made up of α-helix, β-pleated sheet and form collagen triple helix structure.

(ii) Secondary structure of wool α-keratin is the right handed α-helix.

(iii) Three such α-helices form a left handed coil called a protofibril which is stabilized by cross linking disulphide bonds.

(iv) Protofibrils coil into cablelike **microfibrils** of 8 nm diameter.

(v) Several hundred **microfibrils embedded** in are amorphous protein matrix form a **macrofibril.**

(vi) Many macrofibrils together form a fiber of wool.

(vii) When α-keratin is exposed to moist heat and stretched, it converts to β-keratin.

The hydrogen bonds stabilizing the α-helical structure are broken and an extended parallel β-pleated sheet results.

(2) Cytochrome c

(i) Cytochrome c is small, stable and coloured globular protein.

(ii) Single iron atom in cytochrome c molecule alternates between the Fe II and Fe III oxidation states.

(iii) Cyt c has 104-111 amino acid residues, 35 of which are found to be invariant from species to species.

(3) Hemoglobin (Fig. 48.11)

(i) It is the transport protein of RBC.

(ii) Heme iron is bonded to a histidine residue on one side of the heme plane.

(iii) On oxygenation, the iron is liganded to oxygen on the opposite side of the heme in a pocket on the surface of the hemoglobin molecule.

(iv) Heme iron remains in the Fe II state during oxygenation and deoxygenation.

(v) The tetramer $\alpha_2\beta_2$ bind with 4-oxygen molecules.

(vi) Association of α_1-β_1 pair with α_2-β_2 is weaker than within them.

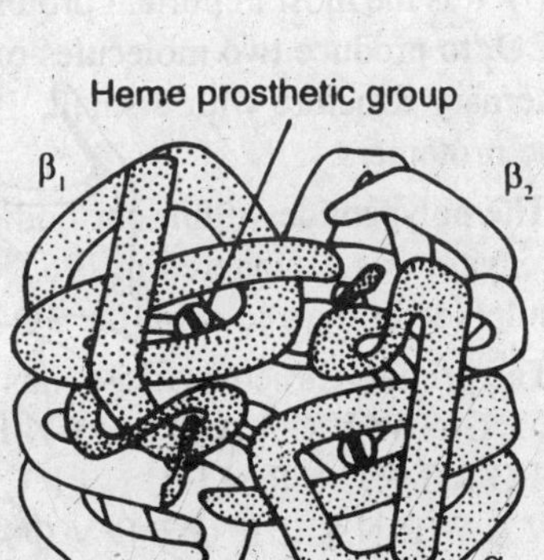

Fig. 48.11. Hemoglobin structure.

(4) Immunoglobin G

(i) It is the most important globular protein related with antibody molecule of immune system.

(ii) It is the serum protein that combines with antigen.

(iii) The quaternary structure of I_gG is made up of 4 polypeptide chains two are **'light'** chains of about 215 amino acid residues and two are **'heavy chains'** of 500 amino acid residues.

(iv) 4 chains are linked by disulfide bridges as form 12 domains. (Fig. 48.12).

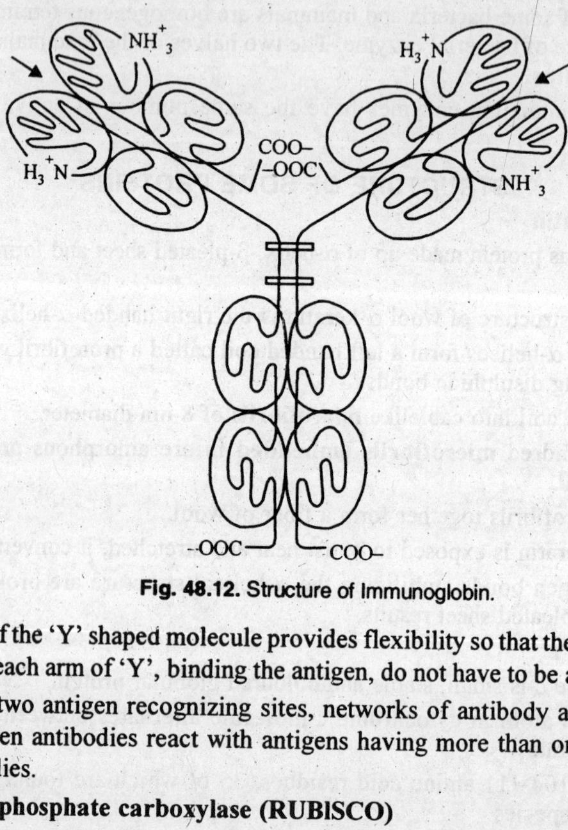

Fig. 48.12. Structure of Immunoglobin.

(v) The joint of the 'Y' shaped molecule provides flexibility so that the two combining sites, located at each arm of 'Y', binding the antigen, do not have to be a fixed distance apart.

(vi) There are two antigen recognizing sites, networks of antibody antigen chains will be formed when antibodies react with antigens having more than one site recognized by the antibodies.

(5) Ribulose bisphosphate carboxylase (RUBISCO)

(i) It is the most important protein in plants catalyzing reaction of ribulose 1, 5, bisphosphate with CO_2 to produce two molecules of 3 phosphoglycerate in CO_2 fixing reaction. RUBISCO has a quaternary structure with 8 small, 100 amino acid residue proteins and 8 large 500 amino acid residue protomers.

The polypeptide within our bodies are folded into complex 3D structures. The 3D structure of **lysozyme** was deciphered in 1965. It is made up of 129 amino acids in linear order and joined by disulphide bridges to form the 3D conformation.

The primary structure of it reads as (NH_2) K V F G R C E L A A A M K R H G L D N Y R G Y S L G N W V C A A K F E S N F N T Q A T N R N T D G S T D Y G I L Q I N S R W W C D N G R T P G S R N L C N I P C S A L L S S D I T A S V N C A K K I V S D G D G M N A W V A W R N R C K G T D V Q A W I R G C R L (COOH).

The linear model of molecule (Fig. 48.13) is some what irregular. If the side chains are stripped away and path of the peptide bond backbone is drawn some regions of protein are ordered in a repeating pattern. It shows both α-helix and β-sheet pattern (Fig. 48.14). In the α-helix region the chain twists around in a spiral while in β-sheet region the polypeptide run alongside each other. In tertiary conformation there can be grooves, clefts and protrusions on the surface of protein where particular amino acid side chains are positioned to form site that bind ligands and catalyse reactions.

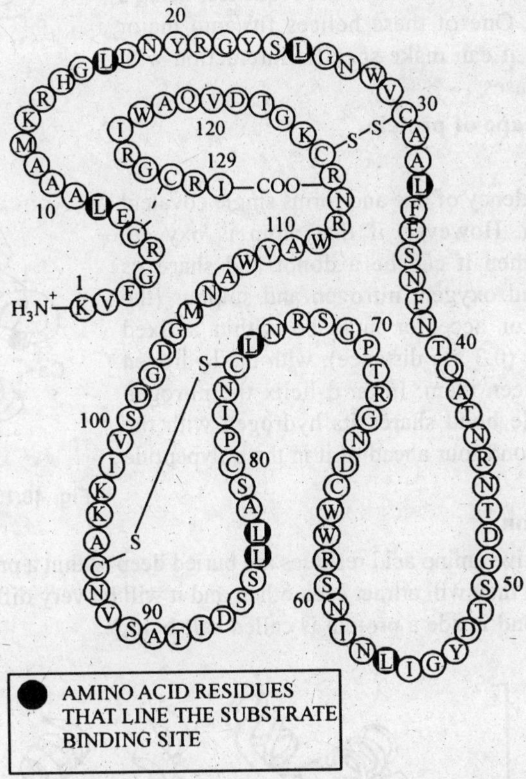

Fig. 48.13. Structure of Lysozyme. (Primary structure).

In calmodulins (calcium binding proteins) the single chain is organized into two domains, joined by only one strand of polypeptide chain. The two domains are very similar and might be originated by duplication of gene (Fig. 48.15).

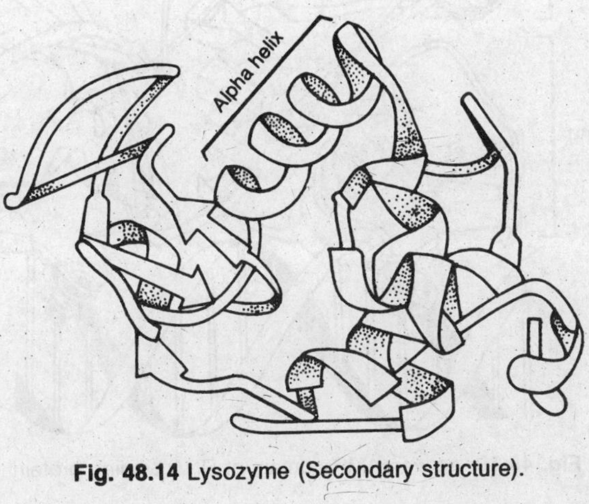

Fig. 48.14 Lysozyme (Secondary structure).

The catabolite activating protein (CAP) of *E. coli* binds to specific sequences of bases on DNA assisting RNA polymerase to bind to its promoters and initiate transcription of *lac* operon. One of the domains of CAP has the job of binding cAMP and another recognizes DNA sequences using a helix turn helix motif. One of these helices fits into major groove of DNA where it can make specific interaction with exposed edges of the bases.

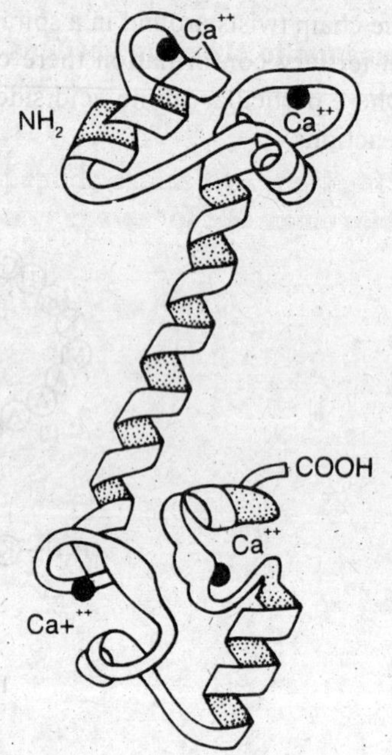

Fig. 48.15. Calmodulin Structure.

Forces molding the shape of protein

Hydrogen bonds

Hydrogen has a valency of one and forms single covalent bond with other atom. However, if that atom is oxygen, nitrogen and sulphur then it can be a donor and share its hydrogen with a second oxygen, nitrogen and sulphur (the acceptor): The donor or acceptor must be within a fixed distance of one another (0.3 nm distance), with the hydrogen on a straight line between them. In an α-helix the nitrogen atom within the peptide bond shares its hydrogen with the oxygen of the peptide bond four ahead of it in the polypeptide chain.

Electrostatic Interaction

If positive or negative amino acid residues are buried deep within a protein, where neither can interact with water, then they will attract each other, and it will be very difficult to tear them apart. Such an electrostatic bond inside a protein is called **salt bridge.**

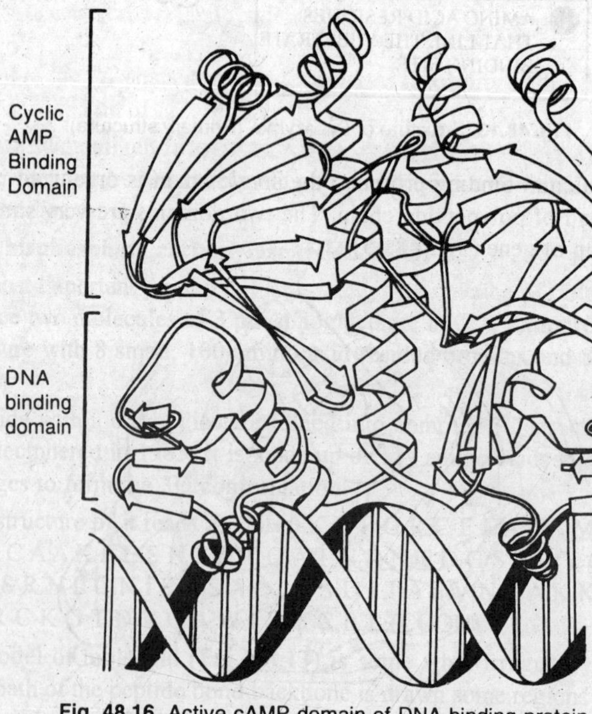

Fig. 48.16. Active cAMP domain of DNA binding protein

Polar groups such as hydroxyl and amide groups are dipoles: they have an excess of electrons on one atom and a compensating deficiency at the other. The partial charges of dipoles will be attracted to other dipoles and to fully charged ions.

van der Wall's Forces

These are relatively weak close range interactions between atoms. These forms are important in the interiors of proteins and membranes and in the specific binding of a ligand to its binding site.

Hydrophobic interactions

A polypeptide with hydrophilic and hydrophobic residues will spontaneously adopt a configuration in which the hydrophobic residues are not exposed to water, either by sitting in a lipid bilayer or by adopting and globular shape in which the hydrophobic residues hide in the centre of the protein.

Disulphide bonds

Extra cellular proteins often have disulphide bonds between specific cysteine residues. These tend to lock the molecules into its conformation. Few proteins contain disulphide bonds. But these are more stable.

SUMMARY

1. Polypeptides are linear polymer of α-amino acids linked by peptide bonds.
2. There are 20 amino acids coded for by the genetic code.
3. Amino acids differ in the properties of their side chains which range from hydrophobic groups to charged and uncharged hydrophilic groups..
4. Proteins are polypeptides that have a complex 3D shape.
5. Proteins have four levels of structures, *i.e.*, primary secondary tertiary and quaternary.
6. Primary structure is the linear sequence of the amino acid monomers.
7. Secondary structures possess regular, repeated folding of the peptide chains.
8. Secondary structures are held together by hydrogen bonds between the carboxyl oxygen and the hydrogen of the amino group of the peptide bonds.
9. Common types of secondary structures are
   (a) α-helix where the chain coils upon itself making a spring like spiral with hydrogen bonds running parallel to the length of the spiral.
   (b) In β-sheets the hydrogen bonds are between extended strands of polypeptide that runs along side of one another.
10. Tertiary structure includes complex foldings.
11. Tertiary structures are stabilized by interactions between side chains.
12. Proteins also have quaternary structures that is an association of subunits, each of which has a tertiary structure.

QUESTION BANK

Very Short Answer Questions

1. Name one imino acid.
2. What is the one letter symbol of leucine and phenyl alanine?
3. Name one example of globular protein.
4. Name one example of triple helix fibrous protein.
5. Who discovered α-helix model of secondary structure of protein.
6. What is RUBISCO?
7. What is immunoglobin G?
8. Draw the structure of immunoglobin G.
9. Draw the structure of α-helix model.
10. Draw the structure of hemoglobin.

Short Answer Questions

1. Explain that proteins are polymers of amino acid.
2. Write notes on primary structure of protein.

3. Write short notes on : (a) amino acids with positively chaged 'R' group. (b) amino acids with polar but unchanged 'R' group.
4. Make structures of Histidine, Lysine,- Aspartic acid and Glutamine.
5. Write three letter and one letter symbols of atleast 10 amino acids.
6. Write short note on forms of protein.
7. Write note on tertiary structure of proteins.
8. Write notes on secondary structures of proteins.
9. Write short notes on : (a) α-helix (b) β-pleated sheets.
10. Write short notes on : (a) Cytochrome c (b) Hemoglobin
11. Write short notes on RUBISCO.

Long Answer Questions

1. Describe the structures of protein.
2. Describe various forces that mold the shape of protein.

Multiple Choice Questions (Objective Type Questions)

1. Which amino acid has no asymmetric carbon atom
   (a) Proline (b) Glycine
   (c) Aspartic acid (d) Cystine
2. Basic unit of protein is
   (a) Polypeptide (b) Peptide
   (c) Amide (d) Amino acid
3. Proteins were named by
   (a) Flemming (b) Sanger
   (c) Berzelius (d) Kunhe
4. α-helix and β-pleated sheats are folding pattern of
   (a) Primary structure of protein (b) Secondary structure of protein
   (c) Tertiary structure of protein (d) Quaternary structure of protein
5. Three dimensional structure of protein is stabilized by
   (a) Ionic bonds (b) Covalent disulphide bonds
   (c) Hydrogen bond (d) All the above
6. Primary structure of protein is maintained by
   (a) Hydrogen bonds (b) Peptide bonds
   (c) Glycosidic linkage (d) Phenolic linkage
7. Hemoglobin and collagen are examples of ...... respectively.
   (a) Fibrous protein and chromoprotein (b) Chromoprotein and fibrous protein
   (c) Nucleoprotein and cell sap protein (d) Chromosomal protein and cytoplasmic protein
8. Which of the two amino acids are essential for growth.
   (a) Arginine and Histidine (b) Cystine and Valine
   (c) Leucine and Isoleucine (d) Arginine and Methionine
9. Histone proteins have large amount of
   (a) Glycine and glutamic acid (b) Protein and phenyl alanine
   (c) Lysine and Arginine (d) Methionine and Arginine
10. Which of the following is imino acid
    (a) Proline (b) Hydroxyproline
    (c)Phenylalanine (d) Both 'a' and 'b'
11. Amino acids are
    (a) Negatively charged (b) Positively charged
    (c) Zwitter ions (d) Cations

ANSWERS

1. (b)	2. (d)	3. (c)	4. (b)	5. (d)	6. (b)
7. (b)	8. (a)	9. (c)	10. (d)	11. (c).	

49

CHAPTER

DNA Damage and Repair

The first demonstration about the damage repair was given by **Kelner** *et al* where they have shown that UV induced effects could be reversed by exposing the cells to visible light in blue spectrum region. The process is called **photoreactivation**. It can only be done before the genetic material is permanently affected.

Setlow and **Setlow** demonstrated that photoreactivation effect primarily depends upon a specific enzyme **photolyase** which can split the pyrimidine dimer produced by UV radiation.

There are also several ways of DNA repair other than that of photoreactivation. The changes in DNA are mainly of 2 types.

(i) Single base change: Which are corrected through DNA replication leading to change in DNA sequence.

(ii) Structural distortions : Resulting from a single stranded nick, removal of a base or covalent link between bases of same or different strands.

In dark repair system, enzyme recognition of the distorted section of a DNA strand, is carried out by excision of defective sequence and replacement of section by new sequence of nucleotides formed as a complement to the undamaged DNA strand. In *E. coli.*, ultraviolet repair (*uvr*) mutation may affect different steps in these processes.

Polymerase I enzyme was used by **Kornberg** *et al* for *in vitro* replication of DNA. This method is used in the dark repair process by **polymerase I mutants** (Pol A) which are extremely sensitive to UV radiation.

In *Micrococus luteus* a different method for elimination of thymine dimer involves a **glycosylase** enzyme which C leaves the glycosylic bond between are of the dimerized thymine bases and its deoxyribose sugar. It leaves a vacant baseless 'AP' (apyrimidinic or apurinic) site on the DNA chain which is then nicked by special AP endonucleases. Glycosylase enzyme is highly specific in helping to remove damaged bases, *e.g.*, hypoxanthine, 7-methyl guanine, deaminated cytosine (uracil), etc., methylnitronitrosoguanidine (MNNG) involves enzymes acting specifically on damaged bases by directly removing methyl or ethyl group from the O^6 position of guanine.

Those DNA aberrations which are not directly repaired by excision and replacement may be more indirectly impaired by enzymes involved in recombination process. The procedure of multiplicity reactivation depends upon recombination events between the various intact parts of damaged phages leading to the formation of recombinant viable phages.

According to **Howard Flanders** the recombination repair process acts mostly on DNA strand carrying gaps caused by intra or interstrand cross links (Fig. 49.1). When such strands replicate gaps remain enabling 'free' end of a single strand section to initiate and exchange with a similarly oriented strand on a homologous DNA molecule. The exchange forms a new unbroken single strand which can serve as template for an enzyme complex to remove and replace the defective DNA part.

Excision repair involves different lengths of DNA as **very short patch repair (VSP); short patch repair** and **long patch repair.** VSP repairs single base damage. VSP also deals with mismatch repair, *e.g.*, **GC—GT** type which gives rise to mutant type.

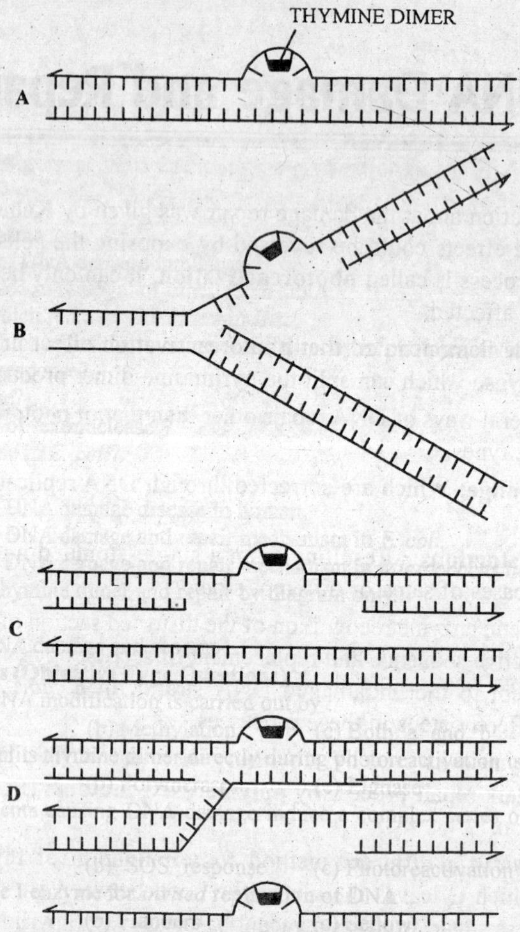

Fig. 49.1. DNA repair (Howard Flanders).

In *E. coli* recombinational repair mechanisms are mostly tied to *rec* A protein which has a dual function catalyzing both **recombination** and **protein cleavage.** The ability of *rec* A to cleave particular proteins regulates its own abundance. The protein repressor (a product of the *lex* A gene) restricts *rec* A gene activity is susceptible to cleavage by *rec* A protein. Due to mutational defects at replication forks an increase in single stranded DNA sections activiting *rec* A proteins to cleave its *lex* A repressor amplifing *rec* A protein production. This type of cellular response was called **"SOS".**

In *E. coli* many treatments carry DNA damage or inhibition of DNA replication, induce a complex series of phenotypic changes are called **SOS response.** It initiates the interaction of *Rec* A protein with **lex** A repressor. It involves damaging treatment activating *Rec* A protein (protease), *Rec* A protease interacts with *Lex* A pressor (**Lex** A gene) all operons to which **Lex** A is bound are induced including many operons or genes, *e.g.*, *din* genes (*uvr* A, *uvr* B, *uvr* C and *hin* A). Each such gene has a **SOS** box. After DNA damaging signal is removed *Rec* A is not induced and *Lex* A accumulates repressing all other genes belonging to SOS system (Fig. 49.2)

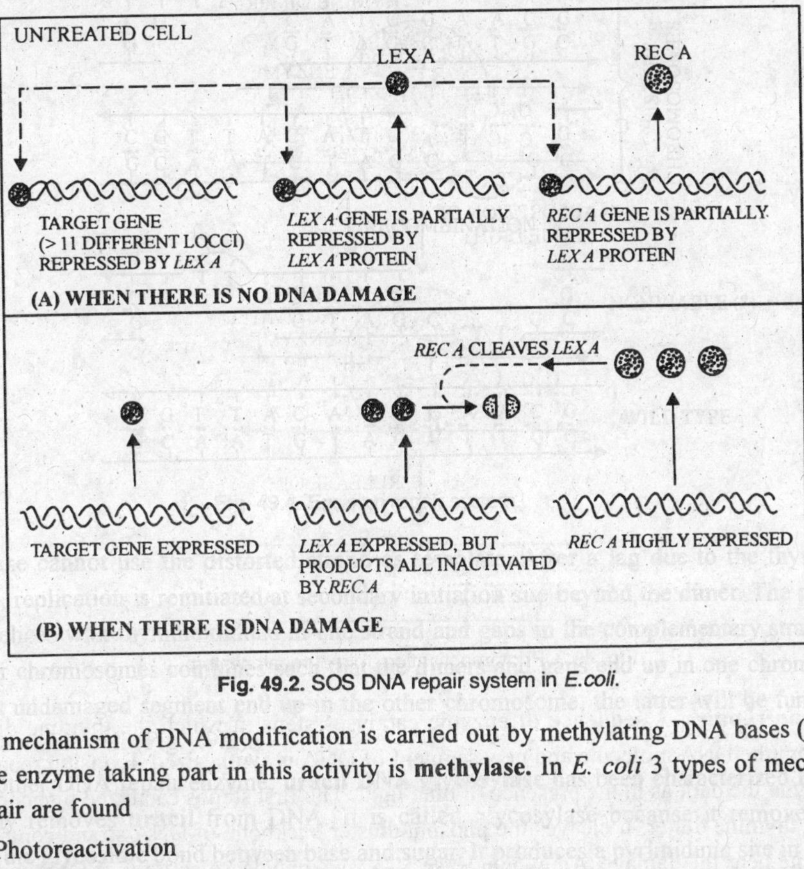

Fig. 49.2. SOS DNA repair system in *E.coli.*

The mechanism of DNA modification is carried out by methylating DNA bases (addition of CH_3). The enzyme taking part in this activity is **methylase.** In *E. coli* 3 types of mechanisms of DNA repair are found

(a) Photoreactivation

(b) Excision repair

(c) Post replication recombination repair

Photoreactivation : It involves an enzyme which splits thymine dimer directly without the removal of nucleotides (Fig. 49.3). Enzyme binds with thymine dimer in DNA in the dark but can not catalyze cleavage of the bonds joining the thymine molecules without energy derived from visible light; light of blue region of spectrum. Enzyme also activates cytosine dimers and cytosine thymine dimers. When ultraviolet light is used as an experimental mutagen, the treatment is usually carried out in the dark to maximize the mutation frequency, photoreactivation is also called **direct repair system** (Fig. 49.3).

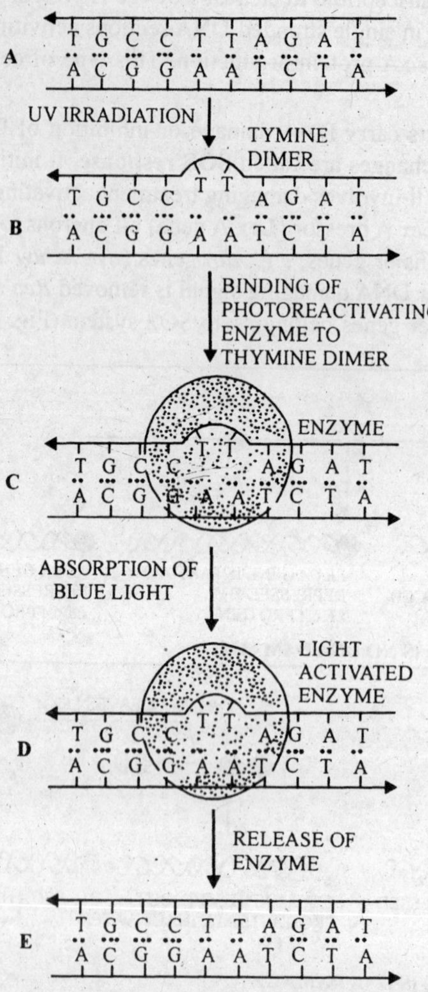

Fig. 49.3. Photoreactivation : direct repair system.

Excision repair : A sequence of enzyme catalyses steps in which the thymine dimers are removed from the DNA molecule and new segment of DNA is synthesized. Excision repair occurs as efficiently in the dark as in the presence of blue light. The first step is catalyzed by endonuclease recognizing thymine dimer; it cleaves the phosphodiester backbone of DNA strand containing the dimer at a site near the damage. An exonuclease 5′→ 3′ exonuclease activity of DNA polymerase I removes the dimer *in vitro.* (Fig. 49.4).

The segment adjacent to the endonuclease is then removed including dimer. DNA polymerase I fills the gap using complementary strand as template. DNA ligase catalyzes covalent closure or phosphodiester linkage between the adjacent nucleotides. Mutant lacking in enzyme DNA polymerase I are defective in excision repair. In addition uvr A, uvr B and uvr C genes are also required.

Post replication recombination: When DNA molecules containing thymine dimers are replicated, gaps are formed in the nascent complementary strand opposite the dimers because DNA

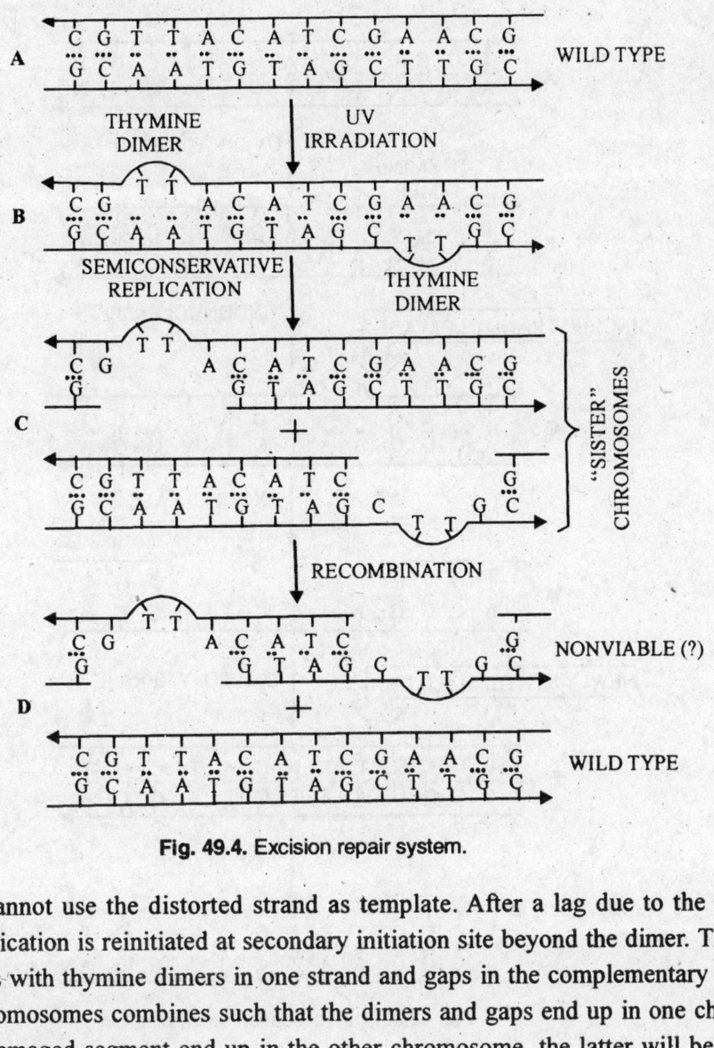

Fig. 49.4. Excision repair system.

polymerase cannot use the distorted strand as template. After a lag due to the thymine dimer blockage, replication is reinitiated at secondary initiation site beyond the dimer. The progeny has double helices with thymine dimers in one strand and gaps in the complementary strand. If these two sister chromosomes combines such that the dimers and gaps end up in one chromosome and the intact undamaged segment end up in the other chromosome, the latter will be functional and produce a viable cell (Fig. 49.5).

Another DNA repair enzyme, **uracil DNA glycosylase** has been characterized in *E. coli.* It efficiently removes **uracil** from DNA. It is called glycosylase because it removes uracil by cleaving the glycosidic bond between base and sugar. It produces a pyrimidinic site in DNA. Such apyrimidinic site undergoes an excision repair mechanism. Uracil is produced in DNA by deamination of cytosine. *E. coli* mutants deficient in uracil DNA glycosylase exhibit increased sensitivity to deaminating agents such as nitrous acid, indicating that enzyme is important in the repair of this type of damage to DNA (Fig. 49.6).

The larger the number of **essential genes** the more sites at which a lethal mutations occur and given equivalent repair process, the greater the probabilities that a given dose of mutagen will cause a lethal mutation.

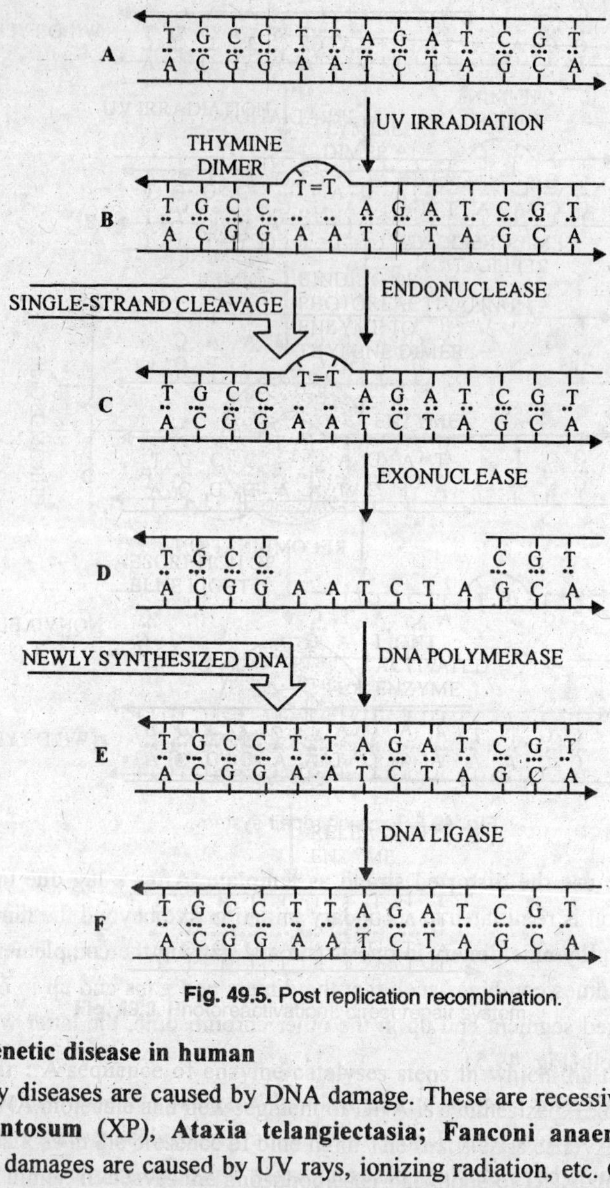

Fig. 49.5. Post replication recombination.

DNA repair and genetic disease in human

In human many diseases are caused by DNA damage. These are recessively inherited, *e.g.*, **Xeroderma pigmentosum** (XP), **Ataxia telangiectasia; Fanconi anaemia,** and **Bloom's syndrome,** etc. The damages are caused by UV rays, ionizing radiation, etc. **Cleaver** has shown that the cells of such affected individuals are unable to repair DNA damage induced by UV.

The best known example of an inherited deficiency in the repairs of radiation damage in human is the autosomal recessive disease **Xeroderma pigmentosum**. The skin of individuals with **Xeroderma pigmentosum** exhibit extreme sensitivity to sunlight resulting in a high frequency of tumors in the epidermal cells of exposed areas such as face. The cells are deficient in the repair of UV induced damage to DNA, such as thymine dimer. Therefore high incidence of skin cancer results directly from the inability of individuals with xeroderma pigmentosum to repair damage in DNA induced by UV in sunlight. The defective enzyme may be endonuclease which recognize as the thymine dimer and excise them. Mutation in about 6 genes can cause the disease

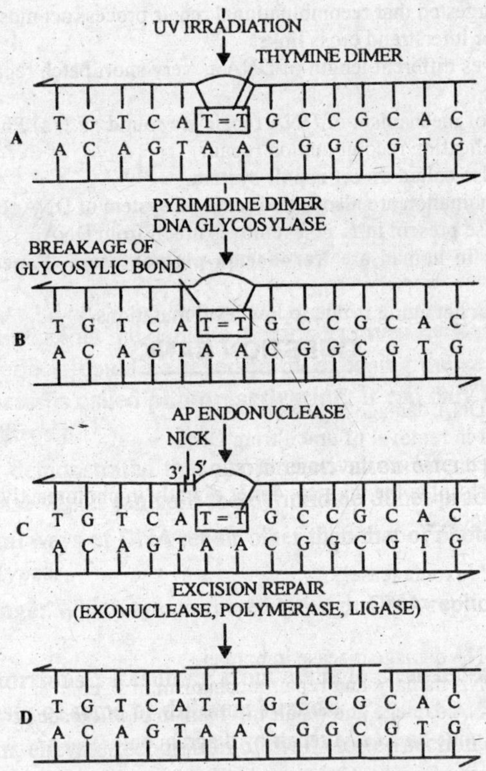

Fig. 49.6. Uracil DNA glycosylase repair system.

Some individuals suffering from this disease may develop neurological abnormalities resulting from premature death of cells. Mutation also contribute in the process of aging. According to **F. M. Burnet,** aging is the consequence of low level mutations **Fanconis anemia, ataxia telangiectasia** and **Bloom's syndrome** are the other three inherited (autosomal recessive) diseases in humans in which the primary defect may be in a DNA repair pathway. Leukemia is found in Fanconis anemia, cells are abnormally sensitive to ionizing radiations in **ataxia telongiectasia.** Individuals with Bloom's syndrome exhibit a high frequency of chromosome breakage resulting in aberration. All the three diseases are due to defect in DNA metabolism, *i.e.*, repair process.

The DNA damage repair system are grouped under

(a) Direct repair of DNA system (or) Photoreactivation

(b) Excision repair system : The damaged section is first excised and then repaired, it involves.

(i) Very short patch repair

(ii) Short patch region

(iii) Long patch repair

(iv) Involves excision of about 1500 bases long segment to 9000 bases long segments.

(c) Mis match repair system : Single base alters due to methylation or deamination.

(d) Retrieval system : Post replication repairs.

(e) Tolerance repair system : Damages blocking normal replication at the damaged site.

SUMMARY

1. The first demonstration about the damage repair was given by **Kelner** *et al* where they have shown that UV induced effect can be reverse by visible light.
2. **Setlow and Setlow** demonstrated that photoreactivity depends upon the enzyme which can split the pyrimidine dimer produced by UV radiation.

3. **Howard Flanders** suggested that recombinational repair process act mostly on DNA strands carrying gaps caused by intra or interstrand cross links.
4. Excision repair involves different length of DNA as very short patch repair (VSP), short patch repair and long patch repairs.
5. In *E.* coli three types of mechanism of DNA repair are found *i.e.*, (a) Photoreactivation (b) Excision repair and (c) Post replication recombination repair.
6. Photoreactivation is also called **direct repair system.**
7. Post replication recombination are also called retrieval system of DNA repair.
8. Uracil DNA glycosylase present in *E. coli* removes uracil from DNA.
9. DNA damage disease in human are **Xeroderma pigmentosum, Fancom's anemia** and **Bloom's syndrome,** etc.
10. According to **F. M. Burnet** aging is due to low level mutations.

QUESTION BANK

Very Short Answer Questions

1. Name two diseases of DNA damage in human.
2. Name the enzyme used in removal of uracil from DNA.
3. Name the enzyme which removes the dimer in *istio.*
4. Name the enzyme which splits the thymine dimer directly in photoreactivation.
5. What is 'SOS' type response?
6. What is endonuclease?
7. What is the function of 'exonuclease'?
8. What is SOS response in *E. coli*?

Short Answer Questions

1. Write short notes on DNA damage disease in human.
2. Write short notes on DNA damage and repair mechanism in *E. coli.*
3. Write short notes on DNA damage and repair mechanism in *Micrococcus luteus.*
4. Explain removal of thymine dimer and repair by diagram only.

Long Answer Questions

1. Write an essay on DNA damage and repair.

Multiple Choice Questions (Objective Type Questions)

1. The mechanism of DNA modification is carried out by :
   (a) Acetylation (b) Methylation (c) Both 'a' and 'b' (d) None
2. The enzyme which splits thymine dimer directly during photoreactivation is :
   (a) Photolyase (b) Polymerase (c) Lignase (d) Invertase
3. In *E. coli* the treatments causing DNA damage induce a complex series of phenotypic changes are called :
   (a) Excise repairs (b) 'SOS' response (c) Photoreactivation (d) All the above
4. Who used polymerase I enzyme for *invited* replication of DNA :
   (a) Kornberg (b) Flanders (c) Setlow (d) Kelner
5. Glycosylase enzyme is specific in helping the removal of damaged bases:
   (a) Hypoxanthine (b) 7-methyl guanine (c) Uracil (d) All the above
6. The disease due to DNA damage in human are :
   (a) Xeroderma pigmentosum (b) Bloom's syndrome (c) Albinism (d) Both 'a' and 'b'
7. Who for the first time observed DNA damaged repair :
   (a) Kelner (b) Stelon (c) Kornberg (d) None of the above
8. Who suggested that aging is due to low level of mutation :
   (a) Kelner (b) Burnet (c) Setlow (d) Flanders
9. Removal of Uracil from DNA is called :
   (a) Mismatch repair system (b) Excision repair system
   (c) Retrieval repair system (d) Tolerance repair system
10. Photoreactivation was demonstrated by :
    (a) Setlow ans Setlow (b) Setlow and Kelner (c) Kelner and Burnet (d) Setlow and Burnet,

ANSWERS

1. (b) 2. (a) 3. (b) 4. (a) 5. (d) 6. (d)
7. (a) 8. (b) 9. (a) 10. (a)